Developments in Petroleum Science, 26

hydraulic proppant fracturing and gravel packing

DEVELOPMENTS IN PETROLEUM SCIENCE
Advisory Editor: G.V. Chilingarian

Volumes 1, 3, 4, 7 and 13 are out of print.

2. W.H. FERTL
ABNORMAL FORMATION PRESSURES

5. T.F. YEN and G.V. CHILINGARIAN (Editors)
OIL SHALE

6. D.W. PEACEMAN
FUNDAMENTALS OF NUMERICAL RESERVOIR SIMULATION

8. L.P. DAKE
FUNDAMENTALS OF RESERVOIR ENGINEERING

9. K. MAGARA
COMPACTION AND FLUID MIGRATION

10. M.T. SILVIA and E.A. ROBINSON
DECONVOLUTION OF GEOPHYSICAL TIME SERIES IN THE EXPLORATION FOR OIL AND NATURAL GAS

11. G.V. CHILINGARIAN and P. VORABUTR
DRILLING AND DRILLING FLUIDS

12. T.D. VAN GOLF-RACHT
FUNDAMENTALS OF FRACTURED RESERVOIR ENGINEERING

14. G. MOZES (Editor)
PARAFFIN PRODUCTS

15A O. SERRA
FUNDAMENTALS OF WELL-LOG INTERPRETATION
1. THE ACQUISITION OF LOGGING DATA

15B O. SERRA
FUNDAMENTALS OF WELL-LOG INTERPRETATION
2. THE INTERPRETATION OF LOGGING DATA

16. R.E. CHAPMAN
PETROLEUM GEOLOGY

17A E.C. DONALDSON, G.V. CHILINGARIAN and T.F. YEN
ENHANCED OIL RECOVERY, I
FUNDAMENTALS AND ANALYSES

18A A.P. SZILAS
PRODUCTION AND TRANSPORT OF OIL AND GAS
A. FLOW MECHANICS AND PRODUCTION
second completely revised edition

18B A.P. SZILAS
PRODUCTION AND TRANSPORT OF OIL AND GAS
B. GATHERING AND TRANSPORTATION
second completely revised edition

19A G.V. CHILINGARIAN, J.O. ROBERTSON Jr. and S. KUMAR
SURFACE OPERATIONS IN PETROLEUM PRODUCTION, I

19B G.V. CHILINGARIAN, J.O. ROBERTSON Jr. and S. KUMAR
SURFACE OPERATIONS IN PETROLEUM PRODUCTION, II

20. A.J. DIKKERS
GEOLOGY IN PETROLEUM PRODUCTION

21. W.F. RAMIREZ
APPLICATION OF OPTIMAL CONTROL THEORY TO ENHANCED OIL RECOVERY

22. E.C. DONALDSON, G.V. CHILINGARIAN and T.F. YEN (Editors)
MICROBIAL ENHANCED OIL RECOVERY

23. J. HAGOORT
FUNDAMENTALS OF GAS RESERVOIR ENGINEERING

24. W. LITTMANN
POLYMER FLOODING

25. N.K. BAIBAKOV and A.R. GARUSHEV
THERMAL METHODS OF PETROLEUM PRODUCTION

Developments in Petroleum Science, 26

hydraulic proppant fracturing and gravel packing

DETLEF MADER

Paul-Hindemith-Ring 4B
D-6054 Rodgau 1 (Jügesheim) F.R.G.

ELSEVIER — Amsterdam — Oxford — New York — Tokyo 1989

ELSEVIER SCIENCE PUBLISHERS B.V.
Sara Burgerhartstraat 25
P.O. Box 211, 1000 AE Amsterdam, The Netherlands

Distributors for the United States and Canada:

ELSEVIER SCIENCE PUBLISHING COMPANY INC.
655 Avenue of the Americas
New York, NY 10010, U.S.A.

ISBN 0-444-87352-x (Vol. 26)
ISBN 0-444-41625-0 (Series)

Printed in The Netherlands

Preface

Hydraulic proppant fracturing is a relatively young technique of hydrocarbon pay stimulation which has been invented only abt. 40 years ago (CLARK 1948, KHRISTIANOVICH & ZHELTOV 1955) and which entered the focus of attention by reservoir and technical engineering particularly during the last ten years since the invention of synthetic intermediate- and high-strength ceramic proppants that can withstand the high closure stresses especially in deep tight gas-bearing sandstone pay formations (COOKE 1976, 1977). The possibility of now treating also high-pressure hydrocarbon-bearing intervals where natural sand crushes as a consequence of elevated closure stress triggered considerable expansion of hydraulic proppant fracturing stimulation which culminated in various MHF campaigns predominantly in USA and Europe where up to 1 Mio. gal of fluids and more than 6 Mio. lbs (3,000 t) of proppants have been pumped into large-scale fractures within a singular jumbo treatment and proppant quantities of 200 - 500 t per job are quite common in large-scale hydraulic reservoir stimulation.

The first field application of sintered bauxite (which was the earliest artificial high-quality proppant) in Western Europe only dates back almost exactly ten years (ERDOEL-ERDGAS AKTUELL 1978 a). Since that time, not only the cumulative quantity of proppants pumped annually downhole was progressively increasing, but also the amount of technical literature on hydraulic proppant fracturing was exponentially growing. Much hope for future booming of hydrocarbon appraisal and development was based on hydraulic proppant fracturing of low-permeability gas and oil reservoirs predominantly during the last five years, and whole exploration and appraisal drilling strategy concepts in petrophysically poorer facies belts of sedimentary basins were founded on the certainty of inevitable stimulation before well completion in the golden years of the hydrocarbon industry in the late 1970`s and early 1980`s due to back-up by a high oil price and a strong US $. The year of publication of this book celebrates the 25th anniversary of hydrocarbon exploration and hydraulic fracturing stimulation in the North Sea which is the oil and gas basin with the greatest potential in Western Europe, the 40th jubilee of commercial execution of hydraulic proppant fracturing treatments in the oil and gas world, and for myself 10 years professional experience as reservoir geologist in hydrocarbon industry.

Hydraulic proppant fracturing and 1986 oil/gas industry crisis

The full significance, sensitivity and fate of hydraulic proppant fracturing for the whole petroleum industry comprising production, service and supply companies, as well as the dependency of exploration and exploitation from the political scenery and its stability, however, was most markedly illustrated by the 1986 crisis when the oil price nearly suddenly collapsed (and effective for Europe in addition the US $ exchange rate almost at the same time catastrophically dropped), and the susceptibility of especially hydraulic proppant fracturing to deteriorations of the general economical framework was violently underlined by the fact that in contrast to drilling and conventional completion which always remain at a certain minimum level, stimulation can at least in some areas be totally suspended if no longer profitable.

The crisis came faster than expected and hit the industry worse than anyone had predicted and thus was a pronounced shock to awake dreaming reservoir engineers, euphoric hydrocarbon marketing people and also partially reality-decoupled service and supply companies by abruptly visualizing that critically changing interactions between several parts of the world`s political network and financial fabric and the resulting consequences for the different economical systems can not only force to seriously modify scientific and technical models and plays as well as to suspend complete strategical concepts of energy supply, but can also turn commercial businesses almost over night from profit into loss and thus also threat the social structure of whole industry branches.

My involvement in the stimulation market

I happened to enter the hydraulic fracturing stimulation scenery as Marketing Manager Proppants Europe just when the 1986 oil industry crisis was almost at its peak which converted the anticipated progressive sound and reasoned approach to my new professional duty into a sudden jump into the cold water of a collapsed drilling and treating market. The necessity of quickly stabilizing and reviving the Western European hydraulic fracturing and gravel packing market for proppant application and the urgency of performing market extension especially towards Eastern Europe, Southern Europe and Northern Africa (parts of which were hitherto almost white patches on the stimulation map) confronted me with the need of rapid collection of large amounts of information during consulting discussions, evaluations of fracturing potential and demand, and recommendations of application of proppant types and grain sizes depending on the requirements of the reservoir types and their geological structures when visiting hydrocarbon production and service companies, as well as reviewing the extensive literature, and fast processing of the assessed material into revised marketing strategy concepts based on geological, reservoir engineering, technical and commercial aspects with the ultimate goal of quickly increasing the areal proppant sales level.

The international marketing activity and the necessary acceleration of direct approach also required me to learn a couple of additional languages in order to have more working instruments for independent action during my professional reorganization and to enable additional transparency with respect to competition, especially when realizing the favourable constellation for expansion of the stimulation market in Eastern Europe at the time of political and economical restructuring (perestroika; GORBATCHEV 1987).

In this critical situation, fortunately the understanding of geological models and reservoir engineering principles as well as exploration and production problems of the main gas-bearing formations with the greatest potential of hydraulic fracturing and thus largest proppant consumption in Western Europe being terrestrial sandstones of Rotliegend, Carboniferous and Buntsandstein was enhanced by the experience which I got from my own independent sedimentological, petrographical and palaeoecological research on continental red bed formations (MADER 1985). Valuable improvement of the scientific and technical knowledge was also achieved by incorporating the background from my earlier practical professional work as Area Production Geologist responsible for an oil- and gas-field region in Germany FRG. Without all this experience, the necessary speedy acquisition of the required knowledge level would certainly not have been possible.

Planning of the present proceedings volume

The more or less evident slow revival of the stimulation activity in Western Europe in 1987 with definite plans for extensive hydraulic proppant fracturing campaigns particularly in the North Sea mainly in 1988 - 1990 and beyond, a profound reorganization of the European marketing strategy of the USA-based major proppant suppliers at the same time, and my independent work as an international consultant for petroleum geology and reservoir engineering triggered the idea to compile a report including review, status and forecast of hydraulic proppant fracturing and gravel packing particularly in Europe and adjoining international markets. This elaboration originally intended to not only sketch the various consequences of the crisis, but also to indicate a suite of possibilities of improvement of the stimulation business in the next years in an outlook focussing on marketing and economical aspects that although being of subjective nature according to my view and assessment, relates nevertheless to the demand of the customers with which I have been confronted during my investigations of the fracturing and particularly the proppant market.

Compiling the report, however, soon revealed that the marketing and economical questions are intimately interwoven with various technical points, and therefore I decided also to give a technical overview of hydraulic proppant fracturing and gravel packing as well as stimulation monitoring by predominantly well logging. This expansion also required to renounce from the areal limitation to Europe and to process information and literature on theories and applications all over the globe, with the emphasis on technological review and summary having been put on the USA which is probably the most mature stimulation market of the world. At the time of beginning to write down my experience and starting to evaluate the literature, however, I could never imagine that this project could ever end up in such a voluminous compilation as presented now to the readership.

Design of my fracbook

Several monographs and textbooks on hydraulic proppant fracturing and related reservoir stimulation have already appeared since the introduction of the treatment technology to the exploration and production industry. While the classical monograph by HOWARD & FAST (1970 a) has been the standard reference for more than three decades since its first edition, in the last years other compilations have been presented which are mainly proceedings volumes and review collections (USATCHOV 1986, ECONOMIDES & NOLTE 1987, GIDLEY 1988). Summaries of technological advances of hydraulic fracturing have not only been published as books, but also as longer journal articles (VEATCH 1983, VEATCH & MOSCHOVIDIS 1986) and congress lectures (FUHRBERG 1983). Most of these compilations, however, have been written by reservoir or technical engineers and concentrate on aspects of fracturing mechanics, fluid rheology, pressure evolution, fracture geometry, treatment modelling and simulation. In addition, marketing approaches hitherto having been carried out by supply and service companies were chiefly performed by commercial representatives with frequently insufficient technical and scientific background and above all not enough understanding of the particular pay intervals and their geological and engineering characteristics.

This is the reason why I became convinced that it would be more than desirable to fill the gap and discuss the subjects of hydraulic fracturing and gravel packing from the standpoint of a reservoir geologist with emphasis being put on the relationships between proppants and hydrocarbon-bearing formations, thereby underlining the significance of proppant type and grain size selection according to the sedimentological and petrographical nature of the pay zone to be treated. Many parts of the discussion thus focus on proppant choice as a function of granulometry, thickness, vertical and lateral continuity, consolidation, cementation and geometrical fabric organization of the reservoir formations that are the targets of hydraulic fracturing and gravel packing. Proppant aspects are the main topic of the present book which has the conception of a proceedings volume instead of the layout of a textbook, and are illustrated not only from scientific and technical view, but also from marketing and economical side. Throughout the technical evaluation, emphasis is put on sketching the suite of reservoir types treated by hydraulic fracturing, comprising low- and high-permeability sandstone, carbonate, chalk, diatomite, coal, shale and crystalline rocks (granite) which all require different fracture geometry as well as proppant quantities, types and grain sizes for technically successful and economically profitable stimulation.

Selection of aspects discussed in the text

The possibilities of proppant application in hydraulic fracturing and gravel packing are very widespread and quite complex. Discussing the topic from all the important viewpoints provides an endless suite of interesting aspects. In order to restrict the present elaboration to a more or less practical, well-organized summary without loosing the red line, no detailed review of the extensive scientific and applied literature in full coverage was intended, but the compiled overview was initially scheduled mainly as an enhanced and upgraded experience report deriving basically from my professional duty as Marketing Manager Proppants responsible for Western and Eastern Europe and resulting from my own assessment of the demand and supply situation of this market. During course of the literature research, however, numerous significant articles on various aspects of hydraulic proppant fracturing and gravel packing have been encountered and evaluated which I decided to present and summarize in an as much as possible condensed form and above all reworked with my own ideas wherever my knowledge permitted, and to bring this interwoven synopsis to the attention of the readership. That is why the size of the volume started to expand progressively, with finally only the deadline for publication having been able to stop the growth.

Certainly the task of limiting the total volume of the book to an economically still acceptable size meant inevitably careful selection of the topics to be discussed, with therefore the suite of items having been restricted to more or less proppant-related subjects, and thus for example aspects of stimulation fluids, fracturing models, matrix acidizing and acid fracturing, reservoir and fracture simulation, fracture geometry and mechanics, pressure evaluation, well testing, flow regimes and production planning in fractured formations, and others could only be treated very briefly in connection with proppant questions and were not examined in more detail, because these topics have been already discussed in more extension in various other compilations published so far to which reference is made. I put the emphasis on such technical and economical aspects of hydraulic proppant fracturing and gravel packing where type and size of proppants play the key role in design and application and which according to my feeling have hitherto not been condensed into actual reviews and summaries.

The most interesting theme for me as a geologist of education was to document significance and necessity of different fracture design for various pay formation types that are prospective for gas, oil or water and which demand hydraulic fracturing stimulation for achievement of economically feasible production rates, with the application spectrum of hydraulic treating including low- and high-permeability sandstone, carbonate, chalk, diatomite, coal, shale and crystalline (granite) reservoirs. The mentioned relationships are the reason why I have to beg the reader for understanding if some special topics could not have been respected in this proceedings volume which according to its title only presents a choice of aspects, but does not want to claim to be regarded as a textbook and also was not written with the intention to provide sufficient coverage to be able to reach the status of a textbook.

Hierarchical structure of the text

The organization of the text in a complex structure required subdivision of the main chapters into various hierarchical orders of subchapters, sections and subsections. As a consequence of comparatively small amounts of illustrations and tables breaking the vast desert of the voluminous text, I have put special emphasis on detailed splitting of the text into numerous paragraphs in a hierarchical manner for the purpose to guide the reader easily through the various chapters and the whole book. Six main text chapters are distinguished comprising proppant selection, marketing and economics, Rotliegend stimulation in Europe, hydraulic proppant fracturing, gravel packing and sand control, and propped fracture and gravel pack monitoring. The beginning of each order of subdivision contains also a short general introduction into the various topics which are discussed in the sections of the hierarchically next inferior level in order to connect the individual parts of different hierarchical stage and to lay a red line through the whole text.

The main table of contents presents only the general organization of the book including the first three hierarchical orders of subdivision, whereas the associated tables of contents assigned to each chapter following the introductory summary provide the full splitting into sections, and while serving as a detailed list of topics, they fulfill at the same time the task of being a sequential subject index by allowing to search for a lot of key words in the sequence of headers and subheaders. Further assistance for rapid access to information on special topics is provided by the alphabetical subject index at the end of the book.

Compilation of literature references

The compilation of the abundant references confronted me with another problem. In contrast to many geological (particularly sedimentological) journals which contain mainly longer articles that treat predominantly broader subjects in greater depth and present a detailed bibliography at the end which can often be quoted as a whole, most of the publications on hydraulic fracturing and gravel packing and related engineering and technical aspects are short SPE (Society of Petroleum Engineers) papers that are split up into very detailed topics and comprise brief experience reports of latest scientific, technical and economical knowledge compiled very actually for special subject symposia and specialists meetings and workshops, and therefore include in the majority of cases only a few references to some very closely related articles. Looking at the flood of papers surging annually from the suite of SPE meetings and visualizing the short time between individual conferences and sharp deadlines for paper submission and acceptance, it is easy to understand that the publications could only be practical status reports and with a very subordinate amount of exceptions, it could not be the intention of the authors to plunge a bit deeper into the technical literature, but limitation had to be made to a few practical references corresponding to the actual working environment.

The pressure on the individual researchers to produce results almost without interruption that is exerted by impatient supervisors and managers, competition between ambitious specialists for job promotion, prestigious behaviour, and above all the bare rockbottom demand to prove effectivity and indispensibility as a base for salary upgrading and job security particularly in the time of catastrophical staff layoff campaigns in a shrinking and restructuring hydrocarbon industry forces the people to maintain continuous output of news in form of numerous internal reports and a reasonable amount of contributions to meetings and workshops which in terms of contents are of the latest knowledge levels, but therefore necessarily in fact do not represent more than brief status reports. Thus the information on hydraulic proppant fracturing and gravel packing is extremely scattered in the literature by the division into a vast amount of individual papers with limited cross-referencing, and a more or less comprehensive compilation of this broad suite of quotations results in an enormous reference list.

Organization of the bibliography

The bibliography, however, is often the most valuable source of information for many readers which are looking for special items that could not have been treated in sufficient detail in the main text. In order to serve as a source of material for further plunging into specialized subjects, a good organization of the reference collection was inevitable, and also the literature compilation was considered to be a major section of the book. This conclusion is particularly valid in view of the present volume being an especially pronounced example of my already repeated experience that after thorough working through the literature, the reference list of the final paper is often longer than the first original draft of the whole text. No one, however, including myself could ever foresee that the voluminous literature compilation could grow so much as to almost reach the size of a little book itself.

Therefore I decided to not only present the literature list in alphabetical order of references and to include therein as much quotations as I felt necessary for proper quoting of the state-of-the-art of the individual topics, but also to process the assemblage for specific evaluation following detailed goals of acquisition. Thus the bibliography at the end of the book includes also a subject key classification to allow its easy consultance independent from the text, with the possibility of accessing the topic-related articles by key words and/or authors names. The overwhelming amount of the papers which are contained in the subject key classification (that in fact represents some type of index to the reference list) is also directly referring to the treated subjects and are therefore also implemented and cited in the text. It is my sincere hope that the reader will not be shocked by the tremendous reference list which exceeds by far what is normal for such a type of proceedings volumes, but that acknowledge is made to a comprehensive source of literature quotations providing a broad suite of specialists with further material for more detailed evaluation.

Choice and presentation of illustrations

Another aim of the enormous reference list and the accompanying subject key classification is to provide a possibility to get access to numerous illustrations and graphical presentations of various relationships. While at least parts of the text could be written in the form of a drastical summary or abstract to include as many individual papers as possible in a highly processed and condensed form, figures needed another approach. In view of the obvious limitation of the present book to a still handy volume (and the size that has turned out at the end is certainly representing the absolute maximum), no reason was seen why valuable printing space should be wasted by repeating illustrations which are contained in numerous earlier publications and have partially already been once or several times copied in other summaries, overviews and textbooks.

Emphasis was therefore put on proper referencing of the text in order to enable the reader to look for figures and further graphical information in the original literature. Advantage could be made in this respect of the excellent storage and access system of the numerated SPE papers which allows the interested individual fast and uncomplicated grasping of the additional information by having an extensive library on stand-by and call at his disposition for fast service by just quoting SPE paper number and credit card number. Reasonable compilations of numerous illustrations relating to most of the discussed topics are also contained in the above mentioned textbooks and monographs as well as overview journal articles. Thus the figures, tables and plates in the present proceedings volume were limited to such illustrations which derived from my own conception and present material and information that previously has not been published in related form, and which is necessary to underline and to further comment the aspects that are discussed chiefly from my experience and knowledge based on own investigations and research.

Processing of the material

As much of the information which I have received orally during my international attention of the stimulation market by consulting and interviewing numerous specialists in various countries is confidential and cannot be published in written or drafted form, many facts could only be sketched in the possible general way. While description in the text can still make reasonable advantage of indirect expression, only very limited obliteration of information for the sake of protecting confidential material is possible in illustrations, and therefore in addition to the reasons discussed in earlier sections, figures and tables in the present proceedings volume had to be restricted to a very small quantity.

In order to address the insider properly and still maintain the confidence barrier to the outsider, many aspects have been formulated in such a way in the text that the insider understands what is expressed, and also the outsider gets his information, although access to confidential material is inhibited. Respecting the confidence level when preparing this summary, however, I wish to express the expectation that the performed analysis will nevertheless be of considerable importance for allowing the reader to get a reasonably quick overview on what is happening where and when in the stimulation scenery, and hopefully the report will also contribute a little bit to promote hydraulic proppant fracturing and gravel packing particularly in Europe and other international markets during the next few years.

According to the relative underrepresentation of illustrations of any type with respect to the tremendous amount of text, I considered it to be indispensible to break the text into as much chapters, subchapters, paragraphs and items (in total a hierarchical system of seven headline orders has been employed) as possible, in order to provide not only a better overview on the list of topics and to convert almost already the table of contents into a sequential subject index of the book (that is assisted by the alphabetical subject index at the end of the volume), but also to facilitate reading of the pages by providing interruptions of the otherwise endless continuing text desert. For the purpose of keeping the red line, each chapter contains a brief outline of the contents of its subchapters, and cross references to other sections in the book discussing related items or other aspects of particular subjects have been implemented as often as possible.

Acknowledgements of sources of practical information

It is my sincere intention to thank all the persons with whom I sat down face to face discussing various problems of hydraulic proppant fracturing and gravel packing in the hydrocarbon exploration and production areas which they represent and are responsible for in their companies, for their hospitality and especially for their generally open mind, confidence and willing cooperation to give me generously the necessary direct practical information for not only making them specific technical and marketing recommendations and sales proposals for solution of their problems and satisfaction of their demand, but also for enabling me to get a rapid overview of the predominantly European stimulation market by putting together the individual pieces of data and splintered meanings to a mosaic which forms the base for the integrated compilation of this review.

In addition to my professional duties, I would like to take also this opportunity to acknowledge the help which I received from numerous individuals by welcoming me for some talking and providing me with material, and to express my gratefulness for the good cooperation for mutual benefit by returning my generalized evaluation and interpretation together with results from my own investigations and research in the form of the present proceedings volume with my very best compliments, thanks and greetings. I do hope that through this technical and marketing review and status report, I can offer some contributions to the solution of various still open questions in order to help to achieve a more widespread distribution and especially considerable expansion of hydraulic proppant fracturing and gravel packing in Western and Eastern Europe in the near future (when hopefully the general economical framework is again more promising; unfortunately I am afraid that except of a few key political people nobody has any possibility to influence the main controlling factors) to the scientific and commercial benefit of all of us together.

Acknowledgements of publication

Looking for a publisher with the aims on fast decision on acceptance and speedy printing in order to respect the very actual nature of marketing status report and prediction as well as technical summary and review which therefore need rapid distribution among the technical and economical specialists in order to be still reasonably up-to-date upon appearance of the book (particularly in view of the permanent possibility of rapidly changing economical framework conditions without expectation or prior notice, and in light of the almost uninterrupted surge of technical publications that brings progress and innovation nearly every quarter of the year), I found invaluable support by Dr. Martin Tanke (Department of Earth Sciences, Elsevier Science Publishers) who very quickly guaranteed me publication of the volume in the series " Developments in Petroleum Sciences " in camera-ready form which enabled me to bring the manuscript on the latest level still very shortly before printing. I am very grateful to Martin for his understanding of my requirements especially in terms of quantity of pages and for having given me all the necessary confidence and freedom to finish the volume along the lines of my intention.

My satisfaction and happyness cannot be overempahsized in view of the fact that although the final size of the book certainly has considerably shocked Martin because it was more than twice as much as he expected and as was confirmed by contract, he did not hesitate to accept the volume for publication in the submitted form and without delay made all the necessary internal preparations for speedy processing. Particular acknowledgement deserves the coordinating engagement of Martin in connection with the good cooperation for realizing my dream to catch the most suitable publication date by filling in the gap between the annual SPE meeting in October 1988 when for the last time a greater amount of brand-new technical information was received for incorporation into the text, and March 1989 when after the winter break the printed book could be presented at the first SPE specialized meetings on low-permeability gas reservoirs and hydraulic proppant fracturing.

Acknowledgements of manuscript preparation

During the stressy period of final editing of the text and preparation of the camera-ready manuscript in the above mentioned slot minus the time required by the publisher for production, I received invaluable support and unrenouncable assistance from my dear friends Elzbieta Zawadzka (Marine Institute, Gdańsk/Poland) and Małgorzata Truszel (Geological Institute, Sosnowiec/Poland) who devoted their whole physical and mental working capacity to me in order to finish the project in the requested time and in a good shape. Without the sacrificing and hearty help of Ela and particularly Małgosia with the time-consuming work of text processing and camera-ready printing during countless hours which they spent patiently and in spite of heavy work with excellent humour at the computer in my office, it would absolutely not have been possible to meet the technically and economically most desirable deadline for delivery of the final manuscript as agreed with the publisher.

As I am alone responsible for all the contents of the present proceedings volume and because I need to keep up-to-date for the preparation of a revision for an eventual second edition, I would appreciate to receive the comments of the readerships through critical discussion and complimentary information. I would also like to express my sincere hope that the present book forms a valuable contribution to the progressive understanding and realization of hydraulic proppant fracturing and gravel packing in various parts of the world and a wide spectrum of hydrocarbon reservoirs.

Rodgau 1 (Jügesheim), January 1989 — Detlef Mader

Table of Contents

Abstract

Hydraulic proppant fracturing and gravel packing are common stimulation and stabilization treatments during completion, testing and exploitation of hydrocarbon reservoirs in oil and gas exploration and production industry. The purpose of natural sand or synthetic proppants of different type and grain size is in hydraulic fracturing to support the crack in order to keep it open against the closure stress acting in pay zone depth and to maintain a highly-conductive drainage path through the tight reservoir rock matrix for oil and gas flowing to the borehole, and in gravel packing to plug the perforation tunnels and to build a gravel mantle along the borehole wall in order to filter the hydrocarbons flowing into the wellbore and prevent pay zone sand from moving. Hydraulic proppant fracturing and gravel packing lead to enhanced oil and gas recovery from low-permeability and weakly-cemented to loose friable sandstones, with the treatments improving hydrocarbon flow rates and minimizing influx of disturbing sand into the borehole, respectively. The present proceedings volume is organized into seven chapters comprising proppant selection, marketing and economics, Rotliegend stimulation in Europe, hydraulic proppant fracturing, gravel packing and sand control, propped fracture and gravel pack monitoring, and bibliography.

Proppant Selection

Proppant type and grain size selection is the key element of both hydraulic fracturing and gravel packing design, because natural sand or synthetic proppants are the only material left in place downhole after termination of the operation and are the critical agents whose performance decides on success or failure of the job.

Proppant types and properties

The five standard proppant types which are available at the moment in the hydrocarbon stimulation market for different prices are natural quartz sand, synthetic intermediate-strength low-density alumina silicate (ceramic) proppant, intermediate-strength high-density alumina oxide and silicate proppant, high-strength bauxite proppant and high-strength zirconia-silicate proppant. The spectrum of currently commercially marketed materials for fracture support is completed by sand and proppants coated by single or double resin pellicles of precured or curable nature. The different proppant types and grain sizes are discussed along the lines of their mineralogical composition and closure stress resistivity, chemical composition and acid solubility, grain sizes and conductivity per US $, and grain shapes. Careful choice of proppant type and grain size according to closure stress of the reservoir, bottom hole temperature, composition of the formation fluids and permeability contrast between pay zone and infilled fracture are economically essential due to the fact that in deep tight gas wells, the proppant cost can reach up to two thirds of the total massive hydraulic fracturing (MHF) expenses.

Aspects requiring consideration for proppant selection are surface morphology and internal microporosity, crushing behaviour and crystalline structure, friction angle and grain shape, fluid salinity and mechanochemical stability, fluid temperature and hot brine aggressivity, equipment abrasion and grain hardness, specific gravity and fluid suspension properties, grain size and embedment, manufacturing process and pellet composition, conductivity discount and proppant mixing. Computer programs which model investment and revenue for various proppant types and grain sizes and give a forecast of the expected production increase and its net present value and discounted cash flow aid in stimulation planning by providing criteria for proppant selection from both technical and economical aspects.

Proppant applications

While traditionally intermediate-strength high-density alumina silicate and oxide proppants as well as high-strength bauxite proppants have been utilized in the cumulative largest amounts for massive hydraulic fracturing of deep tight gas reservoirs, the market is more and more shifting towards application of intermediate-strength low-density alumina silicate proppants which have the advantages of approaching the specific gravity of natural sand and being considerably cheaper than the other synthetic proppant types. As most of the presently stimulated reservoirs are in such depth ranges where the closure stress resistivity of alumina silicate proppants would be still sufficient even in light of maximum pressure drawdown during long-term field production, consumption of the lightest type of synthetic proppants is expected to progressively increase during the next years. High-strength zirconia-silicate proppants represent a technically favourable product if staying reasonably below its stability boundary at which it instantaneously fails to dust in contrast to bauxite-based proppants which break into halves and quarters that retain often sufficient residual conductivity, but the elevated price prevents zirconia-silicate proppants from achieving a larger market share. Synthetic proppants are also used outside of hydrocarbon exploration and exploitation especially as heat-exchange pellets in cupola furnaces in the steel foundry, alumina and glass industry, as dust-absorption or -filtering pellets in the gas purification industry, and for fluid and gas filtration, shot-blasting and grinding, and constituents of high-stability cements.

Marketing and Economics

In terms of marketing and economical aspects, since the invention and first successful field application of synthetic sintered alumina oxide (ceramic) and fused zirconia-silicate proppants abt. 10 years ago, the hydraulic stimulation market as well as the gravel packing scene has been repeatedly influenced in both positive and negative manner by changes of hydrocarbon prices, US $ exchange rate (with respect to DM and £) and proppant prices. Various economical influences on hydraulic proppant fracturing are outlined by comparing oil price evolution with US $ exchange rate history and proppant price development, with special comments offered on present and future consequences of the 1986 oil price collapse for the fracturing market. An outline is also given on the impact of the oil price history on the gas market and its evolution. The major crisis of oil and gas industry in early 1986 hit the hydraulic fracturing market segment in Europe much stronger than that in USA, with the most important stimulation market differences being based on frequency of drilling and stimulation, success/failure ratio and consortial structure, operating mentality and treatment cost, increased cancellation rates of jobs, US $ exchange rate decline, and proppant marketing strategy.

Impact of the oil price collapse and the US $ exchange rate drop

The oil price collapse and the US $ exchange rate drop are expected to lead in future to declining reserves due to suspension of exploration and development (including stimulation), regaining power of the OPEC, and persisting market uncertainty due to of oil price instability and volatility. The most significant present consequences of the oil price crash and the US $ break are changes in fracturing strategy, reduction of capital expenditure, stimulation boat and drilling rig unemployment, and restructuring of operating and service companies along with budget cutting and staff layoff. The impact of the oil price drop on the gas market is at least in parts of Europe considerably weakened or compensated by firm planning of gas delivery which is secured by contracts and thus the hydraulic proppant fracturing activity in the Southern North Sea gasfield belt in going to become the absolute highlight of stimulation since the beginning of treatment campaigns in Europe despite the unfavourable economical framework conditions, and will also represent the culmination of proppant consumption so far in this area.

In comparison to the USA and other parts of the world, the understanding of the areal market is in Europe particularly complicated due to different economical systems and resulting different commercial capacities due to the question of currency convertibility and national organization of companies in relationship to trade offices and ministries, as well as due to various mentalities and languages. The present stimulation market is characterized by reorganization in both Western and Eastern Europe. While in Western Europe some major tight gas development campaigns are based on the inevitability to stimulate most of the wells by massive hydraulic proppant fracturing thereby following a concept which has already been frequently applied in many low-permeability gas patches in the USA, improving economical possibilities and more administrative independence in Eastern Europe along the lines of a major restructurization as well as the shifting emphasis from drilling and fast completion to increasing exploitation of reserves in existing wells allow slowly progressing maturing of hydraulic proppant fracturing also in this area.

Aspects of areal hydraulic proppant fracturing potential

The account of past, present and future areal hydraulic proppant fracturing potential in the European market with comparisons to USA and other parts of the world focusses on massive stimulation of tight gas sandstones in Rotliegend and Carboniferous in British Southern North Sea and Germany FRG which are characterized by the overwhelming share of proppant consumption. Further highlights are massive treatments on Tertiary sediments in Yugoslavia where together with the mentioned formations in Germany FRG the hitherto largest jobs have been carried out in terms of proppant quantity and fluid volume. The impact of the oil price crash is particularly illustrated along the lines of unemployment of drilling rigs and stimulation boats, increased cancellation rates of jobs, staff layoff and company merging, and business loss for proppant suppliers. Tail-in proppant staging represents a major tool of cost containment in stimulation jobs in the British Southern North Sea where the greatest amount of fracturing activity is taking place in the coming years. The highest significance for promotion of hydraulic reservoir stimulation in Eastern Europe is the reorganization (perestroika) in the USSR which together with increasing financial possibilities in the satellite countries such as world bank credits and export revenues enable to realize numerous projects in the near future.

Rotliegend Stimulation in Europe

Rotliegend (and to subordinate amounts also Carboniferous and Buntsandstein) gas reservoirs are high-potential examples of low-permeability (tight) gas sands which have considerable economical significance as gas resources in present and future, and because they represent the centre of massive hydraulic fracturing (MHF) activity in Europe and have seen some or the largest stimulation operations carried out with alumina oxide and silicate proppants worldwide, they are characterized as a case history in more detail. Rotliegend tight gas development by massive treatments in the facies belt between the marginal aeolian dune and fluvial braided-river to sheet-flood sandstones and the central playa and lacustrine mudstones in several parts of the North Sea will be the highlight of hydraulic proppant fracturing during the next five years in Europe and thus also the centre of proppant consumption in this area.

Rotliegend (and also Carboniferous) gas reservoir fracturing is sketched along aspects of distribution of main fields in horizontal and vertical direction, historical development of production and fracturing, proppant consumption, sedimentary-petrophysical and geochemical zonation of fracturing potential, permeability prediction and effective permeability, permeability deterioration and clay mineral distribution, minimum pre-fracturing gas production rate and poroperm properties, differences between exploration and development wells, stimulation cost and fracturing policy, philosophy of off- and onshore exploration well fracturing, implications of fracturing philosophy on proppant selection, appraisal drilling strategy concepts and experience, horizontal and vertical differences of fracturing potential in terms of permeability and gas composition, water sensitivity, water saturation, gas demand and reservoir performance.

The illustration of fracturing performance includes also comments on stimulation boat availability and capacity, spot vs. batch fracturing, exploration/appraisal vs. development wells in terms of proppant selection and cancellation risk, and impact of political-economical systems on stimulation requirement. An outline of sedimentological and petrophysical evolution of various tight gas sandstones is also incorporated into the discussion, and foam fracturing as a special technique for water-sensitive reservoirs is presented more generally. Various sedimentary structures of fluvial, aeolian and lacustrine deposits in Rotliegend and Buntsandstein illustrating distribution of reservoir heterogeneities and natural fractures as well as being of considerable significance for flow patterns of pay fluids and propagation of hydraulically-induced fractures are presented in photographical plates and accompanied by detailed captions that can even be evaluated independent from the main text.

Hydraulic Proppant Fracturing

The account of hydraulic proppant fracturing focussing on technological perspectives of potential extension of the stimulation market in the near future includes illustration of various aspects of containment of fracture propagation in thin sandstones, proppant saturation and conductivity improvement with pumping of up to ultra-high concentrations, targets of unconventional gas stimulation, hydraulic proppant fracturing of carbonate rocks, high-permeability reservoir fracturing, geothermal fracturing, minifracturing (mHF) and other small-scale fracturing, oil-reservoir fracturing, communication fracturing and fracture interference, fracturing of deviated wells, fracture damage, and aggressive fracturing design.

Fracture containment and propagation

Containment of fracture propagation is particularly significant for avoiding the origin of communication through the crack to under- and/or overlying water- or other type of hydrocarbon-bearing horizons, overpressured or other trouble-making beds, and the ground water storey in case of very shallow reservoir depth. Another decisive impact is wasting of stimulation material and energy for opening and plugging of adjoining barren intervals at the expense of propagation and infilling of the crack in the prospective horizon according to design to achieve the necessary length and conductivity for profitable reservoir drainage. The most important aspects of containment of fracture propagation in thin sheet sandstones are single vs. multiple fracturing and diverting techniques, treatment parameter adjustment, horizontal stress differences and fracture geometry, confining stress contrast, multiple-zone fracture migration, selective proppant placement, vertical fracture growth blockage by buoyant or settling diverters, reservoir elasticity vs. plasticity; reservoir heterogeneity from sedimentological, petrophysical and rock mechanical point of view; influence of geological discontinuities on fracture extension and morphology, and combination of hydraulic fracturing and gravel packing.

Proppant saturation and conductivity improvement

Proppant concentration is one of the major tools for conductivity improvement, with multilayer fracture propping with high saturation of material being the only possibility of significant reduction of flow capacity damage by proppant crushing and embedment in many hard and soft reservoir rocks, respectively. Multilayer high-concentration propping also helps to maintain sufficient fracture width and thereby to avoid the occurrence of non-Darcy flow. Ultrahigh fracture conductivity can be achieved apart from optimization of proppant saturation by selecting a higher-quality proppant type and by application of larger proppant grain sizes. The significance of proppant concentration is illustrated by discussing proppant distribution in the fracture, proppant transport and fracture damage, gel filter cakes deposition on fracture walls, fluid viscosity and carrier capacity, proppant density/fluid viscosity and/or density relationship, overpressure calibrated stimulation design, impact of delayed crosslinking, fluid heat capacity and reservoir temperature, fluid leakoff containment, fracture width vs. proppant bridging, mixing of round and angular proppant grains, and uniform vs. changing fracture conductivity.

Targets of unconventional gas stimulation

The main targets of unconventional gas stimulation are tight gas sandstones, geopressured reservoirs, coal seams, shales, high-temperature pay zones, and gas condensate-bearing intervals. Various fracturing stimulation types are outlined for the different reservoir rock types. While creation of long fractures deeply penetrating into the pay and supported by intermediate- to high-strength proppants that are resisting to high closure stresses are the best approach to tight gas sandstones and classical massive hydraulic treatment is the most feasible means of acquisition of gas reserves, explosive and non-proppant nitrogen gas fracturing have proven to be at least partially successive in shale stimulation, although in many cases hydraulic proppant fracturing is also the most superior technique in the latter type of pay formations. Coal seams are characterized by special mechanical properties which lead to elevated treating pressures and frequently restrict the volume of fluid and proppants that can be inserted into short and wide cracks. Fracturing strategies are mainly designed according to thickness and interval spacing of coal seams in the stratigraphical column. High-temperature reservoirs provide particular challenges to thermal fluid and proppant stability. Aspects of other sources of deep and unconventional gas such as abiogenic gas, subducted gas, gas condensate and gas hydrate are also briefly mentioned.

Carbonate and high-permeability sandstone proppant fracturing

Hydraulic proppant fracturing of both sandstone and carbonate rocks is considered to be the most significant stimulation technique in low-permeability reservoirs regardless of their composition. While tight sandstone pay zones are already since long time treated by massive hydraulic fracturing all around the world, low-permeability carbonate reservoirs have not yet been fully understood as profitable targets for hydraulic proppant stimulation. The main reason for the effectivity of proppant fracturing in hydrocarbon-bearing intervals of any type is the conductivity contrast between reservoir rock matrix and proppant package in the fracture. While hitherto mainly acidizing and acid fracturing have been carried out in carbonate reservoirs in Europe, spreading application of proppant fracturing partially in combination with acid fracturing is considered to be the only reasonable possibility to significantly increase the output from this type of hydrocarbon-bearing sections in the next years. Aspects of hydraulic proppant fracturing of carbonate rocks are summarized along the lines of extension of drainage path and conductivity contrast, open vs. propped fractures, combination of natural and artificial fractures, and combination of hydraulic fracturing and gravel packing. The comparative discussion of acidizing stimulation and proppant fracturing also includes comments on advantages and drawbacks of fracture propping, conductivity improvement by acid optimization, enhancement of acid penetration distance, acid retardation, and acid fluid-loss control.

In a similar way as carbonates, also moderate- and high-permeability sandstone reservoirs can be successfully stimulated by hydraulic proppant fracturing by creating a sufficient conductivity contrast between pay zone matrix and proppant wedge in the fracture, with the variables for increasing conductivity contrast being type, grain size and concentration of the proppants used for plugging of the crack.

Chalk fracturing

One major target of carbonate proppant fracturing is the Cretaceous chalk oil reservoir in the North Sea which is illustrated as a case study in more detail. The chalk requires frequently combinations of either acid and proppant fracturing, or fracturing and gravel packing for technically successful and economically profitable stimulation. While some chalk types are stable and extensively naturally fractured, other chalks are soft and flowing as a consequence of overpressuring and undercompaction and belong to the most complicated reservoirs in terms of technically successful and economically profitable stimulation. Special chalk fracturing techniques include tip screenout, interface, dendritic and water fracturing.

The combination of acid and proppant fracturing is further sketched along the lines of viscous fingering acidizing and wormhole growth, whisper fracturing, formation breakdown and closed fracture acidizing. The best solution are frequently short wide precise fractures infilled by high proppant concentrations placed in multilayers. Comments are also offered on reservoir compaction and field subsidence which is a major problem accompanying stimulation and exploitation of chalk fields and which has in the last years attracted considerable attention through spectacular platform jackup operations. An outline of sandstone acidizing is also provided focussing on reaction types and penetration distance of acid. Special methods for effectivity enhancement are delayed acid reaction and in-situ acid generation.

Geothermal fracturing

Geothermal fracturing is chiefly confronted with the problem of fluid stability in the hot reservoirs. While hitherto in many cases both conventional and stem-induced explosive fracturing without insertion of proppants have been applied as stimulation techniques of geothermal reservoirs, improvement of fluid resistivity, temporary formation cooling by pumping of large pre-pad and pad volumes, application of water as fracturing fluid and enhanced proppant placement techniques have allowed to consider proppant fracturing also as a competitive stimulation technique in geothermal reservoirs. Giant-scale water fracturing encompasses the hitherto by far largest amounts of fluid pumped during course of an individual stimulation job. Communication fracturing connecting injection and production wells is of particular significance in geothermal pays in contrast to their undesired nature in oil- and gas-bearing formations. Enhancement of drainage is achieved by fracture system enlargement by heat extraction as a consequence of cold water injection into hot dry rocks. Nuclear fracturing is a possibility of especially large-scale explosive reservoir rubblization and is able to overcome the drawbacks of more standard explosive fracturing techniques comprising short fracture extension and intensive matrix pulverization. An outline is also given on shear stimulation vs. tensile fracturing.

Minifracturing and other small-scale fracturing

Minifracturing (mHF) or comparable smaller-scale treatments are very significant for the initial approach of many marginal and partially still uncommercial reservoirs in the near future and are assessed as being one of the major methods to maintain or even accelerate the technological evolution of hydraulic fracturing despite the presently unfavourable economical constellation. Following definition of mini-, micro- and massive fractures, minifracturing and its importance is discussed along the lines of pre-fracture breakdown test and fracture parameter determination, formation damage around the wellbore, field experiments and mHF vs. MHF, pilot stimulation of marginal reservoirs, application of resin-coated proppants in unstable pay zones, horizontal or drainhole drilling and fracturing, uptake capacity of oil-field water injection and waste-liquid disposal wells, combination of natural and artificial fractures, late-stage refracturing of old wells, combination of acid and proppant fracturing, and fracture length and height optimization.

The main significance of minifracture calibration experiments comprises determination of closure stress and fluid-loss coefficient which are important parameters for reduction of screenout failure rates. Various minifracture tests and microfracture breakdown procedures as well as proppant vs. non-proppant minifracturing are discussed. Short precise propped minifractures are suitable for breakthrough of formation damage haloes around the wellbore, cost containment of stimulation of various reservoir types, and pilot approach of marginal pays. Special techniques are tip screenout fracturing and tail-in of resin-coated proppants for prevention of embedment and flowback. Modified micro- and minifracturing analysis for overpressure calibrated stimulation design are also outlined.

Horizontal borehole and water-injection well fracturing

Horizontal well fracturing is most profitable in combination with rather than as alternative to horizontal hole drilling, and particularly thorough access to the reservoir matrix can be achieved by multiple parallel vertical fracturing of horizontal wellbore trajectories creating a suite of slices intersecting the pay which is most successful if the horizontal borehole is drilled in accordance with the geostress field. Oil-field water-flood and waste-liquid disposal wells are characterized by almost permanent operation above fracture pressure which results not only in reservoir parting, but also in keeping the created cracks open as long as injection at elevated pressures is going on, with therefore in most cases propping of the fractures not being required. Distinction has to be made between controlled fracturing following goals of improvement of well injectivity, and accidental fracturing which can provoke submature fluid breakthrough and thus deteriorate sweep efficiency. Disposal fracturing is in some areas the only possibility to create sufficient uptake capacity for safe underground removal of waste liquids from drilling, completion and hydrocarbon processing.

Naturally-jointed reservoir stimulation

Hydraulic proppant fracturing of naturally jointed reservoirs focusses on interaction between natural and hydraulic cracks. Permeability continuity and stress sensitivity are important parameters controlling interfracture communication. Other aspects which are discussed include fracture interference and superimposition, fracture divergence, temporal natural fracture damage, dual fluid leakoff (comprising standard loss to matrix and accelerated loss to hairline fissures which open up in an advanced stage of the job), and proppant plugging of natural fractures. A special stimulation technique for naturally cracked reservoirs is dendritic fracturing which comprises entering and ballooning of the pre-existing joints by the hydraulic treatment fluids and connecting the whole system to the wellbore. Propping of natural cracks only leads to acceleration of production, but does not enable to increase the recoverable reserves, as no additional drainage area is created. Aspects of tailored pulse loading and shear slippage vs. tensile failure are also discussed.

Refracturing

Refracturing of already earlier stimulated boreholes and original fracturing of infill wells in late stage of reservoir pressure depletion and hydrocarbon exploitation is a suitable means of cheap mobilization of additional oil and gas quantities, because old boreholes are already completely amortized or written-off and the original fracture represents a weakness plane which allows easier reopening at lower injection pressures. Removal of the old proppant package has so far not turned out to be a major problem. Refracturing candidate well choice has to follow the experience of the original stimulation job, and proppant selection for refracturing partially also depends on the aims of acceleration vs. increasing recoverable reserves. The approach of refracturing candidate choice is partially comparable to that of minifracture calibration test evaluation. The main advantage of renewed fracturing following an original treatment independent from the time that has passed in between benefits from recovery enhancement independent from reservoir depletion stage. Special methods are combinations between either hydraulic and explosive or acid and proppant fracturing.

Concerning fracture length and height optimization, the most important parameters influencing design are reservoir permeability, well spacing, fracture conductivity and proppant distribution. Generally long and short fractures are required in low- and high-permeability reservoirs, respectively. Net present value optimization reveals that excessive fracture length does not result in additional investment return, but is rather a waste of capital expenditure, because beyond an optimum fracture length, hydrocarbon production rate and volume remains at a plateau and therefore the payout is deteriorating.

Oil-reservoir fracturing

Oil-reservoir fracturing is in contrast to many cases of gas-pay zone stimulation characterized by higher-permeability rock treating which puts emphasis on achievement of a significant conductivity contrast between reservoir rock matrix and proppant package in the fracture. The higher viscosity of many oils frequently requires to select coarse to very coarse proppants in order to achieve the necessary conductivity and flow capacity. Most oil reservoirs demand stimulation by short wide fractures that are infilled with proppant multilayers and which in some cases represent not very much more than an enhanced gravel pack. A special effect is the possibility of combination of horizontal and vertical fracturing due to changing geostress regime by thermal reservoir expansion in steam-drive enhanced oil recovery operations. Other possibilities of changing fracture orientation from vertical to horizontal or altering fracture azimuth direction to create networks of perpendicular crack elements are disturbing impact of a nearby hydraulic fracture which provokes a stress halo or belt, and thermal reservoir contraction by injection of cold water during secondary water-flood recovery. While fracture orientation changes from vertical to horizontal are common during standard steam-drive operation, fracture azimuth changes can be achieved by staggered steam injection into off-trend well pairs.

Communication fracturing

Communication fracturing with connection of either two superimposed reservoir storeys separated by a tight barrier or two neighbouring wells is an important alternative to sidetrack drilling in field appraisal and to conventional interval-wise fracturing with subsequent commingling of separate layers. Connection of isolated sand bodies in lenticular reservoirs by communication fracturing links much more discrete sand units to the borehole than could be achieved by drilling. Fracture system and well pattern planning has to be made according to crack azimuth as predetermined by the in-situ stress constellation, and sand lens axis orientation. During connection of separated reservoir storeys and field drainage optimization stimulation, caution has to be exercised that unwanted fracture interference is not short-circuiting the planned drainage areas and thus deteriorating the effect of the individual fractures for channelling and delivery of hydrocarbons to the wellbores. In contrast to oil and gas reservoirs, fracture interference and borehole connection is significantly enhancing accessible pay volume and fluid cycling effectivity in geothermal reservoirs.

Deviated-well fracturing

Fracturing of deviated wells suffers from the drawback of limited communication interval between a highly-inclined or horizontal borehole and a vertical fracture intersecting it. The most favourable constellation comprises a wellbore trajectory being oriented parallel to the principal in-situ stress direction which allows generation of an extensive singular vertical fracture that most effectively drains the reservoir along the high-angle to horizontal borehole path. The other extreme is an orientation of the wellbore trajectory perpendicularly to the principal in-situ stress direction which permits creation of multiple vertical fractures cutting orthogonal through the highly-deviated wellbore like a set of parallel slices. All the solutions between these two end members of a continuous spectrum suffer from productivity loss due to limited borehole/fracture contact length.

Other features in highly-inclined boreholes are proppant stratification and fluid layering, and also an increased probability of non-Darcy flow. The drawback of limited communication interval between wellbore and fracture applies equally for highly-deviated to horizontal deep boreholes crossing gently-inclined reservoirs and subvertical shallow wells traversing steeply-dipping pay formations. The best improvement of the indispensibly situation of extended-reach drilling from offshore platforms in a star-like pattern would be achievement of a S-shaped borehole trajectory with again decreasing the inclination prior to intersecting the reservoir. An outline of hydrocarbon production and flow regimes is also given.

Fracture damage

Fracture damage is the worst deterioration of a stimulation job, and factors leading ultimately to treatment failure are outlined by commenting on proppant settling, proppant flowback, proppant crushing and fluid aspects. Proppant settling is influenced by clustered proppant transport and thus hindered segregation of the particles in the slurry. Proppant trapping in the crack can be accelerated by forced fracture closure. The decisiveness of proppant banking depends on grain size distribution and fluid stratification in the reservoir, with distinction having to be made between coarsening- and fining-upwards cyclothems as well as homogeneous pays and formations divided by an oil-water-contact into lower and upper portions. Proppant flowback during well cleanup can be avoided above all by stabilization of the package by a tail of resin-coated proppants which provides an aggregated wedge blocking the fracture mouth. Other possibilities of proppant flowback control are pressure depletion assistance by careful fluid production, limitation of exploitation rates and choke size, and avoidance of excessive pressure drawdown.

Proppant crushing by shock loading and stress corrosion cracking are mainly consequences of unsuitable completion and offtake practices and can be reasonably prevented by operation with low production rates and low flowing pressure. The best insurance against proppant crushing, however, is selection of a proppant type that is able to sufficiently withstand all the possible reservoir pressure drawdowns during field life especially in the immediate surroundings of the wellbore where fluid velocity and pressure depletion culminate. The investment for a better proppant type and thus the insurance against failure of the agent supporting the fracture pays off many times during course of long-term field exploitation. Production cycling by repeated well shut-in leads to irrecoverable fracture conductivity diminution unless a proppant with high enough closure stress resistivity also for peak loading has been chosen and should therefore be avoided.

Aggressive fracturing design

For the sake of the necessary progress in unconventional reservoir stimulation with innovative techniques, an aggressive fracturing design is recommended which, however, has necessarily to consider the risk of some drawback by failures in pilot treatments of marginal reservoirs. The danger of premature screen-out termination of challenging jobs is particularly high when working with excessive proppant concentrations and extraordinarily large proppant grains which, however, are the most effective ways of further increasing the conductivity contrast between reservoir rock and proppant wedge infilling the fracture. Conclusion is made that given an again more stable oil price at a reasonable level and partially also an again stronger US $, there are many growing stimulation and stabilization markets with increasing demand of hydraulic proppant fracturing and gravel packing in the near future in various parts of the world.

Gravel Packing and Sand Control

The spectrum of sand control methods encompasses mechanical gravel packing, chemical reservoir consolidation and mechanical stable sand arch formation. Gravel packing is the most commonly applied and most reliable method of sand control in wells where weakly-cemented or even loose friable sands are not sufficiently stable enough to resist to fluid forces during hydrocarbon production. Distinction has to be made between reservoirs producing sand from the very beginning of hydrocarbon offtake, and pays where sand starts to flow into the wellbore when rising water cut changes fluid viscosity or progressive exploitation leads to declining pressure which both lower integrity and support of the grain fabric of the rock matrix. Some overpressured and undercompacted chalk oil reservoirs also require gravel packing partially in combination with hydraulic fracturing.

Standard gravel packing

The significance of careful selection of type and grain size of the gravel-pack material for mechanical control of pay sands is outlined by discussing aspects of hydrocarbon productivity vs. sand control, reservoir depth and closure stress in perforation tunnels, gravel grain size choice vs. formation grain size, absolute stoppage of moving and flowing reservoir sand, impact of gravel pack hardware, fluid viscosity and gravel concentration, chemical formation consolidation vs. mechanical gravel packing, and combination of gravel packing and hydraulic fracturing. Comments are also offered on gravel-less hardware installation with building of a natural gravel pack around screen or liner by intruding formation sand, triple-wrapped screen utilization where the different wire-wrap layers are scheduled to stop progressively finer grain sizes, and linerless gravel placement. Gravel packing completion fluids and gravel saturation are illustrated along the lines of fluid viscosity and gravel concentration, transport and packing capacity of the slurries, and gravel settling diminution in the carrier fluid.

The effect of a tail-in terminal stage in hydraulic fracturing which places coarser proppants near the crack mouth that keep in place the finer grains of the main lot is comparable to that of a gravel pack, and a row of narrowly-spaced gravel-filled perforation tunnels approaches the significance of a propped minifracture. Reservoir stability and fluid dynamics, and particle transport through perforations are other important topics for further characterization of the significance of gravel packing, with emphasis being on formation pressure and fluid composition, perforation packing and particle bridging, achievement of perforation tunnel plugging, and influence of perforation technique and geometry. Steam-drive enhanced oil recovery requires special adjustment of the gravel-pack material in order to resist to the hot alcaline aggressive milieu, with sintered bauxite and nickel-coated sand having turned out to be sufficiently stable, whereas natural quartz sand is considerably dissolved.

Special gravel packing

Special gravel pack applications are consolidated pack installation, resin-coated gravel placement, gravel packing of deviated wells with the effect of gravel duning, gravel packing with liner vibration, concentrated gravel slurry pumping, density slurry and foam gravel packing, and viscous polymer gravel packing. Consolidated pack installation and resin-coated gravel placement have the advantage of aggregating the perforation tunnel infillings to stable gravel wedges and therefore a gravel mantle seaming the borehole wall and holding the perforation hole plugs in place is not required, with this special type of application thus restricting the operation to outside gravel packing which leaves the wellbore free of hardware and also considerably reduces the required gravel quantity. Consolidated pack installation and resin-coated gravel placement are especially suitable for old production wells and steam-drive injection boreholes.

Gravel compaction and overcoming of the phenomenon of gravel duning in deviated wells can be attained by flow resistance created by flexible baffles and rubber disks, large-diameter washpipes, perforation prepacking, string rotation, liner vibration, low particle density/carrier fluid density ratio, inverted slurry circulation, gravel-less slotted liners and prepacked screens. Liner vibration compensates initial undercompaction of the gravel by triggering settling into a hexagonal package which allows to utilize coarser gravel grain sizes, and the absence of after-pack settling does not necessitate a reserve blank section above screen and perforated interval. Concentrated gravel slurry pumping profits from the tackifying action of the intermixed minor amounts of resin. Viscous polymer gravel packing is associated with buildup of nodular gravel deposits at the perforations and annular sediments on the screen. Comments are also offered on puddle packing which is a method for infilling of borehole wall irregularities generated by earlier sand production or explosive shooting in older wells and thus is a significant repair technique in open borehole sections.

Other sand control methods

Other sand control methods comprise resin consolidation, hydroxy-aluminum stabilization, electroless nickel plating, silica cementation by silicalock and Solder glass methods, warm air coking, stable sand arch formation, and pre-pack liners and multiwire-wrapped screens. The specific features of chemical resin consolidation of the formation sand are illustrated along the lines of activation, catalyzation and curing of the resin. Performance of resin consolidation can be achieved by prepacking, preheating and preflushing of the reservoir. Stable sand arch formation can take place already naturally in formations with sufficient grain friction due to rough particle surfaces or in reservoirs in greater depth with higher overburden stress, but can also be semiartificially introduced or enhanced by packer inflation. Pre-pack liners that are prepared with assistance of vibration and multiwire-wrapped screens can be used for gravel packing and pump protection.

The influence of completion type on gravel packing is outlined by sketching features of inside gravel packing such as significance of perforation tunnel plugging, and open hole gravel packing such as borehole underreaming, port collar implementation and pack washing. Formation damage during gravel packing is introduced by the placement of the mantle itself as well as by trapping of fluid-loss additives at the interface between gravel and reservoir where they are kept in place by the pack. Removal or formation damage by acidizing prior to gravel packing with partial utilization of the gravel itself as diverting agent is usually leading to satisfactory results, whereas acidizing after having installed the gravel pack results in selective acid penetration through the permeable gravel infilling the perforation tunnels as well as to gravel unconsolidation and dissolution and should therefore be avoided whenever possible.

Propped Fracture and Gravel Pack Monitoring

Propped fracture and gravel pack monitoring by well logging, seismical and other methods is not only a means of checking quality of performance of the stimulation and stabilization operations after termination of the job, but is also a technique of long-term behaviour control by repetition runs and gives even important impact on future planning and design of fracturing treatments and gravel pack placements by delivering direct information from the supervision of existing propped fractures and installed gravel packs, respectively.

Propped fracture height monitoring

Fracture detection and determination by well logging aiming chiefly on height monitoring can be carried out by temperature, conventional and spectral gamma-ray, thermal decay time, acoustic and noise logging. Multiple-tracer discrimination by gamma-ray spectroscopy allows to distinguish several treatment stages and enables determination of provenance of radiation from fractured formation, cement sheath or borehole fluid. Temperature logging is the cheapest and simplest fracture height monitoring technique, but is sometimes disturbed by thermal conductivity anomalies of the reservoir which lead to warm noses in the temperature record. The anomalies can be verified and correct fracture height interpretation can be made by combination of temperature with noise or gamma-ray logging.

Radioactivity detection by conventional single- or double-marker gamma-ray logging or advanced multiple-tracer gamma-tay spectroscopy can be performed using three different types of sources comprising naturally radioactive proppants, artificially tagged proppants (including coatings and full grains) and synthetic markers intermixed into fracturing fluid or slurry. Gamma-ray spectroscopy is able to differentiate between several inserted tracers and thus permits to distinguish various fluid and proppant stages as well as seperate layers in multiple-interval treatments if selective tagging of the individual lots is carried out, and is thus the most mature technique of fracture height monitoring. Combination of temperature and conventional gamma-ray logging is the most profitable solution for cost containment in multiwell field campaigns by providing the possibility of mutual calibration in some key wells and later running only one of the registrations. Thermal decay time logging allows also monitoring of acid stimulation due to the acid effect on pulsed neutron capturing. Distinction can be made between propped fracture height as reflected by gamma-ray logging of proppant radioactivity, created fracture height as documented by temperature logging and gamma-ray monitoring of fluid radioactivity, and permeable or effective fracture height as indicated by thermal decay time logging.

Propped fracture azimuth mapping

Possibilities of fracture azimuth mapping are caliper and acoustic logging, borehole televiewer, formation microscanner, and seismical and gravimetrical methods, with some of the means also revealing information on fracture width. Acoustic logging employs mainly reading and processing of the circumferentially travelling Rayleigh and Stoneley waves, with various techniques of waveform modelling and interpretation existing. Caliper logging records borehole ellipticity or ovalization as a response to the in-situ stress distribution, with the major eccentricities indicating the direction of principal horizontal stresses and thus reflecting fracture azimuth. Borehole televiewer and formation microscanner permit direct observation of fractures by visualization in contrast to indirect interpretation of other measurement methods. Distinction can be made between recognition of the natural fracture system in the reservoir rock and evaluation of hydraulically generated fractures intersecting both pay zone matrix and pre-existing joint assemblages. The most suitable techniques for detection of natural fractures are borehole televiewer, formation microscanner, dipmeter, dual laterolog, and sonic logging with full waveform analysis. The suite of different fracture indicators which are all subject to various disturbing influences is best evaluated by integration into a multiple-source fracture probability program. Mechanical properties logging and fracture migration analysis are important predictive tools of fracture modelling and design.

Seismical, gravimetrical and geoelectrical methods mainly aim on fracture azimuth mapping, although some techniques such as vertical seismic profiling can also be applied for fracture height monitoring. Microseismic fracture registration can be performed by active or passive recording, with most of the microearthquake events being created by the closing fracture. Enhanced interpretation can be achieved by combining microearthquake and pressure pulse documentation. Anelastical strain recovery analysis of cores is a suitable method for prediction of fracture azimuth. Distinction of vertical and horizontal hydraulic fractures can be performed by tiltmeter analysis of surface deformations, whereas for natural cracks the separation mode of shallow and deep dual laterolog records is the most indicative instrument. Other geophysical techniques for chiefly fracture azimuth mapping are acoustic emission, hydraulic impedance, hydraulic resonance and magnetometry. Further fracture monitoring can be performed by stimulation data analysis, inflatable impression packers, and photography and imagery. Fracture pressure interpretation does not only allow to calculate fracture height and length, but is also an instrument of premature screenout termination forecast and thus detailed observation of the pressure evolution during course of the treatment enables screenout prevention by permitting remedy actions in time.

Gravel pack evaluation

Gravel pack checking and evaluation can be performed by photon, gamma-ray and neutron logging and can be further checked by utilizing data from the telltale screen. Photon and neutron logging employ the density change generated by gravel insertion into the borehole, whereas gamma-ray logging has to record natural gravel or artificial gravel and/or fluid radioactivity. Calibration with the end members of 0 % gravel in the blank section above the screen and 100 % gravel in the interval of highest instrument count enable to perform quantitative gravel pack evaluation by neutron and gamma-ray logging with expression of percentages of gravel density in the surveyed wellbore section. Special requirements for discrimination of gravel packing stages or levels can be solved by gamma-ray spectroscopy of multiple tracers. Interpretation confidence can be enhanced by combination of neutron and gamma-ray logging, and progressive settling of gravel after installation can be monitored by time-lapse repetition logging. Repair of detected voids can be carried out by the photon-vibrapack combination which allows synthetic improvement of gravel settling by vibration and shaking, with repeated logging and exciting being done until the voids in the pack have disappeared.

The tell-tale screen recognizes completion of gravel insertion once it becomes covered with gravel and thus serves as a reliable indicator when to stop pumping. Mechanical properties logging allows detection of strong and weak sands and helps in candidate screening for the necessity of sand control treatments. While in single-well operations the cost for an integrated sand production forecast analysis could be higher than the expenses for sand control jobs, considerable completion cost could be saved in multiwell field campaigns when mechanical properties logging and evaluation allows to distinguish between stable sands which do not lead to sand influx and unstable sands that would result in sand intrusion into the borehole unless controlled by mechanical gravel packing or chemical formation consolidation.

Bibliography

The extensive literature compilation is processed in a bibliography which consists of two parts. The first section gives a subject key classification of the references into various special topics of mainly technical and reservoir engineering nature, and the second division contains the reference list which includes all the quotations in alphabetical order of the authors names. The main topic headers in the subject key classification are overview of hydraulic fracturing and economical framework of hydrocarbon drilling and stimulation, fracture design and evaluation, fracture features, proppant features; technical, fluid and temperature features; field case studies, special applications of hydraulic fracturing, gravel packing and sand control, and petrophysical evaluation for hydraulic fracturing and gravel packing design. The reference list contains more than 3,500 quotations of papers on technical and economical aspects of hydraulic proppant fracturing and gravel packing as well as associated completion and stimulation techniques.

List of figures

List of tables

List of plates

1. Proppant Selection

1.0. Summary and associated table of contents

Hydraulic proppant fracturing and gravel packing activity in Europe is of different significance in the western and eastern parts of the area, and has again other emphasis in the USA, due to geological (mainly sedimentological-petrophysical), reservoir engineering, technological, marketing and economical reasons. The five basic proppant types which are commercially available at the moment in the hydrocarbon stimulation market for different prices are natural quartz sand, synthetic intermediate-strength low-density alumina silicate (ceramic) proppant, intermediate-strength high-density alumina oxide and silicate proppant, high-strength high-density bauxite proppant, and high-strength low-density zirconia-silicate proppant (the different materials are listed in a sequence of increasing specific gravity, closure stress resistivity and price). Careful selection of type, grain size and to minor amounts also grain shape of the proppant according to closure stress of the reservoir, bottom hole temperature, composition of the formation fluids and permeability contrast between pay zone and infilled fracture are economically essential due to the fact that in deep gas wells, proppant cost can reach up to two thirds of the total massive hydraulic fracturing (MHF) expenses. In deep low-permeability tight gas reservoir fracturing, proppant quantities of 100 - 500 t are frequently needed and some jumbo jobs even require more material in the range of 500 - 1,500 t, with some supergiant treatments needing even record lots up to 3,000 t, and proppant choice is thus the most critical part of designing stimulation operations. Since the first hydraulic proppant fracturing jobs in 1947, the technique has evolved during course of four decades from small operations on single-zone completions to massive treatments on multiple-interval installations, with the main stages of stimulation focus having been formation damage breakthrough and small-scale sand fracturing in the late 1940`s through early 1970`s, tight gas reservoir MHF treatment with progressively increasing amounts of synthetic proppants in the late 1970`s through early 1980`s, and high-permeability pay zone stimulation from the early 1980`s onwards. The evolution of hydraulic fracturing technology is characterized by progressively increasing proppant quantities and crack wing length for low-permeability tight gas reservoir stimulation as well as successively rising proppant concentration and fracture width in moderate-to high-permeability oil and gas pay treatment. Since the innovation and inauguration of hydraulic proppant fracturing shortly after the end of the Second World War, to date almost one million jobs have been carried out worldwide. The major technological breakthrough has been achieved by invention and market introduction of intermediate- and high-strength synthetic proppants in 1976 which allowed hydraulic stimulation of deep high-pressure reservoirs that were prior to availability of closure-stress-resistant proppants unaccessible to fracture treatments. While natural quartz sand fails by crushing of the grains already between 3,000 and 5,000 psi closure stress which restricts its application to relatively shallow reservoirs, synthetic intermediate-strength low-density alumina silicate proppants have almost the same specific gravity as sand, but resist to closure stresses up to 8,000 psi. Intermediate-strength high-density alumina oxide and silicate proppants (stability up to 10,000 psi) and high-strength high-density alumina oxide or low-density zirconia-silicate proppants (resistant up to 15,000 psi) are the suitable materials for infilling of fractures in deep high-pressure reservoirs and have in Europe during the last ten years been frequently applied in MHF operations in tight gas-bearing Rotliegend, Carboniferous and Buntsandstein sandstones in Germany FRG and Netherlands onshore as well as British and Dutch Southern North Sea offshore, and in Tertiary sandstones and conglomerates in Yugoslavia, and have in the USA been used in many tight gas sand reservoirs in various hydrocarbon-prospective basins. Resin-coating of synthetic proppants or natural sand results in aggregation of the individual grains to a stable package in unconsolidated reservoirs and is a suitable method for hydraulic fracturing and gravel packing of unstable soft oil-bearing chalks and sands in the Cretaceous and Jurassic of the Norwegian and subordinately also the British Northern North Sea as well as in Cretaceous and Tertiary in various countries on the European continent. Double resin coating of proppant grains with an outer curable pellicle bonding particles together and an inner precured film providing additional strength of the material combines the benefits of a self-consolidating proppant with the advantages of a high-strength proppant by providing high closure stress resistivity, permanent bonding ability and low density together at reasonable cost. Tail-in of resin-coated proppants stabilizes the grain assemblage in the proximal part of the crack and inhibits fracture damage by evacuation at its mouth. Tail-in design with a smaller lot of coarser (mainly 16/20 or 12/20) and frequently higher-quality proppants representing the terminal stage of a fracture treatment and following a larger main quantity of finer (in most cases 20/40) and often lower-quality proppants has the advantage of ascertaining maximum fracture conductivity in the immediate borehole vicinity where reservoir drawdown is greatest by higher proppant permeability and closure stress resistance at lower overall cost, as well as of forming a pseudo-gravel pack inhibiting proppant flowback from fracture mouth to wellbore and thus preventing stimulation result damage by fracture evacuation. Experimental approaches for possible future innovation of proppant types of again enhanced technical performance and improved economical feasibility in order to adjust to the increasingly diversified spectrum of application focus on prototype materials based on several ceramic oxides, cordierite and silicon carbide, and look mainly into specific gravity lowering by fabrication of partially hollow or particles with high internal microporosity. Concerning the currently commercially available and field-experienced proppant types, crystalline alumina oxide and silicate proppants that are produced by sintering have rough surface and more internal microporosity in comparison to the brittle glassy zirconia-silicate proppants which are manufactured by fusing and have smooth surface and less internal microporosity. On the other hand, the brittle glassy zirconia-silicate proppants tend to fail almost suddenly in a catastrophic manner by breakage into tiny splinters when the boundary of closure stress resistance is reached, whereas the less brittle and partially even ductile crystalline alumina oxide and silicate proppants are split into halves and quarters in a more continuous manner when approaching the boundary of mechanical stability. Various consequences of the different mechanical and chemical properties of the proppant types of different mineralogical composition and crystallographical structure are discussed along the lines of resistivity against closure stress, temperature, and composition of brines. The numerous aspects influencing proppant choice are illustrated along the lines of friction angle and grain shape, fluid salinity and mechanical stability, fluid temperature and hot brine aggressivity, equipment abrasion and grain hardness, specific gravity and fluid suspension properties, grain size and embedment, and manufacturing process and pellet composition. Proppant conductivity as measured during laboratory testing has to be considerably discounted in field applications as a consequence of embedment of brittle proppants into the formation, plugging of pores by fracturing gel residues and fluid-loss filter cakes, discontinuities in crack geometry, and stress corrosion of proppants by leaching and crushing. Embedment into gel filter cakes on fracture walls and precipitation of gel residues in the pore space are the most serious types of proppant permeability deterioration apart from grain crushing. Mixing of proppant types and grain sizes has in most cases considerable deleterious effects on proppant package conductivity, while mixing of different grain shapes can be beneficial for increasing the friction angle of the material. Selection of proppant type and grain size has to be made on the base of calculation of conductivity per US $ vs. cost of the hydraulic fracturing operation and pay-out by increased hydrocarbon production during a certain time, with computer programs for the economical assessment of proppant choice, performance prediction and feasibility modelling being available in the market. Economical profitability analysis of hydraulic proppant fracturing has to be based on net present value calculation, with the most important variables influencing investment return being pre- and post-fracture hydrocarbon production rates as well as proppant type and quantity. The most important technical aspects of proppant selection are closure stress, surface morphology, crushing behaviour, friction angle, fluid salinity, fluid temperature, equipment abrasion and specific gravity, whereas in the economical feasibility evaluation, proppant price has the dominant role particularly in light of the currently quite different pricing of the various materials. In addition to oil- and gas-field applications of proppants for hydraulic fracturing and gravel packing, ceramic proppants are used in other market segments as heat-exchange pellets in cupola furnaces in steel foundry, alumina and glass industry, as dust absorption or filtering pellets in the gas purification branch, for fluid and gas filtration, for shot blasting with hardening of stainless steel by peening rather than cutting, for grinding, as constituents of high-stability cements and for various other purposes. Hydraulic fracturing is in both USA and Europe also in the future remaining to be the most significant application of the different types and grain sizes of the presently commercially available proppants, and experimental approaches carried out in the last years even suggest the possibility of development of new proppant types of enhanced technical performance and better economical feasibility in case of adequate demand by the market.

1.1. Introduction

Hydraulic proppant fracturing of predominantly low-permeability oil and gas reservoirs has become a widely applied technology of hydrocarbon pay stimulation in order to increase recoverable reserves, to accelerate production by increasing rates or to even permit offtake at all. The promotion of hydraulic proppant fracturing to an outstanding technique of well stimulation which allows acquisition of enormous gas reserves in hitherto economically unexploitable deep tight gas sandstone (cf. section 4.4.1.), shale (cf. section 4.4.4.) and coal (cf. section 4.4.3.) pay zones took particularly place during the past dozen of years since synthetic intermediate- and high-strength proppants had been invented and introduced to the petroleum industry in late 1976 (COOKE 1976; COOKE, GIDLEY & MUTTI 1977; ATTEBERRY, TUCKER & RITZ 1979; COOKE & GIDLEY 1979). The significant and decisive impact of high-quality man-made proppants on the booming of hydraulic fracturing is that they could withstand higher closure stresses and therefore allowed to proceed to higher-pressured reservoirs in greater depths than was previously possible when only natural sand as a propping agent was available that starts to crush already in moderate formation depth and thus is unsuitable to support fractures in deep prospective intervals of the geological column.

The following introductory remarks begin with a short characterization of tight gas reservoirs and massive hydraulic proppant fracturing (MHF) stimulation which has turned out during the last 10 years to be the motor of hydrocarbon well stimulation activity and therefore is the key to the whole discussion of hydraulic proppant fracturing by representing the technically and economically most significant application of well treatment. MHF stimulation is also the procedure where the various properties of different proppant types have the highest significance in terms of technical performance and economical feasibility of the operation. After a succeeding outline of evolution of hydraulic fracturing technology and oil price history as well as importance of proppant selection and following some comments on special aspects of the European stimulation market, the organization of the review and status report is briefly sketched.

1.1.1. Tight gas reservoirs and MHF stimulation

Hydraulic proppant fracturing enables to complete hydrocarbon exploration and production wells with an original effective reservoir permeability in the microdarcy range and has thus permitted to satisfy increasing demand for natural gas by focussing attention on unconventional tight gas horizons that were historically considered to be uncommercial, but which are now representing important resources that are already delivering appreciable shares of the total gas production and which will become particularly significant in the future when gas supply from traditional reservoirs declines more and more (RANDOLPH 1974; FAST, HOLMAN & COVLIN 1977; KUUSKRAA, BRASHEAR, DOSCHER & ELKINS 1978; AGARWAL, CARTER & POLLOCK 1979; KUUSKRAA, BRASHEAR, ELKINS & MORRA 1979; KUUSKRAA 1980, BAKER 1981, WALLS 1981; AHMED, SCHATZ, ABOU-SAYED & JONES 1982; BRASHEAR, ELKINS & HAAS 1982; MORRA, BRASHEAR & HAAS 1982; O`SHEA & MURPHY 1982; O`SHEA, ROSENBERG & BRASHEAR 1982; KAZEMI 1983; ROSENBERG, O`SHEA, MERCER, MORRA & BRASHEAR 1983; VEATCH & BAKER 1983; BRASHEAR, ROSENBERG & MERCER 1984; DOE 1984; HAAS, BRASHEAR & MORRA 1985; VEATCH 1986, CLARK 1987, DUDA & HANCOCK 1987, KUUSKRAA & HAAS 1988, NORTHROP 1988; aspects of tight gas reservoirs are also evaluated in sections 3.4.1. and 4.4.1.). The discussion as follows concentrates on significance of tight gas reservoir fracturing for future gas supply, hydrocarbon acquisition expenses and stimulation share, economical incentives of tight gas reservoir fracturing, and proppant quantities applied in MHF stimulation.

1.1.1.1. Significance of tight reservoir fracturing for future gas supply

Massive hydraulic proppant fracturing (MHF; cf. section 4.8.1.2.) is usually the most effective reservoir stimulation treatment in low-permeability hydrocarbon-bearing intervals (ECONOMIDES 1987 b), is in fact the only proven economical development method for tight gas reservoirs to date (VEATCH 1983) and is expected to be able to provide additional gas reserves of 4 - 8 Tcf (Trill. ft3) in the USA if prudently applied (PAI, GARBIS & HALL 1983). Massive hydraulic fracturing comprising large treatment sizes in terms of proppant quantity, fluid volume and crack length has been first applied to deep gas reservoirs in 1975 (RANDOLPH 1974, WORLD OIL 1975 a, SLUSSER & RIECKMANN 1976; cf. sections 1.2.4. and 2.4.1.) and since that time has rapidly become the probably most widely utilized technique in hydrocarbon stimulation especially in USA and Western Europe. The significance of unconventional tight gas sandstone exploitation by MHF stimulation is underlined by the progressively rising contribution of tight reservoir gas to the total gas production in the USA from abt. 5 - 10 % in the 1980`s to abt. 30 % and 50 % forecasted for the years 2000 and 2020, respectively (BAKER 1981; cf. section 2.2.1.6.4.).

Independent from, but certainly considerably promoted by the challenging acquisition of gas in tight reservoirs of sandstone, shale and coal type, the use of hydraulic proppant fracturing for stimulation of wells in sandstone and carbonate pays in oil and gas industry and subordinately also in geothermal heat exploitation (cf. section 4.7.) has risen significantly during the last years and this trend is expected to continue and to even accelerate, because it will become increasingly important to maximize the amount of oil and gas that can be produced from a given well and pay in the near and far future when many of the high-capacity conventional reservoirs become progressively exhausted and depleted (McDANIEL & WILLINGHAM 1978).

1.1.1.2. Hydrocarbon acquisition expenses and stimulation share

Significance of hydraulic proppant fracturing for increasing accessible hydrocarbon reserves and optimization of profitability of oil and gas field exploitation is underlined by the fact that during the last decade, petroleum industry in free world spent annually more than 100 Bill. US $ for oil and gas exploration and production (OEHME 1986; cf. section 2.2.1.2. and 2.2.1.5.2.). It is difficult to estimate the share of stimulation, but it is only logical to postulate that in view of such high expenses every effort should be undertaken to maximize hydrocarbon offtake by application of hydraulic proppant fracturing as a more or less common practice of well completion in fields where the output can be increased through this technique. In the USA which probably represent the most mature oil and gas reservoir stimulation market, hydraulic proppant fracturing treatments have to be carried out in 35 - 40 % of all the wells that are drilled and completed (VEATCH 1983, OIL GAS JOURNAL 1984; cf. section 2.3.).

In the British sector of the North Sea which includes the Southern Rotliegend Basin that is the most important tight gas province in Europe and which is at the moment in the stage of advancing to the centre of hydraulic proppant fracturing in Europe and to represent the highlight of the whole well stimulation history in this area (cf. section 2.4.1.2.), abt. 50 Bill. £ have been invested into exploration and production during 25 years since the discovery of the first commercial gas field in 1965 (OFFSHORE ENGINEER 1986 d), with this capital expenditure containing already reasonable shares of stimulation due to various big campaigns having already taken place in the Rotliegend tight gas belt.

1.1.1.3. Economical incentives of tight gas reservoir fracturing

As a consequence of world-wide increased demand for natural gas and the creation of tight gas incentive price and tax in the USA (HAAS, BRASHEAR & MORRA 1985) in 1978 (thus almost coinciding with the beginning of commercial availability of intermediate- and high-strength proppants; cf. section 1.2.4.1.), low-permeability reservoirs have been recognized as the next major source of future gas supply particularly in the USA and also in several parts of Europe (BAKER 1981; cf. section 2.2.1.6.4.) and have been regarded as having the prospect of offsetting the predictable decline in conventional reserves (HAAS, BRASHEAR & MORRA 1985). The nature of the two economical incentives, providing either higher prices when gas is in short supply or tax credits when it is not, creates a situation which is beneficial for tight gas exploration and exploitation.

Because tight gas formations produce primarily only with low uneconomical rates, the necessary increase of productivity has to be secondarily achieved by hydraulic proppant and/or acid fracturing (cf. section 4.5.1.) or horizontal borehole drilling (SUNG & ERTEKIN 1987; cf. sections 4.8.6.2.1. and 4.8.6.2.2.) or a combination of both. Hydraulic proppant fracturing is generally considered to be the most viable method of improving recovery in tight formations (ECONOMIDES 1987 b), and maximization of the exploitation effectivity can be achieved by installation of multifracture systems in highly-deviated to horizontal drainage wells (cf. section 4.8.6.2.4.) especially in thin reservoirs, whereas for thick pay zones, deeply penetrating normal fractures emanating from vertical wells are still the optimum stimulation technique.

1.1.1.4. Proppant quantities applied in MHF stimulation

Enhancement of the ability to accurately predict and control the geometry of hydraulic fractures increases the chances of achieving optimum propped fracture lengths and heights (cf. section 4.8.11.) and thus leads to an ameliorated economical incentive for development of tight gas reservoirs (HOLDITCH, ROBINSON, WHITEHEAD & ELY 1987). Thick tight reservoir formations (with permeabilities being less than 0.1 md; PAI, GARBIS & HALL 1983; cf. section 3.4.1.2.) containing enormous amounts of gas reserves in USA and Europe that only have to be liberated from the low-permeability matrix require frequently massive hydraulic fracturing treatments (MHF; RANDOLPH 1974; MALONE, STAHL & COULTER 1977; MURPHY & CARNEY 1977; PARKER 1980, 1981; WHITE & DANIEL 1980, 1981; HOLCOMB 1982, PAI & GARBIS 1983 a; PAI, GARBIS & HALL 1983) where proppant quantities of frequently up to more than 400 - 600 t or even abt. 1,000 t are pumped within a single job comprising fluid volumes up to more than 1 Mio. gal, with some of the jumbo stimulations challenging the present technological and operational boundaries.

The world record to date comprises a MHF gas well treatment in the USA with 3,150 t or 6.3 Mio. lbs of 20/40 natural sand as proppants pumped in a total fluid volume of abt. 1.54 Mio. gal into a tight gas reservoir of abt. 300 ft (100 m) pay thickness in abt. 9,300 ft (3,100 m) formation depth in the Wilcox Sand in the Benavides field (Zapata county, South Texas) in 1987 (CONSTIEN, BRANNON & BANNISTER 1988; PITTS 1988). In order to carry out the jumbo job, 18 high-pressure trucks pumped for almost 11 hours and in total abt. 100 pieces of equipment had to be installed at the wellsite. The maximum quantity of synthetic proppants of bauxite base having been injected so far into a large-scale fracture is abt. 1.5 Mio. lbs, with the main reason for the significant discrepancy in world record quantity between natural sand and synthetic proppants not being suspension and transport characteristics within the carrier fluid but being expense of the operation due to the price difference between natural sand and artificial high-quality man-made proppants.

1.1.2. Evolution of fracturing technology and oil price history

The evolution of the various main types and basic concepts of hydraulic fracturing technology (VEATCH 1983; cf. section 1.2.1.2.) and different generations of artificial proppants of various strength and of mainly corundum, mullite and zirconia composition (DOWELL SCHLUMBERGER 1985, WESTERN PETROLEUM 1985; cf. section 1.3. and tabs. 1 - 3) took place parallel to several drastical oil price changes earlier in upwards and later in downwards direction (GRIFFITHS 1986, RUNGE 1986) that were in Europe partially even accentuated by variations of the US $ exchange rate (with respect to DM and £; cf. section 2.2.1.3.).

Planning and execution of hydraulic proppant fracturing jobs has thus been not only influenced by reservoir engineering, geological-petrophysical and technical modelling of optimum hydrocarbon exploitation in terms of rate and duration of production as a base for commitment of gas supply contracts (cf. section 2.2.1.6.2.) and thus discounted cash flow and pay-out of investment, but particularly during the last three years since the major oil price decline (MENGES 1986, OEHME 1986; cf. section 2.2.1.1.2.) also by economical evaluation of the feasibility of carrying-out of the operations in light of the current hydrocarbon price scenario (that was in Europe during the last couple of years negatively influenced by the US $ exchange rate drop; cf. section 2.2.1.3.). Proppant price changes also have significant impacts on fracture operation budgeting and design, because proppant cost comprises the highest share of the total expenses of the hydraulic stimulation job (cf. section 2.2.2.).

1.1.3. Significance of proppant selection

As a consequence of different technical properties, changing operational performance and different prices of the various proppant types, careful selection according to proppant behaviour under reservoir conditions (closure stress, temperature and brine composition) and economical frame is of considerable significance for feasibility modelling of the hydraulic fracture operation, because in deep tight gas wells, proppant cost can reach up to two thirds of the total stimulation expenses (for example the cumulative cost for a MHF treatment in a Rotliegend well in the Söhlingen gas field/Germany FRG in late 1982 where abt. 550 t of proppants had been injected amounted up to 6 Mio. DM or 2.5 Mio. US $, but the jumbo job operation paid off by proving abt. 500 Mio. m3 gas reserves; JOHN 1983, KLOSE & KRÖMER 1983, BLEAKLEY 1984; cf. section 3.3.). Some comments are offered as follows on importance of fracturing for feasibility assessment of hydrocarbon fields, and central role of proppant placement in the stimulation treatment.

1.1.3.1. Importance of fracturing for feasibility assessment of hydrocarbon fields

Hydraulic fracturing operations do not only often represent a large fraction of initial well costs, but also determine the economical viability of a particular borehole or field (NEWBERRY, NELSON & AHMED 1985). A stimulation job being too large can be an unnecessary waste of completion funds, whereas a fracture treatment being too small may result in such inefficient reservoir drainage to make a well unprofitable, and because of this economical double-edged sword, hydraulic fracturing jobs have to be designed in such a way as to optimally deplete the pay formation, with choice of proppant type and grain size playing a key role in this modelling.

After the fracture treatment has been carried out and the fluid has been produced back, the only thing which is left in place in the reservoir is the proppant package supporting the crack, keeping the pathway open thereby providing a high-conductivity channel to enable enhanced flow of oil or gas to the borehole and thus controlling long-term productivity (HOLDITCH 1984, ANDERSON & PHILLIPS

1986, HALL & LARKIN 1986; cf. sections 1.4.12. and 4.13.). This relationship highlights the outstanding significance and critical role of the proppant in the hydraulic fracturing operation and thus the importance of its proper selection.

1.1.3.2. Central role of proppant placement in the fracturing treatment

SIEVERT, WAHL, CLARK & HARKIN (1981) summarize the importance of proppant selection by underlining that all the time and money spent developing and perfecting fracturing fluids, techniques and equipment are peripheral to the central objective of placing proppant in the fracture in order to create a highly conductive flow channel. Economical aspects of hydraulic proppant fracturing are also evaluated by VEATCH (1986).

This high rank of proppant choice as the most critical point of hydraulic fracture treatment design (HOLDITCH 1984) for the different applications in oil, gas and water reservoir management is the reason for the necessity of carrying out a sophisticated multidisciplinary proppant marketing with inclusion of geological (mainly sedimentological-petrophysical) and reservoir engineering concepts into the technical and commercial marketing strategy. Proppant selection requires technical and reservoir engineering recommendations in view of the commercially best net present value per invested US $ (BRIM 1986, ANDERSON & PHILLIPS 1987) which can also be expressed as the need of looking for a proppant to give maximum conductivity per US $ (NORTON-ALCOA PROPPANTS 1986, STANDARD OIL PROPPANTS 1986; cf. sections 1.3.4. and 1.4.12.). The significance of proppant choice is highlighted by the fact that particularly in deep well application, the type of proppant used can make a difference between an economical success or a failure of the whole treatment (McDANIEL & WILLINGHAM 1978).

1.1.4. Special aspects of the European stimulation market

While the geological and reservoir engineering relationships and characteristics of oil and gas pay zones requiring treatment by hydraulic proppant fracturing are basically more or less comparable in different parts of the world, fundamental differences exist in terms of economical and commercial setting between the stimulation markets in Europe and USA. In contrast to the USA which can be regarded as a large more or less homogeneous unit, Europe is not only politically a complex and heterogeneous aggregate of countries speaking a suite of different languages and having a spectrum of various mentalities, with the major distinction concerning economical situation, status of technological development, commercial capacity and foreign trade activity having to be made between Western and Eastern Europe, but also includes a wide range of oil and gas reservoir types in terms of stratigraphical, palaeogeographical, sedimentological and petrophysical characterization. Therefore an assessment of the European areal hydraulic proppant fracturing potential from both technical and marketing point of view (which is presented as a regional case study in chapters 2 and 3 in this compilation) has to summarize the general geological and commercial scene by comparing the different cases and integrating not only scientific, engineering and technological, but also economical and even ideological aspects.

Due to my technical and marketing experience, preferential reference is made within this compilation to examples deriving from Western and Eastern Europe, although the analysis and review (with the exception of the areal market study presented in chapter 2 and the technical-economical example sketch of a tight gas reservoir formation in chapter 3) applies for other parts of the world as well, with most of the case histories and special conclusions being based on experience from the USA which is probably the most important market for hydraulic proppant fracturing and gravel packing in the world both in past and future.

1.1.5. Organization of the review and status report

The organization of the review and status report of hydraulic proppant fracturing and gravel packing in the present proceedings volume is sketched along the lines of evolution of experience and derivation of information as well as subdivision of the book into seven chapters.

1.1.5.1. Experience evolution and derivation of information

The present outline gives a review and discusses status and perspectives of hydraulic proppant fracturing (acid fracturing stimulation without proppants is excluded from this account) and gravel packing of oil and gas reservoirs particularly in Western and Eastern Europe based to considerable amounts on the information which I have collected during my responsibility as Marketing Manager Europe for proppants (MADER 1987). Various comparisons of USA and Europe are used in the chapters concentrating on marketing and economical aspects for giving more general overviews and lead to the mainly technical sections that represent common analyses without reference to specific example areas and are valid for most parts of the world. Concerning the divisions focussing on Europe, many both general and specific comments derive from the experience of developing a marketing strategy for proppants that includes geological and reservoir engineering concepts in various European countries. Abundant technical information has been achieved through extensive literature studies as well as attention of topical symposia and congresses.

In addition, numerous other sedimentological, petrophysical, engineering and technical factors as well as points of finance, mentality and international cooperation which I have compiled during my former professional task as Area Oil and Gas Production Geologist as well as during my own independent scientific research on depositional environmental modelling, palaeogeographical constellation, ecological milieu evolution and diagenetical history of Lower Permian (Rotliegend) and Lower Triassic (Buntsandstein) fluvial, aeolian and lacustrine red bed sediments (part of which form some of the most important natural gas reservoirs onshore Middle Europe and offshore in the North Sea; MADER 1985 a) and violet calcrete palaeosols (MADER 1985 b) are incorporated into this summary report. The main work for establishing this compilation was carried out since I am an independent consultant for international petroleum geology and reservoir engineering.

1.1.5.2. Subdivision of the book into seven chapters

The review and status report of hydraulic proppant fracturing and gravel packing aspects in mainly USA and Europe in the present proceedings volume is divided into seven chapters. The first chapter discusses proppant selection (cf. tabs. 1 - 4), the second section presents marketing and economical aspects, and the third division reports technical and marketing relationships of Rotliegend and Carboniferous stimulation in Europe, thereby forming a transition from the economically oriented parts to the technically dominated chapters. The fourth chapter gives an account of technical and engineering aspects of hydraulic proppant fracturing, and the fifth section focusses on gravel packing. The sixth unit summarizes fracture and gravel pack monitoring. While the second division includes a marketing analysis of Western and Eastern Europe with some comparative assessment of the USA and the third unit represents an areal interdisciplinary case study of the Rotliegend basin being the most important tight gas reservoir province in Europe, the first and fourth to sixth sections are general technical compilations not only being valid for Europe and USA, but also for many other parts of the world, although many of the selected examples are illustrated by experience and progress achieved in Europe and USA.

The evaluation of proppant types and grain sizes is introduced by a brief account of historical development of hydraulic proppant fracturing. The summary of proppant choice offers general comments as well as particular references on the European oil- and gas-bearing formations requiring stimulation. The seventh chapter of the present book contains the bibliography which is divided into subject key classification and general reference list. A preliminary outline of some marketing and economical aspects of hydraulic proppant fracturing has been given by MADER (1987), MADER (1988 a) briefly illustrates a selection of the most important technical and reservoir engineering aspects, and a short interdisciplinary characteristic of the Rotliegend as the most significant tight gas pay horizon in Europe and thus the centre of hydraulic proppant fracturing activity in this area both offshore and onshore is provided by MADER (1988 b).

1.2. Historical development of hydraulic proppant fracturing

Hydraulic proppant fracturing is an established stimulation technique since abt. 40 years from the late 1940`s onwards. Hydraulic fracturing was introduced to the petroleum industry in 1947 - 1949 (CLARK 1948, FARREY 1953, WILSEY & BEARDEN 1954, KHRISTIANOVICH & ZHELTOV 1955, ZHELTOV & KHRISTIANOVICH 1955) and since the first commercial treatments in 1947 (well Klepper 1 in Hugoton gas field in Grant county/Western Kansas/USA; CLARK 1948, HOWARD & FAST 1970 a) through early 1949 (WATERS 1980) already more than one hundred thousand operations have been performed until 1955, more than half a million treatments have been executed during the first 20 years until 1968, and almost one million jobs have been carried out world-wide to date since innovation and inauguration of the method shortly after the end of the Second World War (HUBBERT & WILLIS 1957, WATERS 1980; HICKEY, BROWN & CRITTENDEN 1981; VEATCH 1983; VEATCH & MOSCHOVIDIS 1986). Hydraulic proppant fracturing has evolved during course of four decades from small treatments on single-zone completions to massive operations on multiple-interval installations (GRIFFITH & MADISON 1988). During the peak year of 1955, hydraulic stimulation jobs in the USA have been performed at a rate of 4,500 every month, whereas after 17 years of commercial use in 1965, the rate levelled off to abt. 2,000 actions per month (HOWARD & FAST 1970 a) and later even temporarily declined to abt. 1,500 operations per month.

Hydraulic proppant fracturing is a younger reservoir treatment method than acid fracturing which is common since more than 50 years (HOWARD & FAST 1970 a, WATERS 1980, ECONOMIDES 1986). Before the advent of hydraulic proppant fracturing, well stimulation in sandstone reservoirs was frequently done by detonating different types of explosives in the open hole which resulted in rubblization of the formation and radial propagation of fractures for a short distance from the wellbore, thus effectively increasing borehole diameter (WATERS 1980; cf. section 4.7.3.), and to minor amounts also by acidizing or acid fracturing (cf. section 4.5.1. and 4.5.5.), with all these pre-proppant fracturing methods, however, having had only limited success which at the bottom of the line was in many cases far from satisfying. Since its commercial introduction in 1947/1948, hydraulic fracturing has affected every facet of oil and gas industry by altering pipeline construction, changing production practices, killing the nitroglycerin oil well-shooting business, upsetting drilling companies, revolutioning well service spectrum, and giving rise to re-evaluation of many pay sections that had been previously bypassed and/or neglected (HOWARD & FAST 1970 a).

Hydraulic proppant fracturing is performed in various reservoir depth ranging from very shallow levels from abt. 500 ft (150 m) in case of muddy fine-grained sandstones, shales and chalks bearing oil and/or gas, to very deep storeys in excess of 20,000 ft (6,500 m) in case of tight sandstone, shale and coal seam gas pay horizons. After review of some general points, the following outline mainly focusses on discussion of aspects of limitations of natural sand, restriction of alternative propping materials before the invention of synthetic ceramic proppants, impact of early synthetic ceramic proppants, significance of advanced ceramic proppants, impact of resin-coated proppants, possible

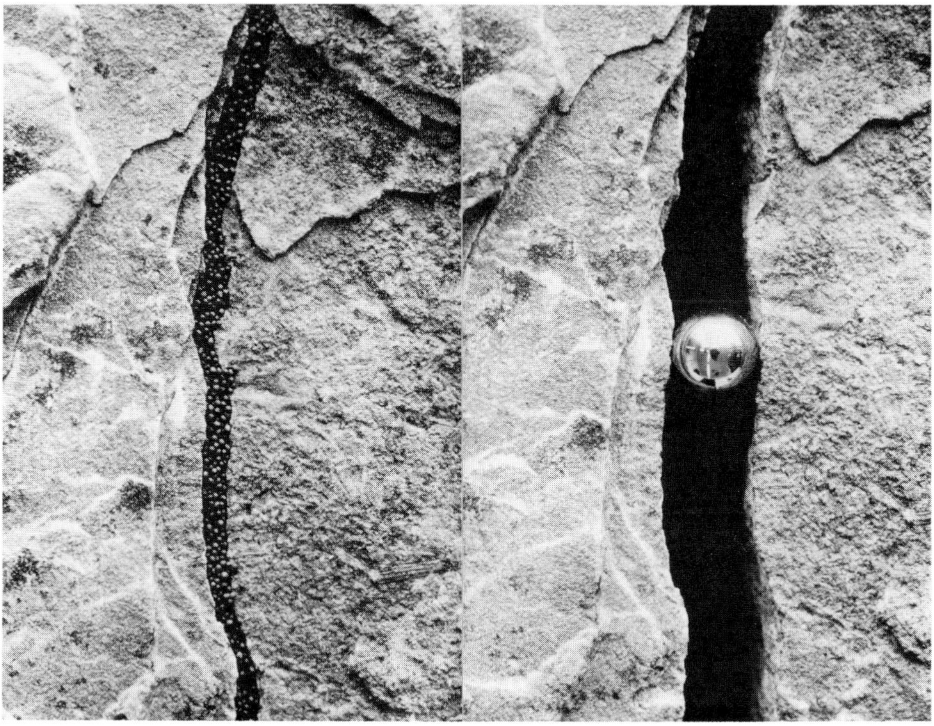

Fig. 1

Schematical illustration of supporting open hydraulically-induced fractures in sandstone reservoirs by infilling the cracks with spherical stress-resistant propping materials which keep the fractures open and prevent them from closing once hydraulic pressure exerted during the stimulation job subsequently declines and closure stress of the formation becomes effective. Typical fracture support by proppant multilayers with application of preferentially round grains of considerably larger grain size than that of the reservoir formation provides high-conductivity drainage paths or superhighways through the tight pay zone along which the hydrocarbons can flow linearly from the outward points of the reservoir to the borehole instead of radially as in undisturbed pays. Proppant closure stress resistivity has to be such that crushing of spherical particles in the fracture upon increasing closure stress with progressive reservoir depletion is avoided, with this goal being predominantly achieved by application of intermediate- and high-strength synthetic proppants instead of natural quartz sand which has only limited pressure resistivity. Another important consideration is resistivity against reservoir temperature and aggressivity of formation water which in some cases may be saturated brines. Selection of type and grain size of the proppants to be pumped into the crack has therefore a key role in hydraulic fracturing design, because in some large-scale treatments, proppant cost can reach up to two thirds of the total stimulation job expenses.

future innovation, and vertical vs. horizontal fracture orientation.

1.2.1. General aspects

Proper understanding of the necessity of application of various proppant types and the significance of proppant selection is facilitated by briefly introducing basic aims and procedures of hydraulic fracturing operations as well as sketching the evolution of main fracturing design concepts as follows.

1.2.1.1. Basic aims and procedures of hydraulic fracturing jobs

Hydraulic proppant fracturing plays a major role in enhancing petroleum reserves and production (VEATCH 1983). The hydraulic fracturing stimulation operation comprises pumping of a slurry consisting of high-viscosity fluid (gel, emulsion or foam) and proppant grains at high rates, various relative concentrations, and high pressures into the formation where due to progressively building up of wellbore pressure finally compressive earth stresses and rock tensile stresses are exceeded and as a consequence of reservoir failure a fracture is hydraulically created that propagates perpendicularly to the least principle earth stress (thus the fracture is usually vertical in wells exceeding 3,000 ft or 1,000 m depth; cf. section 1.2.8.). Following a sketch of the significance of fracture propping, some stimulation goals in view of reservoir productivity enhancement are illustrated.

1.2.1.1.1. Fracture propping

The fracture opened at high hydraulic pressure is kept open by support of hard pressure-resistant proppants which are infilled into the crack and inhibit its closure (cf. fig. 1), with thus a high-conductivity artificial drainage avenue or superhighway originating through which large volumes of hydrocarbons can easily and rapidly flow from the formation extremities to the wellbore in a linear concentrated and funnelled manner rather than previously in a radial dispersed and divided manner (HOWARD & FAST 1970 a, VEATCH 1983). Without support by proppants, hydraulically formed fractures tend to heal or return to their initial state after the parting pressure is released (CARROLL & BAKER 1979).

Hydraulic cracks have two wings extending in opposite directions from the well. Fracturing operations usually consist of successive pumping of four different fluid stages with different rheological properties comprising pre-pad, pad, proppant-laden carrier and displacement (LEE & DANESHY 1985). The whole stimulation treatment can be divided into four phases comprising fracturing, shut-in, clean-up and production (AHMED, SCHATZ, HOLLAND, JONES & GREENFIELD 1982). Marked advancements have been made in materials, equipment and treating techniques during the four decades since hydraulic proppant fracturing was introduced and first applied (WATERS 1980; cf. section 1.2.1.2.).

1.2.1.1.2. Stimulation goals

Hydrocarbon and water wells are fractured in order to increase the rate of fluid offtake and/or recoverable reserves and to decrease the pressure drop around the wellbore (SINCLAIR 1980). Production increase forecast achieved by hydraulic proppant fracturing treatment requires knowledge of the physical characteristics of the propped crack, fracture orientation, and the behaviour of the inserted proppants under closure stress and temperature. Well stimulation by fracturing can be performed either by providing a flow path through a damage zone around the borehole or by altering the current pattern in the reservoir. Small-volume fracture operations overcome wellbore damage (cf. section 4.8.3.) and restore productivity to the borehole by removing the zone of low-permeability or high pressure drop from the flow pattern. Large stimulation treatments provide a much greater surface for fluid offtake (or in other cases also injection; cf. section 4.8.7.), thereby intensifying a well by altering the reservoir flow pattern from radial to linear (cf. section 4.11.3.) which is the key result of a fracture job. Hydraulic proppant stimulation technology requirements are multifacetted and are becoming more complex as the reservoir formations get deeper, hotter and tighter (VEATCH 1983).

1.2.1.2. Evolution of main fracturing design concepts

The historical evolution of main fracturing design concepts is closely linked with the technological progress of treatment fluids and particularly proppants and comprises the successive periods of formation damage breakthrough and small-scale sand fracturing, tight gas reservoir MHF stimulation with synthetic proppants, and high-permeability pay zone fracturing. Some comments on significance of fracture parameter resolution are also offered (the application spectrum of proppants in hydraulic fracturing is summarized in tab. 1).

1.2.1.2.1. Formation damage breakthrough and small-scale sand fracturing

The first commercial hydraulic fracturing treatments in the late 1940`s were small-scale operations with less than 1,000 gal of total fluid quantity and with little amounts of sand up to abt. 5,000 lbs or even no propping material at all (DERBY & SMITH 1979). In the 1950`s, most hydraulic proppant fracturing jobs were carried out in order to break through the formation damage halo around the wellbore and did not aim on true pay stimulation farther penetrating into the virginal reservoir (SMITH 1987). Predominantly short fractures were created with fluid volumes of less than 20,000 gal and propped with small amounts of natural sand not exceeding abt. 10,000 lbs (ELY 1988). After an initial high phase during the first decade of technology availability when hydraulic fracturing received much promotion by the effect of being a brand-new fashion, the activity declined slowly in the 1960`s, partially as a consequence of the low oil price at a time when occasionally high-quality water was more expensive than oil. From the early 1970`s onwards, natural gas began to replace oil and other fuels in the energy supply spectrum (cf. section 2.2.1.6.4.) and tight gas fracturing started due to better understanding of the processes and as a consequence of availability of suitable materials and methods, and also proppant quantities and concentrations quickly increased.

In a similar way as in the early years of hydraulic proppant fracturing, even today many of the stimulation successes with remarkable production increases are due to removal of significant near-wellbore damage having been present prior to the treatment (ELY 1988; cf. section 4.8.3.).

1.2.1.2.2. Tight gas reservoir MHF stimulation with synthetic proppants

The most significant factors triggering the boost of hydraulic stimulation have been the first oil price jumps (cf. section 2.2.1.1.) and the invention of high-strength ceramic proppants (cf. section 1.2.4.1.). Tight gas fracturing comprising creation of long cracks intersecting the low-permeability reservoir and plugging of the fractures with high-conductivity proppants soon became more and more popular and quite quickly reached its golden years already in the late 1970`s/early 1980`s when particularly MHF operations were carried out abundantly especially in USA and parts of Europe. Since this peak the worldwide significance of tight reservoir fracturing declined to various degrees due to changes of the economical framework comprising oil price and US $ exchange rate, consolidation of the market after the initial burst and attractivity of the brand-new fashion, and partial saturation of the market when many treatments had been successfully carried out that had to wait so far until intermediate- and high-strength proppants were available, but is certainly keeping an important role in reservoir development especially with increasing shares of gas in the energy supply spectrum in near and far future.

1.2.1.2.3. High-permeability pay zone fracturing

From the early 1980`s onwards, moderate-permeability pay formation fracturing (cf. section 4.6.) was becoming more common and in the last years, hydraulic stimulation of high-permeability reservoirs with creation of sufficient conductivity contrast between crack and formation by high proppant saturations and wide fractures (cf. sections 4.3.3. and 4.5.1.1.2.) is progressively spreading out. In contrast to microdarcy wells where a 10 - 20 times productivity increase is achieved in optimum cases, millidarcy wells only reach a 2 - 5 times productivity increase which, however, is sufficient for feasible exploitation. While in hydraulic proppant fracturing history the main aim was on creation of long cracks in massive stimulation of tight gas pays, the trend in the future is more and more towards greater width and higher conductivity of shorter fractures particularly in moderate- to high-permeability reservoirs.

There are basically four hydraulic fracturing modes comprising confined vertical height and unrestricted horizontal crack extension (slope 1/4 - 1/3), stable height or fluid loss fracture with uncontrolled leakoff signaling hairline fissure opening, flow restriction leading to slurry dehydration or screen-out by proppant bridging (wellbore storage mode) and unstable height crack (PAI, GARBIS & HALL 1983; SMITH 1987).

1.2.1.2.4. Significance of fracture parameter resolution

The sensitivity of the predicted and actual results of hydraulic proppant fracturing treatments to the quality of the design depends on both relative operation cost and nature of experience in an individual formation (VEATCH 1983). In some areas, it may be the best and at the bottom of the line also cheapest approach to try a number of alternative fluids, proppants, job sizes and injection procedures in several wells of one group of reservoir exposures in order to arrive at a set of standard treatment parameters that provide acceptable results. This approach is particularly suitable to obtain relatively quick and effective results in regions where fracturing operations constitute a relatively small portion of the total drilling and completion expenses, especially applying for high-permeability reservoirs where short cracks are adequate.

In low-permeability formations where deeply penetrating fractures are required, however, resolution of the necessary hydraulic stimulation parameters is very important. In areas where MHF treatments can account for abt. half of the total well costs, the importance of fracturing is equal or greater than that of development drilling for increasing recoverable reserves (cf. section 2.3.), and here a high degree of crack parameter resolution is essential. Aspects of regional evolution of hydraulic proppant fracturing technology and stimulation concepts are also discussed by ROBERTS (1981).

1.2.2. Limitations of natural sand

In the early days of hydraulic reservoir stimulation, only natural quartz sand from different ancient and recent sources was available for propping of the fractures. In the first hydraulic fracturing jobs in the late 1940`s, no propping agents were used at all, while in the early 1950`s, river sand was simply added to the fracturing fluids (FRACFAX 1988 g). In the 1950`s and

1960's, different sand sources were exploited and rescreening of natural sand provided higher conductivities, with Ottawa and Texas sands having become the accepted industry standard fracturing sands since the 1960's (cf. section 1.3.1.1.). Following an outline of the closure stress resistivity interval of natural sand, some aspects of closure stress vs. matrix permeability and production rate are discussed.

1.2.2.1. Closure stress resistivity interval

20/40 mesh grain size sand is preferentially used due to its cheap acquisition as a by-product of glass and foundry industry sands (WATERS 1980). Other more or less common grain sizes depending on reservoir conditions comprise the coarser 16/20, 12/20 and occasionally also 8/12 and 6/10, as well as the finer 30/50, 40/70 and sometimes for special applications also 70/140 and 100 mesh (cf. tabs. 1 and 4 as well as section 1.3.). The low cost, relative abundance, good sphericity and low specific gravity of high-quality, specially selected sands have made it a good proppant for many hydraulic fracturing treatments (CUTLER, JONES, SWANSON & CARROLL 1981).

Crushing of the sand (HUITT, McGLOTHLIN & McDONALD 1959, TUNN 1971) at closure stresses between 3,000 and 5,000 psi (depending on sand quality) seriously limited the application of hydraulic fracturing in greater depths (most commonly in excess of 3,000 m (10,000 ft; CUTLER, JONES, SWANSON & CARROLL 1981) during the first years of industrial operation. Closure stress or compressive earth stress is the numerical difference between the bottom hole treating or fracturing pressure and the reservoir pore pressure or bottom hole flowing pressure, with the latter being the product of fracturing gradient times depth. According to falling reservoir pressure upon drawdown during course of production history, the closure stress increases particularly around the wellbore (ATTEBERRY, TUCKER & RITZ 1979; TUCKER 1979, SARDA 1981, CLARK 1983; cf. section 4.12.4.2.). Closure stress resistivity testing of sand is carried out by TUNN (1971) and SLUSSER & RIECKMANN (1976).

1.2.2.2. Closure stress vs. matrix permeability and offtake rate

The obvious correlation between closure stress and loss in fracture flow capacity in many earlier treatments carried out with natural sand as a propping agent indicates that the created fracture has partially or completely closed due to either embedment or crushing of the proppants (TUNN 1971, TUCKER 1979; cf. section 4.3.3.). The rate of closure stress is functionally related to original matrix permeability and offtake rate. Reservoirs of superior matrix permeability are better able to produce at commercial rates without severe near-wellbore pressure decline, whereas boreholes which exhibit very low flow capacities or where hydrocarbon withdrawal is performed at excessive rates experience rapid bottom hole pressure decline near the wellbore allowing fracture closure due to proppant collapse. The relationship between closure stress and proppant crushing is also reflected by pressure transient testing results. Therefore nowadays the generally accepted practice is to avoid the application of natural sand when closure stresses in excess of 4,000 - 6,000 psi have to be expected at any time of well life (CUTLER, JONES, SWANSON & CARROLL 1981; PAI, GARBIS & HALL 1983; cf. section 1.4.10.4.).

Prior to invention and market introduction of intermediate- and high-strength proppants, deep gas reservoir stimulation performance with natural sand has been improved by pumping high concentrations with the help of high injection rates and high fluid viscosities (HOLDITCH & ELY 1973; CALLANAN, CIPOLLA & LEWIS 1983; cf. section 4.3.2.3.). The pump rates and fluid viscosities necessary to pump high sand saturations, however, have sometimes been detrimental by causing a large pressure drop along the fracture and promoting excessive height growth into unproductive zones (NOLTE 1982).

1.2.3. Restrictions of alternative propping materials before the invention of synthetic ceramic proppants

With progressive technological evolution of hydraulic fracturing stimulation, more and more natural materials other than quartz sand as well as artificial products such as glass beads (stability boundary abt. 6,000 psi), iron and steel shots, broken walnut shells, rounded walnut hulls, alumina pellets, mined garnet and cordierite, and plastic beads or polymer spheres were tried with changing success and each of these novel proppants had its advantages and drawbacks, with only the high-strength annealed glass beads surviving until the early 1970`s (WATERS 1980; HICKEY, BROWN & CRITTENDEN 1981; WESTERN PETROLEUM 1985). Some features of glass beads, metal shots and walnut hulls which were the most widespread alternative propping materials before the invention of synthetic ceramic proppants are outlined as follows.

1.2.3.1. Glass beads

Glass beads have been commercially used for many years (DAVIS 1975), but the strength of the glass decreases greatly in the presence of brine (McGLOTHLIN, HUITT & McKAY 1963; HOWARD & FAST 1970 a, COOKE 1973; HICKEY, BROWN & CRITTENDEN 1981). The brittle glass beads also fail instantaneously into fine powder-size splinters (such as also collapse of zirconia-silicate proppants takes place; cf. section 1.4.2.2.) and not into larger fragments of half-, quarter- or smaller size as does quartz sand when exceeding the crush resistivity boundary of abt. 5,000 psi and crystalline alumina oxide and silicate proppants once their stability border is crossed (cf. section 1.4.2.1.).

Severe crushing of glass beads also occurs in hot water environment as well as in multilayer grain packages (FRACFAX 1988 g). In addition, glass beads can lose much of their compressive strength if being nicked while moving through the pumps, and glass beads which are scratched can be easier crushed than sand (HOLDITCH 1984). Metallized coatings consisting of brass or aluminum on glass beads also did not improve the situation. As a consequence of their catastrophical complete brittle failure in water and brine at elevated closure stress and temperature as well as of their premium price, glass beads were later completely withdrawn from the market in 1975 (SINCLAIR 1980).

1.2.3.2. Metal shots and walnut hulls

Malleable metallic materials are comparatively expensive and as they corrode when contacting high-salinity salt water, their use has been very limited (COOKE 1973, HOLDITCH 1984). Apart from corrosion, metal and steel shots are also too dense to be applied as a proppant in hydraulic fracturing (FRACFAX 1988 g), with transport and suspension properties in most of the fracturing fluids being unacceptable, and excessive segregation and banking in the fracture preventing adequate propping of the crack. Nutshells deform without crushing, but have also not turned out to be a suitable material, because walnut hulls are only deformable under low temperatures and load. Nutshells are also too light to be able to be packed with sufficient density within the fracture without elevated fluid loss which is desirable in gravel packing with lightweight materials (cf. section 5.3.3.), but is untolerable in hydraulic fracturing due to progressively increasing screenout risk. Walnut hulls are today used as fluid-loss additives rather than as proppants (SINCLAIR 1980), in a similar way as applies for 100 mesh sand (cf. section 1.4.11.2.1.). Garnet grains are too brittle, plastic pellets deform too rapidly, and fluidized coke is too weak (FRACFAX 1988 g).

Thus at the bottom of the line none of the aforementioned early-stage synthetic propping agents performed satisfactorily for usage in deep high-pressure reservoir fracturing treatments (COOKE & GIDLEY 1979) and all of them had more

PROPPANT TYPES AND GRAIN SIZES AND THEIR APPLICATION IN PETROLEUM RESERVOIR STIMULATION AND OTHER ENERGY INDUSTRY PROCESSES		TYPE					GRAIN SIZE									SHAPE	
APPLICATION		Quartz sand	Alumina silicate proppants	Alumina oxide and silicate proppants	Alumina oxide proppants	Zirconia-silicate proppants	6/9	8/12	12/20	16/20	20/40	30/50	40/70	70/140	100	Round	Angular
HYDRAULIC FRACTURING	Oil production	⬤	⬤	•			•	•	●	●	⬤					⬤	●
	Water production and injection	⬤	⬤	•			•	●	●	⬤	⬤					⬤	●
	Shallow gas production	⬤	⬤	●			•	•	●	●	⬤	•	•			⬤	●
	Moderately deep gas production	●	⬤	⬤		●			●	⬤	⬤	•	•			⬤	•
	Deep gas production	•	•	⬤	⬤	⬤			•	⬤	⬤					⬤	•
GRAVEL PACKING	Oil production	⬤	●		•	•			•	●	⬤	●	⬤	•		⬤	•
	Water production and injection	⬤	●		•	•		•	●	⬤	⬤	●	●			⬤	•
	Steam-drive enhanced oil recovery	•			⬤			•	●	●	⬤	•	•			⬤	
OTHER TREATMENTS	Fracture fluid-loss additive	⬤													⬤	●	●
	Geothermal well stimulation	•	•	●	⬤	•		•	•	●	⬤					⬤	●
	Coal seam degasification	●	●	●	●					•	⬤	●	●			⬤	●
	Foundry cupola furnace heat transfer				⬤				⬤							⬤	
	Mechanical gas purification				⬤				⬤							⬤	

Tab. 1

disadvantages such as high price, high density and lack of stability which overcompensated the benefits, with thus for many years natural quartz sand remaining to be the only really applicable material for hydraulic fracture support (HICKEY, BROWN & CRITTENDEN 1981).

1.2.4. Impact of early synthetic ceramic proppants

The decisive breakthrough in the technological evolution of hydraulic proppant fracturing was achieved in the late 1970`s with the invention of synthetic intermediate- and high-strength proppants of sintered bauxite (ceramic; COOKE 1976; COOKE, GIDLEY & MUTTI 1977; COOKE & GIDLEY 1979; cf. section 1.3.2.1.) and fused zirconia (SEPR 1983) composition which - in contrast to the low-strength sand - could resist to closure stresses up to more than 10,000 - 15,000 psi (cf. section 1.3.) and thus together with the development of high-capacity modern fracturing fluids opened vast possibilities of stimulation of deep high-pressure reservoirs. Some comments are offered as follows on sintered bauxite invention triggering fracturing boom and importance of proppant testing for quality improvement.

1.2.4.1. Sintered bauxite invention triggering fracturing boom

Sintered bauxite was first applied as propping agent for hydraulic treatment in the field in the USA in 1976 and as a consequence of spectacular stimulation results triggered a boom of deep well fracturing in the following years (HICKEY, BROWN & CRITTENDEN 1981), because it has enabled to achieve stimulation in many deep formations where in high-pressure regimes hydraulic fracturing had been impossible before, and better results of treatments in other shallower horizons (WATERS 1980). The main advantages of sintered bauxite over traditional propping materials are fine particle size, low porosity, high sphericity, good corrosion resistance, and ability to deform slightly under high loads (COOKE 1977; ATTEBERRY, TUCKER & RITZ 1979). Sintered bauxite has excellent conductivity up to 20,000 psi (139 MPa) closure stress (CUTLER, ENNISS, JONES & SWANSON 1985).

While at the beginning of artificial high-strength proppant application mainly angular synthetic bauxite-based proppants were made as by-products of abrasive manufacturing (HARBEN 1978, LARSEN & SMITH 1985; cf. section 1.3.5. and 1.4.11.3.2.), later almost perfectly round corundum/mullite proppants were presented which were produced in special plants, with roundness and sphericity, surface morphology, crystallite size, crush resistivity and size screening having been progressively improved during the years by optimization of grinding of raw materials, intermixing of flux media, sintering and fusing temperatures and control sieving. These ameliorations were the consequence of strongly increasing quality demand by the customers (oil and gas production companies through field service companies) and aggressive competition between the different proppant production companies.

1.2.4.2. Importance of proppant testing for quality improvement

The improvement of proppant properties and their matching with standard parameters was reinforced by quality and performance testing campaigns which were carried out by proppant manufacturers (HICKEY, BROWN & CRITTENDEN 1981; NORMAN, CIPOLLA & WEBB 1983; BROWN & MUCH 1986, COBB & FARRELL 1986, COMALCO 1986, STANDARD OIL PROPPANTS 1986, MUCH 1987, MUCH & PENNY 1987), independent consulting and research laboratories (AHMED, ABOU-SAYED & JONES 1979; SINCLAIR 1980; CUTLER, ENNISS, JONES & CARROLL 1983; DGMK 1983, 1986; CUTLER, ENNISS, JONES & SWANSON 1985; STIM-LAB 1986, PENNY 1987; PURSELL, HOLDITCH & BLAKELEY 1988), hydrocarbon production companies (BECQ, ROQUE & SARDA 1984; CHEUNG 1985; ROODHART, KUIPER & DAVIES 1986) and service companies (McDANIEL & WILLINGHAM 1978;

WENDORFF 1982, KIM & LOSACANO 1985; McDANIEL 1986, 1987, 1988; PARKER & McDANIEL 1987, McDANIEL & PARKER 1988; cf. section 1.4.). Fused zirconia-based proppants suffered in the early years from appendix-like subsize to mini-size round grains growing like noses and horns out of standard-size particles (thereby already a priori including quite some fines), but progressive quality amelioration by optimizing fusing temperatures, reaction time and raw-material processing has overcome this drawback.

1.2.5. Significance of advanced synthetic ceramic proppants

In the late 1970`s and early 1980`s, even two second-generation artificial proppant types (CUTLER, JONES, SWANSON & CARROLL 1981; JONES, CUTLER & SWANSON 1981; CUTLER & JONES 1982; CALLAHAN, CIPOLLA & LEWIS 1983; CUTLER, ENNISS, JONES & CARROLL 1983; SPARLIN & HAGEN 1983, FITZGIBBON 1984; CUTLER, ENNISS, JONES & SWANSON 1985) have been introduced to the hydrocarbon stimulation market which basically consist more or less of mullite and mullite-corundum mixtures, respectively (cf. section 1.3.), and are low-density and high-density intermediate-strength materials resisting to closure stresses up to 8,000 psi and 10,000 psi, respectively. These two intermediate-strength proppant types are filling the gap between low-strength natural sand on the one hand and high-strength synthetic bauxite or zirconia on the other hand in terms of specific gravity, conductivity, closure stress resistivity and price (cf. tabs. 1 - 4).

Both types of intermediate-strength proppants are lighter than high-strength sintered bauxite and thus allow better suspension in and transport by the fracturing fluids and placement in the crack by the carrier medium with less settling. The two advanced synthetic ceramic proppants are therefore permitting to achieve a better performance of the operation at lower cost and thus considerably broadening the spectrum of technical and economical application by providing a wider and more differentiated scope of job-specific proppant choice.

1.2.6. Impact of resin-coated proppants

With time, also the technique of resin-coating (SANTROL 1975, SINCLAIR & GRAHAM 1977; UNDERDOWN, DAS & SPARLIN 1980; GRAHAM, SINCLAIR & BRANDT 1982; UNDERDOWN & DAS 1982; SINCLAIR, GRAHAM & SINCLAIR 1983) of both natural sand and synthetic proppants was developed in order to achieve particular effects in special applications such as formation of consolidated proppant packages in the fracture by cementation of the grains by adhesion of the plastically flowing resin coatings at their contacts to provide stability and to minimize embedment, and relieving high stresses caused by grain-to-grain contacts, thereby improving the load-carrying capacity of the proppant pack (SINCLAIR & GRAHAM 1978, KANAT 1980; HICKEY, BROWN & CRITTENDEN 1981; POPE, WILES & PIERCE 1987).

The first field operations with resin-coated sand and proppants were carried out in 1976 and 1977, respectively (FRACFAX 1988 c, 1988 i). While for many years preference in deep well application was given to intermediate- and high-strength ceramic proppants, the beginning hydrocarbon industry recession in the early 1980`s draw more attention to proppant cost and the less expensive resin-coated sand which had been invented and patented almost at the same time as sintered bauxite and fused zirconia have been discovered and introduced began to be used in much larger quantities. Replacing sintered bauxite by resin-coated sand can result in abt. 50 % total proppant cost savings due to both cheaper pound price and higher pound volume of the latter product (SINCLAIR, GRAHAM & SINCLAIR 1983; cf. also section 1.4.12.4.2.). The following outline illustrates some points of resin types and functions, reactions of resin coatings to closure stress and temperature, and double resin coating combination.

1.2.6.1. Resin types and functions

Resin-coated sand and proppants have deformable resin pellicles that polymerize at elevated temperatures under closure stress to form a cohesive high-strength layer of permeable sandstone (SINCLAIR 1980). Curable material requires time at downhole temperatures to cure the thermosetting resin, whereas pre-cured proppants do not have to rely on downhole conditions to set the resin (CUTLER, ENNISS, JONES & CARROLL 1983). Aspects of reduction of point loading at grain contacts, and specific gravity and proppant price lowering are outlined as follows.

1.2.6.1.1. Reduction of point loading at grain contacts

The resin pellicles surrounding the proppants increase the contact area between individual grains which can thus withstand higher closure stresses before crushing. While uncoated proppants have point-like contacts comprising less than 10 % of grain diameter, resin-coated proppants exhibit broader contact ranges reaching up to 25 - 40 % of grain diameter (SINCLAIR, GRAHAM & SINCLAIR 1983). In addition to reduction of point-to-point loading caused by the plastic flow of the resin films, the resin pellicles bind and contain crushed fragments of the sand grains and thereby decrease the amount of mobile fines which improves conductivity, whereas on the other hand, deformation of the resin pellicles under load decreases conductivity (CUTLER, ENNISS, JONES & SWANSON 1985). In comparison to precured material, curable proppants have much broader contact zones by better flowing of the resin, lower initial conductivity due to deformed resin coatings, lower fines generation and higher conductivity at high closure stress.

1.2.6.1.2. Specific gravity and proppant price lowering

The type of resin which is precured around the sand grains is critical to the final proppant properties, as are amount of resin and extent of cure (UNDERDOWN & DAS 1982). The greatest advantage of resin-coated sand is that its bulk density is much less than that of intermediate- and high-strength proppants, and that its price is considerably competitive to the cost of the latter products. The much lower bulk density of resin-coated sands allows significantly better suspension in many different fracturing fluids than possible for heavier intermediate- and high-strength proppants, and the much cheaper material renders many operations economically more feasible and as a consequence of the elevated closure stress resistivity, resin-coated sand is an acceptable solution for many marginal reservoirs where expensive synthetic proppants are difficult to justify at least in pilot projects. The only artificial proppant type which can compete with resin-coated sand in terms of both technical performance and economical feasibility is intermediate-strength low-density alumina silicate proppant.

1.2.6.2. Reactions of resin coatings to closure stress and temperature

The deformable resin coating on the proppant grain surface spreads crushing and embedment loads over a greater area, thereby preventing crushing and resisting embedment (SINCLAIR, GRAHAM & SINCLAIR 1983). The tough resin film is chemically inert in crude oils, brines and most acids. The thin and tenacious resin pellicle comprises less than 5 % of the total proppant weight and has only a thickness of abt. 0.001 in. Resin-coated proppants consolidate under formation stresses and temperatures both within the reservoir and within laboratory test cells. Various effects of closure stress and temperature and the reactions of resin-coated proppants are illustrated as follows.

1.2.6.2.1. Closure stress

At high rates of application of closure stress, resin-coated proppants tend to be shock-loaded (cf. section 4.12.4.1.), because there is no time for the deformable pellicle to adjust and protect the proppant grains. In field application, however, closure stress returns over a period of hours after the termination of the fracture treatment (cf. section 4.12.2.2.), with this giving the deformable resin film ample time to adjust and protect the grains (SINCLAIR, GRAHAM & SINCLAIR 1983).

At extremely high closure stresses when also resin-coated materials are finally cracked or broken, most fragments are still attached to the tenacious resin envelope, and migration of any fines is minimized since the resin is securely bonded to most of the fragments of the split particles (this effect is much like that of the shatter proof safety glass used in car windshields; SINCLAIR, GRAHAM & SINCLAIR 1983). While resin curing at high reservoir temperatures and pressures in deep wells does not require external catalyzation, resin-coated proppants are also applicable in shallow wells with low formation temperatures and closure stresses once curing is initiated by an activator. Distinction can thus be made between chemical resin bonding at low temperatures and thermal resin bonding at high temperatures above abt. 150 oF.

1.2.6.2.2. Temperature

High temperatures are well tolerated by the resin coating, as it is tightly attached to the proppant surface by a bonding agent and does not come off (SINCLAIR, GRAHAM & SINCLAIR 1983). While high-temperature water or brine will attack fresh siliceous surfaces, the resin pellicle is non-reactive to any reservoir fluid system. Thicker resin films surrounding proppant grains lower specific gravity and increase transport capability of the proppants, with partially total specific gravities significantly lighter than that of natural sand being achieved when enveloping sand grains with resin films. Variations in proppant grain size as well as resin pellicle thickness can be made in order to optimize results of different types of treatments and can be tailored for optimum downhole performance. Tailing-in of resin-coated proppants in the near-wellbore region inhibits fracture evacuation and couples borehole and productive formation (cf. section 4.12.3.3.). Proppant flowback prevention by resin-coated tail-in is made regardless of type or amount of main proppant used in the fracture.

Resin-coated synthetic proppants have lower conductivities than uncoated ones when similar size distributions are tested (CUTLER, ENNISS, JONES & SWANSON 1985). Although containment of fines within the resin pellicles improves conductivity, resin envelope deformation under load decreases conductivity. Flow of resin films, particle rearrangement, elastic deformation, and particle crushing all tend to narrow fracture width as closure stress is increased. Resin-coated proppants are in laboratory testing best evaluated as a function of temperature because of the possible creep of the polymeric pellicles.

1.2.6.3. Double resin coating combination

The latest development is double resin-coating of proppants with an inner precured pellicle and an outer curable envelope (SANTROL 1986). Functions of inner precured and outer curable resin films as well as pressure and temperature reaction of resin pellicles are outlined as follows.

1.2.6.3.1. Functions of inner precured and outer curable films

The inner precured resin film provides additional strength of the material and the outer curable resin coating bonds the particles together, thus giving a combination of the benefits of a high-strength proppant with the advantages of a self-consolidating proppant and providing high strength, low density, and permanent bonding ability together at reasonable cost (SANTROL 1986). The inner resin pellicle has undergone tempering representing advanced polymerization process and imparts great strength to the grain for crush resistance. The outer envelope is a rather thin layer of curable resin that uses downhole temperature and fracture closure stress to form a multilayer proppant package which stays in place.

The curable resin coating also encapsulates and traps any broken grains or fines. The resin pellicle is so thin that it bonds together only upon contact stress in the fracture, in contrast to earlier submature thicker films which sometimes also bonded together in the wellbore. Maximum conductivity and strength of the proppant is achieved by 3 - 4 % coating of the material. Proppants with curable envelopes are the only ones which do not embed into the reservoir formation. In low-temperature pays (below 130 oF), an activator is required to initiate and speed-up bonding of the resin pellicles. Dual resin-coated sand and proppants were introduced to the hydrocarbon stimulation market in 1984 (FRACFAX 1988 i).

1.2.6.3.2. Pressure and temperature reactions of resin pellicles

Comparative closure stress resistivity testing reveals that uncoated ceramic proppants crush up to 6 % and more at 10,000 psi and 300 oF, whereas dual resin-coated proppants of any type at the same conditions do not suffer any crushing and fines generation (FRACFAX 1988 d). Closure stress resistivity testing at 275 oF and 8,000 psi illustrates that double resin-coated proppants of any type are much more stable in monolayers and at proppant loadings of 0.5 - 3.0 lbs/ft2 which correspond to frequent actual industry designs in contrast to the API recommendation of 4 lbs/ft2 for proppant testing (FRACFAX 1988 l). While high-temperature weakens all high- and intermediate-strength ceramic proppants, higher temperatures increase closure stress resistivity of resin-coated proppants.

Double resin-coated material is stronger than almost all uncoated proppant types, and the strength contrast becomes more pronounced as temperature rises (SANTROL 1986). Double resin-coated proppants can be successfully applied up to 25,000 ft reservoir depth and 600 oF bottomhole temperature and bonds together by means of both formation temperature and closure stress. While resin-coated sand can particularly in case of double pellicles successfully replace ceramic proppants up to 12,000 psi closure stress, dual resin-coated ceramic proppants outperform sintered bauxite and all other synthetic types in crush resistance up to closure stresses reached in the hitherto deepest oil and gas wells of the world with highest bottomhole pressures and temperatures.

1.2.7. Possible future innovation

In the last years, continuing quality demand and the need of cost containment triggered experimental approaches of developing even once again new proppant types of enhanced technical performance and better economical feasibility in order to adjust to the increasingly diversified application spectrum (CUTLER, JONES, ENNISS & SWARTZ 1983; CUTLER, ENNISS, JONES & SWANSON 1985; DGMK 1986), with some lines of possible proppant performance improvement including prototype materials based on cordierite, silicon carbide, and several ceramic oxides.

Experimental approaches even looked into the possibility of manufacturing hollow or porous proppants in an attempt to reduce specific gravity of high-quality material below that of sand (CUTLER, JONES, SWANSON & CARROLL 1981; CUTLER & JONES 1982; CUTLER, JONES, ENNISS & SWARTZ 1983), but reviewing the suite of materials and production techniques which have already been tested, it has to be doubted at the moment that it would be possible to develop an ultralight synthetic proppant with fluid suspension properties and certainly sufficient closure stress resistivity even better than sand or low-density intermediate-strength alumina silicate proppants at an acceptable cost (the same applies for similar approaches of gravel packing material; cf. section 5.3.3.2.). The following discussion outlines proppant particle composition types, specific gravity lowering, strength retainment of hollow spheres, and proppant fabrication methods.

1.2.7.1. Proppant particle composition types

CUTLER, JONES, SWANSON & CARROLL (1981) and CUTLER, ENNISS, JONES & SWANSON (1985) give an overview of experimental modelling of lightweight innovative proppant types. An ideal proppant for massive hydraulic fracturing would have a specific gravity of less than 2.0, be able to withstand a closure stress of 20,000 psi (139 MPa), be chemically inert in brine at temperatures up to 200 oC (390 oF), have perfect sphericity, and have a narrow grain size distribution at an acceptable price. Testing of various metals, ceramics, plastics and combinations of these three material groups has revealed that ceramics with or without polymer coatings represent the only sound approach to obtaining a low-cost, corrosion-resistant, low-specific-gravity and high-strength proppant.

In order to meet the specification for low gravity, proppants may be fabricated as solid spheres (fully dense), hollow globules (fully dense except for hollow core), porous spheres (uniform finely dispersed and interconnected microporosity throughout the bead), and composites of ceramic and polymeric materials. For fully dense solid spheres and hollow globules with a small center void, the contact zone stresses are assumed to be controlling, and thus the central empty space decreases specific gravity without affecting fracture pressure. Once a critical void size is exceeded, the limiting stress becomes that at the inside surface, and further increases in empty space diameter thus decrease fracture pressure.

1.2.7.2. Specific gravity lowering

Specific proppant gravity may be lowered significantly by fabricating a hollow bead without any loss in strength (CUTLER, JONES, SWANSON & CARROLL 1981; CUTLER, ENNISS, JONES & SWANSON 1985). Hollow spheres, however, are not always superior to porous globules, which is economically very important, because it is much easier to fabricate proppants with uniform porosity than to make hollow spheres. The excessive cost of raw materials may in many cases exclude the use of high-purity materials, although it is obvious that high-strength high-quality ceramics may be the answer to meeting the design goal of 2.0 g/cm3 specific gravity and 20,000 psi closure stress.

Several candidate proppant materials have been identified which show promise as lightweight proppants in hollow, porous or solid form comprising high-purity alumina, mullite, cordierite and high-strength porcelains, with densities ranging from 1.8 to 2.5 g/cm3. Fracture conductivity results for spray-dried alumina are very encouraging since it appears that the spray-dried - fluidized bed concept produces an easily sintered material which will have a conductivity comparable to sintered bauxite if proppants of similar particle size and shape are tested. Spray drying also permits the economical fabrication of hollow proppants through an agglomeration technique and coating process in the fluidized bed.

1.2.7.3. Strength retainment of hollow spheres

The most economical approach to strength retainment of hollow spheres appears to be spray drying of thin-walled small hollow globules into a fluidized bed, agglomerating several of these spheres together into a larger bead and then coating the agglomerate cluster with a solid envelope. Sintered bauxite produced in this manner would have a specific gravity near that of sand, but retain its high strength, although the process is still in experimental stage. Both dense and hollow alumina spheres have been fabricated by spray fluidization and countercurrent spray drying (CUTLER, ENNISS, JONES & SWANSON 1985).

Spray drying into a fluidized bed is also a viable method for producing porous proppants characterized by disseminated microporosity throughout the grain. Lightweight experimental hollow alumina proppants have lower crush resistance than bauxite but are substantially stronger than sand although lower in density, with non-concentric and multiple voids within individual proppants being the main reasons for the hitherto inadequate strength. Selection of proper spray-drying parameters such as dryer design, nozzle type, additives, temperature and flow rates are important for optimization of hollow proppant fabrication by spray drying.

1.2.7.4. Proppant fabrication methods

A wide variety of fabrication methods can be used to make proppants including mix-pelletization, spray drying, droplet generation and coating (CUTLER, ENNISS, JONES & SWANSON 1985). While traditional proppants are predominantly manufactured by mix-pelletization and occasionally also by droplet generation, the most promising experimental approach comprises coating of cheap ceramic materials and clustering of several smaller spheres to larger particles. Intermediate-stage low-density proppants are mullite proppants fabricated by mix-pelletization and carbon-coated alumina microspheres produced by droplet generation. On laboratory scale, hollow proppants have been fabricated with crush resistance values between sand and intermediate-strength proppants. Manufacturing of either hollow beads or globules with finely dispersed interconnected porosity appears to have the potential for combining adequate strength with low density. It is therefore believed that hollow proppants are the key to obtaining lower specific gravity proppants at competitive costs.

1.2.8. Vertical vs. horizontal fracture orientation

As a consequence of the availability of various proppant types of different mechanical and chemical properties, mineralogical composition and possibilities of industrial application that are reflected in different prices, numerous hydraulic proppant fracture operations have been carried out during the last ten years, with major influencing factors on design and execution being not only proppant performance with respect to their resistivity against closure stress and aggressive formation fluids and their suitability of application in the geological and reservoir engineering framework, but also economical aspects of hydrocarbon price (cf. section 2.2.1.), proppant cost (cf. section 2.2.2.), US $ exchange rate (cf. section 2.2.1.3.) and commercial gas production rates, as well as foreign trade and financing particularly in Eastern European countries (MADER 1987; cf. section 2.4.4.).

Other important points which have to be considered are minimum reservoir properties required for economically profitable stimulations (DIETZEL & ERNST 1987; cf. section 3.6.1.2.). Attention has also to be paid to the question of vertical vs. horizontal fracture propagation direction. The brief outline as follows sketches some aspects of fracture orientation and reservoir depth as well as geotectonical setting, borehole/fracture intersection and reservoir drainage, and coexistence and transition of horizontal and vertical fractures.

1.2.8.1. Fracture orientation and reservoir depth

As a consequence of increasing overburden stress with depth and therefore shifting relationship between horizontal and vertical principal stress, there is a correlation between fracture propagation direction and reservoir depth. The cracks created by hydraulic stimulation treatments are oriented vertically to least principal stress (which is approximately equal to the fracture closure pressure) in the earth`s crust (HUBBERT & WILLIS 1957, DUNLAP 1963, HAIMSON & FAIRHURST 1969, HOWARD & FAST 1970, VEATCH 1983). As a consequence of the geo-stress field, above abt. 3,000 ft (1,000 m) the cracks are generally horizontal, whereas below that boundary depth predominantly vertical fractures exist (SMITH 1965, McGARR & GAY 1978, AGUILERA 1983; ROBINSON, HOLDITCH & WHITEHEAD 1986; other authors assume the borderline between horizontal fracture origin and vertical fracture growth at least in some areas to be already in a depth of abt. 1,000 - 2,000 ft or 300 - 700 m; VEATCH 1983).

1.2.8.2. Fracture orientation and geotectonical setting

Concerning geotectonical setting, the least principal stress is mainly horizontal in tectonically relaxed areas which are only subjected to subsidence and extensional (normal) faulting, and fracturing therefore occurs along vertical planes, whereas in tectonical stress fields where also compressional (inverse or thrust) faulting, shearing and folding takes place, the (externally generated) horizontal stresses may be greater than the overburden stress and due to the least principal stress being vertical, horizontal fractures have to be expected (stress effects on crack orientation and pressures required to produce vertical and horizontal fractures are also discussed by HUBBERT & WILLIS 1957). Some comments are offered as follows on extensional vs. compressional tectonics as well as on comparative significance of horizontal and vertical fractures.

1.2.8.2.1. Extensional vs. compressional tectonics

Natural joint origin is associated with diastrophism during folding and faulting where cracks are generated along the line of the fault which in turn produce a zone of dilatance (cf. section 4.8.1.3.1.); deep erosion of the overburden permitting the upper parts to expand, uplift and fracture through planes of weakness; and volume shrinkage as in case of cooling of igneous rocks (cf. plate VI/8 and IX/5-6) and desiccation of sediments (cf. plate VI/1-7 and section 4.8.8.1.). Thrust faulting is a possibility of creating horizontal and low-angle inclined cracks (cf. plate XI/3) also in shallow depth (AGUILERA 1983). Horizontal fractures, however, often tend to close owing to the effect of overburden. Formation heating by steam injection is a significant means of fundamentally changing the in-situ stress state in shallow reservoirs from a regime favouring origin of vertical cracks in cold environment to a framework supporting propagation of horizontal fractures in hot milieu (HOLZHAUSEN, CARD, RAISBECK & DOBECKI 1985; cf. section 4.9.4.).

1.2.8.2.2. Comparative significance of horizontal and vertical fractures

As most of the hydrocarbon-bearing formations which are stimulated are in depths exceeding the boundary of abt. 3,000 ft (1,000 m) and are in basins with only moderate tectonical stress undergoing still slight subsidence and having been subjected to normal faulting with only minor tilting, shearing and compression, the discussion as follows concentrates on vertical cracks intersecting the reservoir rocks. Aspects of horizontal fractures are emphasized by CRAWFORD & LANDRUM (1956), WAHL (1956, 1965), LANDRUM & CRAWFORD (1957), MORRISSON & HENDERSON (1960), HARTSOCK & WARREN (1961), WAHL & CAMPBELL (1963), LOWE & HUITT (1966), SUN (1969), GRINGARTEN & RAMEY (1974), ERDLE (1981) and SUNG & ERTEKIN (1987). EVANS (1983) and EVANS & HOLZHAUSEN (1983) also comment on development of shallow hydraulic fractures. Ideally, horizontal stress is independent from azimuth in isotropic rocks which, however, are hardly developed in nature. In anisotropic rocks which are the real cases, crack orientation runs perpendicularly to the least horizontal stress (LEICHT 1985). Industrial practice has shown that horizontal fractures constitute a relatively low percentage of the situations experienced to date (VEATCH 1983) and are rather special cases than being more common.

1.2.8.3. Borehole/fracture intersection and reservoir drainage

In terms of reservoir drainage, the effect of horizontal fractures in vertical wells can be compared with that of vertical cracks crossing highly-deviated or horizontal boreholes (cf. chapter 4.11.). The restricted cross-section of the fracture plane being in communication with the wellbore trajectory that perforates the horizontal crack like a slice in contrast to the continuous and full fracture height exposure in case of ideal vertical cracks results in only limited drainage through the fracture and very restricted enlargement of effective wellbore diameter. Therefore except of special applications, preference is always given to vertical cracks for the reason of better reservoir depletion due to longer borehole/fracture communication distance if any influence can be exerted on crack orientation. Some extraordinary indications of the suitability of horizontal fractures, partially in combination with vertical cracks, are outlined in section 4.9.4.2.

1.2.8.4. Coexistence and transition of horizontal and vertical fractures

Horizontal fractures start in many cases to propagate as vertical cracks and then reorient themselves to become horizontal and to run perpendicularly to the least principal stress (DANESHY 1973 b), with initial vertical crack formation being substituted by following horizontal fracture origin which then becomes dominant and thereby severely limits further growth of the vertical crack (WOOD, PARKIN, YOTAM, HANSON, SMITH, ABBOTT, COX & O`SHEA 1983). In some cases, a horizontal fracture and a vertical crack may even keep open and propagate simultaneously, with the rate of growth of the horizontal fracture being greater than that of the vertical crack. Mushroom-type fractures consisting of a primary vertical crack topped by a secondary horizontal fracture like a cap (EVANS, HOLZHAUSEN & WOOD 1982; FITZ-PATRICK, KARR & O`SHEA 1985; SMITH, REN, SORRELS & TEUFEL 1985) also occur occasionally in depth ranges that are transitional for both basic crack types (cf. section 6.2.3.4.) as well as due to changing closure stress conditions by steam-drive enhanced oil recovery operations (cf. section 4.9.4.2.). Networks of orthogonal fractures consisting of two interlocking systems being oriented perpendicularly to each other can be created by altered stress fracturing (cf. section 4.9.4.1.3.1.). Horizontal fractures in shallow depth are recognized by parting gradients greater than 1.

Proppant selection (STEANSON, ELBEL & WENDORFF 1979; CLARK 1983, DOWELL SCHLUMBERGER 1985, MONTGOMERY & STEANSON 1985, NORMAN & FAST 1985, NORTON-ALCOA PROPPANTS 1985, PHILLIPS & ANDERSON 1985, WESTERN PETROLEUM 1985, STANDARD OIL PROPPANTS 1986; cf. section 1.4.) for hydraulic fracture treatments is influenced by all the mentioned features. Its key significance in operation design as a consequence of the considerable share which proppant cost has in the total MHF expenses requires detailed discussion of the individual proppant properties and possibilities of proppant application as follows.

MINERALOGICAL AND CHEMICAL COMPOSITION OF PROPPANTS FOR HYDRAULIC FRACTURING AND GRAVEL PACKING			PROPPANT COMPOSITION													
			MINERALOGY									CHEMISTRY				
			CRYSTALLINE							AMORPHOUS						
PROPPANT TYPE																
PROVENANCE	STRENGTH	TYPE	Quartz	Corundum	Mullite	Cristobalite	Spinels	Others	Zirconium oxide	Silica glassy phase	Other amorphous phases	SiO_2	Al_2O_3	TiO_2	Fe_2O_3	ZrO_2
natural	low-strength	quartz sand	98 - 100	-	-	-	-	-	-	-	-	90 - 100	-	-	-	-
synthetic	intermediate-strength	low-density alumina silicate proppants	-	-	30 - 70	30 - 70	-	>2	-	-	-	46 - 50	46 - 51	1.5 - 2.5	0.7 - 1.1	-
		high-density alumina oxide and silicate proppants	-	30 - 70	30 - 70	-	>8	>2	-	-	>10	12 - 17	73 - 78	2.5 - 4.0	4.5 - 7.0	-
	high-strength	alumina oxide proppants	-	>70	<20	-	<15	<10	-	-	-	3.0 - 6.0	85 - 90	3.0 - 4.0	4.0 - 7.0	-
		zirconia-silicate proppants	-	-	-	-	-	-	41	59	-	44.0	7.5	6.5	0.3	41.5

Tab. 2

1.3. Proppant types and grain sizes

Presently basically five different proppant types for hydraulic fracturing and gravel packing are available in the hydrocarbon stimulation market in various grain sizes and for different prices from several manufacturers : natural quartz sand, synthetic intermediate-strength low-density alumina silicate (ceramic) proppant, intermediate-strength high-density alumina oxide and silicate proppant, high-strength high-density bauxite proppant and high-strength low-density zirconia-silicate proppant (arranged in an order of progressively increasing specific gravity, closure stress resistivity and proppant cost; the main characteristics are compiled in COMALCO 1983, KIM 1983, SEPR 1983, HOLDITCH 1984, DOWELL SCHLUMBERGER 1985, NORMAN & FAST 1985, NORTON-ALCOA PROPPANTS 1985, WESTERN PETROLEUM 1985, BROWN & MUCH 1986, HALLIBURTON 1986, MESSINA 1986, STANDARD OIL PROPPANTS 1986; cf. tabs. 2 - 4). The current proppant prices are increasing in the mentioned sequence from cheap natural quartz sand via moderately expensive synthetic alumina oxide and silicate proppants to expensive artificial zirconia-silicate proppants.

Following an overview of the provenance of natural sand and synthetic proppants and the discussion of the various mechanical and chemical proppant properties, aspects of proppant selection for application under a spectrum of different reservoir conditions are evaluated. The section below includes the outline of mineralogical composition and closure stress resistivity, chemical composition and acid solubility, grain sizes and conductivity per US $, and grain shapes of the presently commercially available proppant types.

1.3.1. Provenance of sand and proppants

Natural sand and synthetic proppants are exploited and produced, respectively, in various places in the world. While traditionally both types of materials chiefly derived from the USA, in the last decade increasing contributions are also coming from other countries and continents.

1.3.1.1. Natural sand

High-quality sands (known as Ottawa Sand and other sands with regional names; WESTERN PETROLEUM 1985) are mined traditionally in the USA, with the main provenances being Jordan Formation in Minnesota, Hickey Formation in Central Texas, St. Peters Formation in Ottawa/Northern Illinois, and Galesville and Ironton Formations in Wisconsin, and smaller quantities also deriving from Eastern Arizona, Central Colorado, Ohio and Northern Carolina (WATERS 1980). Many of the high-quality sands represent ancient aeolian dune sands where effective windblown transport, sorting and deposition has given rise to high quartz content, negligible amount of fines, high grain roundness and sphericity, and very good sorting with narrow range of grain size distribution. Minor and negligible or larger and significant differences exist between excellent quality sands and inferior-quality sands depending on provenance and composition, with the latter group being restricted in application to shallower reservoirs with lower closure stress regimes, whereas the former including first of all Ottawa Sand had in earlier years prior to invention of intermediate- and high-strength ceramic proppants also been applied in deep well fracturing.

Brady Sand or Hickory Sand coming from Texas is the most frequently applied proppant in oil and gas industry, with millions of tons having been pumped during the last decades, because in contrast to recent years, the distribution of the top-quality Ottawa Sand formerly suffered from limited supply (SINCLAIR 1980) and shortages in meeting the enormous demand of the boosting oil and gas industry after the first oil price jump in the early 1970's required the delivery of every sand available almost regardless of quality in the hottest periods of hydrocarbon well stimulation during the golden oil and gas years. Other natural sands of differing quality come from various European countries and are deriving in the Middle and Far East also frequently from local sources.

While the first sand used as fracture proppant in the first treatments in 1947 and following years was just taken out of Arkansas river and pumped downhole, later increasing quality demand required more and more sophisticated mechanical processing of the natural sands for the purposes of hydraulic fracture proppant and especially sand control packing gravel where quality standards demand primarily for narrow grain size distribution, high content of preferably monocrystalline quartz, and excellent roundness and sphericity.

1.3.1.2. Synthetic proppants

Intermediate-strength alumina oxide and silicate (ceramic) proppants are mainly manufactured in the USA, with minor amounts deriving from Brazil and only very limited quantities of high-density material having so far been produced in Hungary that have not yet reached the stage of economical application beyond local field tests of tail-in nature. High-strength sintered bauxite proppants are not only made in the USA, but slowly increasing quantities are coming from Australia and Brazil, and trial runs on more or less laboratory scale without yet reaching field application have been carried out in China (cf. section 2.4.6.4. and tab. 3).

While high-strength bauxite proppants in USA, Brazil and China are exclusively made by mix-pelletization and sintering of the artificially formed pellets, the Australian bauxite proppants derive from screening and conditioning of natural spheres coming from a pisolithic bauxite deposit which following sieving and chemical processing are sintered (cf. section 1.4.9.1.3.). High-strength fused zirconia-silicate proppants finally are manufactured in France. Several other suppliers have so far only submitted laboratory-scale samples for proppant testing. As a result of the wide variety of proppant types that are available today in the market from various producers, the American Petroleum Institute (API) has established recommended test procedures for several proppant properties to distinguish quality and usefulness of each proppant (AMERICAN PETROLEUM INSTITUTE 1985).

1.3.2. Mineralogical composition and closure stress resistivity

Concerning mineralogical composition and closure stress resistivity of proppants, distinction can mainly be made between alumina oxide and silicate proppants and zirconia-silicate proppants (cf. tab. 2).

1.3.2.1. Alumina oxide and silicate proppants

In terms of mineralogical composition, natural sand consists of macrocrystalline quartz which has no crystallographical cleavage, but a shelly breakage into both larger fragments and smaller splinters at higher closure stresses exceeding 3,000 - 5,000 psi. Resin-coated sand may be stable up to 7,000 - 8,000 psi closure stress (cf. section 1.2.6.). Intermediate-strength low-density alumina silicate (ceramic) proppants are mainly built up of micro- to cryptocrystalline mullite and crystobalite deriving from clay as raw material. Intermediate-strength high-density alumina oxide and silicate proppants are composed of micro- to cryptocrystalline corundum and mullite coming from a raw material mixture of bauxite and clay.

High-strength high-density bauxite proppants are almost exclusively consisting of corundum with little quantities of mullite deriving from sintering of nearly pure bauxite (cf. section 1.4.9.1.). As a consequence of composition of interlocking tiny crystals which give much better toughness than a glassy material and provide some deformability under load prior to failure and crushing, the three latter proppant types are predominantly breaking into halves, quarters and successively finer fragments when being subjected to closure stresses beyond 8,000 psi (3 % crush of intermediate-strength low-density alumina silicate proppants); 10,000 psi (4 % crush of intermediate-strength high-density alumina oxide and silicate proppants); and 15,000 psi (5 % crush of high-strength high-density bauxite proppants), with reasonable amounts of finer splinters only originating during advanced failure as a consequence of exposition to excessive closure stresses (cf. section 1.4.2.). Intermediate-strength proppants can be used up to closure stresses of 12,000 - 13,000 psi and since higher pressures are not occurring that frequently, this proppant type suite will be in focus for application during the coming years. Sintered bauxite has excellent conductivity up to 20,000 psi (139 MPa) closure stress (CUTLER, ENNISS, JONES & SWANSON 1985).

Natural sand becomes already unstable at 3,000 - 5,000 psi closure stress depending on quality and grain size, but its resistivity can be extended up to more than 10,000 psi by resin enveloping, and similarly, resin-coated sintered bauxite can tolerate closure stresses up to 30,000 psi (SINCLAIR, GRAHAM & SINCLAIR 1983; cf. section 1.2.6.).

1.3.2.2. Zirconia-silicate proppants

High-strength low-density zirconia-silicate proppants finally are built up of micro- to cryptocrystalline monoclinic and cubic zirconia that is embedded in a modified alumina silicate amorphous or glassy phase originating by fusing of zircon and quartz. The latter brittle fused product (cf. section 1.4.9.2.) behaves like a glass when subjected to higher pressures and thus almost at once catastrophically fails by breakage into exclusively tiny splinters (cf. section 1.4.2.) at closure stresses beyond abt. 15,000 psi (2.5 % crush; all crush values refer to proppants of 20/40 mesh grain size; the data in this chapter are compiled from BJ-HUGHES 1982, COMALCO 1983, KIM 1983, SEPR 1983, NORTON-ALCOA PROPPANTS 1985, BROWN & MUCH 1986; MESSINA 1986, 1987; STANDARD OIL PROPPANTS 1986). Thus collapse of zirconia-silicate proppants is physically similar to that of glass beads which also crush at once into fine powder (cf. section 1.2.3.1.) although taking place at a much more elevated closure stress niveau in case of zirconia spheres. Up to 14,000 psi (97 MPa), there are essentially no conductivity differences between zirconia-silicate proppants and sintered bauxite proppants (CUTLER, ENNISS, JONES & SWANSON 1985).

In contrast to the aforementioned boundary closure stresses, ROBINSON, HOLDITCH & WHITEHEAD (1986) conclude their evaluation of fracture damage and its prevention with recommendation of maximum allowable closure stresses of 4,000 psi for sand, 7,000 psi for intermediate-strength proppants and 10,000 psi for high-strength proppants. In comparison with most results of proppant testing, however, the values for the latter two proppant types are assessed as being considerably too pessimistic, but certainly when designing a suitable proppant type for long production history up to 20 - 30 years, it should be preferred to go absolutely on the safe side and to better select a more expensive proppant type that is resisting to higher closure stresses rather than to risk fracture damage due to proppant failure under conditions of excessive drawdown in the late stage of reservoir depletion.

1.3.3. Chemical composition and acid solubility

Concerning chemical composition and acid solubility, distinction has to be made between natural sand and other proppants containing free silica in larger amounts, and proppant types including only minor shares of free silica or having almost totally non-silica constituents (cf. tabs. 2 and 3).

MECHANICAL PROPERTIES AND PRODUCTION/MANUFACTURING OF PROPPANTS FOR HYDRAULIC FRACTURING AND GRAVEL PACKING			MECHANICAL PROPERTIES				PRODUCTION/MANUFACTURING									
PROPPANT TYPE							TYPE			AREA						
PROVENANCE	STRENGTH	TYPE	Specific gravity (g/cm³)	Bulk density (lbs/ft³)	API crush test : % fines at 10,000 psi (20/40 grain size)	Closure stress resistivity and field application boundary (psi)	Natural	Sintering	Fusing	USA	Europe (France, Hungary, etc.)	Australia	South America (Brazil)	Far East (China)	Africa
natural	low-strength	quartz sand	2.65	96.0 - 103.0	40.8 - 59.0	3,000 - 5,000	●			●	●	•	•	•	●
synthetic	intermediate-strength	low-density alumina silicate proppants	2.70 - 2.75	99.0 - 102.4	4.3 - 9.5	8,000		●		●			•		
		high-density alumina oxide and silicate proppants	3.15 - 3.27	112.5 - 116.8	3.5 - 6.1	10,000 - 12,000		●		●	•		•		
	high-strength	alumina oxide proppants	3.60 - 3.80	130.0 - 144.0	1.5 - 5.0	15,000	•	●		●	•	●	•	•	
		zirconia-silicate proppants	3.15 - 3.17	106.0 - 120.0	0.3 - 4.6	15,000			●		●				

Tab. 3

1.3.3.1. Chemical composition

Natural quartz sand is almost exclusively consisting of SiO_2 and thus has a specific gravity of 2.62 - 2.65 g/cm3. Synthetic intermediate-strength low-density alumina silicate proppants are built up of 46.0 - 50.0 % SiO_2, 46.0 - 51.0 % Al_2O_3, 0.7 - 1.1 % Fe_2O_3 and 1.5 - 2.5 % TiO_2, with the specific gravity being 2.70 - 2.75 g/cm3 which is only slightly more than that of sand. Intermediate-strength high-density alumina oxide and silicate proppants are composed of 12.0 - 17.0 % SiO_2, 73.0 - 78.0 % Al_2O_3, 4.5 - 7.0 % Fe_2O_3 and 2.5 - 4.0 % TiO_2, and have a specific gravity of 3.15 - 3.27 g/cm3 which is considerably higher than that of sand and intermediate-strength low-density alumina silicate proppants and is partially similar or even more elevated than that of high-strength low-density zirconia-silicate proppants, but still much lower than that of almost pure bauxite-based high-strength high-density proppants. Intermediate-strength high-density alumina oxide and silicate proppants are thus representing a transitional proppant type linking several classes of products both in terms of physical properties and practical applications which is underlined by the fact that this proppant type is the hitherto most frequently utilized synthetic proppant for oil and gas reservoir hydraulic stimulation.

High-strength high-density alumina oxide proppants are typically made up of 3.0 - 6.0 % SiO_2, 85.0 - 90.0 % Al_2O_3, 4.0 - 7.0 % Fe_2O_3 and 3.0 - 4.0 % TiO_2, with the specific gravity of 3.6 - 3.8 g/cm3 being the highest of all synthetic proppants. High-strength low-density zirconia-silicate proppants consist typically of 44.0 % SiO_2, 7.5 % Al_2O_3, 0.3 % Fe_2O_3, 6.5 % TiO_2 and 41.5 % ZrO_2, with its specific gravity of 3.15 - 3.17 being even lower than that of some intermediate-strength high-density alumina oxide and silicate proppants (the data are compiled from BJ-HUGHES 1982, COMALCO 1983, KIM 1983, SEPR 1983, NORTON-ALCOA PROPPANTS 1985, BROWN & MUCH 1986, COBB & FARRELL 1986; MESSINA 1986, 1987; STANDARD OIL PROPPANTS 1986).

1.3.3.2. Acid solubility

In contrast to quartz sand and glass beads which are quite soluble in hot formation brines, the synthetic alumina- or zirconia-based proppants are considerably more resistant. The solubility in 12/3 HCl/HF acid (30 min at 150 oF) is for intermediate-strength low-density alumina silicate proppants 1.5 - 2.0 % weight loss, intermediate-strength high-density alumina oxide and silicate proppants 3.3 - 4.1 %, high-strength high-density alumina oxide proppants 2.0 - 3.0 and high-strength low-density zirconia-silicate proppants 2.5 - 3.9 %, in contrast to natural sand where the solubility range is 0.5 - 7.0 % depending on mineralogical and crystallographical composition and structure of the quartz grains, and being higher in alcaline environment (the data are taken from the aforementioned sources. CHEUNG (1985), however, comes to the opposite conclusion that especially alumina oxide and silicate proppants and also sintered bauxite are seriously dissolved and mechanically weakened in acid environment, whereas natural quartz sand is the most resistant material in this particular chemical milieu, thereby having strong impact on planning of acidization of fractures or even joint operation of acid and proppant fracturing (cf. also section 4.5.1., 4.5.5. and 4.8.9.).

1.3.4. Grain sizes and conductivity per US $

The various proppant types are applied according to their different resistivity against crushing in different depths and their ability to withstand aggressive formation brines of different salinity, with proppant selection not only being influenced by aspects of conductivity and permeability as well as long-term stability of the proppant package (which in big fracture jobs is the major share of the whole cost of the operation), but also by the price which raises with increasing closure stress resistivity of the proppant. Careful economical evaluation is therefore necessary in every individual case by an assessment of conductivity per US $ (cf. section 1.4.12.2.) that is obtained when choosing a more expensive proppant type of higher stability and thus better long-term performance in view of permeability. Conductivity is potentially accentuated by selecting a coarser grain size, compared with increasing pumping cost with rising specific gravity and grain size of the proppant, increasing screenout risk when using heavier and coarser proppants, and increasing equipment abrasion by harder proppants (STEANSON, ELBEL & WENDORFF 1979; HOLDITCH 1984, DOWELL SCHLUMBERGER 1985, MONTGOMERY & STEANSON 1985, WESTERN PETROLEUM 1985, HALLIBURTON 1986). Some aspects of conductivity improvement by grain size selection vs. increasing screenout risk and commercially available proppant grain sizes are briefly outlined as follows.

1.3.4.1. Conductivity improvement by grain size selection vs. increasing screenout risk

Grain size selection of proppants also has considerable significance for improving the resulting conductivity in the fractured reservoir. The larger the grain size and the higher the specific proppant gravity, the greater is the risk of material screenout being an only reduced uptake of the formation with respect to design, with a screenout being able to be assessed either negatively as indicating partial failure or positively as testifying to achievement of maximum proppant saturation and uptake potential of the formation. Screenouts may be beneficial by indicating that the fracture is fully packed from the tips backwards in direction to the wellbore and should have a high conductivity, whereas on the other hand, low quantities of injected proppants may reflect uneven proppant distribution in the crack and discontinuous concentration in the surroundings of the well, thereby not matching the attempted fracture length (KOHLHAAS 1982). Aspects of screenout failure are also discussed in sections 4.2.2.2.3. and 6.2.4.2.1.

Some more advanced techniques such as tip screenout fracturing (SMITH, MILLER & HAGA 1985; cf. section 4.5.4.4.1.) even aim a priori on limited and controlled screenout. Together with sophisticated selection of adequate proppant type and grain size, however, also serious attention has to be paid to proper choice of fracturing fluid, because if high-quality proppants are used together with unsuitable fluids, this can ruin considerable amounts of reservoir permeability and fracture conductivity. The significance of high-efficiency stimulation fluids for the successful treatment of high-temperature, low-permeability reservoirs is discussed by BAUMGARTNER, PARKER, WILLIAMS & WOODROOF (1983); HARMS, GOSS & PAYNE (1984) and WALSER (1985). Influences of fracturing fluid properties on proppant transport and crack height behaviour are discussed by MORALES & ABOU-SAYED (1985) and ACHARYA (1986; cf. section 4.3.4.). Larger proppant grain sizes and/or higher-quality proppant types provide greater fracture flow capacity that is not as affected by released formation fines as smaller lower-quality proppants that are crushing due to increasing closure pressure (CROSS & SHELDON 1977).

Conductivity per US $ (or net present value per US $; BRIM 1986, ANDERSON & PHILLIPS 1987) that is obtained when selecting a particular proppant type and grain size has to be calculated against total expenses of the hydraulic fracturing treatment and pay-out by increased cumulative hydrocarbon production within a reasonable time after stimulation (this economical modelling is facilitated by various computer programs for proppant choice and prediction of fracturing operation results; AGARWAL, CARTER & POLLOCK 1979; CHENG 1984; COBB 1985, 1986). Following the guideline of proppant cost per unit conductivity is in fact the most straightforward way of proppant selection (ANDERSON & PHILLIPS 1986; cf. section 1.4.12.).

1.3.4.2. Commercially available proppant grain sizes

The different natural and synthetic propping materials are available in three standard mesh sizes for hydraulic fracturing : 20/40 (the most abundantly used grain size), 16/20 and 12/20. For gravel pack application, sometimes also 30/50, 40/60 or 40/70, 100 and even 70/140 are offered (WESTERN PETROLEUM 1985), and for special tail-in jobs or high-viscosity oil reservoir treatments also 16/30, 10/16 or even 8/12 and 6/10 could be obtained (some aspects of the application of coarser grain sizes are discussed by SCHWARTZ 1969; GRUESBECK, SALATHIEL & ECHOLS 1979; BARBER & THEMIG 1985, ROBINSON 1985; PARKER, ADAMS & LIANKUI 1986; BARBY & BARBEE 1987; cf. section 1.5.1.1.2. and tab. 1). Sand grain sizes other than 20/40 only represent abt. 15 % of the total quantity of proppant sand used (WATERS 1980).

Most of the rarer grain sizes have only been offered from sand suppliers which have the easy way of manufacturing by sieving without further treatment, whereas most of the proppant plants need higher quantities to make economically feasible production runs. This is the explanation why so far 100 mesh intermediate- or high-strength proppant has not been produced although it would have been needed already at least in a couple of cases (cf. section 4.8.8.3.2.3.5.). The common grain sizes of intermediate- to high-strength proppant manufacturing are 12/20, 16/20, 20/40 and 40/60, with all the other finer or coarser grades being exceptions that are partially only available as special order and at a higher price. Exceptions of the latter constellation are only proppant companies that use a natural pisolithic instead of an earthy bauxite raw material (COMALCO 1983, 1986; cf. section 1.4.9.1.3.) or where a large range of grain sizes originates during the electrical fusing process in contrast to the sintering method (SARDA 1981, SEPR 1983, MESSINA 1986; cf. section 1.4.9.2.), with in both cases only sieving being the additional production step for exotic grain sizes.

1.3.5. Grain shapes

In the early stage of high-strength proppant application in deep gas reservoir fracturing, mainly angular proppants were offered at the market which were simple by-products of abrasive manufacturing (HARBEN 1978; cf. tabs. 1 and 4 as well as section 1.2.4.). Increasing conductivity demand by the customers later led to introduction of progressively better rounded particles of both natural quartz sand and synthetic alumina oxide and silicate proppants. The best roundness and sphericity is provided by zirconia-silicate proppants which together with their smooth surface almost represent ideal globules. Proppant strength is generally a function of sphericity for a given size distribution of the same grade of material (STEANSON, ELBEL & WENDORFF 1979).

While during the last ten years large amounts of well-rounded material have been consumed for hydraulic stimulation and gravel packing, LARSEN & SMITH (1985) present a renaissance study of angular fracturing sand by emphasizing new conductivity found in blends of angular and round grains (cf. section 1.4.11.3.). Intermixing of angular grains may also be a suitable means for increasing the friction angle of the proppant pack to avoid its flow out of the fracture at the mouth of the crack near the wellbore (cf. section 1.4.3.) as well as to increase stability of gravel plugging perforation tunnels (cf. section 5.2.4.). At lower closure stresses in shallower reservoirs, angular proppants even tend to have higher conductivities than round ones (WESTERN PETROLEUM 1984).

Concerning natural sand, angular sand has much higher absolute permeability than round sand at low closure stresses up to 2,000 psi, whereas at higher closure stresses above 2,000 psi, round sand is much stronger than angular sand. In the early years of alternate proppant manufacturing and experimentation, va-

SIGNIFICANCE OF PROPPANTS IN PETROLEUM RESERVOIR STIMULATION AND OTHER ENERGY INDUSTRY PROCESSES			PROPPANT APPLICATION													MORPHOLOGY			
PROPPANT TYPE			HYDRAULIC FRACTURING					GRAVEL PACKING			OTHER TREATMENTS					GRAIN SURFACE		GRAIN SHAPE	
PROVENANCE	STRENGTH	TYPE	Oil production	Water production and injection	Shallow gas production	Moderately deep gas production	Deep gas production	Oil production	Water production and injection	Steam-drive enhanced oil recovery	Fracture fluid-loss additive	Geothermal well stimulation	Coal seam degasification	Foundry cupola furnace heat transfer	Mechanical gas purification	Rough	Smooth	Round	Angular
natural	low-strength	quartz sand	⬤	⬤	⬤	●	•	⬤	⬤	•	⬤	•	●			⬤	⬤	⬤	●
synthetic	intermediate-strength	low-density alumina silicate proppants	⬤	⬤	⬤	⬤	•	●	●			•	●			⬤	●	⬤	
		high-density alumina oxide and silicate proppants	•	•	●	⬤	⬤					●	●			⬤		⬤	
	high-strength	alumina oxide proppants					⬤	•	•	⬤		⬤	●	⬤	⬤	⬤		⬤	•
		zirconia-silicate proppants				●	⬤	•	•			•					⬤	⬤	

Tab. 4

rious shapes including oblong, donut- and pellet-size particles apart from round and angular grains have been tested, but except of round and angular grains, no other particle shape has turned out to have any benefit in the hydraulic fracturing proppant application (FRACFAX 1988 g).

1.4. Proppant choice for hydraulic fracturing and gravel packing

As a consequence of the ever increasing development of low-permeability oil and gas reservoirs, hydraulic proppant fracturing has at least in the USA already become one of the most important aspects of well completion (cf. section 2.3.). Because the stimulation treatment can account for 10 - 50 % of the total well cost, significant emphasis should be placed on optimizing the design of the operation, especially proppant selection. Proppant choice is particularly important in view of the fact that the only tangible element of the actual hydraulic fracturing job which will contribute to well productivity increase is the proppant that was pumped into the generated crack, with thus placing correct type, grain size and amount of proppant in the fracture for a given set of reservoir and producing conditions being extremely critical (HOLDITCH 1984, PHILLIPS & ANDERSON 1985; ROBINSON, HOLDITCH & WHITEHEAD 1986) and even representing the key element in hydraulic stimulation operation planning.

In addition to the difference in type of failure between crystalline alumina oxide proppants and glassy zirconia proppants, several other conclusions concerning proppant choice have emerged which comprise aspects and consequences of surface morphology and internal microporosity, crushing behaviour and crystalline structure, friction angle and grain shape, fluid salinity and mechanochemical stability, fluid temperature and hot brine aggressivity, equipment abrasion and grain hardness, specific gravity and fluid suspension properties, and grain size and embedment. Apart from technical proppant selection, proppant price is also a very important factor in terms of economical evaluation. The brief account as follows also includes comments on manufacturing process and pellet composition, as well as conductivity discount and proppant mixing. Finally, the significance of computer programs for proppant choice is outlined, and the numerous results are summarized in some general guidelines for proppant selection. Various aspects of proppant choice are also discussed by RIXE, HOWARD & FAST (1963) and PHILLIPS & ANDERSON (1985). The significance of proppant testing for assessment and improvement of proppant quality and for reliable proppant selection according to reservoir conditions is briefly outlined in the introductory comments.

1.4.0. Significance of proppant testing for quality improvement

In order to optimize fracture design by selection of suitable proppants and to reinforce material quality, laboratory research with experimental testing started in the early days of successful hydraulic treatment performance (SINCLAIR 1980). While in the 1950`s and 1960`s not very many alternative materials to natural sand existed, the situation changed in the 1970`s when finally high-strength synthetic proppants for hydraulic fracturing of deep high-pressure reservoirs were invented, and this diversification of the proppant suite also triggered progressive sophistication of testing procedures along with successively rising quality demand.

During course of quality assessment and improvement, numerous laboratory tests (aiming on proppant selection in long-term studies and on quality control in short-term checks, with the intention of determination of proppant acceptability under various reservoir conditions and particularly closure stress ranges; CUTLER, ENNISS, JONES & SWANSON 1985) have been carried out with the different proppant types and grain sizes under various environmental (pressure, temperature, fluid composition) conditions by independent consultant and research departments (AHMED, ABOU-SAYED & JONES 1979; SINCLAIR 1980; CUTLER, ENNISS, JONES

& CARROLL 1983; DGMK 1983, 1986; CUTLER, ENNISS, JONES & SWANSON 1985; STIM-LAB 1986, PENNY 1987; PURSELL, HOLDITCH & BLAKELEY 1988) as a consequence of orders of fracturing customers, as well as by proppant manufacturers (HICKEY, BROWN & CRITTENDEN 1981; NORMAN, CIPOLLA & WEBB 1983; BROWN & MUCH 1986, COBB & FARRELL 1986, COMALCO 1986, STANDARD OIL PROPPANTS 1986, MUCH 1987, MUCH & PENNY 1987), oil/gas production companies (BECQ, ROQUE & SARDA 1984; CHEUNG 1985; ROODHART, KUIPER & DAVIES 1986) and service companies (McDANIEL & WILLINGHAM 1978, WENDORFF 1982, KIM & LOSACANO 1985; McDANIEL 1986, 1987, 1988; PARKER & McDANIEL 1987, McDANIEL & PARKER 1988). Closure stress resistivity testing of natural sand prior to invention of synthetic proppants has been carried out by TUNN (1971) and SLUSSER & RIECKMANN (1976).

1.4.1. Surface morphology and internal microporosity

As a consequence of the various production processes being sintering and fusing (cf. section 1.4.9.) and of the different internal structure being crystalline and glassy (cf. section 1.4.2.), alumina oxide proppants have a rough strawberry-like surface and a higher internal microporosity (open porosity; BECQ, ROQUE & SARDA 1984) than zirconia-silicate proppants which have a smooth surface and almost no internal microporosity (SEPR 1983, MESSINA 1986; closed porosity; BECQ, ROQUE & SARDA 1984; cf. tab. 4). Comments are offered as follows on laboratory experiments and results as well as field testing and experience.

1.4.1.1. Laboratory experiments and results

Closure stress resistivity is no major difference between both high-strength compositional types of proppants which are both able to mechanically withstand to more than 10,000 psi or 700 bar before starting to be significantly crushed (BROWN & MUCH 1986, COBB & FARRELL 1986, STANDARD OIL PROPPANTS 1986), and apparent enigmatically high differences in permeability and conductivity loss or preservation during long-term flow tests with exposition to brines of various composition (DGMK 1983, 1986) are in most cases not supported by sufficient reproduction runs of the tests or are put into doubt by the use of not properly stainless teflon and hastalloy equipment and debatable methods of fluid passage through the package. Much conflicting conductivity data has been presented in the last years, with frequently the choice of testing parameters or the design of the test cell having been responsible for the discrepancies (McDANIEL 1986).

The amount of internal microporosity is indicated by the tap density of the material (cf. section 1.4.7.2.). Interconnected microporosity reaching to grain surface is generally undesirable because it provides an increase in cumulative particle surface area and thus better possibility for attack by acids and aggressive formation brines (CUTLER, ENNISS, JONES & CARROLL 1983).

1.4.1.2. Field testing and experience

Some of the reported extremely high destructions of conductivity and fluid throughput are even completely contradicted by the long-term performance of various Rotliegend and Carboniferous gas wells where one or several parts of the prospective reservoir complex have been fractured up to 10 years ago (BRINKMANN, FUHRBERG & SCHÖBER 1980; BRINKMANN, KRÖMER & REINICKE 1980; BRINKMANN 1982, SCHWARZ & SCHÖBER 1982, JOHN 1983, LEICHT 1985; REINICKE, BRINKMANN, SCHWARZ & HUENI 1985; cf. sections 2.4.1.1. and 3.3.2.) and are thus pronounced field experiments highlighting that proppants of either type do well perform and no significant reduction of gas flow has been proven during the period of in some cases a full decade, with the expectation in case of the realistic nature of the permeability diminution observed in some of the laboratory tests (DGMK 1983, 1986) requiring that none of the gas wells which were fractured many years ago would be allowed to produce at acceptable rates still today.

Subordinate quantities of alumina silicate and oxide proppants which entered the borehole by partial fracture evacuation and were bailed out of some Carboniferous and Rotliegend wells several years after hydraulic stimulation execution exhibit only negligible mechanical and chemical surface alterations with respect to fresh proppant grains, thus confirming the physical consistency of the proppant wedge in the crack which is also reflected by the satisfactory gas production rates (cf. section 1.4.10.5.).

1.4.2. Crushing behaviour and crystalline structure

Comparability of the results of the long-term flow tests carried out by the various institutions is hampered by differences in size and composition of the test cell, way and period of pressure increase, rate of fluid flux through the proppant package, singular throughput or repeated circulation of the solutions, and temperature and salinity of the fluids. The discussion as follows focusses on failure of microcrystalline and glassy proppant types as well as internal proppant structure.

1.4.2.1. Failure of microcrystalline and glassy proppant types

The difference in crushing behaviour including slow continuous breakage into larger pieces of tougher and partially even ductile crystalline alumina oxide proppants vs. rapid catastrophical failure into tiny splinters of brittle glassy zirconia-silicate proppants seems to favour the latter material at lower closure stresses where preferentially low percentages of the former product start to break into fragments, while zirconia-silicate proppants are still completely stable. The crystalline bauxite proppants, however, are in a much superior situation once the package of glassy zirconia-silicate proppants totally fails into almost dust at the critical closure stress level where the crystalline bauxite proppant package assemblage still has a reasonable conductivity by being crushed predominantly into halves and quarters of grains instead of tiny splinters.

The fine zirconia-silicate proppant dust can be transported and redistributed by fluids, whereas the coarser fragments of bauxite proppants remain in place and form residual packages that still have quite some conductivity which is sufficient to economically drain many tight gas reservoirs by still providing a reasonable conductivity contrast to the low-permeability rock matrix (cf. sections 4.6. and 4.12.4.). Crystalline bauxite proppants thus exhibit a broad breakage curve in contrast to a fairly well-defined and narrow breakage peak of zirconia-silicate proppants (MESSINA 1987).

1.4.2.2. Internal proppant structure

The superior strength characteristics of sintered bauxite are related to its microcrystalline structure which contains many small grain boundaries within each particle separated by an amorphous phase which has subordinate share in total grain composition, with the latter responding to stress by flexing and distributing the load on larger groups of particles, thereby inducing exceptional crush strength and resistance to proppant fragmentation (HICKEY, BROWN & CRITTENDEN 1981). The same applies for other alumina silicate and oxide proppants. The glassy zirconia-silicate proppants, however, consist of monoclinic and cubic zirconia crystallites which are floating in the alumina silicate glassy matrix and therefore due to the rarity to even lack of crystallite contacts, the material has an overall amorphous structure and thus breaks homogeneously into tiny splinters upon failure. Natural quartz sand which has no cleavage is characterized by shattering and irregular breaking, while polycrystalline synthetic sintered bauxite is cracking and splitting into larger parts of the grains (cf. section 1.4.2.1.) instead of failing into powder.

Reviews of proppant performance in field applications with special emphasis being put on assessment of behaviour of sintered bauxite vs. sand particularly in terms of crushing degree and concerning net present value per invested US $ are performed by ATTEBERRY, TUCKER & RITZ (1979); CARROLL & BAKER (1979), KOHLHAAS (1982), CLARK (1983) and BRIM (1986). Detailed comparative effectiveness evaluations of the influence of proppant type and grain size on well productivity are particularly carried out by McDANIEL & WILLINGHAM (1978); HICKEY, BROWN & CRITTENDEN (1981) and NORMAN, CIPOLLA & WEBB (1983).

1.4.3. Friction angle and grain shape

The rough surface of crystalline bauxite proppants gives a much better fluid flow through the proppant package than the smooth surface of the glassy zirconia proppants (cf. also ACHENBACH 1974). The different surface morphology also results in different friction angles of the proppant package. While a lower friction angle of the glassy zirconia proppants may favour better emptying of the silos during blending of proppants with the fluid when carrying out the fracturing stimulation job, with respect to the higher friction angle of the crystalline bauxite proppant, and assuming that even despite dispersion in the fluid there is still an advantage of the lower friction angle of the glassy zirconia proppant for easier rushing into the fracture and for a less tendency of bridging in the crack, the stability of the proppant package within the fracture during subsequent production is another question. Some aspects of proppant backflow out of the fracture mouth and stability enhancement of the proppant package are briefly discussed as follows.

1.4.3.1. Proppant backflow out of the fracture mouth

As production of small amounts of proppants with the gas testifies to proppant loss at the fracture mouth where the flow rate is highest and creates a choke-like effect due to pressure drawdown and current acceleration, the lower friction angle results in diminuished stability of the proppant package and corresponding to the better rushing-in triggers also an enhanced rushing-out of the smooth-surface material from the fracture with respect to the rough-surface material (problems of near-wellbore proppant flowback are discussed by SMITH 1973; ROBINSON, HOLDITCH & WHITEHEAD 1986; cf. section 4.12.3.). Some aspects of proppant transport are also reported by CLARK & QUADIR (1981); SIEVERT, WAHL, CLARK & HARKIN (1981); AHMED (1984); CLARK, HALVACI, GHAELI & PARKS (1985) and ACHARYA (1986, 1987). SHAH & LEE (1986) comment on friction pressures of proppant-laden hydraulic fracturing fluids.

Proppant loss near the crack mouth by backflow induces fracture closure over a certain distance which obviously has a considerably deteriorating effect on proppant package conductivity (cf. section 4.12.3.1.). Fracture evacuation by proppant escape in the immediate wellbore vicinity where pressure drawdown and flow rates are highest is one of the most deleterious types of fracture conductivity damage. In addition, the flow-out of proppants (particularly sintered bauxite) from the crack through the perforations into the hardware-equipped borehole also erodes tubings, valves, chokes and even the wellhead (similar effects are provoked by influx of unconsolidated formation sand into the wellbore through insufficient gravel pack filters or in case of absence of mechanical sand control; cf. section 5.2.1.3.). The general rule is that proppant packages with lower friction angle are less stable but also create less erosion, whereas such with higher friction angle are more stable but cause also more erosion (questions of particle transport through perforations are reviewed by GRUESBECK & COLLINS 1978; cf. also section 5.6.). Proppant backflow is best eliminated by resin pellicles surrounding the individual grains and creating a stable aggregated package (CUTLER, ENNISS, JONES & CARROLL 1983) which prevents fracture evacuation by keeping the coupling of wellbore and productive formation (SINCLAIR, GRAHAM & SINCLAIR 1983; cf. section 4.12.3.3.).

Within a packed bed of proppant spheres filling the width of the fracture, compressive forces tend to arrange the beads in a closely packed structure (CUTLER, JONES, SWANSON & CARROLL 1981). The most typical idealized packing arrangement is the rhombohedral one where each globular particle contacts twelve neighbours in six pairs of diametrical contact loadings (KNUDSEN 1959). While this regularity is likely not to be preserved over long distances, it is believed that the loading pattern is representative of the forces transmitted to each sphere.

1.4.3.2. Stability enhancement of the proppant package

In terms of stability of the proppant package in the fracture, bauxite proppants with rough surface and higher friction angles are thus superior to zirconia proppants with smooth surface and lower friction angles. An even better stability by a higher friction angle could be achieved by intermixing of angular grains into a population of well-rounded particles, with this technique, however, requiring a compromise solution of conductivity reduction vs. stability enhancement (aspects of conductivity of angular blends of fracturing sand are also discussed by LARSEN & SMITH 1985; cf. section 1.3.5.). Flowback is also retarded by application of resin-coated natural sand or synthetic proppants (POPE, WILES & PIERCE 1987; cf. section 4.12.3.3.). Coarse tail-in terminal lots also function like a gravel pack and hold the finer main proppant quantity in place by bridging of the smaller grains at the interface between both populations (cf. section 5.8.8.). Some comments on significance of proppant backflow prevention in hydraulic fracturing and gravel packing as well as prevention of proppant evacuation are offered as follows. The comparative importance of proppant evacuation in hydraulic fracturing and gravel packing is also assessed.

1.4.3.2.1. Proppant backflow prevention in hydraulic fracturing

Fracture evacuation near the wellbore by proppant flowback is one of the most serious types of crack damage (cf. section 4.12.3.) and thus not only optimum type, grain size and concentration of the proppant have to be selected, but also cleanup, flowback and initial hydrocarbon production of the well have to be carefully controlled in order to avoid extreme drawdown which would subject the proppants suddenly to very high closure stresses that might cause significant proppant crushing partially by excessive closure stress, but partially also by shock loading (ROBINSON, HOLDITCH & WHITEHEAD 1986).

In contrast to hydraulic fracturing where proppant flowback can be a serious problem and thus zirconia-silicate proppants with lower friction angle have a technically competitive disadvantage with respect to alumina oxide and silicate proppants with higher friction angle, proppant flowback is only insignificant to negligible in gravel packing.

1.4.3.2.2. Proppant evacuation in fracturing vs. gravel packing

The main reason for the at least partial independence of packing gravel from intergrain stabilization is the fact that the perforation tunnel plugs are hold in place by the internal gravel pack mantle along the borehole wall which in turn is kept in position by the gravel-pack hardware comprising slotted liner or wire-wrapped screen (cf. section 5.2.3.5.). As the angle of repose of gravel packing materials should be as low as possible in order to allow ready infilling of all devices and pockets in the formation in open-hole completions (COBERLY & WAGNER 1938), zirconia-silicate proppants which have the lowest angle of repose of all presently commercially available propping materials should be together with their low specific gravity the ideal gravel type, but unfortunately they are too expensive for most of the sand-control operations (cf. section 5.2.2.3.3.).

On the other hand, however, the smooth surface of zirconia-silicate proppants and glass beads does not allow easy bridging due to lacking friction which is desirable in hydraulic fracturing in order to avoid premature screen-out failure, whereas the rough surface of natural sand as well as alumina oxide and silicate proppants has sufficient friction to permit ready bridging which is supporting perforation tunnel infilling and stabilization in gravel packing, but is unsuitable in hydraulic fracturing as a consequence of increasing screen-out risk.

1.4.4. Fluid salinity and mechanochemical stability

Smoother surface and lower internal microporosity give the glassy zirconia proppants a principally better resistivity against chemical solution than the crystalline alumina oxide proppants, as well as a lower abrasive effect to pumps and equipment than particularly the high-strength bauxite proppants which cause considerable wear and tear of the machinery. Some aspects of brine composition and proppant dissolution, formation water salinity, and formation water aggressivity vs. proppant stability are shortly sketched as follows.

1.4.4.1. Brine composition and proppant dissolution

An inverse performance of the different proppant types with salinity level and pH-value, however, has been recorded, with in low-salinity environment and in alcaline milieu the glassy zirconia proppant being inferior to the crystalline bauxite proppant, whereas the former withstands better in high-salinity environment and in acid milieu than the latter. Some aspects of low and high fluid salinities as well as proppant surface stabilization are outlined as follows.

1.4.4.1.1. Low and high fluid salinities

Proppant testing under different fluid concentrations revealed considerable permeability reductions of the proppant package (in extreme cases up to 10 % residual value) if exposed to high-salinity solutions, whereas only negligible permeability reduction occurs at low fluid salinities, with the reason being mainly proppant corrosion. The aggressive fluids lead to dissolution of parts of the globules, with the effect being partial fluid supersaturation which finally results in reprecipitation of material in other parts of the proppant package thereby forming overgrowths and cements, combined with the effect of collapse of the framework of the residual globule fragments, with the failure of the package being a feature principally comparable to compaction of sediments by also reducing pore space and thus ultimately also permeability (cf. also section 4.5.4.1.6.).

Low salinities do not allow supersaturation to occur and thus there is no reprecipitation partially obstructing or even completely plugging the pores. Permeability reduction in high-salinity environment is triggered by both chemical corrosion and mechanical compaction due to pore space diminution by changing the punctual contacts of the initial globules into more areal contacts of the solution remnants. Finally, pressure solution at the punctual contacts also has its influence.

1.4.4.1.2. Proppant surface stabilization

In addition, if hot water is pumped at once through the proppant layer, a gel forms which plugs the pores and reduces the permeability. If the water first stands within the proppant package and is then flushed through, however, recrystallization takes place and gel formation is prevented by inactivation or protection of the proppant surface (ALBERTSEN 1985). Therefore in practice no danger is seen that such a gel could plug the pores of the proppant package infilling the fracture, as the common policy is to shut in the well after the treatment to allow the high-viscosity fracturing gel to break into simpler components which form a low-viscosity fluid and after a few days start clean-up and production, with the shut-in period being equivalent to the standing time in the test and thus stabilization of the surface during that time prevents gel formation. Aspect of dissolution and condensation of silica in various fluids are discussed by THORNTON & RADKE (1985) and UDELL & LOFY (1985).

1.4.4.2. Formation water salinity

While in the USA formation waters have in most cases only slight to moderate salinity, considerably aggressive brines with almost saturated salinity occur in the Rotliegend and Carboniferous deep tight gas reservoirs of Germany FRG (KLOSE & KRÖMER 1983, ALBERTSEN 1985; cf. section 1.4.10.4.) and surrounding central European countries and areas which require selection of a proppant type that is stable in this high-concentration fluid environment. High salinities of formation waters are also present in Tertiary sandstones in Yugoslavia (ECONOMIDES, CIKES, PFÖRTER, UDICK & URODA 1986) where MHF treatments comparable in size to those in Germany FRG (and partially even larger than the biggest jobs in the latter area) are carried out (cf. section 2.4.3.1.).

Many proppant laboratory stability tests cannot be accurately compared with field conditions due to predominant application of 2 % KCl solution as experimental fluid which matches the composition of prepad and displacement during the fracturing treatment (where it is used as a standard safety precaution for minimization of formation damage and protection of water-sensitive reservoirs by clay stabilization; WATERS 1980, SYDANSK 1984; cf. section 1.4.10.4.2.), but not due to composition of aggressive formation waters which become effective later when the invasion by the injected fluids is compensated and replaced, and also the fracture is flushed with high-salinity formation brines. RENSVOLD (1983) evaluates the stability of sand consolidation resins in hot brines. GRAHAM, MONAGHAN & OSOBA (1959) discuss the influence of propping sand wettability on productivity of hydraulically fractured oil wells.

1.4.4.3. Formation water aggressivity vs. proppant stability

Formation water aggressivity is critical to proppant stability, because even in gas reservoirs with very low water saturation, some amounts of formation water are produced together with the gas and thus flow also through the proppant package. Completely dry gas production without any traces of water represents a theoretically ideal case, but is in practice almost never achieved. As many deep tight gas wells where fractures have been created to drain the surrounding reservoir are scheduled to produce for long time being up to several decades, during such a long period even negligible traces of water within the gas add up to serious chemical corrosion if being of such aggressivity to be able to attack the proppants supporting the fracture. The situation is becoming more accentuated when gas with higher water content (ARENS & BÖHNEL 1977) is produced. Therefore realistic proppant testing and selection has to be based on the highest stress exposure including both maximum reservoir pressure drawdown and formation brine composition.

1.4.5. Fluid temperature and hot brine aggressivity

While most of the synthetic proppants and natural quartz sands have a more or less comparable performance under standard reservoir conditions of temperature according to the regional geothermal gradient and moderate salinity of the formation water (in this context not respecting closure stress that is exerted by hydrostatic reservoir pressure and lithostatic overburden pressure), differences in reaction are exhibited at higher temperatures of the reservoir rocks beyond the normal geothermal gradient and at increased salinities and/or pH-values of the brines beyond the saturation of the pay fluids common in most formations. Some aspects of fluid temperature and hot brine aggressivity in steam-drive enhanced oil recovery and deep tight gas reservoirs are summarized as follows.

1.4.5.1. Steam-drive enhanced oil recovery

In terms of temperatures, the most significant aspect is the behaviour of proppants under conditions of steam-drive enhanced oil recovery in heavy-oil fields when gravel packing and/or hydraulic fracturing is required in the steam injection wells in such formations (cf. section 5.7.). Particularly in case of an alcaline hot environment in the surroundings of the steam injectors, natural quartz sand as well as all the synthetic proppant types containing free quartz (or equivalent crystalline phases or glassy silica) or alumina silicates are unstable due to silica dissolution, whereas sintered bauxite consisting almost exclusively of inert corundum has been proven to be the suitable material due to its stability in this extreme temperature milieu (REED 1980; UNDERDOWN & DAS 1985; cf. also WATKINS, KALFAYAN, WATANABE & HOLM 1985; aspects of steam-flood silica dissolution are also discussed by THOMAS, YOST & CATHEY 1987; cf. section 5.7.1.). Another resistant material which may even perform better than sintered bauxite under steam-drive conditions is nickel-coated sand (SACUTA, NGUYEN & KISSEL 1986; cf. section 5.7.2.).

In contrast to the USA, however, steam-drive enhanced oil recovery is in Europe not yet widely distributed (mainly in Cretaceous sandstones in Germany FRG and Netherlands; LILLIE & SPRINGER 1981; PROYER, CHAZITEODOROU, MUSS & ROSSKAMP 1983; LILLIE & WIETHOFF 1985, REINICKE & WIETHOFF 1987; for further references on steam-drive projects in Europe cf. section 5.7.3.), and due to different methods of steam conditioning, the problem of silica dissolution in an alcaline environment is in Europe not so pronounced as in the USA (cf. section 5.7.5.).

1.4.5.2. Deep tight gas reservoirs

Considerably elevated temperatures in the range of 120 - 240 oC are also present in deep Rotliegend and Carboniferous gas wells in Germany FRG (PHILIPP & REINICKE 1982, JOHN 1983, KLOSE & KRÖMER 1983, BLEAKLEY 1984; cf. fig. 3 and sections 2.4.1.1. and 3.3.) and in deep Tertiary gas and condensate wells in Yugoslavia (ECONOMIDES, CIKES, PFÖRTER, UDICK & URODA 1986; cf. section 2.4.3.1.) which create together with the extraordinarily high formation water salinities (partially saturated brines) and high pressures a very hostile environment and require stable materials to resist not only to the high closure stresses, but also to the chemically aggressive milieu (cf. section 1.4.4.2.). Aspects of thermal stability of resin-coated proppants in hot brines are discussed by RENSVOLD (1983).

1.4.6. Equipment abrasion and grain hardness

The economical feasibility modelling of a hydraulic fracturing operation has in terms of proppant choice not only to include reservoir-relating features, but also to respect handling and maintenance questions. The following outline focusses on erosional behaviour of different proppant types, equipment erosion by proppants, and proppant damage by equipment.

1.4.6.1. Erosional behaviour of different proppant types

The different mechanical properties of the various proppant types of different mineralogical composition result in different hardness and thus abrasive effect which has its influence on tanks, pumps and tubes of the equipment during course of stimulation operation execution. While the more expensive zirconia-silicate proppants (cf. section 1.3.2.2.) have a relatively gentle treating behaviour to the equipment (as a consequence of their smooth surface and their low specific gravity), the cheaper alumina oxide and silicate proppants (cf. section 1.3.2.1.) are considerably abrasive particularly in case of high alumina oxide concentration. Some aspects of sintered bauxite and zirconia-silicate proppants are outlined as follows.

1.4.6.1.1. Sintered bauxite

The most severe destruction of the operating equipment is caused by sintered bauxite (which was originally developed as a by-product of abrasive manufacturing; cf. section 1.2.4.1.), because corundum is one of the mineralogically hardest phases. Careful evaluation has thus to be made in terms of calculation of total treatment cost, with the performance of sintered bauxite in the fractured reservoir having to be compared with the additional expenses by wear and tear of the pumping machinery.

Flowback of sintered bauxite can be extremely hazardous, because it may result not only in cutting out of valves and christmas trees, but also give rise to blowouts and fires when large quantities of sintered bauxite are returned with the well fluids (OSBORNE, McLEOD & SCHROEDER 1981; the same applies for sand influx through insufficient gravel packs or in case of lacking sand control; ANAND & JONES 1977; cf. section 5.2.1.3.1.). Resin-coated sand which is in many applications a suitable replacement for intermediate- and high-strength synthetic proppants (cf. section 1.2.6.) has an excellent operational behaviour by almost not causing any hardware damage (SINCLAIR, GRAHAM & SINCLAIR 1983).

1.4.6.1.2. Zirconia-silicate proppants

Particularly in case of frequent stimulation jobs in deep high-pressure reservoirs, it may be desirable to use the more expensive zirconia-silicate proppants which are (apart from resin-coated sand) the most protecting material for the pumping equipment, rather than applying the cheaper alumina oxide and silicate or sintered bauxite proppants and accept high cost for repair or even replacement of tanks, blenders, channels and strings after almost every individual large-scale operation. Equipment abrasion by hard aggressive proppants has thus significant impact on proppant selection. Low- and high-density alumina silicate and oxide intermediate-strength proppants perform much better and are considerably less destructive to the equipment than high-density bauxite high-strength proppants, but are definitely more affecting the fracturing hardware than the extremely protecting zirconia-silicate proppants or also resin-coated sand.

1.4.6.2. Equipment erosion by proppants

MHF jobs typically require high injection rates and high proppant concentrations pumped at high pressures (PAI, GARBIS & HALL 1983). Larger proppant amounts pumped for longer periods of time can cause serious erosion problems in treatment line, tubing, and in and around the wellhead area. In terms of proppant types, sintered bauxite has the greatest destructive impact and is able to completely ruin equipment when pumped for longer times due to its high abrasiveness. High-density alumina silicate and oxide proppants also occasionally exhibit serious aggressivity, whereas all the other proppant types give only in some cases rise to subordinate damage and generally leave the materials in almost original conditions. The most severe damages occur at diameter changes and threaded connections which are briefly illustrated as follows. Comments are then given on prevention of proppant erosion by hardware adjustments, whereas possibilities of inhibition of material destruction by proppant selection have already been examined in the previous section.

1.4.6.2.1. Diameter changes

Erosion of hardware by abrasive proppants occurs mostly at places having sudden diameter changes or abrupt alterations in flow direction such as elbows, tees, chicksans, and threaded connections where nippled edges restrict fluid flow (PARKER 1980; PAI, GARBIS & HALL 1983). Erosion problems become more severe with increasing proppant concentration and rising fluid velocity. A borehole isolation tool separates the expensive wellhead equipment from high treatment pressures and thus protects it from erosion effects of the proppant slurry, because if erosion at the transition of tubing and borehole head would become too pronounced, the tubing could rupture thereby exposing the high pressure to casing and wellhead equipment (GARBIS, BROWN & MAURITZ 1985) which in the worst case would result in well blowout (OSBORNE, McLEOD & SCHROEDER 1981).

1.4.6.2.2. Threaded connections

Proppant erosion in treatment line and wellhead area requires close watching particularly in places where diameter changes occur and at threaded connections where the edge of the nipple opposes the direction of fluid flow (PARKER 1980). Abrupt diameter changes between borehole head isolation tool and tubing cause turbulence in this area which enhances the erosive effect of fluid and proppants. Other places on the wellhead isolation device which are subject to erosion-related failure are changeover swedge valves and other threaded connections. Changeover swedges are required because most fracturing lines have larger diameters than the borehole head isolation device, with damaging erosion being created by turbulence at diameter changes and threaded connections.

1.4.6.2.3. Prevention of proppant erosion

Problems with valves and threaded connections can be improved by tapering the diameter change with a machined integral swedge, with thus the alteration in diameter being effected over a longer interval thereby reducing turbulence (PAI, GARBIS & HALL 1983). Some aspects of replacement of threaded by integral connections as well as joints and tubulars of higher weight and grade are discussed as follows.

1.4.6.2.3.1. Replacement of threaded by integral connections

In order to further diminuish the danger of proppant erosion, threaded connections should be eliminated in all swedges and valves and should be replaced by integral ones. Monitoring of downhole treating pressures by placing an open ended tubing in the casing without a packer is performed by bottom hole operation pressure recording while pumping through multiple entries into the casing-tubing annulus. As the proppant-laden fluid enters the annulus, the tubing is subjected to an erosive cutting action which can eventually separate it, with blast joints having increased wall thickness being a possible compensation of erosion in this critical area. Choke abrasion can be considerably reduced by using remote-controlled adjustable tungsten carbide elements for well cleanup (OSBORNE, McLEOD & SCHROEDER 1981). Chokes are set on a small opening until presence or absence of sintered bauxite grains deriving from proppant escape due to fracture evacuation can be determined and choke size can possibly be increased. High early drawdown pressures have to be avoided and only gradually increasing choke opening is recommended as the well cleans up and stabilizes.

1.4.6.2.3.2. Joints and tubulars of higher weight and grade

In order to minimize or eliminate proppant erosion, the top one or two joints of tubing should be of higher weight and grade to combat the effects of turbulence-induced erosion (PAI, GARBIS & HALL 1983). The treating line diameter is usually much larger than that of the wellhead isolation device, thus requiring the use of a changeover swedge. When the operation is performed using the triple-entry technique by pumping via tubing and tubing-casing annulus (NALL, CAMPBELL & BONEY 1983; PAI, GARBIS & HALL 1983), the high-concentration proppant-laden slurry entering the annulus at high velocity produces tremendous erosional action on the tubing and may eventually cut it.

Thus high-strength and weight blast joints with greater wall thicknesses should be used in this critical area to withstand erosion caused by impact with high-velocity proppant particles. HOOVER & ADAMS (1988) report utilization of a static tubing string which not only reduces friction and vibration to minimum levels, but also virtually eliminates erosion of tubular and standard wellhead equipment. Similar erosional and destructional effects as provoked by abrasive proppants are triggered by reservoir sand flowing into the well in case of absence or insufficience of gravel packing (cf. section 5.2.1.3.).

1.4.6.3. Proppant damage by equipment

In an inverse approach to the aforementioned, effects of pumping equipment on sand-laden slurries are discussed by ROLL, HIMES, EWERT & DOERKSEN (1986). Aspects of gravel packing sand quality are studied by ZWOLLE & DAVIES (1983; cf. section 5.2.3.5.2.), and WENDORFF, BURKE & GARVIS (1979) emphasize the influence of handling equipment on fracturing sand quality. An only insignificant amount of fines is generated by proppant grain damage when pumping proppant-laden fluids with centrifugal and positive displacement pumps and through crossover tools at low pumping rates such as those used in gravel packing, but significant amounts of fines may originate when proppant-laden slurries pass through small orifices such as chokes at high pump rates and pressures like those used in hydraulic fracturing (ROLL, HIMES, EWERT & DOERKSEN 1986). A general reduction in grain size can occur if poor-quality sands are applied, although this does not necessarily lead to fines generation. Grain breakage can be directly correlated to sand and proppant quality (ZWOLLE & DAVIES 1983) by affecting first chiefly inhomogeneous particles which have inherent microfissures and only secondly at higher stress levels switching to real crushing of originally top-quality grains.

1.4.7. Specific gravity and fluid suspension properties

Proppant selection is in terms of risk of treatment failure by screenout also especially influenced by specific proppant gravity. While resistivity to closure stress has top priority and stability against formation brines is also essential, many cases of application leave the choice between two proppant types of different specific gravity in the advanced stage of proppant selection. Irrespective of factors such as proppant price (cf. section 2.2.2.) and equipment wear and tear by proppant abrasiveness (cf. section 1.4.6.), specific gravity of the propping agents is an important question in hydraulic fracturing and gravel packing. The heavier the proppant, the more viscous carrier fluids are required for reasonable and safe transport and placement of the proppant in the crack, and also the greater the danger of screenout by total plugging of the fracture by proppant bridging and/or slurry sticking before achieving the goal of performance matching design of the operation (cf. section 4.3.4.2.). Some aspects of fluid viscosity and proppant settling as well as specific gravity vs. tap density are illustrated as follows.

1.4.7.1. Fluid viscosity and proppant settling

Heavier proppants are much easier tending to settle in the stimulation fluids before the fracture closes to trap them (cf. section 4.12.2.). The considerably lower specific gravity is one of the major advantages of high-strength zirconia-silicate proppants in deep high-pressure reservoirs where this product can even technically compete with intermediate-strength high-density alumina oxide and silicate proppants because of still being lighter (cf. section 1.3.2. and tab. 3), with the considerable price difference between zirconia- and alumina-based proppants, however, being the other part of the story and representing the economical drawback (cf. section 2.2.2.3.). In terms of the different types of alumina oxide and silicate proppants, preference should always be given to the lightest product if sufficient freedom of choice is provided by the required closure stress resistivity level. The reasons for this are that the lower the specific proppant gravity, the lower the risk of premature screenout failure (cf. sections 4.2.2.2.3. and 6.2.4.2.1.), the better the vertical height containment of the crack by application of lower-viscosity fluids operating with lower pressure (cf. section 4.2.2.3.), and the cheaper the whole operation. Particle transport in viscous and viscoelastic fracturing fluids is also evaluated by ACHARYA (1986, 1987), and proppant settling correlations are discussed by SHAH (1986).

The greatest advantage of decreasing specific proppant gravity is that identical proppant transport with respect to that of heavier proppants in higher-viscosity fluids can take place for lighter proppants in lower-viscosity fluids which are therefore pumpable at lower pressure (CUTLER, ENNISS, JONES & CARROLL 1983). Vertical propagation of the fracture can be limited by reducing fracturing fluid pressure (SIMONSON, ABOU-SAYED & CLIFTON 1978; NOLTE 1982; cf. section 4.2.2.3.). The viscosity range of fracturing fluids makes minimizing of fluid viscosity a much more effective method of controlling pressure than lowering pumping rate, and alternatively, higher proppant volumes can be injected in a given amount of high-viscosity fracturing fluids (cf. section 4.3.4.2.). Lighter proppants can also be inserted in higher concentrations into the fracture than heavier proppants with the same type of carrier fluid, and placement of identical saturations requires only a lower-viscosity transport medium for lighter proppants with respect to that needed for deposition of heavier proppants.

Another economical advantage of lighter proppants is their higher volume per weight unit which results in a two-fold saving of stimulation expenses, comprising first the cheaper base price of lighter proppants and second the requirement of less weight units to meet a designed volumetric quantity of material (SINCLAIR, GRAHAM & SINCLAIR 1983; cf. also sections 1.2.6. and 1.4.12.5.).

1.4.7.2. Specific gravity vs. tap density

In contrast to specific gravity which indicates the characteristics of the whole proppant grains, tap density of the material reflects the amount of interconnected microporosity within the grains (CUTLER, ENNISS, JONES & CARROLL 1983). Interconnected microporosity does not only provide a higher surface area for the attack of acids and aggressive brines, but also affects proppant transport behaviour if the fracturing fluid or its filtrate can also penetrate the microporosity network, because the proppant will then act heavier than indicated by tap density. Intermediate-strength proppants have significantly lower tap densities than sintered bauxite and appear to have adequate strength for the majority of hydraulic fracturing applications (CUTLER, ENNISS, JONES & SWANSON 1985).

1.4.8. Grain size and embedment

The type of rock to be fractured affects the choice of proppant grain size and compressive strength. Following an outline of reservoir composition and proppant requirement, factors influencing proppant embedment are discussed.

1.4.8.1. Reservoir composition and proppant requirement

Soft formations require a large-diameter propping agent that will resist to embedment (cf. section 4.3.3.). Medium- to high-permeability reservoirs need a large-size proppant in order to provide adequate flow capacity with a strong permeability contrast (cf. also sections 4.5.1.1.2. and 4.6.1.), with the optimum being a wide and highly-conductive fracture infilled by proppant multilayers (cf. section 4.3.3.1.). Propping of cracks in hard rocks necessitates a material with high compressive strength to resist to the high closure stress exerted, and since embedment is insignificant in hard rocks with continuous reservoir distribution, small-size proppant can be utilized which allows an increased pack concentration in an oligo- to multilayer fabric counteracting the high stress.

Highly-lithified hard rocks are usually more or less tight and do not require large flow capacities with extreme permeability contrasts, with the best treatment design being long, narrow, highly-packed fractures (CLARK 1983), whereas soft to moderately-consolidated formations are better stimulated with shorter and wider fractures packed by proppant multilayers. If composed of discrete lenticular sand bodies (cf. section 3.4.1.1.1.) that are separated by mudstones, hard sandstone reservoir complexes also suffer from proppant embedment which occurs in the intervening mudstone units.

1.4.8.2. Factors influencing proppant embedment

Proppant embedment is influenced by proppant concentration, proppant size, proppant distribution, rock type and embedment surface, and variables affecting fracture closure are closure pressure, proppant size, proppant distribution, proppant saturation, formation hardness and surface toughness (VOLK, RAIBLE, CARROLL & SPEARS 1981; cf. section 4.3.3.). The rate of fracture closure depends on the proppant amount per unit area or the percent proppant monolayer. Principally fracture closure with a 100 % monolayer is the same for a 50 % monolayer at half the closure pressure and for a 25 % monolayer at one quarter the closure pressure, respectively. Proppant embedment into mudstone is more rapid initially but quickly becomes linearly dependent on closure pressure, with the transition occurring faster than that for sandstone embedment as a consequence of the finer grain size of shales which offers support for the proppant soon after embedment commences. Therefore embedment with a complete monolayer is actually less for shale, with the rate of closure of sandstones dropping at increasing pressure until it is less than shale.

At high closure pressures after significant embedment has occurred, the rubblized sandstone being pushed ahead of the embedding proppant begins to fill the pore spaces in the rock and to distribute the load more uniformly, with the stronger sandstone matrix ultimately resulting in a lower proppant embedment rate than is possible with shale. The rate of fracture closure for sandstones increases as proppant coverage decreases, whereas that for shales is directly related to proppant contact area or number of proppant per unit area. Shales propped by a partial proppant monolayer of 50 % or greater resist to fracture closure about as well as reservoir sandstones, but the interaction with fracturing fluids could significantly alter the rate of fracture closure in shales. As proppant coverage drops below 50 % of a monolayer, the rate of fracture closure is much more rapid than for sandstones. At higher closure stresses, fracture width is almost independent of surface roughness. Embedment aspects of high-strength proppants into low-permeability reservoir rocks are also discussed by MUCH (1987) and MUCH & PENNY (1987), and proppant embedment vs. crushing is discussed along the lines of proppant concentration in the carrier fluids in section 4.3.

1.4.9. Manufacturing process and pellet composition

Raw material composition and manufacturing process have an important influence on proppant pellet microstructure and integrity and thus strength and performance of the proppants. Generally a variety of fabrication methods can be used to make proppants including mix-pelletization, spray drying, droplet generation and coating (CUTLER, ENNISS, JONES & SWANSON 1985; cf. section 1.2.7.), with final proppant hardening of raw agglomerations by sintering or fusing. Distinction has to be made between alumina oxide and silicate proppants and zirconia-silicate proppants. Some comments are also given on natural sand.

1.4.9.1. Alumina oxide and silicate proppants

Alumina oxide and silicate proppants are manufactured by milling or grinding the raw material to a fine powder particle size, forming spherical pellets using a disk pelletizer, and final densification (sintering) of the pellets (COBB & FARRELL 1986) by fluidized bed agglomeration (LUNGHOFER 1984) or high-speed rotary mixing (FITZGIBBON 1984). Crystal phase formation and densification of the pellets are typically carried out in a rotary kiln at a sintering temperature up to abt. 1,500 oC (CUTLER, JONES, SWANSON & CARROLL 1981). Some aspects of grinded powder particle size vs. pellet integrity, sintering temperature and kiln environment, and artificial pellet formation vs. natural pisolite processing are discussed as follows.

1.4.9.1.1. Grinded powder particle size vs. pellet integrity

Changes in ground powder particle size distribution and mean affect other process variables and final pellet integrity. The correct powder particle size distribution is essential for optimal packing in the pelletization process and for more complete densification in the sintering process. Finer grinding of the raw material normally results in a more compact, stronger pellet, with this being beneficial in case of utilization of lower-grade ores where finer grinding is essential to produce strengths similar to those obtained from higher-grade ores with coarser grinding (COBB & FARRELL 1986). Finer particle sizes of the proppants always result from longer grinding before pelletization and not from the difference in pelletization technique (CUTLER, JONES, SWANSON & CARROLL 1981).

1.4.9.1.2. Sintering temperature and kiln environment

In addition to raw material grinding, sintering is also critical. The final crystal sizes depend also on time-temperature relationships as well as atmospherical conditions in the kiln where the pellets are fired for the purpose of mineralogical transformation of bauxite and clay into corundum and mullite. Kiln conditions are also responsible for the occurrence of diffusion, recrystallization and grain growth. Raw material chemistry thus interrelates with manufacturing variables to define proppant microstructure and thus performance of the product (COBB & FARRELL 1986). Sphere diameter increases as sintering temperature decreases since the beads do not shrink as much at lower temperatures (CUTLER, JONES, SWANSON & CARROLL 1981).

1.4.9.1.3. Artificial pellet formation vs. natural pisolite processing

Experience has shown that grinding of raw material, formation of pellets, and sintering of the pellets by firing in the kiln produces much higher crystallite interlocking and thus proppant strength than processing of naturally occurring bauxite pellets in pisolithic deposits (COMALCO 1983, 1986) by cleaning and sieving and subsequent straightforward sintering by firing in the kiln. The reason for the inferior quality of the fired natural pisolites are internal inhomogeneities which provoke a greater percentage of breaking of grains during firing, and even in those grains which pass the quality specification after production remain as weakness fissures that result in earlier crushing of the processed natural pellets with respect to the completely synthetic proppants. Another factor is often the onion-type layer structure of the pisolites which is comparable to that of carbonate ooids, with the suite of concentrical pellicles also including potential instability centres and breaking easier away at elevated closure stresses (cf. section 1.3.1.2.). Artificially formed pellets are characterized by an almost isotropic distribution of powder particles which upon firing in the kiln results in nearly isotropic distribution of corundum and mullite crystallites and thus provides the grain with very high strength due to crystallite interlocking in all directions.

1.4.9.2. Zirconia-silicate proppants

Zirconia-silicate proppants are produced by melting ceramic materials and preparing proppants by using a droplet generator (CUTLER, ENNISS, JONES & CARROLL 1983; CUTLER, ENNISS, JONES & SWANSON 1985). In an effort to minimize surface area, the molten droplets form into fused spheres upon cooling and solidification. This procedure almost eliminates generation of internal microporosity in the amorphous or glassy groundmass in contrast to various amounts of interstitials between interlocking crystallites in alumina silicate and oxide proppants (cf. section 1.4.1.). In addition, various grain sizes are produced much easier than by sintering of alumina oxide and silicate proppants due to the absence of formation of raw-material pellets of predetermined size and difficulty of exact adjustment of a narrow grain size interval by droplet generation, and therefore also extreme grain sizes can be produced with less handling problems by simple sieving of the total material assemblage after manufacturing. On the other hand, the fusing procedure includes much more rejects as a consequence of internal inhomogeneities, external appendices and broken grains, which makes the manufacturing process at the bottom of the line more expensive and is also the reason for the higher price of zirconia-silicate proppants in comparison to alumina oxide and silicate proppants.

In the context of production, packaging, handling and delivering of proppants, an important means to keep standard specifications, to enhance quality maintenance and to reinforce the designed and expected performance level is onsite quality control of fluids and proppants prior to the fracturing job (WENDORFF, BURKE & GARVIS 1979; ROBINSON 1985; SHAH, LEE & JENSEN 1985; HALL & LARKIN 1986).

1.4.9.3. Natural sand

Natural sand produced from the same geological formation could be principally expected to exhibit more or less homogeneous quality distribution and to show only differences in grain size according to depositional environment and facies type. In reality, however, first original variations include also grain composition and stability as a consequence of different palaeogeographical provenance and second artificial discrepancies are introduced by different production, preparation and handling techniques of the various companies. Primary variations of grain composition and stability comprise the number of individual crystals building up quartz sand grains, with distinction having to be made between mono-, oligo- and polycrystalline aggregates. As larger grain boundaries are always possible weakness planes, monocrystalline particles consisting of the rounded and worn fragment of a single quartz crystal have the highest stability, whereas oligocrystalline beads being constructed by a few interlocking quartz crystals may be subjected to breakage into halves or quarters by sliding and parting along the contact planes unless the grain boundaries are sutured or stylolithic instead of concave-convex. Polycrystalline quartz aggregates being built up of numerous tiny interlocking crystallites exhibit very high stability due to similar pressure load redistribution and partial internal equalization as reflected by microcrystalline alumina silicate and oxide proppants. Inclusions such as bubble belts, mica trains and rutile needle girdles in quartz grains may be additional sources of weakness and breakage by sliding and parting.

Sandstone maturity as expressed by percentage of well-rounded (textural maturity) and/or monocrystalline (mineralogical maturity) quartz grains therefore has an important impact on closure stress resistivity of the fracture propping material, and as sandstone maturity is determined by the palaeogeographical and palaeotectonical evolution and to minor amounts also by the palaeoclimatological history of the sandstone sequence, suitability as fracture propping material is directly related to depositional environmental conditions in the sedimentary basin. Secondary effects include processing parameters in the sandstone quarry or sand pit such as material access by digging or blasting, material disintegration by loading, shaking or milling, material separation by floating, and material screening by sieving. As a consequence of possible significant impact of all the mentioned factors, WENDORFF (1982) recommends to test samples provided by all the different companies even when coming from the identical geological formation, because separation, clean-up and sieving procedures give somewhat different particle-size distribution. In a similar way, synthetic proppants manufactured by different companies may also have different surface areas and permeabilities.

1.4.10. Conductivity discount

The most important property for post-operational performance and thus effectivity of hydraulic fracturing stimulation is proppant conductivity (COBB & FARRELL 1986). Crack conductivity is a function of proppant properties (strength, size, composition, roundness, sphericity, distribution, concentration, content of fines), closure stress load on the proppant package, pressure drawdown and fluid flow rate, formation properties (proppant embedment or crushing conditions), potential plugging from fracturing fluid residue or gel filter cake, and resultant propped crack width (MONTGOMERY & STEANSON 1985). Within the last decade, significant advances have been made in analysis of pressure transient data for hydraulically fractured wells (McDANIEL 1988). One of the most meaningful developments has been methods allowing an operator to calculate post-treatment estimates of propped fracture length and crack conductivity. In many instances, however, calculated fracture conductivity values have been considerably lower than those predicted by laboratory proppant testing, thus stimulating industry interest in the use of experimental conditions more closely approaching in-situ parameters.

Realistic proppant conductivity data are essential to both hydraulic fracturing treatment design and accurate forecast of the resultant fractured well production when using a computerized reservoir simulator to model post-stimulation withdrawal capability (McDANIEL 1987, 1988). After discussion of some general aspects including laboratory vs. field conductivity discrepancy and impact of experimental techniques, the implications of long-term proppant stability testing are outlined, and the results are then summarized in recommended conductivity correction factors. Finally, some examples of proppant performance in long-term field application are mentioned.

1.4.10.1. Laboratory vs. field conductivity discrepancy

Post-fracture pressure analysis frequently calculates proppant conductivity and effective crack lengths that fall short of the designed values (McDANIEL 1986, 1987, 1988; MUCH 1987, PARKER & McDANIEL 1987, McDANIEL & PARKER 1988). Treatment schedules therefore need to be modified to take into account a more realistic value of fracture conductivity under in-situ conditions. More accurate laboratory modelling of the proppant bed environment under simulated closure stress and temperature conditions in the presence of gel filter cakes (cf. section 1.4.10.3.5. and 4.3.4.1.4.) which were formed under dynamic conditions has provided a suitable approach to this problem. Concerning laboratory vs. in-situ data, SECCOMBE & ANDERSON (1982) report that standardized experimental data have to be discounted up to 90 %, with the remaining 10 % being the effective conductivity of the proppant package in place in the fracture under reservoir conditions (cf. also DGMK 1983, 1986).

Post-fracture operation performance analysis typically calculates a conductivity value significantly lower than that predicted by laboratory measurements (McDANIEL 1986, 1987, 1988; PARKER & McDANIEL 1987). HICKEY, BROWN & CRITTENDEN (1981) also comment that proppant performance under downhole conditions often does not duplicate the results of laboratory evaluations. Drastical differences of crack and proppant conductivity are already evident when comparing the results of long-term high-temperature and short-term ambient-temperature laboratory testing (COOKE 1973; McDANIEL 1986, 1987). Accurate design of fracturing treatments requires conductivity measurements at simulated reservoir conditions (McDANIEL & PARKER 1988). VEATCH (1977) introduces a computerized method for determining in-situ fracture flow capacity in low-permeability gas-bearing formations, and VLIS, HAAFKENS, SCHIPPER & VISSER (1975) present criteria for proppant placement and fracture conductivity. Some comments are offered as follows on proppant behaviour and reaction to stress as well as on proppant grain size effects.

1.4.10.1.1. Proppant behaviour and reaction to stress

The reasons for the discrepancy in fracture conductivity between laboratory testing and downhole performance include embedment of brittle proppants into the formation, plugging of pores by fracturing gel residues and fluid-loss filter cakes, discontinuities in crack geometry, stress corrosion of proppants by leaching and crushing and other phenomena not typically measured in laboratory experiments (COOKE 1973, ALMOND & BLAND 1984; McDANIEL 1986, 1987, 1988; PARKER & McDANIEL 1987, McDANIEL & PARKER 1988). Entrainment and deposition of fine particles in porous media are also evaluated by GRUESBECK & COLLINS (1982). Significant proppant conductivity discounts (with post-fracture conductivity often differing by a factor 0.1 to 0.5 times the test data) are also proposed by SOLIMAN, VENDITTO & SLUSHER (1984); BRITT (1985), MONTGOMERY & STEANSON (1985) and McDANIEL (1986, 1987, 1988) as a conclusion of laboratory testing and field performance evaluation.

Experimental work showed that when proppants are subjected to concentrated brines, a very rapid permeability decrease occurs, and even 2 % KCl solutions result in a quick rate of proppant permeability decline (COBB & FARRELL 1986, MUCH & PENNY 1987, McDANIEL 1988). In total, numerous laboratory and field experiments have shown that in-situ fracture conductivity is often considerably less than laboratory-measured values (COOKE 1973, 1975; SECCOMBE & ANDERSON

1982; CALLANAN, CIPOLLA & LEWIS 1983; NORMAN, CIPOLLA & WEBB 1983; SOLIMAN, VENDITTO & SLUSHER 1984; REINICKE, BRINKMANN, SCHWARZ & HUENI 1985; LI & ZHU 1986; McDANIEL 1986, 1987, 1988; ROODHART, KUIPER & DAVIES 1986; PARKER & McDANIEL 1987, MUCH & PENNY 1987, POULSEN & SOLIMAN 1987, McDANIEL & PARKER 1988).

1.4.10.1.2. Proppant grain size effects

Comparison of proppant conductivity should be made on the base of as identical grain size distribution as possible, because for example 20/40 mesh material can exhibit considerably different conductivities if granulometrical distribution includes 75 % or 25 % of grains larger than 30 mesh (CUTLER, ENNISS, JONES & CARROLL 1983). Even between several carefully screened samples of the same proppant type, an error band of performance difference of abt. 5 - 10 % can occur, with the major reason for sample-to-sample variation being proppant packing in the test cell, although variations in proppant size distribution still occur even when the proppant assemblage is screened subtly (CUTLER, ENNISS, JONES & SWANSON 1985). As proppant testing results can only be significant and representative if all the examined proppant lots have the same grain size distribution in order to enable identification of material-caused differences, rescreening of production-derived size distributions are frequently necessary before starting experimental work. Fracture width in the testing apparatus also has to be closely monitored, because flow of resin coatings, particle rearrangement, elastical deformations and particle crushing all tend to narrow crack diameter as closure stress is increased and thus various measurements of a test series may not be properly comparable due to not identical base parameters and conditions.

Realistic in-situ proppant conductivity assessment and prediction, however, is essential for cost and performance optimization modelling of hydraulic fracture treatments with realistic crack performance values allowing expense-effective stimulation treatments and maximum return on proppant investment (MUCH & PENNY 1987; cf. section 1.4.12.). Long-term proppant laboratory stability testing has turned out to provide useful information for proper evaluation of proppant field performance, but experimental procedures and results have to be very carefully and critically assessed. Absolute grain size has significant impact on measured proppant conductivity, because crushing of coarser grades starts at lower closure stresses than that of finer fractions (DAVIES & KUIPER 1988) as a consequence of coarser grains having fewer contact points with their neighbours and thus less possibilities of stress distribution (cf. section 4.3.5.1.).

1.4.10.2. Impact of experimental techniques

When reviewing proppant conductivity discount as a result of laboratory stress resistivity experiments performed in radial or linear flow cells, caution has to be exercised, as many reports of catastrophical conductivity deterioration in the past suffer from inadequate choice of testing parameters and improper design of the evaluation cell (McDANIEL 1986, 1987, 1988). After discussion of some general aspects, influences of proppant grain size distribution and proppant concentration are outlined.

1.4.10.2.1. General aspects

Many erroneous procedures involved recirculation of test fluid in contrast to single flow as happening in nature, plumbing of the experiment cell, application of too high oxygen concentrations of the fluid, lacking pre-test silica saturation of the fluid to match conditions in sandstone reservoirs, usage of steel plates instead of reservoir rock core slabs as boundary planes of the experiment cell, and performance of tests only at ambient temperatures, low fluid salinities (in most cases 2 % KCl solutions corresponding to the composition of the pre-pad and displacement stages of the fracturing job are used; cf. SYDANSK 1984; cf. section 1.4.10.4.2.) and short times of closure stress exposition. McDANIEL (1986, 1987, 1988) emphasizes that at least part of the hitherto published drastical conductivity losses of proppant packages are the result of inadequate testing methods that do not sufficiently approach reservoir conditions.

NORMAN, CIPOLLA & WEBB (1983) carry out special damage tests in order to model conductivity decline in the fracture after proppant placement and conclude that the dollar amount of lost revenue due to deterioration of fracture flow capacity by proppant failure increases with formation permeability. While in tight reservoirs the conductivity contrast between crack and formation is not seriously disturbed unless catastrophic proppant crushing occurs, fracture flow capacity can in moderate- to high-permeability rocks already suffer from lower amounts of proppant failure and comparatively small changes in conductivity can already cause significant production losses by considerably reducing conductivity contrast between fracture and reservoir matrix. The effect of proppant performance decline, however, can be controlled by proper material selection, because the proppants with the highest conductivity (which on the other hand are the most expensive ones) show the least revenue loss due to fracture flow capacity damage.

1.4.10.2.2. Proppant grain size distribution

As the stress level at which significant crushing occurs varies with both proppant type and grain size, some obliteration of true conductivity comparison can also arise from the variation of sieve distributions for different proppant types (McDANIEL 1987; cf. section 1.4.10.1.2.). For any given proppant type, the measured conductivity will be higher when the average particle size is larger until proppant crushing becomes significant (cf. section 4.3.5.1.). When for instance two samples of 20/40 mesh proppant have different distributions of particle sizes within the nominal sieve range, measured fracture conductivity will shift accordingly. Short-time test data usually favour the proppant with the larger average particle size, but this is not always the case for long-term conductivity measurements. Wide differences in fracture conductivity between lots of the same proppant material are generally the result of variations in proppant grain size distribution and processing parameters (CUTLER, ENNISS, JONES & SWANSON 1985).

1.4.10.2.3. Proppant concentration

Proppant saturation is also a parameter which needs attention (McDANIEL 1987). For accurate definition of the variation of permeability vs. proppant concentration or fracture width, tests have to be done at proppant saturations both lower and higher than the present industry norm of 2 lbs/ft2 (10 kg/m2), because many fracturing treatments using viscous fluids result in large percentages of the propped crack length with proppant concentrations of 1 lb/ft2 (5 kg/m2) or less, and the lower proppant saturations may be subject to more severe crushing at the same stress level than thicker proppant beds. In terms of multilayer packing, practical experience has shown that proppant concentrations up to 3 - 4 lbs/ft2 (15 - 20 kg/m2) can be placed without technical problems if fluids of sufficient viscosity are used (cf. also sections 4.3.1. and 4.3.4.).

1.4.10.3. Implications of long-term laboratory proppant testing

The economical success of hydraulic fracturing treatments depends on obtaining sufficient fracture conductivity within the proppant bed (McDANIEL & PARKER 1988). Proppant properties experimental modelling can be carried out as short-term testing, long-term testing without gel filter cake, and long-term testing with gel filter cake. Long-term proppant stability laboratory experiments reveal that generally there is a rapid conductivity decline during the first few weeks of measurement interval followed by little or no further decline during the remainder of the test period. The quick decrease in the first phase is interpreted as being the effect of proppant consolidation, compaction and reorientation, whereas the slow further diminution in the second stage is the consequence of stress corrosion. COBB & FARRELL (1986) report a magnitude of proppant conductivity decline of 20 % during the first two weeks and less than 5 % of a long-term experiment run during the remaining nine weeks, indicating that stress corrosion and pressure dissolution do not strongly affect proppant performance after consolidation of the package was achieved.

MUCH & PENNY (1987) outline that the major conductivity decline of sand during the first 20 hours of long-term testing is due to crushing and rearrangement of grains which reduce package porosity and release fines that plug the pore throats in the proppant assemblage. The remainder of the conductivity decline from 20 hours testing time onwards is probably fines migration and redistribution within the proppant assemblage. Proppant pack impairment is caused by embedment into the filter cake deposited on the fracture wall and plugging of the package with formation debris, proppant crush fragments and stimulation fluid residues and precipitates (ROODHART, KUIPER & DAVIES 1986; cf. section 4.3.3.). PARKER & McDANIEL (1987), McDANIEL (1988) and McDANIEL & PARKER (1988) document that gel filter cakes precipitated on fracture walls probably trigger the highest reduction in proppant conductivity due to particle embedment and interstitial plugging in the outer layers of the proppant package within the crack (cf. section 4.3.4.1.4.). Some aspects of gel residue damage, acid dissolution, stress corrosion cracking; effects of high temperature, high brine saturation and high closure stress; and effects of reservoir rock slabs and proppant concentration are briefly outlined as follows. Emphasis is also put on the impact of gel filter cakes on fracture walls.

1.4.10.3.1. Gel residue damage and acid dissolution

Proppant degradation as observed in the field is mainly interpreted to be the consequence of gel residue damage (which can decrease proppant permeability by 20 - 80 %; KIM & LOSACANO 1985) and short-term package consolidation which accounts for abt. 15 - 20 % decline. Lower proppant performance levels than expected from laboratory testing can also be the influence of incomplete sintering of the particles as evidenced by density differences being the result of minor internal additional proppant porosity that cause the grains to be more susceptible to stress corrosion and pressure solution (COBB & FARRELL 1986). Long-term proppant performance under simulated reservoir conditions is also reported by MUCH & PENNY (1987) and PENNY (1987). Crosslinked fracturing fluids leave much more residue than polymer emulsion fluids. While polymer emulsion fluids provide a loose proppant package that is virtually residue-free (cf. section 4.3.4.6.4.), using of crosslinked fluids produces a proppant assemblage in the crack containing a lot of fibrous material between the grains which are then glued together (DAVIES & KUIPER 1988).

During hydraulic proppant fracturing treatments in low-permeability sandstones, stimulation fluid filtrate penetrates into the reservoir formation while a filter cake is deposited on the crack face. Fluid filtrate invasion into pay zone matrix and microfissures causes impairment of permeability and gas flow due to water blocking (cf. sections 3.11.2.4.2. and 4.8.8.3.1.5.), clay swelling (cf. sections 3.5.2. and 3.11.) and dispersion of fine solid particles entrained in the filtrate. Strength reduction of the reservoir rock matrix fabric adjacent to the crack face may initiate collapse and/or plugging of the fracture wall. Filter cake accumulation gives rise to proppant embedment into the draped borehole wall, and dispersion of fine gel residue causes plugging of the pore space of the proppant package in the crack. Proppant conductivity impairment can be provoked by mechanical debris from the collapsed fracture wall, fine fragments from crushed proppant grains, and precipitates from the treatment fluid (ROODHART, KUIPER & DAVIES 1986). GALL & RAIBLE (1985) report on molecular size studies of degraded stimulation fluid polymers, and GHAZALI & WILLHITE (1985) comment on permeability reduction in proppant packages using aluminum citrate/polymer operations.

Various concentrations of HCl/HF acid can dissolve significant amounts of the proppant grains which also seriously impairs conductivity (CHEUNG 1985; cf. sections 4.5.1. and 4.5.5.). Highly-siliceous proppant types also severely degrade in brines at elevated temperatures. Mechanism and temperature of polymer breaking have also an important influence on proppant pack flow impairment, with differences of degree of damage existing for various base fluids (ALMOND & BLAND 1984).

1.4.10.3.2. Stress corrosion cracking

A significant aspect for explanation of the environment-sensitive behaviour of proppants is stress corrosion cracking which occurs in case of contact of water or brine with crystalline or amorphous solids (HILLIG & CHARLES 1965, WIEDERHORN & BOLZ 1970, COOKE 1973; CUTLER, ENNISS, JONES & CARROLL 1983; CUTLER, ENNISS, JONES & SWANSON 1985). Water promotes sandstone fracturing only at elevated temperatures, and the drier the environment, the higher the material strength. The earlier conductivity decrease in water as compared to nitrogen in proppant testing reflects that propagation rates of stress corrosion cracks depend on the environment at the fracture tip, with higher water vapour concentrations increasing crack growth rate. Some aspects of proppant sensitivity to aqueous solutions and grain weakening by stress-intensified corrosion are outlined as follows.

1.4.10.3.2.1. Proppant sensitivity to aqueous solutions

Proppant conductivity testing should therefore be carried out in aqueous environment since this represents a worst possible case situation (CUTLER, ENNISS, JONES & CARROLL 1983; CUTLER, ENNISS, JONES & SWANSON 1985). While sintered bauxite consisting almost completely of corundum is insensitive to aqueous solutions, mullite which is the main constituent of alumina silicate proppants and an important component of alumina silicate and oxide proppants (cf. tab. 2) is susceptible to stress corrosion cracking. Proppant placement within the fractured reservoir should be preferentially carried out with high-quality foamed or energized fluids in order to restrict contact of proppants and aqueous phase to a minimum by guaranteeing immediate invasion of the proppant package by formation gas following quick well cleanup due to the excellent flowback properties of foamed or energized fluids as a consequence of gas drive (cf. section 3.11.2.).

1.4.10.3.2.2. Grain weakening by stress-intensified corrosion

When proppants are held at elevated closure stress and temperature for longer time, the degree of crushing is increased by stress-intensified corrosion which weakens the grains (McDANIEL 1987). As with most corrosion mechanisms, increased temperature environments result in the most stress-intensified corrosion effect. Chemical composition, crystal structure and presence of surface flaws are the primary factors controlling the degree of additional crushing resulting from stress-intensified corrosion at a given closure stress and temperature. Experimental work reveals different permeabilities of proppant packages in different fluids and at different temperatures. Therefore proppant conductivity testing can only approach the reality downhole if carried out under reservoir temperature, pressure and fluid conditions (COOKE 1973).

1.4.10.3.3. Effects of high temperature, high brine saturation and high closure stress

McDANIEL (1986) summarizes research on influence of extended time at different test conditions, variation between ambient and elevated temperature, and effects of brine flow at high temperature and closure stress on proppant conductivity decline. While conventional proppant testing showed that crushing, embedment and plugging by fine solid particles are contributing to in-situ fracture conductivity loss, high temperature coupled with high closure stress and high brine saturations has been proven to have more damaging effects on long-term proppant conductivity than all the combined features previously mentioned together. Drastical proppant conductivity differences are obtained from long-term high-temperature tests and short-term ambient-temperature experiments (COOKE 1973, McDANIEL 1987), with significantly lower proppant conductivities having been measured under realistic reservoir conditions than reported for earlier short-term testing (DAVIES & KUIPER 1988). The following overview focusses on silica saturation and oxygen concentration as well as on reservoir condition spectrum approach.

1.4.10.3.3.1. Silica saturation

Important parameters of proppant conductivity testing are also oxygen content and pre-test silica concentration of the fluid (McDANIEL 1986). At a constant closure stress level, conductivity is irreversibly declining with rising temperature and is not recovered by subsequent cooling down. Closure stress increases the adverse effects of temperature and water-based fluids in many proppants. In actual downhole fracture conditions in a sandstone reservoir, any water present would be saturated with silica due to exposure to the vast amount of silicate surfaces of the reservoir rock and corresponding leaching to the limit of uptake capacity of the fluid.

If a silica-saturated test fluid is used (cf. also MUCH & PENNY 1987), many of the previously reported catastrophical conductivity losses of proppants and sand cannot be reproduced. On the other hand, the brine used in fracturing fluids contains no silica or alumina, while the brine produced from a sandstone formation will be silica- and probably also alumina-saturated. Recommendation can therefore be made that a realistic conductivity test would incorporate a silica/alumina-free brine in the early stage which is then followed by a silica/alumina-saturated brine in the later phases, because the original flowing medium in the reservoir is switching over to a silica/alumina-saturated brine after a period of time (MUCH 1987).

1.4.10.3.3.2. Oxygen concentration

The oxygen level has to be extremely lowered before significant differences in test results are obtained, with resin-coated sand being no longer discoloured or blackened and the frequency of plumbing leakage being drastically reduced. In contrast to hastelloy, stainless steel test cells corrode when exposed to heated brine that contains oxygen, and therefore absolutely minimal amounts of oxygen are essential (MUCH 1987), with even trace quantities of oxygen creating minute corrosion of the stainless steel cell which can provoke dramatical proppant conductivity decline by precipitation of corrosion products within the proppant package. Fracturing fluids carrying the proppants, however, are often supersaturated with oxygen because of the method of blending and proppant addition, while it is assumed that hydrocarbon reservoirs contain no oxygen.

Accurate modelling of a stimulation treatment in a test cell therefore requires heated brine containing oxygen as the initial flowing medium which excludes use of stainless steel cells. Later the flow medium should be reduced to zero oxygen brine after some finite amount of time which would model returned fracturing fluid and formation brine after the oxygen in the solution has reacted with the fracturing gels (oxidizers are commonly added as gel breakers). Thus both oxygen and silica/alumina saturation have to be changed throughout the experiment, with oxygen and silica/alumina concentrations having to be reduced and increased, respectively, during progression of the test and therefore have an inverse correlation.

The oxygen-rich/oxygen-free brine is a very important consideration when evaluating resin-coated proppants which are susceptible to oxidation giving quite different performance results in oxygen-free as opposed to oxygen-rich brines. Hastelloy cells are the ultimate choice for testing proppants containing dissolved oxygen, because hastelloy is almost inert in an oxygen-containing brine as long as fluid pH is controlled. CALLANAN, CIPOLLA & LEWIS (1983) document that exposure to hot brines can reduce proppant conductivity to 50 - 75 %. The exposition to hot brine can cause a two- to three-fold permeability reduction in sand and bauxite proppant packs (COOKE 1973). McDANIEL (1987) concludes that increased temperature environments result in the most stress-intensified corrosion effect which weakens the proppant grains and increases the degree of crushing.

1.4.10.3.3.3. Reservoir condition spectrum approach

McDANIEL (1988) performs proppant conductivity testing under the hitherto broadest and most complete set of environmental conditions approaching in-situ reservoir situation as closely as possible, including downhole temperature, extended test time, core slabs and wafers as cell limitations, gel residue in the proppant pack, gel filter cakes from dynamic fluid loss tests on the simulated crack walls, fracture fluid clean-up period, wet gas as flowing medium, varying closure stresses, and non-Darcy flow effects. Considering the whole suite of parameters, the test results allow more accurately modelling of fracturing treatments within a design simulator and better predicting of post-fracturing production using a reservoir simulator.

1.4.10.3.4. Effects of pay rock slabs and proppant concentration

The traditional conductivity vs. stress evaluation can overestimate proppant performance under downhole conditions (MUCH & PENNY 1987). Conductivity at stress vs. time at temperature provides more meaningful data for estimating well performance. Bulk density and porosity values of proppants assessed over time at closure stress give more accurate information than those at no stress and zero elapsed time. Some aspects of wall slip and metal piston vs. filter cake and core slab as well as fracture porosimetry as evaluated as follows.

1.4.10.3.4.1. Wall slip and metal piston vs. filter cake and core slab

Proppant assessment using hard metal pistons as test cell boundaries may introduce a variable wall slip that can cause measured proppant assemblage conductivity to be greater than absolute proppant package conductivity, whereas if performing experiments with medium-hard sandstone core slabs, proppant embedment becomes a significant factor when exceeding 8,000 psi closure stress and reduces the amount of wallslip (cf. sections 1.4.10.3.4.1. and 4.3.4.1.4.). Filter cakes deposited on the fracture walls when crosslinked polymer-gelled fluids leak off into the reservoir formation can reduce proppant package conductivity at 2 lbs/ft2 (10 kg/m2) concentration by up to 50 %, with filter cake accumulation and subsequent proppant embedment decreasing effective width of the supporting material package regardless of proppant type. The various tests have shown that regardless of the environmental setting, ceramic intermediate-strength proppants can provide conductivities which are over 1000 % greater than those of sand at equal proppant saturations and closure stress conditions.

The proppant grains which are in contact with the experiment cell face or platen are under greatest stress during the test, because each bead has fewer contact points with other spheres than those farther away from the platen surface and toward the fracture center that have more contact points amongst their neighbours depending on their random manner of packing (MESSINA 1987). In steel experiment cells, embedment into the platen in order to release stress is not possible and therefore stress equalization finally takes place by grain crushing, whereas core slabs as bounding faces of the test cell allow embedment of proppant grains and thus reduce the amount of proppant crushing.

1.4.10.3.4.2. Fracture porosimetry

WENDORFF (1982) introduces the fracture porosimeter for fast and simplified determination of crack conductivity at downhole stresses. Permeability, porosity and width of the fracture depend on type and amount of proppants and magnitude of applied stress and are measured between samples of reservoir rock to provide accurate information on crushing and embedment. As closure stress rises, surface area increases, but not nearly as much as permeabilities diminuish. Fracture porosity decreases as load is applied because proppant grains crush and embed in the rock matrix. In competent rock, proppant porosity is usually reduced by crushing, whereas in soft formations, proppant embedment coupled with proppant crushing results in very low fracture porosity. In most propped fractures, both embedment and crushing (cf. section 4.3.3.1.) occur and the resulting porosity is a measure of the combined effect.

1.4.10.3.5. Gel filter cake on fracture wall

PARKER & McDANIEL (1987), McDANIEL (1988) and McDANIEL & PARKER (1988) document the effect of the gel filter cake being deposited on the fracture walls as the fluid leaks off into the formation on conductivity of the proppant package within the crack. The main impacts of the gel filter cake on the fracture wall are proppant embedment and reduction of effective fracture width as well as proppant permeability damage by filter-cake buildup. Comments are also offered on proppant concentration and fluid type as well as on removal of gel filter cake by combined fluid-loss additive and gel breaker. A brief comparison of gel filter cakes in hydraulic fracturing and gravel packing is also provided.

1.4.10.3.5.1. Proppant embedment

Dynamic fluid loss tests indicate that in many cases, average filter cake thicknesses can be very close to proppant grain size for 20/40 mesh particles (PARKER & McDANIEL 1987, McDANIEL 1988). Thus there is a potential to fill the proppant porosity with filter cake and loose the flow along the wall of both fracture faces and for a depth of one or more proppant grains on each side of the crack. As proppants embed into the filter cake when the fracture is closing, they may displace parts of the thick gel residue towards the center of the proppant package within the crack. This embedment, deformation and dispersion of the gel filter cake seriously deteriorates proppant permeability and thus fracture conductivity.

Experimental work has shown that conductivity diminution is least with the base gel alone. Further reduction is observed with addition of liquid hydrocarbon phase for fluid loss control, and the most severe decline occurs when using a solid particulate fluid leakoff additive. Wet gel filter cakes provoke drastical reductions in proppant conductivity even at low closure stress levels. The test results for Ottawa Sand reveal that the range of closure stress for the application of natural sand is significantly lower than that required for many fracturing applications, with the use of intermediate-strength proppants being already necessary from closure stress values of 4,000 psi onwards (McDANIEL & PARKER 1988).

1.4.10.3.5.2. Reduction of effective fracture width

MUCH & PENNY (1987) document that the main mechanism of fracture conductivity deterioration is reduction of the effective width of the proppant package by formation of a filter cake covering the crack walls and embedment of the outer layers of the proppant wedge into the filter cake drape. Testing with reservoir core slabs provides more realistic conditions by allowing fluid leakoff and consequently origin of a filter cake, whereas proppant evaluation using hard metal pistons introduce a wall slip variable that causes the measured proppant conductivity to be greater than the absolute value for the reason of suppression of filter cake development (MUCH & PENNY 1987). Wall slip effects in steady shear viscosity measurements of hydraulic fracturing fluids are also described by GUILLOT & DUNAND (1983) and KNOLL (1985). Filter cake deposition and proppant embedment reduce the effective width of the proppant package regardless of proppant type (cf. also sections 1.4.10.3.4.1. and 4.3.4.1.4.).

Additional effects are generated by dispersion of fine fracturing fluid residue in the proppant package and non-Darcy flow (cf. section 4.3.3.3.) in the restricted pore space, as well as water blocking (cf. sections 3.11.2.4.2. and 4.8.8.3.1.5.) and clay swelling (cf. sections 3.5.2. and 3.11.1.) in the reservoir formation adjoining to the fracture walls (ROODHART, KUIPER & DAVIES 1986). In terms of relative deterioration effect, crosslinked gels show the highest impairment because of formation of very thick filter cakes, and batch-mixed gelled oils exhibit higher deleterious impact than compared with on-the-fly mixed systems. The least damaging fracturing fluid appears to be the polymer emulsion system (ROODHART, KUIPER & DAVIES 1986). The significance of non-Darcy flow effects on the accuracy of proppant conductivity modelling is also emphasized by COOKE (1973); EVANS, HUDSON & GREENLEE (1985); EVANS & EVANS (1986); MALONEY, GALL & RAIBLE (1987) and McDANIEL (1988).

1.4.10.3.5.3. Proppant permeability damage by filtercake buildup

The filter cake forming on the crack walls upon leakoff of high-pressure fracturing fluid consists mainly of polymer and fluid-loss additives (DAVIES & KUIPER 1988). Filter-cake thickness is determined by the particular stimulation fluid used, formation characteristics, fracture-to-reservoir pressure difference, and erosional effects caused by slurry being pumped along the crack faces. During fracture closure, proppants are embedded into the filter cake, making it difficult to remove the crack wall drape during production. While crosslinked fluids with diesel as fluid-loss additives generate thick filter cakes which can almost completely block thin fractures, polymer emulsion fluids do not give significant filter-cake buildup (cf. section 4.3.4.6.).

As proppant permeability damage by crosslinked fluids substantially depends on filter-cake buildup, minimization of the phenomenon can be achieved by using an effective viscosity breaker which is particularly necessary in low-temperature shallow formations. While inadequate breakers lead to virtually complete loss of proppant pack conductivity, aggressive breaker schedules on the other hand can provide high retention factors in excess of 80 %, but can also give rise to excessive proppant settling in the fracture before closure, with therefore a proper balance being required.

1.4.10.3.5.4. Proppant concentration and fluid type

Proppant saturation has a significant influence on remaining fracture conductivity in the presence of a gel filter cake (PARKER & McDANIEL 1987). Low proppant concentrations accentuate permeability deterioration caused by the gel filter cake, and therefore high proppant saturations are essential to achieve the desired fracture conductivity. In the presence of gel filter cakes, the permeability of a propped crack is not constant, and design simulations should be altered to allow input of conductivity data as a function of proppant concentration (McDANIEL & PARKER 1988) and thus variable conductivity from a maximum at the fracture mouth to a minimum at the crack tip. Gel filter cakes cause even lower conductivity than that measured in long-term tests without filter cakes (cf. section 4.3.4.1.4.). Different types of fracturing fluids form filter cakes that result in varying degrees of damage to proppant bed conductivity. High reservoir temperatures may actually lessen the singular effect of gel filter cakes due to more degradation of the gel residue.

Another variable is the type of filter cake present in the proppant pack with fluid-loss additives creating a thinner filter cake on the fracture wall as a consequence of retarded fluid leakoff and thus diminished solid agent segregation and deposition. Less fluid leakoff which means less gelling agent in the filter cake will normally give rise to less proppant bed damage. The degree of gel degradation in the filter cake is also of considerable importance. The results of the studies of McDANIEL (1986, 1987, 1988), PARKER & McDANIEL (1987) and McDANIEL & PARKER (1988) thus clearly show that time, temperature, stress and presence of gel filter cakes are all important factors influencing crack conductivity. Another variable affecting proppant conductivity is the number of phases present. While proppant package conductivity is normally measured with single-phase flow, addition of a second or third phase reduces effective proppant assemblage permeability to the original phase significantly (DAVIES & KUIPER 1988).

1.4.10.3.5.5. Gel filter cakes in fracturing and gravel packing

A comparable phenomenon to gel filter cakes originating on formation surfaces at fracture walls are gel covers created on the reservoir face during gravel packing as a consequence of fluid leakoff and filtration (TORREST & MARCANO 1986). These gel covers can rapidly build up a low-permeability drape that dominates system behaviour and inhibits further fluid loss of the gravel carrier thereby deteriorating gravel placement and pressure response. The layer buildup from concentrated solutions would also tend to give false signals of job completions as pumping pressures rise rapidly. Reduced gravel pack quality and performance could result from failure to account for the effect of surface gel filtration (cf. sections 4.3.4.1.4. and 5.3.1.1.2.2.).

1.4.10.3.5.6. Removal of gel filter cake by combined fluid-loss additive and gel breaker

CANTU & BOYD (1988) present an organic acid polycondensation product being a polyester which acts as combined fluid-loss additive and gel breaker and that has the advantage to disintegrate into soluble monomers at reservoir temperature after termination of the stimulation treatment, thereby secondarily removing the filter cake which has been primarily deposited on the fracture walls and finally leaving no substrate for proppant embedment thus maximizing fracture conductivity. Upon thermal degradation to soluble monomers, the organic ester exposes increasing acid functional groups which break the crosslinked and/or linear fracturing gel used as stimulation fluid as a consequence of decreasing pH (cf. also section 4.3.4.6.4.).

The excellent cleanup of the filter cake with no remaining residue is supported by the concentration of the polyester particulates on the fracture wall in contrast to conventional water-soluble breakers which leak off into the reservoir formation. The particulate polyester has a fluid-loss agent performance at least as good as that of silica flour (cf. section 4.3.4.6.2.) in crosslinked and linear gels as well as in nitrogen foam, and its decisive advantage with respect to classical two- or multicomponent fluid-loss additive/gel breaker systems is that the polyester is only one agent with a double function, with the two successive tasks being fulfilled while changing from solid polyester to soluble monoester upon thermal and/or chemical degradation and thus leaving no particulate residues that could impair proppant conductivity (cf. also sections 4.3.4.6.4. and 4.8.8.3.2.3.5.).

1.4.10.3.6. Gelling agent concentration

Proppant-pack permeability is significantly impaired by the fracturing fluid, with damage degree increasing with rising gelling agent saturation (HAWKINS 1988). In many hydraulic fracture treatments, the gelling agent is often concentrated over ten-fold upon crack closure. Laboratory studies investigating gelling agent saturation within the proppant package, proppant concentration, closure stress, shut-in time, and displacement fluid and its flow rate reveal that even at relatively low gelling agent saturations, significant proppant package permeability impairment occurs. As high displacement rates during well flowback only marginally remove the damage, high breaker concentrations are necessary to significantly reduce permeability impairment. Following discussion of some general aspects, an outline is given on cleanup flow rate and closure stress as well as proppant grain size and breaker saturation. Comments are also offered on hydrocarbon-based cleanup fluids for proppant package permeability improvement.

1.4.10.3.6.1. General aspects

Parameters affecting proppant package permeability have to be divided into parameters operative under all conditions and such applying only when gelling agents are present (HAWKINS 1988). Addition of fracturing fluids to the proppant package greatly complicates both experiments and analysis of results. Factors influencing proppant permeability reduction are gelling agent type and concentration, type and stability of crosslinker, and breaker content. Other important gelling agent properties are molecular weight/size and residue volume after action of a given fracturing fluid breaker or thermal degradation.

Formation of an immobile residue arising from increasing gelling agent saturation due to fluid loss and crack closure and/or as the result of breaking mechanisms is a major source of permeability reduction. As polymeric gelling agents which are not degraded to small molecular size do not penetrate into the matrix of formations which are typically candidates for fracturing treatments, the gelling agent is concentrated due to fluid leakoff during pumping and fracture closure and accumulates within the proppant package with appreciable rigidity. Further complications may arise due to formation of residues as the result of chemical or thermal breaking processes.

1.4.10.3.6.2. Cleanup flow rate and closure stress

Concerning cleanup flow rate, an increase even over a twelve-fold span has little effect on final steady-state permeability or cleanup rapidity (HAWKINS 1988). Higher flow rates even tend to result in somewhat lower permeability recovery, with thus evidently not dislodging restricting species. Over the range of 140 - 200 oF (thus below onset of thermal fluid degradation), temperature has only a minor effect on final permeability or permeability recovery of proppant packages containing concentrated fracturing fluids without breakers, and permeability values are only slightly improved after high cleanup flow rates. With increasing shut-in time at temperature and under closure stress, both final permeability and permeability recovery of the proppant package decrease dramatically. Therefore a minimal shut-in time is maximizing proppant package permeability.

Increasing closure stress also substantially reduces permeability of the proppant assemblage containing fracturing fluid although no unusual flow pattern is present and high flow rates have only a marginal effect. Increasing gelling agent saturation rapidly reduces proppant package permeability, with use of higher proppant loadings increasing retained proppant assemblage conductivity. Filter cakes deposited on fracture walls have increasingly greater influence on permeability, with proppant concentration decreasing to such an amount that filter cake thickness becomes a significant fraction of the cross-sectional flow area.

1.4.10.3.6.3. Proppant grain size and breaker concentration

Investigation of the effect of increasing proppant grain size and narrowing distribution by changing from 20/40 to 20/25 mesh size reveals that proppant package permeability can be significantly increased especially after high cleanup flow rates, with thus increasing proppant grain size being one of the most suitable ways of increasing proppant permeability and fracture conductivity under all possible laboratory and field conditions (cf. section 4.3.5.). The degree of permeability damage resulting from selected fracturing fluids containing no breakers at 175 oF parallels their temperature stabilities which depends from chemical composition, and proppant packages containing fracturing fluids with high gelling agent concentrations require high breaker saturations for reasonable permeability improvement (HAWKINS 1988). At saturations where the breaker is highly effective in reducing fluid viscosity at their original polymer concentration, it is much less effective than the highly-enriched polymer within the proppant package, and therefore a breaker system has to be devised which will maintain fluid properties during the treatment and improve fracture permeability. Caution has to be exercised, however, because high breaker saturations could give rise to extensive fluid degradation during the fracturing treatment.

1.4.10.3.6.4. Hydrocarbon-based cleanup fluids for proppant package permeability improvement

Hydrocarbon-based cleanup fluids cause an initial decrease in apparent permeability as a consequence of intermittent minor emulsion generation which rapidly dissipates followed by a significant increase. Hydrocarbon-based fracturing fluids have the effect of very high proppant package permeability under all conditions, with hydrocarbon-based stimulation fluids clearly causing much less permeability reduction both absolutely and as a percentage than aqueous gelled fluids (HAWKINS 1988). Therefore hydrocarbon-based treatment fluids should be used whenever fracture permeability needs to be maximized. In order to maximize proppant package permeability when aqueous gelled fracturing fluids have to be applied, emphasis has to be put on minimization of use of crosslinking, polymer concentration and shut-in time, and maximization of proppant saturation/loading, proppant grain size, and breaker content of the fluid.

1.4.10.4. Recommended conductivity correction factors

As a result of the outlined testing campaign, McDANIEL (1986) concludes that laboratory measurements of proppant conductivity at ambient temperatures and short times of experiment duration at each closure stress level are typically too optimistic as compared to more realistic in-situ well conditions. Conductivity correction factors of 0.5 - 0.6 for intermediate- and high-strength proppants are recommended to compensate all the detrimental effects by the mentioned test conditions when planning fracture conductivity for an actual stimulation treatment by recalculating laboratory data to in-situ downhole conditions in the fractured reservoir rock. MUCH (1987) even reports examples where laboratory-derived proppant conductivities have to be reduced by 85 - 90 % in order to correlate with history match conductivity values for the reasons of damage by fracturing fluid gel residue, proppant embedment, proppant crushing, and non-Darcy flow in the proppant package.

An important cross-check before the final operation is a minifracture calibration treatment which can verify the correctness of the assumptions and discounting (cf. section 4.8.1.). MONTGOMERY & STEANSON (1985) put forward that in deep well completions, the final recommended conductivity needs to be two to three times higher than the theoretically designed conductivity, with therefore overdesign being necessary to finally account for downhole conditions and results (cf. section 1.4.10.4.3.1.). Further comments on realistic fracture conductivity of proppants particularly as a function of reservoir temperature are given by McDANIEL (1987, 1988), and PENNY (1987) reports on the effects of environmental conditions and stimulation fluids on long-term proppant conductivity. Some aspects of influence of testing equipment, impact of brine composition and conductivity improvement are briefly outlined as follows. Comments are also offered on practical use of proppant conductivity data.

1.4.10.4.1. Influence of testing equipment

ROBINSON, HOLDITCH & WHITEHEAD (1986) put forward that because most of the fracture conductivity tests are done with steel plates, embedment effects are not considered and therefore the measured data are too optimistic. Unless using reservoir core slabs as bounding faces of the simulated fracture, the conductivity values should be reduced by a factor 2 - 3. Proppant conductivity is further diminished by the presence of gel residue, formation fines and other solid materials in the pore space. Although this problem varies according to formation and fluid type and usually is not as serious as crushing and embedment, it has to be considered that solid fine particles can further decline fracture conductivity which should be accounted for by decreasing the values by another factor 2 - 3 in order to properly simulate actual reservoir conditions. Better experimental conductivities than those achieved downhole in the pay formation are also the consequence of wall slip features (cf. section 1.4.10.3.4.1.). Non-Darcy flow phenomena in the propped fracture result in an extra pressure drop which can reduce effective fracture flow capacity by more than a factor 3 (DAVIES & KUIPER 1988).

1.4.10.4.2. Impact of brine composition

Similar or even higher discounts would be necessary for laboratory data obtained from tests in 2 % KCl solutions which match the composition of the prepad and displacement stages in MHF operations and are the base fluid for many crosslinked gels (PAI, GARBIS & HALL 1983; BLEAKLEY 1984), but do not agree with the partially saturated aggressive brines in Rotliegend and Carboniferous reservoir rocks in Germany FRG which become effective after termination of the fracturing treatment when the invasion of the pay horizon caused by the injected fluids has been reversed and also the crack is flushed with high-salinity formation water in sections below the gas-water-contact, as well as during progressive depletion when the water cut in the gas is steadily rising in reservoir intervals above the gas-water-boundary. The outline as follows discusses clay stabilization by KCl water in field stimulation application and KCl solutions as laboratory testing fluids.

1.4.10.4.2.1. Clay stabilization by KCl water in field stimulation application

2 % KCl solutions which are frequently taken as proppant testing fluids are the base composition for the majority of stimulation fluids and are used in the prepad of hydraulic fracturing treatments for the purpose of clay stabilization and thus prevention of excessive formation damage particularly in water-sensitive reservoir rocks as a standard safety precaution (BLACK & HOWER 1965; BLACK, RIPLEY, BEECROFT & PAMPLIN 1979; WATERS 1980; NALL, CAMPBELL & BONEY 1983; SYDANSK 1984, MUCH 1987, MUCH & PENNY 1987; cf. section 3.5.2.2.).

2 % KCl water in combination with clay stabilizers renders the stimulation fluid non-damaging by controlling particle sloughing and migration, and an additional surfactant reduces boundary tension at the formation and fracture fluid interface to facilitate treatment cleanup after sufficient viscosity reduction of the gel by breaking (PAI & GARBIS 1983 b). In addition to stimulation fluids, KCl can also be applied as inhibitor in drilling mud and cement slurries (MESSINA 1986 b). While 2 % KCl water prevents formation damage by filtrate, 5 % KCl water inhibits shale sloughing and helps to maintain good hole conditions, and 2 - 5 % KCl polymer systems provide viscosity and water loss control.

1.4.10.4.2.2. KCl solutions as laboratory testing fluids

2 % KCl solutions have frequently been used as standard test fluids in laboratory proppant conductivity experiments (McDANIEL 1986, 1987, 1988; MUCH & PENNY 1987, PARKER & McDANIEL 1987, PENNY 1987, McDANIEL & PARKER 1988). Since a large number of gas wells which are hydraulically fractured produce mostly gas after initial cleanup, a better approach to in-situ conditions is provided by changing the flow medium from 2 % KCl solution to wet gas after some days in order to simulate the sequence of fracturing fluid backproduction followed by gas withdrawal. Experimental work reveals better permeability preservation when wet gas is used as testing medium after a couple of days than with 2 % KCl water alone. The difference in remaining proppant conductivity is interpreted to be partially the result of gel filter cake dehydration by the wet gas, with this shrinking effect exerted on the gel filter cake increasing proppant package porosity and permeability (McDANIEL 1988). Further investigations have to consider liquid hydrocarbons to check accurate modelling of oil reservoirs in the experiments. Wet gas as flowing medium in tests including gel filter cakes has also been used by ROODHART, KUIPER & DAVIES (1986). Apart from KCl, clay stabilization can also be achieved with KOH (SLOAT & LARSEN 1984, SYDANSK 1984, SLOAT 1988).

Other aspects of stabilization of both dispersed and massive clays and prevention of their migration during fracturing operations and other reservoir treatments are presented by McLAUGHLIN, ELPHINGSTONE & HALL (1976); CLEMENTA (1977); McLAUGHLIN, ELPHINGSTONE, REMINGTON & COATES (1977); PETERS & STOUT (1977), THOMAS & CROWE (1978); COULTER, COPELAND & HARRISBERGER (1979); WILLIAMS & UNDERDOWN (1980); CALLAWAY, ORTIZ & HOLCOMB (1982); HILL (1982), BORCHARDT & BROWN (1984); BORCHARDT, ROLL & RAYNE (1984) and SLOAT & LARSEN (1984).

1.4.10.4.3. Conductivity improvement

POULSEN & SOLIMAN (1987) emphasize that a correction factor of 0.5 is best compensating the up to ten-fold decrease in proppant conductivity from laboratory experiment to field execution conditions, and PHILLIPS & ANDERSON (1985) account for proppant performance damage by a factor of 0.2 - 0.25. KIM & LOSACANO (1985) also propose a correction factor of 0.4 - 0.6 to compensate gel residue plugging of proppant pore space. SECCOMBE & ANDERSON (1982) suggest to multiply short-term proppant testing data by a correction factor of 0.1 to match post-fracturing calculations, and SOLIMAN (1983) determines post-stimulation conductivities of 0.125 - 0.25 of the values predicted from laboratory experiments. REINICKE, BRINKMANN, SCHWARZ & HUENI (1985) find correction factors being in some cases as low as 0.05. Some comments are offered on stimulation treatment overdesign as well as demand of higher proppant quality and saturation.

1.4.10.4.3.1. Stimulation treatment overdesign

Failure to consider the various types of proppant conductivity impairment and to compensate them by correction factors may cause dramatical underdesign of the fracturing treatment, with therefore the actual job having to be overdesigned for the purpose of achieving the desired in-situ conductivity after all the possible flow capacity deteriorations even in the worst case of combined action of all detrimental factors (ROODHART, KUIPER & DAVIES 1986). In terms of layout, a conductivity correction factor of 0.5 means that for constant fracture conductivity, both required proppant concentration at the wellbore and total proppant amount have approximately to be doubled (POULSEN & SOLIMAN 1987). Conductivity improvement when applying correction factors in the range of 0.2 - 0.5 can be achieved by increasing the maximum allowable proppant saturation or by using a higher conductivity proppant of better quality and/or larger grain size.

As many laboratory proppant tests under different conditions have suggested that the final in-situ conductivity of the proppant package within the fracture can be as low as abt. 10 % of the original proppant conductivity (LI & ZHU 1986), proper compensation of the deterioration should be made by overdesigning crack conductivity in such a manner that the fracture flow capacity which is ultimately left downhole is at least equal to the expectation and to the necessary minimum to economically feasibly drain the reservoir, if not higher (ROODHART, KUIPER & DAVIES 1986).

1.4.10.4.3.2. Demand of higher proppant quality and saturation

Using realistic proppant conductivity data in hydraulic stimulation design reveals that in order to account for the necessary conductivity loss compensation, higher proppant concentrations and/or qualities may be required to achieve the desired fracture conductivity in the reservoir environment (McDANIEL 1987). Fracture overdesign should also be made in terms of length, as many low-permeability reservoir treatment jobs have turned out to actually reach only abt. 70 % of the designed propped crack length (HOLDITCH & LEE 1979). An unique and unambiguous history match can only be obtained if the dimensionless fracture conductivity is 10 or greater (cf. section 4.8.11.1.).

The application of realistic and corrected proppant conductivities for fracture stimulation design often results in the necessity to choose considerably higher-strength proppants and/or higher proppant saturations and/or also larger grain sizes than originally planned and expected in order to adequately design and accurately model the fracture operation (McDANIEL 1987). Accurate design of fracturing treatments requires conductivity measurements at simulated reservoir conditions (McDANIEL & PARKER 1988). The comparison of short-term testing, long-term testing without gel filter cakes and long-term testing including gel filter cakes reflects progressive conductivity deterioration as the experimental conditions are more and more approaching the actual downhole reservoir environmental situation. Therefore reliable fracture conductivity modelling and design has to be based on proppant testing carried out under conditions that are as closely as possible approaching those present in the field.

1.4.10.4.4. Practical use of proppant conductivity data

Proppant conductivity data are practically used in two basic applications in oil and gas industry (McDANIEL 1988). The first one comprises offtake increase calculations where an average conductivity value of the resultant propped fracture is used to forecast multifold productivity increase of the well (TINSLEY, WILLIAMS, TINNER & MALONE 1969; SOLIMAN 1983), and the second one includes computerized fractured reservoir simulators which can be modified to incorporate the effects of non-Darcy flow. Some aspects of laminar vs. turbulent flow and conductivity necessity for cleanup vs. production are discussed as follows.

1.4.10.4.4.1. Laminar vs. turbulent flow

Hitherto typically laminar-flow conductivity values have been used and no allowance is made for the increased flow resistance caused by non-Darcy effects as a consequence of turbulences. For gas or oil wells producing at very high rates or exploiting very low-viscosity oil, respectively, however, the use of laminar-flow conductivity data results in overestimating the effectiveness of the planned stimulation treatment, and therefore non-Darcy flow phenomena including turbulences have to be taken under consideration (McDANIEL 1988; cf. section 4.3.3.3.).

Concerning computerized fractured reservoir simulators which can be modified to incorporate the features of non-Darcy flow due to current turbulences, it is essential that conductivity values used in the simulator are measured under test conditions which model realistic reservoir parameters. Attention has to be paid not only on proppant type and size, but also on long-term exposure to stress at in-situ temperature including any variations from theoretical behaviour caused by such effects as presence of gel filter cakes occupying parts of the proppant bed porosity.

1.4.10.4.4.2. Conductivity necessity for cleanup vs. production

In terms of proppant performance modelling, it has to be respected that the optimum fracture conductivity necessary to clean-up the formation may be much higher than that required to produce gas from the reservoir (SOLIMAN & HUNT 1985). Therefore crack parameters should be especially for tight gas reservoirs designed such that a fracturing treatment will leave enough proppant conductivity for efficient cleanup of stimulation fluid. McDANIEL (1987, 1988) and PARKER & McDANIEL (1987) also recommend to place higher proppant concentrations than theoretically necessary in the crack in order to overcome closure effects, to achieve satisfactory in-situ fracture conductivity and thus further improve stimulation results. DAVIES & KUIPER (1988) underline that recent long-term conductivity measurements under realistic reservoir conditions have revealed a technical requirement for stronger, more conductive proppants not only at higher closure stresses in deeper pay zones, but also at lower closure stresses in shallower prospective intervals in order to guarantee optimum in-situ conductivity installation. The latter relationship also applies if coarser proppant grades are used in an attempt to increase fracture conductivity, because crushing of coarser grain sizes starts at lower closure stress levels than that of finer grain sizes.

1.4.10.5. Performance experience in long-term field application

All the experimentally concluded reductions of proppant conductivity, however, are in practical application in the field at least in low-permeability reservoirs in many cases of low significance at the bottom of the line, as the resulting in-situ proppant conductivity is still much higher than formation permeability and thus gives especially in tight sandstones still a marked contrast between reservoir rocks and proppant-filled drainage fracture in a satisfactory way to be the guarantee for long-term gas production at economically sufficient and appreciable rates (as evidenced by numerous Carboniferous and Rotliegend gas wells in Germany FRG which have been treated up to 10 years ago and are still producing at reasonably feasible rates that are not significantly different from those achieved shortly after the operation or are following the predicted natural rate decline with progressive depletion). Some examples from North China are also illustrated.

1.4.10.5.1. Rotliegend and Carboniferous in Germany FRG

Minor amounts of back-produced proppants that were recovered in some deep Rotliegend and Carboniferous gas wells reveal only subordinate mechanical and chemical alterations of the injected proppant grains thus confirming the physical stability of the proppant package in the fracture as evidenced by the persisting satisfactory gas withdrawal rates and also testify to only subordinate or even negligible fracture evacuation by proppant flowback (cf. also section 1.4.1.). Examination of grain surfaces and shapes reflects only minor dissolution and crushing of some parts of the particle population and thus confirm the excellent proppant behaviour which is independently suggested by the promising hydrocarbon offtakes rates. COBB & FARRELL (1986) conclude that the API conductivity data have to be discounted by abt. 20 % if the well conditions are at least as severe as 10,000 psi closure stress, 93 oC (200 oF) reservoir temperature and 2 % KCl fluid saturation (corresponding to the prepad flushing).

REINICKE, BRINKMANN, SCHWARZ & HUENI (1985) observe in Rotliegend and Carboniferous gas wells in Germany FRG a fracture conductivity reduction by a factor two to five which enables performance of a good history match including pressure drop, thus agreeing with GUPPY, CINCO-LEY & RAMEY (1982) that the apparent conductivity obtained from buildup solutions can be up to 85 % lower than the true conductivities. Cross-flow during shut-in times can be expected to become significant if some horizons of the fractured reservoir section are depleted preferentially with increasing exploitation, but no cross-flow was observed in production control surveys during more than 5 years. Concerning field performance interpretation, REINICKE, BRINKMANN, SCHWARZ & HUENI (1985) record a difference of a factor two to five or even greater when comparing computed fracture conductivities with simulation-derived or integral-method-derived values, respectively.

1.4.10.5.2. North China

LI & ZHU (1986) document that the propped fracture conductivity as calculated from transient pressure test data in the range of 2 - 5 months after the stimulation job in moderate- to high-depth oil reservoirs in North China is only 30 - 70 % of the laboratory experiment-derived values, with a minimum of abt. 10 % also being reached in some cases. An analysis of the phenomena indicates that in addition to damage caused by scraped fines from the fracture wall by proppant-carrying fluid, migration of formation fines, embedding or crushing of proppants in the crack, and incompatibility of treatment fluid to formation fluid, the damage generated by degraded fluid residue has significant impact. Fracture conductivity damage can be effectively reduced by increasing proppant concentration. BAKHTAR, BARTON, RAKOP & JONES (1985) discuss modelling of crack permeability around a well during depletion. Long-term performance of hydraulic fracture treatments is also reported by ZAHNER & CRAFTON (1985).

1.4.11. Proppant mixing

The conventional application of proppants in hydraulic fracturing comprises placement of material of unimodal grain size and homogeneous type, but also several possibilities exist for usage of different proppant types and grain sizes in one operation, comprising tail-in and mixing of proppants. Tail-in is performed by pumping of separated lots of proppants of successively increasing closure stress resistivity or conductivity and grain size in order to achieve maximum support in the vicinity of the wellbore where the highest closure stress occurs due to maximum drawdown and highest flow velocities (HOWARD & FAST 1970 a; HICKEY, BROWN & CRITTENDEN 1981; cf. also section 2.4.1.2.). While during course of execution of tail-in fracturing jobs mixing of different proppant types and grain sizes is not intended and can only occur accidentally to minor amounts at the transitions between different treatment stages, mixing of proppants with different physical properties was frequently approached experimentally in order to check its significance for possibly increasing or decreasing proppant conductivity and stability above the level achieved by unimodal homogeneous populations (McDANIEL & WILLINGHAM 1978). Comments are offered as follows on the effects of mixing of proppant types, grain sizes and grain shapes.

1.4.11.1. Mixing of proppant types

The results of most proppant mixing investigations reveal that the best effects in terms of conductivity are obtained with unimodal homogeneous distributions. Mixing of proppant types and grain sizes hardly has an improving impact, if any. Some aspects of mixing of natural sand and synthetic proppants as well as of different sand types are sketched as follows.

1.4.11.1.1. Mixing of natural sand and synthetic proppants

Mixing of natural sand with sintered bauxite or glass beads has been temporarily been common as a consequence of expected possibilities of proppant cost reduction especially in the early 1970`s when glass beads prior to market introduction of sintered bauxite were the only reliable high-strength material and in 1976 - 1978 when the newly invented sintered bauxite was only available in limited quantities at an excessive price which was up to 15 times that for natural sand. Following a general characterization of mixing of sintered bauxite or glass beads with sand, some aspects of conductivity deterioration by proppant mixing are outlined.

1.4.11.1.1.1. Mixing of sintered bauxite or glass bead with sand

Concerning mixing of proppant types of different provenance, McDANIEL & WILLINGHAM (1978) tested mixtures of sand and glass beads or sintered bauxite and found out that except of 50 : 50 (and sometimes also 75 : 25) ratios, all the other mixtures had not more conductivity than sand alone or even less in closure stress regimes up to 8,000 psi, with the permeability of the proppant mixture also rapidly declining with rising closure stress. In the range of 6,000 - 10,000 psi, a mixture of sand and sintered bauxite in a 50 : 50 ratio reveals an average permeability drop of abt. 65 % with respect to pure sintered bauxite populations, and the productivity of other mixture ratios is even more approaching that of sand alone or below in the most unsuitable cases.

1.4.11.1.1.2. Conductivity deterioration by proppant mixing

Mixing of natural quartz sand with synthetic high-performance proppants is thus not an effective method for increasing fracture conductivity at high closure stresses, because the original target of saving treatment cost by replacing parts of the expensive man-made proppant lot by cheap sand is not reached, but in contrast to the expectation, the properties of the artificial high-quality proppants are significantly deteriorated by contamination with much poorer performing sand. Apart from absolutely tremendous drops of conductivity level with respect to pure populations of synthetic proppants, only 50 : 50 (and in some cases also 75 : 25) mixtures have relatively slightly higher permeabilities than sand alone, and all the other mixtures perform only equal or even worse than homogeneous sand populations (McDANIEL & WILLINGHAM 1978, CLARK 1983). In mixtures with inferior material, the better physical properties of a superior proppant type are thus not offset, and proppant selection therefore has to be based on the individual conductivity and closure stress resistivity of unimodal homogeneous proppant populations.

In contrast to the aforementioned results, DAVIS (1975) reports tail-in and mixing of glass beads for achievement of conductivity improvement, high proppant concentration, and proppant cost reduction. Favourable mixtures resulting from field experience are 75 % 20/40 sand and 25 % 12/20 glass beads. Mixing with sand is according to DAVIS (1975) suitable for diminishing glass bead quantity and thus decreasing proppant expenses, and has acceptably good performance in the fracture in reservoir depth (this conclusion, however, is contradicted by the results of McDANIEL & WILLINGHAM 1978 and CLARK 1983).

1.4.11.1.2. Mixing of different sand types

Conductivity deteriorations even occur when mixing of two different sand types is carried out, because also the original sand populations (as screened by sieving after mining in the natural occurrences) have different permeabilities. According to NORMAN, CIPOLLA & WEBB (1983), Hickory Formation sand from Central Texas is more permeable within fracture packages than St. Peter Formation sand from Northern Illinois at lower closure stresses, whereas in higher pressure regimes above abt. 3,500 psi, St. Peter sand performs better than Hickory sand. The explanation of this phenomenon is probably that the St. Peter sand has a higher content of more stable monocrystalline quartz than the Hickory sand and thus less grains crush at higher closure stresses (cf. section 1.4.9.3.), and the Hickory sand contains more frequently oligo- to polycrystalline quartz intergrowths and also some feldspars and rock fragments after processing which are partially better and worse rounded due to improved wearing and inherent cleavage, respectively, than the irregularly breaking quartz grains and thus provide better properties at lower closure stresses until oligo- to polycrystalline quartz, feldspars and rock fragments start to crush at reasonable rates.

1.4.11.2. Mixing of proppant grain sizes

Similarly as mixing of proppant types, combination of several proppant grain sizes has detrimental effects on fracture conductivity unless the different grain sizes are not inserted as an interspersed homogeneous distribution, but as temporally discrete lots that will receive spatially separate placement in the crack. Although it is the aim of quality control during fracturing treatments to avoid unwanted contamination, accidental mixing can occur if solid fluid-loss additives are used or if multi-stage proppant scheduling with coarser or finer tails or heads as compared to the main body are applied, respectively. The most pronounced example of accidental or attempted mixing of grain sizes is 100 mesh sand which is added to fracturing treatments carried out with 20/40 and/or 16/20 mesh proppants for various purposes. The following account sketches possible sources of contamination by outlining the significance of 100 mesh sand and coarser proppants and the drawbacks of the mixing of these two different size classes in hydraulic stimulation jobs. Consequences of mixing of various coarse proppant grain sizes are also analyzed.

1.4.11.2.1. Mixing of coarse proppants and fine 100 mesh sand

During a fracturing operation, it is desirable to minimize fluid leakoff from the crack into the reservoir by adding finely divided solids to the liquid phase (COOKE 1975). The function of particulate fluid-loss additives is to plug the large pore spaces and/or natural cracks and fissures of the pay zone rocks at the fracture walls. The drawback of solid fluid-leakoff additives of insoluble nature (such as 100 mesh sand in contrast to calcium carbonate flour and fine-grained salt; cf. sections 1.4.11.2.1.2.2. and 5.11.4.), however, are significant detrimental effects on proppant permeability in the crack, the problem of displacement of the particles from the fracture face after deposition, and the question of their mobility and freedom to move through the crack when the well is put on production. Solid fine-grained fluid-loss additives can indeed cause very serious decreases in proppant permeability and thus fracture conductivity (PYE & SMITH 1973) particularly in case of mixing with the coarser proppants. Following a short outline of the application spectrum of 100 mesh sand, its permeability deterioration in case of intermixing with fracturing proppants is sketched, with distinction being made between preventing and provoking permeability destruction. Some aspects of 100 mesh sand as exclusive or participant proppant are also discussed.

1.4.11.2.1.1. Application spectrum of 100 mesh sand

Concerning the effect of mixing of different grain size populations, the application of 100 mesh sand as an additive to reduce fluid leakoff through natural cracks in hydraulic proppant fracturing (THOMPSON 1977, McDANIEL & WILLINGHAM 1978; COOPER, NELSON & SCHOPPER 1983; SCHUMACHER & DICKERMAN 1986) and acid fracturing (MILLER & WAREMBOURG 1975, BAILEY & WICKHAM 1984, DOWELL SCHLUMBERGER 1988 d), to scour away near-wellbore damage by perforation erosion (CRAMER 1987), to facilitate slurry insertion by acting as a lead-in proppant at the head of fracturing treatments for plugging of natural and/or hydraulic microjoints and -fissures thereby increasing well productivity (SINCLAIR 1980, GRI 1988), or to limit vertical fracture height propagation by performing as a bridging agent to control downward growth of the crack (HODGES & PAOLI 1982) has to be carried out very carefully, and also good separation of individual stages during tail- in operations is essential.

100 mesh sand has been widely used particularly in the USA und is usually pumped ahead of the sand or proppant lot that is designed to plug the hydraulically created fracture for the purpose of frequently combined fluid leakoff control and supporting solids lead-in as well as microproppant plugging hairline fissures. Hydraulic proppant and acid fracturing treatments using 100 mesh sand as an additive for fluid-loss control and/or hairline crack propping benefit from achieving higher sustained production, reaching higher concentrations of fracture-supporting proppants, fewer screenout failures happening, more treatments staying in zone, and conventional fluid-loss additives not being required (MILLER & WAREMBOURG 1975). Examples of application of 100 mesh sand are reported by GWINN & McMANUS (1977); CAMPBELL, HANOLD, SINCLAIR & VETTER (1981); ROBERTS (1981), McMECHAN & CONWAY (1983), BEGNAUD & CLAIBORNE (1985), ZAHNER & CRAFTON (1985), NIEMEYER & REINART (1986) and NOLTE (1988 b; cf. also sections 4.8.8.3.1.2. and 4.8.8.3.2.3.).

The use of 100 mesh, 40/60 mesh or 70/140 mesh sand in the first 25 % of the fracture treatment representing a lead-in schedule helps in fluid leakoff control through hairline natural fractures and hydraulic fracture initiation (PAI, GARBIS & HALL 1983). The fine sand acts as a secondary proppant and is a highly-conductive means of tip packing in secondary hairline fissures and jointlets which do not accept larger proppant grains, thereby inhibiting excessive loss of high-viscosity fluids (LAMBERT, DOLAN & GALLUS 1983). Some aspects of preventing permeability deterioration in fracture propping, application of 100 mesh sand as exclusive or participant proppant, and provoking permeability destruction in plug setting for height growth limitation are outlined as follows.

1.4.11.2.1.2. Preventing permeability deterioration

Intermixing of 100 mesh sand with coarse fracturing proppants of most commonly 20/40 and/or 16/20 mesh grain size results in drastical proppant package conductivity deterioration. Measures for prevention of deleterious effects are addition of only limited amounts of 100 mesh sand, proper separation of 100 mesh sand by spacer fluid cushions from the main proppant lot, replacement of 100 mesh sand by fine salt which is soluble, and utilization of resin-coated 100 mesh sand or proppants which do not flow out of the hairline fissures into the main crack once placed within the jointlets. The discussion as follows focusses on proppant permeability destruction by fines, 100 mesh sand vs. fine salt, and 100 mesh microproppants for hairline fissures.

1.4.11.2.1.2.1. Proppant permeability destruction by fines

If the fine fluid-loss control sand happens to become intermixed with the larger proppant grains, already a 5 % content of 100 mesh sand within the proppant package can reduce fracture permeability up to 40 %, and at closure stresses above 4,000 psi, contamination of the proppant population by 20 % fine sand can already have the result that permeability of this mixture approaches the value of 100 mesh sand alone which is considerably less than that of the pure proppant population, with thus the effect of the fracturing treatment almost having gone away. A slightly better solution to reduce fluid leakoff might be addition of silica flour which is sufficiently fine-grained not to interfere any more with the coarse-grained proppants in very low concentrations, but also here caution has to be exercised, because 3 - 4 % of fines of any type can in the worst case already destroy 90 % of proppant conductivity (McDANIEL & WILLINGHAM 1978; SMITH 1987).

A safety procedure for effective separation is the pumping of a large spacer pad or cushion between the 100 mesh fluid-loss control sand and the fracture proppants (BLASIO, BAILEY & BERGTHOLD 1985; BARBY & BARBEE 1987). 100 mesh sand has turned out to be a particularly effective fluid-leakoff supervision additive in case of very short crack closure times. 70/140 mesh sand has repeatedly proven to be less damaging to fracture conductivity than silica flour (LAMBERT, DOLAN & GALLUS 1983). In contrast to fine particles which plug the proppant package interstices, larger pieces of debris can possibly obstruct the perforation holes (OSBORNE, McLEOD & SCHROEDER 1981).

Examples of strong permeability reduction by excessive amounts of fine particles invading the proppant pack are also reported by ROLL, HIMES, EWERT & DOERKSEN (1986) who also conclude that in spite of the best efforts to design and perform hydraulic fracturing or gravel packing treatments, excessive fines generation and/or immigration can defeat the purpose of the operation (the detrimental effect of excessive fines is also emphasized by PENBERTHY & COPE 1979; SHRYOCK, DUNLOP & MILLHONE 1979 and KRUEGER 1986).

1.4.11.2.1.2.2. 100 mesh sand vs. fine salt

An alternative to 100 mesh sand as fluid-loss additive for bridging of natural hairline cracks is fine salt of 30/60 and/or 80/120 mesh grain size which packs to lower permeability in the fractures under pressure than silica (SCHRIETER & SHAW 1978), and to some extent also calcium carbonate flour. Salt has also the advantage with respect to sand that it is subsequently dissolved again to permit unrestricted flow from the hairline cracks which thus can contribute to hydrocarbon production, and that it eliminates pump and equipment damage problems provoked by abrasive sand. Calcium carbonate flour and fine-grained salt have the advantage to be able to be later dissolved by hydrochloric acid and water, respectively, in contrast to silica flour and 100 mesh sand which could only be attacked by hydrofluoric acid that also can have detrimental effects on the proppant package (calcium carbonate flour and fine-grained salt are frequently utilized in gravel packing operations because of the advantages mentioned; cf. section 5.11.4.).

1.4.11.2.1.2.3. 100 mesh microproppants for hairline fissures

On the other hand, support of hairline jointlets by 100 mesh microproppants provides additional conductivity to the whole fracture system contributing to production, with open microfissures after pressure return at least partially being able to close again and to deteriorate the effect of the integrated network (WARPINSKI 1988 a). Therefore in many cases it may be desirable in hydraulic fracturing to utilize 100 mesh sand as a combination of fluid-loss agent and microproppant for effectively limiting accelerated fluid leakoff into hairline jointlets and to adequately support the microcracks to keep them open for later reservoir exploitation (cf. sections 4.8.8.3.1.2. and 4.8.8.3.2.3.). As continuous addition of 100 mesh sand guarantees positive fluid-loss control impact throughout the job, in some low-permeability tight gas sandstones it can be feasible to accept minor conductivity reduction of high-quality proppants and to secure fluid leakoff limitation by permanent 100 mesh sand addition until the end of the operation (WARPINSKI, BRANAGAN, SATTLER, CIPOLLA, LORENZ & THORNE 1988).

Such small amounts of 100 mesh sand which do not seriously destroy proppant conductivity, however, have only limited effectivity in fluid-loss restriction and can reasonably slow down accelerated leakoff, but cannot stop it entirely and thus remaining fluid-loss rates are still very high (WARPINSKI 1988 a). In deeper reservoirs with higher closure stress, considerable improvement of dual fluid leakoff control could be achieved by 100 mesh intermediate- to high-strength synthetic proppants as an effective micropropping agent in hairline cracks, but this grain size of artificial high-quality proppants is not yet commercially available (NORTHCUTT, ROBERTSON & HANNAH 1988). Resin-coated 100 mesh sand or proppants have the advantage that once placed within the microfissures, agglomeration of the resin pellicles prevents flowback of the microproppants thus stabilizing them as supporting wedges in the hairline cracks and inhibiting mixing of 100 mesh particles with coarse proppants in the main fracture branch (cf. sections 1.2.6. and 4.8.8.3.2.3.).

1.4.11.2.1.3. 100 mesh sand as exclusive or participant proppant

Fluid leakoff minimization is essential for success of the fracturing treatment, because excessive fluid loss automatically leads to screenout failure (cf. sections 4.2.2.2.3. and 6.2.4.2.1.), but also proppant performance must not be seriously affected, as deteriorating proppant conductivity also results in technical and economical failure of the treatment. EASON (1985) and WARPINSKI (1988 a) describe combined action of 100 mesh sand as fluid-loss additive and as microproppant in naturally fractured reservoirs where it first controls fluid leakoff and later provides adequate permeability and acts as a proppant in the natural and/or hydraulic microcracks once fracture ballooning subsides. When naturally jointed reservoirs are hydraulically treated, a large fracture is created but many small side fissures can open during high-rate injection. These microcracks are prevented from closing by plugging with 100 mesh sand which allows the propped hairline joints to contribute to higher well productivity due to increased surface area and fracture volume. Some comments on application of 100 mesh sand for infilling of hairline fissures in acid and proppant fracturing as well as an outline of the utilization of 100 mesh sand as exclusive proppant are given as follows.

1.4.11.2.1.3.1. Application in acid and proppant fracturing

100 mesh sand as a microproppant is therefore very beneficial in hydraulic proppant stimulation of naturally cracked formations as well as in acid fracturing (FRACFAX 1988 f). 100 mesh both regular and resin-coated sand has abt. 5 - 7 darcy permeability at almost any closure stress which provides still a reasonable contrast to the milli- to microdarcy tight rock matrix. If microproppant flowback has to be inhibited, curable resin-coated 100 mesh sand is kept in place by resin bonding. 100 mesh microproppants help to increase production because they support very fine natural hairline cracks as well as narrow segments or side branches of the acid- or gel-induced fracture. CRENSHAW & JOINES (1977) also emphasize the dual role of 100 mesh sand as fluid-loss additive and microproppant allowing drainage from secondary joints.

In acid fracturing, microproppants are also able to extend stimulated production and increase ultimate recovery by preventing of premature closure or plugging of the etched fracture morphology, vug spaces, hairline side branches and wormhole features. In order to accept propping agent, fracture width must be in the range of 2.5 times the proppant diameter, with the standard proppant grain sizes (16/20 and 20/40) fulfilling this requirement for most of the hydraulic fractures, while 100 mesh sand fits into the hairline microcracks (THOMPSON 1977). 100 mesh microproppant can be regarded as prepack stage ahead of the main stages of various treatments comprising both proppant and acid fracturing (MILLER & WAREMBOURG 1975).

1.4.11.2.1.3.2. 100 mesh sand as exclusive proppant

Apart from its application as fluid-loss additive and/or partial proppant, 100 mesh sand can also be exclusively used as proppant for reservoir stimulation (THOMPSON 1977; HARPER, HAGAN & MARTINS 1985) if the placement of coarser grains would bear high screenout risk, if only moderate crack conductivity is sufficient such as in minifracturing breaking through near-wellbore damage (cf. section 4.8.3.), or if stimulation cost have to be seriously contained for various reasons. 100 mesh as well as 40/60 mesh or 70/140 mesh sand can also be utilized as fine-grained lead-in stage of the proppant assemblage in order to not only plug hairline cracks in the reservoir matrix at the fracture tip, but also to facilitate proppant placement at the extremities of especially very long cracks, with the aims being achieving smooth proppant slurry flow-in, good proppant concentration and avoiding screenout failure. Migration of formation fines and 100 mesh sand particles from microfissures into the coarse proppant package in the main fracture can be prevented by application of resin-coated 100 mesh sand which locks the microproppants together to a stable package that maintains high fracture conductivity.

Generally the amount of 100 mesh sand and other fine solid fluid-loss additives should not exceed abt. 10 %, because in case of higher shares of fines, fracture conductivity is significantly decreasing (LI & ZHU 1986). Fracture conductivity damage created by high-strength bauxite fines is less than that generated by low-strength sand materials, with a fracture conductivity of abt. 70 % and 30 % of the values calculated from field transient pressure tests being reached when using fine bauxite and fine sand, respectively, as fluid-loss additives.

1.4.11.2.1.4. Provoking permeability destruction

In contrast to the general aim of preventing permeability deterioration of proppant assemblages pumped downhole and inserted into hydraulically generated fractures, some special applications which partially are designated to improve proppant placement and thus fracture efficiency look on provoking permeability destruction of some lots of the solid particulates which are driven into the crack by the carrier fluids. Such a special case of desired permeability deterioration is for example the application of mixtures of various proppant grain sizes as buoyant or settling diverters (cf. section 4.2.2.4.) for sealing the upper or lower tip of a fracture extending from the reservoir sandstone into the overlying mudstone barrier and thereby restricting vertical upwards or downwards height growth of the crack, respectively.

NOLTE (1988 b) reports an example where successful blocking of the upper tip of the fracture has been achieved by a tight flow block consisting of a mixture of silica flour and 100 mesh, 20/40 mesh and 10/20 mesh sand which, if properly set in place, has virtually no permeability at all. Careful selective insertion of buoyant diverter for flow block purposes and proppant for fracture support with different fluid stages of different viscosity and with adjusted pumping rates is necessary to avoid interference of both solid loads and to keep the proppants of the later lot which are designated for conductivity creation clean and separated from the grain size mixture which is planned for fracture sealing by permeability deterioration (cf. section 4.2.2.4.1.). Proper design of the fluid spacer separating the two stages of solid material introduction therefore achieves key significance for successful installation of the desired proppant distribution and realization of the attempted fracture height growth control.

Another type of permeability destruction is plugging of formation permeability in thief zones by acrylic epoxy resin systems representing a special application of in-situ chemical reservoir consolidation (KNAPP & WELBOURN 1978; ZAITOUN, KOHLER & GUERRINI 1989; cf. section 5.9.1.7.).

1.4.11.2.2. Mixing of various coarse proppant grain sizes

Accidental mixing of 20/40 and 16/20 or 12/20 proppant grains at the transitions between different stages of tail-in (cf. section 2.4.1.2.) or lead-in multi-proppant hydraulic stimulation operations has the detrimental effect of plugging of the interstitials between the coarser grains by the finer particles of the bimodal distribution. In addition, the four times faster settling velocity of abt. twice larger grains (according to Stokes` law) may lead to vertical separation of grain sizes by falling of the coarser grains faster down to the bottom of the fracture than the finer grains (LOCKETT & AL-HABBOOBY 1973, MIRZA & RICHARDSON 1979), resulting in origin of a vertical proppant grain size zonation due to existence of different layers composing the proppant bank, which is in contrast to the desired effect of lateral succession of belts with proppants of different grain size along the crack. Moreover, gravitational splitting of the populations may result in origin of asymmetrical wedges of coarser material that is not in continuous contact with the reservoir along the full height of the fractured interval and thus the anticipated homogeneous mantle-like distribution of the tail around the wellbore is not accomplished. Aspects of settling velocities of proppants are discussed by DUNAND & SOUCEMARIANADIN (1985), KIRKBY & ROCKEFELLER (1985) and ROODHART (1985), and ACHARYA (1986) evaluates particle transport in viscous and viscoelastic fracturing fluids.

CLARK (1983) summarizes that an intermixing of different proppant quality types results in much less resistance to crushing of the fracture-infilling package with respect to that of homogeneous monoquality proppant suites, whereas an intermixing of various proppant sizes or sometimes also shapes causes less fracture conductivity due to increased packing efficiency of the particles.

1.4.11.3. Mixing of proppant grain shapes

While accidential or planned mixing of proppant types and grain sizes is usually detrimental to performance and gives frequently rise to even worse properties of the mixture than those of some of the ingredients alone, designed mixing of proppant grain shapes can effectively improve stimulation treatment performance by providing higher conductivity and closure stress resistivity to the combination with respect to the characteristics of the individual components alone. An outline of mixing of round and angular sand grains and its significance for conductivity maximization as well as the resulting benefit for the hydrocarbon stimulation industry is given as follows.

1.4.11.3.1. Enhanced conductivity of round-angular sand blends

LARSEN & SMITH (1985) study the influence of angular vs. round grains in 20/40 natural sand. Oil and gas industry generally demands fracturing sand and proppants to meet high roundness and sphericity specifications in order to provide the optimum geometrical pack in the fracture. Some comments on conductivity improvement and package stability as well as crushing behaviour are offered as follows.

1.4.11.3.1.1. Conductivity improvement and package stability

Perfect roundness and sphericity of the grains indeed decrease point pressure and reduce crushing and fines generated in the proppant assemblage, but the optimum geometrical pack for crush resistance does not always offer maximum conductivity (LARSEN & SMITH 1985). While a high grain sphericity has to be maintained because this eliminates proppant particles with elongated shapes and protrusions and prevents their fitting into natural void spaces thereby restricting conductivity, it is not always advantageous for all the sand grains to be either perfectly round and smooth or completely angular. Although totally angular grain populations have the highest porosity and permeability in a loose pack and at very low to low closure stress, they possess surfaces that will break off or crush easily with any significant closure stress because of uneven pressure distribution on the surface points that touch each other and thus at moderate and high closure stresses, entirely round proppant grain populations are superior to wholly angular assemblages.

If a certain percentage of angular grains is interspersed in a very round and spherical sand, however, the benefit of a highly porous angular sand can create a higher conductivity of the mixture than either pure round or homogeneous angular grain populations alone possess. The increased void space of a combination of round and angular sands overcomes crushing susceptibility and enhances overall proppant performance.

1.4.11.3.1.2. Crushing behaviour

Laboratory experiments have revealed that of all mixtures tested, a 50 : 50 round-angular blend of 20/40 mesh sand has on the average the highest relative fracture conductivity through all the closure stress levels up to 6,000 psi (LARSEN & SMITH 1985). Round-angular blends of 20/40 sand have even more conductivity at certain closure stress levels than the exclusively round product which the oil and gas industry has conventionally required. The benefit of the angular grains for improving conductivity of the mixtures results from the internal structure of round and angular quartz grains. Many angular grains have authigenic idiomorphic quartz overgrowths which break off from the parent crystal at low pressures and consequently the generated fines may block effective pore space.

While this may be detrimental at lower closure stresses, the fines may actually surround and support the detrital core at higher closure stresses, and by this protection allow the angular grains to perform better than the more round grains as they pass through their regular crushing sequence. Round grains are frequently monocrystalline and resist to high crushing pressures. When their failure point is reached, however, crushing of monocrystalline round grains is more severe due to their higher brittleness than that of oligocrystalline grains where some deformation can occur by sliding along crystal boundaries. In case of angular grains, a large amount of fracture stress may be relieved when the authigenic euhedral quartz overgrowths break off, allowing the parent grain to retain its internal strength longer.

In addition, angular sands tend to pack tighter than round ones because of the interlocking effect which the rough surfaces have on each other. If the proper percentage of angular grains were interspersed in a population of round grains such that each angular grain is kept from contacting any other, benefit can be taken from the additional void space of the angular grains and the significant reduction of the interlocking effect within the mixture, as well as from the increasing friction providing higher stability of the grain assemblage (cf. sections 1.3.5. and 1.4.3.).

1.4.11.3.2. Benefit for the hydrocarbon stimulation industry

According to the benefit of mixtures of round and angular sand grain populations, a better exploitation of many sandstone quarries and sand pits can be performed and lots that have been separated in the past as being below standard and out of specification can be incorporated into the quantities delivered to the wellsites, thereby securing supply and helping to avoid a repetition of serious fracturing sand shortages (as experienced in the USA during the 1970`s when oil and gas well stimulation boosted after the first oil price jump) in the future when hydraulic stimulation of hydrocarbon reservoirs reaches again such a frequency that all the presently available sources of natural sand and synthetic proppants reach the limits of production and delivery potential (LARSEN & SMITH 1985).

As combinations of round and angular natural sand have turned out to have higher conductivity than round populations alone, the same can be assumed to be valid for synthetic proppants, thereby providing a significant tool for fracture flow capacity maximization and enhancement of the formation-fracture conductivity contrast. This may give rise to a renaissance of angular sintered bauxite which was offered during early availability of synthetic high-strength proppants as a simple by-product of abrasive manufacturing (cf. sections 1.2.4. and 1.3.5.).

1.4.12. Computer programs for proppant selection

Hydraulic proppant fracturing design should ideally be a blend of computer simulation tempered with field experience (PAI & GARBIS 1983 b; GARBIS, BROWN & MAURITZ 1985). The expenses for a hydraulic proppant fracture treatment are composed of proppant, fluid, fixed set-up and hydraulic horsepower costs (ZAHNER & CRAFTON 1985). Due to the fact that proppant expenses can amount up to two thirds of the total stimulation operation cost, proppant selection is not only the key element of fracture design from technical standpoint in terms of closure stress resistivity, but also from economical view concerning job feasibility and profitability. While proppant choice and design optimization in larger hydraulic fracturing campaigns in whole fields have historically been frequently done by trial-and-error evolution, in the last few decades increasing sophistication of computerized modelling has considerably improved the possibilities of accurate planning with economical assessment of payout time and cash-flow rate. In order to obtain highly conductive and deeply penetrating cracks, accurate fracture design is imperative (CHARLES, HUDOCK, UDICK & HALL 1983).

Proppant selection for the individual applications according to reservoir conditions had been traditionally made by type curve analysis and evaluation of proppant characteristics. In the last years, several computer programs have appeared on the market which are based on a type-curve-like algorithm (AGARWAL, CARTER & POLLOCK 1979) and allow fast calculation of various cases of economical feasibility of proppant investment vs. payout by enhanced additional post-fracture hydrocarbon production (COOKE & GIDLEY 1979, HOLDITCH & LANCASTER 1982, CHENG 1984; COBB 1985, 1986; HALLIBURTON 1985; ANDERSON & PHILLIPS 1986, 1987). The programs can be run in small portable computers fitting into brief-cases which allow convenient display at any time and place and thus permit air travelling to customer offices and treatment execution sites. Microcomputer models for the design of hydraulic fractures are also presented by CLEARY, KECK & MEAR (1983).

Following discussion of some aspects of net present value calculation, several points concerning significance of the output for economical feasibility assessment are outlined. The economical importance of proppant selection and fracture design modelling are underlined by an introductory overview of general economical criteria and their relative importance. Comments are also offered on proppant selection based on volume vs. weight.

1.4.12.1. General economical criteria

The most important general economical criterion in hydraulic fracturing design modelling where proppant choice plays a key role is net present value which is present value after treatment minus present value before the operation minus present value of all costs or investments associated with the job (VEATCH 1986). Net present value is the net cash flow resulting from the fracturing treatment, assumes the cost of capital rate, and is the difference between present value of all future cash flows and cost of asset (BARBA 1988). Present value is the cash flow coming from well production and expense streams, and is the actual value of some amount in the future at a specific discount rate (VEATCH 1986). Payout period is the length of time required for the cumulative cash position to reach zero, and discounted payout interval is the period of time for the present value of the cumulative cash position to attain zero. Return on investment is the ratio of cumulative net cash flow over the project life to the maximum cash outlay. Discounted return on investment is the ratio of the present value of cumulative net cash flow over project life to present value of the investment for the treatment. Some aspects of investment return and profitability as well as design optimization and success definition are briefly illustrated as follows.

1.4.12.1.1. Investment return and profitability

Rate of return or profitability index is the compound interest rate whose discount factors will make the present value of a project`s net cash flow equal to zero (VEATCH 1986). Synonyms for profitability index include marginal efficiency of capital, internal rate of return (BAILEY & WICKHAM 1984), true yield, and discounted cash flow method. Other factors are undiscounted profit/investment ratio and benefit/cost ratio. The internal rate of return method balances inflows and outflows, and outputs the cost of capital required to obtain this balance, thereby allowing the operator to compare the performance of the assets and thus being a more universal technique (BARBA 1988). Evaluations involving economical analysis are also sensitive to other parameters such as product price, cost escalation, operating expenses and initial investment (KOZIK, BAILEY & HOLDITCH 1979).

The mentioned terms are used throughout industry for evaluation of the economical viability of projects. While present value is used as the base criterion for economical comparisons, there is no reason why any given item of a set of criteria should be more appropriate than any other, with the choice of points for making economical decisions depending heavily on the financial goals of the investor. Present value calculations can be based on discretized (incremental) or continuous discount factors. It is important to note the significance of early time cash flow due to acceleration on present values as well as differences resulting from different production decline performance. Net present value is the ultimate term used to investigate present value economics coming from hydraulic fracturing operations.

1.4.12.1.2. Design optimization and success definition

The economically optimum fracture design is the one which provides the largest net present value (WAREMBOURG, KLINGENSMITH, HODGES & ERDLE 1985). The economically optimum design is mainly affected by duration of production forecast from which net present value is calculated, net discounted exploitation revenue, and amount of investment required to achieve the design option. The critical variables interact and accuracy is extremely important for production forecasting, stimulation design and economical optimization. The cost to determine more accurate values for the critical variables is usually a very small percentage of the total well cost. Early optimization prevents repetitive errors from occurring in the designs for future wells. The economical optimum can be assessed prior to any expenditure for a fracture stimulation treatment. The success of a hydraulic fracturing operation can be quantified in exact terms regardless of the volumes of fluid and proppant actually pumped. Using all the available technology in a logical engineering sequence decreases the risk associated with any stimulation job and the production results, and the experience factor is accelerated and enhanced.

Another important economical criterion is the definition of success of a stimulation treatment (GIDLEY 1985). Fracture stimulation jobs are usually designed and justified on the folds of increased production based on the steady state offtake rate of the well before fracturing (SLUSSER & RIECKMANN 1976). While in many cases fracturing operations are classified as successes when incremental hydrocarbon production pays out execution cost within a certain time period and satisfies a predetermined profitability index, a better approach is assessing a treatment as success if it resulted in a net positive production rate buildup. Evaluations of stimulation job success can be made on the basis of initial potential of the well, but a much better approach is using actual and projected ultimate recovery of the borehole (HOEL 1988).

1.4.12.2. Net present value calculation

The design of hydraulic fracturing treatments has three basic requirements including determination of oil and gas production rates and cumulative recovery volumes that might be expected from various crack lengths and conductivities for a given reservoir, layout of the fracture operation in order to achieve the desired crack lengths and conductivities, and maximization of economical returns (VEATCH 1983, 1986). Revenue estimates are developed for various fracture parameter constellations from prediction runs with a reservoir performance simulator, whereas a hydraulic fracturing simulator computes treatment volumes, types of materials, and pumping schedules which are necessary to achieve the planned crack geometry. As fracture conductivity is particularly in tight gas reservoirs together with crack length the most important design parameter, careful proppant selection for the benefit of all requirements has a key position in fracture planning, and therefore special computer programs have been developed for proppant choice on both technical and economical background.

The wide range of cost and performance of the various commercially available proppant types has resulted in proliferation of proppant selection models during course of the technological evolution of hydraulic fracturing (PHILLIPS & ANDERSON 1985; ANDERSON & PHILLIPS 1986, 1987). While in earlier years a rather vague relationship between well depth and proppant strength dictated the choice of proppant type, more recently computerized models of varying complexity have been utilized which include net present value calculations by comparing cost and performance of the selected materials and economical revenue of the stimulation job. Simpler non-computerized evaluation techniques are based on either proppant expenses or technical merits of the job design and do not consider productivity or revenue, and include cost/performance comparisons and nomographs. The purpose of computerized proppant selection models is to determine how post-fracture incremental and cumulative productivity is affected by changes of type, size and quantity of the proppant supporting the crack. The technical choices are then expressed in economical terms by application of net present worth calculations, thereby allowing to select the fracturing design which best fits the feasibility performance goals of the well to be treated. The overall guideline is always cost of proppant per unit conductivity or conductivity per US $ (BRIM 1986, ANDERSON & PHILLIPS 1987; cf. section 1.3.4.1.).

1.4.12.3. Simulation techniques and input data

The computer programs for proppant selection require entering of reservoir, hydrocarbon, proppant and economical data and deliver an output of proppant quantity and cost, and revenue from hydrocarbon production after the treatment, thereby enabling to choose the most feasible case out of various runs having been made with different parameters. Initial drilling and completion expenses are not included into the analysis, and effects of income tax and windfall profits are also omitted for simplicity (NORMAN, CIPOLLA & WEBB 1983). The results of proppant selection evaluation by computer programs indicate the effect of fracture permeability on estimated production increase and of proppant choice on the treatment economics (McDANIEL & WILLINGHAM 1978). Some aspects of cost effectiveness of various proppant types, sizes and quantities as well as evolution of post-fracture well performance prediction are outlined as follows.

1.4.12.3.1. Cost effectiveness of various proppant types, sizes and quantities

Assuming that all hydraulic fracturing expenses except proppant cost are constant during several stimulation runs, it is possible to characterize the cost-effectiveness of various types, sizes and amounts of proppants by comparing net present value of future production when different parameter combinations are examined in the program loops (ANDERSON & PHILLIPS 1987), with the optimum proppant being the one giving the highest cumulative net present value. More sophisticated programs do not simulate the fracture as an entire wedge, but as a number of discrete cells, thereby allowing to include also proppant tail-ins and head-ins into the evaluation. The routine applications of computerized reservoir simulators and proppant selection models is particularly important in the current economical state of the hydrocarbon industry and allows fracturing treatment decisions to be based exclusively on well productivity goals of the operator and specific reservoir conditions (ANDERSON & PHILLIPS 1986).

1.4.12.3.2. Evolution of post-fracture well performance forecast

As both computer programs and additional manual techniques have got increasingly sophisticated with time, resulting also in more and more reliable predictions and assessments, there is an evolution of skill in projections of post-fracture well performance (BRIM 1986). In terms of computerized fracture simulation, production and service companies have developed various computational models for offtake forecast (for example numerical simulation; SETTARI 1980, LEICHT 1985). A comparative assessment of various proppant selection models and their significant advantages and drawbacks is carried out by ANDERSON & PHILLIPS (1987). Concerning economical evaluation, production rates vary with time and therefore cumulative withdrawal after some time period should be chosen as a base for comparison (ELBEL & SOOKPRASONG 1985) which is in fact incorporated into many computational models.

1.4.12.4. Significance for economical feasibility modelling

The main advantage of integrated computer programs for proppant selection and production forecast based on entering of detailed reservoir matrix and fluid data is economical cross-checking including control whether the recommended choice of a higher-strength proppant is justified by sufficient reserves and reasonably fast payout of the investment of the stimulation job as designed. It could happen that reservoir conditions would require a higher-quality and thus more expensive proppant which is, however, not supported by overall size, pay thickness and hydrocarbon content of the structure (ROBINSON, HOLDITCH & WHITEHEAD 1986). Comparative evaluation of proppant performance has to include economical analysis, as proppant effectiveness can be best estimated from the resulting productivity index increase and the net present value incremental increase (NORMAN, CIPOLLA & WEBB 1983). Some comments are offered as follows on proppant quality and drainage efficiency differences as well as early and late production history.

1.4.12.4.1. Proppant quality and drainage efficiency differences

If the application of a more expensive proppant type even in low closure stress regimes is economically feasible, it should be carried out because of the higher conductivity of the better-quality proppant and thus more pronounced permeability contrast between fracture and formation (NORMAN, CIPOLLA & WEBB 1983 also report that economical analysis often indicates that the incremental present value obtained by fracturing can be increased by using a high-performance proppant even though proppant expenses are considerably higher).

Comparing the effectivity of sand and intermediate-strength proppants in field applications, ROBINSON, HOLDITCH & WHITEHEAD (1986) found that reservoir zones treated with intermediate-strength proppants clearly had a much larger conductivity than those operated with sand even though job sizes were about the same in terms of proppants pumped, the intermediate-strength proppant was injected into much deeper horizons of the productive column, and sand was pumped in much higher concentrations than intermediate-strength proppants. In cases where it is technically and operationally feasible, the use of larger-sized proppants of the same type may be more cost/performance effective than using a more expensive proppant type of the same grain size (ANDERSON & PHILLIPS 1986).

1.4.12.4.2. Early and late production history

One parameter which fracture simulation computer programs usually do not incorporate is field status during course of exploitation history. HOLDITCH (1984) demonstrates that in early field life, fracturing with bauxite proppants results in higher gas-flow rates and thus a greater percentage of gas production in comparison to the quantities obtained after stimulation with sand. Inversely, however, in late exploitation history, the average rate coming out of a sand-propped crack is higher than that delivered by a bauxite-infilled fracture, because in the latter case, most of the hydrocarbon reserves have already been drained off in early field life and therefore in stages of advanced depletion the already considerably depleted and exhausted reservoir only releases further hydrocarbons at considerably lower rates. This inverse profitability behaviour of sand and bauxite proppants during late and early field life, respectively, needs adequate consideration in designing the propping agent in the range of lower closure stresses where the danger of sand crushing is not yet present and the only decisive question is how to obtain maximum conductivity and production during the whole time of well or field exploitation (cf. also section 4.8.9.4.).

Proppant cost factors which have to be examined for economical feasibility of fracture treatments do not only comprise proppant retail price which is only one facet of the overall expenses of using a specific proppant, but include also number of proppant grains by unit (volume per pound), additional polymer loading of the fracturing fluid that is necessary to transport the denser proppant, additional equipment required on the wellhead such as special flowback manifolds, and additional maintenance costs that are caused by abrasion and erosion (SINCLAIR, GRAHAM & SINCLAIR 1983). Replacing sintered bauxite by resin-coated sand can result in abt. 50 % total proppant expense savings due to both cheaper pound price and higher pound volume of the latter product (cf. sections 1.2.6. and 1.4.7.2.). The change from a high-density to a low-density proppant enables reduction of polymer loading of the carrier fluid up to 30 % without affecting proppant transport or settling characteristics.

1.4.12.5. Proppant volume vs. weight

Technical and economical comparison of different proppant types has to include an evaluation of expenses based on volume vs. weight, as both parameters result in different overall proppant cost (CUTLER, ENNISS, JONES & CARROLL 1983; CUTLER, ENNISS, JONES & SWANSON 1985). Neglecting conductivity differences between various proppant types, fracture design should be performed basing on the concept to pump the same volume of proppants regardless of material density in order to achieve the same volumetrical proppant concentration per unit area within the crack. Identical unit volumes of different proppant types, however, have different weights as a consequence of the different specific gravities of the materials. As proppants are marketed on a weight base, more higher-density material is needed to achieve the same volumetrical proppant saturation as with lower-density material. The price difference of proppants when comparing equal weights is therefore increasing when comparing equal volumes (cf. section 1.4.7.1.).

Neglecting proppant transport and conductivity differences, tap density (cf. section 1.4.7.2.) is the more important parameter for economical evaluation with respect to specific gravity, because it determines proppant cost on a volume basis. A given fracture will be filled by different weights for proppants with different tap densities, with the lower the tap density, the less weight it takes to fill the fracture. Proppant pricing should therefore be compared on a volume basis by multiplying expenses on a weight basis by tap density to determine price per volume.

1.4.12.6. Other aspects

An additional or partially also alternative technique to computerized fracture simulation modelling are hand-applied integral evaluation methods especially for post-treatment performance assessment (BRINKMANN, KRÖMER & REINICKE 1980; REINICKE, BRINKMANN, SCHWARZ & HUENI 1985). Manual and graphical ways of designing and evaluating fracturing stimulation operations are compiled in HALLIBURTON (1971). Type curves provide for many cases a relatively fast and inexpensive method as compared to iterative numerical computerized reservoir simulators (VEATCH 1986). Aspects of pre-fracturing well deliverability prediction are also discussed by SINHA & FURLONG (1979). Combined computer fracture simulation and type curve matching is reported by ROBERTS (1981) and SOLIMAN, VENDITTO & SLUSHER (1984). VEATCH (1977) presents computer studies for determination of in-situ fracture flow capacity in low-permeability gas reservoirs. Some aspects of proppant volume minimization and design optimization as well as proppant selection vs. payout maximization are outlined as follows.

1.4.12.6.1. Proppant volume minimization and design optimization

POULSEN & SOLIMAN (1987) present a numerical procedure for determining proppant schedules for hydraulic fracturing treatments which aims on proppant quantity minimization and optimization based on crack conductivity distribution curves. The procedure is based on modelling of fracture length, fracture width, proppant type and proppant concentration, and includes an economical analysis that is considering cost of each treatment alternative, time-dependent production rate as predicted by a fractured-well simulator, current and expected oil and gas prices, and minimum attractive rate of investment return. The criteria of the operator for making investment decisions are also important parameters, with limited amount of capital available for investment requiring to compare incremental benefit-cost ratios between alternatives in an effort to maximize the return on a given investment, whereas higher investment potential allows comparison of incremental rates of return. The design optimization procedure can be used for either stair-stepped or ramped proppant schedules (cf. also HANDKE 1987; cf. section 4.3.3.5.).

The choice of approach for determining folds of production increase achieved from fracturing and the selection of performance decline behaviour is very important and can be quite critical when doing economical optimization studies (VEATCH 1986). The approaches should be verified with field experience whenever possible. It is often feasible to develop short-cut treatment design and cost estimation methods when assessing large numbers of designs, with development of typical curves for specific formations being a suitable tool. These typical curves can be programmed for automatic interpolation and provide a method for rapidly generating massive sets of estimates at minimal expenses. Fracture design must not be limited to selection of proppant and fluid types and quantities, but a very important and even highly critical factor is well behaviour. Well performance prediction as a consequence of the layout of the fracture treatment has key significance in planning of the operation.

1.4.12.6.2. Proppant selection vs. payout maximization

Sometimes crack design and proppant selection according to reservoir requirement is difficult to match with economical feasibility or at least with maximum payout. OSBORNE, McLEOD & SCHROEDER (1981) present an example of economical unfortunity where wells with weaker gas production capacity due to poorer reservoir facies are the ones in greatest need of a fracturing treatment with exclusively intermediate- or high-strength proppants, because they are the most likely to be subjected to large drawdowns during early well life. On the other hand, wells with strong gas production capacity as a consequence of better reservoir facies are not subjected to such high drawdowns and usually only require a tail-in of intermediate- to high-strength proppants following pumping of natural sand in the main stage, thereby saving considerable proppant expenses. In such cases, the feasibility evaluation has to be made for the whole group of wells or even for the entire field including crestal and marginal areas rather than comparing treatment costs and production revenues for individual wells.

1.4.13. Proppant selection guidelines

Resulting from the discussion in the previous sections, some general guidelines of rule-of-thumb character can be given as a summary for proppant selection for the application in oil and gas well stimulation in hydrocarbon industry. Although the following cum-grano-salis guidelines are helpful for the first approach of proppant choice, caution has to be exercised not to overemphasize these general criteria due to their partially considerable abstraction to even oversimplification which cannot account for all the specific cases (cf. McDANIEL 1987). More precise and sophisticated proppant selection inevitably requires plunging into the detailed assessments which have been briefly outlined in the preceding divisions (the proppant application spectrum in oil and gas reservoir hydraulic fracturing stimulation is summarized in tabs. 1 and 4). The most important characteristics of natural sand, intermediate-strength low-density alumina silicate proppants, intermediate-strength high-density alumina oxide and silicate proppants, high-strength high-density alumina oxide proppants, and high-strength low-density zirconia-silicate proppants are briefly sketched as follows.

1.4.13.1. Natural sand

Natural sand is the cheapest of all proppant types and has always been available in nearly unlimited quantities due to widespread occurrences, uncomplicated accessibility and easy processing. Its application, however, is restricted to shallow wells due to its low closure stress resistivity which is the reason for classifying natural sand as low-strength proppant (cf. section 1.3.1.1.). Having been applied since the very beginning of hydraulic fracturing technology (cf. section 1.2.2.), natural sand is nowadays more and more replaced by synthetic high-conductivity proppants in all the cases where no extreme cost containment is necessary, and also in subcritical closure stress regimes man-made intermediate- and high-strength proppants that give better permeability contrast between fracture and formation can be selected.

The oil price drop in 1986 has temporarily again focussed attention on the cheap sand wherever applicable as a consequence of pronounced to partially even exaggerated fracture treatment cost containment (cf. section 2.2.1.5.), but increasing demand of hydrocarbon reservoir stimulation for progressive production enhancement will automatically push natural sand into a backseat position in the proppant market during the next years. The various intermediate- and high-strength synthetic proppants which during the last ten years already successively replaced natural sand in many sections of the oil and gas market are briefly discussed as follows.

1.4.13.2. Intermediate-strength low-density alumina silicate proppants

Intermediate-strength low-density alumina silicate proppants (cf. section 1.3.2.1.) have the best pumping characteristics of all synthetic proppants due to their low specific gravity which is comparable to that of sand. The higher closure stress resistivity allows the application of this material in shallow to intermediate-depth reservoirs beyond the pressure boundary of natural sand. Due to the availability of intermediate-strength low-density alumina silicate proppants only since a few years, they have much future potential for application in particularly many marginal oil and gas pay zones by replacement of natural sand in order to profit from the higher conductivity of the synthetic material. Intermediate-strength low-density alumina silicate proppants are also the cheapest of all artificial fracture-supporting materials and are harmless to the operational equipment. Effects of proppant settling are still insignificant for a wide variety of carrier fluids and a broad spectrum of proppant concentrations. Thus intermediate-strength low-density alumina silicate proppants are the economically most feasible proppant type in any respect if the boundary of closure stress resistivity is not exceeded.

1.4.13.3. Intermediate-strength high-density alumina oxide and silicate proppants

Intermediate-strength high-density alumina oxide and silicate proppants (cf. section 1.3.2.1.) are mainly applied for hydraulic fracturing of gas reservoirs in moderate to high depth. Being cheaper, lighter and less abrasive than sintered bauxite, they are chosen in all the cases where lightweight synthetic proppants are no longer resisting to the closure stress properly, but high-strength alumina oxide proppants are not yet necessary, and thus both cost premium and disadvantage of even higher particle density can be avoided. As intermediate-strength high-density alumina oxide and silicate proppants are also still not very harmful to the treating equipment, profit of their higher conductivity is frequently also made in boundary cases, and the material is often preferred with respect to natural sand or intermediate-strength low-density proppants in order to go absolutely on the safe side and not to risk any failure of the proppants which have to resist to a broad change of pressure drawdown and buildup during long-term field production history that can comprise up to several decades. The specific gravity is still low enough to allow good pumping behaviour with little risk of screenout, but depending on carrier fluid composition and weight, effects of proppant settling may already become significant.

Experience in the past in Europe has shown that after its invention and introduction to the market, intermediate-strength high-density alumina oxide and silicate proppants have become the standard material for tight gas reservoir stimulation in moderate to high depth (cf. section 2.4.) and have in most applications totally replaced high-strength alumina oxide proppants which were during the early years of deep gas fracturing the exclusive synthetic proppants available in the market (cf. section 1.2.4.1.).

1.4.13.4. High-strength high-density alumina oxide proppants

High-strength high-density alumina oxide proppants or sintered bauxite (cf. section 1.3.2.1.) have been the first synthetic proppants that were introduced to the oil and gas industry. While they have been very commonly applied in the early years of deep reservoir stimulation, their market share has been increasingly taken over in the last years by proppants with lower specific gravity which have a better pumping behaviour and that are often also cheaper. The high specific gravity of sintered bauxite does not only lead to problems of proppant settling in lighter carrier fluids, but also increases the risk of premature screenout termination of the fracturing operation when using heavier transport media in order to minimize or to avoid settling (cf. sections 4.2.2.2.3. and 6.2.4.2.1.). The higher risk of treatment failure can only partially be compensated by special pricing of sintered bauxite which traditionally has been the most expensive of all proppants, but has been offered recently very competitively in order to try to regain some lost market shares or due to the attempts of rivalizing proppant companies to increase their market penetration at the expense of that of their competitors.

Another major disadvantage of sintered bauxite is its considerable abrasiveness to the treatment equipment (cf. section 1.4.6.) which further deteriorates its economical feasibility. The application of sintered bauxite thus seems nowadays to be more or less limited to deep high-pressure reservoir fracturing where all the other materials fail (including boundary cases where it is essential to go on the absolutely safe side for undisturbed long-term field performance), and replacement of lighter proppants by sintered bauxite in shallower reservoirs can only be achieved by a more than competitive pricing probably coupled with an incentive for successful placement downhole in the fracture. A special utilization of sintered bauxite is hydraulic proppant fracturing and gravel packing in steam-injection wells in enhanced oil recovery projects (cf. sections 1.4.5.1. and 5.7.1.).

1.4.13.5. High-strength low-density zirconia-silicate proppants

High-strength low-density zirconia-silicate proppants (cf. section 1.3.2.2.) are an almost ideal material for a wide range of applications as a consequence of their technical properties. The material has been introduced to the hydraulic stimulation market only a couple of years after sintered bauxite has first been presented and successfully applied in the field and thus zirconia-silicate proppants belong together with sintered bauxite to the oldest synthetic high-quality proppants (cf. section 1.2.4.). Low specific gravity and high closure stress resistivity of zirconia-silicate proppants are excellent characteristics for usage in shallow to deep wells without any problems of placement and settling, and the very good conductivity provides in almost all the cases the necessary contrast between formation and fracture in order to allow hydrocarbon flow at economically feasible rates.

The major technical disadvantages, however, are the sudden catastrophical failure of the brittle glassy material into powder-like crushing remnants when the boundary closure stress is exceeded (cf. section 1.4.2.), and the low friction angle which does not only guarantee a better entry of the material into the crack, but also an easier subsequent escape from the fracture by flowback (cf. section 1.4.3.). All the technical benefits of high-strength zirconia-silicate proppants (including their harmless behaviour to the operational equipment), however, are completely compensated and deteriorated by the high price of the material. High-strength low-density zirconia-silicate proppants are the most expensive of all commercially available proppant types and therefore their market penetration is seriously limited despite the favourable technical properties (cf. section 2.2.2.3.).

1.5. Proppant applications

Synthetic proppants of various composition and grain size are applied in the oil and gas industry for hydraulic fracturing and gravel packing of production and injection boreholes in hydrocarbon-bearing reservoirs, as well as in other market segments outside hydrocarbon exploitation such as in the steel foundry industry for heat exchange and in the gas purification industry for dust absorption and filtration. The most significant applications are briefly sketched as follows (a summary of utilization of proppant types and grain sizes in hydraulic fracturing stimulation of oil and gas reservoirs and other markets is provided in tabs. 1 and 4).

1.5.1. Oil- and gas-field applications

The major oil- and gas-field application of natural sand and synthetic proppants is hydraulic fracturing of deep low-permeability gas-bearing sandstones and carbonates and subordinately also shallow or deep low- to moderate-permeability oil-bearing sandstones and chalks (cf. chapter 4). Gas-bearing tight sandstone formations in various depth intervals are typically characterized by 25 - 100 ft net thickness, 3 - 8 % porosity; 1,500 - 15,000 psi reservoir pressure and in-situ permeability of up to 1 md (HOLDITCH 1984), but in most cases being in the microdarcy range (AGARWAL, CARTER & POLLOCK 1979; ATTEBERRY, TUCKER & RITZ 1979; cf. section 3.4.1.).

Proppant selection should generally focus on getting the maximum conductivity or the optimum net present value per invested US $, respectively (BRIM 1986, ANDERSON & PHILLIPS 1987; cf. sections 1.3.4. and 1.4.12.). The second major oil- and gas-field application is gravel packing of primarily poorly-cemented and/or secondarily unstable, more or less shallow oil-bearing sandstones and chalks (cf. chapter 5). The brief outline as follows offers only a few comments on the most important branches of proppant application with particular reference to proppant types and grain sizes (more details with special emphasis on reservoir composition and distribution are presented in chapters 2 and 3). Distinction in the short overview is made between high-density alumina oxide and silicate proppants, low-density alumina silicate proppants and zirconia-silicate proppants.

1.5.1.1. High-density alumina oxide and silicate proppants

In terms of application of high-density alumina oxide and silicate proppants, distinction can be made between standard grain sizes comprising mainly 20/40 and subordinately 16/20 mesh, and coarser grain sizes including 12/20 or even larger.

1.5.1.1.1. Standard grain sizes

The most abundantly applied propping material in oil- and gas-field hydraulic fracturing is still 20/40 low-strength natural sand due to the current dominance of rather shallow wells that require not more than a tail-in of artificial intermediate- to high-strength proppants of abt. 10 - 30 % the quantity of the previously pumped sand, if man-made materials are used at all. Sand of 20/40 and also 30/50 or 40/60 grain size is the dominant material used for gravel packing. Next in the ranking list of cumulative annual consumption are 20/40 and to lesser amount (especially for tail-in applications; cf. section 2.4.1.2.3.) also 16/20 synthetic intermediate-strength high-density alumina oxide and silicate proppants in the predominant part of the cases and high-strength bauxite proppants in the subordinate share of this portion. Most of the past and future massive hydraulic fracturing jobs in tight deep Rotliegend (Lower Permian), Carboniferous and subordinately also Buntsandstein (Lower Triassic) gas-bearing sandstones in Germany FRG, Netherlands, British and Dutch Southern North Sea and adjoining areas have been carried out and will be done with this proppant type (MADER 1987; cf. chapter 2.4.).

Some special applications include pumping of 20/40 low-density intermediate-strength proppants in the early stage and 20/40 and/or 16/20 high-density intermediate-strength and/or high-strength proppants in the late stage of the fracturing operation for the reason of getting a safer lead-in of the lighter proppants with less surface friction that minimize the danger of premature screenout. Other extraordinary cases are tail-in treatments with a larger quantity of 20/40 sand followed by a smaller portion of 20/40 or 16/20 intermediate- or even high-strength proppants in order to maximize closure stress resistivity in the immediate wellbore surroundings where reservoir pressure drawdown reaches its maximum (MADER 1987, 1988 a; cf. chapters 2 and 3). A special utilization is also pumping of exclusively intermediate-strength high-density alumina oxide and silicate or low-density alumina silicate proppants with a smaller quantity of 20/40 material first which is followed by the main lot of 16/20 material (cf. section 2.4.1.2.3.).

1.5.1.1.2. Coarser grain sizes

12/20 intermediate-strength high-density and high-strength proppants are used in some extraordinary reservoirs to achieve a particularly pronounced permeability contrast between formation and proppant package, with the application in most cases being limited to tail-ins during advanced stages of fracturing jobs, because the coarse grain size is considerably increasing the risk of screenout. 6/9, 8/12 and 12/20 material of various type (partially only as a tail-in coupled with finer grain sizes) is already occasionally required (some examples of coarse proppant usage are reported by FAST, HOLMAN & COVLIN 1977; BLACK, RIPLEY, BEECROFT & PAMPLIN 1979; GRUESBECK, SALATHIEL & ECHOLS 1979; LAMBERT, DOLAN & GALLUS 1983; BARBER & THEMIG 1985, ROBINSON 1985, NIEMEYER & REINART 1986; PARKER, ADAMS & LIANKUI 1986; BARBY & BARBEE 1987) or could be an application of high future promotion potential for high-viscosity oil production, oil-field water injection, oil- and gas-field as well as aquifer waste liquid disposal (cf. section 4.8.7.), coal seam degasification (cf. section 4.4.3.), and geothermal heat exploitation fracturing (cf. section 4.7.). Occasionally associated non-oil-field utilizations comprise mineral leaching and production of aqueous solutions by wells (BARBY & BARBEE 1987).

6/9 sand has formerly been extensively used for oil- and water-production well gravel packing, but later more sophisticated design techniques have proven smaller-sized gravel to be superior for effective sand control (SCHWARTZ 1969, SAUCIER 1974; cf. section 5.2.3.). In some cases being coarse reservoir sands of aeolian or fluvial high-energy origin, coarse 12/20 and 16/20 or even 8/12 intermediate- and high-strength alumina oxide and silicate proppants and/or sand are still nowadays required for gravel packing.

1.5.1.2. Low-density alumina silicate proppants

20/40 intermediate-strength low-density alumina silicate proppant (cf. section 1.3.2.1.) which is only commercially available since abt. 5 years is presently greatly expanding and has - in combination with the resin-coated variety - certainly a big future due to the possibility of its application in many marginal reservoirs where so far no suitable stimulation technique has been discovered (such as oil-bearing marginal chalks as well as carbonate shell and quartz sand streaks and sheets interbedded with mechanically and/or chemically unstable mudstones; cf. section 4.8.5.1.). The major advantage of synthetic low-density intermediate- strength alumina silicate proppants is the low specific gravity that is almost comparable to that of natural quartz sand which results in a favourable pumping behaviour, coupled with a considerably higher closure stress resistivity than quartz sand. Thus intermediate-strength low-density alumina silicate proppants are scheduled to replace low-strength natural quartz sand in increasing shares due to their better conductivity and higher closure stress resistivity, if the price difference is not too high for applications where traditionally only natural quartz sand had been used, but a much better fracture performance can be achieved by using intermediate-strength low-density alumina silicate proppants.

1.5.1.3. Zirconia-silicate proppants

High-strength low-density zirconia-silicate proppants (cf. section 1.3.2.2.) have the major disadvantage to be considerably more expensive than alumina oxide (ceramic) proppants (both in terms of volume and weight; cf. section 1.4.12.5.) as a consequence of the different production process with considerably larger amounts of rejects for the reason of grain size being out of specification, and due to termination of governmental support for the manufacturing company some years ago. Thus in contrast to alumina oxide and silicate proppants which have become cheaper during the last years and only very recently were again pushed up in price but still remain below the former level which they had during many years of the golden stimulation period, zirconia-silicate proppants have kept their price or even got more expensive, thereby having an important marketing disadvantage with respect to the competitive products (cf. section 2.2.2.3.).

The application of high-strength low-density zirconia-silicate proppants is therefore more or less limited to some onshore areas (including MHF treatments), but is very difficult to economically justify offshore for more than tail-in jobs (aiming in many cases on the possibility of gamma-ray fracture height control monitoring by well logging of natural proppant radioacticity (cf. section 6.2.1.2.). A temporarily better situation had arisen in Europe in the early years of availability of synthetic high-strength proppants when a high oil price and a strong US $ permitted treatment execution when technically necessary and not when economically most feasible, with in those days repeatedly having occurred that purchase of zirconia-silicate proppants from the local supplier in France turned out to be the better solution at the bottom of the line that to wait for availability of sintered bauxite or to pay the air freight premium to fly in the material from the production plant in the USA to the wellsite in Europe. On the other hand, the major technical advantage of zirconia-silicate proppants is the low specific gravity of that high-strength material which is even lower than that of some of the intermediate-strength high-density proppants (cf. section 1.3.2.).

The high price in combination with the very limited mechanical and chemical stress resistivity also triggered the disappearance of the expensive glass beads (cf. section 1.2.3.1.) in former years when the cheaper and harder ceramic proppants were introduced to the oil and gas market.

1.5.2. Non-oil- and gas-field applications

8/12 and 12/20 high-strength bauxite proppants are also applied outside of the oil and gas market predominantly as heat-exchange (static environment) and heat-transfer (dynamic milieu) pellets in cupola furnaces in the steel foundry market, furnaces in alumina and glass industry, and furnaces for hydrocarbon combustion (aiming on retrieval of residual heat), as well as dust absorption or filtering pellets in the gas purification market. Other non-oil- and gas-field applications of ceramic proppants of various type and grain size are resin, polymer, water and gas filtration; shot blasting by hardening stainless steel by peening rather than by cutting, fluid bed reactions, wear resistance, metal deburring, filling, grinding, insulation and cement constituents, with most of these market segments at the moment in Europe still being in an immature stage, in contrast to the USA where up to a few percent of the total annual ceramic proppant sales is made to industries other than oil and gas companies already since longer time.

As a consequence of their significance in Europe for the development of an important second market segment in the coming years, foundry heat exhange and gas purification applications of sintered bauxite proppant pellets are briefly outlined as follows.

1.5.2.1. Foundry heat exchange

Heat exchange application of sintered bauxite of preferably 12/20 grain size in cupola furnaces for hot wind production in steel foundries has been developed in the last ten years (first introduction to foundry industry in 1976) as a means of effective heat retrieval (KGT 1986). The cupola furnace is composed of an upper gas part and a lower air part. In the upper gas part, soaking is done with considerable gas velocity, whereas in the lower air part, cold air is injected with high pressure. Hot blast-furnace gas comes from the adjoining burning chamber and penetrates a cylindrical tower containing several distribution disks which are perforated by numerous small holes and also contain some larger orifices. Gas velocity is adjusted in a way that the proppants or pellets are kept in suspension as a fluidized bed between the disks, but with time slowly pass downwards through the orifices to the next disk below, whereas the gas stream moves upwards.

Heat transition is optimized by controlled remaining time of the pellets between the distribution disks due to exact calibration of the gas stream. Having taken up sufficient heat during their way downwards, the pellets finally pass via a gas-tight transition from the upper hot gas part into the lower cold air part where they are cooled down by transferring their heat to the cold wind (which is thereby heated up and leaves the chamber as hot wind going into the foundry process system) and then are lifted to the top of the cupola furnace where the circuit starts again with entry of the pellets into the disk tower of the upper gas part. The heated pellets form a gas-tight closure between upper gas part and lower air part even at high differential pressures. Effects of proppant or pellet type and grain size are discussed in the following sections.

1.5.2.1.1. Pellet type

The effectivity of heat retrieval in cupola furnaces through heat exchange by the sintered bauxite pellets is much higher than in conventional recuperators which suffer from the problem of sintering of dust on brick lining and pipes, whereas the pellet system in the cupola furnace includes self-cleaning by later removal of the dust sintered onto the pellet surface during cooling and gravitational segregation of dust and pellets in the lift system, thereby also saving considerable maintenance work and expenses.

Concerning proppant type selection, high-strength alumina oxide proppants (sintered bauxite) are much more suitable than intermediate-strength high-density alumina oxide and silicate proppants due to their mineralogical homogeneity, chemical inertness and higher specific density. The mineralogical homogeneity of sintered bauxite being almost pure corundum provides a better pellet sintering during manufacturing and thus higher mechanical and thermodynamical stability with a lower potential of explosion and cracking of grains during repeated heating and cooling, respectively (experience has shown that the sintered bauxite pellets are sufficiently resistant against breakage in the range of applied changing temperatures which go up to 900 - 1,000 oC). The chemical inertness due to scarcity to almost lack of silica prevents solution in the hot alcaline environment (similarly as in steam-drive areas in oil fields when used for gravel packing and/or hydraulic proppant fracturing; cf. section 5.6.) and thus also enhances pellet stability. The higher specific density ameliorates the effectiveness of gravitational segregation of pellets and dust in the lift system where the dust is separated and removed.

1.5.2.1.2. Pellet grain size

In terms of grain size choice, 12/20 sintered bauxite has so far proven to be the optimum for the foundry heat exchange application. Smaller pellets generally result in better heat transition than larger ones, but if they are too small, however, gas velocity in the cupola furnace is too high and the pellets are transported against the stream upwards to the next disk and are even blown out with the flue gas current (the same happens when pellets of too low specific density are used, as evidenced by accidental field tests with the lighter alumina oxide and silicate proppants instead of the heavier sintered bauxite pellets). Optimization of the process has to be limited to changes of pellet size and specific density, because the gas velocity is more or less fixed to guarantee the required melting capacity of the cupola furnace and is indirectly proportional to the raw iron throughput. Larger and heavier pellets lead to better segregation from dust and also prevent loss of material by blowout at the top.

Concerning the gas-tight transition between upper gas part and lower air part of the cupola furnace, an unimodal pellet grain size distribution such as the narrowly-screened 12/20 sintered bauxite gives satisfactory separation. A bimodal pellet population would result in an even better tightness due to plugging of the interstitials between the coarse grains by the fine grains, but the problem would be gravitational segregation of both grain size classes in the disk tower in the upper gas part before the assemblage reaches the transition at the end of its downwards movement, and also during recycling and cleaning of the grains which are lifted to the top of the disk tower system with different transport velocities due to their different weight.

1.5.2.1.3. Alumina and glass industry heat exchange

A similar heat exchange application of sintered bauxite pellets as in cupola furnaces in the steel foundry industry is performed in alumina melting furnaces, hydrocarbon combustion furnaces, and glass industry furnaces. While the heat exchange application is already common practice in the USA since many years and the whole package of various non-oil- and gas-field proppant applications comprises up to a couple of percent of the total annual proppant sales level, the technique is in Europe still in the initial stages and the market segment is developing within the next few years to probably also reach an important second sector of usage of material with a possibly even higher share of marketing potential than in the USA particularly in light of the expected boosting of the foundry cupola furnace market segment.

1.5.2.2. Mechanical gas purification

Mechanical gas purification is performed in the cement industry with gravel-bed filters (LURGI 1983) where sintered bauxite or glass bead pellets of abt. 6/10 or subordinately also 8/12 or 12/20 grain size are needed. Sintered bauxite has the advantage to be mechanically more resistant and to be considerably cheaper than glass beads. The rough bauxite surface in comparison to the smooth glass surface also gives a better performance particularly during cleaning of the pellets from adhering dust. Distinction can be made between operational and cleaning stage of gravel-bed-filters in mechanical gas purification plants.

1.5.2.2.1. Operational stage

At the end of the heat-exchanging rotary kiln in the cement industry process line, the furnace and then the filter are installed. The filter is a combination between a mechanical separator (precleaning cyclone) and the tube-type gravel-bed filter (granulate-tube-filter; LURGI 1983) where the pellets are enclosed between two wire nets. The gravel-bed filter principally has to fulfill three different functions in the process : purification of the gas by separation of dust, maintenance of transportability in the system (the pellets have to possess sufficient buoyancy in the gas current), and possibility of cleaning the pellets from the adhering dust. The raw gas is cleaned first in the cyclone, with the remaining fine dust then second being trapped in the granulate filter where it sticks to the surface of the pellets preferably in the first quarter of the gravel bed thickness. With time, the gravel-bed filter is too much saturated with dust and then cleaning has to be made.

1.5.2.2.2. Cleaning stage

The whole plant consists of abt. eight moduls which are running separately. The operation includes simultaneous purification of the gas in abt. five to six moduls, and the other moduls are during that time in cleaning stage. Once they are put back on stream, other moduls are taken out from purification and are cleaned. Cleaning is performed by blowing compressed air into the modul and installing a ventilator with a cyclone at the top. The dust-bearing pellet granulate is blown upwards where gravitational separation takes place, with the heavier pellets falling back to the ground on the gravel-bed filter and the lighter dust being blown out.

The grain size of the pellets in the gravel bed is determined by the spacing of the holes in the wire nets that enclose the granulate. Sintered bauxite is more suitable than intermediate-strength high-density alumina oxide and silicate proppants, because the specifically denser material has a better segregability at cleaning stage (similarly as the pellets in the lifting system of the cupola furnace; cf. section 1.5.2.1.). If they are too heavy, the gas current rate has to be readjusted.

In contrast to the dynamical stress in the above described gravel-bed filter, other systems operate with pellets that are subjected to only statical stress.

1.6. Conclusion and outlook

Hydraulic fracturing and gravel packing of oil- and gas-bearing formations with various specific problems can be done with five different proppant types which are available in several grain sizes and at different prices in the market. Depending on mineralogical and chemical composition as well as mechanical and chemical properties, the various proppants can be applied in different pressure, temperature and brine regimes, with closure stress being the most important factor controlling proppant selection. The predominant application in Europe comprises intermediate- to high-strength proppants (and partially also sand in combination with or in some cases even as replacement of synthetic proppants) for hydraulic fracturing of Rotliegend and Carboniferous deep tight gas-bearing sandstones in Germany FRG and Netherlands onshore and British and Dutch Southern North Sea offshore. The main proppant applications in hydraulic fracturing and gravel packing are in the fields of oil production, water offtake and injection, shallow and deep gas withdrawal, and steam-drive enhanced oil recovery. Other proppant utilizations are in the market segments of fracturing fluid-loss additives, geothermal well stimulation and heat extraction, coal seam degasification, foundry cupola furnace heat transfer, and mechanical gas purification.

Based on the present discussion of aspects of proppant choice for various applications from predominantly technical point of view, a more detailed account of marketing and economical aspects of hydraulic fracturing and gravel packing with detailed review, status, forecast and perspectives of stimulation activity in Western and Eastern Europe is given in chapter 2, technical and marketing aspects of Rotliegend and Carboniferous stimulation in Europe are sketched in section 3 as a transition from the economically influenced divisions to the technically concentrated parts, and various technical and reservoir engineering possibilities of promoting and enhancing the fracturing potential, particularly in marginal oil- and gas-bearing formations, are outlined in chapter 4. Section 5 presents an account of gravel packing, and division 6 summarizes hydraulic proppant fracturing and gravel packing monitoring by well logging and other techniques. The bibliography is compiled as subject key classification and as general reference list in chapter 7.

Hydraulic proppant fracturing and gravel packing of oil and gas reservoirs is considered to be a growing branch of the stimulation market in Europe and will also in the future remain the most important market segment of application of the various types of currently commercially manufactured and distributed proppant types and grain sizes, with the largest quantities of material being continued to be consumed by MHF treatments of deep tight gas sands. Experimental approaches during the last years have even demonstrated that provided an adequate demand by the market, some new proppant types of enhanced technical performance and better economical feasibility may be developed in the coming years to adjust to an increasingly diversified spectrum of field application.

2. Marketing and Economics

2.0. Summary and associated table of contents

Hydraulic proppant fracturing and gravel packing in Western and Eastern Europe takes place in different types of reservoirs and under different geological, engineering, technical, marketing, economical and political conditions. In comparison to the USA, general influences on hydraulic fracturing and gravel packing such as changes of hydrocarbon prices and proppant prices have in Europe more serious and also more diversified consequences for the overall stimulation scenery. The impacts of reduction of hydrocarbon prices and proppant prices during the last few years are in Europe even accentuated by the decline of the US $ exchange rate (with respect to other world guide currencies such as DM and £). The technological evolution of manufacturing of various proppant types during the last ten years took place parallel to several major changes of the hydrocarbon price scenario and the US $ exchange rate scenario in either direction. The major oil price drop in late 1985/early 1986 hit the (mainly Western) European stimulation market much stronger than that in the USA due to various reasons comprising amount of drilling rigs, frequency of stimulation jobs, operating mentality and philosophy, proppant price at the bottom of the line, combination with falling US $ value, and proppant marketing activity. The oil price history of the last 20 years comprises the 1973 and 1979 oil price explosions, the 1986 oil price collapse, and the 1987 and 1988 oil price weakness. While prior to the invention and market introduction of high-strength synthetic proppants in 1976 possibilities of stimulation of deep high-pressure reservoirs have been very limited and the first years of availability of artificial high-quality proppants suffered from shortage in supply and elevated prices of the new products, the early 1980`s became the golden years of hydraulic proppant fracturing due to performance of numerous treatments on various scale in the economical framework of a high oil price and a strong US $. The 1986 oil price crash completely reversed oil price evolution and stimulation market potential. In contrast to the USA, many parts of Europe are so far in terms of hydraulic proppant fracturing insufficiently attended markets where a large potential exists with numerous possibilities of stimulation, but complicated geological and engineering reservoir conditions, many different mentalities and administrative interlocking require much more time and activity, with particular necessity being the development of a sound marketing strategy which integrates geological, reservoir engineering, technical and economical concepts. The main differences of the hydraulic proppant fracturing market between USA and Europe comprise absolute and relative frequency of drilling and stimulation, success/failure ratio and consortial structure, operating mentality and treatment cost, integration of stimulation into the total well concept, and proppant marketing strategy. Lacking continuity in marketing attention, missing understanding of the market requirements, and failure to foresee developing market trends has in Europe in the last years repeatedly resulted in serious business losses for proppant companies both to proppant competition and natural sand. Areal market evaluations are given for both Western Europe and Eastern Europe for the periods 1977 - 1985, 1986 (partially including the first half of 1987), and 1987 (comprising mainly the second half) - 1990 (and also extending further into the future), thus containing review, status, forecast and perspectives. Emphasis is also put on economical, commercial, political and ideological aspects of hydraulic proppant fracturing especially in Eastern Europe. In Western Europe, the main targets of stimulation by hydraulic proppant fracturing in past and future are deep low-permeability gas-bearing Rotliegend (Lower Permian) and Carboniferous sandstones in Germany FRG and Netherlands onshore as well as in British and Dutch Southern North Sea offshore, with mainly intermediate- and high-strength proppants being required due to higher closure stresses in the eastern part of the prospective Rotliegend and Carboniferous belt, whereas in its western part, lower closure stresses also permit the application of natural sand in combination with or even in replacement of synthetic proppants. The tail-in fracturing strategy comprising pumping of smaller quantities of intermediate- or high-strength proppants in the second stage of the treatment following injection of larger amounts of sand in the first stage of the operation seems to be the most suitable technical and economical stimulation design for many British Southern North Sea Rotliegend reservoirs, with the advantages being cutting cost and increasing effectivity of hydraulic fracturing by minimizing proppant expenses and maximizing closure stress resistivity in the vicinity of the wellbore, respectively. The Rotliegend and Carboniferous MHF operations are carried out with proppant quantities ranging from 100 t up to 650 t in the largest treatments. Apart from Rotliegend and Carboniferous, Buntsandstein (Lower Triassic) gas reservoirs also require hydraulic proppant fracturing stimulation in parts of the mentioned areas. Small fracturing jobs with 20 - 70 t of sand have in the past been mainly done in shallow oil-bearing Cretaceous sandstones in Germany FRG and Netherlands as well as Tertiary sandstones in Austria, Czechoslovakia, Yugoslavia and Hungary. In the Norwegian and subordinately also British Northern North Sea, unstable oil-bearing Cretaceous chalks and Jurassic sandstones require gravel packing and/or hydraulic fracturing with intermediate-strength low-density proppants and/or sand. Particularly significant are combinations of hydraulic fracturing and gravel packing with both conventional and resin-coated low-density intermediate-strength proppants in some soft flowing chalks in undercompacted and overpressured reservoirs. The dramatical effect of the oil price decline in late 1985/early 1986 (which had an even more severe and tragical impact due to the combination with a serious drop of the US $ exchange rate) was in Western Europe even accentuated by an increased cancellation rate of planned treatments due to several technical reasons resulting from a mismatch of prediction and penetration and also due to partially long drilling times extending over several periods with different general economical framework. Proppant price evolution in the last five years includes the 1985 proppant price lowering and the 1987 proppant price increase. In addition to price reductions and rises enacted by supplying companies, fluctuations in US $ exchange rate provoke alterations of competitiveness of various proppant types. As a consequence of the 1985 proppant price cut and the 1986 oil price drop, a much higher-quality proppant material is now available in significantly larger quantities at a considerably lower price than some years ago, but is seriously limited in application due to a presently poor economical framework particularly in Europe, whereas some years ago, a much lower-quality proppant material that was offered at a markedly higher price than now was virtually unrestricted in distribution due to a strong demand exceeding the previous capacities of supply. Therefore the 1986 oil price collapse completely reversed the stimulation market by changing the situation from good to excellent feasibility of hydraulic fracturing when only insufficient amounts of higher-quality proppants were available at high prices to poor attractivity of hydraulic fracturing when more than enough proppant material is available at low prices. The various price drops and rises for alumina oxide and silicate proppants had also serious impacts on their competitive power with respect to zirconia-silicate proppants which kept almost all tne time their former elevated price position due to more expensive production process, and that are in terms of real price in Europe not influenced by variations of the US $ exchange rate in either direction. In contrast to Western Europe, the different economical system in Eastern Europe leads to a more or less independency from the oil price scenario and also from the US $ exchange rate scenario which means that basically the drilling and stimulation programmes go ahead regardless of general economical crises in Western Europe and USA. In Eastern Europe, the hydraulic proppant fracturing highlight is currently Yugoslavia where MHF treatments focus on deep hot gas- and condensate-bearing Tertiary sandstones and conglomerates, with proppant quantities being needed up to 650 t per well which are thus exceeding partially even the lots pumped during the largest operations in Western Europe. The MHF jobs in Yugoslavia are mainly carried out with smaller amounts of intermediate-strength low-density proppants in the initial stage and larger volumes of high-strength proppants in the main stage of the treatment. In the other Eastern European countries, so far mainly smaller fracture jobs with almost exclusively sand as propping agents in quantities up to 20 - 150 t have been carried out. Major stimulation campaigns have also been performed in Turkey. In terms of economical and financial handling, the treatments are in Eastern Europe often done by own service with purchase of equipment abroad, and hard currency payment is enabled by favourable export volumes or World Bank credits. Considerable improvement of the stimulation potential in Eastern Europe is expected from the political and economical reorganization under the guidelines of reconstruction (perestroika), acceleration (uskorenie) and transparency (glasnost) in USSR and sooner or later also satellite countries, and from the obligations of fulfilling the goals of oil and gas production as fixed in the current five year plan. As oil and gas are the major source of hard currency for the USSR and because both oil price drop and US $ exchange rate collapse result in lower net returns, the losses have to be compensated by selling increasing amounts of hydrocarbons in Western countries. This has been rendered possible by meeting and overshooting of the plan forecast in the last years by achievement of a historical record level of oil production. The partial independence of the USSR from the world market and its efforts to ameliorate hydrocarbon supply security are underlined by the fact that in contrast to considerable budget cutting in the western hemisphere for drilling and completion and resulting seriously less exploration and development activity with respect to original plans before the 1986 oil price crash, the USSR has in 1986 even stepped up drilling in order to rapidly replace maturing oil production from progressively depleted fields. Because of its role as an overwhelming source of hard currency, petroleum development commands top priority in the USSR also after the 1986 oil price drop, whereas at the same time in the USA, petroleum has fallen to the bottom of the governmental priority list. While operators in USA and Europe are reluctant to invest money in new projects, the USSR has boosted its efforts to sustain and expand oil and gas production and exports. The logical result of these enormous efforts regardless of world economical constellations have been the overpassing of the plan goal for the annual oil production quantity in 1987 with breaking the exploitation record so pronouncedly that oil exports reached a maximum. Concerning marketing approach for stimulation and other services, the different logistical and financial situation between Western and Eastern Europe results in a different strategy of treatment operation. While in Western Europe and various other parts of the world oil and gas production companies commonly ask established service companies of mainly North American provenance to carry out the whole stimulation jobs including supply of all the mechanical and chemical additives for the operation, in Eastern Europe the production companies often prefer to buy their own equipment, to purchase most or all of the material directly, and to do the jobs without the support of Western service companies after probably a few introductory, training or pilot treatments. Various stimulation projects in both Western and Eastern Europe in the near future are briefly outlined. The main necessity for many European countries in the near future is to increase domestic hydrocarbon production in order to achieve some more independency from unfavourable changes of the economical and political setting (this applies particularly for Eastern Europe). As a consequence of the regional marketing assessment, the overall conclusion can be made that provided an again more stable oil price at a reasonable level, partially also a stronger value of the US $ (especially applying for Western Europe), and an adequate pricing of proppants and service, both Western and Eastern Europe are growing stimulation markets with increasing demand of proppants for hydraulic fracturing and gravel packing including promotion of treatments in marginal reservoirs in the near future.

2.1. Introduction

Hydraulic proppant fracturing of low-permeability oil- and gas-bearing formations is particularly in Europe (but also in many other parts of the world) not only influenced by technical and reservoir engineering aspects, but also strongly depends on marketing, economical and political relationships. In contrast to the USA, Europe is not only politically a complex aggregate of countries speaking a suite of different languages and having a variety of mentalities, with the major distinction concerning economical situation, status of technological development, commercial capacity, financial relationships, administrative procedure and foreign trade activity having to be made between Western and Eastern Europe, but also includes a wide spectrum of oil and gas reservoir types in terms of stratigraphical, tectonical, sedimentological, petrophysical and engineering characterization. An evaluation of the areal hydraulic fracturing and gravel packing potential for application of proppants from both technical and marketing point of view has to integrate the general situation by comparing the different cases and summarizing not only scientific, engineering and technological, but also economical and even ideological and mentality-related aspects.

Following an introduction of proppant types and grain sizes including assessment of the most significant aspects of composition and properties as a base for proppant selection for various specific applications (chapter 1), the present discussion gives review, status and forecast of hydraulic proppant fracturing in Western and Eastern Europe predominantly from marketing and economical points of view, with emphasis being put on several important factors controlling the general feasibility situation of the oil and gas industry. Comparative evaluation with USA, Middle East and other parts of the world is frequently performed during course of the general and regional assessments. The account of technical and marketing aspects of Rotliegend and Carboniferous stimulation in Europe in division 3 represents an areal case study and forms a transition from the economically influenced sections to the parts concentrating on technical and reservoir engineering points of hydraulic proppant fracturing in chapter 4. Division 5 presents a mainly technical overview and subordinately also marketing outline of gravel packing. Unit 6 summarizes fracture and gravel pack monitoring. Various possibilities of promotion and enhancement of reservoir stimulation by hydraulic fracturing and gravel packing in connection with application of proppants are compiled in chapter 4 that is focussing on technical and reservoir engineering aspects of formation treatment. A compilation of an extensive bibliography with a subject key classification into various special topics and the general reference list are given in section 7.

After an introductory outline of the main economical influences on hydraulic proppant fracturing being changes of hydrocarbon prices and proppant prices and commenting on the major fluctuations of the oil price in either direction during the last 15 years (discussed in combination with the fluctuations of the US $ exchange rate), differences of the stimulation market between USA and Europe are taken as a tool to sketch the specific reactions of both markets to the late 1985/early 1986 oil price drop and thereby induced major crisis of the hydrocarbon production industry. A major part of the report is devoted to an areal marketing assessment of hydraulic proppant fracturing in Western and Eastern Europe, with distinction being made between review (1977 - 1985), status (1986/early 1987) and forecast (late 1987 - 1990 and also following years) for the individual sections of the market.

2.2. Economical influences on hydraulic proppant fracturing

During the last 10 years since hydraulic proppant fracturing (VEATCH 1983, VEATCH & MOSCHOVIDIS 1986) of deep tight reservoirs started to boom with the invention of the first high-strength proppants (COBB 1985, WESTERN PETROLEUM 1985) and their introduction to petroleum industry in late 1976 (COOKE 1976, 1977; ATTEBERRY, TUCKER & RITZ 1979; HICKEY, BROWN & CRITTENDEN 1981; cf. section 1.2.), various events in the general world economy had considerable influences on the hydrocarbon stimulation market and the possibilities of proppant promotion. The most important factors controlling the hydraulic proppant fracturing activity are changes of hydrocarbon prices (BEUDELL 1986 a, 1986 b; GRIFFITHS 1986, HARBEN 1986, MENGES 1986, OEHME 1986, RUNGE 1986, VARZI 1986, MADER 1987) and changes of proppant prices.

As both prices are valid world-wide in US $ (with the exception being only a few proppant types that are sold for other currencies as well), the variations of the exchange rate of the US $ (with respect to other convertible guide currencies such as DM and £) have also an important superimposed significance on the European stimulation market (comparable to their impact on the relative oil price evolution in Europe). Various impacts of hydrocarbon price changes and proppant price changes on hydraulic proppant fracturing potential are discussed as follows.

2.2.1. Influences of hydrocarbon price changes

The technological evolution of manufacturing of various proppant types took place parallel to several major changes of the hydrocarbon price scenario. Following a summary of the oil price evolution in the last decades, various actual consequences of rising and falling oil price for the stimulation market are discussed particularly in view of the tremendous impact of the 1986 oil price drop and the decline of the US $ exchange rate which reached its historical minimum level in the fourth quarter of 1987 and kept being weak throughout the first part of 1988 (the recovery of abt. 20 - 25 % of its value having taken place in the second half of 1988 has at the bottom of the economical line to be classified as being still insufficient, mainly due to the ongoing unreliability of the US $ evolution and the resulting still retarded readiness for investments, and was even already partially again compensated by a renewed drop at the end of 1988). An outlook is also given on possible future consequences of the 1986 oil price collapse. Finally, some comments are offered on the impetus of the oil price evolution for the gas market.

2.2.1.1. Oil price evolution

The oil price evolution in the last 15 years was characterized by several booming and falling events which seriously affected hydrocarbon industry in either way. Changes of the oil price in upwards or downwards direction were also repeatedly accentuated by shifting levels of the US $ exchange rate with respect to other leading world currencies such as DM and £. The oil price history of the last 20 years can be classified into two main periods comprising the 1970`s oil price explosions and the 1980`s oil price drop and crash. The discussion as follows focusses on 1973 and 1979 oil price explosions, 1986 oil price collapse, and 1987 and 1988 oil price weakness. Emphasis is particularly put on the role of the Organization of Petroleum Exporting Countries (OPEC) on the progression of oil price evolution. Some reasons and effects of the 1986 oil price crash are also outlined, and comments are also offered on cyclicities during course of the oil price evolution.

2.2.1.1.1. 1973 and 1979 oil price explosions

The late 1985/early 1986 oil and gas industry crisis is the fifth after the Second World War (earlier crises have been in the middle 1950`s, late 1950`s, early 1960`s and in the late 1960`s/early 1970`s before the first oil price increase shock in late 1973/early 1974) and probably had the worst effect of all the hitherto crises, as the oil and gas market was most severely hit by numerous operations becoming uneconomical almost over night (cf. also LEE 1982). Following the drastical oil price increase in two main steps in late 1973/early 1974 (when the OPEC began to fix and to control the oil price; BILD 1987) and late 1979/early 1980 (OEHME 1986, RUNGE 1986) up to a level that was during extreme periods more than 20 times of that in the early 1970`s, the price rapidly collapsed almost to a niveau that existed in the time shortly after the first oil price shock (at least if calculating not only with the absolute figure, but also incorporating inflation and different exchange rates between US $ and other guide currencies in the world market; cf. section 2.2.1.3.). Some comments are given as follows on oil price trend changes as well as nominal oil price level and impact of oil price jumps.

2.2.1.1.1.1. Oil price trend changes

As a consequence of the stabilization of the situation after the second oil price shock in late 1979/early 1980, the oil price slowly started to leave its certainly much exaggerated level in 1982 and approached more and more a reasonable niveau until late 1985 when the drastical drop was initiated through some precursory fluctuations and soon after the beginning of 1986 almost within a couple of months until early 1986 reversed more than 12 years oil price history (with the oil price decline becoming in Europe at the bottom of the line even more serious by the accompanying decrease of the US $ exchange rate; cf. section 2.2.1.3.). Oil price evolution and the associated periods of crisis and boom are summarized by ATTIGA (1987).

Many of the actual problems are the result of the oil price evolution after the Second World War (SPATSCHEK 1987). In the booming period of 1950 - 1972, low oil prices supported a rapid to saltating increase of oil demand and progressive replacement of other energy media, and in the high-price period 1973 - 1985, the OPEC-dictated oil price explosions enabled the extension of exploration to new frontiers and the development of energy-saving technologies. At least partially this historical oil price and energy supply constellation programmed the 1986 oil price crash.

2.2.1.1.1.2. Nominal oil price level and impact of oil price jumps

The historical oil price evolution includes several major explosionary increases of the level from formerly abt. 1 - 5 US $ per barrel (BISHOP 1988) before 1973 to (partially as a reaction to the Israelian/Egyptian war) 10 - 15 US $ in 1973/1974 (when the OPEC group started to dictate the oil price level by strongly controlling production quantities and their physical distribution as well as by political threatening) and then (partially triggered by the beginning of the Iraqian-Iranian war) up to 35 US $ in 1979/1980 (with a maximum up to 40 US $ or in places such as the spot market even more than 45 US $ having been reached temporarily in 1980/1981).

The consequence of this oil price development was that 1981 has seen the peak of the world-wide hydrocarbon drilling rig count so far (GRIFFITHS 1986, HARBEN 1986, DOUGHERTY 1987; the delay of the climax of rig employment in the North Sea to 1984/1985 is a special effect of retarded approach to this expensive offshore area; OILMAN 1985; cf. section 2.2.1.2.1.). The high oil price niveau guaranteed the OPEC group an additional income of in total more than 5 Bill. DM during the last 13 years (OEHME 1986; cf. section 2.2.1.3.2.). Aspects of European oil and gas supply after the first oil price explosion are discussed by LÜTTRINGHAUS & SCHMITT (1975).

2.2.1.1.2. 1986 oil price collapse

In late 1981 and particularly in 1982, the oil price started slowly to leave its certainly seriously exaggerated niveau (being partially an effect of a generally decreasing energy demand due to the efforts put on conservation; KEG 1986) and gradually declined to 25 - 30 US $ (MINERALÖLWIRTSCHAFT 1983, BEUDELL 1986 c). In late 1985, the decrease had partially already proceeded to real prices in the range of 20 - 25 US $ (although the official OPEC price was still at 28 - 30 US $; BECK 1987, ERDOEL-ERDGAS AKTUELL 1984, ERDÖL UND KOHLE - ERDGAS - PETROCHEMIE 1987 b), thereby announcing the big event which arrived shortly after the beginning of 1986 when the price at once fell down to 10 - 15 US $ and reached temporarily even levels of 5 - 9 US $ (GRIFFITHS 1986, HARBEN 1986, KEG 1986, THOMAS 1986, ALGAR 1987 b, ERDÖL-ERDGAS-KOHLE NACHRICHTEN 1987 c; MABRO 1987, 1988; SCHMIDT 1987, McCANN 1988). Some effects of the 1986 oil price drop are outlined as follows, and comments are also given on return to reasonable oil price level and impact of the crash.

2.2.1.1.2.1. Effects of the 1986 oil price drop

This event represented a short-term reduction of the oil price to less than 50 % of its value during abt. half a year or a 60 - 70 % reduction of the late 1985 price (CROUSE 1987) and was a consequence of the fact that in view of persisting oversupply and declining market shares, the OPEC (ROBERTS 1986, 1987) renounced of its quantity and price control effective end of 1985 and thereby terminated a period of oil price dictation that lasted more than 12 years since its beginning in late 1973, giving way to a free market system due to the lack of a surplus management concept for production to match overall demand made up of actual consumption and stock movements (BEUDELL 1986 c, ATTIGA 1987; cf. sections 2.2.1.2.2. and 2.2.1.2.4.). The oil price of early to mid 1986 ranged in real terms near the low levels just prior to the first oil price explosion in 1973 (THOMPSON & SINGLETON 1988).

In mid to late 1986, the average oil price (the actual price kept fluctuating with a range of a few US $ upwards or downwards) left its minimum below US $ 10 (an average value of 6 - 9 US $ was held in July 1986 at the peak of the crisis; McCANN 1988) and gradually approached a niveau of abt. US $ 13 - 15 (CROUSE 1987). With no agreement between OPEC and non-OPEC to restrain production, the downward pressure on the oil price became irresistible, and finally OPEC was forced to formally abandon its market share policy and to revert to cutting back in output (McCANN 1988).

2.2.1.1.2.2. Return to reasonable oil price level and impact of the crash

Near the end of 1986, regaining consensus and agreement of the OPEC under the force of enormous losses of income and the threat of even some members facing bankruptcy established the average oil price at abt. US $ 18 (BECK 1987, CROUSE 1987; ERDÖL UND KOHLE - ERDGAS - PETROCHEMIE 1987 a, 1987 b; VIELVOYE 1987 a, 1987 b; cf. section 2.2.1.4.3.8.), but could not liquidate numerous minor variations in predominantly downwards direction due to current escalations in the political conflict situation in the Middle East as well as temporal slight changes in the world oil demand and supply constellation.

The instability of the oil price is also caused by the disagreement in the OPEC group on the status of the US $ 18 convention price level (OIL GAS JOURNAL 1987 c). While the key members intend to hold the price at this niveau in order to allow the oil to regain some of its lost market share, other aggressive countries would like to use this base as a springboard to higher prices in 1988 to 1990 (cf. section 2.2.1.4.3.8.). Big national markets such as that in the USA more or less tracked the international oil price evolution, with the changes in the USA having been from US $ 24 - 27 in January 1986 via US $ 13 in the second quarter and US $ 8 - 9 in August to US $ 14 - 15 in December 1986 and US $ 17 - 19 in the second quarter of 1987 (BECK & SMITH 1987, BECK & WILLIAMS 1987, OIL GAS JOURNAL 1987 g, WEST 1987). Evaluating the effect of an US $ 18 oil price before and after the 1986 crash (KINNEY 1987) reveals that while an US $ 18 level was feared by the hydrocarbon industry when the price was still at US $ 25 - 30, after the drop down below US $ 10 a renewed climb to US $ 18 was taken as a sign of market stabilization and triggered slight to moderate pick-up of activity with reasonable improvement of exploration and production cashflow.

2.2.1.1.3. 1987 and 1988 oil price weakness

Throughout 1987, the oil price more or less remained in the range of US $ 15 - 22 with the average being US $ 17 - 18. In early 1988, however, some political and economical effects related to the still ongoing Iranian-Iraqian war and the renewed weakness of the US $ without any hope of near-future reasonable improvement of its exchange rate to other leading world currencies such as DM and £ triggered another at least temporal decline of the real oil price to an average level of US $ 12 - 15 despite the official list price remaining at US $ 18 (AL-CHALABI 1987, ATTIGA 1987, VIELVOYE 1987 a, NARAGHI 1988; OIL DAILY 1988 a, 1988 e) at which the economics of finding and developing new reserves in non-OPEC producing countries are rather marginal (NARAGHI 1988). Short-term fluctuations (daily to monthly) of oil price may include plus or minus US $ 5.

In mid 1988, neither the end of the Iranian-Iraqian war nor the slightly recovering US $ exchange rate had a stabilizing impact on the oil price which remained to be weak and in the longer range unpredictable in terms of its further evolution. In late 1988, an again slightly falling value of the US $ after an intermittent recovery and various political constellations including predominantly the threat of another OPEC price war (WILLIAMS 1988) leading indispensibly to a repetition of the 1986 price collapse with probably even worse consequences had a downwards impact on the oil price which averaged at US $ 12 - 15 (OIL GAS JOURNAL 1988 d) and in short periods even ranged at US $ 10 - 13. Downwards pressure on the US $ exchange rate and thus also the real oil price was also exerted by the presidential election in the USA at the end of 1988 which gave a forecast of the economical policy to be expected for the next couple of years.

2.2.1.1.4. Cyclicities during course of the oil price evolution

Several cyclicities have been of key significance during course of the oil price evolution and had decisive impact on the progression of the suite of hydrocarbon price changes. Oil price history was in summary characterized in 1979 - 1981 by cyclically overheated oil prices triggering an overexpansion of drilling that in turn caused an overbuilding of capacity and inventory (PERKINS 1988). The declining activity due to slowly falling oil price led to cyclical correction of excess inventory in 1982 - 1984, whereas the scenario was disturbed in 1984 - 1985 by a cyclically overvalued US $.

In 1986, cyclical correction of oil prices caused drilling collapse, and additional impact was given by bringing the US $ back to realistic values and even below that. 1987 was a transitional year after the big slide and was characterized by hope and outlook to better possibilities, and the temporary agreement on a convention price (cf. section 2.2.1.4.3.8) gave an artificial stabilization of the situation which many companies took advantage of and consolidated. The 1988 renewed oil price instability and volatility is mainly the result of the indecisiveness of the OPEC as a whole concerning price politics and lacking discipline of various member countries to observe allocated and agreed output quotas (CROUSE 1988 b). Mid to late 1988 has parallels with the great 1986 oil price crash even though netback pricing is discontinued, with the only comfort being that a full rerun of 1986 is not expected (VIELVOYE 1988 b).

2.2.1.1.5. Reasons and effects of the 1986 oil price crash

Reviewing the political and economical constellations, the renouncement of definite production quantities and fixed crude oil prices in late 1985 and the replacement of the former leading aspects by the new goal of achievement of a reasonable share of world oil production by the OPEC consortium had the result that the producing countries had to concentrate on the world market concerning pricing and contract negotiating (ERDÖL-ERDGAS-KOHLE NACHRICHTEN 1987 c). The main influencing factors are netback pricing and market share strategy vs. fixed pricing and production limitation, and oil price defense and market compatibility vs. changing pricing policy.

2.2.1.1.5.1. Netback pricing and market share strategy vs. fixed pricing and production limitation

The consumers utilized the cost-effective supply at netback conditions or on the spot market (BECK & WILLIAMS 1987, OIL GAS JOURNAL 1987 f), with the consequence being that in 1986, almost 90 % of the world oil trade was done based on spot market conditions. This new strategy of netback conditions finally caused the 1986 oil price crash. In late 1986, the OPEC group adjusted themselves to the new situation by renouncing from market share policy and returning to production quantity agreement and discipline (with the latter aim having been at least intended, while achievement was hampered by continuous violation through some members of the cartel), and stabilized the oil price at abt. US $ 18 by deciding to prevent further extension of the price war, stop netback sales by end of 1986 and return to base prices and production limitation (cf. section

2.2.1.4.3.8.).

Thus the increasing oil market share strategy orchestrated by Saudi-Arabia in late 1985 brought about the most precipitous decline in oil prices in the OPEC history within a very short time period (NARAGHI 1988). The major implication of this exercise in contrast to prevailing conventional wisdom was that there existed a short run price elasticity of demand for OPEC oil of substantial magnitude and effectiveness. The immediate task faced by the OPEC was to implement a market-clearing oil price strategy that would balance current economical utility of member countries, would effectively represent the marginal cost of crude oil production by non-OPEC countries, and would curtail substitution of the other competing fuels (cf. sections 2.2.1.2.2. and 2.2.1.2.4.).

2.2.1.1.5.2. Oil price defense and market compatibility vs. changing pricing policy

In the past, OPEC had been able to defend the oil price when its market share had been more or less compatible with its needs and could be distributed among member countries without creating financial hardship (AL-CHALABI 1987). In the golden years, OPEC was able to maintain its role of a swing producer to defend a price from which others benefited in terms of greater investments in oil sources (cf. section 2.2.1.2.2.1.). With declining market share, however, the price defense mechanism became more and more precarious in 1985. Saudi Arabia, assuming the role of swing producer within the OPEC by absorbing the additional fall in demand and allowing the other members to produce at their full quota or even exceed it, was temporarily able to still defend the price. Once demand dropped below the critical rate of 2 Mio. bbl/d production, Saudi Arabia had to change its policy in order to have the possibility to produce its full quota as fixed in late 1984. The market-oriented price mechanism of Saudi Arabia then led to de facto abandonment of the fixed price system which could be defended only by production programming, to more or less market-share pricing policy, which in turn had tremendous impact on the oil price level and caused the worldwide petroleum industry crash.

TEWKSBURY (1977) performed a long-range forecast of oil price evolution which concluded with the prediction of a reasonable oil price reduction in 1985, with this prognosis having been actually met in late 1985/early 1986 almost ten years after the prognosis was made and nearly exactly as it was previewn. The free fall of oil market prices and ensuing market chaos from a more or less stable range of US $ 28 - 34 for many years up to late 1985 and the rapidity with which nominal prices plunged to US $ 6 - 9 in early to mid 1986 is the drawback of the former success of the OPEC in maintaining high oil prices (McCANN 1988). The high prices of the late 1970`s and early 1980`s called forth a rapid increase in non-OPEC oil production which increasingly threatened OPEC`s market share.

2.2.1.2. Effects of oil price changes on hydrocarbon industry

Summarizing the historical evolution, after two oil price explosions which generated an increase in exploration and development activity up to a level being much higher than ever before since the Second World War, the slow and continuous depression from the certainly exaggerated peak to more or less normal positions and somewhat below was suddenly accelerated by a pronounced oil price crash that turned the apparent or artificial undersupply problem into a real underdemand crisis. While the hydrocarbon industry in all parts of the world was more than happy to adjust to the higher oil prices by expansion of the business, the reverse adaptation to the lower prices by cutting back was very painful and hit the whole economical system, especially because it was also more or less accompanied by the most serious slide-down of the US $ exchange rate during the last 40 years.

The considerable consequences of the oil price crash on exploration and exploitation activity have to be understood in light of cost per barrel being the most important economic decision tool (EYCK 1987). The oil price collapse initiated a whole series of fundamental changes in the hydrocarbon industry, ranging from actual sharp cuts of income for governments and companies to a very changed environment with many uncertainties (HOLMES 1987). The outline as follows comments on drilling rig count as well as artificial and natural market forces. Some variables determining oil price level, market forces and oil price evolution, and oil consumption and output are also illustrated.

2.2.1.2.1. Drilling rig count

An overview of the historical oil price evolution including comments on price elasticity and possibilities of prediction of future oil price changes and credibility of such forecasts is also presented by DOUGHERTY (1987). The impact of the upwards and downwards changes of the oil price during the last fifteen years is highlighted by the USA count of active rigs drilling on hydrocarbon targets (that can be considered as being the most prescriptive indicator of the industry`s investment level).

2.2.1.2.1.1. 1986 and 1987

The annual average USA rig count reached from a number below 1,500 before 1973 its climax in 1981 with abt. 4,500 rigs, with this figure being five and a half times higher than both ten years earlier and five years later when in 1986 a total world-wide number of less than 1,000 active rigs was recorded (MILLER & NEWLIN 1987, SCHUBERT 1987) comprising in early 1986 abt. 900 and in mid 1986 abt. 700. In the North Sea representing the most important European petroleum province, the sharp incision caused by the 1986 oil price crash is reflected by a total count of wells drilled of 260 in 1984 and 1985 each vs. 180 in 1986 and 200 in 1987 (QUINLAN 1988). In the whole offshore Europe area comprising North Sea, Irish Sea, Baltic Sea and Adriatic Sea, abt. 50 - 60 rigs were active after the oil price crash in 1986 and 1987, whereas prior to that more than 90 rigs were running (McNALLY 1987). A similar indication is given by the worldwide number of seismical parties active in hydrocarbon exploration which declined from abt. 1,250 in the golden year of 1981 to abt. 400 in early 1987 (ERDÖL UND KOHLE - ERDGAS - PETROCHEMIE 1987 b). In the USA, abt. 12,700 operators drilled in 1982 at least one well on an operator-of-record basis, but in 1987, this number had dropped to only abt. 5,700 operators, with virtually almost all of these 7,000 lost operators having been independents (MOORE 1988 a; cf. section 2.2.1.5.2.). The total number of wells completed in the USA during 1986 was abt. 35,000 which was the lowest amount since 33,000 boreholes were finished in 1974.

With the continuation of low oil prices, exploration and appraisal drilling levels in the North Sea were slashed to the bare minimum in the second half of 1986 and the first half of 1987 (cf. sections 2.2.1.5.2. and 2.4.2.), whereas in the second half of 1987, activity picked up considerably, because companies learned to live with the prospect of a medium-term oil price around US $ 18 (cf. section 2.2.1.4.3.8.), and relatively inexpensive rig and therefore total operating rates started to look attractive (CAPEL 1988). Drilling activity is particularly sensitive to boom/bust cycles (PERKINS 1988). Drilling is inherently cyclical because it is extraordinarily capital-intensive. The high risk, large number of decision makers and high frequency of reinvestment decisions add to the volatility.

Contrasting with all other reports, POPESCU & HODGSHON (1987) put forward that exploratory drilling in Western Europe as a whole was virtually unchanged in 1986 from its 1985 level, and North Sea drilling reduction in 1986 was only abt. 10 %, with, however, the amount of wells having been drilled in the first quarter of 1986 in the British North Sea having been higher than in any other quarter of United Kingdom offshore oil and gas activity. This relationship falsifies the overall picture, as many of the items drilled in the first quarter of 1986 would have been suspended or cancelled if it would not have been too late at the moment when the oil price rapidly dropped. On the other hand, a slight to moderate decrease in drilling frequency in the British North Sea in 1986 is not necessarily a direct consequence of the oil price collapse, but has also to be understood in light of 1984 and 1985 having set records of United Kingdom offshore drilling abundance (HOLBROOK 1987). Aspects of hydrocarbon acquisition activity in an era of low oil prices are also discussed by WOODS & HOLTBERG (1986). The fluctuations of the drilling rig number as a consequence of the economical feasibility scenario changes highlight that the recoverable hydrocarbon reserves are affected by oil and gas prices (GRUY, GARB & HERRON 1980; GARB, GRUY & HERRON 1981).

2.2.1.2.1.2. 1988

In 1988, overall world drilling is rising again considerably in many areas and is exceeding the level of 1986 and possibly even that of 1985 (WORLD OIL 1988 a). Total number of wells drilled worldwide are abt. 60,100 in 1986, abt. 57,100 in 1987 and abt. 60,900 in 1988. By end of 1987, total world oil reserves have reached abt. 800 Bill. bbl, and total world gas reserves have attained abt. 3,860 Tcf. In the USA, oil well completions decreased from abt. 41,000 in 1984 via abt. 26,000 in 1985 to abt. 17,000 in 1986, and gas well completions diminished from abt. 15,700 in 1984 via 10,000 in 1985 to only 6,500 in 1986 (MOORE 1986, 1987 b). Reviewing oil price history with a price escalation of abt. 40 %/year on average over a period of six years (1973 - 1979) reveals that the impetus of the order of magnitude increase in price (which was greatly in excess of the inflation rate) caused reduced consumption through buyer choice, conservation, substitution, enhanced efficiency and conversion to other energy sources, and brought on major increases in non-OPEC oil production capacity.

After the peak of the crash, the USA drilling rig count only slowly improved via a slight further decline until early 1987 reflecting numerous project cancellations and postponements to a little higher figure of abt. 800 - 950 in mid 1987 (BECK 1987), thus still staying at a level of only abt. one fifth of that in the golden years (AL-CHALABI 1987). In the first half of 1988, an increase in active drilling rigs of abt. 20 % was recorded which could bring the annual rig count average by end of 1988 to 1,030 - 1,120, with 1988 thus overall appearing to be another recovery year for the petroleum industry as activity increases for the second straight year since the 1986 oil price collapse (CROUSE 1988 a). It thus temporarily looked like as if the boundary of 1,000 rigs might probably for the first time be exceeded again in 1988 after the depressed years of 1986 and 1987 when following a number of almost 2,000 in 1985 the rig count fell for two years on average below 1,000. The mid to late 1988 oil price decline from US $ 16 - 18 to 12 - 15, however, resulted in the average monthly rig count in the second half of 1988 dropping to 920 - 940 (OIL DAILY 1988 f) which most probably brings the annual average again down below 1,000.

Counting with a cutoff of 45 days activity period, the number of active rigs was in 1988 with abt. 1,530 abt. 10 % higher than in 1987 when abt. 1,390 rigs were active in the USA, indicating that after six consecutive years of declining number of available rigs, heavy rig fleet attrition in the USA is probably coming to an end (FITTS & CROWHURST 1988). Rig utilization which is defined as the ratio of active rigs to available rigs was 56 % in 1988 with respect to 42 % in 1987 which, however, is still considerably lower than the 67 % of 1984 and lags markedly behind the average rate found during the last 35 years.

2.2.1.2.2. Artificial and natural market forces

The unyielding downward pressure of this combination caused the export volume of the OPEC to fall dramatically, finally resulting in capitulation of the OPEC to market forces in late 1985 with abandonment of production restraint in favour of seeking its fair share in the international oil market (RANDOL 1986). Some aspects of swing production for market imbalance compensation as well as OPEC residual market share and price effects are outlined as follows.

2.2.1.2.2.1. Swing production for market imbalance compensation

The main reason for the 1986 oil price crash was the decision of Saudi Arabia to be no longer the swing producer in the OPEC (by swinging, Saudi Arabia for a long time successfully prevented the collapse of the oil price due to market power; ERDÖL UND KOHLE - ERDGAS - PETROCHEMIE 1987 c, OIL GAS JOURNAL 1987 f, VIELVOYE 1988 b), because the country was genuinely fed up with its former self-chosen role of being the swing producer (BECK & WILLIAMS 1987; cf. section 2.2.1.4.2.4.) and single-handedly propping up the sagging price structure of the association while other OPEC members showed no discipline in restraining output and while there was no cooperation from non-OPEC oil exporters to help maintain a stable market.

While the self-restriction of Saudi Arabia`s production to abt. 2.5 Mio. bbl/d in 1985 still effectively compensated the excesses of exploitation by other undisciplined OPEC members that continued to violate their quotas, an increase to abt. 5.0 Mio. bbl/d since the end of 1985 (ERDÖL UND KOHLE - ERDGAS - PETROCHEMIE 1987 c) when Saudi Arabia was forced to pay more serious attention to its own economical and financial balance rather than to subvent brother countries that are unwilling to respect and accept the rules and regulations, respectively, initiated the free fall of the oil price down to a level which reflects its natural market value rather than the artificially lifted position that was characterized by political threatening and dictation. With Saudi Arabia being unwilling to further play the swing producer, no other OPEC nation is likely to be ready or even able to fulfill this function (PETROLEUM REVIEW 1987 a).

In order not to threat the US $ 18 agreement price, Saudi Arabia nevertheless continued to play the role of swing producer in 1987 when necessary (BECK 1987, CROUSE 1988 b). During the last three years, almost all the increases and reductions of total OPEC output within a fluctuation marge of 3 - 4 Mio. bbl/d were performed and backed-up by Saudi Arabia through swinging, with Saudi Arabia therefore having a pivotal role in the world oil market (THOMPSON & SINGLETON 1988). Comments on oil price and demand interactions are also given by LUMSDEN (1988).

2.2.1.2.2.2. OPEC residual market share and price effects

In terms of market share, various oil price explosions and the big oil price crash have all drastically changed the participation of the OPEC group in the total world oil market (BILD 1987). The OPEC group started to fix and to control the oil price and thus to claim a market share in 1973/1974. After the second oil price shock in 1979/1980, the slow downwards orientation of the oil price forced the OPEC bit to bit to concede on price, while at the same time the market share drastically fell and was halved in the period 1981 to 1985. Realistically, the OPEC group always had only the residual market share (PETROLEUM REVIEW 1988 c) in supplying those quantities necessary for delivery which the customer countries could not establish by domestic production or import from somewhere else.

Increasing exploitation of own reserves and considerable effects of energy conservation were the main reasons for the halving of the residual market share of the OPEC. As market share is at the bottom of the line always calculated in quantity times price, following the official renouncement of a fixed price in late 1985 and the give-up of production limitation quotas, the tremendous price fall in combination with slightly increasing quantities could not reverse the downwards orientation of the oil sale revenues for the OPEC group in 1986, and only the return to fixed prices and limited production quantities in late 1986/ early 1987 was able to trigger again a moderate increase in market share and revenues. The net effect of the 1986 oil price crash was a 16 % increase of the OPEC oil production to reach abt. 33 % share of world oil output (ERDÖL-ERDGAS-KOHLE NACHRICHTEN 1987 e).

The 1986 oil price crash has a much bigger impact on company upstream spending that upon actual supply (PETROLEUM REVIEW 1987 a). It has turned out that non-OPEC production costs were lower than was generally assumed, and a further buffer against more drastic indigenous production losses were those output tax regimes in place. If a company is paying a relatively high marginal tax rate, oil price declines can initially impact more severely on the treasury of the government imposing that tax than upon the company itself. Simultaneously, the price fall engendered a strong upstream efficiency drive along with sharply falling oil field equipment and service costs.

2.2.1.2.3. Variables determining oil price level

Some variables determining oil price level are discussed as follows along the lines of coupling of oil production rate and price, current ideal oil price level, oil production costs and oil sector dominance, and oilpolitical triangle USA - Saudi Arabia - Iran.

2.2.1.2.3.1. Coupling of oil production rate and price niveau

Oil price evolution also has shown that the primary determinant of crude oil price is the volume of crude oil offered for sale, with relatively small percentage changes in volume causing relatively large percentage changes in price (for example, a lowering of the quantity offered for purchase by abt. 4 % of the world consumption in late 1986/early 1987 lifted the price by abt. 40 % from 12 to 18 US $; DOUGHERTY 1987, NARAGHI 1988, OIL DAILY 1988 e). Calculating with the official OPEC list price of 29 US $/bbl which was valid still in late 1985, the crash in 1986 brought the price down to abt. 25 - 30 % of the former value at the peak of the crisis and abt. 40 % on average, whereas the regaining consensus of the group and return to production limitations moved the price back to abt. 60 % of its earlier level without, however, reestablishing confidence in its stability that has been lost during the price collapse. In late 1987, oil price was abt. 25 % higher than in late 1986 (CROUSE 1988 b).

At the bottom of the line, the oil price is even established by the rate selected by the major producer of the world, with this position having changed from the USA to Saudi Arabia in the 1960`s. Examining the OPEC efforts to come to an agreement on production rates and prices, and judging the role of Saudi Arabia from their interest in both securing national income and strengthening their market leadership, the 1986 oil price upheaval cannot be interpreted as reflecting the collapse of the OPEC, but rather was the logical unfolding of Saudi Arabia`s plan to bring oil price down to a sustainable level. The effect of Saudi Arabia`s increasing oil market share strategy was plummeting of the average marker oil price from US $ 28 - 30 to US $ 12 - 15, and during the time interval of free market forces, a 1 % increase in global oil supply resulted in a 12 % decline in average oil price (NARAGHI 1988). Other economical aspects of oil price and its changes in past and future are discussed by AL-BLEHED (1987), DOUGHERTY & AL-BLEHED (1987) and LUMSDEN (1988), and the constellations within the OPEC group that led to the oil price crash are evaluated by RANDOL (1986).

SAFER (1977) comments on challenges and opportunities of the world oil market, prediction of oil price evolution, and role of the OPEC group (cf. also TEWKSBURY 1977). WEFA (1988) analyzes boom or bust of the oil market including up- and downswings, and MACKAY CONSULTANTS (1988) present a prediction of the oil and gas industry trends in the next years. KONZELMANN (1976) even assesses oil as being the fate of mankind.

The global oil price is predominantly set by competition among the world`s oil-exporting countries and not by economists of the oil-importing nations (IVANHOE 1988). Oil consumed internally within any producing country sells at the international price set by the exporters. Non-OPEC production increases since the first oil price explosion in 1973 that are used internally within any nation are not available for export and consequently this oil does not affect the international price, with the critical volume thus only being the net amount that is available for export which is usually less than 25 % of the total production of non-OPEC countries.

2.2.1.2.3.2. Current ideal oil price level

The ideal oil price level for at least the OPEC group is at the moment abt. US $ 18, because it is low enough to spur demand but not sufficiently high to head off declines in non-OPEC production (OIL GAS JOURNAL 1988 e). US $ 18 means a level of demand for OPEC oil that yields optimum revenues for OPEC members thus satisfying OPEC needs while taking into account the price elasticity of supply/demand. While US $ 18 is highly desirable, US $ 15 would be the absolute minimum tolerable for the OPEC group. At US $ 18, demand for OPEC oil is abt. 19.5 Mio. bbl/d in 1987/1988. If oil prices decline, OPEC`s revenue shortfall increases geometrically, with the gap expanding at its fastest rate once prices drop under US $ 15. If oil prices were to remain in the US $ 17 - 22 range, OPEC could basically generate the same amount of revenues if it adjusts its production accordingly, whereas for prices of more than US $ 22, OPEC would again be worse off.

Therefore it is expected that in light of own interest, the OPEC group will be attempting to stabilize the oil price at abt. US $ 18 thereby confirming the late 1986 agreement (cf. sections 2.2.1.1.2. and 2.2.1.4.3.8.) by restricting production to abt. 19.5 Mio. bbl/d. US $ 18 can be regarded as the maximum desirable price for the OPEC at the moment, because at this level even denominated in weakening dollars, some new production continues to come on stream outside OPEC, and with higher prices, natural gas and coal begin to replace oil (OIL GAS JOURNAL 1987 n).

2.2.1.2.3.3. Oil production costs and oil sector dominance

In view of the enormous difference in production costs of 0.05 - 0.2 US $/bbl in the OPEC area and 5 - 15 US $ in many of the non-OPEC countries, it would have been a simple operation for OPEC to have driven down the price to a point lower than non-OPEC long-run marginal cost levels (McCANN 1988). For various reasons, however, this was never a viable solution for OPEC. Such a price decline would be politically unacceptable to those OPEC members lacking large reserves and would lead to increasing divisions between countries with major and minor reserves, with low-reserve OPEC members thus being effectively excluded from participating in a long-term strategic price decline.

Another reason is the dominance of the oil sector to OPEC member economies and export revenues as opposed to the relatively minor part played by the oil sector in the total gross national product of many bigger non-OPEC producers (such as Great Britain, Norway and Mexico). Finally, OPEC oil production is relatively price-inelastic especially in the short term, with the desired non-OPEC response to any price fall being lagged, whereas the short-term costs to the cartel would be felt immediately. Thus the restoration of market stability in late 1986 by placing the US $ 18 agreement price (cf. sections 2.2.1.1.2. and 2.2.1.4.3.8.) at more or less a midpoint between the price highs of the early 1980`s and the lows of early to mid 1986 has turned out to be the overall optimum solution.

2.2.1.2.3.4. Oilpolitical triangle USA - Saudi Arabia - Iran

In terms of political relationships, the oil price gyrations of the 1970`s, the oil price collapse of 1986, and the subsequent oil price recovery are the result of the interactions within a triangular relationship with the USA at the top and Saudi Arabia and Iran forming the base of the triangle (PETROLEUM REVIEW 1987 a). It appears from oil price history that whenever USA and either Saudi Arabia or Iran agree on a certain policy towards oil price, the third party has very little influence. In 1973, the Arabian oil embargo led by Saudi Arabia was undermined by the USA which voiced its total opposition and actively encouraged Iran and other non-Arabian countries to take advantage of the situation and to overproduce.

In the late 1970`s, USA supported Saudi Arabia to flood the oil market in order to prevent oil prices from rising even higher. Throughout the 1980`s, the USA was keen to keep oil prices as low as possible in order to reduce energy costs and promote growth, which coincided with the interest of Saudi Arabia to stimulate oil demand that had been badly affected by the price hikes of the 1970`s. The level of oil prices and the influence which the political triangle USA - Saudi Arabia - Iran has upon them will play an important role in future energy security. In addition, without the OPEC, oil prices would decline to levels that would encourage greater dependence on Gulf supplies, thereby possibly harming long-term energy security of the OECD.

2.2.1.2.4. Market forces and oil price evolution

The oil price evolution during the last 20 years was predominantly driven by market forces (PETROLEUM REVIEW 1988 c). Four periods can be distinguished : 1970 - 1973 (leading to the first oil price increase), 1974 - 1979 (guiding to the second oil price uplift), 1979 - 1985 (giving rise to the 1986 oil price plunge) and the current situation 1986 - 1989 and beyond. Comments are also offered on market regulation vs. disequilibrium compensation.

2.2.1.2.4.1. 1970 - 1979

Between 1970 and 1973, world oil demand increased by over 8 Mio. bbl/d mainly due to the declining real oil prices of the previous two decades (PETROLEUM REVIEW 1988 c). As non-OPEC oil supply increased only marginally, demand for OPEC oil increased sharply by almost 7.5 Mio. bbl/d. Therefore the production level came very close to OPEC output capacity and if the situation had continued, demand would have by far exceeded supply. Thus market forces set the stage for the first oil price increase in late 1973, with OPEC`s role having been important in timing, but political factors such as the Arab-Israel war and the magnitude of price increases having been considerably more significant. In the second period, demand for oil despite the now reasonably higher price continued to increase.

While non-OPEC oil supply began to rise, demand for OPEC oil was close to its production capacity, with thus the oil price exploding again in late 1979 in order to create a brake of the excessively progressing demand. The Iranian revolution and the Iranian-Iraqian war were important factors, not so much that they reduced the oil flow from the region, but because they created demand for precautionary and speculative stock, with again market fundamentals playing a crucial role in the two sharp price increases (cf. section 2.2.1.1.1.).

2.2.1.2.4.2. 1979 - 1986

The 1986 oil price collapse was also caused by market forces. The sharp fall in oil demand between 1979 and 1985 and the even greater increase in non-OPEC oil supplies resulted in a world requirement for OPEC oil at a level that was too small for the organization to sustain (PETROLEUM REVIEW 1988 c). The decline of more than 13 Mio. bbl/d in OPEC oil demand led to an OPEC oil production of less than 16 Mio. bbl/d in 1985 and a record minimum of 13.4 Mio. bbl/d in 1987 (compared with the maximum of 31 Mio. bbl/d in 1973; ERDÖL-ERDGAS-KOHLE NACHRICHTEN 1987 d). The long-term exposure to the two oil price explosions has been a significant advance by the industrialized countries in energy conservation measures on the one hand (representing a macroeconomical response to the oil shocks) and a significant improvement in energy efficiency (being a microeconomical response to the price rises; McCANN 1988). As market forecasts in mid to late 1985 indicated further deterioration in demand for OPEC oil in case of absence of major steps, the only option was to float oil prices in order to stop or even reverse the declining trend (PETROLEUM REVIEW 1988 c).

The impact of the 1986 oil price plunge was thus logically a turnaround of the oil market. Demand increased relatively sharply by 1.8 Mio. bbl/d between 1985 and 1987, and the rise in non-OPEC supply was more or less terminated. It was this turnaround in the market that was instrumental in OPEC once again embarking on a regulation of the oil market. Thus while OPEC played a regulating role, market fundamentals set the stage for the two oil price increases of 1973/1974 and 1979/1980 (cf. section 2.2.1.1.1.) as well as for the 1986 price collapse (cf. section 2.2.1.1.2.). The role of OPEC was important in the two price increases in relation to the magnitude of the rise and officializing it, but its role was even more significant between 1982 and 1985 when it managed to prevent a price crash by reducing output.

2.2.1.2.4.3. Market regulation vs. disequilibrium compensation

Historically, the oil market has been regulated by major oil companies until 1970 and by OPEC afterwards (PETROLEUM REVIEW 1988 c). The system of rigid oil pricing, however, will have difficulties in surviving because of the structural changes that have taken place in oil trading. The first rigidity in the OPEC pricing system arose because with given demand, it is not possible to fix both supply and prices, but it could only happen by coincidence if the market is in balance at a given price. If adjustments to the system are not made, however, disequilibrium emerges. One possibility is to have a swing producer to adjust its output to balance the market, but unless overall supply and prices are frequently adjusted, the swing producer will lose its market share.

In the last decade, fundamental changes in the system of oil trading have forced oil prices to become more market responsive. The series of reasons includes increase in spot transactions, emergence of oil future and forward markets, increasing integration of the downstream sector into the oil segment of the producing countries; appearance of new marketing practices such as netback pricing, processing deals and barter trade; increasing number of participants in the oil market, and fall in the OPEC share in global oil supplies.

2.2.1.2.5. Oil consumption and output

Aspects of oil consumption and output are discussed along the lines of OECD demand vs. conservation and substitution as well as OPEC and non-OPEC production vs. demand as follows.

2.2.1.2.5.1. OECD demand vs. conservation and substitution

Oil and total energy consumption per unit of gross national product dropped in most of the Organization for Economic Cooperation and Development (OECD) countries during the 1973 - 1985 interval, but since the oil price collapse in 1986, this trend has reversed (NARAGHI 1988). Due to conservation measures and fuel substitution effects brought about by the 1973/1974 and 1979/1980 price hikes, the OECD countries used 38 % less oil per unit of output in 1985 than in 1973. In 1986, oil consumption levels in all regions of the world have increased for the first time since 1981 as a consequence of now much cheaper oil being available, and global oil consumption in 1986 has reached the same niveau as in 1976. Levelling-off and/or slightly increasing oil fuel share of primary energy consumption in 1986 took place on a world-wide basis.

The consequence of the return to a more elevated list price of US $ 18 in late 1986 means an abt. 40 % income increase with respect to a US $ 12 scenario representing the average level during the peak of the crisis. In 1986, for the first time since 1978, the share of oil in primary energy consumption reversed a declining pattern, and in the free market economies (particularly in the USA), the increasing share of apparent oil consumption in 1986 was accompanied by decreasing shares for gas and coal consumption. If continued in the future, these trends could signal possible reversal of consumer behaviour and conservation measures and provide a basis for the classical debate over the ever-illusive equilibrium price of crude oil.

2.2.1.2.5.2. OPEC and non-OPEC production vs. demand

Although frequently exhibiting no consensus and discipline concerning price and production quantities, the OPEC group has extraordinarily well held together since in 1981/1982 the first losses in revenues occurred due to beginning of declining demand and falling prices (HOLMES 1987). Because in late 1985 as a consequence of lower demand and higher non-OPEC production the absolute level of oil export by the OPEC had fallen down to half the amount of 1980/1981, the OPEC tried to get again more market share by renouncing from fixed price and production quotas which provoked the 1986 oil price crash that had threefold results. The first effect was OPEC`s loss of revenue with very much lower prices, because the income in 1985 was only still half of that in 1980/1981, and the revenue in 1986 was again only half of that in 1985. Nevertheless, the OPEC group retained its cohesion and re-established price target and volume discipline.

The second main result of the 1986 oil price collapse was the curtailment of non-OPEC production. The 1986 output levels were abt. 0.8 Mio. bbl/d less than they might have been expected at prices remaining at 1985 niveau. The bulk of this loss was in the USA where production may have dropped as much as 0.5 Mio. bbl/d. The third important effect of the 1986 oil price slide was the influence of lower oil prices on demand. World demand increased by abt. 0.8 Mio. bbl/d over what it might have been if prices remained at abt. US $ 28. Some of this increased demand in 1986 was substantial stock build especially in Europe, and some was due to substitution of other energy sources by low-priced fuel oil particularly in the USA, thus underlining that oil demand at the margin is strongly influenced by oil price level.

2.2.1.2.6. Other aspects

Effects of the 1986 crisis on oil and gas industry are also discussed by DANCY, DANCY & TAYLOR (1987) and KELKAR (1987), and other oil price aspects are evaluated by ROBERTS (1984), HILL (1987) and POWERS & STEVENSON (1987). CAMPBELL & HUBBARD (1984), ZWICKY (1984) and DOUGHERTY (1987) investigate possibilities of reasonably reliable prediction of the directions of oil price evolution, and KELLY (1987) assesses the question of oil price stability and consolidation. VEATCH (1986) evaluates economical aspects of hydraulic proppant fracturing. Aspects of the role of the OPEC in world oil market and price evolution are also discussed by GATELY (1984), and AL-CHALABI (1987) and ATTIGA (1987) summarize oil price history and associated periods of boom and crisis particularly from the viewpoint of the organization of Arabian petroleum exporting countries (OAPEC). Political prospects within the Arabian area are illustrated by SHARABI (1988), and the relationship between political influences and oil/gas price evolution is sketched by GLASNER (1985). FRANSSEN (1988) comments on political, strategical and resource implications of oil supply dimensions, and LUMSDEN (1988) evaluates oil price and demand interactions. MORSE (1988) illustrates the impact on markets of non-OPEC oil supplies. TARIKI (1985) illuminates the situation of global oil markets of being under stress.

As a consequence of the diminishing revenues from oil sale, more and more OPEC members are pushing into downstream ventures in order to establish integrated production (WILLIAMS 1988; cf. section 2.2.1.4.3.3.). The downstream moves are the most apparent reintegration efforts and are a natural result of their capacity surplus, competition in the oil market, and the desire of the producing countries for more assured outlets given this market share competition (PETROLEUM REVIEW 1987 a). Downstream integration of OPEC states as a compelling competitive edge is also emphasized by PETROLEUM REVIEW (1987 a) and STAUFER (1988).

2.2.1.3. Impact of varying US $ exchange rate on oil price evolution

The low oil price which was in 1986 temporarily considerably below US $ 10 and also throughout 1987 and early 1988 never exceeded an average value of US $ 18 was outside the USA and other dollar-currency countries even accentuated by the accompanying weakness of the US $ (cf. also section 2.3.5.). The exchange rate of the US $ with respect to other world guide currencies such as DM and £ started to decline in mid to late 1985, reached in the fourth quarter of 1987 its historical minimum level after diminishing its value by 40 - 50 % within two years, kept being low throughout the first part of 1988, and only in the second half of 1988 experienced a slight recovery of abt. 20 - 25 % of its value that, however, may be of only temporary nature, as already indicated by an already renewed slight drop after the presidential election in the USA at the end of 1988).

Another major influencing factor is the inflation which had the consequence that the net return from hydrocarbon sale in Europe was in mid 1986 only abt. 40 % of that on average in 1985, and that an oil price in nominal terms of 10 - 15 US $ in 1986 is in real terms equivalent to abt. 5 US $ in 1972/1973. Relative to the DM, the value of the US $ was halved from 3.20 DM in early 1985 to 1.60 DM (and temporarily even below) in late 1987 (THOMPSON & SINGLETON 1988) which in terms of oil price means that an US $ 27 barrel price in early 1985 was equivalent to DM 86, whereas US $ 19 in late 1987 corresponded to only DM 31. Thus the world oil price was only 30 % lower in US $ but 65 % lower in DM, resulting in the fact that on the international market, a barrel of oil was more than 50 % cheaper in late 1987 than in early 1985, with over half of this decrease being directly attributable to the depreciated US $.

The effects of the sharp 1986 oil price drop have been dramatic to even traumatic, leading for production, service and supply companies to a battle for survival of the fittest, with many small companies having been faced with going out of business and various larger companies having tried to escape this fate by merging and mutual corporate takeover (cf. section 2.2.1.5.2.4.) for easier survival until better times come back (GRIFFITHS 1986, MEISTER & CADY 1986). US $ revenues to OPEC have considerably declined both in real and nominal terms due to the depreciating US $ on foreign exchange markets during the last years (McCANN 1988). Thus OPEC was squeezed from two sides recently, with on the one hand lower oil prices having reduced the value of oil export revenues in nominal terms, whereas on the other hand, the depreciating US $ against a basket of other OECD-member currencies led to a decline of purchasing power of the nominal revenues in real terms. The various relationships and impacts are illustrated as follows by outlining some examples comprising import price for crude oil in Germany FRG, OPEC oil sale income, and current vs. constant US $ rate calculation. Impacts of changing US $ rate on the world oil market are also discussed by WEFA (1988).

2.2.1.3.1. Import price for crude oil in Germany FRG

As a consequence of oil price drop and US $ rate collapse, the import price of crude oil at the Germany FRG boundary which reached its historical maximum in March 1985 with 697 DM/t and had an average of 622 DM/t in 1985 fell via 493 DM/t in January 1986 down to the minimum of 120 DM/t in July 1986 (REINICKE, TRENEL, HERBST, KEMPGEN & LINZ 1988) and 178 DM/t in August 1986 (ERDÖL-ERDGAS-KOHLE NACHRICHTEN 1987 c, ERDÖL UND KOHLE - ERDGAS - PETROCHEMIE 1987 b) and probably at least temporarily even further in late 1987/early 1988 when another slight real oil price decline coincided with an extreme weakness of the US $. The minimum of 178 DM/t in August 1986 represents a level that for the last time had been achieved at the end of 1973 before the onset of the OPEC-induced price explosions.

The import price of crude oil as well as the revenues of domestic crude oil thus were in 1986 and 1987 more than 60 % lower than in 1985, with the improvement of the oil price level from 1986 to 1987 having been during most of the time more or less compensated by a further weakening of the US $ (ERDÖL UND KOHLE - ERDGAS - PETROCHEMIE 1988). In 1987, the import price for crude oil increased to abt. 220 - 260 DM/t (ERDÖL-ERDGAS-KOHLE 1988 a) throughout the first half of the year due to the increase of the average oil price from 8 - 12 US $ in mid 1986 to 18 US $ in late 1986 and early 1987 (cf. sections 2.2.1.1.2. and 2.2.1.4.3.8.), whereas the US $ kept being weak and had no strengthening influence.

As a consequence of the diminishing strength of the US $ in later 1987, the revenue for domestic crude oil in Germany FRG fell from 200 - 220 DM/t on average in 1986 and 1987 to 140 DM/t in late 1987/early 1988. In a similar way, the net oil bill of Germany FRG was more than halved from 64.5 Bill. DM in 1985 to 30.8 Bill. DM in 1986 in spite of increased net oil imports (ERDÖL-ERDGAS-KOHLE NACHRICHTEN 1987 c). Impacts of US $ weakness on oil price and exploration/development activity are also sketched by QUINLAN (1988).

2.2.1.3.2. OPEC oil sale income

As a consequence of both energy saving campaigns and oil price drop, the total income of the OPEC group from sale of their oil fell down from the maximum of abt. 280 Bill. US $ in 1980/1981 to abt. 134 Bill. US $ in 1985 and to only abt. 75 Bill. US $ in 1986 (ALGAR 1987 b; ERDÖL-ERDGAS-KOHLE NACHRICHTEN 1987 b, 1987 c; ERDÖL UND KOHLE - ERDGAS - PETROCHEMIE 1987 d, HOLMES 1987). In terms of oil volume, the declining power of the oil market following the second oil price explosion in 1979 and the world petroleum recession which culminated in the 1986 oil price crash triggered a steady reduction of the OPEC oil export quantity from 11 Bill. bbl in 1979 to only 5 Bill. bbl in 1986 when the capacity of other countries had reached such a level that USSR/China exported 2 Bill. bbl and other non-OPEC nations exported 3 Bill. bbl of oil, thereby cutting off considerable business from the OPEC, and the rest was compensated by energy conversion and conservation (IVANHOE 1988).

The combined loss of the OPEC group in 1986 as a consequence of the oil price collapse exceeds 50 Bill. US $ (AL-CHALABI 1987, McCANN 1988), and in 1987, the OPEC oil revenues have been again only abt. one quarter of the peak level of 280 Bill. US $ reached in 1980/1981. The establishment of the US $ 18 agreement price in late 1986 (cf. sections 2.2.1.1.2. and 2.2.1.4.3.8.) brought the 1987 OPEC oil sale revenue back to abt. 97 Bill. US $ which is abt. the mid-point of the levels of 1985 and 1986 (McCANN 1988). Some aspects of oil price convention and defense, OPEC vs. non-OPEC production, and OPEC production ceiling and overhang capacity are illustrated as follows.

2.2.1.3.2.1. Oil price convention and defense

In late 1986, the situation was gradually stabilizing again to an oil price niveau of abt. 18 US $ on average (being the result of a reorientation of the OPEC group towards a consensus forced by the drastical collapse of their income; IBG 1986; ROBERTS 1986, 1987; ALGAR 1987 b, NARAGHI 1988; cf. sections 2.2.1.1.2. and 2.2.1.4.3.8.) which was still more or less valid in mid 1987 (within the range of 15 - 22 US $ according to short-term fluctuations of the real value), with again, however, this nominal- terms level being as a consequence of US $ weakness and inflation in real terms equivalent to only abt. 8 - 10 US $ in 1972/1973 (MENGES 1986, OEHME 1986, RUNGE 1986). The 1986 revenues of the OPEC group amounted only to their 1973 level with inflation assumed to be just 6 % annually in the interim period (OIL GAS JOURNAL 1987 f). These several abrupt price changes during the last 15 years have turned repeatedly oil and general world economics out of equilibrium (SCHAEFER & POWERS 1984, OEHME 1986). The daily OPEC sale volume fell from 31 Mio. bbl in 1979 before the second oil price explosion via 19 Mio. bbl in 1982 to 16 Mio. bbl and less in 1985 prior to the major oil price crash (AL-CHARABI 1987; ERDÖL UND KOHLE - ERDGAS - PETROCHEMIE 1987 c).

The OPEC price defence policy by setting prices at certain levels below which no member country sold its oil and by leaving the production niveau to be regulated by the market had the only drawback that the prices were set at such high levels that the results became self-defeating by drastical drop of demand for oil in general and increasing supplies of oil outside the OPEC, resulting in a decline of the OPEC world oil supply share from abt. 70 % in 1980/1981 to less than 40 % in 1985/1986 (AL-CHARABI 1987). Between 1975 and 1985, some 10 Mio. bbl/d of additional supplies from the non-OPEC producing region were added to non-Communist output, and some 7 Mio. bbl/d were lost in non-Communist world demand as a consequence of energy conservation and fuel substitution.

2.2.1.3.2.2. OPEC vs. non-OPEC production

The tremendous loss in OPEC oil revenues as a result of slowly falling prices in the period 1981 - 1985 and the oil price drop in 1986 is highlighted by comparing OPEC and non-OPEC production (PETROLEUM REVIEW 1988 a). The evolution of OPEC to non-OPEC (non-Communist world) output passed from 30 - 31 to 14 - 15 Mio bbl/d in 1973 through 1977 via 31 to 17.5 in 1979, 22.5 to 18.8 in 1981 and 18.9 to 20.1 in 1981 to 15.4 to 23.1 in 1985 (McCANN 1988), with thus the relationship having been for the first time already reversed in 1981 and having reached a significant disequilibrium for the OPEC in 1985. OPEC share in world energy dropped from one third in 1973 to abt. 10 % in 1987 (SCANLON 1988). Thus by maintaining high prices, OPEC has at the bottom of the line priced itself out of the market and was continuously losing market share (PETROLEUM REVIEW 1988 c).

In 1987 and 1988, as a consequence of increasing demand for the reason of lower prices, OPEC production climbed again to abt. 18 - 20 Mio. bbl/d (OIL GAS JOURNAL 1988 d, VIELVOYE 1988 b) after having reached a record minimum of 13.4 Mio. bbl/d in early 1987 (ERDÖL-ERDGAS-KOHLE NACHRICHTEN 1987 d) and having amounted to 14 - 18 Mio. bbl/d in the period 1983 - 1986 (ERDOEL-ERDGAS AKTUELL 1984). On the other hand, non-OPEC output in 1986 through 1988 stayed more or less constant at 22.5 - 22.6 Mio. bbl/d, with the stagnant situation even already slightly pointing downwards (McCANN 1988). Thus while in 1973 OPEC and non-OPEC counted for 67.5 % and 32.5 % of total non-Communist world oil production, respectively, the ratio had completely turned around in 1986 with non-OPEC accounting for over half of total non-Communist world oil production with 55.2 % and OPEC delivering the balance or the residue of 44.8 %.

2.2.1.3.2.3. OPEC production ceiling and overhang capacity

The average agreed production ceiling of the OPEC group in the last years was in the range of 15 - 17 Mio. bbl/d, with the rest of the actual total output of up to 3 Mio. bbl/d being contributed by uncontrolled and violated overproduction of some member countries which for various reasons did not observe their quotas (PETROLEUM REVIEW 1987 a). The average production ceiling of abt. 16.5 - 17.0 Mio. bbl/d is more or less matching domestic oil consumption in the USA (BECK 1987) which can only be covered by abt. 50 % by own production, with the second half of it having to be imported (CROUSE 1987). The abt. 8.5 Mio. bbl/d import and domestic production volume of the USA, respectively, again is equivalent to abt. two thirds of total USSR domestic output (cf. section 2.4.4.1.). The USA is the biggest oil importer of the world, whereas the USSR is the largest producer and the greatest non-OPEC exporter (PETROLEUM REVIEW 1987 a).

The current overhang production capacity of the OPEC is illustrated by the fact that an annual output potential of 1.5 Bill. t is possible, but at present only 800 - 900 Mio t are produced per year (LÜBBEN & LEINER 1988). Annual OPEC output rose from abt. 440 Mio. t in 1960 continuously up to 1.5 Bill. t in 1973, with this level having been kept more or less constant until 1979 when a gradual slide started that brought annual OPEC production down to 800 Mio. t in the last years. On the other hand, non-OPEC output was abt. 600 Mio. t in 1966 and since that year increased up to 2 Bill. t which was reached some years ago, with this level having so far been kept constant or already slightly declining, because in contrast to the OPEC, the non-OPEC countries do not have any overhang capacity, but instead of this have already reached the maximum limit and the inevitable decline begins as a consequence of faster consumption than reserve addition by new discoveries. Total world annual oil production in the last years amounted to 2.8 - 3.0 Bill. t (ERDOEL-ERDGAS AKTUELL 1982, ERDÖL- ERDGAS-KOHLE NACHRICHTEN 1987 e). OPEC annual oil export in 1974 through 1979 reached 1.45 Bill. t, and due to the price explosions, the income was doubled from 91 Bill. US $ in 1974 to 199 Bill. US $ in 1979 (ERDOEL-ERDGAS AKTUELL 1980 a).

2.2.1.3.3. Current vs. constant US $ rate calculation

YU (1988) gives an outline of the impact of US $ exchange rate variations on oil price evolution by examining the development of oil prices as paid in local currencies. The rise and fall of oil prices in recent years has been dramatical enough based on the US $, but with the fluctuating exchange rate of the world trade base currency, oil price alterations are even more startling when converted into the local currencies as paid by the oil importing countries. From late 1980 to early 1985, the average exchange rate of the US $ increased by almost 60 %. As a result of this, although the world crude oil price in the same period declined from its peak of US $ 45 or more to US $ 30 (representing a drop of 33 % in current dollars or 47 % in 1980 constant dollars), the average price drop margin calculated in the net importing countries 1980 local currencies was consistently much smaller than the former, and in 1984, there was even a rise. In addition, in Western Europe which is the world`s largest oil importing market with a 38 % share of total world oil imports, the average price of imported oil in terms of local 1980 currencies even rose from early 1983 and reached its climax by late 1984 when it was 67 % higher than that of the USA.

Consequently, the effect of price drop in US $ during this period was substantially weakened in stimulating world oil demand and containing high-cost oil production, energy conservation and oil substitution. Since early 1985, the average US $ exchange rate dropped to a level in late 1986 being 28 % below that of early 1985. This decline made the margin of the price fall in constant local currencies of oil-importing countries larger than that calculated in current and constant US $ and narrowed the gap between them. In late 1986, the actual world oil price was 67 % lower than its peak in late 1980. In 1987, the weakening of the US $ continued, with the US $ 18 oil price in late 1986 suffering a further drop by 20 % due to exchange rate fluctuations in 1987. After reaching its historical minimum in late 1987 following a value diminution of abt. 40 - 50 % within two years, the US $ exchange rate slightly improved in mid 1988 by abt. 20 - 25 % which brought the still valid official US $ 18 oil marker price a little back to its reality also in terms of other currencies. In late 1988, however, another downwards trend had already again partially compensated the benefit of the mid 1988 recovery of the US $, and also the oil price was on a renewed downwards movement considerably below the still valid official marker price.

2.2.1.4. Possible future consequences of the 1986 oil price collapse

The most important future consequences of the 1986 oil price collapse (cf. section 2.2.1.1.2.) and the uncertainty in hydrocarbon industry which it created (cf. also section 2.4.2.) and that still had considerable impact on planning and operating during 1987 and 1988 are declining reserves due to suspension and cancellation of exploration and development, regaining power of the OPEC, oil price instability and market uncertainty, and energy supply diversification effects.

2.2.1.4.1. Declining reserves due to suspension of exploration and development

The greatest danger of the 1986 major oil price drop and the resulting uncertainty and frightening of the market is particularly suspension of much of the exploration and appraisal drilling activity as well as stimulation operations (cf. also sections 2.4.2. and 2.3.) in many parts of the world. Although the oil price collapse will not depress North Sea production rates in the short term due to the relatively low operating cost of most fields particularly at full utilization of capacity, it will result in postponement and restructuring of new projects, because at the current oil price level, all of the discoveries coming up for development are either just marginally economical or even not economical at all at least in the British sector (BEUDELL 1986 c). Emphasis in the transitional stage will therefore be on retarded development including partial exploitation of reservoirs through fewer wells. Some aspects of production decline exceeding reserve addition, oil demand vs. supply in the next decades as well as influence of political systems and financial relationships are discussed as follows.

2.2.1.4.1.1. Production decline exceeding reserve addition

Considerable effects are especially visible in the USA where exploration drilling has dropped catastrophically after the 1986 oil price crash (HOLMES 1987; cf. section 2.2.1.5.2.). The USA has always operated on a low reserve basis which has been possible in recent years because of a high level of exploration activity that has tended more or less to replace the produced oil. With such a low amount of activity as in 1986 (cf. section 2.4.2.) and with the severe dent in confidence in the industry in that country, however, it will be difficult to regenerate exploration activity at a sufficiently high level to stem production decline that has been predicted and is inevitable, but has been accelerated by several years by the 1986 oil price collapse. Assuming a slight to moderate oil price rise, many non-OPEC countries will seek to encourage oil exploration and production for reasons of supply security, payment balance, foreign exchange, and local employment almost irrespective of price level.

Nevertheless, the additional oil which can be expected is likely to be more than outweighed by the reduction of output in USA and Great Britain (PETROLEUM ENGINEER 1988 c) where production has already peaked and is inevitably going down. This will result in growing dependence on imported oil in a number of key industrial countries. The low oil price in 1986 particularly illustrated that capital expenditure on oil projects is sensitive to adequate cash flow and price prospects. The significance of additional exploration and appraisal for prevention of another oil crisis is also underlined by LORBACH (1988 b), and the discrepancy between much oil and not enough exploration is illuminated by STAHMER (1979).

2.2.1.4.1.2. Oil demand vs. supply in the next decades

As oil will definitely remain to be the leading primary energy source within the next years (probably even beyond the year 2000) despite a slow decline in relative market share due to substitution by gas and nuclear power (LINDEN 1985, FEHRINGER 1986, OEHME 1986, SCHMIDT 1987), neglectance of extension of existing capacities due to economical forces combined with lack of encouragement and missing optimistic outlook, and the resulting curtailment of exploration and development activities for a couple of years (cf. also HOLBROOK 1987, RENSBURG & MALIK 1987, SCHAEFER & POWERS 1987, WOODS 1987) could mean that within the near future when more and more currently productive wells and fields become exhausted, not enough new reserves might be immediately available for exploitation and the global energy supply will decrease (an analysis of supply and demand for oil and gas and a study of the challenges of the industry is performed by SWEENEY & WYANT 1980, SCHLESINGER 1982, ECK 1983, HILL 1983, PICKENS 1983 and CECIL 1987). Aspects of oil price crises and supply security are also discussed by BAUER (1981), and a prognosis of world energy demand is given by WESTERHAUSEN (1980).

Total world energy demand, however, will significantly increase until the year 2000 predominantly due to rising consumption in the developing countries by adjustment of living standard and industrial growth (HULTIN 1983, LINDEN 1985, DOUGHERTY & AL-BLEHED 1987, OSMAN 1987, SCHMIDT 1987), whereas in the highly industrialized countries, demand will only more or less fluctuate around its present level due to still increasing effectivity of energy conservation, process rationalization, oil substitution chiefly by gas and only slight industrial growth to stagnation or even some shrinkage due to progressive market saturation in many sectors and branches (IBG 1986).

2.2.1.4.1.3. Influence of political systems and financial relationships

Unless the general demand will be reduced by a major economical recession in the near future (OFFSHORE OIL INTERNATIONAL 1988), the delayed exploration of new reserves (cf. also section 2.2.1.5.2.) could result in an established shortage in the coming years. Prognosis is given that following the 1986 oil price crash (cf. section 2.2.1.1.2.), oil production in the British North Sea will remain steady for about another year and begin falling after that unless delayed projects are revived, whereas in the Norwegian North Sea, new fields are able to boost production unless the government continues to limit exports (OIL GAS

JOURNAL 1986). Oil output decline in the USA will gain speed in 1987 due to moribund drilling in 1986, whereas the USSR by contrast stepped up drilling in 1986 in its efforts to rapidly replace maturing production in progressively depleted fields (cf. section 2.4.4.). Risk, stopping rules and economical viability of exploration drilling programs are discussed by RANKIN (1982), and JAMISON (1984) analyzes future petroleum exploration in the USA. NEAL & REED (1981) comment on the impact of taxes on exploration economics and thus feasibility of discovery of additional hydrocarbon reserves. Economical aspects of the low oil price scenario are also investigated by FLATH (1987). GLASNER (1985) discusses the dependency of oil price evolution from political influences.

The consequences of the oil price history on demand and supply are summarized by YU (1988). The low oil price in the first twenty-five post-Second World War years had triggered the rapid growth of world oil demand and the conversion of world energy consumption structure from coal to petroleum. The drastical escalation of the world oil price during the 1970`s and early 1980`s (cf. section 2.2.1.1.1.) had brought about a decline of demand and the present oil glut. Unfortunately in the mid 1970`s and early 1980`s, almost the whole world ignored the impact which the drastical escalation of oil prices had upon energy demand and supply, and failed to foresee the subsequent oil glut and price fall. Therefore the meaning of the present low level of oil prices in the future in terms of demand and supply and the price of oil is a problem which calls for close watch and serious study.

2.2.1.4.2. Regaining power of the OPEC

The evolution as sketched above will at least in in some areas sooner or later enable the Organization of Petroleum Exporting Countries (OPEC) to acquire again a considerably larger oil market share than before (which is the nightmare of all non-OPEC countries), with the first slight renewed market extension already having been achieved in late 1986/early 1987 (OFFSHORE ENGINEER 1986 d, ALGAR 1987 b, ERDÖL-ERDGAS-KOHLE NACHRICHTEN 1987 b, OIL DAILY 1988 a), with market factors moving in OPEC`s favour as demand rises. In 1987/1988, world oil requirement has already increased by abt. 1.8 Mio. bbl/d since 1985, resulting in a 1.6 Mio. bbl/d or 10 % increase in demand for OPEC oil (PETROLEUM REVIEW 1988 c). With the assumption of 1.5 % annual increase in world oil requirement and non-OPEC oil supply remaining more or less at the current level, the demand for OPEC oil will remain below 20 Mio. bbl/d in the near future, but in the second half of the 1990`s, it will reach close to 27 Mio. bbl/d which is near the maximum OPEC production capacity, with this long-term trend, however, not respecting sharp short-term fluctuations in demand.

Historically, demand for OPEC oil has fluctuated by more than 1 Mio. bbl/d on either side of the average annual demand, and particularly these fluctuations have created problems for oil market stability. The regaining power of the OPEC in the future is discussed as follows along the lines of evolution of world oil production and demand, symmetrical boom-crash-boom cycle, reserve life and oil demand increase; netback pricing, swing production and market share recapturing; oil reserve distribution and OPEC capacity utilization, imbalance between oil reserves and consumption, and increasing oil demand exceeding critical OPEC output rate.

2.2.1.4.2.1. Evolution of world oil production and demand

The dependency of the world from OPEC oil in the future is also illustrated by production capacity and reservoir duration in OPEC and non-OPEC groups (LÜBBEN & LEINER 1988). OPEC annual production rose from 440 Mio. t in 1960 continuously to 1.5 Bill. t in 1973 (representing the full output capacity of the association) which was kept until 1979, but later declining demand brought the annual OPEC production down to 800 - 900 Mio. t in the last years. In contrast to this, non-OPEC annual output climbed from 600 Mio. t in 1966 up to 2 Bill. t in recent years which means reaching of the upper limit of the withdrawal potential with no overhang capacity being left, whereas the OPEC group has at the moment abt. 35 - 45 % overhang production potential (cf. section 2.2.1.3.2.3.). As world oil reserves are currently estimated as being abt. 120 Bill. t or 1,750 Bill. bbl (with these reserves being distributed by 59 % to OPEC, 27 % to non-OPEC non-communist and 14 % to communist countries; PETROLEUM REVIEW 1987 a), based on present production levels of 2 Bill. t of the non-OPEC group and 900 Mio. t of the OPEC cartel, the total world reserves would reach abt. 42 years if calculated with total world withdrawal of 2.9 Bill. t (ERDOEL-ERDGAS AKTUELL 1982, ERDÖL-ERDGAS- KOHLE NACHRICHTEN 1987 e). The peak of world oil production has been reached in 1979 and 1980 with abt. 3.1 - 3.2 Bill. t (ERDOEL-ERDGAS AKTUELL 1981 a). The boundary of 3 Bill. t annual world oil production was exceeded for the first time in 1977, with in this year three countries contributing half of the delivery including USSR with abt. 550 Mio. t (cf. section 2.4.4.), USA with abt. 460 Mio. t and Saudi Arabia with abt. 450 Mio. t (ERDOEL-ERDGAS AKTUELL 1978 b).

Current world daily oil output is abt. 53.6 - 55.1 Mio. bbl, of which OPEC association contributes 15.7 - 17.5, non-OPEC non-communist group 22.4 - 22.7, and communist block 15.0 - 17.5. Mio. bbl (BECK 1987). Daily production of the USSR with 12.1 - 12.6 Mio. bbl (cf. section 2.4.4.) contrasts sharply with that of the USA of 8.3 - 9.1 Mio. bbl and the total demand of the USA of 16.2 - 16.5 Mio. bbl, with import thus accounting for abt. 50 % of the USA oil demand (MOORE (1988 d). If evaluating separately with OPEC and non-OPEC production quantities, non-OPEC reserves would last only still 15 years, whereas the OPEC potential is still spanning abt. 100 years. Due to declining output in the non-OPEC world, demand of OPEC oil can be expected to rise to abt. 1.3 Bill. t in the mid to late 1990`s which would most probably have the consequence of an oil price increase to abt. US $ 20 - 30. Estimations of future oil demand are also given by LORBACH (1988 a), and technical and political aspects of prevention of another oil price crisis are discussed by LORBACH (1988 b). World oil reserves and demand are evaluated by LORBACH (1980) and WESTERHAUSEN (1980). SINOR (1986) investigates long-term cycles and prediction of petroleum demand.

2.2.1.4.2.2. Symmetrical boom-crash-boom cycle

When therefore at some time in the future the world is again facing a natural or artificial oil shortage (MINERALÖLWIRTSCHAFT 1983) or a special demand/supply situation is at least apparently generated due to the effectivity of the natural market mechanisms in combination with directed changes or independently from external disturbances (OEHME 1986), prices could be expected to rise again dramatically in a similar way as already happened less then fifteen years ago, with then a full symmetrical cycle of oil price evolution comprising boom, crash and renewed burst having been completed (GRIFFITHS 1986). In terms of market forces, in the next few years already pressure is expected to build up on oil prices (PETROLEUM REVIEW 1988 c).

Temporarily at least balancing of the market by output adjustment can prevent another oil price collapse. The first half of the 1990`s should witness a more or less stable oil market without, however, returning to fixed and rigid prices. Towards the second half of the 1990`s, tightening of supply and demand balances are predicted to culminate in an upwards pressure on oil prices. While the average annual demand for OPEC oil still remains within the production capacity, the short-term demand fluctuations will inevitably result in shortages which may lead to another very sharp increase in oil price.

2.2.1.4.2.3. Reserve life and oil demand increase

In view of the drastical changes of the OPEC oil export volume from 11 Bill. bbl in 1979 to only 5 Bill. bbl in 1986, the critical level at which the OPEC can be expected to be back in the leading position for global oil pricing is abt. 7.3 Bill. bbl per year which corresponds to abt. 20 Mio. bbl/d, because at this volume OPEC is supposed to be able to maintain price discipline within its ranks. In mid to late 1988, OPEC production was still abt. 15 % below this critical export level, and still slightly diminishing demand does at the moment not provide any threat that the market could turn around soon again. As it can be estimated, however, that none of the non-OPEC countries except of Norway and Mexico (and certainly USSR and China) have oil reserves lasting beyond the year 2000 and in light of the diminishing global oil discovery rate since already 1965, it is only a question of time when OPEC will be again in a strong position, because this area of the world is the only one where still new giant oil fields are discovered, whereas in non-OPEC nations new discoveries are no longer keeping up with production.

The overhang of oil oversupply at the moment is highlighted by the fact that due to quota agreements within the group in order to stabilize and defend the oil price level, OPEC is limited to produce abt. 15 - 18 Mio. bbl/d which is abt. half of the actual output capacity (AL-CHALABI 1987) and abt. half of the real production volume in 1973 - 1979 (28 - 32 Mio. bbl/d; HOLMES 1987). With declining output in non-OPEC countries within the next years, the annual increase in demand for OPEC oil is estimated to be abt. 1 Mio. bbl/d in the near future (HOLMES 1987).

The current oil price level does no longer sufficiently cover the increasing exploration and production costs, and the tendency will be that the oil price will rise the faster, the longer the present unequal expense situation is lasting (OEHME 1986). Predictions of future oil price evolution are also made by SEKERA (1986) and THOMPSON & SINGLETON (1988). Geopolitical implications of Middle East oil are discussed by KEPLINGER (1986), RANDOL (1986), ALGAR (1987 b), MIDDLETON (1987) and ROBERTS (1987), and possibilities of reliable forecast of future oil price evaluation are assessed by DOUGHERTY (1987). Political aspects of energy policy are summarized by BARRETT (1986), SCOTT (1986) and FRANSSEN (1988), and CETRON (1986) outlines international political risks and energy project investments. The role of the OPEC group in the world oil market is also highlighted by SAFER (1977) and TEWKSBURY (1977).

2.2.1.4.2.4. Netback pricing, swing production and market share recapturing

RANDOL (1986) describes some reasoning behind the late 1985 decision of Saudi-Arabia to be no longer the swing producer in the OPEC (AL-CHALABI 1987, BECK & WILLIAMS 1987; cf. section 2.2.1.2.2.1.), because the country became fed up with its self-chosen role of single-handedly propping up the sagging price structure of the OPEC, while other OPEC members show no discipline in restraining output and while there was no cooperation from the non-OPEC oil exporters to help maintain a stable market. Netback pricing in 1986 has turned out to be a very effective OPEC instrument to recapture oil market shares on a volume base from non-OPEC producers which had been lost in earlier years and to rebound from record low market shares in 1985, but on the other hand, the flood of OPEC netback oil overburdened an already soft oil market and forced the oil price to tumble below US $ 10 (BECK & WILLIAMS 1987; cf. section 2.2.1.1.2.). The successful recapturing of market shares for the OPEC is highlighted by the fact that the oil imports in Western Europe in 1986 increased by abt. 10 % following a 5 % drop in 1985, with most of this difference coming from OPEC countries (CROLL 1987).

The 1986 oil price crash that was provoked by the OPEC decision to abandon production limitation in favour of seeking the fair share of the international oil market leads to retardation of further development on non-OPEC supplies as a consequence of sharp exploration and production expenditure cutting in response to the earnings impact of plummeting oil prices. The OPEC also speculates on the relationship that the lower the oil price in short-term, the stronger the forces pushing it up again in the long-term (cf. also OEHME 1986), bringing the group then quickly back to the leading role in the world energy market. Aspects of oil supply and demand as well as price evolution from the standpoint of the organization of Arabian petroleum exporting countries (OAPEC) are summarized by ATTIGA (1987).

2.2.1.4.2.5. Oil reserve distribution and OPEC capacity utilization

Over 90 % of world oil reserves are situated in OPEC, USA, USSR and European Economic Community (EEC; NARAGHI 1988). While accounting for in excess of 60 % of the oil demand of the free world, USA and EEC have the highest oil reserve depletion rates along with the lowest proven oil reserves. Assuming that OPEC is the price setter, non-OPEC producers are price takers and supporters, and no major materially adverse political upheavals occur around the Persian Gulf, it can be surmised that the annual weighted average crude oil price will most probably stay in the US $ 18 - 20 range until 1992 - 1994 when the OPEC capacity utilization rate will have risen from its current 65 - 67 % to the historically significant 80 - 85 % level.

Consequent to the 1992 - 1994 period when OPEC is expected to become the major supplier of crude oil again, the oil price will experience with more dramatic annual increases in real terms. Non-OPEC oil production is expected to reach its summit in 1988 to 1990 and then is considered to decline slowly (PETROLEUM REVIEW 1988 a). OPEC produces abt. 40 % of the world sources, and abt. 85 % of the OPEC reserves are situated in the Gulf region (cf. also SCANLON 1988) where production costs per barrel are 0.05 - 0.2 US $ in contrast to 5 - 15 US $ in many parts of the rest of the world (McCANN 1988). As a consequence of all these relationships, it can be predicted that the whole world will depend upon the Gulf OPEC countries concerning oil supply sometime after 1995, with OPEC revival being initiated by cheaper oil currently discouraging energy efficiency and frontier exploration.

2.2.1.4.2.6. Imbalance between oil reserves and consumption

The continuing dependence of the world oil market on the OPEC is underlined by the considerable imbalance between reserves and consumption (OIL GAS JOURNAL 1987 m). The OECD countries representing the non-communist industrial world account for more than 57 % of total world consumption but hold little more than 6 % of world oil reserves, whereas the OPEC group represents little more than 5 % of total consumption but has more than 75 % of total recoverable reserves, with 63 % of the total reserves being concentrated in the Persian Gulf region. Oil production rates are much more evenly distributed than reserves, with the OECD group accounting for nearly 26 % of total output, OPEC 32 %, communist block 28 % and developing countries 14 %.

Reserve lives at the bottom of the line comprise for the OECD abt. 11 years, communist block abt. 14 years, developing countries abt. 27 years, OPEC total abt. 104 years and Persian Gulf OPEC members abt. 131 years. These relationships reflect that despite its current inability to defend an US $ 18 reference price, the OPEC group will remain in control of the oil market for as long as the non-OPEC world needs its oil. Temporary apparent abandonment of the price target by the OPEC seems at least partially to be designed to guarantee that the requirement of OPEC oil persists and grows.

2.2.1.4.2.7. Increasing oil demand exceeding critical OPEC output rate

THOMPSON & SINGLETON (1988) sketch a very pessimistic worst-case scenario and predict considerable increase of the oil price already within the next few years. As a consequence of lower oil prices after the crash, world petroleum consumption increased by 2.5 % from 1985 to 1986 following four years of almost no growth, and similar or even larger increase rates are expected for 1987 and 1988. Therefore the demand for OPEC oil will soon exceed the 20 Mio. bbl/d boundary and can be expected to rise to 25 - 30 Mio. bbl/d in some years. This demand level would bring OPEC into a very strong position to force real price increases, with the model of THOMPSON & SINGLETON (1988) predicting that the oil price is going to rise to the 1980/1981 niveau in real terms and is holding that level through 1994/1995. The high niveau of real oil price will persist longer in the early 1990`s than in the early 1980`s primarily because OPEC`s price-making power will be more heavily concentrated in the Persian Gulf region and USA consumers will not have access to low-cost capital to make energy conservation investments.

GUSTAJTIS (1985) comments already before the 1986 oil price crash and looking out from the 1985 oil price level that the unprecedented rise of world oil price in the last fifteen years has resulted in a massive investment boom in oil and gas exploration and development. Much of the investment made since 1980 is now viewed as an enormous business mistake which leads to a decline of investment capital. As expectations that abt. 70 % of production required in the next decade will come from reserves yet undiscovered will prove to be unattainable unless investments are made, failure to do so will precipitate another oil crisis by catapulting the price considerably upwards.

2.2.1.4.3. Oil price instability and market uncertainty

After the peak of the 1986 crisis, the hydrocarbon market began to recover with the support of an again slowly rising oil price, but in late 1986/early 1987 first some little steps forward were often alternating with other almost compensating steps in backwards direction which gave no security for planning of capital expenditure, second so far an adequate oil price level with sufficient certainty of stabilization which would enable to return to an activity at least more or less comparable to that before the onset of the price fall has not yet been reached, and third the US $ exchange rate (that started to decline in late 1985 and reached its preliminary minimum level in early to mid 1986) kept being very low still in mid 1987, with only minor perspectives of an increase to a reasonable level within the next few months (thus by mid 1987 an again higher absolute oil price which is almost at a niveau similar to that in mid 1985 is in Europe relatively even much lower than in those days due to the now much weaker US $ than two years ago and the progressive inflation since that time).

In late 1987, a slightly stabilizing oil price was compensated by further falling of the US $ exchange rate down to its absolute minimum since the end of the Second World War, and in early 1988, a slightly reviving US $ which stopped its decline and at least stayed more or less constant coincided with another little decrease of the average real oil price (OIL DAILY 1988 a; cf. section 2.2.1.3.). At the end of 1988, the presidential election in the USA confirming the general economical guidelines for the coming years resulted in renewed downwards pressure on the US $ exchange rate.

This constellation of another downwards tendency, although comprising only decline of the real oil price of a few US $ per barrel, created again much uncertainty in the hydrocarbon exploration and production industry and caused not only managements to be even more cautious with long-term development and investment decisions as well as personnel planning, but had also political consequences on the highest level related to the oil scene. The discussion as follows concentrates on cooperation attempts between OPEC and non-OPEC, market shares vs. price stability, downstream integration of oil-producing countries, pricing policy of list vs. flexible prices, possibilities of another oil price war, impact of a possible general economical recession, oil price collapse as result of lacking international cooperation, and different assessment of the US $ 18 agreement price.

2.2.1.4.3.1. Cooperation attempts between OPEC and non-OPEC

For the first time in their history, the OPEC issued a serious alarm for protection of the oil price and prevention of its further downwards orientation not only within the group, but addressed the invitation for a critical session also to various non-OPEC oil producing countries (McCANN 1988, OIL DAILY 1988 a). LUKMAN (1987) emphasizes that cooperation between OPEC and non-OPEC countries is the key to market stability (cf. also OIL GAS JOURNAL 1988 g), and AL-CHALABI (1987) even concludes that trilateral cooperation between consuming nations, OPEC countries and other oil-producers is necessary to defend the oil price which is backed up by the fact that in the long run, non-OPEC oil-producing countries could even suffer more than OPEC from any price collapse given the very high cost of replacement of consumed oil through expensive discoveries of new fields or enhanced development of old fields. Uncertainties about future world oil prices are also discussed by KILGORE (1982). International cooperation between oil producers and consumers aiming on market regulation is expected to be a very positive and constructive step in the direction of preventing further sharp oil price changes (PETROLEUM REVIEW 1988 c).

2.2.1.4.3.2. Market shares vs. price stability

Price stability which is at the moment difficult to achieve by political arrangement and market instruments can in the near future be improved in case of a general economical recession which lowers demand for oil (OFFSHORE OIL INTERNATIONAL 1988). The request for market stability as a base for international oil pricing and decision-making is reviewed by STORY (1986), and DAVID & CARSON (1983) comment on energy financing in times of uncertainty. ROSBACO (1983) gives an overview on investment evaluations in inflationary and unstable scenarios. Aspects of world energy markets coping with instability are also compiled by ROWSE (1987). Further oil price unconsolidation in addition to abandonment of production restraint in favour of seeking fair share in the international oil market are netback pricing deals (RANDOL 1986; cf. section 2.2.1.4.2.4.) which remove the incentive for oil companies to sharply curtail liftings from the OPEC by insulating them from market reality. OPEC strategy for pricing decisions is one of the most important topics for the global energy markets, with rationality and sustainability of OPEC imposed prices, prospect for continued stability of OPEC as a cohesive body, and outlook for oil prices in the future attracting greatest attention (NARAGHI 1988).

2.2.1.4.3.3. Downstream integration of oil-producing countries

In mid 1988, the continuing failure of OPEC to sustain official selling prices in the face of competition by non-OPEC exporters as well as its own inability to rein production has sparked new emphasis on OPEC reintegration (WILLIAMS 1988). Since late 1987, oil markets have remained mostly depressed, OPEC divisiveness has grown worse, and overtures by a group of non-OPEC exporters to share output restraint and thus buoy oil prices have so far failed. More and more OPEC members are therefore pushing into downstream ventures in order to become integrated producers. The downstream activities of large-reserve producers are intended to increase market share and diversify risk, but the result could be continued low oil prices, as these producers try to keep demand high (WORLD OIL 1988 a). In addition, the market share of the large producers comes at the expense of smaller producers that do not have downstream access (cf. section 2.2.1.5.2.) which will have a divisive effect on OPEC unity and will offer little hope that the cartel will be able to improve cooperation and discipline among its members.

The focus on diversification, the foreign downstream push, a possible massive intervention in future markets recently, and the spurning of non-OPEC overtures on production restraint point to the likelihood that market-related pricing will continue to spread and that key members in OPEC will no longer tolerate loss of market share to non-OPEC exporters in order to bolster prices (WILLIAMS 1988). Longer-term reintegration by some OPEC members and non-OPEC exporters could mean depressed oil prices, depressed margins for refiners and marketers, and accelerating oil imports in net consuming nations through the 1990's. Much of the impetus for OPEC reintegration comes from the threat to the market share of the organization by rising production from non-OPEC exporters.

2.2.1.4.3.4. Pricing policy of list vs. flexible prices

Increasing non-OPEC oil supply and lagging demand growth is going to rein OPEC's ability to raise oil prices until the year 2000, because historically, OPEC has been able to jack up oil prices only when the group's production was at 80 % or more of its total capacity (WILLIAMS 1988). Competition between OPEC and non-OPEC producers for market share is expected to persist at least through 1995. Fixed pricing has proven unworkable without sacrifices that key members such as Saudi Arabia are no longer willing to make (cf. section 2.2.1.2.2.1.).

Thus in order to maintain market share, OPEC probably must adopt a flexible pricing policy and eventually abandon official list pricing. The group could then take advantage of immediate increases in spot prices and avoid the role of swing producer. Abt. 85 % of the projected export volume should be sold under contract and abt. 15 % should be brought into the spot market for more than the negotiated prices in the contract. Such an approach would not duplicate the netback-price-related market crash of 1986 (cf. section 2.2.1.1.2.) because there would be adherence to quotas vs. the quota-breaking effort to regain market share in terms of volume, and marginal OPEC oil would sell at a premium instead of at a discount. Spot sales also would be a buffer against supply/demand imbalances, because spot buyers are the first to cut purchases in a soft market and to pay a premium in a tight market.

2.2.1.4.3.5. Possibilities of another oil price war

The downstream push, however, may be another sign that key OPEC producers may be girding for another price war (WILLIAMS 1988). Even in the scenario for flexible pricing, the OPEC members must strictly adhere to quotas, because otherwise the oil price will collapse in an oversupplied market no matter how the supply is made available. Until the main oil exporting countries gain enough share in the downstream market to feel comfortable about their integrated and overall income, they will keep the oil price unstable in order to discourage development of reserves elsewhere. Saudi Arabia seems to be prepared to teach other oil-exporting countries once more the lesson that failure to cooperate in restraining production will have disastrous effects on everyone's income, with a short price war being expected to encourage cooperation and liability. Continuous OPEC overproduction in late 1987 and almost throughout 1988 was progressively undermining the oil price (cf. section 2.2.1.1.3.), and in combination with the prolonged general weakness of the US $, the oil price may again falter if the surge of OPEC output continues (BECK & WILLIAMS 1987).

On the other hand, however, the situation in mid to late 1988 is somewhat different from that in 1986 because in order to move oil in the earlier market share battle, the OPEC countries had to guarantee their customers´ refining margins with netback sales agreements (OIL GAS JOURNAL 1988 c; cf. section 2.2.1.4.2.4.). In 1988, however, the refining markets are strong due to margin widening by the oil price drop, and therefore renewed substantial overproduction and market flooding would not serve the apparent economical interests of any oil-exporting country. A suitable brake for a longer and deeper oil price slump and a reasonable warning for the OPEC to stop lapse of discipline is the memory of the most devastating aspect of the 1986 oil price collapse which was the fear that the oil price could remain below US $ 10 for a longer time, if not in economical terms meaning almost indefinitely.

2.2.1.4.3.6. Impact of a possible general economical recession

In addition to instability generated within the OPEC group, a major external factor in the assemblage of consumers is the fear of recession which has become an important wild card in oil market forecasts (OIL GAS JOURNAL 1987 e). Slower economical growth reduces demand for petroleum and its products and therefore calls on OPEC oil. A recession would seriously complicate efforts to stabilize and restore oil prices, with prospects of recession thus posing a growing threat to the oil market. The weakness of the US $ (cf. section 2.2.1.3.) is one of the strongest signs of unfavourable economical framework and is expected to precipitate also sooner or later a collapse of the oil price. On the other hand, the fear of another oil market crash which at the bottom of the line would seriously hurt all the participants in the game has an offsetting influence which holds associations of rivalizing members despite lacking consensus still together (OIL GAS JOURNAL 1987 f). Oil price volatility also leads to company management restructuring with changing emphasis from technical skill to financial experience and understanding of executives (OIL GAS JOURNAL 1987 g).

2.2.1.4.3.7. Oil price collapse as result of lacking international cooperation

Another possible oil price collapse as the logical continuation of the real price decline by abt. US $ 5 from abt. US $ 18.50 in April 1988 to abt. US $ 13.75 in September 1988 can occur if producers continue to failure to cooperate (OIL GAS JOURNAL 1988 e). The recent downward pressure on prices mainly stems from fears that overproduction will lead to a sharp increase in stocks by year end. The oil price drop of more than US $ 4 in July 1988 was provoked by the refusal of Saudi Arabia and Kuwait to accept production cut offers made by non-OPEC countries, thereby forcing OPEC to focus on internal problems such as output discipline. Many of the underlying factors that brought about the 1986 debacle have reemerged in 1988, and solving the perpetual lack of quota discipline among OPEC member nations remains central to the future health of the hydrocarbon industry (WORLD OIL 1988 a). The non-OPEC proposal of a 5 % production cutback for both OPEC and non-OPEC countries was rejected by some OPEC key members because of their view that demand would be more stimulated with oil prices below US $ 18, whereas any production restrictions could raise the price above US $ 18 (CROUSE 1988 a, McCANN 1988, OIL GAS JOURNAL 1988 g).

2.2.1.4.3.8. Different assessment of the US $ 18 agreement price

In addition, the Gulf members of the OPEC group put forward that OPEC had done enough to curb excess supplies unilaterally since OPEC output was down abt. 40 % from 1979 levels, while non-OPEC countries were producing at capacity (CROUSE 1988 a). Further instability is created by different meanings of the OPEC members on the US $ 18 agreement price (cf. section 2.2.1.1.2.2.), with the Gulf Cooperation Council (GCC) having a fairly optimistic and realistic understanding of demand and not being anxious to see prices sustained above US $ 18 for the reason of fear that rising prices will depress demand, whereas among OPEC members outside the Gulf region, US $ 18 is seen more as the minimum acceptable price (OIL GAS JOURNAL 1988 g). Despite the unsuccessful attempt in early to mid 1988, future cooperation between OPEC and non-OPEC is therefore more rather than less likely to be a valuable policy option for consolidating the world oil market (McCANN 1988). Aspects of oil price instability are also discussed by CROUSE (1987) and OIL GAS JOURNAL (1987 n).

2.2.1.4.4. Energy supply diversification effects

Another international effect is that at oil prices below US $ 15, the world tends to have too few energy options for the long term, whereas at prices above US $ 30, there are too many of them, with an optimum oil price at the bottom of the line thus being in the range of US $ 15 - 30 (OFFSHORE ENGINEER 1986 f). Within the frame of an oil price of US $ 15 - 30, a balance exists between the use of low-cost oil and higher-expense energies including higher-cost oil. On the other hand, a world-wide survey of undeveloped marginal offshore fields reveals that even with an oil price of US $ 15 and in light of existing technology and fiscal regimes, the development of small offshore oil fields around the world will not stop entirely, but for ameliorating its pace, either oil prices must rise above US $ 20 or governments must improve fiscal conditions to provide a better incentive to the industry. Some of the marginal fields currently held by major operators but considered to be uninteresting by them could be attractive to imaginative independent oil companies once they are relinquished.

2.2.1.5. Present consequences of the 1986 oil price drop for the stimulation market

The most important present consequences of the 1986 oil price drop for the European stimulation market are illustrated along the lines of fracturing strategy of deep gas wells before and after the 1986 oil industry crisis, reduction of capital expenditure and organizatory streamlining of exploration and production companies, business deterioration for service companies, and banking relationships and investment strategies.

2.2.1.5.1. Fracturing strategy of deep gas wells before and after the 1986 oil industry crisis

Abt. 15 years ago when the oil price approached its first big increases (RUNGE 1986) thereby ameliorating the feasibility of many hitherto marginal or even uneconomical projects, many deep gas wells in Western Europe und USA were ready for MHF stimulation, but could not be treated, because suitable synthetic intermediate- and high-strength proppants that could withstand the high closure stresses in these depths were not yet developed, and only natural sand existed which crushed under those conditions (TUNN 1971; ATTEBERRY, TUCKER & RITZ 1979; COOKE & GIDLEY 1979; HICKEY, BROWN & CRITTENDEN 1981; KOHLHAAS 1982, CLARK 1983, BRIM 1986). Some aspects of the fracturing potential before and after the 1986 oil price crash are outlined as follows.

2.2.1.5.1.1. Fracturing potential before the 1986 oil price drop

Many economically very attractive fracture jobs could not be carried out many years ago due to lack of suitable propping materials in the early 1970`s. In the golden years of high oil prices in the late 1970`s and early 1980`s (in the latter period in Europe even accentuated by a high US $ exchange rate), supply of intermediate- and high-strength proppants could as a consequence of insufficient manufacturing capacities in relationship to the growing market due to wrong or not enough optimistic forecast and assessment of the requirements temporarily not keep pace with the partially enormous demand.

Thus as a result of preferential service of the USA market, particularly in Western Europe numerous high-rank large-scale fracture jobs had to be delayed until the required quantities of high-quality proppants were available. Expensive waiting times of rigs in those days justified in some cases even payment of air freight for fast delivery of intermediate- or high-strength proppants from the plants in the USA to the well site in Europe or buying a considerably more expensive proppant type than scheduled from local competition in order to get the job rapidly done and make profit from quick pay-out in the scenario of high oil price and high US $ exchange rate (cf. section 2.2.2.3.). Much promotion of the stimulation market, however, was hampered by the undersupply of both European and world markets with proppants in the period 1976 - 1984 comprising the time from invention of intermediate- and high-strength synthetic proppants to beginning of more serious oil price decline and prior to the 1985 proppant price changes in downwards direction along with improving supply capacity (cf. section 2.2.2.1.).

2.2.1.5.1.2. Fracturing potential after the 1986 oil price crash

After the major oil price decline in early 1986 (GRIFFITHS 1986, RUNGE 1986), the situation has reversed, with again numerous deep gas wells that are waiting on treatment (and which partially have originally been drilled in order to be fractured soon afterwards as a result of prediction of tight reservoir facies in the low-permeability zone explored by unconventional gas appraisal in the belt of marginal pay zone development, with the stimulation cost already having been added to the drilling cost in the economical feasibility assessment before spudding of the wells) not being able to be fractured, but this time for economical instead of technical reasons. In contrast to the earlier years, high-quality proppants are now available in almost oversupply, with thus being no limitation of quantities to be delivered straight after order entry, and proppants being also offered for nearly dumping prices with respect to the former situation.

The low oil price (which has in Europe even a more serious impact due to the accompanying weakness of the US $), however, now does hardly give any economical justification for carrying-out of many jobs prior to a reasonable restabilization of the hydrocarbon price scenario at an acceptable level (which would be most suitably reinforced by a rising value of the US $) to be adequate to the investments to be made by executing stimulation (this situation is in parts of Europe still current). Similar difficulties of economical feasibility at lower oil prices as applying for hydraulic proppant fracturing are also touching enhanced oil recovery projects (BRASHEAR, BECKER & BIGLARBIGI 1988). In addition, the most important factor characterizing the hydrocarbon exploration and development scene after the oil price crash in 1986 is lacking confidence for future investment of money despite existing hope or even optimism (KINNEY 1987), and this criterion is the most important reason why the capital expenditure program slashed down and only slowly picks up again. The impact of the confidence loss is so strong that it can be expected that even if the oil market rebounds and enters another boom phase still this century, the lessons of the oil market shocks in the 1970`s and 1980`s may change for ever the guidelines for capital investment and industry management (OIL GAS JOURNAL 1987 g).

2.2.1.5.2. Reduction of capital expenditure and organizatory streamlining in exploration and production companies

The revision of budgets for drilling, completion and exploitation by the hydrocarbon exploration and production companies has not only affected their own capital expenditure, financial flexibility and cash-flow structure, but also had significant impact on service company business, banking relationships and investment strategies. Economical reassessment of the North Sea oil- and gas-field development plans after the 1986 oil price drop (cf. sections 2.2.1.1.2. and 2.4.2.) has resulted in reduction of capital expenditure on current development projects, deferring of investments in new field development projects, preference given to British Southern North Sea Rotliegend gas acquisition and to small-scale developments using subsea completions for tying-in of satellite fields to existing production platforms, severe cuts in planned exploration programs of abt. 25 - 40 % (KEG 1986, OIL GAS JOURNAL 1986, ALGAR 1987 b, HOLBROOK 1987, SCHRÖDER 1987) which will affect future availability of fields for development, and reduction of operating costs for existing producing fields (BEUDELL 1986 b, OFFSHORE ENGINEER 1986 a, THOMAS 1986).

This is more or less reflecting the decreasing earnings and revenues by abt. 34 % and 24 %, respectively, for the group of major oil and gas companies following the 1986 oil price crash (BECK & SMITH 1987). The discussion as follows focusses on development and completion activity, drilling meterage, staff lay-off, company merging, cash management, survival of independents, responsibility transfer by contracting and takeover, and executive skill requirement change.

2.2.1.5.2.1. Development and completion activity

The 1986 oil price collapse led to a 15 % decline in worldwide exploration drilling activity which resulted in loss of reserve replacement of abt. 2.5 - 3.0 Bill. bbl oil (POPESCU & HODGSHON 1987). This was the culmination of a trend which started in 1982 when the zenith was passed and once the nadir in 1986 was reached, worldwide exploration drilling had decreased by almost 30 % since the golden years. In the USA, the effect was even more pronounced and comprised a 60 % reduction in exploration drilling since the peak in 1981 (cf. section 2.2.1.2.1.).

Subsea development of satellite fields forms a veritable complementary set to major traditional developments with platform and pipeline, because it enables small fields to start producing on the basis of existing processing capacities and capabilities that have become available because of the decline of reasonably nearby fields (VALENTIN, VALAIS & CHAMPLON 1988). BARTON (1988) evaluates possibilities of making smaller fields viable, and HARPER (1988) investigates developing cost-effective technology for the future. COX (1984) discusses economical aspects of marginal field development in the North Sea. Adjustments of oil field activities to low revenues are also illustrated by REINICKE, TRENEL, HERBST, KEMPGEN & LINZ (1988). Another point illuminating the severity of the impact of the 1986 oil price crash is the fact that for the first time since drilling began almost 25 years ago, oil production declined in the British North Sea sector in 1986 (McNALLY 1987). In the USA, oil production was 9 % lower in 1987 than in 1985, largely because of shutting-in of stripper wells (THOMPSON & SINGLETON 1988).

2.2.1.5.2.2. Drilling meterage

Germany FRG is an excellent example how drilling meterage fluctuates as a consequence of oil price explosions and collapses. In Germany FRG, drilling in 1986 was less than 50 % of the meterage of 1985, and in 1987 and 1988 further decreases took place (ERDÖL UND KOHLE - ERDGAS - PETROCHEMIE 1988) which can be explained by proper planning of the whole year including cancellation of such marginally feasible items that in 1986 have already been spudded or were ready to start when the oil price dropped. The evolution of total drilled meterage in Germany FRG during the last 15 years is pronouncedly reflecting oil price history (SCHRÖDER 1980, 1987, 1988). From the low in 1973 with 136.4 km total oil and gas drilling, a continuous increase took place to 1977 and 1978 when 243.4 and 264.9 km were drilled, respectively. In 1979, however, the drilling meterage growth stopped, stagnation occurred, and even a slight short reverse evolution was initiated, leading from 253.7 km in 1979 to 236.1 km in 1980.

As a consequence of the second oil price explosion in 1979/1980, drilling meterage drastically increased in the golden years of hydrocarbon acquisition and reached 262.1 km in 1981 and the record of 334.6 km in 1982. In 1983 and 1984, beginning decline of oil prices provoked a fall-back from the 1982 summit to more or less the niveau of 1981, with in total 252.2 km and 265.6 km having been drilled in 1983 and 1984, respectively. A meterage of 246.6 km in 1985 was already the indication of an accelerated slide down which was followed by the complete collapse in 1986 when only 176.9 km were drilled, and due to numerous cancellations not having been immediately effective in 1986 but fully broke through in 1987, the result of 1987 was with 118.9 km even considerably less than in 1986, and 1988 is not coming out better.

In 1986 and following years, continuing oil price instability encouraged only a few very strong American companies to increase again capital expenditure significantly (MOORE 1988 d), while most of the corporations were continuing to cut back in all branches of investment. In the USA, total oil company losses in 1986 and 1987 as a result of the oil price crash amounted to more than 50 Bill. US $ (AL-CHALABI 1987).

2.2.1.5.2.3. Staff layoff

Further consequences of adjustment to the revised situation and company streamlining were cut-backs on general overheads and staff (BEUDELL 1986 c) which resulted in shrinkage or closure of regional offices or even business liquidation, giving rise to several terrible world-wide lay-off actions affecting considerable shares of geoscientists, engineers and administration people (ALGAR 1987 b estimates that world-wide abt. a quarter of a million hydrocarbon industry jobs disappeared due to the 1986 oil price crash). The oil price drop is the main reason why predominantly in USA, Great Britain and Germany FRG thousands of geologists and geophysicists are unemployed and cannot find a qualified job at all, and also reservoir engineers have run into serious problems concerning working place and job security. As a consequence of high layoff rates, some of the oil and gas capitals even partially turned into ghost-like cities as a result of empty office towers and abandoned living districts due to mass exodus of people having become unemployed. One of the world`s largest oil and gas exploration and production corporations adjusted its staff stepwise to the declining market since the slight beginnings of the recession by end of 1981 and up to now reduced its manpower by abt. 45 %.

Most of the oil and gas companies in the USA laid off abt. 25 - 30 % of their employees in 1986 (THOMPSON & SINGLETON 1988). In early 1988, most companies had restructured and were operating with minimum personnel required to perform business (CROUSE 1988 b). Some companies, however, even started another layoff round of abt. 10 % of their staff in 1988 (MOORE 1988 d). Job rotation and loss affected all levels of company employment from lowest workers to highest executives, with top management restructuring aiming on better economical guidance of the corporation which was carried out by changing emphasis from technical skill to financial experience and understanding of executives (OIL GAS JOURNAL 1987 g).

In Germany FRG, abt. 15 % staff reduction since the oil price crash was reached by end of 1987 mainly as a consequence of renouncement of replacing retiring or leaving employees, but partially also due to layoff (ERDÖL UND KOHLE - ERDGAS - PETROCHEMIE 1988). The renewed oil price volatility in mid to late 1988 which brought the price level down to 12 - 15 US $ in real terms and created a situation at the swell of another possible crash led several companies to plan further manpower reduction by 25 - 30 % until 1992 which has to be carried out preferentially by early retirement, but some layoff undoubtedly will not be able to be avoided. SCHOLL (1986) comments on future manpower requirements in a dramatically changing petroleum industry.

2.2.1.5.2.4. Company merging

The dramatical oil price drop in 1986 had particularly severe consequences for smaller production, service and supply companies, many of which were confronted with the fate of having to go out of business due to bankruptcy sooner or later (SCOTTISH PETROLEUM ANNUAL 1987) or being taken over by or merging with larger corporations, including also considerable dismemberment of independents (BEUDELL 1986 c, 1986 d; MEISTER & CADY 1986). The suite of consequences of the economical crisis are loss of jobs, lying idle of capital assets, occurrence of bankruptcies and absence of incentive in the future (SCOTTISH PETROLEUM ANNUAL 1987). Service companies cut staff by 30 % on average to adjust to the considerably diminished demand of well operations and treatments (ERDÖL UND KOHLE - ERDGAS - PETROCHEMIE 1988).

For larger production companies, the crisis was a matter of streamlining by swing and roundabouts including diversification of cash flow outside the North Sea region and even into other integrated activities such as downstream facilities (with partially additional benefit coming from scaling down capacities in manufacturing plants, disposing of excess plants and equipment, and writing down inventories to estimated net realizable value; MILLER & NEWLIN 1987), but for smaller independents without downstream possibilities, at once abt. 50 % or even two-thirds of the income was wiped out (GRIFFITHS 1986). Aspects of vertical integration of up- and downstream branches as well as corporate takeover for investment security and reserve access are outlined as follows.

2.2.1.5.2.4.1. Vertical integration of up- and downstream branches

Vertical integration of up- and downstream branches of international companies was the only chance to stay strong during the peak of the crisis (CROUSE 1987). Larger companies also had to rationalize and reorganize, partially by means of merging with others, with repeated major and minor corporate takeovers and fusions highlighting the restructuring of the hydrocarbon industry (MEISTER & CADY 1986, MILLER & NEWLIN 1987) for long-term survival (HASSELL 1986). Companies with big debt loads have hardly any possibility to fend off takeover bids (OIL GAS JOURNAL 1987 g). Company restructuring is also guided by thinking of financial rather than vertical integration, with joint ventures and acquisitions taking place across industrial and geographical boundaries. Profit centers may develop horizontally rather than vertically. Aspects of company acquisition and merging are also discussed by SIMOES (1985).

2.2.1.5.2.4.2. Corporate takeover for investment security and reserve access

The reasoning behind takeover actions is that at low oil prices, big production companies prefer to obtain reserves by buying their competitors rather than by investing heavily in exploration efforts (OIL DAILY 1988 c). Aspects of company merging and takeover are also mentioned by SCHUBERT (1987). In some outstanding examples, even mega-merging was done, with the managements of the previousy independent companies having completely changed the way to look at one another from competitors and arch rivals to partners (MILLER & NEWLIN 1987, SCHUBERT 1987) which are good example cases of how to survive and prosper in a depressed market. In 1987, companies continued to cut costs and increase efficiency by focussing on core businesses and selling or writing down marginal branches or assets that do no longer meet the requirement of more stringent rate of return (BECK & SMITH 1987).

Leveraged buyouts in the petroleum industry are also illustrated by PAULUS (1988). Various aspects of the restructuring in the oil patch are outlined by MODESITT (1986) and EDGAR (1987), and points of financing mergers and acquisitions are commented by PETRIE (1983) and SOBOTKA (1986). BENTLEY (1988) evaluates company strategies in a changing market, and ANDERSON (1988) reviews the reorientation of the oil and gas industry. Examples of significant corporate deals including takeovers and acquisitions are also reported by CAPEL (1988), with in the North Sea intercompany dealing going to increase and leading to the end result of decreasing number of smaller companies.

2.2.1.5.2.5. Cash management

Independent producers are under increasing pressure to seek out every available dollar of internally generated cashflow from otherwise idle assets or premature financial obligations (CHAN 1985 a, 1985 b). Many petroleum companies can better manage their existing operations to improve their performances through cash management, updated collection policies, cash processing and disbursements, short-term investments, and more accurate planning. Net income is not always indicative of an independent`s viability because actual cashflow may bear no relation to either net income or the traditional definition of cashflow. Operational cashflow adjusts the traditional cashflow for changes in working capital accounts such as inventory, accounts receivable and payable, and others in order to derive net cash income, and non-operational cashflow breaks apart the external sources and uses of funds.

The most fundamental concept involved in the petroleum industry today is obtaining sufficient cash to satisfy debt obligations, drilling commitments, and necessary capital expenditures, and still to be in a position to participate in sound economical acquisitions and development projects when opportunities arise. An increasingly important aspect to competing in the economical environment surrounding today`s petroleum industry is the quality of cash management that an independent producer uses. Effective cash management systems provide efficient methods to gather and disburse cash, forecast working capital needs or cash surpluses, implement short-term investment and borrowing policies, establish reporting and controlling procedures, and direct banking relationships.

2.2.1.5.2.6. Survival of independents

During the prolonged slump of the oil and gas industry following the 1986 oil price crash, the worst damage has been suffered by service companies and independent operators (MOORE 1988 a). While in 1982 in the USA abt. 12,700 operators drilled at least one well on an operator-of-record basis, this number dropped in 1987 to only abt. 5,700 operators, with virtually almost all of the 7,000 lost operators having been independents. Those independents still existing are trying to survive, with outside financial capital being still the lifeline of the independent, but most of it has been used up and competition for what remains is fierce. Therefore independents are looking for less exposure of their capital and for quicker payouts, with many of them returning into existing fields and trying to boost production by reentering old wells. This is a major reason why in the USA in 1987 recompletions for the first time since the 1980`s outnumbered new exploitation installations. Independents which have survived the 1986 oil price crash follow strategies calling for reduction in debt and operating costs, sale of marginal leases, and cautious approach to capital spending (WEST 1987). MANCHESTER (1983) outlines the application of project financing principles to independents, and WYMAN (1978) gives an overview of the role of small independent producers in the rapidly changing oil and gas industry.

In 1987, the slight recovery of oil prices enabled the independent operators in the USA to return to profitability after considerable losses in 1986 (WILLIAMS 1988 b). The absence of hefty writedowns of oil and gas properties which axed earnings in 1986 helped to renew profitability for the independents in 1987. Only a couple of them still suffered from losses compared with more than half of them in 1986, although company performance also in 1987 and 1988 fluctuates with oil and gas price variations. Large and small independents alike returned to profitability in 1987 after taking charges against earnings in 1986 to write down oil and gas book values. The independents also followed the lead of the majors in sending more upstream investment dollars overseas as oil prices cratered and slightly less as oil prices rebounded, with their earnings and operational results tending to reflect the trend.

2.2.1.5.2.7. Responsibility transfer by contracting and takeover

Big companies with sufficient possibilities of diversification are partially in a position to pick up assets for sale world-wide including production as well as acreage for future exploration. The low oil prices are also dissipating the current market premiums of weaker companies thus setting the stage for further rationalization of the independent sector (BATT 1983, RAMSAY 1983, ELTING 1987), because so far companies that have not been expected to survive sooner or later have had premiums in their share prices due to expectations of investors that they will be the subject of a takeover bid (ALGAR 1987 b). Cost savings and streamlining campaigns have also considerable impact on project management or contracting of small- to medium-sized companies which show increasing interest in the scope of transferring responsibilities and risks for the provision and operation of production facilities to the large international service companies with first class credit ratings and strong balance sheets. CHAN (1985) emphasizes the nature of oil-field cashflow as the lifeblood of the independents for their survival, and THOMPSON (1987) illustrates the impact of the new tax law on internal cash flow generation.

2.2.1.5.2.8. Executive skill requirement change

When oil prices boomed in the late 1970`s and early 1980`s, almost every business decision in energy industry and investment banking community (regardless of being positive or negative and good or bad, respectively) turned out to be a financial success simply because the oil price continued to rise (MOORE 1988 a). Survival after the 1986 oil price crash, however, requires much more careful attention on much less capital, and more importantly, it is necessary to anticipate future changes and to react and adapt to them quickly and efficiently.

Therefore the continuing oil price volatility has also serious impact on restructuring of the company managements, with top level reorganization being characterized by a change from emphasis on technical skill to financial experience and economical understanding (OIL GAS JOURNAL 1987 g). While in the early 1980`s during the drilling and completion boom at times of high oil price and strong US $ mainly engineers and geologists ran the businesses and main decision potential of the corporation executives was on technical aspects of hydrocarbon reserves and ways to unlock them under a more or less predictable price regime, the 1986 oil price crash has completely reversed this constellation. In a competitive environment of low oil price and weak US $, financial experts and economical specialists are now constituting the boards of many corporations, because managing cashflow and economical risk is now the chief task of petroleum company executives in a milieu of highly unpredictable price evolutions.

2.2.1.5.3. Business deterioration for service companies

For service and supply companies, less business available means increasing competition, though cut-throat competitors cannot survive for long, and from the operator`s point of view, the cheapest tender is not necessarily the best solution in the long run, because expertise is part of the service offered and screwing-up of jobs is much more costly than approving at once a more expensive and more sound proposal (GRIFFITHS 1986). Some aspects of stimulation boat employment, well service expenditures and total number of treatments, and service cost per well and share of recompletions are discussed as follows.

2.2.1.5.3.1. Stimulation boat employment

Competition was accentuated at least during the peak of the crisis in 1986 and early 1987 by the coexistence of three service vessels of almost equal performance capability (cf. section 3.8.1.2.) which had to share a shrinking or stagnant North Sea stimulation market (FOX 1985, OILMAN 1985 b), with all of the three high-level technology treatment boats having been purpose-built vessels rather than representing converted and adapted supply ships (COOPER & MARSHALL 1984, OFFSHORE ENGINEER 1984; cf. section 3.7.1.1.). The advantages of boat operations over jobs from skids mounted on the platform are reduced rig up/down time, fewer equipment lifts and moves, onsite blending, and the possibility of pumping multiple jobs from one shipload of fluids and proppants (BRANNON, NETTERS & GRIMMER 1987). Only the introduction of stimulation vessels with large proppant and fluid storage capacity has allowed to perform MHF treatments on North Sea reservoirs which have been formerly impossible due to limited space available on the platforms.

On the other hand, disadvantages of vessel treatments vs. skid jobs are partially time- and cost-expensive travel to/from location, and bad weather can cause an operation to be aborted or delayed. Aspects of North Sea stimulation logistics and requirements are outlined by MACADAM (1984). The solution which emerged in the North Sea in 1986/1987 from the treatment fleet overcapacity dilemma as a consequence of underdemand, aggressive to exaggerated competition, establishment or reputation and experience, and various relationships was at least temporarily that due to proposal submission and approval under the influence of service companies and operators, one boat carried out the bulk of the proppant fracturing work, the second vessel took care of the main amount of the acid fracturing jobs, and the third boat spent much of the time unemployed in the harbour. Resulting from the decreasing amounts of orders and the conclusion of business at considerably lower rates due to strong competition which was sometimes even near the break-even point, also service companies had to adjust to the new situation by performance of adaptive shrinkage which was inevitably linked with office reorganization and dismissal of numerous people (the same also applies for drilling contractors). Service company strategies for survival are also evaluated by CLARK (1986).

2.2.1.5.3.2. Well service expenditures and total number of jobs

The amount of business decline for the service companies due to the 1986 oil price crash is highlighted for the USA market by MOORE (1988 b). Well service expenditures by USA operators dropped from the 1981 high of more than 4 Bill. US $ via abt. 3.5 Bill. US $ in 1984 and abt. 2.5 Bill. US $ in 1985 to abt. 1.5 Bill. US $ in 1986 which represents only one third of the level in the peak year of 1981 or a 65 % value drop. Picking up of business in late 1986 brought the well service expenditure level in 1987 back to 2.3 Bill. US $. In terms of executed service jobs, the decline was from 565,000 jobs in 1981 to 480,000 operations in 1985 and 390,000 treatments in 1986, and the improvement of the economical feasibility scene in 1987 lifted the number again to 415,000 jobs. The relationships between total number of operations and cumulative expenditures emphasizes that at low oil prices, first of all considerably cheaper well service jobs are carried out, and the number of jobs decreases only in the second line.

2.2.1.5.3.3. Service cost per well and share of recompletions

The trend as discussed above is further illustrated by the evolution of well service cost on a per-well basis from abt. 5,100 US $ in 1981 via abt. 3,900 US $ in 1983, abt. 3,200 US $ in 1984, abt. 2,750 US $ in 1985 and abt. 1,800 US $ in 1986 to abt. 2,350 US $ in 1987 and abt. 2,500 US $ in 1988 (MOORE 1986, 1987 b). At the bottom of the line, 1986 was the worst year for the service industry since expenditures peaked in 1981, with activity having dropped by more than 50 % after the oil price crash. After a consistent climb of abt. 1 - 4 % annually in the total USA well population over the last decade, total borehole number decreased for the first time again in 1986, because well abandonments outpaced new completions. The disturbed market is highlighted by the fact that operators spent nearly as much money for well servicing in Texas alone during 1981 as they did in the whole USA in 1986.

Completions outpaced recompletions in 1986 by a ratio of only 7 to 5 compared with ratios of 2 to 1 in 1985 and 3 to 1 between 1982 and 1984. A higher percentage of recompletions is an obvious indication that operators are more willing to recomplete older wells with proven reserves and production capabilities than risk money drilling for unproven oil and gas reserves (the same guideline applies for company acquisitions and takeovers, because it is safer to buy known fields with proven reserves than to invest money for exploration and appraisal drilling which could partially result in failures and almost complete loss of capital expenditure). Stimulation jobs in the USA declined by 25 % from abt. 28,800 in 1985 to abt. 21,500 in 1986.

2.2.1.5.4. Banking relationships and investment strategies

Following illustration of the situation of exploration and production as well as service companies, the evaluation of present consequences of the 1986 oil price drop is concluded by illumination of banking relationships, investment strategies and other aspects.

2.2.1.5.4.1. Banking relationships

Considerable stress was also put on the banking sector by the 1986 oil price collapse as a consequence of cash-poor smaller production and service companies seeking postponement of their principal and interest payments on loans and credit facilities for either projects undertaken or acquisitions made earlier (BEUDELL 1986 c). Companies reaching bankruptcy status throws the problem on the banks whether to foreclose or not in view of the underlying assets having been halved when put up for sale, with partially even the situation occurring that the stockmarket quote is more important to either bank or company purchaser than the underlying assets (ALGAR 1987 b). Other casualties will be among those companies owned by conglomerates, banks and financial institutions which will seek to withdraw from North Sea exploration and production. STEWART (1986) comments on secured oil and gas lending in bankruptcies, and MINTZ (1981) evaluates specific requirements and alternatives of reserve-based financing. PELLEY & CONSTABLE (1982) investigate reservoir risks and economical assumptions with respect to bank lending.

Resulting from the severe devaluation of oil, banks are expected not to lend readily anymore for new oil-field developments and also place strict covenants on any corporate borrowing secured on low-priced oil (BEUDELL 1986 c), whereas on the other hand gas projects are strongly supported (cf. section 2.2.1.6.2.). This could force major operators to defer oil projects, because their financially weaker participants may have no cash or insufficient collateral to pay their share of new or incremental development costs. Thus at the bottom of the line, the oil price crash initiated a suite of domino-type effects in the economical scene of the hydrocarbon industry. Perhaps the greatest damage was done to future hydrocarbon production rates and to confidence, because having learned the painful lesson of the sudden and unexpectedly serious 1986 oil price crash, company managements will probably never again automatically assume steady or increasing oil prices and therefore years may pass before investment regains momentum (ALGAR 1987 b). While retarding exploration and new developments which are no longer economically feasible, however, the tumbling oil prices have given enormous incentive to innovative technology which can further help in overall cost cutting. The necessity of effective evaluation of loans to reduce risk is emphasized by McPHAIL (1987), and ROSBACO (1983) comments on investment evaluations in inflationary and unstable scenarios. PELLEY (1982) discusses relationships between banking and energy.

2.2.1.5.4.2. Investment strategies

During the last decade, the petroleum industry in the free world spent annually more than 100 Bill. US $ for oil and gas exploration and production (OEHME 1986). According to such high expenses, it is reasonable to postulate that the share of stimulation should be increased as much as possible in order to achieve maximum profit from the investments. In the British sector of the North Sea, over 50 Bill. £ were invested into hydrocarbon exploration and production since the discovery of the first commercial gas field in 1965 (OFFSHORE ENGINEER 1986 d; cf. sections 2.2.1.6.2. and 3.3.). As a consequence of the high significance of the Rotliegend gas in the Southern North Sea, the hitherto expenses included already considerable shares for stimulation, and the amount of investments for hydraulic proppant fracturing is even going to increase during the major Rotliegend gas development campaigns in 1987 - 1990 and beyond (cf. sections 2.4.5.1. and 3.1.2.). Estimation can be made that the annual stimulation market potential in the North Sea is abt. 20 Mio. £ (OILMAN 1985 b; cf. section 3.8.1.2.).

In 1987, the promising outlook for higher average oil prices as compared with those of 1986 resulted in less severe world-wide capital cutting comprising only abt. 10 - 20 % in contrast to 25 - 40 % in 1986 with respect to the original budgeting made in late 1985, thereby slowing down the slide in hydrocarbon exploration and development (BEUDELL 1987 a; some aspects of picking-up of business in 1987 are also outlined by QUINLAN 1988). Important influences are the benefit of the oil and gas industry from the sharp reduction in cost of drilling, workover, completion and stimulation services, with lower rates being a good reason for budget determination in the light of stabilization of the capital expenditure level.

2.2.1.5.4.3. Other aspects

Although this situation was again slightly undermined by another minor decline of the real oil price in early 1988 as a consequence of the political uncertainties in the Middle East area (which was accepted to be so serious that the OPEC alarmed also non-group countries to participate in a critical session looking for joint efforts to stop the downwards orientation of the oil price; OIL DAILY 1988 a; cf. section 2.2.1.4.3.), the general rising trend of the hydrocarbon exploration and production industry recovering from the 1986 crisis is going on, as also reflected by doubling of the number of rigs being active in the North Sea with respect to 1987 (OIL DAILY 1988 b) following their halving from early to late 1986 when numerous unemployed rigs were stacked along the coastline (SCOTTISH PETROLEUM ANNUAL 1987; cf. section 2.4.2.). The most significant influence on retarded pick-up of activity despite more or less stabilizing oil price level considerably higher that at the peak of the crisis is lacking confidence in future investment of money despite sufficient hope or even optimism (KINNEY 1987). Managements simply do not rely on the situation, and hesitation with postponement if any doubt can be found is characteristic for the scenery. Challenges to industry viability from the viewpoint of the investment community are analyzed by MAXWELL (1986).

The traditional areas of hydraulic proppant fracturing in the world are predominantly the USA and parts of Western Europe (including mainly the British and Dutch Southern North Sea as well as Netherlands and Germany FRG onshore). Eastern Europe and Northern Africa are developing areas where so far insufficient marketing activity was made by proppant and partially also service companies, with major regions of hydraulic proppant fracturing in Eastern Europe being Yugoslavia, Hungary and Turkey. In view of proppant marketing and sales, the 1986 oil price decline had different consequences in USA and Europe (cf. chapter 2.3.), and an important additional feature exclusively affecting the European market was the US $ exchange rate drop (cf. section 2.2.1.1.3.).

2.2.1.6. Oil price evolution and gas market

Oil and gas are different hydrocarbon phases which often occur together (at least more or less gas is dissolved within the oil, or in some cases, the oil reservoir has a gas cap when free gas has exsolved from the oil), but also are frequently distributed in different reservoirs comprising separate storeys of the geological column. In addition to the split natural occurrences, the industrial application and economical significance of oil and gas are in many respects different and thus the market structure and potential of both hydrocarbon phases is not simply comparable. These relationships have an important impact on the reaction and position of the gas market absolutely as well as relatively to that of the oil market in view of the changes of the oil price level in either direction in the past. Aspects of characteristic features of the gas market and its independence from the oil market, influence of the 1986 oil price drop on gas development drilling in the North Sea, significance of the east west energy trade in the European gas market, and evolutionary trends of the world gas market are evaluated as follows.

The outline of various economical and technical factors of the gas market including future trends is presented in some more detail because of the significant role which stimulation plays in enhancement of gas reserves and their producibility. Hydraulic proppant fracturing is already now much more abundant in gas reservoirs than in oil pay zones both in Europe and in North America, and this ratio is expected to shift even more towards gas in the coming decades, in contrast to the past when emphasis has been almost everywhere in the world on the oil sector before the increasing recognition of the economical value of gas started some fifteen years ago (almost coincident with the first political availability threat and oil price jump) and accelerated at the time when synthetic high- and intermediate-strength proppants for deep well stimulation were introduced to the hydrocarbon industry.

2.2.1.6.1. Characteristic features of the gas market and its independence from the oil market

As a consequence of its environmentally less hazardous nature, gas which went at the bottom of the line later on stream for exploitation and large-scale industrial and urban application than oil, has a much better future in comparison to oil at least in Europe due to increasing replacement of oil by gas for all types of heating and combustion systems. Hydrocarbon exploration and development drilling in the next years particularly in the North Sea will predominantly focus on several rich gas provinces which will secure supply until the year 2000 or even beyond (FISHMAN 1986; GREGORY 1986, 1987; THOMAS 1986, ALGAR 1987 a, BAND 1987, POTTER 1987; cf. section 2.4.5. and 3.1.).

Although oil and gas prices are linked in some respect (LINDEN 1985), the gas market has a certain independence and continues to evolve rapidly, whereas the oil market is characterized by a more or less stagnant nature due to still persisting aims on increasing replacement of oil by gas and other alternatives in order to get rid of the political dependence from the OPEC group and to reach the point where demand can be satisfied in large amounts from the North Sea and other European oil provinces. Natural gas has in the past and future years also an important role in the energy trade between Eastern and Western Europe (D`ANDREA 1987) and thus contributes to the progressive approach of the two political complexes with different economical and social structure. Aspects of gas demand and pricing are outlined by FRANSSEN (1982) and JIMISON (1986). HOLTBERG, WOODS & ASHBY (1987) conclude that natural gas will play a major role in a highly competitive energy mix well into the next century. DAHLE (1988) illustrates similarities and differences of gas and oil, and GADONNIEUX (1988) outlines gas challenges in a competitive market. BRIERLEY (1988) summarizes the significance of gas as a global energy source. TECLAW (1988) characterizes natural gas as a strategic fuel for the 1990`s. World total annual gas production is in the range of 1,500 - 2,000 Bill. m3 since several years (ERDOEL-ERDGAS AKTUELL 1981 b).

Because gas is likely to remain a premium fuel in the foreseeable future, demand should continue to increase over the next decade or more despite a short-term erosion of competitiveness following the 1986 oil price collapse (FISHMAN 1986).

2.2.1.6.2. Influence of the 1986 oil price drop on gas development drilling in the North Sea

In terms of the scheduled major development campaigns of Rotliegend gas fields in the British Southern North Sea (cf. sections 2.4.3. and 3.1.), the 1986 oil price drop had only minor impetus on the progress of preparation and acquisition as a consequence of the relative independence of the gas market from the oil market due to long-term planning and arrangements, with most of the gas having already been sold prior to production drilling and stimulation. The influences of the 1986 oil price collapse on development drilling particularly in the British North Sea are illustrated along the lines of gas supply contracts and economical safety, continuity of planned and approved gas-field development, government back-up and supply security, gas-field development vs. gas price and demand evolution, disturbance of oil-field development, and gas import and condensate field development.

2.2.1.6.2.1. Gas supply contracts and economical safety

Gas supply contracts are long-term rather than spot-market-related like oil demand and supply, and include specified prices, minimum offtake quantities, take-or-pay clauses (YURKEWICZ 1986), and strict formulas for price adjustments, with all these measures being necessary to safeguard the large investment in fully dedicated field facilities and infrastructure of export pipeline and import terminal needed to get the gas to the particular market to which it has been contracted (THOMAS 1986). Thus low oil prices have little direct impact on British Southern North Sea gas-field development, independent from the duration of the period of cheap oil and without any connection to the adaptation of oil-field development. While most of the contracts remain unchanged, other negotiations will suffer from a decline in prices up to abt. 30 % as a consequence of some agreement linkage to the now lower gas/oil price ratio (BEU-

DELL 1986 b), with the overall effect, however, being nothing like as severe as that experienced by oil production companies who had to contend with a price slump of some 60 %.

The only crosslinking element is the fact that most of the companies withdrawing gas in the British Southern North Sea are also major oil producers and some of the economy measures being taken in the wake of low oil prices are affecting budgets not just for the oil areas, but to some extent also for gas exploration and development (BEUDELL 1986 c, FISHMAN 1986). In many areas, however, there is a clear shift of activity with promotion of gas projects at the expense of oil plans which are cut back or suspended (BEUDELL 1986 b). On the other hand and in contrast to Europe, the gas market in the USA also suffered considerably from the oil price crash. Gas supply and marketing trends are also outlined by BERGMANN (1982).

2.2.1.6.2.2. Continuity of planned and approved gas-field development

Gas development drilling in the British Southern North Sea (chiefly in the tight Rotliegend gas belt), but to minor amounts also in Netherlands and Germany FRG in Rotliegend and Carboniferous, is therefore more or less independent from the oil price scenario, because supply has to be provided in order to fulfill the negotiated and concluded long-term delivery contracts (cf. sections 2.4.5. and 3.1.). As the gas market is still booming due to continuously increasing replacement of oil by gas particularly in urban heating, development of the tight Rotliegend gas belt has a good future at least for the next five years when the major drilling and fracturing campaigns are going to be carried out (cf. sections 2.4.1.2. and 2.4.5.1.). In addition, development of the planned fields for which the gas had been already sold by contract has been approved by the government and is backed-up against the danger of gas imports from abroad. Proppant demand is also more or less foreseeable, because exploration and appraisal drilling has already shown that in many fields almost all the development wells will have to be fractured (OILMAN 1987 a), and the closure stress in the reservoir depth as well as the long-term field production planning of up to 20 - 30 years leaves no other possibility for going on the safe side than to prop the whole fracture with intermediate-strength proppants and not to use natural sand at all.

Therefore the huge drilling and fracturing campaigns in the Southern North Sea Rotliegend gas belt from 1987 onwards until the early 1990`s (FISHMAN 1986; GREGORY 1986, 1987; OFFSHORE ENGINEER 1986 a, OILMAN 1987 a) comprise the bulk of the European business for the proppant companies in the next years and could even represent one of the major hydraulic fracturing stimulation actions which oil and gas industry has experienced so far. The significance of natural gas in the North Sea is underlined by a strong increase in production in 1987 following the general picking-up of activity after the 1986 oil price crash (QUINLAN 1986). The development campaigns in 1987 - 1990 and beyond are probably the greatest activity since the initial boom that followed gas discovery in the British Southern North Sea in the early 1960`s (VIELVOYE 1988 a) and are certainly the largest Western European gas field development project in the last ten years (McNALLY 1987).

2.2.1.6.2.3. Government back-up and supply security

This upsurge of gas development has been triggered by the need of British Gas to replace supplies from the Frigg field (MURE 1987) which started to produce in 1977 (ERDOEL-ERDGAS AKTUELL 1976 b) and will be depleted by 1992/1993 (initial recoverable reserves of Frigg have been 250 Bill. m3 or 10 Tcf, of which abt. half each is on British and Norwegian North Sea territory; ERDOEL-ERDGAS AKTUELL 1976 b). British Gas has originally planned to replace the Frigg supplies with gas from the Sleipner field in the Norwegian North Sea, but the British government vetoed the import contract in 1985 (abt. 20 % of the total United Kingdom gas supply is coming from the Frigg field; PETROLEUM REVIEW 1988 d). In addition to replacement, gas demand has considerably increased during the late 1970`s and early 1980`s. Therefore still before the 1986 oil price crash huge development plans for Rotliegend gas fields have been made and started to confirm by appraisal drilling which is the base for conclusion of gas delivery contracts that have been signed with British Gas, with the first fields starting supply to British Gas in autumn 1988 (VIELVOYE 1987 c, 1987 d) and the others successively following through 1990 and beyond (CAPEL 1988, OILMAN 1987 a).

Strong governmental support of gas development in the British Southern North Sea gives the operating companies sufficient safety also after the 1986 oil price collapse that all the projects can be finished economically feasibly. There is only little concern that British Gas could stop the gas development boom by starting to contract gas from the Norwegian Sleipner and Troll (GRAY 1987) fields for import from the mid 1990`s (Norway has openly stated that it needs to get into the United Kingdom gas market; OILMAN 1987 a), because although having converted recently to a private company, British Gas still remains the monopolistic gas buyer in Great Britain and needs permission to import natural gas from the government which prefers to back-up domestic companies and to keep out imports, with national hydrocarbon operators also theirselves maintaining pressure on the government to stick to this attitude (VIELVOYE 1988 a). The Troll field has abt. 45 Tcf (1,125 Bill. m3) recoverable gas reserves which make it the tenth largest gas field in the world (McNALLY 1987). The planned putting on stream of the gas field complex Vulcan, Valiant, Vanguard, Ravenspurn South (cf. fig. 6) and Cleeton as the first group of newly developed Rotliegend fields (with the exception of Cleeton having all required considerable amounts of hydraulic proppant fracturing stimulation) still in fall 1988 has been successfully achieved (PETROLEUM REVIEW 1988 d, VIELVOYE 1988 c).

2.2.1.6.2.4. Gas-field development vs. gas price and demand evolution

The development history of the British Southern North Sea Rotliegend gas field belt is an excellent example how gas price controls acquisition activity (PETROLEUM REVIEW 1988 d). After the initial phase of intense exploration and development in the late 1960`s and early 1970`s, a downturn occurred until the beginning of the 1980`s which was triggered by market forces. By that time, British Gas had secured sufficient supplies from the original large fields to meet their requirements and the price they were offering for additional reserves did not warrant further exploration. During this period, the operators turned their attention to the Northern North Sea and the development of oil reserves. Throughout the 1970`s, however, demand for gas trebled, and by the end of the decade, British Gas had to look for new supplies and offered prices that made it attractive for the operators to return to the Southern North Sea and to resume exploration which initiated the gas field drilling, completion and stimulation boom of the 1980`s that will extend at least to the early 1990`s.

2.2.1.6.2.5. Disturbance of oil-field development

In contrast to the gas market situation which is characterized by more or less security, the 1986 oil price collapse had terrible consequences for the oil market, with fields still remaining to be developed having become hopeless uneconomical after the 1986 oil price drop and in mid to late 1987 in a slightly more stabilized situation still remaining marginal except of larger fields. Smaller satellites can be made more feasible only by widespread use of subsea completions with production processed at existing platforms at the big spots nearby (THOMAS 1986; aspects of development of small oil and gas field discoveries in the North Sea are illustrated by MORRISON & JOLLIFFEE 1982 and VOSPER & STEVENS 1982; cf. section 2.2.1.5.2.1.). Thus the activity after the 1986 oil price crash focussed on spur exploitation especially for fields being close to existing infrastructures with spare capacities and cheap tying-in of satellites (ALGAR 1987 b). On the other hand, many new oil field developments are still economically attractive at an oil price of abt. US $ 15, although a higher oil price level and particularly a better back-up by a stronger US $ exchange rate would certainly give a much better incentive.

2.2.1.6.2.6. Gas import and condensate-field development

The 1986 oil price collapse could indeed have a positive impact on the activity in the British Southern North Sea Rotliegend gas belt due to the delayed development of gas condensate fields in the Northern North Sea (FISHMAN 1986 a, GREGORY 1987) which suffer from the problem of large incremental capital investment being required to cycle the gas and of deferral of income from selling the gas throughout the cycling period (ALGAR 1987 b). Gas condensate field development, if taking place at all, is therefore restricted to those patches with exceptionally high liquid content, and even those will need some type of enhanced oil recovery investment incentive to be economically feasible. The exclusion of Norwegian gas so far from import into the United Kingdom (KNOTT 1986) also after the privatization of British Gas which still acts as monopolistic gas buyer, is triggering considerable further upgrading of the British Southern North Sea Rotliegend gas belt (GREGORY 1986), which means again further promotion of the significance of hydraulic proppant fracturing for increasing acquisition of gas for covering rising domestic demand in Great Britain without the support of contracting gas abroad, but exclusively from indigeneous sources.

As it is evident that British Gas has to buy significant new quantities for securing supply of the forecasted demand until the next century, there is a good market for the gas liberated from Rotliegend fields in the British Southern North Sea, and the economical impetus is given particularly by the fact that if the demand of British Gas cannot be satisfied by supply from the United Kingdom North Sea sector, British Gas would have to contract substantial volumes from Norway already in the next years (GREGORY 1987). British Gas itself, however, has a strong interest not to discourage domestic exploration and development by buying abroad earlier than absolutely necessary, but at the bottom of the line has to meet the demand of its customers. Therefore the constellation of the British gas market is considered to be the driving force for exploration and production companies to develop as many gas fields as possible thoroughly which indispensibly incorporates large-scale hydraulic proppant fracturing campaigns for liberation of additional amounts of gas.

2.2.1.6.3. Significance of the east-west energy trade in the European gas market

The importance of gas development drilling including stimulation in the North Sea and other hydrocarbon provinces in Europe in past and future has to be understood with the background of the significance of the east-west energy trade in the European gas market (D`ANDREA 1987). The development of the economical partnership between Eastern and Western Europe was the logical consequence of restructuring of the energy consumption profile during the last decades which was to considerable extent influenced by the oil price evolution. Some comments are offered as follows on evolution of the European gas market supply and demand situation, and international cooperation and competitive challenges, and future evolution of the North Sea gas province.

2.2.1.6.3.1. Evolution of the European gas market

During course of more than 100 years of natural gas acquisition and exploitation history, 1987 was the year of first exceeding 100,000 Bill. m3 (4,000 Tcf) world gas reserves (ERDÖL UND KOHLE - ERDGAS - PETROCHEMIE 1987 d). So far abt. 150,000 Bill. m3 (6,000 Tcf) have been discovered, of which hitherto abt. one quarter has been produced and/or consumed. The biggest gas producers of the world are in descending order USSR, USA, Canada, Netherlands, Algeria, Great Britain, Romania. Indonesia, Mexico and Norway, which together have more than 80 % of the total world gas production. Since 1950, world gas production has increased more than ten times.

Germany FRG is the second largest natural gas importer of the world (after Japan) with a total import quantity of 40 Bill. m3 in 1985 which derived by 33 % from Netherlands, 24 % from USSR and 13 % from Norway and Denmark, with the remaining 30 % coming from domestic production (ERDÖL UND KOHLE - ERDGAS - PETROCHEMIE 1986) of Rotliegend, Buntsandstein, Zechstein and Carboniferous fields in the Northwest German gas-field belt. The diversification of contracted gas among various sources of delivery secures supply, diminishes or even eliminates political dependency, and increases flexibility and competition (ERDOEL-ERDGAS 1978). The following outline focusses on historical changes of energy consumption spectrum as well as political and economical influences on multilateral energy trade.

2.2.1.6.3.1.1. Historical changes of energy consumption spectrum

In the 1950`s, coal and to a lesser amount oil and water power contributed over 95 % of primary energy consumption, with the share of natural gas being almost non-existent (D`ANDREA 1987). Significant changes in energy consumption pattern started to develop in the 1960`s and 1970`s, and the process of reorganization has stabilized in the last years, as a number of countries had achieved a reasonable balance in energy supply diversification. Important gas discoveries in Western Europe and USSR in the late 1950`s and early 1960`s provided various countries the opportunity and desirability of establishing the necessary infrastructure for the use of natural gas. Factors influencing development and growth of gas markets in Europe and world are security of energy supplies and diversification, higher efficiency of natural gas compared to other fuels, availability of natural gas at relatively low costs to consumers, long-term contractual relationship in sale and purchase of gas supplies, capability for transport of natural gas over long distances by large-diameter pipelines and sea-going vessels, development of technology in exploration and production of offshore resources, promotion of inter- and intraregional gas trade, and increased public awareness of the role of natural gas for an improved environment.

2.2.1.6.3.1.2. Political and economical influences on multilateral energy trade

Significance and influence of the above mentioned factors have varied in the different European gas markets due to existence of different economical systems; prevailing political, economical and fiscal relationships; and exposure of gas markets to competing sources of energy (D`ANDREA 1987). In Western Europe, security of energy supply and diversification achieved a high degree of importance following the first major incision into oil price history by the 1973/74 oil price explosion (cf. section 2.2.1.1.1.) when governments implemented various measures to reduce their respective dependence on oil which as a top priority item included an increased usage of natural gas. In light of being no longer constrained by physical limitations for moving gas supplies to markets, natural gas quickly received renewed recognition as a viable energy alternative to economies which had placed primary reliance on coal and nuclear power to alleviate their energy problems.

This situation also caused renewed interest and growth in the overall perspective of energy trade within Europe as well as inter- and intraregional dependency on gas trade. Increased interdependence on gas trade from sources within and outside Europe caused governments, often to the consternation of private commercial enterprises, to be directly involved in negotiating gas trade agreements, as political and economical considerations achieved a new level of strategical importance in bilateral and multilateral energy trade relations. Some aspects of the energy policy in the European community are summarized by CLEUTINX (1986), and characterizations of the European gas market are performed by BERGMANN (1988) and NORVIK (1988). D`ANDREA (1979) comments on the prospects of natural gas supplies and markets in Europe and their influence on the USA. LIESEN (1977) analyzes present and future significance of natural gas for Western Europe.

2.2.1.6.3.2. Supply/demand situation

The gas reserves of Western Europe are much greater than its oil reserves with regard to their energy content, and its gas production covers nearly three quarters of its own needs as opposed to only one third for crude oil (VALENTIN, VALAIS & CHAMPLON 1988). Western European gas reserves are thus an important asset in the security of its energy supplies. The most striking aspect of European gas resources is the offshore concentration which is more pronounced than that of any other continent, with more than 64 % of the proven reserves and probably more than 75 % of the ultimate resources being situated offshore, and with the North Sea having the greatest share due to Norway, Netherlands and United Kingdom there having more than 85 % of the overall reserves. After discussion of some general aspects, country-specific comments on supply and demand are offered for Netherlands, Great Britain, Norway and USSR. Some points of supply security and mutual benefit are also touched.

2.2.1.6.3.2.1. General aspects

Total natural gas consumption in Europe in 1985 amounted to 925 Bill. m3 (37 Tcf) of which 27 % and 73 % were taken in Western and Eastern Europe, respectively, with the USSR having an 85 % share of the Eastern European quantity (D`ANDREA 1987). Source of gas supplies for this consumption level originated essentially from within Europe itself, with only 2.5 % having been imported from Algeria and Libya. In Western Europe, gas consumption was considerably increasing from 1970 to 1979, then declining in the years 1980 - 1982 reflecting a general decrease in energy demand, and since 1983 rising again to reach a level higher than ever before in the last years (KEG 1986).

While several significant producers such as Germany FRG, Italy, France and Romania could not afford to export gas due to domestic energy demand exceeding internal supply and even requiring considerable import, sufficient quantities in Netherlands, Norway and USSR enabled the latter countries to export large volumes of gas, with in addition to that only Denmark supplying small amounts to foreign countries. Great Britain is more or less self-sufficient with neither significant import nor considerable export if the quantities from the Frigg field (which is crossed by the boundary between the British and Norwegian sectors of the North Sea and thus represents a special situation) are neglected.

2.2.1.6.3.2.2. Netherlands

The Netherlands have hitherto been the principal gas exporting country to Western European markets because of early discovery and development of sizeable reserves in the giant Groningen field (JENSEN 1988; cf. fig. 6 and section 3.3.) and other smaller patches producing from partially tight Rotliegend where also some stimulation by hydraulic fracturing was necessary to put the facilities on stream, but recent evaluations of gas reserves and domestic market requirements indicate that exports from the Netherlands will decline in the early 1990`s as part of the long-term strategy of the Netherlands to retain natural gas as the primary source of energy for domestic supply requirements. This concept, however, could also be a significant incentive to push development of further Rotliegend gas fields in the Dutch North Sea to compensate the declining output from Groningen and the other existing fields, which includes a reasonable contribution of hydraulic proppant fracturing for the acquisition of large new reserves in the tight facies belt of the Rotliegend.

2.2.1.6.3.2.3. Great Britain

Great Britain which so far has exported gas only in very limited amounts also needs increasing amounts of its indigeneous gas production for domestic demand, but while it temporarily seemed that the United Kingdom would be forced to contract some quantities to be imported from Norway, at least intermittent governmental decision to renounce from import and to expand the indigeneous facilities has considerably promoted the development potential in the British Southern North Sea as well as the prospects of measuring market penetration of natural gas in the United Kingdom (KNOTT 1986, OILMAN 1987 a). At the moment, 20 - 30 % of the Great Britain domestic gas supply is still coming from the Frigg field (DUFOND & LAFFRONT 1978, GREGORY 1987, MURE 1987) which is cut by the border line between United Kingdom and Norwegian North Sea sectors, but in the early 1990`s, declining delivery from Frigg will be compensated by increasing output from Southern North Sea tight Rotliegend fields following stimulation of the reservoirs by hydraulic proppant fracturing and completion during course of the current and coming major development campaigns (cf. sections 2.4.5.1.).

Therefore the United Kingdom can be expected to maintain self-sufficiency of oil and gas supply beyond the year 2000 (PETROLEUM REVIEW 1988 b) which in terms of gas is primarily ascertained by increasing development of marginal fields and increasing recovery from existing fields chiefly through hydraulic proppant fracturing stimulation (KNOTT 1985). LEWIS-DAVIES (1988) investigates the future gas supply and demand of the United Kingdom, and WHITTING (1979) outlines the British natural gas supply and demand strategy. JOHNSON (1978) presents an economical and financial analysis of the British North Sea sector relating gross national product to investment strategy, and BAYLEY (1984) evaluates market prospects and the gas balance of the United Kingdom.

In terms of distribution of Rotliegend gas reserves, abt. 20 Tcf (500 Bill. m3) had been cumulatively produced from the Southern North Sea through abt. 20 years since offtake started in 1967 (PETROLEUM REVIEW 1988 d). Average daily offshore production in the British sector is abt. twice that of the Dutch sector. The vast Groningen field onshore Netherlands, however, on the other hand produces alone almost as much as the entire United Kingdom continental shelf. According to the large reserves, the British Southern North Sea sector contains enough gas to satisfy the demand of British Gas as well as to look into direct sales to industrial users.

2.2.1.4.3.2.4. Norway

Norway has no domestic markets for gas and therefore is aiming on playing an increasing role in satisfying future Western European gas requirements especially following development of several large fields such as Sleipner and Troll (GRAY 1987) in the last years and in the near future (KNOTT 1986). Norway has so far divided its gas exports between United Kingdom and continental Europe (JENSEN 1988). The development of the giant gas fields Troll and Sleipner in the coming years faces Norway with the fact that its reserve additions in natural gas particularly in the more remote Northern North Sea are in excess of any early market requirements, and therefore Norway has already openly stated that it needs to get into the United Kingdom gas market with exports (OILMAN 1987 a) after termination of exploitation of the Frigg field (crossing the boundary between the British and Norwegian sectors of the North Sea) in a semi-joint venture. Troll is the tenth largest gas field of the world due to recoverable reserves of 45 Tcf (1,125 Bill. m3), and the large amounts of gas enable Norway to become the world`s third biggest exporter of natural gas after USSR and Netherlands (McNALLY 1987).

The future of Norwegian gas is analyzed by AGER-HANSSEN (1982), JOHNSEN (1982) and JACOBSEN (1988), and the general situation of Norway being a country with an energy surplus in the European energy market is illustrated by FROGNES (1981, 1983). Aspects of Norway`s position in the oil market and its relations to the OPEC are outlined by DAHLE (1987), GODEC & ALLARD (1987), NORENG (1987), SOLHEIM (1987) and UDGAARD (1987). Algeria which pioneered liquid natural gas exports to Europe from its giant Hassi R`Mel field lost some market position during the extensive price debates following the second oil price shock (JENSEN 1988).

2.2.1.6.3.2.5. USSR

The dominant gas exporting country in whole Europe is the USSR which holds currently abt. 43 % of the world`s proven gas reserves and ranks first in the world gas production (D`ANDREA 1987; ERDÖL UND KOHLE - ERDGAS - PETROCHEMIE 1986, 1987 d; JENSEN 1988, OIL GAS JOURNAL 1988 f), and is the largest oil and gas producer of the world (CHADWICK, LONG & NISSANKE 1988). The USSR total gas reserves reached abt. 1,500 Tcf (37,500 Bill. m3) by end of 1987 (FUEG 1987) in comparison with abt. 3,860 Tcf (96,500 Bill. m3) worldwide (WORLD OIL 1988 a). Having sufficient potential and being faced with uncertainties of the future prospects of oil production being adequate to meet domestic and export necessities, the USSR has given natural gas a high priority in its overall energy planning activities (cf. section 2.4.4.). Natural gas has provided a reasonable choice for the USSR of diversifying its energy supplies and in contributing an important share in hard currency earnings especially with the recent decline in exporting other fuels.

The USSR accounts together with Iran and Qatar for nearly 80 % of the world`s exportable gas surpluses (JENSEN 1988), and six other countries including Norway, Algeria, Nigeria, Malaysia and Abu Dhabi have ample gas reserves to support any foreseeable world-scale export project. The Ob region in Western Siberia/USSR has the largest concentration of gas reserves in the world. It includes with the supergiants Urengoi (producing sweet gas from Upper Cretaceous Cenomanian reservoirs in 1,100 - 3,100 m depth, and the Jurassic in 4,500 m depth is oil-bearing in some fields) and Yamburg the world`s largest and third largest gas fields which are connected by the Yamal pipeline system of over 5,500 km length and 1,420 mm (56 in) diameter to Western Europe. Other supergiant gas fields which are partially still in the development stage are Astrakhan, Bovanenko, Karachaganak, Medvezhye and Sovietabad (FUEG 1988; cf. tab. 9). Urengoi accounted in 1987 for 42 % of the total gas production in the USSR and had initial reserves more than double than those proven for the entire North Sea, and Yamburg also is believed to have greater production potential than all North Sea fields found to date (OIL GAS JOURNAL 1988 f). Western Siberia is scheduled to reach abt. 70 % of the total USSR gas output in 1990, and the supergiant Urengoi which came on stream in 1978 (OIL GAS JOURNAL 1987 o) and has already cumulatively produced over 40 Tcf (1,000 Bill. m3) of gas is not expected to decline until the late 1990`s (RIGASSI 1986).

Western Siberia contains with abt. 60 - 70 % the bulk of the USSR gas reserves (FISCHER 1981). Total USSR gas production reached the record level of 25.6 Tcf (640 Bill. m3) in 1987 after a rapid evolution from 10.2 Tcf (255 Bill. m3) in 1975 via 15.4 Tcf (385 Bill. m3) in 1980 and 22.7 Tcf (568 Bill. m3) in 1985, and is expected to increase until the year 2000 up to 45 Tcf (1,125 Bill. m3; OIL GAS JOURNAL 1987 o). Therefore by 1989, gas will probably for the first time represent a larger percentage of the USSR fuel balance than oil. Western Siberia`s share in the total USSR gas output increased from 1.3 Tcf (32.5 Bill. m3) in 1975 via 5.6 Tcf (140 Bill. m3) in 1980 and 13.1 Tcf (328 Bill. m3) in 1985 to 16.1 Tcf (403 Bill. m3) in 1987 which represents abt. 63 % of total USSR gas production and is expected to rise to abt. 35.3 Tcf (883 Bill. m3) in 2000 which then would correspond to over 78 % of the total USSR output. The magnitude of the Soviet exportable gas surpluses and the relatively favourable economics of their delivery has rendered import policy of the customers rather than available reserves and product pricing the limiting factor determining the market share which Soviet gas can carve out in Europe. The reserves are so large that the USSR could more than double its gas exports to Western Europe within ten years (STERN 1986). According to the economical necessity of the USSR to earn hard currency from chiefly hydrocarbon sale (cf. section 2.4.4.), there is no reason of fearing commercial or security risks of importing oil and gas from the USSR which is a solid trade partner and delivers reliably and has no interest to exercise economic leverage for political concessions. Gas deliveries to Western Europe have consequently reached hitherto record levels in 1986 and 1987, and plans are made for supply also to other countries than those already contracting gas from the USSR (GORST 1988).

2.2.1.6.3.2.6. Supply security

Up to the year 2000, abt. 60 % of the total European gas consumption can be certainly supplied from sources within Europe, whereas later also various quantities may derive from other countries (KEG 1986). In addition, the gas supply network structure and the related security measures guarantee the compensation of delivery interruptions such as possible embargo situations enacted by one supplier by flexibility to shift to other sources and to switch quantities within the system by increasing gas availability on specific demand. The supply safety is mainly given by the broad diversification of sources of imported volumes without excessive dependency on one individual supplier, reasonable security measures for the case of unforeseen delivery interruptions such as reserve capacities for domestic production and underground storage of peak compensation volumes, and boundary-crossing cooperation between the national gas industries (cf. also LIESEN 1985). The energy scenario history in Western Europe can be summarized as reflecting increasing reliance on natural gas (D`ANDREA 1983). Aspects of the USSR natural gas industry are also discussed by FISCHER (1981).

2.2.1.6.3.2.7. Mutual benefit

Thus the oil price explosions and related energy shocks of the 1970`s have had a profound impact on the energy economy of Europe, because governments implemented policies to reduce their dependence on oil imports, increase their domestic energy production, promote interfuel substitution and increase energy efficiency through conservation and technical innovations (D`ANDREA 1987). An important component of this transformation in the European energy economy was the expansion of the east-west trade. Increased energy imports from the east have allowed industrialized market economies in the west to diversify their sources of energy supply and provide business opportunities, while increased revenues from energy exports have permitted centrally planned economies in the east to import goods and services that are essential for the upgrading of their industrial technologization.

2.2.1.6.3.3. International cooperation and competitive challenges

Some future challenges to the European gas industry are also outlined by LIESEN (1985). Gas is bought and sold on an international market where substantial and often fundamental differences persist among various countries, and the challenges faced differ correspondingly. While in some regions, the development of the gas industry is essentially complete and the strategic objectives are concerned with future supply for the long term or for the consolidation of market positions in an increasingly competitive environment, the cost-effective installation of a gas transmission and distribution system is the target in other areas, and probably in certain countries, natural gas will never be of major importance for a variety of reasons. The necessity to adjust the existing capacity to meet a smaller overall energy demand is the most serious challenge faced on other energy sectors, but does hardly apply to the gas industry which enjoys continuity due to technical and structural factors.

The chief requirement is to maintain and promote a constructive attitude towards gas on the part of consumers and governments. The major technological advantages of gas are the lack of any severe impact on the environment and the easy control. External source diversification coupled with a major indigeneous supply base is the key strategy for energy delivery security, which is backed up by the proven world-wide natural gas reserves exceeding current world gas consumption by a factor of over 50. One of the main arguments for natural gas in Europe is the reduction in unilateral import dependence and the increased security of supplies achieved by the integration of gas in the overall energy market.

Aspects of government policy and international gas trade in the USA are examined by PRICE (1986), and LICHTBLAU (1986) comments on petroleum import and export policies in the USA. KALISCH (1988) gives a gas energy supply outlook until the year 2010 for the USA under different oil and gas price scenarios. NAYLOR & FERRIES (1988) summarize gas price and marketing evolution in the USA. Other relationships in the natural gas industry are illustrated by MAY & MATHEWS (1983). SCHLESINGER (1982) presents an outlook of gas energy demand and competitiveness in the USA, and other aspects of the USA gas market are discussed by ABBOTT (1982), BUSH 1982), KING (1982), MEANS (1982) and HOCHHEISER (1988). KUMAR (1988) comments on restructuring of natural gas markets by regulation and legislation. CUNNINGHAM (1982) emphasizes the role of market competition for shaping of future gas prices. TEK (1979) characterizes the challenge of worldwide natural gas matching supply-limited markets and market-limited supplies. RIETMAN & KIDD (1977) document improvement of ultimate recovery of natural gas by realistic gas prices.

2.2.1.6.3.4. Market shares and reserves distribution

While proven worldwide gas reserves compare favourably to those of oil, many gas discoveries are remote from consuming markets and thus costly to commercialize (JENSEN 1988). Less than half of the world`s total gas reserves are dedicated to specific local or export market outlets, and an equally large quantity of gas may also be classified as exportable surplus being large enough to support an export project, but having no market outlet yet available. Since new export commitments are not growing as rapidly as gas discoveries in major export regions are increasing, the exportable surplus category is steadily becoming larger which also progressively enhances overall supply security. On world-wide scale, oil consumption declined by 3.5 % between 1977 and 1986, whereas gas consumption increased by 29 %, with the gains that natural gas has made in the world energy markets, however, having been regionally very uneven.

In the United States where gas has historically held a high share of energy markets, demand has declined in absolute terms over the mentioned period, whereas its usage in Japan has more than tripled as a consequence of growth from a small base. By far the largest increment of expansion has taken place in the USSR where an aggressive gas utilization program has established one of the world`s most gas-intensive economies (cf. section 2.4.4.). Growth of gas consumption in Europe has been comparatively rapid at an earlier period, but for many parts of Europe, gas has now achieved more or less a level of market maturity and is not significantly increasing its share of energy markets. The reduced growth prospects for gas demand in Europe and North America which have dampened the expectations for world gas trade relative to what was anticipated a decade ago are partially the result of energy conservation and industrial restructuring.

2.2.1.6.3.5. Future evolution of the North Sea gas province

The most important gas province in Western Europe is the North Sea where currently Norway, Netherlands and United Kingdom have more than 85 % of the overall reserves (VALENTIN, VALAIS & CHAMPLON 1988). The nearer and farther future will be characterized by gradually diminishing production sources of easily accessible and retrievable gas. The decline of the presently producing fields as well as those which are developed at the moment can only be compensated in the future by exploration and acquisition of new generations of fields in increasing water depth, more remote geographical location and more unfavourable climatic conditions. Future projects concerning new generations of fields in Western Europe and especially in the North Sea will require larger and larger investments together with technological developments which will be far ahead of the techniques that have already been mastered, with this being the price that Western Europe will have to pay in order to limit the increase in its dependency on imported gas supplies. The frontier gas fields (DELACOUR 1986) which have to be developed are at the limit or even beyond of commercial feasibility given the present state of technology and economical conditions (cf. also VALAIS 1987).

2.2.1.6.4. Evolutionary trends of the world gas market

LINDEN (1985) outlines some evolutionary trends of the world gas market by commenting on the role of gas in a changing technological and competitive environment. Gas currently covers abt. 20 % of the total world energy demand (LIESEN 1985), but this rate is considered to grow substantially during the next decades (LINDEN 1985) which are expected to be characterized by accelerating technological change, growing interfuel competition and continuing structural changes in the industrialized economies, with gas more and more substituting and replacing oil in many branches of application. Because radical changes of the world`s future energy perception have occurred over the relatively short span of the last twenty years, the reliability of a forecast for the next twenty years is mainly depending on future rate of energy demand growth, adequacy of world oil and gas supplies, and outlook for world oil price stability.

BRIERLEY (1988) emphasizes the significance of gas as a global energy source, and GADONNIEUX (1988) illustrates gas challenges in a competitive environment. GUNDERMANN (1982) characterizes natural gas in the energy market of the future, and ERDOEL-ERDGAS (1978) outlines development of production and distribution as well as supply security of natural gas up to the year 2000 and beyond. The summary as follows concentrates on problematic linking of oil and gas prices, energy supply diversification concepts, economical and technical trends, oil substitution by gas, unconventional gas exploitation, and gas marketing strategies.

2.2.1.6.4.1. Problematic linking of oil and gas prices

One of the major problems of the present energy market is the artificial linking of the price structure for gas and oil (LINDEN 1985). Tying gas to oil geologically, economically and conceptually leads to looking at its prospects with tunnel vision, resulting in failure to pursue aggressive gas supply and utilization strategies. Independent from the oil price evolution, the value of gas varies widely depending on where it is sold, at what rate it is taken and at what degree of reliability it is provided. Because gas has an essential, very capital-intensive, geographically constrained and therefore quite appropriately regulated utility function to provide reliable and affordable energy services to residential, commercial and industrial customers that have little or no fuel switching capability, there is no immutable and fundamental rationale for linking the price of gas with that of residual fuel oil. Concerning supply, the obvious abundance of natural gas of both conventional and unconventional type raises the question of the appropriate use of available financial and intellectual resources to increase future gas supply security.

2.2.1.6.4.2. Energy supply diversification concepts

The future supply diversification concept should be based on the least-cost energy service strategy which moves natural gas as the cheapest, cleanest and highest effective energy type into a very favourable position (LINDEN 1985). Although there are national security, monetary and broader geopolitical and geostrategical considerations that require flexibility in terms of energy supply spectrum, the question is whether any of those political and economical concepts are based on mistaken notions of the potential of domestic resources, the inevitability of shortages, the inherent unreliability of foreign sources, and the national security benefits of autarky. Adequate and properly balanced energy research programs are clearly an essential component of sound least- cost strategies for providing energy services which not only have to include solutions to definable problems, but also insurance for a reasonable range of contingencies and opportunities for innovation and non-traditional approaches.

HOLTBERG, WOODS & ASHBY (1987) outline that natural gas will pay a major role in a highly competitive energy mix well into the next century. New supply technologies, particularly advanced techniques for production from unconventional gas resources, will enhance the accentuated position of gas in interfuel competition and assure the reliabilty of long-term supplies.

2.2.1.6.4.3. Economical and technical trends

Concerning perceivable economical and technical trends, the overriding factor is the large increase in cost and risk of financing major energy projects (LINDEN 1985). The unlikeliness of significant mitigation of the underlying causes being abnormally high interest rates, uncontrollable slippage of construction schedules, large cost overruns and general volatility of the energy market, has drastically shifted the relative attractiveness of alternative energy supply and utilization investments. Major oil and gas exploration projects in remote locations and hostile environments have become less attractive, whereas development of even marginal hydrocarbon resources in low-risk milieu and conservation have become more attractive. New exploration, completion and production technologies expanding the available prospects are the guarantee that major commercial oil and gas discoveries will continue well beyond the year 2000. The tendency of making marginal sources of gas economically viable will be enhanced by the growing recognition of the true form value of gas relative to oil, coal and centrally generated electricity with the advent of new and improved end-use technologies. In light of securing supply, some of the fastest-moving energy technologies in the next decades will be more accurate and effective exploration concepts and drilling diagnostics, remotely operated and controlled deep onshore and deep-water offshore drilling, deep-water completion and gas-fired cogeneration.

As there is presently no reason to assume any major upwards oil price move in real terms before the year 2000, the future of gas is characterized by continued tough burnertip competition from fuel oil in stationary heat energy applications where the superior combustion characteristics and non-polluting nature of gas do not offer a substantial competitive advantage. Even without oil competition, the price of gas in low form value applications would be limited by coal and electricity competition, particularly due to the relatively high efficiencies and low investment costs of the latter. The rapid introduction of more efficient and cost-effective gas utilization technologies and the stabilization of natural gas prices have now stemmed the growing electrical inroads into the traditional gas markets and have opened many new markets where gas is of superior competitiveness.

2.2.1.6.4.4. Oil substitution by gas

In areas with abundant gas resources, gas will continue to substitute oil in various applications, thereby further reducing pressure on world oil demand and thus extending the period of relative oil price stability. In areas without sufficient domestic gas resources but within reach of economically competitive export projects, gas consumption will also remain at high levels or even grow (LINDEN 1985). Concerning import, the sources of gas are highly diversified and have so far proven to be more reliable than those of oil, with further increase of energy security being provided by increasing diversification of fuel use and ongoing substitution of oil by gas, thereby increasing the oil export potential. As world-wide gas exploration is still in a relatively early stage, there is a high probability that both proven reserves and remaining recoverable resources will substantially rise over the next decades, particularly with progressive exploration and exploitation of unconventional gas. A gas energy supply outlook until the year 2010 is also given by KALISCH (1988), and JENSEN (1988) comments on world gas supplies, markets and trade. Aspects of areal gas potential are also provided by SHARMA, PATIL, KAMATH & GODBOLE (1988).

2.2.1.6.4.5. Unconventional gas exploitation

The definition of unconventional gas is flexible, because the boundary between what is economically and technologically possible or not has already shifted several times quite rapidly during course of gas acquisition and withdrawal technology evolution (LINDEN 1985). Coproduction of immobile gas in solid hydrates and gas dissolved in geopressured brines (cf. section 4.4.2.) by new techniques looks economically promising even at today`s market prices. The economically and strategically most important unconventional natural gas resources until the year 2000 are tight gas sands (cf. section 4.4.1.), shale gas (cf. section 4.4.4.) and deep unmineable coal seam gas (cf. section 4.4.3.; HAAS, BRASHEAR & MORRA 1985; KUUSKRAA & HAAS 1988), with all of these reservoirs requiring hydraulic proppant fracturing in smaller or larger scale in order to ensure feasible production rates (cf. also section 4.4.). As these three sources together yield already substantial amounts of gas for commercial production in the USA which is by far the most mature gas-producing region of the world, the transition from unconventional to conventional gas resources takes already place, and in the near future, huge marginal gas resources of the mentioned types will come into play in other countries as well.

If massive hydraulic proppant fracturing is prudently applied, a total of more than 600 Tcf (15,000 Bill. m3) of tight gas reserves are waiting on acquisition in the USA, with up to 4 - 8 Tcf (100 - 200 Bill. m3) per year being able to be accessed by 1990 (PAI, GARBIS & HALL 1983).

The economical and strategical importance of adding to natural gas resources that can be recovered at competitive costs is obvious (LINDEN 1985). It would take the construction of synthetic fuel plants at great expense to produce substitute gas from coal, or replace liquids from oil shale or coal, to yield as much useful output over their economical lifetimes as the conversion of additional marginal natural gas resources into feasible reserves. Further improvements in deep exploration and drilling technology will permit extension of the effectivity gains to new deep gas discoveries. At least in the USA, however, the earliest large increment of new supply is clearly that obtainable through improvement in stimulation techniques of both shallow and deep tight reservoirs which will make recovery of a greater portion of these resources profitable at current and projected market-clearing prices, with again hydraulic proppant fracturing playing a key role in this scenario.

2.2.1.6.4.6. Gas marketing strategies

Concerning gas marketing strategy, least-cost service options to energy consumers are key elements of the concept which has to be designed to achieve its essential objectives independent from oil price evolution and competition of gas with other energy types (LINDEN 1985). A strategy flexible enough to cope with the uncertainties of the world energy market must emphasize value of service over volume of service, which in turn demands continued rapid development and commercialization of high-efficiency end-use technologies opening a wide spectrum for gas penetration. These technologies also have to offer attractive investment opportunities with short payback times to present and prospective gas users or outside investors, even if that means temporary or even permanent shrinkage of sales volume. Any downside effect on gas demand caused by the introduction of more efficient end-use technologies will be offset by the increased probability of keeping existing customers instead of losing them to competitive energy service options, and the enhanced ability to further penetrate iexistng markets with relatively cheap gas that has been conserved, as well as to capture new markets. Gas supply and marketing trends are also studied by BERGMANN (1982).

Therefore a marketing strategy based on more efficient end-use is also a concept for maintaining gas industry profitability. Cost-effective end-use efficiency improvements increase rather than decrease the margin between delivered cost of gas and what it is worth to end users, because higher fixed costs per unit sale associated with lower volumes will be more than compensated by the combination of lower average acquisition cost resulting from slower depletion of relatively low marginal cost sources of supply, and higher break-even burner-tip price or form value coming from using less gas to provide the same or even higher-quality energy service in improved equipment whose higher first cost is paid back quickly. The existing fixed investment can therefore be justified on the basis of higher net-added value of the energy services provided rather than merely related to the volume of gas sold. The competitive position of gas is further improved by progressive conservation and cogeneration by cost-effective efficiency ameliorations.

Marketing and economical aspects of gas sources and supplies are also discussed by AL-FARISS (1988), JENSEN (1988), KALISCH (1988), NAYLOR & FERRIES (1988) and SALAMA (1988). Comments on the future market for natural gas and its effect on exploration are also given by GRUB (1984). MAY & WHITE (1981) investigate problems and prospects of natural gas price regulation.

2.2.2. Influences of proppant price changes

Apart from oil price evolution (cf. section 2.2.1.1.) and development of the US $ exchange rate (cf. section 2.2.1.3.), variations in proppant price also have significant influence on the economical feasibility of hydraulic fracturing in the actual oil and gas economical scenario and financial constellation, and thus have important impact on the marketing potential of reservoir stimulation treatments.

Changes of proppant price level took also place almost parallel to major changes in the hydrocarbon price niveau and in a later stage also to the US $ value fluctuations. In the late 1970`s and the early 1980`s, large proppant demand which temporarily even could not be met by supply kept proppant prices high, with the strong to partially even excessive demand having been not only a consequence of the favourable general economical situation during the golden age of hydraulic fracturing, but resulting also from the satisfaction of the consumers that after a long time of frustration, expectation and hope now finally suitable proppants that could withstand the high closure stresses in deep tight reservoirs were available and thus allowed to carry out operations which formerly had to be suspended due to lack of adequate proppants that would have resisted to crushing and thus collapse of the hydraulically-generated fracture in greater depth. While no changes of proppant prices were necessary in the golden age of hydraulic fracturing, the oil price decline beginning in the early 1980`s created another situation.

Proppant price evolution in the last five years includes the 1985 proppant price lowering and the 1987 proppant price increase. Some special impacts of proppant price changes on competition of various proppant types are also illustrated.

2.2.2.1. 1985 proppant price lowering

The expectations from and the consequences of the 1985 proppant price lowering are explained along the lines of marketing considerations and economical results. Aspects of superimposition and accentuation of the 1985 proppant price lowering by the US $ exchange rate drop are also outlined.

2.2.2.1.1. Marketing considerations for 1985 proppant price lowering

The main marketing considerations for the 1985 proppant price lowering were promotion of hydraulic fracturing activity as well as capture of market share from or even elimination of competition. When the hydrocarbon prices started to decline slowly a couple of years ago (being the preceding steps of the anyway unexpected big drop that took place in early 1986, but being partially also the consequence of the increase of the US $ exchange rate with respect to other world guide currencies such as DM and £ in the early 1980`s since its first minimum level in the late 1970`s), competition of the proppant companies manufacturing alumina oxide and silicate proppants as a result of diminishing both absolute and relative market shares, and customer demand required enhanced quality, better availability and lower prices of the various types of proppants offered at the hydrocarbon stimulation market. Another idea was to increase the competitive difference to zirconia-silicate proppants that are sold in currency other than US $ and which moved into a worse feasibility position by loosing governmental financial support for production and marketing (cf. section 2.2.2.3.).

Thus in combination with the installation of new manufacturing capacities by construction of new plants and extension of existing facilities in order to be able to fulfill the expected demand when lowering the price, the traditional marketing concepts were revised and some parties initiated several pulsatory steps of falling prices for alumina oxide and silicate proppants to finally abt. half its original level in mid to late 1985, which including inflation means even almost reduction to abt. a quarter of the former price, whereas the price for zirconia-silicate proppants was kept more or less at its elevated position or even increased due to the termination of governmental protection and support.

2.2.2.1.2. Superimposition by the US $ exchange rate drop

For the European customers, the reduction of prices for alumina oxide and silicate proppants was especially in 1986 and early to mid 1987 even accentuated by the decline of the US $ exchange rate with respect to strong European guide currencies such as DM and Ŀ that after achieving its peak niveau in the very early 1980`s slowly started to go down in the early to mid 1980`s and then accelerated to fall from late 1985 onwards to reach in late 1986/early 1987 for the second time in the finance political history after the Second World War its relative minimum level (and later in the fourth quarter of 1987 even achieving its historical absolute minimum niveau). The favourable impact of the US $ exchange rate drop on the proppant price for the European customers being hydrocarbon production companies at the bottom of the line, however, was more than compensated by its disastrous consequence on the oil price which represents a further devaluation and thus increasing deterioration of the general economical situation completely independent from the oil price crash itself (cf. section 2.2.1.3.).

Given the scenario of high oil prices (in combination with a high US $ exchange rate) in the early 1980`s, this drastical reduction of proppant prices (for Europe favourably combined with the drop of the US $ exchange rate) could really have triggered the beginning of the greatest hydraulic fracturing campaign which oil and gas industry had ever seen, because particularly in deep tight gas reservoir stimulation, the proppant cost may reach up to two thirds of the total expenses of the treatment and an abt. 50 - 75 % reduction of proppant cost (or more if inflation and decline of US $ value are properly included) would have been quite an amelioration of the economical feasibility of the whole job and would have enabled the execution of hydraulic proppant fracturing in such formations of more or less marginal type where its application was formerly far from being commercial (for example the cumulative cost of a MHF operation in a Rotliegend well in the Söhlingen field/Germany FRG in 1982 where abt. 550 t of proppants had been pumped amounted up to 6 Mio. DM; JOHN 1983, KLOSE & KRÖMER 1983, LEICHT 1985; cf. sections 2.4.1.1. and 3.3.).

Many proppant end users therefore heartily welcomed the reduction of prices for alumina oxide and silicate proppants as a very suitable opportunity to consider now more wells for hydraulic proppant fracturing than ever before, because in the past the higher proppant cost rendered many high-potential items marginally economical and particularly frightened smaller operating companies with limited financial possibilities, as a consequence of the fact that for European customers and their way of understanding, fracturing stimulation of a well is often not part of the whole investment for drilling and testing, but is something additional and optional (cf. section 2.3.4.). The economical result of the proppant price reductions, however, was not meeting the expectations due to the disturbing effect of the oil price collapse which took place completely independent from both US $ exchange rate drop and proppant price decrease.

2.2.2.1.3. Economical results of 1985 proppant price decline

Unfortunately, however, shortly after the enormous reduction of the prices for alumina oxide and silicate proppants enacted by some market-leader-type proppant companies came into force in mid to late 1985, the oil and gas market was considerably disturbed in early 1986 by becoming flooded with vast quantities of oil deriving from uncoordinated and violated overproduction of rivalizing OPEC countries (the oil oversupply was partially also the consequence of the Iranian-Iraqian war and of the desperate attempts of Saudi Arabia to get the other OPEC countries under control and to reason, unity and reliability;

cf. section 2.2.1.1.4. and 2.2.1.5.) and resulting in such a terrible lowering of the oil price that for most of the hydrocarbon production companies serious cutting back of the capital expenditure budget including suspension of almost all the stimulation jobs was the only answer which was economically sound as an acceptable solution.

Coupled with a drastical fall of the value of the US $ which started to go down to reach for the second time in the post-Second-World-War history a lower exchange rate (for example in relation to DM and £) than ever before, oil consumers could make double profit of the oversupply and price collapse situation by buying almost unlimited quantities for cheap prices, but the production companies were seriously hit in a two-fold manner, with a few smaller operations being now relatively (with respect to cash backflow) much more expensive and therefore much less economically feasible than several larger jobs before. Some aspects of reversal of fracturing potential and proppant availability as well as capital expenditure cutting and stimulation job suspension are outlined as follows.

2.2.2.1.3.1. Reversal of fracturing potential and proppant availability

The particular tragedy for proppant and service companies is that during the golden years of hydrocarbon drilling and stimulation of the late 1970`s and early 1980`s, abundant fracture jobs were carried out at high proppant and service prices (backed-up by high oil prices and strong US $ guaranteing fast pay-out of investment for the hydrocarbon production companies), whereas since the major oil price decline in early 1986 until at least early to mid 1987, another situation has arrived. Following proppant price reduction, oil price crash (cf. section 2.2.1.1.2.) and US $ exchange rate drop (cf. section 2.2.1.3.), proppants and - due to inevitable adjustment of the service companies (cf. section 2.2.1.5.3.) - service are cheaper than ever before and proppants are nowadays readily available in almost unlimited quantity and even enhanced quality with respect to that a couple of years ago when proppant demand was so large that the manufacturing companies had to produce at maximum capacity, and no time was left for quality improvement and efforts were only made to increase the output given a constant quality.

As a consequence of the 1985 proppant price cuts and the 1986 oil price drop, a much higher-quality proppant material is now available in significantly larger quantities at a considerably lower price than some years ago, but is seriously limited in application due to a presently poor economical framework particularly in Europe, whereas some years ago, a much lower-quality proppant material that was offered at a markedly higher price than now was virtually unrestricted in distribution due to a strong demand exceeding the previous capacities of supply. Therefore the 1986 oil price collapse completely reversed the stimulation market by changing the situation from good to excellent feasibility of hydraulic fracturing when only insufficient amounts of high-quality proppants were available at high prices to poor attractivity of hydraulic fracturing when more than enough proppant material is available at low prices.

2.2.2.1.3.2. Capital expenditure cutting and stimulation job suspension

In 1986 and partially even still in early 1987, the bad general economical situation of an oil price approaching almost the level around or even below the first classical price shock abt. 15 years ago (at least if progression of inflation since that time and the present low US $ value are incorporated into the calculation; cf. section 2.2.1.1.2.) during the worst phase and being in Europe even accentuated by the decline of the US $ exchange rate (with all the effects together triggering in early 1986 within only a few weeks a recurrent incision into the oil price evolution to a level far back in post-Second-World-War history), however, could even given the mentioned proppant- (and also service-) related facts not only not justify an expansion of the stimulation activity as an answer to the considerably lowered proppant and service cost, but also even not provide any convincing arguments for at least maintaining the previous level of fracture treatments, with most of the hydrocarbon production companies thus cutting down their budget of capital expenditure by at least 25 % (OEHME 1986) and postponing almost all their stimulation activity for at least one or two years.

The general situation is only reasonably ameliorating in the major centres of activity since abt. mid 1987, whereas in other common and marginal areas, much of the promising picking-up was in late 1987 again seriously disturbed by the collapse of the US $ exchange rate down to a level which had never been reached before (cf. section 2.2.1.3.), and in early to mid 1988 by an again little slide of the oil price which although being almost negligible in its range, nevertheless hit a very sensible industry that interpreted the fluctuation beyond the average variations as a market signal and became again considerably uncertain concerning further investment decisions. The seriosity of the ongoing crisis constellation is highlighted by the alarm which the OPEC group addressed also to various non-members for a desperate trial to stop further downwards orientation of the oil price (cf. section 2.2.1.4.3.).

2.2.2.2. 1987 proppant price increase

As the 1985 proppant price lowering had not the expected favourable impact on the hydrocarbon exploration and production industry by triggering a renewed boom of stimulation activity as a consequence of market deterioration by the 1986 oil price crash, the proppant suppliers were already anxiously waiting for the earliest possibility to raise the price again in order to both compensate the business losses during the crisis initiated by oil price collapse and US $ exchange rate drop and provoked stimulation level decline, and increase their profit in a time of renewed picking-up of treatment activity and rising proppant consumption. The outline as follows includes comments on the reasoning for renewed proppant price increases, and market reaction and economical consequences of the 1987 proppant price increase.

2.2.2.2.1. Reasoning for renewed proppant price increases

Independent from all economical influences, the future is characterized by need and urgency for unconventional gas (KUUSKRAA 1980) which automatically promises at least some booming of hydraulic proppant fracturing. Effects of price and technology on gas resources in low-permeability reservoirs are also discussed by BAKER (1981). SCHAEFER & POWERS (1984) comment on the impact of rapidly changing economical environment on petroleum exploration and acquisition. Effects of increasing gas prices on known gas reserves are outlined by HICKS & RAHMAN (1984).

Concerning proppant price evolution, preparation of the supply companies for the coming big demand of material during course of the Rotliegend gas field development campaigns in the British Southern North Sea in 1987 - 1990 in Europe (cf. sections 2.2.1.6.2., 2.4.1.2. and 2.4.5.1.) as well as the adaptation to the world-wide picking-up of hydraulic proppant fracturing included more or less carefully planned and timed adjustment of the proppant prices to the expected major businesses and profits. In order to optimize their net return during this probably historical highlight of hydraulic reservoir stimulation in Europe as well as maximizing the payback with progressively improving business perspectives (thereby also partially compensating the income losses generated by the 1985 proppant price lowering which had not the expected impact on sales volumes due to the shortly afterwards following 1986 oil price crash), some parties in mid to late 1987 reversed their 1985 price drop policy and increased the prices again significantly, without, however, reaching the level held in 1985 before the initiation of the price fall.

2.2.2.2. Market reaction and economical consequences of the 1987 price increase

While some slight to moderate increase in proppant price would certainly have been more than justified in order to let also proppant and service companies survive and reconsolidate, the market has reacted with serious business losses for some parties which showed limited flexibility in pricing to major customers and tried to play the card of increased prices even for contracts where bidding was made according to the earlier pricing concept during the transitional phase between old and new list prices, and a few major deals went to competitors thus punishing and disadvantaging some of the initiators of the price rise which did not want to adjust to the demand of the market or simply did not understand the European market and its special features (cf. also section 2.4.2.4.1.).

In view of the expected demand of annually between 5 and 10 million lbs. (2,300 - 4,600 t) of intermediate-strength proppants for the period of three or four years in Europe (comprising chiefly quantities needed for Rotliegend fracturing in the British Southern North Sea), even a price increase of a couple of cents results in case of more or less coordination with the competition in considerable amelioration of the desolate to critical financial situation of the proppant companies which more or less made serious losses during the 1986 crisis. The late 1987 proppant price increase was subsequently in early to mid 1988 from the view of the European customers even accentuated by an again slightly stabilizing US $ exchange rate.

2.2.2.3. Competition of various proppant types

The various price drops and rises for alumina oxide and silicate proppants had also serious impacts on their competitive power with respect to zirconia-silicate proppants which kept almost all the time their former elevated price position due to more expensive production process, and which are manufactured in Europe and sold for currencies other than US $, with thus their pricing in Europe and other non-dollar markets of the world not being influenced by variations of the US $ exchange rate in either direction. The following outline focusses on some aspects of consequences of the 1985 price reduction and the 1987 price increase for the competitiveness of the different proppant types.

2.2.2.3.1. Influence of the 1985 proppant price reduction

While before the first proppant price reduction in mid 1985 competition had to be made through performance and service at a level of more or less comparable prices for all the different products, without anyone being significantly cheaper to allow application of the price instrument as a powerful marketing and sales tool, the price drops for alumina oxide and silicate proppants have moved them into a very advantageous price situation with respect to zirconia-silicate proppants.

As a consequence of the fact that due to the price cuts, the cost for alumina oxide and silicate proppants was finally only between two thirds and half the price for zirconia-silicate proppants, the latter product got almost unmarketable in Europe and lost considerable shares. This effect was even accentuated by the starting decline of the US $ exchange rate which made alumina silicate and oxide proppants in Europe again cheaper, and by termination of governmental financial support for manufacturing and distribution of zirconia-silicate proppants that rendered their pricing even more tough with leaving even less room for downwards adjustments than before. During the time since mid 1985 and particularly after the oil price crash in early 1986, almost no customer was willing to pay such a high premium for zirconia-silicate proppants and thus their application was in Europe virtually limited to a few tail-in operations in order to make benefit of the possibility to perform gamma-ray fracture height monitoring by logging the natural radioactivity of the zirconia-silicate proppants (cf. section 6.2.1.2.1.).

Therefore the companies manufacturing and selling alumina oxide and silicate proppants had a big advantage from their price drop by nearly completely eliminating a serious competitor with whom they had to share the market in former years, but who now particularly after the oil price collapse when cost containment advanced to a top-priority item in most of the operating companies had almost no arguments left to convince clients to take its product. The competitive difference between alumina oxide and silicate proppants and zirconia-silicate proppants was for European customers in 1986 and early 1987 even increased by the falling US $ exchange rate down to its relative minimum in post-Second World War history, thereby providing another proppant price cut beyond the influence of the supplier of the material.

2.2.2.3.2. Influence of the 1987 proppant price increase

The late 1987 proppant price increase, however, was already a step in backwards direction. Although even after the reverse price adjustment in order to ameliorate the net return situation, the difference in cost between alumina oxide and silicate proppants and zirconia-silicate proppants is still sufficient to remain a powerful marketing tool, the danger is there that another price rise for alumina oxide and silicate proppants could turn the market into another direction. The producers of zirconia-silicate proppants, however, have little if any possibilities of adjustment by lowering their prices, because the process of making fused proppants with respect to sintered ones is considerably more expensive and the European proppant company producing zirconia-silicate proppants could not benefit from the automatical additional price decrease that was provided by the decline of the US $ exchange rate.

According to the situation in early to mid 1988 when a slightly again stabilizing US $ exchange rate performed another minor price rise for alumina oxide and silicate proppants independent from any intentions of their suppliers, the producers of the latter materials have to be warned that another company-induced price increase would bring the zirconia-silicate proppants back into their competitive position which they had prior to the mid 1985 proppant price drops and then the expected improvement of net returns for alumina oxide and silicate proppants could be more than compensated by considerable business losses to zirconia-silicate proppants (cf. also section 2.3.6.3.). This danger is particularly serious because a further recovering US $ exchange rate could provide an automatical price rise for alumina oxide and silicate proppants at the bottom of the line for the European customers, with the producers of the latter materials having no influence on an increasing value of the US $ except of once again lowering the proppant price in Europe which then, however, would be certainly at the expense of credibility, confidence and reliability that they receive from their customers.

While the recovering and strengthening US $ in mid to late 1988 provoked already another step towards a possible renewed turnaround of the proppant market, the again declining power of the US $ after the presidential election in the USA at the end of 1988 released at least temporarily this pressure and prevented the suppliers of alumina silicate and oxide proppants from regretting the 1987 proppant price increase. Although the forecast of the evolution of the US $ exchange rate in the very near future does not include any significant son to fear the achievement of such a constellation as sketched above, the whole international multilateral political and economical game contains too much uncertainties that any party can feel their position to be safe, and changes against the prognosis are always able to arrive.

2.3. Stimulation market differences between USA and Europe

With respect to the USA market, the early to mid 1986 crisis (in parts of the area still lasting to late 1986 and even early 1987 or longer) that was initiated by the serious oil price decline hit the (mainly Western) European stimulation market much stronger, resulted in much more cut-back of the activity and took much more time to recover and to readjust to an again better situation (the latter phase is still not terminated and is slowly moving along the rising branch) due to several reasons. The differences outlined as follows include frequency of drilling rigs and stimulation, success/failure ratio and consortial structure, operating mentality and treatment cost, US $ exchange rate decline, and proppant marketing strategy.

2.3.1. General aspects

MARTINS (1987) presents a summarized comparison of the stimulation markets of the tight gas basins in the USA and the Rotliegend/Carboniferous tight gas belt in the Southern North Sea to illustrate general aspects of the different setting of the hydraulic proppant fracturing markets in Europe and USA. Absolutely higher well and fracture cost in Europe are the reason for a relatively lower fracture/well expense ratio of abt. 10 - 20 % in Europe with respect to abt. 40 % in the USA. Well numbers per field reach in the Southern North Sea only 4 - 40 in contrast to 40 - 1,000 wells per field in tight gas basins in the USA. The lower fracture/well cost ratio relatively promotes stimulation in Europe as an additional step of testing or even an integrated part of the whole drilling and completion operation, while the higher well quantity in the USA gives rise to an absolutely higher fracture activity in the latter area.

Fluid efficiency is in Europe less than in the USA, whereas gas flow rates are better in the former area, but are often linked with serious non-Darcy flow which is considered to be unimportant in the latter region. Service boat capacity is often the limiting factor of the fracture treatments in the Southern North Sea where the locations are up to several hundreds of kilometres away from the shore line (cf. section 3.8.1.2.), in contrast to many land tight gas fields in the USA where plenty of material and equipment can be installed without physical limitation of surface facilities (cf. section 1.1.1.4.). ODD (1982) characteristics the economics of oil and gas production in North America as a benchmark for Europe, and MORRISON & JOLLIFFEE (1982) present an outline of the North Sea tax and economical system.

2.3.2. Frequency of drilling and stimulation

Much more drilling rigs are operating in the USA with respect to Europe, and thus as a consequence of the absolutely considerably larger market also a relatively higher potential of stimulation is present in the USA. The difference in rig count also remained during the peak of the crisis, with the difference in proppant demand also relatively staying independent from the absolute change of the sales level in both areas. Some comments are offered as follows on percentages of fracturing necessity for economical production as well as treatment cost and significance of fracturing in well completion.

2.3.2.1. Percentages of fracturing necessity for economical production

The absolute as well as the relative frequency of stimulation jobs is much higher in the USA than in Europe both before the late 1985/early 1986 oil price decline and after the collapse of the hydrocarbon production industry due to deterioration of the general economical situation. In the USA, abt. 35 - 40 % of all the hydrocarbon wells drilled in the last decade have been fractured in order to achieve commercial production, with abt. 25 - 30 % of the total USA reserves having been made economically producible by hydraulic stimulation (VEATCH 1983, OIL GAS JOURNAL 1984), whereas in Europe, the percentage is generally much lower and even in high-potential areas of stimulation often does not exceed 10 - 20 %, if reaching these shares at all.

Percentages of fracturing necessity of 30 - 50 % or even more (in extreme cases up to 100 %) apply in Europe only for well-defined patches of some plays such as a suite of tight Rotliegend gas fields in the Southern North Sea (cf. sections 2.2.1.6.2., 2.4.1.2., 2.4.5.1. and 3.3.), with the overall potential in the whole facies belt, however, being considerably lower. In addition to the relative difference in fracture potential, as a consequence of much more wells being drilled in USA with respect to Europe, also an absolutely larger amount of wells is stimulated according to the mentioned shares. The higher frequency of jobs in the USA results in more often use of the equipment, and thus service charges are also much lower with respect to Europe. Stimulation logistics and requirements in the North Sea in Europe are outlined by MACADAM (1984).

2.3.2.2. Treatment cost and significance of fracturing in well completion

In the USA, the percentage of wells being fractured has considerably increased in the last three years, because treatment cost has decreased due to price cuts by service and proppant companies, with cheaper jobs fitting particularly better into revised budgets of independents (BATT 1983, RAMSEY 1983, ELTING 1987) which only have a rather fixed certain amount of money to be spent for stimulation in their capital expenditure planning period and for which cash management has top priority (cf. section 2.2.1.5.2.5.). CHAN (1985) stresses the nature of oil-field cashflow as the independents` lifeblood for survival, with the independent producers being under increasing pressure to seek out every available dollar of internally generated cashflow from otherwise idle assets or premature financial obligations.

As a consequence of the ever increasing development of low-permeability oil and gas reservoirs, hydraulic proppant fracturing has in the USA already become one of the most important aspects of well completion (WATERS 1980; ROBINSON, HOLDITCH & WHITEHEAD 1986), whereas in Europe, it is still a long way to go to this stage, and presently the evolution of stimulation is in Europe in many areas still at the level of performance only in case of exceptional and indispensible circumstances, with the only exception probably being the thoroughly planned Rotliegend development campaigns in the North Sea in 1987 - 1990 and beyond which are based on the concept that encountering the expected tight facies could mean fracturing almost every well in the suite of production drilling.

2.3.3. Success/failure ratio and consortial structure

In the USA, it is already since a long time well understood that because the search for oil and gas becomes more difficult with increasing prospection and exploration maturity of more and more basins where the big and obvious deposits have been found already much earlier, and nowadays and in the future smaller conventional as well as innovative deposits have to be discovered which require more and more sophisticated strategies and techniques and where the success/failure ratio may be lower, it is increasingly important to fracture as many existing wells as possible to compensate the declining impact of exploration drilling on supply of additional reserves by the impetus of stimulation (KNOTT 1986). Hydraulic proppant fracturing will play an ever increasing role in maintaining the level of oil and gas production in the USA as high as possible (WATERS 1980), and this conclusion also needs understanding in Europe and other areas where it is not yet part of common sense in hydrocarbon reservoir completion and exploitation.

In Europe, the structure of the consortia requires agreement between the partners, with fracture jobs as well as whole drilling projects being often seriously delayed if one of the two to four or even more parties sees no need for investment at the moment, and the still higher prices of materials and operations with respect to the USA have in Europe so far not had a similar triggering and promoting effect, and if there had been one, it was killed by the oil price drop (GRIFFITHS 1986, RUNGE 1986; cf. section 2.2.1.1.) and the decline of the US $ value (cf. section 2.2.1.3.; although at least some North Sea fields are still able to cover operating cost and royalty payments in case of an even still lower oil price; BEUDELL 1986 a). Threatening unwilling partners with sole-risk operations only works in exploration, but cannot be applied in development drilling and stimulation. Financing projects via turnkey contractors is outlined by MARPLE & RULEY (1987).

2.3.4. Operating mentality and treatment cost

The operating mentality and philosophy in the USA is much more direct and decisive than in Europe, being an effect of the generally straightforward American thinking in contrast to the hesitating and well-reasoning European mentality. In Europe, repeated review of the plans, renewed economical feasibility studies, more complicated hierarchical permission procedures with often too many and partially rivalizing deciding functions, another mentality where sometimes subjective interests can seriously influence objective purposes and can hamper project approvals, and certainly sometimes also lacking encouragement for unpopular approaches and missing readyness to consider more risk often lead to alteration, suspension or even total cancellation of planned jobs, whereas in the USA action is made according to the rule "get it done as soon as possible", and also a higher risk is often accepted (NEWENDORP 1983 outlines a strategy for implementing risk analysis) in contrast to Europe where everything should ideally be on the most safe side. Evaluations of economical risk for North Sea offshore field development are carried out by SULLIVAN (1982) and TWEEDIE & KENNY (1982). Some aspects of integration of stimulation into the total well concept, and proppant price and treatment cost are outlined as follows.

2.3.4.1. Integration of stimulation into the total well concept

The fundamentally different stimulation approach between Europe and USA is partially also a consequence of the American concept and understanding that the fracturing treatment is part of the drilling and testing budget and is automatically included therein, whereas according to the traditional European thinking, hydraulic proppant fracturing was for long times considered to be something additional and optional and therefore once decision was to be made, there was always considerable reasoning whether the additional investment should be made or not. Thus at a time after the oil price crash when even well-logging runs were subject to significant streamlining, casing schedules for drilling were changed in order to save money, reservoir coring was reduced to the bare minimum necessary for petrophysical calibration of log interpretation or was even almost totally eliminated, and many expensive stimulation jobs were either suspended and postponed or completely cancelled. The American way of understanding the whole package as an integrated object is only recently spreading in Europe as well, mainly in connection with the huge development drilling campaigns in the British Southern North Sea Rotliegend belt (cf. sections 2.2.1.6.2., 2.4.1.2., 2.4.5.1. and 3.3.) where any other strategy of gas acquisition would be condemned to failure.

The necessity of a more risky operating mentality is particularly in Germany FRG underlined by the phenomena that the last couple of bigger gas strikes in the Carboniferous by exploration wells dates back more than five years (SCHRÖDER 1985) and only a few smaller discoveries were made in 1986/1987 (SCHRÖDER 1987), which is in marked contrast to the fact that most of the several dozens of MHF jobs having been carried out in Germany FRG so far have been performed in the Carboniferous. Discovery rates in the Rotliegend have been considerably higher than in the Carboniferous in Germany FRG in the last years due to a strong and pushed gas exploration and appraisal drilling programme even after the 1986 oil price crash, but an even higher amount of new fields or blocks could have been proven to contain economically accessible gas reserves if hydraulic proppant fracturing stimulation would have had an adequate position as compared to the emphasis on drilling.

2.3.4.2. Proppant price and treatment cost

The total proppant price in Europe is at the bottom of the line almost times 1.5 or even twice the price in the USA as a consequence of sea freight as well as additional handling and warehousing surcharges. The nearly doubled price makes many operations in Europe (with respect to jobs in the USA) much less economically feasible (if at all) and thus is one of the main reasons for the absolutely much lower stimulation activity (in view of hydraulic proppant fracturing, because proppant cost is often the largest share of the total treatment expenses) already in good times and thus especially in the 1986 crisis and the following years. In contrast to up to 100 % cost difference between Europe and central USA, alterations of expenses up to 20 - 30 % can also occur for regional positions within the USA with respect to the middle portion of that area as a result of logistical and environmental differences.

2.3.5. US $ exchange rate decline

The cost for material and operations in Europe was relatively cheaper from late 1985 onwards with respect to the USA also as a consequence of the declining US $ exchange rate (with respect to DM and £) that superimposed in Europe the price reductions of the supply and service companies. At the same time, however, the decreasing value of the US $ deteriorated progressively the general economical situation by giving a lower net return from the hydrocarbon sales (thus in Europe inducing an even stronger reduction of the oil price than effective in the USA; cf. section 2.2.1.3.). This financial political constellation had an important additional impact on the stimulation market in Europe and other non-dollar-currency areas in contrast to its insignificance in the USA.

The cheaper price to be paid for goods and service resulting from price cuts together with US $ devaluation therefore did not lead to an increase of investments in Europe, because the negative effect of the weak US $ on the already unfavourably low oil price did no longer permit to make investments with a sufficiently high discounted cash flow rate even under simplest technical conditions. Thus in many predominantly Western European companies an almost total freeze of the budget of capital expenditure (above certain inevitable obligations set by the mining authority when licensing the concessions and except of such items where contracts had already been signed) was the general strategy to survive until a time with higher oil price and stronger US $ came slowly back by slight stabilization from late 1986 onwards, although having still in mid 1988 not yet reached a level that allows renewed safe planning of significantly higher investments than during the peak of the crisis and shortly after it.

2.3.6. Proppant marketing strategy

As the main proppant producers have both head offices and manufacturing plants in the USA, this area (including Canada) is much better covered by marketing and sales promotion activity and is developed since many years, whereas many parts of Europe were virtually unattended until recently. Sales to Europe were in the last years predominantly made upon spot-like request by the market, with the highlights being the MHF treatments in Germany FRG; British, Dutch and Norwegian North Sea; and Yugoslavia, and orders were obtained by individual proppant companies in some cases by pure chance and luck (sometimes including unexpected backdoor approaches through supply companies which got the business, but were unable to deliver the requested quantities from their own sources), but not as a result of stepwise groundwork by sound marketing and sales promotion concepts based on understanding of geological, engineering and economical relationships.

The much better communication and transparency of the latest technical news concerning application of proppant types and the consulting by the sales representatives in the USA reinforces the potential of hydraulic fracturing in this area and thus helps production and service companies to decide in a direction favourable for the proppant company in a much more effective way than in Europe. Given comparable standard of marketing management, an adequate penetration is also in Europe undoubtedly possible and is only a question of manpower and time. The following discussion summarizes aspects of lacking continuity in marketing attention, business shares of the different proppant companies, and price changes and market turning.

2.3.6.1. Lacking continuity in marketing attention

The hitherto unsatisfying proppant sales level in Europe is therefore the logical and inevitable consequence of lacking continuity and concept of marketing and sales strategy in that area. In contrast to market transparency and illustration, confusion and embarassment was created by various proppant companies by not only neglecting their marketing activity, but in addition to that changing names of proppant products and companies (in some cases repeatedly within a couple of years), alternation of limited periods of time when local technical sales representatives were acting with phases when nobody except the head office in the USA was responsible, employment of marketing managers with soon following premature termination without logical and understandable reasoning and without respecting the requirements of the European market, and finally changing proppant prices repeatedly in downwards and upwards direction. As a consequence of all these shortcomings, Europe is in many parts still nowadays a poorly understood proppant market.

Reviewing the past and current marketing activity of proppant companies in Europe, conclusion has to be made that the generally unsatisfying past and present sales level is the logical and inevitable result of serious neglectance and exaggerated underevaluation of the potential of a market where numerous possibilities of stimulation exist, but complicated reservoir conditions, many different mentalities and administrative interlocking require much more time and activity, coupled with patience, to achieve the goal of the marketing strategy by having the operation of hydraulic proppant fracturing jobs. This target, however, can even today with good intercontinental oral and written communication in scientific, technical and commercial aspects not be achieved without a full-time local marketing management representation within Europe.

Release of the proppant marketing potential in Europe and its conversion into orders can therefore at the moment only be achieved by investment into the market which, however, should in fact having been done already much earlier.

Again it has to be emphasized that continuity of work should be achieved by a longer assignment of marketing personnel rather than hiring and firing policy with premature personnel termination without understandable reasons simply because executives are not willing to wait sufficiently long to observe the market reaction after a reasonable lean time. The market needs in fact quite some time in order to response and to reflect the results of the market attendance efforts in visibly increasing sales levels independent from seasonal fluctuations and variations according to the general economical framework.

2.3.6.2. Business shares of the different proppant companies

Aspects of business shares of the different proppant companies are evaluated along the lines of skewed market share ratio of major proppant suppliers as well as impact of service companies vs. own treatment management as follows.

2.3.6.2.1. Skewed market share ratio of major proppant suppliers

As the service companies being the main proppant customers do not divide the proppant business by 50 : 50 between the major suppliers (a similar relationship applies for the well-logging business provided by the hydrocarbon production companies), but almost always generate an asymmetrical distribution with slight favourization of one supplier with respect to its competition, the individual proppant companies should be inspired to undertake efforts to skew the ratio towards 60 : 40 or even 70 : 30 to their benefit and at the expense of their competition. With the background of identical prices and comparable performance of the various correlating proppant types, this goal can only be achieved by the experience and knowledge of the technical and marketing representative, with geological and reservoir engineering experience being essential to obtain credibility and confidence and being regarded by the customers not simply as a sales agent that tries to promote product no matter how and irrespective of practical demand, but as a technical and scientific expert not only for the commercially offered proppant types, but also for hydraulic fracturing in general and the requirements of the specific reservoirs.

Additional market shares can be obtained by trying to convince service companies and operators to increasing amounts of the benefits of replacing natural sand by ceramic proppants. Talking to all the operators is the only way to make sure that all the relevant information is properly fed to the end user and is not obliterated or drowned in the bulk proposal of the service company.

The skew of the proppant market from 50 : 50 towards 60 : 40 or even 70 : 30 was achieved by one of the major proppant suppliers in Europe for quite some time by good cooperation and confident business relationship with one service company which according to its competitiveness in terms of price and performance won most of the proppant fracturing contracts in the North Sea during the last few years, with the proppant company thereby having had an accidental preferential outlet into the North Sea market (which is the largest share of the European fracturing scene) despite generally insufficient marketing attention. Following the late 1987 proppant price increase, however, this advantage was lost to the competition as a consequence of playing the price card too high due to limited understanding of the market requirements (cf. section 2.4.2.4.1.). This example most significantly highlights the need of improved marketing and sales strategies by experienced local representatives who are also familiar with the particular needs of the European market and the mentality of the people working there.

2.3.6.2.2. Impact of service companies vs. own treatment management

The formerly strong to dominant impact of the service companies on stimulation design and chiefly proppant selection is declining. The reason is that on the one hand, Western European companies have in increasing amounts their own experts who play simulations in the computer with in-house programs and make a design including proppant choice, and then ask the service company to either modify and improve the concept or even almost directly execute the planned treatment. On the other hand, Eastern European companies have frequently own equipment and carry out many jobs (particularly smaller ones) themselves, with even the service companies being prepared to accept direct proppant purchase by the Eastern European operators and trying to focus their business on fluids, additives and other chemicals. The Eastern European mentality is quite different from the Western European and American way of thinking. Friendship has a very high rank also in business life. Once a company representative is acknowledged to be a technical expert and is regarded as a personal friend, the Eastern European specialists will almost always address themselves to him and also give him their business, with in some cases even not talking to competition at all regardless of any pricing structure.

2.3.6.3. Price changes and market turning

Since the 1985 price drops for alumina oxide and silicate proppants which moved the more expensive zirconia-silicate proppants into an almost unmarketable situation, the market in Europe was virtually only shared by the major producers of alumina oxide and silicate proppants. The ratio of 60 : 40 or 70 : 30 which may be achieved by one of the major companies with respect to their competition, however, is only valid as long as the price difference between alumina oxide and silicate proppants on the one hand and zirconia-silicate proppants on the other hand is sufficient to prevent reasonable market penetration of the latter product.

Therefore again a warning has to be issued to the manufacturers of alumina oxide and silicate proppants that following the late 1987 price rise and the recovering US $ exchange rate in early 1988, another company-induced price increase could bring the zirconia-silicate proppants back into their former competitive position which would certainly destroy much of the above sketched marketing concepts, with the probability being very high that in this case, the zirconia-silicate proppants would again achieve at least 20 - 30 % market share as compared to their almost insignificance at the moment comprising not more than 5 - 10 % of the whole proppant business (cf. section 2.2.2.3.2.). The warning not to perform another artificial price rise has to be examined particularly with respect to possible automatical price increases for the European customers by a further regaining US $ exchange rate to which the proppant companies have absolutely no influence at all.

2.4. Areal marketing assessment of hydraulic proppant fracturing

Hydraulic proppant fracturing activity in Europe is of different significance in the western and eastern part of the area due to reservoir engineering, technological and economical reasons. Areal marketing evaluations are given as follows for both Western Europe and Eastern Europe for the periods 1977 - 1985, 1986 (including partially the first half of 1987), and 1987 (comprising mainly the second half) - 1990 (and partially even beyond), thus containing review, status, forecast and perspectives. The first period comprises the range before the major 1986 oil price crash and covers the time from the onset of availability of synthetic high-quality proppants through the golden years of hydrocarbon stimulation to the market collapse. The second interval includes the immediate oil price drop range and the period of reduced and/or retarded activity shortly after the market turning. The third section finally comprises the beginning recovery from the oil price crash and the following years which are more than ever before characterized by realistic project assessment without exaggerated optimism, but on the other hand certainly also not based on excessive pessimism. As a result of the difficulty to prognose and to plan in exact time frames due to many technical and economical uncertainties, the latter period is partially even farther extending into the future as a consequence of some delays and postponements as well as appearance of follow-up projects due to favourable impact of earlier jobs and campaigns.

2.4.0. General aspects

Emphasis is in the present review, status report and prediction also put on economical, commercial, political and even ideological aspects of hydraulic proppant fracturing especially for Eastern Europe. As far as more or less regionally or temporally linked with hydraulic proppant fracturing, comments are also offered on gravel packing which in terms of proppant consumption has generally only a small share of the total proppant quantity delivered to hydrocarbon industry and is in many areas with geologically older and more lithified sandstones even almost of negligible nature in comparison to hydraulic fracturing, whereas other regions with geologically younger and less consolidated sands and chalks are characterized by proppant utilization predominantly for gravel packing with only a subordinate quantity going to hydraulic fracturing (cf. chapter 5). References on general and specific, regional and stratigraphical, reservoir engineering and technical aspects of hydraulic proppant fracturing are compiled in chapter 7, and further marketing details are collected in MADER (1987). An overview map of the centres of activity in Europe is presented in fig. 2, and a summary of characteristic features is exhibited in tab. 5.

The European proppant market can be split into centres of activity of massive hydraulic fracturing (MHF), smaller-scale fracturing and gravel packing (cf. fig. 2). Massive hydraulic fracturing activity in past and future concentrates on British and Dutch Southern North Sea, Netherlands onshore, Germany FRG onshore, and Yugoslavia. Smaller-scale fracturing takes place in the areas of MHF execution and in addition to these in the British and Norwegian Northern North Sea, Germany GDR, Poland, Soviet Union, Czechoslovakia, Hungary, Austria, France, Italy, Bulgaria, Romania, Turkey, Tunisia and Algeria. Gravel packing focusses on Norwegian and British Northern North Sea, Germany FRG, Austria, Czechoslovakia, Italian and Yugoslavian Adriatic Sea, and Soviet Union (cf. figs. 4 - 8). The North Sea is by far the area with highest stimulation potential in past and future and is a mature oil and gas province, with current developments representing already the third generation (THOMAS 1986).

2.4.1. Western Europe 1977 - 1985

When the first intermediate- and high-strength synthetic proppants were introduced into the stimulation market abt. 10 years ago (DOWELL SCHLUMBERGER 1985, WESTERN PETROLEUM 1985; cf. section 1.2.4.), especially in the USA and parts of Western Europe many wells were waiting on hydraulic fracturing that indispensibly required synthetic high-quality proppants which could withstand high closure stresses in deep tight gas reservoirs (cf. section 2.2.2.1.3.). As before the invention of adequate proppants the treatments could frequently not be carried out due to the fact that sand and other traditional materials were not stable enough to prevent crushing, the appearance of intermediate- and high-strength proppants (the first commercial field application of sintered bauxite for propping of stimulated formations took place in 1976; COOKE 1976, 1977; ATTEBERRY, TUCKER & RITZ 1979; HICKEY, BROWN & CRITTENDEN 1981) initiated a boom of hydraulic fracturing which became the golden years of stimulation in the late 1970`s and particularly the early 1980`s when the economical feasibility of the treatments was even enhanced by further rising oil prices and especially in the latter period in Europe backed- up by a high US $ exchange rate (cf. section 2.2.1.3.). The most important areas of fracturing performance are Germany FRG and British Southern North Sea (cf. fig. 6 as well as tabs. 6 and 7) which are discussed in the sections below.

2.4.1.1. Germany FRG

In Western Europe, the main targets of MHF stimulation in the period 1977 - 1985 have been deep tight Rotliegend (cf. chapter 3) and Carboniferous gas-bearing sandstones in Germany FRG onshore, Netherlands onshore and offshore, and British Southern North Sea. In some cases, fracture operations have also been performed in Buntsandstein gas-bearing sandstones in the mentioned areas (sedimentological and diagenetical aspects of Buntsandstein and Rotliegend deposits as a base for reservoir modelling are discussed in PLEIN 1978, LÜTZNER 1981, ROBINSON 1981; DRONG, PLEIN, SANNEMANN, SCHUEPBACH & ZIMDARS 1982; GLENNIE 1983, 1986; LEE 1984; MADER 1981, 1982, 1983, 1984, 1985, 1988 b; MADER & TEYSSEN 1985, MADER & YARDLEY 1985; BIFANI, GEORGE & LEVER 1987; GRALLA 1988, MADER & CHATALOV 1988, MARTIN & EVANS 1988 and RICHARDSON, SANGREE & SNEIDER 1988; cf. section 3.2.1.1.). Some aspects of deep gas and shallow oil fracturing are outlined as follows.

2.4.1.1.1. Deep gas fracturing

The most important fields where MHF stimulation treatments of Rotliegend and Carboniferous deep tight gas reservoirs have been carried out in Germany FRG are in the East Hannover area (field names are presented in alphabetical order in each region) Hamwiede and Wittorf in the Carboniferous and Ebstorf, Munsterlager, Preyersmühle-Hastedt and Söhlingen in the Rotliegend; in the South Hannover area Alfeld-Elze in the Rotliegend and Hildesheimer Wald in the Carboniferous; in the South Oldenburg area Cappeln, Goldenstedt, Syke, Uchte, Varnhorn and Wietingsmoor in the Carboniferous (the Rotliegend is in the latter area not present or not developed in reservoir facies); in the West Emsland area Rütenbrock in the Rotliegend and Emlichheim, Fehndorf, Frenswegen and Rütenbrock in the Carboniferous (accompanied by Coevorden on the immediately neighbouring Dutch territory on the other side of the state boundary); and in the East Friesland area Hamburger Sand, Leer and Leybucht in the Rotliegend and Uphuser Meer in the Carboniferous (cf. maps and explanations in BRINKMANN 1982, JOHN 1983; REINICKE, BRINKMANN, SCHWARZ & HUENI 1985; SCHRÖDER 1986; cf. fig. 4 and tab. 6). Deep gas fracturing in Rotliegend and Carboniferous tight sandstones in Germany FRG is illustrated along the lines of reservoir depth and proppant quantities, fracture length and stimulation ratio, and cost and temporal execution of treatments as follows. The technical and economical fracturing potential in the period 1985 - 1989 is also briefly outlined.

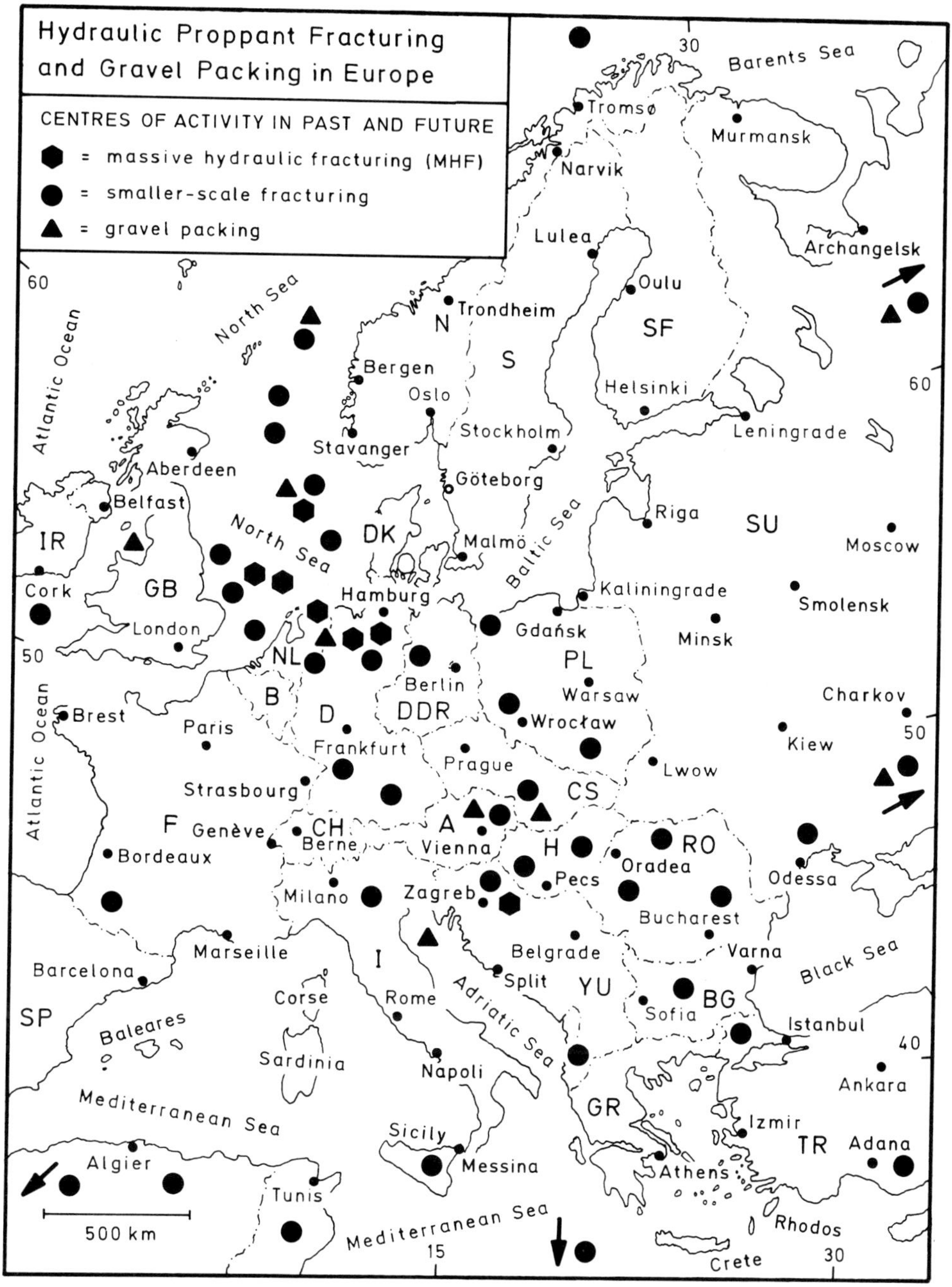

Fig. 2

Overview map of past, present and future hydraulic proppant fracturing and gravel packing potential in Western and Eastern Europe. Distinction can be made in the proppant stimulation market between massive hydraulic fracturing, smaller-scale fracturing, and gravel packing.

2.4.1.1.1.1. Reservoir depth and proppant quantities

Comments on reservoir depth and proppant quantities in hydraulic fracturing treatments in Germany FRG are given for the periods 1975 - 1978 when high-strength synthetic proppants were introduced to the hydrocarbon stimulation market, and 1978 - 1985 comprising the golden years of hydraulic proppant fracturing in both Europe and USA.

2.4.1.1.1.1.1. 1975 - 1978

The demand/supply situation in the European gas market led to an extensive effort to develop tight gas reservoirs in Germany FRG starting in 1977 (BRINKMANN 1982) with some precursor treatments beginning in 1975. Massive hydraulic fracturing in Germany FRG was first applied in 1975 in the well Alfeld-Elze Z 4 (Rotliegend) and 1976 in the well Hildesheimer Wald Z 1 (Carboniferous) where as a consequence of still unavailability of high-strength proppants abt. 80 - 150 t of natural sand have been inserted into the cracks (ERDOEL-ERDGAS AKTUELL 1975, 1976 a). Smaller-scale sand fracturing has already been carried out earlier, with success, however, having been limited by rapid closing of the cracks due to sand crushing in the high-pressure regime of the deep gas reservoirs (TUNN 1971).

In 1978, both sintered bauxite and zirconia-silicate proppants (cf. section 1.3.) were first used in massive hydraulic fracturing operations in Germany FRG when abt. 70 t of sintered bauxite and abt. 140 t of zirconia-silicate proppants were pumped into two different Carboniferous reservoir storeys in the well Hamwiede Z 2 (ERDOEL-ERDGAS AKTUELL 1978 a). The first successful application of high-strength proppants opened suitable perspectives for the economically feasible acquisition of Rotliegend and Carboniferous tight gas reserves, and although in the transitional phase occasionally still natural sand was used (well Cappeln Z 4 a was treated with 250 t of sand in late 1978; ERDOEL-ERDGAS AKTUELL 1979 a), synthetic high-strength proppants were abundantly to exclusively utilized in progressively increasing quantities.

2.4.1.1.1.1.2. 1978 - 1985

The biggest treatments and the largest number of massive hydraulic fracturing jobs in Europe during the earlier part of the period 1977 - 1985 have been carried out in Germany FRG (BRINKMANN, FUHRBERG & SCHÖBER 1980; BRINKMANN, KRÖMER & REINICKE 1980; BRINKMANN 1982, GUNDERMANN 1982, SCHWARZ & SCHÖBER 1982; JOHN 1983, 1987; KLOSE & KRÖMER 1983, BLEAKLEY 1984, RIECKMANN & PUSCH 1984, LEICHT 1985; REINICKE, BRINKMANN, SCHWARZ & HUENI 1985; SOMMER 1987, 1988) where the Rotliegend (cf. chapter 3) and Carboniferous reservoir rocks (cf. fig. 3) are in depths between 3,000 and 6,000 m (equivalent to between 9,000 and 18,000 ft) with bottom hole temperatures between 120 and 180 oC (PHILIPP & REINICKE 1982, JOHN 1983, KLOSE & KRÖMER 1983) and thus in almost all the cases the whole treatment had to be made with intermediate- or high-strength proppants, and only in a few exceptions also sand had been used (cf. section 3.3.2.). Details of Rotliegend and Carboniferous fracture operations are given by KLOSE & KRÖMER (1983); REINICKE, BRINKMANN, SCHWARZ & HUENI (1985) and SOMMER (1987), and aspects of individual massive hydraulic fracture operations are also reported by ERDOEL-ERDGAS AKTUELL (1977, 1980 b) and SCHRÖDER (1980).

As such high reservoir depths of productive sandstones and consequently also closure stresses are nowhere else reached in the Rotliegend and Carboniferous gas-field belt from Great Britain to Poland, and also a similar concentration of prospective patches with hydraulic fracturing potential is not found elsewhere, Germany FRG was for many years the biggest European market for intermediate- and high-strength proppants. Several dozens of large-scale fracture jobs have been carried out during this time in Northwest Germany since 1976/1977 (a couple of operations per year) in the East Hannover, South Oldenburg, West Emsland and East Friesland areas (ALBERTSEN 1985) between the bounding gas fields Groningen/The Netherlands (STÄUBLE & MILIUS 1970, ROSSUM 1975, BEEK & TROOST 1978) in the west and Salzwedel/Germany GDR in the east (JOHN 1983; cf. fig. 4) where the treated wells occupy a broad seam representing a zone of high fracturing potential also for further wells in the future. Proppant quantities consumed in the MHF treatments exceeded in most cases 100 t, with in at least one third of the operations comprising more than 200 t and in the biggest jobs even ranging between 400 and 650 t.

2.4.1.1.1.2. Fracture length and stimulation ratio

Fracture length (as evaluated with actual treatment data) varied between 100 and 550 m, and propped fracture height ranged within the limits of 10 m (thin Carboniferous sandstones) and 115 m (thick Rotliegend sandstones; KLOSE & KRÖMER 1983; REINICKE, BRINKMANN, SCHWARZ & HUENI 1985). Various wells have been fractured in several vertically successive horizons that are separated by thick mudstone intervals from each other, and have been fractured individually and then commingled for pressure build-up analysis and partially also production (REINICKE, BRINKMANN, SCHWARZ & HUENI 1985; with commingling here implying interhorizon communication only through the wellbore and not via the fracture in these multilayer/multifracture systems; BENNETT, REYNOLDS & RAGHAVAN 1982 a, 1982 b; cf. sections 4.2.1.1. and 4.10.1.). While most of the fracturing operations have been great successes resulting in stimulation ratios of 8 - 10, other treatments have been less effective and only gave rise to improvement ratios of 3 - 6, and also several jobs have turned out to be failures mainly due to high water saturation (cf. section 3.12.1.) and soft reservoir rocks (BRINKMANN 1982), with negative items including both premature screenout termination and complete failure, and borderline cases have also come out (SLUSSER & RIECKMANN 1976).

2.4.1.1.1.3. Cost and temporal execution of treatments

Total fracturing expenses in the deep tight formations reached up to 6 Mio. DM or 2.5 Mio. US $ per job (JOHN 1983, KLOSE & KRÖMER 1983, BLEAKLEY 1984, LEICHT 1985). The big Söhlingen Rotliegend gas field has become the highlight of hydraulic proppant fracturing in Germany FRG particularly due to size and success of the executed jobs, but comparably voluminous and beneficial treatments have also been performed in many other fields. Late 1985 has seen some of the largest operations that were ever done in Europe which, however, have so far been also the last jobs in Germany FRG. As a consequence of the serious oil price decline in early 1986 (cf. section 2.2.1.1.2.; that had an even worse effect due to the accompanying US $ exchange rate drop; cf. section 2.2.1.3.), no hydraulic proppant fracturing operation was performed in Germany FRG in 1986 at all, and the same subsequently applied for 1987 and 1988 when in spite of the slight to moderate market recovery only a few small- to mini-scale treatments were carried out in a time of uncertainty of oil price evolution coupled with extremely weak and unreliable US $ (cf. section 2.2.1.1.3.).

The termination of the MHF activity in Germany FRG is the consequence of seriously cutting back of the drilling budget (cf. section 2.2.1.5.2.2.) that comprises in 1987 and 1988 probably not more than one third and/or one half, respectively, of both number of items and total meterage of the niveau achieved in 1985 before the beginning of the oil price drop. The activity level in 1986 is somewhat artificial and disturbed by various influences, as after the worst final stage of the oil price fall in early 1986, all the companies tried to delete or streamline as much objects as possible, but that was no longer able for such items where the drilling contract had already been awarded, the drilling site had been prepared or the well had even already been spudded. Containment of capital expenditure especially affects predominantly development drilling and completion, whereas exploration and appraisal drilling continued to run at comparatively higher levels in 1986 and early 1987, but were in late 1987 and 1988 also progressively reduced to a minimum as a consequence of persisting oil price and US $ weakness. Appraisal and development drilling is currently only performed at a level that is sufficient to secure gas delivery contract obligations in the near future, and to increase proven reserves in order to allow more profitable gas marketing and thus business consolidation.

2.4.1.1.1.4. Technical and economical fracturing potential in 1985 - 1989

While at least some marginal MHF candidates are still available, the only perspective for again a reasonable stimulation activity in Germany FRG comparable to that prior to 1986 would be an oil price of at least US $ 20 with a sufficiently stable average value and an again considerably stronger US $ than at the moment, with the boundary probably being an exchange rate of at least 2 DM. Unfortunately, however, even in mid to late 1988 there is not yet any hope for either amelioration of the economical feasibility of renewed MHF treatments in Germany FRG, and there is despite a slight improvement of the US $ exchange rate evolution in the last third of 1988 almost no hope that the boundary of 2 DM is going to be crossed in the foreseeable future, because at the end of 1988 the US $ was already sliding down again (cf. section 2.2.1.3.).

Therefore since the last jobs having been carried out in late 1985, no massive hydraulic fracturing treatment has been performed for more than three years in the period 1986 - 1988, and according to the currently still depressed oil price and US $ exchange rate situation, it looks like as if also in 1989 no stimulation will take place. While in most of the MHF treatments carried out so far in Germany FRG conventional water- or oil-based gels of various viscosities have been used as fracturing fluids, recently also CO2-foam has been taken as agent transporting and placing the proppants being a pilot application in Europe (SLATTER, RUCKER & CRISP 1986; SOMMER 1987, 1988; cf. section 3.11.). Laboratory testing has revealed that the Carboniferous reservoir rocks exhibit a high degree of plasticity (SLUSSER & RIECKMANN 1976; cf. section 4.2.3.2.3.).

2.4.1.1.2. Shallow oil fracturing

While shallow oil fracturing has been hitherto performed in various European countries almost exclusively with natural sand as propping agent, some reservoirs exist where in terms of adequate closure stress resistivity intermediate-strength synthetic proppants are required.

2.4.1.1.2.1. Natural sand

From the suite of small fracture jobs carried out with abt. 20 - 70 t of sand, several campaigns in Cretaceous shallow sandstone oil reservoirs in the Emsland area in the boundary belt between Netherlands and Germany FRG deserve particular mentioning. In the Tertiary of the Vienna Basin in Austria, sometimes even occasionally very small treatment sizes of abt. 6 - 12 t can be the optimum (DOLAK & PERNTHANER 1976, MURER 1981) and are successful both in new wells and in old wells producing under low formation stress and with good isolation against neighbouring intervals. In the Lower Cretaceous Bentheimer Sandstein (Valanginian) in the Emsland area, numerous fracture treatments as well as varying amounts of gravel pack operations have been carried out on both sides of the state boundary in several fields comprising mainly Rühlermoor and Georgsdorf on the German side and Schoonebeek on the Dutch side (GESEMANN & BADURA 1976; for location maps and field cross sections cf. BOIGK 1980; cf. fig.

Fig. 3

Schematical geological cross-section (without scale) through the gas-bearing reservoir column in Northwest Europe (Germany FRG and adjoining areas). Tight gas sandstones are predominantly Rotliegend and Carboniferous, whereas the Buntsandstein requires only in some regions artificial transmissibility enhancement by stimulation. While Carboniferous, Rotliegend and Buntsandstein pays contain exclusively sweet gas (in the Rotliegend, apart from combustible methane occasionally also inert nitrogen occurs in larger shares), the Zechstein carbonates include sour gas with considerable amounts of the aggressive and toxic hydrogen sulphide. In many cases, the individual storeys of the reservoir pile are characterized by different gas-water-contact levels. Rotliegend anticlines and horst-graben-structures are especially favourably developed below Zechstein salt diapirs, because the thick salt plugs resulted in retarded heat flux and lower reservoir temperature than in the surroundings of the salt domes. The Carboniferous contains not only sandstone reservoirs, but also bears numerous coal seams where the gas for all the four pay horizons originated during course of coalification and subsequently migrated through the cover strata along faults and fissures into the various traps. The thickest reservoirs with the best petrophysical development in the Rotliegend occur in the lower and upper parts of the section, whereas the middle portion of the sequence is in most cases characterized by low-permeability sandstones which require massive hydraulic fracturing stimulation in order to achieve economically feasible gas production rates. The tight Carboniferous needs in most cases fracturing treatments for allowing profitable gas withdrawal, while in the Buntsandstein, hydraulic stimulation has only occasionally to be performed in some areas of the red bed basin.

Legend : 1 = gas-bearing sandstone, 2 = water-containing sandstone, 3 = coal seam, 4 = carbonates, 5 = mudstones and marls; 6 = Zechstein salt layers, cushions and plugs; 7 = gas-water-contact, 8 = tight gas-bearing sandstone reservoirs with various hydraulic proppant fracturing potential, 9 = unconformity.

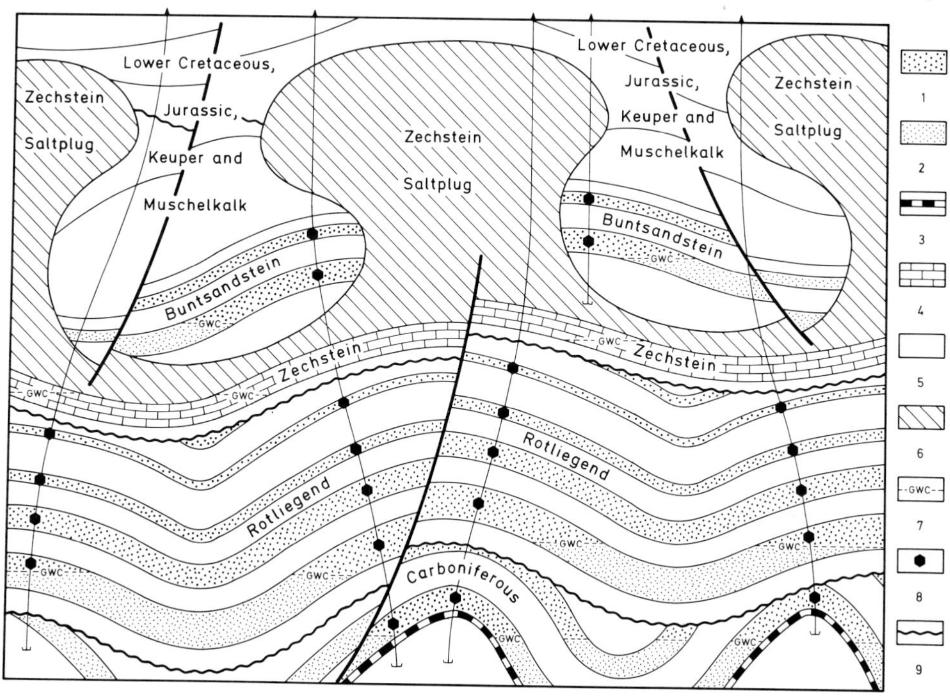

Fig. 5

Overview map of centres of hydraulic proppant fracturing and gravel packing in past, present and future in Western and Eastern Europe. For more details in Germany FRG, Netherlands and North Sea cf. figs. 4 and 6, and for general outline cf. also fig. 2.

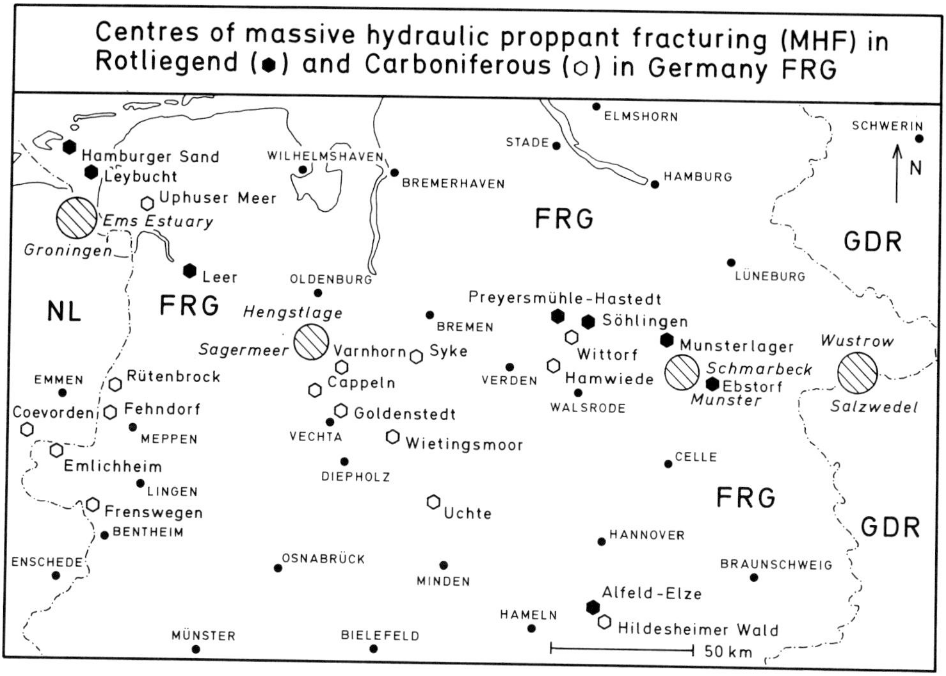

Centres of massive hydraulic proppant fracturing (MHF) in Rotliegend (⬢) and Carboniferous (⬡) in Germany FRG

Fig. 4 (above) and Fig. 6 (below)

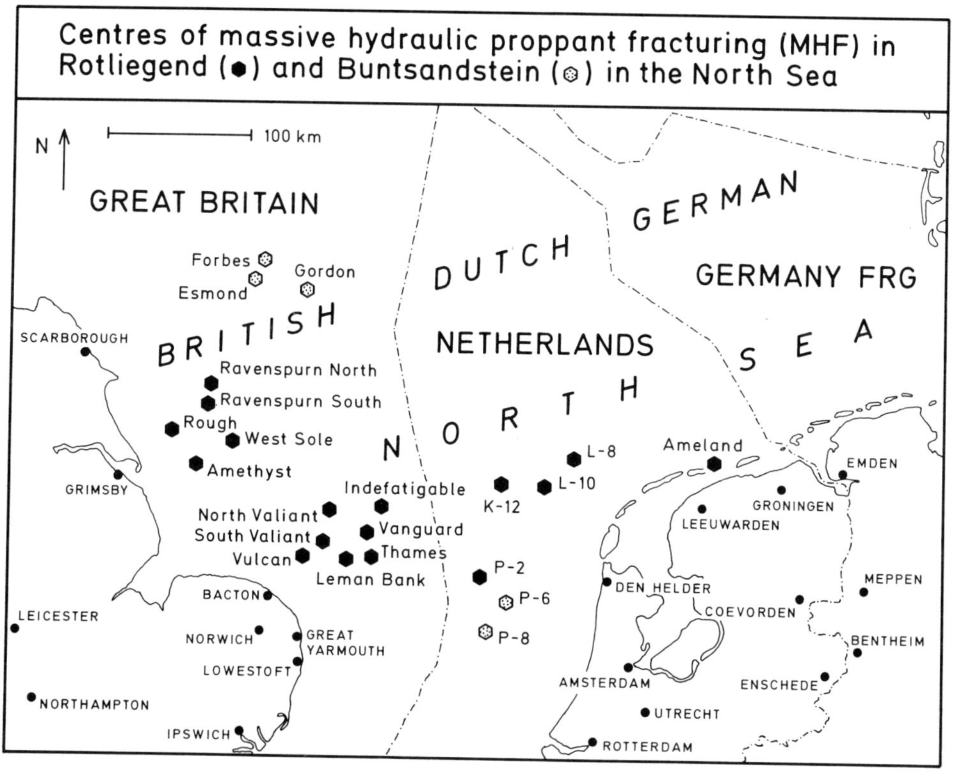

Centres of massive hydraulic proppant fracturing (MHF) in Rotliegend (⬢) and Buntsandstein (⯃) in the North Sea

5 and tab. 6).

Sand fracturing was also common since the early days of oil exploration and production shortly after the Second World War in many parts of the oil field belt in Northwest Germany and to subordinate amounts in the last 20 years also in Southern Germany, with some representative fields being in Northwest Germany Adorf, Emlichheim, Meerdorf and Scheerhorn, and in Southern Germany Hauerz, Mönchsrot and Pfullendorf-Ostrach. To minor amounts, sand fracturing was also done in Tertiary sandstones in the Vienna Basin in Austria (for example in the fields Matzen, Hochleiten, Neusiedel, Stockerau, Tratnach and Zistersdorf; DOLAK & PERNTHANER 1976, MURER 1981) as well as in the Po Valley and the Adriatic Sea in Italy (cf. fig. 5 and tab. 8).

2.4.1.1.2.2. Intermediate-strength synthetic proppants

The low closure stresses in the shallow reservoir depth of abt. 700 - 2,500 m (2,100 - 7,500 ft) did so far not require the application of intermediate-strength synthetic proppants in oil-well fracturing in Germany FRG and other countries. The few deeper wells in Germany FRG where sand would not have been an adequate propping agent were suspended for the time being due to insufficient reserves to economically justify stimulation with expensive proppants in the period of both high oil and proppant prices in the early 1980`s. After the decline of both oil and proppant prices, however, the jobs are in the late 1980`s still far from being feasible, even in light of only small quantities of sand or proppants being needed in oil-well fracturing that do not have such a large share of the total cost of the stimulation treatment in comparison to deep gas wells.

2.4.1.2. British Southern North Sea

In the British Southern North Sea, Rotliegend and partially also Carboniferous reservoir rock depth is much shallower and closure stress consequently is considerably lower than in Germany FRG and Netherlands (ALLISON 1986; cf. section 3.3.), and thus many fracture treatments can be carried out at least partially still with sand in contrast to the other area where operations with almost exclusively intermediate- and/or high-strength proppants are inevitable for successful long- term achievement of the goal of the stimulation job. Following discussion of some general aspects, various points of moderately deep gas fracturing, tail-in fracturing policy, differences between offshore and onshore fracturing, and moderately deep oil fracturing are outlined.

2.4.1.2.1. General aspects

Due to a time lag in exploration and development drilling and completion with respect to the onshore areas, hydraulic proppant fracturing in the British and Dutch Southern North Sea predominantly took place in the later part of the golden stimulation period (cf. section 3.3.), with this being one of the reasons why especially the British Southern North Sea is still a very proliferic area for stimulation in the coming years despite the deteriorated general economical situation as a consequence of much lower oil prices than before (SCHÖNEICH 1983, STEPHENS & MARTINS 1985; cf. section 2.2.1.1.2.) and the additional impact of the considerably low US $ exchange rate since abt. early 1986 (cf. section 2.2.1.3.). An outline of the petroleum potential of the North Sea is also given by HOYLE (1983), and NYSTAD (1981) provides and economical analysis of the North Sea oil and gas region.

Another major reason for the indispensible operation of some large stimulation campaigns in the coming years are already signed gas delivery contracts (that guarantee the sale of the gas which still has to be acquired by field development; OILMAN 1987 a; cf. section 2.2.1.6.2.) which necessitate to fulfill the entered commitments almost regardless of the economical and political framework (certainly backed-up, however, by at least partially still good prices having been fixed in the agreements, as well as by government protection to keep gas imports of considerations; cf. section 2.2.1.6.3.). The Southern North Sea is the traditional centre of gas production in Great Britain and has historically contributed up to abt. 90 % of the total gas supply in the United Kingdom, and it is expected that the Southern North Sea will continue to play its dominant role until the end of the century, because it still holds over 50 % of all United Kingdom proven and probable gas reserves.

The accentuated position of the British Southern North Sea in terms of stimulation potential and hydrocarbon prospectivity within the whole North Sea in comparison to the sectors of other countries is highlighted by the predominant share of this area in the record drilling rig count in 1984/1985 (OILMAN 1985). Economical evaluations of the British Southern North Sea Rotliegend gas belt are also given by FISHMAN (1986), GREGORY (1986), THOMAS (1986), ALGAR (1987 a), BAND (1987) and POTTER (1987). Successful stimulation practices in the Dutch Southern North Sea are summarized by COULTER & PURVIS (1980). Development details for both offshore provinces until the year 2000 are given by OFFSHORE ENGINEER (1985 c, 1986 c). Apart from production wells, some fracture jobs have in this area also been carried out in underground gas storage wells (cf. HOLLIS 1982).

2.4.1.2.2. Moderately deep gas fracturing

Proppant quantities between 100 and 250 t are most common for hydraulic fracturing of Rotliegend gas-bearing sandstones in the British Southern North Sea, with larger jobs requiring 250 - 400 t of propping material being rare exceptions. Some aspects of hydraulic proppant fracturing in Carboniferous and Rotliegend are outlined as follows.

2.4.1.2.2.1. Carboniferous

The Carboniferous has so far in the British Southern North Sea only in a few cases been encountered gas-bearing, as most of the deeper reservoirs are below the gas-water-contact which is similarly as formation depth rising from Germany FRG in the East to the British Southern North Sea in the West (some more Carboniferous discoveries are expected during progressing exploration; OILMAN 1987 e, VIELVOYE 1987 b, CAPEL 1988, PETROLEUM REVIEW 1988 d). In addition, the Carboniferous underlying the Rotliegend (cf. fig. 3) has in many cases been almost totally ignored so far, because the majority of exploration and development wells have been drilled to a target depth just penetrating the top of the Carboniferous (OILMAN 1987 e) as a consequence of the comparatively poor reservoir facies and the high gas-water-contact in this storey. The Buntsandstein has in some fields in the British and especially the Dutch Southern North Sea also repeatedly been fractured.

2.4.1.2.2.2. Rotliegend

Reservoir depth of the Rotliegend sandstones is in most cases abt. 8,000 - 9,000 ft (corresponding to abt. 2,700 - 3,000 m), and only in some patches abt. 6,000 ft (abt. 2,000 m) for the shallower range and abt. 10,000 ft (abt. 3,300 m) for the deeper range (cf. figs. 3 and 6). Summaries of North Sea well stimulation are given by LAFLEUR & JOHNSON (1973) and MARTINS & HARPER (1985). The most important fields where MHF stimulation operations of tight gas reservoirs have already been carried out and/or are still going to be performed are in the British North Sea in the Rotliegend (field names are stated in alphabetical order) Amethyst, Indefatigable, Leman Bank, Ravenspurn, Rough, Thames, Valiant,

Tab. 5

Centres of past and future activity of hydraulic proppant fracturing in Europe

CENTRES OF PAST AND FUTURE ACTIVITY OF HYDRAULIC PROPPANT FRACTURING IN EUROPE	PROPPANT TYPE			HYDROCARBON TYPE			RESERVOIR TYPE				RESERVOIR FORMATION								RESERVOIR LOCATION
	Natural sand	Intermediate-strength proppants	High-strength proppants	Oil	Gas	Condensate	Sandstone	Carbonate	Chalk	Other	Carboniferous	Rotliegend	Buntsandstein	Other Triassic	Jurassic	Cretaceous	Tertiary	Other	Representative hydrocarbon fields
Albania	•	•		•	•		•	•									●		?
Austria	●	•	•	●	•		●	•		•				•			●	•	Matzen, Stockerau, Tratnach, Zistersdorf
Belgium																			
Bulgaria	•	•		•	•		•	•						•	•		•		Badarski, Lukuvit, Tarnak
Czechoslovakia	●	•		●	•		●	•									●		?
Denmark	•	•		•	•		•		•							•	•		Dan, Gorm
Finland																			
France	•	•		•	•		•	•							•	•	•		Lacq, Cerneux, Ville Perdue
Germany FRG	●	●	●	●	●		●		•		●	●	•	•	•	●	•		Söhlingen, Goldenstedt, Hamwiede, Rühle
Germany GDR	•	•			•		•					•							Salzwedel-Peckensen, Winkelstedt, Wenze
Greece																			
Hungary	●	•		•	●	•	●	●									●		Barcs, Belezna, Inke, Mako, Nagykanisza
Ireland	•	•		•	•		•								•	•			?
Italy	•	•	•	•	•		•	●						•			●		Bronte, Gagliano
Netherlands	●	●	•	•	●		●				•	●	•			•			Ameland, Coevorden, B-12, D-5, L-8/10, P-6/12
Norway	•	●		●	•		●		●						•	●	•		Valhall, Haltenbanken, Brage, Ekofisk, Hod
Poland	•	•			•		•					•					•		Międzyzdroje, Uciechów-Bogdaj, Żuchlów, Buk
Portugal																			
Romania	•	•		●	●		●	•								•	●		?
Spain																			
Sweden																			
Switzerland																			
Turkey	●	●		•	•		●	•									●		?
United Kingdom	●	●	•	•	●		●	•	•		•	●	•		•	•			Indefatigable, Ravenspurn, Vulcan, Valiant
USSR	●	●	•	●	●	•	●	•		•	•	•			•	•	•	•	Krasnodarskoe, Orenburgskoe, Astrahanskoe
Yugoslavia	●	●	●	●	●	●	●	•		•							●	•	Molve, Kalinovac, Stari Gradec, Oresac

Fig. 7

Various possibilities of proppant distribution types within hydraulic fractures exist including tail-in, lead-in, banking and pillaring of proppants in order to optimize flow conditions and closure stress resistivity. Legend : 1 = reservoir sandstone surrounding the hydraulic fracture, 2 = bounding mudstones, 3 = fine-grained proppants (40/60 or 50/70), 4 = medium-grained proppants (20/40), 5 = coarse-grained proppants (16/20, 12/20 or 8/12), 5 = boundary of the hydraulic fracture wing, 7 = perforations. Schematically, no scale.

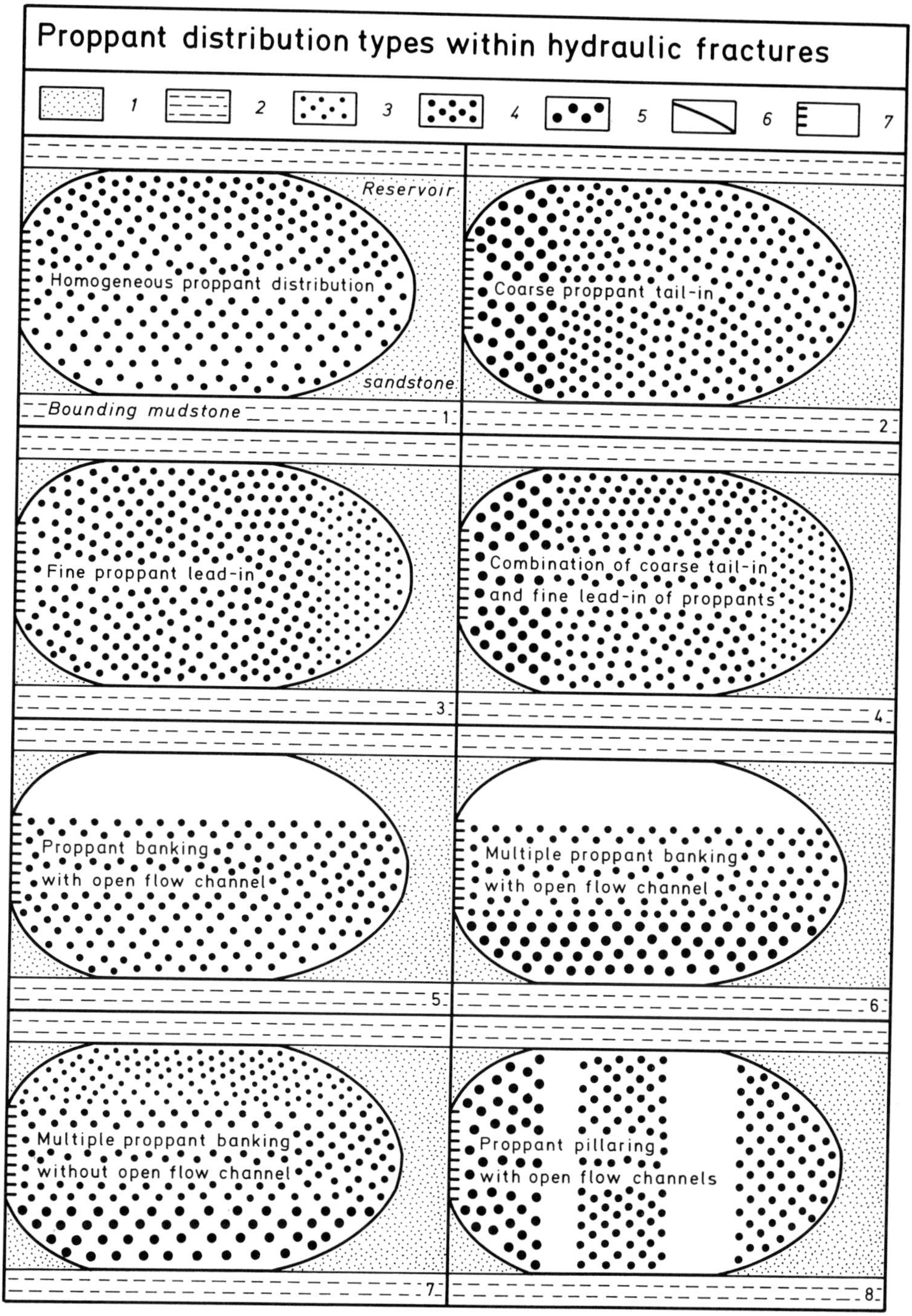
Proppant distribution types within hydraulic fractures
1
2
3
4
5
6
7
Reservoir
Homogeneous proppant distribution
sandstone
Bounding mudstone
1
Coarse proppant tail-in
2
Fine proppant lead-in
3
Combination of coarse tail-in and fine lead-in of proppants
4
Proppant banking with open flow channel
5
Multiple proppant banking with open flow channel
6
Multiple proppant banking without open flow channel
7
Proppant pillaring with open flow channels
8

Fig. 7

Vanguard, Vulcan and West Sole, and in the Buntsandstein Esmond, Forbes and Gordon (cf. figs. 6 and 8 as well as tab. 7). In the Dutch North Sea, emphasis of the activity is in the Rotliegend in Ameland, K-12, L-8, L-10 and P-2, whereas Buntsandstein stimulation focusses on P-6 and P-8 (cf. figs. 6 and 8 as well as tab. 8).

While in the past fracture jobs have been predominantly done with either only sand or exclusively intermediate- to high-strength proppants, in the last years the concept of tailing-in (HOWARD & FAST 1970 a, HOWELL & THOMAS 1978; HICKEY, BROWN & CRITTENDEN 1981) has been more and more applied, with this method commonly including first pumping of 100 - 200 t of sand of 20/40 grain size followed by a tail-in of 25 - 75 t of intermediate-strength proppants of 20/40 or 16/20 grain size. Subordinate applications are three-fold divided jobs with successive pumping of 20/40, 16/20 and finally 12/20 grain size of propping material, or in rarer cases also successions of injection of 40/60, 20/40 and 16/20 or 12/20 proppant grain sizes.

2.4.1.2.3. Tail-in fracturing policy

Tailing-in can comprise sequences of different proppant types and/or grain sizes, with the best results being achieved when varying both parameters (comparative effectivity studies of various proppant successions and mixtures are carried out by McDANIEL & WILLINGHAM 1978; HICKEY, BROWN & CRITTENDEN 1981). The outline as follows includes aims and performance of proppant tail-in schedules, principles of tail-in proppant staging, proppant cost containment (being the major tail-in advantage), proppant mixing and settling (being the most serious tail-in drawbacks), proppant flowback prevention, and tail-in optimization and effectivity. The illustration of the proppant tail-in policy is included into this chapter due to the predominant application of tail-in operations in Rotliegend fracture stimulation of tight gas reservoir sands in the British Southern North Sea.

2.4.1.2.3.1. Aims and performance of proppant tail-in schedules

Tailing-in usually includes succession of discrete quantities that are not intermixed (cf. fig. 7); blending of materials (such as formerly done with sand and glass beads or sintered bauxite; cf. section 1.4.11.1.1.) is nowadays no longer considered, as a mixture of different types and/or grain sizes of proppants leads to equalization and homogenization and partially also compensation of favourable properties of the individual components (McDANIEL & WILLINGHAM 1978 and KOHLHAAS 1982 report that mixtures of sand and glass beads have at high stress not any conductivity more than that of sand alone or have even less depending on mixture ratio, grain size and stress regime; cf. chapter 1.4.11.).

The use of higher-quality proppant types and/or larger proppant grain sizes for tail-ins provides maximum conductivity near the wellbore where it is needed most urgently (CALLANAN, CIPOLLA & LEWIS 1983). Determination when to use a tail-in instead of pumping all high-quality proppant is a function of incremental increase in hydrocarbon production vs. added proppant cost. A special application of tail-in schedules is inserting radioactively-coated proppants in the terminal stage for placement near the wellbore in order to enable gamma-ray fracture height monitoring of natural or artificial proppant radioactivity (HART, ENGI & MORRIS 1983; cf. section 6.2.1.2.).

Apart from proppants, also modern water-based fracturing fluids can be tailored to meet viscosity specifications in a wide range of temperature applications by proper use of primary and secondary gelling agents (ATTEBERRY, TUCKER & RITZ 1979; PAI, GARBIS & HALL 1983; LEE & DANESHY 1985; cf. section 4.3.4.), and similarly, tailored-pulse loading can be used for multiple explosive fracturing of boreholes (WARPINSKI, SCHMIDT, COOPER, WALLING & NORTHROP 1979;

HYDRAULIC PROPPANT FRACTURING AND GRAVEL PACKING IN EUROPE

PAST AND FUTURE STIMULATION POTENTIAL IN IMPORTANT HYDROCARBON FIELDS IN SELECTED EUROPEAN COUNTRIES 1

		STIMULATION TYPE		PROPPANT TYPE		HYDROCARBON TYPE		RESERVOIR TYPE				RESERVOIR FORMATION						
		Fracturing	Gravel packing	Sand	Proppants	Oil	Gas	Sandstone	Carbonate	Chalk	Other	Carboniferous	Rotliegend	Buntsandstein	Jurassic	Cretaceous	Tertiary	Other
	Adorf	●	▲	●		●		●								●		
	Alfeld-Elze	⬢			⬢		⬢	⬢					⬢					
	Cappeln	⬢			⬢		⬢	⬢				⬢						
	Ebstorf	⬢			⬢		⬢	⬢					⬢					
	Emlichheim	⬢	▲	▲	⬢	▲	⬢	⬢				⬢				▲		
	Fehndorf	⬢			⬢		⬢	⬢				⬢						
	Frenswegen	⬢		⬢	⬢		⬢	⬢				⬢						
	Georgsdorf	●	▲	●		●		●								●		
	Goldenstedt	⬢			⬢		⬢	⬢				⬢						
GERMANY	Hamburger Sand	⬢			⬢		⬢	⬢					⬢					
	Hamwiede	⬢			⬢		⬢	⬢				⬢						
	Hauerz	●		●		●		●									●	
	Hildesheimer Wald	⬢			⬢		⬢	⬢				⬢						
	Leer	⬢			⬢		⬢	⬢					⬢					
	Leybucht	⬢			⬢		⬢	⬢					⬢					
	Meerdorf	●		●		●		●							●			
	Mönchsrot	●		●		●		●									●	
	Munsterlager	⬢		⬢	⬢		⬢	⬢					⬢					
	Pfullendorf-Ostrach	●		●		●		●										●
FRG	Preyersmühle-Hast.	⬢			⬢		⬢	⬢					⬢					
	Rühle	●	▲	●		●		●								●		
	Rütenbrock	⬢			⬢		⬢	⬢				⬢						
	Scheerhorn	●	▲	●		●		●								●		
	Söhlingen	⬢			⬢		⬢	⬢					⬢					
	Syke	⬢			⬢		⬢	⬢				⬢						
	Uchte	⬢			⬢		⬢	⬢				⬢						
	Uphuser Meer	⬢		⬢	⬢		⬢	⬢				⬢						
	Varnhorn	⬢			⬢		⬢	⬢				⬢						
	Wietingsmoor	⬢			⬢		⬢	⬢				⬢						
	Wittorf	⬢			⬢		⬢	⬢				⬢						

Tab. 6

SCHMIDT, WARPINSKI & COOPER 1980; SWIFT & KUSUBOV 1980, 1981; CUDERMAN 1982, 1986; McHUGH & KEOUGH 1982; KEOUGH, CURRAN & McHUGH 1984; STOLLER 1985; cf. also sections 4.2.2. and 4.8.8.6.). CHARLES, HUDOCK, UDICK & HALL (1983) report tailoring of proppants in order to obtain the desired conductivity ratio.

2.4.1.2.3.2. Principles of tail-in proppant staging

The most common tail-in schedule is pumping a large amount of 20/40 sand or intermediate-strength proppant followed by a terminal little quantity of 16/20 or even 12/20 intermediate- or high-strength proppant comprising abt. 10 - 40 % of the total proppant quantity (PAI, GARBIS & HALL 1983; GARBIS, BROWN & MAURITZ 1985). The tail-in amount of proppants should be in terms of crack length ideally abt. 100 ft in small fractures, while in larger cracks, the tail-in proppant length should at least equal actual fracture height. On the other hand, also the reverse arrangement representing lead-in strategy is possible, with first pumping of a small amount of 20/40 or even finer sand or intermediate-strength proppant as the head of the treatment and then following with the main body of 20/40 or 16/20 intermediate-strength proppants. This configuration can be even completed by a little final tail-in of 16/20 or 12/20 intermediate- or high-strength proppants.

While 20/40 and 40/70 sand are most commonly used as lead-in proppant, SINCLAIR (1980) describes successful application of 100 mesh sand from Millcreek/ Oklahoma which is more frequently used as fluid-loss additive and retains a large percentage of its permeability under load due to its many support points allowed by the fine grain size (cf. sections 4.8.8.3.1.2. and 4.8.8.3.2.3.). Aspects of tail-in proppant scheduling of various both coarser tail or finer head grain size assemblages and successions are also discussed by DAVIS (1975), KUNDERT & SMINK (1979), LINDLEY & McGHEE (1983), McMECHAN & CONWAY (1983), PAI & GARBIS (1983 b) and GARBIS, BROWN & MAURITZ (1985).

Special divided treatments of lead-in or tail-in character comprise pumping of first smaller or larger quantities of light intermediate- or high-strength proppants and second larger or smaller quantities of heavy intermediate- or high-strength proppants. This concept aims on enabling a safer lead-in by the light proppant which has better suspension properties in the fluid and decreases the risk of early screenout before reasonable amounts of proppants could be placed in the distal segment of the fracture, and on achieving a stronger fracture support by a higher-quality material in the proximal section of the crack where the highest closure stress is effective and screenout failure in the later stage of the operation would not have such a terrible effect, because at least already the light proppant has been successfully inserted (this special combined fracturing strategy has been applied with great benefit in Yugoslavia; cf. section 2.4.3.1.).

2.4.1.2.3.3. Proppant cost containment

The main purpose of the tail-in strategy is to cut down stimulation cost by inserting an expensive synthetic proppant only in the immediate vicinity of the wellbore where closure stress has its maximum particularly upon high drawdown of reservoir pressure in the stage of advanced depletion after many years of production history (AGARWAL 1980), whereas in those parts of the fracture wings that are farther away from the immediate surroundings of the borehole, cheap natural sand is often still sufficient to resist to the comparatively low closure stresses in the British Southern North Sea Rotliegend reservoirs in comparison to those in Germany FRG and Netherlands where depth and thus pressure of most of the gas-bearing formations are slightly to considerably higher and thus intermediate- or high-strength proppants are in most cases indispensible for the whole operation (cf. section 3.3.). Minimization of hydraulic fracturing costs can be particularly achieved by selective application of different proppant types as tail-in lots as a consequence of different prices of natural sand, intermediate- and high-strength proppants (PAI & GARBIS 1983; cf. section 1.3.).

During times of high proppant prices, ceramic proppants were at least three to four times as expensive as sand (in the early days of high-strength proppant introduction and application, limitation of production capacities resulted in a price difference of sintered bauxite and sand up to 15 times or even more) and now in a period of low proppant prices, ceramic proppants are still two to four times as expensive as sand (cf. section 2.2.2.). Therefore the tail-in philosophy in the relatively shallow Rotliegend reservoirs in the British Southern North Sea makes sure that in the most critical section around the wellbore where closure of the fracture would deteriorate the effect of the whole operation (cf. section 4.12.4.2.), a proppant adequate to all possible ranges of closure stress originating during long production history guarantees that the drainage path is kept open in the zone which is with advanced depletion and decline of the reservoir pressure subjected to the highest closure stresses (effects of producing time and drawdown are discussed by AGARWAL 1980).

2.4.1.2.3.4. Proppant mixing and settling

On the other hand, the insertion of sand in the more distant tips of the fracture wings is sufficient to withstand to the comparatively low closure stresses and as a consequence of the volume splitting between sand and intermediate-strength proppants, considerable amounts of stimulation cost can be saved especially in large fracturing campaigns affecting up to several dozens of wells in big tight gas fields. The selection of coarser grain sizes in the nearer vicinity of the wellbore in comparison to finer grain sizes in the fracture wing sections being farther away from the borehole is also independently from proppant type aiming on enhancement of conductivity between fracture and formation (cf. section 4.3.5.) in the zone of greatest flow rate at the fracture mouth.

The major drawbacks of the tail-in policy, however, are the possibility of drastical permeability reduction in the transitional zone between different grain sizes in the proppant package due to at least some mixing (cf. section 1.4.11.) with plugging of the pores between coarser grains by finer grains, and the - according to Stokes` law - abt. four times faster settling velocity of abt. twice larger grains which may lead to creation of asymmetrical wedges of coarser material near the borehole (cf. section 4.2.2.6.) that are not in continuous contact with the reservoir and thus a homogeneous strip-like distribution of the tail around the wellbore is not achieved (aspects of settling velocities of proppants are discussed by LOCKETT & AL-HABBOOBY (1973), MIRZA & RICHARDSON (1979), DUNAND & SOUCEMARIANADIN 1985, KIRKBY & ROCKEFELLER 1985, ROODHART 1985; cf. section 4.12.2.).

Another disadvantage of tail-in proppant size distribution within the fractured reservoir is that an accidental unwanted intermixing of various sizes and shapes of proppants (cf. section 1.4.11.) within the transition zone of several pumping stages would lead to increasing packing efficiency of the grains and partially also enhanced inflow of proppant particles into the wellbore (CLARK 1983).

2.4.1.2.3.5. Proppant flowback prevention

Tailing-in of coarser proppants into the proximal part of the fracture near the wellbore is also an effective means for proppant flowback prevention, because the coarser proppant slug acts like a gravel pack (CROW 1977; cf. section 4.12.3.2.). If the grain size difference between tail-in lot and main proppant quantity has been properly selected, the tail-in segment bridges moving smaller proppant grains at the interface between both proppant stages and stops further migration, thereby stabilizing the proppant population in place and securing persistence of high fracture conductivity. The tail-in of coarser grains retaining the smaller particles in the more distal portions of the crack is comparable to stopping movement of formation sand by perforation tunnel plugging with gravel in mechanical sand control (cf. section 5.8.8.).

2.4.1.2.3.6. Tail-in optimization and effectivity

In light of the above discussed drawbacks of the tail-in treatment technique, in view of the boundary stress resistivity of sand and for the purpose of absolutely going on the safe side to have the security that the propping material will not collapse even when subjected to the highest possible reservoir pressure drawdown and thus closure stress increase during long-term hydrocarbon production (cf. section 4.12.4.) from the offshore fields where any remedy operation causes significantly higher expenses than in onshore settings, several companies finally decided to perform in the 1987 - 1990 fracturing campaigns in the British Southern North Sea only a tail-in concept concerning grain sizes of intermediate-strength proppants (mainly starting with a little volume of abt. 25,000 - 100,000 lbs. (abt. 12 - 50 t) of 20/40 followed by the main quantity of abt. 300,000 - 400,000 lbs. (abt. 150 - 200 t) of 16/20, with in some cases even a little final tail of 12/20 material being possible), but not to use sand at all as a propping agent.

Other intermediate-strength proppant treatment schedules are leading-in abt. 25 - 80 t (abt. 50,000 - 160,000 lbs) of 20/40 that is followed by abt. 100 - 200 t (200,000 - 400,000 lbs) of 16/20 or in some cases also 20/40, with the job being terminated by tailing-in abt. 25 - 50 t (50,000 - 100,000 lbs) of 16/20 or even 12/20 material. Apart from getting higher conductivities of the fractures propped with intermediate-strength material with respect to cracks infilled with sand, this design schedule also gives the guarantee of satisfactory fracture performance throughout field life unaffected from any changes of pressure regime and thus the investment will pay off very well.

Concerning tail-in effectivity, LI & ZHU (1986) report poor performance in moderate- to high-depth oil reservoirs in China where high-strength bauxite tail-ins following pumping of sand have not provided higher conductivities than uniform fracture packages. The quantitative share of the tail with respect to the total proppant load should not be selected too low, because this could cause stratification and/or mixture of proppants at the wellbore which are enhanced by permeability and fracture gradient variations (GARBIS, BROWN & MAURITZ 1985). Aspects of Rotliegend gas reservoir fracturing in the British Southern North Sea are also discussed by ACHARYA & KIM (1987), and COULTER & PURVIS (1980) present experience of Rotliegend and Buntsandstein fracturing in partially water-sensitive reservoirs in the Dutch Southern North Sea.

2.4.1.2.4. Differences between offshore and onshore fracturing

Due to the different logistical situation of onshore and offshore projects (cf. also section 3.8.) and resulting from previously more or less fixed field development plans including commitment of selling the gas prior to the start of production drilling (cf. section 2.2.1.6.2.), the oil price drop in early 1986 (in Europe superimposed by the decline of the US $ exchange rate which reached a minimum almost at the same time; cf. section 2.2.1.3.) has only lead to a more or less cut back of the level of stimulation activity (which was for the service and proppant companies negatively accentuated by the coincidence with higher cancellation rates for technical reasons; cf. section 2.4.2.2.) including postponement of only limited amounts of projects, but has in contrast to the situation in parts of onshore Europe (cf. section 2.4.1.1.1.) not given rise to total suspension and thus also quite some jobs have been carried out in

HYDRAULIC PROPPANT FRACTURING AND GRAVEL PACKING IN EUROPE

PAST AND FUTURE STIMULATION POTENTIAL IN IMPORTANT HYDROCARBON FIELDS IN SELECTED EUROPEAN COUNTRIES 2

Country	Field	STIMULATION TYPE: Fracturing	STIMULATION TYPE: Gravel packing	PROPPANT TYPE: Sand	PROPPANT TYPE: Proppants	HYDROCARBON TYPE: Oil	HYDROCARBON TYPE: Gas	RESERVOIR TYPE: Sandstone	RESERVOIR TYPE: Carbonate	RESERVOIR TYPE: Chalk	RESERVOIR TYPE: Other	RESERVOIR FORMATION: Carboniferous	RESERVOIR FORMATION: Rotliegend	RESERVOIR FORMATION: Buntsandstein	RESERVOIR FORMATION: Jurassic	RESERVOIR FORMATION: Cretaceous	RESERVOIR FORMATION: Tertiary	RESERVOIR FORMATION: Other
GREAT BRITAIN	Amethyst	⬢		⬢	⬢		⬢	⬢					⬢					
	Brae	●		●	●	●		●			●				●			
	Esmond	⬢		⬢	⬢		⬢	⬢						⬢				
	Forbes	⬢		⬢	⬢		⬢	⬢						⬢				
	Gordon	⬢		⬢	⬢		⬢	⬢						⬢				
	Heather	●		●	●	●		●							●			
	Hutton	●		●	●	●		●							●			
	Indefatigable	⬢		⬢	⬢		⬢	⬢					⬢					
	Leman Bank	⬢		⬢	⬢		⬢	⬢					⬢					
	Morecambe		▲	▲			▲	▲						▲				
	Montrose	●		●		●		●							●			
	Ravenspurn	⬢			⬢		⬢	⬢					⬢					
	Rough	⬢		⬢	⬢		⬢	⬢					⬢					
	Tartan	●		●		●		●							●			
	Thames	⬢		⬢	⬢		⬢	⬢					⬢					
	Thistle	●		●		●		●							●			
	Valiant	⬢			⬢		⬢	⬢					⬢					
	Vanguard	⬢			⬢		⬢	⬢					⬢					
	Vulcan	⬢			⬢		⬢	⬢					⬢					
	West Sole	⬢		⬢			⬢	⬢					⬢					
NORWAY	Brage	●		●	●	●		●							●			
	Ekofisk, Eldfisk	●	▲	●		●				●						●		
	Gullfaks		▲	▲		▲		▲							▲			
	Haltenbanken	●			●	●	●	●							●			
	Hod	●	▲	●		●				●						●		
	Statfjord		▲	▲		▲		▲							▲			
	Tommeliten	●	▲	●		●				●						●		
	Tor	●	▲	●		●				●						●		
	Troll		▲	▲		▲		▲							▲			
	Valhall	⬢	▲		⬢	⬢				⬢						⬢		

Tab. 7

1986 in the British Southern North Sea Rotliegend gas belt (in contrast to Germany FRG where nothing took place).

1987 has seen already again some major campaigns and particularly 1988 met the schedule to become the beginning of a pronounced boom of hydraulic proppant fracturing in this area (which extends until 1989 and 1990 or even later), although continuing oil price and US $ weakness caused in some fields quite some delay of the original planning and in some cases even included schedule modifications for the purpose of cost containment. Further aspects of different fracture philosophy and strategy in offshore and onshore areas are discussed in section 3.8. Some differences between offshore and onshore fracturing stimulation treatments are also outlined by ROODHART, KUIPER & DAVIES (1986).

2.4.1.2.5. Moderately deep oil fracturing

In addition to moderately deep gas fracturing in the British Southern North Sea, so far also some moderately deep oil fracturing in the British Northern North Sea has been carried out in a few fields (DESPAX, CHARLEZ, CLINKEMAILLE & ECONOMIDES 1987). The treatments have been performed in the Jurassic Brent Sand group (JOHNSON & KROL 1984) and have been executed either with sand or with intermediate- or high-strength alumina oxide and silicate proppants. While hitherto mainly individual wells have been stimulated due to actual demand, several smaller or larger operation campaigns consisting of a suite of jobs are scheduled for the coming years. The Tertiary Eocene sandstones which have recently been proven to be another major oil-bearing reservoir storey (BEUDELL 1987 b, CAPEL 1988) are expected to rather require gravel packing than hydraulic proppant fracturing. Some of the hydrocarbon fields in the Northern North Sea are considerably overpressured (LINDBERG & FERTL 1980).

The most important fields where moderately deep oil fracturing treatments have already been carried out and/or are still going to be performed are (in alphabetical order) Brae, Heather, Hutton, Montrose, Tartan and Thistle (cf. fig. 8 and tab. 7).

2.4.2. Western Europe 1986

The considerably decreased level of hydraulic proppant fracturing in Western Europe in 1986 (and partially still in the first half of 1987) is not only a consequence of the unexpectedly strong oil price drop (cf. section 2.2.1.1.2.) and the accompanying superimposing impact of the severe US $ exchange rate decline (cf. section 2.2.1.3.), but also results from the coincidence of the deterioration of the general economical situation with increased cancellation rates of stimulation operations due to technical reasons, and is also the effect of the hitherto marketing strategy and activity of the proppant companies. The following summary includes oil price decline, areal fracturing activity, increased cancellation rates of jobs, and proppant marketing activity.

2.4.2.1. Oil price decline

The oil price dropped seriously in late 1985 and especially early 1986 (RUNGE 1986; cf. section 2.2.1.1.2.), with the effect in Europe being even more serious due to the fall of the US $ exchange rate which accelerated in late 1985 and reached its minimum in early 1986 through mid 1987, and improved not sufficiently significant to regain confidence for larger investments until mid to late 1988 (cf. section 2.2.1.3.). The result of the marked changes of the general economical framework, especially at the peak of the crisis in early to mid 1986 when the fresh oil price crash was hurting the whole hydrocarbon industry severely, was that almost all the oil and gas production companies in Western Europe severely revised their budgets of capital expenditure, and cut down investments by at abt. 25 - 40 % (cf. section 2.2.1.5.2.) by drastically reducing or even totally freezing the drilling and stimulation programmes chiefly within the North Sea (BEUDELL 1986 c, GRIFFITHS 1986, HARBEN 1986, OEHME 1986, VIELVOYE 1986, ALGAR 1987 b; MABRO 1987, 1988). Summarizing for Western Europe as a whole, however, indicates that exploratory drilling was virtually unchanged in 1986 from its 1985 level, thereby reflecting primarily the higher interest in onshore exploration where much less stimulation potential exists (POPESCU & HODGSHON 1987).

The streamlining of companies and their capital expenditure also affected general overheads and personnel, giving rise to several major lay-off campaigns in hydrocarbon production and service industry affecting world-wide probably almost a quarter of a million employees and partially causing mass people exodus from oil capitals thus leaving there some ghost-like abandoned quarters (cf. section 2.2.1.5.2.). Nevertheless, the activity of some major production companies that have a better overall position, are backed-up by being owned or share-holded by other profitable businesses, have market protection or financial support by the government and are thus almost of semi-national nature, or are committed to fulfill their gas delivery contracts which had been concluded prior to the crisis (cf. section 2.2.1.6.2.) and thus are forced to go ahead regardless of the economical situation, together with a couple of jobs from smaller companies still maintained a principally reasonable level of market potential in view of hydraulic proppant fracturing.

2.4.2.2. Areal fracturing activity

The areal hydraulic proppant fracturing activity in Western Europe in 1986 is characterized along the lines of Rotliegend gas fracturing in the North Sea as well as unemployment of drilling rigs and stimulation boats as follows.

2.4.2.2.1. Rotliegend gas fracturing in the North Sea

Hydraulic proppant fracturing in Western Europe was in 1986 almost exclusively restricted to the Rotliegend gas fields in the British and Dutch Southern North Sea where intermediate-strength high-density alumina oxide and silicate proppants were used either for the full treatment or as tail-in terminal stages (cf. section 2.4.1.2.3.) following a major early phase of sand pumping. The few exceptions were Cretaceous oil-bearing sandstones in the Dutch Southern North Sea where sand was applied, and Cretaceous oil-bearing chalk in the Norwegian Northern North Sea where both conventional and resin-coated intermediate-strength low-density ceramic proppants were taken (HARTLEY & BOSMA 1985; TANSOE, KING & HOLMAN 1986) for combined hydraulic fracturing and gravel packing in soft flowing undercompacted and overpressured chalks (ATTARD, MATHES & MOWER 1988; cf. section 4.5.4.3.). Some tight Mesozoic sands in the Northern North Sea were also stimulated (DESPAX, CHARLEZ, CLINKEMAILLE & ECONOMIDES 1987). Onshore Germany FRG and Netherlands, nothing took place both in deep gas-well fracturing with synthetic proppants and shallow oil-well fracturing with natural sand (cf. section 2.4.1.1.1.).

The British Southern North Sea Rotliegend gas belt has its own development concept and due to predominant sale of the gas prior to development drilling and hydraulic proppant fracturing (cf. section 2.2.1.6.2.), the province kept its activity more or less unaffected by the 1986 oil price drop and was straightforward continuing its approach to the major development campaigns in 1987 - 1990 (economical aspects are also summarized by FISHMAN 1986, GREGORY 1986, OFFSHORE ENGINEER 1986 a, THOMAS 1986, ALGAR 1987 a, BAND 1987, POTTER 1987). This is reflected by the fact that in the first quarter of 1986, in total more wells had been drilled in the British North Sea than in any other quarter in the history of United Kingdom offshore oil and gas activity, and for 1986 as a whole, only slightly fewer wells have been drilled than the average number having been brought down in the period 1983 - 1985 (POPESCU & HODGSHON 1987). In the Dutch Southern North Sea, projects were shifted from oil spots to gas patches along with general budget cutting as a consequence of the oil price collapse (BEUDELL 1986 b).

2.4.2.2.2. Unemployment of drilling rigs and stimulation boats

The considerably reduced drilling and stimulation activity in the North Sea (GRIFFITHS 1986, VIELVOYE 1986) was best expressed in the second half of 1986 by numerous unemployed jack-up platforms having been lined up like a pearl chain along the British and Norwegian coasts (particularly between Newcastle and Aberdeen on the British side and between Stavanger and Bergen on the Norwegian side; cf. also SCOTTISH PETROLEUM ANNUAL 1987). An indirect indicator was the employment of the stimulation boats in the North Sea where according to the fact that three service vessels (FOX 1985, OILMAN 1985 b) of purpose-built nature (COOPER & MARSHALL 1984, OFFSHORE ENGINEER 1984) had to share a stagnant or shrinking fracturing market, some boats kept waiting for jobs in the harbour of their offshore base up to several months time (cf. sections 3.7.1.1. and 3.8.1.2.).

The total number of active rigs in the North Sea dropped by almost 30 % during 1986, with more than 40 platforms being without job in late 1986 in contrast to only a couple of them one year earlier (OEHME 1986). Up to abt. 70 rigs have been stacked off Northwest Europe in the worst period of 1986 and early 1987, and less than 40 rigs remained working in that period (McNALLY 1987, VIELVOYE 1987 b). In the second half of 1987, recovery of the North Sea market resulted in rise of the active drilling rig number to abt. 65 and fall of the quantity of stacked rigs to abt. 40.

2.4.2.3. Increased cancellation rates

According to Murphy`s law, the crisis that was triggered by the oil price decline was for the European hydraulic proppant fracturing marketing segment even negatively accentuated, as the reduced amount of drilling and stimulation activity in mainly the North Sea during the peak of the crisis coincided with an unusually increased rate of cancellations of planned fracture jobs (with respect to the expectations) in early to late 1986. Most of the reasons for the increased cancellation rates relate to reservoir properties and well type.

2.4.2.3.1. Reservoir properties

The reasons for stimulation job cancellation against expectation and planning are multiple and depending on actual well results. The first possibility is that no gas was found in a water-bearing reservoir in exploration wells when the target formation was encountered deeper than prognosed or the gas-water-contact was met in a higher position than predicted. The second example is that the proven reservoir is too tight to allow economical gas withdrawal even after a hydraulic proppant fracturing treatment (too low pre-fracturing gas offtake rate or completely tight) in contrast to the assumption of presence of low permeabilities in a favourable range for a commerciality of production after fracturing. The third cause is that the reservoir has unexpectedly good permeability and thus a fracturing operation is not necessary, thereby contradicting the expectation of poor reservoir conditions that would have required stimulation. The fourth effect is that the hydraulic treatment is suspended and the well is shut-in to wait for a better general economical situation as an answer to the change of framework conditions during drilling of deeper wells which were often spudded with the expectation of the necessity of later fracturing at a time when the oil price was still promising, but which finally encountered the target formation at a time when the oil price level together with the US $ value

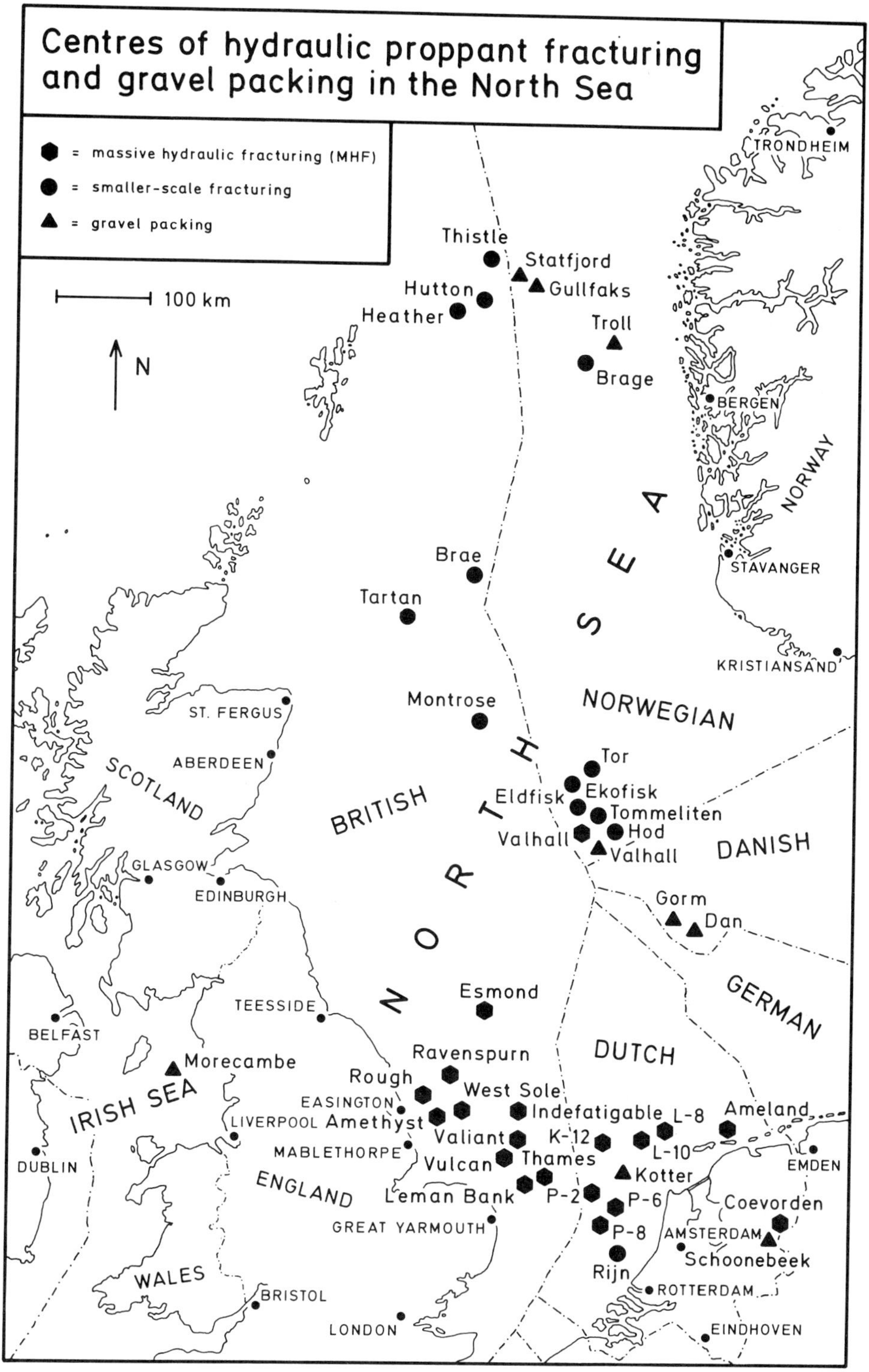
Centres of hydraulic proppant fracturing and gravel packing in the North Sea
= massive hydraulic fracturing (MHF)
= smaller-scale fracturing
= gravel packing
100 km
N
Thistle
Statfjord
Gullfaks
Hutton
Heather
Troll
Brage
TRONDHEIM
BERGEN
NORWAY
STAVANGER
KRISTIANSAND
Brae
Tartan
Montrose
NORWEGIAN
ST. FERGUS
ABERDEEN
SCOTLAND
BRITISH
NORTH SEA
Tor
Eldfisk
Ekofisk
Tommeliten
Valhall
Hod
Valhall
DANISH
GLASGOW
EDINBURGH
Gorm
Dan
GERMAN
Esmond
TEESSIDE
BELFAST
DUTCH
Ravenspurn
Morecambe
IRISH SEA
Rough
West Sole
EASINGTON
Indefatigable
L-8
Ameland
LIVERPOOL
Amethyst
Valiant
K-12
L-10
MABLETHORPE
Vulcan
Thames
EMDEN
DUBLIN
Kotter
ENGLAND
Leman Bank
P-2
P-6
Coevorden
GREAT YARMOUTH
P-8
AMSTERDAM
Schoonebeek
Rijn
WALES
BRISTOL
ROTTERDAM
LONDON
EINDHOVEN

Fig. 8

had so far collapsed that the original plans could no longer be maintained and decision was made to postpone the treatment.

All the mentioned reasons have repeatedly resulted in cancellations of hydraulic proppant fracturing treatments that had been scheduled and designed before, with some of the withdrawals having been come into force almost in the last minute of job preparation predominantly in 1986 and partially also in the first half of 1987. For the individual proppant and service companies, all these effects additionally interfered with loss of some business to competition which, however, is also taking place in good times, and with getting some orders later than expected or even not at all due to intermittent stock liquidation by the service companies enabling them to carry out several operations of various size without in between taking new proppant material (cf. section 2.4.5.5.2.). In the poor economical period, all these factors contribute to hit the industry very seriously. A negative consequence directly related to the streamlining of the budgets in the crisis situation (cf. section 2.2.1.5.2.) was replacement of expensive synthetic intermediate- and high-strength proppants by cheap natural sand in the stimulation design wherever possible according to reservoir conditions, in contrast to sometimes generous application of a higher-quality proppant type even if not indispensibly required for the whole job in the golden years of hydraulic fracturing.

2.4.2.3.2. Well type

The high rate of cancellations in early to late 1986 was mainly caused by the predominance of exploration and appraisal wells where geological and reservoir engineering risk is often very high. For (at least the second half of) 1987 and especially for 1988, the cancellation rate can be expected to have been much lower, as in these years, production (development) drilling was prevailing where the risk is considerably lower, because the structures and their contents are more or less already known. In the first half of 1987, the cancellation rates were already lower than in 1986 also for the reason of reduced original planning of treatments in light of the poor economical framework, with the deletions of the items prior to their appearance on the program list of wells to be spudded mainly affecting higher-risk prospects which could easily be subjected also to later cancellation (in 1986 shortly after the worst stage of the crisis was reached, it was often too late to stop and prevent such items to be drilled).

2.4.2.4. Proppant marketing activity

In view of the total marketing and sales level of proppants in Europe, another reason for the comparatively poor result in 1986 with respect to the expectations is the hitherto underdevelopment of Eastern Europe, Southern Europe and Northern Africa as a consequence of the neglectance of these markets in the past. If having been better attended, hydraulic fracturing campaigns would be now in a much more advanced stage of preparation in Eastern Europe, Southern Europe and Northern Africa and could have partially even already in 1986 and following years resulted in additional business for particularly the proppant companies (that supply in addition to the established service companies which usually operate the jobs as contractors also to those hydrocarbon production companies that carry out smaller or larger jobs themselves without intercalation or intervention of service companies; cf. section 2.4.4.4. and 2.4.5.5.1.) that would have been at the bottom of the line a very welcome compensation of the quantities lost in Western Europe due to the reasons discussed above.

The insufficient marketing attention by the proppant companies during the last years is highlighted by the fact that even in high-potential areas such as Great Britain and Netherlands, there have still recently been encountered some oil and gas companies where the responsible reservoir engineers had never heard before of the product suite and the logistical organization of the major proppant suppliers, although these companies also do have wells that require fracturing with intermediate-strength proppants if adequate layout is performed. On the other hand, the proppant companies so far know very little or almost nothing about the demand of several potential Western and Eastern European countries. Some examples of business loss to proppant competition and natural sand are mentioned as follows.

2.4.2.4.1. Business loss to proppant competition

Business loss to proppant competition occurred repeatedly on large-scale order basis in the last years in Yugoslavia and in the British Southern North Sea. Some particularly spectacular examples where major chances of business conclusion have been lost due to various reasons are illustrated as follows. While some of the cases are the consequence of excessive price pokering or simple neglectance of the market in earlier times, other items are the result of elimination of the natural market forces by short-circuiting due to intracorporation relationships.

2.4.2.4.1.1. Yugoslavia

Following a small-scale precursor treatment in Yugoslavia (cf. section 2.4.2.4.2.1.) which was very successful and opened the way for a big MHF campaign in that country, the insufficient marketing attention in Eastern Europe by the major proppant companies was seriously punished by enormous business loss for some of them in late 1985 when after an unsuccessful approach of various proppant companies by the Yugoslavians for a quotation, in total 5,400 t of proppants were purchased in almost one lot from supply companies which was scheduled to be enough for the 10 - 12 planned jobs of the MHF stimulation campaign throughout a couple of years, and several major proppant companies were only awaking by this event and too late realizing that also at least parts of Eastern Europe have considerable marketing potential. While some of the major proppant companies could still slip into the deal by an unexpected backdoor approach due to the limited delivery capacity of one of the competitors, other parties had absolutely no participation in this action which probably was the historically biggest proppant deal in one movement in whole Europe.

2.4.2.4.1.2. British Southern North Sea

Some case studies of proppant business loss to competition in the British Southern North Sea are illustrated along the lines of price insistence and lacking market understanding as well as internal company structures and short-cut deals.

2.4.2.4.1.2.1. Price insistence and lacking market understanding

Another serious business loss to proppant competition happened during the preparation of one of the big hydraulic fracturing stimulation campaigns in the British Southern North Sea Rotliegend tight gas province. While bidding of the service companies to the operator was made in early to mid 1987, the prices for alumina oxide and silicate proppants were raised effective late 1987 by the suppliers (cf. section 2.2.2.2.) who for a long time had already waited for a good opportunity to increase the price after the unfortunate and subsequently regretted price lowering in late 1985 (cf. section 2.2.2.1.). While for many years one service company performing most of the stimulation work in the North Sea had a preferential relationship to one major proppant supplier, with even frequently taking their material at once without even talking to competition at all, the business negotiations for the big 1988 campaign changed the situation completely.

Becoming upset about the intention of its former standard supplier to charge the new higher proppant price already for the deal that was obtained by the service company based on offering the old lower price, the service company made an agreement with the competition which was willing to listen and to understand the special relationships and precommitments and therefore guaranteed still the old proppant price for the whole campaign comprising several thousands of tons of material, the service company changed its supplier obviously not only for this one deal, but was certainly looking into longer-term fruitful cooperation for mutual benefit similar to the practice done with the competitor before. Thus as a consequence of playing the price card too high due to lacking understanding of the market requirements and the future potential, the proppant outlet into the North Sea was almost completely taken away from the former supplier and given to its competitor, resulting in serious business loss and nearly totally reversing several years of purchase strategy of the service company carrying out most of the hydraulic proppant stimulation work.

These examples illustrate that insufficient understanding of the European stimulation market by remote handling and lacking local knowledge can cost much more money in terms of lost business that would be the expenses for an adequate regional representation. A proper coverage of the whole European area would comprise the implementation of geological and reservoir engineering concepts into the proppant marketing strategy and the operation out of a permanent European marketing representation (the significance of inclusion of geological factors into hydraulic proppant fracture treatment design is also emphasized by BRITT & LARSEN 1986).

2.4.2.4.1.2.2. Internal company structures and short-cut deals

Company and corporation internal structures, however, can in some cases seriously disturb market forces and can result in business approval for certain proppant suppliers only thanks to intra-multinational organization relationships. An example of such action against marketing efforts of competitors and short-circuiting of business conclusion with rendering all the real and open marketing activities of other suppliers and natural market forces a priori useless and a waste of time and money happened also in the preparation of the 1988 proppant fracturing campaigns in the British Southern North Sea. Although having not the best reputation to the service company executing the jobs as a result of inadequate business behaviour and exaggerated price pokering, one of the major proppant companies got the contract to supply material for a large campaign only by intervention of the operator which had formerly together with the proppant manufacturer been under the head of a big multinational corporation, and even after divorce as a consequence of company splitting and sale during course of streamlining measures, a short-cut agreement without involvement of the service company guaranteed the operator a better price for the material and gave the proppant supplier the big deal.

This business in fact represented the first major action of direct proppant sale and purchase, respectively, from supplier to operator without integration and involvement of the service company. It is typical for such type of marketing force invalidation that at the bottom of the line, opportunistic behaviour with securing of former in-house relationships for the mere sake of business conclusion and money-making enabled the proppant manufacturer to agree without hesitation to a price for the whole proppant lot which it had earlier repeatedly rejected from taking into consideration when directly talking to the executing service company.

Conclusion has therefore to be drawn that in some cases, marketing efforts on a sound economical and technical base can be completely spoiled by opportunism and advantage exploitation due to intra-multinational corporation relationships and deals are made all of a sudden which completely diverge from or contradict the commonly exhibited marketing policy of the proppant manufacturer. The outlined examples illustrate that business loss to proppant competition has very diverse aspects which comprise not only technical and economical questions, but include also political and organizatory points that sometimes even incorporate complete elimination of the market forces, with in such cases being at the bottom of the line no difference between satisfactory and engaged marketing attention on the one hand and insufficient recognition and neglectance of the market on the other hand.

2.4.2.4.2. Business loss to natural sand

In addition to business loss to proppant competition, several deals of partially considerable quantities of synthetic proppants fell through because of loss to natural sand for various reasons. Some examples from Yugoslavia, Turkey and British Southern North Sea are outlined as follows. The reasons for the missed opportunities have been similarly as in case of business loss to proppant competition predominantly neglectance of the market and excessive price pokering.

2.4.2.4.2.1. Yugoslavia

The undersupply of the European market with material for many years as a result of wrong planning, unrealistic approach, insufficient attention and lacking understanding had its consequences still in 1985 before the proppant price reduction when neither production nor service company were able to obtain intermediate-strength proppants for a hydraulic fracturing job in Yugoslavia (which for American proppant companies was so far a white patch on the stimulation map) that precursed the MHF campaign in this country, and as a result of the unavailability of the desired product in time that was selected as being suitable, coupled with only limited possibilities of rapid delivery of local proppants, the designed fracture length had to be reduced and the crack had to be infilled with sand followed by a tail-in of high-strength proppants rather than being completely plugged with intermediate-strength proppants as would have been appropriate in the given closure stress regime (ECONOMIDES, CIKES, PFÖRTER, UDICK & URODA 1986).

2.4.2.4.2.2. Turkey

Another example of loosing business as a consequence of lacking marketing activity was the purchase of tremendous amounts of cheap sand by local operators for a major fracturing campaign in Turkey where intermediate-strength proppants would have been technically required at least for tail-in operations. With the background of a huge sand inventory in the warehouse, however, neither proppant nor service companies had any argument to convince the production company in Turkey to change the job design, and a level of technically sound understanding was only reached when the sand pile had finally been completely cleared by pumping downhole. Appropriate marketing attention of the area would certainly have prevented the excessive purchase of sand by the local operator and would have opened the door for considerable intermediate-strength proppant business instead of the sand.

2.4.2.4.2.3. British Southern North Sea

Large-scale business loss of proppant suppliers to natural sand also occurred in the British Southern North Sea during course of Rotliegend gas field development. Although the operator in the example under discussion conventionally used natural sand as the main quantity of the fracturing treatment due to still

HYDRAULIC PROPPANT FRACTURING AND GRAVEL PACKING IN EUROPE — PAST AND FUTURE STIMULATION POTENTIAL IN IMPORTANT HYDROCARBON FIELDS IN SELECTED EUROPEAN COUNTRIES 3		STIMULATION TYPE		PROPPANT TYPE		HYDROCARBON TYPE		RESERVOIR TYPE				RESERVOIR FORMATION						
		Fracturing	Gravel packing	Sand	Proppants	Oil	Gas	Sandstone	Carbonate	Chalk	Other	Carboniferous	Rotliegend	Buntsandstein	Jurassic	Cretaceous	Tertiary	Other
AUSTRIA	Matzen	●	▲	▲		▲	▲	▲									▲	
	Neusiedel	●	▲	▲		▲	▲	▲									▲	
	Prottes	●		●			●		●									●
	Stockerau	●	▲	▲		▲	▲	▲									▲	
	Tratnach	●	▲	▲		▲	▲	▲								▲	▲	
	Zistersdorf	●	▲	▲	●	▲	●	▲									▲	●
DENMARK	Dan		▲	▲		▲				▲						▲		
	Gorm		▲	▲		▲				▲						▲		
GDR	Salzwedel	●		●			●	●					●					
HUNGARY	Barcs	●		●	●		●	●									●	
	Belezna	●		●	●		●	●									●	
	Inke	●		●	●		●	●									●	
	Mako	●		●	●		●	●									●	
	Nagykanisza	●		●	●		●	●									●	
ITALY	Adriatic Sea		▲	▲		▲	▲	▲									▲	
	Bronte	●			●		●	●									●	
	Gagliano	●			●		●	●									●	
	Po Valley	●		●		●	●	●									●	
NETHERLANDS	Ameland	⬢		⬢	⬢		⬢	⬢					⬢					
	Coevorden	⬢		⬢	⬢		⬢	⬢				⬢						
	Groningen	●	▲	●	●		●	●					●					
	K-12	⬢			⬢		⬢	⬢					⬢					
	Kotter		▲	▲		▲		▲								▲		
	L-8, L-10	⬢		⬢	⬢		⬢	⬢					⬢					
	Leeuwarden	⬢			⬢		⬢	⬢					⬢					
	M-10	⬢			⬢		⬢	⬢					⬢					
	P-2	⬢			⬢		⬢	⬢					⬢					
	P-6, P-8	⬢		⬢	⬢		⬢	⬢						⬢				
	P-15 (Rijn)	●		●	●	●		●								●		
	Schoonebeek	●	▲	▲		▲		▲								▲		

Tab. 8

sufficiently shallow reservoir depth (although already near the boundary range of sand stability) and inserted only a comparatively small volume of intermediate-strength synthetic proppants as tail-in for the purpose of providing maximum conductivity and closure stress resistivity in the surroundings of the borehole, replacement of natural sand by low-density alumina silicate proppants was in serious discussion stage and was also heavily promoted by the contracted service company which recognized this opportunity also as a chance of further sharpening of profile and application spectrum. The deal comprising at least 4 Mio. lbs (2,000 t) of material, however, finally fell through and natural sand was preferred, because the proppant supplier could not be convinced that for such a major pilot campaign of utilizing low-density alumina silicate proppants in the whole treatment throughout the field to be developed and to renounce completely from sand (progressive replacement of natural sand by synthetic proppants is one of the major marketing guidelines in the current proppant scene; cf. section 2.2.2.3.), another reduction of the price difference to natural sand would be necessary to obtain the approval from the operator.

As a consequence of lacking understanding of the situation and missing outlook capability, the proppant manufacturer insisted on the standard price for the intermediate-strength low-density alumina silicate proppants even in view of the large quantity to be firmly ordered as one lot and despite the cooperation offered by the service company in marking up the material considerably less than usual to help bridging the price gap close to natural sand. At the bottom of the line, the proppant supplier thus preferred to sell only the tail-in quantity (which amounted to abt. one fifth of the volume under discussion for the whole jobs) at the usual price rather than to deliver a considerably larger lot for a slightly reduced price. This example illustrates that the difference of even a couple of cents per pound in proppant price can be sufficient to kill major possibilities of proppant promotion into new segments of application if lacking encouragement, understanding of the technical necessity, and forecast of stimulating follow-up activity are compensated by rigorous insisting on standard price.

2.4.3. Eastern Europe until 1986

Almost near the end of the period 1977 - 1985 which represents the golden age of hydraulic proppant fracturing at least in Germany FRG (in the British Southern North Sea, the time 1988 - 1992 might become the peak of the activity) in Western Europe, MHF stimulation started to take also place in parts of Eastern Europe. Several partially extremely large fracture jobs have been carried out in Yugoslavia in 1985 and particularly 1986 (ECONOMIDES, CIKES, PFÖRTER, UDICK & URODA 1986), and without the serious drop of the oil price in early 1986, it would have been even more. Comments are offered as follows on Yugoslavia and Hungary, USSR, other Eastern European countries, and Southern Europe and Northern Africa.

2.4.3.1. Yugoslavia and Hungary

Reservoir conditions in Yugoslavia (and also in adjoining Hungary) are quite complicated due to unusually high temperatures of 180 - 240 oC in 3,000 - 4,000 m depth (corresponding to 9,000 - 12,000 ft) which cause problems with the fracturing fluids (cf. sections 4.3.4.4. and 4.7.4.2.) and make the pumping of heavier proppants very difficult. In addition, the reservoir fluids contain H_2S and CO_2 which are aggressive to the equipment. The gas- and condensate-bearing formation (ECONOMIDES, CIKES, PFÖRTER, UDICK & URODA 1986) consists of Tertiary conglomerates, sandstones and limestone breccias. One of the main aims of hydraulic fracturing is to avoid excessive retrograde condensation (CASTELIJNS 1981, KENYON & BEHLE 1987; DANESH, HENDERSON, KRINIS & PEDEN 1988). The fields belong to the Molve-Kalinovac-Stari Gradac complex in the Pannonian Basin (DRAGICEVIC, MARINOVIC, STANKOVIC, BATUSIC & NIKOLIC 1983; IBRAHIMPASIC 1983; RI-

GASSI 1985, 1986; BAUK, SECEN, BABIC & SACER 1986; ENERGIEWIRTSCHAFT 1986 b, GORST 1987) on the Yugoslavian side and to the Barcs-Inke-Mako group on the Hungarian side (cf. fig. 5 and tabs. 8 - 9).

The fracture jobs in Yugoslavia have been carried out in a two-step manner with pumping of 100 - 200 t of intermediate-strength low-density proppants in the early stage and 200 - 450 t of intermediate-strength high-density or high-strength proppants in the late stage, thus being a special type of tail-in and/ or lead-in fracturing strategy (HICKEY, BROWN & CRITTENDEN 1981; cf. section 2.4.1.2.2.). Total proppant quantities per treatment ranged between 300 and 650 t and thus the operations in Yugoslavia included some of the biggest MHF jobs which have ever been carried out in Europe (only in Germany FRG have so far been treatments of comparable magnitude performed; cf. section 2.4.1.1.1.). The design of the jobs in Hungary is smaller due to other reservoir characteristics, technical circumstances and economical conditions with respect to those in Yugoslavia.

2.4.3.2. USSR

The first hydraulic proppant fracturing operations were already carried out in the USSR in 1952. The total annual fracturing activity in the USSR reached its peak in the years 1958 - 1962 with up to 1,500 operations per year, and later declined to a plateau-like level with more or less 100 treatments per year. The centres of hitherto hydraulic fracture operation performance in the USSR are the Krasnodar, Volgo-Ural, Tataria (mainly the Romashkinskoe and Tumaisi fields), Bashkiria, Kujbyshev, Turkmenia, Dagistan and Siberia (mainly the Salimski and Surgutskoe fields) areas. Hydraulic fracture stimulation has been mainly done in oil-production, water-injection and water-disposal wells, but subordinately also in degasification wells of coal-bearing strata and geological investigation wells. MHF treatments have so far not been done in the USSR, and also small-scale fracturing did not play a major role in well completion, because preference was given to drilling and fast exploitation of easily accessible and recoverable reserves. The full significance of stimulation was only lately better realized when oil production lagged behind the goals of the five-year plan (cf. section 2.4.4.1.5.) and the beginning of the major world oil price crisis together with a diminishing strength of the US $ gave the impetus for serious looking into possibilities of enhanced exploitation of existing capacities (EBEL 1986, ORASIANU 1987; CHADWICK, LONG & NISSANKE 1988).

Up to now, only natural sand in quantities of up to 130 t, but in most cases only 20 - 50 t has been used as propping agent in hydraulic fracturing treatments in the USSR. Synthetic high-quality proppants have either not been necessary in shallow reservoirs, or deep pay zones had so far little priority for stimulation. All the hitherto performed jobs have been carried out in own service using high-pressure pumps of American provenance which were bought more than 10 years ago still before the embargo enacted by President Carter in 1978 when American service companies even had offices in Moscow, as well as with domestic and Romanian equipment. The hydraulic fracturing technology is well understood and appreciated in the USSR from where in fact also some of the first inspirations and concepts derived and pioneering work had been done when the stimulation method was born almost 40 years ago (CLARK 1948, WILSEY & BEARDEN 1954, KHRISTIANOVICH & ZHELTOV 1955; cf. section 1.2.).

2.4.3.3. Other Eastern European countries

In most of the other Eastern European countries, similarly as in the USSR so far only smaller fracturing treatments have been done with abt. 20 - 150 t of sand as propping agent (examples comprise mainly Hungary, Poland, Yugoslavia and Germany GDR, and subordinately also Czechoslovakia and Romania, whereas in Bulgaria and Albania, nothing has been done up to now). Intermediate- to high-strength proppants have hitherto only been applied in very limited quantities for a tail-in in Hungary (coming from very restricted domestic manufacturing as by-product in alumina industry; cf. section 1.3.1.2.).

In view of the apparent delay in development of reservoir stimulation in Eastern Europe, it has to be understood that before the 1986 oil price crash, first the priority of domestic oil and gas acquisition was on drilling and completion of easily accessible pay zones which caused no major treatment problems and necessities, and second the USSR delivered sufficient amounts of oil and gas on barter-trade base to its satellites, thereby securing supply and thus not triggering any serious need of increasing domestic hydrocarbon offtakes in the other countries of the socialistic block. The 1986 oil price collapse coupled with the US $ exchange rate drop, however, abruptly changed the whole situation by affecting the total interwoven fabric between the group members on bilateral scale with respect to the USSR as well as on multilateral scale among each of the satellites and from each of them to the USSR.

2.4.3.4. Southern Europe and Northern Africa

Larger fracturing campaigns including cooperation with Western service companies have been made in 1986 and earlier years in Turkey where again almost exclusively sand was used as propping material. Intermediate-strength low-density proppants would have been most suitable at least as a tail-in for a number of operations, but the early purchase of vast amounts of sand inhibited the application of the better, but also more expensive material in the hitherto performed campaigns (cf. section 2.4.2.4.2.).

In Northern Africa, hydraulic proppant fracturing of low-permeability sandstones has so far mainly been carried out in Algeria and in subordinate amounts also in Libya, Egypt, Sudan, Oman, Tunisia, Zaire, Gulf of Guinea and Congo, with in most cases only sand having been pumped, but intermediate- or high-strength proppants also already repeatedly having been applied. Stimulation in the farther Middle East hitherto mainly included India and Pakistan.

2.4.4. Status in Eastern Europe

In contrast to Western Europe, the different economical system in Eastern Europe results in a more or less independency from the oil price scenario and partially also from the US $ exchange rate scenario which means that basically the drilling and stimulation programmes should have gone ahead also during 1986 through 1988 without major changes. In view of market penetration for hydraulic proppant fracturing, however, the Eastern European market (with the exception being Yugoslavia and partially also Hungary; cf. sections 2.4.3.1. and 2.4.6.2.) is seriously underdeveloped or partially still virtually unattended by Western service and proppant companies due to neglectance of marketing activity in the past when basically only the big and urgent demand in USA and parts of Western Europe was satisfied by marketing consultancy and advicing as well as by delivery of material.

Due to the hitherto preferential both exploration and production drilling as well as conventional completion and testing with only little stimulation activity, the Eastern European market is nowadays still in an immature to submature stage of hydraulic proppant fracturing and partially also gravel packing in contrast to Western Europe which is already mature since many years, and to the partially supermature situation of the USA especially during the golden age of stimulation in the late 1970`s and early 1980`s. As a consequence of the interlocking relationship between foreign trade offices and more or less national hydrocarbon production companies as well as various financial complications, negotiation and conclusion of business contracts and application of technology and material take much more time in Eastern Europe than in Western Europe. Comments are offered as follows on oil price and US $ exchange rate, political restrictions in the past, present stimulation potential and economical restructuring, self-servicing treatment policy, economical and financial aspects, and reservoir engineering and gas supply aspects.

2.4.4.1. Oil price and US $ exchange rate

The Eastern European countries (except of the USSR) are in the present situation of low oil prices, increasing devaluation of their currencies with respect to US $, DM and £, and progressive inflation more or less severely forced to increase their domestic oil and gas production by all means, because the USSR has after the 1986 oil price crash (cf. section 2.2.1.1.2.) and the more or less accompanying US $ exchange rate collapse (cf. section 2.2.1.3.) cut back the quantities of hydrocarbon supply to its satellites (ERDOEL-ERDGAS AKTUELL 1985 d, RIGASSI 1985) due to increasing own demand as a source of hard currency by sale to Western countries. The outline as follows presents some aspects of export of USSR oil and gas to Western and Eastern Europe, dependency of Comecon countries on USSR oil and gas, barter trade vs. hard currency sale of oil and gas, role of hydrocarbon exports in the USSR foreign trade, USSR oil production, and currency convertibility and exchange rates.

2.4.4.1.1. Export of USSR oil and gas to Western and Eastern Europe

The gas coming from the giant Siberian gas field Urengoi and others is to large amounts brought in transit through the USSR satellite states to Western Europe where it is sold to various consumers against high prices in hard currency that were fixed by delivery contracts partially still during the time of high oil and gas prices and that are now certainly kept valid by the USSR and are backed-up by reliable delivery, and only balanced and corresponding quantities of the gas are distributed in the Eastern European satellites of the USSR (FEHRINGER 1986). The USSR 1986 gas sales to Western Europe amounted to 1.39 Tcf (abt. 35 Bill. m3), whereas export to the socialistic countries reached still 1.43 Tcf (abt. 36 Bill. m3), but was already on a marked decline (OIL GAS JOURNAL 1987 o). Another reason for the oil supply shortage of the USSR was declining own oil production in the years 1982 - 1985 (RIGASSI 1985), with the cumulative offtake even falling behind the goals set in the five-year plan. In 1986 when the oil price crash took place, oil production in the USSR started to increase again as a consequence of the effectivity of the reorganization politics (perestroika; GORBATCHEV 1987; cf. section 2.4.4.3.2.) and reached 1987 a record level, but because larger quantities were associated with lower prices, the result at the bottom of the line remains the same, and in order to increase hard currency income, oil and gas export have to rise further.

Export of oil and gas account for abt. 60 - 80 % of convertible currency earning of the USSR and therefore have an overwhelming importance in foreign trade (STERN 1986, AL-CHALABI 1987, GORST 1988). This requirement of the USSR to get hard currency from hydrocarbon sale and the large gas reserves which are sufficient for even doubling the supply quantities to Western Europe in the next years are the reason that there is no basis of fearing commercial or security risks of importing gas or oil from the USSR which has no interest to exercise economic leverage for political concessions. The USSR exported in 1987 abt. 12 % and abt. 31 % of its total gas and oil production volume, respectively (OIL GAS JOURNAL 1988 f), with gas and oil representing abt. 10 % and abt. 33 % of the total value of the USSR exports, respectively. Total USSR oil export reached a record high in 1987 with 3.7 Mio. bbl/d (with total production amounting to 12.3 - 12.5 Mio. bbl/d), and its significance for the state budget is reflected by the fact that even in 1985 when the total output fell to the unprecedented low of 11.9 Mio. bbl/d, the export volume achieved 3.4 Mio. bbl/d which is only slightly less than that of the record year of 1987 (OIL GAS JOURNAL 1987

HYDRAULIC PROPPANT FRACTURING AND GRAVEL PACKING IN EUROPE		STIMULATION TYPE		PROPPANT TYPE		HYDROCARBON TYPE		RESERVOIR TYPE				RESERVOIR FORMATION						
PAST AND FUTURE STIMULATION POTENTIAL IN IMPORTANT HYDROCARBON FIELDS IN SELECTED EUROPEAN COUNTRIES 4		Fracturing	Gravel packing	Sand	Proppants	Oil	Gas	Sandstone	Carbonate	Chalk	Other	Carboniferous	Rotliegend	Buntsandstein	Jurassic	Cretaceous	Tertiary	Other
BULGARIA	Badarski	●		●	●	●		●							●			
	Lukuvit	●		●	●	●		●							●			
	Tarnak	●		●	●		●	●							●			
CSSR	Vienna Basin	●	▲	▲		▲	▲	▲									▲	
POLAND	Buk	●		●	●		●	●					●					
	Bukowiec	●		●	●		●	●					●					
	Carpathian Foredeep	●		●	●		●	●									●	
	Grodzisk	●		●	●		●	●					●					
	Klęka	●		●	●		●	●					●					
	Międzyzdroje	●		●	●		●	●					●					
	Uciechów-Bogdaj	●		●	●		●	●					●					
	Ujazd	●		●	●		●	●					●					
	Wierzchowice	●		●	●		●	●					●					
	Załęcze	●		●	●		●	●					●					
	Żuchlów	●		●	●		●	●					●					
ROMANIA	Tertiary Basin	●	▲	●	●	●	●	●									●	
TURKEY	Tertiary Basin	●		●	●	●	●	●									●	●
USSR	Astrahanskoe	●		●	●	●	●	●										
	Borodinskoe	●		●	●	●	●	●										
	Krasnodarskoe	●		●		●		●										
	Orenburgskoe	●		●	●	●	●	●										
	Saratovskoe	●		●	●	●	●	●										
	Shebelinskoe	●		●	●	●	●	●										
	Stavropolskoe	●		●	●	●	●	●										
YUGOSLAVIA	Donya Bukovica	⬢			⬢		⬢	⬢			⬢						⬢	⬢
	Kalinovac	⬢			⬢		⬢	⬢			⬢						⬢	⬢
	Molve	⬢			⬢		⬢	⬢			⬢						⬢	⬢
	Novigrad Podravski	⬢			⬢		⬢	⬢			⬢						⬢	⬢
	Oresac	⬢			⬢		⬢	⬢			⬢						⬢	⬢
	Stari Gradec	⬢			⬢		⬢	⬢			⬢						⬢	⬢

Tab. 9

o). The 1986 oil price crash, however, more than compensated the increasing volume by the considerably dropping price. After rising to 51.6 % in 1984, the oil and gas share in overall value of USSR exports to all countries slided to 49.4 % in 1985 and 43.7 % in 1986.

2.4.4.1.2. Dependency of Comecon countries on USSR oil and gas

At the bottom of the line, all the Comecon (Rat für gegenseitige Wirtschaftshilfe, RGW) countries are depending to larger or smaller degree on the USSR in terms of both oil and gas supply (FISCHER 1978), because due to shortage or lack of hard currency, they cannot afford to buy oil and gas at the world market, and with the exception of Romania, all the socialistic satellites of the USSR cannot cover their whole own demand by domestic production. In addition, pipeline infrastructure does not allow the smaller Comecon countries to renounce from the steady direct supply from the USSR and to access other sources. During the last 20 years, the USSR has sufficiently delivered oil and gas to the satisfaction of the Comecon countries and even sometimes supplied them with larger quantities at the expense of cutting exports to the West, and only the 1986 oil price crash and the accompanying US $ weakness have created another situation.

All the aforementioned technical, economical and political relationships, especially the lower oil price and the weaker US $ with respect to earlier years, highlight again the force lasting on the USSR and partially also the other Eastern European countries to increase their domestic oil and gas production by all possible means (OILMAN 1987 c), with stimulation of reservoirs in existing fields being one of the top-priority items. Particular exceptions from the aforementioned scenario are Czechoslovakia where countertrade with the USSR is still on a mutually sufficient level, and in some extent also the GDR which profits from special trade arrangements with the FRG and from particularly close relationships to the USSR, and takes whenever possible advantage of its geographical position at the boundary of the two big political blocks.

2.4.4.1.3. Barter trade vs. hard currency sale of oil and gas

Oil and gas are the major source of hard currency for the USSR (GRIFFITHS 1986), and as both oil price drop and US $ exchange rate collapse result in lower net returns (cf. section 2.2.1.3.), the losses have to be compensated by selling increasing amounts of hydrocarbons in Western countries which together with an increased domestic demand by enlarging the industry in the USSR are missing for delivery into the socialistic satellite states on barter-trade basis. In addition, many of the Eastern European countries virtually face bankruptcy and are no longer able to purchase even from the USSR which in the present general economical scenario in turn is no longer willing to supply free of charge or even on credit.

Instead of bartering with the USSR, the satellite countries are theirselves more and more forced to get access to hard currency by selling whatever possible in the Western world, and the USSR does the same with products formerly delivered on exchange basis to its group members. On the other hand, however, the USSR has to maintain oil exports to its satellite countries at a reasonable level for both political and economical reasons, because this enables the USSR to obtain products on barter level that it otherwise would have to purchase for US $ on the world market, and permits the satellites to avoid a further increase of their already tremendous debt in hard currency (RIGASSI 1985).

The most pronounced example of bartering is the purchase of Romanian drilling and completion equipment which is paid by oil and gas deliveries (OIL GAS JOURNAL 1987 l, RIGASSI 1986). The significance of this barter trade of equipment vs. hydrocarbons is underlined by the fact that in the period 1981 - 1987,

Romania accounted for 43 - 50 % of the USSR`s total imports of oil and gas field equipment which comprise chiefly heavy drilling rigs, but also substantial amounts of completion and stimulation equipment. The steady increase of import level from Romania in the last years went parallel to the declining oil price that had the consequence of more and more cutting equipment import quantities from Western Europe and Japan (those from the USA were independently slashed by the restrictions on strategic sales; cf. section 2.4.4.2.) by rates of 50 - 80 %.

2.4.4.1.4. Role of hydrocarbon exports in the USSR foreign trade

The significant role of oil exports in the overall USSR balance of payments is also illustrated by CHADWICK, LONG & NISSANKE (1988), whereas GORST (1985) outlines the persistent shortage of oil in the Eastern European satellite countries due to insufficient delivery volumes coming from the USSR. The situation of past and present domestic production, export potential and import requirements in USSR and other Eastern European countries is summarized by OILMAN (1987 c) and ORASIANU (1987).

The 1986 oil price crash was the reason that for the first time since 40 years, the value of the total foreign trade of the USSR slashed by abt. 10 %, with most of the drop stemming from lower oil and gas prices (OIL GAS JOURNAL 1987 h, 1987 l). At the bottom of the line, however, in 1986 reduced exports to Western Europe and other world market countries accounted for the entire decline in the USSR foreign trade value, whereas the USSR trade with the socialistic satellites continued its steady increase in 1986. The reason for this constellation is the transitional phase from high to low oil prices which was of different duration in Western and Eastern Europe. While exports to Western Europe (mainly to Germany FRG) were rapidly reduced because the low oil price after the crash was at once valid there, agreements with its socialistic satellites to base the trade on average world prices during the previous five years (corresponding to plan periods) rather than current world prices enabled the USSR to make still comparatively more profit from either selling or bartering with other Eastern European countries in a prolonged transitional phase which, however, as a consequence of persisting relatively low oil prices until the end of 1988 and beyond, is certainly soon coming to an end.

2.4.4.1.5. USSR oil production

In 1987, the annual plan quantity of oil production was for the first time since 9 years not only met, but also exceeded (RIGASSI 1986, 1987), and benefiting from this, oil export volume rose to a record level, but on the other hand, lower oil price, increasing exploration costs and weaker US $ have in the meantime eroded the profitability of energy trading (GORST 1988). Calculating on global level, 1987 has brought the USSR to the second place after Saudi Arabia in the world league of oil exporting countries. Despite gloomy forecasts following a two-year decline in output in 1984 and 1985 and the reorganization phase in 1986, overshooting of the target quantity for 1987 was established so well that the lower range of the 1990 production goal has already been almost reached.

This again underlines the necessity of stabilization of this level and even expansion of capacity in the next years which certainly has to include a reasonable amount of stimulation. The successful achievement of the record exploitation level also highlights the effectivity of the political and economical reorganization (perestroika; GORBATCHEV 1987; cf. section 2.4.4.3.) which led to major improvements already in a short period of time. The evolution of oil production in the USSR during the last decade is summarized in a discussion of meeting and overshooting of plan forecast, impact of record production for oil exports, and significance of Western Siberia share.

2.4.4.1.5.1. Meeting and overshooting of plan forecast

Oil production level of the USSR in the last years peaked at 12.4 Mio. bbl/d in late 1982 and then gradually decreased via 12.3 in 1983 and 12.2 in 1984 to 11.9 in 1985, but then as a consequence of the effectivity of economical and technical reorganization (perestroika; GORBATCHEV 1987; cf. section 2.4.4.3.) started to rise via 12.3 in 1986 and 12.3 - 12.5 in 1987 to the record of 12.4 - 12.8 in 1988, with the plan forecast of 12.8 in 1990 thus being already touched (RIGASSI 1986, 1987). This overshooting of the plan target terminated an almost fifteen year period of annual falling short of the plan goal, opens the way to further economical and technical improvement of oil exploitation and supply, and underlines the effectivity of the restructuring concept. It is planned to further increase daily oil production to abt. 13.0 - 13.5 Mio. bbl by the year 2000 (OIL GAS JOURNAL 1987 o) and due to the record production of 1986 - 1988, it is almost without doubt that the plan goal of 12.8 Mio. bbl for 1990 will be met.

2.4.4.1.5.2. Impact of record production for oil exports

The record annual oil production of 624.2 Mio. t in 1987 (ERDÖL-ERDGAS-KOHLE NACHRICHTEN 1988 b) enabled the USSR to become the most important oil supplier for the European Economic Community (EEC) group in 1987 and to increase its market share with respect to 1986 considerably at the expense of that of the OPEC and particularly Saudi Arabia (ERDÖL-ERDGAS-KOHLE NACHRICHTEN 1988 a). While in 1986 the EEC net oil imports amounted to 355.5 Mio. t of which 33.4 Mio. t or 9.4 % derived from the USSR, 71.3 Mio. t or 20.1 % came from Saudi Arabia and 246.1 Mio. t or 69.2 % were delivered by the OPEC in total, the trend was completely reversed in 1987 when the 346.9 Mio. t of net oil imports were covered by 41.1 Mio. t or 11.8 % by the USSR, 39.3 Mio. t or 11.3 % by Saudi Arabia, and 212.8 Mio. t or 61.3 % by the OPEC in total.

2.4.4.1.5.3. Significance of Western Siberia share

Western Siberia currently accounts for more than 60 % of the total USSR oil output. The supergiant oil field Samotlor is the second largest in the world and has already produced nearly 20 Bill. bbl of oil (RIGASSI 1986). Other Western Siberian supergiants are Fedorovo and Mamontovo (OIL GAS JOURNAL 1987 o). Western Siberia provided abt. 30 % of the USSR oil output in 1975, and since then the share of the area progressively increased via 52 % in 1980 and 61 % in 1985 to 64 % in 1987, with estimations predicting a further rise to abt. 75 % in 2000. Assessments document that probably only some 30 % of the ultimate geological oil resources have been found so far in the USSR, and current proven and highly probable recoverable reserves reach abt. 100 Bill. bbl (excluding additional reserves that can be produced by enhanced oil recovery techniques), with two thirds of all the known reserves being concentrated in Western Siberia (RIGASSI 1987).

2.4.4.1.6. Currency convertibility and exchange rates

The USSR has a special situation in the world economical system, as the agreed or fixed exchange rates between the non-convertible rouble and the convertible US $, DM and Ŀ remain more or less constant despite any minor or major alterations of exchange rates among Western currencies and other Eastern European currencies. In addition, the USSR is the only Eastern European country which can buy goods in almost unlimited quantity and without unfavourable restrictions of timing against hard currency that can be paid without any problems once Western material, equipment or technology is urgently needed. The only complication is the early budget fixation in the five-year plans which makes major changes and shifts during running plan periods difficult unless supported by exceptional unforeseen situations and backed-up on a high political and technical level.

2.4.4.2. Political restrictions in the past

The impact of political restrictions in the past is characterized by discussing the USA petroleum equipment supply embargo 1978 - 1987 and the status after liquidation of the foreign trade sanctions.

2.4.4.2.1. USA petroleum equipment supply embargo 1978 - 1987

The marketing and sales potential for Western service and proppant companies in the USSR was until recently seriously disturbed by the foreign trade policy sanctions enacted by President Carter in 1978 and later issued in even more serious form in late 1979/early 1980 and late 1981/early 1982 (OIL GAS JOURNAL 1987 p) which practically nearly completely stopped exports of petroleum equipment from the USA to the USSR for almost ten years. The sanctions have only been removed by President Reagan in early 1987 to adjust to the fact that USA companies no longer clearly dominate the world market in oil-field equipment, but comparable material could now be purchased by the USSR from all around the world and especially from Europe and Japan (COTTRILL 1987, ERDÖL-ERDGAS-KOHLE NACHRICHTEN 1987 a, LANDERS 1987, OIL GAS JOURNAL 1987 i), with at the bottom of the line the USA having lost tremendous business opportunities to competitors in other parts of the world rather than achieving the goal of the embargo to hurt the USSR by hindering the growth of its hydrocarbon industry.

Due to the widespread foreign availability of products of comparable technology level particularly in Europe and Japan, the American sanctions on hydrocarbon equipment and technology had totally lost their impact, and in addition to the technical nonsense, they had a significant debilitating effect on the American petroleum industry by resulting in abt. 2 Bill. US $ business loss during the time of the embargo (OIL GAS JOURNAL 1987 i). Trade sanctions, however, still apply for computer technology, advanced seismic products, and high-pressure equipment including stimulation pumps (RIGASSI 1987).

2.4.4.2.2. Status after liquidation of foreign policy sanctions

As the early 1987 oilfield sanctions lift, however, took place after conclusion of the organization of the new five-year plan (1986 - 1990), the renewed marketing approach to the USSR by foreign companies may apart from some exceptions take more time than usually expected, but this is no reason for any delay, because the preparations and negotiations should preferably be concluded early enough prior to the final decisions on the budget and the distribution of expenses for the next five-year plan, thus resulting for the USA at the bottom of the line in another serious business loss to European and Japanese competition. Therefore the ultimate relaxation of the restrictions may have arrived too late for many American companies (OIL GAS JOURNAL 1987 p), and regaining of lost market shares is even complicated by the unreliability of the USA foreign policy which at any time could revert back to sanctions, and the efforts of the USSR to become self-sufficient in oil and gas technology requirements.

The presently tremendous demand of application of stimulation as a viable tool of increasing oil and gas production in the USSR became very prominent during the international oil and gas exhibition in Moscow in October 1987 (Neftegaz; COTTRILL 1987) which was the first major stage of unlimited conversation on all aspects after lifting of the American oil-field embargo and which was consequently used by the Soviets to express their enormous interest in cooperation with Western companies on the base of mutual benefit. Particular interest of American and Western European companies aimed on the offshore market in the USSR which has abt. one fifth of the world`s entire continental shelves, but has so far not very much developed offshore activity (COTTRILL 1987; cf. also section 2.4.4.5.5.).

2.4.4.3. Present stimulation potential and economical restructuring

Hydraulic proppant fracturing in Eastern Europe (with the exception of JUGOSLAWIA) included so far mainly small- to medium-size jobs that were carried out with natural sand as a propping agent and with predominantly Russian or Romanian equipment in own service (such jobs have particularly been done in Hungary, Poland, Germany GDR and the USSR, and subordinately also in Czechoslovakia and Romania, but have so far not been performed in Bulgaria and Albania). The overview as follows illuminates some aspects of supply of synthetic high-quality proppants, political and economical reorganization, and hydrocarbon production goals of the five-year plan. Differences in drilling activity in Western and Eastern Europe after the 1986 oil price crash are also outlined.

2.4.4.3.1. Supply of synthetic high-quality proppants

Domestic production of synthetic intermediate- and high-strength proppants is hitherto limited to manufacturing of very small quantities in Hungary as a by-product in aluminum industry (cf. also section 1.3.1.2.). As a consequence of the urgent need of the USSR to obtain hard currency by petroleum sale in the West and the necessity of increasing domestic supply of oil and gas in the satellite countries of the USSR, stimulation by both acidizing and hydraulic fracturing is considered to have a great potential in Eastern Europe in the coming years, with supply of synthetic proppants and equipment and/or service from Western countries (cf. also sections 2.4.5.5.3. and 2.4.6.3.2.) being indispensible for achieving the goal of acceleration of own hydrocarbon production in the Eastern European states.

2.4.4.3.2. Political and economical reorganization

Acceleration, restructuring and transparency are the current top guidelines of the USSR both domestic and foreign policy (uskorenie, perestroika and glasnost; GORBATCHEV 1987) which certainly will be sooner or later transferred also to the satellite countries, with the only question being the actual timing of the ideological transfer. Due to their validity in all branches of the economical and political systems, the guide slogans of acceleration, restructuring and transparency thus also apply for oil and gas industry development, with enhancement of hydrocarbon production having presently particularly high priority. While restructuring is the major goal of improvement of the political system, speeding-up is most important to enhance the effectivity of the oil and gas industry to enable faster access to known potential by both increasing reserves and accelerating offtake, and transparency is a top item for amelioration of the technological evolution by guaranteing better domestic and international communication and exchange and thus facilitating understanding.

2.4.4.3.3. Hydrocarbon production goals of the five-year plan

According to the new USSR five-year plan (1986 - 1990), oil consumption should be decreased and oil is to be replaced in increasing amounts by gas which means intensification of natural gas production that is certainly including quite some stimulation (ERDOEL-ERDGAS AKTUELL 1985 b, ERDÖL-ERDGAS-KOHLE NACHRICHTEN 1986 a). Further hydraulic fracturing and gravel packing impetus can be expected from the recent unsatisfying oil production which lagged behind the commitments in the last five-year plan (1980 - 1985; EBEL 1986, ERDÖL-ERDGAS-KOHLE NACHRICHTEN 1986 a; cf. also section 2.4.4.1.5.), and therefore something has to be done to compensate the backlog by continuously meeting the goal at the end of the current five-year plan by increasing output from existing fields in addition to new discoveries and appraisals. Another indicator of the enormous stimulation potential is the retardation of field development which is highlighted by the fact that although the USSR is one of the oldest oil-producing countries in the world, over 70 % of the current output comes from only 5 % of the deposits (ORASIANU 1987).

In 1987, for the first time since 9 years the planned annual oil production volume had been fulfilled and even exceeded (RIGASSI 1986, 1987) and a record in exploitation was set, with now the emphasis being on stabilization of this level so that considerable quantities can be diverted for export to Western Europe (OIL GAS JOURNAL 1987 l, ORASIANU 1987, GORST 1988). In 1987 and 1988, the record oil production reached 12.4 - 12.8 Mio bbl/d (FUEG 1988; cf. section 2.4.4.1.5.). CHADWICK, LONG & NISSANKE (1988) give an overview of USSR oil exports and associated trade adjustments, refining constraints and market behaviour, as well as substitution of oil by gas as a consequence of backlogs in oil exploration and exploitation. Past, present and future oil production volumes and potentials of the USSR are summarized by ORASIANU (1987).

2.4.4.3.4. Differences in drilling activity in Western and Eastern Europe after the 1986 oil price crash

The partial independence of the USSR from the world market and its efforts to ameliorate hydrocarbon supply security are underlined by the fact that in contrast to considerable budget cutting in the western hemisphere for drilling and completion and resulting seriously less exploration and development activity with respect to original plans before the 1986 oil price crash, the USSR has in 1986 even stepped up drilling in order to rapidly replace maturing oil production from progressively depleted fields (OIL GAS JOURNAL 1986). Because of its role as an overwhelming source or hard currency, petroleum development commands top priority in the USSR also after the 1986 oil price drop, whereas at the same time in the USA, petroleum has fallen to the bottom of the governmental priority list (KINNEY 1987). While operators in USA and Western Europe are reluctant to invest money in new projects, the USSR has boosted its efforts to sustain and expand oil and gas production and exports. The logical result of these enormous efforts regardless of economical constellations have been the overpassing of the plan goal for the annual oil production quantity in 1987 with breaking the exploitation record so pronouncedly that oil exports reached a maximum (RIGASSI 1987, GORST 1988). In order to maintain or even increase this favourable constellation, it is planned to exceed the exploration and appraisal drilling budget volume of the period 1981 - 1985 by abt. 40 % in 1986 - 1990 (RIGASSI 1986).

2.4.4.4. Self-servicing treatment policy

Concerning marketing approach for stimulation and other services, the different logistical and financial situation between Western and Eastern Europe results in a different strategy of treatment operation. While in Western Europe and various other parts of the world oil and gas production companies commonly ask established service companies of mainly North American provenance to carry out the whole stimulation jobs including supply of all the mechanical and chemical additives for the operation (cf. also PRAST 1980), in Eastern Europe the production companies often prefer to buy their own equipment, to purchase most or all of the material directly and to do the jobs without the support of Western service companies after probably a few introductory, training or pilot treatments (exceptions of this attitude are up to now the MHF operations in Yugoslavia and to a lesser amount also in Hungary on the other side of the state boundary).

While the self-servicing strategy so far mainly included performance of smaller fracturing jobs with predominantly Russian or Romanian equipment, several Eastern European countries (particularly Yugoslavia, subordinately also Poland and USSR) are intending to carry out also bigger treatments theirselves with high- pressure equipment which they are going to buy in the west. The business of Western service and supply companies in Eastern Europe is therefore in many areas restricted to delivery of proppants, fluids, additives and other chemicals, with only in some regions also including introductory service, training, supervision and advice.

2.4.4.5. Economical and financial aspects

In terms of economical systems and financial possibilities, differences have to be made between the individual Eastern European countries. Comments are offered on world bank credits for Yugoslavia, Hungary, Poland and Romania; export revenues in Yugoslavia, Czechoslovakia, Germany GDR and Bulgaria; hard currency acquisition of USSR and Germany GDR, political and economical reorganization and joint ventures.

2.4.4.5.1. World Bank credits for Yugoslavia, Hungary, Poland and Romania

The first group of Eastern European countries includes Yugoslavia, Hungary, Poland and Romania which are members of the International Monetary Fund and thus are able to obtain credits from the World Bank in US $ (cf. also RIGASSI 1985, 1986; SAUNDERS & SHIRAZI 1986) for expansion of domestic hydrocarbon production by drilling, testing and stimulation. World Bank Petroleum Projects including cooperation with Western service and supply companies are already running in Yugoslavia and Hungary (cf. sections 2.4.3.1. and 2.4.6.2.) and are coordinated by the national foreign trade offices. Poland and Romania are at the moment preparing applications for World Bank sponsorship or are already in the stage of negotiation on amount and timing of funding.

2.4.4.5.2. Export revenues in Yugoslavia, Czechoslovakia, Germany GDR and Bulgaria

Other Eastern European countries comprising mainly Yugoslavia, Czechoslovakia, Germany GDR and Bulgaria have sufficient income of hard currency from high export volumes of various goods and services to be independent from World Bank Credits and to be able to finance petroleum drilling and stimulation themselves. The most pronounced example of this category so far was the MHF campaign in Yugoslavia in 1986 (cf. section 2.4.3.1.) which was performed without any support by the World Bank and which is continued in 1987/1988 in addition to an independent World-Bank-sponsored action being part of a larger petroleum project.

2.4.4.5.3. Hard currency acquisition of USSR and Germany GDR

Hard currency for payment of material, equipment and service which cannot be purchased in Eastern Europe and which are indispensible for the decided technological progress and the achievement of the set goals of economical planning is principally no problem for USSR and Germany GDR. The USSR is profiting from large incomes from export of predominantly gas and oil (the role of natural gas in the east-west energy trade in Europe is outlined by D`ANDREA 1987; cf. also section 2.2.1.6.3.). Other USSR convertible money sources are international land and sea cargo and passenger service and export of various other goods, and the GDR is making considerable advantages from the preferential trade with the FRG at special bilateral conditions as well as taking fees from transit travel, obligatory exchange at an expensive artificial rate from foreign citizens coming to the country as visitors as well as participants in conferences, fairs and exhibitions. To a lesser amount, almost the same applies for Romania and Bulgaria which are both working as drilling contractors abroad, with Romania also extensively rebuilding Western oil-field equipment and selling it within Eastern Europe for certainly dumping prices, but for exclusively hard currency or also barter trade of goods which it otherwise would have to purchase for much money on the open world market.

2.4.4.5.4. Political and economical reorganization

The political and economical reorganization which is assessed to be the second revolution in the USSR (GORBATCHEV 1987) is briefly characterized along the guidelines of restructuring, transparency and acceleration (perestroika, glasnost and uskorenie), and the mutual benefit for international cooperation is also stressed.

2.4.4.5.4.1. Restructuring, transparency, acceleration (perestroika, glasnost, uskorenie)

The political and economical reorganization (perestroika; GORBATCHEV 1987) in the USSR gives also increasing freedom and responsibility to exploration and production companies and field operations. While formerly all the subsidiaries were fully dependant on the ministries of oil and gas industry concerning budgeting, approval, negotiations with foreign companies and purchasing decisions, many of them are now already more or less independent and have their own financial budget which they can spend according to their requirements without the necessary back-up by the ministries, and even not everything has to go necessarily through the foreign trade offices which hitherto were almost exclusively responsible for conduction of negotiations and conclusion of contracts for import and export (OILMAN 1987 b).

The economical restructuring requiring individual enterprises to be self-financing is expected to be a suitable motor for energy conservation and reserve maximization in order to avoid closure of companies failing to show profitable operation (GORST 1988). The main guidelines for economical restructuring and administrative reorganization are institution of cost accounting as well as profit and loss criteria in determining efficiency of different industry branches (OIL GAS JOURNAL 1987 o). This new structure characterized by transparency (glasnost) gives much more flexibility for problem analysis and treatment contract conclusion and therefore is one of the ways that will very much allow to expand the volume of well stimulation in the USSR. Direct negotiations with Western supply and service companies as well as improved cash flow based on own budgets are significant elements of acceleration (uskorenie) of the procedures which allow much faster business conclusion than hitherto (OIL GAS JOURNAL 1987 p).

2.4.4.5.4.2. Mutual benefit for international cooperation

Temporally matching with the lifting of the former USA sanctions against the USSR oil and gas industry, the present reorganization (perestroika) does not only affect political relationships, but also the economical pattern by simplifying earlier interwoven relationships, short-circuiting of procedures, and generation of direct communication and approval. The results of the governmental rebuilding of the state structure for the hydrocarbon industry are mutual benefit for international cooperation and joint efforts to master a situation which is critical for either side and where only combined actions with the aim of achieving the optimum goal of the objectives can give rise to the expected success and overall improvement of the indigeneous hydrocarbon supply situation as well as international cash flow and healthyness of service and supply industry (some aspects of the economical restructuring in the USSR are also discussed in OIL GAS JOURNAL 1987 p).

2.4.4.5.5. Joint ventures

Joint ventures are of particular interest for the USSR for prospection and acquisition of hydrocarbon reserves in offshore areas. Comments are offered as follows on Baltic Sea and other USSR offshore areas.

2.4.4.5.5.1. Baltic Sea

The special situation of the USSR is also valid for the joint venture exploration and exploitation of the Baltic Sea by USSR, Poland and Germany GDR (OIL GAS JOURNAL 1987 b), which is the only case within whole Eastern Europe where even cooperation with and coordination by a foreign trade office is not necessarily included into the procedure of practically almost only direct negotiations between the joint venture hydrocarbon production company and Western supply and service companies. 25 wells have been drilled in the Baltic Sea since 1981 and several promising hydrocarbon accumulations have been found (FUEG 1988). Conclusion can thus be made that more or less all the Eastern European countries have possibilities of financing hydraulic proppant fracturing done with Western material, equipment and service, and the only task is to approach these markets with a sound strategical concept and sufficient patience for the long necessary preparation of business conclusion and job execution, and with the mentality of unrestricted cooperation for mutual benefit without any prejudgements.

Regardless of their different hard currency possibilities, all the Eastern European companies are always looking into the matter of counter or barter trade to enhance their financial power by saving convertible money for other purposes. As many Eastern European companies can deliver various raw materials and goods very cheap due to low-expense labour, exchange trade with suitable materials is certainly also of interest to Western suppliers, particularly because an agreement on counter or barter trade is often the fastest way of business conclusion in Eastern Europe.

2.4.4.5.5.2. Other USSR offshore areas

The economical and political reorganization also created some new business possibilities for foreign companies in the USSR by entering join ventures where foreign participants can hold up to 49 % share and have to repatriate profit (FUEG 1988). Such joint ventures have their own independent budget, their self-financing, and are not subject to the national plan (RIGASSI 1987). The joint venture also has priority in USSR allocation of financial and construction resources. Several Western European and American companies are already in advanced negotiation stage with Soviet ministries and firms to bring such joint ventures into life. A joint venture could also emerge as solution of the demarcation dispute between USSR and Norway concerning the Barents Sea (RIGASSI 1986). The special interest of the USSR is concluding joint ventures for exploration and development of remote offshore areas (RIGASSI 1987). The joint ventures generally should have the ability to earn hard currency through exports (OIL GAS JOURNAL 1987 p).

2.4.4.6. Reservoir engineering and hydrocarbon supply aspects

Different necessities and requirements exist for gas and oil supply in the USSR as a consequence of reservoir characteristics and reserve availabilities. While gas delivery in large amounts exceeding the actual demand is no problem at the moment and in the future, considerable improvement of oil supply is essential to enable satisfactory industrial and urban delivery as well as to fulfill the goals of the five-year plan.

2.4.4.6.1. Gas supply

The reason why MHF operations on gas reservoirs have so far not been carried out in the USSR, and also smaller-scale fracturing treatments were frequently performed in individual wells upon special occasions and not in larger planned campaigns affecting whole fields or even groups of patches, is the presence of huge amounts of gas in moderate- to high-permeability sandstone reservoirs particularly in Siberia where stimulation is not necessary, because significant increases of the already primarily high production rates cannot be achieved. In contrast to reservoir treatments which are the dominant challenges in many Western European gas provinces, surface measures caused by permafrost-related problems are the major task in the Siberian gas district. The easily accessible reserves in the Siberian giant gas fields which can be produced cheaply are the main base for the large gas quantities which the USSR exports to Western Europe for the sake of receiving hard currency (cf. section 2.2.1.6.3.), and the potential for satisfying both foreign and domestic demand seems to be almost unlimited at least during the next decades, especially after a long large-diameter pipeline has been built and is already under safe operation without any disturbances since several years.

Therefore in contrast to Middle Europe there is at the moment and probably also in the next few years no need for abundant MHF stimulations on gas reservoirs in the USSR, because the enormous amounts of gas which can be recovered without treatment are sufficient to fulfill the international delivery obligations allowing even still extensions of the commitments in case the Western European market is further growing and demanding and to secure indigeneous supply with again sufficient free capacity to meet potentially rising requirement by domestic industrial enlargement. Given a comparable situation of the North Sea Rotliegend tight gas field belt in the USSR, there is no doubt that in such a case also similar large-scale hydraulic fracturing would take place there, because the technological understanding is well developed and no problems exist concerning financial capability for purchases of equipment, materials and/or service in the world market.

2.4.4.6.2. Oil supply

Oil exploitation, however, needs in contrast to gas production much intensification particularly in the coming years (EBEL 1986, ORASIANU 1987), because also the USSR was hit by the 1986 oil price crash (cf. section 2.2.1.1.2.) and the recent US $ exchange rate collapse (cf. section 2.2.1.3.) and thus has to compensate profit losses from sale of oil on the world market which were not foreseen and planned and thus the equivalents of money and material are lacking in the budgets (cf. section 2.4.4.1.5.). While most of the hitherto performed hydraulic proppant fracturing in the USSR already focussed on oil production and oil-field water injection wells and comprised to minor amounts also unmineable coal seam degasification boreholes, still considerably more oil reservoirs have to be treated in the near future for the purpose of meeting the goals set by the new five-year plan in order to secure domestic supply, delivery to satellite countries by barter trade, and sale in the world market for hard currency at a sufficiently high level.

2.4.5. Western Europe 1987 - 1990

After the slow readjustment and restabilization of the oil price level following the major decline in early 1986, with, however, almost no chance being seen that it would again reach the niveau of the golden late 1970`s and particularly early 1980`s, hydraulic proppant fracturing in Western Europe has already in the first half of 1987 more or less reasonably resumed (this being a time lag to the USA where business started to slowly pick up again already in the last third of 1986) and good expectations exist for predominantly 1988 and 1989. In later 1987, the decline in worldwide exploration drilling finally passed its nadir which was reached in mid 1986 and was more or less kept until early to mid 1987 and that marked the culmination of a progressive slide-down from the zenith in 1981/1982 (POPESCU & HODGSHON 1987).

The readaptation of the oil price to a higher level than during the peak of the 1986 crisis (VARZI 1986), however, is in Europe at the bottom of the line retarded due to the still low US $ exchange rate which kept fluctuating around its preliminary relative minimum throughout mid to late 1986 and even reached in the fourth quarter of 1987 its historical absolute minimum level (with in spite of the temporary slight to moderate improval of the US $ strength in mid to late 1988 even now no perspectives being seen for a renewed reasonable increase of the US $ value within the predictable near future especially in view of the renewed downwards pressure on the US $ value at the end of 1988; cf. section 2.2.1.3.). A general forecast of oil price evolution and US $ rate development representing the basic framework of hydrocarbon exploration and exploitation is given by WEFA (1988), and MACKAY CONSULTANTS (1988) provide a prediction of the oil and gas industry trends.

In Western Europe, an increasing number of gas fields is developed or planned to be developed in the next years in which hydraulic proppant fracturing stimulation is essential to obtain economical production rates, and in many cases, feasible development requires well production maximization by effective fracture treatments with optimum proppant performance (OFFSHORE ENGINEER 1985 c, 1986 c; ROODHART, KUIPER & DAVIES 1986). The most important projects in the British and Dutch Southern North Sea as well as other areas are briefly outlined as follows. A short assessment of the significance of hydraulic fracturing for increasing proven gas reserves is also given, and some comments are offered on the impact of proppant marketing strategy on the stimulation scenery.

2.4.5.1. British Southern North Sea

The outstanding significance of the British Southern North Sea tight gas reserves and their acquisition in the next years are illustrated by discussing as follows role of the major stimulation campaigns in the European gas development scene, cancellation risk of stimulation jobs, self-sufficiency of the United Kingdom gas industry, and significance of gas development activity.

2.4.5.1.1. Role of major stimulation campaigns in the European gas development scene

The main reason for the coming big hydraulic proppant fracturing campaigns preferentially in the British Southern North Sea are definitely planned field developments in the low-permeability zone of the Rotliegend gas belt (ALLISON 1986; GREGORY 1986, 1987) which are reinforced by the fact that the gas has already been sold by means of commitment of supply contracts (cf. sections 2.2.1.6.2. and 2.2.1.6.3.2.) and is one of the major primary energy resources during the next decades even beyond the year 2000 (LÜBBEN 1982, LINDEN 1985, SCHINDEWOLF 1985, STEINMANN 1985, FEHRINGER 1986, BAND 1987, OILMAN 1987 a), and domestic activity is also strongly backed-up by the government which wants to keep imports of foreign gas not being taken into consideration for at least the near future. The Rotliegend development campaigns in the British Southern North Sea in 1987 - 1990 and even later (GREGORY 1986, 1987; OFFSHORE ENGINEER 1986 a, THOMAS 1986, FOX 1987, POTTER 1987, QUINLAN 1988; cf. section 3.3.) are probably the only suite of jobs and/or potential area where similarly as being already common sense in the USA, hydraulic proppant fracturing has become one of the most important aspects of well completion (ROBINSON, HOLDITCH & WHITEHEAD 1986; cf. section 2.3.4.).

Apart from some patches of the Norwegian Northern North Sea where oil-bearing Upper Cretaceous chalk indispensibly requires proppant fracturing, Rotliegend gas development in the British Southern North Sea is the major share of the European hydraulic stimulation activity in early 1987 and will probably more or less keep this role throughout the next years (cf. also section 2.4.1.2.). The Rotliegend campaign is the largest gas development project in the North Sea in the last ten years (McNALLY 1987), and the planned putting on stream of the first group of fields comprising Vulcan, Valiant, Vanguard, Ravenspurn South and Cleeton in late 1988 has been achieved according to schedule (PETROLEUM REVIEW 1988 d; cf. section 2.2.1.6.9.). The inauguration of this field complex in autumn 1988 matches with abt. 50 rigs drilling in the British Southern North Sea which represents the highest level of activity since several years (PETROLEUM REVIEW 1988 d) and again underlines the significance of the Rotliegend belt for gas supply of the United Kingdom.

2.4.5.1.2. Cancellation risk of stimulation jobs

Another reason for an optimistic expectation of the hydraulic fracturing scenery is resulting from the large development plans and includes the shift from mainly exploration drilling in 1986 to predominantly appraisal and development drilling in 1987 through 1990 and beyond. Exploration drilling is often linked with considerable geological and reservoir engineering risk as confirmed by the high cancellation rates (having been above the expectation level just at the peak of the crisis, thereby accentuating it in a Murphy-law-type manner; cf. section 2.4.2.2.). On the other hand, appraisal and production drilling which is starting in the second half of 1987 and lasting until 1989 or even longer has much lower risk, because the structures and their contents are already more or less known, with much lower cancellation rates thus being able to be expected (but not guaranteed due to always present geological uncertainties as a consequence of natural variability even beyond sophisticated human and computerized modelling and interpretation). From appraisal and early development drilling, expectation can be made with sufficient confidence that most of the wells, if in some fields not all of them, will have to be stimulated by hydraulic proppant fracturing in order to establish economically feasible gas flow rates from the tight sandstones (OILMAN 1987 a; cf. sections 3.4.2. and 3.8.).

The centres of MHF stimulation of tight Rotliegend gas reservoirs are Indefatigable, Ravenspurn, Rough, Thames, Valiant, Vanguard and Vulcan (cf. fig. 6 and tab. 7). An outline of the petroleum exploration, development and stimulation potential of the North Sea is also given by HOYLE (1983), and NYSTAD (1981) provides an economical analysis of the North Sea oil and gas region.

2.4.5.1.3. Self-sufficiency of United Kingdom gas industry

The significance of the British Southern North Sea is highlighted by the facts that it is the traditional centre of United Kingdom gas production, has historically contributed up to 90 % of Great Britain gas delivery, still holds over 50 % of the total British proven and probable gas reserves and needs extension of delivery capacity in order to be able to meet winter requirements of daily up to 10 Bill. ft3 (cf. section 2.2.1.6.3.). The circumstance that British Gas as the national distributor is needing significant new supplies of gas in the 1990`s is the major challenge for the operating companies to develop Rotliegend fields in the British Southern North Sea (cf. section 2.2.1.6.2.), because they have an important advantage over the Norwegian competition in that the United Kingdom tax system is far more favourable than the harsh Norwegian tax regime, and that development and production costs are in the Southern North Sea considerably cheaper than in the Northern North Sea.

Including abundant large-scale hydraulic fracturing stimulation, sufficient amounts of gas can be liberated from the Rotliegend field belt to meet the demand of both British Gas and direct industrial users (PETROLEUM REVIEW 1988 d). By end of 1987, the Southern North Sea Rotliegend field belt had cumulatively produced abt. 80 Tcf (2,000 Bill. m3) of gas since the beginning of offtake in 1967. Average daily withdrawal and number of fields being developed in the British sector is twice as high as in the Dutch sector. On the other hand, the vast Groningen field onshore Netherlands produces alone almost as much as the entire United Kingdom continental shelf.

Additional inspiration and feed-back are the improvement of the prospects of obtaining bank finance for gas developments, as at least some banks want to increase loans that are backed by gas with its more stable price than by the unreliable oil (FISHMAN 1986) where much of the earlier investments and credits have been devalidated (cf. section 2.2.1.5.4.). Although the area is mature and production from the major fields is in decline, there is still a suite of smaller accumulations which can be tied into the existing infrastructure to meet and to compensate the projected United Kingdom gas supply shortfall in the next decade (GREGORY 1986).

2.4.5.1.4. Significance of gas development activity

The British Southern North Sea is expected to be the most active European area for gas field development during the next few years (KNOTT 1986), particularly due to so far cancellation of the idea of contracting gas from new Norwegian fields such as Sleipner and Troll in order to compensate the declining delivery from Frigg (which is crossed by the boundary between United Kingdom and Norwegian North Sea sectors) and decision to replace the quantities by increasing offtake from tight Rotliegend gas fields (cf. section 2.2.1.6.2.). The field development campaigns necessary to deliver the gas quantities that have to replace the declining and later ceasing supply from Frigg and to satisfy the increased demand have the deadline to feed first gas volumes into the pipeline network of British Gas by autumn 1988 (which has in the meantime been successfully achieved) and subsequently progressively through 1990 and beyond (OILMAN 1987 a, CAPEL 1988, VIELVOYE 1988 a).

The considerable drilling, completion and stimulation campaigns taking place since mid 1987 represent the most pronounced activity in the British Southern North Sea since the initial boom that followed gas discovery in the early 1960`s (VIELVOYE 1988 a) and mark a major milestone almost matching with the 25 year jubilee of development history of gas fields in the British Southern North Sea (OILMAN 1987 a). A suite of development proposals have already been approved by the government in time to fit into the gas delivery schedule, most of the gas has been sold to British Gas which is still after its privatization the monopolistic gas buyer in the United Kingdom (cf. section 2.2.1.6.2.), and another galaxy of fields is waiting on acceptance of development plans and contraction of gas (OILMAN 1987 a). Although Great Britain is a relatively young hydrocarbon producer and supplier because North Sea exploration and exploitation started only some 20 years ago, it is already self-sufficient in oil and gas supply since many years, and the large-scale acquisition of further gas reserves including significant contributions of hydraulic proppant fracturing stimulation of tight Rotliegend gas helps to secure the United Kingdom self-sufficiency of gas at least until the year 2000 (KNOTT 1985, 1986).

2.4.5.2. Dutch Southern North Sea

In comparison to the British Southern North Sea, the gas prospectivity of the Rotliegend in the Dutch North Sea is slightly to considerably lower and thus also the hydraulic proppant fracturing potential is not so pronounced. Reasonable activity, however, is also going to take place in the Dutch Southern North Sea as a consequence of shifting projects from the oil spots to the gas patches following the 1986 oil price crash, with the main attractivity given to investment and profit levels by the shallower water depth of the Dutch sector and the general proximity to existing pipelines and processing facilities, combined with a reasonable tax regime (BEUDELL 1986 b). Rotliegend hydraulic fracturing stimulation in the Dutch Southern North Sea has been already carried out and/or is still going to be performed mainly in the fields K-12, L-8, L-10 and P-2, and Buntsandstein treatments focussed in the past and/or concentrate in the future on the fields P-6 and P-8 (cf. fig. 6 and tab. 8). Some aspects of Rotliegend gas fields in the Dutch Southern North Sea are discussed by HAAK & ELEWAUT (1988).

In the German North Sea, so far no commercial gas or oil discovery has been made, and due to the sequence of exploration wells including (at least in view of economical feasibility in terms of the offshore situation) so far only failures because of insufficient structural reserves or too high nitrogen content of the gas (cf. section 3.3.4.; with the only exception being two little inshore fields), only very little gas prospectivity (if at all) remains in this area and certainly there is at the moment no fracturing potential unless a future exploration well has a more favourable result. Carboniferous fracturing onshore as well as offshore in the British and Dutch Southern North Sea will certainly be in a better position if gas production with high water cut (ARENS & BÖHNEL 1985) can be put on a more economical base, thereby allowing also to stimulate Carboniferous gas reservoirs with higher water saturation in the near future (cf. section 2.4.1.2.2.1.). Oil and gas field development details for British, Dutch and Norwegian North Sea until the year 2000 are given by OFFSHORE ENGINEER (1986 c).

The coming decades are also in Europe characterized by the need and urgency for unconventional gas (KUUSKRAA 1980) which is underlined by the tight Rotliegend fracturing campaigns. Economical aspects of the British Southern North Sea Rotliegend gas belt and the development of various major fields in this province in the near future are also discussed by FISHMAN (1986), GREGORY (1986, 1987), THOMAS (1986), ALGAR (1987 a), BAND (1987) and POTTER (1987; cf. sections 2.2.1.6.2., 2.2.1.6.3.2. and 2.4.1.2.). An overview of developed and scheduled oil and gas fields in British and Norwegian North Sea is given in OFFSHORE ENGINEER (1985 c, 1986 c) and SCOTTISH PETROLEUM ANNUAL (1987).

2.4.5.3. Other areas

Apart from Rotliegend gas field development in the British Southern North Sea that will in terms of proppant application be either mainly done with 20/40 sand followed by a tail-in of 16/20 intermediate-strength proppants (of abt. 10 - 30 % of the sand quantity) in case of shallower reservoirs (abt. 6,000 - 8,000 ft) or in case of deeper reservoirs (abt. 8,000 - 10,000 ft) exclusively with intermediate-strength proppants of first 20/40 (5 - 20 % of the whole lot) and then second 16/20 material, and which will comprise by far the largest amount of hydraulic proppant fracturing in Europe in the very near future, stimulation activity in (mainly the second half of) 1987 and 1988 (probably still including the following couple of years) will also focus on (classified information is compiled in tabs. 6 - 9, and localities and regions are illustrated in figs. 4 - 8)

* some Rotliegend and Carboniferous sandstone gas wells in Germany FRG (BRINK-

MANN, FUHRBERG & SCHÖBER 1980; BRINKMANN, KRÖMER & REINICKE 1980; BRINKMANN 1982, SCHWARZ & SCHÖBER 1982; JOHN 1983, 1987; LEICHT 1985; REINICKE, BRINKMANN, SCHWARZ & HUENI 1985; cf. fig. 4) which will be treated with intermediate- or high-strength high-density proppants in a reservoir depth between 3,000 and 6,000 m (9,000 - 18,000 ft; cf. section 2.4.1.1.),

* parts of some Rotliegend gas fields and Cretaceous sandstone oil fields in the Dutch Southern North Sea (cf. fig. 6) which will require sand and/or intermediate-strength high-density ceramic proppants in abt. 6,000 - 10,000 ft (2,000 - 3,300 m) reservoir depth (cf. section 3.3.3.),

* some Upper Triassic oil-bearing sandstones in Germany FRG that require intermediate-strength proppants in abt. 1,500 - 2,500 m (4,500 - 7,500 ft) reservoir depth (cf. section 2.4.1.1.2.2. and fig. 5),

* some Mesozoic and Cenozoic oil- and gas-bearing sandstones in Yorkshire and Midlands onshore Great Britain where sand or intermediate-strength proppants will be applied in a reservoir depth of 5,000 - 9,000 ft (1,500 - 3,000 m),

* parts of some Jurassic sandstone (and locally also conglomerate) oil fields in the British and to subordinate amounts also Norwegian Northern North Sea where sand and/or intermediate- to high-strength proppants will be applied (Brent Sand group; cf. JOHNSON & KROL 1984; DESPAX, CHARLEZ, CLINKEMAILLE & ECONOMIDES 1987) in a reservoir depth of 10,000 - 14,000 ft (3,300 - 4,600 m; cf. fig. 8),

* some Jurassic and Tertiary oil fields in the British Northern North Sea where intermediate- or high-strength proppants are required in 9,000 - 11,000 ft (3,000 - 3,300 m) reservoir depth (cf. fig. 8),

* some Cretaceous chalk oil fields in the Norwegian Northern North Sea (HARTLEY & BOSMA 1985; TANSOE, KING & HOLMAN 1986; ATTARD, MATHES & MOWER 1988; cf. section 4.5.4.3.) where conventional and resin-coated intermediate-strength low-density ceramic proppants are needed in a combination of hydraulic fracturing and gravel packing in abt. 2,500 - 3,300 m (8,000 - 10,000 ft) reservoir depth (cf. fig. 8),

* some Tertiary gas- and condensate-bearing sandstones and conglomerates in Yugoslavia (ECONOMIDES, CIKES, PFÖRTER, UDICK & URODA 1986; cf. fig. 5) where the requested material is intermediate-strength low-density ceramic proppants in the early stage and high-strength proppants in the late stage of the operations in a reservoir depth of abt. 3,000 - 4,000 m (9,000 - 12,000 ft),

* parts of some gas fields in Turkey which will be done with sand potentially followed by a tail-in of intermediate-strength low-density ceramic proppant in a reservoir depth of abt. 1,500 - 2,500 m (4,500 - 7,500 ft; cf. fig. 2),

* some Tertiary sandstone gas fields in Sicily/Italy (cf. fig. 5) where intermediate-strength high-density proppants will be needed in abt. 3,500 - 5,000 m (10,500 - 15,000 ft) reservoir depth, and

* probably some Tertiary sandstone and carbonate wells in Hungary, Southern France and Northern Africa which also require intermediate-strength high-density proppants in a reservoir depth of abt. 1,500 - 4,500 m (4,500 - 13,500 ft).

* The oil fracturing with sand in Cretaceous shallow fields in Germany FRG and Netherlands is also scheduled to go ahead during the coming years, with demand of abt. 20 - 70 t of sand per well in a reservoir depth of abt. 700 - 2,500 m (2,100 - 7,500 ft).

2.4.5.4. Significance of fracturing for increasing proven gas reserves

Rotliegend fracturing in Germany FRG has particularly to be assessed in view of the estimation of JOHN (1983) that proven gas reserves (in late 1982) were 120 Bill. m3, but the remaining stimulation potential was at that time at least in the range of 90 - 120 Bill. m3, implying that consequent successful hydraulic fracturing can - in case of verification of all the assumptions - nearly double the hitherto proven Rotliegend gas reserves (cf. section 3.1.2.). REINICKE, BRINKMANN, SCHWARZ & HUENI (1985) also come to the conclusion that abt. 100 Bill. m3 of gas are only able to be accessed by large-scale hydraulic proppant fracturing of the tight pay formations, with the total hitherto proven gas reserves amounting to abt. 300 Bill. m3 (with this quantity, however, including also the Zechstein carbonate sour gas reserves) and another 300 Bill. m3 being predicted for future discoveries, with thus one third of this quantity reaching economical production only through MHF stimulation treatments (this matches in percentage with the situation in the USA where it is estimated that 30 - 35 % of the recoverable hydrocarbon reserves may be attributed to the application of hydraulic fracturing stimulation; WATERS 1980).

Offshore hydraulic proppant fracturing which is so far limited to rig-, platform- and boat-based stimulation jobs will be promoted during the next years by the technological innovation of a subsea stimulation vessel that is equipped with a riser and that is able to carry out operations independent from well heads reaching the water surface, thus opening the possibility of fracturing also wells with subsea completion in the North Sea (STENA 1986). Some operations have also already been carried out from a semisubmersible (HOOD & POWELL 1982).

2.4.5.5. Impact of proppant marketing strategy

In terms of marketing strategy and its impact on proppant sales levels, comments are offered on direct proppant sales to end users, liquidation of service company stocks, and stand-by supply by proppant companies.

2.4.5.5.1. Direct proppant sales to end users

The hydraulic fracturing scenery and particularly proppant choice can to some extent also be influenced by the marketing strategy of the proppant companies. While traditionally sales were almost exclusively made to service companies as intermediate users which then incorporated the proppants into the whole stimulation operation and offered their clients, the production companies representing the end users, the complete job package as an integrated set (including a standard business mark-up range of the proppants), the rapid revision of many budget items for drilling and fracturing after the 1986 oil price drop has opened the door for a considerable change also in this matter. Some aspects of drawbacks of service company stocks and treatment self-performance in Eastern Europe are outlined as follows.

2.4.5.5.1.1. Drawbacks of service company stocks

As a consequence of partially excessive stocks on which service companies kept sitting for quite some time after optimistically ordering enough material to be able to carry out two to three jobs over night in good times of the fracturing scenery and then painfully awaking when after the 1986 oil price crash the expected programmes were suspended or cancelled by the operators, with the purchased material blocking the service company warehouses and remaining in their books, the mentality of both service companies and proppant suppliers is at least in parts of the European stimulation market changing towards direct purchase of the proppants by the end customers, the production companies.

This constellation would certainly give much more market transparency to the proppant companies than before, because it would be now immediately evident who is the end user and how much material he needs for which type of job. The service company would get rid of the risk of sitting on the proppants once an operation is suspended or cancelled, and would compensate the hitherto mark-up of the proppant price by pumping surcharges per volume unit, thus not significantly losing money at the bottom of the line and earning profit almost exclusively through service and own chemicals. The possibility of direct proppant purchase from the manufacturers instead of taking the material through the service companies is also mentioned by CUTLER, ENNISS, JONES & CARROLL (1983).

2.4.5.5.1.2. Treatment self-performance in Eastern Europe

The concept of direct sales would be particularly valid for Eastern Europe where not only the whole economical structure is quite different from that in Western Europe and USA, but many operating companies have own pumping equipment at least for some jobs and the influence of service companies is in some cases even limited to sale of chemicals. In Western Europe, major companies performing larger fracturing campaigns in 1987 - 1990 may wish to seriously look into possibilities of direct purchase of proppants (cf. also section 2.4.2.4.1.2.2.) which can be carried out even in agreement with the service companies. Concerning the still highly competitive situation, it is certainly worthwhile that both parties talk together on all these aspects in order to find a solution which is most beneficial for both of them.

2.4.5.5.2. Liquidation of service company stocks

A temporary drawback for the proppant suppliers in the transitional phase of probably at least partially changing marketing policy and purchase strategy as well as parallel to business pick-up along with recovery from the peak of the oil price crash is the liquidation of the inventories by the service companies. During the golden years of hydraulic proppant fracturing, the service companies could make much money and win contracts against competitors by being able to move towards location immediately after order entry by call without having still to get access to material. The common policy in these days was to keep proppants, fluids and other chemicals in such amounts in the warehouse and on the boat that two or even three standard jobs could be carried out overnight without any restocking.

Having sit on this reserve inventory for quite some time after the sudden oil price drop, the picking up of the activity in late 1986 and throughout 1987 was taken by the service companies as a welcomed occasion to liquidate their excessive stocks not only in the offshore bases, but also in warehouses at several places on the continent, and to get the inventories out of their books. This could be done because the experience has shown during the last years that most of the operations do not have to be carried out at once, or decision and approval are made early enough to allow proper preparation in terms of material acquisition and stimulation vessel loading, and warehouses of many suppliers in strategically suitable locations in Europe allow shipping of material to arrive at the boat base within a couple of days. That is why the proppant companies did not feel the improving business situation immediately by a jump of the sales level, because by liquidating their stocks, the service companies carried out several jobs from their own supply (with even some short-circuit material move between competing service companies having occurred for the purpose of cheaper buying and faster inventory liquidation) before upon clearing of their warehouse purchasing new material for the following operation.

2.4.5.5.3. Stand-by supply by proppant companies

As the production companies as ultimate customers, however, would certainly not risk to repeat the unpleasant experience of the service companies to pile up material which is then not used up for some time once the proppant market turns at least partially to direct purchase by the end users, the marketing and sales policy of the proppant supply companies would have to focus more and more on short-term stand-by, with orders coming in almost in the last minute and the largest competitive advantage being with the proppant company which can deliver the fastest and always in sufficient quantity. The frequent ordering for delivery almost overnight indispensibly requires the presence of local stocks near a central port in Europe from which both North Sea and onshore areas can be supplied very fast, with those companies that hitherto relied on timely orders early enough to bring in the material from the USA having then certainly a big disadvantage which in the worst case can cut them completely off from business.

Stock management improvement has to aim on enhancing the possibilities of reasonable forecasting of coming stimulation jobs and their material demand in terms of proppant type, grain size and quantity. This task needs continuous market survey with close attention to the original sources and definitely requires a regional European contact office as well as frequent travelling for visiting and interviewing stimulation specialists and commercial representatives in production and service companies in order to receive the most actual information for the planning of the physical distribution of the proppant quantities.

While direct sales of proppants to production companies as end users have been performed already in a few cases in Eastern Europe (cf. section 2.4.4.4.), this strategy would be more or less a marketing innovation for Western Europe (except of earlier limited deals with some production companies which have their own in-house service branch for smaller jobs), but for the reasons discussed above, this change of the marketing trend would be more than self-understanding and could considerably help to secure business for all the participating parties by offering competitive solutions for the production company by intensifying the price bidding of the service companies to get the job which, once proppant price mark-up is smaller than it used to be, will have to chiefly or even almost exclusively concentrate on service rates and thus efficiency of work, and feasibility and reliability of equipment.

2.4.6. Eastern Europe 1987 - 1990

In contrast to Western Europe where the expected boom of stimulation in the period 1987 - 1990 in the North Sea is partially the result of the recovery of the hydrocarbon industry after the major breakdown due to the oil price drop in early 1986, and partially was anyway scheduled in the frame of long-term oil- and especially gas-field development plans that were controlled by the extraordinary offshore logistical situation (cf. section 3.8.1.), the more or less independence of the Eastern European market from oil price tragedies and US $ exchange rate crises up to a certain degree (cf. section 2.4.4.) is the reason for the absence of a special boost in this area during the next years. Eastern Europe is characterized in the very near future by continuation of the existing and already firmly planned campaigns and projects as well as by performance of additional jobs as a consequence of successful promotion by both external marketing activity and internal economical force to ameliorate production for enhanced domestic supply particularly in case of renewed severe winters comparable to that in early 1987, although the risk of technical delay and administrative postponement of operations is in parts of Eastern Europe much greater than in Western Europe.

2.4.6.1. General aspects

The main necessity in the near future for many European countries, but particularly for Eastern Europe, is to increase both relative and absolute amount of domestic hydrocarbon production in order to satisfy the increasing urban and industrial demand (which the severe winter 1986/87 has drastically underlined that had particularly terrible effects of energy shortage in Romania and was in fact the worst winter since abt. 40 years; RIGASSI 1985, 1986) and to achieve some more independence from unfavourable changes of the political and economical framework. As exploration drilling has in parts of the areas already reached such a mature stage that spectacular new discoveries seem to be not very likely and the possible increase of producing fields by new exploration hits might be compensated by declining output from older progressively depleted fields, the only really significant impact is seen in major stimulation campaigns in existing fields and hitherto marginal reservoirs.

Unless a comprehensive agreement on limitation of strategical conventional and nuclear arms and a ban of the threat of a space war is reached between USA and USSR reasonably soon, the urgency of the Eastern European countries to ameliorate their own hydrocarbon supply is considered to become even more serious within the next few years (cf. also RIGASSI 1985), thus also increasing the necessity of stimulation in order to achieve as much as possible political autarky and power. Many areas of application of hydraulic fracturing can be expected, because so far in most of the countries drilling and conventional completion had a higher priority. Some of the most evident and partially already concluded projects in Yugoslavia and Hungary, USSR, other Eastern European countries, and Southern Europe and Northern Africa are briefly outlined as follows, and comments are also offered on China.

2.4.6.2. Yugoslavia and Hungary

Yugoslavia probably remains to be the most important area in terms of hydraulic proppant stimulation in Eastern Europe because of the continuation of the MHF campaign (cf. section 2.4.3.1.; reinforced by the fact that investment for the proppants for all the planned jobs has already been made, cf. section 2.4.2.4.1.1., and the wells which originally have been drilled for the purpose to be subsequently fractured, are presently shut-in and are waiting on fracturing to be able to be produced) that was partially suspended and rescheduled following deterioration of the general national economical situation after the oil price decline in early 1986. The best promotion of a rapid execution of the rest of the MHF project is the partially excellent result of the first part of the large-scale treatment campaign in 1986 which gave a fast pay-out of the total stimulation expenses.

Apart from the MHF jobs being financed by the Yugoslavians themselves without any external help, the independent World Bank Petroleum Project (cf. SAUNDERS & SHIRAZI 1988) includes a suite of smaller fracture jobs where also both intermediate-strength low-density and high-strength proppants are going to be used. Various reservoirs are also waiting on small- to medium-scale treatment with sand potentially followed by a tail-in of intermediate- or high-strength proppants, with the latter jobs, however, preferentially being operated by the Yugoslavian companies with own equipment. The scheduled purchase of high-pressure equipment aims on increasing independence enabling the Yugoslavians to carry out also larger fracture jobs without the support of Western service companies.

Hydraulic proppant fracturing is also planned in Hungary in those parts of the field zone which extend from Yugoslavia across the state boundary as well as in other regions in Southwestern and Eastern Hungary, with most of the treatments being included in the current World Bank Petroleum Project and being done with sand and/or intermediate- to high-strength proppants.

2.4.6.3. USSR

Other activity in Eastern Europe is so far difficult to predict. Certainly traditional sand fracturing will continue in the USSR with own high-pressure equipment of American provenance which has been bought still before the sanctions enacted by President Carter in 1978 that have in the meantime finally been lifted by President Reagan in early 1987 (ERDÖL-ERDGAS-KOHLE NACHRICHTEN 1987 a, LANDERS 1987; cf. section 2.4.4.2.) came into force. Some aspects of hydraulic proppant fracturing potential, import vs. domestic manufacturing of high-strength proppants, significance of intensification of oil and gas exploitation, and priority of petroleum development vs. oil price level are discussed as follows.

2.4.6.3.1. Hydraulic proppant fracturing potential

Hydraulic proppant fracturing potential of partially even massive size in the next few years is particularly seen in low-permeability gas-bearing sandstones in the old gas fields Shebelinskoe/Ukraina (3,000 - 4,000 m reservoir depth; cf. tab. 9), Stavropolskoe/Northern Caucasus (2,000 - 3,000 m depth), Saratovskoe (2,000 m depth) and Krasnodarskoe (3,000 - 4,000 m depth). In the new gas fields Orenburgskoe (3,000 - 4,000 m depth) and Karatchaganakskoe or Borodinskoe (4,000 - 5,000 m depth), contracts with North American service companies are already concluded for carrying out fracturing treatments with intermediate-strength proppants still in 1988 or 1989. Potential for larger- or smaller-scale hydraulic fracturing is also present in the old oil field Sachalinskoe (2,000 - 3,000 depth) and the new oil and gas fields Samotlorskoe/Siberia (2,000 m depth; this is a whole complex of fields with many potential stimulation candidates), and Irkutskskoe (4,000 m depth). Other prolific oil and gas fields where hydraulic proppant fracturing will become important in the near future are situated in Astrachan at the northern border of the Caspian Sea, in the area north of the Black Sea, and in the surroundings of Irkutsk and Novosibirsk. Particular problems presently exist in sour gas exploration and production (especially in Orenburgskoe and Astrahanskoe/Lower Volga).

2.4.6.3.2. Import vs. domestic manufacturing of high-strength proppants

The USSR has a large suite of oil and gas reservoirs in various geological formations and in different depth ranges and as there is hitherto either no possibility or no intention of domestic manufacturing, it is only a question of time (and of priority in the governmental planning) when intermediate- and/or high-strength proppants will be purchased from Western suppliers for application in preferentially deep tight oil and gas wells in the USSR.

Although according to the general state of the industry, it has to be assumed that if required the USSR certainly has possibilities of domestic production of synthetic proppants in sufficient quantity, it is almost certain that at least during the next few years, man-made proppants will have to be imported into the USSR for the stimulation treatments in that country, and only later there might be an option for the Russians to proceed to domestic fabrication which, however, might be a more expensive approach including quite some rejects and sub-specification material before the advanced experience of the traditional proppant producers can be met and proppants of comparable quality can be made in sufficient quantity for widespread and unrestricted application according to the requirements (cf. also section 2.4.4.3.1.).

2.4.6.3.3. Significance of intensification of oil and gas exploitation

Intensification of oil and gas exploitation has currently a very high priority in the general state industrial planning (KINNEY 1987, ORASIANU 1987), and the new five-year plan (1986 - 1990) contains the obligation to decrease oil consumption by progressively substituting oil by gas, which results in the necessity of increasing natural gas production that indispensibly includes quite some stimulation (ERDOEL-ERDGAS AKTUELL 1985 b, EBEL 1986, ERDÖL-ERDGAS-KOHLE NACHRICHTEN 1986 a). The economical impetus of increasing oil production by all possible means is the overwhelming importance of oil exports for foreign trade by being together with gas the source of some 60 % of convertible currency earning (STERN 1986; cf. sections 2.4.4.1.1. and 2.4.4.1.4.). While particularly oil production has steadily declined since 1984, the economical requirement of increasing production after the 1986 oil price crash has considerably accelerated exploitation culminating so far by mid 1988 in a hitherto record output, with the progressive improvement of oil production since the 1986 oil price collapse having enabled the USSR to export both in 1987 and 1988 higher quantities of oil than ever before, and also gas production has reached its so far climax level in early 1988 (OIL GAS JOURNAL 1988 b; cf. section 2.4.4.1.5.).

2.4.6.3.4. Priority of petroleum development vs. oil price level

This is the result of petroleum development commanding top priority in the USSR since several years despite the intervening 1986 oil price crash, with the USSR even after the 1986 oil price drop having still again boosted its efforts to sustain and expand oil and gas production and exports, while on the other hand at the same time in the USA, petroleum has fallen to the bottom of the governmental priority list and lacking confidence in future investment of capital leads to postponement of any projects if only slight doubts are cast on their irrevocable economical feasibility in various scenarios (KINNEY 1987). The persisting top priority of petroleum development and the significance of increasing oil and gas production is also underlined by the fact that the oil price crash in 1986 was the reason that for the first time since the Second World War, the total foreign trade value of th USSR slashed by abt. 10 % in 1986, with most of the drop stemming from lower oil and gas prices (OIL GAS JOURNAL 1987 h; cf. section 2.4.4.1.).

2.4.6.4. China

In a similar way as for the USSR, an increasing fracturing potential can be expected in China, because the new Chinese five-year plan (1986 - 1990) also includes the commitment of significantly increasing domestic oil and gas production (REID 1981, EBEL 1986, ENERGIEWIRTSCHAFT 1986 a, WONG 1987), and while certainly much additional capacity is expected to come from successful offshore exploration, economical exploitation will be indispensibly linked with hydraulic proppant fracturing (aspects of fracturing stimulation in China are discussed by LI & ZHU 1986 and PARKER, ADAMS & LIANKUI 1986). The market perspective is increasing, because a restructuring campaign comparable to that in the USSR (cf. section 2.4.4.5.4.) is also taking place in China and is to streamline and decentralize oil and gas industry administration and operations (KHIN 1988). Reorganization of energy industry, relaxation of offshore contract terms, considerable offshore discoveries, and extensive offshore and onshore developments are the principal highlights of activity in China in the first half of 1988, and in 1987, drilling increased for the eighth consecutive year, moving China into the third rank after USA and USSR in drilling.

Some time ago, China even presented its own high-strength proppants which during tests in Western research laboratories have been proven to be of good quality (cf. section 1.3.1.2.), but later not very much was heard about application and while quite some domestic pumping of the material can be assumed, the Chinese proppants so far were never inserted into a fracture in foreign countries. So far both natural sand and high-strength bauxite proppants have already been pumped in hydraulic fracturing treatments in China, with total proppant quantities not exceeding 120 t in many operations (LI & ZHU 1986). Field experience has shown that in various cases, fractures propped with lower concentrations of bauxite have better conductivities than cracks plugged with higher saturations of sand. The use of high-strength proppants as tail-in after the injection of sand as main proppant load gives little superiority in getting better fracture conductivity, with an uniform proppant package therefore providing the optimum result.

2.4.6.5. Other Eastern European countries

Germany GDR which has limited domestic supply by gas production from Rotliegend sandstones that is not meeting the considerable demand (cf. section 3.10.2.2.) and in light of probably no longer receiving sufficient delivery from the USSR is assessed to be sooner or later forced to enhance offtake by larger-scale hydraulic proppant stimulation than done so far. Several oil- and gas-bearing formations in smaller or larger fields which would require stimulation for achievement of adequate production rates are also present onshore and offshore in Poland, Romania (cf. ORASINAU 1986) and Bulgaria and are expected to come to conclusion and job execution within the next few years.

Hydraulic proppant fracturing potential can be reasonably assumed in every Eastern European country, but the operations are expected to take in some states such as Czechoslovakia, Bulgaria and particularly Albania (which has a special organization of foreign affair policy) still some time of preparation for various circumstances of both technical and administrative nature. The discoveries in the Baltic Sea which have so far been made mainly in its eastern part are hitherto not commercial and development would present major environmental hazards (OIL GAS JOURNAL 1987 b), while in the promising western part recently the geological model collapsed, and structural reinterpretation and rethinking is necessary in order to upgrade this part of the oil and gas province.

2.4.6.6. Southern Europe and Northern Africa

In connection with the fracturing potential in the Eastern European complex, also some highlights from Southern Europe and Northern Africa deserve attention. Turkey should be primarily mentioned in this connection as another area of considerable stimulation potential where following the successful hydraulic proppant fracturing campaigns in 1986 and earlier years quite some treatments to be carried out with sand and/or intermediate-strength low-density proppants in several gas fields are scheduled for 1987 and succeeding years (cf. section 2.4.2.4.2.2.).

Other potential areas of hydraulic proppant fracturing with sand and/or intermediate- and high-strength proppants are Algeria, Libya, Oman, Gulf of Guinea (Gabon, Cameroun and Nigeria), Angola, Congo and Zaire in Northern Africa as well as India and Pakistan in the farther Middle East. In the Far East, systematical fracturing is not carried out at the moment, and only occasionally fracturing jobs are performed with predominantly natural sand. In the Middle East, the overwhelming share of carbonates in the suite of reservoirs is the reason why almost exclusively acid fracturing is done in this area. In South America, proppant fracturing focusses on Brazil and Argentina, whereas Venezuela is characterized by acid fracturing. Proppant fracturing has formerly also been carried out in Peru. While in most of the mentioned countries fracture stimulation is chiefly made with natural sand, especially in Brazil and Algeria various types of synthetic proppants are incorporated into the projects of well treatment.

2.5. Concluding assessment of proppant marketing potential

As a consequence of the regional marketing assessment comprising review, status and forecast, and in combination with an outline of technological perspectives of enhancement of the potential applications and thus extension of the market (chapter 4), conclusion can be made that provided an again more stable oil price at a reasonable level (cf. also ERDÖL-ERDGAS-KOHLE NACHRICHTEN 1986 b), partially also a more favourable higher US $ exchange rate (especially in Western Europe) and an adequate pricing of proppants and service based on high technical quality, both Western and Eastern Europe are growing stimulation markets with increasing demand of intermediate- and high-strength proppants for hydraulic fracturing and gravel packing including promotion of treatments in marginal reservoirs in the near future (a stable higher oil price is considered to benefit all producing countries including OPEC members, because during the 1986 crisis, even Saudi Arabia has suspended some of its future capital expenditure programmes; GRIFFITHS 1986). Careful marketing approach with strategies integrating geological, reservoir engineering, technical and economical concepts are essential for developing of new areal markets where so far stimulation has not had such a significance as in parts of Western Europe or in the USA.

Proppant marketing for hydraulic fracturing and gravel packing is in Europe not so much depending on scientific, technical and engineering aspects of proppant choice for the specific application, but requires in comparison with the USA much more diversification due to different economical, financial and mentality-derived aspects in the spectrum of countries speaking not only a suite of different languages, but having also more or less different strategical planning and practical execution of hydrocarbon exploitation. The main task of technically oriented marketing concepts for application of proppants in the near future is promotion along the lines of technological possibilities of enhancement of economical feasibility of marginal oil- and gas-bearing formations (chapter 4) which will have to contribute in increasing amounts to the overall production with continuing decline of the supply from conventional reservoirs.

3. Rotliegend Stimulation in Europe

3.0. Summary and associated table of contents

The Rotliegend (Lower Permian) continental red bed reservoir complex containing large amounts of gas in British and Dutch North Sea as well as Netherlands and Germany FRG onshore has the greatest hydraulic proppant fracturing potential of all European hydrocarbon-bearing formations. Numerous MHF treatments in tight Rotliegend gas sandstone pay horizons have already been carried out during the last ten years and will be performed in the near future, with proppant quantities of up to 650 t per well being injected into the hydraulically induced cracks during stimulation jobs which cost up to several Mio. US $. The economical significance of Rotliegend proppant fracturing is underlined by the fact that abt. 100 Bill. m3 (4 Tcf) and 500 Bill. m3 (20 Tcf) gas reserves in Germany FRG and North Sea, respectively, can only be acquisited and feasibly produced by large-scale stimulation of the tight pay zones. Successful jumbo treatments are able to verify and confirm up to 500 Mio. m3 (20 Bcf) of additional gas reserves, and this fast amortization is the most pronounced justification for the performance of supergiant stimulation operations in Rotliegend and also Carboniferous tight prospective intervals. The hydraulic proppant fracturing potential of Rotliegend and also Carboniferous reservoirs, to minor amounts also Buntsandstein (Lower Triassic) pay zones, is influenced by various geological and petrophysical factors giving rise to a zoned organization of reservoir quality and thus also stimulation necessity. Distinction can be made between primary petrophysical zonation according to reservoir rock permeability, secondary geochemical zonation due to gas composition, and tertiary political zonation of Rotliegend gas reservoirs with fracturing potential differences as a consequence of organization and dynamics of the economical and administrative systems in the territories where the gas fields are located. Considerable shares of inert nitrogen accompanying the combustible methane is the major geochemical drawback of reservoir prospectivity and thus also hydraulic proppant fracturing potential in several areas of the Rotliegend Basin comprising particularly German North Sea and Germany GDR onshore. The Rotliegend in Germany FRG is stratigraphically divided into three pay storeys with different primary petrophysical potential and/or secondary reservoir engineering requirement of hydraulic proppant fracturing within the Weser-Elbe province or East Hannover facies zone (Wechselfolge facies belt). The Schneverdingen-Formation containing the Schneverdingen-Sandstein at the base of the Rotliegend succession in Germany FRG is equivalent to the Lower Slochteren-Formation in Netherlands and Dutch North Sea, and the Emden-Formation including the Hauptsandstein in the middle section of the Rotliegend profile in Germany FRG correlates with the Upper Slochteren-Formation in Netherlands and Dutch North Sea. The Hannover-Formation containing the Wustrow-Sandstein and various other Wechselfolge sandstones at the top of the Rotliegend series in Germany FRG corresponds with the Ten Boer-Formation in Netherlands and Dutch North Sea. The aeolian dune Schneverdingen-Sandstein (abt. 20 - 50 millidarcy) in the lower part and the fluvial Wustrow-Sandstein (abt. 5 - 10 millidarcy) in the upper section of the Rotliegend sequence of the Weser-Elbe facies province in Germany FRG are generally moderate- to high-permeability reservoirs not requiring stimulation, whereas the aeolian and playa sand flat Hauptsandstein (below 1 millidarcy down to abt. 0.5 - 1 microdarcy) in the middle portion is mainly a low-permeability to tight reservoir and Rotliegend MHF treatments are usually carried out in this horizon. In the British North Sea, distinction is made between two laterally intertonguing units in the Rotliegend stratigraphical complex being the Leman Sandstone Formation which consists of clean aeolian and fluvial sandstones at the southern and western basin margins, and the interdigitating Silverpit Formation in the basin centre that is built up of marginal aeolian/playa and lacustrine sandstones, siltstones, mudstones and evaporites. The Rotliegend reservoir complexes in the Mid-European Basin (Southern Permian Basin) are mainly concentrated to the aeolian facies belt separating the marginal alluvial-fan chain and braided river plain from the central playa sand flat and desert lake area. While cross-stratified aeolian dune sands and horizontal-laminated dry sheet sands represent excellent reservoirs usually not necessitating permeability improvement by hydraulic proppant fracturing, tight facies units originate by increasing humidity in the depositional area of a damp to wet playa sand and mud flat in the Wechselfolge facies belt, with the finer-grained and poorer-sorted intercalations chiefly deteriorating vertical permeability. The main primary-depositional features of permeability destruction that are secondarily requiring enhancement of transmissibility by hydraulic proppant fracturing are adhesion of sand and mud during transitional stages between dry, damp and wet playa and floodplain surfaces as well as sedimentation of thin mud drapes as a consequence of interdune and overbank sandflat flooding by fluvial invasion and atmospherical precipitation. Reservoir quality is influenced by facies development and changes as a consequence of palaeoenvironmental, palaeoclimatological and palaeotectonical history of the Rotliegend Basin. In some areas, the facies complex succession also reflects an evolution of fluvial style including progressive expansion of the alluvial braided-river system within the depositional basin as the marginal fan chains and the associated aeolian dune belt shrink in extension or retreat further towards the border of the sedimentary area as a consequence of basin enlargement by stepwise onlap on the fringing highlands with successive infilling of the trough. Progressive diminution of the degree of braiding of the river systems as the erosional highlands in the provenance area are denudated, decreasing palaeoslope gradient, declining amount and grain size of clastic material delivered from the source region, and ceasing transport capacity of the channel network give rise to enlargement of interstream overbank flats where upon desiccation and reasonable persistance of quiet aquatic sedimentary conditions migration of aeolian dunes and accumulation of aeolian sheet sands from the detritus deflated from emerged sand and gravel bars in the shallow to moderately deep braided river channels is becoming more and more widespread, or reversely is diminishing in importance in case of successively increasing frequency and extension of alluvial inundations by flood pulses overtopping the channel banks at high stage in combination with high ground water level as a consequence of higher stream discharge and increasing precipitation rates once climatological conditions are becoming damper and wetter. Apart from the primary depositional facies zonation controlling together with the diagenetical history predominantly of clay minerals the distribution of petrophysical properties, an anisotropical arrangement of hydraulic proppant fracturing potential is also created by the gas composition within the Rotliegend reservoir rocks. This secondary geochemical zonation originates from the admixture of inert nitrogen to the combustible hydrocarbons in some areas of the Rotliegend Basin, with the nitrogen content exceeding a certain boundary level no longer permitting commercial exploitation of the gas accumulations. The impact of the gas composition on fracturing potential can be modified or even compensated by economical and administrative aspects of a tertiary political zonation based on domestic hydrocarbon demand vs. supply and financial capacity. The continental-inland arid aeolian and fluvial Rotliegend is underlain by the Carboniferous terrestrial and marginal-marine humid fluvial and deltaic grey bed facies belt containing coal seams, with this formation having acted both as source of the gas during coalification and as collector by trapping considerable amounts of gas in channel and overbank sandstones. Above the Rotliegend and separated by the Zechstein marine evaporite cycle sequence including mudstones, carbonates, sulphates and salt, the Buntsandstein continental-inland arid fluvial, aeolian and lacustrine red bed complex has also captured gas having migrated from the Carboniferous coal seams through faults, joints and fissures into the fluvial sheet-sand reservoirs. In both Rotliegend and Buntsandstein, the primary depositional facies zonation was in various parts of the basin unfavourably overprinted by secondary tectonical deformation and subsidence. The high-quality reservoir belt in the southern part of the area is still at the surface or within shallow depth and therefore is not able to retain gas, whereas the moderate- to poor-quality pay zone seam in the northern part has been buried in several thousands of metres depth where structural traps are well sealed and allowed the accumulation of large gas fields. This combination of poor reservoir facies buried in suitable depth for hydrocarbon accumulation and trapping in contrast to the good pay facies occurring in shallow depth or even at the surface where gas concentration is not possible due to lack of sealing and trapping underlines the significance of the Rotliegend and Buntsandstein distal-medial tight reservoir facies belt which crosses the northern part of Middle Europe and that represents the most potential and important belt of hydraulic proppant fracturing. Rotliegend gas accumulations in Germany FRG are frequently found beneath Zechstein salt plugs where the lower total geopressure and the retarded heat flux with lower reservoir temperature often preserved better porosities than in the regions between the salt domes. All the four different sedimentary geological formations have been laid down in temporally separate depositional basins which horizontally occupied more or less the same geographical area, resulting in their vertical superimposition in the stratigraphical column during course of burial history. Laterally, the various individual basins are not homogeneous, but in contrast to partially good correlation along strike significant facies differences occur perpendicularly to strike, and also parallel to the basin margins, frequently considerable diversification of depositional environment leading to complex juxtaposition of different genetical sedimentary bodies is encountered. These relationships give rise to vertical and horizontal zonation of different reservoir complexes and together with effects of hydrocarbon generation, migration and accumulation also to segregated arrangement of natural gas deposits within the heterogeneous and anisotropic pay zone networks according to the internal geometrical organization of the latter. The discussion of technical and marketing aspects of Rotliegend stimulation in Europe includes horizontal and vertical distribution of main fields, historical and areal development of production and fracturing, permeability prediction and effective permeability, permeability deterioration and clay mineral distribution, minimum pre-fracturing gas production rate and poroperm properties, stimulation cost and fracturing policy, philosophy of off- and onshore exploration well fracturing, implications of fracturing philosophy on proppant selection, horizontal and vertical differences of fracturing potential, water sensitivity and other aspects. The experience of MHF stimulation of both Rotliegend and Carboniferous sandstones in the last ten years has shown that unless a certain minimum pre-fracturing gas production rate is present, no economical producer can be achieved by carrying out a hydraulic fracturing operation. During course of field development campaigns, distinction can be made between spot fracturing comprising sequentially drilling, testing and stimulation of one well after the other, and batch fracturing meaning first drilling of all the wells one after the other and then second in another round perforating, stimulating and putting on stream of the whole sequence of wells successively. Service boat capacity is often the limiting factor for stimulation in the Southern North Sea. Some parts of the Rotliegend pay series are extremely water-sensitive as a result of the clay mineral constellation, with foamed oil-based fracturing fluids being suitable carriers for treatment of such reservoirs. Carbon dioxide or nitrogen can be used as foaming or energizing agents in such fluids, and successful stimulation can also be performed with utilization of crosslinked acid systems and low-pH methanol systems. Conclusion is made that the case study of Rotliegend tight gas reservoirs reflects the complex problems of terrestrial low-permeability pay zones and helps to understand the suitable approaches of tight gas reserves which will contribute to the total world energy supply in increasing amounts during the coming years. Important sedimentary structures of fluvial, aeolian and lacustrine deposits in Rotliegend and Buntsandstein illustrating distribution and origin of reservoir heterogeneities and natural fractures as well as illuminating the significance of these features for propagation of hydraulically induced fractures and function as flow barriers or preferential current paths are figured in eleven photographic plates that are accompanied by detailed explanations which can be evaluated independent from the main text.

3.1. Introduction

Hydraulic proppant fracturing in Western and Eastern Europe is presently in the status of rethinking and reconstruction. In Western Europe, the major recession of stimulation of oil- and gas-bearing formations provoked by the late 1985/early 1986 oil price drop (cf. section 2.2.1.2.; negatively accentuated by the accompanying US $ exchange rate fall which was partially responsible for the prolongation of the unfavourable general economical constellation until late 1987 when the US $ exchange rate reached its historical minimum level; cf. section 2.2.1.3.) following the golden years of hydraulic proppant fracturing in the late 1970`s/early 1980`s (VEATCH 1983, GRIFFITHS 1986, OEHME 1986, RUNGE 1986, VEATCH & MOSCHOVIDIS 1986, VIELVOYE 1986, MADER 1987) has provided the necessity of reviewing and/or rescheduling and partially also redesigning the MHF treatments in predominantly deep tight Rotliegend and Carboniferous gas-bearing sandstones in Germany FRG and Netherlands onshore as well as British and Dutch Southern North Sea offshore.

In Eastern Europe, the increasing urgency for hard currency coming from export of oil and gas, the rising demand to ameliorate domestic hydrocarbon supply and the political reorganization under the guide slogans of acceleration (uskorenie; GORBATCHEV 1987), reorganization and reconstruction (perestroika) and transparency (glasnost), in combination with the more or less independency from the general world economical situation that is dictated by oil price scenario and US $ value scenario (cf. section 2.4.4.), will already in the near future result in focussing much more on stimulation of oil- and gas-bearing formations than in the past when drilling and fast completion of easier accessible hydrocarbon reserves had higher priority.

Much promotion of the hydraulic proppant fracturing and gravel packing scenery in Western and Eastern Europe in the coming years, however, has to include sound geological, technical and reservoir engineering concepts of development of increasing amounts of hitherto marginal reservoirs, with production from these targets becoming in the individual countries soon more or less important contributions for the compensation of the natural decline of the output from many progressively depleted conventional reservoirs. Rotliegend (Lower Permian) and also Carboniferous, to subordinate amounts also Buntsandstein (Lower Triassic) tight gas sandstone stimulation in Western Europe represent the highlights of hydraulic proppant fracturing in this area both in past and future. After discussion of some general and economical aspects, factors influencing the fracturing potential of Rotliegend gas reservoirs are outlined.

3.1.1. General aspects

Following the preceding outline of proppant selection as a consequence of reservoir conditions and proppant properties (chapter 1) and the summary of marketing and economical aspects of hydraulic fracturing and gravel packing in Western and Eastern Europe (MADER 1987; chapter 2), the present report emphasizes various geological, technical and reservoir engineering possibilities of enhancing Rotliegend stimulation potential which according to my view and understanding will become increasingly significant for achieving the goals of ameliorated supply-demand-relationships independent from any political threatening in the next years. As deep tight gas-bearing Rotliegend and Carboniferous sandstones will remain to be the major targets of massive hydraulic fracturing (MHF) operations in Western Europe in the near future, various factors influencing the fracturing potential of these particular reservoirs are illustrated (cf. fig. 9).

While many important tight gas sands in various basins in the USA have already been extensively described (a recent summary volume has been compiled by SPENCER & MAST 1986), a geological and reservoir engineering synthesis of Rotliegend and Upper Carboniferous which are the most important tight gas sand formations in Europe has hitherto not been presented. The purpose of this outline is to provide an integrated overview of geological, technical and economical aspects of Rotliegend and Carboniferous tight gas sands in North Sea and adjoining onshore areas in Europe where the predominant amount of massive hydraulic proppant fracturing in past and future is concentrated.

The present section is a geological and engineering case study forming the transition between the economically influenced divisions and the technically focussed parts of the review and actual summary of hydraulic proppant fracturing. The discussion in chapter 4 represents a status report of hydraulic stimulation technology and emphasizes numerous methods for enhancing the hydraulic fracturing potential in Europe and other parts of the world. Chapter 5 gives an account of gravel packing as the second application of natural sand and synthetic proppants in oil and gas industry, as well as an overview of other sand control techniques. Unit 6 summarizes hydraulic proppant fracturing and gravel packing monitoring. The extensive literature is included into a general reference list on hydraulic proppant fracturing and gravel packing and is also processed into subject key sections in a bibliography (chapter 7). Technical overviews of the state of the art of hydraulic proppant fracturing and summary experience reports are given by HOWARD & FAST (1970 a), WATERS (1980), SCHOLS (1983), VEATCH (1983), ECONOMIDES (1986), VEATCH & MOSCHOVIDIS (1986), ECONOMIDES & NOLTE (1987) and GIDLEY (1988).

3.1.2. Economical aspects

Several dozens of MHF operations in Rotliegend and Carboniferous gas-bearing sandstones have already been carried out onshore and offshore in the Southern Permian Basin (North Sea and Northwest Europe onshore; ZIEGLER 1982) in the past until 1986, and almost up to hundred treatments or even more are scheduled for the current and coming major offshore (British Southern North Sea) development campaigns (1987 - 1990; cf. section 2.4.1.). In parts of the area, also the Buntsandstein has subordinately been subjected to stimulation operations (mainly in the British and Dutch Southern North Sea) and will also have to be fractured in some more patches in the near future. Some comments are offered as follows on Rotliegend fracturing policy and understanding as well as fracturing job sizes and reserve acquisition potential.

3.1.2.1. Rotliegend fracturing policy and understanding

In view of the current and coming major activity in 1987 - 1990 (cf. sections 2.2.1.6.2. and 2.4.5.1.), the British Southern North Sea Rotliegend belt (cf. fig. 10) is probably the only area in Europe where almost according to the already common sense in the USA, hydraulic proppant fracturing is one of the most important aspects of well completion (ROBINSON, HOLDITCH & WHITEHEAD 1986; cf. section 2.3.), because the development drilling campaigns have been planned with the certainty that in case of encountering the Rotliegend reservoir in the predicted or expected tight facies, almost every well would have to be fractured in the poorest facies patches and every second or third well would require hydraulic stimulation in other sections of the low-permeability facies belt.

In addition, as a consequence of long-term development scheduling and production planning, the British Southern North Sea Rotliegend gas belt is more or less independent from the fluctuations of oil price and general economical framework (cf. section 2.2.1.6.2. and 2.2.1.6.3.2.), with the activity having been only subordinately to negligibly affected by the 1986 oil price drop and having so far straightforward proceeded to the beginning of the major development campaigns in 1987 - 1990 (economical aspects of this Rotliegend gas province are also summarized by FISHMAN 1986, GREGORY 1986, THOMAS 1986, ALGAR 1987, BAND 1987 and POTTER 1987; cf. section 2.4.1.2.).

3.1.2.2. Fracturing job sizes and reserve acquisition potential

The importance of the Rotliegend and Carboniferous reservoirs in terms of both geological-engineering and marketing aspects of stimulation as well as the considerable expenses of MHF treatments require an outline of a suite of factors that are positively or negatively influencing the fracturing potential of Rotliegend and subordinately also Carboniferous gas wells. Some aspects of stimulation treatment volumes and and expenses as well as probably accessible gas reserves are discussed as follows.

3.1.2.2.1. Stimulation treatment volumes and expenses

For example, the total cost of the operation in a Rotliegend well in the Söhlingen field/Germany FRG in late 1982 where abt. 550 t (1.2 Mio. lbs) of proppants have been pumped into the fracture amounted to abt. 6 Mio. DM (2.5 Mio. US $ according to the exchange rate at that time) which was economically justified due to acquisition of abt. 500 Mio. m3 (20 Bcf) gas reserves by the successful jumbo treatment (JOHN 1983, KLOSE & KRÖMER 1983, BLEAKLEY 1984; cf. sections 2.4.1.1. and 3.3.2.2.). The numerous treatments having been carried out with proppant quantities of abt. 100 - 400 t in the period 1977 - 1985 in Germany FRG all have amounted to abt. 1 - 2 Mio. US $ per job (several million DM each with respect to varying US $ exchange rate), and the stimulation operations in the British Southern North Sea can be assessed to have been even more expensive and have reached a couple of million US $ each, with some discount probably having been achieved in case of whole field campaigns with continuous employment of the stimulation boat (cf. section 3.7.1.1.).

Various aspects of MHF stimulations carried out in the last years in Rotliegend and Carboniferous sandstones are reported by TUNN (1971); BRINKMANN, FUHRBERG & SCHÖBER (1980); BRINKMANN, KRÖMER & REINICKE (1980); BRINKMANN (1982), GUNDERMANN (1982), SCHWARZ & SCHÖBER (1982), JOHN (1983, 1987), KLOSE & KRÖMER (1983), BLEAKLEY (1984), RIECKMANN & PUSCH (1984), LEICHT (1985); REINICKE, BRINKMANN, SCHWARZ & HUENI (1985); STEPHENS & MARTINS (1985), ACHARYA & KIM (1987), SOMMER (1987, 1988) and TAN, McGOWEN, LEE & SOLIMAN (1988).

3.1.2.2.2. Probably accessible gas reserves

The significance of hydraulic proppant fracturing is underlined by the estimate that in Germany FRG, the remaining potential of Rotliegend and Carboniferous tight gas sands is abt. 100 Bill. m3 (4 Tcf; this is one third of the expected future discoveries) which can only be produced economically in case of involvement of MHF stimulation treatments (JOHN 1983; REINICKE, BRINKMANN, SCHWARZ & HUENI 1985). The actual proven gas reserves are abt. 250 Bill. m3 (10 Tcf), of which at least 10 % can only be economically acquisited and withdrawn by massive hydraulic fracturing stimulation (SCHRÖDER 1988). ACHARYA & KIM (1987) estimate the total proven and probable Rotliegend gas reserves in North Sea, Netherlands and Germany FRG to be in the range of abt. 85 Tcf (2,125 Bill. m3), and FISHMAN (1986) expects an additional quantity of abt. 20 Tcf (500 Bill. m3) to be able to be obtained by development (including stimulation) during the next years. MARTIN & EVANS (1988) outline that the Rotliegend aeolian dune sand reservoirs in the British Southern North Sea are estimated to contain recoverable reserves of over 800 Bill. m3 (32 Tcf) in fields in production and under development. It can be estimated that at least a quarter of this amount can only be economically liberated by hydraulic fracturing either in past or future.

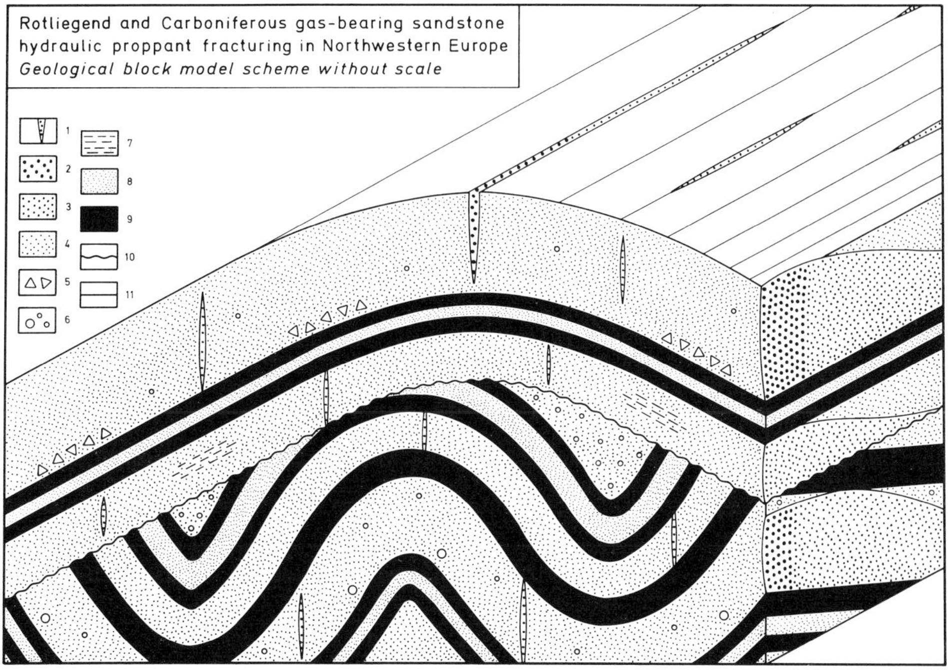

Fig. 9 (above) and Fig. 10 (below)

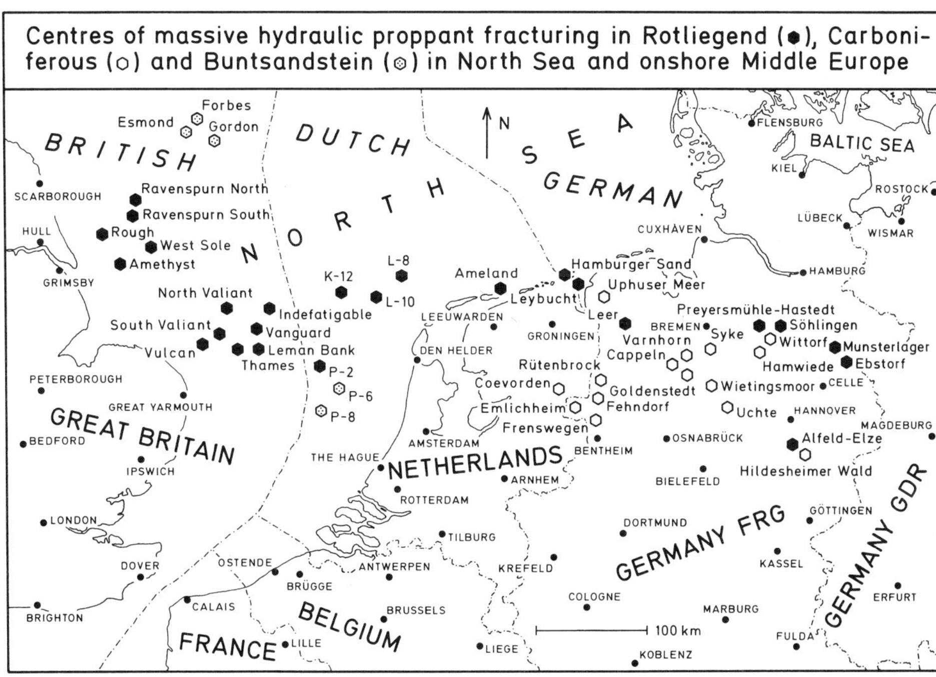

3.1.3. Factors influencing the fracturing potential of Rotliegend gas reservoirs

Deep tight Rotliegend and Carboniferous sandstone gas reservoirs (cf. fig. 11) have been the major targets of the MHF treatments in Great Britain (offshore), in the Netherlands (onshore and offshore) and particularly in Germany FRG (onshore) during the last ten years (with the quantities of pumped proppants ranging between 100 and 650 t per well) and will continue to play the leading roles of hydraulic proppant fracturing in Western Europe (especially in the British Southern North Sea) during the next years in terms of both frequency and size of the jobs (and thus also proppant consumption; MADER 1987, 1988 b).

After an introduction by sketching some general and economical aspects and commenting on distribution of main fields and historical development of production and fracturing, the discussion as follows includes questions of permeability prediction and effective permeability, permeability deterioration and clay mineralogy, water saturation, gas demand, minimum pre-fracturing gas production rates and poroperm properties, reservoir performance, stimulation cost and fracturing policy, philosophy of offshore and onshore exploration well fracturing, implications of fracturing philosophy on proppant selection, appraisal drilling strategy concepts and experience, horizontal and vertical differences of fracturing potential (including permeability and gas composition aspects), and water sensitivity (various references on both general and special topics from numerous fields of application are compiled in chapter 7).

Sedimentological and diagenetical aspects of Buntsandstein and Rotliegend deposits as a base for reservoir modelling are discussed by PLEIN (1978), LÜTZNER (1981), ROBINSON (1981); DRONG, PLEIN, SANNEMANN, SCHUEPBACH & ZIMDARS (1982); GLENNIE (1983, 1986), LEE (1984), MADER (1981, 1982, 1983, 1984 a, 1985 a, 1985 b, 1985 c, 1988 b), MADER & TEYSSEN (1985), MADER & YARDLEY (1985); BIFANI, GEORGE & LEVER (1987); GRALLA (1988), MADER & CHATALOV (1988), MARTIN & EVANS (1988) and RICHARDSON, SANGREE & SNEIDER (1988). An outline of the depositional environment of the Upper Carboniferous is given in OILMAN (1987 e). Important sedimentary structures of fluvial, aeolian and lacustrine deposits in Rotliegend and Buntsandstein illustrating distribution of reservoir heterogeneities and being significant for fluid flow and propagation of hydraulically induced fractures are compiled in photographic plates and are equipped with longer explanations that can be evaluated also independent from the main text (cf. section 3.13.).

3.2. Distribution of main fields

The Mid-European Rotliegend Basin (Southern Permian Basin; ZIEGLER 1982) and the gas-prospective continental-inland arid aeolian, fluvial, alluvial-fan, lacustrine and palaeosol red bed facies belt along its southern margin reaches from England through British and Dutch Southern North Sea, Netherlands, Germany FRG and Germany GDR to Poland and almost to the Russian boundary in Pomerania (PLEIN 1978, GLENNIE 1983, MADER 1985 a). The Rotliegend is underlain by the Carboniferous terrestrial and marginal-marine humid fluvial and deltaic grey bed facies belt containing coal seams, and overlain by the Zechstein marine evaporite cycle sequence including mudstones, carbonates, sulphates and salt, and the Buntsandstein continental-inland arid aeolian, fluvial, lacustrine and palaeosol red bed facies belt (cf. fig. 11).

All these four different sedimentary geological formations have been laid down in temporally separate depositional basins which horizontally occupied more or less the same geographical area, with the more or less spatial coincidence resulting in their vertical superimposition in the stratigraphical column during course of burial history. Laterally, the various individual basins are not homogeneous, but in contrast to partially good correlation along strike, significant facies differences occur perpendicularly to strike, and also parallel to the basin margins, frequently considerable diversification of depositional environment leading to a complex juxtaposition of different genetical sedimentary bodies is encountered. These relationships give rise to vertical and horizontal zonation of different reservoir complexes and together with effects of hydrocarbon generation, migration and accumulation also to segregated arrangement of natural gas deposits within the heterogeneous and anisotropic pay zone networks according to the internal geometrical organization of the latter. The discontinuous arrangement of gas fields in the sedimentary facies belt as a consequence of tectonical overprinting and trap formation also has significant impact on areal distribution of proppant stimulation requirement. Vertical and horizontal distribution of gas fields and hydraulic proppant fracturing potential are discussed in more detail as follows.

3.2.1. Vertical distribution

Within the Rotliegend trend, gas fields have been mainly discovered in the British Southern North Sea, Dutch North Sea, Netherlands onshore, Germany FRG onshore and Western Poland. Only minor amounts of accumulations have been found in Germany GDR and Eastern Poland (cf. figs. 5 and 10). The Carboniferous grey coal-bearing sandstone and mudstone succession underlies the Rotliegend in the whole area of distribution of suitable Permian reservoirs (cf. fig. 11) and has acted both as source formation of the gas during coalification of the various thinner and thicker seams (maturity aspects of the Upper Carboniferous are discussed by KETTEL 1981), and as collecting horizons by trapping considerable amounts of the generated gas in Upper Carboniferous fluvial and deltaic sandstones. Rotliegend and Carboniferous are in many cases separated by a pronounced angular unconformity (cf. figs. 3 and 11 as well as plate I in section 3.13.1.).

Due to migration along non-sealing normal faults and through fissures and joints in the geological column, the gas which was liberated from Upper Carboniferous coal seams not only accumulated in Upper Carboniferous fluvial sandstones of frequently ribbon- and lenticular type and only subordinately blanket nature, and Rotliegend aeolian and fluvial sandstones of in many case blanket character and only to minor amounts also ribbon- and lenticular type, but in some places also reached Lower Triassic Buntsandstein fluvial sheet-sand reservoirs of prevailing blanket geometry. Aspects of depositional environment, tectonical subsidence and gas accumulation, stratigraphy, and gas composition of the suite of reservoir storeys are outlined as follows (illustrations of sedimentary structures of Rotliegend and Buntsandstein facies associations are assembled on photographic plates which together with detailed explanations are compiled in section 3.13. and in MADER 1985 a, 1985 b).

3.2.1.1. Depositional environment

The reservoir complexes in the Rotliegend are mainly associated to the aeolian facies belt (cf. plate II) between the marginal alluvial-fan (cf. plate III) and fluvial gravelly sandstone seam (cf. plates IV and V) and the central damp to wet playa and desert lake mudstone body (cf. plates VI and VII) containing salt intercalations (Wechselfolge facies and marginal facies; GRALLA 1988). Some aspects of reservoir quality and sedimentary facies, palaeoclimatological evolution, and palaeotectonical and palaeoenvironmental evolution are briefly outlined as follows.

Fig. 11

Schematical geological cross-section through the gas-bearing reservoir column in Northwest Germany and surrounding areas. Three tight gas sandstone storeys can be distinguished comprising Carboniferous, Rotliegend and Buntsandstein. The Zechstein carbonates contain sour gas with considerable shares of the aggressive and toxic hydrogen sulphide, whereas the Rotliegend includes sweet gas that only consists of combustible methane and sometimes also of inert nitrogen. All the gas derives from coalification of Carboniferous coal seams and has migrated through the cover horizon sequence along faults and fissures. Carboniferous and Rotliegend tight gas sandstones are in many cases candidates for large-scale hydraulic proppant fracturing stimulation, whereas the Buntsandstein requires only occasionally synthetic permeability improvement. The gas-water-contacts in the different reservoir levels are frequently different, with thereby each of the units representing a separate pay storey.

Legend : 1 = gas-bearing sandstones, 2 = water-containing sandstones, 3 = coal seams, 4 = carbonates, 5 = mudstones and marls, 6 = Zechstein salt, 7 = Muschelkalk salt, 8 = hydraulic proppant fracturing potential, 9 = unconformity.

Fig. 12

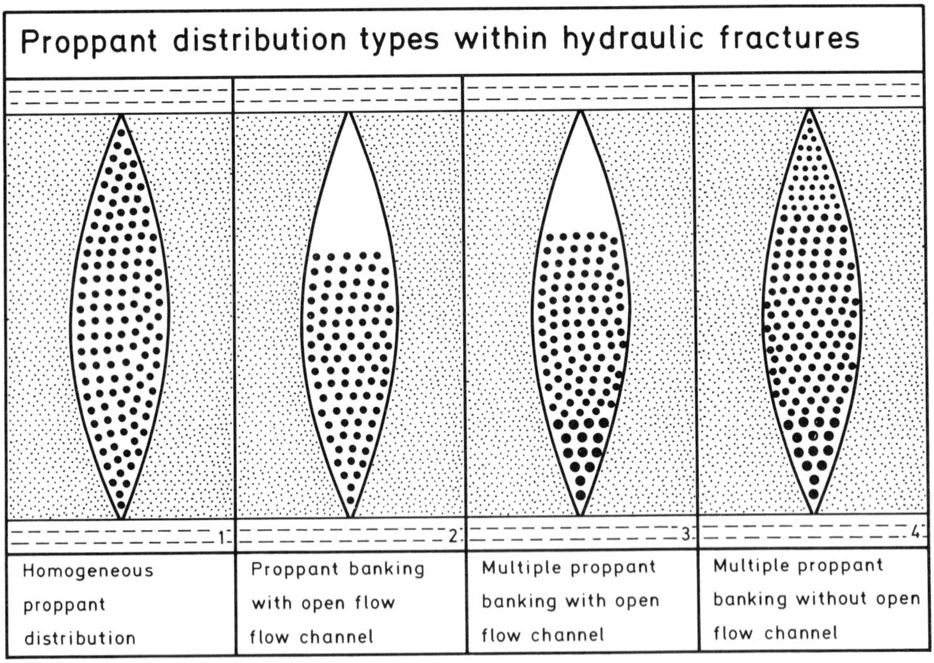

Fig. 11

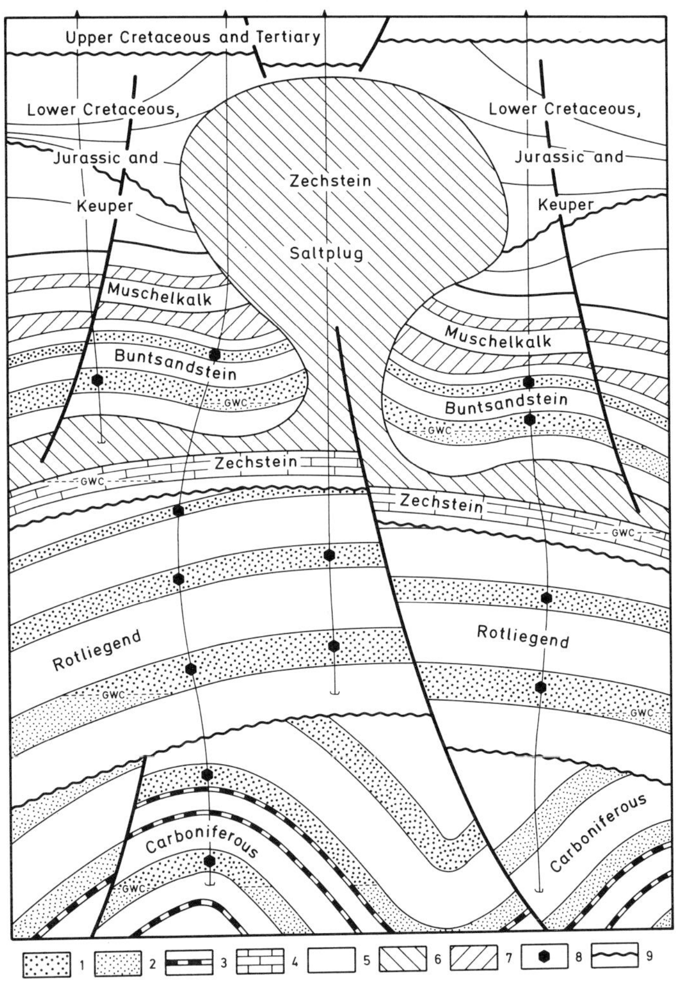
Upper Cretaceous and Tertiary
Lower Cretaceous,
Jurassic and
Keuper
Zechstein
Saltplug
Lower Cretaceous,
Jurassic and
Keuper
Muschelkalk
Buntsandstein
GWC
Muschelkalk
Buntsandstein
GWC
Zechstein
GWC
Zechstein
GWC
Rotliegend
GWC
Rotliegend
GWC
Carboniferous
GWC
Carboniferous
1
2
3
4
5
6
7
8
9

3.2.1.1.1. Reservoir quality and sedimentary facies

The best reservoirs are found in large-scale cross-bedded aeolian dune sands (MADER & YARDLEY 1985; cf. plate II/1-6), whereas poorer facies conditions are linked with small-scale cross-bedded aeolian dune sands and horizontal-laminated aeolian sheet sands (cf. plate VIII/1-5) which frequently contain intercalations of horizontal-stratified damp to wet playa sandstones and mudstones (MADER 1983 a; cf. plate IX) that deteriorate predominantly vertical permeability (cf. also section 3.10.1.1.). Special traps of favourable Rotliegend reservoir facies associations are synsedimentary tectonical features of chiefly graben nature where accelerated subsidence with respect to the surrounding platforms and additional protection from erosion permitted accumulation of particularly thick and sometimes also especially pure aeolian dune sands (the most significant syndepositional Rotliegend graben feature is the Schneverdingen-Graben; DRONG, PLEIN, SANNEMANN, SCHUEPBACH & ZIMDARS 1982; HEDEMANN, MASCHEK, PAULUS & PLEIN 1984; GRALLA 1988). Characterization of continental depositional facies types for hydrocarbon production are given by RICHARDSON, SANGREE & SNEIDER (1986).

3.2.1.1.2. Palaeoclimatological evolution

Throughout the Rotliegend sequence, a palaeoclimatological evolution leading from drier conditions in the lower part where mainly thick cross-bedded aeolian dune sands (cf. plate II/1-6) formed to wetter conditions in the upper part where more and more small-scale cross-stratified aeolian dune sands and horizontal-laminated aeolian sheet sands (cf. plate VIII/1-5) and damp to wet aeolian, adhesion and aquatic playa sediments (cf. plates VI and VII) originated together with increasing amounts of lacustrine mudstones and fluvial sandstones (cf. plate IV) that were laid down in large shallow playa lakes and braided small drainage channels intersecting the sand and mud flat as wells as sheetflood inundations of the continental plain. In some areas, the succession of facies complexes also reflects an evolution of fluvial style (MADER 1983 b, 1984 a, 1985 a, 1985 b; MADER & CHATALOV 1988) including progressive expansion of the alluvial braided-river system within the depositional basin as the marginal fan chains and the associated aeolian dune belt shrink in extension or retreat further towards the border of the sedimentary area as a consequence of basin enlargement by stepwise onlap on the fringing highlands with successive infilling of the trough.

3.2.1.1.3. Palaeotectonical and palaeoenvironmental evolution

Progressive diminution of the degree of braiding of the river systems as the erosional highlands in the provenance area are denudated, palaeoslope gradient decreases, amount and grain size of clastic material delivered from the source region declines, and transport capacity of the channel network ceases, gives rise to enlargement of interstream overbank flats where upon desiccation and reasonable persistence of quiet aquatic sedimentary conditions migration of aeolian dunes (cf. plate II/1-6) and accumulation of aeolian sheet sands (cf. plate VIII/1-5) from the detritus deflated from emerged sand and gravel bars in the shallow to moderately deep braided river channels is becoming more and more widespread (MADER 1983 b, 1984 a, 1985 a, 1985 b), or reversely is diminishing in importance in case of successively increasing frequency and extension of alluvial inundations by flood pulses overtopping the channel banks at high stage in combination with high ground water level as a consequence of higher stream discharge and increasing precipitation rates once climatological conditions are becoming damper and wetter.

3.2.1.1.4. Other aspects

Aspects of aeolian (cf. plates II and VIII), fluvial (cf. plates III - V and VII) and lacustrine (cf. plates VI, VII and IX) depositional environments in the Rotliegend reservoir complex, the spatial and temporal intertonguing of the different sedimentary milieus, the factors controlling sand body geometry and distribution, and the consequences for gas accumulation are discussed by GLENNIE, MUDD & NAGTEGAAL (1978); PLEIN (1978); DRONG, PLEIN, SANNEMANN, SCHUEPBACH & ZIMDARS (1982); GLENNIE (1983), LEE (1984), MADER (1983, 1985, 1988 b), MADER & YARDLEY (1985), GRALLA (1988), HAAK & ELEWAUT (1988), MARTIN & EVANS (1988) and RICHARDSON, SANGER & SNEIDER (1988).

The Carboniferous tight gas sands originated as channel bars and drapes in shallow braided rivers and sheet floods and have predominantly sheet-like geometry (BRINKMANN 1982). The Buntsandstein gas reservoir sands mainly represent also stream bars and sheets (cf. plate IV) having been laid down in watercourses of shallow braided river systems or by sheet floods, but occasionally also aeolian dune and interdune sheet sands (cf. plate VIII/1-5) occur (MADER 1985, MADER & YARDLEY 1985). In all three formations, depositional environment and diagenetical history influence reservoir rock characteristics and thus treatment options.

3.2.1.2. Tectonical subsidence and gas accumulation

In terms of distribution of potential and necessity of hydraulic proppant fracturing for economical gas production, the primary zonation of sedimentary facies belts in both Rotliegend and Buntsandstein in Middle Europe was unfavourably overprinted by secondary tectonical deformation and tertiary hydrocarbon accumulation. Aspects of sedimentary vs. tectonical basin zonation and gas-water-contact level are briefly outlined as follows.

3.2.1.2.1. Sedimentary vs. tectonical zonation

The marginal and proximal depositional facies seams in the southern parts of the Rotliegend and Buntsandstein basins where coarse fluvial and aeolian sands (cf. plates II and VIII) represent high-quality reservoirs (marginal facies; GRALLA 1988) are nowadays either outcropping at the surface or buried in shallow depth without sufficiently isolated structural, stratigraphical and/or facies traps to retain gas. The more distal and central facies realms further to the north, however, which are moderate- to low-quality pay zones containing only smaller amounts of aeolian cross-bedded dune and horizontal-stratified sheet sands (cf. plate VIII), but including mainly fine-grained fluvial channel sand (cf. plate IV) and overbank mud deposits (cf. plates III/1-2 and IX), damp to wet playa sand and mud sediments (cf. plates VI and VII; Wechselfolge facies; GRALLA 1988), and playa-lake mud and salt (basin facies; GRALLA 1988), have undergone subsidence to such a depth where the thick sealing overburden allows the accumulation of huge amounts of gas that only have to be liberated from the poor reservoirs, and tectonical overprinting has generated structural traps of mainly anticlinal and horst/graben nature (flower structures representing inverted graben fillings that are common in the Mesozoic oil-bearing sequence of Jurassic and Cretaceous are not developed in the Palaeozoic gas-bearing part of the geological column).

This combination of poor reservoir facies buried in suitable depth for hydrocarbon accumulation and trapping in contrast of the good pay facies occurring in shallow depth or even at the surface where gas concentration is not possible due to lack of trapping and sealing underlines the significance of the Rotliegend and Buntsandstein distal-medial tight reservoir facies belt which crosses the northern part of Middle Europe (MADER 1985 a) and which represents the most potential and important belt of hydraulic proppant fracturing in the near future. If the sedimentary and tectonical zonations would be of inverse nature, certainly not very much large-scale hydraulic fracturing would be necessary for access of the gas reserves, but according to the situation developed in Rotliegend and Buntsandstein, the most suitable conditions for accumulation and retainment of gas are concentrated in the relatively poorer or even almost worst reservoir rocks where only large-scale expensive stimulation jobs allow economically feasible exploitation.

The most important aspect of salt diapirism tectonics or halokinetics is the effect of retarded diagenesis in the Rotliegend sandstones below the salt plugs as a consequence of the cooling effect of the thick salt mass (temperatures in Rotliegend reservoirs below salt diapirs are on average 12 - 15 oC lower than those in Rotliegend pays between the salt domes; FÄRBER 1984).

3.2.1.2.2. Gas-water-contact level

An additional complication in Rotliegend sequences where thickness and completeness of the sequence has been enhanced by synsedimentary graben tectonics is subsidence of the pay zone column partially below the gas-water-contact. Therefore in many cases in Germany FRG the lower section (the Schneverdingen-Sandstein; cf. sections 3.10.1. and 3.2.1.3.1.) of the Rotliegend succession containing the best reservoirs is either partially or in some cases even completely below the gas-water-contact, whereas the tight middle section (the Hauptsandstein) is in most cases above the gas-water-contact. While certainly much more primary production from the good lower reservoir storey would take place in case of a more favourable gas-water-level, subsidence of the best pay zone interval partially below the gas-water-contact is another criterion giving rise to necessity of large hydraulic fracturing stimulations in the poor middle portion of the reservoir pile.

While in some occasions at least reasonable portions of the Schneverdingen-Sandstein or even the whole Rotliegend sequence including longer sections of the underlying Carboniferous profile are above the gas-water-contacts in Germany FRG, the Carboniferous in the British Southern North Sea is in the overwhelming part of structures below the gas-water-level which in addition to the tight reservoir facies is another important drawback of the gas prospectivity of this geological storey (cf. also section 3.8.1.).

3.2.1.3. Stratigraphy

Concerning Rotliegend stratigraphy in Middle Europe (Southern Permian Basin; ZIEGLER 1982), distinction has to be made between Germany FRG, Netherlands and British Southern North Sea.

3.2.1.3.1. Germany FRG

The Rotliegend section in Germany FRG is particularly well-developed and can be classified into various formations according to recent stratigraphical investigations and correlations (HEDEMANN, MASCHEK, PAULUS & PLEIN 1984; GRALLA 1988). The most important facies province is the Weser-Elbe province (GRALLA 1988) or East Hannover province (HEDEMANN, MASCHEK, PAULUS & PLEIN 1984) where most of the MHF stimulations of Rotliegend tight gas sands have so far taken place (cf. fig. 4) and which is the centre of Rotliegend exploration and production. In this facies province, three storeys can be distinguished in the stratigraphical column. The lower level comprises the Schneverdingen-Formation which reflects the initial stage of basin infilling with central subsidence and is only present in a suite of smaller and larger grabens. The middle section includes the Emden-Formation which represents the main stage of basin infilling with areal coverage of sedimentation. The upper portion comprises the Hannover-Formation or Wechselfolge which devotes its origin to the late stage of basin infilling when large expansion of the depositional area to the south enabled the subsequent extensive Zechstein transgression.

The main reservoir rocks are the Schneverdingen-Sandstein in the Schneverdingen-Formation at the base of the Rotliegend profile, the Slochteren-Hauptsandstein or simply Hauptsandstein in the middle section of the Rotliegend sequence, and the Wustrow-Sandstein (GRALLA, NIEBERDING & SOBOTT 1988) representing the petrophysically best developed one of numerous thin sandstones in the Hannover-Formation or Wechselfolge at the top of the Rotliegend succession. In the East Friesland province, the Hauptsandstein is very reduced, the Schneverdingen-Formation has so far not been encountered by drilling, and the main pay zone in this region are several Wechselfolge sandstones.

3.2.1.3.2. Netherlands

In terms of correlation with the Netherlands adjoining Germany FRG in the west and due to the offshore extension of the Rotliegend basin in the subsurface also connecting the stratigraphy from Germany FRG to Dutch and British North Sea, the Schneverdingen-Formation containing the Schneverdingen-Sandstein in Germany FRG is equivalent to the Lower Slochteren-Formation in Netherlands and westwards adjoining areas. The Emden-Formation including the Hauptsandstein in Germany FRG is correlating with the Upper Slochteren-Formation in Netherlands and North Sea, and the Hannover-Formation containing the Wustrow-Sandstein and other Wechselfolge sandstones in Germany FRG can be connected with the Ten Boer-Formation further westwards.

3.2.1.3.3. British Southern North Sea

In the British Southern North Sea, the Rotliegend section is not subdivided in vertical direction, and distinction is made between two laterally intertonguing units being the Leman Sandstone Formation which consists of clean aeolian and fluvial sandstones in the southern and western basin margins, and the interdigitating Silverpit Formation in the basin centre that is built up of marginal aeolian/playa and lacustrine sandstones, siltstones, mudstones and evaporites (MARTIN & EVANS 1988). Other aspects of Rotliegend stratigraphy in the British Southern North Sea are discussed by GRAY (1975), MARIE (1975), VEEN (1975) and GAGE (1980).

3.2.1.4. Gas composition

In contrast to the formation of sweet gas deposits in Carboniferous, Rotliegend and Buntsandstein consisting only of combustible methane (unless inert nitrogen is associated with the hydrocarbons in greater shares; cf. section 3.10.2.), immigration of coal-seam-deriving gas into the Zechstein carbonates was accompanied by mixing with additional amounts of gas that were generated within the Zechstein and which include also the poisonous and corrosive H2S in smaller or larger amounts, thus particularly in the Hengstlage-Sagermeer complex in the South Oldenburg area (cf. fig. 4) in Germany FRG (but in subordinate amounts also in Poland) creating complications of industrial exploitation due to the necessity of removal of H2S by expensive sour gas purification in large special plants. Hydraulic proppant fracturing has so far never been carried out in the Zechstein carbonates in Middle Europe where stimulation is restricted to matrix acidizing and rarely also acid fracturing (if necessary at all), with all types of reservoir treatment often being hampered and complicated by high H2S-concentrations in the formation gas. Sour gas purification comprising removal of the aggressive and poisonous H2S is also a great problem for exploitation of many gas-bearing carbonate formations in the USSR.

The only drawback in the Rotliegend sweet gas reservoir belt and storey complex is the association of nitrogen with the hydrocarbons in parts of the Lower Permian pay seam, with the nitrogen content reaching from quantitatively negligible admixtures via reasonable shares which require removal in gas purification plants to extraordinarily high amounts which no longer allow economically feasible hydrocarbon production (cf. section 3.10.2.). Helium, quicksilver and higher-homology hydrocarbons are only associated in traces to the methane in Rotliegend reservoirs which do not require attention concerning gas withdrawal, processing and transmission.

3.2.2. Horizontal distribution

In terms of horizontal distribution of hydraulic proppant fracturing potential of Rotliegend and Carboniferous tight gas sandstones, distinction can be made between Germany FRG, British and Dutch North Sea, and Germany GDR and Poland. While in Germany FRG most of the operations have already taken place during the last ten years and current to near-future activity is almost zero as a consequence of the oil price crash and the US $ exchange rate weakness, the predominant amount of stimulation work in the British and Dutch North Sea is carried out at the present time and within the next few years, although also already considerable treatment campaigns have been undertaken in the past (cf. sections 2.2.1.6.2. and 2.4.1.). In Germany GDR and Poland, so far only little stimulation has been performed, but the near future will be characterized by increasing and accelerating necessity to utilize hydraulic proppant fracturing for increasing domestic reserves and supply. Horizontal distribution of gas fields and hydraulic proppant fracturing potential in Germany FRG, British and Dutch North Sea, and Germany GDR and Poland is outlined in more detail as follows.

3.2.2.1. Germany FRG

Typical places of Rotliegend and Carboniferous MHF execution in Germany FRG in the gas field belt between the bounding patches of Groningen/The Netherlands (STÄUBLE & MILIUS 1970, ROSSUM 1975, BEEK & TROOST 1978) in the west and Wustrow (Germany FRG)/Salzwedel (Germany GDR) in the east (PHILIPP & REINICKE 1982, JOHN 1983) have been so far Söhlingen (ERDOEL-ERDGAS AKTUELL 1985 c) and satellites in the area of the Schneverdingen-Graben (DRONG, PLEIN, SANNEMANN, SCHUEPBACH & ZIMDARS 1982; HEDEMANN, MASCHEK, PAULUS & PLEIN 1984; GRALLA 1988) in the Rotliegend of the East Hannover area or Weser-Elbe province (where the hitherto largest hydraulic proppant fracturing operations have been carried out; KLOSE & KRÖMER 1983, BLEAKLEY 1984, LEICHT 1985; cf. figs. 4 and 10).

Other Rotliegend items of considerable MHF significance are Leer, Leybucht and Hamburger Sand in the East Friesland area and facies province, and Alfeld-Elze in the South Hannover area and facies province. The highlights of Carboniferous hydraulic proppant fracturing are Uphuser Meer in the East Friesland area (DIETZEL & HANTELMANN 1985); Cappeln, Goldenstedt, Syke, Uchte, Varnhorn and Wietingsmoor as well as Hamwiede in the South Oldenburg and East Hannover areas, respectively (for location maps cf. JOHN 1983; REINICKE, BRINKMANN, SCHWARZ & HUENI 1985; SCHRÖDER 1986, 1987); and Emlichheim, Fehndorf and Frenswegen in the West Emsland area (SOMMER 1987, 1988; cf. figs. 4 and 10 and tab. 6), with Coevorden adjoining in the Netherlands sector on the other side of the state boundary.

3.2.2.2. British and Dutch North Sea

The activity in the British Southern North Sea concentrates on the Rotliegend field complex around the Sole Pit area (comprising mainly the region of the Amethyst, Indefatigable, Leman Bank (VEEN 1975; GLENNIE, MUDD & NAGTEGAAL

1978; AMOCO 1988), Ravenspurn, Rough, Thames, Valiant, Vanguard, Vulcan and West Sole fields (LAFLEUR & JOHNSON 1973; OFFSHORE ENGINEER 1985 c, 1986 a, 1986 c; WORLD OIL 1985, ALLISON 1986, BRITISH PETROLEUM 1986, FISHMAN 1986, GLENNIE 1986; GREGORY 1986, 1987; THOMAS 1986, ACHARYA & KIM 1987, FOX 1987, OILMAN 1987 a, POTTER 1987, MARTIN & EVANS 1988) offshore Great Yarmouth between Teesside in Great Britain and Den Helder in the Netherlands (for location maps cf. ALLISON 1986, OILFIELD PUBLICATIONS 1986, QUINLAN 1988; cf. figs. 6 and 10 as well as tabs. 7 - 8). In the Dutch North Sea, the main places of Rotliegend hydraulic proppant fracturing stimulation are the K-12, L-8, L-10 and P-2 fields. While most of the hydrocarbon accumulations contain gas, the Argyll field at the margin of the central North Sea graben is nearly the only place where the Rotliegend sandstones contain oil (BIFANI, GEORGE & LEVER 1987).

3.2.2.3. Germany GDR and Poland

In Germany GDR, so far only the Salzwedel field at the FRG (Wustrow)/GDR boundary (cf. fig. 4) and some small satellites around the big patch are productive, because the reservoir contains further to the east (for example near Berlin) almost exclusively inert nitrogen (cf. also section 3.10.2.). In Poland, again sweet combustible gas has accumulated in Rotliegend sandstones, with the most important fields being in the Fore-Sudetic Monocline Bogdaj-Uciechów near Ostrów Wielkopolski; Załęcze, Zuchlów and Wierzchowice; in the area south of Poznań Grodzisk, Ujazd, Bukowiec, Buk and Kłęka; and in the Pomeranian area around Szczecin Międzyzdroje (cf. fig. 5 and tab. 9). Hydraulic proppant fracturing has hitherto been carried out in the Rotliegend in Germany GDR and Poland only occasionally and on smaller scale in contrast to abundant MHF application in Germany FRG.

3.2.3. Other aspects

The Rotliegend is in Middle Europe already since several years the most prospective target horizon of gas exploration (SCHRÖDER 1985) and will keep this position also at least in the near future. The Carboniferous will be of progressively increasing importance, but is expected to always lag behind the Rotliegend in significance, with the main reason being the higher water saturation within the petrophysically poorer reservoirs which are frequently situated close to or even below the gas-water-contact (cf. section 3.8.1.). Laboratory testing has shown that the Carboniferous reservoir rocks exhibit a high degree of plasticity (SLUSSER & RIECKMANN 1976; cf. section 4.2.3.2.3.).

Rotliegend gas fields in Germany FRG are frequently found beneath Zechstein salt plugs where the lower total geopressure and the retarded heat flux often preserved better porosities than in the regions between the salt domes (FÄRBER 1984; cf. section 3.2.1.2.1.). This geological constellation is in contrast to the conditions in the USA where except of overhang drilling, no hydrocarbon production is encountered under salt diapirs, and the penetration of salt in greater thicknesses pointing to a salt plug is even usually the sign to abandon the well (BLEAKLEY 1984).

3.3. Historical development of production and fracturing

The beginning of Rotliegend gas exploration, production and stimulation in Europe only dates back some 25 years (McNALLY 1988). In 1959, the giant Groningen field in the Netherlands (the eastern tip of which extends through the Ems estuary on Germany FRG territory) was discovered (by exploration well Slochteren 1; ACHARYA & KIM 1987), and in 1963 its large size was recognized from appraisal drilling (SCHOONBEEK 1976), with Groningen belonging to the top ten gas fields in the world and containing abt. 2,500 Bill. m3 (100 Tcf) reserves. The Groningen gas strike exactly 20 years ago (only some 10 years after the industrial introduction of hydraulic proppant fracturing; cf. section 1.2.) was the breakthrough for the beginning of intense exploration in British North Sea, Germany FRG, Dutch North Sea, German North Sea and other areas.

3.3.1. British North Sea

The discovery of the giant Rotliegend gas field Groningen (STÄUBLE & MILIUS 1970, ROSSUM 1975, BEEK & TROOST 1978) focussed subsequently the attention of exploration on both onshore and offshore parts of the belt of favourable Rotliegend reservoir facies in the Southern Permian Basin (for palaeogeographical maps cf. ZIEGLER 1982, GLENNIE 1986). Some aspects of treatment size and proppant quantity, offshore fracturing technology and stimulation boats, and significance of North Sea reserves for Great Britain are discussed as follows.

3.3.1.1. Treatment size and proppant quantity

In 1965, West Sole became the first commercial Rotliegend gas discovery in the British Southern North Sea (only one year after the general start of North Sea oil and gas drilling) and soon later also the target of first offshore hydraulic fracturing treatments in the Rotliegend (COOPER & MARSHALL 1984, BRITISH PETROLEUM 1986) which in those days were of comparatively small- to miniature-size and were carried out with abt. 50,000 - 200,000 lbs. (25 - 100 t) of natural sand (abt. 10 years of partially even more before the invention and market introduction of synthetic intermediate- and high-strength proppants). Other 1965/1966 British Southern North Sea Rotliegend gas strikes were Leman, Hewett, Indefatigable and Viking. For comparison, the first commercial North Sea oil discovery was only made in 1969 (thus four years after the first major gas strike; THOMAS 1986). With time, treatment technology evolved progressively to massive hydraulic fracturing with quantities of abt. 100 - 500 t (200,000 - 1 Mio. lbs) of natural sand and/or intermediate-strength synthetic proppants which started to be carried out in a second major series of campaigns in the late 1980`s almost 20 years after the original discovery of Rotliegend gas prospectivity in the North Sea and the onset of proppant stimulation for the amelioration of gas liberation (cf. section 2.2.1.6.2.).

3.3.1.2. Offshore fracturing technology and stimulation boats

Hydraulic proppant fracturing in the Southern North Sea evolved from first major treatments in the mid 1960`s via numerous jobs with skid-installed, platform-mounted equipment between 1965 and 1980 and first ship operations in 1980 by supply vessels that were converted into stimulation boats to treatments carried out by purpose-built fracturing vessels from 1984 onwards (COOPER & MARSHALL 1984), with now three comparable high-level technology stimulation boats being available in the North Sea (FOX 1985, OILMAN 1985 b; cf. also section 3.8.1.2.). The introduction of high-capacity fracturing ships allowed to drastically increase treatment sizes in terms of fluid and proppant volume and only thanks to the stimulation boats it was possibly to achieve the stage of massive hydraulic proppant fracturing in the hostile North Sea offshore environment. Aspects of hydraulic fracturing in partially water-sensitive Rotliegend and Buntsandstein gas reservoirs are also presented by COULTER & PURVIS (1980).

3.3.1.3. Significance of North Sea reserves for Great Britain

The Southern North Sea is the traditional centre of gas production in Great Britain and has historically contributed nearly 90 % of the total gas delivery in the United Kingdom, and there is every reason to believe that the Southern North Sea will keep its dominant role until the year 2000, as it still contains more than 50 % of the total proven and probable gas reserves in the United Kingdom (FISHMAN 1986; cf. also sections 2.2.1.6.2. and 2.2.1.6.3.2.). In the some twenty years from the 1965 start of British offshore gas drilling, stimulation and exploitation, over 50 Bill. Ł have been invested into hydrocarbon exploration and production in the United Kingdom sector of the North Sea (OFFSHORE ENGINEER 1986 d). As a consequence of the significance of Rotliegend gas acquisition including considerable shares of hydraulic fracturing stimulation, Great Britain advanced until 1985 to the world`s fourth biggest gas producer and simultaneously also achieved the rank of being the world`s fifth biggest oil producer. The stimulation market of the North Sea can be estimated to have an annual potential of abt. 20 Mio. Ł (OILMAN 1985 b; cf. section 2.2.1.5.3.).

3.3.2. Germany FRG

The development of Rotliegend and Carboniferous hydraulic proppant fracturing in Germany FRG is summarized in an account of evolution of gas-field discovery as well as treatment size and proppant quantity.

3.3.2.1. Evolution of gas-field discovery

Following the discovery of the giant Groningen field in the Netherlands in 1959 and the first Rotliegend gas fields in the British Southern North Sea in 1965 (West Sole field and satellites; cf. section 3.3.1.), onshore Middle Europe in 1966 the gas field Wustrow/Salzwedel at the FRG/GDR-boundary (cf. fig. 4) was found which formed the third milestone in Rotliegend gas access. The breakthrough in Germany FRG was achieved with the strike of Schmarbeck in 1971 which opened a whole new gas-field province in the East Hannover area, with in subsequent years a large suite of wells having been drilled and the new discoveries culminating so far in 1980 when one of the most important and greatest fields was found in Söhlingen which confirmed various strategical concepts of exploration concerning tectonics, sedimentation and diagenetical history (BRADEL & DRAXLER 1982, PHILIPP & REINICKE 1982) including the play of synsedimentary Rotliegend graben tectonics (DRONG, PLEIN, SANNEMANN, SCHUEPBACH & ZIMDARS 1982; HEDEMANN, MASCHEK, PAULUS & PLEIN 1984; GRALLA 1988) that subsequently gave rise to an entire belt of new gas strikes around the new important discovery spot.

In the last decade, a whole suite of satellite gas fields have been found around the big Söhlingen patch predominantly in a system of synsedimentary grabens (of which the Schneverdingen graben is the most important one) where excellent storage capabilities exist in the Schneverdingen-Sandstein, and in the cover strata pile where chiefly the laterally unlimited Hauptsandstein contains enormous quantities of gas in primarily tight facies that is indispensibly necessitating hydraulic proppant fracturing on larger scale for economical gas liberation. 1987 - 1990 finally is going to see a major campaign of development of a whole suite of tight Rotliegend gas fields in the British Southern North Sea which according to the quantities of accumulated gas and the difficulties to recover it in an economically most feasible way will become one of the highlights in Rotliegend gas acquisition, hydraulic proppant fracturing and exploitation (cf. section 2.2.1.6.2.).

3.3.2.2. Treatment size and proppant quantity

Hydraulic proppant fracturing of Rotliegend gas reservoirs in Germany FRG also started soon after the first discoveries of low-permeability pay zones (with in the early years due to lack of suitable material also mainly smaller operations having been carried out with abt. 25 - 100 t (50,000 - 200,000 lbs) of natural sand; cf. TUNN 1971) and was particularly pronounced in the last ten years since the beginning of availability of synthetic intermediate- and high-strength proppants which were first applied in Germany FRG in 1978 (ERDOEL-ERD-

GAS AKTUELL 1978 a), with the hitherto highlight being the so far biggest operation in the well Söhlingen Z 4 in 1982 where the abt. 550 t (1.2 Mio. lbs) of bauxite proppants injected were temporarily together with the final proppant pumping rate of abt. 9,000 lbs./min. even world record in MHF execution with high-strength proppants, with the result of the successful jumbo job having been the approval of abt. 500 Mio. m3 (abt. 20 Bcf) of gas reserves (KLOSE & KRÖMER 1983, BLEAKLEY 1984).

For comparison, big treatments having been carried out in the USA approached pumping of abt. 3 - 5 Mio. lbs of sand (WATERS 1980), and 1987 has seen the world record sand quantity of 3,150 t or 6.3 Mio. lbs having been pumped into a gas well in South Texas/USA (CONSTIEN, BRANNON & BANNISTER 1988; PITTS 1988; cf. section 1.1.1.4.) which also paid off very quickly due to confirmation of enormous additional gas reserves. Massive hydraulic fracturing with proppant quantities of 100 - 650 t has been frequently carried out in Rotliegend and Carboniferous deep tight gas reservoirs in the period 1977 - 1985 (cf. section 2.4.1.1.).

3.3.3. Dutch North Sea

In the Netherlands, offshore exploration also was the logical consequence of the discovery of the giant Groningen field onshore and within the following years, also important Rotliegend gas fields were found especially in the K-, L- and P-blocks of the Dutch Southern North Sea (for geographical distribution cf. OILFIELD PUBLICATIONS 1986, QUINLAN 1988; cf. fig. 10). Hydraulic proppant fracturing has already been carried out in reasonable amounts in both British and Dutch Southern North Sea, but its climax will be reached in 1987 - 1990 when the major development campaigns of Rotliegend gas fields in the British Southern North Sea (for economical aspects cf. also BEUDELL 1986 b, FISHMAN 1986; GREGORY 1986, 1987; THOMAS 1986, ALGAR 1987 a, BAND 1987, POTTER 1987) require outstanding stimulation activity which might be much more than all the treatments performed hitherto in the whole Rotliegend Basin together. Successful stimulation practices in the Dutch Southern North Sea are summarized by COULTER & PURVIS (1980).

3.3.4. German North Sea

Unfortunately so far no commercial Rotliegend gas strike has been made in the German North Sea (if neglecting the two small fields Emshörn and Leybucht which are only a few km away from the coastline and have to be classified as inshore rather than offshore localities; OFFSHORE ENGINEER 1986 b). Some gas accumulations have been found that could be exploited economically feasible if they would be situated onshore, but the additional transport expenses for up to 300 km from the offshore location to the onshore gas terminal cancels the project due to profitability reasons. Other Rotliegend reservoirs contain too high nitrogen shares (up to 60 % and more) in the gas to be commercially attractive (PLEIN 1975, SCHÖNEICH 1981).

3.3.5. Other areas

The areas of the Danish North Sea as well as of the German, Polish and Russian Baltic Sea contain only reduced Rotliegend sections in unfavourable facies development including only negligible amounts of sands with reservoir quality, or are already outside of the Permian depositional basin on the Fennoscandian Shield. The Irish or Celtic Sea has so far also not delivered economical gas accumulations in the Rotliegend in contrast to some feasible deposits in the Lower Triassic Buntsandstein-facies red beds (such as the Morecambe field). In Germany GDR onshore, apart from Salzwedel (cf. fig. 4) and its small satellites eastwards no other big Rotliegend gas field has been encountered so far with reasonable shares of combustible methane (in the Berlin region, the reservoir contains almost only inert nitrogen), whereas in Poland onshore, commercial Rotliegend gas discoveries were made particularly around Ostrów Wielkopolski south of Poznań and north of Wrocław as well as around Szczecin (cf. fig. 5 and section 3.10.2.1.).

Reservoir subsidence due to framework compaction by pore collapse is negligible in the Rotliegend sandstones and comprises abt. 10 - 25 cm in 20 - 40 years in contrast to significant subsidence in Cretaceous chalk pays which ranges up to several metres (SCHOONBEEK 1976; cf. also section 4.5.4.1.4.).

3.4. Permeability prediction and effective permeability

Within the low-permeability zone of the gas-bearing Rotliegend sandstone belt in the British and Dutch Southern North Sea (and in less pronounced form also in Germany FRG and the Netherlands onshore), prediction of the necessity of hydraulic proppant fracturing is often very difficult, because little changes in permeability in either direction can already result in execution or cancellation of a stimulation job, with this relationship meaning that out of an offshore drilling campaign which comprises several dozens of wells (a couple of such actions are scheduled in the British Southern North Sea in the coming years) where half of the amount of producers is planned to have to be fractured, formation evaluation by logging and testing can result in the necessity to switch to hydraulic proppant fracturing of either only a quarter or less or even almost three quarters or more of the total quantity of the wells, depending upon degree and orientation of the mismatch of prediction and acquisition of permeability in direction of better or poorer facies. Some aspects of tight reservoir permeability range, permeability determination and prediction, and temporary damage and restoration of permeability are briefly sketched as follows.

3.4.1. Tight reservoir permeability range

Tight gas sandstone reservoirs are the major source of future gas production and are predicted to achieve a share of abt. 50 % by the year 2020 in comparison to their contribution of abt. 5 - 10 % in the 1980`s in the USA (BAKER 1981). The maximum recoverable gas reserve in the tight sand resource is over 600 Tcf (15,000 Bill. m3; NORTHROP & FROHNE 1988; cf. section 4.4.1.). Permeability distribution trends are extremely complex and no reasonable and reliable prognosis is possible, with the customary method of exploration and appraisal being pattern drilling on whole trends which have been found by seismic evaluation.

In contrast to continuous or blanket reservoirs, lenticular pay zones must be contacted by drilling wells on a spacing more typical of oil-field than conventional gas-field development, and communication fracturing between the isolated sand bodies to connect them to their neighbours and to the borehole has to be applied frequently (cf. section 4.10.2.). Field boundaries are set only partly by structure and porosity, but to large amounts by permeability, thickness and economics, with interspersed higher-permeability areas occurring as sweet spots. The following outline sketches some special features of Rotliegend and Carboniferous in Europe by commenting on microdarcy pay zone permeability, stratigraphical distribution of permeability, and gas and fluid flow in tight rock matrix. The discussion is introduced by an overview of the general sedimentological and diagenetical evolution of tight gas sands.

3.4.1.1. Sedimentological and diagenetical evolution of tight gas sands

Tight gas sands are characterized by an in-situ permeability to gas of 0.1 md or less and can be differentiated into blanket and lenticular sandstones (SPENCER 1983). While blanket sandstones are predominantly of marine or marginal-marine genesis, lenticular sandstones mainly originated in fluvial and deltaic environments. A third category of tight reservoirs comprises chalk (cf. section 4.5.4.). Sandstones are tight as a consequence of primary fine grain size or due to secondary intense cementation, with the pores being poorly connected by small throats and capillaries (cf. section 3.5.1.1.). Discrete gas-water contacts are absent in lenticular and some of the blanket sandstone reservoirs, and most of the gas occurs in stratigraphical permeability traps. Low porosity, high clay content and high grain densities of tight gas sands render well log analysis difficult (cf. section 3.5.2.3.), and many tight gas basins are either overpressured or underpressured (cf. section 4.4.2.). In contrast to conventional pay complexes, tight gas sands do not exhibit a consistent relationship between porosity and permeability. Some aspects of blanket and lenticular sandstones, porosity and permeability relationships, and reservoir pressure and formation damage are summarized as follows.

3.4.1.1.1. Blanket sandstones

Blanket sandstones in engineering sense have relatively better horizontal continuity than lenticular reservoirs and respond to hydraulic fracturing usually in a somewhat predictable manner. Increasing the proppant volume results generally in increasing well productivity up to certain limits. Some examples of blanket sandstones in the Rocky Mountains tight gas basins in USA include the Lower Cretaceous "J" Sandstone in the Denver-Julesburg Basin, the Upper Cretaceous Almond Formation and Frontier Formation in the Greater Green River Basin, and the Upper Cretaceous Corcoran and Cozzette Sandstones in the Piceance Creek Basin. Although being considerably more extensive in horizontal direction than lenticular reservoirs, also blanket sandstones may exhibit reasonable lateral variability (LORENZ 1983, NORTHROP & FROHNE 1988). A summary of blanket tight gas sands in Texas is given by FINLEY & O`SHEA (1983) and FINLEY (1986).

3.4.1.1.2. Lenticular sandstones

Lenticular reservoirs are fluvial sandstones which are very discontinuous and exhibit many internal permeability variations (cf. section 4.2.4.2.). Geometry and dimensions of the reservoir units are difficult to predict, and therefore the response of lenticular sandstones to hydraulic fracturing is very erratic and generally the stimulation results are poorer than usually possible in massive blanket sandstones. Some examples of lenticular sandstones in the Rocky Mountains tight gas basins in USA include fluvial sandstones in the Upper Cretaceous Mesaverde group and Tertiary sandstones in the San Juan, Uinta, Piceance Creek and Greater Green River Basins. The economical severity of sandstone lenticularity depends on lens geometry, with broad lenses being nearly as commercially attractive as blanket sands (KUUSKRAA, BRASHEAR, ELKINS & MORRA 1979). Economical feasibility of lenticular reservoirs can be considerably improved by fracture penetration through lenses which are not directly in contact with the well by communication fracturing (cf. section 4.10.2.1.).

Chalks are marine microcrystalline limestones of blanket character and are in the Rocky Mountains tight gas basins represented by the Upper Cretaceous Niobrara and Greenhorn formations which contain also blanket tight gas sands (aspects of chalk stimulation are compiled in section 4.5.4.).

3.4.1.1.3. Porosity and permeability relationships

Tight gas reservoirs exhibit two varieties of porosity-permeability relationships. The first type comprises rocks with high porosity but low permeability owing to the small size of pores and throats which are usually well interconnected. This type is represented by chalk (cf. section 4.5.4.) and some marine blanket sandstones. The second variety includes low-porosity sandstones where significant porosity amounts are created by post-depositional dissolution of framework grains and/or matrix cements, with secondary porosity (SCHMIDT & McDONALD 1979) mainly exhibited as small isolated pores or microvugs. Relict grains and particles resulting from incomplete dissolution do also occur. The second type of porosity-permeability relationships is mainly represented by fluvial lenticular and marine blanket sandstones. Secondary microvugs and residuals of primary porosity are scattered through the reservoir and are connected by small ribbon- or sheet-like tortuous capillaries. The small size of the capillaries restricts gas flow and causes generally high capillary pressures which can provoke liquid imbibition during drilling and fracturing that in turn can trigger formation damage.

The reservoirs are very stress-sensitive due to closing of small capillaries or cracks when confining pressure is applied. Many of the microvugs in very tight rocks are partially or completely filled with authigenic clay minerals (cf. section 3.5.1.1.). Non-pore-filling clays are also present as detrital shale clasts and as alteration products of volcanic and other rock fragments (MADER 1981, 1982, 1983 c). Blanket sandstone reservoirs are in some cases higher in percentage of quartz and tend to become tight at slightly greater depths than lenticular sandstones. While partially primary porosity is preserved, considerable amounts of secondary porosity are also developed in many tight gas sands (AL-SHAIEB & WALKER 1986; LAW, POLLASTRO & KEIGHIN 1986). Diagenetic complexity is frequently a function of depositional environment as well as burial and thermal history of the basin.

3.4.1.1.4. Reservoir pressure and formation damage

Reservoir pressure is dependent on organic richness, palaeotemperature, present-day temperature, ability of organic matter to still generate gas, and occurrence of lateral and vertical sealing beds. The main mechanism for overpressuring is active thermal gas generation in source beds associated with low-permeability reservoirs. As a consequence of their pore and capillary character, tight gas reservoirs have a high potential of formation damage. The types of possible formation damage include movement of secondary clays causing plugging of pore throats, swelling of clay minerals, increasing water saturation with consequent reduction of relative gas permeability, leaving of fracturing gel compounds in the pay zone, and precipitation of minerals and compounds caused by chemical additives during acidizing and hydraulic proppant fracturing.

Other geological and engineering aspects of tight gas sands are summarized by KEIGHIN (1979), RICE & SHURR (1980), BROWN & CRAFTON (1981), GAUTIER & RICE (1981), WALLS (1981); FINLEY, GARRETT, HAN, LIN, SAUCIER & TYLER (1983); FINLEY & O`SHEA (1983); FINLEY (1984, 1986); FINLEY, DUTTON, LIN & SAUCIER (1985); JENKINS (1985), AL-SHAIEB & WALKER (1986); BROWN, SMAGALA & HAEFELE (1986); LAW, POLLASTRO & KEIGHIN (1986); LAUGHREY & HARPER (1986), MOSLOW & TILLMAN (1986); PITMAN, ANDERS, FOUCH & NICHOLS (1986); WEIMER, SONNENBERG & YOUNG (1986) and DUTTON & FINLEY (1988).

3.4.1.2. Microdarcy pay zone permeability

Tight reservoirs such as occurring in the European Rotliegend and Carboniferous and in other formations have permeabilities in the microdarcy range up to abt. 1 millidarcy (ATTEBERRY, TUCKER & RITZ 1979; KLOSE & KRÖMER 1983, BLEAK-

LEY 1984, ALBERTSEN 1985; REINICKE, BRINKMANN, SCHWARZ & HUENI 1985). Some aspects of absolute permeability levels and permeability creation are illustrated as follows.

3.4.1.2.1. Absolute permeability levels

According to the definition by the Federal Energy Reserve Commission (FERC) of the USA, tight gas reservoirs have a matrix permeability less than 0.1 mD (HUNT, RAYMOND, HASKETT & PIRIE 1981; HAAS, BRASHEAR & MORRA 1985; PAI, GARBIS & HALL 1985; KUUSKRAA & HAAS 1988). In such tight reservoirs, drawdown of flowing pressure is particularly pronounced during production life and thus a wide range of closure stresses is placed on the proppants especially when field exploitation covers up to several decades (CLARK 1983).

Rotliegend and Carboniferous gas-bearing sandstones in Germany FRG exhibit up to 20 % and 15 % total porosity in extreme cases, respectively, but clay minerals and water saturation considerably reduce effective porosity and thus available permeability for gas flow is only 10 - 50 microdarcy (in comparison to Western and Eastern USA tight gas sands with 3 - 9 % porosity and 1 - 50 microdarcy permeability; GIDLEY, MUTTI, NIERODE & KEHN 1977; ELKINS 1980, JONES & OWENS 1980; McKETTA, KOZIAR & COOK 1980; HUNT, RAYMOND, HASKETT & PIRIE 1981; WALLS 1981; JOHN 1983, 1987; SPENCER 1983, ECONOMIDES 1987; SATTLER, RAIBLE, GALL & GILL 1988; SOEDER & CHOWDIAH 1988) or 1 - 100 microdarcy (BRINKMANN 1982; REINICKE, BRINKMANN, SCHWARZ & HUENI 1985). A comparison of conventional and low-permeability gas reservoirs is given by GAUTIER & RICE (1981), and a review of low-permeability gas sands is provided by KAZEMI (1983).

3.4.1.2.2. Permeability creation

The microdarcy permeability of many tight gas sandstones is caused by considerable reduction of primary porosity up to remnant slot pores between adjoining authigenic quartz overgrowths on detrital sand grains (SOEDER & CHOWDIAH 1988), and much of the pore volume is of secondary nature (sensu SCHMIDT & McDONALD 1979), although flowpaths between the secondary pores often consist of narrow slot pores (RANDOLPH, SOEDER & CHOWDIAH 1984). In some reservoirs, the degree of pore alteration and hence the degree of tightness varies directly as a function of depth (SOEDER & CHOWDIAH 1988), and transitional stages between high-permeability sandstones (up to 10 md or more) containing only slightly altered primary porosity as the one end member and low-permeability sands (down to 1 microdarcy or less) with highly-altered primary porosity characterized by narrow slot pores as the other end member occur within the geological column with progressive burial depth and in relation to thermal history. Fluorescent petrography and mercury porosimetry are useful techniques for rapidly and inexpensively assessing the quality of a tight sandstone reservoir and can quickly reveal which zones contain highly-permeable, grain-supported primary porosity, which intervals are characterized by tight, productive slot pore/solution pore morphology, and which sections include a highly-altered low-permeability pore structure.

Low-permeability gas sandstones can only produce gas from the rock matrix at economical rates if flow to the wellbore is aided by natural or hydraulically-induced fracture systems or a combination of both sets of cracks, but regardless of this, gas production rate is still governed by the rate of gas flow from the matrix into the fractures which depends on detailed matrix pore structure and distribution of water contained therein (WARD & MORROW 1985).

3.4.1.3. Permeability stratigraphy

Effective permeability under reservoir conditions also varies with stratigraphy of the Rotliegend (cf. also section 3.10.1.2.), with the Schneverdingen-Sandstein in the lower part of the Rotliegend succession in the Weser-Elbe facies province in Germany FRG (for Rotliegend stratigraphy cf. section 3.2.1.3.) being the best section with abt. 20 - 50 millidarcy (according to production tests and core measurements), and the Wustrow-Sandstein in the upper part being also reasonable with up to abt. 5 - 10 millidarcy, whereas the Hauptsandstein in the middle part is often very poor with in some cases not exceeding 1 millidarcy, but frequently having much less permeability down to abt. 0.5 - 1 microdarcy (KLOSE & KRÖMER 1983; REINICKE, BRINKMANN, SCHWARZ & HUENI 1985). As economical gas production without fracture treatment requires effective reservoir permeability of at least 0.5 - 1 millidarcy (LEICHT 1985), the middle section of the Rotliegend pay column (the Hauptsandstein) is the major target for MHF operations in Germany FRG. Some rocks, however, are too tight to produce at commercial rates even in case of presence of abundant natural fractures and/or following hydraulic proppant fracturing (SNEIDER, KING, HAWKES & DAVIS 1981; cf. section 3.6.1.2.).

3.4.1.4. Gas and fluid flow in tight rock matrix

The behaviour of tight gas sandstone permeability in response to changing confining pressure can be explained qualitatively by the complex and tortuous pore structure which results from extensive compaction and diagenesis (WALLS 1981). Narrow slot-like apertures exist between pores, with these thin cracks providing the major connectivity as they allow fluid to move when the rock is under effective pressure conditions, but they are easily closed by increasing overburden pressure. At higher closure stress, the contribution of the narrow slot-like apertures to flow is almost eliminated and permeability is provided by the rounder cross-section channels. The effect of partial saturation on gas permeability is also largely controlled by pore structure. Water tends to collect in small pores and cracks due to capillary forces and thereby prevents these channels from transmitting gas unless the pressure gradient is sufficiently large to cause water displacement which may be partially responsible for the failure of some hydraulic fracturing operations to improve well performance. The stimulation fluid itself invades the formation to some extent thereby creating a zone of almost zero gas permeability on each side of the fracture, with this same waterblock phenomenon being on a much larger scale probably the trapping mechanism for huge volumes of gas in some deep basins (MASTERS 1979; cf. section 3.11.2.4.2.).

Gas in place can be estimated for hydraulically fractured wells in tight gas reservoirs within a margin of error acceptable for engineering purposes if the pressure data is collected after pseudo-radial flow regimes (SULLIVAN, POSTON & PIPER 1988).

3.4.2. Permeability determination and prediction

Effective permeabilities are determined by statistical evaluation of laboratory core measurements and recomputation with a correction factor to reservoir conditions (JONES & OWENS 1980, LEICHT 1985). An important complication is permeability anisotropy and heterogeneity in tight formations (ROSE 1982). Some comments are offered as follows on determination, prediction and variations of permeability.

3.4.2.1. Permeability determination

In terms of proper determination, caution has to be exercised for not falsifying the true permeability by unsuitable laboratory handling techniques (FREEMAN & BUSH 1983, SOEDER & DOHERTY 1983, CHOWDIAH 1986, LUFFEL & HOWARD 1987). Some aspects of relative permeability determination are discussed by DELSHAD, MacALLISTER, POPE & ROUSE (1985); WARD & MORROW (1985), FIROOZABADI & AZIZ (1986); GRAVIER, LEMOUZY, BARROUX & ABED (1986); HONARPOUR, KOEDERITZ & HARVEY (1986); PARMESWEAR & MAEREFAT (1986); WATSON, KERIG, RICHMOND & TAO (1986); CIVAN & DONALDSON (1987); DONALDSON, ALAM & CIVAN (1987); MOHANTY & MILLER (1988) and WATSON & ERTEKIN (1988). MILLER & RAMEY (1985) evaluate the effects of temperature and consolidation on relative permeability, and ALI, AL-MARHOUN, ABU-KHAMSIN & CELIK (1987) comment on the effect of overburden pressure on relative permeability.

AGGOUR & MALIK (1987) analyze the effect of cyclic formation pressure changes on permeability, and WATSON (1988) investigates the impact of steep temperature gradients on relative permeability. BAKER (1988) comments on three-phase relative permeability correlations. GRI (1988) presents examples of permeability alterations as a consequence of stress changes. WARD & MORROW (1985) outline that differences between relative permeabilities according to overburden pressure increase with decreasing porosity and increasing sensitivity of permeability to confining pressure. Low-permeability gas sandstones have high surface area per unit volume, and surface adsorption and capillary condensation contribute to retention of significant amounts of water at extremely high capillary pressures. BYRNES, SAMPATH & RANDOLPH (1979) investigate the effect of pressure and water saturation on tight gas sandstone permeability. BERGOSH, LORD & OWEN (1987) introduce the overburden multidirectional permeability technique.

Effective permeabilities of Rotliegend and Carboniferous sandstones are also studied by RIECKMANN (1971). GOGGIN, CHANDLER, KOCUREK & LAKE (1986) outline permeability patterns of aeolian deposits. Vertical permeability can be related to a known distribution of lateral permeability by a method of shale management in field-scale models including the generation of a pattern of deterministically and stochastically distributed shales (HALDORSEN & LAKE 1984).

3.4.2.2. Permeability prediction

In contrast to high-permeability reservoirs where moderate differences of petrophysical properties between expectation and result have often almost no consequences for completion and testing, production planning is frequently seriously affected by only little deviations from the prognosis in low-permeability pays. While porosity mapping may be still more or less reliable in sandstone reservoirs even in the vicinity of the boundary between sand-dominated and mud-dominated facies belts, permeability contouring is often more difficult as a consequence of much more factors with stronger variability and thus less certain possibilities of pre-well assessment influencing permeability with respect to porosity, thereby causing more surprises after penetration of the reservoir formation by the drilling bit.

3.4.2.3. Permeability variations

Permeability changes in tight gas reservoirs can often have dramatic effects on recoverable reserves. KUUSKRAA, BRASHEAR, ELKINS & MORRA (1979) outline an example where a permeability of 0.01 md and a constant fracture design results in a recovery efficiency of 68 % of the gas in place, with permeability changes of one order to magnitude in upwards and downwards direction to 0.1 and 0.001 md leading to recovery efficiencies of 80 % and 25 % of the gas in place, respectively.

Apart from petrophysical effects related to diagenetical clay mineral authigenesis and transformation, permeability is influenced by sedimentological and petrographical complications particularly in conglomerates where varying distribution of gravel clasts of different size, frequency, shape and composition introduce further anisotropy and inhomogeneity factors into the reservoir rock beyond the control exerted by sand grains and mud particles and their original nature and secondary alteration (cf. plates III/5-8, V/1-4, VI/4 and VII/5 in section 3.13.).

3.4.3. Temporary damage and restoration of permeability

In-situ gas permeability in tight sandstone reservoirs is often ten to more than thousand times less than indicated by routine tests owing to the combined effects of overburden pressure, reduced gas slippage, and presence of connate water (JONES & OWENS 1980). Invasion of aqueous drilling or fracturing fluid filtrate has little permanent influence on rock permeability, but clean-up times may be extensive because of the low level of formation permeability (TANNICH 1975), although permeability preservation in the invaded reservoir zone can be assisted by minimization of infiltration (JONES & OWENS 1980). The chemical composition of mud filtrates or fracturing fluids is ordinarily of secondary importance in preventing permeability impairment. Effects of stress, gas slippage and water are correlating with permeability, but not directly with clay content, with tight reservoirs experiencing large effects with both low and high clay contents.

Most of the reservoir formations regain up to 70 - 80 % of the initially undamaged permeability or original productivity after being subjected to treatment fluids (DERBY & SMITH 1979). In tight gas pays, permeability reduction due to fracturing fluid invasion does not appreciably affect productivity due to the insignificant pressure drop across the invaded zone as compared to the pressure drop from reservoir to wellbore (PARROT & LONG 1979). Restoration of gas permeability by removing high fluid saturations adjacent to the fracture face can also be achieved by gas cycling (BRANAGAN, COTNER & GETTMAN 1981; ZAHNER & CRAFTON 1985). Reliable reservoir permeability determination is essential for preventing costly design of MHF treatments (JONES & OWENS 1980). POOLLEN (1957), HOLDITCH (1979) and SOLIMAN & HUNT (1983) also assume temporary permeability damage, whereas others have put forward arguments in favour of permanent productivity damage.

3.5. Permeability deterioration and clay mineral distribution

An important mechanism of permeability deterioration which applies for both Rotliegend and Carboniferous sandstones is the diagenetical transformation of more or less unstable framework grains (mainly rock fragments and feldspars) into aggregates of clay minerals which often have the crystal morphology of rolls and booklets, partially in combination with the additional authigenic growth of such minerals on the surface of framework grains or in the pore space. Although the replacement of isometrical fabric grains by clay mineral packages results in origin of secondary inter-clay mineral microporosity (SCHMIDT & McDONALD 1979) within the interwoven networks, plugging of the former inter-fabric grain pore space by neoformed clay mineral and iron oxide aggregates (MADER 1981, 1982, 1983 c), as well as organization of the much smaller and partially not properly interconnected interstitials in the pore network give rise to much worse flux of water and gas through the reservoir than before clay mineral diagenesis.

In the Carboniferous, this inter-clay mineral microporosity often leads also to a higher water saturation (due to increasing amounts of connate water) than that expected from porosity determination and gas shows, which is one of the main problems of proper petrophysical well log interpretation in this storey of the gas-prospective geological column and based on that decision on running casing, perforating, testing and probably also fracturing, or not to do anything but plug and abandon the well.

Formation damage by permeability deterioration increases with fracturing fluid pH and time of exposure (SIMON & COON 1976, ZAHNER & CRAFTON 1985). Reduction of pH-value can be effectively performed by addition of low amounts of methanol to a crosslinked stimulation gel (HARP 1980). Formation damage by increasing relative water permeability and water imbibition (cf. section 4.5.4.5.) is becoming worse the longer the reservoir is exposed to the treatment fluids as well as the larger the fracture area is subjected to fluid invasion, with thus frequently longer cracks being more seriously affected than shorter ones (ZAHNER & CRAFTON 1985). Aspects of clay mineralogy and morphology as well as permeability deterioration are discussed as follows.

3.5.1. Clay mineralogy and morphology

Different aspects of clay mineralogy and morphology as affecting reservoir sandstone porosity and permeability as well as supporting or inhibiting hydrocarbon flow are illustrated as follows. An outline is also given on the differences between detrital and authigenic clay minerals.

3.5.1.1. Clay morphology

Three main morphological types of dispersed clay minerals in sandstone reservoirs can be distinguished : discrete, pore-lining intergrown and pore-bridging aggregated particles (NEASHAM 1977 a). Discrete particles are scattered throughout the pore system and do not form intergrown crystal frameworks on pore walls or within intergranular pores. They have a more or less random arrangement with respect to one another and affect petrophysical properties mainly by reducing intergranular pore volume and behaving as migrating fines in the pore system. Pore-lining particles are attached to pore walls to form a relatively continuous thin clay mineral coating. The crystals are oriented either parallel or perpendicular to the pore wall surface.

Pore-bridging aggregates extend far into or completely across pore throats to create the bridging effect by extensive development of intergrown and/or intertwined clay minerals within the pore system, creating both microporosity and tortuous fluid-flow pathways. Low-permeability reservoir rocks frequently contain pore-bridging clay mineral types. Distinction should be made between pores which are large spaces surrounded by rock particles, and throats which are the narrow parts between two adjacent particles (GAO, SHEN & TU 1986).

3.5.1.2. Clay mineralogy

Formation clays have a profound reducing effect on matrix permeability (KUKAL, BIDDISON, HILL, MONSON & SIMONS 1983; McLEOD 1984). Different clay minerals precipitate as different morphological forms. Kaolinite is deposited as discrete booklets of clay crystals scattered throughout the pore system, and since kaolinite more commonly occupies intergranular space than it coats or lines pore walls, it is susceptible to migration when conditions of fluid flow prevail. Migrating clays adversely affect permeability by clogging or partially bridging and blocking pore throats. The most commonly observed form of diagenetic clays is as pore coatings or linings. Chlorite, illite and smectite are all deposited as relatively continuous linings on matrix grain surfaces. Smectite and illite tend to grow outward from the grain surface into the pore space, resulting in microporosity and drastically reduced permeability. Smectite swells when in contact with water foreign to the reservoir, thereby rendering permeability reduction during the stimulation treatment a serious problem.

3.5.1.3. Detrital and authigenic clay minerals

Distinction can be made between detrital clay minerals having been incorporated into the sand during deposition within the sedimentary basin below the base level of erosion, and authigenic clay minerals which originated after deposition of the sand through early palaeosol formation (MADER 1983 c, 1984 b), early to advanced ground-water-table influence, advanced burial diagenesis or even late renewed tectonical uplift and weathering impact.

Detrital clays which are originally inserted into the reservoir at the time of deposition are often laid down as horizontal or subhorizontal laminae or lenticles within a sandstone which act as directional permeability barriers limiting vertical permeability greatly while having in most cases little effect on horizontal permeability (unless the clay drapes are parallel to cross-stratification structures where they extend obliquely through the reservoir and can also considerably disturb or inhibit lateral flow). The origin of authigenic clays means formation of secondary porosity (SCHMIDT & McDONALD 1979) at the expense of primary porosity. Primary porosity reduction contributes heavily to the high capillary pressure of the rock and results in high water saturation and low relative gas permeability. Aspects of clay mineralogy and morphology as well as determination of volume, type and distribution of clay minerals from wireline logging records are summarized by FERTL & CHILINGAR (1988). Relationships of Rotliegend porosity and permeability are discussed by FERTL & CAVANAUGH (1973).

3.5.1.4. Other aspects

Morphology and distribution of clay minerals in oil- and gas-bearing pay formations (partially based on scanning electron microscope analysis, energy dispersive spectrometry, image processing and x-ray diffraction analysis) are also studied by TIMUR, HEMPKINS & WEINBRANDT (1971); GAIDA, RÜHL & ZIMMERLE (1973); KIEKE & HARTMAN (1973), STALDER (1973), DULLIEN & DHAWAN (1974); HOLUB, MALY, NOEL & WEINBRANDT (1974); THOMAS, CROWE & SIMPSON (1976); NEASHAM (1977 b), WILSON & PITTMAN (1977), THOMAS (1978), WARDLAW & CASSAN (1978), KEIGHIN (1979), PITTMAN (1979), PITTMAN & THOMAS (1979), COULTER & PURVIS (1980), SCHRANK & HUNT (1981); SNEIDER, KING, HAWKES & DAVIS (1981); WALLS (1981); CROCKER, DONALDSON & MARCHIN (1983); KUKAL, BIDDISON, HILL, MONSON & SIMONS (1983); HARRIS, HAYNES & EGGER (1984); LEE (1984), ORTIZ & McLANE (1984), RUZYLA (1984), BACKFISCH & HELING (1985); GARBIS, BROWN & MAURITZ (1985); McLANE & EDGINGTON (1985); GAO, SHEN & TU (1986); LAUGHREY & HARPER (1986); SCHAIBLE, AKPAN & AYOUB (1986); GIES (1987) and KRUEGER (1988).

Pore structures and authigenic clay minerals in Rotliegend and Carboniferous gas reservoirs are studied by DRONG (1979), COULTER & PURVIS (1980), JOHN (1983, 1987), DEURER & MENZ (1984), DIETZEL & HANTELMANN (1985); in Western USA tight gas sands by WALLS (1981) and SOEDER & RANDOLPH (1987), and in other formations by VITTHAL, GUPTA & SHARMA (1987). WILSON (1981) reviews the origin of clays controlling permeability in tight gas sands. RUHOVETS & FERTL (1982) present volumes, types and distribution of clay minerals in reservoir rocks on the base of well-log evaluation. An assessment of formation properties from stimulation treatments for Appalachian tight gas sands is performed by CHARLES, HUDOCK, UDICK & HALL (1983) and BRANAGAN, CIPOLLA, LEE & WILMER (1985). SPENCER (1983), FINLEY (1985); WARPINSKI, BRANAGAN, SATTLER, LORENZ, NORTHROP, MANN & FROHNE (1985) and JOHN (1987) discuss some geological and petrophysical aspects of tight gas reservoirs. OSTENSEN (1983); RANDOLPH, SOEDER & CHOWDIAH (1984) and SOEDER & RANDOLPH (1987) summarize the knowledge on porosity and permeability of tight sands. An overview of the diagenetical evolution of Buntsandstein red beds is compiled by MADER (1981, 1982, 1983 c).

Experimental petrophysical modelling of tight gas sands is carried out by SATTLER, HECKES & CLARK (1984). CHOWDIAH & SOEDER (1988) present a comparison of pore geometry in high- and low-permeability sandstones. DUTTON & FINLEY (1988) and FRACASSO, DUTTON & FINLEY (1988) give a comprehensive sedimentological and petrographical evaluation of a typical tight gas sandstone formation. DAVIES & ALMON (1981) stress the role of diagenesis in successful formation interpretation, well production, reservoir stimulation and enhanced recovery.

3.5.2. Permeability deterioration

LEONE & SCOTT (1987) summarize the mechanisms by which rock-fluid interactions can lead to permeability damage by migration of fines. Distinction has to be made between mechanical fines migration, water and/or brine sensitivity, and geochemical transformation. Permeability loss by mechanical fines migration occurs when loosely attached particles are mobilized due to drag forces exerted by the flowing fluids and subsequently collect in pore throats in sufficient quantities to cause flow path plugging. Water and/or clay sensitivity leads to swelling of clay minerals, with the particles in the swollen state requiring more space than normally and easier tending to releasing fines. Geochemical transformations comprise dissolution, precipitation and chemical reaction when foreign fluids are injected into the reservoir such as during stimulation or enhanced recovery. Some aspects of clay mineral dispersion and water sensitivity, clay mineral stabilization, clay mineral distribution and log interpretation, negative orientation of permeability changes, and capillary pressure and permeability relationships are discussed as follows.

3.5.2.1. Clay mineral dispersion and water sensitivity

The most important process of permeability deterioration is clay mineral dispersion and migration in sandstone reservoirs (GRAY & REX 1965). Distinction can be made between natural primary-detrital provenance or secondary-authigenic origin of clay mineral crystals which are mobilized by external influences coming from drilling, completion, testing and treating mechanisms, and tertiary-artificial derivation, with the particles being synthetically inserted into and distributed within the reservoir by various acquisition and stimulation processes. The outline as follows focusses on aspects of clay mineral dispersion control as well as clay mineral constellation and permeability damage. Comments are also offered on water- vs. oil-based stimulation fluids for the control of water-sensitive reservoirs.

3.5.2.1.1. Clay mineral dispersion control

Clay mineral dispersion within predominantly water-sensitive reservoirs can take place by hydration of exchangeable cations and particle surfaces, repulsion of interacting clouds of exchangeable cations, desorption or chemical removal of sorbed binding agents, neutralization of positive charges on particle edges, mechanical shear, and thermal (Brownian) motions (HOLCOMB 1986). Options for clay control in damage-potential reservoirs are leaving clays alone, render clays either temporarily or permanently inactive, remove clays by dissolution, or take off water from hydrating or mobilizing forces. KCl or NH4Cl brines should be used for temporary clay control only.

Organic cationic polymers are the most widely used clay-stabilizing additives that form monomolecular films on clay surfaces by cation exchange, with this filming allowing more permanent clay control due to particle bridging over multiple cationic sites along polymer chains. Longer lasting stimulation results can also be provided by anhydrous low-pH methanol prepads or breakdown treatment including fines-suspending surfactant and clay control additive to leak off in the near-wellbore region to partially solubilize, shrink, stabilize and suspend released fines as well as reduce waterblock (cf. section 3.11.2.4.2.) or imbibition (cf. section 4.5.4.5.). In any case, preventing damage to a reservoir formation is always easier than attempting to correct damage already caused.

3.5.2.1.2. Clay mineral constellation and permeability damage

The pore space of Rotliegend sandstones is often dominated by fibrous illite composed of needles and leaves which start radially from the hematite and clay pellicles and form thick coatings around the framework grains in a more outwards position than the earlier surface drapes (JOHN 1983, 1987; DIETZEL & HANTELMANN 1985). Mechanical stress causes illite needles to break into pieces, yielding together with silt-sized grains a large potential of permeability damage by plugging with fines. As a consequence of temperature, depth and burial history, most of the authigenic clay minerals have been transformed to illite in later diagenetical stages, with the total content of dispersed clay minerals varying in lateral and vertical direction. Particularly serious for permeability destruction is the concentration of illite fibres in the pore throats (leaving the centres of large pores either open or closing them also with a microporous fabric) and thus reasonably preserve porosity, but obstruct or plug the connections between adjoining pores, thereby drastically reducing the interconnectedness of the interstitials and thus also fluid flux (JOHN 1983, 1987).

Effects of pore-blocking mechanisms on formation damage are also reported by WOJTANOWICZ, KRILOV & LANGLINAIS (1987), and MUECKE (1978), GAUTIER & RICE (1981), SHARMA & YORTSOS (1986) and PRIISHOLM, NIELSEN & HASLUND (1987) comment on fines migration, blocking and clay swelling as main reservoir deterioration mechanisms. POTTER & DIBBLE (1983) present formation damage by colloid plugging. Restricted pore connections are also indicated by high capillary pressures in relation to total porosity. Vertical permeability is partially very limited, resulting in directional permeability anisotropy up to 80 % (cf. section 4.2.4.2.). PALLATT, WILSON & McHARDY (1984) discuss the relationship between permeability and morphology of diagenetic clay minerals. Aspects of migration of fines and clay minerals in the matrix of partially water-sensitive sandstones are also discussed by GABRIEL & INAMDAR (1983); SHARMA, YORTSOS & HANDY (1985) and SHARMA & YORTSOS (1986). COULTER & PURVIS (1980) analyze clay minerals in the Rotliegend of the Dutch North Sea for the purpose of fracture fluid compatibility determination.

3.5.2.1.3. Water- vs. oil-based stimulation fluids

Although illites are not swelling clays, they exhibit a certain water sensitivity. The pore space is infilled with connate water and initial gas saturation, with the illite needles and leaves being enveloped by connate water (DEURER & MENZ 1984, DIETZEL & HANTELMANN 1985). During low-rate gas flow, the connate water does not move and the rigid illite crystals are sheltered against mechanical forces. Flushing of the pore space with a water-based fluid, however, transmits the current movement to the connate water and thus onto the illite needles, resulting in breaking-off of the rigid illite structures and transporting the debris with the flushing fluid through the pore space. In addition to disintegration of the matrix framework compoundness, plugging of pore throats by the illite pieces may occur.

Using an oil-based fluid, the connate water saturation may be decreased to the irreducible minimum, but the water envelopes still protect the clay crystals against the current movement in the non-wetting flushing fluid, and due to the discrete interface, no axial forces are transmitted through the irreducible water saturation onto the illite needles. This clay mineralogical constellation is the reason for the water sensitivity of the Rotliegend reservoirs in several patches of the basin (DIETZEL & HANTELMANN 1985; SOMMER 1987, 1988), whereas in other spots, the formation is non-water sensitive and common water-based fluids can be applied (KLOSE & KRÖMER 1983; cf. also section 3.11.). Water sensitivity of Rotliegend and Buntsandstein gas reservoirs due to clay minerals is also reported by COULTER & PURVIS (1980).

3.5.2.2. Clay mineral stabilization

Clay stabilizing agents act by permanently holding clay particles together and preventing their movement or dispersion and subsequent particle plugging (DOWELL SCHLUMBERGER 1988 d). They are used to inhibit permeability damage due to foreign water intrusion in water-sensitive reservoirs. Clay stabilizers can be applied after matrix acidizing prior to gravel packing or as a spearhead for hydraulic proppant fracturing with water-based fluids. Clay mineral stabilization in water-sensitive reservoirs can usually be best achieved by 2 % KCl solutions which are used as a standard safety precaution in the prepad and displacement stages of hydraulic fracturing treatments since many years (WATERS 1980; cf. section 1.4.10.4.2.). KCl-bearing water-based fluids can also be used for stimulation if the proper pH-value is chosen and carefully kept constant during the treatment (ORTIZ & McLANE 1984). Some comments are offered as follows on low- and high-pH solutions as well as methanol prepad for illite preconditioning.

3.5.2.2.1. Low- and high-pH solutions

Low-pH (abt. 4) KCl-bearing water-based fluids are very effective in dirty sandstones by removing some of the authigenic clay structures that clog pore throats, thereby reducing the number of sites for potential waterblocking while leaving the detrital cementing clays intact. Higher-pH (abt. 7) solutions, however, are considerably damaging, because although not causing swelling or migration of clay particles, the more or less neutral fluid does not have the capacity to clean-up clay-filled pore spaces and may in fact be trapped in the extremely small cavities of the microporosity network. Other aspects of clay stabilization with KCl solutions are discussed by BLACK & HOWER (1965); McLAUGHLIN, ELPHINGSTONE & HALL (1976); BLACK (1978); BLACK, RIPLEY, BEECROFT & PAMPLIN (1979) and YOUNG & McLAUGHLIN (1979; cf. section 1.4.10.4.2.).

3.5.2.2.2. Methanol prepad for illite preconditioning

Another method of clay mineral stabilization is preconditioning of the fracture wall surfaces with a methanol prepad containing also acetic acid and surfactant additives which reacts with filmy illite bridging the pore throats in such a way that the illite roots still cling to the grain surfaces, thereby helping to maintain matrix competence, whereas the pore space is left unobstructed after flushing (ORTIZ & McLANE 1984). This preconditioning results in enlarging of pore interstitials and reducing of surface tension of the invading water, and thus a following water-based treatment fluid is less likely to be trapped by the microporosity of the authigenic illite webs. Aspects of clay protection chemicals in stimulation fluids are also discussed by CALLAWAY, ORTIZ & HOLCOMB (1982). Formation damage during waterflooding and water-based fracturing in high-clay-content reservoirs is evaluated by ORTIZ & McLANE (1984) and LEONE & SCOTT (1987), and CIVAN & KNAPP (1987) comment on effects of clay swelling and fines migration on formation permeability.

3.5.2.3. Clay mineral distribution and log interpretation

Problems of log interpretation of tight gas sands are assessed by RUHOVETS & FERTL (1982); KUKAL, BIDDISON, HILL, MONSON & SIMONS (1983); MONSON & SIMONS (1983), KUKAL (1984); BUSCH, SOBKOWICH, NEILL, SINHA & FERTL (1985); KUKAL &

SIMONS (1985), RAIBLE & GALL (1985), TSAY & FANG (1986) and KUCHUK (1987). The interpretation of wireline logs in low-porosity and -permeability tight gas sandstones is complicated by time-dependent resistivity changes due to deep filtrate invasion (KUUSKRAA & HAAS 1988). Correct identification of drilling fluid-flushed intervals that have high gas saturation which traditional log evaluation would read as water-bearing, accurate prediction of irreducible water saturation, and permeability calculation can be performed with an integrated log analysis system. Some aspects of porosity evaluation and deep filtrate invasion in tight gas sands are illustrated as follows.

3.5.2.3.1. Porosity evaluation

Porosity interpretation in tight gas sands from geophysical well logs is complicated by several geologically-dependent factors including matrix variations, incomplete fluid invasion, complex reservoir lithologies, and formation clays (KUKAL, BIDDISON, HILL, MONSON & SIMONS 1983). Special complications are formation clay quantification, water resistivity interpretation and water saturation evaluation. The geologically dependent assessment problems are related to the complex depositional, diagenetical and structural histories of the tight gas sands. Other difficulties of tight gas sand log interpretation are discussed by KUKAL (1981) and ROSEPILER (1981).

3.5.2.3.2. Deep filtrate invasion in tight sandstones

Low-permeability low-porosity rocks are generally deeply invaded by mud filtrate during the process of drilling a well, while more porous and permeable formations have principally shallower invasion (KUKAL, BIDDISON, HILL, MONSON & SIMONS 1983). The reason for this is that in more porous rocks which have more space available for liquids, a lesser depth of filtrate invasion occurs before a mudcake builds up on the borehole wall which is essentially halting the invasion process. Conversely, in less porous rocks, the same amount of filtrate will extend further into the formation. Most tight gas sands are compacted and the sonic compaction factor can be disregarded, with shallow or underpressured tight gas sands being rare exceptions.

3.5.2.4. Negative orientation of permeability changes

Permeability variations in sandstones during course of stimulation, acidizing, injection and production have almost always a negative orientation resulting in deterioration as a consequence of the combined effects of swelling and retention of clay particles (CIVAN & KNAPP 1987). Positive orientations of permeability transformation giving rise to enhancement of transmissibility are rather rare exceptions. Water-based substances such as brine and surfactants alter local equilibrium conditions between the porous solid matrix and fluids contained in its pore spaces, giving rise to swelling and enlarging of water-sensitive clay minerals. Being subsequently unable to fit to their original positions, clay particles along the pore surface are captured by the shear force of the flowing fluid and carried downstream, and during movement through the pore system, some of the clay flakes are immobilized by absorption on pore walls, deep-bed filtration or plugging of interstitials and pore throats.

3.5.2.5. Capillary pressure and permeability relationships

AMMER, SAWYER & DROPHIN (1984) and ERTEKIN, KING & SCHWERER (1986) comment on production and flow mechanisms in tight gas reservoirs. HALE (1986) analyzes well production histories in tight gas sands, and KABIR & HASAN (1986) evaluate prefracture testing in tight gas reservoirs. WARDLAW & TAYLOR (1976), WALLS & AMAEFULE (1985) and WARD & MORROW (1985) report on capillary pressure, porosity and permeability relationships in tight gas sands. Capillary pressures can be quite significant especially in low-permeability low-pressure reservoirs and could constitute additional barriers to flow (AMAEFULE & MASUO 1984). THOMAS & WARD (1972) and WALLS, NUR & BOURBIE (1982) emphasize the effect of overburden pressure and water saturation on gas permeability, and JENNINGS, CARROLL & RAIBLE (1981) stress the relationships of permeability and confining pressure. NOMAN & ARCHER (1987) discuss the effect of pore structure on non-Darcy gas flow in some low-permeability rocks. MAZZULLO & MAZZULLO (1985) investigate the clay mineralogy of some low-permeability reservoirs. Other aspects of tight gas sands are evaluated by JONES & OWENS (1980) and HALE, FIRTH, HANSEN, MURPHY & WOODCOX (1987).

3.6. Minimum pre-fracturing gas production rate and poroperm properties

The experience of MHF stimulation of both Rotliegend and Carboniferous sandstones in the last ten years has shown that unless a certain minimum pre-fracturing gas production rate is present, no economical producer can be achieved by carrying out a fracture job especially in the present time of lowered hydrocarbon prices. The consequence of this relationship is that deep tight Rotliegend and Carboniferous gas wells which would have been at once fractured some years ago when log- and core-derived petrophysical interpretation and formation test evaluation of the low-permeability reservoir rocks were still far from their present advanced experience and when major lessons of success and failure in completion and stimulation had still to be learnt, are nowadays carefully evaluated in terms of existing pre-fracturing and expected post-fracturing production rates.

In case of doubt, it is frequently preferred to shut-in the wells or to perform again longer production tests and pressure build-up measurements rather than to carry out the stimulation jobs straight away, or even to collect further information on reservoir composition and behaviour by performing a small-scale or miniature (or even micro-) treatment prior to a later full-scale or even jumbo job. A more advanced result of this changing concept is that even wells which formerly were drilled with the firm pre-spud knowledge that fracturing would be inevitable, are nowadays often postponed in order to await a better general hydrocarbon price scenario (and in Europe also a stronger US $ exchange rate scenario) or are even for the time being totally removed from the drilling budget despite their partially high rank in appraisal drilling strategy studies. Some aspects of pre-fracturing economical feasibility estimation and differences between exploration and development wells are outlined as follows.

3.6.1. Pre-fracturing economical feasibility estimation

Pre-fracturing economical feasibility estimation can be made along the lines of minimum fracture flow capacity and gas offtake rate, minimum reservoir properties, and stimulation rentability classification.

3.6.1.1. Minimum fracture flow capacity and gas offtake rate

Reasonable estimation of economical feasibility can be made in the Rotliegend of Germany FRG with the product of permeability times net thickness where the lower boundary is abt. 2 - 5 mdm (for example a reservoir of 0.1 md permeability and 20 - 50 m net thickness; DIETZEL & ERNST 1987; minimum fracture flow capacities in other formations are indicated by ATTEBERRY, TUCKER & RITZ 1979), and with the pre-fracturing gas production rate which has to be at least 800 - 1,500 m3/h according to hitherto stimulation and gas withdrawal experience, with these values applying at least under the conditions of the present economical scenario. KLOSE & KRÖMER (1983) determined for the well Söhlingen Z 4 which was fractured in 1982 up to 1 mdm and abt. 600 m3/h, and LEICHT (1985) concludes that the result of a fracture treatment is only economical in case of a pre-fracturing gas production rate of at least several hundreds of m3/h. If either value were smaller, commercial production cannot be expected even after an expensive MHF treatment.

3.6.1.2. Minimum reservoir properties for profitable stimulation

DIETZEL & ERNST (1987) also comment on minimum reservoir properties required for profitable stimulations. Rentability of withdrawal from low-permeability gas reservoirs is mainly determined by the increment of production obtainable by hydraulic proppant fracturing as well as by the initial flow capacity of the untreated well. A minimum value of the internal rate of return of 15 % before taxes is considered as profitability boundary. The limit for non-stimulated wells is characterized by the dimensionless equivalent radius rD = 1 which specifies the minimum combination of net height and permeability required for an economically feasible production without stimulation. The range between rD = 1 and rD = 400 in a net height vs. permeability plot covers the minimum parameter combination for a fractured well, with stimulation within this interval being necessary and also profitable in case of sufficient fracture conductivity. Below rD = 400, economically feasible production is not possible even if a technically successful and perfectly designed hydraulic fracture stimulation treatment was performed.

3.6.1.3. Stimulation rentability classification

Experience of past hydraulic proppant fracturing operations allows to carry out a quick-look classification of wells based on test results and log evaluation into wells rentable without stimulation, wells possibly profitable after MHF treatment, and wells not economically feasible in any case. Once a suitable model which is composed according to reservoir and economical boundary conditions has been established for a specific region, the net height vs. permeability curve can be directly used for the exploration risk analysis (DIETZEL & ERNST 1987). Boundary conditions and feasibility requirements of gas recovery from deep layered tight formations are also analyzed by JOHN (1987).

3.6.2. Differences between exploration and development wells

Distinction has further to be made between exploration wells where a lower post-fracturing gas production rate would be still economical due to the possibility of writing off the drilling cost, whereas a development well requires a higher post-fracturing gas production rate, as the drilling expenses cannot be written off (cf. section 3.8.2.1.). The experience during the last ten years has shown that a five- to six-times (in extreme cases also four- to ten-times) increase of the pre-fracturing gas production rate after the stimulation operation is the upper limit of expectation, with this relationship setting the boundary of hope according to the flow capacity of the untreated reservoir. In addition, the highest production rate is achieved immediately after stimulation and clean-up and later progressively declines with increasing offtake. The degree of original production rate improvement as well as the amount of subsequent capacity diminution are decisive for the evaluation of technical and economical success of the operation (ALBERTSEN 1985).

3.7. Stimulation cost and fracturing policy

The considerable to in some cases even almost exaggerated cost containment of oil and gas production companies following the major oil price decline in

1986 (cf. section 2.2.1.5.) and the worsening accompanying US $ exchange rate drop (cf. section 2.2.1.3.; with the limitations of capital expenditure partially even now still not having been lifted again) also influences the philosophy of hydraulic proppant fracturing of offshore and onshore exploration wells (particularly in the British Southern North Sea) which underwent considerable revision according to the changed general economical framework.

Estimating the cost of a hydraulic proppant fracturing treatment to be in the range of 5 - 20 % (onshore) or 10 - 40 % (offshore) of the total drilling expenses (depending on reservoir depth, thickness of the hydrocarbon-bearing section, stimulation job design and treatment performance), in former years an apparently dry tight hole was much more often still subjected to a final trial to convert it into an economical producer by means of stimulation before plugging and abandoning an expensive offshore well and probably burying hopes in smaller spots or larger patches of the facies belt or hydrocarbon play (for comparison, a fracturing treatment can account for 5 - 50 % of the total well cost in the USA; PAI, GARBIS & HALL 1983; ROBINSON, HOLDITCH & LEE 1983; VEATCH 1983; GARBIS, BROWN & MAURITZ 1985; ROBINSON, HOLDITCH & WHITEHEAD 1986; cf. section 2.3.). The following discussion focusses on aspects of fracturing policy, pricing scenario and fracturing treatment sequence strategy.

3.7.1. Fracturing policy

Particularly since the oil price seriously dropped in early 1986 (cf. section 2.2.1.1.2.) and at the same time the US $ exchange rate collapsed (cf. section 2.2.1.3.), however, exploration wells are especially offshore fractured only under exceptional circumstances and are much more frequently than during earlier years left without complete detailed investigation. The more comprehensive assessment is postponed to a later exploration or appraisal well in other parts of the structure or even to the development campaign when the feasibility is improved by first drilling not only one well from one platform location, but drilling a suite of deviated wells in a star- or spider-like geometrical fabric from one rig site that later becomes the place of the production platform. Some aspects of impact of fracturing boats on stimulation strategy as well as reservoir depth and stimulation expenses are discussed as follows.

3.7.1.1. Impact of fracturing boats on stimulation strategy

Secondly, also the basic fixed cost for the stimulation vessel which might be on call for the whole suite of wells is lower with respect to that if ordering the boat to come for only one fracturing job in an exploration well. There are no operational restrictions for a fracturing service vessel being on call or even on standby, because currently three stimulation boats are sharing the still very competitive and limited hydraulic proppant treatment market in the North Sea (FOX 1985; cf. sections 2.2.1.5.3., 3.3.1.2. and 3.8.1.2.), with thus waiting time for boat availability being almost not existing. This latter revised stimulation strategy has been overlain in 1986 by the superimposing effect of increased cancellation rates of fracturing treatments due to technical reasons which even deteriorated the already poor general setting (MADER 1987; cf. section 2.4.2.3.).

3.7.1.2. Reservoir depth and stimulation expenses

The different relationships between drilling and stimulation cost between onshore Germany FRG and Netherlands and offshore British and Dutch Southern North Sea result from the changing reservoir depth that gives rise to relatively lower drilling expenses in the North Sea than in Western Europe onshore if only comparing reservoir depth and net cost to reach the target. On the other hand, offshore drilling is in total generally much more money-consuming than onshore drilling as a consequence of considerably higher fixed gross rig rates. The high cost for hydraulic proppant fracturing is the consequence of thick pay zones which require large quantities of proppants for the treatments. These reservoir and technical engineering differences contribute to the economical feasibility of the stimulation operations and thus inevitably influence the fracturing philosophy.

3.7.2. Pricing scenario

Particularly in the 1986 crisis but still lasting at the moment, the offshore drilling cost scenario is considerably disturbed due to only dumping prices that have to be paid for drilling a well as a consequence of many unemployed rigs waiting for jobs, and also the stimulation expenses are lower than before as a result of serious competition between the service companies in order to get as much operations as possible to keep more or less active the in total three fracturing vessels presently working in the North Sea (cf. section 2.2.1.5.3.).

As both drilling and stimulation cost have remarkably dropped, the relationship between both can be expected to have more or less remained the same. In the second half of 1987, extreme price cutting has no longer been necessary, as the production companies converted from exaggerated cost containment to more appreciation of maintenance of a high technical quality level of the work carried out by the service companies and have arrived at the understanding that a fair pricing of the service companies has to be accepted in order to let them also survive, improve treatment quality and reliability, and give some rentability to all the three purpose-built stimulation boats (COOPER & MARSHALL 1984, OFFSHORE ENGINEER 1984) which have to share the North Sea market (FOX 1985, OILMAN 1985 b; cf. section 3.8.1.2.). With a little more consolidated oil price situation in the second half of 1987 and an increased hydraulic proppant fracturing and particularly acid fracturing activity (which is the most profitable stimulation work for the service companies), there is also no longer an economical base or requirement for extreme price dumping of the stimulation jobs.

3.7.3. Fracturing treatment sequence strategy

During course of field development campaigns, two different fracture treatment sequence strategies have to be distinguished comprising spot and batch fracturing which are characterized in more detail as follows.

3.7.3.1. Spot fracturing

Spot fracturing comprises sequential drilling, testing and stimulation of one well after the other with complete termination of work at one location before moving to the next, thereby allowing to incorporate all the information and experience from earlier operations into design and execution of later treatments. Batch fracturing means first drilling of all the wells one after the other and then second in another round perforating, stimulating and putting on stream of the whole sequence of wells successively. While some companies prefer the first or the second strategical concept throughout the field development campaign, it may be economically more feasible to start with spot fracturing until a reasonable level of experience has been gained and the reservoir and its behaviour after stimulation is fairly well understood, and then later switch over to batch fracturing, in order to have the stimulation vessel continuously operated and thus having lower general treatment expenses by avoiding time- and money-consuming journeys of the boat from base to location and back.

3.7.3.2. Batch fracturing

Batch operation is thus a more mature fracturing strategy that should be favoured once reasonable knowledge of reservoir behaviour and satisfactory experience of treatment performance have been accumulated. Batch execution of jobs is particularly suitable for large development campaigns where sufficient information has already been gained during appraisal well evaluation and early production drilling assessment, whereas spot fracturing is in addition to its submature stage in large fields that are to be developed the most effective strategy in exploration and appraisal stage as well as for development of small fields with only a couple of boreholes. Batch completion aspects of gravel packing in oil wells are discussed by HENG (1987).

The large field campaigns of hydraulic proppant fracturing of Rotliegend tight gas sands in the British Southern North Sea in the period 1987 - 1990 are based on both concepts, and selection of the stimulation strategy is made according to field development maturity stage by drilling and completion. While some campaigns are carried out along the spot guideline, other actions follow the batch strategy.

3.8. Philosophy of off- and onshore exploration well fracturing

In terms of strategical planning and operational execution, distinction has to be made between offshore and onshore exploration well fracturing concerning logistical approach and economical arrangement of job sequences and field developments.

3.8.1. Philosophy of offshore exploration well fracturing

In contrast to Germany FRG and Netherlands onshore where the Carboniferous has been frequently found gas-bearing and has often been fractured during the last ten years (cf. fig. 10), the Carboniferous in the British Southern North Sea has been so far met in most structures water-bearing (as a consequence of its greater reservoir depth of 11,000 - 13,000 ft or 3,700 - 4,400 m, in comparison to the gas-water-contact which is often quite further up in the stratigraphical section) and only in the last years exploration succeeded in encountering more and more gas-bearing patches with favourable closure height and suitable petrophysical properties being above all water saturation. The common strategy in the Carboniferous at the moment is not to immediately fracture exploration wells because time is even in the present low-price scenario still too expensive to let the rig wait for the stimulation treatment including its preparation. Carboniferous development, however, could be considerably enhanced by a couple of experimental fracture jobs in a few selected wells in order to get a better feeling for the potential of the reservoir and to improve production planning. This has particularly been demonstrated by the difficulties to get the petrophysical evaluation of the Carboniferous to a reasonable approach to the reality, with the main problem being the water saturation of the tight Carboniferous reservoir sandstones.

The 1986 oil price crisis has transferred this policy of hesitation of fracturing of exploration wells after a drill stem test also to Rotliegend items, with again, however, the execution of a few selected hydraulic stimulation treatments in every field being essential to back-up the decision of commerciality by getting an understanding of the real capacity of the reservoir. A special application of hydraulic proppant fracturing in the Rotliegend of the British Southern North Sea was stimulation of reservoirs in some wells of underground gas storage fields in the last years. Some comments are offered as follows on well evaluation and fracture design time, and service boat capacity limit.

3.8.1.1. Well evaluation and fracture design time

An important aspect of exploration well fracturing regardless of an offshore or onshore situation is the time that is needed for proper well evaluation and fracture design (cf. BRIM 1986). Although computer programs exist which can provide fast estimation of the optimum solution in terms of choice of proppant type and grain size (CHENG 1984; COBB 1985, 1986; cf. section 1.4.12.), designing of the whole treatment may need too much time to let the rig wait, and if too many rapid guesses are done, the risk of unsatisfactory performance or even failure of the stimulation operation is increased. Fracturing of later development wells has the big advantage that reservoir parameters are already more or less known from earlier exploration and appraisal wells (although many surprises still may occur), but the major drawback of renouncing to stimulate an exploration well is always the danger that based on insufficient reservoir testing without fracturing, decision could be made that exploitation of the hydrocarbon content of the encountered structure is uncommercial and thus the patch never sees development into a field, with thereby probably considerable gas reserves being lost that would have more than paid off the initial stimulation operation.

While many wells within a field which is subjected to a larger hydraulic fracturing campaign can be successfully and profitably stimulated following only one or a few base proposals and test jobs, in other cases more or less individualized fracturing treatment design according to local circumstances in the boreholes is required (MACK & MASON 1985) and much experimental fracturing including submature jobs and also failures are the only way to optimize treatment design in the field campaigns and to achieve the best possible operation layout for the particular reservoir complex.

3.8.1.2. Service boat capacity limit

Service boat capacity is often the limiting factor for stimulation treatments in the Southern North Sea (MARTINS 1987). In the early years of European offshore proppant fracturing, only platform-mounted equipment was available, with treatment quantities having been quite limited (LOUVIERE 1987). In the early 1980`s, the operations got larger when the first self-contained service vessels were developed that at the beginning represented converted supply boats (COOPER & MARSHALL 1984; cf. section 3.3.1.2.). The service ships include pumping equipment, fluid and proppant storage, job control and monitoring, and quality supervision laboratory. Stimulation vessels provide greatly increased versatility in rates, volumes and designs of the treatments, whereas platform stimulations were limited both in rate and volume due to space constraints (SNOW & HOUGH 1988).

Injection rates are often limited by well configuration and friction loss, with the main influencing factors being rate, horsepower and government regulations. In the mainly highly-deviated to even occasionally almost horizontal wells, the longer boreholes limit rate and pressure of fracturing by service boats. The three purpose-built stimulation vessels (COOPER & MARSHALL 1984, OFFSHORE ENGINEER 1984; cf. section 2.2.1.5.3.) currently operating in the North Sea are all comparable in capacity of job performance such as pumping power and material storage facilities (FOX 1985, OILMAN 1985 b; cf. section 3.7.1.1.), with therefore boat employment having been a critical competitive factor during the worst time of the crisis following the 1986 oil price crash (cf. section 2.2.1.3.2.2.). North Sea stimulation logistics and requirements are outlined by MACADAM (1984).

3.8.2. Philosophy of onshore exploration well fracturing

According to the poor general economical situation after the major oil price drop in early 1986 (cf. section 2.2.1.1.2.) and the parallel US $ exchange rate fall (cf. section 2.2.1.3.), also the onshore Rotliegend exploration and appraisal strategy concept in Germany FRG (and partially also in the Netherlands) has been considerably changed. Some aspects of job suspension vs. cost writing-off and changing aims of hydraulic fracturing are briefly illustrated as follows.

3.8.2.1. Job suspension vs. cost writing-off

In contrast to former years where exploration and appraisal wells had been drilled also on such targets that were known before to require expensive large-scale fracturing if containing gas, such prospects are presently postponed and only items are drilled where reasonable production rates without stimulation could be expected (in contrast to this onshore policy, the development campaigns of tight gas fields offshore are quite another story). Thus at the moment fracture jobs in onshore exploration wells are only coming up randomly and by chance when the predicted better facies situation was not encountered accordingly, and when sufficient quantities of gas are found that could only be produced at economical rates after stimulation also in the revised general financial framework. As a consequence of presently sufficient capacity, one of the first drilling program trimming and cancellation actions after the 1986 oil price collapse affected development wells in onshore tight gas fields, and also exploration and appraisal wells in reservoir patches where MHF stimulation would be undoubtedly indispensibly for proper pay zone evaluation have been in many cases at least suspended and postponed, if not partially for the time being completely deleted (cf. sections 2.2.1.5. and 2.2.2.1.3.).

This adjustment to changed economical feasibility framework conditions is also reflected by the fact that in Germany FRG the last MHF stimulation operations on Rotliegend and Carboniferous tight gas reservoirs have been carried out in 1985 as a culmination of the period 1977 - 1985 when several dozens of wells have been fractured in total. According to suspension and cancellation of drilling and stimulation activity after the 1986 oil price crash, however, not a single MHF job has been carried out in the period 1986 - 1988, and the expectations for 1989 are also not very favourable, with only little hope remaining that the former hydraulic proppant fracturing acquisition of Rotliegend and Carboniferous tight gas will be resumed within the next few years (cf. also sections 2.4.1.1. and 2.4.2.).

Apart from the overall feasibility, exploration wells often confront with the necessity of fast decision of either at once plugging and abandoning tight reservoirs, writing off the well as a failure and probably forget the expected reserves in some spots, or trying to get still a positive result by hydraulic proppant fracturing (and thus to confirm geological and reservoir engineering concept and model), whereas development wells cannot be written-off in case of negative result (cf. section 3.6.2.). Thus given the fact of meeting a pre-fracturing gas-flow rate that is so far not commercial but above a certain minimum level which permits to expect a good producing capacity after stimulation, the well is sooner or later fractured.

3.8.2.2. Changing aims of hydraulic fracturing

The general fracturing policy in Western Europe has also changed during the last years. In the past, stimulation of Rotliegend and Carboniferous sandstones was done as much as possible in order to maximize the proven gas reserves. Nowadays the main purpose of stimulation is to increase the general potential of the area and to double the life time of the fields, with demand of increasing the proven gas reserves being no longer so violently challenging than before, because most of the hitherto encountered gas has already been sold through long-term supply contracts and also the market for sale of gas is at least in parts of the area (and probably only temporarily) approaching saturation due to oversupply from various sources and less quick growth and thus more slowly rising demand than expected a few years ago. Unless gas requirement will again become higher in the next years due to again slightly accelerating urban and industrial growth, the latter tendency of the gas market could also seriously affect the hydraulic proppant fracturing market. Some differences between offshore and onshore hydraulic fracturing stimulation treatments are also outlined by ROODHART, KUIPER & DAVIES (1986).

3.9. Implications of fracturing philosophy on proppant selection

In addition, if a hydraulic fracturing job in such exploration wells is carried out at all, it is generally only done for delivery of information on which an extension drilling and field development concept could be based on, and the borehole is often plugged back and abandoned after temporary post-fracturing test production that has the only task to determine the real well capacity. This statement is valid unless the borehole has an extraordinarily favourable result which may lead to straight completion for being later put at once on stream when acquisition of the patch by pipeline or other hydrocarbon transport facilities is made, or the development campaign is carried out with installation of a production and gathering platform).

As a consequence of no previous information in the field when the first exploration well is drilled, there is also often still the risk that the production rate is even uncommercial after the fracturing treatment and thus the borehole would have to be discarded due to economical reasons, in contrast to development wells where the risk is much lower and the boreholes are looking towards a long production history for probably several decades. The various possibilities of fracturing philosophy are summarized as follows by commenting on exploration strategy and stimulation testing, field experience and development concepts, and appraisal drilling strategy concepts and experience.

3.9.1. Exploration strategy and stimulation testing

Concerning proppant marketing, many operators unfortunately believe that short-term test periods for the goal of analysis of real well capacity only require propping of the fractures with cheap sand that can more or less reasonably withstand higher pressures for a couple of days to weeks in temporarily metastable conditions, instead of using expensive intermediate- to high-strength proppants that would be adequate to the closure stress in the pay formation and could also resist to increasing closure stresses with drawdown of the reservoir pressure upon depletion after long-term gas production during up to 20 years or more. Some aspects of proppant cost containment and tail-in policy as well as proper interpretation of reservoir capacity are illustrated as follows.

3.9.1.1. Proppant cost containment and tail-in policy

Exceptions of this cost-containment policy are in some cases minor tail-in quantities of synthetic high-quality proppants in the terminal stages of the treatments in order to create better support in the vicinity of the borehole (HOWARD & FAST 1970 a; HICKEY, BROWN & CRITTENDEN 1981; cf. also sections 2.4.1.2. and 4.2.2.). The tail-in strategy ascertains proper plugging of the induced crack by high-quality proppants at least in the immediate surroundings of the wellbore where flow velocity and pressure drawdown reaches the highest values, and therefore short-term test production may almost approach real reservoir capacity if adequate proppant tail-in is made. It would, however, certainly provide a much better assessment of pay capability in terms of long-term productivity if a full-scale fracturing treatment with corresponding proppant distribution within the crack would be carried out.

Apart from proppants, also fracturing fluids can be tailored to meet viscosity specifications in a wide range of temperature applications by proper use of primary and secondary gelling agents (ATTEBERRY, TUCKER & RITZ 1979; LEE & DANESHY 1985; cf. section 4.3.4.3.), and similarly, tailored-pulse loading can be used for multiple fracturing of boreholes (SWIFT & KUSUBOV 1981, CUDERMAN 1982; cf. section 4.2.2.1.3.3.). In addition to tail-in setting, selective proppant placement within different successive parts of the fracture wing can also be achieved by density control foam application (BARBER & THEMIG 1985; cf. section 4.2.2.3.3.3.) and multiple-fluid treatment (LEE & DANESHY 1986).

3.9.1.2. Proper interpretation of reservoir capacity

The main drawback of a money-saving fracturing layout in exploration and appraisal wells with pumping only sand or restricting synthetic proppants to a little terminal tail-in (cf. section 2.4.1.2.3.) is similar to the danger of totally renouncing of the stimulation operation. Failure of the natural sand to withstand the closure stress sufficiently in its metastable nature in the high-pressure regime during the relatively short testing period, impact of the worse conductivity of sand with respect to synthetic high-quality proppants (which is falsifying or mimicking the real reservoir exploitability), improper fracturing and/or production results or other technical complications can easily lead to misinterpretation of reservoir capacity and potential, and based on such inadequate information, assessment could be made that the deposit is uncommercial and thus the strike would never see its development into a field.

Although the application of high-quality proppants in exploration and appraisal well fracturing is far from being a guarantee for favourable results due to many geological and engineering factors contributing to overall reservoir performance, the possibility that a poor formation will produce at appreciably high rates after stimulation can be maximized by choosing the optimum propping material. Much money can be lost if based on false or insufficient information decision is made that the patch is uncommercial, and a fortune can be gained in case of reasonable results from a properly designed treatment.

3.9.2. Field experience and development concepts

In contrast to exploration and appraisal where in many to in some areas even most cases exclusively sand or a little tail-in of intermediate-strength proppants following a main load of sand are injected into the fracture for the reason of short-term cost containment (and in many instances unfortunately at the expense of long-term reservoir behaviour information), decisions for the major development campaigns in 1987 - 1990 (cf. sections 2.2.1.6.2. and 2.4.1.2.) have been made by several companies to go on the absolutely safe side in view of long-term production history and to choose exclusively intermediate-strength proppants and not to select any quantities of sand for the design of the hydraulic stimulation treatments. Some comments are offered as follows on advantages of intermediate-strength proppant insertion and retardation of Carboniferous strikes.

3.9.2.1. Advantages of intermediate-strength proppant insertion

Concerning all the possible reservoir pressure changes during repeated shut-in and open-up of wells (cf. section 4.12.4.3.) and progressive depletion during long field life (cf. section 4.12.4.2.), the additional investment for the proppant with higher conductivity and stress resistivity performance is certainly paying off very well and in view of improved gas production rates also reasonably fast. In terms of getting adequate field experience, some companies even avoid propping of fractures in exploration wells with cheap sand and prefer the higher investment for taking intermediate-strength proppants in order to carry out a sound field experiment which gives the confirmation that the design of the operations in the later development stage is technically possible to be carried out without major drawbacks and the result is satisfactory.

3.9.2.2. Retardation of Carboniferous strikes

The geological and engineering disadvantage of the present exploration well fracturing philosophy is highlighted by the fact that (with only a few minor exceptions) the last couple of major gas strikes in the Upper Carboniferous in Germany FRG dates back more than five years (SCHRÖDER 1985), although most of the hitherto accomplished several dozens of MHF stimulations in Germany FRG have been carried out in the Upper Carboniferous and plenty of exploration wells have been drilled in the meantime. Straight plugging and abandoning of exploration wells of apparently failure nature, however, is not considered to be an adequate strategy for increasing tight gas reserves and better understanding of the complex reservoir organization and architecture, and hydraulic proppant fracturing for final confirmation of suitability and profitability of well completion is regarded to be essential particularly in the complicated Carboniferous pay assemblages. This relationship underlines the necessity of a more aggressive fracturing policy to try to get some extension of this producing horizon (this conclusion is valid for Germany FRG as well as surrounding countries in both onshore and offshore areas).

3.9.3. Appraisal drilling strategy concepts and experience

The former expectation and estimation of the potential of tight Rotliegend and Carboniferous sandstones in terms of hydraulic proppant fracturing have been considerably revised during the last years particularly in Germany FRG. While previously many low-permeability reservoirs that are worth fracturing to become economical producers were predicted in both formations, in the Rotliegend either very good or extremely poor permeabilities have been frequently encountered which either did not require stimulation or leave no other possibility than to plug the well. The Carboniferous has often been proven to be either completely tight or to be watered.

Adjusting to the new economical situation which according to all the present forecasts will not again lead to a feasibility setting comparable to that during the early 1980`s within the next few years, the gas appraisal drilling strategy has also been changed in various extent in many offshore and onshore parts of Western Europe. In contrast to the golden period in the early 1980`s when during a time of partially excessively high oil price and overreasonably strong US $, big planning was made for unconventional gas exploration and appraisal in such facies belts of especially the Rotliegend Basin (Southern Permian Basin; ZIEGLER 1982) where hydraulic proppant fracturing could already be included into the economical assessment of the well prior to its spudding, these high-risk and -cost items have now been in many cases slightly to considerably postponed in both drilling schedules and appraisal strategy studies, and near-future exploration concentrates on less critical areas in facies seams with basically higher permeability.

This applies particularly for onshore gas exploration in the Rotliegend in Germany FRG where the present focus is on acquisition of petrophysically well-developed intervals of the Schneverdingen-Sandstein (cf. section 3.2.1.3.) in synsedimentary graben systems which do not require fracturing stimulation and that permit to withdraw large quantities of gas over longer periods of time before upon depletion of the high-permeability Schneverdingen-Sandstein the overlying low-permeability continuously-extending Hauptsandstein has to be prepared for production by large-scale hydraulic proppant fracturing. The offshore planning, however, can partially not react so quickly, because contracting of platforms for development drilling and prior to that selling of the proven and parts of the unproven gas reserves result in the necessity of keeping more or less the established appraisal drilling strategy concept almost irrespective of economical deteriorations in the meantime(cf. section 2.2.1.6.2.).

3.10. Horizontal and vertical difference of fracturing potential

The fracturing potential in the Rotliegend Basin (Southern Permian Basin; ZIEGLER 1982) is not homogeneous and isotropic in neither horizontal nor vertical direction in the gas-bearing reservoir column. Horizontally, the distribution of pay zone belts suitable for hydraulic proppant fracturing is controlled by the sedimentary facies and thus reservoir quality zonation within the depositional area, with the seam where stimulation is essential for economical gas production being concentrated to the boundary between the marginal aeolian dune field (cf. plates II and VIII in section 3.13.) and braided fluvial wadi systems (cf. plates III and IV) and the central playa mud flat and desert lake (cf. plates VI, VII and IX) within the main depositional basin (Wechselfolge facies association; GRALLA 1988; cf. also section 3.2.1.1.). The succession of several reservoir storeys in the gas-bearing formation pile (for stratigraphical subdivision of the Rotliegend series cf. section 3.2.1.3.) results in superimposition of similar sedimentary and petrophysical facies zones, or in case of shifting depositional facies belts in the main trough or concentration of some sediments in subbasins or syndepositional tectonical features such as grabens gives rise to vertical stacking of different facies and reservoir quality zones and thus petrophysically variable sediments.

Distinction can be made between primary petrophysical zonation according to permeability of the reservoir rocks, secondary geochemical zonation due to gas composition, and tertiary political zonation of Rotliegend gas reservoirs with fracturing potential as a consequence of the economical and administrative systems in the territory where the gas fields are located. Comments are offered as follows on permeability which exhibits mainly a vertical zonation, and on gas composition which includes predominantly horizontal variations.

3.10.1. Permeability

The Rotliegend in Germany FRG is stratigraphically divided into three reservoir storeys (BRADEL & DRAXLER 1982, PHILIPP & REINICKE 1982, KLOSE & KRÖMER 1983; HEDEMANN, MASCHEK, PAULUS & PLEIN 1984; GRALLA 1988; cf. section 3.2.1.3.) with different primary petrophysical potential and/or secondary reservoir engineering requirement for hydraulic proppant fracturing. The composition of the continental succession of three different palaeoenvironmental and petrophysical complexes of stratigraphical importance is particularly well-developed in the Weser-Elbe facies province (GRALLA 1988) or East Hannover area (HEDEMANN, MASCHEK, PAULUS & PLEIN 1984).

The lower section (Schneverdingen-Sandstein within the Schneverdingen-Formation in Germany FRG or Lower Slochteren-Formation in the Netherlands and Dutch North Sea) and the upper part (Wustrow-Sandstein as the most potential Wechselfolge pay zone within the Hannover-Formation in Germany FRG or Ten Boer-Formation in the Netherlands and Dutch North Sea; GRALLA, NIEBERDING & SOBOTT 1988) are generally moderate- to high-permeability reservoirs not requiring stimulation, whereas the middle portion (Hauptsandstein within the Emden-Formation in Germany FRG or Upper Slochteren-Formation in the Netherlands and Dutch North Sea; Slochteren-Hauptsandstein) is mainly a low-permeability to tight reservoir and usually requires proppant fracturing (in the latter horizon, most of the MHF operations on Rotliegend have been carried out, particularly in the Söhlingen field; KLOSE & KRÖMER 1983). Some points of depositional environment and grain size distribution giving rise to this permeability stratigraphy are discussed as follows. Relationships between sedimentary facies and hydraulic proppant fracturing potential are also outlined.

3.10.1.1. Depositional environment and grain size distribution

The different necessity of hydraulic proppant fracturing mainly relates to the sedimentary and petrophysical facies of the individual storeys of the gas-bearing Rotliegend column (the discussion as follows is based on the Weser-Elbe or East Hannover facies province in Germany FRG where the bulk of Rotliegend exploration, development and fracture stimulation occurs). The aeolian Schneverdingen-Sandstein (consisting frequently of cross-bedded sands having been produced by migration of large and high dunes; cf. plate II/1-6 in section 3.13.) and the fluvial Wustrow-Sandstein (having been deposited by small shallow braided streams or sheet floods; cf. plates III and IV) usually have sufficient primary and/or secondary porosity and thus also permeability, whereas the mainly dry to wet interdune playa sand flat (with in many cases only subordinate amounts of aeolian dunes, but abundant dry aeolian, damp playa and wet lacustrine sheet sands; cf. plates VII, VIII/1-5 and IX) Hauptsandstein has frequently poor permeability due to abundant participation of adhesion of sand and mud in sedimentary accretion, primary or secondary clay infiltration during deposition and/or early diagenesis, and worse grain size distribution.

Depositional environments and diagenetical history of Rotliegend and Buntsandstein sediments as a base for reservoir modelling are discussed by PLEIN (1978), LÜTZNER (1981), ROBINSON (1981); DRONG, PLEIN, SANNEMANN, SCHUEPBACH & ZIMDARS (1982); GLENNIE (1983, 1986), LEE (1984), MADER (1981, 1982, 1983, 1984, 1985, 1988 a), MADER & TEYSSEN (1985), MADER & YARDLEY (1985); BIFANI, GEORGE & LEVER (1987); GRALLA (1988), MADER & CHATALOV (1988), MARTIN & EVANS (1988) and RICHARDSON, SANGREE & SNEIDER (1988; cf. also section 3.2.1.1.).

In some parts of a few fields, however, also the Hauptsandstein can be encountered in a facies of sufficient quality for being a commercial producer without stimulation (in case of larger contributions of cross-bedded aeolian dune sands (cf. plate II/1-6) which represent the highest reservoir quality in the Rotliegend), or the Schneverdingen-Sandstein and the Wustrow-Sandstein might be only economical pay formations after hydraulic proppant fracturing (if the former contains more alluvial-fan breccias (cf. plate III/5-8) built up of volcanic rock clasts instead of aeolian dune sands as a consequence of shifting shares and rates of intertonguing of both facies associations in the marginal portions of graben systems closer to the boundary faults, or if both are more tightly cemented in the vicinity of syndepositional graben faults, above extraordinarily thick volcanic lava flow piles, or in the water-bearing aureole of the gas fields).

3.10.1.2. Permeability stratigraphy

Effective permeabilities under reservoir conditions (as determined from production tests and core measurements) are in the Söhlingen field in the Weser-Elbe or East Hannover facies province in the Schneverdingen-Sandstein 20 - 50 millidarcy, in the Wustrow-Sandstein and other Wechselfolge sandstones up to 10 millidarcy and in the Hauptsandstein in the best cases 0.5 - 1 millidarcy, but frequently much lower permeability values down to only 0.5 microdarcy occur (KLOSE & KRÖMER 1983; for Rotliegend stratigraphy cf. section 3.2.1.3.). This permeability distribution in vertical direction clearly reflects the major significance of the Hauptsandstein as the target of MHF operations, whereas the other Rotliegend sandstones can usually be put on stream without stimulation (the minimum permeability for economical gas production without treatment is abt. 0.5 - 1 millidarcy; LEICHT 1985; cf. also sections 3.4.1.3. and 3.6.1.2.).

3.10.1.3. Sedimentary facies and proppant fracturing potential

In Rotliegend and Buntsandstein continental red bed sequences in Middle Europe, potential and/or necessity of hydraulic proppant fracturing stimulation with the aim of increasing recoverable reserves and/or offtake rates of natural gas are influenced by the sedimentary facies of the reservoir rocks as well as the lateral continuity of the sand bodies. Distinction can generally be made between aeolian, fluvial and alluvial-fan, and lacustrine deposits (characteristics of continental depositional facies types for hydrocarbon production are given by RICHARDSON, SANGREE & SNEIDER 1986). Some aspects of calcrete palaeosols are also mentioned, and an outline is also given on lateral continuity of reservoir units. Detailed sedimentological characterizations and interpretations of aeolian, fluvial and lacustrine deposits as well as palaeosols in Rotliegend and Buntsandstein (including numerous colour photographs) are contained in MADER (1985 a, 1985 b), and a selection of structures are also presented in plates I - XI together with longer explanations that are partially independent from the text (cf. section 3.13.).

3.10.1.3.1. Aeolian sediments

Aeolian dune sands with well-developed cross-stratification (MADER 1985 a, 1985 b; MADER & YARDLEY 1985; cf. plate II/1-6) and weak to moderate diagenetical cementation are the reservoir rocks with the highest permeability in Rotliegend and Buntsandstein and usually allow production of gas at sufficiently high rates without necessitating fracturing treatment. While large-scale cross-stratified dune sands form extensive sand bodies in both horizontal and vertical direction, small-scale cross-bedded dune sands (cf. plate VIII/1-5) are not only limited in thickness, but may also be restricted in lateral continuity and communication between the subunits.

Small-scale cross-stratified aeolian dune sands are frequently vertically interbedded and laterally intertonguing with horizontal-laminated aeolian sheet sands deposited on dry substrates (MADER & YARDLEY 1985) as well as playa and playa-lake sand and mud laid down on damp and wet ground with varying influences of deflation and adhesion of aeolian sand (MADER 1983 a, 1985 b; cf. plates VIII and IX). Damp to wet surface conditions give rise to intercalation of mud laminae and layers into the sand complexes (cf. plate III/1-2) which deteriorate reservoir communication, and also lead to insertion of finer-grained sand which has negative impact on permeability. In contrast to large-scale cross-bedded dune sands which usually do not require hydraulic fracturing stimulation, small-scale cross-stratified dune sands in combination with horizontal-laminated dry aeolian sheet, damp playa and wet playa-lake sediments (MADER 1983 a, 1985 b; MADER & YARDLEY 1985) frequently necessitate reservoir treatment by hydraulic proppant fracturing in order to increase recoverable reserves and to allow to produce the gas at economically feasible rates.

The significance of aeolian dune sands (cf. plate II/1-6) as the highest-permeability gas reservoirs in the Rotliegend of the North Sea is also outlined by HAAK & ELEWAUT (1988), MARTIN & EVANS (1988) and RICHARDSON, SANGREE & SNEIDER (1988). Sedimentary and petrophysical features of aeolian sandstone reservoirs are also illustrated by MADER & YARDLEY (1985), NURMI (1985) and GOGGIN, CHANDLER, KOCUREK & LAKE (1986).

3.10.1.3.2. Fluvial and alluvial-fan deposits

Fluvial (cf. plate IV/3-8) and alluvial-fan (cf. plate III/5-8) deposits are similarly as aeolian dune sands (cf. plate II/1-6) in many cases good to moderate primary reservoir rocks, but while the poroperm properties of dry aeolian sheet, damp playa and wet playa-lake sediments are already primarily deteriorated by insertion of mud and fine sand, permeability of fluvial and alluvial-fan deposits suffers in many cases from secondary diagenetical destruction. Well-sorted fluvial sandstones (cf. plate IV/3-8) and conglomerates (cf. plate V/1-4) originating in shallow to moderately deep braided river channels or partially even unconfined sheet floods (MADER 1983 b, 1984 a, 1985 c; MADER & TEYSSEN 1985, MADER & CHATALOV 1988) have sufficiently coarse grain size to possess good primary porosity and permeability. Intensive quartz authigenesis during course of the diagenetical evolution (MADER 1981, 1982, 1983 c), however, gives rise to frequently tight cementation of the matrix which leads to requirement of hydraulic proppant fracturing for liberation of trapped gas at economically feasible rates.

Poorly-sorted alluvial-fan breccias, conglomerates and sandstones (cf. plates III/5-8, VI/4 and VII/5) have already primarily worse poroperm properties due to dispersion of mud and fine sand in the matrix as well as often irregular shape of sand and gravel grains and discontinuous distribution of gravel clasts, and therefore represent already primarily less suitable pay rocks. Secondary diagenetical lithification is an important contributing factor for further deterioration of reservoir characteristics. If such submature to immature alluvial-fan deposits are gas-bearing, hydraulic fracturing stimulation of frequently large scale is the only way to achieve production at economically feasible rates, but in some cases, reservoir conditions are so poor and/or lateral pay continuity is not sufficient to achieve reasonable gas offtake rates at all.

3.10.1.3.3. Lacustrine sediments and calcrete palaeosols

Lacustrine sediments are mainly mudstones and fine sandstones (MADER 1983 a; cf. plates III/1-2, VI, VII/1-6 and IX) which in Rotliegend and Buntsandstein form rather sealing horizons and layers instead of representing reservoir rocks themselves. If fine-grained homogeneous monolithic or heterogeneous polylithic lacustrine sediments happen to contain gas, economical production rates from these tight reservoirs can only be achieved by hydraulic proppant fracturing stimulation. Reservoir heterogeneity is mainly created by alternating dry, damp and wet conditions at the surface of temporary lakes of playa type with intertonguing and succession of dune and wind ripple migration, adhesion of deflated aeolian sand, aquatic redeposition of windblown sand, and sedimentation of fine sand and mud in the shallow playa lake.

Calcrete palaeosol formation (MADER 1984; cf. plates X and XI/8) by pedogenic overprinting of various types of primary sediments usually comprises authigenesis of clay minerals and iron oxides in the pore space (MADER 1981, 1982, 1983 c) together with growth of carbonate concretions either as isolated nodules or as tubes following fossilized root tubes (cf. plate X/1-5), with in mature stages of caliche formation aggregation of isolated concretions to continuous crusts taking place (MADER 1985). Origin of calcrete palaeosols may transform even high-quality reservoir horizons into low-quality pay zones or even tight barrier layers.

3.10.1.3.4. Lateral continuity of reservoir units

Lateral continuity of reservoir units is an important factor controlling economical success of the stimulation operations. Hydraulic proppant fracturing of continuous pay zone sheets provides optimum results and helps to decrease well number in the field by allowing to increase the drainage area of the individual offtake points. Fracturing stimulation of lenticular formations (cf. plates II/7-8 and V/3-4) where the different reservoir units and subunits have limited horizontal extension and several laterally adjoining pay zone elements must not be connected, but in many cases are separated by mudstone intervals, however, requires proper feasibility modelling and calculation prior to job execution, because lateral discontinuity of gas-bearing sedimentary facies bodies can result in access of only limited reserves and/or production at only restricted rates which soon diminish after initial peak offtake.

Drainage of lenticular pay zones also frequently requires closer well spacing with respect to production from continuous reservoirs which in turn affects the drilling cost budget, and together with the larger number of stimulation candidates careful assessment has to be made whether the efforts pay off or not. A possible solution is connection of isolated sand lenses with neighbouring separate sand bodies and with the wellbore through a large-scale crack by communication fracturing (cf. section 4.10.2.1.).

3.10.2. Gas composition

In addition to the primary petrophysical control of hydraulic proppant fracturing potential in the gas-bearing Rotliegend sandstones in mainly vertical direction, a secondary economical influence in predominantly horizontal direction is geochemically exerted by the gas composition which includes larger amounts of inert nitrogen apart from combustible methane in various parts of the gas field provinces (PHILIPP & REINICKE 1982) that cannot be used, but has to be separated and removed in purification plants. The overview as follows concentrates on nitrogen content and impact of economical systems on gas-field operation.

3.10.2.1. Nitrogen content

Laterally, the nitrogen content increases from Söhlingen and Munster in the west (abt. 20 % or less) to Wustrow/Salzwedel in the east (60 - 90 %) and finally in the Berlin area almost 100 % nitrogen is encountered (cf. fig. 4). Farther to the east, the nitrogen content decreases again to abt. 60 - 80 % near the western boundary of Poland and then further along the Rotliegend facies belt towards the eastern basin margin at the Pomeranian Swell to abt. 20 %. Other areas of high nitrogen content (up to more than 60 %) include the German North Sea (OFFSHORE ENGINEER 1986 b; cf. section 3.3.4.), whereas only very low nitrogen shares occur in Dutch and British North Sea Rotliegend fields (SCHRÖDER & SCHÖNEICH 1986) as well as in Netherlands onshore (abt. 15 % in Groningen; ROSSUM 1975).

To minor amounts (partially abt. 2 - 3 % within 100 m), the nitrogen content is also vertically increasing with depth in the western part of the Rotliegend facies belt from central Northern Germany FRG to Salzwedel/Wustrow at the GDR-FRG-boundary, whereas in Poland around Ostrów Wielkopolski and Szczecin, increasing reservoir depth sometimes provides lower nitrogen contents (cf. fig. 5). Therefore the whole lateral distribution of nitrogen contents in the Rotliegend gas is probably not a simple bisymmetrical arrangement, but it rather seems that Ostrów Wielkopolski and surrounding fields in Poland already belong to another major gas province, with both horizontal and vertical distribution of combustible methane in relation to inert nitrogen being influenced by the stratigraphical and sedimentological composition of the geological record and its palaeotectonical and palaeogeochemical evolution as well as its diagenetical and gas maturation history.

Traces of helium and quicksilver follow the nitrogen trends in normal and inverse orientation, respectively. The nitrogen derives from metamorphosis of deeply buried sediments far below Rotliegend and Carboniferous and has risen along major faults (PHILIPP & REINICKE 1982; aspects of formation of Rotliegend gas deposits are also discussed by HETZER & KATZUNG 1985).

3.10.2.2. Impact of economical systems on gas-field operation

Concerning hydraulic proppant fracturing potential, operations in the western part of the East Hannover area (especially Söhlingen) where the nitrogen content is comparatively low are certainly very feasible, whereas around Wustrow in the eastern part on Germany FRG territory, it would be hardly possible to economically justify an expensive MHF treatment within a reservoir containing predominantly to almost exclusively nitrogen and only subordinate shares of hydrocarbons (and in addition to that having in the meantime arrived at an advanced stage of gas exploitation and reservoir pressure depletion after more than 20 years of production).

The Wustrow/Salzwedel complex (cf. fig. 4) which is divided by the FRG/GDR-boundary into two halves, however, is an excellent example of the effectivity of a tertiary political zonation of hydraulic proppant fracturing potential. On the Germany GDR side in the Salzwedel substructure, the situation is quite different from that in Wustrow on the Germany FRG side. In contrast to the large belt of Rotliegend gas fields in Germany FRG especially in the East Hannover area, with Söhlingen representing the centre of the richest province and the Munster-Schmarbeck complex being another pronounced spot, Salzwedel is the only big Rotliegend gas field on GDR territory and the small satellites surrounding the large patch can deliver only relatively minor additional quantities.

Therefore any effort to increase domestic gas supply in Germany GDR has to take under consideration large-scale fracturing of the Rotliegend reservoir rocks in the Salzwedel field regardless of gas composition, economical payout and partially also stage of gas withdrawal, as in case of own service for the stimulation operations, financing can be made in domestic currency which is non-convertible, whereas purchase of gas abroad has to be made on the basis of barter trade comprising export of goods from Gérmany GDR that are then lacking there, or even by investment of rare hard currency.

3.11. Water sensitivity

Some parts of the Rotliegend pay series are extremely water-sensitive as determined by flow tests on varied fluid damage of both matrix and fracture conductivity (JOHN 1983, DIETZEL & HANTELMANN 1985). Shut-in and clean-up under reservoir conditions results in regained initial permeability and fracture conductivity, with much superior recovery being achieved by foamed in comparison to unfoamed fluid versions. Foamed oil-based fracturing fluids are suitable carriers for treatment of such reservoirs. Application of water-based fracturing fluids leads to complete loss of crack conductivity and severe proppant embedment. The following outline focusses on general aspects, features of foam fracturing stimulation, characteristics of crosslinked acid systems, methanol systems, and liquid carbon dioxide proppant fracturing. The different treatment techniques for water-sensitive reservoirs are discussed in more detail within the chapter on Rotliegend and Carboniferous stimulation in Europe because of their significance in successfully approaching these pay formations which often suffer from considerable water sensitivity and therefore are frequently not accessible to conventional fracturing operations.

3.11.1. General aspects

The general introduction starts with commenting on water sensitivity distribution in the Rotliegend and then highlights the significance of expandable clay minerals as one of the more common reasons for water sensitivity of various reservoir rocks.

3.11.1.1. Water sensitivity distribution in the Rotliegend

The water sensitivity of Rotliegend and partially also Carboniferous sandstones is mainly due to incorporation of water envelopes around authigenic illite needles and leaves in the pore space into the current movement in case of flushing with a water-based fluid, resulting in destruction and breakage of illite crystals which leads to disintegration of matrix compound, framework collapse and plugging of pore throats by illite fragments (DIETZEL & HANTELMANN 1985). Using an oil-based fluid for flushing, the connate water saturation is decreased to the irreducible minimum, but due to the discrete interface between non-wetting oil-based fluid and wetting water pellicles of the illite crystals, the clay mineral needles and leaves do not suffer axial forces and thus the rock fabric remains in its original stable situation (cf. also section 3.5.).

In contrast to water-sensitive Rotliegend sections, there are also patches where the stratigraphical succession is not water sensitive (KLOSE & KRÖMER 1983), indicating that differences in clay-mineralogy and diagenesis probably related to differences in burial history exist in lateral direction between various parts of the Rotliegend basin, with this primary mineralogical zonation in an irregular and locally-influenced manner giving rise to a secondary division into water-sensitive and non-water-sensitive parts of the formation that induce a tertiary distinction of treatment modes with main emphasis on the medium on which fracturing fluids are based.

Many low-permeability sandstone formations are highly sensitive to damage from increased water saturation (SLUSSER & RIECKMANN 1976). Such zones are characterized by high irreducible water saturation caused by the small pore size distribution in the rock, and by high retention and slow release of invaded water above the minimum irreducible value because of the effect of high capillary pressure. This may result in excessively long cleanup times after well completion or fracture stimulation especially if large volumes of aqueous fluids are lost into the zone.

3.11.1.2. Significance of expandable clay minerals

Water sensitivity of reservoir sandstones is characterized by a sharp loss of formation permeability upon contact with water that is substantially less saline than connate water (ECONOMIDES 1986). The most frequent reason of water sensitivity is the presence of expandable clay minerals such as montmorillonite which swell when touching water, thereby pinching off pores and blocking flow channels (SMITH, PAVLICH & SLOVINSKY 1964). Water sensitivity can also occur in sandstones containing only non-expandable clays such as illite or kaolinite and may be caused by changes in pH and salinity (MUNGAN 1965). Permeability reduction is mainly provoked by blocking of the pore passages by dispersed clay minerals, cement particles and other fines. Dispersed clay minerals are present as discrete, pore-lining and pore-bridging clays, with the different clay morphologies significantly affecting sandstone porosity and permeability (NEASHAM 1977, PITTMAN & THOMAS 1978).

Many sandstones contain large amounts of very small, loose solid particles that vary widely in size and chemical composition, with particle wettability and surface/interfacial forces strongly influencing particle mobility when multi-fluid phases are present (MUECKE 1978). Thus water sensitivity is a very complicated phenomenon not only caused by swelling of expandable clays, but dispersion, migration and redeposition of fine particles of various sizes and chemical compositions creating the damage. As brines are not nearly so damaging to sandstone as fresh water is, 1 - 3 % KCl solutions are normally used as the base liquid in fracturing fluids (ECONOMIDES 1986). Potassium ions stabilize clays against water invasion and consequently prevent swelling, and also help neutralizing the negative charges on the clay particle surface which supports to control dispersion tendencies.

Apart from swelling and migrating clays, formation sensitivity problems include emulsifying tendencies with formation fluids, capillary blocks and skin damage caused by fluid residue (DEYSARKAR, DAWSON & ACHARYA 1987). Water sensitivity effects of clay minerals in petroleum reservoir sands are also discussed by DODD, CONLEY & BARNES (1955), HEWITT (1963) and GAUTIER & RICE (1981).

3.11.2. Foam-fracturing stimulation

Water-sensitive clay-bearing parts of the Rotliegend and Carboniferous reservoir column can also be successfully approached with the technique of carbon dioxide foam fracturing (details on the successful European pilot treatment in the Fehndorf field in the Emsland area/Germany FRG are given by SOMMER 1987, 1988; further aspects on carbon dioxide foam fracturing are discussed in SLATTER, RUCKER & CRISP 1986; questions of dynamic fluid loss in foam fracturing are treated by KING 1977 and HARRIS 1982; and aspects of occurrence and acquisition of carbon dioxide are compiled in TRIELOFF 1983). Foam is a low-viscosity, perfect-support medium which provides more rapid and complete treatment fluid recovery, helps to clean the well without swabbing, and reduces formation damage by minimizing the quantity of aqueous fluid which enters the formation (PHILLIPS, COUCHMAN & WILKE 1987). Foam is a particularly excellent medium in reservoirs with low bottomhole pressures such as some shallow gas sands (BALLARD 1977).

Foam has been used since more than 20 years in oil and gas production in all phases of recovery including drilling, acidizing and fracturing stimulation, secondary and tertiary recovery, diverting during workover, and gravel packing (HARMS, SMITH, KING & POSEY 1988). The main favourable mechanical properties of foam are low hydrostatic head, ability to support fines, low leakoff rates, and shear-thinning rheological behaviour, which in combination with low liquid content and energized fluid return facilitate the use of foam in water-sensitive formations in low-pressure gas zone completion and low-permeability reservoir treatment. Aspects of proppant transport capacity and foam stability, carbon dioxide vs. nitrogen as foaming or energizing agent, flow behaviour and foam viscosity, and fluid recovery and well cleanup are discussed as follows.

3.11.2.1. Proppant transport capacity and foam stability

Foam fracturing is carried out in hydrocarbon reservoir stimulation since more than 15 years (HOLCOMB, CALLAWAY & CURRY 1980, 1981; HOLCOMB 1983). Foams used for carrying solid particles are actually three-phase systems, with the proppants grains being suspended in the gas-liquid boundary by surface tension. Particles attached to foam bubbles prevent bubble coalescence and therefore enhance foam stability. Foam bubbles are stabilized at reservoir temperature and pressure conditions by surface tension due to surfactant effect, viscosity due to gel stabilizer effect, and bubble drainage control. Foam breakage during well cleanup is achieved by foam expansion and bubble coalescence as pressure declines, turning the foam into a mist of liquid droplets in gas which has low viscosity and can no longer provide proppant support for the third phase being the proppants. This expansion/coalescence trend dominoes from the fracture region through the perforations into the wellbore.

As the mist begins to climb up the borehole tubulars, turbulence and compression due to back pressure from the choke assembly can trigger foam regeneration which, if taking place after foam breaking and leaving proppant behind in the fracture can be an advantage in maintaining simultaneous recovery of both gas and liquid. Using too much foaming agents, stabilizer or liquid gelling agent can cause overstability of the foam and might not allow foam expansion and bubble coalescence to take place, thereby inhibiting foam breakage and thus causing foam blockage and/or proppant flowback. On the other hand, not enough foaming agent can give rise to premature foam instability that can lead to job screenout or high foam inefficiency especially in terms of proppant carriage. The discussion as follows focusses on impact of proppant addition on foam stability and fluid viscosity, rheological characteristics of foam systems, and wall-building of gel-bearing foams.

3.11.2.1.1. Impact of proppant addition on foam stability and fluid viscosity

The primary limitation of the foam fracturing process is the amount of proppants that can be physically placed into the foam (JENNINGS & DARDEN 1979). If 2 lbs/gal sand concentration is desired in an 80 % quality foam, 10 lbs/gal of sand have to be added to the 20 % aqueous phase prior to dilution with nitrogen gas. Foam is generally a perfect proppant transport medium with slow settling rates, with the proppant essentially not settling as the foam moves down the fracture during the time of treatment (BLAUER & KOHLHAAS 1974). While conventional foam stimulation operations have been frequently inconsistent in placing high proppant concentrations and often reached a pressure limitation which prematurely terminated the job, the constant internal phase technique has overcome previous proppant saturation boundaries and has allowed operations to be pumped with reduced pressure fluctuations (HARRIS, KLEBENOW & KUNDERT 1988). Conventional foam treatments on either carbon dioxide or nitrogen base when pumped at high rates down small-diameter tubing frequently suffer from considerable increase of operational pressure as proppant is added until the maximum allowable pressure is reached as a consequence of high friction pressures.

While proppant addition causes no major change in viscous character of non-foamed fracturing fluids, the two-phase structured foam fluid consisting of gaseous internal phase and liquid external phase is considerably disturbed and proppant addition to high-quality foams gives rise to a significant viscosity increase. In order to maintain a constant-viscosity fracturing fluid, the balance between internal phase vs. external phase must be kept constant. The concept of constant internal phase design keeps both downstream flow rate and viscosity constant by addition of solid proppant while maintaining a constant liquid rate but decreasing gas flow rate sufficiently to equal absolute solid flow rate. In conventional foam treatments, proppant is added to the liquid phase by blending, and the slurry passes then through high-pressure pumps to the wellhead where nitrogen or carbon dioxide are added thus diluting proppant concentration by severalfold. By using constant internal phase design, however, nitrogen or carbon dioxide shares are decreased as proppant saturation is increased, and therefore the dilution effect of the gas is less and higher downhole proppant concentrations may be achieved. Pumping pressure usually rises to a maximum by the end of the pad stage, but then steadily decreases as added proppant increases the hydrostatic weight of the fluid column.

3.11.2.1.2. Rheological characteristics of foam systems

Desirable properties of foam fracturing fluids are high proppant transport capability, low hydrostatic head, low pressure drops due to friction, and no reduction of fracture conductivity due to fluid ingredients (BLAUER & KOHLHAAS 1974, CAWIEZEL & NILES 1987). Foams have extremely good rheological characteristics and fluid loss performance under a fairly wide range of conditions (VEATCH & MOSCHOVIDIS 1986). Carbon-dioxide-foam fracturing fluids also provide a long sustained solution-gas drive due to high solubility of carbon dioxide with reservoir fluids (DEYSARKAR, DAWSON & ACHARYA 1987). Carbon dioxide can be commingled with conventional fracturing fluids as an energizing component of subordinate quantitative share or used as the gaseous phase in formulating foams (GARBIS & TAYLOR 1985).

Energized or foam systems should be considered when the pay zone is underpressured and fracturing fluid recovery is questionable, in case the formation is sensitive to contact by foreign fluids, and if a low-viscosity perfect-support fluid is desired to minimize crack height growth. NORTON & HOFFMAN (1982) report that conventional water- or oil-based fracturing treatments may outperform foam stimulation operations only because of the placement of larger amounts of proppants by the former fluids in the fractured formation. BLAUER, PHILLIPS & CRAIG (1988) comment on rheological and physical differences between foam and emulsion fracturing fluids.

3.11.2.1.3. Wall-building of gel-bearing foams

The types of foams which are used in many hydraulic fracturing treatments contain gel in the liquid phase and are wall-building fluids (HARRIS 1982) which are constructing gel layers on the fracture faces that control fluid leakoff. Foams containing no gel in the liquid are not-wall-building fluids. Foams containing polymers building fracture wall drapes include the potential of damage of original matrix permeability. Since excessive fluid leakoff is one potential cause of premature job termination and also foam fracturing treatments occasionally screen out, adequate knowledge of fluid loss coefficients is essential for proper supervision of foam fracturing jobs, too.

3.11.2.2. Carbon dioxide vs. nitrogen as foaming or energizing agent

High-quality foams were developed into a viable fracturing stimulation tool for oil and gas wells in the 1970`s (HARRIS, KLEBENOW & KUNDERT 1988). Foam fluids typically containing abt. 65 - 80 % nitrogen gas by volume are characterized by low damage to sensitive formations and rapid recovery of treatment fluid. Proppant concentrations in nitrogen foams could be risen from 2 - 3 lbs/gal at the beginning to 5 - 7 lbs/gal in the early 1980`s due to blender and pump design improvement. High-quality carbon dioxide foam fluids were introduced in 1982. Carbon dioxide had been used in shares of 5 - 50 % since the 1950`s to provide gas-assisted fluid recovery. The use of typically 65 - 80 % carbon dioxide caused a significant boost in viscosity for carrying proppant into the formation. The following discussion focusses on fluid viscosity and compatibility, fluid solubility and proppant transport, foam density and recoverability, and rheological nature and transitions of foams and emulsions.

3.11.2.2.1. Fluid viscosity and compatibility

In addition to the other advantages of nitrogen foam fluids, carbon dioxide foam provides better compatibility with formation fluids and lowered interfacial tension, and its higher density gives greater foam stability and allows generally lower pumping pressures during job execution (HARRIS, KLEBENOW & KUNDERT 1988). The greater solubility of carbon dioxide in formation fluids provides gas assist to reservoirs having lower pressures thus enhancing fluid recovery. As for example 70 % carbon dioxide foam treatments can be pumped down casing at significantly lower operational pressures than those needed for similar nitrogen foam jobs, the application of carbon dioxide foams allows treatments to be pumped into deeper and hotter reservoirs than is practical with nitrogen foams.

Simple foams containing water, nitrogen and surfactant are very unstable, because the liquid around the bubbles is thin and drains away easily (ECONOMIDES 1986). Increasing foam stability can be performed with polymer addition (WENDORFF & AINLEY 1981), crosslinking of the polymer in the aqueous phase (WATKINS, WENDORFF & AINLEY 1983) which also improves foam rheology and fluid loss control, and compositional altering of the foam by exchanging nitrogen against carbon dioxide in order to create a denser foam and consequently lower surface treating pressures. Methanol may be used in high concentrations in the aqueous phase to make the foam much less damaging and even easier to cleanup from water-sensitive formations (ECONOMIDES 1986).

3.11.2.2.2. Fluid solubility and proppant transport

Concerning application both as major (foam) or minor (energizer) component, carbon dioxide offers several advantages over nitrogen, because it is more soluble in treating fluids and thus longer shut-in times may be tolerated without excessive loss of gas energy, and the higher density provides increased hydrostatic pressure which may result in lower surface treating pressures, providing the advantage is not offset by higher tubular friction pressures (HARRIS, KLEBENOW & KUNDERT 1988). The higher density of carbon dioxide also has a positive impact on proppant transport capacity. In shallow low-temperature wells, longer shut-in times are required with carbon dioxide than with nitrogen in order to allow it to heat up sufficiently to vaporize. Many deep wells have been treated with carbon dioxide foams simply because the pressure limitations of the tubulars would not allow to carry out nitrogen foam treatments. In contrast to its technical advantages, however, an economical and logistical drawback of carbon dioxide is the fact that nitrogen is usually supplied and pumped by the service companies performing the stimulation treatments, whereas carbon dioxide is normally delivered by an outside contractor (GARBIS & TAYLOR 1985).

3.11.2.2.3. Foam density and recoverability

Foaming additives give better recoverability with carbon dioxide than with nitrogen (NIELL, DOBBS, PRUITT & CRAWFORD 1964). Comparative evaluation of the effects of carbon dioxide and nitrogen as energizing (commingled systems) and foaming agents are also undertaken by HARRIS (1982, 1985, 1987); HARRIS, REIDENBACH & CHISHOLM (1986); REIDENBACH, HARRIS, LEE & LORD (1986) and YOST, OVERBEY, WILKINS & LOCKE (1988). DOSCHER, OYEKAN & EL-ARABI (1983) and WARD (1984) compare the performance of nitrogen vs. carbon dioxide for oil displacement in highly-inclined reservoirs. Another application is energized fracturing with 50 % carbon dioxide for improved hydrocarbon recovery (BLACK & LANGSFORD 1981). Although nitrogen is more expensive than carbon dioxide, it is often preferred as gasifying agent of fracturing fluids in order to enhance fluid return during cleanup as a consequence of its better controlling characteristics of the pH of the fracturing fluid (AHMED, SCHATZ, HOLLAND, JONES & GREENFIELD 1982).

3.11.2.2.4. Rheological nature and transition of foams and emulsions

Foam and emulsion fracturing fluids are used when it is desirable to restrict the volume of either water or solid gelling materials pumped into a formation because of their high apparent viscosity, low fluid loss and good proppant transport capacity that can be achieved with the mixture of two immiscible fluids and a small quantity of surfactant (BLAUER, PHILLIPS & CRAIG 1988). The mechanics responsible for these rheological properties are essentially the same for both foams and emulsions. Foam consists of discontinuous gas bubbles suspended in a liquid, whereas an emulsion is composed of separate droplets of liquid suspended in a second immiscible liquid. The desired fluid properties are achieved when the volume of the discontinuous phase is greater than abt. 52 % when the individual droplets or bubbles begin to interfere with each other during flow which gives rise to increasing viscosity.

Distinction between foam and emulsion is very simple if the average fluid pressure is less than the critical pressure of both liquids within the fluid, with foam or emulsion being present if the temperature is greater or less than

4.1. Introduction

As a consequence of the points discussed in the preceding economically-oriented chapters, the relatively retarded situation in the hydraulic proppant fracturing and gravel packing scenery in Europe and other international markets (with respect to the USA where activity is generally on a higher level) after the oil price drop in 1986 (cf. section 2.2.1.1.2.; partially lasting until 1987 and still not fully recovered in 1988 in some areas, also as a consequence of its accentuation by the accompanying and following US $ exchange rate collapse up to its historical minimum level; cf. section 2.2.1.3.) should be taken by both service and proppant companies as a challenging opportunity to understand the lesson and not to repeat the mistakes of the past, but to lay now patiently the necessary groundwork in hitherto insufficiently attended markets for near-future development or even booming by helping to introduce and to promote various technological possibilities (incorporated into sound marketing concepts) of enhancing the general stimulation potential and thus at the bottom of the line oil and gas supply. After briefly sketching and characterizing possibilities of expanding the hydraulic proppant fracturing market in the near future, an outline of the technical review is given.

4.1.1. Possibilities of expanding the hydraulic proppant fracturing market in the near future

According to my geological, reservoir engineering, technical and marketing experience, a suite of more or less unconventional methods of performance or improval of hydraulic proppant fracturing and gravel packing exist in economically and technically more or less mature status prepared for execution which partially are already more or less familiar in the USA, but have still not reached common sense or widespread application in Europe and probably also other international markets. Other methods are still in the status of technological preparation, enhancement and perfectivation and need introduction to the market by pilot operations supported by triggering attractive innovative offers. Both groups of techniques could include very soon some quite suitable concepts of stimulation of hitherto marginal reservoirs when increasing demand and declining supply in the near future is also in Europe and other areas outside of the USA more and more accentuating the force of enhancement of domestic production within the individual countries (this applies especially for Eastern Europe in light of the present economical, political and financial framework; cf. section 2.4.4.).

Many of the concepts and technologies discussed in this chapter, however, still require some homework before their widespread field utilization in various reservoir types, and success in the particular operations needs improved understanding, skilled selection and careful implementation. The main strategies of market expansion aim on progressive replacement of natural sand by synthetic proppants in applications near the boundary of physical and chemical stability of sand and for the reason of achieving higher conductivity, and on hydraulic stimulation of troublesome reservoirs for which a suitable and successful concept has not yet been found in the past, but can be offered when utilizing all the benefits which are provided by the various types of man-made high-quality proppants.

4.1.2. Outline of the technical review and status report

A few particularly important technological aspects of near-future enhancement of oil and gas production by approaching marginal reservoirs by hydraulic proppant fracturing or by carrying out unconventional stimulation jobs in Europe and other international markets (as far as recognized from my view) comprise containment of fracture propagation in thin sandstones, proppant saturation and conductivity improvement, targets of unconventional gas stimulation, hydraulic proppant fracturing of carbonate rocks, high-permeability reservoir fracturing, geothermal fracturing, minifracturing and other small-scale fracturing, oil-reservoir fracturing, and communication fracturing. These methods and techniques are proposed for consideration by the areal specialists and are briefly presented in summarizing outline and commented as follows including discussion of the most significant literature (numerous references on general and specific technical and reservoir engineering aspects are compiled in both main list and subject key classification in chapter 7; aspects of production geology and associated reservoir engineering are summarized by DICKEY 1981, DIKKERS 1985 and NORTH 1985).

Factors influencing or triggering fracture damage are also reported, and as a general conclusion an aggressive fracturing design is recommended in order not to get stuck with conservative and conventional techniques, but to maintain progress by carrying out unconventional and challenging innovative methods despite of including some risk of drawbacks or even intermittent failures in marginal applications or pilot operations. The account does not aim on any completeness, but focusses on subjects which due to my own assessment according to my experience in the market need preferential attention during the coming years for improving especially the European stimulation potential and to enhance the fracturing activity in this scene.

4.2. Containment of fracture propagation in thin sandstones

While fracture operations within thick sandstone reservoir columns sufficiently far away from gas-water-contact and bounding mudstones are particularly taking place in greater depth where danger of breakthrough to the ground-water storey is no topic for environmental discussion and the threat of connection to water-bearing sandstones within or below the pay column is also generally no item of worrying, treatment of thin sandstone reservoirs within sandwich-type alternating sandstone-mudstone successions with partially different fluid content of the various sandstone horizons creates problems of containment of fracture propagation in vertical direction in order not to achieve unwanted deteriorations of the effect of the fracture jobs (DANESHY 1978; SIMONSON, ABOU-SAYED & CLIFTON 1978; ROSEPILER 1979; WARPINSKI, SCHMIDT & NORTHROP 1980; PRATS 1981, TEUFEL & CLARK 1981, EEKELEN 1982; WARPINSKI, CLARK, SCHMIDT & HUDDLE 1982; SETTARI 1983, WARPINSKI & TEUFEL 1984).

Precise control of fracture height growth and prevention of penetration of the crack through the bounding barriers is essential for avoiding unwanted effects such as water or gas coning and breakthrough after the treatment and thus deterioration of the result of the operation, and for securing optimization of the area drained by the fracture (BEN NACEUR & TOUBOUL 1987). Undesirable zones of fracture extension not only comprise gas caps and water bottoms of oil reservoirs, but generally include all unproductive layers adjoining to the prospective interval, with fracture containment aiming on maximization of crack propagation within the horizon of interest and minimization of excessive upwards and downwards growth. The intrusion of a hydraulic fracture from the pay zone into the horizons lying above and below is a serious concern in crack design, because if the hydraulic fracture is not contained within the producing formation and progagates in both vertical and lateral directions, failure of the stimulation treatment can occur due to substantial loss of fluid and proppants which are used up and wasted for stimulating the unproductive layers adjoining the reservoir (NGUYEN & LARSON 1983, ACHARYA & KIM 1987). Field experience has shown that the most common cause of poor fracture treatment response is excessive fracture height growth (HOLDITCH, ROBINSON & WHITEHEAD 1986), because out-of-control fracturing results in shorter fracture lengths than desired (GREGORCZYK, PAULS, HOLTMYER, CHISHOLM & VENDITTO 1984).

Massive hydraulic proppant fracturing has been historically developed by trial and error and its results are uncertain in many situations (TEUFEL & CLARK 1981; cf. sections 1.1.1. and 4.8.1.2.). While some of these large-scale stimulation efforts have been successful, others have turned out to be extremely disappointing failures mainly for the reason of lack of improved understanding of fracture propagation and containment. The economical significance of fracture extension control is that if the crack is not confined within the producing sandstone and propagates in both vertical and lateral directions in an elliptical shape, then there is an effective wasting of expensive stimulation fluids and proppants which are consumed to fracture and to support barren horizons next to the reservoir complex, respectively. The outline as follows concentrates on areal distribution of potential reservoirs, technical aspects of fracturing operation, fracture propagation, reservoir heterogeneity and anisotropy (including sedimentological, petrophysical and rock mechanical aspects), combination of hydraulic fracturing and gravel packing, and economical significance.

4.2.1. Areal distribution of potential reservoirs

Thin single or multiple, bundle- or sandwich-type sandstone horizons within mudstone sequences (cf. plates IV/4 - 8, VII/1 - 6, VIII and IX) with different thicknesses of both sandstone layers and mudstone intervals vertically separating adjoining sandstone sheets of laterally more or less continuous type and subordinately also overlapping or juxtaposed lenses of various degree of interconnectedness are in Europe particularly present in the upper part of the Rotliegend gas-bearing sandstone succession (Hannover-Formation or Ten Boer-Formation sandstones; Wechselfolge facies association; BRADEL & DRAXLER 1982, PHILIPP & REINICKE 1982, KLOSE & KRÖMER 1983; HEDEMANN, MASCHEK, PAULUS & PLEIN 1984; GRALLA 1988; cf. section 3.2.1.3.) in the British and Dutch Southern North Sea as well as in Netherlands and Germany FRG onshore, Germany GDR and Poland; in the Jurassic oil-bearing sandstone succession in the British and Norwegian Northern North Sea, in some Palaeozoic and Mesozoic oil- and gas-bearing sections in the Baltic Sea offshore Poland and USSR, in Mesozoic and Cenozoic oil- and gas-bearing sandstones onshore and offshore (Black Sea) in Bulgaria, in Cretaceous and partially also Jurassic shallow oil-bearing sandstones and carbonates in the northwestern part of Germany FRG (partially extending across the boundary to Netherlands), in Tertiary to Triassic moderately-deep oil-bearing sandstones in the southern part of Germany FRG, and in Tertiary oil- and gas-bearing sandstones in the Vienna Basin in Austria (cf. KREUTZER 1985, 1986; BRÄUER 1987) and adjoining Czechoslovakia (cf. THON 1985) and to minor amounts also in the Adriatic Basin in Italy and Yugoslavia offshore (cf. DRAGICEVIC, MARINOVIC, STANKOVIC, BATUSIC & NIKOLIC 1983; IBRAHIMPASIC 1983) as well as onshore in Hungary and Yugoslavia (cf. tabs. 5 - 9). Some unstable Cretaceous shallow gas-bearing sandstones of limited thickness are also present in Mozambique in Africa.

4.2.2. Technical aspects of fracturing operation

While in parts of the mentioned formations as well as in other comparable reservoirs in various parts of the world already fracturing jobs with limited vertical propagation (and in many cases consequently also restricted horizontal extension, resulting in consumption of only low to moderate quantities of propping materials) have been successfully carried out, much optimization could still be done by enhancing proppant choice in terms of both type and grain size, amelioration of tail-in operations, accentuation of the permeability contrast between reservoir sandstone and proppant package by shifting to larger proppant grain sizes and varying the relative quantities of different materials in stepwise treatments, and improvement of fracture width by pumping of higher proppant concentrations per fluid unit. Aspects of singular vs. multiple fracturing and diverting techniques, limited-entry fracturing technique, fracture containment in layered reservoirs by treatment parameter adjustment as well as by buoyant and settling diverters and blocking additives, multiple-zone fracture migration concept, selective proppant placement, and special phenomena in coal seams are discussed as follows.

4.2.2.1. Singular vs. multiple fracturing and diverting techniques

Hydraulic proppant fracturing can be performed in either singular or multiple manner depending on reservoir pile composition and completion type including commingling of several separate layers or not. Singular storey-wise fracturing with first stimulation of the lower zone and following its isolation and upwards movement second treating of an upper horizon (cf. sections 2.4.1.1., 3.3. and 4.10.1.) has to be carried out if the specific reservoir conditions require different fracturing fluid types (HOEL 1988) or different proppant types to render the individual jobs technically successful and/or economically profitable. If the same fracturing fluid and the same proppant type can be applied for stimulation of several vertically successive levels of the prospective complex, multiple fracturing representing joint treating of all the zones is the adequate solution provided the in-situ stresses are in a comparable range to allow more or less simultaneous opening of the fractures in all the targeted sections.

CLARK (1987) compares tubing vs. casing as treating medium and evaluates casing preparation for later fracturing stimulation. If the operation is performed through the casing, the well must be killed after fracturing fluid cleanup in order to install the tubing (WEBSTER 1977), with well killing always including considerable risk of crack damage (SLUSSER & RIECKMANN 1976). An alternative to well killing for tubing installation is therefore fracturing down tubing and/or annulus at reduced injection rates and with high friction pressure. The most important questions are diverting agent performance and selection as well as multiple fracturing aspects.

4.2.2.1.1. Diverting agent performance

In terms of operations, the fractures can be carried out separately within every individual sandstone horizon of the sandwich-type interbedding by sealing with bridgeplug and packer (cf. sections 2.4.1.1., 3.3. and 4.10.1), or pumping may in the early stage of the treatment include the whole section of several vertically successive sandstone layers (with the injection points being fixed by the perforated intervals), and in the late stage of the job, balls are dropped or baffles are set to certain levels to control fracture confinement to distinct reservoir segments and to separate the individual production horizons by closing certain perforated intervals (this method has so far particularly been applied for fracturing of water injection wells; CLARK 1968; BALE 1984; cf. section 4.8.7.). In many cases, ball sealer dropping at a rate of one ball per barrel of fluid satisfactorily helps to obtain uniform reservoir breakdown (BLASIO, BAILEY & BERGTHOLD 1985), and normally 50 % excess ball sealers are spaced evenly in the fluid to achieve a ball out and to ensure that all the perforations are opened (PAI & GARBIS 1983 b).

Ball sealers can be used in both proppant and acid fracturing in order to divert the treatment to the next zone between each stage (WEBSTER 1977). ROBINSON, HOLDITCH & LEE (1983) recommend application of ball sealers twice as much as perforation holes are shot. GWINN & McMANUS (1977) report multiple-zone completions using balls and baffles where up to five horizons can be stimulated in one continuous operation. Utilizing the ball and baffle technique, the lowermost zone is perforated and fractured, and is then allowed to flow back immediately. The next zone is perforated as soon as flowback permits entry of the perforating strip into the well. After this zone is perforated, the properly-sized ball is pumped down with breakdown acid and allowed to seat on the baffle ring below the zone to be next stimulated. Each successive zone is treated in a similar manner using larger-diameter fracturing balls and baffles. The immediate flowback after each stage is generally strong enough to unseat and produce back the fracturing balls.

The conventional or modified limited-entry technique (cf. section 4.2.2.2.) using ball sealers offers the best possibilities of diverting stimulation treatments of new well completions (HARRISON 1971). Single-stage multiple-zone stimulation is also discussed by EASON (1985). Selective fracturing with sealing balls as diverting materials is adjusting the requirement of zones with different fracture pressure capacity and increasing the coverage of propped fracture within a long perforation interval (LI & ZHU 1986). The discussion as follows focusses on advantages and drawbacks of ball sealers, injection rate vs. treatment staging, floating and sinking ball sealers, treatment sequence in multilayer reservoirs, and insertion and removal of blocking materials. Some comments are also offered on wax-based pellets.

4.2.2.1.1.1. Advantages and drawbacks of ball sealers

Multistage fracturing with dropping ball diverters and sealers between individual stages is especially suitable for long completion intervals with considerable differences between gross and net thickness resulting in various thin target horizons interspersed within a thick reservoir storey complex (PEARCE 1979). In many cases, however, diverting with ball sealers is not effective due to washing-out of perforations with proppants such that balls do not sufficiently seal off pressure, and also as a consequence of frequent covering of the pay zones with proppants already before ball diverters can be set.

Staged treatments using ball sealers as well as temporary sand, gel or mechanical plugs are also impractical if only thin mudstone intervals separate the successive sandstone storeys and in case of high pressure differences between the individual target zones (BUNDY 1981). On the other hand, sealing balls are particularly effective for complete plugging off of thin reservoir layers separated by preferentially thicker barriers. At least 200 ft of rathole containing clean fluid is required for performing a successful multistage diversion. In addition, unless a reservoir is extremely homogeneous, the probability of fracturing several pay zones together which are separated by mudstone breaks is remote due to stress differences between the various layers (PEARCE 1979; cf. sections 6.2.5. and 6.3.6.).

4.2.2.1.1.2. Injection rate vs. treatment staging

Injection rate affects fracture height and fluid distribution and is very critical to the overall success of the treatment (PAI, GARBIS & HALL 1983). Where interval yield of tubing and/or casing limits injection rate, the operation should be staged using ball sealers. At suitable pumping rates, natural pressure diversion of the treating fluid can also be achieved (NALL, CAMPBELL & BONEY 1983). Multiple-stage jobs can also be performed with a modified limited-entry baffle and ball arrangement, with dropping of ball sealers closing the baffles above the lower formation and isolating the upper formation for stimulation, and subsequent opening of the well permitting the ball sealers to flow out, thereby guaranteing fluid cleanup from both treated zones simultaneously (BLACK, RIPLEY, BEECROFT & PAMPLIN 1979). ERBSTOESSER (1980) comments on improved ball sealer diversion. While settling heavy ball sealers require higher injection rates for satisfactory performance, buoyant ball sealers are operationally difficult in highly-inclined boreholes, because they tend to become stuck downhole and flow back at later times in the life of the well thereby causing problems with the surface equipment (SNOW & HOUGH 1988).

4.2.2.1.1.3. Floating and sinking ball sealers

Similarly as buoyant and settling diverting materials for fracture top and bottom plugging and growth limitation (cf. section 4.2.2.4.), floating and sinking ball sealers can be applied for fracturing treatment diversion (McLEOD 1984). Sinking ball sealers usually require 200 % excess and a high pumping rate. Floating or neutral-density ball sealers have a specific gravity matching that of the fluid, and therefore better ball action takes place as compared to sinking sealers. On the other hand, surface equipment has to be modified to catch the floating ball sealers during flowback (GABRIEL & ERBSTOESSER 1984). In terms of acidizing, ball sealers are restricted in use and should not be applied in long intervals with high perforation density, low-rate treatments and gravel-packed wells (McLEOD 1984). An alternative to ball sealers are retrievable coiled-tubing inflatable packers and bridge plugs which can be used for selective stimulation treatments by isolating a zone from a fracturing job utilizing either a packer to stimulate a lower zone or a plug to treat an upper zone (WALSH & HOLDER 1988).

4.2.2.1.1.4. Treatment sequence in multilayer reservoirs

While in most cases sequential storey-wise fracturing with assistance of ball sealers has to be carried out from bottom to top of the borehole, occasionally an inverted procedure with stepwise stimulation from top to bottom can be performed if the different horizons to be treated are characterized by different reservoir depletion stage (SNOW & HOUGH 1988). The less depleted zones are preferentially stimulated first. If the less depleted layer is lower in the prospective section, it is treated and a plug is set across those perforations while the upper segment is shot and stimulated. If the less depleted zone is higher in the sequence, it is perforated and treated first, and once the lower interval is opened and stimulated, ball sealers are pumped to divert away from the previously treated horizon.

4.2.2.1.1.5. Insertion and removal of blocking materials

Other techniques for simultaneous fracturing of several layers in boreholes with long hydrocarbon-bearing sections include diversion of the fluid to the different zones where treating packers are used or solid chemicals as diverting agents are added to the fluids. Simultaneous injection into several selected pay zone layers can also be performed by pin-point sand fracturing (MURPHY & JUCH 1960; limited-entry fracturing; cf. section 4.2.2.2.). Extremely viscous gels can be applied alone and can be tailored to be almost solid at bottom hole conditions and yet will revert to a low-viscosity fluid after a controlled period of time (cf. section 4.3.4.). In case of operation through perforated casing, high-strength rubber-covered plastic balls are used to seal the perforations taking fluid and divert the fluid to other perforations (WATERS 1980).

In terms of diverting agents, one major difficulty encountered with any blocking material is removing it from the permeable zone after treatment (HARRISON 1971). The perfect blocking material is one that lasts long enough to divert fluid during an operation and then becomes ineffective without having to be dissolved by contact with other agents or by internal breakage (the same applies for fluid-loss additives; cf. section 4.3.4.6.4.). The optimum is a diverting agent that stays intact only for the duration of the treatment, with a solid material degrading immediately being most desirable, whereas diverting agents depending on viscosity alone do not appear to offer positive blocking because of the variation and inpredictability of gels and emulsions. HUCKABEE (1988) reports utilization of rock salt as diverting agent for acid stimulations in carbonate reservoirs, with the water-soluble rock salt particles being one of the most suitable diverting materials due to their subsequent unhindered vanishment once the formation starts to flow or perforation washing is carried out (rock salt has also excellent characteristics as fluid-loss additive for temporary hairline crack plugging followed by dissolution and thus elimination of any blocking effects; cf. sections 1.4.11.2. and 4.8.8.3.).

4.2.2.1.1.6. Wax-based pellets

An efficient diverter used several decades ago are wax-based pellets which improve multiple fracture patterns in single operation and permit better control of acidizing (OIL GAS JOURNAL 1965) so that it works over the entire face of the exposed wellbore. The wax-based beads function as temporary sealing agents which plug any opening in the wellbore through which fluids pass. Wax-based pellets block hydraulic fractures as soon as they are created and propped, thereby permitting to stimulate within the reservoir complex several other times during the same operation. In acidizing, wax-based beads plug any generated holes, thereby preventing the acid from being drained off into the formation at a single point and forcing it to attack other areas of the borehole as well.

4.2.2.1.2. Diverting agent selection

Selection of diverting technique and treatment method for simultaneous multiple fracturing of several sand stringers within one interval depends on the sealing nature of separating mudstones (PARKER, ADAMS & LIANKUI 1986). If the enclosing mudstones act as barriers to vertical crack growth, limited-entry perforating and fracturing techniques are suitable for joint stimulation of a bundle of sand stringers. If the bounding mudstone, however, does not form a barrier to upward crack extension, fracture height is governed by pumping rate and job volume (cf. section 4.2.2.3.), and cluster perforations in a central area combined with high-rate large-quantity treatments may be best for synchronous stimulation of several sand stringers. Cluster perforation fracturing may require additional perforations after the stimulation operation in order to commingle production from the various horizons (cf. sections 4.10.1.1. and 4.10.3.).

While traditionally perforation breakdown prior to stimulation treatments was performed by ball and cup sealers, temporary blocking agents, other diverting means (HARRISON 1971 gives a historical review of diverting agents), and high fluid injection rates, SCOTT (1986) presents precision perforation breakdown by using a hook-wall straddle packer that is operated with minimal pressure and pumping rate to help prevent crack extension or prolongation into unwanted gas- or water-bearing zones or horizons. Multiple fracturing within one well can include creation of a single fracture each in every pay horizon or can comprise formation of several cracks within one reservoir level. In the latter case, the multiple system consists of primary and auxiliary fractures with usually four- to eight-wing radial fractures (CHOO & WU 1987; cf. also sections 4.2.2.1.3. and 4.7.3.). In chalk fracturing, crosslinked polymers are occasionally utilized as diverting agents (HILL, CLEMENT & WARD 1977; cf. section 4.5.4.).

4.2.2.1.3. Multiple fracturing aspects

A very important question of multiple fracturing operations is formation breakdown and ascertainment that all the perforations will accept fluid, which is best ensured by breaking down each zone separately (MARPLE, RULEY, WORLEY & FREEMAN 1987). Fracturing of each zone independently enhances the quality of the stimulation treatment within each layer and decreases the possibility of screenout failure. Comments are offered on fracture morphology, fracture number and stress state, tailored pulse loading, controlled vs. conventional explosive fracturing, simultaneous propagation of multiple fractures, and other techniques and effects.

4.2.2.1.3.1. Fracture morphology

Mineback experiments by physical excavation of hydraulic fractures and observation of their relationships to geological discontinuities reveal that the widely held assumption of ideal smooth planar singular parallel fractures of constant height is probably untenable in many reservoirs (WARPINSKI & TEUFEL 1984). Multiple stranding by offsetting (cf. plate IV/6), crack meandering (cf. plate XI/3), en-echelon fracturing, large-scale surface roughness and waviness, and overlapping stranding are common morphological features of fractures. Some comments are offered as follows on fracture stranding and offsetting as well as fracture roughness, waviness and width variation.

4.2.2.1.3.1.1. Fracture stranding and offsetting

Multiple stranding is predominant in naturally fractured reservoirs, with hydraulic fractures being significantly offset when they cross the joints (WARPINSKI, SCHMIDT & NORTHROP 1980; WARPINSKI 1983 a; cf. section 4.2.3.2.4.). Three types of interactions between hydraulic fractures and preexisting joints occur. The hydraulic fractures either cross the joint, are arrested by opening and dilating the crack (cf. plates VI/1 - 7 and XI/3 + 7) as indicated by fluid flow along the joint, or are stopped by shear slippage of the crack with no dilation and fluid flow along the joint (cf. also section 4.2.3.4.).

Roughness of fracture surface depends on rock type and diminishes with increasing confining stress (BIOT, MEDLIN & MASSE 1984). The effect of variability of fracture shape, path and surface features is an increase in pressure drop along the length of the crack compared with that of the ideal case (WARPINSKI 1983 a), with the main elements provoking larger pressure drop being tortuosity of fracture path, multiple crack strands, sharp turns or corners of the flow path originating at intersections of natural joints and hydraulic fractures, and rock property variations. Friction factors associated with flow in hydraulic fractures are frequently 2 - 3 times greater than predicted by viscous theory as a result of surface roughness, crack offsetting and multiple stranding, with the latter effect occurring regardless of rock type and stress. Extensive hydraulic fracture contorting takes place at natural joints and inclusions (WARPINSKI 1983 a; cf. plate IV/1 - 2).

4.2.2.1.3.1.2. Fracture roughness, waviness and width variation

Other features are fracture waviness (cf. plate XI/3) and width variability. Offsets of the hydraulic crack happen at every point where a natural joint is intersected. Fracture width variation or large-scale fracture roughness have pronounced effects on crack shape, path and surface features, with the main feature being increasing pressure drop along the length of the fracture as compared to the ideal case. Sharp turns such as at fracture offsets (cf. plate XI/3) can have significant effects on pressure losses in high-Reynolds number treatments characterized by high flow rate, low viscosity or small height.

SMITH, ROSENBERG & BOWEN (1982) document small-scale deviations of ideally straight vertical fracture wings on both sides of the wellbore as revealed by optical television (cf. section 6.2.1.9.3.). Such deviations include slight jogs, bifurcations and spalling of the fracture edges, and are presumably related to the dynamic nature of the breakdown phenomenon which is probably including fracture initiation at multiple points in the wellbore.

4.2.2.1.3.2. Fracture number and stress state

In a reservoir complex consisting of various sandstone storeys separated by mudstone barriers, single fractures are created in case of presence of a single gradient shift, whereas multiple fractures with individual cracks contained in each layer are generated in case of several distinct gradient shifts being developed (BUNDY 1981). Higher treatment pressures than usual can be caused by viscous flow through a complex multiple fracture system instead of a singular planar fracture (MEDLIN & FITCH 1983, WARPINSKI & TEUFEL 1984) as well as by a layer of relatively small cracks around a large hydraulic fracture that triggers significant increase of apparent fracture energy with crack growth and thus increase of operational pressures (SHLYAPOBERSKY 1985; cf. section 4.3.4.2.3.). Effects of stress state on hydraulic fractures are also discussed by ZOBACK, ZOBACK & SEEBURGER (1981).

When a fracture penetrates from a low-stress zone into a higher-stress zone, a crack geometry with larger widths is created in the low-stress zone (SMITH, ROSENBERG & BOWEN 1982). For a fracture in a homogeneous formation, the width/pressure slope is greater for points near the center of the crack. Modulus variations and barrier characters of rock layers adjoining the reservoir can trigger fracture width variations. Optical television inspection in some examples reveals a very wide fracture in the upper part of the pay interval and a narrower crack through the shales and lower pay section (cf. section 6.2.1.9.3.). Sharp fracture closures over the bottom of the reservoir are at least partially the consequence of a high-modulus layer below. In addition to these relationships, actual crack widths over the reservoir interval are probably smaller than the measured widths due to spalling and erosion of fracture edges. Aspects of fracture width are also discussed by PERKINS & KERN (1961) and MEDLIN & MASSE (1984).

4.2.2.1.3.3. Tailored pulse loading

Multiple fracture generation by tailored pulse loading is a combination of the effects of hydraulic fracturing and explosive well shooting (SWIFT & KUSUBOV 1981; cf. section 4.7.3.). Tailored pulse loading is particularly suitable for naturally fractured reservoirs (cf. section 4.8.8.5.). Ideally, the optimum stimulation technique is to produce maximum surface area for the least amount of energy expended. Considering the limitations of hydraulic fracturing and explosive shooting, it is apparent that a technique for initiating and sustaining multiple fractures in a wellbore with minimal damage imparted to the borehole wall provides an attractive means for maximizing permeable communication of a well to the surrounding media. Some points of intermediate strain rates and multiple radial cracks are touched as follows.

4.2.2.1.3.3.1. Intermediate strain rates

Tailored pulse loading at intermediate strain rates can create multiple fractures without damaging the wellbore region (WARPINSKI, SCHMIDT, COOPER, WALLING & NORTHROP 1979; SCHMIDT, WARPINSKI & COOPER 1980). The degree of multiple fracturing is quite sensitive to borehole conditions as well as to loading rate and appears to be strongly influenced by transient pore-pressure effects. Fracture threshold-breakdown pressure increases proportionately with loading rate, with this dependence being attributed to transient water intrusion around the borehole. Multiple fracturing by tailored pulse loading couples deformation and failure character of the rock to fluid diffusion through the permeable rock matrix and fluid flow in the cracks (SWIFT & KUSUBOV 1981).

Tailored pulse loading is an intermediate strain rate rock fracturing process generating multiple crack networks which have a higher probability of intersecting far-field natural fractures than cracks created by explosive or conventional hydraulic fracturing (STOLLER 1985; cf. section 4.7.3.). As a consequence of the control by the in-situ stress state of the formation, a singular hydraulic fracture tends to parallel the dominant trend of the existing natural fracture system, thereby reducing the possibility of its intersection (cf. section 4.8.8.1.). Coupling of the hydraulic fracture into the natural crack and joint system is thus dependent upon spacing of the minor or orthogonal fracture trend and extent to which the latter is connected to the dominant natural fracture trend (cf. section 4.2.4.3.).

The merits of a multiple fracture network depend upon creation of cracks of sufficient length to substantially influence production and long-term maintenance of created fracture conductivity. Mechanisms which can maintain conductivity of a multiple crack network despite the lack of inserted proppants are channelling and gouging on the fracture surfaces, and translation of one fracture surface relative to the other when not oriented in the principal stress directions, with shearing stress influence on off-axis cracks being very important. If fracture conductivity can thus be maintained even under high closure stresses, productivity enhancement results through a synergistic combination of increases of effective reservoir permeability and effective wellbore radius.

4.2.2.1.3.3.2. Multiple radial cracks

ZUBER, LEE & GATENS (1987) outline the effectivity of multiple radial fractures created by tailored pulse loading in Devonian shale gas reservoirs (cf. section 4.4.4.6.). The controlled detonation of explosives in the wellbore preferentially by insertion of the charge into natural cracks or into a hydraulically-induced minifracture generates a tailored pulse (cf. section 4.8.9.6.). The main requirements and benefits of this technique are that peak radial stress of the pulse is below flow stress and above tensile strength of the reservoir rock, initial loading rate is large enough to trigger the origin of fractures, and duration of pulse and generated gases from the explosion are sufficient to extend the cracks for significant distances. Multiple radial fracturing created by tailored pulse loading has turned out to be economically much more attractive than standard explosive fracturing (cf. section 4.7.3.), but is of inferior significance compared to hydraulic fracturing which is also in the Devonian shale the most effective stimulation technique (cf. section 4.4.4.6.). Caution has to be exercised concerning orientation of induced fractures with respect to the natural crack system which creates an anisotropical permeability distribution (cf. section 4.2.4.2.) in the shale reservoirs and results in the relationship that a well with an induced fracture perpendicular to the natural crack trend performs much more favourably than a well with an induced fracture parallel to the natural crack trend.

Tailored pulse loading is also a successful technique for microfracturing breakdown for minimization of naturally fractured reservoir damage (BRANAGAN & WILMER 1988; cf. section 4.8.2.2.). Tailored pulse explosive breakdown procedures entail the use of a relatively slow burning propellant designed to generate a multiple set of fractures radiating like spokes from the perforated wellbore. In order to create the correct environment for the tailored pulse explosive to properly initiate multiple fractures, a fluid with a density about that of water but considerably in excess of most available gases is required as a tamping medium in the casing, with liquid carbon dioxide being an ideal wellbore tamping fluid. Tailored pulse loading is also a suitable technology for combination of hydraulic and explosive fracturing (cf. section 4.8.9.6.).

4.2.2.1.3.3.3. Other aspects

A suite of factors controls origin of multiple cracks by tailored pulse loading (SWIFT & KUSUBOV 1981). Threshold breakdown pressure increases proportionately with loading rate, and the onset loading rate for multiple fracturing decreases with increasing water content. Crack number increases with loading rate above the onset loading rate. If fluid intrusion is prevented, only a few fractures occur regardless of loading rate and confinement conditions. The multiple fracturing process at intermediate loading rates is governed by fluid diffusion through the porous solid coupled to the structural deformation of the solid. The mechanism responsible for multiple fracturing appears to be the effect of transient pore pressure buildup in the near vicinity of the wellbore resulting from the diffusive intrusion of the borehole fluid. Brittle fracturing for understanding of high-energy gas stimulation is discussed by KRUGMAN (1986).

Aspects of multiple fracturing by tailored pulse loading, propellants or other techniques are also discussed by WARPINSKI, SCHMIDT, COOPER, WALLING & NORTHROP (1979); SCHMIDT, WARPINSKI & COOPER (1980); SWIFT & KUSUBOV (1980, 1981); AHMED, SCHATZ, HOLLAND, JONES & GREENFIELD (1982); CUDERMAN (1982); McHUGH & KEOUGH (1982), HINN (1983); KEOUGH, CURRAN & McHUGH (1984); CUDERMAN & NORTHROP (1984); FOURNEY, HOLLOWAY & SIMHA (1984); HANSON, SCHMIDT, COOLEY & SCHATZ (1984); AHMED, NEWBERRY & CANNON (1985); STOLLER (1985) and WATSON, BENSON & FILLO (1986). STRUBHAR, FITCH & GLENN (1974) outline multiple vertical fracturing in highly-deviated wellbores as a means to enhance drainage effectivity of inclined boreholes (cf. section 4.8.6.).

4.2.2.1.3.4. Controlled vs. conventional explosive fracturing

In conventional explosive fracturing, firing of a charge in the wellbore generally creates a large number of fractures of short extent (SWIFT & KUSUBOV 1981). The rate of pressurization is very high which causes compressive stresses in the wellbore vicinity to be much higher than the in-situ stress state. This stress environment can cause compaction or pulverization of a finite zone around the wellbore to such a degree that permeability is significantly decreased (SMITH, BASS & TYLER 1978). Thus even though the induced multiple fractures may intersect with many of the natural preexisting cracks (cf. plate IV/1), the damage imparted to the wellbore wall by compaction or pulverization hinders permeable communication to the surrounding media. Decoupling the explosive charge from the wall can lower the applied stress and lessen the amount of damage to the wall region, but the short duration of the explosive impulse does not allow fracture growth to long distances (SWIFT & KUSUBOV 1981; cf. section 4.7.3.).

Tailored pulse loading is a possibility to create multiple fractures of maximum length by controlling the increase rate and therefore preventing formation of a pressure stress cage which is common in standard explosive fracturing (FOURNEY, HOLLOWAY & SIMHA 1984). When detonating ordinary explosives, the pressure rise rates within the borehole are too rapid and create a stress cage by plastic deformation of the rock in the wellbore vicinity. After the passage of outgoing stress waves, the explosive loading results in a compressive residual stress which prevents the high-pressure gases from entering the newly-formed fractures, thereby leading to much shorter fractures than desired and also inhibiting the free flow of gas or oil back into the wellbore. Tailored pulse loading by propellants (SCHMIDT, WARPINSKI & COOPER 1980) or stemming (FOURNEY, BARKER & HOLLOWAY 1981; FOURNEY, HOLLOWAY & SIMHA 1984; YOUNG, BARKER & CLARK 1984), however, can successfully create sufficiently long multiple fractures.

While propellant application is quite expensive and the rate of deflagration of propellants depends strongly on confinement and therefore on pressure, stem-induced fracturing works with highly-decoupled conventional explosive charges that are located at the bottom of a borehole that has been drilled to a depth beyond the pay zone. The well is closed off with a packer or pedal basket located at the top of the pay zone, and several metres of crushed gravel are used to ensure that the stemming column will bridge in the hole when the explosive detonates. Upon firing, air shock and detonation products propagate upwards in the borehole and when they reach the stemming, shock reflecting creates a several-fold pressure increase of favourable magnitude and rate. The absence of stress cage and fine particles plugging the cracks with respect to a charge-area situation prevents quick arrest of the generated multiple radial fracture system which is driven for longer distances by the increased pulse duration. Other aspects of stem-induced tailored-pulse fracturing are discussed in section 4.7.3.3.

4.2.2.1.3.5. Simultaneous propagation of multiple fractures

Concerning simultaneous propagation of multiple fractures, distinction has to be made between vertical and horizontal wells from which the various cracks emanate either into separate layers of the prospective interval and being contained more or less within each storey, or into the same layer like a sequence of slices and being as a whole series confined within this horizon.

4.2.2.1.3.5.1. Vertical wells

BEN NACEUR & ROEGIERS (1988) comment on multiple fracture propagation simultaneously from one vertical wellbore into different vertically successive productive layers of the reservoir complex. Simultaneous growth of fractures can occur with vertical or horizontal cracks extending from different perforation sets. In the first case, the fractures are collinear, whereas they are approximated by parallel cracks in the second one, although the fractures may actually tend to move away from each other. Due to elastic interaction, the widths for two vertical fractures for a given pressure are larger than that for a single crack. Multiple vertical fractures are also more elliptical, with maximum width of the upper fracture being reached above the center of the crack. Since width is proportional to fracture length (cf. section 4.3.3.3.), there is an asymptotic tendency of width ratio, and also merging of fractures occurs. For vertical cracks of different sizes, this effect depends on the relative sizes of the fractures, with modification of crack width profile being larger for a fracture that is influenced by a larger crack size.

4.2.2.1.3.5.2. Horizontal wells

Horizontal wells can be effectively stimulated by creation of a series of parallel vertical fractures to connect the different strata (cf. also section 4.8.6.2.2.). Interaction between several cracks results in smaller fracture widths. During a first phase, only one fracture representing the crack with the lower closure stress can propagate until a boundary stress level is reached. The second fracture starts propagating with the same pressure as the first one, and the pressure tends asymptotically to the lower curve. The injection rate in the first crack is at the beginning constant and then starts declining rapidly to one-half value. Interaction of multiple vertical fractures in both vertical and horizontal wells results in shorter extension of the cracks with respect to length which could be achieved by singular fractures.

Modelling simultaneous propagation of a series of hydraulically-induced fractures thus reveals that the elastic interaction between collinear cracks gives rise to larger fracture widths or lower apparent stiffnesses, whereas for parallel cracks, the results are reverse, and therefore individual fractures are invariably shorter in the first case. In a series of fractures originating from different intervals, the various individual cracks grow independently from each other and even are not knowing the existence of their neighbours. Qualitative estimates of fracture geometries can be made corresponding to incremental injection pressures, and the knowledge of stress distributions prevailing within the perforation intervals is essential for an accurate prediction of fracture sizes.

YOST, OVERBEY, WILKINS & LOCKE (1988) describe simultaneous propagation of multiple hydraulic fractures in shale gas reservoirs traversed by a horizontal borehole. Multiple hydraulic cracks can be induced from a horizontal borehole during a single pumping event, but fracture propagation is strongly influenced by the natural crack system. Low injection rates during a hydraulic fracturing operation allow extension of natural joints with a low-angle relationship to the principal stress orientation, whereas higher injection rates generally result in inducing fractures controlled by the stress field. WANNER, CHESNUT & COX (1979) point out that in thick gross reservoir sections where numerous thin sands are distributed which form only a fraction of the total net pay indicated by logs, the key to successful exploitation of these thin sands lies in the use of relatively small discrete fracture treatments for each sand rather than massive fracturing of a thick gross section (cf. also KNUTSON & WARD 1977).

4.2.2.1.3.6. Other techniques and effects

FREDRICKSON & BROADDUS (1976) and BARBER & THEMIG (1985) introduce the concept of selective placement of fluids in a fracture by controlling density and viscosity (cf. section 4.2.2.6.4.), with separate loading of the individual stages with different proppant types and/or grain sizes representing a divided treatment schedule and resulting in placing of distinct horizontal segments of the whole proppant quantity being ideally head at the fracture tip, main body in the middle section of the wedge, and tail at the crack mouth, or vertically discrete lots at top, middle and bottom of the fracture (cf. also sections 2.4.1.2.3. and 4.3.3.2.). HANNAH, HARRINGTON & ANDERSON (1977) describe possibilities of location of various fluid schedule segments in the fracture after pumping which provides useful information for planning of gelling agent and breaker loading timetables as well as proppant placement.

Case studies of interaction between geological conditions, treatment characteristics and production are reported by McLENNAN, ROEGIERS & MARX (1983). A continuous multistage fracturing technique is outlined by WEBSTER, GOINS & BERRY (1970). MURPHY & JUCH (1960) present pin-point sand fracturing as a method of simultaneous injection into selected sands. Other features of multiple-fractured wells are outlined by WILLIAMS, NIETO, GRAHAM & LEIBACH (1973); CHOO & WU (1987); SMITH, MILLER & HAGA (1987) and SULLIVAN, LEE & HOLDITCH (1987; cf. also section 2.4.1.2.). KULAR, CHHINA & BEST (1988) report multiple hydraulic fracture propagation in oil sands. AL-TAMIMI & EL-MZEIN (1987) comment on stimulation techniques of tight multilayered carbonates. Aspects of single-stage multiple-zone stimulation are discussed by EASON (1985). Design of fracturing treatments in multilayered formations is investigated by BEN NACEUR & ROEGIERS (1988).

4.2.2.2. Limited-entry fracturing technique

Multiple fracturing is commonly permitted by the limited-entry stimulation technique where the number of perforations in a completion interval is limited for promotion of development of perforation friction pressure during a stimulation operation (LAGRONE & RASMUSSEN 1963, HOWARD & FAST 1970, SMALL 1985, CRAMER 1987). The limited-entry technique comprises accomplishing of treating several intervals in a thick pay section with a single-stage operation by limiting number of perforations and providing an injection rate high enough to assure that all the perforations will be taking fluid (LAGRONE & RASMUSSEN 1963). The pumping rate is chosen such that the frictional pressure drop across the perforations will increase pressure in the wellbore to a value above the formation breakdown pressure of each zone perforated, with this concept being an effective and thus widely applied diverting technique. Limited-entry fracturing has originally been introduced to the literature as pin-point sand fracturing (MURPHY & JUCH 1960).

Limited-entry perforation design may be depicted as the ability of a set of perforations to pass a finite volume of liquid at a given differential pressure (EASON 1985). Each perforation acts as a choke diverting fluid to each successive one. With increasing injection rate, all perforations will eventually take up fluid at maximum capacity if a sufficient pressure differential is hold across the perforations. The following overview presents some aspects of simultaneous fluid entry into multiple zones, pressure drop across the perforations, influence of screenouts, and fluid distribution and perforation arrangement.

4.2.2.2.1. Simultaneous fluid entry into multiple zones

The choking effect resulting from the limited-entry operation creates excess pressure in the casing, allowing the simultaneous entry of fracturing fluid into multiple zones of varying in-situ stress states. Having been restricted to acid fracturing and small-scale proppant fracturing treatments in earlier years, limited-entry techniques have recently also been introduced to MHF stimulations (CRAMER 1987). Particular applications include reservoirs with two porosity zones which are separated by a tight barrier (BEGNAUD & CLAIBORNE 1985) where hydraulic communication between the two pay intervals occurs routinely, although proppant remains isolated in the perforated interval due to an apparent width reduction in the portion of the fracture penetrating the bounding interbed.

When both zones are treated separately, the adjacent unperforated pay zone acts as a permeable fluid sink and thus proppant concentration with associated screenout risks will occur. In case of simultaneous stimulation with limited-entry techniques, however, this leakoff source will be eliminated and less total fracturing fluid will be needed to displace a given volume of proppant. Interzonal rate distribution is dictated by coupling of perforation and formation fracture extension pressure characteristics (CRAMER 1987). Material changes at the perforations and in the fractures can cause injection rate and thus treatment volume redistribution among zones which leave some sections of the reservoir complex understimulated and in certain circumstances can contribute to a well screenout. Limited-entry perforation enables multiple-zone completion and helps to control excessive fracture height growth (HOOVER & ADAMS 1988).

4.2.2.2.2. Pressure drop across the perforations

Limited-entry fracturing design frequently requires the use of a minimum perforation pressure drop (BUNDY 1981). Many stimulation jobs designed to treat large intervals fail because the fracturing fluid is not properly distributed over the entire completed interval and when a particular zone of the completed section does not receive enough fluid, it screens out early in the job, because the necessary fracture width has not been created (cf. sections 4.2.3.2.5. and 6.2.4.2.1.). The minimum pressure drop across the perforations forces the fracturing fluid into all the perforations.

Other limited-entry treatments fail because crack design assumes that a single fracture will be generated across the entire completed interval, whereas in reality a multifracture system originates consisting of independent cracks within the individual horizons if sufficiently thick barriers separate the storeys, with each fracture having to receive enough fluid in order to prevent premature screenout termination of the job (cf. section 4.2.2.2.3.). The more independent cracks originate across the completed segment, the more accurate fluid distribution has to be for the purpose of avoiding screenout failure. Some comments on fluid volume and perforation number as well as shear degradation and fluid viscosity are offered as follows.

4.2.2.2.2.1. Fluid volume and perforation number

The amount of fluid that is required for each zone is determined by its height, and the required number of perforations for each interval can be calculated once the optimum pressure drop across the perforations has been established. Maximizing pressure drop ensures that fluid is diverted to all the perforations. Pressure drop across the perforations is constant throughout the completed interval only when formation treating pressure (being fracture gradient times depth) is constant across the section. In case of varying fracture gradients, distribution and number of perforations have to be adjusted in order to compensate different treating pressures in various zones. Minimum perforation pressure drops have to be used if the fracturing rate is limited by tubing and pressure restrictions. Prefracturing injectivity testing and ballout are performed to ascertain that all the perforations are broken down and will take fluid.

4.2.2.2.2.2. Shear degradation and fluid viscosity

As shear degradation of fracturing fluids (cf. sections 4.3.3.3. and 4.3.4.2.) usually takes preferentially place at the perforations, highly limited-entry fracture treatments are characterized by lower fluid viscosities which provide deeper proppant penetration within a long crack with less height growth, and thus connect more net pay to the wellbore (HOEL 1988). Low-viscosity fracturing fluids having undergone shear at the perforations in limited-entry constellations, however, are not able to generate sufficient width to transport larger amounts of proppants. On the other hand, if a higher viscosity fluid were pumped in an attempt to carry large proppant quantities, larger fracture widths could be obtained which are causing more height growth out of zone and less proppant penetration of the central reservoir section. The greatest benefit of limited-entry fracturing is thus improved proppant penetration which compensates for the fact that this goal is achieved at the expense of high proppant concentration.

4.2.2.2.3. Influence of screenouts

Designing effective limited-entry treatments requires a blending of engineering disciplines in a synthesis of conventional fracture modelling with knowlege of the pressure loss characteristics at the perforations. While screenouts can be desirable in certain situations (cf. SMITH, MILLER & HAGA 1987), they have to be strictly avoided in limited-entry applications, because the differential pressure at the offending zone will decrease and cause a redistribution of treatment volume into the other horizons, with the resultant rate loss accelerating the screenout process in that interval (CRAMER 1987). The remaining treatment volume will then be distributed among the open horizons, and the pad volumes in these zones may be insufficient to develop fracture area and establish leakoff control for the additional load, with total treatment screenout being able to occur. All horizons need to be fairly similar in leakoff characteristics for the limited-entry technique to be viable. If one interval has relatively high leakoff characteristics, the overall treatment staging needed to prevent a screenout may leave the other horizons inadequately stimulated, with the possible occurrence of diluted proppant concentrations.

Other aspects of candidate screening are high stress differences between the individual storeys of the reservoir complex (cf. sections 4.2.2.5.3. and 4.2.3.2.2.) which lead to great differences in proppant entry per perforation between the zones. In a large-magnitude treatment, a runaway condition may occur in which injection ceases in the higher-stressed horizon due to the inability to maintain sufficient pressure drop across perforations. Even in completion intervals characterized by small differences in instantaneous shut-in pressure, perforation erosion will eventually outrun the ability to compensate the increasing treating rate. Screenout failures (cf. sections 4.2.3.2.5. and 6.2.4.2.1.)

are frequently provoked by the spikes in stairstep proppant schedules, with smooth ramping schedules bringing in higher proppant concentrations into the fracture with less screenout risk (cf. section 4.3.3.5.).

4.2.2.2.4. Fluid distribution and perforation arrangement

Limited entry permits the fracture to be placed in all desirable parts of a reservoir and gives maximum control of stimulation fluids at each fracture point (AHMED, NEWBERRY & CANNON 1985). In essence, limited entry provides for equal distribution of treating fluids through all perforations by limiting number and size of perforations, and by controlling the differential pressure across the perforations. Correctly applied under the proper well conditions, limited-entry can be extremely effective. Aspects of pseudolimited-entry fracturing are discussed by MATHIAS (1970). Application of ball sealers and diverters represents modified limited entry technique (BLACK, RIPLEY, BEECROFT & PAMPLIN 1979; PAI, GARBIS & HALL 1983; cf. sections 4.2.2.1.1. and 4.2.2.4.).

Limited-entry technique often ascertains that the number of open perforations matches treating rate which can be achieved (CONWAY & HARRIS 1982). If too many perforations exist over a large interval in a multiple-entry design, then too many of the perforations may serve only as fluid thief zones which can cause too short fracture length due to pressure dissipation or can even provoke premature screenout termination. On the other hand, if only a few of the perforations that have been shot are open, then high treating pressures may result which in turn can give rise to screenout failure or non-optimum proppant placement. Limited-entry perforating is often combined with separate acid ball-out in order to open the perforations (WEBSTER 1977).

Another application of limited-entry techniques in a manner comparable to that in hydraulic proppant fracturing is in steam-flood enhanced oil recovery to contain steam override and to minimize the origin of oil pockets or blind spots from insufficient vertical conformance of the injected steam (HUANG & HIGHT 1986; cf. section 4.9.4.). Limited-entry fracturing is a significant means of vertical fracture growth restriction due to the general tendency of the crack to propagate upwards as a consequence of decreasing closure stress (in-situ stress varies with depth; CLIFTON & ABOU-SAYED 1983) with declining overburden, with fracture height control thus commonly being performed by only perforating the lower portion of the interval to be treated (LAMBERT, DOLAN & GALLUS 1983; PAI, GARBIS & HALL 1983). Aspects of limited-entry fracturing technique are also discussed by ODEH (1976), ICHARA & RAGHAVAN (1981), CHANCELLOR & JOHNSON (1988) and ZHOU, QI & XIAO (1988).

4.2.2.3. Fracture containment by treatment parameter adjustment

The dominant factor for arresting or at least restricting vertical growth of fractures is the in-situ minimum horizontal stress contrast between pay zone and barrier (WARPINSKI, SCHMIDT & NORTHROP 1980; TEUFEL & CLARK 1981, NGUYEN & LARSON 1983; cf. sections 4.2.2.5.3. and 4.2.3.2.2.). If shear strength of the interface is small relative to tensile strength and minimum horizontal compressive stress of the boundary layer, vertical fractures may turn into interfacial fractures (cf. section 4.5.4.4.2.) and thus containment is achieved (TEUFEL & CLARK 1981), although this effect may be only temporary, because after the fracture is laterally displaced at the interface, it can still grow into the bounding layer if the frictional properties of the interface change (HANSON, SHAFFER & ANDERSON 1980).

Shales are no absolute barriers to fracture migration, but are only efficient in containing cracks if a certain critical stress contrast is not underpassed (AHMED 1988; cf. sections 4.2.2.5.3. and 4.2.3.2.2.). Weakly-bonded interfaces at shallow depth may stop fracture growth (DANESHY 1978), and also stress discontinuities such as faults may terminate fracture propagation (ROEGIERS & WILES 1981; ROEGIERS, McLENNAN & MURPHY 1982; McLENNAN, ROEGIERS & MARX 1983; cf. section 4.2.3.2.4.). Calculation of vertical fracture containment in layered formations is evaluated by FUNG, VIJAYAKUMAR & CORMACK (1987). Following discussion of some factors influencing fracture containment, emphasis is put on injection rates, fluid properties, propped fracture height vs. reservoir thickness, and perforation interval selection.

4.2.2.3.1. Factors influencing fracture containment

Contrast in material properties and pore pressure generally has only secondary effects on fracture containment, with elastic modulus being more influential than the other material properties, but being only able to reduce the width of the out-of-zone fracture segment and not to stop its ungoverned height propagation (EEKELEN 1982, NGUYEN & LARSON 1983). If the fracturing pressure is high enough, the crack can propagate across any barrier (HUNT, RAYMOND, HASKETT & PIRIE 1981; cf. section 4.2.3.2.4.), with therefore pressure control playing a chief role in fracture containment attempts. High injection rates are necessary to carry the proppant in the fracturing gel and to reduce the probability of a premature screenout during stimulation (PAI & GARBIS 1983; cf. section 6.2.4.2.1.).

In order to contain a fracture within the formation of interest or to minimize its vertical height growth, the excess pressure within the fracture must be minimized (HOLDITCH, ROBINSON, WHITEHEAD & ELY 1988). The excess pressure is almost entirely dependent on the friction pressure drop down the fracture (cf. section 4.3.4.2.2.) which in turn is a function of gel viscosity, injection rate, and friction coefficient between fluid and fracture wall, with gel viscosity and friction coefficient having the greatest influence on friction pressure drop (cf. section 4.3.4.2.). The outline as follows illustrates fracture orientation changes, fracture geometry variations, confining stress contrast, reservoir depletion and stress contrast magnitude, positive and negative fracture containment, and fracture growth termination at interfaces.

4.2.2.3.1.1. Fracture orientation changes

Special influences of screenouts are changes of fracture orientation from horizontal to vertical in shallow reservoir storeys (SCHRAUFNAGEL & LAMBERT 1988). Multiple coal seam fracturing in a stress level where horizontal rather than vertical fractures are more likely to originate can be performed by the limited-entry stimulation technique which provides for multiple horizontally induced fractures and/or complex fractures with both horizontal and inclined components (aspects of composed fractures are also discussed in sections 4.4.3.4.3. and 4.9.4.2.). Perforations serve as pressure chokes that control the rate of fluid injection into each targeted zone. The limited-entry technique requires maintenance of a predetermined wellbore pressure throughout the operation. Perforation diameter grows due to proppant abrasion which causes pressure reductions. Significant pressure drops can indicate screenout of several perforations and initiation of a vertical fracture following termination of propagation of the earlier horizontal crack. Synthetic altered-stress fracturing instead of natural crack orientation changes profits from artificially-induced conditions for alteration of fracture propagation direction (cf. section 4.9.4.1.3.1.).

When the critical treating pressure is reached and the fracture breaks out of zone, the stress concentration is highest in the near wellbore area (ROSE 1988). Therefore uncontrolled fracture height growth can only occur at the crack tip where no proppant slurry is present, with this condition typically resulting in a sudden loss of pad volume and tip screenout. Experimental modelling of the stress gradient affecting the height of vertical hydraulic fractures is carried out by THIERCELIN & LEMANCZYK (1983).

4.2.2.3.1.2. Fracture geometry variations

Fracture geometry can also be influenced by treatment parameters such as fluid viscosity and density, proppant concentration, pumping rate, and perforation placement (NGUYEN & LARSON 1983). By varying fluid properties, proppant saturation and/or injection rate, frictional pressure drop along the fracture faces can be kept below the minimum horizontal stress difference between confining zone and pay horizon. Detection of the possible onset of undesirable growth of fracture height is possible from treating pressures which may allow to undertake corrective actions during the operation (NOLTE & SMITH 1981). Excessive pressure caused by the fluid flowing down the fracture can overcome the confining stresses in adjacent rock layers, resulting in excessive fracture height growth as well as uncontrolled horizontal crack propagation (HOLDITCH, ROBINSON & WHITEHEAD 1986).

4.2.2.3.1.3. Confining stress contrast

The main factors controlling vertical height growth of hydraulic fractures are confining stress contrast, elastic stiffness contrast, fracture toughness, and influence of settled proppant bank or buoyant proppant for downward or upward crack extension (SETTARI 1985). Confining stress contrast is the strongest containment mechanism, whereas stiffness contrast cannot completely arrest height growth. Fracture toughness is important only for cracks created with low-viscosity fluids or after fluid breakdown. A proppant bank at the fracture bottom or buoyant proppants concentrated at the crack top can be an effective barrier to vertical propagation (cf. section 4.2.2.4.), and local temperature changes at the fracture tip are not likely to increase height growth rate seriously (a summary of fracture containment mechanisms is given by SNEDDON & LOWENGRUB 1969).

In-situ stress contrasts between sandstones and mudstones (cf. section 4.2.3.2.2.1.) can in some cases be sufficient to contain the fractures if they are large enough (cf. sections 4.2.2.5.3. and 4.2.3.2.2.). Smaller in-situ stress contrasts can effectively limit fracture propagation if assisted by higher pressure drops in vertical direction due to small crack width or inefficiencies associated with fracture extension across multiple complex beds containing separating mudstone drapes where crack extension energy is progressively diminished (WARPINSKI, BRANAGAN, SATTLER, CIPOLLA, LORENZ & THORNE 1988). Fracture containment by treatment parameter adjustment is generally known as controlled process zone fracturing including optimization of injection rates, pressure, fluid properties and other factors in order to tailor the crack (AHMED, STRAWN, WILSON & SCHATZ 1983). Effects of stress state on hydraulic fractures are also illustrated by ZOBACK, ZOBACK & SEEBURGER (1981).

4.2.2.3.1.4. Reservoir depletion and stress contrast magnitude

Depletion of reservoir pore pressure with progressive hydrocarbon exploration leads to increasing fracture gradient between pay zone and adjacent shales and therefore creates a more effective boundary for crack propagation (HUEBINGER, WEBSTER, CHISHOLM, VENDITTO & HUNT 1988). With a higher confining stress contrast, fractures can be better restricted to the core of the prospective interval, and the resulting reduction of crack height allows to substantially decrease both total proppant quantity and proppant concentration (case studies reveal that proppant volume can be diminished by up to 40 %). This is the reason why refracturing (or alternatively late-stage infill-well stimulation; cf. section 4.8.9.) is frequently comparatively cheaper than original or early-stage treatments and is often still economical even with declining production and falling oil price.

4.2.2.3.1.5. Positive and negative fracture containment

Distinction has to be made between positive and negative fracture containment conditions (WARPINSKI, CLARK, SCHMIDT & HUDDLE 1982). In formations with positive containment conditions, fracture growth out-of-zone is not anticipated, and long deeply penetrating fractures can be effectively used to develop the resource and have a high economical pay-off. In reservoirs with negative containment conditions, however, excessive fracture height growth can only be prevented by renouncing from large-scale treatments which would waste fluids and proppants in adjoining non-prospective formations, and by limitation to small-scale operations resulting in only lower well productivity improvement, because the only other alternative is to reject well and reservoir horizon as unfeasible.

4.2.2.3.1.6. Fracture growth termination at interfaces

Fracture containment may be achieved by termination of fracture growth at a discrete, weak or unbounded interface (ANDERSON 1979, TEUFEL 1979, TEUFEL & CLARK 1981) or restriction in the bounding layer to which the crack propagates and where extensive growth does not take place (WARPINSKI, CLARK, SCHMIDT & HUDDLE 1982; cf. section 4.2.3.). If the fracture penetrates into a region having an adverse stress gradient, continued propagation results in higher stresses on the fracture thus self-limiting growth (SIMONSON, ABOU-SAYED & CLIFTON 1978). Another mechanism of fracture restriction is propagation into a higher modulus region where the decreased width results in increased pressure drop in the crack which might inhibit extensive growth into that region relative to the lower modulus zone. Hydraulic fracturing can be done through tubing in case of smaller treatments in order to avoid killing of the well after termination of the stimulation job and because tubing strength is much greater than casing strength (WEIMER, SONNENBERG & YOUNG 1986).

4.2.2.3.2. Injection rates

Containment of fractures in thin sandstones is in practice generally reached by reduced pumping rates when in a stable system the crack is just running along the sandstone layer and is conducted by the bounding tight rock horizon (following the least resistance for its propagation), with the mudstone seal forming a tight barrier that is limiting vertical fracture extension unless the critical pressure stress of the bounding mudstones to confine the crack is exceeded (cf. also CONWAY, McGOWEN, GUNDERSON & KING 1985; BEN NACEUR & TOUBOUL 1987). Adverse containment conditions being small stress contrasts between adjoining layers are overcome by fracture height control with small injection rates and/or low fluid viscosities in order to avoid exceeding the critical pressure that would lead to propagation out of zone especially during fracture initiation.

Although injecting at a higher rate yields a more effective treatment, it is often necessary to keep the operational pressures below formation critical pressure during the job in order to maximize lateral growth and to control vertical propagation of the fracture (MENG & BROWN 1987). For determination of the maximum allowable pumping rate, pipe friction pressure, shear degradation of the fracturing fluid, and surface treating pressure have also to be considered. The discussion as follows focusses on decreasing pumping rate and total job size, pressure growth and screenout termination, fracture height and fluid distribution, and tubular enlargement and horsepower diminution.

4.2.2.3.2.1. Decreasing pumping rate and total job size

Another possibility to minimize vertical fracture growth is to limit the size of the stimulation treatment along with restricted pumping rates (BARBER & THEMIG 1985). Fracture containment control by decreasing pumping rate and fluid viscosity leads in many cases automatically to a smaller total job size, as the reduced rate causes increasing fluid loss which is limiting cumulative fluid quantity by increasing screenout risk. An alternative combination is increasing fluid density, and lowering pumping rates or reducing fluid pressure (SIMONSON, ABOU-SAYED & CLIFTON 1978; KOZIK, BAILEY & HOLDITCH 1979). Decreasing pumping rates and nevertheless maintaining adequate proppant placement thereby reducing the number of uneconomical stimulation jobs can be best achieved by utilizing a lower-density proppant, with therefore intermediate-strength low-density alumina silicate proppants being of considerable significance (CUTLER, JONES, SWANSON & CARROLL 1981) for assisting vertical fracture containment in thin reservoirs. In some special cases, pay conditions and stability of the boundaries can necessitate to reduce treatment size and pumping rates so significantly to achieve the goal of the crack staying in zone that proppant fracturing is below the minimum economical feasibility volume and cannot compete with acid fracturing and matrix acidizing in carbonates (HUCKABEE 1988).

The viscosity range of existing fracturing fluids makes minimizing fluid viscosity a much more effective method of controlling pressure than lowering pumping rate (NOLTE 1982). Decreasing of specific proppant gravity has the advantage that identical proppant transport to that currently achievable can take place in lower-viscosity fluids, or alternatively higher proppant volumes can be pumped in a given amount of a high-viscosity fracturing fluid (CUTLER, ENNISS, JONES & SWANSON 1985; the same applies for transport and settling of gravel in sand control carrier fluids; cf. section 5.3.3.). Another possibility of adaptation is application of high-strength synthetic proppants at lower concentrations and in narrower fractures instead of higher natural sand saturations in broader cracks, because narrower fractures are slower propagating in vertical direction and can therefore be better controlled than wider cracks (LI & ZHU 1986). This procedure is backed up by the field experience that in many cases, fractures propped with lower concentrations of bauxite have better conductivities than cracks propped with higher saturations of sand.

Boundary zones stopping fracture propagation and inhibiting further crack extension can be regions of increased formation stress, with the high-stress barrier sharply contrasting with the low-stress reservoir, or can be represented by the point at which fluid leakoff becomes equal to pumping rate in permeable columns and thus also further fracture propagation is not possible (HUNT, RAYMOND, HASKETT & PIRIE 1987). Difficulties in crack containment by treating parameter adjustment can occur in depleted reservoirs where fairly severe vertical growth and/or leakoff can easily occur (BEGNAUD & CLAIBORNE 1985). Effects of treatment parameters on fracture geometry are also discussed by DANESHY (1971 b).

4.2.2.3.2.2. Pressure growth and screenout termination

Pressure growth comprising abnormally large increases in treating pressure is mainly caused by fracture branching and is a function of pumping rate and fluid viscosity (MEDLIN & FITCH 1983). Thin fracturing fluids pumped at high rates generate smooth pressure curves which rise parabolically with time to first order, whereas thick gels pumped at low rates create erratic pressure curves that tend to rise more or less linearly with time and show frequently sharp irregularities with abrupt increases and declines. The beginning of screenout appears as a sudden sharp rise in treating pressure on the record regardless of fluid type and pumping rate (cf. section 6.2.4.2.1.).

The screenout is associated with the amount of pressure increase rather than the magnitude of the pressure itself. While pressure growth could be avoided or greatly reduced by pumping thin fracturing fluids at low rates, this approach is impractical due to lacking proppant transport capability. Thin fluids must be pumped at high rates to carry proppants at practical concentrations, with these high rates then in turn again increasing screenout risk. A two-stage treatment is the optimum solution of these problems. The first stage comprises pumping of a thin fluid at low rates without proppants in order to generate a large fracture. After cleanup, the second stage includes pumping of a conventional thick gel treatment with high proppant saturation which simply reopens the existing fracture and thus a pressure increase is not expected.

4.2.2.3.2.3. Fracture height and fluid distribution

Injection rate affects fracture height and fluid distribution and is very critical to the overall success of the treatment (PAI, GARBIS & HALL 1983). Where interval yield of tubing and/or casing limits injection rate, the operation should be staged using ball sealers. It could be favourable for the job procedure to start at high injection rates and low proppant concentrations and then later tapering injection rate and increasing proppant saturation (GIDLEY, MUTTI, NIERODE, KEHN & MUECKE 1979; MATTHEWS, MILLER & SCHLOTTMAN 1982). The high initial injection rate allows for the majority of the ultimate fracture height to be created while pad and low-proppant-concentration slurries are pumped, thereby establishing leakoff control over the entire fracture height while the pad is being inserted and thus reducing screenout risk (PAI, GARBIS & HALL 1983). Diminishing injection rate during later treatment stages decreases the chances of further fracture height growth and also prevents equipment overload during high proppant concentration pumping. On the other hand, however, high injection rates are necessary to carry the proppants in the gel and to reduce the probability of a premature screenout during fracturing (PAI & GARBIS 1983), with thus a sound compromise for satisfying all the purposes having to be found.

AHMED, STRAWN, SCHMIDT, O`SHEA & VEGHTE (1983) comment on stimulating tight sands in the presence of weak stress barriers (cf. section 4.2.3.2.4.) by modified hydraulic fracturing techniques. The effect of bounding formations on vertical propagation of a hydraulically-induced fracture can be ignored for short cracks, but is very prominent and of considerable influence for long fractures (KOZIK, BAILEY & HOLDITCH 1979; cf. section 4.8.11.). The time lag between fracture tip arrival (width response) and fluid arrival (pressure response) provides an estimate of fluid lag (WARPINSKI 1983 a).

4.2.2.3.2.4. Tubular enlargement and horsepower diminution

Decreasing injection rates and consequently diminishing hydraulic horsepower required for the execution of the treatment can also be achieved by a higher-strength casing through which a larger tubing can be run (GARBIS, BROWN & MAURITZ 1985). Reduction of hydraulic horsepower by the larger tubing should offset the extra expenses for the larger tubular goods (cf. section 4.3.4.2.2.1.). Apart from successful stimulation of a thick reservoir zone in a cost-effective manner, another benefit of the larger tubing/casing program is lower surface treating pressure during the fracturing operation which results in safer on-site conditions.

Fracture height containment can also be achieved by creating an artificial barrier with additives (NGUYEN & LARSON 1983), with caution, however, having to be exercised by preventing the crack to grow through the barrier (LUISKUTTY, TOMUTSA & PALMER 1986) by exceeding the boundary stress beyond the fracturing gradient of the adjoining beds with higher elasticity moduls. Field experience of the significance of natural barriers to contain vertical fracture height growth is reported by HOLDITCH, ROBINSON, WHITEHEAD & ELY (1987). Laboratory experiments on fracture propagation and limiting effects are carried out by MEDLIN & MASSE (1984).

4.2.2.3.3. Fluid properties

Fracturing fluid properties also have an influence on proppant transport and variable crack height behaviour (MORALES & ABOU-SAYED 1985, BEN NACEUR & TOUBOUL 1987). Vertical fracture migration out of control occurs when larger fluid injection rates and/or higher-viscosity fluids are used (LI & ZHU 1986). Crack arrest or retardation of fracture propagation can be achieved by altering the viscous properties of the injected fluid within parts of the crack (CLEARY 1980, SETTARI 1985). While heat transfer effects near the fracture tip will generally increase leakoff due to fluid degradation and will not selectively affect the propagation (cf. section 4.3.4.4.), a much greater potential is provided by the injection of light or heavy proppants (BRAUNLICH 1967, NGUYEN & LARSON 1983). SMITH, MILLER & HAGA (1987) propose a comparable tip screenout technique for soft formations in order to increase the fracture opening at the wellbore (cf. section 4.5.4.4.1.). Some comments on proppant density and fluid viscosity, foam and energized fluids, and density-controlled fluids are offered as follows.

4.2.2.3.3.1. Proppant density and fluid viscosity

Heavy particles falling at the bottom of the fracture increase package viscosity, or equivalently vertical transmissibility is decreased by presence of concentrated packages. Selected propping agents that have a density lower than the treating fluid will rise to the fracture top through buoyancy and eventually decrease crack propagation (BEN NACEUR & TOUBOUL 1987; cf. section 4.2.2.4.). When both upward and downward growth of the fracture has to be controlled, the combined use of heavy settling and light buoyant particles is required. The characteristics of the structure of such bi-disperse slurries as well as settling and rising rate differ significantly from those applying for mono-disperse systems due to the presence of streaming columns in the slurry (BATCHELOR & RENSBURG 1986, BEN NACEUR & TOUBOUL 1987).

Vertical propagation of the fracture can be limited by reducing fluid pressure which is much better achieved by minimizing fluid viscosity than by lowering pumping rates (SIMONSON, ABOU-SAYED & CLIFTON 1978). Lower fluid viscosities are possible to be applied in connection with lighter proppants, with almost identical transport conditions being provided for light proppants by low-viscosity fluids and for heavy proppants by high-viscosity fluids (the same applies for placement of gravel by sand-control carrier fluids; cf. section 5.3.3.). The possibility of reducing fluid pressure by fluid viscosity for fracture containment by variation of proppant weight is the greatest advantage of low-density intermediate-strength proppants (CUTLER, ENNISS, JONES & CARROLL 1983).

4.2.2.3.3.2. Foam and energized fluids

Minimization of fracture height growth can also be achieved by application of low-viscosity perfect-support fluids such as systems with carbon dioxide or nitrogen as major (foam) or minor (energizer) component (GARBIS & TAYLOR 1985). Energized and foam systems give enhanced fluid recovery and therefore better control of fracture propagation and proppant placement. The main advantages of the use of carbon dioxide in fracture stimulation treatments are faster and increased load recovery, buffering of aqueous fracturing fluids to a pH of abt. 3.5 thereby controlling clay-swelling in water-sensitive reservoirs, and replacement of some of the water in the operation, thus reducing the liquid volume that has to be recovered (cf. also section 3.11.2.). Carbon dioxide or nitrogen foam systems can also be used for selective proppant placement (BARBER & THEMIG 1985; cf. section 4.4.4.2.).

4.2.2.3.3.3. Density-controlled fluids

MISAK, ATTEBERRY, VENDITTO & FREDRICKSON (1978 a, 1978 b) present density-controlled fracturing including insertion of a higher-density pad followed by lower-density proppant-carrying treating fluids. The higher-density pad occupies the lower portion of the crack and acts as a cushion for the following lower-density proppant-transporting fluids which slide above the heavy pad and move the proppant only into the upper part of the fracture. As the lower section of the crack receives no proppant due to settling inhibition at the interface of the two fluid systems with different specific gravity, fracture closure with falling pressure leaves only the upper segment of the crack propped (cf. also section 4.2.2.6.4.).

The purpose of this special treatment is to exclude fracture growth into the lower water-producing interval of the gas-bearing pay formation due to its close position to the gas-water-contact in a downstructural situation, or if creation of the crack cannot be prevented, the density-controlled fracturing fluid sequence aims on exclusion of propping of the lower section of the crack, thereby provoking its closure and rehealing after termination of the job and thus guaranteeing that gas exploitation from the propped upper section of the fracture will not be disturbed by water production from the lower interval. Comments on vertical fracture propagation containment by controlling stimulation fluid density as well as pumping rates and pressure are also given by KOZIK, BAILEY & HOLDITCH (1979).

4.2.2.3.4. Propped fracture height vs. reservoir thickness

Distinction has generally to be made between propped fracture height and reservoir thickness, with vertical propagation of the crack out of control in thick formations and small-size jobs being much less serious than that occurring in large-size treatments in thin pay zones (LI & ZHU 1986). Fracture height propagates vertically out of interval when larger injection rates and higher-viscosity fluids are used. When pay thickness does not exceed 30 % of the propped fracture height, the stimulation operation is not going to have an actual response, and therefore understanding of the distribution of in-situ stress is important for controlling fracture height and enhancing stimulation result.

Differences in lithological formation properties have decisive impact on differences in fracture pressure capacity (net pressure in the fracture) in different reservoirs which represent differences in minimum in-situ stress between both treated zone and barrier layer. Selection of the right perforation interval can advance to the key element of fracture height containment under certain fracture-pressure capacities according to the distribution of in-situ stress in the zones.

4.2.2.3.5. Perforation interval selection

Perforation interval choice is also important for limiting vertical fracture growth, with perforating only the lower portion of the section to be treated guaranteeing upwards declining hydraulic pressure transmission, because the fractures are generally propagating upwards into the direction of decreasing overburden stress (LAMBERT, DOLAN & GALLUS 1983; PAI, GARBIS & HALL 1983; cf. section 4.2.2.5.3.1.). On the other hand, HUCKABEE (1988) presents an example of fracture extension potential toward the bottom of the reservoir complex, with stimulation operations consequently being initiated in the upper portion of the reservoir. Preferential downward growth of hydraulic fractures is also reported by WARPINSKI, BRANAGAN, SATTLER, CIPOLLA, LORENZ & THORNE 1988).

4.2.2.4. Fracture containment by buoyant and settling diverters and other blocking additives

Apart from treatment parameter adjustment, vertical fracture containment can be achieved by the application of buoyant and settling diverters and other blocking additives in order to limit or stop vertical height growth of the propagating fracture. Some points of fluid viscosity and diverter density as well as pressure diversion are also illustrated as follows.

4.2.2.4.1. General aspects

When a fracture is initiated in a productive interval, it will grow in a vertical radial pattern until it encounters resistance to this growth mode. Such boundaries to propagation are normally barrier rocks or strata that contain higher stress and normally are tougher and less permeable than the productive rock (cf. sections 4.2.2.5.3. and 4.2.3.2.2.). If such barriers exist above and below the prospective interval, vertical growth of the crack will be restrained at these boundaries and the fracture will then continue to grow outward into the hydrocarbon-bearing formation, thereby changing its shape from ideally circular to really elliptical. These conditions provide efficient use of fracturing fluid in creating a deeply penetrating fracture. If the barrier rocks are too weak to withstand the pressure required to propagate the fracture, the boundaries will rupture and vertical growth will continue in downwards and/or upwards direction, and stimulation efficiency is very poor, because much of the fracture area is outside of the economically attractive interval and lateral penetration within the hydrocarbon-bearing section is severely limited due to waste in over- and/or underlying barren zones.

In addition to barren rocks above and below, an oil-bearing section for example might contain water in the bottom of the zone and/or a gas cap at its top. Oil production could be considerably deteriorated in case of fracture breakthrough from the oil-bearing part to the water- and/or gas-containing segments of the sequence. As offtake increases are directly related to fracture area in or crack penetration of the potential interval, containment of the fracture in the productive zone not only increases fracturing efficiency, but also provides greater exploitation increases (DOWELL SCHLUMBERGER 1987).

Instead of reducing pumping rates and fluid viscosity which also lead to cut back of the overall size of the treatment and thus diminuish fracture penetration in the reservoir, crack containment can be achieved by creating artificial upper and lower propagation barriers by inserting light buoyant and heavy settling proppants acting as diverters and controlling upwards and downwards extension of the crack, respectively. The whole system acts as an effective synthetic bracket of top and bottom of the interval to be fractured (bracket stimulation; DOWELL SCHLUMBERGER 1986, 1987).

4.2.2.4.2. Buoyant diverters

Fracture height containment towards the top of the formation can be achieved by creating an artificial barrier with a light solid additive (NGUYEN & LARSON 1983). A nonreactive buoyant diverter is used to control upward propagation of the fracture during the treatment which offers an unique solution of preventing excessive crack height growth independent of formation properties and geological stress environment. The buoyant diverter is pumped as a slurry before the proppant stages of a fracture treatment and accumulates in the uppermost portion of the newly created crack, forming a compact wedge-shaped low-permeability flow block.

Since this artificial barrier does not allow fluid pressure to reach the upper bounding layer, treating pressure can be increased without the risk of having upward fracture propagation. The particles of the buoyant diverter must be small enough so that only negligible mechanical damage occurs during the pumping operation, and must be large enough so that the diverter will not be carried into the proppant bed by production fluids and reduce the available fracture conductivity by invasion of the interproppant porosity (cf. section 1.4.11.2.1.). The rising rate of the diverter decreases with increasing carrier fluid viscosity. In order to be effective, the buoyant diverter has to be introduced in the prepad stage for rapid deposition of the major portion of the blocking additive near the wellbore, as fracture height is greatest in the immediate surroundings of the borehole, because the hydraulic pressure in the very vicinity of the wellbore is greater than at any other point along the length of the fracture.

GABRIEL & ERBSTOESSER (1984) introduce buoyant ball sealers that provide improved diversion possibilities (ERBSTOESSER 1980; cf. also section 4.2.2.1.1.3.). Lightweight buoyant proppants in a prepad migrate upwards thereby creating a high-pressure gradient in the fracture which consequently encourages lateral fracture penetration (McLENNAN, ROEGIERS & MARX 1983). Aspects of fracture packing with lightweight additives which float and create a permeability barrier and thereby reduce upwards crack migration are also discussed by AHMED, STRAWN, WILSON & SCHATZ (1983).

4.2.2.4.3. Settling diverters

Inversely, downward growth of the fracture can be limited by heavy solid additives which fall out of the fracturing fluid and plug the lower tip of the crack wing, thereby retarding or terminating its propagation. The combination of light buoyant diverters at the top and heavy settling diverters at the bottom of the fracture acts as a dual bracketing device which effectively controls vertical height extension of the fracture. The simultaneous application of both components is possible by density-control fluids (cf. section 4.2.2.3.3.3.). Limiting fracture height to within only the pay zone results in diversion from barren zones, bottom water or gas cap, thereby giving rise to improved crack efficiency, deeper penetrating fractures within the target horizons, improved proppant distribution and thus consistently better stimulation results (DOWELL SCHLUMBERGER 1986, 1987).

The method works particularly well if being combined with fluid-viscosity control. Apart from different types of synthetic heavy proppants and other materials, also 100 mesh sand has been successfully used as settling diverter for plugging of the lower fracture tip in order to limit downwards fracture extension (HODGES & PAOLI 1982; cf. sections 1.4.11.2.1. and 4.8.8.3.2.3.). Similarly as buoyant and settling diverter and propping materials in the hydraulic fracture, floating and sinking ball sealers can be applied for stimulation treatment diversion in the borehole (McLEOD 1984; cf. section 4.2.2.1.1.3.).

4.2.2.4.4. Fluid viscosity and diverter density

Field application examples of upwards and downwards limitation of fracture height growth by light buoyant and heavy settling proppants are reported by NOLTE (1988 b). Bridging of the light buoyant material at the entry of the mudstone barrier and performing a tight flow block is achieved by a mixture of silica flour as well as 100 mesh, 20/40 mesh and 10/20 mesh sand, representing a combination where virtually no permeability is left if it is properly set in place (cf. section 1.4.11.2.1.4.). As subsequent fluid cannot penetrate the flow obstacle, slurry dehydration caused by height growth is eliminated, and vertical fracture extension is also arrested because the plug does not allow the active fluid pressure in the main portion of the crack to reach the height extremities.

Improvement of the process is achieved by low fluid loss and low viscosity of the pad in order to minimize vertical fracture propagation already before the buoyant material forming the flow block is placed. The subsequent plug-carrying stage should have relatively high viscosity with an increased injection rate to enhance crack height growth again for the purpose of ameliorating vertical flow and facilitating placement and bridging of the light flow block material in the entrance of the bounding mudstones. The following proppant-transporting fluids should have minimum viscosity needed to move the particles and should be inserted at lower pumping rates in order to reduce subsequent pressures, with pressure government being required to control fracture width in the flow barrier region without unseating the plug.

The indicated changes in fluid viscosity do not only minimize ultimate fracture height, but also enhance fluid displacement down and along the crack through mobility analogy. The higher viscosity of the plug-carrying stage with respect to the pad provides a favourable mobility condition for displacing the pad down the fracture without breakthrough of flow-blocking material. The lower viscosity of the proppant-transporting fluid relative to the plug stage gives an unfavourable mobility condition for displacing the buoyant diverter down the crack and enhances proppant breakthrough which maximizes the amount of flow block material being displaced vertically into the shales and minimizes the quantity of it moved towards the lateral fracture tip. SETTARI (1985) also concludes that a proppant bank at the fracture bottom or buoyant proppants concentrated at the fracture top can provide effective barriers to vertical crack propagation.

4.2.2.4.5. Other blocking additives

Apart from the concept of buoyant diverters (NGUYEN & LARSON 1983), various other additives and methods have been proposed in order to limit vertical fracture extension and to keep the crack within the zone of interest. HANSON & LYONS (1964) present creation of a semipermeable proppant slug in order to close off the paths of least resistance in the fracture by alternating increasing and decreasing injection rate and/or proppant concentration repeatedly. The blocking concept of HANSON (1964) is based on including plastically deformable solids into the fracturing fluid which have densities equal to or different from those of the carrier. PRATER (1968) suggests simultaneous action of two or more fluids of different densities along with a proppant that matches the density of one fluid, with density choice depending on whether downward or upward fracture propagation is to be controlled.

Downward crack growth can also be restricted by properly placing a heavy proppant into the bottom of a newly created fracture which leads to subsequent diversion of the flow of the fracturing fluid upward and laterally (BRAUNLICH 1967). 100 mesh sand can also be successfully used as bridging agent to control the downward crack propagation (HODGES & PAOLI 1982; cf. sections 1.4.11.2.1. and 4.8.8.3.2.3.). NOLTE (1982) reports achievement of fracture height control by pumping a mixture of various sand grain sizes in a high-viscosity fluid immediately after the prepad before the proppant-laden slurry is introduced into the crack, with the sand mixture being designed to bridge in the vertical tips of the fracture to form a flow block and discourage the crack from growing further vertically. Aspects of artificial barriers for fracture height growth limitation are also discussed by LI & ZHU (1986). DOERLER & PROUVOST (1987) comment on the performance of diverting agents and modelling of resultant zone injectivity.

4.2.2.4.6. Pressure diversion

Another possibility of fracturing treatment diversion comprises exploitation of the pressure growth phenomenon which includes abnormally large increases in operating pressure and is a function of pumping rate and fluid viscosity (MEDLIN & FITCH 1983). Thin stimulation fluids pumped at high rates create smooth pressure curves that rise parabolically with time to first order, whereas thick gels pumped at low rates result in erratic pressure curves which tend to rise more or less linearly with time and show frequently sharp irregularities with abrupt increases and declines. The pressure increase leading in most cases to screenout (cf. section 6.2.4.2.1.) can be used as a method of fluid diversion, because the pressure diversion scheme directs the largest fractions of the treatment to the better reservoir zones.

The pressure diversion technique provides a means of fracturing very long intervals containing many productive zones in a single treatment. After perforation and breakdown of each zone by ball sealers (cf. section 4.2.2.1.1.), fluid is diverted from horizon to horizon during the main stimulation operation in accordance with the natural increases in treating pressure. Initially, crack propagation begins in the zone with the lowest fracture gradient. After some increment of pressure increase in this level, fluid diversion begins to the interval of next highest fracture gradient, with this diverting process continuing until all storeys receive fluid in proportion to their fracture gradients. Under ideal conditions, the job could be continued until final screenout starts at maximum operational pressure and each zone contains the largest practical fracture. Proppant transport problems, however, might prevent this ideal condition from being achieved. Proppant fillup in the wellbore has to be avoided by using a crosslinked gel with near-perfect transport properties (cf. section 4.3.4.3.). Another potential problem in proppant bridging of fractures before their width becomes great enough to allow proppant movement.

4.2.2.5. Multiple-zone fracture migration

AHMED, NEWBERRY & CANNON (1985) introduce the fracture migration concept as a technique for optimization of multiple-zone vertical hydraulic stimulation treatments. After outlining of some general aspects and commenting on continuous survey and modelling, examples of application in reservoir complexes with vertical fracturability differences are illustrated.

4.2.2.5.1. General aspects

Conventional methods of designing stimulation of wells with multiple pay zones include limited entry, staging (with the use of diverting agents and ball sealers) and packing-off storeys (including bridge plugs and sand plugbacks) and are based on the assumption that all horizons open up and the fractures all propagate in a similar fashion (AHMED, SCHATZ, HOLLAND, JONES & GREENFIELD 1982; cf. section 4.2.2.1.). All the previous multiple-zone treatment techniques, however, suffer from lack of capability of quantitatively predicting the created fracture characteristics in each interval, mainly as a consequence of missing technology for prediction of fracture migration (CLIFTON & ABOU-SAYED 1981; CLEARY, KAVVADAS & LAM 1983; AHMED, THOMPSON, KELKAR, VEGHTE & HATHAWAY 1984), and the unavailability of reliable and continuous fracture gradient pressure and mechanical properties data has hitherto hindered the commercial development of a fracture migration predictive tool (cf. also section 6.2.5.).

4.2.2.5.2. Continuous survey and modelling

The multiple-zone fracture migration model (AHMED, NEWBERRY & CANNON 1985) provides a continuous survey along the prospective column. Propagation of the crack commences with the intervals averaging the least fracture gradient pressure, and other horizons preferentially open up and propagate as the wellbore pumping pressure increases. Poststimulation production history surveys often reveal that smaller fracture lengths are typically associated with pay zones exhibiting the highest fracture gradient pressure, leading to inefficient depletion, whereas low-stressed intervals give rise to high crack conductivity which is often more than is required to drain the surrounding reservoir, and treatments initiated in structurally deep low-stressed storeys may easily migrate into a nearby aquifer. Hydraulic fracture propagation from one or several points depends on comparison of fracture gradient pressure and rock mechanical properties between source and surrounding points. As the crack grows, the hydrostatic fluid gradient begins to affect the pressure at every point on the fracture tip which in turn affects fracture propagation.

If pressure potential within the crack tips and stress potential within the surrounding rock can be defined, prediction of fracture propagation with increasing treatment volume and pressure within the crack is possible. Fracture propagation at various multiple storeys can therefore be controlled by maneuvering the pressure at interval face, by selectively positioning the perforations, and by using necessary staging and packoff requirements. The multiple-zone fracturing prognosis technique (AHMED, NEWBERRY & CANNON 1985) combines sonic waveform analysis with bulk volume evaluation and linear fracture mechanics with treatment fluid density in order to yield an accurate fracture gradient pressure distribution in and around the hydrocarbon-bearing intervals and to produce an exact prediction of crack migration behaviour in relation to bottomhole treatment pressure (cf. also section 6.2.5.). Fracture height migrates at the wellbore with increases in treatment pressure above the average fracture gradient pressure of the perforated zone. Fracture pressure gradient is a function primarily of overburden stress gradient, pore pressure gradient, and ratio of horizontal to vertical stress (EATON 1969).

4.2.2.5.3. Vertical fracturability differences

If differences in fracturability exist between several zones, such as for example if the lower and middle sections open up simultaneously whereas the upper level breaks down at a higher pressure and thus lags behind, and the upper layer has upwards fracture migration tendency, the middle interval has strong barriers at both sides, and the lower horizon exhibits downward fracture migration trend, adjustments of perforation arrangement have to be carried out if all the three zones are to be stimulated simultaneously. Increasing perforation size and density in the upper level has the consequence that it undergoes one less net pressure step, and placing perforations on the bottom part of the top layer and on the upper part of the base horizon inhibits fracture growth away from the hydrocarbon-bearing interval.

If sufficient pressure differences cannot be created by placing fewer perforations in the top horizon and incremental perforations in the bottom section, then each portion should be treated separately in order to create sufficiently long fractures in both zones. Excessive fracture pressures have to be avoided, because this may lead to fluid loss into boundary lithologies between the pay sand intervals which can trigger premature screenout failure of the job (cf. section 6.2.4.2.1.). Some comments on fracture orientation and movement as well as in-situ stress differences with facies type are offered as follows.

4.2.2.5.3.1. Fracture orientation and movement

The multiple-zone fracture migration concept (cf. section 6.2.5.) only applies to vertical hydraulic stimulation. Horizontal hydraulic fracturing cannot be modelled, but absolute values of fracture gradient pressures permit to discern if a pay zone is amenable to creation of horizontal fractures. The transversely elastical model used to calculate in-situ stress from elastical properties does not include any local tectonical effects. Although most tectonical effects between several intervals cancel out during computation, scaling by direct measurement through minifracturing treatments is recommended in order to improve the multiple-zone crack migration modelling.

Hydraulic fractures usually move upwards due to increasing overburden stresses with depth, although in many instances, the cracks have shown a preponderance of downwards propagation (cf. also section 4.2.2.3.5.). Fracture height is generally related to competence, thickness and hardness of the barrier and is also influenced by the cement bond (PAI, GARBIS & HALL 1983). BLINTEN & AZIZ (1985) comment on stimulation of very long gross intervals.

4.2.2.5.3.2. In-situ stress differences with facies type

The reason for vertical fracturability differences with some reservoir zones opening up earlier than others and restrictions of connectivity of superimposed pay sections through intervening barrier horizons are differences of in-situ stresses with depositional environment and lithology (NORTHROP 1988; cf. section 4.2.2.3.1.3. and 4.2.3.2.2.). Large contrasts of in-situ stress magnitude between reservoir and barren rocks imply a strong tendency for fracture containment and very high pressures are required to break through non-reservoir lithologies. While the horizontal stresses in mudstones are essentially isotropic and on average slightly below the overburden value, horizontal stresses in sandstones and siltstones are anisotropic and have a fairly consistent orientation reflecting a clockwise rotation of the maximum horizontal stress direction with depth as a consequence of stresses associated with the large topographic relief superimposed on the regional stress field of the basin (CLARK 1983; cf. also section 4.2.3.2.2.1.).

As a result of different in-situ stresses in different lithologies, hydraulic fractures do not easily break out of a low-stress sandstone, traverse the high-stress confining rocks and intersect another sandstone (NORTHROP 1988). Therefore intersection, propping and drainage of a large number of remote and isolated reservoir units such as sandstone lenses and ribbons which are not connected to the wellbore are not feasible in the presence of large stress contrasts, but will only pay off in case of small stress contrasts. GRI (1988) reports dense carbonates alternating with tight gas sandstones, with the carbonates being efficient barriers to vertical fracture growth due to their higher in-situ stress.

4.2.2.6. Selective proppant placement

BARBER & THEMIG (1985) present various possibilities of optimization of the results of fracturing treatments by effectively limiting downward fracture growth into the water-bearing section of the reservoir. Many depositional sand bodies are characterized by fining-upwards grain size distribution within the environmental cyclothems (cf. plates III/1 - 2, IV/5, V/1 - 4, VII/3 and IX/1 - 4) which reflect decreasing energy during course of aggradation of the milieu features (such as for example fluvial channel and overbank facies sequences) that gives rise to decreasing permeability from bottom to top of the sedimentary package. Diagenetical cementation and mineral authigenesis often prefers the finer-grained upper portions of the pay which consequently become quite tight in contrast to the coarser-grained lower parts of the reservoir that are left with better permeabilities due to weaker postdepositional lithification. Subsequent fluid stratification after hydrocarbon immigration frequently results in an oil-water-contact crossing the depositional sand body and separating oil in the upper low-permeability section from water in the lower high-permeability storey of the prospective column.

In case of lack of competent intraformational barriers to prevent downward growth of hydraulic fractures out of the upper tight oil-producing interval into the lower permeable water-producing section, attempts have to be made to design the stimulation treatment for minimization of downwards crack propagation in order to maximize effective stimulation of the upper tight reservoir zone. This can be performed by limiting pumping rates and controlling job size, and by selective proppant placement by density-control foam fracturing. In an unbounded pay, high pumping rates will generally promote fracture height growth, because the friction pressure created within the fracture from the wellbore to the fracture tip will rise with increasing rate (EEKELEN 1982).

Therefore low-rate stimulation operations of restricted size are a significant means of limiting vertical fracture extension, provided the pumping rates are not too low, because this would provoke premature screenout termination of the job (cf. sections 4.2.2.2.3., 4.2.3.2.5. and 6.2.4.2.1.). The use of low-viscosity fluids is also effective in restricting vertical fracture propagation, but at the same time the proppant transport capabilities are severely affected (cf. section 4.2.2.3.3.). Aspects of accidental proppant fallout and banking, controlled placement by fluid density regulation, proppant stratification and zonation, and fluid density control and proppant distribution are discussed as follows.

4.2.2.6.1. Accidental proppant fallout and banking

Insufficient carrier ability of low-viscosity fluids may result in proppant fallout during pumping and before closure takes place, giving rise to proppant banking in the lower part of the fractured segment of the reservoir (BARBER & THEMIG 1985). While accidental proppant banking at the bottom of the crack as a consequence of proppant settling in the stimulation fluid prior to closure of the fracture is one type of selective proppant placement and may be desirable in coarsening-upwards sedimentary sand bodies that are fully oil-bearing in order to prop effectively the lower portion of the hydraulic fracture intersecting the tight lower interval of the reservoir in case of absence of an oil-water-contact within the pay zone, it is detrimental to the result of stimulation treatments in fining-upwards sequences if proppant banking provides preferential support of the fracture walls in the lower water-bearing section of reservoirs with fluid stratification.

Therefore the benefit of proppant banking depends on sedimentological composition and/or diagenetical history of the prospective column, and the depositional environment giving rise to the specific trend of grain-size distribution decides together with fluid immigration and displacement history whether proppant banking in hydraulic fractures will be favourable or detrimental for the result of the stimulation job and the enhancement of oil production (aspects of proppant banking are also discussed in sections 4.3.3.2. and 4.12.2.2.).

4.2.2.6.2. Controlled proppant placement by fluid density regulation

Fluid density control giving rise to selective proppant placement is a viable alternative to more conventional types of hydraulic stimulation if fracture height propagation cannot be sufficiently controlled by pumping rate, job size and fluid viscosity (BARBER & THEMIG 1985). Fluids with certain density differences with respect to specific gravity tend to override or underride a preceding fluid in the fracture depending on whether it is lighter or heavier, respectively (FREDRICKSON & BROADDUS 1976). Fluid viscosities also have an effect in controlling density separation, with the higher the density and viscosity differences between both fluids, the more pronounced the density separation effect (cf. section 4.2.2.3.3. and 4.2.2.6.4.).

Proper viscosity modelling of both successive fluids is a critical part of the operation (BARBER & THEMIG 1985). If the viscosity of the preflush is too low, it will be swept away from the wellbore area and quite possibly remain ahead of the fracturing fluid. The preflush must have enough viscosity to provide some resistance to movement within the cracks to allow the fracturing fluid to override the preflush in order to achieve the desired density effect. If the preflush is too viscous, however, it becomes immobile to some extent, and instead of overriding, viscous fingering (cf. section 4.5.4.2.1.) takes place in which the fracturing fluid moves through the preflush in several paths or channels instead of the desired single continuous fluid front.

4.2.2.6.3. Proppant stratification and zonation

Fluid density control allows selective placement of the proppants in the upper or lower portion of the fracture in the cases of overriding or underriding regardless of where the lower or upper bounds of height growth may be, respectively (cf. section 4.2.2.3.3. and 4.2.2.6.4.). The use of foam in this process provides excellent proppant transport and thus eliminates most of the concerns about proppant settling to premature uncontrolled banks during pumping and before fracture closure. The use of nitrogen in foam gives the density difference with potassium chloride water and would eliminate the need for using weighted saltwater brine solutions as is usually required in performing density control treatments.

BARBER & THEMIG (1985) document significant improvement of fracture stimulations in fluid-stratified fining-upwards sand-body reservoirs by fluid density control, with overriding permitting selective proppant placement in the upper tight oil-bearing portion of the pay zone while the lower segment of the crack plane intersecting the lower permeable water-bearing interval of the prospective column is left unpropped and thus water production can be minimized (cf. also MISAK, ATTEBERRY, VENDITTO & FREDRICKSON 1978 b) by allowing the lower portion of the fracture to reheal once closure stress becomes effective after the end of the hydraulic treatment. Selective proppant placement in the upper reservoir portion by overriding created by pumping of a high-density pad followed by lower-density proppant-bearing fluids allows to inhibit connection to the lower water-bearing section of the prospective interval (MISAK, ATTEBERRY, VENDITTO & FREDRICKSON 1978 b).

4.2.2.6.4. Fluid density control and proppant distribution

In a similar way as applying for layering of proppant-supported and unplugged sections of the reservoir, also horizontal stratification of different proppant types can be achieved by inserting two successive proppant-laden fluids of different viscosity into the fracture, resulting in placement of a proppant of different type and/or grain size in the upper portion of the fracture with respect to the lower segment of the crack, thereby permitting selective optimization of the fracture-formation conductivity contrast in reservoirs with pronounced grain size differences between vertically successive units or storeys (cf. also section 4.2.2.3.3.).

Fluid density control, however, can only give rise to different vertical intervals within the fracture and vertical proppant zoning, whereas placement of horizontally differing proppant types and/or grain sizes resulting in lateral coexistence of proppant pillars (cf. section 4.3.3.2.) of different composition has to be achieved by pumping alternating lots of fluids with more or less similar viscosity to avoid over- or underrunning, and being saturated with or without proppants if creation of discrete proppant pillars separated by open columns is desired, or containing successive lots of proppants of different type and/or grain size in case horizontal zoning of proppants in the fracture with tail and/or head being of different composition than the main body is wanted.

Selective proppant placement, however, has to take into consideration that varying conductivity in the fracture from wellbore to tip can significantly affect production rates (BENNETT, ROSATO & REYNOLDS 1981) either in positive or negative direction depending on geometrical arrangement of type and grain size of proppants and effect of possible void spaces between the individual discrete proppant lots. It is therefore very important to determine proppant distribution schedules in such a way that achievement of the appropriate conductivity profile in the fracture is assured (VEATCH & MOSCHOVIDIS 1986).

4.2.2.7. Special phenomena in coal seams

In coal seam gas reservoirs, high treatment pressures may lead to unconfined vertical fracture growth in case of absence of sufficient stress contrasts in the formations bounding the coal or for thin coal seams (JONES, BELL, MORALES & SCHRAUFNAGEL 1987). Modulus contrasts may play an indirect, but important role in affecting fracture containment. Fracture-tip plugging with coal fines prevents uncontained vertical growth of fractures in coal seams by quickly sealing off of narrow cracks created in high-modulus bounding formations (cf. section 4.4.3.3.).

Coal seams interbedded with sandstones and shales can act as thief zones for fracturing slurry when hydraulic stimulation of a sandstone is attempted, because the minimum horizontal stress within the coal is less than within sandstones and shales (DESPAX, CHARLEZ, CLINKEMAILLE & ECONOMIDES 1987). As coal also has a low elastic modulus, connection of the seam with the perforations would lead to preferential fracturing of the coal rather than the sandstone, thus setting the proppant out of the pay zone and resulting in failure of the stimulation job (cf. section 4.4.3.4.).

Therefore containment analysis in such complex interbedded reservoir sequences has to include in-situ stress measurements by minifractures in all relevant layers comprising sandstone, shale and coal. In order for a fracture to stay contained within the formation of interest or in an effort to minimize the vertical height growth, the excessive pressure within the crack has to be minimized. Excess pressure is almost entirely dependant on friction pressure drop down the fracture which in turn is a function of gel viscosity, injection rate, and friction coefficient between fluid and fracture wall (HOLDITCH, ROBINSON, WHITEHEAD & ELY 1987).

4.2.2.8. Other aspects

Hydraulic proppant fracturing of thin sheet-type oil-bearing sandstones is also reported by KOHLHAAS (1982). KELKAR, THOMPSON, HOLLAND, STRAWN, VEGHTE & HATHAWAY (1985) comment on hydraulic crack height extension control using permeability barriers and determination of geological and engineering parameters for fracture design.

Another important parameter affecting height growth and propagation of hydraulically induced cracks is fracture toughness (THIERCELIN, JEFFREY & BEN NACEUR 1987; WANG & CLIFTON 1989). Influences on fracture toughness are exerted by large-scale heterogeneities, confining pressure and pore pressure. Bounding layers with higher crack toughness than that of the pay interval can act to effectively contain a hydraulic fracture between them, and a pay zone with higher crack toughness than bounding layers can lead to fracture growth in the limiting beds at the expense of propagation in the reservoir (cf. also section 4.3.4.2.3.).

Aspects of fracturing in layered reservoirs where more or less containment of crack propagation is required are also discussed by CAMACHO, RAGHAVAN & REY-

NOLDS (1984); SCHULTE (1984); BENNETT, CAMACHO, REYNOLDS & RAGHAVAN (1985); BEN NACEUR & TOUBOUL (1987) and BOUTECA (1987). ABOU-SAYED, SINHA & CLIFTON (1984) and ABOU-SAYED, CLIFTON, DOUGHERTY & MORALES (1984) comment on the influence of in-situ reservoir conditions on the geometry of hydraulic fractures. A quantitative analysis of factors influencing vertical and lateral crack growth is performed by SETTARI (1985). Possibilities of fracture penetration through an interface are evaluated by BIOT, MEDLIN & MASSE (1983; cf. section 4.2.3.2.4.).

4.2.3. Fracture propagation

While in thick sandstones that are only perforated along parts of their thickness according to the limited-entry technique (LAGRONE & RASMUSSEN 1963, HOWARD & FAST 1970, SMALL 1985, CRAMER 1987) the fracture propagates radially through the whole reservoir thickness to the bounding mudstones and then continues along the sealing horizons in lateral direction, thin sandstones are often perforated along almost the whole stratigraphical interval and the fracture is thus nearly instantaneously propagating in lateral direction along the guiding barrier. Particularly in thin sandstones, however, the danger of uncontrolled vertical crack growth and breakthrough to adjoining beds becomes greater with increasing fracture length depending on differences in stability behaviour and stress status of the superimposed formations (KLOSE & KRÖMER 1983) and also on the higher pressures that have to be applied to create longer fractures beyond a certain boundary length.

When the stress intensity factor at the tip reaches its critical value, the hydraulic fracture becomes unstable and begins to move towards the direction where the stress intensity factor has the highest value (LU & YEN 1985). After referring to terminology and nomenclature, various factors controlling fracture propagation are briefly reviewed as follows. An outline of the two different processes of shear slippage and tensile failure of rocks creating open joints is given, the influence of geological discontinuities is sketched, comments are offered on fracture propagation and pressure evolution, and a field example of required seal thickness for successful containment of a large-scale fracture is reported.

4.2.3.1. Terminology and nomenclature

Concerning terminology and nomenclature, the pressure necessary to propagate a fracture is the minimum crack extension pressure which is required at the tip to keep the fracture open (WHITEHEAD, HUNT & HOLDITCH 1987). During injection of fluid, the pressure at the wellbore is the extension pressure plus any excess pressure in the fracture plus pressure losses due to friction through the perforations. After shut-down of pumping, the instantaneous shut-in pressure at the wellbore is equal to the least principal horizontal stress plus any excess pressure in the fracture. As the excess pressure decreases due to fluid leak-off, the crack will close at the in-situ stress which is the resultant pressure. The most common synonyms for in-situ stress are least principal horizontal stress (cf. section 1.2.1.1.), minimum principal stress, fracture closure pressure, closure stress, least compressive stress and fracture gradient. The resultant pressure can be determined by pump-in/flow-back tests, minifracturing and in-situ stress tests (cf. section 4.8.2.), with all of these tests being able to be performed in both open and perforated cased hole.

Fracture propagation concern is also a matter of treatment scale and reservoir thickness, with vertical fracture growth in a thick formation and a small-size job not being so serious as that occurring in a large-size job in a thin formation (LI & ZHU 1986).

4.2.3.2. Factors controlling fracture propagation

Most fractures are oriented more or less in a vertical plane and propagate outward in opposite directions from a wellbore (VEATCH 1983). The fractures are typically extending either in a radial fashion to form penny-shaped cracks, or propagate predominantly in lateral direction along natural formation interfaces and less frequently in vertical direction across bedding planes. Local stress fields and variations in stresses between adjacent reservoirs have dominant effects in controlling fracture orientation and vertical growth tendency, whereas regional stresses may affect the azimuthal trend of the crack. The principal sources of fracture containment are horizontal stress variations with depth and plasticity of bounding shale layers (MEDLIN & FITCH 1983). The outline as follows comments on in-situ stress contrast and other influencing factors, horizontal stress differences and fracture geometry, reservoir elasticity and plasticity, fracture propagation across weakness planes, premature screenout termination, and stress differences and containment.

4.2.3.2.1. In-situ stress contrast and other influencing factors

Fracture propagation is mainly controlled by in-situ stress contrasts between the individual rock layers (WHITEHEAD, HUNT & HOLDITCH 1987), with commonly reservoir sandstones and barrier mudstones being low-stress and high-stress rocks, respectively (NOLTE 1988 a, 1988 b; cf. also sections 4.2.2.3.1.3. and 4.2.3.2.2.). Accurate prediction of the magnitude of vertical fracture growth is imperative for proper design of fracture length, width and height. Apart from fracture propagation upwards or downwards out of the zone of interest, further complications and drawbacks can be multiple fracture closures, extreme perforation damage and communication with other sets of perforations through the crack or behind the pipe. Fracture containment depends mainly on in-situ stress contrasts between formation and adjoining barriers (SIMONSON, ABOU-SAYED & CLIFTON 1978; EEKELEN 1982, MENG & BROWN 1987), but also on rock stiffness and payzone thickness (BOUTECA 1987). Thus the efficiency of a stress barrier has to be considered together with elastic modulus of the rock and reservoir thickness.

Further parameters affecting vertical fracture propagation are in-situ stress gradient (cf. section 4.2.2.3.1.), elastic or plastic reservoir rock behaviour (cf. section 4.2.3.2.3.), toughness and ductility discontinuities (cf. section 4.2.4.3.), natural fractures (cf. section 4.8.8.), leakoff of fracturing fluids (cf. section 4.3.4.6.) and heat-transfer properties (cf. section 4.3.4.4.), as well as variations of in-situ stresses existing in different layers of the geological column (closure stress profile), relative bed thickness of the formations in the vicinity of the fracture, bonding between the rocks, fluid pressure gradients in the fracture, and pore pressure variations from one reservoir section to the other (VEATCH 1983). When a prospective zone contains excellent barriers to fracture growth and well engineered stimulation treatments are performed, expectation can be made to achieve effective crack lengths that are reasonably close to the designed fracture length (ROBINSON, HOLDITCH & LEE 1983).

Other factors influencing vertical and lateral fracture growth in addition to in-situ conditions are rheological properties and flow rates of the stimulation fluid, proppant placement and temperature profiles. The most effective mechanism of fracture containment results from suitable combination of viscosity and injection rate of the fracturing fluid and staging of heavy proppant (BEN NACEUR & TOUBOUL 1987). Aspects of horizontal stress differences, reservoir elasticity vs. plasticity, fracture propagation across weakness planes, and premature screenout termination are outlined as follows.

4.2.3.2.2. Horizontal stress differences and fracture geometry

Containment of the fracture in vertical direction by the over- and underlying tight beds is only guaranteed if the horizontal stresses below and above the target horizon are still during execution of the job and superimposition by the treatment pressure greater than in the reservoir to be stimulated, with otherwise the possibility of undesired crack propagation in penny shape existing (LEICHT 1985). The bounding mudstones are capable of stopping fracture propagation in the sandstone and to conduct further crack extension along the interface (cf. section 4.2.2.3.5.) between both different lithofacies only if there are also reasonable differences in the elasticity modul (KLOSE & KRÖMER 1983 report for Rotliegend mudstones an abt. seven times greater elasticity modul than in the underlying sandstone that was fractured; cf. section 3.3.2.). Reservoirs bounded by weak barriers which have not enough stress to confine an induced fracture in the pay zone at higher stresses can only be treated by smaller jobs involving lower pressures that are still guaranteing fracture termination at weak or unbonded interfaces at low confining stresses (AHMED, STRAWN, WILSON & SCHATZ 1983; cf. section 4.2.3.2.4.).

In case of suitable pressure levels, hydraulic fractures can grow vertically as much as 20 ft even when very low injection rates and total pumped volumes are used (WHITEHEAD, HUNT & HOLDITCH 1987). Given insufficient stress contrast between the net pay zones and the rock layers above and below (cf. section 4.2.2.3.1.3. and 4.2.3.2.1.) can be a great problem to achieve the goal of creating long conductive cracks due to excessive vertical height growth as a consequence of lacking fracture containment (HOLDITCH, ROBINSON, WHITEHEAD & ELY 1987). The effect of the borehole to the stress intensity factor at the fracture tip is negligible when crack length is larger than the radius of the well which is the case in almost all the hydraulic stimulation treatments (BOWIE 1956, YEW & LI 1987). Some aspects of stress variations with lithology, natural fracture distribution in various rocks, in-situ stress contrasts, and fracture instability and movement are discussed as follows.

4.2.3.2.2.1. Stress variations with lithology

Horizontal stress differences between the individual layers of the geological column are in reservoir sections particularly pronounced between sandstones, carbonates and mudstones (BRITT & LARSEN 1986). As carbonates have higher closure stress than sandstones in comparable reservoir depth, selective fracturing of a carbonate horizon within an interbedded sequence has to include proper shut-off of the sandstones above and below the carbonate, because otherwise once communication from the carbonate to the sandstone would be established, only negligible fracture extension would occur in the carbonate zone, and virtually maximum crack length which could be expected in the carbonate interval would be the length already achieved at the time of communication with the sandstone bed . On the other hand, dense carbonates alternating with tight gas sandstones are effective barriers to vertical fracture growth due to their considerably higher in-situ stress (GRI 1988). Apart from lithology, stress level is partially also depending on depositional environment (WARPINSKI, BRANAGAN & WILMER 1985; cf. also sections 4.2.2.3.1.3. and 4.2.3.2.1.).

Shales and impermeable carbonates have higher horizontal stresses than interbedded permeable sandstones particularly after formation pressure drawdown in the latter (LABUDOVIC 1981). In-situ stress can be determined by both open-hole and cased-hole stress testing. Interpretation difficulties may arise from fracture propagation out of the zone of interest, multiple fracture closure, fracture growth around the packer, extreme perforation damage, communication with other sets of perforations through the fracture or behind pipe, or equipment malfunctions requiring surface rather than downhole shut-ins. YEW & CHIOU (1983) discuss effects of in-situ stresses and layer properties on hydraulic fracture containment, and EEKELEN (1981) evaluates the relationship between horizontal stress and depth in sedimentary basins.

4.2.3.2.2.2. Natural fracture distribution in various rocks

Similarly as applying for hydraulically-induced fractures, size and distribution of natural cracks within the reservoir complex depends on differences of in-situ stresses in the individual lithologies, with main distinction being made between low-stress high-permeability sandstones and high-stress low-permeability mudstones (NORTHROP 1988; cf. section 4.8.8.1.). Fractures intersecting sandstone bodies terminate at mudstone contacts both at inter- and intrareservoir boundaries or lithological discontinuities. Therefore the natural fracture system is heterogeneously distributed through non-communicating layers (BRANAGAN, LEE, CIPOLLA & WILMER 1988).

The density of natural fractures is inversely proportional to thickness, with the thinnest layer being the most conductive due to narrow joint spacing, but on the other hand not containing the bulk of producible gas which is concentrated in thicker beds that are comparatively less conductive as a consequence of wide crack spacing. In the absence of material property differences, the minimum horizontal stress differences between pay zone and adjacent layers control the degree of vertical migration of a hydraulic fracture (BEGNAUD & CLAIBORNE 1985). LERCHE & NARR (1984) estimate subsurface crack density in core and discuss effects of variable fracture spacing, and YEW & LODDE (1983) analyze propagation of hydraulically-induced cracks in a layered medium.

4.2.3.2.2.3. In-situ stress contrasts

Concerning the effect of in-situ stress contrasts on hydraulic fractures (cf. section 4.2.2.3.1.), high-stress regions above and below the cracked zone result in a nearly rectangular-shaped fracture for low-viscosity low-flow-rate tests, whereas higher viscosities and flow rates that induce higher fracture pressures result in greater heights and a less rectangular shape (WARPINSKI 1983 a). The pressure near the tip of the crack is the pore pressure and not the minimum in-situ stress value. BIOT, MASSE & MEDLIN (1986) outline that fracture propagation is in most reservoirs controlled mainly by the elastical energy required to force apart the crack faces and not by surface energy effects at the tip. Stress contrasts in the order of 300 - 400 psi are generally sufficient to restrict fracture growth into the higher stress region (AHMED, STRAWN, WILSON & SCHATZ 1983), whereas underpassing of the minimum critical stress contrast has the effect that the adjoining layers are no absolute barriers to fracture growth (AHMED 1988).

Material property interfaces alone have little effect on containment of hydraulic fractures (cf. section 4.2.3.2.4.), and in-situ stresses have the dominant influence on crack behaviour and direction of fracture growth (TEUFEL & CLARK 1981). Crack containment is most likely to occur as a result of an increase of minimum horizontal compressive stress which takes place at the transition from high shear modulus into low shear modulus rocks. Inhibition of vertical fracture growth is also triggered by weak interfacial shear strength of the different layers. LU & YEW (1985) discuss the significance of stress intensity factor at the fracture tips which depends not only on the relative moduli of the layers but also on the relative position of the cracks to the layer interface as well as their relative size.

4.2.3.2.2.4. Fracture instability and movement

In hydraulic fracturing design, two criteria apply in many cases comprising instability of the crack which begins to move when the stress intensity factor at the fracture tip reaches the critical value which is regarded as a material property of the medium, or migration of the crack into the direction at which the stress intensity factor has the highest value (LU & YEW 1985). Upward or downward movement of the fracture can be decided merely by comparing the relative magnitude of the calculated stress intensity factor at the crack tips. Material interfaces or discontinuities have very small effects on the stress intensity factor when the fracture is situated far from the bounding planes. The relative magnitude of the stress intensity factor at the crack tips can be used to indicate the direction of fracture movement. Fracture pressure gradient is a function of overburden stress gradient, pore pressure gradient, and ratio of horizontal to vertical stress (EATON 1969).

4.2.3.2.3. Reservoir elasticity vs. plasticity

Fracture propagation modelling is unrealistic if reservoir rocks are treated as ideally elastic materials as done by all of the commonly used theories, and only becomes reasonable if plasticity and surface energy of the formations are included in the approach (MEDLIN & MASSE 1986). Fracture operations are frequently carried out in brittle sandstone or limestone intervals that are bounded by soft shales which are likely to be very plastic. As crack propagation in plastic mudstones would require much higher propagating pressures than in other layers, it can be inferred that a fracture initiated in a brittle layer with plastic shale boundaries would tend to be confined there by propagating pressure considerations regardless of other effects. Based on the elasticity concept, surface energy and shear modulus differences could also lead to fracture confinement in a layer between dissimilar materials (BIOT, MEDLIN & MASSE 1983).

Rock failure in hydraulic fracturing occurs at the leading edge of the induced crack (SLUSSER & RIECKMANN 1976). If elastic deformation and brittle failure take place, fracture extension requires essentially no increase in pressure at the crack tips. With increasing rock plasticity, the formation tends to balloon before breaking at the fracture tip, thus requiring higher pressures to extend the crack. The increase in applied pressure causes an increase in fracture width and by material balance considerations a shorter crack length. An increase in injection pressure at a constant rate together with an increase in instantaneous shut-in pressure during the treatment may indicate that the formation is behaving in a plastic manner. Comments are offered as follows on distribution of plasticity zones in the reservoir, plasticity vs. non-linear elasticity, and plasticity effects on hydraulic fracturing.

4.2.3.2.3.1. Distribution of plasticity zones in the reservoir

Zones of plastic behaviour are often heterogeneously distributed in thick reservoir intervals (MEDLIN & MASSE 1986). While many prospective rocks behave like brittle materials even at the highest confining stresses of practical interest, other pay formations are brittle at low confining stress, but plastic to some degree at high confining stresses which are typical of deeper wells. Results of tensile stress/strain investigations show that plasticity effects in shales are likely to be more important in controlling fracture height than other factors including far-field stress (MEDLIN & MASSE 1986). Formation plasticity is of considerable practical interest, as the most important factors controlling fracture containment are plasticity of bounding shale layers and horizontal stress variations with depth (MEDLIN & FITCH 1983; cf. section 4.2.2.3.1.). Ductility in reservoir rocks may also lead to abnormally high treating pressures in large operations. Laboratory experiments are carried out by MEDLIN & MASSE (1984), and other aspects of fracture propagation are contributed by BIOT, MASSE & MEDLIN (1986).

Plasticity depends on the relative magnitudes of surface energy and energy dissipated in plastic deformation of the reservoir rock (MEDLIN & MASSE 1986). Plasticity generally shows up in fracturing records as abnormal treating pressures and results in cracks which are shorter and wider than normal for purely brittle behaviour. As energy spent in plastic deformation ahead of the fracture tip increases in proportion to crack length, plasticity effects which might be negligible in small treatments could become significant in very large or massive hydraulic fracturing treatments. Plastic zones, however, usually do not comprise whole reservoir layers and storeys, but are characterized by non-uniform patchy distribution within the prospective formation. Plasticity must generally be more or less continuously distributed throughout an interval to be fractured to have significant effect on stimulation results, because plastic zones of limited extent within such horizons are expected to have only little effect since fractures simply propagate around them.

Plasticity in shales plays an important role in controlling fracture height. Fracture treatments are frequently carried out in brittle sandstone or carbonate intervals bounded by soft shales which are likely to be very plastic. Fractures initiated in brittle sandstones or carbonates with plastic shale boundaries tend to be confined there by propagating pressure considerations regardless of other effects.

4.2.3.2.3.2. Plasticity vs. non-linear elasticity

Typical reservoir rocks may display a wide variety of behaviour. Many behave like brittle materials even at the highest confining stresses of practical interest, while others are brittle at low confining stress but plastic to some degree at higher confining stresses typical of deeper wells. A few pay zones show sound degree of plasticity even at modest confining stress. Distinction has to be made between plasticity and non-linear elasticity which can be done by hysteresis curves that define the amount of recoverable strain. Hysteresis results show that strain beyond a well-defined yield point usually represents plastic deformation. Gradual bending over the stress-strain curve is generally due to non-linear elastic deformation with recoverable strain. Most of the rock materials showing significant bending of the stress-strain curves display considerable creep at fixed elongation.

Limiting vertical fracture height strictly to the pay zone leads to more efficient and more productive cracks which also often extend deeper into the formation. Both upwards and downwards growth of the fracture can be reasonably controlled in order to channel the crack as designed in the direction of the pay zone (DOWELL SCHLUMBERGER 1986). Shales with fracture gradients greater than 1.0 psi/ft do not behave elastically and processes such as creep and possibly fracturing are the dominant mechanisms controlling the stress state (WARPINSKI, BRANAGAN & WILMER 1985). If sandstones and siltstones having much lower stresses with fracture gradients of 0.8 - 0.9 psi/ft are interbedded, crack containment can be expected due to in-situ stress contrasts.

4.2.3.2.3.3. Plasticity effects on hydraulic fracturing

High degrees of plasticity of formation rocks have three main effects on hydraulic fracturing (SLUSSER & RIECKMANN 1976) comprising creation of wider cracks than predicted by conventional designs based on brittle rock failure, establishment of shorter fracture length, and requiring increasing surface treating pressure at constant injection rate during the stimulation job. Wider cracks are a beneficial result, because they increase fracture conductivity especially in the critical area near the wellbore, particularly by reducing the magnitude of fracture damage by proppant crushing and embedment (cf. section 4.3.3.1.1.). The extra width obtained in more plastic formations can be very important in maintaining long-term production improvements which otherwise could be lost by fracture closure.

On the other hand, shorter fracture length is a detrimental effect of reservoir plasticity. Productivity improvements obtained by fracturing increase generally with propped crack length provided the crack has adequate conductivity. A shorter fracture length for a given size stimulation operation, however, may result in significantly less production improvement and could turn the anticipated successful job into an unexpected economical failure. Increasing treating pressures are also an unfavourable drawback, because higher bottom-hole injection pressures required for fracturing of the more plastic formations will also require correspondingly higher surface injection pressure. As a result, injection rates and pressures for deep tight gas fracturing may be limited by the safety pressure ratings on wellhead and tubular goods.

4.2.3.2.4. Fracture propagation across weakness planes

DANESHY (1974) investigates hydraulic fracture propagation in the presence of weakness planes. Large material flaws are unable to drastically change crack orientation. Even if the fracture does not encircle formation defects and overrides their possible effect, the influence of material flaws is local and insignificant on the results of the stimulation treatment. Distinction has to be made between opening, sliding and tearing mode of fracture propagation (IRWIN 1957). Hydraulic fractures are usually induced by the opening mode, possess a volume and are created by normal stresses and material tensile failure, whereas sliding- and tearing-mode fractures originate by shear stresses and have no volume. Small- and medium-scale open or closed material flaws can be easily crossed by propagating hydraulic fractures, with the advantage of weakness existence being a reduction in effective rock surface energy that consequently decreases the energy necessary for crack extension.

Although the hydraulic fracture cannot directly cross natural joints due to material discontinuity, it can encircle the flaw and thus maintain its continuity if its dimensions are much larger than that of the defect. If the plane of weakness is so large that it cannot be encircled by the hydraulic fracture, the crack may intersect the flaw and maintain its original direction, or the hydraulic fracture may reorient itself and propagate along the weakness plane. Once the hydraulic crack has established its orientation, it is unlikely for it to undergo drastical reorientation, because it takes less fluid pressure to continue the fracture in the same plane than to start a new one in a substantially different direction. In severely naturally fractured reservoirs with very high joint permeabilities, it is difficult to predict hydraulic crack propagation due to the mutual interference of the many flaws, but this relationship does not seriously jeopardize the validity of hydraulic fracture extension modelling, since intensively naturally jointed formations are highly productive without requiring hydraulic stimulation for hydrocarbon offtake enhancement (DANESHY 1974; cf. section 4.8.8.).

AHMED, STRAWN, SCHMIDT, O`SHEA & VEGHTE (1983) comment on stimulating tight sands in the presence of weak stress barriers by modified hydraulic fracturing techniques. Weak interfaces and in-situ stress differences are the most likely factors to contain a propagating hydraulic fracture (WARPINSKI, CLARK, SCHMIDT & HUDDLE 1982; cf. sections 4.2.2.3.1. and 4.2.2.5.3.). While weak interfaces are probably effective only at shallow depths, in-situ stress differences are the predominant control in greater depths (cf. section 4.2.3.2.2.3.). Differences in elastic moduli between successive layers may result in stress differences that either enhance or deteriorate crack growth into the bounding non-reservoir strata. Fracture containment may often occur at material property interfaces because the boundary layer is a higher stress region.

4.2.3.2.5. Premature screenout termination

Premature screenout termination of the hydraulic stimulation job interrupts fracture propagation at an unexpected and not designed point and is generally an unwanted drawback by being an early failure not corresponding to the treatment plan. Screenout originates by fracturing material becoming stuck in the crack by proppant bridging and/or fluid sticking due to excessive carrier viscosity or slurry dehydration. Fracture blockage by proppant bridging takes place if crack width is smaller than at least 2.5 to 3 times the proppant grain diameter (BEN NACEUR 1987; cf. sections 1.4.11.2.1. and 4.8.8.3.2.3.). Fluid sticking originates from development of friction between particles and fracture walls if the proppant slurry is subjected to sufficient fluid loss to allow the suspended proppant grains to come into contact, with the resultant solid mass of clustered proppants becoming immobile and forming an obstacle in the crack. Accelerated slurry dehydration by leakage into natural fractures and fissures is a very serious reason leading to screenout failure (cf. sections 4.2.2.2.3., 4.2.3.5.3. and 6.2.4.2.1.). The outline as follows focusses on aspects of gel degradation, insufficient fracture width, and impact of 100 mesh sand lead-in stages.

4.2.3.2.5.1. Gel degradation

The structure of most fracturing gels tends to degrade and lose its proppant transport and suspension ability to some degree with both increasing temperature and time (POULSEN & LEE 1984). Although some gels require elevated temperatures to attain their maximum degree of crosslinking, these fluids also degrade with sufficient temperature and time (cf. section 4.3.4.). Since fluid temperature and residence time in the fracture both increase with distance from the wellbore, it follows that the closer the fluid is to the crack tip, the less proppant-carrying ability it has. Therefore frequently proppant bridging and blocking takes place at the fracture tip provoking premature or submature screenout termination of the job (cf. also sections 4.2.2.2.3., 4.2.3.5.3. and 6.2.4.2.1.).

4.2.3.2.5.2. Insufficient fracture width

Other causes include insufficient fluid volume and viscosity modelling as well as unsatisfactory fracture opening, with the created width not reaching the designed value. Fractures propagating from low-stress sandstone reservoirs into the adjoining high-stress mudstone boundaries suffer from the fact that the relative crack width is very narrow in the mudstone section and therefore top and bottom screenout can occur. On the other hand, proper fracture treatment planning and operation can implement sophisticated techniques for vertical growth containment and lateral extension limiting by crack blockage with light buoyant proppants at the top or heavy settling proppants at the bottom (cf. section 4.2.2.4.) and tip plugging (cf. section 4.8.4.3.), respectively, representing artificial and controlled screenout interruptions of fracture propagation in a determined direction. Premature accidental screenout termination is also caused by failure to use a properly viscosified base fluid with adequate leak-off and foam stability characteristics especially in naturally fractured reservoirs (SATTLER, RAIBLE, GALL & GILL 1988).

In case of a properly executed stimulation job, unexpected screenout termination gives an indication of the optimum treatment size for the individual reservoir and is therefore a positive information being a realistic indicator of natural pay zone geometry limitations and their impact on hydraulic permeability enhancement. Unfortunately, however, the useful aspect of screenouts is often obliterated by job failure due to poor operational control or application of insufficiently clean fluids, with these drawbacks being the insertion point of major efforts of improving stimulation performance.

4.2.3.2.5.3. Impact of 100 mesh sand lead-in stages

In composed treatments consisting of 100 mesh lead-in stage and 20/40 mesh main lot of proppants, radical increases in pressure slope testify to beginning of bridging in the reduced width portion of the fracture in the lower and upper barriers (BEGNAUD & CLAIBORNE 1985). Bridging is due to reduction in width related to the difference in rock properties and horizontal stresses between pay zone and barriers, with natural joints within the pay zone tending to significantly reduce fracture width within the reservoir due to leakoff. Bridging leads to controlled height growth with crack extension occurring with a concurrent increase in net pressure which is conceptually due to the increase in fracture propagation pressure along the crack face.

When the small spacer pad of crosslinked gel run ahead of the 100 mesh stage leaked off, the 100 mesh stage dehydrates at the fracture tip thereby provoking a screenout (aspects of 100 mesh sand lead-in stages are also discussed in sections 1.4.11.2.1. and 4.8.8.3.2.3.). Premature screenout termination can be avoided by moving the 100 mesh stage further back in the pad. Screenout failure can also be due to opening of secondary fractures (cf. also NOLTE & SMITH 1981; cf. section 6.2.4.2.1.). The effect of alleged bridging diminishes proppant concentration in barrier and reservoir thus severely reducing effective conductivity through the barrier if indeed any might exist (BEGNAUD & CLAIBORNE 1985).

4.2.3.2.6. Stress differences and containment

Fracture propagation under stress containment is characterized by two asymptotic regions (THIERCELIN, BEN NACEUR & LEMANCZYK 1985). At the beginning of propagation, crack shape evolution is close to radial extension and the fracture is not contained. After a given time, good crack containment is achieved by decreasing height growth and in turn increasing length extension, especially when fracture height is almost constant for 50 % of crack length. The tip of the extending fracture is frequently of elliptical shape according to the leading edge concept (SETTARI & CLEARY 1986). Fracture height near the wellbore is often somewhat smaller than that farther away from the borehole, which is related to the stress concentration near the wellbore that is always active near the vertical end of the crack (THIERCELIN, BEN NACEUR & LEMANCZYK 1985).

Hydraulic fractures usually tend to move upwards due to increasing overburden stress with depth that limits downwards crack propagation, although in many instances, fractures have shown a preponderance of downwards growth (LAMBERT, DOLAN & GALLUS 1983; PAI, GARBIS & HALL 1983). Fracture height is generally related to competence, thickness and hardness of the mudstone barrier, and is also influenced by the nature of the cement bond. Some comments are offered as follows on fluid breakthrough vs. injection pressure and proppant bridging vs. barrier stability.

4.2.3.2.6.1. Fluid breakthrough vs. injection pressure

BEGNAUD & CLAIBORNE (1985) report examples of fluid breakthrough from one horizon through the bounding seal into another reservoir storey as a consequence of high injection pressures during fracturing operation. Fluid communication existed during course of the stimulation job, but not after the fracturing job due to lack of proppant support of the portion of the crack running through the separating barrier. The breakthrough can be confirmed by post-fracture gamma-ray height monitoring (cf. section 6.2.1.3.) which reveals that varying amounts of proppants reflecting lesser radiation intensity levels were transported through the barrier into the other pay zone level. The effect of alleged bridging diminishes proppant concentration in barrier and reservoir, thereby seriously reducing effective conductivity through the barrier.

4.2.3.2.6.2. Proppant bridging vs. barrier stability

In terms of proppant bridging vs. barrier stability, some bounding seals are only marginally effective in fracture height containment unless proppant bridging prevents crack propagation pressure from reaching the fracture tip (BEGNAUD & CLAIBORNE 1985). In case of complete barrier breakthrough, proppant bridging might also prevent crack growth into the lower stress pay zone on the opposite side of the boundary seal. Early proppant bridging improves containment and helps to prevent fracture propagation through the barrier into the adjacent reservoir storey. If the crack has not completely grown through the barrier at the time the proppant is introduced, then bridging in the reduced width may prevent fracture propagation pressure from reaching the cracked edge in the boundary seal thus inhibiting extension into the neighbouring prospective interval. Regardless of the effectivity of proppant bridging, it seems that two adjoining pay intervals separated by a barrier can usually not be effectively propped with a fracture initiating from only one of them.

4.2.3.2.6.3. Other aspects

Experimental approaches of fracture propagation modelling are also carried out by TEUFEL & CLARK (1981); HANSON, ANDERSON, SHAFFER & THORSON (1982); WARPINSKI, CLARK, SCHMIDT & HUDDLE (1982); AHMED, WILSON & STRAWN (1983); PAPADOPOULOS, NARENDRAN & CLEARY (1983) and MEDLIN & MASSE (1984). Fracture height or width depending on length propagation models are presented, compared and/or discussed by PALMER & CARROLL (1982, 1983); CLEARY, KECK & MEAR (1983); ABOU-SAYED, CLIFTON, DOUGHERTY & MORALES (1984); PALMER & LUISKUTTY (1985) and CROCKETT, OKUSU & CLEARY (1986). Aspects of fracture propagation are also discussed by BARENBLATT (1962), DUNLAP (1963), HOWARD & FAST (1970 b); ABE, MURA & KEER (1976); DANESHY (1976), CLEARY (1978), TRAVIS (1981); CLEARY, NARENDRAN, SETTARI & WONG (1982); BARREE (1983), SETTARI (1985); THIERCELIN, BEN NACEUR & LEMANCZYK (1985) and EL RABAA (1987). RODRIGUEZ, HORNE & CINCO-LEY (1984) report partially penetrating fractures.

4.2.3.3. Shear slippage and tensile failure

Fracture propagation and opening of joints is controlled by two different rock mechanical processes comprising shear slippage and tensile failure (MURPHY & FEHLER 1986; cf. also sections 4.2.3.4.4. and 6.2.2.3.). Interactive coupling of fluid dynamics with rock stresses and deformations reveals two types of stimulation patterns. In the first case, low frictional resistance to shear slippage and large maximum dilatancy to shear slippage gives rise to creation of only a single joint which is in agreement with the conventional hydraulic fracturing theory that a single crack opens at a pressure equal to the minimum earth stress (DANESHY 1973 a). In the second case corresponding to high shear resistance or small dilatancy, however, multiply joint stimulation occurs. Shear slippage along the cracks is accompanied by shear-stress drops, and the interaction of these stress drops with the acting earth stresses results in opening of joints more perpendicular to the maximum stress, thus forming a dendritic or branched pattern (MURPHY & FEHLER 1986; cf. section 4.8.7.3.). Comments on joint deformation principles and influence of fluid viscosity are offered as follows.

4.2.3.3.1. Joint deformation principles

Joint opening or dilation behaviour for shear slippage and separation by tensile failure is quite different. With rising fluid pressure, dilation is first small and simply results from decreasing effective closure stress. At higher pressures, shear slippage of joint surfaces occurs until one large roughness asperity lies atop another and further slippage would allow the largest asperity to slide over and down the other, with this being a natural limit to shear dilation which is maximally in the order of a fraction of a millimeter (BARTON, BANDIS & BAKHTAR 1985). Further increasing joint pressure leads to separation of the crack walls, with dilation being typically in the range of tens of millimeters (PERKINS & KERN 1961, DANESHY 1973 a), thereby being many times that of shear dilation.

In an actual hydraulic fracturing operation, the entire spectrum of joint deformation comprising shear slippage and tensile separation occurs (MURPHY & FEHLER 1986). Near the injection well where the flow passage area is limited, fluid velocities and pressure gradients are large and tensile failure of the rock gives rise to crack separation. Near the tips of the joints far away from the injection well, however, pressures and velocities of the fluid are much reduced, and shear slippage takes place.

4.2.3.3.2. Influence of fluid viscosity

Distinction has to be made between low- and high-viscosity fluids. High-viscosity fluids are typically used in hydrocarbon reservoir stimulation and are injected at high rates. Therefore joint separation is dominant, and if only a few natural cracks are present as frequently happens to be the case in petroleum-bearing formations, actual fracturing with creation of new joint planes by tensile failure of intact rock occurs. Geothermal hot dry rock reservoirs, however, are often intensely naturally jointed, and high downhole temperatures render most of the viscosifying agents of fracturing fluids useless. Thus water is used in geothermal fracturing and this low-viscosity fluid gives rise to dominating shear slippage by moving and separation of already pre-existing cracks without opening of new fractures by actual rupturing of the rock (cf. section 4.7.6.).

4.2.3.4. Influence of geological discontinuities

The widely held assumption that the hydraulic fractures are ideal planar features of usually constant height is probably untenable in many reservoirs because of geological discontinuities (WARPINSKI & TEUFEL 1984). Geological discontinuities such as joints (cf. plates I/1 - 6, IV/4 - 8, VII/3 - 4, VIII/1 - 2 and XI/4 - 6), faults (cf. plate XI/3), bedding planes (cf. plates II, III/1 - 2, IV, VII, VIII and IX) and stress contrasts are ubiquitous features whose effect on propagation of hydraulic fractures depends on ancillary treatment and such reservoir parameters as operating pressure, in-situ stresses, orientation of discontinuities, and permeability.

Geological discontinuities influence the overall geometry and effectiveness of the hydraulic fracture by stopping vertical propagation due to containing action of in-situ stress distribution, material property variations, bounding shale plasticity, and cohesionless or unbonded interfaces (cf. also ANDERSON 1981); arresting lateral propagation across faults (cf. plate XI/3) and sand lens boundaries (cf. plates II/7 - 8, VII/1, VIII/7 and IX/8) where stresses may increase, reducing total length by fluid leakoff, and facilitating origin of multiple parallel-fracture systems, hindering proppant transport and placement because of non-planarity of the singular fracture or multiple crack system, and inducing additional height growth from higher treating pressures. The result may range from negligible to catastrophic depending on the values of the ancillary parameters. The most important aspects are fracture branching, impact of faults and joints, interaction between hydraulic fractures and geological discontinuities, shear slippage vs. tensile failure, and consequences for fracture morphology which are shortly illustrated as follows.

4.2.3.4.1. Fracture branching

Physical excavation of rocks in order to observe the fracture directly by mineback (TEUFEL & WARPINSKI 1983, WARPINSKI & TEUFEL 1984) documents that the in-situ stress distribution has the overriding influence on hydraulic fracture propagation and containment (cf. also sections 4..2.2.3.1. and 4.2.3.2.2.). Fracture containment from a high-stress region is a first-order effect, whereas interfaces, modulus or strength changes, fluid pressure gradients, and most bedding planes have only second-order effects on height growth. After a short characterization of fracture stranding and meandering, some factors influencing crack branching are outlined.

4.2.3.4.1.1. Fracture stranding and meandering

The hydraulic fractures in the mineback sections are frequently observed to diverge considerably from the usual picture of a planar feature, with multiple stranding, fracture meandering (cf. plate XI/3), sharp turns and corners, and large-scale roughness and irregular morphology of fracture surfaces being common occurrences. In the vicinity of geological discontinuities, there is frequently nothing that can be called a planar fracture, and the crack complex can be better described as a zone of multiple fracturing which in extreme cases attains 15 - 30 ft (5 - 10 m) width. Fracture branching therefore is at least in some reservoirs a common phenomenon. Mineback visualization of natural and hydraulically-induced fractures is also carried out by LAYNE & BYRER (1986).

Pressure growth comprising abnormally large increases in treating pressure leads to premature screenout termination (cf. sections 4.2.2.2.3., 4.2.3.2.5. and 6.2.4.2.1.) and therefore has adverse influences on fracture effectiveness (MEDLIN & FITCH 1983). The most important reason for pressure growth is fracture branching at reservoir heterogeneities or natural cracks, with the most prominent response to pressure growth being a large increase in effective fracture width. Fracture branching is considerably altering the normal mechanics of crack propagation. Fracture growth no longer consists of a simple opening of the crack faces against the minimum earth stress, but slip motion is introduced along many of the branch faces which generates friction that increases the energy needed to sustain crack growth.

4.2.3.4.1.2. Factors influencing crack branching

As the additional energy must be supplied by a rise in fluid pressure within the branched crack network, fracture branching leads to a natural increase in treating pressure. Once the initial singular hydraulic fracture has branched repeatedly to form a large network of small fractures, at some distance from the wellbore the density of branched cracks is large enough to approach a more or less homogeneous system when viewed on a larger scale. Under such conditions, fracture extension virtually ceases because of the excessive friction associated with the many branch crack faces, with continued fluid injection then resulting in ballooning of the main fracture with a severe pressure increase.

Fracture branching is a function of pumping rate, with low injection rates minimizing or even totally suppressing crack splitting. Extremely high propulsion rates such as occurring during explosive fracturing (cf. section 4.7.3.) on the other hand give rise to immediate propagation of a complex network of branched small cracks and may completely inhibit formation of a singular main fracture stem. Pressure growth produces only little vertical fracture extension and indicates that crack height seems to be controlled more by perforated interval length than by pressure behaviour. Fractures generally appear to remain well contained within perforated intervals in spite of large pressure increases.

4.2.3.4.2. Impact of faults and joints

In naturally fractured regions, multiple stranding of the cracks is predominant, with singular cracks only seldom having been observed. Fractures are usually offset at joints where sometimes two or three fracture branches are initiated when crossing the natural cracks and becoming offset. Some of the fracture strands die out in a short distance, while others persist for long distances and divide and coalesce many times along the fracture length. Evolution, scale and total effect of the multiple fracture zone depends to a large degree on the ancillary parameters.

Faults also exhibit a significant influence on fracture geometry and morphology. Fractures often terminate near faults, and in case they propagate across the fault, their orientations frequenty change, which can be invariably attributed to a change in stresses across the fault and not to the fault plane itself. Concerning the impact of bedding or parting planes, two effects act in concert. There is sufficient friction on the parting plane to allow the fracture to re-initiate on the other side, but not enough transmitted stress to allow it to propagate easily in the higher stress layer above. Thus the two factors of bedding plane and stress contrast together provide an excellent containment feature.

Although geological discontinuities may considerably alter fracture geometry, their effect on proppant transport may even be more severe. Forcing sand through such features as narrow multistrand fractures or offsets will be much more difficult than through a wider singular hydraulic fracture. The consequence will certainly be higher fracturing pressures and quite often screenout can occur. When local small screenouts start to take place in some branches of the crack system, the increasing frequency of these events can ultimately provoke a complete screenout of the hydraulic fracture complex.

4.2.3.4.3. Interaction between hydraulic fractures and geological discontinuities

Three types of interactions between hydraulic fractures and pre-existing joints as a representative of geological discontinuities have been recognized in laboratory experiments. The hydraulic fractures either cross the joint, are arrested by opening and dilating the joint as indicated by fluid flow along the crack, or are stopped by shear slippage of the joint with no dilation and fluid flow along the crack. The different types of interactions are a function of the horizontal differential stress and the angle of approach of the hydraulic fracture to the pre-existing joint. The hydraulic fractures cross the pre-existing joints only at high horizontal differential stress and at angles of approach of 60 degrees or greater, do not cross the joints at low horizontal differential stress and low angles of approach primarily because the fluid pressure in the hydraulic fracture is sufficient to open and divert fluid along the joints, and are arrested by shear slippage of the pre-existing joint with no opening and diverting of fluid along the joint only at high differential stress and low angles of approach (cf. also BLANTON 1982).

4.2.3.4.4. Shear slippage vs. tensile failure

A complete analysis of the influence of geological discontinuities on fracture propagation is difficult because of a hierarchy of mechanisms available to perturb the fracture (WARPINSKI & TEUFEL 1984). Considering only leakoff, the initial loss mechanism is a result of permeability of the joints, faults or bedding planes. With increasing fluid pressure in the natural cracks, shear slippage may occur, which may increase permeability and give rise to a self-enhancing process, and when fluid pressure becomes great enough to dilate the joints, the result will probably be catastrophic leakoff. The situation for multiple zones of fracturing is even more complex. Good joint permeability may lead to sufficient fluid and pressure in the joint to allow flaws to activate and form a multistrand system. In addition, shear slippage - if occurring - may enhance the permeability, allow more fluid to enter the subsidiary strands and trigger the origin of a wider interval of fracturing.

If the pressure becomes large enough to dilate the joints, an extremely wide zone of fracturing is possible, and additionally these effects may occur as the fracture intersects the discontinuity, or at much later times as fluid leaks off into the joint system. Thus the actual results strongly depend on the ancillary parameters of crack permeability, frictional properties, in-situ stresses, joint spacing and orientation with respect to the stress field, treatment pressure and fluid viscosity. In all cases where the fracture tip is blunted by a joint, the natural crack will dilate because the treatment pressure is greater than the maximum stress, with this probably being the main reason for the origin of a zone of multiple fracturing.

4.2.3.4.5. Consequences for fracture morphology

Geological discontinuities can therefore be the cause of many adverse effects during a fracture treatment. When the differences between in-situ stresses are small, geological discontinuities that are favourably oriented to blunt the crack will also probably be dilated, giving rise to a wide zone of fracturing with many parallel strands propagating side by side. When the differences between in-situ stresses are large, joint dilation is not expected, but shear slippage over an extensive area is possible, leading to enhanced leakoff, which in addition may also be sufficient to cause some further secondary fracturing by activating flaw sites along the crack.

The presence of hydraulic fractures stabilizes the joints in such a way that once a fracture crosses a joint, dilation or shear slippage becomes more difficult even though the increased pore pressure resulting from leakoff may be quite large. Thus fracture branching is probably a crack-tip phenomenon, and joints along the crack length are expected to start opening up when some threshold pressure is reached after the fracture has passed. These effects more or less apply for both vertical crack systems and horizontal planes of weakness.

4.2.3.5. Fracture propagation and pressure evolution

Fracture pressure monitoring and interpretation is a valuable means for determination of crack propagation in predominantly qualitative manner (NOLTE 1988 a, 1988 b). The most important tool is the log-log plot of net pressure vs. time or preferably vs. injected volume if pumping rate varies. Restricted height growth and unlimited lateral propagation are reflected by a straight line with a small positive slope in this crossplot. Nearly constant pressure at constant injection conditions testifies to accelerated fluid loss to opening natural fissures or stable height growth at moderate rates. Increasing pressure on approximately unit slope reflects restricted fracture extension, with the limitation of crack penetration being caused by proppant bridging or excessive fluid viscosity due to slurry dehydration, with both effects leading ultimately to premature screenout termination of the operation (cf. sections 4.2.2.2.3., 4.2.3.2.5. and 6.2.4.2.1.). Decreasing pressure without changing injection conditions indicates unstable height growth of runaway nature resulting from vertical fracture extension into another low-stress zone after crossing of the high-stress barrier interval.

Fracture pressure modelling is particularly significant for height growth analysis through formations with variable horizontal stress in different storeys. Horizontal stresses in mudstones are generally higher than in sandstones.

As fracture stiffness is inversely proportional to initial zone height (which in ideal cases represents gross thickness of the reservoir interval to be fractured) and net pressure is also approximately dependent in the same manner, net pressure is higher in smaller horizons than that in larger sections. The resultant height of the smaller interval is generally greater than that of the larger section, whereas the average width of the smaller zone is usually less than that of the larger segment. As a consequence of negligible fluid loss in bounding mudstones, fracture penetration for smaller sandstone intervals is principally greater than that for longer reservoir zones with the same volume of injected fluid.

4.2.3.6. Field experience of required seal thickness

In terms of necessary or required minimum seal thickness to achieve safe fracture containment, in the MHF operation in the Rotliegend tight gas sandstones in the well Söhlingen Z 4 where abt. 550 t of bauxite proppants were injected into a fracture with abt. 115 m (380 ft) height and abt. 250 m (800 ft) length, the 10 m (30 ft.) thick bounding mudstone above and the 40 m (130 ft) thick fanglomerate below the fractured sandstone, respectively, have proven to have been sufficient for containing and longitudinally conducting even such a large-scale hydraulic fracture (KLOSE & KRÖMER 1983, BLEAKLEY 1984; cf. also sections 2.4.1.1. and 3.3.2.).

4.2.4. Reservoir heterogeneity and anisotropy

Fracture propagation is generally disturbed by lateral and vertical reservoir heterogeneity and anisotropy. While internal discontinuities of sandstone complexes such as boundaries between individual separated or amalgamated sandstone bodies like channels (cf. plates I/7, V/3 - 4, VII/1 and VIII/6), lenses (cf. plates II/7 - 8) and bars (cf. plate III/1 - 2), onlap (cf. plate I/1 - 6) and offlap features, erosional contacts and unconformities may be of only minor significance for the extension of large fractures in thick sandstone series, they are quite seriously affecting fluid flow (WEBER 1982) as well as fracture propagation in thin sheet sandstone layers. Reservoir geometry modelling therefore has to evaluate prior to the treatment whether the target sheet sandstone is more or less homogeneous and isotropic, or whether it is only an apparent sheet body consisting of more or less distinct elements that are only overlapping or merging to form a semicontinuous drape (cf. MADER & YARDLEY 1985).

In order to improve hydraulic fracturing technology and treatment efficiency, it is essential to better understand the operation processes and thus get enhanced knowledge of geometry and facies distribution of the reservoir when hydraulic fracturing is carried out (HOLDITCH, ROBINSON & WHITEHEAD 1987; HOLDITCH, ROBINSON, WHITEHEAD & ELY 1987). Various sedimentological, petrophysical and rock mechanical aspects are discussed as follows, and numerous depositional and tectonical features are illustrated on plates I - XI (cf. section 3.13.).

4.2.4.1. Sedimentological aspects

Depending on how pronounced the interunit boundaries are developed, the fractures may be contained within individual depositional sequences or sandstone bodies and propagation across the boundary may be only possible in case of application of higher pressures and pumping rates which automatically increases the risk of perforating also the sealing mudstone horizon. Intrasheet heterogeneities (cf. plates III/1 - 2, IV/4 - 8), V/1 - 2, VII/3 - 6 and IX/1 - 8) are also important for the communication between several wells in terms of water flooding, chemical injection or steam drive, and in case of clear separation of isolated sandstone bodies (cf. plates II/7 - 8, IV/4 - 5, V/4 and VII/1), probably interbar or transchannel communication fracturing (cf. also section

4.10.) is a suitable solution, if technically possible under the given circumstances.

The presence of mud layers with significant parameter and stress contrast may cause problems such as screenout conditions as the pad volume is spent for fracture initiation in weak sections and subdivision of the pay interval concerning long-term post-fracture reservoir performance prediction as a consequence of differences in closure characteristics between sandstones and mudstones (JOHN 1983). Aspects of inter- and intra-sand body geometry as well as matrix and fracture anisotropy are briefly sketched as follows.

4.2.4.1.1. Inter- and intra-sand body geometry

Many sedimentological mechanisms and features of sand bodies (cf. MADER 1985 a, 1985 b) have not only more or less pronounced influences on fluid flow in reservoirs (WEBER 1982), but have also significant impact on fracture propagation. Important sandstone body aspects for optimum fracture planning respecting possibilities of deflection of smaller cracks in thinner sandstones by internal anisotropies are primary-depositional distribution and secondary-erosional reworking and preservation of sand as well as separation and comparison of sheet-type and non-sheet type sands such as lenticular (cf. plate II/7 - 8) and channelized (cf. plates I/7, V/3 - 4, VII/1 and VIII/6) units in terms of inter- and intra-sand body geometry. Sheet sands are characterized by a continuous distribution in constant or variable thickness across larger areas, whereas non-sheet sands are limited to smaller patches with more rapidly changing thickness. Some aspects of sand body coupling and separation, reservoir heterogeneity and treatment size, and role of mud drapes are illustrated as follows.

4.2.4.1.1.1. Sand body coupling and separation

The most significant features of inter-sand body geometry fabrics are vertical amalgamation to multistorey superimposed complexes (cf. plates I/7, II/1 - 5, III/1, IV/7 - 8, V/1 - 5, VII/1 - 2 and XI/1 - 2), horizontal overlapping to multiphase drapes, and vertical and/or horizontal separation of isolated sand bodies by muddy interventions (cf. plates III/2, IV/4 - 5, VII/3 - 6, VIII/8 and IX/1 - 8) and finer-grained sandy intercalations. Important characteristics of intra-sand body organization are uniphase cyclical composition (cf. plates III/1 - 2, IV/4 - 5 and VI/5 - 8), multiphase microcyclical structure and interbedding of dividing mudstone layers and lenses (cf. plates IV/8, V/1 - 4, VII/5 - 6 and IX/1 - 8).

While barriers for reservoir communication and fracture propagation in sheet sands almost exclusively derive from intra-sand body aspects, considerable control of reservoir distribution and thus fracture effectivity in non-sheet sands is exerted by inter-sand body features which are accentuated by the intra-sand body geometry. On a larger scale, many sedimentary piles are heterogeneous due to the interformational structure comprising ideally anticlines with primary-condensed or secondary-faulted sequences, and intraformational composition of a suite of sand bodies with vertical separation and lateral isolation by limited extension.

4.2.4.1.1.2. Reservoir heterogeneity and treatment size

While such elements of reservoir heterogeneity as stated above may be only of subordinate significance for the MHF stimulation (cf. section 1.1.1.) of thick sandstones where intra- and inter-sand body barriers of fluid flow are crossed by fracturing due to the high pressures applied, the aforementioned elements of pay anisotropy could be decisive for success or failure of mHF treatments (cf. section 4.8.1.) where pressures and volumes are much lower and thus the propagating fracture could be deflected, split into several branches or even already be completely stopped by minor lithofacies features as enumerated above.

4.2.4.1.1.3. Role of mud drapes

Point-bar sand bodies originating in meandering-channel river systems probably represent the reservoir type with strongest internal heterogeneity in terms of sedimentary structures and petrophysical properties. Comparative evaluation of meandering-channel point bars and braided-river mid-channel bars (XUE 1986) reveals a much lower oil displacement efficiency in the former. Semicommunication in the point-bar sand body is mainly due to mud drapes on lateral accretion surfaces which divide the reservoir into oblique slice-shaped subunits. The development of chute bars on top of point bars provides further complication. The significance of intra- and inter-sandbody mud drapes (cf. plates III/1 - 2, IV/4 - 8, VII/5 - 6 and IX/1 - 8) in gas reservoirs is also emphasized by HAAK & ELEWAUT (1988). The occasional presence of mud drapes on top of sands causes disturbances in vertical flow paths, with the severity of disturbance depending on the number of mud drapes and their horizontal extent which is expressed in terms of tortuosity. Aspects of small-scale horizontal and vertical flow barriers and heterogeneities in reservoir rocks on the scale of mud drapes are also discussed by WEBER, PUSCH & MÜLLER (1987) and SILVA, NIKO, BERGH & SANCEVIC (1988).

4.2.4.1.2. Matrix and fracture anisotropy

Aspects of stimulation and effective permeability distribution in heterogeneous reservoirs are also discussed by OTERO, SAEZ & RUSINEK (1987). The significance of reservoir anisotropy for the evaluation of natural fracture systems and the performance of stimulation in naturally cracked reservoirs is summarized by NELSON (1986), and further points of natural fracture system anisotropy are illustrated by BRANAGAN, CIPOLLA, LEE & YAN (1987). The discussion as follows presents aspects of blanket vs. lenticular sand bodies, interference and pulse testing, tracer analysis, and impact of the natural fracture system.

4.2.4.1.2.1. Blanket vs. lenticular sand bodies

HODGES & KNUTSON (1981) comment on channel continuity and directivity and associated flow anisotropy effects in tight gas sands. Continuous blanket sands provide a more extensive drainage area than narrow lens-shaped sand bodies, and in lenticular sands (cf. plates II/7 - 8), the effectiveness of fracture treatments is related to shape and size of the sand body, fracture orientation and wellbore position in the sand body (cf. also GAUTIER & RICE 1981). MEEHAN, HORNE & AZIZ (1988) outline effects of reservoir heterogeneity and fracture azimuth on optimization of fracture length and well spacing. In complicated heterogeneous non-continuous reservoirs, the combination of widely varying permeabilities and different lens sizes seem to control well response partially even more than stimulation treatments (CHARLES, HUDOCK, UDICK & HALL 1983).

Lenticular non-continuous or semi-communicated formations are characterized by EVANS & CARROLL (1981), PETERSON & KOHOUT (1983), LORENZ (1985); SATTLER, RAIBLE & GALL (1985); XUE (1986); BRANAGAN, CIPOLLA, LEE & YAN (1987), EVANS & UBANI (1987) and UBANI & RAY (1988). Blanket-geometry tight gas sands and their lateral variability are reviewed by FINLEY & O`SHEA (1983), LORENZ (1983), FINLEY (1986) and GELLAND, TAYLOR, TESSMAN & LARNER (1986). TYLER & FINLEY (1988) comment on reservoir architecture in heterogeneous pay complexes. MARTIN, EVANS & RAPER (1988) document prediction and modelling of sand-body connectivity. ANDERSON (1982) utilizes repeat formation tester and dipmeter for definition of reservoir heterogeneity and anisotropy.

4.2.4.1.2.2. Interference and pulse testing

BENSON & LAI (1986) analyze interference data in a highly heterogeneous naturally fractured geothermal reservoir. WILSON (1956) presents some aspects of sand-oil fracturing in long heterogeneous sand sections. BENNETT, ROSATO & REYNOLDS (1981) and DUSSAN & SHARMA (1987) discuss various points and features of fracture heterogeneity and anisotropy. Improving performance prediction by more accurate heterogeneity assessment is illuminated by JENSEN & CURRIE (1988), and the influence of critical heterogeneities on production performance of different reservoir units is illustrated by HONARPOUR, TOMUTSA, CARROLL, SZPAKIEWICZ, SHARMA & TILLMAN (1988). Horizontal and vertical pulse and interference testing in multilayered reservoirs with special emphasis on the role of non-communicating layers is investigated by BANDYOPADHYAY (1981), CHU & RAGHAVAN (1981); HUTFILZ, COCKERHAM & McINTOSH (1981); DAKE (1982), CHEN & LESCARBOURA (1983), PRATS (1983), OGBE & BRIGHAM (1984 a, 1984 b), SMITH & BROWN (1984), ROEST (1986), WIJESINGHE & KECECIOGLU (1986 b); BRANAGAN, LEE & WILMER (1988); GRABOWSKI, LOCKE, SARADJI, FEILER & YOST (1988); OPDAL & KOSSACK (1988) and SAMPSON & MARCOU (1988). NAJURIETA, ROBLES & EDWARDS (1986) present interference testing for prediction of early water breakthrough in naturally fractured reservoirs, and SANDAL & RAMEY (1978) comment on interference testing with wellbore storage and skin effect at the production well. SWIFT & BROWN (1976) review interference testing for reservoir definition. Other aspects of interference testing in naturally and hydraulically fractured wells are discussed by LESCARBOURA, WALTHER & WILSON (1975) and RAMEY (1975 b).

KAMAL (1982) and KAMAL & HEGEMAN (1988) review multiple-well interference and pulse testing in homogeneous isotropic reservoirs, homogeneous reservoirs with vertically fractured wells, naturally fractured and multilayered reservoirs with double-porosity behaviour, and heterogeneous reservoirs. STEWART, GUPTA & WESTAWAY (1984) interpret interference tests in reservoirs with sealing and partially communicating faults. Lacking interference may be due to water block effects around the observation wells or it may indicate that there is no or poor connection between the natural fractures intersecting individual wells either because of shale breaks or limited nature of fracture interconnectedness (WARPINSKI, BRANAGAN, SATTLER, LORENZ, NORTHROP, MANN & FROHNE 1985).

4.2.4.1.2.3. Tracer analysis

ROBINSON, TESTER & BROWN (1984) discuss reservoir analysis and sizing by inert and chemically reacting tracers. WAGNER (1977) reviews tracer application for diagnosing interwell reservoir heterogeneities. D`HOOGE, SHEELY & WILLIAMS (1987) comment on interwell tracer analysis of multipay discontinuous reservoirs with up to four tracers for monitoring of differential fluid movement, and other aspects of tracer and pressure test analysis for characterization of areally heterogeneous reservoirs are covered by MISHRA, BRIGHAM & ORR (1988). McLENNAN, ROEGIERS & MARX (1983) report points concerning interaction between geological conditions, treatment characteristics and production. TARIQ, ICHARA & AYESTARAN (1985) and ICHARA & CANNON (1987) investigate the effect of perforation length, density and phasing angle variations on the performance of heterogeneous reservoirs that are characterized by shale-lamination and cross-bedding. LIGHT (1983) outlines the effect of lateral anisotropy on flood pattern dimensions and orientation. Other aspects of tracer testing in heterogeneous formations are discussed by SEETHARAM & DEANS (1986).

4.2.4.1.2.4. Impact of the natural fracture system

GRINGARTEN (1988) analyzes well tests in fissured and multilayered reservoirs with high permeability contrasts between individual beds. JALALI-YAZDI &

ERSHAGHI (1987) comment on heterogeneous naturally fractured reservoirs. BLASINGAME & LEE (1986) report on some homogeneous reservoirs. Heterogeneity aspects of multi-layered reservoirs are discussed by EHLIG-ECONOMIDES & JOSEPH (1985); TANG, LIU & JIANG (1986) and HATZIGNATIOU, OGBE, DEHGHANI & ECONOMIDES (1987). CLEARY (1978 b) gives an outline of primary factors governing hydraulic fractures in heterogeneous stratified formations. SMITH, SAWYER & ESPOSITO (1981) explain the effect of anisotropy and vertical fracture penetration on production decline curves for low-permeability gas reservoirs. Geological and petrophysical factors controlling excess pressures in isolated sand lenses within shales that are characterized by limited reservoir connection are evaluated by OZKAYA (1987).

NORTHROP (1988) provides a detailed characterization of a heterogeneous and anisotropic naturally fractured reservoir that is subjected to hydraulic fracturing stimulation. A synergistic approach including various techniques is necessary to accurately describe the pay section geometry. The depositional environment controls reservoir morphology, size, internal structure and natural fracturing. While in marine blanket sandstones the azimuth of the fracture plane has no influence, the angle between fracture plane and channel direction is very important in fluvial lenticular or ribbon sandstones and can result in well over an order of magnitude difference in the area of intersected pay (cf. also section 4.10.2.).

4.2.4.1.2.5. Other aspects

McELROY (1982) comments on the effect of lateral anisotropy on flood pattern dimensions and orientation. ABBASZADEH-DEHGHANI & BRIGHAM (1984) employ well-to-well tracer flow for determination of pay heterogeneity. ZUBER, LEE & GATENS (1987) document permeability anisotropy in Devonian shale gas reservoirs which is created by the natural fracture system, with this type of prospective interval heterogeneity causing hydraulic or explosive fractures (cf. section 4.4.4.) running perpendicularly to the natural fracture trend being more efficient than those being parallel to it. Reservoir heterogeneity by natural fracture anisotropy is significantly reflected by the absence of pressure interference between closely-spaced neighbouring wells (LORENZ & FINLEY 1987).

In terms of matrix heterogeneity, the success of fracturing treatment depends on distinction of the major reservoir rock types and selective investigation of fluid compatibility in order to eliminate failure causes due to excessive fluid sensitivity (SIMON, KAUL & CULBERTSON 1977). The degree of fluid sensitivity is generally related to rock type, permeability and porosity, with the higher-permeability cleaner sandstones being relatively insensitive to the treatment fluid used. Naturally fractured reservoirs can be regarded as double-porosity media consisting of isotropic heterogeneous primary rock matrix and anisotropic heterogeneous secondary fracture system (NAKORNTHAP & EVANS 1986). Other features of reservoir heterogeneity are discussed by KAMAL (1977), LAMBERT (1981); ARYA, HEWETT, LARSON & LAKE (1985) and LEE & OKUYIGA (1986). HIRASAKI, MORROW & WILLHITE (1984) estimate reservoir heterogeneity from waterflood performance.

4.2.4.2. Petrophysical aspects

As a consequence of the different types and scales of sedimentological discontinuities and anisotropies in reservoir rocks, many pay formations also exhibit heterogeneous distribution of petrophysical parameters, with permeability being most affected by depositional inhomogeneities. Permeability anisotropy in vertical and/or horizontal direction is the result of predominantly grain size differences reflecting energy fluctuations during sedimentation of the sand in various environments with different levels and maintenance of energy in time and space. Apart from its depositional provenance, permeability heterogeneities can also originate during diagenetical history by differential mobilization and/or precipitation of framework grains and/or cements. Aspects of permeability anisotropy as well as intersection of natural fracture systems are briefly characterized as follows.

4.2.4.2.1. Permeability anisotropy

Petrophysical reservoir heterogeneities comprising permeability and conductivity anisotropy can be three-dimensionally diagnosed and monitored by interwell tracer or interference testing (WAGNER & BACKER 1974, WAGNER 1977, ABBASZADEH-DEHGHANI & BRIGHAM 1982, HAGOORT 1982; OHNO, NANBA & HORNE 1985; HEISLER 1986; ONURI, REYNOLDS & RAGHAVAN 1986). Determination of petrophysical reservoir heterogeneities and resulting flow pattern anisotropy has an important impact not only for fracture propagation modelling, but also for optimization of sweep efficiency of water-, steam- and chemical-flooding enhanced oil recovery (OHNO, NANBA & HORNE 1985). Interference and tracer testing should not be regarded as being competitive techniques, but both methods should be applied in a complementary sense in order to achieve optimization of determination of internal reservoir structure (MISHRA & RAMEY 1985, BRIGHAM & ABBASZADEH-DEHGHANI 1986). JOHNSON & KROL (1984), MARTIN & COOPER (1984) and STALKUP & EVANS (1986) give an overview of permeability variations in different depositional environments. EVANS (1988) documents the influence of common sedimentary structures and diagenesis on permeability heterogeneity and anisotropy in selected sandstones.

REVIERE & WU (1986) outline porosity stringerization giving rise to extremely erratic pay continuity that heavily deteriorates the effectivity of waterflooding. Aspects of heterogeneous reservoir composition are also discussed by KNUTSON, HODGES & RIGHTER (1981); BEGG, CARTER & DRANFIELD (1987); RAVENNA, GALLI, MATHIEU, MONTADERT & RUDXIBWICZ (1987); SHARMA, HONARPOUR, SZPAKIEWICZ & SCHATZINGER (1987) and TYLER (1987). ROSE (1982, 1983), BALL & PITTS (1983) and PEACEMAN (1983) comment on permeability anisotropy in tight formations which is mainly based on the fact that vertical permeability is smaller than horizontal permeability (GIGER, REISS & JOURDAN 1984; GIGER 1985, 1987; JOSHI 1986; KARCHER, GIGER & COMBE 1986; KOSSACK & OPDAL (1988); KUCHUK, BRICE, THAMBYNAYAGAM, SHERRARD & GOODE 1988). LEUNG (1986); TAHA, HILL & SEPEHRNOORI (1986) and HOUSEWORTH (1988) report variable directional permeabilities in anisotropic and heterogeneous reservoirs.

TANG, HENG & JIANG (1986) present complications of waterflood effectivity in heterogeneous sand-body arrangements with pronounced inter- and intra-layer permeability differences. HORIE, FIROOZABADI & ISHIMOTO (1988) outline capillary continuity in fractured reservoirs. LEWIS (1988) documents a hierarchical classification of permeability heterogeneity comprising seven orders and emphasizes the need of linking of outcrop and core studies for more accurate three-dimensional prediction of inter-well heterogeneity, with the special requirement being quantification of reservoir heterogeneity developed in different depositional environments. JONES (1987) utilizes the non-Darcy flow coefficient for reservoir heterogeneity characterization. MARTIN & COOPER (1984) document permeability barrier distribution modelling in sedimentologically complex reservoirs.

4.2.4.2.2. Intersection of natural fracture systems

LORENZ, BRANAGAN, WARPINSKI & SATTLER (1986) describe a system consisting of subparallel unidirectional natural fractures originating under conditions of high pore pressures and relatively low differential horizontal stress (cf. also section 4.8.8.1.). Permeability anisotropy is mainly created by the heterogeneous fracture system allowing only marginal fracture inter- and cross-connections, with thus orthogonal permeability being abt. 1 - 2 orders of magnitude lower than primary-direction permeabilities. Although fractures of any scale and opening may increase gross reservoir permeability by more than an order of magnitude over matrix values, the enhanced pay drainage will be highly anisotropic due to the heterogeneous fracture geometry fabric. Aspects of reservoir heterogeneity related to permeability anisotropy are also outlined by WARPINSKI, BRANAGAN, SATTLER, LORENZ, NORTHROP & FROHNE (1985) and LORENZ & FINLEY (1987), with emphasis being put on depositional environment controlling the characteristics of the matrix rock and sedimentary reservoir heterogeneity as well as fracture permeability system.

4.2.4.3. Rock mechanical aspects

In homogeneous formations where no significant stress contrasts can be assumed between reservoir zones selected on the basis of permeability variations, fracture growth is radial and the fracture is always propagating into the layers adjoining the pay zone, with penetration depth being determined by differences in stiffness and horizontal in-situ stress between reservoir and neighbouring beds (cf. sections 4.2.2.3.1.3., 4.2.2.5.3. and 4.2.3.2.2.; particularly in chalk formations; EEKELEN 1982, HARTLEY & BOSMA 1985). A better fracture containment in such homogeneous and poorly stabilized formations can only be achieved during the advanced stage of depletion when different reservoir pressure regimes may have developed in the various individual horizons and thus created a secondary stress zonation in a primarily almost isotropic formation in terms of stress distribution and fracture growth. In-situ stress differences are more effective for crack containment than stiffness contrasts (CLEARY 1979, CLIFTON & ABOU-SAYED 1979; WARPINSKI, SCHMIDT & NORTHROP 1980). Reservoir heterogeneity can considerably influence the success of hydraulic proppant fracturing (WATERS 1980 reports good results in continuous pay horizons, but less favourable performance in discontinuous or lenticular formations). Various points of fracture extension across material discontinuities and other features are outlined as follows.

4.2.4.3.1. Fracture extension across material discontinuities

RUMMEL, HEUSER, MÜLLER & OTTO (1986, 1987) document by fracture propagation modelling based on laboratory experiments that material inhomogeneities, anisotropy and discontinuities have considerable impact on direction and velocity of crack extension (the same applies for pre-existing natural fractures; BLANTON 1982; with fracture connectivity having an important influence; WIJESINGHE 1985, WIJESINGHE & KECECIOGLU 1986; cf. also section 4.8.7.). An outline is given on fracture deflection at bounding planes, material property differences, and interface shear strength as follows.

4.2.4.3.1.1. Fracture deflection at bounding planes

Lithological discontinuities can slow down or speed up fracture growth or can change the direction of crack propagation by deflection, depending on the elastical and fracture-mechanical contrast of the rock units on both sides of the discontinuity. High elasticity moduli and stress inhomogeneities in front of the propagating fracture result in lowering of the stress intensity at the crack tip and therefore slow down fracture growth.

Discontinuities which are inclined with respect to the direction of original fracture propagation lead to deflection of the crack into another direction (JEFFREY, VANDAMME & ROEGIERS 1987 comment on mechanical interactions in branched or subparallel hydraulic fractures). The orientation of radially extending cracks is highly sensible against changes of the orientation of principal stress directions, with the fracture always propagating in the plane which is perpendicular to the least main horizontal stress. In case of characteristical differences in material properties between adjoining rock types, it was possible to extend the crack across the boundary plane by application of high-viscosity fracturing fluids and/or high injection rates.

4.2.4.3.1.2. Material property differences

The rock mechanical results show that containment of fracture propagation by boundary horizons with significantly different material properties can be successfully achieved in case of controlled pumping rates and adjusted treatment design. This conclusion from the experimental investigations is confirmed by the examples of field application (cf. section 4.2.2.3.) where the goal of containing fracture extension to thin reservoir horizons has been successfully reached (cf. KOHLHAAS 1982; CONWAY, McGOWEN, GUNDERSON & KING 1985). BIOT, MEDLIN & MASSE (1983) point out that based on the elasticity concept, surface energy and shear modulus differences could also lead to fracture confinement in a layer between dissimilar materials. Incorporating plasticity and surface energy into fracture propagation modelling, MEDLIN & MASSE (1986) conclude that plasticity effects in shales are more important in controlling fracture height than any other phenomena (cf. section 4.2.3.2.3.). A fracture generated in a brittle sandstone or limestone layer bounded by plastic shales is likely to be contained because of propagating pressure differences between plastic and brittle horizons (cf. also sections 4.2.2.5.3. and 4.2.3.2.2.).

Fractures can propagate across rock interfaces without slip because of friction (TEUFEL 1979) if the pressure between the rocks exceeds a value approximately equal to their tensile strength. In terms of vertical fractures, this implies that slip does not occur at top or bottom of the crack even for perfectly flat and unbounded beds, if the effective vertical pressure (overburden minus pore pressure) exceeds the tensile strength of the rock. For normal gradients, slip does thus not occur in competent rocks in depths below abt. 1,750 ft. or if the overburden pressure exceeds the pore pressure by more than the tensile strength, whereas in shallower depth, slip may take place and result in decreasing pressure to extend a fracture.

4.2.4.3.1.3. Interface shear strength

TEUFEL & CLARK (1981) document that fracture growth in layered rock is controlled by differences in mechanical properties of the formations on either side of the interface between two lithologies of different stress level, changes in horizontal stress state across the interface, and shear strength of the interface. Containment can occur whenever shear strength of the layer interface is sufficiently weak relative to tensile strength and minimum horizontal compressive stress of the bounding layer, with the fracture then more easily becoming an interfacial fracture than extending across the interface into the bounding layer (cf. section 4.5.4.4.2.). A low shear strength interface can occur at low overburden stresses due to the frictional effect of the applied stress on the interface, provided that the overburden stress is sufficiently greater than the horizontal stresses to produce a vertical hydraulic fracture.

Differences in elastical properties of the formations in a dilithological layered rock sequence is the dominant factor in determining horizontal stress state within individual layers, with a possible compressional increase in horizontal stress going from high shear modulus to low shear modulus layers. Containment and propagation of fractures are generally two competing mechanisms at lithological boundaries separating rocks with different stress level.

4.2.4.3.2. Other features

Rock mechanical experiments on fracture propagation in layered rocks are also carried out by TEUFEL & CLARK (1981), BLANTON (1982) and SHAFFER, THORPE, INGRAFFEA & HEUSE (1984). LUISKUTTY, TOMUTSA & PALMER (1986) present a semianalytical model of hydraulic crack growth through weak barriers. EL RABAA (1987) reports features of hydraulic fracture propagation in the presence of stress variations. MOSCHOVIDIS (1987) comments on minifracturing treatments carried out at the interface of two formations, and BIOT, MEDLIN & MASSE (1986) evaluate crack penetration through an interface. LAM & CLEARY (1984) outline slippage and reinitiation of hydraulic fractures at frictional interfaces. JOHN (1987) reports aspects of gas recovery from deep tight formations in light of nonlinear elastical deformation characteristics. THIERCELIN, JEFFREY & BEN NACEUR (1987) and WANG & CLIFTON (1989) outline the influence of fracture toughness on crack geometry.

Rock mechanical aspects of North Sea reservoirs are evaluated by HOLT, INGSOY & MIKKELSEN (1987). EVANS (1983) and EVANS & HOLZHAUSEN (1983) comment on development of shallow hydraulic fractures as examined through the surface deformation field. LABUDOVIC (1984) reviews the effect of Poisson`s ratio on fracture height. LERCHE & NARR (1986) discuss possibilities of estimating subsurface crack density from core evaluations. Mechanical aspects of fracture extension are also outlined by HARRISON, McGUIRE & KIESCHNICK (1954); HUBBERT & WILLIS (1957), HOWARD & FAST (1970 a, 1970 b) and DANESHY (1976).

4.2.5. Combination of hydraulic fracturing and gravel packing

Within many shallow sandstones and carbonate shell layers, the question of fracture containment is accompanied by the problem of insufficient reservoir rock stability that requires often gravel packing in addition to hydraulic stimulation or application of resin-coated proppants (cf. section 1.2.6.) in order to create a stable massive multilayer drainage wedge within the loose formation by cementation of the proppant grains at their contacts due to melting and glueing of the resin films coating the proppant particles and to minimize or even avoid embedment of proppants in the soft formation (cf. sections 4.5.4.3. and 5.3.3.). Many conventional fracturing jobs carried out so far in such horizons failed because of too much proppant embedment in the soft reservoirs (cf. section 4.3.3.), fracture closure due to plastic mobility of poorly consolidated or overpressured, intercalated or adjoining mudstones (cf. plate IX/7), or sand control problems in the wells. The combination of proppant fracturing and gravel packing is particularly in soft sands and chalks a very suitable means for avoiding proppant and formation flowback into the wellbore after hydraulic stimulation of the reservoir (cf. section 4.5.4.3. and 4.12.3.3.). Combined hydraulic fracturing and gravel packing treatments of unconsolidated pay zones are usually performed in depth ranges up to 5,000 ft (1,700 m), but in some cases also formations in greater depth are treated in this manner.

Once resin-coated materials become necessary for hydraulic fracturing and/or gravel packing (SINCLAIR & GRAHAM 1977, 1978; CONSTIEN & MAYER 1978, SAUNDERS & McKENZIE 1979, KANAT 1980; UNDERDOWN, DAY & SPARLIN 1980; POPE, WILES & PIERCE 1987), proppant price is always a critical factor due to the more expensive material, and marketing activity has to take notice of the need of special introductory offers for promoting pilot treatments. Main targets for the approach in this context are marginal chalks, shell-debris and sand-streak intercalations into mudstones in many of the aforementioned geological formations and geographical areas (cf. sections 4.2.1. and 5.5.4.) that are mechanically or chemically insufficiently stable to allow conventional stimulation, and loose pure or muddy sands alternating with mudstones in a sandwich-type manner. Aspects of combination of hydraulic fracturing and gravel packing are also discussed by POOLLEN & MALONE (1959), SPARLIN & COPELAND (1972) and SOLUM (1986; cf. sections 4.5.4.3. and 5.3.3.).

4.2.6. Economical significance

While many specific technological questions of the most suitable methods and materials have still to be solved by field and laboratory studies, I am realizing in hydraulic proppant fracturing (partially in combination with gravel packing) of reservoirs of limited thickness and changing stability considerable potential of extension of both reservoir stimulation and hydrocarbon exploitation markets in both Western and Eastern Europe according to my experience, as most of these horizons become increasingly important during the next years when production from other pay zones which are easy to handle in view of reservoir and technical engineering is declining more and more, and the mentioned horizons where production was carried out so far in reduced rates partially far below the real capacity or was even completely suspended due to the discussed problems will have to contribute in increasing amounts in order to delay and slow down the general decline of annual cumulative domestic oil (and partially also gas) production in the individual countries.

A similar conclusion will apply in the nearer or farther future also to the major Rotliegend and Carboniferous gas-bearing sandstones when new discoveries by exploration reach the stage to be no longer able to keep pace with the still increasing demand of gas, and then the missing quantities have to derive also in these major formations in greater amounts from stimulation of hitherto unconventional tight reservoirs (this being more or less a continuation of the in some areas presently suspended exploration and appraisal drilling strategy concept).

In such big operations in both Europe and USA where the MHF expenses can account for up to half of the total well costs, the importance of fracture stimulation can even be equal to or greater than that of development drilling for increasing recoverable reserves and a high degree of resolution in fracture design is economically very viable (VEATCH 1983).

4.3. Proppant saturation and conductivity improvement

Advances in technology and equipment allowed in recent years an important breakthrough in fracturing treatment effectivity by successively increasing proppant concentration per fluid unit. While conventional fracturing stimulation also includes progressive stepwise increase of proppant saturation during course of the job, the decisive innovation in the last years was the achievement of ultra-high proppant concentrations in the final stages of the operation. In recent years, the trend in hydraulic fracturing is to employ higher injection rates, longer pumping time, higher proppant concentrations, and more efficient fluids (SHAH 1984). After discussion of some general aspects, comments are given as follows on selection of proppant type, proppant embedment and fracture width, proppant transport and fracture damage, ultra-high fracture conductivity achievement, and uniform vs. changing fracture conductivity.

4.3.1. General aspects

Historically, stimulation design has repeatedly changed according to progress in the technological evolution of hydraulic fracturing. In earlier years, proppant concentrations up to abt. 1.5 lbs/gal were used. With improving equipment capacity, stepwise advance was made via saturations of abt. 3 - 5 lbs/gal until recently it has become possible to handle even ultra-high concentrations up to 15 - 20 lbs/gal or even more, and also the present technical barriers are challenged (GREGORCZYK, PAULS, HOLTMYER, VENDITTO & CHISHOLM 1984; MARPLE, RULEY, WORLEY & FREEMAN 1987; similar improvements have been achieved for packing gravel concentration in mechanical sand control; cf. section 5.3.1.). The world record of the highest sand saturation in a hydraulic proppant fracturing treatment is 26 lbs/gal which was pumped in Caddo Parish in Louisiana/USA in 1981 (FRACFAX 1988 k). Following definition of the different types of proppant concentration, the significance of high proppant saturation is outlined.

4.3.1.1. Types of proppant concentration

Two types of proppant saturation have to be distinguished in ultra-high sand concentration treatments (USC jobs; SHAH, SMITH & DONALDSON 1983). Sand concentration on the surface is expressed in pounds of sand added to a gallon of clean fluid (lbs/gal), whereas sand saturation within the crack is referred to as pounds of sand per square foot of fracture area (lbs/ft2). In most of the conventional treatments conducted with proppant-suspending fluids, surface sand concentrations vary from 0.5 to 4 lbs/gal which translates to less than 2 lbs/ft2 (10 kg/m2) in the fracture downhole, whereas ultra-high saturations of 20 lbs/gal at the surface result in placement of 7 lbs/ft2 (35 kg/m2) downhole. The current industry standard ranges more or less between 8 and 15 lbs/gal for most parts of the treatment, and extraordinarily high saturations of 15 - 25 lbs/gal are occasionally reached in the terminal stages of the jobs.

4.3.1.2. Significance of high proppant saturation

The method of progressively rising proppant concentration during the job with finally reaching ultra-high saturations (COULTER & WELLS 1972, DOMSELAAR & VISSER 1974, SCHOLS & VISSER 1974, WATERS 1980; SHAH, SMITH & DONALDSON 1983; GREGORCZYK, PAULS, HOLTMYER, VENDITTO & CHISHOLM 1984; DUNAND & SOUCEMARIANADIN 1985; PAULS, VENDITTO, CHISHOLM, HOLTMYER & GREGORCZYK 1985) in the terminal part of the treatment allows better infilling of the fractures and therefore permits also to increase crack width thus improving fracture flow capacity, and to optimize tail-in policies (cf. section 2.4.1.2.3.) by using different proppant types and grain sizes in the sequence of steps of the procedure. Improval of the result of the stimulation job by increasing proppant saturations is considered to be one of the most important tools of promotion and economically feasible operation of hydraulic fracturing of deep tight Rotliegend and Carboniferous gas wells in Europe within the next years (cf. sections 2.4.1. and 3.3.). The more complete fracture fillup by high proppant concentrations leads to greater sustained fracture flow capacity and higher proppant crushing resistance (cf. section 4.12.4.) which ultimately give rise to a potentially higher sustained production increase after the stimulation (ILSENG & CORTEZ 1985).

4.3.2. Selection of proppant type

Ultra-high proppant concentrations in the fluid together with increasing fracture width sometimes even allow the application of a lighter and thus cheaper proppant (cf. section 1.3.), as some crushing of grains is considered to be negligible and to be compensated by the high proppant saturation where the individual grains are much better supporting each other in a wider fracture and closure stress is distributed on more grains per area unit than in case of lower proppant concentrations and narrower fractures. Ultra-high proppant saturation is a particularly suitable method for enhancing fracture flow capacity and diminishing effects of proppant embedment within soft formations. Proppant density and concentration, significance of intermediate-strength proppants, and benefits of high sand concentrations prior to synthetic proppant invention are discussed as follows.

4.3.2.1. Proppant density and concentration

The application of intermediate-strength low-density proppants is especially attractive, as with lower specific gravity and thus bulk weight, more propping material in terms of volume could be pumped per fluid unit than in case of heavier proppants (cf. sections 1.4.7. and 1.4.12.5.). If closure stress is too high to allow the usage of intermediate-strength low-density proppants, the high-weight proppant type with the lowest density would certainly have an advantage of choice if its price is not too high and thus beyond consideration (cf. sections 1.3. and 2.2.2.3.).

Practical experience has shown so far that sand concentrations up to 16 lbs/gal fluid can be handled without any problems and sand saturations of 8 - 12 lbs/gal are fairly common in modern treatments. Concerning intermediate- and high-strength proppants, concentrations higher than 6 - 8 lbs/gal are usually not needed because of the better flow capacity of the high-quality synthetic proppants and the negligibility to even absence of embedment in deeper formations where man-made proppants are required in contrast to serious embedment problems in shallow reservoirs where natural sand is still frequently pumped as a matter of cost containment. Rising proppant concentration, however, always has to take into account the principally increasing risk of screenout failure by easier bridging and sticking of the saturated slurry in the crack as a consequence of increasing friction (cf. sections 4.2.3.2.5. and 6.2.4.2.1.).

4.3.2.2. Significance of intermediate-strength proppants

Intermediate-strength proppants can be applied up to closure stresses of 12,000 - 13,000 psi with satisfactory performance (cf. section 1.3.2.). As most reservoirs do not exhibit such high pressures, there is not very much requirement for high-strength proppants such as sintered bauxite to justify the cost premium. A much better solution for improvement of fracture conductivity is thus increasing proppant concentration instead of selecting high-strength proppants for intermediate-strength ones. Progressive proppant saturation gives rise to a gradual conductivity increase and lets the fracture respond like having a higher conductivity with less total proppant material.

Economical and technical improvement include two options comprising application of either less proppants to get the same conductivity or using the same amount of proppants to achieve a better result when comparing the performance of two different proppant quality types. In higher concentrations, even a proppant type of inferior quality can perform better than a superior proppant type in low saturations, and therefore most of the hydrocarbon-bearing prospective intervals can be satisfactorily fractured with intermediate-strength low- or high-density proppants, with the light alumina silicate material having the greatest potential due to its specific gravity comparable to sand but much higher closure stress resistivity which is even increased in multilayer packages (cf. section 1.3.2.).

4.3.2.3. Benefits of high sand concentration prior to synthetic proppant invention

Prior to invention and market introduction of intermediate- and high-strength synthetic proppants for deep gas reservoir stimulation (cf. section 1.2.), reasonable resistivity of the proppant package against high closure stresses has been achieved by pumping high to excessive concentrations of natural sand (HOLDITCH & ELY 1973). The advantage of sand multilayering with respect to partial or complete monolayering has proven to be quite significant at least to limited closure stress levels of abt. 6,000 psi, and therefore high concentrations of natural sand have temporarily been even a cost alternative to sintered bauxite (CALLANAN, CIPOLLA & LEWIS 1983). On the other hand, however, pumping rates and fluid viscosities necessary to inject high sand saturations are sometimes detrimental by causing a large pressure drop along the fracture and promoting excessive height growth into unproductive zones (NOLTE 1982; cf. section 4.2.2.3.).

Aspects of high proppant concentrations and large total proppant loads in high- and low-temperature applications are discussed by SANDY, WIGGINS & VENDITTO (1986), and further examples of high proppant saturations in hydraulic fracturing are contributed by SEIDEL & STAHL (1972), HOLDITCH & ELY (1973), CROSS & SHELDON (1977); COULTER, MATTHEWS, SEGLEM & SMITH (1980); PETZET (1982), WATKINS (1982); HARRIS, REIDENBACH & CHISHOLM (1986) and ROSE (1988). DEHLINGER, BROWNE & BUNDRANT (1954) comment on optimum sand concentrations in well treatments. Proppant packing features of hydraulic fractures are also evaluated by ALDERMAN & WENDORFF (1969).

In terms of proppant type and saturation, field experience in moderate- to high-depth oil reservoirs in China has shown that in various cases, fractures propped with lower concentrations of bauxite have better conductivities than cracks propped with higher saturations of sand (LI & ZHU 1986).

4.3.3. Proppant embedment and fracture width

Proppant concentration together with fracture width is an extremely important fracture design parameter in treating both hard and soft rocks. Partial monolayer (single layer) propping (DARIN & HUITT 1960) theoretically provides maximum fracture flow capacity, but is not practical due to embedment and crushing in soft and hard formations, respectively (CLARK 1983). Thus under most conditions, partial monolayer propping is not suitable, because it is very difficult to achieve the desired placement geometry (HOLDITCH 1984). Aspects of multilayer fracture propping, proppant pillaring, fracture width vs. non-Darcy flow, fracture width vs. proppant bridging, and proppant concentration schedules are outlined as follows.

4.3.3.1. Multilayer fracture propping

Multiple layer fracture propping is beneficial in soft rocks because only the outer proppant layers are subjected to embedment, and in hard formations, the increased number of particles results in a greater amount of contact points that acts to minimize fracture deterioration due to closure-stress induced proppant crushing (cf. sections 1.2.1.1. and 4.12.4.). While embedment improves retention of proppant permeability at higher closure stresses because the stress is spread over a larger portion of the particle, it also allows partial fracture closure thereby reducing crack width and conductivity (MONTGOMERY & STEANSON 1985). Embedment depends on diameter and number of proppant grains per unit area of the fracture which matches proppant concentration and is also very sensitive to formation hardness (HOWARD & FAST 1970). Proppant embedment is often combined with fines migration (SHAH, SMITH & DONALDSON 1983). Formation strength which is a controlling factor on embedment can be determined by the mechanical properties log (ANDERSON et al. 1975; cf. sections 6.2.1.10.7. and 6.3.6.).

4.3.3.1.1. General aspects

Aspects of embedment or engulfment of high-strength proppants into low-permeability reservoir rocks are discussed by VOLK, RAIBLE, CARROLL & SPEARS (1981), and HUITT & McGLOTHLIN (1958) and McGLOTHLIN & HUITT (1966) comment on relation of formation rock strength to propping agent strength. ROODHART, KUIPER & DAVIES (1986) outline proppant rock impairment. Fracture width is directly proportional to gel viscosity and injection rate of the stimulation fluid (BARBY & BARBEE 1987), and increasing proppant saturation gives rise to higher gel viscosities thereby triggering the origin of wider fractures. Operating conditions which cause high pressure drop along the fracture such as high injection rates and viscous fluids result in relatively wide cracks, whereas conversely treatment parameters causing low pressure drop along the fracture such as low injection rates and thin fluids give rise to relatively narrow cracks (PERKINS & KERN 1961, BARBY & BARBEE 1987; cf. section 4.2.2.3.). Pay zone thickness has no effect on pressure or width of horizontal fractures, while pressure drop along restricted vertical fractures is large for thin zones but small for thick intervals. On the other hand, crack width at the wellbore is nearly independent of vertical fracture height (cf. section 6.2.2.2.). Reservoir depth has generally little effect on fracture width except of varying rock properties to which fracture width is not particularly sensitive.

High proppant concentrations in order to avoid embedment are also recommended for geopressured reservoirs (ROBINSON, HOLDITCH & LEE 1983; cf. section 4.4.2.). In hard reservoir rocks of continuous sand body distribution, proppant embedment does not occur, but impairment of conductivity is caused by reservoir formation creep at the fracture wall. In hard sandstone pay complexes where isolated sandstone lenses are separated by mudstones (cf. plates II/7 - 8, III/4, VII/3 + 6, VIII/7 - 8 and IX/1), proppant embedment into intervening shales may be a problem. A much larger fracture width than expected in proppant embedment testing indicates occurrence of proppant stacking (VOLK, RAIBLE, CARROLL & SPEARS 1981). The most important aims of multilayer fracture propping are crushing and embedment reduction as well as conductivity improvement which are illustrated below, with emphasis being put on the different properties of soft and hard reservoir rocks and their different reaction to mono- and multilayer fracture propping.

4.3.3.1.2. Crushing and embedment reduction

While some tight reservoirs with permeabilities of abt. 0.001 - 0.05 md can still be sufficiently drained with crushed proppants or sand because the fragments still provide sufficient conductivity contrast (cf. sections 4.6.1. and

4.12.4.), serious conductivity impairment is generated in the proppant package of fractures in moderate- to high-permeability formations by crushing and embedment (HOLDITCH & LEE 1979). A better design solution aiming on higher proppant concentration per unit area of the fracture wall is therefore a multiple-layer pack configuration of proppant particles consequently infilling a wider fracture. After discussion of some general aspects, special applications in soft sands and hard sandstones are outlined.

4.3.3.1.2.1. General aspects

Fracture porosity decreases as load is applied because proppant grains crush and embed into the rock matrix (WENDORFF 1982). While in competent rocks, proppant porosity is usually reduced by crushing, proppant embedment coupled with proppant crushing results in very low fracture porosity in soft formations. In most propped cracks, however, both embedment and crushing occur (cf. section 4.12.4.3.2.) and the resulting porosity is a measure of the combined effect.

High proppant concentrations plugging wider fractures are beneficial in soft rocks because only the outer proppant layers are subjected to embedment and the inner layers remain unaffected, and in hard rocks due to a greater number of contact points of the increased amount of particles minimizing fracture flow capacity deterioration by grain crushing in higher closure stress regimes which is limited to parts of the outer layers. Conductivity of fractures infilled by proppant multilayers is much less depending on reservoir rock hardness than that of those plugged by proppant monolayers or even partial monolayers (DARIN & HUITT 1960, COOKE 1973). Wide fractures and high proppant concentrations resulting in a multilayer package are also the best solution for stimulation of soft tight chalk formations in the North Sea (HARTLEY & BOSMA 1985; TANSOE, KING & HOLMAN 1986; cf. section 4.5.4.).

Apart from monolayers suffering from embedment and crushing in soft and hard rocks, respectively, other factors tending to reduce fracture conductivity are gel residue and fines resulting from incomplete clean-up of broken crosslinked stimulation fluids (cf. section 4.3.4.1.) and improper sieving and screening of the propping agent, respectively (CLARK 1983). Rocks with low embedment pressure are soft, with proppants tending to embed in these type of formations which results in a healing effect of the fracture, and rocks with high embedment pressure are hard and not susceptible to proppant embedment, but rather subjected to proppant crushing. These two different formation types are characterized in more detail as follows.

4.3.3.1.2.2. Soft sands

Multilayer high-conductivity propping of fractures even permits hydraulic stimulation of unconsolidated dirty sands (cf. section 4.8.5.4.) which cannot be treated with conventional techniques due to instantaneous total fracture closure as a consequence of complete embedment or engulfment of proppant monolayers into the soft rock matrix (LAMBERT, DOLAN & GALLUS 1983). Heterogeneous loose sands with considerable movement of formation fines impairing hydrocarbon flow can be successfully stimulated by reduction of fluid flow velocity at and within the rock face through effective wellbore enlargement by formation fracturing with maximizing proppant saturation and minimizing proppant grain size. Higher carrier fluid viscosity provides excellent proppant suspension and permits to use very high proppant concentrations with relatively small pad volumes. Proppant saturation maximization can also be obtained through limiting fracture growth by only perforating the lower portion of the interval to be treated (as the fracture is propagating in upwards direction due to decreasing overburden and closure stress; cf. section 4.2.2.3.1.). Proppant packages infilling wider fractures are less subjected to plugging by fines migration and proppant embedment (SHAH, SMITH & DONALDSON 1983).

100 mesh or 70/140 mesh grain size as a lead-in proppant and fluid-loss agent allows high-conductivity tip packing of secondary hairline fractures which do not accept larger proppant grains (cf. sections 1.4.11.2.1. and 4.8.8.3.2.3.). Flow capacity optimization near the wellbore can be achieved with 8/16 or 10/20 mesh tail-in following conventional pumping of 20/40 mesh proppants (cf. section 2.4.1.2.3.). The multilayer proppant package with high proppant concentration per fracture wall area unit guarantees that a high-conductivity flow channel still remains open after considerable proppant embedment into the weak formation matrix and is the only possibility of stimulating soft unconsolidated sands. In case of proper application, hydraulic multilayer proppant fracturing is much superior to gravel packing in such types of reservoirs, because gravel packing cannot sufficiently minimize formation fluid velocities to stop migration of fine particles within more distal sections of the reservoir and only provides inhibition of fines movement in the proximal pay zone seam around the borehole (cf. section 5.3.5.). The general possibility of hydraulic fracturing of soft unconsolidated sands is frequently highlighted by considerable losses of solids-laden drilling mud (cf. section 5.11.2.) when penetrating such horizons (LAMBERT, DOLAN & GALLUS 1983).

Soft formations require a large-diameter proppant which resists to embedment into the reservoir matrix, and medium- to high-permeability pay zones need a large-size proppant in order to provide adequate flow capacity (CLARK 1983).

4.3.3.1.2.3. Hard sandstones

Crushing reduction independent from proppant concentration can also be achieved by replacement of natural sand by intermediate- or high-strength proppants in hard sandstone reservoirs in both deep and shallow wells (PEARCE 1983; cf. sections 2.3.6. and 2.4.2.4.). Proppant monolayers in hard sandstones not permitting embedment suffer from crushing primarily by point-to-point loading of individual sand grains which are too brittle to deform. Increasing sand concentration in shallow pay zones where pumping has to be done with low injection rates (cf. section 4.2.2.3.) bears the danger of premature screenout termination, but reduction of crushing can be effectively performed by utilizing a higher-strength proppant particularly when proppant placement is carried out in pillars or columns while the fracture is hold open by hydraulic pressure (PUGH & SEGLEM 1977, PEARCE 1983; cf. section 4.3.3.2.). High-quality proppants in multilayer packages together with increased concentration are thus effective methods of treating both very soft and very hard reservoir rocks regardless of their structural depth.

Hard rocks usually have high closure stresses associated with them and therefore require a high compressive strength proppant (CLARK 1983; cf. section 1.2.1.1.). Since embedment is not the major problem in these formations, a small-sized proppant can be utilized which also allows an increased pack concentration that tends to counteract the high closure stresses also in less wide fractures, because hard sandstones are usually low-permeability rocks which do not require large flow capacities within the cracks and are best treated with long narrow highly-packed fractures (cf. also VOLK, RAIBLE, CARROLL & SPEARS 1981). The assumption of negligible embedment and proppant permeability independent of proppant saturation is only valid for proppant multilayers and hard reservoir rocks (CALLANAN, CIPOLLA & LEWIS 1983).

4.3.3.1.3. Conductivity enhancement

As fracture conductivity is defined as crack width times proppant package permeability, wider fractures infilled by proppant multilayers are a significant means of increasing crack flow capacity beyond the technical and economical boundaries of fracture length improvement (COOKE 1973). Proppant multilayers originate when stimulation fluid viscosity is low enough to allow the proppants to settle to the bottom of the vertical crack during fluid injection, as well as when fracturing fluid viscosity is high enough to suspend the proppant in case of sufficiently high proppant concentrations and satisfactorily wide fractures (cf. section 4.3.4.2.).

Special applications comprise high proppant saturations in very small stimulation jobs which are placed close to the wellbore in order to overcome formation damage or to successfully treat unconsolidated formations with high pressure drawdown (in the latter case, the insertion of resin-coated proppants has also proven to be useful; cf. section 4.12.3.3.). Short high-conductivity fractures which are characterized by multilayer propping are a common means of stimulating moderate-permeability sandstone oil reservoirs for both oil production and water injection (cf. sections 4.5.1. and 4.6.), with the necessary high conductivity being achieved by both high concentration and grain size of the proppants (BRITT 1985, BRITT & BENNETT 1985, BRITT & LARSEN 1986), and have also significant application in chalk oil pay zones (cf. sections 4.3.5. and 4.5.4.).

Aspects of conductivity contrast enhancement by multilayer propping of wider fractures are also discussed by MARPLE, RULEY, WORLEY & FREEMAN (1987), with the particular advantages of higher proppant concentrations within the fracture being more complete crack fillup, higher fracture flow capacity, less proppant crushing and improved proppant pack tolerance to fines. High proppant saturations require selection of a fracturing fluid which will have maximum viscosity and proppant transport properties once outside the perforations (cf. section 4.3.4.2.), and it is also imperative that fluid loss within the fracture is held at a minimum (cf. section 4.3.4.6.).

4.3.3.2. Proppant pillaring

Another concept of fracture flow capacity improvement is proppant pillaring which comprises proppant introduction into the crack in such a way that no continuous sheet-type wedge is formed, but open flow channels remain between discrete patches or pillars which provide even higher conductivity than the high-permeability proppant package (TINSLEY & WILLIAMS 1975; FAST, HOLMAN & COLVIN 1977; cf. figs. 7 and 12) if their subsequent closure due to high reservoir stress can be prevented. Some aspects of open flow channels between discrete proppant patches, discontinuous proppant distribution, and fluid density control and selective proppant load are summarized as follows.

4.3.3.2.1. Open flow channels between discrete proppant patches

The proppant pillars have to be spaced sufficiently close to each other in order to avoid fracture deterioration by partial closure of open channels between the supporting proppant patches (TINSLEY & WILLIAMS 1975). In combination with multilayer propping, the technique of pillar propping (cf. figs. 7 and 12) certainly is a very effective means to maximize fracture flow capacity if properly designed and performed. Proppant pillaring has the advantage of creating less difficulties with screenout failures due to alternating pumping of fluid slugs with and without proppant load (cf. sections 4.2.3.2.5. and 6.2.4.2.1.). FAST, HOLMAN & COLVIN (1977) believe that proppant pillaring gives results at least comparable - if not superior - to those of totally proppant packed fractures, because the open crack portions have an enormous flow capacity with respect to the proppant bed or pack (TINSLEY & WILLIAMS 1975). Proppant pillaring can under these circumstances of technically acceptable performance advance to an important means of stimulation expense containment by lowering of proppant cost due to requirement of smaller total proppant quantities than would be necessary for conventional continuous plugging of the crack.

Partial monolayering of proppants has been more successful in treatments where insoluble particles are injected along with the proppant and the mixture is deposited as a multilayer (SINCLAIR 1980). Subsequent dissolution of these additional particles theoretically leaves only the proppant monolayer, but results from these treatments can also be explained by other mechanisms, such as that a multilayered proppant package is placed with an open fracture channel left above it. Such a geometrical arrangement in horizontal division of proppant package and open flow channel represents proppant banking (cf. section 4.12.2.2.), whereas the corresponding vertical distribution matches proppant pillaring (cf. figs. 7 and 12). Proppant banking can be technically achieved by gradually reducing the pumping rate near the end of the treatment in order to screen the borehole out and leave the fracture well-propped near the mouth and close to the bottom with an open high-conductivity channel towards the top (CAMPBELL, HANOLD, SINCLAIR & VETTER 1981), whereas proppant pillaring is triggered by pumping alternating lots of fluid with and without proppant load (cf. also section 4.2.2.6.).

4.3.3.2.2. Discontinuous proppant distribution

The performance of non-uniform proppant packing with open flow channels between isolated proppant islands is underlined by still favourable results in many old wells where formerly technically poor treatments had been carried out and continuous proppant placement could not be achieved (TINSLEY & WILLIAMS 1975; cf. section 4.8.9.). The transition from continuous to discrete proppant distribution in the fracture can be controlled by proppant concentration. As proppant settling is associated with convection cells (cf. section 4.12.2.), higher proppant saturations flatten and dampen out the convection cells in the fluid if sufficient concentration is reached with only little amounts of low-saturated fluid left in between.

Transitional stages between uniform and non-uniform proppant distribution are proppant banks at the lower side of the fracture and isolated proppant patches in the upper part of the crack with open flow channels in between, with some proppant pillars going up from the proppant bank zone into the proppant spot zone. In some cases, also horizontal proppant banking with enrichment of proppants to a layer in the lower part and an open flow channel in the upper section of the fracture is a profitable solution, because upon banking the proppants settle in the created width instead of the fracture closing on a less concentrated and narrower pack as placed by perfect-support fluids (HOEL 1988; cf. figs. 7 and 12).

4.3.3.2.3. Fluid density control and selective proppant load

Proppant banking can also be realized by fluid density control with over- or underriding of two successive fluids of different specific gravity and viscosity (BARBER & THEMIG 1985; cf. section 4.2.2.6.), resulting in selective proppant placement in bank-type distribution in the upper or lower interval of the fracture in case of pumping fluids with and without proppant load, or in stratification of different proppant types and/or grain sizes in the upper and lower reservoir segments in case of injecting two fluids charged with different proppant loads, respectively (cf. section 4.2.2.4.). MALLINGER, RIXE & HOWARD (1964) comment on propping agent spacers in order to increase well productivity.

The importance of selective proppant placement within the fracture is highlighted by the fact that crack conductivity varying from wellbore to tip (cf. section 4.3.6.) can significantly affect production rates (BENNETT, ROSATO & REYNOLDS 1981). It is therefore decisive to determine proppant placement and scheduling programs for a treatment design in order to assure that the appropriate distribution of conductivity in the fracture is achieved (VEATCH & MOSCHOVI-

DIS 1986). Aspects of proppant pillaring are also discussed by PEARCE (1983).

Generation of non-continuous proppant pillars by using spacer fluids between proppant slugs requires that the proppant-carrying fluid is capable of perfectly transporting the proppants during duration of pumping, the spacer volumes finger through the proppant-laden fluid to leave channels throughout the fracture, and the volume of proppant-laden fluid and spacer fluid slugs are such that the pillars are close enough together to ascertain almost complete closure of the void channels when the well is placed on production (PUGH, McDANIEL & SEGLEM 1977). The main economical benefit of pumping alternating volumes of viscous proppant-laden fluid and thin spacer fluid is significant reduction of treatment costs due to lower proppant quantities with respect to continuous fracture infilling.

4.3.3.3. Fracture width vs. non-Darcy flow

Another advantage of multilayer propping of fractures in comparison to monolayer or even partial monolayer support is that the wider the fracture and the larger the proppant grains, the better the preservation of Darcy flow. In narrow fractures propped by monolayers or partial monolayers consisting of small proppant grains, the velocity of the fluid flowing in the fracture may increase so much that turbulences or inertial effects originate which greatly increase the flow resistance of the fracture under conditions of non-Darcy flow (COOKE 1973, COOKE & GIDLEY 1979; cf. section 4.11.3.2.) and thus reduce gas well productivity for low-conductivity fractures by creation of additional pressure losses in the crack which decrease the flowing bottomhole pressure required to withdraw at a desired flow rate if the well is produced against a constant back pressure. The inertial forces are caused by the tortuosity of the flow path as well as by acceleration and deceleration of fluid as it flows through the pore throats in a porous medium.

Inertial gas flow in a hydraulic fracture is substantially affecting the shape of the pressure buildup curve (HOLDITCH & MORSE 1976) since inertial flow effects do not always quickly dissipate when the well is shut in. Fracture conductivity can be considerably affected by the deviation from Darcy flow at high current rates (COOKE 1973), with the additional pressure drop provoking a reduction of effective fracture conductivity by more than a factor of 3 (DAVIES & KUIPER 1988; cf. section 1.4.10.4.). Under such conditions, Forchheimer`s equation rather than Darcy`s law has to be used for correct pressure gradient estimation near the wellbore and consequently flow behaviour of the hydraulically fractured well for high-velocity environments (EZEUDEMBAH & DRANCHUK 1982, PEREZ & KELKAR 1988; PURSELL, HOLDITCH & BLAKELEY 1988). Some aspects of proppant type and grain size, shearing of polymer fluids, highly-deviated wells and secondary fracture narrowing are illustrated as follows.

4.3.3.3.1. Proppant type and grain size

Turbulences leading to non-Darcy flow are also created by small- to micro-scale heterogeneities such as crushed grains in the proppant package (cf. section 1.4.10.), with crushed larger grains being more heterogeneous than intact packs of initially small grains. Gas and low-viscosity oil are most significantly affected in linear flow by turbulences at high current rates. In case of narrow fractures, non-Darcy flow effects are significant and negligible in fracture and formation, respectively (WATTENBERGER & RAMEY 1969; REINICKE, BRINKMANN, SCHWARZ & HUENI 1985).

Non-Darcy flow inhibition of hydrocarbon production is the more serious, the poorer is proppant conductivity, with high-quality synthetic proppants of larger grain size giving rise to much less non-Darcy flow disturbance than natural sand (CALLANAN, CIPOLLA & LEWIS 1983; cf. section 1.4.10.). Both proppant type and grain size are independently and jointly affecting fracture conductivity damage by non-Darcy flow, with higher proppant quality and larger proppant grain size decreasing the heterogeneity effects of non-Darcy flow considerably. Conversely, any proppant-pack conductivity damage increases the non-Darcy flow factor, with greater increases being recorded for higher degrees of damage (DAVIES & KUIPER 1988).

PURSELL, HOLDITCH & BLAKELEY (1988), however, conclude from inertial flow studies that the non-Darcy flow coefficient depends on initial proppant mesh size and is independent of proppant type, with grain size distribution being the most important controlling factor. Permeabilities calculated from the Forchheimer equation exhibit the same trend of decreasing permeability with increasing pore pressure as the Klinkenberg relationship for Darcy flow. Non-Darcy current in porous media is controlled by pore structure (NOMAN & ARCHER 1987), with inertial flow effects being mainly influenced by curvature of streamlines, expansion and contraction losses, throat to pore coordination number, heterogeneities and surface roughness (PURSELL, HOLDITCH & BLAKELEY 1988). The inertial flow coefficient increases with rising pore pressure. Other aspects of non-Darcy flow in proppant packages are discussed by EVANS & EVANS (1986).

4.3.3.3.2. Shearing of polymer fluids

Uncomplexed polymer solutions are frequently in turbulent or transitional flow when crosslinking reactions take place (GARDNER & EIKERTS 1982). The level of turbulence or wall shear stress can be expected to have significant effect upon apparent gel viscosity. Polymer systems are affected by high shear during the crosslinking reaction as well as by level and duration of pumping. The ability of a gel to gain viscosity after a period of high shear flow is strongly affected by level and duration of high shear flow. Proppant addition to a gel increases slurry viscosity at a constant value over that predicted for similar proppant concentrations in Newtonian fluids. Approximation of apparent viscosity of a gel as it enters the fracture at the wellbore requires estimation of the combined effects of shear and proppant on gel viscosity. The effects of temperature and of long periods of low shear rate on gel viscosity after mixing and flowing at high rates are needed to better characterize gels while being in the fracture.

Effects of pore structure on non-Darcy gas flow are reported by NOMAN & ARCHER (1987), and other aspects of non-Darcy gas flow are discussed by RIECKMANN (1970), WONG (1970), JONES (1972, 1987), GEERTSMA (1974), RAMEY (1975), HOLDITCH & MORSE (1976); JONES, BLOUNT & GLAZE (1976); HOLDITCH & LEE (1979), GUPPY (1980); GUPPY, CINCO-LEY & RAMEY (1980); KADI & BROWN (1980); ALLAM, CRICHLOW & SOLIMAN (1981); KADI & BOUTEMY (1981), UBANI & EVANS (1982); CALLANAN, CIPOLLA & LEWIS (1983); EVANS, HUDSON & GREENLEE (1985); NOMAN, SHRIMANKER & ARCHER (1985); EVANS & EVANS (1986), NGUYEN (1986); SALDANA, BRIGHAM & RAMEY (1986); SCHMIDT, CAUDLE & MILLER (1986); LEE, LOGAN & TEK (1987); MALONEY, GALL & RAIBLE (1987); FRAIM & LEE (1988) and PEREZ & KELKAR (1988). Non-Darcy flow is particularly originating in highly-deviated wells (cf. section 4.11.3.2.) and in case of secondary fracture narrowing which are briefly outlined as follows.

4.3.3.3.3. Highly-deviated wells

Non-Darcy flow is a particular problem in highly-deviated wells (MARTINS 1987). Flow has both viscous and inertial components. Consequences of non-Darcy flow are large reduction in effective fracture conductivity (up to 5 - 20 times lower than normal), and well performance is very sensitive to poor crack/wellbore communication as a consequence of limited perforation intervals in highly-inclined boreholes. Fracture flow capacity of a hydraulically stimulated gas well can be significantly reduced by non-Darcy gas flow at high velocity (HOL-

DITCH & MORSE 1976). Excess pressure gradients can be caused by either turbulence or inertial resistance or a combination of the two, and non-Darcy gas flow can also give rise to increasing closure stress which provokes reduction of fracture permeability (cf. also section 4.11.3.2.).

4.3.3.3.4. Secondary fracture narrowing

Non-Darcy flow can also originate in primarily sufficiently wide fractures if they are secondarily narrowed by gel filter cake deposition on the fracture walls (cf. sections 1.4.10.3.5. and 4.3.4.1.4.) as the fluid filtrate leaks off into the reservoir formation and subsequent embedment of the outer proppant layers of a multilayer package into the soft filter cake occurs (ROODHART, KUIPER & DAVIES 1986; McDANIEL 1987, 1988; MUCH & PENNY 1987, PARKER & McDANIEL 1987, McDANIEL & PARKER 1988). Apart from decreasing effective fracture width, an additional effect triggering non-Darcy flow is partial plugging of the pore space of the proppant package by fines deriving from formation collapse, proppant crushing and precipitation from fluids (ROODHART, KUIPER & DAVIES 1986; cf. section 4.12.4.). With decreasing Darcy permeability due to the damaging effects, the non-Darcy flow or Beta factor is correspondingly increasing.

COOKE (1973) was the first to quantify the effects of non-Darcy flow in proppant packages under stress. When no turbulence is present, the Beta factor is zero and simple Darcy conditions for laminar flow exist.

4.3.3.4. Fracture width vs. proppant bridging

In order to accept proppants, fracture width must be at least in the range of 2.5 times the proppant diameter, with this requirement being fulfilled by most of the hydraulically induced fractures for standard proppant grain sizes (16/20 and 20/40), while 100 mesh microproppant fits into hairline cracks and fissures (THOMPSON 1977; cf. sections 1.4.11.2.1. and 4.8.8.3.2.3.). POULSEN & SOLIMAN (1987) outline that the fracture has to be sufficiently wide to prevent bridging of proppants between the crack walls which would result in screenout failure of the job. Some aspects of minimum fracture width for proppant infilling and friction pressure drop are outlined as follows.

4.3.3.4.1. Minimum fracture width for proppant infilling

For the selected created fracture length (which has to be abt. 10 % longer than the desired propped crack length in order to minimize the possibility of a screenout at the fracture tip; cf. section 4.8.11.), crack width at the desired propped fracture length at the time of job completion has to exceed 2.5 times the diameter of the proppant particles for allowing unhindered passage of the proppants throughout the fracture (cf. sections 1.4.11.2.1. and 4.8.8.3.1.2.). Crack width has also to be enough for attaining the required fracture conductivity at the wellbore using a reasonable proppant concentration in the slurry. The amount of chosen proppant per area of fracture face divided by the maximum allowed proppant saturation in the slurry establishes the lower bound of the fracture width which in terms of geometry calculations leads to a lower bound in crack length (cf. section 4.8.11.). Aspects of single- and multiple-particle bridging in fractures are also discussed by LOEPPKE, GLOWKA & WRIGHT (1988).

4.3.3.4.2. Friction pressure drop

Fracture width is essentially controlled by frictional pressure drop in the crack (PERKINS & KERN 1961). Operating conditions that lead to high pressure drop result in relatively wide cracks, whereas small pressure drops produce thin cracks. Proppant concentration changes are sometimes necessary during course of stimulation treatments in order to accommodate proppant bridging or if bottomhole treating pressure is lower than closing pressure (GREENFIELD & AHMED 1983; cf. sections 4.3.3.5. and 6.2.4.3.). Diminishing proppant saturation to avoid bridging is also an important means of screenout failure prevention, because lower proppant concentration permits the fluid flow rate to increase and thus keeps the fracture open to receive further proppants (cf. sections 4.2.3.2.5. and 6.2.4.2.1.).

4.3.3.5. Proppant concentration schedules

Achievement of a continuous fracture conductivity distribution requires successively changing proppant saturation throughout the treatment (POULSEN & SOLIMAN 1987) which can be performed by currently available computer-controlled blenders. Distinction can be made between stair-stepped and ramping proppant staging schedules (CHURCH & PETERS 1987, HANDKE 1987). Some comments on microprocessor control and automation as well as on blending equipment type vs. treatment size are also offered.

4.3.3.5.1. Ramping vs. stair-stepped schedules

In stair-step arrangements, the proppant concentration increment between each treatment stage is large and abrupt and is not restricted to being constant and fluent throughout the fracture operation, every stage of the assemblage typically has a different slurry volume, and the general shape of the system is curvilinear with time. In ramping organizations, the proppant saturation increment between each job stage is usually small and continuous and is rigidly maintained at a constant value throughout the fracturing job, every stage of the schedule has an equal total slurry volume except of the first and last stages which have total fluid volumes each exactly 50 % the size of the other ramp stages, and the general shape of the association is linear increasing or decreasing with time (CHURCH & PETERS 1987, HANDKE 1987, POULSEN & SOLIMAN 1987).

Ramping proppant schedules have the advantage that by continually increasing proppant apportionment rates in small increments throughout the treatment, more control over pressure considerations can be achieved, and the effects of excessive fluid viscosities in the fracture can be eliminated (CHURCH & PETERS 1987; cf. section 4.3.3.3.). The spikes in stairstep schedules are frequently the reason of premature screenout failure of the job (cf. sections 4.2.3.2.5. and 6.2.4.2.1.), with smooth ramping schedules bringing in much higher proppant concentrations with less screenout risk (HOOVER & ADAMS 1988).

4.3.3.5.2. Microprocessor control and automation

The different step changes in proppant saturation between stages of a fracturing operation can be performed automatically and quickly with a blender retrofitted with a microprocessor-controlled proppant delivery system, with the probability of overshooting a new proppant concentration setpoint also being drastically reduced. The capabilities of field-proven fracturing blenders are only enhanced and not restricted once retrofitted with a microprocessor-supervised proppant delivery system. Mixed proppant saturation schedules consisting of an initial stair-step pattern followed by a tail-in of ramping feature are also possible and have in field tests been proven to run very smooth without setpoint overshooting (HANDKE 1987, POULSEN & SOLIMAN 1987). Proppant concentration can be metered with radioactive densimeters (MARPLE, RULEY, WORLEY & FREEMAN 1987). Ramp-controlled blenders with automatic proppant control systems can be directly implemented in new equipment and can be retrofitted in older equipment. The ability of computer-interfaced blending equipment facilitates high proppant concentrations pumped in short stages at high injection rates (CLEARY 1988, HOOVER & ADAMS 1988) as well as high saturations of larger proppant sizes in shorter fluid stages (cf. section 6.2.4.3.).

Pumping high proppant concentrations downhole can lead to excess of proppant transport capacity of the blender (PAI, GARBIS & HALL 1983). In such cases, the job can be divided into three banks with three blenders in a multiple bank system, with all the blenders pumping in parallel. Standby blenders should be rigged up in line and should be close to a proppant source so that minimum changeover time is required and the job may continue as per schedule in case of blender failure. Microprocessor-controlled blenders allowing interaction from the computer van and automatic proppant addition through computer-interfaced densitometers are also described by HOOVER & ADAMS (1988).

4.3.3.5.3. Blending equipment type vs. treatment size

Conventional continuous-mix blending equipment greatly facilitates handling of large volumes of fluid and proppant and enable implementation of massive hydraulic fracturing treatments in low-permeability reservoirs, but is poorly suited to prepare and displace small precise volumes of slurry with the aim of achievement of high proppant concentrations in wide fractures in moderate- to high-permeability pay zones (CRAMER & SONGER 1988; cf. sections 4.5.4.4.1. and 4.6.1.). Some aspects of continuous-mix and batch-mix fracturing are briefly outlined as follows.

4.3.3.5.3.1. Continuous-mix fracturing

Conventional continuous-mix treatments suffer frequently from excessive fracture growth as a consequence of large slurry volumes used, with then only a small portion of the fracture contributing to wellbore transmissibility and the danger of accelerated water breakthrough, hindered vertical and lateral sweep efficiency in waterflooded pay zones, and channelling into vertically adjacent reservoirs existing (CRAMER & SONGER 1988; cf. section 4.2.2.3.). The relatively modest proppant concentrations of conventional continuous-mix operations lead to corresponding small crack width which implies low fracture conductivity that is further deteriorated by gel residue and other proppant pack damage problems (cf. section 4.3.4.1.). Conventional fracturing equipment is poorly suited to mix and discharge small precise volumes of slurry with high proppant saturation. Non-uniform proppant distribution occurs in the often-used high-capacity volumer blender tub apparatus, irregular proppant addition takes place due to fluctuating treatment rate, and inhomogeneous proppant distribution in the fracture can result in low-conductivity bottlenecks and job screenout due to proppant slugging (cf. sections 4.2.3.2.5. and 6.2.4.2.1.).

On the other hand, continuous-mix technology provides additional flexibility to fracturing treatments especially in case of very large jobs (CONSTIEN, BRANNON & BANNISTER 1988). Fluid viscosity can be raised or lowered to accommodate apparent viscosity increases caused by proppant and rapid changes in proppant concentration and polymer saturation can be adjusted in response to downhole pressure changes, with this flexibility economizing materials and increasing job control (cf. section 6.2.4.3.). Fracturing fluids are normally prepared using a combination of batch and continuous-mix processes. Batch mixing consists of formulating several components of the fracturing fluid in large tanks prior to the treatment, and once the operation starts, the remaining components such as crosslinking agents, degradants and proppants are continuously metered into the fluid stream to produce the final fluid composition.

4.3.3.5.3.2. Batch-mix fracturing

In batch-mix fracturing, high proppant concentrations can be pre-blended with gelled liquid in the inventory compartments and slurry density can be verified before the start of the treatment. This assures placement of an uniform proppant distribution in the fracture, because proppant saturations at the surface will not be affected by operation rate fluctuations (CRAMER & SONGER 1988). Therefore incremental improvements in wellbore transmissibility can in many cases be achieved due to creation of significant conductivity contrast between reservoir and fracture. Small precise slurry volumes with high proppant concentrations are particularly suitable for moderate- to high-permeability reservoirs (cf. sections 4.5.1.1. and 4.6.1.) and successfully limit vertical and lateral fracture extension, avoid negative consequences of interzonal communication and waterflood sweep inefficiency, and significantly reduce screenout problems by achievement of uniform proppant distribution and the use of a high pad/slurry ratio.

In most batch-mix fracturing operations, a large percentage of the treatment volume consists of pad fluid (up to 65 - 80 %) insuring leakoff control and sufficient fracture width for passage of the concentrated proppant slurry (cf. section 4.3.4.6.1.). Case studies indicate that even if coarse higher-permeability proppants (8/12 or 12/20 mesh) are used, longer narrowly propped fractures created by conventional continuous-mix blending are less productive and less conductive than shorter wider propped cracks generated by batch-mix fracturing. Batch mixing also has to be carried out for continuous pumping of high-viscosity crosslinked fracturing gels (FREE, FREDERICK & THOMPSON 1975). Aspects of batch-mix and continuous-mix fracturing are also discussed by EVANS & WILLIAMS (1985) and OLSEN & DEBONIS (1988; cf. section 4.3.4.3.1.4.).

4.3.4. Proppant transport and fracture damage

In order to obtain multiple proppant layers, it is necessary to choose a fracturing fluid which provides sufficient viscosity to create wide fractures and to carry the amount of proppant necessary to pack the large crack. Selection of a fracturing fluid is not only a technically, but also an economically critical part of the stimulation treatment design, because fluid and additives (proppants) may together represent more than 50 % of the total cost of the operation (HOLDITCH 1984). The selected fluid should have high enough viscosity to transport high proppant concentrations through the pipe without settling (cf. section 4.12.2.) and yet avoid excessive pipe friction pressure, and once the proppant-laden fluid is in the fracture, sufficient viscosity of the fluid is required to keep the proppants in total suspension in order to avoid settling and banking which would deteriorate conductivity distribution in the crack (SHAH, SMITH & DONALDSON 1983).

The ideal fracturing fluid keeps the proppant particles completely in suspension during both pumping operation and crack closure sequence, thereby minimizing proppant settling (CLARK 1983). The following account gives a brief summary of aspects of breakdown and residue concentration, viscosity and transport ability, impact of delaying, heat capacity and reservoir temperature, viscoelasticity, and leakoff containment of preferentially crosslinked fracturing fluids.

4.3.4.1. Fluid breakdown and residue concentration

Hydraulic fracturing mechanics consists of transmission of energy from surface pumps via a stimulation gel to the formation to be treated (PAI, GARBIS & HALL 1983). Important parameters for treatment fluid selection are sufficient viscosity to transport the proppants downhole, fluid-loss additives, low tubular friction pressure, fracture and formation fluid compatibility, non-damaging nature, ease of recovery after the job, reasonable price and operational safety.

4.3.4.1.1. General aspects

The hydraulic fracturing process typically consists of blending special chemicals to generate the appropriate stimulation fluid, mixing it with proppants, and pumping the slurry into the pay zone at sufficiently high rates and pressures such that the fluid hydraulically wedges and extends a fracture within the formation where the proppants are carried up to the extreme tips of the crack, thereby creating a drainage avenue or superhighway for oil and gas to flow easily from distal portions of the reservoir into the borehole (VEATCH 1983; cf. section 1.2.1.1.). Fluid properties strongly govern fracture propagation behaviour as well as distribution and placement of proppants. Rapid fluid leakoff gives rise to low efficiency in hydraulic wedging and extending of the fracture, and may also leave an undesirable concentration of particulate residue in the crack that could reduce conductivity (cf. sections 4.8.8.3.1.2. and 4.8.8.3.2.3.). Effective fluid viscosity controls internal fracturing pressure and proppant transport characteristics.

Desirable features of stimulation fluids for most hydraulic treatments include low fluid loss in order to obtain the designed penetration with minimum fluid volumes, sufficient effective viscosity to create the necessary fracture width as well as to transport and distribute the proppants in the fracture as required, no excessive friction in the crack, good temperature stability for the particular reservoir to be treated, good shear stability, minimal damaging effects on formation permeability, minimal plugging effects on fracture conductivity, low friction-loss behaviour in the pipe, wettability to leave formation and proppant package water-wet (DERBY & SMITH 1979; GARBIS, BROWN & MAURITZ 1985), low pH to prevent clay swelling, minimum insoluble residue leaving, good post-treatment breaking characteristics and short breaking time, good post-operation clean-up and flowback behaviour, and low cost. While some fracturing fluids exhibit a Newtonian flow behaviour, others are non-Newtonian but essentially behave as power-law fluids, and the crosslinked polymers commonly used for MHF treatments are very complex non-power-law fluids. Many Newtonian fluids degrade with time, and elevated temperature and/or shear accelerate degradation.

Transport of proppants in higher saturations by the carrier fluids is accomplished by increasing the viscosity by adding water-soluble polymers to the water phase and crosslinking them with a metal ion thereby forming chelate complexes (COOKE 1975; AHMED, ABOU-SAYED & JONES 1979; ALMOND & BLAND 1984, KIM & LOSACANO 1985, ACHARYA 1987; cf. section 4.3.4.3.). The following discussion focusses on some aspects of fluid breakdown, residue concentration, gel filter cake on fracture wall, and possibilities of fracturing fluid improvement.

4.3.4.1.2. Fluid breakdown

Viscosity and thus transport capacity of the crosslinked fluid varies with polymer and crosslinker type and concentration, fluid temperature, and method of fluid placement (VEATCH 1983, ALMOND & BLAND 1984). Upon completion of the treatment, removal of the gelled fluid to allow increased production rate is achieved by breakdown of the high-viscosity crosslinked fluids to low-viscosity solutions through thermal degradation of the polymer at reservoir temperature and/or addition of chemical breakers (cf. sections 4.3.4.2.2. and 4.12.2.3.). Some of the broken gels, however, leave unsoluble residues in fracture and formation which may cause considerable flow impairment due to significant reduction of proppant permeability or crack conductivity by blocking or restricting pore spaces. The most suitable fracturing carrier fluids are thus polymers which after completion of the treatment degrade thermally or break chemically without leaving any residue.

Experimental work carried out with sand has shown that sand permeability is continuously decreased with increasing closure stress (cf. section 1.2.2.), but beyond 4,500 psi the permeability change is inconsistent with polymer concentration and closure stress variations, with this divergent behaviour possibly being related to the bilinearity of sand porosity with rising closure stress. Alternatives to crosslinked fluids are discussed by CLARK, HALVACI, GHAELI & PARKS (1985), and ROODHART & DAVIES (1987) report on the revival of polymer emulsions as fracturing fluids.

While thermal fluid breakdown may be satisfactory in high-temperature reservoirs, thermal degradation of fracturing fluids is insufficient in low-temperature reservoirs where only chemical breakers can decompose the fluids.

4.3.4.1.3. Residue concentration

Fracture design with higher proppant saturations thus has to include assessment of fracture conductivity damage by crosslinked gel residue which varies with concentration of the viscosity-increasing polymer vs. incremental conductivity benefit and protection against embedment or crushing from higher proppant saturations in fluid and fracture which require the application of more viscous fracturing fluids that are capable of transporting larger proppant quantities per unit volume (COOKE 1975). Some points of filter cake thickness, polymer and breaker concentration, and residues plugging matrix and proppant permeability are illustrated as follows.

4.3.4.1.3.1. Filter cake thickness

Fluid leakoff from the induced fracture into reservoir matrix and natural cracks has to be minimized in order to avoid premature screenout termination of the hydraulic stimulation operation (cf. sections 4.2.3.2.5. and 6.2.4.2.1.) and to prevent crack damage by deteriorating proppant permeability. Fracture conductivity damage is caused by local concentrations of proppants and fluid remnants as a consequence of proppant settling by gravity, formation and erosion of a filter cake containing polymer residues and/or fluid-loss additives on the crack walls, and interference of the fracture faces with proppant transport along the crack. Some comments are offered as follows on reservoir permeability and fluid type as well as fluid pressure and breaking.

4.3.4.1.3.1.1. Reservoir permeability and fluid type

The filter cake is thicker in high-permeability reservoir intervals where fluid leakoff occurs at higher rates, thereby increasing the local concentration of residue-forming components opposite permeable zones, whereas tight pay sequences are characterized by lower fluid escape rates into the formation (cf. sections 1.4.10.3.5. and 4.3.4.6.4.). Polymer remnants, however, may help to minimize surface area effects of fluid-loss additives which otherwise have a strong influence on fracture permeability. Enveloping of fluid-leakoff agents by polymer residue may minimize fracture conductivity damage by preventing movement of fine particles through pores to plug narrow openings and by decreasing the surface area affecting flow resistance.

Crosslinked fluids leave more residue plugging the porosity of the proppant package than polymer emulsion fluids (DAVIES & KUIPER 1988). While polymer emulsion fluids result in a proppant package that is virtually residue-free, crosslinked fluids produce a proppant package containing a lot of fibrous material between the grains which are then glued together.

4.3.4.1.3.1.2. Fluid pressure and breaking

High-pressure fracturing fluids leaking away into formation matrix and natural cracks provoke generation of a filter cake consisting mainly of polymer and fluid-loss additives (cf. sections 1.4.10.3.5. and 4.3.4.6.4.). Filter-cake thickness is determined by type of stimulation fluid, formation characteristics, fracture-to-reservoir pressure, and erosional effects caused by slurry being pumped along the crack faces. Filter cakes created by crosslinked fluids with diesel as fluid-loss additive can almost completely plug narrow induced fractures, whereas polymer emulsion fluids do not give significant filter-cake buildup.

Proppant permeability damage by crosslinked fluids substantially depends on filter-cake buildup and is minimized by use of an effective viscosity breaker which is particularly necessary in low-temperature shallow formations. While an inadequate breaker leads to virtually complete loss of proppant-package conductivity, aggressive breaker schedules on the other hand can provide high retention factors over 80 %, but can also result in excessive proppant settling in the fracture before closure, with thus a proper balance being required (cf. section 4.12.1.1.). Polymer emulsion fluids are advantageous because conductivity recovery is less sensitive to breaker efficiency, whereas surfactants that are sometimes used in these fluids generate oil-wetness of the proppant package, with retained effective permeability to water being much lower in such cases.

4.3.4.1.3.2. Polymer and breaker concentration

Drastical reductions of fracture conductivity particularly occur in case of low proppant and high fluid residue saturations. In terms of accumulations of high concentrations of fines causing plugging of the proppant package, effects of column length are important (PYE & SMITH 1973) and polymer remnants may even give rise to more serious deterioration of fracture conductivity than fluid-loss additives (cf. section 1.4.10.3.). Factors determining crack conductivity reduction by influences of stimulation fluids are proppant concentration in the carrier medium, amount of residue in the fluid, proppant porosity and quantity share of residue retained in the fracture as treatment fluid leaks off (COOKE 1975). The amount of fluid remnants depends on polymer concentration as well as type and saturation of the breaker cutting the crosslinked gel into simpler constituents which finally achieve a viscosity in the range of that of water.

Local variations in proppant and residue saturations in a crack can explain why fracturing treatments can be successful using stimulation fluids that drastically decrease crack conductivity at normal saturation of injected proppants. Effects of treatment fluid velocity on fluid-loss agent performance are investigated by HALL & DOLLARHIDE (1964). GALL & RAIBLE (1985) report on molecular size studies of degraded fracturing fluid polymers. In case of problems with the original breaker concentration and operational impossibility to wait for complete thermal breakage, remedial breaker treatments by injecting breaker solutions infilling the fracture porosity with excess can be carried out in order to ensure fracture conductivity (WARPINSKI, BRANAGAN, SATTLER, LORENZ, NORTHROP, MANN & FROHNE 1985; cf. section 4.12.2.3.).

Another aspect of minimizing residue concentration and fracture conductivity damage is iron control which applies for both acid and proppant stimulations (DILL 1983). Many hydrocarbon-producing zones contain iron compounds in both matrix and fluids which can generate incompatibilities during treatments, because iron in solution may precipitate and deposit in the formation or the fracture, thus reducing flow capacity. Iron therefore has to be maintained in solution until recovery of the stimulation fluids, because dissolved iron has no blockage potential and production deliverability is at its maximum, in contrast to iron precipitation as crystalline fines or gelatinous mass which seriously hinders flow capacity (cf. section 4.5.1.1.4.).

4.3.4.1.3.3. Residues plugging matrix and proppant permeability

Results of MHF treatments in tight gas sands vary from extremely successful to very disappointing failures (AHMED, ABOU-SAYED & JONES 1979), with the latter failing to create large fractures of appropriate conductivity in the zone of high gas saturation. Achievement of the necessary propped crack geometry is mainly impaired by fracture deviation from the gas-bearing zone, production damage due to stimulation fluid interaction with the host rock matrix, and fracture conductivity damage due to treatment fluid interaction with the proppant package (cf. section 4.12.). Fracturing fluid selection for an effective MHF job depends not only upon fluid effectiveness in fracture creation and proppant transport, but also on minimization of damage to formation permeability and fracture conductivity.

Loss of stimulation fluid to the formation adjacent to the fracture face creates a zone of immobile fracturing liquid and reduces gas flow through the formation face due to diminution of relative gas permeability, and in addition to that, stimulation fluids leave behind solid and liquid residues. Reduction of matrix permeability by fracturing fluid is provoked by swelling of clay minerals and subsequent softening of the fracture face which leads to proppant embedment (cf. section 4.3.3.). Comparative evaluation of different treatment fluids has revealed that foam is the most favourable and successful fracturing medium (cf. section 3.11.2.) and that different stimulation fluids cause various amounts and intensities of damage and interact differently with specific lithologies by leaving different solid and liquid residues. Therefore reduction of fracture conductivity depends on the combination of treatment fluid and reservoir formation and their mutual compatibility.

Diminution of fracture conductivity with stress is principally due to proppant packing density. Conductivity damage of fractures plugged with full monolayers or multilayers is less pronounced than that of partial proppant monolayers. The most important reasons for fracture conductivity diminution at in-situ conditions are proppant packing, proppant embedment due to fracture face softening, proppant crushing, residues from stimulation fluids, release of fines from fracture face, and clay flocculation around proppant embedment sites (cf. section 1.4.10.).

4.3.4.1.4. Gel filter cake on fracture wall

Gel filter cakes accumulating on the fracture walls when fluid leaks off into the reservoir formation have an important deteriorating impact on fracture conductivity unless reasonably compensated by higher proppant concentrations infilling wider cracks (PARKER & McDANIEL 1987; cf. section 1.4.10.3.5.). Some aspects of gel filter cake thickness vs. proppant grain diameter and proppant concentration vs. fracture conductivity are outlined as follows.

4.3.4.1.4.1. Gel filter cake thickness vs. proppant grain diameter

Gel filter cake thickness is in many cases close to average proppant grain diameter for 20/40 mesh particles (PARKER & McDANIEL 1987). This relationship creates a potential to fill the proppant porosity with filter cake, and to loose the flow along the wall of both fracture faces and for a depth of one or more proppant grains on each side of the crack. As proppants embed into the filter cake, they may displace parts of this thick gel pellicle towards the center of the proppant package within the fracture, thereby seriously destroying fracture conductivity unless multilayer propping of the crack is achieved by high proppant concentrations.

4.3.4.1.4.2. Proppant concentration vs. fracture conductivity

Experimental results show that doubling proppant concentration from 1 to 2 lbs/ft2 (5 to 10 kg/m2) gives rise to a five- to ten-fold increase of fracture conductivity, leading to conclusion that in case of presence of gel filter cakes, proppant saturations of 1 lb/ft2 may not be sufficient for adequate production stimulation even at low closure stress levels. In soft reservoir rocks, proppant concentrations of 3 - 4 lbs/ft2 (15 - 20 kg/m2) are recommended to compensate proppant embedment into the formation (GARBIS, BROWN & MAURITZ 1985; McDANIEL & PARKER 1988).

The same applies in unstable collapsing chalk reservoirs where proppant saturation should be at least 2 lbs/m2 in order to avoid embedment deterioration of fracture conductivity (SIMON, COULTER, KING & HOLMAN 1982; cf. section 4.5.4.). McDANIEL & PARKER (1988) even conclude that gel filter cake damage is not universally represented with only data at 2 lbs/ft2. As permeability of a propped fracture is not constant in the presence of gel filter cakes, design simulators have to allow for input of conductivity data being a function of proppant concentration. Aspects of proppant saturation and conductivity in the presence of gel filter cakes are also discussed by ROODHART, KUIPER & DAVIES (1986); PARKER & McDANIEL (1987), PENNY (1987) and McDANIEL (1988).

Similarly as gel filter cakes on formation surfaces at fracture walls deteriorate crack conductivity by proppant embedment, gel surface filtration from polymer solutions during gravel packing creates a low-permeability drape on the reservoir face which upon rapid buildup dominates system behaviour (TORREST & MARCANO 1986) and inhibits further fluid leakoff into the pay matrix, thereby having deleterious effects on gravel placement and thus quality and performance of mechanical sand control (cf. sections 1.4.10.3.5.3. and 5.3.1.1.2.2.).

4.3.4.1.5. Possibilities of fracturing fluid improvement

Improvement of fracture conductivity in terms of fluid composition can be chiefly made by selection of completely degradable fluid-loss additives which are stable at high temperatures and are decomposed without reasonable residue (cf. sections 4.3.4.6.4. and 4.8.8.3.1.), and of delayed-activity breakers which stay dormant until the end of the fracturing job and then quickly split the crosslinked gel into small fragments with again not leaving any significant residue. This is particularly important due to the fact that many of the presently applied breakers act very shortly after their addition to the fluid stream and 60 - 90 % of the original fracture conductivity is lost after placement of the proppant package by residue plugging (cf. also section 1.4.10.). Some aspects of foam utilization and fluid efficiency vs. residue deposition are outlined as follows.

4.3.4.1.5.1. Utilization of foam

The application of foam fracturing fluids (cf. section 3.11.2.) enables retainment of up to 90 % of the original proppant package conductivity due to the better clean-up characteristics of the foam which does not require breaking, but first foams are limited in application due to their temperature sensitivity and second high foam quality of abt. 75 % leaves only 25 % liquid phase for the carriage of proppants, resulting in only lower proppant concentrations to be placed downhole. This effect, however, is compensated by the better fracture conductivity retainment with respect to high proppant saturations placed by crosslinked gels (cf. section 4.3.4.3.) where later much conductivity is destroyed by residue plugging as well as by better clean-up properties which are particularly significant in low-pressure formations where reservoir energy available for fluid flowback after the end of the stimulation treatment is limited. In low-permeability formations, leakoff coefficients for some foams can be lower than those of crosslinked aqueous fracturing fluids (VEATCH & MOSCHOVIDIS 1986).

4.3.4.1.5.2. Fluid efficiency vs. residue deposition

In addition to impairment of fracture conductivity by plugging with residue formed upon breakage or degradation of gelled fluids, permeability of the proppant package can be deteriorated by capture and entrainment of fine particles deriving from proppant crushing (CARROLL & BAKER 1979; cf. section 4.12.4.). The principal mechanisms that cause deposition during transport of particles within a porous medium are sedimentation, direct interception, and surface attractive properties of particles and medium. Plugging of a propped fracture corresponds to removal of suspended particles by contact between the pore walls and the grains when the pores are generally larger than the particles in suspension. Residue-forming material caught in the pores upon release of the parting pressure reduces pore diameter and partially blocks the interconnecting channels (cf. sections 4.3.4.1.3.).

Problems encountered in hydraulic proppant fracturing of deep reservoirs are often created by high bottom hole pressures and temperatures, small diameter tubing, and tubing contraction during stimulation (CONWAY & HARRIS 1982). The high operating pressures often necessitate the use of fluids with minimal pumping friction pressures. In order to minimize contraction in long tubing strings with small diameter, treating fluids must often be preheated (cf. section 4.5.1.), with the high initial fluid temperatures considerably altering fluid properties downhole. Fluid efficiency in most hydraulic fracturing operations varies from 30 - 90 %, with maximum percentage reflecting minimum leakoff into the formation which is desirable for successful execution of the job. It is highly desirable to use fluids which can provide adequate viscosity for proppant transport, but yet are controllable so that neither excessive fracture height nor excessive fracture width do not occur.

4.3.4.2. Fluid viscosity and transport ability

Concerning proppant transport effectivity, low-viscosity fluids (10 - 20 cp) result in settling of proppants to the bottom of the fracture, and bed height can be irregular when particles of different size are used during the treatment. Low-viscosity carrier fluids are therefore only suitable for propping of shorter fractures, with undesired segregation effects having to be taken into consideration in case of transport of proppants farther away from the borehole. Medium-viscosity fluids (50 - 200 cp) provide good but not perfect transport, with proppant beds still possibly being irregular. Medium-viscosity fluids can carry 20/40 or smaller proppants for longer distances in more extensive fractures, but larger grain sizes will settle more rapidly. High-viscosity fluids (over 200 cp) can carry all sizes of proppants with only minimal settling far away from the wellbore within long cracks, and the resulting fracture is usually very uniform (HOLDITCH 1984).

High fluid viscosity reduces friction while moving through the pipe (SHAH 1984 and SHAH & LEE 1986 comment on friction pressures of proppant-laden hydraulic fracturing fluids; cf. section 4.3.4.2.2.1.), permits transport of high proppant saturations and creates wider fractures thereby making it easier to place the propping agent (WATERS 1980). The following discussion focusses on proppant concentration and vertical fracture growth as well as on proppant density/fluid viscosity relationship and fluid volume per reservoir thickness unit. An outline is also given of the concept of overpressure calibrated stimulation design.

4.3.4.2.1. Proppant concentration and vertical fracture growth

The significance of proppant concentration for vertical fracture growth is illustrated along the lines of high- and low-viscosity stimulation fluids as well as shearing of gelled fluids.

4.3.4.2.1.1. High-viscosity fluids

High-viscosity fluids create wider fractures and are better for transporting propping agents, but are also exerting higher treating pressures which have a potential for undesired vertical height growth (MENG & BROWN 1987; cf. section 4.2.2.3.3.). Proppant concentration and pad volume determine the final propped fracture penetration and conductivity. Insufficient pad volume often results in premature screenout due to early pad depletion (cf. sections 4.2.3.2.5. and 6.2.4.2.1.), and short propped penetration is not uncommon when pumping excess pad volume. It is therefore necessary to determine the optimum pad size and proppant schedule in order to ensure that the proppants will be transported to the fracture tip at the end of the injection process (NOLTE 1984, MENG & BROWN 1987). Generally high-viscosity fluids provide improved results from hydraulic fracturing treatments (ROSENE & SHUMAKER 1971).

4.3.4.2.1.2. Low-viscosity fluids

Low-viscosity fluids injected at low pumping rates in treatments of restricted size are an effective means of controlling vertical growth of fracture height, but suffer from limited proppant transport capabilities, with proppant banking occurring as a consequence of proppant settling during fluid insertion into the fracture and before closure of the crack (BARBER & THEMIG 1985; cf. section 4.12.2.2.). Coordinated rather than random and undesired proppant banking or stratification as well as selective placement can be achieved by fluid density control including pumping of two fluids of different specific gravity and viscosity which override or underride each other and which can be charged with different proppant loads for distribution of solid-supporting particles in the upper and/or lower segment of the fracture as horizontal layers (cf. sections 4.2.2.4. and 4.3.3.2.3.).

4.3.4.2.1.3. Shearing of gelled fluids

Viscous gelled water systems feature moderate polymer loadings and low residue on breaking (HANNAH & MATSON 1977). Shearing of gelled fluids results in reduction of viscosity and thus diminution of proppant-carrying capacity (ROLL, HIMES, EWERT & DOERKSEN 1986; cf. section 4.3.3.3.2.). Viscosity decrease of fluids by shearing is highly undesirable, because it leads to proppant fallout during pumping and placement which gives rise to premature screenout failure by proppant bridging (cf. sections 4.2.3.2.5. and 6.2.4.2.1.) in pump, casing or tubing, screen, crossover tool, perforations and other devices. Proppant concentration in the carrier fluid has only little effect on shear degradation of the gel. Fracture height and width is generally greater with viscous fluids (OSBORNE, McLEOD & SCHROEDER 1981). In deep high-pressure wells, the tubular goods often limit injection rate, and shear rates are lower in these fracturing treatments so that apparent viscosity of fluid flowing in these cracks is higher.

Shearing of gelled fluids is predominantly occurring at the perforations in limited-entry constellations (HOEL 1988; cf. section 4.2.2.2.). Crosslinked fluids having undergone shear exhibit lower viscosity which renders them unable to carry and place higher proppant concentrations, but allows deeper proppant penetration with less height growth that results in connection of more net pay to the wellbore (HOEL 1988). If a higher-viscosity fluid were pumped in an attempt to transport large volumes and to place high saturations of proppant, broader fractures could be obtained which cause more height growth out of zone and less proppant penetration within the core of the prospective interval. Case studies reveal that at least for some reservoirs, deep proppant penetration at the expense of high proppant concentration is the better solution of drainage maximization.

4.3.4.2.1.4. Other aspects

High apparent viscosities lead to high pressure drops in the fracture which can cause premature shut-downs because of wellhead pressure limits, or else barrier shales surrounding the sandstone formation are parted and the fluid fractures out of zone. Premature shut-downs give rise to shorter cracks than expected, lower final injected proppant concentrations and poorer fracture conductivities, and stimulating out-of-zone leads to higher but shorter cracks, with in both cases the expected productivity being lower than that desired.

HANNAH, HARRINGTON & ANDERSON (1977) document that the temperature of fluids flowing in a fracture is near the formation temperature over 50 - 75 % of the crack length. The apparent viscosity at the leading edge of the fracture is the most important one since that is where the greatest treatment fluid pressure drop takes place. It is also important that the apparent viscosities of the final fluid stages are not too high under near-wellbore cool-down conditions, as this could cause undesirable fracture height extension into barren shales or nonproductive sand sections.

4.3.4.2.2. Proppant density/fluid viscosity relationship and fluid volume per reservoir thickness unit

The greatest advantage of decreasing specific proppant gravity is the fact that lighter proppants can be transported by lower-viscosity fluids at identical conditions as compared to heavier proppants and higher-viscosity fluids, with pumping of lower-viscosity fluids being able to be carried out at lower pressure (CUTLER, ENNISS, JONES & CARROLL 1983; cf. sections 1.3.2. and 4.2.2.3.3.). Vertical propagation of the crack can be limited by reducing fracturing fluid pressure, and minimizing fluid viscosity by insertion of lighter proppants is a much more effective method of controlling pressure than lowering the pumping rate (SIMONSON, ABOU-SAYED & CLIFTON 1978; NOLTE 1982; cf. section 4.2.2.3.2.). Alternatively, higher proppant volumes can be pumped in a given amount of high-viscosity fracturing fluid.

Fluid volumes per reservoir thickness unit also have an important influence on proppant concentration placed within the fracture and thus post-treatment well performance (NALL, CAMPBELL & BONEY 1983). Field experience has shown that the lowest skin effects, greatest calculated fracture permeabilities, and higher decreased absolute open-flow potential tests were obtained when volumes of 2,000 gal/ft net reservoir thickness were pumped. In terms of higher fluid quantities, the key to the observed productivity increase is the ability to place more proppants in higher saturation in the fracture with the larger fluid amounts, but economics have to be considered on a well-to-well basis to determine the best possible treatment volume for the money spent. The discussion as follows focusses on friction loss in tubular goods, gel degradation, fluid viscosity modification, and fracture height vs. width.

4.3.4.2.2.1. Friction loss in tubular goods

Tubing and casing design also have a large influence on treatment rates and pressures which in turn influence proppant concentration that the formation can accept (PAI & GARBIS 1983 b). Friction calculations for the fluid in the pipe and across the perforations need to account for the increased viscosity and density due to high proppant saturations (SHAH, SMITH & DONALDSON 1983). Because friction loss across the perforations can become excessive as fluid density rises, it is not uncommon to increase the perforation number by 50 or 100 % in an ultra-high proppant concentration job. During course of the operation, the actual bottomhole treating pressure may be gradually rising being a possible indication of a screenout, whereas the surface pressure is dropping due to the large increases in fluid densities.

4.3.4.2.2.2. Gel degradation

The structure of most fracturing gels tends to degrade and lose its proppant transport ability to some degree with both increasing temperature and time (POULSEN & LEE 1984). Although some gels require elevated temperatures to attain their maximum degree of crosslinking, these fluids also degrade with sufficient temperature and time (cf. section 4.3.4.3.). Since temperature and presence time of the fluid in the fracture both increase with distance from the wellbore, it follows that the closer the fluid is to the crack tip, the less proppant-carrying ability it has, with therefore at the fracture tip most easily proppant bridging and blocking occurring which provokes premature or submature stimulation job termination by screenout.

4.3.4.2.2.3. Fluid viscosity modification

When monitoring net pressure evolution during course of the fracturing treatment, the ability to alter fluid viscosity can improve stimulation effectiveness (OLSEN & DEBONIS 1988). Fluid viscosity and pumping rate affect fracture geometry almost equally when altered at the same magnitude, but in most cases, it is operationally easier and more cost efficient to triple fluid viscosity than to triple pumping rate. Therefore modifying fluid rheology seems to be a more viable means of altering fracture geometry in a real-time mode than pumping rate, although both can be used concurrently (cf. section 6.2.4.3.). If net fracturing pressure exceeds critical formation pressure or capacity of the latter, either excessive fluid leakoff or uncontrolled fracture height growth occur which can be both detrimental to fracturing effectiveness (cf. section 4.2.2.3.). If formation pressure capacity for a given reservoir is known, stimulation fluid viscosity can be modified during the treatment to prevent net pressure from exceeding formation pressure capacity.

4.3.4.2.2.4. Fracture height vs. width

Pad or other non-proppant stages are propagating the hydraulically-induced fracture. As this is the first fluid exposed to new rock in the fracture, pad stage often suffers most severely from fluid-loss conditions. Excessive pad fluid loss gives rise to almost complete fluid leakoff into the formation thus bringing proppant-bearing fluid near the fracture face. As proppant-bearing fluids do not propagate hydraulic cracks, fracture penetration is restricted and balloons in width due to the large difference between injection rate and leakoff rate (cf. section 6.2.4.2.1.). At this point, there is no need for fluid viscosity to maintain fracture width, and thus polymer loading could be reduced to the minimum viscosity required for proppant transport which diminishes the amount of polymer used, decreases conductivity damage in this confined fracture area, and also slows down pressure increase. A suitable improvement is considerable enlargement of the pad fluid stage of the purpose of both leakoff control and reservoir cooling (cf. section 4.3.4.4.2.2.).

On the other hand, sudden unexpected and unavoidable occurrence of excessive fracture height can dramatically decrease crack width which in turn can cause proppant bridging in the fracture leading to rapid pressure increase and premature treatment termination by screenout (cf. section 4.2.2.2.3.). The worst case is the combination of high fluid loss and serious width reduction caused by uncontrolled height growth which can provoke proppant bridging fairly close to the wellbore that can result in screenout failure at a very early stage of the treatment. Viscosity increase of the fluid could compensate for the width loss from height growth and help preventing submature screenout failure (cf. sections 4.2.3.2.5. and 6.2.4.2.1.). Field experience reveals that ramped polymer loading jobs have much higher pressure falloff rates after shutdown than conventional constant polymer loading operations.

4.3.4.2.3. Overpressure calibrated stimulation design

Another method of enhancing treatment size and thus proppant quantity and concentration is overpressure calibrated stimulation design (SHLYAPOBERSKY, WONG & WALHAUG 1988) where the net crack pressure measured during mini- or microfracturing jobs (SHLYAPOBERSKY, WALHAUG, SHEFFIELD & HUCKABEE 1988; cf. section 4.8.2.6.) shortly after shut-in (overpressure) is used to determine apparent fracture toughness of the formation which is then taken as an input in a reservoir simulator to design hydraulic fracture stimulation volume and proppant injection schedule. The overpressure is defined as the difference between the corrected instantaneous shut-in pressure and the minimum in-situ stress. The outline as follows focusses on conventional underestimation of net fracture propagation pressure, reasons for elevated net fracture propagation pressures, scale dependency of apparent fracture toughness, and higher proppant volumes and concentrations pumped with overpressure calibrated fracture design.

4.3.4.2.3.1. Conventional underestimation of net fracture propagation pressure

With conventional models, net fracture propagation pressure (defined as fracture propagation pressure minus minimum in-situ stress or closure stress) is always underestimated, with these models thus predicting a narrower and longer propped crack than actually is being created in the field, and therefore the design is not optimized to produce the desired fracture geometry and maximum crack conductivity through high proppant concentrations (SHLYAPOBERSKY, WONG & WALHAUG 1988). Apparent modelling of wider cracks than technically possible as a consequence of underestimation of net fracture propagation pressure can also lead to an unwanted aggressive treatment that has a high potential of submature screenout.

Optimization of the hydraulic stimulation operation thus requires calibration of the model with the field net crack propagation pressure which results in shorter predicted length that is more consistent with the productive response of the well. Overpressure calibrated design enables to perform more aggressive treatments with smaller pad and larger amount and saturation of proppants than with conventional models and permits to significantly diminish the risk of premature screenout termination (cf. sections 4.2.2.2.3., 4.2.3.2.5. and 6.2.4.2.1.). The necessary data for overpressure calibrated stimulation design are measured during mini- and microfracture field tests (SHLYAPOBERSKY 1988; SHLYAPOBERSKY, WALHAUG, SHEFFIELD & HUCKABEE 1988; cf. section 4.8.2.6.).

4.3.4.2.3.2. Reasons for elevated net fracture propagation pressures

Higher net fracture propagation pressures as usual can be created by higher than expected frictional pressures developing in perforations, near the wellbore region and in the crack especially near its tip (SHLYAPOBERSKY, WONG & WALHAUG 1988; cf. section 6.2.4.2.1.). The enhanced friction could be caused by increased slurry viscosity due to leakoff and proppant concentration, fracture surface roughness, turbulent fracture flow under non-Darcy conditions (cf. section 4.3.3.3.), and major current constrictions along the crack caused by orthogonal joints such as natural fractures. Other reasons for elevated net crack propagation pressures are higher fracture toughness than expected due to a scale-dependent crack layer at the fracture tip, and presence of a fluid-nonpenetrating region at the latter point.

Creation of a set of multiple parallel fractures (cf. section 4.8.6.2.4.) instead of a single dominant crack is unlikely to be a reason for higher net fracture propagation pressures, as many successful treatments have been pumped with moderate net crack propagation pressures with high proppant saturations and a viscous slurry. During fluid injection, both frictional and fracture toughness effects are present and difficult to separate, whereas shortly after shut-in when the crack remains open and nearly ceases to propagate with fluid redistribution and pressure instabilization, the frictional effects are minimized, and the net fracture pressure at this moment representing the overpressure can then be primarily attributed to apparent fracture toughness (SHLYAPOBERSKY 1987).

4.3.4.2.3.3. Scale dependency of apparent fracture toughness

Successive field fracturing tests conducted in the same formation with identical rate and fluid viscosity revealed occasionally that the instantaneous shut-in pressure increases with rising injection volumes, whereas closure stress remains nearly unchanged (SHLYAPOBERSKY, WONG & WALHAUG 1988). As proppant bridging cannot occur due to proppant absence in the mini- or microfracture tests, the increase of instantaneous shut-in pressure or overpressure with injection volumes may be a manifestation of the scale dependency of apparent fracture toughness (SHLYAPOBERSKY 1987). In order to predict realistic crack dimensions, hydraulic fracture design therefore has to include apparent crack toughness of the rock in addition to the other interacting processes of viscous fluid flow, elastic rock deformation, and fluid loss through fracture walls.

The energy analysis of hydraulic stimulation model (SHLYAPOBERSKY 1985) incorporates the effect of fracture toughness for uncontained (circular or radial) and contained (rectangular or linear) fractures. The model considers the fluid-filled hydraulic crack as an open thermodynamic system and the quasi-equilibrium propagation of the fracture as the evolution of the system. An approximate solution for such a system is obtained by minimizing total energy dissipation rate as required by the universal evolutionary criterion. Energy dissipation takes place by creation of the new fracture surface, crack-opening by deforming the surrounding reservoir, and viscous dissipation in fluid.

4.3.4.2.3.4. Higher proppant volumes and concentrations pumped with overpressure calibrated fracture design

Field examples of comparative assessment of overpressure calibrated design and conventional modelling show that implementation of the overpressure concept can more than double the treatment volume by permitting creation of wider fractures that accept higher proppant concentrations and larger total proppant quantities (SHLYAPOBERSKY, WONG & WALHAUG 1988). Even with some chosen degree of conservatism, the resulting overpressure calibrated treatments are still much more aggressive in both slurry volume and amount of proppant injected than conventional designs. Thus overpressure calibrated fracture modelling leads to very effective stimulation operations with much larger fluid volumes and considerably more aggressive proppant injection schedules than conventionally designed. The increased job sizes are the confirmation that the standard models underestimate net fracture propagation pressure. The technical and economical benefit of overpressure calibrated fracture design is the possibility of more than doubling total treatment volume and halving pad size with respect to conventional approaches.

Overpressure calibration can also be used to optimize the aggressiveness of the proppant schedule (cf. section 4.13.) and to reduce screenout risk according to field observations and sound engineering judgement. A larger overpressure generally results in a wider fracture that requires a greater treatment volume to produce a desired economical fracture length and allows a more aggressive proppant schedule to optimize treatment productivity, whereas an overpressure smaller than that measured in the field can be used in the design to control the aggressiveness of the proppant schedule and to reduce the risk of premature screenout (cf. section 6.2.4.2.1.). The utilization of overpressure instead of net fracture propagation pressure as a calibration parameter for hydraulic fracture stimulation design thus provides considerable advantages particularly in new fields where experience and uncertainties with the model data have not been fully established, and is an excellent tool for minimization of frictional uncertainties, unreliabilities in using treatment pressure to predict fracture growth, and risk of premature screenout failure. Aspects of overpressure calibrated fracture design are also discussed by HUCKABEE (1988).

4.3.4.3. Impact of delayed crosslinking

The impact of delayed crosslinking on proppant concentration is predominantly the reduced deterioration of proppant saturation by premature material fallout. After some introductory comments on crosslinking types, significance and performance, improvement of proppant placement by retarded crosslinking as well as enhanced fluid stability and transport capacity is outlined.

4.3.4.3.1. Crosslinking types

Crosslinking of fracturing fluids with some type of transition metallic compound such as borate or titanate gives rise to formation of chelate complexes which have considerably enhanced proppant transport and temperature stability characteristics than conventional fracturing fluids (GRONE, BAUMGARTNER & WOODROOF 1983; VEATCH 1983, WALSER 1985; KRAMER, PRUDHOMME, NORMAN & SANDY 1987), and that also aids in fluid-loss control and combats potential screenout failure (NALL, CAMPBELL & BONEY 1983). The crosslinking reaction rate of the organometallic chelate with derivatized cellulose or guar is influenced by a combination of pH and temperature (HARRIS, HARMS & NORMAN 1987). The following overview contains some comments on general benefit of crosslinking as well as conventional and delayed crosslinking. Aspects of viscosity relationships, continuous vs. batch mixing, delayed crosslinked fracturing fluids vs. dual-viscosity gravel packing fluids, and mixing of delayed and rapid crosslinking agents are also illustrated.

4.3.4.3.1.1. General benefit of crosslinking

Crosslinking also increases fluid viscosity and thus proppant transport capacity (GRONE, BAUMGARTNER & WOODROOF 1983). The introduction of crosslinked fracturing gels in the late 1960`s has in many ways been revolutionary, because their unique advantage is the very high viscosity yield for modest polymer saturations which makes them ideal for placing higher proppant concentrations due to inherent high efficiency, ability to create wide fractures, excellent proppant transport characteristics (HANNAH, HARRINGTON & POTTER 1980; PAI & GARBIS 1983 b) and extremely low fluid loss (NORTON & HOFFMAN 1982). As they employ lower polymer concentrations, crosslinked fluids tend to break cleaner and produce less damaging residue than uncrosslinked fluids. Conventional crosslinked fluid systems are stable up to 275 oF, but with enhanced temperature stability mixtures, successful fracture treatments have been carried out at temperatures up to 540 oF (280 oC; WILLIAMS, WOODROOF & BOX 1982; cf. section 4.7.4.2.). Different ligands can be utilized for faster or slower crosslinking of the fluid. Delayed crosslinking systems are formulated to activate after the fluid has been pumped down the tubulars and through the perforations and has been one of the recent significant improvements in fracturing fluid technology (VEATCH & MOSCHOVIDIS 1986).

The only disadvantage of crosslinked fracturing fluids is their unsatisfactory fluid compatibility (NORTON & HOFFMAN 1982), because many jobs using crosslinked gels result in moderate to severe water blockage problems (cf. section 3.11.2.4.2.). While addition of nitrogen to the crosslinked gel helps recovering the treating fluid, the water blockage problems still occur despite presence of nitrogen and use of surfactants. Water blockage, however, also occurs with application of non-crosslinked water-based gels. An unusual special application of crosslinking is connection of an acrylic resin and a liquid epoxy resin for destruction of reservoir permeability by plugging of thief zones (KNAPP & WELBOURN 1978; ZAITOUN, KOHLER & GUERRINI 1989; cf. section 5.9.1.7.).

4.3.4.3.1.2. Conventional and delayed crosslinking

Three types of crosslinker systems have to be distinguished (ZIGRYE, OSBORNE & WESTBROOK 1986). Early time systems begin crosslinking immediately when injected into the base gel. Delayed crosslinker systems start to act some time after injection into the base gel, with the reaction being activated by temperature. The controlled early time crosslinker system is a combination of the two other systems, and when it is injected into the base gel, some crosslinking occurs immediately followed by a major crosslinking reaction initiated at higher temperature of abt. 100 oF (38 oC). The primary advantage of delaying the crosslinking reaction is that viscosity would not increase until shortly before the fracturing fluid reaches the bottom of the wellbore and enters the fracture. Delaying can be achieved by using an inherently delayed crosslinker (WALSER & OTT 1985) or adding a delaying agent to a fast-reacting crosslinker (RUMMO 1982).

This delay in crosslinking minimizes friction pressure in the injection tubing as well as degradation of crosslinked fracturing fluid caused by high shear in the tubing and thus diminishes hydraulic horsepower required for the job (GARBIS, BROWN & MAURITZ 1985) and permits higher injection rates (ECONOMIDES 1986). The delayed crosslinked gel forms when the fluid is far down the tubing string after the high-shear portion of the treatment (HARRIS, HARMS & NORMAN 1987). The small immediate viscosity increase generated by the controlled early time crosslinker system is sufficient to insure proppant transport as the fluid flows into the fracture, but it does not significantly increase tubing friction pressure. HODGE & BARANET (1987) evaluate field methods to determine crosslinking time of fracturing fluids.

4.3.4.3.1.3. Viscosity relationships

The viscosity of delayed crosslinked fracturing fluids is a function of flow geometry, critical stress and time as well as shear and thermal histories (CAMERON, GARDNER & VEATCH 1988). The stress-transformation property of delayed crosslinked gels is the result of gel breaking-down and migrating into the low-viscosity fluid at the walls, and possibly shear-induced flocculation and secondary flows. In low-stress state where the gel shows very low apparent viscosity because of wall slip, perfect proppant transport may be possible since the bulk fluid is completely gelled. Delayed crosslinked gels are generally characterized by non-homogeneous flow (CAMERON 1988). Delaying crosslinking until after exposure to high shear increases gel viscosity (PRUDHOMME, ELLIS, CONSTIEN & KNOLL 1988). The rheology of crosslinked gels is determined by the structure of the gel network which in turn is controlled by crosslinker reaction kinetics and flow conditions during crosslinking. In fast crosslinking systems where the crosslinking process is rapid enough that the gel structure is established during mixing or during exposure to high shear rates, every different configuration of tubing geometry and pumping rate produces a different gel network structure and different fluid rheology, and the same gel chemistry generates different viscosities depending upon magnitude and duration of shear.

Delaying of gel formation until a time when only low levels of shear are experienced by the fluid (for example after the fluid has travelled through the wellbore) allows a more highly structured gel to originate thereby resulting in higher viscosity. The rate at which the gel is chemically formed can be controlled by the addition of chelates which slow the reaction between the polymer molecules and the transition-metal ion (KRAMER, PRUDHOMME, NORMAN & SANDY 1987). The viscosity of the delayed system is several times greater than that of the fast linking gel and maintains a higher viscosity at elevated temperature, thus demonstrating that higher gel viscosity can be obtained for a given chemistry if gel formation occurs after exposure to high shear rates.

4.3.4.3.1.4. Continuous vs. batch mixing

Delayed crosslinked gel viscosity is influenced by shear rate and the degree of base gel hydration (HARRIS, HARMS & NORMAN 1987). Chemical factors influencing viscosity development include base gel concentration, crosslinking agent saturation, fluid pH and ionic strength. Hydration time and pH requirements may differ for specific polymer-crosslinker pairs. For many years, hydraulic fracturing operations have used pregelled or batch-mixed viscosifiers as the standard base fluid for application of crosslinked gelled fracturing fluids. Continuous or on-the-fly mixing of fluids, however, is more economical, but it is essential that base gel polymer and additives are delivered to an aqueous stream so that rapid uniform gelation occurs, and gelation must be sufficiently complete prior to crosslinking agent reaction to give full viscosity development.

The key elements in producing continuously mixed fluids having performance equal to batch-mixed fluids include efficient gel delivery, proper mixing energy level, pH control and utilization of a delayed onset crosslinker. The most rapid base gel hydration is attained with high mixing energy, low pH and minimum salt concentration. Normally tubing transit time is long enough and formation temperature sufficiently high to allow polymer hydration and to initiate the crosslinking reaction, but shallow wells with very short tubing transit time and relatively low formation temperature may require increased precision in use of delayed crosslinked gel fluids. Higher crosslinker concentrations are required for quicker achieving higher crosslinked gel viscosity at low temperatures, whereas lower crosslinker saturations should be used at elevated temperatures.

The most obvious application of delayed crosslinking agents are jobs where it is desirable to reduce polymer loading and yet obtain comparable viscosities to those achieved with conventional systems at higher polymer loadings (CONWAY & HARRIS 1982) as well as better fluid stability (ECONOMIDES 1986). The possibility of delayed crosslinked fluids to transport higher proppant concentrations enables performance of treatments with smaller fluid volumes. Pumping of high-viscosity crosslinked fracturing gels generally requires batch-mixing prior to the job (FREE, FREDERICK & THOMPSON 1975; cf. also section 4.3.7.5.3.).

4.3.4.3.1.5. Delayed crosslinked fracturing fluids vs. dual-viscosity gravel packing fluids

Delayed crosslinking of fracturing fluids can to some extent be compared with dual carrier viscosity modelling of gravel packing fluids (cf. section 5.3.2.2.), although the viscosity change takes place in reverse direction in the latter system. While delayed crosslinked fracturing fluids have lower viscosity at the surface and on their way down the tubulars in order to diminish friction pressure and achieve higher viscosity prior to entrance into the fractured reservoir in order to be able to suspend and transport high proppant concentrations as well as to limit fluid loss, dual-viscosity gravel pack fluids have higher viscosity at the surface for ameliorating mixing and pumping of the slurry and lower viscosity downhole in order to facilitate the necessary high fluid-loss rates to achieve a tight pack.

4.3.4.3.1.6. Mixing of delayed and rapid crosslinking agents

Special applications are mixing of delayed and rapid crosslinking agents (ZIGRYE, OSBORNE & WESTBROOK 1986; HARRIS, HARMS & NORMAN 1987). As delayed crosslinking agents may not produce sufficient crosslinked gel viscosity near the perforations to transport proppant, adding of a rapid crosslinker ensures that the fluid in or near the wellbore has also crosslinked gel viscosity. Another special type of utilization is in-situ crosslinking and subsequent gelation of water-soluble polymer solutions in the vicinity of the wellbore in order to decrease relative permeability to water and to inhibit water coning and channelling through natural fractures at high-permeability streaks in heterogeneous reservoirs (MODY, McKITRICK & SHAHSAVARI 1988).

Aspects of delayed crosslinked fluid systems are also reported by CONWAY & HARRIS (1982), RUMMO (1982), BARANET & AINLEY (1985), LI & ZHU (1986); CAMERON, GARDNER & VEATCH (1988); CONSTIEN, BRANNON & BANNISTER (1988); HOLDITCH, ROBINSON, ELY & RAHIM (1988) and ROSE (1988). Rheological features of crosslinked fracturing fluids are also discussed by PRUDHOMME, ELLIS, CONSTEIN & KNOLL (1988).

4.3.4.3.2. Crosslinking significance and performance

Crosslinked fluids develop high viscosities that exhibit several advantages over non-crosslinked systems being extremely good fluid leakoff control, yield of wider fractures that accept higher proppant concentrations, perfect proppant transport properties, and better temperature stability (PAI & GARBIS 1983 b). Field experience has shown that controlled early time crosslinker systems generally perform much better than delayed crosslinker systems especially when the bottom hole flowing temperature of the fracturing fluid is not high enough to initiate the crosslinking reaction before the fluid reaches the fracture, which would give rise to proppant drop-out resulting in pressure increase and screen-out failure of the stimulation job.

Therefore early time viscosity and retained viscosity are equally important to insure success of a fracturing treatment. Delayed crosslinking considerably enhances fluid viscosity and enables to increase proppant concentration up to 20 lbs/gal fluid at the surface (translating to up to 7 lbs/ft2 or 35 kg/m2 in the fracture; SHAH, SMITH & DONALDSON 1983; cf. section 4.3.1.1.) or even more over long pumping times (GREGORCZYK, PAULS, HOLTMYER, CHISHOLM & VENDITTO 1984). Aspects of friction pressure lowering, screenout rate reduction, friction pressure vs. fracture height growth, fluid viscosity and proppant suspension improvement, and fracture closure and breaker action are briefly discussed as follows.

4.3.4.3.2.1. Friction pressure lowering

Employing delayed crosslinker systems also avoids excessive friction pressure in the tubular goods which enables reduction of the hydraulic horsepower requirements for the fracturing treatment and thus permits to lower the operational costs (SHAH & WATTERS 1986). Correlations of friction pressures of delayed crosslinked fracturing fluids from laboratory measurements, yard tests or field annular flow situations, however, may not be applicable to friction loss prediction for flow down a long vertical tubing string (LORD & McGOWEN 1986). In particular, correlation of laboratory friction diminution measurements made with stimulation gels with or without proppants overpredicts friction loss of these fluids when they are injected down field tubing strings. Delayed crosslinked fracturing fluid systems generally exhibit excellent viscous properties in the crack, good temperature stability, and practically no increase in turbulent friction pressure over base gel (SHAH & WATTERS 1986).

The viscosity obtained from crosslinked fluids is progressively reduced by time and intensity of shear to which the fluid is exposed, and preheating of fluids for the purpose of tubing contraction minimization increases crosslinking rate and shear degradation of fluid viscosity (McMECHAN & CONWAY 1983). Delayed crosslinked fluids with controllable crosslinking time are not subjected to the high shear conditions found in the tubular goods, and the rheology is much more insensitive to treating fluid temperature (CONWAY & HARRIS 1982). The use of temperature-delayed crosslinked fluids results in lower and more consistent friction pressures with respect to conventional organometallic crosslinkers and base linear-gelled fluids, and an excellent friction pressure tolerance and compensation range even leads to absence or negligibility of friction pressure rise upon addition of proppants up to considerably high concentrations (WALSER 1985).

4.3.4.3.2.2. Screenout rate reduction

The much improved proppant transport conditions are underlined by significantly reduced screenout rates (SHAH & WATTERS 1986). Higher average proppant saturations allow more proppant to be placed downhole with less fluid which opens way for to hydraulic horsepower diminution, treatment fluid requirements reduction, and polymer loading lowering, with all of these effects giving rise to significant savings of stimulation expenses. As a consequence of very effective proppant suspension, proppant transport characteristics are somewhat independent of depth, bottom hole static temperature and pumping time. HOLDITCH, ROBINSON & ELY (1988) comment on the effect of viscous fluid properties on excess friction pressures during hydraulic fracturing treatments. WOODROOF & ANDERSON (1977) discuss friction reducers in acid stimulations and the possibility of provoking formation damage. Aspects of friction loss diminution are also discussed by OUSTERHOUT & HALL (1961); PRUITT, SIMMONS, NEILL & CRAWFORD (1965) and SUTTON (1976).

4.3.4.3.2.3. Friction pressure vs. fracture height growth

Higher slurry viscosities resulting from high proppant concentrations will substantially increase the friction pressure in the fracture to the point where it may completely dominate the excess pressure especially when narrower cracks are being created (HOLDITCH, ROBINSON, ELY & RAHIM 1988). As a result of higher friction pressures in the fracture, uncontained height growth can occur even when the excess pressure is rapidly rising. Increasing pressure and thus stress distribution in a reservoir away from the wellbore due to partial depletion gives way to increasing stress at the crack tip as the fracture propagates. Delayed crosslinked fluid systems exhibit very complex non-Newtonian behaviour due to insignificance of shear degradation, retainment of much higher viscosity, and transport of much higher proppant concentrations than with conventional crosslinked gels. The viscosities of delayed crosslinked fluid systems are constantly changing with time and temperature within the fracture (cf. section 4.3.4.3.1.2.) and the friction properties are substantially higher due to greater proppant saturations which is in contrast to conventional crosslinked fluids where by the time they reach the fracture, fluid viscosity has already been considerably reduced by shear degradation and does not change much during the treatment.

As the downhole viscosities of delayed crosslinked fracturing fluids are much larger than those of conventionally crosslinked ones, they can carry much higher proppant saturations which leads to again increasing viscosity, and because proppant concentrations are variable within the fracture in most cases (cf. also section 4.3.6.), slurry viscosities and friction pressure gradients are also variable in the crack. Another influence on friction properties of the fracturing gel is wall slip. In case of higher proppant saturations pumped into narrow cracks, slurry viscosity and resulting friction pressure increase considerably and are seriously affecting the excess pressure. Under such conditions, fracture height is no longer being contained, but height is growing at a much faster rate than length towards the end of the treatment, with the increase in height extension being caused by the rapid increase in excess pressure which is dominated by the friction in the system.

4.3.4.3.2.4. Fluid viscosity and proppant suspension improvement

Since the inception of massive hydraulic fracturing (cf. section 1.1.1.) where the large cracks are supported by multiple proppant layers, viscosities significantly greater than that of water are required for the stimulation fluids (OLSEN & DEBONIS 1988). As proppant settling velocity is inversely proportional to carrier fluid viscosity, the more viscous the fracturing fluid, the farther the proppant can be transported into the crack without settling out of the carrier (cf. section 4.12.2.). Viscous fluids need more energy for moving through a fracture than thin fluids, and the additional energy in form of pressure is applied against the crack walls, thereby causing the fracture to widen which enables placement of more proppants within the crack in order to obtain the desired multiple layers. While too low fluid viscosity is not beneficial, also too high viscosity can be detrimental. Increased fluid viscosity tends to increase the pressure in the fracture which is related to crack height, with increasing fluid viscosity and fracture pressure increasing the danger of uncontrolled height propagation (cf. section 4.2.2.3.).

If fracture height greatly exceeds prospective interval thickness, the stimulation treatment loses effectiveness by wasting fracturing fluid and proppant upon unproductive rock. Proppant embedment in gel filter cakes is a significant contribution to proppant conductivity damage, with higher gel loadings causing greater conductivity deterioration than lower gel saturations (cf. sections 1.4.10.3.5. and 4.3.4.1.4.). It is therefore necessary to establish a balance between too little and too much viscosity to optimize proppant placement in the crack, with the design goal being minimum viscosity and fluid-loss control for accurate proppant installation.

Fluids which are conventionally crosslinked at the surface lose a percentage of the initial proppant transport capacity when being forced through a limited number of perforations at high injection rates particularly during limited-entry fracturing (EASON 1985; cf. section 4.2.2.2.). Delayed crosslinking taking place in the formation after passage through the perforations keeps fluid degradation to a minimum, and preserves proppant suspension and move properties.

4.3.4.3.2.5. Fracture closure and breaker action

In low-temperature reservoirs, the use of a mixture of delayed crosslinker and rapid crosslinker reduces the viscosity increase at elevated temperatures obtained with a delayed crosslinker only (ALMOND & GARVIN 1985; cf. section 4.3.4.3.1.6.). Optimization of the polymer-crosslinker interaction can diminish the amount of base polymer required to achieve a specific viscosity and thus considerably decrease the cost of gelling. Rheological differences between various types of crosslinked fracturing fluids are also investigated by GARDNER & EIKERTS (1982, 1983), CRAIGIE (1983); LESCARBOURA, SIFFERMAN & WAHL (1984); ROGERS, VEATCH & NOLTE (1984); GUILLOT & DUNAND (1985), KNOLL (1985) and SHAH &

WATTERS (1986). Field experience with delayed crosslinked gels is reported by PEARSON, LYNCH, SCHMIDT & McCASLIN (1988). Despite considerable progress made in recent years, rheological characterization of crosslinked fracturing fluids remains a difficult and elusive challenge (VEATCH & MOSCHOVIDIS 1986).

In case of large treatments and efficient fluids, fracture closure would not occur for several hours which would allow ample time for complete proppant settling in water-based low- to moderate-viscosity gels (HANNAH, HARRINGTON & POTTER 1980). Crosslinked fracturing gels, however, can retain the ability to support proppants for several hours at elevated temperatures unless their higher viscosity is degraded to some critical point due to the action of breakers or absolute thermal degradation. Oxidizing breaking agents affect both proppant support characteristics and fluid loss properties of crosslinked gels. If breaker action significantly reduces the time during which perfect proppant support is present and fracture closure is retarded, then imperfect proppant distribution will result and the overall performance may be no better than if a simple gel were used (cf. section 4.12.1.1.).

On the other hand, if breaker action significantly increases fluid loss during closure, then fracture closure time may be reduced and may assist in trapping proppants between the crack walls in a favourable distribution. In some cases, the addition of breakers destroys wall leakoff control and transforms it into viscosity control. Base gel viscosity dictates performance of crosslinked fluids, and partially broken gels perform slightly poorer than would be expected due to their base gel viscosity (CONWAY & HARRIS 1982).

4.3.4.3.3. Improvement of proppant placement

FREEMAN, BILDEN & HOSSAINI (1986) document the superiority of delayed foamed crosslinked gels which allow the placement of much higher proppant concentrations in the fracture than will be possible to insert with foamed linear gels as a consequence of extended fluid-loss properties. The enhanced characteristics of foamed delayed crosslinked gels even allow reduction in foam quality in considerable magnitude without detrimental effects on proppant transport capability. The discussion as follows includes some aspects of proppant transport capacity and fluid-loss additives as well as fracture penetration vs. fluid crosslinking.

4.3.4.3.3.1. Proppant transport capacity and fluid-loss additive

Comparative assessment has shown that solid fluid-loss materials which have to be incorporated with foamed linear gels and result in productivity impairment due to the damage caused by these additives are not necessary to be applied with delayed crosslinked foams because of the higher viscosity of the latter (FREEMAN, BILDEN & HOSSAINI 1986). The greater viscosity of the foamed crosslinked gel also reduces proppant fall rate up to almost zero proppant settling in ideal cases, thereby not only improving proppant saturation which can be successfully placed, but also permitting a delayed start of load recovery after the treatment that considerably decreases the risk of proppant flowback by allowing sufficient fracture closure restraining the proppant package (cf. sections 4.12.2.3. and 4.12.3.1.2.). Rheological aspects of foamed crosslinked fracturing fluids are also discussed by REIDENBACH, HARRIS, LEE & LORD (1983); WATKINS, WENDORFF & AINLEY (1983); HARRIS & REIDENBACH (1984); ROYCE, BECK & RICKARDS (1984) and CRAIGHEAD, HOSSAINI & FREEMAN (1985).

Crosslinked fracturing fluids increase friction drop through tubing which is further increased by the presence of proppants due to abrasion (PAI, GARBIS & HALL 1983). Employment of delayed crosslinkers which prevent crosslinking of fluid until it reaches tubing bottom lowers the treating pressure and enables to save considerable expenses of horsepower, as well as to make the entire operation safer and more efficient by not overloading the equipment (cf. section 4.3.4.3.1.). While crosslinking of fracturing fluids generally aids in leakoff control and combats potential screenout failure, proppant distribution in a non-crosslinked gelled fluid is more effective when increased rates are used to promote proppant suspension because of increased fluid velocity.

4.3.4.3.3.2. Fracture penetration vs. fluid crosslinking

Field experience has shown that crosslinked gelled 2 % KCl water is more evenly distributing the proppants throughout the fracture, whereas the non-crosslinked gelled 2 % KCl water leaves a larger amount of proppants near the wellbore, with this effect being the consequence of the better transporting characteristics of the crosslinked system. Non-crosslinked gelled fluids provide a deeper penetration of the propped fracture due to narrower crack width with respect to crosslinked fluids, but on the other hand, the more uniformly packed fracture created by crosslinked fluids is considered to be more important in sustaining production rates over the life of these wells than the less uniform deeper penetrating fractures generated by non-crosslinked systems.

Delayed crosslinked gels allow higher pumping rates as well as transport and placement of higher proppant saturations with considerably lower rates of screenout failure (KUUSKRAA & HAAS 1988 report examples of successful pumping of sand concentrations up to 12 lbs/gal into reservoirs where saturations up to 3 lbs/gal in conventional fluids led formerly to screenout termination). Other measures to avoid screenout finishing are designing of optimum large pad volumes (cf. sections 4.3.4.6.1.4.).

4.3.4.3.4. Enhanced fluid stability and transport capacity

The main advantages of high-viscosity crosslinked fracturing fluids as compared to the characteristics of non-crosslinked fluids are extremely good leakoff control, yield of wider fractures which accept higher proppant concentrations, perfect proppant transport properties, and enhanced temperature stability (PAI & GARBIS 1983). Apart from neutral fluids, also acids can be gelled and crosslinked (cf. also section 4.5.1.2.) which considerably broadens the application spectrum of these aggressive stimulation fluids. The outline as follows presents some aspects of acid gelling and crosslinking, rheological differences of fracturing fluid stages, and impact of shear during mixing and pumping.

4.3.4.3.4.1. Acid gelling and crosslinking

Apart from more or less neutral water- or polymer-based fluids, also acid can be gelled and crosslinked in both conventional and delayed form (PABLEY & HOLCOMB 1980, 1981, 1982; CHURCH, QUISENBERRY & FOX 1981; PABLEY, EWING & CALLAWAY 1982; McLANE & SCOTT 1983; SMITH, DAWSON & SCOGGINS 1983; JOHNSON, FOX, BURNS & O`MARA 1988; FOX, O`MARA, BURNS & JOHNSON 1989). Some aspects of viscosity and temperature stability as well as low-concentration acid as fracturing base fluid are discussed as follows.

4.3.4.3.4.1.1. Viscosity and temperature stability

As crosslinked acid systems are characterized by high strength, elevated temperature stability and considerable proppant transport capacity, they are an optimum tool for combination of acid and proppant fracturing (cf. sections 4.5.4.2. and 4.8.10.) and therefore compete with polymer- or water-based more or less neutral crosslinked fluids, foam and energized fluids, and emulsified acids. Delayed crosslinked acid systems have very high reduction in tubular friction pressure and leave no insoluble residue, because once the fluid is in the fracture and the acid is spent, the polymer inverts from mildly cationic to anionic thus becoming a brine-soluble polymer which does not precipitate even when becoming completely neutralized in the presence of calcium ions.

The low pH of acid gel systems minimizes clay damage of the formation and thus also permits application of crosslinked acid fluids to water-sensitive sandstone reservoirs. Viscosity stability reaches in acid strengths up to 28 % hydrochloric acid concentration and up to 220 oF temperature. Fluid loss control matching or surpassing that achieved by crosslinked water-based fluids ensures deep penetration of live acid, and an extremely retarded reaction rate allows in many cases the crosslinked acid to be placed into the formation with little or no decrease in acid strength (cf. also sections 3.11.3. and 4.5.1.2.).

Field application of high-strength crosslinked acid gel systems is discussed by COULTER, HARRIS & KLEBENOW (1980); PABLEY, EWING & CALLAWAY (1982); PABLEY & HOLCOMB (1982), McLANE & SCOTT (1983) and PAI & GARBIS (1983 b). MOLON & FOX (1983) and BAILEY & WICKHAM (1984) outline enhancement of fracture penetration and productivity by acid gelling. Temperature stability of acid gelling polymers is investigated by NORMAN, CONWAY & WILSON (1981) and FOX, O`MARA, BURNS & JOHNSON (1989). In a comparable way as for fracturing fluids, extended thermal stability of acid can be achieved for acidizing utilization, with buffer regulation of hydrofluoric acid permitting sandstone acidizing up to 550 oF (280 oC) reservoir temperature (SCHEUERMAN 1988; cf. sections 4.4.5.1. and 4.5.5.).

4.3.4.3.4.1.2. Low-concentration acid as fracturing base fluid

COULTER, HARRIS & KLEBENOW (1980) present the advantages of using low-concentration hydrochloric acid as a base fluid for hydraulic fracturing of some sandstone formations. A gel system with low-residue crosslinked polymer with 3 - 5 % hydrochloric acid has superior fluid-rock compatibility characteristics and eliminates the earlier problems of rapid degradation of most gelling agents with hydrochloric acid. Acid gelling reduces acid reduction rate and therefore increases acid penetration distance (BAILEY & WICKHAM 1984). Other methods of decreasing acid reaction rate and thus enhancing acid penetration distance are addition of chemical retarding agents and acid emulsification in oil, as well as application of gelled water preflushes and utilization of alternating stages of crosslinked polymer fluids and acid (cf. sections 4.5.1.3.2.2. and 4.5.5.3.). Aspects of retardation of reaction rate by acid gelling are also discussed by NIERODE & KRUK (1973); CROWE, MARTIN & MICHAELIS (1981) and JOHNSON, FOX, BURNS & O`MARA (1988).

4.3.4.3.4.2. Rheological differences of fracturing fluid stages

Hydraulic fracturing treatments usually consist of successive pumping of four fluid stages with different rheological properties comprising pre-pad, pad, proppant-laden carrier and displacement (LEE & DANESHY 1985). Rheological properties and predominantly viscosity of each fluid stage can change significantly due to temperature changes in the fracture and the time duration of exposure to each temperature. Shearing in a high-stress environment can destroy a gel after crosslinking (GARDNER & EIKERTS 1982, 1983; SHAH & WATTERS 1984; ROYCE, BECK & RICKARDS 1984), whereas high levels of shear prior to crosslinking have little effect on overall performance. Overshearing leads to breaking of the polymer chain thereby decreasing the average molecular weight which ultimately means loss of polymer functionality (HASHEMI, CAOTHIEN & ROVIG 1988).

4.3.4.3.4.3. Impact of shear during mixing and pumping

The properties of crosslinked polymer fluids in hydraulically-induced fractures depend on intensity and duration of shear to which the fluid was subjected while mixed and pumped from the surface into the fracture (SHAH & WATTERS 1986), and consequently fracture geometry and proppant distribution are also affected. Increased pressure drop in turbulent flow is an indication of crosslinking reaction occurring at a faster rate than shear degradation. Time- and shear-dependent behaviour of crosslinked polymer fluids is also investigated by ROGER, VEATCH & NOLTE (1981) and CRAIGIE (1983). With respect to polymer emulsion fluids, crosslinked gelled water-based fluids cause less formation damage, can carry higher proppant concentrations, are less flammable and less expensive, and have a much faster load recovery time (ROBERTS 1981; cf. section 4.3.4.2.). Temperature is generally activating the crosslinking mechanism (GARDNER & EIKERTS 1983; cf. section 4.3.4.3.1.). Viscosity loss of the carrier stage due to time and temperature can give rise to inability of further suspending the proppants, with stabilization of this fluid stage being able by replacement by a two-stage gelled fluid where the second gelling agent is activated at higher temperature (cf. also section 4.3.4.4.3.).

As crosslinked fracturing fluids are shear-history dependent (cf. section 4.3.4.2.1.), they do not possess unique rheological characteristics (CONWAY & HARRIS 1982). The rheological properties are additionally influenced by the variable conditions encountered during the treatment such as tubing size, pumping rate, surface fluid temperature and pipe residence time.

4.3.4.4. Fluid heat capacity and reservoir temperature

Heat transfer during fracture propagation takes place by convection along the fracture combined with conduction and convection in the reservoir (BEN NACEUR & STEPHENSON 1985). Heat transfer mechanisms play a crucial role in deep well fracturing especially when fluids with high temperature dependency like crosslinked polymers or foams are used. In addition to conduction and convection, also diffusion by thermal exchanges and mechanical leakoff take place in reservoir rocks adjacent to the fracture. Following discussion of some general aspects, an overview is given on heat blockage and formation cooling, primary and secondary gellants, and fracturing fluid viscosity improvement.

4.3.4.4.1. General aspects

The influence of heat transfer is very significant for proppant transport properties due to viscosity variations, as well as for fracturing pressure analysis. At low fluid efficiency, fluid temperature in the crack remains close to injection temperature, whereas with increasing efficiency, the temperature along the fracture tends exponentially to the reservoir temperature (VEATCH 1983). Aspects of heat transfer in hydraulic fracturing and fluid temperature profiles in the crack during the treatment are also discussed by WHEELER (1969), WHITSITT & DYSART (1970), SINCLAIR (1971); CLARK, CLOUD, WAREMBOURG & YUNKO (1978); HARRINGTON, HANNAH & BEIRUTE (1978); CROCKETT, VO & CLEARY (1984) and POULSEN & LEE (1984; cf. also section 4.4.5.1.). Variational temperature analysis in hydraulic fracturing is carried out by BIOT, MASSE & MEDLIN (1984). MOODY (1988) comments on thermal displacement in fractured formations. The outline as follows focusses on temperature sensitivity of fracturing fluid viscosity as well as temperature difference between formation and stimulation fluids.

4.3.4.4.1.1. Temperature sensitivity of fracturing fluid viscosity

Reservoir temperature and heat transfer effects have an important impact on fracturing fluid viscosity and thus also proppant transport and suspension capability (SINCLAIR 1971), particularly in wells with bottomhole temperatures beyond 200 oF. Fracturing fluid viscosity is highly temperature-sensitive. Gel degradation at high temperature takes predominantly place by hydrolysis and oxidation (CONWAY & HARRIS 1982). When the pH is held in the optimum range, the rate of polymer degradation can be slowed down by addition of temperature stabilizers, but little effect on initial apparent viscosity at temperature is achieved, and the rate of viscosity loss for a crosslinked fluid will be no less than that of an equally stabilized base gel. Fluid shear history is the major factor dictating viscosity enhancement at high temperatures. Crosslinking time of high-temperature crosslinking agents controls the ability of the latter to provide viscosity enhancement in the fracture.

4.3.4.4.1.2. Temperature difference between formation and stimulation fluids

The temperature difference between the cool injected stimulation fluid and the hot reservoir rock at the fracture walls (cf. also section 6.2.1.1.) causes heat flow from the pay zone to the crack. Both conduction and convection of heat takes place in vertical fractures (DYSART & WHITSITT 1967, WHITSITT & DYSART 1970, PEACEMAN 1976). Heat blockage is created by cool transpiring carrier fluids leaking out of the crack into the formation, because a leaking-off fluid moving in a direction opposite to that of the conductive heat transfer tends to insulate or block the flow of heat to the fluid remaining in the fracture. Heat flux includes the primary effects of convection within the crack, heat conduction within the fluid, and heat exchange between fluid and reservoir, and the secondary effects of heat conduction and heat generation by energy dissipation in the fluid (CLIFTON & WANG 1988).

4.3.4.4.2. Heat blockage and formation cooling

Conventional water-based gels which are very effective coolants are typified by low apparent viscosities and high leakoff rates, with thus fluid efficiency being low, generated fractures being narrow and heat blockage being very important, because water has a much higher heat capacity than oil. Some aspects of fluid rheology and proppant grain size, enlargement of pre-pad and pad size, and changing temperature and fluid viscosity are briefly outlined as follows.

4.3.4.4.2.1. Fluid rheology and proppant grain size

Since high temperatures have detrimental effects on polymers, additives and acids, precooling of hot formations with water or brine ahead of fracturing treatments are considerably improving the effectiveness of the job (cf. section 4.3.4.4.2.2.). Viscous water-based gels are non-Newtonian fluids that are occasionally even viscoelastical (PAI, GARBIS & HALL 1983; BEN NACEUR & STEPHENSON 1985). Viscous water-based gels exhibit high apparent viscosity at low or moderate shear rates, with these rheological properties providing low tubular friction loss, high fracture friction diminution, low fluid leakoff and good proppant-carrying ability, with viscous water-based gels thus generating wide fractures and allowing utilization of coarse proppants for plugging of the cracks, and at the same time having the high heat capacity of water. Viscous water-in-oil dispersions with very viscous oils permit the routine use of coarse 8/12 mesh proppants in oil-well stimulation and help to create wide fractures (MATTHEWS 1969; cf. section 4.3.5.). Heat blockage by suitable stimulation fluids is an effective means to improve crack performance, because when heat blockage occurs, fractures are wider in the early stages of the treatment and have greater volumes, and the need for fluid-loss additives that could damage reservoir formation and proppant package is reduced or even completely eliminated.

Wider fractures which are favoured by effective heat blockage allow the use of coarser proppants and permit to start proppant injection much earlier than usual during course of the stimulation treatment, leading to a better proppant fill-up of the fracture and enabling a better proppant schedule for plugging of the crack. Substantial ameliorations of fracture geometry and design are the consequence of the importance of heat blockage, giving rise to increasing proppant volumes, earlier beginning of proppant insertion in the job program, reduction or total elimination of fluid-loss additives, and extension of the upper temperature limit of the fracturing fluid. The overall significance of effective heat blockage of the carrier is the creation of wider fractures and the possibility of plugging of the cracks with higher proppant concentrations and coarser proppant grain sizes, thereby maximizing fracture conductivity (cf. section 4.3.5.). Aspects of impact of formation temperature on hydraulic fracturing design and heat transfer in cracks are also discussed by HILL & WAHL (1968) and WHEELER (1969; cf. section 4.4.5.1.).

4.3.4.4.2.2. Enlargement of pre-pad and pad size

Formation cooling by a pre-pad of abt. 100,000 gal of non-crosslinked polymer fluid has the purpose of slowing-down viscosity loss in the fracturing fluid and has the additional effect of establishing a filter cake on the crack wall in order to minimize fracturing fluid leakoff which might result in premature screenout termination (PAI 1982; PAI, GARBIS & HALL 1983). During pre-pad pumping, rates should be maximized in order to create already the total fracture height with the pre-pad and pad fluids instead of the expensive polymer carrier fluid. Formation cooling by pumping of large fluid volumes is particularly significant in high-temperature reservoirs to guarantee extended stability of crosslinked fracturing fluids (GRONE, BAUMGARTNER & WOODROOF 1983; cf. section 4.7.4.2.).

PAI & GARBIS (1983 b) present a pre-pad system consisting of gelled weak hydrochloric acid and carbon dioxide containing surfactants and clay stabilizer in order to cool down the formation to minimize viscosity loss in proppant-laden fluid, condition the reservoir by creating a low-pH environment to minimize clay swelling and sloughing, establish a filter cake on the fracture face and minimize leakoff of crosslinked fluids thus reducing screenout tendency, and decrease clean-up time by carbon dioxide addition. Aspects of cooldown pad pumping are also discussed by SNOW & HOUGH (1988). The following overview characterizes reservoir cooling and fluid saturation as the two main purposes of pad enlargement and comments on some multiple pad schedules.

4.3.4.4.2.2.1. Reservoir cooling and fluid saturation

Similar enlargement of pre-pad and pad stage sizes as applying for reservoir cooling can be performed for fluid saturation of the pay zone complex and diminution of fluid loss. Proper sizing of the pad pumped ahead of the proppant carrier fluid is very important in order to keep the proppant slurry away from the fracture tip region and thus avoiding tip screenout failure (LEE 1987; cf. sections 4.2.3.2.5. and 6.2.4.2.1.). While continuous-mix blending fracturing operates with lower proppant concentrations, batch-mix fracturing aiming on placement of higher proppant saturations commonly is carried out with large percentages of treatment volume consisting of pad fluid (up to 60 - 80 % of the whole job) insuring leakoff control and sufficient fracture width for passage of the concentrated proppant slurry (CRAMER & SONGER 1988; TAN, McGOWEN, LEE & SOLIMAN 1988; cf. sections 4.3.3.5.3. and 4.3.4.3.1.4.).

4.3.4.4.2.2.2. Multiple pad schedules

WARPINSKI, BRANAGAN, SATTLER, CIPOLLA, LORENZ & THORNE (1988) report a special case where three different pads were successively pumped including an initial foamed-water/breaker spearhead, a middle normal foam pad, and a terminal pad with 100 mesh sand for fluid leakoff control (cf. sections 1.4.11.2.1. and 4.8.8.3.2.3.) prior to the proppant-laden fluid stages. HOLDITCH, ROBINSON, WHITEHEAD & ELY (1988) present an example of very low fluid efficiency of only 20 % which has to be compensated by a large pre-pad and a 50 % viscous pad, because without successfully controlling the large amount of leakoff, large hydraulic proppant fracture jobs cannot be pumped due to rapid dehydration screen-out termination (cf. sections 4.2.2.2.3. and 6.2.4.2.1.). In addition to considerably enlarged pre-pad and pad, extra volumes of liquid and solid fluid-loss additives have to be used for leakoff minimization (cf. sections 4.3.4.6. and 4.8.8.3.).

4.3.4.4.2.3. Changing temperature and fluid viscosity

As a consequence of temperature differences between surface and formation, significant reservoir and stimulation fluid cooldown occurs across a portion of the fracture (OLSEN & DEBONIS 1988). As many treatments are designed for a worst-case temperature with at most two fluids of different viscosity to compensate for fluid temperature variance, the result is a fracturing fluid with varying viscosity throughout the crack. On-site microprocessor control (cf. sections 4.3.3.5.3. and 6.2.4.3.) and accurate continuous-mix gelling equipment allows to vary gel loading in order to compensate for time and temperature effects thus yielding a stimulation fluid with a more consistent downhole viscosity while using less gelling agent.

Most fracturing fluids undergo considerable temperature changes during a stimulation operation due to elevated reservoir temperatures, and viscosity also changes, because all modern viscosified fluids are non-Newtonian in nature where viscosity is not constant but is a function of shear rate. Determination of fluid rheology throughout the crack thus requires knowledge of the temperature profile in the fracture. In large treatments, temperature and thus fluid viscosity variations could be compensated to a limited extent by batch mixing of several different polymer loadings. Continuous mixing, however, allows accurate tailoring of polymer loading to meet the time and temperature conditions to which it will ultimately be exposed (cf. section 4.3.3.5.3.).

Ramped viscosity procedures by continuous mixing benefit from cost saving by reduced polymer loading and improved fracture conductivity by less proppant package damage due to diminished polymer loadings, with the lower gel concentration also requiring less breaker near the wellbore. Ramped polymer loading procedures, however, can only compensate the maximum time and temperature which a fluid experiences, with constant viscosity being achieved only until near the end of the operation. High-temperature reservoir fracturing frequently requires viscosity stabilization of the stimulation fluid (THOMAS & ROOT 1979; cf. section 4.7.4.2.).

4.3.4.4.3. Primary and secondary gellants

PAULS, VENDITTO, CHISHOLM, HOLTMYER & GREGORCZYK (1985) comment on mixing problems of fluids and proppants when high sand concentrations should be placed into high-temperature reservoirs. Oil-based fracturing fluids are traditionally gelled by batch mixing at the surface (cf. section 4.3.3.5.3.), with all the gelling agent and activator required being added prior to commencing the treatment. In case of high-temperature formations up to 260 oF (130 oC), the downhole viscosity which conventionally mixed fluids provide is much less than necessary, but if more primary gellant and activator are added by surface batch mixing, the high viscosity of these fluids renders them difficult both to handle with surface equipment and to mix with solid materials such as fluid-loss additives and proppants, especially concerning high concentrations of the latter.

Therefore solid secondary gelling agents may be added to the base gel after passage of the blender on the fly to produce a fluid which is not too viscous to be handled at the surface and yet provides the required viscosity in the fracture downhole, with this high-temperature secondary gellant considerably extending the usable range of gelled oil-base fluids to higher fracture temperatures and longer pumping times, and also allowing to stimulate high-temperature reservoirs with high proppant saturations. This dual addition of primary and secondary gelling agents represents a sequential combination of batch- and continuous-mix fracturing and links the advantages of both separate methods into one joint process.

4.3.4.4.4. Fracturing fluid viscosity improvement

Simultaneous application of primary and secondary gelling agents representing sequentially combined batch- and continuous-mix stimulation comprises the flexibility of formulating a fluid to meet both surface and downhole conditions. Other aspects of improving downhole fracturing fluid viscosity without excessive problems of surface mixing by addition of secondary gellants and/or activators to a base gel downstream of the blender are discussed by BURNHAM, HARRIS & McDANIEL (1978). LOUVIERE (1987) outlines that gel preparation can be done by batch mixing or semi-continuous operation, with the latter technique having no limitation on certain storage capacity and job size being able to be varied during the treatment.

Fracturing fluid recovery is essential for successful stimulation operations and depends to large amounts on gel volume added to the base fluid (PAI, GARBIS & HALL 1983). The addition of gel breakers and low surface tension agents may considerably facilitate clean-up, and the addition of an energizer such as nitrogen or carbon dioxide to the treatment tail may increase the possibility of an early hydrocarbon show. Reasonable cost reductions can be achieved by tapering polymer concentration of the fluid system. As gel weight is reduced, job costs decrease due to diminishing polymer expenses, less horsepower necessary as a result of lower friction pressure, and faster well clean-up as a consequence of less polymer having been pumped downhole (SCHLOTTMAN, MILLER & LUEDERS 1981).

4.3.4.5. Fluid viscoelasticity

ACHARYA (1987) illustrates the significance of viscoelasticity of crosslinked fracturing fluids for proppant transport. Concerning proppant placement along the fracture, proppant movement velocity is a function of fracturing fluid rheology and proppant particle characteristics such as size and density. Crosslinked gels can respond to deformation by viscous flow which dissipates energy and by elastic displacement which stores free energy. When the Reynolds number of the particles is greater than two, proppant settling rate can be influenced by viscous as well as elastic characteristics of the fracturing fluids. The network structure of the crosslinked gel restricts proppant settling, and disruption of the network structure by shear, temperature or chemical breakup only allows the proppant to fall through the gel.

Aspects of proppant transport are also discussed by NOVOTNY (1977), CLARK & QUADIR (1981); SIEVERT, WAHL, CLARK & HARKIN (1981); AHMED (1984); CLARK, HALVACI, GHAELI & PARKS (1985); MORALES & ABOU-SAYED (1985) and LEE & DANESHY

(1986). Effects of pumping equipment on sand-laden slurries are covered by ROLL, HIMES, EWERT & DOERKSEN (1986). HANDKE (1987) reviews quality and versatility improvement of fracturing operations by sand concentration control. Comments on gravel packing fluid viscoelasticity are given by HUDSON & MARTIN (1988; cf. section 5.3.).

4.3.4.6. Fluid leakoff containment

Fluid loss plays a significant role by being one of the more dominant factors controlling hydraulic proppant fracturing treatments (VEATCH & MOSCHOVIDIS 1986, McGOWEN & McDANIEL 1988) and being able to have as much effect on crack penetration as vertical fracture height growth has (VEATCH 1983). The volume of fluid leakoff determines fluid efficiency as well as affects treatment size and proppant scheduling. The rate of fluid loss also impacts fracture closure time and thus final proppant distribution within the crack. While minimization of fluid leakoff is desirable during the pumping operation, it is undesirable to control fluid loss to the point of excessive proppant settling due to very long fracture closure time (cf. section 4.12.2.2.), with thus a sound optimization having to be performed by minifracture calibration tests (cf. section 4.8.1.). Fluid leakoff is generally a function of reservoir properties and filter cake (WILLIAMS 1970).

4.3.4.6.0. Types and impact of fluid loss

Fluid loss comprises leakage of the stimulation fluid out of the main fracture into formation matrix porosity, natural hairline cracks or macrofractures, joints and faults (ECONOMIDES 1986). Hairline cracks are thief fractures which steal fluid from the main fracture and consequently limit its extension (cf. sections 4.8.8.3.1.2. and 4.8.8.3.2.3.). Fluid leakoff behaviour depends on a number of factors including type and quantity of fluid loss additive and gelling agent, pressure differential between fracture and formation, reservoir permeability and porosity as well as fluid flow and compressibility behaviour, fracturing fluid and fluid filtrate viscosity and temperature behaviour, reservoir temperature, and character of the natural fracture system intersecting the pay zone matrix. Other parameters controlling fluid efficiency are fluid properties and injection rate (WAREMBOURG, KLINGENSMITH, HODGES & ERDLE 1985). Fluid leakoff models and evaluations are also presented by SETTARI (1985, 1988); SETTARI, KRY & YEE (1988) and WARPINSKI (1988 a).

Experimental modelling of fluid loss has shown that pressure drop is initially totally dissipated across the formation face and fluid leakoff occurs as if no additive were present (WILLIAMS 1970). This beginning stage is followed by an intermediate stage where a filter cake starts to form and leakoff velocity is lower because some pressure drop occurs across the drape. The final stage comprises achievement of a steady state in which the cake has constant thickness and fluid leakoff velocity is constant. Flow velocity through the steady-state drape is depending on flow velocity across the rock surface, fluid and additive properties, and matrix pore size. Total fluid loss generally consists of spurt loss and fluid-loss or wall-building coefficient. Spurt loss pertains to the instantaneous loss when fluid is first exposed to the fracture face, whereas fluid-loss or wall-building coefficient represents long-term behaviour over the duration of exposure. The outline as follows discusses possibilities of fluid-loss control, natural fractures and particulate additives, mushroom-type fractures and fluid efficiency, and combined fluid-loss control additive and gel breaker. MEDLIN & MASSE (1986 b) analyze fluid-leakoff mechanisms for gels and suspensions.

4.3.4.6.1. Possibilities of fluid-loss control

The impact of fluid leakoff on the result of the hydraulic proppant fracturing treatment is underlined by the possibility to achieve significantly longer fractures with fluids with lower fluid-loss coefficients. Fluid leakoff may be controlled with solid or liquid additives. While solid particulate fluid-loss additives may be very effective for moderate- to high-permeability formations but are much less effective in tight low-permeability reservoirs, liquid fluid-loss additives such as hydrocarbon phases can reduce fluid leakoff significantly in tight formations but have little or no effect in moderate- to high-permeability rocks (cf. section 4.3.4.6.2.). Liquid additives mainly restrict fluid loss into the matrix by viscosity and compressibility differences of reservoir and stimulation fluids, whereas particulate additives principally limit fluid leakoff into natural fractures (McMECHAN & CONWAY 1983) by plugging of cracks and building of filter cakes on the wall.

4.3.4.6.1.1. General aspects

Polymer and crosslinker types have a pronounced effect on fluid leakoff behaviour (McGOWEN & McDANIEL 1988). Lower residue polymers exhibit higher levels of fluid loss, whereas crosslinker type influences crosslinked fluid leakoff. Addition of proppants often seems to have no altering impact on fluid-loss behaviour of crosslinked gels, and also the filter cakes deposited on the fracture walls do not exhibit any signs of erosion which may have resulted from inclusion of proppant to the flow.

Pumping rate obviously compensates for fluid leakoff into the formation if high enough and feasible (VEATCH 1986), but does usually not compete economically with fluid-loss additives to overcome leakoff. High injection rates of suitable fluids can offset fluid loss and ensure adequate proppant transport within the fracture (NORTON & HOFFMAN 1982). In low leakoff situations, optimum rates are generally dominated by friction loss behaviour and hydraulic horsepower costs, whereas in high leakoff cases where volume requirements are affected by pumping rates, fracturing material expenses enter the picture. Distinction can be made between storage-dominated or low fluid-loss fracturing, and fluid-leakoff-dominated or high fluid-loss fracturing (SMITH 1981). Drawbacks of high fluid leakoff rates as well as aspects of solid particle size vs. pore throat diameter, fluid uptake saturation of the formation by pre-pad and pad stages, and fluid viscosity vs. leakoff are discussed as follows.

4.3.4.6.1.2. Drawbacks of high fluid leakoff rates

Fluid leakoff during hydraulic fracturing leads to increasing fluid viscosity due to dehydration and consequently increasing proppant concentration in the crack as a result of volume diminution of the total slurry which may be desirable up to a certain point (because otherwise fluid viscosity decreases by shear and temperature degradation and can reach a point where satisfactory proppant transport and suspension capacity is no longer guaranteed), but at the same time excessive fluid loss considerably increases the danger of premature screenout failure by gel thickening as well as proppant bridging and plugging and thus flow stuck in the fracture (HARRIS & PENNY 1985; PEARSON, ABOU-SAYED & NGUYEN 1985; VEATCH & MOSCHOVIDIS 1986, SMITH 1987; cf. sections 4.2.3.2.5. and 6.2.4.2.1.). In addition, high fluid leakoff can give rise to severe curtailment of production due to formation matrix damage, adverse reservoir fluid interactions or simply altered fracture geometry (PENNY, CONWAY & LEE 1984). As fluid loss can exceed 70 % of the injected volume if not controlled properly, an overwhelming amount of fluid can be necessary to achieve a desired fracture geometry in a MHF treatment and thus fluid efficiency can govern the economical success of the stimulation operation.

Leakoff rates of unbroken fracturing fluids are generally higher than those of broken ones because water is squeezed out, but polymer particles cannot build a filter cake, and therefore only water leaks off. While unbroken fluid cannot form filter cakes at the fracture wall, broken particles can migrate and cover the crack boundary with a filter cake which is then limiting further loss by rate reduction. Absence of a filter cake prior to fluid breaking as well as matrix saturation with water or brine after initial leakoff are the reasons for much higher fluid loss in the early stage prior to breaking than in the late stage after breaking. BENNY & FORD (1985) emphasize the influence of downhole conditions on the leakoff properties of fracturing fluids. Another effect of fluid loss into the near-fracture pore space can be fluctuations in closure stress representing backstress (WARPINSKI, BRANAGAN, SATTLER, LORENZ, NORTHROP, MANN & FROHNE 1985).

4.3.4.6.1.3. Solid particle size vs. pore throat diameter

Small-diameter solids dispersed in fracturing fluids have sufficient concentration and size range to act as fluid-loss additives in formations with small pore-size distributions (CARROLL & BAKER 1979). Bleed-off to the formation from the stimulation fluid prepared with such materials includes a high probability of some damage to reservoir permeability. Bleed-off to the formation from a treatment fluid containing dispersed solids and residue-forming materials gives rise to building of a filter cake upon the formation face before the parting pressure is released, after which the proppants adjacent to the formation face settle into the rather soft filter cake and come to rest at the formation face. Depending upon filter cake thickness, the void volume of the proppant immediately adjacent to the fracture wall could be completely filled.

Experimental research reveals that the possibility of plugging a propped fracture is controlled by particle size distribution in relation to pore size distribution. Residue-forming material in the gels generate a particle size distribution that even in low concentrations will severely damage fracture conductivity. Proppants start crushing or at least releasing fines immediately upon application of overburden pressure, and fines adhere to the proppants which upon liberation in a fracturing fluid may act as a fluid-loss additive. Particle size distributions used to generate fracturing fluids in water may be sufficiently high enough to contribute to both formation and fracture plugging.

4.3.4.6.1.4. Fluid uptake saturation by pre-pad and pad stages

The pre-pad and/or the pad stages of a hydraulic fracturing treatment are often sacrificed for fluid loss into the reservoir in order to have already achieved a saturation level in fluid uptake by the formation once the carrier stage with the suspended proppants is inserted into the fracture (SMITH 1987). Formation fluid leakoff rates can be estimated from pressure decline analysis following injection by minifracture calibration test (NOLTE 1979, NOLTE & SMITH 1981) and from changes of fracturing gradient during pumping based on measurements of increasing instantaneous shut-in pressure (NIERODE 1983).

Concerning fluid-loss modelling by laboratory experiments, dynamic tests can yield quite different results than static runs, and shear rate and history have significant effects (GULBIS 1983, ROODHART 1983; ZIGRYE, WHITFILL & SIEVERT 1983; PENNY, CONWAY & LEE 1984; HARRIS & PENNY 1985, FORD & PENNY 1988). PEARSON, LYNCH, SCHMIDT & McCASLIN (1988) outline improvement of fluid-leakoff control by successively increasing the pad stage volume during course of an evolution of fracturing technology. Increasing pad quantity also aims on screenout risk minimization, because early pad depletion due to high fluid-loss rates increases the danger of premature failure of the job. This applies particularly for bridgeout conditions occurring when the movement of proppant grains at some position in the fracture is restricted by insufficient fracture width.

In energized fracturing treatments, the pre-pad aims on establishment of a filter cake on the fracture face in order to minimize leakoff of crosslinked carrier fluid, thereby reducing the tendency of premature screenout failure (cf. sections 4.2.3.2.5. and 6.2.4.2.1.), as well as cooling down the formation to decrease viscosity loss of the proppant-laden fluid, condition the reservoir to minimize clay swelling and sloughing, and reducing clean-up time by addition of carbon dioxide or nitrogen gas (PAI & GARBIS 1983; cf. section 3.11.2.2.).

4.3.4.6.1.5. Fluid viscosity vs. leakoff

Most viscosifying agents used in fracturing fluids also play an important role in controlling fluid loss (OLSEN & DEBONIS 1988). Reducing viscosity in the latter portions of a treatment thus could increase fluid leakoff, and if needed, other materials could be added to the fluid to maintain an adequate loss rate and yet not affect viscosity. If fluid loss from a wall-building viscosifier is a function of fluid amount that has leaked off across a given rock face, then the filter cake should always be thickest near the wellbore where exposure time has been longest and taper down towards the fracture tip where exposure time is shortest. Concerning proppant embedment in gel filter cakes triggering conductivity deterioration (cf. sections 1.4.10.3.5. and 4.3.4.1), the danger of fracture performance damage particularly near the wellbore also justifies to reduce the amount of fluid-loss material in the latter portion of the treatment. Fluid-leakoff control should be preferentially done in the pad stage of the operation to maintain a more even distribution of filter cake.

The aim of minimizing fluid loss in hydraulic fracturing for the purpose of optimizing proppant transport and prevention of screenout failure due to proppant bridging and/or slurry dehydration (cf. sections 4.2.2.2.3., 4.2.3.2.5. and 6.2.4.2.1.) is in contrast to the goal of high fluid leakoff in gravel packing for the benefit of adequate packing density and proper plugging of the perforation tunnels (cf. section 5.3.1.1.).

4.3.4.6.2. Natural fractures and particulate additives

Fluid leakoff is in an overriding amount influenced by natural fractures and/or high-permeability streaks of the stimulated reservoir. Natural cracks and fissures provide an accelerated fluid depletion mechanism within the propagating crack analogous to a pressure regulator and are principally undesirable, because uncontrolled fluid loss leads to slurry dehydration and proppant bridging which give rise to screenout failure of the treatment (NOLTE 1988 b; cf. sections 4.2.3.2.5. and 6.2.4.2.1.). Once the high leakoff rates in these weakness elements are curbed, fluid loss to the low-permeability matrix is strongly governed by the shear rate within the fracture, particularly in high-rate jobs in the early part of the pad where crack widths are relatively small, which can lead to changing leakoff rates during the treatment.

The characteristics of a particular formation dictate the selection of fluid-loss additives that are required to achieve the necessary leakoff control. Particulate additives such as 100 mesh sand, silica flour, calcium carbonate or oil-soluble resin grains are essential for containing leakoff to high-permeability reservoir streaks (above abt. 5 md permeability) and natural fractures, whereas liquid hydrocarbon additives function well in low-permeability matrix control (PENNY, CONWAY & LEE 1984; SMITH 1987). Hydrocarbon phases mixed with silica flour or other fine-grained particulates and a surfactant can also significantly reduce fluid loss (GULBIS 1983, ROODHART 1983; ZIGRYE, WHITFILL & SIEVERT 1983; PENNY, CONWAY & LEE 1984; HARRIS & PENNY 1985, FORD & PENNY 1988). An overview of fracturing fluid additives is given by HARRIS (1988). WILLIAMSON & ALLENSON (1989) introduce a nondamaging particulate fluid-loss additive. The discussion as follows focusses on liquid and solid fluid-leakoff agents, foams and resins, and impact of fracture geometry and morphology.

4.3.4.6.2.1. Liquid fluid-loss additives

In crosslinked gel systems, diesel oil provides better fluid-loss control than particulate fluid-loss additives, whereas in non-crosslinked linear gel systems, particulate additives provide lower leakoff than diesel oil (OSBORNE, McLEOD & SCHROEDER 1981). The reason for this diametrical behaviour and the superiority of the diesel oil in crosslinked gels is that the latter concentrates in a thin layer along the fracture walls as filtrate is squeezed into the formation pores, resembling a fibrous mat composed of many polymer entanglements, with the diesel oil dispersed into small droplets by suitable surfactants filling in the interstices between these entanglements. CANTU & BOYD (1988) introduce a combination of fluid-loss control additive and gel breaker.

McDANIEL, HOUX & BARRINGER (1976) introduce a new generation of solid-free fracturing fluids which do not need solid additives for leakoff control. In contrast to conventional solid-laden fluids where polymers impart viscosity and fluid-loss additives help to limit fluid leakoff, solid-free fracturing fluids minimize permeability damage to both formation and packed crack. These solid-free stimulation fluids are applicable in high- and low-temperature reservoirs. The high-viscosity water-based fluids are mainly used as spearheads in acidizing treatments and as fracturing fluids for proppant placement.

4.3.4.6.2.2. Solid fluid-loss additives

The application of solid fluid-loss additives as the head of the proppant slurry has to be done with caution in order not to provoke intermixing of proppants and fine sand which would drastically deteriorate the flow capacity of the fracture (cf. section 1.4.11.2.1.). In many naturally jointed reservoirs, additional 100 mesh sand in a concentration of up to 2 lbs/gal fluid is sufficient to reduce leakoff enough to prevent proppant screenout (COOPER, NELSON & SCHOPPER 1977; SCHUMACHER & DICKERMAN 1986; cf. sections 4.8.8.3.1.2. and 4.8.8.3.2.3.). The best effect is achieved when the natural fractures and fissures are opened by maximizing net pressure and sealed with 100 mesh sand before the proppant is placed, with caution, however, having to be undertaken that crack tip does not screen out prematurely (NOLTE 1988 b; cf. section 4.3.4.6.3.). The discussion as follows focusses on comparative evaluation of various agents, 100 mesh sand insertion for dual leakoff control, accelerated leakoff and screenout, and fine sand vs. bauxite.

4.3.4.6.2.2.1. Comparative evaluation of various agents

Field examples of application of 100 mesh sand and silica flour as fluid-loss control additives are also reported by PEARSON, LYNCH, SCHMIDT & McCASLIN (1988) and SATTLER, RAIBLE, GALL & GILL (1988; cf. also section 1.4.11.2.1.1.). PENNY (1982) and WILLIAMSON & ALLENSON (1989) outline non-damaging fluid-loss additives. CRAWFORD (1983) presents proppant schedules and fluid leakoff calculations. SINHA (1976) comments on static and dynamic fluid-loss conditions. HALL & HOUK (1983) investigate fluid-loss control in naturally fractured formations. ZIGRYE, WHITFILL & SIEVERT (1983) evaluate fluid leakoff containment differences of crosslinked and linear fracturing fluids.

DOLAK & PERNTHANER (1976) propose cement as an alternative to 100 mesh sand for plugging of hairline fissures, with pumping of abt. 100 sacks of cement prior to the main fracture fluid stages sufficiently closing the hairline cracks in small-scale sand fracturing treatments. Cement, however, is permanently destroying hairline permeability and does not allow the tiny fissures to later contribute to production, whereas 100 mesh sand is highly-permeable and is the ideal combination of both hairline-crack-plugging and hydrocarbon-draining agent (cf. sections 4.8.8.3.1.2. and 4.8.8.3.2.3.).

4.3.4.6.2.2.2. 100 mesh sand insertion for dual leakoff control

Dual leakoff is a common phenomenon in hydraulic fracturing of tight gas reservoirs (WARPINSKI 1988 a; cf. sections 4.8.8.3.1.2. and 4.8.8.3.2.3.) which is characterized by a significantly increased (up to 50 times) fluid leakoff occurring from a threshold pressure above formation closure pressure onwards. The increased leakoff results in rapid fluid drainage and dehydration, and is a probable cause of early screenouts unless restricted by using 100 mesh sand both as fluid-loss additive and as microproppant infilling the hairline cracks and plugging them open in order to guarantee that they will keep sufficient conductivity to enable gas flow during subsequent exploitation. A relatively flat pressure curve during injection is a characteristic feature of dual leakoff behaviour. Low concentrations of 100 mesh sand that are added to both pad and proppant-laden stages throughout the treatment and which have to be kept limited in order to not seriously destroying proppant conductivity, however, can only reasonably retard accelerated fluid loss and are by no means able to stop it entirely, with leakoff rates still being very high (WARPINSKI, BRANAGAN, SATTLER, CIPOLLA, LORENZ & THORNE 1988).

On the other hand, an exclusive addition of 100 mesh sand during the pad stage may not permit to bring in and to deposit enough material to perfectly seal the hairline cracks such that they are not contributing to fluid leakoff during the following proppant-laden stages. The accelerated secondary loss may start at 50 times the base-level leakoff rate and can be reduced to 20 times shortly after 100 mesh sand enters the perforations and to 5 times by treatment end provided continuous insertion of small amounts of 100 mesh sand, which is certainly a significant retardation of fluid escape into the reservoir through thief hairline cracks (cf. also sections 4.8.8.3.1.2. and 4.8.8.3.2.3.). In order to accept propping agents, fracture width must be in the range of 2.5 times the proppant diameter, with this condition being fulfilled for hydraulically induced fractures by standard proppant grain sizes (16/20 and 20/40), while 100 mesh microproppants fit into hairline cracks and fissures (THOMPSON 1977; cf. section 1.4.11.2.1.).

4.3.4.6.2.2.3. Accelerated leakoff and screenout

In naturally fractured reservoirs with dual leakoff behaviour, solid fluid-loss additives are indispensible, because otherwise early screenouts impair the treatments by resulting in short hydraulic fractures, high pressures and natural crack damage. If control of accelerated or secondary leakoff is only partial due to necessary limitation of 100 mesh sand concentration for the purpose of proppant conductivity preservation, considerable fluid loss still occurs and screenout is imminent at shut-in as revealed by history match results. In such cases, only restricted amounts and limited saturations of proppants can be injected into the reservoir unless a more efficient means of fluid leakoff control can be discovered, because on the other hand additives that plug the natural fractures and are not readily produced back are unacceptable since reasonable withdrawal rates are only achieved through the natural fracture system.

Accelerated leakoff modelling provides also a plausible damage mechanism for foams when little gel and small liquid volumes are injected (WARPINSKI 1988 a; WARPINSKI, BRANAGAN, SATTLER, CIPOLLA, LORENZ & THORNE 1988; cf. sections 4.8.8.3.1.2. and 4.8.8.3.2.3.). If high fluid-loss rates occur due to opening of natural fractures within the reservoir rocks, leakoff of the total foam system into the cracks has to be expected. During post-fracture cleanup, the natural cracks close and production of the liquids trapped within the joints would be nearly impossible, with gel and liquids within the natural fractures then considerably damaging reservoir permeability. Thus breaker pre-pads should be used in order to facilitate cleanup of any gels that are injected past the 100 mesh sand bridges within the hairline cracks, but the main control on damage is the effort to minimize leakoff volumes, with stimulation and production results suggesting that this is a successful strategy.

4.3.4.6.2.2.4. Fine sand vs. bauxite

Fine solid particles of sand and bauxite are effective leakoff additives if the strength compatibility between fine particles and proppants is considered to avoid damage caused by low-strength grains (LI & ZHU 1986). The amount of fine fluid-loss particles should generally not exceed 10 % of the total weight of the proppants because damage of the propped zone by the fine solids increases sharply with rising amount of fines. The damage created by high-strength bauxite is less than that generated by low-strength sand. HUDSON, COFFEY, SAUER & TEOT (1982) discuss non-solid fluid-loss control agents and their performance.

4.3.4.6.2.3. Foams and resins

Oil foams are a reasonable alternative for fluid-loss control in naturally fractured reservoirs (NORTON & HOFFMAN 1982). Fluid leakoff rates of oil foams are very low which reduce the need to use other fluid-loss additives except for naturally fractured pay zones where some additives further support fluid leakoff control. Aspects of oil foam fracturing are also discussed by DRISCOLL, BOWEN & ROBERTS (1980) and GAYELOS & HARRIS (1980). Oil-soluble resins are suitable fluid-loss additives when they are properly sized and have a high enough softening point enabling them to bridge pore spaces, because they can subsequently be dissolved in the produced hydrocarbons and thus do not provoke any formation or proppant pack damage (ECONOMIDES 1986). Oil-in-water emulsions containing little amounts of diesel produce two-phase flow upon fluid loss which drastically reduces leakoff.

4.3.4.6.2.4. Impact of fracture geometry and morphology

Geological discontinuities may considerably alter fracture geometry (cf. section 4.2.3.4.) and thus their effect on proppant transport and fluid leakoff may be serious (WARPINSKI & TEUFEL 1984). Forcing proppants through such features as narrow multistrand fractures or offsets is much more difficult than through a single hydraulic crack, with the consequence being higher fracturing pressure and frequent occurrence of screenouts. If local small screenouts take place in several branches of a multiple fracture system, little additional crack growth is still possible until the abundance of these events becomes too large and ultimately provokes a complete screenout of the hydraulic fracture system (cf. sections 4.2.3.2.5. and 6.2.4.2.1.). Therefore fluid leakoff coefficients may change during course of a treatment as a complicated crack pattern develops.

Successful hydraulic proppant fracturing also requires careful evaluation of the main reservoir rock types in heterogeneous formations, with the classification of fluid compatibility aiming on elimination of possible failure reasons due to excessive fluid sensitivity of some layers of the pay storey complex (SIMON, KAUL & CULBERTSON 1977) with leakoff characteristics again being one of the most important parameters. A summary of fracturing fluid loss is given by POULSEN (1988).

4.3.4.6.3. Mushroom-type fractures and fluid efficiency

A special fluid-loss treatment is necessary in case of mushroom-type fractures consisting of a vertical crack topped by a horizontal fracture like a cap in transitional depth ranges or when injection rates exceed the stress required to lift the overburden (FITZ-PATRICK, KARR & O`SHEA 1985; cf. sections 1.2.8.4., 4.9.4.2. and 6.2.2.4.). Fluid-loss replacement to the horizontal fracture component and maximization of the extension of the vertical fracture component can be achieved by increasing the pad size to almost 100 % and addition of a junk sand stage designed to bridge off the entrance of the narrow horizontal fracture, thereby forcing the trailing proppant to stay in the vertical fracture. The pad is generally a more or less sacrificial fluid quantity which might leak off and thus saturate the primary and secondary porosity of the reservoir formation in order to prevent excessive loss of carrier fluid which would provoke premature screenout failure (SMITH 1987; cf. sections 4.2.3.2.5. and 6.2.4.2.1.), and also serves for creation of sufficient fracture width (cf. section 4.3.4.4.1.2.). As fluid loss is particularly high at the crack tip, the proppant-laden slurry catches up with the fracture tip where it ideally should have arrived at the end of the treatment (cf. section 4.3.4.6.2.2.). Useful information about fluid loss is given by the pressure decline pattern (cf. section 6.2.4.2.).

Fluid efficiency is the fraction of the injected volume that is in the crack at the time injection stops, and the remaining fraction is lost to the formation by fluid loss (NOLTE 1984 a). Because fluid efficiency can be determined for a stimulation treatment independent of any fracture model, the prospect arises that the efficiency for a calibration operation should be used to design a subsequent stimulation job including pad size and optimum proppant schedule completely without the need for a model and such associated parameters as height, loss coefficient and viscosity. This would be especially desirable for treatments in new prospect areas where little if any information may be available to determine what model or associated parameters are appropriate. Fluid efficiency can be determined from pressure-decline analysis for a minifracture calibration treatment performed before the actual stimulation job which provides inferred values of fluid-loss coefficient and fracture width and penetration.

4.3.4.6.4. Combined fluid-loss control additive and gel breaker

CANTU & BOYD (1988) present an organic acid condensation product being a polyester as a combined fluid-loss control additive and gel breaker for fracturing fluids. The organic ester exposes increasing acid functional groups when degraded to soluble monomers at temperatures above 150 oF following the fracturing stimulation treatment, and therefore the degraded additive imposes only minimal formation damage. Some aspects of proppant permeability retainment due to secondary filter cake dissolution, polyester concentration and filter cake resorption, and degradable polyester vs. 100 mesh sand are discussed as follows.

4.3.4.6.4.1. Proppant permeability retainment due to secondary filter cake dissolution

The high-acid-content degradation product breaks the linear and/or crosslinked gel used in the fracturing fluid and effectively cleans the proppant package which significantly increases fracture conductivity, with the polymer breakdown being mainly a consequence of decreasing pH caused by the degraded parts of the organic acid particulates and taking place by backbone cleavage (TYSSEE & VETTER 1981). Therefore secondary filter cake dissolution reinstalls primary proppant permeability which thus is retained in the later stage of the treatment in contrast to conventional operations where gel filter cake precipitation destroys proppant pack permeability by embedment. Replacement of the common silica flour as fluid-loss additive by the polyester can result in more than

200 % increase in fracture conductivity and permeability (CANTU & BOYD 1988) and in leakoff control performance as good as or better than with silica flour in crosslinked and linear gels as well as nitrogen foams.

The particulate polyester is concentrated and retained in the gel filter cake as the fluids leaks off into the formation, with filter cake cleanup thus being much more efficient with a combined fluid-loss control additive and gel breaker, because conventional water-soluble gel breakers leak off into the reservoir matrix more easily. The significance of fluid-loss control is underlined by the fact that fracturing fluids must have a minimum leakoff while transporting proppants in order to achieve an effective treatment.

Fluids containing polyester additives can retain all of the initial proppant permeability, whereas the silica flour gel breaker system may only leave 10 - 15 % of original proppant permeability even with excess breaker concentration in the slurry volume (cf. sections 1.4.10. and 4.12.1.1.). This indicates that in order to effectively break residual gels in fracture systems, it is important to place the gel breakers properly in the filter cake. Conventional silica flour/gel breaker systems can only approach the effectivity of the newly introduced polyester additive if abt. 5 % diesel is added to the mixture, but in any case the polyester is still slightly better performing. Addition of 5 % diesel to the polyester mixture, however, drastically reduces and delays activity and effectivity of the latter.

4.3.4.6.4.2. Polyester concentration and filter cake resorption

The polyester concentration in the filter cake is also important in gel breaking as evidenced by increasing proppant conductivity with higher polyester saturation (CANTU & BOYD 1988). The higher the polyester concentration, the more acid is released as the particles degrade to soluble monomers, and with decreasing pH of the filter cake, polymer breakdown leading to filter cake resorption begins. The fluid buffering capacity also determines the rate at which gel breaking takes place. If the fluid surrounding the polyester particulates has a high buffering capacity, the pH change will be limited due to consumption of the soluble acid by the buffer and thus filter cake cleanup is slower.

Proper fluid conditioning and optimum polyester concentration gives rise to complete cleanup of the filter cake by breaking into soluble monomers, and no filter cake layer is left into which proppant embedment could take place. This removal of the filter cake in an advanced stage of the treatment when it has successfully fulfilled its task to limit fluid loss and is no longer needed is a very effective mechanism to liquidate the formation damage effect which the filter cake had provoked so far and to leave the fracture walls completely unimpaired that also does not create any remaining damage of the proppant package and therefore guarantees maximum fracture conductivity (cf. also section 1.4.10.3.5.4.).

4.3.4.6.4.3. Degradable polyester vs. 100 mesh sand

Gel breaking and filter cake cleanup by the polyester are a function of temperature, concentration and fluid pH. Fluid-loss control by non-ionic degradable platelet additives is described by NORTHCUTT, ROBERTSON & HANNAH (1988). Such degradable fluid-leakoff additives which later are completely dissolved or disintegrated and removed could also be a good approach for dual fluid-loss control in naturally fractured reservoirs (WARPINSKI 1988 a; cf. sections 4.8.8.3.1.2. and 4.8.8.3.2.3.) because in contrast to 100 mesh sand, they do not decrease proppant conductivity (cf. section 1.4.11.2.1.). On the other hand, however, absence of microproppants plugging the hairline cracks could result in their closure with declining pressure and thus no longer satisfactory contribution to gas production.

Therefore it is desirable in deep and high-pressure reservoirs to insert 100 mesh microproppants into the natural jointlets in order to secure flow of gas through the fissures into the borehole, with an improvement in pay zones with high closure stress being application of 100 mesh intermediate- to high-strength synthetic proppants once this grain size becomes commercially available (NORTHCUTT, ROBERTSON & HANNAH 1988; cf. section 4.8.8.3.2.3.). An attractive intermediate solution would be resin-coated 100 mesh sand which similarly as conventional 100 mesh sand has abt. 5 - 7 darcy permeability at almost any closure stress, with aggregation of resin films locking the grain package in place if microproppant flowback from hairline cracks is unacceptable (FRACFAX 1988 f). VARGAS & OHLSON (1986) discuss suitable breakers for gravel-pack completion fluids.

4.3.5. Ultra-high fracture conductivity achievement

Ultra-high fracture conductivities can be achieved with a combination of very high proppant saturations and extra large proppant grain sizes (BARBY & BARBEE 1987) or by mixing of angular and round proppant grains (LARSEN & SMITH 1985).

4.3.5.1. Extraordinarily large proppant grain size

The highest possible conductivity is particularly essential to be reached in high-permeability reservoirs (cf. section 4.6.1.). Large proppant diameters of 8/12 or 6/10 result in negligible embedment even in hard, well-cemented reservoirs. The danger of wellbore screenout failure in such jobs is particularly high if fracture width is not large enough to accept the high concentrations and large sizes of proppants, and in order to perform a smooth treatment, often higher pumping rates are required, with then caution having to be exercised that excessive fracture height growth does not occur (cf. sections 4.2.3.2.5. and 6.2.4.2.1.). Other operational drawbacks are that the consequences of mechanical problems such as equipment failure or engineering errors in the fracture design are usually unforgiving and irreversible in these jobs requiring a high level and standard (BARBY & BARBEE 1987).

The gain in fluid density with increased proppant concentration should cause a significant net reduction in pumping pressure, with the absence of pressure decrease thus early indicating a premature screenout. In terms of proppant flowback, the larger the proppant grain size, the more stable is the package in the fracture and proppant back production is reduced unless coarse abrasive proppants erode the fracture face at the wellbore upon entry or large-volume acidizing annihilates the fracture mouth. Some advantages and drawbacks of ultra-high fracture conductivity achievement with extraordinarily large proppant grain sizes are summarized as follows.

4.3.5.1.1. Advantages

In terms of conductivity maximization, 6/10 intermediate-strength proppants at 1 lb/ft2 (5 kg/m2; cf. section 4.3.1.1.) concentration display greater conductivity than 8/12 particles at 2 lb/ft2 (10 kg/m2) saturation, with thus only half the volume of 6/10 proppants being needed to achieve more conductivity than with 8/12 proppants (BARBY & BARBEE 1987). The lower concentration of 6/10 proppants would be easier on the pumping equipment, leave less proppant in the fracture mouth near the wellbore that could be subjected to flowback, and deliver the equivalent production rate at lower costs. A general guideline for fracture design is that if a formation can accept higher saturations of smaller proppants, it is also possible to place lower concentrations of larger proppants. Examples of field application of 8/12 mesh proppants in oil reservoirs where both long low-conductivity and short high-conductivity fractures have been created are described by BRITT & LARSEN (1986).

Fracture flow capacity and thus proppant conductivity layout should be done based on the application of a conductivity correction factor of 0.5 in order to account for the performance difference of proppants in laboratory and field (POULSEN & SOLIMAN 1987; cf. section 1.4.10.3.). Improvement of fracture flow capacity can be achieved by increasing the maximum allowable proppant concentration and/or by choosing a higher conductivity proppant type and/or grain size. Failure of recognition of the differences between apparent and effective proppant conductivity result in underdesign of the fracturing treatment. Larger grain sizes of higher-quality proppant types provide greater fracture flow capacity that would not be as affected by formation fines which are released as smaller lower-quality proppants are crushing due to increasing closure pressure (CROSS & SHELDON 1977).

4.3.5.1.2. Drawbacks

Drawbacks of extraordinarily large proppant grain sizes are crushing and embedment (cf. section 4.3.5.1.). Closure stress affects first coarser grains of a given type of proppants due to the larger surface of attack and thus crushing of larger particles begins earlier as compared to smaller grains, and embedment in soft formations also starts with coarser particles due to their lower quantity and less dense package with respect to finer grains. Net present value (being defined as extra hydrocarbon revenue by stimulation minus treatment cost; cf. sections 1.4.12.1. and 4.8.11.4.) rises with increasing proppant grain size, proppant concentration and fracture half-length. Larger proppant grain size, however, requires multilayer propping in order to sufficiently decrease deterioration by embedment. In narrower fractures, a better package permeability can be achieved with smaller proppant grains. Other difficulties with large-diameter proppant grains are particle bridging in perforations and fracture resulting in screenout failure of the job (cf. sections 5.6. and 6.2.4.2.1.). NIEMEYER & REINART (1986) document comparative job design evaluation which exhibits significantly higher screenout rates for 12/20 mesh with respect to 20/40 mesh proppants under the same well conditions.

An operational drawback can occur in soft unconsolidated sands, because tails of 8/16 or 10/20 mesh proppants following pumping of 20/40 mesh material might to large amounts remain within the wellbore. The situation of coarse tail-in lots being left in the casing can be avoided by increasing the share of the tail in the whole treatment quantity (LAMBERT, DOLAN & GALLUS 1983; cf. section 2.4.1.2.3.).

4.3.5.2. Mixing of round and angular proppant grains

Maximization of formation-fracture conductivity contrast and fracture flow capacity can further be done by mixing of round and angular proppant or sand grains, because the increased void space of a round-angular grain blend enhances the performance of the combination with respect to that of the ingredients alone (LARSEN & SMITH 1985). Laboratory experiments on 20/40 sand have revealed that of all the mixtures tested, a 50 : 50 round-angular blend has on the average the highest relative fracture conductivity throughout all closure stress levels up to 6,000 psi (cf. section 1.4.11.3.). Therefore angular proppant grains interspersed into round grain populations would increase stability of the package and improve the conductivity of the proppant wedge within the fracture.

Other possibilities of maximizing fracture conductivity are proppant banking and proppant pillaring including creation of horizontal and vertical open flow channels within the fracture, respectively (cf. sections 4.2.2.4. and

4.3.2.2.). PAULS, VENDITTO, CHISHOLM, HOLTMYER & GREGORCZYK (1985) illustrate the beneficial effect of a combination of maximizing fracture length and proppant concentration. The long cracks contact additional reserves and the high proppant saturations not only produce highly-conductive fractures, but also settled bank heights that are capable of infilling the crack both below and through the reservoir.

4.3.6. Uniform vs. changing fracture conductivity

SOLIMAN (1986 b) evaluates the performance effectivity of uniform and changing conductivity fractures and concludes that a changing conductivity crack behaves in many respects as an uniform and average conductivity fracture. The apparent conductivity of the whole fracture depends upon conductivity distribution and not just average conductivity. The reason why decreasing and uniform conductivity fractures are almost equivalent in effect is that the amount of fluid passing through a section of the crack increases nearer to the wellbore, and therefore in case of uniform fracture conductivity, the pressure gradient along the crack increases as it nears the wellbore. Further away from the wellbore, this pressure gradient can be so small that decreasing fracture conductivity would not appreciably increase total pressure drop across the crack. Unless fracture conductivity is changing very quickly, a fracture may appear to be uniform, and well test analysis may never detect changing conductivity inside a crack.

Thus an uniform high conductivity inside the fracture may not be always necessary. Usually crack conductivity near the wellbore is high, and the pressure gradient inside the fracture is much lower than its gradient in the formation. If this crack conductivity near the wellbore is considered to be sufficient to produce a particular formation, then it is not indispensible to maintain such a conductivity throughout the crack. As flow rate through the fracture decreases with distance away from the well, it is sufficient to maintain only the low pressure gradient. This constant pressure gradient is achieved through decreasing crack conductivity, with therefore instead of using a constant conductivity fracture design, application of constant pressure gradient being more efficient. This has significant impact on proppant quantity and thus size and expenses of the stimulation treatment, as a constant pressure gradient fracture may require less than half of the proppant volume needed for a constant conductivity crack. Aspects of changing fracture conductivity and non-uniform distribution are also discussed by GUPPY, CINCO-LEY & RAMEY (1980) and BENNETT, ROSATO, REYNOLDS & RAGHAVAN (1981).

4.4. Targets of unconventional gas stimulation

Targets of unconventional gas stimulation comprise chiefly tight gas sandstone reservoirs (cf. section 3.4.1.) which are already exploited in several parts of the world to various degree, with the current centres being USA and Western Europe. In addition to low-permeability clastic gas-bearing sequences, some other marginal targets for stimulation with particular emphasis on hydraulic proppant fracturing in the coming years include geopressured reservoirs, coal seams, shales, high-temperature reservoirs, and gas condensate reservoirs. Emphasis is put on these various sources of unconventional gas because of their already current importance in parts of the world (LINDEN 1985) and their increasing significance in the energy supply spectrum until the year 2000, with the share of unconventional gas in the world gas delivery being able to be forecasted as ranging in the order of 10 - 30 % (LIESEN 1985, LINDEN 1985) or as being abt. 30 % and 50 % by the years 2000 and 2020, respectively (BAKER 1981). An overview of unconventional natural gas in the USA is given by DOE (1984) and KUUSKRAA & HAAS (1988).

The low permeability of unconventional reservoirs imposes severe restrictions on the ability of the gas to migrate appreciable distances (LAW, POLLASTRO & KEIGHIN 1986). Therefore geological factors such as temporal relationship of hydrocarbon generation and migration with respect to development of structural and stratigraphical traps are not as important as in conventional reservoirs. The more important geological characteristics of unconventional reservoirs are source rock quality and quantity, organic maturation, thermal history, and distribution and nature of porosity and permeability (cf. also SPENCER 1985; cf. section 3.4.1.).

Unconventional gas prospects are mainly characterized by major geological and engineering challenges that need to be faced in translating these potential resources into economically proven reserves (KUUSKRAA & HAAS 1988). Improved advances in geosciences, strategical well placement, and enhanced hydraulic proppant fracturing technology are required to overcome the natural geological discontinuities and permeability barriers impeding the ability of traditional infill drilling to recover the unconventional gas resources. The following discussion includes hydraulic proppant fracturing aspects of tight gas sandstones, geopressured reservoirs, coal seams, shales and high-temperature reservoirs. Some comments are also offered on gas condensate reservoirs.

4.4.1. Tight gas sandstones

Tight gas sandstones are the most prominent unconventional gas reservoirs as well as the largest and most widely recognized of the low-permeability natural gas resources (KUUSKRAA & HAAS 1988) and are already the focus of exploration and production since some time. Tight gas sandstones have provided both in USA and Europe already frequently the stage for execution of spectacular large-scale hydraulic proppant fracturing treatments (cf. section 3.4.1.) which repeatedly set world records in terms of fluid and proppant quantity pumped as well as operation expenses (cf. sections 1.1.1.4. and 2.4.1.1.). The low-permeability (less than 0.1 md) matrix can only liberate the entrapped gas in sufficient amounts if intersected by a deeply penetrating long hydraulic fracture that crosses as much rock matrix as possible in order to maximize gas collection along the high-conductivity pathway of the propped fracture (cf. section 4.8.11.). Large treatment volumes, properly placed and controlled fractures, and satisfactory support of the crack by pressure-resistant proppants are essential for achieving economically attractive rates of gas production. The outline as follows concentrates on tight gas sand body morphology and significance of hydraulic fracturing, distribution of important tight gas sand formations and basins, interrelationship of tight gas sands and coal seams, and overhydrostatic reservoir pressure.

4.4.1.1. Tight gas sand body morphology and significance of hydraulic fracturing

Tight gas sands form blanket-shaped or lenticular bodies (cf. section 3.4.1.), with both types requiring hydraulic generation of ribbon- and wedge-like high-permeability passageways through which the gas can flow from the tight matrix to the wellbore (SPENCER 1983, DOE 1984). Lenticular sand bodies (cf. plates II/7 - 8, III/4, V/4 and VII/3) necessitate in addition linking of various isolated sand lenses together along an extensive fracture in order to sustain sufficient production over the economical life of the project (cf. section 4.10.2.1.). Location of discrete sand lenses can be performed by 3-D reflection seismic section evaluation, vertical seismic borehole profiling and cross-wellbore seismic checking (cf. section 6.2.2.1.1.). Other heterogeneous geological properties such as unconformities and discontinuities (cf. plate I) further complicate the accessibility of natural gas resources in tight sandstones which as a consequence of their definition represent an arbitrary delineation of a natural geological continuous permeability spectrum of the reservoir rocks. Aspects of tight gas sandstone reservoirs are also discussed in sections 1.1.1., 2.4.1. and 3.4. An overview of the potential of tight gas sands is given by HAAS, BRASHEAR & MORRA (1985).

Massive hydraulic fracturing has already in the early 1970`s been identified as a promising alternative technology for economical development of tight gas reservoirs (KUUSKRAA, BRASHEAR, ELKINS & MORRA 1979). Technological improvements which have the largest impact on ultimate recovery and net present value include improved ability to differentiate and characterize the tight pay zones, capacity to stimulate multiple reservoirs for a common wellbore (cf. section 4.2.2.1.3.), improved predictability of fracture performance, increased effectively propped fracture length (cf. section 4.8.11.), ability for fractures to intersect lenses not directly in contact with the well (cf. section 4.10.2.1.), and optimizing field development comprising well spacing (cf. section 4.10.5.1.), well pattern and fracture size.

4.4.1.2. Distribution of important tight gas sand formation and basins

The tight gas sand basins in the USA are expected to have more than 400 Tcf (10,000 Bill. m3) cumulative recoverable reserves (OFFICE OF TECHNOLOGY ASSESSMENT 1985). Some of the most important tight gas sand basins are Northern Great Plains/Williston Basin (138 Tcf or 3,450 Bill. m3 gas in place), Greater Green River Basin/Wyoming and Colorado (135 Tcf or 3,375 Bill. m3 GIP), Piceance Creek Basin/Utah and Colorado (49 Tcf or 1,225 Bill. m3 GIP), Wind River Basin/Wyoming (34 Tcf or 850 Bill. m3 GIP), Cotton Valley Basin/Texas and Louisiana (22 Tcf or 550 Bill. m3 GIP), Uinta Basin/Utah and Colorado (21 Tcf or 525 Bill. m3 GIP), Edwards Line Basin/Texas (14 Tcf or 350 Bill. m3 GIP), Denver Basin/Colorado, Nebraska and Wyoming (13 Tcf or 325 Bill. m3 GIP), Val Verde Basin/Texas (5 Tcf or 125 Bill. m3 GIP), and San Juan Basin/New Mexico and Colorado (3 Tcf or 75 Bill. m3 GIP; KUUSKRAA & HAAS 1988). The maximum technically recoverable gas reserve in the tight gas sandstone resource is over 600 Tcf or 15,000 Bill. m3 (NORTHROP & FROHNE 1988).

Some of the most significant tight gas sands in the USA are in the Pennsylvanian the Atoka Formation in New Mexico, and the Cleveland and Morrow Formations in the Anadarko Basin/Oklahoma and Texas. The Upper Jurassic includes the Cotton Valley Sandstone (Schuler Formation) in the East Texas Basin/Texas and Louisiana, and the Lower Cretaceous contains the Muddy "J" and the Codell Sandstones in the Denver-Julesburg Basin/Colorado, the Travis Peak (Hosston) Formation in the East Texas Basin/Texas and Louisiana, and the Dakota Sandstone in the San Juan Basin/New Mexico. The Upper Cretaceous comprises the Corcoran, Cozzette and Rollins Sandstones in the Mesaverde Group in the Piceance Creek Basin/Colorado and Greater Green River Basin/Wyoming, Utah and Colorado. The Upper Cretaceous Mesaverde Group consists of a lower marginal marine section with mostly blanket reservoirs (Corcoran, Cozzette and Rollins Sandstones) and an upper fluvial to deltaic and paludal part with predominantly lenticular pay zones (JOHNSON & NUCCIO 1986).

Further Upper Cretaceous reservoirs are the Niobrara and Greenhorn Formations in Rocky Mountains and Great Plains area/Colorado, Kansas and Nebraska; the Olmos Formation in the Maverick Basin/Texas, the Frontier and Almond Formations in the Greater Green River Basin/Wyoming, Utah and Colorado; and the Tuscher Formation in the Eastern Uinta Basin/Utah. The Lower Tertiary finally bears the Wasatch and Green River Formations in Piceance Creek Basin/Colorado and Eastern Uinta Basin/Utah (compiled from SPENCER & MAST 1986). The most important tight gas sands in Europe are Rotliegend and Upper Carboniferous in North Sea and onshore Netherlands and Germany FRG (cf. section 2.4.1. and chapter 3).

4.4.1.3. Interrelationship of tight gas sands and coal seams

Tight gas sands are frequently interrelated with coal seams representing the primary sources and reservoirs of natural gas which partially later migrated and accumulated secondarily in sandstone pays (RIGHTMIRE & CHOATE 1986). The most prominent examples include the San Juan and Piceance Creek Basins in the Cretaceous of the USA, and the Upper Carboniferous in North Sea and adjoining onshore areas in Europe (cf. section 3.2.1.). Mass balances reveal that only small portions of the large volumes of gas generated during coalification remained within the coal seams themselves, with large amounts of additional gas having emigrated into adjacent sandstone reservoirs or escaped to the atmosphere. As a consequence of continuity of depositional environment, tight gas sand complexes are ready recipients for the large volumes of excess coal-bed gas.

4.4.1.4. Overhydrostatic reservoir pressure

Many Upper Cretaceous and Lower Tertiary tight gas sands in Colorado, Utah and Wyoming in USA are overpressured (LAW, POLLASTRO & KEIGHIN 1986) as a consequence of gas accumulating at rates greater than that it is depleted and lost (LAW & DICKINSON 1985). During thermal generation of gas, dewatering is accelerated and water saturation is decreased to irreducible levels which is immobile, thereby producing a relatively closed hydrological system in which transportation of dissolved material out of the system is impeded and significant enhancement of permeability and porosity is precluded. As processes leading to amelioration of permeability and porosity are ineffective, other permeability- and porosity-reducing processes continue, thus resulting in a very low-permeability rock sequence in which rates of gas accumulation exceed rates of gas depletion.

The gas accumulations are characterized by the presence of up-dip water-bearing reservoirs and down-dip gas-containing pays, and the top of the overpressured gas-bearing reservoirs cuts across structural and stratigraphical boundaries and is not associated with any particular lithological unit. In the deeper parts of the basin, the relatively closed nature of the system imposes severe restrictions on the ability of the gas to migrate appreciable distances from the interbedded source rocks. The top of overpressuring in the geological column almost matches with the transformation of smectite into illite and is situated near the zone of coexistence of chlorite-smectite in regular interstratification (LAW, POLLASTRO & KEIGHIN 1986; cf. section 3.5.1.).

Other aspects of overpressuring in tight gas sands are reported by LAW, SPENCER & BOSTICK (1980); LAW (1984) and SPENCER (1984). Together with other pay zone types and gas distribution types, overpressured tight gas sands belong to the group of geopressured reservoirs which are discussed in the section below.

4.4.2. Geopressured reservoirs

Apart from thinner or thicker, sheet-type or lenticular tight gas sandstones, another major challenge of necessary approach in the next years is seen in geopressured reservoir stimulation where gas is dissolved in overpressured brines or may also occur as free phase (TUCKER 1979, WEEKS 1979). Geopressured reservoirs are characterized by abnormally high pore fluid pressures exceeding the hydrostatic pressure required to raise a fluid column from the subsurface formation to the surface (DUTTA 1986). Overpressuring occurs within isolated sandstones that are encased into thicker mudstone sequences. Thus the fluid in the saturated buried sediments is trapped within the pore spaces and pressure changes from hydrostatic to lithostatic during progressive burial (JONES 1969; SWANSON, BERNARD & OSOBA 1986). As a consequence of having pressures greater than normal, geopressured reservoirs are capable of production without needing pumps or other equipment to bring gas and water to the surface, in contrast to hydropressured reservoirs with average pressure which contain a vast amount of gas that might be recoverable but will not flow naturally to the surface (GRI 1988 b).

Reservoir engineering aspects of geopressured formations are also discussed by TIMKO & FERTL (1971), ISOKRARI & KNAPP (1976); KNAPP, ISOKRARI, GARG & PRITCHETT (1977); LAW & SPENCER (1981), McPEEK (1981), QUITZAU & BASSIOUNI (1981), SPENCER & LAW (1981); DOHERTY, RANDOLPH, ROGERS & POONAWALA (1982); QUONG, OWEN & LOCKE (1982); ROBINSON, HOLDITCH & LEE (1983); ANDERSON, PETERSON & PARISI (1984); LAW (1984), McMULLAN & BASSIOUNI (1984), ENGINEER (1985), LAW & DICKINSON (1985), SOOKPRASONG & ELBEL (1985), BEGLAND & WHITEHEAD (1986); LAW; POLLASTRO & KEIGHIN (1986); PRASAD (1986); WERES, JUN & TSAO (1986); CHARPENTIER, LAW & PRENSKY (1987); IRELAND & ROBINSON (1987), POSTON & CHEN (1987), PRASAD (1987) and FERTL & LEACH (1988). Case studies of hydraulic proppant fracturing in geopressured reservoirs are carried out by DOSCHER & AZARI (1980) and ROBINSON, HOLDITCH & LEE (1983). A review of technology and economics of methane production from geopressured aquifers is given by DOSCHER, OSBORNE, RHEE, COX & KUUSKRAA (1979), and DOSCHER & AZARI (1980) analyze the effect of critical gas saturation on methane productivity. GEER & COOK (1978) discuss creation, augmentation and exploitation of hydrocarbon pore-volume saturation in geopressured aquifers, and MATTHEWS (1981) reviews possibilities of enhancing gas production from geopressured aquifers. Some comments on geotectonical setting, poroperm properties, ball-out treatments and other sources of deep gas are offered as follows.

4.4.2.1. Geotectonical setting

Geopressured gas accumulations are often deep-basin or basin-centered reservoir and deposit types which are commonly very large, occupy the deeper parts of sedimentary and structural troughs, are frequently downdip from water-bearing rocks and consist nearly always of low-permeability sandstones (CHARPENTIER, LAW & PRENSKY 1987). The abnormal pressures are caused by active gas generation in the younger geological history, with the poor permeability of the sandstone pays not allowing the pressure to equilibrate to hydrostatic levels (MEISSNER 1981; LAW, SPENCER & BOSTICK 1980; LAW & DICKINSON 1985). The occurrence of the unconventional geopressured gas deposits is not completely controlled by structure and stratigraphy, but the more important aspects of these accumulations are temperature, thermal maturity, organic richness, gas compressibility, water saturation and especially permeability (LAW 1984; CHARPENTIER, PRENSKY & LAW 1987). In the USA, geopressured aquifers containing natural gas occur particularly in the Gulf Coast area in Texas and Louisiana (RANDOLPH 1977).

4.4.2.2. Poroperm properties

Geopressured gas reservoirs are often characterized by high secondary porosities (BERG & HABECK 1982). The influence of increasing formation temperature with burial causes the transformation of smectite to illite whereby water is released that causes increasing pore pressure due to its entrapment (cf. section 3.5.1.). Aquathermal pressuring originates from thermal expansion of interstitial water with depth of burial. During normal burial, dewatering of the sedimentary section is unequal as a consequence of restricted drainage in geopressured reservoirs that are surrounded by impervious permeability barriers. Generation of oil and gas from organic material further increases the fluid volume. Secondary porosity is created by dissolution of mineral cements by the flow of water from high-pressure zones to low-pressure patches. The pressure response in geopressured gas reservoirs is adequately explained in a changing pore compressibility scenario (POSTON & CHEN 1987).

In abnormally pressured wells, production rate and flowing tubing pressure decline rapidly following hydraulic fracture stimulation treatments (SOOKPRASONG & ELBEL 1985). Generally the production rate decline depends on many factors such as formation permeability, fracture length, fracture conductivity, reservoir pressure and reservoir size. In deep high-pressure low-permeability wells, closure stress can play an important role on the rate of decline in production when large rapid pressure drawdown is required for economical offtake rates (cf. sections 1.2.1.1. and 4.12.4.). The economical significance of recovery of gas dissolved in geopressured brines is also stressed by LINDEN (1985). Production of the gas dissolved at bubble point pressure can be stimulated by both fracturing and acidizing aiming on creation of a high-permeability zone around the wellbore (DOSCHER & AZARI 1980). Excess pressures of isolated sand lenses within shales representing limitations of reservoir connection are controlled by geological and petrophysical factors (OZKAYA 1987). Flowback and cleanup of fracturing fluids in geopressured reservoirs is generally not a problem due to the abnormally high pore pressures (ROBINSON, HOLDITCH & LEE 1983).

BROWNE & PARKER (1986) report subnormally pressured and fractured reservoirs, and McLENNAN, ROEGIERS & MARX (1983) present underpressured siltstone and shale sequences. In addition to deep high-pressured hard sandstones, also shallow low-pressured soft sandstones may be tight and contain appreciable gas reserves within the low-permeability matrix that has to be liberated by hydraulic fracturing (RICE & SHURR 1978; NYDEGGER, RICE & BROWN 1979; BROWN & CRAFTON 1981, GAUTIER & RICE 1981; KUKAL, BIDDISON, HILL, MONSON & SIMMONS 1983). Some tight gas sandstones are considerably overpressured (McMECHAN & CONWAY 1983, BRANAGAN & WILMER 1985). Combined geopressured and geothermal aquifers with low-salinity water at abnormal pressures and elevated temperature containing dissolved natural gas also occur (KNAPP, ISOKRARI, GARG & PRITCHETT 1977).

4.4.2.3. Ball-out treatments

Many low-permeability geopressured reservoirs are rather soft rocks and should be balled out prior to production testing and fracture treating of the prospective interval, because formation damage can occur if large pressure gradients are imposed (ROBINSON, LEE & HOLDITCH 1983). Ball-out treatments usually change the skin from a positive value caused by drilling mud and cement damage to a slightly negative value due to creating minor fractures around the wellbore, with this amount of stimulation allowing to produce a well at moderate flow rates for short periods of time without imposing large pressure gradients around the wellbore. In addition, during pre-fracturing testing, certainty has to be made that all the net gas pay is contributing production to the well. As a consequence of the geopressure, it is expected that not all the perforation holes will be in effective communication with the formation, and therefore ball-out treatments are necessary to open up all of the perforations so that a more accurate pre-fracturing flow test and pressure buildup test can be performed.

Aspects of limited entry and initial ballout in geopressured reservoirs are also discussed by WEEKS (1979).

4.4.2.4. Other sources of deep gas

Other sources of deep gas can be classified into two categories comprising abiogenic and subducted gas (DOE 1984). Abiogenic gas is speculated to have been formed by non-organic means in the mantle during the primordial creation of the earth, and during course of geological time, it slowly seeped to shallower depths. Subducted gas is of organic origin, but instead of being formed from layers of compacted sediments, the gas is created when marine organisms and vegetation are pushed deep into the earth by tectonical crustal movements. A combination of chemical action, magmatic heat, pressure and time caused the gas to be generated which later either migrated to the surface or accumulated beneath caps of impervious rock. As both abiogenic and subducted gas reservoirs are still far from realistic access and are partially still semihypothetical items, no comments can be made concerning hydraulic fracturing potential of these types of hydrocarbon accumulations.

4.4.3. Coal seams

Methane from coal seams has been produced on various scale since the early 1900`s, but has only in the last decades received larger attention (RIGHTMIRE & CHOATE 1986). Methane found in coal deposits was once regarded as a safety hazard to miners and was flared and vented to the atmosphere, but is today regarded as a potential and valuable energy source (KUUSKRAA & MEYER 1980, DOE 1984, KUUSKRAA & HAAS 1988). In addition to its provenance from secondary enrichments, unconventional gas is also coming in increasing amounts directly from their primary places of generation represented by the coal seams (which are both source and reservoir rocks) where significant quantities are trapped in natural microcracks (sometimes tragically accentuated by explosions of gas which is released during subsurface mining in the galleries) and has been prevented from migration.

The large volumes of gas which are still contained within the coal seams and have not yet emigrated into interrelated tight sandstones are acquisited by coal fracturing in various scale (TREVITS, HANSON & WARD 1982; LASSITER & HOLLINGSWORTH 1984, SARDA 1984, SECCOMBE & SAKASHITA 1985; BOYER, STUBBS & SCHWERER 1986; JONES 1986; JONES, BELL & MORALES 1986; LAYNE & BYRER 1986; LOGAN, SECCOMBE & JONES 1986; MILITZER & SCHWERER 1986, PERLMAN 1986; HANSON, NIELSEN, SORRELS, BOYER & SCHRAUFNAGEL 1987; SCHRAUFNAGEL 1987; REEVES, WALLACE & BEAVERS 1987; ELY, HOLDITCH & CARTER 1988; HOLDITCH, ELY, SEMMELBECK & CARTER 1988; JEU, LOGAN & McBANE 1988; KELAFANT, BOYER & ZUBER 1988; LOGAN, ERWIN & McBANE 1988; ZUBER, KUUSKRAA & SAWYER 1988). The joint significance of coal seams as sources and reservoirs of natural gas and their interrelationships with tight gas sands are emphasized by RIGHTMIRE & CHOATE (1986; cf. section 4.4.1.3.) After illustration of some general aspects, the outline as follows includes relationships of cleat geometry, special mechanical coal properties, fracturing stimulation of coal seams, fracture conductivity maximization, vertical hydraulic fracturing vs. horizontal drilling, reservoir pressure, and relationships between coal seam and bounding strata.

4.4.3.1. General aspects

The methane associated with coal seams was formed as a by-product of the natural biogenic and thermal maturation process of coal and exists both as adsorbed gas on the internal surfaces of the microporous coal structure and as free gas within the pores and in the natural fracture and cleat system of the coalbed (KUUSKRAA & HAAS 1988), and/or as dissolved gas in ground water within the coal bed (RIGHTMIRE & CHOATE 1986) and thus is not simply compressed within the pore spaces of the reservoir (JEU, LOGAN & McBANE 1988). Gas generated in excess of that which can be adsorbed on the coal surfaces is initially present as free gas within the porosity of the coal, most noticeably within the fracture porosity. This gas is available for solution in ground water moving through the bed and migration under hydrodynamic conditions, or it may be retained in the coal bed as trapped free gas under proper hydrogeologic conditions.

Production of methane from coal seams involves removing of water from the coal cleat system, lowering reservoir pressures to reflect gas desorption, and initiating the subsequent flow of gas through the reservoir into a producing wellbore. Following discussion of the importance of the cleat system for gas drainage from coal seams, an account is given on significance of hydraulic fracturing in deep unmineable coal seams for gas exploitation, aims and purposes of coal fracturing, and distribution of important coal basins and gas reserves.

4.4.3.1.1. Importance of the cleat system for gas drainage from coal seams

The mechanism of gas production from coal is substantially different from that in conventional reservoirs (HOLDITCH, ELY, SEMMELBECK, CARTER, HINKEL & JEFFREY 1988). There is in most cases little or no free gas present in the coal, but most of the pore space in the cleat system is water-saturated and most of the gas is adsorbed on the coal surface. In order to allow gas withdrawal, the pressure in the cleat system must be reduced to cause the gas to desorb from the coal surface to the cleat system and to diffuse through the coal matrix. Normally, significant volumes of water must be produced for lowering the pressure in the cleat system so that gas desorption can begin. While fracture intersection of very high-permeability zones would be beneficial to gas offtake in sandstones and carbonates, gas exploitation in coal seams becomes difficult due to the large volumes of water that must be removed, and therefore coal fracturing does not necessarily aim on crossing of the sections with highest cleat permeability (ZUBER, REEVES, JONES & SCHRAUFNAGEL 1988). As gas desorption is the primary source of production, the gas flow rate from a coal seam may increase with time, and it is not uncommon for maximum gas flow rates to occur months or years after starting exploitation. Desorption is also a significant mechanism for gas liberation in shales (cf. section 4.4.4.1.).

Economical gas withdrawal from coal seams requires existence of an extensive cleat system in order to provide the necessary permeability, a high enough gas content to constitute a resource that is worth developing, and connection of the cleat system to the wellbore. The primary reservoir properties which control gas production from coal seams are adsorbed gas content, desorption behaviour characteristics, effective drainage area size, reservoir pressure, and permeability volume and water saturation of the cleat system. Following illustration of some aspects of the significance of hydraulic fracturing in deep unmineable coal seams for gas exploitation, the distribution of important coal basins and gas reserves is sketched.

4.4.3.1.2. Significance of hydraulic fracturing in deep unmineable coal seams for gas exploitation

While hydraulic fracturing has formerly been used as a pure safety procedure in order to accelerate methane removal from coal seams ahead of underground mines which then was flared and vented, during the last decade a growing awareness of the vast economical and industrial potential of hydraulic fracturing for methane exploitation from coal seams has been established (HOLDITCH, ELY, SEMMELBECK, CARTER, HINKEL & JEFFREY 1988). Coal-bed reservoirs are low-pressure low-flow producers because they are at approximately hydrostatic pressure and once the free gas is depleted, production is controlled by methane desorption and diffusion to the fractures. The relative permeability to gas and water are critical to initiating gas withdrawal. Where the coal seams are water-saturated, the water content must be reduced in order to permit desorption and flow of the gas to the wellbore, because high water saturation is seriously limiting gas flow. Stratigraphically, coal is often distributed in relatively thin seams over large intervals which means that for economical recovery, special production practices and multiple completions from the same wellbore are required.

Special emphasis in this connection is put on acquisition of deep unmineable coal seams by drilling and subsequent fracturing, whereas further exploitation of coal seams associated with an active underground coal mine is of less economical significance. Hydraulic fracturing of degasification wells of coal-bearing strata has in the past already repeatedly been carried out in the USSR. An alternative or complementary technique for acquisition of in-situ coal-seam gas is horizontal or lateral drilling (TRACY 1988; cf. sections 4.4.3.6. and 4.8.6.2.1.). LINDEN (1985) emphasizes the importance of deep unmineable coal seam gas for the improvement of the energy acquisition spectrum in the next decades (cf. section 2.2.1.6.4.). Coalbed methane currently represents the fastest growing and potentially the economically most attractive of the low-permeability gas resources (KUUSKRAA & HAAS 1988).

4.4.3.1.3. Aims and purposes of coal fracturing

Hydraulic fracturing in coal attempts to bypass near-wellbore damage (cf. section 4.8.3.), stimulate production and accelerate dewatering, distribute the pressure drawdown to reduce fines production (cf. section 4.2.2.7.), and effectively connect the wellbore to the reservoir (JEU, LOGAN & McBANE 1988). The primary purpose of coal fracturing is to bypass any formation damage that may be present from drilling mud or cement invasion, because formation damage sharply reduces the extent of pressure drawdown that reaches into the reservoir, thereby slowing the dewatering process and delaying methane liberation. Removing or bypassing of formation damage enables early gas production and higher overall exploitation.

Deeply penetrating hydraulic fractures can stimulate higher productivity in lower-permeability coals as in other tight formations (cf. section 4.8.11.). Because of the need to quickly drop reservoir pressure throughout the pay and the benefit of accelerating the dewatering process, high fracture conductivity is paramount especially in the moderate permeability of shallow coals. Fracture stimulations provide better access to the coal reservoir than can be obtained by perforating alone. While open-hole completions should adequately access the individual coal layers, perforating of cased wells may not sufficiently access the alternating structure of thin layers of cleated coal and non-cleated carbonaceous shale. Hydraulic fracturing can help to reach beyond the wellbore to contact coal deposits that the borehole may not actually penetrate (cf. section 4.10.2.1.).

4.4.3.1.4. Distribution of important coal basins and gas reserve

The coal basins in the USA are expected to have more than 400 Tcf or 10,000 Bill. m3 recoverable reserves (BYRER, MROZ & COVATCH 1984; OFFICE OF TECHNOLOGY ASSESSMENT 1985, RIGHTMIRE & CHOATE 1986) or even up to 800 Tcf or 20,000 Bill. m3 production potential (LAYNE & BYRER 1988). Estimates of resource potential in various coal basins and assessments of reserve recoverability are also carried out by CHOATE, JURICH & SAULNIER (1981); CHOATE, LENT & RIGHTMIRE (1982) and ADAMS et al. (1982). Some of the most important coalbed methane basins in the USA are Piceance Creek Basin/Utah and Colorado (84 Tcf or 2,100 Bill. m3 gas in place), San Juan Basin/Colorado and New Mexico (80 Tcf or 2,000 Bill. m3 GIP), Northern Appalachian Basin (61 Tcf or 1,525 Bill. m3 GIP), Central Appalachian Basin (48 Tcf or 1,200 Bill. m3 GIP), Powder River Basin/Montana and Wyoming (39 Tcf or 975 Bill. m3 GIP), Greater Green River Basin/Wyoming and Colorado (31 Tcf or 775 Bill. m3 GIP), Illinois Basin (21 Tcf or 525 Bill. m3 GIP), Warrior Basin/Alabama (20 Tcf or 500 Bill. m3 GIP), Raton Mesa Basin (18 Tcf or 450 Bill. m3 GIP), Arkoma Basin/Oklahoma and Arkansas (4 Tcf or 100 Bill. m3 GIP), Wind River Basin/Wyoming (2 Tcf or 50 Bill. m3 GIP), and Uinta Basin/Utah and Colorado (1 Tcf or 25 Bill. m3 GIP; KUUSKRAA & HAAS 1988).

The greatest current activity is in the Warrior Basin and in the San Juan Basin. In the USA, associate methane from coal seams is already now the primary source of natural gas for Alabama and is rapidly becoming a major source of natural gas in the San Juan Basin in New Mexico and Colorado (HOLDITCH, ELY, SEMMELBECK, CARTER, HINKEL & JEFFREY 1988). Aspects of cleat geometry, special mechanical coal properties, fracture stimulation types of coal seams, fracture conductivity maximization, vertical hydraulic fracturing vs. horizontal drilling, reservoir pressure, and relationships between coal seams and bounding strata are outlined as follows.

4.4.3.2. Cleat geometry

Most of the retained gas within the coal seams (which is much less than the total amount generated during coalification due to considerable emigration into nearer or farther high-permeability reservoirs) is trapped by absorption on the surface of micropores. Smaller amounts are also enclosed as free gas within fractures or pores and as dissolved gas in formation water. The cleat system which provides the primary flow channel and thus represents coal permeability usually consists of two sets of closely-spaced vertical natural fractures which are oriented orthogonal to each other. The primary set of natural fractures are face cleats, and a less defined second set of perpendicular fractures are butt cleats (DECKER & SECCOMBE 1986; LOGAN, SCHWOEBEL & HORNER 1987). The face cleats are continuous and crosscutting, whereas the butt cleats terminate at the face cleats. Coal cleats form generally in response to local structural forces, although the process of coalification and other variables also influence cleat density (McFALL, WICKS, KELSON, SEDWICK & BRANDENBURG 1987; cf. also section 4.2.2.7.).

Coal gas reservoirs typically have very low matrix permeability, with the gas being desorbed from the matrix to the wellbore through a system of cleats. In addition to these natural fractures and fissures, coal seam reservoirs usually require hydraulic fracturing stimulation to achieve economical production as a consequence of their tight nature. Hydraulically-induced cracks commonly propagate parallel to the maximum stress or face cleat direction of the coal seam and therefore may not necessarily adequately access an anisotropic reservoir. Significant improvement can be made by a combination of a horizontal borehole placed perpendicularly to the maximum permeability direction (LOGAN, SCHWOEBEL & HORNER 1987; PETZET 1988) and a hydraulic fracture (cf. sections 4.4.3.6. and 4.8.6.2.1.). Cleat width and spacing control coal permeability and as in-situ stress increases with burial depth, the deeper the coal, the lower the permeability (JEU, LOGAN & McBANE 1988).

Assessment of gas reserves and production potential in deeply buried unmineable coal seams is made by DOE (1983 b); BYRER, MOROZ & COVATOH (1984); RIGHTMIRE, EDDY & KIRR (1984); DECKER & SECCOMBE (1986) and McFALL, WICKS, KELSON, SEDWICK & BRANDENBURG (1987). Comments on the coal-bed degasification process are offered by ANCELL, LAMBERT & JOHNSON (1980) and GRAY (1987).

4.4.3.3. Special mechanical coal properties

Some special mechanical properties of coal seams that are significant for design and execution of stimulation operations are discussed along the lines of treatment pressures and coal chipping as well as impact of hydraulic fracture geometry as follows.

4.4.3.3.1. Treatment pressures and coal chipping

Treatment pressures observed during hydraulic fracturing of coal seams are often different from those of conventional formations (JONES, BELL, MORALES & SCHRAUFNAGEL 1987; cf. section 6.2.4.2.1.). After an early steep pressure rise, overall operation pressures often exceed the lithostatic pressure, but produce surprisingly little vertical fracture extension. The accumulation of coal chips at the wellbore and partial plugging of an advancing crack tip by coal fines carried ahead of or within the fluid operation pad are the responsible mechanisms which account for the anomalous treatment pressure distribution (cf. section 4.2.2.7.). As a consequence of narrow cleat spacing, coal tends to break into small cleat-bounded fragments already at moderate mechanical stress. In addition, the particular properties lead to mechanical destruction of coal around the wellbore with time, with the loose coal causing blockage at the wellbore and partial screenout at the fracture tip. The effects can be partially overcome by jetting the coal with high-pressure water prior to fracturing in order to create a cavity around the borehole and thereby remove any skin damage induced by loose coal fragments and to reduce breakdown or fracture initiation pressures (REEVES, WALLACE & BEAVERS 1987).

The low value of Young`s modulus in coal seams (cf. section 4.4.3.8.3.) creates very wide hydraulic fractures as well as very complex multiple fracture systems propagating in multiple directions (HOLDITCH, ELY, SEMMELBECK, CARTER, HINKEL & JEFFREY 1988). The complex hydraulic fracture behaviour is a consequence of the presence of the cleat system, the complex structure of the coal matrix, and the stratigraphical relationship of the coal seams with respect to the surrounding sediments (JEU, LOGAN & McBANE 1988; cf. section 4.4.3.8.). This generates considerable difficulties to create and to prop long hydraulic fractures. Therefore coal-seam reservoirs with poorly developed cleat systems may not be candidates for development, because they may not be capable of supplying enough gas to the relatively short hydraulic fracture. Variations of stress level in coal seams may be the consequence of faults crossing the gas field which locally cause weakening of the coal matrix so that it will readily fail and spall during the fracture treatment, with spalled coal chips blocking the crack together with high closure stresses resulting in elevated treating pressures (ZUBER, REEVES, JONES & SCHRAUFNAGEL 1988; cf. section 4.2.2.7.). There is also a relationship between borehole stability and coal friability (LOGAN 1988).

4.4.3.3.2. Impact of hydraulic fracture geometry

Reduction of the amount of coal fines produced is a potential benefit of hydraulic fracturing (JEU, LOGAN & McBANE 1988). A highly-conductive fracture can distribute the pressure drop along the crack instead of concentrating it in the immediate wellbore area. Thus the propensity of generating coal fines should be reduced, with utilization of larger proppant grain sizes being able to assist maintenance of fracture conductivity by allowing the smaller coal fines to pass through the proppant package. Another problem of proppant fracturing in coal seams is production of coal grit, proppant and water together with the gas. The application of resin-coated proppants in the tail-in portion of the fracture (cf. section 4.12.3.3.) gives sufficient stabilization to prevent proppant movement and flowback (FRACFAX 1988 e).

The optimum size of fracture half-length depends on well spacing, coal seam properties and stimulation costs (cf. section 4.8.11.), with the highest performance being achieved by very high conductivity and long extension of the fractures (JONES, BELL, MORALES & SCHRAUFNAGEL 1987). When treatment pressure exceeds lithostatic stress, horizontal fractures are initiated which may possibly be effective in contacting higher permeability joints in the formation that may not normally be accessed by vertical fractures. Because of their radial growth, however, horizontal fractures are considerably shorter than confined vertical fractures for the same treatment volumes and therefore will not be as effective in consistently increasing production as long high-conductivity vertical fractures.

4.4.3.4. Fracturing stimulation types of coal seams

Design and execution of stimulation operations in coal seams is not straightforward, but high injection pressures, complex fracture systems, screenouts, and production of proppants and coal fines are typical problems (HOLDITCH, ELY, SEMMELBECK, CARTER, HINKEL & JEFFREY 1988). Hydraulic proppant fracturing is normally inevitable for coal seam gas exploitation. The type of stimulation treatment depends on depth, thickness and stratigraphy of the coal seam. Some different fracturing operation types of coal seams are briefly sketched as follows along the lines of fluid viscosity, multiple interval fracturing, fracturing strategies, and fracture propagation direction and treatment pressure increase.

4.4.3.4.1. Fluid viscosity

ZUBER, KUUSKRAA & SAWYER (1988) compare different hydraulic fracturing types and well spacings for economical recovery of coalbed methane. Evaluation of water, foam and gel fracturing reveals that the best results are achieved with large-scale gel stimulation treatments, with the economical benefit of the costlier gel-based fracturing operations more than compensating for the relatively high treatment expenses. Well spacing and reservoir stimulation effectiveness are thus the two key parameters dictating economical returns from coalbed methane wells and have a dramatic effect on both timing of peak production rates and ultimate well recovery (KUUSKRAA & HAAS 1988). Some comments are offered on gel- vs. water-based stimulation as well as proppant settling and rolling in water fracturing as follows.

4.4.3.4.1.1. Gel- vs. water-based stimulation

HANSON (1988) finds out that low-viscosity fluids such as water can effectively induce fracture height growth through several coal seams via entry at the bottom of the interval, and high-viscosity fluids such as cross-linked gels promote a measure of fracture height containment and lateral extension in layered media. An optimal treatment for many coal seams consists of a large water prepad for the purpose of height growth followed by a crosslinked gel pad for the reason of fracture widening and extension from the wellbore. Slurry stages call for crosslinked gel to effectively carry the proppant and for coarse-grained proppants to enhance fracture conductivity. Casing slotting through coal can reduce high treatment pressure and in part minimize fracture-inlet constrictions.

REEVES, WALLACE & BEAVERS (1987) perform a comparative assessment of different stimulation techniques in terms of cost and production which indicates that wells fractured with water and foam-water treatments produce at higher water and gas rates than those fractured with either gel or foam. In addition, foam-water-stimulated wells exhibit high fluid loss during the first two months of production, whereas water-fractured wells produce more water after the second month. Therefore water fracturing appears to be the most effective treatment in terms of both expenses and offtake. KUUSKRAA & HAAS (1988) find out that gel-based stimulations are superior to water-based treatments, because they achieve peak production rates at earlier times and nearly double the peak rates achieved by water-based jobs. The increase in gas recovery due to the higher fracture conductivity and length of the gel-based stimulation more than compensates for the additional cost of these treatments.

4.4.3.4.1.2. Proppant settling and rolling in water fracturing

In some exceptionally well-cleated and thus high-permeability coal seams, very small- to medium-size fracture stimulations using water with only very low proppant concentrations can already be considerably effective (ZUBER, REEVES, JONES & SCHRAUFNAGEL 1988). The poor proppant transport characteristics of fresh water result in proppant settling near the wellbore during slurry injection, with proppants then being rolled to the head of the bank across its top during flush in order to increase propped fracture length. Although proppant transport is considerably poorer with water than with high-viscosity fluids, the banking phenomena caused by proppant settling can potentially result in a high-conductivity fracture with proppants in the bottom and an open channel in the upper section of the crack crossing the coal seam (cf. LASSITER & HOLLINGSWORTH 1984; cf. section 4.3.3.2.). Water fracturing is also of considerable significance in chalk sequences (cf. section 4.5.4.5.2.) and geothermal reservoirs (cf. section 4.7.3.5.).

4.4.3.4.2. Multiple interval fracturing

SCHRAUFNAGEL & LAMBERT (1988) state that hydraulic proppant fracturing can be successively carried out in various superimposed coal seams and is accomplished through the use of baffles whereby a ball is dropped down the casing to seal off a zone following its treatment, with multiple repetition of this procedure in upwards direction along the prospective section in the borehole resulting in uphole fracturing one level after the other in sequential order (cf. section 4.2.2.1.1.). Another possibility of multiple coal seam fracturing is the limited-entry technique which requires maintenance of a predetermined wellbore pressure throughout the operation (cf. section 4.2.2.2.). Some aspects of reservoir depth of coal seams and limited-entry stimulation are illustrated as follows.

4.4.3.4.2.1. Reservoir depth of coal seams

In shallower reservoir storeys where horizontal rather than vertical fractures are more likely to form (cf. section 1.2.8.), the stimulation technique must provide for multiple hydraulically-induced fractures of horizontal orientation and/or complex fractures with both horizontal and inclined components (SCHRAUFNAGEL & LAMBERT 1988).

Thus in many cases deeper reservoirs require coal seam penetration by highly-inclined to horizontal boreholes from which multiple vertical fractures emanate (cf. section 4.8.6.2.4.), whereas in shallower pay zones, the coal seams are acquisited by vertical boreholes from which multiple oblique drainholes may radiate off (cf. section 4.8.6.1.2.) that are intersected by multiple horizontal fractures in case of single thicker coal seams, or are traversed by highly-deviated to horizontal boreholes from which only singular horizontal hydraulic cracks can be propagated into the coal matrix for the occasion of groups of thinner coal seams.

4.4.3.4.2.2. Limited-entry stimulation

The limited-entry stimulation technique applies for simultaneous treatment of multiple zones by engineering placement, size and number of casing perforations which serve as pressure chokes that control the rate of fluid injection into each targeted zone (SCHRAUFNAGEL & LAMBERT 1988; cf. section 4.2.2.2.). Proppant abrasion leads to perforation diameter growth and consequently pressure reduction. Significant pressure drops in some treatment stages can be caused by screenout of various perforations and initiation of a vertical fracture following termination of propagation of a horizontal crack (cf. sections 1.2.8.4. and 4.9.4.2.). Limited entry has turned out to be a viable approach to stimulating a series of coal seams even at shallow depths where horizontal fractures are anticipated, and at greater depths where the primary crack orientation likely would be vertical, the concept may also prove viable in completing multiple seams within individual coal groups.

4.4.3.4.3. Fracturing strategies

Four basic fracturing strategy scenarios for hydraulic stimulation of coal seams (HOLDITCH, ELY, SEMMELBECK, CARTER, HINKEL & JEFFREY 1988) including horizontal fractures in a singular shallow coal seam, single planar vertical fracture in a series of thin coal seams, multiple complex fracture system in a single thick coal seam, and initial fracture confinement within a single coal seam passing later to vertical crack propagation into the bounding layers. After discussion of some general aspects, an overview is given on single vertical fractures intersecting several thin coal seams and complex multiple fractures contained in a single thick coal seam which are the most frequently developed cases of coal fracturing. Some comments are also offered on injection rates, fluid loss and treatment pressure particularly for generation of complex T-shaped fractures.

4.4.3.4.3.1. General aspects

Concerning the general strategy applicable to any coal seam, the main purpose of hydraulic fracturing is to interconnect the cleat system of the coal to the borehole. Without a well-developed cleat system, creation and propping of a long fracture in coal seams is usually not profitable and commercial at current gas prices (HOLDITCH, ELY, SEMMELBECK, CARTER, HINKEL & JEFFREY 1988). Bottomhole pressure during production should be minimized to accelerate desorption of natural gas. A wide conductive propped fracture system that can withstand the maximum closure stress must be created to minimize the pressure drop down the fracture. As a consequence of high treating pressures (cf. section 4.4.3.8.3.) and complex nature of the fracture systems, real-time changes during the operation are the rule rather than the exception (cf. section 6.2.4.3.). Cutting of proppant insertion and reinitiating the pad or changing pad volume based upon treating pressures may be required to improve the results.

In shallow reservoir depth, individual horizontal fractures within the separate coal seams have to be created by using either limited-entry methods (cf. section 4.2.2.2.) or mechanical diversion (cf. section 4.2.2.1.). Linear fluids with a moderate pad volume should be applied and the bottomhole treating pressure should be in excess of 1 psi/ft. Complex multiple fracture systems may be generated if bottomhole pressures increase substantially during the job.

4.4.3.4.3.2. Single vertical fractures intersecting several thin coal seams

Concerning single vertical fractures intersecting a series of thin coal seams, this scenario is analogous to vertical cracks penetrating layered clastic and/or carbonate reservoirs (HOLDITCH, ELY, SEMMELBECK, CARTER, HINKEL & JEFFREY 1988). Well-developed cleat systems in coal seams may provoke higher fluid leakoff than in conventional pay zones. A common practice of stimulation of thin coal seam complexes is to perforate in a clastic zone adjacent to the coal and to fracture the coal by allowing the crack to grow vertically, with this technique usually resulting in lower operating pressures than if treating the coal exclusively. Excess pressures are moderate and decline throughout most of the job as the fracture continues to grow vertically. Fracture propagation gradients should be less than 1 psi/ft, and pad volumes of abt. 30 - 35 % are appropriate. Viscous shear-stable delayed crosslinked fluids or foams should be used to provide adequate proppant transport and to minimize proppant settling.

4.4.3.4.3.3. Complex multiple fractures contained in a single thick coal seam

The scenario of complex hydraulic fractures contained within a single thick coal seam is unique to coal gas reservoirs and is well documented by minebacks. High treating pressures occur, with the excess pressure in the fracture increasing rapidly when pumping begins and remaining high during the job, as a consequence of creation of multiple vertical or T-shaped fractures (or mushroom-like cracks; cf. sections 1.2.8.4. and 4.9.4.2.) near the wellbore (HOLDITCH, ELY, SEMMELBECK, CARTER, HINKEL & JEFFREY 1988). The tortuous path for fluid flow generated by these multiple fractures results in a zone where the pressure drop is quite large. Other explanations for high treating pressures are slip zones formed in highly-cleated areas immediately ahead of the fracture tip.

Due to stress concentrations near the fracture tip, slip can occur that tends to absorb energy which otherwise would be used to propagate the crack. As a consequence of the high fluid leakoff that can take place in cleated coal, backstresses can increase during the job which also raises injection pressure through poroelastic effects (cf. section 4.4.3.8.3.). If large volumes of coal fines are generated, they can concentrate at the fracture tip and plug it thereby inhibiting further propagation and increasing the operating pressure (cf. sections 4.2.2.7. and 4.4.3.3.). Coal fines may also remain entrained in the fluid and thus cause increasing apparent viscosity which also may give rise to elevated injection pressure (JONES et al. 1987). Several parallel fractures also have considerable impact on treating pressure and crack width (JEFFREY, VANDAMME, HINKEL & HORNER 1987).

4.4.3.4.3.4. Injection rates, fluid loss and treatment pressure

For realization of the strategy of complex hydraulic fractures contained within a single thick coal seam, high injection rates have to be used during the stimulation treatment to combat high fluid loss (HOLDITCH, ELY, SEMMELBECK, CARTER, HINKEL & JEFFREY 1988). In addition to high injection rates, high-viscosity shear-stable delayed crosslinked fluids (cf. section 4.3.4.3.) and bridging fluid-loss additives (cf. section 4.3.4.6.) are required to offset high leakoff. Because the coal has low Young modulus and high formation compressibility and as a consequence of the presence of complex crack systems, hydraulic fractures seldom penetrate more than a few hundred feet away from the wellbore.

As excess pressure rises due to complex fracture geometry (cf. section 4.4.3.8.3.), a vertical component may be initiated into the boundary layers at a weakness point at the border interface. If this happens, fluid escaping to the boundary layer could cause the fractures in the coal seam to decrease in width which could lead to screenout if high proppant concentrations are pumped when vertical excess propagation starts (cf. sections 4.2.3.5.3. and 6.2.4.2.1.). If the sharp drop in treating pressure indicating height growth out of zone occurs near the end of the job, increasing pumping rate may provide the necessary fracture width required to finish the job. If the pressure drop, however, takes place earlier in the operation, reinitiating the pad and pumping of fluid without proppant until sufficient fracture width is obtained may be necessary prior to continuing the proppant stages of the treatment. If the pad is reinitiated, a large enough pad volume must be pumped to develop a fracture that is sufficiently wide to accept proppant before starting back with proppant insertion.

4.4.3.4.4. Fracture propagation direction and treatment pressure increase

Treatment pressure increases occurring shortly after proppants enter the coal seam reflect flow restriction due to near-wellbore proppant blockage in the fracture, and subsequently short periods of pressure decline take place as the proppants are flushed away from the wellbore (JONES 1987, 1988). Simulation runs reveal that the fracture blockage in coal seams frequently contains both proppants and coal chips (cf. sections 4.2.2.7. and 4.4.3.3.). Fracturing with water-based fluids results in shorter propped fracture lengths, whereas higher-viscosity fracturing media such as linear gel and nitrogen foam that have also better proppant transport characteristics than water lead to much longer propped cracks.

Hydraulic fractures in coal seams usually propagate parallel to the maximum stress or face-cleat direction (cf. section 4.4.3.2.) and therefore may not adequately access an anisotropic reservoir (LOGAN 1988). In some cases, a horizontal borehole placed perpendicularly to the maximum permeability direction may therefore be a more effective stimulation technique (cf. sections 4.4.3.6. and 4.8.6.2.1.2.). During coal stimulation treatments, fractures are hydraulically newly induced or pre-existing natural fractures in the coal seam are expanded (LAYNE & BYRER 1986). The complex natural crack system in coal seams and surrounding formation may inhibit the productivity of a stimulation job by diverting fluid paths away from the pay zone fracture channel.

All coal deposits contain methane, although the concentrations vary from seam to seam and sometimes even change within the same seam (DOE 1984). Unlike oil and gas wells where water helps to push the hydrocarbons to the surface, the water in coal seams hinders the gas from moving to the wells and thus first the water has to be pumped out before the gas can be drained off (cf. section 4.4.3.1.). In addition, natural pressures are relatively low in a coal seam, forcing the gas to be either used on-site or compressed in order to enable flow through a longer-distance pipeline. Coal seams can in places be overpressured and water-saturated and can possess high structurally enhanced permeability (McBANE & JEU 1988). Attempts to produce natural gas from coal seams usually encounter significant volumes of water which must be removed in order to reduce reservoir pressure and hence allow desorption of methane from the coal (WAY 1988). Deeper coal seams have overall lower permeabilities than shallower ones, but are nevertheless desirable and economically attractive targets because of their higher rank and higher gas content (McKEE, BUMB & KOENIG 1986; SCHRAUFNAGEL & LAMBERT 1988). A comparative evaluation of hydraulic fracturing techniques for coal seams is carried out by LOGAN, ERWIN & McBANE (1988).

4.4.3.5. Fracture conductivity maximization

In order to maximize fracture conductivity in coal seam stimulation, partially coarse 10/20 mesh proppants have to be applied (cf. also section 4.3.5.) and the total proppant quantity has to be increased over standard designs (JONES, BELL, MORALES & SCHRAUFNAGEL 1987; REEVES, WALLACE & BEAVERS 1987; ZUBER, REEVES, JONES & SCHRAUFNAGEL 1988). The use of low-viscosity banking fluids will allow proppants to settle to the bottom of the fracture and to form a proppant bank which bridges the entire width of the ballooned fracture (cf. sections 4.3.3.2. and 4.12.2.2.). Banking fluids may allow an ultra-high-conductivity open fracture to form above the proppant bank (LASSITER & HOLLINGSWORTH 1984). As a consequence of inefficient proppant transport, the proppant-banking approach using water results in relatively short propped fracture length which is sacrificed for higher conductivity (JONES, BELL, MORALES & SCHRAUFNAGEL 1987).

Thin coal seams are not amenable to vertical drilling and thus highly-deviated to horizontal holes have to follow the coal seam more or less along its dip or strike in order to first render stimulation more effective and second to produce more gas (DOE 1984; cf. section 4.8.6.2.1.). Special techniques for economical acquisition also include birdfoot drilling from one central vertical stem well radially into various directions.

4.4.3.6. Vertical hydraulic fracturing vs. horizontal drilling

Apart from predominantly vertical hydraulic fracturing as a consequence of the mainly greater depth of coal seams, horizontal drilling is another suitable technique for gas drainage from coal reservoirs, with horizontal boreholes for coal-seam degasification having been implemented for the first time in the USA more than 30 years ago (ERTEKIN, SUNG & SCHWERER 1986). The main advantage of horizontal boreholes is that the direction of wellbore trajectory can be controlled with respect to the principal permeability directions of the coal, whereas vertical hydraulic fractures have random orientation to the cleat system which is determined by the geotectonical stress state (cf. section 4.8.6.2.1.5.).

Coal seam thickness has a more pronounced effect on a vertical fracture than on a horizontal borehole, because fracture surface area increases with coal seam thickness in contrast to the exposed surface area of a horizontal borehole which only increases with length of the well. Based on identical exposed surface areas, vertical fractures and horizontal wells create higher gas production rates in early and late exploitation, respectively (cf. also section 4.8.6.2.1.). Therefore the optimum solution for coal seam drainage could be horizontal boreholes from which multiple vertical fractures are propagating, with this combination accessing the largest possible coal seam surface area (cf. also section 4.8.6.2.4.). Horizontal wells in combination with hydraulic fracturing are also significant instruments for gas drainage from shales (cf. section 4.4.4.3.2.).

4.4.3.7. Reservoir pressure

Coal seam reservoirs in various basins are generally characterized by overpressuring, underpressuring or normal pressure regime (McFALL, WICKS, KELSON, SEDWICK & BRANDENBURG 1987). Overpressured areas are basin regions where gas generation within the coal deposits may still be active, where the coals have been uplifted, or where the overburden has been eroded. Underpressured areas are basin regions where the coals are no longer actively generating gas, where the void space may have increased due to natural fracturing or coal desorption, or where the gas may have escaped from the coal deposits to the over- and underlying rock strata. Normally pressured areas are basin regions in which mobile water and residual gas coexist within the cleat and natural fracture system of the coal deposits. In the latter areas, it is assumed that an equilibrium hydrostatic condition has been reached after water has imbibed and resaturated the coal seam.

4.4.3.8. Relationships between coal seams and bounding strata

Fracture design in coal seams must reflect the lithology of the bounding strata, because this affects fracture height growth and crack geometry (JEU, LOGAN & McBANE 1988). Distinction has mainly to be made between sandstones and shales as lithologies next to the coal seam. Some mechanisms of fracture height containment and elevated treating pressures are also illustrated.

4.4.3.8.1. Sandstones

Where coal seams are adjacent to sandstone, hydraulic fracture growth occurs in both sandstone and coal, and stimulation pressures are less than 1.1 psi/ft. Linear fracturing fluids can be used effectively to place proppants in the coal seam if the sandstone overlies the coal (JEU, LOGAN & McBANE 1988). If the sandstone is situated beneath the coal seam, in many cases successful stimulation can be achieved by perforating the sandstone interval and propagating the hydraulic crack from the sandstone upwards into the coal seam. This upwards fracturing technique has the advantage that the guiding sandstone permits a longer fracture extension deeper into the reservoir than would be possible within the coal seam alone and also enables insertion of larger quantities and higher concentrations of proppants which at least in parts of the created crack reach also the coal seam (similar uphole fracturing is performed in chalk sequences, with the crack being initiated in a more competent horizon underlying soft and collapsing storeys where most of the hydrocarbon reserves are contained; cf. section 4.5.4.3.2.).

4.4.3.8.2. Shales

Much higher treating pressures occur when coal is bounded by shale, with the fractures being primarily contained to the coal and the high operational pressures having resulted in part from a complex fracture geometry (JEU, LOGAN & McBANE 1988). High-strength casing is useful in accommodating the higher treating pressures and allows larger jobs to be pumped where many operations had to be terminated prematurely when the burst strength of the tubulars was reached. Higher injection rates and solid fluid-loss material create sufficient width in the complex geometry and thereby increase the likelihood of a successful treatment.

In many cases, the in-situ stress in sandstone is less than that in coal which in turn is less than that in shale (cf. sections 4.2.2.3.1., 4.2.3.2.2. and 4.2.4.3.1.). In coal seams bounded by shales, confined fracture height results in higher treating pressures which causes a complex T-shaped fracture geometry that has both vertical and horizontal components (JEFFREY, VANDAMME & ROEGIERS 1987; cf. sections 1.2.8.4. and 4.9.4.2.). In terms of propagation, the radius of the horizontal fracture quickly exceeds coal seam thickness and is more narrow, whereas the vertical crack is usually very wide due to its confinement (JEU, LOGAN & McBANE 1988). Fracture growth in coal seams bounded by shale is usually contained within the coal and adjacent shales even though pressures far exceed the in-situ stress of the shales. Fracturing pressures frequently are significantly higher than 1.1 psi/ft. Successful fracture designs must ensure that the complex T-shaped crack will have adequate width to accommodate proppant which is achieved by including a greater percentage of pad (cf. section 4.3.4.6.1.4.), higher injection rates, and use of solid fluid-loss agents such as 40/70 or 100 mesh sand (cf. sections 4.8.8.3.1.2. and 4.8.8.3.2.3.).

4.4.3.8.3. Fracture height containment and elevated treating pressures

Potential mechanisms for fracture height containment are plastic deformation of shales, slippage at the coal/shale interface, the nearly tenfold contrast in Young`s modulus between coal and that of shale and sandstone, and limitation of fracture height growth by creation of a horizontal fracture due to the high treating pressure. Magnitude and rate of pressure rise accompanying the start of proppant injection limits crack length, with the pressure rise often reaching the burst strength of the casing and a pressure out causing a premature termination of the job. Accommodation of the abnormally high treating pressures is required to be able to create the very long hydraulic fractures that are needed to tap deep low-permeability coals. High operating pressures generally create shorter and wider fractures, and complex crack geometries cause fracture penetration to be very inefficient.

Possible mechanisms for abnormally high treating pressures are unusual fracture height containment, multiple parallel fractures (cf. section 4.8.6.2.4.), interaction resulting from leakoff inflation of coal cleats (sometimes referred to as poroelastic or backstress effects), coal fines increasing fluid viscosity, coal fines screening off the fracture tip region due to coal failure, plastic coal deformation at the wellbore, partial blockage of the fracture at the borehole due to coal failure (cf. sections 4.2.2.7. and 4.4.3.3.), and increases in effective fracture toughness due to fracture splitting near the tip region.

4.4.4. Shales

In addition to coal seam treatment and tight sandstone stimulation, significant contributions of unconventional gas are further deriving from shale fracturing (SCHRIDER, OVERBEY, KOMAR & PASINI 1977; SCHRIDER & WISE 1980; SWARTZ & AHMED 1982; CHEN, ALAM, VOZNIAK & BLANTON 1984). Shales have a very low matrix permeability (0.1 - 10 microdarcy) and therefore an extensive natural or synthetic fracture system is required in the reservoir to move economical quantities of gas to a wellbore (SOEDER 1988). Shales are rocks made up of clay- to silt-sized particles independent from petrographical composition that are laminated or fissile or both, with the term shale thus not implying that the rock has a significant content of clay minerals (HENSEL 1988). Following an outline of reservoir properties and production mechanisms, aspects of non-proppant fracturing, proppant fracturing, effectivity of proppant vs. nitrogen fracturing, special features of shale gas reservoirs and explosive fracturing are discussed.

4.4.4.1. Reservoir properties and production mechanisms

Shale gas resources include free gas in the natural fracture and porosity system as well as adsorbed gas on the surface of the organic kerogen (KUUSKRAA & HAAS 1988). The gas adsorbed on organic matter desorbs only slowly in response to decreasing reservoir pressure. The gas residing in the natural fractures is in dynamical balance with the adsorbed gas in the matrix of the adjacent wall rock. When a well encounters the gas-filled fractures or is joined to the natural cracks by hydraulic stimulation, gas moves rapidly from the fracture system into the wellbore. During decline of flush production, the fractured gas shale reacts as a conventional porous reservoir. The following outline focusses on gas drainage through natural fractures, significance for gas industry and reserve potential, and impact of horizontal drilling and hydraulic fracturing.

4.4.4.1.1. Gas drainage through natural fractures

Shale wells generally exhibit a fairly rapid initial decline curve as gas is drained from the natural fracture system, followed by a slow gradual decline as gas from the matrix moves to the cracks (KUUSKRAA & HAAS 1988). Gas movement through the low-permeability shale matrix is so slow that only after several years the volume of matrix gas moving into the fractures equals the volume of gas flowing from the cracks into the wellbore. Prior to attaining this steady-state flow, the withdrawal diminution curve shows a steep drop, but soon after the steady-state flow from matrix via fracture to wellbore has been reached, the production decline curve flattens almost to the horizontal and shows only slight decline during the next several decades. Gas drainage through natural fractures is also a significant mechanism for exploitation of coal seams (cf. section 4.4.3.1.).

An increase in the amount of surface area connected to the borehole by hydraulic fractures causes more of the adsorbed gas to be released and produced over the entire life of the well (MERCER, PRATT & YOST 1988). KAMATH & GODBOLE (1985) comment on hot brine stimulation for gas production from natural gas hydrates. OZKAYA (1986) analyzes natural hydraulic fracturing of shales during sedimentation. An assessment of the potential of shale gas reservoirs including outline of effective production techniques is given by KUUSKRAA, WICKS, SAWYER, ESPOSITO & KOMAR (1983); McBANE & THOMPSON (1984), MASON (1987) and VANORSDALE (1987).

4.4.4.1.2. Significance for gas industry and reserve potential

Gas from shale reservoirs actually gave birth to the natural gas industry in the USA when the first commercial gas well was drilled into Devonian shales in Canadaway Creek near Fredonia/New York in 1821 (thus preceding the first oil well drilled in 1859 near Titusville/Pennsylvania by 38 years) and heralded a new energy source (DOE 1984, WITT 1986) which was already abundantly exploited more than 100 years ago. The Devonian shales of the Appalachian, Illinois and Michigan basins in the USA contain in fact one of the largest worldwide concentration of organically rich carbon and gas in place (KUUSKRAA & HAAS 1988) and are estimated to contain more than 100 Tcf or 2,500 Bill. m3 recoverable resources (OFFICE OF TECHNOLOGY ASSESSMENT 1985). The current development activity by drilling and stimulation is concentrated in West Virginia, Kentucky and Ohio.

DOE (1983 a); GATENS, LEE & RAHIM (1985); KOZIAR (1985); KUUSKRAA, SEDWICK & YOST (1985); LEE & GATENS (1985); HAZLETT, LEE, NARAHARA & GATENS (1986); KEMPER, FRANKIE, MOODY & JOHNSTON (1986); WITT (1986); GATENS, LEE, LANE, WATSON, STANLEY & LANCASTER (1987) and GUIDRY, GRAHAM, LANCASTER & CURTIS (1987) present reviews on evaluation, completion and testing of the production potential of Devonian shales. LINDEN (1985) underlines the role which shale gas plays in the gas supply diversification concepts of the near future (cf. section 2.2.1.6.4.). A systematical approach for economical development of shale gas resources is documented by AMERI, AMINIAN, MILLER, DORICICH & YOST (1985). Problems of log interpretation in shale gas reservoirs are summarized by CAMPBELL & TRUMAN (1986); McBANE, CAMPBELL & TRUMAN (1986); GRI (1987 b) and VANORSDALE (1987), with traditional log evaluation having difficulties in distinguishing between apparent and effective porosity (KUUSKRAA & HAAS 1988).

4.4.4.1.3. Impact of horizontal drilling and hydraulic treating

Similarly as applying for coal seams, the gas production potential of shales can be greatly improved by a combination of horizontal well drilling and multiple hydraulic fracturing (YOST, OVERBEY, SALAMY, OKOYE & SARADJI 1987; ZUBER, LEE & GATENS 1987; cf. sections 4.8.6.2.1. and 4.8.6.2.4.). Directional drilling of a horizontal well rectangularly crossing the natural cracks followed by multiple vertical hydraulic fracture stimulation orthogonally to the borehole trajectory is the optimum means to increase the shale surface area in contact with the borehole in order to release more of the adsorbed gas (HORTON, MERCER & SAWYER 1982; SALAMY, SARADJI, OKOYE, MERCER & YOST 1987).

Aspects of interference of natural and hydraulically-induced fractures in shales are also discussed by PASINI & OVERBEY (1969) and BLANTON (1982). Porosity and permeability of gas shales are evaluated by SOEDER (1988). Aspects of shale well stimulation by hydraulic fracturing are outlined by FOSTER (1975), TILLMAN (1983); DUTTON, BEMIS, JOHNSON & WHITE (1984); SHAW, GATENS, LANCASTER, LEE, AVARY & TERRY (1986); LANCASTER, GUIDRY, GRAHAM, CURTIS, SHAW & BLAKE (1987); KUUSKRAA, KELAFANT & HORNER (1988); LAYNE & SIRIWARDANE (1988) and NEARING & STARTZMAN (1988). Shale refracturing is outlined by GUIDRY, GRAHAM, CURTIS, SHAW & BLAKE (1987). Distinction has to be made between non-proppant and proppant hydraulic fracturing as well as explosive fracturing.

4.4.4.2. Non-proppant fracturing

The most important non-proppant fracturing techniques in shale gas reservoirs are nitrogen fracturing, carbon dioxide fracturing and foam fracturing which are briefly outlined as follows.

4.4.4.2.1. Nitrogen fracturing

When pumped at sufficient rates and pressures, gaseous nitrogen alone is successful as a fracturing medium for shale even without proppants and may even outperform other treatment types (ABEL 1982, ABEL et al. 1983). The advantages of nitrogen are non-damaging characteristics, minimum shut-in time, and minimum nominal treatment costs. Liquids can cause clay migration and swelling in the formation which reduces permeability (water sensitivity; cf. section 3.11.2.), and therefore liquid content of fracturing media is extremely crucial. In contrast to fluids, nitrogen is inert, relatively insoluble and compressible, with these properties preventing any damage to shale formations. After an outline of advantages of nitrogen gas fracturing, some comments are offered on wellbore vicinity damage removal and reservoir breakdown.

4.4.4.2.1.1. Advantages of nitrogen gas fracturing

Concerning advantages of nitrogen stimulation, no fluid loss additives are necessary in nitrogen gas treatments. Since nitrogen gas is mobile, fractures that viscous liquids or solids could not penetrate can be connected for optimum deliverability. The shut-in time is also crucial, because the longer the liquids remain in the well, the more severe formation damage provoked especially in shales. Fracturing treatments using only gaseous nitrogen under pressure to create cracks are generally less costly than other hydraulic methods and since no liquids are applied, liquid retention and removal problems do not occur (BRANDENBURG 1988). Because these operations are conventionally performed without proppants, spalling or fracture face irregularities are relied upon for maintaining fracture conductivity.

4.4.4.2.1.2. Wellbore vicinity damage removal and reservoir breakdown

Nitrogen fracturing treatments in shale gas reservoirs mainly remove near-wellbore damage (similarly as in coal-seam stimulation; cf. section 4.4.3.1.3.) and establish communication with previously non-producing perforations and the formation (ELY 1988, LANCASTER 1988). As nitrogen fracturing virtually eliminates the clean-up problems associated with liquid systems (cf. sections 3.11.2.2.), production is not lost and costly swabbing units to recover fluids are not needed. Low initial cost for nitrogen fracturing guarantees rapid pay-back and renders it even economical to repeat the stimulation treatment at a later stage. Nitrogen fracturing of gas shales with partial proppant insertion is outlined by GOTTSCHLING, ROYCE & SCHUCK (1985), and KING (1982, 1983) documents hydraulic proppant stimulation with liquid carbon dioxide. An economical analysis of foam fracturing in shales (sometimes including proppant addition) is performed by LIEBENTHAL, KOMAR, RIEKE & SKILLERN (1979) and STRANG & NORTON (1983). COURSEN (1983) discusses the effect of explosive stimulation with varying charge types on gas productivity of shales.

Nitrogen is also a suitable medium for performing microfracturing breakdown treatments (cf. section 4.8.2.2.) in order to minimize natural fracture system damage in shales and tight sandstones (BRANAGAN & WILMER 1988). Several techniques can be applied including hydraulic nitrogen fracturing using surface pump trucks, perforating with an underbalanced nitrogen column, high-pressure nitrogen impulse perforating/microfracturing combination, nitrogen impulse injection using a tubing pump-out plug, and tailored pulse explosive fracturing.

4.4.4.2.2. Carbon dioxide and foam fracturing

Apart from straight nitrogen fracturing, other non-proppant hydraulic fracturing techniques commonly applied in shale gas reservoirs are carbon dioxide fracturing and foam fracturing, with the latter method particularly frequently also being combined with proppant insertion in order to improve the result of the job beyond simple communication between borehole and reservoir as well as formation damage removal around the wellbore (cf. section 4.8.3.). Some aspects of effectivity of carbon dioxide vs. nitrogen treating, foam stimulation and tailored pulse loading, and auto-propping of cracks created by gas fracturing are discussed as follows.

4.4.4.2.2.1. Effectivity of carbon dioxide vs. nitrogen treating

YOST, OVERBEY, WILKINS & LOCKE (1988) compare the effectivity of different hydraulic stimulation treatments in a horizontal borehole traversing shale gas reservoirs with extensive natural jointing. Carbon dioxide jobs turn out to be more than four times better than nitrogen fracturing initially and to last twice as long prior to closure of the non-propped cracks. The density contrast between nitrogen and carbon dioxide may have caused the propagation of more and wider fractures with carbon dioxide resulting in high improvement ratios. Although stress ratios indicate that the area under study is nearly tectonically relaxed, proppants are still required to maintain permeable flow paths for more than several weeks.

4.4.4.2.2.2. Foam fracturing and tailored pulse loading

In addition to nitrogen and carbon dioxide fracturing, other non-proppant stimulation methods hitherto applied in shale gas reservoirs include foam fracturing and tailored pulse loading, with the latter technique comprising ignition of slow-burning propellants in the borehole (DOE 1984; cf. sections 4.2.2.1.3.3., 4.7.3., 4.8.8.5. and 4.8.9.6.). The force of the burn plus the produced combustion gases crack the rock in several directions at once instead of the usual single-direction fractures created by water or foam. Conventional explosive fracturing has also historically been applied in shale gas reservoirs and results in communication of the borehole with the pay zone and in a relatively small degree of stimulation because the created fractures are multidirectional, unpropped and of limited penetration (BRANDENBURG 1988; cf. section 4.4.4.6.).

4.4.4.2.2.3. Auto-propping of cracks created by gas fracturing

Experience with non-proppant nitrogen fracturing treatments has shown that auto-propping of the cracks by gas erosion generating an etched surface in the formation at the fracture wall, rapid flowback causing pay spalling, or some other mechanisms of reservoir deformation giving rise to an auto-propping conductive flowpath is extremely rare, if existing at all (ELY 1988). Calculation of current rates of nitrogen across the fracture face shows that the possibility of significant erosion is remote, and the fact that considerable amounts of reservoir material are produced with rapid flowback procedures does not prove the installation of a conductive path by self-propping, but rather testifies to failure of the shale under the imposed stress. Auto-propping also is the mechanism to keep open cracks having been generated by acid fracturing, but also in carbonate stimulation, the much better approach is hydraulic proppant treatment with proper infilling and support of the crack (cf. section 4.5.1.1.1.2.).

Therefore much superior results of shale gas stimulation can be achieved by hydraulic proppant fracturing. Aspects of foam fracturing in shale gas reservoirs are also discussed by KOMAR & YOST (1979) and DRISCOLL, BOWEN & ROBERTS (1980).

At the beginning of hydraulic fracturing in gas shales in the early 1950's, only water was used as treating fluid and yielded considerably higher gas flow rates than earlier explosive stimulation (WITT 1986). The later switch to foam and liquefied gas fracturing further increased gas recovery, decreased the time required for the well to return the fracturing fluid, largely prevented water blockage in the cracks (cf. section 3.11.2.4.2.), and permitted more carefully designed procedures for stimulating multiple zones in one well (cf. section 4.2.2.1.3.).

4.4.4.3. Proppant fracturing

Although straight nitrogen fracturing both as small-scale formation and perforation breakdown treatment and as large-scale stimulation operation are very successful in shale gas reservoirs particularly due to their extremely low cost and technical simplicity, long-term stimulation effects can only be achieved by proppant fracturing in different scales. Experience has shown, however, that shale gas reservoirs are very sensitive pay zones which require proper design of the treatments, with emphasis being on achievement of sufficient proppant concentration in the crack (cf. section 4.3.) and rapid cleanup of the well after termination of the job (cf. section 3.11.2.). Some aspects of proppant placement and fluid retrieval as well as vertical vs. horizontal hole drilling and fracturing are discussed as follows.

4.4.4.3.1. Proppant placement and fluid retrieval

LANCASTER, GUIDRY, GRAHAM, CURTIS, SHAW & BLAKE (1987) present proppant fracturing of shale gas reservoirs both for original treatments and for later restimulation. As previous experience has shown that large hydraulic proppant fracturing treatments are effective means of shale stimulation, the optimal design strategy should maximize proppant placement and minimize liquid volumes. Good solutions are achieved with nitrogen foam fracturing fluids carrying high proppant concentrations with application of appropriate chemical additives to assist in efficient cleanup of liquids following the treatment (cf. section 3.11.2.). Proppant fracturing has hitherto comparatively rarely been applied in shale gas reservoirs, but is considered to be of progressively rising significance in the coming years. Aspects of shale proppant fracturing are also discussed by KING (1982, 1983) and GOTTSCHLING, ROYCE & SCHUCK (1985).

Gas production in shale reservoirs is frequently associated with intervals having up to 4 % higher porosity and permeability than the usual shale section, having a substantial portion of the porosity filled with hydrocarbons (the free gas saturation is greater than abt. 2 % of the bulk volume) and having a relatively high kerogen content (greater than 2.5 %). The following discussion focusses on comparative superiority of proppant fracturing as well as combination of gas and foam-proppant fracturing.

4.4.4.3.1.1. Comparative superiority of proppant fracturing

ZUBER, LEE & GATENS (1987) compare the effectivity of different stimulation techniques in Devonian shale reservoirs. Permeability anisotropy in the Devonian shale is present in various degrees and is a result of variation of natural fracture spacing, conductivity, geometry and other heterogeneities. As a consequence of anisotropic permeability caused by natural crack trends, a well with an induced fracture perpendicularly to the natural joint trend will perform much more favourably than a well with an induced fracture parallel to the natural crack trend. Comparative assessment of single hydraulic fracturing, tailored-pulse selective and controlled explosive fracturing creating two or more radial fractures being perpendicular to each other (cf. sections 4.2.2.1.3.3. and 4.7.3.3.), and standard explosive fracturing reveals a ranking of effectivity of the different methods in the mentioned sequence. Hydraulic proppant fracturing therefore has turned out to be the economically most attractive stimulation principle also in Devonian shale gas reservoirs. Other treatment methods comprise nitrogen-foam (KOMAR, YOST & SINCLAIR 1980) or pure nitrogen gas fracturing with or without proppant insertion (FREEMAN, ABEL, KIM & HEINRICH 1981; cf. section 4.4.4.2.1.).

As a consequence of the experience with more advanced techniques in the last years, the historically traditional standard explosive stimulation by borehole shooting with 80 % gelatinated nitroglycerine (cf. section 4.4.4.6.) has been progressively replaced by single hydraulic and multiple radial tailored-pulse fracturing. OVERBEY, YOST & WILKINS (1988) report nitrogen-foam proppant fracturing. Other aspects of shale reservoir stimulation are discussed by KOMAR (1978) and HORTON (1982). OZKAYA (1984) outlines computer simulation of hydraulic fracturing in shales.

4.4.4.3.1.2. Combination of gas and foam-proppant fracturing

LAYNE & SIRIWARDANE (1988) find out from field experience that the most effective fracture design may consist of a hybrid treatment with a carbon dioxide pad and high-quality foam as proppant transport fluid which prevents screenout and formation damage while maintaining post-stimulation fracture conductivity (cf. section 3.11.2.2.). Carbon dioxide may be the preferred base fluid for many shale formations, whereas nitrogen foam has often higher proppant transport capacity than carbon dioxide foam. Fluid entrance into enlarged natural fractures during course of the stimulation treatment takes place as annulus loss and large injection flow into discrete fractures. Leakoff is defined as a small amount of fluid that does not penetrate or significantly deform the reservoir, whereas large injection flow represents a significant rate of fluid penetration that is capable of carrying a proppant and inflating existing fractures to enhance reservoir permeability.

YOST, OVERBEY, WILKINS & LOCKE (1988) conclude that the best results of hydraulic stimulation of a horizontal borehole through shale gas reservoirs which are water-sensitive would be a carbon dioxide pneumatic proppant transport fracturing process that moves sufficient amounts of proppants to maintain crack conductivity without a water base fluid (KING 1983, LEINAN & HARRIS 1983, SINAL & LANCASTER 1987; MERCER, PRATT & YOST 1988) or a mixed treatment consisting of a carbon-dioxide pad followed by proppant-laden nitrogen foam where the liquid phase contains water and methanol (cf. section 3.11.2.2.). As straight carbon dioxide fracturing operations had the highest initial improvement ratio, ultimately a carbon dioxide-based treatment may be preferred. Nitrogen foam can also be applied in massive hydraulic proppant fracturing jobs with placement of proppant concentrations up to 6 lbs/gal at downhole conditions, with liquid volume minimization and proppant transport maximization being the only possibility to achieve a reasonable relationship between treatment cost and effect in the shale gas reservoirs (GATENS 1988). Aspects of proppant-laden nitrogen foam fracturing are also discussed by LANCASTER (1988).

4.4.4.3.2. Vertical vs. horizontal hole drilling and fracturing

In terms of well type, case studies in shale gas reservoirs revealed that four horizontal unstimulated boreholes can produce 1.5 times as much gas as sixteen vertical untreated wells in both infill drilling and virgin reservoir development cases (MERCER, PRATT & YOST 1988). The effect of four unstimulated horizontal boreholes can be achieved by only two horizontal wells if they are hydraulically fractured. On the other hand, BRANDENBURG (1988) concludes that in a number of instances, hydraulic fracturing from vertical wells appears to offer the best opportunity to extract gas reserves from shales in an economically efficient manner.

Horizontal wells crossing multiple natural fractures in the reservoir are required in low-permeability formations such as shales in order to increase natural gas recovery and to reduce the risk of drilling a dry hole (LAYNE & SIRIWARDANE 1988). Pumping of high-pressure fluid down through tubing and annulus gives rise to enlargement of numerous natural joints in the shale sequence intersected by the horizontal well (cf. section 4.8.6.2.). Fracture containment in vertical direction promotes extensive fluid penetration throughout the fracture network and enables induction of highly elongated equivalent cracks through more than a single natural joint orientation (an equivalent fracture has the same crack volume extended into the reservoir and does not follow a single direction of maximum principal stress). In pay zones with numerous fracture intersections, low angles of induced crack approach, and low ratios of maximum and minimum horizontal stresses, interconnection of natural and hydraulic fractures with mutual fluid penetration has a very high potential of occurrence (BLANTON 1986). Intersection and penetration of fluids in a natural fracture tends to impede propagation in the same orientation and may divert the fluid back to the wellbore. Horizontal wells in combination with hydraulic fracturing are also significant instruments for gas drainage from coal seams (cf. section 4.4.3.6.).

4.4.4.4. Effectivity of proppant vs. nitrogen fracturing

Proppant-laden treatments are generally performed with a foam or gelled water carrier to part the formation and transport proppant to support the created fracture and to hold it open (BRANDENBURG 1988). Although many shales are not particularly sensitive to fluids, capillary action may hold water in small pores, thereby reducing their permeability to gas. As reservoir pressures in many shales are insufficient to overcome these capillary forces and to remove the water, foam is often preferred because it reduces the liquid volume of the treatment (cf. section 3.11.2.).

Hydraulic proppant fracturing of shale gas reservoirs in various scales frequently had the surprising result that performance of the wells after the stimulation jobs was not appreciably better than that after straight nitrogen fracture treatments which did little more than establish communication between borehole and reservoir. In addition, well performance after small nitrogen breakdown jobs was often similar to that seen after full-scale nitrogen stimulations (SHAW, GATENS, LANCASTER, LEE, AVARY & TERRY 1986; LANCASTER, GUIDRY, GRAHAM, CURTIS, SHAW & BLAKE 1987). Unsatisfactory stimulation results falling considerably behind the expectations are also reported by LANCASTER (1988). After discussion of some general aspects, a few points of liquid retention and well cleanup, and feasibility of MHF treatments are investigated as follows.

4.4.4.4.1. General aspects

The gas reservoirs are low-pressure and low-permeability layers which are distributed vertically within the shale section, thereby making completions mechanically difficult. Stress contrasts between productive intervals and potential stimulation barriers appear to be minimal (cf. sections 4.2.2.3.1. and 4.2.3.2.2.), with thus vertical fracture growth containment being an important placement consideration. As a consequence of low bottomhole pressure and low permeability, shale gas reservoirs should generally respond very well to the large increase in permeable surface area available for flow that can be achieved with properly conducted fracture treatments (ZUBER, LEE & GATENS 1987; ELY 1988). Unsatisfactory results of many hitherto performed hydraulic proppant fracturing treatments in shale reservoirs can be due to usage of insufficient breaker to degrade the polymer viscosifiers in the proppant-carrying fluids (cf. sections 4.12.1.1.), with the resulting downhole viscosity severely lowering gas permeability in and around the created fracture.

Other reasons are insufficient proppant concentrations having been placed to achieve the necessary fracture conductivity, standard flowback procedures sometimes resulting in little or no proppant being placed across the producing formation, and slow liquid cleanup due to low bottomhole pressure, with unbroken gel or gel lumping being able to cause either failure to stimulate or even damage shale gas wells. Field experience has shown that liquid cleanup in the low-pressure shale gas reservoirs after large-scale fracturing treatments is very slow and will probably never be complete (GATENS 1988), with the retained liquids in the fracture therefore for a longer period of time seriously reducing fracture conductivity.

4.4.4.4.2. Liquid retention and well cleanup

Long cleanup periods following hydraulic proppant fracturing treatments in shale gas wells are due to a combination of large retained liquid volumes, insufficiently broken polymer fluids, and low bottomhole pressure, with foamed fluids therefore being superior solutions of the problem (cf. section 3.11.2.). Another answer to the question of improvement of performance are increased proppant concentrations to be placed in the fracture (cf. section 4.3.), with the higher conductivity achieved by higher proppant saturations enhancing fluid cleanup rates and creating more efficient flowpaths to the wellbore (GATENS 1988). Insufficient stimulation effects following proppant fracturing of small- or large-scale type can also be the result of uncontained crack height growth which may have caused placement of only inadequate proppant concentrations across the pay zone (LANCASTER 1987; cf. section 4.2.2.).

Reduction of fluid retention time can also be obtained by introducing surfactants which are formulated so that minimal liquid absorption occurs, thereby allowing to enhance flowback of low surface tension fluids. An accompanying measure can be treating of the proppants with non-wetting surfactants in order to enhance conductivity of the proppant package. Because of the low permeability of the shale gas reservoir and its inherent low leakoff impact to properly designed fracturing fluids, crack closure may not occur for many hours if fluid flowback is not initiated by pumping or valve opening to artificially accelerate fracture closure or is not enhanced by energizing components or foams. As foam is an excellent proppant transport medium, but not a perfect proppant suspension fluid, there is a possibility that significant proppant settling occurs unless the well is immediately flowed back.

4.4.4.4.3. Feasibility of MHF treatments

GATENS (1988) reviews massive hydraulic proppant fracturing in shale gas reservoirs where treatments have been carried out with up to 1.6 Mio. lbs. (800 t) of sand in order to create long propped cracks. Comparison of performance of MHF-stimulated wells with that of offset wells which were shot with nitroglycerine explosives revealed the surprising fact that the MHF-treated wells are frequently just slightly better than the average shot wells despite the large stimulation jobs pumped. Some aspects of reservoir pressure and proppant concentration are discussed as follows.

4.4.4.4.3.1. Reservoir pressure

A partial explanation of this paradoxon is linked with the much lower reservoir pressure in the MHF-treated wells that were stimulated several decades after the other wells had been almost initially shot with nitroglycerine at a time when reservoir pressure was considerably higher, with thus the MHF-treated wells not having the same amount of reservoir energy available to help bringing the gas to the borehole (cf. section 4.8.9.). If normalized for pressure, the MHF-stimulated wells appear all to be performing similarly to the best shot well, but if normalized for reservoir quality (permeability times net thickness), the MHF-treated wells are each clearly performing better than even the best shot well in the area. Nevertheless it is still surprising why the costly MHF operations do in many cases not result in significantly better performance than rather simple well shooting and nitrogen non-proppant fracturing, and the much higher investment is often not compensated by also a comparatively better post-stimulation production.

Massive hydraulic fracture evaluation by simulation reveals that if the fractures were well contained, propped fracture half-lengths up to 1,000 ft. (330 m) or more could have been obtained, and if the cracks were unconfined, half-lengths of 500 ft (165 m) or more could have been achieved, with thus long propped fractures having been created regardless of existence of good barriers to vertical height growth. Interzone communication behind the pipe indicates that significant vertical height growth has occurred.

4.4.4.4.3.2. Proppant concentration

Economical comparison of small nitrogen breakdown treatments, straight nitrogen fracturing operations of full-scale type, and large-scale proppant fracturing jobs reveals that if properly designed and carried out with emphasis on high proppant concentration and rapid well cleanup after termination of the stimulation job, massive hydraulic fracturing is the ultimately best treatment type in the long-term range of gas field exploitation. It is also attractive to restimulate a well which had originally been treated by straight nitrogen fracturing by small- or large-scale proppant foam fracturing (LANCASTER 1988; cf. also section 4.8.9.). Aspects of massive hydraulic fracturing in shale gas reservoirs are also discussed by CREMEAN, McKETTA, OWENS & SMITH (1979) and GATENS, HOLGATE & LEE (1987). The significance of proppant concentration maximization and liquid volume minimization in large coal and shale fracturing treatments is also emphasized by KUUSKRAA & HAAS (1988).

4.4.4.5. Special features of shale gas reservoirs

The complex flow conditions in shale gas reservoirs can be partially explained by bubbling and seeping of condensate and/or natural gas liquids out of natural cracks along the borehole wall and sublimating into gas, as revealed by direct observation of this process as well as by recording of stained fracture planes by video camera survey (OVERBEY, YOST & YOST 1988; cf. section 6.2.1.9.3.). This phenomenon raises the potential of two-phase flow through the fracture systems and can account for some unusual production patterns. Another noteworthy effect in shale gas reservoirs is gas drainage in an elliptical pattern instead of a circular one common to most conventional wells (DOE 1984), with this feature implying that significant additional quantities of gas could be recovered by special infill drilling.

Other particular features of shale gas reservoirs are that gas is not immediately flowing after perforation, and once gas current is finally established, there is a long period of wellbore storage during both flow and shut-in periods (SCHETTLER, PARMLEY & LEE 1987; LANCASTER 1988). The problem of retarded initiation of gas current can be overcome by breaking down the well with a small nitrogen treatment (cf. section 4.8.2.2.), and elimination of the wellbore storage effect requires improved test designs including both sufficiently long flow and shut-in periods. Small-scale nitrogen operations have not only the task of creating communication with the formation and establishing a gas flow rate, but are also carried out in order to permit a pre-stimulation flow and pressure buildup test.

4.4.4.6. Explosive fracturing

Explosive fracturing was the first stimulation technique in shale gas reservoirs when this resource started to undergo more widespread development more than 50 years ago (WITT 1986). Explosive fracturing was carried out by detonating gelatinated nitroglycerine under a water or gravel tamp (cf. section 4.7.3.). Massive shots with several tons of explosive considerably increased the gas yield, but also often provoked subsequently a massive and prolonged cleanout job because of the large volume of shale rubble that generally collapsed into the wellbore. More than 90 % of the wells in some fields were stimulated by explosive fracturing before the invention of hydraulic fracturing in the late 1940`s (RAY 1976). Explosive stimulation mainly generated permeability pathways that linked the wellbore with extensive natural fracture systems which drain gas from the matrix, with size, extent and geometry of the natural fracture system largely governing productivity. Wells in which shooting failed to intersect extensive natural fracture systems remained dry holes and were failures.

4.4.5. High-temperature reservoirs

Another target is increasing stimulation of very high-temperature tight gas and condensate formations where successful jobs have so far been run in Europe mainly in Yugoslavia (ECONOMIDES, CIKES, PFÖRTER, UDICK & URODA 1986; cf. section 2.4.3.1.) and a suite of further wells is waiting on operation. After discussion of some aspects of fluid stability adjustment to elevated pay temperature, heat transfer in fractures and temperature stability of stimulation fluids, fracturing fluid requirement in hot and cold environment is briefly outlined.

4.4.5.1. Fluid stability adjustment to elevated pay temperature

The decisive necessity of approaching high-temperature reservoirs of either hydrocarbon or geothermal water type (cf. section 4.7.) by hydraulic fracturing operations is extended thermal stability of fracturing fluids which can be reached by both direct and indirect treatments. Direct ways of improvement of temperature resistivity include crosslinking for achieving higher viscosity (cf. section 4.3.4.3.) as well as intermixing of additives that retard thermal breakage of the polymers. Indirect possibilities are formation cooling by pumping large pre-pad and pad stages prior to insertion of the crosslinked polymer carrier stage (cf. section 4.3.4.6.1.4.), thereby mimicking to the latter phase a comparatively colder formation than present in reality. While this heat blocking may work very effectively in not too hot pay zones, some extremely hot geothermal reservoirs can only be stimulated with more or less water as fracturing fluids (cf. section 4.7.3.5.).

On the other hand, high-efficiency fracturing fluids for low-temperature reservoir rocks are reported by ALMOND & GARVIN (1985). Experimental evaluation of thermal stability of fracturing fluids is carried out by LESCARBOURA, SIFFERMAN & WAHL (1984). Various aspects of fracturing fluid selection and performance experience are discussed by BLASIO, BAILEY & BERGTHOLD (1985); SOLIMAN & HUNT (1985) and ROODHART & DAVIES (1987). High-quality water-based fracturing fluids are also described by CHEN, CHEN & ZI (1982). High-temperature effects of proppants, fluids and equipment are also significant for geothermal reservoir stimulation (cf. section 4.7.).

4.4.5.2. Heat transfer in fractures and temperature stability of stimulation fluids

Temperature effects on stimulation design are also discussed by WHITSITT & DYSART (1970) and McDANIEL (1986), and questions of heat transfer in fractures are examined by WHEELER (1969), WHITSITT & DYSART (1970), SINCLAIR (1971); CLARK, CLOUD, WAREMBOURG & YUNKO (1978); HARRINGTON, HANNAH & BEIRUTE (1978); CROCKETT, VO & CLEARY (1984) and POULSEN & LEE (1984; cf. section 4.3.4.4.1.). The significance of high-efficiency fracturing fluids for the successful treatment of high-temperature, low-permeability reservoirs is outlined by ELY, CHATTERJI, HOLTMYER & TINSLEY (1975); THOMAS & ELBEL (1979), ELBEL & THOMAS (1980), WHITE & DANIEL (1980), HSU & CONWAY (1981), WENDORFF & AINLEY (1981), HORTON (1982); BAUMGARTNER, PARKER, WILLIAMS & WOODROOF (1983); GRONE, BAUMGARTNER & WOODROOF (1983 a, 1983 b); HARMS, GOSS & PAYNE (1984); POULSEN & LEE (1984) and WALSER & OTT (1985).

Time-temperature stability of crosslinked stimulation fluids is investigated by CONWAY, PAULS & HARRIS (1980) and HARRIS & REIDENBACH (1984). Increasing high-temperature performance of fracturing fluids is mainly achieved by reducing the rate of viscosity loss which results in more effective proppant transport and allows to replace the flammable methanol (ELBEL & THOMAS 1980). The flexibility of formulating oil-based fracturing fluids to meet both surface and downhole conditions in high-temperature reservoirs in terms of viscosity and handling by addition of secondary gellants and/or activators to a primary base gel on the fly downstream of the blender is investigated by BURNHAM, HARRIS & McDANIEL (1978) and PAULS, VENDITTO, HOLTMYER & GREGORCZYK (1985). The use of viscosity stabilizers in high-temperature fracturing is evaluated by THOMAS & ROOT (1979).

4.4.5.3. Fracturing fluid requirements in hot environment

More and more hydrocarbon reserves are currently found in deep hot reservoirs which demand unique stimulation fluid characteristics (HARMS, GOSS & PAYNE 1984). The ideal fracturing fluid for high-temperature application should contain the lowest polymer loading consistent with sufficient viscosity for proppant transport and minimal fluid leakoff (cf. section 4.3.4.6.). The fluid also has to maintain its velocity and fluid-loss control under shear and prolonged exposure to high temperature in order to extend the fracture and place the proppant, and friction pressure losses should be low for reducing tubing pressures and minimizing pumping energy (cf. section 4.3.4.2.2.). A unique combination of heterogeneously substituted polysaccharide as a key component for gelling aqueous systems, stabilizers and delayed crosslinking chemicals provides the extended thermal stability and low friction pressure necessary for high-temperature MHF stimulation while minimizing the possibility of formation damage (HARMS, GOSS & PAYNE 1984).

GRONE, BAUMGARTNER & WOODROOF (1983) document the chemical evolution of a high-temperature fracturing fluid. Shear and thermal stability at high bottomhole temperatures is provided by high polymer concentrations and/or large fluid volumes which maintain effective fluid viscosities in the fracture, but on the other hand, high polymer saturations lead to higher friction pressure, higher treatment cost and high broken gel residue levels, and large fluid volumes increase the cost of the treatment (cf. section 4.3.4.2.2.). Crosslinked fracturing fluids have been developed for extended temperature stability due to favourable variations of polymer, buffer, gel stabilizer and crosslinker (cf. section 4.3.4.3.). Pumping large volumes of fluid aim on cooling down the formation and can partially be operated with a dual-stage polymer system (cf. section 4.3.4.6.1.4.). Gel stabilizers prevent or retard the chemical degradation of polymer solutions, but many of them interfere with the crosslinking mechanism.

The current temperature records of successful fracturing comprise bottom hole values of 420 oF for a 15,000 ft deep gas well, 540 oF (280 oC; WILLIAMS, WOODROOF & BOX 1982) for a 5,500 ft deep hydrothermal well, and 600 oF for a 13,500 ft deep hot dry rock well (FRACFAX 1988 a). Comments on high-temperature drilling mud and fluid-loss control additives for geothermal reservoirs and deep heat-sensitive formations are given by HILSCHER & CLEMENTS (1982) and CARNEY, GUVEN & McGREW (1982), respectively (cf. also section 4.7.4.2.). Similarly as fracturing fluids, thermal stability extension of acid can be achieved, with buffer regulation of HF acid allowing sandstone acidizing up to 550 oF (280 oC) reservoir temperature (SCHEUERMAN 1988).

4.4.5.4. Fracturing requirements in cold environment

In contrast to extremely high-temperature reservoirs, hydraulic fracture operations have also to be adjusted to extraordinary low-temperature environments resulting from cold weather in Arctic latitudes where currently exploration is boosting. SCHUMACHER & DICKERMAN (1986) report successful hydraulic proppant stimulation in a naturally fractured reservoir in an atmospherical temperature below -50 oF where proppant and additives could only be mixed on the fly and rheological conditions of the fracturing fluids were still adequate to place up to 4 lbs of proppant per gallon of carrier medium into the hydraulic crack intersecting the formation.

4.4.6. Gas condensate reservoirs

Gas condensate reservoirs require in many cases sophisticated stimulation treatments particularly if being tight and abnormally pressured. If the retrograde condensation point (DANESH, HENDERSON, KRINIS & PEDEN 1988) is very close to the initial formation pressure, condensate production can lead to retrograde fallout and increasing water saturation in the vicinity of the wellbore, and two-phase flow results in reduction of the relative permeability to gas and thus flow capacity of the well (ENGINEER 1985). Reservoir pressure can be principally maintained high enough to minimize retrograde liquid condensation by dry gas injection, but in heterogeneous tight reservoirs, gas injectivity is too low and thus the desired effect is not achieved. In such cases, hydraulic fracturing has been shown to create a conductive path which considerably improves the drainage pattern of the reservoir and thereby helps to keep the flowing pressure above the retrograde condensation point. Hydraulic fracturing for avoiding retrograde fallout (DANESH, HENDERSON, KRINIS & PEDEN 1988; cf. section 2.4.3.1.) has been successfully carried out in high-temperature gas condensate wells in Yugoslavia (ECONOMIDES, CIKES, PFÖRTER, UDICK & URODA 1986), and potential is also seen in other fields with comparable conditions.

In addition to the aforementioned types of unconventional gas deposits, natural gas may also be trapped in ice-like crystals below the permafrost or beneath the seafloor, although the potential of such concentrations is still undetermined (DOE 1984). Such methane hydrates were first identified in the 1930's when they formed in natural gas pipelines that were built in cold climates. Methane hydrates have in the recent decades been proven to occur naturally in both onshore and offshore deposits and may in cases even form a natural seal over more conventional gas-bearing strata. Aspects of natural gas hydrates are also discussed by GODBOLE (1985).

4.5. Hydraulic proppant fracturing of carbonate rocks

In Europe and many other parts of the world, so far predominantly sandstone reservoirs are stimulated by hydraulic proppant fracturing. Carbonate formations which are also commonly treated with this method and with encouraging results in the USA (RENSHAW & JOINES 1977, WILLIAMS 1977; MISAK, ATTEBERRY, VENDITTO & FREDRICKSON 1978 b; ATTEBERRY, TUCKER & RITZ 1979; KOZIK, BAILEY & HOLDITCH 1979; ROSEPILER 1979; SCHLOTTMAN, MILLER & LUEDERS 1981; CLARK 1983, BAILEY & WICKHAM 1984, EASON 1985, ILSENG & CORTEZ 1985, MACDONALD & FRANK 1985, BRITT & LARSEN 1986, CRAMER 1987; HUCKABEE 1987, 1988; HOOVER & ADAMS 1988; JOHNSON, FOX, BURNS & O`MARA 1988) are in Europe still almost exclusively subjected to conventional matrix acid or acid fracturing stimulation. Massive hydraulic fracturing (MHF; cf. RANDOLPH 1974, MURPHY & CARNEY 1977, PAI & GARBIS 1983; cf. section 1.1.1.) usually stands for proppant fracturing and is often defined as comprising cracks of at least 1,500 ft (500 m) penetration in each direction from the wellbore (FAST, HOLMAN & COVLIN 1977), whereas in most cases no propping agents are inserted in massive acid fracturing (MAF; McDONALD 1983). Acid treatments can be principally split into matrix acidizing and acid fracturing (KING 1986).

While matrix acidizing with hydrochloric acid in carbonates is already common practice since more than 60 years (LABRID 1975) and with hydrofluoric acid in sandstones (cf. section 4.5.5.) has been introduced over 30 years ago (SMITH & HENDRICKSON 1965), acid fracturing or fracture acidizing in carbonate rocks has been first applied more than 20 years ago (HENDRICKSON & CAMERON 1968) and is therefore a considerably younger procedure than hydraulic proppant fracturing which is an established technique since over 40 years (cf. section 1.2.). BAILEY & WICKHAM (1984) investigate proppant fracturing vs. fracture acidizing, and BEN NACEUR & ECONOMIDES (1988 b) evaluate the effectiveness of acid fracturing. LO & DEAN (1988) model acid fracturing.

Distinction has to be made between acid fracturing and fracture acidizing (BRISCOE 1978), with acid fracturing including the primary process of opening of the crack and subsequently improving its conductivity by acid attack, whereas fracture acidizing comprises chiefly secondary acid injection into pre-existing either natural joints or artificially-created fractures. Aspects of extension of drainage path and conductivity contrast, combination of natural and artificial fractures, and possibilities of application in Europe are outlined as follows. Particular emphasis is given to chalk stimulation which is a very special type of carbonate reservoirs with complicated conditions that require extensive and sophisticated treatments. Comments on sandstone acidizing are also offered.

4.5.1. Extension of drainage path and conductivity contrast

Although in many cases the economically comparatively cheap and technically relatively simple acid jobs have excellent results in carbonate reservoir stimulation and particularly pay-out of the treatment is very fast due to the low cost and also apparently many carbonate formations respond better to acidizing than to hydraulic proppant fracturing (JENNINGS & DARDEN 1979), there are quite some hydrocarbon-bearing carbonate formations in Europe and other parts of the world where hydraulic proppant fracturing would be the best solution in view of reservoir engineering and would give rise to the highest production rates of all stimulation treatments that could be applied. The evaluation as follows focusses on advantages and drawbacks of fracture propping, conductivity improvement by acid optimization, enhancement of acid penetration distance, acid foaming, open vs. propped fractures, and acid fluid-loss control.

4.5.1.1. Advantages and drawbacks of fracture propping

The main reason for the superiority of proppant fracturing also in carbonate rocks is the much larger extension and the more continuous shape of the drainage path through the reservoir in case of a propped fracture as compared to the effect of enlarging natural joints or solving out relatively unstable components of the matrix of carbonate rocks by acidizing. In addition, the drainage radius of a propped fracture is certainly much larger than that of natural joints accentuated by acid solution which is more or less restricted to the closer vicinity of the borehole.

While the advantages comprise mainly generation of a pronounced conductivity contrast between reservoir matrix and propped fracture (cf. also section 4.6.1.), the most important drawbacks of installating propped fractures in carbonate rocks are scale formation in the proppant package and fluid loss (cf. section 4.5.1.6.) in extensive and narrowly-spaced natural crack systems. BAILEY & WICKHAM (1984) report a definite production benefit by proppant fracturing over acid fracturing in carbonate oil reservoirs. The outline as follows focusses on matrix acidizing vs. fracturing, conductivity contrast between reservoir matrix and propped fracture, carbonate scale formation and fluid loss, and iron sequestering and complexing.

4.5.1.1.1. Matrix acidizing vs. fracturing

While matrix acidizing below fracture pressure is still beneficial in high-permeability damaged carbonate formations (over 50 md), acid fracturing is generally used in low-permeability carbonate reservoirs (less than 10 md; in contrast to sandstone proppant fracturing where such values comprise still moderate- to high-permeability targets), because much more effective stimulation is achieved with long acid-etched fractures (McLEOD 1984), and an even better result can be obtained with proppant fracturing. The main purpose of matrix acidizing in sandstones and carbonates is to dissolve formation-damaging clay minerals and other fines and to restore more or less original permeability (DOWELL SCHLUMBERGER 1988 d; cf. section 4.5.5.2.) by flow pattern through the natural matrix permeability below fracture pressure.

Extraordinary high amounts of soluble minerals can also give rise to increase of secondary solution permeability beyond its primary level, with caution, however, having to be exercised that excessive matrix dissolution does not lead to rock framework collapse and destruction of the stimulation effect (cf. section 4.5.5.4.2.). Acid fracturing is applied in order to improve low formation permeability in sandstones and carbonates beyond its primary amount by creation of a conductive crack at injection pressures above fracture gradient. Flow pattern is essentially through the hydraulically-generated crack, although much of the fluid does leak off into the matrix along the fracture faces. HUCKABEE (1988) presents a comparative evaluation of technical and economical effectivity of proppant fracturing, matrix fracturing, and single- and multiple-stage matrix acidizing for a case example. Some aspects of removal of formation damage, auto-propping of rugged fracture walls after etching, fracture conductivity vs. fracture length, and double fracturing in reservoirs with hairline cracks are illustrated as follows.

4.5.1.1.1.1. Removal of formation damage

Matrix acidizing for removal of formation damage around the borehole is only successful in shallow penetration depth, whereas deep solid permeability plugging of the reservoir in the wellbore vicinity has to be corrected by creating a conductive fracture through the damage zone either by acid or by proppant fracturing (cf. section 4.8.3.). On the other hand, there are plenty of cases where acid cannot dissolve the materials causing well and formation damage or where even acid itself may cause damage upon contact with the reservoirs (cf. section 4.5.5.3.). Under such conditions, either chemical treatment with non-acid fluids (BROADDUS 1988) or - in many instances with much better result - hydraulic proppant fracturing are required for damage removal or creation of a convective avenue from the wellbore to the undamaged virginal pay zone, respectively.

Formation damage removal and permeability restoration during matrix acidizing is achieved by either dissolving the damaging material itself or parts of the rock in which the damage exists (DOWELL SCHLUMBERGER 1988 d), with the latter method being most effective in carbonate reservoirs with acid solubilities greater than 50 %. Although fracturing pressure should generally not be exceeded, it may be necessary to create an intermittent fracture in order to open the perforations after which pressure can be reduced to such an extent that the fracture closes again, thereby reinstalling matrix flow pattern (similarly as in gravel packing; cf. sections 5.6.4. and 5.6.6.). In sandstone acidizing performed with a combination of hydrochloric acid and hydrofluoric acid (cf. section 4.5.5.), a spearhead of hydrochloric acid should be injected to react with carbonates before introduction of the HCl/HF mixture if the rock contains abt. 5 - 10 % carbonates. In case of carbonate contents above 20 %, hydrofluoric acid is probably not needed at all except for giving entry through the clay damage mantle around the borehole.

In addition to or as an alternative of large-scale matrix acidizing treatments, small volume operations can be carried out by spotting acid across the perforations or open-hole interval and allowing it to soak (HUCKABEE 1988). Further acid volumes can be squeezed into the reservoir. These techniques can also be combined or can alternate with proppant fracturing on various scale (cf. sections 4.5.1.6.2. and 4.5.4.2.).

4.5.1.1.1.2. Autopropping of rugged fracture walls after etching

In acid fracturing treatments, proppants are in many cases not inevitably required for providing fracture support and conductivity enhancement, because acid reaction with the crack faces of carbonate reservoirs is normally very uneven, with this heterogeneous etching pattern creating roughened surfaces with relief which prevent the fracture from fully closing (McLEOD 1984, DOWELL SCHLUMBERGER 1988 d). Therefore a highly conductive crack may exist after hydraulic pressures are released, but enough acid strength and volume must be used to generate sufficient dissolution for establishment of the required conductivity. On the other hand, the rugged fracture walls do not keep open in such a continuous manner that all the individual etched channels and holes stay in connection after fracture closure, but frequently cavities are separated by protruding pillars where acid attack was slower. This primary permeability deterioration is further enhanced by secondary collapse and crushing of unstable pillars and bridges once closure stress is applied, and irregular and discontinuous auto-propping of parts of the fracture walls (cf. also section 4.5.1.5.1.) with broken carbonate rock further decreases crack conductivity. All these effects can be avoided by adequate infilling of the etched fracture with proppants which keeps the whole space open and provides an ideally connected and conductive flow channel to the borehole.

Although fracture flow capacity resulting from acid reaction is generally very high, it may be considerably decreased when rock embedment strength and/or rock solubility are low or closure stress is high (NIERODE & KRUK 1973). The main advantage of acid fracturing is deep crack penetration into carbonate reservoirs with low-cost fluids and acid replacing expensive proppants (HENDRICKSON & CAMERON 1968). Acid fracturing is therefore very economical, because inexpensive stimulation fluids are used to create the crack area and to propagate it to final penetration including undertaking measures against fluid loss, and then much smaller quantities of acid are following in the second stage of the treatment to etch the desired conductivity at the required depth of penetration in the reservoir. As conventionally no proppants are applied which could bridge in the crack (cf. sections 4.2.3.2.5. and 6.2.4.2.1.), there is no danger of screenout failure of acid fracturing job.

4.5.1.1.1.3. Fracture conductivity vs. fracture length

A comparative evaluation of acidized fracture conductivity vs. acid concentration profile, amount of rock dissolved per unit area, closure stress and formation rock strength has shown that stimulation is generally limited by acidized fracture conductivity and not by acidized crack length (NOVOTNY 1976). Acidized fractures created by hydrochloric acid have limited length but high conductivity, whereas low fluid-loss acid emulsions (cf. section 4.5.1.1.3.2.) generated longer acidized cracks than hydrochloric acid, because the acid emulsion reacts a further distance down the fracture and thus resultant conductivity is lower than that created with hydrochloric acid. When closure stresses on the cracks are high and fracture conductivity limited stimulation, hydrochloric acid treatments provide slightly better results, whereas at low closure stresses and when acidized fracture length limits stimulation, the acid emulsion gives substantially improved treatment effect over plain hydrochloric acid.

4.5.1.1.1.4. Double fracturing in reservoir with hairline cracks

A suitable combination of hydraulic proppant and acid fracturing (cf. section 4.5.4.2.) is double fracturing in reservoirs containing reasonable amounts of carbonates that can be dissolved by acid (DOWELL SCHLUMBERGER 1988 c, 1988 d). Double fracturing consists of alternating stages of non-reactive viscous pad fluid and acid (cf. section 4.5.1.6.2.). Many carbonate formations contain numerous natural hairline cracks and fissures which are extremely narrow or closed and have little effect on overall permeability, but once opened hydraulically or enlarged by acid etching, they can cause excessive fluid leakoff and become the dominant influence on fluid loss (similarly as applying for dual fluid leakoff in sandstone reservoirs; WARPINSKI 1988 a; cf. sections 4.8.8.3.1.2. and 4.8.8.3.2.3.). Using intermittent stages of pad fluid and acid, the acid-enlarged hairline cracks are exposed to non-reactive pad fluid as the hydraulic fracture penetrates outward into the reservoir which keeps the acid in the main fracture and avoids acid loss by further opening of the hairline cracks.

While normal fluid-loss agents do not restrict flow into the hairline fractures, viscous pad stages alternating with acid stages are put against and into hairline cracks as soon as they begin to comprise the main source of leakoff, thereby controlling and subsequently reducing flow into the intersecting enlarged hairline fractures which allows the acid to reach degree of penetration and folds of increase as designed. Even more effective fluid leakoff control can be achieved by bridging the hairline cracks with 100 mesh sand that is suspended in the pad fluid (cf. also MILLER & WAREMBOURG 1975; COULTER, CROWE, BARRETT & MILLER 1976; WARPINSKI 1988 a; cf. sections 4.8.8.3.1.2. and 4.8.8.3.2.3.). An alternative solid fluid-loss agent to 100 mesh sand is fine salt of 30/60 and/or 80/120 mesh grain size which is better than silica because it packs to lower permeabilities in hairline cracks under pressure and subsequently dissolves again out of the fractures, thereby allowing them to stay open and to contribute to production (SCHRIETER & SHAW 1978; cf. sections 1.4.11.2.1.2. and 5.11.4.2.). Salt additive also avoids pump and equipment abrasion caused by silica.

4.5.1.1.2. Conductivity contrast between reservoir matrix and propped fracture

The explanation for the optimum effect of stimulation by hydraulic proppant fracturing of both sandstone and carbonate reservoirs is the permeability contrast that is created between the tight matrix of the hydrocarbon-bearing rocks (regardless of their lithological composition) and the propped fracture (cf. section 4.6.1.). Depending on depth and closure stress, the permeability contrast can be influenced by the possibility of choice between various proppant types which have different conductivities as well as between different grain sizes (cf. section 1.3.) that independently from type also result in changing permeabilities of the proppant package infilling the fracture. Some aspects of fracture flow capacity contrast as well as permeability and conductivity contrast are outlined as follows.

4.5.1.1.2.1. Fracture flow capacity contrast

A flow capacity contrast of at least 100 between formation and fracture is considered to be essential for the achievement of adequate stimulation results, and this approach can only be controlled by selecting a suitable high-quality proppant (WATERS 1980). PAI, GARBIS & HALL (1983) recommend flow capacity contrast ratios between crack and formation of 1,000 - 1,000,000, with fracture conductivity being a function of type, size, strength, quality, and density of the proppant, as well as formation overburden pressure (cf. sections 1.3. and 4.6.1.). Stimulation of carbonate rocks in Europe and in many other parts of the world has so far been mainly done by acidizing due to much lower cost of such treatments with respect to the expenses for hydraulic proppant fracturing that has turned out to be very successful in many operations in the USA, and as a consequence of hitherto more or less satisfactory technical and economical results of acid fracturing or acid washing. Effects of flow capacity contrast between fracture and formation on well productivity are also discussed by FAST, FLICKINGER & HOWARD (1961).

4.5.1.1.2.2. Permeability and conductivity contrast

The permeability contrast between formation matrix and propped fracture is particularly high in tight gas sandstones such as the Rotliegend-Hauptsandstein in some fields in Germany FRG where the reservoir rock matrix permeability of abt. 10 microdarcy contrasts with a permeability of the proppant pack in the fracture of abt. 100 darcy (KLOSE & KRÖMER 1983; cf. sections 3.4. and 3.5.). MONTGOMERY & STEANSON (1985) list effective permeabilities for natural sand in the range of 35 - 80 darcy, for intermediate-strength proppants within the limits of 80 - 900 darcy, and for high-strength proppants within the boundaries of 65 - 1000 darcy (depending on grain size and mineralogical composition of the proppants as well as on closure stress regime; cf. section 1.3.). The effective permeability for 20/40 intermediate- and high-strength proppants which are the most frequently applied synthetic materials is in the range of 180 - 260 darcy, and that of 20/40 natural sand which is still today the most common propping agent is within the limits of 50 - 80 darcy. Absolute permeability and its dependence on confining pressure, pore pressure and temperature are reviewed by GOBRAN, BRIGHAM & RANEY (1981).

4.5.1.1.3. Carbonate scale formation and fluid loss

On the other hand, packages of natural sand or synthetic proppants infilling hydraulic fractures in carbonate rocks (particularly chalks) suffer from secondary permeability deterioration by scale formation due to carbonate cementation of the proppant agglomerations. One of the major drawbacks of carbonate fracturing with respect to sandstone stimulation is the higher density and effectiveness of natural cracks in carbonate reservoirs in comparison to sandstone pays (cf. section 4.5.2.) which give rise to more serious fluid leakoff, thereby creating a higher risk of premature screenout failure of the hydraulic proppant fracturing job (DEES & COULTER 1986; cf. sections 4.2.3.5.3. and 6.2.4.2.1.), and also have a worse influence on hydraulic crack propagation which is subjected to various deflection and refraction events when crossing the abundance of pre-existing joints in the carbonate pay zone. Thus the natural fractures are acting as pronounced fluid thieves which have a serious impact on the effectivity of creation, propagation and plugging of the hydraulic fracture (cf. sections 4.8.8.3.1.2. and 4.8.8.3.2.3.). Some comments on scale precipitation prevention and corrosion inhibition are offered as follows.

4.5.1.1.3.1. Scale precipitation prevention

VETTER, NILSSEN, LANKFORD & SHELTON (1988) investigate scale inhibitors for proppant and acid fracturing. The addition of scale inhibitors to stimulation fluids is essential for prevention of precipitations during well or reservoir operations after the fracturing treatment is completed. In terms of chemical compatibility, numerous scale inhibitors prevent the stimulation fluids from acquiring any gel strength even at very low preventor concentrations, and many inhibitors even provoke rapid complete breakage of crosslinked gel systems. Gel strength in the presence of scale preventors can be provided by increasing crosslinker concentration, with some of the excess crosslinker becoming sacrificial, but on the other hand, the gel becomes extremely unstable upon application of mechanical shear as a consequence of supersaturation, with the possibility of creating formation damage. Chemical and physical incompatibilities between scale inhibitors and stimulation fluids can become serious problems.

Concerning hydraulic proppant fracturing, scale preventors can be added by dissolution within the aqueous phase of the stimulation fluid to the liquid pre-pad during the treatment where they hopefully do not get into contact with the fracturing fluid itself, or to the proppant in solid form where they do not dissolve to any major extent within the stimulation fluid. While some fracturing fluids are rather sensitive to any scale inhibitor addition, others are not. Concerning acid fracturing, scale inhibitors are ineffective at low pH values, because the pH value of produced fluids after an acid stimulation job may stay too low for effective scale prevention. Scale inhibitors may not absorb properly on the internal surface area of the reservoir rock thus drastically reducing the lifetime of a squeeze job during or prior to an acid job.

4.5.1.1.3.2. Corrosion inhibition

DAVIES, LIEVAART & NITTERS (1988) study effects of corrosion inhibitors and mutual solvents on matrix acidizing treatments. Filter-cake-forming inhibitors have a stronger impact on high-permeability rocks, whereas the effects of adsorbing inhibitors are greater in low-permeability rocks. Filtered-out or adsorbed preventors can be removed from the formation by mutual solvents which solubilize damaging inhibitor residues and prevent formation plugging during acid injection. Effects of acid corrosion preventors on matrix stimulation results are also investigated by WOODROOF, BAKER & JENKINS (1975); SMITH, DOLLARHIDE & BYTH (1978) and CROWE & MINOR (1982). Inhibitor adsorption and formation plugging by preventor residues can have a major influence on clean-up rate and ultimate effective permeability of the treated zone. Formation damage can also be provoked by friction reducers in acid stimulation (OUSTERHOUT & HALL 1961; PRUITT, SIMMONS, NEILL & CRAWFORD 1965; SUTTON 1976, WOODROOF & ANDERSON 1977).

4.5.1.1.4. Iron sequestering and complexing

Many carbonate formations contain iron oxides and sequestering agents must be used with acid treatments in these formations (JENNINGS & DARDEN 1979). Iron precipitation has been recognized as a significant problem in acidizing operations for over 30 years. Iron dissolved by acid from reservoir formation or borehole completion remains in solution until the acid is spent (SMITH, CROWE & NOLAN 1968). As pH of the spent acid begins to rise, iron loses its solubility and precipitates, thereby seriously damaging the flow channels just opened by the acid reaction in the pay zone complex. Sequestering agents act by complexing iron ions to inhibit precipitation as hydrochloric acid spends. The presence of iron compounds in a formation is also important when a zone is treated by hydraulic fracturing, since acid is frequently used as a formation breakdown fluid prior to stimulation. After an outline of the most important sources of iron in acidizing and fracturing, problems of compatibility of iron-sequestering and -complexing agents are illustrated.

4.5.1.1.4.1. Sources of iron in acidizing and fracturing

Analysis of iron problems in formations not containing iron-bearing minerals in sufficient quantity reveals that the primary source of iron in acid treatments is pipe dope and mill scale of iron and rust in drilling string, casing and tubing of the well (GOUGLER, HENDRICK & COULTER 1985; cf. section 5.11.5.). Particularly serious sources are new tubular goods where in extreme cases, the mill scale could nearly spend the first portion of acid thereby causing it to be ineffective. Other aspects of iron precipitation during acid treatments and its prevention are examined by SMITH, CROWE & NOLAN (1968); DILL & FREDETTE (1983), COULTER & GOUGLER (1984), CROWE (1984), GDANSKI & PEAVY (1986) and SMOLARCHUK & DILL (1986). HALL & DILL (1988) summarize compatibilities of iron-control additives for limestone and sandstone acidizing of sweet and sour wells. Deposits of iron compounds in tubulars and formation determine amount of iron dissolved by the acid and thus severity of iron problems.

Iron-sequestering additives may present compatibility difficulties and may lose effectiveness because of degradation. The main iron sources are mill scale in new pipe, iron-bearing minerals in sandstone and carbonate formations, and iron in recovered spent acid solutions (SMITH, CROWE & NOLAN 1968). Injection wells usually contain more iron oxide scales than production wells. While hydrochloric acid in carbonate formations does not show compatibility problems with iron sequestering additives, considerable difficulties occur with hydrochloric-hydrofluoric acid solutions in sandstone formations (cf. section 4.5.5.).

In reservoir rocks containing abundant detrital and/or diagenetical iron-bearing minerals such as siderite, ankerite, hematite, limonite and pyrite, hydrochloric acid dissolves the iron compounds which can reprecipitate as the acid completely spends on the carbonate phases. Iron compound precipitation therefore has to be prevented or controlled by iron-sequestering or -complexing agents in the acid (McLEOD 1984).

4.5.1.1.4.2. Compatibility of iron-sequestering and -complexing agents

SMITH, CROWE & NOLAN (1968) present a comparative performance evaluation of various iron-sequestering or -complexing agents, with all of them, however, having to be cautiously applied, as there is principally no iron-control additive that can economically prevent iron precipitation without danger of agent precipitation if iron is not found downhole in the expected quantity, and in some cases, the iron even keeps the sequestering additive in solution in spent acid. A similar effect as iron precipitation following acidizing is iron colloid plugging in sandstone matrix (POTTER & DIBBLE 1985). Flocculation and/or coagulation of the ferric hydroxide leads to formation of filter cake in low-pH cases, whereas colloid and quartz surface interaction produces a more uniform accumulation of colloid at higher pH. In addition to preventing plugging, selective colloid fouling might be used as a mobility buffer in reservoirs with heterogeneous permeability distribution.

Iron-contaminated acid is able to promote precipitation of asphaltenes when acidizing certain oil-bearing reservoirs (NEWBERRY & BARKER 1985, JACOBS & THORNE 1986) even in the presence of moderate amounts of anti-sludging agents. Traditional iron-control additives are ineffective as they sequester iron after the acid is spent. High levels of anti-sludging additives can sometimes form reverse emulsions as a consequence of original iron complexes with non-emulsifying agents and anti-sludging additives. Acid strength has serious impact on the amount of sludge and the inability of typical anti-sludging additives to control it. The use of an improved iron-supervision additive in conjunction with appropriate amounts of anti-sludging agent and corrosion inhibitor can eliminate asphaltene precipitation in moderate-strength acids when measures are taken to keep the iron content in the acid to a minimum. Therefore it is most important to remove iron scale from all piping by back-washing techniques.

4.5.1.2. Conductivity improvement by acid optimization

Acid fracturing is performed by injection of acid with a bottomhole treating pressure exceeding formation fracture pressure which leads to hydraulic opening of the crack and etching of the created fracture faces by the acid, thereby giving rise to an open flow channel (ECONOMIDES 1987 b). The possibilities of acid improvement include acid strength and fluid-loss control, acid gelling and crosslinking, and acid heating.

4.5.1.2.1. Acid strength

Increasing fracture conductivity and length is the most important means of improvement of acid fracturing treatments (NOVOTNY 1976) as well as proppant fracturing operations (cf. sections 1.3., 1.4.10. and 4.8.11.). This can be achieved by increasing the dissolving power of the acid system. Acid-external emulsions create long acidized fractures, but they normally contain only one-third acid and thus the resultant crack conductivity is low. Plain hydrochloric acid gives rise to origin of very conductive fractures in carbonates, but because of its poor fluid-loss characteristics, crack length is short and limits stimulation.

Acid emulsions provide excellent improvements after treatment in shallow reservoirs where closure stresses are low and fracture conductivity is high, as the crack stays more or less open, whereas in deeper reservoirs, unpropped fractures are considerably damaged by closure and thus reasonable amelioration can be only obtained by supporting the cracks with preferably large proppants in order to get maximum conductivity of the propped drainage path that would still be sufficiently higher than a partially closed unfilled fracture. While autopropping of fracture walls (cf. sections 4.5.1.1.1.2. and 4.5.1.5.) can be still quite effective in shallow pay depth, crushing of broken rock fragments causing self-propping in deep reservoirs is a significant factor of conductivity damage in fractures created by acid and not plugged by high-quality synthetic proppants.

4.5.1.2.2. Acid fluid-loss control

Fluid loss can be controlled by using a viscous inert pad ahead of plain hydrochloric acid, but first hydrochloric acid is denser than the pad fluid which would cause the hydrochloric acid to flow along the bottom of the fracture (this problem could be corrected by applying balance-density fluids; FREDRICKSON & BROADDUS 1976; cf. section 4.2.2.6.), and second the poor fluid-loss control of hydrochloric acid allowed it to wormhole through the pad and destroy effect (NOVOTNY 1976). NIERODE & KRUK (1973) also comment on acid fluid-loss additives and acidized fracture conductivity. The use of alternating stages of acid and non-acid fluid (cf. section 4.5.1.6.2.) containing particulate fluid-loss additives can be effective in controlling fluid leakoff to the formation (PARKER, ADAMS & LIANKUI 1986). Viscous acid allows bulk displacement of fluid down the fracture in order to promote exposition of the entire crack to acid, with retardation of acid reaction rate being an additional benefit (aspects of acid fluid-loss control are also discussed in section 4.5.1.6.).

Acid concentrations are determined more by formation mineralogy than by the plugging solid damaging the reservoir (McLEOD 1984). Various acid concentrations dissolve damage particularly if small amounts are critically placed around the perforations, but lower acid saturations reduce precipitation problems in acid-sensitive pay zones. Low hydrochloric acid concentrations (3 - 5 %) are occasionally used for various drilling and completion purposes (SIMON & UNDERWOOD 1977).

4.5.1.2.3. Acid crosslinking

Acid optimization can be effectively achieved by gelling and/or crosslinking the acid (COULTER, HARRIS & KLEBENOW 1980; PABLEY & HOLCOMB 1980, 1981, 1982; CHURCH, QUISENBERRY & FOX 1981; PABLEY, EWING & CALLAWAY 1982; McLANE & SCOTT 1983; SMITH, DAWSON & SCOGGINS 1983; cf. section 4.3.4.3.4.1.). Crosslinked high-strength acid systems are excellent fluids for acid fracturing and are also capable of carrying substantial amounts of proppants, thereby representing the ideal medium for combined acid and proppant fracturing (cf. section 4.5.4.2.). High-strength crosslinked acid systems provide considerable reduction of tubular friction pressure.

Further advantages are the ability to inhibit clay swelling in water-sensitive reservoirs, and therefore the high-strength crosslinked acid system can even compete with foam fracturing fluids and emulsified acids also because of superior proppant transport capacity. Fluid-loss control equalling or exceeding that achieved by crosslinked water-based fluids ensure deep penetration of live acid, and an extremely retarded reaction rate allows in many cases the crosslinked acid to be placed into the formation with little or no decrease in acid strength. Controlled breakout effectively allows the acid to spend as the fluid is moving back toward the wellbore during cleanup (cf. also sections 3.11.3. and 4.3.4.3.).

4.5.1.2.4. Acid gelling

Gelling of acid by polymers provides stable viscosity and continued solubility of the acid at elevated temperatures due to resistance to thermal degradation and hydrolysis of the acid (JOHNSON, FOX, BURNS & O`MARA 1988; FOX, O`MARA, BURNS & JOHNSON 1989; cf. section 4.3.4.3.4.1.). Some polymers are also soluble in the calcium and magnesium brines produced by spending of hydrochloric acid in carbonate formations. The viscosity of the brine is essentially the same as that of the original gelled acid which guarantees continued resistance to fluid loss. Acid retardation (cf. section 4.5.1.3.2.) by producing a stable viscosity restricts convection in the acid. Under dynamic conditions where acid spending rate is heavily influenced by current regime, maintenance of laminar flow is important in prolongation of spending time. Enhancement of production stimulation resulting from matrix or fracture acidizing is due to improved acid penetration, ameliorated acid retardation and reduced fluid loss.

In some special applications, combinations of gelled and non-gelled acids are applied, with gelled acid being inserted first for the purpose of fluid-leakoff control and live acid penetration amelioration, and non-gelled acid being pumped behind the gelled acid to promote fingering of the thin fluid through the more viscous fluid (cf. section 4.5.4.2.1.) and enhancement of a differential etch and dissolution pattern on the fracture face. Lightly gelling of acid is occasionally carried out in order to enhance wormhole penetration (cf. section 4.5.1.6.3.) and to provide sufficient residual viscosity for fines suspension and removal. As a consequence of its increased viscosity, gelled acid has considerably reduced fluid loss and also inhibits migration of ions which retards reaction rate. Viscosity is retained after acid spending which allows insoluble materials to remain suspended and to be returned with flowback.

4.5.1.2.5. Acid heating

Acid heating also is a possibility to improve acid penetration and effectivity (HOCH, WALKER, FREDRICKSON & NORMAN 1986). Heated acid increases the rate of rock dissolution especially on cooler formations, and the increased acid temperature can soften organic deposits frequently found on reservoir faces which hinder the acid-to-rock contact. Heated acid can also minimize reprecipitation of solids in oil used by rapid thermal shock and can prevent excessive tubing contraction.

4.5.1.3. Enhancement of acid penetration distance

When a pad of fluid is used ahead of the acid to achieve a long wide fracture, maximum possible penetration distance is based on the assumption of acid reaction in the fracture geometry existing at the end of the pad, and the minimum possible acid penetration distance will be the dynamic crack length calculated assuming that fluid loss is controlled by the viscosity of reacted acid, with these two conditions being reaction rate limit and fluid-loss boundary, respectively (WILLIAMS & NIERODE 1971). The actual penetration distance probably varies between the maximum and minimum distance during a treatment, because acid reaction produces wormholes leading from the fracture into the formation (cf. section 4.5.1.6.3.) which eliminate fluid-leakoff control by the pad. Following an outline of acid injection and reaction, some aspects of acid retardation are illustrated.

4.5.1.3.1. Acid injection and reaction

The most important variables affecting acid penetration distance are acid injection rate, fracture width created by the pad fluid, temperature and acid concentration. Acid penetration and reaction rate are controlled by flow velocity, existing damage, acid concentration and temperature, and total amount of acid injected, as well as by reaction products, fluid viscosity, and area/volume ratio (BRANNON, NETTERS & GRIMMER 1987; DOWELL SCHLUMBERGER 1988 d). Increasing temperature leads to diminishing acid viscosity and an associate increase in the effective mixing coefficient which altogether give rise to reduction of penetration distance. Temperature controlling acid reaction is affected by acid injection temperature as a major factor and by the heat liberated by the injection itself as a minor factor. As acid spends, reaction rate decreases due to reduced live acid concentration and the effect of dissolved reaction products. Some aspects of acid channelization and fluid property modification are illustrated as follows.

4.5.1.3.1.1. Acid channelization

Acid flow rate can be increased by higher injection rates or by designing a channelized treatment where the acid contacts only a fraction of the total crack height, with the necessary containment being achieved by using a viscous pad fluid so that acid channels through the fluid (cf. section 4.5.4.2.1.) rather than uniformly displacing it (GRAHAM, KERVER & MORGAN 1965). Acid channelization to abt. one fourth of total fracture height significantly improves acid penetration distance (WILLIAMS & NIERODE 1971). The importance of preceding acid injection with a viscous fluid pad is underlined by the increase of fracture width in proportion to fluid viscosity (GEERTSMA & DEKLERK 1969; cf. section 4.5.1.6.2.).

Increasing acid saturation enhances penetration distance since the more concentrated acid is more viscous and has a lower value of the effective mixing coefficient (WILLIAMS & NIERODE 1971), with further increase of penetration distance being able to be achieved by adding more materials which create higher acid viscosity. Advantages of high acid concentrations are also outlined by HARRIS, HENDRICKSON & COULTER (1966). SEVOUGIAN, SCHECHTER & SEPEHRNOORI (1987) comment on optimization of vertical acid fractures in steady-state flow. On the other hand, reduced acid strengths give rise to slower reaction rates that trigger a pH environment which stabilizes the solutions thus minimizing precipitation of spent acid by-products (BRANNON, NETTERS & GRIMMER 1987). This particularly inhibits acid reaction sloughing which may lead to permeability diminution via clay and fines migration and subsequent blockage.

4.5.1.3.1.2. Fluid property modification

Acid penetration distance and rate can be increased by addition of 100 mesh sand in various concentrations and fracturing treatment stages as a fluid-loss control additive in case of high leakoff (BAILEY & WICKHAM 1984; cf. sections 1.4.11.2.1.1. and 4.8.8.3.2.3.). Longer persistence of life acid that can move further outwards into the formation along the fracture is also achieved by reduction or retardation of acid reaction rate by chemical retarding agents, emulsification of acid in oil, and acid gelling.

Further reduction of leakoff and reaction rate can be obtained by various volumes of gelled water preflush, with thinner gelled fluids being designed as cooldown and leakoff modifiers, whereas thicker fluids injected ahead of the acid are used as fluid loss modifiers and present a thick fluid through which the acid can finger (cf. section 4.5.4.2.1.). Another method of increasing acid penetration is using alternating stages of crosslinked polymer fluid followed by acid (cf. section 4.5.1.6.2.). Reduction of surface tension by mutual solvents and elimination of water blocking (cf. section 3.11.2.4.2.) also improves acid penetration and increases clean-up (HARMS, SMITH, KING & POSEY 1988). ALMOND, BRADY & UNDERDOWN (1988) discuss monitoring of acid penetration depth and spending rate by return fluid analysis.

4.5.1.3.2. Acid retardation

Acid effectivity optimization with increasing penetration depth can also be achieved by retardation of reaction rate and thus spending time (DOWELL SCHLUMBERGER 1988 d). Retarded acid travels farther away from borehole and perforations into the formation before becoming spent and is thus able to perform special treatments such as creation of a few long large-diameter channels, deep uniform matrix invasion, deep penetration and etching of natural hairline cracks, and generation and attack of hydraulically-induced fractures.

Acid retarding agents deposit hydrophobic films on carbonate surfaces, or are foaming additives which stabilize a carbon dioxide foam barrier on the rock surface (CROWE, McGOWAN & BARANET 1988). Distinction can be made between chemical retarders, emulsions, foams, gelled acid (cf. sections 4.3.4.3.1.4. and 4.5.1.2.3.) and organic acids. Techniques involving injection of pad fluids to cool the fracture faces (cf. section 4.3.4.6.1.4.) and to deposit a protective filter cake are also employed. In the absence of filter cake development, viscosity imparted by addition of gelling agents has little effect on reaction rate. Emulsification of acid reduces reaction rate, with the greatest retardation provided by oil-external acid emulsions. The summary as follows compiles some points of slowly reacting acids and solid additives, acid emulsification, acid gelling and filter cakes, and acid concentration and friction loss.

4.5.1.3.2.1. Slowly reacting acids and solid additives

Retardation of hydrochloric acid is most effective in fracture acidizing and can be carried out by adding a slower reacting acid such as acetic, formic or propionic acid to the hydrochloric acid. Other possibilities of retardation are using agents to introduce a barrier on the surface of the rock preventing normal contact with the acid, applying additives such as calcium chloride to generate a common ion effect, and increasing acid concentration in order to prolong spending time. With respect to the slowly reacting acetic, formic and propionic acids (DILL & KEENEY 1978), deeper matrix penetration is obtained by the faster reacting hydrochloric acid, because channelling or wormholing (cf. section 4.5.1.6.3.) reduces the area/volume ratio thereby prolonging reaction time. Organic acids are particularly used to treat high-temperature wells (CROWE, McGOWAN & BARANET 1988).

On the other hand, acetic, formic and propionic acids obtain deeper penetration in fractures than hydrochloric acid, but larger volumes would be required to dissolve an equivalent amount of rock and the higher cost may prohibit their extensive use. Formic and acetic acids are particularly characterized by low corrosiveness on metals and compatibility with oil. Incorporation of calcium chloride into hydrochloric acid retards reaction rate and provides higher viscosity of the solution once the acid is spent which may aid in clean-up. Calcium chloride addition further extends the inherent longer spending time of the concentrated hydrochloric acid. Hydrofluoric acid retardation can also be performed by sequential HF liberation (HOLDEN, PRIHODA & HALL 1981; cf. section 4.5.5.4.3.).

4.5.1.3.2.2. Acid emulsification

Emulsification is one of the earliest methods used for acid retardation, with both oil-external and acid-external emulsions having been applied (CROWE, McGOWAN & BARANET 1988). Acid-in-oil emulsions also have retarded reaction rates, because the acid in the emulsion is to some extent prevented from completely contacting the rock surface by presence of an oil film particularly in emulsions with at least 20 % oil as the outer phase. In addition, the higher viscosity of the emulsion tends to restrict flow into the smaller pores, thus minimizing contact with extremely high area/volume ratios. Acid-in-oil emulsions provide increased channelling or wormholing (cf. section 4.5.1.6.3.) in matrix flow. Surfactants result in the presence of oil in a hydrophobic or water-repellent oil-like film on the rock surface which restricts acid-rock contact and is thus also beneficial in reducing acid reaction rate leading to extension of spending time and travel distance. Some oil-external acid emulsions are capable of retarding reaction rate by as much as 98 %, but on the other hand, such an extreme level of acid retardation would lead to inadequate etching of fracture faces (BERGSTROM & MILLER 1975).

Fluid-loss materials and gelling agents being acid thickening additives (cf. section 4.3.4.3.4.1. and 4.5.1.2.3.) also tend to reduce the reaction of hydrochloric acid by film development on rock surfaces and restriction of diffusion (CROWE, MARTIN & MICHAELIS 1981). DAVIS, MANCILLAS & MELNYK (1965) present a procedure using additives in a spearhead fluid ahead of the acid job in order to create new fractures. The specially compounded additives are only slowly dispersible in acid, and their presence on exposed formation and fracture surfaces as a thin sealing film inhibits and postpones acid reaction until the entire treatment volume has entered the reservoir.

4.5.1.3.2.3. Acid gelling and filter cakes

The use of gelled acids has increased due to development of thickening agents having greater stability in hot acid (CROWE, McGOWAN & BARANET 1988). Acid gelling reduces reaction rate by slowing convection and hydrogen ion diffusion, and further diminution of gelled acid reaction rates can be achieved by addition of a foaming agent or a crosslinker (cf. sections 4.3.4.3.4.1. and 4.5.1.2.3.). Although increasing acid viscosity does not necessarily diminuish reaction rate, gelling agents can retard acid reaction by depositing a filter cake on carbonate surfaces (GDANSKI & NORMAN 1983). On the other hand, neither polymer nor surfactant gelled acids exhibit significant retardation in the absence of conditions which cause filter cake formation, but in many cases reaction rates are instead accelerated (CROWE, McGOWAN & BARANET 1988).

Escape of gelled acid into permeable carbonates can result in generation of filter cakes provided that leakoff velocity is not sufficient to cause wormhole development (cf. section 4.5.1.6.3.). As the primary site of acid fluid-loss is from the area around the wormhole tips rather than into the fracture faces, little filter cake development occurs on the crack walls and thus reaction rate is not retarded in case of extensive wormholing. When using gelled acids, it is doubly important to control acid fluid-loss in order to both hydraulically extend the fracture and allow filter cake development (cf. sections 4.3.4.3.1.4. and 4.5.1.2.3.).

4.5.1.3.2.4. Acid concentration and friction loss

High acid concentrations have longer reaction times than lower saturations because there is more acid to react, the additional reaction products further retard reaction rates, and the enlarged flow path with reduced area/volume ratio prolongs spending time and travel distance of a high-concentration acid. NIERODE & KRUK (1973) and MILLER & BERGSTROM (1975) present an acid emulsion system which overcomes the viscosity and friction-loss problem associated with many emulsion systems and provides better live acid penetration. Acid fracturing stimulation can be increased by using effective fluid-loss additives in hydrochloric acid or significantly increase acid viscosity. While acid emulsions have a low fluid-loss rate and are retarded, oil-wetting surfactants usually give no retardation at typical field inspection rates. Retarded hydrofluoric acid reacts slower with clay minerals and feldspars and thus allows deeper penetration of live acid (GDANSKI & PEAVY 1986; cf. section 4.5.5.4.).

The reaction rate of acid on solid carbonate surfaces is also affected by confining pressure (CROWE, McGOWAN & BARANET 1988), with dissolution rates at lower pressures being much higher due to surface agitation by carbon dioxide bubbles, but increasing pressures reducing reaction rate. The impact of various retarding agents on temperature sensitivity of acid reaction rates is very different. Aspects of acid retardation are also discussed by KNOX, LASATER & DILL (1964), NIERODE & KRUK (1973), GDANSKI (1985) and CROWE, McGOWEN & BARANET (1988).

4.5.1.4. Acid foaming

Acid foaming also has significant impact on acid penetration distance and rate (FORD & ROBERTS 1982). Increasing foam quality results in decreasing acid penetration distance, and the higher the foam quality, the lower the acid content of the foam, with the less acid present in the foam, the lower the rock-dissolving power of the foam. The wider the fracture, the longer it takes for hydrogen ions to reach the rock surface and thus the acid travels farther down the crack before spending, with this relationship applying for foamed acids as well as for non-foamed acids. If pumping rate is increased and fracture height remains constant, the distance which foamed acid travels down a fracture before spending is increasing. Acid spending in a crack is governed primarily by the mass transfer of the acid to the fracture wall. Channel etching is due to density effects and may be reduced by using a foam stabilizer which increases gelled fluid viscosity.

Foamed acid is non-damaging and provides unexcelled fluid-leakoff control especially when used in alternating stages of acid and inert pad fluids (SCHERUBEL & CROWE 1978; cf. section 4.5.1.6.2.), and is particularly beneficial in low-pressure, low-permeability, liquid-sensitive formations (FORD 1980; FORD, BURKLECA & SQUIRE 1980) as well as in highly soluble reservoirs that release large amounts of fines. As acid enlarges the flow paths thereby constantly increasing leakoff rate, the best fluid loss control by stable acid foam is achieved by pumping gelled water-based pads ahead of the acid foam in order to reduce escape of gas (SCHERUBEL & CROWE 1978). The use of foamed acid in fracture acidizing treatments has gained widespread acceptance in the last years. Aspects of foamed acids are also discussed by HOLCOMB (1977), HOLCOMB & WILSON (1978), WILSON (1978), PETRYK & GARUK (1979), FORD (1981) and HARMS, SMITH, KING & POSEY (1988).

Aspects of optimization and improvement of acid reaction in matrix and fracture acidizing are also discussed by SUTTON & LASATER (1972); DOMSELAAR, SCHOLS & VISSER (1973); McCUNE, AULT & DUNLAP (1975) and COULTER, CROWE, BARRETT & MILLER (1976).

4.5.1.5. Open vs. propped fractures

Proppant plugging of the acidized fracture is considered to be the most effective way of improving conductivity in acid fracturing treatments of deeper reservoirs where closure stress is too high to allow acidized fractures to stay open with only negligible damage and to provide sufficient conductivity for long-term hydrocarbon production. Some aspects of drawbacks of auto-propping of acidized fracture walls and significance of proppant infilling of the fracture are briefly discussed as follows.

4.5.1.5.1. Drawbacks of auto-propping of acidized fracture walls

Self-propping by crushing of exposed portions of the irregular fracture wall created by acid leaching or bridging of broken fragments of the reservoir rock matrix within the crack (cf. also section 4.5.1.1.1.2.) can be effective, but might also result in serious fracture conductivity deterioration if large amounts of fines are created upon squeezing and crushing of soft ductile and hard brittle carbonate formation matrix, respectively. While smooth fracture walls tend to easier primary closure once sufficient stress is acting, rough crack faces exhibit more resistance to closure stress, and once exceeding the stability boundary, it is possible that breaking of the protrusions into larger pieces leads to auto-propping of the crack at least in metastable nature, but it is much more possible that crushing of these fragments into fine splinters will later result in secondary closure of the fracture.

The possibility of crushing of protrusions and subsequent at least partial fracture closure is especially given in hard brittle fine-grained, well-banked and -jointed sedimentary and bioclastic carbonates, whereas massive biolithic carbonates constructed by reef-building organisms (cf. plate I/8) provide more irregularities both in terms of primary crack opening and secondary selective fracture etching. Partially closed cracks resemble pillar propping (cf. section 4.3.3.2.), but in contrast to infilling with round stable synthetic high-conductivity proppants, support of the fracture by irregular natural carbonate fragments that are subject to considerable crushing due to their unsuitability to withstand the closure stress represents patchy flow obstruction rather than amelioration, with fracture auto-propping by breaking and collapsing carbonate particles therefore being especially in greater reservoir depth highly undesirable. Generally, etched flow channels created by acid fracturing can have significantly lower conductivity than propped cracks because of the much higher likelihood of the open unsupported fracture to close due to formation collapse (ECONOMIDES 1987 b).

4.5.1.5.2. Significance of proppant infilling of the fracture

Fracture wall roughness therefore is an important parameter for preservation or deterioration of crack conductivity created by acid treatments only in shallow to moderately-deep reservoirs. Proppant plugging of the fracture should at least be considered for the immediate wellbore surroundings, as this tail of the fracture wedge is most susceptible to pressure drawdown (cf. sections 2.4.1.2.3. and 4.12.3.3.). Securing the result of an acid fracturing job by applying a proppant of known conductivity for infilling of the crack is especially beneficial as a consequence of the fact that the conductivity delivered by an acid treatment cannot be predicted with certainty (WILLIAMS & NIERODE 1971) in contrast to that provided by a proppant package which can be more or less forecasted (cf. section 1.4.10.). Conductivity obtained from acid fracturing operations is a function of formation type, acid concentration, contact time between acid and rock, and closure stress and thus reservoir depth, whereas proppant conductivity is (within the range of closure stress resistivity of the individual proppant types) independent from all these factors and only is related to the configuration of the proppant package.

In high-permeability reservoirs with low fracture conductivity, gas breakthrough occurs early in time and near the wellbore which reduces fracture effectivity for gas production (SOLIMAN & HUNT 1986). This drawback can be overcome by designing a higher crack conductivity by implementing proppants of larger grain size and better quality (cf. sections 4.3.5. and 4.6.1.). AL-TAMIMI & EL-MZEIN (1987) comment on stimulation techniques of tight multilayered carbonates. Another combination method related to acid and proppant fracturing is chemical fracture-squeeze treatment (TINSLEY, LASATER & KNOX 1967).

4.5.1.6. Acid fluid-loss control

Acid fluid loss is extremely difficult to control and is generally considered to be the major factor limiting the effectiveness of acid fracturing treatments (CROWE, HUTCHINSON & TRITTIPO 1987). Chemical erosion of crack faces and development of wormholes are largely responsible for the reduced efficiency of acid fracturing. The creation of acid wormholes increases the effective area from which leakoff occurs thus reducing the hydraulic efficiency of the acid. Once wormholes form, most acid fluid loss originates from these wormholes rather than penetrating uniformly into the fracture face (cf. sections 4.5.1.2.1. and 4.5.1.6.3.). After an introductory outline of acid reactions during fracturing, some comments are offered on alternating acid and gelled water stages, wormhole growth and acid fluid loss, and possibilities for improving acid fluid-loss control.

4.5.1.6.1. Acid reactions during fracturing

Acid fracturing comprises acid injection into a carbonate formation at pressures sufficient to fracture the reservoir or to open existing natural joints (CROWE, HUTCHINSON & TRITTIPO 1987). As acid flows along the fracture, portions of the crack faces are dissolved and eroded in a non-uniform manner, and thus conductive channels are created which usually remain following fracture closure. The effective length of the crack is determined by the volume of acid used, its reaction rate, and the acid fluid-loss rate, and productivity improvement is largely a function of the length of the etched fracture.

The maximum effectivity distance obtainable by acid fracturing is limited by either spending or fluid leakoff. If the acid spends too rapidly, the etched portion of the crack will be rather short. If acid fluid-loss characteristics are poor, excessive leakoff will terminate fracture growth. In many acid fracturing treatments, operating pressure continually declines and eventually falls below a level required to propagate the crack. In some cases, treating pressure never reaches a level sufficient to initiate a hydraulic fracture, indicating that the acid is expended in a matrix stimulation of the near-wellbore region.

Control of fluid loss during acid fracturing in carbonate formations presents problems which are unique to reactive fluids, because most fluid-leakoff additives and gelling agents used in nonreactive aqueous fracturing fluids are chemically unstable in the aggressive acid. In addition, acid flow along carbonate crack surfaces produces constant chemical erosion, thus making it difficult for wall-building fluids to deposit an effective filter cake. Acid tends also to selectively enlarge pores and natural fractures resulting in wormhole creation and channel generation perpendicularly to the crack face, with these phenomena greatly increasing the effective surface area from which leakoff occurs and generally being believed to adversely affect fluid loss.

4.5.1.6.2. Alternating acid and gelled water stages

While acid- and temperature-resistant additives are either of limited chemical and/or physical applicability or are in many cases too expensive, acid leakoff control can easily be performed by injection of a viscous nonreactive pad preceding the acid (DAVIS, MANCILLAS & MELNYK 1965; COULTER, CROWE, BARRETT & MILLER 1976; CROWE, HUTCHINSON & TRITTIPO 1988). This technique usually employs a water-based fracturing fluid to cool the fracture (cf. section 4.3.4.6.1.4.) and lay down an impermeable filter cake on the crack face. The filter cakes covering fracture walls, however, are relatively ineffective for acid fluid-loss control, because they are quickly penetrated by wormholes (cf. section 4.5.1.6.3.) and then their effect is gone. While the use of the pad probably provides only limited acid fluid-leakoff supervision, it does have other useful functions including cooling of the tubular goods through which acid flows thereby reducing corrosion, cooling the fracture and increasing crack width resulting in reduced acid reaction rate and increased live acid penetration, and promoting acid fingering through its viscosity thus reducing the amount of reactive surface area to which the acid is exposed and improving fracture length and conductivity (cf. also sections 4.5.1.1.1.4. and 4.5.4.2.1.).

Enhanced variations of pad pumping are initial fracture creation by a gelled pad after which alternating stages of acid and additional pad are injected that are designed to enter and seal the wormholes created by the preceding acid. By alternating acid and gel stages, acid leakoff into wormholes and enlarged natural fractures is controlled and treatment efficiency improved. Fine particulate materials such as 100 mesh sand are often added to the pad stages to aid in fluid-loss supervision (cf. section 4.5.1.1.1.4.). Acid fluid leakoff can also be reduced by acid gelling (cf. sections 4.3.4.3.4.1. and 4.5.1.2.3.).

4.5.1.6.3. Wormhole growth and acid fluid loss

Wormhole growth is generated by most of the acid entering a few enlarged pores. Creation and branching of wormholes and enlargement of natural fractures has a major effect on acid fluid loss since it increases the effective amount of surface area from which acid fluid leakoff occurs. Some aspects of wormhole penetration and acid channelling as well as fluid invasion pattern and fracture conductivity development are outlined as follows.

4.5.1.6.3.1. Wormhole penetration and acid channelling

Factors governing wormhole propagation are analogous to those affecting acid spending in a hydraulic fracture with wormholes continuing to grow in length providing that live acid is delivered to the wormhole tip. Increased temperature diminishes wormhole penetration distance, while wormhole depth is lengthened by increased acid concentration and injection rate. Once wormholes reach the maximum spending-limited depth, injection of additional acid mainly acts to enlarge the diameter of existing wormholes. Wormholes grow rapidly in length during early stages of acid injection due to selective acid channelling. In many cases, no single wormholes but rather a branched wormhole network is developed. Thus leakoff of only a small quantity of acid greatly increases the amount of new surface area from which fluid loss can occur, with this being a major reason why acid fluid leakoff is more difficult to control than escape of non-reactive fluids. Factors other than fluid-loss velocity influencing wormhole depth are temperature and acid concentration (DACCORD, TOUBOUL & LENORMAND 1987; HUNG, HILL & SEPEHRNOORI 1987).

4.5.1.6.3.2. Fluid invasion pattern and fracture conductivity development

In addition to increasing acid fluid loss, wormholes also alter the pattern of fluid invasion into the fracture face. Once developed, acid fluid leakoff originates predominantly from existing wormholes with little fluid loss actually occurring at the fracture face. Filter cakes deposited by non-reactive pad fluids have in many cases little influence on subsequent acid fluid loss, because once wormholes penetrate the filter cake, they are responsible for almost all acid fluid leakoff regardless of the presence of the remaining filter cake (cf. section 4.5.1.2.1.). On the other hand, acid reaction rate is retarded by addition of filterable solids to the acid, with the additives depositing filter cakes which act as barriers between acid and fracture surface. While this method of acid retardation is effective under conditions of little or no wormhole development, the existence of deeply penetrating wormholes eliminates most leakoff at the fracture face and prevents filter cakes from being formed (DACCORD, TOUBOUL & LENORMAND 1987; HUNG, HILL & SEPEHRNOORI 1987).

The existence of wormholes also influences the development of fracture conductivity. Since acid fracturing depends upon etching of fracture faces to create flow channels, acid lost to wormholes results in less dissolution of rock from the crack face and thus lower conductivity. It is more beneficial to expend acid in etching fracture walls rather than in creation of wormholes perpendicular to the hydraulic fracture. Localized changes in crack face permeability have a far smaller effect on productivity than does fracture length or conductivity.

4.5.1.6.4. Possibilities for improving acid fluid-loss control

Both fluid-loss and wormhole development can be effectively controlled by injection of multiple stages of acid and gelled water in an alternating manner (cf. sections 4.5.1.6.2. and 4.5.4.2.). Gelled acids have found increased usage in acid fracturing since development of more stable acid gelling agents which, however, reduce both fluid leakoff and acid reaction rate due to acid thickening by forming micellar networks within the acid solution. Viscosity loss at elevated temperature is not caused by gelling agent degradation, but occurs as the result of disruption of the micellar association network, and reaction products formed during acid spending also disrupt the micellar assemblage, thereby causing the spent acid to revert to a thin fluid which aids in recovery after the treatment.

The wormhole problem in acid fluid-loss control can be solved with a gelled acid having unique rheological properties. This gelled acid system has a moderate initial viscosity, but spending of the acid during escape produces a rise in pH which initiates a process creating a very large temporary increase in effective leakoff viscosity. Final spending of the acid triggers another chemical reaction which reverses the temporary viscosity increase, thereby allowing the spent acid to be easily recovered from the formation.

4.5.2. Combination of natural and artificial fractures

Another means of enhancement of the effectivity of carbonate stimulation is the combination of natural jointing and artificial hydraulic fracturing (cf. also section 4.8.8.). As carbonates and sandy carbonates are in many cases intersected by better penetrating and also closer spaced natural cracks than sandstones (cf. section 4.5.1.1.1.3.), they also suffer in a larger degree from the fact that natural fractures being crossed by a hydraulically-induced propagating fracture may deflect, attenuate or even totally stop the growth of the latter (CONWAY, McGOWEN, GUNDERSON & KING 1985; cf. sections 4.2.2.3.1., 4.2.3.2.2. and 4.2.4.3.1.). As gelled hydrochloric acid could help preventing fracture deflection and arrestment if the formation has significant hydrochloric acid solubility (GDANSKI & NORMAN 1983), the combination of acid and proppant fracturing in carbonate rocks is also a very effective means of stimulation of naturally-jointed formations, with emphasis being ideally put on enlargement of existing cracks parallel to their orientation rather than creation of a new independent system of fractures perpendicular to the pre-existing one (cf. also section 4.8.8.).

4.5.3. Possibilities of application in Europe

Various carbonate reservoirs in different geological formations are targets of unconventional gas and oil exploration and production in Europe as well as other areas of the world in the near future. While major formations such as the Zechstein carbonate sour gas belt (reaching from Great Britain to Poland and containing the largest and most important reservoirs in Germany FRG) are only in exceptional cases needing treatments beyond matrix acidizing or acid fracturing (if requiring stimulation at all), particularly Jurassic, Cretaceous and Tertiary carbonate rocks in the Northern North Sea, onshore Germany FRG, Southern France, Yugoslavia, Hungary, Italy, Austria and the Middle East are proliferous goals for optimization of exploitation by hydraulic proppant fracturing.

The most effective help for laying decisive groundwork in this approach, however, would be a rising oil price and increasing US $ value which will considerably increase the attractivity of hydraulic proppant fracturing of carbonate rocks in Europe and other parts of the world, because in the present unfavourable general economical framework, hydraulic proppant fracturing of carbonate formations would be everywhere on the very bottom of the ranking list of planned investments (cf. section 2.2.1.5.). The most promising carbonate reservoir type for hydraulic proppant fracturing (sometimes in combination with gravel packing) is the chalk which is characterized in more detail in the following section.

4.5.4. Chalk stimulation

A special case of carbonate reservoirs is the Upper Cretaceous and partially also Lower Tertiary chalk in the North Sea and parts of onshore Middle Europe as well as in the Rocky Mountain area in Colorado, Nebraska and Kansas/USA.

This high-porosity, but low-permeability and low-stability formation contains considerable oil and occasionally also gas reserves. While the chalk in some deeper fields is more consolidated and sufficiently naturally fractured to have only to be treated by acidizing, the chalk in other shallower fields is soft and flowing and in these cases inevitably has to be approached by sophisticated both hydraulic proppant fracturing and gravel packing for the achievement of economically feasible production rates (various stimulation-related aspects of chalk reservoirs are discussed by MANCILLAS, MATSON & ZIARA 1976; ZIARA, MANCILLAS & MATSON 1977; VLIS, DUNS & FERNANDEZ-LUQUE 1979; BARK & THOMAS 1980, NEWMAN 1981, BETZ 1982; SIMON, COULTER, KING & HOLMAN 1982; WATTS 1983, BREWSTER & DANGERFIELD 1984, PEETERS & HARTLEY 1984, GISTAU 1985, HARTLEY & BOSMA 1985; BREWSTER, DANGERFIELD & FARRELL 1986; JOHNSON & RHETT 1986; TANSOE, KING & HOLMAN 1986; SMITH, MILLER & HAGA 1987; FARRELL 1988, HOWES 1988, SNOW & HOUGH 1988).

Unstable chalk is probably the worst reservoir type to handle in terms of completion and stimulation in North Sea and other areas. The main oil-producing chalk fields in the Norwegian and Danish North Sea are Ekofisk (BARK & THOMAS 1980, 1981; OFFSHORE ENGINEER 1985 b, PEKOT & GERSIB 1987, SNOW & HOUGH 1988), Eldfisk (MICHAUD 1987), Balmoral, Hod (NORBURY 1987), Tor (D`HEUR 1987), Dan and Valhall (TANSOE, KING & HOLMAN 1986; LEONARD & MUNNS 1987; ATTARD, MATHES & MOWER 1988; BISHLAWI & PENG 1988; RUDDY, ANDERSEN, PATTILLO & FOGED 1988), whereas other fields such as Troll and Tommeliten (D`HEUR & PEKOT 1987) still have to be developed. Aspects of petrophysical properties and productiveness, combination of acid and proppant fracturing, combination of hydraulic fracturing and gravel packing, special chalk fracturing techniques, and water imbibition and oil recovery of unstable soft flowing chalk are discussed as follows. Some comments are also offered on horizontal well drilling.

4.5.4.1. Petrophysical properties and productiveness

The chalk is the reservoir rock where the first commercial oil discovery in the North Sea was made in 1969 in Ekofisk in the Norwegian North Sea, with production having started in 1971 (thus four years after the first gas strike and six years after the onset of oil and gas exploration in the North Sea; THOMAS 1986; cf. section 3.3.1.). Ekofisk is also the first giant oil field that was discovered in Western Europe (BARK & THOMAS 1980, 1981; SNOW & HOUGH 1988). In the last years, apart from oil production and reservoir stimulation, the nature of the chalk pay zones in parts of the North Sea has been highlighted by tremendous field subsidence as a consequence of reservoir drainage and depletion particularly in Ekofisk (BLEAKLEY 1986, DIJK & KLÖTERS 1987; THOMAS, DIXON, EVANS & VIENOT 1987; BOADE, CHIN & SIEMERS 1988; JOHNSON, RHETT & SIEMERS 1988; MENGHINI 1988, MES 1988, RENTSCH & MES 1988, SULAK & DANIELSEN 1988) which had to be compensated by a complicated platform jack-up operation to render to equilibrium a vertical difference of more than six meters (MANNESMANN REXROTH 1988, SMITH 1988; SMITH, SMITH & MONSHAUGEN 1988). Subsidence on an even larger scale than in Ekofisk has been detected in the Valhall chalk oil field (OILMAN 1987 d; RUDDY, ANDERSON, PATTILLO, BISHLAWI & FOGED 1988). Oil- and gas-field subsidence problems also occur in other chalks, diatomites and loose finer-grained sands (GEERTSMA 1973, WAAL & SMITS 1988; WENDEL, KUNKEL & SWANSON 1988). MORITA, WHITFILL, NYGAARD & BALE (1988) describe determination of subsidence, reservoir compaction, and in-situ stress induced by reservoir depletion.

4.5.4.1.1. General aspects

Chalk is hitherto hydrocarbon-productive in North Sea, Middle East, Gulf Coast and Rocky Mountain area/USA, and Scotian Shelf/Canada (BLANTON 1981, HARTLEY & BOSMA 1985). Chalk is a massive, rather homogeneous, soft and extremely fine-grained sediment which has a ductile behaviour, with all these features causing or triggering pore collapse and loss of permeability, embedment of proppants (cf. section 4.3.3.) and deterioration of conductivity, evenly etching by acid resulting in low crack conductivity when the pressure is released and an unpropped fracture closes, and lack of significant crack barriers leading to radial fracture development without containment (KING & HOLMAN 1982). As a consequence of the softness, any ridges left after acidization collapse under the stresses imposed by production unless they are multi-layer propped (BLANTON 1981; SIMON, COULTER, KING & HOLMAN 1982). Some rock mechanical characteristics of North Sea reservoirs are also investigated by HOLT, INGSOY & MIKKELSEN (1987). Aspects of stratigraphy, petrography and petrophysics, and natural fracture types are briefly summarized as follows. Comments are also offered on primary chalk properties and stimulation response as well as secondary field subsidence.

4.5.4.1.2. Stratigraphy

North Sea chalk (HARPER & SHAW 1974) exhibits differences in petrophysical properties in stratigraphical succession (TANSOE, KING & HOLMAN 1986) by being a multistorey reservoir with distinct thickness, porosity, permeability and water saturation trends within each layer (BREWSTER, DANGERFIELD & FARRELL 1986). Distinction can be made between Hod, Tor and Ekofisk formations, and in some fields, even a fracture stratigraphy can be established.

4.5.4.1.2.1. Hod, Tor and Ekofisk formations

The Tor Formation (upper part of the Upper Cretaceous Maastrichtian chalk sequence) is a high-porosity (40 - 50 %) naturally fractured chalk with a matrix permeability of abt. 5 md, whereas the underlying Hod Formation (lower part of the Upper Cretaceous Maastrichtian chalk series) is much denser with abt. 30 - 35 % porosity and permeabilities between abt. 2 md and a few microdarcy. A third stratigraphical level is the Lower Tertiary Danian chalk (Ekofisk Formation; FARRELL 1988).

While the Hod chalk at the base is relatively stable and also due to its tighter facies is only subordinately productive, and also the Ekofisk chalk at the top is a thin tight layer generally acting as a hydraulic barrier, the Tor chalk in the middle of the reservoir complex contains the major chalk oil fields in the North Sea, but is often seriously unstable and at least in the Valhall field almost every well needs more or less extensive treatment by both hydraulic proppant fracturing and gravel packing to be able to be produced at economically feasible rates (cf. section 4.5.4.3.). In the Ekofisk field, the chalk is more stable and only conventional hydraulic fracture stimulation, but no gravel pack consolidation of the reservoir formation has to be carried out (FARRELL 1988).

4.5.4.1.2.2. Fracture stratigraphy

In some fields, the correlation of natural crack trends and distribution within the reservoir column results in establishment of a fracture stratigraphy. In the Ekofisk field, the Tertiary Ekofisk Formation is mainly characterized by tectonical fractures of conjugate type, whereas the Cretaceous Tor Formation contains predominantly stylolite-associated cracks which propagate between and terminate at stylolites and form anastomosing networks of nearly vertical extension fractures. The correlation between crack intensity and sedimentary facies of the chalk gives rise to a predictive fracture model (FARRELL 1988).

4.5.4.1.3. Petrography and petrophysics

The Tor chalk consists of coccoliths and calcite crystals and is almost uncemented which apart from overpressuring and undercompaction is one of the main reasons for its instability. The low formation strength is also due to a very low concentration of connate water (less than 5 %) which has decreased the normal effect of pressure solution at grain contacts and thus cementation is severely retarded. The high porosity of the Tor chalk has also been preserved by a very low net confining stress (overburden minus pore pressure) caused by a very high overpressure, resulting in prevention of formation porosity collapse because the fluid in the pore-space is a major supporting element of the overburden. While the matrix permeability of the Tor chalk is only abt. 5 md, pressure buildup evaluation reveals permeabilities up to 100 md which are the effect of an extensive natural fracture network. The soft chalk does not fail and move as individual fine particles, but as a deformable plastic flow of an unit mass (BLANTON 1981, NEWMAN 1981; TANSOE, KING & HOLMAN 1986; ATTARD, MATHES & MOWER 1988). Some aspects of autochthonous and allochthonous chalk as well as fracture closure are discussed as follows.

4.5.4.1.3.1. Autochthonous and allochthonous chalk

Chalk porosity is both stratigraphically and mineralogically controlled (SNOW & HOUGH 1988). The chalk horizons of Hod, Tor and Ekofisk formations consist of alternating layers of allochthonous and autochthonous sediments. The autochthonous chalks were deposited by gradual raining down of calcareous material onto the ocean floor, whereas the allochthonous chalks were laid down rapidly by movements on graben faces. Because of their slower sedimentation rate, the autochthonous chalks are harder, have more cementation and higher silica content, underwent more dewatering, have lower porosity and permeability, have less frequent natural fracturing, and a greater percentage of their natural fractures are cemented than in the allochthonous deposits. These variations in characteristics can cause dramatic changes in porosities across a field and within a wellbore and are also at least partially the source for different fracturing stimulation requirements in various places and storeys.

4.5.4.1.3.2. Fracture closure

Laboratory testing proved the natural fractures to heal and etched open channels to close above 2,000 psi and 800 psi net confining pressure, respectively. Wide fractures and high proppant concentrations in the cracks provide stable fracture flow capacity when confining pressure is increased in contrast to healing of unpacked fractures at moderate confining loads. Excessive pressure drawdown results in failure of the formation and catastrophic loss of production including collapse of cracks created by both proppant and acid fracturing. D`HEUR (1984) documents porosity and hydrocarbon distribution in North Sea chalk reservoirs.

Formations comparable to the North Sea and Rocky Mountain chalks in terms of petrographical composition and hydrocarbon production difficulties are diatomaceous earths and diatomites in California/USA (STRUBHAR, MEDLIN, NABI & ANDREANI 1984; STRICKLAND 1985; WENDEL, KUNKEL & SWANSON 1988).

4.5.4.1.4. Natural fracture types

The North Sea chalk reservoirs are extensively naturally fractured. Distinction can be made between healed, tectonic and stylolite-associated fractures (BREWSTER, DANGERFIELD & FARRELL 1986; FARRELL 1988). The main chalk stimulation targets are also outlined.

4.5.4.1.4.1. Healed and tectonic fractures

Healed fractures are the earliest cracks, because they are cross-cut by all other joint classes (BREWSTER, DANGERFIELD & FARRELL 1986; FARRELL 1988). Healed fractures are single or anastomosing networks and are infilled with fine material being harder than the chalk matrix. No visible porosity can be detected within the healed fractures which presently do not form permeable flow channels, but rather may inhibit fluid flow by acting as microbarriers. The overall geometry and texture of the healed fractures suggests their origin from local brittle deformation of partially lithified sediment, and this type of cracks is intimately associated with reworked chalk.

Tectonic fractures provide the most effective permeability enhancement and are essentially high-angle normal faults commonly forming parallel sets of conjugate cracks. In such gregarious zones, one set of fractures is always dominant and may have an average crack spacing as small as 10 - 15 cm. The fractures are usually planar with smooth surfaces, with slickensides and secondary mineralization of the cracks being unusual. Communication of reservoir fluids between matrix and tectonic fractures is considered to be very good. Displacement across the cracks is commonly in the order of a few cm. Tectonic fractures developed under a near vertical maximum principal stress.

4.5.4.1.4.2. Stylolite-associated fractures

Stylolite-associated fractures also originated under a vertical principal stress system (NELSON 1981, WATTS 1983; BREWSTER, DANGERFIELD & FARRELL 1986). These cracks form adjacent to stylolites and usually parallel the stylolite columns. They are interpreted as tension gashes, although a shear fracture geometry is occasionally seen. The cracks terminate at the stylolite surfaces and never cross them. Stylolite-associated fractures may occur individually in areas with low crack intensities, but are usually seen in anastomosing networks where a fracture spacing of 5 cm is common and 0.5 cm is not unusual. Individual cracks do not extend far into the chalk matrix from the stylolites. Fracture length may reach up to 5 - 15 cm. Stylolite-associated cracks are statistically near vertical, but they form bedding-parallel permeable zones adjacent to individual stylolites. Additional reservoir permeability is in some parts of the chalk provided by fractures along slump surfaces. Outlines of the natural fracture system in North Sea chalk reservoirs are also given by THOMAS, DIXON, EVANS & VIENOT (1987); SYLTE, HALLENBECK & THOMAS (1988) and JENSEN, BRESLING, RASMUSSEN, FOGED & PETERSEN (1989).

4.5.4.1.4.3. Chalk stimulation targets

Besides matrix permeability and net pay thickness, well productivity is highly dependent on natural fracture occurrences (SNOW & HOUGH 1988). Because the natural cracks are short and occur frequently, and due to high matrix permeabilities, dual porosity behaviour is not observed in transient tests, but the natural fractures essentially behave as an increased matrix permeability. Three stimulation targets in chalks of the Ekofisk area can be distinguished : chalks with good matrix permeability and good natural fracturing, chalks with good matrix permeability but poor natural fracturing, and chalks with low matrix permeability and poor natural fracturing. Natural crack occurrence is both locally and regionally controlled. Locally the natural joints appear to radiate from the crest of the structure, whereas regionally they trend NNE-SSW. Conductive natural fractures are influenced by in-situ stresses and may occur in a direction perpendicular to that of minimum stress.

4.5.4.1.5. Primary chalk properties and stimulation response

The various natural fracture systems are especially well-developed in hard chalks that are sufficiently consolidated to allow the cracks to stay open. This is only the case in some fields and stratigraphic horizons, whereas in other patches and storeys, the chalk is soft and poorly cohesive and thus the natural fractures are in most cases rehealed. Ultimate recovery, producing rate and drainage efficiency for each well depend on the success of connecting the natural cracks with the wellbore by a hydraulic fracture (the same applies for very fine-grained hard or soft sandstones to siltstones of different matrix cohesion types which can in terms of rock mechanical properties be compared with chalk; PEARCE 1983; and the same is true for diatomites and diatomaceous earths; STRUBHAR, MEDLIN, NABI & ANDREANI 1984; STRICKLAND 1985; WENDEL, KUNKEL & SWANSON 1988).

The chalk is a difficult stimulation target, because its massive uniform nature results in short radial fracture growth (SNOW & HOUGH 1988). Low hardness and ductile behaviour of the chalk lead to pore collapse and creep causing excessive proppant embedment (cf. section 4.3.3.). The homogeneous nature of the chalk also gives rise to evenly etched crack faces during acidizing and acid fracturing. Low hardness and low yield strength of the chalk can also result in healing of natural cracks and failure of any acid-etched grooves. The chalks also lose mechanical strength when contacted by foreign fluids, thus causing treatments with water-based fluids to actually damage instead of stimulating the formation.

Productivity decline of chalk reservoirs following fracturing and/or acidizing is partially due to two-phase flow and condensate dropout. The majority of the output loss is a consequence of total failure and plastic flow of the chalk, stable creep flow, crushing of the fracture face, brittle failure, breaking off of fines on the crack face thereby clogging up the fracture, and slippage along the joint wall of the chalk. As effective stresses increase, chalk undergoes creep failure and pore collapse, and as this failure occurs on the fracture face, the crack loses some mechanical strength and may close. Fines having been chipped off the crack wall and having filled up the conductive channel are the most important reason for conductivity reduction.

4.5.4.1.6. Primary chalk properties and secondary field subsidence

OEN, ENGELL-JENSEN & BARENDREGT (1988) describe a combination of small-scale faults and microfractures with large-scale faults in a crack system extending throughout the reservoir which provides a communicating network of highly permeable fluid conduits that result in high well productivities and excellent field drainage. The fault and fracture system is quite well supported by the chalk matrix framework in order to stay open even in case of considerable renewed compaction of the reservoir due to field subsidence (SULAK & DANIELSEN 1988).

The latter process is provoked by oil depletion and phase changes due to replacement of the pore volume formerly infilled by oil subsequently by water which alters fluid viscosity and thus rock cohesion, and thereby triggers secondary recompaction of the matrix without, however, significantly affecting fracture porosity and oil productivity. In contrast to the rather negligible impact of field subsidence on petrophysical properties and reservoir engineering, recompaction of the chalk as an adjustment to altered fluid phase distribution has tremendous consequences for technical field operation and requires complicated jack-ups for equilization of subsidence (cf. DIJK & KLÖTERS 1987, MANNESMANN REXROTH 1988; SMITH, SMITH & MONSHAUSEN 1988). The discussion as follows focusses on reservoir compaction and field subsidence, Permian (Rotliegend) sandstone vs. Cretaceous chalk subsidence, and subsidence rates in different chalk fields.

4.5.4.1.6.1. Reservoir compaction and field subsidence

Subsidence is the gradual settling of the land mass or ocean floor over oil and gas reservoirs or aquifers where fluid removal causes pressure reduction in the pore space (BLEAKLEY 1986). Subsidence is triggered by compaction of the pay rock under pressure from the overburden when reservoir pressure is relieved. Effective stress on the pay formation is equal to overburden pressure minus reservoir pore pressure. As the pore pressure declines due to oil and gas production, stress on the pay rock increases and compaction occurs depending on formation compressibility. For large areal reservoir extensions, the overburden cannot be expected to provide support through arching and bridging of grains because of the distances involved regardless of the nature of the overburden rock. Subsidence takes place if the pay zone covers a large area, has great thickness and is compressible, and a large pressure drop is present in the pore space. Subsidence is also more likely to occur over a shallow reservoir than a deep one.

In overpressured and undersaturated chalks, depletion of reservoir pressure with field development causes large changes in net stresses on the formation which result in plastic deformation and compaction (RUDDY, ANDERSON, PATTILLO, BISHLAWI & FOGED 1988). The effects of rock compaction manifest themselves in two important ways being significant contribution to reservoir energy in form of lithic drive, and partial transfer of this compaction through the overburden leading to mudline subsidence. The lithic drive process resulting from rock compaction provides a significant part of the hydrocarbon recovery from overpressured and undersaturated chalks, with rock compressibility and its variation with pore pressure being one of the most important parameters in reservoir modelling. Aspects of reservoir compaction and field subsidence due to pore collapse in chalk formations are also discussed by SMITS, WAAL & KOOTEN (1988). BLANTON (1981) emphasizes the unusual ductile behaviour of the chalk which generally hinders or inhibits the fracturing process. The transition from positive to negative dilatancy with increasing confining pressure reflects a transition from microcracking to pore collapse.

4.5.4.1.6.2. Permian sandstone vs. Cretaceous chalk subsidence

While in sandstones pay formation compaction generally results in permeability reduction as pores are closed and the rock becomes more dense, in chalks occasionally constantly remaining hydrocarbon production rates (SULAK & DANIELSEN 1988) reflect that significant permeability diminution did not occur. On the other hand, the more stable framework of sandstones inhibits major pore collapse which leads to only minor subsidence rates (SCHOONBEEK 1976, BLEAKLEY 1986), whereas in chalks, the grain fabric is much more subjected to pore failure and closure that gives rise to much higher compaction which can be up to 10 - 20 times greater than for sandstones. Water flooding in oil and gas fields suffering from subsidence does not only aim on increasing ultimate recovery, but serves a dual role including maintaining reservoir pressure to mitigate compaction. Waterflooding is the first choice of pressure-maintenance system in oil fields, because gas injection into the reservoir means direct loss of revenue unless inert nitrogen is utilized instead of combustible methane.

The special mechanical behaviour of chalk reservoirs with field subsidence as a consequence of compaction by pore collapse is illustrated by a comparison between Cretaceous (Maastrichtian) chalk oil fields and Permian (Rotliegend) sandstone gas fields (SCHOONBEEK 1976). While subsidence equalization in the chalk oil fields Ekofisk and Valhall in the Norwegian North Sea requires jack-up operations up to 6 m vertical difference for compensation of an annual settling rate of abt. 8 - 16 in. (20 - 40 cm; BLEAKLEY 1986), Rotliegend sandstone subsidence as a result of framework condensation due to gas withdrawal in the Groningen field/Netherlands onshore is in the range of 10 - 25 cm for a production period of 20 - 40 years. This is the consequence of the much more stable framework of Rotliegend sandstones with respect to Cretaceous chalks (SCHOONBEEK 1976). Other hydrocarbon fields with significant reservoir subsidence of abt. 2 m (6 ft) or more are Wilmington and Inglewood in California/USA, Bachaquero and Lagunillas in Venezuela and one field in the USSR (BLEAKLEY 1986).

4.5.4.1.6.3. Subsidence rates in different chalk fields

The Ekofisk chalk field is subsiding at a rate of abt. 20 - 40 cm/year (8 - 16 in.), and from the annual rate, maximum total subsidence from the onset of production in 1971 until the year 2011 when the operating license expires has been calculated and considered as a base for compensation of compaction by the 1987 jack-up operation (cf. DIJK & KLÖTERS 1987, MANNESMANN REXROTH 1988; SMITH, SMITH & MONSHAUSEN 1988). While subsidence so far has ranged from abt. 8 in. via 12 in. on average to 16 in. per year, it is expected that as a consequence of massive waterflood pressure maintenance, future subsidence will be no more than half of that (PETROLEUM ENGINEER INTERNATIONAL 1987 a). The Valhall chalk oil field, however, is even slightly faster subsiding than Ekofisk at a rate of more than 40 cm/year, with present estimates being that the maximum sinkage of Valhall will be between 3 and 5 m over the next ten years (OILMAN 1987 a) instead of throughout 20 - 40 years in Ekofisk.

4.5.4.2. Combination of acid and proppant fracturing

Combination of acid and proppant fracturing is a suitable stimulation technique for chalk reservoirs, because it allows to profit from the advantages of both individual treatment possibilities. The sequence of combined acid and proppant fracturing includes gel injection to initiate fracture growth, acid insertion to etch the fracture walls, again gel pumping for reopening of an even wider fracture due to acid etching, and finally injection of high proppant concentrations into the crack (SNOW & HOUGH 1988; cf. also section 4.5.1.1.1.4.). Aspects requiring attention during joint acid and proppant fracturing are viscous fingering acidizing and wormhole growth, whisper fracturing, and fracture propagation control. The comparative effectivity of proppant and acid fracturing in chalk stimulation is also discussed.

4.5.4.2.1. Viscous fingering acidizing and wormhole growth

Enhancement of chalk stimulation can be achieved by a combination of multilayer propping of short large fractures and viscous fingering acidizing (HARTLEY & BOSMA 1985; DAVIES, BOSMA & VRIES 1987; KING & SCHER 1987; SMITH, MILLER & HAGA 1987), with the latter technique creating widely-spaced etched ridges and thus an irregular fracture wall (SHELL 1979). Viscous fingering acidizing (FREDRICKSON & BROADDUS 1976; DAVIES, BOSMA & VRIES 1987) is to establish fracture geometry with a viscous crosslinked non-acid pad followed by low-viscosity acid (PARKER, ADAMS & LIANKUI 1986). The low-viscosity acid fingers through the viscous pad and creates high-conductivity channels in the crack, in contrast to conventional acid fracturing where the entire crack face is etched and conductivity depends on the differential etching patterns created by the nature of the rock. Widely-spaced perforations can help to get good separation of the acid fingers or channels as they enter the fracture (PARKER, ADAMS & LIANKUI 1986).

PEETERS & HARTLEY (1984) report acid fingering across the crack face and filling of the resulting etched ridges with proppants, with in this technique some parts of the fracture face not being contacted by acid. Other aspects of viscous fingering are covered by DACCORD, NITTMAN & STANLEY (1986); LENORMAND (1986), SARMA (1986), FAYERS & NEWLEY (1987), HOMSY (1987), ALLEN & BOGER

(1988), EWING & RUSSELL (1988), GRADER & O`MEARA (1988), KOSSACK & OPDAL (1988), ODEH & COHEN (1988); EWING, RUSSELL & YOUNG (1989); KU, HIRSH & TAYLOR (1989) and MILLER, MOISSIS & WHEELER (1989). The discussion as follows focusses on whisper fracturing, and viscosity contrast and density difference required for fingering.

4.5.4.2.1.1. Whisper fracturing

Whisper fracturing is a pseudolimited-entry treatment specially designed for chalk reservoirs (HARTLEY & BOSMA 1985). In limited-entry layout, the perforations are restricted enough to create high friction pressures in the perforations which behave like backpressure valves ensuring that high pressures are applied across all perforations and that breakdown of the entire interval occurs (SNOW & HOUGH 1988; cf. section 4.2.2.2.). The high fluid velocities at each perforation during production result in high friction losses, reduction of well productivity, and potentially high perforation erosion and subsequent casing weakening. In addition, only small acid volumes could be injected into hydraulic fractures before they close up. In pseudolimited-entry designs, more perforations are used, the treatments are staged, and the phases are separated by diverters (cf. sections 4.2.2.1.).

In whisper fracturing, the perforations are spaced such that viscous fingering occurs and uneven ridges are etched by the acid. Another pattern is to perforate close to the crack boundary to increase fracture penetration by generating semi-radial instead of radial crack growth. Whisper fracturing treatments can utilize both acid and proppant. If the chalk is unstable, proppant can be injected following the acid. Large proppant volumes may be used to support the entire etched crack, or small proppant quantities could be injected to support fracture conductivity in only the near-wellbore region. For more competent chalks, a proppant job followed by an acid slug to establish high near-wellbore conductivity is also a suitable solution.

4.5.4.2.1.2. Viscosity contrast required for fingering

Promotion of viscous fingering has to be ensured by establishing a viscosity contrast of at least 200 - 300, with the high-viscosity gel also reducing leakoff and reaction rates by creating a wider fracture (SNOW & HOUGH 1988). At least some of the acid must have reduced leakoff to increase acid penetration especially in the high fluid-loss portions of the chalk, with reduced leakoff and increased penetration being achieved by acid gelling (cf. sections 4.3.4.3.4.1. and 4.5.1.2.2.). Aspects of fracture acidizing treatment design are also outlined by BROADDUS, KNOX & FREDRICKSON (1975) and ANDERSON & FREDRICKSON (1987), and mechanical modelling of wormhole growth in carbonate acid fracturing is performed by DACCORD, TOUBOUL & LENORMAND (1987) and HUNG, HILL & SEPEHRNOORI (1987).

The wormhole growth phenomenon (cf. section 4.5.1.6.3.) in matrix acidizing of carbonate formations is analogous to the viscous fingering effect in waterflood studies (ARAKTINGI & ORR 1988, VOGE & CRAFTON 1988). As the conductivity of acid-etched fractures is sensitive to reservoir depletion and any elevated to excessive pressure drawdown associated with production increases the effective stress beyond the stability threshold of the chalk, propping of the fractures is inevitable. While single-layer propped cracks suffer in performance seriously from embedment of grains (SIMON, COULTER, KING & HOLMAN 1982; HARTLEY & BOSMA 1985), multi-layer propping is the optimum solution. Another technique is closed fracture acidizing (FREDRICKSON 1987). In some areas, harder and more permeable chalks achieve economical production rates already with acid fracturing, whereas softer and less permeable chalks can only be feasibly exploited if stimulated by proppant fracturing (ANDERSEN, HANSEN & FJELDGAARD 1988).

4.5.4.2.1.3. Density difference required for fingering

For density differences between acid and pad fluid of less than 0.1 g/cm3, gravity has no effect on viscous fingering (SNOW & HOUGH 1988), whereas for density differences greater than 0.2 g/cm3, gravity over- or underrunning can dominate the flow pattern. In the range between 0.1 and 0.2 g/cm3 density difference, both fingering and gravity segregation occur, with these limits being slightly increased if the gel is crosslinked. Once fingering begins, it continues regardless of the viscosity ratio. Uneven etching develops in homogeneous chalks even without viscosity differences. Fingering is initiated by near-perforation effects and the uneven fracture face, and once initiated, uneven etching becomes a self-propagating process, because as acid etches, the fracture widens, thus causing the acid to continue to flow in the wider section of the crack.

4.5.4.2.2. Fracture propagation control

A critical factor in chalk reservoirs is the control of fracture propagation in vertical direction (EEKELEN 1982, McDONALD 1983, HARTLEY & BOSMA 1985). Fracture growth into neighbouring zones had repeatedly occurred in chalk formations, being a consequence of the special reservoir rock properties of the chalk. The fractures nearly always enter the layers adjoining to the pay zone, and penetration depth is determined by the differences in stiffness and horizontal in-situ stress between pay zone and neighbouring beds. Fracture growth is radial in chalk formations where no significant stress contrast exists between reservoir zones selected on the basis of permeability variations in an otherwise rather homogeneous formation. Although vertical zonation of the chalk complex based on reservoir quality (as chiefly reflected by permeability and framework stability) may not be important for fracture growth in the initial production stage, it may become significant with depletion once different pressure regimes are developed in the various superimposed pay column zones. Some aspects of short wide fractures and high proppant saturation as well as impact of stimulation boats on acid vs. proppant fracturing are discussed as follows.

4.5.4.2.2.1. Short wide fractures and high proppant saturation

Appropriate stimulation solutions according to field and laboratory evidence are short wide fractures that are induced hydraulically and are enlarged by creation of widely-spaced etched ridges by acidizing (SHELL 1979, HARTLEY & BOSMA 1985, KING & SCHER 1987). The resulting wide fracture is propped by a multilayer infilling. Another possibility is combination of hydraulic proppant fracturing and gravel packing with both conventional and resin-coated low-density intermediate-strength proppants where the aggregation of the resin films at grain contacts provides a stable wedge- and mantle-like package in the soft reservoir (TANSOE, KING & HOLMAN 1986; ATTARD, MATHES & MOWER 1988; cf. section 4.5.4.2.). The spectrum of treating techniques carried out in chalk reservoirs also includes whisper fracturing (cf. section 4.5.4.2.1.1.).

Chalk fracturing provides flow avenues from the collapsing matrix to the wellbore if proppant concentration is in excess of 2 lbs/ft2 in order to avoid embedment impairment (SIMON, COULTER, KING & HOLMAN 1982). Oil-based fluids are suggested for stimulating chalk intervals with high oil saturation, because they do not contribute to chalk reservoirs losing strength nearly as much as water-based fluids which provoke fluid phase and viscosity changes that ultimately could give rise to matrix framework collapse in a similar way as reservoir depletion leads to field subsidence.

4.5.4.2.2.2. Impact of stimulation boats on acid vs. proppant fracturing

Prior to invention of stimulation boats, less treatment versatility due to space constraints on platforms resulted in conventional acid stimulation of the chalk (SNOW & HOUGH 1988). Acid treatment has the dual effect of overcoming near-wellbore damage and of putting the natural cracks in direct fluid communication with the borehole. Proppants were originally not inserted, because it was expected that proppants would screen out or flow back rather than stimulate the natural fractures. In areas with poor natural jointing, hydraulic fracturing occurred during acidizing due to pressure parting (cf. section 4.5.1.1.1.1.; similar pressure parting takes occasionally place in gravel packing; cf. sections 5.6.4. and 5.6.6.).

4.5.4.2.3. Comparative effectivity of proppant and acid fracturing

Flow in the near perforation region is extremely turbulent and velocity profiles around the perforation make even acid flow and etching unlikely in this area (SNOW & HOUGH 1988). Proppant fracturing is likely to result in greater crack lengths than acid fracturing from productivity standpoint. Initial crack conductivities are comparable or moderately lower in proppant treatments than in acid jobs primarily due to the potential for a fines problem. Fines are generated both as proppant is flowed into the fracture, upon etching of the fracture face, and during production. Some aspects of longevity of created fractures as well as formation breakdown and closed fracture acidizing are illustrated as follows.

4.5.4.2.3.1. Longevity of created fracture

Proppant treatments, however, are not as effective in stimulating natural fractures as acid jobs and may actually clog them up or otherwise damage them. As the primary mechanisms contributing to the loss of conductivity are the same for both acid and proppant operations, the longevity of the created fracture is not greater for proppant jobs except in case of an ultra-high proppant concentration where higher fracture conductivities may be retained for a longer period of time because the fines problem will not be as severe. As the wells are deep, hot and highly deviated, and at least moderate proppant saturations are required, the possibility of proppant dropout in the wellbore is high. Once the proppant enters the fracture, the risk of screenout is high, because the crack may bend and turn, and since leakoff to natural fractures may be high.

Greater fracture lengths than can be attained with acid are only required in the low-permeability reservoir portions, whereas optimum fracture lengths in the moderate- to high-permeability low-natural crack areas are also obtainable with acids. Thus at the bottom of the line, acid fracturing may be superior in moderate- to high-permeability regions, whereas proppant fracturing is preferential in low-permeability areas of the chalk oil fields. Proppant stimulation should be carried out with moderate concentration and with possibly some acid ahead of the proppant lot to etch the natural fractures depending on chalk characteristics. Pseudolimited-entry acid fracturing (whisper fracturing; cf. section 4.5.4.2.1.1.) has turned out to be the most suitable stimulation technique for the chalk reservoirs in the Ekofisk field.

4.5.4.2.3.2. Formation breakdown and closed fracture acidizing

High-rate acid injection leads to formation breakdown in all parts of the field including areas with narrow spacing of natural cracks (SNOW & HOUGH

1988). While some of the breakdown is caused by natural joint opening, hydraulic fracturing due to pressure parting accounts for much of the breakdown in many regions (cf. sections 4.5.1.1.1.1. and 4.5.4.2.2.). Increased injection rates have increased acid penetration by creating higher pressures in the induced fractures thus keeping them open and widening them, and by limiting leakoff.

The spectrum of chalk acid stimulation also includes closed fracture acidizing comprising opening of the crack by the pad followed by permitting it to close and to stay partially open by auto-propping (cf. sections 4.5.1.1.1.2. and 4.5.1.5.), with then the acid being injected into the partially open crack below fracture pressure and etching much more selectively. There is, however, the risk that the acid may react quickly in the near-wellbore region and equalize the uneven ridges, thereby decreasing rather than increasing fracture conductivity or even contributing to casing failures. In order to avoid casing collapse by large cavities forming immediately around the borehole, it is necessary to maximize acid velocity in the near-wellbore region so as to place the acid far out into the formation and minimize excessive near-wellbore etching.

4.5.4.3. Combination of hydraulic fracturing and gravel packing

Special techniques including combinations of hydraulic fracturing and gravel packing or proppant and acid fracturing as well as various reservoir engineering peculiarities considerably enlarge the possibilities of hydraulic proppant fracturing of carbonate rocks. The most spectacular case of application of combined hydraulic fracturing and gravel packing is the Valhall field (ATTARD, MATHES & MOWER 1988) where the Tor Formation which contains abt. 2/3 of the oil-in-place has extremely high porosity (abt. 40 - 50 %) as a consequence of the overpressured nature of the reservoir. After discussion of some general aspects, the outline as follows concentrates on up-and-under stimulation vs. combined hydraulic proppant fracturing and gravel packing, design and purpose of chalk gravel packing, and wellbore cleanout effectivity.

4.5.4.3.1. General aspects

The inability of formation fluids to escape as the chalk was buried resulted in lack of compaction thereby preserving higher porosity than in other fields and stratigraphical levels (ATTARD, MATHES & MOWER 1988). The high porosity in combination with lacking cementation between the chalk grains gives rise to a very weak matrix which is prone to collapse and to plastic flow when subjected to stresses such as large pressure drawdowns from the pay zone into the wellbore (cf. section 4.5.4.1.3.). In contrast to the Tor Formation, the lower Hod Formation containing the remaining 1/3 of the oil-in-place is a much stronger and more typical limestone reservoir. Despite the high porosity, the permeability of the Tor Formation in the Valhall field is relatively low (abt. 1 - 10 md) due to the very small size of the calcite particles composing the rock. Natural fractures in crestal places can increase permeability up to 100 md.

The only possible completion technique for economical production of the Tor Formation in the Valhall field is hydraulic proppant fracturing followed by gravel packing, with both treatments being carried out with resin-coated alumina silicate proppants in most stages (only the hydraulic fracturing job is initiated with conventional alumina silicate proppants for cost containment purposes). As above certain confining stresses the Tor Formation fails by entering mass flow or plastic flow mode instead of migration of individual particles and because the presence of water results in more rapid failure of the chalk at lower net confining pressures, wide fractures with high proppant concentrations that are aggregated together by resin film bonding (cf. section 1.2.6.) are indispensible to provide stable flow capacity at high drawdowns, and an additional gravel pack mantle guarantees satisfactory and reliable consolidation of the chalk reservoir in the immediate vicinity of the wellbore.

4.5.4.3.2. Up-and-under stimulation vs. combined hydraulic proppant fracturing and gravel packing

Initially developments were fractured up into the Tor Formation through perforations in the Hod Formation, with the reasoning of this up-and-under technique having been that most of the pressure drop through the wellbore during production would be experienced by the more competent Hod Formation and the weaker Tor Formation would not fail and collapse (similar uphole fracturing is performed occasionally in coal-bearing strata, with an underlying sandstone guiding the crack intersecting the coal seam; cf. section 4.4.3.8.1.). High-rate wells, however, suffered subsequently from massive flow of proppants and chalk when drawn down below relatively high wellhead flowing pressures, and newly drilled wells became more and more difficult to complete with Hod to Tor fracturing as the reservoir pressure differential between Tor and Hod formations increased owing to faster depletion of the more productive Tor Formation (ATTARD, MATHES & MOWER 1988).

Therefore the combination of fracturing and gravel packing was introduced including a two-stage gravel pack consisting of fracture/prepack (cf. section 5.6.7.) followed by an annulus pack with screen and liner (cf. section 5.10.1.). High shot density and large average perforation diameter (cf. section 5.6.1.) were selected to provide a large flow area for production and thus minimize the pressure drop across the gravel pack. The hydraulic proppant fracturing treatment is performed to create a wide high-conductivity crack through the Tor Formation with as many of the perforations as possible packed with proppants thereby also acting as a prepack to precede the actual gravel pack job. The desired fracture constellation is achieved by tip screenout operation (cf. section 4.5.4.4.1.). Aggregation of the proppant wedge by resin bonding (cf. section 1.2.6.) had the special aim to minimize proppant flowback during well cleanup and production (cf. section 4.12.3.3.).

4.5.4.3.3. Design and purpose of chalk gravel packing

The purpose of chalk gravel packing was to prevent mass flow of chalk into the wellbore through perforations left unfilled after fracturing and to restrain the fracture/prepack proppant plugs (ATTARD, MATHES & MOWER 1988). 20/40 mesh gravel grain size was chosen because testing had revealed that this size effectively restrains the fractured chalk while retaining substantial hydrocarbon flow capacity. Gravel grain size design following the standard rules (cf. section 5.2.3.) would result in a prohibitively small gravel grain size for ensuring stoppage of the tiny chalk particles, which would be simply a waste of effect because plastic chalk flow occurs as a cohesive mass rather than as single grains.

As a consequence of reservoir depth, borehole deviation and necessity to minimize brine loss, a one-trip circulation slurry gravel pack was selected as appropriate technique, and a total amount of 3,000 - 4,000 lbs (1,300 - 1,800 kg) of alumina silicate gravel was pumped with a maximum gravel concentration of 7.5 lbs/gal (900 kg/m3) in order to avoid gravel bridging in the small clearance between liner and gravel pack assembly (cf. section 5.2.3.4.). According to the high reservoir pressure and the water sensitivity of the chalk, the gravel pack had to be carried out under pressure using a snubbing unit and with diesel as the work fluid. Opening of as much perforations as possible prior to fracturing and gravel packing was achieved by breakdown with acid or gelled diesel and ball sealers for diversion (cf. section 4.2.2.1.1.).

4.5.4.3.4. Wellbore cleanout effectivity

The light alumina silicate proppants (cf. section 1.3.2.1.) were chosen instead of sand to increase fracture conductivity and instead of other heavier synthetic proppants to achieve an effective borehole cleanout prior to running the gravel pack equipment (ATTARD, MATHES & MOWER 1988). Wellbore cleanout with proppant removal can best be performed with more viscous fracturing gel instead of thin diesel. Increasing fluid velocity and improvement of borehole cleanout efficiency was reached by reverse circulation during gravel packing (cf. section 5.3.2.3.). Reperforations and gravel packing under pressure in an already completed interval also resulted in satisfactory productivities. As a consequence of the successful performance of these combined treatments, the wells are producing at the expected flowrates with no signs of borehole damage and at much lower flowing wellhead pressure than without stimulation, resulting in correspondingly higher withdrawal rates.

4.5.4.4. Special chalk fracturing techniques

As the chalk is a peculiar reservoir rock with special properties, also hydraulic fracturing stimulation has to apply unusual methods in order to successfully treat this type of pay formation. Some special chalk stimulation techniques such as tip screenout fracturing, interface fracturing, and dendritic fracturing are outlined as follows.

4.5.4.4.1. Tip screenout fracturing

Tip screenout fracturing is a special technique for creating short wide high-conductivity fractures (SMITH, MILLER & HAGA 1987). The design of this method is based on intentionally screening out the tip of the fracture with sand and then continuing to pump slurry in order to increase fracture width and to fill the crack with proppant multilayers to obtain the necessary high conductivity (cf. section 4.3.3.1.). The fluid-leakoff coefficient of the formation has to be carefully determined by a minifracturing calibration job (cf. section 4.8.2.) before the main treatment, because this special technique involves severe risk of premature screenout (cf. section 6.2.4.2.1.) and failure to achieve the planned tip screenout will not yield the desired amelioration of production rate (cf. also section 4.8.5.3.).

While tip screenout is characterized by plugging of the fracture front with proppants or solid fluid-loss additives, rapid fluid leakoff of proppant-laden stages and corresponding increasing proppant concentrations in the slurry within the fracture give rise to dehydration screenout (WARPINSKI, BRANAGAN, SATTLER, CIPOLLA, LORENZ & THORNE 1988; cf. sections 4.2.3.5.3. and 6.2.4.2.1.). Tip screenout fracturing is especially suitable for the chalk in the Valhall field which is softer, more water-sensitive, and more likely to undergo plastic deformation than the chalk in the Ekofisk field (SMITH, MILLER & HAGA 1987; ATTARD, MATHES & MOWER 1988). In order to obtain high proppant saturations and to help restrain the chalk from flowing in on the propped fracture, an engineered screenout is provoked, resulting in higher pressures within the fracture which cause the crack to widen. As the screenout continues, the wider fracture is packed full of proppant, thereby giving rise to a large well-packed crack (SNOW & HOUGH 1988).

4.5.4.4.2. Interface fracturing

Interface fracturing is a special technique of creating hydraulic cracks near or at the interface of two formations with different leakoff characteristics and mechanical properties for the purpose of stabilization of North Sea chalk completions (MOSCHOVIDIS 1987). The principal oil-bearing zone which is poorly consolidated and creates solids production problems is brought in communication with the underlying more competent chalk zone that is also oil-bearing via a hydraulic fracture in order to minimize inflow of formation fines (cf. also sections 4.5.4.3. and 4.4.3.8.1.). Good treatment results with high-conductivity fractures are achieved by a screenout towards the end of the operation.

In this combined hydraulic fracturing and gravel packing application, the volume curves intersect towards the end of the treatment, but a considerable sand volume has been pumped above the simulated crack capacity, with a possibility of explanation of this relationship being that the simulations do not account for any ballooning of the fracture due to higher treating pressure caused by total or partial crack propagation arrest. Higher operating pressures than normal are observed towards the end of the fracturing jobs and are intended to increase crack width. The intersection points of gravel pack and fracture volume curves determine the slurry volume at which a global screenout occurs, with corrections having to be made for the porosity of the proppant package when dehydrated at screenout conditions, the accessibility of the fracture volume to the proppants, and increasing crack width due to plastic deformation and non-linearity of reservoir behaviour.

4.5.4.4.3. Dendritic fracturing

Another special operational procedure representing a combination of natural and hydraulic fracturing is dendritic fracturing where the treatment is confined to already existing natural cracks by using large volumes of lightly gelled water pumped at high rates and velocities, thereby avoiding creation of new fractures as more viscous fluids would (LOPUS, SEIFERT & SCHEIN 1987; cf. section 4.8.8.4.). Various stages of the treatment incorporate small amounts of proppants in order to aid in abrasion of existing fractures plugged by secondary mineral precipitations as well as fluid-loss control, or bridging agent to cause slight changes in differential pressure within a fracture to assist in diverting into other fractures (cf. section 6.2.1.3.4.5.). After briefly outlining some general aspects, the special constellation of combined dendritic natural and hydraulic fracturing is illustrated. Some refracturing aspects are also sketched.

4.5.4.4.3.1. General aspects

Dendritic fracturing is carried out by injection of a thin water-based fluid at high rates into the formation (KIEL 1976, 1977, 1984), and pumping is then stopped and the well is allowed to flow back, with this injection/flowback cycle being repeated several times. The process causes in-situ material to break off from the fracture face and lodge inside of the crack thereby propping it open (MURPHY & FEHLER 1986). Dendritic fracturing thus includes auto-propping of the crack walls by formation fragments and broken infilling of healed joints (cf. sections 4.5.1.1.1. and 4.5.1.5.1.). Dendritic fracturing (cf. also section 4.8.8.4.) has been successfully carried out in the Austin Chalk (Texas/ USA) and in the Ekofisk chalk (Norwegian North Sea) as well as in various other naturally fractured carbonate reservoirs.

4.5.4.4.3.2. Combination of dendritic natural and hydraulic fracturing

Dendritic hydraulic fracturing is particularly suitable for chalks which are characterized by an already dendritic natural fracture system as a consequence of tectonical forces such as subsidence, uplift and intrusion having been applied at different times both regionally and locally on an altered stress field (HUBBARD & PIERSON 1986; cf. section 4.8.8.1.). In some chalks, the dendritic natural fracture system is the hydrocarbon reservoir with little or no matrix porosity contribution (cf. section 4.5.4.1.4.). As the natural chalk fractures are often only partially conductive and can be plugged beyond repair by drilling mud and cement, and can also be inherently filled with calcite and salt, dendritic hydraulic fracturing is a suitable stimulation technique. Dendritic hydraulic fracturing utilizes large volumes of lightly gelled fresh water pumped down the casing at high velocity (cf. also section 4.8.8.4.). The thin fluid with its high leakoff rate seeks fine hairline cracks rather than propagating and widening major fractures as more viscous fluids would do. Small quantities of proppants are used as a bank-bridging medium.

Since the thin fluid is not a perfect transport system, the proppants are banked in the fractures representing higher-velocity streams until velocity is no longer adequate to transport the proppants. At this point, the proppants act to create a higher pressure differential along exposed crack faces that subsequently permits fluid and proppants to be carried into other fractures as they open. The process is continued until further diversion within the formation is no longer possible with proppants alone. At this stage, temporary diverting agents such as rock salt are introduced (cf. section 4.2.2.1.1.). It is not unusual that more than one temporary plug is required for diversion, and several may be utilized throughout the treatment (cf. section 6.2.1.3.4.5.). Since the purpose of temporary plugs is to achieve diversion away from the wellbore, relatively small-volume plugs are utilized, and the procedure is repeated in stages until changes in bottomhole pressure indicate excessive fracture growth.

4.5.4.4.3.3. Refracturing aspects

Refracturing of a conventionally stimulated well requires only slight modification of the procedure. Standard treatments carry large proppant quantities and attempt to create and pack fractures. Thus an efficient filler system is provided for deposition of eroded fines as well as accumulation and redeposition of salt and calcite. Hydrochloric acid is utilized ahead of each fracturing stage to remove calcite precipitation from the earlier proppant packages, thus allowing the original proppant infilling of the crack to get unconsolidated by destruction and dissolution of the secondary calcite cement and to be transported away from the wellbore which requires somewhat larger high-velocity pads (cf. section 4.8.9.).

4.5.4.4.4. Other aspects of chalk fracturing

Natural fractures in the chalk of the Ekofisk fields are short, occur most frequently in the reworked intervals, and are non-continuous from top to bottom of the sections (BREWSTER, DANGERFIELD & FARRELL 1986; FARRELL 1988, SNOW & HOUGH 1988). Because of their discontinuous nature, hydraulic fracture penetration is often benefiting production beyond anticipation by encountering natural cracks. Fluid leakoff coefficients can be extremely high because of intense natural jointing, large capillary forces, and quick acid-chalk reaction rates. Induced fracture growth is generally vertical and radial, with local changes in stresses not being large enough to confine fracture height, although the cracks tend to grow commonly upwards due to the in-situ stress gradient. Deviations from this ideal fracture geometry occur frequently due to several reasons. As the wells are highly inclined up to 60 degrees in order to access numerous locations by extended-reach drilling (cf. section 4.8.6.1.3.) from the platforms, at least in the region near the wellbore fracture growth will not be ideal, but will either result in multiple fractures or in a continuous or en-echelon bending of the crack to vertical position (STRUBHAR, FITCH & GLENN 1975; SNOW & HOUGH 1988; cf. section 4.11.1.).

4.5.4.5. Water imbibition and oil recovery

Water imbibition plays an important role in oil recovery from chalk formations both through water flooding with injection in one well and production from another through matrix and natural cracks, and water fracturing by pumping water down with a pressure exceeding the reservoir fracture gradient and taking off the fluid from the same well. In a water-wet reservoir such as the chalk, imbibition is the process by which oil representing the non-wetting phase is displaced by water being the wetting phase (OEN, ENGELL-JENSEN & BARENDREGT 1988). Transverse imbibition occurs in fractured reservoirs (double-porosity rocks; BECKNER, ISHIMOTO, YAMAGUCHI, FIROOZABADI & AZIZ 1987; BECKNER, FIROOZABADI & AZIZ 1988). BECH, JENSEN & NIELSEN (1989) comment on gravity-imbibition and gravity-drainage processes. WENDEL, KUNKEL & SWANSON (1988) report spontaneous water imbibition in diatomite reservoirs which are rock-mechanically quite comparable to chalks. Diatomites and diatomaceous earths can be successfully stimulated with high proppant concentrations up to 20 lbs/gal (STRUBHAR, MEDLIN, NABI & ANDREANI 1984; STRICKLAND 1985). The outline as follows focusses on water flooding and water fracturing. Some comments are also offered on foam fracturing.

4.5.4.5.1. Water flooding

Concerning secondary oil recovery, naturally fractured chalk reservoirs will produce under a displacement process during waterflooding, in contrast to sandstones which rely on a piston-like action of the water (OFFSHORE ENGINEER 1985 b). In the very low-permeability chalk, the primary movement of fluids is through the natural cracks rather than the actual rock, allowing oil to be displaced from the rock as the water is preferentially absorbed. Apart from this water imbibition (cf. MATTAX & KYTE 1962, AGUILERA 1975, MENOUAR & KNAPP 1980, CHIERICI 1984, TORSAETER 1984, BALDWIN 1986, WU & PRUESS 1986; THOMAS, DIXON, EVANS & VIENOT 1987; OEN, ENGELL-JENSEN & BARENDREGT 1988), the injected water can move too fast through the fracture system thus bypassing much of the formation and quickly breaking through to the wellbores. These effects also have to be taken under consideration when designing hydraulic proppant and acid fracturing stimulation in the chalk.

In the North Sea, the chalk in the Ekofisk field is more stable and oil recovery can be enhanced by waterflooding through the natural crack system without necessarily involving hydraulic fracturing and gravel packing, whereas the chalk in the Valhall field can only be brought to economically feasible production by the special technique of combined hydraulic proppant fracturing and gravel packing with both conventional and resin-coated low-density intermediate-strength alumina silicate proppants (cf. section 4.5.4.3.).

Water imbibition displacing and expulsing oil as the dominant production mechanism in fractured chalk reservoirs that are subjected to secondary water flooding is largely depending on reservoir wettability being the magnitude of fluid-surface interaction (BALDWIN 1986). Water imbibition can also significantly be affected by the removal of small amounts of tightly bound organic coatings on the chalk surfaces. The higher the level of oxidation of the most tightly bound material, the greater is water imbibition. EHRLICH (1970) discusses the effect of temperature on water-oil imbibition relative permeability. In both chalk and diatomite reservoirs, waterflooding is essential for both oil recovery improvement and reservoir pressure maintenance which is halting oilfield subsidence (WENDEL, KUNKEL & SWANSON 1988; cf. section 4.5.4.1.6.). Other aspects of water imbibition in chalk reservoirs are discussed by SYLTE, HALLENBECK & THOMAS (1988).

4.5.4.5.2. Water fracturing

Water imbibition together with creation of a minifracture is also the reason why small-scale water fracturing in depleted chalk oil fields by pumping water downhole with a pressure exceeding formation tensile stress and producing it back from the same borehole together with incremental oil is a comparatively simple, cheap and effective stimulation technique (JOURNAL PETROLEUM TECHNOLOGY 1987). Such small-scale water fracturing can be principally compared with steam-soak huff and puff treatments and basically cleans up the wellbore and its immediate vicinity, dissolves and removes particles that are plugging the formation, breaks through near-borehole damage haloes and displaces the oil. Injection of water at supercritical pressures gives rise to opening of either singular or multiple fractures including at least partial enlargement of pre-existing natural cracks, and penetration of the water into both the artificial fracture and natural crack systems. The enlargement of the reservoir surface accessible to the injected water is the reason for an ameliorated drainage area for oil, with therefore increased amounts of oil flowing to the well once pumping of water is stopped and back-production of the injected fluid in combination with oil offtake starts.

In contrast to mainly small-scale water fracturing in carbonates and particularly chalk, large-scale non-proppant water fracturing is frequently carried out in geothermal reservoirs where often various fracturing fluids are no longer stable due to high reservoir temperature (cf. section 4.7.3.5.). Water fracturing is also of considerable significance in coal seams (cf. section 4.4.3.4.1.).

4.5.4.5.3. Foam fracturing

Apart from the North Sea and adjoining onshore areas in Europe, large belts of Upper Cretaceous chalk occur in the Rocky Mountain area in Colorado, Nebraska and Kansas in USA. Although oil and gas have been produced from chalk reservoirs in the Rocky Mountains/USA since the early 1900`s (SCHOLLE 1977), only in the 1970`s an upsurge of exploration and development of chalk gas in the Upper Cretaceous Niobrara and Greenhorn formations took place as a consequence of the successful acquisition and exploitation of hydrocarbon reserves in Upper Cretaceous and lower Tertiary chalks in the North Sea in Europe (POLLASTRO & SCHOLLE 1986). Other factors promoting development of the chalk gas boom were increasing gas prices, advanced exploration techniques for low-permeability reservoirs, and major improvements in recovery technology including above all hydraulic proppant fracturing.

Foam fracturing has proven to be the most effective stimulation technique in Niobrara Formation chalks (ROHRET & JONES 1978, LOCKRIDGE & SCHOLLE 1978) because of minimizing fluid volume and significantly reducing formation damage by allowing rapid clean-up after the treatment. Foam fracturing fluids also carry high proppant concentrations and prevent large pressure drops due to friction (cf. section 3.11.2.). The advantage of alcohol in foamed methanol-water systems is rapid and more complete cleanup due to high vapour pressure, high solubility in produced gas, high miscibility in formation waters, and low surface tension properties (ROHRET & JONES 1978; cf. section 3.11.4.). The evolution of fracturing stimulation techniques is summarized by BROWN, CRAFTON & GOLSON (1982). Oil-based foam fracturing in chalks is discussed by BOWEN & DRISCOLL (1980). Aspects of chalk fracturing by different methods and in various formations are also discussed by WEBSTER (1977) and LUPPENS (1980).

4.5.4.6. Horizontal well drilling

Comparative evaluations of horizontal wells vs. vertical fractures (cf. also section 4.8.6.2.1.) in the North Sea chalk have shown that a non-stimulated horizontal well could yield a productivity equal to a successfully propped vertically fractured deviated well, whereas installation cost of the latter borehole is higher (ANDERSEN, HANSEN & FJELDGAARD 1988). While the productivity index improvement for a matrix-acidized horizontal well compared with an optimally fractured vertical well is only marginal, a matrix-acidized horizontal well could be beneficial in multiphase fluid reservoirs because breakthrough of gas and/or water is delayed as a result of drawdown reduction.

4.5.5. Sandstone acidizing

In contrast to carbonates which need proppant fracturing rather than acid fracturing in order to achieve larger production improvements, some sandstones can in addition to proppant fracturing arrive at still ameliorated output by fracturing or matrix acidizing with hydrofluoric acid or HCl/HF-mixtures (ROSS, PIERSON & COULTER 1963; SMITH, ROSS & HENDRICKSON 1965; GATEWOOD, HALL, ROBERTS & LASATER 1970; WILLIAMS & WHITELEY 1971, HOLCOMB 1975; McCUNE, FOGLER, LUND, CUNNINGHAM & AULT 1975). Aspects of sandstone acidizing are also discussed by FARLEY, MILLER & SCHOETTLE (1970); GIDLEY (1970, 1975), SUTTON & LASATER (1972), BURKHILL & PIERRE (1975), HALL (1975); McCUNE, AULT & DUNLAP (1975); WILLIAMS (1975); GIDLEY, RYAN & MAYHILL (1976); LUND, FOGLER & McCUNE (1976); ASELTINE & ELLENBERGER (1977); HILL, LINDSAY, SCHECHTER & SILBERBERG (1977); THOMAS & CROWE (1978 b), PACCALONI (1979); WILLIAMS, GIDLEY & SCHECHTER (1979); SHAUGHNESSY & KUNZE (1981), CROWE & MINOR (1982), SCHECHTER (1982); McLEOD, LEDLOW & TILL (1983); GIDLEY (1985), BURMAN & HALL (1986); SMITH, CLARK, HENDRICKSON & EGBOGAH (1986); TAHA, HILL & SEPEHRNOORI (1986, 1987) and BRANNON, NETTERS & GRIMMER (1987). REN & XIONG (1986) report factors influencing effective acid penetration in fracture systems. Some aspects of acid reaction, acidizing targets, acid penetration distance, and acid types and dissolution possibilities are briefly illustrated as follows.

4.5.5.1. Acid reaction

The commercial development of acidizing as a stimulation technique has become widespread already more than 60 years ago when hydrochloric acid treating of carbonates was introduced (LABRID 1975). Sandstone acidizing was invented over 30 years ago (SMITH & HENDRICKSON 1965) and is based on the unique quality of hydrofluoric acid to attack silica and alumino-silicates. Clay and feldspar dissolution is much faster than that of silica probably because of the larger surface area, and takes place in the first stage by uniform alteration of the crystal lattice and then in the second phase by progressive extraction of aluminum in the form of fluorinated complexes. The main reaction product of silica solubilization is fluorosilic acid accompanied by a small amount of colloidal silica (for thermodynamic and kinetic aspects of acid reactions with various minerals cf. LUND, FOGLER, McCUNE & AULT 1973; LABRID 1975). Chemical precipitation mechanisms are important factors of formation damage during stimulation with hydrofluoric acid and require careful evaluation (WALSH, LAKE & SCHECHTER 1982). Afterflushing and reduction of hydrofluoric acid concentration can be used to combat silica precipitation in sandstone matrix acidizing (SHAUGHNESSY & KUNZE 1980). Some aspects of partial matrix disintegration as well as separation of connate water and acid by preflush are discussed as follows.

4.5.5.1.1. Partial matrix disintegration

The most important element of acid reaction with sandstone is the partial disintegration of the matrix and the release of formation fines in the pore space (GIDLEY 1970). Dissolution of matrix cement and loosening of fine particles, however, also includes the danger of downstream flow of mobilized solid fragments which later tend to plug pore channels. Effective acid treatment of sandstone formations should provide a substantial increase in permeability, no adverse changes in formation wettability, rapid cleanup following the job and minimum operational cost. The most efficient treatment is composed of three stages comprising preflush, acid and afterflush and is conducted at rates that will not open fractures in the formation which did not already exist before acid insertion (GIDLEY, RYAN & MAYHILL 1976 found that the poorest treatment response occurred when acid was injected at pressures exceeding fracturing pressure).

4.5.5.1.2. Separation of connate water and acid by preflush

The preflush consisting of hydrochloric acid should provide a separation between connate water and hydrofluoric acid in order to prevent formation of sodium fluorosilicate, react with the carbonates present in the reservoir rock to avoid dissipation of the more expensive hydrofluoric acid, and maintain a low pH to inhibit deposition of calcium fluoride as product of hydrofluoric acid reacting with carbonates. Concerning amount of acid, caution has to be exercised, as excessive quantities of acid ultimately dissolve matrix cements and destroy sandstone consolidation, resulting in recompaction by overburden stress to a new and lower porosity and a greatly reduced permeability. The better the lithification of the reservoir rock and the more mature its stage in burial diagenesis, the larger the acid volume required for effective sandstone stimulation.

4.5.5.2. Acidizing targets

Concerning acid flow through sandstone matrix, damaged formations are partially plugged by clay and hence have lower porosity and higher concentration of dissoluble minerals (McCUNE, FOGLER, LUND, CUNNINGHAM & AULT 1975). Although the total concentration of dissoluble particles is increased, much of them might be bypassed because of possible channelling, giving rise to a net decrease of minerals accessible to acid, with the overall net effect probably being a decrease in acid capacity number and an increase in required acid volume or concentration for reservoir damage removal. Suitable targets for acid treatment are particularly mineralogically immature and submature sandstones which contain comparatively low amounts of quartz, but have higher shares of detrital feldspars and unstable rock fragments, as well as clastic and diagenetic clay minerals and authigenic carbonate cement. Mature quartz sandstones can only be successfully stimulated by acidizing if containing reasonable quantities of carbonate cement. The following discussion comments on clay and iron-bearing minerals as well as on carbonate minerals.

4.5.5.2.1. Clay and iron-bearing minerals

Larger amounts of clay and iron minerals are usually undesirable characteristics in sandstone acidizing candidates, because such rock compositions are often susceptible to chemical (clay damage) or mechanical (particulate plugging caused by entrained solids in the completion fluid) impairment (BRANNON, NETTERS & GRIMMER 1987) and also subsequent iron precipitation problems (cf. section 4.5.1.1.4.). On the other hand, however, dirty sandstones with excessive contents of clay minerals are hardly stimulation candidates at all because of too poor reservoir properties that in some cases even cannot be sufficiently improved for achievement of economical hydrocarbon production rates by neither hydraulic fracturing nor acidizing. Caution has to be exercised in acidizing deep hot sandstone formations with hydrochloric acid, because deterioration of productivity might occur (BARBY & BARBEE 1987). Field studies have shown that in some reservoirs, high injection rates of water can have a better formation breakdown effect than acid treatments. Acid can cause migration of fines and iron colloids which may plug the pore throats of the reservoir.

4.5.5.2.2. Carbonate minerals

High contents of carbonate cement in quartz sandstones give rise to high acid solubility of the rock. If the cement share is exceeding a certain boundary percentage of total rock volume, carbonate dissolution can result in severe loss of formation strength following acid treatment and thus formation matrix collapse. Therefore in many sandstone reservoirs, hydraulic proppant fracturing is the superior stimulation method and gives in most cases better drainage effects with leaving the matrix fabric intact in comparison to acid treatments which not only on the one hand create pathways for fluid flow, but also on the other hand weaken formation cohesion up to supercritical levels where grain fabric collapse deteriorates the effect of the stimulation operation (PEARCE 1979). A suitable combination of hydraulic proppant fracturing and acidizing would certainly be the optimum solution for improval of both matrix and fracture drainage, but it requires considerable experience to get a satisfactory feeling how much of the matrix can be dissolved without risking framework collapse that would screw up most of the success of the stimulation job.

4.5.5.3. Acid penetration distance

Acid penetration distance in fractures is the way the acid travels before spending to some predetermined degree (ROBERTS & GUIN 1974) which is in most cases abt. 10 % of its original concentration (NOVOTNY 1976). Acid penetration distance is essential for estimating the production improvement obtainable by fracture acidizing. While acid penetration distances in carbonates are limited by high spending ratios due to reaction of the acid with most, if not all, of the reservoir matrix, acid can migrate extremely far in sandstones where commonly only subordinate shares of the framework react with the acid and spending is thus considerably slower in sandstones with respect to carbonates. Some aspects of type of formation damage, acid concentration and reaction rate, low- and high-pH acids, and water blocking are briefly outlined as follows.

4.5.5.3.1. Type of formation damage

Comparing the efficiency of acidizing treatments of undamaged, natural-damaged and mud-damaged sandstone formations, production increases are most significant for interstitial acidizing of damaged formations (GATEWOOD, HALL, ROBERTS & LASATER 1970). In reservoirs with drilling-mud damage resulting from clay particle invasion, volumes of acid sufficient to remove only the clay contained within the narrow damage mantle around the wellbore yield the most economical production increases if no natural damage has occurred as a consequence of contacting of water-sensitive clays by mud filtrate. When treating formations with natural clay damage, the production increase is directly dependent upon the distance which live hydrofluoric acid can be pumped into the reservoir. Penetration depth of live hydrofluoric acid is a function of clay concentration, formation temperature, initial hydrofluoric acid concentration, rate of hydrofluoric acid reaction and pumping rate. Increasing formation temperature and clay concentration both decrease the penetration radius of live or unspent acid.

4.5.5.3.2. Acid concentration and reaction rate

Greater penetration depth of live acid leading to higher production increase is obtained by rising the initial hydrofluoric acid concentration, retardation of the reaction rate of hydrofluoric acid on silicates (cf. section 4.5.5.4.3.), and increasing pumping rate. Enhancement of stimulation results can also be achieved by acid heating (WALKER, FREDRICKSON & NORMAN 1984; cf. section 4.5.1.2.4.). While an improvement of stimulation by increasing acid volumes is desirable and limitation of quantities is usually given by economical factors of job feasibility, acid concentration has to be carefully selected according to formation characteristics, because the higher reaction rate of more concentrated acid also gives rise to increased swelling, detachment and migration of sensitive clay minerals and thus deterioration of permeability by plugging of pore throats (GIDLEY 1971). Buffer regulation of hydrofluoric acid enables extension of thermal stability and application for sandstone acidizing up to 550 oF (280 oC; SCHEUERMAN 1988).

4.5.5.3.3. Low- and high-pH acids

In terms of performance, distinction has also to be made between low-pH (below 4) and high-pH (4 - 6) acid stimulation systems (ABRAMS, LYBARGER, RICHARDSON & NEASHAM 1978; TEMPLETON, RICHARDSON, SCHEUERMAN & ABRAMS 1979; ABRAMS, SCHEUERMAN, TEMPLETON & RICHARDSON 1983). BRISCOE (1978), SCHERUBEL & CROWE (1978) and FORD (1981 a, 1981 b) introduce stabilized foamed acid which provides greater fracture penetration along with increased fracture conductivity throughout fracture length (cf. section 4.5.1.4.). Another improvement is a sequential hydrofluoric acid system which has a retarding effect (HALL & ANDERSON 1977; HALL, TINNEMEYER & UNDERWOOD 1981; cf. section 4.5.5.4.3.). LEE (1985 b) discusses geometry determination for multi-stage acidizing treatments with or without viscous preflush, and CLARK, WONG & MUNGAN (1982) introduce new acid systems for sandstone stimulation.

4.5.5.3.4. Water blocking

Sandstone matrix acidizing often suffers from very slow clean-up of spent acid as a result of water blocking in the critical matrix surrounding the wellbore (McLEOD & COULTER 1966; McLEOD, McGINTY & SMITH 1966; cf. section 3.11.2.4.2.). Addition of alcohol to the acid can often prevent the water-blocking problem thereby speeding up spent acid clean-up. Although water blocking can occur in clean well-consolidated sandstone, it is more likely to be present together with other forms of drainage such as particle blocks or rock wettability changes. Water blocking is most severe in heterogeneous sandstone formations particularly when high- and low-permeability layers alternate such as in many cross-bedded sediments (cf. plate II/5 - 6). Alcohol additives are successful in water-block removal in both matrix acidizing and hydraulic fracturing.

The most critical factors for successful sandstone matrix acidizing are mud acid volume and treating pressure (GIDLEY, RYAN & MAYHILL 1976). The poorest job response occurs when acid is injected at pressures exceeding formation fracture gradient, because the acid is then only channelling along the crack and does not adequately flow through the sandstone matrix.

4.5.5.4. Acid types and dissolution possibilities

Selection of acid types for treatment of both carbonate and sandstone formations depends on the composition of the contaminants which are plugging reservoir permeability in case of formation damage removal necessity (McLEOD 1984). Following discussion of some general aspects, an outline is given on clay mineral type vs. acidizing treatment as well as on delayed acid reaction and in-situ acid generation. Some comments are also given on selective acid stimulation of injection wells.

4.5.5.4.1. General aspects

Hydrochloric acid does not dissolve pipe dope, paraffin or asphaltenes, with treatment of these solids or plugging agents requiring an effective organic solvent. Acetic acid effectively dissolves calcium carbonate scale, but does not dissolve iron oxide scale. Calcium sulphate can be converted to calcium carbonate or calcium hydroxide by treatment with potassium hydroxide or sodium carbonate, and hydrochloric acid can then be used to dissolve the converted scale. Calcium sulphate can also be dissolved in one step with the sodium salt of ethylene diamine tetraacetic acid. Hydrochloric acid does not dissolve formation clay minerals or drilling mud, with hydrofluoric acid having to be used to dissolve these aluminosilicates in rock pores around the wellbore in sandstones, whereas this type of formation damage can be repaired in carbonates also by hydrochloric acid which etches parts of the carbonate rock matrix instead of the damaging materials that are left in place, thereby also increasing permeability.

Mud acid contains hydrochloric and hydrofluoric acids together with additives for proper silicate control, demulsification, and surface tension characteristics which allow the acid to restore and increase permeability near the wellbore. Mud acid can be used for matrix acidizing of sandstone reservoirs, spearheading of hydraulic proppant fracturing treatments and squeeze cementing, loosening stuck drill pipe, and cleaning gravel-packing liners and screens. The primary use of hydrofluoric acid together with intermixed hydrochloric acid is in sandstone matrix treatments to remove clay-particle damage with permeability restoration or to increase original permeability of clay-bearing sandstones. High concentrations of hydrochloric acid in the presence of hydrofluoric acid may improve hydrofluoric acid efficiency (LUND, FOGLER & McCUNE 1976) and speed up hydrofluoric acid reaction (McCUNE, FOGLER, LUND, CUNNINGHAM & AULT 1975).

The use of organic acids such as acetic, formic and propionic acids as stimulation and perforation fluids in carbonate reservoirs has increased in the last years, because they have inherently retarded reaction rates and are much easier to inhibit against corrosion at high temperatures (300 - 400 oF) than hydrochloric acid (DOWELL SCHLUMBERGER 1988 d), whereas in sandstones, their utilization does only occasionally provide technical and economical advantages with respect to mud acid. FRENIER (1989) presents organic chemicals for inhibition of acidizing fluids for high-temperature reservoir stimulation.

4.5.5.4.2. Clay mineral type vs. acidizing treatment

Concerning different clay minerals, caution has to be exercised to select the proper treatment version for different clay mineral types (cf. section 3.5.1.). Kaolinite forms stacks of hexagonal plates which are easily dispersed and can migrate through rock pores bridging in smaller pore throats, and therefore has to be controlled by clay stabilizers in the final overflush after a hydrofluoric acid job. Illite is a fibrous clay mineral that can be broken off or moved by high flow velocities or surges close to the wellbore (cf. section 3.5.2.1.), and thus low acid injection rates and controlled production rates after the treatment are essential. Chlorite is usually iron-rich and partially soluble in hydrochloric acid, with iron being extracted leaving a silica residue behind acid. Therefore chlorite must be acidized cautiously or not at all, with preference to be given to selection of a completion procedure that makes acidizing unnecessary. Smectite can only be successfully acidized if it is sparsely distributed (less than 3 wt. %) in high-permeability formations, whereas in reservoirs with higher smectite contents (over 5 wt. %), any aqueous fluid including acid may cause great damage and well stimulation has to be made with proppant fracturing. Other significant diagenetical minerals usually comprise carbonates which are soluble in hydrochloric acid.

Mud acid has the ability to dissolve clay minerals and is mainly employed for removal of formation damage consisting of plugging of pore throats by clay minerals in the wellbore vicinity (GIDLEY 1985). Gas wells, oil wells and water injection wells each require different treatments to obtain maximum response to mud acid stimulation. Gas wells respond to mud acidizing most predictably with uniform increase in production buildup with increasing acid volume, and with uniform increase in stimulation success with increasing permeability. Oil wells experience maximum in response as a function of the amount of acid used. Treatment success does not appear to increase uniformly with formation permeability, and the afterflush-to-acid volume ratio does not affect buildup uniformly. Results on oil wells seem to be more site-specific and to be more influenced by the nature of the reservoir fluids. Water injection wells sometimes show the odd response of having the greatest buildup to the least acid job.

Enhanced removal of deep clay damage can be performed by in-situ hydrofluoric acid creation (HALL 1978) and self-generating mud acid (TEMPLETON, RICHARDSON, KARNES & LYBARGER 1975).

4.5.5.4.3. Delayed acid reaction and in-situ acid generation

Delayed-action acid systems attribute their success to slow generation of hydrofluoric acid in-situ from hydrolysis of methyl formate which yields methyl alcohol and formic acid, and the latter then reacts with ammonium fluoride to release hydrofluoric acid, with hydrofluoric acid originating downhole in the reservoir in such distances from the borehole that mud acid which is conventionally pumped normally does not reach before spending. In order to achieve optimum result, the wells have to be returned to production by gradually opening the choke over a three-month period and never allowing an excessive drawdown. Another process of in-situ hydrofluoric acid generation is hydrolyzation of fluoboric acid to form hydroxyfluoboric acid and hydrofluoric acid (THOMAS & CROWE 1978 a) which provides a slow release source of hydrofluoric acid that can penetrate deeply before spending, and in addition, the hydroxyfluoboric acid that is created very slowly reacts with clay minerals to form a non-swelling, non-dispersing product that stabilizes clay and even holds fine silica particles in place (cf. also McBRIDE, RATHBONE & THOMAS 1979). On the other hand, fluoboric acid may spend at various formation temperatures at a rapid rate almost similar to that of conventional hydrofluoric acid (KUNZE & SHAUGHNESSY 1980).

Acid return experiments and analyses have revealed that excess hydrochloric acid is almost completely consumed by the hydrofluoric acid chemical process (GDANSKI & PEAVY 1986), with the main mechanisms of hydrochloric acid consumption being spending on carbonates, exchange spending for sodium on ion-exchangeable clays such as mixed layers, and the hydrofluoric acid spending process itself. Lower hydrofluoric acid concentrations should be used when acidizing formations containing more than 15 % feldspars, and higher hydrochloric acid saturations should be applied with hydrofluoric acid to allow maximum formation dissolution (cf. also BRANNON, NETTERS & GRIMMER 1986). PERTHUIS (1989) discusses weak or slowly generating hydrofluoric acid solutions as a choice for sandstone acidizing.

4.5.5.4.4. Selective stimulation of injection wells

Selective acidizing of sandstone reservoirs in injection wells can discretely stimulate ineffective intervals and can significantly improve injection profiles (ELLENBERGER & ASELTINE 1976).

4.6. High-permeability reservoir fracturing

Traditionally hydraulic proppant fracturing was in Europe and other parts of the world almost exclusively limited to the stimulation of low-permeability, more or less tight sandstone reservoirs (less than abt. 5 md reaching down to the microdarcy range; cf. section 3.4.1.). The philosophy of treating also high-permeability hydrocarbon-bearing sandstones (abt. 10 - 30 md or more) which is already common sense in the USA since longer time, however, has so far not been adopted as general procedure in Europe and other parts of the world, with the main reasons being cost containment for stimulation jobs and also comparatively lower necessity of application due to frequently sufficient production rates also without treatment in many formations in Europe. Moderate- and high-permeability reservoir fracturing is considered to have in Europe the greatest impact on enhancement of stimulation of oil-bearing formations (KOHLHAAS 1982, McDONALD 1983; NORMAN, CIPOLLA & WEBB 1983; BRITT 1985). The most important aspects are fracture-formation conductivity contrast, proppant grain size and concentration, proppant/fluid slurry mixing types, and fracture design characteristics which are briefly commented as follows.

4.6.1. Fracture-formation conductivity contrast

Extension of the stimulation potential in the near future, however, is considered to be also in Europe inevitably linked with the application of hydraulic proppant fracturing also in moderate- to high-permeability sandstone reservoirs. High-permeability reservoirs generally require high fracture conductivities but do not need deeply penetrating cracks, whereas low-permeability formations require deeply penetrating fractures but can tolerate lower crack conductivities (VEATCH 1983). While fracture half-length requirements are typically less than 1,000 ft (330 m) in moderate- to high-permeability pay zones, tight gas sands may demand crack half-lengths up to 4,500 ft (1,500 m; cf. section 4.8.11.). Some aspects of reservoir types and fracture flow capacity are briefly outlined as follows.

4.6.1.1. Reservoir types

Similarly as applying for carbonate reservoir fracturing, in high-permeability sandstone stimulation again the critical factor is the conductivity contrast between the propped crack and the matrix of the fractured formation, with various types and grain sizes of proppants allowing to model the optimum permeability contrast within the frame of depth and closure stress (cf. section 4.5.1.1.2.). Enhancement of the contrast between the finite-conductivity fracture and the reservoir can be done by improving type and/or grain size of the proppant inserted for support of the crack. While in very tight reservoirs even a poor fracture permeability provides already sufficient conductivity contrast (CALLANAN, CIPOLLA & LEWIS 1983), optimum proppant properties and distributions have to be selected and installed in order to result in satisfactory conductivity contrasts between fracture and formation in moderate- to high-permeability reservoirs.

Special applications are seen for insufficiently consolidated sandstones and carbonate shell layers that frequently in addition to proppant fracturing require stabilization by gravel packing or by usage of resin-coated proppants (cf. section 1.2.6.) which form a massive hard wedge in the loose reservoir upon interlocking of grains by glueing of the resin films at their contacts (cf. also sections 4.2.5. and 4.12.3.3.), because the high permeability often creates a weak rock fabric that collapses with declining flowing pressure or increasing water cut of the oil or is even unstable from the very beginning of production (cf. section 5.4.), with the propped fracture staying open as a rigid wedge and providing a pathway for the hydrocarbons to the wellbore.

4.6.1.2. Fracture flow capacity

The general rule of the larger the permeability contrast between formation and proppant package in the fracture, the better the flow capacity and rate, equally applies to sandstones and carbonates and is one of the strongest arguments to support hydraulic proppant fracturing of not only sandstone, but also carbonate reservoirs of both low- and high-permeability types, because even the best natural formation permeability is still separated by a pronounced steep gradient from the extremely high-permeability proppant package (cf. section 4.5.1.1.2.). A contrast of at least 100 between the flow capacities of formation and fracture are considered to be essential to obtain adequate stimulation results, and this can only be influenced by selecting a high-quality proppant to fill in the fracture (WATERS 1980), with grain size in addition to and/or in combination with type of proppants being a very important tool for maximizing the conductivity contrast between fracture and formation (cf. section 4.3.5.). High-conductivity short fractures and their economical benefit in moderate-permeability oil reservoirs are illustrated by BRITT (1985; cf. section 4.9.2.).

The necessity of larger proppant grain sizes for effective stimulation of moderate- to high-permeability reservoirs is not only the consequence of the requirement to provide adequate flow capacity, but is also due to the need to resist embedment into the frequently soft to poorly consolidated formations (CLARK 1983; cf. section 4.9.2.).

4.6.2. Proppant grain size and concentration

Ultra-high fracture permeabilities maximizing the conductivity contrast between reservoir and drainage path can thus be best achieved by a combination of very large proppant saturation of the fracturing fluid leading to high-concentration multilayer propping of the fracture and extra large proppant grain sizes (BARBY & BARBEE 1987; cf. sections 4.3.4. and 4.3.5.). The greater the conductivity contrast, the higher the production rates, and cumulative quantities will be up to limits defined by fracture and reservoir geometrical and hydrodynamical properties.

Fracture permeability and thus conductivity can be enhanced by tail-in proppant placement (cf. sections 2.4.1.2.3. and 4.12.3.3.). Ultra-high conductivity fracture stimulation by predominantly grain-size modelling of synthetic proppants is also discussed by BARBY & BARBEE (1987). Moderate- to high-permeability reservoirs can often already be successfully stimulated by microfracturing or minifracturing (CRAMER & SONGER 1988; cf. section 4.8.1.).

4.6.3. Proppant/fluid slurry mixing types

Moderate- to high-permeability reservoirs are best stimulated by pumping of small precise fluid volumes containing high proppant concentrations (CRAMER & SONGER 1988) for creation of short highly conductive fractures. The following account presents a discussion of batch-mix preparation and continuous-mix blending. Comments are also offered on special design for waterflood operations.

4.6.3.1. Batch-mix preparation

While continuous-mix blending equipment greatly facilitates handling of large fluid and proppant volumes and enables implementation and execution of massive hydraulic fracturing treatments in tight reservoirs, it is poorly suited to prepare and displace small precise slurry volumes (CRAMER & SONGER 1988; cf. section 4.3.3.5.3.). Batch-mix fracturing comprising utilization of pre-mixed fluid/proppant slurries is a suitable technique to increase wellbore transmissibility in moderate- to high-permeability reservoirs due to bypass of the wellbore-vicinity damage zone (cf. section 4.8.3.) and extension of the effective borehole radius.

High proppant concentrations can be pre-blended with gelled liquid in the batch-mix inventory compartments and slurry density can be verified before the start of the job. This assures placement of an uniform proppant distribution in the fracture, since proppant saturations at the surface are not affected by treatment rate fluctuations. Incremental improvements in wellbore transmissibility can in many cases be achieved by creation of a significant conductivity contrast between reservoir and fracture. The use of small precise slurry volumes also successfully limits vertical and lateral fracture extension, avoids negative consequences of interzonal communication and waterflood sweep inefficiency, and reduces screenout problems by achievement of uniform proppant distribution and use of a high pad/slurry ratio.

4.6.3.2. Continuous-mix blending

Conventional continuous-mix treatments have inherent limitations in moderate- to high-permeability reservoirs (CRAMER & SONGER 1988; cf. section 4.3.3.5.3.). As relatively large slurry volumes are often used, fracture dimensions can become extensive and this can result in only a small portion of the crack contributing to wellbore transmissibility giving rise to wastefulness, and can lead to achievement of excessive fracture length and height enabling accelerated water breakthrough, hindered vertical and lateral sweep efficiency in waterflooded reservoirs, and channelling into vertically adjacent pay zones. Continuous-mix blending equipment greatly facilitates handling of large fluid and proppant volumes, but is poorly suited to prepare and displace small precise slurry volumes. Non-uniform proppant distribution and irregular proppant addition due to fluctuating treatment rate occurs in the frequently used high capacity/volume blender tub apparatus. Irregular proppant distribution in the fracture can result in low-conductivity bottlenecks and in screenout failure due to proppant slugging (cf. sections 4.2.3.5.3. and 6.2.4.2.1.).

Conventional treatments also often utilize relatively modest proppant concentrations and together with narrow fracture width this leads to low crack conductivity which is further deteriorated by gel residue and other proppant package damage mechanisms. Depending on the viscoelastic characteristics of the stimulation fluid and on crack closure time, proppant settling may occur and enhance fracture conductivity in the lower portions of the crack (cf. sections 4.3.3.2. and 4.12.2.2.) which can lead to preferential water production if a water leg is present or may aggravate flood water cycling if a higher-permeability zone is located at the fracture base. Screenouts often occur when performing conventional treatments in moderate- to high-permeability reservoirs due to severe fluid-loss rates during the job, giving rise to substantial pad fluid leakoff and slurry dehydration at the leading edge of the fracture, with the screenouts further contributing to treatment waste and cost inefficiency (cf. sections 4.2.3.5.3. and 6.2.4.2.1.).

4.6.3.3. Special design for waterflood operations

Special considerations concerning fracture length and conductivity apply to waterflooded moderate- to high-permeability oil reservoirs (cf. also section 4.8.7.1.). Premature injection water breakthrough may result when conductive fractures in excess of 25 % of the drainage radius are created in both injection and production wells (CRAMER & SONGER 1988). Water cycling may be exacerbated when a fracture is extended into a high-permeability rock layer at the expense of lower-permeability horizons or relatedly when proppant is placed preferentially adjacent to the higher-permeability zone due to settling (cf. sections 4.2.2.4. and 4.12.2.2.2.). Both effects decrease sweep efficiency of the flood and necessitate injection of significantly larger water volumes to achieve a given oil recovery. If reservoirs are bounded by rock layers not offering sufficient restriction to vertical crack height growth, fracture communication into aquifers, gas caps or other productive zones as a consequence of a large-scale uncontrolled treatment may cause severe reservoir disruption.

4.6.4. Fracture design characteristics

The primary treatment objectives in moderate- to high-permeability reservoirs are to bypass wellbore vicinity damage (cf. section 4.8.3.), extend effective borehole radius, maximize fracture conductivity (cf. sections 4.3.5., 4.5.1.1.2. and 4.6.1.), avoid excessive fracture height and length (cf. section 4.2.2.), and minimize treatment cost (CRAMER & SONGER 1988). Pad fluid volume and properties should be selected to enable proppant entry and optimize proppant distribution in the fracture. Enough pad should be pumped to keep pace with spurt loss from the fracture while building a low-permeability filter cake at the crack walls thus reducing fluid-loss rate.

In some high-permeability reservoirs and with certain fluid systems, filter cake generation is difficult to achieve at all and the whole fracturing fluid can leak off into the formation matrix. Total loss of pad fluid gives rise to proppant bridging near the fracture tip thereby provoking tip screenout (cf. sections 4.2.3.5.3. and 6.2.4.2.1.), and lack of filter cake generation would result in slurry dehydration and proppant concentrating-type screenout if slurry surface saturations are sufficiently high and slurry residence time is satisfactorily long. Some aspects of fracture width and length as well as high proppant concentration and grain size are briefly outlined as follows.

4.6.4.1. Fracture width and length

Treatment injection rate and the viscous characteristics of the pad combine with pad fluid loss properties to determine hydraulic fracture width. In order to prevent the occurrence of proppant bridging in the fracture, hydraulic width must be at least 2.6 times greater than maximum proppant particle diameter (CRAMER & SONGER 1988). As propped width is proportional to hydraulic width, pad viscosity, pad volume and injection rate can be increased to widen the fractured well beyond the minimum entry width for the purpose of increasing crack conductivity. Proppant selection and scheduling ultimately determine treatment results, with propped fracture length being on average up to 25 % of the drainage radius. Actual propped length can exceed projected length due to post-injection afterflow and variable fluid/slurry velocity profiles in the crack during insertion. In addition to pad injection techniques, fracture conductivity can be maximized by increasing surface proppant concentration in the slurry thus increasing propped fracture width, which can be technically achieved by batch mixing rather than continuous-mix blending without increasing screenout risk (cf. sections 4.3.3.5.3. and 4.6.3.1.).

4.6.4.2. High proppant concentration and grain size

Fracture conductivity can be further enhanced by not adding any fluid-loss material during the slurry stages, because residue development on the crack walls forms primarily after injection cessation and is directly attributable to the remnant characteristics of the slurry (CRAMER & SONGER 1988). High proppant concentrations also hasten fracture closure which minimizes settling and afterflow effects, and also maximizes propped crack height in cases where proppant settling is significant. Conductivity maximization by coarse proppants (such as 8/12 and 12/20 mesh; cf. section 4.3.5.) has to take into account that unless perforation hole diameter is at least six times the average proppant particle size, proppant bridging at the perforations leads to premature treatment failure (cf. section 5.6.1.). The fracture conductivity improvement benefit of higher-quality proppants has to be evaluated against the anticipated incremental production return in order to determine if the use of higher cost/higher performance relationship is warranted.

4.7. Geothermal fracturing

Apart from oil and gas reservoirs, hydraulic proppant fracturing is also a suitable means of stimulation of formations bearing hot water or being hot dry rocks, with the possibility of exploitation of geothermal heat. Some comments are offered as follows on geological aspects as well as areal distribution and exploitation of geothermal reservoirs, explosive and non-proppant fracturing, proppant fracturing, communication fracturing, fracture system enlargement by heat extraction, and shear stimulation vs. tensile fracturing.

4.7.1. Geological aspects of geothermal reservoirs

Permeability and fluid flow in high-temperature geothermal systems is predominantly fracture-controlled (MOORE & NIELSON 1986). The fracture system consists of faults and joints. Geological and geochemical techniques can provide valuable insights into behaviour and evolution of active fracture-controlled geothermal reservoirs. Some comments on distribution of geothermal reservoir types and permeability evolution are offered as follows.

4.7.1.1. Distribution of geothermal reservoir types

Many geothermal systems are hosted by volcanic rock series in structural complexes of calderas and stratovolcanoes, but sandstone and plutonic geothermal reservoirs also occur. Various methods can be applied for evaluation of permeability distribution and fluid flow patterns in different geological environments. In many volcanic geothermal fields, fluid flow is principally confined to steeply dipping normal faults and subsidiary fractures, although interbedded sandstone lenses can also act as stratigraphic aquifers when intersected by the faults (HULEN & NIELSON 1982, 1983).

Tectonical doming can subject the strata in various parts of a geological pile to different stresses (MOORE & NIELSON 1986). In the upper portions of the dome, the rocks are under tension, resulting in normal faulting and open joints along the crestal part of the structure. The upper tensional zone is separated by the neutral plane from the lower compressional zone where conjugate shears are present and the joints are more or less closed. Primary exploration below the depth of the neutral plane is non-productive due to lack of permeability and thus large-scale hydraulic fracturing is necessary to make advantage of the higher reservoir temperature in greater depth.

4.7.1.2. Permeability evolution

It is important in geothermal exploration to differentiate between fossil fluid conduits and those which host the presently active geothermal system. Trace element signature studies can be used to distinguish between fluid flow paths utilized by ancient geothermal systems and the guide channels of the modern system. Chemical and isotopical analyses of fluids can be applied for assessment of flow patterns and residence times. Lack of mixing of several brine types indicates restriction of fluid flow to discrete fault and fracture zones in a low-permeability matrix in some reservoirs. In mountaineous terrains, these conditions favour fluid flow driven by topographically controlled heat differences rather than convective flow common in more highly permeable geothermal fields (ADAMS, MOORE & FORSTER 1985). The importance of topographical control on the fluid flow pattern is in some cases illustrated by the coincidence of the thermal and topographical divide separating areas of different temperature and fluid type.

Vein mineralization and fluid inclusion relationships in sandstone geothermal reservoirs can document the transition from matrix to fracture permeability during course of the structural evolution of geothermal complexes (ELDERS 1979, MOORE & NIELSON 1986) where also geothermal metamorphism of sandstones (McDOWELL & ELDERS 1979) takes place. Distinction can be made between an initial phase with stratified porous flow of thermal fluids in sandstones beneath a largely impermeable mudstone and evaporite caprock complex that forms an aquitard throughout the history of the system, and an advanced stage with upwards migration of fluids of different salinity to different depths along fractures, and then laterally in partially porous rocks at shallow depths where fluid flow becomes increasingly fracture-controlled.

4.7.2. Geothermal reservoir distribution and exploitation

Some points of geothermal reservoir distribution and exploitation are discussed as follows along the lines of geographical distribution of European geothermal reservoirs and worldwide geothermal energy exploitation.

4.7.2.1. Geographical distribution of European geothermal reservoirs

Geothermal sources in crystalline and sedimentary rocks that are partially already naturally fractured are in Europe predominantly present in Iceland (Krafla; STEFANSSON 1981, BODVARSSON et al. 1984; Nesjavellir; BODVARSSON et al. 1986; Svartsengi; GUDMUNDSSON & OLSEN 1985), Italy (Bagnore; ATKINSON, CELATI, CORSI & KUCUK 1980; Larderello; ATKINSON et al. 1978), Southern France and Yugoslavia (JELIC 1982) and are partially already commercially used (RAMEY & GRINGARTEN 1975). Pilot studies are also carried out in Germany FRG (ERDOEL-ERDGAS AKTUELL 1978 c, LEYDECKER 1981, MEIER & ERNST 1981) and other countries. Other initial projects are located in England (PINE & LEDINGHAM 1983). An overview of the European geothermal drilling experience is given by BARON & UNGEMACH (1981), and aspects of geothermal heat exploitation are discussed by ERNST & HUEBINGER (1979).

Distinction has to be made between low- and high-enthalpy reservoirs. Low-enthalpy reservoirs are prolific and dependable in foredeep and intracratonic basins wherever sedimentary multi-aquifer systems are developed (such as Aquitanian and Paris Basins/France, Po Valley/Italy; Wessex, East Yorkshire and Northern Ireland Basins/United Kingdom, Southern Molasse Alpine Foredeep and Northern Munsterland/Germany FRG, Netherlands and Denmark), low- to high-enthalpy formations in continental rift valleys and areas with young but extinct volcanism (such as Limague and Rhône Valley/France, and Rhine Graben and Urach region/Germany FRG). High-enthalpy reservoirs occur in areas with subrecently to recently active volcanism and tectonics which are placed along the Mediterranean belt, comprising chiefly the Tuscanian distensive tectonic system, the Northern Latium and Campanian/Italy Pliocene-Quaternary volcanism that is partially still active at least at post-volcanic stage, and the subduction magmatism of the Aeolian Arc/Greece. High-enthalpy geothermal reservoirs are in Europe restricted to Central and Southern Italy and to Greece (BARON & UNGEMACH 1981). In terms of energy carrier phase, distinction has to be made between vapor-dominated and water-dominated high-temperature geothermal systems (GUDMUNDSSON & OLSEN 1985).

4.7.2.2. Worldwide geothermal energy exploitation

DIPIPPO (1984) presents an overview of worldwide geothermal power development, with the ranking list in terms of heat generating capacity being USA, Philippines, Italy, Japan, Mexico, New Zealand, El Salvador, Iceland, Indonesia, Kenya, Soviet Union, China, Azores/Portugal, Turkey, Nicaragua, Guadeloupe/France, Guatemala, Costa Rica and Greece, with other countries being only of subordinate significance. Most of the geothermal plants in the USA are concentrated in The Geysers field in Northern California and Imperial Valley in Southern California, with other areas of activity including sites in Hawaii, Idaho, Nevada, Oregon, Utah, New Mexico and Texas-Louisiana. Italy operates more individual geothermal power units than any other country, although lagging behind in total cumulative capacity. The most important geothermal fields in Italy are in alphabetic order Bagnore, Bellavista, Castelnuovo, Gabbro, Lagoni Rossi, La Leccia, Larderello, Molinetto, Monte Rotondo, Pianacce, Piancastagnaio, Radicondoli, San Martino, Sasso, Serrazzano, Travale, Tuscan and Vallonsordo (BARON & UNGEMACH 1981; CIGNI, GAUDIO & FABBRI 1981). The Wairakei field in New Zealand is already in service for more than 25 years (STACEY & THAIN 1983). Present state and future prospects of geothermal energy exploitation development are summarized by EDWARDS, CHILINGAR, RIEKE & FERTL (1982) and CATALDI & SOMMARUGA (1986).

In the farther Middle East, geothermal energy is exploited in India. High-temperature proppant testing for particular application in geothermal fracture stimulation is performed by SINCLAIR (1980). Aspects of geothermal well stimulation are also discussed by SINCLAIR, PITTARD & HANOLD (1980); GRANT, DONALDSON & BIXLEY (1982); MORRIS & SINCLAIR (1984) and WALKUP & HORNE (1985). Experience reports on geothermal fracturing are given by ERNST (1977); HUNSBEDT, KRUGER & LONDON (1977); MURPHY & TESTER (1979); WILLIAMS, ROWLEY, NEUDECKER & BRITTENHAM (1979); MORRIS, VERITY, SCHROEDER & GORANSON (1980); CAMPBELL, HANOLD, SINCLAIR & VETTER (1981), and TESTER, MURPHY, GRIGSBY, POTTER & ROBINSON (1986).

4.7.3. Explosive and non-proppant fracturing

Explosive fracturing has in former decades been abundantly used as an alternative to hydraulic fracturing in both hydrocarbon (cf. sections 1.2.1. and 4.4.4.) and geothermal reservoir stimulation, but due to various reasons, its application has declined during the last years, mainly as a consequence of the progressively improving effectivity of hydraulic fracturing. In recent time, however, some revival and renaissance of explosive fracturing can be observed in the field of enhancement of oil, gas and water production, resulting from some innovations which make advanced explosive fracturing competitive to hydraulic fracturing or even allow a combination of both stimulation technology branches (cf. section 4.8.9.6.).

In geothermal reservoirs, stimulation can generally be performed by hydraulic fracturing, explosive fracturing and thermal stress cracking (HUNSBEDT, KRUGER & LONDON 1977). The aim of geothermal reservoir stimulation is to increase vertical and horizontal formation permeability in order to provide adequate fluid circulation and to expose new rock surface to allow heat transfer from the hot rock to the fluid (HUNSBEDT, KRUGER & LONDON 1975). Following the outline of some general aspects and a short review of conventional explosive fracturing and recent innovations, the stem-induced explosive fracturing technique (YOUNG, BARKER & CLARK 1984) is sketched which provides a reasonable means of controlled reservoir stimulation both in hydrocarbon and geothermal fields. Some comments on nuclear and water fracturing are also offered.

4.7.3.1. General aspects

Geothermal stimulation has so far been mainly done by creation of joints in tight formations by conventional explosive fracturing where a more or less multiple radial fracture system (cf. plate IV/1) originates around a detonated cylindrical cavity (in contrast to a hydraulically-induced fracture which frequently is singular or in some cases also consists of a narrow bundle of multiple closely-spaced cracks in a similar way as natural structural faults due to its formation perpendicularly to the least principal stress; SCHOLS 1983), but in several cases self-propping (cf. sections 4.5.1.1.1.2. and 4.5.1.5.1.) of the various cracks has not proven to be the optimum solution.

In addition, the permeability around a detonated cavity following explosive fracturing decreases exponentially with distance, with the effect thus being hardly higher than that obtained by hydraulic proppant fracturing, if not even inferior in result. The stimulation effect frequently suffered from only very limited possibilities to control quantity and size of the propagating fractures, as well as from considerable conductivity deterioration by excessive rubblization of the reservoir rock in the central part of the fractured zone around the borehole. Some aspects of formation damage removal in various reservoir types as well as short fracture extension and reservoir rock rubblization are briefly outlined as follows.

4.7.3.1.1. Formation damage removal in various reservoir types

Explosive fracturing by detonation of chemical or nuclear devices in open holes was before the advent of hydraulic proppant fracturing also a common stimulation technique of sandstone oil and gas reservoirs, with the explosion creating a rubblized formation and fractures radiating out from the wellbore (cf. plate IV/1) thereby increasing its effective diameter (cf. sections 1.2.1. and 4.4.4.). The main purpose of standard explosive fracturing is to increase the size of the wellbore and to remove any near-borehole formation damage resulting from mainly drilling and subordinately also completion (ZUBER, LEE & GATENS 1987; cf. section 4.8.3.). Optimization was also made by placing explosives into natural fractures in the formation, but all these applications had a considerably inferior effect in comparison to hydraulic proppant fracturing (WATERS 1980).

The limited production increase from oil and gas wells stimulated by explosive fracturing could preferentially be attributed to the inadequate or even lacking success of self-propping (cf. sections 4.5.1.1.1.2. and 4.5.1.5.1.) of the created fractures which consequently already prematurely closed or healed. Aspects of explosive fracturing are also discussed by HOWELL & HILLS (1970); RAMEY, KRUGER & RAGHAVAN (1973); CARTER & OLINGER (1976); GRADY, KIPP & SMITH (1979); SCHMIDT, WARPINSKI & COOPER (1980); McHUGH (1983); YOUNG, BARKER & CLARK (1984); WATSON, BENSON & FILLO (1985); BENNETT (1987) and BRANAGAN, LEE, CIPOLLA & WILMER (1988). WINKLE & MIGNOTTE (1963) outline implosive shock treatment of injection wells.

4.7.3.1.2. Short fracture extension and reservoir rock rubblization

Explosive fracturing creates a highly cracked and partially rubblized region around the wellbore through propagation of high-energy stress waves (STOLLER 1985). The extremely high pressures exceed the dynamic compressive strength of the wellbore rock, with the resultant permanent rock compaction producing a very low-permeability barrier at the wellbore. Even without this permeability barrier, effective explosive fracturing is restricted to those formations with a high natural crack density because of the relatively short fractures created, with the main stimulation effect being interconnection of numerous isolated natural cracks by the radially propagating short fractures generated by the explosive treatment (cf. plate IV/1). In terms of operations, conventional explosive fracturing is limited to open holes where the shock wave coming from the detonation of the charge in the wellbore has direct circumferential access to the reservoir formation without any restriction. Selective placement of charges in larger natural fractures or perforation tunnels is a possibility to extend more sophisticated explosive fracturing to cased holes.

BRANAGAN, LEE, CIPOLLA & WILMER (1988) use high-pressure nitrogen gas and explosive fracturing in tight gas sandstones in order to eliminate foreign liquids from entering the reservoir crack system following perforation for the purpose of minimization of external intervention in the natural production process.

4.7.3.2. Conventional explosive fracturing and recent innovation

During the last decade, increasing modification of conventional explosive fracturing techniques in order to develop cost- and performance-competitive alternatives to high-effectivity hydraulic stimulation have led to various innovations which mainly focus on propellant-based charges for well treatment (FOURNEY, BARKER & HOLLOWAY 1983; CUDERMAN & NORTHROP 1984; YOUNG, BARKER & CLARK 1984). Research efforts concentrated on development of explosive- and propellant-founded techniques designed to minimize wellbore damage which is provoked by conventional explosive stimulation treatments, and to maximize the radial distance to which fractures might be propagated into the formation.

The explosive-based techniques include methods for directly displacing a sensitive liquid explosive into a hydraulic fracture system before detonation (SPENCER 1981) which leads to development of star-like secondary cracks radiating from the main fracture that have a similar increasing drainage area effect as a bundle of low-angle to horizontal drainholes emanating from a main subvertical or inclined wellbore, and introduction of a low-detonation-velocity explosive that would be placed directly across the pay zone (WATSON, BENSON & FILLO 1985). Insertion of explosive charges into hydraulically-created fractures and selective firing represents a combination of explosive and hydraulic fracturing (cf. section 4.8.9.4.), but in a similar way also introduction of explosive charges into wider natural joints is possible. This technique has the advantage that either hydraulic fracture or natural crack serve as a guide channel, and the increasing distance of the detonation point away from the borehole leaves the immediate wellbore vicinity intact and avoids formation damage by reservoir rock rubblization in the region closest to the borehole.

Contained explosions and deflagrations represent a new perspective of well shooting (THOMAS & ROOT 1979). The efforts using propellants as chemical energy source have focussed on tailoring propellant burn rate and consequent pressure/time history so that maximum wellbore fracturing and minimum wellbore damage are realized. Explosive-based techniques which are much cheaper and more reliable in field performance, however, may have broader applicability and may be preferable over any propellant-based means. Out of many concepts, the method of stem-induced explosive fracturing (YOUNG, BARKER & CLARK 1984) has proven to be one of the most successful innovations.

4.7.3.3. Stem-induced explosive fracturing

The effectivity of explosive fracturing can be greatly enhanced by the stem-induced technique (YOUNG, BARKER & CLARK 1984) involving placement of the primary explosive charge in an expendable rathole below the target zone in such a manner that the horizon to be fractured is not directly subjected to the high detonation pressures of the explosive in contrast to conventional stimulation where the explosive charge is fired immediately opposite to the target interval. The high-pressure gases generated by the detonation of the explosive, however, are used to fracture the wellbore through the placement of a solid stem of gravel at or near the top of the zone to be treated. Wellbore fracturing is initiated and largely controlled by the pressure/time history of the explosive gases as they reflect off the solid stem after moving up through the open borehole section between the explosive charge and the gravel plug (FOURNEY, BARKER & HOLLOWAY 1981; FOURNEY, HOLLOWAY & SIMHA 1984; YOUNG, BARKER & CLARK 1984).

Stem-induced explosive fracturing is suitable for treatment of only one isolated target horizon or two distinct and separate pay zones, with the multiple-layer stimulation configuration designed to fracture the upper reservoir section by reflection of the explosive gases off the solid gravel stem and to crack the lower pay level by its proximity to the explosive charge. The discussion as follows concentrates on controlled origin of limited multiple fracture system, vertical fracture orientation regardless of geostress state, limitations and possibilities of improvement, and proppant insertion into explosion-generated fractures.

4.7.3.3.1. Controlled origin of limited multiple fracture system

The main advantage of the stem-induced explosive fracturing technique is its controlled performance with origin of only two sets of radial cracks propagating in roughly orthogonal directions as visualized by borehole television and photography (YOUNG, BARKER & CLARK 1984; cf. section 6.2.1.9.3.). The fact that multiple fracturing by stem-induced explosive detonation is so limited spatially and does not exceed four orthogonal fractures even in close proximity to the primary charge attests to the beneficial effects of energy source decoupling and is the main difference to conventional explosive fracturing where a whole bundle of closely-spaced or even almost merged multiple fractures in a frequently star-like pattern (cf. plate IV/1) is created which together with rubblization virtually destroys the reservoir fabric in contrast to smooth intersection of an intact formation matrix by discrete perpendicular fractures.

This superiority of the stem-induced explosive fracturing concept has been verified by comparative experiments with high detonation rates and resulting high loading rates of nitroglycerin charges that led to origin of up to ten vertical cracks forming a multiple bundle even in slim boreholes, with fracture aperture and thus also penetration depth into the formation being quite limited as opposed to stem-induced treatments where good fracture openings indicate reasonable propagation into the reservoir.

4.7.3.3.2. Vertical fracture orientation regardless of geostress state

The most significant characteristic feature of stem-induced explosive fracturing is the dominant vertical orientation and extent of the cracks regardless of the in-situ stress distribution in the reservoir column (YOUNG, BARKER & CLARK 1984). The ability to develop dominant vertical fracturing also in an in-situ stress state that generally dictates the origin of horizontal cracks (for example in hydraulic stimulation of shallow reservoirs) is a significant benefit that can be obtained with the proper application of explosive and/or propellant stimulation techniques and is one of the advantages with respect to hydraulic fracturing where no influence can be exerted on the in-situ stress state and thus orientation of the cracks. This relationship even allows the combination of horizontal hydraulic and vertical explosive fracture creation giving rise to orthogonal crack networking for special reservoir stimulation (cf. section 4.8.9.6.).

4.7.3.3.3. Limitations and possibilities of improvement

The stem-induced explosive fracturing technique has thus been proven to be quite effective in modifying or tailoring the pressure/time histories obtained from a large explosive charge (YOUNG, BARKER & CLARK 1984). The stem-induced approach offers the possibility of fracturing without significant wellbore damage a zone with much more explosive gas energy than could be placed in that reservoir level directly by either decoupled explosive charges or controlled-burn-rate propellants. The only limitation of the stem-induced fracturing technique is the availability of an expendable rathole for placement of the primary explosive charge.

On the other hand, however, the major disadvantage of any explosive fracturing method with respect to hydraulic stimulation is the inability to support the cracks with proppants in order to prevent them from subsequently closing. The lack of plugging material except of random fine formation rubble (auto-propping; cf. sections 4.5.1.1.1.2. and 4.5.1.5.1.) gives rise to rapid fracture flow deterioration by partial closure of the crack upon pressure drawdown during hydrocarbon offtake leading to reservoir depletion. Another negative effect is infilling of parts of the cracks with rubblized or even pulverized rock debris originating from reservoir framework destruction in the immediate surroundings of the detonation point of the explosive charge (cf. plate IV/1).

4.7.3.3.4. Proppant insertion into explosion-generated fractures

Therefore the most challenging task for future explosive fracturing optimization would be the investigation of possibilities of first opening a double set of cracks by explosive treatment and then second following with a hydraulic operation which is to enlarge the already existing fractures, to flush out the fine debris from the cracks, and to infill them with proppants in order to avoid their subsequent closure (such a combination of explosive and hydraulic fracturing resembles in fact an interlocking of preexisting natural and later artificially-created hydraulic fractures; cf. sections 4.5.2. and 4.8.9.4.), with such a linking, however, only being able to be performed in geostress regimes favouring the development of vertical hydraulic fractures due to the exclusive creation of vertical cracks by explosive stimulation. Other aspects of stem-induced tailored-pulse fracturing are discussed in section 4.2.2.1.3.2.

4.7.3.4. Nuclear fracturing

Apart from conventional charges, explosive fracturing of underground formations can also be carried out by detonating nuclear charges (HOWARD & FAST 1970 a). While nuclear fracturing has so far only rarely been applied in geothermal reservoirs but would certainly represent a suitable technique for large-scale cracking of tight rocks, nuclear explosions have more frequently been utilized in oil and gas exploitation. Nuclear detonations can be contained in the zone of interest and are capable of rubbling large rock volumes thereby creating multiple fractures such as originating also during conventional oil-well shooting (cf. section 4.7.3.1.), but on a much larger scale. Some comments on principles and applications are offered as follows.

4.7.3.4.1. Principles

Fractures resulting from underground nuclear explosions differ from those created by hydraulic stimulation in both number of fractures generated and penetration depth into the matrix around the wellbore (HOWARD & FAST 1970 a). Nuclear explosions are expected to generate a cavity of 100 - 300 ft (30 - 100 m) diameter, with numerous fractures radiating from it thus forming a high-permeability zone and considerably enlarging the effective wellbore radius. Upon detonation of the nuclear device, a cavity is generated as a result of vaporization of rocks and saturating fluids. The gases contained within the cavity are initially at extremely high pressure and temperature. Compaction of the rock in the lower hemisphere coupled with upward and lateral rock movement results in a spheroid cavity. The gas pressure temporarily supports the overburden thus preserving the cavity shape. Subsequent heat losses, gas leakoff through the crack system, and vapor condensation reduce the pressure until the fractured rock above the cavity can no longer be supported. Rock collapse into the cavity forms a complex chimney-rubble zone, with most of the molten material and radioactive fission products collecting in the bottom of the zones. Changes in cavity shape from spherical may be due to implosion of cavity walls as well as uplift and dropping of the ceiling.

In areas where hydraulic fracturing is effective, nuclear explosions cannot compete because of high costs, but substantial volumes of hydrocarbons are in place in reservoirs that do not yield to present hydraulic fracturing techniques, and these pay zones may therefore by potential sites for nuclear explosive stimulation.

4.7.3.4.2. Applications

The hitherto main areas of application of nuclear fracturing are New Mexico and Colorado/USA (HOWARD & FAST 1970 a) and Eastern Siberia/USSR (OIL GAS JOURNAL 1987 b, 1987 d). After lowering the nuclear charge into the borehole, the well is cemented to the surface, and only in such a hermetically sealed environment, the nuclear charge is fired, with no radiation being vented to the atmosphere. Nuclear explosions in the USSR have recently proven to be able to interconnect a number of small separate oil-bearing lenses from which oil so far could not be recovered profitably. The explosion results in fracturing beyond the reservoir boundaries and combines the minideposits into a single oil patch which permits maximum recovery, with the nuclear explosion being a special case of communication fracturing to link isolated pay units (cf. section 4.10.2.1.). Earlier attempts to fracture such formations with conventional explosives were only marginally effective. Nuclear blasts also have turned out to be feasible to create underground storage reservoirs in salt formations. Aspects of nuclear fracturing in various reservoir types are also discussed by ATKINSON & LEKAS (1963); COFFER, BRAY, KNUTSON & RAWSTON (1964) and BRAY, KNUTSON, WAHL & DEW (1965).

A probable further application for nuclear fracturing are massive oil shales in Colorado/USA and surroundings which can be retorted in place (HOWARD & FAST 1970 a). The oil is removed in liquid or gaseous form from the cavity created by nuclear explosion, thereby eliminating the need to mine the shale rock and to process it through a retort to extract the hydrocarbons (HOWARD & FAST 1970 a).

4.7.3.5. Water fracturing

As a consequence of high reservoir temperatures which often exceed the thermal stability boundary of many treating fluids, geothermal well stimulation is frequently carried out by simple water fracturing without proppants. Hydraulic fractures connecting natural crack systems providing contact to a fault zone or creating communication to a neighbouring well are generated by injection of water above fracture pressure. Since water has very limited possibilities of proppant transport, such treatments are commonly operated without proppant insertion. Similar non-proppant fracturing includes nitrogen stimulation of shales (cf. section 4.4.4.). While geothermal water fracturing comprises mainly large-scale jobs (cf. section 4.7.4.1.2.), small-scale water fracturing can successfully enhance oil production from carbonates and especially chalk by supporting water imbibition (cf. section 4.5.4.5.2.). Water fracturing is also of considerable significance in coal-seam stimulation (cf. section 4.4.3.4.1.)

4.7.4. Proppant fracturing

Hydraulic fracturing stimulation of geothermal wells presents challenging problems due to formation temperatures in the range 300 - 600 oF which require careful evaluation of the behaviour of fluids, proppants and equipment in this hostile hot brine environment (CAMPBELL, HANOLD, SINCLAIR & VETTER 1981; cf. section 4.4.5.). In order to avoid possible damage to the producing horizon, high-temperature chemical compatibility between in-situ and inserted materials must be verified. The most significant point is that in geothermal wells, the required techniques must be capable of bringing about the production of very large amounts of fluids. This necessity for high flow rates represents a significant departure from conventional petroleum well stimulation and demands the creation of very high near-wellbore permeability and/or fractures with very high flow conductivity. Hydraulic stimulation treatments in geothermal fields may be carried out in formations which produce either hot water or steam from both matrix permeability and from natural fracture systems.

Interesting targets for hydraulic fracture stimulation are geothermal wells that do not intersect nearby major natural crack systems, which can benefit from the establishment of high-conductivity linear flow channels to improve flow capacity from surrounding localized regions of low-permeability formation, which suffered from artificial damage during drilling, completion or workover operations including mud or cement invasion, and that require periodical remedial treatment as a result of fluid-production-related damage. If fracture stimulation can diminish or eliminate the need for new well or sidetrack drilling in the aforementioned situations, the potential for improving geothermal development economics and extending the resource base is certainly substantial. Some aspects of proppants, fluids and treatments in geothermal reservoir stimulation are briefly sketched as follows.

4.7.4.1. Proppant aspects

In geothermal wells, closure stress should in most cases be much lower than in equivalent-depth oil and gas wells because of the single-fluid system (although in some cases a two-phase system occurs if both water and steam are present). Closure stress is more or less in the order of the hydrostatic gradient which is abt. 0.5 psi/ft and thus ranges from 2,000 to 6,000 psi in wells from 4,000 to 12,000 ft in depth (SINCLAIR 1980). Therefore in most cases natural sand could be applied as propping agent (cf. sections 1.2.2. and 1.3.2.) if the high-temperature environment is neglected. According to the hot formation water and in order to meet also boundary conditions in deeper wells, however, sintered bauxite is certainly the superior material for propping of fractures in geothermal wells, because there is a definite need for chemically inert proppants that retain high permeability for long periods of time in the high-temperature reservoir and thus temperature-sensitive proppant types have to be discarded (cf. sections 1.3.2.1. and 5.7.1.). Some comments on high-temperature proppant performance and field experience are offered as follows.

4.7.4.1.1. High-temperature proppant performance

High-temperature proppant testing under geothermal reservoir conditions has shown that sintered bauxite, resin-coated sintered bauxite and resin-coated sand are the most stable materials and are insoluble up to 500 oF (250 oC) and either insoluble or only slightly soluble above 500 oF (SINCLAIR 1980). Conventional natural sand which is widely used in oil and gas industry is not strong enough to withstand the conditions in most geothermal wells, because it is affected by high-temperature water and brine particularly under stress (cf. section 5.7.1.). The main disadvantages of sand are brittle failure by point-to-point loading, weakening by numerous microfractures and faults within the grains, and stress corrosion cracking (cf. section 1.4.10.3.2.) in hot fluids already at low closure stresses, with natural sand therefore being inadequate as proppant in geothermal reservoirs.

Glass beads are not acceptable for geothermal well stimulation due to their brittle failure along with stress corrosion cracking in high-temperature brine (cf. section 1.2.3.1.). Resin-coated sand is the most suitable and probably also the most cost-effective proppant for geothermal wells due to its insensitivity to temperature and load, protective coating of inert resin, and low density (cf. section 1.2.6.). Resin-coated sintered bauxite is the best proppant ever tested showing no temperature sensitivity and permeability decrease under load and would certainly be the first choice for geothermal reservoir fracturing if being less expensive (SINCLAIR 1980; CAMPBELL, HANOLD, SINCLAIR & VETTER 1981). Although slightly crushable, sintered bauxite is much stronger than sand and effectively inert in hot brines also without resin pellicles enveloping the grains.

4.7.4.1.2. Field experience

Experience reports of proppant fracturing of geothermal reservoirs are given by CAMPBELL, HANOLD, SINCLAIR & VETTER (1981) and MORRIS & SINCLAIR (1984). The hottest geothermal well in the world which has been proppant fractured so far is Baca in New Mexico/USA where a high-conductivity crack has been generated by propping with sintered bauxite in a hydrothermal well in a depth of abt. 5,000 ft (1,700 m) and a bottomhole temperature of 540 oF (280 oC; VERITY & MORRIS 1981; WILLIAMS, WOODROOF & BOX 1982; MORRIS & SINCLAIR 1984). Although the stimulation treatment did not ultimately result in a commercial well because productivity declined to an uneconomical level after the fracturing job due to relative permeability reduction associated with two-phase flow effects in the water-steam system that results from restricted inflow as a consequence of the tight reservoir matrix around the fracture, the successful technical performance of the operation promises hydraulic fracturing of geothermal pay zones to be a valid future alternative to sidetracking or new well drilling. Aspects of proppant fracturing of geothermal wells are also treated by CARDEN, NICHOLSON, PETTIT & ROWLEY (1983). Geothermal well stimulation operations are also evaluated by CAMPBELL, MORRIS & VERITY (1981).

Non-proppant geothermal fracturing of hot dry rock reservoirs with chiefly water as fracturing fluid (cf. section 4.7.3.5.) includes the largest hydraulic stimulation operations ever carried out in terms of fluid volume. The current world record comprises a treatment in 13,500 ft reservoir depth with 600 oF bottomhole temperature where over 10 Mio. gal of water plus friction reducer and calcium carbonate have been injected (FRACFAX 1988 b).

4.7.4.2. Fluid aspects

In terms of fluids, high-viscosity gels which are commonly applied in low-temperature oil and gas wells cannot be used in geothermal reservoirs where low-viscosity fluids have to be the rule, because of the extremely high temperatures that degrade the gelled polymer fluids, and also due to stimulation fluid cost. Some aspects of sensitivity and stabilization of fluid viscosity are outlined as follows.

4.7.4.2.1. Fluid viscosity sensitivity

Water-soluble polymers are the main viscosifiers for hydraulic fracturing, but above 250 oF, almost all the polymer systems show a decline in viscosity (CAMPBELL, HANOLD, SINCLAIR & VETTER 1981; cf. section 4.4.5.). Delaying of this decline or degradation in properties can be performed by adding small amounts of methanol to the polymer-water solutions (cf. section 3.11.4.). High-temperature fluid stabilization can also be achieved by adding an oxygen scavenger in order to minimize polymer degradation by dissolved oxygen. Rapid decline in viscosity of the fluid once inserted into the fracture would inevitably result in poor proppant placement or would even provoke premature screenout failure (cf. sections 4.2.3.2.5. and 6.2.4.2.1.) of high-temperature geothermal stimulation treatments. High-viscosity fracturing fluids can only be applied in high-temperature geothermal reservoirs if preceded by a large cooling water pre-pad that gives rise to heat blockage (cf. section 4.3.4.6.1.4.).

4.7.4.2.2. Fluid viscosity stabilization

HARMS, GOSS & PAYNE (1984) present a new generation of fracturing fluids for ultrahigh-temperature application where an unique combination of heterogeneously substituted polysaccharide polymer, stabilizers and delayed crosslinking chemicals provides the extended thermal stability. An innovative crosslinking mechanism enhances the thermal properties of the fluid, and chemically controlled crosslinking reaction allows the maintenance of crosslinking time over a wide range of chemical and/or physical variables such as temperature and polymer loading. The controlled viscosity development gives rise to lower friction pressure and better downhole stability.

GRONE, BAUMGARTNER & WOODROOF (1983) document a crosslinked fracturing fluid system which has extended thermal stability due to favourable variations of polymer, buffer, gel stabilizer and crosslinker and that allowed successful proppant fracturing jobs up to 540 oF (280 oC) reservoir temperature (WILLIAMS, WOODROOF & BOX 1982; cf. also section 4.4.5.2.).

4.7.4.3. Treatment aspects

Geothermal well stimulation utilizes both large-scale conventional hydraulic fracturing of deep low-permeability zones and miniature fracturing treatments of shallow high-permeability reservoirs (CAMPBELL, HANOLD, SINCLAIR & VETTER 1981). Fluid flow within the fracture systems and between wells can be monitored by tracer return analysis and tracer retention modelling (TESTER, BIVENS & POTTER 1982; WALKUP & HORNE 1985) which is particularly useful for proper planning of waste water reinjection into highly-fractured geothermal reservoirs in order not to reduce the enthalpy of the produced water by short-circuiting within the pay zone (PRUESS & BODVARSSON 1984), with a correspondence existing between fast tracer return rates and enthalpy deterioration upon reinjection (HORNE 1982, WALKUP & HORNE 1985).

Aspects of hydraulic fracturing in geothermal energy reservoirs are also discussed by AAMODT et al. (1977), BODVARSSON & GAULKE (1986) and TESTER, MURPHY, GRIGSBY, POTTER & ROBINSON (1986). Effects of non-proppant fracturing treatments are reported by MAHONEY, STUBBS, SCHWERER & DOBSCHA (1980). Testing of geothermal and geopressured formations is reviewed by McCOY, HARTSOCK & DOBSON (1980); DA SIE & HOANG (1985); LEAVER, SAGEEV & RAMEY (1986) and MARCOU & GUDMUNDSSON (1986). BODVARSSON, PRUESS & LIPPMANN (1986) summarize modelling of geothermal systems including evaluation of fracture-dominated flow. Injection and recovery effects in fractured geothermal reservoirs are analyzed by PRUESS & BODVARSSON (1984) and BODVARSSON, PRUESS & SULLIVAN (1985).

4.7.5. Communication fracturing

In terms of geothermal communication fracturing, distinction has to be made between connection of injection and production wells on the one hand and connection of wells and large-scale natural fractures on the other hand.

4.7.5.1. Connection of injection and production wells

Apart from conventional production of hot water from aquifer-type geothermal reservoirs with or without stimulation, a special method in geothermal energy exploitation is the creation of communication fractures (cf. section 4.10.) between neighbouring wells and pumping of cold water from one borehole through the crack to the other well in order to heat it up in the hot dry rock. The direction of fracture propagation can be estimated from determination of the stress field by oriented cores or in-situ measurements by breakout logging (cf. section 6.2.1.8.), and the target well can then be drilled accordingly (possibilities of in-situ stress determination are outlined by ROEGIERS 1974; ABOU-SAYED, BRECHTEL & CLIFTON 1977; WARPINSKI, SCHMIDT & NORTHROP 1980; VOEGELE & JONES 1980, BLÜMLING 1983, TEUFEL 1983; DANESHY, SLUSHER, CHISHOLM & MAGEE 1984; BLANTON & TEUFEL 1986). Enhancement of fracture propagation through the centres of the boreholes can be achieved by orienting the perforations to the directions of the stress field of the reservoir (AHMED, THOMPSON, KELKAR, VEGHTE & HATHAWAY 1984; JARED, SCOTT & EVANS 1984; SCOTT 1986; cf. sections 4.10.3. and 4.10.5.).

As also in hydraulic fracturing of crystalline rocks and tight sedimentary formations the self-propping (cf. sections 4.5.1.1.1.2. and 4.5.1.5.1.) of some pilot fractures has not turned out to be satisfactorily effective and natural cracks of both sandstone and carbonate formations have also not shown to be sufficient pathways, the application of proppants for supporting the cracks will become more and more important with the progress of geothermal energy exploitation in the near future. Field examples of communication fracturing between two wells in geothermal reservoirs are reported by ALBRIGHT & PEARSON (1980), PEARSON (1981); BATCHELOR, BARIA & HEARN (1983); SARDA (1984), TALEBI & CORNET (1985), BAME & FEHLER (1986) and ROBERTS & CRAMPIN (1986).

4.7.5.2. Connection of wells and large-scale natural fractures

A special application of communication fracturing in geothermal reservoirs is to connect the wells with nearby major productive cracks or faults and to achieve the desired high temperatures of the produced fluid (up to 270 oF or more; CAMPBELL, HANOLD, SINCLAIR & VETTER 1981). Dendritic fracturing technique (cf. also sections 4.5.4.4.3. and 4.8.8.4.) offers the best chance of intersecting major natural cracks, whereas a single planar fracture might only parallel and not cross the principal natural cracks. Dendritic fractures are caused by pulsing the formation with reverse flow which provokes formation spalling and diversion of the fracture wings by downhole stress modification. The best results are achieved in naturally jointed formations where major and minor fracture systems already exist but may not have flow capability. Multiple pumping periods are used with each stage of dendritic fracturing utilizing a low-viscosity fluid, sand slugs and several flow-back periods, with high flow rates and friction reduction being applied in these treatments (CAMPBELL, HANOLD, SINCLAIR & VETTER 1981; MURPHY & FEHLER 1986). An alternative to the generation of new fractures can be reactivation of pre-existing joints and fissures in some cases (MEIER & ERNST 1981).

4.7.6. Fracture system enlargement by heat extraction

TESTER, MURPHY, GRIGSBY, POTTER & ROBINSON (1986) document hydraulic fracturing and subsequent increase of accessible reservoir volume and cracked rock surface area by energy extraction operations which causes substantial thermal drawdown in portions of the formation and results in reservoir contraction due to cooling which includes opening of shrinkage joints (in a comparable manner as taking place during solidification and crystallization of volcanic lava flows and pipe fillings; cf. plate XI/5 - 6). Excessive reservoir stresses resulting from heat removal can be released with stress unlocking techniques comprising injection at elevated pressures exceeding the in-situ normal and frictional stresses which hold the fracture faces together, thus giving rise to shear slippage and microseismicity (cf. section 6.2.2.3.2.). Large-scale heterogeneities such as superposition of flows in multiple joints undoubtedly exert great influences on heat transfer behaviour, since the spatial positioning of these low-impedance conduits effectively defines the accessible volume of the rock. In the absence of thermal stress cracking, reservoir growth due to heat extraction is really a thermal contraction effect, with the fractures expanding due to shrinkage of the rocks surrounding the cracks upon thermal drawdown. Overpressurization during injection can accelerate enlargement of the fracture system, and further support is given by remedial pressurization leading to stress unlocking.

BODVARSSON, PRUESS & O´SULLIVAN (1985) also document the beneficial effect of injection for increasing the long-term energy output from wells because of support of heat extraction from the reservoir rocks and maintenance of high flow rates. This relationship is very helpful for geothermal operations where in most high-temperature fields brines have to be reinjected as a consequence of reservoir pressure maintenance, enhancement of energy recovery from hot dry or wet formation rocks, and environmental safety obligations not permitting surface disposal of the brines. Aspects of injection response of geothermal reservoirs are also discussed by LIPPMANN, TSANG & WITHERSPOON (1977); BODVARSSON & TSANG (1982); GUDMUNDSSON, HAUKSSON, THORHALLSSON, ALBERTSSON & THOROLFSSON (1984) and GUDMUNDSSON & OLSEN (1985). Reinjection has to be done with fluids of low gas saturation, because the lower the flowing gas concentration, the more backflow of fluids from the fractures into the matrix of the reservoir rock takes place, thereby ultimately increasing the recoverable geothermal energy reserves (O´SULLIVAN, BODVARSSON, PRUESS & BLAKELEY 1985; BODVARSSON & GAULKE 1986). Success monitoring of reservoir enlargement by heat extraction can also be performed by multiple interference testing of wells which delineates reservoir boundaries (EARLOUGHER & RAMEY 1973; LEAVER, SAGEEV & RAMEY 1986; cf. section 4.2.4.1.2.2.).

4.7.7. Shear stimulation vs. tensile fracturing

Field experience of hydraulic treatments of jointed geothermal reservoirs indicates shear stimulation to be the decisive mechanism in naturally fractured formations rather than conventional tensile failure (MURPHY & FEHLER 1986; cf. sections 4.8.8.4.2. and 6.2.2.3.2.). A coupled rock mechanics and fluid flow model shows that shear slippage along pre-existing joints is more easily induced than conventional tensile failure, particularly when the difference between minimum and maximum earth stresses is large, the joints are oriented at angles between 30 and 60 degrees to the principal earth stresses, and a low-viscosity fluid is injected. Aspects of origin of shear slippage, dilation behaviour, joint deformation spectrum, and dendritic fracturing treatments are outlined as follows.

4.7.7.1. Origin of shear slippage

Most rock masses contain pre-existing fractures or joints (MURPHY & FEHLER 1986). Injection of fluids results in reduction of the effective closure stress. With continuing injection, the fluid pressure can attain a value high enough that the effective closure stress no longer provides sufficient friction in order to resist shearing stresses acting parallel to the joint mode, and shear slippage will occur. Sufficient shear slippage leads to overriding of one surface asperity over or atop another, so that even if the pressure is suddenly reduced, joint opening and permeability are irreversibly increasing, resulting in shear stimulation.

If fluid injection rates are modest, shear stimulation may give rise to sufficient permeability that no further increase in pressure is attainable. If, however, the formation of void space by shearing is insufficient to accommodate the fluid volume injected into the rock joints, the pressure will continue to rise and eventually attain a value equal to the earth stress acting normal to the joint and consequently the effective closure stress is zero, with then the opposing surfaces of the rock that meet at the joint starting to part. This joint separation does not involve actual rupturing of the rock and thus the process should not be called fracturing. If proppants, either purposely injected with the fluid or accidentally broken off the joint surfaces, are trapped within the crack following shut-in, joint stimulation is achieved by irreversible increase of joint opening.

4.7.7.2. Dilation behaviour

The dilation behaviour for slippage and separation of joints is quite different (MURPHY & FEHLER 1986). With increasing fluid pressure, dilation is first small and simply results from the decrease of effective closure stress, but then shear slippage ensues. With continuing joint slippage, the crack surfaces attain a state in which one large roughness asperity lies atop another and further slippage would allow the largest asperity to slide over and down the other, with the maximum shear dilation typically being in the order to a fraction of a millimeter (BARTON, BANDIS & BAKHTAR 1985). Further increasing joint pressure then leads to separation with origin of an opening of typically tens of millimeters (PERKINS & KERN 1961, DANESHY 1973 a). Therefore the key to understanding stimulation lies in a combination of rock mechanics and fluid dynamics and involves both shear slippage and tensile failure (LOCKNER & BYERLEE 1977, MURPHY & FEHLER 1986; cf. sections 4.8.8.4.2. and 6.2.2.1.2.).

Thus while hydraulic fracturing has been conventionally considered to comprise almost exclusively tensile failure of the rock and separation of the crack walls, shear slippage stimulation is also a very important process (LOCKNER & BYERLEE 1977, HAST 1979; SOLBERG, LOCKNER & BYERLEE 1980), particularly in hot dry rock geothermal reservoirs (MURPHY & FEHLER 1986; TESTER, MURPHY, GRIGSBY, POTTER & ROBINSON 1986).

4.7.7.3. Joint deformation system

In an actual hydraulic fracturing operation, the entire spectrum of joint deformation takes place (MURPHY & FEHLER 1986). Near the injection well where the flow passage area is limited, fluid velocities and pressure gradients are large and separation occurs. Near the tips of the fractures far away from the injection well, however, fluid velocities and pressures are much reduced and shear stimulation takes place. Distinction has to be made between rock mechanics and fluid dynamics in hydrocarbon and geothermal reservoirs. In petroleum-bearing formations, very viscous fracturing fluids are normally injected at high rates, with consequently joint separation being dominant, and because in many cases only a limited amount of natural cracks exists, actual jointing of the rock by tensile failure occurs. In geothermal reservoirs, however, natural joints occur frequently, and high downhole temperatures render most viscosifying agents useless, with therefore water being used as fracturing fluid and this low fluid viscosity leads to a dominance of shear stimulation (cf. section 4.7.3.5.).

Microearthquake monitoring of geothermal reservoir stimulation reveals shear slippage along existing joints to be the dominant cause of seismicity and a zone of multiple fractures occurring predominantly by shear failure, thus being in contrast to the conventional hydraulic fracturing theory that predicts the propagation of a single crack due to tensile rock failure at a pressure equal to the minimum earth stress (MURPHY & FEHLER 1986; cf. section 6.2.2.3.2.). Shear stimulation modelling by interactive coupling of fluid dynamics with rock stresses and deformation shows that when the orientations of the pre-existing joints were rotated 30 degrees from the principal stress directions and a low-viscosity fracturing fluid like water is used, two types of stimulation pattern occur.

The first type comprises creation of only a single joint if frictional resistance to shear slippage is low and maximum dilatancy (cf. section 4.8.1.3.1.) due to shear is large. The second type includes multiple joint stimulation corresponding to high shear resistance or small dilatancy. Shear slippage along the joints is accompanied by shear-stress drops, and the interaction of these stress drops with the acting earth stresses results in opening of joints more perpendicular to the maximum stress, with thus a dendritic or branched joint pattern originating. This dendritic stimulation fabric consisting of multiple branches has important implications in reservoir engineering, because volume drainage is more efficient than areal drainage.

4.7.7.4. Dendritic fracturing treatments

Dendritic fracturing treatments have originally been introduced to the hydrocarbon exploitation industry and are usually carried out by repeatedly fracturing a well with a proppant-bearing fluid and then shutting-in and venting it (KIEL 1977; cf. sections 4.5.4.4.3. and 4.8.8.4.). The first cycle of pressurization results in spalling and self-propping (cf. sections 4.5.1.1.1.2. and 4.5.1.5.1.) of the main fracture, and in subsequent cycles, the proppant purposely introduced in the fracturing fluid bridges the spall-proppants, with thus pressure rising and lateral fractures being propagated perpendicular to the main crack. While the mechanism proposed by KIEL (1977) possibly works when the principal stresses in two directions are nearly the same, the model of MURPHY & FEHLER (1986) shows that in case of considerably differing stresses, pressure rise in the blocked main fracture would simply result in further lift-off of the large crack, thereby overcoming the temporary blockage and allowing the main fracture to continue its propagation without opening of secondary lateral cracks.

The key to dendritic fracturing, however, is shear which allows the necessary reduction of the earth stress parallel to the main fracture in order to permit opening of lateral joints (cf. section 4.8.8.4.2.). Dendritic fracturing can thus occur under certain conditions, being that the major joints are not parallel to the principal earth stresses and that flow rate and fluid viscosity within the joints are low enough that shear dilation is still sufficient to transmit the fluid rate without excessive pressure gradients.

4.8. Minifracturing and other small-scale fracturing

Hydraulic proppant fracturing carried out in Europe and other parts of the world so far mainly comprised more or less full-scale larger or smaller treatments with respect to reservoir size and aimed on optimum stimulation (with the achievement of this goal, however, also including high cost of the operation), with in most cases both fracture length and fracture height having been designed near the upper boundaries of technical possibility of performance as well as economical feasibility (cf. section 4.8.11.). Miniature hydraulic proppant fracturing and other small-scale fracturing, however, can not only often be a technical alternative to large-scale stimulation, but can in many cases also represent economically suitable approaches of marginal reservoirs.

Various aspects of minifracturing and other small-scale fracturing including pre-treatment pressure breakdown test and fracture parameter determination, formation damage around the wellbore, field experiments and mHF vs. MHF, pilot stimulation of marginal reservoirs, horizontal well or drainhole drilling and fracturing, uptake capacity enhancement of injection wells, combination of natural and artificial fractures, late-stage refracturing of old wells, and combination of acid and proppant fracturing and other aspects are briefly sketched as follows. Comments are also offered on optimization of fracture length and height which are some of the key parameters for treatment design. The discussion is introduced by a short overview of characteristic features and definitions of minifracture, microfracture and MHF operations.

4.8.1. Minifracture, microfracture and MHF definitions and principles

The account on minifracturing and other small-scale treatments is introduced by briefly defining and outlining minifracture, microfracture and massive fracture operations with characterization of typical features and comparative effectiveness. Some comments on mini-massive fracture are also offered.

4.8.1.1. Minifracture and microfracture

Minifractures are scaled-down treatments which are pumped at more or less the same fluid rates per time interval as the full-size big operation (VEATCH 1983). Minifractures are in most cases performed by injection of fluid at rates of 12 - 15 bbl/min, and the only difference to large-scale fracturing jobs is the limited total pumped volume of the minifracture which is mainly in the range of 5,000 - 40,000 gal (PAI, GARBIS & HALL 1983). The fluid taken for the minifracture calibration treatment (cf. section 4.8.2.1.) is usually the same as anticipated to be applied throughout the major portion of the MHF operation. For proper formation testing, the minifracturing fluid has to be proppant-free in order to allow unrestricted closure of the fracture (cf. section 4.8.2.5.). Microfractures differ from minifractures by comprising reduction of both total volume and pumped fluid rates with respect to full-size operations.

Microfractures are typically performed with very low fluid rates down to 1 gal/min and with very low total fluid amounts up to abt. 500 - 1,000 gal, and are small breakdown fractures comprising hydraulic formation breakdown after bullet or jet perforation (BRANAGAN & WILMER 1988; cf. section 4.8.2.2.). While proppant insertion into cracks created by minifracturing may be quite common, microfracturing is exclusively done without proppants. Minifractures or microfractures for special purposes may comprise volumes as low as 500 - 10,000 gal or 1 - 250 gal, respectively (WARPINSKI, BRANAGAN & WILMER 1985).

Aspects of microfracturing are also discussed by EVANS & ENGELDER (1986), CRAMER & SONGER (1988), HUCKABEE (1988); SHLYAPOBERSKY, WALHAUG, SHEFFIELD & HUCKABEE (1988) and SHLYAPOBERSKY, WONG & WALHAUG (1988). Minifracturing fluid quantity is often designed to be approximately equal to the pad volume of the main treatment (SATTLER, RAIBLE, GALL & GILL 1988). Points of mini- and micro-fracturing are also illustrated by McLENNAN & ROEGIERS (1982) and SHLYAPOBERSKY (1985). Another small-scale stimulation treatment is the bucketfracture where more or less a bucket of natural sand or synthetic proppants is injected into the very near surroundings of the borehole (cf. DETOURNAY, McLENNAN & ROEGIERS 1986; ROEGIERS 1987). CROW (1977) documents miniature massive fracturing as a transitional or intermediate stimulation technique for low-porosity and -permeability zones (cf. section 4.8.1.3.).

4.8.1.2. Massive fracture

Massive fractures are characterized as follows by proppant quantities and fracture wing lengths in comparison to minifractures and microfractures. Some examples of application of massive hydraulic fracturing in tight gas sand stimulation are also outlined.

4.8.1.2.1. Proppant quantities and fracture wing lengths

In comparison to the size of minifractures and microfractures, massive hydraulic fracturing (MHF) treatments comprise fluid volumes in the range of 100,000 -1 Mio. gal or even more (PAI, GARBIS & HALL 1983; VEATCH 1983; up to 3,000 m3; TEUFEL 1981). Proppant volumes from abt. 100 t (200,000 lbs) via frequently 1,000 t (2,000,000 lbs) up to 2,500 t (5 Mio. lbs) are pumped in MHF operations where fracture wings in excess of 1,000 ft (300 m) up to 2,500 ft (800 m) in most applications and in extreme cases up to 4,500 ft (1,500 m) length and over 200,000 ft2 area are created (cf. section 1.1.1.4.). Gross fracture height determining crack volume and net fracture height marking the height over which fluid loss will occur are in the ranges of 100 - 1,000 ft (30 - 330 m) and 30 - 300 ft (10 - 100 m), respectively, in most of the MHF treatments. MHF operations are generally carried out in tight gas sands with less than 0.1 md effective reservoir matrix permeability (cf. section 3.4.1.) and have in the USA so far been successfully performed in many regions of the Rocky Mountains area.

4.8.1.2.2. Applications in tight gas sand stimulation

Some of the key application patches of MHF treatments in the USA comprise San Juan Basin (Dakota reservoir) of Northwest New Mexico, Wattenberg field (Muddy "J") in the Denver Basin of Colorado, Wamsutter field (Mesa Verde) in the Green River Basin in Southern Wyoming, Moxa Arch (Frontier) in Southwestern Wyoming, Cotton Valley Sand and Lime in East Texas, Cotton Valley Formation in Louisiana and Arkansas, Canyon Sand Formation of West Texas, Austin Chalk of South and Central Texas, and Morrow and Atoka Formations of Southeast New Mexico (PAI, GARBIS & HALL 1983; cf. section 3.4.1.1.). Massive hydraulic fracturing in tight gas sands in Europe is chiefly carried out in Rotliegend and Upper Carboniferous in British and Dutch Southern North Sea as well as Netherlands and Germany FRG onshore (cf. sections 2.4.1. and 3.3.) and in the Tertiary of Yugoslavia and Hungary (cf. section 2.4.3.1.).

While most tight gas sands are deep, compacted and normal- to overpressured, occasionally also shallow or underpressured tight gas sands occur (RICE & SHURR 1978; NYDEGGER, RICE & BROWN 1979; BROWN & CRAFTON 1981, GAUTIER & RICE 1981; KUKAL, BIDDISON, HILL, MONSON & SIMONS 1983). Distinction has also to be made between blanket-type and lenticular tight gas sands, with the latter category of low-permeability reservoirs being widely distributed in the Greater Green River, Piceance, Wind River and Uinta basins (WARPINSKI, BRANAGAN, SATTLER, LORENZ, NORTHROP, MANN & FROHNE 1985). Other aspects of tight gas sand formations and basins are discussed in section 3.4.1.

4.8.1.3. Mini-massive fracture

Mini-massive stimulation is an intermediate or transitional technique between massive treatment and minifracturing which combines several methods commonly used to increase fracturing efficiency and to improve results including control of fluid loss into hairline fractures and matrix permeability, supervision of fracture height, and achievement of a deeply-penetrating and highly-packed fracture (CROW 1977). Following some comments on dilatancy of hydraulic fracturing, the design of mini-massive fractures is outlined.

4.8.1.3.1. Dilatancy of hydraulic fracturing

Hydraulic fracturing of underground formations is characterized by dilatancy which comprises generation of numerous tiny cracks in certain directions as rock approaches its breaking point (CROW 1977). Subsurface rocks are triaxially loaded due to the force exerted upon them by the overlying strata. This vertical force or stress imposed on the rock causes a horizontal stress to exist within the rock system. If the rock were not confined within the earth, this vertical force would express itself by causing the rock to deform or expand its lateral dimensions. When pressure is applied from within such a system, the rock ruptures or fractures in a plane perpendicular to the least amount of stress on the system, with these stresses controlling fracture direction and determining whether the crack plane will be horizontal, vertical or inclined.

During hydraulic fracturing stimulation, numerous tiny cracks form as the rock approaches its breaking point due to the applied pressure, and dilatancy then continues to occur along the leading edge of the crack as it progresses outward into the formation. Dilatancy during hydraulic fracturing is the reason for premature screenout in very low-permeability reservoirs (cf. sections 4.2.3.5.3. and 6.2.4.2.1.), failure of the fracture to penetrate the designed length (cf. section 4.8.11.), and failure of well productivity to achieve predicted folds of increase. The explanation is that a portion of the stimulation fluid leaks off into the numerous tiny cracks and the encountered natural fractures, with the fluid having been lost to original and artificial hairline fissures not being available for hydraulic fracture extension (this secondary fluid leakoff into natural and synthetic hairline cracks is also discussed by WARPINSKI 1988 a; cf. section 4.8.8.3.2.3.).

4.8.1.3.2. Design of mini-massive fracturing

Secondary fluid loss into hairline fissures can be satisfactorily controlled with 100 mesh sand as a spearhead in the hydraulic fracturing treatment (CROW 1977; cf. sections 1.4.11.2.1. and 4.8.8.3.1.2.). Dropping of proppants out of the crack prior to its closure and fallout in tubing or casing can be avoided by bleeding the well immediately back after the treatment to allow the fracture to close and to hold the proppants in place (cf. section 4.12.2.2.1.), and further improvement can be made by using high proppant concentration for complete packing of the entire crack (cf. section 4.3.5.). Therefore mini-massive fracturing is designed to tie together several techniques that have been used individually in the past to increase crack efficiency. The sequence of elements in mini-massive treating includes ideally pumping of a pad fluid containing 100 mesh sand in order to prop and to control leakoff into small cracks and natural fractures, injecting a stimulation fluid limiting loss into matrix permeability, using a proppant saturation as high as operationally feasible in order to fill the entire created fracture height, and bleeding the well back as soon and as fast as practical so that the fracture will close and trap the proppant while it is still suspended in the stimulation fluid (cf. section 4.12.3.1.).

Field experience has revealed that the optimum sizing of mini-massive fracture jobs is abt. 10,000 - 50,000 lbs of 100 mesh sand, 50,000 - 100,000 lbs of 20/40 proppants, facultatively up to 10,000 lbs of 16/20 or 10/20 proppants, and 20,000 - 50,000 gal of gelled water or other fracturing fluid. Thus the fluid volume is within the range of conventional minifractures, whereas the proppant quantity is already approaching the lower boundary of massive stimulation treatments, and the bridge from mere calibration test to real reservoir improvement is built by incorporating the 100 mesh sand stage which is a major element of performance amelioration.

4.8.2. Pre-fracturing pressure breakdown test and fracture parameter determination

Miniature hydraulic fracturing jobs (minifractures, mHF; NOLTE 1979, SMITH 1981; CONWAY, McGOWEN, GUNDERSON & KING 1985) have hitherto been mainly done as pre-fracturing breakdown and/or injection tests in order to open the formation without insertion of proppants, with the main purpose being to obtain definite figures that could replace the preliminary and partially still questionable assumptions of fracture gradient, fracture pressure, closure stress (cf. sections 1.2.1.1. and 1.3.2.1.), closure time, fluid efficiency, effective fluid leakoff coefficient, fracture length (cf. section 4.8.11.), fracture height, fracture width, and other important data in the design with the fracturing simulator in order to carry out final optimization runs before the big job is executed (LOCKE & SAWYER 1975; ABOU-SAYED, BRECHTEL & CLIFTON 1977; ZOBACK & POLLARD 1978; NOLTE 1979, 1982; SMITH 1981; VOEGELE, ABOU-SAYED & JONES 1982; AHMED, KELKAR & SCHATZ 1983; VEATCH 1983, CUESTA & ELPHICK 1984; PEARSON, ABOU-SAYED & NGUYEN 1984; CONWAY, McGOWEN, GUNDERSON & KING 1985; ELPHICK 1985; LEE 1985, 1987; MARTINS & HARPER 1985; HOLDITCH, ROBINSON & WHITEHEAD 1986; ROBINSON, HOLDITCH & WHITEHEAD 1986; TANSOE, KING & HOLMAN 1986; MOSCHOVIDIS 1987; SMITH, MILLER & HAGA 1987; TAN, McGOWEN, LEE & SOLIMAN 1988).

The determined parameters are then used for evaluation of presence of natural fractures, pad volume required, best fluid-loss additives to be used, and achievement of optimum fracturing treatment design (TAN, McGOWEN, LEE & SOLIMAN 1988). Some points of minifracture calibration experiments, microfracture breakdown procedures, pressure analysis and design input data collection, size and significance of minifracturing, proppant vs. non-proppant minifracturing, and modified microfracturing and minifracturing analysis for overpressure calibrated fracture design are outlined as follows.

4.8.2.1. Minifracture calibration experiments

Minifractures can therefore also be regarded as calibration experiments for three-dimensional fracture simulators (BRANAGAN, CIPOLLA, LEE & YAN 1987; HOLDITCH, ROBINSON, WHITEHEAD & ELY 1987) and in fact represent scaled-down trial runs of the main fracture treatment (BARBY & BARBEE 1987) or dummy stages prior to the actual fracturing operation (OSBORNE, McLEOD & SCHROEDER 1981) and are in fact data-gathering jobs (HOOVER & ADAMS 1988). The most important aims and purposes of minifracture calibration experiments are diminution of screenout rates and fluid-loss coefficient determination.

4.8.2.1.1. Diminution of screenout rates

If properly performed, minifractures have considerable significance for cutting down screenout rates in the main stimulation treatments, because many screenout reasons are based on insufficient knowledge of formation behaviour and incorrect assumptions of fluid loss (cf. sections 4.2.3.5.3. and 6.2.4.2.1.). Fluid leakoff may be caused by occurrence of natural fractures, failure of fluid-loss additive and higher matrix permeability than expected.

Other common screenout reasons are formation response other than anticipated and less than adequate quality control of fracturing fluids. In addition to better information derived from minifracturing experimental treatments, screenout rates can be decreased by computer monitoring (cf. section 6.2.4.3.) which often enables early screenout prediction during the job and its prevention by on-the-spot changing of the pumping schedule. Improvements of fluids to guarantee their stability in higher temperature for longer time also assist in avoiding screenout failure.

4.8.2.1.2. Fluid-loss coefficient determination

One of the most important fracturing treatment design parameters to be determined by minifracture analysis is the fluid-loss coefficient (TAN, McGOWEN, LEE & SOLIMAN 1988). Loss of stimulation fluid to the formation is a dominant and decisive factor in hydraulic fracture propagation. Local geological variability and presence of natural cracks render laboratory simulations ambiguous and require in-situ fluid-leakoff parameter determination from pressure decline following a small-scale calibrated fracturing schedule. Since the aim of mapping a minifracture by techniques such as acoustic emission interpretation (cf. section 6.2.2.3.2.5.) is to look at the feasibility of large-scale hydraulic fracturing treatments, it is important to measure the evolution of downhole pressure which can be correlated with local stresses (SARDA, PERREAU & DEFLANDRE 1988; cf. section 6.2.4.2.4.).

Applicability and reliability of minifracturing analysis has in the last years been facilitated significantly by introduction of computerized vans equipped with special hardware to more accurately monitor pressure decline and development of adequate software (cf. section 6.2.4.3.). Aspects of minifracture assessment are also discussed by LEE (1986) and SOLIMAN (1986). An additional technique of formation evaluation is pressure decline analysis (NOLTE 1979, NOLTE & SMITH 1981) which gives optimum results in combination with minifracture interpretation. Some points of microfracture breakdown procedures, pressure analysis and design input data collection, size and significance of minifracturing, and proppant vs. non-proppant minifracturing are outlined as follows.

4.8.2.2. Microfracture breakdown procedures

BRANAGAN & WILMER (1988) discuss a series of microfracturing breakdown procedures to minimize natural fracture damage in jointed tight gas sandstone reservoirs. If perforating alone does not provide sufficient production or when perforations yield restricted, uneven or damaged flow paths, microfracturing breakdown is carried out in order to mitigate these undesirable effects. Microfracturing breakdown creates small cracks extending from the wellbore through the perforations into the formation.

The microfracturing process generally consists of pumping of a low-viscosity non-damaging fluid at sufficiently high rates to initiate and propagate a short crack that emanates from the perforations and extends beyond the damage zone. The use of ball sealers or other diverting agents (cf. section 4.2.2.1.1.) during microfracturing provides adequate control of fluid and fracture placement. In the course of drilling and completing a well, numerous mechanisms can degrade the original high conductivity of the natural fractures (cf. section 4.8.8.3.) such as invasion of drilling mud (cf. section 5.11.2.) that may ultimately act as a plug, cements squeezed into the fracture, perforations and near-wellbore stress changes that could tend to close the fracture, and interfacial tensional forces generally described by capillary pressure of invaded completion liquids which limit fracture flow capacity.

Non-aqueous microfracturing breakdown techniques that aim on avoiding water blocking in the natural fracture system (cf. section 3.11.2.4.2.) include hydraulic nitrogen fracturing using surface pump trucks, perforating with an underbalanced nitrogen column, high-pressure nitrogen impulse perforating/microfracturing combination, nitrogen impulse injection using a tubing pump-out plug, and tailored pulse explosive fracturing (cf. section 4.2.2.1.3.3.). Nitrogen alone can be effectively employed in microfracturing breakdown procedures (as well as on larger scale for hydraulic fracturing of shales; cf. section 4.4.4.2.). Perforating in a dry underbalanced nitrogen wellbore column successively connects borehole and highly conductive natural cracks. Nitrogen impulse perforating/microfracturing combination using high-pressure nitrogen to simultaneously create fractures during perforation is also successful in achieving the desired wellbore enhancement. Nitrogen impulse technique is a practical remedial treatment of near-wellbore and natural fracture damage.

4.8.2.3. Pressure analysis and design input data collection

The introduction of minifracture analysis by pressure modelling resulting in crack parameter and geometry determination in 1979 (NOLTE 1979, NOLTE & SMITH 1981; cf. section 6.2.4.2.4.) was the second major breakthrough in hydraulic fracturing technology apart from the invention of synthetic intermediate- and high-strength proppants (SOLIMAN 1986). Minifracture evaluation consists of performing a small fracturing treatment with little or no proppants after which the well is shut in and pressure decline with time is monitored. As the rate of pressure decline with time depends on formation, fluid and fracture parameters, minifracture assessment allows field calculation of important input data for the design of the main treatment. The most important goals of minifracturing are crack diagnostics and containment prediction (SATTLER, RAIBLE & GALL 1985). The discussion as follows focusses on minifracturing test and evaluation types, in-situ stress determination, closure stress vs. fluid type, premature screen-out termination recognition, and fluids and mechanics testing.

4.8.2.3.1. Minifracturing test and evaluation types

The minifracturing procedure often includes conventional step rate test, repeated pump-in/flow-back tests and pump-in/shut-in pressure decline test (FELSENTHAL 1974; SMITH, ROSENBERG & BOWEN 1982; KIM, CHAMPION & COOPER 1985; WARPINSKI, BRANAGAN, SATTLER, LORENZ, NORTHROP, MANN & FROHNE 1985; BRITT & LARSEN 1986, ROSE 1988; TAN, McGOWEN, LEE & SOLIMAN 1988; WAREMBOURG 1988, WARPINSKI 1988 a). The individual elements or stages of minifracturing are briefly characterized as follows.

4.8.2.3.1.1. Step-rate test

The step-rate test is used to determine fracture extension pressure and rates, with the injected fluid usually being an incompressible Newtonian fluid such as water (TAN, McGOWEN, LEE & SOLIMAN 1988). The fracture extension pressure or parting pressure provides an accurate indication of pressure required to extend an existing fracture and thus determine the hydraulic horsepower needed to perform the actual treatment. Following discussion of some general aspects of multiple-step-rate test, the two-step-rate test representing an improved method is introduced.

4.8.2.3.1.1.1. General aspects of multiple-step-rate test

During conventional multiple-step-rate tests, the fluid is injected into the formation at a very low initial rate which is then stepped up in small increments and maintained at each increment for a constant time span which is sufficient for pressure stabilization (TAN, McGOWEN, LEE & SOLIMAN 1988). The step-rate test is particularly beneficial for determination of the maximum injection rate that can be pumped under matrix conditions. Aspects of step-rate tests are also discussed by FELSENTHAL (1974), ALLEN & BAZA (1980), BENNETT & CLARK (1985); WAREMBOURG, KLINGENSMITH, HODGES & ERDLE (1985); PALMER & VEATCH (1987) and SINGH, AGARWAL & KRASE (1988). Step-rate testing is the primary method to define the maximum safe injection pressure without fracturing the reservoir rock, being the formation or fracture parting pressure which initiates a crack in an unfractured well and extends an existing crack in a fractured well.

4.8.2.3.1.1.2. Two-step-rate test

SINGH & AGARWAL (1988) present the two-step-rate test as an advanced procedure for determining formation parting pressure. Compared to a conventional step-rate test which consists of several (usually seven or more) constant-rate injection steps of equal duration, the two-step-rate test requires only two constant-rate steps during which pressures are measured continuously. The two-step-rate test has potential for a large-scale use because of time and cost savings as compared to conventional step-rate tests. The two-step-rate test can be run in at least four different modes. The first mode comprises shut-in of the injection well for the first step followed by injection at a constant rate for the second step to exceed formation or fracture parting pressure. The second mode includes the well being on stabilized low-rate injection, with the first step consisting of the injection rate being stepped up to a constant value below formation or fracture parting pressure, and only the second step exceeding the latter value. The third mode represents reduction of injection rate to a lower constant rate for the first step and increase to a higher constant rate for the second step. In the fourth mode, the well is stabilized at shut-in conditions, with the test being composed of two constant-rate injection steps with progressively increasing rate.

All the four different modes are characterized by the pressures attained for the entire duration of the first step being below fracture or formation pressure, and the injection rate for the second step being chosen such that the injection pressure exceeds the formation or fracture parting pressure during this step. It is important that injection rates are stable and constant before and during each step and the time-rate-pressure data are recorded accurately on a continuous basis during the steps. The results are analyzed by multiple-rate superposition methods. The test should be designed such that the duration of the first step is long enough to be out of wellbore storage effects or at least long enough for the pressure data to be out of the entire wellbore storage dominated flow period. The duration of the first step can also be designed based on the time to reach the radial/pseudo-radial flow regime (cf. section 4.11.3.), or for a fractured well it can be limited to the end of the linear flow regime. If the first step consists of a shut-in or a reduced rate injection period (comprising the first or third modes), equal-time step size can be used for the first and second steps. If the injection rate is increased for each subsequent step (including the second or fourth modes), the first step should be twice as long as the second step.

4.8.2.3.1.2. Pump-in/flow-back test

The pump-in/flow-back test is applied to determine the least principal stress or closure pressure of the formation and is performed by injecting a volume of fluid (normally water) at fracturing rates followed by flowing back the well at a constant rate until closure has been detected (TAN, McGOWEN, LEE & SOLIMAN 1988). Closure pressure is the fluid pressure required to initiate opening of an existing fracture, and is equivalent to and counteracts stress in the rock perpendicular to the crack plane. Closure pressure may be equal or less than breakdown pressure required to initiate an existing fracture, but will always be less than the fracture extension pressure. If the flowback is within the correct range, closure pressure is indicated by the inflection point or change in curvature from concave to convex on the pressure vs. time plot (cf. section 6.2.4.2.), with the inflection point being caused by transition from linear to radial flow as the fracture closes (cf. section 4.11.3.). Closure pressure can be confirmed by the pressure that rebounds from a shut-in after the flowback period, and repetition of the pump-in/flow-back test using different flowback rates gives a more reliable value.

Closure pressure after the main fracturing treatment is in most cases higher than that determined prior to the main job by a minifracturing calibration test. Minifracturing analysis is chiefly performed by conventional type-curve matching and energy-balance equation technique (LEE 1987). Maintaining constant flowback rate is critical to closure pressure evaluation which together with fluid-leakoff coefficient belongs to the most important variables to be obtained in minifracturing analysis that may warrant modification of fluid volume, proppant schedule, pumping rate, necessity of additives and/or requirement of enhanced closure of the main treatment. Multiple injections while changing fluid properties and additives enhance data collection while providing a means of determining whether or not fluid-loss agents are desirable. If fluid leakoff is relatively low, it may not be advantageous to reduce it further for proppant settling characteristics. If the pad fluid without fluid-loss additives is not sufficient for fluid-leakoff control, subsequent injections and declines including fluid-loss agents can be run until fluid leakoff management is acceptable.

4.8.2.3.1.3. Pump-in/shut-in test

The pump-in/shut-in test is probably the most common minifracturing test used and comprises injection of the stimulation fluid chosen for use in the main treatment at the designed rate for the big job for a specific time, with then the well being shut-in for pressure decline monitoring (TAN, McGOWEN, LEE & SOLIMAN 1988). In formations with high leakoff or fracture growth into higher-stress zones, closure pressure determined from pressure decline data may be more appropriate than that measured from pump-in/flow-back tests. During shut-in, the pressure decline approaches a linear relation with the square root of time from shut-in, with the inflection or change of slope from the linear decline indicating closure. Identification of the correct straight line may be enhanced by utilization of additional plots.

Estimation of fracturing gradient and injection rate can also be obtained by a combination of minifracture dummy treatment and ball-out operation (OSBORNE, McLEOD & SCHROEDER 1981; cf. section 4.4.2.3.). SHELLEY & McGOWEN (1986) document how pump-in test correlations can be used for proppant placement predictions. CASTILLO (1987) and WAREMBOURG (1988) outline fracture height determination from pump-in calibration treatments.

4.8.2.3.2. In-situ stress determination

All the pump-in methods have become the most prevalent techniques for measuring in-situ stress with determination of both magnitude and direction (WARPINSKI, BRANAGAN & WILMER 1983). The minimum in-situ stress can be calculated from the inflection points in the observed pressure profiles before and after fracture closure (ECONOMIDES 1987 b). Other techniques for in-situ stress determination are anelastical strain recovery of oriented cores (TEUFEL 1982, 1983, 1985; BLANTON & TEUFEL 1983; cf. section 6.2.2.3.3.4.), acoustic wavetrain measurements (ARON & MURRAY 1978; MAO, SWEENEY, HANSON & COSTANTINO 1984; NEWBERRY, NELSON & AHMED 1985; cf. sections 6.2.1.6.1. and 6.2.1.10.4.), and mechanical properties logging (cf. sections 6.2.1.7.7. and 6.3.6.). WOLFF, BREDEHOEFT, KEYS & SHUTER (1974) perform tectonical stress determination by minifracturing.

While pressure decline monitoring and analysis of minifracturing treatments was hitherto carried out in uniform reservoirs, MOSCHOVIDIS (1987) presents an interpretationn of minifractures initiated near or at the interface of two formations of different leakoff characteristics and mechanical properties and penetrating both sedimentary layers (cf. section 4.5.4.4.2.). Distinction can be made between high-efficiency (storage-dominated) minifractures where the crack area is proportional to pumping time, and low-efficiency (leakoff-dominated) minifractures where the crack area is proportional to the square root of pumping time. Fracturing treatments initiated near or at the interface of two formations have been applied in North Sea chalk oil fields to obtain relatively stable well completions. Similar interface-inserting fracturing is significant in coal-seam stimulation, with an underlying sandstone guiding the crack that extends upwards into the coal seam (cf. section 4.4.3.8.1.).

When hydraulic fracturing operations are conducted in thick reservoir sections, it is common for the perforated interval to be long but yet still significantly shorter than the distance between reliable barriers to crack propagation, or the height of the created fracture is of the same order as the perforated section, with crack propagation being unaffected by confining strata (MARTINS & HARPER 1985). Under such circumstances, fracture shape changes continuously as it evolves from the perforated interval, thus complicating interpretation of pressure decline data measured during pre-stimulation minifracturing tests. A suitable approach is to model the fracture as having evolved as a series of confocal ellipses.

4.8.2.3.3. Closure stress vs. fluid type

As closure stress determinations with acid may be quite near to those obtained with water, the use of moderate acid volumes ahead of shut-in periods does not hinder and indeed aids closure pressure evaluation as a result of perforation cleanup (BEGNAUD & CLAIBORNE 1985; cf. section 5.6.4.2.). Very small volumes pumped in high-resolution methods lead to a wide scatter in measured minimum stresses and introduce vulnerability to anisotropic characteristics. It is therefore imperative to apply sufficiently large volumes for minifracture calibration testing. Another consideration when pumping moderate volumes utilized in the pay zones is the possibility of leakoff-induced backstress increasing closure stress determination. Comparison of closure stresses obtained from microfracture and minifracture calibration jobs often shows good agreement which adds credibility to the hypothesis of early barrier effects. The inferred late occurrence of barrier phenomena in the reduced pore pressure interval might be attributed to less growth into the barrier early on due to lower relative fracture propagation pressure.

4.8.2.3.4. Premature screenout termination recognition

An important application of minifracturing calibration treatments is also recognition of possible premature screenout termination of the main job and collection of information to prevent its occurrence or to reschedule the main operation accordingly (cf. section 6.2.4.2.3.). Another significant utilization are in-situ stress measurements, with creation of small hydraulic fractures being the most practical method to determine the state of stress acting at great depths within the earth (GREENFIELD & AHMED 1983). The in-situ stress measurement values are necessary for calculation of the bottomhole treating pressure required to create and to contain a large-scale hydraulic fracture, with the determined maximum bottomhole treating pressures being indispensibly for avoiding fracture propagation into overlying and underlying strata.

In-situ stress direction and magnitude determination is also essential for planning and design of well pattern and spacing in order to prevent fracture interference (cf. section 4.10.5.) or even manipulations of the pressure regime in order to change the preferential direction of fracture extension (cf. section 4.9.4.1.). Microhydraulic fracturing stress measurements are also carried out by BREDEHOEFT, WOLFF, KEYS & SHUTER (1976). Field experience of minifracturing for parameter determination is also reported by HODGES & PAOLI (1982), BEGNAUD & CLAIBORNE (1985), GRI (1988), NORTHROP (1988); HOLDITCH, ROBINSON, WHITEHEAD & ELY (1988); PEARSON, LYNCH, SCHMIDT & McCASLIN (1988) and WARPINSKI, BRANAGAN, SATTLER, CIPOLLA, LORENZ & THORNE (1988).

4.8.2.3.5. Fluids and mechanics testing

In order to be fully reliable, minifracture analysis has to consider also the rheological properties of the stimulation fluid (LEE 1987). This goal can be achieved by incorporation of an energy-balance equation into minifracture evaluation, allowing geometry prediction from minifracture interpretation to agree well with that deriving from full-scale fracture design programs based on the same or on similar models. The significance of minifracturing and pressure decline analysis for proper full-scale job planning is also outlined by CLEARY (1988).

In addition to measuring during the minifracture, closure stress can be determined from an in-situ stress test and from the actual full-scale fracturing stimulation (ROBINSON, HOLDITCH & WHITEHEAD 1986). In-situ permeability and initial reservoir pressure are obtained from production and pressure build-up tests, and minimum bottom hole flowing pressure can be estimated from field producing conditions. With all the variables obtained from minifracture and other sources, optimum crack length can be determined by projecting recovery efficiency as a function of fracture length and drainage area (HOLDITCH, JENNINGS & NEUSE 1978; ROBINSON, HOLDITCH & LEE 1983; cf. section 4.8.11.). Aspects of optimum fracture length are also discussed by HOLDITCH, JENNINGS, NEUSE & WYMAN (1978) and BENNETT, ROSATO & REYNOLDS (1981).

Minifracture calibration and checking treatments also serve as a mechanical test on fracture line, wellhead protector, tubing, packer, and all other equipment being used, and gives also an idea of average treating pressures (PAI, GARBIS & HALL 1983). Minifracturing jobs are conducted in conjunction with routine perforation breakdown operations (cf. section 4.8.2.2.). Case studies of minifracturing pressure tests and measurements are also reported by BUNDY (1981); KIM, CHAMPION & COOPER (1985); LEE (1987) and WARPINSKI & BRANAGAN (1988). SINGH, AGARWAL & KRASE (1988) document systematic design and analysis of steprate tests for determination of formation parting pressure.

4.8.2.4. Size and significance of minifracturing

While the performance of minifractures for evaluation of the above outlined parameters without plugging of the crack by proppants is already common sense in preparation of large-scale stimulation treatments since quite some time for the reason of determination of unrestricted fracture closure in pre-MHF calibration tests (PAI, GARBIS & HALL 1983), propping of the short cracks that are created during the pressure breakdown test before the main operation with the aim on stimulating the formation already with the scaled-down precursor run, however, has only in the last years become slightly more common. After discussion of some general aspects, proppant quantities and fracture wing length of minifracturing treatments in comparison to massive fracturing operations are illustrated.

4.8.2.4.1. General aspects

Computer programs support the interpretation of pressure decline after breakdown-test minifractures (HALLIBURTON 1985). The significance of pre-fracture injection surveys for a successful main stimulation operation is also stressed by

BUNDY (1981), and LEE (1987) comments on effects of fluid rheology on minifracture analysis.

WYMAN, HOLDITCH & RANDOLPH (1979) report proppant minifracturing for the purpose of production and buildup testing in order to determine reservoir capacity, with the field experimental treatment in addition to this gathering information on the number of perforations actually accepting fluid for the reason of selection of rate and pressure to be used on the large-scale fracturing job.

4.8.2.4.2. Proppant quantities and fracture wing lengths

Miniature hydraulic proppant fracturing jobs are generally performed with abt. 5 - 50 t of proppants, in exceptionally large cases also with volumes up to 100 - 150 t (depending on thickness, lithology and hydrocarbon potential of the reservoir) that are infilling fractures ranging in length predominantly between a few m and several tens of m and only in subordinate cases reaching up to abt. 100 m (in contrast to massive hydraulic fracture operations which include crack lengths between abt. 100 m and most frequently abt. 500 m, but in extreme cases up to more than 1,500 m that are plugged with proppant quantities between abt. 100 t and most abundantly abt. 800 t, but in ultra-large jumbo jobs up to more than 2,500 t; cf. sections 1.1.1.4. and 4.8.1.2.).

The relationships between reservoir capacity, fluid type, deposit size and minifracture design are very important (cf. AGARWAL, CARTER & POLLOCK 1979), because a minifracturing or other small-scale stimulation job carried out in a high-potential deep thick tight gas-bearing sandstone could be considerably larger in view of absolute size than a full-scale conventional fracturing operation in a shallow thin moderate-permeability oil-bearing sandstone.

4.8.2.5. Proppant vs. non-proppant minifracturing

The reason for hitherto mainly carrying out non-proppant minifracturing is to avoid having an obstacle in the crack in case of later performing larger-scale refracturing. Reopening of an unpropped crack is mechanically simpler, and greater safety against premature screenout termination is achieved if the minifracture does not have to be cleared by removing or displacing the small-scale proppant package which keeps it open. On the other hand, however, refracturing of a proppant-filled crack does often perform quite smoothly (cf. section 4.8.9.5.), and proppant minifracturing has the decisive advantage to allow proper assessment of the reservoir capacity due to adequate support of the crack that is limited to the near wellbore surroundings which is the region subjected to the highest closure stresses due to maximum pressure drawdown upon depletion (cf. section 3.9.1.). The drawback of enhanced screenout danger by the proppant plug in the small-scale crack which hinders reentry of the fracture can be eliminated by considerably increasing width of the younger crack up to four times its original size and to fill in the larger fracture with higher proppant saturations (SMITH, MILLER & HAGA 1987).

Another reason for non-proppant minifracturing is to avoid proppant interference with and distortion of the pressure decline curve after shut-in which could affect the calculation of the fluid-loss coefficient (WAREMBOURG, KLINGENSMITH, HODGES & ERDLE 1985) and could make good matching impossible (TAN, McGOWEN, LEE & SOLIMAN 1988). There is also no assurance that the calculated fluid volume inside the fracture after pumping (based on calculated fluid efficiency) is larger than bulk proppant volume. Field examples of proppant minifracturing are reported by TAN, McGOWEN, LEE & SOLIMAN (1988).

Proppant minifracturing and subsequent pressure drawdown testing can therefore also serve as field experiments investigating the suitability of various proppant types for plugging of massive hydraulic fractures, because the short minifractures happen to be situated in the zone of greatest closure stress sensitivity within the pressure depletion cone surrounding the wellbore. The following outline discusses some points of effectivity difference, comparative assessment and short precise propped minifractures.

4.8.2.5.1. Effectivity difference

HARPER, HAGAN & MARTINS (1985) evaluate some case studies of small-scale proppant and non-proppant fracturing treatments. Although proppant-free hydraulic stimulations do provide some conductivity as reflected by productivity increases of the wells (particularly if fluid-loss agents are added to the fracturing fluids, such as 100 mesh sand which has been proven to have a certain propping effect itself; SINCLAIR 1980; CAMPBELL, HANOLD, SINCLAIR & VETTER 1981; WARPINSKI 1988 a; cf. sections 1.4.11.2.1. and 4.8.8.3.2.3.) and almost no evidence for a more rapid decline in well capability after fracturing without proppant relative to proppant stimulation could be observed in the productivity decline curves, the proppant fracturing treatments have an overall markedly better effect than the proppant-free stimulations.

In some cases, non-proppant fracturing can eliminate sand-related operating problems (cf. section 5.3.5.2.), and a reason why fracturing without proppant may give rise to a conductive channel is shear displacement of a magnitude comparable to sandstone grain diameter between opposing crack faces so that a mismatch results when the fracture attempts to close (cf. sections 4.7.7. and 6.2.2.3.2.), in a similar way as large block sliding and dilation of joints occurs occasionally in naturally fractured reservoirs, giving rise to some increased crack roughness (TSANG & WITHERSPOON 1983).

4.8.2.5.2. Comparative assessment

Relatively soft rocks experience more shear displacements leading to increasing fracture conductivity than stiffer rocks, and high stress differences favour relatively high shear displacements (HARPER, HAGAN & MARTINS 1985). The distance for which propagation in the wellbore plane occurs rather than curving into a principal plane might be maximized by high pumping rates during the fracturing treatment. The small-scale shear displacements are likely to arise when cracks initiate from wellbores inclined to principal stress planes with minor deviations of the borehole from the principal stress planes being commonplace, because it is highly improbable that a wellbore will be drilled precisely in a principal stress plane. PEMPER, FLECKER, McWHIRTER & OLIVER (1988) report non-proppant fracturing in order to increase natural gas production by applying fluid pressure for enhancing communication between natural cracks.

The evaluation of earlier reports on proppant-free fracturing treatments (GHAURI 1980, LAMBERT & TREVIS 1980; MAHONEY, STUBBS, SCHWERER & DOBSCHA 1980; FREEMAN, ABEL & KIM 1981) leads HARPER, HAGAN & MARTINS (1985) to the conclusion that at least short-term improvement of well productivity in low-permeability reservoirs can be achieved by stimulation without proppants, even in wells without skin damage or without the potential for access to natural fractures or more permeable zones close to the wellbore. Whenever comparisons are available, however, non-proppant fracturing has been clearly less effective than propped treatments. Aspects of unpropped minifracturing are also mentioned by BRANAGAN, CIPOLLA, LEE & WILMER (1985); BRITT & LARSEN (1986); BRANAGAN, CIPOLLA, LEE & YAN (1987); HOLDITCH, ROBINSON & WHITEHEAD (1987) and HOLDITCH, ROBINSON, WHITEHEAD & ELY (1987). Other types of non-proppant fracturing are nitrogen stimulation of shales (ABEL 1982; cf. section 4.4.4.2.) and water fracturing of geothermal (cf. section 4.1.3.5.) and chalk (cf. section 4.5.4.5.2.) reservoirs as well as waste-liquid disposal well stimulation (cf. section 4.8.7.2.2.).

4.8.2.5.3. Short precise propped minifractures

A special application of proppant minifracturing is creation of short precise hydraulic cracks for distinctive stimulation of sensitive reservoirs (SMITH 1981) aiming on generation of a propped fracture of the required length without regard to the total hydraulic length, and following that on operation of the well below fracture closure pressure. Some aspects of proppant concentration and distribution as well as impact of batch-mix fracturing are evaluated as follows.

4.8.2.5.3.1. Proppant concentration and distribution

In comparable situations, formerly low-rate, low-viscosity, low-proppant concentration approaches have been used to dribble-in a short fracture, but the latter process had the disadvantages that inadequate knowledge and poor predictive capabilities concerning proppant settling made it difficult to design fracture length accurately, low-viscosity fluid resulted in a triangular-shaped proppant distribution, and low proppant saturation gave rise to non-uniform and inadequate fracture conductivity. The basic concept of creation of short precise hydraulic cracks is to pump a relatively large pad thus ensuring adequate width, and then to switch directly to a heavy-concentration proppant slurry. Major problems are the requirement of small exact slurry volumes and the effect of afterflow of the fracturing fluid. After well shut-in, the slurry continues to flow until the crack closes to lock the proppants in place. The length of this afterflow period depends on fluid-loss rate and proppant saturation.

4.8.2.5.3.2. Impact of batch-mix fracturing

Operation of short precise small-volume fracturing treatments placing high proppant concentrations in cracks crossing moderate- to high-permeability reservoirs can be optimized by batch-mix fracturing (CRAMER & SONGER 1988; cf. sections 4.3.3.5.3. and 4.6.3.1.) comprising use of pre-mixed fluid/proppant slurries. Hydraulic fractures in moderate- to high-permeability reservoirs increase borehole transmissibility mainly due to bypass of the wellbore-vicinity damage zone (cf. section 4.8.3.) and extension of the effective borehole radius. Optimal performance in pay zones intersected by closely-spaced wells requires utilization of small precise fluid volumes containing high proppant saturations. As most continuous-mix equipment is poorly suited to prepare and displace small precise slurry volumes, such treatments are preferably carried out by batch mixing where high proppant concentrations can be pre-blended with gelled liquid in the batch-mix inventory compartments and slurry density can be verified before the start of the job.

Batch mixing assures placement of an uniform proppant distribution in the fracture since proppant saturations at the surface will not be affected by treatment rate fluctuations, and incremental improvements in wellbore transmissibility can be achieved in many cases due to creation of a significant conductivity contrast between reservoir and crack. The use of small precise slurry volumes also successfully limits vertical and lateral fracture extension, avoids negative consequences of interzonal communication and waterflood sweep inefficiency, and reduces screenout failure problems due to achievement of uniform proppant distribution and use of a high pad/slurry ratio.

4.8.2.6. Modified micro- and minifracturing analysis for overpressure calibrated fracture design

SHLYAPOBERSKY, WALHAUG, SHEFFIELD & HUCKABEE (1988) propose modified mini- and microfracture tests in order to determine the fracturing parameters required for overpressure calibrated treatment design (SHLYAPOBERSKY, WONG & WALHAUG

1988; cf. section 4.3.4.2.3.). Fracture reopening pressure is defined from low-rate injection tests, and fracture closure pressure is defined from flowback and/or extended falloff tests analyzed by global pressure match which results in more objective determination of minimum in-situ stress and total leakoff coefficient than conventional techniques. The revised model also defines corrected instantaneous shut-in pressure, fracture overpressure and created crack dimensions. The outline as follows focusses on layout and purpose of conventional microfractures and minifractures, modified microfracture and minifracture procedures, and design parameters for optimum micro- and minifracture tests.

4.8.2.6.1. Layout and purpose of conventional microfractures and minifractures

Conventional microfractures are small hydraulic fracturing treatments performed in a selected interval and are typically used to determine a minimum in-situ stress profile or vertical stress variation for predicting vertical crack growth (SHLYAPOBERSKY, WALHAUG, SHEFFIELD & HUCKABEE 1988). Conventional microfracture tests assume that the created crack is small and is nearly closed at the moment of shut-in. Minifractures are larger hydraulic fracturing treatments and are used for determining leakoff coefficient, representative values of gross fracture interval, and created crack dimensions of length and average width, with these parameters being obtained from pressure decline analysis after shut-in (cf. section 4.8.1.). Unlike the microfracture test where the minimum in-situ stress is obtained from the instantaneous shut-in pressure, it is derived in the minifracture test from either an extended pressure falloff or a constant-rate flowback. As field experience has revealed that the net fracture pressure is frequently much higher than that predicted by classical design, overpressure calibrated fracture modelling (SHLYAPOBERSKY, WONG & WALHAUG 1988; cf. section 4.3.4.2.3.) is introduced and modifications of micro- and minifracture test analysis (SHLYAPOBERSKY, WALHAUG, SHEFFIELD & HUCKABEE 1988) are proposed. Other aspects of the revised micro- and minifracturing sequence are discussed by HUCKABEE (1988).

Discrepancies between net fracture pressures predicted by conventional models and those measured in the field are the consequence of the assumption of the classical models that the apparent fracture toughness of the formation rock is a material constant, whereas in reality it is a scale-dependent parameter (cf. section 4.3.4.2.3.3.). Therefore apparent fracture toughness must be determined from field measurements instead of laboratory evaluations, and fracture pressure decline analysis has to include the effect of field apparent fracture toughness (SHLYAPOBERSKY 1987).

4.8.2.6.2. Modified microfracture and minifracture procedures

The approach for measuring fracture parameters as proposed by SHLYAPOBERSKY, WALHAUG, SHEFFIELD & HUCKABEE (1988) involves a selected sequence of reopening, propagation, shut-in and flowback procedures which are similar for both micro- and minifracture tests and differ only in the magnitude of rates and volumes pumped and the types of fluid used. Before crack initiation, a filtration test can be performed in order to determine the matrix leakoff properties of an unfractured formation and the compressibility of the borehole fluid by pumping of water or brine at a very low rate for pressure buildup without exceeding the expected minimum in-situ stress. The filtration test also helps to identify an adequate reopening rate and the fracture breakdown injection rate in permeable formations.

During the first injection at a rate sufficient to create a single dominant fracture, the formation is broken down for the purpose of initiating a single dominant crack from the borehole. In low-permeability pays, injection rate is normally much larger than filtration rate, whereas in moderate- to high-permeability reservoirs and naturally fractured sequences, injection rate has to be much greater than maximum leakoff rate. Following the breakdown, a series of reopening, propagation, shut-in and flowback cycles can be conducted. Some aspects of reopening, propagation, shut-in and flowback cycles as well as global vs. local pressure decline match are discussed as follows.

4.8.2.6.2.1. Reopening, propagation, shut-in and flowback cycles

During course of reopening, propagation, shut-in and flowback cycles, the fracture is propagated to a desired size, and shut-in and flowback procedures are performed to identify crack closure pressure, overpressure, fluid-loss coefficient, fracture dimensions and other crack-specific parameters (SHLYAPOBERSKY, WALHAUG, SHEFFIELD & HUCKABEE 1988). Fracture closure and opening pressures can measure the minimum in-situ stress only if most of the created crack surface is perpendicular to the minimum stress component. As the hydraulic fracture may be initiated at some angle to the plane perpendicular to the far-field minimum in-situ stress, and after propagating away from the well the crack surface will be aligned with this direction, the fracture has to be propagated far enough from the well to minimize the effect of initial crack orientation at the borehole.

Because viscous frictional pressures can be very high even for low-viscosity fluids pumped at small rates as a consequence of small fracture aperture and amplification of the effect by waviness and roughness of the real crack surface, and since the pressure measured at the borehole may differ significantly from the average pressure in the fracture, and crack opening and closure pressure determined from reopening and flowback procedures may be very sensitive to flow rates, the fracture should be sized large enough, but not too large to risk breaking into adjacent test intervals. The accurate design of the injected volume for creating a crack of desired dimension which is confined in the tested zone has to account for the field-determined overpressure and the corresponding apparent fracture toughness as well as for the leakoff coefficient.

4.8.2.6.2.2. Global vs. local pressure decline match

In later stages of reopening, propagation, shut-in and flowback cycles, the fracture is reopened at a constant pumping rate to determine crack reopening pressure. After the designed volume has been pumped, the well is shut in again and extended shut-in and/or flowback tests are performed to determine fracture closure pressure (SHLYAPOBERSKY, WALHAUG, SHEFFIELD & HUCKABEE 1988). A more accurate analysis of the crack pressure decline can be performed if the theoretical pressure decrease is matched with field pressure data for a time interval (global match) rather than matching the pressure derivative (local match) at one point only. Several reopening, propagation, shut-in and flowback cycles should be performed at different reopening and flowback rates in order to identify the bounds of fracture closure and opening pressures which are considered to be the estimates of minimum in-situ stress. Constant-rate flowback procedures are used to accelerate and accentuate fracture closure.

The distinct flowback pressure character is due to the increase of frictional pressure in the fracture and/or the decrease of crack compliance during continuous fracture aperture reduction before complete mechanical closure occurs. Crack reopening at a constant low rate may identify an upper bound estimate for the minimum in-situ stress. As during injection into the fracture the measured bottom-hole pressure is higher than the pressure in the crack because of viscous frictional losses, the fracture reopening pressure gives an upper limit for the minimum in-situ stress. A global pressure match analysis of the extended falloff test can also be used to determine minimum in-situ stress or fracture closure pressure and total leakoff coefficient in more objective and reliable manner than with conventional techniques (SHLYAPOBERSKY 1988).

4.8.2.6.3. Design parameters for optimum micro- and minifracture

Design of micro- and minifracture tests involves appropriate selection of injection fluid, injection rate and volume, pressure falloff time after shut-in, and flowback rate and time (SHLYAPOBERSKY, WALHAUG, SHEFFIELD & HUCKABEE 1988). Some comments are offered as follows on test fluid rates and viscosities as well as shut-in time and flowback rate.

4.8.2.6.3.1. Test fluid rates and viscosities

The microfracture test fluid should ideally be a low-viscosity penetrating fluid which minimizes near-wellbore and fracture frictional effects. Minifracture tests often use two or more different fluids, with a fluid representative of the full-scale treatment having to be used to determine the leakoff coefficient properly which is required for design calibration (cf. section 4.8.2.1.). Injection rates for reservoir breakdown and fracture propagation must be adequate to ensure creation and extension of a single dominant crack and to generate sufficient fracture volume for subsequent extended shut-in and flowback. Microfracture tests should be pumped at rates as low as practical in order to minimize friction and to propagate the fracture in a quasi-static manner, while minifracture tests should be pumped at the planned full-scale treatment rates in order to evaluate potential fracture containment barriers and to determine the rate sensitivity of the crack overpressure (cf. section 4.8.1.1.). Fracture reopenings should be performed at several low injection rates in order to minimize the uncertainties of friction in perforations and near the wellbore, as crack reopening pressure can be very sensitive to the reopening rate because of the pressure-dependent fracture permeability.

The combination of injection rate and volume must create a crack with adequate fluid volume remaining in the fracture at shutdown to provide a stable constant-rate flowback before crack closure. The objective of a microfracture test is to generate a small radial crack limited to the interval of interest, whereas the objective of a minifracture is to provoke a crack of adequate dimensions as to test all fracture parameters to be encountered in the full-scale treatment, and the pumped volume should be enough to propagate the fracture to the expected containment boundaries for the main job.

4.8.2.6.3.2. Shut-in time and flowback rate

Shut-in time between stop of injection and start of flowback should be long enough to allow the pressure inside the fracture to equalize and is in practice often at least 50 % of the total injection time, although considerable deviations of this rule may be required by special circumstances. Flowback rate yielding the desired pressure character depends strongly on the actual test conditions and may vary between 5 % and 20 % of the injection rate. Minimum and maximum flowback rates are determined by leakoff rate and volume of fluid remaining in the fracture at the beginning of flowback. Minimum flowback rate must be greater than leakoff rate at the start of the flowback to generate the characteristic increase in pressure decline rate at fracture closure. Maximum flowback rate must provide a stable pressure decline before crack closure which is dependent on maintaining adequate fracture volume at the start of flowback. On-site modifications are critical for successful testing because crack imperfections may dominantly appear in reopenings and flowbacks, and the proposed procedures offer three independent ways of estimating minimum in-situ stress from fracture reopening and closure pressures.

4.8.3. Formation damage around the wellbore

Minifracturing including insertion of proppants thus generally represents relative limitation of fracture propagation to the nearer surroundings of the borehole (with respect to full-scale treatments) and has therefore hitherto in most cases been used only to overcome formation damage around the wellbore by breaking through the belt with deteriorated permeability seaming the borehole into the virginal reservoir (some of the proppant minifractures can be compared with enhanced gravel packs; DAVIES & HAGELAARS 1981; cf. section 5.3.5.), with this special workover- or repair-type application being independent from permeability of the virginal formation and thus having been carried out in both low- and high-permeability reservoirs. Following some remarks on origin and types of formation damage, breakthrough methods by minifracturing are briefly sketched. Comments are also offered on formation damage removal by acidizing.

4.8.3.1. Origin of formation damage

Reservoir damage comprises any impairment of productivity or injectivity within wellbore, perforations, formation adjacent to the wellbore, or fractures communicating with the wellbore that ultimately reduce the natural capacity to produce or inject fluids or gases (ALLEN & ROBERTS 1982, HOLCOMB 1986; KRUEGER 1986, 1988). While most of the minifracturing treatments to overcome formation damage are carried out in the early stage of well life after drilling, cementing and completion, a future tendency will be to focus also on older wells in late stage of production history in order to get still considerable amounts of hydrocarbon reserves out which otherwise would remain within the pay zone (cf. section 4.8.9.).

One of the most important features of formation damage is dispersion and migration of clay minerals in sandstones (GRAY & REX 1965), with the particles being of either natural primary-detrital provenance or secondary-authigenic origin (cf. section 3.5.1.) and being mobilized by external influences deriving from drilling, completion, testing and/or treating, or being tertiary-artificial and result from being synthetically inserted into and distributed within the reservoir by various acquisition and stimulation processes. Some aspects of formation exposure during well operation and fluid loss into the reservoir are illustrated as follows.

4.8.3.1.1. Formation exposure during well operation

Formation damage comprises physical, chemical and/or biological reduction of reservoir permeability in the immediate surroundings of the wellbore due to various different drilling and completion influences (GIDLEY 1970) at any time during life of the well (SCHAIBLE, AKPAN & AYOUB 1986). Generally every well operation disturbs the natural chemical, hydrodynamical and thermodynamical equilibrium and may thus potentially damage the formation (ALEGRE & DOUGHERTY 1988). Permeability deterioration around the perforations can occur as a consequence of matrix crushing and compaction caused by the shaped charge or by gun debris (WHITE, WALTER & DIEBOLD 1965).

Any operation such as drilling, cementing, perforating, completion, stimulation, gravel packing, injection and workover that exposes the formation to an alien fluid may result in adverse interactions between wellbore fluids and formation fluids and/or the formation itself. Solids suspended in the wellbore fluids may plug the formation, and organic and/or inorganic chemical species may precipitate out of either fluid group and reduce formation permeability (HOUCHIN & HUDSON 1986). Clay minerals within the reservoir framework may in response to chemical and/or mechanical forces change location and configuration and thus reduce fluid transmissibility (cf. section 3.5.2.).

4.8.3.1.2. Fluid loss into the reservoir

Loss of completion fluids, filtrates or particles from drilling mud may cause clay swelling (MONAGHAN, SALATHIEL, MORGAN & KAISER 1959), particle plugging by dispersed formation fines, particle invasion (GLENN & SLUSSER 1957) or adverse changes in fluid saturation, fluid viscosity and formation wettability. Highly alcaline filtrates occurring with the cementing operation may reduce effective permeability near the wellbore, and permeability deterioration may also occur during hydrocarbon production such as precipitation of scales, asphaltenes and other insolubles as formation pressure and temperature are lowered. Many of the adverse permeability effects correlate directly to swelling, dispersion and plugging of clay particles in the fine-grained fraction of the reservoir rock matrix (GIDLEY 1970).

Most of the effects causing formation damage are provoked by clogging of pores with fines (RIESE, SCHECHTER & RIESE 1988). The ultimate net result of all the different types of formation damage is reservoir permeability reduction. A progress report on the application of expert systems for the diagnosis of formation damage problems is given by ALEGRE & DOUGHERTY (1988). An overview of formation damage by alteration of physical characteristics of reservoir rocks in near-wellbore zones is provided by MICHAJLOV (1987).

4.8.3.2. Types of formation damage

Distinction can be made between mud damage and natural damage of reservoir formations (GATEWOOD, HALL, ROBERTS & LASATER 1970). Mud damage originates by drilling mud permeating the pore spaces of the formation adjacent to the wellbore. The degree of damage is usually dependent upon drilling time, pay zone permeability, mud type, and number of trips made during the drilling operation, with maximum penetration of drilling mud particles into the interstices of the formation being shallow. Additional damage may result when mud filtrate leaks into the reservoir and contacts naturally occurring formation clays. In pay zones with high percentages of water-sensitive clays, filtrate damage could be as severe as damage caused by mud particle invasion. Naturally damaged reservoirs exhibit a reduction in virgin permeability due to swelling or migration of sensitive clays to plug the formation flow channels. This can occur after a well has been exposed to fresh water, brines, acid or completion fluids, or after mud filtrate leakoff. Another type of damage is water blocking in fracture and formation (HOLDITCH 1979, LIN & HUANG 1988; cf. section 3.11.2.4.2.).

Aspects of formation damage are discussed by HOLDITCH (1979); McCORRISTON, DEMBY & PEASE (1981); POTTER & DIBBLE (1983); ALLEN, RILEY & STRASSNER (1984); AMAEFULE & MASUO (1984); AMAEFULE, PADILLA, McCAFFERY & TEAL (1984); GOODE, BERRY & STACY (1984); MARX & RAHMAN (1984), POTTER (1984); SOMERTON, CHEN, SCHUH & YUEN (1984); TODD, SOMERVILLE & SCOTT (1984); TYLER, METZGER & TWYFORD (1984); DEURER, SAHR & UJMA (1985); MEYERS, SKILLMAN & HERRING (1985); HOLCOMB (1986), MORGENTHALER (1986); AMAEFULE, AJUFO, PETERSON & DURST (1987); BARKER (1987), LEONE & SCOTT (1987); PRIISHOLM, NIELSEN & HASLUND (1987); VITTHAL, GUPTA & SHARMA (1987) and WOJTANOWICZ, KRILOV & LANGLINAIS (1987). The role of petrographical analyses in the design of non-damaging treatment programs is reviewed by DERSEY (1986). KING & HOLMAN (1986) report on hydrocarbon solvents as an alternative to acid for removing some formation damage. MALY (1976) stresses the importance of paying close attention to the smallest job details for minimizing formation damage, and COLLE (1978) introduces a completion technique for overcoming formation damage. SARKAR & SHARMA (1988) review formation damage by fines migration in two-phase flow.

4.8.3.3. Formation damage breakthrough by minifracturing

Minifracturing in order to overcome formation damage has frequently shown very promising results and as a consequence of the gathered experience, recently also some proppant minifractures were made as the first step preceding an optional following second expensive step being the MHF treatment. Some points of small proppant volumes for plugging of short cracks as well as damage halo breakage by short high-conductivity fractures are discussed as follows.

4.8.3.3.1. Small proppant volumes for plugging of short cracks

Formation damage breakthrough by minifracturing requires only very small proppant volumes in order to plug the short cracks extending from the borehole into its immediate surroundings. As most of the reservoir deterioration zones represent radial mantles enveloping the wellbore and hardly exceeding a couple of decimetres thickness, singular or multiple fractures in the order of a few metres length would be sufficient for optimization of the drainage. Caution has to be exercised concerning selection of suitable proppants, because the small-scale fractures in the closest vicinity of the borehole are subjected to maximum pressure drawdown upon depletion of the reservoir during course of field production (cf. sections 2.4.1.2.3. and 4.12.4.2.).

As it has been proven that in many cases already the minifracture could result in a reasonable increase of flow rate which is sufficient for converting a previously uncommercial well into an economical producer and thus the MHF treatment that was scheduled to follow was no longer necessary to be carried out, considerable stimulation cost could be saved if minifractures with proppant infilling of the cracks become a more common tool of formation evaluation prior to planning and decision of execution of expensive MHF treatments. Technical performance of small-scale fracturing has not necessarily to be done exclusively by hydraulic pumping and injection, but methods have also been developed during the years for fracturing by vibration (METZGER 1958). Field examples of small-scale hydraulic stimulation for breaking through wellbore-surrounding formation damage are reported by GRIFFIN (1985).

4.8.3.3.2. Damage halo breakage by short high-conductivity fractures

NIEMEYER & REINART (1985) document complex formation damage which is caused by interactions between mud filtrate, cement filtrate and connate water, giving rise to calcium carbonate and barium sulphate precipitation, water block from completion fluids (cf. sections 3.11.2.4.2. and 4.8.4.2.), clay dispersion by a drilling fluid additive and mobile formation fines. The small conductivity contrast attainable in the moderate-permeability oil reservoir made it uneconomical to create fractures any longer than necessary to bypass near-wellbore damage and establish reasonable withdrawal rates by flow efficiency improvement. Short high-conductivity fractures have turned out to be a satisfactory means to break through the formation damage halo around the borehole and to access the reserves in the pay zone.

Excessive fluid leakoff into the moderate-permeability matrix is compensated by pumping a large pad comprising more than half of the total job volume (cf. section 4.3.4.6.1.4.) and addition of silica flour and 100 mesh sand (cf. sections 1.4.11.2.1. and 4.8.8.3.2.3.). Minifracturing with reduction of both injection rate and total quantity of the treatment permits successful crack height limitation and prevents vertical fracture breakthrough into adjoining other pay zones in cases where seal thickness is insufficient to contain a standard-size job and fracture emigration is considered to be undesirable due to other fluid and pressure regimes in the overlying next reservoir storey. Aspects of well stimulation and scale inhibitors aiming on prevention of formation damage are discussed by VETTER & LANKFORD (1988), and PRASAD (1985) evaluates productivity improvement of a damaged well by fracturing.

4.8.3.4. Formation damage removal by matrix acidizing

Formation damage removal has often been carried out by matrix acidizing or acid fracturing instead of proppant fracturing for the reason of cost containment (cf. section 4.5.1.1.1.). Matrix acidizing, however, only successfully takes away formation damage in shallow depth around the borehole, whereas deep solid permeability plugging has to be corrected by creating a conductive fracture through the damage zone by either acid or proppant fracturing (McLEOD 1984). On the other hand, formations can be easily damaged by improper use of hydrofluoric acid, because spent hydrofluoric acid will precipitate silica, calcium fluoride and other compounds, especially when not enough hydrochloric acid preflush is used to remove calcium carbonate in the reservoir prior to hydrofluoric acid pumping (WALSH, LAKE & SCHECHTER 1982; McLEOD 1984; cf. section 4.5.1.1.). Fluid buffers have to be used to isolate formation fluids from acid in order to avoid undesired reservoir damage.

4.8.4. Field experiments and mHF vs. MHF

Particularly in light of the oil price crisis in 1986 (cf. section 2.2.1.1.2.) which is in parts of Europe and also in other sections of the world still strongly influencing the budget of capital expenditure (cf. section 2.2.1.5.2.), much promotion of stimulation activity can be expected in the next years from an abundant application of cheap proppant minifractures which could frequently replace expensive MHF treatments or at least postpone their execution without renouncing too much on the desired effect of acceleration of production and thus save at least the cost for the difference in proppant quantity which can comprise up to a couple of hundred tons of additional material (for which the expenses may amount up to 1 Mio. US $) for a MHF instead of a mHF. Some aspects of fracturing expenses vs. effectivity and fluid leakoff vs. water blocking are briefly outlined as follows.

4.8.4.1. Fracturing expenses vs. effectivity

Apart from proppant and fluid cost, the other expenses often remain more or less the same, because for offshore jobs the stimulation vessel has to go anyway on location and for onshore treatments also most parts of the fracturing truck assembly have to be installed (cf. section 3.8.). Proppant cost, however, is the largest share of the whole operation particularly in deep gas wells (cf. section 2.3.4.), and thus the application of only abt. 10 - 25 % of the proppant quantity and abt. 20 - 50 % of the fluid volume necessary for a MHF job in a mHF treatment is really a challenging alternative if the latter operation is sufficient for achievement of an economical production rate from the reservoir after the operation. The significance of a mini- or microfracture treatment before the main job with the option of possibly the small-scale operation already giving a satisfactory result leading to suspension or cancellation of the expensive big job is also emphasized by BRANAGAN & WILMER (1988).

Hydraulic proppant minifracturing is significant predominantly for unconventional reservoirs where minifractures can also be regarded as field experiments to get reliable information upon which then especially designed stimulation projects could be based on. The popularization of cheap minifractures is thus assumed to be a very important tool for promotion of stimulation by hydraulic proppant fracturing during the next years, because the economical feasibility of smaller treatments is much better than that of larger jobs and much expenses can be saved if the minifracture leads already to an acceptable solution in terms of increase of production rate at least for a certain time.

An interesting example is reported by NORTHCUTT, ROBERTSON & HANNAH (1988) from an intended MHF stimulation of tight gas reservoirs. As a consequence of underestimation of fluid leakoff in the naturally fractured pay zone, the treatment suffered from very early dehydration screenout, but although technically being a failure with only having achieved a job size comparable to that of a larger minifracture, it was a considerable commercial success by several times multiplying the pre-fracturing gas flow rate, thus emphasizing that the main effect of the operation was already achieved after breaking through the damage zone in the immediate borehole vicinity with a comparatively small to very small treatment. The near-wellbore formation deterioration responsible for flow inhibition without fracture job is the result of fines migration, scaling, closure of natural gas reservoir flowpaths or a combination of these factors.

4.8.4.2. Fluid leakoff vs. water blocking

Fluid leakoff which is very significant for proper design of the job depends on fracture fluid viscosity, reservoir rock permeability, viscosity and compressibility of the hydrocarbons in the pay formation, and the ability of the stimulation fluid to deposit a mud cake at the borehole wall, with fluid loss being best determined by a minifracture or pressure breakdown test (cf. section 4.8.2.1.2.). If too much fracturing fluid filtrate leaks off during the treatment into the formation matrix and is there fixed by capillary and interface forces, water saturation can considerably increase up to residual gas concentration near the fracture wall, and effective permeability can be considerably deteriorated by water blocking inhibiting initiation of gas flow especially in tight reservoirs with high capillary pressure (HOLDITCH 1979; cf. section 3.11.2.4.2.). Mitigation of water blocking can be achieved by using a solvent in the pad fluids with a surfactant in the rest of the fluid, thereby reducing water blocking in the fracture as well as significantly enhancing fracturing fluid recovery and production (PHILLIPS & WILSON 1984), because otherwise considerably high excess pressures are necessary to break through the water-blocking flow inhibition (ROODHART, KUIPER & DAVIES 1986).

Results from pump-in/shut-in tests (cf. section 4.8.2.3.1.3.) can be used to determine the maximum recommended job size and critical proppant concentrations for low-permeability reservoirs where leakoff into natural fractures occurs (SHELLEY & McGOWEN 1986; TAN, McGOWEN, LEE & SOLIMAN 1988) if the formation has sufficiently low matrix permeability that natural fractures predominantly control fluid loss.

4.8.4.3. Other aspects

Stimulation data analysis on-site (cf. section 6.2.4.2.4.) after the minifracture is a suitable means of evaluation of the treatment and can help to decide on necessity, design and sequence of further operation stages (ELPHICK 1985, PALMER & VEATCH 1987). Closure stress can be checked by step-rate test or pump-in/flow-back test (cf. section 4.8.2.), and fluid leakoff is given by an unpropped minifracture, with closure stress depending on formation pore pressure which varies during stimulation due to fluid loss (cf. also section 4.4.10.).

An important aspect for the planning of mHF vs. MHF, as well as optimum length of the fracture wings independent from the size of the job, is the fact that every given fracture conductivity has its associated optimum length, with excessively long fractures then having no further effect except of causing unnecessary additional cost (BAUMGARTNER 1983, LEICHT 1985). This means that fracture length is ultimately also depending on proppant selection and the conductivity of the designed propping material (cf. section 4.8.11.). Aspects of excessive fracture length are also discussed by CIPOLLA & LEE (1987).

4.8.5. Pilot stimulation of marginal reservoirs

Little cheap fracturing jobs are also quite attractive as pilot projects for both better investigation and introductory trial treatments (which have to be reinforced by the service and proppant companies by special promotion pricing) in marginal reservoirs. The following discussion gives an outline of areal distribution of potential reservoirs, application of resin-coated proppants in unstable pay zones, tip screenout fracturing, and dirty sandstone stimulation.

4.8.5.1. Areal distribution of potential reservoirs

Potential reservoirs of marginal type which can be successfully approached by minifracturing as the first step of stimulation technology are for example

* single or multiple sandwich-type sandstone sheets and lenses,
* unconsolidated to weakly-lithified fine-grained flasery or muddy-silty sandstones and soft unstable chalks (which occur in North Sea, Gulf Coast, Rocky Mountain region and Middle East; BLANTON 1986),
* individual or grouped carbonate shell (bioclastic) layers in mudstone sequences,
* low-permeability sandstone belts or sheets horizontally or vertically separating high-permeability zones, and
* isolated shoe-string or lenticular sandstone bodies within mudstone formations.

In many of these applications, intermediate-strength low-density alumina silicate proppants in both conventional and resin-coated form (SINCLAIR & GRAHAM 1977, 1978; COSTEIN & MAYER 1978, SAUNDERS & McKENZIE 1979, KANAT 1980; UNDERDOWN; DAY & SPARLIN 1980; POPE, WILES & PIERCE 1987; cf. section 1.2.6.) will become increasingly significant during the next years, as many earlier trial runs with sand have not led to promising results and heavier proppants are not an adequate material in the relatively shallow depth in most of the cases due to settling problems in the carrier fluids (cf. section 4.12.2.) and the higher price of the material (ATTARD, MATHES & MOWER 1988; cf. sections 1.3.2. and 2.2.2.3.).

Suitable target formations for such pilot minifracturing of marginal reservoirs in Europe are mainly Middle Jurassic and Lower Cretaceous sandstones, Upper Jurassic or Lower Cretaceous carbonate shell layers in mudstone sequences, and Upper Cretaceous and partially also Lower Tertiary chalk in Germany FRG and the Netherlands; Jurassic sandstones and Upper Cretaceous to Lower Tertiary chalk in parts of the Norwegian and British Northern North Sea; and Tertiary sandstones in Austria (cf. DOLAK & PERNTHANER 1976, KRÖLL 1984; KREUTZER 1985, 1986; BRÄUER 1987), Bulgaria, Czechoslovakia (cf. THON 1985), Italy, Hungary and Yugoslavia (cf. DRAGICEVIC, MARINOVIC, STANKOVIC, BATUSIC & NIKOLIC 1983). Required proppant quantities for such small-scale jobs may be as low as 6 - 12 t, but may also comprise higher or even also lower volumes in case of problems to meet the average figures (DOLAK & PERNTHANER 1976).

4.8.5.2. Application of resin-coated proppants in unstable pay zones

The main significance of the application of resin-coated proppants in a suite of the aforementioned cases is to provide a stable aggregated package in soft to insufficiently consolidated formations with low embedment pressure where conventional proppants tend to embed, thereby resulting in a healing effect to occur in the fracture (in contrast to crushing of grains in a fractured hard lithified rock with high embedment pressure; CLARK 1983; cf. section 4.3.3.). Some aspects of proppant wedge stabilization and embedment prevention as well as tail-in propping for flowback inhibition are discussed as follows.

4.8.5.2.1. Proppant wedge stabilization and embedment prevention

While in many soft to moderately-cemented formations, the choice of a proppant with a larger diameter is already enough for satisfactorily limiting proppant embedment, other rock types like the above enumerated ones have to receive an artificial stabilization at least within the fracture infill wedge and/or the gravel pack mantle by sticking together of resin-coated grains (cf. section 4.12.3.3.). Thermal stability of resin-coated proppants and their consolidation properties are investigated by RENSVOLD (1983).

Resin coating can be carried out for any commercially available proppant by adding a thin pellicle of phenolic resin to the individual grains (POPE, WILES & PIERCE 1987). When temperatures in excess of 140 oF and closure stress are applied to the curable resin-coated proppants in the reservoir formation, the resin softens, starts to cure and chemically bonds together to generate a consolidated permeable package (cf. section 1.2.6.).

4.8.5.2.2. Tail-in propping for flowback inhibition

Concerning tail-in propping policy (cf. section 2.4.1.2.3.), at least one third to half of the amount of the pumped proppants have to be resin-coated in order to get an aggregated rigid plug of the fracture which is withstanding collapse (cf. section 4.12.3.3.), whereas the outer parts of the crack near the tips may be infilled with conventional proppants if a multi-layer package is guaranteed. In unstable formations, however, any cost containment achieved by the partial replacement of resin-coated proppants by conventional ones can be more than compensated by serious conductivity deterioration by proppant embedment and fracture closure (cf. section 4.3.3.). FRIEDMAN, SURLES & KIEKE (1986) comment on high-temperature sand consolidation. Tailing-in of resin-coated proppants is a particularly successful method for controlling proppant flowback near the wellbore and thus avoiding fracture damage (cf. section 4.11.3.) by creating a stable consolidated wedge holding the proppant in place irrespective of flow rates and velocities of the produced fluids.

4.8.5.3. Tip screenout fracturing

Another technique for the stimulation of weak, unstable formations is tip screenout fracturing (SMITH, MILLER & HAGA 1987; cf. section 4.5.4.4.1.) which principally aims on creating short high-conductivity fractures in soft reservoir rocks such as chalks, diatomaceous earths and unstable sands. Some principles and applications of tip screenout fracturing are outlined as follows.

4.8.5.3.1. Principles

The design is based on intentionally screening out the tip of the fracture with proppants and then continuing to pump slurry to increase crack width and to pack the fracture with proppant multilayers in order to obtain high conductivity (SMITH, MILLER & HAGA 1987). Because this involves severe risk of premature screenout (cf. section 6.2.4.2.1.) and as failure to achieve tip screenout will not yield the desired stimulation, it is indispensible to perform special prefacture tests in order to exactly determine design variables (particularly fluid leakoff coefficient).

Tip screenout fracturing can also be performed in multifracture systems with for example perforating the lower interval and carrying out a tip screenout fracturing treatment to create a high-conductivity conduit from the upper zone to the perforations (SMITH, MILLER & HAGA 1987; interface fracturing; cf. section 4.5.4.4.2.). As widely-propped cracks are necessary to maintain fracture conductivity in soft formations, conductivity loss resulting from proppant embedment into weak unstable reservoirs would make the flow conduit from the upper interval to the perforations ineffective. Aspects of perforating in one zone and fracturing into another horizon are also discussed by HOLM (1975); AHMED, WILSON & STRAWN (1983) and ATTARD, MATHES & MOWER (1988).

4.8.5.3.2. Applications

Tip screenout fracturing has the drawback of significant pressure drop and loss of productivity through the crack which might occur at high hydrocarbon offtake rates (SMITH, MILLER & HAGA 1987). If tip screenout fracturing is not successfully executed, hydrocarbon production rates reach a point where further wellhead pressure reduction has very little or no effect on well performance. In order to eliminate this choking effect, fracture conductivity has to be increased and/or distance between upper zone and perforations has to be decreased. Increasing fracture conductivity can be achieved by redesigning the operations such as to reflect triple or quadruple width instead of primary double-width design philosophy, along with increasing maximum proppant concentration. In addition, detrimental pressure drops through the crack can be avoided by carrying out a multifracture completion instead of a single stimulation job. Multifracture completion may be executed as separate tip screenout operations in discrete layers which are later commingled (cf. section 4.10.1.1.), or as joint multiple fracturing including creation of communication pathways between the vertically superimposed reservoir storeys if necessary and desirable (cf. section 4.10.1.2.). Tip screenout fracturing is a particularly suitable technique in water-sensitive soft formations where acidizing cannot be done (cf. also section 4.5.4.3.1.).

Minifracturing is also considered to be a suitable pilot stimulation means for the successful approach of other unconventional targets such as geopressured reservoirs, coal seams, shales and high-temperature reservoirs (cf. also section 4.4.).

4.8.5.4. Dirty sandstone stimulation

Minifracturing as precursor- or experimental-type treatment is also a suitable method for the approach of dirty sandstones which comprise 80 % of the world´s clastic petroleum reservoirs, but only 15 % of the world´s oil fields (DAVIES 1980). While hydraulic proppant fracturing of clean sandstones is practiced since longer time, dirty reservoir stimulation has only moved into the focus of treating schedules in the last decades. Particularly as a consequence of the necessity of increasing oil reserves in spite of production exceeding additional discoveries (cf. section 2.2.1.5.2.), hydraulic proppant fracturing of dirty sandstones will become progressively more important in the near future. With properly designed stimulation procedures, 80 % of the world`s prospective sandstone potential become highly prospective exploration and development targets.

4.8.6. Horizontal well or drainhole drilling and fracturing

In addition to conventional vertical drilling and although being quite expensive as well as technically complicated, potential of enhanced hydrocarbon recovery is also seen for hydraulic proppant smaller- or larger-scale fracturing of horizontal or drainhole wells (GIGER 1983, STEENBOCK 1985, VICANEK 1985, JOSHI 1987) in various reservoirs of predominantly thin sheet-like sandstone type. The most common procedures for acceleration of oil production are hydraulic fracturing, horizontal drilling and vertical infill drilling (PLAHN, STARTZMAN & WATTENBARGER 1987). Various aspects of horizontal hole drilling and fracturing are briefly sketched as follows.

4.8.6.1. Horizontal hole drilling

Horizontal or drainhole drilling has already repeatedly been performed in Europe in France (Lacq and Castera Lou fields; WINNOCK & PONTALIER 1970; MONTIGNY, SORRIAUX, LOUIS & LESSI 1988), Germany FRG (Georgsdorf, Berkhöpen and Lehrte fields; PREVEDEL 1985, 1987), Italy (Rospo Mare field; GIGER 1983, BEUDELL 1985 a, ERDOEL-ERDGAS AKTUELL 1985 e, REISS 1985, FOX 1986, OFFSHORE ENGINEER 1986; DUSSERT, SANTORO & SOUDET 1988), Netherlands (MURPHY 1988), England (HARDMAN 1986; BORNER, HALDORSEN, HARRISON, HOPWOOD & MACDONALD 1988; MACDONALD 1988; OIL GAS JOURNAL 1988 b, PETROLEUM REVIEW 1988 e), Denmark (WENANDE 1987; ANDERSEN, HANSEN & FJELDGAARD 1988), Norway (FANTOFT, KROGH & POLLEN 1988), India (KHIN 1988) and in the USA (where horizonal well drilling has been applied for speeding-up of oil production since over 40 years; STORMONT 1935, RANNEY 1941, LANDRUM & CRAWFORD 1955, ROEMERSHAUSEN & HAWKINS 1955, DIAMOND & OYLER 1980, MOORE 1980, STIEGLER 1982, GORODY 1984, DICKINSON & DICKINSON 1985, BOSIO 1986; DICKINSON, ANDERSON & DICKINSON 1986; LI, CHANDELLE & BRYCH 1986; WILKIRSON, SMITH, STAGG & WALTERS 1986; SHERRARD, BRICE & MacDONALD 1986; OVERBEY, YOST & WILKINS 1988) as well as in the USSR (where the technology is also applied since also more than 40 years; BORISOV 1964, GRIGORIAN 1969, BUTLER 1984, REISS 1985, ERDÖL UND KOHLE - ERDGAS - PETROCHEMIE 1987 e) without major technical difficulties.

4.8.6.1.0. Historical development and status of horizontal wells

While early horizontal drilling in both USA and USSR in the 1950`s to early 1970`s lead to conclusion that horizontal or highly-deviated wells are technically feasible but economically disappointing (AUSTIN, ROSE & SCHUH 1988), the development of improved directional drilling methods resulting from experience gained in Gulf of Mexico and North Sea in the late 1970`s and 1980`s turned the key and gave way to economical attractivity of highly-inclined and horizontal boreholes for profitable reservoir exploitation (REISS, JOURDAN & GIGER 1984).

In the last decade, abt. 150 important horizontal boreholes have been drilled in the world (PETZET 1988), and currently abt. 30 horizontal wells are successfully producing oil worldwide at economically feasible rates (JOSHI 1986 a, 1987 a). Some estimates even place the current number of horizontal wells between 300 and 600 (AUSTIN, ROSE & SCHUH 1988). Horizontal drainhole lengths up to 2,000 m (6,000 ft) in reservoir depth up to 3,000 m (9,000 ft) have been achieved (ARMESSEN, JOURDAN & MARIOTTI 1988). Drilling failures of horizontal wells are comparatively rare, penetration rates approach those of vertical wells, drilling of accurate trajectories has proven to be completely feasible, and drilling costs also are not substantially higher than those of conventional wells if compared on a base of drilling progress unit. Following an outline of the main targets, the discussion concentrates on aspects of distinction of horizontal wells and drainholes, horizontal boreholes vs. extended-reach drilling, inverted drilling, significance for reservoir drainage, and horizontal hole completion and stimulation.

4.8.6.1.1. Main targets of horizontal wells

The main targets of horizontal well drilling are tight gas formations and heavy oil reservoirs which are partially also treated by steam injection flooding (BUTLER & STEPHENS 1980, RIAL 1984, GOODE & THAMBYNAYAGAM 1985, HUANG & HIGHT 1986; cf. section 4.9.4.), reservoirs with gas and water coning problems (CHAPERON 1986; KUSSACK, KLEPPE & AASEN 1987), naturally fractured pay zones with vertical cracks (MERCER, PRATT & YOST 1988; cf. section 4.8.6.1.5.1.), thin prospective sections (JOSHI 1986 a, 1987 a; FANTOFT, KROGH & POLLEN 1988), and edge water or gas drive reservoirs (BOSIO & GIANNESINI 1988, PETZET 1988). Subordinate applications are coal-seam degasification wells (ERTEKIN, SUNG & SCHWERER 1986; KING, ERTEKIN & SCHWERER 1986; SUNG, ERTEKIN & SCHWERER 1986; LOGAN, SCHWOEBEL & HORNER 1987; cf. section 4.4.3.6.) and gas shale exploitation holes (SALAMY, SARADJI, OKOYE, MERCER & YOST 1987; YOST, OVERBEY, SALAMI, OKOYE & SARADJI 1987; cf. section 4.4.4.3.2.).

In oil fields, horizontal wells can be also used to control and to diminuish severe coning problems (REISS 1985), to accelerate gravity-drainage oil production (BEZAIRE & MARKIW 1979), to enhance offtake from solution-gas-drive reservoirs (PLAHN, STARTZMAN & WATTENBARGER 1987), and to reduce drawdown (MACDONALD 1988). Horizontal drilling also improves oil and gas production from reservoirs with gas caps and bottom water (PETZET 1988; FANTOFT, KROGH & POLLEN 1988). Other candidates are pay zones with sparsely distributed vertical fractures that a normal well has little chance of intersecting (JOSHI 1986 a). Short-radius lateral holes can serve as hydraulic fracture replacement in chalk and other naturally fractured formations. Horizontal drilling is not only suitable for oil and gas reservoir exploitation, but also profitable for condensate pay zone depletion (HOWES 1988).

Summaries of horizontal well drilling are compiled by REISS, JOURDAN, GIGER & ARMESSEN (1984); JOURDAN, REISS & JOLY (1986) and JOSHI (1987 a), and a status report of operational aspects is given by PETZET (1988). Horizontal boreholes aim on production from plane or parallel flow with uniform pressure drop in contrast to vertical wells which operate under radial circular flow with concentrated pressure drop (HOWES 1988), and for achievement of plane flow, both high angle and long reservoir intersection path are critical, with horizontal hole trajectory through the pay zone being preferably at least 10 times the reservoir thickness (cf. section 4.11.3.).

4.8.6.1.2. Horizontal wells vs. drainholes

Distinction has to be made between horizontal wells and drainholes (JOSHI 1987 a). Horizontal wells are normally new boreholes reaching a length of the horizontal section of 1,000 - 3,000 ft (300 - 1,000 m) and are drilled from the surface by starting off vertically and then at the kick-off point progressively increasing inclination by building-up deviation angle up to finally subhorizontal and horizontal. Drainholes or lateral holes are sidetracks from existing vertical wells and attain a length of 100 - 700 ft (30 - 250 m). Either single or multiple drainholes can be drilled through a single vertical well, with multiple drainholes emanating from the same or from different elevations of the parent well and generating a birdfoot pattern. Drainholes represent secondary high-angle to horizontal holes which are drilled radially from primary vertical or low-angle holes (TRACY 1988). The larger the diameter of the host vertical well, the more high-angle or horizontal drainholes can be drilled out of it (PETZET 1988). Most of the horizontal boreholes drilled so far have been installed as drainholes out of vertical wells that have already been reasonably good producers (AUSTIN, ROSE & SCHUH 1988). Following some comments on radius for borehole trajectory turning, applications of drainhole drilling are outlined.

4.8.6.1.2.1. Radius for borehole trajectory turning

Classification of horizontal well and drainhole drilling techniques can be made into four broad categories depending on the radius required to turn the borehole trajectory from vertical to horizontal direction or the rate of angle building, thus distinguishing ultra-short-, short-, medium- and long-radius drainholes (EDLUND 1987, JOSHI 1987; AUSTIN, ROSE & SCHUH 1988; HOWES 1988, JÜRGENS 1988, LOGAN 1988, MAHONEY 1988, TRACY 1988, ZALESKI & SPATZ 1988). DICKIN-

SON, ANDERSON & DICKINSON (1986) outline multiple radial drilling. The primary difference between conventional or extended-reach directional drilling and shorter-radius drainhole techniques is build angle, with a considerably higher build angle for shorter-radius drainholes resulting in less borehole trajectory and thus less cost with respect to conventional directional drilling (LOGAN 1988).

4.8.6.1.2.2. Drainhole drilling applications

Drainhole drilling is a promising technology enhancing oil recovery and improving production efficiency and economics (TRACY 1988). Its major applications in oil and gas industry are irregular reservoirs which can be reached without additional vertical wellbores, limitation of invasion of unwanted formation fluids, penetration of vertical fractures, and maximization of production of low-permeability and low-energy reservoirs by increasing formation exposure. Drainhole drilling can also be used in mining industry to increase recovery efficiency and control in solution mining, to improve methane removal from coal prior to mining start-up, and to facilitate underground gasification in coal deposits.

4.8.6.1.3. Horizontal wells vs. extended-reach drilling

Another type of highly-deviated to horizontal borehole application is extended-reach drilling which has the objective to create a drainage point in the reservoir as far as possible from the surface outlet (BOSIO & GIANNESINI 1988). Current technology allows to drill lateral displacements of the extended-reach sections up to 10 km (6 mi) in a true vertical total depth of abt. 4 km (HOWES 1988, OFFSHORE ENGINEER 1988) or abt. 32,000 ft lateral displacement at a true vertical depth of 12,000 ft at an inclination of 72 - 75 degrees (PETROLEUM ENGINEER INTERNATIONAL 1988 a). Some comments are offered on horizontal displacement and total accessible area as well as extended- vs- limited-reach drilling as follows.

4.8.6.1.3.1. Horizontal displacement and total accessible area

The current world record of successful extended-reach drilling is abt. 15,100 ft (5,000 m) lateral displacement at a true vertical depth of abt. 8,000 ft (2,600 m) at an angle of abt. 70 degrees. The hitherto maximum achievement in Germany FRG is abt. 2,100 m (6,300 ft) lateral displacement at a true vertical depth of abt. 1,400 m (4,200 ft) at a maximum angle of 65.5 degrees (GREBE 1981). Horizontal borehole sections up to 3,000 ft (1,000 m) in extended-reach wells have been frequently achieved (FERTL & NICE 1988).

Extended-reach drilling is generally carried out with inclination angles between 60 and 90 degrees and usually aims at horizontal displacements in excess of 2 miles (HOWES 1988). The reason for extended-reach drilling is to access remote targets from a central point thus reducing the number of high-cost platforms or subsea wells that are necessary for development of the field usually up to 50 % or in extreme cases even up to 75 %. If drilling from a platform is increased from a maximum angle of hole inclination from 40 to 80 degrees, the total accessible area is rising by a factor of 30 (MOORE 1987 a). Extended-reach drilling in field appraisal is illustrated by LOCKE, JOHNSON & JEWKES (1988).

4.8.6.1.3.2. Extended- vs. limited-reach drilling

Extended-reach drilling is characterized by two main building curves that are initiated by two kick-off points and which are connected by a long straight section between the kick-off marks, with the upper buildup curve being often situated fairly high in the well, and the lower connecting the oblique straight section with the horizontal or subvertical landing trace depending on depletion and stimulation requirements (JOURDAN, ARMESSEN & POUSSELET 1988). Limited-reach drilling is in contrast to this characterized by only one buildup curve and a J-type shape of the borehole trajectory, with the single kick-off point being situated at a fairly deep level. For better reservoir penetration geometry especially in case of vertical hydraulic fracturing and gravel packing, a S-shaped borehole trajectory with one or two kick-off points is preferential, with this shape being created by reducing again the wellbore inclination prior to propagation through the pay complex (BRUIST, JEFFERIS & BOTTS 1983).

4.8.6.1.4. Inverted drilling

A special type of high-angle to horizontal drilling is inverted drilling and completion (PETZET 1988) which involves creation of borehole trajectories in excess of 90 degrees. Inverted wells start vertically and then gradually build up an inclination up to 90 degrees which is then in the terminal phase exceeded up to abt. 100 - 110 degrees by turning again upwards, thus propagating in inverse direction in the final stage. Inverted drilling may leave a greater distance or standoff between the gas-oil-contact in a reservoir and the perforations which extends long-term well productivity and practically eliminates gas channelling. Examples of inverted drilling with terminal angles in excess of 90 degrees are also reported by GREENER, SAUVAGEAU & PASTERNACK (1988), and STEENBOCK & MÜLLER-LINK (1987) present alternative drainhole drilling with an articulated spring for inverted drilling up to 110 degrees hole inclination. Examples of inverted drilling with final angles of abt. 100 degrees are also documented by WESTERN OIL WORLD (1988).

4.8.6.1.5. Significance for reservoir drainage

Horizontal drilling aims on radically changing of the flow conditions in the reservoir by creation of a planar flow pattern instead of the conventional radial circular flow network (BOSIO & GIANNESINI 1988) which is the controlling factor of productivity and/or recovery improvement (cf. section 4.11.3.). The most fundamental parameter of horizontal drilling is therefore the length of the drainage section traversing the reservoir horizontally, or more specifically the relative length of the drainage interval in relation to reservoir thickness and drainage area dimensions. Some comments are offered as follows on communication path between wellbore trajectory and reservoir, improved natural fracture intersection, choke effect on vertical fractures, distance from fluid contacts, reduction of coning, selection of drainage and injection direction, reservoir geometry and natural fracture distribution, and wellbore instability.

4.8.6.1.5.1. Communication path between wellbore trajectory and reservoir

Higher production rates and in many cases also higher ultimate recovery results from a long communication path between wellbore trajectory and reservoir in a horizontal borehole in contrast to a vertical well where the length of the pay formation open to the borehole is more or less limited to prospective zone thickness (PETZET 1988). In this connection, however, attention has to be paid to the fact that the reservoir drainage effect of horizontal boreholes traversing slightly-dipping formations is equivalent to that of vertical wells crossing steeply-inclined beds providing the most suitable constellation of maximum communication path of wellbore trajectory and pay zone, whereas the inverse combination of horizontal boreholes intersecting steeply-dipping strata and vertical wells propagating through gently-inclined reservoirs representing the worst possible case of shortest contact length of wellbore and pay interval (cf. also section 4.11.1.2.2.).

Exploitation of incremental reserves by horizontal boreholes is frequently the consequence of reduced coning, improved pressure maintenance, and larger drainage area. Even if they do not raise ultimate recovery, horizontal wells may improve economics by allowing the boreholes to produce at rates that payout faster. The effectivity and profitability of horizontal wells strongly depends from permeability anisotropy in the reservoir (JOSHI 1988, MUKHERJEE & ECONOMIDES 1988; cf. section 4.2.4.2.).

4.8.6.1.5.2. Improved natural fracture intersection

An important advantage of horizontal wells is their much better and more effective intersection of systems of vertical natural fractures in the reservoir (REISS 1985) which allows much greater production profit from the joint fabric than in vertical wells. As according to the rock stress distribution, vertical or highly-inclined natural fractures are present in most of the oil and gas reservoirs which can be better developed with horizontal well drilling, this acquisition and completion technique has considerable significance for enhanced exploitation of both matrix- and fracture-porosity pay systems.

Another advantage is the larger spread of the formation interval intersected by the borehole which permits selective production of head or tail of the penetrated reservoir segment in a much more efficient way than would be possible for vertical wells. The presence of open natural fractures in the reservoir formation is usually indicated by severe mud losses up to total leakoff of circulation during the drilling phase. Intersection of numerous vertical natural fractures by the horizontal well leads to drastical increase of well productivity, allowing in turn an offtake under very limited flowing pressure which gives rise to only gentle deformation of the oil-water-contact, with this effect being one of the keys for a very satisfactory oil sweep efficiency in horizontal boreholes (cf. also section 4.8.6.2.4.2.).

4.8.6.1.5.3. Choke effect on vertical fractures

The contact between horizontal boreholes and vertical fractures results in a choke being a special skin effect (MUKHERJEE & ECONOMIDES 1988). Hydraulically-fractured horizontal wells can be regarded as choked vertical cracks because of the limited contact between borehole and fracture (cf. section 4.11.1.). This choke effect is negligible in the ideal constellation of vertical fracture plane orientation being parallel to horizontal borehole trajectory for the case of a single fracture, whereas its maximum is reached for the geometrical relationship of multiple vertical cracks intersecting the horizontal well orthogonally to the orientation of its trajectory. On the other hand, multiple fractures propagating perpendicularly to the horizontal borehole aid in the best coverage of the drainage area. The choke-type skin effect results in an additional pressure drop due to the radial flow generated around the entry point.

As the skin effect is equal to zero when the well is vertical or inclined along the fracture direction, deviated holes should be preferentially drilled such that they are turned vertically downward for the intersection of the reservoir in order to minimize contact skin effect (cf. also section 4.8.6.1.3.). In terms of reservoir exploitation, the performance of a fractured horizontal well which has been drilled along the crack plane cannot be better than that of a fractured vertical hole, whereas horizontal well productivity can substantially benefit via multiple fracture completion despite the occurring skin effects in the latter case in comparison to almost zero skin in the former constellation.

4.8.6.1.5.4. Distance from fluid contacts

Other advantages of horizontal wells include longer completed intervals resulting from increased direct contact between borehole and pay zone and therefore increased production rates in the longer wellbores, increased standoff from fluid contacts in oil reservoirs overlying an aquifer or located under a gas cap improving production rates without causing or with at least deferring coning, reduction of drawdown for a given production rate by the longer wellbore, field development with fewer boreholes because of the larger drainage area of each well, and allowing development of some fields where conventional vertical boreholes would be uneconomical (SHERRARD, BRICE & MACDONALD 1986) or the surface area is unaccessible due to environmental reasons. The critical coning rate for a horizontal well is higher than that for a corresponding vertical borehole (MACDONALD 1988).

If coning is inevitable, horizontal wells still provide improved sweep efficiency, because the shape of conventional cones in vertical boreholes results in less displacement of hydrocarbons towards the producer than the tent-shaped cone which is characteristic for horizontal wells. The ratio of vertical to horizontal permeability is very important in delaying water breakthrough in horizontal boreholes (GIGER 1986).

4.8.6.1.5.5. Reduction of coning

Concerning reduction of gas- and water-coning tendencies, horizontal wells exhibit better performance than vertical boreholes, as it requires much smaller pressure drawdown to produce oil at the same rate and because it has almost a linear pressure gradient from the wellbore to the drainage radius, resulting in a steady gas dip or water rise over the long producing horizontal-well length (SHERRARD, BRICE & MACDONALD 1986). In contrast, in a vertical borehole a log-linear pressure gradient from wellbore to drainage radius, along with a conical flow at a single point, accelerates coning problems. Further promising aspects are the ability to complete several to various highly-deviated or horizontal wells from a single location covering a large drainage area which is essential in environmentally sensitive or hostile areas as well as offshore where preparation of drilling sites is a major expense (HUANG & HIGHT 1986).

Particularly feasible applications of horizontal wells are thin oil zones which are sandwiched between gas cap and aquifer where production rates in vertical boreholes are severely limited by gas and water coning (FANTOFT, KROGH & POLLEN 1988; GILHUUS 1988). The main advantages of horizontal wells in such a setting are lower fluid velocities near the wellbore due to longer producing section and longer distance to fluid contacts, whereas a potential disadvantage is the inability to shift the vertical depth of the producing interval to compensate for movements of the phase contacts. Horizontal boreholes located close enough to the gas-oil-contact would have early gas breakthrough rather than water breakthrough, and following recompletion for gas, production can continue with little or no water cut.

The horizontal well has to penetrate all permeability barriers in order to ensure high layer sweep efficiency and maximum oil recovery. While vertical well productivity is limited by reservoir thickness, horizontal well exploitability can be improved by increasing its length (MACDONALD 1988). Water breakthrough can be delayed by placing the standoff of the horizontal section above the oil-water contact as great as possible (GIGER 1986). Another application of horizontal wells or drainholes is connection of vertical shafts or horizontal gallery systems to improve the effectivity of oil mining (BEST, CORDELL & HASTON 1987; ERDÖL-ERDGAS-KOHLE 1988 b, MOORE 1988 c). This technique aims on gravity drainage of heavy oil sands via horizontal wells emanating from the underground tunnel system.

4.8.6.1.5.6. Selection of drainage and injection direction

Horizontal wells also allow to inject and/or produce fluids orthogonal to those from a vertical borehole, thereby providing the potential of improving sweep efficiency of a flood and thus recovery efficiency. The decisive advantage of horizontal wells with respect to vertical boreholes and/or vertical fractures (cf. section 4.8.6.2.1.) is the possibility of drilling the highly-deviated borehole in every direction that is desirable from reservoir drainage point of view, in contrast to the more or less predetermined direction of hydraulic fracture propagation (also altered-stress fracturing does not allow to turn the crack into every possible direction, but only permits to shift the orientation of the fracture in several lines more or less normally to each other; cf. section 4.9.4.1.3.1.) and the standard perpendicular penetration of vertical wells.

Injection of steam from a horizontal borehole eventually provides a more homogeneous steam front and better injectivity indices (RIAL 1984; SUNG, ERTEKIN & SCHWERER 1986; cf. section 4.9.4.2.1.2.). Horizontal wells produce on average around four times more than vertical boreholes, and as they produce in a wider area, it is not only the productivity, but also the total exploitation that can be multiplied by a factor of abt. four (DUDA & KUMAR 1988, HOWES 1988). A greater production is achieved with an increased recovery factor and wider drainage area. Comparative evaluations of productivity and feasibility of horizontal vs. vertical wells are carried out by HUANG & HIGHT (1986); KOSSACK, KLEPPE & AASEN (1987); DUDA & KUMAR (1988) and MUKHERJEE & ECONOMIDES (1988).

4.8.6.1.5.7. Reservoir geometry and natural crack distribution

In summary, the main fields of application of horizontal drilling are thus pay zones with vertical fractures that are sparsely distributed and cannot be accidentally crossed by vertical wells or vertical hydraulic cracks alone, reservoirs with coning problems where a horizontal borehole can be positioned away from the fluid contact and allows a lower drawdown of the unwanted fluid, in thin pay zones where too many vertical wells would be required, in edge water or gas drive reservoirs where a horizontal borehole combines the effect of a more efficient gravity drainage and a wider drainage area, secondary or tertiary recovery processes where horizontal wells can improve injectivity and areal sweep efficiency, evaluation wells after a discovery is made in order to study the lateral evolution of the facies and to provide valuable information before the final development decision is made, to help solving environmental problems by reducing the space occupied by drilling sites and production lines, and to reduce the number of required offshore platforms (BOSIO & GIANNESINI 1988).

An important application of both horizontal wells and vertical hydraulic fractures is penetration of isolated sand bodies of lenticular geometry (MACDONALD 1988). The amelioration of reservoir continuity with intersection of more sand bodies provides access to larger hydrocarbon volumes (cf. section 4.10.2.1.).

4.8.6.1.5.8. Wellbore instability

Highly-deviated to horizontal boreholes are subjected to considerable problems of wellbore instability which provoke frequently borehole collapse and formation fracturing (FUH, WHITFILL & SCHUH 1988). Borehole collapse pressure gradient increases with hole angle, whereas fracture gradient remains the same. As a consequence of this relationship, the working range between the two gradients is narrower at higher hole angles and thus with increasing deviation angle of the borehole, the potential for unwanted accidental formation fracturing during drilling increases. Optimum mud weight for wellbore stability in highly-inclined to horizontal holes ranges between borehole collapse pressure gradient representing the lower limit and formation fracture gradient being the upper boundary. Aspects of instability of highly-deviated to horizontal wells are also discussed by BRADLEY (1979); HOTTMAN, SMITH & PURCELL (1979); AADNØY & CHENEVERT (1987), HSIAO (1987), MAURY & SAUZAY (1987) and YEW & LI (1987).

4.8.6.1.6. Horizontal hole completion and stimulation

REISS (1985) gives an experience report of horizontal well production after five years, and HUANG & HIGHT (1986) and PETZET (1987) evaluate steamflood processes in horizontal boreholes. A comparison of exploitability of horizontal and vertical wells is performed by KOSSACK, KLEPPE & AASEN (1987). Hydraulic fracturing of highly-deviated wells is investigated by YEW & LI (1987), and gravel packing of horizontal boreholes in analyzed by DICKINSON & ANDERSON (1987) and SPREUX, GEORGES & LESSI (1988; cf. also section 5.8.3.). JOSHI (1987 a) presents a review of horizontal well and drainhole technology.

Logging methods in horizontal boreholes are evaluated by BRATOVICH, BELL & KAAZ (1977); BARON & WITTRISCH (1982), DANIEL & FERTL (1984); JOLY, DORMIGNY, CATALA & PINSON (1985); REISS (1985); DICKINSON, ANDERSON & DICKINSON (1986); ANDERSON & LATOS (1987), FERTL & HOTZ (1987), FERTL & MARTIN (1987); SPREUX, LOUIS & ROCCA (1987); FERTL & NICE (1988); MONTIGNY, SORRIAUX, LOUIS & LESSI (1988); MACEWEN (1988) and SPREUX, GEORGES & LESSI (1988).

WEIRICH, ZALESKI & MULCAHY (1987) summarize perforation techniques for horizontal boreholes. Productivity of horizontal wells is evaluated by BABU & ODEH (1988), JOSHI (1988) and MUKHERJEE & ECONOMIDES (1988). Horizontal well completion types comprise open hole, slotted or perforated liner or casing, external casing packers with tubing-operated subs, cemented liner or casing, prepacked liner, and screen with external gravel packing completions (AUSTIN, ROSE & SCHUH 1988).

4.8.6.1.7. Other aspects

Other aspects of horizontal wells and drainholes are discussed by JOURDAN, REISS, BARON & WITTRISCH (1982); PUGH (1982); JOURDAN, REISS, BARON, GIGER & WITTRISCH (1983); GIGER, REISS & JOURDAN (1984); HYLAND (1984), JOURDAN & BARON (1984), GIGER (1985, 1987), GOODE & THAMBYNAYAGAM (1985), CLONTS & RAMEY (1986), JOSHI (1986); KARCHER, GIGER & COMBE (1986); SHERRARD, BRICE & MACDONALD (1986); DIETRICH (1987), FINCHER (1987), HSIAO (1987); LOGAN, SCHWOEBEL & HORNER (1987); MOORE (1987); PLAHN, STARTZMAN & WATTENBARGER (1987); SALAMY, SARADJI, OKOYE, MERCER & YOST (1987); WEIRICH, ZALESKI & MULCAHY (1987); WILSON & SABINS (1987) ; YOST, OVERBEY, SALAMY, OKOYE & SARADJI (1987); FUH, WHITFILL & SCHUH (1988); GREENER, SAUVAGEAU & PASTERNACK (1988); KERR & BROUTHERS (1988), KUCHUK & GOODE (1988), OZKAN & RAGHAVAN (1988), PETZET (1988); REILEY, STAGE & VIDRINE (1988); SHAW & SUTHERLAND (1988), STEWART & WILLIAMSON (1988), VRIELINK & HIPPMANN (1988) and WINTERS, LUCIA & JOHNSON (1988).

A spectacular highly-deviated to horizontal drilling operation for the Grand Canyon waterline project is presented by LATTIMORE, CARDEN & FISCHER (1987). HERNANDEZ & CADE (1988) discuss screening criteria for horizontal wells. A more sophisticated technique is short-radius lateral drilling where various short horizontal radial holes are made from a vertical wellbore (DICKINSON & DICKINSON 1985, PENDLETON & RAMESH 1985; DICKINSON, ANDERSON & DICKINSON 1986; PARSONS & FINCHER 1987). The building-up of the high inclination angle up to finally horizontal position can in shallow depths be supported by slant drilling rigs with mast inclinations of 30 - 45 degrees which allows to access targets that could not be reached by conventional directional drilling (LOXAM 1982, BLEAKLEY 1985, MOORE 1987 a, EDWARDS & STRELKOV 1988, HOWES 1987). REHM (1987) reports horizontal drilling in slim holes. COOPER & THRONCOSO (1988) give an overview of horizontal well completion technology.

4.8.6.2. Horizontal hole fracturing

Most of the aforementioned reservoir examples would also be suitable targets for horizontal or drainhole drilling including subsequent minifracturing and other smaller-scale hydraulic stimulation. The combination of horizontal well drilling and vertical or horizontal fracturing is considered to be the only available method to enable economically feasible production rates to be achieved from some thin sheet reservoirs. Comments are offered as follows on horizontal hole drilling vs. fracturing, horizontal hole drilling in combination with fracturing, preperforated liner completion and proppant requirements, and multiple vertical fracturing of highly-inclined boreholes.

4.8.6.2.1. Horizontal hole drilling vs. fracturing

Horizontal wells and drainholes can be modelled as representing stimulated boreholes where fracture height is limited by the wellbore diameter (JOSHI 1987 a; cf. section 4.11.1.). A properly designed horizontal borehole would therefore be equivalent to a vertical well with a fully penetrating fracture and in fact represents a long controlled tube-like instead of sheet-shaped vertical crack. After discussion of some general aspects of alternative or combination, the outline as follows concentrates on tight gas reservoir drainage, identical exposed surfaces in coal seams, single- vs. multiple-layer reservoirs, and orientation to natural fractures and permeability anisotropies.

4.8.6.2.1.1. Alternative or combination

While in most hydraulic fracturing stimulation jobs it is difficult to obtain infinite crack conductivity which even decreases over time, a horizontal borehole offers an almost permanent infinite-conductivity fluid flow path. In addition, in reservoirs where the bottom water or top gas cap renders fracturing difficult, a horizontal well offers an alternative to obtain high production rates without gas or water coning. A horizontal well therefore provides a viable completion option and will compete with hydraulic fracturing in the future (JOSHI 1987 a).

Horizontal holes can be approached as representing and behaving like thin long tube-like infinite-conductivity induced fractures (SNOW & HOUGH 1988). Horizontal wells perform satisfactorily in thin reservoirs with good vertical permeabilities, whereas in thick and/or layered pays or in case of low vertical permeability, several horizontal stems are required. In such occasions, a high-angle well of more than 60 degrees inclination has the advantages of both exposing more borehole length to the formation and of penetrating several layers of the reservoir. In case of effective stimulation by hydraulic fracturing, there is an increased likelihood of highly-deviated wells to intersect multiple horizons.

As the cost of stimulation is usually only a fraction of the drilling expenses (cf. sections 2.3.4. and 3.7.), horizontal or drainhole wells should not be regarded as an alternative to hydraulic fracturing (VICANEK 1985), but emphasis should be put on optimum reservoir depletion by a combination of horizontal or drainhole drilling and stimulation especially for the enhancement of production from low-permeability pay formations. SUNG & ERTEKIN (1987) give a performance comparison of vertical and horizontal fractures and horizontal boreholes in low-permeability reservoirs. Aspects of low-permeability reservoir development with horizontal wells are also discussed by GIGER (1987). DUDA & KUTSKA (1988) study gas production from tight sandstone reservoirs via horizontal boreholes by simulation runs.

4.8.6.2.1.2. Tight gas reservoir drainage

KRUYADIJK & NIKO (1988) discuss horizontal hole drilling vs. massive hydraulic fracturing for draining tight naturally jointed gas reservoirs. The break-even point for a massive hydraulic fracture intersecting a naturally jointed zone with respect to a horizontal well is determined by the balance between extra fracture surface and pressure drop over the crack wing intersecting the fractured interval. Short cracks require very high conductivities in order to be competitive with horizontal holes, and operational disadvantages and lack in fracture direction control are major concerns in MHF scenarios. As horizontal boreholes can be oriented according to reservoir necessity which therefore allows selection of a favourable direction with respect to hydrocarbon-water-contact and/or vertical fracture pattern (cf. section 4.8.6.1.5.), horizontal wells are a real alternative to hydraulic fracturing in the development of tight naturally jointed gas reservoirs. In view of the large contrast between fractured zone and reservoir permeability, wellbore length is of no importance, and borehole deviation angle is significant only with regard to the wellbore surface intersecting the fractured interval.

Extended effective fracture length by linking of natural cracks by the hydraulic fracture gives rise to prolonged bilinear flow (cf. section 4.11.3.), with then fractured section conductivity having only a minimal effect on pressure response during this flow period. Horizontal drilling as sidetracks out of existing vertical holes can in some cases also be comparatively more feasible at the bottom of the line than remedial gravel packing (MURPHY 1988). Horizontal boreholes of optimum length can occasionally produce abt. twice as much gas as a vertical well in combination with a vertical hydraulic fracture in blanket sandstones, whereas in lenticular sands, the larger intersection plane of the hydraulic fracture may be more profitable than the limited trajectory of the horizontal borehole (DUDA & KUMAR 1988). In the case of an idealized vertical isotropic medium, horizontal wells appear to be preferable to vertical wells in almost all cases, whereas for reasonable vertical anisotropy and low permeability, even an extended-reach horizontal well does not outperform a vertical well intersected by a vertical hydraulic fracture (MUKHERJEE & ECONOMIDES 1988).

4.8.6.2.1.3. Identical exposed surfaces in coal seams

ERTEKIN, SUNG & SCHWERER (1986) and LOGAN (1988) discuss the comparative effectivity of vertical hydraulic fractures and horizontal boreholes in coal seams which are characterized by a dual porosity system consisting of matrix pores and natural fractures comprising face and butt cleats. While face cleats are continuous throughout the reservoir, butt cleats are discontinuous and end at the face cleats. This relationship creates the anisotropical permeability nature of the coal seam macropore structure, with permeability in face cleat direction being considerably larger than that in butt cleat direction. Some aspects of cleat system orientation and multiple-wing boreholes are outlined as follows.

4.8.6.2.1.3.1. Cleat system orientation

With respect to a vertical hydraulic fracture that has more or less random orientation to the cleat system which is governed by the stress pattern in the geological sequence, horizontal boreholes in coal seams have the decisive advantage that their orientation to face and butt cleats is controllable, and also penetration length of horizontal boreholes can be much easier varied than propagation distance of hydraulic fractures (ERTEKIN, SUNG & SCHWERER 1986; LOGAN 1988). Vertical hydraulic fractures usually extend parallel to the maximum stress or face cleat direction and therefore may not adequately access an anisotropic reservoir, and a horizontal borehole placed perpendicular to the face cleats representing the maximum permeability direction may be a more effective stimulation technique due to providing maximum access to primary flow channels. Short- and medium-radius directional drilling techniques (cf. section 4.8.6.1.2.) are frequently used to successfully place horizontal drainholes into coal seams. Horizontal boreholes drilled perpendicularly to the face cleat direction provide the highest gas production potential (ERTEKIN, SUNG & SCHWERER 1986).

4.8.6.2.1.3.2. Multiple-wing boreholes

Further improvement of horizontal well drilling in thin coal seams can be achieved by multiple boreholes consisting of several wings, with a two-wing wellbore originally being able to influence a drainage area that is two times larger than that affected by a four-wing borehole during early exploitation, whereas in late production, the four-wing borehole system is characterized by a more uniform movement of pressure transient due to its position in the center of a square reservoir (ERTEKIN, SUNG & SCHWERER 1986; LOGAN 1988). Coal seam thickness is a more critical factor influencing vertical crack performance than that of a horizontal borehole, because fracture surface area increases as coal seam thickness increases, whereas the exposed surface area for a given-diameter borehole increases only when its length is increased.

Comparative modelling based on identical exposed surface areas reveals that hydraulic fractures initially produce gas at higher rates than horizontal boreholes, because the distance for the gas to travel is shorter in a fully contained crack than in the horizontal well. In later exploitation, depletion of the cleat system and movement of pressure transients to distances farther away from vertical fracture and horizontal borehole results in wellbore length becoming the governing parameter and the horizontal borehole then performing better than the vertical fracture, with the difference in gas production rates before and after crossover also rising with increasing coal seam thickness.

4.8.6.2.1.4. Single- vs. multiple-layer reservoirs

While horizontal wells are very effective and superior in performance to both unstimulated and vertically fractured vertical boreholes in thin single-layer reservoirs, pay zones with high vertical permeability and prospective intervals with gas- and water-coning problems, vertical wells of either fractured or unstimulated nature are better solutions for thin multiple-layer reservoirs and pay zones with low vertical permeability where horizontal boreholes are unsuitable (JOSHI 1986 a). Vertical wells which are stimulated by vertical fractures are generally more effective than unfractured boreholes, and multiple vertical fracturing of horizontal wells can be a competitive procedure in intermediate reservoirs. If the length of a horizontal borehole is significantly larger than pay thickness, then horizontal well production is in the long time the same as that obtained from a fully penetrating infinite-conductivity vertical fracture.

Vertical permeability distribution has considerable impact on productivity augmentation through horizontal wells, with low and high vertical permeability reducing and enhancing horizontal borehole productivity, respectively. A horizontal well replaces more vertical borehole in a thin-layer reservoir than in a thick-bed pay zone, whereas in contrast to horizontal wells, slant boreholes (cf. section 4.8.6.2.2.2.) are more productive in thick reservoirs than in thin pay zones. Thus maximization of hydrocarbon production can be performed by selection between vertical, slant and horizontal wells depending on reservoir thickness and anisotropy.

4.8.6.2.1.5. Orientation to natural fractures and permeability anisotropies

In terms of the geometrical relationship between natural fractures and permeability anisotropies in the reservoir formation, horizontal wells can be very easily drilled in any direction to adjust to the optimum configuration of normal penetration of the wellbore trajectory through the natural joints. The propagation of hydraulically-induced fractures is in most cases directionally determined by the existing geostress field, and without thermal influence, only altered-stress fracturing (WARPINSKI & BRANAGAN 1988; cf. section 4.9.4.1.3.1.) can give rise to different orientations of crack extension and permit the option to model the angle of intersection of natural cracks and hydraulic fractures. Altered-stress fracturing may ultimately even lead to interconnected networks consisting of cracks propagating in alternating standard direction and in that perpendicular to it. The orientation of both hydraulic fractures and/or horizontal wells in hydrocarbon reservoirs is also important with respect to tectonical dip direction as well as sand body geometry and extension.

4.8.6.2.2. Horizontal hole drilling in combination with fracturing

SUNG & ERTEKIN (1987) present a performance comparison of vertical and horizontal hydraulic fractures and horizontal boreholes in low-permeability reservoirs. The results of a suite of numerical simulation exercises show that for equal surface area, the production from horizontal boreholes is expected to be greater than that from hydraulic fractures in vertical wells. Production rates from both horizontal wells and hydraulic fractures rise with increasing reservoir penetration length, but achievement of greater intersection lengths is technically more feasible in case of horizontal boreholes. Some comments are given as follows on reservoir thickness and permeability, and wellbore angle buildup technology and slant holes.

4.8.6.2.2.1. Reservoir thickness and permeability

Concerning reservoir thickness and permeability, the superiority of horizontal wells becomes more apparent in thinner and tighter formations, whereas the effectivity of hydraulic fractures improves in thicker pay zones. In case of anisotropic reservoirs, horizontal boreholes should be drilled and hydraulic fractures should be propagated perpendicularly to the largest permeability direction. Under comparable conditions, the production rate from a horizontal crack system is greater than that from a vertical fracture system, because a more uniform pressure decline over the entire reservoir caused by the horizontal crack helps achieving better drainage.

The best, but certainly also the most expensive solution, however, is a combination of horizontal drilling and hydraulic fracturing that provides the maximum drainage area in the reservoir. Particularly suitable targets for this dual approach are oil-bearing chalks (cf. section 4.5.4.) and gas-bearing shales (cf. section 4.4.4.). MERCER, PRATT & YOST (1988) document that two hydraulically fractured or four unstimulated horizontal wells can produce 1.5 times as much gas as 16 untreated vertical wells in case of both infill drilling and virgin reservoir development (cf. section 4.4.3.6.). Comparative performance evaluation of horizontal and vertical wells is carried out by BORNER, HALDORSEN, HARRISON, HOPWOOD & MACDONALD (1988); KING & ERTEKIN (1988) and WOOTEN, MUKHERJEE & MACH (1988). Special features of hydraulic fracturing of highly-deviated or horizontal boreholes are discussed in chapter 4.11. Field aspects of hydraulic fracturing in horizontal well sections are also discussed by GIGER, REISS & JOURDAN (1984); YOST, OVERBEY, SALAMY, OKOYE & SARADJI (1987); PEMPER, FLECKER, McWHIRTER & OLIVER (1988); SOLIMAN, HUNT & EL RABAA (1988) and SPREUX, GEORGES

& LESSI (1988).

DOSCHER, YING & KOSTURA (1988) discuss the utilization of specially fractured horizontal wells to improve steam-drive enhanced recovery of high-viscosity hydrocarbons. Properly positioned and fractured horizontal production wells permit commercial exploitation of many heavy oil and bitumen reservoirs which otherwise would be completely economically unfeasible.

4.8.6.2.2.2. Wellbore angle buildup technology and slant holes

Apart from vertical drilling at the beginning of horizontal well generation, building of angle in shallow slant holes can be accelerated by inclined rig position (VRIELINK & HIPPMAN 1988). Using slant well technology, the required well spacing unit is achieved with minimum surface land usage, optimum hole quality and minimum environmental impact, with slant well drilling therefore providing an attractive alternative to conventional directional multiwell pad projects. Slant borehole drilling is particularly suitable for shallow reservoirs with reduced interwell spacing, high facilities and tie-in cost per borehole head, high operating and maintenance cost per location, and requirement to drill deviated straight holes in order to minimize tubular wear (RICHARDSON 1985). Modelling of spud inclination, spud azimuth and surface location can be done for optimization of surface distance, wellbore curvature and downhole separation. Vertically spudded wells are not to be cross-drilled in order to eliminate the collision and interpretation danger. Rig inclination in slant hole drilling is usually abt. 10 - 30 degrees, but can in extreme cases also reach 45 - 60 degrees.

4.8.6.2.3. Preperforated liner completion and proppant requirement

Long horizontal wells are typically completed with preperforated liners to preserve hole integrity (YOST, OVERBEY, WILKINS & LOCKE 1988). The disadvantage of this type of completion is the associated risk of pulling the liner at a later stage of production history and re-running and cementing a casing string such that selective placement of fracturing fluids can be accomplished. An alternative approach is zone isolation performed by installation of external casing packers and port collars as an integral part of a casing string run along the horizontal section. Such a completion arrangement provided stimulation intervals with ready-made perforations injecting fracturing fluids into an openhole stimulation condition behind pipe. The latter method avoids the problems of formation damage associated with cementing and eliminates the need for tubing-conveyed perforating of numerous treatment intervals.

Comparison of different fracture operation designs in horizontal wells traversing shale gas reservoirs indicates that carbon dioxide stimulation can be more than four times better than nitrogen fracturing initially and can last twice as long prior to crack closure (cf. section 4.4.4.2.). The density contrast between nitrogen and carbon dioxide may have caused the propagation of more and wider fractures with carbon dioxide resulting in high improvement ratios. Proppants are definitely required for fracturing of low-stress shale gas reservoirs to sustain gas production beyond baseline conditions.

4.8.6.2.4. Multiple vertical fracturing of highly-inclined boreholes

A suitable technique to maximize drainage effectivity of highly-inclined to horizontal boreholes is multiple vertical fracturing (STRUBHAR, FITCH & GLENN 1974; STRUBHAR & GLENN 1974; cf. section 4.11.) which integrates extension of geometry and flow capacity of existing cracks as well as creation of new fractures (OVERBEY, YOST & WILKINS 1988). Using multiple fractures in a horizontal well, theoretically the same amount of total production will be recovered at an accelerated rate and thus economical payout of the borehole will be obtained in a shorter period of time (AUSTIN, ROSE & SCHUH 1988). Aspects of sets of parallel fracture planes, natural fracture intersection, multiple- vs. limited-entry fracturing, fracture morphology and orientation, and influence of injection rates are briefly outlined as follows.

4.8.6.2.4.1. Sets of parallel fracture planes

The object of the multifracture completion concept is to obtain greater fracture area through closer crack spacing than could be achieved from the same number of wells with single vertical fractures (PASINI & OVERBEY 1969). Optimization of multiple stimulation in highly-deviated boreholes requires the azimuth of the wellbore to be at high angle (preferentially perpendicular) to the fracture azimuth. Some aspects of angular relationship between wellbore trajectory and induced cracks as well as fracture spacing vs. drainage effectivity are outlined as follows.

4.8.6.2.4.1.1. Angular relationship between wellbore trajectory and induced cracks

There are all stages of transition of distances of the individual fractures and angular relationship to the wellbore trajectory between the end points of a singular vertical crack propagating parallel to the horizontal borehole trajectory in case of orientation of the latter parallel to fracture azimuth, and multiple parallel vertical cracks extending perpendicularly to the horizontal borehole trajectory in case of orientation of the latter orthogonal to fracture azimuth. Prediction of the induced fracture azimuth can be made from strain relaxation of oriented cores (cf. section 6.2.2.3.3.4.), borehole breakout elongation analysis (cf. section 6.2.1.8.), or directional determination of cracks generated during micro- or minifracture calibration treatments in adjoining or preceding vertical wells (AUSTIN, SCHUH & ROSE 1988; cf. section 4.8.2.).

4.8.6.2.4.1.2. Fracture spacing vs. drainage effectivity

The different fractures have to be kept parallel and separate in order not to deteriorate the drainage effect by crack interference and intersection, with the latter drawback being able to be avoided by sufficient lateral spacing of the individual fractures and careful control of their azimuth, inclination and length for making sure that contact of two independent cracks does not occur (ROEGIERS & DETOURNAY 1988). Several vertical independently producing cracks crossing a single deviated wellbore are much more economical than individual vertical fractures in separate vertical boreholes and represent the most feasible combination of horizontal well drilling and hydraulic fracturing, and certainly although being the most expensive reservoir treatment, definitely represent the most effective means of drainage maximization.

Field examples of multiple vertical fractures in horizontal boreholes and enhanced productivity modelling in comparison to single-fractured vertical wells are presented by STRUBHAR, FITCH & GLENN (1974); GIGER, COMBE & REISS (1983); GIGER (1985); KARCHER, GIGER & COMBE (1986); ANDERSEN, HANSEN & FJELDGAARD (1988) and OVERBEY, YOST & WILKINS (1988).

4.8.6.2.4.2. Natural fracture intersection

Similar aspects apply for optimization of drainage of naturally fractured reservoirs where effective depletion is reached when a horizontal well is drilled perpendicularly to the orientation of the natural cracks thereby intersecting a whole bundle of parallel pre-existing fractures (JOSHI 1987; cf. section 4.8.6.1.5.1.). Influences of natural cracks on horizontal borehole performance are also discussed by CARVALHO & ROSA (1988). Insights to hydraulic fracturing of horizontal wells in naturally jointed formations are also given by LAYNE & SIRIWARDANE (1988). Multiple vertical fractures propagating from horizontal boreholes are also the optimum solution for gas drainage from coal seam reservoirs because of suitable combination of both elements, as the former and the latter give rise to higher production rates in early and late exploitation stages, respectively (ERTEKIN, SUNG & SCHWERER 1986; cf. section 4.4.3.6.).

Optimization of number and size of multiple hydraulic fractures intersecting horizontal wells can be performed with reservoir simulators based on the net present value concept (MENG & BROWN 1987; BALEN, MENG & ECONOMIDES 1988; MUKHERJEE & ECONOMIDES 1988; cf. sections 1.4.12.2. and 4.8.11.4.). In horizontal wells, the sum of the maximum net present values of the individual multiple fractures must be substantially larger than the net present value of a vertical crack intersecting a vertical borehole in order to be comparatively more profitable.

4.8.6.2.4.3. Multiple- vs. limited-entry fracturing

AUSTIN, ROSE & SCHUH (1988) present simultaneous multiple-entry hydraulic fracturing in horizontal wells. Limited-entry perforation at several crack initiation points allows equal placement of proppants and fluids in one stimulation treatment, and in carbonates, proppant fracturing can also be replaced by acid fracturing if economically feasible. The number of crack initiation points can be maximized if the horizontal borehole is oriented perpendicularly to the azimuth of the induced fractures. Unless the specific reservoir requires something else, the individual crack initiation points should generally be spaced 250 - 300 ft apart. Before the actual treatment is carried out, each set of perforations at each fracture initiation point should be broken down and a stable injection rate established.

4.8.6.2.4.4. Fracture morphology and orientation

Similarly as multiple vertical hydraulic fractures emanating from highly-inclined to horizontal boreholes represent substantial reservoir drainage improvement with respect to singular cracks fractures intersecting deviated wells or even vertical boreholes in some cases, success probability of exploration and appraisal of naturally fractured pay zones can be considerably increased if directional wells are drilled perpendicularly to the orientation of the natural cracks instead of normal boreholes (AGUILERA 1983; cf. sections4.8.6.1.5.1. and 4.8.6.2.4.2.).

BEN NACEUR & ROEGIERS (1988) investigate simultaneous propagation of multiple vertical hydraulic fractures. Distinction can be made between collinear sets emanating from vertical boreholes and parallel sets intersecting horizontal wellbores. The elastic interaction between collinear fractures in vertical wells results in larger fracture widths or lower apparent stiffnesses, whereas the consequences for parallel fractures in horizontal wells are opposite. Therefore individual fractures are invariably shorter in the first case (cf. also section 4.2.2.1.3.4.).

Selective multiple fracturing with introduction of several equidistant planparallel fractures from horizontal drainholes can considerably improve productivity index (ANDERSEN, HANSEN & FJELDGAARD 1988). While in the early stage of exploitation there is no interference between the cracks, after some period of production the fracture planes start interfering, with the degree of interference steadily increasing.

4.8.6.2.4.5. Influence of injection rates

OVERBEY, YOST & WILKINS (1988) and YOST, OVERBEY, WILKINS & LOCKE (1988) study multiple vertical hydraulic fracture origin in horizontal boreholes traversing shale gas reservoirs that are extensively naturally jointed. Multiple hydraulic fractures can be propagated in a wellbore which has been completed to provide adequate access to multiple natural cracks. Low injection rates during a hydraulic fracturing operation allow the propagation of natural cracks with a low-angle relationship to the principal stress orientation, whereas higher injection rates generally result in inducing fractures controlled by the stress field. Some comments are offered as follows on dendritic fracturing and stress field relationships as well as inflation and ballooning of natural cracks.

4.8.6.2.4.5.1. Dendritic fracturing and stress field relationships

Low injection rates thus lead to development of a dendritic fracture pattern by preferentially entering and ballooning pre-existing natural cracks (cf. sections 4.5.4.4.3. and 4.8.8.4.), whereas high pumping rates give rise to abandonment of the original joints created by the palaeostress field and generation of new fresh fractures that are in accordance to the actual stress field in case of changes of palaeomagnetical and palaeotectonical stress situation between creation of natural cracks and induction of hydraulic fractures. In the special case of neither rotation nor tilting of the crustal blocks between formation of natural joints and triggering of hydraulic fractures, also higher injection rates can only result in lateral and longitudinal expansion of pre-existing natural cracks including some breakthroughs between individual joints which also leads to dendritic fracturing.

Distribution of the propagated fractures is more random at low than at high injection rates, with the latter seeming to trigger auto-selection by the formation concerning spacing and distribution of cracks. Multiple hydraulic vertical fractures can be induced from a horizontal wellbore during a single pumping event. Even though stress ratios indicate that the investigated area is nearly tectonically relaxed, proppants are still needed to maintain permeable flow paths for more than several weeks. Different closure pressures having been recorded in different stimulation treatments can be the consequence of induction of different fractures having near-adjacent angles, and subsequent pressure drops indicate that some of the created cracks were closing off.

4.8.6.2.4.5.2. Inflation and ballooning of natural cracks

Improvement of gas recovery efficiency of highly-inclined or horizontal boreholes in naturally fractured reservoirs by extension of geometry and flow capacity of existing cracks and creation of new hydraulically-induced fractures can be best accomplished by stimulating the natural joints which exist in the reservoir by inflating them with a non-damaging fluid and propping the ballooned fractures to maintain the enhanced flow capacity and to induce additional fractures by increasing injection rate (OVERBEY, YOST & WILKINS 1988; cf. also section 4.8.8.6.). Multiple natural crack sets are generated by both normal or block faulting and thrusting. Field experience reveals that during stimulation operations, two sets of natural fractures were opened and propagated, and induced cracks were created along a third direction controlled by the regional stress field. Multiple-oriented hydraulic fractures can be generated from a horizontal wellbore under openhole conditions where the orientations of natural and induced cracks form an angle of abt. 15 degrees.

4.8.7. Uptake capacity enhancement of injection wells

Minifracturing and other smaller-scale stimulation are also regarded to be suitable means to increase the surface of the formation surrounding the borehole for enhanced uptake of water, tensid, polymer, acid or steam in injection wells, and to ameliorate the conductivity of naturally jointed reservoirs by smaller- or larger-scale artificial fracturing. The most significant items of application of injection-well uptake capacity improvement by hydraulic fracturing are oil-field water-injection wells and waste-liquid disposal wells.

4.8.7.1. Oil-field water-injection wells

The most abundant utilization of uptake capacity amelioration of injection wells by hydraulic proppant small- or large-scale fracturing is the stimulation of oil-field water-injection wells. Distinction can be made between controlled fracturing by a properly designed hydraulic stimulation treatment and accidental fracturing by injection operation above formation fracture gradient.

4.8.7.1.1. Controlled fracturing

In many cases, the permeability of sandstone reservoirs is poorer in the aquifer zone surrounding the oil field than in the oil-bearing column due to early oil immigration into embryonic structures which resulted in retarded diagenesis and thus better permeability preservation in the oil-saturated belt with respect to the water-containing marginal seam that was often more tightly cemented. Therefore in many fields the oil-production wells do not or only to subordinate amounts require stimulation, whereas the water-injection wells need hydraulic fracturing in order to ameliorate their uptake capacity and thus enhance reservoir pressure improvement and maintenance by increasing injection rates. Apart from enhanced reservoir depletion by secondary recovery, water flooding amelioration is an important measure for counteracting pay recompaction and resulting oil-field subsidence (cf. section 4.5.4.1.6.). After discussion of some general aspects, a few points of proppant size and reservoir pressure as well as decreasing in-situ stress by cold water injection, and sweep efficiency vs. injectivity are outlined.

4.8.7.1.1.1. General aspects

An artificial fracturing operation of smaller or larger size performed in already naturally jointed reservoirs can also considerably enhance the conductivity of the latter and can be technically handled much easier by just enlarging and propping the already existing zones of weakness (aspects of naturally fractured reservoirs are discussed by PINE & BATCHELOR 1982, RUMMEL 1985, BLANTON 1986; GATENS, OLAREWAJU & LEE 1986; GUEVARA & TYLER 1986, HARRISON & TILDEN 1986, NELSON 1986; TANSOE, KING & HOLMAN 1986; RUMMEL, HEUSER, MÜLLER & OTTO 1987; cf. section 4.8.8.). Comments on hydraulic fracturing of injection wells are given by RILEY (1959), JONES (1960), HOWELL & THOMAS (1978), ILSENG & CORTEZ (1985), KONING (1985), KONING & NIKO (1985), PERKINS & GONZALEZ (1985); BIOT, MASSE & MEDLIN (1986); CHEN (1986), REVIERE & WU (1986), HOLZHAUSEN & EGAN (1987) and WILLIAMS, SHERRARD & LIN (1987).

As a consequence of injection well fracturing, injection rate may be increased at a near-constant injection pressure, or if necessary injection pressure may be reduced while a constant injection rate is maintained (GUERREDO 1958, POWELL & JOHNSTON 1960, HARTSOCK & SLOBOD 1961). With increasing water input, the floodout time for a particular reservoir would be reduced (HOWARD & FAST 1970 a). Large-volume fracture stimulations in a waterflood operation are reported by ILSENG & CORTEZ (1985). Fracturing for assistance of steam-flooding is evaluated by BRITTON, MARTIN, LEIBRECHT & HARMON (1983) and STANG & SONI (1987; cf. also section 4.9.4.). Fluid flow and leakoff properties from fractures induced in injection operations are discussed by SETTARI (1988). Hydraulic stimulation of oil-field water-injection wells has in the past already frequently been performed in the USSR.

4.8.7.1.1.2. Proppant size and reservoir pressure

HOOVER & ADAMS (1988) outline that proppant selection for waterflood projects has to be based on size and ultimate performance of injection operation. Switching from 20/40 to 12/20 mesh proppants is often favourable in order to avoid or to diminuish filter cake buildup on the fracture face which affects proppant performance and is in essence a reduction in ultimate propped crack width. Pumping of large quantities of large-sized proppants partially offsets the diminution of propped fracture width attributed to stimulation fluid-caused filter cake remnants. The damage aspects are a combination of fluid retention, fines present in the proppant package, fines having migrated from the formation, and embedment due to closure stress.

The pay horizons to be treated by more or less minifracturing and other small-scale hydraulic stimulation include both hydropressured and geopressured formations (questions of geopressured reservoirs are reported by SAMUELS 1979, TUCKER 1979, WEEKS 1979; McCOY, HARTSOCK & DOBSON 1980; AMIRIJAFARI, SHARER & RIGBY 1982; HAMMERSHAIMB & KUUSKRAA 1982, ROSE & DOHERTY 1982; cf. also section 4.4.2.), with normal hydrostatic pressures being abt. 0.45 psi/ft and hyperhydrostatic pressures reaching up to 0.9 psi/ft or even more (BRIM 1986). KUO, HANSON & BRISAY (1984) evaluate prediction of fracture extension during waterflood operation, and GARON, YIN & DUNAYEVSKY (1988) simulate thermally-induced waterflood fracturing.

4.8.7.1.1.3. Decreasing in-situ stress by cold water injection

Special features in water-, thermal- or chemical-injection projects are decreasing in-situ stress or fracture closure pressure with declining pore pressure when finally closure pressure in sandstone reservoirs is much lower than in the surrounding shales (SMITH 1981), with such an in-situ stress contrast providing a strong barrier to vertical growth, and penetration of the fracture tip beyond the relatively pressurized area of the injection project into regions of low reservoir pressure and/or closure stress. Some aspects of reservoir pressure distribution and fracture penetration as well as coupled formation cooling and fracture elongation are outlined as follows.

4.8.7.1.1.3.1. Reservoir pressure distribution and fracture penetration

Increase and redistribution of reservoir pressure by injection of liquids for pressure maintenance or enhanced recovery has the result that as the fracture penetrates beyond the increased pressure region, it would be entering a lower closure stress area thus possibly producing excessive afterflow of the fracturing fluid. The latter effect takes place when after shut-in of the well, the slurry continues to flow until the fracture closes to lock the proppant in place (cf. section 4.12.2.2.1.), with the length of this afterflow period depending on fluid-loss rate and proppant concentration. This special problem of pressure variation in reservoirs where injection project areas are active needs serious attention for interpreting and recognizing these synthetic pay zone pressure distribution correctly and to adjust the stimulation treatment to it.

In-situ stress is also significantly affected by fluid leakoff during fracturing.

Injection of cool water can reduce earth stresses around uptake wells considerably, causing them to fracture at pressures substantially lower than would be expected in the absence of the thermoelastic effect (PERKINS & GONZALEZ 1985). Depending on the shape of the cool region and the difference between minimum and maximum in-situ horizontal earth stresses, minor fractures perpendicular to the main fracture eventually open, thus creating a jointed system of primary and secondary fractures (cf. also section 4.9.5.).

4.8.7.1.1.3.2. Coupled formation cooling and fracture elongation

Reservoir behaviour associated with cold water injection into a hot formation are very complex because of the coupled nature of the mechanisms involved (GARON, LIN & DUNAYEVSKY 1988). Water injection cools the reservoir by conduction and convection. Prior to fracturing, flow distribution is controlled by reservoir characteristics, and temperature field is governed by flow pattern. In-situ stress distribution is subsequently determined by temperature and pressure fields, and fracture gradient may be decreased until at some injection pressures, formation parting occurs. Flow pattern is strongly influenced by any fractures developing which control subsequent temperature and stress distributions. Reductions of stress near the fracture tips result in fracture elongation and further stress diminution, with this coupling effect being able to give rise to rapid fracture extension (cf. section 4.9.5.). Other aspects of thermally-induced waterflood fracturing are discussed by PERKINS & GONZALEZ (1984); MORALES, ABOU-SAYED, JONES & AL-SHAFFAR (1986) and WILLIAMS, SHERRARD & LIN (1987).

4.8.7.1.1.4. Sweep efficiency vs. injectivity

HOWELL & THOMAS (1978) emphasize the role of injection well fracturing for maximization of sweep efficiency. Field experience has shown that injectivity of shot holes is inferior to all other types of stimulation, and in particular fracturing of shot holes is less effective than fracturing in gauge holes. Some wells treated with low-viscosity fluid have greater decline before stabilizing than those stimulated with high-viscosity fluids, whereas high-viscosity fluid jobs exhibit fewer mechanical screenout problems than low-viscosity ones. Shot holes may provide only small open fractures.

In addition to or as an alternative of hydraulic proppant fracturing, selective acidizing of sandstone reservoirs in injection wells can discretely stimulate ineffective intervals and significantly improve injectivity profiles (ELLENBERGER & ASELTINE 1976; cf. section 4.5.3.). Stimulation of water injection wells is also discussed by CLEMENTZ (1982).

4.8.7.1.2. Accidental fracturing

Injection wells can already be accidentally stimulated by exceeding the formation fracture gradient by too high injection rates and pressures. Such accidental unplanned fracturing in water-flood or -disposal projects have to be avoided in many cases for reservoir engineering and environmental safety reasons. Some of the main drawbacks of accidental injection-well fracturing are illustrated as follows along the lines of fluid breakthrough vs. sweep efficiency, and open cracks above fracture pressure and non-proppant hydraulic stimulation.

4.8.7.1.2.1. Fluid breakthrough vs. sweep efficiency

Premature fracturing of injection wells may result in bypass of reservoir regions with significant oil and gas reserves (AMAEFULE & MASUO 1984), thereby deteriorating sweep pattern and efficiency as well as recovery factor. Excessive injection pressures can also give rise to water channelling in the cement sheath between casing and borehole wall or water passage through the natural crack system in the pay sandstone into vertical fractures and horizontal bedding planes of overlying boundary shales, with the water causing swelling of clay minerals which leads to serious casing damage (CHEN 1986).

Accidental injection-well fracturing may lead to premature breakthrough of injected fluids, poor sweep efficiency, reduced recovery and loss of costly injection fluids (SINGH & AGARWAL 1988). In some reservoirs, fracturing is not recommended at all, because radial flow has a more favourable impact on waterflood performance than the linear flow paths which usually result from fracturing (SLOAT 1988; cf. section 4.11.3.). Aspects of accidental injection-well fracturing by injection pressure exceeding formation or fracture parting pressure (fracture extension/propagation pressure) are also discussed by DICKEY & ANDRESEN (1945), JONES (1960), WILLIAMS (1977); HAGOORT, WEATHERHILL & SETTARI (1980); HAGOORT (1981), KONING & NIKO (1985); TANG; LIU & JIANG (1986) and PRADO, KRUYSDIJK, NIKO, SANCEVIC & RODRIGUEZ (1988).

4.8.7.1.2.2. Open cracks above fracture pressure and non-proppant hydraulic stimulation

Injection well fracturing, however, has to take always into consideration that in case of continuous injection operation above fracture pressure, open cracks exist which are kept open by the permanent overpressure and conductivity might be severely deteriorated if the fractures are infilled with proppants (BARBY & BARBEE 1987). If an injection well has insufficient uptake capacity at or above fracture pressure, it may be possible to increase injectivity by hydraulic stimulation of the reservoir using a high-rate, low fluid-loss, slick-water treatment without proppant.

The high rate and large volume may extend vertical and horizontal communication to more permeable layers of the reservoir. Once the barriers are broken through, the regular injection above the fracture pressure should keep the fractures open, thereby not necessitating a proppant to support the crack and thus not requiring to reduce the maximum flow capacity of an open fracture by infilling with proppants. Non-proppant hydraulic reservoir stimulation is also significant in some minifracturing treatments of sandstones and carbonates (cf. section 4.8.2.5.), water fracturing of chalks (cf. section 4.5.4.5.2.) and geothermal reservoirs (cf. section 4.7.3.5.), and nitrogen fracturing of gas shales (cf. section 4.4.4.2.).

4.8.7.1.2.3. Other aspects

Accidental fracturing of water-injection wells has unforeseen consequences on waterflood sweep efficiency (DIKKEN & NIKO 1987). Fracturing often causes cooling of hot formations by cold injection water. Another undesired effect is crack wall impairment as a result of filter cake build-up due to poor-quality injection water. On the other hand, injection operation far below the formation or fracture parting pressure leads to injection volumes much lower than the allowable maximum and thereby result in a considerably reduced rate of oil recovery. A reliable estimate of the fracture or formation parting pressure is thus of critical importance in conducting secondary and tertiary recovery projects (SINGH & AGARWAL 1988; cf. section 4.8.2.3.1.). WILLIAMS (1977) evaluates the influence of induced vertical cracks on overall reservoir performance in accidentally fractured water-injection wells as a consequence of operation above fracture pressure.

Prior to introduction of hydraulic fracturing to the hydrocarbon industry some 40 years ago, accidental fracturing as a consequence of injection pressures exceeding formation fracture gradient occurred during acidizing (HOWARD & FAST 1970 a), water injection (GREBE 1943, DICKEY & ANDERSON 1945, YUSTER & CALHOUN 1945) and squeeze cementing (TORREY 1940, REISTLE 1945, TEPLITZ & HASSEBROEK 1946, HOWARD & FAST 1950) as reflected by acceptance of considerably larger volumes by the well than predetermined by reservoir evaluation and calculation. Formation lifting and pressure parting resulted in much higher uptake capacity of the pay than usual.

4.8.7.2. Waste-liquid disposal wells

Cheap minifractures may also be the only feasible way for some special applications of stimulation such as to open zones for take-up of disposed waste liquids deriving from drilling, gas production and purification (mainly sour-gas treating), oil and water separation in fields producing with higher water cuts, steam-drive enhanced oil recovery (excess water), and other industrial processing, in order to remove them safely into properly sealed underground formations (unless they are stored within salt caverns) within preferentially isolated depleted oil and gas reservoirs. After commenting on safe removal of waste liquids by underground disposal, some aspects of proppant vs. non-proppant disposal fracturing are outlined.

4.8.7.2.1. Safe removal of waste liquids by underground disposal

The experience of looking for suitable disposal horizons that could take up reasonable quantities of waste liquids of different viscosity and with varying amounts of dispersed solids has shown that there are not many geologically safe and unobjectionable harmless formations for this application, and the few possible target layers predominantly require permeability enhancement by fracturing in order to avoid rapid plugging of the injection wells by particles filtrated out of the disposed fluids.

It is therefore assumed that hydraulic proppant fracturing of underground waste-liquid disposal wells will change from being unconventional to more or less popular in light of the increasing need of safe clearance of various fluids in the near future when conventional injection capacities are approaching exhaustion (this applies particularly for waste liquids that cannot be disposed into aquifers surrounding producing oil and gas fields without risking to deteriorate the hydrocarbon quality or to seriously disturb oil and gas drainage). Fracture containment in vertical direction is in this special application of particular significance for the reason of environmental and reservoir engineering safety (cf. section 4.2.2.). Aspects of so far common practices of disposal of excess water are discussed by SPRINGER & MÜLLER (1985), and economical questions of disposal methods for North Sea oil and gas fields are studied by FISHMAN (1986). WILSON & KISER (1987) also comment on aspects of hazardous waste-liquid disposal by injection into subsurface formations, and SUN & MONGAN (1974), DOSCHER & EL-ARABI (1983) and McBANE & JEU (1988) report examples of waste water disposal at pressures above fracture gradient or breakdown pressure.

4.8.7.2.2. Proppant vs. non-proppant disposal fracturing

It has been recognized very early in hydraulic fracturing history that stimulation of disposal boreholes can establish a low-pressure, high-fluid-injection capacity well almost everywhere in the hydrocarbon field complex (HOWARD & FAST 1970 a), with thus disposal fracturing being an easy solution of lacking waste disposal capacities in reasonably close situation to oil and gas production and processing. In terms of fracture propping, the hydraulically-induced crack may be left open if sufficiently supported by permanent injection above fracture pressure (cf. section 4.8.7.1.2.2.), because proppant packages plugging the crack may be subjected to rapid permeability deterioration by invasion of fine solids dispersed within the disposed waste fluids into the proppant interstitials and thus destroying uptake capacity.

If waste-liquid disposal is not continuously carried out at sufficient overpressure to keep the fracture open or to enable unproblematic repeated reopening of the crack upon return to injection operation, the fracture should be plugged with coarse to very coarse proppants in bank or pillar distribution (cf. section 4.3.3.2.) instead of continuous infilling, and wide cracks should be supported by proppant multilayers in order to provide as much porosity and permeability as possible and to enable maximum fluid acceptance capacity.

4.8.8. Combination of natural and artificial fractures

Smaller- or larger-scale hydraulic proppant fracturing within naturally jointed reservoirs is particularly significant, as the pre-existing crack network can be successfully employed for enhancement of the effectivity of the artificial stimulation operation. Naturally jointed (also called double-porosity) pays are sedimentary formations which were after deposition fractured by tectonical or chemical processes, and therefore have a primary intergranular matrix porosity which is controlled by sedimentary environment and diagenetical evolution, and a secondary fracture porosity that is governed by fracturing or jointing as a result of failure during mechanical deformation (BARUA, KUCUK & GOMEZ-ANGULO 1985; HENSEL 1988). Naturally jointed reservoirs can be thus regarded as double-porosity media consisting of isotropic heterogeneous primary rock matrix and anisotropic heterogeneous secondary fracture system (HOUZE, HORNE & RAMEY 1984; NAKORNTHAP & EVANS 1986). ABDASSAH & ERSHAGHI (1986) introduce a triple porosity model for representing naturally cracked pays. Some important naturally fractured oil and gas reservoirs include Monterey Formation and overthrust sandstone formations in the western USA, chalk formations in the North Sea in Europe (cf. section 4.5.4.), and carbonate formations in the Middle East (BERGOSH, LORD & OWEN 1987).

Unfortunately some confusion occurs due to simultaneous application of the term secondary porosity in sedimentary petrography to describe renewed origin of void spaces by mineral transformation and dissolution within the rock matrix during course of burial diagenesis independent from any influence of fractures (SCHMIDT & McDONALD 1979 a, 1979 b; AL-SHAIEB & WALKER 1986), and in petrophysics for characterization of reservoirs including primary matrix and secondary crack porosity. Therefore fracture porosity in petrophysical sense should better named tertiary porosity to allow for proper distinction of all the possible porosity types (the same applies for permeability). The various phenomena arising from coexistence of natural and hydraulic fractures are summarized in sections commenting on fracture interference and superimposition, fracture divergence, temporary natural fracture damage, dendritic fracturing technique, tailored pulse loading, proppant plugging of natural fractures, and other aspects.

4.8.8.1. Fracture interference and superimposition

Interference and superimposition of natural and hydraulic fractures are the results of carrying out artificial stimulation treatments in originally jointed formations. Aspects of distribution and generation of natural fractures, interaction between natural and hydraulic fractures, and interfracture communication are outlined as follows.

4.8.8.1.1. Distribution of natural fractures

Two types of natural crack systems can be distinguished comprising multiple fracture sets associated with structurally deformed strata and single regional crack sets (LORENZ & FINLEY 1987, NORTHROP & FROHNE 1987). Fractures of the first type are caused by local faulting or folding and commonly cut indiscriminantly across lithological boundaries. Fractures of the second type are caused by regional stresses of much less magnitude in conjunction with high pore pressures and occur in structurally undeformed formations. Some aspects of stress differences and depositional environment as well as reservoir permeability and communication are outlined as follows (various types and distributions of natural fractures are illustrated on plates I - XI in section 3.13.).

4.8.8.1.1.1. Stress differences and depositional environment

The distribution of regional fractures is generally controlled by subtle stress differences of the various lithologies, with thus depositional environment that governs the sedimentary heterogeneity of the reservoir also supervising the distribution of regional fractures within the pay zone (LORENZ & FINLEY 1987, NORTHROP & FROHNE 1987). Therefore changes in crack-related production from reservoirs with regional fractures can be correlated with changes in depositional milieu, and some sedimentary environments have a better prospectivity for crack-controlled hydrocarbon production than others. Fracturing style often changes with depth from shear fracturing near the surface to extension fracturing at depth in the hydrocarbon reservoir (CLARK 1983; cf. also section 6.2.2.3.). Shear cracks are associated with movement parallel to the plane with no perpendicular motion, whereas extension fractures are created as the walls move apart (HENSEL 1988). AGUILERA (1987) distinguishes tectonical, regional and contractional natural fractures.

4.8.8.1.1.2. Reservoir permeability and communication

Reservoir permeability is significantly enhanced by the natural fracture system, but the irregularly distributed original cracks allow only imperfect communication across the pay and represent an unidirectional accomodating permeability enhancement only in a preferred direction (LORENZ & FINLEY 1987, NORTHROP & FROHNE 1987). Therefore hydraulic proppant fracturing stimulation is required to fully connect all reservoir segments to the wellbore, but the fracture permeability system is easily damaged by treatment fluids and stresses. Crack-dominated permeability systems are thus an important consideration in tight gas reservoir modelling (cf. section 3.4.1.), with in certain cases factors such as reservoir pressure, sizes and heterogeneities providing secondary controls on fracture-dominated production.

Typical naturally fractured reservoirs are composed of brittle rock with low intergranular porosity (HENSEL 1988). Occurrence and extent of natural joints is governed by initial shear strength, internal friction, rock ductility, effective confining stress, temperature, depth of burial, rock type, bed thickness, and adjacency to other layers. Crack porosity is mainly a function of block size and fracture width. Fractures have such a dramatic impact on reservoir performance that their hydrocarbon storage capacities tend to be inflated.

4.8.8.1.2. Generation of natural fractures

Natural fractures are generated by diastrophism in case of folding (cf. plate I/1 - 7) and faulting (cf. plate XI/3 - 4) where cracks are created along the line of the fault which in turn produce a zone of dilatance, deep erosion of the overburden that permits the upper parts to expand, uplift and fracture through planes of weakness (frequently in connection with origin of unconformities; cf. plate I/1 - 7), and volume shrinkage as in case of cooling of igneous rocks (cf. plate XI/5 - 6) and desiccation of sediments (cf. plates VI/1 - 7 and XI/7; AGUILERA 1983). As the probabilities of intercepting vertical fractures with a vertical hole are very slim, exploration and appraisal success in naturally fractured reservoirs would be greater if directional wells are drilled perpendicular to the orientation of the natural cracks (cf. sections 4.8.6.1.5.2. and 4.8.6.2.4.2.).

The common offshore exploration strategy of drilling a vertical well in the crestal area of anticlinal structures may be a double pitfall, because first the vertical hole might not intercept vertical or high-inclination natural fractures and second, in contrast to popular belief, the crestal area of asymmetrical anticlines might contain only little fracture porosity. The discussion as follows concentrates on impact of reservoir heterogeneities, fracture spacing vs. bed thickness, and fracture mineralization and stress sensitivity.

4.8.8.1.2.1. Impact of reservoir heterogeneities

LORENZ, BRANAGAN, WARPINSKI & SATTLER (1986); LORENZ & FINLEY (1987), NORTHROP (1988) and NORTHROP & FROHNE (1988) describe reservoirs containing a system of unidirectional subparallel natural fractures which occur principally in sandstones and siltstones and terminate at mudstone contacts both in vertical and horizontal direction as well as both at interreservoir and intrareservoir lithological discontinuity planes. Thin mudstone interbeddings in sandstone complexes such as overbank floodplain members of fluvial braided-river cyclothems (cf. plates III/1 - 2, IV/3 - 5, VII/3, VIII/3 and IX/1 - 3), interdune pond deposits in aeolian dune and sheet sand as well as playa sequences (cf. plate VIII/1 - 2), and low-water drapes in fluvial channel sand units (cf. plates IV/6 - 8, VII/5 - 6, VIII/8 and IX/4 - 8) are already sufficient mechanical property and rock stress contrasts to stop natural fracture propagation.

Discontinuities such as unconformities between folded basement and flat-lying cover strata (cf. plate I/1 - 7) or massive carbonate reef domes interspersed in bedded carbonate sediments (cf. plate I/8) are also areas of fracture extension termination and change of geometrical constellation of the cracks at the transition between two different joint systems of different genesis. Natural fractures occur in a wide spectrum of lengths, widths, spacings and densities, and the cracks and their interconnections are often narrow and/or mineralized, resulting in a stress-sensitive, easily damaged system.

4.8.8.1.2.2. Fracture spacing vs. bed thickness

The average fracture spacing in heterogeneous reservoirs is always significantly less than would be suggested by a linear relationship between joint spacing and bed thickness. The natural cracks formed under conditions of high pore pressures and relatively low horizontal stresses (LORENZ, BRANAGAN, WARPINSKI & SATTLER 1986) which were provided by tectonical compression of the basin. Fracturing occurred at high in-situ pore pressures that were caused by gases released during maturation of organic material in the formation which decreased the average effective stress sufficiently to allow failure to take place at relatively low horizontal stresses (NORTHROP 1988).

4.8.8.1.2.3. Fracture mineralization and stress sensitivity

In many tight sands, gas production is principally controlled by a network of open and partly mineralized natural fractures (PITMAN & STRUNT 1986) which formed in response to high pore-fluid pressures that developed during hydrocarbon generation as well as to widespread tectonical stress with late-stage basin subsidence and sediment loading associated with periods of uplift and erosion.

Spacing, orientation and mode of cracks are related to reservoir lithology (various examples of geometry and distribution of natural fractures are illustrated on plates I - XI in section 3.13.).

Fractures are either open and non-mineralized or are partly to completely occluded by fine- to coarse-crystalline calcite (LORENZ, BRANAGAN, WARPINSKI & SATTLER 1986; LORENZ & FINLEY 1987). Incomplete calcite crystal growths along joint walls are very common and may keep the cracks propped open which helps to maintain the increased permeability produced by fracturing. On the other hand, joints completely cemented by calcite may bar fluid flow at depth. Vertical fractures in sandstones predominantly result from extension, because they are in most cases open with reasonable wall separation and show little if any evidence of lateral displacement. Cracks in shales are numerous and tend to be non-mineralized and closed. Their highly irregular surfaces which are inclined at low angles to bedding and which are commonly polished and slickensided indicate that the cracks originated by shear, and permeability along these fractures under overburden stress is negligible.

According to the stress sensitivity of the reservoir, production can be essentially shut off by reducing formation pressure which effectively squeezes the cracks closed, whereas conversely high-pressure gas injection gives rise to dramatically increased pay permeability due to joint opening (RANDOLPH, SOEDER & CHOWDIAH 1984; NORTHROP & FROHNE 1988). Other natural fractures which appear to be discontinuous on core scale are believed to enhance matrix permeability and increase permeability anisotropy rather than act in the manner of cracks associated with the dual-porosity system (BELFIELD 1988).

4.8.8.1.3. Interaction between natural and hydraulic fractures

Natural cracks have an important influence on modification or alteration of propagation of hydraulically induced fractures (cf. also section 4.5.1.). Preexisting natural joints are secondarily enlarged first by drilling mud and later by stimulation fluids (KULANDER & DEAN 1988; NURMI, WATERHOUSE, KHOBAR & AKBAR 1988). A natural crack which is crossed by a hydraulically-generated fracture behaves like a high-permeability streak or a reasonable lithological discontinuity (cf. also section 4.2.4.2.) and is able to deflect or complete arrest its growth (cf. section 4.2.4.3.). A significant increase in bottom hole treating pressure is often required to reinitiate fracture propagation and results in formation of a multiple crack network with very short effective penetration into the reservoir (BLANTON 1980; CONWAY, McGOWEN, GUNDERSON & KING 1985). Connectivity of the system of natural fractures is a significant parameter (WIJESINGHE 1985, WIJESINGHE & KECECIOGLU 1986 a).

Many naturally jointed formations are fault-related, and single and intercepting sealing faults have an important impact on reservoir communication (AGUILERA 1985). Aspects of combination of natural and hydraulic fracturing are also discussed by BEN NACEUR (1988), and experimental modelling of the interaction between hydraulically-induced and pre-existing fractures is performed by BLANTON (1982). Natural cracks control the permeability of many tight gas reservoirs and can create tens-of-millidarcy pay zones out of submicrodarcy rocks (NORTHROP 1988; cf. section 3.4.1.). Following discussion of some general aspects, orientation of natural and hydraulic fractures as well as relationship between ancient and actual stress system are outlined.

4.8.8.1.3.1. General aspects

Most tight reservoirs are naturally fissured, with the cracks either being very small and irrelevant to flow thus reducing the pay to a homogeneous and likely isotropic medium, or with production being significantly affected by intersection of wellbore and cracks in massive naturally jointed formations (BEN NA-

CEUR & ECONOMIDES 1988). During execution of hydraulic fracturing treatments, the presence of fissures and cracks affects induced fracture geometry, because major natural cracks lead to a change of hydraulic fracture path (JEFFREY & VANDAMME 1987; cf. section 4.2.4.3.) thereby causing problems for proppant transport, and the high permeability of fissures results in their action as thief zones where fracturing fluid-loss is enhanced. Depending on crack magnitude, it is possible that leakoff control without suitable diverting agents does not work because filter cakes from polymer residues may be impossible to originate. In limestone formations and during acid fracturing, fissuring may enhance wormhole growth (BEN NACEUR & ECONOMIDES 1988; cf. section 4.5.4.2.1.). Influences of natural cracks on propagation of hydraulic fractures are also discussed by TEUFEL & WARPINSKI (1988).

4.8.8.1.3.2. Orientation of natural and hydraulic fractures

JEFFREY, VANDAMME & ROEGIERS (1987) comment on mechanical interactions in branched or subparallel hydraulic fractures. Hydraulic stimulation of naturally jointed reservoirs can result in generation of complex geometries such as offsets at crosscutting natural fractures. Pumping fluid into a cracked rock formation can inflate the natural joints, and the resulting hydraulic fracture may propagate as a system of parallel to subparallel fractures with both opening and sliding movement across the crack surfaces (PINE & BATCHELOR 1984). Some aspects of cross-cutting and multiple branching of induced fractures as well as hydraulic fracture propagation parallel to natural cracks are outlined as follows.

4.8.8.1.3.2.1. Cross-cutting and multiple branching of induced fractures

If the natural joints are more widely-spaced, the hydraulic fracture may propagate as a single fracture but interact with natural cracks one by one as it approaches them, with offsetting and branching of the hydraulic fracture resulting from such interactions (JEFFREY, VANDAMME & ROEGIERS 1987; cf. section 4.2.4.3.). Slip occurring on weakness planes after the hydraulic fracture crosses them leads to a reduction in crack width at the site of offset which may cause proppant bridging.

The study of the propagation of multiple branched fractures and interaction of a hydraulic fracture with a cross-cutting natural crack reveals that multiple branched fractures are a quite common result of such interferences. The interaction between the two (or more) branches of the fracture gives rise to increasing treating pressure as compared with the single crack. Hydraulic stimulation of sandstone reservoirs characterized by subparallel unidirectional natural fractures suffer in effectivity from the fact that the hydraulic cracks propagate parallel to the natural fracture trend (LORENZ & FINLEY 1987).

4.8.8.1.3.2.2. Hydraulic fracture propagation parallel to natural cracks

As hydraulically-induced fractures are generally parallel to natural and to the principal horizontal stress of the system, the artificial fractures consequently often fail to intercept the main natural fractures (AGUILERA 1983). Conjugate cracks might be crossed and connected by hydraulic fractures, but have lower transmissibilities than the main natural joints which guide the propagation direction of hydraulically-induced fractures to extend more or less parallel to them. In addition, since conjugate cracks are under compressive stress, they probably tend to close as the pressure within the hydraulic fractures is reduced, thereby provoking sharp declines in production rates. Another stimulation problem arises from the usually large thickness of naturally fractured reservoirs without plastic layers in between. Even if only a short interval is perforated and treated according to the limited-entry technique, the hydraulic fracture grows upwards and downwards until it reaches the plastic barriers bounding the reservoir complex, and consequently the extent of the fracture away from the wellbore in lateral direction is limited (cf. section 4.2.3.2.).

4.8.8.1.3.3. Relationship between ancient and actual stress system

The combination of impacts of depositional environment, natural cracks, and in-situ stresses places constraints on the effectiveness of hydraulic fracturing as a means of stimulating naturally jointed reservoirs (NORTHROP 1988, NORTHROP & FROHNE 1988). If palaeostress state having created the natural crack system and actual stress state guiding the origin of hydraulic fractures have the same orientation, the hydraulic fracture will parallel the natural joints and thus intersect relatively few of the more conductive paths, with thus short hydraulic fractures perpendicular to the natural crack direction being just as effective as a much larger hydraulic fracture which parallels the natural joint system (BRANAGAN, CIPOLLA, LEE & CHEN 1987). Accessibility of gas reserves in such naturally cracked reservoirs therefore sometimes even renders artificial alteration of the stress state feasible in order to trigger hydraulic fracture propagation perpendicular to the natural joint system (WARPINSKI & BRANAGAN 1988; cf. section 4.9.4.1.3.1.).

If natural fractures are oriented in an acute angle to the principal stress direction, they can be inflated during stimulation and can propagate further before the treating pressure is inducing failure in the stress-determined direction (OVERBEY, YOST & YOST 1988). The sequential development of this pattern is described by YOST, OVERBEY, WILKINS & LOCKE (1988) as a succession of first opening and widening of natural fractures at lower injection pressures in a low-angle position to the principal stress and second creation of newly induced cracks in the predetermined direction at higher pressures. EASON (1985) also points out that a successful hydraulic stimulation comprises lateral penetration of the induced fracture away from the wellbore and intersection of a large number of natural cracks.

4.8.8.1.4. Interfracture communication

LORENZ, BRANAGAN, WARPINSKI & SATTLER (1986) describe a system of unidirectional subparallel or anastomosed natural fractures that originated under conditions of high pore pressures and relatively low differential horizontal stresses rather than by structural deformation which control the evolution of secondary permeability in many tight lenticular non-marine flat-lying clastic reservoir rocks. The cracks are subparallel with infrequent low-angle intersections allowing only limited interfracture communication, and orthogonal connecting joints are rare. Some aspects of permeability continuity and stress sensitivity are outlined as follows.

4.8.8.1.4.1. Permeability continuity

Permeability continuity across the fracture pattern within reservoirs does exist, but requires that fluids travel along the joints and zig-zag back and forth through low-angle intersections of adjacent subparallel cracks, with these paths being an order of magnitude or more longer than the direct route (LORENZ, BRANAGAN, WARPINSKI & SATTLER 1986). Co-linear fractures are connected by en-echelon offsets that may coalesce to become an apparent singular crack during erosional stress relief, and joints occur in a spectrum of lengths and spacings. The en-echelon connections commonly create effective fracture lengths that are bounded only by the limits of the reservoir or by large-scale internal heterogeneities.

4.8.8.1.4.2. Stress sensitivity

Most fractures terminate vertically within reservoir sandstones at minor lithological discontinuities such as mudstone interbeds and lenses in fluvial channel sands (cf. plates II/7 - 8, III/1 - 2, IV/7 - 8, VII/5 - 6 and IX/4 - 8) as well as mudstone contacts bounding the sandstone pays (cf. plates IV/4 - 5, VII/3, VIII/6 - 8 and IX/1 - 3), with therefore fractures within different reservoir storeys being poorly connected to cracks within the over- and underlying beds which results in restricted vertical fracture continuity (LORENZ, BRANAGAN, WARPINSKI & SATTLER 1986). As a consequence of their mainly narrow width, most of the joints are considered to be highly stress-sensitive due to closure of gaps between asperities in the same way that low-permeability matrix rocks are stress sensitive to microcracks, and many of the flowpaths may be sufficiently narrow to create large capillary forces thereby inhibiting liquid flow which emphasizes the non-linear relationship of fracture width and transmissibility.

High treatment pressures during hydraulic stimulation of such naturally jointed reservoirs indicate complex fracturing due to interaction with the pre-existing cracks. Decreasing flow rate and average formation capacity of the naturally jointed pay despite the presence of a new artificial high-conductivity fracture reflects damage of the pre-existing cracks having been intersected by the hydraulic fracture (cf. section 4.8.8.3.).

4.8.8.2. Fracture divergence

The initially parallel elements of the fracture bundle diverge from one another at an angle that diminishes with increasing injection rate, fluid viscosity, field stress, and initial spacing (LORENZ, BRANAGAN, WARPINSKI & SATTLER 1986). Such uniform divergence and simultaneous propagation are not expected to be observed in actual hydraulic fractures, because slight difference between the two branches would likely lead to one branch dominating and continuing to grow at the expense of the other. Branching is nevertheless expected to affect the width of the dominant fracture near the division by possibly causing local restrictions that might give rise to proppant bridging and screenout (BLANTON 1982, 1986; JEFFREY, VANDAMME & ROEGIERS 1987). Abnormal treating pressures in hydraulic fracturing operations are also reported by MEDLIN & FITCH (1983).

Thus a combination of natural and artificial fracturing should preferentially aim on enlargement of the natural cracks parallel to them rather than creation of an independent system of synthetic fissures perpendicular or oblique to them, thus implying that pre-fracturing determination of the geostress field by wellbore breakout elongation interpretation (cf. section 6.2.1.8.1.) and probably also natural crack monitoring by borehole television (FLECKENSTEIN 1984, CLERKE & AKKEREN 1986, DARILEK 1986; cf. sections 6.2.1.9. and 6.2.1.10.1.) to be necessary for optimum design and performance of artificial hydraulic fracturing in a naturally cracked or jointed reservoir. Aspects of interaction between pre-existing (primary) and hydraulically-induced (secondary) fractures are also discussed by BLANTON (1982). HOUZE, HORNE & RAMEY (1984); LANCASTER & GATENS (1986), MURPHY & FEHLER (1986); BRANAGAN, CIPOLLA, LEE & CHEN (1987) and BEN NACEUR & ECONOMIDES (1988) also comment on hydraulic fracturing of naturally jointed formations. In terms of orientation of natural fractures within the geostress field, DULA (1981) comments on correlation between deformation lamellae microfractures, macrofractures and in-situ stress measurements.

4.8.8.3. Temporary natural fracture damage

Due to the large hydrocarbon reserves held by naturally fractured pays, their exploitation has become a challenging task for the petroleum industry (PRADO & PRAT 1987) and much potential is seen for the near future. Artificial enhancement of natural fracture systems is expected to play a major role for increasing productivity of these reservoirs. There is, however, the danger that hydraulic fracturing may result in at least temporary subsequent decrease of the flow capacity of the natural joint system near the wellbore and the artificially induced crack, thus limiting production capacity of the well (BRANAGAN, CIPOLLA, LEE & YAN 1987). Natural fracture damage by water and stimulation fluids can be up to such an extent that post-treatment offtake is less than pre-job exploration rates (SATTLER, RAIBLE & GALL 1985). Aspects of interactions between natural fractures and stimulation fluids, fluid loss, limited fracture interconnectedness, elevated treatment pressure, and stress sensitivity behaviour and asperity shearing are outlined as follows.

4.8.8.3.1. Interactions between natural fractures and stimulation fluids

Natural fracture damage can be the result of residual treatment liquids and/or stimulation fluid blockage (similar as water blocking due to capillary pressure; HOLDITCH 1979; cf. sections 3.11.2.4.2.) impairing the permeability of the natural joints intersected by the artificial crack, but is often of transitory nature and can be overcome by an extended shut-in period to enable significant clean-up. As liquid impairment of natural fractures could severely reduce their relative gas permeability, the high capillary pressures of the matrix rock cause during an extended shut-in period the liquid in the natural cracks to imbibe into the matrix and thus the almost total liquid saturation of the natural joints decreases with time.

Fracturing fluid residue blocking also degrades due to increasing molecular weight of the polymers with time and temperature, with thereby both processes leading to increasing gas permeability. The worst case of natural crack damage occurs in anisotropic systems where a direction of the subsequently induced hydraulic fracture parallel to the maximum-permeability natural joints can magnify the deterioration effects and minimize production enhancement (BRANAGAN, CIPOLLA, LEE & YAN 1987).

GALL, MALONEY, SATTLER & RAIBLE (1988) report permeability deterioration of natural fractures caused by stimulation fluid polymers, and SATTLER, RAIBLE, GALL & GILL (1988) describe suitable fracturing fluids for naturally cracked tight gas sandstones in order to minimize natural joint damage. In many tight gas sandstones, natural fractures dominate pre-stimulation production, and a very important question of post-stimulation offtake is how well the natural cracks remain open. Interactions between natural fractures and treatment fluids are reflected by high operating pressures, accelerated or dual leakoff (cf. section 4.8.8.3.2.3.), and highly increased values of accessible reservoir surface upon reversal of gas flow. The summary as follows includes comments on natural crack width, dual fluid leakoff, capillary forces, impact of breaker systems and water blocking.

4.8.8.3.1.1. Natural crack width

The narrowness of many natural cracks makes them susceptible both to fluid damage during stimulation and to changes in the local stress field as a result of hydraulic fracturing or drawdown from production. Very high pore capillary pressures may tend to keep the narrow natural joints relatively free of water. Abnormally high treatment pressures above the closure stresses or even above the containment barriers as determined from in-situ stress testing and also occurring during screenouts (cf. section 6.2.4.2.1.) suggest that the natural fractures open up under such high treating pressures or that stimulation gel is forced into them or both. Fracturing fluid once forced into the narrow natural cracks would be difficult to recover at existing reservoir pressure.

4.8.8.3.1.2. Dual fluid leakoff

Opening up of natural fractures with treatment fluid injection at higher stimulation pressures is also indicated by dual fluid leakoff consisting of normal leakoff at lower pressures and accelerated leakoff at a pressure high enough to trigger the natural cracks to open up (WARPINSKI 1988 a). If gel stabilized by methanol and containing little breaker is pumped into the natural fracture system, post-stimulation workover operations involving large quantities of water may suppress formation temperatures and retard degradation of the gel trapped in the natural crack network. 100 mesh sand does not increase permeability damage in the presence of a breaker, but seems to increase permeability deterioration without a breaker, and breaker added to biogel may mitigate permeability damage (SATTLER, RAIBLE, GALL & GILL 1988). The discussion as follows focusses on 100 mesh sand as fluid-leakoff additive and microproppant, conventional fluid-loss additives vs. 100 mesh sand, and flip-flop insertion of 100 mesh sand.

4.8.8.3.1.2.1. 100 mesh sand as fluid-loss additive and microproppant

100 mesh sand is very effective as a combination of fluid-loss additive and microproppant by reducing fluid leakoff due to hairline fissure plugging, but installing a sufficiently permeable supporting wedge within the microcracks to allow them later to effectively contribute to reservoir drainage (WARPINSKI 1988 a; cf. also section 4.8.8.3.2.3.). 100 mesh sand diminishes secondary fluid loss through intersecting hairline fissures, but does not restrict fluid travel down the principal fracture and thus permits the subsequent stimulation fluid to extend the main crack to the desired distance into the reservoir (THOMPSON 1977; cf. section 1.4.11.2.1.). 100 mesh sand is primarily used as an added conditioning step in the pad fluids. In partially depleted zones which have been heavily acidized, it is necessary to use large volumes of 100 mesh sand in order to fill all the previously created voids so that the new treatment can enter earlier unstimulated portions of the interval. Hairline fissures thieving off some of the fracturing fluid and thus preventing achievement of the predicted penetration are either naturally occurring or are stress fractures caused during the process of hydraulic fracturing and associated dilatancy (cf. section 4.8.1.3.1.).

4.8.8.3.1.2.2. Conventional fluid-loss agents vs. 100 mesh sand

The particle size of conventional fluid-loss additives is too small to effectively bridge off and limit fluid leakoff in the miniature cracks (WARPINSKI 1988 a). The hairline fissures are actually propagated with fluid only and serve as escape routes for the fluid to leave the primary crack and to limit its extension. Fluid loss through the hairline fissures causes a buildup of proppant concentration in the principal fracture which aggravates the situation, and when the pressure is released after the treatment, the microfractures containing only fluid-loss additives can only very little contribute to hydrocarbon production.

100 mesh sand perfectly bridges off the fissures before travelling very far down the hairline and builds up a secondary granular matrix within the cracks which serves as a base for other fluid-leakoff additives to work against, thereby further limiting fluid loss into the hairlines and keeping the rest of the

fluid within the principal fracture. In combination with 100 mesh sand, only soluble fluid-leakoff agents should be used, because invasion into the 100 mesh sand may cause such a degree of plugging that a low-pressure formation cannot kick out. 100 mesh sand permits the main fracture to penetrate deeper into the reservoir which in turn results in connection of more and more hairlines to the principal crack that ultimately leads to more production avenues causing both higher initial offtake, longer lasting well productivity, and greater cumulative exploitability. 100 mesh sand plugging of hairline cracks can also help to contain vertical fracture propagation (cf. section 4.2.2.4.).

4.8.8.3.1.2.3. Flip-flop insertion of 100 mesh sand

In terms of operations, the flip-flop method has turned out to be the best technique, consisting of alternating slugs of pad fluid and 100 mesh sand slurry in the same fluid being pumped ahead of the acid or proppant slurry (WARPINSKI 1988 a). These spacers serve as make-up fluid for that fluid lost to the hairline fractures in the bridging process and maintain an even flow of constant 100 mesh sand concentration. The overall result of the fracturing treatment seems to be better if the total volume of 100 mesh sand is abt. equal to the total proppant volume in smaller jobs or if at least 50,000 lbs of 100 mesh sand are used in larger operations. 100 mesh sand can be applied in various concentrations in the range of 1 - 6 lbs/gal.

In many cases, using 100 mesh sand pads allows to place larger saturations and higher quantities of proppant into the main fracture at lower pumping rates and with less efficient stimulation fluids than in case of treatment without 100 mesh sand pads. Effective trapping of 100 mesh sand in the hairline cracks when pressure is released at the end of the job leaves a propped microfracture that is fully capable of feeding accelerated hydrocarbon production into the main fracture.

4.8.8.3.1.3. Capillary forces

Natural fracture conductivity damage by fracturing fluids is mainly controlled by initial pay permeability and polymer gel type (GALL, SATTLER, MALONEY & RAIBLE 1988). Capillary forces could cause redistribution of liquids within the reservoir and result in removal of water from the natural cracks, and polymer residues in the fracture could decompose at pay temperatures of 200 oF. As gelled polymer fluids can enter narrow natural joint flow channels and leave behind residues which could not be removed by gas flow, at least parts of the natural fracture conductivity damage caused by polymer remnants has to be assumed to be permanent. Increased leakoff rates with polymer solutions containing breakers suggest that buildup of polymer residues in the crack is slower than with non-breaker-bearing gels. Polymer remnants deposited in natural cracks intersected by hydraulic fractures result in restricted gas flow, with the overall increase after the stimulation treatment being considerably less than expected. The addition of breaking agents at low concentrations may be of some help in reducing polymer residues and damage to gas flow in narrow fractures. As crack permeability is also sensitive to stress changes, pressure drawdown in a naturally jointed reservoir could cause reductions in gas permeability.

As installation of a conductive hydraulic fracture is often accompanied by diminution of flow capacity of the natural joints that are in contact with the induced crack, effectiveness of hydraulic fracture stimulation in tight lenticular naturally jointed gas sands is dependent on preserving the flow capacity of the existing natural fracture systems (BRANAGAN, CIPOLLA, LEE & WILMER 1985).

4.8.8.3.1.4. Impact of breaker systems

A breaker system generally should allow the viscosity of a stimulation fluid to remain reasonably high and stable over the shorter term to carry proppants and fine-grained fluid-loss additive sand out into the hydraulically-created fracture, whereas in the longer term, the viscosity of the stimulation fluid system should degrade to facilitate cleanup and recovery of the polymer (GALL, SATTLER, MALONEY & RAIBLE 1988). Damage from treatment fluid polymers is responsible for most of the early post-fracturing production problems, and other forms of mechanical deterioration as well as closure due to drawdown can make the natural cracks even more narrow and can compound polymer damage which might be mitigated by use of a prepad consisting of a breaker solution.

This approach would put additional breaker into the formation and would entail little risk of premature viscosity degradation, because the gel has no contact with the additional breaker until being in the reservoir itself. Field and laboratory evidence indicate that fluid systems including a breaker prepad are apparently satisfactory for stimulations of naturally fractured tight gas sandstone reservoirs, whereas failure to use a properly viscosified base fluid with adequate leakoff and foam stability characteristics would risk an uphole or downhole screenout.

BRANAGAN, LEE, CIPOLLA & WILMER (1988) also emphasize the extreme sensitivity of natural crack systems in some tight gas sandstones to invading completion and hydraulic stimulation liquids. Although reservoir matrix damage is considered to be detrimental, the more serious problem is clearly maintaining the highly productive flow paths of the natural cracks in their original state. The deterioration mechanisms to the joints are so severe that fracture flow capacity could be diminished to the point of rendering them useless as current conduits (BRANAGAN & WILMER 1988).

4.8.8.3.1.5. Water blocking

Water blocking of fracture and/or formation occurs if the drawdown pressure gradient in the reservoir near the crack face does not exceed the rock capillary pressure sufficiently for enabling gas to flow (HOLDITCH 1979; ABRAMS & VINEGAR 1985; cf. section 3.11.2.4.2.). Water blocking is generally not a serious problem in most tight formations because drawdown pressure and gas mobility are usually high enough for efficient displacement of stimulation fluid from the reservoir, but may develop if pay pressure gradients in the near fracture-face region are low or if fluid mobilities are considerably reduced by formation or crack damage. Capillary pressure increases when capillaries are small, surface and interfacial tensions are high, and the contact angle between liquid and rock is less than 90 degrees (BROADDUS 1988). When capillary pressure exceeds bottom-hole pressure, the interval is water-blocked.

The best solution to avoid water blocking is performing the stimulation operation with only partially aqueous or even non-aqueous treatment media such as foam, energized fluids, nitrogen or carbon dioxide (cf. sections 3.11. and 4.8.8.3.2.1.). Prevention can also be made by adding a surfactant to treating fluids which imparts low surface tension and changes the angle at which fluid contacts the rock face.

4.8.8.3.2. Fluid loss

Damage of flow and uptake capacities of naturally fractured intervals of reservoir complexes may also be caused during drilling when serious lost circulation problems occur, especially when complete mud losses take suddenly place and leakoff rates exceed rig pumping capabilities as a consequence of pressures within the disturbed zone being commonly well below that of a fresh water gradient (CANSON 1985; cf. section 4.8.8.6.). Partial plugging of the natural fracture system is achieved by blind drilling or insertion of bridging solids into the fluid stream in order to control the borehole and to allow further drilling progress, or even by repeated cementing of the intervals taking up excessive amounts of mud. Penetration of a low-pressure naturally fractured rock interval is usually signaled by a sudden loss of returns at a high rate accompanied by a notable increase in rotary torque and relative drilling roughness which are a consequence of crossing openings that are large enough for bit teeth to fall into or hang up in. Some aspects of breakdown fracturing procedures, capillary pressure and surface tension, and dual leakoff behaviour are outlined as follows.

4.8.8.3.2.1. Breakdown fracturing procedures

BRANAGAN & WILMER (1988) discuss breakdown procedures designed to minimize naturally fractured reservoir damage (cf. section 4.8.2.2.). In the course of drilling and completing a well, a number of mechanisms degrades the original high conductivity of natural cracks, including invasion of drilling mud that may ultimately act as a plug, cements squeezed into the fracture, perforations and near-wellbore stress changes that could tend to close the crack, and the interfacial tensional forces generally described by capillary pressure of invaded completion liquids which limit fracture flow capacity. Some aspects of fracture aperture restriction by water and non-aqueous techniques for avoiding of water blocking are discussed as follows.

4.8.8.3.2.1.1. Fracture aperture restriction by water

Liquids present in the natural joint system restrict the available fracture aperture through which gas can flow. If water is used in the breakdown process, it will undoubtedly fill most of the nearby cracks. When the well is later put on production, the injected water must either be removed from any fracture through the borehole or it must be allowed to imbibe into the matrix, because otherwise gaseous flow will be almost totally impeded due to the inability of gas to flow horizontally through the water (this phenomenon is known also from hydraulically-induced fractures as water blocking; HOLDITCH 1979; cf. sections 3.11.2.4.2. and 4.8.8.3.1.5.).

4.8.8.3.2.1.2. Non-aqueous techniques for avoiding of water blocking

Non-aqueous microfracturing breakdown treatments avoiding water blocking comprise hydraulic nitrogen fracturing using surface pump trucks, perforating with an underbalanced nitrogen column, high-pressure nitrogen impulse perforating/microfracturing, nitrogen impulse injection using a tubing pump-out plug, and tailored pulse explosive fracturing.

Nitrogen impulse perforating/microfracturing simultaneously perforates the well and creates cracks that extend from the borehole and intersect the natural joint system in the wellbore vicinity. Dynamically, nitrogen impulse fracturing is more explosive rather than injective, as the compressional energy stored in the high-pressure nitrogen column is almost instantaneously released into the perforations in the moment when they are created. Tailored pulse explosive breakdown procedure (cf. section 4.2.2.1.3.3.) entails the use of a relatively slow burning propellant designed to create a multiple set of fractures radiating like spokes from the perforated wellbore. In order to create the correct environment for the tailored pulse explosive to properly initiate multiple fractures, a fluid with a density about that of water but considerably in excess of most available gases is required as tamping medium in the casing, with liquid carbon dioxide being an ideal wellbore tamping fluid. Pulsed explosives can be successfully applied for creation of multiple sets of unobstructed fractures that connect wellbore and natural crack system. The same effect can be achieved by perforating in a dry underbalanced wellbore column of nitrogen.

4.8.8.3.2.2. Capillary pressure and surface tension

Enhanced stimulation fluid recovery can also be achieved by improving proppant pack drainage and minimizing capillary pressure in the formation through judicious application of surface active agents which is applicable particularly in tight gas reservoirs and low-pressure oil wells and may be used in all aqueous fracturing treatments including slick and gelled water as well as foamed and crosslinked systems (PHILLIPS & WILSON 1984 a). The technique involves utilization of a solvent in the pad fluids and a surfactant system (CLARK, PIKE & RENGEL 1979, 1980) throughout the entire fracturing treatment, with this combination improving proppant pack drainage and minimizing capillary water blocks in the invaded formation while maintaining the water-wet condition of both proppant pack and formation (PHILLIPS & WILSON 1984 a). This dual combination accounts for the complex relationship of effective oil permeability and thus oil recovery being highest when the substrate is strongly water-wet and water saturation is low, and conversely a strongly oil-wet substrate and low oil saturation resulting in high relative permeability to water and enhanced load water recovery (OWENS & ARCHER 1971).

Another important effect minimizing water damage is that ultra-low surface tension provided by the surfactants helps reducing capillary imbibition of injected water into the formation at the fracture face (PHILLIPS & WILSON 1984 a), and minimization of capillary imbibition of the initial leakoff fluid is achieved by adding methanol to the pre-pad and pad stages (cf. section 3.11.4.). Reservoirs with low closure stress and greater development of natural fractures require an energized system for improved return of stimulation fluids, with up to 40 % nitrogen within delayed crosslinked fluids forming a stable foam structure inside the fluid and supplying adequate energy for the return of load water (EASON 1985; cf. section 3.11.2.).

Diminution of fluid loss during the proppant-laden main fracturing treatment stages can be similarly as for reservoir cooling be achieved by formation saturation with large amounts of fluid due to pre-pad and pad enlargement (cf. section 4.3.4.6.1.4.), with the cheap and simple pre-pad and pad fluids being sacrificed to save the expensive and complex main stage fluids. The purpose of running extremely large pads is to allow ample leakoff into natural fractures and matrix in order to block the pre-existing cracks and to avoid further fluid loss into the tectonical and sedimentary discontinuity planes and fissures during the proppant-laden main treatment stages. In exceptional cases, the pad stage can comprise up to 60 - 80 % of the whole treatment quantity (TAN, McGOWEN, LEE & SOLIMAN 1988).

4.8.8.3.2.3. Dual leakoff behaviour

Naturally fractured tight gas reservoirs with more or less well interconnected joint systems are characterized by dual leakoff behaviour of fracturing fluids during course of hydraulic stimulation treatments (WARPINSKI 1988 a). At relatively low injection pressures, the base fluid-loss coefficient is very low and has a magnitude that is consistent with effective reservoir permeabilities of tens of microdarcies. Above some critical pressure, however, leakoff accelerates and becomes as much as 50 times greater than the base level. When proppant-laden stages enter the perforations under these elevated secondary leakoff conditions, the slurry rapidly suffers significant increase in viscosity and proppant concentration due to fluid dissipation, and dehydration screenout failure occurs. The associated characteristic pressure behaviour is rapid flattening of the pressure curve above some critical pressure value.

As excessive fracture height growth can be excluded due to similar compliances of all the rocks in the investigated interval of the case studies, high leakoff is the likely candidate for this behaviour, although the pressure decline pattern has invariably yielded a low fluid-loss coefficient even at early times. Thus an accelerated leakoff above a threshold pressure takes place through the natural fracture system. Some aspects of mechanisms of primary and secondary fluid loss, insertion of 100 mesh sand, impact on reservoir stimulation, 100 mesh sand vs. other fluid-loss additives, high-quality 100 mesh microproppants, and degradable fluid-leakoff additives are outlined as follows.

4.8.8.3.2.3.1. Mechanisms of primary and secondary fluid loss

Dual fluid leakoff can be triggered by two mechanisms. The first possibility is opening of natural fractures within the sandstone pays, with threshold pressure then being dependent upon differences in horizontal in-situ stresses and other reservoir and treatment factors (WARPINSKI 1988 a). The second way is due to fracture height growth into adjacent siltstones and sandstones containing wide natural cracks, with threshold pressure being related to the pressure at which enough height propagation occurs for these open joints to be intersected. In some tight gas formations, stresses and other parameters have such magnitudes that it is difficult to determine which of the two modes is the really decisive one.

In case of recognition of the problem being fluid leakoff and not fracture height growth, complex stimulation or some other process, fluid-loss control must be effected through some additive or change in the treatment schedule to keep injection pressure below threshold value. As it has turned out to be impossible to lower the pumping pressure significantly and still transport reasonable amounts and concentrations of proppants, leakoff additives have to be selected which are not entirely plugging the natural fractures that represent the primary production paths. 100 mesh sand is an optimum additive, because it bridges off excessive fluid leakoff, but also leaves propped conductive flow paths for subsequent production (cf. also sections 1.4.11.2.1. and 4.8.8.3.1.2.).

4.8.8.3.2.3.2. Insertion of 100 mesh sand

Experimental stimulation work has revealed that 100 mesh sand can reduce the leakoff during pumping by a factor of 3.5 from 70 times the base level to only 20 times the initial rate if added only in small concentrations. 100 mesh sand also helps to bridge-off fracture height growth at least while it is actively injected, and also gives rise to increasing fluid-loss coefficient during pressure decline (WARPINSKI 1988 a; cf. also section 4.8.8.3.1.2.). This behaviour is triggered by propping of the natural joints by the 100 mesh sand, thereby providing an enhanced flow path for hydrocarbon production into the hydraulic fracture.

Optimum performance of 100 mesh sand is given by continuous addition of small saturations through almost the entire treatment with the exception of an initial pad, with the quantities, however, having to be limited to such an amount that conductivity deterioration as a consequence of intermixing of 20/40 proppants and 100 mesh sand (cf. section 1.4.11.2.1.) is still subordinate and insignificant. A fracture conductivity reduction by abt. a factor of two which can be accepted in most of the low-permeability reservoirs because still a very high conductivity contrast is left permits to add already such concentrations of 100 mesh sand that the dual task of micropropping the hairline cracks and controlling fluid leakoff rate can be successfully carried out.

4.8.8.3.2.3.3. Impact on reservoir stimulation

Dual fluid-loss modelling and control is a suitable mechanism to improve stimulation results in low-permeability highly-anisotropic naturally fractured tight gas reservoirs which are otherwise unable to be treated without encountering an early screenout that gives rise to only short propped fracture lengths and high pressures above closure (WARPINSKI 1988 a). Dehydration screenout is a result of high leakoff conditions at elevated injection pressures and can be triggered by either opening of natural fractures within the pay or height propagation into nearby layers containing many wide open cracks (cf. section 6.2.4.2.1.). If the fractures in the reservoir dilate at pressure conditions above some threshold, the damage is probably due to injection of large amounts of stimulation fluid into the natural fractures, and cleanup of these cracks would be difficult because of occurring under low-pressure closed conditions. Recognition of the dual leakoff phenomenon requires analysis of both pressure decline after shut-in and injection pressure evolution in conjunction with a history match of the treatment pressure.

4.8.8.3.2.3.4. 100 mesh sand vs. other fluid-loss additives

While in some tight gas reservoirs accelerated fluid leakoff due to opening of natural fractures can be avoided by simply keeping treatment pressures below the critical level, in other formations economical stimulation with corresponding low injection rates is not possible and fluid-loss additives are required in order to allow to raise operating pressures to the level needed for profitable treatment (WARPINSKI 1988 a). As the natural fractures are also the primary gas production pathways, fluid leakoff additives totally plugging the hairline cracks such as silica flour cannot be employed because there is no guarantee that complete cleanup of the fissures including ejection of the blocking materials could be effected. 100 mesh sand is able to control both fluid loss into natural fractures and provide enhanced permeability paths by acting as a plugging microproppant in the narrow joints (cf. also 4.8.8.3.1.2.). Some enhanced bridging of natural cracks may even result from interaction with the 20/40 mesh main proppants.

As addition of 100 mesh sand does not completely stop accelerated fluid leakoff at concentrations which do not yet seriously destroy proppant conductivity (cf. section 1.4.11.2.1.), retarded fluid loss still limits the amount of proppants that can be placed into such reservoirs without suffering from an early screenout. Underestimation of fluid-leakoff coefficient giving rise to premature screenout termination of the stimulation job is also reported by NORTHCUTT, ROBERTSON & HANNAH (1988). The use of 100 mesh sand in naturally fractured reservoirs as a transitional stage between pad and proppant-laden fluid (cf. also HOLDITCH, ROBINSON, WHITEHEAD & ELY 1987) serves as fluid-loss additive and as initial proppant and allows significant increase of proppant concentration due to effective leakoff control.

4.8.8.3.2.3.5. High-quality 100 mesh microproppants and degradable fluid-leakoff additives

In deeper reservoirs with higher closure stress, it would be desirable to use 100 mesh intermediate- to high-strength synthetic proppants as an effective microproppant for hairline cracks, but unfortunately this grain size of artificial high-quality proppants is not yet commercially available (cf. sections 1.3.4. and 1.5.1.1.). A suitable alternative would be resin-coated 100 mesh sand which similarly as conventional 100 mesh sand has abt. 5 - 7 darcy permeability at almost any closure stress (FRACFAX 1988 f), with resin bonding creating a stable microproppant package if no flowback can be tolerated (cf. section 1.2.6.).

Another approach to dual fluid leakoff control is a degradable additive which later after fulfilling of its function of retarding or even preventing fluid escape is removed by dissolution or disintegration and ultimately leaves the natural fractures open in order to enable them to contribute to production without any restriction (CANTU & BOYD 1988; NORTHCUTT, ROBERTSON & HANNAH 1988; cf. section 4.3.4.6.4.). On the other hand, however, primarily plugged and secondarily reopened hairline cracks as a consequence of later degradation of the originally inserted fluid-loss additive may result in closure of the open fissures in high-pressure reservoirs as a consequence of absence of support against the compressive stress, with the natural fractures thus no longer being able to contribute to gas production. Micropropping of hairline cracks with materials that can adequately resist to the closure stress is therefore the preferential solution of dual fluid leakoff control. The significance of secondary fluid-loss into natural and synthetic hairline fractures and the successful control of dual fluid leakoff by 100 mesh sand is also discussed by CROW (1977).

4.8.8.3.3. Limited fracture interconnectedness

LORENZ, BRANAGAN, WARPINSKI & SATTLER (1986) also illustrate temporal damage of a complex natural crack system due to influence by liquids during hydraulic fracture stimulation. Natural joints are common in numerous geological settings where major structural movements are important, but they also form in provinces of little or no structural deformation if sufficient in-situ pore pressures and some horizontal stress anisotropy exist. These conditions give rise to a fracture network consisting of unidirectional subparallel cracks that are aligned with the maximum horizontal stress and which are limited vertically by irregular lithological discontinuities in the reservoir. Fractures rarely extend across intervening shales between major sandstones and thus each reservoir sandstone storey represents an isolated stratigraphical trap if surrounded by mudstone also in terms of crack distribution. With sufficient time of shut-in after the hydraulic fracturing operation, liquids in the natural joints could then be imbibed into the even smaller pore spaces of the matrix rock, thus allowing gas flow in the natural cracks to return to the original undamaged condition.

Naturally fractured reservoirs with limited crack interconnectedness are stress-sensitive (NORTHROP 1988). Production can be essentially shut-off by reducing the bottomhole pressure, with the increasing effective stress being sufficient to squeeze the natural fractures shut. Conversely injection of gas at high pressures results in a dramatically increased volume of accessible reservoir surface due to opening of the fractures. Depending on stability or mobility of stress directions during course of the geological history of the petroleum-prospective basin, hydraulically induced fractures will propagate in a direction parallel or at an angle to the orientation of the natural cracks (cf. section 4.8.8.1.3.2.). The most favourable situation for interconnection of natural joints by the hydraulic fracture is a rotation of the principal stress direction after formation of the natural cracks to enable an acute angle between natural fractures and intersecting hydraulic fracture to occur.

In case of still parallel stress direction, the hydraulic fracturing process either only includes inflation and enlargement of existing natural cracks, or the stress state has to be artificially altered by mechanical or thermal influences (cf. section 4.9.4.). GRI (1988) describes tight gas sandstones in an area of tectonical quiescence where natural fractures, drilling- and coring-induced cracks and hydraulic fractures have all the same trend in the field.

4.8.8.3.4. Elevated treatment pressure

Temporary natural fracture damage is best expressed by abnormally high treatment pressures being up to four times greater than predicted. As the hydraulic fracture intersects subparallel cracks that are deviated at some angle, it is likely that some or many of these natural joints open up and initiate secondary hydraulic fracture strands, giving rise to a wide zone of fracturing. The extra tortuosity and additional wall surfaces of the many crack strands increase fluid friction and thus give rise to considerably higher injection pressures. The damage of the natural fracture system is a self-driven mechanism which temporarily impairs productivity by eliminating the most favourable depletion mechanism of the tight gas sands.

The higher treatment pressures triggered by increasing friction in the natural joint system result in greater fluid leakoff into the natural fractures, larger shear stresses and higher pore pressures to induce shear slippage, and greater effective propped widths to elevate proppant-induced stresses. Increasing operating pressure in response to the presence of the natural fracture system gives also rise to increasing deterioration. The damage scenario is most easily seen for small hydraulic fractures. Larger treatments contact more extensive matrix areas and thus lost production through the natural fractures is compensated somewhat by increased matrix flow. With small jobs, productivity loss is a clear signal of deleterious operation.

Concerning damage of narrow natural crack systems by stimulation fluids, high fracturing pressures either open the intersecting natural joints or force fluid and its associated organic materials into them under high differential pressure (NORTHROP 1988, NORTHROP & FROHNE 1988). Opening of the crack system is indicated by significantly increased leakoff coefficients above a threshold pressure. Other reasons of very high treatment pressures are high stresses in the confining lithologies, backstresses, multistrand fracturing, or presence of thin high-stress stringers in the reservoir. Apart from damage of the natural crack system, another deleterious effect of high stimulation pressures is the creation of shorter and wider fractures with less effective drainage. Effectiveness of hydraulic fracturing is limited by an anisotropic natural joint system, high fracturing pressures, accelerated or dual fluid leakoff (cf. section 4.8.8.3.2.3.), and damage to the natural crack network. Breakdowns and extended cleanup and test times are required for accurate reservoir assessment, and insitu stress behaviour varies with depth and lithology, is complex, and significantly affects stimulation and gas production.

4.8.8.3.5. Stress-sensitivity behaviour and asperity shearing

Damage mechanisms of natural fracture systems by hydraulic stimulation treatments include apart from liquids problems also stress-sensitivity behaviour and asperity shearing with subsequent fracture closure. Liquids problems comprise gel residue phenomena, but may also cover capillary pressure effects (cf. section 4.8.8.3.2.2.) of liquids in the narrow passageways of the natural cracks. If many of the passages through the fractures are extremely small, water movement becomes difficult and cleanup very slow. Capillary effects are exacerbated by large molecular gel residues in the fracture.

Stress-sensitivity problems originate if the propped hydraulic fracture imparts a large stress on adjacent natural cracks thus reducing their effective widths. As this stress cannot be greater than the treatment injection pressure above closure stress, this mechanism may not entirely shut down flow in the natural fractures, but may also act in concert with gel or capillary pressure problems (cf. section 4.8.8.3.1.5.). Asperity-shearing along natural fractures and the resultant closure and trapping of fines, gels or water is caused by fluid leakoff and/or induced treatment stresses, with unknown parameters being extent of the shear slippage region and amount of closure representing damage severity. Aspects of temporary natural fracture deterioration are also discussed by BRANAGAN, CIPOLLA, LEE & YAN (1987).

4.8.8.4. Dendritic fracturing technique

A special hydraulic treatment method for various naturally jointed reservoirs is the dendritic fracturing technique (KIEL 1977; LOPUS, SEIFERT & SCHEIN 1987). This process is to overcome the problems associated with gaining vertical selectivity and connecting or opening the natural crack network, whereas conventional stimulation in the past could only affect a modest portion of the pay interval. Aspects of connection of natural and hydraulic fractures as well as shear slippage vs. tensile failure ae summarized as follows.

4.8.8.4.1. Connection of natural and hydraulic fractures

The dendritic fracturing procedure confines the operation to already existing natural joints by the use of large volumes of lightly gelled fresh water pumped at high rates and velocities rather than creating new cracks as more viscous fluids would (KIEL 1977; LOPUS, SEIFERT & SCHEIN 1987). The high rates effectively dissolve fracture-bridging calcite, salt and other secondary mineral deposits to improve their permeability. Various stages of the treatment incorporate small amounts of proppants in order to aid in abrasion of existing fractures as well as fluid-loss control, or bridging to cause slight changes in differential pressure within a crack to assist in diverting into other joints. The fracture gradients measured during both conventional and dendritic stimulation treatments are considerably lower than the unfractured matrix gradient, which indicates that crack closure in the standard sense does not occur, but the reasons for the well response to the dendritic stimulation process include dissolution of secondary mineral deposits from fracture openings and fracture walls, thereby providing a consequential increase in fracture conductivity. Insoluble materials plugging the cracks are eroded by the proppant-laden fluid stream and transported away from the wellbore during treatment.

Conventional fracturing operations in various formations suffer from the behaviour of the proppant wedge in the crack as a filler causing the solubilized minerals in the return treatment fluids to reprecipitate salt and calcite within the proppant package, with cementation being possible to the point that fluid flow is severely restricted and a reservoir of limited extension and rapid depletion is mimicked by the low bottom hole fracturing pressure. In such situations, hydrochloric acid utilized in the dendritic stimulation process with high fluid velocity has successfully helped in removal of this damage. Some points of diverting stages and reverse-flow pulse stimulation are sketched as follows.

4.8.8.4.1.1. Diverting stages

Past attempts at limited entry and selective perforating (cf. section 4.2.2.2.) including the use of ball sealers (cf. section 4.2.2.1.1.) have proven to be ineffective in ameliorating zonal coverage, because regardless of fluid control inside or directly behind the pipe, the interconnecting fracture network resulted in only the well-developed portions of the pay zone accepting most of the treating fluid (KIEL 1977; LOPUS, SEIFERT & SCHEIN 1987). The dendritic fracturing technique provides diverting stages using a gelled fluid containing rock salt and proppants pumped at sufficient rates to achieve diversion away from the borehole. Small variations in surface and bottom-hole treating pressures during the diverting stages are the consequence of low gradients in many naturally fractured formations, with therefore the magnitude of pressure changes required for good diversion being low and often appearing insignificant.

Independent of number of diverter stages and amount of solid diverter used, it is imperative during treatment to maintain the highest possible velocity within the fractures, with the only reason for reduction of rate during the operation being to mix and pump the diverter stages at proper concentrations (HARRISON 1971 gives a historical review of the use of various diverting agents in oil- and gas-field operation). The high fluid velocity creates a mechanism for carrying the sand used for its bridging and abrasive characteristics. During retreatments of conventionally fractured wells, high fluid velocity is necessary to carry the initial proppant quantity away from the wellbore (cf. section 4.8.9.5.), with hydrochloric acid having to be included in these instances to break the proppant package free of calcite and salt cementing it together.

If continued pumping permits the rate to be increased with no significant rise in surface operation pressure, this could indicate a change in injectivity because of reduced diversion effectiveness or the result of pumping into previously untreated fractures creating less flow restriction. If the slope of the pressure versus time graph for the various stimulation stages usually shows a positive trend, this reflects restriction or resistance to flow caused by the new fracture entry, or that the carrier fluid has minimally efficient proppant transport capabilities leading to formation of a sand bank that, as it builds, will act to create pressure differential across the crack causing fluid to enter other secondary fractures. Aspects of dendritic stimulation of naturally jointed reservoirs are also discussed by HANNAH (1976) and HUBBARD & PIERSON (1986). Application of rock salt as diverting agent for stimulation block stages is also reported by HUCKABEE (1988).

4.8.8.4.1.2. Reverse-flow pulse stimulation

Dendritic fracturing is a reverse-flow stimulation technique with respect to conventional or normal planar hydraulic fracturing and is very suitable for geothermal well stimulation, because it offers the best chance of intersecting major natural cracks, whereas a single planar fracture might only parallel and not cross the principal natural joints (CAMPBELL, HANOLD, SINCLAIR & VETTER 1981; cf. section 4.8.8.1.3.2.). Dendritic fractures are provoked by reverse-flow pulsing of the reservoir, thereby inducing formation spalling and diversion of the fracture wings by downhole stress modification. Multiple injection periods are used with each stage utilizing low-viscosity fluids, sand slugs, and several flow-back times, with high flow rates and friction reduction being applied in these treatments. The best results of dendritic fracturing are achieved in naturally jointed formations where major and minor crack systems already exist but may not have flow capability.

4.8.8.4.2. Shear slippage vs. tensile failure

The origin of a multiple branched joint system or dendritic fracturing is not controlled by tensile rock failure as assumed by KIEL (1977), but is rather governed by shear slippage which results in local redistribution of stresses (MURPHY & FEHLER 1986; cf. section 4.7.7.). If frictional resistance to shear slippage is low and maximum dilatancy due to shear is large, only a single joint is stimulated. In case of high shear resistance or small dilatancy (cf. section 4.8.1.3.1.), however, multiple crack stimulation occurs. Shear slippage along the joints is accompanied by shear-stress drops, and the interaction of these stress drops with the acting earth stresses results in opening of cracks more perpendicular to the maximum stress, thereby giving rise to formation of a branched or dendritic pattern.

Dendritic fracturing in hydrocarbon reservoir stimulation (KIEL 1977) is technically performed by repetitive fracturing with proppant-bearing fluids, shutting-in and venting. The first cycle of pressurization results in spalling and self-propping of the main fracture (cf. section 4.7.7.4.), and in the subsequent cycles, the proppant purposely introduced in the treating fluid bridges the spall-proppants, thereby leading to pressure rise and propagation of lateral fractures perpendicularly to the main crack (KIEL 1977). This model, however, does not work when considerable stress differences occur, because the pressure rise in the blocked main fracture then simply results in further lift-off of the large crack, overcoming the temporary blockage and allowing the main fracture to continue its propagation (MURPHY & FEHLER 1986).

Dendritic fracturing is rather controlled by shear which permits the necessary reduction of the earth stress parallel to the main fracture to initiate opening of lateral joints. Aspects of shear slippage giving rise to asperity shearing along natural fractures and temporary damage of the pre-existing crack system due to hydraulic reservoir stimulation are also discussed by LORENZ, BRANAGAN, WARPINSKI & SATTLER (1986). Field experience of dendritic fracturing is also reported by McKETTA, KOZIAR & COOK (1980).

4.8.8.5. Tailored pulse loading

Tailored pulse loading comprises intermediate strain rate rock fracturing processes that are capable of generating multiple crack networks which have a much higher probability of intersecting natural joint systems than standard explosive and conventional hydraulic fracturing (STOLLER 1985; cf. sections 4.2.2.1.3.3. and 4.7.3.). Conductivity maintenance of the non-propped multiple fractures can be achieved by channelling and gouging which occur on the crack surfaces, and by translation of one fracture surface relative to the other if the cracks are not oriented in the principal stress directions, with the shearing stress influence on the off-axis fractures being very important.

In terms of creation of multiple fractures, distinction has to be made between early stage when fracture-fracture interactions must be considered to allow all the individual cracks to grow, and late stage when optimization of the pressure pulse is required to achieve maximum fracture lengths. The crack geometry created by tailored pulse loading tends towards rectilinear as opposed to radial, which results from the more dynamic nature of the fracturing process that masks out the influence of static effects such as in-situ stresses.

4.8.8.6. Proppant plugging of natural fractures

Hydraulic stimulation treatments of naturally fractured formations often suffer from the uncertainty whether the reservoir has been really broken down with creation of new crack planes or whether only pre-existing natural joints have been enlarged and infilled with proppants. In many naturally fractured formations, fluid and proppants cannot be put down fast enough to avoid leakoff into the pre-existing crack system or even into the rock matrix. This is already highlighted by the fact that naturally jointed reservoirs frequently cause serious drilling problems due to lost circulation during penetration, with mud loss rates often exceeding rig pumping capabilities and requiring temporary progress by blind drilling until the zone is crossed (CANSON 1985; cf. section 4.8.8.3.2.). In terms of hydraulic stimulation of extensively naturally fractured reservoirs, two possibilities exist comprising plugging of natural cracks with proppants and origin of inclined fractures by Coulomb failure. The discussion as follows includes aspects of production acceleration by propping of natural cracks, fresh reservoir volume acquisition and recovery enhancement by artificial fracturing, and natural propping of sedimentary shrinkage cracks.

4.8.8.6.1. Production acceleration by propping of natural cracks

Natural fracture propping, however, does not result in additional hydrocarbon exploitation, because the initial peak production is later compensated by faster decline of output rates and cumulative offtake volume. The reason for this effect is the lack of newly created fracture planes which could access incremental hydrocarbon quantities that are not reached by the existing drainage system. Only hydraulic fracturing can give rise to substantial productivity improvement by exposing fresh surfaces and generating new intersections through the tight matrix along which hitherto inaccessible reserves can be acquired, with then natural cracks that are crossed by the induced fracture and thus tied into the active system connected directly to the borehole also achieving considerable significance for production contribution, and in this case propping of natural fractures ascertaining that they stay open also with returning closure stress and keep being permeable and conductive hydrocarbon delivery paths to the main hydraulic fracture stem and thereby ultimately to the wellbore.

Natural fracture propping results in simple acceleration by faster flow of the hydrocarbons that are already within the reach of the existing drainage pattern to the wellbore as a consequence of increasing conductivity of the flow paths. Concerning comparative assessment of effectivity, hydraulic fractures have the advantage of being fresh surfaces, whereas many natural cracks can be clay-covered or mineralized which may considerably reduce their permeability and/or conductivity. In addition, the distribution of natural fractures in the geological column is heterogeneous, with tight gas sands containing very few natural cracks, whereas high-permeability sandstones and carbonates have a lot of natural fractures.

4.8.8.6.2. Fresh reservoir volume acquisition and recovery enhancement by artificial fracturing

Enhancement of the drainage effectivity of natural joints together with improvement of cumulative productivity can only be achieved by interconnecting natural and hydraulic or explosive fractures. Artificially created cracks propagating at high rates run through natural fractures and continue to extend into the matrix, whereas those travelling at low rates turn into the direction of natural joints once a critical velocity of propagation is no longer exceeded. In terms of explosive fracturing, however, high propagation rates require high detonation rates which result in pulverization of the rock and creation of a lot of fines, unless the explosive charge is selectively put into natural fractures and ignited therein (cf. section 4.7.3.). A suitable combination of hydraulic and explosive fracturing (cf. section 4.8.9.6.) would be the performance of a hydraulic minifracture into which the explosive charge is inserted.

If long-term production maximization is desired to be achieved, hydraulic or explosive fractures should be interconnected with the smallest natural crack system for recovery factor optimization. On the other hand, if opening-up of the formation and high initial production rates are wanted, hydraulic or explosive fractures should be interconnected with the largest natural crack system. This guarantees good flow rates, but as a consequence of only large matrix blocks being included in the coarse network, recovery factors are low, whereas enclosure of small matrix blocks by the former technique results in good areal drainage and high recovery factors.

ZUBER, LEE & GATENS (1987) review various stimulation techniques for Devonian shale gas reservoirs. The historically abundantly applied standard explosive fracturing method by detonating gelatinated nitroglycerine in the borehole (well shooting; cf. section 4.4.4.6.) has in the last years been more and more replaced by single hydraulic fracturing with or without proppant insertion as well as multiple radial tailored-pulse fracturing representing selective and controlled explosive stimulation (cf. section 4.7.3.). The natural fracture system intersecting the shale creates an anisotropical permeability distribution. This reservoir heterogeneity and the access of additional gas volumes only by opening of new cracks rather than enlarging pre-existing joints are the reason for a much better performance of wells with induced fractures running perpendicular to the natural fracture trend than such with artificially created cracks being parallel to the existing natural joint system.

4.8.8.6.3. Natural propping of sedimentary shrinkage cracks

Sedimentary shrinkage cracks of subaerial desiccation or subaqueous synaeresis origin in fluvial and lacustrine mudstones are excellent examples to illustrate a natural propping mechanism of fractures (cf. plates VI/1 - 7 and XI/7). Contraction of the mud surface in quiet environment either by atmospheric drying or subaqueous shrinkage gives rise to opening of extension cracks. Subsequent high-energy fluvial or aeolian depositional events lead to return of sand accumulation which does not only cover the sedimentary surface, but also plugs the open syndepositional cracks. The infilling of the shrinkage fractures by sand of different grain size according to the granulometrical spectrum supplied by the provenance area and transported by wind or water represents natural propping of natural fractures which results in keeping the cracks open and ascertaining their function of high-permeability avenues also in post-depositional history.

4.8.8.7. Other aspects

KOHLHAAS (1982) describes a reservoir with lateral change of patches with and without natural fractures. The zones with natural cracks have better pre- and post-treatment results than the spots without natural joints. Injection of fluid and proppants leads to enlargement of natural fractures and partial plugging with proppants in addition to creation of new hydraulically induced cracks. This example underlines the effectivity of a combination of natural and artificial fracturing for the successful stimulation of hydrocarbon reservoirs of limited thickness and extension.

Reservoir management aspects in naturally fractured or fissured pay formations, especially in terms of gravity drainage, are discussed by JACQUIN, LEGAIT, MARTIN, NECTOUX, ANTERION & RIOCHE (1987) and NECTOUX (1987). NELSON (1986) presents a summarizing geological analysis of naturally fractured formations with emphasis on strategies of reservoir management, detection and prediction of fracture orientation and spacing, reservoir anisotropy and statistical modelling, and possibilities of stimulation of naturally jointed pays. HENNINGTON (1980) reviews unconventional exploration for natural fractures, and RUOTSALA & WILLIAMS (1982) stress the significance of reflection seismology as an exploration tool for fractured zones. ABDASSAH & ERSHAGHI (1986) define naturally cracked reservoirs as triple-porosity systems, whereas WARREN & ROOT (1963) idealized the naturally jointed pay as consisting of two distinct systems being matrix and fractures. NAJURIETA (1979) and DERUYCK, BOURDET, PRAT & RAMEY (1982) comment on the significance of interference and pulse testing for the evaluation of communication within naturally fractured reservoirs.

BOURDET & GRINGARTEN (1980) assess determination of fissure volume and block size in naturally fractured formations. ULDRICH & ERSHAGHI (1979) present estimations of the interporosity flow parameter. Well log evaluation of naturally jointed reservoirs is discussed by AGUILERA (1976) and AGUILERA & ACEVEDO (1982), and REYNOLDS, CHANG, YEH & RAGHAVAN (1985) comment on wellbore pressure response. Various points of naturally fractured pays are summarized by AGUILERA (1980). Aspects of naturally cracked reservoirs are also treated by AGUILERA (1975, 1976, 1985), KOSTURA & RAVENSCROFT (1977); BRANAGAN, COTNER & LEE (1984); BLASINGAME & LEE (1986); BRANAGAN, CIPOLLA & LEE (1987) and EVANS & UBANI (1987).

4.8.9. Late-stage refracturing of old wells

Another mHF or related small-scale stimulation application, but occasionally also larger-scale treatment to even MHF utilization, is refracturing of particularly shallow oil wells comprising renewed fracturing in horizons which had already been fractured once long time ago (or equivalently original fracturing in advanced depletion stages of reservoirs intersected by boreholes that had never been stimulated before). After an outline of general refracturing significance, the discussion as follows comments on refracturing candidate well choice, areal distribution of potential reservoirs, mobilization of additional hydrocarbon reserves, proppant selection for refracturing, fracture reopening and proppant placement during refracturing, and combination of hydraulic and explosive fracturing.

4.8.9.0. General refracturing significance

Refracturing can be very successful by reopening the old fracture (that has in the meantime become completely closed due to crushing of nearly all the formerly inserted propping material) which acts as a pre-existing zone of weakness and is infilled again with proppants, with refracturing becoming increasingly important for workover of old oil fields. Since the advent of hydraulic proppant fracturing, periodical restimulation has been employed as a viable means of economically increasing production rates and capturing recoverable reserves, and in many instances, retreatment of existing wells has become a feasible alternative to new exploration or drilling additional development wells. In some geographical areas, production increases from restimulation have demonstrated that recompletion of targeted wells can be one of the best investment opportunities available to the producer (HALLIBURTON 1987). Refracturing of wells having already passed considerable production history and having been earlier stimulated shorter or longer time ago has become particularly interesting after the 1986 oil price crash as a consequence of being a comparatively cheap investment leading to rapid payout in comparison to much more expensive fracturing of new wells with a much more unfavourable discounted cash flow rate (cf. section 2.2.1.5.3.).

Refracturing is a remedy operation in order to improve productivity of such wells that have already been fractured some time ago with poor technical performance, insufficient reservoir response, or inadequate stimulation designs (MARTINS 1987) or in a more unfavourable economical environment. Consistent success can be achieved by refracturing if the original treatment has been performed some time ago and both technical and economical framework have changed, whereas if recently a good large-scale job has been carried out with satisfactory results, an enhancement by refracturing may be difficult to achieve. If the former crack is too densely propped, the new stimulation treatment probably does not ameliorate the rate and could even by poorer than the old operation. There is a complete range between just breaking down the perforations or pumping the whole old treatment away when carrying out refracturing.

The environment for refracturing of already earlier stimulated boreholes as well as initial fracturing of late-stage infill production points is depleted reservoir pore pressure which increases the confining stress contrast between pay zone and shale barriers, thereby allowing better fracture height growth control (cf. section 4.2.2.) which leads to lower proppant quantity being required to plug the smaller crack (HUEBINGER, WEBSTER, CHISHOLM, VENDITTO & HUNT 1988).

4.8.9.1. Refracturing candidate well choice

In the current unfavourable economical climate caused by uncertain oil price (cf. section 2.2.1.1.3.) and a curtailed gas market (cf. section 2.2.1.6.3.; in Europe and other parts of international markets also unfavourably superimposed by the simultaneous US $ weakness; cf. section 2.2.1.3.), restimulation of existing wells to increase production rates and/or economically recoverable reserves has become an alternative to exploration or drilling development wells (CONWAY, McMECHAN, McGOWEN, BROWN, CHISHOLM & VENDITTO 1985; cf. section 2.2.1.5.3.). Production increases realized from refracturing in several areas have demonstrated that recompletion of certain wells can be one of the best investment opportunities presently available. Improvement of the effect of the original stimulation by the renewed operation can be achieved because of recent advances in fracturing technology employing more efficient fluids that are capable of transporting higher proppant concentrations, with in combination with modified surface equipment now treatments being able to be performed that in terms of fluid and proppant volume as well as proppant saturation were previously not possible.

Periodical refracturing has been employed as a method for increasing production rates and/or economically recoverable reserves since the advent of hydraulic fracturing, and focusses particularly on wells which formerly have been underestimated by utilization of too low proppant concentrations and too small jobs. SINHA & ELBEL (1983) comment on partial rehealing of fractures as pressure declines with progressive depletion of the reservoir, with an advanced stage of closure of the original crack being an indicator for refracturing provided satisfactory production rates have been obtained after the primary operation. Well performance analysis after original stimulation as a base for further treatment planning includes a comparable approach for refracturing candidate selection and minifracture calibration test evaluation (cf. section 4.8.2.).

As a consequence of progressively increasing experience from earlier treatments and resulting design optimizations for successive operations both in connection with refracturing of old wells long time after the original stimulation jobs and in relationship with fracturing campaigns in new fields with short periods of time between the individual treatments, an evolution of fracture operation design and performance (TINDELL, NEAL & HUNTER 1981) takes often place from early more or less experimental stages through advanced routine phases up to a mature high-technology level. HOWARD & FAST (1970 a) summarize the purpose of refracturing as being extension of the existing fracture system, reopening of previously generated cracks, washing of fracture faces, replenishing of embedded proppants, and opening of new cracks in previously unstimulated intervals. The outline as follows illustrates well performance after original stimulation treatment and characterizes also the comparable approach of refracturing candidate selection and minifracture calibration test evaluation.

4.8.9.1.1. Well performance after original stimulation treatment

Concerning refracturing candidate selection, previously fractured wells with sufficient remaining reserves for consideration can be divided into four categories. Following discussion of some general aspects, success vs. failure of original fracture job is analyzed.

4.8.9.1.1.1. General aspects

The first class comprises wells with insufficient production increase from the first hydraulic stimulation job due to premature screenout failure. Analysis of treating pressure from the first operation is necessary to identify the problem and to modify the second treatment such as to overcome the formation characteristics that cause the first job to fail (cf. CONWAY, McGOWEN, GUNDERSON & KING 1985). The second group includes boreholes with insufficient production increase following successful pumping of the first fracture operation. A modified refracturing procedure has to be set up which is promising in terms of both technical and economical aspects.

The third class contains wells that were formerly already successfully fractured, but production has fallen below projected decline levels in the meantime. Severity and time frame of the abnormally reduced production help to determine the degree of modification of the original design which is necessary for successful restimulation. The fourth assemblage bears boreholes with successful original treatment that are still producing at projected decline rates. In this case, the main question is whether improvements in fracturing technology have made it possible to achieve a large enough production increase over that from the first stimulation job to justify the cost of the refracturing treatment (CONWAY, McMECHAN, McGOWEN, BROWN, CHISHOLM & VENDITTO 1985).

4.8.9.1.1.2. Success vs. failure of original fracture job

If a well responded favourably to original stimulation, it is probably a good candidate for refracturing, because boreholes drilled in areas of high potential but treated with conventional design show in many cases satisfactory early production that later declines rapidly (VENDITTO, CHISHOLM, WIGGINS & CONWAY 1986). Where the initial fracturing job screened out, good potential may exist for refracturing after identification of the screenout mode and solution of the problem which caused the premature failure (cf. section 6.2.4.2.1.). Sometimes a stimulation operation is a technical success, but an economical failure due to disappointing resulting production. If it can be determined what the original treatment accomplished and if it was defective in some way, refracturing may be economically justified. Restimulation is also the appropriate solution where a well was successfully fractured, but exploitation decline indicates that the crack is no longer effective. Wells having been stimulated and producing at projected decline rates may even be candidates if they were originally fractured using conventional loads pumped at low proppant concentrations. Refracturing of still producing wells with massive treatments and high proppant saturations has high economical feasibility potential.

Older wells that were never stimulated but have good production histories can often be rejuvenated by late-life fracturing. Restimulation is generally a viable way to reach untapped hydrocarbon reserves. While refracturing by MHF technology is highly successful, also small-scale fracturing and acidizing have given excellent production increase returns. Other refracturing candidates are wells where formerly bad and damaging fluids, insufficient proppant concentrations or inadequate proppant types have been used. CROWELL & JENNINGS (1979) analyze refracturing candidate selection in low-permeability gas reservoirs and underline the definite need of well performance diagnosis prior to restimulation. An important source of information for refracturing decision is pressure development obtained after original stimulation (TUCKER 1979).

4.8.9.1.2. Comparable approach of refracturing candidate selection and minifracture calibration test evaluation

In some tight gas fields, wells which had initially been successfully completed with small-volume fracture treatments can be successfully restimulated by MHF operations, whereas wells that responded poorly to original small-volume fracture jobs will not sufficiently react to refracturing on larger scale to pay out the expenses of the restimulation treatment (PARROT & LONG 1979). In order to find out these relationships as a tool of containing fracturing expenses and to test the formation in reality rather than relying completely on more or less theoretical planning, minifracturing with insertion of small quantities of proppants (cf. section 4.8.2.3.) should be done before MHF treatments especially in areas with little experience from other wells in the field. The significance of proppant minifracturing of tight gas sands requiring MHF stimulation for ultimate economical production is not only fracture pressure gradient, fluid-leakoff coefficient, and closure stress determination by calibration testing (cf. section 4.8.2.1.), but also feasibility checking by probing formation response, and thus comprehensive MHF treatment planning including preceding minifracture can be compared to large-scale refracturing of wells having been formerly subjected to small-scale stimulation jobs.

4.8.9.2. Areal distribution of potential reservoirs

High-rank targets for refracturing in Europe are seen especially in Lower Cretaceous oil-bearing sandstones in the large oil fields in the boundary belt between Germany FRG and the Netherlands as well as in various other older production patches in the seam of Jurassic and Lower Cretaceous oil fields in the Lower Saxony Basin in the northern part of Germany FRG (for location maps cf. BOIGK 1980, SCHRÖDER 1986), and in Tertiary oil- and gas-bearing sandstones in the big fields in the Vienna Basin in Austria (DOLAK & PERNTHANER 1976) and Czechoslovakia. A high potential for refracturing is also considered to be present in large oil fields in the USSR such as for example around the Caspian Sea and in Siberia.

Another restimulation goal are gas-bearing Rotliegend and Carboniferous sandstones in the Mid-European Basin (Southern Permian Basin; cf. section 2.4.1. and 3.3.) which had been already fractured before the advent of intermediate- and high-strength ceramic proppants and where still sufficient reserves are left to justify an improvement of the unsatisfying earlier treatment by performing a modern high-technology operation.

4.8.9.3. Mobilization of additional hydrocarbon reserves

The mobilization of additional hydrocarbon reserves by refracturing of earlier stimulated boreholes or even first fracturing in the late stage of well and field history is supported by an ameliorated feasibility due to already amortization of the original drilling and completion investment through prolonged production. The technical execution of the restimulation treatment is easier than that of an original job because of the already existence of a weakness plane leading to formation opening at lower injection pressures than usual (cf. section 4.8.9.5.), and the creation of new drainage paths improves flow conditions and acquisition area and thus leads to recovery enhancement independent from reservoir depletion stage. Some comments are also offered on production decline offsetting and reservoir types suitable for refracturing.

4.8.9.3.1. Ameliorated feasibility due to well amortization

The main significance of refracturing or partially also even first stimulation in the late stage of hydrocarbon exploitation and reservoir pressure depletion is to mobilize parts of the remaining hydrocarbon quantities from tighter streaks and layers in order to optimize the recovery factor and thus more effectively drain particularly huge shallow oil fields containing enormous amounts of reserves (at least in terms of relative view when comparing the size of all the domestic production patches within the individual countries) prior to plugging and abandoning of the wells. Late-stage refracturing of horizons that have already been exploited for partially more than half a century can be in places combined with stimulation of marginal reservoirs over- or underlying the main producing storey in little distance which had been previously not perforated due to the much larger reserves in the main target formation and the much better output rates from the perforated intervals therein, with the marginal layers therefore having been put on a backseat position in the ranking list as long as offtake from the main high-capacity reservoir was satisfactory.

Once the main producer is more or less depleted, however, the feasibility situation for refracturing of senile horizons along with first treatment of juvenile marginal pays is very favourable, because the only expenses for then accessing the oil in the marginal reservoir are the cost for recompletion and stimulation, as the well is already there and would afterwards anyway have to be plugged with cement, and also after up to several decades of production from the main target horizon there is nothing left for still writing off, and long-term exploitation has rather not only paid off all the previous investment, but also given rise to considerable profit. Aspects of late depletion stage and end of field life economics in North Sea hydrocarbon patches are outlined by KEMP, KELLAS & ROSE (1987).

4.8.9.3.2. Production decline offsetting

GRIFFITH & MADISON (1988) outline field examples where production declines have been partially offset by refracturing. It is not uncommon for a well to have been refractured several times throughout its producing life. While acid stimulations have traditionally resulted in short-lived production increases with limited effects on the ultimate recovery of the borehole, refracturing has led to a greater production increase sustained over a longer period of time. Extension of propped fracture life and prevention of early conductivity destruction by mobile fines and scale deposition can be achieved with larger fracture width and increasing proppant saturation, and therefore during course of stimulation evolution, small multi-staged low-proppant-concentration fracturing treatments were replaced by larger single-stage limited-entry high-proppant-saturation fracturing jobs. The benefits of the single-stage limited-entry fracture operation include larger propped fracture width, better areal coverage of net pay zone, and stimulation of previously unfractured pay zones (cf. section 4.2.2.2.).

4.8.9.3.3. Easier technical treatment performance by presence of weakness plane

Restimulation options include refracturing, reacidizing, reperforating, heated acid treatments (cf. section 4.5.1.2.4.), closed fracture acidizing, water fracturing (cf. section 4.5.4.4.2.), subzone redevelopment, and horizontal well or drainhole drilling (cf. section 4.8.6.), of which refracturing is by far the most effective method (JOURNAL PETROLEUM TECHNOLOGY 1987). While fracturing of a depleted well for the first time is much more difficult than stimulating a virgin well because it requires a much higher fracture initiation pressure, refracturing is facilitated by a crack already existing representing a weakness plane in the rock fabric which does no longer require to overcome the tensile stress of the formation (cf. section 4.8.9.5.). Therefore less pressure is needed to reopen the old fracture. The proppant package plugging the original crack is already suspended, and in many cases, there is an efficient barrier at the bottom of the fracture as a consequence of proppant banking which acts as a block limiting downwards fracture propagation (cf. section 4.2.2.4.3.) and thus prevents the crack from migrating into the underlying water zone.

WYMAN, HOLDITCH & RANDOLPH (1979) report special displacement techniques for proppants within old fractures at the beginning of restimulation operations. In order to ensure that a screenout would not result from pickup of the proppant inserted during the earlier treatment by the leading edge of the pad during the refracturing job, combined and alternated nitrogen, methanol and water mixture injection is carried out. In the middle of this displacement procedure, a small foam slug is pumped into the annulus for the purpose of further scouring any old proppant from the fracture walls, with the following nitrogen breaking it up and allowing any entrained proppant to settle to the bottom.

4.8.9.3.4. Recovery enhancement independent from reservoir depletion stage

Late-stage hydraulic proppant stimulation also of depleted wells is supported by the fact that fracturing is the most effective treatment technique by superimposing a newly-created high-permeability structure within the reservoir which is in communication with the wellbore, although the overall pay permeability cannot be changed (ECONOMIDES 1987 b), and this increase of formation surface which is accessible from the borehole through the fracture system and that can contribute to drainage of hydrocarbons from the rock matrix to the borehole through the crack network is the reason for recovery enhancement in pay sections regardless of their depletion stage. In tight gas reservoirs, fluid load recovery in original treatments can require a few weeks to several years, and for refracturing, it may need even longer time because of declined reservoir pressure due to production (PARROT & LONG 1979).

Refracturing of wells which were already once treated before is also reported by KOHLHAAS (1982) who also describes successful first fracturing in a mature oil field in late stages of depletion, resulting in access to additional remaining reserves that would never have come out of the reservoir without stimulation. Renewed fracturing of a gas and condensate well after a first smaller treatment by a second larger operation is outlined by ECONOMIDES, CIKES, PFÖRTER, UDICK & URODA (1986). WATERS (1980) also emphasizes the necessity of many shallow stripper wells to be fractured in order to remain productive. Aspects of refracturing are also discussed by COULTER & MENZIE (1973), CROWELL & JENNINGS (1978), PARROT & LONG (1979), HUNTER (1985) and SAUCIER (1988), and SINHA & ELBEL (1983) investigate stimulation of old producing wells. Recompletion proppant fracturing for improving productivity of multipay intervals is also evaluated by DARR & CARLTON (1988). GRIFFITH & MADISON (1988) outline optimization of proppant concentration and profitability for refracturing. DOLAK & PERNTHANER (1976) document that also in old highly-watered reservoirs with very low formation pressure, comparatively cheap and relatively small-scale fracturing treatments can lead to production of significant amounts of additional oil.

4.8.9.3.5. Reservoir types suitable for refracturing

While most of the refracturing work takes place in sandstone reservoirs, also gas-bearing shales can be an economically attractive and technically challenging target for proppant restimulation (LANCASTER, GUIDRY, GRAHAM, CURTIS, SHAW & BLAKE 1987; cf. section 4.4.4.). Coal refracturing is outlined by JEU, LOGAN & McBANE (1988; cf. section 4.4.3.). Restimulation can also comprise acid non-proppant fracturing especially if gelled or crosslinked acid is applied, and in some cases, such a treatment can follow an unsuccessful hydraulic proppant treatment that was terminated by screenout failure in carbonate reservoirs (JOHNSON, FOX, BURNS & O`MARA 1988). BRINKMANN (1982) describes an example of unwanted accidental refracturing of an earlier stimulated interval during course of a MHF campaign carried successively out in several horizons within one well.

4.8.9.4. Proppant selection for refracturing

Important guidelines for proppant selection for advanced-stage first or repeated fracturing of oil- or gas-bearing reservoirs are the relationships between discounted operating cash income, time and proppant type (HOLDITCH 1984). Some aspects of synthetic proppants vs. natural sand as well as reservoir pressure vs. proppant concentration are outlined as follows.

4.8.9.4.1. Synthetic proppants vs. natural sand

Examination of field examples reveals that discounted income during the first few years is greater when using synthetic bauxite proppants instead of natural sand due to higher initial gas flow rates as a consequence of greater fracture conductivity which results in production of a greater percentage of gas during early field history and better acceleration of cumulative gas recovery. On the other hand, however, the average production rate of a fracture propped with sand is usually slightly higher in comparison to plugging of the crack with bauxite proppants during the later production periods as a consequence of less pronounced acceleration, with the slightly higher rate produced through the sand-infilled fracture in comparison to that coming still out of the bauxite-plugged crack (which has already far more strongly depleted the reservoir with respect to that treated with sand) being the explanation for the higher discounted profit generated during late field life for the case of a sand-propped fracture.

Some contributing factors which give rise to low well productivity and characterize refracturing candidates are low proppant concentration, polymer degradation or filter cake plugging, crushed proppants or formation particles, and embedment and flowback of proppants (FRACFAX 1988 h). Curable resin-coated proppants are frequently a suitable material for refracturing (cf. section 1.2.6.). When they are injected into an old fracture filled with conventional proppants, the viscous pad pushes back the old proppant, and curable resin-coated particles are then pumped in sufficient quantity to cover the perforated interval up to a desired fracture length, with resin bonding locking the coated particles in place at the fracture face together to form a stable mass which prevents flowback of unstabilized old proppants as well as guarantees invalidization of new proppants (cf. section 4.12.3.3.).

4.8.9.4.2. Reservoir pressure vs. proppant concentration

In terms of treatment size and proppant saturation, refracturing usually takes place in an environment of depleted reservoir pore pressure which creates more effective boundaries between pay zone and adjacent shales because of the increased fracture gradient and in-situ stress difference (cf. sections 4.2.2.3.1. and 4.2.3.2.2.) between the two media (HUEBINGER, WEBSTER, CHISHOLM, VENDITTO & HUNT 1988). The increased fracture gradient following somewhat depletion of reservoir pore pressure allows to design fracturing operations that will be more contained within the prospective interval. With respect to the original stimulation jobs, both less total proppant quantity and less proppant saturation are required to fill in a shorter fracture during retreatment which is concentrated to the core of the pay zone, and thus restimulation jobs turn out to be considerably cheaper than initial treatments when reservoir pore pressure was at its virgin level (case studies show that proppant volume can be reduced up to abt. 40 % to achieve a comparable effect).

This is the reason why refracturing can still be economical even with declining production and falling oil price. The other possibility opened by higher fracture gradient is propagation of a larger crack infilled by larger amounts of proppants than would have been safely enabled by initial reservoir pressure, thereby giving rise to deeper pay penetration by a more conductive fracture which considerably increases hydrocarbon drainage and cumulative offtake, but also requires higher initial capital expenditure resulting in longer amortization time.

Proppant pumping in refracturing should follow an aggressive schedule (VENDITTO, CHISHOLM, WIGGINS & CONWAY 1986; cf. also section 4.13.). The best way to achieve long fractures and high volumes of proppant deposition is to progress quickly from treatment pad to heavy proppant saturation which propagates the crack rapidly and helps to ensure proppant delivery sufficiently fast to the fracture limits. In terms of choice of sand vs. high-strength proppants, field experience has shown that high sand concentrations can be more effective than moderate saturations of intermediate- or high-strength proppants that have been kept limited due to economical reasons.

4.8.9.5. Fracture reopening and proppant placement during refracturing

Proppant selection for refracturing has to be combined with treatment design. An important point of consideration is the residual proppant package in the earlier crack which has to be effectively removed when reopening the fracture in order to prevent premature screenout failure of the refracturing operation by getting stuck at the older plug within the crack (cf. section 4.8.9.3.3.). The following summary contains evaluation of increasing fracture width and proppant concentration as well as possibilities of reinjection complications.

4.8.9.5.1. Increasing fracture width and proppant concentration

Fracture mouth clearance by displacement of the earlier inserted supporting material can be achieved by considerably increasing fracture width and proppant concentration, with the optimum shifting effect of the high-density carrier slurry being obtained if the mixture travels by plug flow. Transport capacity of the fluid is certainly enhanced by choosing a light proppant type (cf. section 4.3.4.), plug flow behaviour can be improved by adding resin as grain pellicles or dispersed particles (cf. section 1.2.6. and 5.8.5.), and placement amelioration and embedment minimization of the proppant grains can be secured by resin films surrounding the particles which give rise to artificial cementation of the proppant package to a stable rigid wedge supporting the crack. The following outline focusses on aspects of arrangement of primary and secondary proppant load, refracturing treatment size, pressure and stress distribution, and perforated interval configuration.

4.8.9.5.1.1. Arrangement of primary and secondary proppant load

Practical experience has shown that the original fractures generally do not disturb the restimulation operation. As the old crack is still a weakness plane and opens earlier than an unfractured formation, the fracture gradient is lower than usual, but on the other hand fluid loss is higher due to the thief effect of the existing fracture. Displacement of the old proppant plug consisting of either still intact or already crushed grains is principally no problem provided sufficient width of the newly enlarged fracture. Arrangement of primary and secondary proppant load is therefore generally very good, and only occasionally some subordinate reinjection complications occur (cf. section 4.8.9.5.2.).

Field examples of large-scale refracturing with high proppant concentrations in reservoirs that have formerly been stimulated with small-scale treatments and low proppant saturations reveal a significant increase in production rate and comparatively rapid pay-out even if the fracturing job expenses approach 50 % of the total well costs (BRINKMANN 1982; GREGORCZYK, PAULS, HOLTMYER, VENDITTO & CHISHOLM 1984). Refracturing does not only apply for renewed treatments months or years after the original operations, but also for continuation of jobs after several days in case of necessary interruption of the pumping schedule due to bad weather conditions or severe mechanical problems with the equipment (PEARSON, LYNCH, SCHMIDT & McCASLIN 1988).

4.8.9.5.1.2. Refracturing treatment size

Refracturing can be carried out as small-scale jobs with abt. 10,000 lbs (5 t) of proppants through all stages of conventional and massive operations up to jumbo treatments comprising 3 Mio. lbs (1,500 t) of proppants (VENDITTO, CHISHOLM, WIGGINS & CONWAY 1986). In many cases, massive hydraulic refracturing treatments with high proppant concentrations are carried out in wells having experienced small-scale stimulation with low proppant saturation in a period of 1 - 40 years ago. The benefits of massive high-concentration refracturing include reduction of effects of fines on fracture flow capacity as a consequence of the proppant bed being subjected to less plugging by fines migration, more effective drainage of stringer sands, leaving of the created fracture propped for a much higher percentage of its entire height and length, and improvement of both conductivity and length of fracture thus ameliorating production rate and total recoverable hydrocarbons in very tight reservoirs.

4.8.9.5.1.3. Pressure and stress distribution

When low-permeability gas wells are fractured upon initial completion, reservoir pressure and horizontal stress are uniform throughout the pay formation (HOLDITCH, ROBINSON, ELY & RAHIM 1988). If the borehole is produced prior to the stimulation treatment or if later-stage refracturing is carried out, however, to a greater or lesser amount both a pressure distribution and stress gradient are developed in the reservoir, with stress in the rock being larger in the areas away from the wellbore where pressure has not been affected by production. In such a situation, the stress at the fracture tip and the resulting bottomhole treating pressure increase as the crack propagates into reservoir areas not having been drained so far by the existing wellbore. Stress changes at the fracture tip therefore deserve serious attention during refracturing design.

4.8.9.5.1.4. Perforated interval configuration

In many cases, hydraulic proppant or acid fracturing has to be performed by limited-entry technique in order to avoid unwanted excessive fracture extension in either upward or downward direction into non-productive horizons (HUCKABEE 1988; cf. section 4.2.2.2.). While in new completions this goal can be achieved by selective perforation placement according to reservoir requirements, restimulation is often confronted with the fact that a long perforated section exists in the borehole and communication occurs with a much thicker segment of the pay complex than can be allowed for fracture initiation. Refracturing operations therefore frequently have to apply mechanical diversion methods such as sand plugbacks, straddle packer assemblies, and retrievable bridge plugs and packers. On the other hand, the higher in-situ stress contrast between pay and barrier as a consequence of depleted reservoir pressure diminishes the danger of uncontrolled excessive fracture height growth and thus limited-entry technique which is necessary to inevitably in original stimulation at virgin reservoir pressure is frequently not required in late-stage refracturing.

4.8.9.5.2. Possibilities of reinjection complication

While in most cases no difficulties have been encountered during reopening of old fractures, replacement of the original proppant infill and insertion of the new proppant load during course of restimulation operations, BEGNAUD & CLAIBORNE (1985) report the occasional occurrence of pressure-out complications due to interference of earlier and current proppant lots, especially in case of new treatments breaking through barriers to get in communication with old fractures in adjoining horizons. Such interferences can result in proppant bed redistribution which may have the effect that a gain of productivity in one storey is jeopardized by a reduction of exploitability in the other layer.

Occasional higher fracture reopening pressures in coal seam gas reservoir refracturing can be created by accumulation of coal fines near the wellbore which bridge the crack and constrict the borehole vicinity (ZUBER, REEVES, JONES & SCHRAUFNAGEL 1988; cf. sections 4.2.2.7. and 4.4.3.3.). Coal seams can become particularly friable if loosening of the fabric is exacerbated by shear stresses imposed by faults in the surroundings of the borehole. In rare events, it can be imagined that pushing of the old proppant package infilling the original crack in front of the pad into the now wider opening and deeper penetrating fracture can take place faster than crack extension and thus reaching of the old proppant lot at the front of the restimulation treatment to the fracture tip could provoke screenout termination (cf. section 6.2.4.2.1.). As most of the refracturing jobs, however, perform smoothly without screenout problems, propagation of the crack during refracturing probably takes place at a rate faster than that of proppant-laden fluid advance. Other complications may occur in reservoir rocks with multiple natural joint systems that are partially healed (DOLAK & PERNTHANER 1976). In such cases, one of the main problems is opening up of additional fissure systems for uptake of additional amounts of proppants.

4.8.9.6. Combination of hydraulic and explosive fracturing

Late-stage refracturing of wells where an earlier hydraulic stimulation treatment has been performed some time ago can considerably profit from explosive fracturing superimposed on the older hydraulic treatment (YOUNG, BARKER & CLARK 1984). This is particularly due to the independency of orientation of cracks generated by stem-induced hydraulic fracturing which are vertical even in reservoir columns where the in-situ stress state would dictate the origin of horizontal cracks. The intersection of earlier horizontal hydraulic fractures by later vertical explosive fractures therefore is a very effective means of increasing the drainage area by combining several distinct crack systems (cf. also section 4.9.4.2.1.4.).

In addition, fractures generated by explosive detonation may often also intersect larger portions of the pay zones that were not adequately tapped by the earlier hydraulic stimulation treatments. The large flush in production resulting from explosive refracturing of wells that had been previously hydraulically stimulated underlines the significant benefit of stem-induced explosive retreatments in order to acquire new parts of the reservoir that have so far not been drained by the hydraulically-induced crack. Combination of hydraulic and explosive fracturing are partially reviving oil-well shooting by firing explosive charges which has been abundantly carried out prior to introduction of hydraulic fracturing to the hydrocarbon market, with well shooting aiming on fracturing or rubblizing of the reservoir in order to increase both initial flow and ultimate recovery of oil, water and gas wells (GRANT, DUVALL, OBERT, ROUGH & ATCHISON 1950).

4.8.10. Combination of acid and proppant fracturing

Miniature hydraulic proppant fracturing could also be a promising improvement of acid fracturing by having a combination of the effect of dissolution by the acid and propping of the fracture that was opened by the pad, and in case of good results could be the initial step for changing towards larger-scale hydraulic proppant fracturing of carbonate formations in Europe. Some combined proppant-acid-fracturing jobs have already been carried out in Cretaceous chalk in the Norwegian and Danish North Sea (TANSOE, KING & HOLMAN 1986; cf. section 4.5.4.2.) and in Middle Triassic carbonates in Austria. A particularly important marketing potential of minifractures carried out with intermediate- to high-strength proppants is seen at least in parts of Eastern Europe where administrative inhibitions of fast approval of major investments would favour to perform some initial treatments as minifractures with own equipment and service (cf. section 2.4.4.4.) and thus only comparatively little amounts of proppants would have to be purchased in the west. Aspects of proppant dissolution, treatment schedule and acid improvement, and special applications are briefly outlined as follows.

4.8.10.1. Proppant dissolution

Concerning combination of acid and proppant fracturing (cf. section 4.5.4.2.), attention has to be paid to the fact that under acid conditions, natural quartz sand has the lowest solubility in HCl/HF acid (CHEUNG 1985, COBB & FARRELL 1986). Sintered bauxite and alumina oxide and silicate proppants dissolve more or less quickly when exposed to HCl/HF-acid by leaching and formation of pits in the particles. The rate of solubility increases with higher HF concentrations, longer exposure time to acid, and agitation of the fluid (cf. section 1.3.3.2.). HCl/HF-mixtures in the ratio 12 : 3 attack all the proppants much more rapidly than either hydrochloric acid or hydrofluoric acid individually. The presence of corrosion inhibitors (cf. section 4.5.1.1.3.2.) which are to protect the tubings only slow down, but do not stop the dissolution rate of alumina oxide-containing proppants.

In addition to chemical dissolution, the alumina oxide-based proppants lose considerable percentages of their mechanical strength, and sintered bauxite even becomes weaker than natural quartz sand that has been similarly treated. Therefore combination of proppant and acid fracturing has to focus on application of natural quartz sand as propping agent when HCl/HF-mixtures are used as acid phase, and for deep wells propped with alumina oxide-based proppants, extreme caution should be taken in the use of HCl/HF-acid and preferably only hydrochloric acid should be applied which does not significantly dissolve or change the physical properties of gravels and proppants.

4.8.10.2. Treatment schedule and acid improvement

In terms of treatment schedule, distinction has to be made between joint or separated operation of acid and proppant fracturing. As acid alone is only a poor fracturing fluid with insufficient proppant transport properties, its effectivity for proppant carriage has to be frequently improved by gelling and/or crosslinking of the acid (cf. sections 4.3.4.3.4.1. and 4.5.1.2.). Some comments on suitable successions of stages are also offered.

4.8.10.2.1. Joint or separated operation

Combination of acid and proppant fracturing has to distinguish between temporally joint or separated operation. CHEUNG (1985) is favouring acidizing after gravel packing or hydraulic fracturing, but this can include only acid washing or matrix acidizing. Acid and proppant fracturing always should be executed more or less simultaneously, because in case of first proppant fracturing and second acid stimulation of carbonates or carbonate-bearing sandstones, the crack would be enlarged by acid leaching and the proppant distribution would be no longer suitable to support the fracture and prevent its closure.

Inversely, first acidizing or acid fracturing and second hydraulic proppant fracturing could result in propagation of the large proppant-infilled crack in another direction or along other predetermined cleavage planes than the acid-induced void spaces, which would mean that no combination effect is achieved, but the two separate steps of operation rather act individually and spatially distinct. Simultaneous proppant and acid fracturing guarantees that after the main reaction period of the acid, the hydraulically induced and chemically enlarged fracture is infilled by the proppants which support the whole final geometry and shape of the crack.

4.8.10.2.2. Acid gelling and crosslinking

On the other hand, acid is a very unsuitable fracturing or gravel packing fluid which performs much worse than water- or oil-based gels due to poor leak-off control and low viscosity unless upgraded (cf. also section 4.11.). Acid improvement, however, can be done by gelling or crosslinking, or by using alternating stages of acid and gel or fluid-loss additive (ELPHICK 1985; cf. also sections 4.3.4.3.4.1. and 4.5.1.2.2.). High-strength crosslinked acid systems with elevated temperature stability (PABLEY & HOLCOMB 1982; SMITH, DAWSON & SCOGGINS 1983) have excellent proppant transport capacity and are therefore the optimum medium for a combination of acid and proppant fracturing. Crosslinking of the acid can be done both in conventional and delayed manner (cf. section 4.3.4.3.) which is considerably increasing the possibility of reducing tubular friction pressure.

The low pH of acid gel systems minimizes clay damage and thus allows crosslinked acid systems to be a reasonable alternative to foam and energized fracturing fluids in water-sensitive sandstone reservoirs. Fluid-loss control identical or larger than that obtained by crosslinked water-based fluids insures deep penetration of live acid, and an extremely retarded reaction rate allows in many cases the crosslinked acid to be placed into the formation with little or no decrease in acid strength. Aspects of effective acid penetration in fracture systems are discussed by REN & XIONG (1986), and high-temperature stability of acid gelling polymers is investigated by NORMAN, CONWAY & WILLIAMS (1981) and NORMAN & CONWAY (1983).

4.8.10.2.3. Succession of stages

Concerning technical realization of combined acid and proppant fracturing, it may be in many cases the optimum solution to pump acid first without proppants, open the fracture hydraulically with the acid pad, and follow then with injection of a gelled carrier fluid that is saturated with proppants once the reaction of the acid in the fractured formation is almost terminated. Caution has to be exercised concerning proppant choice and acid selection, because while hydrochloric acid is generally not hazardous to any proppant type, HCl-HF-mixtures severely corrode proppants consisting of alumina oxide and silicate, whereas natural quartz sand is the chemically most stable propping material in acid environment (CHEUNG 1985, COBB & FARRELL 1986) (aspects of sandstone acidizing are discussed in section 4.5.5.).

4.8.10.3. Special applications

A special type of combination of acid and proppant fracturing has to be applied in some chalk formations where short wide cracks are induced hydraulically and subsequently enlarged by creation of widely-spaced etched ridges by acidizing in order to make the fractures sufficiently wide for infilling with proppant multi-layers that minimize embedment of grains into the soft formation (SHELL 1979, HARTLEY & BOSMA 1985; TANSOE, KING & HOLMAN 1986; cf. section 4.5.4.2.2.). A suitable technique for such weak unstable reservoirs is tip screenout fracturing (SMITH, MILLER & HAGA 1987; cf. sections 4.5.4.4.1. and 4.8.5.3.).

From the discussed points, conclusion can thus be made that minifracturing and related small-scale stimulation is a high-rank means to expand the hydraulic proppant fracturing market in both Western and Eastern Europe by not only allowing to cut stimulation cost, but also permitting to better justify the approach of unconventional oil and gas reservoirs at least at an experimental stage where so far no sound technical and economical stimulation concept has been developed.

4.8.11. Fracture length and height optimization

A discussion of mHF vs. MHF and other smaller- or larger-scale proppant stimulations automatically has to focus on the significance of fracture length (expressed as half length from the wellbore to the tip of one wing) and fracture height and their controlling factors and influences. Hydraulic fracturing treatments do not only represent a large fraction of initial well expenses (cf. sections 2.3.4. and 3.7.), but also determine the economical viability of a particular borehole or field (NEWBERRY, NELSON & AHMED 1985). A stimulation job being too large can be an unnecessary waste of completion funds, whereas a fracturing operation being too small may result in such inefficient reservoir drainage as to render a well unprofitable.

This economical double-edged sword necessitates that hydraulic fracturing treatments have to be designed in such a way that optimum depletion of the pay zone is reached, with proppant type and grain size selection being one of the key elements of this procedure (cf. section 1.4.12.), but an almost equivalent control exerted by choice of adequate fracture length which is the item governing total job size given more or less constant fracture height within the limits of reservoir thickness between the upper and lower bounding seals. The outline as follows focusses on reservoir permeability and well spacing vs. fracture length, fracture conductivity vs. fracture length, fracture flow capacity vs. proppant distribution, net present value optimization and fracture height. Some field examples are also mentioned.

4.8.11.1. Reservoir permeability and well spacing vs. fracture length

Reservoir permeability and well spacing are important factors influencing fracture length selection. Long and short fractures have to be chosen in low- and high-permeability pay zones, respectively, and the tighter the reservoir, the closer the boreholes have generally to be spaced. The following discussion also includes comments on reservoir and fracture parameters vs. economical aspects, pressure decline and fluid flow pattern, and fracture extension vs. reservoir shape.

4.8.11.1.1. Long and short fractures in low- and high-permeability reservoirs

Generally, fracture length is similarly as well spacing determined by the drainage radius in the hydrocarbon-bearing formation which is in turn a function of reservoir permeability (MONTGOMERY & STEANSON 1985, POULSEN & SOLIMAN 1987). This relationship already leads to conclusion that long fractures in connection with MHF treatments are required in low-permeability pay horizons and particularly tight gas sands, whereas moderate- and high-permeability formations can be successfully approached by shorter cracks of high conductivity (cf. sections 4.5.4.2.2. and 4.6.) and smaller job sizes and thus are suitable targets for minifracturing and its application in the various examples outlined in the preceding sections.

As fracture length is also linked with fracture flow capacity or conductivity (cf. section 1.4.10.), this means that support of long fractures in tight reservoirs has to be provided by intermediate- and high-strength synthetic proppants, whereas infilling of shorter cracks in moderate- to high-permeability pay horizons may also include natural sand in many cases, although a better conductivity contrast is always achieved by the application of a high-performance man-made proppant. Shorter fractures in moderate- to high-permeability reservoirs require also efficient multilayer propping to avoid embedment (cf. section 4.3.3.1.). The following discussion concentrates on reservoir permeability and fracture length, near-wellbore damage zone bypass, and fracture flow capacity and length optimization.

4.8.11.1.1.1. Reservoir permeability and fracture length

As low-permeability reservoirs with broad drainage radii respond favourably to creation of long propped fractures which is achieved by large fluid and proppant volumes, fracturing research has for years concentrated on fracture length improvement (CRAMER & SONGER 1988). This focus has unfortunately led for a long time to belief that well productivity is directly proportional to treatment volume and fracture length regardless of reservoir characteristics. A different approach, however, is required for moderate- to high-permeability pay zones where optimal results are achieved by creation of short high-conductivity fractures (cf. sections 4.5.4.2.2. and 4.6.). In hydrocarbon-bearing intervals intersected by closely-spaced wells, this is most efficiently accomplished by using a relatively small slurry volume containing high proppant concentration. Wellbore transmissibility is increased by borehole vicinity damage zone bypass (cf. section 4.8.3.) and extension of effective wellbore radius.

4.8.11.1.1.2. Near-wellbore damage zone bypass

Damage zone bypass is achieved by creation of a flow path with sufficient conductivity contrast from wellbore to virginal reservoir formation and is normally responsible for most of the productivity improvement attained from fracture stimulation of moderate- to high-permeability pay zones (CRAMER & SONGER 1988; cf. section 4.8.3.). Extension of effective wellbore radius also requires existence of a large conductivity contrast between propped fracture and reservoir pore network (cf. sections 4.5.1.1.2. and 4.6.1.). With all other things being constant, the benefits of conductivity contrast and resulting transmissibility continually diminish with increasing fracture length. Within the range of achievable fracture conductivities, optimal fracture half-length varies from 10 - 30 % of the well drainage radius. The primary treatment objectives in moderate- to high-permeability reservoirs are to bypass wellbore damage, maximize fracture conductivity, avoid excessive fracture height and length, and minimize operation cost.

Lenticular tight gas sands (cf. section 3.4.1.) require more complicated solutions, because the relationship between sand body long axis orientation and fracture azimuth as a consequence of horizontal stress direction has important impact on the possibility of fracture interference and thus stimulation effect deterioration (cf. section 4.10.5.1.). Given favourable conditions, square drainage patterns with relatively short fractures are more efficient in reservoirs with higher permeability (over 0.1 md), whereas rectangular drainage configurations with long fractures are the optimum possibility for pay zones with lower permeability (below 0.1 md; KUUSKRAA, BRASHEAR, ELKINS & MORRA 1979).

4.8.11.1.1.3. Fracture flow capacity and length optimization

Based on known reservoir permeability, it is important to optimize fracture length and conductivity by comparing treatment cost and expected production revenue (MONTGOMERY & STEANSON 1985). Pressure drop along a propped fracture (of finite conductivity; infinite-conductivity cases are hardly, if at all, developed in nature; cf. REINICKE, BRINKMANN, SCHWARZ & HUENI 1985) that has an insufficient flow capacity will seriously limit well productivity, whereas a fracture with excessive flow capacity is not effective and is rather a waste of money, because the incremental investment is not compensated and returned by further additional production (this applies for excess of fracture length, whereas additional crack height may have a positive impact on well productivity; CIPOLLA & LEE 1987).

When the effective in-situ reservoir permeability is greater than abt. 0.1 md, the desired fracture length is generally abt. 1,000 ft (abt. 300 m) or less, whereas in tight pay horizons with less than 0.1 md, production can be almost directly proportional to crack length before boundary conditions are reached and with adequate fracture flow capacity, the general rule is the longer the crack, the higher the producing rate. In tight reservoirs, thus fracture length is usually abt. 2,500 - 4,500 ft (abt. 750 - 1,500 m; BAKER 1981, VEATCH 1983), with longer fractures particularly being necessary in lenticular instead of continuous blanket sandstone pay segments (cf. section 3.4.1.). MEEHAN, HORNE & AZIZ (1988) outline effects of reservoir heterogeneity and fracture azimuth on optimization of fracture length and well spacing.

4.8.11.1.2. Reservoir and fracture parameters vs. economical aspects

Fracturing stimulation treatment optimization has three basic steps comprising evaluation of increased income which might be expected from oil- or gas-producing performance resulting from various fracture lengths and conductivities, determination of costs required to achieve the different lengths and conductivities, and assessment of net revenue vs. fracture length to pick up the treatment design which yields the maximum payback (VEATCH & MOSCHOVIDIS 1986). Although fracture length requirements depend greatly on reservoir permeability, fracture conductivity, fracture penetration and formation permeability, other factors such as fracture height and net pay thickness can become important considerations in fracture economics by having significant incremental effects. Some comments on fracture penetration vs. net pay thickness and other aspects are offered as follows.

4.8.11.1.2.1. Fracture penetration vs. net pay thickness

Optimum fracture penetration gets longer as net pay thickness increases for moderate- to high-permeability reservoirs, but remains rather constant for low-permeability hydrocarbon-bearing intervals (VEATCH & MOSCHOVIDIS 1986). Optimum crack length can therefore vary widely for a given pay permeability and fracture conductivity depending on net pay magnitude. Fracture height can have significant impact on optimum economical crack penetration which in turn affects well spacing, because the distance between neighbouring boreholes can be increased if fracture length and/or height increase, thereby replacing relatively high drilling expenses by lower stimulation costs. Optimization of fracture length and fracture height therefore has important consequences on planning of well spacing and thus field development expenses.

4.8.11.1.2.2. Other aspects

Additional factors requiring consideration for fracturing stimulation treatment optimization are duration of production forecast from which net present worth (cf. sections 1.4.12.2. and 4.8.11.4.) is calculated, net discounted production revenue, and amount of investment required to achieve the design option (WAREMBOURG, KLINGENSMITH, HODGES & ERDLE 1985), as well as hydrocarbon price (cf. section 2.2.1.), interest (discount) factors, technology level and risk (ROSENBERG, O`SHEA, MERCER, MORRA & BRASHEAR 1983; BRASHEAR, ROSENBERG & MERCER 1984). BRITT (1985) documents economical benefits of short high-conductivity fractures in moderate-permeability oil reservoirs (cf. section 4.6.) and documents the impact of different fracture lengths and flow capacities on incremental present worth.

In reservoirs where fracturing is an integral part of the development program, well spacing and appropriate fracture length requirements are paramount (VEATCH 1986; cf. section 2.3.4.). The impact of late time well interference from overlapping drainage areas of inefficient well patterns on economics is considerably influenced by fracture azimuth which has a prominent control on well spacing in order to prevent unwanted crack connection and cutting off drainage paths. BARBA (1988) comments on reservoir justification for maximizing hydraulic fracture length.

4.8.11.1.3. Pressure decline and fluid flow pattern

Low-permeability reservoirs intersected by long hydraulic fractures are characterized by rapid decline rates during the first months of production after the treatment (HOLDITCH & LEE 1979). Unsuccessful stimulation results as a consequence of extremely quick rate diminution beyond normal ranges are the result of an extraordinarily low formation permeability (less than 0.001 md), insufficient fracture length which is usually caused by fracturing out of zone or by poor proppant transport, insufficient fracture conductivity due to proppant crushing and embedment, or insufficient gas-in-place. In very tight reservoirs, longer fractures can be created by pumping larger treatment volumes, and conductivity improvement can be made by employing higher gel concentrations, smaller proppant grain sizes, and density-controlled fluids (cf. section 4.2.2.6.).

In many cases, however, the average treatment achieves only abt. 70 % of the designed fracture length, implying that the actual crack is wider and shorter and fluid loss is larger than predicted by conventional design techniques, and the barriers controlling fracture height might be underestimated. While in reservoirs with permeabilities of 0.001 - 0.05 md and with fracture lengths of 500 - 1,500 ft (150 - 500 m), even crushed sand or proppants usually provide sufficient conductivity to adequately drain the pay zone (cf. section 4.12.4.), crushing and/or embedment of proppants can create serious problems in other hydrocarbon-bearing intervals.

Low fracture conductivity gives rise to an essentially radial flow pattern in the reservoir (cf. section 4.11.3.), whereas high fracture conductivity triggers an unique flow pattern, and an unique history match is only obtained when the dimensionless fracture conductivity is 10 or greater (cf. section 4.8.11.3.). As the majority of propped fracture lengths appears to have reached only abt. 70 % of the designed lengths (HOLDITCH & LEE 1979, SCOTT 1979), it may be advisable to overdesign a stimulation treatment in terms of length, especially when selection of fracture barriers is difficult. Concerning excessive crack length, large fracture surface areas do not necessarily produce proportionally more, with offtake from any given formation also being controlled by nonhomogeneous permeability distribution, capillary pore size, crushed proppant beds and fluid interfacial tension (DERBY & SMITH 1979).

4.8.11.1.4. Fracture extension vs. reservoir shape

Fracture length modelling has considerable significance in drainage of lenticular tight gas sands where crack length also depends on the angular relationship between horizontal stress direction and sand lens axis orientation (KUUSKRAA, BRASHEAR, ELKINS & MORRA 1979). Distinction has to be made between fracture propagation parallel to long and short axes of the sand lens as well as intermediate oblique extension. The optimum case is fracture orientation parallel to the long axis of the lenticular sand body where maximum fracture length can be achieved in order to allow for efficient and feasible gas recovery. When regional tectonical forces are essentially perpendicular to lens direction, fractures intersecting isolated lenses have to be kept short in order to avoid waste of fracture extension in surrounding barrier rocks. Fracture length can only be maximized if lens spacing is so narrow that several separate lenses can be connected together and to the borehole by a long fracture. If the relationship between crack azimuth and sand lens long axis is random, optimum fracture extension is also limited by the lens boundary unless other adjoining lenses can be reached by the same crack.

Thus fracture length in lenticular tight gas sandstones is mainly a function of lens diameter, crack orientation with respect to lens axis, and frequency and spacing of lenses in comparison to economically feasible fracture length. In contrast to lenticular tight gas sands, crack azimuth has no significance in blanket sandstones except of well pattern planning for avoiding fracture interference (cf. section 4.10.5.1.), and fracture length can be designed exclusively according to permeability distribution and recoverable gas reserves based on net present value optimization (cf. section 1.4.1.2.2. and 4.8.11.4.).

Connection of isolated sand lenses does not only include acquisition of strictly laterally adjoining lenses, but incorporates also obliquely displaced lower and higher lenses in underlying and overlying reservoir levels. Therefore drainage optimization in lenticular tight gas sands also includes fracture height modelling which automatically also has its influence on fracture length. The significance of maximization of hydraulic crack extension to the reservoir limits of sand lenses is also emphasized by PETERSON & KOHOUT (1983).

4.8.11.2. Fracture conductivity vs. fracture length

The relationship of fracture conductivity and crack length can be summarized in the requirement of shorter higher-conductivity fractures for moderate- to high-permeability reservoirs and longer fractures for tight pay intervals which do not necessarily have to be of high-conductivity nature, but can also be of absolute low-conductivity type, as even in the latter case still a significant relative conductivity contrast between the low-permeability rock matrix and the proppant package in the fracture is created (cf. sections 4.5.1.1.2. and 4.6.1.). The following outline also focusses on impact of proppant type, fracture flow capacity optimization, fracture productivity and fluid flow pattern, acceleration vs. ultimate recovery enhancement, and differences between created, propped and effective fracture length.

4.8.11.2.1. Impact of proppant type

Fracture length is linked with crack conductivity, with longer fractures in tighter formations justifying the application of a better-performing proppant in order to maximize drain-off rates of hydrocarbons to the wellbore. This is particularly due to the fact that most of the tight formations which have to be stimulated by MHF treatments involving long fractures are situated in greater depth where higher closure stresses indispensibly require the application of intermediate- or high-strength synthetic proppants, and even in case of doubt of their necessity because of the closure stress still being in the range of the boundary stability of sand, it is for the reason of securing long-term production history up to 20 - 30 years recommended to go on the absolutely safe side and choose a proppant type that would be able to resist to any possible pressure drawdown during course of progressive depletion of the reservoir (cf. sections 1.3.2. and 2.4.1.).

Fracture length has to be optimized by variation of crack conductivity, but also caution has to be exercised to contain the fracture within the planned vertical interval for reservoir engineering and environmental safety reasons (cf. section 4.2.2.). As a rule of thumb, high-permeability reservoirs require high fracture conductivities but do not need deeply penetrating cracks, whereas low-permeability formations demand deeply penetrating fractures but can tolerate lower crack conductivities (VEATCH 1983).

4.8.11.2.2. Fracture flow capacity optimization

Deeply penetrating fractures are usually recommended for low-permeability reservoirs, but the incremental improvement on well production diminishes with crack length due to the fact that the dimensionless fracture conductivity or flow capacity decreases with increasing crack length, with thus a balance between fracture characteristics and reservoir properties having to be achieved in order to optimize pay deliverability (MENG & BROWN 1987). In reality, however, the selection of an optimum propped fracture length based on reservoir conditions is influenced to great extent by non-technical aspects such as operator budget and common area practices (PHILLIPS & ANDERSON 1985; ANDERSON & PHILLIPS 1986, 1987), and based on these considerations as well as on pay properties, penetration distance and proppant concentration profile are obtained using fracture design programs (cf. section 1.4.12.). Therefore the optimum dimensionless conductivity or flow capacity of at least 10 in tight reservoirs can not always be economically or operationally obtained.

In terms of fracture conductivity vs. length, attention has also to be paid to conductivity variations along the extension of the fracture wings away from the borehole (cf. section 4.3.6.). Fracture conductivity is in practice a decreasing function of distance from the wellbore (BENNETT, ROSATO, REYNOLDS & RAGHAVAN 1981). If fracture conductivity decreases monotonically with distance from the borehole, then at late production time the variable fracture conductivity solutions behave like a constant conductivity crack with conductivity equal to the arithmetic conductivity average. Decreasing fracture conductivity along the crack extension can be due to changing proppant types and/or grain size as a consequence of tail-in operations, or due to narrowing fracture width away from the wellbore in direction of the crack tip. In case of presence of a lower conductivity zone in the fracture reach, the crack limits the capacity of higher conductivity intervals beyond it to supply fluid or gas to the well. Fracture conductivity near the borehole can be determined only if data are recorded during the bilinear flow period.

4.8.11.2.3. Fracture productivity and fluid flow pattern

A wide and varied assortment of methods of both graphic and computerized nature are available to estimate effects of fracture length and conductivity on reservoir productivity, and various crack geometry models exist which may predict different fracture lengths for a given amount of injected fluid at a given rate with all the other parameters being the same, with the different results being a consequence of the basic premises used to develop the models (VEATCH 1983). In high-permeability reservoirs where steady-state or semi-steady state flow is dominant, less than a three-fold production increase is available regardless of fracture length and conductivity, whereas in low-permeability pay sections with chiefly unsteady (transient) performance behaviour, the more deeply penetrating fractures enhance the folds of production rate increase, but conductivity-function values of 5 or more are required to realize significant offtake enhancement, and in order to maximize exploitation rate, a dimensionless fracture conductivity in the range of 100 - 500 should be achieved.

CALLANAN, CIPOLLA & LEE (1983) outline that dimensionless fracture conductivities greater than 10 imply enough permeability to handle all the gas which the formation can supply, whereas values in excess of 100 are representing infinitely conductive fractures. DAVIES & KUIPER (1988) argue that a dimensionless fracture conductivity of 15 is an appropriate design value for pseudo-steady-state flow conditions, but tight reservoirs with high initial transient production rates require higher values because of the prolonged duration of the transient rates (cf. section 4.8.11.3.).

4.8.11.2.4. Acceleration vs. ultimate recovery enhancement

In high-permeability reservoirs with steady-state or semi-steady state flow, hydraulic fracture stimulation increases early-life production rates which increases cash flow, but has little or no impact on ultimate recovery and thus basically represents acceleration of reservoir depletion (cf. section 4.6.). In low-permeability formations with unsteady-state (transient) performance behaviour, however, deeply penetrating fractures can significantly improve ultimate recovery, with advanced technology such as long and conductive fractures being able to increase recoverable reserves from tight gas sands by 40 - 75 % (BAKER 1981, VEATCH 1981). LANCASTER, GUIDRY, GRAHAM, CURTIS, SHAW & BLAKE (1987) carry out fracture length optimization in shale gas reservoirs and conclude that while longer fractures lead to increased gas production at some point of time, the increased exploitation may become insufficient to offset the increased costs associated with pumping of larger treatments.

4.8.11.2.5. Created, propped and effective fracture length

The created crack length always has to be abt. 10 % larger than the subsequently propped fracture length in order to minimize the possibility of a screenout at the crack tip (POULSEN & SOLIMAN 1987). For any fracture length, crack width has to be sufficiently large for prevention of bridging of proppants between the fracture walls. Crack width at the desired propped fracture length at the time of job completion has to exceed twice the diameter of the proppant particles in order to allow unhindered passage of the proppants throughout the crack.

Formation permeability, fracture conductivity and crack penetration in terms of length of the drainage path are affecting production rate folds of increase and recoverable reserve increases (VEATCH 1986). Cash flow from rate acceleration has a significant impact on economics even without any reserve increase. Distinction has to be made not only between fluid-created and actually-propped fracture length (cf. section 6.2.1.5.2.), but also between propped and effective fracture length (ZAHNER & CRAFTON 1985). Effective fracture length is the portion of propped fracture length which contributes to production and is smaller than total propped fracture length due to formation damage adjoining to the crack wall by increasing relative water permeability and water imbibition which are rising with both time and surface area of exposure to invading fracturing fluids. Reduction of crack area by increasing relative water permeability and water imbibition thus represents reduction of fracture length at the bottom of the line if crack height is considered to be constant.

4.8.11.3. Fracture flow capacity vs. proppant distribution

In terms of dimensionless fracture conductivity or flow capacity, the optimum length-to-width ratio in high-permeability reservoirs exceeding 1 md permeability or in case of cumulative production during long-term field operation is 1.26 (PRATS 1961). For tight formations, however, the early-time transient production rates prior to reaching pseudo steady-state conditions are important when considering the economics or pay-out of a fracturing treatment (MORSE & GONTEN 1972), and therefore higher values of fracture flow capacity are more feasible. ELBEL (1985) documents that in the permeability range of 0.1 - 0.001 md, a dimensionless fracture flow capacity of 5 should be considered when the proppant volume has been predetermined, and HOLDITCH & LEE (1979) and PHILLIPS & ANDERSON (1985) point out that an unique history match can only be obtained when the dimensionless fracture flow capacity is 10 or greater.

Varying proppant distribution for a given proppant volume and crack length can affect production rates depending on fracture flow capacity (cf. section

4.3.6.). CINCO-LEY, SAMANIEGO & DOMINGUEZ (1978) and AGARWAL, CARTER & POLLOCK (1979) recommend as a rule-of-thumb to design fracture flow capacity to be at least 10 which according to the investigations of ELBEL (1985) has obvious benefits, but on the other hand requires eight times the amount of proppant than for a value of 1.26. Optimization modelling can be performed giving production forecasts for various equal proppant volume fracture geometries with various formation permeabilities, constant fracture length with different proppant volumes, and equal fracture length and proppant volume with varying proppant distributions. Square drainage simulation can be replaced by more appropriate rectangular drainage modelling in case of long fractures in tight reservoirs (HOLDITCH, JENNINGS, NEUSE & WYMAN 1978), and if fracture azimuth (SMITH, HOLMAN, FAST & COVLIN 1978) is known and the wells can be spaced for rectangular drainage, then it may be possible to economize the treatment by designing for lower fracture flow capacities. Crack conductivity increase can be further achieved by improvement of gas permeability within the fracture possibly by removal of stimulation fluid or by fluid segregation, as well as by sustained production (SCOTT 1979).

Proppant transport effectivity is decisive for the success of a hydraulic fracturing job, because fracture conductivity is controlled by proppant distribution (NOVOTNY 1977). As the slurry travels away from the wellbore along the crack, proppants and fluid are heated and the reservoir rock is cooled (cf. section 4.3.4.6.1.4.). Because carrier fluid is continuously lost to the formation by leakoff, proppant concentration and slurry viscosity increase and fluid velocity diminishes. Fracture width also decreases away from the wellbore which alters fluid velocity. As is cannot be assumed that even a completely suspending fluid carries the proppants to the tips at the end of the fracture, length of the proppant-fluid slurry is usually abt. 2/3 of the total fracture length.

4.8.11.4. Net present value optimization

Net present value (being defined as hydrocarbon revenue by stimulation minus treatment cost; cf. section 1.4.12.2.) drastically increases with rising fracture half length, increasing proppant concentration and increasing proppant grain size. After some distance of steep rise, net present value achieves a plateau and does not increase further when fracture length is continued to be progressively increased. The unfeasibility of excessive crack length is best reflected by the net present value even again declining after the plateau section, because the creation of successively longer fractures is linked with again increasing treatment expenses that are not compensated by rising additional hydrocarbon recovery and thus the increasing operational costs give rise to diminution of the difference defining net present value. Many stimulation jobs, however, remain in the pre-plateau stage of the fracture length vs. net present value plot which testifies to lack of achievement of optimum fracture length (ECONOMIDES & NOLTE 1987).

It is in practice certainly not easy to reach exactly the optimum point of the net revenue curve at which cost to achieve longer fractures starts to exceed the revenue generated by production from the additional fracture length (VEATCH 1983, 1986), but this optimum solution can be closely approached especially in large field-wide campaigns affecting up to several dozens of wells in one big patch where sufficient comparison of treatment success and economical payback can be done between the individual wells. ZAHNER & CRAFTON (1985) also emphasize that there is no correlation between performance and propped fracture length beyond a maximum and consequently optimum value of the latter.

In terms of operations, it is easier to create long fractures in thin reservoirs than in thick pay formations (JOSHI 1987). While generation of short fractures of abt. 100 - 200 ft (30 - 70 m) length is quite easy and cheap, creation of long fractures of abt. 3,000 ft (1,000 m) length is both difficult and expensive (FAST, HOLMAN & COVLIN 1977). Aspects of fracture treatment optimization by net present value modelling are also discussed by COCIANCIG, GRIFFITHS & SCHÜLER (1980).

4.8.11.5. Field examples

MONTGOMERY & STEANSON (1985) present an example showing that in the given drainage radius and thus also fracture length, an unstimulated well would produce 8 % of the initial gas in place in 20 years, whereas in case of creating a crack of abt. 1,300 ft (abt. 400 m) length, the same well would deplete 34 % of the initial gas in place in the same amount of time. The initial rates for the fractured well would be higher than those for the unstimulated one and ultimate recovery would also be greater. These two factors are key points in determination of the economics of fracture length and well spacing. Aspects of fracture length are also discussed by POOLLEN, TINSLEY & SAUNDERS (1958); McGUIRE & SIKORA (1960), PRATS & LEVINE (1963); BENNETT, ROSATO & REYNOLDS (1981); HOLDITCH, LEE & GIST (1981); CLARK (1983) and CAMACHO, RAGHAVAN & REYNOLDS (1984). BARBA (1988) evaluates reservoir justification for maximizing hydraulic fracture length, and BARBA (1986) comments on improving return on hydraulic treatment investment with wireline inputs. The importance of vertical fracture height on steady-state production increase is commented by TINSLEY, WILLIAMS, TINNER & MALONE (1969).

The MHF operations in Rotliegend and Carboniferous gas-bearing formations in Germany FRG which are partially the largest stimulation jobs carried out in Europe within the most important gas reservoirs in this area (cf. section 2.4.1. and 3.3.) have been performed in pay horizons with permeabilities of less than 0.1 md and in extreme cases as low as 0.5 microdarcy (REINICKE, BRINKMANN, SCHWARZ & HUENI 1985). Proppant quantities up to 550 t (1,100,000 lbs) have been injected into fractures reaching in length between 200 and 550 m (600 and 1,600 ft) and in height between 10 and 115 m (30 and 345 ft; KLOSE & KRÖMER 1983). FAST, HOLMAN & COVLIN (1977) outline MHF treatments in tight gas sands in Colorado/USA where proppant volumes up to 500 t have been pumped into fractures up to 3,000 ft (1,000 m) in length.

4.8.11.6. Fracture height

In contrast to fracture length which beyond an optimum will be ineffective with respect to further increase of the productivity index and therefore is economically not justified (cf. section 4.8.11.4.), excess crack height may have a positive impact on the result of the stimulation job (CIPOLLA & LEE 1987). While conventional fracture design is based on constant crack height being equal or less than reservoir thickness (PERKINS & KERN 1961, GEERTSMA & DEKLERK 1969; TINSLEY, WILLIAMS, TINNER & MALONE 1969; NORDGREN 1972), more recent fracture modelling indicates that hydraulic fractures are often not limited to reservoir thickness (CLEARY 1980, PALMER & CARROLL 1982, SETTARI & CLEARY 1982; ABOU-SAYED, SINHA & CLIFTON 1984; MEYER 1985).

Fracture height selection is the most important variable in hydraulic treatment design, because it dictates volumes and rates required for execution of the operation (GARBIS, BROWN & MAURITZ 1985). Fracture height can be divided into net height representing the amount of zone where fluid leakoff occurs, and gross height being the distance between zone barriers.

Three-dimensional reservoir simulation studies indicate that propped fracture height in excess of the productive thickness of the reservoir can add to well productivity, particularly when fracture conductivity is relatively low compared to formation permeability (CIPOLLA & LEE 1987). The magnitude of the productivity increase is a function of the dimensionless crack flow capacity and excess height. Fracture height extending above and below the reservoir into non-productive strata can significantly affect transient pressure behaviour of vertically fractured wells. In terms of proppant selection and crack design, excess fracture height can result in requirement of less proppant quantity for a given crack length (cf. section 4.3.), necessity of a less conductive (and therefore expensive) proppant type (cf. section 1.3.2.), increase of the optimum economic fracture length (cf. section 4.8.11.) and/or increased estimated post-stimulation production. TINSLEY, WILLIAMS, TINNER & MALONE (1969) present curves showing production increase as a function of fracture height, fracture length, reservoir height, drainage radius, fracture flow capacity, formation permeability and wellbore radius.

4.9. Oil-reservoir fracturing

In contrast to the USA where both oil- and gas-reservoir fracturing is common sense and hydraulic proppant stimulation has already evolved to the stage of being one of the most important aspects of well completion (WATERS 1980; ROBINSON, HOLDITCH & WHITEHEAD 1986; cf. section 2.3.4.), in Europe and other parts of the world preferentially gas wells are treated. After discussion of some general points, several aspects are outlined which illuminate the benefit of oil-reservoir fracturing with high-performance synthetic proppants in view of conductivity contrast and economical feasibility. Comments are also offered on steam-drive effects on fracture propagation and cold water injection effects on hydraulic fracturing.

4.9.1. General aspects

The outline of some general introductory aspects comprises sketching of gas vs. oil reservoir stimulation as well as proppant application principles.

4.9.1.1. Gas vs. oil reservoir stimulation

In comparison to the stimulation of gas-bearing formations, oil-reservoir fracturing is in Europe and other parts of the world only performed in a considerably lower share and in smaller scale and is predominantly restricted to special cases where apart from a few exceptions, almost exclusively natural sand is used as propping agent due to shallow depth reasonably below the critical pressure stability of sand and as a consequence of serious cost containment for economical reasons of investment limitation. Although the general feasibility situation in the years after the 1986 oil price crash (cf. section 2.2.1.1.2.) does not allow introduction of new techniques, considerable improvement of oil production from marginal reservoirs can be achieved in the near future by increasing the stimulation activity in oil-bearing pay horizons and performing the operations in rising amounts with synthetic proppants instead of using sand, with the effect of obtaining a more pronounced conductivity contrast between fracture and especially moderate- to high-permeability formations and thus a better drainage of the oil (cf. sections 4.5.1.1.2. and 4.6.1.).

While several aspects of oil-reservoir fracturing have already been discussed in the divisions on chalk stimulation (section 4.5.4.), high-permeability reservoir fracturing (section 4.6.) and late-stage refracturing of old wells (section 4.8.9.), some points are compiled here to outline in a summarizing manner the advantages of oil-pay fracturing with placement of synthetic high-conductivity proppants in the crack as a perspective of near-future enhanced oil recovery.

4.9.1.2. Proppant application principles

The effectiveness of hydraulic fracturing for improving well productivity is a function of fracture length and fracture conductivity (TINSLEY, WILLIAMS, TIN-

NER & MALONE 1969; cf. sections 1.4.10. and 4.8.11.) and in order to achieve the most cost-effective stimulation, particularly optimization of fracture flow capacity is essential which is primarily governed by type and concentration of the proppants (NORMAN, CIPOLLA & WEBB 1983; cf. section 1.3.2.). In moderate- to high-permeability oil reservoirs, high proppant concentrations plugging wide cracks in multilayers are one way to improve stimulation results (COULTER & WELLS 1972; SHAH, SMITH & DONALDSON 1983; cf. also section 4.3.), but a more significant amelioration can be obtained by varying proppant type and partially also grain size. While intermediate- and high-strength proppants are widely applied in deep gas reservoirs where their usage is inevitable due to corresponding closure stresses, so far comparatively little field experience exists on the potential of these high-conductivity proppants in low- to moderately-stressed relatively shallow oil-bearing pay horizons (KOHLHAAS 1982; NORMAN, CIPOLLA & WEBB 1983). The following outline particularly emphasizes conductivity contrast and economical feasibility.

Aspects of shallow to moderate-depth oil-reservoir fracturing are also discussed by BRITT (1985), BRITT & BENNETT (1985), BRITT & LARSEN (1986), LI & ZHU (1986) and MARPLE, RULEY, WORLEY & FREEMAN (1987). GRAHAM, MONAGHAN & OSOBA (1959) outline the influence of propping sand wettability on productivity of hydraulically fractured oil wells. RICE & SHURR (1978); BLACK, RIPLEY, BEECROFT & PAMPLIN (1979) and NYDEGGER, RICE & BROWN (1979) comment on stimulation of low-permeability, low-pressure shallow gas reservoirs. SINHA & ELBEL (1983) present productivity increase of many old wells by properly designed stimulation treatments.

4.9.2. Conductivity contrast

Intermediate- and high-strength proppants have not only the advantage of resisting to higher closure stresses than sand, but they also offer up to ten times the conductivity of sand even at the low closure stress levels encountered in most oil wells as a consequence of the combination of better roundness and sphericity, less crushing and narrower sizing in comparison to sand (NORMAN, CIPOLLA & WEBB 1983). The higher bulk density of the ceramic proppants together with their better conductivity with respect to sand enable to reduce the propped fracture widths and thus to streamline the stimulation operation. Comments are offered as follows on proppant type and grain size.

4.9.2.1. Proppant type

In many moderate-permeability oil reservoirs, productivity increases and optimum fracture length from the individual wells are limited by fracture conductivity. Generally, longer fractures require more conductive, high-performance proppants in order to achieve increasing profitability (NORMAN, CIPOLLA & WEBB 1983). Increasing fracture length can significantly ameliorate hydrocarbon production when proppant effectiveness is moderate to high (cf. also section 4.8.11.). As the ultimate controlling factor of the success of the stimulation job is the conductivity contrast between fracture and formation (cf. sections 4.5.1.1.2. and 4.6.1.; which according to WATERS 1980 has to be at least 100 times conductivity difference, and PAI, GARBIS & HALL 1983 recommend flow capacity contrast ratios between fracture and reservoir matrix of 1,000 - 1,000,000, with fracture conductivity being a function of type, size, strength, quality, and density of the proppant as well as formation overburden pressure), a higher-quality intermediate- or high-strength proppant has to be selected instead of sand particularly in case of higher pay interval permeability in order to accomplish the necessary conductivity contrast for production increases maximization.

In terms of comparative effectiveness of natural sand and synthetic ceramic proppants, the performance differences are increasing and diminishing at higher and lower closure stresses, respectively. The proppants with the highest conductivity also show the least revenue loss due to fracture damage by proppant crushing, embedment, flowback or plugging (NORMAN, CIPOLLA & WEBB 1983; cf. also section 4.12.).

4.9.2.2. Proppant grain size

Creation of sufficient conductivity contrast between the moderate- to high-permeability shallow oil sand and the fracture infilling can also be performed by selecting coarser proppant grain sizes such as 10/20 or even 8/12 (MARPLE, RULEY, WORLEY & FREEMAN 1987; cf. section 4.3.5.1.). Additional benefit in conductivity can be achieved by increasing fracture width and placement of higher proppant concentrations per unit area of the fracture walls by multilayer deposition (cf. section 4.3.3.1.). Case studies show that modified treatments using one-third less total volume and twice as much proppants with respect to the original operation design give rise to significant improvements in well productivity. The performance seen on newly completed boreholes even suggests that refracturing of older wells with the revised job schedule may have economical viability (cf. section 4.8.9.).

NIEMEYER & REINART (1986) report effective bypass of near-wellbore damage and reservoir stimulation by short high-conductivity fractures in moderate-permeability oil zones. Complex formation damage in these prospective intervals resulted from interactions between mud filtrate, cement filtrate and connate water leading to precipitation of calcium carbonate and barium sulphate, water block from completion fluids (cf. sections 3.11.2.4.2. and 4.8.8.3.1.5.), clay dispersion by a drilling mud additive and mobile formation fines (cf. section 3.5.2.). In order to compensate for excessive leakoff into the moderate-permeability matrix, a large pad exceeding half of the total job volume had to be pumped (cf. section 4.3.4.6.1.4.), and silica flour and 100 mesh sand were also included to restrict fluid loss (cf. sections 1.4.11.2.1. and 4.8.8.3.2.3.).

While fracturing with coarse proppants of 12/20 mesh size has turned out to be a problem due to both perforation and formation screenout failure, satisfactory conductivity contrasts have been achieved with 20/40 proppants in various stimulation fluids. Small-scale treatments with reduced rates and cumulative quantities have been performed with successful fracture height limitation in wells where the reservoir is only capped by a thin seal of insufficient thickness to prevent breakthrough to another pay storey with different fluid and pressure regime. Some aspects of conductivity improvement by coarse proppant grain sizes in oil reservoirs are also discussed by LAMBERT, DOLAN & GALLUS (1983).

4.9.3. Economical feasibility

Examining the comparative effectiveness of natural sand and synthetic proppants in a variety of oil reservoirs in different depths and pressure regimes, economical analysis indicates frequently that the incremental present value obtained by fracturing can be ameliorated by using a high-performance proppant even though proppant cost is higher, because ceramic proppants offer significantly more permeability than the same size sand even at low closure stresses as encountered in most oil wells (NORMAN, CIPOLLA & WEBB 1983). Some aspects of synthetic proppants vs. natural sand as well as fracture design and plugging are briefly outlined as follows.

4.9.3.1. Synthetic proppants vs. natural sand

Simulations indicate that if a high-performance proppant is used instead of sand, greater recovery efficiencies are obtained in a shorter period of time and up to 25 % greater returns can be achieved. KOHLHAAS (1982) records in an intermediate-depth oil field return-on-investment ratios from the additional income of 5.6 : 1 and 23 : 1 when using natural sand and man-made bauxite proppants for plugging the fracture, respectively.

High-performance synthetic proppants are economically desirable for formation permeabilities of 1 - 10 md or more even at closure stress levels not exceeding 2,000 psi, and below 1 md reservoir permeability, the use of manufactured proppants can be economically justified at closure stresses of abt. 4,000 psi or more. As the application of intermediate- or high-strength proppants enables to plug longer, more feasible fractures and because of the much higher conductivity than sand even at low closure stresses (cf. section 1.3.2.), NORMAN, CIPOLLA & WEBB (1983) recommend to consider man-made proppants for use in oil reservoirs even below the crushing point of sand. KOHLHAAS (1982) emphasizes the profitability of such an investment not only in early-stage field production, but also in late-stage offtake with advanced reservoir depletion when still considerable quantities of oil can be drained through a high-conductivity fracture that would otherwise never come out of the pay horizon (cf. section 4.8.9.).

4.9.3.2. Fracture design and plugging

BRITT (1985) documents the economical benefits of short high-conductivity fractures even in moderate-permeability oil reservoirs undergoing both primary and secondary depletion. LI & ZHU (1986) present comparative performance evaluation of sand and bauxite proppants in moderate- to high-depth oil reservoirs. Field experience has shown that in various cases, fractures propped with lower concentrations of bauxite have better conductivities than those propped with higher saturations of sand. The use of high-strength proppants as a tail-in after the injection of sand as main proppant load gives little superiority in obtaining better fracture conductivity, with the optimum being provided by an uniform proppant package. Moderate- to high-permeability oil reservoirs can frequently already be successfully stimulated by microfracturing or minifracturing (CRAMER & SONGER 1988; cf. chapter 4.8.).

4.9.4. Steam-drive effects on fracture propagation

Steam-drive operation is a common means of enhanced oil recovery in heavy-oil reservoirs and has already been abundantly applied in many parts of the world. Steam-flooding results in heating of the pay zone that formerly belonged to the regional geothermical temperature regime up to an artificial level according to the temperature of steam or condensed hot water entering the injection well and migrating towards the production borehole. In many heavy-oil sand reservoirs, injection of significant amounts of steam often requires creation of fractures, and monitoring of initiation, propagation and collapse of fractures may provide important information about the distribution of injected fluids within the reservoir (SINGHAI & CARD 1988).

Apart from shallow-depth heavy-oil reservoirs, steam-flooding can also be successfully applied in intermediate- and high-depth light-oil pay zones where steam distillation is more important and gravity override is less significant (SARATHI, ROARK & STRYCKER 1988). Steam-drive is carried out in vertical, inclined and horizontal wells (JAIN & KHOSLA 1985, PETZET 1987). Tertiary thermal oil recovery frequently requires fracture-assisted steam-flood technology (SONI & HARMON 1986), because conventional steam-flooding in thin reservoirs generally suffers from excessive heat losses to the over- and underburden, and high-rate steam injection is often constrained by limited reservoir injectivity. Fracture-assisted steam-drive allows high rates of steam injection into shallow pay sands through hydraulically-created horizontal fractures which limits heat losses by shortening steamflood project life.

Heating of the reservoir formation associated with steam-drive operations can provoke a fundamental change of the in-situ stress state of the geological column, leading from a geostress regime favouring propagation of vertical fractures at the cold geothermical temperature level before steam injection onset to a geostress framework supporting creation of horizontal cracks at the hot temperature niveau that was artificially created by steam-flooding (HOLZHAUSEN, CARD, RAISBECK & DOBECKI 1985). Following an outline of the changing stress regime by thermal reservoir expansion, the concept of combination of horizontal and vertical fractures for steam-drive reservoir stimulation is illustrated. Comments are also offered on changes of fracture azimuth orientation.

4.9.4.1. Changing stress regime by thermal reservoir expansion

Basically two types of hot water and steam injection processes including fluid displacement are in use in enhanced oil recovery (BRITTON, MARTIN, LEIBRECHT & HARMON 1983). The first type includes drive or matrix flow process in which hot water or steam or some intermediate mixture is continuously injected into a reservoir at relatively low rates and pressures to heat and displace oil in a modified water flooding manner. This technique works satisfactorily if oil at natural pay conditions is sufficiently mobile to be moved at practical rates by hot fluid injection without pressure pasting the reservoir or uncontrolled fingering. The second type comprises conduction heating of the pay by hot fluid passing through a highly permeable zone such as a horizontal fracture, a gas cap at the top of the prospective section, or a relatively thin interval within or adjacent to the main reservoir complex such as a water horizon at the bottom of the deposit. Hydraulic fracturing for enhanced uptake of injected steam is particularly significant for the latter type of enhanced oil recovery operation.

The reservoir section adjacent to the highly permeable zone is heated by vertical conduction of heat from steam or hot water in the channel and also by convection of steam or hot water which may have leaked from the channel. When steam is injected into a pay containing viscous immobile fluids, steam tends to either channel and push its way through the immobile fluids towards the low-pressure sink, or steam flow stops momentarily and the pressure builds up until the formation parts so that fluid injection can continue (GESHELIN, GRABOWSKI & PEASE 1981; SETTARI & RAISBECK 1981). As such fractures seldom develop at the best location and with the most suitable orientation, it is helpful to hydraulically stimulate the wells prior to steam injection, with mainly horizontal fractures originating in the shallow reservoir depth of most heavy-oil sands (REYNOLDS, SCOTT, POPHAM & COFFER 1961; MARTIN, ALEXANDER, DEW & TYNAN 1972; DOSCHER, LABELLE, SAWATSKY & ZWICKY 1964; CLOSMANN 1971, SETTARI & RAISBECK 1981, CHHINA 1986; cf. section 1.2.8.). Other aspects of fracture preheating in thermal oil recovery are discussed by BRITTON, MARTIN, McDANIEL & WAHL (1981) and RESTINE (1983).

Low steam injectivity in many heavy-oil reservoirs necessitates fracturing of the formation for enhanced uptake in order to achieve commercial rates of steam injection, with energy distribution and oil depletion following the fracture trends (VITTORATOS 1986). The change in geostress state is caused by thermal expansion of the oil pay during steam injection. The following outline concentrates on stress inversion due to anisotropic reservoir expansion as well as on cyclic fracture collapse and opening. Comments are also offered on other types of changing stress direction.

4.9.4.1.1. Stress inversion due to anisotropic reservoir expansion

With progressive heating, some vertical expansion of the formation material should be possible in spite of restriction by overburden and lithostatic pressure, whereas horizontal enlargement may be suppressed by the confining effect of the surrounding rocks which are compacted and lithified by overburden stress and diagenetical cementation, respectively (HOLZHAUSEN, CARD, RAISBECK & DOBECKI 1985). Suppression of lateral expansion would give rise to an increase in the magnitude of horizontal in-situ stress components, leaving the vertical stress largely unaffected.

Therefore horizontal fractures start to originate when horizontal stresses exceed the vertical in-situ compression, with more than two weeks of steam injection being required for triggering this stress distribution modification. Injected steam and hot water may initially enter the formation through vertical fractures that have been hydraulically created prior to the onset of steam flooding, whereas with increasing reservoir temperature, the vertical fractures may close due to thermal expansion of the oil sands unless propped with suitable materials, and horizontal cracks open if enhanced oil recovery operation is performed above formation fracturing pressure which guide the injected steam and hot water away from the borehole into the formation.

Horizontal fracture generation during steam-drive operations often takes place in cycles comprising propagation and collapse (HOLZHAUSEN, CARD, RAISBECK & DOBECKI 1985). As fluid entry into the heavy-oil reservoir is restricted by the high viscosity of the pore fluid, pressure rapidly builds up with continuous steam injection after the beginning of the operation. Once a critical pressure is reached, formation breakdown results in propagation of a horizontal fracture away from the wellbore.

4.9.4.1.2. Cyclic fracture collapse and opening

The volume increase of the system due to the aforementioned processes causes fluid pressure to drop which in turn gives rise to flashing of some water in the wellbore into steam (VITTORATOS 1986). Aspects of water flashing, thermal reservoir expansion and plasticity changes, steam injectivity improvement by reservoir fracturing, interwell communication and fracture interference, and gravity drainage through the fractures are outlined as follows.

4.9.4.1.2.1. Water flashing

The pressure drop in the newly created fracture finally declines to a value that is too low to continue to lift the overburden, with the crack then collapsing and closing (HOLZHAUSEN, CARD, RAISBECK & DOBECKI 1985; VITTORATOS 1986). Some pressure loss before collapse may be the consequence of steam condensation after the fracture propagated into cooler portions of the oil sand. Rehealing of the crack after collapse provokes again restricted fluid entry, pressure buildup and fracture generation at the beginning of the next cycle. Tiltmeter monitoring reveals that interruptions of steam injectivity correspond to opening and closing episodes of fractures (SINGHAI & CARD 1988; cf. section 6.2.2.3.4. and 6.2.2.4.1.). Cyclic fracture opening and rehealing depending on steam injection pressure is also reported by ZIEGLER (1988).

Water flashing takes place if the pressure in the formation heated by steam injection falls below the steam saturation pressure (VITTORATOS 1986). This occurs in the vicinity of the stimulated wellbore as the pressure declines at the end of steam injection, and also further out in the reservoir at lower temperatures if the pressure drops sufficiently due to high fluid production rates.

Flashing often occurs in conjunction with convective flow which both give rise to regions of uniform reservoir temperature. GUDMUNDSSON, MENZIES & HORNE (1983) analyze steamtube relative permeability functions for flashing steam-water flow in fractures.

4.9.4.1.2.2. Thermal reservoir expansion and plasticity changes

The change of the geostress state by thermal expansion during steam-drive operation can also be explained by changing plasticity of the reservoir rock and decreasing matrix stress contrast between various lithotypes (CONSTANT & BOURGOYNE 1986). In geothermical temperature situations, the cohesive resistance or internal friction of a sandstone is greater than that of many mudstones, thereby permitting a greater difference between vertical and horizontal matrix stress to develop (cf. sections 4.2.2.3.1. and 4.2.3.2.2.). With rising temperature due to artificial heating by steam injection, the formations become more plastical (cf. sections 4.2.3.2.3. and 4.9.4.2.1.1.) and thus there are less differences between horizontal and vertical matrix stress. Comments on in-situ stress changes during steam-drive stimulation of oil reservoirs are also given by CLOSMANN & PHOCAS (1976), DUSSEAULT & SIMMONS (1982) and WONG (1986).

4.9.4.1.2.3. Steam injectivity improvement by reservoir fracturing

Cyclic steam stimulation effects in multi-well systems (instead of single-well systems as usually assumed and modelled) are discussed by VITTORATOS, SCOTT & BEATTIE (1988) involving considerable interwell communication or interference upon formation failing (cf. section 4.10.5.1.4.). Steam injectivity is generally achieved by reservoir fracturing (BUCKLES 1979, VITTORATOS 1986). There is essentially no water or steam injectivity unless formation failure occurs (VITTORATOS, SCOTT & BEATTIE 1988). Both vertical and horizontal fracturing take place at crack initiation, with horizontal fractures being more common near areas that have undergone intensive steam injection. Extensive interwell interactions are the result of pay failure that is required in order to achieve commercial levels of steam injectivity. Reservoir fracturing can extend to distances equal to that of well spacing or more and thus overlapping of the failed portions of the pay zone may occur. The disturbed intervals associated with the wells create conduits of enhanced pressure and fluid propagation between the boreholes, with the conduits embedded in a matrix with significantly slower pressure and fluid propagation.

4.9.4.1.2.4. Interwell communication and fracture interference

Inferences on the shape of the interwell communication region can be made from the heat to water breakthrough ratio, with cylindrical channels having values of 2 - 3, whereas fractures have values of 6 and more, because their larger surface for a given reservoir volume permits more heat to be transferred into the pay zone (VITTORATOS, SCOTT & BEATTIE 1988). Since fracturing is necessary for steam injectivity, interwell communication and crack interference can only be avoided by increasing well spacing, or by placing the production and injection wells in favourable geometrical patterns according to the geostress system (cf. section 4.10.5.1.). Another method is to lay out the system in such a way that the field is dissected into communication couplets between injection and production wells.

The highest recovery, however, is achieved with zero overlap of fractures extending from neighbouring boreholes, but in practice, the high interwell communication associated with no overlap leads to high well downtime because of sand-related pump failures. Therefore a 50 % overlap appears to be the optimum for spreading the heat between the wells, increasing off-trend movement of steam, retarding the tendency of the reservoir to form couplets, and minimizing operationally disruptive borehole interactions.

4.9.4.1.2.5. Gravity drainage through fractures

During fracture-assisted steam-flooding, the oil drains by gravity into the open horizontal fracture where it is rapidly transported as dispersion or entrainment to the production well via the crack, thus bypassing the relatively cold matrix between injector and offtake point (SONI & HARMON 1986). Growth of the steam-heated zone above the horizontal fracture is substantially more than that below (CLOSMANN & SMITH 1983), reflecting a horizontal crack which rises slowly by slumping of the unconsolidated sand above the fracture as a gravity-dominated heat transfer in which the movement of the upper boundary of the vertically growing steam zone is limited by counter-current fluid redistribution. As a consequence of this upwards growth, original installation of the horizontal crack by hydraulic fracturing should be performed low in the pay.

4.9.4.1.2.6. Other aspects

Comparable thermal expansion of the reservoir rock as in steam-drive operations is known from in-situ combustion enhanced oil recovery where partial closing of the natural fractures intersecting the pay zone matrix may occur due to formation heating and may ameliorate reservoir drainage and sweep efficiency (SCHULTE & VRIES 1985). Cyclic steaming of tar sands through hydraulically-induced fractures is discussed by DIETRICH (1983), and WATKINS, KALFAYAN & BLASER (1987) outline cyclic steam stimulation of tight clay-rich reservoirs. MILHEM & AHMED (1987) and SYLVESTER & CHEN (1988) report field experience of cyclic steam stimulation projects. Improvement of steam stimulation performance by chemical emulsion viscosity reduction is assessed by CURRENT (1988). Aspects of cyclic steam stimulation above fracture pressure are also discussed by ITO (1984), KENNEDY (1984); PETHRICK, SENNHAUSER & HARDING (1986), LIN (1988) and TORTIKE & ALI (1989).

As a consequence of the shallow reservoir depth of most heavy-oil sand deposits, predominantly horizontal fractures originate (REYNOLDS, SCOTT, POPHAM & COFFER 1961; DOSCHER, LABELLE, SAWATSKY & ZWICKY 1064; WAHL 1965, CLOSMANN 1971; MARTIN, ALEXANDER, DEW & TYNAN 1972; SETTARI & RAISBECK 1981, CHHINA 1986; cf. section 1.2.8.). Wells are notched and horizontally fractured and then stimulated with high-pressure steam in order to provide a heated target (BRITTON, MARTINS, LEIBRECHT & HARMON 1982). Horizontal fractures are then created from the injection well to the production boreholes, and steam is injected at very high rates and pressures to hold the fracture open, preheat the formation and mobilize the heavy oil.

4.9.4.1.3. Other possibilities of changing stress direction

Apart from steam-drive reservoir heating, expansion and fracturing, other possibilities exist to change the stress direction in the rock complex and to exploit this advantage of manipulation of main formation orientation for distinctive hydraulic fracturing. Some techniques are briefly illustrated as follows by commenting on altered-stress fracturing as well as structural control vs. stress control.

4.9.4.1.3.1. Altered-stress fracturing

WARPINSKI & BRANAGAN (1988) present altered stress fracturing which is also at least partially taking advantage of artificial changes of the geostress pattern. The purpose of altered-stress fracturing is to create hydraulic fractures propagating perpendicularly or obliquely to the natural crack system and across the pre-existing joints in tight gas sandstones instead of parallel to them (cf. section 4.8.8.1.3.), thereby opening up a much larger surface of the reservoir and connecting numerous natural fractures to the hydraulically-induced drainage path. Altered-stress fracturing is a concept whereby in most cases a hydraulic fracture in one well is reoriented by another hydraulic crack in a nearby location. While traditionally most efforts to control fracture height growth and azimuth propagation have been modifications of stimulation design and drilling pattern with emphasis on well spacing and configuration, altering the stress field to a more favourable orientation or magnitude is a suitable method particularly in older fields where borehole distribution is given and can be neither changed nor supplemented by renewed drilling. As significant stress perturbations are induced by any process that changes reservoir pressure or fractures the rock, a hydraulic crack is an effective mechanism for altering the stress field because of the high pressures needed for fracture dilation and the large surface area created.

In order to overcome coalignment of natural and hydraulic fractures, earlier approaches included propellant fracturing and directional drilling. Propellant fracturing or tailored pulse loading (CUDERMAN & NORTHROP 1984; cf. sections 4.2.2.1.3.3. and 4.8.8.5.) induces multiple radially oriented cracks, but fracture length that can be achieved is very limited. Directional high-angle or even horizontal drilling in lenticular sands intersects more natural joints if the hole is oriented perpendicular to natural fracture azimuth (cf. sections 4.8.6.1.5.1. and 4.8.6.2.4.2.), but if drilling damages the natural cracks or if too few natural fractures are encountered in a restricted sand body, conventional stimulations in a directionally drilled well are not more effective than in vertical boreholes. Altered-stress fracturing is discussed as follows for two-well and one-well applications. Some comments are also offered on stress contrast and leakoff control.

4.9.4.1.3.1.1. Two-well application

Altered-stress fracturing is an attempt to reorient the stress field in a region by conducting a hydraulic fracture nearby (SHUCK 1977). While the original concept of stress alteration using multiple wells is somewhat limited because of borehole spacing criteria, recent innovation allowed the technique to be also applicable in a single deviated hole (UHRI 1987), and some stress modifications can also be carried out in vertical wells (WARPINSKI & BRANAGAN 1988). An offset well is used to create a hydraulic fracture that alters the stress field around the borehole under discussion, with the offset well being fractured first and then being continued to be treated with high pressure while the production borehole is fractured.

If the stress difference is not too large, the wells are relatively close, and treatment pressure and fracture size in the offset well are sufficiently large, enough stress can be added to the virgin minimum horizontal in-situ stress to convert it to the maximum horizontal stress. Pumping has to be maintained in the offset well fracture, because stress perturbation will decay as pressure declines. If a treatment in the production borehole is started under these conditions, the hydraulic fracture in the production well will propagate perpendicular to the usual direction of propagation and extends a certain distance before it is turning into the original orientation because of weakening conversion pressure farther away from the pressurized fracture in the offset well. Thus altered-stress fracturing enables to create an interlocking network of orthogonally-oriented induced crack systems for optimum intersection and drainage of the reservoir.

As stress perturbation decays rapidly with distance in both horizontal and vertical direction representing fracture length and height, the two wells should be as close together as possible, and also the difference between the two horizontal stresses must not be large, because stress perturbation is proportional to pressure in the offset well fracture which has operational limits. High treatment pressures in the offset well can be obtained by reservoir features such as good stress contrasts or by induced features such as screenouts. Stress contrasts are also important because fracture width in high-stress regions will be small so that less strain will be induced and the effect of extra height growth will be less than in a homogeneous formation. High stresses in surrounding mudstones reduce the magnitude of the stress change, with the effect increasing with crack size, while low-stress layers of course enhance the induced stress change. Lower and higher anisotropy ratios reduce and increase, respectively, the advantage of the altered-stress fracture.

4.9.4.1.3.1.2. One-well application

As applicability of altered-stress fracturing in separate wells is limited, emphasis has to be put on carrying out the operation in a single directionally drilled borehole (UHRI 1987). The first treatment is conducted down the annulus while the second job is performed down the tubing, resulting in much closer distances than could be obtained in multiple wells and therefore profiting from less pressure decline. In addition, if the first operation to alter the stress is initiated in a high-stress muddy barrier (cf. sections 4.2.2.3.1. and 4.2.3.2.2.), high treating pressures and large height are facilitated so that much greater stress changes occur. Altered-stress fracturing cannot be used to reverse the horizontal stress in a vertical well when overburden stress is also vertical, but if dipping strata, buried topography (cf. plate I/1 - 7) or other effects are misaligned with the normal direction, altered-stress fracturing gives feasible results (WARPINSKI & BRANAGAN 1988).

If carried out in one well, however, it may not be possible to guarantee sufficient pressure in the offset fracture to reverse the stress field, and also the complexity of the operations can be a problem. Unusual fracture designs for such cases are tip screenout (cf. section 4.5.4.4.1.) for the first fracture which ensures high pressures as well as making maintenance of those easy by allowing for relatively slow pumping of only liquids after the tip screenout has formed. It may also be possible to induce a tip screenout and pack enough proppants into the fracture so that no additional pumping is needed, with the propped fracture width and the strain it produces being equivalent to the pressure required to create that width. As stress perturbation at some distance from a fracture is independent of modulus, the compliance of natural fractures is not attenuating the decay distance.

4.9.4.1.3.1.3. Stress contrast

Another possible application of the stress-modification concept is for alteration of the vertical distribution of the minimum horizontal in-situ stress, because a hydraulic fracture not only creates a large compressive stress in the layer it resides, but also induces large tensile stresses above and below which could be used to advantage if hydraulic cracks are propagating into undesirable zones such as aquifers (WARPINSKI & BRANAGAN 1988). Changing of insufficiently low to satisfactorily high stress contrast can be performed in a vertical well, a deviated well or two adjacent wells. Altered-stress fracturing does not change fracture orientation from vertical to horizontal as provoked by steam flooding of the reservoir and resulting pay expansion due to heating, but rather gives rise to changes of crack azimuth in a direction running perpendicularly to that determined by the undistributed geostress field. The effect of altered-stress fracturing can be even accentuated by staggered steam injection which combines geostress pattern disturbance by pressure impact from a hydraulic fracture and from reservoir expansion as a consequence of increasing pay temperature in steam-drive operations (cf. section 4.9.4.3.2.).

If an altered-stress concept is used to reorient the fracture, the initial breakdown technique in the production well should be carefully considered. Normal breakdown or ball-out (cf. section 4.4.2.3.) initiates hydraulic fractures in the usual orientation, and afterwards attempts to start an altered-stress crack may be difficult and stress perturbations may be mitigated. It is therefore better to break down under reversed stress conditions so that the fracture is oriented correctly from the start. An alternate breakdown technique is using propellant fracturing (CUDERMAN & NORTHROP 1984) through perforations with at least 90 degrees phasing, with cracks initiating out of each set of perforations giving reasonable initiation sites for the follow-on treatment (WARPINSKI & BRANAGAN 1988).

4.9.4.3.1.4. Leakoff control

Another important consideration for altered-stress operations is leakoff control. If the hydraulic fracture is altered to cross the most permeable natural cracks rather than running parallel to them, then fluid leakoff may be much greater than normally found in the reservoir. Height containment conditions may also suffer particularly if maximum and minimum stress have different vertical distribution, which occurs if viscoelastic shales are present (WARPINSKI & TEUFEL 1987) and if tectonic or other strains are induced differently in the two principal directions. Accurate fracture height estimations can be made if stress measurements are carried out in an offset well which enables to relatively easily determine the beginning of significant fracture height growth.

Another possibility to create vertical fractures in shallow reservoir storeys where usually horizontal cracks originate is limited-entry stimulation (cf. section 4.2.2.2.) of multiple layers until proppant abrasion of perforation tunnels results in their enlargement and consequently pressure drop followed by screenout of various perforations, as well as leading to initiation of a vertical fracture after termination of propagation of the earlier horizontal crack (SCHRAUFNAGEL & LAMBERT 1988). The change of orientation from an original horizontal fracture to a subsequently vertical crack due to pressure rise at screenout stage is particularly pronounced when proppant bridging and sticking leads to considerably increasing treatment pressure early during proppant injection (HANSON 1988).

4.9.4.1.3.2. Structural control vs. stress control

McLENNAN, ROEGIERS & MARX (1983) even conclude that there is little argument that hydraulic fracturing is a stress-controlled phenomenon wherein the crack tends to open perpendicularly to the smallest principal stress. In addition to stress government, structural supervision in the tectonical framework of the oil and/or gas deposit exerts a modifying influence on fracture orientation. Therefore significant anisotropy due to layering and often consequently higher horizontal permeability can enhance the likelihood of horizontal fracture generation at slightly greater depths (KOMAR & FROHNE 1973, SUN & MONGAN 1974). The structural control is most important when vertical stress and minimum horizontal stress are approximately equal (McLENNAN, ROEGIERS & MARX 1983).

This constellation frequently results in development of a zone with horizontal fractures followed by a transitional region where horizontal and vertical fractures may coexist (or may be generated vice versa) that is grading into an interval with exclusively vertical cracks in greater depth (cf. section 1.2.8.). In considerably tectonically deformed areas, fracture propagation trend may parallel the faulting trend, with either the faults themselves controlling crack extension direction or the regional stress field being dominant. Under some circumstances, it may be feasible to alter in-situ stress contrasts by pore-pressure manipulation, fluid viscosity and rate control, perforation placement, and prenotching to provoke generation of vertical fractures instead of horizontal ones in shallower depth storeys (CLEARY 1980 b; McLENNAN, ROEGIERS & MARX 1983).

4.9.4.2. Combination of horizontal and vertical fracturing

Steam-drive enhanced oil recovery can be considerably supported by utilizing the changing geostress regime (cf. section 4.9.4.1.) for an extraordinary design of special hydraulic stimulation treatments comprising a combination of vertical and horizontal fractures which would be impossible in other reservoirs not subjected to artificial heating and thus controlled thermal expansion. Aspects of creation, management and determination of horizontal and vertical fractures are briefly outlined as follows.

4.9.4.2.1. Fracture creation

Creation of a vertical fracture by hydraulic stimulation prior to the onset of steam injection provides a base for improved entry of steam into the heavy-oil-bearing pay zone. Aspects of thermal reservoir expansion and increasing closure stress, horizontal fractures and horizontal vs. vertical boreholes, multiple horizontal and vertical fracture propagation, and combination of hydraulic and explosive fracturing are outlined as follows.

4.9.4.2.1.1. Thermal reservoir expansion and increasing closure stress

If a vertical crack is generated hydraulically before the start of steam flooding, caution has to be exercised for adequate propping of the vertical fracture, because thermal expansion of the reservoir due to heating by the steam gives rise to increasing closure stresses as a consequence of an additional pressure component coming from volume enlargement of the rock matrix. Proppant choice therefore has to respect the necessity of higher resistivity than according to depth and formation pressure to avoid collapse of the crack by proppant failure due to thermal expansion stress, and also has to be made according to the subsequent high-temperature environment in terms of thermal stability of the material supporting the fracture walls, and resistivity against aggressive alcaline solutions (similarly as applying for steam-injection well gravel packing; cf. section 5.7.).

Another aspect which has to be taken into consideration is increasing plasticity of the reservoir rock with rising temperature (cf. section 4.2.3.2.3. and 4.9.4.1.2.2.) which provokes increasing embedment of the proppants, and thus multilayer propping of the crack has to be planned (cf. section 4.3.3.1.). Concerning proppant type, all the mentioned adverse conditions in the high-temperature high-alcalinity steam-injection wells can only be satisfactorily met by sintered bauxite and nickel-coated sand (cf. section 5.7.).

4.9.4.2.1.2. Horizontal fractures and horizontal vs. vertical wells

Horizontal fractures intersecting vertical wells suffer considerably from the very limited communication interval between wellbore and fracture plane in a similar way as applying for vertical cracks crossing highly-deviated or horizontal boreholes (cf. section 4.11.1.2.). The only difference in thermal enhanced oil recovery projects is the steam-induced combination with vertical fractures that gives rise to a crossing pattern of two fractures perpendicular to each other and representing a special case which cannot be achieved by either horizontal cracks intersecting horizontal boreholes or vertical fractures cutting vertical wellbores.

The effectivity of horizontal fractures generated by stress regime inversion due to steam flooding (cf. section 4.9.4.1.), however, can be considerably improved with horizontal steam-injection wells where the horizontal fractures have a long communication interval with the wellbore comparable to that of vertical fractures intersecting vertical boreholes. Thus horizontal wells in steam-flood operations provide the unique opportunity to increase the direct contact between wellbore and pay zone, between wellbore and fracture, and between fracture and pay zone, with this combination being the optimum solution and leading to maximum hydrocarbon recovery from the reservoir. Other impacts of horizontal wells on steam-drive processes are outlined by BUTLER & STEPHENS (1980), RIAL (1984), GOODE & THAMBYNAYAGAM (1985) and HUANG & HIGHT (1986; cf. also section 4.8.6.).

Field examples of hydraulic creation of a vertical fracture prior to steam injection and subsequent formation of a horizontal crack as a result of steam stimulation are reported by SINGHAI & CARD (1988). Aspects of fracturing of heavy-oil sand reservoirs during course of steam injection are also discussed by BABCOCK (1975); GESHELIN, GRABOWSKI & PEASE (1981); GESHELIN & GRABOWSKI (1982), ABAD & HENSLEY (1984) and PEAKE (1984). It has to be emphasized that only the combined effect with the vertical cracks renders the horizontal fractures sufficiently effective, because they alone would be hampered in performance by limited cross-section with the borehole in a similar way as applying for vertical cracks crossing highly-deviated or horizontal boreholes (cf. section 4.11.1.2.). The very small communication interval between perforated zone and fracture plane has a suite of negative impacts on treatment performance and effect which, however, in the special case of combined design are compensated by the coexisting vertical fracture.

4.9.4.2.1.3. Multiple horizontal and vertical fracture propagation

Developing interwell communication between injectors and producers in early steam-flood operation is very important for effective enhanced oil recovery (KULAR, CHHINA, BEST & MACKENZIE 1988). For more effective reservoir heating, the preferred interwell communication path is either low or in the middle of the pay zone, with fracture geometry playing an important role. Principles and significance of multiple horizontal and vertical fracture propagation are outlined as follows.

4.9.4.2.1.3.1. Principles

Achieving horizontal fractures at depth where in-situ stress is in near-isotropic state is uncertain because of pore-pressure increase due to fluid leakoff, thermal expansion and fracture loading which may alter the stress field, thereby inducing a change in fracture orientation or resulting in multidirectional cracks (KULAR, CHHINA, BEST & MACKENZIE 1988). Heat injection and various fracture loadings induce changes of in-situ stress regime. Crack loading which is defined as difference between fracture extension pressure and formation closure stress has significant influences on changes of in-situ stress (CHHINA & AGAR 1985, WONG 1986). In case of smaller fracture loadings, horizontal cracks may stay horizontal and may only be moderately inclined, whereas for greater fracture loadings, an initially horizontal crack may progress into multiple fractures with varying orientations which is caused by stress reversals occurring in a near-isotropic stress environment. Qualitatively high crack loadings can induce changes in the horizontal to vertical stress ratios to favour multiple fracturing.

While immediately adjacent to the fracture the initial total stress ratio does not change significantly because both horizontal and vertical total stress increase, some distance away from the crack face the stress ratio has changed, with a reversal of the minimum stress direction having occurred from originally vertical to now horizontal. If a propagating horizontal fracture or high fluid pressure reaches the vertical fracture zone or stress reversal interval, a conjugate vertical crack starts propagating. Initiation and extension of this secondary vertical fracture depends on sand failure mechanism and pore pressure distribution off the crack face (aspects of composed T-shaped or mushroom fracture complexes are also discussed in sections 1.2.8.4., 4.4.3.4.3., 4.9.4.2.3.2. and 6.2.2.4.4.). Pore pressure can increase in the formation due to leakoff or due to fracture loading under undrained conditions. As oil sands are characterized by high fluid loss and low tensile strength, hydraulic fractures propagate by both shear and tensile failure (DUSSEAULT 1985), with shear failure generally occurring prior to tensile failure.

4.9.4.2.1.3.2. Significance

Shear failure region, high fluid leakoff, or pore pressure generation due to undrained fracture loading are some of the mechanisms by which the horizontal crack could penetrate into the vertical fracture zone past the buffer zone where the initial stress ratio did not change. Pressure drop is primarily affected by stimulation fluid vicosity and only secondarily by fluid injection rate (KULAR, CHHINA, BEST & MACKENZIE 1988). For unconsolidated formations, treatment fluid viscosity is a critical fracture design parameter due to its influence on crack loading which in turn dictates fracture orientation. Apparent slurry viscosity tends to increase crack loading as proppant injection is initiated.

Multiple horizontal fractures emanating in different reservoir levels and from different wells which are interlinked with a conjugate vertical crack are efficient means of pay zone and borehole connection through a network of fractures of different orientation. This geometrical combinatory constellation provides an important possibility of building bridges between various reservoir segments in horizontal and vertical direction and thus much more efficient pay complex intersection than could be achieved by horizontal or vertical fractures alone.

4.9.4.2.1.4. Combination of hydraulic and explosive fracturing

The secondary high-temperature environment of steam-drive projects in primary low-temperature reservoirs and the associated change in geostress state which leads to inversion of fracture propagation orientation from originally vertical to subsequently horizontal also enables profitable combination of hydraulic and explosive fracturing. While the cracks generated by explosive detonation in the borehole are usually propagating in vertical direction, the primary geostress state in many shallow oil reservoirs and the secondary geostress state in steam-flood operations trigger hydraulically-generated cracks to penetrate the pay interval in horizontal orientation. This relationship permits in both mentioned cases a suitable additional permeability improvement of the reservoir by horizontal hydraulic fracturing superimposed by vertical explosive stimulation, enabling enhanced oil offtake in production wells (cf. section 4.8.9.6.) and ameliorated steam uptake capacity in thermal injection boreholes.

4.9.4.2.2. Fracture management

Once a suitably equipped vertical hydraulic fracture has been installed within the reservoir, steam injection at high rates will first lead to fluid entry preferentially via the vertical crack. Increasing injection rates exceeding uptake capacity of the drainage path will then give rise to pressure buildup, and achievement of a critical pressure level results in generation of horizontal fractures intersecting the earlier vertical cracks. Continuous steam injection with sufficiently high rates keeps the horizontal fracture open and thus allows fluid entry through two perpendicular crack systems. As proppant insertion into the horizontal fractures is not possible in the hot high-pressure steam-drive environment which also must not be interrupted or disturbed in order to achieve an optimum effect of oil mobilization, the horizontal cracks can only be kept open by maintaining injection pressure at a sufficiently elevated niveau, or cyclical generation and collapse of horizontal fractures according to the model of HOLZHAUSEN, CARD, RAISBECK & DOBECKI (1985) has to be accepted.

In very shallow reservoir depth above 3,000 ft (1,000 m), an inverse procedure is possible with first hydraulic generation of a horizontal fracture which is stable in this low-overburden environment, and second either enlarging the existing fracture by steam pressure or creation of new cracks running preferentially parallel to the older fracture. Vertical cracks would in such shallow depth only originate when the formation is split into sealed fault blocks or the reservoir sand is present as discontinuous lenses and channels pinching out within thicker mudstone complexes, with thus lateral expansion of the pay zone upon heating being considerably restricted and in some cases according to the shallow depth being even more limited than vertical expansion. In addition, perforation friction occurs due to limited communication between wellbore and a horizontal fracture in shallow depth, and this effect can cause a small vertical fracture to open which breaks down the wellbore, but then soon turns into the proper direction. Aspects of fracture-assisted steam-flood technology are also discussed by BRITTON, MARTIN, LEIBRECHT & HARMON (1983) and SONI & HARMON (1986).

4.9.4.2.3. Fracture determination

The method most commonly used to determine whether a hydraulic fracture is vertical or horizontal is comparison of the pressure required to propagate the fracture (which is represented by the instantaneous shut-in pressure) to the vertical in-situ stress at the injection depth (HOLZHAUSEN, CARD, RAISBECK & DOBECKI 1985). If the pressure is less than the vertical stress, the fracture cannot be horizontal because the pressure is insufficient to lift the overburden. The assumption that the vertical stress is a principal stress leads to conclusion that the fracture is vertical, because it should form in a plane perpendicular to the least compressive principal stress which must be horizontal.

On the other hand, if the pressure in the fracture exceeds the vertical stress, the crack plane is taken to be horizontal under the assumption that the vertical stress is the least compressive principal stress. Thus in general only either vertical or horizontal fractures can originate in a given reservoir, and the generation of both types of cracks one after each other to finally coexist in the formation and make up two perpendicular intersecting drainage paths is not possible unless special conditions are present, with the most important means of changing geostress regime being thermal expansion of the rock column as discussed above. Other methods of discrimination between horizontal and vertical cracks are tiltmeter monitoring (cf. sections 6.2.2.3.4. and 6.2.2.4.1.) and temperature profiling (VITTORATOS 1986). Some aspects of fracture gradient vs. overburden stress gradient as well as reservoir expansion and changing in-situ stress are illustrated as follows.

4.9.4.2.3.1. Fracture gradient vs. overburden stress gradient

In terms of crack determination in shallow-depth reservoirs, fracture gradients exceeding the overburden stress gradient may reflect vertical cracks, whereas fracture gradients approximately equivalent to the overburden stress gradient are interpreted as indicating creation of horizontal cracks (STOLLER 1985). Temperature profile monitoring is a suitable instrument to determine fracture shape and size, thermal diffusivity of the formation, and fluid flow velocities in convective regions (VITTORATOS 1986). Tiltmeter survey can distinguish vertical and horizontal fractures which produce a radial symmetrical uplift of the ground and a trough at the surface parallel to fracture strike, respectively (HOLZHAUSEN, CARD, RAISBECK & DOBECKI 1985; cf. sections 6.2.2.3.4. and 6.2.2.4.1.).

4.9.4.2.3.2. Reservoir expansion and changing in-situ stress

Steam-drive operations in heavy-oil fields thus provide the unique possibility of reservoir stimulation with a combination of horizontal and vertical fractures as a consequence of the changing in-situ stress regime by thermal pay expansion which would be impossible in other formations that are not subjected to temperature-related enhanced oil recovery. Aspects of horizontal fractures in shallow heavy-oil reservoirs stimulated by steam flooding are also discussed by CLOSMANN & PHOCAS (1976); REDDY, PROWSE & REDFORD (1980); CLOSMANN & SMITH (1983) and VITTORATOS (1986), and DUSSEAULT & SIMMONS (1982) analyze injection-induced stress and fracture orientation changes.

CLOSMANN & PHOCAS (1976) conclude that favourable conditions for horizontal fractures can be created by heating along two parallel vertical cracks. DUSSEAULT & SIMMONS (1982) document that pore pressure has more significant effects than temperature on inducing stress changes, and that the stress field is altered enough from fluid injection alone to allow horizontal fractures to be created. CLEARY, CROCKETT, MARTINEZ, NARENDRAN & SLUTSKY (1983) put forward that fluid leakoff increases the pore pressure around the crack and that the induced backstresses act to close the fracture subsequently. CROCKETT, VO & CLEARY (1984) show that thermally-induced stresses are small compared with those generated by fluid loss over the time scale of typical hydraulic fracturing in tight reservoirs.

Combinations of vertical and horizontal fractures originate also in reservoirs with normal geotemperature without influence of artificial formation heating by opening of a primary vertical crack which secondarily turns into a horizontal fracture in shallow depth that breaks out at the top of the vertical crack (WOOD, PARKIN, YOTAM, HANSON, SMITH, ABBOTT, COX & O`SHEA 1983; cf. section 6.2.3.). T-shaped fractures consisting of vertical and horizontal components also originate in coal seams that are bounded by shales due to abnormally high treating pressures (JEU, LOGAN & McBANE 1988; cf. section 4.4.3.8.). Other aspects of composed mushroom-shaped fracture assemblages are discussed in sections 1.2.8.4., 4.4.3.4.3., 4.9.4.2.1.3. and 6.2.2.4.4.

4.9.4.3. Changes of fracture azimuth orientation

In-situ stress changes during steam stimulation of heavy-oil sand reservoirs do not only give rise to generation of horizontal fractures in original geostress regimes favouring formation of vertical cracks without thermal influence, but allow also changes of vertical fracture orientation with respect to the azimuth of the crack preferentially in deeper reservoir storeys where horizontal fractures are difficult to create also under artificially elevated pay zone temperatures and pressures (WONG 1986). This has important consequences for planning of borehole pattern and spacing as well as utilization of fracture interference and creation of cracks perpendicularly to each other for enhanced hydrocarbon exploitation. The outline as follows focusses on on-trend and off-trend well patterns, staggered steam injection into off-trend well pairs, and mechanisms of fracture azimuth orientation changes.

4.9.4.3.1. On-trend and off-trend well patterns

Analyzing two wells, distinction can be made between on-trend pattern when both boreholes are aligned along the fracture direction according to the undisturbed geostress field, and off-trend arrangements when both wells are placed in such a way that the connection line runs perpendicularly to the crack direction determined by the inherent geostress distribution (WONG 1986). In terms of steam stimulation, discrimination can be made between simultaneous and staggered steam injection of two neighbouring boreholes. If two on-trend wells are steamed simultaneously or in a staggered manner, the fractures which are created are oriented initially in the natural crack direction where they do not interact with each other before they start to grow considerably in length, with the amount and rapidity of interaction by interference and connection depending on well spacing and injection pressure. Continued growth of the fractures in on-trend direction with progressive steam injection occurs because the preferred crack orientation ahead of the fracture does not change.

In contrast to the on-trend case, the injection sequence has significant impact on the orientations of fractures in two off-trend wells. Simultaneous steam injection into two off-trend boreholes results again in crack orientation and growth in the virginal fracture direction as long as they do not interact with each other, and once a fracture length has been reached to allow for connection and interference, the cracks also continue to propagate in the original fracture orientation.

4.9.4.3.2. Staggered steam injection into off-trend well pairs

Remarkable changes in fracture orientation, however, are provoked by staggered steam injection into two off-trend wells (WONG 1986). If steam stimulation is staggered in such a way that injection at the first borehole has generated a crack of length in the order of well spacing, then injection at the second well causes a fracture growing perpendicularly to the first one and propagating in off-trend direction as a consequence of significant changes in the principal orientations along the off-trend direction from a crack.

Staggered steam injection of borehole pairs placed in an off-trend pattern is therefore a possibility to make use of in-situ stress distribution changes by reservoir heating in order to create fabrics of two fracture systems running perpendicularly to each other which allow to drain the pay zone in much wider extent than could be done by a number of parallel fractures, because additional quantities of hydrocarbons can be accessed with fewer wells by normal connections of parallel fractures and generation of orthogonally linked or approaching cracks instead of a suite of parallel fractures. Staggered steam injection of off-trend well pairs thus represents a special method of altered-stress fracturing by being a combination between geostress distribution disturbance and transformation by the first hydraulic fracture and the pressure necessary to generate it and to keep it open (cf. section 4.9.4.1.3.1.), and the additional impact of reservoir heating by steam flooding which provokes changes of geostress pattern due to pay expansion.

4.9.4.3.3. Mechanisms of fracture azimuth orientation changes

Steam stimulation therefore effects significant changes in magnitudes and orientations of stresses (WONG 1986). Steam injection basically increases the in-situ stresses, while the stress alterations are recovered mainly early in the production cycle. At the end of an injection cycle, most of the stress changes are around the middle of the fracture in the off-trend direction, whereas ahead of the crack, only small alterations in the principal orientations occur. Fracture loading is the dominant mechanism resulting in changes of both magnitude and direction of stress, because the magnitude of fracture loading is relatively high compared with the average pressure/temperature alteration. Pressure and temperature changes simply alter the stress magnitudes, but leave the principal stress orientations almost unaffected.

The maximum alterations take place around the tip of the disturbed zones of pressure/temperature changes where the largest gradients occur. At the end of production cycles, the in-situ stresses return essentially to their initial states as a consequence of linear-elastic behaviour and small average pressure/temperature changes, but residual stresses and deformations are expected once nonlinear behaviour is taken into account. From the analysis of stress alterations during cyclic steam stimulation, fracture orientation can be manipulated to a certain extent by choosing an appropriate combination of well layout and operating schedule, thereby optimizing hydrocarbon production. Stress changes induced by pore fluid pressure and/or temperature changes are also investigated by CLOSMANN & PHOCAS (1976), DUSSEAULT & SIMMONS (1982); CLEARY, CROCKETT, MARTINEZ, NARENDRAN & SLUTSKY (1983); CROCKETT, VO & CLEARY (1984) and PERKINS & GONZALEZ (1984, 1985).

An alternative technology to steam-drive mobilization of heavy oil is mining of the high-viscosity deposits (HARDING & ALI 1984; LYMAN, PIPER & RIDDELL 1984) which, however, is frequently of very limited economical feasibility in comparison to thermal enhanced oil recovery. The effect of oil mining can be enhanced by horizontal wellbores connecting vertical shafts and partially replacing horizontal galleries or to create orthogonal communication tubes between parallel galleries (BEST, CORDELL & HASTON 1987, ERDÖL-ERDGAS-KOHLE 1988 b, MOORE 1988 c).

4.9.5. Cold water injection effects on hydraulic fracturing

Apart from steam drive comprising tertiary-recovery hot steam and water injection, also cold water injection by conventional secondary-recovery water flooding into oil reservoirs can have significant effects on hydraulic fracturing by modifying or changing the initial in-situ stress distribution in the pay interval (PERKINS & GONZALEZ 1985). For typical water flooding of a moderately deep oil reservoir, horizontal earth stresses may be reduced by several hundred psi. If the injection pressure is too high or if suspended solids in the water plug the formation face at the perforations, the pay will be fractured hydraulically. As the crack grows, the flow system evolves from an essentially circular geometry in the plan view to one characterized more nearly as elliptical. Aspects of in-situ stress and fracture pressure reduction, reservoir pressure and fracture gradient reduction, formation parting and stress reduction at fracture tip, primary and secondary fractures; and reservoir pressure, injectivity and fracture length are outlined as follows.

4.9.5.1. In-situ stress and fracture pressure reduction

Temperatures of injected fluids are typically lower than in-situ reservoir temperatures, and therefore a region of cooled rock forms around each injection well and grows as additional fluid is injected (PERKINS & GONZALEZ 1985). Formation rock within the cooled region contracts, and this leads to decreasing horizontal earth stress near the injection borehole. In addition, waters normally available for injection contain suspended solids, and consequently plugging of formation rock occurs thereby possibly severely limiting injectivity. In field operations, injection often simply continues at a bottomhole pressure that is high enough to initiate and extend hydraulic fractures which then enable the injected fluid to leakoff readily through the large crack face area (cf. section 4.8.7.1.2.). Because of the lowering of horizontal earth stresses that result from cold fluid injection, hydraulic fracturing pressures can be much lower than would be expected for an ordinary low-leakoff hydraulic stimulation treatment.

In terms of initial fracture orientation, moderately deep formations are typically in a state of triaxial compression with the largest principal stress oriented vertically, with therefore vertical hydraulic cracks tending to initiate and propagate perpendicularly to the minimum horizontal stress (cf. section 1.2.8.). When cold fluid is injected into an unfractured well, the region of cooled rock might be considered radially symmetrical with respect to the borehole. Superimposition of the thermoelastic stress thus will decrease the maximum and minimum horizontal stresses by an equal amount. Although the well can be fractured hydraulically at a reduced pressure, there will be no effect of a radially symmetrical cooled region on the initial orientation of an induced crack.

4.9.5.2. Reservoir temperature and fracture gradient reduction

Considerable reservoir temperature reductions in waterflood operations lead to significant decline of fracture gradient and requires diminution of injection pressures to avoid accidental formation parting (GARON, LIN & DUNAYEVSKY 1988). Thermally induced stress reduction and fracturing facilitation is of considerable importance for proper waterflood design and modelling, because changes in reservoir temperature and pore pressure fields resulting from cold water injection can cause changes in in-situ stress levels. Decreasing reservoir temperature results in to reduction of compressive stress, while rising pore pressure increases in-situ stress. Uncontrolled vertical and lateral fracture growth could seriously impact vertical and areal sweep efficiencies, whereas reduction of injection pressures could provoke severe underinjection in many waterflood patterns. Reservoir behaviour associated with cold water injection into a hot formation are very complex because of the coupled nature of the mechanisms involved. Water injection cools the pay by conduction and convection. Prior to fracturing, flow distribution is controlled by reservoir characteristics, and temperature field is governed by flow pattern. The in-situ stress distribution is subsequently determined by temperature and pressure fields, and fracture gradient may be decreased.

4.9.5.3. Formation parting and stress reduction at fracture tip

At some injection pressure, formation parting occurs. Flow pattern is obviously strongly influenced by any fractures which develop and this controls the subsequent temperature and stress distributions. Stress diminution near the crack tips results in fracture elongation and further stress reduction, with this coupling effect being able to give rise to rapid fracture extension (PERKINS & GONZALEZ 1985). The maximum level of stress reduction increases and the region of stress distribution expands as the cooling region grows with continuing water injection. The maximum stress decline occurs at some distance from the wellbore because of the combination of thermal and pore-pressure effects.

Simulation of fracture opening and continuous crack propagation can be performed by utilization of a pressure-dependent permeability multiplier which below fracture pressure has a constant value of one and above crack pressure equals the assigned fracture permeability divided by the product of original reservoir permeability and relative permeability to water and residual oil saturation. Vertical permeability barriers often have a higher undisturbed minimum horizontal stress than the adjacent pay sandstones and are less affected by cooling, thereby providing a relative barrier to inhibit vertical fracture growth. Other aspects of thermally induced waterflood fracturing are also discussed by PERKINS & GONZALEZ (1984); MORALES, ABOU-SAYED, JONES & AL-SAFFAR (1986) and WILLIAMS, SHERRARD & LIN (1987).

4.9.5.4. Primary and secondary fractures

Different injection conditions can be distinguished. If the injection rate exceeds slightly the ability of the unfractured formation to accept the fluid, a short crack extends from the wellbore, and the surrounding region of cooled rock is nearly circular in shape and only slightly elliptical, with therefore the thermoelastic reduction of horizontal stresses being nearly uniform in all directions (PERKINS & GONZALEZ 1985). If the injection rate is much larger or some fracture-face damage has occurred, the crack extends a greater distance from the well, and the cooled region becomes more elongated in shape. As the cooled area elongates, the thermoelastic reduction in stress parallel to the fracture exceeds the thermoelastic stress diminution perpendicularly to the crack which tends to decrease the difference between stresses within the cooled region that are parallel and perpendicular to the fracture, respectively.

If continued injection causes the fracture to extend further, a flatter shape of the cooled area causes the stresses parallel to the crack to become less than those across the fracture, and thus minor cracks or joints perpendicular to the main fracture would open preferentially. The shape of the jointed region then presumably adjusts itself continually to maintain open fractures in both directions as the size of the cooled area expanded. The opening of secondary cracks perpendicular to the main fracture system is also known from geothermal well stimulation where also cold water is injected into a hot reservoir (BAZANT 1979, BARR 1980; cf. section 4.7.).

4.9.5.5. Reservoir pressure, injectivity and fracture length

Fracture extension pressures are a function of reservoir pressure, permeability and size of the region cooled by injection water. Low pay pressure enables the fluid to leak off through a relatively short crack, whereas higher reservoir pressure requires a longer fracture to accommodate fluid at the injected rate. Injection of cool water can reduce earth stress around injection wells substantially, causing them to fracture at pressures considerably lower than would be expected in the absence of the thermoelastic effect. Depending on the shape of the cool region and the difference between minimum and maximum in-situ horizontal earth stresses, minor fractures perpendicular to the main crack open thus creating a jointed fracture system consisting of primary and secondary fractures. Short cracks resulting from low injection rate, high permeability or low reservoir pressure do not significantly change the shape of the waterflood front from the circular shape that would be expected for an unfractured well, whereas longer cracks deriving from high injection rates, low permeability, high pay pressure or poor water quality cause the waterflooded region to become noticeably elliptical in shape.

4.10. Communication fracturing

Conventional hydraulic proppant fracturing is confined to one singular hydrocarbon-bearing storey in the prospective column. When several stimulation jobs have been carried out so far in one well in Europe, there was generally a clear temporal succession of first performing the treatment in the lowermost sand, afterwards testing and/or producing it until depletion in case of success of the stimulation job or up to declaration as a failure, then partially plugging back the borehole by cementing, sanding-up and/or setting bridges sealing off the lower horizon which is to be abandoned, and subsequently fracturing separately the next potential sandstone higher up in the section. Following an outline of conventional storey-wise fracturing in discrete time steps, some aspects of application of simultaneous communication fracturing in production and appraisal drilling as well as fracture interference are discussed. Comments are also offered on influence of perforation arrangement.

4.10.1. Conventional storey-wise fracturing

Conventional hydraulic fracturing by storey-wise treatment of vertically successive separate reservoir subunits and intervals in discrete time steps has been applied very widespread in Europe and other parts of the world during the last decades. Starting at the bottom of the borehole and climbing up the staircase of pay zones requires sealing off of the inferior unit and perforating of the superior unit after plugging of the borehole floor in each stimulation step. Some aspects of repeated stepwise fracturing in Rotliegend and Carboniferous of Germany FRG, commingling of separate layers after individual fracturing, multilayer reservoirs with and without crossflow, and connection of separate layers by joint fracturing are illustrated as follows.

4.10.1.1. Repeated stepwise fracturing in Rotliegend and Carboniferous of Germany FRG

Examples of such repeated stepwise fracturing with following exploitation in vertically successive reservoirs are predominantly the gas-bearing sandstone columns in the Upper Carboniferous and subordinately also in parts of the Rotliegend (cf. sections 2.4.1.1. and 3.3.2.) of Northwest Germany (mainly the Wechselfolge in the Hannover-Formation or Ten Boer-Formation at the top where similarly as in the Upper Carboniferous the more or less thin individual sandstone horizons are separated by thicker or thinner mudstone intervals; BRINKMANN 1982; REINICKE, BRINKMANN, SCHWARZ & HUENI 1985; cf. section 3.2.1.3.).

Separation of the individual treatments in Rotliegend and Carboniferous wells was achieved by sand plugging. In some cases, too small intervals between perforations and insufficient cement bond caused communication behind casing and reduced the stimulation ratio. Multistage treatments including perforating, fracturing, setting a sand plug to isolate the interval, and moving uphole for the next stage have also been performed in the Cotton Valley Formation (Upper Jurassic) in East Texas and Louisiana/USA (JENNINGS & SPRAWLS 1975, HUFFT 1977) and in multiple coal seam sequences (SCHRAUFNAGEL & LAMBERT 1988; cf. section 4.4.3.). Successive multistage fracturing can be conducted from one horizon to the next in upwards direction by mechanical diversion with sand plugback, straddle packer assemblies, and retrievable bridge plugs and packers (HUCKABEE 1987; cf. section 4.2.2.1.3.).

Conventional storey-wise fracturing with isolation of the lower zone after termination of the treatment and upwards movement for the next job is unavoidable if vertically successive horizons have to be stimulated with different fracturing fluid types (HOEL 1988) or cracks within them have to be infilled with different proppant types due to specific reservoir conditions that are considerably changing within the prospective complex. If the same fracturing fluid and the same proppant type can be utilized for several successive layers, then multiple fracturing in one run without discrete handling of individual horizons is possible (cf. section 4.2.2.1.3.) provided the in-situ stresses in all the levels is comparable to allow more or less simultaneous opening of the separate fractures in all the zones (cf. section 6.2.5.). In case of significant in-situ stress differences between the individual horizons, conventional storey-wise stimulation instead of multistage fracturing has to be performed.

4.10.1.2. Commingling of separate layers after individual fracturing

Many of the Rotliegend and Carboniferous wells in Germany FRG were stimulated individually in several horizons and then commingled for pressure build-up analysis and partially also production (thus representing multilayer/multifracture systems; BENNETT, REYNOLDS & RAGHAVAN 1982 a, 1982 b; cf. sections 2.4.1.1. and 3.3.2.). Commingled systems comprise multistorey reservoirs where the individual layers do not communicate in terms of fluid flow through the formation, but although crossflow is absent, they may be completed together and produced by the same wellbore (KUCHUK & WILKINSON 1988). Many different types of commingled systems exist, with the borehole being vertical, horizontal, inclined, fractured or partially penetrated, and the individual reservoirs being homogeneous, heterogeneous or fractured, with different outer boundary conditions such as infinite extent, constant pressure or no-flow being developed.

In the Rotliegend and Carboniferous formations in Germany FRG, the mudstone beds between the different fractured horizons are thicker than necessary for containment of the cracks, with communication between the horizons thus only being expected to occur through the wellbore (REINICKE, BRINKMANN, SCHWARZ & HUENI 1985), with therefore no true commingling by contact and exchange only via the fracture in contrast to vertical communication everywhere in the reservoir (BENNETT, REYNOLDS & RAGHAVAN 1982 b) being given (commingling comprises communication of several layers of the reservoir storey complex only through the wellbores; ONUR, REYNOLDS & RAGHAVAN 1986).

4.10.1.3. Multilayer reservoirs with and without crossflow

Aspects of multiple-zone fracturing are also discussed by AHMED, NEWBERRY & CANNON (1985); EASON (1985), CHOO & WU (1987) and SULLIVAN, LEE & HOLDITCH (1987), and some points of commingled multilayered reservoirs are reported by COBB, RAMEY & MILLER (1972); LARSEN (1981), GREENFIELD & AHMED (1983); ONUR, REYNOLDS & RAGHAVAN (1986); CHEN (1987); HATZIGNATIOU, OGBE, DEHGHANI & ECONOMIDES (1987) and SMITH, MILLER & HAGA (1987). SULLIVAN, LEE & HOLDITCH (1987) present a special case of multilayer reservoirs where small thin high-permeability channel sands of very limited areal extent are already naturally commingled with much larger thicker lower-permeability blanket sands of relatively large areal extent.

DEANS & GAO (1983), GAO (1983), GAO & DEANS (1983), PRATS (1983); PRIJAMBODO, RAGHAVAN & REYNOLDS (1985); WIJESINGHE & KECECIOGLU (1986 b) and LEE & OLAREWAJU (1988) outline multilayer reservoirs with crossflow, and EARLOUGHER, KERSCH & KUNZMAN (1974); RAGHAVAN, TOPALOGLU, COBB & RAMEY (1974); OTUOMAGIE & MENZIE (1976); KUCUK, KARAKAS & AYESTARAN (1984); KUCUK, SHAW, AYESTARAN & NICHOLSON (1986); FETKOVICH, BRADLEY, WORKS & TRASHER (1988) and KUCHUK & WILKINSON (1988) comment on multilayer pays without crossflow. OLAREWAJU & LEE (1988) discuss combinations of crossflow and commingling. LEUNG (1986 b) presents dual-porosity and -permeability aquifers as well as two interconnected single-porosity aquifers.

4.10.1.4. Connection of separate layers by joint fracturing

Joint fracturing of several vertically superimposed sandstones by injection of proppants and fluids through several perforated intervals (cf. section 4.2.2.) within one job (BENNETT, REYNOLDS & RAGHAVAN 1982 b), with the individual fractures remaining contained within the individual horizons, have hitherto not been carried out in Europe beyond the point of experimental stage.

In all the cases, so far creation of a vertical or horizontal communication fracture between several similar or different reservoir facies zones through a separating barrier has generally not been attempted in European hydrocarbon production wells and has only in a few cases been made in water-injection wells near the margin of the field in order to increase the water uptake, with the design of fracture propagation having generally aimed on confinement within one horizon in order not to break through the bounding seals of the reservoir. Limited communication fracturing between two vertically superimposed pay intervals has been performed in the North Sea chalk (SMITH, MILLER & HAGA 1987; cf. section 4.5.4.) where only one zone has been perforated and the fracture has been extended into the other horizon by interface fracturing (cf. section 4.5.4.4.2; some aspects of this technique are also discussed by HOLM 1975; AHMED, WILSON & STRAWN 1983 and BRITT & LARSEN 1986). Multiple-stage fracturing and communication treatments, however, have already been frequently carried out in the USA and occasionally also in other parts of the world. Some comments are offered here concerning application of communication fracturing in production and appraisal drilling as well as fracture interference as follows.

4.10.2. Application in production drilling

Production problems in hydrocarbon-bearing formations with complicated geometry of sand bodies in horizontal and/or vertical direction, however, support at least for the future amelioration of exploitation (given an economically more favourable oil price scenario which is expected to be reached sooner or later; OEHME 1986; and in Europe also necessarily enhanced by an again stronger US $; cf. section 2.2.1.3.) the possibility of partially being solved with unconventional communication fracturing where horizontal or vertical connection is made between several facies zones, thereby adapting to procedures which are already common in the USA since quite some time. Some aspects of connection of isolated sand bodies and separated reservoir storeys are briefly outlined as follows.

4.10.2.1. Connection of isolated sand bodies

Examples of application of communication stimulation comprise fracturing from one isolated channel sandstone into another in horizontal direction in order to improve water- and steam-flood projects or to increase the production radius, fracturing from one high-permeability zone to another in horizontal or vertical direction through a separating low-permeability sandy or muddy barrier (where some operational difficulties of self-containment of the fracture within one sandstone body still have to be overcome by finding the right pressure distribution for performing the communication fracture across lithofacies boundaries, but nevertheless keeping horizontal and vertical crack propagation reasonably under control) in order to enhance the drainage radius or to allow joint instead of selective completion in one well, and fracturing from one borehole to another for implementing a guide channel for the propagation of tensids, polymers or steam during tertiary oil recovery in marginal reservoirs or for allowing exploitation of geothermal energy by pumping cold water through a hot dry formation where it heats up (cf. chapter 4.7.). Some aspects of vertical interference testing across a low-permeability zone which is partially based on communicating fractures are discussed by EHLIG-ECONOMIDES & AYOUB (1986). The outline as follows focusses on magnitude of in-situ stress contrast, fracture width vs. lens axis orientation, and sand body size and spacing.

4.10.2.1.1. Magnitude of in-situ stress contrast

Communication fracturing for the purpose of connection of many isolated sandstone units of lenticular or ribbon-type character to the wellbore, however, is only feasible in case of presence of small to almost negligible (in terms of rock mechanics) in-situ stress contrasts between different lithologies (NORTHROP 1988, NORTHROP & FROHNE 1988). In case of large in-situ stress differences, a hydraulic fracture does not easily break out of a low-stress sandstone, traverse the high-stress confining rocks in horizontal or vertical direction, and cross another sandstone of continuous or isolated nature. Thus intersection, propping and drainage of a large number of remote reservoir units is hardly feasible in the presence of large stress contrasts. Models of fracture growth through lithological or tectonical barriers are presented by LUISKUTTY,

TOMUTSA & PALMER (1986). Reservoir communication can be checked by multiple interference testing which enables detection of pay limits and extension of the corresponding blocks (EARLOUGHER & RAMEY 1973; LEAVER, SAGEEV & RAMEY 1986; cf. section 4.2.4.1.2.2.).

4.10.2.1.2. Fracture azimuth vs. lens axis orientation

Connection of isolated sand bodies by communication fracturing is very important in lenticular tight gas sandstones where recovery efficiency and thus economical feasibility increases with crack penetration through separate lenses that are not in direct contact with the borehole (KUUSKRAA, BRASHEAR, ELKINS & MORRA 1979; cf. section 3.4.1.1.2.). At a rate of one additional lens accessed by the hydraulic fracture for each lens encountered at the wellbore, the initial ten-year gas recovery could be up to 70 % higher than for the single-lens case.

Fracture azimuth as a consequence of horizontal stress direction in relationship to long axis orientation of sand lenses is a decisive factor in controlling fracture length, and spacing and diameter of sand lenses in horizontal and vertical direction also influences crack height. Maximum fracture length can be achieved if the crack propagates parallel to the long axis of the isolated sand body. If the fracture is oriented perpendicular or oblique to sand lens long axis, fracture length has to be kept shorter unless neighbouring sand lenses can be accessed by the same crack and can also be connected to the wellbore. The importance of fracture extension to the reservoir limits of lenticular sand bodies is also stressed by PETERSON & KOHOUT (1983), and MEEHAN, HORNE & AZIZ (1988) outline effects of reservoir heterogeneity and crack azimuth on optimization of fracture length and well spacing.

4.10.2.1.3. Sand body size and spacing

Communication fracturing between separated sand lenses depends in design on size and spacing of the isolated bodies which is a function of primary depositional environment and secondary erosional condensation of the sedimentary sequence (cf. MADER 1985 a). Limited natural interconnection of original sand bodies may even still persist in multistorey and multilateral complexes of vertically and horizontally amalgamated polylenticular aggregates due to separating mud drapes (cf. plates III/1 - 2, IV/4 - 8, VII/5 - 6 and IX/1 - 8) and grain size fluctuations in sand-dominated cyclothems. Therefore even fracture intersection of such heterogeneous sand lens assemblages cannot completely remove the phenomenon that some reservoir intervals are quickly drained while other nearby segments are only slowly depleted (WARPINSKI, BRANAGAN, SATTLER, CIPOLLA, LORENZ & THORNE 1988). Insufficient natural original sand body interconnection by primary-depositional or secondary-erosional juxtaposition has to be artificially compensated by communication fracturing in order to allow for profitable reservoir acquisition and drainage.

4.10.2.2. Connection of separated reservoir storeys

Communication fracturing from one reservoir storey to another aiming on later joint completion and production is only reasonable if both horizons are not too far away in vertical direction and their thickness is ideally each close to marginal, with therefore single completion and production being hardly feasible. While two superimposed hydrocarbon-bearing strata can also be exploited by two pumps in dual borehole completion, multiple layer-cake-type oil- and gas-bearing sections with critical thicknesses, petrophysical properties and thus also productivities of the individual sheets can only be economically exploited by communication fracturing and then following joint depletion.

Considerable significance of communication fracturing is seen for creating a connection path between laterally or diagonally coexisting similar facies bodies that are separated by tight interventions, in order to allow proper performance of secondary and tertiary oil recovery by guaranteing the arrival of the injected fluids or gases at the margin of the drainage cone of the production well and not being stuck at facies anisotropies in heterogeneous reservoirs (the experience of oil exploitation in Middle Europe in the last decades has shown that frequently failures of water flooding were caused by intercalated impermeable facies units which prevented communication between injection and production wells).

A special type of communication fracturing is generation of connections from steam injection wells to neighbouring production boreholes via a horizontal fracture according to the shallow reservoir depth of most heavy oil deposits (BRITTON, MARTIN, LEIBRECHT & HARMON 1982; cf. section 4.9.4.2.). An alternative method to communication fracturing is connection of vertical wellbores by lateral boreholes or drainholes which resemble hydraulic fractures that are reduced from a plane to a line and condensed in the wellbore tube (FINCHER 1985; cf. section 4.8.6.1.). This method, however, is only suitable for thin reservoir sands, whereas connection of boreholes through thick pay zone bodies can only be performed by large-scale hydraulic proppant fracturing.

4.10.3. Influence of perforation arrangement

The success of communication fracturing between several boreholes can be optimized by an orientation of the line connecting the boreholes parallel to the geostress field by appropriate well location prior to drilling of the target boreholes, and by improving fracture propagation through both well axes by oriented precision perforations (JARED, SCOTT & EVANS 1984). Aspects of perforation placement optimization are also discussed by McLEOD (1978); AHMED, THOMPSON, KELKAR, VEGHTE & HATHAWAY (1984); TARIQ, ICHARA & AYESTARAN (1985); HUSHBECK (1986), SCOTT (1986) and ICHARA & CANNON (1987). CHENEVERT (1985) comments on perforation stability in low-permeability gas reservoirs, and TODD & BRADLEY (1986) illustrate the influence of perforation geometry on well productivity and fracture performance. AHMED & KELLER (1984) document the significance of perforation placement techniques for modification of hydraulic fracturing treatments. The following outline discusses some aspects of perforation orientation and stress direction, angular distribution of perforations, and breakdown of perforations.

4.10.3.1. Perforation orientation and stress direction

Concerning relationships between perforations and hydraulic fractures, the influence of perforations on the created hydraulic crack depends on stress distribution around the wellbore, because cracks originate whenever the maximum tensile stress induced at the borehole wall exceeds that of the formation (DANESHY 1973 b). In terms of arrangement, a helical distribution of the perforations enables better matching of the reservoir stress directions than linear rows where coincidence with the orientation of formation stress may be difficult to achieve, because perforation lines may not be in a direction close to the fracture plane. In comparison to open holes, the existence of a perforated casing greatly increases breakdown pressure to almost twice its value.

With progressive deviation of the perforations from the expected fracture plane, the chances that cracks initiate from perforations decreases, with thus hydraulic fractures not necessarily inserting at the perforations. BRANAGAN & WILMER (1988) illustrate the two extreme cases of large- or small-scale fractures running parallel or orthogonal to perforation tunnels. Cracks propagating parallel to perforation holes are the continuous further extension of them away from the wellbore in case of perforation tunnel direction matching with geostress orientation (with then gravel-filled perforation holes providing a smooth transition from mechanical sand control to reservoir fracturing stimulation; cf. section 5.3.5.). Propagation of the large- or small-scale cracks perpendicularly to the perforation tunnels and inserting at the end of it, however, provides only very limited contact to the perforation hole in casing and cement sheath and thus is considerably impairing the effect of the stimulation or breakdown treatment in case of perforation tunnel direction being normal to geostress orientation.

4.10.3.2. Angular distribution of perforations

Experiments with two diametrically opposite lines of perforations in the casing showed that for an angle between perforations and expected fracture plane up to 30 degrees, most of the cracks start to propagate from perforations on both sides of the hole, whereas for an angle up to 60 degrees, the fractures usually extend only from one side and ignore the perforations on the other side (DANESHY 1973 b, BRANAGAN & WILMER 1988). For an angle up to 90 degrees, the fracture mostly ignores all the perforations and initiates on the borehole wall perpendicularly to the least lateral principal stress, with stimulation fluid entering the perforations and travelling between formation and outer casing wall before reaching the fracture. This fluid migration can easily become the source of screenout failure due to proppant bridging in the narrow path which causes plugging of the crack by slurry sticking and proppant blocking. Helical configuration of perforation arrangement assures that at least some perforations are always very close to the fracture plane in terms of angular relationship.

4.10.3.3. Perforation breakdown

Breakdown pressure differences show that if the fracture orientation is known with sufficient certainty, it is better to perforate the well along a line in the crack plane than to perforate helically. Shorter perforations have a lower breakdown pressure than longer ones, and larger perforations lead to a lower tensile strength of the formation than smaller ones.

Precision perforation breakdown helps to increase the effectivity of fracturing stimulations (HUSHBECK 1986). The benefits of this technique include cleaning out of perforations and slight formation breakdown in order to give the treatment an opening to begin an acid job or hydraulic fracturing operation. In addition, the status of perforations is established, which is important due to the sensitivity of pumping rates and pressures to number and diameter of perforations. Limited-entry stimulation (cf. section 4.2.2.2.) is considerably aided by detailed knowledge of perforation status.

4.10.4. Application in appraisal drilling

Communication fracturing from a poor facies belt into the adjoining better zone of reservoir development could support and improve the appraisal drilling strategy concepts of fields in complicated formations with facies changes that are difficult to predict and in situations where hydrocarbon-bearing sandstones lap on and off a basement with pronounced topographical highs and lows (such as associated with unconformities and palaeoreliefs; cf. plate I/1 - 7). Conventional field appraisal so far included plugging back and possible sidetracking of a well that encountered a non-reservoir facies belt or a basement high, with the deviation of the borehole aiming on reaching a better facies zone or meeting the reservoir rocks in depressions between the basement elevations (aspects of palaeotopography on exploration and appraisal well siting and consequently also hydraulic proppant fracturing planning are discussed by THOMAS, SONNENBERG, LANGENBAHN & MURPHY 1988).

The purpose of reservoir facies belt bridging and appraisal drilling improvement could at least in some cases be reached considerably cheaper by smaller- or larger-scale hydraulic proppant fracturing instead of comparatively expensive directional sidetrack drilling, with the partially big difference in investment at least justifying to try to turn the present failure into a commercial producer by carrying out a communication fracturing treatment before in case of negative result of the stimulation, partial plugging back and deviation drilling is still a remaining possibility before finally cementing and abandoning the whole well. Thus also communication fracturing is expected to be in various cases a suitable method to reduce the overall development expenses particularly in light of necessary cost containment when the general economical situation restricts the drilling budget.

Drilling of appraisal wells in fractured reservoirs and subsequent aiming on connection of the borehole section to a nearby large-scale natural crack or fault by a hydraulic fracturing treatment can be considerably optimized by immediately drilling the well on intersections of major natural fractures as revealed from photogrammetry and imagery (REYNOLDS 1987; cf. section 6.2.7.).

4.10.5. Fracture interference

While communication fracturing from one borehole to another includes only one hydraulic stimulation operation in only one source well, with the crack then propagating to the target borehole, interference of several fractures can occur when a couple of neighbouring wells in gas or oil fields are separately treated and if their connection line is in accordance with the geostress system. In contrast to the desired effect of communication fracturing of several boreholes by one crack, fracture interference by overlapping or connection of several cracks propagating from different wells may be unwanted in oil and gas exploitation, as this effect seriously reduces the attainable productivity improvement for each borehole by diminishing drainage radius and offtake area of the individual well (VISSER & DOMSELAAR 1974, SCHOLS 1983, LEICHT 1985).

Careful planning of borehole spacing, fracture length (which depends on drainage radius and is also linked to fracture conductivity; McDANIEL & WILLINGHAM 1978, MONTGOMERY & STEANSON 1985; cf. section 4.8.11.) and well orientation has thus to be made prior to job execution in order to optimize hydrocarbon production within fields of limited extension and resources. The significance of fracture azimuth prognosis for the improvement of hydrocarbon recovery is analyzed by HASSAN (1982). Fracture interference assessment is different in hydrocarbon and geothermal reservoirs.

4.10.5.1. Oil and gas reservoirs

In oil and gas fields with more or less narrow spacing of wells within pressure- and flow-homogeneous reservoir blocks, caution has not only to be exercised for avoiding direct fracture interference by merging of several individual cracks to a dendritic system (cf. section 4.8.8.4.), but also for preventing interference of cracks emanating from the individual boreholes with the drainage cones of other wells. Therefore the tip-to-tip length of the fractures does not only have to be reasonably in tolerance with the well-to-well distances, but should be designed in such a way that intersection of neighbouring drainage cones does not occur. The following discussion illustrates aspects of fracture length and well drainage radius, fracture orientation prediction and interference avoidance, lenticular tight gas sands, and steam-drive operation.

4.10.5.1.1. Fracture length and well drainage radius

In tight gas reservoirs of microdarcy permeability range, fracture lengths greater than normal well drainage radius are often necessary to achieve economically feasible production (SMITH, REN, SORRELS & TEUFEL 1985). In these cases, it is necessary to determine fracture azimuth prior to infill drilling in order not to destroy the effectivity of the long linear drainage paths by premature termination of the flow channels at neighbouring boreholes. Fracture azimuth monitoring techniques (cf. section 6.2.2.3.) represent invaluable support for well pattern planning and stimulation design in order to avoid interference with pre-existing wells and to prevent interference as field development proceeds (OVERBEY, YOST & YOST 1988). Knowledge of azimuthal orientation of induced fractures in hydraulically stimulated boreholes in tight gas reservoirs can increase recoverable gas reserves by as much as one third through optimal spacing of production wells within the gas field.

4.10.5.1.2. Fracture orientation prediction and interference avoidance

Even with massive hydraulic fracturing stimulation, much of the gas reserves in tight sandstone reservoirs may be left unrecovered because the drainage pattern around a fracture-treated well is elliptical rather than radial in low-permeability pay zones (TEUFEL 1981), with the elliptical drainage network being even accentuated and enhanced as the fracture increases in length. Therefore borehole location and spacing based on radial drainage may not be the optimum economical gas recovery procedure in a particular field due to overlapping drainage at crack tips as a consequence of fracture interference and due to dead zones between crack flanks. Thus reliable methods of fracture orientation prediction are of great importance to gas industry, as knowledge of crack direction enables to avoid fracture interference and allows to achieve a higher depletion stage in reservoir exploitation (techniques for fracture azimuth prognosis and determination are discussed in sections 6.2.1.8. and 6.2.2.3., respectively).

Another reason for fracture orientation forecast and determination is because massive hydraulic cracks propagate parallel to the maximum horizontal stress direction, knowledge of this orientation prior to stimulation permits optimal placement of seismic sensors which can track position and advance of the fracture (CLARK 1983). Effects of stress state on fracture orientation are also discussed by ZOBACK, ZOBACK & SEEBURGER (1981). FORGOTSON, FORGOTSON & BROWN (1980) illustrate prediction of hydraulic fracture orientation.

4.10.5.1.3. Lenticular tight gas sands

Achievement of optimum reservoir drainage and avoidance of fracture interference is particularly important in lenticular tight gas sands where well spacing and crack length are a function of horizontal stress orientation controlling fracture propagation direction and azimuth of sand body long axis having significant impact on crack length and height modelling. The geometrical relationship between sand lens orientation, fracture azimuth and well pattern governs drainage configuration where depending on prevention of crack interference and controlled by permeability level, either square drainage networks with relatively short fractures or rectangular patterns with long fractures are the more efficient and feasible solution (KUUSKRAA, BRASHEAR, ELKINS & MORRA 1979).

4.10.5.1.4. Steam-drive operations

VITTORATOS, SCOTT & BEATTIE (1988) report fracture interference in steam-drive operations in heavy oil sand reservoirs. Interwell communication with fracture breakthrough from injector to producer and formation of numerous of such borehole couplets considerably reduces sweep efficiency and pay zone exploitation (cf. section 4.9.4.1.2.). The extensive interwell interactions are a result of reservoir failure that is required in order to achieve commercial levels of steam injectivity. As failure can extend up to distances equal to borehole spacing, overlapping of the failed pay portions occurs, and the disturbed zones associated with the wells create conduits of enhanced pressure and fluid propagation between the boreholes, with the paths embedded in a matrix with significantly slower pressure and fluid propagation. Apart from well spacing and fracture orientation optimization by proper geometrical planning of the enhanced oil recovery operations according to the geostress pattern in order to avoid fracture interference and steam expansion shortcutting along borehole connection couplets created by fracture breakthrough, steaming of rows of wells sequentially with 50 % overlap in injection time between adjacent rows significantly reduces water transfer between boreholes and increases the conformance of injected heat.

4.10.5.2. Geothermal reservoirs

On the other hand, interference fracturing may be another means to increase the possibilities of geothermal heat exploitation (cf. chapter 4.7.) by having flow from and to the borehole from several directions and thus optimizing pumping of the injected cold water through a hot dry formation, with in case of a suitable orientation of the wells to the geostress field even allowing to save some injection wells and put more investment in production facilities and stimulation design.

Fracture interference modelling in either negative manner to avoid it to happen or in a positive way to achieve this effect, however, has to take into account that irrespective of the general rule that fracture orientation takes place according to the regional stress field, local deviations up to 15 degrees are possible as a consequence of patchy compressional influences particularly in anticlinal structures.

4.11. Fracturing of deviated wells

With increasing offshore development of tight gas fields where large stimulation campaigns have to be carried out and in some patches nearly all the wells have to be treated, hydraulic proppant fracturing of moderately- to highly-deviated holes is becoming increasingly significant. This is a consequence of the standard offshore development strategy to drill directionally in all possible ways from a central platform from which a lot of production wells emanate in a star- or spider-like pattern. One of the major examples of offshore hydraulic proppant fracturing campaigns in an area with numerous highly-deviated wells is the Rotliegend reservoir belt in the British Southern North Sea (cf. sections 2.4.1.2. and 3.7. to 3.9.). The discussion as follows focusses on fracture propagation and orientation, proppant stratification and fluid layering, hydrocarbon production and flow regimes, and improvements of fracture effectivity in deviated wells.

4.11.1. Fracture propagation and orientation

The effects of fracture propagation and orientation in highly-deviated to horizontal boreholes are illustrated by commenting on wellbore axis being not parallel to in-situ principal stress direction, and limited communication interval between fracture plane and wellbore section.

4.11.1.1. Wellbore axis not parallel to in-situ principal stress direction

The key feature of deviated wells is that their axis is generally not aligned with the direction of the in-situ stress components in the rock medium, but is oriented at an angle other than normal and/or parallel to it (YEW & LI 1987) unless oriented highly-deviated to horizontal drilling is made along or orthogonal to the geostress pattern. Therefore the trajectories of deviated boreholes are not parallel to the least principal stress (BUCHANAN 1987), and the medium surrounding the well is under the combined action of normal and shear stresses and is thus in a three-dimensional stress state (BRADLEY 1979, YEW & LI 1987). Hydraulic fractures initiate when the maximum tensile stress induced on the wellbore wall exceeds the tensile strength of the formation rock (KIM, CHAMPION & COOPER 1985), and propagate in a plane perpendicular to the least in-situ principal stress. In vertical wells, one of the in-situ principal stresses is usually parallel to the borehole axis, and the resulting fracture plane is horizontal in shallower and vertical in deeper formations (cf. section 1.2.8.).

In deviated wells, none of the in-situ principal stresses is parallel or normal to the borehole axis, and therefore initiation of a fracture takes place from two opposite points along the circumference of the wellbore. The crack sweeps the entire perforated or open zone in the deviated well only in case of the borehole azimuth being very near to that of the fracture plane. As a consequence of the geometrical constellation, the best case of crack orientation is represented by coincidence of crack plane and well trajectory azimuth (especially if the borehole is deviated to horizontal), whereas the worst case comprises a wellbore being normal to the fracture plane (BUCHANAN 1987). It is also possible to create several parallel fractures in the same wellbore (cf. section 4.8.6.2.4.).

On the other hand, AADNOY (1987, 1988) postulates that in deviated wells, vertical hydraulic fractures are always initiated along the wellbore axis regardless of borehole inclination, which means that fractures cutting the hole at an angle are not expected, with therefore the length of the cracked well section determining stimulation success. As they propagate away from the borehole, the fractures change direction and orient themselves normal to the least in-situ stress. SNOW & HOUGH (1988) report that in highly-deviated wells, at least in the near-wellbore region crack growth is not ideal, but will either result in multiple fractures or in a continuous or en-echelon bending of the crack to vertical position (cf. also STRUBHAR, FITCH & GLENN 1988).

4.11.1.2. Limited communication interval between fracture plane and wellbore section

Unless the azimuthal orientations of fracture plane and well trajectory coincide, there is only a very small communication interval between perforated borehole section and crack plane in moderately- to highly-deviated wells (BUCHANAN 1987) in comparison to a vertical fracture in a vertical borehole where after a successful stimulation treatment, the entire vertical extent of the formation should be in communication with the perforations through the fracture (HOLDITCH & LEE 1979). Horizontal wells and drainholes even are characterizedby very limited fracture height intersection which is equal to wellbore diameter (JOSHI 1987). After discussion of some drawbacks in operation and performance of vertical fractures in highly-deviated to horizontal boreholes traversing slightly dipping reservoirs, the comparable situation of vertical wells crossing steeply dipping prospective formations is illustrated.

4.11.1.2.1. Drawbacks in operation and performance

The unfavourable geometrical relationship of vertical fractures intersecting highly-deviated to horizontal wells with only a very restricted contact plane results in high pressure losses during treatment, low stimulation rates, small fractures of limited extension away from the borehole intersection, and wellbore plugging and perforation leakoff. Multiple fracturing through different perforations may occur as a consequence of repeated onset of the crack at several spatially separated places along the oblique borehole section, because at any of these points, the boundary conditions for fracture initiation are simultaneously fulfilled and the individual cracks propagate each in the direction normal to the least principal stress, giving rise to various parallel travelling smaller fracture planes instead of one large coherent crack face in vertical wells. Some comments are offered as follows on wellbore instability and fracture height determination by well logging.

4.11.1.2.1.1. Wellbore instability

Twisting effects occur as the fracture reorientates itself from the deviated wellbore in case it started to propagate in different directions. In highly-deviated and horizontal wells, stimulation boat capacity is the limiting factor for offshore fracturing, with the longer boreholes restricting pumping rate and pressure that can be achieved and maintained by the vessels (LOUVIERE 1987; cf. section 3.8.1.2.). Fracturing in highly-deviated boreholes even bears considerable risk of wellbore screenout due to high fluid leakoff through the perforations. With increasing borehole inclination, wellbore instability problems increase and give more frequently rise to borehole collapse and formation fracturing (WILSON & WILLIS 1986; FUH, WHITFILL & SCHUH 1988; cf. section 4.8.6.1.5.8.). It is therefore essential to predict any significant stress changes around the wellbore when the hole angle increases under certain geological and in-situ stress conditions. Borehole collapse pressure gradient increases with well angle, whereas fracture gradient remains the same, and as a result, the working range between the two gradients is narrower at higher hole angles.

The same relationship as applying for highly-inclined to horizontal wells traversing gently-dipping to flat-lying formations is also true for vertical boreholes intersecting steeply-dipping rocks (cf. section 4.11.1.2.2.). Optimum mud weight for wellbore stability in highly-inclined holes ranges between borehole collapse pressure gradient representing the lower limit and formation fracture gradient being the upper boundary (cf. section 6.2.3.2.6.). Aspects of instability of highly-deviated boreholes are also discussed by BRADLEY (1979); HOTTMAN, SMITH & PURCELL (1979); AADNOV & CHENEVERT (1987), HSIAO (1987), MAURY & SAUZAY (1987) and YEW & LI (1987).

4.11.1.2.1.2. Fracture height determination by well logging

The limited communication interval between vertical fracture and highly-deviated wellbore trajectory has also negative impact on the effectivity of fracture height determination by well logging techniques (PEETERS & HARTLEY 1984; cf. section 6.2.1.10.8.5.). As all methods of fracture height detection by wireline logging have only a small radius of investigation and cannot register crack height farther away from the wellbore, monitoring of fracture height is progressively deteriorated with increasing borehole inclination and successively shorter contact intervals between wellbore trajectory and crack plane. Hydraulic fracture direction determination in highly-deviated wells is also discussed by EL RABAA (1988).

4.11.1.2.2. Comparable situation in vertical wells traversing steeply dipping reservoirs

A geometrical situation comparable to horizontal drilling traversing flat-lying to gently-inclined beds can be generated by vertical wells penetrating steeply-dipping formations (ELBEL & SCHOTTLER 1977). This constellation also gives a long borehole intersection with the reservoir and represents the most favourable stimulation situation in vertical wells by enabling a very long contact interval between vertical fracture, highly-inclined pay formation and wellbore trajectory, in contrast to horizontal drilling of flat-lying or low-angle dipping strata where the vertical fracture in greater depth allows only a limited communication path between borehole and crack and a favourable geometrical arrangement comparable to vertical fractures intersecting steeply-dipping reservoir rocks in vertical wells can only be reached by horizontal fractures crossing slightly-inclined pay zones in highly-deviated to horizontal boreholes in shallower depth storeys. The same geometrical constellation of very small communication interval between perforated section and fracture plane as applying for vertical cracks crossing highly-deviated or horizontal boreholes is valid for horizontal fractures traversing vertical wells in shallow reservoir depth.

In order to improve the geometrical constellation of wellbore trajectory and vertical hydraulic fracture, extended-reach drilling should whenever technically possible and economically feasible not be carried out in a J-shaped manner, but in case of necessity of subsequent vertical hydraulic fracturing or gravel packing, preference should be given to a S-shape which is generated by again reducing borehole inclination to an angle below 45 degrees prior to penetration of the pay complex (BRUIST, JEFFERIS & BOTTS 1983; cf. section 4.8.6.1.). Such adjustments, however, are not possible in case of horizontal drainholes that are installed parallel to the reservoir boundaries for the purpose of hydrocarbon exploitation improvement by prolongation of the contact path between wellbore trajectory and pay stratification.

4.11.2. Proppant stratification and fluid layering

Proppant transport in highly-deviated wells follows tortuous paths. Dehydration of the proppant-fluid-slurry is very high due to extensive fluid loss through the perforations. Proppant stratification during pumping also occurs, because segregation into high-density slurry on the lower side and low-density slurry at the upper side of tubing or casing takes place which gives higher friction pressure and looses hydrodynamic head. The heavier the proppant grains, the more pronounced is the settling and bedding effect (a similar banking and bedding of proppants occurs during gravel pack operations in highly-deviated wells; cf. section 5.8.3.1.). Extreme cases of stratification might even lead to origin of a settled bank of proppants. Gravity layering of fluids in deviated wells during displacement flushing also occurs.

4.11.3. Hydrocarbon production and flow regimes

The influences of the special geometrical relationship between highly-deviated boreholes and intersecting hydraulic fractures are highlighted by discussing aspects of reduction of long-term well productivity by changing flow pattern, and non-Darcy flow and skin factor as follows.

4.11.3.1. Reduction of long-term well productivity by changing flow pattern

Concerning production in formations where the wellbore is not in communication with the created fracture over its total height, but is restricted to only a small interval of it as a consequence of the deviated borehole (cf. section 4.11.1.2.), major drawbacks are high-pressure losses during exploitation, high velocities in the fracture near the wellbore intersection, diminished fracture conductivity because of non-Darcy flow, and reduced overall well productivity (LEE & HOLDITCH 1979, SCHULTE 1984). If the fracture is in communication with the wellbore over its total height in vertical boreholes, a bilinear flow regime is developed consisting of transient linear horizontal flow within the formation towards the fracture in combination with transient linear horizontal flow within the fracture towards the wellbore. The different flow interval types in vertical and horizontal wells are outlined as follows.

4.11.3.1.1. Flow interval types in vertical wells

Distinction can be made between four different flow intervals for drawdowns of vertically fractured wells comprising fracture-linear, bilinear, formation-linear and pseudoradial flow (CINCO-LEY & SAMANIEGO 1981). If the communication between wellbore and fracture is limited in deviated boreholes, a linear/radial flow pattern is developed which comprises transient linear horizontal flow within the formation towards the fracture in combination with transient radial current within the crack towards the interval of the fracture intersecting the wellbore (LEE & HOLDITCH 1979, SCHULTE 1979). In a semiradial fracture, the flow regime following linear/radial current is transient linear flow in the formation and steady-state current in the crack, whereas in a very elongated fracture, the linear/radial flow regime is followed by the bilinear current regime. In both cases, pressure loss resulting from the limited inflow interval at the wellbore is seen as a skin factor. An intermediate behaviour is pseudo-radial current (RUSSELL & TRUITT 1964).

CINCO-LEY & MENG (1988) introduce two additional flow periods being trilinear and formation bilinear current, thereby increasing the number of flow intervals to be distinguished to six. Trilinear current is present instead of the bilinear flow regime during the stabilization period for small volumes of dimensionless fracture storativity, and formation bilinear current occurs during the pseudolinear flow period when most of the fluid expansion is provided by the matrix in linear current regime under the same conditions.

4.11.3.1.2. Flow intervals in horizontal wells

In horizontal wells, the typical infinite acting radial flow pattern of vertical wells is complicated by fluid current in the wellbore vicinity perpendicular as well as parallel to the sedimentary bedding (KUCHUK, GOODE, WILKINSON & THAMBYNAYAGAM 1988). As a consequence of reservoir anisotropy, horizontal wells start with an elliptic cylindrical flow pattern which later gives way to radial current that in turn is followed by a hemi-cylindrical flow period. An intermediate time linear current may develop once the pressure transient reaches the upper and lower boundaries that pass into intermediate time radial flow after the effects of top and bottom boundaries reach steady-state.

4.11.3.2. Non-Darcy flow and skin factor

The limitation of the communication between wellbore and fracture to only a part of the fracture height in highly-deviated boreholes seriously affects the long-term productivity of the well. The flow in the formation follows Darcy`s law, whereas inside the fracture, non-Darcy flow effects have to be taken into account which may significantly increase the skin factor as a result of partial communication. The deteriorating influence can be so large that the semi-steady state flow is only one third of that where wellbore and crack are in full communication. In addition, the apparent fracture conductivity may be an order lower than the real value as a consequence of non-Darcy flow effects. Marked layering of the pay zone is also contributing to the overall results, because the various beds start to deliver hydrocarbons to the crack at different times and possibly also with different rates. The deterioration of long-term productivity due to a well inflow interval being much smaller than total created fracture height must be taken into serious consideration particularly for the design of cracks from highly-deviated or even horizontal wells and drain holes into tight gas reservoirs (other aspects of non-Darcy flow are discussed in section 4.3.3.3.).

4.11.4. Improvement of fracture effectivity in deviated wells

Hydraulic fracturing in highly-deviated wells can be considerably improved if the pay zone could be penetrated by an almost vertical borehole section which, however, causes high additional directional drilling expenses. The borehole trajectory should be planned in such a way that at least across the reservoir interval, azimuth of wellbore axis and fracture plane are more or less coinciding, because this gives maximum intersection of the borehole by the fracture. An example of such fracture-suitable directional drilling modelling is reported by OIL GAS JOURNAL (1988 b) where although maximum inclination angle of the borehole was 76 degrees, angular wellbore deviation was reduced to 27 degrees for penetration through the pay zone which is reasonably subvertical to considerably diminuish or even cancel the problems that would have been occurred in the high-angle hole.

On the other hand, an azimuth of the wellbore axis perpendicular to that of the crack plane gives rise to multiple parallel fractures crossing the borehole like a set of slices which might be desirable in some special applications (cf. section 4.8.6.2.4.). Other important measures for improving stimulation are increasing effective wellbore diameter, using multiple deep-penetrating perforation charges, enlarging perforations and planning a staged fracturing treatment (cf. section 4.2.2.1.3.). Perfect proppant transport can be achieved with shear-independent fluids (cf. section 4.3.4.2.1.).

4.12. Fracture damage

Following the outline of various possibilities of technological improvement of the predominantly European fracturing potential by more or less unconventional stimulation techniques in marginal reservoirs, the discussion comprising a suite of geological, reservoir engineering and technical aspects will be concluded in this section by reviewing several important points which are controlling amelioration or deterioration of fracture effectivity. The summary is presented in the final division of the review and status report on hydraulic proppant fracturing because of its significance also for crosslinking the chapter on proppant selection with the account of the technological perspectives. For effective proppant choice and application as well as for successful performance of the stimulation treatment, it is essential to consider aspects of possible fracture damage after the end of the operation and ways of its minimization or even complete prevention which are illustrated as follows. After discussion of some general aspects, the significance of proppant settling, proppant flowback and proppant crushing are outlined. The impact of fluids is also briefly sketched.

4.12.1. General aspects

Important points for illustration of stimulation result deterioration are fracture conductivity vs. formation permeability damage as well as proppant crushing and embedment vs. closure stress.

4.12.1.1. Fracture conductivity vs. formation permeability damage

Fracture conductivity damage can occur by proppant flowback into the wellbore, embedment or crushing of proppant grains (cf. section 4.3.3.1.2.), and plugging of the proppant porosity by formation fines, insoluble fluid-loss additives and gel residues (ROBINSON, HOLDITCH & WHITEHEAD 1986) and has considerably detrimental effects on well performance (COOKE 1973, HOLDITCH 1979). In comparison, damage of the formation around the fracture could be caused by swelling clays (cf. section 3.5.2.), mechanical movement of solid fines, chemical precipitations in the pore space, plugging of interstitials by unbroken gel, and various other physical or chemical reactions between the injected synthetic materials and the formation and its natural constituents. Some aspects of contrast magnitude, relative significance, and breaker type and concentration are illustrated as follows.

4.12.1.1.1. Contrast magnitude

Fracture damage is much more important than formation damage around the crack, because deterioration of proppant porosity is much stronger diminishing the permeability contrast between fracture and formation than destruction of parts of reservoir rock matrix porosity (ROBINSON, HOLDITCH & WHITEHEAD 1986 estimate that it requires a 1000 fold decrease in formation permeability before the effects are comparable to only a 40 fold decrease in fracture conductivity). The only major exception are probably low-pressure water-sensitive reservoirs where formation damage can severely curtail gas or oil production.

Other types of fracture deterioration are related to non-achievement of the ideal result of a highly-conductive crack intersecting the reservoir matrix in its original configuration and in full vertical thickness of the pay interval (ECONOMIDES 1986). Damaged fractures are surrounded by an ellipsoidal or cylindrical zone of formation matrix deterioration as a consequence of stimulation fluid leakoff into the reservoir matrix or water blocking in the pay zone adjacent to the fracture (cf. section 3.11.2.4.2.). Choked cracks suffer from reduced conductivity and permeability in the immediate wellbore vicinity as a result of proppant flowback or crushing due to either unsuitable closure time or inadequate cleanup and production start-up procedures (CINCO-LEY & SAMANIEGO 1981 b). Partially penetrating fractures fail to span the entire reservoir thickness in height due to insufficient volume and pressure during treatment execution.

4.12.1.1.2. Relative significance

Proppant pack permeability damage minimization is considerably more important than formation deterioration minimization, because reservoir matrix damage has a relatively short-term effect on productivity, whereas proppant pack deterioration has a very prolonged deleterious effect (ECONOMIDES 1986). Fracturing fluids become more concentrated as the treatment progresses, because polymers and solid fluid-loss additives are filtered out on the rock surface. While earlier studies assumed that the filter cake is redispersed and distributed evenly throughout the proppant package (STEPHEN-PYE & SMITH 1973), later investigations have shown that this model is incorrect and the filter cake progressively builds on the fracture walls (cf. sections 1.4.10.3.5. and 4.3.4.1.4.) and gives rise to proppant embedment (McDANIEL 1987, 1988; cf. section 4.3.3.1.2.). Reduction of fracture conductivity is caused by a combination of proppant embedment into filter cake and/or formation, proppant crushing, proppant compaction and pack configuration changes, and proppant interstitial plugging by formation fines and residue from fracturing fluids (ECONOMIDES 1986).

4.12.1.1.3. Breaker type and concentration

Fracture conductivity can be correlated to molecular size of the fluid residue, and complete breakage of the fluid is essential to achieve good conductivity (ALMOND 1982). Excessively high injection pressures are required to pump partially broken fracturing fluids through the crack. Higher breaker concentrations increase flow capacity, whereas flow reduction increases as crosslinker and/or gelling agent saturation increases. For the purpose of studying fracture conductivity damage, each stimulation fluid must be evaluated as a system composed of polymer, gelling agent, crosslinker, breaker and additives. Enzyme oxidizer and thermal breaking mechanisms have different degradation efficiency for different polymer fluid systems (ALMOND & BLAND 1984, GALL & RAIBLE 1985, KIM & LOSACANO 1985). The lower the proppant concentration, the higher the fracture conductivity damage (COOKE 1975). Residue saturation in the fracturing fluid is a function of gelling agent, breaker type and breaker concentration, and factors influencing fracture conductivity damage are proppant porosity and permeability, residue saturation in the fluid, and retained residue in the fracture.

4.12.1.2. Proppant crushing and embedment vs. closure stress

Fracture damage often may occur by proppant crushing or embedment. In hard formations, proppants tend to crush, while in soft formations, they are likely to embed into the reservoir rock (cf. section 4.3.3.1.2.). Simultaneous severe proppant crushing and embedment, however, can take place in deep soft reservoirs (such as the Vicksburg Formation in South Texas; ROBINSON, HOLDITCH & WHITEHEAD 1986). In this respect, an important consideration is proppant concentration and its succession during course of the treatment. High proppant saturations result in wide fractures infilled with proppant multilayers and thus fracture conductivity is proportionally increased and the effects of crushing and embedment are not so critical.

Similarly as applying for hydraulically-induced fractures, the normal gradual closure behaviour of pre-existing natural cracks around a borehole during various stages of depletion when the effective normal stress or closure stress (being equivalent to total pressure minus pore pressure) is progressively rising has a major impact on changes of fracture permeability and subsequent well production (BAKHTAR, BARTON, RAKOP & JONES 1985). While deterioration of natural cracks (cf. section 4.8.8.3.) cannot or only to minor extent be influenced, the damage of hydraulically-induced fractures can be controlled by both layout of the proppant package and design of clean-up and early production schedules.

4.12.2. Proppant settling

Hydraulic fracturing treatments are generally performed to increase well productivity by enlarging the surface area available for flow from the formation and by providing a more permeable conduit for current from the extremities of the drainage area to the wellbore (POULSEN & LEE 1984). Proppants are placed in the fracture by mixing them with the stimulation fluid, thereby generating a slurry which is pumped into the borehole to create the fracture, with ideally on completion of the treatment the proppant being distributed such that it covers the entire exposed producing interval. As propping agents are generally denser than the fluid transporting them, they may settle toward the fracture bottom throughout the job and continue settling until the crack closes, thereby possibly leaving the upper portion of the fracture unsupported as a consequence of banking in the lower part of the fracture. Proppant settling velocity is mainly controlled by flow behaviour of the transporting fluid which is altered through selection of gelling agent, crosslinker, stabilizing additives and their respective concentrations. For a given fluid, flow behaviour as well as proppant transport and suspension ability change with time and temperature.

Since fracturing treatments often take hours to complete and fluid temperature can change radically from wellbore to crack tip, both time and temperature are important factors in stimulation operation design.

Proppant settling during the time required for fracture closure after a hydraulic stimulation job can have an important bearing on proppant distribution in the crack and may determine success or failure of the operation (NOVOTNY 1977). In a hydraulic stimulation treatment, a fracture is created by rupturing the formation at high pressure by means of a special fluid transmitting the hydraulic energy, and proppants transported by the carrier fluid are placed in the induced fracture channel to prevent the crack from closing and rehealing after pressure release (ROODHART 1985).

Productivity improvement is mainly determined by proppant dimensions and distribution which in turn are largely controlled by proppant settling velocity in the fluid, with high settling velocities resulting in triggering of proppant banking at the fracture bottom, whereas a very low settling velocity permits the proppant to remain in suspension distributed over the total fracture height. Some aspects of post-fracturing pressure decline, proppant transport and banking, carrier fluid viscosity, pad and carrier fluid volumes, and slurry consistency and proppant clustering are briefly outlined as follows.

4.12.2.1. Post-fracturing pressure decline

Analysis of stimulation injection pressures during the job and pressure decline after the treatment (NOLTE 1979, NOLTE & SMITH 1981; ROBINSON, HOLDITCH & WHITEHEAD 1986) is important for determination of the fracture closure pressure which is approximately equal to the least principal stress. The control by pressure evaluation is essential, because the crack must have already closed before the well is opened for clean-up and the timing of fracture closure must be prior to beginning of gel breakage in order to trap the proppants before they can settle in the decomposing fracturing fluid. The rate of post-fracturing pressure decline depends on fluid leakoff which is a function of formation permeability (both primary matrix and secondary joint permeability; cf. section 4.8.8.1.), with in high-permeability reservoir rocks the pressure falling off rapidly, thereby allowing the fracture to close in a relatively short period of time, whereas in tight pay horizons, fluid leakoff may be retarded and thus the fracture will stay open for some longer time.

4.12.2.2. Proppant transport and banking

Proppant transport prediction models are generally based on Stokes` law for laminar or creeping flow of Newtonian fluids at low Reynolds numbers as well as on Newton`s law for turbulent flow at high to very high Reynolds numbers (CLARK & QUADIR 1981, SHAH 1982, VEATCH 1983). While some fracturing fluids exhibit a Newtonian flow behaviour, others are non-Newtonian fluids but essentially behave as power-law fluids, and the crosslinked polymers commonly used for MHF treatments are very complex non-power-law mixtures (cf. section 4.3.4.3.). Proppant settling velocities in Newtonian fluids are a function of gravitational acceleration, fluid density, particle density, particle diameter, fluid viscosity and surface roughness. In slurries, settling takes place in a somewhat different manner because of particle interference and/or clumping (CLARK & GULER 1983) and clustered proppant transport which is hindering gravity segregation (SIEVERT, WAHL, CLARK & HARKIN 1981).

Aspects of proppant transport and settling are also discussed by LOCKETT & AL-HABBOOBY (1973), BARNEA & MEDNICK (1975), NOVOTNY (1977), DANESHY (1978 b); HARRINGTON, HANNAH & WILLIAMS (1979); MIRZA & RICHARDSON (1979), ZIGRANGE & SYLVESTER (1981), SHAH (1982, 1986), PEDEN & LUO (1987) and CLIFTON & WANG (1988, 1989). Proppant settling, proppant banking and formation grain size distribution, fracture closure and fluid viscosity, and other aspects are briefly discussed as follows.

4.12.2.2.1. Proppant settling

While proppant is injected into the fracture, it travels along with the fluid away from the wellbore and settles downward at a rate that depends on fluid properties and surrounding conditions (NOVOTNY 1977). Movement paths of proppants are often tortuous due to secondary flow caused by irregular morphology of the crack walls (VEATCH & MOSCHOVIDIS 1986). As carrier fluid is lost through the permeable walls of the fracture, the crack starts to close at the end of the treatment, with proppant continuing to settle until a proppant bank is formed at the bottom of the crack, the proppant concentration in the slurry becomes so high that it can no longer settle, or the fracture closes on the slurry thereby trapping the proppant.

Proppant banking beneath the productive reservoir sand interval before closure of the fracture stops settling by trapping of the proppants between the walls of the crack, and this process can result in such fracture deterioration that almost all the stimulation effect is gone. As it is not unlikely that the crack closes instantaneously at the end of proppant injection, proppant settling during the time the fracture stays open has to be reduced as much as possible by modifying fluid viscosity and proppant concentration in order to improve proppant suspension in the carrier medium. Some aspects of clustered proppant transport and hindered settling, forced fracture closure, and wellbore vicinity and tail-in are outlined as follows.

4.12.2.2.1.1. Clustered proppant transport and hindered settling

In non-Newtonian fluids, proppant grains are transported in clusters rather than as single particles (SIEVERT, WAHL, CLARK & HARKIN 1981). When the flow is stopped, the clusters fall at a much faster rate than single particles to the bottom of the crack, with the suspended proppants thus settling in a very short time. If grains fall through a liquid containing other particles, the terminal settling velocity is different than that of a single grain falling through the same liquid as a consequence of hindered settling (CLARK & QUADIR 1981; cf. also section 4.12.2.5.1.). Terminal settling velocities of proppants in hydraulic fracturing treatments are governed by fluid viscosity, fluid velocity, fracture width, particle size and particle density, and are also a function of gravitational and drag forces. An absence of external shear on the fluid produces lower settling velocities and hence longer proppant packages. Apparent proppant settling velocities decrease with increasing proppant concentration (SIEVERT, WAHL, CLARK & HARKIN 1981).

Hindered settling effects which account for the influence of interactions between particles in a slurry on the free-fall velocity of the grains are not sufficiently pronounced to explain the whole magnitude of deviation between apparent and actual settling velocities. If the proppant particles move at a horizontal rate less than that of the carrying fluid, proppant concentration in the slurry increases which in non-Newtonian fluids in turn increases pack growth rate, while Newtonian fluids do not exhibit a clear-cut influence of proppant saturation. On the other hand, doubling of the viscosity of Newtonian fluids has only a small effect on transport behaviour in these systems, whereas in non-Newtonian systems, a severalfold increase in transport and suspension capability takes place. Impingement of proppant grains against fracture walls during transport and settling is not considered to be a major influencing factor, because most of the particles stay in the central high-velocity zone and are not tending to concentrate in the marginal low-viscosity zone.

4.12.2.2.1.2. Forced fracture closure

ROSE (1988) presents a technique allowing after placement of the proppants in the fracture to induce closure of the crack so that the proppants are trapped in the producing interval instead of settling to the bottom of the fracture. Simple methods of ascertaining proppant deposition within the prospective interval and avoiding significant and unacceptable settling are separate treatment of each interval of the reservoir column (cf. section 4.10.1.) and limited-entry perforation and fracturing (cf. section 4.2.2.2.). As crack height may still be excessive or not completely contained within the perforated interval, a forced fracture closure technique can be applied to force the crack to shut before proppant settling can occur.

Forced fracture closure can be accomplished by controlling flowback rate of treating fluids thus reducing the pressure differential near the wellbore and allowing fracture faces to close and to trap the high proppant concentration, thus stopping flowback of earlier proppant stages and resulting in better well productivity. Rapid return of carbon dioxide foam fluids supports forced crack closure by quick pressure drop within wellbore and fracture during fluid recovery, thereby allowing the crack walls to shut upon proppant suspended within the perforated interval.

4.12.2.2.1.3. Wellbore vicinity and tail-in

The most significant proppant settling occurs at the place where fracture width is largest which is in the immediate surroundings of the wellbore (ELY 1988). As partial fracture rehealing as a consequence of proppant settling is particularly detrimental in the most proximal section of the crack, assistance has to be given to fracture closure by immediate flowback of the well after termination (cf. section 4.12.3.1.) of the treatment or to enhancement of cleanup by energizing components or foams (cf. section 3.11.2.). Excessive proppant settling either in pipe or fracture during pumping may result in premature screenout termination of the job (SHAH, SMITH & DONALDSON 1983; cf. section 6.2.4.2.1.). Fracture closure can be observed by monitoring surface pressure following a treatment. By insuring that the crack is completely closed before returning the well to hydrocarbon offtake, proppant production can be mitigated.

Proppant settling does not stop when injection is terminated, but continues until the proppant either settles on top of the existing proppant bed in the lower part of the crack or becomes trapped between the walls as the fracture closes (POULSEN & LEE 1984). While near the wellbore the proppants are nearly perfectly suspended and only very little settling occurs, the suspended proppant height drops more rapidly closer to the leading edge of the proppant-laden fluid farther outwards within the crack, thereby indicating fluid degradation which provokes higher settling velocities. If large portions of the proppant load are deposited at the fracture bottom, much of the upper interval of the crack is left unpropped. Near the wellbore where conditions of time and temperature are least severe, the largest amount of proppant is suspended when the fracture closes at that location. This effect even occurs although crack width is initially greatest next to the wellbore and the fracture takes the longest time to close there.

As fluid stability in the immediate borehole surroundings is very good, however, the pronounced fracture width and extended time of crack closure are not effective for giving rise to enhanced proppant settling. Two means exist for increasing propped fracture height at closure comprising decreasing proppant settling rate by selecting a more stable or viscous fluid, or increasing fluid loss to provoke sooner crack closure. Since fluid leakoff affects the fluid portion available for fracture creation, the latter solution may require some compromise as to what fluid loss characteristics are desirable. Tail-in proppant scheduling (cf. sections 2.4.1.2.3. and 4.12.3.3.) increases proppant settling in the wellbore vicinity due to higher density or larger grain size of the tail-in material, with the tailed-in proppant lot therefore also requiring greater gel strength and stability to keep it suspended.

4.12.2.2.1.4. Other aspects

Fracture temperature calculations which aim on proppant settling estimation have to employ more accurately represented flow rates and fluid-loss rates. Fracture closure calculations are determining proppant distribution in the closed crack. Treatment costs are reduced by staging gel, stabilizer and breaker concentrations based on fluid behaviour in the fracture.

Fluid loss to the formation by leakoff decreases fluid velocity and increases proppant saturation and slurry viscosity, thereby diminishing proppant settling possibilities. Drag of fluid and proppant towards the fracture walls is negligible for most stimulation conditions. Settling velocity of the proppant-fluid slurry is influenced by the combination of fracture geometry, fluid loss, fluid heating and reservoir formation cooling (BABCOCK, PROKOP & KEHLE 1967; NOVOTNY 1977).

4.12.2.2.2. Proppant banking and formation grain size distribution

Proppant banking in the fracture decreases the cross-sectional area for the flow which increases fluid velocity. Since proppant bank permeability is much greater than formation permeability, fluid loss can be assumed to occur over the entire permeable reservoir sand even if it is covered by a proppant bank. With continuing building of the proppant bank, fluid velocity across its top may become so large that no more proppant can be deposited, with the proppant bank then reaching its equilibrium height. Some aspects of fining- and coarsening-upwards grain-size sequences as well as hydrocarbon distribution in granulometrical cyclothems are discussed as follows.

4.12.2.2.2.1. Fining- and coarsening-upward grain-size sequences

Proppant banking may be desired to create a particularly dense proppant package in the lower portion of the fracture in such reservoirs where most of the hydrocarbon production is coming from the bottom of the prospective interval (LEE & DANESHY 1985), such as for example in fining-upwards grain size sequences where the highest permeabilities occur at the base of the pay zone. In the reverse case of coarsening-upwards granulometrical successions, however, proppant banking with concentration of the material supporting the fracture at the bottom of the reservoir would cause considerable deterioration of exploitation possibilities, because the best section at the top of the pay zone would hardly receive any proppant plug. Proppant banking is also investigated by SCHOLS & VISSER (1974).

Similarly as applying for coarsening-upwards clastic reservoirs with full oil saturation, proppant settling and banking is detrimental to sandstone pay intervals of either grain size distribution and organization that are divided into an upper oil-bearing section and a lower water-bearing portion by the oil-water-contact. Proppant settling prior to fracture closure would lead in such reservoirs to reduced proppant concentrations in the upper interval with high oil saturations and to increased proppant package density adjacent to the lower water-productive section (CRAMER & SONGER 1988).

4.12.2.2.2.2. Hydrocarbon distribution in granulometrical cyclothems

In ideal cases representing fully hydrocarbon-bearing fining-upwards or coarsening-upwards sandstone sequences, proppant banking can be the optimum solution if fracture propagation in downwards direction stops at the reservoir base to allow upwards bank growth from the actual beginning of the net pay interval and not wasting proppant material by settling into the barren section below the target zone. In such cases, banking results in proppant settling into the full created fracture width instead of the crack closing on a less concentrated and narrower pack such as set in place by perfect-transport fluids (HOEL 1988). If high-quality proppants are used, a sufficient conductivity contrast is always given everywhere in the fracture regardless of whether the proppant banks are opposite finer-grained low-permeability or coarser-grained high-permeability basal members of coarsening- or fining-upwards cyclothems, respectively, or whether the open flow channel is situated next to the corresponding upper coarse and fine units.

4.12.2.2.3. Fracture closure and fluid viscosity

Crosslinked fracturing fluids have the unique advantage of yielding high viscosity for modest polymer concentration which makes them ideal for placing higher proppant saturations due to inherent high efficiency, the ability to create wide fractures, and excellent proppant transport and suspension characteristics (HANNAH, HARRINGTON & POTTER 1980; cf. section 4.3.4.3.). In case of large treatments, fracture closure does not occur for several hours which would allow ample time for complete proppant settling in water-based gels, whereas crosslinked fracturing gels can retain the ability to support proppants for several hours at elevated temperatures unless their viscosity is degraded to some critical point by breaker action or absolute thermal degradation.

If breaker action significantly reduces the time during which perfect proppant support is present and fracture closure is retarded, then imperfect proppant distribution will result and the overall performance may be no better than if a simple gel were used. On the other hand, if breakers significantly increase fluid loss during closure, then fracture closure time may be reduced and thus a favourable proppant distribution may be safely trapped between the shutting crack walls. In low-viscosity fracturing fluids, proppants may settle rapidly, thereby leaving a short crack with low bed height which could result in only the bottom few feet of the zone of interest actually being propped with a conductive fracture (KUNDERT & SMINK 1979). Fracture closure usually occurs at the end of linear flow (KULAR, CHHINA, BEST & MACKENZIE 1988; cf. section 4.11.3.).

4.12.2.2.4. Other aspects

Principally proppant settling is a function of fluid shear rate, but when particles are flowing through a fracture, dragging near the crack walls hinders grain settling, with wall roughness also being an important factor (ACHENBACH 1974). Proppant saturation probably has the greatest effect on settling velocity of the solid particles, with progressive fluid loss during course of the hydraulic stimulation treatment resulting in successively higher fluid viscosity and proppant concentration and thereby proppant settling occasionally being severe during initial stages, but being inhibited more and more towards the end of the operation by better suspending and supporting conditions in the more saturated and more viscous carrier medium (NOVOTNY 1977). Static and dynamic proppant settling velocities are different in shear-sensitive fluids.

Experimental work by BIOT & MEDLIN (1985) and MEDLIN, SEXTON & ZUMWALT

(1985) for thin uncrosslinked fluids has revealed four regions of different proppant transport phenomena. The lowest zone is the settled bank where concentration is a function of proppant characteristics. In upwards direction, the following segments of the flow profile comprise bed load being a fluidized layer of relatively small height, zone of viscous drag transport where proppant saturation is more or less constant, and interval of turbulent transport through which concentration declines to zero.

Aspects of proppant transport are also discussed by KERN, PERKINS & WYANT (1959), LOWE & HUITT (1966); CLARK & QUADIR (1981); SIEVERT, WAHL, CLARK & HARKIN (1981); CLARK & GULER (1983); GOTTSCHLING, ROYCE & SHUCK (1983); AHMED (1984); CLARK, HALVACI, GHAELI & PARKS (1985); ROODHART (1985) and ACHARYA (1986, 1987). Similar effects of transport and settling as exhibited by proppants in fracturing stimulation fluids apply for rock cuttings in drilling mud (PEDEN & LUO 1987).

4.12.2.3. Carrier fluid viscosity

Optimization of proppant distribution within the fracture and minimization of proppant settling can be best achieved by using high proppant saturations (cf. section 4.3.) and shortening closure time of the fracture (BARBY & BARBEE 1987). The following overview presents some aspects of crosslinking types, proppant transport capacity, and fluid rheology stability.

4.12.2.3.1. Crosslinking types

Proppant settling rates are particularly critical in low-viscosity carrier fluids and with long fracture closure times. Concerning viscosity modelling, different types of crosslinkers can be used, and shear forces caused by pumping of the viscous crosslinked gel down the tubulars could provoke reduction of the effective viscosity (cf. section 4.3.4.3.). Conventional crosslinking agents lead to immediate crosslinking, whereas a delayed crosslinker could be timed to crosslink as moving downhole, thereby building up viscosity during the entire pumping time. Dual-viscosity crosslinkers form at once a slight crosslinking and then increase viscosity by further crosslinking while being displaced. Delayed and dual-viscosity crosslinkers should not be as extensively affected by the shearing force in the tubulars as the conventional crosslinker (GARDNER & EIKERTS 1982).

4.12.2.3.2. Proppant transport capacity

Concerning fluid viscosity vs. proppant settling, low-viscosity fluids (10 - 20 cp) lead to gravitational differentiation and concentration of much proppant material to the bottom of the fracture, and bed height can be irregular if different particle sizes are used during the treatment (cf. section 2.4.1.2.3.). Medium-viscosity fluids (50 - 200 cp) provide good, but not perfect proppant transport, and proppant beds can still be irregular. Medium-viscosity fluids carry only 20/40 or smaller proppants for long distances, whereas larger grains will settle more rapidly (the dependency of settling rate from particle size is also illustrated by LOCKETT & AL-HABBOOBY 1973 and MIRZA & RICHARDSON 1979). High-viscosity fluids can transport all proppant grain sizes with only minimal settling, and the resultant fracture usually is very uniform (HOLDITCH 1984).

Aspects of proppant distribution in vertical fractures are also reported by CLARK & QUADIR (1981); SIEVERT, WAHL, CLARK & HARKIN (1981) and BABCOCK, PROKOP & KEHLE (1987). Proppant settling is discussed by KIRKBY & ROCKEFELLER (1985), ROODHART (1985) and SHAH (1986). CLARK, HALVACI, GHAELI & PARKS (1985); DUNAND & SOUCEMARIANADIN (1985) and ACHARYA (1986, 1987) comment on particle transport in viscous and viscoelastic fracturing fluids. SHAH & LEE (1986) analyze friction pressures of proppant-laden hydraulic fracturing fluids.

4.12.2.3.3. Fluid rheology stability

The rheological properties of the carrier fluid can vary significantly due to temperature changes in the fracture and the time duration of exposure to each temperature (LEE & DANESHY 1985). Viscosity loss of the carrier due to time and temperature can lead to inability of further proppant suspension, but proppant settling can be prevented by replacement by two-stage gelled fluids where the second gelling agent is activated at higher temperature. Many fluids, however, can maintain their viscosity even up to temperatures approaching 300 oF and thus proppant suspension at least for several hours is guaranteed. Tail-in pumping schedules with insertion of first stronger and second weaker gels or vice versa are also possible. Drilling and fracturing fluids are generally representing power-law-type fluids, and the problem of rock cutting settling out from drilling mud and of proppants from fracturing fluids is complicated by the shear-dependent characteristics of the fluids (PEDEN & LUO 1987).

4.12.2.4. Pad and carrier fluid volumes

The danger of excessive proppant settling can also be minimized by favourable variations of pad and carrier fluid volumes (LEE & DANESHY 1985). When fluid efficiency is low, either a large amount of primary pad is required to complete the treatment, or if the leading proppant-laden elements lose their suspended proppants, they can serve as secondary pad and thus the quantity of primary pad can be reduced. Determination of an optimum amount of primary pad has to be based on the joint consideration of fluid loss and proppant settling. If the leading proppant does not settle out, propped fracture length can be controlled by varying the primary pad volume. The less pad remaining at the end of this stage, the longer the propped fracture will be. As fracture width at the tip is narrow and a little extra volume of pad will occupy a long region, propped fracture length may thereby be reduced. Too little pad, however, will increase the chance of proppant reaching the fracture tip, thereby creating high pressure. Alternative treatments include two-stage gelled fluid pumping where the second gelling agent is activated at higher temperatures.

In tight formations with low fluid leakoff rates, fracture closure time is inevitably very high, and thus it is practically impossible to keep proppant from settling to the fracture bottom at closure. In order to provide an adequate amount of proppant in the pay zone, proppant concentration and volume should be as high as possible without risking premature screenout failure (cf. section 6.2.4.2.1.).

Fracture geometry is most influenced by the fluid viscosity at the tip of the crack, with the higher the viscosity, the shorter the fracture length for a given injected volume. Therefore in order to create narrow long fractures, fluid viscosity at the tip of the crack has to be maintained low, whereas for wider fractures, fluid sequences resulting in higher viscosity at the crack tip have to be selected.

4.12.2.5. Slurry consistency and proppant clustering

KIRKBY & ROCKEFELLER (1985) analyze proppant settling velocities in non-flowing slurries, because static sedimentation behaviour is important in proppant transport not only prior to closure when the placed slurry is standing in the crack, but also during pumping and fluid insertion into the fracture, as a high percentage of the proppant slurry resides in regions of low-shear rate. Sedimentation behaviour of fracturing fluids is dominated by particle clustering which gives rise to average static slurry settling velocities many times faster than that of single particles (CLARK, HARKIN, WAHL & SIEVERT 1977; SIEVERT, WAHL, CLARK & HARKIN 1981; cf. section 4.12.2.2.1.), with the consequence of clustering being an inhomogeneous slurry and a time-dependent distribution of settling velocities (TORY & PICKARD 1977). Some comments are offered as follows on proppant clustering and fluid viscosity variation as well as slurry sedimentation and proppant cluster formation.

4.12.2.5.1. Proppant clustering and fluid viscosity variation

Proppant clustering occurs in all fracturing fluids at typical proppant concentrations, with settling velocity depending on proppant saturation in a much different manner than assumed by conventional proppant transport models (NOVOTNY 1977). Proppant clustering can therefore significantly increase proppant settling velocity independent of fluid type (KIRKBY & ROCKEFELLER 1985).

Fracturing fluid viscosity varies greatly within the crack due to wide variations in local shear rate (KIRKBY & ROCKEFELLER 1985), with the viscosity of some of the fracturing fluids being extremely sensitive to shear rate (CLARK 1979). The proppant plug flowing in the center of the crack resides in a region of very low shear rate (NOLTE 1982). As a consequence of viscosity differences at low shear rates and static settling velocity, rheological characterization of the fluids at low shear rates is essential for differentiating suspension performance (cf. also CLARK, HALVACI, GHAELI & PARKS 1985).

4.12.2.5.2. Slurry sedimentation and proppant cluster formation

The slurry sedimentation behaviour indicates significant differences between Newtonian and fracturing fluids (KIRKBY & ROCKEFELLER 1985). Newtonian fluids exhibit relatively uniform concentration in the slurry phase as well as defined fluid/slurry and slurry/proppant bank interfaces, whereas fracturing fluids display multiple concentration phases without distinct interfaces. As the primary slurry body settles, a slow-settling dilute-slurry phase remains above. Dense proppant banks form very slowly, with a slow-settling, less dense slurry phase appearing above the bank. Some fracturing fluids deposit dense proppant banks that are highly fluid and can readily be redispersed, whereas Newtonian fluids lay down immobile proppant banks.

An important aspect of clustered sedimentation is the dependence of slurry-settling velocity on distance, time and local concentration (TORY & PICKARD 1977, KIRKBY & ROCKEFELLER 1985). Rapid settling does not occur until the particles have traversed sufficient distance to allow the formation of clusters. The natural instability of the deposition process leads to inhomogeneous proppant concentration (THACKER & LAVELLE 1978). If flow within a fracture is highly non-uniform, average settling velocity is determined by the suspension behaviour of the slow-moving zones which tend to have higher viscosities. Slow-moving horizontal fluid bands with high apparent yield stress can greatly inhibit proppant settling, thereby acting as a sedimentation barrier.

4.12.3. Proppant flowback

Proppant flowback after completion of the hydraulic stimulation can seriously damage fracture conductivity by removing the support of the crack to stay open and thus allowing the fracture to partially close and to disconnect the drainage path. Proppant flowback into the wellbore is thus generally undesirable since the primary goal of hydraulic fracturing is to place proppant in the created fracture to increase crack conductivity and thereby ameliorate reservoir productivity via a greater effective wellbore radius (HOLCOMB 1983). Apart from fracture conductivity damage especially in the proximal seam around the borehole, proppant flowback causes wellbore fill which must be circulated or bailed out before the borehole can be produced efficiently, and erosional and plugging damage to tubulars, wellhead assemblies, valves, chokes and flowlines, with this equipment destruction being a serious deterioration of both technical performance and economical feasibility. Proppant distribution and thus also overall fracture quality are severely affected by proppant inflow into the wellbore (CLARK 1983).

In addition, flowback of highly abrasive sintered bauxite proppants can be extremely hazardous, with not only cutting out of valves and christmas trees occurring where large quantities of sintered bauxite proppants are returned with well fluids, but also in extreme cases blowouts and fires taking place (OSBORNE, McLEOD & SCHROEDER 1981; cf. section 1.4.6.). The most important procedures of controlling and avoiding proppant flowback are careful cleanup of the well after the fracturing job, pumping of a tail-in of curable resin-coated sand in the terminal stage of the hydraulic stimulation job, utilization of foam and energized fracturing fluids, and shut-in time adjustment.

4.12.3.1. Well cleanup

Well cleanup has a very important impact on fracture performance due to its control on proppant particle adjustment during crack closure with declining pressure level. The influences of borehole cleanup are summarized by briefly commenting on fracture closure and proppant trapping, proppant flowback control and diminution, and effects of fracturing fluid and capillary pressure on well cleanup.

4.12.3.1.1. Fracture closure and proppant trapping

The effectivity of a hydraulic fracturing treatment is mainly controlled by final pumped proppant concentration and crack width at the wellbore, with these two factors predominantly determining optimum well productivity (OSBORNE, McLEOD & SCHROEDER 1981). Offtake capability of the fractured well is thus often governed by the connection of the last volume of injected proppants to the wellbore. Therefore not only crack evacuation has a considerably detrimental effect on fracture performance, but also proppant overdisplacing can be disastrous when proppant is disconnected from the borehole (RAYMOND & BINDER 1967, SMITH 1973). In the latter case, fluid must flow out of the fracture and through the low formation permeability into the wellbore over the disconnected distance. Some aspects of pressure depletion and fluid leakoff as well as pressure decline assistance by controlled fluid production are illustrated as follows.

4.12.3.1.1.1. Pressure depletion and fluid leakoff

In low-permeability reservoirs, 12 - 24 hrs may be required before the pressure depletes sufficiently to permit fracture closure and trapping of the proppants between the crack walls, whereas in high-permeability pay horizons, fracture closure may already be achieved in a time between a quarter to a couple of hours. As 12 - 24 hours is also the time range when most of the added gel breakers start to become active and begin to decompose the gel including viscosity lowering, it may be necessary to assist fracture closure by slowly and carefully flowing the well back on low rates and through a narrow choke in order to bleed off pressure from the crack. Production rates, however, have to be adjusted to keep proppant flowback limited to a minimum in order not to deteriorate proppant plugging of the fracture particularly in the wellbore surroundings. After fracture closure is detected or when pressure has fallen down below the known value of closure stress, the well has to be shut in to allow the gel to break. For fracturing and closure pressure prediction and recognition as well as for design of mud, casing and cementing programs, fracturing technique and proppant type, accurate values of in-situ stress are needed.

Closure time which is very critical for proper well cleanup modelling can be understood as the period necessitated for either complete fluid leakoff or total proppant settling and trapping (LEE & DANESHY 1985). The time required for all suspended proppants either to settle down or to be trapped between the fracture walls (POULSEN & LEE 1984) gives a better approach, because the period needed for the created fracture volume to lose all of its fluid when pumping stops (NOLTE 1979) can be excessively high in case of slow fluid leakoff in tight reservoirs.

4.12.3.1.1.2. Pressure decline assistance by controlled fluid production

It is certainly a critical decision to produce fluid back in order to assist pressure decline which automatically involves flowback of at least little quantities of proppants from the fracture into the wellbore, but it is often the lesser of two evils, because proppant settling down to the bottom of a persistently open wide crack and even potentially out of zone is the worst possible case of fracture damage and would deteriorate the effect of the whole treatment (ROBINSON, HOLDITCH & WHITEHEAD 1986). Therefore everything has to be undertaken to ensure that the crack heals before the gel breaks and allows proppant settling, and cleanup and production should not be started before the proppant is trapped in the closed fracture.

Special care has to be taken to prevent proppant disconnection in case it is necessary to shut down pumping operations to repair leaks or to work on equipment malfunction (OSBORNE, McLEOD & SCHROEDER 1981). Shut-down has to be made with crosslinked gel and proppants still being in tubing and blender until repairs are made and then enough volume of base gel without any fluid-loss additive, crosslinker or proppants is pumped to re-establish rate and pressure which is followed by starting of simultaneously pumping crosslinker, fluid-leakoff additive and proppant at the same concentration as before shut-down. It is not advantageous to start up again with a lower proppant concentration than previously pumped or running any crosslinked gel without proppant. The thin noncrosslinked gel in the intermediate stage fingers into the more viscous crosslinked gel and by this as well as by fluid leakoff will be absorbed to the point where no disconnection will occur between the proppant-laden elements.

4.12.3.1.2. Proppant flowback control and diminution

Well flowback and cleanup after closure of the crack has to be performed very carefully, because in the first hours to days after the fracture operation, the proppant particles are adjusting and packing due to the closure stress which is exerted on the proppant. If the fracturing fluids are produced back too fast at too high rates, proppant failure or flowback into the borehole can occur. Some points of fluid flow velocity and pressure drop, and fluid viscosity and well cleanup are illustrated as follows.

4.12.3.1.2.1. Fluid flow velocity and pressure drop

Broken gel still retains some viscosity and thus imposes a much greater drag on the proppants than does dry gas, and therefore adjusting and packing proppants can be carried away from the fracture by the produced fluids. As pressure drop and fluid velocity reach their maximum in the immediate vicinity of the wellbore, proppant flowback and crushing is greatest in the very surroundings of the borehole, and this effect can only be avoided by carefully adjusting cleanup to very low rates and maximizing back-pressure against the formation. Proppant flowback can also be retarded by intermixing angular grains into the population of round particles (LARSEN & SMITH 1985; cf. sections 1.3.5. and 1.4.3.) or the application of resin-coated material (POPE, WILES & PIERCE 1987; cf. section 4.12.3.3.).

Fluid flow velocity reaches its maximum in the nearest zone of the wellbore because of funnelling of the flow lines from the whole fracture height to the perforated interval which (according to the limited-entry technique; HOWARD & FAST 1970 a, SMALL 1985, CRAMER 1987) comprises usually only parts of the former and pressure drawdown becomes particularly critical due to concentration to the narrow fracture where linear flow instead of radial flow in unstimulated reservoirs occurs (cf. section 4.11.3.) and focusses the effect of increasing closure stress to some small sections of the pay horizon (such pressure distribution funnelling is illustrated for Rotliegend and Carboniferous gas wells in Germany FRG by BRINKMANN, KRÖMER & REINICKE 1980 and REINICKE, BRINKMANN, SCHWARZ & HUENI 1985; cf. sections 2.4.1.1. and 3.3.2.).

4.12.3.1.2.2. Fluid viscosity and well cleanup

Improved well cleanup with significantly decreased risk of proppant flowback can be provided by delayed foamed crosslinked fracturing gels which in comparison to conventional linear gels have much better proppant transport capability and suffer much less from fluid loss due to their higher viscosity and better developed two-phase interlocking structure (FREEMAN, BILDEN & HOSSAINI 1986; cf. section 4.3.4.3.). Conventional linear foamed gels have the disadvantage that in order to minimize nitrogen or carbon dioxide leakoff for the purpose of maximizing liquid recovery and limiting proppant settling, well cleanup has to start already a couple of hours after termination of pumping, with energy and velocity of foam returns combined with the relatively short shut-in time often giving rise to excessive proppant flowback and recovery. The higher viscosity and stability of delayed foamed crosslinked gels enables better support and profile of the proppant package within the fracture, and because phase separation or gas leakoff is significantly reduced with a crosslinked foam, the start of load recovery by well cleanup can be delayed which allows for sufficient crack closure restraining the proppant package. The significance of initial production via a small choke at restricted rates for maintenance of optimal fracture performance is also underlined by KUNDERT & SMINK (1979).

OSBORNE, McLEOD & SCHROEDER (1981) recommend adding of extra breaker to the crosslinked gel in the last fluid pumping stage which helps to obtain lower fluid viscosity after fracture closure in the section adjoining to the wellbore, thereby improving fluid recovery and well cleanup. Assistance can be given by taking samples from the last fluid injection stage into the laboratory and supervising breaker reaction and viscosity evolution in order to better determine the most suitable time for termination of well shut-in and beginning of fluid cleanup.

4.12.3.1.3. Effects of fracturing fluid and capillary pressure on well cleanup

SOLIMAN & HUNT (1985) illustrate some effects of fracturing fluid and capillary pressure on well cleanup which also have implications on proppant performance in the fracture. After the total stimulation treatment volume has been pumped, fluid loss will continue while the pressure in the fracture dissipates. Imbibition may cause treatment fluid to penetrate even deeper into the formation after the fluid pressure in the crack has equalized with the pore pressure of the reservoir. As a result of the fracturing operation, stimulation fluid exists inside the crack and for a certain depth within the formation around the fracture. The invaded zone has a reduced relative permeability to gas due to the presence of both gas and liquid. With beginning offtake, initially mostly treatment fluid is produced thus cleaning up the pay zone and leading to increasing gas saturation in the invaded interval, with the cleanup pattern depending on formation and crack properties. Aspects of dimensionless fracture conductivity and capillary pressure are briefly outlined as follows.

4.12.3.1.3.1. Dimensionless fracture conductivity

The dimensionless fracture conductivity or flow capacity has particularly significant impact on well cleanup (SOLIMAN & HUNT 1985), with special emphasis being on time and location of gas breakthrough in the invaded zone. At low dimensionless fracture conductivities, the gas breakthrough into the crack occurs near the wellbore, implying that except for the borehole vicinity, gas flow into the fracture is obstructed by the presence of stimulation fluid thus greatly reducing crack effectiveness. If pressure tests are conducted under such conditions, the fracture will behave as if being of much shorter length comprising only abt. 20 - 30 % of the actual length.

As the dimensionless fracture conductivity increases, the change in treatment fluid saturation inside the crack becomes more gradual, and at a value of abt. 50, stimulation fluid concentration in the fracture gets almost uniform. At higher dimensionless fracture flow capacities between 200 and 1,000, a mild gas breakthrough occurs at the crack tip. In order to help avoiding early and severe gas breakthrough at the wellbore, the dimensionless fracture conductivity should be at least 50. The optimum dimensionless fracture flow capacity is in the neighbourhood of 100 or abt. three times the value required to produce the formation fluid efficiently and quickly (cf. also section 4.8.11.1.1.).

4.12.3.1.3.2. Capillary pressure

In tight gas reservoirs, capillary pressure can be very high thus restricting fracturing fluid flow and requiring a very high pressure drawdown for cleanup to take place (SOLIMAN & HUNT 1985). High capillary pressure reduces the effectiveness of a fracture to produce stimulation fluids by preventing water from flowing easily, with a minimum differential pressure being required before any water can begin to flow. High capillary pressure also inhibits early gas breakthrough at the wellbore, with gas breakthrough occurring in areas with least water saturation which are usually situated at the crack tip. With increasing capillary pressure, cumulative fracturing fluid production at a certain time declines. High capillary pressure means less water production which in turn results in less reservoir cleanup.

Concerning water block effects on gas offtake and well cleanup, the phenomenon has no serious consequence as long as no permanent damage to the formation had occurred (HOLDITCH 1979; cf. sections 3.11.2.4.2. and 4.8.8.3.1.5.). Waterblock is also of no significance when drawdown pressure gradient in the near fracture face region of the matrix is several hundred psi/in. higher than the capillary entry pressure (ABRAMS & VINEGAR 1985). As a consequence of all these effects, it is recommended to design crack conductivity sufficiently high to adequately cleanup the stimulation fluids (SOLIMAN & HUNT 1985). In some instances, the optimum fracture flow capacity necessary to clean up the formation properly and reasonably quickly may be much higher than that required for gas production from the reservoir. It is also desirable to attempt to reduce capillary pressure effect by any means possible.

4.12.3.2. Proppant grain size

Generally the larger the proppant grain size, the better the proppant would stay in the fracture mouth and flowback can be also minimized by selecting larger proppants (BARBY & BARBEE 1987). Additional benefit could be achieved by careful selective overflushing of the middle and/or upper part of the perforated interval, with slight overflushing sometimes being able to save tremendous expenses for pump repair and wellbore cleanout. Even with large proppant grain size, however, some embedment must occur to hold the proppant in place. If the fracture face at the wellbore is severely eroded by highly-abrasive large proppants or annihilated by large volumes of acid, excessive proppant flowback may occur. These effects can also be avoided by tailing-in resin-coated proppant in order to form a stable wedge near the fracture mouth (cf. section 4.12.3.3.).

Proppant grain size stabilization can already be achieved to some extent by conventional tail-in comprising insertion of coarser proppant grains in the final stage than in the main part of the treatment (cf. section 2.4.1.2.3.4.). The larger proppant grains near the fracture mouth are themselves staying better within the crack due to higher drag resistance and friction, but in addition to that, the coarse proppant tail-in acts like a gravel pack on the finer grains of the medial to distal portions of propped fracture length (cf. section 5.8.8.). In case of proper grain size selection, the coarser tail-in quantity grains bridge the finer main-lot particles at the interface of both proppant stages and prevent their further movement towards the wellbore and out of the fracture mouth. The effect of the coarser proppant tail on the finer proppant main stage is comparable to the impact of perforation tunnel fillings or borehole wall linings with gravel on reservoir formation sand in mechanical sand control (cf. section 5.6.1.). Aspects of proppant return minimization by tail-in function like a screen are also discussed by CROW (1977).

4.12.3.3. Resin-coated proppant tail-in

Proppant package stability improvement by application of resin-coated grains is a common and efficient means of prevention of fracturing operation failure by proppant flowback. The most significant aspects of resin-coated proppant tail-in are outlined as follows.

4.12.3.3.1. Fracture operation failure by proppant flowback

Different possibilities of fracture operation failure by proppant flowback can be distinguished comprising damage shortly after the hydraulic stimulation, shortly after water-flood response, shortly after acid treatment and at random (POPE, WILES & PIERCE 1983). Most of the failures occur within 100 days of the fracturing operation and can be attributed to low closure pressure and increased fluid velocity through the perforations in the shallow oil wells. The high producing fluid level in boreholes that respond well to the fracturing treatment corresponds to a high bottomhole flowing pressure and consequently a low closure pressure which can be insufficient to keep the proppant in place, and increased fluid velocity through the perforations can displace the proppant into the wellbore. Higher fluid velocities and rates through the perforations are also reasons for proppant flowback after response to water flooding and acidizing. Acidizing can even destroy a calcite scale which acts as a cement holding the proppant in place in the surroundings of the wellbore.

Proppant flowback does not only result in fracture damage by removal of the support of the crack and thereby triggering its partial closure, but similarly as does sand production also leads to borehole operational problems such as stuck pump plungers, eroded pump valves, sucker rod failures, stuck tubing and proppant fill (POPE, WILES & PIERCE 1987; cf. section 5.2.1.3.). Proppant flowback can also fill surface equipment with sand which gives rise to inaccurate fluid measurements and eventually requires extensive cleaning. In many cases, the cost to return a well to oil production after failure by proppant flowback has been so excessive that the well had to be prematurely abandoned.

In terms of fracture evacuation by proppant flowback in the near-wellbore area, RAYMOND & BINDER (1967) document that nearly all of the potential stimulation is lost if the unfilled part of the crack is only 0.5 % of the total fracture length and if this portion is adjacent to the borehole (cf. also SINCLAIR, GRAHAM & SINCLAIR 1983). In addition, crack evacuation by proppant flowback causes fracture diameter narrowing to that width where crack healing occurs, and even if complete closure does not take place, a narrow fracture can be easily plugged by spalled pieces and fine formation particles.

4.12.3.3.2. Package stability improvement by resin-coated proppants

Several alternative methods have been used to control proppant flowback following fracture stimulation comprising raising the pumping equipment above the perforations, using special sand pumps, installing a gravel pack, and running a tubing-conveyed sand filter, but sufficient success was not achieved with any of these procedures. Satisfactory control of proppant flowback, however, could be obtained by performing the fracturing treatment with a tail-in of curable resin-coated sand (cf. section 1.2.6.) of abt. 10 - 20 % of the total proppant quantity. The following discussion summarizes some points of proppant flowback control as well as tail-in length and additional gravel packing stabilization.

4.12.3.3.2.1. Proppant flowback control

Under conditions of reservoir temperature and pressure, the resin chemically bonds together to form a consolidated permeable barrier against proppant flowback once carried out properly (cf. section 1.2.6.). Resin-coated proppant tail-in effectively prevents fracture evacuation in the near-wellbore area, because the consolidated proppant package firmly couples borehole and crack thus leading to optimum drainage of the productive formation (SINCLAIR, GRAHAM & SINCLAIR 1983). In addition to inhibition of fracture evacuation, proppant flowback prevention greatly reduces well maintenance, wear, erosion and abrasion problems.

Failure rates in fractures with a wedge of resin-coated proppants in the terminal interval near the wellbore are much lower than in conventionally equipped boreholes. Failures of the resin-coated sand to control proppant flowback can be attributed to failure of resin bonding and failure of the resin-coated sand to cover the whole perforated interval. If the resin cures before the gel breaks and thus before the proppant grains come into contact, the ability of consolidation of the proppant package is lost. If the proppant is being banked near the wellbore during stimulation, the curable resin-coated sand in the tail-in will override the top of the banked proppant and cover only the perforations that are taking fluid.

Subsequent acidizing does not affect the stability of the consolidated package due to resin inertness. Productivity losses due to fracture evacuation near the wellbore by flowback of uncoated proppants have turned out to be much higher than reductions of exploitation due to slightly lower flow efficiencies of resin-coated proppant consolidated packages with respect to those of cracks propped with conventional sand, thus underlining that a tail-in of resin-coated sand is an effective method of controlling proppant flowback following fracture stimulation. Proppant flowback diminution or elimination by resin-coated proppant tail-in in combination with small-choke cleanup is also outlined by GARBIS, BROWN & MAURITZ (1985).

4.12.3.3.2.2. Tail-in length and additional gravel packing stabilization

In terms of length of the resin-coated proppant tail-in, the optimum is abt.

100 ft in most treatments of standard or average size, because in case of shorter tails some proppant flowback can still occur due to nonuniform proppant distribution and longer tails do not give a superior effect, but are only a premium investment (cf. section 2.4.1.2.3.). Proppant flowback minimization after fracturing treatments by application of resin-coated tail-in is also discussed by JEU, PEREZ & WAY (1988). Similarly as applying for total proppant quantity, tail-in volume and length have to be adjusted to total crack length in order to be effective. Wider fractures near the borehole require larger tail-in quantities to generate the multilayer packing to the designed length, and may also demand longer tail-ins in order to provide additional package stability.

Resin bonding of tail-in lots for proppant stabilization and creation of a consolidated wedge in the borehole vicinity is particularly significant in wider cracks where at least in the central part some more instability than usual can occur. Shorter tail-ins than according to standard design can be sufficient if their primary purpose is to carry radioactive markers either within the proppant crystal lattice or as surficial pellicles for fracture height monitoring by gamma-ray logging (cf. section 6.2.1.3.). In extreme cases, tail-in length may reach from abt. 1 m (3 ft) in very small-scale treatments of mini- or micro-size to abt. 1,000 ft (330 m) in jumbo-type massive hydraulic stimulation jobs where total fracture length can be up to 4,500 ft (1,500 m). In terms of comparison of job size, proppant quantity and fracture length portion required for a tail-in of a MHF operation can be larger than those for a whole smaller-scale hydraulic stimulation treatment.

Particularly in poorly-consolidated formations, both proppant and formation sand flowback into the well can be effectively controlled by setting a gravel pack after the termination of the hydraulic fracturing treatment (this applies especially for soft sands and chalks). Gravel pack stabilization of both reservoir sand and proppant package in the fracture is the safest way of maximum conductivity and effectivity preservation in unconsolidated pay zones, because properly set gravel pack mantle and perforation tunnel plugs hold back everything which tries to move (aspects of combination of hydraulic proppant fracturing and gravel packing are also discussed in sections 4.5.4.3. and 5.3.5.).

4.12.3.4. Foam and energized fracturing fluids

Proppant flowback prevention can also be achieved with utilization of foam and energized fracturing fluids (HOLCOMB 1983; cf. section 3.11.2.). If properly carried out, well cleanup after foam or energized stimulation treatments provides maximum fluid load recovery and minimum proppant flowback into the borehole or to the surface. Caution has to be exercised, however, on velocity of well cleanup, because too rapid fluid production could cause foam regeneration which would then again be able to carry proppant from the fracture mouth into the borehole. While some proppant flowback is associated with almost any type of fracturing treatment and causes little harm if kept to a minimum, the added influence of gas assist from a foam fracturing operation can increase the magnitude of this problem significantly if cleanup is not properly carried out. Proppant flowback can be minimized by adequately flushing the proppants at least to the perforations, slight foam overflushing in order to anticipate some regeneration, and utilization of an adjustable wellhead choke assembly with a high steel alloy or tungsten carbide material in the valve seating. Proper procedure for foam flowback allows the foam to break back uniformly as it leaves the fracture through the perforations and travels up the wellbore tubulars to the surface, thus leaving the proppant in the fracture.

Although it is generally assumed that once crack closure is achieved, it is irreversible, it could be possible for gas expansion within the fracture to reopen the crack and allow some proppant flowback, with continuous cleanup monitoring helping to prevent the occurrence of this effect. Therefore it is desirable to maintain the flowback pressure at a level approximately 200 psi below the pressure necessary to keep fracture closure stress and thereby leave the proppant trapped within the apparently healed crack system.

4.12.3.5. Shut-in time adjustment

Significant influence on well cleanup and proppant flowback can also be achieved by shut-in time adjustment (PAI & GARBIS 1983 b). Productivity as related to fluid retention is a function of shut-in period after treatment, and substantial productivity impairment occurs if the well is left shut-in after stimulation for long periods of time. If for some reason the well has to be left shut-in, it should be stimulated just prior to opening. In order to expedite cleanup, the well should be opened within abt. 30 min after an acid treatment and within abt. 2 - 4 hrs after a fracturing operation, with fracturing stimulations having to be left shut-in longer in order to allow for the gel to break and the crack to heal. Excessively long shut-in times lead to dissipation of stimulated gas into the formation, giving rise to substantial energy and pressure loss which is then lacking in fluid recovery.

4.12.4. Proppant crushing

Apart from or in addition to proppant flowback, more or less severe proppant crushing can occur during cleanup and back-production as a consequence of large pressure drawdown and resulting high closure stress (ROBINSON, HOLDITCH & WHITEHEAD 1986). Following termination of the hydraulic stimulation operation, there is always a pressure drop in the fracture due to its finite conductivity (CINCO-LEY, SAMANIEGO & DOMINGUEZ 1978; CINCO-LEY & SAMANIEGO 1981), as infinite fracture conductivities (GRINGARTEN, RAMEY & RAGHAVAN 1974, 1975) where no pressure drop occurs in the crack are hardly, if at all, developed in nature (REINICKE, BRINKMANN, SCHWARZ & HUENI 1985). Some aspects of proppant shock loading, pressure drawdown and hydrocarbon production rates, and production cycling by repeated well shut-in are briefly outlined as follows.

4.12.4.1. Proppant shock loading

Field experience has shown that because the maximum closure stress did not reach the boundary level of general proppant stability in a number of cases, the explanation of abundant crushing is that large rapid increases in closure stress lead to shock loading of the proppants thus causing greater failure than usually observed with gradual loading (ROBINSON, HOLDITCH & WHITEHEAD 1986). Application of high-strength proppants minimizes proppant crushing by shock loading, but in wells where the fracture is infilled with low- to intermediate-strength proppants, shock loading with excessive closure stresses can be severely detrimental to borehole performance, and significant crushing of proppants in the vicinity of the wellbore where pressure drawdown and fluid flow rates are greatest is seriously deteriorating the effectiveness of the entire created fracture (McDANIEL & WILLINGHAM 1978, CARROLL & BAKER 1979). Aspects of sand stresses around the borehole are also discussed by RISNES, BRATLI & HORSRUD (1982). Some comments are offered as follows on bottom-hole flowing pressure and choke size as well as stress corrosion cracking.

4.12.4.1.1. Bottom-hole flowing pressure and choke size

Fracture damage can also occur long time after the fluids have been cleaned up. Installation of large chokes on wells in order to increase oil or gas production can also cause rapidly rising closure stress and shock loading of the proppants. Therefore it is very critical to avoid exerting quick and excessive closure stresses on the proppants by minimizing drawdown during both fracturing fluid cleanup and later hydrocarbon production. Flowing the boreholes too hard in order to attempt to produce more oil or gas often only leads to fracture deterioration, because in most reservoirs, the additional pressure drawdown does not result in a substantial increase in exploitation, as illustrated by gas deliverability relationships. If the bottom-hole flowing pressure of the well is equal to 50 % of the average reservoir pressure, the borehole will produce abt. 75 % of the absolute open flow. Reducing the bottom-hole flowing pressure below this point will not result in reasonable amelioration of production, but may cause fracture damage (ROBINSON, HOLDITCH & WHITEHEAD 1986).

Only gradual loading of the proppants occurs if the well is produced on a small choke, and the proppant will never be exposed to the high closure stresses which might occur if a large choke is used. Concerning application of low- to intermediate-strength proppants, only a minimal amount of proppant crushing may occur when producing with a small choke, whereas irreparable damage could take place when using a large choke. Severe proppant crushing due to high closure stresses is a major problem when excessive pressure drawdown occurs either in early or late stage of production, because proppant crushing reduces fracture conductivity to the point where well performance is seriously affected and ultimate recovery is dramatically reduced (ROBINSON, HOLDITCH & LEE 1983).

4.12.4.1.2. Stress-corrosion cracking

Another aspect is stress corrosion cracking of proppant grains depending on the fluid medium (CUTLER, ENNISS, JONES & CARROLL 1983; cf. section 1.4.10.3.2.). The earlier decrease in conductivity in water as compared to that in nitrogen in proppant testing is ascribed to stress corrosion cracking, indicating that fracture propagation rates depend on the environment at the crack tip and higher water vapor concentrations increase fracture growth rate. Therefore proppant conductivity testing should be carried out in aqueous environments since this represents the worst possible case situation. Proppant placement in the fracture in the reservoir should thus be performed with foamed fluids of high quality to guarantee that following quick cleanup of the foam with high content of gaseous phase, the proppant package is immediately invaded by formation gas and therefore has only limited possibilities of contacting aqueous phases.

4.12.4.2. Pressure drawdown and hydrocarbon production rates

Aspects of pressure drawdown and hydrocarbon production rates are discussed along the lines of careful slow starting of well production after hydraulic fracturing and gravel packing, effect of crushing on conductivity of sand and bauxite, fracture conductivity degradation vs. residual preservation, and proppant crushing and closure stress resistivity intervals.

4.12.4.2.1. Careful slow starting of well production after hydraulic fracturing or gravel packing

A shut-in or producing well should therefore never be opened on a large choke instantaneously. Starting with small chokes, increasing choke size in gradual increments in reasonably wide time steps should be performed up to a carefully selected optimum and equilibrium level of hydrocarbon production and fracture damage prevention. In terms of crack width vs. proppant type with respect to crushing ratio, MONTGOMERY & STEANSON (1985) report that even though fractures propped with natural sand often are wider owing to the lower specific gravity of sand, the conductivity of cracks supported by bauxite proppants is up to abt. 30 times higher because of the higher strength and very limited crushing of bauxite proppants. Experience reports of proppant crushing by shock loading and careful slow production to minimize fracture deterioration are given by ROBINSON, HOLDITCH & LEE (1983). Well cleanup problems including repeated blowdowns can often have the consequence that effective stress acting on the proppant package approaches total closure stress (WARPINSKI, BRANAGAN, SATTLER, LORENZ, NORTHROP, MANN & FROHNE 1983). When combined with cyclic fatigue loading, proppant crushing can considerably reduce fracture conductivity.

Similarly as applying for hydraulic fracture conductivity damage prevention, gravel pack permeability deterioration can be inhibited by carefully and slowly beginning well production after termination of the sand control treatment (HALLIBURTON 1988). Total fluid rate and formation drawdown have to be kept to a minimum during the first weeks of production, with offtake starting out with small choke sizes and possibly also reduced volumes of lift gas. Such a procedure should minimize initial gravel movement in wellbore and perforation tunnels and allow the borehole to clean up with minimum drawdown across the newly placed gravel pack. Any rapid readjustment of the freshly installed pack is likely to cause formation fines to invade perforation tunnels and pack mantle and intermix with gravel which would lead to permeability loss. Care should be taken during swabbing not to overstress the gravel pack.

Caution has particularly to be exercised in deep gas wells where the most severe closure stress conditions happen in early production history, because the amount of stress required to open the fracture is at the highest value. If the borehole is allowed to produce at maximum rate or is swabbed hard during initial completion, an extremely low pressure will be temporarily created within the fracture and will cause an unduly high closure stress with the danger of severe conductivity damage by proppant crushing (MONTGOMERY & STEANSON 1985).

4.12.4.2.2. Effect of crushing on conductivity of sand and bauxite

While in some tight gas reservoirs even crushed proppants may provide sufficient conductivity to reasonably drain the hydrocarbon reserves (HOLDITCH & LEE 1979; HUEBINGER, WEBSTER, CHISHOLM, VENDITTO & HUNT 1988), proppant crushing generally causes stratification of the proppant bed which in turn will behave like an inhomogeneous formation and limit fluid flow (DERBY & SMITH 1979). Proppant crushing is evidenced by correlations between closure stress and loss in fracture flow capacity, pressure transient testing relationships, and direct observation of collapsed grains which have been produced through tubing to the surface equipment or have been bailed out of the well during workover (TUCKER 1979). Proppant crushing is also the reason for different production histories exhibited by deep gas wells that have been fractured with natural sand and sintered bauxite (CLARK 1983).

Comparison of performance reveals that boreholes which utilized natural sand as propping medium possess a much more rapid decline in production than wells of equal closure stress that have been treated with sintered bauxite. As sand-fractured boreholes which experience lesser values of closure stress do not show such an abnormal decline of flowrate, the comparative assessment clearly reveals that the more severe decline of some wells is the consequence of sand crushing in contrast to bauxite stability. Similarly as applying for proppant crushing, choke size choice is a very critical aspect also in view of proppant flowback. Under normal circumstances, increasing choke size leads to increasing production rate and decreasing flowing pressure due to larger drawdown, but this occurs only as long as the additional closure stress exerted on the proppants does not reduce fracture conductivity by proppant flowback and/or crushing.

4.12.4.2.3. Fracture conductivity degradation vs. residual preservation

Fracture conductivity degradation as a consequence of proppant crushing and subsequent migration of fines can occasionally occur even though flowing pressure drawdowns were minimized and closure stresses maintained within the established limits for the originally chosen proppant type (HUEBINGER, WEBSTER, CHISHOLM, VENDITTO & HUNT 1988). Case studies, however, reveal occasionally that some wells having suspected fracture degradation have limited post-refracturing histories and only little improvement with respect to the original fracture (cf. section 4.8.9.), and other boreholes have production histories that show no signs of fracture damage by proppant crushing and appear to be capable of draining the reservoir efficiently without requiring a refracturing treatment.

Natural sand instead of intermediate- or high-strength proppants can be selected because the economics of higher-quality proppants in virgin-pressured wells cannot be supported because of the large volumes required for an effective fracture, and closure stresses in partially depleted infill boreholes are low enough that the likelihood of crushing is not significant. These relationships outline that crushing of natural sand or crystalline alumina oxide and silicate proppants which leads to formation of halves and quarters leaves in some cases still sufficient conductivity for profitable draining of the reservoir, although certainly the optimum is avoiding of any proppant damage.

4.12.4.2.4. Proppant crushing and closure stress resistivity intervals

TUNN (1971) reports fracture damage by sand crushing from deep stimulated gas wells prior to commercial introduction of synthetic high-strength proppants. The most important reason for comparatively low productivity increase indexes is the insufficient pressure resistivity of the sand, with rapid crushing of the sand resulting in quick partial or almost complete closure of the cracks. The reduction of the hydrocarbon flow rate after initial improvement back to more or less the original level took place in a few days to a couple of weeks time. The explanation possibilities of this phenomenon comprise lacking pressure resistivity of the sand to formation closure stress after decline of the treatment injection pressure, insufficient pressure resistivity of the reservoir rock which covers, surrounds and includes the undestroyed proppant grains, and lacking sand grains in the proximal part of the fracture near the wellbore, with the consequences of all these mechanisms being fracture closure and conductivity destruction.

As a conclusion of the evaluation of fracture damage and possibilities of its prevention, ROBINSON, HOLDITCH & WHITEHEAD (1986) recommend maximum closure stresses of 4,000 psi for sand, 7,000 psi for intermediate-strength and 10,000 psi for high-strength proppants. In comparison to most results of proppant testing, however, these values are considered to be too pessimistic, but in view of careful planning for long production history up to 20 - 30 years, it should be preferred to go absolutely on the safe side and to better take a more expensive higher-performance proppant rather than risking fracture damage by excessive pressure drawdown in late stages of reservoir depletion.

4.12.4.3. Production cycling by repeated well shut-in

Considerable fracture conductivity damage by both proppant crushing and embedment (cf. section 4.3.3.1.) can also be caused by production cycling due to repeated borehole shut-in and again starting offtake of hydrocarbons (KIM & WILLINGHAM 1987). Shut-in of gas wells having been stimulated by hydraulic proppant fracturing is frequently performed for shorter or longer duration due to demand restrictions, completion workover or other reasons. The stabilized gas production rate following each shut-in period, however, is lower than that of the previous offtake cycle at the same wellbore pressure (REINICKE, BRINKMANN, SCHWARZ & HUENI 1985; VENTURA 1985), because fracture conductivity continues to decrease following the subsequent shut-ins as reflected by pressure buildup analysis.

Repeated production and shut-in of the fracture-stimulated borehole causes cyclic loading of the proppant package in the fracture, resulting in progressive closure of the crack as a consequence of both proppant crushing and embedment and thus giving rise to successively decreasing width of the propped fracture with cycles (KIM & WILLINGHAM 1987). Some comments are offered as follows on irrecoverable fracture conductivity diminution as well as combination of proppant embedment and crushing. Comments are also offered on impact of shut-in time.

4.12.4.3.1. Irrecoverable fracture conductivity diminution

Experimental work has shown that the conductivity of the propped fracture remains at the value attained when maximum closure stress is experienced even if the load is later reduced from the maximum to a lower level. Thus the reduction of crack conductivity is an irrecoverable process, and conductivity diminution due to sustained loading at a constant closure stress below the anticipated maximum closure stress is not significant for estimating ultimate propped fracture conductivity. Closure stress acting on the propped fracture continues to increase at a slow rate with progressive reservoir depletion and is approaching maximum load. Upon shut-in, closure stress is slowly released without, however, having a recovery effect on fracture conductivity, and reopening of the well repeats the process of cyclic loading, thereby resulting in further fracture conductivity reduction.

4.12.4.3.2. Combination of proppant embedment and crushing

The overall decrease in fracture conductivity upon production cycling with repeated well shut-in is the combined effect of proppant embedment into the formation rock at the fracture walls and of proppant crushing which are both permanent damage effects (cf. section 4.3.3.1.). The significance of proppant embedment is reflected by the relationship that the harder the reservoir rock, the lesser is the cyclic deterioration effect on fracture conductivity. Larger proppants of 12/20 mesh size perform better in soft formation rock than in hard reservoirs under cyclic loading, whereas at high closure stresses above 6,000 psi, smaller proppant of 20/40 mesh size show less cyclic loading damage effect in hard pay intervals. High proppant concentrations leading to multilayer propping have an additional benefit by minimizing cyclic loading deterioration of fracture conductivity, because proppant embedment which frequently causes more damage to fracture flow capacity than proppant crushing is significantly diminished in multilayer proppant packages.

As fracture conductivity cannot be regained by releasing closure stress, deterioration of fracture flow capacity should be kept at a minimum by avoiding unduly high closure stress during initial completion and early production, and limiting well shut-in periods in advanced offtake stages to absolutely necessary events, because every cycling of the production process is irrevocably linked with at least some fracture conductivity damage.

4.12.4.3.3. Impact of shut-in time

Substantial well productivity impairment can occur in case of long-time shut-in after fracturing stimulation (PAI & GARBIS 1983). Although a certain time of well shut-in is necessary to allow for gel breakage and fracture healing, excessively long shut-in times give rise to dissipation of stimulation gas in energized treatments into the formation leading to substantial loss in pressure and energy that is required in load recovery. Drainage of foam during too long well shut-in builds up a high hydrostatic fluid column which after recovery of a gas cap causes the flowback to cease (HOLCOMB 1983). Gas in the foam located in the fracture also dissipates into the reservoir as foam life is exhausted, which negates the advantage of gas assist to recover the liquid from the farthest reaches of the created fracture system. On the other hand, when wells are opened too much and too quickly after a foam fracturing treatment, casing collapse might occur due to negative pressures caused by the rapid venting of gas from the wellhead overriding the expansion pressure factor due to friction heating and bottom-hole temperature.

Various aspects of fracture damage are also discussed by AHMED, ABOU-SAYED & JONES (1979); CARROLL & BAKER (1979); BRANAGAN, CIPOLLA, LEE & YAN (1987) and HAWKINS & KING (1988).

4.12.5. Fluid aspects

SOLIMAN & HUNT (1986) summarize the effect of stimulation fluid and its cleanup on well performance and crack properties. Hydraulic fracturing usually involves injection of a large volume of proppant-laden fluid into the crack induced within the reservoir. With continuing injection, the fracture enlarges and part of the fluid leaks off into the matrix. Fluid loss will continue after the termination of the job while fracture pressure dissipates. Imbibition may cause treatment fluid to penetrate even deeper into the formation after the fluid pressure in the fracture has equalized with the pore pressure of the reservoir. Due to infiltration, the invaded pay zone will have a reduced relative permeability to gas prior to production of both gas and water during cleanup. The high water saturation within fracture and formation represents a temporary damage effect which is usually reversible by careful cleanup of the well (cf. also section 4.8.8.3.). Fracture length, fracture conductivity and capillary pressure are important influences on the cleanup pattern.

An important mechanism damaging mainly formation and subordinately also fracture is water blocking (HOLDITCH 1979, BROWER & MORROW 1983; PENNY, SOLIMAN, CONWAY & BRISCOE 1983; ABRAMS & VINEGAR 1985; cf. sections 3.11.2.4.2. and 4.8.8.3.1.5.). SOLIMAN & HUNT (1986) find a direct relationship between fracture conductivity and stimulation fluid cleanup, which means that effective fracture flow capacity increases with progressive cleanup of the crack, and thus proper cleanup removes temporary fracture deterioration. The optimum crack conductivity for sound cleanup might be higher than that necessary for economically feasible gas production. In very tight reservoirs, the cleanup of fracturing fluids might be largely controlled by the capillary pressure effect. The existence of a mobile water saturation in the crack representing an apparent fracture damage has a pronounced effect on the buildup pressure response which must be recognized in order to avoid misinterpretation of the buildup data (SOLIMAN & HUNT 1986).

4.13. Aggressive fracturing design

The successful improvement of the European hydrocarbon stimulation potential by application of many of the concepts outlined above and being already more or less common in the USA and other parts of the world also strongly depends on the operating mentality and the encouragement to take some risk into consideration, too. Many marginal reservoirs will never be able to become approached in an adequate manner unless a challenging procedure is carried out which automatically involves some risk. It may partially be a difficult decision especially in the poor economical situation in the years after the 1986 oil price crash (cf. section 2.2.1.1.; in Europe and other international markets even accentuated by the US $ weakness; cf. section 2.2.1.3.), but similarly as many great successes in the past have been performed by risky pilot treatments, it is also in the future essential to open new possibilities by encouraged solutions and an aggressive design of jobs. Some comments are offered as follows on progress instead of routine and risky approaches of marginal reservoirs.

4.13.1. Progress instead of routine

An operating mentality of aggressive fracturing design includes the philosophy that nothing must become routine, but there has always to be progress. In this connection, however, it is indispensible that successes are accompanied by failures, and it depends on skill and sometimes also a bit on luck whether the ratio is at the bottom of the line economically feasible in total or not. On the other hand, much profit can be missed by a continuously too conservative design. It is nice to see everything running smoothly, but one always has to ask the question if there would not be a design still improving the performance. WAREMBOURG, KLINGENSMITH, HODGES & ERDLE (1985) emphasize that a successful treatment is not an operation that has been pumped without problems, but rather a job that provides the production performance enhancement predicted by the design process or even exceeding the forecast.

HOLDITCH (1984) presents a suitable example to illustrate the above suggestions. While abt. 30 - 50 % of the total cost of a fracturing treatment involves equipment and manpower that is used during the job and removed after the operation is finished, the overwhelming part of the expenses of most of the jobs is for proppants and fluids. As fluid leaks off and is produced back, the only thing that remains within the fracture and determines the fate of the whole treatment being either success or failure is the propping agent which was pumped downhole and that is supporting the crack to stay open and to act as a drainage pathway (cf. also PHILLIPS & ANDERSON 1985, HALL & LARKIN 1986). Therefore a wise approach to stimulation is to get as much grains of the designed proppant quantity into the fracture as possible. If, however, every job goes perfectly with no problems of putting the proppants away, then the fracture design is too conservative. Either money is being wasted by using too large pads, gel, fluid-loss additives or others, or the proppant concentration being pumped is not sufficient to pack the fracture. Attempting to determine the optimum proppant saturation for a zone, the possibility of screenouts and pressure-outs increases with approaching the physical boundary of uptake and/or placement capability (SHAH, SMITH & DONALDSON 1983). It is certainly not necessary to provoke a screenout in every treatment, but money should not be wasted by overdesigning the operation.

GREGORCZYK, PAULS, HOLTMYER, CHISHOLM & VENDITTO (1984) report field experience where best stimulation results have been obtained when an aggressive pumping schedule was used. CONWAY & HARRIS (1982) also emphasize that optimum results of designed treatments and major advantages of innovative and conventional components can only be determined from field trials and experiments which automatically have to include some risk.

4.13.2. Risky approach of marginal reservoirs

Thus at the end of the discussion of technological possibilities of enhancing the hydraulic proppant fracturing potential in Europe and other parts of the world, the general recommendation is given to perform more aggressive fracturing designs and also to carry out risky approaches to marginal oil- and gas-bearing formations. Aggressive strategical concepts are the only possibility of achieving the necessary progress with access to hitherto unconventional reserves which are indispensible for the satisfaction of the increasing primary energy demand and for the security of some domestic independence by more or less own hydrocarbon supply in the coming decades. NEWENDORP (1975, 1983), MEGILL (1984, 1985), BLESSING (1988) and GARB (1988) outline strategies for implementing risk analysis in various hydrocarbon exploration and development projects.

Some specialized techniques for application in marginal or unconventional reservoirs even aim principally on screenout events in some stages of the treatment or phases of fracture propagation (such as tip screenout fracturing; SMITH, MILLER & HAGA 1987; cf. section 4.5.4.4.1.).

4.14. Conclusion

The outline of various technological perspectives of enhancement of the hydraulic proppant fracturing and gravel packing potential in Europe in the present review and status report, as well as the areal marketing and economical assessment comprising review, status and forecast of technical, engineering and financial aspects of proppant stimulation in Western and Eastern Europe (MADER 1987; cf. chapter 2) and the evaluation of proppant selection as a consequence of reservoir conditions and proppant properties (cf. chapter 1) permit the conclusion that numerous possibilities of hydraulic proppant fracturing and gravel packing exist in Europe and other international markets in the next years. In case of the necessary back-up of the activity by an again more constant oil price at a reasonable niveau, partially also supported by a stronger US $, and by adequate pricing of proppants and service based on high technical quality and marketing requirements, both Western and Eastern Europe are considered to be expanding stimulation markets with increasing demand of hydraulic proppant fracturing including promotion of treatments in marginal formations in the near future.

It is, however, essential that the technical recommendations by an integrated marketing strategy incorporating also geological and reservoir engineering concepts are supported by a suitable approach of the individual mentalities in the suite of different European countries not only speaking different languages, but also comprising a variety of reservoir rocks with specific problems, and are not again interrupted by another major disturbance of the oil- and gas-producing industry by a renewed dramatical change of US $ exchange rate and oil price.

5. Gravel Packing and Sand Control

5.0. Summary and associated table of contents

Gravel packing and hydraulic fracturing are the main applications of natural sand and synthetic proppants in oil and gas reservoir stimulation and completion. Gravel packing is the most common sand control method comprising mechanical stabilization of perforation tunnels and borehole wall against invasion of loose formation sand. Sand control methods comprise mechanical bridging installations such as gravel packs, slotted liners, wire-wrapped screens or prepacks; chemical consolidation by injection of solidifying fluids (mainly plastic resins) into the formation to provide an in-situ grain-to-grain cementation, and maintenance of a high level of grain-to-grain stress in the reservoir sand within the borehole wall and its near surroundings through sand arching in the pay zone grain fabric, with triggering of the geometrical arrangement being supported or enhanced by using long inflatable packers. Gravel packing is by far the most distributed and popular sand control method due to its comparatively simple technical procedure, good economical feasibility, high success rate and widespread applicability. Although many pay zones may produce up to a certain limit rate without sand influx or with technically tolerable sand invention levels, profitable reservoir exploitation can only be performed if the formation sand is sufficiently stabilized by either natural or artificial means. Gravel packing includes running of a slotted liner or wire-wrapped screen into the open or cased borehole and surrounding it with a gravel mantle that covers the wellbore wall and supports the perforation tunnels which are also filled with gravel in cased holes or prevents the reservoir face section from collapse in open holes. Gravel packing design primarily includes selection of gravel type according to reservoir depth, temperature and brine composition; choice of gravel grain size in order to completely retain movement of formation sand, selection of screen size and/or liner slot openings for the purpose of preventing both gravel and reservoir fines to enter the gravel-pack hardware, gravel pack thickness aiming on achievement of maximum filtration capacity and minimum hydrocarbon flow rate impairment, and perforation tunnel length for guaranteeing stable perforation holes and creating a sufficient capture radius for the effluent hydrocarbons, as well as borehole size and gravel placement technique. While most of the inside gravel packs within the wellbore are not undergoing higher physical stresses and therefore with except of steam-drive enhanced oil recovery operations natural sand is still suitable as gravel material, the grains plugging the perforation tunnels may be subjected to considerable differential stresses and prevention of perforation hole collapse requires in some cases selection of high-quality synthetic proppants with sufficient closure stress resistivity. The main reasons for the predominant to almost exclusive application of natural sand as gravel packing material are its ready nearly unlimited availability and its cheap price, with payment of a premium for a high-quality synthetic proppant type instead of natural sand in most of the shallow reservoir depths where gravel packing is carried out not being technically and economically justified. In boundary cases when deeper pay zones require gravel packing or shallower unconsolidated sands are overpressured, the sections of the gravel pack which are most susceptible to deterioration by grain crushing due to supercritical closure stress are the plugged perforation tunnels and the gravel nodes at the distal ends of the perforation tubes where the more or less full lithostatic pressure is released and the effective closure stress also rises with reservoir pressure decline during progressive depletion. As the perforation tunnels are situated in the immediate vicinity of the wellbore where closure stress upon pressure drawdown reaches its maximum, attention has to be paid to design the material infilling hydraulic fracture path as well as gravel-packed perforation tunnels in the nearest surroundings of the wellbore in such a manner that the plugging is adequate to the highest possible closure stress throughout borehole life. Gravel grain size choice is the key item in gravel pack design and has a much greater importance than in hydraulic fracturing modelling where proppant type is the most decisive point that in turn is of comparatively subordinate significance in gravel packing. In order to completely bridge formation grains migrating to the sand/gravel interface and for the purpose of totally stopping invasion of sand into the borehole, the mean size of the gravel should be abt. five to six times larger than that of the sand. For most of the reservoirs requiring sand control by gravel packing, gravel grain size selection can be summarized that a ratio of gravel median diameter/ formation sand median diameter between 5 and 6 is the ideal range providing absolute stoppage of sand movement and unrestricted initial gravel permeability. Ratios below 5 give rise to limited hydrocarbon production, because proper bridging at the gravel interface is not possible, but yet no invasion of the gravel pack by formation sand occurs. Ratios between 6 and 11 provoke intrusion of reservoir sand into the gravel pack which drastically reduces gravel permeability, but still provides effective sand control once the gravel pack is completely plugged with fine particles and further motion out of the pay zone is inhibited. At ratios in excess of 11, the pores in the gravel package become so large that formation sand is partially migrating through the gravel into the wellbore, thereby increasing gravel permeability again, but already progressively deteriorating hydrocarbon exploitation by having only one effect of slowing down of sand production. At a ratio of abt. 15, original gravel pack permeability is reached again which reflects that at ratios above 15, formation sand can invade the borehole by passage through the gravel pack without any restriction, with in this case the gravel pack being completely useless. Although the gravel pack itself represents some type of formation damage by causing pressure losses across the perforations and fluid flow restriction through the cylindrical gravel mantle, the positive effect is still better than its negative influence, and pressure drop and flow area can be improved by increasing perforation density and diameter or even switching to open-hole completion if technically possible. Gravel-pack completions with plugging of long deeply penetrating perforation tunnels of wide diameter and narrow spacing along the borehole wall are approaching morphology and effectivity of hydraulic proppant minifracturing and thus significantly increasing the effective borehole cross-section. In some reservoirs, slotted liner or wire-wrapped screen can even be run without inserting gravel around it, because if formation sand grain size is sufficiently coarse to provide permeability, influx of pay zone sand would creates a natural gravel pack around the sand control hardware installed in the open borehole. In terms of carrier fluids, low-viscosity fluids can only transport low gravel concentrations and have to be pumped at high velocities which can provoke intermixing of gravel and formation sand, whereas high-viscosity fluids can suspend high gravel saturations and can be injected at lower velocities thus eliminating grain contamination. The most important aspect of gravel packing carrier fluids is sufficient leakoff into the formation in order to achieve the necessary tight packing of the gravel, which is the second major difference to hydraulic fracturing where fluid loss has to be minimized in order to avoid proppant bridging and premature screenout failure. Gravel packing results are improving with progressive fluid leakoff and the rheological properties of sand control transport fluids therefore have to be adjusted such that satisfactorily high rates of fluid leakoff are achieved. Fluid loss into liner or screen and/or formation during gravel packing is necessary to take place in high rates in order to allow deposition of the gravel on the borehole wall where it forms a mantle coating the reservoir/wellbore interface, spearheads plugging the perforation tunnels, a belt bridging the annulus between borehole boundary face and screen or liner, or a plug infilling the whole diameter of the wellbore. Steam-drive enhanced oil recovery is a particularly hostile environment, because the high-pH solutions attack formation sand and packing gravel and dissolve it, thereby deteriorating the sand control installation. Sintered bauxite and nickel-coated sand have been proven to represent suitable materials which are mechanically and chemically stable under these harsh conditions, and while sintered bauxite has already been abundantly applied in the field since many years, nickel-coated sand is still in experimental stage. Special gravel pack applications include consolidated pack installation and resin-coated gravel placement where the individual gravel grains aggregate to a stable matrix due to the cementing action of the resin pellicles, resulting in origin of a synthetic sandstone. The main advantage of these systems is that only an outside gravel pack infilling the perforation tunnels has to be performed, because backflow of gravel from the perforation holes is not possible and therefore support of the casing wall by a gravel mantle is not necessary, thus leaving the borehole free and unobstructed by any sand control hardware. Gravel packing in deviated wells is mainly influenced by the phenomenon of gravel duning which causes incomplete annulus plugging and can even provoke premature annulus bridging. Adequate gravel packing in highly-inclined to horizontal boreholes can be achieved by flexible baffles creating flow resistance, large-diameter washpipe, string rotation and perforation prepacking. Liner vibration during gravel packing changes the pack configuration from cubic to hexagonal and allows to utilize considerably coarser gravel grain sizes with the same resultant packing density than applying for finer gravel without vibration, with the advantage being much higher permeability in case of shaking. In terms of execution of gravel pack treatments during course of production history of predominantly oil and subordinately also gas wells, three cases have to be distinguished : primarily insufficiently cohesive pay zones, as well as secondarily unstable hydrocarbon-bearing sections due to reservoir pressure decline, and water cut increase and corresponding fluid viscosity decrease. The first group of formations comprises completely unconsolidated sands and chalks which require stabilization by gravel packing from the very beginning of production. The main pays in this group are silty and argillaceous, mainly medium- to fine-grained sands as well as soft flowing chalks and diatomaceous earths which according to their fine grain size, insufficient binding nature and lacking cementation have not enough internal cohesion to be satisfactorily stable without artificial improvement. The second assemblage incorporates weakly-cemented sands of various grain size including the whole spectrum from almost silt to nearly gravel where the problems of sand influx into the borehole start with progressive depletion due to decreasing reservoir pressure that is so far maintaining a more or less self-supporting framework of grains, with the formation failing at once a critical boundary pressure is reached and underpassed. Early pressure maintenance methods such as water or gas injection which are very common in oil-field exploitation can help to prevent the sand from becoming soft and flowing, or can at least decelerate the cohesivity transformation, but the changing phase composition and viscosity of the reservoir fluids also has its impact. The third association includes slightly-lithified sands of different grain size comprising a broad range throughout the granulometrical interval of sand where gravel packing becomes necessary due to reservoir breakdown and disintegration with increasing water cut in the oil during advanced production with rising oil-water-contact accompanying pressure depletion, as a consequence of excessive water injection due to pressure maintenance operations, or as a result of water coning and breakthrough due to unfavourable differences in relative permeability and preferred water conductivity along some high-permeability streaks or through the joint system. Changing fluid viscosity results in different cohesion of the grain fabric, and generally induces less stability than before and thus ultimately causes the sand to flow. The main members of the second and third groups of sand-problem reservoirs are weakly-cemented friable sands which might be originally still sufficiently stable, but with falling reservoir pressure and/or changing fluid viscosity, internal support is taken away and the grain fabric starts to collapse and to release more and more loose sand grains which move towards the least restricted spot that is represented by the borehole. Other sand control methods mainly comprise resin consolidation of the reservoir sand which is performed by injection of resin into the pay zone where it hardens and forms an artificial cement of the sand. Resin consolidation and gravel packing representing chemical and mechanical sand control treatments, respectively, have each their preferential applications, but can in some cases also be combined, particularly in case of consolidated pack installation with accompanying additional formation stabilization. Other chemical sand control operations are hydroxy-aluminum consolidation, electroless nickel plating, silica cementation and warm air coking. The construction of stable sand arches around the perforation holes is the explanation why with increasing pay depth, sand production troubles are more and more diminishing and finally vanishing. In shallow reservoirs, stable sand arches can be artificially created by agitating the grain fabric by inflatable packers or vibrating strings. Open-hole gravel packing provides the greatest inflow area for the hydrocarbons and the best filtration effect of the sand plug, with the effective borehole diameter being increased by underreaming. The most critical point in cased-hole gravel packing is adequate infilling of all the perforation tunnels which have to be absolutely clean prior to gravel insertion, with perforation preparation being possible by washing, surging or underbalanced shooting.

5.1. Introduction

Gravel packing for sand control is the second application of both natural sand and synthetic proppants in oil and gas well stimulation and completion (ROGERS 1971; SUMAN 1974, 1975 a; ANAND & JONES 1977; GURLEY, COPELAND & HENDRICK 1977; McLEOD & CRAWFORD 1983; SUMAN, ELLIS & SNYDER 1983; LEDLOW, SAUER & TILL 1985; NEAL & CARROLL 1985, PENBERTHY 1985, SPARLIN & HAGEN 1985; COULTER, MARTINEZ & FISCHER 1987; GURLEY 1988) and is carried out in oil-, gas- and water-production wells; water-, steam- and chemical-injection wells; underground gas-storage wells and occasionally also mineral-leaching and geothermal heat-exploitation wells (SOLUM 1984).

Gravel packing was introduced to the hydrocarbon industry more than 60 years ago and has been applied in water wells since more than 100 years (SOLUM 1986). Since that time, it has progressively evolved and is already since many decades the most common sand control technique. The gravel pack system basically consists of a screen or liner to prevent gravel production, and a gravel mantle to stop formation sand invasion. Gravel packing aims on reduction or controlling of sand production, minimization of formation damage, and increase of hydrocarbon production. The introductory remarks as follows include comments on significance and necessity of gravel packing for sand stabilization and productivity maintenance as well as organization of the present review and status report. An outline is also given on the reservoir composition spectrum requiring support by gravel packing.

5.1.1. Reservoir composition spectrum requiring gravel packing

Gravel packing is done in order to artificially stabilize mainly originally insufficiently consolidated, subordinately also secondarily and tertiarily destrengthened sands and other related grainy sedimentary rocks. Insufficiently consolidated (due to poor carbonate, silica or other cementation) and lithified (due to only slight pressure solution) or not enough cohesive (as a consequence of only little binding by only small amounts of dispersed mud) to even almost loose friable sands of various grain size are the most common trouble-making deposits in terms of sand production.

Apart from unconsolidated coarse, medium or fine, clean or dirty sands, to minor amounts also silty and muddy sands, sand-mud sandwich-type interbeddings, carbonate shell layers and soft chalk require stabilization treatments in the surroundings of the wellbore in order to prevent influx of sand, mud or carbonate particles into the borehole and to stop production of solid formation fragments (mainly sand) together with the hydrocarbons. While gravel packing in sandy reservoirs is quite abundantly applied and has been performed throughout the world since many decades, gravel packing in chalk (ATTARD, MATHES & MOWER 1988; cf. section 4.5.4.3.) and coal (JEU, LOGAN & McBANE 1988) has so far only occasionally been utilized.

5.1.2. Sand stabilization and productivity maintenance

The gravel pack acts as a filter for fines migrating out of the formation and inhibits the fines to plug the slots in the liner or the interwire spaces in the screen, but allows reservoir fluids to flow easily into the wellbore. The permeable gravel pack filter permits maximum production rates while it mechanically props the formation and prohibits the passage of sand into the wellbore. Bridging of the formation sand against the gravel pack in order to allow only natural fluid flow is the key to sand movement control (SCHWARTZ 1969).

Limited amounts of sand influx into the borehole can also be at least temporarily handled by cutting back oil or gas production rates (STEIN, ODEH & JONES 1974) which results in reduction of the tendency of the sand to move (especially in weak sands containing high-viscosity oil; cf. section 5.2.1.4.), but larger quantities of sand intervention can only be stopped by gravel packing or other sand control methods. PEDEN & YASSIN (1986) investigate the effect of perforation conditions upon maximum sand-free production rate in friable sandstones. Although gravel packing as the most common sand control treatment is quite expensive, it is not nearly as costly as losing a production well due to sand problems (cf. section 5.2.1.3.). The economical implications of the damage which flowing formation sand can create are critical enough to require continuous improvements in sand control techniques and sand production prediction methods (GHALAMBOR, KOLIBA, HAYATDAVOUDI & ALCOCER 1988).

5.1.3. Organization of the review and status report

As this review and status report mainly focusses on aspects of proppant application in oil- and gas-well stimulation and completion, the discussion of sand control techniques concentrates on gravel packing where questions of type and grain size of natural sand or synthetic proppants are very decisive for adequate performance of the operation. Gravel packing is still the most effective, practical, successful, reliable, straightforward, and in many cases also cheapest sand control treatment, as underlined by the fact that 60 - 90 % of all sand-stabilization jobs in the USA are carried out by gravel packing (ALLEN & ROBERTS 1982, SUMAN & SNYDER 1982; SUMAN, ELLIS & SNYDER 1983). For the sake of completeness and comparison with gravel packing, the other sand control methods are also briefly discussed, with special emphasis on the role of natural sand or synthetic proppants in some of these applications.

After some general comments, aspects of application of proppant or gravel types and grain sizes, gravel packing completion fluids and gravel saturation, reservoir stability and fluid dynamics, areal distribution of potential reservoirs, particle transport through perforations, steam-drive enhanced oil recovery and other techniques are discussed in connection with gravel pack questions. An account of special gravel pack applications forms a transition to an outline of other sand control methods, and finally some comments are offered on the influence of completion type on gravel packing and formation damage during gravel packing.

5.2. Gravel types and grain sizes

Similarly as proppant type and grain size selection is the key element of hydraulic fracturing design, gravel type and grain size choice is a very critical and central part of gravel packing design. Gravel grain size has to be carefully determined in order to satisfy the dual requirements of the pack to be permeable to formation fluids (primarily hydrocarbons) and well completion fluids, and to act as a filter preventing reservoir sand from moving into the wellbore. The term gravel used in sand control is different from that in sedimentary petrography in not relating to granulometrical assemblages coarser than sand, but comprising the whole grain size spectrum of sand and probably also coarser fractions that are used as packing gravel.

The synthetic intermediate- and high-strength pellets consisting of low- or high-density alumina oxide and/or silicate or low-density zirconia-silicate (cf. sections 1.3. and 5.2.2.3.) are called proppants in hydraulic fracturing to express their function of propping the open crack and providing support against its closure, whereas the same pellets are named gravel in mechanical sand control even if their task is partially approaching that of proppants in hydraulic fracturing when plugging perforation tunnels in addition to composing a cylindrical gravel mantle along the borehole wall or between casing and screen or liner. After a brief general outline of sand production and its control, aspects of gravel type, gravel grain size, and gravel grain surface and shape are discussed as follows.

5.2.1. Sand production and its control

Sand production is one of the oldest oil-field problems (ALLEN & ROBERTS 1982) and is apart from increasing water cut with progressive reservoir depletion the most significant factor deteriorating net oil exploitation rates. In contrast to water invention shifting the relationship of oil to water into an uneconomical direction, sand influx into the borehole is not only an oil production problem, but also provokes considerable damage of the wellbore and surface installations which in the worst case may lead to complete loss of the well. The brief overview as follows introduces the sand-production trouble as a base of the succeeding discussion of gravel packing techniques and materials. After some general aspects, the outline focusses on reasons of sand production, well deterioration by sand production, critical hydrocarbon offtake rate, and necessity and aims of sand control treatments.

5.2.1.1. General aspects

Sand grains composing arenitic reservoir rocks are stabilized by compressive forces due to the weight of the overburden, by capillary cohesion and by particle cementation (ALLEN & ROBERTS 1982, BJ-HUGHES 1986). Causes of sand production above a critical hydrocarbon offtake rate are related to dislodging of grains by drag or frictional forces of flowing fluid which increase with higher current rates and higher fluid viscosity, diminution in formation strength often associated with water production due to dissolution or dispersion of cementing materials between sand grains, reduction in capillary forces with increasing water saturation, declined relative permeability to oil due to rising water saturation which increases reservoir pressure drawdown for a given oil withdrawal rate, diminishing cohesion due to fines mobilization and wettability alterations as a consequence of changing fluid type from oil to water, and declining formation pressure with advancing reservoir depletion which increases overburden stress and thus compaction forces and may disturb cementation between grains. Sand production can be basically controlled by three mechanisms : reducing drag forces by improving well completion practices, mechanical bridging of sand by gravel packing, and increasing formation strength by chemical sand consolidation.

5.2.1.2. Reasons of sand production

Sand production in oil and gas reservoirs can be caused by totally unconsolidated formations, high hydrocarbon offtake rates, onset of water production, subsidence due to reservoir depletion, and improper well completion practices

(HALLIBURTON 1988). Totally unconsolidated formations react to any attempt to withdraw oil or gas with large amounts of sand inflow into the borehole. Other reservoirs produce sand-free if offtake rates are restricted, but increasing gross production rates create excessive stresses on insufficiently lithified pay zones and may exceed the ability of weak cements to bond sand grains together. Clay minerals and silt which are slightly stabilizing other loose sandstones may be seriously affected by water and when water production starts, the bonding is weakened or destroyed and formation sand is flowing into the borehole. Reduction of reservoir pressure may cause the overburden to subside, crush a poorly consolidated reservoir and lead to sand production and serious casing damage.

Improper well completion practices include misuse of acid for stimulation which may remove the small amount of calcareous bonding material in certain weakly-consolidated pay zones giving rise to sand production. Swabbing a well too hard or bringing the borehole to desired offtake capacity too quickly can cause excessive stress on slightly lithified reservoirs and provoke sand flow. Instantly opening some wells to maximum production rate can trigger water to be drawn into borehole and adjacent formation matrix, with such well completion practices being able to trigger premature sand influx in boreholes that might have operated sand-free for longer periods of time.

5.2.1.3. Well deterioration by sand production

Various possibilities of well deterioration by sand production exist which range from damage of only a few parts of downhole and surface installation to complete destruction of the equipment which ultimately can have the consequence that for technical and/or economical reasons the well has to be abandoned. The outline as follows concentrates on damage spectrum caused by sand influx, and economical and technical loss of wells.

5.2.1.3.1. Damage spectrum caused by sand influx

Sand flow from unconsolidated formations into the borehole creates various problems in hydrocarbon exploitation such as premature failure of artificial lift equipment; sand fill-up and bridging in perforations, casing, tubing and/or flowlines thereby shutting-off production; failure and collapse of casing or liner as a consequence of removal and migration of surrounding formations which leads to supercritical differential pressure and reservoir compaction or subsidence, erosion and abrasion of downhole and surface tubular equipment; cutting of pipes, chokes, valves, and artificial lift tools; filling of separator tanks and transport facilities, downhole casing and formation damage that sometimes causes premature abandonment of completions, and handling and disposal of produced reservoir materials (SUMAN, ELLIS & SNYDER 1983). Some comments are offered on equipment failure and sacrification of hydrocarbon exploitation time as follows.

5.2.1.3.1.1. Equipment abrasion

Sand production from unconsolidated reservoirs as well as proppant evacuation from insufficiently stabilized fracture infillings (cf. section 4.12.3.) lead to borehole operational problems such as stuck pump plungers, eroded pump valves, sucker rod failures, stuck tubing and proppant fill, and above all to drastical deterioration of the effectivity of reservoir stimulation and well completion for hydrocarbon production. Damage of surface and downhole equipment by sand erosion and impingement can also be the reason of fire or blowout (ANAND & JONES 1977, HALLIBURTON 1988). Reservoir compaction due to sand influx into the borehole leads to load deformation of casing and gravel-pack screens (WOOLEY & PRACHNER 1984). In addition to formation sand influx, deteriorating effects on wellbore hardware are also caused by proppant flowback from fracture mouths in hydraulic reservoir stimulation (cf. section 4.12.3.) and also during pumping of very abrasive proppant types downhole during fracturing treatments (cf. section 1.4.6.).

5.2.1.3.1.2. Sacrification of hydrocarbon exploitation time

Hydrocarbon exploitation time must be sacrificed to empty separators or to replace damaged completions components which causes high additional unnecessary cost in terms of damaged equipment, expensive workovers and lost oil and gas production, and in many cases, the cost to return a well to hydrocarbon exploitation after extensive sand influx or proppant flowback can be so excessive that the well has to be prematurely abandoned (POPE, WILES & PIERCE 1987) or can only be saved by sidetracking (MURER & STRAUCH 1980, MURER 1981). In addition to the expenses for the sand problems, unexpected pipe leaks or equipment failures represent serious safety and spill hazards particularly in offshore and inland water locations (GHALAMBOR, KOLIBA, HAYATDAVOUDI & ALCOCER 1988). Much money has to be invested every year to combat sand production in petroleum industry (DURRETT, GOLBIN, MURRAY & TIGHE 1977).

5.2.1.3.2. Economical and technical loss of wells

Apart from economical loss of the well for the reason of unfeasible remedy expenses, the ultimate danger of serious sand flow into the borehole can be technical loss of the well by casing failure in the producing interval due to non-uniform lateral loading as the sediment is flowing and pushing and/or high axial compressive load due to slumping of the overlying formations in initially fairly competent reservoir rocks, or due to serious pay zone compaction subjecting the entire casing string to unusual load conditions in initially unconsolidated pays (DEAN 1939; JARDINS 1959). Large-scale sand influx can even result in tremendous reservoir settling and consequently oil-field subsidence (ALLEN 1969, 1973; SULAK & DANIELSEN 1988, WAAL & SMITH 1988), with pay zone shortening also being transmitted to casing, tubing, liner and screen which can be plastically deformed at loads exceeding the elastical limit of steel, leading to severe buckling when lateral restraint is lost due to sand production.

5.2.1.3.3. Other aspects

Deliberate disturbance of the reservoir framework by sand production can also cause significant decrease in pay zone permeability although porosity may only slightly change (HALLIBURTON 1988), with this type of formation damage making successful sand control more difficult. In addition, sand which is produced during early well life may be the supporting matrix for shale streaks which upon loss of support could fall across some of the perforations and prevent satisfactory placement of remedial sand control materials. Slumping and intermixing of overlying shales with reservoir sand can severely restrict fluid flow because permeability losses up to 95 % may occur (RENSVOLD 1982).

The necessity of gravel packing hydrocarbon production wells in order to avoid their destruction by sand flow is particularly urgent in subsea installations where many wells are tied to the central template by subsea flowlines, because the risk of leaving wells unpacked in a subsea completion is considerably higher than normal due to the extraordinary expenses involved (SCRIMGEOUR & SHEPHERD 1988).

5.2.1.4. Critical hydrocarbon offtake rate

Slight sand influx and formation of sand bridges in casing or tubing can at least temporarily be handled by reducing oil or gas production rates in order to decrease the tendency for sand to enter the borehole (effects of flow rate on stability of unconsolidated sands are also discussed by TIPPIE & KOHLHAAS 1973), by increasing produced-fluid rising velocities in tubulars to minimize sand settling by utilizing small tubing strings or by prematurely applying gas lift, or by increasing rising velocities by recycling part of the flow back into the well. Some aspects on toleration level of joint hydrocarbon and sand production, and influence of perforations are outlined as follows.

5.2.1.4.1. Toleration level of joint hydrocarbon and sand production

Although the technically and economically optimum solution is always zero sand production rate, relatively weak formations containing high-viscosity oil may allow hydrocarbon offtake inspite of some sand influx which does not cause problems if the flow rate is sufficiently low (WEISSENBURGER, MORITA, MARTIN & WHITFILL 1987). On the other hand, comparatively strong reservoirs may produce sand once the effective in-situ stress is increased due to declining reservoir pressure by progressive depletion. Following the outline of some examples of exceptionally high sand contents in oil production, comments are offered on sand production vs. hydrocarbon exploitation rate.

5.2.1.4.1.1. Exceptionally high sand contents in oil production

CHEN, ZHOU & LIU (1986) report some flowing wells that occasionally produce oil even with a sand content up to 40 % at least for some time of exploitation (MULLINS, BALDWIN & BERRY 1974; MORSE 1975 and FOSTER & LINVILLE 1979 present methods of detection and monitoring of sand production in flowing wells, and CHAMBERLAIN & FORSYTH 1982 also comment on sand removal from gas streams), but generally the practical limit on sand influx is abt. 0.1 % or even less (ALLEN & ROBERTS 1982). VONDE (1980) documents that with specially designed pumping and treating equipment, crude oil containing up to 70 vol% of entrained sand can be withdrawn in extraordinary cases, and that even higher offtake rates than with conventional rod pumps and sand control completions can be achieved, but on the other hand, excessive sand production leads to reservoir damage.

In the exceptional case study reported by VONDE (1980), one limitation to the application of the special pumps is the assurance that adequate velocity will be maintained in tubing and wellbore to eliminate sand settling. While sand exclusion completions have resulted in severe wellbore plugging and limited oil productivity, significant offtake increases have been obtained by allowing unrestrained sand entry into the borehole. The only long-term effective and generally applicable treatment against sand flow, however, is gravel packing or another sand control operation (SUMAN, ELLIS & SNYDER 1983).

5.2.1.4.1.2. Sand production vs. hydrocarbon exploitation rate

Sand production is a rate-sensitive process, with a critical hydrocarbon offtake rate existing below which no or only negligible amounts of sand will be produced (BJ-HUGHES 1986). This sand-free rate, however, is often uneconomical and therefore the reservoir sand has to be controlled in order to make oil exploitation feasible. The critical hydrocarbon offtake rate above which sand production becomes excessive and untolerable in oil-field maintenance can be determined by beaning-up fluid flow rates (ALLEN & ROBERTS 1982). As fluid flow rate is increased stepwise, sand concentration jumps at each increment and then tapers off to the original saturation.

The surge effect apparently breaks unstable sand bridges which rebuild at the higher rate in subcritical flow regimes. Under supercritical conditions when the tolerance boundary has been exceeded, sand bridges do not reform due to overpassing of the strength of the structure and sand inflow continues at higher rates, with hydrocarbon offtake rate then having to be reduced significantly below the critical range to allow reinstallation of sand bridges. Bean-ing-up of oil production has the side benefit that the gradually increasing rates provide ideal clean-up of induced and inherent fines in the pore channels around the wellbore, thereby often resulting in significantly higher productivity indexes. PEDEN & YASSIN (1986 b) discuss determination of optimum completion conditions for sand-free oil production.

As a consequence of extremely high flow velocities, in gas wells even a small amount of sand for a very short duration can result in complete erosion of steel pipe at critical areas (ANAND & JONES 1977), whereas in oil wells due to lower velocity and higher viscosity, higher concentrations of sand can be tolerated over longer periods of time at least in some cases. In a few very unstable oil sands, the critical rate at which sand starts flowing is less than 1 m3 of oil per day (MURER 1981).

5.2.1.4.2. Influence of perforations

For some reservoirs suffering from sand emigration, maximum hydrocarbon offtake rate without excessive sand production is proportional to the length of the perforated interval, with thereby some sand influx problems also being able to be solved by opening an increased length of pay zone section, or improving perforation density in case interval length has reached its technical boundary by being identical to net reservoir thickness. Following discussion of some general aspects, the advantage of selective perforation of hard layers in interbeddings of soft and hard sands is illustrated.

5.2.1.4.2.1. General aspects

Other factors affecting maximum offtake rate are productivity index and formation strength or drawdown factor. The significance of inertial effects (cf. section 4.3.3.3.) on sand production can be shown by fluctuating current rate experiments where either a sharp increase or a sudden decrease in flow rate causes temporary increase of sand influx (SAUCIER 1974). Given uniform current rate after the change, sand production decreased, thereby apparently indicating reestablishment of bridging effects. Rate changes may thus (independent of the absolute offtake levels) be more detrimental to reservoir stability than magnitude of a certain flow rate once it is kept more or less constant. PEDEN & YASSIN (1986 a) analyze the effect of perforation conditions upon maximum sand-free production rate in friable sandstones.

5.2.1.4.2.2. Selective perforation of hard layers in interbeds

In addition to or alternatively to reduced hydrocarbon offtake rates, selective perforation can be already an effective means of sand control in thick reservoirs unless poorly-consolidated material is dominant (WEISSENBURGER, MORITA, MARTIN & WHITFILL 1987). This applies particularly for interbedded sequences of soft and lithified horizons. While the absence of induced formation damage and a possible significant improvement of productivity index by acceptable mild sand flow are advantages, selective perforation requires strict adherence to operational guidelines with sophisticated perforation interval choice based on a mechanical properties log that has been calibrated by cores, because any errors could result in serious sand problems. Leaving out the weak unstable horizons from being perforated in intimate interbeddings of soft and hard sands provides hydrocarbon offtake through the consolidated layers, with the formation sand loosened within the unstable intervals already being bridged at the interface to the hard interventions and thus even not reach the surroundings of the perforation tunnels.

5.2.1.5. Necessity and aims of sand control treatments

Reservoir engineering planning of field exploitation usually does not have any necessity and understanding for handling sand flow problems by decreasing hydrocarbon production rates, but in many fields the economical framework requires progressively accelerated oil or gas offtake quantities (cf. also STRACKE 1987), and moreover secondary and tertiary oil-recovery methods become senseless once the necessary large to tremendous fluid volumes with considerable excess water cannot be managed including their reinjection for the purpose of pressure maintenance, reservoir fluid pushing and environmentally safe and hazardless underground disposal (cf. also SCHWARTZ 1969).

Therefore performance of gravel pack or another sand control operation is a widely applied means to optimize offtake quantities by stabilizing the grain framework of the sandy pay formation and to prevent its collapse which would disturb regular and/or enhanced reservoir depletion. Comments are offered as follows on sand control method spectrum, formation consolidation vs. gravel packing, and timing of sand control installation.

5.2.1.5.1. Sand control method spectrum

Sand control methods comprise mechanical bridging installations such as gravel packs, slotted liners, wire-wrapped screens or pre-packs; chemical consolidation by injection of solidifying fluids (mainly plastic resins) into the formation to provide an in-situ grain-to-grain cementation (cf. section 5.9.1.), and maintenance of a high level of grain-to-grain stress in the reservoir sand within the borehole wall and its near surroundings through sand arching in the pay zone grain fabric (cf. section 5.9.5.), with triggering of the geometrical arrangement being supported or enhanced by using long inflatable packers. Gravel packing is by far the most distributed and popular sand control method due to its comparatively simple technical procedure, good economical feasibility, high success rate and widespread applicability. Gravel packing is particularly suitable for wells which have already produced quantities of sand, thereby having caused either voids behind the casing and/or development of a zone of dilated pay zone around the borehole (RENSVOLD 1982).

Sand control treatments aim on retention of load-bearing solids which upon movement and emigration would weaken the rock fabric and ultimately give rise to pay zone collapse independent from their detrimental effect within the wellbore itself and the technical installations therein (ALLEN & ROBERTS 1982). Fine solids associated with framework interstitials or formation fluids are not part of the mechanical structure and are always produced in small amounts, with this being in fact beneficial as otherwise they would eventually block the flow channels.

5.2.1.5.2. Formation consolidation vs. gravel packing

While the chemicals for synthetic sand framework cementation are relatively expensive and thus consolidation is normally more costly than gravel packing (cf. section 5.9.1.), the performance of gravel packing requires in any case rig or hoist operation which is not inevitably necessary for consolidation by chemical injection that can even be run through coiled tubing (LANMON & GILL 1966 even present a wire-line completion technique for sand control). As artificial glueing and cementing of loose formation sands can completely ruin the poroperm properties if the job fails, however, gravel packing has turned out during the last decades to be the most effective, practical and successful sand control method since its introduction to water-production industry over 100 years ago (API 1961, CASHION & MARKIE 1970, SOLUM 1986) and its first applications in oil and gas fields over 60 years ago (SCOTT 1928, SAWDON 1937, COBERLY & WAGNER 1938, ECKEL & REISTLE 1940, HILL 1941, WEST 1951).

5.2.1.5.3. Timing of sand control installation

In terms of timing of installation of sand control measures, early well life is in many cases a more suitable period for investment than late production history (HALLIBURTON 1988). Delay of sand control treatments bears the risk that once sand inflow commences with its associated disadvantages if not inhibited initially, cost of remedial actions may by far exceed that of primary sand control, and repair operations may fail due to reservoir damage in contrast to primary control which is much surer and does not only save cost for handling of sand invasion problems, but also allows faster hydrocarbon exploitation and thus drilling and completion amortization. While initial cost may be fairly accurately estimated, the expenses of remedial work are less predictable.

5.2.2. Gravel type

Gravel packing design primarily includes selection of gravel type according to reservoir depth, temperature and brine composition; choice of gravel grain size in order to completely retain movement of formation sand, selection of screen size and/or liner slot openings for the purpose of preventing both gravel and formation fines to enter the gravel-pack hardware, gravel pack thickness aiming on achievement of maximum filtration capacity and minimum hydrocarbon flow rate impairment, and perforation tunnel length for guaranteing stable perforation holes and creating a sufficient capture radius for the effluent hydrocarbons, as well as borehole size and gravel placement technique. Selection of gravel in terms of type and grain size is the key element of sand control design treatment. Some comments on natural sand and synthetic proppants are offered as follows.

5.2.2.1. General aspects of gravel type choice

Both natural sand and synthetic proppants can be applied for hydraulic fracturing as well as gravel packing. The most important criterion for proppant or gravel type selection is closure stress which in hydraulic fracturing affects the whole proppant wedge infilling the crack, whereas in gravel packing, only the proppants plugging the perforation tunnels and composing gravel nodes at the distal ends of the perforation tubes are subjected to more or less full closure stress, but the gravel mantle either along the borehole wall or between casing and screen or liner does not have to resist to the full pressure load due to the absence of the lithostatic component in the open wellbore.

The distal gravel nodes represent a more or less ball-shaped replacement infilling of the rubblized zone at the end of the perforation tunnel where using up of the charge resulted in further formation penetration, and must not be confused with the gravel nodes at the proximal ends of the perforation tunnels which are deposits originating after filling of the shot holes on the casing wall where they tend to bridge the annulus by meeting with annular sediments covering the outer face of screen or liner (cf. section 5.8.7.). Additional more irregular gravel nodes originate at the cement/formation interface or even at the casing/cement boundary as a consequence of infilling of voids that derive from incomplete cement distribution with holes and channels in the cement sheath. Significance, suitability and necessity of natural sand and synthetic proppants as gravel in mechanical sand control are outlined as follows.

5.2.2.2. Natural sand

As deep tight gas-bearing sandstones and moderately-deep oil-bearing sandstones are usually sufficiently cemented and do not disintegrate into loose grains during testing and production even throughout long exploitation history, sand control treatments are in most cases restricted to comparatively shallow or even very shallow weakly-lithified or nearly uncemented formations containing mainly oil and subordinately also gas reservoirs. In other areas, however, sand problems may be encountered up to reservoir depths of 12,000 ft (4,000 m) or more (ALLEN & ROBERTS 1982, BJ-HUGHES 1986).

Production drilling on ground water also has abundantly to include gravel packing of the unstable, often geologically quite young sandstones and conglomerates in frequently very shallow depth particularly in case of conventional drinking water; mineralized or thermal waters commonly derive from deeper storeys of the local geological column which are better lithified. Aspects of gravel type selection are summarized as follows along the lines of reservoir depth and closure stress in perforation tunnels. An overview of the provenance of gravel packing sand is also provided.

5.2.2.2.1. Reservoir depth and closure stress in perforation tunnels

As a consequence of the low closure stresses in the generally shallow depth of the oil, water or gas reservoirs requiring sand control, natural sand is in most cases the adequate material for gravel packing in view of demanded mechanical resistivity, with the closure stress regime in most of the pay formations with sand influx problems being subcritical for sand. Sand crushing may generally only occur in limited amounts during surface handling and downhole pumping, but not at the bottom of the well as a consequence of excessive closure stress in most of the reservoir depth ranges where gravel packing is carried out (cf. section 1.2.2.).

The main reasons for the predominant to almost exclusive application of natural sand as gravel packing material in mechanical sand control are its ready nearly unlimited availability and its cheap price, with payment of a premium for a high-quality synthetic proppant type instead of natural sand in most of the shallow reservoir depths where gravel packing is carried out not being technically and economically justified. In boundary cases when deeper pay zones require gravel packing or shallower unconsolidated sands are overpressured, the sections of the gravel pack which are most susceptible to deterioration by grain crushing due to supercritical closure stress are the plugged perforation tunnels and the gravel nodes at the distal ends of the perforation tubes where the more or less full lithostatic pressure is released and the effective closure stress also rises with reservoir pressure decline during progressive depletion (similarly as applying for propped fractures in hydraulic stimulation; cf. section 4.12.4.). The differential pressure lasting on the gravel in the perforation tunnels is in any case considerably higher than that acting on the gravel mantle in the wellbore.

As the perforation tunnels are situated in the immediate vicinity of the wellbore where closure stress upon pressure drawdown reaches its maximum, attention has to be paid not only in hydraulic fracturing, but also in gravel packing, to go on the absolutely safe side and design the material infilling hydraulic fracture path as well as gravel-packed perforation tunnels in the nearest surroundings of the wellbore in such a manner that the plugging is adequate to the highest possible closure stress throughout the life of the borehole. Perforation collapse (ANTHEUNIS, VRIEZEN, SCHIPPER & VLIS 1976) in friable shallow sandstones and weakly-cemented reservoirs in greater depth has to be avoided by early propping of the perforation tunnels with gravel of suitable type and grain size.

5.2.2.2.2. Provenance of gravel packing sand

High-quality sand of both American and European provenance has been commonly used for gravel packing, with the utilization of cheaper material from local sources even helping to reduce the cost of the operations in case quality is acceptable (examples where expensive import of sand from remote sources could be avoided due to local presence of sand in suitable quality and sufficient quantities were in the last years particularly Libya, Indonesia, and USSR, and the same certainly applies for other parts of Africa and the Far East). In terms of quality demand, the expensive top-class Ottawa Sand is frequently not inevitably necessary as gravel packing material and can be replaced by cheaper sands from areas closer surrounding the oil-field operation, thereby saving both product and transport cost (the same applies for small- to moderate-scale sand fracturing in shallow to intermediate reservoir depth; cf. section 1.3.1.).

Sand production difficulties usually increase with shallowing formation depth where the reservoir rocks are in less mature stages of burial diagenesis and thus lithification and the grain fabric is also poorer supported by lower lithostatic and hydrostatic pressures, as well as with rising percentage of fines (RIKE 1970). This relationship clearly underlines that in most of the cases of sand control demand, reservoir depth and thus bottomhole pressure are so low that natural sand is still enough stable with sufficient security to be used as gravel packing material.

5.2.2.3. Synthetic proppants

Intermediate- and high-strength low-density or high-density alumina oxide and silicate (ceramic) or low-density zirconia-silicate proppants (cf. section 1.3.) have so far only been applied in a few exceptions for gravel packing, with their wider distribution having been mainly limited by the considerably higher prices of the high-quality expensive synthetic materials (cf. section 2.2.2.3.). Sintered bauxite has been particularly utilized in steam-flood injection-well gravel packing (cf. section 5.7.1.), and nickel-coated sand has been recently introduced for the same purpose because of being even superior to sintered bauxite in the thermal high-alcalinity environment (cf. section 5.7.2.).

Resin-coated sand and proppants have also repeatedly been selected for application in the gravel packing market segment (cf. section 5.8.2.). Low-density intermediate-strength alumina silicate proppants are the most suitable material for optimum pumping and placement, because their specific gravity is almost identical to that of sand which allows excellent suspension and transport without settling in most of the gravel packing fluids at the benefit of the much higher conductivity of the high-quality synthetic material. Recently an ultra-light artificial gravel packing material has been introduced for special application in highly-deviated wells (cf. sections 5.3.3. and 5.8.3.). The following outline focusses on sand consolidation and closure stress transfer, perforation tunnel and borehole wall stability, and angle of repose and gravel infilling.

5.2.2.3.1. Sand consolidation and closure stress transfer

For the purpose of securing unlimited success of gravel packing operations, recommendation is made to favour the more expensive high-quality synthetic proppants in all cases where at any stage during long-term field operation and hydrocarbon production history, effective closure stress in the immediate surroundings of the wellbore might upon the highest possible reservoir pressure drawdown exceed the resistivity boundary of natural sand, because crushing of the material infilling the perforation tunnels and forming the gravel nodes at the distal ends of the perforation tubes would be the worst deterioration of the effectivity of the gravel pack job.

Closure stress becomes critical on the proppants in the perforation tunnels particularly in moderately-cemented sandstones in greater reservoir depth. In shallow unconsolidated sands, closure stress is not fully exerted on the proppants in the perforation tunnels because of compensation of the stress by compaction of the soft rock and embedment of the proppant particles. The weight is transferred from the poorly-lithified layers to under- or overlying intensely-cemented horizons such as carbonates and consolidated sandstones. An important application of intermediate- and high-strength proppants is prevention of perforation tunnel collapse in low-permeability overpressured gas sandstones where elevated pressure drawdown leads to closure of the horizontal perforation holes and impedes gas production (CHENEVERT & THOMPSON 1985).

5.2.2.3.2. Perforation tunnel and borehole wall stability

Perforating into the side of a borehole sets up a very high-stress environment which is very critical for perforation tunnel stability, and collapse of the horizontal holes can best be prevented by infilling them with pressure-resistant proppants which plug the voids and guarantee the applicability of high pressure drawdowns for reservoir-adequate offtake rates instead of keeping drawdown very limited in order to avoid collapse of unplugged perforation tunnels and to renounce from economical gas production rates (aspects of perforation stability under high stress are also discussed by PATTILLO & SMITH 1982). STEIN (1976) uses conventional well log data for selection of gravel packing materials for sand control.

5.2.2.3.3. Angle of repose and gravel infilling

The angle of repose of the gravel-pack material should be small in order to allow ready infilling of all crevices and pockets in the formation in barefoot-completed wells (COBERLY & WAGNER 1938). According to their very low angle of repose, high-strength zirconia-silicate proppants should also in terms of their low specific gravity be the ideal gravel-pack material, but unfortunately they are far too expensive for most of the applications (cf. sections 1.4.3. and 2.2.2.3.). Low density and cheap price of intermediate-strength alumina silicate proppants characterize this type of gravel to be the most suitable material to replace natural sand in all the utilizations where higher closure stress resistivity and higher pack conductivity of the gravel is required.

The low specific gravity of low-density alumina silicate proppants results in a pumping behaviour of this material similar to that of natural sand while all the advantages of the high-quality proppant are obtained. The high specific gravity of sintered bauxite is a considerable drawback for the application of gravel material apart from its elevated price, but as a consequence of its performance being superior to all competitive materials in some special applications such as steam-drive enhanced oil recovery (cf. section 5.7.1.), the utilization of the expensive and heavy sintered bauxite is in some cases inevitable for achieving successful sand control.

5.2.3. Gravel grain size

The most extensively used sand-control technique consists of placing a screen in the wellbore and packing gravel around it (GRUESBECK, SALATHIEL & ECHOLS 1979). The screen is sized to hold back the gravel which in turn is sized to retain the formation sand. Grain size selection of the gravel is therefore a function of mean particle size of the reservoir sand, and screen openings have to be adjusted to chosen gravel grain size. Selection of gravel grain size in sand control is therefore of much greater significance than gravel type choice, which is in contrast to hydraulic fracturing where emphasis is on selection of proppant type and although being also considerably important for conductivity optimization, proppant grain size choice is in most cases only of subordinate priority (cf. sections 1.3. and 1.4.).

Formation particle size distribution data are the basis for gravel size and screen opening specifications and have also pronounced effects on completion configuration (ECONOMIDES 1986), because finer reservoir sands requiring smaller gravel for proper control need more perforations than normal or even openhole completion in order to provide an adequate flow area. After the outline of some general features, the brief account as follows concentrates on aspects of grain size selection vs. formation grain size, packing gravel/formation sand size ratios, width of gravel mantle, impact of gravel pack hardware, formation grain size evaluation, grain size intermixing of gravel and formation sand, and hydrocarbon productivity vs. sand control. Some aspects of using screen or liner without gravel pack are also discussed.

5.2.3.1. General aspects of gravel grain size selection

Careful selection of sand or proppant grain size for sand control is essential, as too fine material leads to plugging of the reservoir, whereas too coarse material gives rise to grinding of the liner by allowing formation particles to pass the ineffective filter and to be produced together with the hydrocarbons. Gravel size reduction may diminish the height of the gravel-packed interval and leave some zones without sand control, create voids within perforations and cause their collapse, change gravel size distribution and decrease sand control efficiency, and generate fines and plug the gravel pack, whereas for hydraulic fracturing, proppant size reduction may decrease fracture conductivity and generate fines (CHEUNG 1985). Some aspects of common gravel quantities and grain sizes, prevention of formation sand intrusion, and gravel quality demand are outlined as follows.

5.2.3.1.1. Common gravel quantities and grain sizes

As the quantities of sand that are needed for gravel packing are abt. 1 - 20 t per well and in the smaller-scale applications often only amounts between a few hundreds of kg and some tons are necessary, the total demand of material for gravel packing is much less than that required for hydraulic stimulation and is in most areas cumulatively for fields or complexes of patches only abt. 5 - 20 % and for the individual job only abt. 1 - 10 % of the quantity used for fracturing.

In addition, common grain sizes for gravel packing are often 30/50 and 40/60 mesh instead of predominantly 20/40 and 16/20 in hydraulic fracturing, whereas in other reservoirs, also 12/20 and 20/40 are usual grain sizes for gravel packing instead of or in addition to the finer material. In very coarse pay zones or in case of reasonable gravel compaction by liner vibration (cf. section 5.8.4.), also 8/12 and 6/10 sizes have been occasionally applied (cf. FLANIGAN 1979), whereas extremely fine sands to silts, diatomaceous earths and chalks may require even finer gravel of 40/70 or 70/140 size unless special design has to be made such as in case of mass-flowing chalks with motion of pellets composed of numerous individual microparticles (cf. section 4.5.4.3.).

5.2.3.1.2. Prevention of formation sand intrusion

Choice of gravel size has to follow the guidelines that an effective gravel pack is to completely prevent the intrusion of formation sand both into the wellbore and the gravel pack itself, to provide maximum productivity by minimizing the pseudo-skin associated with the additional flow resistance to the hydrocarbons generated by the gravel pack, and to minimize permeability impairment within the formation (PEDEN, RUSSELL & OYENEYIN 1985). The objective of a good gravel pack is to connect the interstitial tubes of the formation to those of the gravel in such a way that there is no movement of solids at all when the produced hydrocarbons pass through the pack (SOLUM 1984).

Gravel size selection relative to the size of the formation particles is one of the most important aspects of designing gravel pack for maximum productivity (GURLEY, COPELAND & HENDRICK 1977). Gravel should be principally sized to prevent invasion of the finest formation sand within the perforated hydrocarbon-bearing interval of the geological column in case of multiple-horizon completions in order to assure effective sand control and longer-lived gravel packs, with the desired result being stabilized grain bridging and therefore absolute stoppage of sand production without undue limitations of flow capacity and thus exploitation rates (SAUCIER 1974). Coarser pay zone sands are relatively easy to control, but considerable difficulties occur in finer reservoir sands with larger amounts of silt and clay particles. Smaller gravel may require a larger perforation area or even open hole completion for its installation, and is more susceptible to permeability reduction caused by mixing with formation debris and contamination with reservoir sand during placement.

5.2.3.1.3. Gravel quality demand

In terms of quality demand, a minimum of 96 % of the gravel should fall between the designated sieve mesh sizes, and not more than 2 % should be smaller than the last mesh size of the granulometrical interval. The gravel should consist of individual grains and not of clusters of two or more grains. Sand containing more than 1 % of acid-soluble contaminates is not recommended for gravel packing because of representing a source of fines and grain fracturing which reduces gravel pack life. Gravel containing more than 0.5 % of clay-sized fine particles passing 325 mesh screen should also be rejected, as fines quickly destroy pack permeability.

5.2.3.2. Gravel grain size selection vs. formation grain size

In terms of granulometry, experience during the last decades has shown that mean size of the gravel pack sand or proppants should be generally five to six times the mean size of the formation sand expressed by the 50 % diameter of the grain size curve (SCHWARTZ 1969, SAUCIER 1974, SPARLIN 1974; GURLEY, COPELAND & HENDRICK 1977; PENBERTHY & COPE 1979) which is considered to be the optimum for complete prevention or stoppage of movement of formation sand into or through the gravel with little or no effect on initial gravel permeability. The earlier guidelines of selecting a mean size of the gravel pack material of eight to ten times the 10 % coarse percentile of the cumulative sieve analysis curve of the formation sand (COBERLY & WAGNER 1938, HILL 1941) or four to thirteen times that grain size percentage of the granulometrical distribution (SCHWARTZ 1969; WILLIAMS, ELLIOTT & WEAVER 1972) or other rules of thumb (WINTERBURN 1947, TAUSCH & CORLEY 1958) have turned out to result in too many failures (SAUCIER 1974). Gravel pack material/formation sand ratios of five to six have particularly been proven to be successful in wells with severe flow disturbance where bridging is unsatisfactory under the same conditions where still good results are achieved for uniform undisturbed flow.

Oversized gravel can be invaded and plugged by formation fines during production due to insufficient bridging properties (SPARLIN & COPELAND 1972, SAUCIER 1974). Properly sized gravel should be placed in such a manner that perforation tunnels are gravel-filled and even a gravel node is created at the distal ends of the perforation tubes (BELL 1982), because otherwise pay zone sand may fill the perforations and cause severe productivity restriction (SUMAN, ELLIS & SNYDER 1983). Aspects of gravel size selection and gravel-sand size ratios are also discussed by COBERLY & WAGNER (1938), TAUSCH & CORLEY (1959) and STEIN (1983), and COBERLY (1937) outlines questions of screen wire and liner slot openings. OYENEYIN (1987 b) presents formation sand size prediction from well log correlations.

5.2.3.3. Packing gravel/formation sand size ratios

As reservoir flow conditions may change during hydrocarbon withdrawal time and altered fluid forces cause pay instabilities to occur, gravel bridges tend to breakdown and more sand is produced until new bridges form under new conditions of stability (SAUCIER 1974). This is the reason why in contrast to earlier gravel pack design aiming on bridging and reasonably slowing down of formation sand migration with larger gravel-sand mean diameter ratios in the range of 6 - 14, tendency has been during the last decade more and more towards smaller gravel-sand diameter ratios of abt. 5 - 6 in order to absolutely stop formation sand moving and flowing into or through the gravel (SPARLIN 1969; FLANIGAN 1979; SUMAN, ELLIS & SNYDER 1983). Other points which have to be considered are gravel permeability and packing density, formation permeability and granulometry, and impact of formation and gravel fines.

5.2.3.3.1. Absolute stoppage of moving formation sand

Although absolute stoppage of sand intrusion may be desirable from the sand control point of view (SAUCIER 1974), an overdesign of sand particle restraint bears the danger of potential reduction in well productivity (PEDEN, RUSSELL & OYENEYIN 1985), with thus a sound compromise being necessary between maximum restriction of sand movement and highest possible hydrocarbon offtake rate. In contrast to gravel-pack sand sizing, wrapped-screen wire spacing or liner slot width has only a minimal, generally almost negligible detrimental effect on well productivity and thus screen and liner opening sizing exclusively aims on absolute containment of gravel and formation sand (cf. section 5.2.3.5.).

Although the initial permeability of the gravel pack is in case of a material/formation sand size ratio of five to six considerably less than for a ratio of abt. ten, this effect is significantly compensated by much less subsequent gravel pack impairment by pore bridging and plugging in case of smaller ratios with respect to larger relationships (SAUCIER 1974). If gravel pack grain size is too large, formation sand particle bridging fails and entering reservoir sand grains may even reduce effective pack permeability to a value less than the initial one of a finer gravel pack that undergoes less impairment. The lowest gravel pack deterioration is achieved when gravel size and formation sand size are equivalent, but for this constellation, also well productivity is negligible and approaches its minimum. For any gravel size, the permeability of the reservoir sand/gravel pack interface is substantially the same as that of the formation itself or larger (COBERLY & WAGNER 1938).

5.2.3.3.2. Gravel permeability and packing density

Smaller gravel-sand mean diameter ratios below 5 usually create problems due to the low absolute gravel permeability and may only be desirable under seriously disturbed reservoir flow conditions (SAUCIER 1974) and under high-velocity current situations (ALLEN & ROBERTS 1982), because they usually provide sand control only at the expense of hydrocarbon productivity (HIMES & RUIZ 1988) especially in cased-hole completions where smaller, less permeable perforation packages cause an increased pressure drop across them. Some comments are offered as follows on possible range and ideal level of packing gravel/formation sand size ratio.

5.2.3.3.2.1. Possible range of pack gravel/pay sand size ratio

For very high oil offtake rates or in case of expected water production, even a gravel-sand ratio of 2 may be helpful (STEIN 1983). Substantially larger gravel diameter for controlling the same size of formation sand can be chosen if hexagonal instead of cubic gravel grain packing can be achieved by liner vibration (SOLUM 1984; cf. section 5.8.4.). With 8/12 mesh gravel in the hexagonal pack, vibrating will give the same interstitial tube size as 20/30 mesh gravel in the conventional cubic arrangement through flow-type placement, and also the width of liner slots and wrapped-screen wires can be accordingly increased. On the other hand, GULATI & MALY (1974) present evidence arguing for selection of smaller grain sizes in gravel packing. Smaller gravel-sand size ratios are also required when using loose instead of tight packing fabrics (ALLEN & ROBERTS 1982).

In some formations of preferably coarser grain size, gravel-sand size ratios of 6 - 14 allow sand invasion into the pack and ratios over 14 even permit unrestrained sand production through the pack (FLANIGAN 1979, PENBERTHY & COPE 1979). In such reservoirs, complete stoppage of sand movement can be reached with gravel-sand size ratios of 3 - 6. The more uniform the gravel material, the higher the permeability of the gravel pack, and avoiding excess fines makes twice screening of the gravel feasible.

5.2.3.3.2.2. Ideal level of pack gravel/pay sand size ratio

For most of the reservoirs requiring sand control by gravel packing, gravel grain size selection can be summarized that a ratio of gravel median diameter/formation sand median diameter between 5 and 6 is the ideal range providing absolute stoppage of sand movement and unrestricted initial gravel permeability (ECONOMIDES 1986). Ratios below 5 give rise to limited hydrocarbon production, because proper bridging at the gravel interface is not possible, but yet no invasion of the gravel pack by reservoir sand occurs. Ratios between 6 and 11 provoke intrusion of formation sand into the gravel pack which drastically reduces gravel permeability, but still provide effective sand control once the gravel pack is completely plugged with fine particles and further motion out of the pay zone is inhibited.

At ratios in excess of 11, the pores in the gravel package become so large that reservoir sand is partially migrating through the gravel into the wellbore, thereby increasing gravel permeability again, but already progressively deteriorating hydrocarbon exploitation by having only one effect of slowing down of sand production. At a ratio of abt. 15, original gravel pack permeability is reached again which reflects that at ratios above 15, formation sand can invade the borehole by passage through the gravel pack without any restriction, with in this case the gravel pack being completely useless.

5.2.3.3.3. Formation permeability and granulometry

Gravel grain size selection depends on both formation permeability and granulometry. While coarser gravel grain sizes are required for high-permeability reservoirs being clean medium- to coarse-grained sands, smaller gravel grain sizes are needed for low-permeability pay zones being fine-grained sands containing dispersed clay particles. As fines migration is the worst impairment in low-permeability sands, smaller gravel grain sizes have to be selected for bridging and holding back of clay minerals, with better results even being achieved if accepting lower gravel pack permeability with respect to that obtained from larger gravel grain sizes. In many cases, 40/60 or 50/70 mesh gravel pack material is applied for controlling fine-grained low-permeability sands. Although such dirty reservoirs can be exploited without stabilization by reducing production rates, enough inlet area is necessary to achieve reasonable offtake rates. Creation of sufficient inflow area and diminution of flow velocity in high-pressure oil wells may require installation of larger amounts of perforations or open-hole completion, and impairment can be counteracted by increasing inlet area and reducing velocity.

As most of the wells with gravel packing potential are nowadays completed in low- to moderate-permeability gas reservoirs, abt. 80 % of the jobs are carried out using 40/60 mesh or even smaller gravel packing material in contrast to the situation abt. 10 years ago when preferentially 20/40 mesh gravel has been applied for high-permeability oil reservoir sand control. 40/60 mesh gravel has turned out to perform much better in preventing formation fines invasion than 20/40 mesh gravel. Low-permeability sands requiring gravel packing are particularly widespread in the Gulf of Mexico.

5.2.3.3.4. Impact of formation and gravel fines

Gravel pack performance is even more susceptible to formation and gravel fines than hydraulic fracturing behaviour. Contents of clay minerals and other fine materials above 1 % in the sand to be controlled create problems unless clay inhibitors or stabilizers are included, and 2 - 4 % of clay minerals in a sand can screw up a gravel pack job if freshwater systems are introduced. This sensibility of gravel packing against even slight impurifications of fines also implies that gravel pack sand or proppants have to be much narrower screened than the material used in hydraulic fracturing. Although the preparation of specially sieved gravel pack sand is more costly than the production of fracturing sand, much more money can be wasted once the inferior product impairs gravel pack permeability and hence oil production. Therefore even multiple screening of gravel to extract fines below a tolerable minimum residue pays off very well.

The aforementioned guidelines for gravel grain size selection as a function of reservoir granulometry all refer exclusively to sandy pay zones. In chalk formations, these rules are not applicable, and another approach has to be undertaken. Deviation from conventional gravel pack design criteria in the chalk is caused by the extremely fine grain size of this reservoir rock type and its unique flow behaviour in case of lacking consolidation and occurrence of internal overpressure (ATTARD, MATHES & MOWER 1988; cf. section 4.5.4.3.). Instead of a prohibitively small gravel grain size which would have been required to ensure individual formation particle stoppage according to the standard guidelines, 20/40 grain size has turned out to retain substantial flow capacity of the oil and to hold back the chalk matrix which is characterized by mass flow behaviour once becoming unstable and mobile.

5.2.3.4. Width of gravel mantle

A gravel pack width of 5 grain diameters will already create an effective bridge (SAGE & LACEY 1942), but the accepted minimum gravel pack thickness for oil-field applications is 3 in. (SCHWARTZ 1969) according to the problem of placing gravel and fluctuating flow conditions. Therefore open holes have generally to be underreamed to provide 3 in. or even more between screen and formation at the borehole wall (ALLEN & ROBERTS 1982). Water-well completion has shown that a pack thicker than 8 in. cannot be properly developed.

The ideal gravel pack thickness for protecting a high-rate secondary-recovery oil-production well is 6 in., but tool capabilities and casing or screen schemes often limit gravel belt thickness to 5 in. Thicker gravel packs commonly permit higher production rates per unit length of reservoir interval. In cased holes, screen diameter should be as large as possible, but adequate room for gravel packing should be left in the annulus (aspects of gravel mantle width in open and cases holes are also discussed in sections 5.10.1.2. and 5.10.2.1.). MURER (1981) recommends a gravel pack width of at least 100 - 150 mm to create a sufficiently stable sand annulus. LEWIS (1985) argues that at least in some cases, gravel pack thickness plays key function in sand control.

5.2.3.5. Impact of gravel pack hardware

In addition to gravel size selection, considerable improvement of the gravel pack performance can be made by various other technical features (GURLEY, COPELAND & HENDRICK 1977). The greatest pressure drop in a gravel-packed well occurs in the perforation tunnel (WILLIAMS, ELLIOTT & WEAVER 1972, SAUCIER 1974), with minimum pressure drawdown being achieved by increasing the inlet area by increasing density or diameter of the perforations, or even by eliminating the perforations with open-hole completion (SUMAN 1972). Some aspects of slotted liners vs. wire-wrapped screens and packing hardware influence on gravel quality are summarized as follows.

5.2.3.5.1. Slotted liners vs. wire-wrapped screens

While coarser gravel can be sufficiently operated with a slotted liner, wire-wrapped screens are much more suitable for finer gravel, because the wire spacing can be mechanically much better adjusted to be smaller than the diameter of the smallest gravel than could be made adequate slot spacing in a slotted liner, and in addition to that wire-wrapped screens have higher flow capacities than slotted liners with the same slot width as the spacing of the wire. Other benefits of wire-wrapped screens with respect to slotted liners are clog resistance and durability against corrosion, whereas the only competitive advantage of slotted liners over wire-wrapped screens is cost and therefore slotted liners may be feasible in boreholes where very long intervals have to be gravel packed (HALLIBURTON 1988). Following discussion of some general aspects, comments are offered on slot and wire spacing selection as well as on mechanical limitation of slot and liner width.

5.2.3.5.1.1. General aspects

In earlier years, many sand production problems were simply handled by indicious selection of opening width in perforated liners (COBERLY 1937, COBERLY & WAGNER 1938). In many fields, however, this method of reduction or elimination of sand trouble does not lead to satisfactory results because of too fine grain size of the formation sand, poor compaction of the reservoir, low bottom-hole pressure and other adverse conditions, and gravel pack installation in combination with slotted liners or wire-wrapped screens has turned out to be the optimum solution.

Slotted liners are particularly successful in reservoirs where grain size distribution and framework cohesion is such that formation sand filters out and due to bridging tendency generates its own pack. Wire-wrapped screens include wire-wrapped flush with the pipe face with no spacers between wire and pipe, and wire-wrapped over-ribs holding the wire wrapping away from the pipe. The latter stancliff type of wire-wrapped screens has a greatly increased area available for flow into the wellbore which reduces flow velocity in the borehole vicinity.

5.2.3.5.1.2. Slot and wire spacing selection

Liner slots and wire spacings must have a width equal to or less than the diameter of the smallest grains of the gravel pack (COBERLY & WAGNER 1938) or formation sand depending on whether the screen is run in combination with gravel or not (ALLEN & ROBERTS 1982). While a slot with a width twice the grain size at the ten percentile of the gravel pack would hold the gravel in the accepted sense, the small quantity of gravel which is produced during the period when the slot bridge is being built up might be enough to cause the gravel pack to settle, thereby allowing production of sufficient formation sand to brand the installation a failure.

Slot width should be as narrow as possible to retain sand grains, but on the other hand as large as possible for not restricting flow of fluids and interstitial fines (ALLEN & ROBERTS 1982). In order to go on the absolutely safe side, wire spacing or slot width of half or two thirds the diameter of the smallest gravel pack grain is desirable (HALLIBURTON 1988). Absolute stoppage of formation sand and optimum gravel retention require a maximum number of liner slots with a width less than the diameter of the smallest gravel used (NINI & OWEN 1983). Undersizing has proven to reduce the tendency of near-size gravel to plug slots during gravel placement (SHRYOCK & MILLHONE 1979). LEDLOW & SPARLIN (1980) and SOLUM & RAMEZAMI (1987) report downhole protection of sand control screens. Good gravel pack insertion tools are the base for satisfactory gravel pack installation and a penny saved for better hardware could result in loss of thousands of dollars once gravel pack performance is not appropriate due to poor placement.

5.2.3.5.1.3. Mechanical limitation of slot and wire width

As cutting of liner slot and wrapping of wire-screen width is mechanically limited, it may be in some very fine formation sands difficult to keep the rule that the maximum opening should not be larger than the smallest gravel used (FLANIGAN 1979). Although on average, screen opening should be abt. half or two thirds the gravel size, small slot widths are particular problems if screen or liner has to be coated with polymers in order to prevent corrosion. Screen or liner length should be abt. 10 ft. longer than the perforated interval and should extend above and below the perforations where it passes into blank liner (HALLIBURTON 1988). Screen or liner diameter should be such that a clearance of at least 3/4 in. completely around the gravel pack hardware is provided when positioned in the casing in order to allow room for gravel to flow freely, and provide extra gravel reserve for after-pack settling and adequate space for fishing tools (NINI & OWEN 1983). Increasing screen or liner diameter is more important in higher-rate wells especially in case of gas because of turbulence, and less important in low-rate wells (LEDLOW, SAUER & TILL 1985).

5.2.3.5.2. Packing hardware influence on gravel quality

Due to the high cost of accurately screened gravel and the importance of not breaking or crushing it, treatment and placement of gravel with surface equipment is a primary factor in proper sand control (SOLUM 1986). Satisfactory results of gravel packing operations strongly depend on high gravel quality, accurate control of the fluid-gravel ratio, reduced gravel break-up and well-filtered carrier fluids. SOLUM (1986) presents a historical overview of introduction and modification of various surface and downhole equipment and methods during course of the technological evolution of gravel packing. Gravel breakup causes liner plugging and reduces pack permeability. Some aspects of potential gravel breakup points and improvement as well as fluid viscosity and velocity are outlined as follows.

5.2.3.5.2.1. Potential gravel breakup points and improvement

The most potential gravel breakup points are high-angle bends in the manifolding as well as centrifugal pump, restricted flow points, crossover tool and casing wall. Considerable minimization of gravel breakup has been achieved during the last years by introduction of a spherical ball valve in the fluid pump. Although gravel breakup can never be eliminated completely during surface and downhole handling, the effects of most of the mentioned trouble points can be sufficiently minimized by quality improvement of equipment and technology.

Gravel crushing and erosion of casing and formation face can occur in case of application of slim-size crossover parts (FLANIGAN 1979). Enlarged crossover parts for utilization of viscous slurries diminuish significantly the tendency of gravel crushing. Pumping of gravel-laden water mixtures through a narrow crossover part causes fracturing of some of the grains, with the blasting effect of sand hitting the casing walls being able to produce enough fines to completely plug liner slots (HALLIBURTON 1988). In contrast to water-based carrier fluids, gelled polymers require less velocity to move the gravel efficiently through the crossover tool, the cushioning effect of the gel reduces grain-to-grain abrasion which can also generate fines, and virtually no plugging of slots occurs.

5.2.3.5.2.2. Fluid viscosity and velocity

Thin low-viscosity packing fluids have to be pumped at high rate and velocity in order to be able to carry gravel (HALLIBURTON 1988). Low-viscosity gravel slurries, however, have a very severe scouring effect on work strings and cause the pack to collect debris. High fluid velocity provokes grain crushing during gravel placement, and low-viscosity fluids carrying gravel at high pump rates can cause intermixing of gravel and formation sand due to jetting effects. When using thin packing fluids, grain crushing in pumps can be avoided by downstream sand injectors.

5.2.3.6. Use of screen or liner without gravel pack

In addition to gravel packing where the screen holds the gravel in place that in turn retains the reservoir sand, formation sand control is also possible by stabilizing the pay zone with the screen alone without gravel or vice versa (ALLEN & ROBERTS 1982). Comments are given as follows on gravel-less hardware installation and linerless gravel placement.

5.2.3.6.1. Gravel-less hardware installation

The economical benefit of gravel-less hardware installation is that artificially-introduced gravel is a major cost factor in sand control operations which can be saved particularly in coarse clean reservoir sands where flowing formation sands construct a natural gravel pack around the wire-wrapped screen or slotted liner. Ribbed wire-wrapped screens cost two or three times more than slotted pipes, but for the same slot width, the former have perhaps eight to ten times the open area. All-weld screens have twice the open area as ribbed screens, but cost nearly twice as much. Wire-wrapped screens also have the advantage of more erosion- and corrosion-resistant materials. Vertically slotted pipe has more axial strength and bending strength than horizontally slotted pipe. The following outline illuminates some aspects of slot/wire width and natural gravel pack building as well as triple-wrapped screens.

5.2.3.6.1.1. Slot/wire width and natural gravel pack building

The use of properly sized wire-wrapped screens or slotted liners without gravel pack can be a low-cost means of controlling sand for low production rates per unit length of reservoir section. Slot size should be more or less twice the 10 percentile sand size for non-uniform sands, whereas slot width probably should not be greater than the 10 percentile size for uniform sands (COBERLY 1937, ALLEN & ROBERTS 1982). Selection of wire-wrapped screen or slotted liner once the slot size has been determined depends on well conditions. Sawcut slots are cheaper, whereas wire-wrapped screens permit the application of harder, more erosion-resistant metal. Screens set inside casing usually reduce productivity since fine-grained formation sand moving through the perforations fills the annulus between screen and casing, with thus optimization being reached by using the largest possible screen diameter.

This applies particularly for open holes in order to prevent caving of shale laminations, with under these conditions underreaming of the wellbore being neither necessary nor desirable. Such prepacking of liners with borehole collapse onto the liner and generation of natural gravel prepacking is also applicable in highly-deviated to horizontal wells as a completion technique in unconsolidated sands which create considerable borehole instability (AUSTIN, ROSE & SCHUH 1988). Installation of screen or liner alone which results in building of a natural gravel pack consisting of formation sand around the hardware, however, may promote reservoir disturbances and permeability damage (HALLIBURTON 1988; cf. section 5.9.6.2.).

5.2.3.6.1.2. Triple-wrapped screens

Improved hardware for gravel-less installation are triple-wrapped screens (LIKWARTZ 1976). Triple-wrapped screens for self-packing of formation sand around them are designed such that the multiple wraps bridge the largest reservoir sand grains at the outer screen and the smallest particles at the innermost screen. Triple-wrap screens are especially suitable for wells where bottomhole pressure is too low to allow gravel packing (cf. also SPURLOCK & DEMSKI 1972).

5.2.3.6.2. Linerless gravel placement

The inverse solution to screen or liner installation without gravel is linerless gravel packing (HALLIBURTON 1988) where abt. 20 - 25 % of coarse gravel has to be uniformly mixed into the main gravel in order to provide a bridging effect on the perforations, with round particles bridging a hole three times as great as the diameter of one of the particles. The main advantage of linerless gravel packing is complete absence of any hardware obstruction in the hole. Linerless gravel packing, however, should never be attempted in cased holes unless every effort has been undertaken to open up every perforation, because one untreated or insufficiently packed perforation may serve as entrance for formation fines which then plug the gravel cylinder and ultimately result in sand control failure. In order to keep this drawback possibility relatively low, total perforation number should not exceed that necessary to provide adequate production. Spherical grains form stable bridges on openings larger than grain diameter, and grain angularity and shape do not materially affect the opening size for a stable bridge, but do increase bridging range (COBERLY 1937).

In open holes, linerless gravel placement can profit from the absence of hardware in the wellbore by enabling to insert the gravel with higher operating pressure which permits better infilling and plugging of all the cavities and irregularities created at the borehole wall of the unstable sand by underreaming. Linerless gravel packs represent in fact complete infilling of the total inner borehole diameter with a dense cylindrical gravel pack which at the same time is an excellent filter for the oil that flows out at the gravel top in contrast to a tube-shaped gravel mantle between screen or liner and casing or borehole wall generated by conventional gravel packing with hardware utilization.

5.2.3.7. Formation grain size evaluation

Formation grain size evaluation has to be carried out with maximum accuracy, because its proper determination is inevitably necessary for selection of the suitable gravel grain size for the sand control treatment. Formation grain size assessment is also the greatest possibility of erroneous gravel pack design and thus a very potential source of failure of the sand control operation if not leading to a result that is more than closely approaching the reality downhole. Some comments are offered as follows on grain size variations in heterogeneous reservoirs and adjustment of gravel grain size to finest pay beds.

5.2.3.7.1. Grain size variations in heterogeneous reservoirs

While grain size determination may be rather simple in homogeneous reservoirs, many pay zones exhibit considerably heterogeneous composition (cf. section 4.2.4.1.) with grain sizes changing horizontally and/or vertically at random or in an ordered manner on various scales, and additional complications are given by the different effectivity of various borehole sampling techniques. The decisive drawback is that significant grain size changes can take place on sample scale in very anisotropic and heterogeneous pay zones such as cross-stratified small aeolian dune sands (cf. plate II/2 and 6 in section 3.13.), horizontal-laminated aeolian interdune and playa sediments (cf. plate VIII/1-5) and fluvial braided-river channel bar deposits (cf. plates III/1-2, V/1-8, VII/5-6 and IX/1-8).

Typical successions of progressively changing grain sizes in a more or less ordered manner are coarsening- and fining-upwards sequences which are characteristic features of deltaic and fluvial sedimentary complexes, respectively (cf. also section 4.2.2.6.). Microcyclic repetitions of several members of these grain-size trend cyclothems represent heterogeneities on a lower hierarchical order, and intercalations of mud drapes into sand bodies (cf. plates III/1-2, VII/5-6 and IX/1-8) are smaller-scale anisotropies of grain size distribution and depending on extension and thickness of hydrocarbon-bearing intervals can quite seriously disturb correct grain size determination for mechanical sand control design.

5.2.3.7.2. Adjustment of gravel grain size to finest pay beds

Therefore in many heterogeneous and anisotropic reservoirs, formation grain size may vary considerably over the interval to be gravel packed (SAUCIER 1974). In order to achieve maximum effectivity, gravel pack design and grain size selection should be based on the finest sand-size segment of the productive formation that is perforated. This procedure ascertains that even in the worst case, complete stoppage of all grain sizes of moving formation sand is achieved. Although this may result in choice of a smaller gravel grain size than would be suitable for many intervals of the prospective section and thus hydrocarbon offtake capacity is sacrificed due to a higher pressure drop through a less permeable gravel pack tunnel plug and mantle, it is better to prevent drastical gravel permeability destruction by invading formation sand.

Pay zone grain size determination has to be carried out with considerable accuracy and caution including compensation and correction of disturbances and contamination in order not to lead to improper gravel pack layout (MALY & KRUEGER 1971). Caution has to be exercised to restrict sand production in order not to trigger destruction of the natural stratification of the reservoir (SPARLIN & COPELAND 1972). If the original bedding fabric collapses due to excessive sand influx into the borehole, nearby shale lenses and layers will mix with the unconsolidated sand when shifting towards the wellbore. Such mixtures can become severe permeability barriers if stopped and held in place by the gravel pack. Especially when formation grain size varies considerably in the investigated interval in horizontal and vertical direction, attention has to be paid in gravel-sand size ratio design to the smaller reservoir sand grains and the finer pay zone interbeds, particularly in case of higher flow velocity, more non-uniform sand, fluctuating flow rate and high gas-oil ratio (ALLEN & ROBERTS 1982).

5.2.3.8. Grain size intermixing of gravel and formation sand

Concerning grain size mixing, similarly as applying for conductivity deterioration in hydraulic proppant fracturing (cf. section 1.4.11.2.), intermixing of relatively small amounts of formation sand (abt. 10 - 25 %) being considerably finer than the sand-control medium lowers gravel mixture permeability to nearly that of the contaminating pay zone sand alone or even less (MORROW, HUPPLER & SIMMONS 1969; SPARLIN 1974). Therefore intermixing of gravel and formation sand has to be minimized during performance of the treatment, which can be achieved by tackifying agents in the fluid that cause the gravel to flow as a plug or a consistent slurry, or by high gravel concentration and high fluid viscosity.

5.2.3.8.1. High gravel concentration and high fluid viscosity

Significant means of reducing intermixing of gravel pack material and reservoir sand are suspension of gravel at high concentrations (abt. 15 lbs of gravel added to 1 gal of carrier fluid or even more) in high-viscosity fluids and pumping the slurry through perforations at very slow rates (SPARLIN & COPELAND 1972; SUMAN, ELLIS & SNYDER 1983), which also requires less carrier fluid and thus cuts rig time, minimizes formation plugging, reduces screen erosion during placement, and decreases gravel crushing at the crossover part and plugging of slotted liner or wire-wrapped screen (UNDERDOWN, DAS & NGUYEN 1984). The danger of mixing of reservoir sand and gravel pack material is particularly high if the gravel is squeezed outside the casing and injected with high velocities (SPARLIN 1969, SPARLIN & BOND 1969, SPARLIN & COPELAND 1972).

5.2.3.8.2. Tackifying action of resin in the slurry

The amount of mixing of gravel with formation sand can be significantly reduced by suspending high gravel concentrations in a viscous oil as carrier fluid and add resin as a tackifying agent to the slurry in order to make the gravel grains slightly sticky. The slurry is then squeezed into the perforations and a screen is washed in place after the concentrated gravel slurry has been injected. Minimization of gravel and pay zone sand mixing is mainly achieved due to pushing of the loose reservoir sand away from the wellbore by the high-viscosity fluid, the promotion of plug flow of the gravel by the high saturation and the cohesive force of the tackifying agent thereby bringing large amounts of gravel to the formation at the same time, and the slow pumping rate preventing any blasting, fingering or fracturing of the reservoir.

5.2.3.9. Hydrocarbon productivity vs. sand control

The main question when treating sand problems in oil wells is how to perform optimum sand control without restricting hydrocarbon production (SPARLIN & COPELAND 1972). As most operators are now more critically than ever before reviewing the wells in light of possibilities to increase offtake rates at the lowest possible investment and maintenance expenses and to drastically lower withdrawal cost without loosing too much output which has to be remained at an optimum feasibility level, improvement of gravel packing techniques as one potentially profitable way to ameliorate well productivity in combination with giving better sand control is of progressively greater significance. Sand retention vs. hydrocarbon flow capacity and other aspects are briefly outlined as follows.

5.2.3.9.1. Sand retention vs. hydrocarbon flow capacity

Although design of the sand-retention gravel is based on the finest formation material present in the productive interval, gravel selected according to sand stabilization criteria could adversely affect well flow capacity. High success ratios of the gravel pack operation, however, are frequently obtained when carefully considered adjustments are made in design conditions. Gravel plugging with formation sand may be intentionally permitted opposite some finer low-permeability portions of multi-pay intervals to allow highest productivity from the coarser good reservoir sections to which gravel size is optimally adjusted. Alternative solutions in case of low absolute gravel permeability with high pressure drop across the perforation tunnels are increased perforation density and/or diameter, and open-hole gravel pack if technically allowed by the distribution of reservoir types and fluid compositions in the geological column penetrated by the borehole since the last casing was run. In any case, however, gravel-sand diameter ratios that allow continued sand production and lead to possible liner erosion have to be avoided (SUMAN, ELLIS & SNYDER 1983).

The most prevalent theory attributes sand production to friction and resultant pressure drop as reservoir fluids pass through the pores of the formation sand body (RIKE 1970). When the pressure drop is high and cementing materials between sand grains are weak or absent, the individual particles of the sand body are dislodged and carried into the wellbore. Another explanation suggests that environmental water which was present during original deposition of the sand grains is chemically different from the water contained in the aquifer surrounding the oil field. Water production can actually chemically dissolve a portion of the cement between the sand grains, thereby loosening the grain fabric. Another concept assumes that the reservoir compacts as formation pressure declines with progressive depletion, and the changing load tends to shift sand grains and mechanically shear existing cementing materials.

5.2.3.9.2. Other aspects

An important application of gravel packing in combination with hydraulic fracturing is prevention of proppant and formation sand flowback into the borehole after termination of hydraulic stimulation treatments particularly in soft sands and chalks. Double insurance against grain flowback can be made by tailing-in resin-coated proppants in the terminal stage of the hydraulic fracture job (cf. section 4.12.3.3.2.) and performing a combined outside and inside gravel pack operation with resin-coated gravel (cf. section 5.8.2.). In case of proper grain size design, the coarser tail-in of the proppant wedge filling the hydraulically-generated crack acts itself as a gravel pack by bridging the finer proppants of the main fracture-plugging lot at the interface of both grain-size classes (cf. section 5.8.8.). Drawbacks of any type of gravel packing are mainly linked with soft high-permeability thick reservoir intervals which are subject to mechanical formation damage when being coated with the gravel mantle, and horizons where mobile fines or high-viscosity waxy crude oil may create deposits near the wellbore or within the gravel pack that could lower productivity (WEISSENBURGER, MORITA, MARTIN & WHITFILL 1987).

Experience of productivity differences as a consequence of gravel grain size variations shows that higher offtake rates are obtained with coarser gravel due to better permeability of the pack which allows easier hydrocarbon flow through the sand-control filter (STRACKE (1987). GURLEY, COPELAND & HENDRICK (1977) outline designing and planning of gravel-pack operations for maximum productivity. Relationships between gravel pack design and well productivity are also examined by SCHWARTZ (1969) and LYBARGER, SCHEUERMAN & WILLARD (1974).

5.2.4. Gravel grain surface and shape

In terms of surface morphology, application of glass beads or zirconia-silicate proppants which have both perfectly round grain shapes and smooth grain surfaces (cf. sections 1.3.2. and 1.4.1.) requires reduction of the gravel-sand diameter ratio in order to promote bridging on the even surface of the globular or spherical grains (SUMAN, ELLIS & SNYDER 1983) or intermixing of some subspherical or angular grains with rough surface of similar size in order to provide the necessary friction. Following discussion of some aspects of round vs. angular sands for mechanical sand control, differences in grain shape requirement between hydraulic proppant fracturing and gravel packing are illustrated.

5.2.4.1. Angular vs. round gravel in mechanical sand control

Rough-surface natural sand and synthetic alumina silicate and oxide proppants have better bridging properties than smooth-surface artificial glass beads and zirconia-silicate proppants (cf. sections 1.3.2. and 1.4.1.). Angular gravel-pack sand is even less likely to be invaded because sand bridges more readily on the irregular shapes and uneven surfaces, but more consistent packing is obtained with rounded gravel with irregular rough surface (GULATI & MALY 1974). Rounded gravel pack grains result on average in lower pressure drops and apparently show less packing variance than more angular materials as a consequence of inertial effects of flow (SAUCIER 1974; cf. section 4.3.3.3.). Angular sand has a higher tendency to form premature bridges than round sands (HALLIBURTON 1988). Angular sand, however, is also more subject to chipping and fragmenting during placement which reduces pack permeability and can clog slots and screens.

5.2.4.2. Differences between fracturing and gravel packing

The important difference of surface morphology significance between hydraulic proppant fracturing and gravel packing is that in hydraulic stimulation, bridging of proppants in the crack is undesirable because it leads to premature screenout failure, and therefore round smooth particles are principally better suited for hydraulic fracturing. In gravel packing, however, bridging is necessary for the success of the operation, with thus irregular gravel surfaces that provide friction being more favourable than smooth surfaces where intermixing of other grains is necessary to trigger bridging. Stability improvement of all gravel types and grain sizes can be best achieved by liner vibration (SOLUM 1984; cf. section 5.8.4.) which gives rise to origin of a hexagonal particle framework instead of the standard cubic fabric that is created by conventional flow-type gravel placement. Consistency of the hexagonal grain arrangement is also increasing with more pronounced surface roughness or shape angularity of the grains and subordinately also with larger size of the gravel.

5.3. Gravel packing completion fluids and gravel saturation

Gravel packing transport fluids are in many cases more viscous than the hydraulic fracturing carrier fluids, while other mechanical sand control applications are also performed with low-viscosity transport fluids. Viscosity and thus suspension capacity of fluids for gravel packing, however, is in comparison to hydraulic fracturing a much smaller concern in terms of proppant or gravel dispersion, because as a consequence of the mainly low closure stress in the prevailing amount of reservoirs, predominantly natural quartz sand and light, low-density synthetic proppants (either alumina-silicate or zirconia-silicate proppants according to the presently commercially available spectrum; cf. chapter 1.3.) are selected for sand-control treatments. High-density proppants are only used in special applications such as steam-drive projects (cf. section 5.7.1.), because the operation causes difficulties of proper placement by particularly strong settling effects of heavy alumina oxide and silicate proppant grains in the fluids (aspects of gravel-pack carrier fluids are discussed by GURLEY, COPELAND & HENDRICK 1977; CONSTIEN & MAYER 1978, COSTEIN & WALLACE 1980, TORREST 1982, SCHEUERMAN 1984, SKAGGS 1984; UNDERDOWN, DAS & NGUYEN 1984; LEDLOW & SAUER 1985; MURPHEY, ROLL & WONG 1985; PENBERTHY 1985).

Gravel packing should aim on setting the gravel without contamination at the proper location as tightly as possible and then holding it in place for the lifetime of the well (ALLEN & ROBERTS 1982). The most important aspect of gravel packing carrier fluids is sufficient leakoff into the formation in order to achieve the necessary tight packing of the gravel, which is the second major difference to hydraulic fracturing where fluid loss has to be minimized in order to avoid proppant bridging and premature screenout failure. Gravel packing results are improving with progressive fluid leakoff and the rheological properties of sand control transport fluids therefore have to be adjusted such that satisfactorily high rates of fluid leakoff are achieved. The most important aspects are fluid viscosity and gravel concentration, transport and packing capacity of the slurries, gravel settling diminution in the carrier fluid, gel surface filtration from polymer solutions and screenout mimicking, and combination of gravel packing and hydraulic proppant fracturing.

5.3.1. Fluid viscosity and gravel concentration

Distinction can generally be made between conventional circulation gravel packing where low-viscosity fluids are used to carry low gravel concentrations (up to abt. 1.5 lbs/gal which is equivalent to abt. 180 kg/m3) into the well at high circulation rates and large total treatment quantities (high volume/rapid displacement method), and high-density gravel packing where high-viscosity, polymer-gelled fluid systems are applied that transport high gravel saturations (up to abt. 15 lbs/gal which is equivalent to abt. 1,800 kg/m3) to achieve packing with low total volume handled (SPARLIN & COPELAND 1972; SUMAN, ELLIS & SNYDER 1983) by the slurry pack method (PEDEN, RUSSELL & OYENEYIN 1985) or high-density squeeze packing technique where the sequence of prepad, gravel slurry and postpad is successively pumped (ELSON, DARLINGTON & MANTOOTH 1984). High-viscosity carrier fluids and high gravel concentrations result in less tendency of intermixing of gravel and formation sand (GAJDOSIK & WILLINGHAM 1976).

Low-viscosity fluids with low gravel concentration are mainly used in circulation packing in open holes or deviated cased holes where the perforations have been prepacked by viscous slurries transporting high gravel saturations, whereas high-viscosity fluids with high gravel concentrations are preferentially utilized in squeeze packing in cased vertical holes. High gravel concentrations in high-viscosity fluids do not plug the screen so severely as do low gravel saturations in low-viscosity fluids and are also much better promoting adequate perforation packing. A recent outstanding innovation are surfactant-bearing water-based high-viscosity gravel packing fluids which combine the favourable aspects of high fluid loss rates of conventional low-viscosity water-based fluids and high gravel transport capacity of traditional high-viscosity polymer-based fluids (DOWELL SCHLUMBERGER 1988, NEHMER 1988; cf. section 5.3.1.1.2.). Some comments on achievement of sufficient fluid leakoff and prevention of excessive fluid loss are given as follows.

5.3.1.1. Achievement of sufficient fluid leakoff

High-viscosity fluids are thus able to carry abt. ten times the gravel quantity per fluid unit which can be transported by low-viscosity fluids. The viscosity of the carrier fluid is directly related to the designed gravel concentrations for the slurry (PEDEN, RUSSELL & OYENEYIN 1985). High-viscosity fluids characterized by high gravel transport capacity and thus result in shorter gravel placement times, and are therefore an effective means for eliminating blank sections in the wellbore provided pack dehydration can be achieved by fluid leakoff through liner or screen (ELSON, DARLINGTON & MANTOOTH 1984). Low-viscosity fluids may provide tighter gravel placement in the annulus (ALLEN & ROBERTS 1982), but do not promote adequate perforation packing.

While the most significant advantages of high gravel saturations in high-viscosity fluids are excellent perforation packing, less impairment of screen flow capacity, and quick performance thereby saving rig time, several drawbacks also exist. After-pack gravel slumping may create voids in the packs which is a particularly severe problem in deviated wells (cf. section 5.8.3.). The high-density gravel-laden slurry may fall through the lighter previously pumped pad fluid, with the rate of fallthrough becoming greater when large-diameter work strings are used. If circulation is taken above the packer and not through a screen, the roping effect (cf. also section 5.9.1.3.3.) has to be considered to prevent getting gravel on top of the packer.

In contrast to hydraulic fracturing where extraordinarily high fluid loss to the formation can result in premature screenout termination of the treatment by fluid dehydration, proppant bridging and slurry sticking (cf. also sections 4.2.2.2.3. and 6.2.4.2.1), and excessive fluid leakoff rates are therefore undesired, gravel pack operations require high rates of fluid loss into the formation in order to transport and deposit large amounts of gravel and thus to enable high plugging efficiency of the perforation tunnels (NEHMER 1988). Influences of carrier fluid viscosity as well as improvement by surfactant-bearing water-based fluids are briefly illustrated as follows. Some comments are also given on the impact of formation permeability.

5.3.1.1.1. Influence of carrier fluid viscosity

Gravel packing can be performed with thin low-viscosity or thick high-viscosity fluids (HALLIBURTON 1988). Thin low-viscosity fluids are able to carry sand in lower concentrations only at higher pump rates which may cause intermixing of gravel and formation sand with associated permeability damage, and some grain breakage and fines generation may occur due to pumping of sand slurries at high rates through the tools as a consequence of impact. In extremely long productive intervals, the application of low-viscosity fluids may be dictated by economical considerations. High-viscosity fluids allow gravel packing at low injection rates which reduces intermixing of gravel and formation sand when the slurry is pumped through the perforations. Aspects of gravel placement and fluid breaking, gel concentration vs. slurry dehydration, fluid viscosity and formation damage, and fluid viscosity and leakoff rates are briefly outlined as follows.

5.3.1.1.1.1. Gravel placement and fluid breaking

High fluid-loss rates of low-viscosity gravel pack carrier media can be achieved by building a particle bridge on the formation face against which hydrostatic pressure can be applied (ALLEN & ROBERTS 1982). High-viscosity transport media place the gravel more quickly and with less fluid leakoff to the formation which bears also less possibilities of reservoir damage. Increasing carrier-fluid viscosity minimizes gravity segregation of the gravel during placement and thereby minimizes plugging of the liner slots with gravel fines which develop in the setting process. Concerning application of high-viscosity water-based fluids, type and concentration of viscosity breaker must be determined depending on formation temperature, and the sump volume below the screen has to be minimized in order to reduce further settling of the gravel after the viscosity has broken, thereby loosening the pack fabric.

Effectivity and time of breakage of the carrier fluid gel have a direct bearing upon gravel transport efficiency, compaction and settling of the gravel pack, and possible impairment of gravel pack permeability by fluid residues (ALMOND & BLAND 1984, SCHEUERMAN 1984; UNDERDOWN, DAS & NGUYEN 1984). Minimum carrier fluid density required depends on reservoir pressure which is a function of structural depth (PEDEN, RUSSELL & OYENEYIN 1985).

5.3.1.1.1.2. Gel concentration vs. slurry dehydration

Intermediate gel concentrations have higher leakoff at the same pressure than high gel saturations which aids in slurry dehydration both into the formation and through the screen (LEDLOW, SAUER & TILL 1985) if there is no skin and no pressure loss through the perforations. The pad volume in a straight squeeze pack should be decreased to a minimal amount such that less fluid has to leak off in order to achieve a tight pack, whereas in a circulating pack, the pad volume should be increased in order to allow some pad leakoff into the formation and favour circulation up the wash pipe. Increasing pump rate increases packing efficiency and prevents premature screenout or roping.

5.3.1.1.1.3. Fluid viscosity and formation damage

A special type of high-viscosity carrier fluids with high gravel saturations are concentrated gravel slurries (SPARLIN & BOND 1969, SPARLIN & COPELAND 1972) where a resin additive is intermixed as a tackifying agent in order to make the individual gravel grains sticky and to promote plug flow of the slurry (cf. section 5.8.5.). In the last years, the trend of application is more and more towards high-viscosity fluids and high gravel concentrations for the purpose of achieving safe gravel placement in a higher packing density, because experience has shown that low-viscosity systems with low gravel saturations effectively control formation sand, but also restrict hydrocarbon production.

Low-viscosity fluids carrying gravel at high pumping rates can cause intermixing of the coarser packing gravel and the finer formation sand due to the jetting effect (HALLIBURTON 1988) which can provoke severe permeability damage. The highest amount of damage occurs at a ratio of abt. 70 vol% packing gravel and 30 vol% formation sand, with the greater the ratio between gravel and reservoir permeability, the greater the damage potential. High pumping rates also result in generated fines due to gravel shattering during placement which may plug the slots of the screen (NINI & OWEN 1983).

5.3.1.1.1.4. Fluid viscosity and leakoff rates

Fluid leakoff rate to the reservoir is not the most decisive aspect for establishing gravel packs through the perforations (RENSVOLD & DECKER 1978). It is, however, essential that a void already exists behind the casing or that the slurry displaces formation sand to create room for the pack. Thick high-viscosity carrier fluids which have a considerably lower fluid-loss rate to the pay than thin low-viscosity transport media can more effectively displace reservoir sand outside the perforations to generate space for the gravel pack than thin low-viscosity fluids which are characterized by high leakoff rates, but which have insufficient ability to compress loose reservoir sand in the immediate vicinity of the perforation tunnels and therefore can only manage to install a less compact gravel plug than high-viscosity carriers.

In contrast to polymer fluids which suffer from limited leakoff and cause formation damage by microgels and fisheyes, high-viscosity water-based gel systems (which are outlined in the following section) do not contain any polymer and therefore have excellent leakoff and gravel suspension characteristics (DOWELL SCHLUMBERGER 1988 a).

5.3.1.1.2. Improvement by surfactant-bearing water-based fluids

Improvement of fluid leakoff into the reservoir in order to achieve satisfactory perforation tunnel packing and enhancement of gravel suspension properties which are particularly decisive in highly-deviated boreholes can be reached with a viscoelastic surfactant-base gravel-pack carrier fluid which has distinct advantages over conventional polymer gelled fluid systems (NEHMER 1988) and represents a major recent innovation. Comments on gel structure, microgel breaking and removal, and gel filter cakes in fracturing and gravel packing are offered as follows.

5.3.1.1.2.1. Gel structure

Surfactants develop viscous gels by formation of rod-shaped micelles which interact with each other at high concentrations, thereby hindering movement of the micelles and triggering an increase of fluid viscosity, with thus the rod-shaped micelles acting similar to polymer molecules. A salt-induced transition from spherical to rod-shaped micelles is a major mechanism for increasing viscoelasticity of the fluid. The viscoelastic surfactant-base carrier fluid does not cause formation damage in contrast to many polymer-gelled fluids where fish eyes and/or microgels which are even not removed by breakers plug pore throats and thus lead to impaired fluid leakoff. Intense manipulation of polymer-gelled fluids using shear and filter devices helps removing the fisheye and/or microgel particles, but makes the treatment more expensive in terms of equipment and manpower, and excessive shear degrades the polymer, giving rise to poor gravel suspension.

The viscoelastic surfactant gelled fluid system therefore represents a superior solution because of absence of formation damage as well as presence of favourable fluid leakoff, gravel suspension, gravel transport and gravel placement options (DOWELL SCHLUMBERGER 1988 a). The surfactant-bearing water-based system combines the favourable attributes of high fluid-loss rates of traditional low-viscosity water-based fluids and high gravel transport capacity of conventional high-viscosity polymer-based fluids. Another promising alternative is the low particle density/carrier fluid density system introduced by HUDSON & MARTIN (1988; cf. section 5.3.3.2.).

5.3.1.1.2.2. Microgel breaking and removal

The main significance of viscoelastic surfactant gelled fluid systems is fast leakoff of carrying fluids which assists the gravel to penetrate all the way to the ends of the perforation tunnels, because in order to achieve minimum pressure drawdown during production, screen/casing annulus and perforations must be completely packed with gravel to keep out formation sand (DOWELL SCHLUMBERGER 1988 a). Rapid fluid loss is especially desirable in squeeze packs (cf. section 5.3.2.4.2.), because the entire prepad must be leaked into the formation before the slurry can even reach the perforations.

Aspects of fluid quality enhancement by microgel breaking and removal in polymer-gelled gravel packing fluid operations predominantly conducted under constant flow rate are discussed by HASHEMI, CAOTHIEN & ROVIG (1988). Both mechanical and hydraulic shearing break and remove the partially hydrated microgels, with the resulting gels having higher fluid leakoff for a tighter gravel pack. Polymer shear degradation occurs in addition to chemical, biological and thermal decomposition and has to be taken into account. Gel shearing as a step to break up fish-eyes or micelles and augment their hydration can ultimately cause gels to lose the vital property of gravel suspension, and therefore must be monitored closely prior to mixing a gravel pack slurry (COLE, SHAW, CAVENY & BELLENGER 1988).

5.3.1.1.2.3. Gel filter cakes in fracturing and gravel packing

Similarly as gel filter cakes originating on the formation face at fracture walls deteriorate crack conductivity due to proppant embedment, gel surface filtration from polymer solutions during gravel packing covers the reservoir face with a low-permeability drape and deteriorates further leakoff of gravel carrier fluids into the pay matrix, thereby having deleterious effects on gravel placement and thus quality and performance of mechanical sand control (TORREST & MARCANO 1986; cf. sections 1.4.10.3.5.3. and 4.3.4.1.4.). Filtration of microgels during polymer insertion quickly builds up a relatively impermeable surface layer, and the pressure drop for water flow through this pellicle is linear in rate. Once formed, the gel layers can undergo changes in characteristics with water flow at sufficient rates. The layer buildup tends to give false signals of job completions as pumping pressures rise rapidly by mimicking screenout termination of the operation, and reduced gravel pack quality and performance could result from failure to account for the effect of surface gel filtration.

5.3.1.1.3. Impact of formation permeability

Fluid loss during gravel packing is strongly dependent from reservoir formation permeability. High-permeability sands are characterized by good fluid leakoff and thus allow high-quality gravel packing, whereas low-permeability sands permit only poor fluid loss and are therefore difficult to gravel pack with reasonable effectivity. Following a brief characterization of formation permeability classification for gravel packing in comparison to hydraulic proppant fracturing, some aspects of reservoir permeability and composition are outlined.

5.3.1.1.3.1. Permeability classification for gravel packing

In view of gravel packing applications, low-permeability sands are reservoirs with permeabilities in the range of abt. 20 - 75 md (which would have to be characterized as high-permeability pays in light of hydraulic proppant fracturing; cf. section 4.6.), and high-permeability sands are formations with permeabilities in the range of abt. 75 md - 4 d. While polymer fluids have high viscosity and bridge off in low-permeability sands thereby resulting in only limited effectivity of gravel packing, surfactant-based non-polymer fluids have lower viscosity and thus exhibit good leakoff rates also in low-permeability sands, therefore enabling satisfactory gravel placement (DOWELL SCHLUMBERGER 1988 a) and avoiding problems caused by other carrier fluids such as permeability loss from formation damage, low pack efficiency due to slow fluid leakoff, and rapid settling-out of gravel from the slurry.

5.3.1.1.3.2. Reservoir permeability and composition

While high-permeability reservoirs with gravel packing potential are clean medium- to coarse-grained unconsolidated sands, low-permeability pay zones requiring stabilization are fine-grained sands containing dispersed clay mineral particles. Especially complicated situations for gravel packing in view of sufficient fluid leakoff rates are interbedded sequences of high- and low-permeability sands, with again proper fluid loss only being able to be achieved by surfactant-based non-polymer fluids with low viscosity (alternating low- and high-permeability reservoir storeys also have a negative impact on fluid leakoff in hydraulic fracturing which is difficult to limit in contrast to the necessity of its promotion in gravel packing; cf. section 4.3.4.6.).

5.3.1.2. Prevention of excessive fluid loss

High fluid-loss rates during gravel placement are necessary to achieve good pack qualities, but high to excessive fluid leakoff prior to gravel packing can cause serious problems if soluble fluid-loss agents had to be added, because invasion of formation pores, bridging at the entrance of the reservoir to build up a filter cake, and filling of perforation tunnels may prevent complete gravel placement and reduce well productivities (SPARLIN & HAGEN 1986). Excess fluid-loss additives inhibit complete coverage of an interval with gravel due to either inadequate leakoff to the formation or partial plugging of the screen.

Fine grinds of calcium carbonate or silica flour are the most common solid leakoff additives to gelled fluids and form low-permeability filter cakes and packs as they cover reservoir face and perforation tunnels, respectively, but these fine particles can cause serious formation damage, as they are difficult to remove (cf. section 5.11.4.). Due to the fact that any perforations filled with fluid-loss additives will not accept gravel and where fluid flow is restricted, gravel will not be packed sufficiently tightly, all the fluid-leakoff agents used before setting the gravel pack must be removed from wellbore, perforations and near-borehole formation to ensure a complete pack. Some drawbacks of post-packing acidizing as well as a few points of solid fluid-loss additives and partial formation pre-packing are illustrated as follows.

5.3.1.2.1. Drawbacks of post-packing acidizing

Acidizing after a gravel pack is in place should be avoided if possible, because fluid injected into the gravel mantle may disturb the pack and push the gravel away from the wellbore, and where material is dissolved, the gravel will move into the voided space, resulting in holes in the pack and/or gravel and sand mixing together. As acid can be injected through perforations filled with gravel much more easily than through those plugged with calcium carbonate fluid-loss additive as a consequence of permeability and pressure drop contrast, the acid will bypass the calcium-carbonate filled perforations and therefore not completely remove the fine plugging material. It is thus far better to circumvent the need for an acid job after gravel pack installation by removing all previously inserted fluid-leakoff additives by a pre-gravel-pack acid treatment which provides higher well productivity and more positive sand control (other aspects of acidizing prior to and after gravel pack installation are discussed in section 5.11.4.).

5.3.1.2.2. Solid fluid-loss additives and formation pre-packing

The most practical solution is to use a combination of gravel and viscous fluids as a diverting system for the acid which helps to prevent excessive fluid loss ahead of the gravel slurry and partially prepacks the formation with gravel. Acid has to be diverted to all perforations in order to allow for complete fluid-loss additive removal and to prevent acid bypass that leaves some damaged patches in the formation. The gravel diverter stages should be designed such that not more than 10 % of the perforations are filled with gravel in each stage, as the effect of the gravel-filled perforations on acid injection pressure depends on amount of perforations packed by gravel, number of perforations cleaned by acid, severity of plugging of other perforations, undamaged injectivity of the completed interval, downhole dynamic viscosity of the gelled brine, gravel permeability and injection rate.

Injection of calcium carbonate leakoff additives into a gravel pack is often necessary to control excessive fluid loss while preparing the well for production after a gravel pack or while installing a gravel pack in an upper zone, but can lead to serious problems. As these fluid-leakoff additives are often designed such that the particles flow through the gravel and build a filter cake on the formation face at the boundary between reservoir sand and packing gravel, the pores of the gravel in some of the perforation tunnels become plugged and the material is very difficult to remove due to selective dissolution by the acid followed by acid bypassing of perforations filled with fine-grained solids due to differences in pressure drops. Therefore the median particle size of the fluid-loss additive should approximately equal the median pore size of the gravel which minimizes invasion of calcium carbonate into perforation tunnels and formation and still adequately controls fluid loss rate.

5.3.2. Transport and packing capacity of the slurries

Transport efficiency of high-density gravel pack slurries generally improves with increasing fluid velocity, gravel concentration and annular speed (SKAGGS 1984). With increasing gravel saturation, less fluid viscosity is required to maintain a given level of transport efficiency which improves placement by better leakoff rates and reduced pressure drop through the perforations while being packed. Transport efficiency is predominantly controlled by slurry consistency, with similar conditions being able to be achieved with low-viscosity fluids and high gravel concentrations or high-viscosity fluids and low gravel saturations within the range of gravel suspension capacity of the different fluid types, thereby giving ample possibilities of design. The overview as follows includes comments on suspension symmetry, dual carrier fluid viscosity, evolution of gravel packing techniques, and gravel volume and placement.

5.3.2.1. Suspension symmetry

Distinction can generally be made between symmetrical and asymmetrical suspensions for slurry definition (WAHLMEIER & ANDREWS 1988). Symmetrical suspensions in which the solids are uniformly distributed throughout the liquid apply for slurries having extremely high solid concentrations, high velocity, fine particle size, elevated fluid viscosity, and/or small density differences between solids and liquid. Asymmetrical suspensions exhibit a pronounced concentration gradient normal to flow direction and are represented by slurries usually having lower saturations and/or containing larger particle sizes than those of symmetrical suspensions. Asymmetrical suspensions representing heterogeneous flow can be subdivided into the stages of flow within a distorted concentration profile and no particulate deposit, sedimentation within a moving bed of particulates, deposited stationary bed with saltation, and complete settling of the particles from the liquid. The critical or equilibrium velocity of the slurry is the velocity for given liquid and/or flow conditions where a bed of solids begins to form.

5.3.2.2. Dual carrier fluid viscosity

Increasing gravel pack carrier fluid viscosity facilitates mixing, pumping and transport of slurries with high gravel concentrations, improves gravel movement through perforations, stabilizes perforation cavities and tunnels in the formation, decreases erosion and screen plugging by fines, and inhibits gravity segregation of particle sizes and settling of gravel in the slurry (GRUESBECK & COLLINS 1978; SHRYOCK, DUNLOP & MILLHONE 1979; SHRYOCK & MILLHONE 1979). On the other hand, however, increasing viscosity can cause incomplete packing due to decreasing fluid leakoff rates and increasing tendency for gravel layering on the screen and node building at the perforations (SHRYOCK 1980, SCHEUERMAN 1984; cf. section 5.8.7.).

As the fluid viscosity needed for mixing and pumping of gravel slurries at the surface is higher than that required for optimum performance downhole, the necessary dual viscosity is attained by addition of breakers to degrade the viscosifying polymer (SCHEUERMAN 1984). This dual viscosity modelling of gravel packing fluids in order to achieve the required downhole viscosity not before contacting the reservoir can be to some extent compared with delayed crosslinking of fracturing fluids which also provides a useful combination, although in reverse orientation, by having lower viscosity upon entrance to the flowlines thereby reducing tubular friction, and reaching higher viscosity downhole thus being able to suspend and transport high proppant concentrations into the fracture (cf. section 4.3.4.3.).

5.3.2.3. Evolution of gravel packing techniques

In the early years of gravel packing since the late 1930`s when this method became a viable means of controlling sand production in oil and gas wells, gravel was placed downhole in the initial submature phase by simple dry dumping, and then with beginning amelioration of techniques and accumulation of knowledge and experience mainly by the normal circulation method which comprises pumping of the gravel down the tubing string and through a float collar at the bottom of the liner and back up the annular space (SOLUM 1986). Later reverse circulation, squeeze packing and four-position longstroke operation were successfully improving gravel placement effectivity. Following discussion of some general aspects, some comments are given on normal and reverse circulation packing, squeeze packing, combined circulation and squeeze packing, four-position longstroke packing, wash down and pressure packing, and other techniques.

5.3.2.3.1. General aspects

After the early straightforward beginnings, sand control technology has later progressively evolved with both local adaptation and increasing sophistication (MURPHEY, BILA & TOTTY 1974). Clogging of the float collar and the inability to pack the liner top were major drawbacks of the normal circulation method. During the last decades, the reverse circulation technique was developed which consists of pumping of the gravel down the annular space and around the outside of the liner. As the gravel reaches its setting point, the carrying fluids would continue downhole and return to the surface through a perforated nipple at the bottom of the liner. Earlier drawbacks such as bridging off due to poor downhole tools and surface procedures have been overcome during course of the technological evolution, with reverse circulation in combination with squeeze packing being the currently most frequently applied gravel placement method (SOLUM 1986). The original normal circulation technique has recently been revived and utilized in an improved version for gravel packing of highly-deviated wells (STILES 1985; STILES, COLOMB & FARLEY 1986; cf. section 5.8.3.5.).

5.3.2.3.2. Normal and reverse circulation packing

The simplest sand control measure is installation of screen or liner alone without gravel, with influx of formation sand into the borehole then resulting in building of a natural gravel pack (HALLIBURTON 1988; cf. section 5.2.3.6.). No influence can be exerted on gravel grain size distribution in such natural packs arising from pay collapse along the borehole wall which is given by reservoir granulometry, and sand entering the wellbore may cause formation disturbance and permeability damage. Gravity packing is the most primitive means of gravel insertion where a screen is run into the borehole and gravel is slowly poured down the annulus. Care must be taken to avoid gravel bridging around the tubing collars and it is very unlikely that any gravel is forced through the perforations, with consequently hydrocarbon flow into the wellbore being restricted by formation fines invading the pack.

Circulating packs are generally placed in a well by pumping fluids down the workstring through a crossover tool and returning to the surface (LEDLOW, SAUER

& TILL 1985). A circulating pack is best for gravel placement in the screen-casing annulus but has poor performance in perforation tunnel plugging (cf. section 5.6.3.). It is therefore chiefly applied in open-hole completions in vertical wells (cf. section 5.10.2.) where gravel squeezing into perforations is not required and in wells where perforation prepacking (cf. section 5.6.7.) has been performed. For circulation packing, lower tell-tale screen and seal sub are eliminated, and an upper tell-tale screen is placed between shear safety sub and blank pipe (cf. section 6.3.5.). Circulation passes through the primary screen until it is covered with gravel and is then diverted through the upper tell-tale screen, with gravel falling by gravity to fill the blank liner annulus until the upper tell-tale screen is covered and an abrupt pressure increase occurs.

Circulation gravel packing is usually carried out with low-viscosity fluids with low gravel concentrations. Reverse-circulation packing enables to place some gravel through the perforations by shutting off the returns at the surface and pumping continuously until a pressure increase indicates that gravel has covered the tell-tale screen. If no positive pressure is exerted, only gravel packing inside the wellbore is likely to be achieved. Two or three joints of unslotted tubing are generally run above the screen to provide space for gravel to fill the annulus above the screen which normally prevents fluid and gravel movement up the annulus and provides a reserve pack if the gravel settles (cf. sections 5.3.2.4.3. and 5.8.4.1.).

5.3.2.3.3. Squeeze packing

Squeeze packing is suitable for vertical wells with short perforated intervals (HALLIBURTON 1988). No tell-tale screen is implemented at all, and the tool is first placed in reversing position, but instead of reversing, the fluids are circulated taking returns above the packer. Once the first fluids to be squeezed into the formation reach a point just above the packer, the tool is cycled to the squeeze position and gravel is squeezed into the formation until a predetermined pressure is attained and the tool is cycled to reverse position to remove excess gravel. Squeeze gravel packing is usually carried out with high-viscosity fluids with high gravel concentrations. A squeeze pack is placed into a well by ultimately pumping all fluids into the formation with no returns to the surface (LEDLOW, SAUER & TILL 1985). Squeeze packing is most efficient in placing gravel in the perforation tunnels and least effective in placing gravel in the screen-casing annulus. Unless carrier fluid viscosity is decreased or the pressure differential is increased, a straight squeeze pack often results in poor gravel placement.

5.3.2.3.4. Combined circulation and squeeze packing

As a consequence of their inverse effectivity, a combination of circulation and squeeze gravel pack is often the best solution for reasonable packing of both perforation tunnels and screen-casing annulus (HALLIBURTON 1988). Gravel is placed in the well by circulating slurry down until a screenout is obtained on the lower tell-tale screen, and when the slurry crosses the screen, returns are shut off and fluid is pumped into the formation. Gravel is circulated to the lower tell-tale screen until an abrupt pressure increase indicates that the lower tell-tale screen is covered thereby restricting circulation. At this point, the tool is cycled into squeeze position and pumping is continued until a predetermined pressure is reached, with pumping proceeding at the highest practical rate which assists in perforation packing.

The combination of circulation and squeeze gravel packing is particularly suitable for wells with less than 45 ft of completion interval or less than 45 degrees of hole deviation. The combined method also allows for pressure depletion by temporary circulation when gel-out or screenout during squeeze packing is creating pressure-up cycles and mimicking high pressure losses downhole which could cause premature termination of the job.

5.3.2.3.5. Four-position longstroke packing

Four-position longstroke packing is performed with adding an upper circulation position to the combined circulation and squeeze tool. The four-position instrument is operated by vertical workstring movements into squeeze (no returns), lower circulation (returns only through lower tell-tale screen), upper circulation (returns through main screen) and reverse circulation (DOWELL SCHLUMBERGER 1988 b). Slurry is first circulated down to screenout on the lower tell-tale screen, and then second the tool is lowered into squeeze position which closes the circulating part thereby preventing returns from going to the surface and instead of that injecting into the reservoir.

After screenout on the formation, the instrument is raised to the upper circulating position in the third stage and the remaining slurry is dehydrated in the screen-casing annulus, and after all the positions have been tested to verify a screenout in each position, the excess slurry left in the tubing is reversed out. The four-position longstroke crossover instrument with reversing ball in place offers the flexibility to successfully gravel pack most wells including those with longer zones and higher degrees of hole deviation.

5.3.2.3.6. Wash down and pressure packing

Wash down packing consists of depositing gravel to a predetermined height above perforations followed by running screen and liner assembly with a wash pipe and a circulating-type shoe (HALLIBURTON 1988). The screen is washed down through the gravel, and when the shoe is on bottom, gravel is allowed to fall back around screen and liner. This approach has minimum provision for compacting the gravel in the screen-casing annulus and no provision for squeezing gravel through the perforations. The well, however, may be pressure-packed prior to washing down the screen, and there may be some segregation of gravel sizes during the wash down.

Pressure gravel packing enables to fill the perforation tunnels with gravel in order to keep them unclogged by formation sand and fines, to fill void spaces behind the casing, and to restress the reservoir. Circulating pressure pack permits slurry injection down the tubing which may prevent scouring of mud, rust, pipe dope and scale from the tubing-casing annulus that could damage pack permeability. In order to place gravel through the perforations, surface returns may be stopped and slurry squeezed against the pay zone before the perforations are covered with slurry.

5.3.2.3.7. Other techniques

Bullhead or bradenhead pressure packing operates without tell-tale screen or wash pipe because no returns are taken (HALLIBURTON 1988). Pumping pressures increase as more perforations are covered with gravel, and after final sandout, the packer is released from screen and blank pipe at the crossover tool. Combination of several methods can enhance the prospect of achieving a tight uniform pack with gravel occupying each perforation tunnel and restressing all areas of the sand-producing formation. Bullhead pressure packing followed by circulation pressure packing can be beneficially combined to form an outstanding composite design. Reasons for unsatisfactory sand control performance are mainly annulus bridging, intermixing of gravel and formation sand, and mixing of borehole fluids with the gravel packing slurry (RENSVOLD & DECKER 1978).

5.3.2.4. Gravel volume and placement

The volume required for annular and perforation packing depends on placement techniques as well as on the fact whether previous sand production has occurred or not in old or new boreholes, respectively (ECONOMIDES 1986). Different gravel volumes are required in circulation and squeeze packing, and a reserve blank section above screen and perforations has to be included in order to compensate gravel settling.

5.3.2.4.1. Circulation packing

In case of simple circulation pack with no intention of squeezing gravel into the formation, two times the annular volume of screen and blank liner above the zone are usually an adequate gravel volume (ECONOMIDES 1986). In case of squeezing, gravel quantity can be very different depending on formation characteristics, because screenout or pressure-out preventing further injection is highly dependent on reservoir permeability. In sections with clean highly permeable pay sand, gravel slurry dehydration occurs early in the squeeze, thereby reducing the gravel requirement, whereas in dirty low-permeability reservoir intervals, fracturing may take place as a consequence of retarded leakoff and pressure buildup (cf. sections 5.6.4. and 5.6.6.3.), and a screenout may have to be forced by reducing injection rates to a bare minimum. Long zones take less gravel per ft than short intervals, because there is more area available for slurry dehydration. Long sections with highly-permeable portions at the top of the interval may screenout early due to dehydration into that zone, thereby causing bridging which prevents injection into the lower part.

5.3.2.4.2. Squeeze packing

Squeeze placements in general require more or less one time the annular volume of screen and blank liner above the zone plus 0.5 - 1.5 ft3 of perforations (ECONOMIDES 1986). The differences between circulation and squeeze packing in terms of necessary amount of gravel of abt. one time of annular volume of screen and blank liner can be mainly explained with more efficient compaction of the gravel by squeezing under higher pressures where predominantly the perforations are properly filled and then the annulus is also plugged, whereas circulating always is problematical concerning perforation filling and packing density is lower in the annulus than achieved by squeezing.

5.3.2.4.3. Reserve blank section above screen and perforations

Apart from packing the whole length of perforated interval and slotted liner or wire-wrapped screen, a reserve annular section of gravel above the zone has to be placed against a blank liner (ECONOMIDES 1986). The reserve allows excess gravel to compensate for settling after screenout occurs (cf. also section 5.8.4.1.), and it also prevents gravel fluidization during hydrocarbon production. In injection wells, the reserve is available to replace gravel which is pumped away while injecting. When gravel is placed in near vertical wells by circulating, gravel is carried in a low-viscosity fluid and reserve gravel is stored in the blank pipe annulus below the upper tell-tale screen and above the main screen. In vertical wells, circulation provides complete filling of the blank pipe annulus.

After the primary screen is covered, circulation continues through the upper tell-tale screen and gravel settles out until it is covered as well, with reserve gravel being simply a function of blank liner length and its annular volume. High-viscosity fluids have restricted flow through the upper tell-tale screen and bridging occurs before gravel placement is complete. Using high-viscosity fluids, reserve gravel can only be derived by settling of gravel out of the slurry column, with the higher the gravel saturation, the more gravel being available to form the reserve volume. The length of blank liner required is affected not only by gravel concentration, but also by screen length. In open-hole completions, the open-hole section may be considerably larger as a consequence of underreaming than the cased hole diameter where the reserve is located. Due to difficulties of placing reserve gravel below the upper tell-tale screen in highly-deviated wells, continuous screen should extend above the perforations to allow insertion of reserve gravel.

5.3.3. Gravel settling diminution in the carrier fluid

The success of a gravel-pack completion depends on placement of gravel in the perforation tunnel area and the screen-casing annulus (HUDSON & MARTIN 1988). As wellbore deviation increases, gravitational forces and carrier fluid limitations may result in gravel settling or duning (cf. section 5.8.3.) and incomplete packing of much of the perforation tunnel area as well as premature screenout of the liner-casing annulus. Factors influencing gravel placement effectivity include fluid viscosity, fluid density, pumping rate, gravel size, gravel concentration, gravel density, wellbore deviation and downhole assembly. The discussion as follows focusses on pressure drop minimization, decreasing particulate density/transport fluid density ratio, high-density fluids vs. low-density particles, and application spectrum of low-density particles.

5.3.3.1. Pressure drop minimization

In order to maximize productivity, the pressure drop across the gravel pack filter must be minimized which requires choice of the largest gravel particle that prevents movement of the formation material (HUDSON & MARTIN 1988; cf. section 5.2.3.3.). Poor gravel placement with voids in perforation tunnels and screen-casing annulus decreases well exploitability. Low productivity as a consequence of insufficient gravel insertion is a common problem especially associated with highly-deviated wells, because settling of gravel in the slurry (cf. section 5.8.3.1.) as it flows down the annulus causes a gravel concentration gradient in the fluid from the high to the low side of the flow area. When the carrier fluid leaking off through the perforations has only low gravel saturation or even does not contain gravel as a consequence of settling, it is difficult to completely pack the perforations, if possible at all. Therefore settling of gravel in the carrier fluid must be reduced as much as possible in order to improve gravel pack efficiency (cf. also section 5.8.3.4.).

The pressure drop across a gravel pack is rather insensitive to the annular clearance between casing and screen (YILDIZ & LANGLINAIS 1988) and is also independent of the perforation phasing angle, indicating that flow from nearby perforations does not interfere. Shot density and perforation diameter have the greatest influence on pressure drop, while media anisotropy has only subordinate contribution to total pressure drop.

5.3.3.2. Decreasing particle density/carrier fluid density ratio

Gravel placement is generally improved by decreasing the ratio of particulate density to carrier fluid density (HUDSON & MARTIN 1988). Low particulate density/carrier fluid density ratios are obtained using either gravel of various type in high-density brines or low-density gravel-pack materials in commonly used carrier fluids. Regardless of carrier fluid density, the settling rate of the gravel grains decreases as the particle density/carrier fluid density ratio decreases. Addition of a viscosifier also reduces the settling rates of gravel grains at each particle density/carrier fluid density ratio, and therefore particle transport improvement can be achieved with either lower particle density/carrier fluid density ratios or higher fluid viscosities. Particle settling is further influenced by gravel concentration in the carrier fluid, with settling rate decreasing with increasing particle saturation.

Gravel settling is thus reduced with use of high-viscosity carrier fluids and systems with low particle density/carrier fluid density ratios. High-viscosity carrier fluids, however, do not provide the suspension properties required in surface mixing, pumping equipment or under downhole conditions. Polymer concentrations required to yield high viscosity lead to impaired fluid leakoff and false screenouts as well as reduction in formation permeability. The low particle density/carrier fluid density system utilizes buoyancy forces to counter the gravitational forces on the gravel particulate rather than high viscosity to reduce particle settling (HUDSON & MARTIN 1988). Because the polymer saturation can be reduced, it does not have the disadvantages of the high polymer concentration/high viscosity system. Another alternative are surfactant-bearing water-based high-viscosity fluid systems which are characterized by high fluid leakoff rates and high gravel suspension and transport capacity (DOWELL SCHLUMBERGER 1988 a, NEHMER 1988; cf. section 5.3.1.1.2.).

5.3.3.3. High-density fluids vs. low-density particles

Low particle density/carrier fluid density ratios can be obtained using either standard gravel material in high-density brine fluids or low-density particles in commonly applied carrier fluids. The use of high-density fluids introduces additional drawbacks, because the overbalance created by these fluids can lead to severe fluid loss and well control problems. A gravel pack system design utilizing conventional fluids carrying low-density particulate material therefore improves particle transport without the problems associated with high polymer loading or high-density fluids, whereas with conventional gravel, the required fluid density results in high slurry density and high hydrostatic pressures.

Concerning highly-deviated wellbores, HUDSON & MARTIN (1988) therefore recommend the use of light particles of preferably 1.5 - 1.8 g/cm3 specific density as a replacement for sized gravel to yield the required particle density/carrier fluid density ratio when combined with conventional carrier fluids. Experimental work has shown that under these conditions, complete packing of the perforation tunnels also at the upper side of the highly-deviated boreholes can be achieved (cf. also section 5.8.3.4.). The low specific gravity of the ultralight gravel packing material is achieved by coating an inner core with plastic, with this composition having very low acid solubility and providing good permeability at 4,000 psi closure stress.

5.3.3.4. Application spectrum of low-density particles

The high cost of the ultralight gravel packing material (HUDSON & MARTIN 1988) restricts its application to horizontal and highly-inclined boreholes where the achieved maintenance of annular flow by low friction pressure drop and improved packing density pays off, whereas the material is too expensive and does not have sufficient closure stress resistivity for utilization in vertical-hole gravel packing and hydraulic fracturing where (in contrast to only little amounts required for gravel packing) larger quantities of material are needed. In addition, proper deposition of the ultralight material in satisfactory concentrations and packing density can only be achieved by very high fluid-loss rates which are desirable in gravel packing, but unacceptable in hydraulic fracturing where excessive fluid leakoff automatically leads to premature screenout failure of the treatment due to proppant bridging and slurry sticking as a consequence of fluid dehydration (cf. sections 4.2.2.2.3. and 6.2.4.2.1.). Increasing slurry density increases friction pressure which leads to bridging at the end of the screen. Buoyant low-density gravel pack material, however, does not allow gravity settling of the solid particles, with therefore packing only being able to be achieved through fluid leakoff or staged treatments with section-wise placement.

5.3.4. Gel surface filtration from polymer solutions

Gravel packing with viscous aqueous polymer solutions allows rapid placement with suspensions of high solids concentrations (TORREST & MARCANO 1986). High-viscosity fluids generally facilitate infilling of perforations and placement of gravel deposits beyond the perforations while minimizing mixing with formation sand. Viscous aqueous polymer solutions, however, create a relatively impermeable surface layer while flowing through porous media such as gravel pack and formation sand, giving rise to anomalous pressure drops and inlet build-up which is not removable with subsequent water flushing. Surface layer generation is mainly due to microgel plugging of large pore diameters at low current rates, with abt. 4 % of the total polymer volume contributing to the build-up of the gel layer. The pressure drop for water flow through this layer is linear in current rate at least during the initial stages, and the gel layers once formed can undergo changes in characteristics with water flow at sufficient rates. Screenout mimicking by surface layer-induced pressure rise and other aspects are briefly discussed as follows.

5.3.4.1. Screenout mimicking by face layer-induced pressure rise

The significance of surface impairment and retention is that unexpected but substantial filtration at the formation face with viscous polymer solutions during gravel packing is manifested largely as high pressures which is clouding the interpretation of placement behaviour and affected gravel pack quality (TORREST & MARCANO 1986). As pressure increases often associated with premature screenout during packing could thus also be caused by the rapid build-up of a gel surface layer at the formation face, mistaking such false signals as indicator for screenout termination and based on this decision being made to stop pumping leaves the flow lines full of gravel and therefore the pack is building in a way very different from that originally intended.

Rapid microgel layer build-up with placement of very small solution volumes could therefore act as a profile control during gravel packing with concentrated polymer solutions in the wellbore flowing through the perforations and into the formation. If the rising injection pressures are erroneously interpreted as indicating screenout termination long before the gravel pack has actually formed, stopping of the job based on rapid pressure build-up during gravel placement leads to incomplete gravel pack generation and reduced gravel pack quality as a consequence of failure to account for the effect of surface gel filtration. It is therefore essential to be able to recognize such screenout mimicking caused by surface gel layer deposition and to continue pumping of gravel-saturated slurry until the real end of the job.

5.3.4.2. Other aspects

On the other hand, maintenance of high pumping pressures could give rise to fracturing of the reservoir and strongly uniform gravel placement predominantly within the pressure-parted crack (cf. sections 5.3.5. and 5.6.4.2.). Although viscosity breakers can in principle be chosen to allow pumping away of material to the formation, the fine tuning needed to retain high viscosities for concentrated suspensions and low viscosities into the reservoir is not always possible. In view of the apparent successes of gravel packing with viscous aqueous polymer solutions, the kinds of behaviour caused by gel surface filtration and layer generation suggest that other factors including more careful completions and choices for application as well as general improvement in fluid preparation and cleanliness may be important contributions.

In contrast to generation of the surface gel layer, adsorption from polymer molecules in solution or molecular entanglement in concentrated solutions are not significant factors in gravel packing applications in contrast to their partially considerable importance in hydraulic proppant fracturing (cf. section 4.3.4.). Discussion of aspects of polymer adsorption are also carried out by CHAUVETEAU (1981), GRAMAIN & MYARD (1981) and COHEN & METZNER (1982). CHAUVETEAU (1981) comments on hydrodynamic retention, and BURCIK (1969), KOHLER & CHAUVETEAU (1981) and CHAUVETEAU & KOHLER (1984) illustrate microgel plugging.

5.3.5. Combination of gravel packing and proppant fracturing

Viscous gravel pack carrier fluids have an increased drag effect on both gravel and formation sand, allowing better restressing of the reservoir, origin of a more tightly packed gravel mantle, and to help pushing formation sand out of the perforation tunnels. Knowledge of the quality of the formation hydrocarbon-bearing sand enables prediction of the pay zone acceptability for gravel (GURLEY, COPELAND & HENDRICK 1977). Dirty reservoirs containing larger amounts of silt and clay almost always take up more gravel before screenout than previously expected and planned, as the permeabilities of these low-maturity sands are poor and the fluid carrying the gravel will not leak off at a high rate, with the formation thus becoming fractured and the gravel being transported away from the wellbore into the pressure-parted crack in order to prop it in an unwanted hydraulic fracturing operation unless an artificial screenout is triggered by decreasing pumping rates. Inversely, clean and highly-permeable reservoirs are characterized by rapid fluid leakoff and quick screenout after injection of only small amounts of gravel, and consequently a low probability of triggering the opening of a hydraulic fracture during the gravel pack operation.

Pressure parting with opening of a hydraulically-induced fracture can also occur in various situations of gravel prepacking of perforations independent from whether the formation has produced sand prior to gravel prepacking or not, and whether perforation washing or surging is carried out, with the only controlling parameter being operating pressure (PENBERTHY 1985; cf. sections 5.6.4.2. and 5.6.6.). Advantages of combined gravel packing and fracturing as well as formation strength support by hydraulic fracturing alone are briefly discussed as follows.

5.3.5.1. Advantages of combined gravel packing and fracturing

In some cases, it may be desirable from reservoir engineering point of view to carry out a combined hydraulic fracturing and gravel packing treatment in one trip by provoking excess of the fracture gradient, opening of the formation, plugging of the crack by proppants or gravel, and then triggering a tip screenout by lowering pumping rates in order to stop further propagation of the fracture, prop it completely from the wing tips up to the wellbore, fill the perforation tunnels with gravel in cased holes, and subsequently line the borehole wall along the reservoir in open holes or the perforated casing face with the gravel pack mantle. The discussion as follows focusses on adequate stabilization of loose sands and chalks, proper infilling of perforation tunnels and cement sheath defects, and operational sequence of hydraulic proppant fracturing followed by gravel packing.

5.3.5.1.1. Adequate stabilization of loose sands and chalks

Combined small-scale hydraulic fracturing and gravel packing treatments of controlled and not random or accidental technical performance are very important in various low-permeability unstable formations (such as shell debris, chalk and silty sands) and will become increasingly important in the near future with progressive approach of more and more marginal reservoirs (cf. also section 4.5.4.3.). Stabilization of loose sands and chalks can even be enhanced by tailing-in of resin-coated gravel in the terminal stage of the hydraulic fracturing treatment (cf. section 4.12.3.3.2.) and also performing the gravel pack with resin-coated grains (cf. sections 4.5.4.3. and 5.8.2.).

Hydraulic fracturing of the pay zone to be gravel packed has repeatedly occurred during sand control treatments, as indicated by the consumption of unusually high volumes of gravel in some wells and also by considerably increased post-operational production rates (SPARLIN & COPELAND 1972). The experience has shown that if fracturing has taken place during gravel pack jobs, it has apparently not reduced the effectivity of the sand control measure, thus contradicting the earlier fear that in case of fracturing, the gravel would all go into one place by plugging the crack and would not allow a proper solution of the sand problem. Aspects of combined hydraulic fracturing and gravel packing are also discussed by POOLLEN & MALONE (1959) and SOLUM (1986).

5.3.5.1.2. Proper infilling of perforation tunnels and cement sheath defects

LEDLOW, SAUER & TILL (1985) present a positive assessment of hydraulic fracturing during gravel packing operations. In partially depleted reservoirs, often the hydrostatic pressure generated by a column of gravel pack slurry is already sufficient to exceed formation fracture pressure (CIRIGLIANO & LEIBACH 1967, PENBERTHY & COPE 1979, TRAHAN & SPIES 1981). Achievement of pay zone fracturing, infilling of the crack with gravel and subsequently packing perforations and annulus with gravel is in many cases the only possibility to obtain a good gravel pack which properly infills all the perforations and voids and sets a decent inside mantle. The benefit of hydraulic fracturing and - following formation breakdown - complete filling of the wellbore with gravel is that if pumping is stopped too early, the fracture closes and the borehole is not full of sand because equilibrium bed height has not sufficiently risen, the incompletely filled wellbore exhibits little or no stimulation and is susceptible to sand production.

Even fracturing into an underlying water-bearing zone is more positive than negative, because the cement sheath in many wells is defective and of insufficient quality, and if a pressure differential breaks down the cement and creates a channel, the only possibility to plug all the voids in both formation and cement mantle is to infill them with gravel until terminal screenout indicates that maximum packing has been achieved. Exceeding fracture pressure does not prevent gravel screenout, but in fact leads to higher screenout pressures which together with multiple pressure cycles dehydrate gravel pack slurries better.

5.3.5.1.3. Operational sequence of hydraulic proppant fracturing followed by gravel packing

Concerning operational sequence, either the hydraulic fracturing treatment or the gravel packing job can be carried out first. Starting with hydraulic fracturing in new wells has the advantage that a continuous gravel mantle can be installed later which then is no longer intersected by perforation shooting and fracture extension. In old wells which have already been equipped with a gravel pack, hydraulic fracturing stimulation in a more advanced stage of depletion requires crossing of the gravel pack by both perforations and propagating hydraulic fracture.

As new perforation holes have to be shot for allowing the fracture to initiate, optimum completion includes performance of renewed gravel packing for the purpose of plugging of the perforation holes which were newly shot for preparation of the fracture job (aspects of reperforation and re-gravel packing are mentioned by BONOMO & YOUNG 1983). Therefore the ideal sequence is starting with hydraulic proppant fracturing and, upon termination of infilling of the crack with a tight proppant package, continuing with plugging of perforation tunnels and setting of the inside gravel pack mantle. Field examples of hydraulic fracturing in older wells equipped with an earlier gravel pack are reported by DARR, BROWN & MURPHY (1985) and DARR & CARLTON (1988). Aspects of combinations of hydraulic fracturing and gravel packing are also outlined by SOLUM (1986).

5.3.5.2. Formation strength support by proppant fracturing alone

Creation of a conductive path some distance into the reservoir formation by means of a propped hydraulic fracture is also without combination with gravel packing a suitable means of supporting formation strength maintenance or even amelioration of reservoir stability as a consequence of reduction of drag or frictional forces on the pay zone by increasing flow area (ALLEN & ROBERTS 1982). Small-scale hydraulic fracturing together with improving perforation geometry and distribution can already effectively reduce an otherwise serious sand control problem and thus in some cases help to avoid the necessity of gravel packing, thereby leaving a full open wellbore and not giving resistance to fluid flow through perforation tunnel plugs and borehole-seaming gravel mantle. Such fracture packing can be effective for stacked thin reservoir intervals, inclined holes and unstable wells that would be difficult to gravel pack (WEISSENBURGER, MORITA, MARTIN & WHITFILL 1987). Fracture packing, however, requires solid formation adjacent to the poorly-consolidated zones of interest, and thin soft layers are better suited because of the relatively high flow restriction in the fracture.

5.4. Reservoir stability and fluid dynamics

Reservoir stability and fluid dynamics are mutually depending factors. The following overview demonstrates how changing fluid dynamics and composition can affect reservoir consistency and how decreasing formation pressure and shifting hydrocarbon/water ratios can create sand control problems in previously stable pay zones. Comments are also given on rock mechanics, perforation tunnel morphology, and water cut in the hydrocarbons. Some differences between hydraulic fracturing and gravel packing are also outlined.

5.4.1. Formation pressure and fluid composition

In terms of execution of gravel pack treatments during course of the production history of predominantly oil and subordinately also gas wells, three cases have to be distinguished. The suite of possibilities comprises primarily not enough cohesive pay zones as well as secondarily unstable hydrocarbon-bearing sections due to reservoir pressure decline, and water cut increase and corresponding fluid viscosity decrease. The spectrum of cases is summarized in a discussion of insufficient consolidation and declining reservoir pressure, and changing fluid composition and viscosity as follows. Some comments are also offered on achievement of the critical stage of formation collapse.

5.4.1.1. Insufficient consolidation and declining pay pressure

The first group of reservoirs comprises completely unconsolidated sands and chalks which require stabilization by gravel packing from the very beginning of production. The main pay formations in this group are silty and argillaceous, mainly medium- to fine-grained sands as well as soft flowing chalks and diatomaceous earths which according to their fine grain size, insufficient binding nature and lacking cementation have not enough internal cohesion to be satisfactorily stable without artificial improvement. Some loose schill carbonates also belong to this division of reservoir types.

The second assemblage incorporates weakly-cemented sands of various grain size including the whole spectrum from almost silt to nearly gravel where the problems of sand influx into the borehole start with progressive depletion due to decreasing reservoir pressure that is so far maintaining a more or less self-supporting framework of grains, with the formation failing once a critical boundary pressure is reached and underpassed. Early pressure maintenance methods such as water or gas injection which are very common in oil-field exploitation can help to prevent the sand from becoming soft and flowing, or can at least decelerate the cohesivity transformation, but the changing phase composition and viscosity of the reservoir fluids also has its impact. VAZIRI (1986) recognizes changes of permeability and rock properties around the wellbore to be the main reasons triggering sand production.

5.4.1.2. Changing fluid composition and viscosity

The third association includes slightly-lithified sands of different grain size comprising a broad range throughout the granulometrical interval of sand where gravel packing becomes necessary due to reservoir breakdown and disintegration with increasing water cut in the oil during advanced production with a rising oil-water-contact accompanying pressure depletion, as a consequence of excessive water injection due to pressure maintenance operations, or as a result of water coning and breakthrough due to unfavourable differences in relative permeability and preferred water conductivity along some high-permeability streaks or through the joint system.

The changing fluid viscosity leads to different cohesion of the grain fabric, generally induces less stability than before and thus ultimately causes the sand to flow. The main members of the second and third groups of sand-problem reservoirs are weakly-cemented friable sands which might be originally still sufficiently stable (although sand production may occur from the very beginning of being put on stream), but with falling reservoir pressure and/or changing fluid viscosity, internal support is taken away and the grain fabric starts to collapse and to release more and more loose sand grains which move towards the least restricted spot that is represented by the borehole.

5.4.1.3. Achievement of the critical stage of formation collapse

In all the three different cases as outlined above, solid control problems commonly start with more or less unprevented migration of sand, mud and chalk that is not seriously disturbing hydrocarbon production, but later the stage is reached when plugging of reservoir pores and liner slots as well as infilling of the tubing and excessive production of solids stops oil or gas flow and ruins the equipment. In some cases, sand production problems are originally absent and are only secondarily created by insufficient or non-adequate completion practices (ALLEN & ROBERTS 1982).

In some very unstable oil sands, the critical offtake rate at which sand starts to flow is less than 1 m3 of oil per day (MURER 1981). Distinction has to be made between more or less constant sand concentration in the produced oil as a consequence of slow and gradual sand influx, and pulsatory collapse of channels and caverns as well as intermittent arrival of large amounts of loose sand in the borehole alternating with periods of apparent sand stability.

5.4.2. Fluid dynamics

Gas evolution from the flowing fluid also has significant effects on sand production (SAUCIER 1974), and the same applies to flow disturbances caused by rate changes and surging (such as taking place during artificial lift). Reduction of fluid rate leads to very large gas/liquid ratios and results in declining sand production which is not the consequence of pack fluidization, as evidenced by the occurrence of the same features in consolidated packs in full-scale model verification tests. Rate changes affect gravel pack behaviour more significantly than does the magnitude of the flow rate. Sand bridge construction and breakdown as well as influence of production rates are outlined as follows.

5.4.2.1. Sand bridge construction and breakdown

For a given set of reservoir flow conditions, primary and secondary sand bridges form which are stable for the existing geometries and hydrodynamic forces. If fluid forces are altered, instabilities occur, bridges break down, and more sand is produced until new bridges originate under new conditions of stability (cf. section 5.9.5.). High flow rates at the beginning of hydrocarbon exploitation cause higher initial quantities of sand production and flow impairment, thus necessitating to start exploitation of oil reservoirs with sand control problems with low to moderate offtake rates. Hydrodynamic factors affecting gravel pack design are also discussed by SAGE & LACEY (1942).

Concerning special treatments, steam-soak or huff and puff wells often suffer from sand influx, because the cyclical repetition of steam injection and hydrocarbon, water and steam-condensate production results in continuously changing swelling and shrinking of the formation due to alternating heating and cooling, which additionally to the already originally friable nature further loosens the grain fabric (TOMA, LIVESEY & HEIDRICK 1986). In order to maximize relative permeability to oil, the gravel has to be water-wet before it is added to the placement fluid particularly in case of oil-based carrier media (ALLEN & ROBERTS 1982).

5.4.2.2. Influence of production rates

Similarly as applying for hydraulic fracturing (cf. section 4.11.), damage of the gravel pack after its installation has to be avoided by carefully putting the well on stream. The formation pores around the wellbore are loaded with fines carried in by fluids filtering into the reservoir or by inherent fines made movable by fluid filtrate effects. This assembly of fines should be started to be transported back towards the wellbore as soon as possible after completion of the gravel packing job, but at a low rate in order to minimize plugging due to high velocity. Production rates should be gradually increased in order to promote bridging of formation sand on the gravel. Migration of fines into the gravel pack can be minimized by the use of clay-stabilizing organic polymers in the placement fluid which are strongly absorbed by clays, feldspars and other negatively charged silicates, thereby extending gravel pack life.

Water-wet formations produce hydrocarbons without influx of sand if there is no turbulent flow in the reservoir and if water saturation is less than the critical concentration (COLLE 1988). Surface tension of the water holds sand grains, shale particles and other fines in place during non-turbulent flow, whereas turbulent flow overcomes surface tension and causes the water to move with corresponding transport of fines and sand.

Sand influx may not be a serious problem during primary gas production, but may only show up after conversion of wells and reservoir for secondary gas storage (ANAND & JONES 1977), especially if planned gas offtake rates from the underground store are considerably greater than primary gas withdrawal rates.

5.4.3. Rock mechanics

In terms of rock mechanical approach of reservoir and particularly perforation tunnel stability vs. collapse, two types of failure have to be distinguished (MORITA, WHITFILL, FEDDE & LØVIK 1987). Shear failure occurs if well pressure is too low, whereas tensile failure takes place if the normalized pressure gradient is too high. Both shear and tensile failure envelopes are affected by cyclic loading due to repeated shut-in and restarting flow of the well. Sand production is controlled by two independent factors being well pressure or total drawdown, and local pressure gradient around the perforation cavity or completion pressure loss. The outline as follows focusses on shear and tensile failure, in-situ stress state and reservoir depletion, capillary bonding, sand bridging, perforation tunnel stability, and pressure drop and flow resistance in perforation tunnels.

5.4.3.1. Shear and tensile failure

Shear failure suffers from the weakening effect due to plastic fatigue, whereas the shift of the tensile failure envelope depends on the degree of shearing. Once plasticized, the perforation cavity surface cannot recover the deformation, and the stress is released at the cavity surface which is susceptible to tensile failure, as there is not enough stress to bind the sand grains. Therefore the degree of shearing and thus the pressure drawdown should be limited during the production cycle preceding well shut-in. In addition, shear failure usually occurs more abruptly than tensile failure. The unloading effect of the plastic zone around the perforation cavities is serious when production restarts after termination of the shut-in period of the well, and tensile failure occurs within a relatively small flow rate if the drawdown during the previous offtake cycle is too large.

5.4.3.2. In-situ stress state and reservoir depletion

The sand problems are sensitive to the in-situ stress state, with in case of high in-situ stress shear failure tending to occur, whereas the possibility of tensile failure is diminished. Reservoir pressure depletion increases the effective in-situ stress, and therefore a tensile-dominant sand problem may be reduced, but a shear-dominant one becomes serious in weak formations with declining reservoir pressure. Since shear failure is usually of catastrophical nature, reservoir pressure should be maintained or drawdown should be limited in case of perforating and producing a weak pay zone.

Perforation tunnel stability is also sensitive to deformation and strength characteristics of the rock. Two types of failure envelopes usually apply comprising octahedral shear stress vs. effective hydrostatic stress for brittle rocks and plastic strain vs. effective hydrostatic stress for malleable rocks. The shear stability envelope varies with critical plastic strain, although the tensile failure envelope remains unchanged. In highly deformable reservoirs, perforation tunnels are distorted due to fluid forces and thus the cavities are destabilized. Tensile strength improves cavity stability when the drawdown pressure is small; however, a rock normally looses its tensile strength once it is sheared. If the cementation material is not brittle, a certain degree of tensile strength still remains between the sand grains.

5.4.3.3. Capillary bonding

When the water saturation in the reservoir sand around the perforation cavities approaches the irreducible connate concentration during efflux of hydrocarbons into the wellbore, the capillary pressure rapidly increases (MORITA, WHITFILL, FEDDE & LØVIK 1987). Capillary bonding is not weakened during the shearing process and is therefore not affected by cyclical flow. Thus even if a previous drawdown is close to the critical shear failure drawdown, no sand problems occur with a significantly high flow rate after shut-in of the well, because the flow rate to break capillary bonding is very high since the plastic zone is very thin with respect to cavity size.

5.4.3.4. Sand bridging

Sand bridging is also important if grain size is large compared with the perforation tunnel radius and has a significant effect if capillary bonding also exists (MORITA, WHITFILL, FEDDE & LØVIK 1987). If stress around a cavity is released, microcracks appear around it. Unless capillary bonding exists, displacement at the cavity surface is not uniform and crack growth occurs at one side when stress is released. If displacement exceeds a critical fraction of grain size, the sand grain is dislodged which is followed by sand flow until another stable sand arch is formed (cf. section 5.9.5.). Capillary bonding tends to give more uniform deformation and to prevent dislodging of reservoir sand particles. Cohesive strength of the sand grain package increases with wetting phase saturation to a maximum at abt. 80 % concentration, but beyond this point, any increase in wetting phase saturation causes a rapid decrease in cohesive strength (DURRETT, GUBIN, MURRAY & TIGHE 1977).

5.4.3.5. Perforation tunnel stability

Perforation tunnels in friable sandstones that are consolidated to a certain extent are stable when the formation is initially perforated (YASSIN & PEDEN 1988). Failure of the tunnels can occur in an advanced stage if the grain load is increased enough to break the cementation bonds, which is associated with forces generated by fluid depletion through the permeable rock and is amplified with restriction to flow caused by formation damage due to drilling and perforiatng operations. If the plastic zones expand to reach the midpoint distance between adjacent perforation tunnels, instability or failure occurs by large deformation or strain resulting in sand production. Thus the relationship of the plastic zone to a specified distance from the centre of the perforation tunnel represents the stability ratio of the grain fabric. Other aspects of perforation tunnel stability are discussed by VRIEZEN, SPIJKER & VLIS (1975); ANTHEUNIS, VRIEZEN, SCHIPPER & VLIS (1976); NORDGREN (1977) and CHENEVERT & THOMPSON (1985).

5.4.3.6. Pressure drop and flow resistance in perforation tunnel

Gravel-packed perforation tunnels usually offer little flow resistance when perforating density is adequate (McLEOD 1984). When gravel is not packed tightly enough during placement, gravel shifting in perforation tunnels occurs (TORREST 1982, STEIN 1983). Damage to formation sand before gravel installation causes premature pressure outs resulting from viscous fluids entering damaged or reduced permeability near the perforations (McLEOD 1984). As a consequence of high pressures, pumping may be halted before the gravel has concentrated adequately in the perforation holes and if thus the tunnels are only partially filled with gravel, reservoir sand enters the holes upon well production, thereby bridging on the gravel inside the tunnel and packing the partially void hole with formation sand which is much lower in permeability than gravel. If reservoir sand fills the perforation tunnels, pressure drop through the completion increases and flow rate declines.

If gravel is placed correctly in the perforation tunnel and little or no gravel is inserted outside the hole, pay zone sand abuts the tunnel entrance at the cement-reservoir and gravel-formation interfaces. If this prospective interval sand is clean and permeable, the pressure drop caused by spherical flow through the pay sand to the tunnel will be small, but if any damage exists from polymer residue, pipe dope or formation fines, the pressure drop can be substantial and flow greatly reduced. Successful remedy by acidizing depends on damage severity and solvent choice. If damage is moderate (less than 90 % permeability loss), acid can usually dissolve the damage, whereas in case of severe damage (more than 99 % loss in permeability), acid may not enter the perforations fast enough to dissolve the damage.

5.4.4. Perforation tunnel morphology

Perforation tunnels are usually long and of cylindrical nature immediately after creation (MORITA, WHITFILL, FEDDE & LØVIK 1987). Spherical cavities are generally more stable than cylindrical tunnels for shear failure, whereas almost no differences exist for extension failure. The normalized pressure gradient has more effect upon the failure envelope shape of a cylindrical hole than that of a cylindrical cavity.

Therefore a high sand-free hydrocarbon flow rate can be achieved by enlarging long cylindrical perforation tunnels into spherical cavities. Although a long perforation hole is less stable than a short one, it has less completion pressure drop, and thus the use of excessively short or long perforations should be avoided in weak formations. As a consequence of cavity enlargement during production, the completion pressure drop is relatively small in a weak reservoir. Nevertheless the use of excessively long perforation tunnels for reduction of completion pressure loss should not be overemphasized, even though long perforation holes are preferred for strong formations.

Three types of perforation tunnel instability have to be distinguished. Shear failure occurs in weak formations if effective in-situ stress is high or well pressure is low. Tensile failure during loading takes place for an abnormally high pressure gradient around a cavity due to highly damaged permeability or a sudden increase in flow rate such as following well shut-in. Tensile failure for an unloaded zone occurs with a relatively low flow rate if a previous drawdown before a well shut-in exceeds the critical drawdown.

5.4.5. Water cut in the hydrocarbons

Sand problems by shear or tensile failure of the grain fabric are enhanced by appearing and rising water cut in the hydrocarbons, because capillary pressure holding sand grains together is lost when water saturation increases reasonably above the irreducible connate concentration, and flow friction significantly increases due to lowering of relative hydrocarbon permeability by rising water saturation and lowered reservoir pressure when water cut appears in the oil and/or gas (MORITA, WHITFILL, FEDDE & LØVIK 1987; GHALAMBOR, KOLIBA, HAYATDAVOUDI & ALCOCER 1988).

In addition, in order to maintain an economically acceptable hydrocarbon flow rate, the well is inclined to be produced with a higher total flow rate which causes low pressure in the well and high pressure gradient at the cavity surface. Other points triggering increasing sand problems with appearing and rising water cut are increasing plugging rate when water spills fine particles away which previously adhered to sand grains, enlargement of cavity shape resulting in cavity interaction, and cement dissolution between sand grains by the flushing water. As shear failure and cyclical tensile failure sand problems are serious for weak formations, low well pressure should be avoided by using secondary maintenance, and total flow rate of the well should be decreased after appearance of water cut in the hydrocarbons.

5.4.6. Difference between proppant fracturing and gravel packing

In contrast to hydraulic proppant fracturing which concerning reservoir rock can be successfully carried out in all stages of production history, but is often performed in early well evolution for economical reasons based on absolute hydrocarbon reserves and relative oil or gas saturation with respect to that of water in the gas zone, sand control operations have in most cases to be done before production starts or shortly after beginning of exploitation of the hydrocarbon-bearing interval for technical reasons. Gravel packing is only optimally effective when installed before the reservoir rock is seriously disturbed by sand removal, with further sand flow being very difficult to control as the volume of produced sand increases. Experience indicates that initial sand control installations are far more successful than remedial treatments which may even considerably to seriously impair productivity (SUMAN, ELLIS & SNYDER 1983).

In contrast to hydraulic fracturing where fluid leakoff has to be limited to a minimum in order to prevent excessive proppant concentration in the fracture that ultimately leads to a screenout (cf. section 4.3.4.6.), fluid loss into liner or screen and/or formation during gravel packing is desired and necessary to take place in high rates in order to allow deposition of the gravel on the borehole wall where it forms a mantle coating the reservoir/wellbore interface, spearheads plugging the perforation tunnels, a belt bridging the annulus between borehole boundary face and screen or liner, or a plug infilling the whole diameter of the wellbore (depending on type of well completion and gravel pack installation; cf. section 5.10.).

5.5. Areal distribution of potential reservoirs

The most important areas of past performance and future potential of gravel packing are briefly outlined in a discussion of general aspects, centres of gravel packing, potential in Europe and impact of resin-coated sand and proppants as follows.

5.5.1. General aspects

Unstable reservoirs requiring sand control treatments are particularly widespread in younger geological formations which are still in an early stage of the diagenetical evolution and also partially have not yet achieved the stage of full mechanical compaction. Sandy sediments can, however, also preserve a soft nature throughout considerable sections of the geological history if reasonable cementation has been prevented by preferably early oil and subordinately also early gas immigration or by persistent shallow burial under low compressive stress, or if undercompaction is conserved by internal fluid overpressuring. Secondarily, instability can be generated if existing in-situ stresses are altered by drilling and/or completion operations such that the rock matrix is weakened by movement of the borehole wall. A formerly hard sandstone matrix can also become friable again in case of tectonical shearing and faulting of the sedimentary column (this particularly applies for narrower or broader perturbation or even mylonite zones along graben boundary faults or main nappe overthrusts). Tertiarily, weakening of the grain fabric by loosing support can originate during hydrocarbon exploitation, with the most important reservoir engineering factors responsible for onsetting sand production problems being declining formation pressure as a consequence of progressive depletion and changing fluid composition and thus viscosity due to increasing water cut in the oil (cf. section 5.4.1.).

5.5.2. Centres of gravel packing

Reviewing world-wide distribution of gravel packing centres, sand production problems occur most commonly in Tertiary and Cretaceous deposits such as in Gulf Coast (Texas and Louisiana) and California in USA, Gulf of Mexico, Niger Delta and Ivory Coast, Tar Sand Belt in Canada, and in Indonesia, Trinidad and Venezuela (SUMAN, ELLIS & SNYDER 1983; OYENEYIN 1988). The current highlights of gravel packing operations are Gulf of Mexico and Western Africa. Offshore gravel packing can be performed from both platforms with mounted rigs and floating vessels such as semi-submersibles and drill ships (ZALESKI & DONOVAN 1986). The main offshore gravel packing areas in the last years comprise North Sea, Gulf of Mexico, offshore California, Indonesia and Ivory Coast. Most of the reservoirs in Europe requiring gravel packing also are of Tertiary and subordinately of Cretaceous and Jurassic age. The horizons needing sand-control treatments are mainly oil-bearing formations, but gas-bearing pay zones are also affected (ANAND & JONES 1977), with the problem in gas wells often becoming even more pronounced as a consequence of high pressure and high flow velocity which are both serious factors contributing to easy sand production from insufficiently consolidated reservoirs (OYENEYIN 1988).

5.5.3. Gravel packing potential in Europe

The main target reservoir rock types for gravel packing in Europe are sandstones and chalks, with in many cases combinations of hydraulic proppant fracturing and gravel packing having to be carried out with both conventional and resin-coated grains in the chalk, whereas most of the sandstones require only standard gravel packing (cf. tabs. 5 and 11). Coal and shale gravel packing as occasionally already having been carried out at least in experimental stage in the USA has in Europe so far not been considered.

5.5.3.1. Sandstones

Gravel packing in Europe has so far been predominantly done (and will also be necessary in the future) in Tertiary oil- and partially also gas-bearing sands in Austria (fields Matzen and Hochleiten; SAMHABER 1977, MURER 1981), Czechoslovakia, Italy and Yugoslavia (offshore Adriatic Sea; cf. also BEUDELL 1985 a), and Yugoslavia and Hungary onshore. In the Norwegian Northern North Sea, some Jurassic oil-bearing sands and Cretaceous oil-bearing chalks need gravel packing (MASSIE, NYGAARD & MORITA 1987), whereas sand control treatments in the British Northern North Sea are so far only necessary in a few exceptions. While the Jurassic oil-bearing sandstones in the latter area are in most cases sufficiently stable, the Tertiary Eocene oil-bearing sandstones could require sand control once they are going to be developed (BEUDELL 1987 b).

Gravel packing (partially as an alternative or replacement of small-scale sand-oil-fracturing) is also done in some Lower Cretaceous oil-bearing sandstones in Germany FRG (fields Georgsdorf and Rühlermoor; GESEMANN & BADURA 1976, LILLIE & WIETHOFF 1985) and Netherlands (field Schoonebeek; HARMSEN 1979, CHUDY 1981), and occasionally also in Lower Cretaceous and Middle Jurassic sandstones in various parts of the Northwest German oil-field belt (for maps cf. BOIGK 1980). The shallow Lower Cretaceous oil-bearing sandstones in Germany FRG and Netherlands have been frequently the target of open-hole gravel packing by sanding-up the borehole following underreaming (cf. also sections 4.11.2. and 5.10.2.). Generally almost no sand control problems occur in the gas-bearing Rotliegend section in the British and Dutch Southern North Sea as well as in Netherlands and Germany FRG onshore (in the latter areas with local exceptions such as the Groningen field in the Netherlands; DAVIES, ZWOLLE & MEIJS 1983;

CENTRES OF PAST AND FUTURE ACTIVITY OF GRAVEL PACKING IN EUROPE	PROPPANT TYPE			HYDROCARBON TYPE			RESERVOIR TYPE				RESERVOIR FORMATION								RESERVOIR LOCATION
	Natural sand	Intermediate-strength proppants	High-strength proppants	Oil	Gas	Condensate	Sandstone	Carbonate	Chalk	Other	Carboniferous	Rotliegend	Buntsandstein	Other Triassic	Jurassic	Cretaceous	Tertiary	Other	Representative hydrocarbon fields
Albania	•	•		•	•		•										●		?
Austria	●	•		●	•		●										●		Matzen, Stockerau, Tratnach, Zistersdorf
Belgium																			
Bulgaria																			
Czechoslovakia	●	•		●	•		●										●		
Denmark	•	•		●			●		●							●	•		Dan, Gorm
Finland																			
France																			
Germany FRG	●	•	•	●			●		•						•	●	•		Emlichheim, Georgsdorf, Rühle, Adorf
Germany GDR																			
Greece																			
Hungary	●				●		●										●		Hajduszboszlo
Ireland																			
Italy	●	•		●	•		●	•									●		?
Netherlands	●	•		●			●									●			Schoonebeek
Norway	●	●		●	•		●		●						●	●	•		Ekofisk, Gullfaks, Statfjord, Valhall, Troll
Poland																			
Portugal																			
Romania	●	•		●	•		●									•	●		?
Spain																			
Sweden																			
Switzerland																			
Turkey																			
United Kingdom	●	•		●			●								●	•			?
USSR	●	•		●	•		●								•	•	●		?
Yugoslavia	●	•		●	•		●										●		?

Tab. 11

STROBL & PLACKETT 1984; and some wells in Germany FRG; JOHN 1983), and also the Buntsandstein in the mentioned area is sufficiently stable. Other aspects of sand control in oil fields in Germany FRG and Austria are discussed by SAMHABER (1977).

5.5.3.2. Chalks

Gravel packing in combination with hydraulic proppant fracturing is the only possibility to achieve economical production in some soft flowing Upper Cretaceous chalks in the Norwegian Northern North Sea (MANCILLAS, MATSON & ZIARA 1976; VLIS, DUNS & FERNANDEZ-LUQUE 1979; HARTLEY & BOSMA 1985; mainly in the Valhall field; TANSOE, KING & HOLMAN 1986, MOSCHOVIDIS 1987; ATTARD, MATHES & MOWER 1988) that are characterized by undercompaction and overpressuring (cf. section 4.5.4.3.). Minimization of solids production can be best achieved by initiating a fracture at the interface of the highly unconsolidated upper chalk layer and the more competent underlying chalk zone and communication of both horizons through the fracture (MOSCHOVIDIS 1987). In this extremely unstable high-porosity but low-permeability formation, a combination of resin-coated and conventional intermediate-strength low-density alumina silicate proppants has turned out to be the most suitable solution for fracturing stimulation and gravel packing stabilization.

Resin-coated light ceramic proppants and resin-coated sand (SINCLAIR & GRAHAM 1977, 1978; CONSTIEN & MAYER 1978, SAUNDERS & McKENZIE 1979, KANAT 1980; UNDERDOWN, DAY & SPARLIN 1980) are also considered to be the adequate material for gravel packing (partially together with hydraulic fracturing) in order to achieve economical production from unconsolidated Tertiary sand-mud-interbeddings in Austria (cf. KREUTZER 1985, 1986; BRÄUER 1987) and surrounding countries, and carbonate shell layers in Lower Cretaceous mudstones and marginal Upper Cretaceous chalks in Germany FRG and in places also in the Netherlands (probably similarly as in the Norwegian North Sea chalk partially in combination with conventional material).

5.5.4. Impact of resin-coated sand and proppants

The application of resin-coated sand and proppants (cf. section 1.2.6.) is a common procedure in gravel packing and hydraulic fracturing when artificial stabilization of the gravel mantle or proppant wedge has to be achieved by limited cementing of the grains without destroying too much permeability of the sand or proppant package. Aspects of resin bonding and aggregation, application spectrum of resin-coated material, and potential for promotion of marginal sands and chalks are illustrated as follows.

5.5.4.1. Resin bonding and aggregation

Resin-coating includes covering of the grains with thin films of organic resin that is partially polymerized (SANTROL 1975, KANAT 1980; GRAHAM, SINCLAIR & BRANDT 1982; POPE, WILES & PIERCE 1987). At room temperature, the mantled grains do not adhere to each other, but within a certain range of higher temperatures such as encountered in the reservoirs, the resin will first soften and melt to form a viscous liquid and then under proper conditions will set to generate a solid. Curing of the resin can be initiated and/or speeded up by an activator and/or accelerator if formation temperature is too low in shallow reservoirs, with distinction thus having to be made between thermal and chemical resin curing and setting (cf. sections 1.2.6.2. and 5.8.2.1.2.). If the sand or proppant particles are in contact when the resin liquifies and moves, surface tension causes the resin to migrate and to form pendular rings at contact points. Hardening of the resin then cements the grain package and provides a bonding and cushioning effect that distributes applied forces more evenly over the sand grains, thus retarding crushing of grains at point contacts under applied pressure (HICKEY, BROWN & CRITTENDEN 1981). Aspects of resin-coated sand for gravel packing are also discussed by MURPHEY, ROLL & WONG (1985).

5.5.4.2. Application spectrum of resin-coated material

Resin-coated sand is particularly suitable for outside gravel packing within the perforation tunnels which creates aggregated stable spearhead packages that do not need support by an inside gravel mantle and are thus the most economical gravel pack completion of technically advanced maturity (cf. section 5.8.2.). If for the reason of insufficient success of perforation cleaning (cf. section 5.6.4.2.) the plugging of perforation tunnels with resin-coated sand is not possible in acceptable quality, resin-coated sand is also an adequate material for inside gravel packing in the annulus between casing and screen or liner in order to effectively prevent backflow of outside pack gravel in the perforation tunnels as well as proppants in associated hydraulic fractures (cf. section 4.12.3.3.2.) into the wellbore. Resin-bonded inside gravel packs do not move, settle or mix at high flow rates or by opening and shutting of the well, whereas loose conventional gravel may be subjected to some displacement. In terms of operations and borehole conditions, preferential applications of resin-coated gravel are wells with bad casing at or near the pay zone, wells with very long productive intervals and variable non-uniform reservoir permeability, slim-hole completions and multiple zones separated by packers.

5.5.4.3. Potential for promotion of marginal sands and chalks

The artificial aggregation of the gravel pack grains by glueing together of the resin films at the grain boundaries is assessed to be one of the main tools for promotion of gravel packing in marginal formations comprising predominantly loose and soft sands and chalks in various regions of Western and Eastern Europe in the coming years. In contrast to pumping of resin-coated sand or proppants where aggregation of the resin pellicles only at grain contacts preserves the intergrain permeability very well, consolidation by secondary injection of resin (cf. also section 5.9.1.) into a conventional gravel pack or directly into the pay zone sand has in several Western European formations so far in most cases been a failure due to partial or even complete plugging of the pores in occasionally such an extent that almost no more hydrocarbon flux was possible.

Areal case studies of reservoirs requiring gravel packing for economical production and experience reports of successful operations are provided for China by CHEN, ZHOU & LIU (1986). Special aspects of offshore gravel packing are outlined for the Gulf of Mexico by STRACKE (1987). Summary reviews of sand control field experience are given by JARDINS (1959).

5.6. Particle transport through perforations

Sand-controlled wells may be cased and perforated or completed open-hole (cf. also section 5.10.). For perforated wells, the preferable sand-control methods are outside and inside gravel pack operation or chemical-injection consolidation treatment, whereas in open holes, the proper methods are restraining liners or open-hole gravel pack. Cased and perforated wells are normally less productive than open-hole completions when comparable levels of technology are applied (cf. also HARRIS 1966, McLEOD 1978, LOCKE 1980). The use of cemented and perforated casings, however, is frequently required due to multiple and/or thin productive sections in the reservoir column, the necessity to exclude interbedded water or gas horizons or undesirable mobile shale layers or streaks, and the possibility of setting a deeper completion in the well at a future date (SUMAN, ELLIS & SNYDER 1983).

In cased holes, the success of a gravel pack with respect to formation sand restraint implies that effective packing of both screen-casing annulus and perforation tunnels is achieved (PEDEN, RUSSELL & OYENEYIN 1985) which requires wide and clean perforations (ALLEN & ROBERTS 1982). Aspects of perforation packing and particle bridging, impact of fluid viscosity, achievement of perforation tunnel plugging; influence of perforation technique, perforation geometry, reservoir section and treatment variables, gravel prepacking of perforation tunnels and other points are outlined as follows.

5.6.1. Perforation packing and particle bridging

Ideal gravel packing thus includes complete filling of the perforation tunnels (in order to prevent their plugging with reservoir sand), creation of a gravel node at the distal ends of the shot holes by compression of the rubblized zone surrounding the spearhead, and generation of gravel enrichments at the proximal end of the perforation tunnels at the formation-cement interface, in order to optimally prevent reservoir sand from entering perforations and borehole (BELL 1982; cf. also section 5.8.7.). When particles are transported from casing through perforations during gravel pre-packing and hydraulic fracturing, inertial and gravity forces cause particle slurries to increase in concentration downstream of the perforations, giving rise to incomplete packing of voids outside the casing because of particle bridging in the entrance of perforation tunnels or lack of transport of the slurry through the perforations with forming a bed in the casing until an equilibrium velocity is reached. The effect of void and filled perforations on wellbore productivity is discussed by DIETRICH & BONDOR (1976).

Tight packing of perforation tunnels in order to minimize impairment by reservoir sand influx can be best achieved by squeeze packing techniques (SAUCIER 1974). Tube effects may predominate in cases where the cement sheath does not fracture and the perforation tunnels through the cement mantle do not bell out or enlarge (cf. CRAMER 1987). Aspects of inside and outside gravel packing, perforation diameter/particle diameter ratio, number and size of perforations, and tubing-conveyed underbalanced perforating are briefly summarized as follows.

5.6.1.1. Inside and outside gravel packing

Distinction can be made between outside and inside gravel packing. Outside gravel packing or pressure packing includes pumping of gravel through the perforation holes in the casing outside of the borehole into perforation tunnels crossing cement sheath and extending more or less deeply into the formation, and into cavities behind casing deriving from insufficient cementation and earlier production. Inside gravel packing includes infilling of the annulus between wire-wrapped screen or slotted liner and casing with a gravel mantle that restrains the outside pack from moving back into the borehole or stabilizes wellbore walls in open holes against collapse. Gravel grain sizes for outside and inside packing are usually identical. Density and diameter of perforations are important in all types of sand control. High perforation density leads to lower fluid velocity which helps to control screen erosion and keeps the pack gravel from shifting, and also allows to inject gravel easier in a tight close pack without exceeding fracture pressures. High perforation density together with large perforation hole diameter and deep perforation tunnel penetration also provide a transition between a suite of discrete perforation tunnels separated by barrier intervals and a continuous wedge-shaped crack formed by hydraulic fracturing (cf. section 4.5.4.3.).

5.6.1.2. Perforation diameter/particle diameter ratio

In order to prevent particle bridging within the perforation tunnels and thus blocking of the tubular pathways connecting pay zone and borehole in both hydraulic fracturing and gravel packing, the ratio perforation diameter/average gravel or proppant particle diameter should be 6/1 or larger (SAUCIER 1974). At 2/1 or 3/1 size difference, bridging occurs already at very low gravel or proppant concentrations, whereas at 6/1 size difference, even very high gravel or proppant saturations do not lead to bridging, but give rise to complete filling of the whole perforation tunnels and even origin of a gravel node at the distal end of the cement mantle. In order to prevent formation of a particle bank in the casing opposite to open perforations, high particle transport efficiency should be designed by favourably varying pumping rate and fluid viscosity (GRUESBECK & COLLINS 1978). High pressure drops across gravel-filled perforation tunnels can become significant with the current industry trend towards use of smaller gravel sizes in order to prevent invasion of formation sand into the gravel (WILLIAMS, ELLIOTT & WEAVER 1972; SAUCIER 1974).

5.6.1.3. Number and size of perforations

In terms of proppant bridging, the selection of perforations of smaller size usually excludes the use of proppants coarser than 20/40 mesh size prior to significant perforation erosion (CRAMER 1987). On the other hand, reducing number and/or size of perforations allows to maintain a given level of initial pressure drop at a lower injection rate and offers the potential to keep pace with perforation erosion by increasing injection rate from a significantly lower base rate. Comments on bridging particle size distribution are also offered by MAHAJAN & BARRON (1980). Some aspects of pressure losses in the perforation tunnels as well as perforation and screen erosion by gravel surging are outlined as follows.

5.6.1.3.1. Pressure losses in the perforation tunnels

While open-hole completions provide minimum pressure loss and from that point of view should be considered whenever possible in order to optimize hydrocarbon production, the perforation tunnels in cased holes are the primary source of pressure loss which depends on diameter and density of the perforations (ECONOMIDES 1986), while perforation penetration depth into the reservoir formation has only subordinate influence. Acceptable pressure losses for adequate hydrocarbon production can be obtained through high-density large-diameter perforations which with increasing density and diameter progressively approach conditions being realized in open holes or hydraulic fractures emanating from perforation rows in cased holes. Pressure declines in the perforation tunnels have to be related to hydrocarbon type, because pressure losses being suitable for high-viscosity oil reservoirs may be unacceptable for high-rate gas-bearing pay zones or vice versa. Pressure decrease through the gravel pack is directly proportional to the inlet area through the gravel pack, gravel permeability, and fluid viscosity for the case of laminar flow. Once production rate reaches turbulent flow conditions, pressure loss increases sharply.

Pressure loss also varies for single- vs. multiple-phase flow. In high-rate oil wells, open-hole completions may be required in order to obtain a desirable productivity index and favourable pressure loss, with small gravel with relatively low permeability being needed to control the reservoir sand. Pressure drop is also influenced by grain size and resulting permeability of the gravel infilling the perforation tunnels, with therefore gravel pack permeability damage having a severe impact on perforated completions due to causing increasing pressure loss up to excessive and untolerable values. Aspects of pressure drop across perforations are also discussed by TARIQ (1987) and PEREZ & KALKAR (1988).

5.6.1.3.2. Perforation and screen erosion by gravel surging

Gravel placement comprises perforation tunnel plugging and annular packing around screen and blank pipe section above it (ECONOMIDES 1986). Perforation infilling is the most critical part of gravel packing operations, because if the perforations are not packed by gravel, they are plugged with invading formation sand which is considerably impairing productivity. Pressure drop across the perforations may reach the point where most of the production is concentrated to a few perforations, thereby resulting in high current rapidity and surging of the gravel, and the subsequent high flow velocity of the formation particles may cause screen erosion. Therefore purging clean perforations before gravel placement is a key condition to success.

Concerning stability of perforation tunnels, infilling by gravel packing should be carried out shortly after the perforation job and proppants of suitable closure stress resistivity and adequate size have to be used in order to prevent collapse of the perforation holes especially in harder and softer reservoir types, respectively (ALLEN & ROBERTS 1982; cf. section 5.2.2.), with the perforation tunnel collapse mechanisms in moderately-deep pay zones being similar to proppant crushing and embedment in hydraulic fractures (cf. sections 4.3. and 4.12.). PEDEN & YASSIN (1986) investigate the effect of perforation conditions upon maximum sand-free production rate in friable sandstones without gravel pack completion, and SOLUM, RAMEZANI & MIMS (1988) present new technology for gravel packing through perforation tubes.

5.6.1.4. Tubing-conveyed underbalanced perforating

Perforation type and completion method are very important for minimization of pressure losses exerted by the gravel pack itself which represent some type of formation damage provoked by the sand control treatment (LEDLOW, SAUER & TILL 1985; cf. section 5.11.). Underbalanced perforating is carried out with a hydrostatic head less than the reservoir and with the surface valves open to allow controlled flow (ECONOMIDES 1986). Perforating while formation pressure is greater than the pressure inside the casing generates a beneficial backsurge that brings completion debris, drilling mud, cement products, crushed rock and other contaminants out of flow channels (COLLE 1988).

In terms of perforating technique, tubing-conveyed underbalanced perforation is one of the biggest developments supporting and improving gravel packing in the last ten years, because this method gets all the perforations open and clean and gives zero or negative skin damage. Tubing-conveyed perforating comprising placement of the guns inside the casing via attachment to the bottom of the production tubing has been invented in the 1940`s, but has only been refined and developed to actual standard from the 1970`s to the present. Overbalanced perforation by wireline requires subsequent acidizing for the achievement of zero skin damage and is also operationally more complicated. Some aspects of underbalanced vs. overbalanced perforating and advantages of tubing-conveyed operation are illustrated as follows.

5.6.1.4.1. Underbalanced vs. overbalanced perforating

Overbalanced perforation followed by washing frequently creates high positive skins, with subsequently the reservoir section often having to be acidized prior to or after gravel packing (cf. section 5.11.4.). On the other hand, overbalanced perforating followed by washing can sometimes lead to the same average percentage of skin reduction as underbalanced perforating (BRANNON, NETTERS & GRIMMER 1987).

SEANARD (1986) introduces underbalanced perforating in a closed system which allows the use of high pressure differentials to surge the perforations while at the same time limiting sand influx into the wellbore and allowing to bleed trapped pressure, to flow the well in a controlled environment and to remove damage. Underbalanced perforating can also be carried out after installation of the gravel pack for the purpose of increasing shot density and/or opening of additional production zones (cf. section 5.3.5.1.3.). Gravel packed completions require that the well is killed after perforating and the guns are pulled out of the hole. For execution in gravel packed wells, the guns are run below a retrievable packer that allows circulation after it is set, and the completion fluid in the workstring is then displaced using nitrogen to achieve the desired underbalanced condition plus safety factor.

Tubing-conveyed underbalanced perforation has turned out to provide the best results for gravel packing. When perforating with a tubing-conveyed deep penetrating casing gun, formation pressure is instantly released into the wellbore at maximum differential which pushes fluids into the tubing, carrying with it mud filtrate, cement, contamination, and perforating debris, with this cleaning action allowing the formation to produce at its natural capacity (PAI & GARBIS 1983 b). An underbalance of 500 - 750 psi for oil wells and 1,000 - 1,500 psi for gas wells has turned out to be suitable for giving rise to successful perforating with the desired result, and after perforating, the well is flowed in order to remove the perforating debris (LEDLOW, SAUER & TILL 1985; REGALBUTO & RIGGS 1985, 1986). The debris can be prevented from falling back by using an annulus pressure valve and a recloseable unloader sub. Very high pressure differentials, however, can cause excessive sand influx into the wellbore (SEANARD 1986).

5.6.1.4.2. Advantages of tubing-conveyed operation

The main benefits of underbalanced perforating are overcoming of formation damage without acidizing or fracturing in many cases, yielding higher injection and withdrawal rates, accomplishment without sanding-up of the wellbore, perforating of long intervals and cleaning without washing, and yielding of flow efficiency near 100 % especially in combination with high shot density (COLLE 1988). Tubing-conveyed operation provides deeply penetrating perforations, and the disc method even combines the advantages of perforating with a casing gun and the enhanced clean-up achieved with a tubing-conveyed gun. In the disc method, first the well is perforated with differential pressure into the formation using a deeply penetrating casing gun, and tubing is then run dry into the wellbore with a sub and shear-tested disc assembly installed in the tubing near the packer. With tubing and packer set, a surface weight is dropped to rupture the disc, thus creating an instantaneous surge into the borehole, and filtrate and debris is forced into the wellbore as formation pressure is exposed to near atmospheric conditions, with this system allowing for maximum size and penetration of shots, and once the disc is ruptured, the tubing remains open full-gauge for stimulation operations.

Aspects of tubing-conveyed underbalanced perforation are also discussed by WORLD OIL (1975 b), PERRY & SMITH (1980), BONOMO & YOUNG (1983), BELL (1984), YOUNG & ZALESKI (1985); KING, ANDERSON & BINGHAM (1985); REGALBUTO & RIGGS (1985, 1986), SEANARD (1986), HALLECK & DEO (1987); WEIRICH, ZALESKI & MULCAHY (1987), BOWLER (1988) and ZALESKI & WEIRICH (1988). The advantages of underbalanced perforating for gravel pack installation are also emphasized by McLEOD & CRAWFORD (1982), BONOMO & YOUNG (1983) and PENBERTHY (1985).

5.6.2. Impact of fluid viscosity

The main impacts of fluid viscosity on gravel pack performance are direction of gravel buildup in the annulus and flow pattern of the carrier medium. Some aspects of vertical vs. radial packing of the gravel as well as plug flow behaviour of concentrated slurries are illustrated as follows.

5.6.2.1. Vertical vs. radial packing of the gravel

High-viscosity fluids should not be used for placing gravel between screen and casing when a circulating technique is applied. In low-viscosity fluids, the settling velocity of the gravel is sufficiently high that the particles fall to the floor of the borehole when the fluid enters the screen (NOVOTNY 1977). The gravel pack thus forms from bottom to top of the wellbore by positive gravel dehydration in upwards direction which has a high density (abt. 90 % of maximum achieved after mechanical packing) and offers the most effective sand control for inside gravel packs (provided the perforation tunnels have already been infilled by prepacking, because circulation of low-viscosity fluids includes only little chance of transporting gravel also into the perforation holes; SUMAN, ELLIS & SNYDER 1983; cf. section 5.6.6.1.).

When high-viscosity fluids are used to carry the gravel or if the tail pipe is not lowered all the way down to the bottom of the wellbore, the particles follow the fluid trajectories, and as the fluid enters the screen, the pack forms from the outside of the screen radially outward to the casing. This type of inside-out radial packing is less favourable than the bottom-to-top vertical packing, because large pressure gradients are generated across pack and screen and fluid circulation is stopped, and also the danger of irregular pack building with inclusion of voids is much higher (cf. section 5.3.1.1.1.). Low-viscosity carrier fluids for optimum gravel packing also require low-density proppants and/or sand because of the better performance of the gravel pack if the job is technically properly done (cf. section 5.3.3.). High-viscosity fluids are preferentially suitable for squeeze packing of predominantly perforation tunnels, with the applicability spectrum encompassing exclusive squeezing, combined circulation and squeezing, and four-position longstroke packing (cf. section 5.3.2.3.).

5.6.2.2. Plug flow behaviour of concentrated slurries

Improvement of gravel placement technique with concentrated slurries carried in viscous oil- or water-based fluids results also in good sand control and productivity (SPARLIN & COPELAND 1972; LYBARGER, SCHEUERMAN & WILLARD 1974; SCHURTZ, BREINER & COMEAUX 1975), with concentrated slurries allowing the application of finer gravel and thus ameliorating performance. Smaller gravel size permits better formation control due to minimization of reservoir sand and packing gravel mixing by the plug flow behaviour of the concentrated gravel slurry (cf. section 5.8.5.), thereby enabling the gravel to retain most of its ideal permeability after being in place. As a consequence of improved gravel transport, the viscous fluids provide good radial and vertical gravel distribution. Gravel packing can sometimes also be enhanced by increasing perforation radius at the expense of formation penetration (STRACKE 1987). Liner vibration fluidizes the gravel and is therefore an excellent tool for packing gravel through perforations by elimination of fluid loss to the reservoir and squeeze packing (SOLUM 1984; cf. section 5.8.4.). Increasing size and density of perforations generally lead to increasing well productivity (WILLIAMS, ELLIOTT & WEAVER 1972; SAUCIER 1974).

SOLUM (1986) comments on effects of gravel blending and pumping techniques on accurate control and protection of the gravel during the operation. McLEOD (1986) gives an outline of productivity decline in gravel packed wells after acidizing. Gravel packing sand quality is studied by ZWOLLE & DAVIES (1983). BURGER (1987) reports on development of a chemical process for sand control. BONOMO & YOUNG (1983) discuss aspects of perforation cleanup.

5.6.3. Achievement of perforation tunnel plugging

Techniques for infilling the perforation tunnels with gravel vary in effectiveness (ALLEN & ROBERTS 1982). Comments are given as follows on two-stage sand-oil squeeze in long zones, Wikker Dikker technique in short zones, pressure packing and cavity filling, sequence of circulation and squeeze steps, and wash-down and through-tubing packing.

5.6.3.1. Two-stage sand-oil squeeze in long zones

The two-stage sand-oil squeeze probably provides best results in long zones (ALLEN & ROBERTS 1982). The first stage consists of a small hydraulic fracturing job with gravel sized to control the formation sand. Ball sealers dropped randomly or degradable diverting agents may be used to insure that a high percentage of the perforations accepts gravel. During the second stage, a wire-wrapped screen or slotted liner assembly (with a cup type of crossover device at the top and a double check valve washdown shoe at the bottom) is run and washed down through gravel remaining inside the casing. Additional gravel is placed around the liner through the crossover until a surface pressure increase indicates that gravel has covered the telltale slots.

5.6.3.2. Wikker Dikker technique in short zones

For shorter zones, the Wikker Dikker technique temporarily plugs the top of the liner and permits gravel to be placed at high rates and at pressures above the fracture gradient around the outside of the liner and through the perforation tunnels (ALLEN & ROBERTS 1982). The Wikker Dikker method has the advantages that after gravel placement, it is not necessary to run down through the perforated zone again to disturb or plug the gravel, and that mechanical procedures are simplified.

5.6.3.3. Pressure packing and cavity filling

The pressure pack method attempts to carry gravel through open perforations to fill a possible cavity behind the pipe and the perforation tunnel using crossover equipment (ALLEN & ROBERTS 1982). During placement, fluid carries gravel down the annulus between the inside of the casing and outside the screen or slotted liner, with gravel being deposited and fluid moving through the slots to the bottom of the liner assembly and back up through the wash pipe. Return flow into the wash pipe is periodically restricted, thus forcing fluid through the perforations and into the formation below fracture pressure, thereby carrying gravel into the perforation tunnels. The pressure pack technique, however, is not considered to approach the effectiveness of the two-stage sand-oil squeeze or Wikker Dikker methods.

5.6.3.4. Sequence of circulation and squeeze steps

The sequence of steps is extremely critical for achievement of a good result of perforation tunnel plugging with gravel. The normal sequence of lower circulation, squeeze position and upper circulation gives high-quality perforation packing, whereas the inverse sequential order comprising lower circulation, upper circulation and squeeze position does not provide a good chance for proper infilling of perforation holes and tunnels (cf. section 5.3.2.3.). Squeezing at the end of the job leaves frequently no room for filling of perforation tunnels, because the solid dehydrated sand in the annulus cannot move any more and may have also already prematurely bridged the annulus thereby even preventing generation of a continuous gravel mantle. Plugging of perforation holes results in origin of a gravel node before the outlet of the perforations in the annulus which is building towards the wash pipe, is bridging the annular space and can therefore block it off, and layered deposits on the screen propagate into the opposite direction and meet with the gravel nodes (cf. section 5.8.7.). The combined circulate-squeeze process has no need for blank pipe to exceed three joints above the productive zone.

5.6.3.5. Wash-down and through-tubing packing

The wash-down technique is applied in short low-productivity zones (ALLEN & ROBERTS 1982). Gravel is placed through open-ended tubing and then a screen is washed down through the gravel. Circulation is stopped in order to allow the gravel to settle back around the liner. Equipment is also available for gravel packing two zones in a dual completion with maintenance of pressure separation between zones.

Through-tubing gravel packing is performed with small-diameter wire-wrapped screens which can even be run through 2 3/8 or 2 7/8 in. tubing. Therefore also tubingless completions can be gravel packed in order to control sand, and small screens can be gravel packed inside casing below 2 3/8 in. or larger tubing using concentric through-tubing methods. Gravel placement is done by forcing all fluid and gravel into the perforations similar to the Wikker Dikker technique.

5.6.4. Influence of perforation technique

Casing and cement sheath perforation in oil- and gas-well completion is usually done by shaped charges, hydrojets or bullets (DANESHY 1973 b). Conventional bullet or jet perforation creates a dimpling or irregular hole, respectively, through casing and cement sheath into the formation without removing any material (cf. section 5.6.5.). Purging of gun, mud cake, formation and cement debris by underbalanced perforating and flowing, backsurging, and perforation washing and breakdown provides more space for the gravel and reduces the danger of bridging (McCAULEY 1969, SUMAN 1972, CROWE & CRYAR 1975, PENBERTHY 1983, STRACKE 1987). Comments are offered as follows on formation damage by perforation debris, perforation cleaning and proppant-induced perforation erosion.

5.6.4.1. Formation damage by perforation debris

Any debris created by the perforating process as well as sand or mud infilling the perforation channels which are not removed prior to gravel placement can severely choke hydrocarbon production (SPARLIN & COPELAND 1972). Other undesirable solids also must not be flushed into the formation by the carrier fluids, because contaminating solids coming from unfiltered fluids, storage tanks, blending equipment and tubular goods will damage the most permeable zones first and thus after the gravel pack is set and well exploitation starts, the most productive sands are suffering worst from pore plugging by invaded solids, and once they have penetrated into the formation prior to gravel packing, solid particles are trapped within the pay zone by the gravel pack which prevents their renewed emigration from the reservoir during course of subsequent hydrocarbon production (cf. section 5.11.4.).

The cup type of perforation wash tool can provide excellent cleaning of the perforation tunnels, perhaps including breaking away some of the primary cement in order to reduce the length of the hole and creating a cavity behind the casing to permit a greater thickness of the gravel with formation of a node (ALLEN & ROBERTS 1982).

Clean tubing is indispensible for preparing a well for gravel packing and is assuring clean open perforations instead of blocked ones. A string of good plastic-coated tubing is most desirable to prevent injection of rust, dope, scale and mill varnish into the formation. If plastic-coated tubing is not available, a clean string of new or used tubing can be applied, provided any rust, dope, scale and mill varnish is removed by pumping a solvent and a rubber plug down the tubing prior to packer setting. Pipe dope can be kept from the inside of the pipe by applying it sparingly to the pin instead of the collar, because in this way, the pipe dope would be squeezed to the outside when the joints are made up (cf. section 5.11.5.).

5.6.4.2. Perforation cleaning

Cased-hole gravel packs have generally lower productivity indices than open-hole gravel packs which is mainly due to fluid entry through perforations that expose a small fraction of reservoir sand to the well, in contrast to open-hole gravel packs where the whole borehole wall is covered by the gravel plug (PENBERTHY 1985; cf. section 5.10.). If the perforations are partially or completely filled with reservoir sand, lower productivity can result unless proper completion procedures are implemented, because perforations plugged with formation sand lead to high completion pressure drawdown and short completion life. Cased-hole gravel pack productivity can be considerably improved by large-diameter high-density perforation and subsequent cleaning. When no perforation cleaning is carried out, prepacking of the perforation tunnels is the optimum possible procedure without resorting to high pumping rates and pressures. Following discussion of some general aspects, perforation washing and surging are briefly outlined.

5.6.4.2.1. General aspects

Perforations plugged by debris can neither be filled with gravel nor allow penetration of consolidation fluids into the reservoir matrix and thus severely deteriorate sand control effectivity (PENBERTHY 1985). Perforation cleaning by debris removal does not only enhance gravel packing and resin consolidation, but also recovers mud cakes and mud pockets from the cement-formation interface which create injectivity problems. Perforation cleaning prior to gravel packing or resin consolidation also avoids the danger that as a consequence of altered permeability and flow geometry after the treatment, the plugging material in the perforation hole may be ejected and sand production may resume. Perforation cleaning methods include backflow, backsurge, perforation washing and chemical washing by chiefly acidizing.

Perforating is sometimes conducted in mud, with an overbalance followed by backflow to establish injectivity. Backflow without underbalance perforating, however, may not clean up all the perforations as the differential required to purge debris and mud from some perforations may not be reached. Perforation washing is a positive method of removing perforation debris, mud and formation sand from the perforation tunnel and from behind the casing as well as performing squeeze packs. The wash tools are fitted with tail pipe and mule shoe on the bottom. Acids or other clean fluids can be injected through the isolated perforations to establish communication through all perforations. Backsurging exposes perforations instantaneously to greatly lowered pressure to draw out a limited volume of perforating debris, mud, formation sand and reservoir fluid. Perforation washing and surging have different effects in relation to perforation prepacking with gravel.

5.6.4.2.2. Perforation washing

Perforation washing is achieved by straddling a small increment of the perforated interval with a special tool and injecting fluids into some perforations while taking returns from other perforations (DOWELL SCHLUMBERGER 1988 c). The fluid circulates outside the casing and back through the perforations nearest to the tool seals, thereby removing debris and formation sand from the perforations and from behind the casing. The tool is moved in increments equal to seal spacing until the entire perforated interval is washed. Fluid is pumped through one set of perforations through the casing-formation annulus and out another group of holes (NINI & OWEN 1983). The success of perforation washing and the amount of formation sand washed out depends on fluid viscosity and pumping rate (PENBERTHY 1985). Low-viscosity brines pumped at high rates remove larger amounts of reservoir sand than viscosified fluids do, because low-viscosity allows to achieve turbulent flow much more easily, whereas the polymer in high-viscosity fluids produces a filter cake that apparently reduces the possibility to remove sand.

Pumping rate per perforation has important impact, because turbulence and pressure drop through the circulation path causes more debris to be removed. In highly-deviated wells with more than 70 degrees inclination and long intervals to be gravel packed, high circulation rates have to be utilized requiring high pumping pressures, because otherwise formation sand would settle in the casing-drillpipe annulus and stick the washtool (BRUIST, JEFFERIS & BOTTS 1983). In general, however, experience reveals that most of the pay sand settles out in the rathole (cf. section 5.9.1.3.3.) and that regardless of hole angle, no sticking problems occur with the standard washtool arrangement. Some aspects of pressure parting of reservoir sand and comparative effectivity of perforation washing are illustrated as follows.

5.6.4.2.2.1. Pressure parting of reservoir sand

The subsequently resulting gravel prepack geometry is often a pressure part, with appearance and orientation of the fracture being similar to cases without previous sand production or without perforation washing (PENBERTHY 1985; cf. section 5.6.6.3.). Formation sand permeability also influences the amount of sand removed by perforation washing. Pay sands with lower permeability contain a higher percentage of fines and clay minerals which create a natural filter cake, and therefore the capillary pressure in these sands is greater than in higher-permeability reservoirs, with the latter thus being more easily washed out.

At high perforation density, the individual perforations communicate with each other along the casing during washing. Experimental work has shown that the amount of sand removal is greater with low perforation densities because of higher wash rates per perforation hole than with high perforation densities when the path of least resistance is between the perforations along the casing wall, with fluid travelling in this manner not being able to remove formation sand. Breakdown of perforations during washing may partially take place in a pair-wise manner. The lower the pumping rate, the less potential exists for breaking down all perforations during washing. Therefore perforation washing should be conducted at the highest practical rate to increase the pressure drop in order to maximize the number of open perforations, but below fracturing pressure in order to avoid opening of cracks in the reservoir. The element spacing on wash tools should be as close as possible for supporting achievement of high perforation wash rates. The voids which are washed into the formation between the perforations can be subsequently prepacked with gravel (cf. section 5.3.1.2.2.).

5.6.4.2.2.2. Comparative effectivity of perforation washing

Perforation washing may not be as effective as underbalanced perforating or backsurging, because all of the perforations are not connected (ECONOMIDES 1986). Flow takes the path of least resistance, and once a flow avenue is established, all flow proceeds through that channel leaving other perforations plugged.

5.6.4.2.3. Perforation surging

Perforation surging is an alternative to perforation washing and is capable of removing about the same quantity of formation sand per reservoir interval unit as perforation washing. Some aspects of perforation opening and sand removal as well as long and short reservoir intervals are outlined as follows.

5.6.4.2.3.1. Perforation opening and sand removal

Perforation surging can be performed with surge tools and by underbalanced perforating. Backsurging involves exposing of an already perforated interval to a sudden pressure underbalance. Loose material and gun debris from perforation tunnels are dislodged by sudden exposure of the perforated zone to an open chamber at atmospheric pressure, with the created differential pressure causing formation fluids to surge through the perforations into the casing, thereby flushing the perforation tunnels.

Perforation surging may not open all the perforations, with the amount of formation material removed per pay zone unit probably differing in an unpredictable manner, the surge geometry being erratic, and the quantities of surged sand being proportional to the pressure difference. High-surge differential pressures are more effective in removing material from the perforations than low pressures. While intermixing of prepack gravel and formation sand does not occur after perforation washing, gravel prepacking subsequently to surging tends to cause intermixing of gravel and formation sand (cf. section 5.2.3.8. and 5.6.7.). Surging also appears to remove sand more unevenly from the reservoir than washing does. Underbalanced perforating followed by perforation washing seem to provide the best chance to assure clean perforations prior to gravel packing.

5.6.4.2.3.2. Long and short reservoir intervals

Underbalanced perforating and surging have a common disadvantage in long intervals, because energy created by the underbalance is concentrated at the top of the interval and is at a minimum at the bottom of the section (RASHID 1984). Simultaneous surging of an upper and a lower zone can have the effect that the lower perforations are not purged and it may be impossible to squeeze gravel into the perforations when the pack is installed. Prime considerations for selection of perforation cleaning method are interval length, reservoir pressure and competency (ECONOMIDES 1986).

Perforation washing requires adequate reservoir pressure to allow circulation, whereas stability of the formation against collapse is decisive for underbalanced perforating and backsurging which are best suited to relatively short intervals where adequate energy is available to purge the perforations. Longer intervals are better suited to perforation washing using high pumping rates. Backsurging or underbalanced perforating may be the only feasible options in underpressured reservoirs where circulation is difficult if not impossible, whereas in very unconsolidated formations, perforation washing should be preferred (NINI & OWEN 1983). Perforation washing is also a superior technique in long intervals to be gravel-packed and in case of presence of high-viscosity oil (BRUIST, JEFFERIS & BOTTS 1983).

5.6.4.3. Proppant-induced perforation erosion

Concerning proppant-induced perforation erosion (CRUMP & CONWAY (1986), perforation friction pressure is reduced by both inlet rounding and progressive perforation diameter increase. The rate of perforation erosion increases with higher sand concentrations, higher differential pressure and smaller perforations, and is also influenced by casing grade and weight, characteristics of perforation through the cement sheath, and proppant type. The mechanical configuration of the perforations apparently continually changes during a proppant-laden hydraulic fracturing or gravel packing job, resulting in steadily increasing perforation flow capacity at a given pressure drop. If therefore treating rate is not progressively increasing during an operation to keep up with increasing flow capacity, pressure drop across the perforations will decrease and this could cause interzonal rate redistribution. Proppant erosion, however, does not transform perforations into perfect flow nozzles.

Other factors exerting control over perforation friction pressure and interval rate distribution are perforation density, injection rate and fluid density (CRAMER 1987). Perforation erosion can be generally a predictable junction of proppant volume, but the erosion process will blunt its own growth after a certain volume of proppant is displaced. Above a certain critical sand concentration, a particular completion system perforated with holes of uniform characteristics will exhibit an erosion rate that is linear with respect to proppant flowthrough rate. Aspects of erosion of perforation tunnels in gas wells are discussed by VRIEZEN, SPIJKER & VLIS (1975).

5.6.5. Influence of perforation geometry

Increasing width, length and density of perforations through the existing pay zone help to increase the flow area which is an effective means of reducing drag or frictional forces and thus diminishing effects damaging the grain fabric, thereby supporting stabilization of the reservoir framework against excessive sand production (ALLEN & ROBERTS 1982). Good completion practices, use of clean fluids and careful selection of perforating charges and conditions can often effectively reduce an otherwise serious sand control problem. Increasing the perforation hole area or the permeability of the material infilling the perforation tunnel is a primary approach for reducing pressure drop for a given flow rate through a single perforation.

Bullet perforators exhibit distinct advantages over jet perforators in terms of hole characteristics. Bullets produce constant-diameter round holes irrespective of gun clearance or casing grade and thickness, although perforation penetration performance is poor in hard rocks. Jet perforators are affected by gun clearance, casing grade, and shaped charge conditions, and also exhibit considerable randomness (SMALL 1985; KING, BINGHAM & KIDDER 1986; CRAMER 1987). Jet perforations may be required if the drilled hole is in a washed-out condition in the completion intervals and cement sheath thickness exceeds abt. 4 in. (CRAMER 1987). In formations with low compressive strength, bullet guns are preferable to small-hole jets (ALLEN & ROBERTS 1982).

Optimum perforation penetration is equal to flushed zone radius, but a minimum of 8 in. has been experimentally found to be the critical tunnel length required for maximum productivity (OYENEYIN 1987, 1988). The skin effects imposed by gravel pack, screen, perforations and other items are very important factors in the analysis of well performance (HIMMATRAMKA 1981).

5.6.6. Influence of reservoir section and treatment variables

The success of perforation packing with gravel is primarily a function of perforation hole cleanliness and is in addition to that influenced by interval length, single- vs. multiple-zone completions, permeability variations; screen, liner and washpipe characteristics; squeezing and/or circulating technique, and viscosity and leakoff rate of carrier fluid (ECONOMIDES 1986). The discussion as follows focusses on formation permeability variations and fluid viscosity, multiple intervals, and fluid velocity and pressure parting.

5.6.6.1. Formation permeability variations and fluid viscosity

Interval length affects the amount of gravel which can be packed in the perforation tunnels because of fluid leakoff across the area of exposed reservoir, with short perforated intervals suffering to a lesser degree from fluid loss and thus being more efficiently packed per section unit than longer ones (ECONOMIDES 1986). As the carrier fluid dehydrates into the formation, gravel may bridge off in the screen-casing annulus thereby isolating some of the perforations. Variations in pay zone permeability may be even more susceptible to gravel bridging if the high-permeability zone is at the top of the perforated interval, with permeability variations therefore being particularly detrimental to uniform perforation packing. Nodular gravel buildups at the inlet to the perforation tunnels are caused by viscous drag of the carrier fluid dehydrating through the packed perforations (cf. section 5.8.7.). The node remains intact until flow velocity is great enough to shift the nodes. If permeability variations result in reduced flow below the gravel node, it will continue to buildup until bridging occurs.

Low-viscosity carrier fluids do not cause buildup of gravel nodes, but are on the other hand less capable of transporting gravel (TORREST 1982). High-viscosity fluids are more applicable in prepacking of perforations prior to running of screen (GRUESBECK & COLLINS 1978), but should not be used in circulating packs. In high-viscosity fluids, the gravel particles follow fluid trajectories as the fluid enters the screen, and the pack forms from outside the screen radially toward the casing thereby leading to bridging and incomplete packing (cf. section 5.6.2.1.). Considerable permeability variations accelerate this process as gravel nodes are building in opposite direction from the perforated casing towards the screen, thereby enabling meeting of annular deposits on screen surface and gravel nodes in the middle of the annulus which means earlier bridging (cf. section 5.8.7.).

5.6.6.2. Multiple intervals

Multiple intervals commingled in a single gravel pack create special perforation packing problems particularly if they are separated by considerable distance. If an upper sand is much more permeable than a lower sand, single-stage gravel packing results in gravel bridging in the upper zone whereas the lower zone is not packed at all. Different screen diameter can increase the difficulty of properly gravel packing the whole multiple-zone interval. In terms of screen and liner configurations, using a small diameter blank liner above the screen promotes bridging (RASHID 1984), whereas a blank liner of identical outside diameter as the screen prevents bridging and enables settling of reserve gravel.

Blank liners between or below screens as well as bow spring centralizers promote voids in the gravel pack, and the use of an upper tell-tale screen (cf. section 6.3.5.) in conjunction with viscous fluids also promotes bridging at that point. Wash pipe outer diameter to screen inner diameter ratio has an effect on perforation packing, because the greater the restriction to flow inside the screen, the higher the screen annular velocity, with high velocity in this area tending to erode gravel nodes on the perforations thus retarding the bridging effects.

5.6.6.3. Fluid velocity and pressure parting

Carrier fluid velocity should be high enough to transport the gravel into the perforations, but low enough to dehydrate once it has entered the perforations. On the other hand, the optimum carrier viscosity for perforation packing may not be the best for annular packing. Too high fluid injection rates into the perforations result in pressure parting if the fracture gradient is exceeded (cf. section 5.6.4.2.2.1.). Orientation of the single pressure-parting planes is always normal to the direction of least stress when the perforations are not purged. In case of perforation purging, pressure parting does not always occur in a single plane, but multiple partings progress irregularly through the formation which appear to be controlled by the stress state and in some cases are perpendicular to each other, thus probably indicating stress state changes as the slurry is injected. When fluid leakoff is low due to low reservoir permeability, formation fracturing results in pumping of large amounts of gravel into the pay zone crack without adequate dehydration to pack the tunnels. On the other hand, fracturing may aid in perforation packing by providing more area for slurry dehydration (cf. section 5.3.5.).

The optimum procedure for perforation tunnel packing is squeeze packing with the screen in place when the perforated interval is relatively short (cf. section 5.3.2.3.3.), whereas in case of long intervals, the perforations should be purged in shorter intervals and prepacked in stages (cf. section 5.6.7.). There must be adequate leakoff to carry the gravel out of the casing into the perforations. Increased carrier fluid density and viscosity support gravel transport, but increased viscosity also reduces the dehydration process and may promote bridging due to node buildup.

5.6.7. Gravel prepacking of perforation tunnels

Gravel prepacking is a common procedure for enhancement of sand control efficiency and for support of perforation cleaning operations (PENBERTHY 1985). Prepacking is a special stage pumped in advance of the main gravel-pack treatment for the sole purpose of placing gravel into perforation tunnels before the annular gravel mantle is installed (HALL, MATHERNE, KERLIN & PACE 1988). The main advantage of gravel prepacking is high-permeability perforation tunnel filling.

5.6.7.1. High-permeability perforation tunnel filling

Prepacking ensures that the perforation tunnels are more completely filled, thereby maintaining high permeability from screen through perforation hole to formation face. Leakoff and gravel suspending properties of the carrier fluid, however, are equally important for prepacking and for main packing. In case of too slow fluid loss into the reservoir, only insufficient amounts of gravel enter the perforation tunnel through the relatively short duration of prepacking pumping, and thus only a commensurately small gravel volume is carried into the hole to be deposited there. If the perforation tunnel is thus only sparsely packed, subsequent production practices can result in hole collapse and permeability reduction. The combined permeability of a collapsed perforation tunnel partially filled with gravel and formation sand can be 10 to 100 times less than the permeability of annular gravel, and to overcome this increased resistance would require a drawdown 10 to 100 fold greater than that needed for a properly gravel-packed well.

Sufficient fluid leakoff gives rise to complete plugging of the perforation tunnels with gravel thereby guaranteeing optimum stability and maximum permeability. Good results can be achieved by displacing alternating stages of an acid treatment and gravel slurry, because running acid ahead of the gravel prepack stimulates the formation and increases fluid leakoff which aids in packing perforation tunnels with highly permeable gravel. The improved permeability reduces pressure drops across the pack which improves completion life and lessens the danger of screen failures. Prepacking with the gravel-pack screen in place minimizes surging of prepack gravel out of the perforations and also saves one half of rig time needed for the treatment. Distinction has to be made concerning mechanical formation stability whether sand production has already taken place from the reservoir into the well or not, and in terms of perforation cleaning between gravel prepacking subsequent to perforation washing or surging.

5.6.7.2. No sand production prior to gravel prepacking

In case of no prior sand production, planar pressure parting occurs in the unconsolidated formation sand when critical injection pressure level is exceeded, with gravel layer thickness being proportional to amount of gravel pumped (PENBERTHY 1985). Orientation of pressure parting is normal to the least principal stress as is typical for hydraulically-induced fractures, with gravel prepacking of reservoirs that so far have not produced sand leading to a combination of hydraulic fracturing and sand control (cf. section 5.3.5.). Intermixing of gravel and formation sand does not take place except of the immediate perforation vicinity, and the interface between gravel and formation wall is sharp.

The behaviour of pressure parting applies for both unconsolidated and consolidated reservoirs, because fracturing only requires that the minimum parting pressure is exceeded, whereas on the other hand if pressure parting does not occur, gravel builds up against the perforations until a screenout takes place. The pressure parting develops instead of an expected spherical gravel-mass geometry outside the perforations (cf. section 5.8.7.), because the formation sand is not equally free to move in any direction, but there is always a least principal stress direction prominently developed even when there has been no prior sand production.

5.6.7.3. Sand production prior to gravel prepacking

If the formation has already produced sand prior to gravel prepacking and a stable cavity has originated outside of the perforation, gravel placement leads to infilling of the voids (PENBERTHY 1985). If stable cavities do not form as a result of sand production, gravel prepack placement precipitates pressure parting in the formation sand (cf. sections 5.6.4.2.2.1. and 5.6.6.3.), with the orientation of the fracture planes depending on the stress state of the reservoir at any particular location in the perforation vicinity. In contrast to the case of no prior sand production, pressure parting in pay zones that have already produced sand does not always take place in a single plane, and there is also considerable intermixing of gravel and formation sand in the immediate perforation vicinity where consequently lacking prepacked gravel continuity occurs. While in some cases multiple pressure partings connected at the perforations appear, in other instances planar cracks originate which are distinctly perpendicular to each other. This reflects that under low initial stress-state conditions, the stress changes as prepack gravel is forced through the perforations.

When no perforation cleaning has been conducted, prepacking of the perforation tunnels is theoretically the optimum procedure possible without resorting to high pump rates and pressures. On the other hand, gravel packing a well that was earlier completed only by perforating is indispensibly linked with moving the prepack gravel through the perforations from which sand production has already occurred, with therefore prepack geometry depending on size and shape of any void around the perforations as well as stress state of formation sand.

5.6.8. Other aspects

Perforation collapse in very friable shallow sandstones (ANTHEUNIS, VRIEZEN, SCHIPPER & VLIS 1976) and in weakly-cemented reservoirs in greater depth where closure stress is higher has to be avoided in critical pay zones by carrying out of the gravel-pack operations shortly after the perforation job and by adequately propping the perforation tunnels to keep them open.

Aspects of particle transport through perforations are also discussed by HAYNES & GRAY (1972), TORREST & SAVAGE (1975), SKAGGS (1984) and CRAMER (1987), and some points of proppant transport are treated by CLARK & QUADIR (1981); SIEVERT, WAHL, CLARK & HARKIN (1981); AHMED (1984) and MORALES & ABOU-SAYED (1985). TORREST (1982 a) evaluates deposit buildup during gravel packing. DAVIES, HAGELAARS & ROBERTS (1984) comment on some aspects of near-wellbore placement of sand consolidation treatments. GRUESBECK & COLLINS (1982) discuss entrainment and deposition of fine particles in porous media, and ACHARYA (1986) investigates particle transport in viscous and viscoelastic fracturing fluids. Possibilities for improving gravel pack placement are outlined by MALY, ROBINSON & LAURIE (1974), and new design concepts for optimization of sand control are illustrated by ELLIS, SNYDER & SUMAN (1981). SHAH & LEE (1986) contribute to friction pressures of proppant-laden hydraulic fracturing fluids. SUMAN (1972) stresses the importance of cleaning perforations from debris and removal of mud pockets at the cement-formation interface for facilitating proper and uniform sand control placement.

HARRIS (1966), SUMAN (1972), McLEOD (1978), BELL (1982) and BONOMO & YOUNG (1983) summarize the effects of various perforation techniques on well performance and productivity ratios. KLOTZ, KRUEGER & PYE (1974) emphasize the necessity of deep clean perforations for productivity maximization in damaged formations. DANESHY (1973 b) reviews particle transport through perforations in connection with hydraulic fracturing. Various aspects of gravel packing of perforation tunnels and screen-casing annulus are summarized by HAYNES & GRAY (1972), GRUESBECK & COLLINS (1978); GRUESBECK, SALATHIEL & ECHOLS (1979); SHRYOCK, DUNLAP & MILLHONE (1979); SHRYOCK (1980), HODGE (1982), SKAGGS (1984); PEDEN, RUSSELL & OYENEYIN (1985) and PENBERTHY (1986).

5.7. Steam-drive enhanced oil recovery

Many formations containing heavy oil or tar are composed of unconsolidated sands which are unstable once the high-viscosity hydrocarbons are mobilized by steam, with reservoir sand then entering the borehole (RENSVOLD 1982) and therefore gravel packing of such pay zones being often indispensible. Two failure mechanisms affect the use of natural sand as gravel pack material under certain well and reservoir conditions, comprising chemical dissolution of silica by hot high-pH fluids (particularly in steam-flood areas, but also in-situ combustion projects) and mechanical grain crushing which produces pack-plugging fines. Two materials are commonly regarded as the best alternatives of sand in thermal gravel packing. While sintered bauxite has already been frequently applied since many years, nickel-coated sand has only recently been introduced. The discussion as follows includes also application potential of gravel packing in steam-drive projects in Europe, gravel density, and alcalinity reduction of the aqueous injected phase.

5.7.1. Sintered bauxite

Sintered bauxite seems to be the adequate material for gravel packing as well as for hydraulic fracturing in steam-soak (or huff and puff) and steam-drive (or steam-flood) areas as well as in-situ combustion (or fireflood) enhanced recovery projects in heavy-oil fields. Natural quartz sand is chemically unstable in the hot alcaline environment and dissolves more or less rapidly, and also resin-coated sand is seriously affected at high fluid alcalinity, whereas sintered bauxite has been shown to be a perfectly resisting material in many investigations (REED 1980; McCORRISTON, DEMBY & PEASE 1981; WATKINS & WATANABE 1984, UNDERDOWN & DAS 1985; WATKINS, KALFAYAN, WATANABE & HOLM 1985; STANDARD OIL PROPPANTS 1986; GROSSMANN, HÖFLING & MARX 1988; aspects of steam-flood silica dissolution are also discussed by BIRD & BOON 1987 and THOMAS, YOST & CATHEY 1987). Fused zirconia-silicate proppants are also subjected to silica exsolution and thus destruction of the fabric.

BECQ, ROQUE & SARDA (1984), however, come to the conclusion that none of the presently commercially available proppants (including sintered bauxite) resists satisfactorily to very alcaline pH and high temperature such as can be expected to occur in wells subjected to steam flooding, and thus for this special application the search for the ideal material has to take a look at new structures of alumina or zirconium oxide. High-temperature proppant testing for the particular purposes of geothermal fracture stimulation is carried out by SINCLAIR (1980) and RENSVOLD (1982).

5.7.2. Nickel-coated sand

SACUTA, NGUYEN & KISSEL (1986) introduce nickel-coated sand as suitable gravel-packing material for thermal wells as an alternative to sintered bauxite. In contrast to UNDERDOWN & DAS (1985) who report weight losses by dissolution of abt. 46 % for Ottawa Sand and abt. 4 % for sintered bauxite at 300 oC and pH = 11 (the range of these dissolution rates is independently confirmed by the test results of BECQ, ROQUE & SARDA 1984 who, however, in contrast to UNDERDOWN & DAS 1985, consider the documented destruction share of sintered bauxite not to be sufficiently low for undisturbed application of the heavy proppants as steam-drive gravel pack material), SACUTA, NGUYEN & KISSEL (1986) document under the same experimental conditions weight losses by dissolution of abt. 73 % for Ottawa Sand and abt. 39 % for sintered bauxite. SACUTA, NGUYEN & KISSEL (1986) conclude from these experimental results that sintered bauxite is not a generally suitable gravel pack material for elevated alcalinities, and the same applies for zirconia-silicate proppants which exhibit a weight loss by dissolution of abt. 29 %. Reasonable stability at 300 oC and pH = 11 is only exhibited by cement clinker and nickel-coated sand with weight losses by dissolution of abt. 6 % and less than 1 %, respectively. Very low dissolution rates of less than 1 % are also reported by BECQ, ROQUE & SARDA (1984) for some experimental alumina-zirconium oxides and zirconium oxides. Silicon carbide has also been considered (CHEN, ZHOU & LIU 1986).

From the commercially accessible materials, nickel-coated sand is thus the gravel type which is best resistant to degradation under severe thermal and hydrolytic stresses and is by far superior to all the other gravel pack materials tested (including sintered bauxite), and is at least according to the results of SACUTA, NGUYEN & KISSEL (1986) recommended for hydraulic proppant fracturing and gravel packing in steam-drive areas and related heat-application enhanced oil recovery projects such as for example in-situ combustion. Aspects of coating sand grains with nickel pellicles by an electroless nickel plating process or other hydrometallurgical methods are discussed by STRICKLAND, RICHARDSON, HAMBY & TORREST (1975) and TORREST (1975), and a more complicated technique is presented by HANSBERGER (1978).

In terms of high-temperature performance, nickel-coated sand as well as sintered bauxite should also be suitable materials for propping of hydraulic fractures in geothermal reservoirs where temperatures up to 500 oC can be reached (cf. section 4.7.4.1.). In contrast to steam-drive enhanced oil recovery where gravel pack or fracture infilling can be set prior to steam injection in a still cold pay zone interval in the yet untreated borehole and later heating of the well by steam flooding only affects the gravel being already in place, the main problem of geothermal reservoir hydraulic fracturing and gravel packing is the selection of suitable carrier fluids which are stable in the hot environment, because formation temperature is at an elevated level all the time and cannot be reasonably lowered for the performance of the operation (cf. section 4.7.4.). Sand control requirement in geothermal reservoirs is mentioned by BARON & UNGEMACH (1981).

5.7.3. Application potential in Europe

Steam-drive enhanced oil recovery has in Europe so far mainly been applied in Cretaceous sandstones in Germany FRG and Netherlands since 1975 (HARMSEN 1979, PROYER 1980, LILLIE 1981, LILLIE & SPRINGER 1981; GRAUTE, KURTH, SCHLEMM & THIES 1982; ROSSKAMP 1982; PROYER, CHAZITEODOROU, MUSS & ROSSKAMP 1983; LILLIE & WIETHOFF 1985, REINICKE & WIETHOFF 1987) in larger extent, has been run also in France (SAHUQUET & FERRIER 1980; COUDERC, MONFRIN, QUETTIER & SAHUQUET 1987; COUDERC, VERPEAUX, MONFRIN & QUETTIER 1987; COURBOT 1987, SAHUQUET & SITBON 1987), Hungary (FARKAS & VOLL 1987), Turkey (GÜMRAH & OKANDAN 1987) and Soviet Union, and has been performed in smaller pilot runs in Yugoslavia and Austria. As a consequence of differences in steam conditioning and effective reservoir temperatures between USA and Europe, and because of applications resulting in a more or less neutral rather than an alcaline milieu, the use of sintered bauxite and/or nickel-coated sand as alternative gravel pack materials in steam-injection wells has in Europe so far not been necessary in contrast to the widespread use of sintered bauxite in the USA where numerous earlier sand gravel packs later failed by quartz dissolution, but frequently excellent results were obtained with sintered bauxite gravel packing (UNDERDOWN & DAS 1985).

Future applications of sintered bauxite may comprise for example the exploitation of heavy oil sand deposits by steam-drive wells equipped with gravel packs and probably also enhanced by hydraulic proppant minifracturing in the Orinoco belt in Venezuela and in the Athabasca- and Peace-River area north of Edmonton in Alberta/Canada (ERDÖL UND KOHLE - ERDGAS 1983, ERDOEL-ERDGAS AKTUELL 1985 a) as well as in-situ combustion methods.

Similarly as applying for hydraulic fracturing, multiple thin-sand reservoirs are expected to become more and more widespread targets of enhanced oil recovery which in many cases also includes the need of gravel packing (steamflood evaluations of such types of pay formations are carried out by POWERS, DODSON, GHASSEMI & MOORE 1985). Field case studies of steam-flood projects are also evaluated by CHU (1985). Effects of acids on gravel packs are discussed by CHEUNG (1985). Some problems of sand production in heavy oil reservoirs are reported by SMITH (1986). TOMA, LIVESEY & HEIDRICK (1986) outline gravel packing in horizontal steam-injection wells. Aspects of application of sand control technology in thermal recovery systems are also discussed by RENSVOLD (1982). FRIEDMAN, SURLES & KIEKE (1986) present a method of high-temperature sand consolidation which is also resisting to the hot steam-drive environment.

5.7.4. Gravel density

The gravel pack generally becomes more stable when a heavier material is applied (particularly in high-performance wells). On the other hand, the frequently more viscous gravel pack fluids with respect to the fracturing fluids are much better transporting and putting in place lighter sand or proppants (cf. section 5.3.1.). In addition, considerably larger quantities (in terms of weight) of high-density alumina oxide proppants are required to cover the same gravel pack interval by the same volume of material with respect to low-density ceramic proppants or natural sand, with sintered bauxite utilization for gravel packing thus being more expensive both due to the higher unit price of the material and the need of larger tonnages due to higher weight per unit volume (the same applies for hydraulic fracturing; cf. sections 1.4.7. and 1.4.12.5.).

All these drawbacks, however, are more than compensated by the destruction of every presently commercially available propping material except of sintered bauxite and nickel-coated sand in the aggressive alcaline environment of steam-drive enhanced oil recovery, with thus leaving almost no alternative to the heavy bauxite proppants in this special application of gravel packing and hydraulic fracturing (cf. section 4.9.4.). The lower specific gravity could turn out to be the greatest advantage of nickel-coated sand and could facilitate its progressive expansion and distribution by successively replacing sintered bauxite, particularly as the study of SACUTA, NGUYEN & KISSEL (1986) has shown that the stability of nickel-coated sand under steam-drive conditions can even be superior to that of sintered bauxite.

Gravel dissolution in the presence of high-pH high-temperature steam affects both unconsolidated and resin-coated gravel packs (RENSVOLD 1982). As the resin films do not completely cover every gravel grain in the pack, the exposed surfaces are therefore susceptible to dissolution by steam thus leading to failure of the resin bond.

5.7.5. Alcalinity reduction of the aqueous injected phase

Retardation or even prevention of quartz sand dissolution in gravel pack material and reservoir formation sand can also be achieved by the use of ammonium chloride and nitrate as additives which neutralize the alcalinity yielded by bicarbonate decomposition and thus reduce the pH of generator effluents in steam-drive operations (WATKINS & WATANABE 1984; WATKINS, KALFAYAN, WATANABE & HOLM 1985). Most generators produce a steam quality of abt. 80 %, with the remaining aqueous aggressive phase being injected together with the steam into the pay horizon (UNDERDOWN & DAS 1985). Formation sand vs. gravel pack sand dissolution, cyclic steam injection and downhole pH value, mechanical gravel pack stability and other aspects are briefly outlined as follows.

5.7.5.1. Formation sand vs. gravel pack sand dissolution

Neutralization of the hot alcaline water is the only possibility to prevent the reservoir formation sand from chemical dissolution in thermal steam-flood projects, because in contrast to gravel packing material, the pay zone sand cannot be exchanged against substituting resisting products, but has to remain in place in its original composition and thus is continuously exposed to the hostile high-temperature environment. As therefore the reservoir material cannot be adjusted to the fluids in the wellbore as can be done with gravel pack material and also hydraulic fracturing proppants, the reverse adjustment has to be made by selection of a proper composition of the injected fluids in order not to damage the pay zone too seriously (FLANIGAN 1979 also points out that neither the geological environment of the reservoir complex can be changed, nor is it possible in offshore operations to renounce from high-angle directional drilling). As silica dissolution increases with bicarbonate content in the feedwater, the amount of ammonium salts required for compensation of alcalinity and thus reduction or stoppage of silica dissolution downhole is roughly proportional to the bicarbonate content.

UNDERDOWN & DAS (1985) document that pH is the most important factor affecting the stability of all gravel packing materials. As quartz dissolution continues even when pumping neutral water after alcaline brines and because a persistently elevated pH during steam-drive operations leads also to some dissolution and grain surface erosion of sintered bauxite, recommendation is made that in order to prolong the life of gravel packing material in high-temperature steam-injection wells, pH should be maintained at 9.0 or less by treating the water before it enters the steam generator in order to remove all the bicarbonate ions or by adding an acid to the feedwater for lowering of the pH (UNDERDOWN &

DAS 1985). This neutralization also helps restricting reprecipitation of silica and/or alumina silicate minerals in gravel pack, screen and/or reservoir formation as a consequence of supersaturation of the brines following extensive silica dissolution (TOMA, LIVESEY & HEIDRICK 1986). WALDORF (1965); McCORRISTON, DEMBY & PEASE (1981); AMAEFULE, PADILLA, McCAFFERY & TEAL (1984) and SOMERTON, CHEN, SCHUH & YUEN (1984) also comment on formation damage by steam condensate and chemical treatments for injectivity improvement. An overview of steam injection theories in an unified approach is given by ALI (1982), and FAROUQ & MELDAU (1979) present a summary of steamflood technology.

5.7.5.2. Cyclic steam injection and downhole pH value

CHEN, ZHOU & LIU (1986) point out that the decisive question is whether the injected phase arrives with high pH at the bottom of the thermal well. The higher pressure downhole with respect to the surface facilities is thought to lead to decreasing pH as a consequence of reversal of bicarbonate decomposition, as suggested by much less quartz gravel dissolution than predicted according to the aggressivity of the fluid under surface conditions in various field case studies in China. Therefore CHEN, ZHOU & LIU (1986) conclude that given sufficient reservoir depth and thus bottomhole pressure, high-quality quartz sand might still be an effective gravel packing material in steam-injection wells, which would be a significant aspect due to the much lower price of quartz sand with respect to synthetic gravel packing materials. The purer the sand in terms of quartz content and the better the stability of the preferentially mono- to oligocrystalline quartz grains, the longer it will resist to the harsh steam environment (UNDERDOWN & DAS 1985).

The gravel packing material also has to withstand cyclical steam-drive operation in which periods of high injection rate and low production rate alternate with periods of low injection rate and high production rate (cf. section 4.9.4.1.). The resulting cycles of reservoir pressure buildup and blowdown in the formation are improving thermal recovery (DOSCHER, LABELLE, SAWATSKY & ZWICKY 1963; HEISLER 1986), but are also increasing the thermal stress acting on the gravel pack material by alternating expansion and contraction by heating and cooling with changing operation mode.

5.7.5.3. Mechanical gravel pack stability

Maximum efficiency of the gravel pack depends on its mechanical stability (RENSVOLD 1982). Once placed, the gravel pack should not shift or move, because this would allow intermixing of formation sand with packing gravel during production, thus leading to permeability deterioration. Gravel dissolution by steam is the most serious process giving rise to pack instability and movement because of physical and chemical removal of material. In addition to dissolution, other factors responsible for pack movement after placement are expansion and contraction of casing and screen assembly during cyclic steam injection, flow reversal during steaming and production, and formation compaction.

5.7.5.4. Other aspects

SHRYOCK, AHMED, MELOY & KENT (1988) outline reduced density slurry packing, intermediate viscosity slurry packing, and foam gravel packing for sand control in steam-drive areas in shallow underpressured sands. Foam has turned out to be a viable fluid for gravel packing where lost circulation prevents the use of higher density sands. Minimization of packing gravel and formation sand dissolution can also be achieved by steam separation techniques in order to remove the high-pH boiler water, with the ionic strength of the condensed steam then being increased by continuous addition of KCl (RENSVOLD 1982).

Comments on high-temperature sand consolidation are also offered by FRIEDMAN, SURLES & KIEKE (1986). Sand control filters for thermal recovery wells are presented by TOMA, LIVESEY & HEIDRICK (1986). The stability of resin-coated proppants in hot brines is investigated by RENSVOLD (1983).

5.8. Special gravel pack applications

Following the discussion of common practices of gravel packing operation, this section assembles various special techniques of gravel packing which are necessary to carry out according to well conditions or which may represent an economically more feasible and/or technically less complicated alternative of standard gravel packing. The account includes aspects of consolidated pack installation, resin-coated gravel placement, gravel packing of deviated wells, gravel packing with liner vibration, concentrated gravel slurry pumping, density slurry and foam gravel packing, viscous polymer gravel packing and deposit buildup, prevention of proppant flowback by fracture tail-in, and puddle packing.

5.8.1. Consolidated pack installation

Consolidated pack installation is an intermediate technique between conventional gravel packing (cf. sections 5.2. - 5.7.) and chemical formation consolidation (cf. sections 5.9.1. - 5.9.4.; TAUSCH & CORLEY 1959; SPARLIN 1967, 1971; HAMBY & RICHARDSON 1968; YOUNG, COOK & DONALDSON 1969; COPELAND & McAULEY 1974; MURPHEY, BILA & TOTTY 1974; KNAPP, PLANTY & VOILAND 1977; COPELAND & CONSTIEN 1978) and to some extent includes a combination of both traditional methods (HALLIBURTON 1988). Consolidated pack installation involves blending of the slurry at the surface and pumping of the mixture of gravel and glueing agent through the perforations where the added plastic cures and forms a stable gravel pack which behaves like an artificially cemented gravel mantle between screen and casing or borehole wall as well as within the perforation tunnels. Injection of the slurry is stopped in order to screen out the gravel on the formation and to pack the material in the wellbore including covering the perforations and filling them with consolidating gravel.

Stabilized pack installation is particularly suitable for wells where notches behind the casing have already originated by sand production or where cavities accepting gravel nodes can be artificially created by washing through perforations (ALLEN & ROBERTS 1982; cf. section 5.8.7.) or where the casing sheath is defective and contains holes and channels. The brief outline as follows concentrates on aspects of perforation tunnel plugging, differences between consolidated and loose gravel packs, and artificial stable borehole wall.

5.8.1.1. Perforation tunnel plugging

The advantage of consolidated pack installation is that given proper placement in cased holes, only the perforation tunnels have to be kept filled with gravel, whereas remnants of the pack in the wellbore can be washed or drilled out after consolidation of the gravel pack and thus do not leave any restrictions or obstructions to further borehole manipulations, with the absence of screen or liner enabling better investment in production equipment such as larger pump and wider liner (KNAPP, PLANTY & VOILAND 1977). This is in contrast to conventional gravel packing which results in origin of a loose gravel mantle between borehole wall or casing and screen, and the annulus is lost for production manipulation purposes. Some comments on application spectrum and exclusive outside gravel packing are offered as follows.

5.8.1.1.1. Application spectrum

The absence of a gravel pack mantle seaming the borehole wall also eliminates further flow velocity reduction of hydrocarbons that exit the gravel-packed perforation tunnels, therefore resulting in higher well productivity and easier influx of oil into the borehole (cf. section 5.10.2.). Consolidated pack installation can provide a sand control completion with good productivity without a slotted liner or wire-wrapped screen (CONSTIEN & MAYER 1978, SINCLAIR & GRAHAM 1978 b), and because of its low cost due to reduced gravel and hardware demand, the system can be also considered for long interval sand stabilization purposes of both single- and multiple-type where both conventional gravel packing and standard formation consolidation might only be marginally feasible. Consolidated pack installation enables effective sand control in multiple pay zone complexes, with the upper horizons being able to produce sand-free without equipment in the casing that would interfere with offtake from lower levels (RENSVOLD 1982).

5.8.1.1.2. Exclusive outside gravel packing

Consolidated pack installation therefore comprises in ideal case only setting of an outside gravel pack beyond the casing within the perforation holes, with an inside gravel pack not being necessary, because the stabilized gravel aggregates are not able to flow back into the borehole. As the treatment is only successful if every perforation takes up gravel, caution has to be exercised that all the perforations are open and clean. Pumping of the slurry has to be performed below 80 % of the fracture pressure all the time. Prevention of gravel swabbing into the wellbore while tripping out when the consolidated mass may not yet be fully stabilized can be achieved by filling the borehole with gravel which involves washing sand after the hardening period, placing a viscous overflush high in the tubing and immediately beginning the trip out with tubing and annulus open which does not require a trip to wash out gravel after the hardening period, and washing in an open-ended screen with no packer on top so that all perforations are protected and every particle is locked into place after thermal or chemical hardening.

5.8.1.2. Differences between consolidated and loose gravel packs

Resin-coated gravel sized to control the formation sand is injected into the perforations at below fracture pressure until a screenout stops the operation (ALLEN & ROBERTS 1982). With an external catalyst system, the sand remaining inside the casing is washed back with oil, and the resin-coated sand placed in perforation tunnels and cavities is consolidated by injecting the catalyst. With an internal catalyst system, the resin-coated sand is allowed time to consolidate and is then drilled out of the casing, with only the infillings of the perforation tunnels and nodes being left.

In remedial operations, a consolidated pack may be preferable with respect to a conventional loose gravel pack, because the bonded gravel is not likely to be lost to the unstressed formation. Consolidated packs are less expensive than standard gravel packs followed by a chemical consolidation treatment once the loose gravel pack later turns out to be not sufficient and has to be enhanced by a subsequent injection operation. Significant strength degradation of all consolidated pack systems, however, occurs with exposure to hydrofluoric acid, with thus this type of acidizing not being able to be applied any more after installation of the consolidated pack, whereas hydrochloric acid acts as a catalyst that can speed curing (cf. sections 4.5.5.4. and 5.11.4.).

5.8.1.3. Artificial stable borehole wall

Consolidated pack installation has in the last decades frequently been performed in China to create an artificial stable borehole wall (CHEN, ZHOU & LIU 1986). The cured mass forms a hardened pack which contains tortuous capillaries that allow the formation fluid to flow into the wellbore and to simultaneously bridge the reservoir sand in place. Both cement and resin have been successfully used as bonding materials. The strength of the artificial borehole wall is primarily controlled by the ratio of the different materials, with higher strength, but also lower permeability, given by a higher proportion of bonding agent.

Other influencing factors are gravel size, with larger particle diameter giving higher consolidation strength. Experience has shown that the best results are obtained when pumping a gravel quantity equivalent to three times the amount of sand which has been produced so far. Permeability of the consolidated pack can be enhanced by mixing diesel oil with additive-containing cement-gravel-slurry to form an oil-in-water emulsion. After having been squeezed into the formation, the slurry is hardened by hydration under reservoir conditions, with the dispersed diesel and remaining water forming many intercommunicated channels in the bonded sand control barrier. Satisfactory results have also been achieved in oil wells producing with high water cuts and in water-injection boreholes.

5.8.1.4. Other aspects

Proper blending of slurry components is essential, because compressive strength is sensitive to relatively minor variations in resin content. Reduced resin content may lead to inadequate strength, while excess resin in internally catalyzed systems may cause formation plugging due to invasion of plastic droplets. Consolidated gravel packs usually do not require screen or liner to be held in place, but can also be carried out as terminal tail of conventional loose packs in order to prevent sand or proppant flowback from the perforations in a comparable way as resin-coated proppants can support the proximal fracture tip near the wellbore (cf. section 4.12.3.3.2.). Shifting of gravel cannot occur because the grains are bonded together in a coherent permeable mass (RENSVOLD 1982) which is a considerable advantage when large and reversible pressure differentials occur during well production.

5.8.2. Resin-coated gravel placement

Stimulation results are improved by using resin-coated grains not only as proppants in hydraulic fracturing of hard rocks, but also as packing gravel in sand control of soft rocks (CONSTIEN & MAYER 1978, SINCLAIR & GRAHAM 1978 b; SINCLAIR, GRAHAM & SINCLAIR 1983; cf. section 1.2.6.). The resin-coated materials provide stabilization of fracturing proppants and packing gravel and prevent movement or flowback of the inserted grain populations. Resin-coated gravel placement is a combination of mechanical and chemical sand control (HALLIBURTON 1988), because gravel and resin are placed in one operation, and gravel becomes coated with resin before reaching the formation, with this treatment thus being similar to conventional gravel packing except of the pack becoming resin-coated. In addition to the resin-coated packing gravel, extra resin can be injected to stabilize a portion of the formation sand adjacent to the gravel, with this extra resin being placed after washing out or drilling out of the excess packing gravel and before pumping of the acid catalyst (MURPHEY, BILA & TOTTY 1974), thus in full sequence performing a mechanical gravel pack followed by chemical gravel and formation consolidation.

Under normal conditions, almost exclusively 40/60 mesh grain size is used for resin-coated gravel packing, because this grain size leads to a consolidated pack with high mechanical strength and high permeability that will not materially hinder well productivity (RENSVOLD 1982). The pumping of natural sand or synthetic proppants that are coated with uncured resin films which harden and bond when the materials are heated either naturally by the reservoir temperature in intermediate to deep wells or artificially by hot water or steam flush in shallow pay zones is an advanced utilization of gravel packing. Resin-coated gravel can be used for regular or inside gravel pack, outside gravel pack, primary gravel pack in new wells, repair gravel pack in old wells, minifracturing, patching of existing gravel-packed screens and liners, and prebonded resin-coated gravel inner liners (SANTROL 1986 b) in oil- and gas-producing wells and water-, steam- and chemical-injection wells. Gravel mantle stabilization by resin cementation and other aspects are briefly outlined as follows.

5.8.2.1. Gravel mantle stabilization by resin cementation

Gravel mantle stabilization by resin cementation is based on the concept that the resin-coated gravel absorbs heat after placement, and the uncured resin pellicles soften, melt and diffuse to form necks between adjacent sand grains which act like a synthetic gravel cement (cf. section 1.2.6.). In addition, the individual resin-coated grains are slightly tacky and tend to stick to neighbouring grains as they first contact them, although the equivalent of a true tap density is probably only infrequently achieved (MURPHEY, ROLL & WONG 1985). After a period of time, the resin thermosets, forming a permeable consolidated pack by chemical bonding (cf. also section 5.8.1.). If formation temperature is too low in shallow wells, curing can also be initialized or accelerated by addition of an activator which is an alcohol-surfactant blend (POPE, WILES & PIERCE 1987). The outline as follows focusses on aspects of external and internal catalyzation, thermal and chemical curing, application in sand control and fracture stimulation, and consolidation of both gravel and formation sand.

5.8.2.1.1. External and internal catalyzation

The function of the resin system is to consolidate the gravel into a strong permeable filter (ECONOMIDES 1986). In order to achieve this goal, the resin system must first coat the gravel and chemically bond to the gravel surface. Then after the resin-coated gravel is pressure packed against the formation, the resin must cure to form a hard impervious film covering each gravel particle and connecting them at the contact points. A minimum amount of resin is used to achieve a balance between strength and permeability. Once the resin system has coated the gravel, the coupling agent determines the ultimate strength which the aggregated pack obtains. Without coupling agent, compressive strengths in the order to 50 - 300 psi result, whereas with the coupling agent present, compressive strengths up to 2,500 - 3,500 psi can be achieved (in comparison, resin consolidation of formation sand may result in compressive strengths up to 3,000 - 4,000 psi; cf. section 5.9.1.2.3).

The reason for this tremendous influence of the coupling agent on compressive strength is related to the point of failure in the resin-gravel matrix when placed under excessive compressive loads. Compressive failure at excessive pressures occurs at the resin-gravel interface rather than through the resin glue line between particles or through the particles themselves. This is an adhesion failure mechanism and is controlled by the coupling agent rather than by the epoxy resin which forms the coating around the particles and the connecting resin matrix with other particles.

5.8.2.1.2. Thermal and chemical curing

Similarly as applying for resin consolidation of formation sand, placement of resin-coated gravel can be performed by curing agent employment in two different ways comprising thermal and chemical curing (RENSVOLD 1982). The resin system consists of resin, curing agent, coupling agent and cure accelerator (the latter is only required if bottomhole temperature is less than 54 oF). External catalyzation uses a spacer following the resin-coated gravel slurry which in turn is followed by a curing solution, with the resin films on the gravel grains reacting with the curing solution as it is pumped through the permeable mass. The resin hardens rapidly, and fluid flow in the pore space during this process helps to insure that the gravel pack retains a very high permeability. Another advantage is that the resin does not harden prematurely, thereby preventing the gravel from being properly placed. Internal catalyzation involves adding of curing agent to the resin before the resin-coated gravel is pumped into the well which results in uniform curing of the resin on the gravel grains.

5.8.2.1.3. Application in sand control and fracture stimulation

Similarly as stabilization of gravel packs with avoiding proppant flowback from the perforation tunnels and the borehole gravel mantle and thus sand control failure can be achieved by application of resin-coated gravel, aggregation of a tail-in of hydraulic fracturing treatments with proppants precoated with a thermosetting phenolic resin which softens and cures under reservoir conditions to form a stable grain package is a very effective means to prevent crack evacuation by proppant flowback from the fracture mouth and thus partial closure of the crack and disconnection of several sections of the hydraulically-induced drainage path (POPE, WILES & PIERCE 1987; cf. section 4.12.3.3.). In some soft formations, high pressure drawdowns cause the near-wellbore formation to shear and to breakdown which results in spalling and eventually production loss due to well plugging, with resin-coated gravel helping to spread the high drawdown stress and to resist to embedment (SINCLAIR, GRAHAM & SINCLAIR 1983). Concerning operation, the hardening agent may be added to the resin during mixing at the surface or it may be applied as an afterflush once the gravel has been injected (HALLIBURTON 1988). Some comments on compressive strength of resin-coated aggregates as well as single and multiple gravel pack completions are offered as follows.

5.8.2.1.3.1. Compressive strength of resin-coated aggregates

After injection of the resin-coated gravel into the perforation tunnels and subsequent curing of the resin to produce a consolidated gravel mass, the aggregated gravel inside of the casing is drilled out and the well is returned to production. The resin-coated gravel aggregates infilling the perforation tunnels attain a compressive strength of 1,500 and 2,500 psi for 20/40 and 40/60 mesh gravel, respectively, and in some favourable cases can even reach compressive strengths up to 3,500 psi (which is on average still slightly lower than those being able to be acquired in chemical formation consolidation by resin cementation of the reservoir sand ranging between 3,000 and 4,000 psi; cf. section 5.9.1.2.3.). According to this high compressive strength, the perforation tunnel plugs are sufficiently stabilized to be unable to flow back, thus not requiring a gravel mantle along the perforated casing to support them (ECONOMIDES 1986).

5.8.2.1.3.2. Single and multiple gravel pack completions

The high compressive strength is retained over extended periods of time, and the resin-coated gravel packs have excellent resistance to production fluids and most treating chemicals. Resin-coated gravel packing is applicable for single and multiple completions. The high gravel concentration is carried in a viscous gel that improves gravel placement in long perforated sections thereby enhancing job success (cf. section 5.3.1.1.). Well cleanup time is shortened and formation damage is lessened because a relatively small volume of fluid is used to place a large volume of gravel network. While in most cases placement of resin-coated gravel gives way to much better sand control effectivity with respect to setting of conventional gravel, SCRIMGEOUR & SHEPHERD (1988) present an inverse example where resin-coated gravel insertion for internal gravel packing resulted in too high productivity restriction and therefore was in later stages of the completion campaign replaced by conventional gravel.

5.8.2.1.4. Consolidation of both gravel and formation sand

Resin-coated gravel placement allows simultaneous consolidation of gravel pack and formation sand (HALLIBURTON 1988) if sufficient excess resin is included, while elimination of excess resin results in stabilization of only the gravel pack. When dispersing resin and low gravel saturations in a thin carrier fluid in a mixing tank, some of the resin automatically coats the gravel during mixing, with the excess resin remaining in the carrier fluid. As the mixture is pumped, the resin-coated gravel screens off against the formation, and as the carrier fluid moves into the reservoir, tight grain-to-grain contact is attained in the gravel pack, and the excess resin in the fluid coats the pay sand.

Using high gravel concentrations in thick carrier fluids, all of the resin coats the gravel completely already during mixing, with no excess resin remaining dispersed in the transport fluid. The mixture is then pumped into the well, but little or no carrier fluid is lost to the formation. In addition to the aforementioned spatially coinciding combination of gravel packing and reservoir consolidation in the same prospective interval, multiple-zone sand control can be performed as spatially separate association of gravel packing in lower sections and resin stabilization in upper portions of the hydrocarbon-bearing sequence (THOMAS & JOHNSON 1976).

Resin-coated gravel packing was earlier executed with an oil-based carrying medium because of the ease with which the resin coats the gravel in an oil medium, but with development of modern surfactants, it became possible to use aqueous transport fluids and still allow the resin to bond the gravel (ECONOMIDES 1986).

5.8.2.2. Special applications

Resin-coated gravel placement is more sophisticated and also frequently more successful than conventional mechanical sand control where the gravel grains are not mantled and aggregated by resin pellicles. While resin-coated gravel packing could be theoretically suitable for all wells suffering from sand problems, economical feasibility often restricts its application to such reservoirs where conventional loose gravel pack would not properly solve the well troubles. As this relationship between technology and economics affect all types of boreholes, feasible application of resin-coated gravel placement has to be carefully selected on the basis of pay characteristics. The following overview discusses mainly aspects of old production boreholes and steam-drive projects as well as transitional applications between gravel packing and hydraulic fracturing. Comments are also offered on carrier fluid and formation.

5.8.2.2.1. Old production wells

Resin-coated gravel packs are especially beneficial to older wells that have already more or less extensively produced sand, because the viscous resin-coated gravel tends to restress the formation by occupying voids created by sand production. Other advantages of resin-coated gravel are requirement of only small volume and short job time, possibility to treat multiple zones, and unnecessity of screen or liner to hold the pack in place. Resin-coated gravel packing is generally applicable for long and short intervals, cased- and open-hole completions, single and multiple completions, clean and dirty sands as well as almost shale-like reservoirs and chalks, and low- and high-temperature pay zones (up to 400 oF).

Packing of perforation tunnels with resin-coated gravel aggregates and leaving the inner part of the borehole empty due to absence of both gravel mantle and screen or liner is particularly suitable for long multiple completions where the spacing between the separated intervals economically and technically discourages the use of screens (DOWELL SCHLUMBERGER 1988 c). Special applications of resin-coated gravel packing is tailing-in after hydraulic fracturing treatments to retain proppants in the fracture (cf. sections 2.4.1.2.3. and 4.12.3.3.), and gravel packing of unconsolidated sands using fracturing pressures to achieve stimulation with a sand-control operation, with the proppant package being stabilized in place in a similar way as the formation sand. Resin-coated sand slurries can also be used to repair damaged slotted liner or wire-wrapped screen in conventional gravel packs without removing liner or screen from the well by utilizing either concentric or coiled tubing with a small workover hoist (MURPHEY, ROLL & WONG 1985).

5.8.2.2.2. Steam-drive projects

The use of resin-coated sand for gravel packing in thermal wells has been growing in popularity during the last years, because improvements in resin chemistry and manufacturing processes have helped to increase the thermal stability of this material and to prolong its effectiveness in the harsh environment of steam-flood enhanced-recovery boreholes (UNDERDOWN & DAS 1985). Aggregation of the resin films creates a consolidated permeable pack that cannot be displaced by the high-velocity steam which is injected into the reservoir. Resin-coated sand in steam-drive oil fields, however, only performs reasonable up to moderate alcalinities. At high alcalinities, sintered bauxite and nickel-coated sand are superior materials (cf. section 5.7.). Thermal resistance of resin-coated gravel is also investigated by COPELAND & BASS (1976).

5.8.2.2.3. Transition between gravel packing and fracturing

Although the main application of resin-coated gravel is outside packing of perforation tunnels without accompanying inside gravel mantle thereby leaving the borehole completely unobstructed by installing neither screen or liner nor gravel sheath, other possibilities of utilization of resin-coated gravel include inside regular gravel pack, minifracturing, patching of damaged existing gravel-packed screens or liners through coiled or concentric tubing, and prebonded resin-coated gravel inner liner (SANTROL 1986 b). Following discussion of some general aspects, the main transitional stages between gravel packing and hydraulic fracturing are illustrated along the lines of a comparison of minifractures and rows of perforation tunnels.

5.8.2.2.3.1. General aspects

In a regular inside gravel pack mantle in the annulus between casing and screen or liner, resin-coated gravel forms a permanent downhole filter preventing movement or mixing with reservoir particles when opening or shutting-in the well. Similarly as applying for outside gravel packing of perforation tunnels, propping of short minifractures in unconsolidated formations requires preventive maintenance by resin bonding in order to inhibit embedment, particle mixing or contamination, proppant or formation flowback, and susceptibility to extreme pressure drawdown around the wellbore.

5.8.2.2.3.2. Minifractures vs. perforation tunnel rows

Minifractures can be compared in effect with a series of deep perforation tunnels (cf. also plate XI/8) which are concentrated within one plane and where the distances between the discrete tunnels are reduced to zero, with both types representing enlargement of effective wellbore diameter. The two-wing minifracture is morphologically almost equivalent to a double series of perforation tunnel rows in 180 degrees phasing. Gravel plugging of the perforation tunnel rows and proppant infilling of the minifracture has the same drainage and stabilization effect on the reservoir sand (cf. section 5.3.5.2.). Patching of existing gravel-packed screens and liners is a simple repair method where resin-coated gravel is injected into the holes of a screen or liner. After the particles bond together and those that are left inside the screen are removed, flowing of formation particles through screen or liner is prevented and holes containing damaged gravel packs are effectively repaired. Prebonded resin-coated gravel inner liners between screens represent good downhole filters for formation material by providing good protection against particle abrasion and/or erosion.

Aspects of placement of resin-coated gravel are also discussed by SCHROEDER & TUCKER (1974), SANTROL (1975), SINCLAIR & GRAHAM (1977, 1978), CONSTIEN & MAYER (1978); GRIFFIN, CONSTIEN & COLAVECCHIO (1978); SAUNDERS & McKENZIE (1979), KANAT (1980); UNDERDOWN, DAY & SPARLIN (1980); HICKEY, BROWN & CRITTENDEN (1981); GRAHAM, SINCLAIR & BRANDT (1982); RENSVOLD (1982); WILSON, NELSON & STADLEMAN (1983); MURPHEY, ROLL & WONG (1985); POPE, WILES & PIERCE (1987) and FAN (1988).

5.8.2.2.4. Carrier fluid and formation

Successful placement of resin-coated gravel in the perforation tunnels can only be achieved if the carrier fluid is adequately adjusted to formation fluid and matrix in order to prevent unwanted side directions that could disturb or deteriorate coating, migration and setting of the resin downhole (ECONOMIDES 1986). In order to create an environment in which the resin system will coat the gravel, a chelating agent and a surfactant have to be added to the water-based carrier fluid. Some aspects of interfacial fluid property adjustment and formation water cation chelation are outlined as follows.

5.8.2.2.4.1. Interfacial fluid property adjustment

The surfactant is necessary to adjust surface and interfacial properties of the aqueous fluid to the hydrocarbon environment as well as to coat the surface of the gravel grains prior to application of the resin system, because the epoxy resin system would not otherwise wet the sand in the aqueous milieu (ECONOMIDES).

5.8.2.2.4.2. Formation water cation chelation

Most fresh-water supplies contain small amounts of di- and trivalent cations which must be chelated in order to prevent them from competing with the cationic surfactant for sites on the negatively charged gravel grains and from inhibiting effective resin coating of the gravel grains which would lead to reduced strength of the aggregated pack and resin suspension in the carrier fluid (ECONOMIDES 1986). The chelating agent has to be used at the minimum level necessary, since it is possible for chelating materials to undergo cation exchange with formation clays and thus render them susceptible to migration. Chelating agent concentration has to be properly tested on field experiments in order to ascertain that all multivalent cations in the carrier fluid are chelated, but excessive amounts of chelating materials that could lead to fines movement in the reservoir are avoided.

Conventional application of resin bonding in external and/or internal gravel packs aims on gravel stabilization and aggregation partially together with additional formation sand consolidation, with the main goal being permeability retention for hydrocarbon productivity maximization. In some special cases, however, a reverse approach may be required including local patchy permeability destruction by invasion of plastic into the gravel pack for the reason of shutting-off intervals producing water instead of gas or oil (cf. section 5.10.6.).

5.8.3. Gravel packing of deviated wells

While traditionally gravel packing has been performed in vertical wells under ideal gravity conditions, progressive offshore oil and gas field development with a suite of production wells emanating from a central platform in a spider- or star-like pattern and including considerable deviation angles and distances with technically partially difficult and expensive directional drilling (cf. sections 3.8. and 4.8.6.1.) confronts the operators increasingly with the need of gravel packing operations in highly-inclined to sometimes even nearly horizontal well sections (WALLING & STABLER 1978; PEDEN, RUSSELL & OYENEYIN 1985; STRACKE 1987; SPREUX, GEORGES & LESSI 1988). As particularly in offshore operations, but also in environmentally protected or otherwise inaccessible areas (such as wildlife reservates and large towns), high-angle directional drilling is the only possibility of economically developing a field by a star- or spider-like radial borehole assemblage inserting at a central platform, gravel packing completions have to be adjusted to this unrenouncable technical feature of much larger financial scale.

5.8.3.0. General aspects

The main problem in deviated well sections is unstable or incomplete placement of gravel around the screen (GRUESBECK, SALATHIEL & ECHOLS 1979). The high operational costs offshore and in remote land locations give considerable incentive to develop gravel packing techniques for complicated inclined boreholes that have both high initial success ratios and the ability to provide sand-free production for sustained periods of time, in order to save expensive remedy jobs of premature failures. The most common troubles causing completion failure in deviated wells are premature screenout and after-pack settling of gravel (ELSON, DARLINGTON & MANTOOTH 1984).

Variables affecting gravel packing efficiency in highly-deviated holes are grain size, density and concentration of gravel, carrying fluid viscosity, slurry velocity, and directional control of slurry flow (RENSVOLD 1982). Circulation-type gravel packing models have shown that packing efficiency increases with reduced hole angle, lower gravel concentration, decreasing specific gravity of gravel particles, decreasing gravel particle diameter, higher fluid density, high flow rate and increasing resistance to fluid flow in the screen-tailpipe annulus (SUMAN, ELLIS & SNYDER 1983). Aspects of gravel duning, gravel compaction, perforation packing and string rotation, gravel settling, inverted gravel slurry circulation, viscous gravel packing slurries, and gravel-less slotted liners and pre-packed screens are sketched as follows.

5.8.3.1. Gravel duning

Up to hole angles of abt. 45 - 50 degrees, conventional gravel packing is still efficient due to gravity effects on the material, but the performance is already slightly disturbed due to no longer perfect vertical gravity slippage und sliding of the gravel (MALY, ROBINSON & LAURIE 1974; SUMAN, ELLIS & SNYDER 1983). In contrast to wells with borehole angles up to 45 degrees from vertical where gravel packing progress is from bottom to top under predominantly gravity influence thereby giving rise to a more or less continuous pack without voids because annular bridging is inhibited by gravitation, gravel packing proceeds from top to bottom at hole inclinations greater than 45 degrees (HALLIBURTON 1988) which bears increasing risk of void generation and premature annulus bridging due to partially insufficient gravitational effects in the inclined borehole. Annular gravel packing efficiency starts decreasing at abt. 45 degrees hole deviation and becomes a serious problem beyond 60 degrees wellbore inclination. Some aspects of extended-reach drilling shape, dune growth and migration, carrier fluid viscosity and gravel dune dynamics, gravel placement and equilibrium bank position, equilibrium velocity and gravel bank height, fluid velocity and dune movement, and fluid velocity and gravel transport are outlined as follows.

5.8.3.1.1. Extended-reach drilling shape

During gravel placement in highly-deviated wellbores with more than 45 degrees inclination, a gravel dune forms near liner top or crossover to the screen-casing annulus near the inlet of the wellbore by settling of solids in the annulus and as a consequence of bridging with either high- or low-viscosity carrier fluids (GRUESBECK & COLLINS 1978; GRUESBECK, SALATHIEL & ECHOLS 1979). As the degree of deviation increases, the percent pack in the annulus decreases, because gravitational forces tend to cause the gravel to prematurely settle out near the upper part of the zone to be gravel packed, with the result being dune formation at the upper end of the zone (STILES 1985).

Therefore extended-reach drilling should in case of necessity of subsequent gravel packing and also hydraulic proppant fracturing not be carried out in J-shaped manner, but if technically possible, preference should be given to a S-shape by decreasing inclination prior to penetration of the reservoir such that the borehole trajectory through the pay complex does not exceed 45 degrees deviation from vertical (BRUIST, JEFFERIS & BOTTS 1983; cf. sections 4.8.6.1. and 4.11.4.).

5.8.3.1.2. Dune growth and migration

The gravel dune restricts the area open to flow and increases the velocity of the slurry in the open channel above the bank. Dune height increases until fluid velocity in the open channel above the bank is high enough to transport gravel over the dune, to deposit it farther towards the bottom of the wellbore and to stop further net sedimentation of solids in the bank. The gravel dune enlarges until restriction causes fluid velocity to increase so that gravel is transported over the dune and towards its bottom. Fluid escaping from the gravel through the screen eventually lowers the velocity enough so that the channel bridges off completely, with gravel then being only able to be deposited above the bridge resulting in partial filling.

As height of the dune increases and the dune continues to descend down the wellbore, pressure drop along the open channel above the gravel bank may become high enough to force carrier fluids through the gravel into the screen-tailpipe annulus. When transport fluid is lost through the screen, fluid velocity in the open channel above the bank decreases, and gravel is deposited in the open channel, thereby bridging it off totally. Gravel can then be only sedimented upstream of the bridge, leaving the lower section exposed. The lower portion of the screen-casing annulus may thus not be packed, resulting in formation of an incomplete gravel pack with large void spaces in the lower interval.

5.8.3.1.3. Carrier fluid viscosity and gravel dune dynamics

Significant gravel duning and carrier fluid diversion into the screen-tailpipe annulus can be reasonably restricted, if not almost completely prevented, by using high transport fluid viscosities and high gravel saturations which could yield pressure gradients that limit the slurry flow rate (PEDEN, RUSSELL & OYENEYIN 1985). High-viscosity carrier fluids with high gravel concentrations provide good gravel transport and do not give rise to dune formation that originate only in low-viscosity transport media with low to moderate gravel saturations, but are practically unsuitable for use in completion intervals in wells with angles over 80 degrees from vertical (ELSON, DARLINGTON & MANTOOTH 1984). High-viscosity slurries with high gravel concentration also suffer from afterpack slumping which creates voids on the top side of gravel packs in deviated boreholes. Satisfactory transport and improved packing are achieved with lower carrier viscosity and saturation in conjunction with special liner/tailpipe (washpipe) geometrical constellations.

5.8.3.1.4. Gravel placement and equilibrium bank position

The final amount of gravel that can be placed in a deviated wellbore depends on the height of the equilibrium bank or dune formed under the chosen design conditions (GRUESBECK, SALATHIEL & ECHOLS 1979). If the equilibrium bank height is less than the inside diameter of the casing, gravel can be transported over the dune through the upper flow channel and deposited farther toward the bottom of the borehole until migration of the gravel stops at the lower end of the well. When the equilibrium bank extends all the way to the bottom of the wellbore, the carrier fluid enters the screen/tailpipe annulus and gravel deposits in the open channel upwards until the entire annulus is filled. Complete gravel filling of deviated wellbores, however, can only be accomplished if the fluid-loss rate is small, the pressure gradient in the tailpipe annulus is sufficient to force enough fluid to flow in the casing-screen annulus to build an equilibrium bank, and the gravel dune has an open channel at the top through which gravel can be transported all the way down to the bottom of the wellbore.

If the fluid-leakoff rate is excessive or if the pressure gradient in the tailpipe annulus is not sufficient, the equilibrium gravel bank will be equal to casing diameter, and formation of a gravel bridge causes incomplete gravel packing. Equilibrium velocity is the velocity required to carry particles over the top of the dune bank (ECONOMIDES 1986, GRUESBECK & COLLINS 1978). The velocity is high enough to prevent longitudinal concentration changes, but is not necessarily high enough to inhibit changes in saturation perpendicular to the direction of flow.

Above a stabilized bank, distinction can be made between a lower zone with high concentration of particles that jump and slide along the top of the dune, and an upper interval with low saturation of grains that decreases to zero at the top of the stream. As long as the system is in equilibrium and bank height is less than inside casing diameter, the gravel particles are carried over the dune and deposited further downhole until the bank extends all the way to bottom where the fluid enters the screen and gravel is deposited in the open channel from the bottom upward until the entire wellbore annulus is filled.

5.8.3.1.5. Equilibrium velocity and gravel bank height

If equilibrium velocity is disturbed, bank height is changing, with increasing velocity resulting in decreased bank height until a new equilibrium point is reached. If the flow rate is increased sufficiently, complete dispersion of the particles may take place with uniform distribution throughout the current area. As the dune grows towards the bottom of the well, fluid is being lost continually to screen-washpipe annulus and formation, and dune height increases with distance into the screen-casing annulus. If dune height increases to the same height as the casing, the pack bridges and filling then starts back up the screen-casing annulus leaving a void below that point.

Equilibrium velocity depends on particle concentration as well as on viscous, inertial, gravitational, frictional and buoyant forces that act on the grains during their transport over the equilibrium bank. The critical velocity decreases with diminishing hole deviation until an angle is reached at which the particles roll down the inclined bank and the critical velocity is zero, with this angle depending on angularity and effective weight of the particles in the fluid and being in many cases abt. 45 - 55 degrees from the vertical. Similar effects of proppant stratification and settling as observed during gravel packing of inclined boreholes occur during hydraulic fracturing of highly-deviated wells (cf. section 4.11.2.).

5.8.3.1.6. Fluid velocity and dune movement

While simple gel systems are suitable gravel carrier fluids in 90 % of the wells requiring mechanical sand control, highly-deviated boreholes with inclinations of 60 degrees or more from vertical are characterized by the drawback that simple gel systems do not give rise to complete support of gravel (HALLIBURTON 1988). This is most severe at slow pumping rates and with large pipe diameters, with fluid velocities being low and gravel settling out on the bottom side of the tubular goods forming dunes which can roll down the inside of the placement string until they form a bridge and sand off the string. In other cases, the dunes move up to the crossover tool where the high gravel concentration plugs the crossover ports and again sands off the placement string.

Crosslinked gel systems can considerably improve gravel placement in highly-deviated wells, because they are much more viscous than simple gels (cf. section 4.3.4.3.) and can fully support high gravel concentrations even in a static state, and as no gravel settling occurs, the chance of premature screenout is greatly reduced. Another advantage of crosslinked gels is that they add stability to a completion especially in open holes in highly-unconsolidated sands. On the other hand, crosslinked fluids provide only little fluid loss to the formation and therefore do not generate such tight gravel packs as do simple gel systems.

5.8.3.1.7. Fluid velocity and gravel transport

Restricting the clearance between liner and wash pipe increases the amount of gravel that can be placed in the casing-liner annulus in highly-deviated boreholes with low-viscosity carrier fluids, but high fluid velocities are required to maintain the gravel in suspension (RENSVOLD & DECKER 1978). These rates may not be attainable in highly-deviated open-hole completions when fluid losses to the fully exposed formation along the borehole wall are taken into account. High fluid velocity also gives rise to intermixing of gravel and formation sand which considerably reduces pack permeability (cf. section 5.3.1.1.1.), particularly in highly-deviated open-hole completions when a gravel dune builds up and develops a high-velocity flow channel above the dune in the borehole-liner annulus.

The scouring effect of this high slurry velocity on the exposed formation could easily entrain sufficient reservoir sand in the pack to seriously reduce its permeability. On the other hand, low pumping rates which are enabled by thick high-viscosity carrier fluids containing high gravel concentrations would not only diminish the scouring action of the slurry, but would also reduce fluid loss from slurry to formation and help to maintain a more uniform slurry velocity throughout the annulus interval being packed.

5.8.3.2. Gravel compaction

In addition to gravel duning, the inability to adequately compact gravel causes problems in highly-deviated wells. In oblique boreholes, the blank segment above the slotted liner or wire-wrapped screen and the perforated pay zone or open-hole formation wall, respectively, does not provide reserve gravel storage for later settling, because gravity will not move gravel along a highly-inclined to even almost horizontal wellbore, and settling gravel thus lies on the lower borehole wall and by moving not towards the well floor, but segregating and compacting towards the casing mantle gives rise to origin of open channels along the top side of the screen-casing annulus, in contrast to vertical holes where gravel settling leads to compensation of the 15 % additional blank liner section above the perforated interval by downhole movement of the compacting gravel mantle which thus exposes the upper reserve segment and enables it to start slumping.

Annular and perforation packing efficiencies have little direct interaction, because perforation packing depends primarily on fluid-leakoff rate, phase location (top or bottom of the casing in deviated holes) and gravel concentration, whereas annular packing depends predominantly on fluid transport efficiency (viscosity and density), gravel size and concentration, mechanical configuration, and slurry flow rate (ECONOMIDES 1986). Annular packing efficiency increases with smaller gravel sizes, higher carrier fluid density, larger wash-pipe, and increased screen-casing annular flow velocity. Comments on flow resistance created by flexible baffles, large-diameter washpipes and other aspects, fluid viscosity and casing-liner relationship, and gravel pack quality improvement are offered as follows.

5.8.3.2.1. Flow resistance created by flexible baffles

Sufficient compaction of gravel can be obtained throughout a highly-deviated interval when flexible rubber baffles are spaced along the washpipe inside the screen which create a flow resistance on fluid moving in the liner-washpipe annulus by restricting the clearance, thereby forcing sand-laden fluid in the open hole to flow to the bottom of the screen. Placing a tight bladder seal inside the screen or deformable baffles spaced at regular intervals along the tailpipe can divert fluid flow inside the annulus. Some aspects of baffle spacing and stiffness as well as rubber discs are outlined as follows.

5.8.3.2.1.1. Baffle spacing and stiffness

The baffles maintain velocity in the screen-open hole annulus and distribute fluid flow between the tailpipe-screen annulus and over the gravel dune which allows a stabilized bank to propagate all the way to the bottom of the wellbore, thus preventing formation of bridges that stop downward progression of gravel (MALY, ROBINSON & LAURIE 1974; GRUESBECK, SALATHIEL & ECHOLS 1979; SUMAN, ELLIS & SNYDER 1983). The flow in the channel over the dune is maintained and the current in the screen-washpipe annulus is restricted (MALY, ROBINSON & LAURIE 1974; SCHROEDER 1987). Baffle spacing and stiffness are critical for successful gravel packing. If the baffles are spaced too far apart, a gravel bridge can form which leads to incomplete gravel packing, and if the baffles are spaced too close together or are too stiff, excessive pressures can develop which may result in loss of circulation of the carrying fluid and bridging of the gravel in the wellbore. Crossover tools also help to reduce sand placement problems in high-angle boreholes especially in terms of gravel segregation (SAUCIER 1974).

5.8.3.2.1.2. Rubber discs

RENSVOLD & DECKER (1978) investigate the effects of carrying fluid and borehole deviation in relation to liner-wash pipe annulus clearance or placement of rubber discs in the liner-wash pipe annulus. Rubber disc spacing in borehole inclinations up to 60 degrees can be considerably varied in the range of a couple of feet without having major differential influences. Gravel packing efficiency is also improved when the slurry is directed to the lower end of the interval by confining fluid returns through a bottom tell-tale screen and tail pipe (RENSVOLD 1982; cf. section 6.3.5.).

5.8.3.2.2. Large-diameter washpipes and other aspects

Another mechanical means for achieving satisfactory gravel pack results in highly-deviated wells is the use of an oversized large-diameter washpipe (GRUESBECK, SALATHIEL & ECHOLS 1979; SHRYOCK 1980; ELSON, DARLINGTON & MANTOOTH 1984) which also aims on increasing the resistance of fluid flow in the annulus, thereby maintaining more efficient gravel transport. The diameter of the tailpipe needed to gravel pack deviated boreholes depends upon wellbore geometry, fluid and solid properties and flow rate. The discussion as follows focusses on aspects of fluid velocity and viscosity as well as prevention of fluid channelling through the screen.

5.8.3.2.2.1. Fluid velocity and viscosity

Restricting clearance between slotted liner and wash pipe materially increases the amount of gravel that can be placed in the casing-liner annulus in highly-deviated boreholes when a thin carrying fluid is used (HALLIBURTON 1988). High fluid velocities, however, are required to maintain gravel in suspension in thin fluids which may not be attainable in deviated open-hole completions if considerable fluid losses to the formation occur.

High fluid viscosity has also a detrimental effect on intermixing of formation sand and gravel in open-hole completions (cf. section 5.3.1.1.1.), because as the gravel dune builds up in a highly-deviated hole, a high-velocity flow channel develops above the dune in the borehole-liner annulus. The scouring effect of this high slurry velocity on the exposed formation could easily entrain sufficient reservoir sand in the pack to seriously reduce its permeability. On the other hand, low pumping rates which are made possible by using a thickened carrier fluid containing high gravel concentrations would not only reduce the scouring action of the slurry, but would also reduce fluid loss from slurry to formation and help to maintain a more uniform slurry velocity throughout the annulus interval being packed.

5.8.3.2.2.2. Prevention of fluid channelling through the screen

A large outer-diameter tailpipe is desirable to prevent fluid channelling through the screens (ZALESKI & SPATZ 1988). The same applies for a large inner-diameter washpipe in order to minimize pressure on the formation while circulating. Insufficient radial clearance between the screen outer diameter and open hole or casing inner diameter is a primary cause of premature bridging. Therefore a screen and washpipe assembly is required that provides sufficient clearance to prevent bridging and yet does not impose too much backpressure through the washpipe that will increase leakoff inordinately. Screen and washpipe should be selected before the well is drifted to ensure dimensional and hydraulic compatibility with the hole and reservoir being drilled.

5.8.3.2.3. Fluid viscosity and casing-liner relationship

In order to perform an adequate gravel packing job in highly-inclined wells, special liner-tailpipe geometrical constellations are required in conjunction with the optimal slurry properties. Small dimensions of liner-casing annulus can magnify the negative effects of gravel settling, because very little compaction is needed to expose the liner. Extensive settling can also occur by ineffective slurry dehydration at screenout or sandoff followed by carrier-fluid viscosity breakback. Baffled or oversized washpipes as mechanical means for gravel pack enhancement, however, have in most cases been used with only limited success.

Better elimination of gravel duning and premature screenout is accomplished by crosslinked polymer gravel packing fluids with almost zero gravel fall rates (UNDERDOWN, DAS & NGUYEN 1984). Given sufficiently high viscosity, crosslinked polymer fluids transport high concentrations of gravel through the tubing at any angle with only little or no settling of gravel even in deep and/or highly-deviated wells. Adequate fluid loss insures compact packing of gravel against the formation face. Zero gravel fall rate of the gelled fluid is ascertained by its thixotropic nature.

5.8.3.2.4. Gravel pack quality improvement

Gravel pack quality in highly-deviated wells improves with decreasing gravel concentration, decreasing particle diameter, decreasing particle density, increasing fluid density, increasing flow rate and decreasing current capacity between slotted liner and washpipe, with the resistance to flow in the washpipe annulus being the most important factor (GRUESBECK, SALATHIEL & ECHOLS 1979). The impact of wash pipe diameter on gravel placement as a function of borehole inclination is also investigated by RENSVOLD & DECKER (1978). Viscous slurry packing is a suitable technique for high-angle wells and long perforated intervals, although high-rate pressure packing can create costly pack contamination or result in premature bridging (WALLING & STABLER 1978).

Increasing gravel concentration increases flowing pressure gradient above the gravel dune and causes more current in the washpipe-screen annulus thereby increasing the tendency for preventive bridging (ECONOMIDES 1986). Increased screen length results in greater flow inside the screen and greater leakoff into the perforations which cause increasing dune height and may trigger bridging.

5.8.3.3. Perforation packing and string rotation

Particular attention in deviated wells has to be paid to packing of perforation tunnels. While the perforation holes at the lower side of the inclined wellbore are still reasonably filled, considerable problems occur at the upper side. Improvements of tunnel plugging with gravel can be made by adjusting diameter and length of the perforations (cf. section 5.6.1.) as well as specifically selecting a suitable firing technique (cf. sectinon 5.6.4.). Carrier fluid viscosity also has an important impact, with perforation packing appearing to be controlled by supply of gravel and magnitude of gravitational forces on the gravel particles at all but the highest leakoff rates. Annular and perforation packing conditions interact through screen size which is dictating gravel particle trajectory and speed at the entrance of the perforation tunnels, but principally perforation packing is primarily controlled by the perforation characteristics and not by the design conditions for the annular pack (PEDEN, RUSSELL & OYENEYIN 1985). Comments are offered as follows on perforation phasing, perforation prepacking, and liner vibration and drill pipe rotation.

5.8.3.3.1. Perforation phasing

Perforation phasing is also a critical issue in highly-deviated to horizontal boreholes to be gravel packed (ZALESKI & SPATZ 1988). The two basic approaches are perforating of either the entire or the lower half circumference of the casing. Full phasing creates a better leakoff path while placing a gravel pack thus resulting in better completion, but a major disadvantage of shooting the entire circumference is formation settling from the high-side perforations which could stick the guns, prevent a gravel pack assembly from getting to bottom, and fill the low-side perforations thereby negating their existence (NINI & OWEN 1983). Some aspects of single- vs. double-side perforating as well as packing efficiency in top-side perforations are illustrated as follows.

5.8.3.3.1.1. Single- vs. double-side perforating

As a consequence of the aforementioned drawbacks, the conservative approach of perforating the low side only is often preferred, with in most cases shooting of two rows of holes at 60 degrees to the left of the low-side centerline being carried out. While shot density and gravel fillup may be reduced, the perforation tunnels will not be subjected to fill from top-side holes and will not be plugged with surged debris either, because the latter material will collect at the very bottom of the casing. Other reasons for restricting perforation holes to the low side of the casing are avoiding of mixing of formation sand and packing gravel which would destroy permeability, top-side channels that may develop will not endanger the gravel pack success, and elimination of reservoir sand falling into the borehole due to gravity and bridging (NINI & OWEN 1983).

5.8.3.3.1.2. Packing efficiency in top-side perforations

Acidizing often gives rise to productivity decline in gravel packed wells due to dispersion and plugging with formation fines and as a consequence of voids caused by incomplete perforation packing or acidizing (McLEOD 1986; cf. section 5.11.4.). Experimental work on perforation orientation angle and perforation pack efficiency has shown that very few of the top-side perforations in deviated wells remain packed, because following an original plugging to some degree, the gravel is falling out of these perforations into voids in the annulus after screenout termination of gravel implementation (SCHROEDER 1987). Perforation packing within inclined boreholes is primarily affected by an angular relationship other than perpendicular or parallel between perforation tunnels or cavities other than spherical and the principal in-situ stress axes (WEISSENBURGER, MORITA, MARTIN & WHITFILL 1987).

5.8.3.3.2. Perforation prepacking

Perforation prepacking with gravel and gel is another method for reducing excessive fluid leakoff (ZALESKI & SPATZ 1988; cf. section 5.6.7.). The gel invades the formation, slows down, increases in viscosity, and the perforations are filled with gravel. Perforation prepacking is particularly beneficial in high-permeability zones of depleted reservoirs where overbalance and subsequently fluid loss are very high. A less desirable alternative of reducing leakoff is to introduce solids into the perforations at the formation face, but while fluid loss is easily controlled, the perforations become difficult to fill, and the removal of the solids from the entire interval is difficult and expensive.

5.8.3.3.3. Liner vibration and drill pipe rotation

Gravel packing with liner vibration (cf. section 5.8.4.) also is a suitable means of gravel pack improvement in deviated wells (SOLUM 1984). Rotating the drill pipe throughout the packing operation in high-angle holes prevents gravel from laying on the bottom side of the drill string and building up, causing bridges and early pack-off of the crossover tools as well as between liner and casing. Liner vibration requires rotation of the drill pipe during the whole placement operation, and all the gravel inserted in the fluid stream at the top of the well stays in the fluid flow to reach the producing zone.

5.8.3.4. Gravel settling

Factors affecting gravel placement and settling include fluid viscosity, fluid density, pumping rate, gravel grain size, gravel concentration, gravel density, wellbore deviation, and downhole assembly (cf. also section 4.12.2.). In addition to string rotation, perforation tunnel packing in highly-inclined wellbores can be enhanced by diminishing gravel settling rates in the transport fluids. Gravel settling in the carrier fluid can be a serious problem in gravel packing of highly-deviated boreholes, because a particular concentration gradient from the high to the low side of the screen-casing annulus leads to insufficient plugging of the upper perforations (HUDSON & MARTIN 1988). Gravel settling and gravel placement can be reduced and increased, respectively, by decreasing the particulate density/transport fluid density ratio (cf. section 5.3.3.). Low particulate density/carrier fluid density ratios are obtained by using either gravel in high-density brines or low-density gravel packing materials in commonly applied transport fluids.

Gravel settling in the slurry as it flows down the screen-casing annulus causes a saturation gradient from the high to the low side of the flow area. When fluid leaking off through the perforations has only low gravel concentration or almost no gravel at all such as occasionally occurring on the top side of the annulus in highly-deviated wells, it is very difficult to completely pack these perforations. Since packing efficiency in the perforation tunnels area is greater if gravel saturation in the slurry flowing into the perforations is higher, it is important to reduce gravel settling and to maintain a high leakoff rate as the carrier fluid flows through the formation. Some comments are offered as follows on low particulate density/carrier fluid density ratio and impact of low-density gravel packing materials.

5.8.3.4.1. Low particulate density/carrier fluid density ratio

Regardless of carrier fluid density, the settling rate of the grains decreases as the particle density/transport fluid density ratio decreases, viscosity of the slurry increases, and grain concentration increases (cf. section 5.3.3.). Gravel settling is thus reduced with high-viscosity carrier fluids and systems with low particulate density/fluid density ratios. High-viscosity transport fluids do not give the suspension properties required in mixing/pumping equipment or under downhole conditions, and polymer saturations required to yield high viscosity lead to impaired fluid leakoff and false screenouts as well as a reduction in formation permeability.

Low particulate density/fluid density ratios can be obtained by using high-density fluids or low-density particles. High-density fluids create overbalance which leads to severe fluid loss and well control problems. The low particle density/carrier fluid density system utilizes buoyancy forces to counter gravitational forces on gravel particulate rather than high fluid viscosity to reduce grain settling (HUDSON & MARTIN 1988). Experimental work shows that complete packing of perforation tunnels throughout the screen-casing annulus in highly-deviated boreholes can be achieved by using light particles with a preferential specific density of 1.5 - 1.8 g/cm3 in combination with conventional gravel-pack carrier fluids (cf. section 5.3.3.).

5.8.3.4.2. Impact of low-density gravel packing materials

The low-density particles considerably effect friction pressure in turbulent current, while their effect in laminar flow is only subordinate. Improved suspension given by low-density grains enables to use lower fluid viscosity and lower pumping rates during placement which decreases the friction loss generated by the slurry flow and diminuish the pressure gradient between the screen-casing annulus and the screen-washpipe annulus. A lower pressure drop across the screen reduces slurry dehydration through the screen and similarly to using a larger washpipe, the liquid flow rate through the screen during placement is reduced.

As slurry dehydration through the screen increases gravel concentration in the screen-casing annulus until screenout occurs, slowing of this dehydration process results in a more uniform slurry saturation in the annulus during placement thereby effectively reducing premature screenout. The gravel pack system utilizing conventional fluids with low-density grains improves particle transport without the problems associated with high polymer loading or high-density fluids and is therefore the preferred method for improving gravel packing efficiency in highly-deviated and horizontal boreholes.

5.8.3.5. Inverted gravel slurry circulation

Gravel placement, settling and compaction in deviated wells can be considerably improved by inverted gravel slurry circulation from bottom to top similarly to the procedure of casing cementing (STILES 1985; STILES, COLOMB & FARLEY 1986). In case of normal gravel slurry circulation from top to bottom, the percent pack in the annulus decreases as the degree of deviation increases, because gravitational forces tend to cause the gravel to prematurely settle out near the upper part of the zone to be gravel packed, thus resulting in dune formation at the upper end of the interval.

As the dune enlarges and descends down the annulus, more and more of the carrier fluid is diverted through the screen by fluid flowing over the dune which causes slurry velocity to decline that leads to additional gravel settling, with this process continuing until the dune completely blocks flow to the lower portion of the annulus. When this shut-off occurs, the slurry fluid is diverted through the top section of the screen and no more slurry can be placed in the annulus which will adequately pack the upper section but leave a void in the lower portion. The discussion as follows illustrates some aspects of conventional vs, inverted gravel packing, void prevention with inverted slurry circulation, and comparison to normal and reverse circulation packing.

5.8.3.5.1. Conventional vs. inverted gravel packing

In contrast to conventional or normal gravel packs where gravel is pumped from the top of the interval to the bottom, inverted gravel packing reverses the flow and gravel is circulated from the bottom of the interval to the top (STILES 1985; STILES, COLOMB & FARLEY 1986). Thus the gravity component relative to the fluid component is reversed. In conventional gravel packs, the vertical component of the fluid is a negative vector pointing downwards, and gravity effect on particles is also a negative or downward vector. As the slurry is pumped into the screen-casing annulus, the negative vector from the fluid flow is added to the negative gravity effect, with this combination of forces accelerating gravel settling rate and indirectly causing occurrence of the duning phenomenon. When the slurry path is reversed, the vertical fluid component is now an upwards-oriented or positive vector which tends to cancel out the downwards-directed negative vector created by gravity and therefore slows settling rate.

The degree of this slowing process depends on hole angle, fluid viscosity and density, gravel concentration, and gravel grain size. Reduction of the settling rate diminishes or completely eliminates gravel dune formation. As the gravel is being constantly bumped up the hole by fluid force, it is not allowed to settle out, but is kept in the fluid. Instead of settling out and being deposited on the low side of the casing, the gravel moves completely to the top of the screen in plug flow and does not accumulate at the bottom of the zone.

5.8.3.5.2. Void prevention with inverted slurry circulation

Inverted slurry circulation attacks the problem of void formation in the gravel pack also in low-deviation boreholes. Movement through the interval in plug flow enables the gravel to reach all parts of the section before a premature screenout from a more permeable zone could occur (STILES 1985; STILES, COLOMB & FARLEY 1986). In addition, more gravel can be placed in the perforation tunnels independent from borehole inclination. In conventional packs with normal circulation, the gravel is constantly falling ahead of the slurry and is deposited at the bottom of the interval. As this gravel builds up, the pressure exerted through the gravel and out into the perforations is reduced.

Inverted circulation does not include gravel falling ahead of the slurry and accumulating at the bottom of the interval which keeps the slurry mobile for a longer time and allows it to be squeezed into the perforations. As the gravel is moving from the bottom of the zone to the top, more complete dehydration exists across the entire screen length, thus eliminating the need for several joints of blank tubing above the production screen. This technique is also useful in lifting any formation material that might have sloughed in around the screen while running in the hole. The formation material would be circulated up the screen-casing annulus until it was above the productive interval, and the higher-permeability gravel pack then would be circulated around the production screen.

5.8.3.5.3. Comparison to normal and reverse circulation packing

Inverted gravel slurry circulation (STILES 1985; STILES, COLOMB & FARLEY 1988) is quite similar to the normal circulation method which was the first technique of gravel packing when mechanical sand control was introduced to the hydrocarbon industry more than 60 years ago (SOLUM 1986; cf. section 5.3.2.3.2.). Normal circulation was performed by pumping gravel down the tubing string, through a float collar at the bottom of the liner, and back up the annular space. Clogging of the float collar and the inability to pack the liner top were major drawbacks of normal circulation, and therefore with time reverse circulation was initiated which consists of pumping the gravel down the annular space and around the outside of the liner.

As the gravel reaches its setting point, the carrier fluids would continue downhole and return to the surface through a perforated nipple at the liner bottom. After correction of early discrepancies such as bridging off due to defective downhole tools and surface techniques, reverse circulation became the model of current gravel placement procedures. The inverted gravel slurry circulation for the special application of deviated-well gravel packing (STILES 1985; STILES, COLOMB & FARLEY 1986) thus at the bottom of the line represents revival of the original normal circulation method and its utilization in an improved version (cf. section 5.3.2.3.2.).

5.8.3.6. Viscous gravel packing slurries

Good gravel suspension in highly-deviated wells can be achieved by application of viscous gravel packing slurries which are composed of high-viscosity water-based polymer gels containing high gravel concentrations (FLANIGAN 1979).

Pad spacers both precede and follow the slurry in order to prevent contamination of the slurry with wellbore or formation fluids. The slurry often exhibits a roping effect (cf. section 5.9.1.3.3.) by taking the shape of the smallest restriction it passes through as it proceeds down the drill pipe and into the opened hole. In high-angle wells, conventional gravel packing tends to produce looser packs yielding lower calculated fill-ups, whereas the viscous slurry packs provide the tendency of the gravel to restress the formation and the packs become tighter than before. The use of a viscous slurry also reduces fluid turbulence and increases gravel carrying capacity which decrease erosion and essentially pack from the bottom up (cf. also sections 5.8.5 and 5.8.7.). Gravel settling to bottom of drill pipe and gel passing over the gravel make a screenout almost impossible.

5.8.3.7. Gravel-less slotted liners and pre-packed screens

Slotted liners without gravel pack (cf. section 5.2.3.6.1.) and pre-packed screens (cf. section 5.9.6.1.) are frequently preferred with respect to gravel packing in highly-deviated to horizontal boreholes (ZALESKI & SPATZ 1988). Wellbore collapse, however, is generally considerably reducing productivity although formation sand influx will still be effectively stopped. The use of slotted liners without gravel pack and pre-packed screens to control sand and to maintain good production must be carefully applied because the consequences of fines migration and plugging will be difficult if not impossible to reverse. On the other hand, the use of pre-packed screen assemblies eliminates failure due to minor voids in the pack, whereas conventional wire-wrapped screens optimize dimensional criteria but are susceptible to total failure if even a small section is left uncovered.

5.8.3.8. Other aspects

Aspects of gravel pack operations in highly-deviated wells are also discussed by MALY & ROBINSON (1972); LYBARGER, SCHEUERMAN & WILLARD (1974); MALY, ROBINSON & LAURIE (1974); HOLMAN (1975), NOVOTNY & MATSON (1975), FARMIGIANO (1976), ANDERSON & HUTCHINSON (1978), RENSVOLD & DECKER (1978), WALLING & STABLER (1978); GRUESBECK, SALATHIEL & ECHOLS (1979); SHRYOCK (1980), HODGE (1982); PEDEN, RUSSELL & OYENEYIN (1982, 1984, 1985); TORREST (1982 b); BALLEGOOYEN, GIAP & SENG (1983); NINI & OWEN (1983); ELSON, DARLINGTON & MANTOOTH (1984); HAILEY (1984), NADERI (1984); UNDERDOWN, DAS & NGUYEN (1984); OYENEYIN & PEDEN (1985) and SCHROEDER (1987). TOMA, LIVESEY & HEIDRICK (1986); DICKINSON & ANDERSON (1987) and GROSSMANN, HÖFLING & MARX (1988) comment on gravel packing in horizontal boreholes, and hydraulic fracturing in highly-inclined wells is analyzed by YEW & LI (1987; cf. also section 4.11.). STILES, COLOMB & FARLEY (1986) describe a gravity-assisted gravel packing system. In contrast to the conventional method of first perforation and second gravel pack, HAILEY, DONOVAN & SICKLE (1985) introduce an integrated system requiring only one trip into the borehole for performance of both operations. SCRIMGEOUR & SHEPHERD (1988) report successful gravel packing from a semisubmersible even in highly-deviated wells in the North Sea.

5.8.4. Gravel packing with liner vibration

SOLUM (1984) presents sand control optimization by performance of gravel packing with liner vibration which provides a substantial difference in gravel compaction. Standard gravel packing results in cubic aggregation of the particles, whereas the liner vibration technique leads to hexagonal packing of the gravel which gives a denser and more stable pack with abt. 15 - 25 % more gravel in the same volume. The outline as follows focusses on compensation of initial undercompaction by blank liner section, prevention of after-pack gravel settling, achievement of optimum packing density by liner vibration, and gravel and perforation washing.

5.8.4.1. Compensation of initial undercompaction by blank liner section

Conventional flow-type gravel packing which is widely applied in hydrocarbon exploitation since over 50 years is characterized by an after-pack settling of the gravel due to its initial undercompaction which is commonly compensated by infilling the annulus all along the perforated interval or open-hole segment plus a reserve blank section above it of abt. 10 - 15 % of the liner length or barefoot interval (gravel settling can be in extreme cases up to 25 %), in order to guarantee complete coverage of the perforated section by gravel also after diminution of original gravel height as a consequence of subsequent settling by compaction (cf. also section 5.3.2.4.3.). Extra gravel above the wire-wrapped screen or slotted liner is also necessary to compensate for a possible loss of gravel from the annulus through the perforations during production (RENSVOLD 1982).

A circulating pack requires 90 - 120 ft of minimum blank section in order to yield a 60 ft reserve interval of gravel above the screen which is considered as the minimum desirable extra interval (LEDLOW, SAUER & TILL 1985). The required reserve height increases substantially if a large-diameter open hole is underreamed (cf. section 5.10.2.). A straight squeeze pack also needs more blank section to allow for incomplete slurry dehydration or gel-out. A combination of squeezing and circulating permits better placement and minimizes the length of the required blank section. The number of times the screen length necessary for the blank interval is a function of gravel concentration in the slurry (cf. also section 5.3.2.4.3.).

5.8.4.2. Prevention of after-pack gravel settling

After-pack gravel settling provides disturbance of the inside gravel pack by rotation and movement of grains, slipping of the package along borehole wall or perforated casing, and sliding along slots and interwire spacings in screen and liner (LEDLOW, SAUER & TILL 1985). Excessive after-pack gravel settling can provoke partial efflux of gravel from the perforation tunnels if their mouths are not continuously tightly covered by the inside gravel mantle, but are subjected to slight underbalance as a consequence of gravel slumping which may last only for short time, but nevertheless may be already sufficient to enable the hydrocarbon stream to spill out some material, with gravel evacuation of the perforation tunnels thus being the worst possible gravel pack damage (cf. sections 5.6.1. and 5.6.4.).

The wellbore wall in open holes may be loosened surficially by sliding gravel, and formation sand may become intermixed with the moving and rotating gravel grain population thereby deteriorating its permeability. Slots and interwire spacings in screen or liner may be mechanically damaged due to grinding by the slumping gravel. As all these effects are undesirable, gravel settling after installation of the pack and after start-up of hydrocarbon production has to be reduced to a minimum. The effect of after-pack settling, however, can only be prevented by converting the grain fabric from cubic to hexagonal which is achieved by liner vibration with sufficient amplitude during the placement period that eliminates subsequent compaction (SOLUM 1984).

5.8.4.3. Achievement of optimum pack density by liner vibration

Liner vibration during gravel pack installation and settling reduces pore size or interstitial tube size of the gravel which creates stability and gives substantially more control with the reservoir formation sand than conventional cubic pack (SOLUM 1984). When true hexagonal packing is achieved by enhanced compaction due to liner vibration, there is absolutely no movement of the gravel after the flow pack is completed because further compaction is not possible, with thus the primary gravel pack not settling at a later date in the production history of the well. Compaction depends on the effective vertical length of liner vibration and on the rotating type to transmit the vibration energy through the fluid to the gravel. Some aspects of liner vibration in gravel packing as well as for pump protecting pre-packed liners and in primary cementing are outlined as follows.

5.8.4.3.1. Gravel packing

SOLUM, RAMEZANI & MIMS (1988) investigate effects of vibration on degree of compaction and rate of fluid loss to the formation during the gravel placement process. Vibration substantially reduces the degree of fluid loss required to fill the perforation tubes with the packing agent, and an additional 10 % amount of gravel can be placed in the perforation tunnel with the help of vibration. At the minimum fluid loss rate required to fill the perforation hole with vibration, no gravel could be placed in it without vibration.

While original liner vibration aims on improvement of packing density of the primary gravel pack during course of its installation prior to onset of hydrocarbon production, wireline tools capable of vibration are suitable instruments for secondary application for repair purposes at any later stage of well life between post-completion quality monitoring by logging and workover operations in case of subsequent problems. The wireline vibration tools transmit agitation to the fluid by reciprocation which passes vibration pulses to the annulus, and excitation of the gravel together with gravitational forces enables settling and compaction of the gravel (DRESSER ATLAS 1987). Some instruments are even able to perform controlled explosions to emit stronger pulses in order to shake down the pack (SCHLUMBERGER 1988 a; cf. section 6.3.1.2.).

5.8.4.3.2. Pump protecting pre-packed liners

Another application of liner vibration in sand control is achievement of hexagonal gravel packing in pump protectors representing a special type of pre-packed liners for trapping and retention of flour-like formation fines (SOLUM & RAMEZANI 1987; cf. section 5.9.6.1.). The pump protectors consist of a slotted outer sleeve and a slotted base pipe which are gravel packed as a filter medium, with interstitial tube size being selected such that it is smaller than the powder-size material. Generation of hexagonal packing by vibration prior to pump protector installation in the well reduces interstitial gravel tube size and thus permits choice of larger gravel grain diameters.

5.8.4.3.3. Primary cementing

Similarly as gravel settling and compaction can be enhanced by liner vibration, primary cementing can be improved by casing vibration during cement curing time (COOKE, GONZALEZ & BROUSSARD 1985). The effect of vibration is to decrease or eliminate gel strength of the curing cement slurry which then flows downward to compensate for shrinkage and reestablishes the hydrostatic head of the cement. Casing vibration can prevent pore fluid entry into the annulus before the cement cures and increase the radial stress at the interface between formation and cured cement.

5.8.4.4. Gravel and perforation washing

Conventional methods of increasing gravel compaction by settling include perforation washing and subsequent gravel repacking, but this treatment always results in productivity deterioration by reduction of flow rates, because the washing of the pack commingles the gravel with the formation sand and decreases the permeability of the pack (cf. also section 5.2.3.8.). Liner and gravel pack washing for improving gravel settling can mix the gravel with the formation sand and any void areas which may have occurred due to settling may be filled with formation sand and fines (FLANIGAN 1979). Both water and jet perforation washing cannot convert the cubic grain fabric into a hexagonal framework and thus remain very ineffective in terms of gravel settling, let alone the detrimental consequence on hydrocarbon production. As washing also results in fluffing of the gravel which may even lead to decreasing compaction, perforation washing should be avoided after vibration gravel placement.

Gravel pack washing can even result in reversing of the gravel out of the packed interval through the crossover (ELSON, DARLINGTON & MANTOOTH 1984). Comments on perforation cleanup are also offered by BONOMO & YOUNG (1983). Proper gravel settling is also very important for wells where water or steam injection pressure is exceeding overburden pressure, because unless the gravel is completely consolidated, it would break down. All these aspects argue for widespread application of liner vibration in order to optimize downhole performance of the gravel pack (SOLUM 1984).

5.8.5. Concentrated gravel slurry pumping

SPARLIN & BOND (1969) and SPARLIN & COPELAND (1972) introduce the concept of adding resin as tackifying agent to high-viscosity slurries with high gravel saturation in order to promote plug flow of the slurry by making the individual gravel grains sticky. Some comments are offered as follows on better gravel transport and tighter gravel packing as well as tackifying action of intermixed resin.

5.8.5.1. Better gravel transport and tighter gravel packing

In addition to minimization of mixing of gravel and formation sand (cf. section 5.2.2.2.), other advantages of using high-viscosity transport media are the tighter packing of the gravel as it is squeezed through the pack and the better carrying properties which provide good radial and vertical gravel distribution (SPARLIN & BOND 1969, SPARLIN & COPELAND 1972). The greater fluid drag on the gravel particles with respect to low-viscosity fluids is much more effectively pushing the loose reservoir sand back from the wellbore, and thus tighter packing the gravel in place, much better suspending the gravel, and thereby reducing the amount of fluid loss to the formation. Fluid viscosity has to be high enough to suspend the gravel but also sufficiently low to allow for pumping of the slurry. If the viscosity of the slurry is too low, a screenout inside the casing could be the result, and if the viscosity is too high, too much friction pressure might occur.

5.8.5.2. Tackifying action of intermixed resin

In contrast to consolidated pack installation or artificial borehole wall placement (cf. section 5.8.1.) as well as pumping of resin-coated gravel (cf. section 5.8.2.), the amount of resin intermixed into the concentrated gravel slurry is so low that the resin does not solidify and is washed out slowly with production fluids, but exerts its stabilizing role during downhole pumping of the slurry where sufficient cohesion is given to secure plug flow (SPARLIN & BOND 1969, SPARLIN & COPELAND 1972). The tackifying resin is not easily washed off of the gravel during blending and placement, but holds the gravel in position while the screen is being washed in and helps to restrict water but not oil production after completion of the job.

The resin is added in bulk to the slurry consisting of carrier fluid and suspended conventional gravel that is not surrounded by resin pellicles, in contrast to curable resin-coated sand or proppants where the individual gravel grains are enveloped by resin films that start to become active by thermoreaction once placed in the reservoir, which in turn is in contrast to the concentrated gravel slurry where the intermixed resin component begins its tackifying action already during injection of the mixture into the well. Gravel concentrations up to 20 lbs/gal of fluid have been successfully handled by concentrated slurry operation (cf. also section 5.3.1.1.). The slurry is pumped with slow rates until screenout occurs. Aspects of slurry packing of deep high-pressure wells are outlined by CLAWSON & SPARLIN (1975).

5.8.6. Density slurry and foam gravel packing

SHRYOCK, AHMAD, MELOY & KENT (1988) present reduced density slurry and foam gravel packing in steam-drive operations in shallow underpressured unconsolidated sands. Foam is a viable fluid for gravel packing where lost circulation prevents the use of higher-density fluids. Foam gravel packing is also outlined by ELSON & ANDERSON (1982).

5.8.7. Viscous polymer gravel packing and deposit buildup

Gravel packing with viscous polymer solutions is in many ways different from packing with suspensions in water (TORREST 1982 a). Polymer suspensions are much more concentrated so that gravel can be placed with less fluid and pumping time. Dense concentrated slurries exhibit complex dispersal and settling patterns within the borehole. During placement, interactions with the formation are different and deposits originate within the wellbore at the perforations and on the wire-wrapped screen. The slurry is pumped away through the perforations into the reservoir, and after the gravel bridges against the formation material and fills the perforations, deposits are generated within the borehole. Viscous slurry packing has particularly good results in high-angle wells and long perforated intervals (WALLING & STABLER 1978), but on the other hand, high-rate pressure packing can create costly pack contamination or give rise to premature bridging.

The best inside gravel pack is one that can be placed with reasonable procedures and will maintain its integrity during production, with such a gravel pack not having internal voids or excessive porosity and preventing invasion of formation sand or development of open flow channels from reservoir to screen.

5.8.7.1. General aspects

Deposit buildup inside the wellbore is a prominent characteristic during placement when viscous polymer solutions are used for gravel packing, with the sediments growing from the perforations back to the wire-wrapped screen or slotted liner as well as in opposite direction, with thereby in the worst case deposits growing from either side meeting in the middle of the annulus (WALLING & STABLER 1978, TORREST 1982 a). As the sediments enlarge in the annulus, resistance to fluid flow increases and some of the fluid will be diverted through the wire-wrapped screen. Because this fluid carries gravel in suspension, an annular deposit forms on the screen. With increasing fluid velocity in the annulus due to sediment buildup, there is a tendency for deposit growth to be self-levelling and even for parts of the sediments to be washed away. With steady injection of a viscous suspension, a more or less complicated depositional pattern evolves in the wellbore annulus that is transient in time and place.

The nature of the sediment buildups and interaction may lead to variations in annular pack porosity with subsequent adverse consequences for sand control and/or well productivity. In some cases, bridging of the deposits prior to complete filling may lead to erroneous screenout indications and inadequate gravel placement. The apparently simple process of viscous gravel packing is in reality a complex sequence of sediment buildup and readjustment that is transient in both time and space. Gravel nodes are formed inside the wellbore at the perforations and annular deposits appear on the wire-wrapped screen.

In contrast to viscous polymer fluids, gravel packs fill from borehole annulus bottom upwards when using water or other low-viscosity fluids. The outline as follows focusses on nodular and annular deposit buildup at the perforations and on the screen, respectively, as well as on benefits of high carrier fluid viscosity.

5.8.7.2. Nodular deposit buildup at the perforations

With water at typical perforation flow rates, deposit buildup at the perforations inside the wellbore is insignificant and has little influence on downward gravel suspension flow on the annulus (WALLING & STABLER 1978, TORREST 1982 a). With polymer solutions, the nodular sediments are much less symmetrical than with water and far more top heavy. Gravel nodes laid down by viscous polymer fluids have high porosity and permeability as evidenced by the relatively low pressure drops in contrast to nodes placed by water flow which display low porosity and permeability as reflected by high pressure drops. When the deposit reaches the wire-wrapped screen, the solution flow will tend to follow this shortest path regardless of how much larger the node sediment would grow up and down the wellbore annulus.

Thus even though fluid is pumped away into the formation with no intended current through the wire-wrap, bridging of nodes across the annulus will assure some flow into the wire-wrap during packing. Even with uniform perforation current, the extent of node buildup varies along the interval because of varying downflow rates. Node buildups at adjacent perforations are soon merging and may lead to bridging at some depth of the perforated interval where the downflow rate has been sufficiently reduced.

5.8.7.3. Annular deposit buildup on the screen

Similarly as polymer slurry builds nodes at the perforations, annular deposits form on the screen with flow through the wire-wrap. At elevated current rates through the screen, the thick annular layer is rapidly peeling away into the channel with only a very thin layer of beads remaining on the screen (WALLING & STABLER 1978, TORREST 1982 a). For several different initial annulus deposit thicknesses, the bulk fall-away always occurs at more or less the same fluid rate level, and presumably with a high enough downflow rate, a higher screen flow rate is required to retain the annular deposit.

With node buildup and interaction at some distance into the perforated interval, the annular flow is partially diverted into the screen thereby building a deposit on it which interacts with nodes higher in the pack that had not yet reached the screen. This process could lead to a substantial bridge in the upper portion of the pack, with the annulus being only partially filled below, or the pressure buildup with flow through the annular sediment in the upper section could lead to channelling through the previously bridged deposit at the perforations. With all the perforations taking some flow, sediments will build back from perforation throats to the screen and then enlarge. The suspension flow must now follow a winding path with some partial bypass through the screen, thereby building a deposit there. Perforation sediments soon merge, and with the gravel-coated screen reducing the flow cross-sectional area in what is left of the annulus, a narrow spiral channel develops along the wellbore with a high velocity erosive flow tending to keep it open under steady conditions.

In terms of geometry of node buildup and interaction, the deposit has a non-uniform porosity distribution during placement in the borehole, as sediments build from perforation to screen and enlarge laterally and from above. Filling in of the open zone just below the perforation node occurs at relatively low velocity and suggests higher porosity.

5.8.7.4. Benefits of high carrier fluid viscosity

Viscous carrier fluids in gravel packing have often the advantage that the high viscosity minimizes turbulence beyond the perforation tunnel and therefore reduces mixing of gravel and formation sand (WALLING & STABLER 1978, TORREST 1982 a). On the other hand, however, high fluid viscosity limits fluid leakoff to the formation (cf. also section 5.3.1.1.). Without gravel placement out beyond the perforation tunnel to provide a reserve during production, any extra porosity in the annular pack will allow production flows to push gravel back into the annulus which in turn can leave the perforation tunnel only partly full of gravel, thus triggering or provoking subsequent filling with formation sand, permeability impairment and loss of productivity. In more severe cases, the perforation tunnel may be left open and a channel formed from perforation to screen with sand production and screen erosion.

With viscous polymer solutions, nodular gravel deposits grow from the perforation tunnel inlet back into the annulus at low leakoff rates. These sediments are retained with low pressure drop and the extent of their growth depends on the interplay between perforation and annular flow rate as well as solution viscosity. Most of the pressure drop associated with the deposit occurs in the hemisphere just beyond the inlet to the perforation tunnel. Gravel nodes readily grow back to the screen, and then increased resistance to annular downflow causes some fluid to be diverted through the screen thereby building an annular gravel sediment. The overall process of gravel pack buildup with viscous solutions is one of continual deposit growth, interaction, rearrangement and erosion which is transient in both time and position. The localized and non-uniform sediment buildup leads to variations in annular pack porosity. Gravel pack response to fluid production will be sensitive to these variations and their consequences.

5.8.8. Prevention of proppant flowback by fracture tail-in

Hydraulic fracturing performance frequently suffers from subsequent crack conductivity deterioration due to proppant flowback out of the fracture mouth into the wellbore (cf. section 4.12.3.). Tailing-in of coarser proppants in the proximal section of the fracture following pumping of finer proppants into the main portion of the crack comprising the medial and distal parts of the wing is a suitable technique to hold back the proppants and prevent their migration in a comparable manner to gravel packing inhibiting advance of reservoir formation sand (cf. section 2.4.1.2.3.4.). In case of proper grain size selection of proppants for main fracture load and tail-in termination, the coarser tail-in proppants act as a gravel pack and result in bridging of the finer proppants at the interface between both grain size classes, thereby stabilizing the proppant quantity in place and reserving fracture conductivity (CROW 1977). The effect of the coarse tail-in on the fine proppants is similar to the impact of gravel in perforation tunnels or as a mantle seaming the borehole wall on reservoir formation sand.

5.8.9. Puddle packing

Puddle packing is a combination of puddle cementing and gravel packing and is used for infilling of large open-hole voids left by nitroglycerine well shooting many decades ago and recently threatening secondary and tertiary oil recovery potential (SLATON 1988). Infilling of the shot cavities by resin-coated sand which is later drilled out from the borehole centre converts the multiple-shot open-hole completions into cased-hole completions once a liner is set after drill-out. Puddle packing has proven to be an economically viable and attractive alternative to plugging of shothole wells and drilling of new injectors.

Puddle packing provides injectors with zonal isolation between shot sections and controls loss of injection fluid to non-pay horizons. In production boreholes, puddle packing permits selective stimulation by isolating individual zones, allows more effective beam lifting, eliminates the need of cleanout of wellbore fill, and enables to run interpretable injection profile logs. Casing and cementing the open hole without puddle packing suffered from failure of perforating the thick cement sheaths in the shothole sections and from zonal isolation problems. The optimum operational sequence is adding 70/170 mesh sand to a gelled carrier fluid, adding resin to coat the sand grains, and adding graded silica flour that attaches itself to the resin-coated sand as well as becoming itself resin-coated.

5.9. Other sand control methods

Apart from conventional and special gravel packing, other sand control methods are common practice in the oil and gas industry. Their market share, however, is still relatively subordinate, as illustrated by the fact that abt. 85 % of all sand-control operations in the USA are done by gravel packing (SUMAN & SNYDER 1982). From the various techniques tried since the beginning of sand flow treatment, resin consolidation, hydroxy-aluminum and electroless nickel stabilization, silica cementation, warm air coking consolidation, stable sand arch formation, and pre-pack liners and multiwire-wrapped screens have been most frequently applied. These methods of non-gravel-packing sand control are outlined as follows.

5.9.1. Resin consolidation

In-situ chemical plastic formation consolidation treatments (plasticization) as an alternative for mechanical gravel packing have already been developed and introduced to the hydrocarbon industry over 40 years ago (WRIGHTSMAN 1949; MURPHEY, BILA & TOTTY 1974; SINCLAIR & GRAHAM 1978 b) and are designed to bind the loose reservoir sand grains and/or an artificial filler material by interstitial resin deposition into a strong matrix in order to stabilize the borehole wall and to maintain the permeability for oil and gas of the surrounding wellbore area (RIKE 1970). The basic objective of chemical resin consolidation is to increase the strength of the reservoir sand around the wellbore such that sand grains are not dislodged by the drag forces of the flowing fluids and pressure drawdown at the desired production rate (ALLEN & ROBERTS 1982). After an outline of some general aspects, the discussion of resin consolidation as a means of chemical formation stabilization as follows emphasizes resin viscosity and activity, sources of consolidation failure, advantages and problems, possibility of imposed resin consolidation performance, formation consolidation vs. gravel packing, and permeability destruction and plugging of thief zones.

5.9.1.1. General aspects

The resin consolidation type of strengthening of the formation in the vicinity of the borehole eliminates sand production, with successful resin operations being often significant improvements over the customary use of screen and gravel pack for sand control. Since the first formation stabilization treatments in 1946, ameliorated field performance has been achieved during course of the technological evolution of sand consolidation by development of new resin systems of progressively increasing effectivity. The most common resin types are phenolic, phenol-formaldehyde epoxy, furan, and phenol-furfuryl resins, and during the years, also other stabilizing materials such as crude oil coking and nickel plating (cf. sections 5.9.2. and 5.9.4.) have been tried (HALLIBURTON 1988).

Resin consolidation has the advantage that no mechanical obstructions are left in the well and the treatments can be accomplished in tubingless wells and without pulling the completion string (POSTON, TURNER, TOBOLA & BARGER 1986). Resin stabilization of sandstone pays can be carried out in vertical, inclined and horizontal wells (SPREUX, GEORGES & LESSI 1988) and is in most cases performed in short intervals with limited sand-producing tendency and high reservoir pressure. Some comments are offered on mechanical gravel packing vs. chemical formation consolidation, resin mobilization and migration, and resin curing and catalyzation.

5.9.1.1.1. Mechanical gravel packing vs. chemical formation consolidation

While mechanical gravel packing with resin-coated natural sand or synthetic proppants has in the last years been proven to be the most suitable method of stabilization of many loose formations and will certainly be of rising significance for sand control in various marginal reservoirs in the near future (cf. sections 5.5.2. and 5.5.4.), chemical consolidation by injection of liquid polymerizable resins into the pay zone has been performed as an alternative to gravel packing for many years, although the success ratios have not always been as expected. Although being principally a simple solution by introducing an artificial cement between the sand grains of the friable reservoir, the method includes high risk of completely destroying the poroperm properties if the resin is too much obstructing the pore space. Even successful resin precipitation at the grain contacts results in reduction of the relative permeability to oil, because in any case the resin as an artificial cement occupies a portion of the original pore space and the resin surface is oil-wet (ALLEN & ROBERTS 1982).

Although the process of resin consolidation may be quite straightforward, problems arise from the mechanics of placement which may not work properly. Therefore nowadays in most cases gravel packing is preferred in spite of being technically more complicated, because it is at the bottom of the line more successful than resin stabilization. During many years, both mechanical gravel packing and chemical sand consolidation did have preferred applications and as such the methods were complimentary to one another (PEDEN, RUSSELL & OYENEYIN 1985). Resin-coated gravel placement with excess resin in the solution is even a combination of mechanical gravel packing and chemical reservoir consolidation (cf. section 5.8.2.1.3.).

5.9.1.1.2. Resin mobilization and migration

Sand consolidation efficiency depends on placement and strength of the bonding agent (FRIEDMAN, SURLES & KIEKE 1986). Concerning placement, the binding film must surround the grains in such a way to guarantee the desired stabilization as well as to create the minimum impediment to flow of fluid. In terms of strength, the consolidated pack must be able to endure both natural and artificial reservoir temperature during heat-application enhanced oil recovery (cf. section 5.7.) and to resist to the mechanically erosive and chemically aggressive effects of the injected and produced fluids.

Formation sand stabilization usually involves injection of proper preflushes (in order to push away oil and reservoir water and to clean the pay sand belt around the wellbore), resin solutions as plastic consolidation material, and overflushes into an unstable formation so that the reservoir particles are properly cleaned and then chemically cemented to prevent sand flow (SUMAN, ELLIS & SNYDER 1983). Overflushing with special fluids sometimes containing activators and/or accelerators aims on establishing desired permeability and/or speeding-up curing time, improving sweep efficiency, controlling thickness of the plastic film and thereby compressive strength and permeability, and displacing all but a residual resin saturation at the grain-to-grain contact points. The sand to be stabilized can be natural reservoir sand or clean graded sand which has been inserted as a prepacking step and in fact already represents an outside gravel pack, with resin consolidation thus being applicable in both new and old wells without or with sand production history, respectively.

5.9.1.1.3. Resin curing and catalyzation

The resin cures radially from the wellbore in outwards direction as the time of exposure to formation temperature increases and/or as the injected activator or accelerator works its way from the borehole to the interior of the formation. In many applications, catalysts for initialization of the polymerization (polycondensation) are also required (FRIEDMAN, SURLES & KIEKE 1986). The resin cures and bonds the grains together to generate a synthetic sandstone with artificial cement. Sand consolidation resins penetrate the formation and immobilize particles further away from the wellbore which stops grain movement to a point beyond where reservoir fluid velocity can cause particle migration. Sand consolidation resins also build much more strength into the formation than is needed to withstand the stresses of overburden and fluid velocity forces, and the pay is therefore durable while maintaining a high rate of production. Distinction can be made between external and internal catalyzation where the curing agent is added to the resin mixture prior to pumping and injected into the formation after resin placement in the desired saturation in the reservoir by overflushing, respectively (BROOKS, MUECKE, RICKEY & KERVER 1974; SHAUGHNESSY, SALATHIEL & PENBERTHY 1977).

5.9.1.2. Resin viscosity and activity

Low-viscosity resins are to be preferred, because high-viscosity resins cause rising pumping pressures and increase the risk of generating a channel through an unconsolidated sand (SHAUGHNESSY, SALATHIEL & PENBERTHY 1977) by opening of a fracture upon formation parting due to exceeding of breakdown bounding pressure (cf. section 5.9.1.5.4.2.; the same applies for gravel prepacking of unstable pays at too elevated injection pressures; cf. section 5.9.1.5.1.1.). The optimal residual cured resin saturation is 35 % of the original pore volume. Higher residual resin concentrations lower reservoir permeability after consolidation without substantially improving compressive strength, whereas lower saturations contain too little resin to effectively coat and bond together all the sand grains.

5.9.1.2.0. General aspects

Sand control by chemical consolidation comprises injection of chemicals into the naturally stabilized formation to provide grain-to-grain cementation at the contact points and to create a matrix, with excess plastic material being displaced from pore spaces by overflush fluid (HALLIBURTON 1988). Two types of resin consolidation systems can be distinguished (ALLEN & ROBERTS 1982) being internally and externally activated systems (the latter are also called overflush systems).

In both types of resin systems, careful preflushing has to be done in order to clean and condition the reservoir sand for resin penetration (ECONOMIDES 1986). Crude oil saturation must be removed or reduced to residual levels, and paraffinic deposits have to be dissolved. Sand grains must be water-wet to enable proper contact with the resin coupling agent, and skin damage caused by drilling fines has to be eliminated as well. Clay minerals should be removed in order to prevent the catalyst from being prematurely depleted from the oil carrier fluids and calcareous materials must also disappear to exclude reactions with the acidic catalyst. Comments are offered as follows on internally and externally activated resin systems as well as on resin yield and formation permeability retention.

5.9.1.2.1. Internally activated systems

Internally activated systems are composed of resin, diluent, coupling agent and either curing agent or catalyst and accelerator (ALLEN & ROBERTS 1982). They contain a low percentage of active resin which hardens if properly attracted to the grain contact points and thus provides the stabilization effect. The remaining inert material of the subjected mixture fills the center portion of the pore space in order to ensure that sufficient permeability is retained. Curing agents or catalysts are added at the surface prior to pumping in an amount according to formation temperature. Very accurate control of displacement is required to insert the resin through the perforations sufficiently effective, but without over-displacement. Some processes employ phase separation which includes using a relatively dilute solution of resin in hydrocarbon solvent. Combined with an activator (curing agent or catalyst), the liquid resin phase separates from the solvent after a period of time and solidifies, with the still liquid resin after separation being drawn by surface tension to the grain contact points and hardening there.

5.9.1.2.2. Externally activated or overflush systems

Externally activated or overflush systems contain a high percentage of active resin which occupies most of the pore space in the initial step (ALLEN & ROBERTS 1982). Permeability must be subsequently re-established by displacing further into the formation all but a residual resin saturation which in case of proper sand wettability remains at the sand grain contacts. Curing agent is usually contained in the overflush fluid, but can also be added to the initial resin solution. Accurate control of displacement is not as critical, but all sections not overflushed will be plugged. Overflushing is designed to govern plastic film thickness which in turn is decisive for compressive strength and permeability. In some cases, a two-step overflush is used, with the first phase restoring permeability and the second stage introducing a cure activator or an accelerator.

External catalyzation allows unlimited pumping time, because the resin does not harden until contacted with the overflush catalyst solution. Unlimited pumping time also permits reduced pumping rate if necessary for low-permeability formations in order to stay below fracturing pressure. When the activator is already in the resin solution, overflush may still be needed in order to establish or insure permeability, but is not required to ascertain that the plastic will eventually cure. An accelerator in the overflush may speed-up curing and minimize waiting time. External catalyzation provides significant operational advantages, because internally catalyzed resins are limited in pumping time to the settling period and the chemical must be mixed at the well site (SHAUGHNESSY, SALATHIEL & PENBERTHY 1978). Aspects of externally catalyzed resin systems are also discussed by BROOKS, MUECKE, RICKLEY & KERVER 1974).

5.9.1.2.3. Resin yield and formation permeability retention

Perforation density is significant to improve resin distribution and to reduce pressure drawdown. Injection of the resin solution has to be done at low rates in order to promote uniform resin coverage and has to be carried out exclusively below fracture pressure in order to avoid pressure parting of the reservoir and opening of a high-permeability crack which would absorb the overwhelming amount of injected resin solution thereby drastically deteriorating the possibilities of matrix cementation. Some aspects of reservoir composition and resin yield as well as consolidation strength and dirty pay stabilization are discussed as follows.

5.9.1.2.3.1. Reservoir composition and resin yield

Resin yield depends very much on reservoir composition, with the main distinction having to be made between clean and dirty sands. In dirty sands, high-yield resin solutions are required to accommodate the large surface area of the silt and clay particles. Some resin systems have inadequate yield and can only be used in clean sands (HALLIBURTON 1988). Higher shares of clay minerals necessitate large amounts of resin or may extract the curing agent from the resin, and therefore acid preflushing in order to shrink clays is recommended. As the resin leaves only an extremely thin coating on the sand, up to 90 % permeability retention is possible in clean sands after resin consolidation in case of suitable performance of the chemical sand control treatment. Compressive strength of the resin-stabilized formation sand may in ideal case approach 3,000 - 4,000 psi, thus being partially even higher than that which can be achieved by consolidated pack installation and resin-coated gravel placement (cf. section 5.8.2.1.1.).

5.9.1.2.3.2. Consolidation strength and dirty pay stabilization

Maximum consolidation strength occurs in the immediate perforation vicinity where it is needed most while farther away from the borehole surroundings, secondary-artificial cohesivity of the pay sand is progressively diminishing and finally completely fading out. Various chemicals can be pumped through the consolidated reservoir with no damage at all, with the resin being very resistant to wash-off by flushing solutions even when high injection rates are used. Some resins have the ability to permanently shrink several clay types which may result in increased permeability in some clay-bearing dirty formations, thus helping in stabilizing clay-containing reservoirs which are usually not susceptible to consolidation by other sand control treatments and even improving the petrophysical properties beyond the natural state, thereby giving rise to the double success of stopping sand production and enhancing hydrocarbon offtake capability by opening of wider flow capillaries than existed before.

5.9.1.3. Sources of consolidation failure

An important point for satisfactory performance of resin consolidation are solids-free completion fluids, because dispersed solid particles would inevitably bridge and plug perforations and pores, respectively (RIKE 1970). Further improvement of performance can be made by paying attention to the manner of perforating and the continuous use of clean fluids from the time of perforation to the moment of overflushing the formation after placement of the plastic (fluid and material cleanliness are also essential for success of gravel packing; cf. section 5.11.5.). Following an outline of requirements for successful resin consolidation, comments are given on unsuitable drilling and completion practices as well as intermixing of rathole fluid and sand control fluid.

5.9.1.3.1. Requirements for successful resin consolidation

Consolidation of reservoirs by chemical injection requires more or less short intervals, lack of previous sand production (because uniform chemical distribution is difficult when cavities, washholes or other unstressed formation parts exist), limited sand-producing tendency, high reservoir pressure, and high-quality pay sand with sufficient vertical permeability. STROHM, MANTOOTH & PRIESTER (1970) discuss an improved placement system with controlled resin injection for extension of plastic consolidation to sands with horizontally and vertically variable permeability and dirty silty or argillaceous sands, as well as to long and/or discontinuous completion intervals.

The one-trip coverage of several separate pay zones with a multi-injection packer significantly reduces completion cost as compared to a series of individual settings of conventional packers when treating each storey of the prospective pile individually. Failures are most often attributed to contaminated chemicals and improper placement in unstable formations. Principally, the resin is drawn by capillary forces to the grain-to-grain contact points (BEZEMER & MEIJS 1966), but field experience has shown that unwanted pore throat blocking also frequently occurs. RIKE (1970) outlines from field experience that many unsuccessful jobs can be corrected by repetition runs during subsequent treatments.

THOMAS & JOHNSON (1976) investigate reperforating of resin-consolidated zones for capacity improvement. In most wells, sand control by resin stabilization is not lost after reperforation, including boreholes with two-stage sand control completions comprising gravel packing in lower zones and resin stabilization in upper sections.

5.9.1.3.2. Unsuitable drilling and completion practices

The most frequent cause of failure of any sand control process is improper planning and well preparation prior to the formation stabilization treatment (HALLIBURTON 1988). Reservoir permeability damage by invasion of drilling mud and completion fluids disturbs penetration of sand control fluids and gives rise to non-uniform placement of consolidation chemicals or gravel. Caution has to be exercised that drilling fluids do not disperse or swell water-sensitive clays contained in formations adjacent to the wellbore in order not to reduce effective permeability. Washouts through the productive interval due to poor drilling practices may result in perforations which fail to penetrate adequately to the prospective formation, since an excessively thick cement sheath may be present in the enlarged hole section. Some aspects of cementation and perforation are discussed as follows.

5.9.1.3.2.1. Cementation

Larger borehole diameters and casing sizes reduce shear forces of fluid movement into the wellbore, thereby lessening the burden on weak formations that produce sand (HALLIBURTON 1988). Proper cementing is imperative to achieve best results from any type of sand control treatment. Uncemented channels and off-centered casing strings give rise to poor resin consolidation and gravel packing performance. As the off-center casing problem increases considerably with borehole deviation, minimization of uncemented channels and off-center pipe has to be achieved by the borehole being as near to gage as possible, the casing string being adequately centralized, the cement slurry being pumped in turbulent flow, and sufficient spacer being used between mud and cement.

5.9.1.3.2.2. Perforation

Maximum unstimulated well productivity is obtained from open-hole completion with maximum wellbore diameter and no formation damage, but in many instances, casing has to be run and cemented for orderly reservoir control which subsequently requires perforating in order to get access to the prospective interval and to exploit the hydrocarbons (HALLIBURTON 1988; cf. section 5.10.1.). In terms of different perforation techniques, alleviation of reservoir damage can be achieved by placing a filtered fluid across the interval, perforating with underbalance or differential pressure toward the wellbore, allowing formation to produce after perforating with underbalance, cleaning of perforations and reservoir face by backsurging or perforation washing, chemical washing by acids, and hydraulic perforating with low-volume high-velocity jetstream of sand fluid slurry.

5.9.1.3.3. Intermixing of rathole fluid and sand control fluid

An important consideration in performing resin sand consolidation is preventing the lighter rathole fluid from being displaced by the heavier resin and entering the pore space of the reservoir, because this can be a source of failure of the stabilization treatment (SHAUGHNESSY, SALATHIEL & PENBERTHY 1977). Intermixing with the rathole fluid cannot only screw up resin consolidation processes, but can also deteriorate the effect of gravel packing and acidizing operations. The significance of rathole content separation and isolation in these different well treatments is outlined as follows.

5.9.1.3.3.1. Resin consolidation

The objective of preflushing is to remove reservoir fluids which might contaminate the resin, and to oil-wet sand grain surfaces in order to allow generation of continuous resin coatings (ALLEN & ROBERTS 1982). Water entering the formation along with the resin has a very detrimental effect on compressive strength (PENBERTHY, SHAUGHNESSY, SALATHIEL & GRUESBECK 1978), because the preflush ahead of the resin removes water from the pore space, but any water entering the sand along with the resin will not be removed prior to catalysis and curing and thus the development of continuous resin pellicles around the grains is disturbed or even seriously prevented. Circulation of the rathole with a dense saltwater is cumbersome and cannot be applied to bullhead treatments which are carried out through production tubing without a rig. A better solution for operation by either bullhead or workstring technique is to drop in an organic liquid having a greater density than that of the resin and being immiscible with the rathole brine, with the organic liquid falling from the bottom of the perforated interval to the rathole floor and thus also eliminating the need of circulation (SHAUGHNESSY; SALATHIEL & PENBERTHY 1977).

5.9.1.3.3.2. Gravel packing

Similarly as applying for resin consolidation, intermixing of rathole fluid with sand control fluid is also one of the most serious reasons for unsatisfactory performance of gravel packing (RENSVOLD & DECKER 1978). High-viscosity slurries carrying gravel are rapidly falling through the less dense pad or brine being present in annulus and rathole by the roping effect. Bullhead gravel packing displaces rathole fluid upward in the annulus, thereby mixing it with the slurry passing through the perforations. This could result in fines from the rathole debris and residue becoming entrained in the gravel pack slurry, causing gravel contamination which leads to lowered pack permeability.

In addition, fluid losses from the slurry through the tell-tale screen (cf. section 6.3.5.) above the perforations provoke rapid formation of an annular gravel bridge. At high pumping rates and pressures, flow channels are generated through the gravel bridge allowing slurry to reach the perforations, but low pumping rates and pressures may give rise to establishment of a gravel bridge in the annulus sufficiently long to enable misinterpretation of the increasing pressure as testifying to screenout. On the other hand, positioning of the tell-tale at the bottom of the rathole offers several advantages including production of the rathole fluid through the washpipe, thereby preventing contamination of slurry going through the perforations, and promotion of more complete annular fill-up by the gravel bridge opposite to the tell-tale in the annulus. Apart from this, with increasing gravel column height in the annulus, the resultant rise in differential pressure in the annulus below the perforations directs more slurry through the perforations. Placement of the tell-tale screen below the perforations allows abt. 25 % more gravel to be packed than positioning it above.

5.9.1.3.3.3. Acidizing

Concentric tubing strings allow to circulate brine in order to clean out the rathole below the perforated interval before stimulation fluid insertion (McLEOD 1984). Fluids and solids in the rathole below the lowest perforation affect both gravel packing and acidizing. A rathole fluid that is denser than the acid and a fluid above the top perforation which is lighter than the acid should be used, as otherwise the acid could end up in the rathole rather than in the formation. Aspects of such density segregation and separation are discussed by HONG & MILHONE (1977). The use of coiled tubing has the advantages that the rathole can be circulated clean if necessary (BRANNON, NETTERS & GRIMMER 1987).

5.9.1.4. Advantages and problems

If properly applied under the right conditions, chemical resin consolidation has inherent technical advantages with respect to other mechanical and mechanochemical sand control methods including suitability for through-tubing and small-diameter casing operation, maintenance of full open wellbore without obstructing screen, suitability for multiple reservoir completions, ready applicability in abnormally pressured wells, good performance in fine-grained sands which are difficult to control with mechanical gravel packing, and achievement of high compressive strength of the artificially cemented reservoir sand (ALLEN & ROBERTS 1982). The following outline illustrates some aspects of multistage processes vs. uniform perforation coverage as well as special conditioning of dirty sands.

5.9.1.4.1. Multistage processes vs. uniform perforation coverage

On the other hand, however, the problems of chemical resin consolidation are multiple. The basic problem is to increase physical formation strength uniformly throughout the completion zone without excessive reduction in permeability or too patchy distribution of cemented portions which can render overall reservoir performance even worse than originally. Many practical problems evolve from the fact that all sand consolidation techniques utilize multistage processes in which several fluids that are carefully formulated for the specific well conditions must be uniformly injected sequentially into a perforated interval. Uniform coverage of each perforation is a critical requirement. Fingering of one fluid through another must be controlled by low injection rates and by designing each fluid so that its viscosity is similar or slightly greater than that of the fluid which it displaces. Low fluid viscosities are desirable in order to obtain reasonable injection rates at low injection pressures, with time allowed for injection usually being limited.

5.9.1.4.2. Special conditioning of dirty sands

Sands containing more than abt. 10 % dispersed mud present a problem to phase separation systems, since these systems have only a small percentage of active material in the resin solution. The much higher internal surface area of dirty sands robs the limited amount of resin from the sand grain contacts and reduces the strength of the consolidation. In dirty sands, a preflush stimulation treatment designed to increase permeability may be required to improve uniformity of resin placement and to provide needed productivity following stabilization. If sand efflux has created cavities behind the perforations, these holes and notches have to be infilled and plugged with 20/40 or 40/60 mesh prepacking gravel before either phase separation or overflush resin consolidation systems can be used.

5.9.1.5. Possibility of improved resin consolidation performance

Several possibilities exist for improvement of resin stabilization performance in case of prestressed or not sufficiently permeable formations. Other techniques are purely applied to ascertain better resin distribution in the reservoir. Some comments are offered as follows on prepacking, preheating and preflushing of the formation, and on two- and three-stage sand consolidation.

5.9.1.5.1. Formation prepacking

If multiple intervals are to be treated with stabilizing fluids, they should be isolated with packers and/or bridge plugs if spacing permits (HALLIBURTON 1988). If more than one interval is treated without isolation, small differences in injection pressures may interfere with uniform fluid placement. In case of consolidating a reservoir in an older well that has produced an appreciable quantity of sand, the well should be packed with clean gravel prior to stabilization. The benefit of such a prepacking is that it might be difficult to achieve uniform resin distribution into the formation when pumping through a cavity or loose zone. Another advantage of prepacking is restressing of the reservoir. Some comments are offered on pressure parting and gravel placement in fracture as well as on influence of perforation washing and surging.

5.9.1.5.1.1. Pressure parting and gravel placement in fracture

Gravel prepacking in unconsolidated formations gives rise to pressure parting and gravel placement in the fracture as long as the critical minimum pressure is exceeded (PENBERTHY 1985, HALLIBURTON 1988; the same applies for resin stabilization of the reservoir sand at injection pressures exceeding the breakdown boundary of the pay; cf. section 5.9.1.2.), and the interface between gravel and formation sand is sharp, with intermixing only occurring in the immediate perforation vicinity. If the reservoir has produced sand before gravel prepacking and a firm cavity has originated outside of the perforations, gravel prepacking infills the voids and thus supports the formation wall (cf. section 5.6.7.). If stable cavities are not developed, prepack placement precipitates pressure parting in the formation sand, with fracture orientation depending on the stress state in the reservoir at any particular location in the vicinity of the perforation. Washing of perforations can create voids in the formation which can be subsequently prepacked with gravel. The amount of pay zone sand removed by perforation washing increases with pumping rate, and low-viscosity fluids are more effective in washing perforations than high-viscosity fluids.

5.9.1.5.1.2. Influence of perforation washing and surging

Perforation wash volume geometry is dependent on reservoir permeability, with high-permeability sands being removed with less difficulty than low-permeability sands (HALLIBURTON 1988). Perforation washing can also lead to pressure parting of the formation if the fracture gradient is exceeded. On a gross surged-interval basis, perforation surging is capable of removing about the same quantity of formation sand per reservoir interval as perforation washing. Perforation surging may not open all the perforations, with the amount of pay zone material being removed probably differing unpredictably. High surge-differential pressures are more effective than low pressures in removing sand from the perforations. Gravel prepacking subsequent to surging causes intermixing of gravel and formation sand (further aspects of gravel prepacking are discussed in section 5.6.6.).

5.9.1.5.2. Formation preheating

Apart from prepacking, a better effect of resin consolidation can be achieved by pre-heating of the formation sand by in-situ combustion and subsequent cooling down to abt. 400 oF (205 oC) before resin pumping into the pay zone (RENSVOLD 1982). Formation permeability is maintained by injecting inert gas after resin insertion or by entraining the resin in the gas. An alternative method is preheating the formation to 300 - 700 oF (150 - 370 oC) by steam injection. The resin is introduced followed by additional steam which cures the resin and increases its radial penetration into the formation. Heat-resistant furfuryl alcohol resins could in fact be one of the most feasible approaches to sand control in steam-drive operations, and steam injection for reservoir preheating is considered to be a more economical method and a technique being better able to be supervised than in-situ combustion.

FRIEDMAN, SURLES & KIEKE (1986) present a method of high-temperature sand consolidation which is even applicable in steam injection and fireflood (in-situ combustion) wells, with the resin resisting up to 500 oF temperature (WEAVER & MORGAN 1988). As a consequence of its temperature stability, resin is also widely used as core binder in steel foundries (MURPHEY, BILA & TOTTY 1974). CARNES & MIXON (1971) comment on consolidation and gravel packing in one step. One-trip gravel packing systems are also presented by BERCEGEAY & RICHARD (1974; cf. also section 5.10.1.2.).

5.9.1.5.3. Formation preflushing

CHEN, ZHOU & LIU (1986) document field experience of resin consolidation in China and conclude that stabilization within a radius of 0.5 m from the wellbore has to be achieved for successful operations. Preflushing should be made with fluid quantities equivalent to three times the pore volume in order to allow for proper formation cleaning. Hydrochloric acid has turned out to be a catalyst to accelerate the curing and solidifying reaction under reservoir temperature.

Field experience has shown that suitable resin-consolidated borehole vicinity zones range up to 6 ft diameter of the stabilized cylinder, with the cylindrical matrix providing casing support and reservoir strength as well as minimizing particle migration. The greater the diameter of the cylinder, the lower the flow velocity at its extremities, with particle transport tendency being less at lower velocities. Ideal cylindrical consolidation, however, cannot always be achieved due to formation inconsistencies that interfere with uniform resin penetration.

5.9.1.5.4. Two- and three-stage sand consolidation

In case of highly variable in-situ permeability distribution in the reservoir sand, consolidation treatments often are difficult to be carried out in an uniform manner, because the variable permeability provokes fluid channelling through the high-permeability streaks (POSTON, TURNER, TOBOLA & BARGER 1986). The outline as follows presents aspects of two-stage sand stabilization, formation fracturing during resin consolidation, and three-stage sand stabilization.

5.9.1.5.4.1. Two-stage sand stabilization

A more uniformly consolidated interval is provided by a two-stage operation where initial high injection rate forces most of the fluids to the zone of highest permeability, with this first stage reducing the effective in-situ permeability of the high-permeability streaks to approximately that of the low-permeability horizons (POSTON, TURNER, TOBOLA & BARGER 1986). The second stage comprises a similar stabilization job which is conducted at a diminished injection rate in order to uniformly treat the entire interval. Caution has to be exercised that injection rate of the second stage is not too high, because otherwise high pressure differentials are created at the formation face which induce hydraulic fracturing of the second-stage job through the first-stage treatment.

5.9.1.5.4.2. Formation fracturing during resin consolidation

If pressure parting of the reservoir occurs, the resin is forced into the pay zone only along the fractures and not through the matrix which gives rise to an uneven resin distribution (cf. section 5.9.1.2.). Fractures are particularly easily developing at the interface between high- and low-permeability layers if the fracture pressure gradient is exceeded. The risk of fracturing the formation is also increased by high-viscosity epoxy resins which lead to increasing pumping pressure (SHAUGHNESSY, SALATHIEL & PENBERTHY 1977). Aspects of formation fracturing during resin consolidation are also discussed by HOWER & BROWN (1961). Low injection pressures during the second stage prevent resin channelling and allow its uniform flow into the sand matrix. Increased reservoir temperature reduces viscosity of treating fluids which results in thinner resin coatings on the sand grains, more uniform fluid displacement, and higher compressive strength of the consolidated formation.

5.9.1.5.4.3. Three-stage sand stabilization

TREADWAY, BRANDT & PARKER (1966) even introduce a three-step process consisting of first injection of high-viscosity epoxy resin followed by second diesel oil to re-establish permeability, and finished by third injection of an activator to harden the resin. This process is applicable over a wide temperature range, provides adequate strength characteristics, and affords significant permeability retention. Similarly as applying for resin consolidation, two-stage and one-trip operations are also possible for gravel packing (cf. section 5.10.1.2.).

5.9.1.5.5. Other aspects

Aspects of sand consolidation by resin injection are also discussed by CARDWELL (1947), SMITH & POLK (1947), BEZEMER & MEIJS (1966); TREADWAY, BRANDT & PARKER (1966); YOUNG (1967), JENNINGS & MUIRHEAD (1968), SMITH (1968), RICHARDSON & HAMBY (1970), RIKE (1970); STROHM, MANTOOTH & PRIESTER (1970); BROOKS (1971), MUECKE (1974), MABIE (1975), COPELAND & CONSTIEN (1978); PENBERTHY, SHAUGHNESSY, GRUESBECK & SALATHIEL (1978); MURPHEY (1980), RENSVOLD (1983) and WEAVER & MORGAN (1988). New consolidation approaches with application of other materials are presented by TERWILLIGER, SMITH & GOODWIN (1964); FITZGERALD, STEPHENS & TERWILLIGER (1966); HAMBY & RICHARDSON (1968); HARRISBERGER, COULTER & GURLEY (1971); MURPHEY, BILA & TOTTY (1974); STRICKLAND, RICHARDSON, HAMBY & TORREST (1975); RIKE (1975), TORREST (1975); SHAUGHNESSY, SALATHIEL & PENBERTHY (1977); DAVIES, ZWOLLE & MEIJS (1980) and BURGER, GADELLE & MARRAST (1987).

5.9.1.6. Formation consolidation vs. gravel packing

Selection of mechanical gravel packing or chemical formation consolidation or even a combination of both methods for sand control depends on technical and economical considerations (HALLIBURTON 1988). The most important factors include initial sand control cost, expected reliability, effect on productivity, completion repair cost, formation sand quality, presence of multiple thin productive sections, exclusion of interbedded water or gas, level of pressure depletion, presence of undesirable shale streaks, and history of sand production. As consolidation chemicals are relatively expensive, formation stabilization by resin cementation is normally more costly per foot than gravel packing, with the total consolidation expenses therefore being significantly influenced by interval thickness. On the other hand, most gravel pack installations require a rig or a hoist, whereas formation stabilization may sometimes also be performed through coiled tubing, with the latter technique permitting more economical well treatments and reducing production losses resulting from unavailability of workover rigs.

Formation consolidation is advantageous in case of short prospective intervals, lack of previous sand production in new wells, single-horizon pay zones or upper portions of multiple-layer completions where no obstruction is left in the wellbore, and clean sand with sufficient permeability. If these conditions are not given, gravel packing or a combination of mechanical and chemical sand control should be preferred. The main advantages of gravel packing are high productivity in many cases and rather simple installation. The special benefits of chemical formation consolidation are restoration of production at minimum cost in some situations, applicability to relatively short intervals, successful stabilization of fine sands with proper design, and leaving the wellbore unencumbered for future workover operations. Combinations of mechanical gravel packing and chemical consolidation often result in high productivity after the treatment, can be relatively cheap in some simple situations, and are suitable for reservoir zones of older wells which may have produced some sand and would need pre-packing before effective stabilization could be done (cf. section 5.9.1.5.1.).

5.9.1.7. Permeability destruction and plugging of thief zones

While generally resin consolidation of hydrocarbon-bearing formations aims on maximum permeability preservation and optimum matrix stabilization in order to enable profitable oil and gas production, some special applications follow the inverse goal. Similarly as occasionally proppant grain size mixing has the task of provoking permeability destruction for plugging of fracture tips propagating downwards or upwards and thus for limitation of fracture height extension (cf. sections 1.4.11.2.1.4. and 4.2.2.4.), acrylic epoxy gel systems have been developed for plugging of reservoir permeability in thief zones (KNAPP & WELBOURN 1978).

The application spectrum of these delayed gelation fluids includes profile control for water injection wells, shutoff of water influx zones in production boreholes, and plugging of sections stealing fracturing fluid. The gel is generated by crosslinking of acrylic resin and liquid epoxy resin contained in a water-based continuous-phase emulsion. The resin droplets are small enough to flow into typical formation pores so that in-depth plugging of thief zones can be obtained. Various techniques of polymer injection for either stimulation or plugging of horizons in gas, oil or water wells are also discussed by LAKATOS, KRETZSCHMAR, CZOLBE, BITTKOW & WASSERMANN (1988) and ZAITOUN, KOHLER & GUERRINI (1989).

5.9.2. Hydroxy-aluminum and electroless nickel consolidation

Apart from resin stabilization (cf. section 5.9.1.), further possibilities of chemical sand consolidation as an alternative to mechanical gravel packing and other mechanochemical sand control methods comprise among others hydroxy-aluminum stabilization and electroless nickel plating which are characterized as follows.

5.9.2.1. Hydroxy-aluminum stabilization

Sand consolidation with hydroxy-aluminum solutions has been introduced over 15 years ago (VELEY 1969; REED 1972, 1977; REED & COPPEL 1972; BELVIN, COTTON & DUGAS 1973; COPPEL, JENNINGS & REED 1973; HASKIN, REED & COPPEL 1975; HASKIN 1976, BANIK & SHARMA 1984). Hydroxy-aluminum treatments are particularly significant for stabilization of water-sensitive formations. The polymeric inorganic cations are positively charged and are thus tightly absorbed on the negatively charged clay mineral surfaces. Due to an almost identical octahedral structure of some clay minerals and hydroxy-aluminum polymers, once tightly held on the clay surface the polymers are non-exchangeable and the clays become quite resistant to structural expansion and dispersion. Hydroxy-aluminum consolidation, however, does not work in clean sands with only traces of clay minerals or no fines at all (ANAND & JONES 1977).

The effectiveness of the hydroxy-aluminum polymer increases in case of moderate ageing in fresh water after the treatment, because the polymers grow to form larger and more perfectly crystalline sheets on the clay mineral surfaces, whereas excessive ageing leads to recrystallization into considerably larger molecules which then drastically reduce the permeability. Once the interlayer spaces of the clay sheets are filled up with tightly absorbed hydroxy-aluminum ions, the clays can no longer attract water and hence permeability damage by water sensitivity (cf. sections 3.5.2. and 3.11.1.) is no longer possible (VELEY 1969, BANIK & SHARMA 1984). Comments on decreasing formation sand entry by in-place clay stabilization are also given by VELEY & CASEY (1973). Another technique is calcium hydroxide treating of loose dirty sands which stabilizes the clay minerals and creates some type of artificial muddy cement between the framework grains (CHARLES 1976).

5.9.2.2. Electroless nickel plating

Formation sand can also be consolidated by electroless nickel plating from fluids injected into the sand (STRICKLAND, RICHARDSON, HAMBY & TORREST 1975; TORREST 1975). The mechanical strength of this consolidation type is greatest near the borehole. Compressive strength up to 40,000 psi and temperature stability above 500 oF (260 oC) can make electroless nickel plating superior to resin consolidation if properly executed downhole. On the other hand, while nickel plate on the formation sand grains would have the necessary heat resistance, dissolution of the reservoir sand by high-pH steam inserting at grain contact points where coverage by nickel is incomplete could lead to failure, with electroless nickel plating in terms of thermal enhanced oil recovery being more suitable for steam-drive than for cyclic steam-soak stimulation (RENSVOLD 1982).

5.9.3. Silica cementation

In addition to the methods by resin and hydroxy-aluminum injection as discussed in the above sections, other possibilities of chemical reservoir stabilization exist. Chemical sand consolidation techniques by silica cementation are introduced by ASLESEN, SHORT & TERWILLIGER (1981) and DAVIES, ZWOLLE & MEIJS (1983) and comprise Solder glass method and Silicalock concept, respectively, which are briefly outlined as follows.

5.9.3.1. Silicalock concept

The Silicalock concept (DAVIES, ZWOLLE & MEIJS 1983) is a special treatment for gas wells. Liquid silica tetrachloride is vaporized into a stream of high-pressure nitrogen and injected into the formation through regular or coiled tubing. Immobile connate water present in the treated zone reacts with the vaporized chemical, and silica cement is precipitated which connects the grains and thus stabilizes the framework package of loose grains. Silicalock operations have been successfully performed for sand control consolidation in the giant Rotliegend gas field Groningen/The Netherlands.

5.9.3.2. Solder glass method

The Solder glass method (ASLESEN, SHORT & TERWILLIGER 1981) is similar to the Silicalock concept, but amorphous glass is precipitated at the grain contacts instead of microcrystalline quartz. The Solder glass method offers an alternative to heat limitations of other bonding chemicals such as most of the resins in steam-stimulated wells. A special glass consisting of lead oxide, zinc oxide, boric oxide and silica is powered and solubilized in a water/sodium hydroxide/sodium silicate solution and squeezed into a formation sand that has been cleaned and dried by injection of air and downhole forward in-situ combustion. Injected with air immediately upon cessation of burning, the solution contacts and wets the sand grains. Additional hot air subsequently drives off the water and thus triggers the crystallization of glass at the contact points between the grains due to undersaturation of the solution by drying and restriction. Clays are also dehydrated in this process for an additional benefit. The Solder glass method has been successfully applied in the Fruitvale field in California/USA where previously severe and uncontrollable sanding problems existed.

The Solder glass solution is injected into the formation sand through a single perforation, and simultaneous injection of unheated air disperses the solution through the zone that has been previously cleaned by in-situ combustion. Air injection is then continued, with air temperature being increased in carefully staged levels, thereby evaporating the water from the Solder glass solution. Finally air temperature is raised to 800 - 1,200 oF (430 - 650 oC) and maintained at this level for eight hours which triggers devitrification of the Solder glass that has been deposited on the sand grains, thereby forming a strong bond between the grains. The major reason for some unsuccessful treatments is casing failure due to thermal shock (RENSVOLD 1982), whereas the major advantage of Solder glass consolidation is the high temperature stability. On the other hand, the Solder glass bond is dissolved by highly alcaline solutions and hydrofluoric acid. Therefore Solder glass bonds have to be protected from steam dissolution in thermal enhanced oil recovery projects in a similar ways as gravel packs (cf. section 5.7.).

5.9.4. Warm air coking consolidation

Warm air coking is another process of formation sand consolidation which is most suitable for low-gravity oil reservoirs (TERWILLIGER, SMITH & GOODWIN

1964; FITZGERALD, STEPHENS & TERWILLIGER 1966). A downhole heater supplies hot air to the formation causing controlled oxidation of the crude oil around the borehole, with the resulting coke being an infusible hard material which binds the sand grains together. Combustion of the oil is prevented by careful supervision of air temperature and injection rates. Warm air coking consolidation has turned out to be a particularly applicable reservoir stabilization process in steam-drive operations up to 460 oF (240 oC; RENSVOLD 1982). Avoiding ignition during the coking process may be difficult. Variations in formation permeability could result in uneven distribution of heated air in the reservoir which gives rise to differences in coking degree and subsequent resistance to sand production.

5.9.5. Stable sand arch formation

Particle bridging in order to stabilize loose sandy reservoirs and to prevent grains from moving can be achieved by synthetic gravel packing enveloping the pay zone as well as combinations of mechanical sand control and chemical formation consolidation, but is also possible to be reached by semi-natural sand grain arching within the reservoir rock matrix itself (HALL & HARRISBERGER 1970, SUMAN 1975; CLEARY, MELVAN & KOHLHAAS 1979; RISNES, BRATLI & HORSRUD 1982 b). After an outline of some general aspects, comments are given on sand arching combined with resin consolidation, as well as on grain surface roughness and fluid mixing relationships which are the most important factors controlling sand arch stability.

5.9.5.1. General aspects

An unconsolidated pay sand behind perforated casing can have a natural resistance to flow if the grains form a stable dome-shaped arch over the entry hole at the distal end of the cement mantle and/or the perforation tunnel or around the wellbore. The stability of the sand arch is a function of how tightly packed the grains are within the natural matrix of the reservoir sand, and of synthetic enhancement of the original fabric. Although sand arching being formation of a stable grain fabric spanning an opening can be a completely natural feature which is favoured by a suitable granulometrical distribution and grain shape of the reservoir sand and may not require any synthetic manipulating influences (but is at least to some degree also a function of pay depth), the technique of sand control by arching includes abundantly artificial stress loading of the sand by operation of a long inflatable packer in order to provide the necessary grain packing (SUMAN, ELLIS & SNYDER 1983). Improvement of sand arch stability can also be achieved by injection of a low-strength plastic in a combination of mechanical and chemical formation consolidation (HALL & HARRISBERGER 1970). Some aspects of cavity face permeability vs. dilation, sand package disintegration vs. grain interlocking, and wellbore stress state control are outlined as follows.

5.9.5.1.1. Cavity face permeability vs. dilation

In mechanically weak or unconsolidated formations, high earth stresses cause shear failure around spherical and/or cylindrical cavities located within the solid since boundary stresses are zero at its face (PERKINS & WEINGARTEN 1988). If fluid flows towards the cavity, additional stresses are induced which generally cause the outer boundary of the shear failure region to expand as current rate increases. As fluid flows towards the cavity, tensile stresses tend to be induced near its face which pull the sand pack apart thereby causing it to dilate.

In order to preserve mechanical stability, the permeability of the sand at the cavity face must progressively increase by dilation as flow rate rises.

When cavity face porosity has increased sufficiently, mechanical interaction of grains is insufficient to keep the pack from becoming fluidized and producing particles into the cavity. The disaggregated region with high porosity and permeability adjacent to the cavity is surrounded by a cemented zone at the limit of shear stability which is in turn enclosed by a region which behaves elastically. The outer radius of both the disaggregated area and the interval at the limit of shear stability increase with rising flow rate.

5.9.5.1.2. Sand package disintegration vs. grain interlocking

Disintegration of the sand package is probably controlled by shear strain which increases as sand flows plastically towards the cavity if a roughly constant dilation occurs as a result of shearing, with weakly-cemented materials becoming disaggregated when a critical amount of shear strain is exceeded. At the cavity face, small cohesive strengths are sufficient to maintain stability because of low stresses. When fluid is flowing through the disaggregated material toward the cavity, the disintegrated solids expand in order to preserve mechanical consistency which allows a small amount of sand to be produced while preserving overall stability.

In disaggregated materials, apparent unit cohesive strengths are manifested because of particle interlocking which permits mechanical stability to exist at a cavity boundary when fluid is flowing. If fraction porosity, however, increases to a critical value, mechanical interaction of the solids is insufficient to keep the particles from fluidizing and being transported together with the fluid. Other aspects of stability and failure of spherical and/or cylindrical cavities in unconsolidated sand and weakly-lithified sandstone are discussed by TIPPIE & KOHLHAAS (1973, 1974), FAHRENTHOLD (1984), PERKINS & GONZALEZ (1984) and FAHRENTHOLD & CHEATHAM (1986).

5.9.5.1.3. Wellbore stress state control

Wellbore stress state can be controlled by either application of elevated drilling fluid weight in order to prevent dilation and/or shear crushing of sand grains surrounding the borehole, or by establishing the proper stress state for stable sand arches by utilization of one or more inflatable external casing packers opposite the intervals to be perforated (SUMAN 1975 b). Four ranges of sand arch stability can be distinguished. Radial effective stress or arch load generally increases with depth. Formation sands surrounding boreholes in an extremely shallow depth do either not arch over perforations or would generate tenuous arches, and sand could readily flow by rolling and sliding through the perforations. At intermediate depth, rate-sensitive sand arches originate, with sand production depending upon flow rate and restraint being provided by interfacial tension and sand angularity. In greater depth, arch load may be sufficient to provide for high rates of flow without failure, and for a given sand quality, the radial effective stress may reach the level where finally arch failure occurs through grain crushing. In this case, however, arch loading may be self-adjusting by shifting to a lower but stable arch load. Some comments on pore pressure changes and arch load alterations as well as radial effective stress and imposing techniques are offered as follows.

5.9.5.1.3.1. Pore pressure changes and arch load alterations

Increasing pore pressure decreases arch load and thereby shifts load conditions from higher to lower ranges, whereas decreasing pore pressure has the inverse effect (SUMAN 1975 b). Increasing pore pressure can be generated by shut-in of producing wells, injection of stimulation or consolidation treatments, or injection of secondary- or tertiary-recovery fluids. Reduction in arch load causes conditions around the well to shift from no arch formation (crushing region) toward stable arch (crushing ragion), rate-sensitive arch (dilatant region) and no arch generation (dilatant region), whereas rising arch load triggers changes of stability levels in reverse direction.

In terms of application of these concepts, increases in pore pressure during secondary and tertiary recovery operations could lead to failures in production and injection wells where so far no sand control problems occurred. Attic pressuring procedures are particularly susceptible to a shift of sufficient magnitude to cause difficulties. An increased likelihood of sand-up is also provoked by pore pressure elevation around wells that are shut-in for extended periods of time. Decreasing pore pressure is also the consequence of depletion and/or drawdown. Depending upon the original condition, sand production likelihood may actually decrease with increased drawdown pressure if the stress rise causes arch loading to shift to a more stabilized condition. The conditions following increase of pore pressure depend upon the original arch load and the magnitude of pore pressure changes.

5.9.5.1.3.2. Radial effective stress and imposing techniques

Radial effective stress is equal to mud or cement weight minus pore pressure (SUMAN 1975 b). During drilling, the radial effective stress must be maintained at a value below the minor horizontal effective principal stress in order to avoid or minimize the likelihood of reservoir fracturing. Inflatable external casing packers can be used for purging of mud from the casing/formation annulus and imposing load upon the borehole wall up to fracturing pressure. Drilling mud weight has to be high enough for prevention of formation dilation, but caution has also to be exercised to inhibit productivity reduction by mud damage in the reservoir. The radial effective stress imposed during cementation establishes later arch load conditions. The radial effective stress could range from the minimum to prevent dilation up to that which would fracture the formation. Actual arch load which is necessary to be imposed might depend upon original stress state, depth of zone, and expected pore pressure history. Other variables which are important to the degree of obtained arch stability are sand grain angularity and clay content.

5.9.5.2. Sand arching combined with resin consolidation

If a sand arch builds and later fails by grain crushing and fabric collapse, it may reform with relieved stresses in the dilatant range or at the critical constant-value stress. A sand consolidation treatment by resin injection at this stage can provide ample cohesiveness to stabilize the formation sand at a loading level in its dilatant range, but if stresses later increase enough due to pressure drawdown or other causes to reach the crushing stress of the sand, the arch will fail. Fully effective arch stabilization by resin injection thus requires that the sand is able to remain in the dilatant range throughout the borehole life.

The latter necessity can be assured by performing a combination of mechanical and chemical sand control by linking gravel packing, formation consolidation and semi-natural sand arching, including injection of a sand whose critical stress for constant-volume shearing is well above that of the reservoir sand. In case of injection so as to achieve a dense enough packing to be dilatant, such a sand can form very stable arches even when using a plastic with only moderate cohesiveness for its consolidation (HALL & HARRISBERGER 1970) in order to ascertain maximum original permeability retention. The effect of compressive strength on sand stabilization by natural grain arching is already evident in the field where sand flow tendency decreases with increasing well depth as a consequence of the resulting higher grain-to-grain stress by progressive overburden (cf. section 5.9.5.1.3.).

5.9.5.3. Grain surface roughness

Natural sand arch stability requires some surface roughness or angular shape of the sand grains (similarly as synthetic gravel pack and proppant package stability; cf. sections 5.2.3. and 1.4.3., respectively) to provide the necessary cohesion and friction (HALL & HARRISBERGER 1970). Well-rounded sand grains with smooth surface do not generate stable arches. Fluid effects also add to grain-to-grain cohesion, because the residual wetting phase forms pendular rings around grain contact points which provide cohesiveness through interfacial tension. Lacking sufficient grain-to-grain stress and if connate water were the only cohesive force present, arch stability may be lessened by increased saturation of the wetting fluid.

In case of insufficiently rough grain surface, triaxial test results emphasize considerable differences between strong-grained round sands and weaker-grained angular sands with respect to the stress levels for grain crushing and thus contraction by arch collapse (HALL & HARRISBERGER 1970). Arch failure under load occurs at lower stresses for angular sands, with grain crushing usually being quite evident, whereas rounded-grain sands are more stable. Arch stability thus requires dilatancy and cohesiveness or some other restraint on the surface grains of the aggregate, because loosely packed sand does not arch. Shearing which causes crushing and collapse cannot propagate in a confined arch structure in a sand loaded within its dilatant range.

5.9.5.4. Fluid mixing relationships

The formation of a stable load-carrying sand arch spanning the production cavity and stabilizing weakly-cemented reservoirs against further sanding after some initial solids production is facilitated by the presence of two different fluids such as oil and water in the pay zone which may contribute cohesive forces between the sand grains if the wetting-phase saturation is near irreducible (TIXIER, LOVELESS & ANDERSON 1973). The cohesive forces result from the interfacial tension between the two fluids where their surface contacts the grains, and the effect is to pull the sand particles together. The cohesive forces are not effective at lower oil saturations where the hydrocarbons exist as separated globules and there is no contact of the oil-water-interface with the sand grains, as indicated by the observation that many wells do not produce sand until the water cut rises considerably.

Effects of producing rate on sand arch formation in unconsolidated and weakly-lithified reservoirs are investigated by TIPPIE (1973) and TIPPIE & KOHLHAAS (1973, 1974), and the impact of fluid properties on sand arch generation and stability is illustrated by CLEARY (1978 c) and CLEARY, MELVAN & KOHLHAAS (1979). Influence of grain size on sand arch resistance is analyzed by WOOD (1979), and MELVAN (1978); CLEARY, MELVAN & KOHLHAAS (1979) and BENTON (1984) examine the role of different stresses in sand arch stability. Other aspects of sand arch consistency are discussed by AYERS (1965); GETZLER, KOMORNIK & MAZURIK (1968); STEIN & HILCHIE (1972); DURRETT, GUBIN, MURRAY & TIGHE (1977); CLEARY, MELVAN & KOHLHAAS (1979); LASAKI (1980), BRATLI & RISNES (1981); RISNES, BRATLI & HORSRUD (1982 a, 1982 b); MORITA, WHITFILL, FEDDE & LØVIK (1987) and YASSIN & PEDEN (1988).

5.9.6. Pre-pack liners and multiwire-wrapped screens

Advanced sand-control hardware systems with preinstalled or simulated gravel pack material are represented by pre-pack liners and multiwire-wrapped screens which are characterized as follows.

5.9.6.1. Pre-pack liners

Pre-pack liners and multiwire-wrapped screens can even be run without a gravel pack belt between sand-control hardware and borehole wall or casing, whereas slotted liners and conventional wire-wrapped screens of wrapped-on-pipe, grooved, ribbed or all-welded type have to be generally installed in combination with a gravel pack mantle in order to facilitate bridging of formation sand (the only exception is a suitable grain size of the reservoir sand that can create a natural gravel pack around an inserted screen or liner upon pay failure and sand influx into the borehole; cf. section 5.2.3.6.). Some aspects of pre-pack liners in gravel packing and pump protection are illustrated as follows.

5.9.6.1.1. Gravel packing

Pre-pack liners for gravel packing are constructed with an integral surrounding gravel bed designed to restrain the pay formation in accordance to size selection criteria applying for loose gravel. Multi-wrapped screens have several layers of wire wrap, with each succeeding layer from outside-in having smaller openings between wires. The multiple wire wraps are to bridge larger sand grains at the outer layer, intermediate-size particles in successive layers, and smaller sizes at the innermost layer, thus simulating a selectively sorted or zoned gravel pack mantle of the reservoir formation sand within the screen (SPURLOCK & DEMSKI 1972).

Pre-pack liners and multiwire-wrapped screens are normally not applied inside casing because of the need to gravel-pack perforation tunnels and beyond in order to prevent severe productivity impairment. In open hole, pre-pack liners and multiwire-wrapped screens necessarily have to be somewhat smaller than the borehole diameter which permits the reservoir formation that is not withhold by a casing to shift when the well is produced. All these drawbacks favour the installation of a gravel pack between screen and casing or borehole wall as the best solution of sand control treatment in any respect (cf. also section 5.10.1.). Pre-pack liners that are covered with a gravel layer which is cemented by resin have been repeatedly successfully applied in Cretaceous oil sands in Germany FRG (GESEMANN & BADURA 1976) and Tertiary oil sands in Austria (MURER 1981).

5.9.6.1.2. Pump protection

A special type of pre-packed liners are downhole pump protectors which aim on trapping and retention of flour-like formation fines in various concentrations, because the powder-like material cannot be stopped with conventional gravel packs (SOLUM & RAMEZANI 1987; cf. also section 5.8.4.3.2.). The pump protectors consist of a slotted outer sleeve and a slotted base pipe which contain a gravel pack as filter medium. Gravel is selected such that the maximum pore size of the gravel pack is smaller than the size of the fine particles in order to stop the flour-like material completely. As hexagonal arrangement of the pack does not only give stability, but also reduces pore size or interstitial tube size of the gravel (cf. section 5.8.4.1.), hexagonal packing is achieved by vibrating the liner during gravel compaction into the pump protector at the manufacturing plant. For optimum result, the pump protector is assembled, packed and vibrated in vertical position.

The pump protector is installed in the well similarly as conventional prepacked liners, and can be pulled from the hole when and if the filter elements become plugged with flour-like fines. The pump protector can then be disassembled, cleaned and repacked with vibration. The pump protectors can be used repeatedly, as only the filter gravel must be replaced and outer and inner slots cleaned, with the pump protectors thus increasing life span of the pumps and reducing production downtime. When a shutdown occurs due to surface failure, a back flushing effect sends the fines bridged off on the outer sleeve and some fines from the filter pack to bottom, thereby diminishing the pressure drop across the pump protector. In addition, when surface shutdowns occur, the fluid remaining in the tubing is not sand-laden, thus eliminating pump failures due to the pump locking up due to sand after a shutdown.

5.9.6.2. Multiwire-wrapped screens

LIVESEY & TOMA (1984) and TOMA, LIVESEY & HEIDRICK (1986) present a pre-packed liner as a suitable sand-control filter which consists of stainless steel wool sandwiched between coarse steel cloth within the annular space of a dual-walled pipe. This filter was designed for the special application in horizontal steam-injection wells where gravel packing had so far been unsatisfactory. The metallic wool-cloth filter has a higher open area and thus a much higher porosity than a conventional screen-gravel pack filter. The steel cloth serves to confine and compress the steel wool and in addition improves its filtering ability by assisting the bridging of solid particles at the steel wool/ steel cloth interface.

The internal capturing element of the filter consists of a dense network of criss-crossing filaments with irregular cross sections. The capture of particles to the fibrous structure by various adhesion forces is strongly dependant on the probability of getting the solid grains in close or direct contact with the fibers which increases with filter tortuosity that is a function of compression. Field tests have revealed that this special filter can not only in light of technical performance, but also in view of economical feasibility represent a reasonable alternative to gravel packing in the harsh steam-drive environment (cf. section 5.7.). CHEN, ZHOU & LIU (1986) comment on field experience with pre-packed liners in China. Epoxy resin is the most frequently applied bonding agent for the gravel mantle of the liner.

Another means of sand control and also proppant flowback handling in hydraulic fracturing are tubing-conveyed sand filters which, however, are generally inferior in performance with respect to other techniques (POPE, WILES & PIERCE 1987). Wire-wrapped screens and slotted liners have frequently also been installed without gravel pack which allows the well to initially produce formation sand that creates a gravel-pack-type mantle around the hardware (AUSTIN, ROSE & SCHUH 1988; HALLIBURTON 1988). This method, however, promotes formation disturbances and permeability damage, includes the risk of failure of pack establishment, and bears the danger of screen plugging and severe production loss (cf. section 5.2.3.6.).

5.10. Influence of completion type on gravel packing

Borehole conditions have an important influence on both design and performance of gravel packing. Two basic completion types can be distinguished comprising cased and open holes. While the latter may be most desirable from the point of well productivity, it is in most cases necessary to run casing and to perforate it subsequently due to reasons of hole stability, mechanical and/or chemical protection of the reservoir, or separation of the pay zone from other layers such as water- and gas-bearing sands or plastically mobile shales. Some technical aspects of different gravel packing layout in cased and open holes are outlined as follows. Factors improving gravel pack efficiency are briefly summarized, and comments are also given on liner placement, wellbore damage and water control in gravel-packed holes.

5.10.1. Inside gravel packing

The installation of inside or internal gravel packing in cased holes is inevitably associated with some loss of productivity, because apart from the gravel pack, flow resistance through perforated cement and casing is already primarily greater than that in a barefoot well (MUSKAT 1949). The chosen thickness of the gravel pack in cased holes is always a compromise between effective sand stoppage and productivity decline due to increased flow resistance (PEDEN, RUSSELL & OYENEYIN 1985). Commonly a minimum annular gravel width of 1 in. is considered to be desirable in order to provide effective sand bridging, with a much better performance, however, being achieved with gravel mantle thicknesses of 4 - 5 in. (cf. section 5.2.3.3.; similar gravel widths between reservoir wall and sand control hardware are required at least in some open-hole completions, although the average diameter in the latter application is slightly broader; cf. section 5.10.2.2.).

Pressure loss is a function of perforation diameter, with large-diameter high-density perforations being an adequate means to obtain acceptable pressure losses which in case of being still excessive require change to open-hole completion where pressure loss is minimal (ECONOMIDES 1986). Pressure loss cannot be calculated absolutely, but has to be related to hydrocarbon type, as suitable pressure losses in high-viscosity oil reservoirs may be beyond discussion for high-rate gas pay zones and vice versa. Annular packing efficiency can also be ameliorated by increasing screen diameter which results in rising annular slurry transport velocity, and by increasing fluid leakoff rate. Some aspects of significance of perforation tunnel packing as well as two-stage and one-trip systems are offered as follows.

5.10.1.1. Significance of perforation tunnel packing

Inside gravel packing of cased holes with placement of the gravel between screen and casing requires perforation hole packing with viscous carrier fluids. Infilling of the perforation tunnels with compacted gravel is achieved by spotting and squeezing of the fluids into the formation with the open-ended tubing technique (with or without reciprocation) or the packer squeeze method (RAWLINGS 1958, HAYNES & GRAY 1974). High gravel concentrations in viscous fluids may be preferable to minimize intermixing of gravel and reservoir sand. Inside gravel packing can be performed in two stages, with first pre-packing the perforations and second circulating gravel into place in the screen-casing annulus by washdown, reverse circulation or crossover methods, or can be carried out in only one borehole trip (BERCEGEAY & RICHARD 1974; cf. section 5.10.1.2.). The significance of perforation tunnel plugging for successful inside gravel packing is also outlined in sections 5.2.2.1.1., 5.4.3.5., 5.6.1., 5.8.1.1. and 5.11.1.

5.10.1.2. Two-stage and one-trip systems

Gravel compaction is enhanced by using relatively low gravel concentrations and low-viscosity placement fluids. Dense viscous fluids with high sand saturations, however, may also be utilized for advantages such as reduced fluid volume and less screen cutting, and particularly in case of application of the packer squeeze method in single-stage gravel packing (BERCEGEAY & RICHARD 1974). Two-stage gravel packing has the benefit that gravel is more efficiently distributed to the many perforation tunnels if squeezed separately, which is an important consideration in remedial-type operations where the formation has to be restressed in order to restore mechanical stability. This can be best achieved by multiple squeezing, with the repeated squeeze runs being performed to screenout that is reached when no further gravel can be put away at less than original fracture pressure (SUMAN, ELLIS & SNYDER 1983). In the last years, repeatedly one-trip multiple-interval gravel pack systems have been presented which have the advantage of minimization of formation damage and operating time. Some aspects of interval length vs. reservoir permeability are also outlined.

5.10.1.2.1. Minimization of formation damage and operating time

HAILEY, DONOVAN & SICKLE (1985) present an integrated one-trip perforating and gravel packing system. Provided enough rathole available in the well to accept the spent hollow carriers of the combined treatment string when the assembly is moved down to the gravel packing position after firing of the perforation guns, this system does not only save at least one roundtrip of rig time, but elimination of an additional roundtrip also greatly reduces completion-fluid loss to the formation by diminishing exposure time and thus minimizing reservoir permeability damage which is benefitting the long-term productivity of the well. For usual gravel pack operations, a rathole of 5 to 10 ft length is required in order to allow some sand slumping before the gravel is in place (SPARLIN & COPELAND 1972). WEIRICH, ZALESKI & TYLER (1988) introduce a one-trip multiple-zone gravel packing technique for low-pressure shallow wells.

Gravel packing should be ideally performed in one continuous operation instead of repeating many separated short-interval gravel packs (ZALESKI & SPATZ 1988). One-trip multiple-zone gravel packing (BRUIST, JEFFERIS & BOTTS 1983; JEFFERIS, BRUIST & BOTTS 1983) allows very long sections to be gravel packed in short stages without tripping pipe and running individual screen assemblies, and if any phase is unsuccessful, it is also not necessary to remove equipment from previously packed intervals. In terms of interval length, the primary consideration is carrying capability and leakoff tendency of the slurry (ZALESKI & SPATZ 1988). Although gravel transport can be easily accomplished with many types of carrier fluids, leakoff into a high-permeability formation during circulation increases gravel concentration and decreases slurry velocity until a bridge is created, and therefore high fluid-loss rates of gravel transport medium are desirable for generating a tight pack (cf. also section 5.3.1.1.).

5.10.1.2.2. Interval length vs. reservoir permeability

Based on existing carrier fluids and pumping equipment, it seems feasible to attempt gravel packing intervals of 400 ft for low-permeability formations (less than 100 md), whereas reservoirs with very high permeability (more than 1,000 md) should probably be limited to 200 ft intervals unless leakoff rates can be controlled. Regardless of section length and open or cased-hole completion, it will be necessary to circulate the slurry to the end of the screen and take returns through a concentric tailpipe, with a large tailpipe being desirable to prevent channelling of fluids through the screens. For proper gravel placement in long high-permeability intervals, high-viscosity carrier fluids are required to transport gravel to bottom, and pumping rates must be sufficient to allow formation of a stable bank of gravel over which the slurry must continue to pass.

In contrast to conventional stacking of inside perforated casing gravel packs, one-trip multiple-interval gravel packing allows sequential packing of all zones through their port collars and return to lower sections for repacking if required (BRUIST, JEFFERIS & BOTTS 1983; JEFFERIS, BRUIST & BOTTS 1983). Additional benefits of the one-trip multiple-zone gravel pack system are that downhole rotation during the gravel packing operation is eliminated, a full column of gravel-laden slurry can be reversed out after screenout without pressuring other intervals, each horizon can be repacked during completion life, and by installing production sealing units with sliding sleeves inside the screen and liner assemblies, any combination of isolated sections can be produced or selected for water injection.

5.10.1.3. Other aspects

Aspects of productivity of inside casing gravel-pack completions are also discussed by WILLIAMS, ELLIOTT & WEAVER (1972). YILDIZ & LANGLINAIS (1988) present calculation of pressure losses across gravel packs. Special aspects of water-injection well gravel packing are reported by WADE (1967). Another critical area of gravel packing in cased holes is the width of the screen-casing annulus. If the difference between outside screen diameter and inside casing diameter is abt. 1 in., the job is problematical, with better results being achieved if the clearance is 1.5 - 2 in. (cf. section 5.2.3.4.). An annulus of 50 mm allows easy overshooting and overwashing of the liner in case of necessary workover (MURER 1981). Well completion geometry in friable sands also has important impact on maximum sand-free production rates without gravel packing (STEIN, ODEH & JONES 1974). SOKOLL & ROGERS (1982) describe a technique for gravel packing of short intervals. POSTON, TURNER, TOBOLA & BARGER (1986) introduce a technique for two-stage sand consolidation by resin cementation (cf. section 5.9.1.5.4.).

5.10.2. Open-hole gravel packing

Open-hole gravel packing is widely applied where formation characteristics allow barefoot completions and where the sand control installation must permit maximum reservoir fluid flow for higher profitability (SUMAN, ELLIS & SNYDER 1983). The higher productivity with respect to inside gravel packing and also chemical sand consolidation is achieved by the elimination of restrictive casing perforations and improvement of radial flow into the well by enlarging the drainage area of the borehole by underreaming the wellbore which allows the installation of a broader gravel pack mantle between borehole wall and screen or liner than in cased and perforated wells. Following discussion of some advantages and drawbacks of open-hole gravel packing, an outline is given on borehole underreaming as well as port collar implementation and pack washing.

5.10.2.1. Advantages and drawbacks of open-hole gravel packing

The much greater area open to flow of fluids into the wellbore and the corresponding reduction in current velocity provides maximum productivity of open-hole gravel packing which therefore should always be selected if compatible with other completion considerations (ALLEN & ROBERTS 1982). Reverse circulation and crossover gravel packing techniques are commonly used in relatively short open-hole intervals or where wells are not greatly deviated and separation of zones is not necessary. Washing-in liners or screens in open holes bears the danger of formation sand loosening and intermixing with the gravel (MALY & VOZELINEK 1978). Open-hole completions are especially suitable for coarser sands that have some inherent natural stabilization due to grain friction, whereas for very fine and very unconsolidated sands, openhole gravel packing should not be considered (NINI & OWEN 1983).

Open-hole gravel packs are always less expensive than comparable cased-hole packs because of savings on casing, cementing and perforating (ZALESKI & SPATZ 1988). On the other hand, open-hole gravel packs offer less flexibility in plugbacks, selective completions, and remedial treatments for damage removal or premature water production. In comparison with cased-hole completions, it requires four perforations per foot penetrating casing and cement sheath and extending reasonably deep into the formation in order to equal open-hole exploitability. As such a perforation shot density almost resembles a fracture and may in practice be near the boundary of technical performance capability, an opening or cavity into the reservoir beyond the main perforation tunnel is needed to approach open-hole productivity, with such a space having to be held open in unconsolidated sands by gravel or artificial stabilization of the surrounding formation. Open-hole completion exploitability can be considerably increased by underreaming in order to expand the hole to a larger size.

5.10.2.2. Borehole underreaming

Underreaming is frequently carried out during open-hole gravel-pack completion for the purpose of ameliorating reservoir exploitation (HILL 1941) by increasing gravel sheath thickness between wellbore and screen (HALLIBURTON 1988). Open-hole completion by sanding-up following underreaming has been frequently applied for gravel packing of shallow Lower Cretaceous oil-bearing sandstones in Germany FRG and Netherlands (cf. section 5.5.1.). High gravel placement efficiency is ascertained by an uniform diameter throughout the underreamed hole (to be verified by caliper logging; cf. sections 6.2.1.8. and 6.3.7.), avoiding a corkscrewed wellbore profile by underreaming at moderate speed, and thorough circulation of hole opening debris out of the well prior to running liner, screen and gravel.

Experience has shown that frequently good results are obtained by underreaming the well to give a gravel thickness of 3 - 4 in. or even more (for desirable width of the gravel belt or diameter of the gravel plug cf. sections 5.2.3.4. and 5.10.1.), and for the purpose of sufficient safety, the open-hole interval should be underreamed 4 - 6 in. on the diameter to provide the necessary gravel thickness between reservoir wall and sand control hardware, and to remove drilled solids and mud cake from the face of the open hole and the immediately adjoining infiltrated, invaded and deteriorated formation zone (ALLEN & ROBERTS 1982), with maintaining both mechanical borehole stability and fluid compatibility by using properly designed non-damaging completion fluids (SUMAN, ELLIS & SNYDER 1983). Underreaming of open-hole intervals prior to gravel packing, however, can in some cases give rise to serious deterioration of the effect by provoking formation damage by gravel packing which leads to production rate reduction (PRACHNER 1976).

5.10.2.3. Port collar implementation and pack washing

Gravel placement is also enhanced by implementation of a port collar within liner or screen. If insufficient gravel is packed in the annulus, the port collar is closed and the screen is washed while lowering the combination tool. This supposedly dislodges any gravel bridges that have caused voids in the pack, and the port collar may then be reopened to complete the packing operation. Pack washing, however, has to be executed with caution, because added gravel input likely does not come from voids, but from additional hole volume caused by loosened formation sand invading the pack, with this being an effect which is highly undesirable (MALY & VOZELINEK 1978; cf. section 5.8.4.4.). In long open-hole intervals where casing and perforations would be too restrictive, but where pay zones must be separated, external packers and port collars allow selective multiple gravel packing and separation by cementing. Instead of using port collars, also variations of the hydraulic jar can be applied to vibrate the sand into place (MURPHEY, ROLL & WONG 1985).

JONES & THORP (1977) investigate the performance of both inside and open-hole gravel packs with respect to the computed expectation and discuss various reasons for discrepancies between design and result. Open-hole completions have the advantage that the gravel pack is not exerting formation damage by providing an additional pressure loss through the perforations (LEDLOW, SAUER & TILL 1985). As in open-hole gravel packs the wellbore may tend to cave in during or before gravel placement, a longer lower telltale screen allows partial coverage with the slurry being still circulated in place. GOTTSCHLING & LEGAN (1981) characterize long-interval open-hole gravel packing.

5.10.3. Factors improving gravel pack efficiency

STRACKE (1987) outlines the influence of various factors during completion on the success of gravel packing operations and well productivity. Generally, there is no single factor to ensure high-efficiency gravel pack completion, but a combination of factors is involved. The association of underbalanced perforation with a surge chamber and subsequent perforation washing and breakdown produces superior completion results. Acidizing after gravel packing can often yield good results as immediate flow back is permitted but has to be carried out with considerable caution (cf. section 5.11.4.), whereas acidizing during gravel packing should principally be avoided unless there is special demand and experience. The effectivity of the gravel pack treatment can be further improved by increased perforation radius at the expense of formation penetration. Aspects of ameliorating gravel packing techniques for high-rate completions and maximum productivity are also discussed by GURLEY, COPELAND & HENDRICK (1977) and McLEOD & CRAWFORD (1983).

5.10.4. Liner placement

The most commonly used technique for placing the liner into the borehole during gravel packing operations is to wash the liner in after the gravel has been set (SPARLIN & COPELAND 1972), but this should preferably be performed in one continuous job. An alternative method is installation of the liner in the borehole before the gravel is inserted, with this technique being particularly suitable for applications in wellbores with long producing intervals or with low bottom-hole pressures. The gravel is placed by pumping it through a crossover tool (SPARLIN 1969).

The primary advantages of the method of first liner placement and second gravel insertion are avoiding the danger to get the liner stuck in long intervals when adding a joint of pipe during the procedure of washing-in of the liner which is then interrupted by a shut-down, the possibility of performance in wells which do not possess sufficient bottom-hole pressure to support a column of fluid during the liner wash-in operation, elimination of one roundtrip with the tubing, no disturbance of the gravel that is packed inside the wellbore after the formation has ceased to accept gravel, and saving a lot of time on boreholes that require more than one batch of slurry by continuously mixing and pumping the slurry. A combination of both methods of liner placement also has been used successfully, with the hole being bailed clean after the slurry is in place, the liner then being run into the well, and a small gravel slurry volume being pumped down the annulus (SPARLIN & COPELAND 1972). SHURTZ & BREINER (1975) present through-tubing gravel packing techniques.

5.10.5. Wellbore damage in gravel-packed holes

One of the most difficult wellbore cleanup problems is encountered in gravel-packed boreholes (CRAWFORD, COULTER & OSBORN 1980). The gravel pack presents an additional barrier to overcome when bringing mud and solids back into the wellbore, and particularly lost drilling mud is difficult to flow through a gravel pack. An acid system containing effective complexing and dispersing agents has turned out to be an optimum solution to various cleanup problems of wellbore damage in gravel-packed holes (CRAWFORD, COULTER & OSBORN 1980). Another possibility is the application of oil-continuous fluids all the time during open-hole gravel pack completions, with oil-based mud effectively preventing formation deterioration and providing hole stability during underreaming and gravel placement. STADALMAN, NOVOTNY & HOUCHIN (1985) emphasize that understanding wellbore pressures during changing borehole conditions improves sand control longevity. Other aspects of wellbore and formation damage during gravel packing are discussed in chapter 5.11.

5.10.6. Water control in gravel packed completions

The design of high-volume gravel packed completions has considerably evolved during the last 15 years, with the main advances having been increasing of the area open to flow and optimization of wellbore transmissibility of fluids (CARROLL & BULLEN 1988). High formation fluid transmissibility in the completion near the borehole, however, renders it often difficult to control unwanted water production. Setting a plug or a nipple at predetermined intervals in the screen assembly is unsuccessful, because water production bypasses the plug in the ribbed section of the screen or in the high-permeability gravel in the screen-casing annulus. An alternative approach is to locate the water production interval and place material above or within it (cf. section 5.9.1.7.) in order to seal the flow area in the screen and to eliminate gravel permeability at this patch.

A basal plug of sand or cement is set, with depth depending on water entry, perforations, and formation permeability barriers. Plastic is spotted on the base, and plastic set time controlled by additives and reservoir temperature has to be adjusted such as to allow enough time for fluid displacement in the gravel packed completion and total radial penetration of the plastic into the gravel pack which must be achieved in order to restrict vertical movement of produced fluids. Water-production zone determination can be performed by pulsed neutron logging (cf. section 6.3.3.4.). This water-control technique represents secondary gravel consolidation by resin cementation which in contrast to the conventional applications of resin bonding does not have the purpose of gravel stabilization with permeability preservation, but follows the aim of local permeability destruction of the formerly highly porous gravel pack in order to shutoff intervals producing water instead of gas or oil (cf. also section 5.9.1.7.). Improved polyacrylamide treatments for water control in producing wells are outlined by ZAITOUN, KOHLER & GUERRINI 1989).

5.11. Formation damage during gravel packing

In contrast to hydraulic fracturing where almost no earlier damage can affect the freshly generated crack prior to its opening and all phenomena of formation and fracture deterioration are directly linked to the hydraulic stimulation process itself, the open or cased borehole wall to be gravel packed can already have been subjected to various types of formation damage prior to insertion of the sand control filter which consequently later affect the performance of the gravel pack in addition to the features newly originating when setting the sand supervision mantle in place.

Moreover, the design of grain size and thus permeability of the gravel pack follows other criteria than that of grain size and thereby conductivity of the proppant package within the hydraulic fracture (cf. section 1.3.2. and 1.4.10.), with the gravel pack grain size also being limited in upper direction by formation grain size to be withheld, and therefore impairment of transmissibility not being able to be compensated by considerable increases in grain size as could be done in hydraulic fracturing provided the coarser particles could be safely transported to their destination within the crack (cf. section 4.3.5.). The discussion as follows includes perforation tunnel plugging, drilling mud loss, completion fluid loss, formation damage removal by acidizing, and fluid and hardware cleanliness.

5.11.1. Perforation tunnel plugging

Perforation tunnel plugging is the most significant part of the whole gravel packing operation, because unless the perforation holes are properly infilled with tightly packed gravel, the goal of reservoir sand stabilization and pressure drop minimization through the gravel pack cannot be achieved. Prior to commenting on specific aspects of necessity of clean open perforation holes for tight gravel infilling and impact of perforation hole diameter, some general comments are offered on reservoir impairment by the gravel pack as well as covering of damaged formation by the gravel pack.

5.11.1.1. Reservoir impairment by the gravel pack

Gravel packing is also itself representing productivity impairment, because it always provides an additional pressure loss to the well system (LEDLOW, SAUER & TILL 1985). Gravel packs are especially susceptible to impairment, because the gravel is trapping the damage in place and at least complicates, if not completely prevents its removal by mechanical or chemical means (SPARLIN & HAGEN 1986). Formation deterioration prior to or during course of hydraulic fracturing is in most cases equalized by the deep penetration of the crack which breaks through the comparatively narrow damage zone surrounding the borehole and exposes such a long section of reservoir matrix along the high-permeability drainage avenue that pay deterioration at the fracture walls is compensated by the large accessible area of the long crack that in addition is characterized by a very high conductivity contrast to the formation face.

5.11.1.2. Covering of damaged formation by gravel packing

Gravel packing, however, is only covering and infilling the pre-existing borehole geometry and volume, and if parts of the original surface area are deteriorated, this has at once serious impact on gravel pack performance and in the worst case can completely deteriorate its effectivity. Thus formation damage is a much more serious concern in gravel packing than in hydraulic fracturing and special attention is necessary during drilling, testing and completion to keep influences deteriorating reservoir properties as limited as possible.

Many basically successfully gravel packed wells suffer from reduced productivity as a result of formation damage induced by gravel pack completion practices, and gravel packing of even an only slightly deteriorated reservoir can already result in long-term detrimental effects on production (HOUCHIN, DUNLAP & HUTCHINSON 1988). The primary objective of gravel packing completions is limiting pressure drawdown across perforations and sand control filter.

5.11.1.3. Necessity of clean open perforation holes for tight gravel infilling

The most serious problem is that because the perforation tunnels are not completely open prior to gravel packing, optimum fluid leakoff is not attained and dehydration of the gravel carrier fluid is restricted, the perforation tunnels cannot be tightly packed, voids originate in the screen-casing annulus, and therefore the goal of little or no pressure drawdown across the perforations is not reached. Fluid leak-off during gravel packing may be optimized by simultaneous processing of the gravel pack slurry with low-strength acid, whereas subsequent injection of acid may result in creating or expanding existing tunnel voids and pack bridges in the screen-casing annulus (McLEOD 1986). The areas of interest in which formation damage can be limited during gravel packing include perforation, filtration, carrier fluid selection, lost circulation control and workstring hygiene (HOUCHIN, DUNLAP & HUTCHINSON 1988).

5.11.1.4. Impact of perforation hole diameter

Concerning reduction of formation deterioration during perforation (McLEOD 1978, LOCKE 1980), selection of perforation geometry design which includes effective shot density, penetration depth, gun phasing and tunnel diameter can further influence well productivity through additional pressure drops (BELL 1982). As gravel packing involves another pressure drop experience across the filled perforations (McLEOD & CRAWFORD 1983), larger tunnel diameters are required to compensate or to minimize this additional pressure drop (cf. also section 5.6.1.). Large-diameter shots, however, cannot be chosen in highly washed-out zones where the cement sheath is so thick that the perforation charge cannot penetrate it, with tunnel diameter selection thus having to be based on borehole quality and integrity verification by caliper log interpretation (cf. section 6.2.1.8.).

Because perforation area is sacrificed as a function of penetration depth, considerable improvement can be achieved with large-diameter tunnels if they can be properly installed (HOUCHIN, DUNLAP & HUTCHINSON 1988). Damage removal in completely or partially plugged perforation holes can be achieved by washing, surging and acidizing (cf. section 5.11.4.). Underbalanced perforation techniques have the benefit that the crushed material particularly accumulating at the end of the perforation tunnel is at least partially removed by the surge effect (cf. section 5.6.1.4.). Additional cleanup of the rock powder can be achieved with low-strength mud acids and with brines.

Serious damage in gravel packing can be caused when high-density gel/gravel slurry is pumped down into the perforations (McLEOD, LEDLOW & TILL 1983). Pumping this gel/gravel slurry down dirty pipe will squeeze pipe dope, mill scale and other contaminants into the perforations (cf. section 5.11.5.).

5.11.2. Drilling mud loss

Productivity problems in many gravel packing completions are attributable to inadequate removal of lost circulation additives from the drilling period (SPARLIN & HAGEN 1986). This drawback, however, is inherent in the purpose of a lost circulation system which is to create formation damage in order to control and to stop excessive drilling mud leakoff. If deterioration caused by lost circulation material is not removed prior to gravel packing, sufficiently high leakoff rates of the gravel carrier fluid which are essential for perforation tunnel plugging and slurry dehydration will be prohibited (McLEOD & CRAWFORD 1983).

Various mechanical and chemical methods of lost circulation fighting exist, with particulate solids being most common for controlling severe drilling mud leakoff. Solids achieve fluid-loss containment by forming a filter cake along the borehole wall at the reservoir face, invasion of pore space and absolute filling of perforation tunnels. Each of the particulate solids as well as chemical agents having been inserted for the purpose of lost circulation control must be removed prior to gravel packing in order to optimize carrier fluid leakoff and gravel placement (HOUCHIN, DUNLAP & HUTCHINSON 1988). In extreme cases, particulate bridging can be so tight that complete removal is impossible or chemicals that have to be applied for cleaning have in turn a negative effect on formation permeability.

5.11.3. Completion fluid loss

In addition to compensation of the effects created by drilling mud loss control techniques, gravel pack completion fluid leakoff should be restricted to the necessary volumes during actual placement, and further fluid escape into the formation prior to or after setting of the gravel mantle should be avoided, as this unnecessary fluid loss may also cause reservoir damage. After discussion of some general aspects, the significance of solid fluid-loss additives as well as their insertion and subsequent removal is illustrated. Comments are also offered on the application of gravel as diverting agent.

5.11.3.1. General aspects

Density control of gravel packing completion fluids is a viable leakoff government technique particularly in low-pressure formations and should be reinforced by pore pressure data, because drilling mud weight correlations can be inaccurate and give almost always too high values. Fluid density control may be achieved with nitrogen-generated foam which is capable of maintaining its stability throughout the completion operations. Fluid density control is also a valuable technique for amelioration of selective proppant placement in hydraulic fracturing stimulation (cf. sections 4.2.2.3.3.3. and 4.2.2.6.). The following outline illustrates some aspects of mechanical fluid-leakoff control and types of fluid-loss additives.

5.11.3.1.1. Mechanical fluid-leakoff control

Crosslinked polymers have to date provided the best success due to their ability to maintain foam quality for extended periods of time in a broad temperature and pressure range. Mechanical fluid-loss control is accomplished by carrying out a one-trip completion with a combined perforating/gravel packing downhole assembly which limits the amount of fluid leakoff during completion by reducing the time of formation exposure, but is not applicable when perforation surging or washing is required. Completion fluid loss to the formation following perforation cleaning by washing, surging or acidizing has to be limited as much as possible, because this creates additional damage in the perforation tunnels which prohibits proper leakoff of the gravel packing carrier fluids.

5.11.3.1.2. Types of fluid-loss additives

Fluid-loss additives are frequently used in completion fluids and pills in order to reduce fluid costs and formation damage, but many of these agents in fact create more deterioration than they prevent (SPARLIN & HAGEN 1986). Therefore common recommendations are to use fluid-leakoff additives that will be dissolved by the produced fluids, acid or water, and thus can be readily removed after a job, with typical additives being oil-soluble resin, calcium carbonate particles and salt suspensions. Silica flour is more difficult to dissolve and requires hydrofluoric acid utilization which includes the danger of packing gravel attack and dissolution that gives rise to pack slumping and therefore should be avoided.

While it is particularly important that soluble additives are used for wells that are to be gravel packed since gravel is trapping insoluble solids in the borehole, even some soluble materials may remain in perforation tunnels and impair well productivity. Excess fluid-loss additives also prevent complete coverage of a zone with gravel either due to inadequate leakoff to the formation or partial plugging of the screen. It is therefore imperative that these agents are removed by acid water or solvents preceding the gravel slurry, but if their removal causes circulation loss, the results are equally bad.

5.11.3.2. Application of gravel as diverting agent

Thus the most practical solution is using a combination of gravel and viscous fluids as a diverting system for the liquids that removes the additives without creating additional reservoir damage (SPARLIN & HAGEN 1986). When this system is used just ahead of a gravel slurry, it not only helps to remove the fluid-loss agents, but also helps to prevent excessive fluid leakoff and partially prepacks the formation with gravel. Excessive fluid-loss rates which are common in weak or unconsolidated high-permeability sandstones not only hinder complete zone coverage by gravel packs, but also may lead to blowouts. The combination of solids and gelled fluids effectively controls fluid leakoff in most wells.

5.11.3.3. Insertion of solid fluid-loss additives

The most commonly used fluid-loss additives are calcium carbonate particles which are acid-soluble, resin dispersions that are oil-soluble, and saturated salt brine that can be removed by water. Silica flour which is only soluble in hydrofluoric acid is replaced by calcium carbonate that can be dissolved by hydrochloric acid whenever possible in order to prevent pack destabilization by attack of gravel and sand by hydrofluoric acid (cf. also section 5.11.4.3.). Calcium carbonate controls fluid loss by invasion of formation pores, bridging at the reservoir entrance by filter cake generation, and filling of perforation tunnels.

Smaller grinds supervise fluid leakoff more effectively than larger grinds, but potentially cause formation damage as they are more difficult to remove. Larger and smaller grinds are needed for reservoirs with higher and lower permeability having broader and narrower pores, respectively. If too fine grinds are used, the particles are invading deeply into the pay zone matrix and do not effectively reduce fluid-loss rate unless high concentrations or large volumes of particles are applied, whereas if too coarse grinds are utilized, only few of the particles migrate far into the formation, but the high-permeability filter cake is less effective in reducing fluid leakoff.

5.11.3.4. Removal of solid fluid-loss additives

Removal of solid fluid-loss additives is most commonly performed by acidizing if the components are soluble. Incomplete elimination and withdrawal of fluid-leakoff agents prevents proper gravel placement in perforation tunnels and deteriorates the whole gravel packing operation.

5.11.3.4.1. Acidizing

In terms of removal of fluid-loss additives by acid, the net effect of the calcium carbonate-filled formation pores, filter cake, and plugged perforation tunnels at equilibrium is to cause the acid to initially flow uniformly into all zones, but uneven distribution in reservoir sections with different permeability leads to preferential calcium carbonate dissolution in low-permeability intervals that are less likely to have been filled before cleaning starts in the high-permeability portions that have accepted larger volumes of solids. Once acid has removed the calcium carbonate from some of the low-permeability zones, acid will then follow this path of least resistance and bypass the remaining damaged intervals, and the acid front will also become neutralized, with the only way that additional reaction can take place being to increase pressure and to force additional acid into the plugged zones which, however, may give rise to exceeding the formation fracture gradient and resulting pressure parting of the reservoir. On the other hand, if gravel is pumped immediately after acid injection, the gravel traps the undissolved calcium carbonate in perforation tunnels and formation pores.

5.11.3.4.2. Problems of incomplete fluid-loss additive removal

Incomplete removal of fluid-loss additives may cause considerable gravel placement problems. If solids settle out of a pill and pack around the lower telltale screen (cf. section 6.3.5.) or lower part of the main screen, acid bypasses this packed calcium carbonate and is either injected into the formation or circulated through the upper part of the main screen, with in either case being no assurance that gravel will be packed around the lower screen interval. If calcium carbonate-laden fluid is circulated through the screen before the acid has arrived on bottom, any particle sizes equal to or larger than the screen openings are plugging preferentially the lower sections of the screen as they follow the fluid flow. If circulation is through the lower telltale screen, it may become completely obstructed and prevent acid from contacting it, and could be very difficult to clean without pulling it out of the hole. Any perforations that are filled with calcium carbonate do not accept gravel, and where fluid flow is restricted, gravel will not be tightly packed. Therefore all of the fluid-loss agents used prior to gravel packing must be removed from wellbore, perforations and near-borehole formation to assure a complete gravel pack.

Another important item requiring attention is pipe dope that can seriously impair gravel pack and formation. Unless the pipe dope is handled very carefully, other safety precautions are almost useless, because pipe dope damage of formation and/or gravel pack can screw up the whole sand control job. Pipe thread components contain metallic additives to lubricate and help threads seal which are not soluble in acids and other economically available materials, with the best handling approach being to eliminate closing the box end of the tubing and to paint the pin with a small brush.

5.11.4. Formation damage removal by acidizing

Formation damage during gravel packing can further result from dirty workstrings when pipe dope deposits can become serious, dirty completion fluids which are not properly filtered and leave plugging residues in the pore space of both reservoir and gravel pack, and insufficient breaker performance in the carrier fluid. Acidizing is a very common remedy action for various types of deterioration. Distinction can be made between acidizing prior to and after gravel packing. Comments are also offered on gravel unconsolidation and dissolution by acidizing.

5.11.4.1. Acidizing prior to gravel packing

If any potential of formation damage exists prior to gravel packing, acidizing should be carried out as remedial treatment (CRAWFORD, COULTER & OSBORN 1980; McLEOD & CRAWFORD 1983; LEDLOW, SAUER & TILL 1985). The design of a pre-gravel pack acid job, however, is critical, because in case of poor performance, the acid can create more deterioration than previously existed. Low-strength acid reduces the possibility of precipitating acid by-products (McLEOD 1984), and the acid should be diverted to achieve as close to uniform placement as possible (SPARLIN & HAGEN 1986). In order to prevent over-acidization of particular portions of the perforated interval and to promote uniform acid distribution, the size of the stages should be kept to a minimum and sand-laden pills should be injected between the acid stages. The following outline illustrates some aspects of gravel and other materials as diverting agents, diverting agent injection and distribution, and formation wetness and clay stabilization.

5.11.4.1.1. Gravel and other materials as diverting agents

The combination of carrier fluid viscosity and absolute perforation filling by gravel provides the diverting mechanism. The acid should be overflushed into the formation, because the spent acid will remain in the reservoir for an extended period of time before it can be produced. Overflushing the acid a maximum distance from the wellbore eliminates the potential of productivity loss due to acid-induced damage. An additional benefit of the gravel-laden pill diverting technique is perforation pre-packing which is particularly valuable in a circulating gravel pack (cf. section 5.6.7.). Previous pre-packing techniques have involved filling of casing and perforations with gravel pack sand prior to running the gravel-pack installation assembly downhole (PENBERTHY & COPE 1979), with the gravel inside the casing later being washed out. This technique, however, allows for substantial intermixing of gravel and formation sand and is therefore not advisable (HOUCHIN, DUNLAP & HUTCHINSON 1988).

Acid diversion to all perforations during a pre-gravel pack treatment is best accomplished by using the gravel itself as diverter (SPARLIN & HAGEN 1986). The acid treatment is divided into several stages, and gravel carried by gelled brine is injected between the acid stages. Gravel concentration in the gelled brine should not be excessive and is usually in the range of 1 - 5 lbs/gal. The gravel volume in the first and second diverter slurries should be designed to fill abt. 10 % of the perforations each. In wells that have produced a lot of sand before the job, the volume of diverter gravel must be increased. When the diverter gravel slurry enters the perforations, operating pressure increases, because the effect of gravel filling perforation tunnels and voids outside the perforations increases the resistance to flow of the gelled fluid as the postpad is squeezed into the gravel-packed perforations. The effect of gravel-filled perforations on injection pressure depends on number of perforations that are gravel packed and/or cleaned by acid, severity of plugging of other perforations, undamaged injectivity of the completed interval, downhole dynamic viscosity of the gelled brine, gravel permeability and injection rate.

5.11.4.1.2. Diverting agent injection and distribution

The ideal situation is when acid is being lost to the formation at a rate and pressure equal to the injectivity into the undamaged completion interval. The diverting system for pre-gravel pack acidizing using gravel and gelled brine increases the pressure drop through empty perforations and forces acid into calcium carbonate-filled perforations which improves well cleanup and provides a means of maintaining fluid returns during the gravel packing process. Other suitable diverters are oil-soluble resins and nitrogen foam (CROWE 1971, NIERODE & KRUK 1973; BRANNON, NETTERS & GRIMMER 1987). Continuous injection of the diverter at high concentrations helps to establish acid distribution to the whole interval, because the success of well stimulation depends on proper placement of treating fluids into damaged or low-permeability zones and away from higher-permeability layers. Diverting agent added continuously to treating fluid equalizes the flow in all intervals, and diverting agent injected in small stages ahead of treating fluids can cause a highly skewed distribution of the injected fluid.

Oil-soluble resin can also be used as matrix diverting agent at low concentrations for acidizing through gravel packs (NIERODE & KRUK 1973, KING & HOLLINGSWORTH 1979) and is an effective fluid-loss additive in fracture acidizing of carbonates at high saturations. As oil-soluble resin builds a very thin efficient filter cake that is easily and rapidly soluble in hydrocarbons, its application as both diverting agent and fluid-loss additive also through gravel packs is without any danger of damaging well productivity.

5.11.4.1.3. Formation wetness and clay stabilization

Sands with high concentration of mobile water-wet clay minerals bear the danger that the fines migrate towards the wellbore once water production starts and invade the gravel pack near the perforations where bridging can result in productivity impairment (THOMAS & CROWE 1978; McBRIDE, RATHBONE & THOMAS 1979). Fluoboric acid generating hydrofluoric acid in-situ can both stimulate and stabilize clays in poorly consolidated formations, with the treatment being able to be carried out both before and after gravel packing operations. Thickened versions of fluoboric acid can be used also as transport fluids for high-density gravel packing (CONSTIEN & WALLACE 1980), with the system retaining the properties of the acid being retarded hydrofluoric acid generation (cf. section 4.5.5.4.3.) as well as clay stabilization by borosilicate formation and also developing adequate viscosity to transport gravel up to 15 lbs/gal.

Continuation of injection of diverter slurries and acid until all the perforations are filled with gravel is not practical, as it is not possible to count on all the perforations being open to accept the gravel. If too much gravel is injected, some of it will cover the lower telltale screen and interfere with gravel placement. Therefore as a practical guideline the combined volume of gravel in the diverter slurries should not exceed 30 % of the total gravel volume estimated to be packed outside of the casing during gravel packing.

Pre-gravel packing acidizing, however, is in many cases not the best solution, as it is more beneficial and economical to first bring the well on production, evaluate the damage and then second administer a full-scale acid job through a coiled-tubing unit (BRANNON, NETTERS & GRIMMER 1987).

5.11.4.2. Acidizing after gravel packing

Acidizing after installation of the gravel pack in order to remove calcium carbonate fluid-loss additives must be done very carefully and should be avoided if possible (SPARLIN & HAGEN 1986). Fluid injected into the gravel pack may disturb it and push the gravel away from the wellbore into the void spaces generated by carbonate dissolution in perforation tunnels or outside of the casing as well as into cavities created by dissolution of native reservoir material. The undesirable results of this procedure are formation of holes in the gravel pack and/or intermixing of packing gravel and reservoir sand. In particularly severe cases, the gravel column can settle considerably in the annulus and can even expose the upper part of the perforated section as a consequence of destruction of pack integrity by acidizing. Some aspects of selective acid penetration vs. perforation tunnel content as well as difficulty of fluid-loss agent removal through the gravel pack are outlined as follows.

5.11.4.2.1. Selective acid penetration vs. perforation tunnel content

Although there may be dramatical increases in production rates by post-gravel pack acid jobs dissolving residual calcium carbonate fluid-loss additives, multiple acid treatments are often required to approach an undamaged production rate, because it is very difficult for the acid to contact all of the calcium carbonate in a gravel packed well, as acid can be injected through perforations filled with gravel much more easily than through those plugged with calcium carbonate due to considerable permeability differences. Perforations cleaned by pre-gravel pack acidizing accept the post-gravel pack acid much more readily than those which remain filled with calcium carbonate (SPARLIN & HAGEN 1986). if there is adequate leakoff rate through some unplugged perforations, the acid does not remove calcium carbonate packed in the perforations, because there is not enough pressure drop across them to force unspent acid through them. In any case, it is thus far better to circumvent the need for an acid job after gravel pack installation by removing all fluid-loss additives by pre-gravel pack acidizing which provides higher hydrocarbon productivity and more positive sand control.

5.11.4.2.2. Difficulty of fluid-loss agent removal through the gravel pack

An even more difficult problem is removal of calcium carbonate that has been injected into a gravel pack for the purpose to control excessive fluid leakoff while preparing the well for production after a gravel pack or while installing a gravel pack in an upper zone. If the particle sizes of the additives in such fluid-loss pills are such that the particles flow through the gravel and generate a filter cake on the formation face, the calcium carbonate would be most difficult to remove as it fills gravel porosity in some of the perforation tunnels. Every effort should be made to prevent killing a borehole that has been gravel packed with a fluid-leakoff pill by using tools that isolate the gravel pack, but if a calcium carbonate pill is inevitable, the median particle size of the fluid-loss additive should approximately equal the median pore size of the gravel. This will minimize calcium carbonate invasion into perforation tunnels and formation matrix and should nevertheless adequately control fluid loss rate. Complete calcium carbonate removal, however, may also require an alternate fluid-loss agent in the acid treatment.

Improvement of matrix acidizing through gravel packs and prevention of subsequent productivity decline can be performed with fluoboric acid which generates hydrofluoric acid in-situ that can deeply penetrate into the formation and which stabilizes reservoir fines by fusing them (McBRIDE, RATHBONE & THOMAS 1979; KUNZE & SHAUGHNESSY 1980; cf. section 4.5.5.4.3.). SCRIMGEOUR & SHEPHERD (1988) report successful post-pack acidizing of gravel-packed wells with half-strength mud acid for the purpose of obtaining the target productivities and injectivities, whereas ROGERS (1976) documents formation and pack damage by post-gravel packing acidizing. TORREST (1982 d) outlines injectivity testing for gravel pack quality determination. MURER (1981) reports good results of post-gravel packing acidizing with mixtures of hydrofluoric and hydrochloric acid which removed migrating fines and calcite scale that had given rise to reservoir and gravel permeability damage and which enabled in abt. 50 % of the cases achievement of considerable additional oil production.

5.11.4.3. Gravel unconsolidation and dissolution by acidizing

While common processes of reservoir and gravel damage after acidizing are dispersion and plugging with fines and clay minerals, other effects of acidizing are gravel unconsolidation and dissolution (McLEOD 1986). Gravel may be flushed from perforation tunnels during acidizing into dissolved formation voids or acidized gravel cavities in the screen-casing annulus, or perforation tunnel voids may have existed prior to acidizing as a result of poor gravel placement. Production start-up rate controls how reservoir sand enters partially filled voids after acidizing, with moderate offtake rates allowing formation sand to enter the tunnel above the gravel so that a more conductive parallel alignment of sand and gravel in the tunnel is created. After an outline of gravel and reservoir plugging by migrating fines, gravel dissolution and unconsolidation together with formation sand movement are illustrated.

5.11.4.3.1. Gravel and formation plugging by migrating fines

Productivity impairment and restriction after gravel packing which cannot be satisfactorily explained by gravel damage in the perforation tunnels can be reasonably modelled by reservoir sand deterioration adjacent to the gravel packed holes. Concerning movement and trapping of fines following gravel pack acidizing, an entrapping mechanism is necessary for the dispersed fines to cause plugging near perforations. Apart from clay stabilization additives, hydrofluoric acidizing with sufficient overflushing of the spent acid has a consolidating effect itself. When fines move easily through formation sand that has been cleaned by acidizing, the interface between reservoir sand and packing gravel offers the first restriction for fines entrapment. The acid reaction front creates trash along with possible oil sludge that could cause plugging during return flow if rates are too high. Slow startup rates allow some of this material to separate and to be produced through the gravel pack, whereas fast startup rates could dislodge the trash and send it at once to the sand/gravel interface where it can be trapped and form a filter cake that in turn will trap other fines moving out of the formation.

5.11.4.3.2. Gravel dissolution

Apart from fines concentration, gravel pack productivity decline after acidizing is caused by unconsolidation and dissolution of gravel. Once hydrofluoric acid has cleaned formation sand from the clay-sized fines, it works more slowly on the remaining larger silica grains being packing gravel and reservoir sand (GATEWOOD, HALL, ROBERTS & LASATER 1970). Thus some reduction in volume of the gravel pack will occur as a consequence of hydrofluoric acid dissolution especially at higher temperature in deeper pay zones (CHEUNG 1985). Some comments are offered on gravel grain size and acid concentration as well as mobility of loose packing gravel and pay sand.

5.11.4.3.2.1. Gravel grain size and acid concentration

The gravel has a better chance to stay intact with decreasing mesh size. As gravel dissolution and the resulting framework unconsolidation are serious deteriorations of gravel pack structure and effectivity, acid jobs should be designed to dissolve not more than 5 % of the gravel. This can be achieved by reducing effective acid concentration, using weaker hydrofluoric acid, or applying retarded acids with lower available hydrofluoric acid saturation to limit acid attack on the gravel (McLEOD 1986; cf. section 4.5.5.4.3.). Thus the utilization of silica flour as solid fluid-loss control additive which can only be removed by hydrofluoric acid application should be avoided if anyhow possible, because silica flour diminution and withdrawal is automatically linked with packing gravel and formation sand attack by hydrofluoric acid which provokes framework destabilization of reservoir rock and gravel pack.

5.11.4.3.2.2. Mobility of loose packing gravel and pay sand

Excessive gravel dissolution creates a loose pack where gravel can be displaced from perforation tunnels into the wellbore during hydrocarbon production after acidizing, thereby partially emptying the perforations from their supporting gravel plug. Moderate gravel dissolution is generally compensated by first falling of gravel reserve in the annulus downwards to equalize voids before gravel is moving from the perforation tunnels back into the wellbore, but the risk always exists that overacidization of the gravel generates dangerous voids which will not all be filled by slumping of the gravel reserve above the screen in the annulus (CHEUNG 1985).

Temperature and hydrofluoric acid concentration are the most important parameters controlling gravel dissolution (McLEOD 1986). Formation sand size reduction by dissolution is generally no problem, because even if the large particle sizes are diminished, acid removes the fines thereby leaving the same average particle size as before acidizing. At high temperatures when a lot of sand is dissolved, however, the large fraction may be sufficiently decreased in size to invade the gravel packed perforation tunnels.

In addition, reservoir sand always has an entry into tunnels where original gravel insertion has not been complete. If perforation tunnels are only partially filled with gravel due to formation damage before and during gravel pack placement, subsequent flow can produce hydrocarbons at reasonable rates until critical pressure drawdown is exceeded which gives rise to collapse of stable open perforations and entering of the tunnels by reservoir sand (McLEOD & CRAWFORD 1982). Formation sand which has been completely unconsolidated by hydrofluoric acid is free to enter partially filled perforation tunnels (McLEOD 1986) and restricts well flow even if the reservoir sand in the tunnels is a high-quality acid-cleaned material which has a higher permeability than originally due to removal of clay minerals by acidizing.

5.11.4.3.3. Gravel unconsolidation

Flow velocities in the perforation tunnels are sufficient to flush part or all of the gravel from the holes if a void exists outside the tunnel or is created by acidizing at high injection pressures. If acid coverage is poor in case of no diversion, only a fraction of the perforations may be taking acid, thereby resulting in high velocities and flushing gravel out of those tunnels while the others remain tightly packed. When hydrocarbon flow resumes, formation sand enters the holes or mixes with the gravel as it is pushed back into the tunnels. If acid coverage is complete, all perforations are open with equal flow rates per perforation, but flushing of gravel still takes place in all perforations to a lesser extent, and gravel will be slowly displaced out until the velocity above the remaining gravel falls below the critical velocity for saltation flow, with the result being that all perforations are partially filled with gravel.

When perforations are partially flushed, high startup rates can move gravel and sand into the tunnel where both are packed in series with sand on the outside position. At low startup rates, the larger gravel is still lying, and the void above the gravel will be gradually filled with sand as production rate is increased, with the latter condition giving rise to parallel flow paths through formation sand and gravel in the tunnel. As more flow resistance is present when sand and gravel are in series, slow startup after acidizing and limiting pressure drops across the completion are suitable instruments to restrict the amount of reservoir sand coming into the perforation tunnels or shifting the gravel to allow more influx of formation sand.

5.11.5. Fluid and hardware cleanliness

One of the most critical steps in designing and executing a sand control completion is wellbore preparation (LEDLOW, SAUER & TILL 1985). Both tubing work strings which are frequently used for cement squeeze jobs and new tubing strings containing mill scale and rust are undesirable. The most efficient solution is water-blasting the tubing or have an extra string of tubing that is exclusively used for sand control. When water-blasting is not feasible, the tubing can be cleaned with aromatic solvents and acid, with a good procedure being to run the tubing into the hole prior to perforating and circulating an aromatic solvent to spot, and following reverse-out spotting acid which is then also reversed out just before the completion fluid is placed (ECONOMIDES 1986).

Pipe dope should be minimized, with the pin ends being doped sparingly with a fine brush only when required while tripping in the hole in order to minimize plugging of the screen slots and completion fluid contamination (NINI & OWEN 1983). The completion fluid should be designed to have an overbalance of abt. 100 - 150 psi for oil wells and abt. 300 psi for gas wells to control the formation which has to be lower than drilling mud overbalance that is usually abt. 300 - 500 psi, because lack of solids in the completion fluids allow free flow into the reservoir and thus higher overbalance results in excessive fluid loss. In most cases, the drilling fluid left in the casing is heavier than the completion fluid, with the density and viscosity difference between both fluids favouring reverse circulating when swapping out these fluids.

6. Propped Fracture and Gravel Pack Monitoring

6.0. Summary and associated table of contents

Planning and execution of hydraulic proppant fracturing and gravel packing require close monitoring in terms of prediction, supervision and retroactive evaluation by various wireline well logging, seismic and other techniques in order to control mainly height and azimuth of the cracks and to improve the design of current and future treatments. The most important logging methods for fracture height determination are temperature, gamma-ray, acoustic and noise recording. Temperature logging aims on distinction of the cold fracturing fluid pumped into the crack from the warm formation fluid and is a simple and cheap technique if carried out up to 24 hrs after termination of the hydraulic stimulation operation prior to thermal equalization of the borehole surroundings, but can be negatively influenced by anomalous thermal reservoir conductivities. Conventional gamma-ray logging of naturally radioactive zirconia-silicate proppants provides propped fracture height in contrast to created (fluid-penetrated) fracture height as evidenced by temperature display and is also a straightforward and affordable technique, but suffers from frequently inhomogeneous radioactivity and the high price of zirconia-silicate proppants. Thermal decay time monitoring provides permeable fracture height which comprises the section taking up stimulation fluid as well as contributing to hydrocarbon production and therefore represents effective fracture height in terms of reservoir exploitation. Synthetic tracering of fracturing materials by coating of proppants with radioactive pellicles, intermixing of radioactive grains into the proppant lot, and addition of solid or liquid markers to the fracturing fluids has the advantage of insertion of controlled concentration and intensity of radiation which can be easily distinguished from the natural formation background radioactivity. While conventional single- and double-marker monitoring has hitherto been commonly applied either alone or in combination with temperature logging as simple and cheap methods, the possibilities of interpretation are limited. Although being considerably more expensive, multiple-tracer discrimination by gamma-ray spectroscopy is the most sophisticated, advanced and unambiguous fracture height monitoring technique, because different markers can be added to separate proppant and fluid lots in staged and multi-horizon treatments, and distinction of natural and/or artificial radiation coming from fractured formation, and synthetic radiation deriving from cement sheath around the wellbore and borehole fluids can also be performed. Multiple-tracer discrimination by gamma-ray spectroscopy should therefore be applied in all hydraulic fracturing treatments where the results are decisive for the design of other wells of the stimulation campaign and where several discrete fluid and/or proppant steps have to be distinguished. Future innovation of oriented gamma-ray spectrometry could provide an unique combination of horizontal and vertical fracture determination. Distinction of practically up to four tracers can be made based on different energy levels and/or half-life times, with in the latter case also time-lapse monitoring allowing to control changes of fracture height due to tip closure as a consequence of proppant crushing with increasing closure stress during progressive reservoir depletion and pressure drawdown. Fracture determination by acoustic logging is mainly carried out by analysis of sonic wave transformations and particularly exploits attenuation and absorption of circumferentially travelling Rayleigh and Stoneley fluid-formation interface or tube waves. Noise logging records fluid movement in the fracture and is mainly used for verification of temperature anomalies which can come from thermal formation conductivity anisotropies or from the fluid-filled fracture. Caliper logging with borehole eccentricity evaluation is the most important technique of fracture azimuth prediction because of the intimate relationship of geostress field orientation and wellbore ovalization. The long axis of the elliptical borehole cross-section is usually parallel to the direction of least principal formation stress which is perpendicular to fracture orientation, with the fracture thus propagating normally to the long axis of the ellipse. Borehole television and formation microscanning are together with spinner surveying important techniques for direct observation and measurement of fracture height in contrast to other logging methods which are based on inferences due to indirect interpretation and calculation of recorded data curves. Acoustical and electrical imaging of the wellbore wall by televiewer and formation microscanner allow detailed visualization of sedimentary and structural reservoir features which can only be matched or surpassed in terms of quality by expensive coring of the pay sequence. In addition to determination of hydraulically-induced fractures, detection and evaluation of natural fractures can be made by various well logging techniques such as formation microscanner and borehole televiewer (these methods also allow distinction of natural and drilling-induced cracks), dipmeter logging, dual laterolog recording, and sonic logging with full waveform analysis. As many different open-hole logging methods are affected by fluid-filled fractures which cause anomalous readings, a fracture probability program can be developed where various individual crack indicators are simultaneously used to compute the likelihood of a zone of being naturally jointed. The composite fracture index log assigns a level of plausibility to each response and produces a log that reflects the cumulative probability of all fracture indicators. The mechanical properties log which is computed from sonic and density input logs is a predictive tool for retroanalysis of drilled wells and prognosis derivation from the results of an earlier borehole for the better and more effective operation of a subsequent well. Mechanical properties computation enables hydraulic fracture propagation modelling and migration analysis and is particularly significant for proper design of multi-horizon treatments by discrimination of different formation fracturabilities. Seismical and gravimetrical methods aim in contrast to most well-logging techniques not or not only on fracture height determination, but are mainly applied in order to detect crack azimuth. Fracture height monitoring can be performed by in-situ well-seismic detection with an accelerometer that operates principally like a geophone, with either active excitation of the formation or passive evaluation of microseismic events produced by the closing crack. Fracture azimuth has major implications on reservoir exploitation strategy by being in the direction of maximum horizontal in-situ stress and thus being significant information for well pattern and spacing modelling in order to most suitably drain the reservoir with as few boreholes and fractures as possible including avoiding of crack interference. Fracture azimuth mapping is mainly carried out by microseismic event monitoring coming from the closing crack with registration and interpretation of shear and compressional waves and can be combined with pressure pulse registration. Other techniques of fracture azimuth mapping are observation of associated surface deformation by tiltmeter analysis, strain relaxation evaluation of oriented cores, electrical potential field recording, and magnetometrical measuring. Tiltmeter analysis is also able to distinguish horizontal and vertical fractures. Interpretation of stimulation data such as pressures, injection rates, densities, concentrations and volumes is important for pre-treatment scheduling, real-time job supervision and post-operation assessment for the purpose of enhanced control and design of current and future hydraulic fracturing performance. Real-time analysis capabilities enable simultaneous execution and evaluation of the operation by computer-controlled supervision, thus permitting to make design changes during the treatment which are particularly significant for early recognition and avoidance of premature screenout termination of the job. Screenout prediction is the most important task of fracture pressure interpretation, with pressure evolution reflecting propagation of the crack. Quality control is a special type of stimulation data analysis and job performance supervision and basically aims on improving success rates of the treatments by reducing premature screenout failure which may occur due to various reasons. Fracture migration modelling is a pre-stimulation forecast of formation fracturability and prognosis of differences in fracture gradient and thus open-up behaviour between various hydrocarbon-bearing zones of the reservoir column. Fracture migration analysis is a combination of sonic waveform interpretation with bulk volume evaluation and linear fracture mechanics with treatment fluid density and can also be used for fracture height containment prediction. Inflatable impression packers are mechanical monitoring devices which record imprints of propped hydraulic fractures left on the soft pliable coating when the packer is pressed against the borehole wall upon inflation. Stereopair photography and computerized satellite imagery including false-colour display of various processed images are excellent tools for detecting and trend mapping of large natural geomorphological fractures and fracture systems. The main significance of photography and imagery is reduction of the number of dry holes and improvement of well productivity by picking-up of appraisal and offset drilling locations at major fracture intersections which permits also considerably faster prospect generation. Gravel pack checking and evaluation aims on inspection of pack density and continuity as well as detection of void spaces within the pack, and on determination of the top of the cylindrical gravel column in the wellbore in order to verify its proper coverage of the producing zone. Gravel pack effectivity monitoring in qualitative and quantitative manner is possible by gravel density examination and extension control by compensated neutron, gamma-ray and photon (which is a special type of gamma-ray) logging. Photon logging is a shallow reading device which similarly as the other techniques records changing gravel pack density due to the fact that formation and fluid density are more or less constant along the surveyed interval, and therefore the chief factor causing varying log response is gravel pack density. Voids within the gravel pack represent low-density pockets and bridges in an otherwise uniform and continuous medium, and marked density changes also occur at top and bottom of the gravel pack. An enhanced technique is photon/vibrapack combination which is a complete system for identification of deficiencies in gravel packs and immediate in-situ repair by shaking which provokes improved density by settling of the gravel. Gamma-ray assessment of gravel pack density can be performed in quantitative manner by assuming constant formation and fluid density and thus correlating gamma-ray count rates to percentages of gravel pack or only fluid-filled annular sections. Top and bottom of the gravel pack can be identified by sudden increases in count rate, and void spaces of the interval along the screen can also be picked up by their anomalous count rate. Neutron logging is based on the fact that the presence of gravel in the borehole considerably changes the apparent porosity recorded by the tool. Combination of neutron and gamma-ray logging with placement of radioactively-coated gravel and/or -tagged fluid enables cross-checking of the neutron record which is particularly significant in gas-bearing formations where the reading of the neutron tool can be disturbed by the gas influence. Time-lapse gravel pack monitoring is important for determination of progressive gravel settling with increasing compaction unless final packing density is immediately achieved by vibration during insertion. While the various logging methods aim on quality monitoring of the gravel pack once it has been installed between screen and casing or borehole wall, the function of the tell-tale screen is to determine the end of the gravel pack operation by indicating the moment when the treatment is completely performed. Mechanical properties logging is carried out prior to planning and design of sand control treatments in order to check the stability of the various hydrocarbon-bearing sands in the prospective column and to predict prior to onset of exploitation or increase of offtake rates whether the grain fabric would be strong enough to allow to start oil and gas production or to increase pumping rates without the danger of provoking sand problems in the wellbore, or whether the rock structure is too weak and gravel packing and other sand-control measures have to be carried out in order to permit undisturbed hydrocarbon offtake. The mechanical properties log as a predictive tool is thus an aid for screening of gravel packing candidates and has considerable significance in early recognition of potential sand-trouble-making horizons and the necessity of performance of sand-control treatments prior to wellbore damage by onsetting sand production, as well as in money saving and thus budget containment by picking up of horizons which will resist to all the possible drag forces and pressure drawdowns during hydrocarbon offtake without sand production and therefore do not need any artificial stabilization. The integrated sand production forecast system synergistically uses geological description, core testing, log analysis and reservoir simulation to predict perforation cavity stability at initial conditions and throughout the life of the well. Caliper logging in open holes can be used as an aid in estimating the amount of material required for gravel packing, thereby being a tool for optimizing design of the sand control treatment.

6.1. Introduction

Hydraulic proppant fracturing has achieved considerably progressing importance during the last ten years since the invention of intermediate- and high-strength synthetic proppants with regard to improving well productivity, increasing reserves and ameliorating drainage of producing intervals (cf. section 1.2.). Successful technological innovations are an important economical factor by permitting to reduce the number of wells drilled in oil and/or gas fields. In order to optimize the effectivity of hydraulic proppant fracturing, azimuth, length and height of the crack have to be determined by downhole or surface measurements (WITTRISCH & SARDA 1987). Assessment of the effectiveness of the induced fracture requires knowledge of dimensions, orientations and spatial characteristics of the crack in relation to the oil- or gas-bearing formation (HART, ENGI & MORRIS 1983). A comprehensive characterization of fracture propagation not only aids in stimulation treatment design for the reservoir, but can also help specify well spacings which will minimize interference of drainage patterns and thus maximize economical recovery potential.

Hydraulic proppant fracturing and gravel packing operations have to be closely monitored and supervised prior to the job, during course of the treatment and after termination of the borehole manipulation in order to secure optimum results of the stimulation and stabilization treatments especially in numerous problematical reservoirs which cannot produce at commercial rates without artificial improvements of permeability and consolidation. Fracture and gravel pack monitoring basically comprise prediction, supervision and retroactive evaluation. Prediction has immediate impact on the design of the current job, supervision helps to secure safe performance of the operation and provides opportunities to change the schedule during course of the procedure if necessary particularly in light of avoiding premature screenout termination of the treatment, and retroactive evaluation is important for improving the design of follow-up operations by implementing the experience received from the execution of earlier jobs and incorporating the knowledge of the reaction of the reservoir to the treatment and thus assessing match or discrepancy of anticipation and result of the fracturing stimulation. An important task of fracture height monitoring is checking of separating barrier stability in order to avoid undesired breakthroughs to higher or lower horizons.

Following the discussion of various technological aspects of hydraulic proppant fracturing (chapter 4) as well as gravel packing and sand control (chapter 5), this division summarizes methods of fracture detection and determination as well as gravel pack checking and evaluation.

6.2. Fracture detection and determination

Control of predicted vs. achieved height of the hydraulically-induced fracture by geophysical cable (wireline) well-logging (AGNEW 1965; DOBKINS 1979, 1981; PEARCE 1979, FERTL 1980, VOEGELE & JONES 1980; HUNT, HASKETT, RAYMOND & PIRIE 1981; McGHEE, LESSO & WATERFILL 1981; JULIAN 1982, PEETERS & HARTLEY 1984, FRISINGER & COOPER 1985, DARILEK 1986; SORRELLS, GRANT, MAHRER, MAUK, MULCAHY, STUMP & SWANSON 1986; AHMED 1987; NOBLETT, FERTL & GUY 1987) will become increasingly important particularly in such horizons where fracture containment in vertical direction is inevitable due to environmental safety obligations by mining authority or natural protection department, or reservoir engineering requirements such as danger of breakthrough to water-bearing horizons in shallow to moderate depth (cf. section 4.2.).

The progressive improvement of reliability and effectivity of fracture height monitoring is also an important means of promotion of not only well-logging jobs, but ultimately also fracturing treatments. Therefore also cooperation between well-logging companies on the one hand and service and proppant companies on the other hand is necessary to trigger increasing performance of stimulations and following post-job result monitoring which has been facilitated and ameliorated by some big merges and takeovers in the last years that brought in several cases logging and pumping divisions or companies into one corporation. Other important points of fracture monitoring and determination are crack azimuth mapping and fracture width estimation. The possibilities of checking of theoretical planning vs. practical outcome focus on post-job quality control monitoring, but significant contributions can also be made by pre-treatment modelling and prediction. In some instances, also supervision during course of the operation can provide important contributions for enhancement of the results.

6.2.0. General aspects of fracture height monitoring

Determination of achieved fracture height is an important step to evaluate the effectiveness of the stimulation treatment, to estimate the subsequent production behaviour of the well, and to check the accuracy of fracture design and height prediction models that were used to plan the hydraulic job (AHMED 1987). Knowledge of the actual crack height is also required for the calculation of an exact fracture length and is therefore an important input parameter for geometry modelling of the crack and hydrocarbon exploitation forecast. Fracture height determination can be performed by direct measuring with formation microscanner and borehole televiewer, and by indirect inference from various well-logging methods. Direct measurement techniques can only be applied in open holes, whereas the basic limitation of the indirect methods is their small radius of investigation, with the exception of seismical techniques, dual laterolog monitoring, and partially also multiple-tracer gamma-ray spectroscopy.

Several techniques of fracture height determination by well-logging after the hydraulic fracture treatment exist which are mainly based on temperature and radioactivity effects. A less common method is the noise log (DOBKINS 1979) which has its main significance in clarification of anomalous readings of temperature logs. Comments are offered as follows on well logging methods focussing on fracture height control, seismical and gravimetrical methods including fracture azimuth determination, and magnetometrical methods. An outline of the significance of stimulation data analysis is also given. The review and status report also includes a compilation of techniques for detection of natural cracks which play an important role in many reservoirs for both hydrocarbon production and prognosis of propagation direction of hydraulic fractures during stimulation jobs, with acoustical and electrical imaging methods as well as dipmeter, sonic logging and dual laterolog recording being the most suitable means for this purpose. Finally the significance of the fracture migration model as a predictive tool is illustrated. The outline as follows concentrates on monitoring of hydraulically-induced cracks in vertical or moderately-inclined wells, with some special aspects of fracture orientation determination in highly-deviated wells also being discussed.

6.2.1. Well logging methods

Cable (wireline) well logging methods of hydraulic fracture monitoring and determination chiefly aim on crack height control and to subordinate amounts also on fracture azimuth prediction and/or checking. Some information on crack width could also be gained. The most important tools for hydraulic fracture height inspection are temperature, single- and multiple-tracer gamma-ray, acoustic and noise logging instruments. Caliper logging is the most significant tool for in-situ stress pattern determination and hence fracture azimuth prediction, whereas borehole television and formation microscanner also reveal some insight into crack width and morphology. Natural fracture detection can mainly be performed by acoustical and electrical borehole imaging as well as dipmeter, waveform sonic measurements and dual laterolog registration. The mechanical properties log is together with fracture migration analysis (cf. section 6.2.5.) an important predictive tool for fracture geometry modelling before the hydraulic stimulation job. Comments on other well logging methods are also briefly offered.

6.2.1.1. Temperature logging

The simplest and cheapest method of wireline fracture height monitoring is recording the temperature contrast between the cold pumped stimulation fluid within the crack and the warm surrounding formation by running a temperature log up to abt. 24 hrs after termination of the treatment (temperature logging is also possible at any later stage provided preparation of the well by renewed injection of cold fluid into the fracture). The temperature log monitors the vertical extension of the interval that has taken up the injected fracturing fluid which corresponds to created fracture height (cf. section 6.2.1.5.2.). The well has to be shut in for temperature logging, because flowing conditions would seriously disturb the temperature distribution to be recorded. The highly susceptible logging tools are also available in 1 11/16 " slim size fitting through the normal 3 1/2 " tubing and thus saving round-trip time for pulling and installation of the tubing.

6.2.1.1.1. General aspects

Temperature logging is together with conventional gamma-ray recording the oldest technique of height monitoring of hydraulic fractures and is applied in stimulation surveys in the field since more than 25 years. Temperature logging is able to almost unmistakably document the disturbance of the generally linear geothermal temperature gradient by marked breaks and level changes at the upper and lower boundaries of the fracture, as the non-permeable over- and underburden reflecting a temperature distribution according to the geothermal gradient are in pronounced contrast to the cooled fractured interval that is still affected by the injection of stimulation fluid with more or less surface temperature. The best results are obtained when running the temperature log between 18 and 24 hrs after termination of the hydraulic proppant fracturing treatment when only the major cooling effects remain and the minor fluctuations have died away or have been equalized (PEARCE 1979). After 18 hrs, the cased borehole as well as any small channels behind pipe have returned to near the static temperature gradient, and only those intervals where large amounts of fluid have entered during course of the stimulation operation exhibit still a cooling effect.

As temperature logging only records the distribution of the injected cold stimulation fluid within the created fracture height intersecting the warm reservoir (cf. section 6.2.1.5.2.), the method can be applied for post-treatment monitoring of both proppant and acid fracturing in sandstones and carbonates.

At least 20 ft (6 m), but preferably 50 ft (15 m) or more of clean hole between proppant fill and deepest zone of interest is necessary to get good interpretation of both temperature and radioactive tracer surveys. Temperature logging should go upwards above the zone of interest at least 100 % of reservoir interval thickness, and logging speed should be in the range of 20 ft/min. The following outline comprises gradient and differential temperature recording, influence of anomalous thermal formation conductivity, and improvement of interpretation confidence.

6.2.1.1.2. Gradient and differential temperature recording

Distinction has to be made between temperature gradient and differential temperature recording. Gradient documentation gives only the absolute formation temperature, whereas differential reading provides a relative comparison between successive points and allows more distinct recognition of anomalies. Temperature log evaluation can be enhanced by plotting the temperature difference between two points spaced a couple of feet in vertical direction along a multi-recorder tool, by electronic recomputation of the survey by a single-recorder instrument instead of the absolute temperature, or by the time lapse procedure with subtraction of one value from the other, with the differential temperature being constant both above and below the crack, but being variable within the fractured section. The method is environmentally without any hazards, because neither source nor recorder contain radioactive elements. Temperature logging is still the most reliable fracture height determination method due to the elimination of disturbing influences such as fluctuating stimulation chemical and/or formation radioactivity in gamma-ray logging, and is also the economically most feasible and technically most simple means among the various available techniques for fracture monitoring (DOBKINS 1979).

While the absolute cost of a temperature record is comparable or even slightly lower than that of a gamma-ray measurement, sometimes further containment of logging expenses can be achieved by the fact that a post-fracturing temperature log is not requiring a pre-stimulation reference log which is indispensible in case of running a gamma-ray log for fracture height monitoring. Temperature logging is independent from proppant type and its mineralogical composition as well as from admixture of radioactive tracers to proppants and fluids and their environmental hazards. Improvement of conventional temperature gradient or differential surveys can be made with the radial differential temperature log (COOKE 1979, DOBKINS 1979) where the temperature difference is recorded around the casing for some revolutions followed by stopping the rotation of the tool and measurement of the static temperature.

Temperature logging can only record the distribution of the fracturing fluid within the crack, and its interpretation is therefore restricted to created fracture height (cf. section 6.2.1.5.).

6.2.1.1.3. Influence of anomalous thermal formation conductivity

Interpretation of temperature logging surveys can be complicated by low temperature contrasts, flow back from the formation before and/or after the treatment, or fluid movement behind the casing (PEMPER, FLECKER, McWHIRTER & OLIVER 1988) as well as thermal conductivity differences, secondary flow in the fracture, and convection currents (PEETERS & HARTLEY 1984). The most abundant complication of crack height monitoring by temperature logging is development of an anomalous warm nose at the fracture top (DOBKINS 1979, VEATCH 1983) which may be caused by warmed fluid in the fracture flowing back past the wellbore from one portion of the crack to another after pumping has ceased at the end of the stimulation job (PEETERS & HARTLEY 1984). Probably continued fracture extension after shut-in can create a redistribution of fluid in the crack carrying heat back across the wellbore thereby triggering development of the warm nose. Many of the unusual readings, however, are caused by anomalous thermal formation conductivity in some patches of the reservoir interval.

The appearance of a warm anomaly at the top or above the treatment zone (DOBKINS 1979) is a geographically widespread problem in the USA and is partially the consequence of different thermal conductivities of the rocks and also of operational procedures. The overview as follows concentrates on significance and distribution of warm anomalies, clarification of thermal conductivity anomalies, and other aspects.

6.2.1.1.3.1. Significance and distribution of warm anomalies

Thermal conductivity (EVANS 1977) varies depending on the mineralogical and petrographical composition of the rocks, with formations with high thermal conductivity changing temperature more slowly than those with low thermal conductivity. Thus high thermal conductivity zones cool less quickly down than low thermal conductivity zones do when a cold fluid is pumped down a hot wellbore, with thus high thermal conductivity zones showing up as warm anomalies on temperature surveys carried out soon after the stimulation operation, whereas low-conductivity intervals quickly adjust to the changed environmental conditions after insertion of the cool treatment fluids. If the thermal conductivity characteristics of the rock pile are determined by a pre-fracturing cold water circulation treatment followed by a reference temperature survey, thermal conductivity anomalies can be distinguished from crack anomalies in the post-fracturing record. In case of such a reference log being present, the best results of crack height interpretation are obtained when running the post-fracturing temperature log as soon as possible after the end of the stimulation operation.

As a consequence of the occurrence of anomalous high thermal conductivity zones along the fracture height predominantly in the upper part of the section, determination of crack bottom is often quite easily performed from the depth at which successive temperature logs converge or cross, whereas fracture top is frequently more difficult to realize and confident interpretation has in many cases to include cross-checking with gamma-ray intensity distribution (cf. section 6.2.1.4.2.), evaluation of changes in gradients of the temperature profile (PEARSON, ABOU-SAYED & NGUYEN 1985) or noise assessment (cf. section 6.2.1.7.).

Difficulties of interpretation of temperature logs may also arise from prefracture existences of temperature anomalies in thermal or chemical enhanced oil recovery projects (SMITH 1981). Such anomalies may result from the normal injection of hot water into a cooler formation and by the falling fluid level in the wells after shut-in which continued to carry warm fluid into the fractured zone for several hours, thereby obscuring the development of temperature anomalies. While injection of fracturing fluids that are colder than the natural reservoir fluids give rise to an overall cooling effect, insertion of hot water or steam during tertiary oil recovery projects has the inverse consequence of heating the formation up above the geothermical temperature level, with these two reverse temperature changes being able to trigger considerable interpretation difficulties in case of interference and superimposition.

6.2.1.1.3.2. Clarification of thermal conductivity anomalies

A suitable means of clarification whether the recorded temperature anomalies belong still to the fracture or already to the untreated formation is noise or sound logging (cf. section 6.2.1.7.). Pre-fracturing temperature surveying establishing the temperature gradient of the unstimulated formation is also a possibility of proper assessment of anomalous temperature readings (HART, ENGI &

MORRIS 1983) and is often performed after circulating fluid in order to simulate wellbore conditions during pumping. During the fracturing operation, large volumes of cool fluid enter the reservoir thereby locally reducing the temperature. Anomalies generally originate when the temperature of fractured zones decays at a lesser rate than that of unstimulated zones.

Post-fracturing temperature logs frequently show an inflection point roughly opposite the bottom of the perforated interval (PEETERS & HARTLEY 1984). This is probably because the borehole fluid beneath the perforations is not disturbed during the injection of stimulation fluids and only minor cooling occurs as a result of mixing and radiation.

Temperature logging may in some cases give only a pessimistic view of gross fracture height (BEGNAUD & CLAIBORNE 1985). The flow rate during crack propagation within the pay may be 30 - 80 times greater than that through an area of the fracture only 1/3 as wide which can cause proppant bridging and thus produce more cooling over the wide portion of the interval. The pessimistic growth value from temperature logs is in part influenced by the stagnant fluid below the perforations. Although this number may be conservative, usually the height difference missed is in a region of very small width and in most cases would not be propped effectively at all (cf. section 6.2.1.5.2.).

6.2.1.1.3.3. Other aspects

Another problem of obtaining a good temperature survey is that in many cases the bottom is tagged somewhere above or within the perforated interval which does not allow to obtain a good value of the fracture bottom. In order to get a temperature log below the perforations for the purpose of crack bottom determination, the well is flowed back at a moderate rate when screening out and being not fully flushed. A clean bottom-up is recorded after a wellbore volume of fluid has been produced which is to be expected following a screenout, thus indicating that the proppant within the formation is not flowing back. Sometimes it may be difficult to determine the lower boundary of the fracture because of complications in lowering the tools below the perforations due to the accumulation of sand or proppants (DOBKINS 1981 a; HART, ENGI & MORRIS 1983). As the temperature log usually indicates little other than more or less pronounced cooling over the wide portion of the fracture in the main pay, presumably for a normal-cased completion the insulating effect of the cement plus the fact that all the fluid must leave the perforations could create a cool anomaly over the perforated interval. Since this section coincides with the reservoir, interpretation could be erroneously made that confined height growth occurred when in fact excessive height propagation has taken place.

WARPINSKI, BRANAGAN, SATTLER, CIPOLLA, LORENZ & THORNE (1988) report the phenomenon that good results of temperature logging in their case study could only be obtained when running the logs in upward direction, whereas downward runs gave flat temperature profiles as a consequence of tool problems or formation and wellbore behaviour.

6.2.1.1.4. Improvement of interpretation confidence

Improvement of temperature logging can be made by reference runs both before and after the treatment where the cooling effect of the injected fracturing fluid can be clearly picked up and thermoconductivity anomalies become easily distinguishable. Formation thermal conductivity fluctuations and variations in cement annulus thickness affect the pre-operation temperature survey in a similar manner as the post-job record, and therefore the differences in both readings are the influence of the stimulation treatment (DOBKINS 1981 a). The use of a lithology log together with the temperature survey also enhances the interpretation and may even eliminate the need of a pre-operation reference temperature log in every well of the field. In shallow boreholes where the difference between bottomhole temperature and injected fracturing fluid temperature is small, the temperature log may not be able to reliably identify the anomaly caused by the stimulation treatment, whereas in deep wells, the higher the temperature contrast between formation and injected stimulation fluids, the clearer the monitoring of fracture height by temperature logging.

The use of post-operation temperature surveys at frequent intervals and comparison of the results can allow recognition of the zones that took up the majority of the fluid which correspond to the widest fracture segments, with this relationship therefore enabling estimation of the qualitative crack width throughout the created fracture height. Flowback from the formation before and after the stimulation job and fluid movement behind the casing can obscure the temperature log readings and provide interpretation difficulties (AHMED 1987). Improvement of temperature log evaluation can also be made by performing several runs with sufficient time between them in order to pick up the subtle changes necessary for accurate interpretation (WAREMBOURG, KLINGENSMITH, HODGES & ERDLE 1985).

Various aspects of temperature logging for fracture height monitoring are also outlined by WITTERHOLT & TIXIER (1972), SMITH & STEFFENSON (1975); HARRINGTON, HANNAH & BEIRUTE (1978); SMITH, HOLMAN, FAST & COLVIN (1978); WAGES (1982), GREENFIELD & AHMED (1983), MORRIS & SINCLAIR (1984), PEETERS & HARTLEY (1984), JORDEN & CAMPBELL (1985) and PEARSON, ABOU-SAYED & NGUYEN (1985). Improved temperature log interpretation guidelines are summarized by SMITH & STEFFENSON (1973). Field examples of fracture evaluation by temperature logging are presented by PEARCE (1979); AHMED, SCHATZ, HOLLAND, JONES & GREENFIELD (1982); BEGNAUD & CLAIBORNE (1985); KIM, CHAMPION & COOPER (1985); LI & ZHU (1986); NIEMEYER & REINART (1986), NOLTE (1986), LEE (1987), HUCKABEE (1988) and TAN, McGOWEN, LEE & SOLIMAN (1988).

6.2.1.2. Gamma-ray logging of natural proppant radioactivity

Gamma-ray logging aims on recording of formation and proppant radioactivity and is independent from the geothermic gradient, and thus is a suitable alternative to temperature logging for hydraulic fracture height monitoring, with both methods being able to act together in a complementary manner (cf. section 6.2.1.4.). Gamma-ray spectroscopy of multiple radioactive tracers (cf. section 6.2.1.3.) is probably the most sophisticated, advanced and unambiguous fracture height evaluation technique developed to date, and its higher cost with respect to the expenses for the method is in many cases well paid-off by a better retrospective fracture control and positive impact on design of future stimulation operations. The most important aspects of gamma-ray logging are natural proppant radioactivity, artificial proppant or fluid radioactivity, and combination of temperature and gamma-ray logging. Comments are also offered on thermal decay time recording which represents a combination of neutron and gamma-ray logging, and a brief sketch of different fracture height types as registered by various logging techniques is also given.

Gamma-ray logging of natural radioactivity of zirconia-silicate proppants is apart from temperature determination another simple, cheap and straightforward technique of fracture height monitoring if all the emitting and recording conditions are favourable which, however, has in the past turned out not to have always been the case. Due to registration of radiation emitted by the proppants supporting the infilled fracture height, gamma-ray logging of natural proppant radioactivity is limited to proppant fracturing, with monitoring of acid fracturing having to be carried out by insertion of artificial tracers. Following a sketch of the measurement principles, some of the recording and interpretation drawbacks are outlined.

6.2.1.2.1. Measurement principles

Apart from temperature measurements, another standard technique for fracture height determination is running a conventional gamma-ray log in case of infilling of the crack with radioactive proppants and/or fluids. Prior to discussion of environmental effects, it has to be emphasized that this method is often more complicated and does not necessarily deliver clear and undoubted results due to in various cases not constant natural or artificial radioactivity of proppants and/or fluids, changing natural radioactivity of the formation, and more difficult evaluation by the disturbing effect of the diversified background radioactivity of the lithological column.

While recording of a spectral gamma-ray that could eliminate the latter drawback is certainly a difficult economical point of discussion and decision for mono-isotope natural proppant radioactivity display because of considerably higher logging cost and is only reasonably applied for multiple-tracer monitoring in wells where several different zones have been fractured or various types and/or grain sizes of proppants have been injected (cf. section 6.2.1.3.4.), also the expenses for a conventional gamma-ray could be higher (with respect to the cost of a temperature log) unless a reference gamma-ray had been anyway recorded prior to stimulation for the purpose of petrophysical formation evaluation, geological correlation, sedimentological interpretation or completion depth determination, and is still of sufficient quality and actuality to allow an unmistakable match with the post-fracturing gamma-ray run for treatment effectivity monitoring. Some recording and interpretation drawbacks of conventional gamma-ray logging of natural proppant radioactivity are sketched as follows.

6.2.1.2.2. Recording and interpretation drawbacks

Concerning radioactivity of proppants and/or fluids, the environmentally least hazardous and thus also the administratively least complicated and inhibited way is infilling of the fracture with high-strength zirconia-silicate proppants where traces of radioactive cations are diadochously replacing zirconia cations in the lattices of the zirconia microcrystals dispersed within the glassy aluminasilicate matrix of the proppants and derive from naturally implemented radioactive impurities in the crystal lattice of the zircon raw material of primary or secondary magmatic or metamorphic provenance. The advantages of the straightforward and apparently unambiguous technique, however, are deteriorated or even compensated by some drawbacks comprising proppant radioactivity and proppant price.

6.2.1.2.2.1. Proppant radioactivity

The experience in the past has shown that as a result of the use of several zircon raw materials of different stratigraphical, palaeogeographical and geotectonical provenance and thereby different petrographical and geochemical composition and consequently different spectrum and level of radioactivity, production of zirconia-silicate proppants does not necessarily have to lead to homogenization of the radioactivity during the fusing process. Differences in the niveau of radioactivity may even occur between samples taken from different parts of smaller lots of delivered proppants, with field measurements revealing that in the worst cases, distinction of proppant-induced relative radioactivity from background radioactivity coming from the formation may be almost impossible, or even the zirconia-silicate proppants are hardly radioactive at all. In case of segregation effects during proppant processing when using unusually pure raw materials, proppant radioactivity may be below the critical boundary level that is essential for undoubted recognition and distinction from natural reservoir rock radioactivity beyond the obliterating range of technical measurement deviation.

WARPINSKI, BRANAGAN, SATTLER, LORENZ, NORTHROP, MANN & FROHNE (1985) report an analogous example of gamma-ray surveying showing only sporadic traces of radioactive sand in the perforated interval, thus giving no reliable indication of fracture height. MEEHAN & PENNINGTON (1981) ascertain proper radioactive sand distribution by injection of the tracer-bearing proppants throughout the job.

6.2.1.2.2.2. Proppant price

In addition to the drawback of possibly inhomogeneous or insufficient radioactivity, the high-strength zirconia-silicate proppants are much more expensive than intermediate- and high-strength alumina oxide and silicate ceramic proppants, with the higher proppant cost seriously limiting the application due to economical reasons of fracture expense containment (particularly offshore) and often restricting it to a tail-in of smaller quantities in the vicinity of the borehole just for immediate logging purposes, but not for fracture-supporting reasons in the primary line (cf. section 2.2.2.3.). As opposed to the pricing situation, the problem of level and constancy of the radioactivity, however, can be reasonably controlled by limitation of the zircon raw material to only one source where the niveau of radioactivity is sufficiently and reproducibly high. In contrast to fracture height, reasonable results on fracture width have so far not been obtained with gamma-ray logging of natural proppant radioactivity.

The pricing situation can only be rendered more favourable for zirconia-silicate proppants (which are basically a very good product, but are only difficult to market due to their high selling price as a consequence of elevated production costs) by price increases for alumina silicate and oxide proppants up to a level reasonably comparable with that of zirconia-silicate proppants. This price approaching or almost equalization can be achieved either directly by actions undertaken by the proppant companies in terms of price correction (cf. section 2.2.2.2.), or indirectly by a strengthening US $ and as a result of this rising of its exchange rate with respect to other world guide currencies such as DM and Ł (cf. section 2.2.1.1.3.).

Conventional gamma-ray logging of natural proppant radioactivity is limited to monitoring of proppant distribution within the infilled and supported crack, with its interpretation thus providing propped fracture height (cf. section 6.2.1.5.).

6.2.1.3. Gamma-ray logging of artificial proppant or fluid radioactivity

Gamma-ray tracer logging has been utilized since over 25 years for fracture height determination (ZHUVAGIN & AKCHASYANOV 1960; WEBSTER, GOINS & BERRY 1964; PEMPER, FLECKER, McWHIRTER & OLIVER 1988). As artificial tracers can be added to any type of fluid regardless of composition if they are stable in the specific chemical environment, radioactive marker monitoring of fluids for delineation of created fracture height can be applied in both proppant and acid fracturing. The discussion of gamma-ray logging of artificial proppant or fluid radioactivity as follows focusses on tracer aspects, conventional single- or double-tracer monitoring, administrative and environmental aspects of tracer monitoring, and multiple tracer discrimination by spectral gamma-ray logging.

6.2.1.3.1. Tracer aspects

Gamma-ray conventional and spectroscopy monitoring of synthetic tracers that have been inserted into borehole, fracture and/or formation for special purposes and with detailed aims requires understanding of marker behaviour and dynamics. The following summary discusses the most important points along the lines of tracer types and tagging possibilities, marker stability and dispersion, tracer requirements and composition, marker injection and placement, and tracer distribution and recording.

6.2.1.3.1.1. Tracer types and tagging possibilities

In terms of artificial radioactivity of proppants and/or fluids, three different types of tracering can be distinguished. The first type comprises coating of natural sand or synthetic proppants of any chemical and mineralogical composition with radioactive pellicles. The second type is intermixing of little amounts of radioactive grains of any composition into a lot of standard man-made proppants or common sand. The third type includes addition of radioactive tracers in liquid or solid form to the fracturing fluids or the fluid-proppant-slurry (TESLJUK 1957; WHEELER, PARSONS, CONCHIE & DURHAM 1985; PRO-TECHNICS 1987). Differentiation has further to be made between single- and/or double-tracer conventional gamma-ray logging and multiple-tracer gamma-ray spectroscopy, depending on whether only one or two tracers have been added to proppants and/or fluids, whether various proppant and/or fluid stages have been individually tagged by different identifiers, or whether even substages of solid and/or liquid fracturing treatment constituents have been separately marked by distinguishable radioactive isotopes.

Discrimination can further be made between inert and temperature-sensitive chemically reacting tracers (ROBINSON, TESTER & BROWN 1984). While hydraulic fracture height monitoring in most cases requires inert markers with good absorption capabilities, the application of chemically reacting tracers in special situations can provide further information particularly on marker dispersion within the cracked formation. In multiple fracture systems, tracer dispersion is controlled by a combination of three mechanisms being large-scale flow heterogeneities such as cracks of different size and varying current impedance, crossflow between different main fractures as well as side crack branches, and flow dispersion within individual cracks. Various single-fracture dispersion sources exist comprising roughness of fracture surface, hold-up of tracer in dead-end secondary fractures branching off the one in which the flow is occurring, different arrival times of the fluid caused by the velocity profile, and molecular diffusion between different streamlines of the current (HORNE & RODRIGUEZ 1983, ROBINSON & TESTER 1984). NGUYEN & STEVENS (1988) use inert gas radioactive markers for steam injection profiling. Mechanisms of tracer dispersion and migration are also discussed by NERETNIEKS & RASMUSON (1984).

6.2.1.3.1.2. Marker stability and dispersion

Coating of proppant grains with radioactive pellicles, fusing of marker substances on proppant grain surfaces, and intermixing of proppant-sized radioactive grains into the proppant population are the technically most secure and environmentally most safe methods, because it almost guarantees that once set in position within the fracture, the radioactively-tagged proppants remain in place and unless chemical dissolution by aggressive brines or mechanical transport of crushed fines occur, no marker redistribution will happen which enables not only constant-record repetition logging in terms of tracer concentration (regardless of declining marker intensity with time), but also prevents any identifier flowback into the wellbore and contamination of fluids which may either directly or indirectly reach the surface.

In case of intermixing of liquid or solid tracers into the fracturing fluid, emphasis has to be put on either rapid vanishing of radiation by application of radionuclides with very short half-life time (which, however, is seriously limiting the possibilities of time-lapse monitoring), or on sufficient chemical adsorption of crystallographical bonding of the marker isotopes to proppants, formation sand grains, reservoir clay minerals or fracture wall filter cake in order to fix the tracer in place and prevent its redistribution which would not only impair the comparability of repetition logging runs, but also represent environmental hazard due to the possibility of marker transport with the fluids during flowback in well cleanup or later production to the borehole and from there to the surface.

6.2.1.3.1.3. Tracer requirements and composition

Important factors controlling selection of specific isotopes and concentrations of radioactive tracers (PRO-TECHNICS 1987) are relative gamma-ray intensities, fluid volume to expected fracture height ratio, proppant quantities to expected crack height ratio, expected time between stimulation operation and logging, and isotope half-life. Other factors involving chemical and physical nature of the isotopes are critical to accurately tracing the constituents of the treatment. The selected isotopes must possess sufficiently differing gamma energy spectra that can be discriminated with logging tools and software. Liquid tracers must have the ability to be transported into the fracture without significantly absorbing onto the tubulars through which they are pumped, and conversely they must be able to be sufficiently absorbed onto the formation rock so that when the spent stimulation fluids are flowed back, the isotopes will not contaminate wellbore and surface equipment. It is further desirable that solid tracers used to tag proppant lots possess the same size, density and compressive strength as the proppant in order to allow stimulation to achieve maximum accuracy. Solid tracer materials also must possess low washability so that the isotopes they employ do not end up in the fluid phase.

VETTER & CRICHLOW (1979) outline frequently large marker losses due to undesired adsorption in the reservoir matrix. To prevent disappointing results, a better solution than a single tracer is a cocktail of multiple markers where each individual element and chemical compound has predetermined adsorption isotherms, with these isotherms in addition to material balances and variations in retention times allowing good reservoir modelling and tracer record interpretation. Aspects of radioactive marker adsorption in reservoir rocks are also discussed by GRAUE, KOLLTVEIT, LIEN & SKAUGE (1988).

6.2.1.3.1.4. Marker injection and placement

Radioactive slurry configuration is generally a liquid or solid proppant radioisotope held in a viscous carrier fluid (PRIEST 1988). Injection of proppant-entrained radioisotopes at the latest time before fracturing fluid enters the wellbore following flow through the surface equipment chain and high-pressure downhole pumps minimizes personnel exposure to radiation and hazards of transporting radioactive material. Radioisotopes have been hitherto added manually or by mechanical means to the upstream side of high-pressure downhole pumps including the mixing tank on the blending equipment or a self-contained low-pressure injection unit used before or after the blending equipment. Direct injection of the radioisotope in the high-pressure side near the wellhead at the downstream side of the high-pressure pumps can reduce the physical volume of the radioisotope and can greatly minimize the problems of contamination. Radioisotope concentration should take into account the maximum quantity of radioisotope that can be placed in a certain volume of carrier fluid and still be injected in consistent manner. Aspects of choice, implementation and safety of tracers are also discussed by McINTYRE, POLKOWSKI, BRON & POW (1988). Tracer placement techniques for improved radioactive marker logging are also outlined by HILL, BOEHM & AKERS (1988).

6.2.1.3.1.5. Tracer distribution and recording

Gamma-ray logging of injected fluid that has been marked by a tracer reveals created fracture height (cf. section 6.2.1.5.). Interpretation can be influenced by relative porosity and fluid uptake capacity of various zones, because the radioactively-tagged fluid leaks off into the formation not necessarily in homogeneous distribution across the perforated or open interval. Propped fracture height is obtained by gamma-ray monitoring of the distribution of radioactively-coated proppants. For proper fracture height recording, the proppants covered with radioactive pellicles have to be added at the blender ideally throughout the job, because the first amount of proppants does not necessarily enter the same zones as the last lot. The flush volume should be accurately calculated, because overflushing would push the proppants too far into the formation away from the wellbore where the emitted radiation becomes insensitive to the measuring tool, and underflushing would keep some proppant material within the borehole, thereby delaying the logging until the wellbore is cleaned (AHMED 1987) or even falsifying the fracture monitoring record.

Liquid radioactive tracers injected for gamma-ray logging have generally to be overdisplaced in order to remove all the radioactive residue from the wellbore, because the tool can still pick up the marker when it is overdisplaced due to tracer leaking into the reservoir (BUNDY 1981). They also have to adhere to formation sand grains or have to be bound within sheets of clay mineral crystals in the pore space in order to keep more or less constant position and not to be produced back during well cleanup and fluid offtake. Radioactively-tagged proppants, however, cannot be recorded by the instrument if too much overdisplaced into the created fracture. Failure to overdisplace liquid and/or solid tracer will leave radioactive residue in the borehole which can obliterate the readings and in the worst case ruin the log. In connection with fracture height monitoring, gamma-ray logging can also be used to find mechanical well problems such as channelling or casing splits which is indicated by tracer material being recorded in the wrong place. Especially large radioactive peaks registered at the bottom of the treated interval are an indicator of preferential downwards growth of the fracture (WARPINSKI, BRANAGAN, SATTLER, CIPOLLA, LORENZ & THORNE 1988).

Tracer placement techniques for improved radioactive marker logging are discussed by HILL, BOEHM & AKERS (1988). Aspects of marker injection, distribution and recording are also evaluated by WILEY & COCANOWER (1975), ANTHONY & HILL (1984) and HILL & SOLARES (1985). Distinction can principally be made between tracer loss, velocity shot and two-pulse methods for improved marker logging interpretation (HILL & SOLARES 1985; HILL, BOEHM & AKERS 1988). Logging preparation after tracer injection often necessitates casing string decontamination (LABUDOVIC 1981).

6.2.1.3.2. Conventional single- or double-tracer monitoring

As an introduction to an extensive review and status report of gamma-ray spectroscopy recording and interpretation, principles and applications of conventional single- or double-tracer gamma-ray monitoring are outlined as follows.

6.2.1.3.2.1. Principles

The radioactive tracers are in most cases added in the terminal stage of proppant-fluid-slurry pumping to ensure their deposition close to the wellbore where they can be easily recorded by the logging tool and show up as pronounced anomalies considerably above the natural background radiation level even at relatively low-energy activity peaks of the artificial markers (LEICHT 1985). The other advantage of near-wellbore tracer placement in the final stage of fluid or proppant pumping is that only comparatively little amounts of artificially radioactive proppants or fluid are required for tagging of the proppant-fluid-slurry. Recording and processing is identical to that discussed for monitoring of natural proppant radioactivity (cf. section 6.2.1.2.). In terms of gamma-ray spectroscopy (cf. section 6.2.1.3.4.), there is no difference between spectrometrical discrimination of the natural background radioactivity of the reservoir formation and various synthetic tracers which have been artificially introduced into hydrocarbon-bearing interval matrix, fracture, gravel pack mantle and/or borehole.

Conventional gamma-ray fracture height monitoring comprises labelling of proppants or fluids with a single radioactive tracer and logging after completion of the stimulation treatment. Double-tracer application has hitherto been limited to the usage of isotopes with different decay rates that can be discriminated through multiple-run time-lapse conventional gamma-ray logging, because gross gamma-ray counting tools are sensitive only to the overall presence of gamma rays and not to the individual gamma-ray energy signatures characteristic of different tracers (GADEKEN, SMITH & SEIFERT 1987). Conventional fracture height determination by single- or double-tracer gamma-ray logging provides created fracture height in case of fluid tagging and propped fracture height for the occasion of proppant marking, or alternatively tagging of two parts of the whole fracturing material assembly with markers of different half-life leads to determination of both fracture height types in case of proppant and fluid spotting and time-lapse monitoring of isotope decay (cf. sections 6.2.1.4.4. and 6.2.1.5.2.).

6.2.1.3.2.2. Applications

Field performance examples of conventional single- or double-tracer gamma-ray monitoring of hydraulic fractures are reported by PEARCE (1979). The surveys permit to recognize whether large amounts of proppants were dropped out in the wellbore while pumping and whether fractures were created below the top of proppant fill initially (prior to tagging of proppant-laden fluid with radioactive sand) or by natural diversion while pumping with no radioactive sand being placed in the crack (these relationships become only evident after cleaning-out of the wells). Standard gamma-ray and temperature logging for fracture height monitoring is also carried out by HOWELL & THOMAS (1978), MEDLIN & FITCH (1983); GREGORCZYK, PAULS, HOLTMYER, VENDITTO & CHISHOLM (1984); HOLDITCH, ROBINSON, WHITEHEAD & ELY (1987); HUEBINGER, WEBSTER, CHISHOLM, VENDITTO & HUNT (1988); KULAR, CHHINA, BEST & MACKENZIE (1988); NORTHCUTT, ROBERTSON & HANNAH (1988) and TAN, McGOWEN, LEE & SOLIMAN (1988). HOOVER & ADAMS (1988) are able to distinguish between minifracturing data-gathering and calibration treatment and main stimulation operation by differential tagging.

Temperature and single-tracer conventional gamma-ray logging are also viable methods for determination of fluid uptake zones when encountering severe drilling mud losses during penetration of extensively naturally fractured reservoir or barren intervals (CANSON 1985). Lost circulation sections can also be monitored by noise logging as well as by spinner and flowmeter surveys (cf. section 6.2.1.7.). An enhanced gamma-ray logging application is selective downhole marker injection and subsequent recording of the travel path of the fluid contaminated by the inserted tracer.

6.2.1.3.3. Administrative and environmental aspects of tracer monitoring

The technically favourable indications of the suitability of tracer utilization for gamma-ray conventional or spectroscopy monitoring of hydraulic fracture height are quite prominent, but the procedure of field application is often complicated by administrative, environmental and economical aspects. Some points including restrictions and obligations of radioactive marker handling as well as cost containment vs. advanced radioactive tracer monitoring are briefly summarized as follows. Comments are also offered on intermediate safety procedures.

6.2.1.3.3.1. Restrictions and obligations of radioactive tracer handling

While all three methods of artificial introduction of radioactive markers into fracturing fluids and proppants have the advantage of good control of homogeneous distribution of the tagging material within the total proppant and fluid package infilling the fracture as well as maintenance of constant concentration of sufficiently high intensity of radioactivity thereby guaranteeing reproducibility, one major drawback are the administrative complications of handling of these environmentally hazardous materials. Safety obligations by mining authorities, labour departments and national environmental offices are particularly restrictive in Great Britain and Germany FRG and unfortunately often do not allow to take artificial radioactive tracering of proppants and/or fluids into consideration due to the fact that the time needed from application to obtainment of permission and the partially required preparation of the material (being not able to be carried out before approving decision has been made by the authority) is frequently too long and costs a lot of money if a well which is ready for treatment has to wait (this is particularly valid for exploration and appraisal wells where the rig is standing on the drilling site).

Tracer application can under the current handling and safety regulations in Europe almost only meet the obligations set by various authorities in development campaigns where proper planning of marker utilization and performance of the administrative procedure can be done reasonably in advance, with everything being well set once the borehole is ready for stimulation treatment and identifier injection. As particularly in the North Sea in Europe where most of the hydraulic proppant fracturing activity is going to take place during the next years (cf. sections 2.2.1.6.2. and 2.4.5.), field development is performed in well-prepared campaigns affecting a suite of boreholes in different fields, no practical restriction of radioactive tracer application for hydraulic fracture height monitoring by conventional or advanced gamma-ray logging techniques is seen for the predominant amount of stimulation jobs in the near future in Europe.

As the advantages of the exactly known and controlled level of radioactivity which can be used for proper calibration of measuring and recording devices is unfortunately frequently compensated by organizatory drawbacks, at the bottom of the line preference has especially in the past often been given to the simple and cheap methods of temperature logging of any proppant type (cf. section 6.2.1.1.) and gamma-ray logging of naturally radioactive proppants infilling the fracture (cf. section 6.2.1.2.) for not only technical and economical, but also administrative reasons. In the future, however, based on early planning of tracer utilization and performance of obligatory procedures, multiple marker scanning by gamma-ray spectroscopy which is the most sophisticated and advanced technique of fracture monitoring is expected to be rapidly spreading out to increasingly diversified applications in hydraulic well stimulation.

6.2.1.3.3.2. Intermediate safety procedures

An intermediate safety procedure in order to meet both technical and administrative requirements and to perform an environmentally less hazardous treatment is renouncing of the common practice to insert artificially radioactively-coated grains into the final stage of proppant pumping. Injection of tracer-tagged particles with the finer-grained lead-in lot and the normal-grained main quantity of the proppant load gives rise to a good distribution of radioactive material within the fracture, and pumping of conventional normally- or coarse-grained proppants in the terminal or tail-in stage (HART, ENGI & MORRIS 1983) with only little marker tagging under enhanced control or even without any tracer intermixing guarantees that any remaining sand which is circulated out of the borehole and also material that flows back out of the fracture mouth into the wellbore is not radioactively contaminated.

Therefore the operational safety is considerably increased by limitation of marker-tagged proppants to the fracture sections located farther away from the borehole (DRESSER ATLAS 1982). Caution has to be exercised, however, that the radioactive proppant or fluid is not transported too far away from the fracture mouth to become undetectable by the gamma-ray tool (cf. section 6.2.1.3.1.3.), because unless sufficiently confident distinction of natural background radioactivity and artificial proppant and/or fluid radioactivity can be done, the radioactive fracture height monitoring approach is useless. MEEHAN & PENNINGTON (1981) ascertain optimum distribution of radioactively-coated sand within the fracture by continuous injection throughout the stimulation job.

6.2.1.3.3.3. Cost containment vs. advanced radioactive tracer monitoring

Another influencing factor in Europe is the often exaggerated cost containment of the operating companies which particularly limits the more abundant application of advanced gamma-ray tracer logging techniques such as gamma-ray spectroscopy. Service and supply companies always run into great troubles when trying to convince many operators that following drilling of an expensive hole down to terminal depth, a little additional sum of money should be spent on proper investigation of potential productive horizons by logging, testing and stimulation (cf. sections 2.2.1.5.2., 2.3.4. and 2.4.2.). Although the latter expenses are together only a fraction of the drilling cost, in many cases these items are the first which are killed by notorious cost-cutting managers, with objectively no justification being able to be seen for such a procedure. In the USA, first environmental obligations of tracer handling are more favourable and second the mentality is more directed towards acquisition of additional information on the reservoir even if linked with having to pay a premium, and therefore gamma-ray tracer logging of any type is a continuously expanding market there especially after the successful application of gamma-ray spectroscopy in various critical reservoir constellations in the last years.

Thus the major marketing effort of service and supply companies in Europe and other parts of the world outside the USA has in the next years to be put on convincing hydrocarbon production companies that the considerable expenses for drilling are almost impossible to justify if a small additional amount of abt. 10 - 20 % of the total sum is not invested for proper evaluation of the column of geological formations penetrated by the drilling bit and exposed along the borehole wall, and adequate testing and stimulation of all the horizons which might contain important oil and gas reserves that, however, are not sufficiently or unmistakably evident on logging records.

The only possibility to increase recoverable reserves in the framework of more and more difficult exploration strategies and decreasing discovery rates vs. increasing consumption is provenance of additional quantities from hydraulic fracturing treatments of marginal reservoirs, and the satisfactory performance of stimulation has to be secured by sophisticated and unambiguous monitoring of the crack. In this context, two aspects require major emphasis comprising proppant selection and monitoring technique choice. Proppant selection has to focus on application of the most suitable type and grain size for the given reservoir regardless of possibly having to pay a premium for a better proppant, and monitoring technique choice has to concentrate on utilization of the method with highest resolution and accuracy.

6.2.1.3.4. Multiple tracer discrimination by spectral gamma-ray logging

Distinction has to be made between single- and multiple-tracer recording by gamma-ray logging tools. While traditionally single-tracer applications have commonly been fully sufficient for fracture evaluation and double-tracer treatments with isotopes of different decay rates that have to be discriminated by multiple-run time-lapse conventional gamma-ray logging have been subordinate exceptions, nowadays many fracturing operations are more complex. Standard single-isotope recording cannot distinguish between radioactively-tagged material which is located inside the wellbore, in voids or channels of the cement annulus, or actually within the created or propped fracture, with the total gamma-ray activity being a combination of all the emissions from the various sources. Multiple zones are fractured, treatment fluids and proppants are injected in several stages, and diverting agents are used between the individual stages to separately route stimulation materials into the desired intervals. Necessity also exists for distinction of the radioactive material within the fractured formation and within the borehole. It is therefore often desirable to inject and monitor multiple tracers (GADEKEN & SMITH 1986, AHMED 1987, PRO-TECHNICS 1987).

After discussing some general aspects of gamma-ray spectroscopy, the outline of multiple tracer discrimination by spectral gamma-ray logging as follows focusses on distinction of radiation provenance from fractured formation and borehole fluid, spectrometrical deconvolution of composite radiation profiles, and gamma spectroscopy applications in hydraulic fracturing as well as in combinations of natural and hydraulic fracturing.

6.2.1.3.4.1. General aspects of gamma-ray spectroscopy

Multiple tracer discrimination by spectral gamma-ray logging (gamma-ray spectroscopy; gamma-ray spectrometry; compensated spectral natural gamma ray system; WICHMANN, McWHIRTER & HOPKINSON 1975; FERTL & RIEKE 1979; FERTL, STAPP, VAELLO & VERCELINO 1980; FERTL & OLIVER 1983; SMITH, ROBBINS, ARNOLD, GADEKEN & DEATON 1983; GADEKEN, ARNOLD & SMITH 1984; ANDERSON, PEARSON, ABOU-SAYED & MYERS 1986; CRAIGHEAD, McDOUGALL & HOTZ 1987; WILLIAMS & McCARTHY 1987; multiple radioactive tracer scanning; GADEKEN & SMITH 1986; GADEKEN, SMITH & SEIFERT 1987; GADEKEN, SMITH, ARNOLD & ROBBINS 1988; precision radioactive isotope spectral measurement; PEMPER, FLECKER, McWHIRTER & OLIVER 1988; WESTERN ATLAS 1988 b) comprises break-up of the radiation into the individual isotope contributions of various tracers by spectrum unfolding and weighted-square data processing (WESTERN ATLAS 1988 b).

In most cases up to four and in extremely favourable cases even up to six different tracers with half-life times of preferentially abt. 30 - 90 days each can be inserted into various fluid and proppant stages (PRO-TECHNICS 1987) and can be recorded and discriminated in one run. Practical limitation to four different isotopes, however, is recommended, because with a greater amount of components, the decrease in number of degrees of freedom makes it likely that statistical variations could be mistaken for fluctuations in tracer concentrations, especially when both borehole and formation components are selected (GADEKEN, SMITH & SEIFERT 1987), with in the latter case even two or three different markers being sometimes the upper limit for unambiguous interpretation of the results of the tracer scanning record. A higher amount of different markers to be discerned also bears the potential danger that some of the identifier shapes to be discriminated look in fact quite similar.

As the tracer isotopes have different energy peaks and/or half-life times, and fluid and proppants can be differently tagged (PRO-TECHNICS 1987), gamma-ray spectroscopy allows assessment of created vs. propped fracture height (cf.

section 6.2.1.5.2.) by energy spectrum distinction and/or time-lapse differentiation of the individual tracers in contrast to standard gamma-ray logging which gives only propped fracture height, with created vertical crack extension in the latter case having to be determined by temperature logging. Gamma-ray spectroscopy can also be utilized for analysis and deconvolution of the natural radioactivity profile of the reservoir formation (HEFLIN & NETTLETON 1980, FERTL 1982).

6.2.1.3.4.2. Distinction of radiation provenance from fractured formation and borehole fluid

Although the spectrum unfolding or deconvolution technique reveals the vertical position of tracer material along a wellbore corresponding to depth, it is not always apparent whether the detected radiation is coming from formation, cement channels behind casing, or within the borehile itself inside of the casing (PEMPER, FLECKER, McWHIRTER & OLIVER 1988). In case of treatment of two separate intervals of the pay zone complex, the radiation intensity curves may clearly define the major fracture sections, but also show a low level amount of tracer material in the region below, between or above the fractured segments, pointing to radiation deriving from borehole or cement channels behind the casing. As it is generally not uncommon for fluid and proppant to channel behind casing and to create additional fractures in adjacent zones, distinction of various possibilities of radiation provenance is very significant. Basic discrimination techniques and special differentiation methods are briefly outlined as follows.

6.2.1.3.4.2.1. Basic discrimination techniques

Precision radioactive isotope spectral measurements (WESTERN ATLAS 1988 b) permit also distinction of material within the borehole and within the fractured formation by gamma-ray peak-to-downscatter ratios as well as comparative evaluation of Compton downscattering and photoelectrical absorption effects based on the shapes of the tracer spectra, and sensitivity coefficients accounting for absolute magnitude differences and relative shape changes from high to low energies due to Compton scattering effects (GADEKEN, SMITH & SEIFERT 1987; PEMPER, FLECKER, McWHIRTER & OLIVER 1988). Another possibility of distinction of radioactive material within the borehole and within the formation behind the casing is the spectral gamma-ray crossplot of total photo sum vs. background window count rate (ANDERSON, PEARSON, ABOU-SAYED & MYERS 1986). A weighted-least-squares spectrum unfolding algorithm calculates the radioactive intensities as a function of depth.

6.2.1.3.4.2.2. Special differentiation methods

Distinction of gamma-rays coming from borehole and/or fractured formation by photoelectrical effect is based on the fact that the chemical element with the highest atomic number dominates absorption (GADEKEN, SMITH & SEIFERT 1987; PEMPER, FLECKER, McWHIRTER & OLIVER 1988). In cased wells, usually iron is most significant in the low-energy spectrum. Gamma-rays deriving from the fractured formation have to traverse the casing and thus suffer strong absorption by iron, whereas the spectrum of gamma-rays emanating from the fluid-filled wellbore is not changed. In contrast to the photoelectrical effect, Compton scattering is no function of the atomic number of the chemical elements, but is a function of density and thickness of the material traversed by gamma-rays. After Compton downscattering, different shapes of tracer spectra are left as a consequence of different provenance of the radiation, and mathematical type curve matching based on weighted least squares can in ideal cases even distinguish radiation coming from fractured formation, cement sheath around the wellbore, and borehole fluid. The energy spectrum with casing (representing the formation case) shows fewer gamma-rays in the peak region and more gamma-rays in the downscatter window than the spectrum accumulated without casing between instrument and radioactive source (comprising the borehole case). In order to improve the interpretation, either the natural radioactivity has to be subtracted or the tracer concentration has to be increased for allowing clearer modelling of spectra shapes.

The high-energy counts/low-energy counts ratio is a quick indicator of the distance of the source of radiation within the fractured formation, although some ambiguity is left with this downscattering ratio, because for example an exclusive tracer enrichment in the cement sheath might look as if being distributed half each in reservoir and borehole fluid. Another hint to radiation provenance is provided by repeated logging runs. If the gamma-ray spectrum changes with repetition recording, the tracer probably comes from the borehole where it is moving in the fluid, whereas more or less constant radiation profiles point to derivation from the fractured formation where the tracer is either mechanically fixed or chemically bound or adsorbed.

6.2.1.3.4.3. Spectrometrical deconvolution of composite radiation profiles

By means of gamma-ray spectroscopy, the observed total composite radioactivity spectrum can be deconvolved into standard spectra of the natural potassium, thorium and uranium isotopes which compose the background radiation, and of the synthetic tracers that were added to proppants and/or fluids in the hydraulic fracturing operation. Some aspects of elemental separation by spectral stripping, common reference spectra or isotope fingerprints, and discrimination of different fracturing job stages are discussed as follows.

6.2.1.3.4.3.1. Elemental separation by spectral stripping

Elemental separation is achieved by using spectral stripping techniques or a weighted-least-squares algorithm for spectrum unfolding (GADEKEN & SMITH 1986; PEMPER, FLECKER, McWHIRTER & OLIVER 1988). This spectrum unfolding or deconvolution is a method of spectral fitting to obtain elemental yields. Spectrum unfolding or deconvolution has also been proven to be useful in a variety of other well logging applications such as natural (MARETT, CHEVALIER, SOUHAITE & SUAU 1976; SERRA, BALDWIN & QUIREIN 1980; SMITH, ROBBINS, ARNOLD, GADEKEN & DEATON 1983) and pulsed neutron (HERTZOG 1980; GRAU & SCHWEITZER 1987) gamma-ray spectroscopy. In the weighted-least-squares spectrum unfolding technique, the assumption is made that each measured spectrum is composed of a linear combination of the individual spectra from radioactive materials found in formation and borehole environment.

JACKSON & GARTNER (1988) present a special stripping technique for computation of quantitative concentrations for all radioactive isotopes in fracture and formation once being properly calibrated. Apparent distance to the source from the borehole wall is inferred from the reconstructed gamma-ray spectrum which permits discrimination of tracers being in solution in the borehole or collecting in the fractures in the reservoir. Accurate concentrations allow to reconstruct the original gross gamma-ray level, thereby possibly eliminating the need of the pre-fracturing reference log and thus enabling well evaluation in a single logging run. As the data is independent of the measurement system, a comparison of operations performed in different wells in the same area is meaningful. The elemental spectral analysis program can compute concentrations of up to seven isotopes based on up to 50 different spectral windows. The suite of isotopes can be examined before the fracturing operations in order to determine the optimum selection of tracer isotopes. OVERBEY, YOST & WILKINS (1988) develop special procedures for distinguishing effects of various flow rates and pressures during the treatment by selective tracer monitoring.

6.2.1.3.4.3.2. Common reference spectra or isotope fingerprints

Standard reference spectra from each tracer representing isotope fingerprints are obtained via measurements in known test conditions and are then used as inputs along with the observed composite spectra in the deconvolution process which includes mathematical type-curve matching. The best results are achieved if tracers with separated gamma-ray lines are carefully selected, because overlapping of several gamma-ray peaks coming from different isotopes within a relatively narrow energy region can cause problems at least in stripping processing.

6.2.1.3.4.3.3. Discrimination of different fracturing job stages

The decisive effectiveness of gamma-ray spectroscopy of several different markers is the separation of different stages of a fracturing job by selective recording and discrimination of multiple tracers. The observation of the specific gamma-ray signature of each tracer in the fracture or cement material requires only one borehole trip with the logging instrument (GADEKEN & SMITH 1986). Due to high tool resolution, the uphole computer software can be tailored to process virtually any combination of multiple tracers. Gamma-ray spectroscopy is therefore particularly indicated for proper evaluation of multistage proppant and fluid schedules within the hydraulic fracturing treatment where all the individual constituents of the pumping program can be separately marked with distinguishable isotopes. It can also be determined whether fluid and proppant have channelled behind the casing and created additional fractures in adjacent zones by evaluation of material being inside or outside of formation and borehole, respectively. Gamma-ray spectroscopy is able to distinguish between natural formation isotopes and synthetic inserted tracers without necessitating a pre-marker-injection reference log (PEMPER, FLECKER, McWHIRTER & OLIVER 1988). In most cases, the intensity of the tracer material in the fracture is abt. one to two orders of magnitude greater than that of the natural reservoir background.

Gamma-ray spectroscopy tools are available in 3 5/8 " and 1 11/16 " outer diameter (PEMPER, FLECKER, McWHIRTER & OLIVER 1988; WELEX 1988). Although the larger version of the tool has greater counting rate capabilities due to the bigger thallium-activated sodium iodide scintillation crystal implemented, the 1 11/16 " slimhole instrument is more popular, because its utilization does not necessitate removal of tubing from the wellbore. The counting rate capability difference between the two systems can be compensated by reducing the logging speed of the smaller instrument. The spectral standards employed for data analysis have to be matched to composition and dimensions of the crystal for either tool.

6.2.1.3.4.4. Gamma-ray spectroscopy application in hydraulic proppant fracturing

Gamma-ray spectroscopy evaluation of multiple tracers added to the materials inserted into the borehole is a successful monitoring technique in hydraulic proppant fracturing, acid fracturing, cementing, water or chemical flooding, and even steam-drive operations. The main applications of multiple-tracer discrimination by gamma-ray spectroscopy are determination of effectiveness of diverting agents used in staged fracturing treatments, evaluation of fracture height in multistage proppant or acid stimulations, differentiation between proppant and fluid placement, distinction of insertion of resin-coated and conventional proppants or different proppant types in tail-in operations, discrimination of created and propped fracture height, assessment of the placement of different proppant concentrations, appraisal of relative locations of lead and tail slurries in primary cementing operations, evaluation of effectiveness of squeeze cementing and gravel pack procedures, monitoring of injection well performance in secondary and tertiary enhanced oil recovery projects, determination of interwell or interstorey communication or isolation and thus control of the degree of reservoir heterogeneity, survey of injected hazardous wastes in disposal wells, identification of hydraulic contact between treated and thief zones, determination of centres of drilling mud and completion fluid loss, localization of casing or tubing leaks, control of perforation breakdown status, and monitoring of gravel pack height and density. Particularly interesting aspects are distinction between created and propped fracture height as well as joint fracture azimuth and height determination.

6.2.1.3.4.4.1. Created vs. propped fracture height

Distinction between created and propped fracture height by different tagging of fluids and proppants can easily and reliably be performed by gamma-ray spectroscopy both in single- and multiple-horizon stimulation treatments. Fluid labelling and monitoring has the advantage that in addition to helping to indicate fracture segments which may have closed after the end of the operation, also communication paths to nearby water-bearing intervals or bridges between two superimposed zones through the separating barrier can be discerned (GADEKEN & SMITH 1986). In terms of cementing applications, gamma-ray spectroscopy can locate different types of cements and measure the average radial cement thickness (GADEKEN, SMITH & SEIFERT 1987).

WILLIAMS & McCARTHY (1987) comment on the application of multiple radioactive tracers to optimize stimulation design. Field experience of fracture height monitoring by logging radioactivity of multiple artificial tracers is reported by AHMED, NEWBERRY & CANNON (1985); BARBER & THEMIG (1985); PEARSON, ABOU-SAYED & NGUYEN (1985); HOLDITCH, ROBINSON, WHITEHEAD & ELY (1987) and YOST, OVERBEY, WILKINS & LOCKE (1988). CAMPBELL, HANOLD, SINCLAIR & VETTER (1981); TESTER, BIVENS & POTTER (1982); GUDMUNDSSON, HAUKSSON, THORHALLSSON, ALBERTSSON & THOROLFSSON (1984); ROBINSON, TESTER & BROWN (1984) and ROBINSON & TESTER (1985) describe chemical and radioactive tracering for performance monitoring of hydraulic fractures in geothermal reservoirs.

6.2.1.3.4.4.2. Joint fracture azimuth and height determination

An advanced application of multiple-tracer gamma-ray spectroscopy would be oriented recording aiming on fracture azimuth determination in addition to crack height monitoring. The achievement of good results, however, would require perforation orientation according to the geostress field in order to secure optimum fracture propagation and full communication with the borehole trajectory (cf. also section 4.10.3.). This technique which is currently still in experimental stage would represent an unique combination of fracture analysis in both horizontal (fracture azimuth) and vertical (fracture height) direction and would be the most sophisticated method delivering a maximum of results from only one instrument and one borehole run and would therefore be in a very competitive position once commercially available.

Gamma-ray spectroscopy can also be run through tubing with up to 1 11/16 " slimhole tools and can be combined with temperature and casing collar locator instruments in one run. In contrast to standard gamma-ray logging, gamma-ray spectroscopy eliminates pre-treatment surveys for determination of background radiation by spectral deconvolution of the recorded data. Enhanced natural gamma-ray spectroscopy can be utilized for mineralogical and geochemical well logging (SCHLUMBERGER 1987 c).

6.2.1.3.4.5. Gamma-ray spectroscopy application in a combination of natural and hydraulic fracturing

Possibilities of application of gamma-ray spectroscopy in combinations of natural and hydraulic fracturing are illustrated by some examples of natural crack widening by hydraulic treatment, multiple hydraulic fracturing in horizontal wells, and coexistence of two natural fracture systems as follows.

6.2.1.3.4.5.1. Natural fracture widening by hydraulic treatment

LOPUS, SEIFERT & SCHEIN (1987) report identification of conducting natural fractures utilizing multiple radioactive isotope tracering techniques and gamma-ray spectroscopy. The use of multiple radioactive markers in conjunction with a dendritic fracturing technique (cf. section 4.8.8.4.) has been applied for effectivity monitoring of the various diverter stages during the treatment and thus selective vertical distribution check of the individual fracturing pulses. Gamma-ray spectroscopy analysis of the distribution of several tracers has verified the restriction of the hydraulic stimulation to the pre-existing pattern of natural cracks by the dendritic hydraulic fracturing procedure (cf. also section 4.5.4.4.3.).

In the example outlined by LOPUS, SEIFERT & SCHEIN (1987), four tracers were inserted in different stages of the operation. The first marker was used for indication of the most conductive fractures, the second isotope was applied for monitoring of less conductive or blocked fractures after the high-conductivity cracks have been temporarily diverted with proppants, the third tracer was inserted after pumping of another diverter to reflect still less conductive fractures, and following a large volume of fresh water flush in order to minimize the effect of the diverters, the last tracer aimed on monitoring of the gross vertical interval which was effectively treated, thus representing total fracture height checking. Soluble diverters such as rock salt were used to seal off the major natural fractures in order to force the fluid into smaller cracks and to create artificial fractures, and a final large flush had the purpose to dissolve the salt out of the blocked positions.

Fracture height determination by special gamma-ray analysis is also carried out by ANDERSON, PEARSON, ABOU-SAYED & MYERS (1986). Aspects of gamma-ray spectroscopy for multiple radioactive tracer well logging are also discussed by GADEKEN, SMITH & SEIFERT (1987). An advanced technique is pulsed neutron induced gamma-ray spectroscopy (MYERS 1988). Digital sonic logging in combination with high-resolution dipmeter and induced gamma-ray spectroscopy provides an unique opportunity for retroanalysis of fracturing operations and producibility of tight gas sands as well as predictive mechanical properties modelling aiming on crack boundary forecast (HUNT, RAYMOND, HASKETT & PIRIE 1981). Fracture evaluation by quantitative concentrations derived from the natural gamma-ray spectrum is outlined by JACKSON & GARTNER (1988).

6.2.1.3.4.5.2. Multiple hydraulic fracturing in horizontal wells

OVERBEY, YOST & WILKINS (1988) evaluate multiple hydraulic fracturing in horizontal wellbores intersecting naturally jointed reservoirs. Special discrimination procedures allow to separately examine the impacts of various flow rates and pressures which existed during the treatments. Radioactive isotope spectroscopy can be effectively used to trace flow paths to adjacent intervals, because the isotopes could not have appeared in these zones without travelling back to the wellbore via fractures of an orientation considerably different than that of the principal stress direction. The distribution of fluid entry points indicates that natural cracks are selected at low injection rates while induced fractures are chosen at higher pumping rates. Gamma-ray spectroscopy of multiple radioactive tracers in conducting multiple injection tests is thus very beneficial in understanding natural vs. induced preferred fluid paths.

6.2.1.3.4.5.3. Coexistence of two natural fracture systems

LAYNE & SIRIWARDANE (1988) perform gamma-ray spectroscopy of multiple tracers in reservoirs intersected by two distinct natural fracture systems. Marker monitoring revealed that during the first phase of the stimulation job, fluids propagated into the first crack system and entered a fault zone which crosses the middle portion of the reservoir complex. In the second phase of the operation when injection rates and pressures were increased, fluids penetrated the second fracture network and travelled back to the wellbore as indicated by tracer enrichment in the lower segment of the pay. These relationships testify to existence of a highly-connected natural fracture system in the prospective interval which promotes multiple paths of expanded natural cracks from a single stimulation treatment. Fracture diagnostics can even be expanded to identify the relative amount of fluid that entered each crack in a semi-quantitative manner. Two forms of fluid entrance can be recognized including annulus leakoff and large injection flow into discrete fractures.

6.2.1.4. Combination of temperature and gamma-ray logging

The optimum combination in order to reproduce the results, to calibrate the individual logs and to cross-check the reliability of the recorded data as well as to pay attention to the economical need of proppant and logging cost containment is carrying out fracturing jobs with high-strength zirconia-silicate proppant tail-ins (cf. section 2.4.1.2.3.) in some carefully selected wells in the field and running therein both temperature and conventional gamma-ray logs (both tools can be combined and may thus be recorded in one run even in 1 11/16 " slim size) shortly after the stimulation treatment. The feasibility of this combination is particularly high for field-wide monitoring campaigns in tight gas reservoirs where most of the wells have to be fractured and gamma-ray spectroscopy is beyond consideration for economical reasons due to relatively simple fracturing operation design.

It has to be emphasized, however, that any stimulation job requiring discrimination of more than one component of synthetic tracer and/or natural material radioactivity can only be satisfactorily monitored with gamma-ray spectroscopy. The following brief outline includes aspects of logging cost containment in fracturing campaigns, clarification of temperature anomalies, distinction of created and propped fracture height, and time-lapse monitoring repetition logging.

6.2.1.4.1. Logging cost containment in fracturing campaigns

The combination of temperature and gamma-ray logging does not only allow comparison of the results delivered by the different methods in the individual well, but using the existing temperature logs for calibration and reference, it is fully sufficient to run only standard temperature logs in the other wells of the field to obtain the requested information, with infilling of the fractures with sand and/or intermediate- to high-strength proppants of any composition keeping treatment cost within a reasonable frame, and temperature logging being even also slightly cheaper than gamma-ray recording, thereby drastically containing the logging expenses for large fracturing campaigns affecting up to several dozens of wells in big fields in case of unambiguous temperature readings in formations with normal thermal conductivity. In case of hardwork at the bottom of the hole not permitting the tool to get completely down in the situation of short ratholes, at least the upper boundary of the fracture can be clearly picked up.

This type of combination and replacement of temperature and conventional gamma-ray logging in field-wide campaigns with recording of both logs in a few selected wells and later measuring only still either gamma-ray or temperature logs with interborehole calibration, comparison and correlation, however, is only possible in case of not too significant differences between created and propped fracture height on the one hand (cf. section 6.2.1.5.2.) and reservoir thickness setting the stage for absolute fracture height on the other hand, thus guaranteing reasonably matching readings of the two different determination techniques (otherwise the only reliable but also expensive solution is multiple-tracer gamma-ray spectroscopy; cf. section 6.2.1.3.4.).

6.2.1.4.2. Clarification of temperature anomalies

In terms of warm anomalies at the top of the fracture as reflected by temperature logging (cf. section 6.2.1.1.3.), distinction between thermal conductivity anomalies and fracture anomalies can also be made by checking a parallel gamma-ray log that can show whether the warm zone is also infilled by radioactive proppant or fluid material and thus still belongs to the crack or not (DOBKINS 1979). In some cases, however, the gamma-ray log does not agree with the temperature survey, with the temperature log being assumed to be more accurate, because the fluid could enter and cool a fracture regardless of its size, whereas distribution of radioactive proppants in the fracture may have been restricted by proppant grain size and crack width, with probably only little material having entered the fracture at the narrower upper and lower terminal tips which then might not show up on the gamma-ray record due to possibly being below background level of radiation.

On the other hand, in such cases the gamma-ray log shows the effective fracture height, because the upper and lower tips reflected by the temperature log do not contribute to reservoir drainage if not sufficiently propped to become distinguishable on the gamma-ray record. The differences between created crack height as reflected by fluid monitoring with the temperature log and propped fracture height as displayed by proppant surveying with the gamma-ray log (cf. section 6.2.1.5.2.) can only be clarified by comparative evaluation of both records, and in case of doubtful results of these standard techniques, the only reliable possibility is gamma-ray spectroscopy (cf. section 6.2.1.3.4.).

6.2.1.4.3. Distinction of created and propped fracture height

The combination of temperature and conventional gamma-ray logging therefore also allows to determine and to distinguish both created and propped fracture height (cf. section 6.2.1.5.2.) by separate fluid recording (temperature logging) and proppant detection (gamma-ray logging). Temperature and gamma-ray surveys can be further combined with pre-fracturing injectivity testing to make sure that the entire completed interval will be treated during the stimulation job, and if not so, the logs act as a road map to determine where additional perforations are required (BUNDY 1981). Aspects of combination of temperature and conventional gamma-ray logging are also discussed by GREGORCZYK, PAULS, HOLTMYER, CHISHOLM & VENDITTO (1984) and BEGNAUD & CLAIBORNE (1985).

A more sophisticated, although in most cases also a more expensive technique of distinction of created and propped fracture height is gamma-ray spectroscopy (cf. section 6.2.1.3.4.). While a standard combination run of conventional gamma-ray and temperature tools is cheaper than gamma-ray spectroscopy, time-lapse recording by several runs of conventional gamma-ray logging in order to separate fluid and proppant radiation comprising tracers with different half-life times could be more expensive than gamma-ray spectroscopy at the bottom of the line, with the major advantage of the latter sophisticated technique being the unambiguity of results due to clear segregation of different tracers both in terms of energy peak level and decay rates.

6.2.1.4.4. Time-lapse monitoring repetition logging

Repetition of either temperature or gamma-ray logging runs during course of long-term production history can also document progressive closure at the upper and lower ends of the fracture due to crushing of the propping material as a consequence of increasing stress with successive reservoir depletion and pressure drawdown which first affects the narrow upper and lower fracture tips where mono- or oligo-layer propping is present in contrast to multi-layer propping of the middle portions of the crack, with consequently a successively shorter fracture appearing on the records with progressing hydrocarbon exploitation. Some comments are offered on temperature and conventional gamma-ray time lapsing as well as spectral gamma-ray time lapsing.

6.2.1.4.4.1. Temperature and conventional gamma-ray time lapsing

Temperature logging at a later stage, however, requires preparation of the well by renewed injection of cold fluids into the fracture (cf. section 6.2.1.1.), with thus time-lapse monitoring of changes of crack dimension (similarly as time-lapse determination of progressive rise of the hydrocarbon-water-contact during course of depletion by thermal neutron decay time logging; CLAVIER, HOYLE & MEUNIER 1971; SCHLUMBERGER 1974) being in terms of operations much easier and cheaper performed by gamma-ray recording than by temperature measurement. The simple possibility of continuously monitoring crack geometry by repeating gamma-ray logging from time to time during production history also argues for carrying-out of combined gamma-ray and temperature logging shortly after the fracturing operation at least in some wells of a group of treated boreholes in fields where stimulation is an important improvement of the hydrocarbon output.

The only limitation of time-lapse monitoring by conventional gamma-ray repetition logging is the declining intensity of proppant and/or fluid radioactivity. While most of the synthetic tracers are relatively short-lived with half-life times of the isotopes of abt. 30 - 90 days and thus are ideal means for short-term time-lapse monitoring, the radioactive impurities in natural zircon by diadochous cation replacement in the crystal lattice (cf. section 6.2.1.2.) have longer half-life times up to geochronologically exploitable periods. The radioactive impurities in the zircon raw material are more or less homogenized and redistributed during the fusion process of zirconia-silicate proppant manufacturing, with this proppant type therefore representing the most suitable material for long-term time-lapse monitoring of fracture height changes during progressive reservoir exploitation by conventional gamma-ray logging.

6.2.1.4.4.2. Spectral gamma-ray time lapsing

Time-lapse gamma-ray spectroscopy is a suitable means of distinction of several tracers with different half-life times in addition to or as an alternative to segregation of markers with different energy peaks (cf. section 6.2.1.3.4.). It is also a possible, but certainly also more expensive application for monitoring of progressive fracture closure at the upper and lower ends of the crack wings, although this goal can also be achieved by conventional gamma-ray logging. Summaries of recent advances and experience of technical well-logging for production monitoring are given by BRADEL (1980), BRADEL & DRAXLER (1982) and NOBLETT, FERTL & GUY (1987), and problems of log interpretation of tight gas sands are reviewed by RUHOVETS & FERTL (1982); KUKAL, BIDDISON, HILL, MONSON & SIMONS (1983); MONSON & SIMONS (1983), KUKAL (1984), KUKAL & SIMONS (1985), TSAY & FANG (1986) and KUCHUK (1987). An overview of log measurement and interpretation techniques is provided by SERRA (1985, 1987) and ELLIS (1987).

6.2.1.5. Thermal decay time logging

An alternative to temperature and gamma-ray logging for hydraulic fracture height determination is thermal decay time recording (PEETERS & HARTLEY (1984) which in fact is a combination of neutron and gamma-ray logging. After discussing some general aspects of thermal decay time logging, distinction of several fracture height types is made. Comments are also offered on evaluation of acid stimulation.

6.2.1.5.1. General aspects

The thermal decay time tool sends bursts of high-energy neutrons into the reservoir where they are rapidly slowed down to thermal velocities and are then captured by nuclei with a corresponding emission of gamma rays (PEETERS & HARTLEY 1984). Detection of the number of gamma rays as a function of time is a measure of the change in the thermal neutron population in the formation. Because chlorine is a strong thermal neutron absorber, the response of the thermal decay time log is primarily determined by the chlorine present in reservoir brine. Inducing a hydraulic fracture leads to replacement of formation brine by stimulation fluid, and thus comparison of thermal decay time logs run before and after the treatment can give an indication of crack height.

As the thermal decay time instrument has a very limited depth of investigation, consequently the fluid in the perforations has in many cases a greater influence on the tool response than the fracture. In addition to the perforation effect possibly overshadowing that of the crack, the tool is very sensitive to changes in apparent borehole salinities which are usually unknown when completion fluids are displaced by stimulation fluids. A dual-burst thermal decay time logging system is introduced by OLESEN, MAHDAVI, STEINMANN & YVER (1987) and STEINMAN, ADOLPH, MAHDAVI & PREEG (1988). Pulsed neutron logging with analysis of capture cross-section of gamma-rays resulting from thermal decay of neutrons is also applied for gravel pack height and density evaluation (cf. section 6.3.3.).

Reviewing several drawbacks of various fracture height determination logging methods, PEETERS & HARTLEY (1984) conclude that crack height interpretation should be regarded as an art, and in order to improve confidence in modelling, a combination of a number of different techniques should be utilized.

6.2.1.5.2. Fracture height types

Attention has to be paid to the fact that different logging tools respond to different fracture heights. Distinction can be made between created, propped and permeable fracture height (PEETERS & HARTLEY 1984).

6.2.1.5.2.1. Created and propped fracture height

As stimulation fluid enters the fracture and cools the whole length of the crack which was originally opened, temperature logging responds to total created fracture height where fluid had a chance to penetrate (PEETERS & HARTLEY 1984; cf. section 6.2.1.1.). In rare cases when particularly large fluid volumes were injected and the cooling effect is extending upwards and/or downwards beyond the crack tips due to further advance of the cold front beyond the fracture boundaries by conduction, crack height reflected by temperature logging could be even more than the true created one, but this effect is in most cases of negligible magnitude. Proppant distribution in the fracture is influenced by settling occurring during crack closure and grain size limitations for entering the narrow upper and lower tips of the fracture, with therefore propped crack height as measured by the gamma-ray tool in combination with radioactive proppants (cf. section 6.2.1.2.) being in most cases less than created fracture height as registered by temperature logging, with particularly propped crack top being lower in the borehole than created fracture top as a consequence of proppant settling and banking (cf. section 4.12.2.).

In low-viscosity fracturing fluids, the proppant bank is growing from the bottom of the crack in upwards direction by progressive vertical proppant aggradation, and therefore radioactive proppants added only in the late stage of the treatment are not reflecting the true propped fracture height. Thus radioactive grains should be intermixed into the proppant population preferably throughout the job for the purpose of reflecting total propped fracture height upon gamma-ray logging, with addition being stopped just before the after-flush in order to avoid accumulation of radioactive proppant in the tubulars.

Created and propped fracture height as reflected by fluid entry and proppant distribution can also be termed upper and lower crack height (GRI 1988) as a consequence of the fact that created fracture height is generally larger than propped fracture height for the reason of the inability of proppant insertion into the upper and lower narrow terminal tips of the crack. Fracture height obtained from pressure decline evolution of minifracture calibration treatments (CASTILLO 1987) is usually greater and smaller than that recorded by gamma-ray and temperature well logging, respectively (WAREMBOURG 1988). Therefore pressure analysis requires input of a leakoff or fluid-loss crack height which is termed net fracture height (in contrast to gross fracture height being overall crack height) in the calibration job assessment. Fluid-loss or net fracture height is generally greater than the net hydrocarbon pay zone because of permeable intervals above and/or below the net hydrocarbon pay zone.

6.2.1.5.2.2. Permeable fracture height

Liquid radioactive tracers recorded by gamma-ray logging as well as thermal decay time monitoring reveal permeable fracture height which is likely to be different from created crack height as displayed by temperature logging and propped fracture height as registered by gamma-ray logging of radioactive proppants (PEETERS & HARTLEY 1984). Leak-off of the stimulation fluid into the permeable interval of the fracture transports the liquid tracer exclusively into that section of the crack which later is also active in hydrocarbon production. Permeable fracture height therefore represents effective crack height concerning hydrocarbon exploitation. The thermal decay time log responds to the amount of chlorine in the formation brine and therefore also displays the permeable fracture interval where reservoir brine is replaced by stimulation fluids.

Created crack height as reflected by temperature logging could approach permeable fracture height measured by thermal decay time monitoring if no closure after generation of the crack occurs, formation fluid is replaced by fracturing fluid along the full length of the crack, and laterally and vertically progressive reservoir cooling by conduction is negligible. In such cases, the interval where warm reservoir fluids have been replaced by cool stimulation liquids is equal to the section where thermal decay time logging registers a salinity change. If parts of the created fracture are subsequently closing again, the tips may be still reflected by the temperature log due to the cooling effect of the injected treatment fluids, but replacement of formation liquids does not necessarily have to occur and thus salinity checking by thermal decay time logging gives another result.

The existence of three different fracture heights comprising created, propped and permeable crack height which are monitored by different logging techniques is the reason why comparison of several determination methods can never come to agreement concerning fracture height modelling, and therefore proper understanding of all the different crack heights has to be achieved by a combination of a number of independent techniques.

6.2.1.5.3. Evaluation of acid stimulation

In addition to conventional time lapse monitoring of water saturations, pulsed neutron capture logging (cf. also section 6.3.3.4.) is an excellent tool in the evaluation of acid stimulation in chalk and other carbonate reservoirs (CORWITH & MENGEL 1988). Other applications include assessment of pilot-scale water injection projects, water and gas saturation changes, and compaction monitoring. Some aspects of acid effect on pulsed neutron capture logging, gel-pad-created vs. acid-etched fracture height, and influence of well deviation and interval thickness are discussed as follows.

6.2.1.5.3.1. Acid effect on pulsed neutron capture logging

Calculation of the acid effect is based on the change in formation capture cross-section due to acid which gives a good indication of the effectiveness of the diverter stages of a treatment consisting of a pad of stimulation fluid containing either ball sealers or chemical bridging agents which function to temporarily plug off the more permeable zones to allow successive stimulation stages to treat less permeable intervals (CORWITH & MENGEL 1988). Compaction monitoring to control field subsidence by pulsed neutron capture logging has insufficient vertical resolution for the desired calculation precision, with radioactive markers giving better results.

The acid effect on pulsed neutron capture logging is caused by local increases in porosity and/or retention of $CaCl_2$ reaction products by the formation matrix (AL-SAIF, COCHRANE, EDMUNDSON & YOUNGBLOOD 1975). In case of predominant acid fracturing, the porosity increase in the near-wellbore area is not significant, with the acid effect then primarily being triggered by an increase in chloride content in the interstitial water due to $CaCl_2$ reaction product (CORWITH & MENGEL 1988). Therefore the acid effect is expected to dissipate after the well begins to produce water within the oil or after installation of waterflood secondary recovery.

Acid stimulation monitoring is always performed by comparing pre- and post-treatment records, with invasion of drilling fluids influencing the pre-treatment log not disturbing determination of the acid effect. The acid effect results in reduction of the total counts recorded by pulsed neutron capture tool. The counts are also affected in a similar fashion by increased water saturation, but the magnitude of decrease appears to be much greater for acid stimulation than for rising water cut. A comparison of background-corrected counts provides at least a qualitative indication of where acid effect remains.

6.2.1.5.3.2. Gel-pad-created vs. acid-etched fracture height

Thermal decay time and temperature logging in chalk reservoirs occasionally reveals that gel-pad-created fracture height is significantly different from acid-etched crack height (SNOW & HOUGH 1988). While gel-pad-created cracks are generally radial, acid is confined within a zone bounded by low-porosity layers, because the fracture thins at these low-porosity sections thereby restricting acid flow to the wide segments of the interval. Fracture thinning occurs both because the low-porosity zones tend to be harder and more cemented with a higher modulus and since they have slightly higher in-situ stresses. Acid effect on thermal decay time logs and flow contribution on spinner surveys is usually observed from all perforation sets with the cluster perforation technique.

6.2.1.5.3.3. Influence of well deviation and interval thickness

Where wellbore inclination is less than 20 degrees and the perforations are closer than abt. 20 ft together, one perforation set usually does not receive acid and does not contribute to flow, because higher stresses are created in the vicinity of the induced cracks thus increasing the local stresses near the unfractured perforations and restricting crack growth there (SNOW & HOUGH 1988). If the perforated interval is more than 15 ft long and borehole inclination is more than 20 degrees, thermal decay time logging reflects formation of multiple fractures within the same perforated section. Where several hundred feet of perforations are placed, acid effect on thermal decay time log is usually recorded over the entire interval even though flow is usually only from a limited area. The uneven current is due either to permeability variations or poor diversion, with in the latter case the primary flow interval having been hydraulically fractured while the other portions have been only matrix acidized.

6.2.1.6. Acoustic logging

Conventional acoustic logging devices have for a long time been used for locating predominantly natural fracture systems which intersect the wellbore in a horizontal or near-horizontal plane (NOBLETT, FERTL & GUY 1987; cf. section 6.2.1.10.4.). Such tools typically employ shear wave detection techniques applied to full-wave acoustic signatures. Loss of shear wave amplitude, coupled with increasing transit times in the presence of a fluid-filled crack system, is used to determine fracture location. As the direction of the cracks shifts towards the vertical plane with respect to the intersecting wellbore, conventional acoustic devices become unreliable for fracture detection purposes. Acoustic logging generally measures the effect of the cracks on transmission of the acoustic energy through the formation, but the fracture system has to be such that it influences the acoustic energy (HEFLIN 1979).

The Circumferential Acoustilog (SETSER 1981, VOGEL & HEROLZ 1981, BOTTER 1982, GUY & FERTL 1984; GUY, FERTL & OLIVER 1986; GUY, CORLEY, FERTL & FROST 1987; NOBLETT & FERTL 1987; NOBLETT, FERTL & GUY 1987; Circumferential Microsonic Tool; BOTTER 1982, SCHLUMBERGER 1982; Full Wave Sonic Logging; MINEAR 1986, WELEX 1987) employs the principle of generating and detecting circumferentially instead of longitudinally and transversely propagating acoustic waves of boundary character and is based on a circumferentially oriented acoustic transmitter-receiver array for detection of vertical and near-vertical fracture systems such as occurring in reservoirs in greater depth. Auxiliary instruments run in combination with the Circumferential Acoustilog sonde include a natural gamma-ray detector and an orientation section, with the latter measuring deviation of the borehole from vertical plus azimuth and relative bearing of the receiver-containing pad. The data are displayed in variable density log and/or signature waveform format. Following an overview of recognition of fractures from wave transformations, an outline of fracture orientation interpretation is given.

6.2.1.6.1. Recognition of fractures from wave transformations

The main application of the Circumferential Acoustilog is azimuth mapping of natural fracture systems or joint bundles, although the determination of the directional orientation of hydraulically-induced cracks is also possible. Careful calibration and recording should also allow vertical height monitoring of prominent large-scale natural and hydraulically-induced fractures by following continuous trends of more or less constant azimuth readings of major cracks that are separated by zones reflecting only little jointing (if at all) representing boundary seal sections of the fractured pay interval. The brief summary as follows focusses on sonic wave types and their propagation, sonic wave attenuation as fracture indicators, and additional information for fracture detection.

6.2.1.6.1.1. Sonic wave types and their propagation

The omnidirectional compressional wave generated in the fluid by the radial mode excitation of the transmitter ensures that acoustic energy will arrive at the fluid-formation interface at the borehole wall at the critical angle to produce refracted-diffracted bulk compressional and bulk shear waves in varying lithologies (GUY, FERTL & OLIVER 1986). Two types of boundary or interface waves are also created at the fluid-reservoir interface comprising Rayleigh wave and Stoneley wave which propagate in formation and fluid, respectively, circumferentially along the borehole wall in the finite medium of the wellbore, in contrast to longitudinal and transverse travelling of compressional and shear waves, respectively, in an infinite medium. Borehole curvature causes multiple reflection paths to arise at low grazing angles. Another wave is the direct compressional fluid wave, and the order of arrival at the receiver is bulk compressional wave, bulk shear wave, Rayleigh wave, direct fluid compressional wave and guided fluid Stoneley wave (MINEAR 1986).

The most significant wave for fracture detection is the Rayleigh wave which as a consequence of being a boundary wave travelling along the curvature of the borehole is not greatly affected by geometrical spreading. The motion of the particle is normal to the wellbore wall and therefore will be highly attenuated by vertical discontinuities such as fractures intersecting the borehole and lying in the propagation path between source and receiver. The guided fluid Stoneley wave as another boundary wave is damped due to radial energy emission from the fluid into the formation as the wave propagates around the wellbore. The magnitude of damping of the Stoneley wave is to a large degree a function of the contrast between the acoustic impedances of fluid and reservoir. Digital interpretation techniques analyze the circumferential acoustic waveform information as to the first arrival time of the various wave components as well as amplitude measurements within selected time windows (GUY, CORLEY, FERTL & FROST 1987).

6.2.1.6.1.2. Sonic wave attenuation as fracture indicators

Interpretation of the Circumferential Acoustilog is based on the fact that an open vertical or near-vertical fracture intersecting the borehole between transmitter and receiver results in significant attenuation of the Rayleigh wave which will typically not propagate across the crack (WALKER 1962, PAILLET 1981; GUY, CORLEY, FERTL & FROST 1987; NOBLETT, FERTL & GUY 1987). The absence of Rayleigh waves from the sonic records therefore indicates the presence of fractures in the formation. As the tool employs a quadrature measurement technique, disappearance of the Rayleigh wave in two quadrants and not in the other two is interpreted as an indication of the existence of a fracture system. To a lesser degree, the guided fluid Stoneley wave will suffer some attenuation in those quadrants where the crack is encountered and can therefore be used to detect fracturing in lithologies that will not support a good Rayleigh wave due to high crack and matrix porosities (BOTTER & ARKEL 1982), with the measurement being enhanced by the elimination of the direct fluid compressional wave by a baffle system. In contrast to the Rayleigh wave, the guided fluid Stoneley wave propagates across the fracture and only experiences some energy loss.

The orientation section provides the capability for determining directional subsurface trends of major detected fracture systems by measuring azimuth and relative bearing of the reference acoustic pad and deviation angle of the borehole from vertical. The instrument rotates during the logging run in order to allow a more precise estimation of the general direction of major trends by additional azimuth readings in fractured intervals with respect to an axis-related stationary upwards movement in the wellbore. In major vertically fractured intervals, resolution is increased by slowing down or even totally stopping tool rotation.

6.2.1.6.1.3. Additional information for fracture detection

In addition, two independent caliper pairs at the Circumferential Acoustilog instrument provide borehole size measurements. Fractured intervals may show up on the caliper record by thick mud-cake buildup particularly when using lost circulation material or heavily weighted mud, and by wellbore cross-sectional elongation observed preferentially in the main direction of subsurface natural fracture trends. The fracture-related washout/breakout indications result from chipping while drilling those cracked intervals (for evaluation of borehole elongation by caliper logging cf. also section 6.2.1.8.).

The Circumferential Acoustilog or Microsoniclog waveshape, however, can respond to factors other than fractures, including borehole conditions, lithology, porosity and microstratigraphy (SCHLUMBERGER 1982). Distortion can occur near laminae and bed boundaries particularly when dip is large, and low wave amplitudes can result from poor pad contact in rugose boreholes and washouts as well as in sections of shales and high-porosity sands. Given uniform lithology and wellbore conditions and one or more quadrants showing high-amplitude records, low-amplitude readings are indicative of fractures. A better understanding of amplitude attenuation is achieved by comparison of the actual amplitudes with the values reconstructed from porosity and lithology indicators by a non-parametric regression technique, with fractures being likely to account for measured amplitudes being lower than the reconstructed ones.

6.2.1.6.2. Interpretation of fracture orientation

In terms of fracture orientation interpretation, the first arrival wave may be attenuated by several formation and borehole conditions, and attenuation may be observed in one to four measurement quadrants. Attenuation in opposite or in any two adjacent quadrants best defines vertical fractures and the direction of subsurface stress planes, whereas other combinations of attenuation are also the consequence of variations in lithology, porosity and permeability in the generally more or less heterogeneous reservoirs. The variable density log format is most suitable for a quicklook on-site analysis, whereas a more detailed evaluation is accomplished by using the signature waveform presentation.

Field examples of crack evaluation by Circumferential Acoustilog recording are outlined by NOBLETT, FERTL & GUY (1987). Fracture identification and subsurface trend mapping define the direction of properly selected offset wells. Localization of natural cracks also is an important aid for the selection of target zones for stimulation, and recognition of fracture orientation trends can also assist in defining future well locations. Detection of major faults that greatly affect subsurface fracture and stress plane direction results in identification of both drainage pattern and travel path of stimulation fluids and therefore contribute important input for the design of hydraulic fracturing treatments. In other formations, the discovery that shale streaks interbedded with the reservoir storeys are also naturally cracked permits their incorporation into perforation and stimulation including propping of the vertically communicating fractures and thus commingling of several pay zone intervals, giving rise to improved drainage patterns of hydraulically-induced cracks and hence an overall enhanced result of the stimulation job. The Circumferential Acoustilog is very sensitive to high-angle fractures (HEFLIN 1979), and can also be successfully applied in oil-based drilling muds where many other current fracture detection methods do not work (GUY, CORLEY, FERTL & FROST 1987).

SARDA, PERREAU & DEFLANDRE (1988) comment on acoustic emission interpretation for estimating hydraulic fracture extent (cf. also section 6.2.2.3.2.5.). Other aspects of sophisticated acoustic logging with utilization of waveform analysis are discussed by PAILLET & WHITE (1982) and ZEMANEK, ANGONA, WILLIAMS & CALDWELL (1984).

6.2.1.7. Noise logging

Noise or sound logging (McKINLEY, BOWER & RUMBLE 1973; Sonan logging; DRESSER ATLAS 1982) records fluid movement in the fracture or in channels behind the pipe in case of insufficient cement bond. The tool records the sound patterns that are associated with familiar downhole problems such as casing or tubing leaks and flow in the cement annulus behind casing. Simulation of these downhole effects in a surface flow loop facility and recording of their sound patterns generates a set of model responses enabling problem diagnosis. Because of the high degree of sensitivity of the sound-recording microphone, the most reliable data are acquired when wireline and instrument are stationary, thereby greatly reducing or in ideal cases even almost completely eliminating the undesirable background noise that is otherwise generated by the movement of the tool. Therefore sound or noise logging has to be carried out by stop-and-go recording in order to eliminate cable, tubing, casing and borehole fluid noise and to be able to recognize fracture and formation fluid flow behind the casing (BRITT 1977). The sound logging instrument can therefore also only be combined with other passive tools such as temperature log, but not with active mechanical devices. Some comments are offered as follows on verification of temperature anomalies and fluid movement detection by sound recording.

6.2.1.7.1. Verification of temperature anomalies

Running the noise log in combination with a temperature sensor, intervals of increasing noise level at all frequencies matching with temperature anomalies reflect the zones of fluid movement and also help to identify warm noses representing anomalies of thermal conductivity of formation or crack (cf. section 6.2.1.1.2.) as belonging still to the fractured interval. In order to significantly reduce or even almost totally eliminate disturbing background noise, sound logging should be preferentially performed in a step-wise manner with spot-like stationary recording between phases of tool advancement. The most important patches for noise logging are bottom and top of the fracture as well as warm anomalies shown by the temperature log. As a consequence of the operational complication of the stationary recording in critical intervals, noise logging has its most important application in verification of uncertainties left by temperature surveys by checking of limited patches and specific spots, but should generally not be carried out as an exclusive base documentation of fracture height.

In formations which have turned out already in several wells of a field to create anomalies in the temperature record that do not allow straightforward fracture height determination without additional noise logging or conventional gamma-ray recording, replacement of temperature logging and additional interpretation help such as noise registration by gamma-ray spectroscopy (cf. section 6.2.1.3.4.) is recommended and could also be the cheaper solution at the bottom of the line for getting straightforward unambiguous results without any investment of additional time and money in case of complications.

6.2.1.7.2. Fluid movement detection by sound recording

Fluid movement behind the casing is also indicated by the radial differential temperature log which allows interpretation of flux in some type of channel in the annulus where the cement bond has been damaged by the fracturing treatment. Within the formation, fluid flow recorded by noise logging occurs indispensibly within cracks in case of a low-permeability reservoir rock (DOBKINS 1979). During course of testing after the fracturing operation, it is also possible to execute other production control logging which record temperature, flow rates and distribution of specific gravity in the borehole.

Noise logging works particularly well in open holes, but does not measure fracture height growth into intervals where no fluid influx into the wellbore occurs. In a similar way as noise logging, Spinner surveys indicate fluid-entry points into the borehole, but also cannot deliver results in segments where no fluid is moving. Although Spinner surveys can also be performed in cased holes, the tool only identifies that set of perforations which was actually fractured and is effectively contributing to exploitation (AHMED 1987). Spinner surveys indicating large decreases in general productivity in some zones testify together with enhanced injectivity of fluid due to acception of extraordinarily large volumes by the formation, core evaluation, and rapid pressure response between adjacent wells during interference tests to intervals with very well-developed natural joint systems of high permeability (BELFIELD 1988, SNOW & HOUGH 1988). Field examples of combined temperature and noise logging are also reported by PENNEBRAKER & WOODY (1977), KENNEDY & McELHINEY (1979) and WYMAN, HOLDITCH & RANDOLPH (1979). Special aspects of noise logging are also discussed by McKINLEY & BOWER (1977), and borehole audio tracer survey methods are also illustrated by BRITT (1977).

All the discussed methods do only permit to determine the fracture height in the almost immediate surroundings of the wellbore. Fracture height farther away from the borehole, however, cannot yet be monitored with hitherto available techniques (LEICHT 1985) and can only more or less be estimated in an indirect manner by fracture pressure evaluation and interpretation (cf. section 6.2.4.2.).

6.2.1.8. Caliper logging

While temperature, gamma-ray, acoustic and noise logging aim on fracture height monitoring, it is also important to determine the azimuth of the crack, in order to avoid interference of fractures extending from neighbouring wells in tight gas fields (cf. section 4.10.5.). Caliper logging generally locates borehole wall rugosity associated with fractures. Apart from usage in other fields of drilling and completion, two applications of caliper logging exist in hydraulic fracturing design and execution comprising fracture detection and insitu stress determination (from which fracture azimuth prediction is made). Comments are offered as follows on borehole ellipticity as well as breakout vs. creep deformation.

6.2.1.8.1. Borehole ellipticity

A very useful predictive indication of the direction of fracture propagation can be obtained by four-arm caliper logging prior to setting of casing and interpretation of the elliptical elongation (ovalization) of the wellbore cross-section (COX 1983). Caliper logging reflects the radial stress distribution provided that the rock in the neighbourhood of the borehole remains linear elastic and no excessive fluid movement is allowed in or out of the formation (DETOURNAY & FAIRHURST 1982, GOUGH & BELL 1982). In the latter case, erosion plays a non-negligible role and quickly overshadows the elastic deformation field. Aspects of ovalization and stress orientation as well as major and minor eccentricities are outlined as follows.

6.2.1.8.1.1. Ovalization and stress orientation

If the stress distribution is anisotropical as usual, the borehole cross-section is elliptical and the long axis of the ellipse is parallel to the direction of least principal stress which is perpendicular to fracture orientation, with the crack thus propagating normally to the long axis of the ellipse (cf. also GRIFFIN 1985, LEICHT 1985). The erosion pattern, however, is biased by the microcracking due to the in-situ stress field around the borehole, and the resulting ellipticity can be related to stress orientation (COX 1983).

A more sophisticated technique is breakout logging which records the regional tectonical stress direction (SCHLUMBERGER 1983; ZOBACK, MOOS, MASTIN & ANDERSON 1985). A sufficient difference between the principal horizontal stresses causes a circular wellbore to spall and to form breakouts, thereby creating an elliptical borehole with the long axis being at right angle to the expected hydraulic fracture azimuth (GOUGH & BELL 1981, COX 1983, PLUMB & HICKMAN 1985). In-situ stress determination from wellbore elongation is discussed by BLANTON & TEUFEL (1985). TEUFEL (1982, 1985) and TEUFEL & WARPINSKI (1984) outline in-situ stress calculation and prediction of hydraulic fracture azimuth by the anelastic strain recovery method. As breakout evaluation for in-situ stress modelling requires consideration of not only stress, but also failure mode, the only conclusion which can be derived from an elastical model is the link between the location of such breakouts and the principal in-situ stress direction (DETOURNAY & ROEGIERS 1986). Borehole ovality can also be determined by open-hole running of microacoustic cement evaluation tools.

6.2.1.8.1.2. Major and minor eccentricities

Very weak formation spalling resulting only in little borehole eccentricities which are almost below the resolution boundary of the caliper tool can also be a consequence of factors other than stress field orientation and should therefore be only considered for interpretation if matching with data from more pronounced wellbore ellipticity in the same reservoir interval (GRIFFIN 1985). While in many cases the minor eccentricities reflect the stress orientation of the formation and coincide with the alignment of the major ellipticities, the former seem in other cases to result from rock material inhomogeneities and are oriented at oblique angles to the latter, or even testify to the minor stress axis and are directed perpendicularly to the large-scale ovalizations. The major ellipticities, however, are always oriented in the direction of the stress field. The degree of formation spalling leading to borehole ovalization is influenced by bit weight as well as velocity, viscosity and time of circulation.

6.2.1.8.2. Breakout vs. creep deformation

Determination of in-situ stress at reservoir depth from caliper measurements of a deformed wellbore (BLANTON & TEUFEL 1985) is based on the fact that in brittle rock, wellbore deformation results in spalling or breakout of the borehole wall, whereas in ductile rock, creep deformation gives rise to wellbore closure. Equations derived from viscoelasticity relate the minimum and maximum radii of a deformed wellbore to the maximum and minimum stresses perpendicular to the axis of the borehole. For a vertical wellbore drilled perpendicularly to a principal stress plane, the orientation of the greatest horizontal principal stress will be parallel to the direction of maximum shortening of wellbore diameter, and the orientation of the least stress will be parallel to the direction of minimum shortening. Creep closure of a wellbore may also be associated with borehole breakouts. The formation of breakouts may also be time-dependent, with extension of the breakout growing with time in the direction of the minimum stress (PLUMB & HICKMAN 1985). In other cases, wellbore elongation is most likely due to washouts in naturally cracked and induced fracture zones rather than a response to the regional tensional stress regime (OWEN, TORONTO & PETERSON 1988). Other aspects of relationships between borehole breakouts, in-situ stress direction and fracture azimuth are reported by BELL & GOUGH (1978), BLANTON (1983), BLANTON & TEUFEL (1983); FORDJOR, BELL & GOUGH (1983); TEUFEL (1985 b) and ZOBACK, MOOS, MASTIN & ANDERSON (1985).

Borehole breakout determination can be performed by caliper logging of elongation or ellipticity as well as by acoustical televiewer recording (PAILLET & KIM 1985; ZOBACK, MOOS, MASTIN & ANDERSON 1985). Distinction has to be made between ovality and breakout in boreholes. Breakout reflects the stress direction, whereas ovality is mainly coming from the pipe influence. Breakout orientation determination of the in-situ stress pattern is not possible in highly-deviated wells.

6.2.1.9. Borehole televiewer

Borehole television and formation microscanning are together with spinner surveying significant techniques for direct observation and measurement of fracture height in contrast to other logging methods which are based on inferences as a consequence of indirect interpretation and calculation of measured data curves (AHMED 1987). Borehole television can be performed in two ways. The conventional method comprises acoustic television by a specialized sonic logging tool, while the second technique is simple optical television or videocamera recording. An alternative technique of even higher resolution is the electrical formation microscanner which, however, can only be utilized in conductive mud.

6.2.1.9.1. Acoustic borehole television

The conventional borehole televiewer is an acoustic logging tool that has a fine vertical resolution which is extremely useful for detecting fracture orientation, bedding dip direction, thin high-permeability streaks and lithological changes (ZEMANEK, CALDWELL, GLENN, HOLCOMB, NORTON & STRAUS 1969; WILEY 1980, PASTERNACK & GOODWILL 1983, TAYLOR 1983; RAMBOW 1984, 1985; PAILLET, KEYS & HESS 1985; CLERKE & AKKEREN 1986; PLUMB 1988). A borehole televiewer log is comparable to a picture of a continuous core and may yield even more information since it is an image of the core host environment (TAYLOR 1983). Some points of recording mechanism, advantages and drawbacks, and image types are illustrated as follows. Comments are also given on improvement by digital recording, simultaneous monitoring of fracture opening and closing, and effects of formation lithology, borehole wall conditions, and borehole geometry on televiewer log quality. The impact of fracture widening during drilling is also characterized.

6.2.1.9.1.1. Recording mechanism

The well televiewer tool acquires a detailed acoustic image of the entire cylindrical inner surface of the borehole or casing wall with a vertical resolution of abt. 1/3 in. (8 mm) and an angular resolution of abt. 1.5 degrees. Individual acoustic pulse reflections from the rock face are a result of the successive firings of a transverse mounted pulsed acoustic transducer which rotates inside the centralized tool and emits a pulsed ultrasonic beam. The reflected signal is recorded both in terms of amplitude and travel time. A quasi-optical acoustic image is composed from the reflections of the ultrasonic pulses from the borehole wall and displayed in black-white representation, and a flux gate magnetometer rotating with the piezoelectric transducer crystal is used to detect the earth`s magnetic north on each rotation. Fracture orientation is determined by triggering the oscilloscope sweep so that the photographic record is split on the north side of the borehole cylinder, with directional determination of the provoking point being maintained by the flux-gate magnetometer.

The image shows the borehole wall as if it were split vertically and laid flat. Vertical fractures appear as straight lines, and cracks dipping between vertical and horizontal appear as sinusoidal traces with strike and dip of the fractures being determined from amplitude and phase of the sinusoids. The tool is very conclusive in terms of interpretation, but is also very sensitive to hole rugosity and mud conditions which restricts acoustic coupling (HEFLIN

1979).

An important improving modification of the borehole televiewer is an automatic gain control which allows good images to be obtained even in out-of-round boreholes or when the tool is off-centered (RAMBOW 1984). HEARD (1981) presents a system capable of operation in high-temperature environments.

6.2.1.9.1.2. Advantages and drawbacks

The borehole televiewer can not only identify presence and orientation of cracks, vugs and other features, but also reveals the irregularity of the fracture face where it intersects the borehole, such as for example changing aperture in different sections of the crack. The televiewer can also be used to infer fracture orientation by determining the major axis of ellipticity as well as the main axis of wellbore spalling (GRIFFIN 1985), and due to the high sampling rate, the televiewer can be utilized to describe borehole shape with an accuracy and resolution not possible with conventional calipers (TAYLOR 1983). Wellbores are commonly considered to be round or elliptical if two independent calipers read similarly or differently, respectively, whereas televiewer logs reveal that noncircular boreholes are usually elongated, but are not truly elliptical in shape. Transit time measurements generally show wellbores to be eroded or elongated on opposite sides with no erosion indicated around the remaining circumference. The borehole televiewer is also a useful and unique dipmeter device, because magnitude and direction of inclination are calculated from bedding planes and fractures that are visible on the log.

The major drawbacks of the televiewer are that in elongated, rugose or collapsing boreholes it works only poorly or even not at all (SCHLUMBERGER 1987 a). Good pictures require near-perfect centering of the sonde in a nearly circular wellbore, a low content of solids in the borehole fluid, and a constant slow logging speed. The televiewer data still need considerable manipulation even in smooth-walled wellbores. Factors controlling transducer resolution and detection limits, influence of borehole shape and sonde position on the appearance of planar fractures, and effect of wellbore cross section and sonde position on the return-signal amplitude are summarized by GEORGI (1985). Fracturing gel leaking off from the crack face into the borehole after completion of the hydraulic stimulation treatment squelches and absorbs the sonic signal and thus renders the televiewer useless until the bottom of the wellbore is cleaned out and backflow of fracturing gel from the crack into the borehole has ceased.

6.2.1.9.1.3. Image types

The borehole televiewer delivers three different types of images comprising polar plot, amplitude log and time-of-flight log (TAYLOR 1983, CLERKE & AKKEREN 1986). The polar plot is a cross-sectional view of the wellbore at a given depth composed of 250 successive measurements of the round-trip travel time (time of flight) of sonic pulses from transducer to borehole wall and back during one revolution and therefore can be regarded as an acoustic caliper log. The amplitude log is composed from the amplitude of the reflected sonic pulse which is controlled by acoustic impedance of lithology, attenuation length and reflection geometry due to borehole size and shape, and acoustic properties of the mud. The log image is coded in gray shades and is presented as a cylinder that has been sliced open along the north edge and laid flat. The time-of-flight log reflects the travel time of the sonic pulses from transducer to wellbore wall and back, and can be thought of as a relief map of the borehole surface with white ridge tops, black valleys and contours of gray shades (transit time log; TAYLOR 1983). Sensitivity of time contours can be varied according to desired resolution and intended mapping purposes. A fourth real-time log presentation is the average amplitude curve which highlights rapid vertical variations in the wellbore and facilitates comparison with conventional log curves.

Experience reports of fracture detection with the borehole televiewer are given by KEYS (1979), KEYS & SULLIVAN (1979), PAILLET (1981, 1983), KAMP (1983), FLECKENSTEIN (1984), GRIFFIN (1985), CLERKE & AKKEREN (1986), DARILEK (1986), HACKBARTH & TEPPER (1988) and LAUBACH, BAUMGARDNER, MONSON & MEADOR (1988). CAMPBELL, HANOLD, SINCLAIR & VETTER (1981) describe fracture performance monitoring by borehole televiewer surveys in geothermal reservoirs.

6.2.1.9.1.4. Improvement by digital recording and processing

The analog evaluation of borehole televiewer logging can be considerably improved by digital recording and processing (PASTERNACK & GOODWILL 1983). Simultaneous digitizing of acoustic reflectance and travel time from televiewer waveforms recorded on videotape and manipulating the digital data allows both qualitative and quantitative utilization of the tool. Enhancements of logging technology comprise processing of digitized acoustic reflectance images (WILEY 1980) and scrolling raster scan display, image processing capability, simultaneous recording of reflectance and acoustic travel time images, and tilted polar scan display (BRODING 1981). Other improvements include image enhancement by source deconvolution and pattern recognition (TAYLOR 1983), display of range of flight data in addition to acoustic amplitude of the reflected signal (RAMBOW 1984) and control of depth of investigation by varying source frequency (BRODING 1984).

Independent processing of digital data obtained from analog wavetrains recorded in the field (PASTERNACK & GOODWILL 1983) permits various interpretation improvements including application of digital televiewer data as a high-resolution caliper log by considering travel time pictures as caliper images, estimation of degree and direction of borehole ellipticity from the televiewer travel time data, thin bed resolution and examination of casing defects and perforations.

6.2.1.9.1.5. Simultaneous monitoring of fracture opening and closing

Enhanced application technology has permitted to pressurize the well with the televiewer in the borehole (RAMBOW 1984) and thus allows to simultaneously monitor opening and closing of cracks during the hydraulic process. The wireline tool is run through a packoff. Opening of fractures while the well is under pressure can be detected by recording the crack interceptions and estimating their width. Closing of the fractures similarly is performed by comparative modelling of increasing and/or decreasing apparent crack width in the different stages of the job and after termination of the treatment. Observation of progressively increasing vertical fracture height by successive extension of the crack in upwards and downwards direction enables to supervise fracture containment and in case of necessity to reduce pumping rate or to stop injection completely. The system works satisfactorily even in 12 lbs/gal inverted emulsion oil mud, thereby providing an excellent control on generation and extension of hydraulic fractures during course of stimulation operations.

6.2.1.9.1.6. Effects of formation lithology and borehole wall conditions on televiewer log quality

The quality of the acoustical television log depends considerably on formation lithology and borehole wall conditions (PAILLET, KEYS & HESS 1985). Fracture faces at the wellbore wall have undergone stress concentrations and release during drilling and subsequent washout, and mechanical breakage during circulation and multiple trips of the drill string. The way in which drilling and subsequent stresses affect the fracture-borehole intersection significantly controls determination of crack image character obtained from the acoustical televiewer log. Massive crystalline rocks provide an uniformly reflective borehole wall which represents an ideal background for fracture identification. The majority of natural cracks in crystalline rocks are artificially widened near the borehole during drilling by stress concentration at the fracture-borehole intersection during passage of the drilling bit.

Fracture interpretation in sedimentary rocks can be difficult because of the generally less sonic reflectivity, variable background reflectivity, and extensive drilling damage of friable bedded deposits. Vertical variations in borehole wall reflectivity are mainly a function of changing mud content in the formation. Comparison of acoustical televiewer logs and core sequences reveals that many fractures appearing closed in cores show up open on the televiewer log, thereby underlining their secondary opening during the drilling process.

6.2.1.9.1.7. Impact of fracture widening during drilling

Crack aperture and permeability can be at least qualitatively determined by acoustical borehole television, with fractures appearing widest and darkest representing the most permeable ones, whereas cracks with only a faint discontinuous trace being almost tight (PAILLET, KEYS & HESS 1985). An open fracture produces an image because no signal is returned to the logging sonde, and a plugged crack creates a picture if there is sufficient sonic contrast between the filling material in the joint and the host rock to produce a weaker reflected signal (TAYLOR 1983). Distinction of open and filled fractures can be optimized by combining amplitude and transit time logs, with an image on both displays indicating an open fracture, whereas a picture only on the amplitude log characterizes a filled crack. Drawbacks of the fracture permeability estimation approach are linked with the width of the acoustic beam provided by the source which indicates that the small crack width has been convolved with a broad source function. As also fracture opening at the borehole wall may have been substantially widened during drilling, apparent crack width on televiewer logs may not be a reliable indicator of relative and/or effective permeability.

Drilling techniques can greatly affect acoustical televiewer logs because of borehole wall roughness and non-circular wellbore cross-section as well as borehole breakouts. Fracture aperture widening occurs during drilling by hydraulic opening due to invasion of high-pressure mud (not mud filtrate), or by erosion of soft alteration products out of the crack which originally have been present in the vicinity of the fracture plane around the borehole. In terms of televiewer resolution combined with widening by drilling, fractures less than 1/32 in. diameter and even hairline cracks often erode and become wide enough at the wellbore surface to be detected by the televiewer (TAYLOR 1983).

6.2.1.9.1.8. Effects of hole geometry on televiewer log quality

Vertical round boreholes allow optimum results of acoustical television logging, whereas deviations from ideal wellbore geometry trigger deteriorations of log quality (GEORGI 1985). The borehole televiewer utilizes an ultrasonic transducer in a conventional pulse-echo mode. The televiewer images are constructed from the amplitude of specularly reflected acoustic energy. As transducer energy is strongly collimated, it is essential that the sonic beam is nearly normal to the wellbore wall, and the constructed images are adversely affected when the tool is decentralized or the borehole cross-section is not round. Some aspects of hole roundness and instrument centralization, correction in elliptical hole and eccentered tool, and signal attenuation by solid-laden drilling mud are illustrated as follows.

6.2.1.9.1.8.1. Hole roundness and instrument centralization

In circular boreholes with centered instrument, the fracture signatures are simple sinusoids, but in elliptical wellbores with centralized tool or in round boreholes with off-centered instrument, the crack signatures can be severely distorted and fragmented by compression and extension of different parts of the sinusoid, because the wellbore radius is no longer independent of the azimuthal angle (GEORGI 1985). When the angle of incidence diverges from 10 degrees, little or no energy is reflected back towards the transducer, and therefore when the televiewer is eccentered or tilted in a circular borehole, the resulting images exhibit two prominent vertical stripes representing low-amplitude returns (TAYLOR 1983). When the tool is centered in elliptical wellbores, acoustic returns are received from four sectors, centered on the major and minor axes, giving rise to four prominent black and white stripes on gray-scaled images (WILEY 1980).

6.2.1.9.1.8.2. Correction in elliptical hole and eccentered tool

Tool centering in non-circular borehole cross-sections can also distort the classical sine wave signature of dipping planar fractures. Little or no distortion occurs for horizontal and vertical features, whereas maximum distortion takes place for obliquely oriented planar phenomena (GEORGI 1985). For small wellbore ellipticities and centralized tools, the distortion is negligible and little error will be introduced into the estimated fracture dips and strikes. Interpretation problems due to noncircular boreholes and eccentered logging sondes are easily overcome by using the combination of amplitude and transit time logs (TAYLOR 1983). Tool position and wellbore shape cannot be determined from variations in signal strength which are recorded on the conventional amplitude log, but can be determined from interval transit time which varies only as the time required for a signal to travel from the logging sonde to the borehole wall. A white speckled appearance of the transit time log reflecting alternating no signal and far signal typifies a rugose wellbore surface due to erosion or elongation.

6.2.1.9.1.8.3. Signal attenuation by solid-laden drilling mud

Another negative effect in boreholes is attenuation of ultrasonic borehole televiewer signals by suspended solids in freshwater drilling mud which can be up to 10 dB/in. (RAMBOW 1984). In elliptical wellbores, this intensity attenuation depends on the azimuthal angle being largest along the major axis and least along the minor axis. Signal loss also occurs due to a divergence of incident and reflected signals (GEORGI 1985). As this angular separation increases, the nature of the observed signal changes. For small angles, the amplitude of the return signal is dominated by specularly reflected energy, whereas for larger angles, the observed amplitudes are dominated by back-scattered energy.

6.2.1.9.2. Electrical formation microscanning

The formation microscanning instrument (EKSTROM, CHEN, ROSSI, LOCKE & ARON 1986; EKSTROM, DAHAN, CHEN, LLOYD & ROSSI 1986; LLOYD, DAHAN & HUTIN 1986; PLUMB & LUTHI 1986, SCHLUMBERGER 1986 b) and the high-resolution dipmeter tool (SCHLUMBERGER 1982, 1983), however, have much higher vertical and horizontal resolution than the wellbore televiewer and are thus superior means for fracture recognition (PLUMB 1988), mainly as a consequence of different recording techniques. The borehole televiewer is an active acoustic device and behaves poorly in heavy mud where signal transmission is impaired, but works in all types of drilling mud regardless of their conductivity nature. Formation microscanner and dipmeter are passive resistivity tools which, however, do not work in oil-based muds that are non-conductive. In contrast to the quality drawback, the wellbore televiewer sees the full circumference of the borehole wall, whereas the formation microscanner only records sections of it not exceeding half of the total circumference, and an improvement of coverage requires repeated logging runs at different azimuths (as both borehole televiewer and formation microscanner have their chief application in natural fracture detection and picturing, more details on both instruments and their recording techniques are presented in section 6.2.1.10.1.).

6.2.1.9.3. Optical borehole television and videocamera recording

Apart from acoustic-logging borehole television, also direct optical television and photography systems have been developed and successfully applied in fracture height, azimuth, width and morphology inspection (BRIDGES 1964; SMITH, ROSENBERG & BOWEN 1982; YOUNG, BARKER & CLARK 1984). The outline as follows focusses on image resolution and borehole fluid clarity and possibilities of enhanced interpretation.

6.2.1.9.3.1. Image resolution and borehole fluid clarity

High-resolution direct observation of the wellbore wall can also serve to establish whether significant damage by crushing and/or permanent deformation and/or enlargement of the borehole has occurred during explosive fracturing (cf. section 4.7.3. and plate IV/1), and can also provide information on fracture width at its mouth and on benefits obtained from various wellbore-notching techniques. Some of the optical television and photography systems can be run in a fluid-filled borehole if it is an adequately clear water solution, but bailing the well dry has also turned out to deliver good results of observation and documentation. Poor water clarity created by fluid redistribution or fluid movement in the wellbore can deteriorate any optical television fracture inspection (SMITH, ROSENBERG & BOWEN 1982). With the logging cable being inside of coiled tubing, it is possible to enter any well with downhole videocamera and to be confident to see the borehole structures by ensuring that a nitrogen cushion is surrounding the camera while in hole (MACEWEN 1988). By adjusting nitrogen pumping pressure and wellhead pressure accordingly, it is possible to move the nitrogen cushion up and down the well with the camera. Small quantities of nitrogen can be used at the same time in the form of small jets to keep the camera lens clear of wellbore fluids.

Experience of television camera borehole surveys are also reported by FITZPATRICK, KARR & O`SHEA (1985) and SMITH, REN, SORRELS & TEUFEL (1985). Downhole optical television logging and viewing is the simplest way of fracture azimuth determination, because there is no interpretation involved (SMITH, ROSENBERG & BOWEN 1982). Optical pictures of hydraulic fractures in wellbores by photography are also shown by WATERHOUSE, CHARARA & NURMI (1987). OVERBEY, YOST & WILKINS (1988); OVERBEY, YOST & YOST (1988) and YOST, OVERBEY, WILKINS & LOCKE (1988) analyze natural fractures in a horizontal well by borehole videocamera imaging (with the camera having been drill-pipe conveyed) in a clean, air-drilled wellbore.

6.2.1.9.3.2. Possibilities of enhanced interpretation

Special video enhancement and playback equipment can be used to examine the video tape and to orient features in the wellbore. Determination of top and/or bottom of the borehole on the video image enables orientation of natural fractures because wellbore azimuth is known. The angle of interception of a crack with the borehole can be derived from the focal length of the camera as an input for estimating fracture entry point lengths along the wellbore combined with borehole diameter in the tangent function, and from development of a series of models of cracks intersecting the wellbore at low angles and using them as multiplexed refreshed signals to the video monitor for direct comparison to actual wellbore images by superimposition on the same screen.

The knowledge of the orientation of fractures and faults from a video survey can be utilized to make decisions about where to stimulate in order to avoid interference with existing wells and to prevent interference as field development proceeds. Observing cracks stained with natural gas liquids gives proof of an added dimension of complexity of flow through anisotropic shale gas reservoirs. Fracture monitoring by downhole videocamera survey is also performed by LAU (1980), LAYNE & SIRIWARDANE (1988); MERCER, PRATT & YOST (1988); OVERBEY, YOST & WILKINS (1988) and SCHRAUFNAGEL & LAMBERT (1988). Comparison of acoustical and optical borehole television is carried out by PAILLET, KEYS & HESS (1985).

6.2.1.10. Natural fracture detection by various logging methods

In contrast to determination and evaluation of hydraulically-induced and proppant-infilled fractures of predominantly larger scale which is discussed in above (cf. sections 6.2.1.1. - 6.2.1.8.), the smaller-scale networks of natural cracks and joints within the formation can also be assessed by various wireline well logging techniques. Localization and characterization of natural fractures is important to aid in selection of target zones for hydraulic stimulation procedures, and determination of the orientation of natural fractures can assist in choice of sites for drilling of further wells (GUY, FERTL & OLIVER 1986). In addition, natural fractures theirselves may be the key to economical hydrocarbon production in formations with low matrix permeability, and their detection becomes increasingly significant with the progression of exploration and development to successively deeper storeys of the geological column where the petrophysical properties of the rock matrix are increasingly poorer, and especially primary matrix permeability is progressively lower, with secondary fracture permeability (cf. section 4.8.8.1.) becoming decisive for reservoir prospectivity.

As cracks are anomalies in the rock due to loss of cohesion in the formation itself, they cause anomalous responses in wireline logs with respect to readings in undisturbed geological sequences (SCHLUMBERGER 1982). The evaluation of natural fracture permeability of rocks is not only of considerable importance for oil and gas production, but also of significance for enhanced petroleum recovery, geothermal exploration, radioactive waste disposal and earthquake prediction (PAILLET 1981).

6.2.1.10.0. General aspects of natural fracture recognition

The existence of fractures may be derived from logs because of the response of the tools to physical crack properties such as visibility (MYUNG & CALDWELL 1971), porosity (PICKETT & REYNOLDS 1969), permeability (HEFLIN, NEILL & DAVIES 1976), resistivity (MINNE & GARTNER 1979) and acoustic impedance contrast (BRODING 1981). Other indications on logs from which the presence of fractures may be inferred are caliper response to hole-size reduction or elongation, and anomalous response from temperature and noise logs (MYUNG 1976).

Information on the geostress field which determines the direction of longitudinal fracture propagation can be obtained by evaluation of natural crack orientation by borehole televiewer (ZEMANEK, CALDWELL, GLENN, HOLCOMB, NORTON & STRAUS 1969; KAMP 1983, FLECKENSTEIN 1984, CLERKE & AKKEREN 1986, DARILEK 1986), high-resolution dipmeter analysis (especially wave-form interpretation; PERRIER & QUIBLIER 1973, SHIELDS 1974, GILREATH 1977; SCHLUMBERGER 1982, 1983) with four- or even six-arm tools (MORRISON & THIBODAUX 1984; GOETZ, TYSON & GARAT 1987), formation microscanner (EKSTROM, CHEN, ROSSI, LOCKE & ARON 1986; EKSTROM, DAHAN, CHEN, LLOYD & ROSSI 1986; PLUMB & LUTHI 1986; SCHLUMBERGER 1986 b, 1987 a; HANSEN 1988) and fracmeter interpretation (CHARLEZ & DESPAX 1987). Summaries of fracture detection with well logs are given by AGUILERA (1976);

SHANKS, KWON, VRIES & WICHMANN (1976); HEFLIN (1979), SUAU & GARTNER (1979); HIRSCH, CISAR, GLASS & ROMANOWSKI (1981); SCHLUMBERGER (1982, 1987 a) and PLUMB, BRIE & HSU (1985).

Natural fractures are either open and conductive or closed and tight by formation debris and/or mineral precipitation. They are defined by strike, dip, aperture, spacing and conductivity. The simplest and most direct possibility of investigation of the natural crack system is visual inspection of cores. Borehole measurement systems designed for fracture characterization are generally based on one of three principles comprising electrical conductivity of fluid-filled fractures, acoustic attenuation during propagation across fractures, and borehole wall imaging (PAILLET 1981). Fracture detection with wireline logs comprises indirect visualization by formation microscanner and borehole televiewer which are electrical and acoustical imaging devices, and measuring of various formation parameters and their interpretation. The discussion as follows outlines formation microscanner and borehole televiewer, dipmeter logging, other aspects of resistivity logging techniques, dipmeter, sonic logging with full waveform analysis, special methods for horizontal fractures, fracture probability program, mechanical properties logging and other techniques.

6.2.1.10.1. Formation microscanner and borehole televiewer

Formation microscanner and borehole televiewer are passive resistivity and active acoustic tools, respectively, for visual inspection of the borehole and imaging of natural fractures (cf. also section 6.2.1.9.). The formation microscanner can be regarded as an enhanced extension of the high-resolution dipmeter (cf. section 6.2.1.10.2.). Both formation microscanner and dipmeter have the advantage that the caliper orients itself with one pair of pads along the fracture and breakout plane (cf. section 6.2.1.8.). The formation microscanner allows recognition of open cracks, determination of their strike and dip, and distinction between natural fractures and drilling-induced cracks or borehole damage (SCHLUMBERGER 1986 b). Another borehole visualization tool producing high-quality images is the stratigraphic high-resolution dipmeter (SHDT; HACKBARTH & TEPPER 1988; cf. section 6.2.1.10.2.). The discussion as follows concentrates on borehole visualization and formation imaging vs. reservoir coring, distinction of natural and drilling-induced fractures, coring-induced fractures indicating hydraulic fracture propagation, electrical microscanning vs. acoustical television, and acoustical television vs. dipmeter resistivity recording.

6.2.1.10.1.1. Borehole visualization and formation imaging vs. reservoir coring

While cores provide the most direct information on fractures and their relation to the reservoir rock, core recovery in jointed and cracked rock is commonly poor and core orientation methods may fail, with thus geophysical logs that image the borehole being able to provide information on some critical aspects of reservoir fractures (LAUBACH, BAUMGARDNER, MONSON, HUNT & MEADOR 1988). Particularly the formation microscanner produces high-quality reservoir images which are rivaled in detail only by core samples (SCHLUMBERGER 1986 b, 1987 a). The formation microscanner thus represents the closest approach to the ultimate goal of sophisticated wireline logging to provide an in-situ description of the pay that could otherwise only be obtained by the expensive and more time-consuming procedure of full-hole coring.

Borehole visualization by microscanner and televiewer permits also fracture mapping on the wellbore wall and determination of fracture geometry (strike, dip, length, width), morphology, distribution, frequency, spacing, aperture and open vs. closed nature, in contrast to other logging techniques of indirect nature which reveal only the existence of cracks and probably permit to discriminate open and closed fractures, or even suffer from the fact that responses ascribed to cracks can also have been caused by bedding, vugs, mud cake, lithology changes, washouts or even moderate wellbore elongation (which can be caused by breakout collapse of the borehole wall from release of natural strain, or can be an artifact from drilling due to excessive or insufficient mud pressure or damage of the borehole wall by the drilling string).

Comparative evaluation of borehole televiewer logs, formation microscanner logs and reservoir cores reveals that vertical extension fractures are usually visible on both log types, whereas low-angle natural shear cracks cannot be imaged (LAUBACH, BAUMGARDNER, MONSON, HUNT & MEADOR (1988). Both wellbore televiewer and formation microscanner represent fracture shape and distinguish cracks from borehole breakouts. Existing commercial borehole televiewing and reservoir microscanning techiques, however, do not give a quantitative measure of crack aperture. Fracture orientation is readily obtained for inclined cracks from either wellbore televiewer or formation microscanner logs, but the direction of vertical fractures is commonly ambiguous on both logs. Borehole televiewer and reservoir microscanner are generally useful adjuncts to core-based crack studies for evaluation of naturally jointed pays.

6.2.1.10.1.2. Distinction of natural and drilling-induced cracks

Formation microscanner and borehole televiewer can distinguish natural cracks having been present in the rock before drilling from fractures related to wellbore damage by hydraulic formation breakdown or mechanical rock breakout as a direct result of drilling and/or coring. The following discussion focusses on morphological and genetical crack type spectrum, relationship to well trajectory and rock type, hackle-plume geometry and fracture orientation, and conductivity contrast between fracture-filling minerals and host rock.

6.2.1.10.1.2.1. Morphological and genetical crack type spectrum

Induced cracks can be caused by hydraulic fracturing, thermal fracturing and unloading stresses (WATERHOUSE, CHARARA & NURMI 1987). Hydraulic double-winged cracks are created when the wellbore pressure exceeds the combined reservoir pore pressure and rock tensile strength, and the induced fractures are oriented vertically along the lines of least principal stress across the borehole wall. Thermal fracturing occurs during drilling and coring when large thermal gradients are generated between hot rock and cooler drilling fluid, and give rise to tensile cracks around the well in a direction perpendicular to the temperature gradient. Unloading fractures appear as a result of the reduction of overburden pressure near the surface of stress removal and propagate perpendicular to the unloading direction. Numerous unloading fractures also occur in cores as a consequence of removing the samples from the subsurface stress environment.

6.2.1.10.1.2.2. Relationship to well trajectory and rock type

Natural fractures having been present prior to drilling usually cut diagonally across the borehole and are better developed in sandstone intervals of a clastic series, whereas drilling-induced hydraulic, thermal or mechanical fractures run parallel to the borehole trajectory over longer distances, are best developed in mudstone sections and tend to terminate as they enter sandstone interbeds, and are (with the exception of unloading cracks) not found in corresponding cores (cf. also LORENZ & FINLEY 1987, 1988). The possibility of discrimination between natural and drilling-induced fractures is the most significant advantage of formation microscanner and borehole televiewer with respect to other indirect well logging techniques which respond to all types of open cracks encountered along the borehole trajectory. Distinction of natural and drilling-induced fractures is only possible by direct imaging, because the physical record of both types of cracks by indirectly measuring tools is the same.

The borehole televiewer can also discriminate open and sealed fractures by comparing time of flight and reflected amplitude information (TAYLOR 1983, RAMBOW 1984; cf. section 6.2.1.9.1.6.).

6.2.1.10.1.2.3. Hackle-plume geometry and fracture orientation

KULANDER & DEAN (1988 a, 1988 b) comment on identification of drilling-induced fractures in cores. Hackle plume components including crack plane morphology and frequency of closely-spaced induced fractures show a unique geometrical relationship to core boundaries not developed with natural joints. Hackle plume geometry and fracture orientation reflect principal stress directions existing during propagation at any point of the crack face. Induced fracture frequency is commonly influenced by mechanical properties of contrasting lithologies. Properly logged drilling-induced cracks may predict preferred hydraulic fracture stimulation directions and rock behaviour to diverse treating procedures. In contrast to drilling-induced fractures, natural joints show no geometrical relationship to core parameters. Criteria for distinction of natural and drilling-induced fractures in cores are also compiled by SANGREE (1969) and LORENZ & FINLEY (1987, 1988).

6.2.1.10.1.2.4. Conductivity contrast between fracture-filling minerals and host rock

LAUBACH, BAUMGARDNER, MONSON, HUNT & MEADOR (1988) conclude from comparative evaluation of borehole televiewer, formation microscanner and reservoir cores that differentiation of natural and drilling-induced fractures cannot be performed on the base of reliable criteria (cf. also LORENZ & FINLEY 1987, 1988 for results of macroscopical core inspection analysis). Formation microscanning can only distinguish natural from drilling-induced cracks if a sufficient conductivity contrast exists between fracture-filling minerals and host rock. Primarily quartz-filled natural cracks do not contrast in conductivity with the host quartz arenite. Healed fractures are detectable by borehole televiewer if the material filling the crack has an acoustic impedance that is significantly different from that of the host rock (such as calcite-filled cracks intersecting quartzites and quartzitic sandstones). The characteristic concave-downward curve of drilling-induced petal-centerline fractures (LAUBACH & MONSON 1988) can sometimes be used to discriminate between drilling-induced and natural cracks. Inclined drilling-created petal fractures commonly curve or hook into vertical natural cracks thereby generating composite fractures with shapes similar to drilling-provoked petal-centerline cracks.

6.2.1.10.1.3. Coring-induced fractures indicating hydraulic fracture propagation direction

The strike of extension fractures created in front of the drilling bit during coring is a potentially useful indicator of horizontal stress anisotropy that is easily obtained from oriented core (LAUBACH & MONSON 1988). Coring-induced extension cracks have distinctive geometry and surface structures, and are parallel maximum horizontal stress unless near-wellbore stress perturbations or mechanical anisotropy in the rock have a significant effect on fracture strike. Coring-induced cracks can be recognized by the location of their origins near or within the core, characteristic fracture shapes, and absence of mineralization (KULANDER, BARTON & DEAN 1979). Fracture origins can be identified by plume structures and arrest lines. Azimuth of maximum horizontal stress can be determined from strain recovery measurements on oriented core, wellbore breakouts observed with dipmeter and borehole televiewer logs, and strike of cracks created in open-hole stress tests. The outline as follows illuminates some aspects of petal and petal-centerline fracture morphology and propagation as well as stress field perturbations near the wellbore.

6.2.1.10.1.3.1. Petal and petal-centerline fracture morphology

Petal and petal-centerline fractures are two common types of coring-induced cracks which tend to have a preferred orientation that is controlled by either contemporaneous stress or pre-existing rock fabric (KULANDER, BARTON & DEAN 1979). Petal and petal-centerline fractures characteristically curve into the core in downhole direction. Petal cracks are short inclined fractures that are concave downward with various initiation angles and a wide range of curvatures. Petal-centerline fractures are single cracks that are composed of a short smoothly curving concave downward petal segment which gradually merges in downcore direction with a planar centerline fracture segment that may bisect the core. The lower and upper terminations of petal and petal-centerline fractures occur inside and outside of the core, respectively, and the cracks commonly are composed of multiple strands rather than single fracture planes which do not result in complete separation of the core. Petal cracks also frequently intersect and reactivate natural fractures, but coring-induced cracks are more common and longer where natural joints are absent.

6.2.1.10.1.3.2. Petal and petal-centerline fracture propagation

Petal and petal-centerline fractures propagate in front of the core bit as indicated by the edges of petal-centerline cracks being commonly scoured by the bit or crosscut by grooves created by scribe knives located inside the core barrel assembly (KULANDER, BARTON & DEAN 1979). Palaeomagnetical results also testify that petal-centerline fractures predate horizontal spinoff cracks which form after the core enters the barrel. Petal and petal-centerline fractures are hydraulic cracks which follow stress trajectories in front of the core bit that reflect regional stress anisotropy and directly below the bit near the petal fracture origin, with the reduced vertical load being caused by the presence of the borehole. In other cases, high shear stresses created by the rotating core bit can also locally reorient principal horizontal stresses and provoke petal strike to deviate from parallelism with regional stress directions (LORENZ & FINLEY 1987, 1988). Coring-induced petal and petal-centerline fractures are usually subparallel to the strike of natural extension cracks if the modern and palaeostress trajectories are similar (LAUBACH & MONSON 1988), and natural fracture anisotropy does not control the strike of coring-induced cracks.

6.2.1.10.1.3.3. Stress field perturbations near the wellbore

Fractures created in small-scale stress tests are composed of single crack planes or multiple anastomosing fracture strands. Locally, cracks generated in stress tests have dips of 70 - 80 degrees which is not as steep as is typical for near-vertical petal-centerline fractures which may indicate that the dip of the latter type of cracks does not record the precise orientation of the vertical principal stress. Similarly as for retrospective evaluation of the stress field, coring-induced petal and petal-centerline fractures can be used for prediction of hydraulic fracture propagation azimuth. Accuracy of directional forecast, however, can occasionally be deteriorated if petal and petal-centerline cracks reflect perturbations of the stress field near the wellbore instead of regional stresses. Shear stresses caused by bit rotation could provoke coring-induced fractures to develop a strike that deviates in a clockwise direction from the regional maximum horizontal stress (LORENZ & FINLEY 1987, 1988).

Such perturbations can only be evaluated with higher resolution data on regional and local stress directions. Moreover, within-well fluctuation in coring-induced fracture strike may reflect subtle differences in the orientation of horizontal stresses within the formation or imprecise core orientation (NEL-

SON, LENOX & WARD 1987). Coring-induced fractures and natural cracks occurring together in the same core which are not precisely parallel to each other indicate that also hydraulically-induced fractures do not propagate exactly parallel to natural cracks (cf. section 4.8.8.1.). RAO, ADVANI, CHANG, LEE & DEAN (1979) comment on in-situ stress determination based on fracture responses associated with coring operations, and DANESHY, SLUSHER, CHISHOLM & MAGEE (1986) discuss in-situ stress measurements during drilling. Aspects of core-induced fractures are also analyzed by OWEN, TORONTO & PETERSON (1988).

6.2.1.10.1.4. Electrical microscanning vs. acoustical television

The formation microscanner measures electrical conductivity by mapping the borehole wall with a dense array of sensors giving rise to a high-resolution electrical image of the reservoir (SCHLUMBERGER 1986 b, 1987 a). The tool records microconductivity changes caused by bedding surfaces and fracture planes intersecting the wellbore. The arrays of microresistivity sensors are set on two orthogonal pads to give two oriented image slabs of the borehole wall, while all four pads contain electrodes for the standard high-resolution dipmeter recording. A triaxial accelerometer and three single-axis magnetometers provide accurate orientation information and enable speed corrections to be made on the acquired data.

Picturing coverage of the borehole wall is usually 20 % and can be increased up to 50 % by repetition runs which is in contrast to the 100 % coverage by the well televiewer (cf. section 6.2.1.9.). The conductivity changes detected in the formation can be displayed by data processing as a variable-intensity gray scale or colour image. Deteriorations of image quality are caused by excessive stickiness causing extremely irregular tool movement along the borehole trajectory which cannot be completely handled by speed correction algorithms, and out-of-gauge conditions in rugose and washed-out boreholes, but as a consequence of wellbore improvement across pay zones due to mud cake build-up protecting the formation, the danger of potential image loss across reservoir sections is very low. On the other hand, thick filter cakes draping the borehole wall may obliterate visibility of small-scale sedimentary and tectonical structures by triggering accentuation or smoothening of electrical resistivity differences, but such mud-cake impact is in most cases of only negligible order. Some aspects of tool sensitivity and fracture aperture as well as instrument resolution and borehole wall coverage are outlined as follows.

6.2.1.10.1.4.1. Tool sensitivity and fracture aperture

The chief advantage of the formation microscanner over the borehole televiewer is that it can detect a range of crack apertures from those several centimetres across to microfractures which are open only a fraction of a millimeter and can even escape detection in visual examination of core samples (SCHLUMBERGER 1986 b, 1987 a). Electrical imaging by reservoir microscanner is much more sensitive than acoustical picturing by borehole televiewer, because a water- or mud-filled fracture can be up to 1,000 times more conductive than surrounding hydrocarbon-bearing rocks. In comparison, the acoustic impedance contrast between most fluids and rocks is only abt. four to one.

The high resolution of the formation microscanner also allows monitoring of small-scale sedimentary structures such as cross-bedding and horizontal-stratification (PLUMB & LUTHI 1986) and thus enables detailed compositional analysis of the pay interval. The sensitivity of the reservoir microscanner is almost independent from logging speed, whereas the borehole televiewer produces satisfactory results only at slow tool pulling rates. The ideal way is using well televiewer and formation microscanner conjointly for fracture detection because of the 360 degrees borehole coverage of the former and the high sensitivity of the latter (SCHLUMBERGER 1987 a). Field experience of fracture identification by reservoir microscanner application is reported by McNAUGHTON, CASARTA & BORNEMANN (1988), and successful utilization of the borehole televiewer for the same purpose is described by WALBE (1986).

6.2.1.10.1.4.2. Instrument resolution and borehole wall coverage

LAUBACH, BAUMGARDNER, MONSON, HUNT & MEADOR (1988) perform comparative evaluation of borehole television, formation microscanning and reservoir coring. The 360 degrees wellbore coverage of the televiewer permits reliable fracture identification, orientation, and estimation of crack abundance. Borehole breakouts from which principal stress directions can be deduced are also more easily recognized with complete wellbore coverage. The borehole televiewer, however, is limited by lower resolution and by the vertical striping that occurs in elliptical or enlarged holes or when the tool is off-center. The higher sampling rate of the formation microscanner and conductivity contrast being larger than the acoustical contrast between fractures and matrix enables the microscanner to find more cracks (DENNIS, STANDEN, GEORGI & CALLOW 1987), whereas in other cases the televiewer can pick up more fractures because its coverage of the borehole wall is more complete than that of the microscanner.

The high resolution of the reservoir microscanner makes it valuable for detecting small features and for distinguishing little cracks from wellbore spalls. Analysis of fracture origin and relation of cracks to bedding which is useful for evaluating the relation between matrix and fracture permeability is augmented by high-resolution microscanner images of crack shape and sedimentary structures. Conductive joints, however, appear to be wider on microscanner logs than in core. Detection of small fractures and low-angle cracks depends on vertical sampling rate and scale of image more than on tool physics.

6.2.1.10.1.5. Acoustical television vs. dipmeter resistivity recording

Acoustical borehole television has also an important advantage with respect to dipmeter resistivity recording (RAMBOW 1984). While dipmeter tools read with four or six arms and thus obtain only information on limited sections of the wellbore wall (cf. section 6.2.1.10.2.), the televiewer registers the full circumference of the borehole and can therefore lead in various cases to safer interpretation (cf. section 6.2.1.9.). Improvement of wellbore acoustical television technology enables also logging of soft reservoirs which flow and heal most marks within a short period of time. Borehole geometry analysis by dipmeter in soft formations may give evidence of a pronounced ellipticity, but the televiewer reveals that really a key seat is developing along some faces of the wall because of the softness of the formation. Dipmeter evaluation could thus result in misinterpretation of wellbore elongation and thus error in earth stress direction determination, whereas borehole televiewer examination provides the correct information.

Apart from its classical applications of fracture detection and casing inspection, the wellbore televiewer can also itself be used as a high-resolution dipmeter. In many respects, the borehole televiewer surpasses dipmeter performance because it can detect and differentiate between fractures and bedding, folds and faults, and also realize smaller wellbore features such as vugs and burrowing. The televiewer also excels as a borehole geometry tool because of its essentially continuous coverage of the wall circumference (TAYLOR 1983; cf. section 6.2.1.9.1.2.).

6.2.1.10.2. Dipmeter logging

Dipmeter analysis being a high-resolution microresistivity tool interpretation is also a good means for detection of open fluid-filled natural fractures (SHIELDS 1974, GILREATH 1977; DELHOMME, PANCI, LESLIE, CIGNI & SPILA 1981; SCHLUMBERGER 1982 a, 1983; DELHOMME, PILENKO, CHERUVIER & CULL 1986; DRESSER ATLAS 1987 b; ENDERLIN, EDMISTON & EPPS 1988). Some aspects of recording principles, dipmeter and caliper response to fractures, dipmeter evaluation for fracture detection, other reasons than fractures causing dipmeter curve separation, and borehole spalling and dipmeter response are outlined as follows.

6.2.1.10.2.1. Recording principles

The dipmeter sonde is a four-pad hydraulic unit with the pads spaced 90 degrees apart around the borehole and is designed to determine angle and direction of dip of bedding and fracture planes where they intersect the borehole (SCHLUMBERGER 1982, 1983; DRESSER ATLAS 1987 b). A button on each pad records a microfocused conductivity curve with high resolution, and a second button on two pads reads another curve for speed corrections. High inclinometer accuracy is achieved by a triaxial accelerometer and three single-axis magnetometers supplying information on tool deviation and azimuth which are necessary to convert the computed correlations to true formation dip and to perform speed corrections in case of instrument sticking in poor boreholes. The short spacing between the two measure electrodes on each of the four pads means that the side-by-side curve likeness is better than that from pad-to-pad, permitting short correlation intervals to be used. The cross-linking of opposite arms guarantees self centralization of the sonde in the borehole which is necessary for keeping all electrodes co-planar. An advanced version of the tool operates with six pads (MORRISON & THIBODAUX 1984; GOETZ, TYSON & GARAT 1987).

6.2.1.10.2.2. Dipmeter and caliper response to fractures

If the microresistivity recordings are made with less sensitivity, resistivity anomalies related to bedding are removed which allows the suspected fracture to give an anomalous conductive kick when detected by any of the four pads (SCHLUMBERGER 1982, 1983; DRESSER ATLAS 1987 b). A high-angle conductive anomaly such as a fluid-filled open crack within an otherwise resistive geological unit with low to moderate matrix permeability produces pronounced reading, with the variations depending on anomaly geometry and instrument rotation. Strike and dip orientation analysis of open fluid-filled fractures can also be performed. Healed cracks are normally not of interest except of possible indications of planes of weakness which may be used to plan a hydraulic fracturing job. The major advantage of the dipmeter tool with respect to laterolog instruments are the four sensors which provide a greater chance of finding crack anomalies than with a single one, and the additional directional information permitting fracture azimuth determination. Vertical cracks filled with conductive mud or mud filtrate are normally seen as conductive anomalies on a single pad, on two adjacent pads or on two opposite pads.

Fracture detection by dipmeter evaluation is enhanced by combination with the four-arm caliper recording which reflects elongate borehole geometry in the jointed interval. In the breakout plane, instrument rotation normally stops, with the tool trajectory through the fractured zone following a fixed path. The high-resolution dipmeter log is also sensitive to hairline cracks. As many wellbore conditions other than fractures cause responses similar to those related to cracks, fracture indication by dipmeter readings should be compared and corroborated with clues from other log responses such as circumferential acoustic-log (GUY, FERTL & OLIVER 1986; GUY, CORLEY, FERTL & FROST 1987; NOBLETT, FERTL & GUY 1987), and also inspection of reservoir cores should be made if available.

6.2.1.10.2.3. Dipmeter evaluation for fracture detection

Dipmeter fracture evaluation can be performed by comparison of pairs of microresistivity curves from adjacent buttons of the same pad (conductivity anomaly) or from adjacent pads (fracture identification; SCHLUMBERGER 1987 a). Separation of the pairs of curves usually indicates a conductivity anomaly associated with fractures. Comparing resistivity anomalies for different pad pairs can reveal crack strike. Not all the anomalies, however, can be attributed to fractures. Sources of confusion that also trigger dipmeter curve separation include shale breaks, vugs, borehole irregularities, loss of pad contact, tool pulls, floating pads, electrode insulator failures, incomplete response to thin beds, clay clasts and quartz pebbles, and wellbore spalling which can be due to break-out at the apexes of an elliptical hole from earth stress or irregularity of large fractures.

6.2.1.10.2.4. Other reasons than fractures causing dipmeter curve separation

As dipmeter curve separation is not an unique signal of fracturing, the most effective use of the dipmeter is as a means to condemn those zones where no cracks provoking no separation are evident and to high-grade those intervals where possibly fractures occur as evidenced by separation of dipmeter curves. Dipmeter evaluation for crack detection is limited to relatively thick massive formations where the borehole is in good condition (GEORGI 1986). This is because the dipmeter sonde in a round hole tends to rotate and twist from cable torque which increases the likelihood of the pads crossing vertical or steeply dipping fractures. In elliptical wellbores, the tool will stop twisting and the pads may miss cracks that do not cross the borehole. Tool rotation does not change the probability of detecting inclined fractures, but when the sonde stops rotating, only those vertical cracks with strikes parallel to the major or minor diameter of the wellbore will be recorded. Nevertheless the dipmeter is able to at least subdivide the borehole section into probably fractured and unfractured zones.

In rugose wellbores, frequently much curve separation takes place due to alternately sticking and pulling free of the sonde, with this type of separation often being able to be recognized from the tension curve or from the character of the resistivity curves themselves. Separations can also be created by small indentations in the borehole wall caused by the drill string pipe joints, electrical anisotropies of the formation, or can be artifacts due to unequal pad pressures. It is often difficult to establish the origin of the observed curve separation and to eliminate separations not attributable to open fractures. Separation will occur whenever one of the electrode buttons passes over a localized electrical heterogeneity. By comparing the various resistivity curves, it is usually possible to establish whether the separation is due to a drilling-fluid-invaded open crack that gives rise to a low-resistivity anomaly or due to a feature which leads to a high-resistivity anomaly such as pebbles, boulders, chert nodules and cemented fractures.

Simple optical inspection of the tadpole processing display of the dipmeter record computation reveals fractures as isolated high-angle dips in an environment of more or less constant low- to moderate-angle inclinations corresponding to stratigraphical and sedimentary bedding planes (SCHLUMBERGER 1982, 1983; DRESSER ATLAS 1987 b). Comparing pairs of microresistivity curves, separation between curves coming from opposite or adjacent pads also indicates locations of high-angle fractures, because low- to moderate-angle stratification boundaries appear as sharp spikes which correlate on all four dipmeter curves.

6.2.1.10.2.5. Borehole spalling and dipmeter response

Fractures preferentially occur in brittle rocks and jointed reservoirs are often located in tectonically active regions where the horizontal components of the in-situ stress field are unequal (GEORGI 1986). In brittle rocks when the two in-situ stress components in the plane normal to the borehole axis are unequal, the wellbore spalls in a predictable manner (cf. ZOBACK, MOOS, MASTIN & ANDERSON 1985), with the borehole elongating in the direction of minimum stress (cf. section 6.2.1.8.). The unevenness of the spalled wellbore wall causes the dipmeter pads to lift off the borehole wall and gives rise to anomalously low resistivity readings and separation on the dipmeter fracture log, with such a separation having to be discounted. If borehole spalling is extensive and the wellbore becomes elliptical, the diameter caliper data can be used to identify separation due to borehole spalling, and the sonde orientation data can be used to infer the direction of wellbore elongation. It is sometimes necessary, however, to infer the orientation of the spalling from minor inflections in the orientation data or from the general tectonical setting.

The fracture identification log is a microresistivity dipmeter version with a specially formatted output (PAILLET 1981), with each of the different traces representing the overlay of microresistivity curves from adjacent measurement quadrants (cf. also BROWN 1978). Field examples of fracture detection and measurement by dipmeter logging are presented by BABCOCK (1978).

6.2.1.10.3. Other aspects of resistivity logging techniques

Resistivity logs generally compare resistivity values of differing depth of formation investigation in order to obtain indications of invasion of open fractures by drilling fluid which would point to a naturally jointed zone (HEFLIN 1979). After some general aspects, possibilities of fracture detection by dual laterolog recording are illustrated.

6.2.1.10.3.1. General aspects

The vertical resolution of each measurement should be considered in the interpretation. Porosity tools such as acoustic, density and neutron logging compare total porosity to matrix porosity, but fracturing often does not significantly increase total porosity with respect to the resolution of the porosity measuring devices (HEFLIN 1979). The shallow correction curve on compensated density recording is sensitive to minor borehole effects due to fracturing, but other wellbore features such as rugosity give the same response as fracturing and it is often difficult to distinguish cracks from other anomalies. Dipmeter evaluation of fracture orientation comprises comparison of resistivity variations obtained from four or six different orientations in the borehole, but maintaining pad contact is critical in using these with repeated passes to improve the surveyed area of the wellbore.

6.2.1.10.3.2. Dual laterolog recording

Dual laterolog measurements are usually made in three different depths of reservoir investigation (SCHLUMBERGER 1987 a). While the deep and shallow devices read within the virginal formation or distal parts of the invaded interval, the microspherically focussed log analyzes the flushed zone which is the section of the mud-filtrate-penetrated belt closest to the borehole wall. Although the microspherically focussed log response sometimes fluctuates dramatically in front of fractures, its sensitivity to poor wellbore conditions often precludes its use in crack detection. The electromagnetic propagation time log is another very shallow measurement which reacts to horizontal fractures and has a high vertical resolution. Deep and shallow laterologs respond to fractures if the resistivity of the fluid within the crack differs sufficiently from the resistivity of the formation (RASMUS 1981, BOYELDIEU & WINCHESTER 1982). In the presence of fractures, deep and shallow resistivity tools read too high, with the deep instrument recording higher than the shallow detector, whereas the microspherically focussed tool reads much lower than both deep and shallow instruments (SCHLUMBERGER 1982 b).

In general, the dual laterolog does best detecting fractures in low-porosity and high-resistivity formations, and performs worst detecting horizontal cracks in high-porosity and low-resistivity reservoir sections. Laterolog behaviour in a fractured reservoir can be modelled by a finite-element program which delineates that the curve separation between deep and shallow laterologs may be proportional to ratio of crack aperture and conductivity of fracture fluid (SIBBIT & FAIVRE 1985). While oil-based muds preclude use of the laterolog, the induction tool can be applied in oil- or water-base mud, but is not sensitive to horizontal cracks (HOWARD 1981). Resistivity logging is also a possibility of hydraulic fracture height determination (WITTRISCH & SARDA 1987).

6.2.1.10.3.3. Dual laterolog interpretation

Large separations are usually observed between shallow and deep laterolog curves in fractured rocks (SIBBIT & FAIVRE 1985). In case of vertical cracks, the deep curve reads higher than the shallow curve, whereas for horizontal fractures, the deep curve registers lower than the shallow curve (thus dual laterolog evaluation can be utilized for distinction of vertical and horizontal cracks; cf. section 6.2.1.10.5.2.). Only open fractures can be detected, as only these are invaded by drilling mud, but these are exactly those cracks which contribute to hydrocarbon production and thus represent effective secondary crack permeability, and tight fractures which have no effect on reservoir drainage are not displayed by the dual laterolog record. The dual laterolog response is controlled by resistivity of formation blocks, resistivity of invading fluid, extent of invasion and fracture opening. While many crack indicators from well logs are highly influenced by near-borehole effects which can be mistaken for or can mark anomalies caused by cracks, the dual laterolog sees deeper into the reservoir and can reveal the extension of open fractures in the formation which enables assessment of their productivity, in contrast to many cases of spectacular near-borehole fracturing which later often turns out to contribute little to well productivity because of limited extension into the reservoir.

Separation between deep and shallow laterolog curve even allows computation of fracture porosity (BOYELDIEU & WINCHESTER 1982). While density and orientation of cracks are best determined by high-resolution tools such as dipmeter, televiewer and microscanner, the combination of these surveys with apertures derived from dual laterolog permits estimation of porosity and permeability of fractured zones (GOLF-RACHT 1985).

6.2.1.10.4. Sonic logging with full waveform analysis

Array sonic tools equipped with multiple receivers permitting full sonic waveform analysis are becoming increasingly important for natural fracture detection (MORRIS, LITTLE & LETTON 1984; SCHLUMBERGER 1985, 1987 b; MINEAR 1986, PAILLET 1986, WELEX 1986), and sophisticated sonic log interpretation based on waveform analysis is probably the most commonly used fracture detection technique after borehole imaging and dipmeter evaluation. The instruments record longitudinally and transversely propagating as well as circumferentially travelling sonic waves (cf. also section 6.2.1.6.1.).

Sonic waveforms (WU 1984) enable examination of energy and frequency spectra associated with compressional, shear, Rayleigh and Stoneley waves. The attenuation of the sonic waveform`s shear energy in relation to its compressional energy indicates possible vertical natural fracturing in the formation. This information combined with traditional crack indicators such as borehole ellipticity (cf. section 6.2.1.8.), differential resistivity, formation density, bulk density compensation, photoelectric absorption, spontaneous potential, gamma-ray spectrometry (cf. section 6.2.1.3.4.) and in-situ stress analysis provide a fracture probability index for each zone (SCHLUMBERGER 1985; cf. also section 6.2.1.10.6.). The use of various different crack indicators offsets the effects of anomalous readings of the individual logs. The outline as follows focusses on mode conversion attenuation of sonic waveforms, amplitude spike analysis, crisscross analysis, and Stoneley wave evaluation.

6.2.1.10.4.1. Mode conversion attenuation of sonic waveforms

Sonic waveform analysis has the advantage that it can be performed in mud of any weight and type (cf. also section 6.2.1.6.1.). When an acoustic wave reaches a fluid-filled fracture, part of it reflects back into the rock and part changes to a fluid wave in the crack by mode conversion (SCHLUMBERGER 1985, 1987 b). When the fluid wave reaches the opposite wall of the fracture, a second mode conversion divides it into compressional, shear, Rayleigh and Stoneley wave components. The presence of cracks is indicated by the energy loss caused by these mode conversions (MORRIS, GRINE & ARKFELD 1964). Energy loss due to wave attenuation, however, can also occur from poor borehole conditions and changes in formation lithology (particularly interbedded soft shales). The main applications of sonic waveform analysis are natural fracture evaluation, mechanical rock properties determination, formation assessment (especially porosity measurement behind casing), reservoir strength analysis with respect to sand production, seismic correlation and cement evaluation (SCHLUMBERGER 1987 b).

Only the existence, but not the orientation of natural fractures (joints) can be monitored by long-spacing sonic log analysis (particularly variable density log (VDL) or signature wave-form evaluation; SCHLUMBERGER 1974; MORRIS, LITTLE & LETTON 1984; KREIJ 1986).

6.2.1.10.4.2. Amplitude spike analysis

Apart from direct sonic waveform analysis by mode conversion attenuation evaluation, various other acoustic logging and interpretation methods exist (SCHLUMBERGER 1987 a). Amplitude spike analysis relies on enhanced acoustic coupling which represents strengthening of the transmitted or received signal right at the fracture opening. This causes two momentary peaks in compressional wave amplitude as the transducers pass a crack (CHEUNG 1984). The signal enhancement occurs for only a foot or so near the fracture and thus provides a fairly good vertical resolution. Amplitude spike analysis gives a more crack-specific response and distinguishes fractures from uncracked bed boundaries and large washouts.

6.2.1.10.4.3. Crisscross analysis

Crisscross analysis (CHEUNG 1984) examines patterns representing differences in the arrival times of the various wave components which result from mode conversions presumed to occur at fractures. The compressional-shear crisscross is of greatest interest since it is less cluttered by direct arrivals. The vertical resolution of crisscross analysis is less than that of amplitude spike evaluation. Crisscross assessment shows a distinct energy peak indicating a fracture even when the eye fails to detect a pattern in the variable density display (SCHLUMBERGER 1987 a).

6.2.1.10.4.4. Stoneley wave evaluation

More recent sonic logging and interpretation techniques take advantage of the Stoneley wave (HORNBY, JOHNSON, WINKLER & PLUMB 1987; SCHLUMBERGER 1987 a; BRIE, HSU & ECKERSLEY 1988; PLUMB & HORNBY 1988). Fractures attenuate low-frequency Stoneley waves differently from compressional, shear and Rayleigh waves. Very little Stoneley wave attenuation is from mode conversion; most of it is caused by the wave pumping fluid in and out of the crack. The magnitude of this Stoneley wave energy drop can be related to fracture aperture and is independent from crack dip. The presence of the Stoneley wave is easily identified in the recorded waveforms, because its motion is slower than that of compressional, shear and Rayleigh waves, has the lowest frequency and suffers least attenuation in undisturbed formations, thus appearing generally as a large-amplitude event (WATERHOUSE, CHARARA & NURMI 1987).

Open fractures as well as interconnected pores give rise to loss of energy and speed of the Stoneley wave, with cracks generally producing sharper and more spiked disruptions of the sonic waveforms than communicating grain interstices. Fracture modelling from the Stoneley wave includes estimation of the effective crack width. Various possibilities of Stoneley wave record analysis and fracture evaluation are discussed by HUANG & HUNTER (1981); BEYDOUN, CHENG & TOKSÖZ (1983), LIU (1984, 1985), PAILLET (1980, 1983, 1985); PAILLET, CHENG, HESS & HARDIN (1986) and HSU, BRIE & PLUMB (1987). A quite simple means of quicklook fracture recognition is sometimes pronounced cycle skipping on the sonic log record (SCHLUMBERGER 1982).

6.2.1.10.5. Special methods for horizontal fractures

Most of the previously mentioned methods work better in vertical fractures. For apparent horizontal cracks in deep highly-inclined wellbore sections and true horizontal fractures in shallow reservoir depth, acoustic and density techniques are more important. After discussion of some general aspects, distinction of horizontal and vertical fractures by dual laterolog recording is outlined.

6.2.1.10.5.1. General aspects

Horizontal fracture assessment is particularly successful by determination of barite infillings of open cracks in case the drilling mud is weighted by barite (SCHLUMBERGER 1987 a). Poor borehole conditions in the shallow depth where horizontal fractures occur have often negative influence on evaluation. Performing acoustic methods, wavelet processing allows measurement of reflections from fracture planes. Azimuth orientation analysis of apparent horizontal cracks in highly-deviated wells aim on directional determination of the optimum wellbore trajectory, with the goal being to drill perpendicularly to the fracture intersections in order to penetrate a maximum of crack bundles.

In terms of distinction of vertical and horizontal fractures, acoustic tools are generally sensitive to horizontal cracks, whereas resistivity instruments mainly record vertical fractures. The extent of the anomalies on the log also gives some information on the vertical or horizontal nature of the fracture. Horizontal cracks on formation microscanner images, however, may be confused with bedding planes, because the physical apparition is more or less the same.

6.2.1.10.5.2. Distinction of vertical and horizontal fractures by dual laterolog recording

Dual laterolog records can be utilized for distinction of horizontal and vertical fractures (SIBBIT & FAIVRE 1985; cf. section 6.2.1.10.3.3.). Open cracks are deeply invaded by mud during the drilling process, and this conductive fluid causes anomalous readings. For vertical fractures, shallow and deep laterolog curves separate, with the deep curve recording higher than the shallow curve, and the conductivity difference being proportional to the product of fracture opening and conductivity of the invaded fluid. In case of deep invasion, the resistivity of the formation between cracks may be computed which in turn may be used to determine the saturation of the reservoir blocks.

For horizontal fractures, both laterolog curves decrease sharply over an interval of abt. 0.8 m of the well representing total width at the inflection points, and the separation between both curves may reverse, with the deep curve reading lower than the shallow curve. At the shoulders of horizontal cracks, the curves register the resistivity of the surrounding medium. The level of resistivity opposite the fracture is again related to the product of crack opening and conductivity of the invaded fluid. Intersecting horizontal and vertical fractures behave as if individual horizontal and vertical cracks were superimposed.

6.2.1.10.6. Fracture probability program

A satisfactory characterization of fractures often requires integrating data from several imageless types of records in a multilog approach. As many different open-hole logging methods are affected by fluid-filled cracks which cause anomalous readings, a fracture probability program can be developed where various (up to 16 or more) individual fracture indicators are simultaneously used to compute the likelihood of an interval being naturally jointed (GROVES & WHITTAKER 1985, SCHLUMBERGER 1985, SENECAL & MULHERN 1985). Although the reliability of the individual indicators varies as the tools respond to factors other than fluid-filled cracks, the composite fracture index log results from analyzing how each standard tool responds to joints and benefits from enhancing the inspection of individual logs based on local experience and understanding of the analyst by computer-aided independent comparative evaluations. The composite index integration assigns a level of plausibility to each response and produces a grouped log that reflects the cumulative probability of all fracture indicators (SCHLUMBERGER 1982 b, BOYELDIEU & MARTIN 1984, GROVES & WHITTAKER 1985).

As most of the different fracture-recording logs only tend to be affected by open and thus effective cracks, the results of the fracture probability analysis provide usually reliable information for further reservoir and production engineering planning. HARRISON & TILDEN (1986) document a case study of successful application of composite crack identification logs for field-wide fracture predictive modelling following calibration of core data. Field utilization of crack identification logging is also reported by OLIVE (1988).

Natural fracture determination and evaluation by various logging methods always has to respect that fractures can be enhanced or even induced by drilling (criteria for distinction of natural and artificial cracks are discussed by SANGREE 1969, KULANDER & DEAN 1988, LAUBACH & MONSON 1988; cf. also section 6.2.1.10.1.2.), the response of logging tools to fractures may be similar to their reaction to other phenomena such as washouts and bed boundaries, no existing logging tool can unequivocally determine location and orientation of fractures but a qualitative assessment can be made from a combination of records, and logging instruments investigate the borehole and its immediate vicinity only and hence the findings may not be representative of the entire reservoir including deeper sections farther away from the wellbore (BOTTER 1982).

6.2.1.10.7. Mechanical properties log

The mechanical properties log which is computed from sonic and density input logs (cf. also section 6.3.6.) is a predictive tool for retroanalysis of drilled wells and prognosis derivation from the results of an earlier borehole for the better and more effective operation of a subsequent well. Its main purposes are evaluation of the natural fracture system together with assessment of the dynamical elastical properties to provide input data for the design of hydraulic stimulation and sand control treatments. Other tasks are mud loss control and avoidance of pipe sticking. Some principles and applications of mechanical properties logging are briefly summarized as follows.

6.2.1.10.7.1. Principles

Mechanical properties analysis testifies natural fractures to be open as soon as the mud pressure is raised enough to push the tangential stress into tension and towards negative values. The opening of cracks when the tensile strength reaches zero may cause sudden mud losses during drilling, with the computed fracture pressure gradient in the mechanical properties analysis indicating the critical mud pressure gradient. The purpose of hydraulic fracturing is to deliberately induce a tensional tangential force exceeding the tensile strength of the rock. In order to ensure success in the stimulation operation, the tensional stress must overcome the maximum tensile strength existing in the interval to be treated. In the mechanical properties analysis, the pressure required to achieve a successful hydraulic fracture is computed by assuming maximum tensile strength and initial shear strength to be equal. Fracturing then will start if the wellbore fluid pressure is raised to a level such that the tangential stress is exceeding the initial shear strength. Another parameter is the toughness indicator, with an increase in toughness representing a greater resistance to fracturing. With increasing local field experience, the toughness curve may be calibrated to provide a prediction of crack containment.

Mechanical properties computation from the input logs is very important because of the major effect which in-situ stress profiles have on fracture propagation geometry and due to the fact that in-situ stress can significantly vary between adjacent formations even over small vertical intervals (WARPINSKI, BRANAGAN & WILMER 1983). In-situ stress determination can also be performed by minifracture test pressure decline analysis (cf. section 4.8.2.1.), acoustic wavetrain interpretation (cf. section 6.2.1.6.) and anelastical strain recovery of oriented cores (VEATCH & MOSCHOVIDIS 1986; cf. section 6.2.2.3.3.4.). Minifracture testing and mechanical properties modelling are particularly significant for calibration of in-situ formation stress or closure pressure in order to have realistical input values for fracture design which helps to reduce the rate of premature screenout termination of the jobs (cf. section 6.2.4.2.1.).

6.2.1.10.7.2. Applications

The main applications of mechanical properties analysis in terms of fracturing are avoidance of accidental pressure parting during drilling and cementing, assisting in mud-loss control for subsequent nearby wells, differential pressure requirement for hydraulic fracture initiation, zone isolation for selective fracturing, identification of potentially weak zones and resistance to fracturing, and containment of vertical extent of crack propagation (SCHLUMBERGER 1982 b). The significance of mechanical properties determination from log processing is also emphasized by CLEARY (1988), with desirable further developments being calculation of in-situ stresses from the mechanical properties log and an earth-consolidation model to relate horizontal to well-known vertical stresses. Mechanical properties computation is in terms of reservoir stimulation and consolidation significant for both hydraulic fracture propagation modelling and formation sand production forecast (cf. sections 6.2.5. and 6.3.6.). Some comments on fracture pressure evaluation and modelling as well as continuous mechanical properties log vs. isolated minifracturing calibration test data are offered as follows.

6.2.1.10.7.2.1. Fracture pressure evaluation and modelling

Three steps can be distinguished in fracture pressure evaluation from full wave sonic logs : estimation of elastic moduli, assessment of least principle horizontal stress and leakoff coefficient, and application of least principal horizontal stress for prediction of fracture height (MINEAR 1986). An important aim of mechanical properties modelling is picking up of boundaries to vertical fracture growth (HUNT, RAYMOND, HASKETT & PIRIE 1981). The impetus of mechanical properties modelling from log records for in-situ stress and formation properties determination is also outlined by TIXIER, LOVELESS & ANDERSON (1973); ARON, MURRAY & SEEMAN (1978); ROSEPILER (1979), COATES & DENOO (1981); SCHLOTTMAN, MILLER & LUEDERS (1981); SCHLUMBERGER (1982 b); NEWBERRY, NELSON & AHMED (1985); HOLDITCH, ROBINSON & WHITEHEAD (1986); AHMED (1988) and LANCASTER, GATENS, STANLEY, HARRISON & GUIDRY (1988). A case study of pre-fracturing analysis logging is documented by GRI (1987, 1988). ANDERSON, INGRAM & ZANIER (1973) comment on determination of fracture pressure gradients from well logs. Mechanical properties modelling is also an important contributing element for prediction of fracturing pressures in wildcat wells (DAINES 1980).

6.2.1.10.7.2.2. Continuous mechanical properties log vs. isolated minifracturing calibration test data

The primary benefit of using wireline data for fracture migration analysis from determination of in-situ stress, tensile strength and elastical constants of the formation is obtaining of continuous data in contrast to isolated and sporadic values delivered by cores and minifracturing treatments (NEWBERRY, NELSON & AHMED 1985; AHMED 1988). In addition, minifracturing data are usually confined to the pay intervals and give no information on location and quality of the barriers. Economically, the cost of routinely coring long sections and minifracturing numerous levels is prohibitive. The wireline-based hydraulic fracture height prediction model uses only stress differences and not absolute values, with the result that any consistent shift has little effect on the calculations. Core and minifracturing data can be incorporated for calibration of the output in order to yield accurate absolute pressures. Mechanical properties logging and evaluation has also significant application in drilling technology for prediction of formation fracture gradients (SIMMONS & RAU 1988).

6.2.1.10.8. Other techniques

In addition to the aforementioned methods, further impact on natural fracture detection and characterization can be gained from gamma-ray detection of uranium-mineralized joints, formation factor equation exponent value and plots, radar pulsing, and miscellaneous means. Some comments are also offered on fracture height logging in vertical and deviated wells.

6.2.1.10.8.1. Gamma-ray detection of uranium-mineralized joints

Gamma-ray spectroscopy can pick-up natural fractures if they are filled with uranium mineral deposits which occurs quite common in some formations. The technique of picking up high uranium enrichments vs. low potassium and thorium concentrations helps to build evidence for cracks, but is far from confirmatory.

In hard carbonates drilled with barytic mud, fractures are often clearly recognized by a sharp peak on the photoelectric factor curve if the borehole is in good condition (SCHLUMBERGER 1987 b). In other formations, however, zones with high uranium concentrations are in places phosphatic or tuffaceous rather than representing open natural fractures intersecting sedimentary rocks (SENECAL & MULHERN 1985).

6.2.1.10.8.2. Formation factor equation exponent value and plots

In the porosity-resistivity crossplot which is often used to obtain the apparent value of the exponent m in Archie`s formation factor equation, m is frequently found to drop well below the normal value of 2 and ranges usually around 1.4 in fractured formations. Tool azimuth in jointed intervals is often constant, reflecting interruption of instrument rotation when passing along the cracked zone (SCHLUMBERGER 1987 a). Other natural fracture indications are differential response to conventional porosity measurements such as partitioning of intergranular and fracture porosity components on the basis of crossplots, and magnitude of apparent power law exponent relating formation factor to porosity (PAILLET 1981). Fracture identification from the cementation exponent is also discussed by AGUILERA (1976), and differential response indications are examined by PICKETT & REYNOLDS (1969) and BECK, SCHULTZ & FITZGERALD (1977). The utilization of lithoporosity crossplots and spontaneous potential curve readings for detection of fractures is analyzed by POUPON, HOYLE & SCHMIDT (1970).

6.2.1.10.8.3. Radar pulsing

Radar pulsing is another method of natural fracture detection which has a wider registration radius and can also recognize cracks that are farther away from the borehole (OIL GAS JOURNAL 1987 k). Directional fracture mapping with this tool which is still in experimental stage is made by emission of short radar pulses that are reflected by rock discontinuities such as preferentially fractures and return to the receiving antenna of the instrument, with the delay of the return signal indicating the distance of the cracks from the borehole. The high resolution capacity of the tool is provided by the use of short electromagnetic pulses instead of continuous wave transmission, and quick rise time and short duration of the radar pulses are accomplished by using the special characteristics of avalanche transistors.

6.2.1.10.8.4. Miscellaneous

Other methods of borehole logging fracture detection and determination are discussed by ARDITTY, GRAS & STARON (1987); BRIE, HSU & PLUMB (1987); PAILLET (1987), SCHLUMBERGER (1987 a), WITTRISCH (1987) and PLUMB (1988). Naturally jointed reservoirs can also be analyzed by repeat formation testing (STEWART, WITTMAN & GOLF-RACHT 1981; BARUA, KUCUK & GOMEZ-ANGULO 1985). Inference of fracture properties can also be made from indications of fluid motion as recorded by flow-metering of fluid streams into the well (AGUILERA 1979, KEYS 1979). Some examples from well testing and production, and drilling and coring are sketched as follows.

6.2.1.10.8.4.1. Well testing and production

The presence of fractures is also exhibited if during drill stem testing or production testing a flow capacity is found which is greater than that what was to be expected on the basis of matrix permeability alone (BOTTER 1982). Determination of length of waterflood-induced fractures (cf. section 4.8.7.1.) by high-precision pressure-falloff testing is documented by KONING & NIKO (1985) and PRADO, KRUYSDIJK, NIKO, SANCEVIC & RODRIGUEZ (1988). BELFIELD (1988) stresses the significance of productivity data as indicators of natural fractures in light of the economical infeasibility of coring every well or performing an exhaustive pressure transient program. Production data also measure the true in-situ flow capacity of the formation. Interference tests are able to reflect the direction of maximum permeability which matches with the azimuthal orientation of the natural fractures, and pulse tests indicating very high formation transmissivities are an additional confirmation. Good results of interference tests also document that the open natural cracks are well interconnected across the field.

6.2.1.10.8.4.2. Drilling and coring

MACPHERSON & GUENARD (1988) present delineation of fractured pay zones using drilling engineering and mud logging data. Recognition of fractured intervals is accomplished through integration of drilling response modelling and pulsed nuclear magnetic resonance measurements on drilled cuttings. MYUNG (1976) comments on possibilities deriving from production index log and sibilation record. The comparative significance of various techniques is summarized by AGUILERA (1976), SUAU & GARTNER (1979), SCHLUMBERGER (1982, 1987 a) and BOYELDIEU & MARTIN (1984). Fractography is an important method for core and outcrop fracture investigations (KULANDER, BARTON & DEAN 1979). AMERI & RIEKE (1981) use well logs to provide formation damage design values for hydraulic fracture treatments. AADNOY (1988) describes an inversion technique to determine the in-situ stress field from fracturing data.

LAUBACH & MONSON (1988) characterize coring-induced cracks (cf. section 6.2.1.10.3.3.) as indicators of hydraulic fracture propagation in naturally jointed reservoirs. BELANI & JALALI-YAZDI (1988) illustrate estimation of matrix block size distribution and crack intensity from pressure transient data. PLUMB & HORNBY (1988) characterize fractures by integrating wellbore images and borehole Stoneley wave analysis. KLEINBERG, CHOW, PLONA, ORTON & CANADY (1984) outline reliability and sensitivity of crack detection techniques for borehole application. YOST, FROHNE, KOMAR & AMERI (1982) present methods for determination of natural and induced fracture relationships in shale reservoirs. LERCHE & NARR (1984) estimate subsurface density in core and discuss effects of variable fracture spacing.

6.2.1.10.8.5. Fracture height logging in vertical and deviated wells

All wireline logging methods of detecting fracture height have very limited depth of investigation and do not allow to measure crack height farther away from the wellbore (PEETERS & HARTLEY 1984). Only the deep laterolog curve has some more penetration in contrast to all the other devices which are registering shallow to extremely shallow. Therefore efforts to record fracture height are best restricted to nearly vertical wells where normally vertical cracks are created. In highly-deviated and horizontal wells, the vertical fracture is usually in contact with the well only over a limited interval (SCHULTE 1984; cf. section 4.11.1.) and thus the possibilities of properly monitoring fracture height are progressively deteriorated with increasing wellbore inclination (the same applies for horizontal cracks intersecting vertical wells in shallow reservoir depth; cf. section 1.2.8.).

The degree of wellbore inclination can have a very significant effect on interpretation of propped fracture height especially for large intervals (BEGNAUD & CLAIBORNE 1985). The limited extent of investigation can cause a small change in borehole inclination from vertical to limit the amount of fracture height recorded by the log.

EL RABAA (1988) proposes on-site strain relaxation measurements of oriented cores as the most reliable method of fracture azimuth determination in highly-deviated or even horizontal wells. The process is based on the assumption that a sphere of rock separated from a formation and subjected to unequal stress field will relax and deform anelastically into an ellipsoid, the axes of which are aligned with the directions of the released stresses. The analysis can also account for cross-anisotrophy or orthotropy in formation properties caused by bedding.

6.2.2. Seismical and gravimetrical methods

Seismical and gravimetrical methods aim in contrast to most well-logging techniques not or not only on fracture height determination, but are mainly applied in order to detect crack azimuth and in some cases also get information on fracture width, and deliver occasionally also additionally some data on crack height. The following account therefore assembles aspects of fracture height monitoring, fracture width determination and fracture azimuth mapping. Some points of distinction of vertical and horizontal fractures as well as fracture permeability estimation are also outlined.

6.2.2.1. Fracture height monitoring

Apart from various borehole logging methods (cf. section 6.2.1.), another technique of fracture height monitoring is in-situ well-seismic detection with an accelerometer that works principally like a geophone. From the surface, cracks can be evaluated both in terms of direction (azimuth) and distance of extension from the borehole (but not concerning height) by gravimetry, with these horizontal-dimension parameters not being able to be determined by cable well logging that can only investigate the vertical dimension. Various seismical and gravimetrical methods of fracture monitoring and detection are discussed by ALBRIGHT & PEARSON (1980), LEYDECKER (1981), HARDT (1982), DOBECKI (1983), WOOD (1983), WIERCZEYKO (1985), HOLZHAUSEN & EGAN (1986), MAUK & MAHRER (1986) and THORNE & MORRIS (1987). An assessment of borehole seismic fracture diagnostics is given by THORNE & MORRIS (1988). Distinction can be made between spherical crack height and azimuth determination during the actual hydraulic stimulation process with passive borehole instruments and surface tiltmeters, and after termination of the operation with shear-wave polarization and splitting (AHMED 1987). Comments are offered as follows on active vs. passive borehole seismics as well as on seismical fracture height determination.

6.2.2.1.1. Active vs. passive borehole seismics

Distinction can generally be made between active borehole seismics where energy is emitted into the fractured formation, and passive wellbore seismics where the closing crack itself produces microseismic events (DOBECKI 1983). In both cases, calibration of the downhole seismic tool by surface shooting or vibration prior to fracture detection is essential (HOLDITCH, ROBINSON, WHITEHEAD & ELY 1987). While no borehole preparation is necessary immediately after the hydraulic stimulation operation when the fracture is still closing and thus is pressing and rubbing thereby creating microseismic energy, later repetition runs and height checking are only possible by artificial excitation of crack and surrounding formation by minifracture injection tests (cf. section 4.8.2.).

Concerning areal geophysical exploration for natural fractures or field surveys to check sets of hydraulically-induced cracks and to map their geometrical pattern in multi-well hydrocarbon patches, fractures are generally identified as noise zones on certain seismic horizons (REYNOLDS 1987). Conventional seismological techniques are only effective on intermediate and deep jointed intervals. Special tools and methods are required for fracture detection in shallow formations. Areal extension of fractures is very difficult and expensive to determine by conventional seismology, and it is almost impossible to detect crack intersections and to evaluate their horizontal magnitude.

In addition to fracture azimuth and height monitoring, seismic techniques have a second important application in tight gas reservoirs comprising location of isolated sand lenses (DOE 1984). The determination of the spatial position of discrete sand bodies in lenticular pay complexes can be performed by 3-D reflection seismic profile interpretation, vertical seismic wellbore profiling and cross borehole seismic analysis. Seismic borehole tomography for imaging formations around the wellbore is an inverse vertical seismic profiling technique, with the drilling bit being the source and the receiver geophones being located at the surface instead of seismic energy created by shallow hole shooting and recording with borehole geophones (GRI 1988).

6.2.2.1.2. Seismical fracture height determination

THORNE & MORRIS (1987, 1988) summarize of advances of borehole seismic fracture diagnostics in terms of height and azimuth of the crack. The tool includes a synthetic generator of microseismic events which produces sinusoidal signals of specified amplitude, frequency and phase. The following overview concentrates on seismic waves and orientations, seismic event generation and interpretation, and vertical seismic profiling and other aspects.

6.2.2.1.2.1. Seismic waves and orientations

Complex waveforms resulting from the microseismic events can be analyzed and processed also in highly anisotropic geological formations. Acoustic emissions from a hydraulic fracture are thought to be caused by shear failures (cf. section 4.7.7.) induced by the open, pressurized crack, the localized high pore pressure zone surrounding the fracture, and the presence of planes of weakness in the rock around the fracture (DOBECKI 1983). Detection and localization of these acoustic emissions can be used to infer azimuth and height of the crack, because they should occur in a tight band near the hydraulic fracture face (ALBRIGHT & PEARSON 1980).

The method of localization of microseismic events depends on the detection of primary (compressional) and secondary (shear) wave arrivals to compute the distance of the tool to the microseismic event. Determination of orientation by a triaxial geophone set is based on the fact that a primary wave is polarized in the direction of propagation, so that a three-dimensional vector with components proportional to the output of the three geophone axes points either towards or away from the direction of arrival, and also depends on the assumption that the direction of primary wave arrival is the direction to the seismic source. While fracture azimuth recording gives usually very good results, crack height cannot always be established due to lack of microseismic events located at the bottom of the fracture (THORNE & MORRIS 1987, 1988).

6.2.2.1.2.2. Seismic event generation and interpretation

Fracture azimuth, height and extent can be inferred during the initial stages of stimulation from seismic signals received during non-proppant pumping times (HART, ENGI & MORRIS 1983). The seismic events are assessed to be caused by shear features induced by high pore-fluid pressure occurring on planes of weakness in the rock surrounding the large hydraulic fracture (cf. section 4.7.7.). Although these events do not seem to be caused by tensile failure at the expanding edge of the hydraulic crack, their location can still be used to infer fracture azimuth and growth because the shear events are believed to cluster in a tight band near the hydraulic face (ALBRIGHT & PEARSON 1980; cf. section

6.2.2.3.2.). Angular interpretation ambiguity can be overcome by fielding two borehole seismic packages comprising one in the fractured well and another in the offset well (HART, ENGI & MORRIS 1983). Since the borehole seismic packages can only be downhole during non-proppant times, a wellbore seismic tool will be employed in the fractured borehole during the initial stages of the stimulation treatment only, but a package is listening from an offset well as long as events are being received.

6.2.2.1.2.3. Vertical seismic profiling and other aspects

Fracture detection techniques for borehole application are also discussed by KLEINBERG, CHOW, PLONA, ORTON & CANADY (1984). HSU, BRIE & PLUMB (1987) evaluate crack identification by array sonic tools. LACY (1987) compares different hydraulic fracture orientation methods. HARDIN & TOKSÖZ (1985) explain detection and characterization of fractures from generation of tube waves by the presence of open cracks in the formation (HUANG & HUNTER 1982; BEYDOUN, CHENG & TOKSÖZ 1985) by seismic primary waves incident on a fracture intersecting the borehole which is common in vertical seismic profiling. The behaviour is the result of fluid exchange between crack and wellbore, and the observed efficiency of conversion is related to fracture permeability. Crack orientation can be obtained from multiple-source offset vertical seismic profiling surveys, with three offsets being sufficient to geometrically resolve fracture orientation. The spectral method for comparison of incident primary wavelet and generated tube waves is effective given sufficient signal strength, yielding values for fracture parameters which are in reasonable agreement with independent observations.

From the well-logging techniques, even radial differential temperature recording is not suitable for fracture azimuth determination (DOBKINS 1979; cf. section 6.2.1.1.), and caliper or breakout logging (cf. section 6.2.1.8.) can only give a predictive indication of the direction of fracture propagation prior to running casing and performing the stimulation job, but is not able to subsequently monitor the result of the treatment.

6.2.2.2. Fracture width determination

Fracture width monitoring in addition to height determination and orientation checking has so far been more or less successfully approached by the borehole acoustical televiewer (KAMP 1983, SMITH 1983, FLECKENSTEIN 1984, CLERKE & AKKEREN 1986, DARILEK 1986; cf. section 6.2.1.9.1.), closed-circuit optical television camera (SMITH, ROSENBERG & BOWEN 1982; cf. section 6.2.1.9.3.) and advanced litholog tools such as the micro-electron scanning tool (MEST; cf. sections 6.2.1.9.2. and 6.2.1.10.1.) and specialized sonic logging instruments, but is not yet developed in such a reliable and sophisticated manner than the methods of crack height monitoring. WARPINSKI (1983 a) discusses possibilities of fracture width measurement during propagation of the crack by mineback direct inspection. Fracture width estimation can be made by inversion of the formation microscanner image and manipulation of sonic measurements.

Apparent fracture width recorded by the acoustic borehole televiewer is generally one full order of magnitude greater than measured core fracture widths (PAILLET 1981) as a consequence of widening by the drilling process and high-pressure mud invasion. Microresistivity dipmeter acoustic waveform amplitude recording system and circumferential acoustic probes show recognizable and correlate responses for one set of identified core fractures, whereas smaller cracks with very narrow opening are beyond the limits of resolution of these tools.

6.2.2.3. Fracture azimuth mapping

Fracture azimuth has major implications on reservoir exploitation strategy by being in the direction of maximum horizontal in-situ stress and thus being significant information for well pattern and spacing modelling (ECONOMIDES 1987 b). Fracture azimuth and proppant distribution mapping (SMITH, HOLMAN, FAST & COVLIN 1978; SMITH, LOGAN & WOOD 1980; HANSON & OWEN 1982, CLARK 1983; WOOD, PARKIN, YOTAM, HANSON, SMITH, ABBOTT, COX & O`SHEA 1983; FITZ-PATRICK, KARR & O`SHEA 1985; SMITH, REN, SORRELS & TEUFEL 1985; SORRELS & MULCAHY 1986) has particular significance in very tight formations where almost every well has to be stimulated and therefore in accordance to the geostress field the borehole locations have to be carefully selected in order to minimize interference of the long fractures extending from neighbouring wells (LEICHT 1985; cf. also section 4.10.5.). Aspects of crack azimuth determination are also discussed by TEUFEL, HART, SATTLER & CLARK (1984). Field experience of fracture azimuth monitoring with triaxial-geophone seismic recording is outlined by GRIFFIN (1985) and HOLDITCH, ROBINSON, WHITEHEAD & ELY (1987). Fracture azimuth mapping is the most important application of gravimetrical and geoelectrical methods (HARDT 1982, WIERCZEYKO 1985).

Fracture density modelling and variable fracture spacings based on core evaluation are discussed by LERCHE & NARR (1986). HOLZHAUSEN & EGAN (1987) comment on detection and control of fractures in water-injection wells. WESTERMARK & SCHOLES (1987) report fracture identification in open hole using data from cement evaluation tools. The outline of fracture azimuth mapping as follows focusses on geoelectrical and gravimetrical recording, microseismic fracture monitoring, combination of microearthquake and pressure pulse registration, surface tiltmeter measuring, hydraulic impedance analysis, and hydraulic resonance analysis.

6.2.2.3.1. Geoelectrical and gravimetrical recording

Concerning fracture azimuth monitoring, geoelectrical methods (HARDT 1982) aim on registration of the disturbance of the electrical potential field by the fracture infilled with fluid and proppants which acts as an additional electrical conductor. Measuring the electrical potential in the ground by a network of recorders that is radialsymmetrically oriented to the well before and after the stimulation operation, the geometry of the deflection of the potential field after the job enables the determination of fracture orientation (cf. also LEICHT 1985). Following a characterization of geoelectrical fracture detection, other methods of fracture azimuth mapping are illustrated.

6.2.2.3.1.1. Geoelectrical fracture detection

Geoelectrical fracture detection is performed from the surface and utilizes the casings of several neighbouring wells as current electrodes (HARDT 1982). By sending current into the ground through the casing, a strong artificial electrical field is created, and potential differences are registered at the surface in sensors which are radialsymmetrically oriented around the borehole. Prior to hydraulic fracturing, only the casing of the well acts as an electrode and represents a vertical linear electrode, whereas after the stimulation job, the crack is filled with highly-conductive fracturing fluid which represents an additional horizontal electrode. The disturbance of the equipotential line pattern by the fracture can be calculated and plotted to reflect crack orientation and extension. In addition to that, spontaneous potential of the cover strata above the fracture is anomalously changed during fluid injection, and permanent registration of spontaneous potential provides another possibility of determination of fracture propagation direction. Optimization of reliability is achieved by simultaneous utilization of active potential gradient and passive spontaneous potential geoelectrical registration methods.

6.2.2.3.1.2. Other methods of fracture azimuth mapping

Other methods of fracture mapping and azimuth assessment are observation of associated surface deformation (EVANS, HOLZHAUSEN & WOOD 1980; EVANS 1982; cf. section 6.2.2.3.4.), strain relaxation evaluation of oriented cores (TEUFEL 1981; cf. section 6.2.2.3.3.4.) with palaeomagnetical core orientation, microelectric scanning tool recording and interpretation (SCHLUMBERGER 1986 b, 1987 a; HOLDITCH, ROBINSON, WHITEHEAD & ELY 1987; cf. section 6.2.1.10.1.), monitoring of potential gradients (BARTEL, McCANN & KECK 1976) and tracing and evaluation of microseismic events caused by the stimulation job (MORRIS & SINCLAIR 1984; cf. section 6.2.2.3.2.). VOEGELE & JONES (1980) present a wireline tool for determination of in-situ stress contrasts. The relative magnitude of the stress intensity factor at the fracture tips can be used to indicate the direction of crack movement (LU & YEW 1985). Various case studies of fracture diagnostics and proppant distribution mapping are discussed by FITZ-PATRICK, KARR & O`SHEA (1985) and SMITH, REN, SORRELS & TEUFEL (1985). Fracture mapping in geothermal reservoirs by geological and geochemical techniques is reported by MOORE & NIELSON (1986). Methods of crack diagnostics are also discussed by HART, ENGI & MORRIS (1983); HART, ENGI, FLEMING & MORRIS (1984) and GRIFFIN (1985). Production control measurements become important in wells where inflow is limited to part of the fracture height (SCHULTE 1984).

Fracture length is an important input parameter for determination of locations of additional wells and thus drilling pattern planning (DOE 1984). In order to monitor crack length, electrically conductive salt water is pumped into the crack, and electrical fracture resistivity as compared to natural formation resistivity is recorded by both surface sensors and downhole probes. The induced surface electrical potential method used for hydraulic crack characterization measures variations in resistivity contrasts resulting from the flow of conductive fracture fluid into the earth (HART, ENGI & MORRIS 1983). If electrical potential measurements are taken both at the surface and downhole before, during and after the fracturing treatment and are thus compared, diagnostic information about the crack is obtained, and fracture azimuth can be inferred from the raw data.

6.2.2.3.2. Microseismic fracture monitoring

TESTER, MURPHY, GRIGSBY, POTTER & ROBINSON (1986) document microseismic fracture monitoring in geothermal reservoirs. Pressurization of the formation with injected fluids beyond the in-situ normal and frictional stresses holding the fracture faces together gives rise to shear slippage and creation of microseismic events that could be monitored (cf. section 4.7.7.). Examining of the microseismic events originating from either tensile or shear failure depending on the alignment of the in-situ horizontal principal stresses with the joint structure and on the level of anisotropy of the two principal horizontal stresses allows to develop reservoir geometry models. The primary objective of acquiring and analyzing borehole microseismic data is to determine hydraulic fracture azimuth (GRI 1988).

Microseismic events resulting from a hydraulically-induced crack are considered to be caused predominantly by shear failures induced by the open pressurized fracture and the localized high pore pressure zone surrounding the crack (THORNE & MORRIS 1987, 1988). Although these microseismic events are generally not thought to be caused by tensile failure as the hydraulic fracture propagates through the gas reservoir, they should occur in a tight band near the hydraulic fracture face which enables to use their location to infer azimuth, height and wing length of the crack (ALBRIGHT & PEARSON 1980; cf. section 6.2.2.1.2.2.). Microseismic event determination depends on detection of primary and secondary wave arrivals to compute the distance from the receiver to the pulse.

Thus in total azimuth, length and height of the fracture can be determined from oscillations of seismic energy trapped in a low-velocity zone surrounding the crack, and characteristics of fluid waves in the wellbore that are generated at the perforations by motions in the fracture outside the borehole. Crack azimuth and inclination can also be estimated from frequency-domain analysis of highly-polarized components of noise particle motions using the continuous microseismic radiation technique. The discussion as follows focusses on active and passive recording, shear and compressional wave registration, wave record interpretation and signal discrimination, shear-wave polarization and acoustic emission interpretation.

6.2.2.3.2.1. Active and passive recording

Passive microseismic monitoring focusses on mapping of events induced by pressurization during fracturing, heat extraction and overpressurization, whereas active techniques include detonator and piezo-electric acoustic sources in hole-to-hole transmission and reflection tests to study acoustic attenuation caused by the presence of water-filled fractures contained between wells. Results of the survey show that major production and injection zones occur near the intersection of a series of en-echelon secondary cracks with the wellbores, with the secondary fractures permitting lateral fluid communication between the primary vertical joints. In the absence of thermal stress cracking, reservoir growth due to heat extraction is really a thermal contraction effect, with the fractures expanding due to shrinkage of the rock surrounding the cracks by thermal drawdown. Microseismic event mapping during the early stages of fracturing to enlarge the active reservoir showed a distinct growth upward and outward from the injection region. Microseismic fracture monitoring aspects are also contributed by AKI, FEHLER et al. (1982) and ALBRIGHT & PEARSON (1982).

Recording of data is usually performed with a triaxial seismometer which is lowered into the borehole (DOBECKI 1983, GRIFFIN 1985). It contains three orthogonally mounted seismic detectors and is oriented using a coaxially mounted gyroscope. The orthogonal configuration of the detectors allows the seismic signals to be sorted into compressional and shear events, and the compressional events can be oriented to determine the direction of their source being the fracture plane with respect to the tool. If no seismic signals are received from the formation after the hydraulic stimulation job, this indicates that the fracture has already completely closed, with the walls no longer moving and thus no longer producing microseismic events. Very shortly after major fracturing operations, many events may be unusable due to excessive noise, weak signal or poor clamping effects giving rise to questionable tool orientation. In comparison to other techniques, the triaxial borehole seismometer has sometimes turned out to produce the most reliable and precise measurements of fracture azimuth, as it directly assesses the seismic disturbances emanating from the crack face.

SARDA, WITTRISCH, PERREAU & DEFLANDRE (1987) introduce a microseismic fracture monitoring system which enables determination of azimuth, length and height of hydraulically-induced cracks. Microseismic events are generated during opening and closing that occurs after the injection of treatment fluids. Fracture length can also be determined by interpretation of pressure buildup after the stimulation operation. Besides three accelerometers, the monitoring tool includes an inclinometer or magnetic compass to indicate instrument position. Interpretation starts with determination of the coordinates of the hypocenters of the microseismic events, and then a statistical and mechanical analysis of these coordinates is used to evaluate fracture development. Aspects of microseismic crack monitoring are also discussed by MAJER & DOE (1986), and SARDA, PERREAU & DEFLANDRE (1988) comment on acoustic emission interpretation for estimating hydraulic fracture extent.

Active borehole seismics is mainly performed after termination of the fracturing treatment when the crack has closed and is no longer creating microseismic events by wall movements. Microseismic signals recorded during pressure decline and flowback are interpreted to reflect rearrangement of the proppants within the fracture and therefore indicate propped crack length on the mapped distribution of events (WARPINSKI, BRANAGAN, SATTLER, CIPOLLA, LORENZ & THORNE 1988).

6.2.2.3.2.2. Shear and compressional wave registration

Microearthquake hydraulic fracture height, length and shape mapping by recording of shear and compressional waves with borehole geophones and surface seismometers is reported by MURPHY & FEHLER (1986). The distribution of microearthquake locations reflects a dendritic multiple-branch fracture pattern which is mainly a consequence of intense shear slippage along pre-existing natural fractures without creation of new cracks by actual rock rupturing due to tensile failure (cf. section 4.7.7.). Mapping of microearthquakes often does not document a single planar fracture as predicted by the conventional hydraulic treatment theory (HUBBERT & WILLIS 1957), but instead frequently depicts a zone of stimulation distributed through the hot dry rock volume of the geothermal reservoir that is abt. 0.8 km high, abt. 0.8 km wide and abt. 0.25 km thick, with the precision of microseismic event positioning being abt. 30 m. The first motions of microearthquakes and fault plane solutions determined by the surface array of seismometers indicate shear-slip along pre-existing joints, and tensile fracturing is considered to generate only very weak seismic signals that could not be detected by the surface recording devices (HOUSE, KEPPLER & KAIEDA 1985; MURPHY & FEHLER 1986). The predominance of shear slip movement is also reflected by a larger amplitude of the shear wave than that of the compressional wave.

In terms of detection of hydraulic fractures with borehole geophones, the recorded microseismic events are probably generated by shear slippage along pre-existing joints or bedding planes as fluid leakoff increases the pressure and reduces net normal stress (WARPINSKI & TEUFEL 1984; cf. section 4.7.7.). In gas reservoirs, the joints are probably filled mostly with gas, and the evacuated crack case is probably a good representation of the pressure distribution. Therefore the seismic signals are expected to lie along a narrow band adjacent to the fracture which corresponds to the geophysical survey results (HART, ENGI, FLEMING & MORRIS 1984). In geothermal reservoirs, the fluids in joint and treatment are the same and thus pressure coupling can extend far away from the hydraulic fracture if the joints are long or interconnected, giving rise to a cloud-like distribution of seismic events encircling the borehole (BATCHELOR 1984).

6.2.2.3.2.3. Wave record interpretation and signal distinction

The continuous recording of microseismic event data enables compilation of hodograms from cross correlation which exhibit crack azimuth orientation. Asymmetrical fractures can also be determined if one wing is horizontally longer than the other due to more vertical propagation of the other in contrast to preferential horizontal extension of the one. The possibilities of evaluation of the fracture depend on the type of waves which are recorded. If only the compressional wave is obtained, the assessment is restricted to crack azimuth orientation. If the shear wave is also recorded, fracture height determination can be done as well. Propagation distance calculation of the crack wings can also be made from the shear wave record. For fracture azimuth modelling, the geophone or accelerometer tools are preferentially lowered in nearby monitor wells and not directly into the borehole to be fractured itself, because in the latter environment, measuring instruments can only be inserted during minifracturing operations, but not during MHF treatments.

Natural fractures existing in the formation where the hydraulic crack is generated may disturb microseismic fracture azimuth mapping (DOBECKI 1983). Microseismic events associated with the hydraulic stimulation process normally occur near the plane of the vertical fracture, whereas natural cracks that do not intersect the hydraulic fracture can generate microseismic events along almost any radial direction away from the wellbore. Discrimination of signals coming from hydraulic and natural fractures has to include separation of compressional and shear waves and determination of their angle of polarization and difference in arrival time which can then be converted into distances from the source.

6.2.2.3.2.4. Shear-wave polarization

Other microseismic fracture detection methods are based on shear-wave polarization (SCHLUMBERGER 1987 a). As a consequence of their preferred orientation, fractures affect the two shear waves with approximately orthogonal polarization in a different manner, causing them to arrive at different times due to shear-wave splitting. Analysis of the arrival time difference leads to estimation of cross fracture density in the subsurface. Concerning shear-wave polarization, shear-wave particle displacement is in isotropic material normal to the direction of propagation, whereas in anisotropic rocks, particle displacement deviates from normal to the direction of propagation. This indicates the direction of anisotropy and in turn orientation and direction of the in-situ stress responsible for fracture alignment (CRAMPIN & ATKINSON 1985). Field examples of crack azimuth determination from shear-wave polarization and splitting based on measuring of delays between faster and slower split shear waves are presented by MARTIN (1987). Based on the azimuthal anisotropy concept, similar methods can also be applied to calculate fracture height and geometry (AHMED 1987).

6.2.2.3.2.5. Acoustic emission interpretation

Acoustic emission interpretation is quite similar to microearthquake evaluation, but can be applied on various scales not only comprising large-scale water injection in geothermal hot dry rock projects, but also small-scale minifracture calibration tests preceding main stimulation jobs in oil and gas fields (SARDA, PERREAU & DEFLANDRE 1988). The acoustic acitivity generated by hydraulic fracturing is mainly characterized by multiple burst emissions and can be detected in observation wells or in the treated borehole itself. An anchoring pressure being above the downhole pressure is necessary to get good acoustic coupling. Some aspects of acoustic event triggering by shear failure, natural crack solicitation by fluid injection, and refraction events and fracture solicitation are discussed as follows.

6.2.2.3.2.5.1. Acoustic event triggering by shear failure

The acoustic events are mainly triggered by shear failures induced by high pore pressure which occur on planes of weakness in the vicinity of the developed tensile fracture (ALBRIGHT & PEARSON 1980, PEARSON 1981, TALEBI & CORNET 1985). Prediction of fracture orientation before injection begins can be made by shear-wave polarization (ROBERTS & CRAMPIN 1986). Long-period events result from tensile opening of a fluid-driven crack and can be correlated with fracture dimensions and with rock and fluid characteristics (BAME & FEHLER 1986, CHOUET 1986). Evaluation of burst polarity being first motion shows that all the sensors have the same polarity when the crack propagates, with this polarity corresponding to a compression for the sensor and hence to a tensile fracture at the emitting point (MAJER & DOE 1986; SARDA, PERREAU & DEFLANDRE 1988).

6.2.2.3.2.5.2. Natural crack solicitation by fluid injection

Natural fractures can be solicited by the injection of a stimulation fluid, with then difficulties arising for identification of the nature of events emitted by hydraulic fractures and natural cracks (SARDA, PERREAU & DEFLANDRE 1988). There is generally a strong correlation between acoustic activity and pressure evolution. All the sensors have the same polarity when the fracture propagates which corresponds to a compression for the sensor and hence to a tensile fracture at the emitting point. Determination of the source direction can be easily done by the hodogram method practiced on selected signals. Other aspects of acoustic emission interpretation are discussed by SCHUCK & KEECH (1975); BATCHELOR, BARIA & HEARN (1983); TALEBI & CORNET (1985), MAJER & DOE (1986) and BOUTECA (1987). In case of microcracking by anelastical strain relaxation of cores, horizontal stress direction can be determined by acoustic velocity anisotropy evaluation (CLARK 1983).

6.2.2.3.2.5.3. Refraction events and fracture solicitation

Independent confirmation of the results of acoustic emission interpretation by evaluation of the geological structure of the reservoir and particularly the direction of natural fractures (LACY 1987) is an important cross-checking technique (SARDA, PERREAU & DEFLANDRE 1988). The proportion of refraction events is greater during renewed solicitation of the newly created crack by injecting further small fluid volumes in order to generate new acoustic pulses than during the main injection. The main frequency of the acoustic bursts often corresponds to the resonance frequency of the tool along its axis or to the resonance of the accelerometers which enables to detect sources located far away from the tool. Formation characteristics are important for the quality of the results, with the method being more effective when hydraulic stimulation is carried out in lower permeability reservoirs and/or in greater depth. The natural fractures in the pay zone have to be considered for estimating the possible occurrence of shear failures.

6.2.2.3.3. Combination of microearthquake and pressure pulse registration

FITZ-PATRICK, KARR & O`SHEA (1985) present a comprehensive fracture geometry and proppant mapping experiment consisting mainly of remote and borehole seismical and magnetometrical fracture monitoring methods. Seismical crack detection can be performed with low-frequency, high-frequency and long-period systems. High-frequency seismic monitoring attempts to detect microseismic activity associated with a propagating hydraulic fracture, with the microearthquakes originating by small shear motions associated with increased pore pressure in the vicinity of the fracture and the spatial distribution of the microseismic sources providing an approximate map of crack propagation as an expanding cloud of seismic events. Low-frequency seismic monitoring and associated high-resolution dynamic pressure recording of pressure variations or pulsations emanating from the fracture is scheduled to lead to seismic imaging of the crack using adaptive-beam forming techniques by combining the pressure data from the injections with the geophone data. Influences of pressure pulsing on seismic monitoring, possibilities of improvement, potential and limitations of seismic monitoring, anelastic strain recovery analysis, and other geophysical techniques are outlined as follows.

6.2.2.3.3.1. Influence of pressure pulsing on seismic monitoring

The principle of the method is that the pump surges cause periodical variations in fracturing fluid pressure and/or rate, thereby creating hydraulic waves which propagate into the fracture, and once being inside of the crack, these pressure variations vibrate the fracture walls, radiating seismic waves in a process analogous to a loudspeaker cone radiating acoustic waves (FITZ-PATRICK, KARR & O`SHEA 1985). Measuring free pressure oscillations by pulsing is based on the concept that the presence of the fracture will modify the water hammer or tube waves generated by these pulses, with the character of the oscillations and their amplitude decay rates being able to be related to injected fluid volumes and thus fracture geometry parameters. Long-period seismic monitoring aims on measuring of velocity and acceleration signatures of fracture azimuth extension in this passband, and additional information is achieved by combination with tiltmeters (cf. section 6.2.2.3.4.). It is much easier to perform retroactive measurements of hydraulic geometry and final propped geometry than accurately monitor actual proppant transport in a propagating fracture. Aspects of microseismic activity, mapping and fracture azimuth determination are also discussed by BATCHELOR, BARIA & HEARN (1983); CARDEN, NICHOLSON, PETTIT & ROWLEY (1983); HART, ENGI, MORRIS & FLEMING (1984); SMITH, REN, SORRELS & TEUFEL (1985); SARDA, WITTRISCH, PERREAU & DEFLANDRE (1987) and LOCKE, YOST, OVERBEY & GRABOWSKI (1988).

6.2.2.3.3.2. Possibilities of improvement

Fracture detection and performance evaluation is to some extent also possible by assessment of changes of fluid transmissibility during waterflooding experiments (MORALES, ABOU-SAYED, JONES & AL-SAFFAR 1986) and pressure interference tests (ELKINS & SKOV 1960). POWER, SCHUSTER, HAY & THOMAS (1976) comment on detection of hydraulic fracture orientation and dimensions, and determination of areal extent and direction of cracks is also analyzed by REYNOLDS, SCOTT, POPHAM & COFFER (1961). SMITH (1979) discusses some aspects of effect on fracture azimuth on well and field production. MARTIN & DAVIS (1987) outline shear-wave modelling of microfractures which is based on the concept of shear-wave birefringence in anisotropic media (CAMERON 1988).

On-site strain relaxation, differential strain analysis and sonic velocity evaluation are limited in significance by the fact that the measurements are taken on discrete core samples (GRIFFIN 1985). Variations of stress orientation throughout the interval are thus missed and can lead to large errors in the overall orientation estimates. Interference or pulse pressure testing is only applicable if the half-length of the hydraulically-induced cracks is at least exceeding half a well spacing. In terms of deviated boreholes, fracture pressure decline modelling can assist geophysical and logging techniques in determination of crack orientation relative to the azimuth of the inclined well at least in certain areas (KIM, CHAMPION & COOPER 1985).

CLEARY (1988) concludes that the present downhole geophone-based seismic mapping techniques provide only a vague image at all depths and may give a good estimate of fracture orientation. Desirable future innovation would be a wellbore tool (in order to avoid signal fading with depth) which truly responds to the outlines of the crack perimeter instead of giving fracture volume and orientation. Promising techniques include detailed interpretation of short-term pressure transients or surges, downhole transmission and sensing of Stoneley/Rayleigh stress waves reflected from the fracture perimeter, and use of instrumental casing.

6.2.2.3.3.3. Potential and limitations of seismic monitoring

The use of remote seismic methods which require surface-installed sensor systems to monitor hydraulic fracturing treatments at commercially interesting depths, however, is likely to be successful only in those special cases where the shear stresses in the formation undergoing treatment are much larger than those caused by the overburden load (SMITH, REN, SORRELS & TEUFEL 1985). Since this condition is not likely to occur on a widespread scale in oil- and gas-bearing formations, such methods cannot be expected to play a significant role in routinely determining azimuth and dimensions of hydraulic fractures. On the other hand, borehole seismic methods which use sensor systems deployed in the treatment well have the potential for providing reliable estimates of fracture azimuth on a routine basis and also estimates of other hydraulic crack parameters of interest.

Determination of fracture azimuth, length and height by microseismic event monitoring and interpretation with application of triaxial dynamic accelerometers is also presented by WITTRISCH & SARDA (1987). Propagation of a hydraulically-induced crack is usually accompanied by mechanical vibrations or seismic events in the acoustic frequency range. Fracture development determination is performed by statistical and mechanical analysis of the microseismic hypocenter coordinates. Excessive noise can be a serious problem of borehole seismic fracture monitoring and in unfavourable situations only a few analyzable seismic signals are obtained (WARPINSKI, BRANAGAN, SATTLER, LORENZ, NORTHROP, MANN & FROHNE 1985).

Concerning fracture azimuth determination, distinction can be made between direct observation utilizing inflatable impression packers (cf. section 6.2.6.) and borehole televiewer (cf. sections 6.2.1.9. and 6.2.1.10.1.), indirect measurements comprising triaxial wellbore seismometer, and predictive inference including borehole ellipticity monitoring from caliper logging (cf. section 6.2.1.8.), on-site strain relaxation (cf. section 6.2.2.3.3.4.), differential strain analysis, sonic velocity assessment, and fault and structural mapping (cf. section 6.2.7.). Aspects of seismical mapping of geometry and orientation of hydraulic fractures are also discussed by HART, ENGI & MORRIS (1983); HART, ENGI, FLEMING & MORRIS (1984); TEUFEL, HART, SATTLER & CLARK (1984) and THORNE & MORRIS (1987, 1988). Another microseismic technique of fracture azimuth orientation interpretation is frequency-domain analysis of highly-polarized components of noise particle motions (OWEN, TORONTO & PETERSON 1988).

6.2.2.3.3.4. Anelastic strain recovery analysis

Determination of minimum horizontal stress, maximum horizontal stress and fracture azimuth by differential strain curve analysis is based on microcracks resulting from stress relief which occurs during the coring process (BEGNAUD & CLAIBORNE 1985, TEUFEL 1988). Comments are offered as follows on core evaluation for anelastic strain recovery, reliability of horizontal in-situ stress record, and comparison of different strain measuring methods.

6.2.2.3.3.4.1. Core evaluation for anelastic strain recovery

Core expansion is related to the change in stresses, and as laboratory reloading of the core is performed, the strain in several directions is measured and then related to stress conditions downhole. In anelastic strain recovery or expansion, the direction of maximum horizontal core recovery after its retrieval from the subsurface indicates the direction of maximum horizontal stress (LAUBACH & MONSON 1988). The accuracy of differential strain curve analysis is limited in instances where the rock has experienced stress-induced microcracks over geological time, and is enhanced at greater depths due to an increased tendency to develop microcracks. Differential strain curve analysis is based on anisotropic contraction of core when hydrostatically loaded in the laboratory, with the direction of maximum core contraction being inferred to represent the direction of maximum horizontal stress. Wellbore breakouts develop parallel to minimum horizontal stress.

6.2.2.3.3.4.2. Reliability of horizontal in-situ stress record

Anelastic strain analysis is a technique of determining directions and ratio of maximum and minimum horizontal strains (stresses) from oriented core and thus predict fracture azimuthal direction (TEUFEL 1981). When the ambient in-situ stress that exists at depth is reduced appreciably by coring a section of rock, relaxation of the locked-in grains occurs. Total strain relaxation consists of an instantaneous recovery of elastic strains and a continued time-dependent (creep) relaxation of the rock core that tends asymptotically to a final permanent set. These relaxations are measured along known directions of an oriented core and can then be used to calculate the principal directions of strain relief.

Strain relaxations are assumed to be uniform in time, thereby suggesting that strain orientations determined from a segment of the time-dependent portion of the strain relaxation curve may provide an accurate evaluation of the horizontal main in-situ stress directions. Strain relaxation results from microcracking of the core after the in-situ stress is relieved (CLARK 1983, TEUFEL 1988). OWEN, TORONTO & PETERSON (1988) comment on reliability of anelastic strain recovery estimates for stress orientation. Analysis of anelastic strain recovery data provides estimates of the relative magnitudes of the two horizontal stresses in sandstones, whereas those in mudstones are essentially isotropic and on the average slightly below the overburden value (WARPINSKI & TEUFEL 1987). Aspects of fracture determination and in-situ stress measurements by anelastic strain recovery evaluation are also discussed by TEUFEL (1981, 1982, 1983), TEUFEL & WARPINSKI (1984), EL RABAA & MEADOWS (1986) and WARPINSKI & TEUFEL (1986).

6.2.2.3.3.4.3. Comparison of different strain measuring methods

OWEN, TORONTO & PETERSON (1988) comment on the reliability of anelastic strain recovery estimates for stress orientation. Comparative evaluation of various independent techniques being orientation of natural fractures and microcracks, coring-induced petal and petal-centerline fractures (cf. section 6.2.1.10.1.3.), fluid inclusion planes, and stress-test cracks, and borehole ellipticity (cf. section 6.2.1.8.), three-dimensional microseismic surveys, differential strain analysis and anelastic strain recovery have revealed that the results of anelastic strain recovery evaluation are consistent with those from the other independent methods which predict or measure orientation of natural and induced fractures. Pre-existing natural joints or microfissures in recovered oriented core samples may influence horizontal stress directions recorded by anelastic strain expansion of the core. Anelastic strain recovery assessment is advantageous if stress orientation estimates are needed immediately after coring.

6.2.2.3.3.5. Other geophysical techniques

Other, more complicated means of fracture azimuth predictive modelling are directional analysis of the geometrical fabrics in oriented cores by anelastical strain and creep recovery (LOGAN & TEUFEL 1978; TEUFEL 1981, 1982, 1983; BLANTON 1983, BLANTON & TEUFEL 1983, REN & HUDSON 1985), differential strain evaluation (SIMMONS, SIEGFRIED & FEVES 1974; SIEGFRIED & SIMMONS 1978, REN & ROEGIERS 1983, STRICKLAND & REN 1983, WARPINSKI & TEUFEL 1987), differential thermal expansion analysis (REN & ROEGIERS 1983), solid earth tidal strain evaluation (OWEN 1982) or core recompression (WITTRISCH & SARDA 1987).

Further fracture azimuth mapping techniques of indirect predictive inference nature are surface tilt measurements (WOOD, POLLARD & RALEIGH 1976; SMITH, LOGAN & WOOD 1980; DAVIS 1983, EVANS 1983, EVANS & HOLZHAUSEN 1983; HOLZHAUSEN, CARD, RAISBECK & DOBECKI 1985; cf. section 6.2.2.3.4.), electrical potential recording (POWER, SCHUSTER, HAY & TWOMBLEY 1975; SMITH, LOGAN & WOOD 1980), pressure transient analysis (CINCO-LEY & SAMANIEGO 1977), interference and pulse testing (PIERCE, VELA & KOONCE 1974; CINCO-LEY & SAMANIEGO 1977; EKIES, HADINOTO & RAGHAVAN 1977; URAIET, RAGHAVAN & THOMAS 1977; ABOBISE & TIAB 1982, KAMAL 1982, EVANS & HOLZHAUSEN 1983, KAMAL & HEGEMANN 1988), sonic velocity analysis (LACY 1984), and fault and structural mapping (CLARK 1983; TEUFEL, HART, SATTLER & CLARK 1984; cf. section 6.2.7.) or fault-plane solutions (BUCHANAN 1987).

EVANS & ENGELDER (1986) outline three-dimensional stress mapping by microfracturing. FORGOTSON, FORGOTSON & BROWN (1980) illustrate prediction of hydraulic fracture orientation, and TYLER & VOLLENDORF (1975) document mapping of cracks resulting from hydraulic fracturing for in-situ stress measurements.

6.2.2.3.4. Surface tiltmeter measuring

Fracture azimuth mapping can also be performed by surface tiltmeter measuring (SMITH 1979, WOOD 1981, LACY 1984). After discussion of some general aspects, anomalous tiltmeter readings as well as first-order levelling are briefly outlined.

6.2.2.3.4.1. General aspects

Tiltmeters indirectly measure the perturbations in underground deformation fields (such as caused by hydraulic fracturing) by the induced changes in formation tilt on or near the surface (SMITH 1979, WOOD 1981, LACY 1984). Using an array of continuously recording shallow borehole tiltmeters and a suitable inversion technique enables to infer orientation and extent of fractures. Depth resolution can be improved by using an array of at least 12 - 18 surface tiltmeters and the inclusion of the influence of surface noise effects such as vegetation, precipitation and temperature. The tilt signals associated with fracture propagation in greater depth are very small and are difficult to extract when compared to earth tides, surface and site-dependent effects, with the accuracy being considerably improved by the independent readings of numerous individual instruments in the tiltmeter chain. Such tiltmeter arrays can be used for mapping and monitoring of fracture growth dynamics, because they are sensitive enough to detect characteristic surface deformation patterns which relate to fracture orientation and geometry (SUN 1969, POLLARD 1976, POLLARD & HOLZHAUSEN 1979). While tiltmeters cannot be used offshore (BUCHANAN 1987), the current onshore tiltmeter technology allows reliable interpretation in most cases only for shallow depth (CLEARY 1988).

The physical quantity measured at each tiltmeter site of an array of tiltmeters is the spatial gradient of vertical displacement (HART, ENGI & MORRIS 1983). Assuming a homogeneous crust, the measured surface deformation field produced by a fracture is a function only of crack geometry and the Poisson`s ratio of the medium, and the surface tiltfield is dependent only on the well displacements defining the fracture once Poisson`s ratio is known (EVANS, HOLZHAUSEN & WOOD 1982).

Aspects of fracture determination by tiltmeter and borehole seismical methods are also discussed by WOOD, POLLARD & RALEIGH (1976); SMITH, HOLMAN, FAST & COVLIN (1978); SMITH, LOGAN & WOOD (1980); NOLTE & SMITH (1981), DAVIS (1982); EVANS, HOLZHAUSEN & WOOD (1982); DAVIS (1983), EVANS (1983), EVANS & HOLZHAUSEN (1983); WOOD, PARKIN, YOTAM, HANSON, SMITH, ABBOTT, COX & O`SHEA (1983); FITZ-PATRICK, KARR & O`SHEA (1985); HOLZHAUSEN, CARD, RAISBECK & DOBECKI (1985); SMITH, REN, SORRELS & TEUFEL (1985); KULAR, CHHINA, BEST & MACKENZIE (1988); OVERBEY, YOST & WILKINS (1988) and PALMER & VEATCH (1988). Surface deformation evaluation for fracture determination is also outlined by WOOD, POLLARD & RALEIGH (1976); FAST, HOLMAN & COVLIN (1977); SMITH, LOGAN & WOOD (1980) and EVANS, HOLZHAUSEN & WOOD (1982).

6.2.2.3.4.2. Anomalous tiltmeter readings

Some anomalous tiltmeter readings include determination of both primary and secondary azimuth of a large-scale fracture in contrast to corresponding cracks in the same field which have been created by smaller treatments and exhibit only the primary system in the tiltmeter survey. The secondary fracture indication can be interpreted as reflecting the opening of natural cracks and fissures which result in accelerated fluid-leakoff depletion and increase the risk of premature screenout failure of the stimulation operation (NOLTE 1988 b). Drawback of tiltmeters which detect short range or sudden episodes are the requirement of several weeks of baseline data and a large number of instruments in appropriate arrays (SINGHAI & CARD 1988). While tiltmeters end usually up with high-quality results, the costs involved are often excessive compared with the value of the results obtained. Minor vertical fracturing events may be too sluggish to be picked up by tiltmeters which work best with abrupt changes in surface tilts.

6.2.2.3.4.3. First-order levelling

In contrast to tiltmeters which monitor the changes in tilt of near-surface beds induced by underground deformations, first-order levelling attempts to measure the resultant elevation changes of the ground surface (SINGHAI & CARD 1988). In ideal circumstances, tell-tale troughs or radially symmetrical swelling could be expected, but because of extraneous influences such as inhomogeneity of deformation properties and differential ground settling, it may not always be possible to obtain highly-reliable interpretations of tiltmeter data. For monitoring elevation changes, several permanent monuments of helical screw type have to be installed around the wellbores. A more sophisticated version of first-order levelling is laser ranging.

6.2.2.3.5. Hydraulic impedance analysis

The basic principle of the hydraulic impedance method is that by opening and closing a bleed-off valve in a pressurized wellbore, a downgoing pressure wave is initiated (GRI 1986, 1988). This wave reflects from the bottom of the borehole back to the surface where it is reflected downwards again until this free oscillation of wellbore pressure is damped by fluid friction and energy losses into the fracture. Some aspects of free oscillation period and crack growth as well as fracture compliance vs. impedance magnitude are discussed as follows.

6.2.2.3.5.1. Free oscillation period and fracture growth

The fracture is the key element in causing reflections as the wave passes by, and therefore if the downhole wellbore geometry is known prior to stimulation, a numerical borehole model can prognose the resulting free oscillation that would occur. By adding an element which models the hydraulic behaviour of a fracture to the wellbore model, free oscillation waveforms can be predicted in a borehole with a fracture of any dimension. Matching the model forecasts with data recorded after stimulation leads to determination of a best fit of the data sets, with the crack dimensions represented by the matching model then being used as an estimate of the actual fracture dimensions.

The growth of a hydraulic crack increases the period of free oscillations in a well, and simultaneously, the decay rate of free oscillations decreases (HOLZHAUSEN & GOOCH 1985). The properties of forced oscillations in a borehole also change during fracture growth. Crack impedance can be determined directly by measuring the ratio of downhole pressure and flow oscillations, or indirectly from wellhead measurements using impedance transfer functions. Because impedance is a function of fracture dimensions and the elasticity of the surrounding rock, crack impedance analysis enables fracture geometry evaluation based on the dynamics of wave propagation in the well and the effect which the hydraulic crack has on oscillatory pressures and flows. Therefore fracture impedance analysis is the logical extension of earlier pressure interpretation techniques for hydraulic crack assessment.

6.2.2.3.5.2. Fracture compliance vs. impedance magnitude

Continuous monitoring of impedance during hydraulic stimulation also reveals whether fracture growth is dominated by radial expansion with uncontrolled and unrestricted vertical propagation, or whether vertical crack containment is established and maintained by either adjusted treatment conditions or sufficient confining capacity of boundary seals. The high compliance of an internally pressurized fracture leads to high capacitance and low impedance that result in doubling of the period of free oscillations as observed after breakdown. Because compliance changes as the crack grows, it produces alterations in oscillatory pressure behaviour regardless of the details of hydraulic fracture geometry. As crack impedance is dominated by wavespeed, impedance changes can be detected in both open-hole and perforated-casing completions.

6.2.2.3.6. Hydraulic resonance analysis

The objective of hydraulic resonance research is to develop a real-time technique for determining fracture length (GRI 1986, 1988). Pressure sensors are deployed to measure surface casing and tubing pressures for data acquisition. During the pumping process, these pressure signals are recorded as are tachometer signals from all the pumps. Each pump piston stroke causes a pressure pulse to occur at the outlet of the pump, and an aggregation of pumps simultaneously acting causes a complicated pattern of pressure pulses to be applied to the pump discharge line, giving rise to a pumping signal with a large bandwidth character that is detected by the annular pressure sensor.

The pressure pulses travel as acoustic waves in the discharge line down the annulus, through the perforation tunnels, and out into the induced fracture. The waves in the crack reflect from irregularities and from the ends of the fracture wings to create echoes which re-enter the borehole through perforation tunnels. Once again in the wellbore, these waves travel up the annulus and up the tubing, thereby allowing their detection by the tubing pressure sensor. Noting the arrival times of the various reflections, it is possible to identify the reflections in the fracture and to distinguish them from reflections in the borehole. With knowledge of acoustic speed in wellbore and crack, the arrival times can be used to calculate fracture length.

6.2.2.4. Distinction of vertical and horizontal fractures

Tiltmeter monitoring of hydraulic fractures is based on the earth responding more or less elastically to the opening of the crack, with the surface of the earth deforming in a predictable manner (DAVIS 1983; SMITH, REN, SORRELS & TEUFEL 1985; cf. section 6.2.2.3.4.). Tiltmeter fracture recording and evaluation works very well in shallow depth (WOOD, POLLARD & RALEIGH 1976; WOOD, PARKIN, YOTAM, HANSON, SMITH, ABBOTT, COX & O`SHEA 1983), whereas the success in greater depth is limited (LACY 1987). Examples of tiltmeter recordings indicating original horizontal fracture propagation and changing orientation to accomodate a vertical or inclined component later in the job are reported by HANSON (1988) and SCHRAUFNAGEL & LAMBERT (1988). The possibilities of distinction of vertical and horizontal fractures is outlined as follows along the lines of tiltmeter surveys, reservoir fluid flow modelling, dual laterolog curve separation, and combination of horizontal and vertical fractures.

6.2.2.4.1. Tiltmeter surveys

Monitoring of the surface deformation field with high-resolution surface and borehole tiltmeters that record rotations in two perpendicular vertical planes allow distinction of vertical and horizontal fractures that are generated by hydraulic stimulation treatments (HOLZHAUSEN, CARD, RAISBECK & DOBECKI 1985; cf. section 6.2.2.3.4.). A radial symmetrical uplift of the ground surface (as reflected by the tiltmeters rotating radially away from the wellbore) can be interpreted as reflecting the formation of predominantly horizontal fractures centered roughly around the borehole. Backwards tiltmeter rotation radially towards the wellbore indicates subsequent collapse of the horizontal crack (WOOD, PARKIN, YOTAM, HANSON, SMITH, ABBOTT, COX & O`SHEA 1983). In contrast, growth of vertical or steeply-dipping hydraulic fractures produces a trough at the surface that is parallel to crack strike. Distinction of vertical and horizontal fractures by surface deformation field measurements is also carried out by EVANS, HOLZHAUSEN & WOOD (1980); DAVIS (1983), EVANS (1983) and EVANS & HOLZHAUSEN (1983). Estimation of crack radius as a function of elastic properties can be performed by modelling a penny-shaped horizontal fracture by a planar discontinuity in the displacement field. If the crack is horizontal, its areal extent can be inferred, while if the fracture is vertical, its azimuth can be estimated (HART, ENGI & MORRIS 1983).

KULAR, CHHINA, BEST & MACKENZIE (1988) report an example where the early proppant-free stage of a stimulation treatment in a shallow heavy-oil reservoir is reflected by tiltmeter readings as propagating horizontally, while the survey of the late proppant-laden stage testifies to further vertical extension. Comparison of the entire stimulation operation with the theoretical tilts indicates multiple fracture geometry, with both vertical and horizontal crack components increasing during proppant injection. The main fracture is thus initially horizontal and changes its orientation to vertical at the beginning of proppant insertion (aspects of combinations of horizontal and vertical fractures are also discussed in sections 1.2.8.4., 4.4.3.4.3., 4.9.4.2. and 6.2.2.4.4.).

6.2.2.4.2. Reservoir fluid flow modelling

Distinction of vertical and horizontal fractures can also be performed by modelling reservoir fluid flow. SINGHAI & CARD (1988) illustrate an example from steam-soak stimulation of heavy oil sands where matrix flow is characterized by a slow change in salinity (from that of steam condensate to interstitial water), whereas severe water coning or vertical cracks would be highlighted by a rapid transition and horizontal fractures would be indicated by a salinity change in between the two (GESHELIN, GRABOWSKI & PEASE 1981). By comparing data on watercuts of effluents and on salinity changes, it is possible to infer the dominant fluid-flow mechanism within the reservoir (whether taking place through matrix or fractures or through communications with water-bearing formation).

Analyses of injection pressure and flow-rate charts provide additional clues regarding existence of fractures (SINGHAI & CARD 1988). If flow rates and pressure remained steady and the latter is less than vertical depth stress, an absence of fracturing is indicated which in some cases is also confirmed by a multiple-injection-rate/pressure relationship. The presence of a vertical crack is indicated by an injection pressure exceeding the vertical depth stress and then showing a significant pressure drop followed by a fairly rapid increase to a value near pre-breakdown pressure. For horizontal fractures, the minimum breakdown pressure would be close to the vertical depth stress, and both pressure drop and recovery would be much more gradual in comparison to short events associated with vertical fractures.

6.2.2.4.3. Dual laterolog curve separation

Other possibilities of distinction of vertical and horizontal fractures include orientation of separation of shallow and deep dual laterolog curves (cf. section 6.2.1.10.3.3.), with in case of vertical cracks the deep curve reading higher than the shallow one and for horizontal fractures the deep curve reading lower than the shallow one (SIBBIT & FAIVRE 1985; cf. section 6.2.1.10.5.2.).

6.2.2.4.4. Combination of horizontal and vertical fractures

Mushroom-type fractures consisting of a primary vertical crack topped or bottomed by a secondary horizontal fracture like cap or stand, respectively, occur occasionally in depth ranges that are transitional for both basic fracture types (FITZ-PATRICK, KARR & O`SHEA 1985), as reflected by tiltmeter evaluation and bottomhole pressure analysis (cf. section 6.2.2.3.4.). An alternative interpretation of this feature, however, is the tilt being caused by fluid leakoff during injection spreading rapidly through the high-permeability upper reservoir interval causing swelling of the pay, with the radially symmetrical deformation in this case being an effect of pore pressure diffusion (SMITH, REN, SORRELS & TEUFEL 1985). Creation of a horizontal fracture sitting above the vertical crack is only possible if the injection pressure exceeds the overburden stress and thus lifting of the cover strata becomes possible (aspects of composite fractures consisting of horizontal and vertical elements are also outlined in sections 1.2.8.4., 4.4.3.4.3., 4.9.4.2. and 6.2.2.4.1.).

WOOD, PARKIN, YOTAM, HANSON, SMITH, ABBOTT, COX & O`SHEA (1983) describe formation of such mushroom-type fracture systems consisting of a primary vertical crack and a secondary horizontal fracture which broke out from the top of the vertical crack. T-shaped fractures consisting of horizontal and vertical components also originate in coal seams bounded by shales due to abnormally high treating pressures (JEU, LOGAN & McBANE 1988; cf. section 4.4.3.8.). Mushroom-type fractures are also reported by EVANS, HOLZHAUSEN & WOOD (1982).

6.2.2.5. Fracture permeability estimation

In addition to fracture geometry modelling by determination of height, azimuth and length by well logging or seismical methods, important characterization of natural and/or hydraulic cracks can be made by permeability estimation (PAILLET, HSIEH & CHENG 1987). Fracture permeability is mainly assessed by analysis of tube-wave-amplitude attenuation in full-waveform acoustic logs recording viscous fluid dissipation in crack openings (PAILLET 1980, 1983, 1985; HARDIN, CHENG, PAILLET & MENDELSON 1987) and interpretation of tube waves generated in vertical seismic profiling making use of fluid pulses in the borehole produced by seismic waves intersecting the permeable fracture (LEVINE, CYBRIWSKY & TOKSÖZ 1984; HARDIN & TOKSÖZ 1985; HARDIN, CHENG, PAILLET & MENDELSON 1987).

Independent information on fracture permeability can be obtained from packer-isolation-flow tests with pumping or injection into an interval isolated by packers for registering fracture flow into or away from the isolated section (DAVISON, KEYS & PAILLET 1982), and crosshole pumping tests using a slow-velocity heat-pulse flowmeter which identifies inflow and outflow during pumping (HESS 1986; PAILLET, CHENG, HESS & HARDIN 1986; PAILLET & HESS 1986; PAILLET, HESS, CHENG & HARDIN 1987). The existence of permeable fractures is also indicated by acoustic reflectivity imaging of the borehole wall by television (ZEMANEK, CALDWELL, GLENN, HOLCOMB, NORTON & STRAUS 1969; KEYS 1979; PAILLET, KEYS & HESS 1985; cf. section 6.2.1.9.) and conventional well log response to fractures (KEYS 1979; NELSON, MAGNUSSON & RACHIELE 1983; cf. section 6.2.1.10.).

6.2.3. Magnetometrical methods

Mapping of transport and final distribution of magnetized proppants within a hydraulic fracture can be performed by measuring with superconducting quantum interference magnetometers laid out at the surface and high-sensitivity borehole fluxgate magnetometers placed in observation wells very near the anticipated strike of the fracture (FITZ-PATRICK, KARR & O`SHEA 1985). Magnetite of various mesh size can be used as a suitable proppant and is magnetized just downstream of the pump trucks. Magnetite, however, is heavier than all the traditional proppants, has different frictional properties than quartz sand and ceramic proppants, and therefore behaves different in the proppant transport system. The weight difference can even give rise to magnetite fingering through quartz sand in the hopper. Surface magnetometer methods are a viable research tool for magnetic proppant mapping in shallow depth, but have very limited effects or even do not work at all in deeper reservoirs.

WOOD, PARKIN, YOTAM, HANSON, SMITH, ABBOTT, COX & O`SHEA (1983) document fracture proppant mapping by using a surface superconducting magnetometer. A surface array of sensitive magnetometers was deployed to map a shallow horizontal crack plugged by a magnetically tagged proppant. Magnetometer survey was supported and improved by tiltmeter recording (cf. section 6.2.2.3.4.), borehole television (cf. section 6.2.1.9.), and stimulation pressure/flow data analysis (cf. section 6.2.4.2.), and all the methods applied revealed a complex vertical-horizontal fracture formation process. Highly-sensitive broad-bandwidth cryogenic superconducting magnetometers detect the anomalous spatial magnetic field of a magnetically tagged proppant.

6.2.4. Stimulation data analysis

Another possibility of fracture treatment effectivity check is stimulation data analysis (pressures, injection rates, densities etc.) including pre- and post-job pressure transient interpretation and logging, with currently in the North Sea in Europe being much emphasis put on treatment evaluation leading to improved design and execution (ELPHICK 1985). After the discussion of some aspects on significance of treatment parameter evaluation, some points of fracture pressure interpretation and screenout prediction, computerized real-time job supervision, and quality control are outlined.

6.2.4.1. Significance of treatment parameter evaluation

From the assessment of the operating parameters, fluid characteristics such as friction gradients and leakoff coefficients can be calculated, and the response of various formations to fluids can be controlled. The following outline illustrates some aspects of optimization of current and future jobs as well as simulation vs. field evidence.

6.2.4.1.1. Optimization of current and future jobs

The ability to obtain fluid-loss parameters from both main and miniature fracturing treatments support design of a subsequent job in the same well or stimulation operations in nearby boreholes having the same formation characteristics (TAN, McGOWEN, LEE & SOLIMAN 1988). Real-time analysis capabilities enable simultaneous execution and evaluation, thus allowing to make design decisions during the treatment. Analysis of bottom-hole fracturing pressure (NOLTE & SMITH 1979) can indicate the mode of propagation and may give evidence of adverse conditions such as uncontrolled height growth or screenout early enough to modify the pumping schedule. Fluid leakoff characteristics can be calculated through regular pressure decline analyses.

Techniques of measuring parameters affecting fracture design during the treatment and crack height determination from analysis of the fracturing pressure are discussed by PEARSON, ABOU-SAYED & NGUYEN (1985). Aspects of on-site quality control of stimulation operations are also reported by HALL & LARKIN (1986) and JOPLING & KETCHER (1987). HOLDITCH, ROBINSON & WHITEHEAD (1986) comment on the significance of pre- and post-fracturing formation evaluation. The support of real-time monitoring and analysis of hydraulic fracturing operations by computerized field systems is outlined by CLEARY, BUHARALI, CROCKETT & SALEHI (1986); CROCKETT, OKUSU & CLEARY (1986) and CLEARY (1988). WEIGHTMAN & ROBERTS (1987) present a suitcase-size data-recording system for stimulation operations improvement, and BUHARALI, WRIGHT & WILLIS (1987) introduce real-time hydraulic fracture treatment supervision with a portable microcomputer.

PARKER & McDANIEL (1987) present some aspects of fracturing stimulation design improvement by conductivity measurements under in-situ conditions. PALMER & VEATCH (1987) discuss abnormally high fracturing pressures in step-rate tests. PEARSON (1987) comments on development and application of an operator`s stimulation monitoring system.

6.2.4.1.2. Simulation vs. field evidence

Although fracture simulators are very useful aids for treatment design selection, they provide but a simplistic look at the fracturing process (CRAMER & SONGER 1988). In some complex downhole environments, fracture simulators can furnish misleading information. Therefore it is always important to obtain post-treatment feedback from all available sources such as actual jobs, well logs, pressure transient tests, and production/injection histories so that projections can be compared with actualities. In this way, the usefulness of fracture simulators can be evaluated and treatment options may be selected with greater confidence.

6.2.4.2. Fracture pressure interpretation and screenout forecast

An important part of stimulation data analysis is fracture pressure interpretation in order to determine geometrical and dynamical crack parameters from the pressure decline behaviour after the treatment (NOLTE 1979, 1988 a, 1988 b; NOLTE & SMITH 1981). The mode of fracture propagation can be estimated from the excess pressure evolution. Pressure decline analysis is based on a constant relation between pressure and crack volume, and in each case, it is important to determine the validity of this assumption for a formation with varying properties. Fracture treating pressure interpretation permits identification of periods of confined-height extension, uncontrolled height growth, and recognition of reservoir critical pressure which may lead to significant reduction of lateral fracture extension probably followed by screenout or undesired crack height growth (MENG & BROWN 1987). The following account presents aspects of pressure evolution and screenout prevention, pressure evolution and fracture extension, fracture height and reservoir thickness, minifracturing evaluation and three-dimensional simulation, fracture pressure and geotectonical setting, and prevention of pressure-out situations.

6.2.4.2.1. Pressure evolution and screenout prevention

Screenout failure represents premature achievement of the situation that the carrier fluid can no longer transport the proppants towards the lateral-distal tip of the fracture because of proppant bridging, increasing fluid viscosity due to excessive filtrate leakoff and slurry dehydration, or thermal degradation of the fracturing fluid (ECONOMIDES 1987 b; cf. sections 4.2.2.2.3., 4.2.2.3.2.2., 4.2.3.2.5. and 4.8.2.3.4.). Some aspects of screenout types, occurrences, indications and remedy actions, avoidance of screenout failure by pumping rate and volume adjustment, high- and low-stressed zones, other reasons of abnormally high treatment pressures, and screenout forecast possibilities are discussed as follows.

6.2.4.2.1.1. Screenout types and occurrences

Distinction can be made first of all between fracture screenout when the material becomes stuck within the crack itself, and wellbore screenout when immobility blocking occurs in borehole and surface equipment such as pump, casing or tubing, screen, crossover tools, perforations and other devices (ECONOMIDES 1987 b). Other reasons of premature job termination can be related to equipment failure, tubular goods failure, flowline or wellhead leakage, weather, and erroneous treatment design (PAI, GARBIS & HALL 1983). If the operation has to be shut down for any reason other than screenout, the well should be flushed whenever possible and the proppants should be pumped into the formation instead of risking their settling in the borehole.

If the problem is minor and can be corrected in a short time, pumping should continue as per original schedule, because in case of flushing and later resuming of the treatment, there will be a volume of fluid pumped into the fracture system that contains no sand and consequently will allow that portion of the crack to close due to lacking proppant support (PARKER 1980). Aspects of prediction, prevention and remedy of screenouts in hydraulic fracturing are also discussed by PAI (1982). Discrimination is also possible between tip screenout and dehydration screenout (WARPINSKI, BRANAGAN, SATTLER, CIPOLLA, LORENZ & THORNE 1988; cf. also sections 4.2.2.2.3., 4.2.2.3.2.2., 4.2.3.2.5. and 4.8.2.3.4.) which are briefly characterized as follows.

6.2.4.2.1.1.1. Tip screenout

Tip screenout originates by travelling of proppant-laden fluid stages along the crack at a rate faster than that of fracture propagation, with thus proppants or solid fluid-loss particles finally reaching the crack tip and plugging it. This contact of proppants and fracture tip which is also enabled by high leakoff rates of the preceding pad or by fingering of the proppant stages through the pad as a consequence of density differences or excessive pumping pressures is the reason for stopping fracture propagation and provoking the tip screenout, with further continuing pumping then only resulting in widening or ballooning of the crack. In some special applications of creation of wide and short cracks, tip screenout fracturing can therefore be applied as a designed technique of premature termination provocation in order to achieve the desired ballooning of the crack (SMITH, MILLER & HAGA 1987; cf. section 4.5.4.4.1.).

6.2.4.2.1.1.2. Dehydration screenout

Dehydration screenout occurs when high fluid-loss rates in naturally jointed reservoirs cannot be sufficiently decreased by solid fluid-leakoff additives (cf. sections 4.8.8.3.1.2. and 4.8.8.3.2.3.) and consequently proppant concentration increases up to a point where the high-viscosity slurry can no longer move due to proppant bridging and/or fluid sticking, thereby triggering the dehydration screenout (cf. section 4.2.2.3.2.2.). It is often difficult to avoid the latter type of premature treatment failure, because high saturations of fluid-loss additives such as 100 mesh sand would effectively limit fluid leakoff and thus prevent dehydration screenout, but would totally ruin the result of the stimulation operation by complete destruction of proppant conductivity (cf. section 1.4.11.2.1.). On the other hand, insertion of only minor amounts of 100 mesh sand that do not damage proppant conductivity beyond subordinate and negligible levels is not sufficient to combat fluid leakoff and therefore creates a high risk of dehydration screenout. An appropriate intermediate solution could at least in some cases be the combined fluid-loss and breaker system which upon disaggregation of solid polyesters into soluble monomers does not leave any damaging materials and results in subsequent dissolution and removal of the previously deposited filter cake (CANTU & BOYD 1988; cf. section 4.3.4.6.4.).

Screenouts may also occur by overriding or fingering of the injected carrying fluids with accompanying proppant fallout at elevated temperatures without total movement of the heavier proppant concentrations (WEEKS 1979). Excessive proppant settling either in pipe or in fracture during pumping can also give rise to premature screenout termination of the job (SHAH, SMITH & DONALDSON 1983). Higher treating pressures than usual can be the result of viscous flow through a complex multiple crack system instead of a singular planar fracture (MEDLIN & FITCH 1983, WARPINSKI & TEUFEL 1984) or can be triggered by a layer of relatively small cracks around a large hydraulic fracture which provokes a significant increase of the apparent crack energy with fracture growth and thus increase of operational pressures (SHLYAPOBERSKY 1985). It is therefore desirable to study pressure screenouts not only as sandout phenomena, but also as features of crack layer evolution during hydraulic fracturing treatments.

6.2.4.2.1.2. Screenout indications and remedy actions

Blender failures during pumping can provoke screenout termination of the job or changes in fluid rheology that may detrimentally alter the way the formation fractures from that point on (PARKER 1980). Failure of the crosslinking mechanism of the treating fluid is usually indicated by pressure drop and can result in screenout termination, because proppant transport capabilities are diminished and fracture width is reduced with decreasing viscosity. Significant pressure changes caused by the formation are usually the result of stimulating out of zone or screenout. Fracture growth out of zone is usually accompanied by sharp pressure drop because the confining barriers have failed, and with increased crack height, fracture width is decreased if the injection rate remains constant, with screenout failure being able to occur because the reduced crack width may give rise to proppant bridging (McMECHAN & CONWAY 1983). Sudden pressure increases usually indicate a screenout near the wellbore as a consequence of pumping of unusually high proppant concentrations or viscosity loss due to crosslinking failure (PARKER 1980).

These sudden pressure increases bear only little warning time and minimal remedial action that can be taken. If a near-wellbore screenout occurs early in a treatment or at a point where the fracture design would be critically compromised, the borehole should be flowed back and pumping should be resumed if possible which may be successful if the block is superficial and not tightly packed. Gradually increasing pressure in the early stage followed by more rapidly rising pressure testifies to screenout taking place at some distance from the wellbore when the proppants become immobile and start filling the fracture from that point back to the borehole. Such screenout types are also indicated by a general upward trend in surface treating pressure with possibly one or several sharp breaks which precedes the steep pressure increase. The slow pressure increase includes ample warning time and thus remedial action can lead to screenout prevention.

A possible alternative is decreasing proppant concentration for the reason of permitting the fluid flow rate to increase and thus to keep the fracture open to receive further proppants (GREENFIELD & AHMED 1983), but without the prescribed proppants, crack conductivity may be seriously reduced which may be equally as damaging as the screenout itself, and also the possible pressure relief provided it will take some time to reach the trouble spot having to be displaced from the surface to the troubled areas which may be too late to avoid the screenout to occur (PARKER 1980). A more favourable remedy action is increasing injection rate to provide added fracture width that may relieve the restriction preventing proppant transport, with the effect being felt immediately and possibly being adequate to re-establish proppant mobility.

6.2.4.2.1.3. Avoidance of screenout failure by pumping rate and volume adjustment

On the other hand, however, increasing injection rate may in some instances not be a reliable alternative because excessive fracture heights may be the result which can itself give rise to screenout failure by corresponding crack width narrowing (PARKER 1980). If a screenout does occur, it is best to cease pumping and not pressure up again in an attempt to flush casing or tubing, and pumping should be stopped if the injection rate has been decreased to less than half and the pressure is still rising sharply, because if pumping is continued in such a case, the proppants are packed and crushed which leads to serious conductivity damage. The rate of pressure drop when shutting down reflects the difference, with rapid pressure drops testifying to retaining conductivity of fracture and wellbore, whereas if the pressure does not drop but remains high, the proppants are tightly packed in the borehole area and the resulting conductivity is drastically lowered.

Another reason of fluctuating treating pressures that may ultimately give rise to screenout failure is flow restriction through perforations, with an unexpectedly high pressure drop indicating that not all the perforations are accepting fluid. This leads to a surface treating pressure higher than anticipated and may result in an undesirable distribution of the stimulation job. High operating pressures necessitate the use of fluids with minimal pumping friction pressures in order to reduce the screenout risk (McMECHAN & CONWAY 1983), and for the purpose of minimizing contraction in the long tubing strings, stimulation fluids must be often preheated.

Screenout avoidance can also be enhanced by increasing the pad volume to abt. 50 % or more of the gel volume based on fluid leakoff values obtained from experience of earlier jobs terminated by screenout failure in order to prevent excessive fluid loss (cf. section 4.3.4.6.1.4.) which causes proppant to reach the fracture tip in a premature stage or to attain a concentration that is too high for continued transport in the crack. Treating pressure can be reduced by application of delayed crosslinking gel systems which also provides a higher fluid viscosity in the fracture. Fluid retention in the crack also significantly contributes to screenout failure risk as well as to poor post-fracturing hydrocarbon production.

6.2.4.2.1.4. High- and low-stressed zones

In terms of pressure evaluation, propped fractures are likely giving higher shut-in values than unpropped cracks as a consequence of the inability of the propped fracture to close down and the resultant locking of stress into the crack vicinity (McLENNAN, ROEGIERS & MARX 1983). There is no consistent relationship between fracture gradients determined after an acid breakdown and after a large-scale proppant fracturing treatment, with some of the discrepancies being due to the crack moving into higher- or lower-stressed zones during a propped operation. Higher fracture gradient differences lead to increasing production due to wider propped cracks as reflected by stresses locked into the formation by incomplete fracture closure and due to restricted vertical growth, because if the crack grows into a lower-stressed zone, lower shut-in levels may be recorded. On the other hand, increased stress levels are caused solely by permeation over a long time period, and lower fracture conductivity is provoked by higher closure stress across the crack.

6.2.4.2.1.5. Other reasons of abnormally high treatment pressure

Abnormally high treatment pressures do not necessarily have to indicate premature screenout failure of the job, but can also have various other reasons (WARPINSKI, BRANAGAN, SATTLER, LORENZ, NORTHROP, MANN & FROHNE 1985) such as complex fracturing (MEDLIN & FITCH 1983; cf. section 4.2.3.4.). Another possibility is that the fracture reached the boundary of a lenticular sand body so that treatment pressures need to increase for major fracturing to occur in the laterally adjacent and/or vertically over- and/or underlying higher-stress shales. Back stresses due to fluid leakoff giving rise to changing closure stress and presence of high-stress stringers between low-stress zones (cf. sections 4.2.2.3.1. and 4.2.3.2.2.) which reduce fracture width are also effects provoking job pressure increases. Bottomhole treating pressures remain occasionally constant during pad and then gradually decrease during the proppant-laden stages (TAN, McGOWEN, LEE & SOLIMAN 1988). This may indicate either uncontrolled fracture height growth or lower perforation friction pressure during proppant-laden stages due to perforation erosion by proppants. Some aspects of fracture branching and high-viscosity slurry friction are illustrated as follows.

6.2.4.2.1.5.1. Fracture branching

Abnormally high treating pressures leading to screenout at maximum practical operating pressure are frequently caused by fracture branching which is a function of fluid viscosity and pumping rate (MEDLIN & FITCH 1983; cf. section 4.2.3.4.1.). Fracture branching as a response to pressure growth occurs at reservoir heterogeneities and natural cracks. When fracture branching starts, the normal mechanics of crack propagation are considerably altered. Fracture growth no longer consists of a simple opening of the crack faces against the minimum earth stress, because slip motion is introduced along many of the branch faces which generates friction that increases the energy needed to sustain crack growth.

As the additional energy must be supplied by a rise in fluid pressure within the branched crack network, branching leads to a natural increase in treating pressure. Successive branches form as the initial main fracture propagates away from the wellbore, and at sufficient distance, this process produces a fan-shaped volume of rock with multiple fractures. Low pumping rates can considerably reduce or even completely inhibit crack branching, whereas excessive pressuring such as occurring during explosive fracturing (cf. section 4.7.7.) almost immediately generates a broad set of multiple small cracks and may completely prevent the origin of an initial large main singular fracture stem.

6.2.4.2.1.5.2. High-viscosity slurry friction

Another reason of increasing treating pressures is friction of high-viscosity slurries transporting high proppant concentrations in narrow fractures (HOLDITCH, ROBINSON, ELY & RAHIM 1988; cf. section 4.3.4.2.). Instead of representing crack containment upon reaching of the barrier or screenout termination, considerably rising excess pressures could simply be the consequence of markedly increasing friction especially in case of delayed crosslinked stimulation gels which have much higher viscosity downhole and that do not much change their viscosity in the fracture. As a result of pronounced friction, increasingly higher injection pressures are required to pump the thick proppant-laden gel down the crack and thus excess pressure increases with time. Gel friction properties can in some cases control excess pressure evolution at least as much as fracture geometry does, and therefore accurate prediction of crack dimensions requires detailed measuring or modelling of fluid slurry behaviour in the fracture. Thus unconfined height growth can occur even when the excess pressure in the crack is rapidly increasing, with the primary reason of this behaviour being that slurry viscosity at high proppant concentrations increases friction pressure (other aspects of abnormally elevated treatment pressures are discussed in sections 4.2.2.3.2. and 4.3.4.4.3.).

6.2.4.2.1.6. Screenout forecast possibilities

JOPLING & KETCHER (1987) present real-time decision making possibilities through use of a log/log plot of net pressure above closure vs. time during a fracturing job for the sake of early recognition and prevention of a premature screenout. The computing equipment in the treatment monitoring van (cf. section 6.2.4.3.) permits successive follow-up of the progress of the stimulation treatment, with the continuous plotting displaying lateral extension height growth, excessive fluid leakoff, width growth or approaching of screenout conditions, and through early observation giving time to make changes in job design as the situation requires. Screenout prediction by fracture pressure interpretation was originally introduced by NOLTE & SMITH (1981) as hand evaluation, but online assistance has since then considerably improved the effectivity of pressure supervision during the stimulation operation. The use of on-site computers has allowed the continuous gathering, processing, calculating, recording and displaying of data in real-time.

6.2.4.2.2. Pressure evolution and fracture extension

NOLTE (1988 a, 1988 b) gives a summary of fracture monitoring and determination by pressure analysis which is very similar to the assessment of reservoir pressures. The pressure decline after the hydraulic stimulation operation can be used to estimate design parameters and the resultant dimensions for a crack. Pressures during fracturing provide a basis for interpreting the stimulation job either qualitatively by the slope of log-pressure/log-time plots or quantitatively for past or future treatments by comparisons with predictions from computer models. The basic interpretive tool for pressures during hydraulic fracturing is a log-log plot of net pressure (total fluid pressure minus closure pressure or minimum in-situ formation stress) vs. time or equivalently vs. injected volume (the latter preferably if injection rate varies). Fracture stiffness is the ratio of net pressure to average crack width. Some points of constant pressure rise reflecting fracture growth, pressure increase types and corresponding significances, and other aspects are briefly discussed as follows.

6.2.4.2.2.1. Constant pressure rise reflecting fracture growth

Constant pressure rise with increasing time or volume indicates progressive opening of the fracture and ideal performance of the job with restricted vertical height growth and unlimited horizontal extension of the fracture (NOLTE 1988 a, 1988 b). There are, however, numerous deviations from the ideal pressure rise being displayed as a straight line with small positive slope in the plot. Nearly constant pressure at constant injection conditions is interpreted as indicating accelerated fluid loss to opening natural fissures or moderate stable height growth which starts at formation pressure capacity that is the maximum pressure at which the reservoir can generally contain fluid and maintain efficient fracture extension. Excess of formation pressure capacity during stimulation gives rise to excessive fluid loss and ineffective crack extension into adjacent non-reservoir intervals.

6.2.4.2.2.2. Pressure rise types and corresponding significance

Increasing pressure on an approximate unit slope is interpreted as reflecting a restriction to fracture penetration caused by proppant bridging or excessive fluid viscosity due to slurry dehydration, with both effects ultimately leading to premature screenout termination of the job (NOLTE 1988 a, 1988 b). The log-log unit slope for the compressible system is the consequence of storage of increasing quantities of fluid in a restricted volume and testifies to limited flow into both wings of the fracture. If one wing subsequently becomes blocked, the doubling of flow into the remaining and active wing will cause a double log-log slope. A decrease in pressure without changing injection conditions is interpreted to give evidence of unstable height growth of runaway nature as a result of vertical growth into a relatively low-closure-stress zone (generally sandstones and mudstones differ from each other in alternating sequences by being low-stress and high-stress intervals, respectively; cf. sections 4.2.2.3.1. and 4.2.3.2.2.). Excessive fracture growth in vertical direction could lead to sharp pressure decline which is perhaps followed by a smoother trend later on as the resistance to height extension progressively increases (ECONOMIDES 1987 b).

In combined horizontal and vertical fracturing in shallow heavy-oil sands, immediate pressure reduction with proppant insertion into the crack reflects unstable height growth either due to a horizontal fracture migrating upwards to the formation top where in-situ vertical stress is lower or due to initiation of a vertical fracture. On the other hand, such pressure drops can also be caused by proppant erosion around the wellbore (KULAR, CHHINA, BEST & MACKENZIE 1988).

6.2.4.2.2.3. Other aspects

The application of pressure transient analysis techniques to fracture design allows evaluation of current stimulation methods as well as past completion and treatment practices, and leads to better design and utilization of future operations (CHARLES, HUDOCK, UDICK & HALL 1983). Pressure buildup tests are routinely carried out after hydraulic fracturing treatments to determine length and conductivity of the created crack (SOLIMAN & HUNT 1983). Fracture treating pressure vs. stimulation time also provides a means by which crack growth in terms of both height and length can be inferred (HART, ENGI & MORRIS 1983). The dynamic downhole pressure when compared with fracture design model predictions can also be used to estimate the amount of out-of-zone stimulation in terms of excessive fracture height achievement. The pressure measurement is most accurately performed bottom-hole because of varying proppant concentrations, friction losses, and flow rates which make calculation of the fluid head difficult.

Fracture length can also be evaluated by type curve matching (BRINKMANN 1982). BELANI & JALALI-YAZDI (1988) outline estimation of matrix-block-size distribution and fracture intensity from pressure transient data. Pressure performance in stimulation jobs and its application is outlined by JIANG, LIU, YU, SONG & SHAN (1988). Bottom-hole pressure analysis is an important instrument for treatment design optimization (DAVIS, NICKELL, LENT & ALLAM 1980). MURPHY & BROWN (1976) outline investigation of hydraulic fracture geometry and downhole connection conditions with flow testing. VONGVUTHIPORNCHAI & RAGHAVAN (1987) discuss pressure falloff behaviour in vertically fractured wells.

6.2.4.2.3. Fracture height and reservoir thickness

In addition to screenout situations reflecting proppant bridging and/or slurry sticking in the fracture, also sliding resistance (partially due to progressive proppant staging) and bank building are potentially dominant contributors to fracture pressure rise in the late stage of the operation (CLEARY, BUHARALI, CROCKETT & SALEHI 1986). In terms of monitoring and determination, decline analysis of net pressure at shut-in can be used to estimate fracture length and width, time required for the crack to close, and fluid efficiency. Some aspects of fracture height and treatment pressure magnitude as well as fracture width and reservoir stress level are illustrated as follows.

6.2.4.2.3.1. Fracture height and treatment pressure magnitude

The magnitude of net treatment pressure depends primarily on initial height and not on ultimate height of the fracture if unstable vertical growth with penetration of other low-stress zones is not significant (CLEARY, BUHARALI, CROCKETT & SALEHI 1986). For large height increases, the nearly constant pressure results primarily from the decrease in effective flow rate. Pressure control can be primarily achieved with fluid viscosity and only secondarily with injection rate.

The final fracture height is greater for a thin pay section than for a thicker reservoir interval (NOLTE 1988 a, 1988 b). Stiffness is inversely proportional to initial interval thickness or zone height. If two horizons are the same except that their initial heights differ by a factor of 2, the net pressure of the smaller interval will be approximately twice that for the larger zone at the same penetration. The resultant relative (with respect to reservoir thickness) fracture height for the smaller interval will generally be greater than that for the larger zone.

6.2.4.2.3.2. Fracture width and reservoir stress level

Because fracture width in bounding or separating mudstone intervals is relatively narrower than in the initial sandstone complex (cf. section 4.2.2.4.), the average width for the smaller reservoir horizon is generally less than that for the larger zone (CLEARY, BUHARALI, CROCKETT & SALEHI 1986). As negligible fluid loss occurs in mudstones, fracture penetration for the smaller interval is also generally greater than that for the larger horizon with the same volume of injected fluid. The presence of an interbedded mudstone representing a high-stress section in a low-stress sandstone reservoir complex which is to be fractured can significantly increase net pressure and resultant height over that for a homogeneous interval of the same height.

Fluid pressure during hydraulic stimulation governs fracture geometry and potentially also treatment efficiency. The most realistic means of controlling the pressure is by using fluids with minimal viscosity required to transport proppant and to control fluid loss effectively. Although minimal fluid viscosity is obviously desirable for reducing pressure below formation capacity, the use of minimal viscosity magnifies other aspects of a successful treatment relative to the use of excessive viscosity.

6.2.4.2.4. Minifracturing evaluation and 3-D simulation

Screenout termination prediction and prevention can also be assisted by minifracturing calibration treatment evaluation and three-dimensional simulation (CLEARY, BARR & WILLIS 1988). The following discussion outlines some aspects of proppant bridging and fluid loss provoking screenout, simulation modelling and screenout candidate recognition, and fracture height calculation and checking by well logging.

6.2.4.2.4.1. Proppant bridging and fluidloss provoking screenout

Pressure monitoring reveals dramatic pressure drops indicating breakout into a low-stress, high-permeability and high fluid leakoff reservoir zone. Kicking-up of pressure testifies to opening of a dominant pathway for proppant injection (CLEARY, BARR & WILLIS 1988). Proppant is then accepted until bridging develops further away from the injection point after which the pressure curve develops into a classical quadratical backfill shape which leads to final screenout. Bridging and screenout is evident both in the simulation model and in the minifracturing calibration treatment, enabling to prepare measures preventing screenout to actually occur. The alternative explanation of unprecedented degeneration in fluid rheology is quite likely as a consequence of on-site viscosity measurements under crosslinking conditions in the reservoir.

The main reason for screenout termination is the presence of a thief zone of low stress, high permeability and high fluid uptake rate. The deleterious effect of the thief zone cannot be overcome by high rates and high viscosity of the fluid, but can only be reasonably controlled by leakoff additives and/or diverting fluids. In reservoir rocks with dense natural fracture systems, excessive shattering of near-wellbore rock may occur if the number of perforations is too great and therefore numerous narrow pathways may originate conveying fluid from borehole to main fracture further out, with dominant avenues (which are allowing lots of proppants to enter) only opening up at higher pressures. Shifting of the fracture center off the primary injection point can lead to proppant insertion into a narrow corner of the crack such as occurring at its bottom which makes a screenout even more likely, and in addition, any nearby barriers can pinch the fracture closed further.

6.2.4.2.4.2. Simulation modelling and screenout candidate recognition

Simulation runs for the case of an effective upper barrier but absence of a lower boundary indicate that when a low-viscosity low-rate treatment is pumped, the fracture grows down but is restrained by the upper barrier, resulting in an asymmetrical crack with the risk of proppant injection into the top corner (CLEARY, BARR & WILLIS 1988). In case of greater flowrate or higher viscosity, the fracture grows through both boundaries into a relatively circular shape and the screenout risk is reduced, with little change in fracture efficiency.

An important consequence of simulation modelling is recognition of some intervals that represent automatic screenout candidates into which therefore only small jobs can be pumped. In such particular cases where trouble to be expected is indicated prior to the main fracturing treatment, valuable additional information and knowledge of reservoir response can be gained from minifracturing calibration operations (cf. section 4.8.2.). Screenout failure risk can also be minimized by increasing the pad volume for avoiding early pad depletion in case of high fluid-loss rates (PEARSON, LYNCH, SCHMIDT & McCASLIN 1988). Fluid depletion is particularly critical in bridgeout conditions when the movement of the proppant grains is restricted by insufficient fracture width.

6.2.4.2.4.3. Fracture height calculation and checking by well logging

CASTILLO (1987) and WAREMBOURG (1988) outline crack height determination from pressure decline analysis following minifracture calibration treatments. Evaluation of pressure evolution allows calculation of net pressure, fracture half-length and efficiency. In addition to pressure history assessment, crack height is controlled by gamma-ray and temperature logging after the calibration operation. An estimate has to be made for those cases where wireline surveys are unreliable or where they were not run at all, with a major advantage of the pressure analysis technique being the ability to obtain a value for fracture height without cable-tool survey. The estimate of crack height, however, is only as accurate as are the input values for Young`s modulus and fracturing fluid viscosity.

Fracture height obtained from pressure analysis following calibration treatments is usually greater than that determined by gamma-ray surveying (WAREMBOURG 1988). This can be easily explained for cracks being inclined in the wellbore and therefore showing only the best case from the survey. In contrast, fracture height obtained from pressure evolution interpretation is usually less than that recorded by temperature logging (cf. section 6.2.1.1.). Pressure analysis requires input of a leakoff or fluid-loss height which is termed net fracture height in contrast to gross fracture height being overall fracture height (cf. also section 6.2.1.5.2.).

Fluid-loss height is generally greater than net hydrocarbon pay zone because of permeable intervals above and/or below the net hydrocarbon reservoir horizon. Well logging fracture height surveys should always be run to confirm pressure evaluation determination of crack height, because cable-tool surveys show the best case with respect to inclined vs. true vertical fractures, and pressure simulators require more data than are usually available for satisfactory results. Fracture height from well logging surveys should be compared to the height calculated from pressure analysis also as check on other variables such as Young`s modulus and viscosity, although cable-tool survey heights are probably in most cases optimistic. Leakoff coefficient calculations are greatly influenced by the ratio of net to gross fracture height.

6.2.4.2.5. Fracture pressure and geotectonical setting

Fracture pressure is not constant in whole fields or areas, but varies with geotectonical setting (LI & ZHU 1986). Fracture pressure is generally lower in regions with relaxed tectonical stress than in areas where compressive tectonical stress is acting (cf. also section 1.2.8.). For wells in certain fields, fracture pressure is affected by borehole location representing the actual site of tectonical stress, reservoir pore pressure and formation lithology. Fracture pressure gradient in the top part of some anticlines is lower than that on the flank part of the dome-shaped structure which in turn is still lower than that on the synclinal part of the structure which is saturated with water.

6.2.4.2.6. Prevention of pressure-out situations

While screenout sensu strictu implies blockage at the perforations and an inability to flush down, the term pressure-out is used as opposed to screenout for cases when it is possible to flush down to the top perforation at a low rate and maximum pressure, thereby indicating an obstruction at some point within the fracture itself (OSBORNE, McLEOD & SCHROEDER 1981). Pressure-out situations can occur even with acceptable fluid rheology and good fluid-loss control. An evaluation of proppant concentration in the fracture at any time and distance during the treatment shows that the fluid volumes needed ahead of each proppant concentration change are determined by fluid-leakoff coefficient and pumping time (KASPEREIT 1979; OSBORNE, McLEOD & SCHROEDER 1981).

Experimental work has evidenced that even with a fairly low fluid-loss coefficient, it is possible to bridge out in the fracture tip if insufficient pad volume was run ahead and that bridging out is possible at any proppant saturation change if insufficient fluid has been run ahead of the pad and earlier stages. Pressure-out situations usually occur when the pad has leaked off and first proppant-laden element concentrates to a point where the proppant will no longer flow in the fracture. Pressure-out, however, does not necessarily have to occur only at this point, but can also take place in later proppant-laden elements if the pumped proppant saturation is increased too soon.

Too high proppant concentrations in the fracture can cause different effects. The first possibility is that the wellhead pressure increases to the point that the safe boreholehead operating pressure boundary is reached and the treatment has to be shut down, thus representing a pressure-out. The second result is that the downhole pressure builds to a point where it is relieved by extending the fracture into or through overlying shales or other impermeable barriers, as evidenced by a steady increase in pressure indicating a storage phenomenon followed by a pressure drop testifying to fracture height extension (NOLTE & SMITH 1979). This may be followed by a further pressure increase and fracture extension pressure drops until either the treatment is finished or a final sharp pressure increase reaches the maximum allowable wellhead pressure (OSBORNE, McLEOD & SCHROEDER 1981). Proppant schedules should therefore be carefully evaluated or calculated to achieve an adequate proppant distribution in the fracture and to prevent a pressure-out situation. Fluid-loss control is important even in tight sandstones, with fluid-leakoff coefficients being able to be estimated by evaluating past treatments where a screenout or a pressure-out has occurred.

6.2.4.3. Computerized real-time job supervision

After some general aspects of on-line treatment monitoring, comments are offered on special technology for computerized job supervision as well as conservative and progressive fracture operation control.

6.2.4.3.1. General aspects of on-line treatment monitoring

CLEARY (1988) reviews the presently available computerized systems for real-time monitoring, analysis and control of hydraulic fracture stimulation operations which allow on-site calculations immediately as the job proceeds in the field and permit adjustments to be made as needed. The process of history matching the observed and predicted pressures in the early stage of the treatment, during the minifracture calibration run, and/or the pre-pad and pad phases of the main job, enables determination of critical parameters such as variations in confining earth stresses that dominate fracture growth along and out of the pay zone. In case of shadow-fracturing when both previous minifracture operation and the pad phase of the main treatment actually already achieve most of the hydraulically-created fracture area, the parameter values determined in the early stages are essentially valid for later phases of crack widening as the cross-linked and proppant-laden stages are pumped in and are mainly used for interpretation of rheological and proppant transport behaviour, with particular emphasis on predicting and avoiding premature screenout by excess fluid leakoff and/or failure to achieve a sufficiently wide fracture with good proppant-carrying fluid.

As the hydrostatic head pressure of incompressible fluids is only affected by proppant concentration, computer on-site monitoring allows the hydrostatic head pressure of the slurry to reflect changes in proppant addition rate and proppant density (BAUMGARTNER & MACK 1988). Fluid friction pressure is also affected by injection rate and proppant saturation. As hydrostatic head pressure of conventional incompressible oil- and water-based fluids on the one hand changes insignificantly, and of commingled and foam fluids on the other hand changes significantly, respectively, with pressure and temperature gradients encountered in the treating string, corrections of hydrostatic head pressure have to be made.

6.2.4.3.2. Special technology for computerized job supervision

Further improvements of on-site stimulation data analysis and modification for the benefit of better job supervision and screenout failure prevention in the near future are development of automatical feedback control from computer to pumps, blenders and valves in order to achieve a smoother performance of the actual treatment. The transition to automatical computerized survey systems probably passes from operator-activated feedback signals from a central control unit on-site gradually to increasing computerization, but operator observation and override switches will always remain implemented.

CLEARY (1988) concludes that once automatical computerized job control and data analysis is achieved, hydraulic fracturing technology will finally have reached a reasonable level of maturity. The significance of computerized on-site data gathering and monitoring equipment is also highlighted by VEATCH & MOSCHOVIDIS (1986) who emphasize that the computer age has already started for hydraulic fracturing. Other aspects of computerized job supervision, process control and automation are discussed by COOPER, NELSON & SCHOPPER (1983); HANNAH, HARRINGTON & LANCE (1983); HARRINGTON (1985), MACK & BAUMGARTNER (1986), BAUMGARTNER & MACK (1988), OLSEN & DEBONIS (1988) and TOMLINSON, BYRD & BERG (1988). AGUILERA, HOUSTON, PODIO & SONG (1988) describe an automatic pressure buildup data acquisition and interpretation system allowing unattended operation by a portable microcomputer.

6.2.4.3.3. Conservative and progressive fracture job control

CLEARY, BUHARALI, CROCKETT & SALEHI (1986) summarize the necessity of computerized field systems for real-time monitoring and analysis of hydraulic fracturing jobs by stating that among the many operations of the petroleum production process, hydraulic proppant fracturing may be the most complex to conduct and the most evasive to verify in a scientific and engineering manner. The success of the best-designed treatment can often be a victim of the vagaries of nature and there is very little direct information as to exactly how effective was the stimulation work, or what went wrong when production shows poorly. As a consequence of these shortcomings, hydraulic fracturing technology has developed in an understandable conservation fashion by well operations wishing to avoid risk involved in blind experimentation, and subject to technology cost limitations, service companies have not been able to muster the many components needed to develop, verify or control the procedures which might have resolved some of the difficulties. In a sense, the necessarily competitive nature of the industry may have militated against the coordinated effort needed to upgrade the technology.

These drawbacks can be effectively overcome by monitoring, analysis and potential control of hydraulic fracturing treatments in real-time during the field operation by a suite of instruments ranging from trailor-mounted computers via super-miniature computers to entirely portable systems. Comprehensive analysis of earlier treatments and evaluation of the results for the reason of progressive optimization of subsequent operations leads to development of an evolutionary sequence of fracturing technology with successive improvement of the steps advancing towards more mature stages (PEARSON, LYNCH, SCHMIDT & McCASLIN 1988).

6.2.4.4. Quality control

A special type of stimulation data analysis and job performance evaluation is quality control which basically aims on improving success rates of the treatments by reducing premature screenout terminations which may occur due to various reasons (PARKER 1980; OSBORNE, McLEOD & SCHROEDER 1981; PAI, GARBIS & HALL 1983; BEGNAUD & CLAIBORNE 1985, DEES & COULTER 1986, MARTINS 1987; for screenout failure cf. also section 6.2.4.2.1.). Quality control of fracturing treatments is essential in order to obtain benefits of maximum achievable productivity and to reduce time and money required for remedy of problematical or failure operations (DEES & COULTER 1986). The highest frequency of fracturing stimulation problems is related to well conditions and formation properties rather than to service company performance and job execution. Significant variables for stimulation operation quality assessment are treatment fluid type, reservoir lithology, pay depth, maximum proppant concentration, proppant type and grain size, and ratio of zone thickness to injection rate. CONWAY & HARRIS (1982) emphasize that very careful monitoring and quality control of every phase of well drilling and completion is required to make commercial recovery of reserves possible. EVANS & WILLIS (1986) underline that understanding and optimization of numerous variables involved in fracturing job execution improves the effectiveness of stimulation treatments and produces better and more realistic fracturing operations.

BAUMGARTNER & MACK (1988) report on-site computer monitoring of foamed stimulation fluids, and HARRIS, KLEBENOW & KUNDERT (1988) document how constant internal phase design improves treatment results. OLSEN & DEBONIS (1988) discuss real-time optimization of fracturing fluid viscosity to meet changing stimulation conditions. Items where quality control is critical for successful performance of MHF operations are reviewed by PARKER (1980). The significance of quality control is also emphasized by SAUCIER (1988).

6.2.5. Fracture migration analysis

In contrast to all the aforementioned post-treatment monitoring techniques, fracture migration modelling is a pre-stimulation prediction of formation fracturability and prognosis of differences in fracture gradient and thus open-up behaviour between various hydrocarbon-bearing zones of the reservoir column (AHMED, NEWBERRY & CANNON 1985; NEWBERRY, NELSON & AHMED 1985; CORLEY & KLAUTT 1986; SCHLUMBERGER 1986 a, 1987 b; WELEX 1986; BALL, CHACE & FERTL 1987). Fracture migration analysis is a combination of sonic waveform interpretation with bulk volume evaluation and linear fracture mechanics with treatment fluid density (cf. also sections 4.2.2.5.1. and 6.2.1.10.7.). Enhanced formation strength interpretation can be done from recording of the entire wavetrain and its subsequent breakdown into various components comprising compressional, shear and Stoneley waves during processing (SCHLUMBERGER 1985, 1987 b; WELEX 1986). Fracture migration modelling aims on fracture height propagation control in order to prevent the crack from extending away from the perforated hydrocarbon-bearing interval (CORLEY & KLAUTT 1986). Aspects of fracturability prediction and fracture height containment analysis are outlined as follows.

6.2.5.1. Fracturability prediction

Fracture migration analysis is based on calculation of the dynamical elastical rock properties of the formation from compressional and shear wave velocity measurements which are then integrated into two models including prediction of pressure necessary to initialize the fracture and to keep it subsequently open during propagation, respectively. Important formation fracturability indicators are relative horizontal stress and fracture toughness modulus. Some principles and applications are briefly sketched as follows.

6.2.5.1.1. Principles

Distinction has to be made between radial and tangential stress levels as well as fracture and closure pressure gradients. In order to fracture a formation, matrix tangential stress must overcome formation tensile stress, with breakdown occurring when the sum of these two is zero (CORLEY & KLAUTT 1986). The additional pressure which is required to match the value of matrix tangential stress and formation tensile stress is a prediction of the borehole pressure necessary to fracture the reservoir. The dynamical elastical rock properties combined with bulk volume analysis provide a picture of in-situ stress in and around the hydrocarbon-bearing zones. These data plus treatment fluid density are used to predict the fracture migration behaviour relative to bottom-hole treatment pressure. Total crack height at the wellbore for a given injection pressure can be calculated and used as a constant for determining fracturing fluid volumes, and crack geometry can be predicted using a pseudo-three-dimensional model which also allows the simultaneous calculation of fracture height, length and width (SCHLUMBERGER 1985, 1986 a, 1987 b).

The fracture migration model shows containment and relative differences in breakdown between individual zones and formations. If the breakdown pressure of two horizons is more or less identical, both are simultaneously opened once fluid is injected. If the breakdown pressure of two layers is different, singular fracturing has to be done by first opening one level, and then with rising injection pressure the other interval is also breaking up, or pressure differences have to be eliminated through perforations. The fracture migration model shows which zones can be commingled at once and which intervals have to be sealed off and treated separately, and is thus a predictive tool of horizon opening and crack propagation. The results of fracture migration analysis indicate in which zone of the reservoir complex crack opening is going to start and which are the bounding pressures for fracture containment between the enclosing barrier intervals, or which is the minimum pressure to break through a mudstone layer separating two sandstone pay horizons.

6.2.5.1.2. Applications

Assessment of formation behaviour is made by comparing fracture height and increase in net treatment pressure which is the bottom-hole treatment pressure in excess of the fracture extension pressure in the perforated interval (cf. also COATES & DENOO 1981). The fracture migration model assumes balanced horizontal formation stresses. In some areas, the geological column is not tectonically relaxed and the horizontal stress imbalance has to be compensated by empirical adjustment of the crack migration model by adequate fracture treatment data (CORLEY & KLAUTT 1986). The qualitative aspects of crack migration analysis are much more accurate than the quantitative ones.

Fracture prediction modelling can be done by log evaluation of formation properties alone, but always can be carried out more effectively and confidently if backed-up by minifracturing determination treatments. In order to get full control on the reservoir column and to provide the necessary calibration, minifracturing analysis is required not only in the pay zones theirselves, but also in the barriers between the hydrocarbon-bearing permeable sections.

A comparable forecast can be done for sand production modelling in gravel packing. As shear values of the formation are not known, an empirical method based on the Mohr circles has to be used to predict the compressional/shear relationship. In contrast to fracture migration analysis which looks at differences in fracturability level, sand production modelling investigates possible sanding of the formation and is a more quantitative means of reservoir assessment. Radial stress and tangential stress are the normal stresses in a horizontal plane at the borehole wall prior to production. Those reservoirs with the lowest initial tangential stress and greatest initial radial stress will fracture most easily, but will typically have the lowest probability of potential sanding problems, and vice versa (cf. also sections 5.8.6. and 6.3.6.).

6.2.5.2. Fracture height containment analysis

In addition to fracturability prediction, the crack migration model can be used for fracture height containment analysis (CORLEY & KLAUTT 1986). Formation fracture pressure which is required to initiate the crack is lower than reservoir closure pressure that is necessary to extend the crack once formation breakdown has been achieved. For vertical fracture migration to occur and to continue, treatment pressure must be increased to overcome the closure pressures of any reservoirs that the vertical crack encounters during propagation. Thus barriers impeding or preventing vertical fracture migration are those formations which have closure pressures greater than that of the perforated interval. The delta pressure curve indicates the vertical propagation of the fracture away from the perforations for any specific increase in treatment pressure. If a specific maximum crack height is desired, delta pressure will indicate the maximum increase in treatment pressure which can be applied. Therefore pressure modelling by fracture migration analysis permits proper design of the hydraulic stimulation job in order to make sure that fracture height containment is successfully achieved by active control of vertical crack propagation (AHMED, NEWBERRY & CANNON 1985; NEWBERRY, NELSON & AHMED 1985; SCHLUMBERGER 1985, 1987 b).

As fracture migration determination is only dependant on differential closure pressures, errors in the absolute value of the predicted closure pressure have little effect on delta pressure or the resulting crack propagation assessment (cf. also section 4.2.2.5.1.). The fracture migration program can also be run in a microcomputer environment (BALL, CHACE & FERTL 1987).

Aspects of minimum horizontal stress difference determination in pay zones and bounding layers from sonic waveform records in combination with open-hole density logs are discussed by BEGNAUD & CLAIBORNE (1985). Deviations of the measured values from the expected ones can be due to the presence of active tectonics which results in a shift in values for all zones from depth to surface. The substantial values of the maximum horizontal stress indicate a possible tectonical influence, whereas additional compressive forces affecting the fracture pressure cause the difference between sonic-derived stress values which are usually greater and the in-situ values to become smaller. Somewhat insignificant stress ratios are caused by the presence of tectonical forces parallel to the fracture which is supported by the crack azimuth being in general agreement with the direction of areal tectonical forces.

6.2.6. Inflatable impression packers

Hydraulic fracture detection can also be performed with inflatable impression packers which are bladders with a soft pliable coating (HAIMSON 1973, TAM INTERNATIONAL 1986, SCHLUMBERGER 1987 a, WALSH & HOLDER 1988). The following section discusses various aspects along the lines of mechanical fracture imprint study, fracmeter approach, and impression packer vs. borehole television.

6.2.6.1. Mechanical fracture imprint study

Inflatable impression packers are lowered in place and inflated until the soft coating presses against the wellbore, conforming to the rugosity of the hole. The packer is then deflated, withdrawn and examined for the mechanical impression of fractures. The application of the technique is limited by the fact that many hydraulically or naturally fractured formations characteristically have large or irregular wellbores which can result in overinflation and blowout of the packer.

Inflatable impression packers are also a possibility of direct observation of hydraulic fractures apart from borehole imaging (GRIFFIN 1985). The expandable production packers are equipped with a malleable rubber jacket over leaf-type flexible steel elements, are run on tubing and are oriented by landing a wireline-conveyed gyroscope into an orienting profile just above the packer. Once a plug has been wireline set in the bottom of the string, the packers are inflated with tubing pressure. Several packers can be stacked to cover a long interval. The flexible steel leaves cause the packers to retract when tubing pressure is released so that the tools can be pulled out of the hole.

Impression packers can document morphology and azimuth of the hydraulic fracture intersecting the reservoir interval that has been stimulated. Even with a series of packers stacked together above each other, however, significant intervals of information are lost at each coupling. Tailpipe should be run below the bottom packer to minimize plugging of the latter with rathole sediments. It should be ascertained that the packers are not inflated across large vugs or washouts because the steel leaves will lock in the overinflated position which seriously damages the packer when the string is pulled out of the borehole. Field experience of inflatable packer application for fracture monitoring is also reported by BREDEHOEFT, WOLFF, KEYS & SHUTER (1976) and WALSH & HOLDER (1988).

6.2.6.2. Fracmeter approach

A special inflatable packer is introduced as a fracmeter by CHARLEZ, SALEH & DESPAX (1987). The fracmeter is a straddle inflatable packer with a spacing filled with radial displacement transducers. When a hydraulic fracture is created, the displacement field around the wellbore becomes anisotropic with a maximum value perpendicular to the crack plane. The multiplication of the number of equations relating the radial displacement of the borehole and the pressure allows to determine with a single test by an appropriate method the two horizontal geostatic stresses, the elastic constants of the rock, and fracture azimuth.

The rockmeter (CHARLEZ, SALEH, DESPAX & JULIEN 1986) is based on the same combination of two essential elements of the device being uniform radial pressurization of the borehole above the rupture point using a rubber inflatable packer and measurement of the associated deformation of the wellbore for different azimuths using displacement transducers fixed on the packer itself. While the rockmeter is limited to moderate depth of application due to mechanical reasons, the fracmeter (CHARLEZ, SALEH & DESPAX 1987) can also be operated in great depth.

6.2.6.3. Impression packer vs. borehole television

Inflatable impression packers provide in some cases higher resolution of tracing of hydraulically-induced fractures than acoustical borehole television (PAILLET, KEYS & HESS 1985). In addition, impression packers work in any borehole environment, whereas television is depending on mud type and composition and is also influenced by wellbore geometry. Therefore in many applications impression packers provide the primary means for identification of hydraulically-induced fractures, whereas acoustical television is more used for recognition of intervals suitable for hydraulic stimulation. Mirror images of impression packer tracing can be used to account for the fact that the acoustical televiewer looks at the borehole from center to wall, whereas the impression packer presents an image from wall to center.

6.2.7. Photography and imagery

Fracture patterns significantly control accumulation, distribution and flow of hydrocarbons in any given reservoir regardless of age or lithology (REYNOLDS 1987). The number of dry holes can be considerably reduced and well productivity may be importantly increased by drilling on fracture intersections. In some fields, cracks are intimately associated with faults and represent the entire porosity, permeability and trapping mechanism. Fracture pattern identification and location staking on crack intersections can play the main role in hydrocarbon discovery and can give in some cases much superior results than seismic line evaluation. Fracture mapping by stereopair photography and computerized satellite imagery including false-colour display of various processed images are excellent tools for detecting large geomorphological fractures and crack systems. A map with selected intersections of long fractures with well-defined anomalies or structures and highlighted topographical lows and highs may permit to generate prospects abt. 10 times faster than normally.

In fields with excellent productivity, a fracture pattern map can assist in finding out how the crack influenced the best producers and where additional wells could be staked and drilled with the best possible drainage patterns thus avoiding random choice of offset locations. The effectivity of photogrammetry and imagery in recognition and mapping of fractures of various scale for well location determination at crack intersections can be enhanced by combination with seismology, geochemistry, radiometrics, magnetotellurics, microgravimetry, micromagnetics and gravity, especially if direct and indirect image sensing methods are deteriorated by heavy forest covers or thick aeolian sand blankets masking the surface.

Mapping fractured trends facilitates searching for areas of more intense shattering (AGUILERA 1983). The combination of photogeology from airplanes and spacecrafts with structural mapping, residual stress distribution and geochemical information provides significant tools for discovering fractured reservoirs. Techniques which have proven particularly effective are colour normal aerial photography, infrared photography, thermal infrared imagery and side-looking radar imagery (RYAN 1976, AGUILERA 1983). Mapping fracture trends by aerial photography is also documented by ALPAY (1969), and PIRSON (1967) outlines crack development mapping from well logs. CLARK (1983) also emphasizes the significance of directional mapping of lineaments, faults and fractures at the earth surface for determination of horizontal in-situ stress orientations and prediction of azimuth of hydraulic fracture propagation. Distinction has to be made between ancient and actual stress directions for not mistaking palaeopatterns as being indicative of recent conditions and vice versa.

6.3. Gravel pack checking and evaluation

Gravel pack filters between borehole wall or perforated casing and screen or liner (cf. chapter 5) are only fully effective if they are properly placed and settled without containing void spaces and if they cover the whole perforated or underreamed interval. Gravel pack quality monitoring therefore aims on inspection of pack density and continuity as well as detection of void spaces within the pack, and on determination of the top of the cylindrical gravel column in the wellbore in order to verify its proper coverage of the producing zone and of screen and perforated interval.

It is of particular significance to recognize and possibly improve void bridges within the gravel pack filter, insufficient vertical extension of the gravel pack mantle and if possible also unsatisfactory perforation tunnel plugging by gravel spearheads before the onset of hydrocarbon production (cf. section 5.8.1. and 5.9.1.5.1.), because otherwise immigration of sand through internal blank spaces and open interval tops causes premature failure and expensive remedial work. Various well logging techniques are applicable for void detection within and top determination of the gravel pack mantle. The discussion as follows comprises photon, gamma-ray, neutron, caliper, and other gravel pack logging techniques, as well as tell-tale screen and mechanical properties log.

6.3.1. Photon logging

Gravel pack effectivity monitoring in qualitative and quantitative manner is possible by gravel density examination and extension control by compensated neutron, gamma-ray and photon (which is a special type of gamma-ray) cable (wireline) well-logging (NEILL et al. 1977, CHUDY 1981, NEAL 1983, BOYCE & CARROLL 1984, NEAL & CARROLL 1985, SOLLEE 1985). Gravel height checking can also be done by various methods. A recently introduced enhanced technique is the photon/vibrapack integrated system (DRESSER ATLAS 1987). Combinations of neutron and gamma-ray logging are also discussed, and the function of the tell-tale screen as an indicator of the termination of the gravel-pack operation is also outlined.

The mechanical properties log as a means of screening sand control treatment candidates by assessing reservoir lithification and thus forecasting pay zone stability is also introduced, and the use of caliper logging for predicting and estimating gravel quantities for packing open-hole sections is illustrated. STEIN (1976) uses conventional log data for selection of gravel packing materials for sand control. After some introductory comments on gravel pack monitoring vs. hydraulic fracturing inspection, the principles of photon logging are outlined. The discussion then focusses on conventional photon logging and photon/vibrapack combination.

6.3.1.1. Gravel pack monitoring vs. hydraulic fracturing inspection

The following outline on gravel pack monitoring in comparison to hydraulic fracturing inspection focusses on temperature and neutron logging which are the most suitable techniques to illustrate the differences between geophysical evaluation of fractures and gravel packs. After a discussion of general aspects of temperature and neutron logging, the density change generated by insertion of gravel into the borehole and proppants into the hydraulically-induced crack intersecting the reservoir in documented in more detail.

6.3.1.1.1. Temperature and neutron logging

Gravel pack monitoring techniques are different from hydraulic proppant fracturing examination methods (cf. section 6.2.), because the gravel pack in most cases covers a much shorter interval, is in its total extent limited to the immediate surroundings of the wellbore, and has a cylindrical mantle shape encompassing the whole circumference of the borehole (cf. chapter 5), in contrast to mainly longer vertical sections, deep penetration into the formation, and double-wedge shape of hydraulic fractures infilled by proppants (cf. chapter 4). While temperature logging is a very useful tool for hydraulic fracture detection due to pumping of large volumes into the crack (cf. section 6.2.1.1.), it is difficult to apply unambiguously in gravel packing because of utilization of much smaller fluid volumes which do not give such a pronounced cooling effect of the reservoir, if noticeable with certainty at all.

On the other hand, neutron logging is especially suitable for gravel packing (cf. section 6.3.3.) due to exploitation of the additional density effect created by the gravel pack mantle which represents an annulus of additional material installed in the borehole, whereas the proppant infilling of hydraulic fractures only replaces wedge-shaped bodies of formation sand due to pressure parting of the crack walls and generally does not represent an additional proppant body which would result in serious changes in density distribution in the borehole.

6.3.1.1.2. Density change generated by gravel and proppants

Exceptions are probably pre-existing natural fractures in originally jointed formations that if being enlarged and plugged by proppants would then also represent an additional volume of material inserted into the borehole. In contrast to gravel pack mantles which surround the borehole along its total circumference and effectively narrow the actual wellbore diameter, with almost all the gravel pack material being concentrated in the drape seaming the borehole wall, the additional material introduced into a hydraulic fracture is restricted to the width of the crack, and the large proppant quantity filling long deeply penetrating fractures cannot be detected by density-susceptible shallow-recording logging instruments.

Therefore the effect of proppant plugs in hydraulic fractures is very limited due to its bounding to the more or less narrow cross-sections of the two fracture wings in contrast to the continuous circumferential gravel pack mantle in addition to the perforation tunnel infillings. Higher gravity of synthetic proppants (cf. section 1.3.) is in many cases compensated by looser packing density than the reservoir sand due to higher porosity and as a consequence of lacking compaction by overburden with respect to pay matrix, thereby partially equalizing the effect of different specific gravity. The decisive difference between hydraulic fracturing and gravel packing in terms of evaluation by well logging is the installation of a distinctly additional volume by inserting the gravel pack mantle which can be exploited for quantitative determination of pack density, whereas due to the integrated nature of hydraulic fractures within the surrounding formation, plugging of the crack by proppants can only be qualitatively analyzed.

6.3.1.2. Principles of photon logging

Photon logging is based on a non-focussed gamma-gamma-device with a specific source strength and source-detector-spacing that is run in centralized position (NEILL et al. 1977, CHUDY 1981). The two modes of particle and energy interaction which contribute to the response are compton scattering whereby the photon gives off some energy to the electron with which it collides, and photoelectric effect whereby all the energy is given up during collision. Pair production is eliminated by choosing an adequate energy level. The reduction of intensity which is recorded in the detector is proportional to the electron density present and thus in most cases also to bulk density, with the greater the bulk density, the less radiation being detected. Penetration depth is very low, with the tool thus successfully evaluating gravel pack density with no sizeable effect from the reservoir formation characteristics. Centralization guarantees repeatable records of the photon logging device if bulk density is not changed.

The photon logging instrument is adjusted for gravel pack evaluation in such a manner that the formation itself is least effective upon the response of the tool (DRESSER ATLAS 1982). This is accomplished by choosing detector source strength and source-detector spacing such that the radius of investigation does not extend beyond the underreamed or perforated formation interval (cf. section 5.10.). Therefore the response of the instrument is principally affected by fluid, pipe and gravel pack density. As fluid and pipe density are constant throughout the examined section, the chief factor causing varying log response is gravel pack density. Voids within the gravel pack represent low-density pockets and bridges in an otherwise uniform and continuous medium, and marked density changes also occur at bottom and top of the gravel pack. The high number of counts recorded in the detector almost eliminates statistical variation which gives rise to a high degree of repeatability of photon logging provided constant borehole conditions. Distinction can be made between conventional photon logging and photon/vibrapack combination.

6.3.1.3. Conventional photon logging

Conventional gravel pack quality checking by running a photon log after the operation does not necessarily need a pre-pack reference log (NEILL et al. 1977, CHUDY 1981). Similarly as applying for the fracture monitoring instruments (cf. chapter 6.2.), most of the tools for gravel pack evaluation can be run in 1 11/16 " slim size (fitting also through narrow slotted liners, wire-wrapped screens and/or tubings) in combination with a casing collar locator and other production-control logging tools. The record from the instrument´s survey can be interpreted in terms of vertical extension assessment of the gravel pack as well as quality checking including recognition of deficiencies such as gravity settling and bridging of the pack which causes premature screenout during the job.

In addition to gravel pack quality inspection and height determination, the photon log can be used to check whether sand has been produced behind casing or tubing, and detection of cavities in weakly-consolidated sandstones. Photon logging in order to determine the top of the gravel mantle behind the screen in cased or open holes and to locate voids in the pack caused by bridging during placement has been frequently successfully performed in the shallow Lower Cretaceous oil-bearing sandstones in the Schoonebeek field in the Netherlands (CHUDY 1981), and STROBL & PLACKETT (1984) outline gravel-pack logging in Rotliegend gas-bearing pay zones in the Groningen field in the Netherlands (cf. also DAVIES, ZWOLLE & MEIJS 1983; cf. section 5.5.3.1.).

In addition to gravel pack density evaluation, photon logging can also be used for assessment of natural sand and synthetic proppant packages in artificial fractures in the formation by recording of density changes between fractured and unfractured reservoir intervals (in case of development of sufficient density differences; cf. section 6.3.1.1.), as well as for location of fluid interfaces in production wells, determination of cavity tops in storage wells, detection of the position of stuck pipe, and location of cement top and discontinuities within the cement mantle. If properly applied, photon logging could even be the best tool for volumetrical cement evaluation.

6.3.1.4. Photon/vibrapack combination

A more advanced technique is the photon/vibrapack combination (DRESSER ATLAS 1987 a) which is a complete system for identification of deficiencies in gravel packs and immediate in-situ repair. The photon/vibrapack integrated assemblage can be used for location of low-density sections and only partially packed zones and voids within the gravel pack by evaluation of relative density of gravel and fluid in the annulus. The instrument transmits mechanical energy through the fluid to the annulus during revolution, and the vibration together with gravitational settling improves gravel compaction, thus enabling to avoid damage of casing or liner by string-shot explosives and perforation or formation collapse associated with high differential pressures during swabbing, and yet exceeding the energy transmission of frequency-generation techniques. Aspects of application of liner vibration with gravel packing for effective sand control are also discussed by SOLUM (1984; cf. section 5.8.4.).

Repeated alternating logging and shaking of the gravel pack can be done in order to improve gravel mantle density. Inhomogeneous gravel settling as a response to shaking and vibration, however, can result in creation of a void bridge between two packed zones, with multiple and repeated shaking being able to move the bridge upwards. The high degree of repeatability of the photon log (if bulk density is not changed) results in reading of the curves on top of each other in the shoulder areas between several underreamed or perforated and packed intervals and thus allows many times of alternating cross-checking by logging and mechanical improvement of pack quality by shaking. Additional calibration and reliability checking during repetition runs between several gravel pack vibration remedial treatments can be achieved by logging of a standard section in the blank liner zone above the perforated or underreamed interval where borehole conditions remain absolutely constant.

Accurate depth correlation is verified by combination with a casing collar locator. Gravel pack vibration can be run at all borehole deviation angles from vertical to horizontal. Wireline devices for gravel pack repair are also presented by CARROLL, NEAL & PARKER (1986) and SCHLUMBERGER (1988 a). If voids are present in the gravel pack, reciprocation of the repair tool agitates the well fluid in the pack area, and the resultant turbulence breaks up bridges and eliminates voids. If a bridge is not dislodged by fluid turbulence, little controlled explosions can be carried out in order to shake and pulse down the pack.

6.3.2. Gamma-ray logging

Apart from the photon log which is a special method of gamma-ray recording and interpretation, other possibilities of gravel pack evaluation by gamma-ray tools exist. Distinction can be made between nuclear fluid density meter which operates with one source and one detector, and dual-spacing gamma-ray which has two detectors that provide the possibility of looking at different depths of investigation (NEAL 1983, NEAL & CARROLL 1985, KREIJ 1986). Both qualitative and quantitative gamma-ray gravel pack evaluation is possible, and the instrument records both natural and artificial radioactivity of the gravel-pack material. Some comments are also made on gamma-ray spectroscopy of multiple radioactive tracers.

6.3.2.1. Qualitative gamma-ray gravel pack evaluation

In both logging instruments, gamma rays that are emitted by the source into the borehole and its surroundings are partially scattered back to the detector. Number of gamma rays returning to the detector and thus attenuation or absorption factor of the emitted radiation is an indication of the density of the material through which the gamma rays travelled (WAHL, TITTMAN, ALGER & JOHNSTONE 1964). Higher-density materials cause more gamma rays to be slowed down and absorbed, while lower-density materials allow more gamma rays to be returned to the detector. In a gravel-packed well, everything remains constant except of the annular space between screen and casing which can be completely filled with gravel, partially filled with gravel, or have void areas containing no gravel. In each of these cases, the area without gravel is filled with some type of fluid, and since gravel density is different from fluid density, it is possible to identify the zones which are not packed with gravel.

A special application is the control of the degree of uniformity of gravel packing from bottom to top of the column by radioactive pellet introduction into the gravel lot and subsequent gamma-ray recording (FLANIGAN 1979). If no correlations could be drawn between the sequence of placement of radioactive pellets in the slurry and the succession of their appearance on the logs, this constellation indicates that the gravel does not pack in a perfectly uniform bottom-to-top sequence. Some ambiguity, however, might originate from uneven distribution of the radioactive pellets in the gravel pack slurry due to primary mixing and/or secondary segregation effects of the subordinate amount of grains interspersed within the bulk of the gravel pack material. A better solution of the approach, although being more expensive and also more complicated in terms of safety handling, would be insertion of exclusively radioactively tagged gravel.

6.3.2.2. Quantitative gamma-ray gravel pack evaluation

In contrast to most of the other gravel pack monitoring wire-line logging tools which only allow qualitative assessment (with an exception being compensated neutron log; cf. SOLLEE 1985; cf. section 6.3.3.3.), a single-source single-detector gamma-ray logging instrument can also be used for quantitative gravel pack evaluation (NEAL & CARROLL 1985). The annulus between screen and casing is in all possible ways filled with some type of fluid of known density in those places where gravel is absent, and because gravel density is different from fluid density, it is possible to correlate gamma-ray count rates to percentages of gravel pack or only fluid-filled annular sections, respectively. The top of the gravel pack can be identified by a sudden increase in count rate, and similarly also void spaces of the interval along the screen can be picked up by their anomalous count rates. Calibrating with the extremes of totally fluid-filled annulus and optimum gravel pack density, gamma-ray count rate can be recomputed into gravel pack density or void space percentage which can even be expressed in a curve for visual presentation and inspection.

6.3.2.3. Natural and artificial radioactivity logging

Similarly as height monitoring in hydraulic fracturing can be performed by recording either natural or artificial radioactivity (cf. section 6.2.1.3.), also gravel pack density and length evaluation can be done by tracing either original or synthetic radioactivity of sand or proppants as well as artificial radioactivity of tagged fluid residues. Independent from screen extension and thickness of the perforated zone, calibration of gamma-ray logging tools or other instruments for registration of radioactivity can be done by setting some radioactive bullets into the pay zone interval and locating them on control runs. In combination with a casing collar locator, radioactive bullet logging is not only a suitable means for monitoring of gravel and/or formation compaction, but can also be applied for oil-field subsidence monitoring (ALLEN 1969, 1984; RUDDY, ANDERSON, PATTILLO, BISHLAWI & FOGED 1983). RUEDRICH, PERKINS & O`BRIEN (1974) comment on precise joint length determination using a multiple casing collar locator tool.

6.3.2.4. Gamma-ray spectroscopy of multiple radioactive tracers

Gamma-ray logging of radioactive markers in gravel packing applications mainly aims on recording of synthetic tracers added to gravel material and/or carrier fluids, because utilization of naturally radioactive zirconia-silicate proppants (cf. section 6.2.1.2.) is in most cases beyond consideration for gravel packing due to their extraordinarily high price (cf. sections 1.4.3., 2.2.2.3. and 5.2.2.3.). Gravel pack height by radioactive marker monitoring can be best obtained when coating the gravel grains with radioactive pellicles and recording the tracer spectrum. While gravel pack height control usually requires only a single marker which can be registered by a conventional tool, additional tagging of gravel pack fluids can demonstrate the zones of preferential and/or retarded leakoff of gravel carriers which can independently from the gravel record confirm the effectivity of the sand control treatment. Several gravel stages which are brought in discretely between solid-free fluid lots (particularly applying for highly-deviated to horizontal borehole gravel packing) can be distinguished by gamma-ray spectroscopy of multiple tracers (cf. section 6.2.1.3.4.), with special applications being visualization of prepacking of perforation tunnels, inspection of upper blank or reserve section, monitoring of gravel settling efficiency during vibrapack improvement and remedy operations, and control of gravel transport, settling and packing density at the upper and lower sides of highly-inclined to horizontal wells.

In combined mechanical gravel packing and chemical reservoir consolidation treatments, different tagging of gravel and excess resin solution enables to verify perforation penetration of the fluid and reflects the zones of additional formation matrix stabilization. Discrete marking of gravel lots injected in stages during multiple-interval sand control treatments allows to distinguish the different completion sections and permits to recognize proper selective gravel pack installation. Utilization of different particle coatings and/or fluid additives in multiple-level consolidated pack insertion or resin-coated gravel placement is able to monitor after drillout of the inside aggregates whether all the perforations in each interval are properly filled or whether repair has to be made in order to avoid evacuation of perforations and invasion of formation sand into the borehole as a consequence of absence of wellbore-lining gravel mantle. Tagging of fluid-loss agents in pills or pre- and/or post-gravel pack acids monitors distribution of solid additives on the gravel/formation interface or penetration of acid within the borehole-surrounding reservoir matrix, thus being able to recognize whether all the perforations are highly-permeable or whether some of them are bypassed due to plugging with formation sand.

Radioactive density monitoring is also a suitable means for controlling metering of the proppant concentration in the blending and delivery equipment during hydraulic fracturing and gravel packing operations (MARPLE, RULEY, WORLEY & FREEMAN 1987). Field experience of gamma-ray gravel pack monitoring is also reported by ATTARD, MATHES & MOWER (1988).

6.3.3. Neutron logging

Neutron logging with a borehole-compensated tool (NEAL 1983) uses one neutron source and two detectors being located at the near and far ends of the instrument. Neutron logging records the hydrogen index of a formation by emitting fast neutrons and counting those that are slowed down to the thermal energy level. The following outline focusses on operational principles, qualitative and quantitative neutron gravel pack evaluation, and other applications.

6.3.3.1. Operational principles

As hydrogen has the greatest energy-attenuating effect on neutrons, the compensated neutron log reflects primarily the fluid-filled part of the formation and is thus conventionally used as a porosity log for reservoir evaluation (ALGER et al. 1971). Gravel pack logging with neutron instruments utilizes the fact that the two detectors are affected differently by the area between tool and formation. When the annulus between screen and casing is filled with fluid, the instrument is looking at a zone that has apparently 100 % porosity, whereas in case of presence of gravel, the apparent porosity is greatly reduced. Proper calibration and scaling of the near and far count rates cause them to read on top of each other in zones that are 100 % packed, and to separate in zones of less than 100 % gravel pack density, with separation being greatest in zones which have no pack at all (NEAL 1983, KREIJ 1986).

In contrast to all the other tools that can be run in 1 11/16 " slim size fitting also through narrow screens, liners and tubings, the compensated neutron tool is only available in 3 3/8 " diameter and is therefore limited to larger-scale wider gravel packing and production installations. Improvement can probably be done by application of thermal multigate decay pulsed tools (OLIVER, WYATT & SMITH 1986).

6.3.3.2. Qualitative neutron gravel pack evaluation

Pulsed neutron logging in a thermal multigate decay system is a combination between neutron and gamma-ray logging by using a tool containing a pulsed neutron generator and a gamma-ray detector (SCHULTZ, SMITH, VERBOUT, BRIDGES & GARCIA 1983; SMITH, ARNOLD & PEELMAN 1983; BUCHANAN, CLEARMAN, HEIDBRINK & SMITH 1984; QUIREIN, VIGNE & CHAPMAN 1987; WYATT & SMITH 1987). The system measures the thermal neutron capture cross-section of the formation. The composite decay curve from both reservoir and borehole capture gamma rays is registered and separated into two individual components. The application of pulsed neutron tools for qualitative gravel pack monitoring is based on the fact that the presence of gravel changes the neutron absorption properties in the borehole by reducing neutron capture cross-section and increasing diffusion length in the wellbore, and also depends on increasing formation gamma-ray attenuation when traversing the higher-density gravel pack.

Neutron logging can be improved by combination of two curves, with the count rate at the near detector reading conditions around the screen and the far curve recording formation properties (BRUIST, JEFFERIS & BOTTS 1983; JEFFERIS,

BRUIST & BOTTS 1983). Both curves should approximately register the same and track each other when the annulus is filled with gravel, whereas if the annulus is filled with fluid, the readings differ resulting in curve separation. Near and far detector curves of the neutron logging tool, however, do not differentiate between gravel and fill.

6.3.3.3. Quantitative neutron gravel pack evaluation

The decay of gamma rays produced after neutrons die away is proportional to gravel pack density, and under ideal conditions, similar calibration between the end members of 100 % fluid-filled annulus and maximum gravel pack density enables quantitative gravel quality checking. Lacking gravel saturation is depicted by increases in the log value as a result of high-sigma brine filling the casing/tubing annulus instead of low-sigma gravel. For gravel pack evaluation in gas-bearing reservoirs in both qualitative and quantitative manner, correction has to be made for the gas influx on the neutron record (SCHLUMBERGER 1972). Thus in contrast to hydraulic proppant fracturing monitoring where only qualitative assessment of the proppant assemblage infilling the induced fracture or even only fluid penetration within the generated crack is possible, gravel pack mantle distribution and density can also be assessed in a quantitative manner by neutron and gamma-ray (cf. section 6.3.2.2.) evaluation.

6.3.3.4. Other applications

Further applications of pulsed neutron capture logging in a thermal multigate decay system in addition to gravel pack quality checking and evaluation are measuring and quantifying hydrocarbon saturation behind casing which allows utilization of pulsed neutron capture recording for locating bypassed hydrocarbons, observing movement of hydrocarbon-water-contacts, monitoring reservoir depletion, determining residual oil saturation in log-inject-log processes, and measuring efficiency of secondary and tertiary recovery projects (WYATT & SMITH 1987). Aspects of thermal multigate decay pulsed neutron logging in an unusual downhole environment are summarized by AJAM & RAHAL (1986), and RANDALL, GRAY, CRAIK & HOPKINSON (1985) present a new digital multi-scale pulsed neutron logging system.

Field applications of pulsed neutron capture logging are reported by ITOH & SCHOPPER (1977); SCHENEWERK, KNAPP & FERTL (1982); OLIVER & FERTL (1986); VASILEV, RANDALL, OLIVER & FERTL (1987); CORWITH & MENGEL (1988) and WOZNIAK (1988). Thermal decay time logging of gamma rays originating after slowdown of high-energy neutrons emitted as bursts to the formation is also applied for hydraulic fracture height monitoring (PEETERS & HARTLEY 1984; cf. section 6.2.1.5.4.). Gravel-pack top evaluation with dual-burst pulsed neutron logging is reported by SCRIMGEOUR & SHEPHERD (1988). Pulsed neutron capture logging is also a suitable method for assessment of acid stimulation (CORWITH & MENGEL 1988; cf. section 6.2.1.5.3.).

Compensated neutron, nuclear fluid density meter and dual-spacing gamma-ray tools all respond well to changes in gravel density in the annulus between screen and casing (NEAL 1983). Although a straightforward interpretation may be possible in case of presence of sufficient field experience, the evaluation is always improved by comparison with a reference log that is recorded after having set the screen (representing the gravel pack hardware) and before placing the gravel. Aspects of gravel pack evaluation by neutron logging are also reported by ELSON, DARLINGTON & MANTOOTH (1984).

6.3.4. Other gravel pack logging techniques

In addition to the aforementioned possibilities of gravel pack checking and evaluation, further useful information can be obtained by combination of neutron and gamma-ray logging, other methods, and time-lapse monitoring of gravel settling.

6.3.4.1. Combination of neutron and gamma-ray logging

A useful combination of neutron and gamma-ray logging is the placement of radioactively-coated gravel and/or tagged fluid and cross-checking the neutron log record by a simple gamma-ray survey which can very easily detect unpacked zones by the absence of the additional radiation coming from the coated gravel and/or residues of infiltrated tagged fluid. This combination or alternatively the application of the other above described gravel pack monitoring methods is particularly useful in gas-bearing formations where the reading of the neutron log tool can be disturbed by the gas influence (SCHLUMBERGER 1972).

Comparative evaluation of the applicability of gamma-ray and neutron logging for gravel pack monitoring reveals that both qualitative and quantitative interpretation are possible with either tool (SOLLEE 1985; cf. sections 6.3.2.2. and 6.3.3.3.). Gamma-ray instruments predict more reliably the packing percentage around the screen, but are more sensitive to lack of circulation than neutron tools. Both instruments are affected by brine density, with increasing density reducing the gamma-ray count and giving rise to measuring of reduced hydrogen by the neutron record. Calibration with 0 % gravel can be performed in the blank liner section above the gravel top. Although both tools may measure quite small voids, a minimum void length of 3 ft is required for a full count response.

6.3.4.2. Other methods

Principally also acoustic velocity and formation density tools can be used for the investigation of the annulus between screen and casing aiming on gravel pack quality evaluation in terms of density and length, but field experience has shown that the methods mentioned in this chapter give the best and most reliable results (NEAL 1983). Sonic bond logging is sometimes also useful in detecting open sections within the gravel pack (ALLEN & ROBERTS 1982). Checking of gravel pack completion results can also be done by production logging using spinner flow meter (cf. section 6.2.1.7.), gradiomanometer and temperature measuring device (ECONOMIDES 1986). Transient pressure testing is also a valuable tool for assessing gravel pack completion efficiency. Production flowmeters discover gravel voids and associated lack of flow (RASHID 1984).

6.3.4.3. Time-lapse monitoring of gravel settling

Similarly as height monitoring of hydraulic fractures can be performed repeatedly in a time-lapse manner during the lifetime of a well in order to check whether fracture height has diminished as a consequence of proppant crushing and crack closure (cf. section 6.2.1.4.4.), renewed gravel pack quality inspection can be done aiming on determination of progressive gravel settling in case of conventional flow-type placement. Gravel settling with time following placement of the pack is a common effect due to the instability of cubic packs and is usually accounted for by packing also a blank section of abt. 15 % of the liner length above the perforated interval (cf. section 5.3.2.4.3.). Resolution of gravel settling inspection can be considerably enhanced by gamma-ray spectroscopy of multiple radioactive tracers (cf. section 6.2.3.4.) which have been added to several successive lots of the gravel material inserted one after the other into the borehole. Gravel settling after the original installation of the pack can only be avoided by introducing hexagonal fabric instead of cubic framework through liner vibration (SOLUM 1984; cf. section 5.8.4.) which then does not necessitate subsequent control logging due to absence of any settling after initial completion.

6.3.5. Tell-tale screen

While the above discussed logging methods aim on quality monitoring of the gravel pack once it has been installed between screen and casing or borehole wall, the function of the tell-tale screen is to determine the end of the gravel pack operation by indicating the moment when the treatment is - regardless of its achieved quality - completely performed. The tell-tale screen is usually applied in reverse-circulation well equipments (cf. section 5.3.2.3.).

The tell-tale screen is part of the completion string and is detecting the height up to which gravel has been packed in the screen-casing annulus (SUMAN, ELLIS & SNYDER 1983). The tell-tale screen may be a short section of appropriately sized holes or slots placed just below the crossover parts (upper tell-tale screen) or may be a short piece of screen placed below the perforation bottom and just above the sump packer or other false bottom in the well (lower tell-tale screen). Surface pressure builds as the production screen becomes covered, fluid is diverted through the tell-tale, and sudden pressure increase occurs when gravel envelopes the tell-tale and covers the openings thereby restricting fluid return and resulting in pressure increase, with the packing operation then being completed. Following illustration of the function of the tell-tale screen in lower or upper position, an outline is given on its action related to different gravel packing stages.

6.3.5.1. Lower and upper tell-tale screens

The tell-tale screen is above the primary screen to indicate gravel fill after extensive circulation especially for conventional low-density, low-viscosity fluid application (SUMAN, ELLIS & SNYDER 1983). If high-viscosity gelled gravel carrier fluids are used, the tell-tale screen should be placed below the primary screen, with its main purpose not being to serve as an indicator when successful gravel packing is completed, but to give some assurance that premature bridging does not occur with resultant voids in the pack which could give rise to sand production and gravel settling. In deviated holes, the bottom isolated tell-tale screen aids in preventing a premature sandout because of bridging regardless of the fluid used.

Fluid losses from the gravel packing slurry through the tell-tale screen above the perforations cause rapid formation of an annular gravel bridge (RENSVOLD & DECKER 1978). While at high pumping rates and pressures a flow channel forms through the gravel bridge thereby allowing slurry to reach the perforations, low pumping rates and pressures can give rise to establishment of an annular gravel bridge sufficiently long so that the increasing pumping pressure at the surface could be misinterpreted as a sandout. In contrast, positioning the tell-tale screen at the bottom of the rathole allows production of rathole fluid through the washpipe thus preventing contamination of slurry going through the perforations, and promoting more complete fill-up of the annulus by the gravel bridge opposite to the tell-tale screen. Placement of the tell-tale screen below the perforations permits abt. 25 % more gravel to be packed than positioning it above by predominantly improved gravel compaction.

6.3.5.2. Tell-tale screen action and gravel packing stage

While two-stage operations often include a lower tell-tale screen, one-trip tools commonly incorporate an upper tell-tale screen to indicate when gravel fills the annular space. In addition to monitoring of treatment termination, the lower tell-tale screen has in combination with lower sealbore and washpipe the task of assisting in moving the gravel slurry down to the bottom of the screen before excessive leakoff and gravel deposition occurs (PEDEN, RUSSELL & OYENEYIN 1985). The bottom of the washpipe is seated in a lower sealbore below which the lower tell-tale screen is located (FLANIGAN 1979).

During initial gravel packing, all the wellbore fluids and solids are circulated through the lower tell-tale screen only up the washpipe and out the cross-over tool. During the main phase of gravel packing, prepad and slurry tend to rope down the low side of the hole (cf. section 5.9.1.3.3.) and pack the lower tell-tale first. As the pack moves uphole, less fluid is circulated through the lower tell-tale screen and some of the gel is squeezed into the formation as the gravel bridges up tight against it. Checking of gravel fill-in can also be performed already during the job by reversing the annular flow until the pressure increases, or circulation of gravel into the annulus until pressure increase occurs depending on the state of job completion (SPARLIN & COPELAND 1972).

6.3.6. Mechanical properties log

In contrast to all the above discussed logging techniques which aim on direct gravel pack evaluation after completion of the sand control operation and thus represent performance monitoring procedures and retroanalyzing tools, the mechanical properties log (TIXIER, LOVELESS & ANDERSON 1973; COATES & DENOO 1981, SCHLUMBERGER 1982 b) has only indirect relationships to the gravel pack job itself. The mechanical properties log is run prior to planning and design of mechanical, chemical or combined sand control treatments in order to check the stability of the various hydrocarbon-bearing sands in the prospective column and to predict prior to onset of exploitation or increase of offtake rates whether the grain fabric would be strong enough to allow to start oil and/or gas production or to increase pumping rates without the danger of provoking sand problems in the wellbore, or whether the rock structure is too weak and gravel packing or other sand control operations have to be carried out in order to permit undisturbed hydrocarbon withdrawal.

The mechanical properties log as a predictive tool is thus an aid for screening of gravel packing candidates and has considerable significance in early recognition of potential sand-trouble-making horizons and the necessity of performance of sand-control treatments prior to wellbore damage by onsetting sand production, as well as in money saving and thus budget containment by picking up of horizons which will resist to all the possible drag forces and pressure drawdowns during hydrocarbon offtake without sand production and therefore do not need any artificial stabilization. Following an outline of relationships between formation strength and sand production, some aspects of reservoir stability prediction are illustrated. An overview of an integrated sand production forecast system, and comments on single- vs. multiple-well feasibility aspects are also given.

6.3.6.1. Relationship between formation strength and sand influx

Sand production problems are created by destructive forces acting on the reservoir such as pressure gradient near the perforations, flow rate, and scrubbing action of the produced fluid. The ability of the sand to withstand these destructive forces is determined by two main factors being intrinsic formation strength and capability of the reservoir to generate stable sand arches around the perforations or the borehole (cf. section 5.9.4.). The outline as follows focusses on formation permeability and consolidation as well as corrections and improvements.

6.3.6.1.1. Formation permeability and consolidation

Intrinsic formation strength is governed by the state of confining stress (as determined by the difference between overburden stress and pore pressure), grain shape and sorting, and grain cementation including shaliness. Intrinsic formation strength correlates well with reservoir ability to produce hydrocarbons at high flow rate (TIXIER, LOVELESS & ANDERSON 1973). Consistently lower permeabilities of stronger sands limit the magnitudes of flow rates to which they are exposed and thus prevent them from action of excessive forces that could be destructive to the grain fabric, whereas weaker sands have relatively higher permeabilities and are thereby capable of producing at high flow rates even with small drawdowns, with the soft sands being susceptible to grain framework loosening and thus bearing the potential danger of sanding problems. Effects of flow rates on stability of unconsolidated sands are also discussed by TIPPIE & KOHLHAAS (1973).

The mechanical properties log provides a quantitative means for distinction of sands that are strong enough to produce hydrocarbons without any form of sand control from reservoirs where sand problems have to be expected. The method is based on correlation of in-situ strength with dynamic elastic moduli computed from sonic and density logs, with formation strength being inferred from the values of shear modulus and bulk compressibility, because there is a good correlation between intrinsic formation strength and the dynamic elastic constants (STEIN & HILCHIE 1972; TIXIER, LOVELESS & ANDERSON 1973). Shear modulus and bulk compressibility can principally be determined from sonic-compressional and shear velocities as well as bulk density, but in friable geologically young sediments where the sanding problem is most acute, the shear wave is highly attenuated, with shear velocity approaching fluid velocity.

An alternative technique is relating the elastic constants to sonic compressional-wave velocity, bulk density and shaliness of the sand. MONTMAYEUR & GRAVES (1986) discuss prediction of mechanical and elastical properties in consolidated and unconsolidated sands from sonic logging. Consolidated formations having sand grains cemented or compacted so that they remain in place and do not move even if there is some turbulent flow of fluids in pore spaces can be recognized by sonic log travel times in adjacent shales below 100 microseconds/ft and bulk densities in adjacent shales above 2.40 g/cm3 (COLLE 1988).

6.3.6.1.2. Corrections and improvements

Similarly as in neutron logging, correction of the gas influence has to be carried out in mechanical properties analysis in reservoirs containing light hydrocarbons (SCHLUMBERGER 1972), with normalization and calibration of sonic and density logs being best achieved by combination with neutron and induction-electrical logs. An enhancement of formation strength determination is the full wave sonic system (SCHLUMBERGER 1985, 1987 b; WELEX 1986) where the entire acoustic wavetrain is recorded and during digital processing broken down into various components comprising compressional, shear and Stoneley waves (cf. section 6.2.1.6.).

6.3.6.2. Reservoir stability prediction

In terms of practical interpretation and application of the mechanical properties log, experience has shown that a good correlation exists between sanding problems and the determined dynamic elastic moduli (TIXIER, LOVELESS & ANDERSON 1973). If the investigated reservoir sand exhibits both strong and weak sections, completions without sand control can be attempted if perforations are restricted to strong intervals. Estimation of the maximum sand-free hydrocarbon production rate have to be based on transient test and log data, and the maximum sand-free oil or gas offtake rate also determines the maximum drawdown permissible for a given well. Problems in evaluation may occur in boreholes with poorly cemented casing mimicking barefoot completion and misleading stress estimations, and in case of drillstem tests having been made without water cushions that expose the formation to high initial drawdowns. The overview as follows focusses on significance of sand stability forecast, gravel packing candidate recognition, optimum mud weight determination, and maximum hydrocarbon offtake rate estimation.

6.3.6.2.1. Significance of sand stability forecast

The most rewarding finding is that by running the mechanical properties log, various formations have turned out to be sufficiently strong (which was later confirmed by undisturbed open-hole production) in cases when weakness and consequently sanding problems were feared (TIXIER, LOVELESS & ANDERSON 1973). On the other hand, however, if both weak and strong sections are commingled in a longer interval and are exposed by a continuous chain of perforation holes, as well as in case of ambiguous interpretation of the mechanical properties log for any reason, emphasis should be put on economically going on the absolutely safe side and better gravel packing also such reservoir sands that obviously would not have needed it according to the mechanical properties interpretation, than renouncing of gravel packing of doubtful pay sands which later turn out to be really unstable and deteriorate the whole feasibility of the well by excessive sand invasion into the borehole.

Aspects of sand production forecast are also discussed by STEIN, ODEH & JONES (1974); STEIN (1975), COATES & DENOO (1981), BARROW & LASSEIGNE (1984), MASSIE & NYGAARD (1987); MORITA, WHITFILL, FEDDE & LØVIK (1987); MORITA, WHITFILL, MASSIE & KNUDSEN (1987); WEISSENBURGER, MORITA, MARTIN & WHITFILL (1987) and GHALAMBOR, KOLIBA, HAYATDAVOUDI & ALCOCER (1988). YASSIN & PEDEN (1987) present sand stability prediction on a microcomputer. The boundary reservoir drawdown pressure at which sand influx starts to occur is obtained by comparing the existing shear stress around the borehole to the failure shear stress, with a Mohr`s circle failure analysis being used in the comparison (SCHLUMBERGER 1985, 1987 b; WELEX 1986).

6.3.6.2.2. Gravel packing candidate recognition

Radial stress and tangential stress are the normal stresses in a horizontal plane at the borehole wall prior to hydrocarbon production (WESTERN ATLAS 1988 a). Those formations with the lowest initial tangential stress and greatest initial radial stress will fracture most easily, but typically have the lowest probability of potential sanding problems. Consequently sand production and thus gravel packing candidates are reservoirs with high initial tangential stress and low initial radial stress. Initial shear strength is a prediction of internal strength for soft compressible formations and is an essential part of sand production prognosis, whereas tensile stress is applicable to more brittle reservoirs that will fracture. Zones where the critical wellbore pressure is higher than the anticipated flowing well pressure are predicted to produce sand.

6.3.6.2.3. Optimum mud weight determination

Minimum mud weight is a forecast of the borehole pressure allowing sand production to occur, with natural sanding of the formation during drilling taking place if the actual mud weight does not exceed this critical pressure value (WESTERN ATLAS 1988 a). Minimum mud weight also expresses the radial and tangential stress constellation which is necessary to create the Coulomb failure condition that causes sand influx. Drilling is most effective when radial and tangential stresses are equal, because at that condition, there is no shear stress acting along a horizontal axis, with this therefore being ideal mud weight. Maximum mud weight is given by the possible breakdown of the formation when the total of internal tensile strength and tangential stress equal zero. The maximum flowing pressure that can be tolerated without sand production is formation pressure minus drawdown pressure (which is the difference between static formation pressure and the borehole pressure at which sanding occurs).

Minimum, ideal and maximum mud weights which result from sand production analysis and forecast are valuable input parameters for the design of drilling programs of additional wells in the same field. On a sand strength display composed of several curves, potential sanding zones may be observed wherever the difference between pore pressure gradient and hydrostatic head is smaller than the reservoir pressure drawdown. The main applications of mechanical properties evaluation in terms of sand stability or weakness recognition are prediction of wellbore pressures at which shear failure and thus sand influx can occur, limit of safe drawdown for sand-free production, identification of potential sanding zones for a given wellbore pressure and drawdown, and selection of perforating intervals to exclude the weakest zones (SCHLUMBERGER 1982 b). Mechanical properties log assessment for determination of formation strength for drilling and completion modelling and design is also discussed by ELKINGTON, STOUTHAMER & BROWN (1982); KAZI, ZEKAI & BAHAA-ELDIN (1983) and ONYIA (1987, 1988).

6.3.6.2.4. Maximum hydrocarbon offtake rate determination

Sonic logging can also be a suitable means for estimation of maximum hydrocarbon offtake rates before excessive sand production starts (ALLEN & ROBERTS 1982). Field experience has shown that within an area, correlation of formation strength or drawdown factor with sonic interval transit time is possible. Measurement of sonic interval transit time and productivity index at a low offtake rate in new wells could therefore together with the interval length provide an estimate of maximum oil production rate before the onset of sand problems. Other possibilities of assessing formation strength and determining maximum hydrocarbon offtake rates by sonic logging are discussed by STEIN & HILCHIE (1972).

The mechanical properties log is also an important tool for prediction of the three-dimensional shape of hydraulic fractures and to properly design a stimulation treatment in order to control height growth of the crack (HOLDITCH, ROBINSON & WHITEHEAD 1986). The most important application of mechanical properties data is to determine stress profile and varying properties in multilayered systems which are the typical formation type of low-permeability reservoirs. The knowledge of an accurate stress profile is inevitable for design of a fracture treatment that has the maximum opportunity of being contained within the productive interval (cf. also section 6.2.5.).

6.3.6.3. Integrated sand production forecast system

The following outline of the integrated sand production forecast system focusses on external calibration of mechanical properties logs, early appraisal drilling vs. late infill drilling, and synergistic evaluation.

6.3.6.3.1. External calibration of mechanical properties logs

As a consequence of their indirect nature and inherent reliance on simplifying assumptions, mechanical properties logs require external calibration (WEISSENBURGER, MORITA, MARTIN & WHITFILL 1987). Mechanical properties logs are affected by input sonic and density readings, empirical conversions, assumed simplified stress/strain behaviour, and limited perforation cavity failure theory. In mature areas, mechanical properties logs can be calibrated against hydrocarbon production history and sand control experience. In newer areas, mechanical properties logs can be calibrated by some extent against drill-stem tests, but most effectively against core strength tests and rock-mechanics-based cavity stability models. Mechanical properties logs are continuous data in contrast to restricted core samples and therefore inescapably provide the link between point-wise core-based rock mechanical sand production forecasts and uncored intervals or new wells. Mechanical properties logs and drill-stem tests are useful predictors of relative formation strength during initial phases of production, but are not sufficiently quantifiable throughout the anticipated reservoir life without absolute calibration against cores.

6.3.6.3.2. Early appraisal drilling vs. late infill drilling

While comparative mechanical properties modelling during early appraisal drilling might be still quite ambiguous due to lateral changes of facies associations in the reservoir complex, mechanical properties assessment and sand stability forecast may become quite reliable tools in early to late infill drilling if the pay sequence is not too heterogeneous and if facies changes occur in an expressible trend rather than in an abrupt and unpredictable manner. In addition to formation grain size and cementation which remain constant throughout the exploitation period, depletion of reservoir pressure with progressive exhausting of the hydrocarbon reserves introduces a factor of uncertainty that cannot be evaluated by mechanical properties analysis unless late-stage renewed logging and interpretation runs are performed.

6.3.6.3.3. Synergistic evaluation

The integrated sand production forecast system therefore synergistically uses geological description, core testing, log analysis and reservoir simulation to prognose perforation cavity stability at initial conditions and throughout the life of a well. This is of particular significance in newer development areas of especially offshore situation where sand stability prediction has to be made without analogous reservoir performance as a guide, in contrast to many mature producing areas where sand inflow forecast can be based on empirical relationships and historical precedence. The cost of positive sand control measures and deferred revenue from possibly reduced rates of initial hydrocarbon exploitation provide the economical incentive for improved sand production prognosis methodologies and optimized completion planning.

Determination of horizontal stresses and mechanical properties are in addition to sand production also used to analyze other borehole phenomena and problems such as fracturing collapse and lost circulation (AADNOY 1988). For deviated holes, mechanical evaluation provides a potential for minimization of borehole instability by planning of trajectory and direction accordingly.

6.3.6.4. Single- vs. multiple-well feasibility aspects

The potential impact of completion design determines whether a sand production forecast study and the necessary special data measurements are justified. For a single well of suspect strength, the cost of a conservative gravel pack could be less than a quantitative sand prediction study. Alternatively, knowledge of safe operating limits for natural completions in a multiwell field by a sound sand stability prognosis could recover the cost of a sand prediction study many times over by avoiding the expenses of standard gravel packing in every well, and requiring only selective gravel packing in those wells where sand invasion problems really have to be anticipated.

Other aspects of the integrated sand production forecast system are discussed by MASSIE, NYGAARD & MORITA (1987); MORITA, WHITFILL, FEDDE & LØVIK (1987) and MORITA, WHITFILL, MASSIE & KNUDSEN (1987). GHALAMBOR, KOLIBA, HAYATDAVOUDI & ALCOCER (1988) present an enhanced prognosis technique which enables also prediction modelling of gas wells with free-water contributions. Examples of field normalization of formation mechanical properties for use in sand control management are evaluated by JØRANSON, SPURLIN & EDWARDS (1988).

6.3.7. Caliper logging

Caliper logging in open holes can be used as an aid in estimating the amount of material required for gravel packing by depicting the irregularity of the shape of the borehole wall which has been caused by breakout and sanding-off during drilling, thereby being a tool for optimizing design of the sand control treatment by checking the quantity of gravel needed for mechanical reservoir stabilization (ALLEN & ROBERTS 1982). Caliper logging is of considerable significance for predictive modelling of hydraulic fracture propagation direction from borehole ellipticity orientation (cf. section 6.2.1.8.).

7. Bibliography

7.0. Summary and associated table of contents

As a complementary source of detailed specific information, in addition to the discussions in the text a bibliography on hydraulic proppant fracturing and gravel packing as well as associated stimulation, stabilization and monitoring technologies is presented which is divided into two parts. The first section gives a subject key classification of the references into various special topics of mainly technical and reservoir engineering nature. Within each subject chapter and its subdivisions, the references are cited in alphabetical order of the authors names and are illustrated by a few key words. The second section contains the general reference list which includes all the quotations in alphabetical order of the authors names that are contained in both subject key classification of the bibliography and preceding text (including the chapters proppant selection, marketing and economics, Rotliegend stimulation in Europe, hydraulic proppant fracturing, gravel packing and sand control, and propped fracture and gravel pack monitoring). The bibliography is only a selected compilation of most significant references according to their practical usefulness as judged from my experience. The main topic headers in the subject key classification are overview of hydraulic fracturing and economical framework of hydrocarbon drilling and stimulation, fracture design and evaluation, fracture features, proppant features; technical, fluid and temperature features; field case studies, special applications of hydraulic fracturing, gravel packing and sand control, and petrophysical evaluation for hydraulic fracturing and gravel packing design. Following the introductory general section assembling overview of hydraulic fracturing, and economical framework of hydrocarbon drilling and stimulation, the subject header fracture design and evaluation compiles references on general fracture design, special fracture design, tailored-pulse and tail-in fracturing, computer programs for fracture modelling, reservoir engineering aspects of fracturing, pressure analysis in fractured reservoirs, in-situ stress determination and significance, relationships between fracturing and well productivity, and horizontal fractures. The topic division fracture features includes references on geometry, parameters, conductivity and permeability, propagation, containment, damage, simulation, prediction, monitoring and detection, and performance evaluation, and orientation changes of fractures. The subject header proppant features assembles references on selection, testing, concentration and settling of proppants, as well as on proppant and fine particle transport, and resin coating of proppants and gravel. The topic branch technical, fluid and temperature features comprises references on well completion and stimulation operation, fluid loss in fractures, viscosity effects of crosslinked fracturing fluids, water-sensitive formations and foam application, temperature effects and heat transfer in fractures, protection of core samples, combination of acid and proppant fracturing, sandstone acidizing, combination of natural and hydraulic fracturing, formation damage, horizontal or drainhole drilling, explosive fracturing, reservoir heterogeneity; interwell tracer, interference and pulse testing; multilayered reservoirs, multiple-fracture systems, clay stabilization, and reservoir compaction and field subsidence. Examples of field case studies from Europe, USA and Asia are reviewed with special emphasis on aspects of tight gas sands. The subject header special applications of hydraulic fracturing compiles references on minifracturing and pressure injection tests, moderate-permeability oil-reservoir fracturing, chalk fracturing, naturally fractured reservoirs, coal fracturing and degasification, shale fracturing, geopressured reservoirs, geothermal fracturing, water and steam injection well fracturing and gravel packing, refracturing, non-proppant fracturing, high proppant concentrations, and highly-deviated well fracturing. Gravel packing and sand control is discussed along the lines of type and grain size, performance and design, monitoring and evaluation, transport and settling, checking and handling of gravel; highly-deviated well gravel packing, prediction of sand production, special gravel pack applications, formation consolidation, pre-pack liners and wire-wrapped screens, sand arching, and summary reviews and field experience reports. The main aspects of petrophysical formation evaluation for hydraulic fracturing and gravel packing design are relative and absolute permeability determination, permeability deterioration, tight reservoir log interpretation, and clay mineral morphology and distribution. The reference list contains over 3,500 entries of quoted papers on technical and economical aspects of hydraulic proppant fracturing and gravel packing as well as associated completion, stimulation and consolidation technologies.

7.1. Introduction

In the preceding six text chapters, hydraulic proppant fracturing and gravel packing have been discussed along the lines of proppant selection (chapter 1), marketing and economics (chapter 2), Rotliegend stimulation in Europe (chapter 3), hydraulic proppant fracturing (chapter 4), gravel packing and sand control (chapter 5), and propped fracture and gravel pack monitoring (chapter 6). The material was compiled both from my practical market experience and from the extensive literature. As the collection of references is also of considerable importance for anyone entering the field of hydraulic proppant stimulation, the literature is processed here in a bibliography that is self-explanatory and partially independent from the preceding text in case the reader needs to get hand of more details very fast. The significance of the literature compilation is briefly sketched along the lines of subject key classification and reference list, and aims and understanding of the bibliography as follows.

7.1.1. Subject key classification and reference list

The bibliography contains selected references of papers which enable introductory information and allow an approach to further and more specialized literature through the general reference list at the end of the assemblage of cited contributions that are sorted into topic divisions. In order to give the possibility of getting a rapid thematical overview, the bibliography is divided into two parts. The first section presents a classification of the entries into various special topics of mainly technical and reservoir engineering nature, with the references being listed under the individual subject headers in alphabetical order of the authors` names and being characterized by a few key words. According to their thematical diversification, many papers are not only assigned to one topic column, but are cited under several subject keys. The second portion contains the main reference list which includes all the quotations in alphabetical order of the authors` names. This twofold combination allows fast access to the individual references with the help of the authors` names that have been found in the subject classification section or which have already been previously known. For sake of brevity, the key words from the subject headers are in many cases not repeated again in the individual reference lines within each topic division.

7.1.2. Aims and understanding of the bibliography

The bibliography of selected references does only compile those contributions from the vast literature on stimulation which I have assessed to be most significant for the reasonable quick and thorough approach of the individual subjects when preparing the present overview and status report as well as during technical, reservoir engineering and economical consulting when working in the European proppant fracturing market. Therefore the bibliography does by far not claim to be more or less complete, but only represents a selected compilation which has proven to be of practical benefit (otherwise the bibliography could have easily matured to such a size to become a little book in itself). Further access to the literature can often also be made by looking on other papers of the same author following the quotations in the reference list. More literature sources are also contained and discussed in the corresponding text chapters to which cross reference is made at the beginning of the divisions of the bibliography.

The subject key classification is the result of my experience with difficulties of rapid literature search on specialized topics and therefore is scheduled to allow the fast reader who has to compile an urgent report a quick entry to the theme by giving a selection of suitable introductory references. The bibliography is not an alternative to the text, but can be of complementary service to the evaluation of individual text sections. As new papers are continuously coming up, but once a deadline for publication of this review and status report had to be fixed, subject classification and reference list in this bibliography are permanently kept up-to-date on-line.

7.2. Subject key classification

The contributions entered in the reference list are assembled here in a thematical classification into various special topics (an overview of the spectrum of subjects is given in the associated table of contents following the summary in section 7.0.). The references are grouped in alphabetical order of the authors` names under each topic header, and a couple of key words provide further information on the content of the individual papers. The full bibliographical quotation can then be found under the authors` names in the following reference list (section 7.3.).

The main subject classification keys for subdivision of the vast literature into special topics are overview of hydraulic fracturing and economical framework, fracture design and evaluation, fracture features, proppant features; technical, fluid and temperature features; field case studies, special applications of hydraulic fracturing, gravel packing and sand control, and petrophysical evaluation for hydraulic fracturing and gravel packing design.

7.2.1. Overview of hydraulic fracturing and economical framework

Under this topic header, references on historical review, state-of-the-art, proppant invention and development, introduction to stimulation, experience reports, technology proceedings, keynote addresses, and monographies of hydraulic proppant fracturing and other well treatments are listed in the section on overview of hydraulic fracturing. The division of economical framework of hydrocarbon drilling and stimulation contains references on oil price evolution, US $ exchange rate variations, exploration and production reviews, significance of natural gas in the near future, importance of tight gas acquisition and exploitation, hydrocarbon supply and demand, political sanctions, and general energy economics.

7.2.1.1. Overview of hydraulic fracturing

The compilation of quotations on overview of hydraulic fracturing provides selected references to topics discussed mainly in sections 1.1. and 1.2. of the text.

AHMED et al. (1982) : state-of-the-art, Western tight sand, stimulation, frac
ALLEN & ROBERTS (1982) : production operations, fracture, gravel packing, log
API (1961) : history of petroleum engineering, review, evaluation
BROEK (1978) : elementary fracture mechanics, introduction, overview, status
CALLANAN, CIPOLLA & LEWIS (1983) : second-generation proppant, significance
CLARK (1949) : hydraulic fracturing, industry introduction, well stimulation
CLEARY (1988) : fracture engineering, future technology, state-of-the-art
COBB (1985) : overview, proppant application, history, state, problem, solution
COMPRESSED AIR MAGAZINE (1986) : general outline, oil-well persuading
COOKE (1976), COOKE & GIDLEY (1979) : high-strength proppant, deep fracturing
COOKE, GIDLEY & MUTTI (1977) : high-strength proppant, deep well fracturing
CUTLER et al. (1981) : new proppants, deep stimulation, high closure stress
DANIEL & WHITE (1980) : fundamentals of fracturing, mechanics, chemistry, state
ECONOMIDES (1986) : technology conference, hydraulic fracturing, gravel pack
ECONOMIDES & NOLTE (1987) : reservoir stimulation, proceedings volume, review
FUHRBERG (1983) : well completion, well stimulation, review, advance, progress
GIDLEY (1988) : recent advances, hydraulic fracturing, monograph, technology
HANSON, ANDERSON, SHAFFER & THORSON (1982) : stress, friction, fluid flow
HASSEBROEK & WALTER (1964) : 15 year review, advances, progress, summary
HICKEY, BROWN & CRITTENDEN (1981) : comparative effectivity, proppant impact
HOWARD & FAST (1970) : monograph, overview, hydraulic fracturing, proppants
HUBBERT & WILLIS (1957) : fracture and rock mechanics, overview, principles
KHRISTIANOVICH & ZHELTOV (1955) : hydraulic fracturing, introduction, methods
KHRISTIANOVICH et al. (1959) : theoretical principles, oil reservoir fracture
LAFLEUR & JOHNSON (1973) : North Sea, well stimulation, overview, fracturing
MADER (1987) : marketing, economics, Europe, oil price and US $ rate crises
MADER (1988) : technical, reservoir engineering, Europe, market extension
MARTINEZ, STEANSON & COULTER (1987) : formation fracturing, summary, handbook

MENDELSSOHN (1984) : fracture modelling, 2-D and 3-D models, height growth
MURPHEY & CARNEY (1977) : MHF, second look, review, evaluation
NELSON (1987) : fractured reservoir, turning knowledge into practice
PASINI & OVERBEY (1969) : natural and induced fracture systems, exploitation
PRATER (1968) : hydraulic fracturing, U.S. patent, method, application, example
REISS (1980) : reservoir engineering aspects of fractured formations
SARDA (1981) : high-strength proppants, deep reservoir stimulation, composition
SCHOLS (1983) : special topics, seminar, fracture mechanics, fluid relationship
STEPHENS & MARTINS (1985) : North Sea, experience, review, stimulation impact
USATCHOV (1986) : hydraulic fracturing, engineering textbook, overview
VEATCH (1983) : overview, design, treatment, technology, summary report, review
VEATCH & MOSCHOVIDIS (1986) : recent advances, technology, overview, summary
WATERS (1980, 1981) : introduction, hydrocarbon reservoirs, historical review
WESTERN PETROLEUM (1984) : proppants, selection, application, illustration
WILSEY & BEARDEN (1954) : reservoir fracturing, introduction, tight formations
ZHELTOV & KHRISTIANOVICH (1955) : hydraulic fracturing, oil-bearing beds

7.2.1.2. Economical framework of hydrocarbon drilling and stimulation

References on the economical framework of hydrocarbon drilling and stimulation (as discussed in chapter 2 of the text) are classified into oil price history, oil price collapse 1986 and consequences, oil production and trade, gas supply and demand, hydrocarbon acquisition feasibility vs. price, political relationships and conflicts, OPEC impact on oil price scenario, market structures and reactions, future hydrocarbon exploration and reserves, field development and maintenance, energy balance in specific areas; company acquisition, takeover, merging; risk analysis and amortization time, and other aspects.

7.2.1.2.1. Oil price history

The references on oil price history correspond mainly to aspects discussed in section 2.2.1.1. of the text.

ALGAR (1987 b) : OPEC, politics and economics, consequence of 1986 oil crisis
ATTIGA (1987) : the Arabs and the oil crises 1973 - 1986
BECK (1987) : improving oil prices, demand decline, production changes
CECIL (1987) : analysis of supply and demand for crude oil
DOUGHERTY (1987) : oil price evolution, prediction, direction, uncertainties
EBEL (1986) : oil in Soviet Union and China, problems, prospects, potential
ECK (1983) : world and USA outlook for oil and gas
EYCK (1987) : cost per barrel as an economic decision tool
GATELY (1984) : ten-year retrospective, OPEC and world oil market
GLASNER (1985) : politics, prices, petroleum, political influences, evolution
HILL (1987) : oil futures, options, importance increase, industry conditions
HOLMES (1987) : OPEC and oil market prospects today
IBG (1986) : OPEC, price control, limit price, orientation price, reorientate
KELLY (1987) : oil price stability, dream or reality, stabilization
KILGORE (1982) : uncertainties about future world oil prices
KONZELMANN (1976) : oil, fate of mankind, dependency, politics
LEE (1982) : cycles of crisis, seasons of glut, oil price evolution
LORBACH (1980) : world oil reserves and demand
LORBACH (1988) : future oil demand, oil crisis prevention, price evolution
LÜBBEN & LEINER (1988) : world oil reserves, OPEC/non-OPEC production capacity
LUKMAN (1987) : oil market stability key, cooperation, OPEC/non-OPEC countries
McCANN (1988) : OPEC and the future of oil
MOORE (1987 b) : well servicing, decline, increase, oil price
MORSE (1988) : impact on markets of non-OPEC oil supplies
OEHME (1986) : energy economics, change, oil price evolution
PETROLEUM REVIEW (1987 a) : OPEC oil, pressures on US $, demand/supply

POWERS & STEVENSON (1987) : perceptions, key to oil prices, review, evaluation
RANDOL (1986) : OPEC, oil price war, Saudi Arabia, Rambo
ROBERTS (1984) : crude oil price direction, prediction, uncertainty
RUNGE (1986) : oil price evolution, exploration, production, oil/gas, results
SAFER (1977) : world oil market, OPEC role, oil price evolution prediction
SCHMIDT (1987) : energy supply, oil industry, future, forecast
SEKERA (1986) : oil prices, 1921 - 2121, forecast, review, analysis
TEWKSBURY (1977) : prediction of 1985 oil price drop, OPEC role, market impact
VARZI (1986) : oil price, 1987 outlook
WEFA (1988) : boom or bust, recession, up- and down-swing, prognosis
YU (1988) : oil prices as paid in national currencies, evolution

7.2.1.2.2. Oil price collapse 1986 and consequences

The references on oil price collapse 1986 and consequences correspond mainly to aspects discussed in sections 2.2.1.1.2. and 2.2.1.5. of the text.

ALGAR (1987 b) : OPEC, politics and economics, consequence of 1986 oil crisis
ATTIGA (1987) : the Arabs and the oil crises 1973 - 1986
BAKER (1981) : effect of price and technology, gas resources, low-permeability
BECK (1987) : improving oil prices, demand decline, production changes
BENTLEY (1988) : company strategies in a changing market
BEUDELL (1986 c) : learn to live with low-priced oil, 1986 oil price crash
BILD (1987) : OPEC, whatever happened to market share
BRASHEAR et al. (1988) : enhanced oil recovery, lower oil prices, potential
CAMPBELL & HUBBARD (1984) : price forecasting, project evaluation, prediction
CAPEL (1988) : Western Europe, learning to live with low oil prices
CETRON (1986) : international political risks and energy project investments
CHAN (1985) : oil-field cashflow, independents` lifeblood for survival
CLARK (1986) : survival strategy, service industry
COX (1984) : economical aspects, marginal field development, North Sea
DAVID & CARSON (1983) : energy financing in times and uncertainty
ECK (1983) : world and USA outlook for oil and gas
EDGAR (1987) : restructuring in the oil patch
ELTING (1987) : new financial world for the independent petroleum producer
EYCK (1987) : cost per barrel as an economic decision tool
FISHMAN (1986 a) : North Sea, Southern Basin, independence from oil price drop
GARB, GRUY & HERRON (1981) : recoverable reserves are affected by oil/gas price
GLASNER (1985) : politics, prices, petroleum, political influences, evolution
GRUY, GARB & HERRON (1980) : recoverable reserves are affected by oil/gas price
GUSTAJTIS (1986) : dropping world oil price forces next oil crisis
HARBEN (1986) : oil price collapse, drilling reduction, rock bottom
HASSELL (1988) : restructuring for long-term survival
HOLMES (1987) : OPEC and oil market prospects today
KELKAR (1987) : current oil crisis, effects on oil and gas lease, court decide
KELLY (1987) : oil price stability, dream or reality, stabilization
KEMP, KELLAS & ROSE (1987) : end of field life economics, North Sea, oil/gas
KUMAR (1988) : restructuring of gas markets by regulations and legislation
LUKMAN (1987) : oil market stability key, cooperation, OPEC/non-OPEC countries
MABRO (1987) : netback pricing, 1986 oil price collapse, reasons, consequences
MABRO (1988) : oil price crisis, 1986, economical effects, policy responses
MADER (1987) : marketing, economics, oil price, proppant price, US $ rate, area
MENGES (1986) : OPEC crisis, impact on production, oil price drop
MINERALÖLWIRTSCHAFT (1983) : oil price, world economy factor, rise/drop, lack
NARAGHI (1988) : OPEC`s current US $ 18 price policy, future implications
OTTO (1986) : low crude oil prices could trigger serious gas shortages
PETROLEUM REVIEW (1987 a) : OPEC oil, pressures on US $, demand/supply
PETROLEUM REVIEW (1988 r) : regulate to avert a 4th price desaster, oil prices
ROBERTS (1984) : crude oil price direction, prediction, uncertainty
ROBERTS (1986, 1987) : OPEC and Adam Smith, review, evaluation
ROSBACO (1983) : investment evaluations in inflatory and unstable scenarios

SCHOLL (1986) : future manpower requirements, changing petroleum industry
SOLHEIM (1987) : oil market, OPEC, Norwegian foreign policy, oil company view
STAUFER (1988) : downstream integration by OPEC states, compelling competition
STORY (1986) : market stability request, oil-pricing and decision-making
TARIKI (1985) : global oil markets under stress
TEWKSBURY (1977) : prediction of 1985 oil price drop, OPEC role, market impact
WEFA (1988) : boom or bust, recession, up- and down-swing, prognosis
WOODS & HOLTBERG (1986) : hydrocarbon activity in an era of low oil prices
YU (1988) : oil prices as paid in national currencies, evolution

7.2.1.2.3. Oil production and trade

The references on oil production and demand correspond mainly to aspects discussed in sections 2.2.1.2., 2.2.1.4. and 2.2.1.5. of the text.

BAND (1987) : North Sea, production prospects, 2000, development, exploration
BARTON (1988) : how to make smaller fields viable
BAYLEY & COX (1982) : small discovery development conditions, North Sea
BECK (1987) : improving oil prices, demand decline, production changes
BRASHEAR et al. (1988) : enhanced oil recovery, lower oil prices, potential
CECIL (1987) : analysis of supply and demand for crude oil
CHADWICK, LONG & NISSANKE (1988) : USSR oil exports, market behaviour, trade
CHAN (1985) : oil-field cashflow, independents` lifeblood for survival
DAHLE (1987) : international oil markets, Norway`s role
DAHLE (1988) : gas versus oil, similarities and differences
DOUGHERTY & AL-BLEHED (1987) : world total primary energy demand, oil share
EBEL (1986) : oil in Soviet Union and China, problems, prospects, potential
ECK (1983) : world and USA outlook for oil and gas
FISCHER (1978) : Comecon dependency on USSR oil and gas supply
FISCHER (1981) : USSR natural gas industry, potential, requirements
FROGNES (1981) : Norway, oil policy, economic outlook, energy surplus country
FUEG (1988) : Eastern Europe, USSR sets another record
HARPER (1988) : developing cost-effective technology for the future
HILL (1987) : oil futures, options, importance increase, industry conditions
HOLMES (1987) : OPEC and oil market prospects today
KEMP, KELLAS & ROSE (1987) : end of field life economics, North Sea, oil/gas
KONZELMANN (1976) : oil, fate of mankind, dependency, politics
LORBACH (1980) : world oil reserves and demand
LORBACH (1988) : future oil demand, oil crisis prevention, price evolution
LÜBBEN & LEINER (1988) : world oil reserves, OPEC/non-OPEC production capacity
LUMSDEN (1988) : oil price and demand interactions
MENGES (1986) : OPEC crisis, impact on production, oil price drop
MIDDLETON (1987) : political and economical context of oil industry
MINERALÖLWIRTSCHAFT (1983) : oil price, world economy factor, rise/drop, lack
MORSE (1988) : impact on markets of non-OPEC oil supplies
NORENG (1987) : Norway, OPEC, rivalry, joint action, oil strategy, marginal
NYSTAD (1981) : North Sea oil and gas region, economical analysis, potential
ODD (1987) : oil and gas production economics, USA, benchmark for Europe
OFFSHORE ENGINEER (1985 c, 1986 c) : North Sea oil/gas fields, development plan
OSMAN (1987) : worldwide oil and gas supply and demand
PETROLEUM REVIEW (1987 a) : OPEC oil, pressures on US $, demand/supply
PICKENS (1983) : oil and gas industry challenges in the 1980`s
QUINLAN (1988) : North Sea, oil-field survey, new concern over oil prices
REID (1981) : oil exploration possibilities in China, review, outlook
RIGASSI (1985) : USSR, Eastern Europe, production troubles, enhancement need
RUNGE (1986) : oil price evolution, exploration, production, oil/gas, results
SCHMIDT (1987) : energy supply, oil industry, future, forecast
SCOTTISH PETROLEUM ANNUAL (1987) : oil and gas field development, North Sea
SOLHEIM (1987) : oil market, OPEC, Norwegian foreign policy, oil company view
STERN (1986) : Soviet oil and gas exports to the West, commercial transactions
SULLIVAN (1982) : economic risk evaluation, offshore development, North Sea

TARIKI (1985) : global oil markets under stress
WOODS (1987) : current low drilling level, effects on oil and gas production
WOODS & HOLTBERG (1986) : hydrocarbon activity in an era of low oil prices

7.2.1.2.4. Gas supply and demand

The references on gas supply and demand correspond mainly to aspects discussed in section 2.2.1.6. of the text.

AGER-HANSSEN (1982) : future of Norwegian gas
AL-FARISS (1988) : Saudi natural gas, future prospects
BAND (1987) : North Sea, production prospects, 2000, development, exploration
BERGMANN (1982) : gas supply and marketing trends
BERGMANN (1988) : gas market structures in Western Europe
BRASHEAR et al. (1982) : tight gas research, industry perspective, development
BRASHEAR et al. (1984) : tight gas resource, technology, economics, sensitivity
BRIERLEY (1988) : gas as a global energy source
BUSH (1982) : natural gas markets, order restoration by policy makers
CUNNINGHAM (1982) : market competition, shaping of future gas prices
DAHLE (1988) : gas versus oil, similarities and differences
DANCY et al. (1987) : natural gas marketing, declining market, legal duty
D`ANDREA (1979) : Europe, prospects of gas supply and market, influence on USA
D`ANDREA (1983) : prospect of increased reliance by West Europe on natural gas
D`ANDREA (1987) : role of gas, East-West energy trade, Europe, politic impact
DELACOUR (1986) : frontier gas, challenge, technical answers
ECK (1983) : world and USA outlook for oil and gas
FEHRINGER (1986) : natural gas, primary energy, present distribution, future
FISCHER (1978) : Comecon dependency on USSR oil and gas supply
FISCHER (1981) : USSR natural gas industry, potential, requirements
FISHMAN (1986 a) : North Sea, Southern Basin, independence from oil price drop
FISHMAN (1986 b) : North Sea, economics, oil and gas fields, disposal methods
FRANSSEN (1982) : gas demand and pricing
FRANSSEN (1988) : supply dimension, political, strategic, resource, implication
FROGNES (1983) : Norway, energy market, import, export, international energy
FUEG (1988) : Eastern Europe, USSR sets another record
GADONNIEUX (1988) : gas challenges in a competitive market
GAUTIER & RICE (1981) : conventional, low-permeability, gas reservoirs, compare
GREGORY (1986) : North Sea, Southern Basin, development, economical framework
GRUB (1984) : future market for natural gas, effect on exploration
GUNDERMANN (1982) : natural gas, role in the energy market of the future
HAAS, BRASHEAR & MORRA (1985) : tight sands, historic trends, current product
HICKS & RAHMAN (1984) : increasing gas prices, effects on known gas reserves
HOLDITCH (1974) : tight gas pay, economical production, status, outlook, future
JACOBSEN (1988) : Norway as a gas supplier
JENSEN (1988) : world gas supplies, markets and trade, energy surpluses
JIMISON (1986) : natural gas regulation and market, USA, rules changing, game
KALISCH (1988) : gas energy supply, outlook 1987 - 2010
KAZEMI (1983) : low-permeability gas sands, review
KUMAR (1988) : restructuring of gas markets by regulations and legislation
KUUSKRAA (1980) : unconventional gas, need, urgency
KUUSKRAA & HAAS (1988) : natural gas, low-permeability formations
KUUSKRAA & MEYER (1980) : world resources of unconventional gas, review
KUUSKRAA et al. (1978) : unconventional gas sources, vast potential, feasible
KUUSKRAA et al. (1979) : gas recovery, tight formations, technology, economics
LEWIS-DAVIES (1988) : future United Kingdom gas supply and demand
LIESEN (1977) : natural gas, present and future significance
LIESEN (1985) : European gas industry, future challenges, transmission, supply
LINDEN (1985) : gas, technology, competition, environment, conservation, share
LÜBBEN (1982) : natural gas, exploration, production, forecast, assessment
MAY & MATHEWS (1983) : natural gas industry problems, causes, responses
MAY & WHITE (1981) : gas price regulation, problems, prospects, perspectives

MEANS (1982) : gas industry future, Federal Energy Regulatory Commission (FERC)
NAYLOR & FERRIES (1988) : managing gas supplies, deregulated 1980`s markets
NORTHROP (1988) : natural gas production, low-permeability reservoirs
NORVIK (1988) : the European gas market
NYSTAD (1981) : North Sea oil and gas region, economical analysis, potential
ODD (1987) : oil and gas production economics, USA, benchmark for Europe
OFFICE OF TECHNOLOGY ASSESSMENT (1985) : U.S. gas supply, 2000, availability
PICKENS (1983) : oil and gas industry challenges in the 1980`s
PRICE (1986) : government policy and international gas trade
RIETMAN & KIDD (1977) : ultimate gas recovery improvement, realistic gas price
ROSENBERG et al. (1983) : sensitivity analysis, tight gas study, significance
SALAMA (1988) : natural gas supply, federal regulations, tight reservoirs
SCHINDEWOLF (1985) : natural gas, distribution, forecast, risk
SCHLESINGER (1982) : United States. gas, energy demand, competitiveness
SHARMA et al. (1988) : natural gas potential, Alaskan Arctic, outlook
STEINMANN (1985) : natural gas supply, transmission system, Europe, economics
STERN (1986) : Soviet oil and gas exports to the West, commercial transactions
TECLAW (1988) : natural gas, strategic fuel for the 1990`s
TEK (1979) : challenge of worldwide gas, market, supply, limitation
VALENTIN, VALAIS & CHAMPLON (1988) : gas production outlook, European problems
VEATCH (1981) : technology challenge, tight gas reservoir, recovery improvement
WHITTING (1979) : gas supply and demand strategy, Great Britain

7.2.1.2.5. Hydrocarbon acquisition feasibility vs. price

The references on hydrocarbon acquisition feasibility vs. price correspond mainly to aspects discussed in sections 2.2.1.4. and 2.2.1.5. of the text.

ALGAR (1987 b) : OPEC, politics and economics, consequence of 1986 oil crisis
ANDERSON & PHILLIPS (1986) : economic well performance, design optimization
BAKER (1981) : effect of price and technology, gas resources, low-permeability
BAND (1987) : North Sea, production prospects, 2000, development, exploration
BECK (1987) : improving oil prices, demand decline, production changes
BEUDELL (1986 c) : learn to live with low-priced oil, 1986 oil price crash
BRASHEAR et al. (1988) : enhanced oil recovery, lower oil prices, potential
CAMPBELL & HUBBARD (1984) : price forecasting, project evaluation, prediction
CAPEL (1988) : Western Europe, learning to live with low oil prices
CETRON (1986) : international political risks and energy project investments
CUNNINGHAM (1982) : market competition, shaping of future gas prices
DOUGHERTY (1987) : oil price evolution, prediction, direction, uncertainties
FISHMAN (1986 a) : North Sea, Southern Basin, independence from oil price drop
FISHMAN (1986 b) : North Sea, economics, oil and gas fields, disposal methods
FRANSSEN (1982) : gas demand and pricing
GARB, GRUY & HERRON (1981) : recoverable reserves are affected by oil/gas price
GLASNER (1985) : politics, prices, petroleum, political influences, evolution
GREGORY (1986) : North Sea, Southern Basin, development, economical framework
GRUY, GARB & HERRON (1980) : recoverable reserves are affected by oil/gas price
GUSTAJTIS (1986) : dropping world oil price forces next oil crisis
HICKS & RAHMAN (1984) : increasing gas prices, effects on known gas reserves
HOLDITCH (1974) : tight gas pay, economical production, status, outlook, future
KEMP, KELLAS & ROSE (1987) : end of field life economics, North Sea, oil/gas
KUUSKRAA et al. (1979) : gas recovery, tight formations, technology, economics
LINDEN (1985) : gas, technology, competition, environment, conservation, share
LUMSDEN (1988) : oil price and demand interactions
MADER (1987) : marketing, economics, oil price, proppant price, US $ rate, area
MIDDLETON (1987) : political and economical context of oil industry
NYSTAD (1981) : North Sea oil and gas region, economical analysis, potential
ODD (1987) : oil and gas production economics, USA, benchmark for Europe
OTTO (1986) : low crude oil prices could trigger serious gas shortages
POTTER (1987) : North Sea, Southern Basin, development plans, economical base
RENSBURG & MALIK (1987) : exploration change, economical implication, technique

ROSBACO (1983) : investment evaluations in inflatory and unstable scenarios
RUNGE (1986) : oil price evolution, exploration, production, oil/gas, results
SCHAEFER & POWERS (1984) : economics, changes, exploration/acquisition impact
SULLIVAN (1982) : economic risk evaluation, offshore development, North Sea
THOMAS (1986) : North Sea, field development, historic costs, future trend
UKOOA (1984) : North Sea, economics, feasibility, forecast, reserve decrease
WESTERHAUSEN (1980) : world energy demand prognosis, exploration potential

7.2.1.2.6. Political relationships and conflicts

The references on political relationships and conflicts correspond mainly to aspects discussed in sections 2.2.1.1. and 2.2.1.4. of the text.

ABBOTT (1982) : prospects of regulatory reform in natural gas
ALGAR (1987 b) : OPEC, politics and economics, consequence of 1986 oil crisis
ATTIGA (1987) : the Arabs and the oil crises 1973 - 1986
BARRETT (1986) : politics of energy policy, congressional perspective
BILD (1987) : OPEC, whatever happened to market share
CETRON (1986) : international political risks and energy project investments
CHADWICK, LONG & NISSANKE (1988) : USSR oil exports, market behaviour, trade
CLEUTINX (1986) : energy policy in the European community
D`ANDREA (1983) : prospect of increased reliance by West Europe on natural gas
D`ANDREA (1987) : role of gas, East-West energy trade, Europe, politic impact
EBEL (1986) : oil in Soviet Union and China, problems, prospects, potential
FISCHER (1978) : Comecon dependency on USSR oil and gas supply
FISCHER (1981) : USSR natural gas industry, potential, requirements
FRANSSEN (1988) : supply dimension, political, strategic, resource, implication
FROGNES (1981) : Norway, oil policy, economic outlook, energy surplus country
FUEG (1988) : Eastern Europe, USSR sets another record
GATELY (1984) : ten-year retrospective, OPEC and world oil market
GLASNER (1985) : politics, prices, petroleum, political influences, evolution
GODEC & ALLARD (1987) : Norwegian foreign policy, role of oil
IBG (1986) : OPEC, price control, limit price, orientation price, reorientate
JAMISON (1984) : future petroleum exploration, USA, perils of politics
KEG (1986) : energy in Europe, energy politics, demand, supply, development
KELKAR (1987) : current oil crisis, effects on oil and gas lease, court decide
KEPLINGER (1986) : geopolitical implications, Middle East oil
KING (1982) : industry`s view of United States energy policy
KNOTT (1985) : energy balance, decline into self-sufficiency, review, analysis
KONZELMANN (1976) : oil, fate of mankind, dependency, politics
KUMAR (1988) : restructuring of gas markets by regulations and legislation
LANDERS (1987) : oil-field sanctions, USA vs. USSR, lift, trade possibilities
LICHTBLAU (1986) : petroleum import and export, politics, USA
MABRO (1988) : oil price crisis, 1986, economical effects, policy responses
MAY & WHITE (1981) : gas price regulation, problems, prospects, perspectives
MEANS (1982) : gas industry future, Federal Energy Regulatory Commission (FERC)
MIDDLETON (1987) : political and economical context of oil industry
NARAGHI (1988) : OPEC`s current US $ 18 price policy, future implications
NAYLOR & FERRIES (1988) : managing gas supplies, deregulated 1980`s markets
NEAL & REED (1981) : impact of taxes on exploration economics in the USA
NORENG (1987) : Norway, OPEC, rivalry, joint action, oil strategy, marginal
POWERS & STEVENSON (1987) : perceptions, key to oil prices, review, evaluation
PRICE (1986) : government policy and international gas trade
RANDOL (1986) : OPEC, oil price war, Saudi Arabia, Rambo
RIGASSI (1985) : USSR, Eastern Europe, production troubles, enhancement need
SAUNDERS & SHIRAZI (1986) : world bank role in development countries, gas, oil
SCOTT (1986) : politics of energy policy, corporate perspective
SHARABI (1988) : Arabian political prospects, inter-Arabian politics
STERN (1986) : Soviet oil and gas exports to the West, commercial transactions
THOMPSON (1988) : impact of new tax law on internal cash flow generation

7.2.1.2.7. OPEC impact on oil price scenario

The references on OPEC impact on oil price scenario correspond mainly to aspects discussed in sections 2.2.1.1. and 2.2.1.4. of the text.

ALGAR (1987 b) : OPEC, politics and economics, consequence of 1986 oil crisis
ATTIGA (1987) : the Arabs and the oil crises 1973 - 1986
BILD (1987) : OPEC, whatever happened to market share
DOUGHERTY (1987) : oil price evolution, prediction, direction, uncertainties
GATELY (1984) : ten-year retrospective, OPEC and world oil market
HOLMES (1987) : OPEC and oil market prospects today
IBG (1986) : OPEC, price control, limit price, orientation price, reorientate
KEPLINGER (1986) : geopolitical implications, Middle East oil
KILGORE (1982) : uncertainties about future world oil prices
KONZELMANN (1976) : oil, fate of mankind, dependency, politics
LEE (1982) : cycles of crisis, seasons of glut, oil price evolution
LÜBBEN & LEINER (1988) : world oil reserves, OPEC/non-OPEC production capacity
LUKMAN (1987) : oil market stability key, cooperation, OPEC/non-OPEC countries
McCANN (1988) : OPEC and the future of oil
MENGES (1986) : OPEC crisis, impact on production, oil price drop
MIDDLETON (1987) : political and economical context of oil industry
MORSE (1988) : impact on markets of non-OPEC oil supplies
NARAGHI (1988) : OPEC`s current US $ 18 price policy, future implications
NORENG (1987) : Norway, OPEC, rivalry, joint action, oil strategy, marginal
PETROLEUM REVIEW (1987 a) : OPEC oil, pressures on US $, demand/supply
RANDOL (1986) : OPEC, oil price war, Saudi Arabia, Rambo
ROBERTS (1984) : crude oil price direction, prediction, uncertainty
ROBERTS (1986, 1987) : OPEC and Adam Smith, review, evaluation
SAFER (1977) : world oil market, OPEC role, oil price evolution prediction
SHARABI (1988) : Arabian political prospects, inter-Arabian politics
SOLHEIM (1987) : oil market, OPEC, Norwegian foreign policy, oil company view
STAUFER (1988) : downstream integration by OPEC states, compelling competition
UDGAARD (1987) : Norway, IEA, OPEC, oil prices, foreign policy

7.2.1.2.8. Market structures and reactions

The references on market structures and reactions correspond mainly to aspects discussed in section 2.2.1. of the text.

BAKER (1981) : effect of price and technology, gas resources, low-permeability
BAND (1987) : North Sea, production prospects, 2000, development, exploration
BATT (1983) : planning and financial controls of independent producers
BAYLEY (1984) : market prospects, United Kingdom, gas balance
BAYLEY & COX (1982) : small discovery development conditions, North Sea
BECK (1987) : improving oil prices, demand decline, production changes
BERGMANN (1982) : gas supply and marketing trends
BERGMANN (1988) : gas market structures in Western Europe
BUSH (1982) : natural gas markets, order restoration by policy makers
CUNNINGHAM (1982) : market competition, shaping of future gas prices
DAHLE (1987) : international oil markets, Norway`s role
DANCY et al. (1987) : natural gas marketing, declining market, legal duty
D`ANDREA (1979) : Europe, prospects of gas supply and market, influence on USA
FROGNES (1983) : Norway, energy market, import, export, international energy
GATELY (1984) : ten-year retrospective, OPEC and world oil market
GRUB (1984) : future market for natural gas, effect on exploration
GUNDERMANN (1982) : natural gas, role in the energy market of the future
HOCHHEISER (1988) : North American natural gas markets
JENSEN (1988) : world gas supplies, markets and trade
JIMISON (1986) : natural gas regulation and market, USA, rules changing, game
KUMAR (1988) : restructuring of gas markets by regulations and legislation

LUKMAN (1987) : oil market stability key, cooperation, OPEC/non-OPEC countries
MABRO (1987) : netback pricing, 1986 oil price collapse, reasons, consequences
MABRO (1988) : oil price crisis, 1986, economical effects, policy responses
MADER (1987) : marketing, economics, oil price, proppant price, US $ rate, area
MODESITT (1986) : petroleum industry restructuring, environmental regulations
MORSE (1988) : impact on markets of non-OPEC oil supplies
NAYLOR & FERRIES (1988) : managing gas supplies, deregulated 1980`s markets
NORVIK (1988) : the European gas market
ROWSE (1987) : world energy markets, coping with instability
RUNGE (1986) : oil price evolution, exploration, production, oil/gas, results
SAFER (1977) : world oil market, OPEC role, oil price evolution prediction
SAUNDERS & SHIRAZI (1986) : world bank role in development countries, gas, oil
SCHLESINGER (1982) : United States, gas, energy demand, competitiveness
SOLHEIM (1987) : oil market, OPEC, Norwegian foreign policy, oil company view
STERN (1986) : Soviet oil and gas exports to the West, commercial transactions
STORY (1986) : market stability request, oil-pricing and decision-making
TARIKI (1985) : global oil markets under stress
TEK (1979) : challenge of worldwide gas, market, supply, limitation
TEWKSBURY (1977) : prediction of 1985 oil price drop, OPEC role, market impact

7.2.1.2.9. Future hydrocarbon exploration and reserves

The references on future hydrocarbon exploration and reserves correspond mainly to aspects discussed in section 2.2.1. of the text.

AGER-HANSSEN (1982) : future of Norwegian gas
ALGAR (1987 a) : North Sea, review, forecast, exploration, development
AL-FARISS (1988) : Saudi natural gas, future prospects
BAKER (1981) : effect of price and technology, gas resources, low-permeability
BAND (1987) : North Sea, production prospects, 2000, development, exploration
BRASHEAR et al. (1982) : tight gas research, industry perspective, development
BRASHEAR et al. (1984) : tight gas resource, technology, economics, sensitivity
FISHMAN (1986 a) : North Sea, Southern Basin, independence from oil price drop
GADONNIEUX (1988) : gas challenges in a competitive market
GAUTIER & RICE (1981) : conventional, low-permeability, gas reservoirs, compare
GREGORY (1986) : North Sea, Southern Basin, development, economical framework
GRUB (1984) : future market for natural gas, effect on exploration
GRUY, GARB & HERRON (1980) : recoverable reserves are affected by oil/gas price
GUNDERMANN (1982) : natural gas, role in the energy market of the future
GUSTAJTIS (1986) : dropping world oil price forces next oil crisis
HAAS, BRASHEAR & MORRA (1985) : tight sands, historic trends, current product
HILL (1987) : oil futures, options, importance increase, industry conditions
HOYLE (1983) : continuing North Sea exploration, expectation, potential, risk
JAMISON (1984) : future petroleum exploration, USA, perils of politics
KALISCH (1988) : gas energy supply, outlook 1987 - 2010
KEMP (1984) : North Sea, future, forecast, new development, reserve increase
KUUSKRAA & MEYER (1980) : world resources of unconventional gas, review
KUUSKRAA et al. (1978) : unconventional gas sources, vast potential, feasible
LEWIS-DAVIES (1988) : future United Kingdom gas supply and demand
LORBACH (1988) : future oil demand, oil crisis prevention, price evolution
OFFICE OF TECHNOLOGY ASSESSMENT (1985) : U.S. gas supply, 2000, availability
PICKENS (1983) : oil and gas industry challenges in the 1980`s
SEKERA (1986) : oil prices, 1921 - 2121, forecast, review, analysis
TECLAW (1988) : natural gas, strategic fuel for the 1990`s
UKOOA (1984) : North Sea, economics, feasibility, forecast, reserve decrease
VALENTIN, VALAIS & CHAMPLON (1988) : gas production outlook, European problems

7.2.1.2.10. Field development and maintenance

The references on field development and maintenance correspond mainly to aspects discussed in sections 2.2.1., 2.4.1. and 2.4.2. of the text.

BARBA (1986) : improving return, fracture treatment investment, wireline input
BARTON (1988) : how to make smaller fields viable
BAYLEY (1984) : market prospects, United Kingdom, gas balance
BAYLEY & COX (1982) : small discovery development conditions, North Sea
CAMPBELL & HUBBARD (1984) : price forecasting, project evaluation, prediction
COX (1984) : economical aspects, marginal field development, North Sea
GARB, GRUY & HERRON (1981) : recoverable reserves are affected by oil/gas price
GREGORY (1986) : North Sea, Southern Basin, development, economical framework
GRUY, GARB & HERRON (1980) : recoverable reserves are affected by oil/gas price
HARBEN (1986) : oil price collapse, drilling reduction, rock bottom
HARPER (1988) : developing cost-effective technology for the future
JOHNSON (1978) : North Sea and Great Britain, economy, investment strategy, GNP
KEMP (1984) : North Sea, future, forecast, new development, reserve increase
KEMP, KELLAS & ROSE (1987) : end of field life economics, North Sea, oil/gas
MACADAM (1984) : North Sea stimulation logistics and requirements, marketing
MANCHESTER (1983) : project financing principles, independents, application
MINTZ (1981) : reserve-based financing, specific requirements and alternatives
MODESITT (1986) : petroleum industry restructuring, environmental regulations
OILMAN (1985) : North Sea, 1984/1985, record drilling, exploration/development
POTTER (1987) : North Sea, Southern Basin, development plans, economical base
QUINLAN (1988) : North Sea, oil-field survey, new concern over oil prices
ROSBACO (1983) : investment evaluations in inflatory and unstable scenarios
SCOTTISH PETROLEUM ANNUAL (1987) : oil and gas field development, North Sea
THOMAS (1986) : North Sea, field development, historic costs, future trend
TWEEDIE & KENNY (1982) : North Sea development risk, historical analysis
UKOOA (1984) : North Sea, economics, feasibility, forecast, reserve decrease
WOODS & HOLTBERG (1986) : hydrocarbon activity in an era of low oil prices
WORLD OIL (1985) : Western Europe, North Sea, activity, development

7.2.1.2.11. Energy balance in specific areas

The references on energy balance in specific areas correspond mainly to aspects discussed in sections 2.2.1. and 2.4.4. of the text.

AGER-HANSSEN (1982) : future of Norwegian gas
AL-FARISS (1988) : Saudi natural gas, future prospects
BAYLEY (1984) : market prospects, United Kingdom, gas balance
BERGMANN (1982) : gas supply and marketing trends
CHADWICK, LONG & NISSANKE (1988) : USSR oil exports, market behaviour, trade
CLEUTINX (1986) : energy policy in the European community
DAHLE (1987) : international oil markets, Norway`s role
D`ANDREA (1979) : Europe, prospects of gas supply and market, influence on USA
D`ANDREA (1983) : prospect of increased reliance by West Europe on natural gas
D`ANDREA (1987) : role of gas, East-West energy trade, Europe, politic impact
EBEL (1986) : oil in Soviet Union and China, problems, prospects, potential
ECK (1983) : world and USA outlook for oil and gas
FISCHER (1978) : Comecon dependency on USSR oil and gas supply
FISCHER (1981) : USSR natural gas industry, potential, requirements
FISHMAN (1986 a) : North Sea, Southern Basin, independence from oil price drop
FROGNES (1981) : Norway, oil policy, economic outlook, energy surplus country
FROGNES (1983) : Norway, energy market, import, export, international energy
FUEG (1988) : Eastern Europe, USSR sets another record
HOCHHEISER (1988) : North American natural gas markets
HOLTBERG, WOODS & ASHBY (1987) : U.S. energy supply and demand, baseline plan
KEG (1986) : energy in Europe, energy politics, demand, supply, development
LEWIS-DAVIES (1988) : future United Kingdom gas supply and demand

NORENG (1987) : Norway, OPEC, rivalry, joint action, oil strategy, marginal
NORVIK (1988) : the European gas market
NYSTAD (1981) : North Sea oil and gas region, economical analysis, potential
OFFICE OF TECHNOLOGY ASSESSMENT (1985) : U.S. gas supply, 2000, availability
RIGASSI (1985) : USSR, Eastern Europe, production troubles, enhancement need
SCHLESINGER (1982) : United States, gas, energy demand, competitiveness
SOLHEIM (1987) : oil market, OPEC, Norwegian foreign policy, oil company view
STERN (1986) : Soviet oil and gas exports to the West, commercial transactions
WHITTING (1979) : gas supply and demand strategy, Great Britain

7.2.1.2.12. Company acquisition, takeover, merging

The references on company acquisition, takeover, and merging correspond mainly to aspects discussed in section 2.2.1.5.2. of the text.

BATT (1983) : planning and financial controls of independent producers
BENTLEY (1988) : company strategies in a changing market
CAPEL (1988) : Western Europe, learning to live with low oil prices
CHAN (1985) : oil-field cashflow, independents` lifeblood for survival
CLARK (1986) : survival strategy, service industry
EDGAR (1987) : restructuring in the oil patch
ELTING (1987) : new financial world for the independent petroleum producer
HASSELL (1988) : restructuring for long-term survival
MACADAM (1984) : North Sea stimulation logistics and requirements, marketing
MANCHESTER (1983) : project financing principles, independents, application
MARPLE & RULEY (1987) : financing projects via turnkey contractors
MAXWELL (1986) : changes to industry viability, investment community viewpoint
MEISTER & CADY (1986) : corporate takeovers, petroleum industry restructuring
MILLER & NEWLIN (1987) : corporate takeovers and fusions, mega-merging, change
MINTZ (1981) : reserve-based financing, specific requirements and alternatives
MODESITT (1986) : petroleum industry restructuring, environmental regulations
MOORE (1987 b) : well servicing, decline, increase, oil price
MOORE (1988 b) : well servicing, activity bounds back
ODD (1987) : oil and gas production economics, USA, benchmark for Europe
PAULUS (1987) : leveraged buyouts in the petroleum industry
PELLEY (1982) : banking, energy, Federal role
PETRIE (1983) : petroleum mergers and acquisition, perspective, analysis
RAMSAY (1983) : project financing for small independent company
SCHOLL (1986) : future manpower requirements, changing petroleum industry
SIMOES (1985) : company merging and acquisition, petroleum industry
SOBOTKA (1986) : financing mergers and acquisitions
WYMAN (1978) : independent producer, rapidly changing hydrocarbon industry

7.2.1.2.13. Risk analysis and amortization time

The references on risk analysis and amortization time correspond mainly to aspects discussed in section 2.2.1.5.2. of the text.

ANDERSON & PHILLIPS (1986) : economic well performance, design optimization
BARBA (1986) : improving return, fracture treatment investment, wireline input
CAMPBELL & HUBBARD (1984) : price forecasting, project evaluation, prediction
CETRON (1986) : international political risks and energy project investments
COX (1984) : economical aspects, marginal field development, North Sea
GARB (1988) : risk assessment, hydrocarbon reserve estimation
KEMP, KELLAS & ROSE (1987) : end of field life economics, North Sea, oil/gas
MANCHESTER (1983) : project financing principles, independents, application
MEGILL (1984, 1985) : risk analysis, evaluation, management
MINTZ (1981) : reserve-based financing, specific requirements and alternatives
NEWENDORP (1975) : risk analysis, petroleum exploration/development, simulation
NEWENDORP (1983) : strategy for implementing risk analysis

PELLEY & CONSTABLE (1982) : reservoir risks, economical assumptions, bank lend
RANKIN (1982) : risk, stepping rules, economic viability, exploration drilling
ROSBACO (1983) : investment evaluations in inflatory and unstable scenarios
SULLIVAN (1982) : economic risk evaluation, offshore development, North Sea
VOSPER & STEVENS (1982) : development options for small gas fields
WEFA (1988) : boom or bust, recession, up- and down-swing, prognosis

7.2.1.2.14. Other marketing and economical aspects

The references on other marketing and economical aspects correspond to various aspects discussed in chapter 2 of the text.

ABBOTT (1982) : prospects of regulatory reform in natural gas
BARBA (1986) : improving return, fracture treatment investment, wireline input
BATT (1983) : planning and financial controls of independent producers
BENTLEY (1988) : company strategies in a changing market
CHAN (1985) : oil-field cashflow, independents` lifeblood for survival
CLARK (1986) : survival strategy, service industry
COOPER & MARSHALL (1984) : purpose-built stimulation vessel, North Sea usage
DAHLE (1988) : gas versus oil, similarities and differences
DAVID & CARSON (1983) : energy financing in times and uncertainty
ELTING (1987) : new financial world for the independent petroleum producer
KING (1982) : industry`s view of United States energy policy
KNOTT (1985) : energy balance, decline into self-sufficiency, review, analysis
MACADAM (1984) : North Sea stimulation logistics and requirements, marketing
MANCHESTER (1983) : project financing principles, independents, application
MARPLE & RULEY (1987) : financing projects via turnkey contractors
MAXWELL (1986) : changes to industry viability, investment community viewpoint
McPHAIL (1988) : effective evaluation of oil loans to reduce risk
MINTZ (1981) : reserve-based financing, specific requirements and alternatives
MODESITT (1986) : petroleum industry restructuring, environmental regulations
MORRISON & JOLLIFFEE (1982) : North Sea tax and economics
NEAL & REED (1981) : impact of taxes on exploration economics in the USA
ODD (1987) : oil and gas production economics, USA, benchmark for Europe
OTTO (1986) : low crude oil prices could trigger serious gas shortages
PELLEY (1982) : banking, energy, Federal role
PETROLEUM REVIEW (1988 r) : regulate to avert a 4th price desaster, oil prices
QUINLAN (1988) : North Sea, oil-field survey, new concern over oil prices
RANDOLPH (1974) : MHF treatments, potential of doubling U.S. gas supply, status
RAMSAY (1983) : project financing for small independent company
ROSBACO (1983) : investment evaluations in inflatory and unstable scenarios
ROSENBERG et al. (1983) : sensitivity analysis, tight gas study, significance
ROWSE (1987) : world energy markets, coping with instability
SCHOLL (1986) : future manpower requirements, changing petroleum industry
SINOR (1986) : long-term cycles, prediction of petroleum demand
STAUFER (1988) : downstream integration by OPEC states, compelling competition
STEWART (1986) : secured oil and gas lending in bankruptcies
STORY (1986) : market stability request, oil-pricing and decision-making
THOMPSON (1988) : impact of new tax law on internal cash flow generation
WEFA (1988) : boom or bust, recession, up- and down-swing, prognosis
WESTERHAUSEN (1980) : world energy demand prognosis, exploration potential
WYMAN (1978) : independent producer, rapidly changing hydrocarbon industry
YURKEWICZ (1986) : economical impact of take-or-pay settlements
ZWICKY (1984) : price forecasting, project evaluation, prediction, discussion

7.2.2. Fracture design and evaluation

This subject classification key includes a variety of different topics comprising general fracture design, special fracture design, MHF aspects, tailored-pulse and tail-in fracturing, computer programs for fracture modelling, reservoir engineering aspects of fracturing, pressure analysis in fractured reservoirs, in-situ stress determination and significance, relationships between fracturing and well productivity, and horizontal fractures. The topics related to the reference compilations are mainly discussed in chapter 4 of the text.

The section on general fracture design comprises references on laboratory and field evaluation, parameter studies, optimization of stimulation techniques, key factors of design, systematical approaches, and treatment technology. The portion on special fracture design contains references on MHF and mHF design technology, pressure analysis and testing as base information for design, formation stress, fracture fluid response, and parameter determination and optimization. The account on computer programs for fracture modelling quotes references on COMPUPROP, PROPTRAN, mathematical base for computer simulation and prediction, and proppant selection program examples. The division on tailored-pulse and tail-in fracturing comprises references on selective proppant placement, wellbore surroundings, multiple fracturing, loading rates, propellants, and fracture flow capacity enhancement.

The branch on reservoir engineering aspects of fracturing presents references on status reports, experience reports, performance prediction and evaluation, determination of fracture parameters, overview of new developments, influences on productivity increase, stimulation problems, planning and operation, and forecast vs. field results. The division on pressure analysis in fractured reservoirs comprises references on transient pressure, finite conductivity, formation response prediction, fracture pressure behaviour, buildup curves, unsteady state pressure distribution, proppant and fluid schedules, and pressure interpretation. The section on in-situ stress determination and significance includes references on principal and in-situ stress monitoring, influence on fracture containment, field experience reports, and laboratory investigations. The division on relationships between fracturing and well productivity discusses references on improved drainage, stabilized flow, productivity increase, performance evaluation, well spacing and fracture length, and test evaluation. The branch on horizontal fractures includes references on proppant transport, penetration, pressure distribution, tension distribution, reservoir thickness and well performance.

7.2.2.1. General fracture design

The references on general fracture design correspond mainly to aspects discussed in sections 1.2.1. and 4.2. of the text.

ABOU-SAYED, AHMED & JONES (1981) : systematical approach, fracture job design
ANDERSON & PHILLIPS (1986, 1987) : fracture design optimization, economic value
CLEARY (1980) : fracture shape, mechanisms and procedures
CLIFTON & ABOU-SAYED (1979) : computation, 3-D
CONWAY & HARRIS (1982) : laboratory and field evaluation, stimulation technique
COOPER, NELSON & SCHOPPER (1984) : design improvement, on-site computer system
CRITTENDON (1959) : frac treatment, mechanics of design and interpretation
CROCKETT, OKUSU & CLEARY (1986) : complete integrated model, real-time analysis
DANESHY (1973) : vertical fractures
DANESHY et al. (1971) : treatment parameters, effect on geometry
DETOURNAY, McLENNAN & ROEGIERS (1986) : poroelastic concepts, fracturing modes
DYES, KEMP & CHANDLE (1958) : sweep-out pattern, fracture effect
ELBEL (1986) : efficient recovery design
HAIMSON & FAIRHURST (1969) : hydraulic fracturing in porous-permeable materials
HANSON, ANDERSON & SHAFFER (1980) : geometry, effects of various parameters
HOLDITCH & MORSE (1976) : non-Darcy flow, fracture behaviour
HOLDITCH et al. (1978) : well spacing, fracture length, optimization, model
IRWIN & WIT (1983) : summary of fracture mechanics concepts
JAEGER & COOK (1979) : fundamentals of rock mechanics
KECK, CLEARY & CROCKETT (1984) : lumped numerical model, frac design

MENG & BROWN (1987) : geometry requirement, production forecast, design optimum
MEYER (1986) : design formulae, parametric studies, model comparison, 2-D, 3-D
OSBORNE, McLEOD & SCHROEDER (1981) : hydraulic fracturing, problem, analysis
PAI & GARBIS (1983) : MHF treatment design, definition, layout
PAI, GARBIS & HALL (1983) : MHF concepts, execution, review, experience
PEARSON, ABOU-SAYED & NGUYEN (1984) : parameters affecting design, measurement
POULSEN & SOLIMAN (1987) : optimal treatment design, procedures, hydraulic frac
PRATS (1961) : effect of vertical fractures on reservoir behaviour
SINSON, OGBE & DEHGHANI (1988) : stimulation strategy optimization, oil/gas
SMITH (1965) : hydraulic fracture stimulation design
STRUBHAR & JENNINGS (1977) : trends in massive hydraulic fracturing, examples
VEATCH (1983) : overview, design, treatment technology
WAREMBOURG (1980) : professional design
WAREMBOURG et al. (1985) : fracture stimulation design and evaluation
WHITE & DANIEL (1981) : key factors, MHF design

7.2.2.2. Special fracture design

The references on special fracture design are classified along the lines of conductivity significance, fluid importance, special techniques, design technology, and other aspects.

7.2.2.2.1. Conductivity significance

The references on conductivity significance correspond mainly to aspects discussed in sections 1.4.10. and 4.2. of the text.

ALDERMAN & WENDORFF (1969) : proppant packing of hydraulic fractures
BENNETT, REYNOLDS & RAGHAVAN (1982 a) : finite conductivity, single-layer pay
BENNETT, REYNOLDS & RAGHAVAN (1982 b) : finite conductivity, multilayer rocks
COULTER & WELLS (1972) : high proppant concentration, stimulation advantages
DERBY & SMITH (1979) : regained gas permeability, design improvement
HICKEY, BROWN & CRITTENDEN (1981) : tail-in, comparative effectiveness of props
KASPEREIT (1979) : suspended sand packs, fracture length
NOLTE (1984) : proppant and fluid schedule determination, pressure decline
POULSEN & SOLIMAN (1987) : prop schedule, prop quantity, minimization, economy
SMITH, MILLER & HAGA (1987) : tip screen-out fracturing, soft unstable rocks
TINSLEY & WILLIAMS (1975) : prop pillaring, open-crack sections, discrete patch
WAHL (1965) : horizontal fractures, propped fracture area

7.2.2.2.2. Fluid importance

The references on fluid importance correspond mainly to aspects discussed in section 4.3.4. of the text.

ADVANI et al. (1986) : fluid flow, structural response modelling, frac mechanic
ALMOND & GARWIN (1985) : high-efficiency, fracturing fluids, low temperatures
BARBER & THEMIG (1985) : fluid density control, selective prop placement, foam
BLACK & LANGSFORD (1981) : energized fracturing, 30 % CO_2, improved recovery
FORD, PENNY & BRISCOE (1986) : enhanced water recovery improves stimulation
FRISINGER & COOPER (1985) : optimum design, formation stress, fluid response
HANNAH, HARRINGTON & ANDERSON (1977) : fluid segment location, breaker schedule
HANSON et al. (1981) : stress friction, fluid flow, effects on hydraulic frac
KAMATH & GODBOLE (1985) : hot brine stimulation, natural gas hydrates, assess
KING (1982) : liquid carbon dioxide, sand fracturing
McLEOD (1983) : treatment design, high-viscosity cross-linked fluids
MISAK et al. (1978) : density-controlled fluids, selective proppant placement
NOLTE (1984) : proppant and fluid schedule determination, pressure decline
PHILLIPS & WILSON (1985) : new stimulation technique, fracture fluid recovery

POULSEN & LEE (1984) : fluid properties, time/temperature dependance, design
SMITH, ROSS & HENDRICKSON (1965) : hydrofluoric acid stimulation, sandstones
WHITSITT & DYSART (1970) : temperature, effect on stimulation design

7.2.2.2.3. Special techniques

The references on special techniques correspond mainly to aspects discussed in sections 2.4.1.2.3., 4.2. and 4.5.4. of the text.

AHMED, NEWBERRY & CANNON (1986) : multiple zone fracturing, treatment design
BARBER & THEMIG (1985) : fluid density control, selective prop placement, foam
BENNETT, ROSATO & REYNOLDS (1981) : fracture heterogeneity, winglength, reply
BLACK & LANGSFORD (1981) : energized fracturing, 30 % CO2, improved recovery
BLINTEN & AZIZ (1985) : very long gross interval stimulation
CUDERMAN (1982) : multiple fracturing, propellant, borehole aspects
CUDERMAN (1986) : tail-in, cased and perforated holes
CUDERMAN & NORTHROP (1984) : propellant-based technology, multiple wellbores
EASON (1985) : single-stage multiple fracturing, Niobrara Formation, effective
EVANS & HOLZHAUSEN (1983) : shallow hydraulic fractures, surface deformation
HANSON & LYONS (1964) : frac method, semipermeable prop plug, growth control
HANSON et al. (1984) : multiple fracturing, controlled pulse pressurization
HARRIS, BAILEY & EVERTZ (1984) : foam fracturing, evaluation, low-permeability
HICKEY, BROWN & CRITTENDEN (1981) : tail-in, comparative effectiveness of props
HOOD & POWELL (1982) : pilot fracturing treatment, semisubmersible, North Sea
KING (1982) : liquid carbon dioxide, sand fracturing
MACK & MASON (1985) : individualized fracture treatment design, benefit, sand
McHUGH & KEOUGH (1982) : pulse tailoring, permeability enhancement
MISAK et al. (1978) : density-controlled fluids, selective proppant placement
SHLYAPOBERSKY et al. (1988) : overpressure calibrated fracture design
SMITH, ROSS & HENDRICKSON (1965) : hydrofluoric acid stimulation, sandstones
SMITH, MILLER & HAGA (1987) : tip screen-out fracturing, soft unstable rocks
STOLLER (1985) : tailored pulse loading, new stimulation approach, perspective
SWIFT & KUSUBOV (1981) : multiple fracturing, tailored-pulse loading
TINSLEY & WILLIAMS (1975) : prop pillaring, open-crack sections, discrete patch
WAHL (1965) : horizontal fractures, propped fracture area
WARPINSKI & BRANAGAN (1988) : altered-stress fracturing, connectivity
WESTCOTT (1985) : multiple radial fracturing technique, tight reservoir

7.2.2.2.4. Design technology

The references on design technology correspond mainly to aspects discussed in section 4.2. of the text.

AL-KHATIB, KING & WILSON (1984) : hydraulic fracture design and evaluation
BARBER & THEMIG (1985) : fluid density control, selective prop placement, foam
BLINTEN & AZIZ (1985) : very long gross interval stimulation
BOYER, STUBBS & SCHWERER (1986) : frac design, coalbed properties, methane
BRANAGAN & COTNER (1983) : pre-frac well test, improvement procedure, analysis
CLARK, HARKIN, WAHL & SIEVERT (1977) : proppant transport model, design
CLEARY (1980) : comprehensive design formulae
CUESTA & ELPHICK (1984) : pressure test analysis, role in design/evaluation
DAVIS, NICKELL, LENT & ALLAM (1980) : bottom-hole pressure analysis, design
DERBY & SMITH (1979) : regained gas permeability, design improvement
EHLIG-ECONOMIDES & AYOUB (1986) : vertical interference testing across tight
ELBEL & SOOKPRASONG (1985) : cumulative production, type curve use
FRISINGER & COOPER (1985) : optimum design, formation stress, fluid response
HANNAH, HARRINGTON & ANDERSON (1977) : fluid segment location, breaker schedule
HANSON et al. (1981) : stress friction, fluid flow, effects on hydraulic frac
HICKEY, BROWN & CRITTENDEN (1981) : tail-in, comparative effectiveness of props
HOLCOMB (1982) : treatment choice, MHF vs. conventional frac, selection, size

HOLDITCH, JENNINGS, NEUSE & WYMAN (1978) : well spacing, fracture length
HOOD & POWELL (1982) : pilot fracturing treatment, semisubmersible, North Sea
KIEL (1970) : hydraulic fracturing, new process, dendritic, widening of cracks
LENT, NICKELL, DAVID & ALLAM (1980) : bottom hole pressure analysis, design
MACK & MASON (1985) : individualized fracture treatment design, benefit, sand
McLEOD (1983) : treatment design, high-viscosity cross-linked fluids
NOLTE (1984) : proppant and fluid schedule determination, pressure decline
PARKER & McDANIEL (1987) : design improvement, in-situ conductivity measuring
SATTLER et al. (1986) : lenticular tight gas, design/analysis, assessment
SCHLOTTMAN, MILLER & LUEDERS (1981) : MHF design, Cotton Valley Sands, Texas
SCHLUMBERGER (1986) : FracHite log, design technology
SHLYAPOBERSKY et al. (1988) : overpressure calibrated fracture design
SMITH (1981) : mHF, design
TEUFEL (1983) : design application, anelastic strain recovery, oriented cores
VOEGELE, ABOU-SAYED & JONES (1983) : design optimization, in-situ stress data
WHITSITT & DYSART (1970) : temperature, effect on stimulation design

7.2.2.2.5. Other aspects

The references on other aspects of special fracture design correspond to various items discussed in chapter 4 of the text.

AGARWAL (1980) : drawdown curves, producing-time effects, pressure buildup
ALMOND & GARWIN (1985) : high-efficiency, fracturing fluids, low temperatures
BENNETT, ROSATO & REYNOLDS (1981) : fracture heterogeneity, winglength, reply
BRANAGAN & COTNER (1983) : pre-frac well test, improvement procedure, analysis
CUESTA & ELPHICK (1984) : pressure test analysis, role in design/evaluation
EHLIG-ECONOMIDES & AYOUB (1986) : vertical interference testing across tight
ELBEL & SOOKPRASONG (1985) : cumulative production, type curve use
FORD, PENNY & BRISCOE (1986) : enhanced water recovery improves stimulation
HANNAH, HARRINGTON & ANDERSON (1977) : fluid segment location, breaker schedule
HANSON & LYONS (1964) : frac method, semipermeable prop plug, growth control
HANSON et al. (1981) : stress friction, fluid flow, effects on hydraulic frac
HARRINGTON, HANNAH & BEIRUTE (1978) : post-fracturing temperature recovery
HARRIS, BAILEY & EVERTZ (1984) : foam fracturing, evaluation, low-permeability
HOLDITCH, JENNINGS, NEUSE & WYMAN (1978) : well spacing, fracture length
JARED, SCOTT & EVANS (1984) : precision perforation, uniform stimulation
JEU (1983) : MHF optimization, fracture azimuth, reservoir data
KAMATH & GODBOLE (1985) : hot brine stimulation, natural gas hydrates, assess
McHUGH & KEOUGH (1982) : pulse tailoring, permeability enhancement
PAI (1982) : screenout termination, prediction, prevention, remedy
PERKINS & KRECH (1968) : energy-balance concept of hydraulic fracturing
ROBINSON (1985) : on-site inspection and testing, treatment improvement
SIMMONS (1983) : STRESSFRAC, new dimension in well treatment
TEUFEL (1983) : design application, anelastic strain recovery, oriented cores
THIERCELIN & LEMANCZYK (1983) : stress gradient, fracture height

7.2.2.3. MHF aspects

The references on MHF aspects correspond mainly to points discussed in sections 1.1.1., 3.4.1. and 4.8.1.2. of the text.

CONSTIEN, BRANNON & BANNISTER (1988) : world record, MHF stimulation, volume
FAST, HOLMAN & COVLIN (1975) : MHF application, Muddy "J", Wattenberg field
FITCH & MEDLIN (1978) : Piceance Basin, Rio Blanco County, Colorado, MHF
GIDLEY et al. (1979) : MHF, low-permeability gas formations, well performance
HANNAH, MATSON & CRAWFORD (1977) : MHF, Fort Worth Basin, status report
HOLCOMB (1982) : fracture treatment choice, MHF vs. conventional
KOZIK, BAILEY & HOLDITCH (1979) : MHF, Cotton Valley, Fallon and Personville
MALONE, STAHL & COULTER (1977) : MHF from planning to performance

MATTHEWS, MILLER & SCHLOTTMAN (1982) : record MHF job, Cotton Valley, Texas
MURPHY & CARNEY (1977) : MHF technology, review, prospects, applications
PAI & GARBIS (1983 a) : MHF treatments, design, evaluation, review, experience
PAI, GARBIS & HALL (1983) : MHF from concept to execution
PARKER (1980, 1981) : MHF, logistics, operational considerations, problems
PEARSON et al. (1988) : MHF, improvement, Cotton Valley Formation, East Texas
RANDOLPH (1974) : MHF stimulation, reserve doubling, economical significance
SCHLOTTMAN, MILLER & LUEDERS (1981) : MHF design, Cotton Valley Sands, Texas
SHORT (1978) : tight gas sands, Uinta Basin, massive fracturing treatments
STRUBHAR, FITCH & MEDLIN (1980) : MHF demonstration, Piceance Basin, Colorado
VEATCH & CROWELL (1980) : MHF technology acceleration, joint research-operation
WENDORFF & AINLEY (1980) : MHF, high-temperature wells, stable frac foams
WHITE & DANIEL (1980, 1981) : MHF design, key factors, Cotton Valley, Texas

7.2.2.4. Tailored-pulse and tail-in fracturing

The references on tailored-pulse and tail-in fracturing correspond mainly to aspects discussed in sections 2.4.1.2.3., 4.2.2.1.3. and 4.8.8.5. of the text.

AHMED, NEWBERRY & CANNON (1985) : tailored pulse loading, intermediate strain
AHMED et. al. (1982) : tailored pulse loading, propellant, intermediate strain
BARBER & THEMIG (1985) : selective prop placement, density control, foam frac
CALLANAN, CIPOLLA & LEWIS (1983) : tail-in scheduling, type and grain size
CHARLES, HUDOCK, UDICK & HALL (1983) : proppant tailoring, desired conductivity
CUDERMAN (1982) : multiple fracturing, propellants, borehole considerations
CUDERMAN (1986) : tail-in, cased and perforated holes
CUDERMAN & NORTHROP (1984) : propellant-based technology, multiple wellbores
DAVIS (1975) : tail-in scheduling, proppant type and grain size, selective
FREDRICKSON & BROADDUS (1976) : selective fluids, density/viscosity control
FOURNEY, HOLLOWAY & SIMHA (1984) : borehole pressure, distribution model study
GARBIS, BROWN & MAURITZ (1985) : tail-in scheduling, type and grain size
HOWARD & FAST (1970 a) : tail-in scheduling, proppant type and grain size
HICKEY, BROWN & CRITTENDEN (1981) : tail-in scheduling, type and grain size
KEOUGH, CURRAN & McHUGH (1984) : tailored pulse loading, gas well stimulation
KRUGMAN (1986) : brittle fracture theory, high-energy gas fracture, understand
KUNDERT & SMINK (1979) : tail-in scheduling, proppant type and grain size
LINDLEY & McGHEE (1983) : tail-in scheduling, type and grain size, selective
McHUGH & KEOUGH (1982) : fracture permeability enhancement, tailored-pulse load
McMECHAN & CONWAY (1983) : tail-in scheduling, proppant type and grain size
PAI & GARBIS (1983 b) : tail-in scheduling, proppant type and grain size
PAI, GARBIS & HALL (1983) : tail-in scheduling, type and grain size, selective
RISNES, BRATLI & HORSRUD (1982) : sand stress, wellbore surroundings, arching
SCHMIDT, WARPINSKI & COOPER (1980) : load concept, in-situ evaluation, tailored
STOLLER (1985) : tailored pulse loading, new stimulation approach, perspective
SWIFT & KUSUBOV (1980, 1981) : multiple fractures, intermediate loading rates
WARPINSKI et al. (1979) : high-energy gas frac, multifracturing in one well
WATSON, BENSON & FILLO (1986) : tailored pulse loading, explosion control
ZUBER, LEE & GATENS (1987) : tailored pulse load, radial explosive, hydraulic

7.2.2.5. Computer programs for fracture modelling

The references on computer programs for fracture modelling correspond mainly to aspects discussed in section 1.4.12. of the text.

AGARWAL, CARTER & POLLOCK (1979) : base for program, prediction, performance
BUHARALI, WRIGHT & WILLIS (1987) : real-time simulation, portable microcomputer
CLEARY (1988) : computerized real-time job control, data analysis, frac model
CLEARY, KECK & MEAR (1983) : microcomputer, hydraulic fracturing design
COBB (1985) : overview, proppant selection program, COMPUPROP, example, result
COBB (1986) : proppant selection program, COMPUPROP

COOPER, NELSON & SCHOPPER (1983) : on-site computer system, frac design improve
CHENG (1984) : proppant selection program
CLIFTON & ABOU-SAYED (1979) : computation, 3-D geometry
HALLIBURTON (1985) : proppant selection, PROP, MFD, stimulation computer, model
HANNAH, HARRINGTON & LANCE (1983) : real-time calculation, accurate pressure
HARRINGTON (1985) : computer, microprocessor, valuable well completion tools
HOLDITCH & LANCASTER (1982) : PROPTRAN, proppant distribution, numerical model
LOWE, McGLOTHLIN & HUITT (1966) : horizontal case, fracture design program
ROBERTS (1981) : combined computer frac simulation and type curve matching
SOLIMAN et. al. (1984) : combined computer frac simulation, type curve match

7.2.2.6. Reservoir engineering aspects of fracturing

The references on reservoir engineering aspects of fracturing are classified along the lines of general aspects, non-Darcy flow, transient fluid flow, and post-treatment production and pressure behaviour.

7.2.2.6.1. General aspects

The references on general aspects of reservoir engineering of hydraulic proppant fracturing correspond to various items discussed mainly in section 4.2. of the text.

ADVANI et al. (1982) : fluid flow, fracture mechanics, modelling, frac jobs
AGARWAL (1977), AGARWAL et al. (1980) : evaluate, predict, monitoring, analysis
BAKHTAR et al. (1985) : frac permeability model, depletion, well surroundings
BARENBLATT, ZHELTOV & KOCHINA (1960) : seepage of homogeneous liquids, fissures
BARGAS & YANOSIK (1988) : vertical fractures, effects on areal sweep efficiency
BENNETT & FORGERSON (1975) : reserve prediction, flow rate forecast, tight gas
BRINKMANN (1982) : status report, Germany, deep low-permeable
BRINKMANN, FUHRBERG & SCHÖBER (1980) : status, forecast, Germany
BRINKMANN, KRÖMER & REINICKE (1980) : fracspecific parameters, treat gas wells
GOLF-RACHT (1985) : fundamentals of fractured reservoir engineering, textbook
GRINGARTEN (1978) : reservoir limits testing, fractured wells
JEFFREY, VANDAMME & ROEGIERS (1987) : branched, subparallel frac, interaction
JOHN (1983) : stimulation problems, deep tight gas, Rotliegend, Germany
KLOSE & KRÖMER (1983) : fracture planning, fracture operation, Rotliegend, FRG
LEICHT (1985) : reservoir engineering aspects, MHF, Germany FRG
LORENZ, BRANAGAN, WARPINSKI & SATTLER (1986) : pay behaviour, stress sensitive
McGLOTHLIN & HUITT (1966) : formation rock strength, propping agent strength
MILLHEIM & CICHOWICZ (1968) : test, fracture analysis, low-permeability gas
NIERODE (1983) : design, field results, comparison
OLAREWAJU & LEE (1988) : crossflow, commingling, wellbore storage, layer skin
PRATS (1961) : vertical fractures, effect on pay behaviour, incompressible case
RAYMOND & BINDER (1967) : formation damage, productivity, vertical fractures
REINICKE, BRINKMANN, SCHWARZ & HUENI (1985) : buildup data interpretation, MHF
REISS (1981) : reservoir engineering aspects of fractured formations, review
RIECKMANN & PUSCH (1984) : reservoir engineering, new developments
SCHWARZ & SCHÖBER (1982) : MHF, deep tight gas
SENER (1986) : relative diffusion flux, fracture to matrix gas transfer
SLUSSER & RIECKMANN (1976) : low-permeable gas
SMITH (1973) : production ratio, incomplete frac fill-up, wellbore vicinity
TINSLEY, WILLIAMS, TINER & MALONE (1969) : frac height, effect on productivity
VERBEEK (1982) : production test analysis, tight solution gas drive
WATERHOUSE, CHARARA & NURMI (1987) : focus on fractures, significance, offtake
WONG (1970) : non-Darcy flow, turbulence factors, liquid saturation, gas-liquid
WOOD & JUNKIN (1970) : stress, displacement

7.2.2.6.2. Non-Darcy flow

The references on non-Darcy flow correspond mainly to aspects discussed in section 4.3.3.3. of the text.

ALLAM, CRICHLOW & SOLIMAN (1981) : non-Darcy gas flow, turbulence, closure
BROADDUS (1988) : water blocking, formation damage, removal with non-acid fluid
CALLANAN, CIPOLLA & LEWIS (1983) : non-Darcy flow, tight gas, prop conductive
EVANS & EVANS (1986) : non-Darcy flow, influence of saturation, compressible
EVANS, HUDSON & GREENLEE (1985) : liquid saturation, non-Darcy flow coefficient
FRAIM & LEE (1988) : non-Darcy flow, long-term gas production, pay properties
GEERTSMA (1974) : inertial resistance coefficient, porous media, fluid flow
GUPPY (1980) : non-Darcy flow, finite conductivity, vertical fracture
GUPPY, CINCO-LEY & RAMEY (1980) : non-Darcy flow, constant pressure production
HOLDITCH & LEE (1979) : non-Darcy flow, fracture evaluation, low-permeability
HOLDITCH & MORSE (1976) : non-Darcy flow, fraced well behaviour, gas reservoir
JONES (1972, 1987) : non-Darcy flow, unsteady state, Klinkenberg permeameter
JONES, BLOUNT & GLAZE (1976) : non-Darcy flow, turbulence, performance forecast
KADI & BOUTEMY (1981) : non-Darcy well test analysis, solution gas production
KADI & BROWN (1980) : non-Darcy flow, dissolved gas-drive reservoirs, inertial
LEE, LOGAN & TEK (1987) : turbulence, non-Darcy flow, transient flow, porosity
MALONEY et al. (1987) : partial saturation, gel damage, stress, non-Darcy flow
MARTINS (1987) : non-Darcy flow, North Sea tight gas stimulation, comparison
McDANIEL (1988) : non-Darcy flow, gas, oil, low viscosity, high production rate
NGUYEN (1986) : non-Darcy flow, experimental study, flow through perforations
NOMAN & ARCHER (1987) : effect of pore structure on non-Darcy flow
NOMAN, SHRIMANKER & ARCHER (1985) : inertial resistance, non-Darcy flow, rate
OLAREWAJU & LEE (1988) : crossflow, commingling, wellbore storage, layer skin
PEREZ & KELKAR (1988) : non-Darcy flow, two-phase pressure drop prediction
PRATS (1961) : vertical fractures, effect on pay behaviour, incompressible case
RAMEY (1975) : non-Darcy flow, wellbore storage, pressure buildup, drawdown
RIECKMANN (1970) : non-Darcy flow, inertial effects, turbulences, gas flow
SALDANA, BRIGHAM & RAMEY (1986) : wellbore inertial and frictional effects
SCHMIDT, CAUDLE & MILLER (1986) : gas well decline analysis, real gas behaviour
UBANI & EVANS (1982) : non-Darcy compressible flow of real gases, propped frac
WONG (1970) : non-Darcy flow, turbulence factors, liquid saturation, gas-liquid

7.2.2.6.3. Transient fluid flow

The references on transient fluid flow correspond mainly to aspects discussed in section 4.2. of the text.

ADVANI et al. (1982) : fluid flow, fracture mechanics, modelling, frac jobs
BARKER & RAMEY (1978) : transient flow, finite conductivity, vertical fracture
BENNETT & FORGERSON (1975) : reserve prediction, flow rate forecast, tight gas
LEE & BROCKENBROUGH (1983) : analytic solution, finite-conductivity, vertical
LEE, LOGAN & TEK (1987) : turbulence, non-Darcy flow, transient flow, porosity
McDANIEL (1988) : non-Darcy flow, gas, oil, low viscosity, high production rate
OLAREWAJU & LEE (1988) : crossflow, commingling, wellbore storage, layer skin
URAIET, RAGHAVAN & THOMAS (1976) : vertical fracture height, transient flow

7.2.2.6.4. Post-fracture production and pressure behaviour

The references on post-fracture production and pressure behaviour correspond mainly to aspects discussed in section 4.2. of the text.

BAUMGARTNER (1983) : fracture and production behaviour, treated wells, Germany
BRIM (1986) : post-audit fracture stimulation, overpressure, net present value
BRINKMANN, KRÖMER & REINICKE (1980) : fracspecific parameters, treat gas wells
FRAIM & LEE (1988) : non-Darcy flow, long-term gas production, pay properties

GRINGARTEN (1978) : reservoir limits testing, fractured wells
LEICHT (1985) : reservoir engineering aspects, MHF, Germany FRG
LORENZ, BRANAGAN, WARPINSKI & SATTLER (1986) : pay behaviour, stress sensitive
MILLHEIM & CICHOWICZ (1968) : test, fracture analysis, low-permeability gas
REINICKE, BRINKMANN, SCHWARZ & HUENI (1985) : buildup data interpretation, MHF
SMITH (1973) : production ratio, incomplete frac fill-up, wellbore vicinity
TINSLEY, WILLIAMS, TINER & MALONE (1969) : frac height, effect on productivity

7.2.2.7. Pressure analysis in fractured reservoirs

The references om pressure analysis in fractured reservoirs are compiled along the lines of pressure buildup evaluation, pressure decline analysis, pressure regime and derivative, pressure distribution and fall-off behaviour, transient pressure evolution, and other aspects.

7.2.2.7.1. Pressure buildup evaluation

The references on pressure buildup evaluation correspond mainly to aspects discussed in section 6.2.4.2. of the text.

AGARWAL (1979) : pressure buildup, real gas pseudo time, MHF gas, well analysis
AGARWAL (1980) : pressure buildup, drawdown curves, producing-time effects
AGARWAL et al. (1980) : postfrac performance, combined analysis, buildup, MHF
AHMED, KUCUK & AYESTERAN (1985) : pressure buildup analysis, short transient
BOSTIC, AGARWAL & CARTER (1980) : pressure buildup, postfrac well performance
CHEN (1984) : pressure buildup, vertically fractured, low-permeability pay zone
CHEN & KRY (1981) : pressure buildup, vertical fracture, low-permeability pay
CRAWFORD, HAGEDORN & PIERCE (1976) : pressure buildup, naturall fractured pay
EVERDINGEN & MEYER (1971) : buildup curves, post-treatment
GUPPY, CINCO-LEY & RAMEY (1982) : buildup analysis, high flow rates, crossflow
LEE (1967) : pressure buildup test, fracture analysis
ONUR, SERRA & REYNOLDS (1988) : pressure buildup analysis, multi-well system
OTUOMAGIE & MENZIE (1976) : pressure buildup, double layer, no crossflow
RAHIM & LEE (1986) : iterative technique, buildup data, fracture properties
REINICKE et al. (1985) : buildup data, interpretation, MHF, integral methods
ROSATO et al. (1981) : short-time buildup data, finite conductivity, analysis

7.2.2.7.2. Pressure decline analysis

The references on pressure decline analysis correspond mainly to aspects discussed in section 6.2.4.2. of the text.

CASTILLO (1987) : pressure decline analysis, modify, pressure-dependent leakoff
CUESTA & ELPHICK (1984) : pressure test analysis, design/evaluation implication
FETKOVICH (1973) : decline curve analysis, type curves
FRAIM, LEE & GATENS (1986) : advanced decline curve analysis, type curves
LEE (1985 a) : pressure decline analysis, geometry model of fracturing
NOLTE (1979) : pressure decline, fracture parameters, determination
NOLTE (1984) : proppant and fluid schedule determination, pressure decline
NOLTE (1986) : frac pressure decline, general analysis, 3 models, application
NOLTE (1988) : application of fracture design based on pressure analysis
NOLTE & SMITH (1981) : fracture pressure interpretation
SHLYAPOBERSKY et al. (1988) : overpressure calibrated design, stimulation
STRIGHT & GORDON (1983) : decline curve analysis, low-permeability, gas wells

7.2.2.7.3. Pressure regime and derivative

The references on pressure regime and derivative correspond mainly to aspects discussed in section 6.2.4.2. of the text.

ANDERSON, INGRAM & ZANIER (1973) : fracture pressure gradient, well log measure
BOURDET, AYOUB & PIRARD (1984) : pressure derivative, well test interpretation
CHEN (1987) : pressure drawdown, layered reservoir, linear, boundaries
CHEN, YEH, RAGHAVAN & REYNOLDS (1984) : pressure response, observation wells
CINCO-LEY, RAMEY & MILLER (1975) : unsteady-state pressure, inclined fracture
CONWAY, McGOWEN, GUNDERSON & KING (1985) : formation response, pressure regime
CUESTA & ELPHICK (1984) : pressure test analysis, design/evaluation implication
GODBEY & HODGES (1958) : pressure measurements during formation frac operations
GRINGARTEN, RAMEY & RAGHAVAN (1975) : applied pressure analysis
KUCUK & AYESTARAN (1983) : sandface flow rate, simultaneously measured pressure
LENT, NICKELL, DAVID & ALLAM (1980) : bottom hole pressure, design optimization
LORD & McGOWEN (1986) : real-time treating pressure analysis, new correlation
MEDLIN & FITCH (1983) : abnormal treating pressures, MHF
NOLTE (1982) : fracture design considerations, pressure analysis
NOLTE & SMITH (1981) : fracture pressure interpretation
OZKAN, OHAERI & RAGHAVAN (1987) : constant pressure production, unsteady flow
POE & JENNINGS (1988) : backpressure test analysis, stabilized absolute flow
RAGHAVAN & OHAERI (1981) : constant pressure production, unsteady flow
RAHIM & LEE (1986) : iterative technique, buildup data, fracture properties
TIAB & PUTHIGAI (1988) : pressure-derivation type curves, vertical fracture
WONG, HARRINGTON & CINCO-LEY (1984) : pressure derivative function, transient

7.2.2.7.4. Pressure distribution and fall-off behaviour

The references on pressure distribution and fall-off behaviour correspond mainly to aspects discussed in section 6.2.4.2. of the text.

CHEN, YEH, RAGHAVAN & REYNOLDS (1984) : pressure response, observation wells
CINCO-LEY, RAMEY & MILLER (1975) : unsteady-state pressure, inclined fracture
DAVIAU, MOURONVAL, BOURDAROT & CURUCHET (1985) : horizontal wells, pressure
ERDLE et al. (1981) : hydraulic frac treatment, BHP analysis, Peru, results
GRINGARTEN & RAMEY (1974) : unsteady state pressure, partial penetration
GRINGARTEN, RAMEY & RAGHAVAN (1974) : unsteady pressure, infinite conductivity
HATZIGNATIOU et al. (1987) : interference pressure behaviour, multilayered
KUCUK, SHAH, AYESTARAN & NICHOLSON (1986) : multilayer testing and analysis
LEE & MAO (1984) : post-fracture pressure analysis, reservoir interpretation
MEDLIN & FITCH (1983) : abnormal treating pressures, MHF
NOLTE & SMITH (1981) : fracture pressure interpretation
OZKAN, OHAERI & RAGHAVAN (1987) : constant pressure production, unsteady flow
POE & JENNINGS (1988) : backpressure test analysis, stabilized absolute flow
RAGHAVAN & OHAERI (1981) : constant pressure production, unsteady flow
SHLYAPOBERSKY et al. (1988) : overpressure calibrated design, stimulation
SINHA & SCHLITTLER (1977) : pressure fall-off data, highly-dipping formations
SWANSON & MEEKEN (1981) : frac pressure analysis, South Belridge, Lost Hills
VANDAMME, JEFFREY & CURRAN (1986) : pressure distribution, 3-D fractures
VONGVUTHIPORNCHAI & RAGHAVAN (1987) : pressure fall-off behaviour

7.2.2.7.5. Transient pressure evolution

The references on transient pressure evolution correspond mainly to aspects discussed in section 6.2.4.2. of the text.

BINGLE (1987) : transient rate and pressure, TRAP tool, test analysis, logging
BOSTIC & GRAHAM (1979) : pressure transient testing, prefrac, Cotton Valley
CINCO-LEY (1982) : transient pressure, fracture evaluation

CINCO-LEY & SAMANIEGO (1977 b) : wellbore storage and damage, transient flow
CINCO-LEY & SAMANIEGO (1981) : transient pressure, finite conductivity, damage
CINCO-LEY, SAMANIEGO & DOMINGUEZ (1978) : transient pressure, finite conduct
CINCO-LEY, SAMANIEGO & KUCUK (1985) : multiple block size, pressure transient
CLARK (1968) : transient pressure, water injection well
EHLIG-ECONOMIDES & ECONOMIDES (1985) : transient pressure, elongated line flow
EHLIG-ECONOMIDES & RAMEY (1979) : transient rate decline, constant pressure
ERTEKIN et al. (1987) : pressure transient behaviour, finite-conductivity
HOLDITCH & LEE (1979) : transient pressure, frac evaluation, field experience
KENNEDY & McELHINEY (1979) : pressure transient analysis, pre- and post-frac
KUCHUK & WILKINSON (1988) : transient pressure behaviour of commingled pays
KUCHUK et al. (1988) : horizontal well, radial flow, pressure transient
KUMAR, GUPPY & KAGAWAN (1985) : pressure transient analysis, fractured well
LEE & HOLDITCH (1979) : transient pressure, frac evaluation, theory background
MAVOR & CINCO-LEY (1979) : transient pressure behaviour, naturally fraced pay
NUMBERE, JOSEPH & ZOUKAGHE (1985) : transient pressure behaviour, irregular
OLIVER (1984) : pressure transients caused by fracturing
PRAT et al. (1984) : pressure transient testing, fraced reservoir evaluation
ROSS (1984) : pressure transient data analysis, hydraulically fractured wells
RUSSELL & TRUITT (1964) : transient pressure, vertical fracture
SARABIAN (1980) : transient, two-layered heterogeneous system, vertical frac
SATMAN (1988) : pressure transient analysis, composite naturally fractured pay
SCOTT (1963) : transient pressure behaviour, effect of vertical fractures

7.2.2.7.6. Other aspects

The references on other aspects of pressure analysis in fractured reservoirs correspond to various items discussed in section 6.2.4. of the text.

BARUA, KUCUK & GOMEZ-ANGULO (1985) : computer application, fractured rocks
BINGLE (1987) : transient rate and pressure, TRAP tool, test analysis, logging
BLASINGAME & LEE (1986) : homogeneous reservoirs, natural cracks, decline curve
BRINKMANN, KRÖMER & REINICKE (1980) : integral method, hand-applied, specific
CINCO-LEY & MENG (1988) : double porosity reservoir, finite conductivity
CINCO-LEY, RAMEY & MILLER (1975) : unsteady-state pressure, inclined fracture
CINCO-LEY & SAMANIEGO (1977 b) : wellbore storage and damage, transient flow
CROCKETT, WILLIS & CLEARY (1986) : prediction improvement, history matching
CUESTA & ELPHICK (1984) : pressure test analysis, design/evaluation implication
DAVIAU, MOURONVAL, BOURDAROT & CURUCHET (1985) : horizontal wells, pressure
ERDLE et. al. (1981) : hydraulic frac treatment, BHP analysis, Peru, results
GODBEY & HODGES (1958) : pressure measurements during formation frac operations
HANNAH, HARRINGTON & LANCE (1983) : real-time calculation, surface measurement
HATZIGNATIOU et al. (1987) : interference pressure behaviour, multilayered
HOLDITCH, LEE & GIST (1981) : frac length, conductivity, improved technique
JALALI-YAZDI & ERSHAGHI (1987) : heterogeneous, naturally fractured reservoirs
KAZEMI (1969) : naturally fractured reservoirs, uniform fracture distribution
KUCUK & AYESTARAN (1983) : sandface flow rate, simultaneously measured pressure
KUCUK, KARAKAS & AYESTARAN (1984) : commingled zones, no crossflow, well test
KUCUK, SHAH, AYESTARAN & NICHOLSON (1986) : multilayer testing and analysis
LEE et al. (1984) : single-point flow test, permeability estimation
LENT, NICKELL, DAVID & ALLAM (1980) : bottom hole pressure, design optimization
LORD & McGOWEN (1986) : real-time treating pressure analysis, new correlation
OZKAN, OHAERI & RAGHAVAN (1987) : constant pressure production, unsteady flow
OZKAN, RAGHAVAN & JOSHI (1987) : horizontal wells, pressure analysis
POE & JENNINGS (1988) : backpressure test analysis, stabilized absolute flow
SWANSON & MEEKEN (1981) : frac pressure analysis, South Belridge, Lost Hills
TUCKER (1979) : practical pressure analysis, proppant selection, geopressured

7.2.2.8. In-situ stress determination and significance

The references on in-situ stress determination and significance are compiled along the lines of in-situ stress state and distribution, in-situ stress measurement, stress relationship to depth and lithology, differential strain analysis, anelastic strain recovery, and well logging and fracture azimuth prediction.

7.2.2.8.1. In-situ stress state and distribution

The references on in-situ stress state and distribution correspond mainly to aspects discussed in sections 4.2.2.3.1., 4.2.3.2.2. and 4.2.3.2.6. of the text.

ABOU-SAYED (1982) : in-situ stress contrast, laboratory investigation, deep
BLANTON (1983) : recovery deformation, in-situ stress magnitudes, relationship
CLEARY (1979) : structure sensitivity, fracturing rate, fluid saturation
CLEARY et al. (1983) : induced stress, fluid flow, integral schemes, surround
CORNET & VALETTE (1984) : hydraulic geostatic injection test, in-situ stress
DULA (1981) : deformation lamellae, microfractures, macrofractures, stresses
MORITA, WHITFILL, NYGAARD & BALE (1988) : depletion, compaction, subsidence
PINE & BATCHELOR (1982) : in-situ stress, joints, granite, hydraulic behaviour
VOEGELE, ABOU-SAYED & JONES (1983) : stimulation design optimum, in-situ stress
VOIGHT (1968) : virgin state, partial anelastic strain, tensor measurement
WARPINSKI et al. (1982) : frac containment effect, laboratory investigation
WARPINSKI & TEUFEL (1987) : low-permeability, non-marine rocks, in-situ stress
WONG (1986) : in-situ stress changes during steam stimulation, Cold Lake field

7.2.2.8.2. In-situ stress measurement and prediction

The references on in-situ stress measurement and prediction correspond mainly to aspects discussed in sections 6.2.1.8. and 6.2.2.3. of the text.

ABOU-SAYED, BRECHTEL & CLIFTON (1977) : mechanics, approach, determination
BLANTON & TEUFEL (1985) : in-situ stress determination, wellbore elongation
BLANTON & TEUFEL (1986) : recovery deformation, critical evaluation, determine
CHARLEZ, SALEH & DESPAX (1987) : fracmeter, stress determination, crust tension
CHARLEZ, SALEH, DESPAX & JULIEN (1986) : rockmeter, geostatic stress determine
CORNET & VALETTE (1984) : hydraulic geostatic injection test, in-situ stress
DANESHY, SLUSHER, CHISHOLM & MAGEE (1984) : measurement during drilling, stress
DULA (1981) : deformation lamellae, microfractures, macrofractures, stresses
HAIMSON (1978) : hydrofracturing, stress measurement, field results
HAST (1979) : stress measurements, limits, shear slippage, tensile failure
JOHNSON & ALBRIGHT (1985) : crosswell acoustic data, in-situ physical property
REN & HUDSON (1985) : in-situ stress state prediction, wave velocity analysis
ROEGIERS (1974) : determination, field method, development and evaluation
ROSEPILER (1979) : principal stress determination, fracture confinement
STRICKLAND & REN (1981) : prediction, deep wells, differential strain curve
TEUFEL (1982, 1983) : wellbore elongation, breakout, in-situ stress measuring
TEUFEL & WARPINSKI (1984) : wellbore elongation, breakout, in-situ stress state
VOEGELE & JONES (1980) : wireline hydraulic frac, stress contrast determination
WARPINSKI (1983 b) : accuracy, reliability, perforated cased holes
WARPINSKI (1985) : elastic and viscoelastic stress calculation, sediment basin
WARPINSKI, BRANAGAN & WILMER (1985) : measurements, experiment site
WHITEHEAD, HUNT, FINLEY & HOLDITCH (1986) : log-derived value, field measure
ZOBACK & HAIMSON (1982) : hydraulic fracturing, in-situ stress measurements

7.2.2.8.3. Stress relationship to depth and lithology

The references on stress relationship to depth and lithology correspond mainly to aspects discussed in sections 4.2.2.3.1., 4.2.3.2.2. and 4.2.3.2.6. of the text.

ABOU-SAYED (1982) : in-situ stress contrast, laboratory investigation, deep
BLANTON (1983) : recovery deformation, in-situ stress magnitudes, relationship
BLÜMLING (1983) : well logging, relations to stress and strain field
BRECKELS & EEKELEN (1982) : horizontal stress, depth, relationship, basin
CLEARY (1979) : structure sensitivity, fracturing rate, fluid saturation
EEKELEN (1981) : relationships, horizontal stress, reservoir depth, basin
HAST (1979) : stress measurements, limits, shear slippage, tensile failure
MORITA, WHITFILL, NYGAARD & BALE (1988) : depletion, compaction, subsidence
WARPINSKI & TEUFEL (1987) : low-permeability, non-marine rocks, in-situ stress
WHITEHEAD, HUNT & HOLDITCH (1987) : lithology, pay pressure, effect on stress

7.2.2.8.4. Differential strain analysis

The references on differential strain analysis correspond mainly to aspects discussed in section 6.2.2.3.3. of the text.

REN & ROEGIERS (1983) : differential strain curve analysis, in-situ stress
SIEGFRIED & SIMMONS (1978) : differential strain analysis, oriented cracks
SIMMONS, SIEGFRIED & FEVES (1974) : differential strain analysis, frac examine
STRICKLAND & REN (1980) : differential strain curve, in-situ stress prediction
STRICKLAND & REN (1981) : prediction, deep wells, differential strain curve
WARPINSKI & TEUFEL (1987) : differential strain curve analysis, in-situ stress

7.2.2.8.5. Anelastic strain recovery

The references on anelastic strain recovery correspond mainly to aspects discussed in section 6.2.2.3.3. of the text.

BEGNAUD & CLAIBORNE (1985) : vertical fracture growth, anelastic strain state
BLANTON & TEUFEL (1986) : recovery deformation, critical evaluation, determine
EL RABAA & MEADOWS (1986) : strain relaxation method, laboratory and field
LAUBACH & MONSON (1988) : anelastic strain recovery, coring-induced fractures
TEUFEL (1982) : fracture azimuth prediction, anelastic strain recovery, cores
TEUFEL (1983) : anelastic strain recovery, oriented cores, design applications
TEUFEL (1985) : frac azimuth forecast, anelastic strain recovery, insitu stress
TEUFEL & WARPINSKI (1984) : anelastic strain recovery, oriented core, in-situ
VOIGHT (1968) : virgin state, partial anelastic strain, tensor measurement
WARPINSKI (1985) : elastic and viscoelastic stress calculation, sediment basin
WARPINSKI & TEUFEL (1986) : anelastic strain recovery, viscoelastic constitute

7.2.2.8.6. Well logging and fracture azimuth prediction

The references on well logging and fracture azimuth prediction correspond mainly to aspects discussed in sections 6.2.1.8. and 6.2.2.3. of the text.

BELL & GOUGH (1978) : borehole breakout, fracture azimuth, in-situ stress
BLANTON (1983) : relation, recovery deformation and in-situ stress magnitude
BLANTON & TEUFEL (1985) : in-situ stress determination, wellbore elongation
BLÜMLING (1983) : well logging, relations to stress and strain field
FORDJOR, BELL & GOUGH (1983) : fracture azimuth prediction, well logging ways
GOUGH & BELL (1982, 1983) : stress orientation, borehole wall fractures, Canada
JOHNSON & ALBRIGHT (1985) : crosswell acoustic data, in-situ physical property
MAO, SWEENEY, HANSON & COSTANTINO (1984) : sonic technique, estimation, stress

REN & HUDSON (1985) : in-situ stress state prediction, wave velocity analysis
TEUFEL (1985) : frac azimuth forecast, anelastic strain recovery, insitu stress
TEUFEL & BLANTON (1985) : wellbore elongation measurements, stress direction
WARPINSKI (1983 b) : accuracy, reliability, perforated cased holes
WHITEHEAD, HUNT, FINLEY & HOLDITCH (1986) : log-derived value, field measure
ZOBACK, MOOS, MASTIN & ANDERSON (1985) : wellbore breakout, in-situ stress

7.2.2.9. Relationships between fracturing and well productivity

The references on relationships between fracturing and well productivity correspond mainly to aspects discussed in section 4.2. of the text.

AGARWAL, CARTER & POLLOCK (1979) : evaluation, performance prediction, MHF
ANDERSON & PHILLIPS (1986, 1987) : economic well performance, design optimum
ARASTOPOUR (1987) : gas flow analysis, fractured tight reservoir production
CHASE & WILLIAMS (1986) : IPR curves, performance prediction, fractured wells
CHURCH & PETERS (1987) : improved frac technique, production potential increase
DIETZEL & ERNST (1987) : pay property requirement, profitable stimulation
FAST, FLICKINGER & HOWARD (1961) : frac-formation flow capacity contrast
GRAHAM, MONAGHAN & OSOBA (1959) : proppant wettability, productivity influence
HOLDITCH, JENNINGS, NEUSE & WYMAN (1978) : well spacing, frac length, optimize
HUNT (1986) : production systems analysis, fractured wells, performance
KIM & WILLINGHAM (1987) : flow response, repeated production cycles, propped
LORENZ & FINLEY (1987) : frac characteristics, related production, difference
MALLINGER, RIXE & HOWARD (1964) : propping agent spacers, productivity increase
McDONALD (1983) : production test evaluation, oil wells, offshore Qatar, MHF
McGUIRE & SIKORA (1960) : vertical fractures, effect on well productivity
MISAK et al. (1978) : minimizing water production, fracturing technique
MORSE & GONTEN (1972) : stabilized flow, vertical fractures
NIERODE (1983) : design, field results, comparison
NIERODE (1987) : well productivity, modelling, hydraulic fracturing
NOVOTNY (1977) : proppant transport, proppant settling, well productivity
PHILLIPS & WILSON (1984) : improved drainage, fluid recovery enhance, produce
POOLLEN (1957) : permeability damage vs. productivity
POOLLEN, TINSLEY & SAUNDERS (1958) : fracture flow capacity, well productivity
PRATS & LEVINE (1963) : vertical frac effect, reservoir behaviour, oil/gas flow
RAGHAVAN, CADY & RAMEY (1972) : well-test analysis, vertical fracture
RAYMOND & BINDER (1967) : formation damage, vertical fractures
SCHULTE (1984) : inflow limitation, part of frac height, production evaluation
SINHA & FURLONG (1979) : deliverability, prediction, gas wells, tight pay zone
SMITH (1979) : fracture azimuth, effect on production
SOLIMAN & HUNT (1985) : fracturing fluid, clean up, well performance
SOLIMAN, VENDITTO & SLUSHER (1984) : well performance, evaluation, type curves
TINSLEY, WILLIAMS, TINER & MALONE (1969) : frac height, effect on production
VISSER & DOMSELAAR (1974) : interference, productivity, improvement
YOUSEF & EVANS (1986) : type-curve techniques, lenticular formations

7.2.2.10. Horizontal fractures

The references on horizontal fractures correspond mainly to aspects discussed in sections 1.2.8. and 4.9.4. of the text.

BRITTON, MARTIN, McDANIEL & WAHL (1981) : fracture pre-heating, oil recovery
CHHINA (1986) : horizontal fracture test, Athabasca oil sands
CLOSMANN (1971) : horizontal fracturing, subsurface earth formations, US patent
CRAWFORD & LANDRUM (1956) : horizontal fractures, effect on production capacity
DANESHY (1974) : horizontal and vertical fractures, propagation, initiation
DUSSEAULT (1977) : stress rate, hydraulic fracturing, Athabasca oil sands
DUSSEAULT & SIMMONS (1982) : steam injection, stress/frac orientation changes
ERDLE et al. (1981) : BHP analysis, Peru, frac treatment result, horizontal

EVANS (1983), EVANS & HOLZHAUSEN (1983) : shallow fracs, surface deformation
EVANS et al. (1985) : horizontal/vertical frac distinction, surface deformation
GRINGARTEN & RAMEY (1974) : partial penetration, restricted entry, pressure
HARTSOCK & WARREN (1961) : horizontal fracturing, effect on well performance
HUBBERT & WILLIS (1957) : hydraulic fracturing mechanics, initiation pressure
LANDRUM & CRAWFORD (1957) : thick reservoirs, effect on pattern conductivity
LOWE & HUITT (1966) : proppant transport, horizontal fractures
MARTIN, ALEXANDER, DEW & TYNAN (1972) : thermal recovery, Wyoming, horizontal
MORRISSON & HENDERSON (1960) : oil gravity drainage, large horizontal fracs
PERKINS & KERN (1961) : frac width, horizontal, vertical, friction, pressure
REDDY, PROWSE & REDFORD (1980) : steam injection, horizontal frac, simulation
REYNOLDS, SCOTT, POPHAM & COFFER (1961) : areal extent, orientation, determine
SETTARI & RAISBECK (1981) : analysis, numerical modelling, cyclic steam inject
SUN (1969) : surface deformation, uplift, horizontal fracture size, evaluation
SUNG & ERTEKIN (1987) : vertical and horizontal frac, horizontal well, compare
VITTORATOS (1986) : steam injection, horizontal frac, temperature profiling
WAHL (1956) : effect on well productivity, electrolytical model study, horizon
WAHL (1965) : propped fracture area, horizontal fracture design
WAHL & CAMPBELL (1963) : sand movement, horizontal fracturing

7.2.3. Fracture features

This topic header assembles a suite of different subjects including geometry, parameters, conductivity and permeability, propagation, containment, damage, simulation, prediction, monitoring and detection, performance evaluation, and orientation changes of fractures.

The section on fracture geometry comprises references on fracture shape, treatment parameters and their effect on fracture geometry, stress distribution and its influence, optimum design, variational approach, 2-D and 3-D modelling, and geometry determination. The division on fracture parameters contains references on measurement during treatment, stress gradient, determination of parameters, as well as papers on various individual fracture parameters such as mainly width, length, azimuth and flow capacity. The portion on fracture conductivity and permeability presents references on effects of fracturing fluids, temperature and stress, proppant mixtures, fluid-loss additives, and particle size distribution, and also on determination. The branch on fracture propagation collects references on layered formations, geological discontinuities, growth barriers, stress variations, interface penetration, plasticity, deflection, manipulation and pore pressure. The chapter on fracture containment includes references on controlling factors, layered formations, stress distribution, influence of in-situ stress, burial history, consequences for design, and quantitative analysis.

The section on fracture damage assembles references on flowback procedures, well productivity, fluid interaction, proppant crushing, impairment minimization, gel residues, water blocking and closure stress. The division on fracture simulation contains references on stress distribution, finite difference simulation, finite capacity, numerical simulation, type curve analysis and history match. The portion on fracture prediction collects references on formation response, fracture pressure behaviour, history match, variational approach, proppant distribution, numerical model, stress gradient, design technology and predictive modelling. The chapter on fracture monitoring and detection includes references on well logging, temperature survey, radioactive recording, geoelectrical and seismic methods, design vs. measurement, proppant mapping, borehole television, and optimization of design. The division on performance evaluation assembles references on well test analysis, quality control, stimulation data evaluation, flow characteristics, type curves and performance modelling. The segment on fracture orientation changes includes references on stress changes, temperature impact, pore pressure, thermal expansion and contraction, horizontal and vertical fractures, and on- and off-trend well patterns.

7.2.3.1. Fracture geometry

The references on fracture geometry correspond mainly to aspects discussed in section 4.2. of the text.

ABOU-SAYED, SINHA & CLIFTON (1984) : in-situ reservoir conditions, geometry
ABOU-SAYED, SINHA, DOUGHERTY & MORALES (1984) : influence on fracture geometry
ADVANI, KHATTAB & LEE (1985) : geometry modelling, prediction, comparison
ADVANI et al. (1985) : geometry prediction, early time production modelling
AHMED, WILSON & STRAWN (1983) : stress distribution, effect on frac geometry
BOWIE (1956) : infinite plate, radial cracks, internal circular hole boundary
CLEARY (1980) : fracture shape, mechanics and procedures
CLIFTON & ABOU-SAYED (1979) : computation, 3-D
CLIFTON & ABOU-SAYED (1981) : 3-D geometry prediction, variational approach
COCIANCIG, GRIFFITHS & SCHÜLER (1980) : NPV optimization, height, length
DANESHY (1971 b) : effect of treatment parameters on fracture geometry
DANESHY, WILLIAMS & TINSLEY (1971) : treatment parameters, effect on geometry
DOUGHERTY & ABOU-SAYED (1984) : in-situ reservoir conditions, geometry effect
EEKELEN (1982) : fracture containment, layered formations
ELBEL (1985) : fracture geometry, optimum design, considerations
HANSON, ANDERSON & SHAFFER (1980) : effects of various parameters
HOLDITCH, JENNINGS & NEUSE (1978) : fracture length, well spacing, tight gas
HOLDITCH, ROBINSON & WHITEHEAD (1986) : formation evaluation, 3-D shape
LEE (1986) : minifrac analysis, ellipsoidal geometry, model evaluation
LEE & DANESHY (1985) : frac geometry, proppant transport, multiple-fluid treat
MAUK & MAHRER (1986) : frac monitor, long-period seismometer, induced geometry
MEEHAN, HORNE & AZIZ (1988) : reservoir heterogeneity, fracture length, azimuth
MENG & BROWN (1987) : treatment scheduling, production forecasting, geometry
MEYER (1985) : 3-D fracture modelling, parameters, geometry, shape, programs
MORITA, WHITFILL & WAHL (1985) : stress intensity factor, cross-section shape
PALMER & CRAIG (1984) : asymmetrical vertical growth, elongated hydraulic frac
PALMER & LUISKUTTY (1985) : frac model, elongated vertical fracs, comparison
PESLJAK (1985) : circular fractures, description, modelling, calculation
SETTARI (1982), SETTARI & CLEARY (1986) : pseudo-3-D model, development, test
SHI & CHA (1974) : three-dimensional crack problems, fracture criterion
THIERCELIN, JEFFREY & BEN NACEUR (1987) : frac toughness, influence on geometry
TREVITS et al. (1982) : parameters controlling geometry, identify, evaluation
TYLER & VOLLENDORF (1975) : mapping, fracture morphology, physical observation
WOOD, POLLARD & RALEIGH (1976) : in-situ geometry determination, tiltmeter use

7.2.3.2. Fracture parameters

The references on fracture parameters correspond mainly to aspects discussed in sections 4.2. and 4.8.11. of the text.

BARBA (1988) : maximizing fracture length, reservoir justification, Spraberry
BENNETT, ROSATO & REYNOLDS (1981) : wing length, frac heterogeneity, well react
BRINKMANN, KRÖMER & REINICKE (1980) : determination, fracspecific parameters
CIPOLLA & LEE (1987) : excess propped frac height, well productivity, effect
DOBKINS (1979, 1981) : fracture height, determination, improvement
FRISINGER & COOPER (1985) : formation stress, fracture fluid response, quantify
HOLDITCH, LEE & GIST (1981) : conductivity, length, pressure buildup, improve
JENNINGS & FORD (1976) : fracture flow capacity, key to production
LABUDOVIC (1984) : fracture height, effect of Poisson`s ratio
LERCHE & NARR (1986) : frac density, variable frac spacing, core evaluation
MEYER (1986) : parametric studies, model comparison, vertical fracs, 2-D, 3-D
NOLTE (1979) : fracture parameter determination, pressure decline
PALMER & CARROLL (1983) : fracture height, numerical solution, elongation
PEARSON, ABOU-SAYED & NGUYEN (1984) : frac design, measurement during treatment

PERKINS & KERN (1961) : fracture width, determination possibilities
POOLLEN, TINSLEY & SAUNDERS (1958) : fracture flow capacity, well productivity
RAGHAVAN, URAIET & THOMAS (1978) : vertical fracture height
SHLYAPOBERSKY (1985) : fracture toughness, screenout, mini- and microfractures
SHLYAPOBERSKY et al. (1988) : fracture toughness, mini- and microfractures
SMITH, ROSENBERG & BOWEN (1982) : fracture width, design vs. measurement
THIERCELIN & LEMANCZYK (1983) : stress gradient, fracture height, effects
TINSLEY et al. (1969) : frac height, effect on steady-state production increase
TIRANT & DUPUIS (1967) : fracture dimensions, oil reservoir fracturing
WARPINSKI (1983) : width, pressure, measurement, propagating fracture

7.2.3.3. Fracture conductivity and permeability

The references on fracture conductivity and permeability correspond mainly to aspects discussed in sections 1.4.10. and 4.3. of the text.

ALDERMAN & WENDORFF (1969) : proppant packing of fractures
BARBY & BARBEE (1987) : ultra-high frac conductivity, stimulation, case history
CARROLL & BAKER (1979) : particle size distribution, proppant crushing, effect
CINCO-LEY et al. (1987) : low-conductivity vertical fracture, well behaviour
COOKE (1973) : multiple proppant layers, fracture width, applications, benefit
COOKE (1975) : effect of fracturing fluids, proppant package, fracture width
COOKE (1977) : high-strength proppants, applications, benefits, deep treating
CUTLER et al. (1985) : conductivity, comparison, proppants, composition, type
CUTLER, ENNISS & JONES (1983) : conductivity comparison, closure stress, props
DAVIES & KUIPER (1988) : hydraulic fracture stimulation, fracture conductivity
HARPER, HAGAN & MARTINS (1985) : non-proppant frac, shear displace, mismatch
McDANIEL (1986) : conductivity testing, high temperature and stress, influence
McDANIEL (1987) : realistic frac conductivity, reservoir temperature, testing
McDANIEL & WILLINGHAM (1978) : frac permeability, proppant mixtures, effect
SOLIMAN (1986 a) : fracture conductivity distribution study
SOLIMAN (1986 b) : uniform or changing conductivity, constant pressure gradient
STEPHEN-PYE & SMITH (1973) : fluid-loss additive, prop conductivity, permeable
TSANG & WITHERSPOON (1983) : frac roughness, fluid flow parameter, mechanics
VEATCH (1977) : in-situ fracture flow capacity, computer studies, determination
VLIS, HAAFKENS, SCHIPPER & VISSER (1975) : proppant placement, conductivity
WENDORFF (1982) : frac porosimeter, downhole stress, conductivity determinate

7.2.3.4. Fracture propagation

The references on fracture propagation are compiled along the lines of fracture growth and rock properties, fracture extension and fluid interaction, impact of discontinuities and natural cracks, and three-dimensional modelling.

7.2.3.4.1. Fracture growth and rock properties

The references on fracture growth and rock properties correspond mainly to aspects discussed in sections 4.2.2. and 4.2.3. of the text.

ABE, MURA & KEER (1976) : penny-shaped crack, growth rate, propagation
ANDERSON (1981) : unbonded rock interfaces, friction effects, fracture growth
BARENBLATT (1962) : equilibrium cracks, brittle fracture, mathematical theory
BEGNAUD & CLAIBORNE (1985) : vertical frac growth, Ratcliffe, Mission Canyon
BIOT, MEDLIN & MASSE (1983) : fracture penetration through interface
BIOT, MEDLIN & MASSE (1984) : fracture propagation, laboratory experiments
CLEARY (1978) : moving singularities, elasto-diffusive solids, frac propagation
CLEARY et al. (1982) : unsymmetric unidirectional frac propagation, model
DANESHY (1976) : rock properties controlling hydraulic fracture propagation
EL RABAA (1987) : stress variations, fracture propagation

HOLZHAUSEN et al. (1985) : frac growth, steam stimulation, single-well test
LUISKUTTY, TOMUTSA & PALMER (1986) : weak lithological barriers, frac growth
SETTARI (1985) : fracture growth, vertical, lateral, quantitative analysis
SMITH et al. (1987) : tip screenout, vertical communication, multi-frac system
TESTER et al. (1986) : fracture growth, geothermal reservoir, heat extraction
YEW & LODDE (1983) : layered medium, hydraulic fracture propagation

7.2.3.4.2. Fracture extension and fluid interaction

The references on fracture extension and fluid interaction correspond mainly to aspects discussed in sections 4.2.2., 4.2.3. and 4.3.4. of the text.

BARENBLATT (1962) : equilibrium cracks, brittle fracture, mathematical theory
BARREE (1983) : 3-D fracture propagation, heterogeneous, numerical simulator
BEN NACEUR, THIERCELIN & TOUBOUL (1987) : fluid flow simulation, 3D-propagation
BIOT, MEDLIN & MASSE (1983) : fracture penetration through interface
BIOT, MEDLIN & MASSE (1984) : fracture propagation, laboratory experiments
CLEARY (1978) : moving singularities, elasto-diffusive solids, frac propagation
CLEARY et al. (1982) : unsymmetric unidirectional frac propagation, model
DANESHY (1976) : rock properties controlling hydraulic fracture propagation
DUNLAP (1963) : factors controlling orientation and direction of hydrofractures
EL RABAA (1987) : stress variations, fracture propagation
HARRISON, McGUIRE & KIESCHNICK (1954) : frac induction and extension, mechanics
HOWARD & FAST (1957) : fracture extension, optimum fluid characteristics
HOWARD & FAST (1970 b) : fracture extension, controlling factors, monograph
LABUZ, SHAH & DOWDING (1985) : crack, propagation, granite, experiment, analyze
LAM & CLEARY (1984) : slippage, re-initiation, frictional interfaces, mechanics
MEDLIN & MASSE (1976) : fracture initiation pressure, orientation, experiments
MEDLIN & MASSE (1984) : laboratory experiments, fracture propagation, stress
MEDLIN & MASSE (1986) : plasticity, brittle, propagating pressure, confinement
PINE & BATCHELOR (1984) : downward migration, jointed rock, hydraulic injection
RODRIGUEZ et al. (1984) : partially penetrating fractures, transient pressure
RUMMEL et al. (1986, 1987) : modelling, deflection, penetration, anisotropy
SALZ (1977) : propagation pressure, pore pressure
SNEDDON & LOWENGRUB (1969) : crack propagation, elasticity theory
TEUFEL & CLARK (1981) : propagation containment, experimental study, layering
THIERCELIN et al. (1985) : 3-D-propagation, vertical fracture simulation
TOUBOUL, BEN NACEUR & THIERCELIN (1986) : 3-D-propagation, variation, simulate
TRAVIS (1981) : fluid-driven fracture, propagation, modelling, permeable rock
WARPINSKI & TEUFEL (1984) : geological discontinuities, propagation influence
YEW & LODDE (1983) : layered medium, hydraulic fracture propagation

7.2.3.4.3. Impact of discontinuities and natural cracks

The references on impact of discontinuities and natural cracks correspond mainly to aspects discussed in sections 4.2.3.2., 4.2.3.4. and 4.2.4.3. of the text.

ANDERSON (1981) : unbonded rock interfaces, friction effects, fracture growth
BIOT, MEDLIN & MASSE (1983) : fracture penetration through interface
BOONE, WAWRYZNEK & INGRAFFEA (1986) : frac process zone, simulation, hydrofrac
DANESHY (1976) : rock properties controlling hydraulic fracture propagation
DANESHY (1978) : layered formations
DUNLAP (1963) : factors controlling orientation and direction of hydrofractures
EL RABAA (1987) : stress variations, fracture propagation
LAM & CLEARY (1984) : slippage, re-initiation, frictional interfaces, mechanics
LAMONT & JESSEN (1963) : existing natural cracks, effect on hydrofrac extension
LUISKUTTY, TOMUTSA & PALMER (1986) : weak lithological barriers, frac growth
MEDLIN & MASSE (1986) : plasticity, brittle, propagating pressure, confinement
PALMER & CARROLL (1981, 1982) : 3-D, stress variations, stochastic model

PINE & BATCHELOR (1984) : downward migration, jointed rock, hydraulic injection
RODRIGUEZ et al. (1984) : partially penetrating fractures, transient pressure
ROEGIERS, McLENNAN & MURPHY (1982) : pre-existing discontinuities, influence
ROEGIERS & WILES (1981) : pre-existing discontinuities, extension influence
RUMMEL (1985) : manipulation, naturally fractured rock
RUMMEL et al. (1986, 1987) : modelling, deflection, penetration, anisotropy
TEUFEL & WARPINSKI (1983) : in-situ stress variations, mineback test, layering
TEUFEL & WARPINSKI (1988) : natural cracks, influence on hydraulic propagation
WARPINSKI & TEUFEL (1984) : geological discontinuities, propagation influence
YEW & LODDE (1983) : layered medium, hydraulic fracture propagation

7.2.3.4.4. Three-dimensional modelling

The references on three-dimensional modelling correspond to various aspects discussed in chapter 4 of the text.

BARREE (1983) : 3-D fracture propagation, heterogeneous, numerical simulator
BEN NACEUR, THIERCELIN & TOUBOUL (1987) : fluidflow simulation, 3-D-propagation
BOUTECA (1984) : 3-D analytical model, theory and field test
PALMER & CARROLL (1981, 1982) : 3-D, stress variations, stochastic model
THIERCELIN et al. (1985) : 3-D-propagation, vertical fracture simulation
TOUBOUL, BEN NACEUR & THIERCELIN (1986) : 3-D-propagation, variation, simulate

7.2.3.4.5. Other aspects

The references on other aspects of fracture propagation correspond to various items discussed in sections 4.2.2. and 4.2.3. of the text.

ABE, MURA & KEER (1976) : penny-shaped crack, growth rate, propagation
BOONE, WAWRYZNEK & INGRAFFEA (1986) : frac process zone, simulation, hydrofrac
CLEARY (1978) : moving singularities, elasto-diffusive solids, frac propagation
DUNLAP (1963) : factors controlling orientation and direction of hydrofractures
EVANS (1982) : growth, consolidation, shallow fractures, surface deformation
LAM & CLEARY (1984) : slippage, re-initiation, frictional interfaces, mechanics
LEE (1988) : pressure decline analysis, Penny-shaped, Perkins-Kern geometry
MEDLIN & MASSE (1976) : fracture initiation pressure, orientation, experiments
PINE & BATCHELOR (1984) : downward migration, jointed rock, hydraulic injection
SALZ (1977) : propagation pressure, pore pressure
SMITH et al. (1987) : tip screenout, vertical communication, multi-frac system
WONG & CLEARY (1985) : circular fracture, numerical simulation, unsteady flow
ZOBACK & POLLARD (1978) : pressure-time record, in-situ stress determination

7.2.3.5. Fracture containment

The references on fracture containment correspond mainly to aspects discussed in section 4.2.2. of the text.

AHMED et al. (1983) : weak stress barrier, modified frac, tight gas stimulation
BARBER & THEMIG (1985) : low-rate pumping, low fluid viscosity, low job size
BEN NACEUR & TOUBOUL (1987) : fracture height growth, layered media, controls
BRAUNLICH (1967) : fracture control during well treatment
BOUTECA (1987) : fracture geometry, 3-D-model, containment, analysis, testing
DOWELL SCHLUMBERGER (1986) : bracketfrac, fracture height, limitation, control
DOWELL SCHLUMBERGER (1987) : inverta-/divertafrac, bracketfrac, growth control
EEKELEN (1982) : fracture geometry, layered formations
FUNG, VIJAYAKUMAR & CORMACK (1987) : layered formations, containment
HANSON (1964) : plastically deformable solids, subterranean treatment, patent
HANSON & LYONS (1964) : semipermeable prop plug, height growth control, patent
HODGES & PAOLI (1982) : 100 mesh sand, fracture bridging, downward control

KELKAR et al. (1985) : fracture height growth control, permeability barriers
KOHLHAAS (1982) : thin sandstones, fracture stimulation, hydrophobic, Wyoming
LARSON & NGUYEN (1984) : height containment, artificial barrier creation
LU & YEW (1985) : bonded half-planes, containment, hydraulic induction, stress
NGUYEN & LARSON (1983) : height containment, artificial barrier creation
PRATS (1981) : horizontal stress differences, burial history, material property
ROSEPILER (1979) : principal stress determination, fracture confinement
SETTARI (1983) : quantitative analysis, controlling factors
SIMONSON, ABOU-SAYED & CLIFTON (1978) : MHF containment
SMITH (1981) : mHF, design, application, advantages, procedures
TEUFEL & CLARK (1981) : frac propagation, fracture containment, experiments
WARPINSKI, SCHMIDT & NORTHROP (1980) : in-situ stress, containment influence
WARPINSKI et al. (1982) : in-situ stress, containment influence, effect, test
YEW & CHIOU (1983) : in-situ stress, layer properties, containment effect

7.2.3.6. Fracture damage

The references on fracture damage correspond mainly to aspects discussed in section 4.12. of the text.

AHMED, ABOU-SAYED & JONES (1979) : fluid interaction, propped fractures, tight
BRANAGAN et al. (1987) : damage effects, frac performance, natural fractures
CARROLL & BAKER (1979) : proppant crushing, effect on frac conductivity, size
COOKE (1973) : fracture conductivity, multiple proppant layers, damage
GALL, RAIBLE & SATTLER (1988) : permeability damage, fluid polymers, joints
HAWKINS (1988) : proppant-pack permeability damage, typical treatments
HOLDITCH (1979) : effective factors, water blocking, gas flow, fractured wells
KIM & LOSACANO (1985) : conductivity damage, crosslinked gel residue, closure
LARSEN & SMITH (1985) : angular prop blends, friction angle increase, flowback
POOLLEN (1957) : permeability damage, productivity
POPE, WILES & PIERCE (1987) : curable resin-coated sand, prop flowback control
RAYMOND & BINDER (1957) : well productivity, formation damage, vertical frac
ROBINSON, HOLDITCH & LEE (1983) : fracture damage, case study, tight gas
ROBINSON, HOLDITCH & WHITEHEAD (1986) : minimizing damage, controlled flowback
SECCOMBE & ANDERSON (1982) : proppant selection, tight gas, bauxite
SOLIMAN & HUNT (1986) : fracture fluid cleanup, effect on well performance

7.2.3.7. Fracture simulation

The references on fracture simulation correspond mainly to aspects discussed in section 4.2. of the text.

ACHARYA & KIM (1987) : treatment design simulation, Rotliegend, North Sea
ADVANI (1980) : finite element model simulation, hydraulic frac association
AHMED, WILSON & STRAWN (1983) : stress distribution, effect on frac geometry
BARR, WILLIS, CLEARY & ROGERS (1988) : real-time model enhancement, full 3-D
BEN NACEUR, THIERCELIN & TOUBOUL (1987) : fluid flow, 3-D, propagation impetus
BUHARALI et al. (1987) : real-time simulation, portable microcomputer
CHEN, WASSERMAN & FITZMORRIS (1987) : thermal simulation, natural fracturing
CLARK (1983) : computer simulation, proppant selection, pack configuration
CLEARY (1988) : automatic job supervision, computerized on-site simulation
CLEARY, KAVVADAS & LAM (1983) : 3-D fracture simulator
CLEARY, LAM & BARR (1986) : complete 3-D simulator, hydraulic fracturing test
CRAFTON & HARRIS (1976) : finite difference simulation, finite capacity
DEAN & LO (1986) : natural fracture simulator, development, application
EVANS & CARROLL (1981) : stochastic modelling, noncontinuous lenticular pays
GIDLEY et al. (1979) : MHF, well performance, low-permeability, pay simulation
GILMAN & KAZEMI (1983) : naturally fraced reservoirs, simulation, improvement
HAAK & ELEWAUT (1988) : heterogeneity modelling, 3-D simulation improvement
HOSSAIN, CADY & HONARPOUR (1984) : finite-conductivity, real gas flow, simulate

JOHN (1982) : numerical simulation, computation, model development, modelling
JONES, BELL & MORALES (1986) : coal fracing, empirical relationships, computer
KARASAKI, LONG & WITHERSPOON (1988) : fracture-dominated reservoir, analysis
LAM, CLEARY & BARR (1986) : 3-D simulator, fracturing design
LEE & TAN (1987) : multiple porosity/permeability simulation, fractured pay
LOCKE & SAWYER (1975) : pressure injection, numerical, type curve, historymatch
MENDELSOHN (1984) : fracture modelling, 2-D, 3-D, general concepts, review
MORALES, JONES & ABOU-SAYED (1985) : quasi-3-D, fluid properties, proppant move
NGHIEM (1983) : infinite conductivity, vertical frac modelling, source-sink
NGHIEM, FORSYTH & BEHIE (1984) : fully implicit model, hydraulic fracturing
PAPADOPOULOS et al. (1983) : laboratory simulation, hydraulic fracturing
SAWYER, LOCKE & OVERBEY (1974) : finite capacity, vertical fracture
SETTARI & CLEARY (1982) : 3-D simulation
SETTARI & PRICE (1980) : simulation, low-permeability reservoir
THOMAS, DIXON & PIERSON (1983) : fractured reservoir simulation
WANG & CLIFTON (1989) : elastic modulus, toughness, simulation, influence

7.2.3.8. Fracture prediction

The references on fracture prediction correspond mainly to aspects discussed in sections 4.2. and 6.2.1.10.7. of the text.

AADNØY & LARSEN (1987) : fracture gradient prediction, vertical and inclined
AGARWAL, CARTER & POLLOCK (1979) : performance predict, evaluation, MHF, tight
AHMED (1988) : fracture height prediction, modelling, logging, rock properties
BROWN, FORGOTSON & FORGOTSON (1980) : frac orientation, Cotton Valley, Texas
CLARK (1983) : fracture azimuth, prediction, geology, cores, analysis
CLIFTON & ABOU-SAYED (1981) : 3-D geometry prediction, variational approach
CONSTANT & BOURGOYNE (1986) : fracture gradient prediction, offshore wells
CONWAY, McGOWEN, GUNDERSON & KING (1985) : formation response, pressure regime
CROCKETT, WILLIS & CLEARY (1986) : prediction improvement, pressure history
DAINES (1980) : fracture pressure prediction, wildcat well, mechanical property
EATON (1969) : fracture gradient prediction, oilfield operation application
FORGOTSON, FORGOTSON & BROWN (1980) : fracture orientation, prediction
GEERTSMA & DEKLERK (1969) : width, extent, rapid method of prediction
GEERTSMA & HAAFKENS (1979) : width, extent, comparison of prediction theories
HARRINGTON, WHITSITT & HANNAH (1973) : fluid interface location and movement
HARRISON & TILDEN (1986) : predictive model, fracture distribution, intensity
HASSAN (1982) : azimuth prediction, improved hydrocarbon recovery, implication
HOLDITCH & LANCASTER (1982) : prop distribution, finite difference numerical
LOGAN & TEUFEL (1978) : MHF prediction, oriented core analysis
NEWBERRY, NELSON & AHMED (1985) : fracture migration, compressional and shear
PALMER & CARROLL (1983) : fracture height, leakoff, numerical solution
SCHLUMBERGER (1986) : FracHite log, design technology, prediction
STRUBHAR, FITCH & UHRI (1977) : simplified performance prediction, vertical
THIERCELIN & LEMANCZYK (1983) : stress gradient, effect on fracture height
THOMPSON (1981) : performance prediction, constant pressure, tight, capacity

7.2.3.9. Fracture modelling

The references on fracture modelling correspond mainly to aspects discussed in section 4.2. of the text.

ABOU-SAYED, CLIFTON, DOUGHERTY & MORALES (1984) : 3-D, frac geometry, examples
CLEARY, BARR & WILLIS (1988) : full 3-D simulation, real-time model enhancement
CLEARY, KECK & MEAR (1983) : microcomputer model, hydraulic fracture design
CROCKETT, OKUSU & CLEARY (1986) : complete integrated model, real-time, design
DANESHY (1973) : vertical hydraulic fracture design, modelling
GEERTSMA & DEKLERK (1969) : width, extent, prediction, modelling
KEAT, ANNIGERI & CLEARY (1988) : surface integral, finite element, hybrid

NORDGREN (1982) : vertical hydraulic fracture propagation, modelling
PALMER & CARROLL (1983) : numerical solution, hydraulic fracture height
PALMER & LUISKUTTY (1985) : elongated vertical frac, model comparison, apply
PERKINS & KERN (1961) : hydraulic fracture width, modelling
ROEGIERS & ISHIJIMA (1983) : coupled frac model, hydraulic frac application
SETTARI & CLEARY (1986) : pseudo-3-D model, development, testing
SETTARI, PUCHYR & BACHMAN (1987) : decoupled modelling, fracturing processes
SWAAN (1982) : three-phase model, fluid segregation, fractured reservoirs

7.2.3.10. Fracture monitoring and detection

The references on fracture monitoring and detection (corresponding to chapter 6 of the text) are classified along the lines of general aspects and summaries, temperature logging, conventional gamma-ray logging, gamma-ray spectroscopy, acoustic logging, neutron logging, borehole elongation and caliper logging, laterolog and spontaneous potential, dipmeter recording, borehole television, formation microscanner, seismical methods and microearthquake detection; tiltmeter, gravimeter and magnetometer; strain relaxation, core evaluation and in-situ stress; mechanical properties logging, fracture azimuth and width determination, fracture permeability estimation, noise logging, inflatable impression packers; interference, tracer and pulse testing; and other aspects.

7.2.3.10.1. General aspects and summaries

The references on general aspects and summaries of fracture monitoring and detection correspond to various aspects discussed in chapter 6 of the text.

AGUILERA & ACEVEDO (1982) : computerized log interpretation, natural fractures
AHMED (1987) : fracture height prediction, post-treatment measurements, review
BOYELDIEU & MARTIN (1984) : fracture detection and evaluation, summary, review
DENNIS et al. (1987) : frac identification, productivity prediction, carbonate
DOBKINS (1979, 1981) : frac height, improved methods, determination, well log
DRESSER ATLAS (1982) : interpretive methods for production well logs
ELLIS (1987) : well logging for earth scientists, textbook
FERTL (1980) : fracture evaluation, well logging
FITZ-PATRICK, KARR & O`SHEA (1985) : proppant mapping, fracture geometry, test
FRISINGER & COOPER (1985) : well logging, formation stress, frac fluid response
HEFLIN (1979) : log responses, fracture detection, summary, West Coast pay zone
HIRSCH, CISAR, GLASS & ROMANOWSKI (1981) : log fracture detection, techniques
HUNT et al. (1981) : new well logs, application to fracturing, Cotton Valley
JULIAN (1982) : fracture detection techniques, chalk formations
KLEINBERG et al. (1984) : borehole application techniques, sensitive, reliable
McGARR & GAY (1978) : fracture orientation, vertical vs. horizontal, stress
MYUNG (1976) : frac detection, well logging, shale, temperature, sibilation
NOBLETT, FERTL & GUY (1987) : fracture evaluation, recent advances
PAILLET (1985 b) : fractured reservoir evaluation, problems, solutions, review
PIRSON (1967) : mapping fracture development from well logs
PLUMB (1988) : natural fracture characterization from well logs, summary
POWER, SCHUSTER, HAY & TWOMBLY (1976) : orientation, dimension, detection, case
REYNOLDS, SCOTT, POPHAM & COFFER (1961) : frac extent and orientation detection
SCHLUMBERGER (1972, 1974) : log interpretation, application principles
SERRA (1985, 1987) : well-log interpretation, fundamentals, textbook, summary
SMITH et al. (1978 a) : fracture azimuth determination, deep penetrating crack
SMITH, LOGAN & WOOD (1980) : fracture azimuth, shallow experiment
SMITH, REN, SORRELS & TEUFEL (1985) : fracture azimuth, fracture diagnostics
SMITH, ROSENBERG & BOWEN (1982) : fracture width, design vs. measurement
SORRELLS et al. (1986) : hydraulic fracture diagnostics research
SUAU & GARTNER (1979) : fracture detection from logs, summary, evaluation
THORNE & MORRIS (1987) : borehole seismic fracture diagnostics, advances
WARPINSKI (1983) : width, pressure, measurement, propagating fracture

WELEX (1986) : formation strength analysis, wavetrain, frac pressure, sanding
YOST, FROHNE, KUMAR & AMERI (1982) : shale fracture, natural/hydraulic relation

7.2.3.10.2. Temperature logging

The references on temperature logging correspond mainly to aspects discussed in section 6.2.1.1. of the text.

AGNEW (1965) : temperature survey, fracture evaluation, well logging
BIOT, MASSE & MEDLIN (1984) : temperature analysis, hydraulic fracturing
COOKE (1979) : differential temperature logging, flow behind casing, detection
DOBKINS (1979, 1981) : frac height, improved methods, determination, well log
EVANS (1977) : thermal conductivity, temperature logging, North Sea, property
FAGLEY et al. (1982) : water injection wells, temperature log interpretation
JORDEN & CAMPBELL (1985) : temperature log, rock properties, borehole environs
KIM, CHAMPION & COOPER (1985) : deviated wells, temperature survey, evaluation
PEARCE (1979) : tracer, temperature survey, evaluation, fracture treatment
SMITH et al. (1978 b) : temperature logging, fracture height determination
SMITH & STEFFENSON (1973, 1975) : temperature logging, improved interpretation
WAGES (1982) : post-fracture temperature survey interpretation
WITTERHOLT & TIXIER (1972) : temperature logging, injection wells

7.2.3.10.3. Conventional gamma-ray logging

The references on conventional gamma-ray logging correspond mainly to aspects discussed in sections 6.2.1.2., 6.2.1.3. and 6.3.2. of the text.

AHMED (1987) : fracture height prediction, post-treatment measurements, review
ALLEN (1984) : radioactive bullet monitoring, compaction, subsidence, supervise
DOBKINS (1979, 1981) : frac height, improved methods, determination, well log
GESINK et al. (1985) : gamma-ray emitting tracer, preferential flow zones
HILL & SOLARES (1985) : radioactive tracer injection profile logging, improve
MEEHAN & PENNINGTON (1981) : radioactive sand, tracer gamma-ray logging, height
PEARCE (1979) : tracer, temperature survey, evaluation, fracture treatment
ROBINSON, TESTER & BROWN (1984) : inert, chemically reacting tracers, sizing
SMITH, LOGAN & WOOD (1980) : fracture azimuth, shallow experiment
SMITH, REN, SORRELS & TEUFEL (1985) : fracture azimuth, fracture diagnostics
SMITH, ROSENBERG & BOWEN (1982) : fracture width, design vs. measurement
SMITH & STEFFENSON (1973, 1975) : temperature logging, improved interpretation
SORRELS & MULCAHY (1986) : fracture azimuth estimation, microseismic method
SUAU & GARTNER (1979) : fracture detection from logs, summary, evaluation
TAM INTERNATIONAL (1986) : inflatable packer, well testing, workover, fracture
TAYLOR (1983) : borehole televiewer survey, travel time, amplitude, interpret
TESLJUK (1957) : radioactive isotope application
WHEELER, PARSONS, CONCHIE & DURHAM (1985) : radioactive tracers, oil reservoir
WILLIAMS & McCARTHY (1987) : multiple radioactive tracers, stimulation optimize
ZHUVAGIN & AKCHASYANOV (1960) : radioactive tracer control of hydraulic fracs

7.2.3.10.4. Gamma-ray spectroscopy

The references on gamma-ray spectroscopy correspond mainly to aspects discussed in sections 6.2.1.3.4. and 6.3.2.4. of the text.

ANDERSON et al. (1986) : spectral gamma ray analysis, frac height monitoring
ANTHONY & HILL (1984) : two-pulse tracer logging, radioactive injection survey
FERTL (1982) : natural gamma-ray, spectral logging, North America, experience
FERTL & OLIVER (1983) : multiple stage fracturing evaluation, U.S. patent
FERTL & RIEKE (1979) : gamma-ray spectroscopy, fractured shale, source rock
FERTL et al. (1980) : spectral gamma-ray, Austin Chalk, Texas, spectroscopy

GADEKEN, ARNOLD & SMITH (1984) : compensated spectral natural gamma ray, tracer
GADEKEN & SMITH (1986) : gamma-ray spectroscopy, multiple radioactive tracers
GADEKEN, SMITH & SEIFERT (1987) : gamma-ray spectroscopy, multiple tracering
GADEKEN, SMITH, ARNOLD & ROBBINS (1988) : gamma spectroscopy, natural isotopes
GRAU & SCHWEITZER (1987) : gamma-ray spectroscopy, NaI (Tl) crystals, apply
GRAU et al. (1988) : capture gamma-ray spectroscopy, borehole correction model
HEFLIN & NETTLETON (1980) : natural gamma-ray spectroscopy, application, case
HILL, BOEHM & AKERS (1988) : tracer placement techniques, improved marker logs
HILL & SOLARES (1985) : radioactive tracer injection profile logging, improve
JACKSON & GARTNER (1988) : quantitative gamma-ray concentrations, evaluation
LOPUS, SEIFERT & SCHEIN (1987) : natural frac conductivity, gamma spectroscopy
MARETT, CHEVALIER, SOUHAITE & SUAU (1976) : gamma-ray spectrometry, shaly sand
MYERS (1988) : pulsed neutron, induced gamma-ray spectroscopy, application
NGUYEN & STEVENS (1988) : inert gas radioactive tracers, steam injection profil
PEMPER et al. (1988) : multiple radioactive tracers, gamma-ray spectroscopy
PRIEST (1988) : radioisotope injection at wellhead, fracture operation, improve
ROBINSON & TESTER (1984) : tracer-determined flow, residence time, dispersion
ROBINSON, TESTER & BROWN (1984) : inert, chemically reacting tracers, sizing
SERRA, BALDWIN & QUIREIN (1980) : natural gamma-ray spectroscopy, application
SMITH et al. (1983) : gamma-ray spectroscopy, compensation, multi-function
WESTERN ATLAS (1988 b) : PRISM, gamma-ray spectroscopy, multiple tracer logging
WHEELER, PARSONS, CONCHIE & DURHAM (1985) : radioactive tracers, oil reservoir
WICHMANN, McWHIRTER & HOPKINSON (1975) : natural gamma-ray spectroscopy, field
WILLIAMS & McCARTHY (1987) : multiple radioactive tracers, stimulation optimize

7.2.3.10.5. Acoustic logging

The references on acoustic logging correspond mainly to aspects discussed in sections 6.2.1.6. and 6.2.1.10.4. of the text.

AGUILERA (1974) : sonic, resistivity, naturally fractured reservoirs, detection
ALBRIGHT & PEARSON (1982) : acoustic emission, fracture location tool, diagnose
BEYDOUN, CHENG & TOKSÖZ (1983) : tube wave, Stoneley wave, sonic, frac detect
BEYDOUN, CHENG & TOKSÖZ (1985) : vertical seismic profiling, open frac detect
BOTTER (1982) : borehole circumferential acoustic waves, fracture delineation
BOTTER & ARKEL (1982) : circumferential propagation, acoustic boundary waves
BRIE, HSU & ECKERSLEY (1988) : Stoneley normalized differential energy, analyze
BRIE, HSU & PLUMB (1987) : Stoneley normalized differential energy, evaluation
CHEUNG (1984) : sonic logging, fracture detection, naturally cracked reservoirs
GUY, CORLEY, FERTL & FROST (1987) : circumferential acoustic, vertical fracture
GUY & FERTL (1984) : circumferential acoustic measurements, natural fractures
GUY, FERTL & OLIVER (1986) : circumferentially propagated acoustic logging
HORNBY et al. (1987) : borehole Stoneley wave, acoustic, natural fractures
HSU, BRIE & PLUMB (1987) : fracture identification, array sonic tools, method
HUANG & HUNTER (1981) : Stoneley wave, tube wave, sonic, open fracs, fluids
KOERPERICH (1978) : acoustic boundary waves, interference pattern, frac detect
KOERPERICH (1980) : shear-wave velocity, long- and short-spaced acoustic device
LIU (1984) : Stoneley-wave-derived dt-shear log, natural fracture detection
LIU (1985) : Stoneley wave, borehole sonic velocity, fracture evaluation
MORRIS, GRINE & ARKFELD (1964) : acoustic amplitude, compressional, shear, frac
MORRIS, LITTLE & LETTON (1984) : sonic array tool, full waveform logging, frac
NIELSEN & KOHLHAAS (1979) : rock mechanical properties, acoustic and biaxial
NOBLETT & FERTL (1987) : circumferential acoustic log, vertical fractures
PAILLET (1985 a) : borehole acoustic methods, application in rock mechanics
PAILLET (1987) : acoustic waveform, VSP, tube wave, frac permeability, testing
PAILLET & WHITE (1982) : acoustic wave propagation, rock property relationship
PLUMB & HORNBY (1988) : borehole Stoneley wave analysis
SARDA, PERREAU & DEFLANDRE (1988) : acoustic emission, frac extent estimation
SCHLUMBERGER (1985, 1987 b) : array-sonic service, natural fracture detection
SETSER (1981) : circumferential acoustic logging, natural fracture detection
VOGEL & HEROLZ (1981) : circumferential acoustic device, natural fractures

WELEX (1986) : formation strength analysis, wavetrain, frac pressure, sanding
WU (1984) : sonic velocity, acoustic waveforms, digital and analog technique

7.2.3.10.6. Neutron logging

The references on neutron logging correspond mainly to aspects discussed in section 6.3.3. of the text.

AJAM & RAHAL (1986) : thermal multigate decay pulsed neutron logging, use
BRUIST, JEFFERIS & BOTTS (1983) : near and far detectors, curve combination
BUCHANAN, CLEARMAN, HEIDBRINK & SMITH (1984) : thermal multigate decay neutron
CLAVIER, HOYLE & MEUNIER (1971) : thermal neutron decay time log, GWC monitor
CORWITH & MENGEL (1988) : pulsed neutron capture logging, application spectrum
ELSON, DARLINGTON & MANTOOTH (1984) : neutron logging, gravel pack evaluation
HERTZOG (1980) : inelastic neutron scattering, capture gamma-ray spectrometry
ITOH & SCHOPPER (1977) : pulsed neutron capture well logging techniques
JEFFERIS, BRUIST & BOTTS (1983) : near and far detectors, curve combination
KREIJ (1986) : neutron logging, curve separation, gravel density, packing rate
MYERS (1988) : pulsed neutron, induced gamma-ray spectroscopy, application
NEAL (1983) : neutron logging, borehole-compensated tool, gravel pack checking
OLESEN et al. (1987) : dual-burst thermal decay time logging, overview, example
OLIVER & FERTL (1986) : pulsed neutron capture logging, applications, advantage
OLIVER, WYATT & SMITH (1986) : thermal multigate decay pulsed tools, neutron
PEETERS & HARTLEY (1984) : induced frac height detection, thermal decay time
RANDALL, GRAY, CRAIK & HOPKINSON (1985) : thermal multigate decay pulsed system
SCHENEWERK, KNAPP & FERTL (1982) : pulsed neutron capture logging, application
SCHULTZ et al. (1983) : combination of neutron and gamma-ray logging, multigate
SMITH, ARNOLD & PEELMAN (1983) : pulsed neutron generator, multigate decay
VASILEV et al. (1987) : pulsed neutron capture logging, Middle East reservoirs
WELEX (1986) : formation strength analysis, wavetrain, frac pressure, sanding

7.2.3.10.7. Borehole elongation and caliper logging

The references on borehole elongation and caliper logging correspond mainly to aspects discussed in section 6.2.1.8. of the text.

BELL & GOUGH (1978) : borehole breakout, in-situ stress direction, frac azimuth
BLANTON (1983) : borehole breakout, fracture azimuth, in-situ stress direction
BLANTON & TEUFEL (1985) : in-situ stress determination, deformed wellbore
CHARLEZ, SALEH & DESPAX (1987) : fracmeter, stress state determinate, tension
COX (1983) : caliper logging, borehole long axis, ovalization, geostress
DETOURNAY & FAIRHURST (1982) : radial stress distribution, linear elasticity
FORDJOR, BELL & GOUGH (1983) : borehole breakout, fracture azimuth, stress
GOUGH & BELL (1982) : radial stress distribution, linear elastic borehole wall
GRIFFIN (1985) : minor and major borehole eccentricities, significance, analyze
OWEN, TORONTO & PETERSON (1988) : wellbore elongation, washouts, fracture zone
PLUMB & HICKMAN (1985) : borehole elongation, dipmeter, television, relation
SMITH et al. (1978 a) : fracture azimuth determination, deep penetrating crack
SMITH, LOGAN & WOOD (1980) : fracture azimuth, shallow experiment
SMITH, REN, SORRELS & TEUFEL (1985) : fracture azimuth, fracture diagnostics
TEUFEL (1985 b) : wellbore breakout, in-situ stress relation, natural fracture
TEUFEL et al. (1984) : fracture azimuth determination, multiwell experiment
VOEGELE & JONES (1980) : wireline tool, in-situ stress contrast determination
ZOBACK, MOOS, MASTIN & ANDERSON (1985) : wellbore breakout, in-situ stress

7.2.3.10.8. Laterolog and spontaneous potential

The references on laterolog and spontaneous potential correspond mainly to aspects discussed in section 6.2.1.10.3. of the text.

AGUILERA (1974) : sonic, resistivity, naturally fractured reservoirs
BOYELDIEU & WINCHESTER (1982) : dual laterolog, frac porosity, hard carbonates
HOWARD (1981) : induction logging, vertical fractures, borehole fluid presence
POUPON, HOYLE & SCHMIDT (1970) : porosity, crossplot, spontaneous potential
RASMUS (1981) : deep and shallow laterolog, fracture response, curve separation
SCHLUMBERGER (1987 a) : investigation depths, deep, shallow, microspherical
SIBBIT & FAIVRE (1985) : dual laterolog response, horizontal, vertical, fracs
WITTRISCH & SARDA (1987) : resistivity logging for fracture height detection

7.2.3.10.9. Dipmeter recording

The references on dipmeter recording correspond mainly to aspects discussed in section 6.2.1.10.2. of the text.

DELHOMME et al. (1981) : dipmeter, multipad sensors, circumferential microsonic
DELHOMME et al. (1986) : reservoir applications of dipmeter logs
DRESSER ATLAS (1987) : diplog analysis, fundamentals, fracture detection
ENDERLIN, EDMISTON & EPPS (1988) : dip analysis, single sedimentary event
GEORGI (1986) : dipmeter interpretation, fracture detection, guidelines, review
GILREATH (1977) : dipmeter, principles, application
GOETZ, TYSON & GARAT (1987) : six-arm dipmeter, interpretation, new dimension
MORRISON & THIBODAUX (1984) : six-arm dipmeter, borehole cross-section, natural
PERRIER & QUIBLIER (1973) : dipmeter, fracture detection, orientation
PLUMB & HICKMAN (1985) : borehole elongation, dipmeter, television, relation
SCHLUMBERGER (1982, 1983) : dipmeter, fracture orientation, fracture detection
SCHLUMBERGER (1987 a) : dipmeter, televiewer, microscanner, fracture detection
SHIELDS (1974) : dipmeter, sedimentary environment, geological correlation

7.2.3.10.10. Borehole television

The references on borehole television correspond mainly to aspects discussed in sections 6.2.1.9. and 6.2.1.10.1. of the text.

BELL & GOUGH (1979) : borehole televiewer, in-situ stress determination, oil
BRIGGS (1964) : downhole television camera, development, application, describe
BRODING (1981, 1984) : borehole televiewer, volumetric scan logging, cement
CLERKE & AKKEREN (1986) : borehole televiewer, completion, improvement, Permian
DARILEK (1986) : borehole television, gas production, Devonian shales, poroperm
FLECKENSTEIN (1984) : borehole television, Rotliegend, Germany, fractures, gas
FORDJOR, BELL & GOUGH (1983) : borehole television, in-situ stress, breakout
GEORGI (1985) : borehole televiewer images, geometric aspects, shape influence
GRIFFIN (1985) : borehole televiewer, inflatable packer, seismic accelerometer
HACKBARTH & TEPPER (1988) : formation microscanner, dipmeter, thin bedding
HEARD (1981) : high-temperature acoustic borehole televiewer
KAMP (1983) : downhole telemetry, evaluation, practical application
KEYS & SULLIVAN (1979) : borehole television, geothermal reservoir, mapping
KIERSTEIN (1984) : borehole television, frac orientation correction, mapping
KULANDER, BARTON & DEAN (1979) : fractography, outcrop and core frac studies
LAU (1980, 1983) : optical borehole television, fracture detection
LAUBACH et al. (1988) : formation microscanner, comparison, fracture detection
MYUNG & CALDWELL (1971) : frac visibility, logging techniques, natural crack
OVERBEY, YOST & YOST (1988) : borehole video camera, horizontal well, natural
PAILLET & KIM (1985) : borehole television, wall breakout, mapping, distribute
PASTERNACK & GOODWILL (1983) : digital borehole televiewer logging, improvement
PLUMB & HICKMAN (1985) : borehole elongation, dipmeter, television, relation
RAMBOW (1983) : borehole televiewer, experience report, field examples
RAMBOW (1985) : borehole televiewer, wellbore visual inspection
SCHLUMBERGER (1987 a) : dipmeter, televiewer, microscanner, fracture detection
TAYLOR (1983) : borehole televiewer survey, travel time, amplitude, interpret

WALBE (1986) : borehole televiewer, open and cased hole logging, Appalachian
WILEY (1980) : borehole televiewer, review, application, interpretation
ZEMANEK et al. (1969) : borehole televiewer, fracture localization, inspection

7.2.3.10.11. Formation microscanner

The references on formation microscanner correspond mainly to aspects discussed in sections 6.2.1.9. and 6.2.1.10.1. of the text.

BRODING (1981, 1984) : borehole televiewer, volumetric scan logging, cement
EKSTROM et al. (1986) : formation microscanner, imaging, microelectric arrays
HACKBARTH & TEPPER (1988) : borehole televiewer, dipmeter, thin bedding, sand
HANSEN (1988) : imaging, formation microscanner, Permian Basin, application
LAUBACH & MONSON (1988) : borehole televiewer, coring-induced fractures
LAUBACH et al. (1988) : borehole televiewer, comparison, fracture detection
LLOYD, DAHAN & HUTIN (1986) : microscanner, formation imaging, high-resolution
McNAUGHTON & CASARTA (1988) : fracture identification, Lisburne carbonate
PLUMB & LUTHI (1986) : formation microscanner, aeolian sands, borehole imaging
SCHLUMBERGER (1986 b) : formation microscanner, resistivity, natural fractures
SCHLUMBERGER (1987 a) : dipmeter, televiewer, microscanner, fracture detection

7.2.3.10.12. Seismical methods and microearthquake detection

The references on seismical methods and microearthquake detection correspond mainly to aspects discussed in section 6.2.2.3. of the text.

AKI, FEHLER et al. (1982) : seismic data, geothermal, Fenton Hill, New Mexico
ALBRIGHT & PEARSON (1980) : microseismic techniques, hydraulic frac location
BAME & FEHLER (1986) : long-period earthquakes accompanying hydraulic fracture
BATCHELOR, BARIA & HEARN (1983) : microseismic event location, frac monitoring
BEYDOUN, CHENG & TOKSÖZ (1985) : vertical seismic profiling, open frac detect
CARDEN et al. (1983) : microearthquake fracture detection, geothermal prop frac
CHOUET (1986) : dynamics of fluid-driven crack, finite difference methodology
CRAMPIN & ATKINSON (1985) : microseismic fracture detection, microcrack, crust
DOBECKI (1983) : passive borehole seismics, hydraulic fracture orientation
HART et al. (1983) : frac diagnostics, field instrumentation, comprehensive
HOUSE, KEPPLER & KAIEDA (1985) : microearthquake map, geothermal MHF, seismic
IVERSON (1988) : acoustic impedance, crosswell logging, connection, cracking
KLEINBERG et al. (1984) : borehole application techniques, sensitive, reliable
KOERPERICH (1980) : shear-wave velocity, long- and short-spaced acoustic device
LEYDECKER (1981) : seismic fracture detection, geothermic project, hot dry rock
LOCKE, YOST, OVERBEY & GRABOWSKI (1988) : foam fracturing, dynamic pressure
MAJER & DOE (1986) : high-frequency seismic monitoring, hydrofracture study
MAUK & MAHRER (1986) : frac monitor, long period seismometer, induced geometry
MAUK & MAHRER (1988) : long-period seismometer, induced crack geometry, frac
McGARR & GAY (1978) : fracture orientation, vertical vs. horizontal, stress
MOORE & NIELSEN (1986) : frac map, geothermal reservoir, geology/geochemistry
MORALES et al. (1986) : fracture detection, water flooding, transmissibility
MURPHY & FEHLER (1986) : microearthquake mapping, geothermal MHF, shear slip
PEARSON (1981) : microseismics, high pore pressures, fracturing experiments
REYNOLDS (1987) : frac identification, behaviour, seismics, tracer, photography
ROBERTS & CRAMPIN (1986) : shear-wave polarization, hot dry rock geothermal pay
SARDA, PERREAU & DEFLANDRE (1988) : acoustic emission interpretation, extension
SARDA, WITTRISCH, PERREAU & DEFLANDRE (1987) : downhole fracture monitoring
SCHUCK & KEECH (1975) : acoustic emission monitoring, propagating fractures
SCHUSTER (1978) : seismic signals, wellbore detection, fracture monitoring
SMITH et al. (1978 a) : fracture azimuth determination, deep penetrating crack
SMITH, LOGAN & WOOD (1980) : fracture azimuth, shallow experiment, diagnostics
SMITH, REN, SORRELS & TEUFEL (1985) : fracture azimuth, fracture diagnostics
SORRELS & MULCAHY (1986) : fracture azimuth estimation, microseismic method
TALEBI & CORNET (1985) : seismo-acoustic activity, fluid injection generation

TEUFEL et al. (1984) : fracture azimuth determination, multiwell experiment
THORNE & MORRIS (1987) : borehole seismic fracture diagnostics, advances
WIERCZEYKO (1985) : fracture detection, seismic, geoelectrical, surface-derived
WITTRISCH & SARDA (1987) : SIMFRAC, direction of hydraulic fracture

7.2.3.10.13. Tiltmeter, gravimeter and magnetometer

The references on tiltmeter, gravimeter and magnetometer correspond mainly to aspects discussed in sections 6.2.2.3.1., 6.2.2.3.4. and 6.2.3. of the text.

BARTEL, McCANN & KECK (1976) : fracture mapping, potential gradients, character
DAVIS (1983) : surface deformation field, geostress regime, frac orientation
EVANS, HOLZHAUSEN & WOOD (1982) : tiltmeter analysis, fracture geometry analyze
EVANS et al. (1980) : frac mapping, associated surface deformation, observation
EVANS et al. (1985) : surface deformation, horizontal, vertical, orientation
FITZ-PATRICK, KARR & O`SHEA (1985) : proppant mapping, fracture geometry, test
GRIFFIN (1985) : borehole televiewer, inflatable packer, seismic accelerometer
HANSON & OWEN (1982) : fracture orientation analysis, solid earth tidal strain
HARDT (1982) : geoelectrical fracture detection, equipotential pattern, disturb
HOLZHAUSEN & EGAN (1986) : fracture diagnostics, hydraulic impedance method
McGARR & GAY (1978) : fracture orientation, vertical vs. horizontal, stress
PALMER & VEATCH (1988) : tilt, uplift, pressure transient, fracturing, detect
POLLARD (1976) : tiltmeter survey, mechanical interaction, fracture stability
POLLARD & HOLZHAUSEN (1979) : tiltmeter, mechanical interaction, frac stability
WOOD, POLLARD & RALEIGH (1976) : in-situ geometry determination, tiltmeter
WOOD et al. (1983) : proppant mapping, surface superconducting, magnetometer

7.2.3.10.14. Strain relaxation, core evaluation, in-situ stress

The references on strain relaxation, core evaluation and in-situ stress correspond mainly to aspects discussed in sections 6.2.1.10.1. and 6.2.2.3.3. of the text.

BLANTON & TEUFEL (1983) : strain recovery method, in-situ stress determination
FORDJOR, BELL & GOUGH (1983) : borehole television, in-situ stress, breakout
FRISINGER & COOPER (1985) : well logging, formation stress, frac fluid response
HANSON & OWEN (1982) : fracture orientation analysis, solid earth tidal strain
KULANDER, BARTON & DEAN (1979) : fractography, outcrop and core frac studies
LAUBACH & MONSON (1988) : coring-induced fractures, borehole televiewer, detect
LERCHE & NARR (1984) : subsurface fracture density, variable spacing, core
LERCHE & NARR (1986) : frac density, variable frac spacing, core evaluation
NIELSEN & KOHLHAAS (1979) : rock mechanical properties, acoustic and biaxial
OWEN, TORONTO & PETERSON (1988) : anelastic strain recovery, reliability, test
PIERCE, VELA & KOONCE (1974) : pressure transient, pulse test, frac azimuth
RAO et al. (1979) : in-situ stress, coring-induced fracturing, determination
TEUFEL (1981) : strain relaxation, fracture azimuth, oriented core prediction
TEUFEL (1982, 1985 a) : fracture azimuth prediction, anelastic strain recovery
TEUFEL (1985 b) : wellbore breakout, in-situ stress relation, natural fracture
TEUFEL (1988) : anelastic strain recovery, microcrack genesis, petrophysics
TEUFEL & WARPINSKI (1984) : anelastic strain recovery measuring, crack stress
URAIET, RAGHAVAN & THOMAS (1977) : interference test, vertical frac orientation
VOEGELE & JONES (1980) : wireline tool, in-situ stress contrast determination
ZOBACK, MOOS, MASTIN & ANDERSON (1985) : wellbore breakout, in-situ stress

7.2.3.10.15. Mechanical properties logging

The references on mechanical properties logging correspond mainly to aspects discussed in sections 6.2.1.10.7. and 6.3.6. of the text.

AHMED (1988) : fracture height prediction, modelling, rock properties, sonic
ANDERSON et al. (1975) : formation strength, mechanical properties estimation
BALL, CHACE & FERTL (1987) : well data system, frac migration model, computer
CHARLEZ, SALEH & DESPAX (1987) : fracmeter, stress state determinate, tension
COATES & DENOO (1981) : frac migration model, mechanical properties, prediction
CORLEY & KLAUTT (1986) : frac migration model, fracturability prediction
ELKINGTON, STOUTHAMER & BROWN (1982) : rock strength, wireline logs, prediction
FRISINGER & COOPER (1985) : well logging, formation stress, frac fluid response
KAZI, ZEKAI & BAHAA-ELDIN (1983) : uniaxial compressive strength, sonic log
LANCASTER et al. (1988) : acoustic log, in-situ stress tests, stress profiles
MONTMAYEUR & GRAVES (1986) : mechanical properties prediction, sonic logging
NIELSEN & KOHLHAAS (1979) : rock mechanical properties, acoustic and biaxial
ONYIA (1987) : geology drilling log, computer database system
ONYIA (1988) : formation strength, drilling strength, mechanical properties
OYENEYIN (1987 b) : formation sand size prediction, well log correlations
PAILLET & WHITE (1982) : acoustic wave propagation, rock property relationship
WELEX (1986) : formation strength analysis, wavetrain, frac pressure, sanding

7.2.3.10.16. Fracture azimuth and width determination

The references on fracture azimuth and width determination correspond mainly to aspects discussed in sections 6.2.1.8., 6.2.2. and 6.2.7. of the text.

ALPAY (1969) : aerial photography, fracture mapping, infrared, radar
BARTEL, McCANN & KECK (1976) : fracture mapping, potential gradients, character
CINCO-LEY & SAMANIEGO (1977) : finite conductivity, orientation, pressure test
DAVIS (1983) : surface deformation field, geostress regime, frac orientation
ELKINS & SKOV (1960) : frac orientation determination, pressure interference
FITZ-PATRICK, KARR & O`SHEA (1985) : proppant mapping, fracture geometry, test
HANSON & OWEN (1982) : fracture orientation analysis, solid earth tidal strain
MOORE & NIELSEN (1986) : frac map, geothermal reservoir, geology/geochemistry
MORALES et al. (1986) : fracture detection, water flooding, transmissibility
PIRSON (1967) : mapping fracture development from well logs
POWER, SCHUSTER, HAY & TWOMBLY (1976) : orientation, dimension, detection, case
RYAN (1976) : remote sensing, fracture mapping, West Virginia, Kentucky
SMITH et al. (1978 a) : fracture azimuth determination, deep penetrating crack
SMITH, LOGAN & WOOD (1980) : fracture azimuth, shallow experiment
SMITH, REN, SORRELS & TEUFEL (1985) : fracture azimuth, fracture diagnostics
SMITH, ROSENBERG & BOWEN (1982) : fracture width, design vs. measurement
SORRELS & MULCAHY (1986) : fracture azimuth estimation, microseismic method
TEUFEL (1981) : strain relaxation, fracture azimuth, oriented core prediction
TEUFEL (1982, 1985 a) : fracture azimuth prediction, anelastic strain recovery
TEUFEL et al. (1984) : fracture azimuth determination, multiwell experiment
TYLER & VOLLENDORF (1975) : fracture mapping, in-situ stress measurements
URAIET, RAGHAVAN & THOMAS (1977) : interference test, vertical frac orientation
WARPINSKI (1983) : width, pressure, measurement, propagating fracture
WIERCZEYKO (1985) : frac detection, seismic, geoelectrical, surface-derived
WITTRISCH & SARDA (1987) : SIMFRAC, direction of hydraulic fracture

7.2.3.10.17. Fracture permeability estimation

The references on fracture permeability estimation correspond mainly to aspects discussed in section 6.2.2.5. of the text.

DAVISON, KEYS & PAILLET (1982) : borehole geophysical logs, hydrologic tests
HARDIN & TOKSÖZ (1985) : tube wave generation, frac characterization, permeable
HARDIN et al. (1987) : attenuation and generation of tube waves, frac permeable
HESS (1986) : low-velocity borehole flowmeter, fracture permeability
KEYS (1979) : borehole geophysics, igneous, metamorphic, fracture permeability
LEVINE, CYBRIWSKY & TOKSÖZ (1984) : hydraulic conductivity, vertical profiling

NELSON, MAGNUSSON & RACHIELE (1983) : borehole geophysics, frac permeability
PAILLET (1980, 1983, 1985 b) : acoustic propagation, fracture permeability
PAILLET & HESS (1986) : borehole geophysics, fracture permeability
PAILLET, HSIEH & CHENG (1987) : acoustic waveform, vertical seismic profiling
PAILLET, KEYS & HESS (1985) : lithology effect, televiewer quality, analysis
PAILLET et al. (1986, 1987) : tube waves, vertical flow, high resolution

7.2.3.10.18. Noise logging

The references on noise logging correspond mainly to aspects discussed in section 6.2.1.7. of the text.

BELFIELD (1988) : spinner surveys, injection tests, natural fractures, cores
BRITT (1977) : audio tracer survey, sound or noise, stationary, stop-and-go log
DOBKINS (1979) : fluid flow within cracks, noise-logging record, tight rock
KENNEDY & McELHINEY (1979) : combination of temperature and noise logging
McKINLEY & BOWER (1977) : noise logging, special applications
McKINLEY, BOWER & RUMBLE (1973) : noise log, fluid flow behind casing, cement
PENNEBAKER & WOODY (1977) : temperature and noise combination, problem wells
SARDA, PERREAU & DEFLANDRE (1988) : acoustic emission, frac extent estimation
WYMAN, HOLDITCH & RANDOLPH (1979) : combination of temperature and noise log

7.2.3.10.19. Inflatable impression packers

The references on inflatable impression packers correspond mainly to aspects discussed in section 6.2.6. of the text.

BREDEHOEFT, WOLFF, KEYS & SHUTER (1976) : inflatable impression packer survey
CHARLEZ, SALEH & DESPAX (1987) : fracmeter, special inflatable packer, imprint
CHARLEZ, SALEH, DESPAX & JULIEN (1986) : rockmeter, radial displacement, rubber
GRIFFIN (1985) : borehole televiewer, inflatable packer, seismic accelerometer
HAIMSON (1973) : impression packer, earthquake related stresses, Colorado
PAILLET, KEYS & HESS (1985) : comparative resolution, borehole television
TAM INTERNATIONAL (1986) : inflatable packer, well testing, workover, fracture
WALSH & HOLDER (1988) : inflatable packers, production application

7.2.3.10.20. Interference, tracer and pulse testing

The references on interference, tracer and pulse testing correspond mainly to aspects discussed in section 4.2.4.1.2. of the text. The quotations assembled here are predominantly in relation to fracture height as well as azimuth detection and estimation; references on sand body geometry evaluation and reservoir continuity checking with the mentioned test methods are compiled in section 7.2.5.14.

ABOBISE & TIAB (1982) : pulse testing, fracture orientation, pay permeability
BELFIELD (1988) : spinner surveys, injection tests, natural fractures, cores
MORALES et al. (1986) : fracture detection, water flooding, transmissibility
PIERCE, VELA & KOONCE (1974) : pressure transient, pulse test, frac azimuth
ROBINSON & TESTER (1984) : tracer-determined flow, residence time, dispersion
ROBINSON, TESTER & BROWN (1984) : inert, chemically reacting tracers, sizing
TESTER, BIVENS & POTTER (1982) : interwell tracer analysis, geothermal, fracs
URAIET, RAGHAVAN & THOMAS (1977) : interference test, vertical frac orientation

7.2.3.10.21. Other aspects

The references on other aspects of fracture monitoring and detection correspond to various items discussed in chapter 6.

AADNØY (1988) : inversion technique, in-situ stress field, fracturing data
AGUILERA & ACEVEDO (1982) : computerized log interpretation, natural fractures
ALPAY (1969) : aerial photography, fracture mapping, infrared, radar
ARDITTY, GRAS & STARON (1987) : EVA data, open and cased hole, permeability
BALL, CHACE & FERTL (1987) : well data system, frac migration model, computer
BELANI & JALALI-YAZDI (1988) : fracture intensity, pressure transient data
BROWN (1978) : fracture identification log, microresistivity, application
CINCO-LEY & SAMANIEGO (1977) : finite conductivity, orientation, pressure test
CRAIGHEAD, McDOUGALL & HOTZ (1987) : pre-frac design, post-frac diagnosis, two
DAVIS (1983) : surface deformation field, geostress regime, frac orientation
EL RABAA (1988) : fracture direction determination in deviated wells
EVANS & ENGELDER (1986) : stress mapping, microfracturing, wireline
FERTL & OLIVER (1983) : multiple stage fracturing evaluation, U.S. patent
FITZ-PATRICK, KARR & O`SHEA (1985) : proppant mapping, fracture geometry, test
GRIFFIN (1985) : borehole televiewer, inflatable packer, seismic accelerometer
HAIMSON (1973) : impression packer, earthquake related stresses, Colorado
HANSON & OWEN (1982) : fracture orientation analysis, solid earth tidal strain
HERTZOG (1980) : inelastic neutron scattering, capture gamma-ray spectrometry
HOWARD (1981) : induction logging, vertical fractures, borehole fluid presence
HUNT et al. (1981) : new well logs, application to fracturing, Cotton Valley
JIANG et al. (1988) : pressure performance in fracturing jobs, application
KOERPERICH (1980) : shear-wave velocity, long- and short-spaced acoustic device
KULANDER, BARTON & DEAN (1979) : fractography, outcrop and core frac studies
MACPHERSON & GUENARD (1988) : drilling engineering and mudlogging data
McGARR & GAY (1978) : fracture orientation, vertical vs. horizontal, stress
MOORE & NIELSEN (1986) : frac map, geothermal reservoir, geology/geochemistry
MORALES et al. (1986) : fracture detection, water flooding, transmissibility
MURPHY & BROWN (1976) : fracture geometry, connection downhole, flow testing
MYUNG & CALDWELL (1971) : frac visibility, logging techniques, natural crack
OWEN (1982) : solid earth tidal strain method, fracture orientation analysis
PALMER & VEATCH (1987) : step-rate tests, abnormally high fracturing pressures
PARKER & McDANIEL (1987) : design improvement, in-situ conductivity measures
PEARSON (1987) : operator`s stimulation monitoring system, application, develop
PENNEBAKER & WOODY (1977) : temperature and noise combination, problem wells
PIERCE, VELA & KOONCE (1974) : pressure transient, pulse test, frac azimuth
PLUMB & HORNBY (1988) : integrating wellbore images, fracture characterization
POUPON, HOYLE & SCHMIDT (1970) : porosity, crossplot, spontaneous potential
REYNOLDS (1987) : frac identification, behaviour, seismics, tracer, photography
REYNOLDS, SCOTT, POPHAM & COFFER (1961) : frac extent and orientation detection
RYAN (1976) : remote sensing, fracture mapping, West Virginia, Kentucky
SANGREE (1969) : distinction of natural and induced fractures in cores
SARDA, PERREAU & DEFLANDRE (1988) : acoustic emission, frac extent estimation
STEINMAN, ADOLPH, MAHDAVI & PREEG (1988) : dual-burst thermal decay time log
TOMLINSON, BYRD & BERG (1988) : fracturing process control and automation
URAIET, RAGHAVAN & THOMAS (1977) : interference test, vertical frac orientation
WESTERMARK & SCHOLES (1987) : cement evaluation tool usage, open-hole test
WIERCZEYKO (1985) : frac detection, seismic, geoelectrical, surface-derived

7.2.3.11. Fracture performance evaluation

The references on fracture performance evaluation correspond mainly to aspects discussed in section 4.2. of the text.

BENNETT, CAMACHO, REYNOLDS & RAGHAVAN (1985) : layered reservoir production
BRINKMANN, KRÖMER & REINICKE (1980) : type curves, fracspecific parameters
CHARLES et al. (1983) : stimulation treatments, evaluation, Appalachian tight
CLEARY et al. (1986) : computerized field system, real-time monitoring, analyze
DOWDLE & HYDE (1977) : well test analysis, fractured gas sands, evaluation
ECONOMIDES (1987) : fractured well testing, evaluation, recommendations
EVANS & CARROLL (1981) : stochastic modelling, fractured wells, noncontinuous

FLEMING & GONZALEZ (1987) : performance modelling, pattern floods, curvilinear
GROVE & WHITTAKER (1985) : frac probability determination, continuous, Monterey
HALL & LARKIN (1986) : on-site quality control, fracture treatments
HOLBERT et al. (1987) : signature approach, water-driven reservoirs, post-frac
HOLDITCH & LAUFER (1978) : fractured gas well analysis, reservoir simulation
HOLDITCH & LEE (1979) : pressure transient tests, low permeability, field
HOLDITCH et al. (1981) : improved technique, conductivity, fracture length
HOLZHAUSEN & GOOCH (1985) : fracture impedance, measurement, closure pressure
HURST (1985) : model, fracture evaluation, reservoir limit, determination
JOPLING & KETCHER (1987) : real-time decision making, log/log plot, treatment
LEE & HOLDITCH (1979) : pressure transient tests, low permeability, theory
MEAD (1987) : finite system buildup analysis, investigation, evaluation
MOUSLI et al. (1982) : vertical fractures, intercepting active and observation
NURMI, CHARARA & WATFA (1987) : Middle East, evaluation, new techniques
PEARSON, ABOU-SAYED & NGUYEN (1985) : fracture design parameters, measurement
PRADO & PRAT (1987) : unsteady liquid flow, uniformly fractured reservoir
REINICKE et al. (1985) : buildup data, type curves, integral methods, hand
ROBINSON, HOLDITCH & LEE (1983) : MHF, tight gas sand, geopressured evaluation
SCOTT (1979) : flow characteristics, determination, fractured wells, new method
SEARS & PHILLIPS (1987) : fraced reservoirs evaluation, Monte Carlo techniques
SOLIMAN et al. (1984) : type curves, fractured well performance, evaluation
THOMPSON (1981) : constant pressure, finite capacity, type curves, performance
WYMAN, HOLDITCH & RANDOLPH (1979) : frac evaluation, analysis, field example

7.2.3.12. Fracture orientation changes

The references on fracture orientation changes correspond mainly to aspects discussed in section 4.9.4. of the text.

CHHINA & AGAR (1985) : in-situ stress changes, fracture extension pressure
CLEARY et al. (1983) : stress changes, temperature, pore fluid pressure, direct
CLOSMANN & PHOCAS (1976) : heating along parallel vertical cracks, horizontal
CROCKETT, VO & CLEARY (1984) : stress changes, temperature, pore fluid pressure
CONSTANT & BOURGOYNE (1986) : changing reservoir plasticity, stress contrast
DUSSEAULT & SIMMONS (1982) : stress changes, pore fluid pressure, temperature
HOLZHAUSEN et al. (1985) : cyclic fracture steaming, generation and collapse
KOMAR & FROHNE (1973) : factors influencing fracture initiation and orientation
KULAR, CHHINA, BEST & MACKENZIE (1988) : thermal expansion, fluid loss, loading
McLENNAN, ROEGIERS & MARX (1983) : pore-pressure manipulation, pre-notching
PERKINS & GONZALEZ (1984, 1985) : stress changes, temperature, pore fluid press
SCHRAUFNAGEL & LAMBERT (1988) : combination of horizontal and vertical cracks
SHUCK (1977) : altered-stress fracturing, stress reorientation, nearby crack
SINGHAI & CARD (1988) : horizontal and vertical fractures, steam injection
SUN & MONGAN (1974) : disposal fracturing, shale, West Valley, New York, uptake
VITTORATOS (1986) : fracture creation and collapse, overburden, compression
VITTORATOS, SCOTT & BEATTIE (1988) : fracturing for steam injectivity improve
WARPINSKI & BRANAGAN (1988) : altered-stress fracturing, perpendicular network
WONG (1986) : on- and off-trend well pairs, staggered steam injection, azimuth

7.2.4. Proppant features

This subject classification key collects aspects on various different topics including selection, testing, concentration and settling of proppants, as well as proppant and fine particle transport and resin coating of proppants and gravel. Corresponding aspects are mainly discussed in chapters 1 and 4 of the text.

The section on proppant selection comprises references on first- and second-generation proppants, criteria and guidelines for selection, comparative effectiveness, technical specifications, behaviour under extreme conditions, strength and density, conductivity comparison, and application examples of proppants. The chapter on proppant testing contains references on material properties, long-term and short-term stability, behaviour under extreme conditions, recommended testing practices, and fluid interaction of proppants. The branch on proppant concentration presents references on ultra-high saturation and bank buildup of proppants. The division on proppant settling assembles references on incomplete fracture fill-up, dynamic proppant fall, settling velocity, non-Newtonian fluids, and effects on productivity ratio. The section on proppant and fine particle transport collects references on particle size distribution, settling velocity, transport model, numerical solution, and controlling factors. Some experiences of resin coating of proppants and gravel are reported under the last topic line.

7.2.4.1. Proppant selection

The references on proppant selection are classified along the lines of sintered bauxite proppants, zirconia silicate proppants, lightweight proppants, and other aspects.

7.2.4.1.1. Sintered bauxite proppants

References on sintered bauxite proppants correspond mainly to aspects discussed in sections 1.3.2.1. and 1.4. of the text.

COOKE (1975, 1977) : proppant type, fluid type, effect on conductivity
COOKE (1976), COOKE & GIDLEY (1979) : high-strength proppant, sintered bauxite
COOKE, GIDLEY & MUTTI (1977) : high-strength proppant, sintered bauxite
COOKE, HEDDEN & CHARD (1978) : sintered bauxite, hydraulic fracturing, patent
COMALCO (1983) : Duraprop, natural bauxite pisolites, technical specifications
CUTLER, ENNISS, JONES & CARROLL (1983) : conductivity comparison, commercial
CUTLER, ENNISS, JONES & SWANSON (1985) : conductivity comparison, ceramic prop
KOHLHAAS (1982) : bauxite proppants, intermediate depth, oil reservoir fracture
LUNGHOFER (1984) : sintered bauxite, U.S. patent
MESSINA (1987) : proppants, bauxite, technical specifications, specific gravity
MONTGOMERY & STEANSON (1985) : proppant selection, key function, significance
NORTON-ALCOA PROPPANTS (1985) : bauxite proppant products, technical specifics
RIXE, FAST & HOWARD (1963) : proppant selection for hydraulic fracturing
SPARLIN & HAGEN (1983) : intermediate-strength proppant
STANDARD OIL PROPPANTS (1986) : prop for all reasons, technical specifications
WESTERN PETROLEUM (1984) : proppants, description, illustration, selection

7.2.4.1.2. Zirconia silicate proppants

References on zirconia silicate proppants correspond mainly to aspects discussed in sections 1.3.2.2. and 1.4. of the text.

BECQ, ROQUE & SARDA (1984) : extreme conditions, high-strength proppant
COL (1983) : Zirprop, zirconia-silicate high-strength low-density proppants
MESSINA (1986) : Zirprop, technical specifications
SARDA (1981) : high-strength proppant, zirconium oxide
SEPR (1983) : Zirprop, technical specifications

7.2.4.1.3. Lightweight proppants

References on lightweight proppants correspond mainly to aspects discussed in sections 1.3.2.1. and 1.4. of the text.

CALLANAN, CIPOLLA & LEWIS (1983) : second-generation proppant, tight gas pays

CUTLER et al. (1981) : new proppants, deep gas well stimulation, conductivity
CUTLER & JONES (1982), CUTLER et al. (1983) : lightweight proppants, deep gas
CUTLER, ENNISS, JONES & CARROLL (1983) : conductivity comparison, commercial
CUTLER, ENNISS, JONES & SWANSON (1985) : conductivity comparison, ceramic prop
FITZGIBBON (1984) : light proppants, U.S. patent
JONES, CUTLER & SWANSON (1981) : lightweight proppants, deep gas wells
MESSINA (1987) : proppants, bauxite, technical specifications, specific gravity
SANTROL (1975) : resin-coated sand, fracturing, gravel packing, performance
STANDARD OIL PROPPANTS (1986) : prop for all reasons, technical specifications
UNDERDOWN & DAS (1982) : new proppant, lightweight, deep hydraulic fracturing

7.2.4.1.4. Other aspects

References on other aspects of proppant selection correspond to various points discussed in sections 1.2. to 1.4. of the text.

ATTEBERRY, TUCKER & RITZ (1979) : closure stress, low-permeability gas, Texas
BROWN & MUCH (1986) : proppant evaluation
CLARK (1983) : selection method, computer simulation, prop type, pack configure
COBB (1985) : overview, history, examples
COOKE (1975, 1977) : proppant type, fluid type, effect on conductivity
CROSS & SHELDON (1977) : grain-size choice, coarser material, less residue plug
CUTLER, ENNISS, JONES & CARROLL (1983) : conductivity comparison, commercial
CUTLER, ENNISS, JONES & SWANSON (1985) : conductivity comparison, ceramic prop
DAVIS (1975) : proppant mixing, glass beads, cost containment, conductivity
DGMK (1983, 1986) : proppant testing, new proppant model type
DOWELL SCHLUMBERGER (1985) : proppant selection guide, natural sand, synthetic
ECONOMIDES (1986) : technology conference, proppant fracturing, gravel packing
FAST & MASE (1965) : propping agents, effective in gas well fracturing
HARBEN (1978) : abrasives, review
HICKEY (1983); HICKEY, BROWN & CRITTENDEN (1981) : comparative effectiveness
HOLDITCH (1984) : proppant selection criteria
HUITT & McGLOTHLIN (1958) : proppant embedment susceptibility, frac propping
HUITT, McGLOTHLIN & McDONALD (1959) : proppant crushing, fracture propping
KERN, PERKINS & WYANT (1961) : propping fractures with aluminum particles
KIM (1983) : proppants, permeability, conductivity
LARSEN & SMITH (1985) : angular blend, fracturing sand, new conductivity
McDANIEL & WILLINGHAM (1978) : frac permeability, effect of proppant mixtures
McGLOTHLIN & HUITT (1966) : formation rock strength, propping agent strength
MILLER & WAREMBOURG (1975) : 100 mesh sand, prepack technique, frac improvement
MONTGOMERY & STEANSON (1985) : proppant selection, key function
NEAL, PARMLEY & COLPAYS (1977) : deep well treatment, oxide ceramic proppants
NORMAN & FAST (1985) : proppant nomograph
NORMAN, CIPOLLA & WEBB (1983) : oil well, moderate permeability, prop choice
NORMAN & FAST (1985) : proppant monograph, proppant selection, properties
NORTON-ALCOA PROPPANTS (1985) : bauxite proppant products, technical specifics
PEARCE (1983) : high-strength proppants, shallow and hard formations
PHILLIPS & ANDERSON (1985) : fracture optimization, modelling, low-permeability
RIXE, FAST & HOWARD (1963) : proppant selection for hydraulic fracturing
ROBINSON, HOLDITCH & WHITEHEAD (1986) : frac damage, flowback, crushing, shock
SACUTA, NGUYEN & KISSEL (1986) : nickel-coated sands, gravel pack material
SANTROL (1975) : resin-coated sand, fracturing, gravel packing, performance
SECCOMBE & ANDERSON (1982) : proppant selection, tight gas, bauxite vs. sand
SPARLIN & HAGEN (1983) : intermediate-strength proppant
STANDARD OIL PROPPANTS (1986) : prop for all reasons, technical specifications
STEANSON, ELBEL & WENDORFF (1979) : proppant selection
THOMPSON (1977) : 100 mesh sand, stimulation effectivity, Permian Basin rocks
TUCKER (1979) : practical pressure analysis, geopressure, closure stress, crush
VOLK et al. (1981) : high-strength props, embedment, low-permeability formation
WENDORFF (1978) : fracture sand control
WENDORFF (1979) : sand handling equipment, proppant sand quality

WESTERN PETROLEUM (1984) : proppants, description, illustration, selection

7.2.4.2. Proppant testing

The references on proppant testing correspond mainly to aspects discussed in section 1.4.10. of the text.

AHMED, ABOU-SAYED & JONES (1979) : fracture fluid interaction, propped fracture
ALBERTSEN (1985) : status report, energy technology, geotechnics, proppant test
AMERICAN PETROLEUM INSTITUTE (1985) : testing, recommended practices
BECQ, ROQUE & SARDA (1984) : extreme conditions, high-strength, acid, steam
BROWN & MUCH (1986) : long-term, stress corrosion, simulated pay conditions
CHEUNG (1985) : acid effect, gravel, proppants, dissolution
COBB & FARRELL (1986) : long-term proppant stability, testing, evaluation
COMALCO (1986) : long-term conductivity testing
COOKE (1973) : conductivity of fracture proppants in multilayers
DGMK (1983, 1985) : material properties, model investigations, proppant model
HAWKINS (1988) : gelling agent concentration, permeability damage, fluids
HICKEY, BROWN & CRITTENDEN (1981) : flow tests, long-term, field campaign
HILLIG & CHARLES (1965) : stress-dependent surface reactions, strength, test
KIM & LOSACANO (1985) : conductivity damage, crosslinked gel residue, closure
McDANIEL (1986) : prop conductivity, high temperature and stress, brine flow
McDANIEL (1987) : realistic fracture conductivity, reservoir temperature, test
McDANIEL (1988) : wet gas flow, gel filter cake, long-term conductivity testing
McDANIEL & PARKER (1988) : conductivity measurement, simulated reservoir state
MUCH (1987) : proppant evaluation, testing methods, improvement
MUCH & PENNY (1987) : long-term performance, simulated reservoir conditions
PARKER & McDANIEL (1987) : conductivity measurement under in-situ conditions
PENNY (1987) : environmental conditions, frac fluids, long-term conductivity
PURSELL, HOLDITCH & BLAKELEY (1988) : inertial flow, high-strength proppants
ROBINSON (1985) : on-site inspection and testing, treatment improvement
ROODHART, KUIPER & DAVIES (1986) : proppant rock impairment, hydraulic fracture
SINCLAIR (1980) : high-temperature proppant testing, geothermal application
SOLIMAN (1983) : production increase calculation modification, fractured well
STANDARD OIL PROPPANTS (1986) : long-term prop stability, testing, evaluation
STIM-LAB (1986) : environmental conditions, fracturing fluids, testing
SYDANSK (1984) : clay stabilization, potassium hydroxide
THORNTON & RADKE (1985) : silica dissolution, condensation, alcaline fluids
UDELL & LOFY (1985) : permeability reduction, silica dissolution
VOLK et al. (1981) : proppant embedment, low-permeability reservoir
WEINBRANDT, RAMEY & CASSE (1975) : temperature effect, permeability
WENDORFF (1978) : fracture sand control, quality testing
WIEDERHORN & BOLZ (1970) : stress corrosion, static fatigue, glass, experiment
ZWOLLE & DAVIES (1983) : gravel packing sand quality, quantitative study

7.2.4.3. Proppant concentration

The references on proppant concentration correspond mainly to aspects discussed in section 4.3. of the text.

COULTER & WELLS (1972) : high proppant concentration, stimulation advantages
COULTER et al. (1980) : ultra-high concentration, theory, mechanics, aspects
DARIN & HUITT (1960) : partial monolayer of propping agent, fracture flow
DEHLINGER, BROWNE & BUNDRANT (1954) : optimum sand concentration, well treating
DOMSELAAR & VISSER (1974) : fractures, viscous gel creation, shape, concentrate
DUNAND & SOUCEMARIANADIN (1985) : proppant concentration, settling velocity
DUNLAP (1965) : propping agents, influence on packed fractures
FAST, HOLMAN & COVLIN (1977) : proppant pillaring, discontinuous proppant place
GREGORCZYK et al. (1984) : high sand concentration, production boosting, Olmos
HANNAH, HARRINGTON & POTTER (1980) : prop distribution, frac closure, behaviour

HARRINGTON & HANNAH (1975) : selected prop concentration, perfect fluid support
HARRIS, REIDENBACH & CHISHOLM (1986) : carbon dioxide and nitrogen stimulation
HOLDITCH & ELY (1973) : high proppant concentration, successful deep well frac
PAULS et al. (1985) : high proppant concentration, viscosity management, heat
PETZET (1982) : high sand concentration, hydrocarbon output hiking
PUGH, McDANIEL & SEGLEM (1977) : proppant pillaring, Dean Sand, column, isolate
SANDY, WIGGINS & VENDITTO (1986) : prop concentration, load, low-temperature
SCHOLS & VISSER (1974) : proppant bank buildup
SEIDEL & STAHL (1972) : gas well stimulation, viscous water-based fluids
SHAH, SMITH & DONALDSON (1983) : ultra high sand concentration, case history
SMITH (1975) : incomplete fracture fill-up at wellbore, effect on productivity
TINSLEY & WILLIAMS (1975) : proppant pillaring, discontinuous proppant place
WATKINS (1982) : high sand concentrations, East Texas

7.2.4.4. Proppant settling

The references on proppant settling correspond mainly to aspects discussed in section 4.12.2. of the text.

ACHARYA (1986, 1987) : prop transport, viscoelasticity of crosslinked fluids
BABCOCK, PROKOP & KEHLE (1967) : proppant distribution, vertical fractures
BARBER & THEMIG (1985) : proppant banking, selective prop placement, settling
BATCHELOR & RENSBURG (1986) : bi-disperse sedimentation, structure formation
DUNAND & SOUCEMARIANADIN (1985) : concentration effects, settling velocity
HANNAH, HARRINGTON & ANDERSON (1978) : dynamic prop fall rates, measurement
HARRINGTON, HANNAH & WILLIAMS (1979) : dynamic experiments, crosslinked fluids
KIRKBY & ROCKEFELLER (1985) : settling velocity, nonflowing slurries
LOCKETT & AL-HABBOOBY (1973) : differential size settling, two particle species
MIRZA & RICHARDSON (1979) : suspension settling, particles of two or more sizes
NOVOTNY (1977) : prop settling, prop transport, well productivity, prediction
PEDEN & LUO (1987) : settling velocity, particle shape, fracturing fluids
ROODHART (1985) : non-Newtonian fracturing fluids, proppant settling, velocity
ROODHART, KUIPER & DAVIES (1986) : prop rock impairment, hydraulic fracturing
SMITH (1973) : incomplete fracture fill-up, wellbore, productivity rate effect
SCHOLS & VISSER (1974) : proppant bank buildup, fracture without fluid loss
SHAH (1982, 1986) : settling correlations, non-Newtonian fluid, static, dynamic
SWANSON (1967) : free settling velocity, formula for direct determination
THACKER & LAVELLE (1978) : settling stability, suspended sediments
TORY & PICKARD (1977) : three-parameter Markov model for sedimentation
ZIGRANGE & SYLVESTER (1981) : particle settling velocity, liquid-solid systems

7.2.4.5. Proppant and fine particle transport

The references on proppant and fine particle transport correspond mainly to aspects discussed in sections 4.3.4. and 4.12.2. of the text.

ACHARYA (1986) : particle transport, viscous, viscoelastic, fracturing fluids
ACHARYA (1987) : viscoelasticity, crosslinked fluids, proppant transport
ACHENBACH (1974) : surface roughness, tunnel blockage, fluid flow past spheres
AHMED (1984) : fluid flow and leakoff, frac geometry, modelling, prop transport
BARNEA & MEDNICK (1975) : minimum fluidization velocity, correlations
BIOT & MEDLIN (1985) : sand transport in thin fluids, theory, evaluation
CARROLL & BAKER (1979) : particle size distribution, prop crushing, conductive
CLARK & GULER (1983), CLARK & QUADIR (1981) : prop transport, settling velocity
CLARK & QUADIR (1981) : particle settling velocity, proppant transport, review
CLARK et al. (1977) : prop transport, model, design, flow type, prop distribute
CLARK, HALVACI, GHAELI & PARKS (1985) : crosslinked fluid alternatives
CLARK, MANNING, QUADIR & GÜLER (1981) : proppant transport, vertical fractures
CLIFTON & WANG (1988) : thermal effects, 3-D simulation, proppant transport
CLIFTON & WANG (1989) : proppant transport, fluid tracking, 3-D simulation

DANESHY (1978) : sand transport, numerical solution
DONALDSON, BAKER & CARROLL (1977) : particle transport in sandstones
GABRIEL & INAMDAR (1983) : fines migration, porous media, experiment
GOTTSCHLING et al. (1985) : nitrogen gas transport, sand placement, shales pay
GRUESBECK & COLLINS (1978) : particle transport through perforations, effect
GRUESBECK & COLLINS (1982) : fine particles, porous media, entrainment, deposit
HAYNES & GRAY (1972) : particle transport, perforated casing, effectivity
KERN, PERKINS & WYANT (1959) : sand transport, frac mechanisms, sand settling
LEE & DANESHY (1985) : fracture geometry, multiple-fluid treatment, computation
LOEPPKE, GLOWKA & WRIGHT (1988) : particle bridging, single, multiple, width
LOWE & HUITT (1966) : proppant transport, horizontal fractures
MAHAJAN & BARRON (1980) : bridging grain size distribution, non-damaging fluid
MEDLIN, SEXTON & ZUMWALT (1985) : sand transport in thin fluids, experiments
MEYER (1985) : proppant placement analysis
MORALES, JONES & ABOU-SAYED (1985) : frac fluid property, variable height
MUECKE (1978) : formation fines, movement, controlling factors, porous media
NOVOTNY (1977) : proppant transport
SENGUPTA et al. (1982) : migrating formation fines, shear strength, fluids
SHAH & LEE (1986) : friction pressures, proppant-laden hydraulic frac fluids
SHELLEY & McGOWEN (1986) : pump-in test correlation predicts proppant placement
SIEVERT et al. (1981) : proppant transport, large vertical symposium
TANSOE, KING & HOLMAN (1986) : chalk flow, unit mass, failure, collapse, Norway
TORREST & SAVAGE (1975) : particle collection, vertical suspension flow, place
WAHL & CAMPBELL (1963) : sand movement, horizontal fractures

7.2.4.6. Resin coating of proppants and gravel

The references on resin coating of proppants and gravel correspond mainly to aspects discussed in sections 1.2.6. and 4.12.3.3. of the text.

CARDWELL (1947) : consolidation of incompetent formations by plastic
CUTLER, ENNISS, JONES & SWANSON (1982) : resin pellicle deformation, conductive
GRAHAM, MUECKE & COOKE (1975) : resin-coated grains, U.S. patent, applications
GRAHAM, SINCLAIR & BRANDT (1982) : resin-coated proppants, well treatment
JEU, PEREZ & WAY (1988) : proppant flowback minimization, curable resin coating
KANAT (1980) : resin coating of proppants, effect on permeability
POPE, WILES & PIERCE (1987) : prop flowback control, curable resin-coated sand
RICHARDSON & HAMBY (1970) : silty sand consolidation, epoxy resin, overflush
RIKE (1970) : sand consolidation, experience, South Louisiana
SANTROL (1975) : resin-coated sand, application, significance
SANTROL (1986) : double resin coating, inner precured and outer curable films
SAUNDERS & McKENZIE (1979) : performance review, phenolic resin
SINCLAIR & GRAHAM (1977, 1978) : resin-coated gravel
SINCLAIR, GRAHAM & SINCLAIR (1983) : improved well stimulation, resin-coated
SMITH & POLK (1947) : plastics for consolidation of loose sand in wells
STROHM, MANTOOTH & PRIESTER (1970) : sand consolidation, plastic, control
TREADWAY, BRANDT & PARKER (1966) : sand consolidation, plastics, new process
UNDERDOWN & DAS (1982) : resin-coated gravel placement
UNDERDOWN, DAY & SPARLIN (1980) : plastic pre-coated gravel

7.2.5. Technical, fluid and temperature features

This topic header comprises several different subjects (including some miscellaneous branches) which encompass well completion and stimulation operation, fluid loss in fractures, viscosity effects of crosslinked fracturing fluids, water-sensitive formations and foam application, temperature effects and heat transfer in fractures, protection of core samples, combination of acid and proppant fracturing, sandstone acidizing, combination of natural and hydraulic fracturing, formation damage, horizontal or drainhole drilling, explosive fracturing, reservoir heterogeneity; interwell tracer, interference and pulse testing; multilayered reservoirs, multiple-fracture systems, clay stabilization, and reservoir compaction and field subsidence.

The section on well completion and stimulation operation contains references on damage minimization, flowback procedures, handling equipment, perforation techniques, fluid recovery, controlled fracture placement, and liquid removal. The chapter on fluid-loss in fractures collects references on dynamic fluid leakoff, fluid loss additives, measurement, and interaction with the reservoir. The branch on viscosity effects of crosslinked fracturing fluids presents references on treatment design, conductivity damage, dynamical experiments, and effects of shear and proppants. The portion on water-sensitive formations and foam application discusses some aspects of foam fracturing especially with carbon-dioxide foam, and dynamic fluid loss. The division on temperature effects and heat transfer comprises references on high-temperature formations, analytical calculation of heat transfer, conductivity testing and water-sensitive formations. Some methods and materials for preservation of core samples are discussed. The section on combination of acid and proppant fracturing focusses on references on acid reaction rates and thermodynamic relationships, viscous fingering, whispering and wormhole growth. The division on sandstone acidizing highlights design, quality control and experience. The assembly on combination of natural and hydraulic fracturing includes references on fracture interaction, dendritic fracturing, double-porosity reservoirs, and vertical crack interception.

The chapter on formation damage assembles references on water blocking, infiltration, leakoff and visualization of formation damage effects. The branch on horizontal or drainhole drilling presents references with emphasis on drilling techniques and experience, field examples, and success evaluation. The division on explosive fracturing assembles references on well shooting, explosion containment, stem induction, borehole pressure distribution, and radial fractures. The section on reservoir heterogeneity includes references on sedimentary structure and permeability anisotropy, lenticular and blanket sand bodies, channel continuity, multilayered reservoirs, and pay architecture. The chapter on interwell tracer, interference and pulse testing contains references on the mentioned testing techniques. The section on multilayered reservoirs brings together references on commingled zones, crossflow, pressure buildup and drawdown, and reservoir evaluation. The chapter on multiple-fracture systems assembles references on loading rates, tailored-pulse technique and wellbore features. The division on clay stabilization includes references on water-sensitive reservoirs, chemical treatments, metal reactions, adsorption effects, and completion impact. The association on reservoir compaction and field subsidence comprises various aspects of these phenomena in predominantly chalk oil fields.

7.2.5.1. Well completion and stimulation operation

The references on well completion and stimulation operation are classified along the lines of perforation aspects, diverting and sealing agents, water blocking and non-Darcy flow, and other aspects.

7.2.5.1.1. Perforation aspects

The references on perforation aspects correspond mainly to items discussed in sections 4.2.2.2., 4.10.3. and 5.6. of the text.

AHMED & KELLER (1984) : perforation placement technique, frac modification
AHMED et al. (1984) : perforation, placement, optimize, fracture modification
ANTHEUNIS et al. (1976) : perforation collapse, friable sandstones, failure
BELL (1982) : perforating techniques, well productivity maximization
BONOMO & YOUNG (1983) : underbalanced perforating, perforation cleanup
CHENEVERT & THOMPSON (1985) : perforation stability, low-permeability gas pays
CRUMP & CONWAY (1986) : perforation entry friction, bottom-hole treating test
CUDERMAN (1986) : tailored pulse loading, cased and perforated boreholes
GRUESBECK & COLLINS (1978) : particle transport through perforations
HARRIS (1966) : perforating, effect on well productivity
HOLM (1975) : fracture propagation, perforation, performance improvement
HUSHBECK (1986) : precision perforation breakdown, stimulation effectivity
ICHARA & CANNON (1987) : perforating effect, shale-laminated, crossbedding
JARED, SCOTT & EVANS (1984) : precision perforation, uniform stimulation
KING et al. (1986) : perforating charge, performance, post-plug conditions
McCAULEY (1969) : perforations, purging, increasing performance
McLEOD (1978) : perforating conditions, effects on well performance
McLEOD & CRAWFORD (1982) : underbalanced perforation, high-rate completions
MURPHY & JUCH (1960) : pin-point sand-fracing, simultaneous injection, select
PATTILLO & SMITH (1982) : formation flow, effect on perforation stability
PENBERTHY (1985) : underbalanced perforating, perforation cleaning, gravel pack
SCOTT (1986) : perforation breakdown, stimulation improvement, Permian Basin
SEANARD (1986) : underbalanced perforating, closed system
SMALL (1985) : perforations, limited-entry technique, steam injection
SUMAN (1972) : perforations, performance problems, gravel packing
TARIQ, ICHARA & AYESTARAN (1985) : perforated completions, pay heterogeneity
TODD & BRADLEY (1986) : perforation geometry, skin effects, well productivity
VRIEZEN, SPIJKER & VLIS (1975) : perforation tunnel erosion, gas wells
YOUNG & ZALESKI (1985) : tubing-conveyed underbalanced perforating, design

7.2.5.1.2. Diverting and sealing agents

The references on diverting and sealing agents correspond mainly to aspects discussed in section 4.2.2.1. of the text.

BALE (1984) : buoyant ball sealers, matrix acidizing, Saudi Arabia
CRAMER (1987) : limited-entry technique, MHF treatments
DOERLER & PROUVOST (1987) : diverting agents, modelling of zone injectivity
ERBSTOESSER (1980) : improved ball sealer diversion
GABRIEL & ERBSTOESSER (1984) : design of buoyant ball sealer treatments
GWINN & McMANUS (1977) : multihorizon fracturing, ball and baffle method
HARRISON (1971) : diverting agents, history and application
LAGRONE & RASMUSSEN (1962) : controlled fracture placement, limited-entry
WEBSTER (1977) : ball sealer, acid, proppant, multistage

7.2.5.1.3. Water blocking and non-Darcy flow

The references on water blocking and non-Darcy flow correspond mainly to aspects discussed in sections 4.3.3.3. and 4.8.8.3. of the text.

EZEUDEMBAH & DRANCHUK (1982) : Forchheimer equation, non-Darcy flow, high speed
HOLDITCH (1979) : water blocking, gas flow, formation damage
KADI & BROWN (1980) : non-Darcy flow, dissolved gas-drive reservoirs
KADI & BOUTEMY (1981) : non-Darcy flow, well test analysis, solution gas drive
LENORMAND (1986) : viscous fingering, capillary effects, immiscible displace

LIN & HUANG (1988) : water blocking, rock wettability, miscible displacement
NGUYEN (1986) : non-Darcy flow through perforations, experimental study
SALDANA, BRIGHAM & RAMEY (1986) : inertial, frictional, non-Darcy flow, joints
SCHMIDT, CAUDLE & MILLER (1986) : non-Darcy flow, decline analysis, real gas
UBANI & EVANS (1982) : non-Darcy flow, propped fractures, real gases

7.2.5.1.4. Other aspects

References on other aspects of well completion and stimulation operation correspond to various points discussed in chapter 4 of the text.

BECKNER, FIROOZABADI & AZIZ (1988) : transverse imbibition, double porosity
CHAO & WU (1987) : multiple-fractured gas wells, transient pressure behaviour
CLARK (1987) : tubing vs. casing as fracture treating medium
CLARK, PIKE & RENGEL (1979, 1980) : surfactant stimulation agents, watersoluble
CONSTIEN, BRANNON & BANNISTER (1988) : continuous-mix technology, job flexible
CRAMER (1987) : limited-entry technique, MHF treatments
CRUMP & CONWAY (1986) : perforation entry friction, bottom-hole treating test
CUDERMAN (1986) : tailored pulse loading, cased and perforated boreholes
DEES & COULTER (1986) : quality control, job problems, success improvement
EHLIG-ECONOMIDES & AYOUB (1986) : vertical interference test across tight zone
FOX (1985) : three stimulation boats operate in North Sea, capacity, available
GIDLEY & HANSON (1973) : central terminal upset prevention
GRUESBECK & COLLINS (1978) : particle transport through perforations, placement
HANDKE (1987) : sand saturation control, fracture quality/versatility improve
HUSHBECK (1986) : precision perforation breakdown, stimulation effectivity
KNAPP & VLIS (1967) : oil-field subsidence, reasons, prevention, example study
LAGRONE & RASMUSSEN (1962) : controlled fracture placement, limited-entry
LENORMAND (1986) : viscous fingering, capillary effects, immiscible displace
MATTHEWS (1969) : superfrac, field use, new hydraulic fracturing technique
MARTINS (1987) : quality control, fracturing problems, job failure, success
McCAULEY (1969) : perforations, purging, increasing performance
McLENNAN et al. (1983) : interaction, geologic conditions, treatment character
MENG et al. (1982) : production system analysis, vertical fractures
METZGER (1958) : fracturing by vibration, new method for well stimulation
MISAK et al. (1978) : minimizing water production, fracturing technique
MURPHY & JUCH (1960) : pin-point sand-fracing, simultaneous injection, select
ONUR, REYNOLDS & RAGHAVAN (1986) : commingled multilayered pays, interference
OSBORNE, McLEOD & SCHROEDER (1981) : fracturing problems, quality control, test
PATTILLO & SMITH (1982) : formation flow, effect on perforation stability
PENNY, SOLIMAN, CONWAY & BRISCOE (1983) : enhanced load-water, stimulation
PHILLIPS & WILSON (1984) : improved sand pack drainage, fluid recovery improve
PRIJAMBODO et al. (1985) : multilayered reservoirs, crossflow, production
RAWLINGS (1958) : use of hydraulic fracturing equipment for pay sand control
ROBINSON, HOLDITCH & WHITEHEAD (1986) : frac damage minimize, flowback control
ROLL, HIMES, EWERT & DOERKSEN (1986) : pumping equipment, sand-laden slurries
SAUCIER & LANDS (1977) : simulated in-situ perforation tests
SOLIMAN & HUNT (1985) : fracturing fluid, cleanup, well performance
SULLIVAN, LEE & HOLDITCH (1987) : multilayer fractured gas pays, transient
TORSAETER (1984) : water imbibition, Ekofisk field chalk, experiments
VEATCH & CROWELL (1982) : joint research/operation, fracture technics speeding
WATERS (1980) : hydraulic fracturing, historical review, proppants, fluids
WATSON, BENSON & FILLO (1986) : liquid propellant stimulation, Appalachians
WEBSTER, GOINS & BERRY (1970) : continuous multistage fracturing technique
WEEKS (1979) : revised completion procedures, geopressured
WEIGHTMAN & ROBERTS (1987) : data recording system, fracture operation improve
WENDORFF (1979) : sand handling equipment, proppant sand quality
WHITE, WALTER & DIEBOLD (1965) : gas well completion, higher deliverability
WILSON & KISER (1987) : hazardous, liquid waste injection, design, site select

7.2.5.2. Fluid loss in fractures

The references on fluid loss in fractures correspond mainly to aspects discussed in sections 4.3.4. and 4.8.8.3. of the text.

AHMED (1984) : leakoff, proppant transport, fracture geometry, modelling
AHMED, ABOU-SAYED & JONES (1979) : fracture fluid interaction, propped cracks
BROWER & MORROW (1983) : fluid flow, cracks, low-permeability sands
CANSON (1985) : lost circulation treatments, naturally fractured reservoirs
CANTU & BOYD (1988) : combined fluid-loss control additive and gel breaker
COOPER, NELSON & SCHOPPER (1984) : leakoff rate, determination, computerized
CRAWFORD (1983) : proppant scheduling, fluid-loss calculation, fracturing
CRENSHAW & JOINES (1977) : carbonate fracturing, 100 mesh, microcrack propping
CROWE & CRYAR (1975) : oil-soluble resin mixtures, workover, completion, gravel
FORD & PENNY (1988) : leakoff property of fracturing fluid, downhole condition
GULBIS (1983) : dynamic fluid loss
HALL & DOLLARHIDE (1964) : fluid velocity, effect on fluid-loss agent perform
HALL & DOLLARHIDE (1968) : fluid-loss agent, performance, dynamic conditions
HALL & HOUK (1983) : natural fracturing, Buda Formation, fluid-loss control
HARRIS (1982) : dynamic fluid loss, nitrogen foam fracturing fluids, character
HARRIS (1987) : dynamic fluid loss, carbon dioxide foam fracturing fluids
HARRIS (1988) : fracturing fluid additives
HARRIS & PENNY (1985) : temperature and shear history, influence, fluid effect
HUDSON, COFFEY, SAUER & TEOT (1982) : liquid-thickened completion/workover
KING (1977) : dynamic fluid leakoff, foam fracturing fluids
McDANIEL, DEYSARKAR, CALLANAN & KOHLHAAS (1981) : simulated fracture conditions
McGOWEN & McDANIEL (1988) : fluid preconditioning, dynamic fluid-loss data
MEDLIN & MASSE (1986 b) : fluid loss mechanism for gels and suspensions
NAYBERG & LINAFELTER (1984) : cement circulation loss, fractured formations
NIERODE (1983) : fracture design methods, comparison, modelling, field result
NOLTE (1979), NOLTE & SMITH (1981) : fluid-loss determination, pressure decline
PALMER & CARROLL (1983) : fluid loss, elongated hydraulic fracture, numerical
PENNY (1982) : fluid-loss additives, non-damaging
PENNY, CONWAY & LEE (1985) : fluid leakoff, control, modelling
PENNY & FORD (1985) : leakoff properties, fracturing fluids, downhole condition
POULSEN (1988) : fracturing fluid loss, comprehensive theoretical treatment
PYE & SMITH (1973) : fluid-loss additives, proppant conductivity reduction
ROODHART (1983) : measurement, dynamic conditions, fluid loss modelling
SETTARI (1983) : fluid-loss model, general aspects, hydraulic fracturing
SETTARI (1988) : fluid leakoff, injection operation, fracture induction, model
SINHA (1976) : fluid leakoff, dynamic and static conditions, equipment
STEPHEN-PYE & SMITH (1973) : fluid-loss additive, proppant conductivity
THOMPSON (1970) : 100 mesh sand, stimulation effectivity, Permian Basin rocks
WARPINSKI (1988 a) : dual fluid leakoff, 100 mesh sand, acceleration, hairlines
WILLIAMS (1970) : fluid-loss, hydraulically induced fractures, evaluation
WILLIAMSON & ALLENSON (1989) : non-damaging particulate fluid-loss additives
ZIGRYE, WHITFILL & SIEVERT (1985) : control differences, crosslinked, linear

7.2.5.3. Viscosity effects of crosslinked fracturing fluids

The references on viscosity effects of crosslinked fracturing fluids correspond mainly to aspects discussed in section 4.3.4.3. of the text.

ACHARYA (1986, 1987) : viscoelasticity, proppant transport, crosslinked fluids
ACHARYA & DEYSARKAR (1987) : fracture fluid rheology, low-shear conditions
ALMOND & BLAND (1984) : break mechanism, gelling agent, residue, flow impairing
BARANET & AINLEY (1985) : delayed crosslinking, fracturing fluid improvement
BATCHELOR & RENSBURG (1986) : bi-disperse sedimentation, structure formation
BAUMGARTNER et al. (1983) : high-efficiency fracturing fluids, low-permeability
BURNHAM, HARRIS & McDANIEL (1978) : high temperature, viscosity management
CAMERON, GARDNER & VEATCH (1988) : rheological behaviour, delayed crosslinking

CHHABRA & UHLHERR (1979) : polymer solutions, zero shear viscosity estimation
CLARK (1979) : stimulation fluid, rheology, viscosity, crosslinking, approach
CLOUD & CLARK (1980) : crosslinked gels, alternative fluid models, rheology
CONWAY, PAULS & HARRIS (1980) : time-temperature stability, instrumentation
CONWAY et al. (1983) : rheological behaviour, chemical model, crosslinked fluid
CRAIGIE (1983) : rheology, shear history, crosslinking, determination, simulate
ELY et al. (1987) : field measures, in-situ conditions, fluid viscosity, shear
FREE, FREDERICK & THOMPSON (1975) : high-viscosity, crosslinking, batch mixing
FREEMAN, BILDEN & HOSSAINI (1986) : delayed crosslinked gel, foam fracturing
GARDNER & EIKERTS (1982) : effects of shear and proppant
GARDNER & EIKERTS (1983) : rheological character, rapid, delayed, crosslinking
GRONE, BAUMGARTNER & WOODROOF (1983) : high-temperature fluid, chemical analyze
GUILLOT & DUNAND (1985) : fracturing fluids, rheology, laser anemometry
HANNAH, HARRINGTON & POTTER (1980) : frac closure, prop distribution, gel
HARRINGTON, HANNAH & WILLIAMS (1979) : dynamic experiments, proppant settling
HARRIS, HARMS & NORMAN (1987) : continuously mixed crosslinked fluids, flowloop
HARRIS & REIDENBACH (1987) : foam fracturing fluids, high-temperature rheology
HODGE & BARANET (1987) : crosslinking time determination, fracturing fluids
HOEL (1988) : crosslinked gelled water, Spraberry Trend, West Texas
HSU & CONWAY (1981) : fracturing fluids, deep hot formations
JOHNSON, FOX, BURNS & O`MARA (1988) : acid gelling, stimulation improvement
KIM & LOSACANO (1985) : conductivity damage, crosslinked gel residue, closure
KNOLL (1985) : wall slip evaluation, steady shear viscosity measurement, fluid
LESCARBOURA et al. (1984) : fluid stability, heat, pressure, flow loop, test
McLANE & SCOTT (1983) : high-strength crosslinked acid system, Mission Canyon
McLEOD (1983) : treatment design, high-viscosity
MODY, McKITRICK & SHAHSAVARI (1988) : in-situ crosslinking, wellbore vicinity
OLSEN & DEBONIS (1988) : real time optimization, changing fracture conditions
PAULS et al. (1985) : viscosity managing, high-temperature pay, prop saturation
PILEHVARI & CLARK (1985) : rheology, wall slip, viscosity measurement, fluids
PRUDHOMME et al. (1988) : reproducible rheological measurements
ROEGIERS, BEN NACEUR & TOUBOUL (1986) : fluid rheology, frac propagation, 3-D
ROGERS, VEATCH & NOLTE (1984) : fracturing fluid rheology, pipe viscosimeter
ROODHART & DAVIES (1987) : polymer emulsions, revival, fracturing fluids, test
ROSENE & SHUMAKER (1971) : high-viscosity fluids, improved results, fracturing
ROYCE, BECK & RICKARDS (1984) : adjustable crosslinked fracturing fluids
RUMMO (1982) : delayed crosslinking, gel additives
SHAH (1984) : friction pressure, fracturing gel, correlations, prediction
SHAH & WATTERS (1986) : rheological properties, crosslinked fluids, time, shear
SINCLAIR, TERRY & KIEL (1974) : polymer emulsion fracturing
SOLIMAN & HUNT (1985) : fracturing fluid, effect of cleanup, well performance
VARGAS & OHLSON (1986) : viscosity breakers for sand pack completion fluids
WATKINS, WENDORFF & AINLEY (1983) : new crosslinked foamed fracturing fluid
ZIGRYE, OSBORNE & WESTBROOK (1986) : delayed action, crosslinked fluids, flow

7.2.5.4. Water-sensitive formations and foam application

The references on water-sensitive formations and foam application are classified into general foam fracturing, carbon dioxide, nitrogen and other aspects. The assemblage is corresponding to section 3.11. in the text.

7.2.5.4.1. General foam fracturing

The references on general foam fracturing correspond mainly to aspects discussed in section 3.11.2. of the text.

AINLEY & CHARLES (1982) : stabilized foam pad, fracturing technique
BALLARD (1977) : shallow gas fracturing, foam successful application
BLAUER & KOHLHAAS (1974) : foam fracturing, formation application, outline
BLAUER, PHILLIPS & CRAIG (1988) : foam, emulsion fluids, rheological difference

CAWIEZEL & NILES (1987) : foam fracturing fluid, rheological property, downhole
FREEMAN, BILDEN & HOSSAINI (1986) : foam fracturing, delayed crosslinked gels
GAYELOS & HARRIS (1980) : foam fracturing, theories, procedures, results
GOELITZ & EVERTZ (1982) : foam fracturing, Uinta Basin, Dakota and Wasatch
GRUNDMANN & LORD (1983) : foam stimulation, overview
HARRIS (1985) : rheology of foam fracturing fluids, effects of texture
HARRIS, BAILEY, EVERTZ & REEVES (1984) : foam fracturing, stimulation results
HARRIS & REIDENBACH (1984) : high-temperature rheology, foam fracturing fluids
HARRIS et al. (1984) : evolution to foam fracturing, stimulation results
HOLCOMB (1978) : low-quality foamed gel, fracturing efficiency, cleanup, damage
HOLCOMB (1983) : foam stimulation, well cleanup, proppant flowback prevention
HOLCOMB, CALLAWAY & CURRY (1980 a) : foam stimulation, water-sensitive pay
HOLCOMB, CALLAWAY & CURRY (1980 b) : rheology, physico-chemics, aqueous foams
HOLCOMB, CALLAWAY & CURRY (1981) : aqueous stimulation foam, rheology, property
KING (1985) : foam stimulation, nitrified techniques, treatment possibilities
KOMAR & YOST (1979) : foam fracturing, Devonian Shale
MACK & BAUMGARTNER (1986) : foamed stimulation fluids, friction pressure, test
SHEA & BUCHER (1982) : foam fracturing, field study
SIMON & COON (1976) : fluid pH effects, low-permeability sandstone, pay damage

7.2.5.4.2. Carbon dioxide

The references on carbon dioxide correspond mainly to aspects discussed in section 3.11.2.2. of the text.

BLACK & LANGSFORD (1981) : energized CO2 fracturing, improved oil/gas recovery
BULLEN & LILLIES (1983) : carbon dioxide fracturing process, technology, patent
DEYSARKAR et al. (1987) : CO2-laden fracturing fluid, water-sensitive formation
GARBIS & TAYLOR (1985) : CO2, energizing component, frac fluid, effectivity
HARRIS (1987) : dynamic fluid loss, carbon dioxide foam frac fluids, character
HARRIS, HAYNES & EGGER (1984) : CO2-base fracturing fluids, Anadarko Basin
HARRIS, REIDENBACH & CHISHOLM (1986) : carbon dioxide, nitrogen, stimulation
KING (1983) : liquid CO2, low-permeability reservoir stimulation
LEINAN & HARRIS (1983) : carbon dioxide/aqueous mixtures, fracturing fluids
LILLIES (1982) : proppant fracturing, liquid carbon dioxide application
LILLIES & KING (1982) : liquid carbon dioxide, proppant, benefit, technology
NEILL et al. (1964) : carbon dioxide, nitrogen, stimulation, field, laboratory
PHILLIPS, COUCHMAN & WILKE (1987) : high-temperature rheology, CO2 foam frac
REIDENBACH, HARRIS, LEE & LORD (1985) : nitrogen, carbon dioxide, foam rheology
SINAL & LANCASTER (1987) : liquid CO2 fracturing, advantages and limitations
SLATTER, RUCKER & CRISP (1986) CO2-foam, clay-bearing sandstone
SOMMER (1987) : CO2-foam, MHF, Carboniferous, Emsland, Fehndorf, pilot project
WARD (1984) : nitrogen, carbon dioxide, oil-field application, stimulation
WARNOCK, HARRIS & KING (1985) : CO2-foam, Arkansas-Louisiana-Texas, field test

7.2.5.4.3. Nitrogen

The references on nitrogen correspond mainly to aspects discussed in sections 3.11.2.2. and 4.4.4.2.1. of the text.

ABEL (1982), ABEL et al. (1983) : nitrogen gas fracturing of shales, no prop
BRANAGAN & WILMER (1988) : nitrogen breakdown microfracturing, perforations
BRANDENBURG (1988) : nitrogen gas fracturing of shales, non-proppant technique
DOSCHER, OYEKAN & EL-ARABI (1983) : comparison of nitrogen and carbon dioxide
ELY (1988) : nitrogen gas fracturing of shales, non-proppant stimulation
GOTTSCHLING, ROYCE & SCHUCK (1985) : nitrogen gas fracturing of shales, prop
HARRIS (1982) : dynamic fluid loss, nitrogen foam fracturing fluids, character
HARRIS (1985, 1987) : comparative evaluation of nitrogen and carbon dioxide
HARRIS, REIDENBACH & CHISHOLM (1986) : carbon dioxide, nitrogen, stimulation
KING (1985) : foam stimulation, nitrified techniques, treatment possibilities
LANCASTER (1988) : nitrogen gas fracturing of shales, non-proppant stimulation

NEILL et al. (1964) : carbon dioxide, nitrogen, stimulation, field, laboratory
REIDENBACH, HARRIS, LEE & LORD (1985) : nitrogen, carbon dioxide, foam rheology
WARD (1984) : comparative evaluation of nitrogen and carbon dioxide
YOST, OVERBEY, WILKINS & LOCKE (1988) : comparison, nitrogen and carbon dioxide

7.2.5.4.4. Other aspects

The references on other aspects of water-sensitive reservoirs and foam fracturing correspond to various items discussed in sections 3.11.1., 3.11.3. and 3.11.4. of the text.

BLAUER, PHILLIPS & CRAIG (1988) : foam, emulsion fluids, rheological difference
CAWIEZEL & NILES (1987) : foam fracturing fluid, rheological property, downhole
CRAIGHEAD, HOSSAINI & FREEMAN (1985) : delayed crosslinked foamed gels, usage
CRAIGHEAD, HOSSAINI & WATSON (1983) : foamed anhydrous methanol stimulation
CREMA & ALM (1985) : foaming of anhydrous methanol, well stimulation usage
DIETZEL & HANTELMANN (1985) : water-sensitive, high-temperature, oil gel
DRISCOLL, BOWEN & ROBERTS (1980) : oil foams, Niobrara Shale Formation, usage
FREEMAN, BILDEN & HOSSAINI (1986) : foam fracturing, delayed crosslinked gels
HARRIS (1985) : rheology of foam fracturing fluids, effects of texture
HARRIS & REIDENBACH (1984) : high-temperature rheology, foam fracturing fluids
HARRIS et al. (1984) : evolution to foam fracturing, stimulation results
HOLCOMB (1978) : low-quality foamed gel, fracturing efficiency, cleanup, damage
HOLCOMB, CALLAWAY & CURRY (1980 a) : foam stimulation, water-sensitive pay
HOLCOMB, CALLAWAY & CURRY (1980 b) : rheology, physico-chemics, aqueous foams
HOLCOMB, CALLAWAY & CURRY (1981) : aqueous stimulation foam, rheology, property
JOHN (1983) : water sensitivity, oil-based foamed fluids, Rotliegend, Germany
KHILAR & FOGLER (1983) : water sensitivity of sandstones
KHILAR, FOGLER & AHLUWALIA (1983) : sandstone water sensitive, particle capture
KING (1977) : dynamic fluid leakoff, foam fracturing fluids
KOHLHAAS (1982) : water sensitivity, hydrophobic, oil reservoir, Steinle Ranch
MATHIAS (1970) : thick low-pressure water-sensitive sand, pseudolimited entry
McLANE & EDGINGTON (1985) : low-pH methanol, stimulation method, case study
NORTON & HOFFMAN (1982) : oil foams, fractured California reservoirs
ORTIZ & McLANE (1984) : low-pH methanol, alternative, dirty sandstone
SIMON & COON (1976) : fluid pH effects, low-permeability sandstone, pay damage
VELEY (1969) : water sensitivity control, clay stabilization, metal reaction
WALDORF (1965) : effect of steam on permeability, water sensitivity
WATKINS, WENDORFF & AINLEY (1983) : new crosslinked foamed fracturing fluid
WENDORFF & AINLEY (1981) : MHF, high-temperature wells, stable foam fracturing
WENDORFF & EARL (1983) : foam fracturing laboratory, experimental approach

7.2.5.5. Temperature effects and heat transfer in fractures

The references on temperature effects and heat transfer in fractures correspond mainly to aspects discussed in section 4.3.4.4. of the text.

BEN NACEUR & STEPHENSON (1985) : heat transfer model, hydraulic fracturing
CROCKETT, VO & CLEARY (1984) : heat transfer, fluid flow, induced stress
CULLENDER & SMITH (1956) : large temperature gradients, gas-flow, calculation
DIETZEL & HANTELMANN (1985) : water-sensitive, high temperature, oil gel
DYSART & WHITSITT (1967) : fluid temperatures in fractures
HARRINGTON, HANNAH & BEIRUTE (1978) : post-fracturing temperature recovery
HILL & WAHL (1968) : formation temperature, effect on hydraulic fracture design
HOLMAN (1976) : heat transfer, monograph, review
McDANIEL (1986) : conductivity testing, high temperature and stress
MEYER (1987) : heat transfer, hydraulic fracturing
MOODY (1988) : thermal displacement in fractured reservoir
NEMAT-NASSER & OHTSUBO (1978) : fluid flow, heat transfer, hot dry rock masses
PEACEMAN (1976) : convection, vertical fracture, matrix-fissure heat transfer

PRUESS (1983) : heat transfer, fractured geothermal reservoirs
PRUESS & NARASIMHAN (1985) : fluid and heat flow modelling, fractured reservoir
SINCLAIR (1971) : heat transfer effects, deep well fracturing
WEINBRANDT, RAMEY & CASSE (1975) : permeability
WHEELER (1969) : heat transfer, analytical calculations
WHITSITT & DYSART (1970) : temperature, effect on stimulation design

7.2.5.6. Protection of core samples

AUMAN (1986) : core preservation material, laboratory evaluation
HUNT & COBB (1986) : core preservation package, high performance
SOEDER & DOHERTY (1983) : laboratory drying techniques, effect on permeability
THOMAS, HSING & MENZIE (1984) : core damage, oil-based drilling fluid, analyze

7.2.5.7. Combination of acid and proppant fracturing

The references on combination of acid and proppant fracturing are classified along the lines of general carbonate fracturing, viscous fingering and wormhole growth, and other aspects.

7.2.5.7.1. General carbonate fracturing

The references on general carbonate fracturing correspond mainly to aspects discussed in sections 4.5.1. and 4.5.4. of the text.

AL-TAMIMI & EL-MZEIN (1987) : tight multilayered carbonates, stimulation types
ANDERSON & FREDERICKSON (1987) : fracture acidizing design, dynamic etching
BAILEY (1984) : sand fracturing vs. fracture acidizing
BRITT & LARSEN (1986) : carbonate fracturing, minifracturing, oil, water
CRENSHAW & JOINES (1977) : carbonate fracturing, 100 mesh sand, fluid-loss
CROWE, HUTCHINSON & TRITTIPO (1987) : fluid-loss control, acid fracturing, key
DOMSELAAR, SCHOLS & VISSER (1973) : acidizing process, acid fracturing, analyze
HENDRICKSON & CAMERON (1968) : acid fracturing, penetration, etch, conductivity
HOOVER & ADAMS (1988) : carbonate reservor fracturing, Howard Glassock field
HUCKABEE (1987) : carbonate stimulation optimization, fracture geometry design
HUNG et al. (1987) : wormhole growth, carbonate acid fracturing, mechanistics
LO & DEAN (1988) : modelling of acid fracturing
MACDONALD & FRANK (1985) : proppant fracturing, carbonates, Slave Point rocks
MISAK et al. (1978) : carbonate fracturing, density-controlled fluids
NIERODE & KRUK (1973) : acidized fracture conductivity, retardation, fluid loss
NOVOTNY (1976) : stimulation prediction, acid fracturing, finite-conductivity
SMITH et al. (1987) : tip screenout, wide short fractures, communication, chalk

7.2.5.7.2. Viscous fingering and wormhole growth

The references on viscous fingering and wormhole growth correspond mainly to aspects discussed in sections 4.5.1.6.3. and 4.5.4.2.1. of the text.

ALLEN & BOGER (1988) : viscous fingering, mobility control, rheology, impact
ARAKTINGI & ORR (1988) : viscous fingering, heterogeneous porous media
DACCORD, NITTMAN & STANLEY (1986) : viscous fingering acidizing, selective
DACCORD et al. (1987) : wormhole growth, quantitative study, carbonate acidize
DAVIES, BOSMA & VRIES (1987) : viscous fingering, acid fracturing, large-scale
EWING & RUSSELL (1988) : viscous fingering, heterogeneity, miscible displacing
EWING, RUSSELL & YOUNG (1989) : viscous fingering, heterogeneity, displacement
FAYERS & NEWLEY (1987) : viscous fingering, gravity effects, empirical model
FREDRICKSON & BROADDUS (1976) : viscous fingering acidizing, selective solving
GRADER & O`MEARA (1988) : viscous fingering, dynamic displacement, fluid move

HARTLEY & BOSMA (1985) : viscous fingering acidizing, selective dissolution
HOMSY (1987) : viscous fingering, porous media, fluid movement, displacement
HUNG et al. (1987) : wormhole growth, carbonate acid fracturing, mechanistics
KING & SCHER (1987) : viscous fingering acidizing, probability, simulation
KOSSACK & OPDAL (1988) : viscous fingering, heterogeneous reservoir, injection
KU, HIRSH & TAYLOR (1989) : viscous fingering, pseudospectral matrix simulation
LENORMAND (1986) : viscous fingering acidizing, selective matrix dissolution
MILLER, MOISSIS & WHEELER (1989) : viscous fingering, gravity, heterogeneity
ODEH & COHEN (1988) : viscous fingering, simulation, fluid displacement, move
SARMA (1986) : viscous fingering acidizing, flood efficiency, reservoir matrix
VOGE & CRAFTON (1988) : carbonate acidizing, wormhole growth, waterflooding

7.2.5.7.3. Other aspects

The references on other aspects of combination of acid and proppant fracturing correspond to various points discussed in sections 4.5.1. and 4.5.4. of the text.

ANDERSON & FREDERICKSON (1987) : fracture acidizing design, dynamic etching
ASELTINE & ELLENBERGER (1977) : selective acid stimulation, vertical efficiency
BECH et al. (1989) : gravity-imbibition, gravity-drainage, diffusion
BROADDUS, KNOX & FREDRICKSON (1975) : dynamic etching, acid treatment planning
COULTER et al. (1976) : leakoff control, fracture acidizing, stage alternation
CROWE & MINOR (1982) : corrosion inhibitor adsorption, matrix stimulation
DOMSELAAR, SCHOLS & VISSER (1973) : acidizing process, acid fracturing, analyze
FREDRICKSON (1987) : closed fracture acidizing, carbonate stimulation
FREDRICKSON & BROADDUS (1976) : selective fluid placement, density, viscosity
GDANSKI & NORMAN (1983) : acid reaction rates, frac propagation, natural fracs
GRAHAM, KERVER & MORGAN (1965) : acidizing, corrosion inhibition, U.S. patent
HARRIS, HENDRICKSON & COULTER (1966) : high acid concentration, improvement
HENDRICKSON & CAMERON (1968) : acid fracturing, penetration, etch, conductivity
KING (1986) : acidizing concepts, matrix vs. fracture acidizing
LEE & ROBERTS (1980) : temperature distribution, acid penetration, heat effect
NORMAN & CONWAY (1983) : high-temperature gelled acid, development, field test
NORMAN, CONWAY & WILSON (1981) : temperature-stable acid-gelling polymers
PABLEY & HOLCOMB (1982) : crosslinked acid system, proppant transport, strength
REN & XIONG (1986) : effective acid penetration, temperature field effects
SEVOUGIAN et al. (1987) : vertical acid fracture, optimization, steady state
SHAUGHNESSY & KUNZE (1981) : sandstone acidizing, improved field practice
SMITH, DAWSON & SCOGGINS (1983) : crosslinked acid system, proppant transport
SMITH & HENDRICKSON (1965) : hydrofluoric acid stimulation, sandstone reservoir
THOMAS, DIXON, EVANS & VIENOT (1987) : water imbibition, waterflooding, joints
WILLIAMS, GIDLEY & SCHECHTER (1979) : acidizing fundamentals, monograph
WU & PRUESS (1986) : water imbibition.

7.2.5.8. Sandstone acidizing

The references on sandstone acidizing are compiled along the lines of acid retardation; acid gelling, crosslinking and foaming; precipitation inhibition, and other aspects. The literature quotations correspond to the evaluation of items in sections 4.5.1. and 4.5.5. of the text.

7.2.5.8.1. Acid retardation

The references on acid retardation correspond mainly to aspects discussed in sections 4.5.1.3.2. and 4.5.5.4. of the text.

BERGSTROM & MILLER (1975) : acid retardation, acid-in-oil emulsion
CROWE, McGOWAN & BARANET (1988) : retarded acid, effectivity, potential benefit

GDANSKI (1985) : HF retardation for more effective stimulation
HALL (1978) : in-situ HF generation, deep clay damage removal, new technique
HALL & ANDERSON (1977) : new retarded sandstone acidizing system, field result
HALL, TINNEMEYER & UNDERWOOD (1981) : retarded HF, sequential acid system.
HOLCOMB (1977) : foamed acid, extending retardation, penetration, spending
HOLDEN, PRIHODA & HALL (1981) : HF retardation, sequential HF liberation
KNOX, LASATER & DILL (1964) : acid retardation, acidizing improvement
McBRIDE, RATHBONE & THOMAS (1979) : fluoboric acid treatment, multiple-rate
NIERODE & KRUK (1973) : acid retardation, fluid-loss additives, conductivity
PERTHUIS (1989) : weak or slowly generating HF solutions, choice for acidizing
SCHEUERMAN (1988) : buffer-regulated HF acid, 550 oF stability, sand acidizing

7.2.5.8.2. Acid gelling, crosslinking and foaming

The references on acid gelling, crosslinking and foaming correspond mainly to aspects discussed in sections 4.3.4.3.4. and 4.5.1.2. of the text.

BRISCOE (1978) : stabilized foamed acid, acid fracturing, fracture acidizing
BURMAN & HALL (1986) : matrix sandstone stimulation, foam, diverting technique
CHURCH, QUISENBERRY & FOX (1981) : acid gelling, carbonate formations, testing
COULTER, HARRIS & KLEBENOW (1980) : acid base gel system, stimulation process
DEYSARKAR et al. (1984) : crosslinked acid gel
FORD (1981) : foamed acid, fracture acidizing
FORD, BURKLECA & SQUIRE (1980) : foamed acid stimulation, experience
FORD & ROBERTS (1982) : effect of foam on surface kinetics, fracture acidizing
FOX, O`MARA, BURNS & JOHNSON (1989) : thermally stable acid gelling agents
GIDLEY (1970) : sandstone stimulation, acid-mutual solvent method
HALL (1975) : role of mutual solvents in sandstone acidizing
HARMS, SMITH, KING & POSEY (1988) : acid foaming, workover, prolonged stability
HOLCOMB (1977) : foamed acid, extending retardation, penetration, spending
HOLCOMB & WILSON (1976) : foamed acidizing, selective diverting, stable foam
McLANE & SCOTT (1983) : high-strength crosslinked HCl system, Mission Canyon
McLEOD & COULTER (1966) : alcohol additive, waterblock prevention, removal
McLEOD, McGINTY & SMITH (1966) : deep well stimulation, alcoholic acid, glykol
MOLON & FOX (1983) : fracture penetration, productivity increase, xanthan gel
PABLEY, EWING & CALLAWAY (1982) : crosslinked HCl, performance, Rocky Mountains
PABLEY & HOLCOMB (1980, 1981) : high-strength crosslinked acid gel
PABLEY & HOLCOMB (1982) : crosslinked high-strength HCl system, Rocky Mountains
PETRYK & GARUK (1979) : foamed acid, fracture acidizing, experience
SCHERUBEL & CROWE (1978) : foamed acid, fracture acidizing, new concept
SCHEUERMAN (1988) : buffer-regulated HF acid, 550 oF stability, sand acidizing
SUTTON & LASATER (1972) : acid additive selection, sandstone acidizing
WILSON (19789 : foamed acid stimulation, San Andres, area study

7.2.5.8.3. Precipitation inhibition

The references on precipitation inhibition correspond mainly to aspects discussed in section 4.5.1.1. of the text.

COULTER & GOUGLER (1984) : tubing, main source, iron precipitation, acidizing
CROWE (1984 a) : iron precipitation, prevention, spent treating acid, analysis
CROWE (1984 b) : hydrated silica precipitation from spent acid, inhibition
DILL & FREDETTE (1983) : iron control, Appalachian Basin, acid stimulation
GOUGLER, HENDRICK & COULTER (1985) : source and magnitude of iron problems
HALL & DILL (1988) : iron control additives, limestone and sandstone acidizing
JACOBS & THORNE (1986) : asphaltene production, oil sand acidizing, anti-sludge
SMITH, CROWE & NOLAN (1968) : secondary deposition of iron compounds, acidizing
WALSH, LAKE & SCHECHTER (1982) : chemical precipitation, formation damage, HF
WOODROOF, BAKER & JENKINS (1975) : corrosion inhibition, HCl/HF mixtures

7.2.5.8.4. Other aspects

The references on other aspects of sandstone acidizing correspond to various items discussed in section 4.5.5. of the text.

ARMENDARIZ & BARRINGER (1981) : complex weak acid, Canyon Sand stimulation
AULT et al. (1977) : sandstone acidizing, field application, case study
BRANNON, NETTERS & GRIMMER (1987) : sandstone acidizing, quality control, case
CLARK, WONG & MUNGAN (1982) : new acid systems for sandstone stimulation
CROWE (1971) : oil-soluble resin, diverting agent, matrix acidizing
ELLENBERGER & ASELTINE (1976) : selective acid, injection wells, improvement
FARLEY, MILLER & SCHOETTLE (1970) : matrix stimulation, HF-HCl, design criteria
FRENIER (1989) : high-temperature milieu, acid inhibition, organic chemicals
GATEWOOD, HALL, ROBERTS & LASATER (1970) : sandstone acidizing
GDANSKI & PEAVY (1986) : well return analysis, HCl consumption, HF reaction
GIDLEY (1970) : sandstone stimulation, acid-mutual solvent method
GIDLEY (1985) : sandstone acidizing, recent experience, summary report
GIDLEY, RYAN & MAYHILL (1976) : sandstone acidizing, field application study
HALL (1975) : role of mutual solvents in sandstone acidizing
HALL (1978) : in-situ HF generation, deep clay damage removal, new technique
HILL, LINDSAY, SCHECHTER & SILBERBERG (1977) : sandstone acidizing, design
HOCH et al. (1986) : acid heating, increasing dissolution rate
HOLCOMB (1975) : sandstone acidizing, low surface tension, HCl/HF mixtures
KING (1986) : acidizing concepts, matrix vs. fracture acidizing
KING & HOLLINGSWORTH (1979) : oil-soluble resin, diverting agent, acidizing
LABRID (1975) : argillaceous sandstone acidizing, thermodynamics, kinetics
LEE (1985 b) : multi-stage acidizing treatment, viscous preflush, geometry
LUND, FOGLER & McCUNE (1976) : sandstone acidizing, HCl/HF-mixtures
LUND, FOGLER, McCUNE & AULT (1973) : kinetic rate expression, acid reactions
McBRIDE, RATHBONE & THOMAS (1979) : fluoboric acid treatment, multiple-rate
McCUNE, AULT & DUNLAP (1975) : physical and chemical changes, new model, acid
McLEOD & COULTER (1966) : alcohol additive, waterblock prevention, removal
McLEOD, LEDLOW & TILL (1983) : sandstone acidizing, planning, performance
MILLER & BERGSTROM (1975) : acid-in-oil emulsion, acid retardation, stimulation
McLEOD, McGINTY & SMITH (1966) : deep well stimulation, alcoholic acid, glykol
PACCALONI (1979) : matrix stimulation, planning
ROBERTS & GUIN (1974) : acid penetration distance in fractures, prediction
ROSS, PIERSON & COULTER (1963) : matrix acidizing, formation damage correction
SHAUGHNESSY & KUNZE (1981) : sandstone acidizing, improved field practice
SMITH, CLARK, HENDRICKSON & EGBOGAH (1986) : dirty sandstones, buffered HF
SMITH, DOLLARHIDE & BYTH (1978) : acid corrosion inhibitors, effectivity
SMITH, ROSS & HENDRICKSON (1965) : hydrofluoric acid stimulation, sandstone
SUTTON & LASATER (1972) : acid additive selection, sandstone acidizing
TAHA, HILL & SEPEHRNOORI (1986) : sandstone acidizing, heterogeneous reservoir
TAHA, HILL & SEPEHRNOORI (1987) : sandstone acidizing, generalized model
TEMPLETON et al. (1975) : self-generating mud acid
TEMPLETON et al. (1979) : higher pH acid stimulation systems
THOMAS & CROWE (1978 b) : new acid system, fines migration control, matrix
WALKER, FREDRICKSON & NORMAN (1984) : acid heating, improved stimulation result
WILLIAMS (1975) : HF reactions with sandstone formations
WILLIAMS & WHITELEY (1971) : hydrofluoric acid reaction, porous sandstone

7.2.5.9. Combination of natural and hydraulic fracturing

The references on combination of natural and hydraulic fracturing correspond mainly to aspects discussed in section 4.8.8. of the text.

BEN NACEUR & ECONOMIDES (1988) : vertical fracture interception, fissured pay
BLANTON (1982) : fracture interaction, pre-existing, hydraulically induced
BLANTON (1986) : propagation, hydraulic, dynamic, natural fractures

BRANAGAN, CIPOLLA, LEE & CHEN (1987) : hydraulic fracturing, natural joints
HANNAH (1986) : performance improvement, fracturing technique, Mississippian
HOUZE, HORNE & RAMEY (1984) : infinite conductivity, double porosity, vertical
HUBBARD & PIERSON (1986) : dendritic fracturing, Austin Chalk, productivity
JEFFREY, VANDAMME & ROEGIERS (1987) : mechanical interaction, subparallel
LANCASTER & GATENS (1986) : well test analysis, dual-porosity pay, hydrofrac
LOPUS, SEIFERT & SCHEIN (1987) : dendritic fracturing, natural crack flushing
MURPHY & FEHLER (1986) : jointed formations, hydraulic fracturing, problems
NELSON (1986) : stimulation, anisotropy, failure, sedimentary structures, test
PINE & BATCHELOR (1984) : downward migration, jointed rock, hydraulic injection

7.2.5.10. Formation damage

The references on formation damage correspond mainly to aspects discussed in section 4.8.3. of the text.

ABRAMS & VINEGAR (1985) : impairment mechanism, water blocking, Vicksburg
ALEGRE & DOUGHERTY (1988) : expert system application, pay damage diagnosis
ALMOND (1982) : gelling agent residue, low-temperature conditions, controls
AMAEFULE & MASUO (1984) : capillary pressure, rapid evaluation, stimulation
AMAEFULE et al. (1984) : steam condensate, injectivity improvement, chemical
AMAEFULE et al. (1987) : understanding, improved relative permeability analysis
BARKER (1987) : hot oiling, formation damage
BOON et al. (1983) : reaction between rock matrix and injected fluids, oil sand
BROADDUS (1988) : removal with nonacid fluids, water blocking, well damage
BROWER & MORROW (1983) : fluid flow in cracks, tight gas sands, water blocking
COLLE (1978) : completion technique for overcoming formation damage
DERSEY (1986) : non-damaging treatment design, role of petrographical analysis
DEURER, SAHR & UJMA (1985) : physical damage, visualization, fluorescence
DILL (1983) : iron control precipitation prevention, flow-capacity restriction
DONALDSON (1985) : capillary pressure curves, production wells, damage analysis
GLENN & SLUSSER (1957) : particle invasion into porous media, well productivity
GOODE, BERRY & STACY (1984) : remedial treatments, oil-phase drilling mud
GRAY & REX (1965) : clay dispersion and migration, sandstone, formation damage
HOLCOMB (1986) : reservoir damage, minimization, fluid additive performance
HOLDITCH (1979) : water blocking, formation damage, improvement
HOUCHIN, DUNLAP & HUTCHINSON (1988) : gravel-pack completions, formation damage
HOUCHIN & HUDSON (1986) : prediction, evaluation, treatment, organic deposition
KATSANIS et al. (1983) : scale formation, precipitation, geological systems
KING & HOLMAN (1986) : hydrocarbon solvents, acid alternative, damage removal
KRUEGER (1986, 1988) : formation damage, well productivity, oil-field operation
KRUEGER, VOGEL & FISCHER (1967) : clay- or silt-blocked sandstone, cleanup
LEONE & SCOTT (1987) : high clay-content pay, waterflooding, damage control
MALY (1976) : damage minimization, performance improvement, cleanup effect
MARX & RAHMAN (1984) : evaluation, drilling fluid-caused, pressure-reduced pay
McCORRISTON, DEMBY & PEASE (1981) : heavy oil formations, steam injection
MEYERS, SKILLMAN & HERRING (1985) : inhibitor squeeze treatment, control
MICHAJLOV (1987) : physical characteristics alteration, near-wellbore zones
MONAGHAN et al. (1959) : clay-bearing sandstones, formation damage, laboratory
MORGENTHALER (1986) : high-density brine completion fluids, damage tests
NEWBERRY & BARKER (1985) : paraffine/asphaltene deposition control, prevention
ORTIZ & McLANE (1984) : dirty sandstones, water blocking, water sensitivity
POTTER (1984) : brine corrosion inhibitors, formation damage
POTTER & DIBBLE (1983) : colloid plugging, formation damage
PRASAD (1985) : fracturing improves productivity of damaged wells
PRIISHOLM et al. (1987) : fines migration, clay swelling, blocking, geothermal
REED (1977) : permeability damage, mica alteration, carbonate dissolution
RIESE, SCHECHTER & RIESE (1988) : origins and mechanisms of formation damage
SARKAR & SHARMA (1988) : fines migration, two-phase flow, formation damage
SCHAIBLE, AKPAN & AYOUB (1986) : identification, evaluation, treatment
SHARMA & YARTSOS (1986) : permeability impairment, fines migration

SHARMA, YARTSOS & HANDY (1985) : clay release and deposition in sandstones
SOMERTON et al. (1984) : evaluation, screening tests, EOR operations
STOUT & PETERS (1976) : prevention, fracturing, Clinton Formation
TODD, SOMERVILLE & SCOTT (1984) : injectivity decline, depth, measurement
VETTER & LANKFORD (1988) : stimulation, scale inhibition, damage prevention
VOLK, GALL, RAIBLE & CARROLL (1985) : formation damage by fracturing fluids
WOJTANOWICZ et al. (1987) : pore-blocking mechanism, effect on formation damage

7.2.5.11. Horizontal or drainhole drilling

The references on horizontal or drainhole drilling are classified along the lines of horizontal drilling technology, horizontal well completion, and horizontal well fracturing. Corresponding aspects are discussed in section 4.8.6. of the text.

7.2.5.11.1. Horizontal drilling technology

The references on horizontal drilling technology correspond mainly to aspects discussed in section 4.8.6.1. of the text.

AADNOY & CHENEVERT (1987) : borehole stability, highly-inclined wells, fracture
ANDERSEN, HANSEN & FJELDGAARD (1988) : horizontal drilling and completion
BLEAKLEY (1985) : slant hole drilling, England, Morecambe, experience, example
BORISOV (1964) : oil production using horizontal and multiple deviation wells
CHEN & OLYNYK (1985) : tilted horizontal wells
DECH, HEARN, SCHUH & LENHART (1986) : medium-radius, horizontal drilling method
DETMERING (1984) : drainhole drilling update, Empire Abo
DIAMOND & OYLER (1980) : coalbed drainage hole, long horizontal drilling
DICKINSON, ANDERSON & DICKINSON (1986) : second-generation horizontal drilling
DICKINSON & DICKINSON (1985) : horizontal radial drilling system
EDLUND (1987) : medium-curvature, Spraberry Trend area, USA, application
EDWARDS & STRELKOV (1988) : slant-hole drilling, expansion, Canada, experience
ERDÖL-ERDGAS-KOHLE (1988 b) : oil mining, horizontal drilling, alternative
FUH, WHITFILL & SCHUH (1988) : borehole stability analysis, high-angle holes
GIGER (1983, 1987) : horizontal drilling, well number reduction, developing
GIGER, COMBE & REISS (1983) : horizontal wells, significance, oil/gas offtake
GIGER, REISS & JOURDAN (1984) : horizontal drilling, reservoir engineering
GORODY (1984) : horizontal drilling, TEDSI, technology
GREENER, SAUVAGEAU & PASTERNACK (1988) : horizontal drilling, aeolian Nugget
HERNANDEZ & CADE (1988) : screening criteria for horizontal wells
HOTTMAN, SMITH & PURCELL (1979) : borehole instability, formation fracturing
HSIAO (1987) : horizontal wellbore failure, study, experience report
HYLAND (1984) : drainhole drilling, overview, evaluation
JOSHI (1986) : well productivity augmentation, slant and horizontal wells
JOSHI (1987) : horizontal wells, drainhole technology, review
JOURDAN, REISS & JOLY (1986) : state-of-the-art 1986, horizontal drilling
KERR & BROTHERS (1988) : systems approach to horizontal drilling
KRUYADIJK & NIKO (1988) : horizontal drilling vs. MHF stimulation, joints
LI, CHANDELLE & BRYCH (1986) : lateral drilling, promising new application
LOCKE, JOHNSON & JEWKES (1988) : extended-reach appraisal drilling, case study
LOGAN, SCHWOEBEL & HORNER (1987) : coalbed methane recovery, horizontal well
LOXAM (1982) : slant hole drilling, unique horizontal wells, Canada
MACDONALD (1988) : horizontal well application, British North Sea, case studies
MAHONY (1988) : horizontal drilling increase, techniques, effects, review
MAURY & SAUZAY (1987) : borehole instability, rock mechanics approach, cases
MOORE (1980) : drainhole, reservoir placement, ARCO, experience, case studies
PARSONS & FINCHER (1987) : short-radius lateral drilling, alternatives
PETZET (1987, 1988) : horizontal drilling, distribution, application
PREVEDEL (1985) : horizontal drilling, pilot test, optimization, Berkhöpen
PREVEDEL (1987) : horizontal and drainhole drilling, Germany FRG, optimization

REHM (1987) : horizontal drilling in slim holes
REILEY, STAGG & VIDRINE (1988) : horizontal well case histories in Alaska
REISS, JOURDAN & GIGER (1984) : offshore horizontal drilling, deep and prolific
REISS, JOURDAN, GIGER & ARMESSEN (1984) : off- and onshore European horizontal
SHAW & SUTHERLAND (1988) : high-angle wells, innovative approach, Bass Strait
SHERRARD, BRICE & MacDONALD (1986) : Prudhoe Bay, horizontal well application
STEENBOCK (1985) : horizontal drilling, drainhole technique
STRIEGLER (1982) : horizontal, drainhole, ARCO, experience, case studies
VRIELINK & HIPPMAN (1988) : slant well drilling, optimization
WILKIRSON, SMITH, STAGG & WALTERS (1986) : horizontal drilling techniques
WILSON & WILLIS (1986) : high-angle drilling, borehole instability, fracturing
WINTERS, LUCIA & JOHNSON (1988) : large medium-radius horizontal hole drilling

7.2.5.11.2. Horizontal well completion

The references on horizontal well completion correspond mainly to aspects discussed in section 4.8.6.1. of the text.

ANDERSEN, HANSEN & FJELDGAARD (1988) : horizontal drilling and completion
ARMESSEN, JOURDAN & MARIOTTI (1988) : positive/negative factors, review
BORISOV (1964) : oil production using horizontal and multiple deviation wells
BORNER et al. (1988) : vertical vs. horizontal well performance, North Sea
BOSIO, FINCHER, GIANNESINI & HATTEN (1987) : new production method, horizontal
CARVALHO & ROSA (1988) : transient pressure behaviour, naturally fractured pay
CHAPERON (1986) : coning toward horizontal and vertical wells, anisotropic pay
CLONTS & RAMEY (1986) : pressure transient analysis, horizontal drainhole well
COOPER & THRONCOSO (1988) : overview of horizontal well completion techniques
DIAMOND & OYLER (1980) : coalbed drainage hole, long horizontal drilling
DIETRICH (1987) : horizontal-well steam pilot, Kern River, case study
DUDA & KUTSKA (1988) : gas production, tight sands, simulation study, drainage
DUSSERT, SANTORO & SOUDET (1988) : Rospo Mare, Italy, development, drainage
ERTEKIN, SUNG & SCHWERER (1986) : production performance, coal seam, degasify
ESCARON (1983) : deviated wells, logging with standard tools, technique
FANTOFT, KROGH & POLLEN (1988) : thin oil column, recovery enhancement, Oseberg
GIGER (1985) : horizontal well production techniques, heterogeneous reservoirs
GIGER, COMBE & REISS (1983) : horizontal wells, significance, oil/gas offtake
GIGER & JOURDAN (1984) : four horizontal wells producing in Western Europe
GILHUUS (1986) : oil production from thin zones, Midgard field, Norway
GOODE & THAMBYNAYAGAM (1985) : pressure drawdown, buildup, anisotropic media
HSIAO (1987) : horizontal wellbore failure, study, experience report
HUANG & HIGHT (1986) : steamflood processes, horizontal wells
JAIN & KHOSLA (1985) : steam flooding, horizontal wells, Athabasca
JOLY, DORMIGNY, CATALA & PINSON (1985) : production logging, horizontal wells
JOSHI (1986) : well productivity augmentation, slant and horizontal wells
KARCHER, GIGER & COMBE (1986) : horizontal well behaviour prediction
KING & ERTEKIN (1988) : horizontal vs. vertical wells, coal seams, comparison
KOSSACK, KLEPPE & AASEN (1987) : horizontal vs. vertical wells, oil production
KRUYADIJK & NIKO (1988) : horizontal drilling vs. MHF stimulation, joints
KUCHUK & GOODE (1988) : inflow performance, pressure transient analysis
KUCHUK et al. (1988) : horizontal and vertical permeability, skin factor
MACEWEN (1988) : horizontal well logging
MONTIGNY et al. (1988) : reservoir appraisal enhancement, horizontal well logs
MURPHY (1988) : Helder field, Netherlands, horizontal well performance, example
PLAHN et al. (1987) : horizontal well performance, solution gas-drive pays
REILEY, STAGG & VIDRINE (1988) : horizontal well case histories in Alaska
REISS (1986) : horizontal well production, five-year experience report
RIAL (1984) : steam-flood, 3-D thermal stimulation, horizontal injection well
SALAMY et al. (1987) : Devonian shale, recovery, efficiency aspects, horizontal
VICANEK (1985) : horizontal wells, oil production, advantages, comparison, frac
WEIRICH et al. (1987) : perforation, design, techniques, experience report
WILSON & SABINS (1987) : cementing, laboratory investigation, recommendation

WOOTEN, MUKHERJEE & MACH (1988) : horizontal and vertical well performance
YOST et al. (1987) : Devonian shale, wellsite selection, design, experience

7.2.5.11.3. Horizontal well fracturing

The references on horizontal well fracturing correspond mainly to aspects discussed in section 4.8.6.2. of the text.

AUSTIN, ROSE & SCHUH (1988) : simultaneous multiple-entry fracturing treatment
DOSCHER, YING & KOSTURA (1988) : special fracturing, gas injection, steam drive
HOTTMAN, SMITH & PURCELL (1979) : borehole instability, formation fracturing
KING & ERTEKIN (1988) : horizontal vs. vertical wells, coal seams, comparison
KOSSACK, KLEPPE & AASEN (1987) : horizontal vs. vertical wells, oil production
KRUYADIJK & NIKO (1988) : horizontal drilling vs. MHF stimulation, joints
LAYNE & SIRIWARDANE (1988) : hydraulic fracturing, natural joints, drainhole
OVERBEY, YOST & WILKINS (1988) : inducing multiple fractures, horizontal well
SOLIMAN, HUNT & EL RABAA (1988) : horizontal hole fracturing
SPREUX, GEORGES & LESSI (1988) : fracturing, gravel packing, log, consolidate
STRUBHAR, FITCH & GLENN (1975) : multiple vertical fractures, highly-deviated
SUNG & ERTEKIN (1987) : vertical and horizontal fractures, performance, compare
WILSON & WILLIS (1986) : high-angle drilling, borehole instability, fracturing

7.2.5.12. Explosive fracturing

The references on explosive fracturing correspond mainly to aspects discussed in section 4.7.3. of the text.

CARTER & OLINGER (1976) : explosive fragmentation of oil shale
GRANT, DUVALL, OBERT, ROUGH & ATCHISON (1950) : oil well shooting
FOURNEY, BARKER & HOLLOWAY (1981) : explosive well stimulation, techniques
FOURNEY, BARKER & HOLLOWAY (1983) : propellant charges, well stimulation study
FOURNEY, HOLLOWAY & SIMHA (1984) : borehole pressure distribution, model test
GRADY, KIPP & SMITH (1980) : explosive fracturing studies on oil shale
HOWELL & HILLS (1970) : explosive detonation, hydraulically fractured gas well
HUNSBEDT, KRUGER & LONDON (1975, 1977) : geothermal frac, thermal stress crack
McHUGH (1983) : crack extension, internal gas pressure, explosive fracturing
RAMEY, KRUGER & RAGHAVAN (1973) : explosive stimulation, hydrothermal reservoir
SCHMIDT, BOADE & BASS (1979) : well shooting, contained explosion, deflagration
SPENCER (1981) : astrofrac, cost-effective explosive fracturing, tight pay
WATERS (1980) : principles, application, distribution
WATSON, BENSON & FILLO (1985) : liquid propellant stimulation, shallow oil/gas
WINKLE & MIGNOTTE (1953) : implosive shock treatment of injection wells
YOUNG, BARKER & CLARK (1984) : stem induction, field tests, explosive fracture
ZUBER, LEE & GATENS (1987) : radial fracture, explosive vs. hydraulic, compare

7.2.5.13. Reservoir heterogeneity

The references on reservoir heterogeneity are classified along the lines of sedimentary structures and sand body geometry, natural fractures, permeability and other aspects. The literature quotations correspond to various points discussed in section 4.2.4. of the text.

7.2.5.13.1. Sedimentary structures and sand body geometry

The references on sedimentary structures and sand body geometry as examples of reservoir heterogeneity correspond mainly to aspects discussed in sections 3.13. and 4.2.4.1. of the text.

BRANAGAN et al. (1985) : lenticular, tight gas sands, pre- and post-frac test
CLEARY (1978 b) : heterogeneous stratified formations, fracture propagation
EHLIG-ECONOMIDES & JOSEPH (1985) : individual layer property determination
EVANS (1988) : permeability heterogeneity and anisotropy, sedimentary structure
EVANS & CARROLL (1981) : non-continuous, lenticular, stochastic modelling
EVANS & UBANI (1987) : lenticular, multiphase flow modelling, permeability play
FINLEY & O`SHEA (1983) : blanket geometry, tight gas sand, engineering, geology
GELFAND et al. (1986) : river-channel reservoir, seismic lithology modelling
GRINGARTEN (1988) : multilayered reservoirs, high permeability contrast, joint
HODGES & KNUTSON (1981) : channel continuity, directivity, tight gas sands
HOLDITCH, ROBINSON & WHITEHEAD (1987) : lenticular, blanket, complex, examples
ICHARA & CANNON (1987) : perforating effect, cross-bedding, skin, damage, shale
KNUTSON (1976) : noncontinuous sandstone reservoir modelling, Fort Union
LEE & OKUYIGA (1986) : microheterogeneity, shape/size distribution, dispersion
LORENZ (1983) : blanket sandstones, lateral variability
LORENZ (1985) : lenticular reservoir size and orientation prediction, Mesaverde
MARTIN, EVANS & RAPER (1988) : sand body connection, prediction, modelling
NELSON (1986) : natural fracturing, anisotropy, sedimentary structures
OZKAYA (1987) : isolated sand lenses within shales, excess pressures, control
PETERSON & KOHOUT (1983) : lenticular, non-continuous lateral extension, close
RAVENNE et al. (1987) : geometry, sand body, fluvio-deltaic reservoir, internal
REVIERE & WU (1986) : stringerization, erratic pay continuity, waterflooding
ROSE (1982, 1983) : permeability anisotropy, tight reservoirs, bedding impact
SATTLER, HECKES & CLARK (1984) : tight gas sands, lenses, pressure core
SATTLER, RAIBLE & GALL (1985) : lenticular sands, stimulation interpretation
STALKUP & EBANKS (1986) : permeability variation, depositional environment
TANG, LIN & JIANG (1986) : multi-layered heterogeneous sandstone, field improve
TARIQ, ICHARA & AYESTARAN (1985) : anisotropy, laminae, fracturing, perforation
TYLER & FINLEY (1988) : reservoir architecture, extended conventional recovery
WEBER (1982) : sedimentary structures, fluid flow, reservoir models, influence
WILSON (1956) : long heterogeneous sand sections, sand-oil-fracturing
XUE (1986) : point bar facies, reservoir model, semi-communicated sandbody

7.2.5.13.2. Natural fractures

The references on natural fractures as examples of reservoir heterogeneity correspond mainly to aspects discussed in sections 3.13. and 4.2.4.1. of the text.

BENSON & LAI (1986) : interference data analysis, natural fractures, geothermal
BLASINGAME & LEE (1986) : homogeneous reservoirs, natural fracturing, decline
BRANAGAN et al. (1987) : anisotropy, natural fracture systems, post-treatment
EHLIG-ECONOMIDES & JOSEPH (1985) : individual layer property determination
JALALI-YAZDI & ERSHAGHI (1987) : heterogeneous, natural fracturing, type curve
NELSON (1986) : natural fracturing, anisotropy, sedimentary structures
TARIQ, ICHARA & AYESTARAN (1985) : anisotropy, laminae, fracturing, perforation

7.2.5.13.3. Permeability

The references on permeability as an example of reservoir heterogeneity correspond mainly to aspects discussed in sections 3.4. and 4.2.4.2. of the text.

BALL & PITTS (1983) : permeability heterogeneity, simulation, laboratory flood
EVANS (1988) : permeability heterogeneity and anisotropy, sedimentary structure
EVANS & UBANI (1987) : lenticular, multiphase flow modelling, permeability play
GOBRAN, BRIGHAM & RAMEY (1981) : absolute permeability, pore pressure
GRINGARTEN (1988) : multilayered reservoirs, high permeability contrast, joint
HOUSEWORTH (1988) : permeability heterogeneity, characterization, experimental
KNUTSON, HODGES & RIGHTER (1981) : poroperm properties, structural analysis
LEE & OKUYIGA (1986) : microheterogeneity, shape/size distribution, dispersion

LEUNG (1986) : anisotropic, heterogeneous variable directional permeability
MARTIN & COOPER (1984) : permeability distribution modelling, flow barriers
OTERO, SAEZ & RUSINEK (1987) : effective permeability
PEACEMAN (1985) : permeability anisotropy, reservoir simulation, numerical
ROSE (1982, 1983) : permeability anisotropy, tight reservoirs, bedding impact
STALKUP & EBANKS (1986) : permeability variation, depositional environment
TAHA, HILL & SEPEHRNOORI (1986) : sandstone acidizing, permeability anisotropy

7.2.5.13.4. Other aspects

The references on other aspects of reservoir heterogeneity correspond to various points discussed in section 4.2.4. of the text.

ABBASZADEH-DEHGHANI & BRIGHAM (1984) : well-to-well tracer flow, heterogeneity
ANDERSON (1982) : heterogeneity, anisotropy, dipmeter, formation tester, define
ARYA, HEWETT, LARSON & LAKE (1985) : dispersion and reservoir heterogeneity
BEGG et al. (1987) : simulator grid-block parameters, effective value assign
CLEARY (1978 b) : heterogeneous stratified formations, fracture propagation
DUSSAN & SHARMA (1987) : anisotropy, formation testing, pressure transient
EHLIG-ECONOMIDES & JOSEPH (1985) : individual layer property determination
HAAK & ELEWAUT (1988) : heterogeneity modelling, 3-D simulation improvement
HIRASAKI, MORROW & WILLHITE (1984) : waterflood performance, heterogeneity
HONARPOUR et al. (1988) : critical heterogeneities, influence on production
JENSEN & CURRIE (1988) : heterogeneity assessment, performance prediction
KAMAL (1977) : use of pressure transient, reservoir heterogeneity description
LAMBERT (1981) : statistical study of reservoir heterogeneity
LEE & OKUYIGA (1986) : microheterogeneity, shape/size distribution, dispersion
LEUNG (1986) : anisotropic, heterogeneous variable directional permeability
LIGHT (1983) : lateral anisotropy, effect on flood pattern orientation and size
McELROY (1982) : lateral anisotropy, effect on flood pattern orientation/size
MISHRA, BRIGHAM & ORR (1988) : tracer and pressure test analysis, heterogeneity
OPDAL & KOSSACK (1988) : condensate recovery, methane and nitrogen injection
REVIERE & WU (1986) : stringerization, erratic pay continuity, waterflooding
SEETHARAM & DEANS (1986) : tracer testing, model selection, simulation
SHARMA et al. (1987) : critical heterogeneity, waterflood influence, chemical
SMITH & BROWN (1984) : tracer test response, heterogeneous reservoir interpret
STRELTSOVA & McKINLEY (1984) : buildup pattern, flow time, heterogeneity
TAHA, HILL & SEPEHRNOORI (1986) : sandstone acidizing, permeability anisotropy
TANG, LIN & JIANG (1986) : multi-layered heterogeneous sandstone, field improve
TARIQ, ICHARA & AYESTARAN (1985) : anisotropy, laminae, fracturing, perforation
TYLER & FINLEY (1988) : reservoir architecture, extended conventional recovery
WIJESINGHE & KECECIOGLU (1986 b) : interference testing, unsteady crossflow
WILSON (1956) : long heterogeneous sand sections, sand-oil-fracturing

7.2.5.14. Interwell tracer, interference and pulse testing

References on interwell tracer, interference and pulse testing (corresponding to the discussion in section 4.2.4. in the text) are compiled as follows.

7.2.5.14.1. Interference testing

The references on interference testing correspond mainly to aspects discussed in section 4.2.4.1.2.2. of the text.

AL-MARHOUN (1985) : interference testing, new analysis, approach
BANDHYOPADHYAY (1981) : reservoir heterogeneity determination, interference
BRANAGAN, LEE & WILMER (1988) : naturally jointed tight pay, pre-frac testing
BREMER, WINSTON & VELA (1983) : test across low-permeability zones, modelling
CHEN & LESCARBOURA (1983) : anisotropic reservoirs, five-spot data, analysis

CHU & RAGHAVAN (1981) : effect of noncommunicating layers, interference testing
EHLIG-ECONOMIDES & ANSARI (1983) : constant pressure production, interference
GRADER & HORNE (1988) : circular impermeable or compressible subregion
LESCARBOURA, WALTHER & WILSON (1975) : natural and hydraulic fracturing
NAJURIETA, ROBLES & EDWARDS (1986) : water breakthrough prediction, jointing
OGBE & BRIGHAM (1984 a) : wellbore storage, skin effects, interference testing
ONUR, REYNOLDS & RAGHAVAN (1986) : commingled multilayered pays, interference
RAMEY (1975 b) : anisotropic formations, interference analysis
SAGEEV (1986) : geothermal field, multiple interference testing
SAMPSON & MARCOU (1988) : heterogeneity, fractured carbonates, interference
SANDAL & RAMEY (1978) : wellbore storage, skin effect, interference testing
STEWART, GUPTA & WESTAWAY (1984) : sealing and communicating faults, analysis
SWIFT & BROWN (1976) : reservoir definition, review, interference testing
URAIET, RAGHAVAN & THOMAS (1977) : vertical fracture orientation determination
WIJESINGHE & KECECIOGLU (1985) : natural fractures, unsteady interporosity flow

7.2.5.14.2. Pulse testing

The references on pulse testing correspond mainly to aspects discussed in section 4.2.4.1.2.2. of the text.

AYOUB, BOURDET & CHAUVEL (1988) : impulse testing
CINCO-LEY & SAMANIEGO (1977) : interference and pulse test, vertical fractures
DAKE (1982) : vertical and horizontal pulse testing, repeat formation testing
EKIES, HADINOTO & RAGHAVAN (1977) : pulse testing, vertically fractured wells
EVANS & HOLZHAUSEN (1983) : interference and pulse test, shallow hydrofracture
GRABOWSKI et al. (1988) : highly anisotropic dual-porosity pay, pulse testing
HUINONG (1984) : interference and pulse testing, Kenli carbonate, case history
HUTFILZ, COCKERHAM & McINTOSH (1981) : high-permeability milieu, pay modelling
KAMAL (1982) : interference and pulse testing, review
OGBE & BRIGHAM (1984 b) : pulse testing, wellbore storage, skin effects
PRATS (1983) : crossflow between contiguous layers, interpretation, pulse test
ZHUANG (1982) : interference and pulse testing, geological characteristics

7.2.5.14.3. Tracer testing

The references on tracer testing correspond mainly to aspects discussed in sections 4.2.4.1.2.3. and 6.2.1.3.1. of the text.

ABBASZADEH-DEHGHANI & BRIGHAM (1984) : well-to-well tracer flow analysis
COOKE (1983) : areal sweep efficiency, steam-drive, chemical tracers, analysis
D`HOOGE et al. (1981) : fluid movement monitoring, discontinuous multipay
HAGOORT (1982) : response in watered-out reservoir, interwell tracer testing
HEISLER (1986) : steamdrive project continuity, communication, interwell tracer
HORNE & RODRIGUEZ (1983) : fractured geothermal systems, tracer flow dispersion
OHNO, NANBA & HORNE (1985) : reservoir heterogeneity, sweep efficiency, tracer
PRO-TECHNICS (1987) : fracturing, flooding, gravel packing, tracer application
ROBINSON & TESTER (1984) : tracer analysis, dispersed fluid flow, jointed pay
ROBINSON, TESTER & BROWN (1984) : inert and chemically reacting tracers
WAGNER (1977) : interwell reservoir heterogeneity diagnosis, tracer application
WAGNER, BAKER & SCOTT (1974) : multifluid multiwell injection program, tracer
WILEY & COCANOWER (1975) : injectivity profile determination, tracer evaluation

7.2.5.15. Multilayered reservoirs

The references on multilayered reservoirs correspond mainly to aspects discussed in sections 4.2.2. and 4.10. of the text.

BEN NACEUR & ROEGIERS (1988) : fracture treatment design, multilayered pays

COBB, RAMEY & MILLER (1972) : commingled zones, well test analysis
DEANS & GAO (1983) : crossflow, pressure transients, single-phase flow
EARLOUGHER, KERSCH & KUNZMAN (1974) : pressure buildup, no crossflow, bounding
GAO (1983) : crossflow behaviour, two-layer reservoir, reservoir evaluation
GAO & DEANS (1983) : two-layer reservoir, crossflow, pressure drawdown, buildup
LARSEN (1981) : commingled zones, unequal initial pressure and pay properties
ONUR, REYNOLDS & RAGHAVAN (1986) : commingled multilayered reservoir, testing
RAGHAVAN et al. (1974) : commingled zones, unequal thickness, test analysis
WILLIAMS, NIETO, GRAHAM & LEIBACH (1973) : staged treatment for multiple pays

7.2.5.16. Multiple-fracture systems

The references on multiple-fracture systems correspond mainly to aspects discussed in sections 4.2.2.1.3., 4.8.6.2.4. and 4.8.8.4. of the text.

AHMED et al. (1982) : multiple fracturing, state-of-the-art, Western tight gas
CAMPBELL, HANOLD, SINCLAIR & VETTER (1981) : interaction, hydraulic, natural
CUDERMAN (1982) : propellants, borehole considerations, multiple fracturing
HANNAH (1976) : dendritic fracturing, widening and ballooning of natural cracks
HINN (1983) : multizone tight gas completion, Blocker Cotton Valley field
HUBBARD & PIERSON (1986) : dendritic fracturing, connection of natural cracks
KIEL (1977) : dendritic fracturing, primary main and secondary lateral cracks
KNUTSON & WARD (1977) : surface fracture patterns, sand continuity, multifrac
KULAR, CHHINA, BEST & MACKENZIE (1988) : multiple frac propagation, oil sands
LOPUS, SEIFERT & SCHEIN (1987) : dendritic fracturing, natural crack widening
MURPHY & FEHLER (1986) : dendritic fracturing, geothermal MHF, shear slippage
NARENDRAN & CLEARY (1983) : multiple hydraulic cracks, growth and interaction
PEARCE (1979) : multiple fracturing, ball sealers and diverters, monitoring
SWIFT & KUSUBOV (1980, 1981) : multiple fractures, intermediate loading rates
STOLLER (1985) : multiple fracture systems, tailored pulse loading
STRUBHAR, FITCH & GLENN (1974) : multifrac concept in singular deviated well
WANNER, CHESNUT & COX (1979) : discrete fracturing, individual stimulation
WARPINSKI et al. (1979) : multiple fracing, one wellbore, high-energy gas frac

7.2.5.17. Clay stabilization

The references on clay stabilization correspond mainly to aspects discussed in sections 3.5.2. and 5.9.2. of the text.

BORCHARDT & BROWN (1984) : EOR injection rate improvement, clay stabilization
BORCHARDT, ROLL & RAYNE (1984) : mineral fines stabilization, well completions
BORCHARDT, YOUNG & McLAUGHLIN (1979) : effectiveness in high-temperature steam
BLACK (1978) : water-sensitive reservoir, clay stabilization, tight sandstone
BLACK & HOWER (1965) : KCl solutions, water-sensitive formations, stabilization
BLACK, RIPLEY, BEECROFT & PAMPLIN (1979) : shallow gas reservoirs, coarse props
CALLAWAY, ORTIZ & HOLCOMB (1982) : clay protection, stimulation fluids, test
CLEMENTA (1977) : adsorption effects, clay stabilization in sandstones
COULTER, COPELAND & HARRISBERGER (1979) : laboratory study, clay stabilizers
HILL (1982) : best performance criteria, clay stabilization
McLAUGHLIN, ELPHINGSTONE & HALL (1976) : aqueous polymers, clay treating
McLAUGHLIN, ELPHINGSTONE, REMINGTON & COATES (1977) : pay damage correction
PALUMBO et al. (1989) : hydratable clay inhibition, polymers, development
PETERS & STOUT (1977) : hydrolyzable zirconium salts, fracturing treatment
REED (1977) : hydroxy aluminum solutions, sand consolidation, clay stabilizing
ROEHL & HACKETT (1982) : shale-swelling inhibitors, laboratory screening method
SLOAT (1988) : lower residual oil saturation, alkaline potassium solutions
SLOAT & LARSEN (1984) : injectivity improvement, clay stabilization
SYDANSK (1984) : potassium hydroxide, clay stabilization
THOMAS & CROWE (1978) : chemical treatment, stimulation and clay control, sand
THOMAS, CROWE & SIMPSON (1976) : chemical treatment of formation clays, SEM

VELEY (1969) : water sensitivity control, metal reaction, clay stabilization
WILLIAMS & UNDERDOWN (1980) : permanent clay stabilization treatment, polymer
YOUNG, McLAUGHLIN & BORCHARDT (1979) : high-temperature steam milieu, effects

7.2.5.18. Reservoir compaction and field subsidence

The references on reservoir compaction and field subsidence correspond mainly to aspects discussed in section 4.5.4.1.6. of the text.

BLEAKLEY (1986) : reservoir subsidence, compaction, reasons, significance
BOADE, CHIN & SIEMERS (1988) : Ekofisk reservoir compaction, forecast, simulate
DIJK & KLÖTERS (1987) : reservoir subsidence, Ekofisk chalk field, jackup job
GEERTSMA (1973) : land subsidence above compacting oil and gas reservoirs
JOHNSON, RHETT & SIEMERS (1988) : rock mechanics, subsidence evaluation, chalk
MANNESMANN REXROTH (1988) : Ekofisk jackup, review, evaluation
MENGHINI (1988) : compaction monitoring, Ekofisk area chalk fields
MES (1988) : measuring offshore field subsidence, seabed pressure gauges
MORITA et al. (1988) : subsidence, reservoir compaction, depletion, stress
OILMAN (1987 a) : Valhall, chalk oil-field, subsidence rates, compaction
RENTSCH & MES (1988) : measurement of Ekofisk subsidence
RUDDY et al. (1988) : subsidence, chalk, North Sea, Valhall, high-porosity
SCHOONBEEK (1976) : reservoir compaction and subsidence, Rotliegend, Groningen
SMITH (1988) : Ekofisk subsidence, jackup project management
SMITH, SMITH & MONSHAUSEN (1988) : Ekofisk subsidence, jackup design, concept
SULAK & DANIELSEN (1988) : reservoir aspects of Ekofisk subsidence
THOMAS, DIXON, EVANS & VIENOT (1987) : Ekofisk waterflood pilot, compaction
WAAL & SMITS (1988) : reservoir compaction, surface subsidence, prediction
WENDEL, KUNKEL & SWANSON (1988) : waterflood potential, diatomite, compaction

7.2.6. Field case studies

In addition to the other thematical subject headers, papers that explicitely refer to special field areas where the results have been obtained and/or for which the presented interpretation is valid are grouped together here under the title of field case studies. Distinction as follows is made between examples from Europe, USA and Asia. In addition to field case studies, references are given also on field experience associated to hydraulic fracturing.

The field case studies from Europe mainly include Rotliegend and Carboniferous MHF stimulation in Northwest Germany FRG, together with some examples from British and Norwegian North Sea as well as from Yugoslavia. Overviews and summaries on reservoir rock facies analysis in selected formations are also given. The field case studies from USA predominantly comprise tight gas sands in Texas, but also examples from gas and oil fields in various other areas such as Oklahoma, Wyoming, Colorado and Virginia. The section on field experience associated to hydraulic fracturing assembles references on well logging and its interpretation, reservoir rock distribution and facies, oil- and gas-field distribution (including maps), gas production, reservoir mineralogy and diagenesis, and sedimentological and petrographical genesis modelling.

While the references on field case studies Europe are mainly associated to chapters 2 and 3 of the text, the quotations on field case studies USA and Asia are generally not directly corresponding to the text, but represent a complementary assemblage containing numerous example histories that provide more details than could be evaluated and reworked in the text. The references on field experience associated to hydraulic fracturing and gravel packing are mainly assigned to the discussions in chapters 2 and 3 of the text, whereas aspects of tight gas sands are assessed in various parts of chapters 1 through 4.

7.2.6.1. Field case studies Europe

ACHARYA & KIM (1987) : Rotliegend, North Sea, treatment design, simulation
BLEAKLEY (1984) : Söhlingen, Germany, Rotliegend, MHF, world record, tight gas
BRINKMANN (1982) : Germany, status report, deep low-permeability
BRINKMANN, FUHRBERG & SCHÖBER (1980) : Germany, tight deep gas
BRINKMANN, KRÖMER & REINICKE (1980) : Germany, fracspecific parameters, wells
COULTER & PURVIS (1980) : offshore Holland, successful stimulation practices
DESPAX et al. (1987) : Northern North Sea, tight gas sand, stimulation, result
DEURER & MENZ (1984) : Germany, Friesland, Rotliegend, pore structures, clays
DIETZEL & HANTELMANN (1985) : Germany, Friesland, Uphuser Meer, water-sensitive
ECONOMIDES, CIKES, PFÖRTER, UDICK & URODA (1986) : Yugoslavia, high temperature
HARTLEY & BOSMA (1985) : North Sea, chalk, completion, stimulation
JELIC (1982) : Yugoslavia, geothermal energy
JOHN (1983) : Germany, Rotliegend, stimulation problems, deep tight gas, status
KLOSE & KRÖMER (1983) : Germany, Rotliegend, Söhlingen, fracture planning, job
LAFLEUR & JOHNSON (1973) : North Sea, well stimulation, summary, experience
LEICHT (1985) : Germany, Rotliegend, Carboniferous, MHF
LEYDECKER (1981) : Germany, Falkenberg, geothermic project, seismic detection
MADER (1985) : Europe, Buntsandstein, red sandstone, facies architecture, model
MADER (1987) : Europe, proppant, marketing, economics, Rotliegend/Carboniferous
MADER (1988 b) : Europe, proppant, technical, reservoir engineering, Rotliegend
MANCILLAS, MATSON & ZIARA (1976) : Norway, Ekofisk, chalk, North Sea
MASSIE & NYGAARD (1987) : Gullfaks, North Sea, gravel packing, sand production
McDONALD (1983) : offshore Qatar, production test evaluation, oil well, MAF
NEWMAN (1981) : North Sea, chalk, effect of water chemistry, compression
OILFIELD PUBLICATIONS (1986) : North Sea, map, field locations, oil/gas fields
REINICKE, BRINKMANN, SCHWARZ & HUENI (1985) : Germany, build-up data analysis
SCHRÖDER (1985 - 1988) : Germany, maps, field locations, oil and gas fields
SCHWARZ & SCHÖBER (1982) : Germany, deep tight gas, MHF
SIMON et al. (1982) : North Sea, chalk, completion, stimulation
SMITH et al. (1987) : North Sea, chalk, tip screenout, short wide fractures
SOMMER (1987) : Germany, Emsland, Fehndorf, Carboniferous, CO2-foam, pilot job
STEPHENS & MARTINS (1985) : North Sea, experience, stimulation activity, review
TANSOE, KING & HOLMAN (1986) : Norway, Valhall, chalk, high-porosity, treatment
TUNN (1971) : Germany, Rotliegend, Carboniferous, stimulation, natural gas

7.2.6.2. Field case studies USA

The references on field case studies USA are regionally classified along the lines of Anadarko Basin (Oklahoma), Denver-Julesburg Basin (Colorado), Piceance Basin (Colorado), Texas, Wyoming and other areas.

7.2.6.2.1. Anadarko Basin (Oklahoma)

BRUNETTE (1983) : Red Fork Formation, Anadarko Basin, stimulation techniques
DAVIS (1975) : Morrow Sandstone, Anadarko Basin, deep reservoir fracturing
DAVIS, LENT, NICKELL & ALLAM (1980) : Anadarko Basin, Oklahoma, optimization
HARRIS, HAYNES & EGGER (1984) : Red Fork Formation, Anadarko Basin, Oklahoma
HICKEY (1983) : Red Fork Formation, Anadarko Basin, comparative effectiveness
HICKEY, BROWN & CRITTENDEN (1981) : Red Fork Formation, proppant, comparative
LENT, NICKELL, DAVID & ALLAM (1980) : Anadarko Basin/Oklahoma, design, optimize
McMECHAN & CONWAY (1983) : Fletcher field, Deep Anadarko Basin, stimulation
NORTHCUTT et al. (1987) : Morrow Formation, Anadarko Basin, state of the art
SIMON, KAUL & CULBERTSON (1977) : Anadarko Basin, Morrow-Springer Sandstone

7.2.6.2.2. Denver-Julesburg Basin (Colorado)

BLASIO, BAILEY & BERGTHOLD (1985) : Denver-Julesburg Basin, Codell, fluids
CARNEY & WIELAND (1973) : Rocky Mountain area, tight gas stimulation
FAST, HOLMAN & COVLIN (1977) : Wattenberg field, Colorado, Muddy "J" Formation
HARP (1980) : Muddy "J" Sand, Wattenberg field, stimulation treatment evolution
McLANG & EDGINGTON (1985) : Denver-Julesburg Basin, Codell, low-pH methanol
ROBERTS (1981) : Wattenberg field, Colorado, Muddy "J" Formation, optimization
SMITH et al. (1978) : Wattenberg, frac azimuth determination, deep penetration
ZAHNER & CRAFTON (1985) : Wattenberg field, Muddy "J" sand, long-term perform

7.2.6.2.3. Piceance Basin (Colorado)

BRANAGAN et al. (1987) : Paludal Sandstones, Mesaverde Group, natural jointing
BREDEHOEFT et al. (1976) : Piceance Basin, Colorado, regional in-situ stress
LOGAN & SECCOMBE (1986) : Piceance Basin, Colorado, deeply buried coal, frac
LORENZ & FINLEY (1987) : Mesaverde Formation, Colorado, fracture characteristic
RICE & SHURR (1978) : Northern Great Plains, shallow tight gas, resources
STRIGHT & GORDON (1983) : Piceance Basin, decline curve analysis, tight gas
STRUBHAR et al. (1980) : Piceance Basin, Colorado, MHF, demonstration
WARPINSKI, BRANAGAN & WILMER (1985) : Mesaverde Group, Rifle/Colorado, multiple

7.2.6.2.4. Texas

ATTEBERRY, TUCKER & RITZ (1979) : South Texas, sintered bauxite, deep treating
BARBA (1988) : Spraberry trend, West Texas, fracture length, pay justification
CLARK (1983) : Cotton Valley proppant selection, sand crush, stimulation method
GARBIS, BROWN & MAURITZ (1985) : Wilcox Formation, South Texas, completion type
GUEVARA & TYLER (1986) : Central Spraberry Trend, Midland Basin, natural joint
HOLDITCH, ROBINSON & WHITEHEAD (1987) : Travis Peak, East Texas, heterogeneity
HOLZHAUSEN & EGAN (1986) : East Texas, Western Colorado, fracture diagnostics
HOOVER & ADAMS (1988) : Howard Glasscock field, West Texas, carbonate fracture
HUFFT (1977) : Cotton Valley Formation, Texas, fracturing technique evolution
JENNINGS & SPRAWLS (1975) : Cotton Valley Sandstone, tight, stimulation success
KOZIK, BAILEY & HOLDITCH (1979) : Cotton Valley, Fallon, Parsonville, limestone
PEARSON, LYNCH, SCHMIDT & McCASLIN (1988) : Cotton Valley Formation, East Texas
ROSEPILER (1979) : Cotton Valley, principal stress determination, confinement
SAUCIER (1988) : McAllen Ranch, Olmos fields, development program, optimization
SCHLOTTMAN, MILLER & LUEDERS (1981) : Cotton Valley, East Texas, MHF design
SMITH & MURPHY (1974) : Sutton county, Texas, gas well stimulation, evaluation
TINDELL, NEAL & HUNTER (1981) : Cotton Valley Sands, Oak Hill field, evolution
WEEKS (1979) : Wilcox Sands, Southwest Texas, revised completion
WHITEHEAD, HUNT, FINLEY & HOLDITCH (1986) : Travis Peak Formation, East Texas

7.2.6.2.5. Wyoming

AHMED et al. (1981) : Pinedale field, Wyoming, optimized stimulation, tight
HARRISON & TILDEN (1986) : Wertz and Lost Soldier, Wyoming, natural fracturing
KOHLHAAS (1982) : Steinle Ranch/Wyoming, intermediate depth oil, bauxite
MARTIN (1987) : Silo field, Wyoming, three-component seismic investigation
SECCOMBE & ANDERSON (1982) : Wamsutter area/Wyoming, proppant selection
WANNER & COX (1979) : Mesaverde group, Wyoming, tight gas production

7.2.6.2.6. Other areas

AMINIAN, AMERI & BRIGHT (1988) : Big Injun Formation, West Virginia, optimize
BRIM (1986) : Vicksburg Formation, post-audit, fracture stimulations, bauxite
BRITTON et al. (1983) : Street Ranch, fracture-assisted steam-flood, pilot test

BROWNE & PARKER (1986) : Trenton/Black River, subnormal pressure, Stoney Point
CHARLES et al. (1983) : Appalachian tight gas sands, stimulation treatment
CRADDOCK, GOZA & BISHOP (1983) : Morrow, Southern Blaine, Western Canadian
CRENSHAW & FLIPPEN (1968) : deep Ellenburger, Delaware Basin, acid fracturing
CRENSHAW, FLIPPEN & PAULEY (1968) : deep Ellenburger, Delaware Basin, acid frac
ELBEL, HOWARD, TALLEY & McLAUGHLIN (1984) : Cottage Grove Formation, treatment
GREENFIELD & AHMED (1983) : Western tight sand reservoir, fracturing treatment
GRIFFITH & MADISON (1988) : North Ward Estes field, stimulation optimization
HARRIS, BAILEY, EVERTZ & REEVES (1984) : Wasatch Formation, foam fracturing
HARRIS, WARNOCK & KING (1983) : CO2 foam fracturing, Ark-La-Tex region, field
HARRIS et al. (1984) : Wasatch Formation, stimulation results, evolution, foam
HOLDITCH & LEE (1979) : transient pressure tests, fracture evaluation
JENNINGS & DARDEN (1979) : Eastern USA, gas well stimulation
KOZIK, BAILEY & HOLDITCH (1979) : Cotton Valley, Fallon, Parsonville, limestone
KUNDERT & SMINK (1979) : Escondido Sandstone, improved stimulation
LAYNE & BYRER (1986) : Warrior Basin, Alabama, coal fracturing
MACDONALD & CASWELL (1983) : Medicine Hat Formation, fracture job analysis
MACK & MASON (1985) : Clinton Sands, individualized treatment design, benefit
MARPLE et al. (1987) : Carter County, Oklahoma, modified fracturing, shallow
MITCHELL & CHAMBERLAIN (1983) : Randolph Sand, Semitropic field, Kern County
MNISZEWSKI & WESTCOTT (1986) : Devonian shale, high-energy chemical sources
NALL, CAMPBELL & BONEY (1983) : Abo Formation, Pecos Slope field, New Mexico
NYDEGGER, RICE & BROWN (1979) : Bowdoin Dome area, Montana, shallow tight gas
PAI & GARBIS (1983 b) : Morrow Formation, New Mexico, completion practices
PIERCE et al. (1983) : deep Smackover Formation, Johns field, Rankin county
POINTE, BELFIELD & HELWIG (1984) : Monterey Formation, Santa Barbara channel
RASMUS (1981) : Yellow Creek field, Twin Creek Formation, fluid evaluation
RICE & SHURR (1978) : Northern Great Plains, shallow tight gas, resources
ROBINSON, HOLDITCH & LEE (1983) : Wilcox (Lobo) trend, Webb and Zapata counties
SAWOLO et al. (1983) : Yakin field, development, sand control, gravel packing
SAUCIER (1988) : McAllen Ranch, Olmos fields, development program, optimization
SCHRIDER (1985) : Qinton Sandstone, fracturing, development drilling
SCHUMACHER & DICKERMAN (1987) : Lisburne Carbonate, Prudhoe Bay Unit, Alaska
SENECAL & MULHERN (1985) : Monterey Formation, California, frac identification
SHEA & BUCHER (1982) : Benson Formation, Central West Virginia, foam fracturing
SHREDER (1984) : Fryburg Formation, Dakota, acid stimulation, case study
SMITH & SCHMID (1984) : Ferguson Sandstone, Powder River Basin, alternative
SNEIDER et al. (1981) : Deep Basin Gas area, Western Canada, pay detection
STRICKLAND (1985) : Belridge oil field, California, production improvement
STRUBHAR et al. (1974) : Caddo-Pine Island field, multifrac concept, deviated
STRUBHAR et al. (1984) : Belridge oil field, diatomaceous earth, California
SWANSON, GOTTSCHLING & MEEKEN (1980) : Belridge, Lost Hills, California, test
SWARTZ & AHMED (1982) : Eastern Devonian shales, fracture optimization, method
THORNOCK & COVLIN (1977) : Spinal field, Colorado, Upper Cretaceous sandstones
TSAY & FANG (1986) : Hartselle Sandstone, North Alabama, log interpretation
TUCKER (1979) : McAllen Ranch (Vicksburg), geopressured, proppant selection
VENTURA (1985) : Slator Ranch, fracture optimization
WYMAN (1985) : Elmworth, Alberta/Canada, fracturing, drainholes, tight sand
WYMAN, HOLDITCH & RANDOLPH (1979) : Elmworth, Alberta/Canada, fracture analysis

7.2.6.3. Field case studies Asia

CHEN (1986) : Fuyu Sandstone, China, natural fracturing, water injection
CHEN, ZHOU & LIU (1986) : Shengli oilfield, China, sand control techniques
DANIEL (1954) : Middle East, fractured reservoirs, review
HENG (1987) : Malaysia, gravel packing, oil wells, batch completion
LI & ZHU (1986) : North China, deep well fracturing, problems, Moxi oil field
MINNE & GARTNER (1979) : Middle East, fracture detection
PARKER, ADAMS & LIANKUI (1986) : Chinese formations, stimulation studies

7.2.6.4. Field experience associated to hydraulic fracturing and gravel packing

ARENS & BÖHNEL (1985) : Germany, Carboniferous, gas production, high water cut
BACKFISCH (1984) : Rotliegend, Germany, diagenesis, sedimentology, environment
BIFANI, GEORGE & LEVER (1987) : Rotliegend, Argyll field, sedimentology, milieu
BRADEL (1980) : well logging, cased holes, technical and production, experience
BRADEL & DRAXLER (1982) : logging, computer evaluation, Rotliegend, poroperm
BRÄUER (1987) : Austria, Vienna Basin, 12. Tortonian reservoir, CO2-flooding
BROOKS & GLENNIE (1987) : petroleum geology of Northwest Europe, proceedings
COUDERC et al. (1987) : France, Emeraude, offshore steam project, performance
COURBOT (1987) : France, Marienbronn, steam injection, pilot test
DRAGICEVIC et al. (1983) : Yugoslavia, Tertiary, geology, potential, gas.
DRONG (1979) : Rotliegend, Germany, diagenesis, cementation, porosity, burial
DRONG et al. (1982) : Rotliegend, Germany, graben, syndepositional tectonics
DUTTON & FINLEY (1988) : Travis Peak Formation, East Texas, sedimentology
ECONOMIDES (1987) : fractured well testing, evaluation, recommendations
ELKINS (1980) : USA; Western tight gas sand, research requirements
FARKAS & VOLL (1987) : Hungary, Demjén-West, steam-drive, oil recovery
FARRELL (1988) : Ekofisk, North Sea, chalk, natural fractures, distribution
FLECKENSTEIN (1984) : borehole television, Rotliegend, Germany, natural cracks
GRALLA (1988) : Rotliegend, Germany FRG, stratigraphy, facies, correlation
GÜMRAH & OKANDAN (1987) : Turkey, steam-drive, heavy oil recovery, CO2
HETZER & KATZUNG (1985) : Rotliegend, Europe, gas deposits, origin, distribute
HODGES & KNUTSON (1981) : Green River Basin, Wyoming, channel continuity
HOLT et al. (1987) : North Sea, reservoir formations, rock mechanical analysis
JONES & OWENS (1978) : USA, tight gas sands, poroperm, laboratory study
KNUTSON et al. (1981) : reservoir heterogeneity, poroperm properties, structure
KOSSACK, KLEPPE & AASEN (1987) : Troll, Norway, horizontal vs. vertical wells
KREUTZER (1985, 1986) : Austria, Tertiary, reservoir description, layering
KUKAL et al. (1983) : tight gas sands, log interpretation, accuracy, criticism
LÜTZNER (1981) : Rotliegend, Variscan molasse, sedimentology, environment
MADER (1985) : Buntsandstein, Rotliegend, Middle Europe, models, sand geometry
MARTIN & EVANS (1988) : Rotliegend, North Sea, sedimentology, environment
NURMI, WATERHOUSE & AKBAR (1988) : Middle East, natural fractures, distribution
OFFSHORE ENGINEER (1985 a) : North Sea, oil fields, gas fields, status, active
OFFSHORE ENGINEER (1985 b) : North Sea, Ekofisk, chalk, water injection, frac
OFFSHORE ENGINEER (1986 c) : development details, oil/gas fields, until 2000
PHILIPP & REINICKE (1982) : Germany, East Hannover, Rotliegend, gas composition
PLEIN (1978) : Germany, Rotliegend, facies distribution, facies architecture
RICHARDSON, SANGREE & SNEIDER (1988) : Rotliegend, North Sea, aeolian dunes
RIECKMANN (1971) : Rotliegend, Carboniferous, Germany, effective permeability
ROBINSON (1981) : Rotliegend, North Sea, facies types, reservoir quality
RUHOVETS & FERTL (1982) : clay mineral distribution, well-log evaluation
RUOTSALA & WILLIAMS (1982) : fracture zones, reflection seismology, exploration
SCHRÖDER (1985 - 1988) : Germany, maps, field locations, oil, gas, North Sea
THOMAS et al. (1988) : palaeotopography, well siting, communication fracturing
THON (1985) : Czechoslovakia, oil, gas, Tertiary, Palaeozoic, distribution
WALLS (1981) : Western tight gas sands, USA, pore structures, clay mineralogy

7.2.6.5. Aspects of tight gas sands

The references on aspects of tight gas sands are classified along the lines of porosity and permeability, capillary pressure and water saturation, pre- and post-fracturing performance, and regional examples.

7.2.6.5.1. Porosity and permeability

The references on porosity and permeability of tight gas sands correspond mainly to aspects discussed in sections 3.4. and 3.5. of the text.

AL-SHAIEB & WALKER (1986) : Morrow Sandstone, Anadarko Basin, secondary pores
BYRNES, SAMPATH & RANDOLPH (1979) : pressure, water saturation, permeability
DOE (1984) : unconventional natural gas, tight gas sandstone reservoirs
FINLEY (1984) : geology, engineering, selected tight gas sands, characteristic
FINLEY (1985) : reservoir properties, gas productivity, Corcoran, Cozette
FINLEY (1986) : blanket geometry, low permeability, Texas, overview
FRACASSO, DUTTON & FINLEY (1988) : Travis Peak Formation, petrophysics, tight
FREEMAN & BUSH (1983) : permeability measurements, nonsteady state, convention
JENKINS (1985) : low-permeability, sand characteristics, selected case studies
JENNINGS, CARROLL & RAIBLE (1981) : permeability, confining pressure, relation
JONES & OWENS (1980) : laboratory study, tight gas sands
KAZEMI (1983) : low-permeability gas sands, review
KUKAL (1984) : tight gas sands, effective log analysis, systematic approach
KUKAL & SIMONS (1985) : log analysis, permeability quantification, microdarcy
KUKAL et al. (1983) : tight gas sands, log interpretation, accuracy, problems
LUFFEL & HOWARD (1987) : porosity, laboratory measurement, reliability, tight
OSTENSEN (1983) : microcrack permeability, tight gas sands
RANDOLPH, SOEDER & CHOWDIAH (1984) : tight gas sands, porosity, permeability
SAMPATH & KEIGHIN (1982) : gas slippage, affecting factors, Uinta Basin
SATTLER, HECKES & CLARK (1984) : pressure core measurements, petrophysical test
SATTLER et al. (1988) : naturally fractured tight sands, stimulation fluids
SOEDER & CHOWDIAH (1988) : pore geometry, low- and high-permeability sandstones
SOEDER & RANDOLPH (1987) : poroperm, pore structure, Mesaverde Sandstone
SULLIVAN, POSTON & PIPER (1988) : reserve estimation, buildup tests, pressure
THOMAS & WARD (1972) : overburden pressure, water saturation, permeability
WALLS, NUR & BOURBIE (1982) : pressure, water saturation, gas permeability
WARD & MORROW (1985) : capillary pressure, gas-relative permeability, tight
WEI, MORROW & BROWER (1986) : absolute permeability, temperature, pressure
WILSON (1981) : USA, origin of clays, permeability control, Western tight gas

7.2.6.5.2. Capillary pressure and water saturation

The references on capillary pressure and water saturation correspond mainly to aspects discussed in section 3.4. of the text.

BYRNES, SAMPATH & RANDOLPH (1979) : pressure, water saturation, permeability
HOLDITCH, LEE, LANCASTER & DAVIS (1983) : mud filtrate invasion, productivity
KUKAL (1984) : tight gas sands, effective log analysis, systematic approach
KUKAL & SIMONS (1985) : log analysis, permeability quantification, microdarcy
KUKAL et al. (1983) : tight gas sands, log interpretation, accuracy, problems
THOMAS & WARD (1972) : overburden pressure, water saturation, permeability
WALLS & AMAEFULE (1985) : capillary pressure, permeability relationships, tight
WALLS, NUR & BOURBIE (1982) : pressure, water saturation, gas permeability
WARD & MORROW (1985) : capillary pressure, gas-relative permeability, tight
WARDLAW & TAYLOR (1976) : capillary pressure, behaviour, pore structure check

7.2.6.5.3. Pre- and post-fracturing performance

The references on pre- and post-fracturing performance of tight gas sands correspond mainly to aspects discussed in section 2.4.1. and chapter 3 of the text.

BRANAGAN et al. (1985) : pre- and post-fracture performance modelling, lenticle
BRANAGAN et al. (1988) : pre-frac interference testing, natural joints, tight
BROWN & CRAFTON (1981) : low-pressure tight gas reservoir, development

DOE (1984) : unconventional natural gas, tight gas sandstone reservoirs
DUDA & HANCOCK (1987) : development potential, South Rulison area, Garfield
ERTEKIN, KING & SCHWERER (1986) : dynamic gas slippage, flow mechanism, dual
FINLEY (1984) : geology, engineering, selected tight gas sands, characteristic
FINLEY (1985) : reservoir properties, gas productivity, Corcoran, Cozette
FINLEY (1986) : blanket geometry, low permeability, Texas, overview
FINLEY et al. (1983) : primary and secondary tight gas sand objectives, geology
GIDLEY, MUTTI, NIERODE & KEHN (1977) : massive hydraulic fracture, performance
HALE (1986) : Rocky Mountains, well production histories, analysis
HALE et al. (1987) : Rocky Mountains, post-buildup drawdown analysis, tight gas
HOLDITCH, LEE, LANCASTER & DAVIS (1983) : mud filtrate invasion, productivity
JENKINS (1985) : low-permeability, sand characteristics, selected case studies
JOHN (1987) : gas recovery, deep layered pays, non-linear elastic deformation
KABIR & HASAN (1986) : pre-fracture testing in tight gas reservoirs
KUUSKRAA & HAAS (1988) : natural gas from low-permeability formations, tight
NORTHROP (1988) : insights into natural gas production, tight gas sandstones
O`SHEA & MURPHY (1982) : GRI program for tight gas sands research
O`SHEA, ROSENBERG & BRASHEAR (1982) : national petroleum council study, review
RAIBLE & GALL (1985) : formation damage studies, Western tight gas sands
SATTLER et al. (1988) : naturally fractured tight sands, stimulation fluids
SPENCER & MAST (1986) : geology of tight gas sand reservoirs
SULLIVAN, POSTON & PIPER (1988) : reserve estimation, buildup tests, pressure
UBANI & RAY (1988) : performance of non-continuous tight gas sands
VEATCH & BAKER (1983) : U.S. tight gas potential, technology, price influences
WARPINSKI (1988 a) : dual leakoff behaviour, hydraulic fracture, 100 mesh sand
WARPINSKI et al. (1985) : lenticular, paludal, fracturing, testing, case study

7.2.6.5.4. Regional examples

The references on regional examples of tight gas sands correspond mainly to aspects discussed in sections 3.4.1. and 4.8.1.2. of the text.

AL-SHAIEB & WALKER (1986) : Morrow Sandstone, Anadarko Basin, secondary pores
BROWN, SMAGALA & HAEFELE (1986) : Piceance Basin, Cozzette, Corcoran, Rollins
DUDA & HANCOCK (1987) : development potential, South Rulison area, Garfield
FINLEY (1985) : reservoir properties, gas productivity, Corcoran, Cozette
FINLEY (1986) : blanket geometry, low permeability, Texas, overview
FINLEY et al. (1985) : Travis Peak (Hosston) Formation, geology, engineering
FRACASSO, DUTTON & FINLEY (1988) : Travis Peak Formation, petrophysics, tight
HALE (1986) : Rocky Mountains, well production histories, analysis
HALE et al. (1987) : Rocky Mountains, post-buildup drawdown analysis, tight gas
JOHNSON & NUCCIO (1986) : Piceance Creek Basin, Mesaverde Group, West Colorado
KNUTSON, YOST & AMMER (1985) : Clinton Sandstones, Ohio, reservoir geology
LAUGHREY & HARPER (1986) : Upper Devonian, Lower Silurian, Pennsylvania
LAW, POLLASTRO & KEIGHIN (1986) : Green River Basin, Wyoming, Colorado, Utah
LIN & FINLEY (1985) : Travis Peak Formation, East Texas Basin, pay engineering
McKETTA, KOZIAR & COOK (1980) : Eastern tight gas sands, MHF, dendritic frac
MORROW et al. (1986) : Western tight gas sands, fracture flow analysis, matrix
MOSLOW & TILLMAN (1986) : Frontier Formation, Wyoming, reservoir characteristic
PEARSON et al. (1988) : Cotton Valley Formation, MHF, improvement, East Texas
PITMAN, ANDERS, FOUCH & NICHOLS (1986) : Uinta Basin, Utah, Upper Cretaceous
POLLASTRO & SCHOLLE (1986) : chalk, Niobrara Formation, Rocky Mountains
RAIBLE & GALL (1985) : formation damage studies, Western tight gas sands
SOEDER & RANDOLPH (1987) : poroperm, pore structure, Mesaverde Sandstone
SPENCER (1983) : Rocky Mountain region, tight gas reservoirs, geological aspect
SPENCER & MAST (1986) : geology of tight gas sand reservoirs, case studies
TINSON et al. (1982) : Cotton Valley, East Texas, tight gas play, infill wells
WEIMER & SONNENBERG (1982) : Wattenberg field, Denver Basin, Muddy "J" sand.
WEIMER & SONNENBERG (1983) : Codell Sandstone, Denver Basin, exploration play
WEIMER, SONNENBERG & YOUNG (1986) : Wattenberg field, Denver Basin, Colorado

7.2.7. Special applications of hydraulic fracturing

This subject classification key includes a variety of different topics comprising minifracturing and pressure injection tests, moderate-permeability oil-reservoir fracturing, chalk fracturing, naturally fractured reservoirs, coal fracturing and degasification, shale fracturing, geopressured reservoirs, geothermal fracturing, water and steam injection well fracturing and gravel packing, refracturing, non-proppant fracturing, high proppant concentrations, and highly-deviated well fracturing.

The section on minifracturing and pressure injection tests contains references on formation response, pressure test analysis, minifracture design, formation evaluation, and pressure behaviour. Examples of application of synthetic proppants in moderate-permeability oil-reservoir fracturing as well as in high-porosity, but low-permeability and low-stability chalk fracturing are presented. The branch on chalk fracturing assembles also references on rock properties, water imbibition, natural cracks, acidizing and diatomites. The chapter on naturally fractured reservoirs comprises references on dynamical and hydraulical fracture induction, fracture-controlled production, distribution and intensity of fractures, predictive modelling, geological analysis and manipulation, and formation testing. The section on coal fracturing and degasification presents references on coal spalling, cleat systems, single and multiple fracturing, dewatering, gas recovery potential, fracture modelling and strategies, and unconventional gas resources. The division on shale fracturing brings together references on hydraulic and explosive, nitrogen and foam, and proppant and non-proppant fracturing, and horizontal vs. vertical wells.

The branch on geopressured reservoirs includes references on production potential, hydrodynamics of overpressuring, energy resources, stimulation and gas generation. The chapter on geothermal fracturing assembles references on hot dry rock stimulation, combined explosive and hydraulic fracturing, dendritic fracturing, thermal breakdown, high-temperature fluid stability, and heat transfer. The division on water injection well fracturing and gravel packing incorporates references on transient pressure, water-flood induction, deposit build-up, sand control design and viscous polymer solutions. The portion on steam injection well fracturing and gravel packing includes references on stability of proppants and gravel, thermal well completion, steam-drive operation, steam stimulation, and reservoir damage. The association on refracturing presents references on candidate selection, treatment design, and proppant choice and concentration. The group on non-proppant fracturing contains references on shale nitrogen fracturing, sandstone fluid fracturing, and comparative evaluation. References on high proppant concentration and highly-deviated well fracturing are also classified and presented.

7.2.7.1. Minifracturing and pressure injection test

The references on minifracturing and pressure injection test correspond mainly to aspects discussed in sections 4.8.1. and 4.8.2. of the text.

AHMED, KELKAR & SCHATZ (1983) : minifrac, in-situ stress test, permeability
ALLEN & BAZA (1980) : step-rate injectivity test, geothermal reservoir fracture
BENNETT & CLARK (1985) : nitrogen step-rate test, Anschutz Ranch, injection
BUNDY (1981) : pre-frac injection survey, successful frac, necessity
CONWAY, McGOWEN, GUNDERSON & KING (1985) : formation response, pressure behave
CRAMER & SONGER (1988) : microfracturing, moderate-permeability sandstone pays
CROW (1977) : mini massive frac, fluid-loss control, fracture height, penetrate
CUESTA & ELPHICK (1984) : pressure test analysis, role in design and evaluation
FELSENTHAL (1974) : step-rate tests, safe injection pressure level, flooding
HOLDITCH, ROBINSON & WHITEHEAD (1986) : formation evaluation, 3-D shape
KIM, CHAMPION & COOPER (1985) : minifrac, deviated wells, pressure management

LEE (1985) : pressure decline analysis, geometry models, minifrac analysis
LEE (1986) : minifrac analysis, ellipsoidal geometry, model evaluation
LEE (1987) : fluid rheology, effects on minifrac analysis
LOCKE & SAWYER (1975) : pressure injection test, numerical simulation
MARTINS & HARPER (1985) : minifrac pressure decline, long perforated intervals
MOSCHOVIDIS (1987) : pressure decline, minifrac treatment, formation interface
NOLTE (1979) : parameter determination, pressure decline, minifrac introduction
PARRISH, STEVENS & TURNER (1981) : in-situ frac experiments, final results
SHELLEY & McGOWEN (1986) : pump-in test correlation predicts proppant placement
SHLYAPOBERSKY et al. (1988) : mini- and microfractures, testing procedures
SINGH, AGARWAL & KRASE (1988) : systematic design/analysis, parting pressure
SMITH (1981) : mHF, short precise fractures, design
SOLIMAN (1986) : minifrac analysis, fluid compressibility, temperature changes
TAN, McGOWEN, LEE & SOLIMAN (1988) : treatment design, improvement, minifracs
VOEGELE, ABOU-SAYED & JONES (1981) : in-situ stress determination, stimulation
WAREMBOURG (1988) : pump-in tests, fracture parameter determination, analysis
WOLFF, BREDEHOEFT, KEYS & SHUTER (1974) : tectonic stress determination

7.2.7.2. Moderate-permeability oil-reservoir fracturing

The references on moderate-permeability oil-reservoir fracturing correspond mainly to aspects discussed in section 4.6. of the text.

BARBY & BARBEE (1987) : ultrahigh conductivity, very coarse proppants
BRITT (1985) : oil well, moderate permeability, optimization
BRITT & BENNETT (1985) : fracture conductivity, bilinear flow concepts
BRITT & LARSEN (1986) : fracture characterization, multilayered reservoirs
CRAMER & SONGER (1988) : mini- and microfracturing, batch-mix fracturing
GRAHAM, MONAGHAN & OSOBA (1959) : proppant wettability, productivity influence
KOHLHAAS (1982) : intermediate depth, bauxite proppants, field study
LAMBERT, DOLAN & GALLUS (1983) : poorly-consolidated oil reservoir fracturing
MARPLE et al. (1987) : modified stimulation techniques, shallow oil sands
McDONALD (1983) : production test evaluation, MHF, offshore Qatar
NIEMEYER & REINART (1986) : Kuparuk River unit, moderate-permeability oil pay
NORMAN, CIPOLLA & WEBB (1983) : synthetic proppants, application, simulation

7.2.7.3. Chalk fracturing

The references on chalk fracturing correspond mainly to aspects discussed in section 4.5.4. of the text.

ATTARD, MATHES & MOWER (1988) : gravel packing, fracturing, Valhall, pressure
BALDWIN (1986) : water imbibition, North Sea, chalk reservoir characterization
BARK & THOMAS (1981) : Ekofisk, North Sea, giant oil field, chalk reservoir
BISHLAWI & PENG (1988) : Valhall, Norway, rock compaction, pay management
BLANTON (1981) : ductility, dilatancy, microcracking, pore collapse, deforming
BOWEN & DRISCOLL (1980) : oil-based foam fracturing, Niobrara formation
BREWSTER & DANGERFIELD (1984) : chalk fields, geology, reservoir characterize
BREWSTER, DANGERFIELD & FARRELL (1986) : Ekofisk, natural fractures, imbibition
BROWN, CRAFTON & GOLSON (1982) : Niobrara chalk, evolution of fracing methods
FARRELL (1988) : Ekofisk, natural fractures, hydraulic fractures, waterflooding
GISTAU (1985) : chalk stimulation methods and practices
HARTLEY & BOSMA (1985) : chalk completion, fracturing, acidizing, combination
HILL, CLEMENT & WARD (1977) : Austin Chalk, crosslinked polymer, diverter
JOHNSON & RHETT (1986) : chalk compaction, reservoir completion, stress
LOCKRIDGE & SCHOLLE (1978) : Niobrara chalk, foam fracturing, Colorado, Kansas
LUPPENS (1980) : Greater Ekofisk area, chalk completion techniques
MANCILLAS, MATSON & ZIARA (1976) : Ekofisk, North Sea, stimulation
McDONALD (1983) : offshore Qatar, stimulation, production, evaluation, MHF
NEWMAN (1981) : effect of water chemistry, compression, poroperm, North Sea

ROHRET & JONES (1978) : foamed methanol water, Niobrara Chalk stimulation
SCHOLLE (1977) : chalk oil and gas production, Upper Cretaceous, USA
SHELL (1979) : widely-spaced etched ridges, WISPER, European patent
SIMON et al. (1982) : North Sea, completion, stimulation, laboratory study
SNOW & HOUGH (1988) : Ekofisk area, chalk stimulation, field and laboratory
TANSOE, KING & HOLMAN (1986) : chalk production, high porosity, North Sea
VLIS, DUNS & FERNANDEZ-LUQUE (1979) : increasing well productivity, tight chalk
WATTS (1983) : microfractures, poroperm properties, Albuskjell field, North Sea
WEBSTER (1977) : ball sealer, acid, proppant, multistage
WENDEL, KUNKEL & SWANSON (1988) : diatomite, waterflooding, water imbibition
ZIARA, MANCILLAS & MATSON (1977) : chalk stimulation, Ekofisk area

7.2.7.4. Naturally fractured reservoirs

The references on naturally fractured reservoirs are classified along the lines of formation testing, transient flow and pressure behaviour, fracture characterization, and other aspects. The quotations assembled in this division correspond mainly to items discussed in section 4.8.8. of the text.

7.2.7.4.1. Formation testing

The references on formation testing correspond mainly to aspects discussed in sections 4.2.4.2.1. and 4.8.8.3. of the text.

AGUILERA (1975) : influx, imbibition, oil recovery, prediction, natural joints
AGUILERA (1985) : well-test analysis, naturally fractured reservoirs
AGUILERA et al. (1988) : multiphase natural fracturing, pumping, test analysis
BACON, KEMP & THORNE (1984) : water-flood oil recovery, directional drilling
BRANAGAN, CIPOLLA & LEE (1987) : design, evaluation, treatments, natural joint
BRANAGAN, COTNER & LEE (1984) : interference testing, natural fracturing
BRANAGAN & WILMER (1988) : breakdown procedures, microfracturing, injection
CANSON (1985) : lost circulation, fluid uptake, temperature/gamma-ray logging
CASTELIJNS & HAGOORT (1984) : retrograde condensate recovery, gas-condensate
CHEN (1986) : Fuyu Sandstone, development, water injection, fracturing, China
CHEN, YEH, RAGHAVAN & REYNOLDS (1984) : pressure response, observation wells
DERUYCK et al. (1982) : interference tests, double-porosity reservoirs
DOUGLAS et al. (1987) : flow simulation, naturally fractured reservoirs
ERSHAGHI & AFLAKI (1985) : characterization problems, well test data
EVANS (1982) : multiphase flow modelling, naturally fractured reservoirs
EVANS & UBANI (1987) : permeability concept, multiphase flow modelling, lenses
FISHER & CLUTTERBUCK (1986) : arctic fractured carbonate, testing, evaluation
GUEVARA & TYLER (1986) : hydrocarbon recovery, facies architecture, impact
JACQUIN et al. (1987) : fissured reservoirs, gravity drainage, non-equilibrium
LAI, BODVARSSON, TSIANG & WITHERSPOON (1983) : well test data analysis
MORRIS & CAMPBELL (1985) : rate-sensitive flow, well analysis, pressure model
NAJURIETA (1979) : interference and pulse testing, naturally fractured pays
NECTOUX (1987) : North Sea, oil/gas drainage, equilibrium, gravity, velocity
OWUSU & ERSHAGHI (1986) : reserve estimation, recovery performance, complex
SCHULTE & VRIES (1985) : heavy oil reservoirs, in-situ combustion, cracking
SOBBI & STEWART (1988) : well test interpretation, naturally fractured pays
STEWART, WITTMAN & GOLF-RACHT (1981) : repeated formation testing, evaluation
ULDRICH & ERSHAGHI (1979) : interporosity flow parameter estimation, jointing
VELEY & JOLY (1987) : water influx reduction, fractured reservoirs
WALDREN & CORRIGAN (1985) : simulation, geological, engineering, problems
WIJESINGHE (1985) : unsteady flow problems, arbitrary fracture connectivity
WIJESINGHE & CULHAM (1984) : single-well pressure test, arbitrary, connectivity
WIJESINGHE & KECECIOGLU (1985) : interference pressure test, interporosity flow
WIJESINGHE & KECECIOGLU (1986) : unsteady flow problem, arbitrary connectivity
WITHJACK (1985) : bottom water drive analysis, offshore Philippine, Palawan
ZHU (1986) : double porosity reservoir, well testing, China

7.2.7.4.2. Transient flow and pressure behaviour

The references on transient flow and pressure behaviour correspond mainly to aspects discussed in section 6.2.4.2. of the text.

BINGLE (1987) : transient rate and pressure data logging, TRAP tool, analysis
CHEN, RAGHAVAN, SERRA & REYNOLDS (1985) : bounded reservoir, pressure transient
DOUGLAS et al. (1987) : flow simulation, naturally fractured reservoirs
JACQUIN et al. (1987) : fissured reservoirs, gravity drainage, non-equilibrium
KUCUK & SAWYER (1980) : transient flow, Devonian gas shales, natural fractures
MAVOR & CINCO-LEY (1979) : transient pressure behaviour, naturally jointed pay
MORRIS & CAMPBELL (1985) : rate-sensitive flow, well analysis, pressure model
SAMANIEGO & CINCO-LEY (1986) : pressure transient analysis, field case, gas pay
SERRA, REYNOLDS & RAGHAVAN (1983) : pressure transient analysis, natural joint
SOBBI & STEWART (1988) : well test interpretation, naturally fractured pays
SONIER (1987) : naturally fractured reservoir simulation, new simulator
STRELTSOVA (1983) : well pressure behaviour, naturally fractured reservoir
WIJESINGHE (1985) : unsteady flow problems, arbitrary fracture connectivity
WIJESINGHE & KECECIOGLU (1986) : unsteady flow problem, arbitrary connectivity

7.2.7.4.3. Fracture characterization

The references on fracture characterization correspond mainly to aspects discussed in sections 4.5.4.1. and 4.8.8.1. of the text.

ABDASSAH & ERSHAGHI (1986) : triple-porosity systems, representation
AGUILERA (1976) : log analysis, conventional, naturally fractured reservoirs
AGUILERA (1980) : naturally fractured reservoirs, summary, review
AGUILERA (1982) : relative permeability concepts, performance prediction
AGUILERA & ACEVEDO (1982) : computerized well log interpretation
AGUILERA & POOLLEN (1978) : geological aspects, naturally fractured reservoirs
BERGOSH et al. (1987) : core analysis, new developments, natural joints
BLASINGAME & LEE (1986) : homogeneous reservoirs, hydraulic fracture, decline
BOURDET & GRINGARTEN (1980) : fissure volume, block size, type-curve analysis
BRANAGAN & WILMER (1988) : breakdown procedures, microfracturing, injection
CANSON (1985) : lost circulation, fluid uptake, temperature/gamma-ray logging
CHEN, RAGHAVAN, SERRA & REYNOLDS (1985) : bounded reservoir, pressure transient
EVANS & UBANI (1987) : permeability concept, multiphase flow modelling, lenses
GRINGARTEN (1982) : fissured multilayered pays, double-porosity behaviour, test
GUEVARA & TYLER (1986) : hydrocarbon recovery, facies architecture, impact
HARRISON & TILDEN (1986) : fracture distribution, intensity, predictive model
HENNINGTON (1980) : unconventional exploration, natural fractures
HENSEL (1988) : perspective look at fracture porosity
JACQUIN et al. (1987) : fissured reservoirs, gravity drainage, non-equilibrium
JALALI-YAZDI & ERSHAGHI (1987) : naturally fractured heterogeneous reservoirs
JENSEN et al. (1989) : backstripping, finite-element stress, distribution
KOSTURA & RAVENSCROFT (1977) : fracture-controlled production, natural cracks
KUCUK & SAWYER (1980) : transient flow, Devonian gas shales, natural fractures
KULANDER & DEAN (1988) : drilling-induced fractures, distinction, morphology
LEE & TAN (1987) : multiple porosity/permeability simulator, evaluation, test
LEUNG (1986 b) : dual-porosity/permeability, interconnected single-porosity bed
MÄKEL (1987) : South Oman, fractured reservoir origin, implications
MURPHY & FEHLER (1986) : jointed formations, hydraulic fracturing problems
NELSON (1986) : geological analysis, width, spacing, stress, failure, managing
NELSON (1988) : geologist/engineering vs. management view, difference, conflict
PRADO & PRAT (1987) : uniformly fractured zone around the well, unsteady flow
QI & ZHANG (1986) : parameter estimation, cybernetic methods, natural joints
REYNOLDS (1987) : fracture identification, behaviour, photogeology, tracering
RIEPE, PAPE & SCHOPPER (1985) : fractures, microfissures, logging, detection

RUMMEL (1985) : fracture propagation, manipulation, naturally fractured rocks
RUMMEL et al. (1987) : fracture propagation, modelling, experiments, hydraulic
RUOTSALA & WILLIAMS (1982) : reflection seismology, exploration tool, fracture
SHAFFER et al. (1984) : fracture propagation, jointed rock, physical/numerical
TANSOE, KING & HOLMAN (1986) : chalk, North Sea, fracture healing, propping
WARREN & ROOT (1963) : naturally fractured reservoir, behaviour
WIJESINGHE (1985) : unsteady flow problems, arbitrary fracture connectivity
WIJESINGHE & CULHAM (1984) : single-well pressure test, arbitrary, connectivity
WIJESINGHE & KECECIOGLU (1985) : interference pressure test, interporosity flow
WIJESINGHE & KECECIOGLU (1986) : unsteady flow problem, arbitrary connectivity

7.2.7.4.4. Other aspects

The references on other aspects of naturally fractured reservoirs correspond to various points discussed in section 4.8.8. of the text.

AGUILERA (1982) : relative permeability concepts, performance prediction
BACON, KEMP & THORNE (1984) : water-flood oil recovery, directional drilling
BINGLE (1987) : transient rate and pressure data logging, TRAP tool, analysis
BLANTON (1986) : propagation, hydraulic induction, dynamical induction
BRANAGAN, CIPOLLA & LEE (1987) : design, evaluation, treatments, natural joint
CANSON (1985) : lost circulation, fluid uptake, temperature/gamma-ray logging
CASTELIJNS & HAGOORT (1984) : retrograde condensate recovery, gas-condensate
CHEN (1986) : Fuyu Sandstone, development, water injection, fracturing, China
CHEN, ALAM, VOZNIAK & BLANTON (1984) : gas shale production, multiwell testing
CHEN, WASSERMAN & FITZMORRIS (1987) : thermal simulator, naturally jointed pay
DOUGLAS et al. (1987) : flow simulation, naturally fractured reservoirs
EVANS (1982) : multiphase flow modelling, naturally fractured reservoirs
FISHER & CLUTTERBUCK (1986) : arctic fractured carbonate, testing, evaluation
GATENS, OLAREWAJU & LEE (1986) : integrated reservoir description method
HENNINGTON (1980) : unconventional exploration, natural fractures
HENSEL (1988) : perspective look at fracture porosity
JACQUIN et al. (1987) : fissured reservoirs, gravity drainage, non-equilibrium
KULANDER & DEAN (1988) : drilling-induced fractures, distinction, morphology
LEE & TAN (1987) : multiple porosity/permeability simulator, evaluation, test
LITVAK, SATTER & ETEBAR (1988) : performance analysis, novel pay simulator
MORRIS & CAMPBELL (1985) : rate-sensitive flow, well analysis, pressure model
NAKORNTHAP & EVANS (1986) : numerical simulation, reservoir characterization
NECTOUX (1987) : North Sea, oil/gas drainage, equilibrium, gravity, velocity
OWUSU & ERSHAGHI (1986) : reserve estimation, recovery performance, complex
QI & ZHANG (1986) : parameter estimation, cybernetic methods, natural joints
RUOTSALA & WILLIAMS (1982) : reflection seismology, exploration tool, fracture
SATTLER, GALL, RAIBLE & GILL (1988) : fluid systems, tight gas sandstones, MWX
SCHULTE & VRIES (1985) : heavy oil reservoirs, in-situ combustion, cracking
SHAFFER et al. (1984) : fracture propagation, jointed rock, physical/numerical
SONIER (1987) : naturally fractured reservoir simulation, new simulator
THOMPSON (1977) : 100 mesh sand, microproppant, hairline cracks
ULDRICH & ERSHAGHI (1979) : interporosity flow parameter estimation, jointing
VELEY & JOLY (1987) : water influx reduction, fractured reservoirs
WALDREN & CORRIGAN (1985) : simulation, geological, engineering, problems
WARPINSKI (1988 a) : 100 mesh sand, microproppant, hairline crack, dual leakoff
WARREN & ROOT (1963) : naturally fractured reservoir, behaviour
WITHJACK (1985) : bottom water drive analysis, offshore Philippine, Palawan

7.2.7.5. Coal fracturing and degasification

The references on coal fracturing and degasification correspond mainly to aspects discussed in section 4.4.3. of the text.

ADAMS et al. (1982) : potential coal gas recovery, Appalachian Basin, geology

ANCELL et al. (1980) : coal-bed degasification process, Warrior Basin, Alabama
BELL, JONES & MORALES (1987) : spalling, hydraulic fracturing strategy, coal
BOYER, POPOVICH & SCHRAUFNAGEL (1986) : multiple coal seam completion project
BOYER, STUBBS & SCHWERER (1986) : coal properties, frac design, methane offtake
BYRER, MROZ & COVATCH (1984) : production potential, coal-bed methane, USA
CHOATE, JURICH & SAULNIER (1981) : potential coal gas recovery, Piceance Basin
CHOATE, LENT & RIGHTMIRE (1982) : potential coal gas recovery, San Juan Basin
DECKER & SECCOMBE (1986) : production control, deep coal seam, geological data
ERTEKIN, SUNG & SCHWERER (1986) : production performance analysis, horizontal
GRAY (1987) : coal seam reservoir engineering, gas storage and movement
HANSON (1988) : fracture models, coal seam methane drainage, development
HANSON, MAVKO & NIELSEN (1988) : hydraulic fracturing models, coal drainage
HANSON et al. (1987) : thin multiple coal seams, stimulation design, production
HOLDITCH et al. (1988) : fracturing strategies, enhanced coal gas recovery
JEU, LOGAN & McBANE (1988) : different fracturing techniques, Piceance Basin
JONES (1986, 1987) : fracture design, methane recovery, coal seam, techniques
JONES (1988) : hydraulic fracture strategy, spalling, coal seam development
JONES, BELL & MORALES (1986) : empirical relationships, computer simulation
KELAFANT, BOYER & ZUBER (1988) : Central Appalachian Basin, coalbed potential
KING, ERTEKIN & SCHWERER (1986) : numerical simulation, transient behaviour
KUUSKRAA & MEYER (1980) : world resources of unconventional gas, review
LASSITER & HOLLINGSWORTH (1984) : coal-seam degasification method
LAYNE & BYRER (1986) : coal fracturing, Warrior Basin, Alabama, stimulation
LAYNE, SIRIWARDANE & BYRER (1988) : gas production potential, coal seams
LOGAN, ERWIN & McBANE (1988) : different fracturing techniques, deep coal
LOGAN, SECCOMBE & JONES (1986) : fracturing, open hole testing, deep burial
MAHONEY et al. (1980) : non-proppant, stimulation job, coal seam degasification
McBANE & JEU (1988) : Western Cretaceous, coal seam fracturing
McFALL et al. (1987) : coal-bed methane resources, Piceance Basin, Colorado
MILITZER & SCHWERER (1986) : economical assessment, commercial potential
PERLMAN (1986) : coal-seam fracturing, method, review
REEVES, WALLACE & BEAVERS (1987) : reservoir properties, geologic setting, test
RIGHTMIRE, EDDY & KIRR (1984) : coal-bed methane resources, USA, assessment
SARDA (1984) : coal degasification, communication fracturing, orientation
SCHRAUFNAGEL & LAMBERT (1988) : limited entry, prop frac, screenout, orientate
SECCOMBE & SAKASHITA (1985) : economical assessment, deep coalbed exploitation
SUNG, ERTEKIN & SCHWERER (1986) : comprehensive coal seam degasification model
TRACY (1988) : in-situ coal degasification, lateral drilling technology
TREVITS et al. (1982) : fracture geometry, controlling parameters, identifying
WAY (1987) : hydrologic characterization, optimal dewatering, methane drainage
ZUBER, KUUSKRAA & SAWYER (1988) : optimizing well spacing, fracture design

7.2.7.6. Shale fracturing

The references on shale fracturing correspond mainly to aspects discussed in section 4.4.4. of the text.

ABEL (1982) : nitrogen fracturing, non-proppant fracturing, Ohio Shale
AMERI et al. (1985) : economical development, systematic approach, Devonian
BRANDENBURG (1988) : stimulation as key to unlocking shale gas reserves
CHEN, ALAM, VOZNIAK & BLANTON (1984) : gas shales, natural fracturing, offtake
COURSEN (1983) : explosive fracturing, charge types, stimulation, production
CREMEAN, McKETTA, OWENS & SMITH (1979) : MHF stimulation, Lincoln, Virginia
DUTTON, BEMIS, JOHNSON & WHITE (1984) : shale fracturing, Britt Formation
ELY (1988) : hydraulic fracturing, design considerations, treatment volume
FOSTER (1975) : shale fracturing, Devonian, overview, outline
GATENS (1988) : MHF stimulation, Devonian shale, economical analysis
GATENS, HOLGATE & LEE (1987) : MHF stimulation, Lincoln, Virginia, Devonian
GATENS, LEE & RAHIM (1985) : analytical model, history match, Devonian shales
GATENS et al. (1987) : Devonian shales, production data analysis, evaluation
GOTTSCHLING, ROYCE & SCHUCK (1985) : nitrogen gas, sand, new stimulation method

GRI (1987 b) : Devonian shale, specific log interpretation, summary, evaluation
GUIDRY et al. (1987) : Devonian shale, evaluation, completion, testing
HAZLETT et al. (1986) : production data analysis, type curves, Devonian shales
HORTON (1982) : shale reservoirs, multiple fracturing, 95 jobs in 63 wells
HORTON, MERCER & SAWYER (1982) : infill drilling, vertical wells, hydrofrac
KEMPER et al. (1986) : Eastern Kentucky, Letcher and Knott counties, production
KIRR, EDDY, LOCKE & SMITH (1986) : production potential, geology, Rhinestreet
KOMAR (1978) : Devonian shales, stimulation design
KOMAR, YOST & SINCLAIR (1980) : foam fracturing, Devonian shale
KOZIAR (1985) : production data analysis, Southwestern Virginia, Devonian shale
KUUSKRAA, KELAFANT & HORNER (1988) : Devonian, Mississippian, Virginia
KUUSKRAA, SEDWICK & YOST (1985) : technically recoverable reserves, Devonian
KUUSKRAA et al. (1983) : production potential and strategies, effectivity
LANCASTER et al. (1987) : evaluation, completion, testing, stimulation, wells
LANCASTER (1988) : stimulation evaluation, reservoir analysis, Devonian shale
LEE & GATENS (1985) : Eastern Devonian shales, gas production data analysis
MASON (1987) : reserve estimation, high-rate shale oil and gas completion
McBANE & THOMPSON (1984) : exploration/production studies, Devonian gas shales
MNISZEWSKI & WESTCOTT (1986) : high-energy chemical sources, Devonian shale
NEARING & STARTZMAN (1988) : Eastern Devonian shales, stimulation effects
OZKAYA (1984) : computer simulation, hydraulic fracture, primary migration
OZKAYA (1986) : natural hydraulic fracturing of shales during sedimentation
SCHETTLER, PARMLEY & LEE (1987) : gas storage and transport, Devonian shales
SCHRIDER et al. (1977) : natural gas from Eastern USA shales, rate augmentation
SHAH et al. (1986) : reservoir and stimulation analysis, fracturing, Devonian
STRANG & NORTON (1983) : foamed sand, improved stimulation results, Devonian
SWARTZ & AHMED 1982) : Eastern Devonian shales, fracture optimization, method
TILLMAN (1983) : fracture-enhanced reservoir development, Devonian shales
VANORSDALE (1987) : Devonian shale gas reservoirs, evaluation
ZUBER, LEE & GATENS (1987) : explosive, radial, hydraulic fracture, comparison

7.2.7.7. Geopressured reservoirs

The references on geopressured reservoirs correspond mainly to aspects discussed in section 4.4.2. of the text.

AMIRIJAFARI et al. (1982) : flow characteristics, geopressured reservoirs
ANDERSON, PETERSON & PARISI (1984) : enhanced production, water drive
BEGLAND & WHITEHEAD (1986) : depletion, performance, geopressured gas reservoir
BERG & HABECK (1982) : Lower Vicksburg, McAllen Ranch field, Texas, porosity
CHARPENTIER, LAW & PRENSKY (1987) : quantitative modelling, overpressuring
DOHERTY et al. (1982) : methane production from geopressured aquifers
DOSCHER & AZARI (1980) : critical gas saturation, numerical simulation, methane
DOSCHER et al. (1979) : technology, economics, methane production, review
DUTTA (1986) : geopressured reservoirs, proceedings volume, abnormal pore-fluid
ENGINEER (1985) : tight gas condensate, abnormally pressured, case study
FROHNE & MERCER (1984) : reservoir performance study, offset well, interference
GEER & COOK (1978) : creation, augmentation, exploitation, pore volume saturate
HAMMERSHAIMB & KUUSKRAA (1982) : reinjection, economic potential, geopressured
HORTON, MERCER & SAWYER (1982) : infill drilling, vertical wells, hydrofrac
IRELAND & ROBINSON (1987) : reserve prediction, production testing, geopressure
ISOKRARI & KNAPP (1976) : geothermal geopressured aquifers, gas production
JONES (1969) : hydrodynamics, geopressure, Northern Gulf, Mexico Basin
KNAPP, ISOKRARI, GARG & PRITCHETT (1977) : geothermal geopressured aquifers
LAW (1984) : thermal maturity, overpressuring, gas generation and accumulation
LAW & DICKINSON (1985) : conceptual model, origin, low-permeability
LAW & SPENCER (1981) : abnormally high-pressured, low-permeability gas pays
LAW, SPENCER & BOSTICK (1980) : overpressure, gas generation, temperature
MATTHEWS (1981) : enhancing gas production, geopressured reservoirs, possible
McCOY et al. (1980) : wells-of-opportunity, geothermal, geopressured, testing
McMULLAN & BASSIOUNI (1984) : maximum flow rates, prediction, geopressured pay

McPEEK (1981) : giant gas supply, deep overpressured reservoirs, development
MEISSNER (1981) : abnormal pressures, hydrocarbon generation, maturation
MNISZEWSKI & WESTCOTT (1986) : high-energy chemical sources, Devonian shale
POSTON & CHEN (1987) : formation compressibility, gas-in-place, geopressuring
PRASAD (1986) : fracturing, gamma-ray logging, remedial operations
PRASAD (1987) : case studies, abnormally pressured gas reservoirs, importance
QUITZAU & BASSIOUNI (1981) : geopressure energy resource impact, exploration
QUONG, OWEN & LOCKE (1982) : methane extraction, high temperature and pressure
RANDOLPH (1977) : geopressured reservoirs containing natural gas, Gulf Coast
ROBINSON, HOLDITCH & LEE (1983) : reservoir evaluation, tight gas, completion
ROSE & DOHERTY (1982) : data collection, instrumentation, geopressured pays
SAMUELS (1980) : geopressured energy resource, Louisiana, Texas, evaluation
SOOKPRASONG & ELBEL (1985) : hydraulic fracture optimization, geopressured pay
SPENCER & LAW (1981) : overpressured low-permeability gas reservoirs
SWANSON, BERNARD & OSOBA (1986) : geothermal, geopressured, Gulf Coast, testing
TIMKO & FERTL (1971) : hydrocarbon accumulation, geopressure, economic impact
TUCKER (1979) : practical pressure analysis, proppant selection, geopressured
WEEKS (1979) : revised completion, geopressured
WERES, JUN & TSAO (1986) : 3-phase equilibrium model, interpretation, offtake

7.2.7.8. Geothermal fracturing

The references on geothermal fracturing are classified along the lines of geothermal drilling, testing and completion, geothermal stimulation, geothermal field operation, high-temperature fracturing fluid stability, and other aspects. The quotations in this division correspond mainly to items discussed in section 4.7. of the text.

7.2.7.8.1. Geothermal drilling, testing and completion

The references on geothermal drilling, testing and completion correspond to various aspects discussed in sections 4.7.1. and 4.7.2. of the text.

ALLEN & BAZA (1980) : step-rate injectivity test, fracturing, liquid-dominated
BARON & UNGEMACH (1981) : European geothermal drilling experience, case studies
BODVARSSON & TSANG (1982) : injection, thermal breakdown, fractured geothermal
BODVARSSON et al. (1982) : reservoir capacity and longevity, Baca/New Mexico
CATALDI & SOMMARUGA (1986) : present state, future prospects, development
DA SIE & HOANG (1985) : well testing, geothermal field, Heber
DIPIPPO (1984) : worldwide geothermal power development, overview and update
GUDMUNDSSON et al. (1984) : injection and tracer testing, Svartsengi, Iceland
HORNE (1982) : geothermal reinjection experience, Japan
LEAVER, SAGEEV & RAMEY (1986) : multiple interference testing, Ohaaki reservoir
LIPPMANN, TSANG & WITHERSPOON (1977) : injection/production response
MARCOU & GUDMUNDSSON (1986) : developmental model, geothermal reservoirs
MEIER & ERNST (1981) : Urach, Germany FRG, hot dry rock, geothermal pilot test
MURPHY & TESTER (1979) : heat production, fracturing, hot dry rock, results
RAMEY & GRINGARTEN (1975) : high-volume, geothermal steam well
SON, DZIALOWSKI & LOFTIN (1983) : gelatinous oil-base fluids, insulator usage
TESTER, BIVENS & POTTER (1982) : interwell tracer analysis, geothermal fracture
WALKUP & HORNE (1985) : tracer retention, tracer transport, fractured reservoir
WILLIAMS et al. (1979) : equipment for drilling and fracturing hot granite well

7.2.7.8.2. Geothermal stimulation

The references on geothermal stimulation correspond to various aspects discussed in sections 4.7.3. to 4.7.6. of the text.

ALLEN & BAZA (1980) : step-rate injectivity test, fracturing, liquid-dominated

BARIA & GREEN (1986) : viscous stimulation, hot dry rock fracturing, seismicity
BATCHELOR (1982) : combined explosive and hydraulic fracturing, hot dry rocks
BATCHELOR (1984) : hot dry rock reservoir stimulation, explosive, hydraulic
BATCHELOR, PEARSON & HALLADAY (1980) : combined explosive and hydraulic fracs
BODVARSSON & GAULKE (1986) : fractured, geothermal, non-condensible, recovery
BODVARSSON, PRUESS & LIPPMANN (1986) : geothermal system modelling, fracture
BODVARSSON, PRUESS & O`SULLIVAN (1985) : fractured geothermal reservoirs
BODVARSSON & TSANG (1982) : injection, thermal breakdown, fractured geothermal
CAMPBELL, MORRIS & VERITY (1981) : geothermal well stimulation, evaluation
CAMPBELL et al. (1981) : normal, reverse, dendritic, large-scale, miniature
CARDEN et al. (1983) : geothermal prop fracturing, microearthquake detection
CINCO-LEY et al. (1979) : vertical fractures, parallelepiped model, steam
ERNST (1977) : hot dry rock fracturing, single well, hydraulic stimulation
GRONE, BAUMGARTNER & WOODROOF (1983) : proppant fracturing, 540 oF, geothermal
HUNSBEDT, KRUGER & LONDON (1975, 1977) : fracturing types, fluid circulation
MORRIS & SINCLAIR (1984) : bottomhole treatment evaluation, stimulation
MORRIS et al. (1980) : geothermal well stimulation experiments and evaluation
MURPHY & FEHLER (1986) : dendritic fracturing, microearthquakes, shear/tensile
MURPHY & PINE (1984) : dendritic hydraulic fracturing, geothermal energy pays
MURPHY & TESTER (1979) : heat production, fracturing, hot dry rock, results
MURPHY et al. (1977) : geothermal energy reservoir, hydraulic fracturing, test
PINE (1983) : large-scale hydraulic injections, Carnmenellis granite, England
PINE (1987) : medium-viscosity large-volume hydraulic jobs, hot dry stimulation
PINE & BATCHELOR (1984) : hydraulic injections, downward shear migration
PINE & LEDINGHAM (1983) : England, geothermal stimulation, potential
PINE et al. (1986) : combined explosive and hydraulic fracturing, geothermal
PRUESS (1983) : heat transfer, fractured geothermal reservoir
PRUESS (1985) : vapor-dominated reservoirs, heat pipes in fractured rocks
PRUESS & BODVARSSON (1984) : geothermal reservoirs, major fractures, injection
ROWLEY et al. (1983) : fracturing operations, dry geothermal reservoir
SINCLAIR (1980) : geothermal frac simulation, high-temperature proppant testing
SINCLAIR, PITTARD & HANOLD (1980) : geothermal well stimulation
SOLBERG, LOCKNER & BYERLEE (1980) : hydraulic fracturing, shear slippage
TESTER et al. (1986) : fracture growth, geothermal reservoir, heat extraction
THOMAS & ROOT (1979) : viscosity, stabilization, high-temperature fracturing
VERITY & MORRIS (1981) : geothermal fracturing, Baca/New Mexico, stimulation
WILLIAMS, WOODROOF & BOX (1982) : proppant fracturing, 540 oF, geothermal well

7.2.7.8.3. Geothermal field operation

The references on geothermal field operation correspond mainly to aspects discussed in section 4.7.2. of the text.

ADAMS, MOORE & FORSTER (1985) : fluid flow, volcanic terrain, geothermal system
ATKINSON et al. (1980) : Bagnore/Italy, behaviour, field study, geothermal pay
BARON & UNGEMACH (1981) : European geothermal drilling experience, case studies
BODVARSSON et al. (1984) : Krafla/Iceland, geothermal field, case study
BODVARSSON et al. (1986) : Nesjavellir/Iceland, geothermal field, case study
CIGNI, GAUDIO & FABBRI (1981) : Italy, deep geothermal drilling, experience
DA SIE & HOANG (1985) : well testing, geothermal field, Heber
DIPIPPO (1984) : worldwide geothermal power development, overview and update
ELDERS (1979) : geological background, geothermal fields, Salton trough, USA
GUDMUNDSSON & OLSEN (1985) : water influx, Iceland, Svartsengi, geothermal
GUDMUNDSSON et al. (1984) : injection and tracer testing, Svartsengi, Iceland
HORNE (1982) : geothermal reinjection experience, Japan
JELIC (1982) : geothermal energy, Yugoslavia
LEAVER, SAGEEV & RAMEY (1986) : multiple interference testing, Ohaaki reservoir
McDOWELL & ELDERS (1979) : geothermal metamorphism, Salton Sea, pay system
MEIER & ERNST (1981) : Urach, Germany FRG, hot dry rock, geothermal pilot test
STACEY & THAIN (1983) : 25 years of geothermal heat exploitation, Wairakei
STEFANSSON (1981) : Krafla/Iceland, geothermal field, case study

VERITY & MORRIS (1981) : geothermal fracturing, Baca/New Mexico, stimulation

7.2.7.8.4. High-temperature fracturing fluid stability

The references on high-temperature fracturing fluid stability correspond mainly to aspects discussed in sections 4.3.4.4., 4.4.5. and 4.7.4.2. of the text.

BARIA & GREEN (1986) : viscous stimulation, hot dry rock fracturing, seismicity
BAUMGARTNER, PARKER, WILLIAMS & WOODROOF (1983) : high-efficiency fluids
BODVARSSON & TSANG (1982) : injection, thermal breakdown, fractured geothermal
CARNEY, GUVEN & McGREW (1982) : high-temperature fluid-loss additives, fluids
ELBEL & THOMAS (1980) : viscosity stabilizers, high-temperature fracturing
ELY, CHATTERJI, HOLTMYER & TINSLEY (1975) : high-temperature fracture fluids
GRONE, BAUMGARTNER & WOODROOF (1983) : proppant fracturing, 540 oF, geothermal
HARMS, GOSS & PAYNE (1984) : ultrahigh temperature, fracturing fluids
HILSCHER & CLEMENTS (1982) : high-temperature drilling fluid, deep sensitive
HORNE (1982) : geothermal reinjection experience, Japan
HORTON (1982) : high-temperature reservoirs, fracturing fluids
HUNSBEDT, KRUGER & LONDON (1975, 1977) : fracturing types, fluid circulation
SINCLAIR (1980) : geothermal frac simulation, high-temperature proppant testing
SON, DZIALOWSKI & LOFTIN (1983) : gelatinous oil-base fluids, insulator usage
THOMAS & ROOT (1979) : viscosity, stabilization, high-temperature fracturing
VERITY & MORRIS (1981) : geothermal fracturing, Baca/New Mexico, stimulation
WALSER (1985) : high-temperature fracturing fluids, field study, South Texas
WILLIAMS et al. (1979) : equipment for drilling and fracturing hot granite well
WILLIAMS, WOODROOF & BOX (1982) : proppant fracturing, 540 oF, geothermal well.

7.2.7.8.5. Other aspects

The references on other aspects of geothermal fracturing correspond to various points discussed in section 4.7.3. of the text.

ADAMS, MOORE & FORSTER (1985) : fluid flow, volcanic terrain, geothermal system
BARTON, BANDIS & BAKHTAR (1985) : conductivity coupling, shear slippage
BODVARSSON & GAULKE (1986) : fractured, geothermal, non-condensible, recovery
BODVARSSON, PRUESS & LIPPMANN (1986) : geothermal system modelling, fracture
BODVARSSON et al. (1982) : reservoir capacity and longevity, Baca/New Mexico
DIPIPPO (1984) : worldwide geothermal power development, overview and update
ELDERS (1979) : geological background, geothermal fields, Salton trough, USA
GRANT (1983) : geothermal reservoir modelling
GRANT, DONALDSON & BIXLEY (1982) : geothermal reservoir engineering, textbook
HULEN & NIELSON (1982, 1983) : stratigraphic permeability, volcanic series
LEYDECKER (1981) : seismic detection, geothermics
LIPPMANN, TSANG & WITHERSPOON (1977) : injection/production response
LOCKNER & BYERLEE (1977) : shear slippage, high confining pressure, stress
McCOY et al. (1982) : wells-of-opportunity, geopressuring, geothermal, testing
McDOWELL & ELDERS (1979) : geothermal metamorphism, Salton Sea, pay system
MURPHY & TESTER (1979) : heat production, fracturing, hot dry rock, results
PRUESS (1983) : heat transfer, fractured geothermal reservoir
PRUESS (1985) : vapor-dominated reservoirs, heat pipes in fractured rocks
RAMEY & GRINGARTEN (1975) : high-volume, geothermal steam well
SON, DZIALOWSKI & LOFTIN (1983) : gelatinous oil-base fluids, insulator usage
STACEY & THAIN (1983) : 25 years of geothermal heat exploitation, Wairakei
TESTER, BIVENS & POTTER (1982) : interwell tracer analysis, geothermal fracture
WILLIAMS et al. (1979) : equipment for drilling and fracturing hot granite well

7.2.7.9. Water injection well fracturing and gravel packing

The references on water injection well fracturing and gravel packing correspond mainly to aspects discussed in section 4.8.7. of the text.

BRITT & LARSEN (1986) : oil reservoir, multilayering, minifracturing, carbonate
CHEN (1986) : water injection, fracturing, development, Fuyu Sandstone, China
CLARK (1968) : transient pressure, water injection well
CRAWFORD & COLLINS (1953) : vertical fractures, effect on secondary recovery
DICKEY & ANDERSEN (1945) : water-injection well fracturing
DIKKENS & NIKO (1987) : waterflood-induced fractures, accidental, propagation
DOSCHER & EL-ARABI (1983) : waste water disposal above fracture gradient
DOWELL & JOHNSTON (1960) : waterflooding, injection well fracturing
GARON, YIN & DUNAYEVSKY (1988) : thermally induced waterflood fracturing
GREBE (1943) : water-injection well fracturing
GUERREDO (1958) : waterflooding, injection well fracturing
HAGOORT (1981) : waterflood-induced hydraulic fracturing, accidental
HAGOORT, WEATHERHILL & SETTARI (1980) : waterflood-induced hydraulic fractures
HARTSOCK & SLOBOD (1961) : waterflooding, injection well fracturing
HOLZHAUSEN & EGAN (1987) : detection, control, water injection well fracturing
HOWARD & FAST (1950) : injection fracturing, squeeze cementing
HOWARD & FAST (1970 a) : disposal fracturing, injector installation
HOWELL & THOMAS (1978) : fluid viscosity, screenout, well shooting, gauge hole
ILSENG & CORTEZ (1985) : waterflood, West Texas, large-volume frac stimulation
JONES (1960) : accidental stimulation, water-flood project, Milham Sand
KOHLHAAS (1982) : refracturing, late-stage of depletion, mature, hydrophobic
KONING (1985) : water-injection well fracturing, propagation, modelling
KONING & NIKO (1985) : water-injection well fracturing, pressure fall-off test
KUO, HANSON & BRISAY (1984) : fracture extent prediction, waterflood operation
McBANE & JEU (1988) : waste-water disposal well fracturing, coal gas offtake
PERKINS & GONZALEZ (1985) : thermoelastic stresses, effect on injector fracture
PRADO et al. (1988) : waterflood-induced fracture, high-precision fall-off test
REISTLE (1945) : injection well fracturing, squeeze cementing
REVIERE & WU (1986) : water-flood, infill drilling, fracturing, evaluation
RILEY (1959) : water-flood operations, Kermit, Cherrykirk, Pecos Valley fields
SCHWARTZ (1969) : sand control design, high productivity oil and water wells
SINCLAIR (1980) : geothermal, fracture simulation, high-temperature proppant
SUN & MONGAN (1974) : waste disposal fracturing, shales, West Valley, New York
TEPLITZ & HASSEBROEK (1946) : injection well fracturing, squeeze cementing
TORREST (1982) : deposit buildup, viscous polymer solutions, water
TORREY (1940) : injection well fracturing, squeeze cementing
WILLIAMS (1977) : accidental waterflood-induced fractures, overfrac pressure
WILLIAMS, SHERRARD & LIN (1987) : fracture induction, reservoir management
WILSON & KISER (1987) : hazardous, liquid waste injection, design, site choice
YUSTER & CALHOUN (1945) : water-injection well fracturing

7.2.7.10. Steam injection well fracturing and gravel packing

The references on steam injection well fracturing and gravel packing correspond mainly to aspects discussed in sections 4.9.4. and 5.7. of the text.

ABAD & HENSLEY (1984) : effect of fractures and dip on steam sweep efficiency
ALI (1982) : steam injection theories, unique approach, formation reactions
AMAEFULE et al. (1984) : steam condensate, formation damage, chemical treatment
BABCOCK (1975) : fracture phenomena, Athabasca oil sands, Alberta
BIRD & BOON (1987) : silica transport, steam injection, oil sands, dissolution
BLAIR, SCRIBNER & STOUT (1982) : cyclic steam injection, chemical enhancement
BRITTON et al. (1982) : fracture-assisted steamflood technology, Street Ranch
BUCKLES (1979) : steam stimulation fracturing, heavy oil recovery, Cold Lake
CHU (1985) : steam-flood field projects, state-of-the-art review
CLOSMANN & PHOCAS (1976) : thermal stress, heated fracture, pattern changes

CLOSMANN & SMITH (1983) : steam-heated fracture, steam-zone rise, temperature
COUDERC et al. (1987 a, 1987 b) : steam pilot, offshore, Emeraude Vapeur
CURRENT (1988) : steam stimulation improvement, chemical emulsion, viscosity
DIETRICH (1983) : cyclic steaming, tar sands, hydraulically-induced fractures
DOSCHER, LABELLE, SAWATSKY & ZWICKY (1963) : steam-drive recovery, Athabasca
DUSSEAULT & SIMMONS (1982) : injection-induced stress and fracture orientation
FRIEDMAN, SURLES & KIEKE (1986) : high-temperature sand consolidation
GESHELIN & GRABOWSKI (1982) : oil sands steam stimulation, fracture model
GESHELIN, GRABOWSKI & PEASE (1981) : steam stimulation, fractured reservoir
GROSSMANN, HÖFLING & MARX (1988) : ceramic proppant, high-angle well, steam
GUDMUNDSSON et al. (1983) : flashing steam-water flow in fractures, steamtube
HARMSEN (1979) : Schoonebeek, Netherlands, steamflood operation, waterdrive
KENNEDY (1984) : cyclic steam stimulation, Wilmington field, Los Angeles/Calif.
LILLIE & SPRINGER (1981) : steam-drive operation, Emsland, Germany, status
LILLIE & WIETHOFF (1985) : steam drive operation, Emsland, experience report
LIN (1988) : cyclic steam stimulation above fracture pressure
McCORRISTON, DEMBY & PEASE (1981) : reservoir damage, heavy oil
MELDAU, SHIPLEY & COATS (1981) : cyclic gas/steam stimulation, heavy-oil field
MILHEM & AHMED (1987) : cyclic steam stimulation project, performance, Kuwait
PEAKE (1984) : hydraulic fracture and steam stimulation, diatomite reservoir
PETHRICK, SENNHAUSER & HARDING (1986) : cyclic steam stimulation, Wolf Lake
PERKINS & GONZALEZ (1984) : earth stress changes, radial symmetry, temperature
PERKINS & GONZALEZ (1985) : thermoelastic stress, effect on injection fracture
POWERS et al. (1985) : multiple thin-sand reservoir, steam-flood, commerciality
PROYER et al. (1983) : steam-drive operation, Emsland, pilot project, results
REDDY, PROWSE & REDFORD (1980) : steam injection, horizontal fracture, simulate
REED (1980) : steam injection, gravel and sandstone dissolution
REINICKE & WIETHOFF (1987) : steam-drive, Emsland, Germany, forecast, results
RENSVOLD (1982) : sand-control technology, thermal recovery application
SACUTA, NGUYEN & KISSEL (1986) : nickel-coated sands, gravel pack, stability
SAHUQUET & SITBON (1987) : Lacq, France, steam-drive, result
SHRYOCK, AHMAD, MELOY & KENT (1988) : steam-drive gravel packing, foam, slurry
SINGHAI & CARD (1988) : steam stimulation monitoring, fracturing, Athabasca
SOBOCINSKI, LESKOVA & GREEBE (1984) : steam-flood simulation, Georgsdorf, FRG
SOMERTON et al. (1984) : formation damage evaluation, screening, EOR operation
SONI & HARMON (1986) : fracture-assisted steamflood pilot, Saner Ranch
SPRINGER & MÜLLER (1985) : excess water, disposal, possibilities, steam drive
STANG & SONI (1987) : fracture-assisted steam-flood, technology, Saner Ranch
SYLVESTER & CHEN (1988) : cyclic steam stimulation, pressure depletion, model
THOMAS, YOST & CATHEY (1987) : steam-flood silica dissolution, silica removal
TOMA, LIVESEY & HEIDRICK (1986) : sand control filters, thermal recovery wells
TORTIKE & ALI (1989) : fracture-based cyclic steam stimulation, tar sands
UNDERDOWN & DAS (1985) : thermal wells, gravel pack stability
UNDERDOWN & KAMALENDU (1983) : thermal wells, gravel pack stability
VITTORATOS (1986) : steam injection, horizontal fractures, temperature profile
WATKINS & WATANABE (1984) : silica dissolution inhibition during steam drive
WATKINS et al. (1985) : prevention of gravel pack and formation dissolution
WONG (1986) : in-situ stress changes, steam stimulation, Cold Lake oil sands
ZIEGLER (1988) : steamflood fracturing, fracture opening/closing, California

7.2.7.11. Refracturing

The references on refracturing correspond mainly to aspects discussed in section 4.8.9. of the text.

CONWAY et al. (1985) : refracturing, recoverable reserve expansion, oil pays
COULTER & MENZIE (1973) : refracturing design, restimulation, subsurface pay
CROWELL & JENNINGS (1979) : diagnostic technique, restimulation candidates
DARR & CARLTON (1988) : recompletion, prop frac, multipay intervals, improving
GREGORCZYK et al. (1984) : high prop concentration, MHF refrac, Olmos Formation
GRIFFITH & MADISON (1988) : production decline offset by refracturing

HALLIBURTON (1987) : refracturing, well candidate selection, reserve recovery
HOLDITCH (1984) : proppant selection for early- vs. late-stage fracturing
HUEBINGER et al. (1988) : proppant concentration, optimizing program, Olmos
HUNTER (1985) : refracturing, Oak Hill, Cotton Valley, case history
JOURNAL PETROLEUM TECHNOLOGY (1987) : stimulation of old wells, refracturing
KOHLHAAS (1982) : refracturing examples
LANCASTER et al. (1987) : targets for restimulation, attractivity, feasibility
PARROT & LONG (1979) : MHF, refracturing, tight, Muddy "J" Formation
PAULS et al. (1984) : refracturing payout, treatment size, proppant saturation
PEARSON, LYNCH, SCHMIDT & McCASLIN (1988) : time between original and refrac
SAUCIER (1988) : refracturing examples
SINHA & ELBEL (1983) : old producing well stimulation, design, evaluation
TUCKER (1979) : pressure development after original stimulation, refracturing
VENDITTO et al. (1986) : refracturing, cost effectiveness, feasibility, impact
WYMAN, HOLDITCH & RANDOLPH (1979) : displacement of old proppant assemblage
YOUNG, BARKER & CLARK (1984) : combination of hydraulic and explosive fracture

7.2.7.12. Non-proppant fracturing

The references on non-proppant fracturing correspond mainly to aspects discussed in sections 4.4.4. and 4.8.2.5. of the text.

ABEL (1982), ABEL et al. (1983) : shale nitrogen fracturing, non-proppant
BRANAGAN, CIPOLLA, LEE & WILMER (1985) : non-proppant minifracturing, effect
BRANAGAN, CIPOLLA, LEE & YAN (1987) : hydraulic fracturing, naturally jointed
BRANAGAN & WILMER (1988) : microfracturing, nitrogen breakdown
BRITT & LARSEN (1986) : non-proppant minifracturing, significance, feasibility
ELY (1988) : nitrogen gas shale fracturing, non-proppant fracturing, importance
FREEMAN, ABEL & KIM (1981) : stimulation technique using only nitrogen
GHAURI (1980) : high-rate oil backflushing, well stimulation, non-proppant
GOTTSCHLING, ROYCE & SCHUCK (1985) : nitrogen gas shale fracturing, significant
HARPER, HAGAN & MARTINS (1985) : fracturing without proppant, comparison, test
HOLDITCH, ROBINSON & WHITEHEAD (1987) : pre- and post-fracture evaluation
LAMBERT & TREVITS (1980) : feasibility, methane recovery, coal bed drainage
LANCASTER (1988) : nitrogen gas shale fracturing, non-proppant fracturing
MAHONEY et al. (1980) : non-proppant stimulation, treatment, coal gas offtake

7.2.7.13. High proppant concentration

The references on high proppant concentration correspond mainly to aspects discussed in section 4.3. of the text.

COULTER & WELLS (1972) : high proppant concentration, advantages in stimulation
DOMSELAAR & VISSER (1974) : proppant concentration, fracture shape, viscous gel
DUNAND & SOUCEMARIANADIN (1985) : settling velocity, proppant slurry, saturate
GREGORCZYK et al. (1984) : high proppant concentration, production boosting
PAULS et al. (1985) : Olmos Formation, successful stimulation, oil-base fluids
SANDY, WIGGINS & VENDITTO (1986) : large total proppant load, high saturation
SCHOLS & VISSER (1974) : proppant bank buildup, vertical fracture, no fluidloss
SHAH, SMITH & DONALDSON (1983) : ultra-high proppant concentration, stimulation
WATERS (1980) : hydrocarbon reservoir stimulation, overview, possibilities

7.2.7.14. Highly-deviated well fracturing

The references on highly-deviated well fracturing correspond mainly to aspects discussed in section 4.11. of the text.

AADNOY (1986) : stability of deviated boreholes, stress removal failure
AADNOY (1987, 1988) : vertical fracture initiation, borehole axis inclination

AADNOY & CHENEVERT (1987) : borehole instability, wellbore collapse, deviation
BRADLEY (1979) : failure of inclined boreholes, stress removal collapse
BUCHANAN (1987) : deviated well stimulation, proppant transport, stratification
EL RABAA (1988) : fracture height determination, wellbore intersection, angle
FUH, WHITFILL & SCHUH (1988) : borehole instability, wellbore collapse, angle
HOTTMAN, SMITH & PURCELL (1979) : borehole instability, wellbore collapse
KIM, CHAMPION & COOPER (1985) : deviated well stimulation, temperature logging
MAURY & SAUZAY (1987) : borehole instability, wellbore collapse, inclination
PEETERS & HARTLEY (1984) : fracture height determination, wellbore intersection
SCHULTE (1984) : limited well inflow, part of fracture height
SNOW & HOUGH (1988) : multiple fractures, en-echelon crack bending, orientation
STRUBHAR, FITCH & GLENN (1988) : multiple parallel fractures, singular well
WILSON & WILLIS (1986) : highly-deviated wells, borehole instability, collapse
YEW & LI (1987) : deviated well fracturing, elasticity

7.2.8. Gravel packing and sand control

This topic header discusses aspects of gravel type and grain size; performance, design, and monitoring and evaluation of gravel packing; highly-deviated well gravel packing, gravel transport and settling, gravel checking and handling, prediction of sand production, special gravel pack applications, formation consolidation, other sand control methods, sand arching, and summary reviews and field experience reports.

The section on gravel type and grain size contains references on gravel stability, transport, composition, processing and application. The branch on gravel pack design brings together references on completion type, interval length, gravel concentration, pack thickness, impact on well productivity, and fluid properties. The division on gravel pack performance includes references on placement improvement, gravel transport, pack repair, screen effectivity, slurry density, recovery enhancement, and completion types. The chapter on gravel pack monitoring and evaluation comprises references on well logging, density and photon logging, and quantitative approach of evaluation. The branch on highly-deviated well gravel packing presents references on hole inclination impact, gravel density, packing effectivity, liner vibration, gravel duning, gravel bridging, and placement techniques. The assemblage on gravel transport and settling includes references on perforation passage, perforation cleanup, influence of borehole deviation, fluid viscosity and density, and fluid leakoff. Some references on gravel checking and handling are also compiled.

The division on prediction of sand production bears references on mechanical properties, Mohr`s circle, and sand-free production rates. The segment on special gravel pack applications comprises references on consolidated pack installation, resin-coated gravel placement, thermal resistivity, liner vibration, and concentrated gravel slurry. The branch on formation consolidation agglomerates references on resin, hydroxy aluminum and silica stabilization, and warm air coking. The division on other sand control methods contains references on pre-pack liners, multiple-wrap screens, and pressure packing. The section on sand arching comprises references on arch stability, near wellbore stress behaviour, sand-free oil production, and effect of offtake rate on arch formation. The references on summary reviews and field experience reports correspond to case studies in various areas and stratigraphical horizons.

7.2.8.1. Gravel type and grain size

The references on gravel type and grain size correspond mainly to aspects discussed in section 5.2. of the text.

BECQ, ROQUE & SARDA (1984) : steam-flood, sand, bauxite, zirconia, stability
BURDYLO (1977) : gravel pack guide

CHEUNG (1985) : effect of acids, gravels and proppants
COBERLY & WAGNER (1938) : selection, installation, oil-well gravel pack, design
CROWE & CRYAR (1975) : fluid loss, resin mixtures, workover, completion
ECONOMIDES (1986) : technology conference
FLANIGAN (1979) : gravel grain size, screen coating, old field enhancement
GRUESBECK & COLLINS (1978) : particle transport through perforations, drawback
GULATI & MALY (1974) : smaller gravel, permeability evidence, selection
GURLEY, COPELAND & HENDRICK (1977) : size selection, significance for offtake
HALL & HARRISBERGER (1970) : sand arch stability, sand control, key, influence
HAYNES & GRAY (1972) : particle transport, perforated casing, effectivity
HENG (1987) : gravel selection, gravel transport, sand production, placement
HUDSON & MARTIN (1988) : low-density gravel pack material, placement efficiency
KANAT (1980) : resin-coated gravel
MALY & KRUEGER (1971) : formation sampling, gravel size selection, improperties
McCORRISTON, DEMBY & PEASE (1981) : sintered bauxite, superiority, steam-flood
MURPHY, ROLL & WONG (1985) : resin-coated sand, damaged liner repair
PEDEN, RUSSELL & OYENEYIN (1985) : sand intrusion prevention, impair prevention
REED (1980) : steam injection, dissolution
SACUTA, NGUYEN & KISSEL (1986) : nickel-coated sand, stability in steam-drive
SAUCIER (1974) : design considerations, size selection, improvement
SCHWARTZ (1969) : sand control design, high-rate oil and water wells, success
SINCLAIR (1980) : high-temperature proppant testing, geothermal fracturing
SINCLAIR & GRAHAM (1977, 1978) : resin-coated gravel
SOLUM (1984) : liner vibration, gravel packing, new technique, sand control
SOLUM (1986) : gravel blending, pumping techniques, accurate control, protect
SPARLIN (1969) : fight sand with sand, realistic approach to gravel packing
SPARLIN (1974) : permeability, sand, gravel, design study
SPARLIN & COPELAND (1972) : concentrated gravel slurry, pressure packing
STEIN (1976) : gravel material selection, use of conventional log data
STEIN (1983) : changing well conditions, gravel pack design
STRACKE (1987) : offshore gravel packing, productivity differences, gravel size
SUMAN (1975) : gravel packing, sand control, review, status report
TAUSCH & CORLEY (1959) : gravel size selection
THOMAS, YOST & CATHEY (1987) : steam-flood silica dissolution
UNDERDOWN & DAS (1985) : sintered bauxite, alternative for sand in steam-flood
UNDERDOWN, DAY & SPARLIN (1980) : plastic pre-coated gravel
UNDERDOWN, DAS & NGUYEN (1984) : gravel crushing, slot/wire plugging, choice

7.2.8.2. Gravel pack design

The references on gravel pack design correspond mainly to aspects discussed in sections 5.3. to 5.6. of the text.

BALLEGOOYEN, GIAP & SENG (1983) : long perforated intervals, deviated holes
BURDYLO (1977) : gravel pack guide
BURGER (1987) : development of chemical process for sand control
CASHION & MARKLE (1940) : oil-well gravel packing, Gulf Coast fields
COBERLY (1937) : screen opening selection, unconsolidated sands, recommendation
COBERLY & WAGNER (1938) : selection, installation, oil-well gravel pack
CONSTIEN & MAYER (1978) : water-carried resin-coated gravel, screen
CROWE & CRYAR (1975) : fluid loss, resin mixtures, workover, completion
ECKEL & REISTLE (1940) : gravel packing, hardware, technology, well equipment
ECONOMIDES (1986) : technology conference
ELLIS, SNYDER & SUMAN (1981) : design concepts, sand control optimization
GAJDOSIK & WILLINGHAM (1976) : high gravel concentration, high-viscosity fluid
GOTTSCHLING & LEGAN (1981) : long-interval open-hole viscous gravel packing
GURLEY, COPELAND & HENDRICK (1977) : design, plan, execution, maximum offtake
HAILEY, DONOVAN & SICKLE (1985) : one trip perforate and gravel pack, system
HENG (1987) : gravel selection, gravel transport, sand production, placement
HILL (1941) : factors affecting use of gravel in oil wells, gravel selection
HIMES & RUIZ (1986) : new sidewall coring, gravel pack design improvement, test

LEDLOW & SAUER (1985) : design, placement, evaluation, improved performance
LEWIS (1985) : gravel pack thickness, key role in sand control
LYBARGER et al. (1974) : viscous systems, high productivity, water-base, Gulf
McCAULEY (1969) : perforation, purging, gravel placement, backsurging
McLEOD (1986) : productivity decline, post-acidizing, gravel-packed wells
McLEOD & CRAWFORD (1983) : high rate completions
MURPHY, ROLL & WONG (1985) : resin-coated sand, damaged liner repair
NELSON & SMITH (1969) : new development in gravel packing
OYENEYIN (1987) : computer programs, oil-well gravel pack design
OYENEYIN (1988) : gravel packing effect on gas well productivity, analysis
PEDEN, RUSSELL & OYENEYIN (1984) : effective sand control, deviated wells
PENBERTHY & COPE (1979) : design and productivity of gravel-packed completions
POOLLEN & MALONE (1959) : stimulation, sand control, combination
RASHID (1984) : selective completion, improved recovery, packerless, Cognac
SAUCIER (1974) : design considerations, size selection, improvement
SAWDON (1937) : gravel packing in oil wells
SCHEUERMAN (1984) : carrier fluid properties, high permeability, viscous pack
SCHROEDER (1987) : high-pressure wellbore model, gravel-pack studies, fullscale
SCHURTZ, BREINER & COMEAUX (1975) : through-tubing gravel pack, new technology
SCHWARTZ (1969) : sand control design, high-rate oil and water wells, success
SCOTT (1928) : well completion, Gulf Coast, USA, sand control experience
SINCLAIR & GRAHAM (1977, 1978) : resin-coated gravel
SKAGGS (1984) : high-density slurries, transport efficiency
SMITH (1986) : sand production, fluid flow, heavy oil reservoirs, solution gas
SOLUM (1984) : liner vibration, gravel packing, new technique, sand control
SOLUM (1986) : gravel blending, pumping techniques, accurate control, protect
SPARLIN (1969) : fight sand with sand, realistic approach to gravel packing
SPARLIN (1974) : permeability, sand, gravel
SPARLIN (1987) : gravel pack innovations, new developments
SPARLIN & COPELAND (1972) : concentrated gravel slurry, pressure packing
SPARLIN & HAGEN (1985) : gravel packing requires planning
STEIN (1983) : gravel pack design, changing well conditions
STEIN & HILCHIE (1972) : production rate estimation without sand control
STRACKE (1987) : offshore gravel packing, methods, success, experience report
SUMAN (1975) : gravel packing, sand control, review, status report
SUMAN & SNYDER (1982) : sand control design criteria, gravel packing concepts
TAUSCH & CORLEY (1958) : sand exclusion in oil and gas wells
TORREST (1982) : deposit build-up, viscous polymer solutions, water
TORREST & SAVAGE (1975) : particle collection, vertical suspension flow
UNDERDOWN & DAS (1983, 1985) : thermal wells, gravel pack stability
UNDERDOWN, DAY & SPARLIN (1980) : plastic pre-coated gravel
WAHLMEIER & ANDREWS (1988) : gravel placement mechanics, design and evaluation
WATKINS et al. (1985) : steam injection, solution prevention, gravel pack
WEST (1951) : gravel pack completion, water and gas exclusion
WILLIAMS, ELLIOTT & WEAVER (1972) : inside-casing, productivity

7.2.8.3. Gravel pack performance

The references on gravel pack performance correspond mainly to aspects discussed in sections 5.3. to 5.6. of the text.

CARROLL, NEAL & PARKER (1986) : gravel pack repair with wireline devices
CHEN, ZHOU & LIU (1986) : sand control techniques, China, experience report
CHEUNG (1985) : effect of acids, gravels and proppants
COBERLY & WAGNER (1938) : selection, installation, oil-well gravel pack, design
COSTEIN & WALLACE (1980) : performance, carrying fluid design, improvement
CROWE & CRYAR (1975) : fluid loss, resin mixtures, workover, completion
ELSON, DARLINGTON & MANTOOTH (1984) : high-angle completion, experience
HAYNES & GRAY (1972) : particle transport, perforated casing, effectivity
HENG (1987) : gravel selection, gravel transport, sand production, placement
HIMMATRAMKA (1981) : productivity reduction, non-Darcy flow, true skin

LEDLOW & SAUER (1985) : design, placement, evaluation, improved performance
MALY, ROBINSON & LAURIE (1974) : improving gravel pack placement, new tool
MALY & VOZELINEK (1978) : pressure washing of open-hole gravel pack, criteria
RASHID (1984) : selective completion, improved recovery, packerless, Cognac
SAGE & LACEY (1942) : gravel screen effectivity, hydrodynamics, flow conditions
SAUNDERS & McKENZIE (1979) : performance review, phenolic resin
SCHROEDER (1987) : high-pressure wellbore model, gravel-pack studies, fullscale
SCHURTZ, BREINER & COMEAUX (1975) : through-tubing gravel pack, new technology
SCHWARTZ (1969) : sand control design, high-rate oil and water wells, success
SHURTZ & BREINER (1975) : through-tubing gravel packing techniques
SKAGGS (1984) : high-density slurries, transport efficiency
SMITH (1986) : sand production, fluid flow, heavy oil reservoirs, solution gas
SOLUM (1984) : liner vibration, gravel packing, new technique, sand control
SPARLIN (1969) : fight sand with sand, realistic approach to gravel packing
SPARLIN & COPELAND (1972) : concentrated gravel slurry, pressure packing
STADALMAN, NOVOTNY & HOUCHIN (1985) : sand control longevity improvement
STRACKE (1987) : offshore gravel packing, methods, success, experience report
SUMAN (1972) : perforations, performance, problems, gravel placement, improving
SUMAN (1974, 1975) : gravel packing, sand control, review, status report
TORREST (1982) : deposit buildup, viscous polymer solutions, water
TORREST & MARCANO (1986) : gel surface filtration, polymer solution, screenout
TORREST & SAVAGE (1975) : particle collection, vertical suspension flow, place
UNDERDOWN & DAS (1983, 1985) : thermal wells, gravel pack stability
WATKINS et al. (1985) : steam injection, solution prevention, gravel pack, pay
WEIRICH, ZALESKI & TYLER (1988) : one-trip multiple-zone system, low-pressure
WILLIAMS, ELLIOTT & WEAVER (1972) : inside-casing, productivity
ZALESKI & DONOVAN (1986) : gravel packing operations from a floating vessel

7.2.8.4. Gravel pack monitoring and evaluation

The references on gravel pack monitoring and evaluation correspond mainly to aspects discussed in section 6.3. of the text.

AJAM & RAHAL (1986) : thermal multigate decay, pulsed neutron logging usage
ALLEN (1969) : casing collar locator, radioactive bullet logging, subsidence
BOYCE & CARROLL (1984) : logging tool, mathematical modelling
BUCHANAN et al. (1984) : pulsed neutron log, thermal multigate decay, capture
CHUDY (1981) : photon log, monitoring, Schoonebeek, oil well, Netherlands
DRESSER ATLAS (1987) : photon/vibrapack combination, evaluation, repair
JØRANSON, SPURLIN & EDWARDS (1988) : mechanical properties, sand management
KREIJ (1986) : cased-well logging, gravel pack log, density evaluation, control
NEAL (1983) : evaluation, qualitative, neutron, gammaray, nuclear fluid density
NEAL & CARROLL (1985) : evaluation, quantitative approach, gamma-ray
NEILL et al. (1977) : photon log, monitoring, evaluation
OLIVER, WYATT & SMITH (1986) : thermal multigate decay neutron log, monitoring
RANDALL et al. (1985) : new digital multiscale pulsed neutron logging system
RUEDRICH, PERKINS & O`BRIEN (1974) : multiple casing locator, joint length
SCHULTZ et al. (1983) : pulsed neutron capture logging, thermal multigate decay
SMITH, ARNOLD & PEELMAN (1983) : pulsed neutron capture logging, thermal decay
SOLLEE (1985) : logging experiments, gravel pack monitoring, methodology
STROBL & PLACKETT (1984) : gravel-pack logging, Groningen gas field, Netherland
SUMAN, ELLIS & SNYDER (1983) : tell-tale screen, termination of job, height
TIXIER, LOVELESS & ANDERSON (1973) : mechanical properties log, sand stability
WELEX (1986) : formation strength analysis, boundary drawdown, frac profile
WESTERN ATLAS (1988 a) : sand production analysis, mechanical properties, mud
WYATT & SMITH (1987) : pulsed neutron logging, thermal multigate decay

7.2.8.5. Highly-deviated well gravel packing

The references on highly-deviated well gravel packing correspond mainly to aspects discussed in section 5.8.3. of the text.

ANDERSON & HUTCHINSON (1978) : deviated-hole gravel packing
BALLEGOOYEN, GIAP & SENG (1983) : long perforated intervals, deviated holes
DICKINSON & ANDERSON (1987) : horizontal boreholes, gravel packing problems
ELSON, DARLINGTON & MANTOOTH (1984) : high-angle completion, oversized washpipe
GROSSMANN, HÖFLING & MARX (1988) : ceramic proppants, steam drive, high-angle
GRUESBECK, SALATHIEL & ECHOLS (1979) : deviated wellbore, gravel pack design
HAILEY (1984) : long perforated intervals, deviated holes, gravel packing
HODGE (1982) : deviated-hole gravel packing, gravel transport
HOLMAN (1975) : deviated-hole gravel packing
HUDSON & MARTIN (1988) : low-density particulates, gravel settling reduction
MALY & ROBINSON (1972) : inclined well gravel packing, method, U.S. patent
MALY, ROBINSON & LAURIE (1974) : gravel bridging, baffled pipe, velocity
NADERI (1984) : long perforated intervals, deviated holes, experience, review
NINI & OWEN (1983) : high-angle gravel packing, successful techniques
NOVOTNY & MATSON (1975) : deviated-hole gravel packing
OYENEYIN & PEDEN (1985) : effective gravel packing of deviated wells, influence
PEDEN, RUSSELL & OYENEYIN (1984) : effective sand control, deviated wells
RENSVOLD & DECKER (1978) : full-scale gravel packing model studies
SCHROEDER (1987) : full-scale high-pressure wellbore, gravel pack studies
SHRYOCK (1980) : deviated model wellbore, oversized washpipe, compaction
SOLUM (1984) : liner vibration, hexagonal packing, pipe rotation, compaction
SPREUX, GEORGES & LESSI (1988) : horizontal wells, fracturing, consolidation
STILES (1985) : bottom-up technique, gravity assistance, gravel transport
STILES, COLOMB & FARLEY (1986) : bottom-up technique, gravity assistance
STRACKE (1987) : offshore gravel packing, methods, success, experience report
SUMAN, ELLIS & SNYDER (1983) : gravel duning, flow resistance, compaction
TOMA, LIVESEY & HEIDRICK (1986) : steel cloth-wool screen filter, gravel pack
WALLING & STABLER (1978) : deviated-hole gravel packing
UNDERDOWN, DAS & NGUYEN (1984) : highly-deviated wells, crosslinked polymer
ZALESKI & SPATZ (1988) : horizontal wells, perforation, sand control

7.2.8.6. Gravel transport and settling

The references on gravel transport and settling correspond mainly to aspects discussed in sections 5.3., 5.6. and 5.8.3. of the text.

BONOMO & YOUNG (1983) : perforation cleanup methods, analysis, evaluation
BURCIK (1969) : microgel plugging, partially hydrolyzed polyacrylamide
CHAUVETEAU (1981) : coiled polymer solutions, flow through porous media
CHAUVETEAU & KOHLER (1984) : microgel plugging, polysaccharide solutions
CLARK & QUADIR (1981) : proppant transport, particle settling velocity equation
COHEN & METZNER (1982) : absorption effects, polymer solutions, capillary flow
CONSTIEN & MAYER (1978) : water-carried resin-coated gravel, screen
COSTEIN & WALLACE (1980) : performance, carrying fluid design, improvement
CROWE & CRYAR (1975) : fluid loss, resin mixtures, workover, completion
DOWELL SCHLUMBERGER (1988 a) : Permpac, improved gravel suspension, leakoff
GRAMAIN & MYARD (1981) : flexible polymers desorbed in porous media, elongation
GRUESBECK & COLLINS (1978) : particle transport through perforations, drawback
GURLEY, COPELAND & HENDRICK (1977) : maximum productivity, design, planning
HALL & HARRISBERGER (1970) : sand arch stability, sand control, key, influence
HAYNES & GRAY (1972) : particle transport, perforated casing, effectivity
HENG (1987) : gravel selection, gravel transport, sand production, placement
HODGE (1982) : gravel transport, deviated wellbores, principles, models
HOUCHIN, DUNLAP & HUTCHINSON (1988) : formation damage during gravel packing
KOHLER & CHAUVETEAU (1981) : polysaccharide plugging, porous media flow
LEDLOW & SAUER (1985) : design, placement, evaluation, improved performance

LYBARGER et al. (1974) : viscous systems, high productivity, water-base, Gulf
McCAULEY (1969) : perforation, purging, gravel placement, backsurging
MURPHY, ROLL & WONG (1985) : resin-coated sand slurry, damaged liner, repair
NEHMER (1988) : viscoelastic gravel-pack carrier fluid, gravel suspension
PEDEN, RUSSELL & OYENEYIN (1985) : fluid viscosity, gravel concentration
PENBERTHY (1985) : gravel placement, perforation cleaning, gravel transport
SCHEUERMAN (1984) : carrier fluid properties, evaluation, advantages
SCHURTZ, BREINER & COMEAUX (1975) : through-tubing gravel pack, new technology
SHRYOCK, DUNLOP & MILLHONE (1979) : full-scale gravel packing studies, results
SIEVERT, WAHL, CLARK & HARKIN (1981) : proppant transport
SKAGGS (1984) : high-density slurries, transport efficiency
SOLUM (1986) : sand production, fluid flow, heavy oil reservoirs, solution gas
SOLUM, RAMEZANI & MIMS (1988) : gravel packing through perforation tubes
TORREST (1982 a) : deposit build-up, viscous polymer solutions, water
TORREST (1982 b) : placement conditions influence, gravel pack response
TORREST (1982 c) : flow of viscous polymer solutions, gravel packing, porosity
TORREST & MARCANO (1986) : gel surface filtration, polymer solutions, packing
TORREST & SAVAGE (1975) : particle collection, vertical suspension flow, place
YILDIZ & LANGLINAIS (1988) : calculation of pressure losses around gravel pack

7.2.8.7. Gravel checking and handling

The references on gravel checking and handling correspond mainly to aspects discussed in sections 5.2. and 5.3. of the text.

BOULET (1978) : gravel blending, gravel pumping, techniques, influences
FOSTER & LINVILLE (1979) : sand production monitoring, flowing well streams
MULLINS, BALDWIN & BERRY (1974) : surface flowline sand detection
SOLUM (1986) : gravel blending, pumping techniques, accurate control, protect
SUMAN (1975) : gravel packing, sand control, review, status report
ZWOLLE & DAVIES (1983) : gravel packing sand quality

7.2.8.8. Prediction of sand production

The references on prediction of sand production correspond mainly to aspects discussed in section 6.3.6. of the text.

BARROW & LASSEIGNE (1984) : Mohr`s circle technique, sand strength prediction
COATES & DENOO (1981) : mechanical properties log, Mohr`s circle, borehole
GHALAMBOR et al. (1988) : Gulf coast gas wells, sanding forecast, free water
MASSIE, NYGAARD & MORITA (1987) : offshore gravel packing, field experience
MORITA et al. (1987) : realistic prediction, numerical and analytical approach
STEIN, ODEH & JONES (1974) : sand-free production rates, different completions
TIPPIE & KOHLHAAS (1973) : flow rate, effect on sand stability, unconsolidated
TIXIER, LOVELESS & ANDERSON (1973) : mechanical properties log, candidates
WEISSENBURGER et al. (1987) : engineering approach, sand production prediction
YASSIN & PEDEN (1987) : sand stability prediction on microcomputer

7.2.8.9. Special gravel pack applications

The references on special gravel pack applications correspond mainly to aspects discussed in section 5.8. of the text.

BERCEGEAY & RICHARD (1974) : single-stage gravel packing, packer squeezing
BJ-HUGHES (1986) : viscous gravel packing, slurry consolidation, sand stability
CHEN, ZHOU & LIU (1986) : consolidated pack installation, China, experience
CLAWSON & SPARLIN (1975) : slurry packing of deep high-pressure wells
CONSTIEN & MAYER (1978) : water-carried resin-coated gravel, consolidated pack
COPELAND & BASS (1976) : thermal resistance of resin-coated gravel, steam drive

COPELAND & McAULEY (1974) : consolidated pack installation
COPELAND & CONSTIEN (1978) : consolidated gravel pack formation, methodology
CRAWFORD, COULTER & OSBORN (1980) : wellbore damage removal, acidizing, gravel
ELSON & ANDERSON (1982) : foam gravel packing
FAN (1988) : resin-coated gravel packing, sand control technology, evaluation
GRIFFIN, CONSTIEN & COLAVECCHIO (1978) : resin-coated gravel placement
HAILEY, DONOVAN & SICKLE (1985) : one-trip perforation and gravel packing
HAMBY & RICHARDSON (1968) : dirty sands, consolidated pack installation
HICKEY, BROWN & CRITTENDEN (1981) : resin-coated gravel placement
JEFFERIS, BRUIST & BOTTS (1983) : multizone, one-trip, selective, field-proven
KNAPP, PLANTY & VOILAND (1977) : consolidated pack installation
MALY & VOZELINEK (1978) : open hole gravel packing
MURPHEY, BILA & TOTTY (1974) : consolidated pack installation, water-based
MURPHEY, ROLL & WONG (1985) : resin-coated gravel placement
POPE, WILES & PIERCE (1987) : resin-coated gravel placement
SAUNDERS & McKENZIE (1979) : resin-coated gravel placement, performance review
SCHROEDER & TUCKER (1974) : resin-coated gravel placement
SINCLAIR & GRAHAM (1977, 1978) : resin-coated gravel, consolidated pack setting
SLATON (1988) : puddle packing, filling of explosive shotholes, stabilization
SOKOLL & ROGERS (1982) : short interval gravel packing technique, applications
SOLUM (1984) : liner vibration, hexagonal packing, compaction, stability
SOLUM & RAMEZANI (1987) : downhole pump protectors, flow-like fines trapping
SPARLIN (1967) : consolidated pack installation
SPARLIN (1971) : pressure packing, consolidated pack, unconsolidated sand
SPARLIN & BOND (1969) : concentrated gravel slurry, gravel rock placement
SPARLIN & COPELAND (1972) : concentrated gravel slurry, pressure packing
SPREUX, GEORGES & LESSI (1988) : gravel packing in horizontal wells
TAUSCH & CORLEY (1958) : consolidated pack installation
UNDERDOWN, DAY & SPARLIN (1980) : resin-coated gravel placement
WEIRICH, ZALESKI & TYLER (1988) : one-trip multiple-zone packing, low pressure
WILSON, NELSON & STADLEMAN (1983) : resin-coated slurry pack, gas delivery
YOUNG, COOK & DONALDSON (1969) : consolidated pack installation

7.2.8.10. Formation consolidation

The references on formation consolidation are classified along the lines of resin consolidation and other chemical stabilization methods. The quotations in the subject key division correspond mainly to aspects discussed in section 5.9. of the text.

7.2.8.10.1. Resin consolidation

The references on resin consolidation correspond mainly to aspects discussed in section 5.9.1. of the text.

BEZEMER & MEIJS (1966) : resin consolidation, grain cementation, resin migrate
BROOKS (1971) : dirty sand consolidation, resin injection
BROOKS, MUECKE, RICKEY & KERVER (1974) : external resin catalyzation, control
BURGER, GADELLE & MARRAST (1987) : chemical sand control process, development
CARDWELL (1947) : sand consolidation by resin injection
CARNES & MIXON (1971) : one-step gravel packing and consolidation
CHEN, ZHOU & LIU (1986) : sand control techniques, China, experience report
CONSTIEN & MAYER (1978) : water-carried resin-coated gravel, screen
COPELAND & CONSTIEN (1978) : sand consolidation by resin injection
DAVIES & HAGELAARS (1981) : sand consolidation, near wellbore placement
DAVIES, HAGELAARS & ROBERTS (1984) : sand consolidation, near-wellbore placing
GURLEY, COPELAND & HENDRICK (1977) : design, plan, execution, maximum offtake
HAMBY & RICHARDSON (1968) : new sand control process, dirty sands
HARRISBERGER, COULTER & GURLEY (1971) : sand consolidation mechanics, advances
HOWER & BROWN (1961) : sand consolidation, large-scale laboratory investigation

JENNINGS & MUIRHEAD (1968) : permeable consolidation, heat-curable resin
KNAPP & WELBOURN (1978) : permeability destruction, thief zone plugging
MABIE (1975) : sand consolidation, successful installation, review
MANTOOTH & TREADWAY (1965) : new sand consolidation method
MUECKE (1974) : plastic sand consolidation, influencing factors, determination
MURPHEY (1980) : gelled water epoxy sand consolidation system
MURPHEY, BILA & TOTTY (1974) : sand consolidation systems placed with water
PEDEN, RUSSELL & OYENEYIN (1984) : effective sand control, consolidation
PENBERTHY et al. (1978) : sand consolidation by resin injection
POSTON, TURNER, TOBOLA & BARGER (1986) : two-stage sand consolidation technique
RICHARDSON & HAMBY (1970) : epoxy resin overflush, silty sand consolidation
RIKE (1970) : sand consolidation, experience review, South Louisiana
SAUNDERS & McKENZIE (1979) : performance review, phenolic resin
SHAUGHNESSY, SALATHIEL & PENBERTHY (1977) : consolidation, new approach
SMITH (1968) : permeable consolidation of incompetent sands, heat-curable resin
SMITH (1969) : sand consolidation through production tubing
SMITH & POLK (1947) : sand consolidation by resin injection
SPARLIN & COPELAND (1972) : concentrated gravel slurry, pressure packing
STROHM et al. (1970) : sand consolidation, plastic-controlled injection, resin
SUMAN (1975) : gravel packing, sand control, review, status report
THOMAS & JOHNSON (1976) : resin consolidation, reperforation, maintenance
TREADWAY, BRANDT & PARKER (1966) : sand consolidation by resin injection
UNDERDOWN, DAY & SPARLIN (1980) : plastic pre-coated gravel
WEAVER & MORGAN (1988) : resin consolidation, workover replacement, gas storage
WRIGHTSMAN (1949) : sand consolidation method, technology, U.S. patent
YOUNG (1967) : consolidation, clayey sands, flow-rate boosting

7.2.8.10.2. Other chemical stabilization methods

The references on other chemical stabilization methods correspond mainly to aspects discussed in sections 5.9.2. to 5.9.4. of the text.

ASLESEN, SHORT & TERWILLIGER (1981) : silica consolidation, Solder Glass method
BANIK & SHARMA (1984) : hydroxy aluminum, sand stabilization, water sensitivity
BELVIN, COTTON & DUGAS (1973) : hydroxy aluminum consolidation, acid-stimulated
BJ-HUGHES (1986) : viscous gravel packing, slurry consolidation, sand stability
CARNES & MIXON (1971) : one-step gravel packing and consolidation
CHARLES (1976) : calcium hydroxide treatments, sand control effectivity.
COPPEL, JENNINGS & REED (1973) : hydroxy aluminum, well treatment, stabilizing
DAVIES, ZWOLLE & MEIJS (1983) : Silicalock, novel sand-control process, gas
FITZGERALD, STEPHENS & TERWILLIGER (1966) : warm air coking, consolidation
HASKIN (1976) : hydroxy aluminum treatment, review
HASKIN, REED & COPPEL (1975) : hydroxy aluminum well treatment
REED (1972) : clay stabilization, hydroxy aluminum solutions, sand consolidate
REED & COPPEL (1972) : hydroxy aluminum solutions, sand stabilization
SPARLIN & COPELAND (1972) : concentrated gravel slurry, pressure packing
STRICKLAND et al. (1975) : nickel-coating, consolidation, injection
TERWILLIGER, SMITH & GOODWIN (1964) : warm air coking, formation consolidation
TORREST (1975) : nickel-coating, consolidation, injection
VELEY (1969) : water sensitivity, hydroxy-aluminum treatment, clay reaction
VELEY & CASEY (1973) : in-place clay stabilization, pay sand entry cutting
ZAITOUN, KOHLER & GUERRINI (1989) : polyacrylamide, water control, plugging

7.2.8.11. Pre-pack liners and wire-wrapped screens

The references on pre-pack liners and wire-wrapped screens correspond mainly to aspects discussed in section 5.9.6. of the text.

BURGER (1987) : development of chemical process for sand control
CHEN, ZHOU & LIU (1986) : pre-pack liner, field experience, epoxy resin bond

COSTEIN & MAYER (1978) : water-carried resin-coated gravel, screen
LANMON & GILL (1966) : wireline completion technique for sand control
LIKWARTZ (1976) : triple-wrap screen, self-packing, sand bridging
LIVESEY & TOMA (1984) : pre-pack liner filter, steel wool/steel cloth inlay
PEDEN & YASSIN (1986) : perforation conditions, maximum sand-free production
SAUNDERS & McKENZIE (1979) : performance review, phenolic resin
SCHURTZ, BREINER & COMEAUX (1975) : through-tubing gravel pack, new technology
SPARLIN (1967) : new sand control technique for old sand problems
SPARLIN (1971) : pressure packing technique, unconsolidated sand control
SPARLIN & COPELAND (1972) : concentrated gravel slurry, pressure packing
SPURLOCK & DEMSKI (1972) : pre-pack liner, multiwire-wrapped screen, selective
STILES (1985), STILES, COLOMB & FARLEY (1986) : gravity-assisted gravel packing
SUMAN (1975) : gravel packing, sand control, review, wellbore stress, status
THOMAS & JOHNSON (1976) : dual completion, lower gravel pack, upper resin
TOMA, LIVESEY & HEIDRICK (1986) : pre-pack filter, steel wool/steel cloth inlay

7.2.8.12. Sand arching

The references on sand arching correspond mainly to aspects discussed in section 5.9.5. of the text.

AYERS (1965) : sand arching, plastic consolidation, treatment design, mechanics
BENTON (1984) : near wellbore stress behaviour, sand arch formation and failure
BRATLI & RISNES (1981) : sand arching, stability and failure
CLEARY (1978 c) : effect of fluid properties on sand arch stability
CLEARY, MELVAN & KOHLHAAS (1979) : stable arch concept
DURRETT, GUBIN, MURRAY & TIGHE (1977) : sand arching, sand control, solution
FAHRENTHOLD (1984) : elastic and plastic stress, spherical and cylindrical cave
FAHRENTHOLD & CHEATHAM (1986) : incipient yielding, solids production, casing
GETZLER et al. (1968) : model study, sand arching above buried structures
HALL & HARRISBERGER (1970) : sand arch stability, sand control, key, influence
LASAKI (1980) : sand arching, stability failure
MELVIN (1978) : effect of overburden stress on arch formation and stability
PEDEN & YASSIN (1986 b) : sand-free oil production, optimum completion design
PERKINS & WEINGARTEN (1988) : sand arching, stability and failure, cavities
RISNES, HORSRUD & BRATLI (1982) : sand arching, stability and failure
STEIN & HILCHIE (1972) : sand arching, friable sandstones, maximum production
TIPPIE (1973) : effect of producing rate on sand arch formation, unconsolidated
WOOD (1979) : effect of sand size on arch stability

7.2.8.13. Summary reviews and field experience reports

In contrast to the preceding divisions, the references on summary reviews and field experience reports do not correspond to specific sections of the text where items contained within the quoted literature are discussed and reworked, but rather represent an independent collection of case studies that illustrate many of the general evaluations and conclusions on gravel packing and sand control as presented in chapter 5 of the text in more detail.

ANAND & JONES (1977) : gas-well sand production evaluation, sand toleration
ATTARD, MATHES & MOWER (1988) : Valhall, Norway, gravel packing, fracturing
BRUIST (1981) : Beta-type gravel packing completions
CASHION & MARKLE (1940) : Gulf Coast fields, oil-well gravel packing
CIRIGLIANO & LEIBACH (1967) : Venezuela, gravel packing, experience report
COULTER, MARTINEZ & FISCHER (1987) : sand control, overview
DEAN (1939) : Wilmington oil field, drilling and production problems
GESEMANN & BADURA (1976) : Rühlermoor, Cretaceous, FRG, sand trouble, gravel
GURLEY (1988) : gravel packing, review, application
HENG (1986) : gravel packing completion, overview, Sarawak, Shell
JARDINS (1959) : thirty years sand control experience, Black Bayou field

JONES & THORP (1977) : Seria field, Brunei, performance analysis, statistics
LIKWARTZ (1976) : history of sand control, Teak field
MANTOOTH & WILLIAMS (1980) : offshore Louisiana, shallow sands, oil-base fluids
MONROE & PENBERTHY (1980) : high-volume water-supply wells, gravel packing
MOWER & SOVERALL (1980) : Trinidad, sand control challenge, applied technology
MURER (1981) : Matzen, Pirawarth, Hochleiten, Vienna Basin, Tertiary, control
ROGERS (1971) : sand control in oil and gas wells, summary, review
SAMHABER (1977) : sand control, FRG, Austria, oil wells, problems, experience
SCRIMGEOUR & SHEPHERD (1988) : Balmoral, productivity maintenance, sand control
SUMAN, ELLIS & SNYDER (1983) : sand control handbook, summary, review
VAZIRI (1986) : sand production mechanics, permeability changes, rock property
WADE (1967) : California, water-injection wells, treatment, completion methods
WILSON, STALLCUP & GLAZE (1978) : High Island field, high-rate wells, control
WINTERBURN (1947) : Wilmington oil field, sand control, unconsolidated

7.2.9. Petrophysical evaluation for hydraulic fracturing and gravel pack design

This subject classification key includes aspects of relative and absolute permeability determination, permeability deterioration, tight reservoir log interpretation, and clay mineral morphology and distribution.

The section on relative and absolute permeability determination contains references on overburden pressure, formation damage, temperature effects, capillary pressure, and pore water distribution. The division on permeability deterioration comprises references on formation pressure changes, clay swelling and migration, pore structure, and temperature gradient. The branch on tight reservoir log interpretation assembles references on water and gas saturation, clay mineral distribution, permeability quantification, and parameter estimation. The chapter on clay mineral morphology and distribution presents references on pore geometry, particle visualization, reservoir properties, effect of chemical treatments, and porosity nature.

7.2.9.1. Relative and absolute permeability determination

The references on relative and absolute permeability determination correspond mainly to aspects discussed in sections 3.4. and 3.5. of the text.

ALI et al. (1987) : overburden pressure, effect on relative permeability
AMAEFULE et al. (1987) : formation damage, understanding, improved evaluation
BAKER (1988) : three-phase relative permeability correlations
CHIERICI (1984) : drainage, imbibition, novel relations, relative permeability
CHOWDIAH (1986) : pore water distribution, stress laboratory measurements
CIVAN & DONALDSON (1987) : unsteady-state displacements, relative permeability
DELSHAD, MacALLISTER, POPE & ROUSE (1985) : multiphase dispersion, experiments
DONALDSON, ALAM & CIVAN (1987) : simulated reservoir condition, relative
FIROOZABADI & AZIZ (1986) : centrifuge data, relative permeability, measuring
GRAVIER et al. (1986) : gas-condensate, whole core, reservoir conditions
HALDORSEN & LAKE (1984) : relative vertical and lateral permeability, shale
HONARPOUR, KOEDERITZ & HARVEY (1986) : relative permeability of petroleum pays
McKEE et al. (1988) : stress-dependent permeability and porosity, various rocks
MILLER & RAMEY (1985) : temperature effect, oil/water relative permeability
MOHANTY & MILLER (1988) : unsteady relative permeability, mixed-wet reservoir
OYENEYIN (1985) : formation sand size prediction from well log correlations
PARMESWEAR & MAEREFAT (1986) : three-phase relative permeability data, compare
POLIKAR, ALI & PUTTAGUNTA (1988) : high-temperature relative permeability
PURCELL (1949) : capillary pressures, mercury application, permeability usage
QADEER, DEHGHANI, OGBE & OSTERMANN (1988) : correction, oil/water permeability
WARD & MORROW (1985) : capillary pressure, water saturation, relative permeable
WARD & THOMAS (1972) : overburden pressure, water saturation, gas permeability

WATSON & ERTEKIN (1988) : steep temperature gradient, relative permeability
WATSON et al. (1986) : displacement experiments, improved method, estimation
WEI, MORROW & BROWER (1986) : absolute permeability, tight sandstone, effects

7.2.9.2. Permeability deterioration

The references on permeability deterioration correspond mainly to aspects discussed in section 3.5. of the text.

AGGOUR & MALIK (1987) : cyclic formation pressure changes, permeability impact
CIVAN & KNAPP (1987) : clay swelling, fines migration, effect on permeability
GRAY & REX (1965) : clay dispersion and migration, formation damage, sandstone
NOMAN & ARCHER (1987) : pore structure, effect on non-Darcy gasflow, tight
PALLATT, WILSON & McHARDY (1984) : permeability, clay mineral morphology
VITTHAL, GUPTA & SHARMA (1987) : clay distribution, morphology, damage
WATSON (1988) : effect of steep temperature gradient on relative permeability
WOJTANOWICZ et al. (1987) : pore-blocking mechanisms, effects on pay damage

7.2.9.3. Tight reservoir log interpretation

The references on tight reservoir log interpretation correspond mainly to aspects discussed in section 3.4. of the text.

BUSCH et al. (1985) : log-derived evaluation, gas-bearing tight sandstones
FERTL & CAVANAUGH (1973) : Rotliegend, North Sea, porosity, permeability
KUCHUK (1987) : parameter estimation, low-permeability reservoirs, new methods
KUKAL (1981) : fluid-corrected porosity, shallow invasion profiles, tight gas
KUKAL (1983) : variable unflushed gas saturation, low-permeability gas sands
KUKAL (1984) : effective log analysis, systematic approach, tight gas sands
KUKAL & SIMONS (1985) : log analysis, permeability quantification, microdarcy
KUKAL et al. (1983) : tight gas sands, log interpretation, accuracy, problems
PIRSON (1983) : geologic well log analysis
ROSEPILER (1981) : water saturation, calculation, significance, shaly gas sand
RUHOVETS & FERTL (1982) : clay mineral distribution, well-log interpretation
TSAY & FANG (1986) : core-log study, Hartselle Sandstone, North Alabama, tight

7.2.9.4. Clay mineral morphology and distribution

The references on clay mineral morphology and distribution correspond mainly to aspects discussed in section 3.5.1. of the text.

CHOWDIAH & SOEDER (1988) : pore geometry comparison, high- and low-permeability
COULTER & PURVIS (1980) : Rotliegend, Dutch North Sea, fluid compatibility
CROCKER, DONALDSON & MARCHIN (1983) : reservoir rocks, related clays, compare
DULLIEN & DHAWAN (1974) : quantitative photomicrography, mercury porosimeter
FERTL & CHILINGAR (1988) : volume, type, distribution, mineralogy, logging
GAIDA, KESSEL, VOLZ & ZIMMERLE (1987) : EOR, reservoir sandstone parameters
GAIDA, RÜHL & ZIMMERLE (1973) : SEM, pore space, sandstones
GAO, SHEN & TU (1986) : pore space study, image processing
GIES (1987) : micropore system visualization, polarizing microscope, improving
HOLUB et al. (1974) : SEM reservoir monitoring, increasing oil production
MAZZULLO & MAZZULLO (1985) : Morrow Formation, New Mexico, clay mineralogy
NEASHAM (1977) : morphology, dispersed clay, fluid flow properties
PITTMAN (1979) : porosity, diagenesis, productive capability
PITTMAN & THOMAS (1979) : SEM, reservoir rock study, pore morphology
RUZYLA (1984) : pore space characterization, quantitative image analysis
SCHMIDT & McDONALD (1979) : secondary porosity in sandstones, texture, role
SCHRANK & HUNT (1981) : SEM, improved reservoir evaluation, pore systems
STALDER (1973) : crystallographic habit, aggregate structure
THOMAS (1978) : diagenetic sequence, low-permeability, argillaceous sandstone
THOMAS, CROWE & SIMPSON (1976) : effect of chemical treatments upon pay clays
TIMUR, HEMPKINS & WEINBRANDT (1971) : SEM, pore systems

7.3. Reference list

AADNØY, B.S. (1986) : Mechanics of wellhole stability in deviated boreholes. 24th Nat. Congr. Assoc. Petrol. Engin. Mexico, Acapulco/Mexico, 07. - 11.05. 1986.

AADNØY, B.S. (1987 a) : Modelling of the stability of highly inclined boreholes in anisotropic rock formations. SPE Paper, **16526**. SPE Drill. Engin., 3 (1988) : 259 - 268.

AADNØY, B.S. (1987 b) : Continuum mechanics analysis of the stability of inclined boreholes in anisotropic rock formations. Ph.D. Thesis, Univ. of Trondheim/Norway.

AADNØY, B.S. (1988) : Inversion technique to determine the in-situ stress field from fracturing data. SPE Paper, **18023**.

AADNØY, B.S. & CHENEVERT, M.E. (1987) : Stability of highly inclined boreholes. SPE/IADC Paper, **16052**. SPE Drill. Engin., **2** : 364 - 374.

AADNØY, B.S. & LARSEN, K. (1987) : Method for fracture gradient prediction for vertical and inclined boreholes. SPE Paper, **16695**.

ABAD, B.P. & HENSLEY, C.A. (1984) : The effect of stratification, fractures, and dip on steam sweep efficiency and heat requirements of steamflood operations. SPE Paper, **12748**.

ABBASZADEH-DEHGHANI, M. & BRIGHAM, W.E. (1982) : Analysis of unit mobility ratio well-to-well tracer flow to determine reservoir heterogeneity. Dept. Energy Rept., **DE-AC03-815F11164**.

ABBASZADEH-DEHGHANI, M. & BRIGHAM, W.E. (1984) : Analysis of well-to-well tracer flow to determine reservoir layering. J. Petrol. Technol., **36** : 1753 - 1762. SPE Paper, **10760** (1982).

ABBOTT, C.G. (1982) : The prospects of regulatory reform in natural gas. SPE Paper, **11264**.

ABDASSAH, D. & ERSHAGHI, I. (1986) : Triple porosity systems for representing naturally fractured reservoirs. SPE Form. Eval., **1** : 113 - 127. SPE Paper, **13409** (1984).

ABE, H., MURA, T. & KEER, L.M. (1976) : Growth rate of a penny-shaped crack in hydraulic fracturing of rocks. J. Geophys. Res., **81** : 5335 - 5340.

ABEL, J.C. (1982) : Application of nitrogen fracturing in the Ohio Shale. SPE Paper, **10378**.

ABEL, J.C., FREEMAN, E.R., KIM, C.M. & HEINRICH, C. (1983) : A stimulation technique using only nitrogen. J. Petrol. Technol., **35** : 2165 - 2174. SPE Paper, **10129** (1981).

ABOBISE, E.O. & TIAB, D. (1982) : Determining fracture orientation and formation permeability from pulse testing. SPE Paper, **11027**.

ABOU-SAYED, A.S. (1982) : Laboratory evaluation of in-situ stress contrast in deeply buried sediments. SPE Paper, **11069**. Terra Tek Rept.

ABOU-SAYED, A.S., AHMED, U. & JONES, A. (1981) : Systematic approach to massive hydraulic fracturing treatment design. SPE Paper, **9877**.

ABOU-SAYED, A.S., BRECHTEL, C.E. & CLIFTON, R.J. (1977) : In situ stress determination by hydrofracturing. A fracture mechanics approach. Terra Tek Rept., **TR 77-60**. J. Geophys. Res., **83** (1978) : 2851 - 2862.

ABOU-SAYED, A.S., CLIFTON, R.J., DOUGHERTY, R.L. & MORALES, R.H. (1984) : Evaluation of the influence of in-situ reservoir conditions on the geometry of hydraulic fractures using a 3-D simulator : Part 2 - Case studies. SPE/DOE/GRI Paper, **12878**.

ABOU-SAYED, A.S., SINHA, K.P. & CLIFTON, R.J. (1984) : Evaluation of the influence of in-situ reservoir conditions on the geometry of hydraulic fractures using a 3-D simulator : Part 1 - Technical approach. SPE/DOE/GRI Paper, **12877**.

ABRAMS, A. & VINEGAR, H.J. (1985) : Impairment mechanism in Vicksburg tight gas sands. SPE/DOE Paper, **13883**.

ABRAMS, A., LYBARGER, J.H., RICHARDSON, E.A. & NEASHAM, J.W. (1978) : The development and application of a high pH acid stimulation for a deep Mississippi gas well. SPE Paper, **7565**.

ABRAMS, A., SCHEUERMAN, R.F., TEMPLETON, C.C. & RICHARDSON, E.A. (1983) : Higher-pH acid stimulation systems. J. Petrol. Technol., **35** : 2175 - 2184. SPE

Paper, **7892** (1979).
ACHARYA, A. (1986) : Particle transport in viscous and viscoelastic fracturing fluids. SPE Product. Engin., **1** : 104 - 110. SPE Paper, **13179** (1984).
ACHARYA, A. (1987) : Viscoelasticity of crosslinked fracturing fluids and proppant transport. SPE Paper, **16221** (1987), **15937** (1986).
ACHARYA, A. & DEYSARKAR, A.K. (1987) : Rheology of fracturing fluids at low-shear conditions. SPE Paper, **16917**.
ACHARYA, A. & KIM, C.M. (1987) : Hydraulic fracture treatment design simulation for the Rotliegendes formation. SPE/DOE Paper, **16414**. J. Petrol. Technol., **40** (1988) : 139 - 142.
ACHENBACH, E. (1974) : The effects of surface roughness and tunnel blockage on the flow past spheres. J. Fluid Mechan., **65** : 113 - 125.
ADAMS, M.A. et al. (1982) : Geologic overview, coal resources, and potential methane recovery from coalbeds, northern Appalachian coal basin, Pennsylvania, Ohio, Maryland, West Virginia, and Kentucky. Prepared for U.S. Dept. Energy, Morgantown Energy Technology Center, by TRW Coalbed Methane Program, 179 pp.; McLean/Virginia.
ADAMS, M.C., MOORE, J.N. & FORSTER, C. (1985) : Fluid flow in volcanic terrains - hydrogeochemistry of the Meager Mountain thermal system. Geotherm. Resourc. Counc. Trans., **9** : 377 - 382.
ADVANI, S.H. (1980) : Finite element model simulations associated with hydraulic fracturing. SPE/DOE Paper, **8941**. Soc. Petrol. Engin. J., **22** (1982) : 209 - 218.
ADVANI, S.H., KHATTAB, H. & LEE, J.K. (1985) : Hydraulic fracture geometry modeling, prediction, and comparisons. SPE/DOE Paper, **13863**.
ADVANI, S.H., KHATTAB, H., MENDELSOHN, D.A. & LEE, J.K. (1985) : Hydraulic fracture geometry prediction and associated early time production modelling. SPE Paper, **13863**.
ADVANI, S.H., LEE, J.K., HAMID, M.S., GURDOGAN, O. & KHATTAB, H. (1982) : Fluid flow, structural, and fracture mechanics modeling associated with hydraulic stimulation operations. SPE/DOE Paper, **10846**.
ADVANI, S.H., LEE, J.K., KHATTAB, H.A. & GURDOGAN, O. (1986) : Fluid flow and structural response modeling associated with the mechanics of hydraulic fracturing. SPE Form. Eval., **1** : 309 - 318. SPE Paper, **10846** (1982).
AGARWAL, R.G. (1979) : Real gas pseudo-time - a new function for pressure buildup analysis of MHF tight gas wells. SPE Paper, **8279**.
AGARWAL, R.G. (1980) : A new method to account for producing time effects when drawdown curves are used to analyze pressure buildup and other test data. SPE Paper, **9289**.
AGARWAL, R.G., CARTER, R.D. & POLLOCK, C.B. (1979) : Evaluation and performance prediction of low-permeability gas wells stimulated by massive hydraulic fracturing. J. Petrol. Technol., **31** : 362 - 372, 651 - 654. SPE Papers, **6838** (1977), **8145** (1980).
AGER-HANSSEN, H. (1982) : The future of Norwegian gas. EUR Paper, **342**.
AGGOUR, M.A. & MALIK, S.A. (1987) : Effect of cyclic formation pressure changes on permeability. SPE Paper, **15728**.
AGNEW, B.G. (1965) : Evaluation of fracture treatments with temperature surveys. SPE Paper, **1287**. J. Petrol. Technol., **18** (1966) : 892 - 898.
AGUILERA, R. (1974) : Analysis of naturally fractured reservoirs from sonic and resistivity logs. J. Petrol. Technol., **26** : 1233 - 1238. SPE Paper, **4398** (1973).
AGUILERA, R.N. (1975) : Graphical solution of imbibition equations used to predict oil recovery by water influx in naturally fractured reservoirs. J. Petrol. Technol., **27** : 1526 - 1528. SPE Paper, **5418**.
AGUILERA, R.N. (1976) : Analysis of naturally fractured reservoirs from conventional well logs. J. Petrol. Technol., **28** : 764 - 772. SPE Paper, **5342** (1975).
AGUILERA, R. (1980) : Naturally fractured reservoirs. 703 pp.; Penn Well, Tulsa/Oklahoma.
AGUILERA, R. (1982) : Relative permeability concepts for predicting the performance of naturally fractured reservoirs. J. Canad. Petrol. Technol., **1982**/ 9 - 10 : 41 - 48.

AGUILERA, R. (1983) : Exploring for naturally fractured reservoirs. Trans. 24th Soc. Prof. Well Log Analysts Annual Logging Symp., Paper **C**; 27. - 30.06. 1983.
AGUILERA, R. (1985) : Well test analysis of naturally fractured reservoirs. SPE Paper, **13663**. SPE Form. Eval., **2** (1987) : 239 - 252.
AGUILERA, R. (1987) : Multiple rate analysis for pressure buildup tests in reservoirs with tectonic, regional and contractional natural fractures. SPE Form. Eval., **2** : 253 - 260. SPE Paper, **13827** (1984).
AGUILERA, R. & ACEVEDO, L. (1982) : FCL - a computerized well log interpretation process for evaluation of naturally fractured reservoirs. J. Canad. Petrol. Technol., **1982**/1-2 : 31 - 37.
AGUILERA, R., HOUSTON, M., PODIO, A.L. & SONG, S.J. (1988) : Well test analysis of pumping wells in multiphase naturally fractured reservoirs. SPE Paper, **17545**.
AGUILERA, R. & POOLLEN, H.K.v. (1978 a) : Geologic aspects of naturally fractured reservoirs explained. Oil Gas J., 18.12.**1978** : 47 - 51.
AGUILERA, R. & POOLLEN, H.K.v. (1978 b) : How to evaluate naturally fractured reservoirs from various well logs. Oil Gas J., 25.12.**1978** : 202 - 208.
AHMED, U. (1984) : A practical hydraulic fracturing model simulating necessary fracture geometry, fluid flow and leakoff, and proppant transport. SPE/DOE/GRI Paper, **12880**.
AHMED, U. (1987) : Fracture-height predictions and post-treatment measurements. In : ECONOMIDES, M.J. & NOLTE, K.G. (Editors), Reservoir stimulation. Schlumberger Educational Services, 10-1 - 10.13. Dowell Schlumberger Document, **TSL-2612**. Schlumberger Document, **SMP-7018**. Houston/Texas.
AHMED, U. (1988) : Fracture height prediction. J. Petrol. Technol., **40** : 813 - 815. SPE Paper, **18338**.
AHMED, U., ABOU-SAYED, A.S. & JONES, A.H. (1979) : Experimental evaluation of fracturing fluid interaction with tight reservoir rocks and propped fractures. SPE Paper, **7922**.
AHMED, U., KELKAR, S. & SCHATZ, J. (1983) : Mini-frac : an aid to formation insitu stress and permeability measurements. Proc. 24th Symp. Rock Mechan., 319 - 326.
AHMED, U. & KELLER, S. (1984) : Perforation placement technique : a modified hydraulic fracturing technique. SPE Paper, **12841**.
AHMED, U., KUCHUK, F. & AYESTERAN, L. (1985) : Short term transient-rate and pressure-buildup analysis of low-permeability wells. SPE Paper, **13870**. SPE Form. Eval., **2** (1987) : 611 - 617.
AHMED, U., NEWBERRY, B.M. & CANNON, D.E. (1985) : Hydraulic fracture treatment design of wells with multiple zones. SPE/DOE Paper, **13857**.
AHMED, U., SCHATZ, J.F., ABOU-SAYED, A.S. & JONES, A.H. (1982) : A step by step approach to hydraulic fracture treatment design, implementation, and analysis for tight gas sands. SPE/DOE Paper, **10829**.
AHMED, U., SCHATZ, J.F., GREENFIELD, H. & JONES, A.H. (1981) : Optimized stimulation of tight sands in the Pinedale Field, Wyoming. Proc. 16th Intern. Soc. Energy Convers. Engin. Conf., ASME, **2** : 1249 - 1254; New York.
AHMED, U., SCHATZ, J.F., HOLLAND, M.T., JONES, A.H. & GREENFIELD, H. (1982) : State-of-the-art hydraulic fracture stimulation treatment for a Western tight sand reservoir. SPE Paper, **11184**.
AHMED, U., STRAWN, J., SCHMIDT, R., O`SHEA, P. & VEGHTE, R. (1983) : Stimulating tight sands in the presence of weak stress barriers by modified hydraulic fracturing techniques : a status report. SPE Paper, **11836**.
AHMED, U., THOMPSON, T.W., KELKAR, S.M., VEGHTE, R. & HATHAWAY, S. (1984) : Perforation placement optimization : a modified hydraulic fracturing technique. SPE/DOE/GRI Paper, **12841**.
AHMED, U., WILSON, M.D. & STRAWN, J. (1983) : Effect of stress distribution on hydraulic fracture geometry : a laboratory simulation study on one-meter cubic blocks. SPE/DOE Paper, **11637**.
AINLEY, B.R. & CHARLES, J.G. (1982) : Fracturing using a stabilized foam pad. SPE/DOE Paper, **10825**.
AJAM, S.O. & RAHAL, V.E. (1986) : Applications of thermal multigate decay pulsed neutron logs in unusual downhole logging environments. 6th Offshore

Southeast Asia Conference and Exhibition, Singapore, 26.-31-01.1986, pp. 437 - 446.
AKI, K., FEHLER, M. et al. (1982) : Interpretation of seismic data from hydraulic fracturing experiments at the Fenton Hill, New Mexico, hot dry rock geothermal site. J. Geophys. Res., **87** : 936 - 944.
ALBERTSEN, M. (1985) : Stimulation geringpermeabler tiefer Gaslagerstätten (Phase III). Statusbericht zum Forschungsvorhaben, **03E-3068-B** : 483 - 494. Programm Energieforschung und Energietechnologien, Statusreport 1985 Geotechnik und Lagerstätten (KFA Jülich).
AL-BLEHED, M.A. (1987) : An engineering-economic approach to modeling free world total energy demand and oil´s market share. Ph.D. Thesis, Dept. of Petrol. Engin., Univ. of Southern California.
ALBRIGHT, J.N. & PEARSON, C.F. (1980) : Location of hydraulic fractures using microseismic techniques. SPE Paper, **9509**.
ALBRIGHT, J.N. & PEARSON, C.F. (1982) : Acoustic emissions as a tool for hydraulic fracture location : experience at the Fenton Hill hot dry rock site. Soc. Petrol. Engin. J., **22** : 523 - 530. SPE Paper, **9509** (1980), **10898** (1982).
AL-CHALABI, F.J. (1987) : The prospects for oil producers from an OPEC point of view. Petrol. Rev., **1987**/1 : 15 - 17.
ALDERMAN, E.N. & WENDORFF, C.L. (1969) : Prop-packed fractures - a reality on which productivity increase can be predicted. SPE Paper, **2452**.
ALEGRE, L. & DOUGHERTY, E.L. (1988) : Applicability of expert systems to diagnose formation damage problems : a progress report. SPE Paper, **17460**.
AL-FARISS, T.F. (1988) : Future prospects of Saudi natural gas. SPE Paper, **17749**.
ALGAR, P. (1987 a) : North Sea review : pessimism is misplaced. Petroleum Management, **9**/7 : 47 - 52.
ALGAR, P. (1987 b) : OPEC politics and economics. Petroleum Management, **9**/6 : 58 - 62.
ALGER, R.P., LOCKE, S., NAGEL, W.A. & SHERMAN, H. (1971) : The dual spacing neutron log - CNL. SPE Paper, **3565**. J. Petrol. Technol., **24** (1972) : 1073 - 1083.
ALI, H.S., AL-MARHOUN, M.A., ABU-KHAMSIN, S.A. & CELIK, M.S. (1987) : The effect of overburden pressure on relative permeability. SPE Paper, **15730**.
ALI, S.M.F. (1982) : Steam injection theories - a unified approach. SPE Paper, **10746**.
AL-KHATIB, A.M., KING, A.R. & WILSON, M.S. (1984) : Hydraulic fracturing design and evaluation : a case history. SPE Paper, **12482**.
ALLAM, A.M., CRICHLOW, H.B. & SOLIMAN, M.Y. (1981) : Numerical simulation of heterogeneous fractured gas reservoir systems with turbulence and closure stress effects. SPE Paper, **10119**.
ALLEN, D.R. (1969) : Collar and radioactive bullet logging for subsidence monitoring. Soc. Prof. Well Log Analysts Annual Logging Meeting Trans., Houston/Texas, 26.05.1969.
ALLEN, D.R. (1973) : Oilfield subsidence : susceptibility and monitoring techniques. Amer. Assoc. Petrol. Geol., Annual Meeting, Anaheim/California, **1973**/5.
ALLEN, D.R. (1984) : Developments in precision casing joint and radioactive measurements for compaction monitoring. J. Petrol. Technol., **36** : 805 - 810. SPE Paper, **9933** (1981).
ALLEN, E. & BOGER, D.V. (1988) : The influence of rheological properties on mobility control in polymer-augmented waterflooding. SPE Paper, **18097**.
ALLEN, F.L., RILEY, S.M. & STRASSNER, J.E. (1984) : Initial study of temperature and pressure effects on formation damage by completion fluids. SPE Paper, **12488**.
ALLEN, T.O. & ROBERTS, A.P. (1982) : Production operations. **1** : 280 pp.; **2** : 250 pp. Oil and Gas Consultants International, Tulsa/Oklahoma.
ALLEN, T.S. & BAZA, J.R. (1980) : The role of step-rate injectivity testing in liquid dominated, highly fractured geothermal reservoirs. SPE Paper, **9275**.
ALLISON, K. (1986) : BP`s Villages Southern North Sea project - development of Cleeton and Ravenspurn. Hydrocarbons `86, Conference and Exhibition, Great

Yarmouth/England, 01.10.1986.
AL-MARHOUN, M.A. (1985) : Interference testing : a new analysis approach. SPE Paper, **13732**.
ALMOND, S.W. (1982) : Factors affecting gelling agent residue under low temperature conditions. SPE Paper, **10658**.
ALMOND, S.W. & BLAND, W.E. (1984) : The effect of break mechanism on gelling agent residue and flow impairment in 20/40 mesh sand. SPE Paper, **12485**. CIM Paper, **84-35-30**.
ALMOND, S.W., BRADY, J.L. & UNDERDOWN, D.R. (1988) : Return fluid analysis from the Sadlerochit Formation, Prudhoe Bay, Alaska : field study - part I. SPE Paper, **18223**.
ALMOND, S.W. & GARVIN, T.R. (1985) : High-efficiency fracturing fluids for low-temperature reservoirs. SPE Paper, **13617**.
ALPAY, O.R. (1969) : Application of aerial photographic interpretation to the study of reservoir natural fracture systems. SPE Paper, **2567**.
AL-SAIF, A.S., COCHRANE, J.E., EDMUNDSON, H.N. & YOUNGBLOOD, W.E. (1975) : Analysis of pulsed-neutron-decay time logs in acidized carbonate formations. Soc. Petrol. Engin. J., **15** : 453 - 466.
AL-SHAIEB, Z. & WALKER, P. (1986) : Evolution of secondary porosity in Pennsylvanian Morrow Sandstones, Anadarko Basin, Oklahoma. In : SPENCER, C.W. & MAST, R.F. (Editors), Geology of tight gas reservoirs. Amer. Assoc. Petrol. Geol., Stud. Geol., **24** : 45 - 68.
AL-TAMIMI, Y.K. & EL-MZEIN, N. (1987) : Stimulating tight, multilayered carbonates in the Upper Zakum. SPE Paper, **15771**.
AMAEFULE, J.O. & MASUO, S.T. (1984) : The use of capillary-pressure data for rapid evaluation of formation damage or stimulation. SPE Paper, **12475**. SPE Prod. Engin., **1** (1986) : 131 - 142.
AMAEFULE, J.O., PADILLA, P.C., McCAFFERY, F.G. & TEAL, S.L. (1984) : Steam condensate : formation damage and chemical treatments for injectivity improvement. SPE Paper, **12499**.
AMAEFULE, J.O., AJUFO, A., PETERSON, E. & DURST, K. (1987) : Understanding formation damage processes : an essential ingredient for improved measurement and interpretation of relative permeability data. SPE Paper, **16232**.
AMERI, S., AMINIAN, K., MILLER, J.A., DORICICH, D. & YOST, A.B. II (1985) : A systematic approach for economic development of the Devonian Shale gas resources. SPE Paper, **14504**.
AMERI, S. & RIEKE, H.H. III (1981) : Use of well logs to provide formation damage design values for hydraulic fracture treatments. SPE Paper, **10367**.
AMERICAN PETROLEUM INSTITUTE (1985) : API RP 56, recommended practices for testing sand used in hydraulic fracturing operations. 1st ed. 1983. Tentative third draft of recommended practices for evaluating proppant conductivity. Dallas/Texas.
AMINIAN, K., AMERI, S. & BRIGHT, D. (1988) : The evaluation and optimization of hydraulic fracturing treatments for the Big Injun Formation in West Virginia. SPE Paper, **18540**.
AMIRIJAFARI, B., SHARER, J.C. & RIGBY, F.A. (1982) : Flow characteristics of geopressured reservoirs. SPE/DOE Paper, **10834**.
AMMER, J.R., SAWYER, W.K. & DROPHIN, M.J. (1984) : Practical methods for detecting production mechanisms in tight gas reservoirs. SPE/DOE Paper, **12864**.
AMOCO (1988) : Continuing development of Leman and Indefatigable fields. Hydrocarbons `88 Conference, Great Yarmouth/England, 06.10.1988.
ANAND, J. & JONES, B.F. (1977) : Evaluation and control of sand production in gas wells. SPE Paper, **6531**.
ANCELL, K.L., LAMBERT, S. & JOHNSON, F.S. (1980) : Analysis of the coalbed degasification process at a seventeen well pattern in the Warrior Basin of Alabama. SPE/DOE Paper, **8971**.
ANDERSEN, S.A., HANSEN, S.A. & FJELDGAARD, K. (1988) : Horizontal drilling and completion : Denmark. SPE Paper, **18349**.
ANDERSON, A. (1988) : World oil trends 1987 - 88. 128 pp.; Arthur Anderson & Co., Surrey/England.
ANDERSON, B.R. (1982) : Use of the repeat formation tester and dipmeter in defining reservoir heterogeneity and anisotropy, Anschutz Ranch East field.

SPE Paper, **11091**.
ANDERSON, G.D. (1981) : Effects of friction on hydraulic fracture growth near unbonded interfaces in rocks. Soc. Petrol. Engin. J., **21** : 21 - 30. SPE Paper, **8347** (1979).
ANDERSON, G.W. & HUTCHINSON, S. (1978) : How to efficiently wash sand from deviated wellbores. World Oil, **1978**/12.
ANDERSON, H.A. & LATOS, J. (1987) : Logging and completion of highly deviated wells using a coiled tubing system. SPE Paper, **17038**.
ANDERSON, J.A., PEARSON, C.M., ABOU-SAYED, A.S. & MYERS, G.D. (1986) : Determination of fracture height by spectral gamma log analysis. SPE Paper, **15439**.
ANDERSON, L.L., PETERSON, K.P. & PARISI, W.A. (1984) : Enhanced production from a slightly geopressured water-drive gas condensate field. SPE/DOE/GRI Paper, **12866**.
ANDERSON, M.S. & FREDRICKSON, S.E. (1987) : Dynamic etching tests aid fracture acidizing treatment design. SPE/DOE Paper, **16452**.
ANDERSON, R.A., INGRAM, D.S. & ZANIER, A.M. (1973) : Determining fracture pressure gradients from well logs. J. Petrol. Technol., **25** : 1259 - 1268. SPE Paper, **4135** (1972).
ANDERSON, R.W. & PHILLIPS, A.M. (1986) : Practical application of economic well-performance criteria to the optimization of fracturing treatment design. SPE Paper, **14982**. J. Petrol. Technol., **40** (1988) : 223 - 228.
ANDERSON, R.W. & PHILLIPS, A.M. (1987) : Use of single-well simulators and economic performance criteria to optimize fracturing treatment design. SPE Paper, **16236**.
ANNIGERI, B.S. & CLEARY, M.P. (1984) : Surface integral finite element hybrid (SIFEH) method for fracture mechanics. Intern. J. Numer. Methods Engin., **20** : 869 - 885.
ANTHEUNIS, D., VRIEZEN, P.B., SCHIPPER, B.A. & VLIS, A.C.v.d. (1976) : Perforation collapse : failure of perforated friable sandstones. SPE Paper, **5750**.
ANTHONY, J.L. & HILL, A.D. (1984) : Two-pulse tracer logging : a new method for radioactive tracer injection logging. SPE Paper, **13396**. SPE Product. Engin., **1** (1986) : 117 - 124.
API (1961) : History of petroleum engineering. American Petroleum Institute, Dallas/Texas.
ARAKTINGI, U. & ORR, F.M. (1988) : Viscous fingering in heterogeneous porous media. SPE Paper, **18095**.
ARASTOPOUR, H., CHEN, S.T. & MARIRI, H. (1987) : Analysis of flow of gas and water in a fractured & non-fractured low-permeability reservoir under production. SPE Paper, **16948**.
ARDITTY, P., GRAS, C. & STARON, P. (1987) : Permeability, fracture detection, and reservoir analysis from EVA data in open and cased hole. Trans. 28th Soc. Prof. Well Log Analysts Annual Logging Symp., London/England, 29.06. - 02.07.1987.
ARENS, K. & BÖHNEL, F. (1985) : Gasproduktion bei hoher Verwässerung. Erdoel-Erdgas, **101** : 200 - 205.
ARMENDARIZ, V.M. & BARRINGER, D.K. (1981) : Use of complex weak acid in Canyon Sand stimulation. SPE Paper, **9706**.
ARMESSEN, P., JOURDAN, A.P. & MARIOTTI, C. (1988) : Horizontal drilling has negative and positive factors. Oil Gas J., 23.05.**1988** : 37 - 40.
ARON, J., MURRAY, J. & SEEMAN, B. (1978) : Formation compressional and shear interval-transit-time logging by means of long spacings and digital techniques. SPE Paper, **7446**.
ARYA, A., HEWETT, T.A., LARSON, R. & LAKE, L.W. (1985) : Dispersion and reservoir heterogeneity. SPE Paper, **14364**.
ASELTINE, R.J. & ELLENBERGER, C.W. (1977) : Selective acid stimulation to improve vertical efficiency in injection wells - a case history. J. Petrol. Technol., **29** : 25 - 29. SPE Paper, **5869** (1976).
ASLESEN, K.S., SHORT, C.J. & TERWILLIGER, P.L. (1981) : A new method for sand control and well stimulation in unconsolidated dirty sands. SPE Paper, **10172**.
ATKINSON, C.H. & LEKAS, M.A. (1963) : Atomic-age fracturing may soon open up Stubborn reservoir. Oil Gas J., 02.12.**1963** : 154 - 156.

ATKINSON, P. et al. (1978) : Analysis of reservoir pressure and decline curves in Serrazzano Zone, Larderello geothermal field. Geothermics, **7** : 133 - 144.
ATKINSON, P.G., CELATI, R., CORSI, R. & KUCUK, F. (1980) : Behavior of the Bagnore steam/CO2 geothermal reservoir, Italy. Soc. Petrol. Engin. J., **20** : 228 - 238. SPE Paper, **7132** (1978).
ATTARD, M., MATHES, R.A. & MOWER, L.N. (1988) : Gravel packing in an abnormally pressured chalk reservoir : a Valhall field case study. SPE Paper, **18226**.
ATTEBERRY, R.D., TUCKER, R.L. & RITZ, J.W. (1979) : Application of sintered bauxite proppants to stimulation of low permeability South Texas gas reservoirs. SPE Paper, **7924**.
ATTIGA, A.A. (1987) : The Arabs and the oil crisis 1973 - 1986. 473 pp.; Organization of Arab Petroleum Exporting Countries, Safat/Kuwait.
AULT, J.W., McCUNE, C.C. & DUNLAP, R.G. (1977) : Discussion of a study of the field application of sandstone acidizing. J. Petrol. Technol., **29** : 122 - 123. SPE Paper, **6507**.
AUMAN, J.B. (1986) : A laboratory evaluation of core preservation materials. SPE Paper, **15381**.
AUSTIN, C.E., ROSE, R.E. & SCHUH, F.J. (1988) : Simultaneous multiple entry hydraulic fracture treatments of horizontally drilled wells. SPE Paper, **18263**.
AYERS, W.M. (1965) : Mechanics of sand consolidating plastic treatments. SPE Paper, **1135**.
AYOUB, J.A., BOURDET, D.P. & CHAUVEL, Y.L. (1988) : Impulse testing. SPE Form. Eval., **3** : 534 - 546.
BABCOCK, E.A. (1975) : Fracture phenomena in the waterways and the McMurray Formations, Athabasca oil sands region, Northeast Alberta. Bull. Canad. Petrol. Geol., **23** : 810 - 826.
BABCOCK, E.A. (1978) : Measurement of subsurface fractures from dipmeter logs. Bull. Amer. Assoc. Petrol. Geol., **62** : 1111 - 1126.
BABCOCK, R.E., PROKOP, C.L. & KEHLE, R.O. (1967) : Distribution of propping agents in vertical fractures. Producers Monthly, **1967**/11 : 11 - 18. API Drill. Product. Pract., **1967** : 207 - 217.
BABU, D.K. & ODEH, A.S. (1988) : Productivity of a horizontal well. SPE Paper, **18298**.
BACKFISCH, S. (1984) : Die Petrographie der Rotliegendsandsteine Südwestdeutschlands: ein Beitrag zur Diagenese vulkanoklastreicher Sandsteine. Ph. D. Thesis, 235 pp., University of Heidelberg/Germany FRG (unpubl.).
BACKFISCH, S. & HELING, D. (1985) : Diagenese der Oberrotliegendsandsteine Südwestdeutschlands. Erdöl und Kohle - Erdgas - Petrochemie, **38** : 109 - 110.
BACON, J.R. & KEMPTHORNE, R.H. (1984) : Waterflood oil recovery in fractured reservoirs with directionally drilled wells. Soc. Petrol. Engin. J., **24** : 375 - 381. SPE Paper, **11024** (1982).
BAILEY, D.E. & WICKHAM, J.F. (1984) : Sand fracturing vs. fracture acidizing. SPE Paper, **12898**. Western Oil Reporter, **1984**/2 : p. 51 ff.
BAKER, L.E. (1988) : Three-phase relative permeability correlations. SPE/DOE Paper, **17369**.
BAKER, O. (1981) : Gas resources in low permeability formations and the effect of price and technology. SPE/DOE Paper, **9897, 10359**. Proc. ASME 16th Intersoc. Energy Convers. Conf., Atlanta/Georgia.
BAKHTAR, K., BARTON, N.R., RAKOP, K. & JONES, A.H. (1985) : Modeling fracture permeability around a well during depletion. SPE Paper, **13671**.
BALDWIN, B.A. (1986) : Water imbibition and characterization of North Sea chalk reservoir surfaces. SPE Paper, **14108**. SPE Form. Eval., **3** (1988) : 125 - 130.
BALE, G.E. (1984) : Matrix acidizing in Saudi Arabia by using buoyant ball sealers. J. Petrol. Technol., **36** : 1748 - 1752. SPE Paper, **11500** (1983).
BALEN, R.M., MENG, H.Z. & ECONOMIDES, M.J. (1988) : Applications of the net present value (NPV) in the optimization of hydraulic fractures. SPE Paper, **18541**.
BALL, J.T. & PITTS, M.J. (1983) : Simulation of reservoir permeability heterogeneities with laboratory corefloods. SPE Paper, **11790**.
BALL, M.S., CHACE, D.M. & FERTL, W.H. (1987) : The Well Data System (WDS) : an advanced formation evaluation concept in a microcomputer environment. SPE Paper, **17034**.

BALLARD, J.L. (1977) : Successful application of foam frac to shallow gas sands. SPE Paper, **6377**.
BALLEGOOYEN, J.v., GIAP, T.K. & SENG, T.K. (1983) : Experience in gravel packing long, perforated intervals in deviated holes. J. Petrol. Technol., **35** : 2079 - 2086. SPE Paper, **10456** (1982).
BAME, D. & FEHLER, M. (1986) : Observations of long periodic earthquakes accompanying hydraulic fracturing. Geophys. Res. Letters, **13** : 149 - 152.
BAND, G.C. (1987) : U.K. North Sea production prospects to the year 2000. J. Petrol. Technol., **39** : 64 - 70. SPE Paper, **13009** (1984).
BANDYOPADHYAY, P. (1981) : A practical short-term interference testing method to determine reservoir heterogeneities. SPE Paper, **9396**.
BANIK, D.K. & SHARMA, H.K. (1984) : Clay stabilization and sand consolidation : use of hydroxy-aluminum. Petrol. Asia J., **7** : 101 - 109.
BARANET, S.E. & AINLEY, B.R. (1985) : Delayed crosslinking improves performance of fracturing fluids in deep formations. CIM Paper, **85-36-39**.
BARBA, R.E. (1986) : Improving return on hydraulic fracture treatment investment with wireline inputs. Proc. Southwestern Petrol. Short Course, Lubbock/Texas.
BARBA, R.E. (1988) : Optimizing hydraulic fracture length in the Spraberry Trend of West Texas. SPE Paper, **17280**.
BARBER, P.L. & THEMIG, D.J. (1985) : The use of density control foam fracturing for selective proppant placement in the Bartlesville Formation. SPE Paper, **13797**.
BARBY, B.G. & BARBEE, W.C. (1987) : Ultra-high conductivity fracture stimulations : a case history. SPE Paper, **16222**.
BARENBLATT, G.I. (1962) : The mathematical theory of equilibrium cracks in brittle fracture. Adv. Appl. Mechan., **7** : 55 - 129.
BARENBLATT, G.I., ZHELTOV, Y.P. & KOCHINA, I.N. (1960) : Basic concepts in the theory of seapage of homogeneous liquids in fissured rocks. J. Appl. Mathem., **24** : 1286 - 1303.
BARGAS, C.L. & YANOSIK, J.L. (1988) : The effects of vertical fractures on areal sweep efficiency in adverse mobility ratio floods. SPE Paper, **17609**.
BARIA, R. & GREEN, A.S.P. (1986) : Seismicity induced during a viscous stimulation at the Camborne School of Mines hot dry rock geothermal energy project in Cornwall, England. 8th Intern. Conf. on Acoustic Emission, 23 pp., Tokyo/Japan, 10.1986.
BARK, E.v.d. & THOMAS, O.D. (1980) : Ekofisk : first of the giant oil fields of Western Europe. In : HALBOUTY, M.T. (Editor), Giant oil and gas fields of the decade 1968 - 1978. Amer. Assoc. Petrol. Geol. Mem., **30** : 195 - 224.
BARK, E.v.d. & THOMAS, O.D. (1981) : Ekofisk : first of the giant oil fields in Western Europe. Bull. Amer. Assoc. Petrol. Geol., **65** : 2341 - 2363.
BARKER, B.J. & RAMEY, H.J. (1978) : Transient flow to finite conductivity vertical fractures. SPE Paper, **7489**.
BARKER, K.M. (1987) : Formation damage related to hot oiling. SPE Paper, **16230**.
BARNEA, E. & MEDNICK, R.L. (1975) : Correlations for minimum fluidization velocity. Trans. Inst. Chem. Engin., **53** : 278 - 281.
BARON, G. & UNGEMACH, P. (1981) : European geothermal drilling experience - problem areas and case studies. Proc. Intern. Conf. Geotherm. Drill. Complet. Technol., Sandia National Laboratories Rept., **SAND 81-0036 C** : 5-1 - 5-22; Albuquerque/New Mexico.
BARON, G. & WITTRISCH, C. (1982) : A new logging method for horizontal wells : SIMPHOR. Petrol. Technol., **1982**/12 (294) : 42 - 44.
BARR, D.T. (1980) : Thermal cracking in nonporous geothermal reservoirs. M.Sc. Thesis, Mass. Inst. Technol., Cambridge/Massachusetts.
BARR, D.T., WILLIS, R.M., CLEARY, M.P. & ROGERS, L.A. (1988) : Enhancement of real-time hydraulic fracturing models using full 3-D simulation. SPE Paper, **17713**.
BARREE, R.D. (1983) : A practical numerical simulator for three-dimensional fracture propagation in heterogeneous media. SPE Paper, **12273**.
BARRETT, M.F. (1986) : The politics of energy policy : the congressional perspective. SPE Paper, **14628**.
BARROW, D.C. & LASSEIGNE, C.A. (1984) : A field evaluation of the Mohr´s circle

technique for predicting sand strength. SPE Paper, **13087**.
BARTEL, L.C., McCANN, R.P. & KECK, L.J. (1976) : Use of potential gradients in massive hydraulic fracture mapping and characterization. SPE Paper, **6090**.
BARTON, N.R., BANDIS, S. & BAKHTAR, K. (1985) : Strength, deformation and conductivity coupling of rock joints. Intern. J. Rock Mechan., Min. Ser., Geomechan. Abstr., **22**.
BARTON, W.C. (1988) : How to make smaller fields viable. Offshore Northern Seas Conference, Stavanger/Norway, 23. - 26.08.1988.
BARUA, J., KUCUK, F. & GOMEZ-ANGULO, J. (1985) : Application of computers in the analysis of well tests from fractured reservoirs. SPE Paper, **13662**.
BATCHELOR, A.S. (1982 a) : The stimulation of a hot dry rock geothermal reservoir in the Cornubian granite, England. Proc. 8th Workshop Geothermal Reservoir Engineering, Stanford Univ., Stanford/California, pp. 237 - 248.
BATCHELOR, A.S. (1982 b) : The creation of hot dry rock systems by combined explosive and hydraulic fracturing. Intern. Conf. on Geothermal Energy, Florence, 05.1982, BHRA Fluid Engineering, Bedford, pp. 321 - 342.
BATCHELOR, A.S. (1984) : Hot dry rock reservoir stimulation in the UK : an extended summary. European Geothermal Update, Third Intern. Seminar, EUR **8853** EN, Munich/Germany FRG, 29.11. - 01.12.1984.
BATCHELOR, A.S., BARIA, K. & HEARN, K. (1983) : Monitoring effects of hydraulic stimulation by microseismic event location : a case study. SPE Paper, **12109**.
BATCHELOR, A.S., PEARSON, C.M. & HALLADAY, N. (1980) : The enhancement of the permeability of granite by explosive and hydraulic fracturing. 2nd Intern. Seminar on results of the EEC geothermal energy research, Dordrecht/Netherlands. EUR, **6862** : 28 pp.
BATCHELOR, G.K. & RENSBURG, J.v. (1986) : Structure formation in bi-disperse sedimentation. J. Fluid Mechan., **166** : 379 - 408.
BATT, R.E. (1983) : Planning and financial controls for the independent producer. SPE Paper, **11324**.
BAUER, L. (1981) : Erdöl und Erdgas als zukunftsbestimmende Faktoren. Erdoel-Erdgas, **97** : 236 - 240.
BAUK, A., SECEN, J., BABIC, D. & SACER, D. (1986) : Approach to the development of the Kalinovac gas-condensate field. SPE Paper, **15526**.
BAUMGARTNER, S.A. & MACK, D.J. (1988) : On-site computer monitoring of foamed stimulation fluids. SPE Paper, **17531**.
BAUMGARTNER, S.A., PARKER, C.D., WILLIAMS, D.A. & WOODROOF, R.A. (1983 a) : High efficiency fracturing fluids for high-temperature, low-permeability reservoirs. SPE/DOE Paper, **11615**.
BAUMGARTNER, S.A., PARKER, C.D., WILLIAMS, D.A. & WOODROOF, R.A. (1983 b) : High efficiency fracturing fluids for high temperature reservoirs. SPE Paper, **11566**.
BAUMGARTNER, W. (1983) : Analyse des Frac- und Produktionsverhaltens von Massive-Hydraulic-Fracturing (MHF)-behandelten Gaslagerstätten. M. Sc. Thesis, University of Hannover/Germany FRG (unpubl.).
BAYLEY, C.H. (1984) : Market prospects and the U.K. gas balance. SPE Paper, **13010**.
BAYLEY, C.H. & COX, T.F. (1982) : Conditions for the development of smaller discoveries in the North Sea. EUR Paper, **300**.
BAZANT, Z.P. (1979) : Stability and post-critical growth of a system of cooling or shrinkage cracks. Intern. J. Fract., **15** : 443 - 456.
BECH, N., JENSEN, O.K. & NIELSEN, B. (1989) : Modelling of gravity-imbibition and gravity-drainage processes with the diffusion equation - analytic and numerical solutions. SPE Paper, **18428**.
BECK, J., SCHULTZ, A. & FITZGERALD, D. (1977) : Reservoir evaluation of fractured Cretaceous carbonates in South Texas. Trans. 18th Soc. Prof. Well Log Analysts Annual Logging Symp., Houston/Texas, 06.1977.
BECK, R.J. (1987) : Improving oil prices slow changes in U.S. demand, production. Oil Gas J., 27.07.**1987** : 33 - 43.
BECK, R.J. & SMITH, G.E. (1987) : Oil, gas price collapse cuts OGJ group`s profits one third in 1986. Oil Gas J., 25.05.**1987** : 19 - 21.
BECK, R.J. & WILLIAMS, B. (1987) : OPEC restraint trims world oil production in first half. Oil Gas J., 14.09.**1987** : 21 - 23.

BECKNER, B.L., FIROOZABADI, A. & AZIZ, K. (1988) : Modeling transverse imbibition in double porosity simulators. SPE Paper, **17414**.
BECKNER, B.L., ISHIMOTO, K., YAMAGUCHI, S., FIROOZABADI, A. & AZIZ, K. (1987) : Imbibition-dominated matrix-fracture fluid transfer in dual porosity simulators. SPE Paper, **16981**.
BECQ, D.F., ROQUE, C. & SARDA, J.P. (1984) : High-strength proppants behavior under extreme conditions. SPE Paper, **12487**.
BEEK, F.v. & TROOST, P.J.P.M. (1978) : The Groningen gas field : a case study on the development of a giant gas field. SPE Paper, **7423**.
BEGG, S., CARTER, R.R. & DRANFIELD, P. (1987) : Assigning effective values to simulator grid-block parameters in heterogeneous reservoirs. SPE Paper, **16754**.
BEGLAND, T.F. & WHITEHEAD, W.R. (1986) : Depletion performance of volumetric high-pressure gas reservoirs. SPE Paper, **15523**.
BEGNAUD, W.J. & CLAIBORNE, E.B. (1985) : Vertical fracture growth considerations in the Mission Canyon/Ratcliffe formations of the North Alexander area. SPE Paper, **14375**.
BELANI, A. & JALALI-YAZDI, Y. (1988) : Estimation of matrix block size distribution and fracture intensity from pressure transient data. SPE Paper, **18171**.
BELFIELD, W.C. (1988) : Characterization of a naturally fractured carbonate reservoir : Lisburne field, Prudhoe Bay, Alaska. SPE Paper, **18174**.
BELL, G.J., JONES, A.H. & MORALES, R.H. (1987) : Spalling and the development of a hydraulic fracturing strategy for coal. Gas Research Institute, Semiannual Rept., 101 pp.
BELL, J.S. & GOUGH, D.I. (1979) : Northeast-southwest compressive stress in Alberta - evidence from oil wells. Earth Planet. Sci. Letters, **45** : 475 - 482.
BELL, W.T. (1982) : Perforating techniques for maximizing well productivity. SPE Paper, **10033**.
BELL, W.T. (1984) : Perforating underbalanced - evolving techniques. J. Petrol. Technol., **36** : 1653 - 1662. SPE Paper, **13413**.
BELVIN, T.R., COTTON, W.R. & DUGAS, E.J. (1973) : OH-Al treatments sustain acid stimulated production. World Oil, **177**/2 : 26 - 29.
BEN NACEUR, K. (1988) : Production enhancement of naturally fractured reservoirs. Amer. Assoc. Petrol. Geol., Annual Convention, Houston/Texas, 20. - 23.03.1988.
BEN NACEUR, K. & ECONOMIDES, M.J. (1988 a) : Production from naturally fissured reservoirs intercepted by a vertical hydraulic fracture. SPE Paper, **17425**.
BEN NACEUR, K. & ECONOMIDES, M.J. (1988 b) : The effectiveness of acid fractures and their production behaviour. SPE Paper, **18536**.
BEN NACEUR, K., DUNAND, A. & STEPHENSON, P. (1985) : Coupling between fluid-loss and heat exchange during fracturing. 4th Intern. Conf. Numer. Methods in Thermal Problems, Swansea/England.
BEN NACEUR, K. & ROEGIERS, J.C. (1988) : Design of fracturing treatments in multi-layered formations. SPE Paper, **17712**.
BEN NACEUR, K. & STEPHENSON, P. (1985) : Models of heat transfer in hydraulic fracturing. SPE/DOE Paper, **13865**.
BEN NACEUR, K. & TOUBOUL, E. (1987) : Mechanisms controlling fracture height growth in layered media. SPE/DOE Paper, **16433**.
BEN NACEUR, K., THIERCELIN, M. & TOUBOUL, E. (1987) : Simulation of fluid flow in hydraulic fracturing : implications on three-dimensional propagation. SPE Paper, **16032**.
BENNETT, C.O., CAMACHO, R.G.V., REYNOLDS, A.C. & RAGHAVAN, R. (1985) : Approximate solutions for fractured wells producing layered reservoirs. Soc. Petrol. Engin. J., **25** : 729 - 742. SPE Paper, **11599** (1982).
BENNETT, C.O. & CLARK, T.J. (1985) : Design and analysis of a nitrogen step-rate test, Anschutz Ranch East nitrogen injection project. SPE Paper, **14456**.
BENNETT, C.O., REYNOLDS, A.C. & RAGHAVAN, R. (1982 a) : Performance of finite conductivity vertically fractured wells in single-layer reservoirs. SPE Paper, **11029**. SPE Form. Eval., **1** (1986) : 399 - 412.
BENNETT, C.O., REYNOLDS, A.C. & RAGHAVAN, R. (1982 b) : Analysis of finite conductivity fractures intercepting multilayer commingled reservoirs. SPE Pa-

per, **11030**. SPE Form. Eval., **1** (1986) : 259 - 274.
BENNETT, C.O., REYNOLDS, A.C. & RAGHAVAN, R. (1983) . A study of the response of fractured wells in layered reservoirs. SPE Paper, **11714**.
BENNETT, C.O., ROSATO, N.D., REYNOLDS, A.C. & RAGHAVAN, R. (1981) : Influence of fracture heterogeneity and wing length on the response of vertically fractured wells. SPE/DOE Paper, **9886**. Soc. Petrol. Engin. J., **23** (1983) : 219 - 230.
BENNETT, E.N. & FORGERSON, D.C. (1975) : Predicting reserves and forecasting flow rates of relatively tight gas wells using limited performance data. J. Petrol. Technol., **27** : 585 - 591. SPE Paper, **5058** (1974).
BENSON, S.M. & LAI, C.H. (1986) : Analysis of interference data in a highly heterogeneous naturally fractured geothermal reservoir. SPE Form. Eval., **1** : 236 - 248. SPE Paper, **13252** (1984).
BENTLEY, F.W.P. (1988) : Company strategies in a changing market. Offshore Northern Seas Conference, Stavanger/Norway, 23. - 26.08.1988.
BENTON, J.H. (1984) : Near wellbore stress behaviour in an unconsolidated sand during sand arch formation and failure. Thesis, **T-3014**; Colorado School of Mines, Golden/Colorado.
BERCEGEAY, E.P. & RICHARD, C.A. (1974) : A one-trip gravel packing system. SPE Paper, **4771**.
BERG, R.B. & HABECK, M.F. (1982) : Abnormal pressures in the Lower Vicksburg, McAllen Ranch field, South Texas. Trans. Gulf Coast Assoc. Geol. Soc., **32** : 247 - 253.
BERGMANN, B. (1982) : Gas supply and marketing trends. EUR Paper, **341**.
BERGMANN, B. (1988) : Gas market structures in Western Europe. Offshore Northern Seas Conference, Stavanger/Norway, 23. - 26.08.1988.
BERGOSH, J., LORD, G. & OWEN, L.B. (1987) : New developments in the analysis of cores from naturally fractured reservoirs. SPE Paper, **16805**.
BERGSTROM, J.M. & MILLER, B.D. (1975) : Results of acid-in-oil emulsion stimulation of carbonate formations. SPE Paper, **5648**.
BERTELSMANN (1980) : Neuer Weltatlas. 319 pp.; Gütersloh (Bertelsmann).
BEST, D.A., CORDELL, G.M. & HASTON, J.A. (1987) : Underground test facility : shaft/tunnel laboratory for horizontal well technology. J. Petrol. Technol., **39** : 603 - 612.
BETZ, C.A. (1982) : The Austin Chalk - drilling and completion techniques. SPE Paper, **11122**.
BEUDELL, M. (1985) : Oil remains elusive in Italian energy search. Offshore Engin., **1985**/8 : 32 - 38.
BEUDELL, M. (1986 a) : Big UK fields still on at $ 5 barrel, but ... Offshore Engin., **1986**/2 : p. 13.
BEUDELL, M. (1986 b) : Dutch go for gas as oil loses favour. Offshore Engineer, **1986**/11 : 60 - 64.
BEUDELL, M. (1986 c) : Learning to live with low-priced oil. Offshore Engineer, **1986**/5 : 45 - 51.
BEUDELL, M. (1986 d) : Fewer, fitter independents in prospect next year. Offshore Engin., **1986**/8 : 41 - 42.
BEUDELL, M. (1986 e) : Marginal review points to continuing development. Offshore Engin., **1986**/10 : p. 69.
BEUDELL, M. (1987 a) : Growing optimism for late ´87 industry pick-up. Offshore Engin., **1987**/2 : p. 21.
BEUDELL, M. (1987 b) : Eocene oil strikes raise UK North Sea hopes. Offshore Engin., **1987**/9 : 59 - 60.
BEYDOUN, W.B., CHENG, C.H. & TOKSÖZ, M.N. (1983) : Detection of subsurface fractures and permeable zones by the analysis of tube waves. 8th European Form. Eval. Symp., Paper **I**, London/England, 14. - 15.03.1983.
BEYDOUN, W.B., CHENG, C.H. & TOKSÖZ, M.N. (1985) : Detection of open fractures with vertical seismic profiling. J. Geophys. Res., **1985**.
BEZAIRE, G.E. & MARKIW, I.A. (1979) : Esso Resources horizontal hole project at Cold Lake. CIM Paper, **79-30-10**.
BEZEMER, C. & MEIJS, F.H. (1966) : Development and field application of the process for sand consolidation employing EPOSAND resin. SPE Paper, **1591**.
BIFANI, R., GEORGE, G.T. & LEVER, A. (1987) : Geological and reservoir charac-

teristics of the Rotliegend sandstones in the Argyll field. In : BROOKS, J. & GLENNIE, K.W. (Editors), Petroleum Geology of North West Europe, **1** : 509 - 522; Graham & Trotman, London/England.
BILD, P. (1987) : OPEC : Whatever happened to market share? Petrol. Rev., **1987**/4 : p. 6.
BINGLE, C.F. (1987) : Combining transient rate and pressure data with production logging for enhanced well test analysis. SPE Paper, **16224**.
BIOT, M.A., MASSE, L. & MEDLIN, W.L. (1984) : Temperature analysis in hydraulic fracturing. SPE Paper, **13228**. J. Petrol. Technol., **39** (1987) : 1389 - 1397.
BIOT, M.A., MASSE, L. & MEDLIN, W.L. (1986) : A two-dimensional theory of fracture propagation. SPE Product. Engin., **1** : 17 - 30. SPE Paper, **11067** (1982).
BIOT, M.A. & MEDLIN, W.L. (1985) : Theory of sand transport in thin fluids. SPE Paper, **14468**.
BIOT, M.A., MEDLIN, W.L. & MASSE, L. (1983) : Fracture penetration through an interface. Soc. Petrol. Engin. J., **23** : 857 - 869. SPE Paper, **10372** (1981).
BIOT, M.A., MEDLIN, W.L. & MASSE, L. (1984) : Laboratory experiments in fracture propagation. Soc. Petrol. Engin. J., **24** : 256 - 268. SPE Paper, **10377** (1981).
BIRD, G.W. & BOON, J.A. (1987) : Silica transport during steam injection into oil sands : 1. Dissolution and precipitation kinetics of quartz - new results and review of existing data. Chem. Geol., **1987**.
BISHLAWI, M. & PENG, C.P. (1988) : Reservoir management of Valhall field, Norway : a fractured high porosity chalk with rock compaction. SPE Paper, **18352**.
BISHOP, J. (1988) : Oil prices and the "need to know". Petrol. Rev., **1988**/10 : 27 - 29.
BJ-HUGHES (1982) : Fracture permeability of various proppants. Fract. Engin. Bull., **1982**/2 (BJ-469).
BJ-HUGHES (1986) : Viscous gravel packing, slurry consolidation, sand consolidation. Brochure.
BLACK, H.N. (1978) : Criteria for fracturing low permeability, water-sensitive hydrocarbon producing formations. SPE Paper, **7121**.
BLACK, H.N. & HOWER, W.F. (1965) : Advantageous use of potassium chloride water for fracturing water-sensitive formations. API Paper, **851-39-F**.
BLACK, H.N. & LANGSFORD, R.W. (1981) : Energized fracturing with fifty percent carbon dioxide for improved hydrocarbon recovery. SPE Paper, **9705**. J. Petrol. Technol., **34** (1982) : 135 - 140.
BLACK, H.N., RIPLEY, H.E., BEECROFT, W.H. & PAMPLIN, L.O. (1979) : Drilling and fracturing fluid improvements for low-permeability gas wells in Canada. SPE Paper, **7926**. J. Petrol. Technol., **33** (1981) : 26 - 32.
BLAIR, C.M., SCRIBNER, R.E. & STOUT, C.A. (1982) : Chemical enhancement of oil production by cyclic steam injection. J. Petrol. Technol., **34** : 2757 - 2762.
BLANTON, T.L. (1981) : Deformation of chalk under confining pressure and pore pressure. Soc. Petrol. Engin. J., **21** : 43 - 50. SPE Paper, **8076** (1978). EUR Paper, **41** (1978).
BLANTON, T.L. (1982) : An experimental study of interaction between hydraulically induced and pre-existing fractures. SPE/DOE Paper, **10847**.
BLANTON, T.L. (1983) : The relation between recovery deformation and in-situ stress magnitudes. SPE Paper, **11624**.
BLANTON, T.L. (1986) : Propagation of hydraulically and dynamically induced fractures in naturally fractured reservoirs. SPE Paper, **15261**.
BLANTON, T.L. & TEUFEL, L.W. (1983) : A field test of the strain recovery method for determining in situ stress in Devonian Shale. SPE Paper, **12304**.
BLANTON, T.L. & TEUFEL, L.W. (1985) : In-situ stress determination from wellbore elongation measurements. SPE/DOE Paper, **13877**.
BLANTON, T.L. & TEUFEL, L.W. (1986) : A critical evaluation of recovery deformation measurements as a means of in-situ stress determination. SPE Paper, **15217**.
BLASINGAME, T.A. & LEE, W.J. (1986 a) : Properties of homogeneous reservoirs, naturally fractured reservoirs, and hydraulically fractured reservoirs from

decline curve analysis. SPE Paper, **15018**.
BLASINGAME, T.A. & LEE, W.J. (1986 b) : Variable-rate reservoir limits testing. SPE Paper, **15028**.
BLASIO, C.J.d., BAILEY, D.E. & BERGTHOLD, J.L. (1985) : Various fracturing fluids used in the Codell Formation in the Denver-Julesburg Basin of North-eastern Colorado. SPE/DOE Paper, **13890**.
BLAUER, R.E. & KOHLHAAS, C.A. (1974) : Formation fracturing with foam. SPE Paper, **5003**.
BLAUER, R.E., PHILLIPS, A.M. & CRAIG, D.P. (1988) : Rheological and physical differences between foam and emulsion fracturing fluids. SPE Paper, **18214**.
BLEAKLEY, W.B. (1980) : Mitchell Energy foam fracs tight gas zones. Petrol. Engin. Intern., **1980**/12 : 24 - 28.
BLEAKLEY, W.B. (1984) : Mobil AG scores with massive frac. Petrol. Engin., **1984**/1 : 72 - 83.
BLEAKLEY, W.B. (1985) : British Gas meets Morecambe field challenges. Petrol. Engin. Intern., **1985**/9 : 30 - 36.
BLEAKLEY, W.B. (1986) : Subsidence : what it is, what it causes, and what it can cost. Petrol. Engin. Intern., **1986**/11 : 31 - 36.
BLESSING, E.W. (1988) : Risk management or Russian roulette. Gulf Coast Oil World, **6**/10 : 21 - 23.
BLINTEN, J.S. & AZIZ, R.M. (1985) : Stimulating very long gross intervals. SPE Paper, **13709**.
BLÜMLING, P. (1983) : Bohrlochmessungen und ihre Beziehung zum regionalen Spannungsfeld. Technical University Karlsruhe (unpubl.).
BODVARSSON, G.S. & GAULKE, S. (1986) : Effects of noncondensible gases on fluid recovery in fractured geothermal reservoirs. SPE Paper, **15121**. SPE Reserv. Engin., **2** (1987) : 335 - 342.
BODVARSSON, G.S., PRUESS, K. & LIPPMANN, M.J. (1986) : Modeling of geothermal systems. J. Petrol. Technol., **38** : 1007 - 1021. SPE Paper, **13613** (1985).
BODVARSSON, G.S., PRUESS, K. & O`SULLIVAN, M.J. (1985) : Injection and energy recovery in fractured geothermal reservoirs. Soc. Petrol. Engin. J., **25** : 303 - 312. SPE Paper, **11689** (1983).
BODVARSSON, G.S. & TSANG, C.F. (1982) : Injection and thermal breakthrough in fractured geothermal reservoirs. J. Geophys. Res., **87** : 1031 - 1048.
BODVARSSON, G.S. et al. (1982) : Preliminary estimation of the reservoir capacity and the longevity of the Baca geothermal field, New Mexico. Water Resourc. Res., **18** : 1713 - 1723.
BODVARSSON, G.S. et al. (1984) : The Krafla geothermal field. 2 : The natural state of the reservoir. 3 : The generating capacity of the field. Water Resour. Res., **20** : 1531 - 1559.
BODVARSSON, G.S. et al. (1986) : Natural state model of the Nesjavellir geothermal field, Iceland. Proc. 11th Geothermal Reservoir Engineering Workshop, Rept., **SGP-TR-93**; Stanford Univ., Stanford/California.
BOIGK, H. (1981) : Erdöl und Erdölgas in der Bundesrepublik Deutschland. 330 pp.; Enke, Stuttgart/Germany FRG.
BONOMO, J.M. & YOUNG, W.S. (1983) : Analysis and evaluation of perforating and perforation cleanup methods. SPE Paper, **12106**. J. Petrol. Technol., **37** (1985) : 505 - 510.
BOOGAERT, H.A.v.A., HEDEMANN, H.A., MAKURIN, A., SCHOTT, W. & ZHABREV, I.P. (1984) : International map of natural gas fields in Europe 1 : 2 500 000. 9 sheets; Hannover (Bundesanstalt für Geowissenschaften und Rohstoffe).
BOON, J.A., HAMILTON, T. et al. (1982) : Reaction between rock matrix and injected fluids in Cold Lake oil sands - potential for formation damage. CIM Paper, **82-33-37**. J. Canad. Petrol. Technol., **1983** : 55 - 66.
BOONE, T.J., WAWRYZNEK, P.A. & INGRAFFEA, A.R. (1986) : Simulation of the fracture process zone in rock with application to hydrofracturing. Intern. J. Rock Mechan., Miner. Sci., Geomechan. Abstr., **23** : 255 - 265.
BOSSIE-CODREANU, D., BIA, P.R. & SABATHIER, J.C. (1985) : The "checker model", an improvement in modeling naturally fractured reservoirs with tridimensional, triphasic, black-oil numerical model. Soc. Petrol. Engin. J., **25** : 743 - 756. SPE Paper, **10977** (1982).
BORCHARDT, J.K. & BROWN, D.L. (1984) : Clay stabilizers improve EOR injection

rates. Oil Gas J., 10.09.**1984** : 150 - 156.
BORCHARDT, J.K., ROLL, D.L. & RAYNE, L.M. (1984) : Use of a mineral fines stabilizer in well completions. SPE Paper, **12757**.
BORCHARDT, J.K., YOUNG, B.M. & McLAUGHLIN, H.C. (1979) : Clay stabilization agents - their effectiveness in high temperature steam. CIM Paper, **79-30-12**.
BORISOV, J.P. (1964) : Oil production using horizontal and multiple deviation wells. Nedra, Moscow/USSR (in Russian). English edition by R & D Library Translation (edition and translation by J. STRAUSS & S.D. JOSHI); Phillips Petrol. Co., Bartlesville/Oklahoma (1984).
BORNER, D.L., HALDORSEN, H.H., HARRISON, P.F., HOPWOOD, A.P. & MACDONALD, C.J. (1988) : Performance of vertical and horizontal wells in reservoirs containing discontinuous shales; numerical studies of two North Sea fields. SPE Paper, **17600**.
BOSIO, J. (1986) : Horizontal wells are now used for industrial developments. 6th Offshore Southeast Asia Conference and Exhibition, Singapore, 26. - 31. 01.1986, pp. 77 - 81.
BOSIO, J.C., FINCHER, R.W., GIANNESINI, J.F. & HATTEN, J.L. (1987) : Horizontal drilling - a new production method. 12th World Petrol. Congr., Houston/Texas, Session 4, Topic 11.
BOSIO, J. & GIANNESINI, J.F. (1988) : Horizontal wells cut offshore production costs. OSEA Paper, **88127**. Petrol. Inform., **1988**/3 : p. 88.
BOSIO, J.C. & REISS, L.H. (1988) : Site selection remains key to success in horizontal-well operations. Oil Gas J., 21.03.**1988** : 71 - 76.
BOSTIC, J.N., AGARWAL, R.G. & CARTER, R.D. (1980) : Combined analysis of post-fracturing performance and pressure buildup data for evaluating an MHF gas well. J. Petrol. Technol., **32** : 1711 - 1719. SPE Paper, **8280** (1979).
BOSTIC, J.N. & GRAHAM, J.A. (1979) : Prefracturing pressure transient testing : East Texas Cotton Valley tight gas play. SPE Paper, **7941**.
BOTTER, B.J. (1982) : Circumferential acoustic waves in boreholes for the delineation of vertical fractures. Trans. 23rd Soc. Prof. Well Log Analysts Annual Logging Symp., Paper **S**, 06. - 09.07.1982.
BOTTER, B.J. & ARKEL, J.v. (1982) : Circumferential propagation of acoustic boundary waves in boreholes. J. Acoust. Soc. Amer., **1982**/4 : 790 - 795.
BOULET, D.P. (1978) : Gravel for sand control : a study of quality control. SPE Paper, **7002**. J. Petrol. Technol., **31** (1979) : 164 - 168.
BOURDET, D., AYOUB, J.A. & PIRARD, Y.M. (1984) : Use of pressure derivative in well test interpretation. SPE Paper, **12777**.
BOURDET, D. & GRINGARTEN, A.C. (1980) : Determination of fissure volume and block size in fractured reservoirs by type-curve analysis. SPE Paper, **9293**.
BOUTECA, M.J. (1984) : 3D analytical model for hydraulic fracturing : theory and field test. SPE Paper, **13276**.
BOUTECA, M.J. (1987) : Hydraulic fracturing model based on a 3D closed form : tests, analysis of fracture geometry, and containment. SPE/DOE Paper, **16432**.
BOWEN, J.G. & DRISCOLL, P.L. (1980) : Oil base foam fracturing applied to the Niobrara Shale Formation. SPE Paper, **9335**.
BOWIE, O.L. (1956) : Analysis of an infinite plate containing radial cracks originating at the boundary of an internal circular hole. J. Math. Phys., **25** : 60 - 71.
BOWLER, G.V. (1988) : Tubing-conveyed perforating : operating experience - an approach to better gas zone evaluation and completion. OSEA Paper, **17667**.
BOYCE, J.R. & CARROLL, J.F. (1984) : Mathematical modelling of a gravel pack logging tool. SPE Paper, **13138**. SPE Product. Engin., **1** (1986) : 125 - 130.
BOYELDIEU, C. & MARTIN, C. (1984) : Fracture detection and evaluation. Proc. 8th Intern. Symp., Société pour l`Avancement de l´Interprétation des Diagraphies, Paper **21**, Paris/France, 24. - 26.10.1984.
BOYELDIEU, C. & WINCHESTER, A. (1982) : Use of the dual laterolog for the evaluation of the fracture porosity in hard carbonate formations. SPE Paper, **10464**.
BOYER, C.M. II, POPOVICH, G.S. & SCHRAUFNAGEL, R.A. (1986) : The Rock Creek methane from multiple coal seams completion project. SPE Paper, **15259**.
BOYER, C.M. II, STUBBS, P.B. & SCHWERER, F.C. (1986) : Measurement of coalbed

properties for hydraulic fracture design and methane production. SPE Paper, **15258**.

BRADEL, E. (1980) : Probleme und Erfahrungen bei petrophysikalischen und technischen Kabelmessungen in verrohrten Bohrlöchern. Erdoel-Erdgas, **96** : 295 - 300.

BRADEL, E. & DRAXLER, H. (1982) : Erweiterte Computerauswertungen von Bohrlochmessungen in Rotliegendsedimenten Norddeutschlands (Ost-Hannover). Erdoel-Erdgas, **98** : 16 - 24.

BRADLEY, W.B. (1979) : Failure of inclined boreholes. J. Energy Resourc. Technol., **101** : 232 - 239. Trans. ASME, **1979**.

BRÄUER, L. (1987) : Investigations for a CO2-flooding process in the 12. Tortonian reservoir of the Vienna Basin. 4th Europ. Symp. Enhanced Oil Recovery, Miscible and immiscible gas flooding session II, Hamburg/Germany FRG.

BRANAGAN, P.T., CIPOLLA, C.L. & LEE, S.J. (1987) : Designing and evaluating hydraulic fracture treatments in naturally fractured reservoirs. SPE/DOE Paper, **16434**.

BRANAGAN, P., CIPOLLA, C.L. & LEE, S.J. (1988) : Fracture design considerations in naturally fractured reservoirs. SPE Paper, **17607**.

BRANAGAN, P.T., CIPOLLA, C., LEE, S.J. & CHEN, J. (1987) : Designing and evaluating hydraulic fracture treatments in naturally fractured reservoirs. SPE/DOE Paper, **16434**.

BRANAGAN, P.T., CIPOLLA, C.L., LEE, S.J. & WILMER, R.H. (1985) : Comprehensive well testing and modeling of pre- and post-fracture well performance of the MWX lenticular tight gas sands. SPE/DOE Paper, **13867**.

BRANAGAN, P.T., CIPOLLA, C.L., LEE, S.J. & YAN, L. (1987) : Case history of hydraulic fracture performance in the naturally fractured paludal zone : the transitory effects of damage. SPE/DOE Paper, **16397**.

BRANAGAN, P. & COTNER, G. (1983) : Procedures to improve pre-frac well testing and analysis. SPE Paper, **11856**.

BRANAGAN, P.T., COTNER, G. & GETTMAN, G. (1981) : Evidence of permeability enhancement through cyclic dry gas injection. SPE/DOE Paper, **9854**.

BRANAGAN, P., COTNER, G. & LEE, S.J. (1984) : Interference testing of the naturally fractured Cozzette Sandstone : a case study at the DOE MWX site. SPE/DOE/GRI Paper, **12869**.

BRANAGAN, P., LEE, S.J., CIPOLLA, C.L. & WILMER, R. (1988) : Pre-frac interference testing of a naturally fractured, tight fluvial reservoir. SPE Paper, **17724**.

BRANAGAN, P. & WILMER, R. (1988) : Breakdown procedures designed to minimize naturally fractured reservoir damage. SPE Paper, **17716**.

BRANDENBURG, C.F. (1988) : Stimulation - the key to unlocking Devonian Shale reserves. Eastern Devonian Gas Shales Technol. Rev., **5**/1 : 3 - 6.

BRANNON, D.H., NETTERS, C.K. & GRIMMER, P.J. (1987) : Matrix acidizing design and quality-control techniques prove successful in Main Pass Area Sandstone. J. Petrol. Technol., **39** : 931 - 942. SPE Paper, **14827** (1986).

BRASHEAR, J.P., BECKER, A.B. & BIGLARBIGI, K.H. (1988) : Incentives, technology, and EOR : potential for increased oil recovery at lower prices. SPE Paper, **17454**.

BRASHEAR, J.P., ELKINS, L.E. & HAAS, M.R. (1982) : Tight gas research and development : the industry perspective. SPE/DOE Paper, **10807**.

BRASHEAR, J.P., ROSENBERG, J.I. & MERCER, J. (1984) : Tight gas resource and technology appraisal : sensitivity analyses of the National Petroleum Council estimates. SPE/DOE/GRI Paper, **12862**.

BRATLI, R.K. & RISNES, R. (1981) : Stability and failure of sand arches. Soc. Petrol. Engin. J., **21** : 236 - 248. SPE Paper, **8427** (1979).

BRATOVICH, M.W., BELL, W.T. & KAAZ, K.D. (1977) : Improved techniques for logging high-angle wells. SPE Paper, **6818**.

BRAUNLICH, F. (1967) : Controlling fractures during well treatment. U.S. Patent, **3,335,697** (15.08.1967).

BRAY, B.G., KNUTSON, C.F., WAHL, H.A. & DEW, J.N. (1965) : Economics of contained nuclear explosions applied to petroleum reservoir stimulation. J. Petrol. Technol., **17** : 1145 - 1152.

BRECKELS, J.M. & EEKELEN, H.A.M.v. (1982) : Relationship between horizontal

stress and depth in sedimentary basins. J. Petrol. Technol., **34** : 2191 - 2199. SPE Paper, **10336** (1981).
BREDEHOEFT, J.D., WOLFF, R.G., KEYS, W.S. & SHUTER, E. (1976) : Hydraulic fracturing to determine the regional in-situ stress field, Piceance Basin, Colorado. Bull. Geol. Soc. Amer., **87** : 250 - 258.
BREMER, R.E., WINSTON, H. & VELA, S. (1983) : Analytical model for vertical interference tests across low-permeability zones. SPE Paper, **11965**. Soc. Petrol. Engin. J., **25** (1985) : 407 - 418.
BREWSTER, J. & DANGERFIELD, J. (1984) : Chalk fields along the Lindesnes Ridge. Marine Petrol. Geol., **1** : 238 - 278.
BREWSTER, J., DANGERFIELD, J. & FARRELL, H. (1986) : The geology and geophysics of the Ekofisk field waterflood. Marine Petrol. Geol., **3** : 139 - 169.
BRIE, A., HSU, K. & ECKERSLEY, C. (1988) : Using the Stoneley normalized differential energies for fractured reservoir evaluation. Trans. 29th Soc. Prof. Well Log Analysts Annual Logging Symp., San Antonio/Texas, 05. - 08.06.1988.
BRIE, A., HSU, K. & PLUMB, R.A. (1987) : Use of the Stoneley normalized differential energies for fractured reservoir evaluation. Trans. 28th Soc. Prof. Well Log Analysts Annual Logging Symp., London/England, 29.06. - 02.07.1987.
BRIERLEY, C.W. (1988) : Gas as a global energy source. Offshore Northern Seas Conference, Stavanger/Norway, 23. - 26.08.1988.
BRIGGS, R.O. (1964) : Development of a downhole television camera. Proc. 5th Soc. Prof. Well Log Analysts Annual Logging Symp., Midland/Texas.
BRIGHAM, W.E. & ABBASZADEH-DEHGHANI, M. (1986) : Tracer testing for reservoir description. SPE Paper, **14102**. J. Petrol. Technol., **39** (1987) : 519 - 527.
BRIM, H.B. (1986) : A post-audit of fracture stimulations in the Vicksburg Formation of South Texas. SPE Paper, **15508**.
BRINKMANN, F. (1982) : Status report on fracturing of deep and low permeable formations in West Germany. SPE/DOE Paper, **9852**.
BRINKMANN, F., FUHRBERG, H.D. & SCHÖBER, K. (1980) : Frac-Behandlungen in tiefen, geringpermeablen Gaslagerstätten - derzeitiger Stand und weitere Aussichten. Erdoel-Erdgas, **96** : 37 - 44.
BRINKMANN, F., KRÖMER, E. & REINICKE, K.M. (1980) : Methoden zur Bestimmung fracspezifischer Parameter aus dem Förderverhalten von behandelten Gasbohrungen. Erdoel-Erdgas, **96** : 236 - 245.
BRISCOE, H.R. (1978) : Stabilized foamed acid, a new stimulation technique for carbonate gas reservoirs. SPE Paper, **8141**.
BRITISH PETROLEUM (1986) : Village fields development. BP`s Southern North Sea gas project. The West Sole field. Hydrocarbons `86, Exhibition and Conference, Great Yarmouth/England, 01. - 03.10.1986.
BRITT, E.L. (1977) : Theory and applications of the borehole audio tracer survey. SPE Paper, **6552**. Trans. 17th Soc. Prof. Well Log Analysts Annual Logging Symp., 09. - 12.06.1976.
BRITT, L.K. (1985) : Optimized oilwell fracturing of moderate-permeability reservoirs. SPE Paper, **14371**.
BRITT, L.K. & BENNETT, C.O. (1985) : Determination of fracture conductivity in moderate-permeability reservoirs using bilinear flow concepts. SPE Paper, **14165**.
BRITT, L.K. & LARSEN, M.J. (1986) : Fracture characterization of multilayered reservoirs. SPE Paper, **15511**.
BRITTON, M.W., MARTIN, W.L., LEIBRECHT, R.J. & HARMON, R.A. (1983) : The Street Ranch pilot test of fracture-assisted steamflood technology. J. Petrol. Technol., **35** : 511 - 522. SPE Paper, **10707** (1982).
BRITTON, M.W., MARTIN, W.L., McDANIEL, J.D. & WAHL, H.A. (1981) : Fracture preheat oil recovery process. U.S. Patent, **4,265,310** (05.05.1981). Canadian Patent, **1,122,113** (20.04.1982).
BROADDUS, G. (1988) : Well- and formation-damage removal with nonacid fluids. J. Petrol. Technol., **40** : 685 - 687. SPE Paper, **17111**.
BROADDUS, G.C., KNOX, J.A. & FREDRICKSON, S.E. (1975) : Dynamic etching tests and their use in planning acid treatments. SPE Paper, **2362**.
BRODING, R.A. (1981) : Volumetric scan well logging. Trans. 22nd Soc. Prof. Well Log Analysts Annual Logging Symp., Paper **B**, Mexico City/Mexico, 23. - 26.06.1981.

BRODING, R.A. (1983) : Application of the sonic volumetric scan log to cement evaluation. Trans. 24th Soc. Prof. Well Log Analysts Annual Logging Symp., Paper **JJ**, Calgary/Alberta, 27. - 30.06.1983.

BROEK, D. (1978) : Elementary fracture mechanics. Sijhoff & Noordhoff Publ., The Netherlands.

BROOKS, F.A. (1971) : Consolidation of dirty sands by phenol-formaldehyde plastic. J. Petrol. Technol., **23** : 934 - 938.

BROOKS, F.A., MUECKE, T.W., RICKEY, W.P. & KERVER, J.K. (1974) : Externally catalyzed epoxy for sand control. J. Petrol. Technol., **26** : 589 - 594. SPE Paper, **4034** (1972).

BROOKS, J. & GLENNIE, K.W. (1987) : Petroleum geology of Northwest Europe. 1,219 + pp.; Graham & Trotman, London. ISBN 0-86010-703-5.

BROWER, K.R. & MORROW, N.R. (1983) : Fluid flow in cracks as related to low permeability gas sands. SPE/DOE Paper, **11623**. Soc. Petrol. Engin. J., **25** (1985) : 191 - 201.

BROWN, A. & MARRIOTT, F.T. (1987) : Use of tracers to investigate drilling fluid invasion and oil flushing while coring. SPE Paper, **16352**.

BROWN, C.A. & CRAFTON, J.W. (1981) : Exploration and development of a low pressure, low permeability gas reservoir. SPE Paper, **10304**.

BROWN, C.A., CRAFTON, J.W. & GOLSON, J.G. (1982) : The Niobrara gas play : exploration and development of a low-pressure low permeability gas reservoir. J. Petrol. Technol., **34** : 2862 - 2870. SPE Paper, **10304** (1981).

BROWN, C.A., SMAGALA, T.M. & HAEFELE, G.R. (1986) : Southern Piceance Basin model - Cozzette, Corcoran and Rollins Sandstones. In : SPENCER, C.W. & MAST, R.F. (Editors), Geology of tight gas reservoirs. Amer. Assoc. Petrol. Geol., Stud. Geol., **24** : 207 - 220.

BROWN, R.O. (1978) : Application of fracture identification logs in the Cretaceous of North Louisiana and Mississippi. Trans. Gulf Coast Assoc. Geol. Soc., **28** : 75 - 91.

BROWN, R.O., FORGOTSON, J.M. & FORGOTSON, J.M.Jr. (1980) : Predicting the orientation of hydraulically created fractures in the Cotton Valley Formation of East Texas. SPE Paper, **9269**.

BROWN, W. & MUCH, M. (1986 a) : An evaluation of four commonly used proppants. Norton-Alcoa Proppants Publ.

BROWN, W.E. & MUCH, M.G. (1986 b) : Long term stability of proppants and stress corrosion effects under simulated petroleum reservoir conditions. Talk given at the Institut für Erdölgewinnung und Tiefbohrkunde, Technical University Clausthal, 20.06.1986.

BROWN, W.E. & MUCH, M.G. (1986 c) : Performance of proppants under simulated downhole conditions - stress corrosion effects. Norton-Alcoa Proppants Rept., 06.1986.

BROWNE, B.L. & PARKER, E.F. (1986) : Situational drilling and completion practices in the Stoney Point field : a subnormally pressured and fractured Trenton/Black River reservoir. SPE Paper, **15918**.

BRUNETTE, P.T. (1983) : Stimulation techniques in the Red Fork Formation in the Anadarko Basin. SPE Paper, **11574**.

BRUIST, E.H. (1974) : Better performance of Gulf Coast wells. SPE Paper, **4777**.

BRUIST, E.H. (1981) : Beta type gravel pack well completions. SPE Paper, **10175**.

BRUIST, E.H., JEFFERIS, R.G. & BOTTS, T.M. (1983) : Well completions in the Beta field, offshore California. SPE Paper, **11696**.

BUCHANAN, A. (1987) : Fracturing deviated wells. SPE Continuing Education, Hydraulic Fracturing Seminar, London/England, 01.12.1987.

BUCHANAN, J.C., CLEARMAN, D.K., HEIDBRINK, L.J. & SMITH, H.D. (1984) : Applications of TMD pulsed neutron logs in unusual downhole logging environments. Trans. 25th Soc. Prof. Well Log Analysts Annual Logging Symp., Paper **KKK**, New Orleans/Louisiana, 10. - 13.06.1984.

BUCKLES, R.S. (1979) : Steam stimulation heavy oil recovery at Cold Lake, Alberta. SPE Paper, **7994**.

BUECHLEY, T.C., & LORD, D.L. (1973) : What to learn about hydraulic fracturing fluids. Oil Gas J., 17.09.**1973** : 84 - 88.

BUHARALI, A.M., WRIGHT, T.B. & WILLIS, R.M. (1987) : Real-time hydraulic fracturing simulation with a portable microcomputer. SPE Paper, **16486**.

BUI, H.D. (1975) : Application des potentiels élastiques à l´étude des fissures planes de forme arbitraire en milieu tridimensionnel. Centre Res. Acad. Sci., Ser. A, **280** : 1157 - 1160.

BULLEN, R.S. & LILLIES, A.T. (1983) : Carbon dioxide fracturing process and apparatus. U.S. Patent, **4,375,545** (22.02.1983).

BUNDY, T.E. (1981) : Pre-fracture injection surveys : a necessity for successful fracture treatments. SPE/DOE Paper, **9883**. J. Petrol. Technol., **34** (1982) : 995 - 1001.

BURCIK, E.J. (1969) : The mechanism of microgel formation in partially hydrolyzed polyacrylamide. J. Petrol. Technol., **21** : 373 - 374.

BURDYLO, L. (1977) : Dowell Schlumberger gravel pack guide II.

BURGER, J., GADELLE, C.P. & MARRAST, J.R. (1987) : Development of a chemical process for sand control. SPE Paper, **15410**.

BURKILL, G.C.C. & PIERRE, M.L. (1975) : Successful matrix acidizing of sandstones requires a reliable estimate of wellbore damage. SPE Paper, **5590**.

BURMAN, J.W. & HALL, B.E. (1986) : Foam as a diverting technique for matrix sandstone stimulation. SPE Paper, **15575**.

BURNHAM, J.W., HARRIS, L.E. & McDANIEL, B.W. (1978) : Developments in hydrocarbon fluids for high-temperature fracturing. SPE Paper, **7564**. J. Petrol. Technol., **32** (1980) : 217 - 220.

BUSCH, E.A., SOBKOWICH, K.N., NEILL, B.E., SINHA, A.K. & FERTL, W.H. (1985) : Log-derived evaluation of gas-bearing Cherokee, Red Fork, and Morrow Formations, Custer County, Oklahoma. SPE Paper, **13808**.

BUSH, N.J. (1982) : Natural gas markets : can policy makers restore order? SPE Paper, **11267**.

BUTLER, R.M. (1984) : The potential of horizontal wells. 5th Annual Conference on Advances in Petroleum Recovery and Upgrading Technology, Calgary/Alberta, 14. - 15.06.1984.

BUTLER, R.M. & STEPHENS, D.J. (1980) : The gravity drainage of steam-heated heavy oil to parallel horizontal wells. 31st Annual Technical Meeting of the Petroleum Society of CIM, Calgary/Alberta, 25. - 28.05.1980.

BYRER, C.W., MROZ, T.H. & COVATCH, G.L. (1984) : Production potential for coalbed methane in U.S. basins. SPE/DOE/GRI Paper, **12832**. J. Petrol. Technol., **39** (1987) : 821 - 834.

BYRNES, A.P., SAMPATH, K. & RANDOLPH, P.L. (1979) : Effect of pressure and water saturation on permeability of Western tight sandstones. SPE Paper (1979 Enhanced Oil Recovery Symp., Tulsa/Oklahoma, 22. - 24.08.1979). Proc. 5th Annual Symp. Enhanced oil and gas recovery and improved drilling methods, **3** : L 5/1 - L 5/16.

CALLANAN, M.J., CIPOLLA, C.L. & LEWIS, P.E. (1983) : The application of a new second-generation high-strength proppant in tight gas reservoirs. SPE/DOE Paper, **11633**.

CALLAWAY, R.E., ORTIZ, J. & HOLCOMB, D.L. (1982) : Clay protection chemicals : a practical evaluation and application technique for their use in stimulation fluids. SPE Paper, **10663**.

CAMACHO, R.G.V., RAGHAVAN, R. & REYNOLDS, A.C. (1984) : Response of wells producing layered reservoirs ; unequal fracture length. SPE/DOE/GRI Paper, **12844**. SPE Form. Eval., **2** (1987) : 9 - 28.

CAMERON, J. (1988 a) : Shear waves in an anisotropic earth. Amer. Assoc. Petrol. Geol., Annual Convention, Houston/Texas, 20. - 23.03.1988.

CAMERON, J.R. (1988 b) : Viscometry of nonhomogeneous flows and the behaviour of a titanium-crosslinked hydroxypropyl guar gel in Couette flow. J. Rheol., **1988**.

CAMERON, J.R., GARDNER, D.C. & VEATCH, R.W. (1988) : New insights on the rheological behavior of delayed crosslinked fracturing fluids. SPE Paper, **18209**.

CAMPBELL, D.A., HANOLD, R.J., SINCLAIR, A.R. & VETTER, O.J. (1981) : A review of the geothermal reservoir well stimulation program. Proc. Intern. Conf. Geotherm. Drill. Complet. Technol., Sandia National Laboratories Rept., **SAND** 81-0036 C : 21-1 - 21-18; Albuquerque/New Mexico.

CAMPBELL, D.A, MORRIS, C.W. & VERITY, R.V. (1981) : Geothermal well stimulation experiments and evaluation. SPE Paper, **10316**.

CAMPBELL, J.M. & HUBBARD, R.A. (1984) : Price forecasting and project evalua-

ORDER NUMBER

PO NYAATP08

BOOK NUMBER

5952523

BAY NUMBER

7114-11

AUTHOR	Mader, D.
TITLE	Hydraulic Proppant Fracturing and Gravel Packing (Volume 26)

tion in the 1980`s. J. Petrol. Technol., **36** : 817 - 825, 1979 - 1980. SPE Paper, **11315** (1983).

CAMPBELL, R.L. & TRUMAN, R.B. (1986) : Formation evaluation in the Devonian Shale. SPE Paper, **15212**.

CANNON, D.E. (1979) : Log evaluation of a fractured reservoir Monterey Shale. Trans. 20th Soc. Prof. Well Log Analysts Annual Logging Symp.

CANSON, B.E. (1985) : Lost circulation treatments for naturally fractured vugular or cavernous formations. SPE/IADC Paper, **13440**.

CANTU, L.A. & BOYD, P.A. (1988) : Laboratory and field evaluation of a combined fluid loss control additive and gel breaker for fracturing fluids. SPE Paper, **18211**.

CAPEL, J. (1988) : Western Europe. Learning to live with low prices. World Oil, **207** (1988/8) : 61 - 73.

CARDEN, R.S., NICHOLSON, R.W., PETTIT, R.A. & ROWLEY, J.C. (1983) : Unique aspects of drilling and completing hot dry rock geothermal wells. IADC/SPE Paper, **11373**. J. Petrol. Technol., **37** (1985) : 821 - 834.

CARDWELL, P.H. (1947) : Plastic used to consolidate incompetent formations. Trans. AIME, **170** : 174 - 179.

CARNES, J.D. & MIXON, A. (1971) : Consolidate and gravel pack in one step. SPE Paper, **3593**.

CARNEY, L.L., GUVEN, N. & McGREW, G.T. (1982) : Investigation of high-temperature fluid loss control agents in geothermal drilling fluids. SPE Paper, **10736**.

CARNEY, M.J. & WIELAND, D.R. (1973) : Stimulation of low permeability gas wells in the Rocky Mountain area. SPE Paper, **4396**.

CARROLL, H.B. & BAKER, B.H. (1979) : Particle size distributions generated by crushed proppants and their effects on fracture conductivity. SPE Paper, **7923**.

CARROLL, J.F. & BULLEN, B. (1988) : Successful water control examples in Gulf of Mexico gravel packed completions. SPE Paper, **18228**.

CARROLL, J.F., NEAL, M.R. & PARKER, W.B. (1986) : Gravel pack repair with wireline devices. SPE Paper, **15274**.

CARTER, W.J. & OLINGER, B.W. (1976) : Explosive fragmentation of oil shale. SPE Paper, **6070**.

CARVALHO, R.d.S. & ROSA, A.J. (1988) : Transient pressure behavior for horizontal wells in naturally fractured reservoirs. SPE Paper, **18302**.

CASHION, M.L. & MARKLE, F.B. (1940) : Gravel-packing oil wells in Gulf Coast fields. Oil Gas J., 04.04.**1940** : 48 - 55.

CASTELIJNS, J.H.P. (1981) : Modelling of retrograde condensate drainage in condensate-rich gas caps or gas reservoirs. SPE Paper, **10285**.

CASTELIJNS, J.H.P. & HAGOORT, J. (1984) : Recovery of retrograde condensate from naturally fractured gas-condensate reservoirs. Soc. Petrol. Engin. J., **24** : 707 - 717. SPE Paper, **11199** (1983).

CASTILLO, J.L. (1987) : Modified fracture pressure decline analysis including pressure-dependent leakoff. SPE/DOE Paper, **16417**.

CATALDI, R. & SOMMARUGA, C. (1986) : Background, present state and future prospects of geothermal development. World Energy Conference, Cannes/France.

CAWIEZEL, K.E. & NILES, T.D. (1987) : Rheological properties of foam fracturing fluids under downhole conditions. SPE Paper, **16191**.

CECIL, R.E. (1987) : An analysis of supply and demand for crude oil. SPE Paper, **16291**.

CETRON, M.J. (1986) : International political risks and energy project investments. SPE Paper, **14645**.

CHADWICK, M., LONG, D. & NISSANKE, M. (1988) : Soviet oil exports : trade adjustments, refining constraints and market behavior. 263 pp.; Penn Well, Tulsa/Oklahoma.

CHAMBERLAIN, N.P. (1982) : Filtration of gas on oil and gas platforms with special reference to sand removal and slug catching. SPE Paper, **10984**.

CHAN, S.A. (1985 a) : Oilfield cashflow : the independents lifeblood for survival. SPE Paper, **14154**.

CHAN, S.A. (1985 b) : Secured petroleum financing for the independent producer. SPE Paper, **13790**.

CHANCELLOR, R.E. & JOHNSON, R.C. (1988) : Geologic and engineering implications of production history from five Mesaverde wells in Central Piceance Creek Basin, Northwest Colorado. SPE Form. Eval., **3** : 307 - 314. SPE Paper, **15237** (1986).
CHANG, Y.P., KONG, C.S. & CHEN, D.J. (1973) : The use of fundamental Green`s functions for the solutions of problems of heat conduction in anisotropic media. Intern. J. Heat Mass Transfer, **16** : 1905 - 1918.
CHAPERON, I. (1986) : Theoretical study of coning toward horizontal and vertical wells in anisotropic formations : subcritical and critical rates. SPE Paper, **15377**.
CHARLES, J.G., HUDOCK, K.D., UDICK, T.H. & HALL, D.R. (1983) : Evaluation of stimulation treatments in Appalachian tight gas sands. SPE/DOE Paper, **11601** (1983), **11257** (1982).
CHARLES, S.R. (1976) : Results of calcium hydroxide treatments for sand control, Kern River field. SPE Paper, **6179**.
CHARLEZ, P., SALEH, K. & DESPAX, D. (1987) : The fracmeter : a new way to measure the state of stress in the earth crust. SPE Paper, **15773**.
CHARLEZ, P., SALEH, K., DESPAX, D. & JULIEN, P. (1986): A new way to determine the state of stress and the elastic characteristics of rock massive. Intern. Symp. on Rock Stress and Rock Stress Measurements, Stockholm/Sweden, 01. - 03.09.1986, Centeck Publishers.
CHARPENTIER, R.R., LAW, B.E. & PRENSKY, S.E. (1987) : Quantitative model of overpressured gas resources of the Pinedale Anticline, Wyoming. SPE/DOE Paper, **16404**.
CHASE, R.W. & WILLIAMS, M.A.T. (1986) : Dimensionless IPR curves for predicting the performance of fractured gas wells. SPE Paper, **15936**.
CHAUVETEAU, G. (1981) : Molecular interpretation of several different properties of flow of coiled polymer solutions through porous media in oil recovery conditions. SPE Paper, **10060**.
CHAUVETEAU, G. & KOHLER, N. (1980) : Influence of microgels in xanthan polysaccharide solutions on their flow behavior through various porous media. SPE Paper, **9295**.
CHEN, C.C., ALAM, J., VOZNIAK, J.P. & BLANTON, T.L. (1984) : Evaluation of naturally fractured gas shale production utilizing multiwell transient tests : a field study. SPE/DOE/GRI Paper, **12872**.
CHEN, C.C., SERRA, K., REYNOLDS, A.C. & RAGHAVAN, R. (1985) : Pressure transient analysis methods for bounded naturally fractured reservoirs. Soc. Petrol. Engin. J., **25** : 451 - 464. SPE Paper, **11243** (1982).
CHEN, C.C., YEH, N.S., RAGHAVAN, R. & REYNOLDS, A.C. (1984) : Pressure response at observation wells in fractured reservoirs. Soc. Petrol. Engin. J., **24** : 628 - 638. SPE Paper, **10839** (1982).
CHEN, D.T., ZHOU, B.Q. & LIU, Z.Z. (1986) : Sand control techniques in Shengli oil-field. SPE Paper, **14840**.
CHEN, H.G. & LESCARBOURA, J.A. (1983) : Interference analysis of an anisotropic reservoir using five-spot data. SPE Paper, **11967**.
CHEN, L.D., CHEN, G.Y. & ZI, X.X. (1982) : A high quality water-base fracturing fluid. SPE Paper, **10568**.
CHEN, S.M. & OLYNYK, J. (1985) : Sweep efficiency improvements using horizontal wells or tilted horizontal wells in miscible floods. CIM Paper, **85-36-62**.
CHEN, T. (1987) : Pressure drawdown in a layered reservoir with linear boundaries. SPE Paper, **16767**.
CHEN, T. & KRY, R. (1981) : Pressure buildup from a vertically fractured low-permeability formation. SPE/DOE Paper, **9889**. J. Petrol. Technol., **34** (1982) : 917 - 924.
CHEN, W.H., WASSERMAN, M.L. & FITZMORRIS, R.E. (1987) : A thermal simulator for naturally fractured reservoirs. SPE Paper, **16008**.
CHEN, Y.Z. (1986) : Development of Fuyu fractured sandstone reservoirs by water injection. SPE Paper, **14843**.
CHENEVERT, M.E. & THOMPSON, T.W. (1985) : Perforation stability in low-permeability gas reservoirs. SPE/DOE Paper, **13902**.
CHENG, M.C.A. (1984) : Program selects proppants. Oil Gas J., 30.04.**1984** : 67 - 69.

CHEUNG, P. (1984) : Fracture detection using the sonic tool. Proc. 8th Intern. Symp., Société pour l'Avancement de l'Interprétation des Diagraphies, Paper **42**, Paris/France, 24. - 26.10.1984.

CHEUNG, S.K. (1985) : Effect of acids on gravels and proppants. SPE Paper, **13842**. SPE Product. Engin., **3** (1988) : 201 - 204.

CHHABRA, R.P. & UHLHERR, P.H.T. (1979) : Estimation of zero shear viscosity of polymer solutions from filling sphere data. Rheol. Acta, **18** : 593 - 599.

CHHINA, H.S. (1986) : Horizontal fracture test in the Athabasca oil sands. SPE Paper, **15366**.

CHHINA, H.S. & AGAR, J. (1985) : Potential use of fracture technology for recovery of bitumen from oil sands. 3rd Intern. Conf. Heavy Crude and Tar Sands, Long Beach/California, 22. - 31.07.1985.

CHIERICI, G.L. (1984) : Novel relations for drainage and imbibition relative permeabilities. Soc. Petrol. Engin. J., **24** : 275 - 276. SPE Paper, **10165** (1981).

CHOATE, R., JURICH, D. & SAULNIER, G.J. (1981) : Geologic overview, coal deposits, and potential for methane recovery from coalbeds, Piceance Basin, Colorado. Prepared for U.S. Dept. Energy, Morgantown Energy Technology Center, by TRW Energy Engineering Division, 184 pp., McLean/Virginia.

CHOATE, R., LENT, J. & RIGHTMIRE, C.T. (1982) : San Juan Basin report - Upper Cretaceous geology, coal and the potential for methane recovery from coalbeds, San Juan Basin, Colorado and New Mexico. Prepared for U.S. Dept. Energy, Morgantown Energy Technology Center, by TRW Coalbed Methane Program, 149 pp., McLean/Virginia.

CHOO, Y.K. & WU, C.H. (1987) : Transient pressure behavior of multiple-fractured gas wells. SPE/DOE Paper, **16398**.

CHOUET, B. (1986) : Dynamics of a fluid-driven crack in three dimensions by the finite difference method. J. Geophys. Res., **91** (B 14) : 13,967 - 13,992.

CHOWDIAH, P. (1986) : Effects of pore water distribution and stress on the laboratory measurement of tight sandstone properties. SPE Paper, **15210**.

CHOWDIAH, P. & SOEDER, D.J. (1988) : Comparison of pore geometry in high and low permeability sandstones, Travis Peak Formation, East Texas. SPE Paper, **17729**.

CHU, C. (1985) : State-of-the-art review of steamflood field projects. J. Petrol. Technol., **37** : 1887 - 1902. SPE Paper, **11733** (1983).

CHU, W.C. & RAGHAVAN, R. (1981): The effect of non-communicating layers on interference test data. J. Petrol. Technol., **33** : 370 - 382. SPE Paper, **8390** (1979).

CHUDY, S. (1981) : Photon Log. Trans. 22nd Soc. Prof. Well Log Analysts Annual Logging Symp., Paper **GG**, Mexico City/Mexico, 23. - 26.06.1981.

CHURCH, D.C. & PETERS, B.A. (1987) : Improved fracturing technique yields increased production potential. SPE Paper, **17045**.

CHURCH, D.C., QUISENBERRY, J.L. & FOX, K.B. (1981) : Field evaluation of gelled acids for carbonate formations. SPE Paper, **9753**. J. Petrol. Technol., **33** : 2471 - 2474.

CIGNI, U., GAUDIO, P.d. & FABBRI, F. (1981) : Italian experience and problems in deep geothermal drilling. Proc. Intern. Conf. Geotherm. Drill. Complet. Technol., Sandia National Laboratories Rept., **SAND 81-0036 C** : 30-1 - 30-16; Albuquerque/New Mexico.

CINCO-LEY, H. (1982) : Evaluation of hydraulic fracturing by transient pressure analysis methods. SPE Paper, **10043**.

CINCO-LEY, H., BRIGHAM, W.E., ECONOMIDES, M.J., MILLER, F.G., RAMEY, H.J., BARELLI, A. & MANNETTI, G. (1979) : A parallelepiped model to analyze the pressure behavior of geothermal steam wells penetrating vertical fractures. SPE Paper, **8231**.

CINCO-LEY, H. & MENG, H.Z. (1988) : Pressure transient analysis of wells with finite conductivity vertical fractures in double porosity reservoirs. SPE Paper, **18172**.

CINCO-LEY, H., RAMEY, H.J. & MILLER, F.G. (1975) : Unsteady-state pressure distribution created by a well with an inclined fracture. SPE Paper, **5591**.

CINCO-LEY, H., RAMEY, H.J. & SAMANIEGO, F.V. (1987) : Behavior of wells with low-conductivity vertical fractures. SPE Paper, **16776**.

CINCO-LEY, H. & SAMANIEGO, F.V. (1977 a) : Determination of the orientation of a finite conductivity vertical fracture by transient pressure analysis. SPE Paper, **6750**.
CINCO-LEY, H. & SAMANIEGO, F.V. (1977 b) : Effect of wellbore storage and damage on the transient pressure behavior of vertically fractured wells. SPE Paper, **6752**.
CINCO-LEY, H. & SAMANIEGO, F.V. (1981 a) : Transient pressure analysis for fractured wells. J. Petrol. Technol., **33** : 1749 - 1766. SPE Paper, **7490** (1978).
CINCO-LEY, H. & SAMANIEGO, F.V. (1981 b) : Transient pressure analysis : finite conductivity fracture case versus damaged fracture case. SPE Paper, **10179**.
CINCO-LEY, H., SAMANIEGO, F.V. & DOMINGUEZ, N.A. (1978) : Transient pressure behavior for a well with a finite-conductivity vertical fracture. Soc. Petrol. Engin. J., **18** : 253 - 264. SPE Paper, **6014** (1976).
CINCO-LEY, H., SAMANIEGO, F.V. & KUCUK, F. (1985) : The pressure transient behavior for naturally fractured reservoirs with multiple block size. SPE Paper, **14168**.
CIPOLLA, C.L. & LEE, S.J. (1987) : The effect of excess propped fracture height on well productivity. SPE Paper, **16219**.
CIRIGLIANO, J.A. & LEIBACH, R.E. (1967) : Gravel packing in Venezuela. Proc. 7th World Petrol. Congr., Sect. **III** : 407 - 418.
CIVAN, F. & DONALDSON, E.C. (1987) : Relative permeability from unsteady-state displacements : an analytical interpretation. SPE Paper, **16200**.
CIVAN, F. & KNAPP, R.M. (1987) : Effect of clay swelling and fines migration on formation permeability. SPE Paper, **16235**.
CLARK, E.H. (1986) : Strategy for survival : service industry. SPE Paper, **15848**.
CLARK, G.J., WONG, T.C.T. & MUNGAN, N. (1982) : New acid systems for sandstone stimulation. SPE Paper, **10662**.
CLARK, H.B., PIKE, M.T. & RENGEL, G.L. (1979) : The use of fluorochemical surfactants in non-aqueous stimulation fluids. SPE Paper, **7894**.
CLARK, H.B., PIKE, M.T. & RENGEL, G.L. (1980) : Water soluble fluorochemical surfactant well stimulation additives. SPE Paper, **9008**.
CLARK, H.C. (1983) : A method for selection of optimum proppant in the Cotton Valley. SPE/DOE Paper, **11617**.
CLARK, H.C. (1987) : Mechanical design considerations for fracture-treating down casing strings. SPE Drill. Engin., **2** : 116 - 126. SPE Paper, **14370** (1985).
CLARK, J.A. (1982) : Glacial loading : a cause of natural fracturing and a control of the present stress state in regions of high Devonian shale gas production. SPE/DOE Paper, **10798**.
CLARK, J.A. (1983) : The prediction of hydraulic fracture azimuth through geological, core, and analytical studies. SPE/DOE Paper, **11611**.
CLARK, J.B. (1949) : Hydraulic process for increasing productivity of wells. Trans. AIME, **186** : 1 - 8. Soc. Petrol. Engin., Annual Fall Meeting (1948).
CLARK, J.B., FAST, C.R. & HOWARD, G.C. (1952) : A multiple-fracturing process for increasing productivity of wells. API Drill. Product. Pract., **1952** : p. 104 ff.
CLARK, K.K. (1968) : Transient pressure testing on fractured water injection wells. J. Petrol. Technol., **20** : 639 - 643.
CLARK, P.E. (1979) : Stimulation fluid rheology - a new approach. SPE Paper, **8300**.
CLARK, P.E., CLOUD, J.E., WAREMBOURG, P.A. & YUNKO, C.G. (1978) : Importance of fluid temperature calculations in fracture treatment design for high temperature wells. SPE Paper, **7757**.
CLARK, P.E. & GÜLER, N. (1983) : Proppant transport in vertical fractures : settling velocity correlations. SPE/DOE Paper, **11636**.
CLARK, P.E., HARKIN, M.W., WAHL, H.A. & SIEVERT, J.A. (1977) : Design of a large vertical prop transport model. SPE Paper, **6814**.
CLARK, P.E., HALVACI, M., GHAELI, H. & PARKS, R.F. (1985) : Proppant transport by xanthan and xanthan-hydroxypropyl guar solutions : alternatives to crosslinked fluids. SPE/DOE Paper, **13907**.

CLARK, P.E., MANNING, F.S., QUADIR, J.A. & GULER, N. (1981) : Prop transport in vertical fractures. SPE Paper, **10261**.
CLARK, P.E. & QUADIR, J.A. (1981) : Proppant transport in hydraulic fractures : a critical review of particle settling velocity equations. SPE/DOE Paper, **9866**.
CLAVIER, C., HOYLE, W. & MEUNIER, D. (1971) : Quantitative interpretation of thermal neutron decay time logs : Part I. Fundamentals and techniques. Part II. Interpretation example, interpretation accuracy, and time-lapse technique. J. Petrol. Technol., **23** : 743 - 755, 756 - 763. SPE Paper, **2658** (1969).
CLAWSON, F.L. & SPARLIN, D.D. (1975) : Slurry packing deep, high pressure wells. SPE Paper, **5378**.
CLEARY, M.P. (1978 a) : Moving singularities in elasto-diffusive solids with applications to fracture propagation. Intern. J. Solids Struct., **14** : 81 - 97.
CLEARY, M.P. (1978 b) : Primary factors governing hydraulic fractures in heterogeneous stratified porous formations. Paper **78-PET-47**, ASME Energy Technology Conference and Exhibition, Houston/Texas.
CLEARY, M.P. (1978 c) : The effect of fluid properties on arch stability in unconsolidated sands. Thesis, **T-2072**; Colorado School of Mines, Golden/Colorado.
CLEARY, M.P. (1979) : Rate and structure sensitivity in hydraulic fracturing of fluid saturated porous formations. Proc. 20th U.S. Symp. Rock Mechan., 127 - 142, Austin/Texas.
CLEARY, M.P. (1980 a) : Comprehensive design formulae for hydraulic fracturing. SPE Paper, **9259**.
CLEARY, M.P. (1980 b) : Analysis of mechanisms and procedures for producing favourable shapes of hydraulic fractures. SPE Paper, **9260**.
CLEARY, M.P. (1988) : The engineering of hydraulic fractures - state of the art and technology of the future. J. Petrol. Technol., **40** : 13 - 21. SPE Paper, **17260**.
CLEARY, M.P., BARR, D.T. & WILLIS, R.M. (1988) : Enhancement of real-time hydraulic fracturing models with full 3-D simulation. SPE Paper, **17713**.
CLEARY, M.P., BUHARALI, A.M., CROCKETT, A.R. & SALEHI, I.A. (1986) : Computerized field system for real-time monitoring and analysis of hydraulic fracturing operations. SPE Paper, **14087**.
CLEARY, M.P., CROCKETT, A.R., MARTINEZ, M., NARENDRAN, V.M. & SLUTSKY, S. (1983) : Surface integral schemes for fluid flow and induced stresses around fractures in underground reservoirs. SPE/DOE Paper, **11632**.
CLEARY, M.P., KAVVADAS, M. & LAM, K.Y. (1983) : Development of a fully three-dimensional simulator for analysis and design of hydraulic fracturing. SPE/DOE Paper, **11631**.
CLEARY, M.P., KECK, R.G. & MEAR, M.E. (1983) : Microcomputer models for the design of hydraulic fractures. SPE/DOE Paper, **11628**.
CLEARY, M.P., LAM, K.Y. & BARR, D.T. (1986) : A complete 3 D simulator for analysis and design of hydraulic fracturing. SPE Paper, **15266**.
CLEARY, M.P., MELVAN, J.J. & KOHLHAAS, C.A. (1979) : The effect of confining stress and fluid properties on arch stability in unconsolidated sands. SPE Paper, **8426**.
CLEARY, M.P., NARENDRAN, V.M., SETTARI, A. & WONG, S.K. (1982) : General solutions for unsymmetric unidirectional hydraulic fracture propagation in reservoir structures. Mass. Inst. Technol., Resource Extraction Laboratory, Rept., **REL-82-10**.
CLEMENTZ, D.M. (1977) : Clay stabilization in sandstones through adsorption of petroleum heavy ends. J. Petrol. Technol., **29** : 1061 - 1066. SPE Paper, **6217** (1976).
CLEMENTZ, D.M. (1982) : Stimulation of water injection wells in the Los Angeles Basin using sodium hypochlorite and mineral acids. SPE Paper, **10624**. J. Petrol. Technol., **34** : 2087 - 2096.
CLERKE, E.A. & AKKEREN, T.J.v. (1986) : Borehole televiewer improves completion results in a Permian Basin San Andres reservoir. SPE Paper, **15033**. SPE Product. Engin., **3** (1988) : 89 - 95.
CLEUTINX, C. (1986) : Energy policy in the European community. SPE Paper,

14638.
CLIFTON, R.J. & ABOU-SAYED, A.S. (1979) : On the computation of the three dimensional geometry of hydraulic fractures. SPE Paper, **7943**.
CLIFTON, R.J. & ABOU-SAYED, A.S. (1981) : A variational approach to the prediction of the three-dimensional geometry of hydraulic fractures. SPE/DOE Paper, **9879**.
CLIFTON, R.J. & WANG, J.J. (1988) : Multiple fluids, proppant transport and thermal effects in three-dimensional simulation of hydraulic fracturing. SPE Paper, **18198**.
CLIFTON, R.J. & WANG, J.J. (1989) : Fluid tracking and proppant transport in three-dimensional simulation of hydraulic fracturing. SPE Paper, **18455**.
CLONTS, M.D. & RAMEY, H.J. (1986) : Pressure transient analysis for wells with horizontal drainholes. SPE Paper, **15116**.
CLOSMANN, P.J. (1971) : Process for horizontally fracturing subsurface earth formations. U.S. Patent, **3,313,785** (10.1971).
CLOSMANN, P.J. & PHOCAS, D.M. (1976) : Thermal stresses near a heated fracture in transversely isotropic oil shale. SPE Paper, **6183**.
CLOSMANN, P.J. & SMITH, R.A. (1983) : Temperature observations and steam-zone rise in the vicinity of a steam-heated fracture. Soc. Petrol. Engin. J., **23** : 575 - 586. SPE Paper, **9898** (1981).
CLOUD, J.E. & CLARK, P.E. (1980) : Stimulation fluid rheology III. Alternatives to the Power law fluid model for crosslinked gels. SPE Paper, **9332**.
COATES, G.R. & DENOO, S.A. (1981) : Mechanical properties program using borehole stress analysis and Mohr´s circle. Trans. 22nd Soc. Prof. Well Log Analysts Annual Logging Symp., Paper **DD**, Mexico City/Mexico, 23. - 26.06.1981.
COBB, S.L. (1985) : An overview of the use of proppants in hydraulic fracturing. Standard Oil Proppants Co., Intern. Rept., 9 pp.
COBB, S.L. (1986) : Compuprop - a proppant selection program. SPE Microcomp. News, **1986** : 6 - 7.
COBB, S.L. & FARRELL, J.J. (1986) : Evaluation of long-term proppant stability. SPE Paper, **14133**.
COBB, W.M., RAMEY, H.J. & MILLER, F.G. (1972) : Well-test analysis for wells producing commingled zones. J. Petrol. Technol., **24** : 27 - 37. SPE Paper, **3014** (1970).
COBERLY, C.J. (1937) : Selection of screen openings for unconsolidated sands. API Drill. Product. Pract., **1937** : 189 - 201.
COBERLY, C.J. & WAGNER, E.M. (1938) : Some considerations in the selection and installation of gravel pack for oil wells. AIME Techn. Publ., Petrol. Technol. (Class G, Petrol. Divis., **64**), **960** : 1 - 20. SPE Reprint Series, Well Completion, **5** (1970) : 156 - 175.
COCIANCIG, B., GRIFFITHS, J. & SCHÜLER, S. (1980) : Optimierung von Frac-Behandlungen. Erdoel-Erdgas, **96** : 373 - 377.
COFFER, H.F., BRAY, B.G., KNUTSON, C.F. & RAWSON, D.E. (1964) : Effects of nuclear explosions on oil reservoir stimulation. J. Petrol. Technol., **16** : 473 - 480.
COHEN, Y. & METZNER, A.B. (1982) : Adsorption effects in the flow of polymer solutions through capillaries. Macromolecules, **15** : p. 1425 ff.
COL, V. (1983) : Zirprop 126. SEPR - Messina Worldwide, Paris/France.
COLE, C., SHAW, S., CAVENY, B. & BELLENGER, B. (1988) : Monitoring HEC gel shearing to optimize improvements. SPE Paper, **17480**.
COLLE, E.A. (1978) : A completion technique for overcoming formation damage. SPE Paper, **7009**.
COLLE, E. (1988) : Increase production with underbalanced perforation. Petrol. Engin. Intern., **1988**/7 : 39 - 42.
COMALCO (1983) : Duraprop. Calcined bauxite proppant. Technical specifications. Melbourne/Australia.
COMALCO (1986) : Long term conductivity tests on Dura-Prop. Industr. Miner. Techn. Rept. Proppants, RIM TR **1/86** : 8 pp.; Melbourne/Australia.
COMPRESSED AIR MAGAZINE (1986) : Persuading an oil well to produce. Compressed Air Magazine, **1986**/6 : 25 - 29.
CONSTANT, W.D. & BOURGOYNE, A.T. (1986) : Fracture gradient prediction for offshore wells. SPE Paper, **15105**. SPE Drill. Engin., **3** (1988) : 136 - 140.

CONSTIEN, V.G., BRANNON, H.D. & BANNISTER, C.E. (1988) : Continuous mix technology adds new flexibility to frac jobs. Oil Gas J., 06.08.**1988** : 49 - 54.
CONSTIEN, V.G. & MAYER, M.H. (1978) : What! No screen? Gravel packing with water-carried resin coated gravel. SPE Paper, **7003**.
CONSTIEN, V.G. & WALLACE, J.L. (1980) : Designed carrying fluid improves gravel pack performance. SPE Paper, **8802**.
CONWAY, M.W., ALMOND, S.W., BRISCOE, J.E. & HARRIS, E. (1983) : Chemical model for the rheological behavior of cross-linked fluid systems. J. Petrol. Technol., **35** : 315 - 320. SPE Paper, **9334** (1980).
CONWAY, M.W. & HARRIS, L.E. (1982) : A laboratory and field evaluation of a technique for hydraulic fracturing stimulation of deep wells. SPE Paper, **10964**.
CONWAY, M.W., McGOWEN, J.M., GUNDERSON, D.W. & KING, D.G. (1985) : Prediction of formation response from fracture pressure behavior. SPE Paper, **14263**.
CONWAY, M.W. & McMECHAN, D. (1983) : Recent development in hydraulic stimulation treatments in the Fletcher field in the Deep Anadarko Basin. SPE/DOE Paper, **11604**.
CONWAY, M.W., McMECHAN, D.E., McGOWEN, J.M., BROWN, D., CHISHOLM, P.T. & VENDITTO, J.J. (1985) : Expanding recoverable reserves through refracturing. SPE Paper, **14376**.
CONWAY, M.W., PAULS, R.W. & HARRIS, L.E. (1980) : Evaluation of procedures and instrumentation available for time-temperature stability of crosslinked fluids. SPE Paper, **9333**.
COOKE, C.E. (1973) : Conductivity of fracture proppants in multiple layers. J. Petrol. Technol., **25** : 1101 - 1107. SPE Paper, **4117** (1972). SPE Reprint Series, Well Completions, **5 a**/II (1978) : 220 - 226.
COOKE, C.E. (1975) : Effect of fracturing fluids on fracture conductivity. J. Petrol. Technol., **27** : 1273 - 1282. SPE Paper, **5114** (1974). SPE Reprint Series, Well Completions, **5 a**/II (1978) : 227 - 236.
COOKE, C.E. (1976) : Hydraulic fracturing with a high-strength proppant. SPE Paper, **6213**. J. Petrol. Technol., **29** (1977) : 1222 - 1226.
COOKE, C.E. (1977) : Use of high-strength proppant for fracturing deep wells. SPE Paper, **6440**. Fracturing with a high-strength proppant. J. Petrol. Technol., **29** : 1222 - 1226.
COOKE, C.E. (1979) : Radial differential temperature (RDT) logging - a new tool for detecting and treating flow behind casing. J. Petrol. Technol., **31** : 676 - 682. SPE Paper, **7558** (1978).
COOKE, C.E. & GIDLEY, J.L. (1979) : High-strength proppant extends deep well fracturing capabilities. Proc. 10th World Petrol. Congr., 89 - 96, Panel Discussion 7, Paper 5; Bucharest/Romania.
COOKE, C.E., GIDLEY, J.L. & MUTTI, D.H. (1977) : Use of high-strength proppant for fracturing deep wells. SPE Paper, **6440**.
COOKE, C.E., GONZALEZ, O.J. & BROUSSARD, D.J. (1985) : Primary cementing improvement by casing vibration during cement curing time. SPE Paper, **14199**. SPE Product. Engin., **3** (1988) : 339 - 345.
COOKE, C.E., HEDDEN, W.A. & CHARD, W.C. (1978) : Hydraulic fracturing method using sintered bauxite propping agent. U.S. Patent, **4,068,718** (17.01.1978).
COOKE, R.W. (1983) : Areal sweep efficiency determination in steamdrive projects utilizing chemical tracers. SPE Paper, **11805**.
COOPER, G.D., NELSON, S.G. & SCHOPPER, M.D. (1983) : Improving fracturing design through the use of an on-site computer system. SPE Paper, **12063**.
COOPER, M.P., NELSON, S.G. & SCHOPPER, M.D. (1984) : Comparison of methods for determining in-situ leak-off rate based on analysis with an on-site computer. SPE Paper, **13223**.
COOPER, R.E. & MARSHALL, L.B. (1984) : The first purpose-built stimulation vessel for North Sea application. SPE Paper, **12993**.
COOPER, R.E. & THRONCOSO, J.C. (1988) : An overview of horizontal well completion technology. SPE Paper, **17582**.
COPELAND, C.T. & BASS, D.M. (1976) : A study of the thermal resistance of resin-coated gravel. SPE Paper, **6180**.
COPELAND, C.T. & CONSTIEN, V.G. (1978) : Method for forming a consolidated gravel pack. U.S. Patent, **4,074,760** (21.02.1978), **3,867,986**.

COPELAND, C.T. & McAULEY, J.D. (1974) : Controlling sand with an epoxy-coated, high-solids-content gravel slurry. J. Petrol. Technol., **26** : 1215 - 1220. SPE Paper, **4553** (1973).

COPPEL, C.P., JENNINGS, H.Y. & REED, M.G. (1973) : Field results from wells treated with hydroxy-aluminum. J. Petrol. Technol., **25** : 1108 - 1112. SPE Paper, **3998** (1972).

CORLEY, B.H. & KLAUTT, W.R. (1986) : Prediction of fracture migration using elastic rock properties. Intern. Symp. on Petroleum Exploration in Carbonate Areas, Nanjing/People`s Republic of China, 12. - 19.11.1986.

CORNET, F.H. & VALETTE, B. (1984) : In situ stress determination from hydraulic injection test data. J. Geophys. Res., **89**: 11527 - 11537.

CORWITH, J.R. & MENGEL, F. (1988) : Applications of pulsed neutron capture logs in the Ekofisk area fields. SPE Paper, **18147**.

COSTEIN, V.G. & WALLACE, J.L. (1980) : Designed carrying fluid improves gravel pack performance. SPE Paper, **8802**.

COTTRILL, A. (1987) : Showing the Soviet Union the North Sea`s offshore strengths. Offshore Engin., **1987**/10 : 61 - 62.

COUDERC, B., MONFRIN, D., QUETTIER, L. & SAHUQUET, B. (1987) : Emeraude Vapeur, an offshore steam pilot. 4th Europ. Symp. Enhanced Oil Recovery, Thermal Recovery Session IV, Hamburg/Germany FRG.

COUDERC, B., VERPEAUX, J.F., MONFRIN, D. & QUETTIER, L. (1987) : Emeraude Vapeur : a steam pilot in an offshore environment. SPE Paper, **16723**.

COULTER, A.W., COPELAND, C.T. & HARRISBERGER, W.H. (1979) : A laboratory study of clay stabilizers. Soc. Petrol. Engin. J., **19** : 267 - 269. SPE Paper, **7016**.

COULTER, A.W., CROWE, C.W., BARRETT, N.D. & MILLER, B.D. (1976) : Alternate stages of pad fluid and acid provide improved leak-of control for fracture acidizing. SPE Paper, **6124**.

COULTER, A.W. & GOUGLER, P.D. (1984) : Field tests indicate tubing is main source of iron precipitation in the wellbore. Oil Gas J., 03.09.**1984** : 87 - 88.

COULTER, A.W., MARTINEZ, S.J. & FISCHER, K.F. (1987) : Remedial cleanup, sand control, and other stimulation treatments. In : BRADLEY, H.B. (Editor), Petroleum Engineering Handbook, 56-1 - 56-9; Soc. Petrol. Engin., Richardson/ Texas. ISBN 1-55563-010-3.

COULTER, G.R., HARRIS, L.E. & KLEBENOW, D.E. (1980) : Successful stimulation in the Rocky Mountain area using a new acid base gel system. SPE Paper, **9032**.

COULTER, G.R., MATTHEWS, B.A., SEGLEM, R.L. & SMITH, J.E. (1980) : The theoretical and mechanical aspects of ultra high sand concentrations in hydraulic fracturing. Proc. 27th Annual Meeting, Southwest Petroleum Shortcourse, **1980**/4 : p. 97 ff.; Texas Techn. Univ., Lubbock/Texas.

COULTER, G.R. & MENZIE, D.E. (1973) : The design of refrac treatments for restimulation of subsurface formations. SPE Paper, **4400**.

COULTER, G.R. & PURVIS, S.B. (1980) : Successful stimulation practices - Offshore Holland. SPE Paper, **10934**. J. Petrol. Technol., **34** (1982) : 1211 - 1218. EUR Paper, **263**.

COULTER, G.R. & WELLS, R.D. (1972) : The advantages of high proppant concentration in fracture stimulation. J. Petrol. Technol., **24** : 643 - 650. SPE Paper, **3208** (1971).

COURBOT, A. (1987) : Steam injection pilot test on Marienbronn field (France) to produce a 11 o API oil. 4th Europ. Symp. Enhanced Oil Recovery, Thermal Recovery Session IV, Hamburg/Germany FRG.

COURSEN, D.L. (1983) : Early production from some Devonian shale gas wells stimulated with several kinds of explosive charge. SPE Paper, **12322**.

COX, J.W. (1983) : Long axis orientation in elongated boreholes and its correlation with rock stress data. Trans. 24th Soc. Prof. Well Log Analysts Annual Logging Symp., Paper **J**, Calgary/Alberta, 27. - 30.06.1983. The Technical Review, **30** (1982) : 15 - 24.

COX, T.F. (1984) : The economic aspects of marginal field development in an offshore environment. SPE Paper, **12583**.

CRADDOCK, D.L., GOZA, B.T. & BISHOP, J.C. (1983) : A case history : fracturing the Morrow in Southern Blaine and Western Canadian Counties in Oklahoma. J.

Petrol. Technol., **35** : 507 - 511. SPE Paper, **10131** (1981), **11567** (1983).

CRAFTON, J.W. & HARRIS, C.D. (1976) : A direct finite-difference simulation of a gas well with a finite capacity vertical fracture. SPE Paper, **5736**.

CRAIGIE, L.J. (1983) : A new method for determining the rheology of crosslinked fracturing fluids using shear history simulation. SPE/DOE Paper, **11635**.

CRAIGHEAD, M.S., HOSSAINI, M. & FREEMAN, E.R. (1985) : Foam fracturing utilizing delayed crosslinked gels. SPE Paper, **14437**.

CRAIGHEAD, M.S., HOSSAINI, M. & WATSON, R.W. (1983) : Foamed anhydrous methanol stimulation. SPE Paper, **12315**.

CRAIGHEAD, M.S., McDOUGALL, J.G. & HOTZ, R.F. (1987) : New two-part logging program to aid in prehydraulic fracture design and postfracture diagnosis. Trans. 11th Meeting Canad. Well Logging Soc.

CRAMER, D.D. (1987 a) : The application of limited-entry techniques in massive hydraulic fracturing treatments. SPE Paper, **16189**.

CRAMER, D.D. (1987 b) : Limited entry extended to massive hydraulic fracturing. Oil Gas J., 14.12.**1987** : 40 - 45.

CRAMER, D. & SONGER, M. (1988) : The use of microfracturing techniques in moderate permeability sandstone reservoirs. Batch mix fracturing : an effective method of stimulating moderate permeability reservoirs. SPE Paper, **17500**.

CRAMPIN, S. & ATKINSON, B.K. (1985) : Microcracks in the Earth`s crust. First Break, **1985**/3 : 16 - 21.

CRAWFORD, D.L., COULTER, A.W. & OSBORN, F.E. III (1980) : Removal of wellbore damage from highly permeable formations and naturally fractured reservoirs. SPE Paper, **8796**.

CRAWFORD, G.E., HAGEDORN, A.R. & PIERCE, A.E. (1976) : Analysis of pressure buildup tests in a naturally fractured reservoir. J. Petrol. Technol., **28** : 1295 - 1300. SPE Paper, **4558** (1973).

CRAWFORD, H.R. (1983) : Proppant scheduling and calculation of fluid lost during fracturing. SPE Paper, **12064**.

CRAWFORD, P.B. & COLLINS, R.E. (1953) : Estimated effect of vertical fractures on secondary recovery. TP Paper, **3879**.

CRAWFORD, P.B. & LANDRUM, B.L. (1956) : Do horizontal fractures affect production capacity? World Oil, **1956**/5.

CREMA, S.C. & ALM, R.R. (1985) : Foaming of anhydrous methanol for well stimulation. SPE Paper, **13565**.

CREMEAN, S.P., McKETTA, S.F., OWENS, G.L. & SMITH, E.C. (1979) : Massive hydraulic fracturing experiments of the Devonian Shale in Lincoln County, West Virginia. Columbia Final Rept. to the U.S. Dept. Energy, Contract, **E (46-1)** 8014, 01.1979.

CRENSHAW, P.L. & FLIPPEN F.F. (1968) : Stimulation of the deep Ellenburger in the Delaware Basin. SPE Paper, **2075**.

CRENSHAW, P.L., FLIPPEN, F.F. & PAULEY, P.O. (1968) : Stimulation treatment design for the Delaware Basin Ellenburger. SPE Paper, **2375**.

CRENSHAW, P.L. & JOINES, B.R. (1977) : Novel stimulation technique proven effective in the world`s biggest deep gas producer. SPE Paper, **6434**.

CRITTENDON, B.C. (1959) : The mechanics of design and interpretation of hydraulic fracture treatments. J. Petrol. Technol., **11**/10.

CROCKER, M.E., DONALDSON, E.C. & MARCHIN, L.M. (1983) : Comparison and analysis of reservoir rocks and related clays. SPE Paper, **11973**.

CROCKETT, A.R., OKUSU, N.M. & CLEARY, M.P. (1986) : A complete integrated model for design and real-time analysis of hydraulic fracturing operations. SPE Paper, **15069**.

CROCKETT, A.R., VO, H. & CLEARY, M.P. (1984) : Studies of fluid flow, heat transfer and induced stresses in and around underground fractures. Mass. Inst. Technol., Resourc. Extract. Lab. Rept., **REL-84-1** (1984/3), **REL-83-10** (1983).

CROCKETT, A.R., WILLIS, R.M. & CLEARY, M.P. (1986) : Improvement of hydraulic fracture predictions by real-time history matching on observed pressures. SPE Paper, **15264**.

CROLL, D.O. (1987) : Western Europe - oil imports up on low oil prices in 1986. Petrol. Econ., **54**/8 : 305 - 306.

CROSS, R.R. & SHELDON, L. (1977) : Fracturing of the Miocene Sand in South

Louisiana. SPE Paper, **6447**.
CROUSE, P.C. (1987) : Opec reorganization could spell relief. World Oil, **206** (1987/2) : 25 - 28.
CROUSE, P.C. (1988 a) : The focus is at OPEC. World Oil, **207** (1988/8) : 27 - 28.
CROUSE, P.C. (1988 b) : The U.S. industry in `88. A better year depends on oil prices and the economy. World Oil, **207** (1988/2) : 27 - 33.
CROW, W.L. (1977) : Mini massive frac - a stimulation technique for low-porosity, low-permeability zones. SPE Paper, **6627**.
CROWE, C.W. (1971) : Evaluation of oil-soluble resin mixtures as diverting agents for matrix acidizing. SPE Paper, **3505**.
CROWE, C.W. (1984 a) : Evaluation of agents for preventing precipitation of ferric hydroxide from spent treating acid. SPE Paper, **12497**. J. Petrol. Technol., **37** (1985) : 691 - 695.
CROWE, C.W. (1984 b) : Precipitation of hydrated silica from spent hydrofluoric acid - how much of a problem is it? SPE Paper, **13083**. J. Petrol. Technol., **38** (1986) : 1234 - 1240.
CROWE, C.W. & CRYAR, H.B. (1975) : Development of oil-soluble resin mixtures for control of fluid loss in water-base workover and completion fluids. SPE Paper, **5662**.
CROWE, C.W., HUTCHINSON, B.H. & TRITTIPO, B.L. (1987) : Fluid loss control : the key to successful acid fracturing. SPE Paper, **16883**.
CROWE, C.W., MARTIN, R.C. & MICHAELIS, A.M. (1980) : Evaluation of acid-gelling agents for use in well stimulation. SPE Paper, **9384**. Soc. Petrol. Engin. J., **21** (1981) : 415 - 424.
CROWE, C.W., McGOWAN, G.R. & BARANET, S.E. (1988) : Investigation of retarded acids provides better understanding of their effectiveness and potential benefits. SPE Paper, **18222**.
CROWE, C.W. & MINOR, S.S. (1982) : Effect of acid corrosion inhibitors on matrix stimulation results. SPE Paper, **11119**. J. Petrol. Technol., **37** (1985) : 1853 - 1860.
CROWELL, R.F. & JENNINGS, A.J. (1978) : A diagnostic technique for restimulation candidate selection. SPE Paper, **7556**.
CRUMP, J. & CONWAY, M. (1986) : Effects of perforation entry friction on bottomhole treating analysis. SPE Paper, **15474**.
CUDERMAN, J.F. (1982) : Multiple fracturing experiments - propellant and borehole considerations. SPE/DOE Paper, **10845**.
CUDERMAN, J.F. (1986) : Tailored-pulse fracturing in cased and perforated boreholes. SPE Paper, **15253**.
CUDERMAN, J.F. & NORTHROP, D.A. (1984) : A propellant-based technology for multiple fracturing wellbores to enhance gas recovery : application and results in Devonian shale. SPE/DOE/GRI Paper, **12838**. SPE Product. Engin., **1** (1986) : 97 - 103.
CUESTA, J.F. & ELPHICK, J.J. (1984) : Pre- and post-stimulation pressure test analysis and its role in the design and evaluation of hydraulic fracture treatments. SPE Paper, **12992**.
CULLENDER, M.H. & SMITH, R.V. (1956) : Practical solution of gas-flow equations for well and pipelines with large temperature gradients. Trans. AIME, **207** : 281 - 287.
CUNNINGHAM, J.R. (1982) : Market competition will shape future gas prices. SPE Paper, **11096**.
CURRENT, S.P. (1988) : Chemical emulsion viscosity reduction to improve steam stimulation performance. SPE Paper, **17481**.
CUTLER, R.A., ENNISS, D.O., JONES, A.H. & CARROLL, H.B. (1983) : Comparison of the fracture conductivity of commercially available and experimental proppants at intermediate and high closure stresses. SPE/DOE Paper, **11634**.
CUTLER, R.A., ENNISS, D.O., JONES, A.H. & SWANSON, S.R. (1985) : Fracture conductivity comparison of ceramic proppants. Soc. Petrol. Engin. J., **25** : 157 - 170.
CUTLER, R.A. & JONES, A.H. (1982) : Lightweight proppants for deep gas well stimulation. DOE Second Annual Report, Terra Tek Inc., 04/1982. Bartlesville Energy Technology Center Rept., **10038-22** (04/1982).

CUTLER, R.A., JONES, A.H., ENNISS, D.O. & SWARTZ, G.C. (1983) : Lightweight proppants for deep gas well stimulation - Third Annual Report. Bartlesville Energy Technology Center Rept.

CUTLER, R.A., JONES, A.H., SWANSON, S.R. & CARROLL, H.B. (1981) : New proppants for deep gas well stimulation. SPE/DOE Paper, **9869**.

DACCORD, G., NITTMANN, J. & STANLEY, H.E. (1986) : Radial viscous fingers and diffusion-limited aggregation : fractal dimension and growth sites. Phys. Rev. Lett., **56** : 336 - 339.

DACCORD, G., TOUBOUL, E. & LENORMAND, R. (1987) : Carbonate acidizing : a quantitative study of the wormholing phenomenon. SPE Paper, **16887**.

DAHLE, Ø. (1987) : The international oil markets and Norway`s role. In : Norway, OPEC and IEA - global dialogue and price stabilizing measures in the oil market? Intern. Politics, **1987**/5-6; Norwegian Institute of International Affairs.

DAHLE, Ø. (1988) : Gas versus oil - similarities and differences. Offshore Northern Seas Conference, Stavanger/Norway, 23. - 26.08.1988.

DAINES, S.R. (1981) : The prediction of fracture pressures for wildcat wells. SPE Paper, **9254**.

DAKE, L.P. (1982) : Application of the repeat formation tester in vertical and horizontal pulse testing in the Middle Jurassic Brent Sands. EUR Paper, **270**. SPE Paper, **12584** (1984).

DANCY, J. & DANCY, V.A. (1987) : Legal duty to market gas in a deteriorating market. SPE Paper, **16307**.

DANCY, J.R., DANCY, V.A. & TAYLOR, L.S. (1987) : Legal duty to market natural gas in a declining market. SPE Paper, **16239**.

D`ANDREA, L. (1979) : The prospects of natural gas supplies and markets in Europe and their influence on the U.S. SPE Paper, **8292**.

D`ANDREA, L.A. (1983) : The prospect of increased reliance by Western Europe on natural gas and its significance. SPE Paper, **11295**.

D`ANDREA, L. (1987) : The role of natural gas in the European gas market and East-West energy trade. SPE Paper, **16855**.

DANESH, A., HENDERSON, G.D., KRINIS, D. & PEDEN, J.M. (1988) : Experimental investigation of retrograde condensation in porous media at reservoir conditions. SPE Paper, **18316**.

DANESHY, A.A. (1971 a) : True and apparent direction of hydraulic fractures. Soc. Petrol. Engin. J., **11** : 149 - 163.

DANESHY, A.A. (1971 b) : Effect of treatment parameters on the geometry of a hydraulic fracture. SPE Paper, **3507**.

DANESHY, A.A. (1973 a) : On the design of vertical hydraulic fractures. J. Petrol. Technol., **25** : 83 - 93. SPE Paper, **3654** (1971).

DANESHY, A.A. (1973 b) : Experimental investigation of hydraulic fracturing through perforations. J. Petrol. Technol., **25** : 1201 - 1206. SPE Paper, **4333**. SPE Reprint Series, Well Completions, **5 a**/II (1978) : 246 - 251.

DANESHY, A.A. (1973 c) : A study of inclined hydraulic fractures. Soc. Petrol. Engin. J., **13** : 61 - 68. SPE Paper, **4062** (1972).

DANESHY, A.A. (1974) : Hydraulic fracture propagation in the presence of planes of weakness. SPE Paper, **4852**. SPE Reprint Series, Well Completions, **5 a**/II (1978) : 252 - 259.

DANESHY, A.A. (1976) : Rock properties controlling hydraulic fracture propagation. SPE Paper, **5752**.

DANESHY, A.A. (1978 a) : Hydraulic fracture propagation in layered formations. Soc. Petrol. Engin. J., **18** : 33 - 41. SPE Paper, **6088**.

DANESHY, A.A. (1978 b) : Numerical solution of sand transport in hydraulic fracturing. SPE Paper, **5636**. J. Petrol. Technol., **30** : 132 - 140.

DANESHY, A.A., SLUSHER, G.L., CHISHOLM, P.T. & MAGEE, D.A. (1984) : In-situ stress measurements during drilling. SPE Paper, **13227**. J. Petrol. Technol., **38** (1986) : 891 - 898.

DANESHY, A.A., WILLIAMS, J.R. & TINSLEY, J.M. (1971) : Effect of treatment parameters on the geometry of a hydraulic fracture. SPE Paper, **3507**.

DANIEL, E.J. (1954) : Fractured reservoirs of Middle East. Bull. Amer. Assoc. Petrol. Geol., **38** : 774 - 815.

DANIEL, G. & WHITE, J. (1980) : Fundamentals of fracturing. SPE Paper, **9064**.

DANIEL, W.L. & FERTL, W.H. (1984) : Logging high angle, long-reach boreholes. Oil Gas J., 12.**1984** : 103 - 108.
DARILEK, G.T. (1986) : A color borehole television system for developing gas production from Devonian shales. SPE Paper, **15219**.
DARIN, S.R. & HUITT, J.L. (1960) : Effect of a partial monolayer of propping agent on fracture flow capacity. Trans. AIME, **219** : 31 - 37.
DARR, C.D., BROWN, E.K. & MURPHY, J.R. (1985) : A method to convert shot open holes into cased holes with zonal isolation. SPE Paper, **15009**.
DARR, C.D.K. & CARLTON, J.C. (1988) : PUDDLE-PACK recompletion propped fracturing techniques to improve productivity of multipay producing wells. SPE Paper, **17544**.
DA SIE, W.J. & HOANG, V.T. (1985) : The well testing program at the Heber geothermal field. SPE Paper, **13612**.
DAVARZANI, J. & SLOAN, M.L. (1987) : Analysis of geothermal wells using simultaneous logging instruments. SPE Paper, **16818**.
DAVIAU, F., MOURONVAL, G., BOURDAROT, G. & CURUTCHET, P. (1985) : Pressure analysis for horizontal wells. SPE Paper, **14251**.
DAVID, A. & CARSON, G.P. (1983) : Energy financing in times of uncertainty. SPE Paper, **11286**.
DAVIES, D.K. (1980) : Reservoir stimulation of dirty sandstones. SPE Paper, **8795**.
DAVIES, D.K. & ALMON, W.R. (1981) : The role of diagenesis in successful formation interpretation, well production, stimulation, and enhanced recovery. SPE Paper, **9715**.
DAVIES, D.R., BOSMA, M.G.R. & VRIES, W.d. (1987) : Development of field design rules for viscous fingering in acid fracturing treatments : a large-scale model study. SPE Paper, **15772**.
DAVIES, D.R., HAGELAARS, A.M.P.M. & ROBERTS, D.L. (1984) : Near-wellbore placement of small-volume (sand consolidation) treatments. J. Petrol. Technol., **36** : 1905 - 1916. SPE Paper, **10177** (1981).
DAVIES, D.R., LIEVAART, L. & NITTERS, G. (1988) : The effects of corrosion inhibitors and mutual solvents on matrix acidizing treatments. SPE Paper, **17153**.
DAVIES, D.R. & KUIPER, T.O.H. (1988) : Fracture conductivity in hydraulic fracture stimulation. J. Petrol. Technol., **40** : 550 - 552. SPE Paper, **17655**.
DAVIES, D.R., ZWOLLE, S. & MEIJS, F.H. (1983) : "Silicalock" - a novel sand-control process for gas wells. J. Petrol. Technol., **35** : 2087 - 2094. SPE Paper, **9423** (1980).
DAVIS, J.J., MANCILLAS, G. & MELNYK, J.D. (1965) : Improved acid treatments by use of the spearhead film technique. SPE Paper, **1164**.
DAVIS, P.M. (1983) : Surface deformation associated with a dipping hydrofracture. J. Geophys. Res., **88** : 5826 - 5834. Trans. Amer. Geophys. Union, **63** (1982) : p. 1119 ff.
DAVIS, W.E. (1975) : Considerations for fracture stimulation of the deep Morrow in the Anadarko Basin. SPE Paper, **5391**.
DAVIS, W.E., NICKELL, R.L., LENT, J.T. & ALLAM, A.M. (1980) : The use of bottom hole pressure analysis in optimizing frac design - Anadarko Basin of Western Oklahoma. SPE Paper, **9262**.
DAVISON, C.C., KEYS, W.S. & PAILLET, F.L. (1982) : Use of borehole geophysical logs and hydrologic tests to characterize crystalline rock for nuclear waste storage, Whiteshell Nuclear Research Establishment, Manitoba and Chalk River Nuclear Laboratory, Ontario/Canada. U.S. Dept. Energy Rept., **ONWI-418** : 103 pp.
DEAN, C.J. (1939) : Drilling and production problems in the Wilmington field. API Drill. Product. Pract., **1939**.
DEAN, R.H. & LO, L.L. (1986) : Development of a natural fracture simulator and examples of its use. SPE Paper, **14110**.
DEANS, H.A. & GAO, C. (1983) : Pressure transients and crossflow in a multilayer reservoir : single-phase flow. SPE Paper, **11966**. SPE Form. Eval., **3** (1988) : 438 - 448.
DECH, J.A., HEARN, D.D., SCHUH, F.J. & LENHART, B. (1986) : New tools allow medium-radius horizontal drilling. Oil Gas J., 14.07.**1986** : 95 - 99.

DECKER, A.D. & SECCOMBE, J.C. (1986) : Geologic parameters controlling natural gas production from a single deeply buried coal reservoir in the Piceance Basin, Mesa County, Colorado. SPE Paper, **15221**.

DEES, J.M. & COULTER, G.R. (1986) : Fracturing treatment quality control : results of a three-year survey. SPE Paper, **15512**.

DEHLINGER, P., BROWNE, W.H. & BUNDRANT, C.O. (1954) : Optimum sand concentrations in well treatments. Trans. AIME, **201** : 273 - 278.

DELACOUR, J. (1986) : Frontier gas - the challenge and technical answers. Gastech Conference, Hamburg/Germany FRG, 25. - 28.11.1986.

DELHOMME, J.P., PANCI, G., LESLIE, D., CIGNI, M. & SPILA, M. (1981) : An integrated approach to fracture detection from multiple sensors. SAID Sociéte pour l´Avancement de l´Interprétation des Diagraphies, Paris/France, 10.1981.

DELHOMME, J.P., PILENKO, T., CHERUVIER, E. & CULL, R. (1986) : Reservoir applications of dipmeter logs. SPE Paper, **15485**.

DELSHAD, M., MacALLISTER, D.J., POPE, G.A. & ROUSE, B.A. (1985) : Multiphase dispersion and relative permeability experiments. Soc. Petrol. Engin. J., **25** : 524 - 534. SPE Paper, **10201** (1981).

DENNIS, B., STANDEN, E., GEORGI, D. & CALLOW, G. (1987) : Fracture identification and productivity prediction in a carbonate reef complex. SPE Paper, **16808**.

DERBY, J.A. & SMITH, P.L. (1979) : Regained gas permeability - an aid to improved hydraulic fracture design. SPE Paper, **7947**.

DERSEY, D.G. (1986) : The role of petrographic analyses in the design of non-damaging drilling, completion, and stimulation programs. SPE Paper, **14089**.

DERUYCK, B.G., BOURDET, D.P., PRAT, G.da & RAMEY, H.J. (1982) : Interpretation of interference tests in reservoirs with double-porosity behavior - theory and field examples. SPE Paper, **11025**.

DESBRANDES, R. (1985) : Encyclopedia of well logging. 584 pp.; Graham & Trotman, London/England; Gulf Publ., Houston/Texas.

DESPAX, D., CHARLEZ, P., CLINKEMAILLE, J. & ECONOMIDES, M.J. (1987) : On the stimulation of a Northern North Sea tight sand. SPE/DOE Paper, **16428**.

DETMERING, T.J. (1984) : Update on drainhole drilling - Empire Abo. Proc. 31st Annual Southwestern Petroleum Short Course, Texas Techn. Univ., Lubbock/Texas, pp. 25 - 41.

DETOURNAY, E. & FAIRHURST, C. (1982) : Generalization of the ground reaction curve concept. Proc. 23rd Symp. Rock Mechan., Berkeley/California.

DETOURNAY, E., McLENNAN, J.D. & ROEGIERS, J.C. (1986) : Poroelastic concepts explain some of the hydraulic fracturing mechanisms. SPE Paper, **15262**.

DETOURNAY, E. & ROEGIERS, J.C. (1986) : Comment on " Wellbore breakouts and in-situ stress " by Mark D. Zoback, Daniel Moos, Larry Maskin and Roger N. Anderson. J. Geophys. Res., **91** (B 14) : 14161 - 14162.

DEURER, R. & MENZ, D. (1984) : Pore structures of Rotliegend sandstones in the Ostfriesland region. Erdoel-Erdgas, **100** : 240 - 243.

DEURER, R., SAHR, M. & UJMA, K.H. (1985) : Die Visualisierung von physikalischer Trägerschädigung. Erdoel-Erdgas, **101** : 372 - 376.

DEYSARKAR, A.K., DAWSON, J.C. & ACHARYA, A.R. (1987) : Experience with CO2-laden fracturing fluid for stimulating water-sensitive formations. SPE Paper, **16193**.

DEYSARKAR, A.K., DAWSON, J.C., SEDILLO, L.P. & KNOLL-DAVIS, S. (1984) : Crosslinked acid gel. J. Canad. Petrol. Technol., **1984** : 26 - 32.

DGMK (1983) : Stimulation of low permeable deep gas-reservoirs (phase II). DGMK-Project, 167-3, Final Report. DGMK Forschungsbericht, **167-3** : 156 pp. (in German).

DGMK (1986) : Material properties and model investigations related to frac propagation during stimulation tests in low permeability sediments of North-German gas formations. Final Report - Part 1. DGMK-Project, 167/3 : Stimulation of low permeable deep gas reservoirs (phase III). DGMK Forschungsbericht, **167-4** : 137 pp. (in German).

D`HEUR, M. (1984) : Porosity and hydrocarbon distribution in the North Sea chalk reservoirs. Marine Petrol. Geol., **1** : 211 - 238.

D`HEUR, M. (1987) : Tor. In : SPENCER, A.M. et al. (Editors), Geology of the

Norwegian oil and gas fields, 129 - 142; Graham & Trotman, London/England.
D`HEUR, M. & PEKOT, L.J. (1987) : Tommeliten. In : SPENCER, A.M. et al. (Editors), Geology of the Norwegian oil and gas fields, 117 - 128; Graham & Trotman, London/England.
D`HOOGE, J.A., SHEELY, C.Q. & WILLIAMS, B.J. (1981) : Interwell tracers - an effective reservoir evaluation tool : West Sumatra field results. J. Petrol. Technol., **33** : 779 - 782. SPE Paper, **8434** (1979).
DIAMOND, W.P. & OYLER, D.C. (1980) : Drilling long horizontal coalbed methane drainage holes from a directional surface borehole. SPE/DOE Paper, **8968**.
DICKEY, P.A. (1981) : Petroleum development geology. 2nd edition, 428 pp.; 3rd edition (1986), 530 pp.; Penn Well, Tulsa/Oklahoma. ISBN 0-87814-307-6.
DICKEY, P.A. & ANDRESEN, K.H. (1945) : Behavior of water input wells. Oil Weekly, 10.12.**1945**. API Drill. Product. Pract., **1945** : 34 - 58.
DICKINSON, W. & ANDERSON, R.R. (1987) : Gravel packing of horizontal wells. SPE Paper, **16931**.
DICKINSON, W., ANDERSON, R.R. & DICKINSON, R.W. (1986) : A second-generation horizontal drilling system. SPE/IADC Paper, **14804**.
DICKINSON, W. & DICKINSON, R.W. (1985) : Horizontal radial drilling system. SPE Paper, **13949**.
DIETRICH, J.K. (1983) : Cyclic steaming of tar sands through hydraulically induced fractures. SPE Paper, **11684**. SPE Reserv. Engin., **1** (1986) : 217 - 229.
DIETRICH, J.K. (1987) : The Kern River horizontal-well steam pilot. SPE Paper, **16346**. SPE Reserv. Engin., **3** (1988) : 935 - 944.
DIETRICH, J.K. & BONDOR, P.L. (1976) : The effect of void and filled perforations on well productivity. SPE Paper, **6178**.
DIETZEL, H.J. & ERNST, P.L. (1987) : Reservoir properties required for profitable stimulations. SPE/DOE Paper, **16412**.
DIETZEL, H.J. & HANTELMANN, G.A.v. (1985) : Stimulating water-sensitive formations at high reservoir temperature. SPE Paper, **13875**.
DIJK, J.v. & KLÖTERS, W. (1987) : Lifting the platforms in the Ekofisk field. Erdöl und Kohle - Erdgas - Petrochemie, **40** : 415 - 418.
DIKKEN, B.J. & NIKO, H. (1987) : Waterflood-induced fractures : a simulation study of their propagation and effects on waterflood sweep efficiency. SPE Paper, **16551**.
DIKKERS, A.J. (1985) : Geology in petroleum production. Developm. Petrol. Sci., **20** : 240 pp., Amsterdam (Elsevier).
DILL, W.R. (1983) : Iron control in fracturing operations. Proc. 30th Annual Southwestern Petroleum Short Course, Texas Techn. Univ., Lubbock/Texas, 27. - 28.04.1983, pp. 59 - 63.
DILL, W.R. & FREDETTE, G. (1983) : Iron control in the Appalachian Basin. SPE Paper, **12319**.
DILL, W.R. & KEENEY, B.R. (1978) : Optimizing HCl-formic acid mixtures for high temperature stimulation. SPE Paper, **7567**.
DIPIPPO, R. (1984) : Worldwide geothermal power development. An overview and update. Geothermal Resources Council Bull., **1984**. Geothermal World Directory, **1983/84** : 129 - 141.
DITZHUIJZEN, P.J.D. & WAAL, J.A.d. (1984) : Reservoir compaction and surface subsidence in the Central Luconia gas-bearing carbonates offshore Sarawak, East Malaysia. SPE Paper, **12400**.
DOBECKI, T.L. (1983) : Hydraulic fracture orientation using passive borehole seismics. SPE Paper, **12110**.
DOBKINS, T.A. (1979) : Methods to better determine hydraulic fracture height. SPE Paper, **8403**.
DOBKINS, T.A. (1981 a) : Improved methods to determine hydraulic fracture height. J. Petrol. Technol., **33** : 719 - 729. SPE Paper, **8403**.
DOBKINS, T.A. (1981 b) : Procedures, results and benefits of detailed fracture treatment analysis. SPE Paper, **10130**.
DODD, C.G., CONLEY, F.R. & BARNES, P.M. (1955) : Clay minerals in petroleum reservoir sands and water sensitivity effects : clays and clay minerals. Proc. 3rd Conf. National Acad. Sci., National Res. Counc. Publ., **395** : 221 - 238.
DOE (1983 a) : Evaluation of the Devonian shale resource in the Eastern United

States. DOE Document, **DE-83-00-8749**.
DOE (1983 b) : Methane recovery from coalbeds : a potential energy source. DOE Document, **DOE/METC/83-76**.
DOE (1984) : Unconventional gas. DOE/FE Paper, **0046** : 12 pp.; Washington/DC.
DOERLER, N. & PROUVOST, L.P. (1987) : Diverting agents : laboratory study and modeling of resultant zone injectivities. SPE Paper, **16250**.
DOHERTY, M.G., RANDOLPH, P.L., ROGERS, L.A. & POONAWALA, N.A. (1982) : Methane production from geopressured aquifers. J. Petrol. Technol., **34** : 1591 - 1599. SPE Paper, **9468** (1980).
DOLAK, E. & PERNTHANER, P. (1976) : Faziesuntersuchungen und Stimulationsarbeiten im Flysch der Lagerstätte St. Ulrich - Hauskirchen. Erdoel-Erdgas, **92** : 292 - 296.
DOMSELAAR, H.R.v., SCHOLS, R.S. & VISSER, W. (1973) : An analysis of the acidizing process in acid fracturing. Soc. Petrol. Engin. J., **13** : 239 - 250. SPE Paper, **3748** (1972).
DOMSELAAR, H.R.v. & VISSER, W. (1974) : Proppant concentration in and final shape of fractures generated by viscous gels. Soc. Petrol. Engin. J., **14** : 531 - 536. SPE Paper, **4820**.
DONALDSON, E.C. (1985) : Use of capillary pressure curves for analysis of production well formation damage. SPE Paper, **13809**.
DONALDSON, E.C., ALAM, W. & CIVAN, F. (1987) : Relative permeabilities at simulated reservoir conditions. SPE Paper, **16970**.
DONALDSON, E.C., BAKER, B.A. & CARROLL, H.B. (1977) : Particle transport in sandstones. SPE Paper, **6905**.
DOSCHER, T.M. & AZARI, M. (1980) : The numerical simulation of the effect of critical gas saturation and other parameters on the productivity of methane from geopressured aquifers. SPE Paper, **8891**.
DOSCHER, T.M. & EL-ARABI, M.A. (1983) : Waste water disposal above fracture gradients in California. SPE Paper, **11680**.
DOSCHER, T.M., LABELLE, R.W., SAWATSKY, L.H. & ZWICKY, R.W. (1963) : Steam drive - a process for in-situ recovery of oil from the Athabasca Oil Sands. 2nd Athabasca Oil Sands Conference, Edmonton/Alberta, 30. - 31.10.1963.
DOSCHER, T.M., LABELLE, R.W., SAWATSKY, L.H. & ZWICKY, R.W. (1964) : Steam drive successful in Canada`s oil sands. Petrol. Engin. Intern., **1964**/1 : 71 - 78.
DOSCHER, T.M., OSBORNE, R.H., WILSON, T., RHEE, S.W., COX, D. & KUUSKRAA, V.A. (1979) : The technology and economics of methane production from geopressured aquifers. J. Petrol. Technol., **31** : 1502 - 1514. SPE Paper, **7827**.
DOSCHER, T.M., OYEKAN, R. & EL-ARABI, M.A. (1983) : A controversial laboratory study of the mechanism of crude oil displacement by carbon dioxide : Part III - nitrogen vs. carbon dioxide in dipping models. SPE Paper, **11678**.
DOSCHER, T.M., YING, A. & KOSTURA, J.A. (1988) : Improving the steam drive : physical model studies of inert gas injection and horizontal wells. SPE Paper, **17448**.
DOUGLAS, J., PAES LEME, P.J., ARBOGAST, T. & SCHMITT, T. (1987) : Simulation of flow in naturally fractured reservoirs. SPE Paper, **16019**.
DOUGHERTY, E.L. (1987) : Oil prices : can we predict where they are going? SPE Paper, **16850**.
DOUGHERTY, E.L. & AL-BLEHED, M.S. (1987) : Free world total primary energy demand and oil share : an econometric approach. SPE Paper, **16290**.
DOUGHERTY, R.L. & ABOU-SAYED, A.S. (1984) : Evaluation of the influence of in-situ reservoir conditions on the geometry of hydraulic fractures using a 3D simulator. SPE Paper, **13275**.
DOWDLE, W.P. & HYDE, P.V. (1977) : Well test analysis of fractured gas sands. SPE Paper, **6437**.
DOWELL SCHLUMBERGER (1985) : Proppant selection guide. 29 pp.
DOWELL SCHLUMBERGER (1986) : Bracketfrac controls vertical fracture height and improves your payout potential. Dowell Schlumberger Document, **TSL-2594** (06855 M).
DOWELL SCHLUMBERGER (1987) : Divertafrac Invertafrac. Services to control vertical fracture growth. Brochure, **SL-2578**; Houston/Texas.
DOWELL SCHLUMBERGER (1988 a) : PERMPAC : a better suspender to keep your per-

meability up. Advertisement in J. Petrol. Technol.
DOWELL SCHLUMBERGER (1988 b) : Model 21 gravel-packing system. Brochure, **TSL-6052** (088510 M); Houston/Texas.
DOWELL SCHLUMBERGER (1988 c) : Stimulation services and technology. Sand control services and technology. Brochure, **TSL-0536** (09874 M); Houston/Texas.
DOWELL SCHLUMBERGER (1988 d) : Acidizing. Status of the art. Brochure, **TSL-2038** (10861 M); Houston/Texas.
DRAGICEVIC, B., MARINOVIC, D., STANKOVIC, S., BATUSIC, V. & NIKOLIC, D. (1983) : Geologic properties and oil- and gas-bearing capacity of deeper parts of the Pannonian Basin in Yugoslavia. Nafta, **34** : 369 - 379.
DRESSER ATLAS (1982) : Interpretive methods for production well logs. 2nd edition, 159 pp., Houston/Texas.
DRESSER ATLAS (1987 a) : Photon/Vibrapack combination. Brochure; Houston/Texas.
DRESSER ATLAS (1987 b) : Fundamentals of diplog analysis. 210 pp., Houston/Texas.
DRISCOLL, P.L., BOWEN, J.G. & ROBERTS, M.A. (1980) : Oil base foam fracturing applied to the Niobrara Shale Formation. SPE Paper, **9335**.
DRONG, H.J. (1979) : Diagenetische Veränderungen in den Rotliegend Sandsteinen im NW-Deutschen Becken. Geol. Rdsch., **68** : 1172 - 1183.
DRONG, H.J., PLEIN, E., SANNEMANN, D., SCHUEPBACH, M.A. & ZIMDARS, J. (1982) : Der Schneverdingen-Sandstein des Rotliegenden - eine äolische Sedimentfüllung alter Grabenstrukturen. Z. dt. geol. Ges., **133** : 699 - 725.
DUDA, J.R. & HANCOCK, J.S. (1987) : An assessment of tight sand development potential in the South Rulison area, Garfield County, Colorado. SPE/DOE Paper, **16426**.
DUDA, J.R. & KUMAR, H.K. (1988) : Production forecasts for alternative completion methods in low-permeability gas reservoirs : simulation of high-angle and horizontal wells. SPE Paper, **18291**.
DUDA, J.R. & KUTSKA, H.M. (1988) : Gas production from Western tight sands via horizontal wellbores : a simulation study. SPE Paper, **17530**.
DUFOND, R. & LAFFONT, M. (1978) : Frigg, the first giant gas field in the Northern North Sea. EUR Paper, **110**.
DULA, W.F. (1981) : Correlation between deformation lamellae, microfractures, macrofractures and in situ stress measurements, White River uplift, Colorado. Geol. Soc. Amer. Bull., **92** : 37 - 46.
DULLIEN, F.A.L. & DHAWAN, G.K. (1974) : Characterization of pore structure by a combination of quantitative photomicrography and mercury porosimetry. J. Colloid Interface Sci., **47** : 337 - 349.
DUNAND, A. & SOUCEMARIANADIN, A. (1985) : Concentration effects on the settling velocities of proppant slurries. SPE Paper, **14259**.
DUNLAP, I.R. (1963) : Factors controlling the orientation and direction of hydraulic fractures. J. Inst. Petrol., **49** : p. 477 ff.
DUNLAP, I.R. (1965) : How propping agents affect packed fractures. Petrol. Engin. Intern., **1965**/11 : 80 - 90.
DURRETT, J.L., GOLBIN, W.T., MURRAY, J.W. & TIGHE, R.E. (1977) : Seeking a solution to sand control. J. Petrol. Technol., **29** : 1664 - 1672. SPE Paper, **6210** (1976).
DUSSAN, E.B. & SHARMA, Y. (1987) : An analysis of single probe formation tester pressure transient response in an anisotropic formation. SPE Paper, **16801**.
DUSSEAULT, M.B. (1977) : Stress state and hydraulic fracturing in the Athabasca oil sands. In : The oil sands of Canada-Venezuela 1977, CIM Spec. Vol., **17** : 27 - 35.
DUSSEAULT, M.B. (1985) : Fluid injection design to preclude induced shear failure of formations. CIM Paper **85-36-6**.
DUSSEAULT, M.B. & SIMMONS, J.V. (1982) : Injection-induced stress and fracture orientation changes. Canad. Geotechn. J., **19** : 483 - 493.
DUSSERT, P., SANTORO, G. & SOUDET, H. (1988) : A decade of drilling developments pays off in offshore Italian oil field. Oil Gas J., 29.02.**1988** : 33 - 39.
DUTTA, N. (1986) : Geopressure. Geophysics Reprint Series, **7** : 390 pp.; Soc. Explor. Geophys., Tulsa/Oklahoma: ISBN 0-931830-43-5.
DUTTON, K.D., BEMIS, R., JOHNSON, R. & WHITE, J.L. (1984) : Shale production

from the Britt Formation in Northwest Caddo County. SPE Paper, **12621**.
DUTTON, S.P. & FINLEY, R.J. (1988) : Controls on reservoir quality in tight sandstones of the Travis Peak Formation, East Texas. SPE Form. Eval., **3** : 97 - 104. SPE Paper, **15220** (1986).
DYES, A.B., KEMP, C.E. & CHAUDLE, B.H. (1958) : Effects of fractures on sweep-out pattern. Trans. AIME, **213** : 245 - 249. TP Paper, **8033**.
DYSART, G.R. & WHITSITT, N.F. (1967) : Fluid temperatures in fractures. SPE Paper, **1902**.
EAKIN, J.L., JOHANSEN, R.T., HOPKINS, A.D. & TALIAFERRO, R.W. (1962) : How chemical treatment gets rid of gas well water blocks. Oil Gas J., 26.11.**1962** : 83 - 89.
EARLOUGHER, R.C., KERSCH, K.M. & KUNZMAN, W.J. (1974) : Some characteristics of pressure buildup behavior in bounded multiple-layered reservoirs without crossflow. J. Petrol. Technol., **26** : 1178 - 1186. SPE Paper, **4499** (1973).
EARLOUGHER, R.C. & RAMEY, H.J. (1973) : Interference analysis in bounded systems. J. Canad. Petrol. Technol., **1973** : 33 - 45.
EASON, C.R. (1985) : Single-stage multiple zone stimulation : an effective method of stimulating the Niobrara Formation. SPE/DOE Paper, **13889**.
EATON, B.A. (1969) : Fracture gradient prediction and its application in oilfield operations. J. Petrol. Technol., **21** : 1353 - 1360. SPE Paper, **2163** (1968).
EBEL, R.E. (1986) : Oil in the Soviet Union and China : problems and prospects. SPE Paper, **14639**.
ECK, T. (1983) : World and U.S. outlook for oil and gas. SPE Paper, **11665**.
ECKEL, J.E. & REISTLE, C.E. (1940) : An apparatus for gravel packing in wells. U.S. Patent, **2,224,538** (10.12.1940).
ECONOMIDES, M.J. (1986) : Dowell Schlumberger Technology Conference. 217 pp.
ECONOMIDES, M.J. (1987 a) : Observations and recommendations in the evaluation of tests of hydraulically fractured wells. SPE/DOE Paper, **16396**.
ECONOMIDES, M.J. (1987 b) : How to engineer a fracturing treatment. J. Petrol. Technol., **39** : 1343 - 1345. SPE Paper, **17176**.
ECONOMIDES, M.J., CIKES, M., PFÖRTER, H., UDICK, T.H. & URODA, P. (1986) : The stimulation of a tight, very high-temperature gas condensate well. SPE Paper, **15239**.
ECONOMIDES, M.J. & NOLTE, K.G. (1987) : Reservoir stimulation. Schlumberger Educational Services; Dowell Schlumberger Publ., **TSL-2612**; Schlumberger Publ., **SMP-7018**; Houston/Texas.
EDGAR, A.L. (1987) : Restructuring in the oil patch. SPE Paper, **16300**.
EDWARDS, L., CHILINGAR, G., RIEKE, H. & FERTL, W. (1982) : Handbook of geothermal energy. 614 pp.
EDLUND, P.A. (1987) : Application of recently developed medium-curvature horizontal drilling technology in the Spraberry trend area. SPE/IADC Paper, **16170**. J. Petrol. Technol., **40** (1988) : 1178 - 1182.
EDWARDS, R.D. & STRELKOV, G. (1988) : Slant-hole drilling finds expanding role in Canada. Petrol. Engin. Intern., **1988**/2 : 20 - 23.
EEKELEN, H.A.M.v. (1981) : Relationship between horizontal stress and depth in sedimentary basins. SPE Paper, **10336**. J. Petrol. Technol., **34** (1982) : 2191 - 2199.
EEKELEN, H.A.M.v. (1982) : Hydraulic fracture geometry : fracture containment in layered formations. Soc. Petrol. Engin. J., **22** : 341 - 349. SPE Paper, **9261** (1980).
EHLIG-ECONOMIDES, C.A. & ANSARI, J.A. (1983) : Interference analysis for wells produced at constant pressure. SPE Paper, **11715**.
EHLIG-ECONOMIDES, C. & AYOUB, J.A. (1986) : Vertical interference testing across a low-permeability zone. SPE Form. Eval., **1** : 497 - 510. SPE Paper, **13251** (1984).
EHLIG-ECONOMIDES, C.A. & ECONOMIDES, M.J. (1985) : Pressure transient analysis in an elongated linear flow system. Soc. Petrol. Engin. J., **25** : 839 - 847. SPE Paper, **12520** (1984).
EHLIG-ECONOMIDES, C.A. & JOSEPH, J.A. (1985) : A new test for determination of individual layer properties in a multilayered reservoir. SPE Paper, **14167**.
EHLIG-ECONOMIDES, C. & RAMEY, H.J. (1979) : Transient rate decline analysis for

wells produced at constant pressure. SPE Paper, **8387**.
EHRLICH, R. (1970) : The effect of temperature on water-oil imbibition relative permeability. SPE Paper, **3214**.
EKIES, S., HADINOTO, N. & RAGHAVAN, R. (1977) : Pulse-testing of vertically fractured wells. SPE Paper, **6751**.
EKSTROM, M.P., CHEN, M.Y., ROSSI, D.J., LOCKE, S. & ARON, J. (1986) : High resolution microelectrical borehole wall imaging. 10th European Symp. Soc. Prof. Well Log Analysts, Aberdeen/Scotland, 23. - 24.04.1986.
EKSTROM, M.P., DAHAN, C.A., CHEN, M.Y., LLOYD, P.M. & ROSSI, D.J. (1986) : Formation imaging with microelectric scanning arrays. Trans. 27th Soc. Prof. Well Log Analysts Annual Logging Symp., Paper **BB**, Houston/Texas, 09. - 13. 06.1986.
ELBEL, J.L. (1985) : Considerations for optimum fracture geometry design. SPE/ DOE Paper, **13866**. SPE Product. Engin., **3** (1988) : 323 - 327.
ELBEL, J.L. (1986) : Designing hydraulic fractures for efficient reserve recovery. SPE Paper, **15231**.
ELBEL, J.L., HOWARD, R.L., TALLEY, G.R. & McLAUGHLIN, B.H. (1984) : Stimulation study of Cottage Grove Formation. J. Petrol. Technol., **36** : 1199 - 1205. SPE Paper, **11564** (1983).
ELBEL, J.L. & SOOKPRASONG, P.A. (1985) : The use of cumulative production type curves in fracture design. SPE Paper, **14510**. SPE Product. Engin., **2** (1987) : 191 - 198.
ELBEL, J.L. & THOMAS, R.L. (1980) : The use of viscosity stabilizers in high temperature fracturing. SPE Paper, **9036**.
ELDERS, W.A. (1979) : The geological background of the geothermal fields of the Salton Trough. In : ELDERS, W.A. (Editor), Geology and geothermics of the Salton Trough. Campus Museum Contrib., **5** : 1 - 19; Univ. of California, Riverside/California.
EL-KHATIB, N.A.F. (1985) : The effect of crossflow on waterflooding of stratified reservoirs. Soc. Petrol. Engin. J., **25** : 291 - 302.
EL-KHATIB, N.A.F. (1987) : A new approach to interference test analysis. SPE Form. Eval., **2** : 609 - 610.
ELKINGTON, P.A.S., STOUTHAMER, P. & BROWN, J.R. (1982) : Rock strength predictions from wireline logs. Intern. J. Rock Mechan., Min. Sci., Geomechan. Abstr., **19** : 91 - 97.
ELKINS, L.E. (1980) : Western tight sands major research requirements. First Intern. Gas Research Conference, Chicago/Illinois, 09. - 12.06.1980.
ELKINS, L.F. & SKOV, A.M. (1960) : Determination of fracture orientation from pressure interference. J. Petrol. Technol., **12** : 301 - 304. TP, **8137**.
ELLENBERGER, C.W. & ASELTINE, R.J. (1976) : Selective acid stimulation to improve vertical efficiency in injection wells. SPE Paper, **5869**.
ELLIS, D.V. (1987) : Well logging for earth scientists. 450 pp.; Elsevier, Amsterdam/Netherlands.
ELPHICK, J. (1985) : Stimulation data analysis and its application to North Sea hydraulic fracturing. Offshore Northern Seas Conference, Aberdeen/Scotland.
EL RABAA, A.W. (1987) : Hydraulic fracture propagation in the presence of stress variations. SPE Paper, **16898**.
EL RABAA, A.W. (1988) : Determination of hydraulic fracture direction in deviated wells. SPE Paper, **17482**.
EL RABAA, A.W.M. & MEADOWS, D.L. (1986) : Laboratory and field application of the strain relaxation method. SPE Paper, **15072**.
ELSON, T.D. & ANDERSON, G.W. (1982) : Foam gravel packing. SPE Paper, **11013**.
ELSON, T.D., DARLINGTON, R.H. & MANTOOTH, M.A. (1984) : High-angle gravel pack completion studies. J. Petrol. Technol., **36** : 69 - 78. SPE Paper, **11012** (1982).
ELTING, W.W. (1987) : The new financial world for the independent petroleum producer. SPE Paper, **16305**.
ELY, J.W. (1988) : Design considerations for hydraulic fracturing treatments in the Devonian Shale. Eastern Devonian Gas Shales Technol. Rev., **5**/1 : 7 - 11.
ELY, J.W., CHATTERJI, J., HOLTMYER, M. & TINSLEY, J.M. (1975) : Compositions for fracturing high temperature well formations. U.S. Patent, **3,898,165**.
ELY, J.W., HOLDITCH, S.A. & CARTER, R.H. (1988) : Improved hydraulic fracturing

strategy for Fruitland Formation coal-bed methane recovery, San Juan Basin. In : Geology and coal-bed methane resources of the Northern San Juan Basin, Colorado and New Mexico. Rocky Mountain Assoc. Geol. Guidebook.
ELY, J.W., HOLDITCH, S.A. & HASKETT, S. (1987) : Field measurement of fracturing fluid viscosity at in-situ conditions of temperature and shear. SPE Paper, **16915**.
ELY, J., McDOW, G. & TURNER, J. (1982) : Stimulation techniques used in the Austin Chalk. Southwest Petrol. Short Course, Texas Techn. Univ., Lubbock/ Texas, 21. - 22.04.1982.
ENDERLIN, M.B., EDMISTON, C.S. & EPPS, D.S. (1988) : DASSEL: Dip analysis by single sedimentary event location. Amer. Assoc. Petrol. Geol., Annual Convention, Houston/Texas, 20. - 23.03.1988.
ENERGIEWIRTSCHAFT (1986 a) : Chinas Aussichten in der Erdölexploration und -produktion. Erdöl und Kohle - Erdgas - Petrochemie, **39** : 304 - 305.
ENERGIEWIRTSCHAFT (1986 b) : Jugoslawien. Energieprobleme mit fossilen Brennstoffen. Erdöl und Kohle - Erdgas, **39** : 350 - 351.
ENGINEER, R. (1985) : Cal Canal field, California : case history of a tight and abnormally pressured gas condensate reservoir. SPE Paper, **13650**.
ERBSTOESSER, S.R. (1980) : Improved ball sealer diversion. J. Petrol. Technol., **32** : 1903 - 1910. SPE Paper, **7569** (1978).
ERDLE, J.C., BEN NACEUR, K., BELL, J. & BEZIER, C. (1981) : Results of hydraulic fracturing treatment BHP analysis in Peru. SPE Paper, **10390**.
ERDOEL-ERDGAS (1978) : Erdgas auf dem Weg ins nächste Jahrhundert. Versorgungssicherheit und Zukunftsentwicklung des Erdgases in Westeuropa. Erdoel-Erdgas, **94** : 71 - 75.
ERDOEL-ERDGAS AKTUELL (1975) : Mobil fracte erfolgreich. Erdoel-Erdgas, **91** : p. 338.
ERDOEL-ERDGAS AKTUELL (1976 a) : Hildesheimer Wald Z 1 gasfündig. Erdoel-Erdgas, **92** : p. 133.
ERDOEL-ERDGAS AKTUELL (1976 b) : Frigg-Gas ab 1977. Erdoel-Erdgas, **92** : p. 32.
ERDOEL-ERDGAS AKTUELL (1977) : Groß-Frac in Südoldenburg. Erdoel-Erdgas, **93** : 191 - 192.
ERDOEL-ERDGAS AKTUELL (1978 a) : Frac-Erfolge in Hamwiede. Erdoel-Erdgas, **94** : p. 347.
ERDOEL-ERDGAS AKTUELL (1978 b) : Mehr als 3 Mrd t Erdöl gefördert. Erdoel-Erdgas, **94** : p. 34.
ERDOEL-ERDGAS AKTUELL (1978 c) : Geothermische Bohrung Urach im Kristallin. Erdoel-Erdgas, **94** : p. 70.
ERDOEL-ERDGAS AKTUELL (1979) : Frac in der Bohrung Cappeln Z 4 a. Erdoel-Erdgas, **95** : p. 3.
ERDOEL-ERDGAS AKTUELL (1980 a) : OPEC 200 Mrd $ Einnahmen aus dem Ölexport. Erdoel-Erdgas, **96** : p. 209.
ERDOEL-ERDGAS AKTUELL (1980 b) : Fracbehandlung der Bohrung Syke Z 1. Erdoel-Erdgas, **96** : p. 249.
ERDOEL-ERDGAS AKTUELL (1981 a) : 3066 Millionen Tonnen Erdöl gefördert. Erdoel-Erdgas, **97** : p. 6.
ERDOEL-ERDGAS AKTUELL (1981 b) : Immer mehr Erdgas. Erdoel-Erdgas, **97** : p. 39.
ERDOEL-ERDGAS AKTUELL (1982) : Welt 2859 Millionen Tonnen Erdöl gefördert. Erdoel-Erdgas, **98** : p. 36.
ERDOEL-ERDGAS AKTUELL (1984) : OPEC Ministertreffen : alles bleibt beim Alten. Erdoel-Erdgas, **100** : p. 234.
ERDOEL-ERDGAS AKTUELL (1985 a) : Teersand-Dampfflut-Projekt in Alberta. Erdoel-Erdgas, **101** : p. 371.
ERDOEL-ERDGAS AKTUELL (1985 b) : Neuer Fünfjahresplan mit stagnierender Ölförderung. Moskau forciert Ölsuche. Erdoel-Erdgas, **101** : p. 371, p. 178.
ERDOEL-ERDGAS AKTUELL (1985 c) : Frac-Behandlung der Erdgasbohrung Söhlingen Z-5. Erdoel-Erdgas, **101** : p. 337.
ERDOEL-ERDGAS AKTUELL (1985 d) : Hohe Energie-Abhängigkeit von der UdSSR. Erdoel-Erdgas, **101** : p. 337.
ERDOEL-ERDGAS AKTUELL (1985 e) : Horizontalbohrungen im "Rospo Mare"-Ölfeld. Erdoel-Erdgas, **101** : p. 217.
ERDÖL-ERDGAS-KOHLE (1988 a) : Die Mineralölwirtschaft der Bundesrepublik

Deutschland 1987. Erdöl-Erdgas-Kohle, **104** : 309 - 311.

ERDÖL-ERDGAS-KOHLE (1988 b) : Erdölbergbau : Konkurrenz für die Horizontalbohrtechnik? Erdöl-Erdgas-Kohle, **104** : 58 - 59.

ERDÖL-ERDGAS-KOHLE NACHRICHTEN (1986 a) : Unbefriedigende Erdölproduktion in Westsibirien. Erdgas übertrifft Erdöl als Energieträger. Erdöl-Erdgas-Kohle, **102** : p. 11.

ERDÖL-ERDGAS-KOHLE NACHRICHTEN (1986 b) : Ist die nächste Ölpreisexplosion schon in Sicht? Erdöl-Erdgas-Kohle, **102** : p. 541.

ERDÖL-ERDGAS-KOHLE NACHRICHTEN (1987 a) : US-Servicefirmen drängen wieder auf den sowjetischen Markt. Erdöl-Erdgas-Kohle, **103** : p. 107.

ERDÖL-ERDGAS-KOHLE NACHRICHTEN (1987 b) : Opec steigert 1986 ihren Marktanteil. Erdöl-Erdgas-Kohle, **103** : p. 258.

ERDÖL-ERDGAS-KOHLE NACHRICHTEN (1987 c) : Die Mineralölwirtschaft der Bundesrepublik Deutschland 1986. Günstige Ertragssituation war von kurzer Dauer. Erdöl-Erdgas-Kohle, **103** : 324 - 327.

ERDÖL-ERDGAS-KOHLE NACHRICHTEN (1987 d) : OPEC-Ölförderung auf Rekordtief. Erdöl-Erdgas-Kohle, **103** : p. 160.

ERDÖL-ERDGAS-KOHLE NACHRICHTEN (1987 e) : OPEC steigert 1986 ihren Marktanteil. Erdöl-Erdgas-Kohle, **103** : p. 258.

ERDÖL-ERDGAS-KOHLE NACHRICHTEN (1988 a) : UdSSR wichtigster Öllieferant für die EG. Erdöl-Erdgas-Kohle, **104** : p. 247.

ERDÖL-ERDGAS-KOHLE NACHRICHTEN (1988 b) : UdSSR 624,2 Mio t in 1987. Erdöl-Erdgas-Kohle, **104** : p. 57.

ERDÖL UND KOHLE - ERDGAS (1983) : 11. Welt-Erdöl-Kongreß London. Erdöl und Kohle - Erdgas, **36** : 503 - 507.

ERDÖL UND KOHLE - ERDGAS - PETROCHEMIE (1986) : Bundesrepublik zweitgrößter Erdgasimporteur der Welt. Erdöl und Kohle - Erdgas - Petrochemie, **39** : p. 211.

ERDÖL UND KOHLE - ERDGAS - PETROCHEMIE (1987 a) : OPEC : Neuer Richtpreis und neue Förderquoten. Erdöl und Kohle - Erdgas - Petrochemie, **40** : p. 59.

ERDÖL UND KOHLE - ERDGAS - PETROCHEMIE (1987 b) : Aktivitäten der deutschen Erdöl- und Erdgasgewinnungsindustrie 1986. Erdöl und Kohle - Erdgas - Petrochemie, **40** : 148 - 150.

ERDÖL UND KOHLE - ERDGAS - PETROCHEMIE (1987 c) : Entwicklung der Rohölpreise. Erdöl und Kohle - Erdgas - Petrochemie, **40** : 195 - 196.

ERDÖL UND KOHLE - ERDGAS - PETROCHEMIE (1987 d) : Erdgas : Reserven - Funde - Förderung in der Welt. Erdöl und Kohle - Erdgas - Petrochemie, **40** : 241 - 242.

ERDÖL UND KOHLE - ERDGAS - PETROCHEMIE (1987 e) : Horizontalbohrung und Nuklearsprengungen in Westsibirien. Erdöl und Kohle - Erdgas - Petrochemie, **40** : p. 512.

ERDÖL UND KOHLE - ERDGAS - PETROCHEMIE (1988) : Aktivitäten der deutschen Erdöl- und Erdgasgewinnungsindustrie 1987. Erdöl und Kohle - Erdgas - Petrochemie, **41** : 186 - 187.

ERIKSEN, T., HELLE, M., HENDEN, J. & ROGNEBAKKE, A. (1987) : Gullfaks. In : SPENCER, A.M. et al. (Editors), Geology of the Norwegian oil and gas fields, 273 - 286; Graham & Trotman, London/England.

ERNST, P.L. (1977) : A hydraulic fracturing technique for dry, hot-rock experiments in a single borehole. SPE Paper, **6897**.

ERNST, P. & HIEBLINGER, J. (1979) : Aspekte der Nutzung geothermischer Energie. Erdoel-Erdgas, **95** : 412 - 420.

ERSHAGHI, I. & AFLAKI, R. (1985) : Problems in characterization of naturally fractured reservoirs from well test data. Soc. Petrol. Engin. J., **25** : 445 - 450. SPE Paper, **12014** (1983).

ERTEKIN, T., CICEK, O., ADEWUMI, M.A. & DAUD, M.E. (1987) : Pressure transient behavior of non-Newtonian/Newtonian composite systems with a finite-conductivity vertical fracture. SPE Paper, **17053**.

ERTEKIN, T., KING, G.R. & SCHWERER, F.C. (1986) : Dynamic gas slippage : a unique dual-mechanism approach to the flow of gas in tight formations. SPE Form. Eval., **1** : 43 - 52. SPE Paper, **12045** (1983).

ERTEKIN, T., SUNG, W. & SCHWERER, F.C. (1986) : Production performance analysis of horizontal drainage wells for the degasification of coal seams. SPE Pa-

per, **15453**. J. Petrol. Technol., **40** (1988) : 625 - 632.
ESCARON, P.C. (1983) : A technique to evaluate deviated wells with standard logging tools. SPE Paper, **12180**.
EVANS, D. & WILLIS, K. (1986) : Fracture execution - an essential part of every fracture design. CIM Paper, **86-37-70**.
EVANS, E.V. & EVANS, R.D. (1986) : The influence of an immobile or mobile saturation upon non-Darcy compressible flow of real gases in propped fractures. SPE Paper, **15066**. J. Petrol. Technol., **40** (1988) : 1343 - 1351.
EVANS, K. (1982) : The growth and consolidation characteristics of shallow hydraulic fractures as viewed through the surface deformation field. SPE/DOE Paper, **10851**.
EVANS, K. (1983) : On the development of shallow hydraulic fractures as viewed through the surface deformation field. Part 1 - Principles. J. Petrol. Technol., **35** : 406 - 410. SPE Paper, **10851** (1982).
EVANS, K.F. & ENGELDER, J.T. (1986) : A study of stress in Devonian Shale : Part I - 3 D stress mapping using a wireline. SPE Paper, **15609**.
EVANS, K. & HOLZHAUSEN, G. (1983) : On the development of shallow hydraulic fractures as viewed throughout the surface deformation field. Part 2 - case histories. J. Petrol. Technol., **35** : 411 - 420. SPE Paper, **11766** (1982).
EVANS, K.F., HOLZHAUSEN, G.R. & WOOD, M.D. (1980) : The mapping of nitrogen gas induced hydraulic fracture in Devonian shale by observation of the associated surface deformation. SPE/DOE Paper, **8933**.
EVANS, K.F., HOLZHAUSEN, G.R. & WOOD, M.D. (1982) : The geometry of a large-scale nitrogen gas hydraulic fracture formed in Devonian Shale : an example of fracture mapping with tiltmeters. Soc. Petrol. Engin. J., **22** : 755 - 763.
EVANS, R.D. (1982) : A proposed model for multiphase flow through naturally fractured reservoirs. Soc. Petrol. Engin. J., **22** : 669 - 680. SPE Paper, **9940** (1981).
EVANS, R. (1988) : An investigation into the influence of common sedimentary structures and diagenesis on permeability heterogeneity and anisotropy in selected sands and sandstones. SPE Paper, **17130**.
EVANS, R.D. & CARROLL, H.B. (1981) : Stochastic modeling of fractured gas wells completed in low permeability noncontinuous lenticular reservoirs. SPE/DOE Paper, **9884**.
EVANS, R.D., HUDSON, C.S. & GREENLEE, J.D. (1985) : The effect of liquid saturation on the non-Darcy flow coefficient in porous media. SPE Paper, **14206**.
EVANS, R.D. & UBANI, E.A. (1987) : An equivalent cell block porosity and permeability concept for modeling multiphase flow in naturally fractured, lenticular reservoirs. SPE/DOE Paper, **16430**.
EVANS, S. & WILLIAMS, M.E. (1985) : Batch treatment of offshore gas wells in the East Cameron and Vermilion areas. SPE Paper, **14566**.
EVANS, T.R. (1977) : Thermal properties of North Sea rocks. Log Analyst, **1977** : 3 - 12.
EVERDINGEN, A.F.v. & MEYER, L.J. (1971) : Analysis of buildup curves obtained after well treatment. J. Petrol. Technol., **23** : 513 - 524. SPE Paper, **2864** (1970).
EWING, R.E. & RUSSELL, T.F. (1988) : A model of heterogeneity and viscous fingering in miscible displacement based on fine-grid stimulation. SPE/DOE Paper, **17336**.
EWING, R.E., RUSSELL, T.F. & YOUNG, L.C. (1989) : A coarse-grid dispersion model of heterogeneity and viscous fingering in two-dimensional miscible displacement derived from fine-grid simulation. SPE Paper, **18441**.
EYCK, D.K.t. (1987) : Cost per barrel as an economic decision tool. SPE Paper, **16994**.
EZEUDEMBAH, A.S. & DRANCHUK, P.M. (1982) : Flow mechanism of Forchheimer`s cubic equation in high-velocity radial gas flow through porous media. SPE Paper, **10979**.
FAGLEY, J., FOGLER, H.S., DAVENPORT, C.B. & MILLHONE, R.S. (1982) : An improved simulation for interpreting temperature logs in water injection wells. Soc. Petrol. Engin. J., **22** : 709 - 718.
FAHRENTHOLD, E.P. (1984) : Elastic and plastic stresses in a porous medium containing spherical and cylindrical cavities. Ph.D. Thesis, Rice University,

Houston/Texas, 05.1984.
FAHRENTHOLD, E.P. & CHEATHAM, J.B. (1986) : Incipient yielding and solids production for perforated casing completions. J. Energy Resourc. Technol., **108** : 321 - 325.
FAN, C. (1988) : Sand control technology with plastic-coated sand grain. SPE Paper, **17843**.
FANTOFT, S., KROGH, P.K. & POLLEN, S. (1988) : Evaluation of oil recovery from a thin oil column by a horizontal well in the Oseberg Gamma North reservoir. SPE Paper, **18341**.
FÄRBER, A. (1984) : Die Temperaturverteilung im Bereich des Salzstockes Dethlingen. Erdoel-Erdgas, **100** : 276 - 279.
FARKAS, E. & VOLL, L. (1987) : A preliminary study of Demjén-West steam injection project. 4th Europ. Symp. Enhanced Oil Recovery, Thermal Recovery Session, Hamburg/Germany FRG.
FARLEY, J.T., MILLER, B.M. & SCHOETTLE, V. (1970) : Design criteria for matrix stimulation with hydrochloric-hydrofluoric acid. J. Petrol. Technol., **22** : 433 - 440. SPE Paper, **2621** (1969).
FARMIGIANO, J.M. (1976) : Multizone gravel-pack completion method works in high-angle holes. Oil Gas J., 12.01.**1976**.
FAROUQ, S.M.A. & MELDAU, R.F. (1979) : Current steamflood technology. J. Petrol. Technol., **31** : 1332 - 1342. SPE Paper, **7183** (1978).
FARRELL, H.E. (1988) : Fracture characterization of Greater Ekofisk chalks from core analysis. Amer. Assoc. Petrol. Geol., Annual Convention, Houston/Texas, 20. - 23.03.1988.
FARRIS, R.F. (1953) : Hydraulic fracturing, a method for increasing well productivity by fracturing the producing formation and thus increasing the well drainage area. U.S. Patent, **23733** (10.11.1953).
FAST, C.R., FLICKINGER, D.H. & HOWARD, G.C. (1961) : Effect of fracture-formation flow capacity contrast on well productivity. API Drill. Product. Pract. **1961** : 145 - 151.
FAST, C.R., HOLMAN, G.B. & COVLIN, R.J. (1977) : The application of massive hydraulic fracturing to the tight Muddy "J" Formation, Wattenberg field, Adams and Weld Counties, Colorado. J. Petrol. Technol., **29** : 45 - 51. SPE Paper, **5624** (1975). SPE Reprint Series, Well Completions, **5 a**/II (1978) : 267 - 273.
FAST, C.R. & MASE, G.P. (1965) : Propping agent effective in gas well fracturing. Oil Gas J., **63**/39 (27.09.**1965**) : 97 - 102.
FAYERS, F.J. & NEWLEY, T.M.J. (1987) : Detailed validation of an empirical model for viscous fingering with gravity effects. SPE Paper, **15993**.
FEHRINGER, A. (1986) : Beitrag des Erdgases zur Deckung des Primärenergiebedarfes in Westeuropa - Ausblick bis zum Jahr 2000. Erdöl-Erdgas-Kohle, **102** : 536 - 538.
FELSENTHAL, M. (1974) : Step-rate tests define safe injection pressures in floods. Oil Gas J., 28.10.**1974** : 49 - 54.
FERTL, W.H. (1980) : Evaluation of fractured reservoir rocks using geophysical well logs. SPE/DOE Paper, **8938**.
FERTL, W.H. (1982) : Experiences with natural gamma ray spectral logging in North America. SPE Paper, **11145**.
FERTL, W.H. & CAVANAUGH, R.J. (1973) : North Sea experience aids sandstone evaluation. World Oil, **1973**/4 : 77 - 80.
FERTL, W.H. & CHILINGAR, G.V. (1988) : Determination of volume, type, and distribution modes of clay minerals from well logging data. SPE Paper, **17145**.
FERTL, W.H. & HOTZ, R.F. (1987) : Efficiently log and perforate 60o plus wells with coiled tubing. World Oil, **206** (1987/7) : 32 - 35.
FERTL, W.H. & LEACH, W.G. (1988) : Economics of hydrocarbon reserves in overpressured reservoirs below 18,000 ft in South Louisiana. SPE Paper, **18146**.
FERTL, W.H. & MARTIN, J.R. (1987) : Well logging technology for highly deviated and horizontal wellbores. J. Petrol. Sci. Engin., **1** : 83 - 90.
FERTL, W.H. & NICE, S.B. (1988) : Well logging in extended-reach and horizontal boreholes. OTC Paper, **5828**. SPE Paper, **17897**.
FERTL, W.H. & OLIVER, D.W. (1983) : Method and apparatus for evaluating multiple stage fracturing of earth formations surrounding a borehole. U.S. Pa-

tent, **4,415,805**.
FERTL, W.H. & RIEKE, H.H. (1979) : Gamma ray spectral evaluation techniques identify fractured shale reservoirs and source rock characteristics. SPE Paper, **8454**. J. Petrol. Technol., **32** (1980) : 2053 - 2062.
FERTL, W.H., STAPP, W.L., VAELLO, D.B. & VERCELLINO, W.C. (1980) : Spectral gamma-ray logging in the Texas Austin Chalk trend. J. Petrol. Technol., **32** : 481 - 488. SPE Paper, **7431** (1978).
FESTOEY, S. & GOLF-RACHT, T.D.v. (1987) : Gas gravity drainage in fractured reservoirs through new dual continuum approach. SPE Paper, **16980**.
FETKOVICH, M.J. (1973) : Decline curve analysis using type curves. SPE Paper, **4629**.
FETKOVICH, M.J., BRADLEY, M.D., WORKS, A.M. & TRASHER, T.S. (1988) : Depletion performance of layered reservoirs without crossflow. SPE Paper, **18266**.
FINCHER, R. (1985) : Vertical wellbore communication via lateral wellbores - an alternative to hydrofracturing. Solution Mining Research Institute Meeting, Houston/Texas, 15.10.1985.
FINCHER, R.W. (1987) : Short-radius lateral drilling, a completion alternative. Petrol. Engin. Intern., **59**/2 : 29 - 35.
FINLEY, R.J. (1984) : Geology and engineering characteristics of selected low-permeability gas sandstones, a national survey. Univ. of Texas at Austin, Bur. Econ. Geol., Rept. Invest., **138** : 220 pp.
FINLEY, R.J. (1985) : Reservoir properties and gas productivity of the Corcoran and Cozzette tight sandstones, Colorado. SPE/DOE Paper, **13852**.
FINLEY, R.J. (1986) : An overview of selected blanket-geometry, low-permeability gas sandstones in Texas. In : SPENCER, C.W. & MAST, R.F. (Editors), Geology of tight gas reservoirs. Amer. Assoc. Petrol. Geol., Stud. Geol., **24** : 69 - 86.
FINLEY, R.J., DUTTON, S.P., LIN, Z.S. & SAUCIER, A.E. (1985) : The Travis Peak (Hosston) Formation : geologic framework, core studies, and engineering field analysis. Univ. of Texas at Austin, Bur. Econ. Geol., Rept. (Contract), **5082-211-0708** : 230 pp.
FINLEY, R.J., GARRETT, C.M., HAN, J.H., LIN, Z.S., SAUCIER, A.E. & TYLER, N. (1983) : Geologic analysis of primary and secondary tight gas sand objectives. Phase A : selective investigation of six stratigraphic units. Univ. of Texas at Austin, Bur. Econ. Geol., Rept. (Contract), **5082-211-0708** : 287 pp.
FINLEY, R.J. & O`SHEA, P.A. (1983) : Geologic and engineering analysis of blanket-geometry tight gas sandstones. SPE/DOE Paper, **11607**.
FIROOZABADI, A. & AZIZ, K. (1986) : Relative permeabilities from centrifuge data. SPE Paper, **15059**.
FISCHER, A.K. (1978) : Die Mineralöl- und Erdgasabhängigkeit der RGW-Länder von der Sowjetunion. Erdoel-Erdgas, **94** : 248 - 251.
FISCHER, A.K. (1981) : Die sowjetische Erdgasindustrie. Erdoel-Erdgas, **97** : 9 - 14.
FISHER, D.W. (1986) : Testing and evaluation of an arctic fractured carbonate : the Lisburne Formation, Prudhoe Bay. SPE Paper, **15100**.
FISHMAN, D. (1986 a) : Southern Basin sheltered from low oil prices. Offshore Engineer, **1986**/9 : 43 - 44.
FISHMAN, D.M. (1986 b) : The economics of disposal methods for North Sea oil and gas fields : a study in comparative advantage. J. Petrol. Technol., **38** : 570 - 576. SPE Paper, **13780** (1985).
FITCH, J.L. & MEDLIN, W.L. (1978) : Demonstration of massive hydraulic fracturing, Piceance Basin, Rio Blanco County, Colorado. Proc. 4th Annual Symp. on Enhanced oil and gas recovery and improved drilling methods, **2** : F 7/1 - F 7/7; Petrol. Publ. , Tulsa/Oklahoma.
FITTS, R.L. & CROWBURST, M.E. (1988) : 36th annual Reed rig census. World Oil, **207** (1988/10) : 39 - 42.
FITZGERALD, B.M., STEPHENS, G. & TERWILLIGER, P.L. (1966) : Warm air coking sand consolidation - field results in viscous oil sands. J. Petrol. Technol., **18** : 35 - 42. SPE Paper, **1239** (1965).
FITZGIBBON, J.J. (1984) : Sintered spherical pellets containing clay as a major component useful for gas and oil well proppants. U.S. Patent, **4,427,068** (24.01.1984).

FITZ-PATRICK, R.P., KARR, G.K. & O`SHEA, P.A. (1985) : A comprehensive fracture geometry and proppant mapping experiment : Part I - An overview. SPE/DOE Paper, **13893**. SPE Product. Engin., **1** (1986) : 411 - 422.

FLANIGAN, M.J. (1979) : Smaller gravel and coated screens enhance 50-year old field. SPE Paper, **7973**. J. Petrol. Technol., **32** (1980) : 757 - 763.

FLATH, K. (1987) : Niedrige Ölpreise - Segen oder Verhängnis? Schriftenreihe der Gesellschaft für Energiewissenschaft und Energiepolitik, 166 pp.; TÜV Rheinland, Köln/Germany FRG.

FLECKENSTEIN, M. (1984) : BHTV interpretation in a fractured Rotliegend reservoir in the southern Hannover area. Shell Intern. Petrol. Maatsch., Group Production Geology/Petrophysics Conference; The Hague/Netherlands.

FLEMING, G.C. & GONZALEZ, J.A. (1987) : Modeling the performance of fractured wells in pattern floods using orthogonal, curvilinear coordinates. SPE Paper, **16973**.

FORD, W.G.F. (1981 a) : The use of foamed acid in fracture acidizing. SPE Paper, **9652**.

FORD, W.G.F. (1981 b) : Foamed acid - an effective stimulation fluid. SPE Paper, **9385**. J. Petrol. Technol., **33** (1981) : 1203 - 1210.

FORD, W.G.F., BURKLECA, L.F. & SQUIRE, K.A. (1980) : Foamed acid stimulation : success in the Illinois and Michigan Basins. SPE Paper, **9386**.

FORD, W.G.F. & PENNY, G.S. (1988) : Influence of downhole conditions on the leakoff properties of fracturing fluids. SPE Product. Engin., **3** : 43 - 51. SPE Paper, **14016** (1985).

FORD, W.G.F., PENNY, G.S. & BRISCOE, J.E. (1986) : Enhanced water recovery improves stimulation results. SPE Paper, **15851**.

FORD, W.G.F. & ROBERTS, L.D. (1982) : The effect of foam on surface kinetics in fracture acidizing. SPE Paper, **11120**.

FORDJOR, C.K., BELL, J.S. & GOUGH, D.I. (1983) : Breaks in Alberta and stress in the North American plate. Canad. J. Earth Sci., **20** : 1445 - 1455.

FORGOTSON, J.M., FORGOTSON, J.M. & BROWN, R.O. (1980) : Predicting the orientation of hydraulically created fractures in the Cotton Valley Formation of East Texas. SPE Paper, **9269**.

FOSTER, C.R. & LINVILLE, T.W. (1979) : A method of monitoring sand production in a flowing well stream. SPE Paper, **8214**.

FOSTER, J.M. (1975) : A "new" gas supply - the Devonian shales. SPE Paper, **5451**.

FOURNEY, W.L., BARKER, D.B. & HOLLOWAY, D.C. (1981) : Model studies of explosive well stimulation techniques. Intern. J. Rock Mechan., Min. Sci., Geomechan. Abstr., **18** : 113 - 127.

FOURNEY, W.L., BARKER, D.B. & HOLLOWAY, D.C. (1983) : Model studies of well stimulation using propellant charges. Intern. J. Rock Mechan., Min. Sci., Geomechan. Abstr., **20** : 91 - 101.

FOURNEY, W.L., HOLLOWAY, D.C. & SIMHA, K.R.Y. (1984) : Model investigations of borehole pressure distribution. SPE/DOE/GRI Paper, **12883**. SPE Product. Engin., **2** (1988) : 243 - 249.

FOX, C. (1985) : New vessels stimulate North Sea action. Offshore Engin., **1985**/10 : 42 - 43.

FOX, C. (1986) : Rospo Mare - ELF keeps faith with horizontal wells. Offshore Engin., **1986**/9 : 57 - 58.

FOX, C. (1987) : V fields drilling team climbs leaning curve. Offshore Engin., **1987**/11 : 59 - 60.

FOX, K.B., O`MARA, E.M., BURNS, L.D. & JOHNSON, D.E. (1989) : Laboratory and field study of a thermally stable gelling agent. SPE Paper, **18470**.

FRACASSO, M.A., DUTTON, S.P. & FINLEY, R.J. (1988) : Depositional systems and diagenesis of Travis Peak tight gas sandstone reservoirs, Sabine Uplift area, Texas. SPE Form. Eval., **3** : 105 - 115. SPE Paper, **14666** (1986).

FRACFAX (1988 a) : Frac records - highest temperature (BHT) frac job. Fracfax, **1** (02.1988); Santrol, Houston/Texas.

FRACFAX (1988 b) : Frac record - largestvolume frac-job. Fracfax, **2** (03.1988), **4** (05.1988); Santrol, Houston/Texas.

FRACFAX (1988 c) : Super sand passes 10th year anniversary. Fracfax, **2** (03. 1988); Santrol, Houston/Texas.

FRACFAX (1988 d) : South Texas frac results. Which is strongest? Fracfax, **3** (04. 1988), **4** (05.1988); Santrol, Houston/Texas.
FRACFAX (1988 e) : Methane production from coal. Fracfax, **4** (05.1988); Santrol, Houston/Texas.
FRACFAX (1988 f) : Micro proppants - technology for the future. Fracfax, **5** (06. 1988); Santrol, Houston/Texas.
FRACFAX (1988 g) : For a good job - get a good prop! Fracfax, **5** (06.1988); Santrol, Houston/Texas.
FRACFAX (1988 h) : Refracs solve problems. Fracfax, **7** (08.1988); Santrol, Houston/Texas.
FRACFAX (1988 i) : First family of proppants. Fracfax, **8** (09.1988); Santrol, Houston/Texas.
FRACFAX (1988 k) : World record sand concentration. Fracfax, **9** (10.1988); Santrol, Houston/Texas.
FRACFAX (1988 l) : Proppant crushing update. Fracfax, **9** (10.1988); Santrol, Houston/Texas.
FRADKIN, L.J., SOREY, M.L. & McNABB, A. (1981) : On identification and validation of some geothermal models. Water Resourc. Res., **1981** : 929 - 936.
FRAIM, M.L. & LEE, J.W. (1988) : Determination of formation properties from long term gas well production data affected by non-Darcy flow. SPE Paper, **16934**.
FRAIM, M.L., LEE, W.J. & GATENS, J.M. III (1986) : Advanced decline curve analysis using normalized-time and type curves for vertically fractured wells. SPE Paper, **15524**.
FRANSSEN, H. (1982) : Gas demand and pricing. EUR Paper, **343**.
FRANSSEN, H.T. (1988) : Supply dimensions - political, strategic and resource implications. Offshore Northern Seas Conference, Stavanger/Norway, 23. - 26. 08.1988.
FREDRICKSON, S.E. (1987) : Stimulating carbonate formations using a closed fracture acidizing technique. SPE Paper, **14654**.
FREDRICKSON, S.E. & BROADDUS, G.C. (1976) : Selective placement of fluids in a fracture by controlling density and viscosity. J. Petrol. Technol., **28** : 597 - 602. SPE Paper, **5629** (1975).
FREE, D.L., FREDERICK, A.F. & THOMPSON, J.E. (1975) : Fracturing with a high viscosity crosslinked gel - continuous fracturing technique. SPE Paper, **5640**.
FREEMAN, D.L. & BUSH, D.C. (1983) : Low-permeability laboratory measurements by nonsteady-state and conventional methods. Soc. Petrol. Engin. J., **23** : 928 - 936. SPE Paper, **10075** (1981).
FREEMAN, E.R., ABEL, J.C., KIM, C.M. & HEINRICH, C.C. (1981) : A stimulation technique using only nitrogen. SPE Paper, **10129**. J. Petrol. Technol., **35** (1983) : 2165 - 2174.
FREEMAN, E.R., BILDEN, D.M. & HOSSAINI, M. (1986) : Delayed crosslinked gels : their role in aqueous foam fracturing. SPE Paper, **15070**.
FRENIER, W.W. (1989) : Acidizing fluids used to stimulate high temperature wells can be inhibited using organic chemicals. SPE Paper, **18468**.
FRIEDMAN, R.H., SURLES, B.W. & KIEKE, D.E. (1986) : High temperature sand consolidation. SPE Paper, **14093**. SPE Product. Engin., **3** (1988) : 167 - 168.
FRISINGER, M. & COOPER, R.E. (1985) : Techniques for quantifying formation stress and fracture fluid response for optimum hydraulic fracture design. SPE Paper, **14015**.
FROGNES, G. (1981) : Norway, oil policy and economic outlook - the study of a country in energy surplus. SPE Paper, **9577**. J. Petrol. Technol., **34** (1982) : 46 - 54.
FROGNES, G. (1983) : Norway - the dilemma of a small country torn between national and international energy interests. SPE Paper, **11313**.
FROHNE, K.H. & MERCER, J.C. (1984) : Fractured shale gas reservoir performance study - an offset well interference field test. J. Petrol. Technol., **36** : 291 - 300. SPE Paper, **11224** (1982).
FROST, H.J. & RAJ, R. (1982) : Limiting densities for dense random packing of spheres. J. Amer. Ceram. Soc., **65** : C 19.
FUEG, J.C. (1988) : Eastern Europe. The USSR sets another record. World Oil,

207 (1988/8) : 74 - 79.
FUH, G.F., WHITFILL, D.L. & SCHUH, P.R. (1988) : Use of borehole stability analysis for successful drilling of high-angle holes. IADC/SPE Paper, **17235**.
FUHRBERG, H.D. (1983) : Well completion and stimulation. 11th World Petrol. Congr., London/England, Review Paper, **5**.
FUNG, R.L., VIJAYAKUMAR, S. & CORMACK, D.E. (1987) : Calculation of vertical fracture containment in layered formations. SPE Form. Eval., **2** : 518 - 522. SPE Paper, **14707** (1985).
GABRIEL, G.A. & ERBSTOESSER, S.R. (1984) : The design of buoyant ball sealer treatments. SPE Paper, **13085**.
GABRIEL, G.A. & INAMDAR, G.R. (1983) : An experimental investigation of fines migration in porous media. SPE Paper, **12168**.
GADEKEN, L.L., ARNOLD, D.M. & SMITH, H.D. (1984) : Applications of the Compensated Spectral Natural Gamma tool. Trans. 25th Soc. Prof. Well Log Analysts Annual Logging Symp., Paper **JJJ**, New Orleans/Louisiana, 10.-13.06.84.
GADEKEN, L.L. & SMITH, H.D. (1986) : TracerScan - a spectroscopy technique for determining the distribution of multiple radioactive tracers in downhole operations. Trans. 27th Soc. Prof. Well Log Analysts Annual Logging Symp., Paper **ZZ**, Houston/Texas, 09. - 13.06.1986. Log Analyst, **28** (1987) : 27 - 36.
GADEKEN, L.L., SMITH, H.D., ARNOLD, D.M. & ROBBINS, C.A. (1988) : Natural gamma ray spectroscopy applied to borehole measurements. Log Analyst, **29**.
GADEKEN, L.L., SMITH, H.D. & SEIFERT, D.J. (1987) : Calibration and analysis of borehole and formation sensitivities for gamma ray spectroscopy measurements with multiple radioactive tracers. Trans. 28th Soc. Prof. Well Log Analysts Annual Logging Symp., Paper **V**, London/England, 29.06. - 02.07.1987.
GADONNIEUX, P. (1988) : Gas challenges in a competitive market. Offshore Northern Seas Conference, Stavanger/Norway, 23. - 26.08.1988.
GAGE, M. (1980) : A review of the Viking gas field. In : HALBOUTY, M.T. (Editor), Giant oil and gas fields of the decade : 1968 - 78. Amer. Assoc. Petrol. Geol. Mem., **30** : 39 - 54.
GAIDA, K.H., KESSEL, D.E., VOLZ, H. & ZIMMERLE, W. (1987) : Geologic parameters of reservoir sandstones as applied to enhanced oil recovery. SPE Form. Eval., **2** : 89 - 96.
GAIDA, K.H., RÜHL, W. & ZIMMERLE, W. (1973) : Rasterelektronenmikroskopische Untersuchungen des Porenraumes von Sandsteinen. Erdoel-Erdgas, **89** : 336 - 343.
GAJDOSIK, W.C. & WILLINGHAM, J.R. (1976) : Improved method of gravel packing in California. SPE Paper, **5861**.
GALL, B.L. & RAIBLE, C.J. (1985) : Molecular size studies of degraded fracturing fluid polymers. SPE Paper, **13566**.
GALL, B.L., MALONEY, D.R., SATTLER, A.R. & RAIBLE, C.J. (1988) : Permeability damage to natural fractures caused by fracturing fluid polymers. SPE Paper, **17542**.
GAO, C. (1983) : Crossflow behavior in a partially perforated two layer reservoir : the evaluation of reservoir parameters by transient well tests. SPE Paper, **11875**.
GAO, C. & DEANS, H.A. (1983) : Single-phase flow in a two-layer reservoir with significant crossflow - pressure drawdown and buildup behavior when both layers are completed at a single well. SPE Paper, **11439**. Soc. Petrol. Engin. J., **24** (1984) : 97 - 106.
GAO, C.T. & DEANS, H.A. (1988) : Pressure transients and crossflow caused by diffusivities in multilayer reservoirs. SPE Form. Eval., **3** : 438 - 448.
GAO, Y., SHEN, P. & TU, F. (1986) : A study of pore structure by image processing method and its application. SPE Paper, **14872**.
GARB, F.A. (1988) : Assessing risk in estimating hydrocarbon reserves and in evaluating hydrocarbon-producing properties. J. Petrol. Technol., **40** : 765 - 778. SPE Paper, **15921** (1986).
GARB, F.A., GRUY, H.J. & HERRON, E.H. (1981) : Recoverable reserves are affected by oil and gas prices. J. Petrol. Technol., **33** : 33 - 38. SPE Paper, **9360** (1980).
GARBIS, S.J., BROWN, B.J. & MAURITZ, S.J. (1985) : Review of the completion practices in the Wilcox Formation in South and South Central Texas. SPE Pa-

per, **13900**.
GARBIS, S.J. & TAYLOR, J.L. III (1985) : The utility of CO2 as an energizing component for fracturing fluids. SPE Paper, **13794**. SPE Product. Engin., **1** (1986) : 351 - 358.
GARDNER, D.C. & EIKERTS, J.V. (1982) : The effects of shear and proppant on the viscosity of crosslinked fracturing fluids. SPE Paper, **11066**.
GARDNER, D.C. & EIKERTS, J.V. (1983) : Rheological characterization of cross-linked and delayed crosslinked fracturing fluids using a closed-loop pipe viscometer. SPE Paper, **12028**.
GARON, A.M., LIN, C.Y. & DUNAYEVSKY, V.A. (1988) : Simulation of thermally induced waterflood fracturing in Prudhoe Bay. SPE Paper, **17417**.
GATELY, D. (1984) : A ten-year retrospective : OPEC and the world oil market. J. Econ. Literat., **22** : 1100 - 1114.
GATENS, J.M. III (1988) : A review of massive hydraulic fracturing experiments in the Devonian shales in Lincoln County, West Virginia. Eastern Devonian Gas Shales Technol. Rev., **5**/1 : 12 - 22.
GATENS, J.M. III, HOLGATE, K.E. & LEE, W.J. (1987) : Evaluation of massive hydraulic fracturing experiments in the Devonian shales in Lincoln County, West Virginia. S.A. Holditch & Assoc. Inc., Topical Rept. to Gas Research Institute, GRI Contract, **5084-213-0980**, 02.1987.
GATENS, J.M., OLAREWAJU, J.S. & LEE, W.J. (1986) : An integrated reservoir description method for naturally fractured reservoirs. SPE Paper, **15235**.
GATENS, J.M. III, LEE, W.J., LANE, H.S., WATSON, A.T., STANLEY, D.K. & LANCASTER, D.E. (1987) : Analysis of Eastern Devonian gas shales production data. SPE Paper, **17059**.
GATENS, J.M. III, LEE, W.J. & RAHIM, M.Z. (1985) : Application of an analytical model to history match Devonian shales production data. SPE Paper, **14509**.
GATEWOOD, J.R., HALL, B.E., ROBERTS, L.D. & LASATER, R.M. (1970) : Redicating results of sandstone acidizing. J. Petrol. Technol., **22** : 693 - 700. SPE Paper, **2622** (1969). SPE Reprint Series, Well Completions, **5 a**/II (1978) : 191 -198.
GAUTIER, D.L. & RICE, D.D. (1981) : Comparison of conventional and low-permeability reservoirs of shallow gas in the Northern Great Plains. SPE/DOE Paper, **9846**. J. Petrol. Technol., **34** (1982) : 1600 - 1608.
GAYELOS, J.S. & HARRIS, P.C. (1980) : Foam fracturing : theories, procedures and results. SPE/DOE Paper, **8961**.
GDANSKI, R.D. (1985) : AlCl3 retards HF acid for more effective stimulations. Oil Gas J., 28.10.**1985** : 111 - 115.
GDANSKI, R.D. & NORMAN, L.R. (1983) : The effect of filterable solids on acid reaction rates. Using the hollow-core test to determine acid reaction rates. SPE Paper, **12151**. SPE Product. Engin., **1** (1986) : 111 - 116.
GDANSKI, R.D. & PEAVY, M.A. (1985) : Well return analysis causes re-evaluation of HCl theories. SPE Paper, **14825**.
GEER, E.C. & COOK, H.L. (1978) : The creation, augmentation and exploitation of hydrocarbon pore-volume saturation in geopressured aquifers. SPE Paper, **7541**.
GEERTSMA, J. (1973) : Land subsidence above compacting oil and gas reservoirs. J. Petrol. Technol., **25** : 734 - 744. SPE Paper, **3730** (1972).
GEERTSMA, J. (1974) : Estimating the coefficient of inertial resistance in fluid flow through porous media. Soc. Petrol. Engin. J., **14** : 445 - 450.
GEERTSMA, J. & DEKLERK, F. (1969) : A rapid method of predicting width and extent of hydraulically induced fractures. J. Petrol. Technol., **21** : 1571 - 1581. SPE Paper, **2458**.
GEERTSMA, J. & HAAFKENS, R. (1979) : A comparison of the theories for predicting width and extent of vertical hydraulically induced fractures. J. Energy Resources Technol., **101** : 8 - 19. Trans. ASME, **101** : 8 - 19.
GELFAND, V.A., TAYLOR, G., TESSMAN, J. & LARNER, K.L. (1986) : Delineation of river-channel reservoirs by seismic lithologic modeling of 3-D seismic data in Southern Alberta. SPE Paper, **14118**.
GEORGI, D.T. (1985) : Geometrical aspects of borehole televiewer images. Trans. 26th Soc. Prof. Well Log Analysts Annual Logging Symp., Paper **O**, Dallas/Texas, 17. - 20.06.1985.

GEORGI, D.T. (1986) : Guides for the interpretation of dipmeter fracture logs. Trans. 27th Soc. Prof. Well Log Analysts Annual Logging Symp., Paper **TT**, Houston/Texas, 09. - 13.06.1986.

GESEMANN, K.H. & BADURA, K.R. (1976) : Sandschwierigkeiten im Erdölfeld Rühlermoor. 4. Gemeinschaftstagung ÖGEW/DGMK, Salzburg/Österreich, 04. - 06.10. 1976; Kompendium, **76/77** : 153 - 171.

GESHELIN, B.M. & GRABOWSKI, J.W. (1982) : Source distributed fracture model for the numerical modelling of oil sands steam stimulation. SPE Paper, **10508**.

GESHELIN, B.M., GRABOWSKI, J.W. & PEASE, E.C. (1981) : Numerical study of transport of injected and reservoir water in fractured reservoir during steam stimulation. SPE Paper, **10322**.

GESINK, J.C.J., BERGEN, E.A.v.d., MONCHY, A.R.d., RIJNDERS, J.P. & SOET, J. (1985) : Use of gamma ray-emitting tracers and subsequent gamma ray logging in an observation well to determine the preferential flow zones in a reservoir. J. Petrol. Technol., **37** : 711 - 719.

GETZLER, Z., KOMORNIK, A. & MAZURIK, A. (1968) : Model study on arching above buried structures. Proc. ASCE, **94** (No. SM5) : 1123 - 1141.

GHAURI, W.K. (1980) : Production technology experience in a large carbonate waterflood, Denver unit, Wasson San Andres field. J. Petrol. Technol., **32** : 1493 - 1502. SPE Paper, **8406** (1979).

GHAZALI, H.A. & WILLHITE, G.P. (1985) : Permeability modification using aluminum citrate/polymer treatments : mechanism of permeability reduction in sandpacks. SPE Paper, **13583**.

GIDLEY, J.L. (1970) : Stimulation of sandstone formations with the acid-mutual solvent method. SPE Paper, **3007**. J. Petrol. Technol., **23** (1971). SPE Reprint Series, Well Completions, **5 a**/II (1978) : 146 - 153.

GIDLEY, J.L. (1985) : Acidizing sandstone formations : a detailed examination of recent experience. SPE Paper, **14164**.

GIDLEY, J.L. (1988) : Recent advances in hydraulic fracturing. SPE Monograph, in preparation.

GIDLEY, J.L. & HANSON, H.R. (1973) : Prevention of central terminal upsets related to stimulation and sand consolidation treatments. SPE Paper, **4551**.

GIDLEY, J.L., MUTTI, D.H., NIERODE, D.E. & KEHN, D.M. (1977) : Tight gas stimulation by massive hydraulic fracturing - a study of well performance. SPE Paper, **6867** (1977).

GIDLEY, J.L., MUTTI, D.H., NIERODE, D.E., KEHN, D.M. & MUECKE, T.W. (1979) : Stimulation of low-permeability gas formations by massive hydraulic fracturing - a study of well performance. J. Petrol. Technol., **31** : 525 - 531. SPE Paper, **6867**.

GIDLEY, J.L., RYAN, J.C. & MAYHILL, T.D. (1976) : Study of the field application of sandstone acidizing. J. Petrol. Technol., **28** : 1289 - 1294. SPE Paper, **5693**. SPE Reprint Series, Well Completions, **5 a**/II (1978) : 175 - 180.

GIES, R.M. (1987) : An improved method for viewing micropore systems in rocks with the polarizing microscope. SPE Form. Eval., **2** : 209 - 214. SPE Paper, **13136** (1984).

GIGER, F. (1983) : Réduction du nombre de puits par l`utilisation de forages horizontaux. Rev. Inst. Franç. Pétrol., **38**/3 : 351 - 360.

GIGER, F.M. (1985) : Horizontal wells production techniques in heterogeneous reservoirs. SPE Paper, **13710**.

GIGER, F.M. (1986) : Analytic 2-D models of water cresting before breakthrough for horizontal wells. SPE Paper, **15378**.

GIGER, F.M. (1987) : Low-permeability reservoirs development using horizontal wells. SPE/DOE Paper, **16406**.

GIGER, F., COMBE, J. & REISS, L.H. (1983) : L`interêt du forage horizontal pour l`exploitation des gîsements d`hydrocarbures. Rev. Inst. Franç. Pétrol., **38**/3 : 329 - 350.

GIGER, F.M. & JOURDAN, A.P. (1984 a) : The four horizontal wells producing oil in Western Europe. Proc. 5th Annual Conf. on Advances in Petroleum Recovery and Upgrading Technology, AOSTRA, Calgary/Alberta, 14. - 15.06.1984.

GIGER, F.M. & JOURDAN, A.P.L. (1984) : Offshore and onshore European horizontal wells. OTC Paper, **4791**.

GIGER, F.M., REISS, L.H. & JOURDAN, A.P. (1984) : The reservoir engineering aspects of horizontal drilling. SPE Paper, **13024**.

GILHUUS, T. (1986) : A study of possible oil production from the thin oil zone of the Midgard field. Paper presented at the Norwegian Petroleum Directorate seminar on recovery from thin oil zones, Stavanger/Norway, 04.1986.

GILMAN, J.R. & KAZEMI, H. (1983) : Improvements in simulation of naturally fractured reservoirs. Soc. Petrol. Engin. J., **23** : 695 - 707. SPE Paper, **10511** (1982).

GILREATH, J.A. (1977) : The dipmeter. In : ROY, L.W.le (Editor), Subsurface geology in petroleum mining and construction. 4th ed.; Colorado School of Mines, Golden/Colorado.

GISTAU, M.J. (1985) : Stimulation methods and practices on chalk reservoirs. Chalk Symposium, Stavanger/Norway, 21. - 22.05.1985.

GLASNER, D. (1985) : Politics, prices and petroleum. The political economy of energy. 297 pp.; Pacific Institute for Public Policy Research, San Francisco/California. ISBN 0-88410-953-4.

GLENN, E.E. & SLUSSER, M.L. (1957) : Factors affecting well productivity. II. Drilling fluid particle invasion into porous media. Trans. AIME, **210** : 132 - 139.

GLENNIE, K.W. (1983) : Early Permian (Rotliegendes) palaeowinds of the North Sea. Sedim. Geol., **34** : 245 - 265.

GLENNIE, K.W. (1986) : Introduction to the petroleum geology of the North Sea. 2nd ed., 288 pp.; Blackwell, Oxford/England.

GLENNIE, K.W., MUDD, G.C. & NAGTEGAAL, P.J.C. (1978) : Depositional environment and diagenesis of Permian Rotliegendes sandstones in Leman Bank and Sole Pit areas of the UK southern North Sea. J. Geol. Soc. Lond., **135** : 25 - 34.

GOBRAN, D.B., BRIGHAM, W.E. & RAMEY, H.J. (1981) : Absolute permeability as a function of confining pressure, pore pressure and temperature. SPE Paper, **10156**. SPE Form. Eval., **2** (1987) : 77 - 84.

GODBEY, J.K. & HODGES, H.D. (1958) : Pressure measurement during formation fracturing operations. Trans. AIME, **213** : 65 - 69.

GODBOLE, S.P. (1985) : Natural gas hydrates in Alaska : quantification and economic evaluation. SPE Paper, **13593**. SPE Form. Eval., **3** (1988) : 263 - 266.

GODEC, P. & ALLARD, A. (1987) : Oil in Norwegian foreign policy : a vital and inevitable component. In : Norway, OPEC and IEA - global dialogue and price stabilizing measures in the oil market. Intern. Politics, **1987**/5-6; Norwegian Institute of International Affairs.

GOELITZ, R. & EVERTZ, G.E. (1982) : Foam fracturing in the Uinta Basin - a field study of the Dakota and Wasatch formations. SPE Paper, **10883**.

GOETZ, J.F., TYSON, J. & GARAT, J. (1987) : New angles and dimensions in dipmeter interpretation opened by the six arm dipmeter. Trans. 28th Soc. Prof. Well Log Analysts Annual Logging Symp., Paper **JJ**, London/England, 29.06. - 02.07.1987.

GOGGIN, D.J., CHANDLER, M.A., KOCUREK, G.A. & LAKE, L.W. (1986) : Patterns of permeability in eolian deposits. SPE/DOE Paper, **14893**. SPE Form. Eval., **3** (1988) : 297 - 306.

GOLF-RACHT, T.D.v. (1985) : Fundamentals of fractured reservoir engineering. Developm. Petrol. Science, **12** : 710 pp.; Elsevier, Amsterdam/Netherlands.

GOODE, D.L., BERRY, S.D. & STACY, A.L. (1984) : Aqueous-based remedial treatments for reservoir damaged by oil-phase drilling muds. SPE Paper, **12501**.

GOODE, P.A. & THAMBYNAYAGAM, R.K.M. (1985) Pressure drawdown and buildup analysis of horizontal wells in anisotropic media. SPE Paper, **14250**. SPE Form. Eval., **2** (1987) : 683 - 697.

GORBATCHEV, M. (1987) : Perestroika. 344 pp.; Harper & Row, New York (in English); Droemer Knaur, München/Germany FRG (in German).

GORODY, A.W. (1984) : TEDSI develops horizontal drilling technology. Oil Gas J., 01.10.**1984** : 118 - 126.

GORST, I. (1985) : Persistant shortage of oil. Petrol. Econ., **52**/9 : 319 - 321.

GORST, I. (1987) : Yugoslavia - energy shortages impede growth. Petrol. Econ., **54**/7 : 265 - 267.

GORST, I. (1988) : Soviet Union - oil exports rise to record level. Petrol. Econ., **55**/2 : 43 - 44.

GOTTSCHLING, J.C., ROYCE, T.N. & SCHUCK, L.Z. (1985) : Nitrogen gas and sand :

a new technique for stimulation of Devonian shale. J. Petrol. Technol., **37** : 901 - 907. SPE Paper, **12313** (1983).
GOUGH, D.I. & BELL, J.S. (1981) : Stress orientations from oil well fractures in Alberta and Texas. Canad. J. Earth Sci., **18** : p. 638 ff.
GOUGH, D.I. & BELL, J.S. (1982) : Stress orientations from borehole wall fractures with examples from Colorado, East Texas and Northern Canada. Canad. J. Earth Sci., **19** : 1358 - 1370.
GOUGLER, P.D., HENDRICK, J.E. & COULTER, A.W. (1985) : Field investigation identifies source and magnitude of iron problems. SPE Paper, **13812**.
GRABOWSKI, M.A., LOCKE, C.D., SARADJI, B., FEILER, J. & YOST, A.B. II (1988) : Field results of pulse tests in a highly anisotropic dual-porosity gas reservoir. SPE Paper, **17764**.
GRADER, A.S. & HORNE, R.N. (1988) : Interference testing : detecting a circular impermeable or compressible subregion. SPE Form. Eval., **3** : 420 - 428.
GRADER, A.S. & O`MEARA, D.J. (1988) : Dynamic displacement measurements of three-phase relative permeabilities using three immiscible liquids. SPE Paper, **18293**.
GRADY, D.E., KIPP, M.E. & SMITH, C.S. (1980) : Explosive fracture studies on oil shale. Soc. Petrol. Engin. J., **20** : 349 - 356. SPE Paper, **8215** (1979).
GRAHAM, J.W., KERVER, J.K. & MORGAN, F.A. (1965) : Method of acidizing and introducing a corrosion inhibition into a hydrocarbon producing formation. U.S. Patent, **3,167,123** (26.01.1965).
GRAHAM, J.W., MONAGHAN, P.H. & OSOBA, J.S. (1959) : Influence of propping sand wettability on productivity of hydraulically fractured oil wells. Trans. AIME, **216** : 324 - 329.
GRAHAM, J.W., MUECKE, T.W. & COOKE, C.E. (1975) : Resin-coated particles. U.S. Patent, **3,929,191** (30.12.1975).
GRAHAM, J.W., SINCLAIR, A.R. & BRANDT, J.L. (1982) : Method of treating wells using resin coated particles. U.S. Patent, **4,336,842** (29.06.1982).
GRALLA, P. (1988) : Das Oberrotliegende in NW-Deutschland - Lithostratigraphie und Faziesanalyse. Geol. Jb., **A 106** : 3 - 59. Nachr. dt. geol. Ges., **39** : 21 - 22.
GRALLA, P., NIEBERDING, F. & SOBOTT, R. (1988) : Der Wustrow-Sedimentationszyklus des Oberrotliegenden im Bereich einer NW-deutschen Erdgaslagerstätte. Nachr. dt. geol. Ges., **39** : 22 - 23.
GRAMAIN, P. & MYARD, P. (1981) : Elongational deformation by shear flow of flexible polymers adsorbed in porous media. Macromolecules, **14** : p. 180 ff.
GRANT, B.F., DUVALL, W.I., OBERT, L., ROUGH, R.L. & ATCHISON, T.C. (1950) : Research on shooting oil and gas wells. API Drill. Product. Pract., **1950** : p. 303.
GRANT, M.A. (1983) : Geothermal reservoir modelling. Geothermics, **12** : 251 - 263.
GRANT, M.S., DONALDSON, I.G. & BIXLEY, P.F. (1982) : Geothermal reservoir engineering. Academic Press, New York City.
GRAU, J.A., ROSCOE, B.A. & TABANOU, J.R. (1988) : A borehole correction model for capture gamma ray spectroscopy logging tools. SPE Form. Eval., **3** : 62 - 68.
GRAU, J.A. & SCHWEITZER, J.S. (1987) : Prompt gamma-ray spectral analysis of well data obtained with NaI(Tl) and 14 MeV neutrons. Nucl. Geophys., **1** : 157 - 165.
GRAUE, A., KOLLTVEIT, K., LIEN, J.R. & SKAUGE, A. (1988) : Imaging fluid saturation development in long coreflood displacements. SPE Paper, **17438**.
GRAUTE, J., KURTH, M., SCHLEMM, F. & THIES, F. (1982) : Erfahrungen bei der Durchführung von Dampfflutverfahren in den Feldern Georgsdorf und Rühlermoor. Teil 1. Erdoel-Erdgas, **98** : 300 - 309.
GRAY, D.H. & REX, R.W. (1965) : Formation damage in sandstones caused by clay dispersion and migration. Proc. 14th Clay Minerals Conference, Clay and Clay Miner., **1966** : 355 - 366.
GRAY, D.I. (1987) : Troll. In : SPENCER, A.M. et al. (Editors), Geology of the Norwegian oil and gas fields, 389 - 402; Graham & Trotman, London/England.
GRAY, I. (1975) : Viking gas-field. In : WOODLAND, A.W. (Editor), Petroleum and the continental shelf of North-West Europe, **1** : 241 - 248; Applied Science

Publ., London/England.
GRAY, I. (1987) : Reservoir engineering in coal seams : Part 1 - The physical process of gas storage and movement in coal seams. Part 2 - Observations of gas movement in coal seams. SPE Reserv. Engin., **2** : 28 - 34, 35 - 40.
GREBE, D. (1981) : Extreme Ablenkungen von Bohrungen am Beispiel von Produktionsbohrungen des Ölfeldes Schwedeneck. Erdoel-Erdgas, **97** : 60 - 63.
GREBE, J.J. (1943) : Tools and aims of research. Chem. Engin. News, **21** (10.12. **1943**) : p. 2004 ff.
GREENER, J.M., SAUVAGEAU, K.A. & PASTERNACK, I. (1988) : Horizontal drilling of eolian Nugget Sandstone in the Overthrust Belt. SPE Paper, **17527**.
GREENFIELD, H.S. & AHMED, U. (1983) : Fracture stimulation treatment of a Western tight sand reservoir : a case history. SPE Paper, **11825**.
GREENFIELD, H.S., AHMED, U., SCHATZ, J. & JONES, A.H. (1982) : A successful stimulation research program for a tight sand reservoir. SPE/DOE Paper, **10828**.
GREENHORN, R.C. & LI, E. (1985) : Investigation of high-phase volume liquid CO2 fracturing fluids. CIM Paper, **85-36-34**.
GREGORCZYK, W.A., PAULS, R., HOLTMYER, M., VENDITTO, J.J. & CHISHOLM, P. (1984) : High frac sand concentrations boost production in Texas Olmos Formation. Oil Gas J., **82** (17.09.**1984**): 105 - 108.
GREGORY, J. (1986) : Southern Basin shapes up as saviour. Offshore Engineer, **1986**/9 : 36 - 38.
GREGORY, J. (1987) : Southern Basin prepares to meet gas future. Offshore Engin., **1987**/11 : 56 - 57.
GRI (1986) : In focus - tight gas sands, **3**/2.
GRI (1987 a) : In focus - tight gas sands, **4**/1.
GRI (1987 b) : Devonian shale - specific log interpretation. Eastern Devonian Gas Shales Technol. Rev., **4**/1 (10.1987).
GRI (1988 a) : A synopsis of technological advancements used in evaluating the Travis Peak Formation. In focus - tight gas sands, **5**/1 : 48 pp.
GRI (1988 b) : GRID - Gas Research Institute digest. Summer 1988, 44 pp.
GRIFFIN, K.W. (1985) : Induced fracture orientation determination in the Kuparuk reservoir. SPE Paper, **14261**.
GRIFFIN, T.J., CONSTIEN, V.G. & COLAVECCHIO, G.P. (1978) : Water carried resin-coated gravel solves varied sand control problems in West Coast wells. SPE Paper, **7122**.
GRIFFITH, P.J. & MADISON, J.S. (1988) : Optimization of fracture stimulation within the North Ward Estes field. SPE Paper, **17281**.
GRIFFITHS, J. (1986) : Oilwell drilling. Survival of the fittest. Industrial Miner., **230** (1986/11) : 21 - 36.
GRIGORIAN, A.M. (1969) : Drilling of productive layers by horizontal wells and multiple completion. 190 pp. (in Russian); Nedra, Moscow/USSR. Inst. Franç. Pétrol. Publ., **27746** (1980; in French); Paris/France.
GRINGARTEN, A.C. (1978) : Reservoir limits testing for fractured wells. SPE Paper, **7452**.
GRINGARTEN, A.C. (1982) : Interpretation of tests in fissured reservoirs and multilayered reservoirs with double porosity behaviour : theory and practice. SPE Paper, **10044**. J. Petrol. Technol., **36** (1984) : 549 - 564.
GRINGARTEN, A. (1988) : Analysis of well tests in fissured reservoirs and multilayered reservoirs with high permeability contrast between layers : a review. SPE Paper, **17350**.
GRINGARTEN, A.C., BOURDET, D.P., LANDEL, P.A. & KNIAZEFF, V.J. (1979) : A comparison between different skin and wellbore storage type-curves for early time transient analysis. SPE Paper, **8205**.
GRINGARTEN, A.C. & RAMEY, H.J. (1974) : Unsteady-state pressure distributions created by a well with a single horizontal fracture, partial penetration, or restricted entry. Soc. Petrol. Engin. J., **14** : 413 - 426.
GRINGARTEN, A.C., RAMEY, H.J. & RAGHAVAN, R. (1974) : Unsteady-state pressure distributions created by a well with a single infinite-conductivity vertical fracture. Soc. Petrol. Engin. J., **14** : 347 - 360. SPE Paper, **4051** (1972).
GRINGARTEN, A.C., RAMEY, H.J. & RAGHAVAN, R. (1975) : Applied pressure analysis for fractured wells. J. Petrol. Technol., **27** : 887 - 892. SPE Paper, **5496**.

GRONE, C.C.la, BAUMGARTNER, S.A. & WOODROOF, R.A. (1983) : The chemical evolution of a high-temperature fracturing fluid. SPE Paper, **11794**. Soc. Petrol. Engin. J., **25** (1985) : 623 - 628.

GROSSMANN, U., HÖFLING, B. & MARX, C. (1988) : Untersuchungen zur Gravel-Pack-Komplettierung mit keramischen Materialien für thermische Bohrungen. DGMK-spring meeting, Celle/Germany FRG, 10.06.1988.

GROVES, G.P. & WHITTAKER, J.L. (1985) : Continuous fracture probability determination as applied to the Monterey Formation. SPE Paper, **13652**.

GRUB, D. (1984) : The future market for natural gas and its effect on exploration. SPE Paper, **13379**.

GRUESBECK, C. & COLLINS, R.E. (1978) : Particle transport through perforations. SPE Paper, **7006**. Soc. Petrol. Engin. J., **22** (1982) : 857 - 865.

GRUESBECK, C. & COLLINS, R.E. (1982) : Entrainment and deposition of fine particles in porous media. Soc. Petrol. Engin. J., **22** : 847 - 856. SPE Paper, **8430** (1979).

GRUESBECK, C., SALATHIEL, W.M. & ECHOLS, E.E. (1979) : Design of gravel packs in deviated wellbores. J. Petrol. Technol., **31** : 109 - 115. SPE Paper, **6805** (1977). SPE Reprint Series, Well Completions, **5 a**/II (1978) : 84 - 94.

GRUNDMANN, S.R. & LORD, D.L. (1983) : Foam stimulation. J. Petrol. Technol., **35** : 597 - 602. SPE Paper, **9754** (1981).

GRUY, H.J., GARB, F.A. & HERRON, H. (1980) : Recoverable reserves are affected by oil and gas prices. SPE Paper, **9360**.

GUDMUNDSSON, J.S. & OLSEN, G. (1985) : Water influx modeling of Svartsengi geothermal field, Iceland. SPE Paper, **13615**. SPE Reserv. Engin., **2** (1987) : 77 - 84.

GUDMUNDSSON, J.S., HAUKSSON, T., THORHALLSSON, S., ALBERTSSON, A. & THOROLFSSON, G. (1984) : Injection and tracer testing in Svartsengi geothermal field, Iceland. Proc. 6th New Zealand Geothermal Workshop, 175 - 180; Auckland/New Zealand.

GUDMUNDSSON, J.S., MENZIES, A.J. & HORNE, R.N. (1983) : Steamtube relative permeability functions for flashing steam/water flow in fractures. SPE Paper, **11686**. SPE Reserv. Engin., **1** (1986) : 371 - 377.

GUERRERO, E.T. (1958) : Fracturing can help secondary recovery. World Oil, **1958**/7 : p. 126 ff.

GUEVARA, E.H. & TYLER, N. (1986) : Influence of facies architecture on hydrocarbon recovery from naturally fractured submarine fan reservoirs, Central Spraberry Trend (Permian), Midland Basin, Texas. Bull. Amer. Assoc. Petrol. Geol., **70** : p. 597.

GUIDRY, F.K., GRAHAM, R.L., LANCASTER, D.E. & CURTIS, J.B. (1987) : A case study of the evaluation, completion, and testing of a Devonian shale gas well. SPE Paper, **17037**.

GUILLOT, D. & DUNAND, A. (1985) : Rheological characterization of fracturing fluids by using laser anemometry. Soc. Petrol. Engin. J., **25** : 39 - 45. SPE Paper, **12030** (1983).

GULATI, M.S. & MALY, G.P. (1974) : Thin-section and permeability studies call for smaller gravels in gravel packing. SPE Paper, **4773**. J. Petrol. Technol., **27** (1975) : 107 - 112.

GULBIS, J. (1983) : Dynamic fluid loss of fracturing fluids. SPE Paper, **12154**. CIM Paper, **82-33-18** (1982). J. Canad. Petrol. Technol., **1984** : 33 - 36.

GÜMRAH, F. & OKANDAN, E. (1987) : The use of CO2 with steam for heavy oil recovery. 4th Europ. Symp. Enhanced Oil Recovery, Thermal Recovery Session, Hamburg/Germany FRG.

GUNDERMANN, H.P. (1982) : Erdgas in Energiemarkt der Zukunft. Erdoel-Erdgas, **98** : 152 - 155.

GUPPY, K.H. (1980) : Non-Darcy flow in wells with a finite-conductivity vertical fracture. Ph.D. Thesis, Stanford Univ., 06.1980. Soc. Petrol. Engin. J., **22** (1982) : 681 - 698. SPE Paper, **8281** (1979).

GUPPY, K.H., CINCO-LEY, H. & RAMEY, H.J. (1980) : Effect of non-Darcy flow on constant pressure production of fractured wells. SPE Paper, **9344**. Soc. Petrol. Engin. J., **21** (1981) : 390 - 400.

GUPPY, K.H., CINCO-LEY, H. & RAMEY, H.J. (1982) : Pressure buildup analysis of fractured wells producing at high flow rates. J. Petrol. Technol., **34** : 2656

- 2666. SPE Paper, **10178** (1981).
GURLEY, D.G., COPELAND, C.T. & HENDRICK, J.O. (1977) : Design, plan, and execution of gravel-pack operations for maximum productivity. J. Petrol. Technol., **29** : 1259 - 1266. SPE Paper, **5709** (1976).
GURLEY, D.G. (1988) : Gravel packing. In preparation.
GUSTAJTIS, K.A. (1985) : Dropping world oil prices will force next oil crisis. CIM Paper, **85-36-19**.
GUY, J.O., CORLEY, B., FERTL, W.H. & FROST, E. (1987) : Oriented Circumferential Acoustic Logging evaluates vertically fractured reservoirs. Trans. 28th Soc. Prof. Well Log Analysts Annual Logging Symp., London/England, 29.06. - 02.07.1987.
GUY, J.O. & FERTL, W.H. (1980) : Circumferential acoustic measurements - the CA Log. SAID 7th European Logging Symposium, Société pour l´Avancement de l´Interprétation des Diagraphies, Paper **30**, Paris/France, 21. - 23.10.1984.
GUY, J.O., FERTL, W.H. & OLIVER, D.W. (1986) : The use of circumferentially propagated acoustic waves in well logging. SPE Paper, **15484**.
GWINN, J.E. & McMANUS, G.R. (1977) : Current gas well completion practices in Appalachian gas sands. SPE Paper, **6621**.
HAAK, A.M. & ELEWAUT, E.F.M. (1988) : Modeling heterogeneity to improve 3-D stimulation results : a case history on a North Sea gas field. SPE Paper, **17763**.
HAAS, M.R., BRASHEAR, J.P. & MORRA, F. (1985) : Historical trends and current production of gas from tight formations. SPE/DOE Paper, **13897**. J. Petrol. Technol., **39** (1987) : 77 - 88.
HACKBARTH, C.J. & TEPPER, B.J. (1988) : Examination of BHTV, FMS, and SHDT images in very thinly bedded sands and shales. SPE Paper, **18118**.
HAGOORT, J. (1981) : Waterflood-induced hydraulic fracturing. Ph.D. Thesis, 227 pp.; Delft Techn. Univ., Delft/Netherlands.
HAGOORT, J. (1982) : The response of interwell tracer tests in watered-out reservoirs. SPE Paper, **11131**.
HAGOORT, J., WEATHERHILL, B.D. & SETTARI, A. (1980) : Modeling the propagation of waterflood-induced hydraulic fractures. Soc. Petrol. Engin. J., **20** : 293 - 303. SPE Paper, **7412** (1978).
HAILEY, T.T. (1984) : Further discussion of experience in gravel packing long perforated intervals in deviated holes. J. Petrol. Technol., **36** : p. 1352. SPE Paper, **13309**.
HAILEY, T.T., DONOVAN, J.F. & SICKLE, E.W.v. (1985) : One-trip perforate and gravel pack system. SPE Paper, **13648**.
HAIMSON, B.C. (1973) : Earthquake related stresses at Rangely, Colorado. Proc. 14th Annual Rock Mechan. Symp., Amer. Soc. Civil Engin., 689 - 708; University Park/Pennsylvania.
HAIMSON, B.C. (1978 a) : The hydrofracturing stress measurement method and recent field results. Intern. J. Rock Mech., Min. Sci., Geomechan. Abstr., **15** : 167 - 178.
HAIMSON, B.C. (1978 b) : Crustal stress in the continental United States as derived from hydrofracturing tests. In : HEACOCK, J.C. (Editor), The Earth`s crust. Geophys. Monograph Ser., **20**; Amer. Geophys. Inst.
HAIMSON, B. & FAIRHURST, C. (1969 a) : Hydraulic fracturing in porous-permeable materials. J. Petrol. Technol., **21** : 811 - 817. SPE Paper, **2354** (1968).
HAIMSON, B. & FAIRHURST, C. (1969 b) : Initiation and extension of hydraulic fractures in rocks. Soc. Petrol. Engin. J., **9** : 310 - 318.
HALDORSEN, H.H. & LAKE, L.W. (1984) : A new approach to shale management in field-scale models. Soc. Petrol. Engin. J., **24** : 447 - 457. SPE Paper, **10976** (1982).
HALE, B.W. (1986) : Analysis of tight gas well production histories in the Rocky Mountains. SPE Product. Engin., **1** : 310 - 322. SPE/DOE Paper, **11639** (1983).
HALE, B.W., FIRTH, C.H., HANSEN, R.C., MURPHY, M.J. & WOODCOX, S.C. (1987) : Post-buildup drawdown analysis of tight gas wells in the Rocky Mountains. SPE Paper, **16804**.
HALL, B.E. (1975) : The effort of mutual solvents on absorption in sandstone acidizing. J. Petrol. Technol., **27** : 1439 - 1442. SPE Paper, **5377**.

HALL, B.E. (1978) : A new technique for generating in-situ hydrofluoric acid for deep clay damage removal. J. Petrol. Technol., **30** : 1220 - 1224. SPE Paper, **6512** (1977).
HALL, B.E. & ANDERSON, B.W. (1977) : Field results for a new retarded sandstone acidizing system. SPE Paper, **6871**.
HALL, B.E. & DILL, W.R. (1988) : Iron control additives for limestone and sandstone acidizing of sweet and sour wells. SPE Paper, **17157**.
HALL, B.E. & HOUK, S.G. (1983) : Fluid loss control in the naturally fractured Buda Formation. SPE Paper, **12152**.
HALL, B.E. & LARKIN, S.D. (1986) : On-site quality control of fracture treatments. SPE Paper, **15014**.
HALL, B.E., MATHERNE, B., KERLIN, B.A. & PACE, J. (1988) : Perforation prepacking increases production in gravel pack wells. In press.
HALL, B.E., TINNEMEYER, A.C. & UNDERWOOD, P.J. (1981) : Stimulation of the North Coles levee field with a retarded HF-acid. SPE Paper, **9934**.
HALL, C.D. & DOLLARHIDE, F.E. (1964) : Effects of fracturing fluid velocity on fluid-loss agent performance. J. Petrol. Technol., **16** : 555 - 560.
HALL, C.D. & DOLLARHIDE, F.E. (1968) : Performance of fracturing fluid-loss agent under dynamic conditions. J. Petrol. Technol., **20** : 763 - 768. SPE Paper, **1904** (1967).
HALL, C.D. & HARRISBERGER, W.H. (1970) . Stability of sand arches : a key to sand control. J. Petrol. Technol., **22** : 821 - 829. SPE Paper, **2399** (1969). SPE Reprint Series, Well Completion, **5** : 202 - 210.
HALLECK, P.M. & DEO, M. (1987) : The effects of underbalance on perforation flow. SPE Paper, **16895**.
HALLIBURTON (1971) : The Fracbook. Design/data manual for hydraulic fracturing. 111 pp.; Duncan/Oklahoma.
HALLIBURTON (1985) : Planen Sie mit diesem excellenten Halliburton-Fachmann Ihre nächsten Stimulationsarbeiten. Erdoel-Erdgas, **101**/12 : A2 - A3.
HALLIBURTON (1986) : Proppants. Brochure, **F-3235**; Duncan/Oklahoma.
HALLIBURTON (1987) : Refrac. Brochure, **F-3233**; Duncan/Oklahoma.
HALLIBURTON (1988) : Sand control customer manual. Duncan/Oklahoma.
HAMBY, T.W. & RICHARDSON, E.A. (1968) : New sand control process works well in dirty sands. World Oil, **1968**/6 : 84 - 86.
HAMMERSHAIMB, E.C. & KUUSKRAA, V.A. (1982) : The economic potential of reinjection into geopressured aquifers. SPE/DOE Paper, **10833**. J. Petrol. Technol., **36** (1984) : 1164 - 1170.
HANDKE, W.A. (1987) : Microprocessor control of sand concentration improves the quality and versatility of fracturing operations. SPE Paper, **16190**.
HANNAH, R.R. (1976) : New fracturing technique leads to improved performance in the Mississippian trend. J. Petrol. Technol., **28** : 859 - 864. SPE Paper, **5628** (1975).
HANNAH, R.R., HARRINGTON, L.J. & ANDERSON, R.W. (1977) : Stimulation design applications of a technique to locate successive fluid segments in fractures. SPE Paper, **6815**.
HANNAH, R.R., HARRINGTON, L.J. & ANDERSON, R.W. (1978) : Measurement of dynamic proppant fall rates in fracturing gels using a concentric cylinder tester. SPE Paper, **7571**. J. Petrol. Technol., **33** (1981) : 909 - 913.
HANNAH, R.R., HARRINGTON, L.J. & LANCE, L.C. (1983) : The real-time calculation of accurate bottomhole fracturing pressure from surface measurements using measured pressures as a base. SPE Paper, **12062**.
HANNAH, R.R., HARRINGTON, L.J. & POTTER, J.S. (1980) : Post fracturing behavior of fracturing gels and its influence on fracture closure and proppant distribution. SPE Paper, **9331**.
HANNAH, R.R. & MATSON, B.G. (1977) : A family of viscous gelled water systems featuring moderate polymer loadings and low residue on breaking. SPE Paper, **6380**.
HANNAH, R.R., MATSON, B.G. & CRAWFORD, A. (1977) : Status report on MHF in the Fort Worth Basin. Proc. MHF Symp., 28.02. - 01.03.1977, Univ. of Oklahoma.
HANSBERGER, B.G. (1978) : Method of forming gravel packs. U.S. Patent, **4,102,298** (25.07.1978).
HANSEN, S.M. (1988) : Applications for the formation micro scanner imaging tool

in the Permian Basin. SPE Paper, **18112**.
HANSON, A.W. (1964) : Plastically deformable solids in treating subterranean formations. U.S. Patent, **3,159,217** (01.12.1964).
HANSON, A.W & LYONS, O.W. (1964) : Method of fracturing formations. U.S. Patent, **3,151,678** (06.10.1964).
HANSON, J.M. & OWEN, L.B. (1982) : Fracture orientation analysis by the solid earth tidal strain. SPE Paper, **11070**.
HANSON, J.M., SCHMIDT, R.A., COOLEY, C.H. & SCHATZ, J.F. (1984) : Multiple fracture stimulation using controlled pulse pressurization. SPE/DOE/GRI Paper, **12839**.
HANSON, M.E. (1988) : Development of hydraulic fracture models for application to coal seam methane drainage. Quart. Rev. of Methane from Coal Seams Technology, **5**/3-4 : 46 - 48; **6**/1 : 39 - 41.
HANSON, M.E., ANDERSON, G.D. & SHAFFER, R.J. (1980) : Effects of various parameters on hydraulic fracturing geometry. SPE/DOE Paper, **8942**. Soc. Petrol. Engin. J., **21** (1981) : 435 - 443.
HANSON, M.E., ANDERSON, G.D., SHAFFER, R.J. & THORSON, L.D. (1982) : Some effects of stress, friction, and fluid flow on hydraulic fracturing. Soc. Petrol. Engin. J., **22** : 321 - 332. SPE/DOE Paper, **9831** (1981).
HANSON, M.E., MAVKO, B.B. & NIELSEN, P.E. (1988) : Development of hydraulic fracturing models for application to coal seam methane drainage. Gas Research Institute, Quart. Rept. (01. - 03.1988), 25 pp.
HANSON, M.E., NIELSEN, P.E., SORRELS, G.G., BOYER, C.M. & SCHRAUFNAGEL, R.A. (1987) : Design, execution and analysis of a stimulation to produce gas from thin multiple coal seams. SPE Paper, **16860**.
HANSON, M.E. & SHAFFER, R.J. (1980) : Some results from continuum mechanics analyses of the hydraulic fracturing process. Soc. Petrol. Engin. J., **20** : 86 - 94.
HARBEN, P. (1978) : Abrasives - taking the rough with the smooth. Industrial Miner., **1978**/11 : 49 - 73.
HARBEN, P. (1986) : $ 20 oil forces drilling to rock bottom. Industrial Miner., **1986**/2 : 60 - 61.
HARDIN, E.L., CHENG, C.H., PAILLET, F.L. & MENDELSON, J.D. (1987) : Fracture characterization by means of attenuation and generation of tube waves in fractured crystalline rock at Mirror Lake, New Hampshire. J. Geophys. Res., **1987**.
HARDIN, E.L. & TOKSÖZ, M.N. (1985) : Detection and characterization of fractures from generation of tube waves. Trans. 26th Soc. Prof. Well Log Analysts Annual Logging Symp., Paper **II**, Dallas/Texas, 17. - 20.06.1985.
HARDING, T.G. & ALI, S.M.F. (1984) : Mine-assisted heavy oil recovery technology. SPE Paper, **12787**.
HARDMAN, P. (1986) : Beckingham 36 horizontal well. SPE Paper, **15895**.
HARDT, I. (1982) : Geoelektrische Frac-Ortung. Erdoel-Erdgas, **98** : 309 - 312.
HARMS, S.D., GOSS, M.L. & PAYNE, K.L. (1984) : New generation fracturing fluid for ultrahigh-temperature application. SPE Paper, **12484**.
HARMS, S.D., SMITH, J.M., KING, G.E. & POSEY, K. (1988) : Novel workover treatment proves effective in Permian Basin : laboratory and field results. SPE Paper, **17296**.
HARMSEN, G.J. (1979) : Steamflooding in a water drive reservoir in the Schoonebeek field in the Netherlands. 10th World Petrol. Congr., Bucharest/Romania.
HARP, L.J. (1966) : Do not overlook fractured zones. World Oil, **1966**/4 : 119 - 123.
HARP, L.J. (1980) : Another significant step in the evolution of stimulation treatments in the Muddy "J" Sand - Wattenberg field. SPE Paper, **9037**.
HARPER, J.L. (1988) : Developing cost effective technology for the future. Offshore Northern Seas Conference, Stavanger/Norway, 23. - 26.08.1988.
HARPER, M.L. & SHAW, B.E. 1974) : Cretaceous-Tertiary carbonate reservoirs in the North Sea. Offshore North Sea 1974, Explor. Geol. Geophys. Conf., Pap., **G-IV/4**.
HARPER, T.R., HAGAN, J.T. & MARTINS, J.P. (1985) : Fracturing without proppant. SPE/DOE Paper, **13858**.
HARRINGTON, L.J. (1985) : Computers and microprocessors - valuable tools in

well completions. Oil Man, Offshore Europe Exhibition, Aberdeen/Scotland, 10. - 13.09.1985.
HARRINGTON, L.J. & HANNAH, R.R. (1975) : Fracturing design using perfect support fluids for selected fracture proppant concentrations in vertical fractures. SPE Paper, **5642**.
HARRINGTON, L.J., HANNAH, R.R. & BEIRUTE, R. (1978) : Post-fracturing temperature recovery and its implication for stimulation design. SPE Paper, **7560**.
HARRINGTON, L.J., HANNAH, R.R. & WILLIAMS, D. (1979) : Dynamic experiments and proppant settling in crosslinked fracturing fluids. SPE Paper, **8342**.
HARRINGTON, L.J., WHITSITT, N.F. & HANNAH, R.R. (1973) : Prediction of the location and movement of fluid interfaces in a fracture. Southwestern Petroleum Short Course Association, Lubbock/Texas, 26. - 27.04.1973.
HARRIS, M.H. (1966) : The effect of perforating on well productivity. Trans. AIME, **237** : 518 - 528.
HARRIS, O.E., HENDRICKSON, A.R. & COULTER, A.W. (1966) : High-concentration hydrochloric acid aids stimulation results in carbonate formations. J. Petrol. Technol., **18** : 1291 - 1296. SPE Paper, **1654**.
HARRIS, P.C. (1982) : Dynamic fluid-loss characteristics of nitrogen foam fracturing fluids. SPE Paper, **11065**. J. Petrol. Technol., **37** (1985) : 1847 - 1852.
HARRIS, P.C. (1983) : Successful field applications of CO2 foam fracturing fluids in the Ark-La-Tex region. SPE Paper, **11932**.
HARRIS, P.C. (1985) : Effects of texture on rheology of foam fracturing fluids. SPE Paper, **14257**.
HARRIS, P.C. (1987) : Dynamic fluid-loss characteristics of CO2-foam fracturing fluids. SPE Product. Engin., **2** : 89 - 94. SPE Paper, **13180** (1984).
HARRIS, P.C. (1988) : Fracturing fluid additives. SPE Paper, **17112**.
HARRIS, P.C., BAILEY, D.E., EVERTZ, G.L. & REEVES, T.L. (1984) : Stimulation results in the low-permeability Wasatch Formation - an evolution to foam fracturing. SPE Paper, **11837**. J. Petrol. Technol., **36** : 1545 - 1551.
HARRIS, P.C., HARMS, W.M. & NORMAN, L.R. (1987) : Study of continuously mixed cross-linked fracturing fluids with a flow loop viscosimeter. SPE Paper, **17044**.
HARRIS, P.C., HAYNES, R.J. & EGGER, J.P. (1984) : The use of CO2-base fracturing fluids in the Red Fork Formation in the Anadarko Basin, Oklahoma. J. Petrol. Technol., **36** : 1003 - 1008. SPE Paper, **11575** (1983).
HARRIS, P.C., KLEBENOW, D.E. & KUNDERT, D.P. (1988) : Constant internal phase design improves stimulation results. SPE Paper, **17532**.
HARRIS, P.C. & PENNY, G.S. (1985) : Influence of temperature and shear history on fracturing fluid efficiency. SPE Paper, **14258**.
HARRIS, P.C. & REIDENBACH, V.G. (1984) : High-temperature rheological study of foam fracturing fluids. SPE Paper, **13177**. J. Petrol. Technol., **39** (1987) : 613 - 619.
HARRIS, P.C., REIDENBACH, V.G. & CHISHOLM, P.T. (1986) : Method for stimulation of wells with carbon dioxide or nitrogen based fluids containing high proppant concentration. U.S. Patent, **4,627,495** (09.12.1986).
HARRISBERGER, W.H., COULTER, A.W. & GURLEY, D.G. (1971) : Application of recent advances in the mechanics of sand consolidation. SPE Paper, **3589**.
HARRISON, E., McGUIRE, W.K. & KIESCHNICK, W.F. (1954) : The mechanics of fracture induction and extension. Trans. AIME, **201** : 252 - 263.
HARRISON, H.L. & TILDEN, J. (1986) : Predictive model for fracture distribution and intensity in sandstone and carbonate reservoirs, Wertz and Lost Soldier Fields, Wyoming. Bull. Amer. Assoc. Petrol. Geol., **70** : p. 599.
HARRISON, N.W. (1971) : Diverting agents - history and application. SPE Paper, **3653**. J. Petrol. Technol., **24** (1972). SPE Reprint Series, Well Completions, **5 a**/II (1978) : 199 - 205.
HART, C.M., ENGI, D. & MORRIS, H.E. (1983) : A comprehensive fracture diagnostics instrumentation fielding program. SPE/DOE Paper, **11810**.
HART, C.M., ENGI, D., FLEMING, R.P. & MORRIS, H.E. (1984) : Fracture diagnostics results for the Multiwell Experiment`s paludal zone stimulation. SPE/DOE/GRI Paper, **12852**. SPE Form. Eval., **2** (1987) : 320 - 326.
HARTLEY, R. & BOSMA, M.G.R. (1985) : Fracturing in chalk completions. J. Pe-

trol. Technol., **37** : 73 - 79, p. 1498. SPE Paper, **11901** (1983).
HARTSOCK, J.H. & SLOBOD, R.L. (1961) : The effect of mobility ratio and vertical fractures on the sweep efficiency of a five-spot. Prod. Monthly, **1961**/9 : p. 2 ff.
HARTSOCK, J.H. & WARREN, J.E. (1961) : The effect of horizontal hydraulic fracturing on well performance. J. Petrol. Technol., **13** : 1050 - 1056.
HASHEMI, R., CAOTHIEN, S. & ROVIG, J. (1988) : Field measurement of viscosified fluids. SPE Paper, **17462**.
HASKIN, C.A. (1976) : A review of hydroxy-aluminum treatments. SPE Paper, **5692**.
HASKIN, C.A., REED, M.G. & COPPEL, C.P. (1975) : Here`s when and how to use hydroxy-aluminum well treatment. Oil Gas J., 10.11.**1975** : 217 - 220.
HASSAN, D. (1982) : A method for predicting fracture azimuth and the implications thereof to improve hydrocarbon recovery. CIM Paper, **82-33-19**.
HASSEBROEK, W.E. & WALTER, A.B. (1964) : Advancements through 15 years of fracturing. J. Petrol. Technol., **16** : 760 - 764.
HASSELL, R.L. (1986) : Restructuring : survival or suicide. SPE Paper, **15553**.
HASSELL, R.L. (1988) : Restructuring for long-term survival. SPE Paper, **16851**.
HAST, N. (1979) : Limits of stress measurements in the earth`s crust. Rock Mechan., **11** : 143 - 150.
HATZIGNATIOU, D.G., OGBE, D.O., DEHGHANI, K. & ECONOMIDES, M.J. (1987) : Interference pressure behavior in multi-layered composite reservoirs. SPE Paper, **16766**.
HAWKINS, G.W. (1988) : Laboratory study of proppant-pack permeability reduction caused by fracturing fluids concentrated during closure. SPE Paper, **18261**.
HAYNES, C.D. & GRAY, K.E. (1972) : Sand particle transport in perforated casing. SPE Paper, **4031**. J. Petrol. Technol., **26** (1974) : 80 - 84.
HAZLETT, W.G., LEE, W.J., NARAHARA, G.M. & GATENS, J.M. III (1986) : Production data analysis type curves for the Devonian Shales. SPE Paper, **15934**.
HEARD, F.E. (1981) : A high temperature acoustic borehole televiewer. Proc. High Temperature Electronics and Instrumentation Conference, Houston/Texas, 12.1981.
HEARST, J.R. et al. (1976) : Fractures induced by a contained explosion in Kemmerer coal. Intern. J. Rock. Mechan. Miner. Sci., Geomechan. Abstr., **13**.
HEDEMANN, H.A., MASCHEK, W., PAULUS, B. & PLEIN, E. (1984) : Mitteilung zur lithostratigraphischen Gliederung des Oberrotliegenden im nordwestdeutschen Becken. Nachricht. dt. geol. Ges., **30** : 100 - 107.
HEFLIN, J.D. (1979) : Fracture detection in West Coast reservoirs using well logs. Soc. Petrol. Engin. Publ., **7976** : 203 - 217.
HEFLIN, J.D., NEILL, B.E. & DAVIES, M.R. (1976) : Log evaluation in the California Miocene formations. SPE Paper, **6160**.
HEFLIN, J.D. & NETTLETON, K.A. (1980) : Formation evaluation utilizing gamma ray spectral analysis. SPE Paper, **9042**.
HEISLER, R.P. (1986) : Interpretation of radioactive tracer results in a steamdrive project. SPE Paper, **15092**. SPE Reserv. Engin., **3** (1988) : 281 - 287.
HELM, D.C. (1984) : Field based computational techniques for predicting subsidence due to fluid withdrawal. Geol. Soc. Amer., Rev. Engin. Geol., **6** : 1 - 22.
HENDRICKSON, A.R. & CAMERON, R.C. (1968) : New fracture acid technique provides efficient stimulation of massive carbonate sections. Annual Technical Meeting of the Petroleum Society of CIM, Calgary/Alberta, 05.1988.
HENG, H.B. (1986) : An overview of gravel packed completions in Sarawak Shell Berhad (SSB) and Sabah Shell Petroleum Company (SSPC). SPE Paper, **14600**. SPE Product. Engin., **2** (1987) : 81 - 88.
HENG, H.B. (1987) : An overview of gravel-packed completions. SPE Production Engineering, **2** : 81 - 88. SPE Paper, **14600** (1986).
HENNINGTON, W.M. (1980) : Unconventional exploration for natural fractures. SPE/DOE Paper, **8956**.
HENSEL, W.M. (1988) : A perspective look at fracture porosity. SPE Paper, **16806**.
HERNANDEZ, I. & CADE, R. (1988) : Screening criteria for horizontal wells. IADC/SPE Paper, **17239**.
HERTZOG, R.C. (1980) : Laboratory and field evaluation of an inelastic neutron

scattering and capture gamma ray spectrometry tool. Soc. Petrol. Engin. J., **20** : 327 - 340. SPE Paper, **7430** (1978).
HESS, A.W. (1986) : Identifying hydraulicly-conductive fractures with low-velocity borehole flowmeter. Canad. Geotechn. J., **23** : 69 - 78.
HETZER, H. & KATZUNG, G. (1985) : Zur Bildung der Erdgaslagerstätten im Rotliegenden der Mitteleuropäischen Senke. Z. angew. Geol., **31** : 285 - 290.
HEWITT, C.H. (1963) : Analytical techniques for recognizing water sensitive reservoir rocks. J. Petrol. Technol., **15** : 813 - 818.
HICKEY, J.W. (1983) : The comparative effectiveness of propping agents in the Red Fork Formation of the Anadarko Basin. SPE Paper, **11577**.
HICKEY, J.W., BROWN, W.E. & CRITTENDEN, S.J. (1981) : The comparative effectiveness of propping agents in the Red Fork Formation of the Anadarko Basin. SPE Paper, **10132**.
HICKS, J.N. & RAHMAN, H. (1984) : Effect of increasing gas prices on known reserves in domestic gas fields. SPE Paper, **12601**.
HILL, A.D., BOEHM, K.E. & AKERS, T.J. (1988) : Tracer placement techniques for improved radioactive tracer logging. SPE Paper, **17317**.
HILL, A.D., LINDSAY, D.M., SCHECHTER, R.S. & SILBERBERG, I.H. (1977) : Sandstone acidizing : the development of design methods. SPE Paper, **6607**.
HILL, A.D. & SOLARES, J.R. (1985) : Improved analysis methods for radioactive tracer injection profile logging. J. Petrol. Technol., **37** : 511 - 520. SPE Paper, **12140** (1983).
HILL, D.G. (1982) : Clay stabilization - criteria for best performance. SPE Paper, **10656**.
HILL, J. (1983) : Key issues in determining the contour of future energy demand. SPE Paper, **11325**.
HILL, J. (1987) : Oil futures and options increase in importance in volatile industry conditions. SPE Paper, **16311**.
HILL, K.E. (1941) : Factors affecting the use of gravel in oil wells. API Drill. Prod. Pract., **134** : 134 - 143. Oil Weekly, 26.05.**1941** : 13 - 20.
HILL, O., CLEMENT, C. & WARD, J. (1977) : Austin Chalk fracturing design utilizing a crosslinked natural polymer as a diverting agent. SPE Paper, **6869**.
HILL, W.L. & WAHL, H.A. (1968) : The effect of formation temperature on hydraulic fracture design. API Paper, **851-42-1**.
HILLIG, W.B. & CHARLES, R.J. (1965) : Surfaces, stress-dependent surface reactions and strength. In : ZACKAY, V.F. (Editor), High-strength materials. p. 682 ff.; Wiley, New York.
HILSCHER, L.W. & CLEMENTS, W.R. (1982) : High-temperature drilling fluid for geothermal and deep sensitive formations. SPE Paper, **10737**.
HIMES, R.E. & RUIZ, S.J. (1986): New sidewall core analysis techniques improve gravel pack design. SPE Paper, **14813**.
HIMMATRAMKA, A.H. (1981) : Productivity reduction due to non-Darcy flow and true skin in gravel-packed wells. SPE Paper, **10084**.
HINN, R.L. (1983) : Case history : a multizone tight gas completion technique for the Blocker Cotton Valley field, East Texas. SPE Paper, **11934**.
HIRASAKI, G.J., MORROW, F. & WILLHITE, G.P. (1984) : Estimation of reservoir heterogeneity from waterflood performance. SPE Paper, **13415**.
HIRSCH, J.M., CISAR, M.T., GLASS, S.W. & ROMANOWSKI, D.A. (1981) : Recent experience with wireline fracture detection logs. SPE Paper, **10333**.
HOCH, O., WALKER, M., FREDRICKSON, S. & NORMAN, L. (1986) : Heated acids for improved stimulation results. CIM Paper, **86-37-68**.
HOCHHEISER, H.W. (1988) : North American natural gas markets. Symposium on Energy, Finance, and Taxation Policies, Washington/D.C., 19. - 20.09.1988.
HODGE, R.M. (1982) : Gravel transport in deviated wellbores. SPE Paper **10654**.
HODGE, R.M. & BARANET, S.E. (1987) : Evaluation of field methods to determine crosslink time of fracturing fluids. SPE Paper, **16249**.
HODGES, J.E. & PAOLI, B.F. (1982) : Modern techniques and thorough evaluation yield more consistent stimulation results in Frontier Formation. SPE Paper, **10892**.
HODGES, L.T. & KNUTSON, C.F. (1981) : Tight gas sandstone channel continuity and directivity, Upper Cretaceous and Paleocene, Greater Green River Basin, Wyoming. SPE/DOE Paper, **9844**.

HOEL, M. (1988) : Crosslinked gelled water is optimum fracture fluid for Spraberry trend in West Texas. Oil Gas J., 15.08.**1988** : 60 - 69.
HOLBERT, D.R., ARONOFSKY, J.S., CLINE, W.B. & ADAMS, C.L. (1987) : The signature approach to the analysis of fractured, water-driven reservoirs. SPE Paper, **16979**.
HOLBROOK, P. (1987) : Is the current oil price the reason for the reduction in exploration drilling? No. Petrol. Rev., **1987**/1 : 20 - 21.
HOLCOMB, D.L. (1975) : Low surface tension hydrochloric - hydrofluoric acid mixtures in low porosity, low permeability sandstones. SPE Paper, **5411**.
HOLCOMB, D.L. (1977) : Foamed acid as a means for providing extended retardation. SPE Paper, **6376**.
HOLCOMB, D.L. (1978) : Low quality foamed gel systems provide fracturing efficiency, improved clean-up and less formation damage. SPE Paper, **7562**.
HOLCOMB, D.L. (1983) : Correct flowback procedure - key to successful foam stimulation. Proc. 30th Annual Southwestern Petroleum Short Course, Texas Techn. Univ., Lubbock/Texas, 27. - 28.04.1983, pp. 75 - 82.
HOLCOMB, D.L. (1986) : Minimizing reservoir damage by insuring proper additive performance in stimulation fluids. SPE Paper, **14991**.
HOLCOMB, D.L., CALLAWAY, R.E. & CURRY, L.L. (1980 a) : Foamed hydrocarbon stimulation of water sensitive formations. SPE Paper, **9033**.
HOLCOMB, D.L., CALLAWAY, R.E. & CURRY, L.L. (1980 b) : The chemistry, physical nature and rheology of an aqueous stimulation foam. SPE Paper, **9530**.
HOLCOMB, D.L., CALLAWAY, R.E. & CURRY, L.L. (1981) : Chemistry, physical nature, and rheology of aqueous stimulation foams. Soc. Petrol. Engin. J., **21** : 410 - 414.
HOLCOMB, D.L. & WILSON, S.C. (1978) : Foamed acidizing and selective diverting using stable foam for improving acid stimulation. Proc. 25th Southwest Petrol. Short Course, Texas Techn. Univ., Lubbock/Texas.
HOLCOMB, W. (1982) : How to choose a hydraulic frac treatment : MHF versus conventional. Petrol. Engin. Intern., **1982**/8 : 108 - 116.
HOLDEN, W.W. III, PRIHODA, C.H. & HALL, B.E. (1981) : Successful stimulation of Fordoche field with a retarded HF acid. J. Petrol. Technol., **33** : 1485 - 1490. SPE Paper, **9389** (1980).
HOLDITCH, S.A. (1974) : Economic production of tight gas reservoirs looks better. Oil Gas J., 04.02.**1974** : 99 - 102.
HOLDITCH, S.A. (1979) : Factors affecting water blocking and gas flow from hydraulically fractured gas wells. J. Petrol. Technol., **31** : 1515 - 1524. SPE Paper, **7561** (1978).
HOLDITCH, S.A. (1984) : Criteria for propping agent selection. Sovereign Engin. Intern., 2nd. ed., 165 pp.; 3rd ed. (1987), 221 pp.; Norton-Alcoa Proppants, Dallas/Texas.
HOLDITCH, S.A. & ELY, J. (1973) : Successful stimulation of deep wells using high proppant concentration. J. Petrol. Technol., **25** : 959 - 964. SPE Paper, **4118** (1972).
HOLDITCH, S.A., ELY, J.W., SEMMELBECK, M.E., CARTER, R.H., HINKEL, J. & JEFFREY, R.G. (1988) : Enhanced recovery of coalbed methane through hydraulic fracturing. SPE Paper, **18250**.
HOLDITCH, S.A., JENNINGS, J.W., NEUSE, S.H. & WYMAN, R.E. (1978) : The optimization of well spacing and fracture length in low permeability gas reservoirs. SPE Paper, **7496**.
HOLDITCH, S.A. & LANCASTER, D.E. (1982) : PROPTRAN - a finite difference numerical model for predicting proppant distribution in vertical fractures. S.A. Holditch and Assoc. Docum.
HOLDITCH, S.A. & LAUFER, W.L. (1978) : The analysis of fractured gas wells using reservoir simulation. SPE Paper, **7473**.
HOLDITCH, S.A. & LEE, W.J. (1979) : Fracture evaluation with pressure transient tests in low-permeability gas reservoirs. Part II : field examples. SPE Paper, **7930**.
HOLDITCH, S.A., LEE, W.J. & GIST, R. (1981) : An improved technique for estimating permeability, fracture length and fracture conductivity from pressure buildup tests in low permeability gas wells. SPE/DOE Paper, **9885**. J. Petrol. Technol., **35** (1983) : 981 - 990.

HOLDITCH, S.A., LEE, W.J., LANCASTER, D.E. & DAVIS, T.B. (1983) : Effect of mud filtrate invasion on apparent productivity in drillstem tests in low-permeability rock. J. Petrol. Technol., **35** : 299 - 305. SPE Paper, **9842** (1981).
HOLDITCH, S.A. & MORSE, R.A. (1971) : Large fracturing treatments may unlock tight reservoirs. Oil Gas J., 29.05.**1971** : 57 - 60; 05.04.**1971** : 84 - 89.
HOLDITCH, S.A. & MORSE, R.A. (1976) : The effects of non-Darcy flow on the behavior of hydraulically fractured gas wells. J. Petrol. Technol., **28** : 1169 - 1178. SPE Paper, **5586** (1975).
HOLDITCH, S.A., ROBINSON, B.M., ELY, J.W. & RAHIM, Z. (1988) : The effect of viscous fluid properties on excess friction pressures measured during hydraulic fracture treatments. SPE Paper, **18208**.
HOLDITCH, S.A., ROBINSON, B.M. & WHITEHEAD, W.S. (1986) : Pre-fracture and post-fracture formation evaluation necessary to characterize the three-dimensional shape of a hydraulic fracture. SPE Paper, **14086**. SPE Form. Eval., **2** (1987) : p. 523 ff.
HOLDITCH, S.A., ROBINSON, B.M. & WHITEHEAD, W.S. (1987) : The analysis of complex Travis Peak reservoirs in East Texas. SPE/DOE Paper, **16427**.
HOLDITCH, S.A., ROBINSON, B.M., WHITEHEAD, W.S. & ELY, J.W. (1987) : The GRI staged field experiment. SPE/DOE Paper, **16429**. SPE Form. Eval., **3** (1988) : 519 - 533.
HOLLIS, A.P. (1982) : Some petroleum engineering considerations in the changeover of the Rough gas field to the storage mode. EUR Paper, **295**. SPE Paper, **12932** (1984). J. Petrol. Technol., **36** (1984) : 797 - 804.
HOLM, A.E. (1975) : Perforating at base of or below sand and fracturing improves well performance. SPE Paper, **5412**.
HOLMAN, G.B. (1975) : Evaluation of control techniques for unconsolidated silty sands. SPE Paper, **5656**. J. Petrol. Technol., **28** (1976) : 979 - 984.
HOLMAN, J.P. (1976) : Heat transfer. 4th edition, 224 pp.; McGraw-Hill.
HOLMES, P.F. (1987) : OPEC and oil market prospects today. Petrol. Rev., **1987**/7 : 28 - 32.
HOLT, R.M., INGSOY, P. & MIKKELSEN, M. (1987) : Rock mechanical analysis of North Sea reservoir formations. SPE Paper, **16796**.
HOLTBERG, P.D., WOODS, T.J. & ASHBY, A.B. (1987) : 1987 GRI baseline projection of U.S. energy supply and demand to 2010. Gas Research Institute, 79 pp.
HOLUB, R.W., MALY, G.P., NOEL, R.P. & WEINBRANDT, R.M. (1974) : Scanning electron microscope pictures of reservoir rocks reveal ways to increase oil production. SPE Paper, **4787**.
HOLZHAUSEN, G.R., CARD, C.C., RAISBECK, J.M. & DOBECKI, T.L. (1985) : Hydraulic fracture growth during steam stimulation in a single-well test. SPE Paper, **13619**.
HOLZHAUSEN, G.R. & EGAN, H.N. (1986) : Fracture diagnostics in East Texas and Western Colorado using the hydraulic-impedance method. SPE Paper, **15215**.
HOLZHAUSEN, G.R. & EGAN, H.N. (1987) : Detection and control of hydraulic fractures in water injection wells. SPE Paper, **16362**.
HOLZHAUSEN, G.R. & GOOCH, R.P. (1985) : Impedance of hydraulic fractures : its measurement and use for estimating fracture closure pressures and fracture dimensions. SPE/DOE Paper, **13892**.
HOMSY, G.M. (1987) : Viscous fingering in porous media. Ann. Rev. Fluid Mech., **19** : 271 - 311.
HONARPOUR, M., KOEDERITZ, L. & HARVEY, A.H. (1986) : Relative permeability of petroleum reservoirs. 143 pp.; CRC Press, Boca Raton/Florida.
HONARPOUR, M.M., SZPAKIEWICZ, M.J., SCHATZINGER, R.A., TOMUTSA, L., CARROLL, H.B. & TILLMAN, R.W. (1988) : Integrated geological/engineering model for barrier island deposits in Bell Creek field, Montana. SPE/DOE Paper, **17366**.
HONG, K.C. & MILLHONE, R.S. (1977) : Injection profile effects caused by gravity segregation in the wellbore. J. Petrol. Technol., **29** : 1657 - 1663. SPE Paper, **6129** (1976).
HOOD, B.F. & POWELL, S.E. (1982) : Case history of a pilot fracture treatment carried out on the Clair field, west of Shetland, from a semisubmersible. EUR Paper, **309**.
HOOVER, R.S. & ADAMS, R. (1988) : Advances in fracturing techniques in carbonate reservoirs in the Howard Glasscock field in West Texas. SPE Paper,

17282.
HORIE, T., FIROOZABADI, A. & ISHIMOTO, K. (1988) : Capillary continuity in fractured reservoirs. SPE Paper, **18282**.
HORNBY, B.E., JOHNSON, D.L., WINKLER, K.W. & PLUMB, R.A. (1987) : Fracture evaluation from the borehole Stoneley wave. 57th Annual Intern. Meeting, Soc. Explor. Geophys., Expanded abstracts with biographies, 688 - 691, New Orleans/Louisiana, 11. - 15.10.1987.
HORNE, R.N. (1982) : Geothermal reinjection experience in Japan. J. Petrol. Technol., **34** : 495 - 503. SPE Paper, **9925** (1981).
HORNE, R.N. & RODRIGUEZ, F. (1983) : Dispersion in tracer flow in fractured geothermal systems. Geophys. Res. Letters, **10** : p. 289 ff.
HORTON, A.I. (1982) : 95 stimulations in 63 wells. Northeast Oil Reporter, **1982**/2 : 63 - 73.
HORTON, A.I., MERCER, J.C. & SAWYER, W.K. (1982) : Infill drilling for shale gas development : a field case study. SPE Paper, **11182**.
HORTON, R.L. (1982) : Fracturing fluids for high-temperature reservoirs. Drilling, **1982**/12 : 72 - 78.
HOSSAIN, M., CADY, G.V. & HONARPOUR, M.M. (1984) : Simulation of real gas flow through finite-conductivity fractures. SPE Paper, **12919**.
HOTTMAN, C.E., SMITH, J.H. & PURCELL, W.R. (1979) : Relationship among earth stresses, pore pressure, and drilling problems, offshore Gulf of Alaska. J. Petrol. Technol., **31** : 1477 - 1484. SPE Paper, **7501** (1978).
HOUCHIN, L.R., DUNLAP, D.D. & HUTCHINSON, J.E. (1988) : Formation damage during gravel-pack completions. SPE Paper, **17166**.
HOUCHIN, L.R. & HUDSON, L.M. (1986) : The prediction, evaluation, and treatment of formation damage caused by organic deposition. SPE Paper, **14818**.
HOUSE, L., KEPPLER, H. & KAIEDA, H. (1985) : Seismic studies of a massive hydraulic fracturing experiment. Proc. Geotherm. Resourc. Counc. Annual Meeting, Kona/HA, 26. - 30.08.1985.
HOUSEWORTH, J.E. (1988) : Characterizing permeability heterogeneity in core samples from standard miscible displacement experiments. SPE Paper, **18329**.
HOUZE, O.P., HORNE, R. & RAMEY, H.J. (1984) : Pressure-transient response of an infinite-conductivity vertical fracture in a reservoir with double-porosity behavior. SPE Paper, **12778**. SPE Form. Eval., **3** (1988) : 510 - 518.
HOWARD, A.Q. (1981) : Induction logging for vertical structures in the presence of a borehole fluid. Geophysics, **46** : 68 - 75.
HOWARD, G.C. & FAST, C.R. (1950) : Squeeze cementing operations. Trans. AIME, **189** : 53 - 64.
HOWARD, G.C. & FAST, C.R. (1957) : Optimum fluid characteristics for fracture extension. API Drill. Product. Pract., **1957** : 261 - 268.
HOWARD, G.C. & FAST, C.R. (1970 a) : Hydraulic fracturing. SPE Monograph Series, **2** (Henry L. Doherty Series) : 203 pp. Revised edition (1st ed. 1957). Soc. Petrol. Engin. of AIME, Dallas/Texas and New York (ISBN 0-89520-201-8). SPE Paper, **30402**.
HOWARD, G.C. & FAST, C.R. (1970 b) : Factors controlling fracture extension. SPE Reprint Series, Well Completions, **5** : 257 - 264. Canad. Miner. Metall. Bull. Trans., **64** (1961) : 101 - 108.
HOWELL, J.C. & THOMAS, B.D. (1978) : Evaluation of injection well stimulation as applied to a large micellar-polymer project. SPE Paper, **7180**.
HOWELL, W.D. & HILLS, J.B. (1970) : Explosive detonation tested in hydraulically fractured gas wells. J. Petrol. Technol., **22** : 403 - 408. SPE Paper, **2605** (1969).
HOWER, W.F. & BROWN, W. (1961) : Large scale laboratory investigation of sand consolidation techniques. J. Petrol. Technol., **13** : 1221 - 1229.
HOWES, J. (1988) : Horizontal and extended reach drilling come of age. Oilman, **1988**/4 : 15 - 39.
HOYLE, R.S. (1983) : Should we continue to explore in the North Sea? SPE Paper, **11305**.
HSIAO, C. (1987) : A study of horizontal wellbore failure. SPE Paper, **16927**.
HSU, C.H. & CONWAY, M.W. (1981) : Fracturing fluids for deep hot formations. J. Petrol. Technol., **33** : 2213 - 2218. SPE Paper, **9707** (1981).
HSU, K., BRIE, A. & PLUMB, R.A. (1987) : A new method for fracture identifica-

tion using array sonic tools. J. Petrol. Technol., **39** : 677 - 683. SPE Paper, **14397** (1985).
HUANG, C.F. & HUNTER, J.A. (1981) : The correlation of "tube wave" events with open fractures in fluid-filled boreholes : current research. Geol. Surv. Canada Paper, **81-1A** : 361 - 376.
HUANG, W.S. & HIGHT, M.A. (1986) : Evaluation of steamflood processes using horizontal wells. SPE Paper, **14130**.
HUBBARD, D. & PIERSON, N.O. (1986) : New frac treatment boosts Austin Chalk productivity. World Oil, **1986**/9 : 31 - 34.
HUBBERT, M.K. & WILLIS, D.G. (1957) : Mechanics of hydraulic fracturing. Petrol. Trans. AIME, **210** : 153 - 168. Paper TP, **4597** (Petroleum Branch). SPE Reprint Series, Well completions, **5** (1970) : 241 - 256.
HUCKABEE, P.T. (1987) : Carbonate stimulation optimization using state-of-the-art fracture geometry design. SPE Paper, **16884**.
HUCKABEE, P.T. (1988) : Carbonate stimulation optimization using hydraulic fracturing field testing. SPE Paper, **18224**.
HUDSON, T.E., COFFEY, M.D., SAUER, C.W. & TEOT, A.S. (1982) : Fluid loss control through the use of a liquid-thickened completion and workover brine. SPE Paper, **10652**.
HUDSON, T.E. & MARTIN, W. (1988) : Use of low-density gravel-pack material improves placement efficiency. SPE Paper, **17169, 18227**.
HUEBINGER, T., WEBSTER, D., CHISHOLM, P., VENDITTO, J. & HUNT, J. (1988) : Optimizing program increases field`s profits. Oil Gas J., 29.08.**1988** : 35 - 39.
HUFFT, H.F. (1977) : The evolution of a fracturing technique for the Cotton Valley. SPE Paper, **6868**.
HUINONG, Z. (1984) : Interference testing and pulse testing in the Kenli Carbonate oil pool - a case history. J. Petrol. Technol., **36** : 1009 - 1017. SPE Paper, **10581** (1982).
HUITT, J.L. & McGLOTHLIN, B.B. (1958) : The propping of fractures in formations susceptible to propping-sand embedment. API Drill. Product. Pract., **1958** : 115 - 123.
HUITT, J.L., McGLOTHLIN, B.B. & McDONALD, J.F. (1959) : The propping of fractures in formations in which propping sand crushes. API Drill. Product. Pract., **1959** : 120 - 130.
HULEN, J.B. & NIELSON, D.L. (1982) : Stratigraphic permeability in the Baca geothermal system, Redondo Creek area, Valles Caldera, New Mexico. Geotherm. Resourc. Counc. Trans., **6** : 27 - 30.
HULEN, J.B. & NIELSON, D.L. (1983) : Stratigraphy of the Bandelier Tuff and characterization of high-level clay alteration in borehole B-20, Redondo Creek area, Valles Caldera, New Mexico. Geotherm. Resourc. Counc. Trans., **7** : 163 - 168.
HULTIN, S.O. (1983) : The world`s energy supply problems. 11. World Petrol. Congr., London/England, Review Paper, **2**.
HUNG, K.M., HILL, A.D. & SEPEHRNOORI, K. (1987) : A mechanistic model of wormhole growth in carbonate matrix acidizing and acid fracturing. SPE Paper, **16886**.
HUNSBEDT, A., KRUGER, P. & LONDON, A.L. (1975) : Recovery of energy from fracture stimulated geothermal reservoirs. SPE Paper, **5875**. J. Petrol. Technol., **29** (1977) : 940 - 946.
HUNSBEDT, A., KRUGER, P. & LONDON, A.L. (1977) : Laboratory studies of fluid production from artificially fractured geothermal reservoirs. SPE Paper, **6553**.
HUNT, E.R., HASKETT, C.E., RAYMOND, D.H. & PIRIE, R.G. (1981) : Application of new well logs and geology to fracturing and producibility in tight gas sands, Cotton Valley Group. SPE/DOE Paper, **9832**.
HUNT, J.L. (1986) : Production systems analysis for fractured wells. SPE Paper, **15931**.
HUNT, P.K. & COBB, S.L. (1986) : A new high performance core preservation package. SPE Paper, **15382**.
HUNTER, J.C. (1985) : A case history of refracs in the Oak Hill (Cotton Valley) field. SPE Paper, **13878**.

HURST, W. (1985) : Model evaluates hydraulic fractures, determines reservoir limit. Petrol. Engin. Intern., **1985**/2 : 36 - 42.
HUSHBECK, D.F. (1986) : Precision perforation breakdown for more effective stimulation jobs. SPE Paper, **14096**.
HUTFILZ, J.N., COCKERHAM, P.W. & McINTOSH, J.R. (1981) : Pulse testing for reservoir description in a high permeability environment. SPE Paper, **9604**. J. Petrol. Technol., **34** (1982) : 2179 - 2190.
HYLAND, C. (1984) : Drainhole drilling - an old idea whose time is now. SPE Paper, **12792**.
IBG (1986) : OPEC Kompromiß in Genf im Oktober. Quo vadis ohne Yamani. Erdöl und Kohle - Erdgas - Petrochemie, **39** : 536 - 537.
IBRAHIMPASIC, I. (1983) : Hydrocarbon exploration and exploitation in Yugoslavia with special survey of INA-NAFTAPLIN. Nafta, **34** : 417 - 428.
ICHARA, M.J. & CANNON, D.E. (1987) : The effect of perforating on the performance of shale-laminated and crossbedded formations. SPE Paper, **16205**.
ICHARA, M.J. & RAGHAVAN, R. (1981) : Pressure behavior in vertically fractured wells subject to limited entry and bottom water drive. SPE Paper, **10180**.
ILSENG, J.R. & CORTEZ, M.J. (1985) : Case history of large-volume fracture stimulations in a West Texas waterflood. SPE Paper, **13795**.
IRELAND, M.M. & ROBINSON, J.B. (1987) : Reserve predictions from production testing on geopressured gas reservoirs. SPE Paper, **16958**.
IRWIN, G.R. (1957) : Analysis of stresses and strains near the end of a crack traversing a plate. J. Applied Mechan., **24** : 361 - 364.
IRWIN, G.R. & WIT, R.d. (1983) : A summary of fracture mechanics concepts. J. Test. Eval., **11** : 56 - 65.
ISOKRARI, O. & KNAPP, R.M. (1976) : Natural gas production from geothermal geopressured aquifers. SPE Paper, **6037**.
ITO, Y. (1984) : The introduction of the microchanneling phenomenon to cyclic steam stimulation and its application to the numerical simulator (sand deformation concept). Soc. Petrol. Engin. J., **24** : 417 - 430.
ITOH, T. & SCHOPPER, J.R. (1977) : Pulsed neutron well logging techniques. Erdoel-Erdgas, **93** : 131 - 136.
IVANHOE, L.F. (1988) : Future crude oil supply and prices. Oil Gas J., 25.07. **1988** : 111 - 112.
IVERSON, W.P. (1988) : Crosswell logging for acoustic impedance. J. Petrol. Technol., **40** : 75 - 82.
JACKSON, C.E. & GARTNER, M.L. (1988) : A technique for evaluation of fractures using quantitative concentrations derived from the natural gamma ray spectrum. SPE Paper, **18152**.
JACOBS, I.C. & THORNE, M.A. (1986) : Asphaltene precipitation during acid stimulation treatments. SPE Paper, **14823**.
JACOBSEN, B. (1988) : Norway as a gas supplier. Hydrocarbons `88 Conference, Great Yarmouth/England, 05.10.1988.
JACQUIN, C., LEGAIT, B., MARTIN, J.M., NECTOUX, A., ANTERION, F. & RIOCHE, M. (1987) : Gravity drainage in a fissured reservoir with fluids non in equilibrium. 4th Europ. Symp. Enhanced Oil Recovery, Modern Reservoir Management Session III, Hamburg/Germany FRG.
JAEGER, J.C. & COOK, N.G.W. (1979) : Fundamentals of rock mechanics. Chapman & Hall, New York.
JAIN, S. & KHOSLA, A. (1985) : Predicting steam recovery of Athabasca oil through horizontal wells. CIM Paper, **85-36-28**.
JALALI-YAZDI, Y. & ERSHAGHI, I. (1987 a) : A unified type curve for the analysis of heterogeneous naturally fractured reservoirs. SPE Paper, **16778**.
JALALI-YAZDI, Y. & ERSHAGHI, I. (1987 b) : Pressure transient analysis of heterogeneous naturally fractured reservoirs. SPE Paper, **16341**.
JAMISON, H.C. (1984) : Future U.S. petroleum exploration and the perils of politics. SPE Paper, **12728**.
JARDINS, P.E.d. (1959) : Thirty years of sand-control experience, Black Bayou field, Cameron Parish, Louisiana. API Paper, **926-4-C**.
JARED, C.B., SCOTT, T.A. & EVANS, R.T. (1984) : Precision perforation helps achieve uniform stimulation. Oil Gas J., 19.11.**1984** : 98 - 103.
JEFFERIS, R.G., BRUIST, E.H. & BOTTS, T.M. (1983) : A field proven system for

selective multizone, one-trip gravel packing. SPE Paper, **11697**.

JEFFREY, R.G., VANDAMME, L., HINKEL, J.J. & HORNER, D.M. (1987) : Propagation of fractures in coal : equivalent modulus and parallel fracture effects. Proc. Coalbed Methane Symp., Tuscaloosa/Alabama, pp. 113 - 116.

JEFFREY, R.G., VANDAMME, L. & ROEGIERS, J.C. (1987) : Mechanical interactions in branched or subparallel hydraulic fractures. SPE/DOE Paper, **16422**.

JELIC, K. (1982) : Geothermal energy in the Pannonian Basin in Croatia. Proc. Symp. Energy, Zagreb/Yugoslavia, **1982** : 132 - 146 (in Serbocroatian).

JENKINS, R.E. (1985) : Some characteristics of selected gas reservoirs with low permeability. SPE Paper, **13938**.

JENNINGS, A.R. & FORD, D.L. (1976) : Fracture flow capacity - a key to sustained production after hydraulic fracturing. SPE Paper, **6127**.

JENNINGS, A.R. & DARDEN, W.G. (1979) : Gas well stimulation in the Eastern United States. SPE Paper, **7914**.

JENNINGS, A.R., DARDEN, W.G., WENZEL, R.W., FOSTER, J.M. & SHRUT, R. (1977) : Massive hydraulic fracturing in the Eastern United States. SPE Paper, **6866**.

JENNINGS, A.R. & SPRAWLS, B.T. (1975) : Successful stimulation in the Cotton Valley Sandstone - a low permeability reservoir. SPE Paper, **5627**. J. Petrol. Technol., **29** (1977): 1267 - 1276.

JENNINGS, J.B., CARROLL, H.B. & RAIBLE, C.J. (1981) : The relationship of permeability to confining pressure in low permeability rock. SPE/DOE Paper, **9870**.

JENNINGS, J.W., McGREGOR, D.S. & MORSE, R.A. (1988) : Simultaneous determination of capillary pressure and relative permeability by automatic history matching. SPE Form. Eval., **3** : 322 - 328.

JENNINGS, J.W. & MUIRHEAD, J.F. (1968) : Method for permeably consolidating an incompetent formation with heat-curable resin. U.S. Patent, **3,373,813** (19.03.1968).

JENSEN, J.L. & CURRIE, I.D. (1988) : Improving performance prediction by more accurate heterogeneity assessment. SPE/DOE Paper, **17364**.

JENSEN, J.T. (1988) : World gas supplies, markets, and trade. SPE Paper, **17751**.

JENSEN, O.K., BRESLING, S., RASMUSSEN, F.O., FOGED, N. & PETERSEN, K. (1989) : Natural fracture distribution in reservoirs modelled by backstripping and finite element stress analysis. SPE Paper, **18429**.

JEU, S.J. (1983) : A different approach to optimizing MHF using fracture azimuth and reservoir data. SPE Paper, **12190**.

JEU, S.J., LOGAN, T.L. & McBANE, R.A. (1988) : Exploitation of deeply buried coalbed methane using different hydraulic fracturing techniques in the Piceance Basin, Colorado and San Juan Basin, New Mexico. SPE Paper, **18253**.

JEU, S.J., PEREZ, F.D. & WAY, G.A. (1988) : Curable resin-coated proppant minimizes proppant flowback after fracturing treatment. SPE Paper, **17283**.

JIANG, D., LIU, L., YU, S., SONG, W. & SHAN, W. (1988) : Analysis of pressure performance in fracturing job and its application. SPE Paper, **17835**.

JIMISON, J.W. (1986) : U.S. natural gas regulation and markets : changing the rules to fit the game. SPE Paper, **14631**.

JØRANSON, H., SPURLIN, J. & EDWARDS, D. (1988) : Field normalization of formation mechanical properties for use in sand control management. Trans. 29th Soc. Prof. Well Log Analysts Annual Logging Symp., San Antonio/Texas, 05. - 08.06.1988.

JOHN, P.W. (1982) : Hydraulic fracturing Simulator-Rechenmodellentwicklung zur numerischen Simulation der hydraulischen Rißbildung. DGMK-Forschungsbericht, **167-2**.

JOHN, P.W. (1983) : Stimulation problems of deep, tight, layered Rotliegend gas sands in North West Germany. Oil Gas, **9** : 21 - 36.

JOHN, P.W. (1987) : Gas recovery from deep millidarcy layered gas formations with nonlinear elastic deformation characteristics. SPE/DOE Paper, **16425**.

JOHNSEN, A. (1982) : Norwegian gas - reserves and markets. Gastech 1982, 9th Intern. LNG/LPG-Conference and Exhibition, Paris/France, 05. - 08.10.1982.

JOHNSON, C.L.M. (1978) : The North Sea and the UK economy - a financial analysis. EUR Paper, **101**.

JOHNSON, D.E., FOX, K.B., BURNS, L.D. & O`MARA, E.M. (1988) : Carbonate production decline rates are reduced through improvement in gelled acid technology. SPE Paper, **17297**.

JOHNSON, H.D. & KROL, D.E. (1984) : Detailed geological modeling of a heterogeneous sandstone reservoir : Lower Jurassic Statfjord Formation, Brent field (North Sea, U.K. sector). SPE Paper, **13050**.

JOHNSON, J.P. & RHETT, D.W. (1986) : Compaction behavior of Ekofisk chalk as a function of stress. SPE Paper, **15872**.

JOHNSON, J.P., RHETT, D.W. & SIEMERS, W.T. (1988) : Rock mechanics of the Ekofisk reservoir in the evaluation of subsidence. OTC Paper, **5621**. SPE Paper, **17854**.

JOHNSON, P.A. & ALBRIGHT, J.N. (1985) : In-situ physical properties using crosswell acoustic data. SPE Paper, **13881**.

JOHNSON, R.C. & NUCCIO, V.F. (1986) : Structural and thermal history of the Piceance Creek Basin, Western Colorado in relation to hydrocarbon occurrence in the Mesaverde Group. In : SPENCER, C.W. & MAST, R.F. (Editors), Geology of tight gas reservoirs. Amer. Assoc. Petrol. Geol., Stud. Geol., **24** : 165 - 206.

JOLY, E.L., DORMIGNY, A.M., CATALA, G.N. & PINSON, F.P. (1985) : A new production logging technique for horizontal wells. SPE Paper, **14463**. SPE Product. Engin., **3** (1988) : 328 - 332.

JONES, A.H. (1986) : Hydraulic fracture design rationale for the recovery of methane from coal seams. Quart. Rev. Methane Coal Seams Technol., **4** : 38 - 40.

JONES, A.H. (1987) : Hydraulic fracturing of coal seams. Pittsburgh Coalbed Methane Forum, 05. - 06.10.1987, Morgantown/West Virginia.

JONES, A.H. (1988) : Spalling and the development of a hydraulic fracturing strategy for coal. Quart. Rev. of Methane from Coal Seams Technol., **5**/3-4 : 48 - 49; **6**/1 : 41 - 43.

JONES, A.H., BELL, G.J. & MORALES, R.H. (1986) : Coalbed hydraulic fracture treatment empirical relationships and computer simulation. SPE Paper, **15242**.

JONES, A.H., BELL, G.J., MORALES, R.H. & SCHRAUFNAGEL. R.A. (1987) : Examination of potential mechanisms responsible for the high treatment pressures observed during stimulation of coalbed reservoirs. SPE Paper, **16421**.

JONES, A.H., CUTLER, R.A. & SWANSON, S.R. (1981) : Lightweight proppants for deep gas well stimulation - First Annual Report. Bartlesville Energy Technology Center Rept., **10038-19** (10.1981).

JONES, A.H. et al. (1987) : The influence of coal fines/chips on the behavior of hydraulic fracture stimulation treatments. Proc. Coalbed Methane Symp., Tuscaloosa/Alabama, 93 - 102.

JONES, F.O. & OWENS, W.W. (1980) : A laboratory study of low permeability gas sands. J. Petrol. Technol., **32** : 1631 - 1640. SPE Paper, **7551** (1979).

JONES, F.T. (1960) : History of a Milham Sand waterflood following accidental stimulation. J. Petrol. Technol., **12** : 33 - 35.

JONES, L.G., BLOUNT, E.M. & GLAZE, O.H. (1976): Use of short term multiple rate flow tests to predict performance of wells having turbulence. SPE Paper, **6133**.

JONES, P.H. (1969) : Hydrodynamics of geopressure in the Northern Gulf of Mexico Basin. J. Petrol. Technol., **21** : 803 - 810. SPE Paper, **2207** (1968).

JONES, R.E. & THORP, G. (1977) : A method of analyzing performance of gravel pack completions in Seria Field, Brunei, using a simple statistical method. SPE Paper, **6806**. J. Petrol. Technol., **32** (1980) : 496 - 504.

JONES, S.C. (1972) : A rapid accurate unsteady-state Klinkenberg permeameter. Soc. Petrol. Engin. J., **12** : 383 - 397. SPE Paper, **3535** (1971).

JONES, S.C. (1987) : Using the inertial coefficient, beta, to characterize heterogeneity in reservoir rock. SPE Paper, **16949**.

JOPLING, M.W. & KETCHER, N.W. (1987) : Real time decision-making through use of a log/log plot during a frac job. SPE Paper, **16223**.

JORDEN, J.R. & CAMPBELL, F.L. (1985) : Well logging I - Rock properties, borehole environment, mud and temperature logging. SPE Monograph Series, **9** : 175 pp. SPE Paper, **30409**.

JORDEN, J.R. & CAMPBELL, F.L. (1986) : Well logging II. SPE Monograph Series, **10** : 192 pp. SPE Paper, **30410**.

JOSHI, S.D. (1986 a) : Augmentation of well productivity using slant and hori-

zontal wells. SPE Paper, **15375**. J. Petrol. Technol., **40** (1988) : 729 - 739.
JOSHI, S.D. (1986 b) : A review of horizontal well technology. Proc. 1986 Tar Sand Symp., DOE/METC Rept., **87/6073** : 112 - 131.
JOSHI, S.D. (1987 a) : A review of horizontal well and drainhole technology. SPE Paper, **16868**.
JOSHI, S.D. (1987 b) : A review of thermal oil recovery using horizontal wells. In situ, **11**/2-3 : 211 - 259.
JOURDAN, A.P., ARMESSEN, P. & ROUSSELET, P. (1988) : Elf has set up rules for horizontal drilling. Oil Gas J., 09.05.**1988** : 33 - 40.
JOURDAN, A.P. & BARON, G. (1984) : Horizontal well proves productivity advantages. Petrol. Engin. Intern., **1984**/10 : 23 - 25.
JOURDAN, A.P., REISS, L.H., BARON, G.A., GIGER, F. & WITTRISCH, C. (1983) : Review of horizontal drilling progress. Petrol. Inform., **1585** (25.03.1983) : 27 - 33.
JOURDAN, A.P., REISS, L.H., BARON, G.A. & WITTRISCH, C. (1982) : One, two and three horizontal wells in France and Italy. Petrol. Inform., **1567** (15.04. 1982) : 17 - 23.
JOURDAN, A.P., REISS, L.H. & JOLY, E. (1986) : Horizontal wells : the worldwide state-of-the-art, as of 1986. SPE Paper, **14067**.
JOURNAL PETROLEUM TECHNOLOGY (1987) : Stimulation activity of old wells increases in new economic setting. J. Petrol. Technol., **39** : 1552 - 1556.
JULIAN, P.J. (1982) : Fracture detection techniques in the Georgetown and Austin chalk formations. SPE Paper, **11035**.
JÜRGENS, R. (1988) : Gegenwärtiger Stand unterschiedlicher Bohrverfahren für die Horizontal-Bohrtechnik. DGMK Spring Meeting, Celle/Germany FRG, 10.06. 1988.
KABIR, C.S. & HASAN, A.R. (1986) : Prefacture testing in tight gas reservoirs. SPE Form. Eval., **1** : 128 - 138, p. 631. SPE Paper, **12918** (1984).
KADI, K.S. & BOUTEMY, Y. (1981) : Non-Darcy well test analysis in wells producing by solution gas. SPE Paper, **10225**.
KADI, K.S. & BROWN, K.E. (1980) : Non-Darcy flow in dissolved gas-drive reservoirs. SPE Paper, **9301**.
KALISCH, R.B. (1988) : Gas energy supply outlook : 1987 - 2010. SPE Paper, **17750**.
KAMAL, M. (1977) : Use of pressure transient to describe reservoir heterogeneity. SPE Paper, **6885**.
KAMAL, M. (1982) : Interference and pulse testing - a review. SPE Paper, **10042**. J. Petrol. Technol., **35** (1983) : 2257 - 2270.
KAMAL, M.M. & HEGEMAN, P.S. (1988) : New developments in multiple-well testing. SPE Form. Eval., **3** : 159 - 168. SPE Paper, **14100** (1986).
KAMATH, V.A. & GODBOLE, S.P. (1985) : Evaluation of hot-brine stimulation technique for gas production from natural gas hydrates. SPE Paper, **13596**. J. Petrol. Technol., **39** (1987) : 1379 - 1388.
KAMP, A.W. (1983) : Downhole telemetry from the user`s point of view. J. Petrol. Technol., **35** : 1792 - 1796. SPE Paper, **11227** (1982).
KANAT, J. (1980) : Effects of resin coatings on proppant permeability. Norton Co. Intern. Res. Rept.
KARASAKI, K., LONG, J.C.S. & WITHERSPOON, P.A. (1988) : A new analytic model for fracture-dominated reservoirs. SPE Form. Eval., **3** : 242 - 250. SPE Paper, **14171** (1985).
KARCHER, B.J., GIGER, F.M. & COMBE, J. (1986) : Some practical formulas to predict horizontal well behavior. SPE Paper, **15430**.
KASPEREIT, D.H. (1979) : Fracture design for suspended sand packs. SPE Paper, **8343**.
KATSANIS, E.P., KRUMRINE, P.H. & FALCONE, J.S. (1983) : Chemistry of precipitation and scale formation in geological systems. SPE Paper, **11802**.
KAZEMI, H. (1969) : Pressure transient analysis of naturally fractured reservoirs with uniform fracture distribution. Soc. Petrol. Engin. J., **9** : 451 - 462.
KAZEMI, H. (1983) : Low-permeability gas sands. SPE Paper, **11330**.
KAZI, A., ZEKAI, S. & BAHAA-ELDIN, H.S. (1983) : Relationship between sonic pulse velocity and uniaxial compressive strengths of rocks. Proc. 24th U.S.

Symp. Rock Mechan., 409 - 419, Texas Univ., 20. - 23.06.1983.
KEAT, W.B., ANNIGERI, B.S. & CLEARY, M.P. (1988) : Surface integral and finite element hybrid method for two- and three-dimensional fracture mechanics analysis. Intern. J. Fract., **36** : 35 - 53.
KECK, R.G., CLEARY, M.P. & CROCKETT, A. (1984) : A lumped numerical model for the design of hydraulic fractures. SPE/DOE/GRI Paper, **12884**. SPE Paper, **13862**.
KEG (1986 a) : Energie in der Gemeinschaft - Aussichten bis zum Jahre 2000 und darüber hinaus. Energie in Europa, **1986**/6 : 27 - 1; Kommission der Europäischen Gemeinschaften, Generaldirektion Energie, Luxemburg.
KEG (1986 b) : Energie in Europa. Energiepolitiken und Entwicklungen in der Europäischen Gemeinschaft. **6** (1986/12) : 73 pp.; Kommission der Europäischen Gemeinschaften, Generaldirektion Energie, Luxemburg.
KEIGHIN, C.W. (1979) : Influence of diagenetic reactions on reservoir properties of the Neslen, Farrer, and Tuscher Formations, Uinta Basin, Utah. SPE Paper, **7919**.
KELAFANT, J., BOYER, C. & ZUBER, M. (1988) : Production potential and strategies for coalbed methane in the Central Appalachian Basin. SPE Paper, **18550**.
KELKAR, B.G. (1987) : The effect of current oil crisis on oil and gas lease in light of recent court decisions. SPE Paper, **16238**.
KELKAR, S., THOMPSON, T., HOLLAND, M., STRAWN, J., VEGHTE, R. & HATHAWAY, S. (1985) : Hydraulic fracture height growth control using permeability barriers and determination of geologic and engineering parameters for fracture design. SPE Paper, **13856**.
KELLY, M. (1987) : Oil price stability : an impossible dream? SPE Paper, **16293**.
KEMP, A. (1984) : Operators take a view on future prospects. Oilman, **1984**/9 : 21 - 22.
KEMP, A.G., KELLAS, G. & ROSE, D. (1987) : End of field life economics in the North Sea. SPE Paper, **16559**.
KEMPER, J.R., FRANKIE, W.T., MOODY, J.R. & JOHNSTON, I.M. (1986) : Hydrocarbon production from the Devonian Shale in Letcher and Knott Counties, Eastern Kentucky. SPE Paper, **15932**.
KENNEDY, J.H. & McELHINEY, J.E. (1979) : Pressure transient analysis of tight western gas sands - pre- and post-Fracturing. SPE Paper, **7933**.
KENNEDY, M. (1984) : Cyclic steam stimulation in the waterflooded Ranger Sand of the Wilmington oil field, Los Angeles, California. SPE Paper, **12658**.
KENYON, D.E. & BEHLE, G.A. (1987) : Third SPE comparative solution project : gas cycling of retrograde condensate reservoirs. J. Petrol. Technol., **39** : 981 - 997. SPE Paper, **12278** (1983).
KEOUGH, D.D., CURRAN, D.R. & McHUGH, S.L. (1984) : Applicability of tailored-pulse loading for gas well stimulation. SPE/DOE/GRI Paper, **12837**.
KEPLINGER, H.F. (1986) : Geopolitical implication of Middle East oil. J. Petrol. Technol., **38** : 1241 - 1245. SPE Paper, **14625**.
KERN, L.R., PERKINS, T.K. & WYANT, R.E. (1959) : The mechanics of sand movement in fracturing. Trans. AIME, **216** : 403 - 405.
KERN, L.R., PERKINS, T.K. & WYANT, R.E. (1961) : Propping fractures with aluminum particles. J. Petrol. Technol., **13** : 583 - 589.
KERR, D. & BROUTHERS, S. (1988) : A systems approach to horizontal drilling. SPE Paper, **17618**.
KETTEL, D. (1981) : Maturitätsberechnungen für das nordwestdeutsche Obercarbon - ein Test verschiedener Methoden. Erdoel-Erdgas, **97** : 395 - 404.
KEYS, W.S. (1979) : Borehole geophysics in igneous and metamorphic rocks. Trans. 20th Soc. Prof. Well Log Analysts Annual Logging Symp., Paper **00**, Tulsa/Oklahoma.
KEYS, W.S. & SULLIVAN, J.K. (1979) : Role of borehole geophysics in defining the physical characteristics of the Raft River geothermal reservoir, Idaho. Geophys., **44** : 1116 - 1141.
KGT (1986) : Windtemperaturen bis 900 oC für Kupolöfen mit KGT-ECONO-THERM-Rekuperatoren. 13 pp.; KGT Gießereitechnik, Düsseldorf/Germany FRG.
KHILAR, K.C. & FOGLER, H.S. (1983) : Water sensitivity of sandstones. Soc. Petrol. Engin. J., **23** : 55 - 64. SPE Paper, **10103** (1981).
KHILAR, K.C., FOGLER, H.S. & AHLUWALIA, Y. (1983) : Sandstone water sensitivi-

ty : existence of a critical rate of salinity decrease for particles capture. Chem. Engin. Sci., **38** : 789 -800.

KHIN, J.A. (1988) : Far East. It`s starting to warm up out here. World Oil, **207** (1988/8) : 101 - 116.

KHRISTIANOVICH, S.A. & ZHELTOV, Y.P. (1955) : Formation of vertical fractures by means of highly viscous liquid. Proc. 4th World Petrol. Congr., Sect. II, **2** : 579 - 586; Rome/Italy.

KHRISTIANOVICH, S.A., ZHELTOV, Y.P., BARENBLATT, G.I. & MAXIMOVICH, G.K. (1959) : Theoretical principles of hydraulic fracturing in oil strata. Proc. 5th World Petrol. Congr., Sect. II, Paper 23, **2** : p. 298 ff.; New York.

KIEKE, E.M. & HARTMAN, D.J. (1973) : The influence of microporosity on formation evaluation and the use of the scanning electron microscope to gain new insight into pore geometry. SPE Paper, **4531**.

KIEL, O.M. (1970) : A new hydraulic fracturing process. J. Petrol. Technol., **22** : 89 - 96. SPE Paper, **2453** (1969).

KIEL, O.M. (1973) : Method of fracturing subterranean formations using oil-in-water emulsions. U.S. Patent, **3,710,865** (16.01.1973).

KIEL, O.M. (1976) : Hydraulic fracturing process using reverse flow. U.S. Patent, **3,933,205**.

KIEL, O.M. (1977) : The Kiel process - reservoir stimulation by dendritic fracturing. SPE Paper, **6984**.

KIEL, O.M. (1984) : Hydraulic fracturing process. Norwegian Patent, **149,285** (Patent Application, **773,871**); Oslo/Norway.

KIERSTEIN, R.A. (1984) : True location and orientation of fractures logged with the acoustic televiewer (including programs to correct fracture orientation). U.S. Geol. Surv., Water Resour. Invest. Rept., **83-4275** : 71 pp.

KILGORE, W.C. (1982) : Uncertainties about future world oil prices. SPE Paper, **11095**.

KIM, C.M. (1983) : Proppants : permeability and conductivity. BJ-Hughes Brochure, 20.05.1983.

KIM, C.M., CHAMPION, J.H. & COOPER, G.D. (1985) : Evaluation of fracturing results in deviated wellbores through onsite pressure measurement and post-fracture temperature survey. SPE Paper, **14373**.

KIM, C.M. & LOSACANO, J.A. (1985) : Fracture conductivity damage due to cross-linked gel residue and closure stress on propped 20/40 mesh sand. SPE Paper, **14436**.

KIM, C.M. & WILLINGHAM, J.R. (1987) : Flow response of propped fracture to repeated production cycles. SPE Paper, **16912**.

KING, G.E. (1977) : Factors affecting dynamic fluid leakoff with foam fracturing fluids. SPE Paper, **6817**.

KING, G.E. (1985) : Foam and nitrified fluid treatments - stimulation techniques and more. SPE Paper, **14477**.

KING, G.E. (1986) : Acidizing concepts - matrix vs. fracture acidizing. J. Petrol. Technol., **38** : 507 - 508. SPE Paper, **15279**.

KING, G.E., ANDERSON, A. & BINGHAM, M. (1985) : A field study of underbalance pressures necessary to obtain clean perforations using tubing-conveyed perforating. SPE Paper, **14321**. J. Petrol. Technol., **38** (1986) : 662 - 668.

KING, G.E., BINGHAM, M.D. & KIDDER, R.W. (1986) : Factors affecting perforating charge performance and relationship to port plug condition. SPE Product. Engin., **1** : 379 - 387. SPE Paper, **13128** (1984).

KING, G.E. & HOLLINGSWORTH, F.H. (1979) : Evaluation of diverting agent effectiveness and cleanup characteristics using a dynamic laboratory model - high permeability core. SPE Paper, **8400**.

KING, G.E. & HOLMAN, G.B. (1982) : Laboratory study of factors affecting chalk flow in high porosity chalk. SPE Paper, **11124**.

KING, G.E. & HOLMAN, G.B. (1986) : Hydrocarbon solvents : an alternative to acid for removing some formation damage. SPE Paper, **14136**.

KING, G.R. & ERTEKIN, T. (1988) : Comparative evaluation of vertical and horizontal drainage wells for the degasification of coal seams. SPE Reserv. Engin., **3** : 720 - 735. SPE Paper, **13091** (1984).

KING, G.R., ERTEKIN, T. & SCHWERER, F.C. (1986) : Numerical simulation of the transient behavior of coal seam degasification wells. SPE Form. Eval., **1** :

165 - 183. SPE Paper, **12258** (1983).

KING, M.J. & SCHER, H. (1987) : Viscous fingering utilizing probabilistic simulation. SPE Paper, **16708**.

KING, S. (1982) : Sand fracturing with liquid carbon dioxide. SPE Paper, **11341**.

KING, S.R. (1983) : Liquid CO2 for the stimulation of low-permeability reservoirs. SPE/DOE Paper, **11616**.

KING, W.C. (1982) : Industry`s view of U.S. energy policy. SPE Paper, **11269**.

KINLEY, R.M. & BOWER, F.M. (1977) : Specialized applications of noise logging. SPE Paper, **6784**.

KINNEY, G.T. (1987) : OPEC plan working. Oil Gas J., 23.02.**1987** : p. 25.

KIRKBY, L.L. & ROCKEFELLER, H.A. (1985) : Proppant settling velocities in non-flowing slurries. SPE Paper, **13906**.

KIRR, J.N., EDDY, G.E., LOCKE, C.D. & SMITH, J.B. (1986) : Geology and production potential of the Devonian Rhinestreet Shale : a case study of three wells. SPE Paper, **15256**.

KLEINBERG, R.L., CHOW, E.Y., PLONA, T.J., ORTON, M. & CANADY, W.J. (1984) : Sensitivity and reliability of two fracture detection techniques for borehole application. J. Petrol. Technol., **36** : 657 - 663. SPE Paper, **11036** (1982).

KLOSE, G. & KRÖMER, E. (1983) : Frac-Planung und -Behandlung Söhlingen Z-4. Erdoel-Erdgas, **99** : 181 - 188.

KLOTZ, J.A., KRUEGER, R.F. & PYE, D.S. (1974) : Maximum well productivity in damaged formations requires deep, clean perforations. SPE Paper, **4792**.

KNAPP, R.H., PLANTY, R. & VOILAND, E.J. (1977) : A gravel-coating aqueous epoxy emulsion system for water-based consolidated gravel packing : development and application. SPE Paper, **6177**. J. Petrol. Technol., **29** : 1489 - 1496. SPE Reprint Series, Well Completions, **5 a**/II (1978) : 76 - 83.

KNAPP, R.H. & WELBOURN, M.E. (1978) : An acrylic/epoxy emulsion gel system for formation plugging : laboratory development and field testing for steam thief zone plugging. SPE Paper, **7083**.

KNAPP, R.M., ISOKRARI, O.F., GARG, S.K. & PRITCHETT, J.W. (1977) : An analysis of production from geopressured geothermal aquifers. SPE Paper, **6825**.

KNAPP, W.v.d. & VLIS, A.C.v.d. (1967) : On the cause of subsidence in oil-producing area. Proc. 7th World Petrol. Congr., **3** : 85 - 95, Mexico City/Mexico.

KNOLL, S.K. (1985) : Wallslip evaluation in steady shear viscosity measurements of hydraulic fracturing fluids. SPE Paper, **13904**.

KNOTT, T. (1985) : Energy in the balance - decline into self-sufficiency. Offshore Engin., **1985**/11 : 28 - 31.

KNOTT, T. (1986) : Morecambe go-ahead buoys up hydrocarbons conference. Offshore Engin., **1986**/10 : p. 21.

KNOX, J.A., LASATER, R.M. & DILL, W.E. (1964) : A new concept in acidizing utilizing chemical retardation. SPE Paper, **975**.

KNUDSEN, F.P. (1959) : Dependence of mechanical strength of brittle polycrystalline specimens on porosity and grain size. J. Amer. Ceram. Soc., **42**.

KNUTSON, C.F. (1976) : Modeling of noncontinuous Fort Union and Mesaverde Sandstone reservoirs, Piceance Basin, Northwestern Colorado. Soc. Petrol. Engin. J., **16** : 175 - 188. SPE Paper, **5024** (1974).

KNUTSON, C.F., HODGES, L.T. & RIGHTER, S.B. (1981) : Permeability, petrography and small scale structural element analysis of Upper Cretaceous channel sandstones from the Rock Springs Uplift and Wind River Basin, Wyoming. SPE/DOE Paper, **9874**.

KNUTSON, C.F. & WARD, D.C. (1977) : Implications of surface fracture patterns and sand continuity in relation to stimulation of tight gas sands, Eastern Uinta Basin, Utah. SPE Paper, **6740**.

KNUTSON, C.F., YOST, A.B. II & AMMER, J.R. (1985) : Reservoir geology of some tight Clinton Sandstones, Eastern Ohio. SPE Paper, **13853**.

KOERPERICH, E.A. (1978) : Investigation of acoustic boundary waves and interference patterns as techniques for detecting fractures. J. Petrol. Technol., **30** : 1199 - 1207. SPE Paper, **6820** (1977).

KOERPERICH, E.A. (1980) : Shear wave velocities determined from long- and

short-spaced borehole acoustical devices. Soc. Petrol. Engin. J., **20** : 317 - 326. SPE Paper, **8237** (1979).
KOH, W.I. (1977) : Non-Darcy flow of gas in propped fractures. Ph.D. Thesis, Texas A & M Univ., 09.1977.
KOHLER, N. & CHAUVETEAU, G. (1981) : Xanthan polysaccharide plugging behavior in porous media - preferential use of fermentation broth. J. Petrol. Technol., **33** : 349 - 358. SPE Paper, **7425** (1978).
KOHLHAAS, C.A. (1982) : Field study - Steinle Ranch, an intermediate depth oil field, shows significant benefit from bauxite proppants. SPE Paper, **10875**.
KOMAR, C.A. (1978) : Development of a rationale for stimulation design in the Devonian shales. SPE Paper, **7166**.
KOMAR, C. & FROHNE, K.H. (1973 a) : Analysis of factors influencing fracture initiation and orientation in oil reservoir sandstone. U.S. Bur. Mines, Rept. Invest., **7813**.
KOMAR, C.A. & FROHNE, K.H. (1973 b) : Factors controlling fracture orientation in sandstone. SPE Paper, **4567**.
KOMAR, C.A. & YOST, A.B. (1979) : Practical aspects of foam fracturing in the Devonian Shale. SPE Paper, **8345**.
KOMAR, C.A., YOST, A.B. & SINCLAIR, A.R. (1980) : Foam fracturing the Devonian Shale. World Oil, **1980**/7 : 119 - 134.
KONING, E.J.L. (1985) : Fracturing water-injection wells - analytical modelling of fracture propagation. SPE Paper, **14684**.
KONING, E.J.L. & NIKO, H. (1985 a) : Fractured water injection wells - a pressure fall-off test for determining fracture dimensions. SPE Paper, **14458**.
KONING, E.J.L. & NIKO, H. (1985 b): Application of a special fall-off test in a fractured North Sea water injector. SPE Paper, **16392**.
KONZELMANN, G. (1976) : Öl - Schicksal der Menschheit? 364 pp.; Sigloch Service Edition, Künzelsau/Germany FRG.
KOSSACK, C.A., KLEPPE, J. & AASEN, T. (1987) : Oil production from the Troll field : a comparison of horizontal and vertical wells. SPE Paper, **16869**.
KOSSACK, C.A. & OPDAL, S.T. (1988) : Recovery of condensate from a heterogeneous reservoir by the injection of a slug of methane followed by nitrogen. SPE Paper, **18265**.
KOSTURA, J.R. & RAVENSCROFT, J.H. (1977) : Fracture-controlled production. Amer. Assoc. Petrol. Geol., Reprint Ser., **21** : 221 pp.
KOZIAR, G. (1985) : Devonian Shale production data analysis : Southwestern West Virginia. SPE Paper, **14505**.
KOZIK, H.G., BAILEY, B.G. & HOLDITCH, S.A. (1979) : A case history for massive hydraulic fracturing the Cotton Valley lime matrix, Fallon and Personville fields - Limestone County, Texas. SPE Paper, **7911**. J. Petrol. Technol., **33** (1981) : 229 - 244.
KRAMER, J., PRUDHOMME, R.K., NORMAN, L.R. & SANDY, J.M. (1987) : Characteristics of metal-polymer interactions in fracturing fluid systems. SPE Paper, **16914**.
KREIJ, P.d. (1986) : Service-Katalog für Arbeiten im verrohrten Bohrloch. 58 pp.; Schlumberger, Diepholz/Germany FRG.
KREUTZER, N. (1985) : Probleme der Lagengliederung von Öl- und Gashorizonten. Erdoel-Erdgas, **101** : 11 - 14.
KREUTZER, N. (1986) : Die Ablagerungssequenzen der miozänen Badener Serie im Feld Matzen und im zentralen Wiener Becken. Erdöl-Erdgas-Kohle, **102** : 492 - 503.
KRÖLL, A. (1984) : Die Erdöl- und Erdgasregion Matzen/Schönkirchen aus geologischer Sicht. Erdoel-Erdgas, **100** : 185 - 195.
KRUEGER, R.F. (1986) : An overview of formation damage and well productivity in oil-field operations. J. Petrol. Technol., **38** : 131 - 188. SPE Paper, **10029** (1982).
KRUEGER, R.F. (1988) : An overview of formation damage and well productivity in oilfield operations : an update. SPE Paper, **17459**.
KRUEGER, R.F., VOGEL, L.C. & FISCHER, P.W. (1967) : Effect of pressure drawdown on clean-up of clay- or silt-blocked sandstone. J. Petrol. Technol., **19** : 397 - 403.
KRUGMAN, D.G. (1986) : A brittle fracture theory for understanding high energy

gas fracturing. SPE Paper, **15680**.
KRUYADIJK, C.P.J.v. & NIKO, H. (1988) : Alternatives for draining tight naturally fractured gas reservoirs : horizontal drilling vs. massive hydraulic fracturing. SPE Paper, **18339**.
KU, H.C., HIRSH, R.S. & TAYLOR, T.D. (1989) : Simulation of viscous fingering by the pseudospectral matrix element method. SPE Paper, **18443**.
KUCHUK, F.J. (1987) : New methods for estimating parameters of low permeability reservoirs. SPE/DOE Paper, **16394**.
KUCHUK, F., BRICE, B., THAMBYNAYAGAM, M., SHERRARD, D. & GOODE, P. (1988) : Estimation of horizontal and vertical permeabilities, skin factor and reservoir pressure for horizontal wells : a field study. SPE Paper, **18308**.
KUCHUK, F. & GOODE, P.A. (1988) : Pressure transient analysis and inflow performance for horizontal wells. SPE Paper, **18300**.
KUCHUK, F.J., GOODE, P.A., WILKINSON, D.J. & THAMBYNAYAGAM, R.K.M. (1988) : Pressure transient behavior of horizontal wells with or without gas cap or aquifer. SPE Paper, **17413**.
KUCHUK, F.J. & WILKINSON, D.J. (1988) : Transient pressure behavior of commingled reservoirs. SPE Paper, **18125**.
KUCUK, F. & AYESTARAN, L. (1983) : Analysis of simultaneously measured pressure and sandface flow rate in transient well testing. SPE Paper, **12177**.
KUCUK, F., KARAKAS, M. & AYESTARAN, L. (1984) : Well test analysis of commingled zones without crossflow. SPE Paper, **13081**. SPE Form. Eval., **1** (1986) : 342 - 354.
KUCUK, F. & SAWYER, W.K. (1980) : Transient flow in naturally fractured reservoirs and its application to Devonian gas shales. SPE Paper, **9397**.
KUCUK, F., SHAH, P.C., AYESTARAN, L. & NICHOLSON, B. (1986) : Application of multilayer testing and analysis : a field case. SPE Paper, **15419**.
KUKAL, G.C. (1981) : Determination of fluid corrected porosity in tight gas sands and in formations exhibiting shallow invasion profiles. SPE Paper, **9856**.
KUKAL, G.C. (1983) : Log analysis in low-permeability gas sand sequences - correcting for variable unflushed gas saturation. Trans. 24th Soc. Prof. Well Log Analysts Annual Logging Symp., Paper **F**, 27. - 30.06.1983.
KUKAL, G.C. (1984) : A systematic approach for the effective log analysis of tight gas sands. SPE/DOE/GRI Paper, **12851**.
KUKAL, G.C. & SIMONS, K.E. (1985) : Log analysis techniques for quantifying the permeability of sub-millidarcy sandstone reservoirs. SPE/DOE Paper, **13880**. SPE Form. Eval., **1** (1986) : 609 - 622.
KUKAL, G.C., BIDDISON, C.L., HILL, R.E., MONSON, E.R. & SIMONS, K.E. (1983) : Critical problems hindering accurate log interpretation of tight gas sand reservoirs. SPE/DOE Paper, **11620**.
KULANDER, B.R., BARTON, C.C. & DEAN, S.L. (1979) : Application of fractography to core and outcrop fracture investigations. U.S. Dept. Energy, Morgantown Energy Technology Center, Rept., **METC/SP-79/3** : 174 pp., Morgantown/West Virginia.
KULANDER, B.R. & DEAN, S.L. (1988 a) : Drilling-induced fractures : their identification and analysis. Amer. Assoc. Petrol. Geol., Annual Convention, Houston/Texas, 20. - 23.03.1988.
KULANDER, B.R. & DEAN, S.L. (1988 b) : Analysis of fractures and tectonic structures in core. Amer. Assoc. Petrol. Geol., Annual Convention, Houston/Texas, 20. - 23.03.1988.
KULAR, G.S., CHHINA, H.S., BEST, D.A. & MACKENZIE, W.T. (1988) : Multiple hydraulic fracture propagation in oil sands. SPE Paper, **17534**.
KUMAR, J. (1988) : Restructuring of natural gas markets by regulations and legislations. SPE Paper, **18107**.
KUMAR, S., GUPPY, K.H. & KAGAWAN, V.D. (1985) : Pressure transient analysis for fractured wells producing at constant pressure. SPE Paper, **13629**. SPE Form. Eval., **3** (1988) : 169 - 178.
KUNZE, K.R. & SHAUGHNESSY, C.M. (1980) : Acidizing sandstone formations with fluoboric acid. SPE Paper, **9387**. Soc. Petrol. Engin. J., **23** (1983) : 65 - 72.
KUNDERT, D.P. & SMINK, D.E. (1979) : Improved stimulation of the Escondido

Sandstone. SPE Paper, **7912**.
KUO, M.C.T., HANSON, H.G. & BRISAY, C.L.d. (1984) : Prediction of fracture extension during waterflood operations. SPE Paper **12769**.
KUUSKRAA, V.A. (1980) : The need and urgency for unconventional gas. SPE/DOE Paper, **8921**.
KUUSKRAA, V.A., BRASHEAR, J.P., DOSCHER, T.M. & ELKINS, L.E. (1978) : Vast potential held by four "unconventional" gas sources. Oil Gas J., **76** : 47 - 54.
KUUSKRAA, V.A., BRASHEAR, J.P., ELKINS, L.F. & MORRA, F. (1979) : Gas recovery from tight formations : a function of technology and economics. SPE Paper, **7916**. J. Petrol. Technol., **33** (1981) : 1545 - 1556.
KUUSKRAA, V.A. & HAAS, M.R. (1988) : Natural gas from low permeability formations. SPE Paper, **17746**.
KUUSKRAA, V.A., KELAFANT, J.R. & HORNER, D.M. (1988) : Gas production strategies for Devonian/Mississippian clastics of the Appalachian Basin : a case study in the Plateau region of Virginia. SPE Paper, **18546**.
KUUSKRAA, V.A. & MEYER, R.F. (1980) : Review of world resources of unconventional gas. Proc. IIASA Conference on Conventional and Unconventional World Natural Gas Resources, Laxenburg/Austria, 30.06. - 04.07.1980, pp. 27 - 43.
KUUSKRAA, V.A., SEDWICK, K. & YOST, A.B. II (1985) : Technically recoverable Devonian Shale gas in Ohio, West Virginia and Kentucky. SPE Paper, **14503**.
KUUSKRAA, V.A., WICKS, D.E., SAWYER, W.K., ESPOSITO, P.R. & KOMAR, C.A. (1983) : Production potential from Devonian gas shales : effective production strategies. SPE Paper, **12316**.
LABO, J. (1987) : A practical introduction to borehole geophysics. Geophys. Refer., **2** : 336 pp.; Soc. Explor. Geophys., Tulsa/Oklahoma. ISBN 0-931830-47-8.
LABRID, J.C. (1975) : Thermodynamic and kinetic aspects of argillaceous sandstone acidizing. Soc. Petrol. Engin. J., **15** : 117 - 128. SPE Paper, **5156** (1974). SPE Reprint Series, Well Completions, **5 a**/II (1978) : 163 - 174.
LABUDOVIC, V. (1984) : The effect of Poisson`s ratio on fracture height. J. Petrol. Technol., **36** : 287 - 290. SPE Paper, **10307** (1981).
LABUZ, J.F., SHAH, S.P. & DOWDING, C.H. (1985) : Experimental analysis of crack propagation in granite. Intern. J. Rock Mechan., Miner. Sci., Geomechan. Abstr., **22** : 85 - 98.
LACY, L.L. (1987) : Comparison of hydraulic fracture orientation techniques for deep wells. SPE Form. Eval., **2** : 66 - 76. SPE Paper, **13225** (1984).
LAFLEUR, K.K. & JOHNSON, A.K. (1973) : Well stimulation in the North Sea : a survey. SPE Paper, **4315**.
LAGRONE, K.W. & RASMUSSEN, J.W. (1962) : Better completion by controlled fracture placement, limited-entry technique. API No. **906-7-B**.
LAGRONE, K.W. & RASMUSSEN, J.W. (1963) : A new development in completion methods - the limited entry technique. J. Petrol. Technol., **15** : 693 - 702.
LAI, C.H., BODVARSSON, G.S., TSIANG, C.F. & WITHERSPOON, P.A. (1983) : A new model for well test data analysis for naturally fractured reservoirs. SPE Paper, **11688**.
LAKATOS, I., KRETSCHMAR, H.J., CZOLBE, P., BITTKOW, P. & WASSERMANN, I. (1988) : Polymerinjektionen zur Stimulierung oder Abdichtung von Öl-, Gas- oder Wasserbohrungen. Erdöl-Erdgas-Kohle, **104** : 19 - 24.
LAM, K.Y. & CLEARY, M.P. (1984) : Slippage and re-initiation of hydraulic fractures at frictional interfaces. Intern. J. Numer. Analyt. Meth. Geomechan., **8** : 589 - 604.
LAM, K.Y., CLEARY, M.P. & BARR, D.T. (1986) : A complete three-dimensional simulator for analysis and design of hydraulic fracturing. SPE Paper, **15266**.
LAMBERT, M.E. (1981) : A statistical study of reservoir heterogeneity. M.Sc. Thesis, Univ. of Texas at Austin, Austin/Texas.
LAMBERT, S.A., DOLAN, R.T. & GALLUS, J.P. (1983) : Fracturing poorly consolidated sandstone formations. Proc. 30th Annual Southwestern Petroleum Short Course, Texas Techn. Univ., Lubbock/Texas, 27. - 28.04.1983, pp. 84 - 88.
LAMBERT, S.W. & TREVITS, M.A. (1980) : The feasibility of no-proppant stimulation to enhance removal of methane from the Mary Lee Coalbed. U.S. Dept. of Energy, Rept. Invest., **1980**/4.
LAMONT, N. & JESSEN, F.W. (1963) : The effects of existing fractures in rocks

on the extension of hydraulic fractures. J. Petrol. Technol., **15** : 203 - 209.
LANCASTER, D.E. (1988) : Reservoir and stimulation evaluation in the Devonian Shales. Eastern Devonian Gas Shales Technol. Rev., **5**/1 : 23 - 36.
LANCASTER, D.E. & GATENS, J.M. III (1986) : Practical well test analysis methods for hydraulically fractured wells in dual-porosity reservoirs. SPE Paper, **15924**.
LANCASTER, D.E., GATENS, J.M. III, STANLEY, D.K., HARRISON, C.W. & GUIDRY, F.K. (1988) : Using in-situ stress tests and acoustic logs to determine mechanical properties and stress profiles in the Devonian shales. SPE Paper, **18523**.
LANCASTER, D.E., GUIDRY, F.K., GRAHAM, R.L., CURTIS, J.B., SHAW, J.S. & BLAKE, T.H. (1987) : A case study of the evaluation, completion, and testing of a Devonian Shale gas well. SPE Paper, **17037**.
LANCASTER, G. (1986) : Advantages and limitations of liquid CO2 fracturing. CIM Paper, **86-37-69**.
LANDERS, J. (1987) : Oilfield equipment sanctions lifted. The News, Washington Bureau, **1987**.
LANDRUM, B.L. & CRAWFORD, P.B. (1955) : Effect of drainhole drilling on production capacity. J. Petrol. Technol., **7** : 45 - 47.
LANDRUM, B.L. & CRAWFORD, P.B. (1957) : Estimated effect of horizontal fractures in thick reservoirs on pattern conductivity. Trans. AIME, **210** : 399 - 401.
LANMON, C.P. & GILL, H.F. (1966) : A wireline completion technique for sand control. SPE Paper, **1487**. J. Petrol. Technol., **19** (1967) : 741 - 744.
LARSEN, D.G. & SMITH, L.J. (1985) : New conductivity found in angular blends of fracturing sand. SPE Paper, **13814**.
LARSEN, L. (1981) : Wells producing commingled zones with unequal initial pressures and reservoir properties. SPE Paper, **10325**.
LARSON, D.B. & NGUYEN, H.X. (1984) : Fracture height containment by creating an artificial barrier. CIM Paper, **84-35-107**.
LASAKI, G.O. (1980) : Arch stability and failure behaviour in unconsolidated natural sands. Thesis, **T-2299**; Colorado School of Mines, Golden/Colorado, 05.1980.
LASSITER, P.A. & HOLLINGSWORTH, F.C. (1984) : Method of coal degasification. U.S. Patent, **4,471,840** (18.09.1984).
LATTIMORE, G.M., CARDEN, R.S. & FISCHER, T. (1987) : Grand Canyon directional drilling and waterline project. SPE/IADC Paper, **16169**.
LAU, J.S.O. (1980) : Borehole television surveys. Proc. 13th Canad. Symp. Rock Mechan., CIM Spec. Vol., **22** : 204 - 210.
LAU, J.S.O. (1983) : The determination of true orientations of fractures in rock cores. Canad. Geotechn. J., **20** : 221 - 227.
LAUBACH, S.E., BAUMGARDNER, R.W., MONSON, E.R., HUNT, E. & MEADOR, K.J. (1988) : Fracture detection in low-permeability reservoir sandstone : a comparison between BHTV and FMS logs to core. SPE Paper, **18119**.
LAUBACH, S.E. & MONSON, E.R. (1988) : Coring-induced fractures : indicators of hydraulic fracture propagation in a naturally fractured reservoir. SPE Paper, **18164**.
LAUGHERTY, C.D. & HARPER, J.A. (1986) : Comparisons of Upper Devonian and Lower Silurian tight formations in Pennsylvania - geological and engineering characteristics. In : SPENCER, C.W. & MAST, R.F. (Editors), Geology of tight gas reservoirs. Amer. Assoc. Petrol. Geol., Stud. Geol., **24** : 9 - 44.
LAW, B.E. (1984) : Relationships of source-rock, thermal maturity, and overpressuring to gas generation and occurrence in low-permeability Upper Cretaceous and lower Tertiary rocks, Greater Green River basin, Wyoming, Colorado, and Utah. In : WOODWARD, J., MEISSNER, F.F. & CLAYTON, J.L. (Editors), Hydrocarbon source rocks of the Greater Rocky Mountain region, 469 - 490; Rocky Mountain Assoc. Geol.
LAW, B.E. & DICKINSON, W.W. (1985) : Conceptual model for origin of abnormally pressured gas accumulations in low-permeability reservoirs. Bull. Amer. Assoc. Petrol. Geol., **69** : 1295 - 1304.
LAW, B.E., POLLASTRO, R.M. & KEIGHIN, C.W. (1986) : Geologic characterization of low-permeability gas reservoirs in selected wells, Greater Green River

Basin, Wyoming, Colorado, and Utah. In : SPENCER, C.W. & MAST, R.F. (Editors), Geology of tight gas reservoirs. Amer. Assoc. Petrol. Geol., Stud. Geol., **24** : 253 - 270.
LAW, B.E. & SPENCER, C.W. (1981) : Abnormally high-pressured, low-permeability, Upper Cretaceous and Tertiary gas reservoirs, northern Green River basin, Wyoming. Bull. Amer. Assoc. Petrol. Geol., **65** : p. 948.
LAW, B.E., SPENCER, C.W. & BOSTICK, N.H. (1980) : Evaluation of organic matter, subsurface temperature, and pressure with regard to gas generation in low-permeability Upper Cretaceous and lower Tertiary sandstones in Pacific Creek area, Sublette and Sweetwater Counties, Wyoming. Mountain Geol., **17** : 23 - 35.
LAYNE, A.W. & BYRER, C.W. (1986) : Analysis of coalbed methane stimulations in the Warrior Basin, Alabama. SPE Paper, **15218**. SPE Form. Eval., **3** (1988) : 663 - 669.
LAYNE, A.W. & SIRIWARDANE, H.J. (1988) : Insights to hydraulic fracturing of a horizontal well in a naturally fractured formation. SPE Paper, **18255**.
LAYNE, A.W., SIRIWARDANE, H.J. & BYRER, C.W. (1988) : Assessment of gas production potential from coalbeds and adjacent strata. SPE Paper, **17765**.
LEAVER, J.D., SAGEEV, A. & RAMEY, H.J. (1986) : Multiple interference testing in the Ohaaki geothermal field. SPE Paper, **15122**. SPE Form. Eval., **3** (1988) : 429 - 437.
LEDLOW, L.B., SAUER, C.W. & TILL, M.V. (1985) : Recent design, placement and evaluation techniques lead to improved gravel pack performance. SPE Paper, **14162**.
LEDLOW, L.B. & SPARLIN, D.D. (1980) : Downhole protection of sand control screens. SPE Paper, **8803**.
LEE, B. & MAO, M. (1984) : Post frac pressure analysis tells a story. 5th Offshore Southeast Asia Conference and Exhibition, Session **8** (Petrol. Engin.), Singapore, 21. - 24.02.1984.
LEE, B.Y.Q. & TAN, T.B.S. (1987) : Application of a multiple porosity/permeability simulator in fractured reservoir simulation. SPE Paper, **16009**.
LEE, J.C. & KEER, L.M. (1987) : Study of a three-dimensional crack terminating at an interface. J. Applied Mechan., **1967**.
LEE, J.E. (1982) : Cycles of crisis, seasons of glut ... where do we go from here? SPE Paper, **11003**.
LEE, M.C. (1985) : Diagenesis of the Permian Rotliegendes sandstone, North Sea : K/Ar, 018/016, and petrologic evidence. Ph.D. Thesis, 365 pp.; Case Western Reserve Univ., Diss. Abstr. Intern., Sect. B, **45**/10 : 3184-b.
LEE, M.H. & ROBERTS, L.D. (1980) : Effect of heat of reaction on temperature distribution and acid penetration in a fracture. Soc. Petrol. Engin. J., **20** : 501 - 507.
LEE, R.L., LOGAN, R.W. & TEK, M.R. (1987) : Effect of turbulence on transient flow of real gas through porous media. SPE Form. Eval., **2** : 108 - 120. SPE Paper, **14205** (1985).
LEE, S.T. & BROCKENBROUGH, J. (1983) : A new analytic solution for finite conductivity vertical fractures with real time and Laplace space parameter estimation. SPE Paper, **12013**. SPE Form. Eval., **1** (1986) : 75 - 88.
LEE, S.T. & BROCKENBROUGH, J.R. (1986) : A new approximate analytic solution for finite-conductivity vertical fractures. SPE Form. Eval., **1** : 75 - 88. SPE Paper, **12013** (1983).
LEE, S.T. & OKUYIGA, M. (1986) : Analysis of dispersion in layered porous medium with micro heterogeneity of arbitrary shapes and size distribution. SPE Paper, **15387**.
LEE, W.J. (1967) : Analysis of hydraulically fractured wells with pressure buildup test. SPE Paper, **1820**.
LEE, W.J. & GATENS, J.M. III (1985) : Analysis of Eastern Devonian gas shales production data. SPE Paper, **14506**.
LEE, W.J. & HOLDITCH, S.A. (1979) : Fracture evaluation with pressure transient testing in low-permeability gas reservoirs. Part I : Theoretical background. SPE Paper, **7929**. J. Petrol. Technol., **33** (1981) : 1776 - 1792.
LEE, W.J., KUO, T.B., HOLDITCH, S.A. & McVAY, D.A. (1984) : Estimating formation permeability from single-point flow data. SPE/DOE/GRI Paper, **12847**.

LEE, W.J. & OLAREWAJU, J.S. (1988) : The rate performance of a layered reservoir with unsteady-state interlayer crossflow. SPE Paper, **18544**.

LEE, W.S. (1985 a) : Pressure decline analysis with the Christianovich and Zheltov and Penny-shaped geometry model of fracturing. SPE/DOE Paper, **13872**.

LEE, W.S. (1985 b) : Geometry determination for multistage acidizing treatments with or without viscous preflush. SPE Paper, **14515**.

LEE, W.S. (1986 a) : Minifrac analysis based on ellipsoidal geometry. SPE Paper, **15369**.

LEE, W.S. (1986 b) : New method of mini-frac analysis offers greater accuracy and enhanced applicability. SPE Paper, **15941**.

LEE, W.S. (1987) : Study of the effects of fluid rheology on minifrac analysis. SPE Paper, **16916**.

LEE, W.S. (1988) : Fracture propagation theory and pressure decline analysis with Lagrangian formulation for Penny-shaped and Perkins-Kern geometry models. SPE Paper, **17151**.

LEE, W.S. & DANESHY, A.A. (1985) : Fracture geometry and proppant-transport computation for multiple-fluid treatment. SPE Paper, **14511**. SPE Product. Engin., **2** (1987) : 257 - 266.

LEICHT, H. (1985) : Lagerstättentechnische Aspekte bei großvolumigen Frac-Behandlungen. gwf-gas/erdgas, **126** : 154 - 160.

LEINAN, A.B. & HARRIS, P.C. (1983) : Carbon dioxide/aqueous mixtures as fracturing fluids in Western Canada. CIM Paper, **83-34-36**. J. Canad. Petrol. Technol., **1985** : 45 - 48.

LENORMAND, R. (1986) : Scaling laws for immiscible displacements with capillary and viscous fingering. SPE Paper, **15390**.

LENT, J.T., NICKELL, R.L., DAVIS, W.E. & ALLAN, A.M. (1980) : Use of bottom hole pressure analysis in optimizing frac design - Anadarko Basin of Western Oklahoma. SPE Paper, **9262**.

LEONARD, R.C. & MUNNS, J.W. (1987) : Valhall. In : SPENCER, A.M. et al. (Editors), Geology of the Norwegian oil and gas fields, 153 - 164; Graham & Trotman, London/England.

LEONE, J.A. & SCOTT, E.M. (1987) : Characterization and control of formation damage during waterflooding of a high-clay-content reservoir. SPE Paper, **16234**.

LERCHE, I. & NARR, W. (1986) : Estimating subsurface fracture density in core : effects resulting from variable fracture spacing. SPE Form. Eval., **1** : 249 - 258. SPE Paper, **13300** (1985).

LESCARBOURA, J.A., SIFFERMAN, T.R. & WAHL, H.A. (1984) : Evaluation of fracturing fluid stability by using a heated, pressurized flow loop. Soc. Petrol. Engin. J., **24** : 249 - 255. SPE Paper, **10962** (1982).

LESCARBOURA, J.A., WALTHER, H.C. & WILSON, P.L. (1975) : Design and analysis of interference tests. SPE Paper, **5314**.

LEUNG, W.F. (1986 a) : A tensor model for anisotropic and heterogeneous reservoirs with variable directional permeabilities. SPE Paper, **15134**.

LEUNG, W.F. (1986 b) : A new pseudosteady-state model for dual-porosity/dual-permeability aquifers and two interconnected single-porosity aquifers. SPE Reserv. Engin., **1** : 511 - 520. SPE Paper, **12277** (1983).

LEVINE, E.N., CYBRIWSKY, Z.A. & TOKSÖZ, M.N. (1984) : Detection of permeable rock fractures and estimation of hydraulic conductivity and yield by 3-D vertical seismic profiling. Proc. National Water Well Assoc. Conference on Surface and borehole geophysical methods in ground water investigations, 853 - 876, San Antonio/Texas.

LEWIS, J.J.M. (1988) : Outcrop-derived quantitative models of permeability heterogeneity for genetically different sand bodies. SPE Paper, **18153**.

LEWIS, K. (1985) : Gravel pack thickness plays key role in sand control in the project III steamflood. SPE Paper, **14046**.

LEWIS-DAVIES, G. (1988) : Future UK gas supply and demand. Hydrocarbons `88 Conference, Great Yarmouth/England, 05.10.1988.

LEYDECKER, G. (1981) : Seismische Ortung hydraulisch erzeugter Brüche im Geothermik Frac-Projekt Falkenberg. Bundesanst. Geowiss. Rohstoffe Rept., **86549**; Hannover/Germany FRG.

LI, G.C. & ZHU, Z.M. (1986) : A discussion on some problems relating to the

fracturing of deep wells in North China region. SPE Paper, **14863**.
LI, T.K., CHANDELLE, V. & BRYCH, J. (1986) : Lateral drilling : a new application shows promise. World Oil, **1986**/6 : 68 - 71.
LICHTBLAU, J.H. (1986) : U.S. petroleum import and export policies. SPE Paper, **14629**.
LIEBENTHAL, A.M., KOMAR, C., RIEKE, H.H. & SKILLERN, C.R. (1979) : Economic analysis of foam fracturing in the Devonian Shales : preliminary report. SPE Paper, **8738**.
LIESEN, K. (1977) : Erdgas, seine gegenwärtige und zukünftige Bedeutung für Westeuropa. Erdoel-Erdgas, **93** : 412 - 416.
LIESEN, K. (1985) : Future challenges to the European gas industry. Oil Gas, **11**/2 : 12 - 13.
LIGHT, S.M. (1983) : The effect of lateral anisotropy on flood pattern dimensions and orientation. SPE Paper, **10759**.
LIKWARTZ, D.J. (1976) : A history of sand control in Amoco Trinidad`s offshore Teak field. J. Petrol. Technol., **28** : 972 - 978. SPE Paper, **5657** (1975).
LILLIE, W. (1981) : Thermische Erdölgewinnung im Feld Georgsdorf. Erdoel-Erdgas, **97** : 47 - 53.
LILLIE, W.H.E. & SPRINGER, F.P. (1981) : Status of the steam drive pilot in the Georgsdorf field, The Federal Republic of Germany. J. Petrol. Technol., **33** : 173 - 180. SPE Paper, **8385**.
LILLIE, W. & WIETHOFF, B. (1985) : 10 Jahre Dampffluten im Emsland - Eine lagerstättenkundliche Betrachtung für die Felder Georgsdorf und Rühlermoor. Erdoel-Erdgas, **101** : 384 - 390.
LILLIES, A.T. (1982) : Sand fracturing with liquid carbon dioxide. CIM Paper, **82-33-23**.
LILLIES, A.T. & KING, S.R. (1982) : Sand fracturing with liquid carbon dioxide. SPE Paper, **11341**.
LIN, C.Y. (1988) : A new approach for simulation of cyclic steam stimulation above fracture pressure by modifying a thermal simulator. SPE Paper, **18077**.
LIN, E.C. & HUANG, E.T.S. (1988) : The effect of rock wettability on water-blocking during miscible displacements. SPE Paper, **17375**.
LIN, Z.S. & FINLEY, R.J. (1987) : Reservoir engineering properties and production characteristics of selected tight gas fields, Travis Peak Formation, East Texas Basin. SPE/DOE Paper, **13901**.
LINDBERG, P.H. & FERTL, W.H. (1980) : Origin, occurrence, and distribution of overpressures in the Northern North Sea area. SPE Paper, **9339**.
LINDEN, H.R. (1985) : Gas in a changing technological and competitive environment. Oil Gas, **11**/2 : 7 - 12.
LINDLEY, B.W. & McGHEE, B.F. (1983) : An investigation of a high-strength proppant tail-in at McAllen Ranch field. SPE Paper, **11935**.
LIPPMANN, M.J., TSANG, C.F. & WITHERSPOON, P.A. (1977) : Analysis of the response of geothermal reservoirs under injection and production procedures. SPE Paper, **6537**.
LITVAK, B.L., SATTER, A. & ETEBAR, S. (1988) : An analysis of naturally fractured reservoir performance using a novel fractured reservoir simulator. SPE Paper, **17615**.
LIU, O.Y. (1984) : Stoneley wave-derived dt shear log. Trans. 25th Soc. Prof. Well Log Analysts Annual Logging Symp., New Orleans/Louisiana, 10. - 13.06. 1984.
LIU, O.Y. (1985) : Fracture evaluation using borehole sonic velocity measurements. SPE Paper,**14399**.
LIVESEY, D.V. & TOMA, P. (1984) : Filter for separating discrete solids elements from a fluid stream. U.S. Patent, **4,434,054** (28.02.1984).
LLOYD, P.M., DAHAN, C. & HUTIN, R. (1986) : Formation imaging with microelectrical scanning arrays : a new generation of stratigraphic high resolution dipmeter tool. Trans. 10th European Logging Symp., Soc. Prof. Well Log Analysts, Paper **L**, Aberdeen/Scotland, 23. - 24.04.1986.
LO, K.K. & DEAN, R.H. (1988) : Modeling of acid fracturing. SPE Paper, **17110**.
LOCKE, C.D. & SAWYER, W.K. (1975) : Constant pressure injection test in a fractured reservoir - history match using numerical simulation and type curve analysis. SPE Paper, **5594**.

LOCKE, C.D., YOST, A.B. II, OVERBEY, W.K. & GRABOWSKI, M.A. (1988) : Field measurements of dynamic pressure and seismic activity during foam fracturing. SPE Paper, **18555**.

LOCKE, H.A., JOHNSON, J.B. & JEWKES, D.C. (1988) : Extended reach appraisal well : a case history. SPE Paper, **17564**.

LOCKE, S. (1980) : An advanced method for predicting the productivity ratio of a perforated well. SPE Paper, **8804**. J. Petrol. Technol., **33** (1981) : 2481 - 2488.

LOCKETT, J.M. & AL-HABBOOBY, H.M. (1973) : Differential settling by size of two particle species in a liquid. Trans. Inst. Chem. Engin., **51** : p. 281 ff.

LOCKNER, D. & BYERLEE, J.D. (1977) : Hydrofracture in Weber sandstone at high confining pressure and differential stress. J. Geophys. Res., **82** : 2018 - 2026.

LOCKRIDGE, J.P. & SCHOLLE, P.A. (1978) : Niobrara gas in eastern Colorado and northwestern Kansas. In : PRUIT, J.D. & COFFIN, P.E. (Editors), Energy resources of the Denver Basin. Rocky Mountain Assoc. Geol. Guideb., 35 - 49.

LOEPPKE, G.E., GLOWKA, D.A. & WRIGHT, E.K. (1988) : Design and evaluation of lost circulation materials for severe environments. SPE Paper, **18022**.

LOGAN, J.M. & TEUFEL, W. (1978) : The prediction of massive hydraulic fracturing from analyses of oriented cores. Proc. 19th U.S. Rock Mechan. Symp., 340 - 344, Reno.

LOGAN, T.L. (1987) : Application of medium radius horizontal drainhole drilling technology for underground coal degasification, Hanna Basin, Wyoming. In : MARTIN, J.W. & BARONE, S.P. (Editors), Proc. 13th Annual Underground Coal Gasification Symp., Laramie/Wyoming, 08.1987. U.S. Dept. Energy Rept., DOE/METC, **88/6095** : 417 - 426.

LOGAN, T.L. (1988) : Horizontal drainhole drilling techniques used for coal seam resource exploitation. SPE Paper, **18254**.

LOGAN, T.L., ERWIN, T.M. & McBANE, R. (1988) : Exploitation of deeply buried coalbed methane using different hydraulic fracturing techniques in the Piceance Basin, Colorado. SPE Paper, **18253**.

LOGAN, T.L., SECCOMBE, J.C. & JONES, A.H. (1986) : Hydraulic fracture stimulation and open-hole testing of a deeply buried coal seam in the Piceance Basin, Colorado. SPE Paper, **15251**.

LOGAN, T.L., SCHWOEBEL, J.J. & HORNER, D.M. (1987) : Application of horizontal drain-hole drilling technology for coalbed methane recovery. SPE/DOE Paper, **16409**.

LOHRENZ, J. (1988 a) : Bidding optimum bonus for federal offshore oil and gas leases. SPE Paper, **15992**.

LOHRENZ, J. (1988 b) : Profitabilities on federal offshore oil and gas leases : a review. J. Petrol. Technol., **40** : 760 - 764. SPE Paper, **16313** (1987).

LOHRENZ, J. & DOUGHERTY, E.L. (1983) : Federal offshore oil and gas lease bonus bid reactions : viewpoints of bidders and owners. Rejection policy : past and present. SPE Paper, **11309**. J. Petrol. Technol., **35** (1983) : 1951 - 1955.

LOPUS, T.A., SEIFERT, D.J. & SCHEIN, G.W. (1987) : Production improvement through identification of conducting natural fractures utilizing multiple radioactive isotope technology. SPE Paper, **16192**.

LORBACH, M. (1980) : Welterdöl-Reserven und Bedarf. Erdoel-Erdgas, **96** : 5 - 12.

LORBACH, M. (1988 a) : Gedanken zum zukünftigen Erdölbedarf. Erdöl-Erdgas-Kohle, **104** : 146 - 149.

LORBACH, M. (1988 b) : Technik und Politik zur Verhinderung einer nächsten Erdölkrise. Erdöl-Erdgas-Kohle, **104** : 2 - 5.

LORD, D.L. & McGOWEN, J.M. (1986) : Real-time treating pressure analysis aided by new correlation. SPE Paper, **15367**.

LORENZ, J.C. (1983) : Lateral variability in the Corcoran and Cozzette blanket sandstones and associated Mesaverde rocks, Piceance Creek Basin, Northwestern Colorado. SPE/DOE Paper, **11608**.

LORENZ, J.C. (1985) : Predictions of size and orientations of lenticular reservoirs in the Mesaverde Group, Northwestern Colorado. SPE/DOE Paper, **13851**.

LORENZ, J.C., BRANAGAN, P., WARPINSKI, N.R. & SATTLER, A.R. (1986) : Fracture characteristics and reservoir behaviour of stress-sensitive fracture systems in flat-laying lenticular formations. SPE Paper, **15244**.

LORENZ, J.C. & FINLEY, S.J. (1987) : Differences in fracture characteristics and related production of natural gas in different zones of the Mesaverde Formation, Northwestern Colorado. SPE Paper, **16809**.

LORENZ, J.C. & FINLEY, S.J. (1988) : Significance of drilling and coring-induced fractures in Mesaverde core, northwestern Colorado. Sandia National Labor. Rept., **SAND 87-1111** : 29 pp.; Albuquerque/New Mexico.

LOUVIERE, W. (1987) : Practical requirements for fracture stimulation operations. SPE Continuing Education, Hydraulic Fracturing Seminar, London/England, 01.12.1987.

LOWE, D.K., McGLOTHLIN, B.B. & HUITT, J.L. (1966) : A computerized program for designing fracturing treatments (horizontal case). SPE Paper, **1541**.

LOXAM, D.C. (1982) : Texaco Canada completes unique horizontal drilling program. Petrol. Engin. Intern., **1982**/9 : 40 - 52.

LU, C.K. & YEW, C.H. (1985) : Bonded half-planes containing two arbitrarily oriented cracks : a study of containment of hydraulically induced fractures. Soc. Petrol. Engin. J., **25** : 55 - 66. SPE Paper, **11871** (1983).

LÜBBEN, H. (1982) : Erdgas aus heimischen Quellen - Gegenwart und Zukunft. Erdoel-Erdgas, **98**. BEB Mosaik, **3/82** : 6 - 13.

LÜBBEN, H. & LEINER, J. (1988) : Öl-Perspektiven im Upstream-Bereich. Erdöl-Erdgas-Kohle, **104** : 194 - 198.

LUFFEL, D.L. & HOWARD, W.E. (1987) : Reliability of laboratory measurement of porosity in tight gas sands. SPE/DOE Paper, **16401**.

LUISKUTTY, C.T., TOMUTSA, L. & PALMER, I.D. (1986) : A semianalytical model of hydraulic fracture growth through weak barriers. SPE Paper, **15012**.

LUKMAN, A.R. (1987) : Co-operation between OPEC and non-OPEC countries is the key to oil market stability. In : Norway, OPEC and IEA - global dialogue and price stabilizing measures in the oil market? Intern. Politics, **1987**/5-6; Norwegian Institute of International Affairs.

LUMSDEN, G.Q. (1988) : Oil price and demand interactions. Offshore Northern Seas Conference, Stavanger/Norway, 23. - 26.08.1988.

LUND, K., FOGLER, H.S. & McCUNE, C.C. (1976) : On predicting the flow and reaction of HCl/HF acid mixtures in porous sandstone cores. Soc. Petrol. Engin. J., **16** : 248 - 260. SPE Paper, **5646** (1975).

LUND, K., FOGLER, H.S., McCUNE, C.C. & AULT, J.W. (1973) : Kinetic rate expression for reactions of selected minerals with HCl and HF mixtures. SPE Paper, **4348**.

LUNGHOFER, E.P. (1984) : Process for the production of sintered bauxite spheres. U.S. Patent, **4,440,886** (03.04.1984).

LUPPENS, J.C. (1980) : Completion techniques for Albuskjell, Edda and Eldfisk fields, Greater Ekofisk area. SPE Paper, **10872**. J. Petrol. Technol., **34** (1982) : 471 - 476. OTC Paper, **3707**.

LURGI (1983) : The new granulate-tube-filter. Frankfurt/Main.

LÜTTRINGHAUS, G. & SCHMITT, D. (1975) : Die Erdöl- und Erdgasversorgung Westeuropas im Jahr nach der Energiekrise. Erdoel-Erdgas. **91** : 5 - 14.

LÜTZNER, H. (1981) : Sedimentation der variszischen Molasse im Thüringer Wald. Schriftenr. geol. Wiss., **17** : 217 pp.

LYBARGER, J.H., SCHEUERMAN, R.F. & WILLARD, R.G. (1974) : Water-base viscous gravel-pack system results in high productivity in Gulf Coast completions. SPE Paper, **4774**. SPE Reprint Series, Well Completions, **5 a**/II (1978) : 69 - 75.

LYMAN, T.J., PIPER, E.M. & RIDDELL, A.W. (1984) : Heavy oil mining technical and economical analysis. SPE Paper, **12788**.

MABIE, G.J. (1975) : Sand consolidation - an example of how it succeeded. SPE Paper, **5658**.

MABRO, R. (1987) : Netback pricing and the oil price collapse of 1986. Oxford Institute of Energy Studies, Paper, **WPM 10** : 56 pp., Oxford England; 45 pp.; Penn Well, Tulsa/Oklahoma.

MABRO, R. (1988) : The 1986 oil price crisis : economic effects and policy responses. 270 pp.; Penn Well, Tulsa/Oklahoma.

MACADAM, J.M. (1984) : North Sea stimulation logistics and requirements. SPE Paper, **12999**.

MACDONALD, C.J. (1988) : Horizontal wells : an application in the North Sea.

SPE Paper, **18342**.
MACDONALD, R.G. & CASWELL, G.R. (1983) : Design and analysis of fracture treatments on the Medicine Hat Formation using a hydraulic fracture simulator. CIM Paper, **83-34-35**.
MACDONALD, R.G. & FRANK, J. (1985) : Sand fracturing the Slave Point Carbonate. CIM Paper, **85-36-36**.
MACEWEN, H. (1988) : Coiled-tubing-conveyed logging systems. SPE Paper, **18350**.
MACK, D.J. & BAUMGARTNER, S.A. (1986) : Friction pressure of foamed stimulation fluids evaluated with an on-site computer. SPE Paper, **15631**.
MACK, D.J. & MASON, R.A. (1985) : Individualized fracturing treatment design benefits Clinton Sands. SPE Paper, **14513**.
MACKAY CONSULTANTS (1988) : World offshore oil and gas industry 1986 - 1990. 515 pp.; Mackay Publ., Inverness/Scotland.
MACPHERSON, J. & GUENARD, E. (1988) : Delineation of fractured pay zones using drilling engineering and mudlogging data. SPE Paper, **17490**.
MADER, D. (1980) : Zur Substratgebundenheit von Nestbauten der solitären Urbiene Colletes daviesanus (Hymenoptera : Colletidae). Entom. Gen., **6** : 57 - 63.
MADER, D. (1981) : Diagenesis of the Buntsandstein (Lower Triassic) in Western Eifel (Germany). N. Jb. Miner. Abh., **142** : 1 - 26.
MADER, D. (1982 a) : Petrographie und Diagenese fluviatiler Sedimente im Buntsandstein der Westeifel. Teil II. Diagenese und Diskussion. Z. geol. Wiss., **10** : 181 - 205.
MADER, D. (1982 b) : Aeolian sands in continental red beds of the Middle Buntsandstein (Lower Triassic) at the western margin of the German Basin. Sedim. Geol., **31** : 191 - 230.
MADER, D. (1982 c) : Sedimentologie und Genese des Buntsandsteins in der Eifel. Z. dt. geol. Ges., **133** : 257 - 307.
MADER, D. (1983 a) : Lakustrische Sedimentation im Mittleren Buntsandstein der Südeifel. Mainzer geowiss. Mitt., **12** : 217 - 260.
MADER, D. (1983 b) : Evolution of fluvial sedimentation in the Buntsandstein (Lower Triassic) of the Eifel (Germany). Sedim. Geol., **37** : 1 - 84.
MADER, D. (1983 c) : Primäre und sekundäre Eisenoxide im Buntsandstein der Westeifel - ein Beitrag zur Genese der Rotfärbung in der germanischen Trias. Teil I. Verteilung der Eisenoxide und Merkmale der Rotfärbung. Teil II. Deutung der Pigmentgenese. Z. geol. Wiss., **11** : 603 - 637, 729 - 767.
MADER, D. (1983 d) : Aeolische und fluviatile Sedimentation im Mittleren Buntsandstein. Westeifel, Südeifel, Nordeifel. N. Jb. Geol. Paläont. Abh., **160** (1980) : 1 - 41; **161** (1981) : 354 - 407; **165** (1983) : 254 - 302.
MADER, D. (1983 e) : Aeolian sands terminating an evolution of fluvial depositional environment in Middle Buntsandstein (Lower Triassic) of the Eifel, Federal Republic Germany. In : BROOKFIELD, M.E. & AHLBRANDT, T.S. (Editors), Eolian sediments and processes. Developm. Sedim., **38** : 583 - 612; Elsevier, Amsterdam/Netherlands.
MADER, D. (1984 a) : Entstehung der fluviatilen Sedimente in der grobklastischen Marginalfazies im Oberen Buntsandstein von Luxemburg. N. Jb. Geol. Paläont. Abh., **168** : 23 - 86.
MADER, D. (1984 b) : Fluviatile Sedimentation im Wechsel mit Pedogenese in der Marginalfazies der Zwischenschichten im Oberen Buntsandstein von Luxemburg. Oberrhein. geol. Abh., **33** : 15 - 66.
MADER, D. (1984 c) : Beiträge zur Substratgebundenheit von Nestbauten der solitären Urbiene Colletes daviesanus (Hymenoptera : Colletidae) im Rotliegenden und Tertiär von Rheinhessen, Bundesrepublik Deutschland. Zool. Jb. Syst., **111** : 145 - 173.
MADER, D. (1984 d) : Stratigraphische, paläogeographische und palökologische Signifikanz der Fossilien im mitteleuropäischen Buntsandstein. Zbl. Geol. Paläont., Teil I, **1983** : 1165 - 1217.
MADER, D. (1985 a) : Beiträge zur Genese des germanischen Buntsandsteins. 630 + X pp.; Sedimo-Verlag, Hannover/Germany FRG. ISBN 3-9801149-0-2.
MADER, D. (1985 b) : Aspects of fluvial sedimentation in the Lower Triassic Buntsandstein of Europe. Lecture Notes in Earth Sciences, **4** : 626 + VIII pp.; Springer-Verlag Berlin - Heidelberg - New York - Tokyo. ISBN 3-540-

13984-2.
MADER, D. (1985 c) : Fluvial conglomerates in continental red beds of the Buntsandstein (Lower Triassic) in the Eifel (F.R. Germany) and their palaeoenvironmental, palaeogeographical and palaeotectonic significance. Sedim. Geol., **44** : 1 - 64.
MADER, D. (1985 d) : Beiträge zur Substratgebundenheit von Nestbauten der solitären Urbiene Colletes daviesanus (Hymenoptera : Colletidae) im Buntsandstein des Solling und im Dogger der Deisterumgebung. Zool. Jb. Syst., **112** : 495 - 523.
MADER, D. (1985 e) : Aspekte der Stratigraphie und Ablagerungsgeschichte des Buntsandsteins in der Eifeler Nord-Süd-Zone (Deutschland und Luxemburg). Jber. Mitt. oberrhein. geol. Ver., N.F., **67** : 199 - 242.
MADER, D. (1987) : Marketing and economical aspects of hydraulic proppant fracturing in Western and Eastern Europe. Oil Gas, **13**/2 : 34 - 41.
MADER, D. (1988 a) : Technical and reservoir engineering aspects of hydraulic proppant fracturing in Western and Eastern Europe. Oil Gas, **14**/1 : 19 - 23, 46 - 50.
MADER, D. (1988 b) : Geological and petrophysical influences on hydraulic proppant fracturing potential in European Rotliegend gas reservoirs. Oil Gas, **14**/2 : 26 - 31.
MADER, D. (1989 a) : Palaeoecology, palaeophytogeography and palaeophytosociology of the lycopod Pleuromeia and other xerophytes in the Buntsandstein flora. In preparation for publication.
MADER, D. (1989 b) : Aspects of evolution of fluvial style and depositional modelling in Buntsandstein (Lower Triassic) and Rotliegend (Lower Permian) in Europe. In preparation.
MADER, D. & TEYSSEN, T. (1985) : Palaeoenvironmental interpretation of fluvial red beds by statistical analysis of palaeocurrent data : examples from the Buntsandstein (Lower Triassic) of the Eifel and Bavaria in the German Basin (Middle Europe). Sedim. Geol., **41** : 1 - 74.
MADER, D. & YARDLEY, M.J. (1985) : Migration, modification and merging in aeolian systems and the significance of the depositional mechanisms in Permian and Triassic dune sands of Europe and North America. Sedim. Geol., **43** : 85 - 218.
MADER, D. & CHATALOV, G. (1988) : Fluvial depositional evolution of Buntsandstein of the Teteven Anticlinorium (Central Fore-Balcan, Northern Bulgaria). Geol. Balcan., **18** : 3 - 30.
MAHAJAN, N.C. & BARRON, B.M. (1980) : Bridging particle size distribution : a key factor in the designing of non-damaging completion fluids. SPE Paper, **8792**.
MAHONEY, J.V., STUBBS, P.B., SCHWERER, F.C. III & DOBSCHA, F.X. (1980) : Effects of a no-proppant foam stimulation treatment on a coal seam degasification borehole. SPE/DOE Paper, **8931**. J. Petrol. Technol., **33** (1981) : 2227 - 2235.
MAHONY, B.J. (1988) : Horizontal drilling use on the rise : why and how. World Oil, **207** (1988/10) : 45 - 57.
MAJER, E.L. & DOE, T.W. (1986) : Studying hydrofractures by high frequency seismic monitoring. Intern. J. Rock Mechan., Min. Sci., Geomechan. Abstr., **23** : 185 - 199.
MÄKEL, G.H. (1987) : Fractured clastic reservoirs in South Oman : possible modes of origin and implications. SPE Paper, **15700**.
MALLINGER, M.A., RIXE, F.H. & HOWARD, G.C. (1964) : Development and use of propping agent spacers to increase well productivity. API Drill. Product. Pract., **1964** : 88 - 99.
MALONE, W.T., STAHL, E.J. & COULTER, G.R. (1977) : MHF from planning to performance. Proc. MHF Symp., Univ. of Oklahoma, 28.02. - 01.03.1977.
MALONEY, D.R., GALL, B.L. & RAIBLE, C.J. (1987) : Non-Darcy gas flow through propped fractures : effects of partial saturation, gel damage, and stress. SPE Paper, **16899**.
MALY, G.P. (1976) : Close attention to the smallest job details vital for minimizing formation damage. SPE Paper, **5702**.
MALY, G.P. & KRUEGER, R.F. (1971) : Improper formation sampling leads to impro-

per selection of gravel size. J. Petrol. Technol., **23** : 1403 - 1414. SPE Paper, **3041** (1970).
MALY, G.P. & ROBINSON, J.P. (1972) : Apparatus for gravel packing inclined wells. U.S. Patent, **3,637,010**.
MALY, G.P., ROBINSON, J.P. & LAURIE, A.M. (1974) : New gravel pack tool for improving pack placement. J. Petrol. Technol., **26** : 19 - 24. SPE Paper, **4032** (1972).
MALY, G.P. & VOZELINEK, J. (1978) : Visual model study shows when and when not to pressure wash open hole gravel packs. SPE Paper, **7001**.
MANCHESTER, C.M. (1983) : Applying the principles of project financing to the independent oil and gas producer. SPE Paper, **12023**.
MANCILLAS, G., MATSON, R.P. & ZIARA, B.A. (1976) : Stimulation of the Ekofisk. SPE Paper, **6125**.
MANNESMANN REXROTH (1988) : Ekofisk pack-up. RIQ Special "Ekofisk", **RE 09 634** (10.1987), 15 pp.; Mannesmann Rexroth, Lohr/Germany FRG.
MANTOOTH, M.A. & WILLIAMS, R.H. (1980) : Sand control by application of special fluids and techniques in a shallow sand, offshore, Louisiana : a case history. SPE Paper, **8800**.
MANTOOTH, W.A. & TREADWAY, B.R. (1965) : New sand consolidation method. Oil Gas J., 18.01.**1965**.
MAO, N.H. & SWEENEY, J.J. (1986) : Estimation of in-situ stresses from ultrasonic measurements. SPE Form. Eval., **1** : 532 - 538. SPE Paper, **12850** (1984).
MAO, N.H., SWEENEY, J.J., HANSON, J.M. & COSTANTINO, M.S. (1984) : Using a sonic technique to estimate in-situ stresses. SPE/DOE/GRI Paper, **12850**.
MARCOU, J.A. & GUDMUNDSSON, J.S. (1986) : Development model for geothermal reservoirs. SPE Paper, **15119**.
MARETT, G., CHEVALIER, P., SOUHAITE, P. & SUAU, J. (1976) : Shaly sand evaluation using gamma ray spectrometry, applied to the North Sea Jurassic. Trans. 17th Soc. Prof. Well Log Analysts Annual Logging Symp., Paper **DD**, Denver/Colorado, 09. - 12.06.1976.
MARIE, J.P.P. (1975) : Rotliegendes stratigraphy and diagenesis. In : WOODLAND, A.W. (Editor), Petroleum and the continental shelf of North-West Europe, **1** : 205 - 211; Applied Science Publ., London/England.
MARPLE, G.T. & RULEY, G.E. (1987) : Financing projects via turnkey contractors. SPE Paper, **16302**.
MARPLE, G.T., RULEY, G.E., WORLEY, H.W. & FREEMAN, E.R. (1987) : Modified fracture stimulation techniques for shallow oil sands in Carter County, Oklahohoma. SPE Paper, **16220**.
MARTIN, J.H. & COOPER, J.A. (1984) : An integrated approach to the modeling of permeability barrier distribution in a sedimentologically complex reservoir. SPE Paper, **13051**.
MARTIN, J.H., EVANS, A.J. & RAPER, J.K. (1988) : Reservoir modeling of low-sinuosity channel sands : a network approach. SPE Paper, **18364**.
MARTIN, J.H. & EVANS, P.F. (1988) : Reservoir modelling of marginal aeolian/sabkha sequences, Southern North Sea (U.K. sector). SPE Paper, **18155**.
MARTIN, M.A. (1987) : Three-component seismic investigation of a fractured reservoir, Silo field, Wyoming. Ph.D. Thesis, Colorado School of Mines, Golden/Colorado.
MARTIN, M.A. & DEAN, T.L. (1987) : Shear wave modelling of microfractures. The Leading Edge, **1987**.
MARTIN, W.L., ALEXANDER, J.D., DEW, J.N. & TYNAN, J.W. (1972) : Thermal recovery at North Tisdale field, Wyoming. J. Petrol. Technol., **24** : 606 - 616. SPE Paper, **3595** (1971).
MARTINS, J.P. & HARPER, T.R. (1985) : Mini-frac pressure decline analysis for fractures evolving from long perforated intervals and unaffected by confining strata. SPE/DOE Paper, **13869**.
MARTINS, P. (1987) : Fracturing in the Southern North Sea. SPE Continuing Education, Hydraulic Fracturing Seminar, London/England, 01.12.1987.
MARTINEZ, S.J., STEANSON, R.E. & COULTER, A.W. (1987) : Formation fracturing. In : BRADLEY, H.B. (Editor), Petroleum engineering handbook. 55-1 - 55-12; Soc. Petrol. Engin., Richardson/Texas. ISBN 1-55563-010-3.
MARX, C. & RAHMAN, S.S. (1984) : Evaluation of formation damage caused by dril-

ling fluids specifically in pressure-reduced formations. SPE Paper, **12494**.
MASON, R.A. (1987) : Estimating reserves for high-rate Devonian shale oil and gas completions with limited production data. SPE Reserv. Engin., **2** : 394 - 400. SPE Paper, **13369** (1984).
MASSIE, I., NYGAARD, O. & MORITA, N. (1987) : Field experiences from the subsea wells on Gullfaks : an operator`s practical implementation of a new computer based sand production prediction model. SPE Paper, **16893**.
MASTERS, J.A. (1979) : Deep basin gas trap, Western Alberta. Amer. Assoc. Petrol. Geol. Bull., **63** : 152 - 181.
MASTROJANNIS, E.N., KEER, L. & MURA, T. (1980) : Growth of planar cracks induced by hydraulic fracture. Intern. J. Numer. Methods Engin., **15** : 41 - 54.
MASTROJANNIS, E.N., KEER, L.M. & MURA, T. (1985) : Stress intensity factor for a plane crack under normal pressure. Intern. J. Fract., **1985**/6.
MATHIAS, J.P. (1970) : Successful stimulation of a thick, low-pressure, water-sensitive gas reservoir by pseudolimited entry. SPE Paper, **2903**.
MATTAX, C.C. & KYTE, J.R. (1962) : Imbibition oil recovery from fractured, water-drive reservoir. Soc. Petrol. Engin. J., **2** : 177 - 184.
MATTHEWS, B.A., MILLER, W.K. II & SCHLOTTMAN, B.W. (1982) : Record MHF treatment pumped in East Texas Cotton Valley sands. Oil Gas J., 04.10.**1982** : 94 - 98.
MATTHEWS, C.S. (1981) : Possibilities of enhancing gas production from geopressured aquifers. SPE Paper, **9733**.
MATTHEWS, T.M. (1969) : Field use of Superfrac - a new hydraulic fracturing technique. SPE Paper, **2625**.
MAUK, F.J. & MAHRER, K.D. (1986) : Monitoring hydraulic fracture stimulations with long-period seismometers to extract induced fracture geometry. SPE Paper, **15214**.
MAUK, F.J. & MAHRER, K.D. (1988) : Use of long-period seismometers to determine induced fracture geometry. SPE Product. Engin., **3** : 192 - 200.
MAURY, V.M. & SAUZAY, J.M. (1987) : Borehole instability : case histories, rock mechanics approach, and results. SPE/IADC Paper, **16051**.
MAVOR, M.J. & CINCO-LEY, H. (1979) : Transient pressure behavior of naturally fractured reservoirs. SPE Paper, **7977**.
MAXWELL, C.T. (1986) : Current challenges to industry viability : investment community viewpoint. SPE Paper, **15847**.
MAY, H.S. & MATHEWS, C.D. (1983) : Problems in the natural gas industry - causes and responses. SPE Paper, **11307**.
MAY, H.S. & WHITE, C.A. (1981) : Natural gas price regulation - its problems and prospects. SPE Paper, **9555**.
MAZZULLO, L.J. & MAZZULLO, J. (1985) : Geology and clay mineralogy of the Morrow Formation, Southeastern New Mexico. SPE/DOE Paper, **13849**.
McBANE, R.A., CAMPBELL, R.L. & TRUMAN, R. (1986) : Comparison of diagnostic tools for selecting completion intervals in Devonian shale wells. SPE Paper, **15948**.
McBANE, R.A. & JEU, S.J. (1988 a) : Deep coal seam project. Quart. Rev. of Methane from Coal Seams Technol., **5**/3-4 : 31 - 32.
McBANE, R.A. & JEU, S.J. (1988 b) : Western Cretaceous coal seams project. Quart. Rev. of Methane from Coal Seams Technol., **6**/1 : 25 - 29.
McBANE, R.A. & THOMPSON, T.W. (1984) : Exploration/production studies of the Devonian gas shales. SPE/DOE/GRI Paper, **12833**.
McBRIDE, J.R., RATHBONE, M.J. & THOMAS, R.L. (1979) : Evaluation of fluoboric acid treatment in the Grand Isle offshore area using multiple-rate flow test. SPE Paper, **8339**.
McCANN, K. (1988) : OPEC and the future of oil. Petrol. Rev., **1988**/8 : 16 - 20.
McCAULEY, T.V. (1969) : Backsurging and abrasive perforating to obtain improved perforations performance. J. Petrol. Technol., **21** : 1207 - 1212. SPE Paper, **3449** (1971).
McCORRISTON, L.L., DEMBY, R.A. & PEASE, E.C. (1981) : Study of reservoir damage produced in heavy oil formation due to steam injection. SPE Paper, **10077**.
McCOY, R.L., HARTSOCK, J.H. & DOBSON, R.J. (1980) : Preliminary results of the wells-of-opportunity geopressured-geothermal testing program. SPE/DOE Paper, **8958**.

McCUNE, C.C., AULT, J.W. & DUNLAP, R.G. (1975) : Reservoir properties affecting matrix acid stimulation of sandstones. J. Petrol. Technol., **27** : 633 - 640. SPE Paper, **4552** (1973).
McCUNE, C.C., FOGLER, H.S., LUND, K., CUNNINGHAM, J.R. & AULT, J.W. (1975) : A new model of the physical and chemical changes in sandstone during acidizing. Soc. Petrol. Engin. J., **15** : 361 - 370. SPE Paper, **5157**. SPE Reprint Series, Well Completions, **5 a**/II (1978) : 181 - 190.
McDANIEL, B.W. (1986) : Conductivity testing of proppants at high temperature and stress. SPE Paper, **15067**.
McDANIEL, B.W. (1987) : Realistic fracture conductivities of proppants as a function of reservoir temperature. SPE/DOE Paper, **16453**.
McDANIEL, B.W. (1988) : Use of wet gas flow for long-term fracture conductivity measurements in the presence of gel filter cakes. SPE Paper, **17543**.
McDANIEL, B.W. & PARKER, M.A. (1988) : Accurate design of fracturing treatments requires conductivity measurements at simulated reservoir conditions. SPE Paper, **17541**.
McDANIEL, R.R., HOUX, M.R. & BARRINGER, D.K. (1976) : A new generation of solid-free fracturing fluids. SPE Paper, **5641**.
McDANIEL, R.R. & WILLINGHAM, J.R. (1978) : The effect of various proppants and proppant mixtures on fracture permeability. SPE Paper, **7573**.
McDANIEL, R.R., DEYSARKAR, A.K., CALLANAN, M.J. & KOHLHAAS, C.A. (1981) : An improved method for measuring fluid loss at simulated fracture conditions. SPE Paper, **10259**. Soc. Petrol. Engin. J., **25** (1985) : 482 - 490.
McDONALD, S.W. (1983) : Evaluation of production tests in oil wells stimulated by massive acid fracturing offshore Qatar. J. Petrol. Technol., **35** : 496 - 506. SPE Paper, **9641** (1981).
McDOWELL, S.D. & ELDERS, W.A. (1979) : Geothermal metamorphism of sandstone in the Salton Sea geothermal system. In : ELDERS, W.A. (Editor), Geology and geothermics of the Salton Trough. Campus Museum Contrib., **5** : 70 - 76; Univ. of California, Riverside/California.
McELROY, S.L. (1982) : The effect of lateral anisotropy on flood pattern dimensions and orientation. SPE Paper, **10759**.
McFALL, K.S., WICKS, D.E., KELSON, B., SEDWICK, K. & BRANDENBURG, C. (1987) : An analysis of the coal and coalbed methane resources of the Piceance Basin, Colorado. SPE Paper, **16418**.
McGARR, A. & GAY, N.C. (1978) : State of stress in the earth`s crust. Annual Rev. Earth Planet. Sci., **6** : 405 - 436.
McGHEE, B.F., LESSO, W.G. & WATERFILL, D. (1981) : The use of cased hole logging to evaluate hydraulic fracture height and proppant placement. SPE Paper, **10309**.
McGLOTHLIN, B.B. & HUITT, J.L. (1966) : Relation of formation rock strength to propping agent strength in hydraulic fracturing. J. Petrol. Technol., **18** : 377 - 384. SPE Paper, **1131** (1965).
McGLOTHLIN, B.B., HUITT, J.L. & McKAY, J.D. (1963) : A high strength glass pellet as a propping agent. API Paper, **851-37-C**. World Oil, **157** : p. 57 ff.
McGOWEN, J.M. & McDANIEL, B.W. (1988) : The effects of fluid preconditioning and test cell design on the measurement of dynamic fluid loss data. SPE Paper, **18212**.
McGUIRE, W.J. & SIKORA, V.J. (1960) : The effect of vertical fractures on well productivity. Trans. AIME, **219** : 401 - 403. J. Petrol. Technol., **12** : 72 - 74.
McHUGH, S. (1983) : Crack extension caused by internal gas pressure compared with extension caused by tensile stress. Intern. J. Fract., **21**.
McHUGH, S.L. & KEOUGH, D. (1982) : Fracture and permeability enhancement with tailored pulse loading. SPE/DOE Paper, **10844**.
McINTYRE, F.J., POLKOWSKI, G.A., BRON, J. & POW, M.J. (1988) : Radioactive tracers monitor solvent spreading in rainbow vertical hydrocarbon miscible flood. SPE Reserv. Engin., **3** : 273 - 280. SPE Paper, **14440** (1985).
McKEE, C.R., BUMB, A.C. & KOENIG, R.A. (1988) : Stress-dependent permeability and porosity of coal and other geologic formations. SPE Form. Eval., **3** : 81 - 91. SPE Paper, **12858** (1984).
McKETTA, S.F., KOZIAR, G. & COOK, L.S. (1980) : Eastern tight sands, a stimula-

tion assessment. SPE Paper, **9536**.
McKINLEY, R.M., BOWER, F.M. & RUMBLE, R.C. (1973) : The structure and interpretation of noise from flow behind cemented casing. J. Petrol. Technol., **25** : 329 - 338. SPE Paper, **3999** (1972).
McLANE, J.E. & EDGINGTON, L. (1985) : A stimulation method for the Codell Sandstone, Denver-Julesburg Basin of Colorado, using low pH methanol. SPE/DOE Paper, **13888**.
McLANE, J.E. & SCOTT, M.A. (1983) : Evaluation of the use of high-strength cross-linked hydrochloric acid system for Mission Canyon stimulation : SEM/EDS analysis. SPE Paper, **11839**.
McLAUGHLIN, H.C., ELPHINGSTONE, E.A. & HALL, B. (1976) : Aqueous polymers for treating clays in oil and gas producing formations. SPE Paper, **6008**.
McLAUGHLIN, H.C., ELPHINGSTONE, E.A., REMINGTON, R.E. & COATES, S. (1977) : Clay stabilizing agent can correct formation damage. World Oil, **1977**/5 : 58 - 60.
McLENNAN, J.D. & ROEGIERS, J.C. (1982) : How instantaneous are instantaneous shut-in pressures? SPE Paper, **11064**.
McLENNAN, J.D., ROEGIERS, J.C. & MARX, W.P. (1983) : The Mancos Formation : an evaluation of the interaction of geological conditions, treatment characteristics and production. SPE/DOE Paper, **11606**.
McLEOD, H.O. (1982) : The effect of perforating conditions on well performance. SPE Paper, **10649**. J. Petrol. Technol., **35** (1983) : 31 - 39.
McLEOD, H.O. (1983) : A simplified approach to design of fracturing treatments using high viscosity cross-linked fluids. SPE/DOE Paper, **11614**.
McLEOD, H.O. (1984) : Matrix acidizing. J. Petrol. Technol., **36** : 2055 - 2069. SPE Paper, **13752**.
McLEOD, H.O. (1986) : Productivity decline in gravel packed wells after acidizing. SPE Paper, **14824**.
McLEOD, H.O. & COULTER, A.W. (1966) : The use of alcohol in gas well stimulation. SPE Paper, **1663**.
McLEOD, H.O. & COULTER, A.W. (1969) : The stimulation treatment pressure record - an overlooked formation evaluation tool. J. Petrol. Technol., **21** : 951 - 960. SPE Paper, **2287** (1968).
McLEOD, H.O. & CRAWFORD, H.R. (1983) : Gravel packing for high rate completions. SPE Paper, **11008**.
McLEOD, H.O., LEDLOW, L.B. & TILL, M.V. (1983) : Planning, execution and evaluation of acid treatments in sandstone formations. SPE Paper, **11931**.
McLEOD, H.O., McGINTY, J.E. & SMITH, C.F. (1966) : Deep well stimulation with alcoholic acid. SPE Paper, **1558**.
McMECHAN, D.E. & CONWAY, M.W. (1983) : Recent development in hydraulic stimulation treatments in the Fletcher field in the deep Anadarko Basin. SPE/DOE Paper, **11604**.
McMULLAN, J.H. & BASSIOUNI, Z. (1984) : Prediction of maximum flow rates from geopressured aquifers. J. Petrol. Technol., **36** : 503 - 509. SPE Paper, **10282** (1981).
McNALLY, R. (1987) : North Sea operators respond to challenge. Petrol. Engin. Intern., **1987**/8 : 18 - 22.
McNALLY, R. (1988) : North Sea jubilee. Petrol. Engin. Intern., **1988**/7 : p. 65.
McNAUGHTON, D., CASARTA, L.J. & BORNEMANN, E. (1988) : Fracture identification from formation microscanner in Alaska`s Lisburne Carbonate. SPE Paper, **17435**.
McPEEK, L.A. (1981) : Eastern Green River basin : a developing giant gas supply from deep, overpressured Upper Cretaceous sandstones. Bull. Amer. Assoc. Petrol. Geol., **65** : 1078 - 1098.
McPHAIL, W.R. (1987) : Effective evaluation of loans to reduce risk. SPE Paper, **16303**.
MEAD, H.N. (1987) : Using finite-system buildup analysis to investigate fractured, vugular, stimulated, and horizontal wells. SPE Paper, **16379**. J. Petrol. Technol., **40** (1988) : 1361 - 1371.
MEANS, R. (1982) : Future of the natural gas industry : the Federal Energy Regulatory Commission perspective. SPE Paper, **11263**.
MEDLIN, W.L. & FITCH, J.L. (1983) : Abnormal treating pressures in MHF treat-

ments. SPE Paper, **12108**. J. Petrol. Technol., **40** (1988) : 633 - 642.
MEDLIN, W.L. & MASSE, L. (1976) : Laboratory investigation of fracture initiation pressure and orientation. SPE Paper, **6087**.
MEDLIN, W.L. & MASSE, L. (1984) : Laboratory experiments in fracture propagation. Soc. Petrol. Engin. J., **24** : 256 - 268. SPE Paper, **10377** (1981).
MEDLIN, W.L. & MASSE, L. (1986 a) : Plasticity effects in hydraulic fracturing. J. Petrol. Technol., **38** : 995 - 1006. SPE Paper, **11068** (1982).
MEDLIN, W.L. & MASSE, L. (1986 b) : Fluid loss mechanism for gels and suspensions. SPE Paper, **15630**.
MEDLIN, W.L., SEXTON, J.H. & ZUMWALT, G.L. (1985) : Sand transport experiments in thin fluids. SPE Paper, **14469**.
MEEHAN, D.N., HORNE, R.N. & AZIZ, K. (1988) : Effects of reservoir heterogeneity and fracture azimuth on optimization of fracture length and well spacing. SPE Paper, **17606**.
MEEHAN, D.N. & PENNINGTON, B.F. (1981) : Numerical simulation results in the Carthage (Cotton Valley) field. SPE/DOE Paper, **9838**.
MEGILL, R.E. (1984) : An introduction to risk analysis. 2nd edition, 274 pp.; Penn Well, Tulsa/Oklahoma. ISBN 0-87814-257-6.
MEGILL, R.E. (1985) : Evaluating & managing risk. A collection of readings. 152 pp.; Sci Data Publ., Penn Well, Tulsa/Oklahoma. ISBN 0-87814-283-5.
MEIER, U. & ERNST, P.L. (1981) : Drilling and completion of the Urach III HDR test well. Proc. Intern. Conf. Geotherm. Drill. Complet. Technol., Sandia National Laboratories Rept., **SAND 81-0036 C** : 6-1 - 6-24; Albuquerque/ New Mexico.
MEISSNER, F.F. (1981) : Abnormal pressures produced by hydrocarbon generation and maturation and their relation to processes of migration and accumulation. Bull. Amer. Assoc. Petrol. Geol., **65** : p. 2467.
MEISTER, P.M. & CADY, G.V. (1986) : Corporate takeovers and petroleum industry restructuring. SPE Paper, **15169**. J. Petrol. Technol., **38** : 549 - 556.
MELDAU, R.S., SHIPLEY, R.G. & COATS, K.H. (1981) : Cyclic gas/steam stimulation of heavy-oil wells. J. Petrol. Technol., **33** : 1990 - 1998.
MELVAN, J.J. (1978) : The effect of overburden stress on the formation and stability of arches in unconsolidated sands. Thesis, **T-2071**; Colorado School of Mines, Golden/Colorado.
MENDELSOHN, D.A. (1984) : A review of hydraulic fracture modeling. I : General concepts, 2D models, motivation for 3D modeling. II : 3D modeling and vertical growth in layered rock. J. Energy Resources Technol., **106** : 369 - 376, 543 - 553.
MENG, H.Z. & BROWN, K.E. (1987) : Coupling of production forecasting, fracture geometry requirements, and treatment scheduling in the optimum hydraulic fracture design. SPE/DOE Paper, **16435**.
MENG, H.Z., PROANO, E.A., BUHIDMA, I.M. & MACH, J.M. (1982) : Production systems analysis of vertically fractured wells. SPE/DOE Paper, **10842**.
MENGES, U.R. (1986) : Opec-Krise setzt neue Öl- und Erdgasperspektiven. Börsen-Z., 26.04.**1986**. BEB Mosaik, **2/86** : 7 - 8.
MENGHINI, M.L. (1988) : Compaction monitoring in the Ekofisk area chalk fields. OTC Paper, **5620**. SPE Paper, **17853**.
MENOUAR, H. & KNAPP, R.M. (1980) : Numerical simulation of the imbibition process in fractured reservoirs. SPE Paper, **9370**.
MERCER, J.C., PRATT, H.R. III & YOST, A.B. II (1988) : Infill drilling using horizontal wells : a field development strategy for tight fractured formations. SPE Paper, **17727**.
MES, M.J. (1988) : Measuring offshore subsidence with seabed pressure gauges. SPE Paper, **18182**.
MESSINA (1986 a) : Z-Prop. A high temperature-fusion produced zirconium ceramic. An outstanding frac proppant. Brochure; Messina Worldwide, Dallas/Texas.
MESSINA, M.L. (1986 b) : Oilfield chemicals and fluid systems. Oil Asia, **1986** : 74 - 76.
MESSINA (1987) : Proppants. Messina Techn. Bull., 12 pp., 5 pp., Dallas/Texas.
METZGER, H.A. (1958) : Preliminary report on fracturing by vibration - a new method of stimulation wells. J. Petrol. Technol., **10** : 13 - 16.

MEYER, B.R. (1985) : Frac model in 3 D. 1 : New simulator makes fracture design routine. 2 : Proppant placement analyzed. 3 : Hydraulic fracturing simulator`s capabilities examined. 4 : Graphics module outputs fracturing results. Oil Gas J., 17.06.**1985** : 87 - 92; 01.07.**1985** : 62 - 65; 22.07.**1985** : 83 - 92; 29.07.**1985** : 132 - 136.

MEYER, B.R. (1986) : Design formulae for 2-D and 3-D vertical hydraulic fractures : model comparison and parametric studies. SPE Paper, **15240**.

MEYER, B.R. (1987) : Heat transfer in hydraulic fracturing. SPE Paper, **17041**.

MEYERS, K.O., SKILLMAN, H.L. & HERRING, G.D. (1985) : Control of formation damage at Prudhoe Bay, Alaska, by inhibitor squeeze treatment. J. Petrol. Technol., **37** : 1019 - 1034. SPE Paper, **12472** (1984).

MICHAJLOV, N.N. (1987) : Alteration of physical characteristics of reservoir rocks in near wellbore zones. 152 pp. (in Russian), Nedra, Moscow/USSR.

MICHAUD, F. (1987) : Eldfisk. In : SPENCER, A.M. et al. (Editors), Geology of the Norwegian oil and gas fields, 89 - 106; Graham & Trotman, London/England.

MIDDLETON, J.A. (1987) : The political and economical context of the oil industry. SPE Hydrocarbon Economics and Evaluation Symposium, Plenary Session; Dallas/Texas, 02. - 03.03.1987.

MILHEM, M.M. & AHMED, K.N. (1987) : Performance of a pilot cyclic steam stimulation project in Kuwait. SPE Paper, **15733**.

MILITZER, M.R. & SCHWERER, F.C. (1986) : Preliminary economic assessment potential for producing coalbed methane for multiple coal seams completion project at Rock Creek. Gas Research Institute Rept., **GRI-86/0046**; Chicago/Illinois.

MILLER, B.D. & BERGSTROM, J.M. (1975) : Results of acid-in-oil emulsion stimulations of carbonate formations. SPE Paper, **5648**.

MILLER, B.D. & WAREMBOURG, P.A. (1975) : Pre-pack technique using fine sand improves results of fracturing and fracture acidizing treatments. SPE Paper, **5643**.

MILLER, C.A., MOISSIS, D.E. & WHEELER, M.F. (1989) : Simulation of miscible viscous fingering using a modified method of characteristics : effects of gravity and heterogeneity. SPE Paper, **18440**.

MILLER, J.A. & NEWLIN, T.W. (1987) : Mega-merger case study. Petrol. Managem., **9**.

MILLER, M.A. & RAMEY, H.J. (1985) : Effect of temperature on oil/water relative permeabilities of unconsolidated and consolidated sands. Soc. Petrol. Engin. J., **25** : 945 - 953. SPE Paper, **12116** (1983).

MILLHEIM, K.K. & CICHOWICZ, L. (1968) : Testing and analyzing low-permeability fractured gas wells. J. Petrol. Technol., **20** : 193 - 198. SPE Paper, **1768** (1967).

MINEAR, J.W. (1986) : Full wave sonic logging : a brief perspective. Trans. 27th Soc. Prof. Well Log Analysts Annual Logging Symp., Paper **AAA**, Houston/Texas, 09. - 13.06.1986.

MINERALÖLWIRTSCHAFT (1983) : Ölpreis : ein Faktor in der Weltwirtschaft. Erdöl und Kohle - Erdgas, **36** : 189 - 190.

MINNE, J.C. & GARTNER, J. (1979) : Fracture detection in the Middle East. SPE Paper, **7773**.

MINTZ, F. (1981) : Reserve based financing - specific requirements and alternatives. SPE Paper, **9578**.

MIRZA, S. & RICHARDSON, J.F. (1979) : Sedimentation of suspension particles of two and more sizes. Chem. Engin. Sci., **34** : p. 447 ff.

MISAK, M.D., ATTEBERRY, R.D., VENDITTO, J.J. & FREDRICKSON, S.E. (1978 a) : A fracturing technique to minimize water production. SPE Paper, **7563**.

MISAK, M.D., ATTEBERRY, R.D., VENDITTO, J.J. & FREDRICKSON, S.E. (1978 b) : Edwards Limestone completions improved by new stimulation technique. SPE Paper, **7115**.

MISHRA, S., BRIGHAM, W.E. & ORR, F.M. (1988) : Tracer and pressure test analysis for characterization of areally heterogeneous reservoirs. SPE/DOE Paper, **17365**.

MISHRA, S. & RAMEY, H.J. (1985) : A comparison of pressure transient and tracer concentration-time data for layered reservoirs under injection. SPE Paper,

14170.
MITCHELL, M.D. & CHAMBERLAIN, C.R. (1983) : An analysis of stimulation treatments of the Randolph Sand in the Semitropic field, Kern County, California. SPE Paper, **11709**.
MNISZEWSKI, K.R. & WESTCOTT, P.A. (1986) : The characterization of high-energy chemical sources for fracturing Devonian shale. SPE Paper, **15254**.
MODESITT, L.E. (1986) : Environmental regulations : leading the way toward restructuring the petroleum industry. SPE Paper, **14636**. J. Petrol. Technol., **39** (1987) : 1113 - 1118.
MODY, B.G., McKITRICK, R.S. & SHAHSAVARI, D. (1988) : Proper application of crosslinked polymer decreases operating costs. SPE Paper, **17288**.
MOHANTY, K.K. & MILLER, A.E. (1988) : Factors influencing unsteady relative permeability of a mixed-wet reservoir rock. SPE Paper, **18292**.
MOLON, J.P. & FOX, K.B. (1983) : Increased fracture penetration and productivity using xanthan gelled acid in massive carbonate formations. SPE Paper, **11502**.
MONAGHAN, P.H., SALATHIEL, R.A., MORGAN, B.E. & KAISER, A.D. (1959) : Laboratory studies of formation damage in sands containing clays. J. Petrol. Technol., **11** : 209 - 215.
MONROE, S.A. & PENBERTHY, W.L. (1980) : Gravel packing high-volume water supply wells. J. Petrol. Technol., **32** : 2097 - 2102.
MONTGOMERY, C.T. & STEANSON, R.E. (1985) : Proppant selection : the key to successful fracture stimulation. J. Petrol. Technol., **37** : 2163 - 2172. SPE Paper, **12616** (1984).
MONTIGNY, O.d., SORRIAUX, P., LOUIS, A.J.P. & LESSI, J. (1988) : Horizontal-well drilling data enhance reservoir appraisal. Oil Gas J., 04.07.**1988** : 40 - 48.
MONTMAYEUR, H. & GRAVES, R.M. (1986) : Prediction of static elastic/mechanical properties of consolidated and unconsolidated sands from acoustic measurements : correlations. SPE Paper, **15644**, **14159** (1985).
MOODY, J.D.G. (1988) : Thermal displacement in a fractured reservoir. SPE Paper, **17090**.
MOORE, J.N. & NIELSON, D.L. (1986) : Geological and geochemical techniques for mapping fractures and fluid flow patterns in geothermal reservoirs. SPE Paper, **15125**.
MOORE, S.D. (1986) : Well servicing expenditures, activity drop substantially. Petrol. Engin. Intern., **1986**/7 : 20 - 26.
MOORE, S.D. (1987 a) : High-angle drilling comes of age. Petrol. Engin. Intern., **59**/2 : 18 - 22.
MOORE, S.D. (1987 b) : Well servicing : nowhere to go but up. Petrol. Engin. Intern., **59**/7 : 16 - 22.
MOORE, S.D. (1988 a) : Change with the changing times? Petrol. Engin. Intern., **60**/7 : p. 6.
MOORE, S.D. (1988 b) : Well servicing activity bounds back. Petrol. Engin. Intern., **60**/7 : 16 - 20.
MOORE, S.D. (1988 c) : Alberta pushes ahead with underground drilling. Petrol. Engin. Intern., **60**/2 : 28 - 30.
MOORE, S.D. (1988 d) : Trends point in both directions. Petrol. Engin. Intern., **60**/2 : p. 6.
MOORE, W.D. III (1980) : ARCO drills horizontal drainhole for better reservoir placement. Oil Gas J., **1980** : 139 - 148.
MORALES, R.H. & ABOU-SAYED, A. (1985) : The effects of frac fluid properties on proppant transport and variable height fracture behavior. SPE Paper, **13616**.
MORALES, R.H., ABOU-SAYED, A.S., JONES, A.H. & AL-SAFFAR, A. (1986) : Detection of a formation fracture in a waterflooding experiment. J. Petrol. Technol., **38** : 1113 - 1121. SPE Paper, **13747** (1985).
MORALES, R.H., JONES, A.H. & ABOU-SAYED, A.S. (1985) : Analysis of the effects of frac fluid properties on proppant transport and variable height fracture behavior by using a quasi 3-D simulator. SPE Paper, **13616**.
MORGENTHALER, L.N. (1986) : Formation damage tests of high-density brine completion fluids. SPE Product. Engin., **1** : 432 - 436.
MORITA, N., WHITFILL, D.L., FEDDE, O.P. & LØVIK, T.H. (1987) : Parametric stu-

dy of sand production prediction : analytical approach. SPE Paper, **16990**.

MORITA, N., WHITFILL, D.L., MASSIE, I. & KNUDSEN, T.W. (1987) : Realistic sand production prediction : numerical approach. SPE Paper, **16989**.

MORITA, N., WHITFILL, D.L. & WAHL, H.A. (1985) : Stress intensity factor and fracture cross-sectional shape predictions from a 3D model for hydraulically induced fractures. SPE Paper, **14262**.

MORITA, N., WHITFILL, D.L., NYGAAD, O. & BALE, A. (1988) : A quick method to determine subsidence, reservoir compaction, and in-situ stress caused by reservoir depletion. SPE Paper, **17150**.

MORRA, F., BRASHEAR, J.P. & HAAS, M.R. (1982) : Comparative economics of gas production from conventional, tight, and deep reservoirs. SPE/DOE Paper, **10864**.

MORRIS, C.F., LITTLE, T.M. & LETTON, W. III (1984) : A new sonic array tool for full waveform logging. SPE Paper, **13285**.

MORRIS, C.M. & CAMPBELL, D.A. (1979) : Geothermal reservoir energy recovery : a three-dimensional simulation study of the East Mesa field. SPE Paper, **8229**.

MORRIS, C.M. & CAMPBELL, D.A. (1985) : Analysis of wells in naturally fractured formations with rate-sensitive flow. SPE Paper, **14169**. SPE Form. Eval., **2** (1987) : 567 - 572.

MORRIS, C.W. & SINCLAIR, A.R. (1984) : Evaluation of bottomhole treatment for geothermal well hydraulic fracture stimulation. J. Petrol. Technol., **36** : 829 - 836. SPE Paper, **11157** (1982).

MORRIS, C.W., VERITY, R.V., SCHROEDER, R.C. & GORANSON, C. (1980) : Geothermal well stimulation experiments and evaluations. SPE Paper, **9277**.

MORRIS, R.L., GRINE, D.R. & ARKFELD, T.E. (1961) : Using compressional and shear acoustic amplitudes for the location of fractures. J. Petrol. Technol., **13** : 623 - 632.

MORRISON, D.D. & JOLLIFFEE, V. (1982) : North Sea tax and economics. EUR Paper, **299**.

MORRISON, R. & THIBODAUX, J. (1984) : The six-arm dipmeter, a new concept by Geosource. Trans. 25th Soc. Prof. Well Log Analysts Annual Logging Symp., Paper **MMM**, New Orleans/Louisiana, 10. - 13.06.1984.

MORRISSON, T.E. & HENDERSON, J.H. (1960) : Gravity drainage of oil into large horizontal fractures. Trans. AIME, **219** : 7 - 15.

MORROW, N.R., HUPPLER, J.D. & SIMMONS, A.B. (1969) : Porosity and permeability of unconsolidated, Upper Miocene sands from grain-size analysis. J. Sedim. Petrol., **39** : 312 - 321.

MORROW, N.R., WARD, J.S., BROWER, K.R. & CATHER, S.M. (1986) : Rock matrix and fracture analysis of flow in Western tight gas sands. 1985 Annual Rept. **DOE/** MC/21179-2032, **NTIS/DE86001055**, 68 pp. Quart. Techn Progr. Rept., 07. - 09.1986, New Mexico Petroleum Recovery Research Center, 34 pp.; Socorro/New Mexico.

MORSE, E.L. (1988) : The impact on markets of non-OPEC oil supplies. Symposium on Energy, Finance and Taxation Policies, Washington/D.C., 19. - 20.09.1988.

MORSE, J.V. (1975) : Field use of turbine meters in sand-laden crude oil. SPE Paper, **5606**.

MORSE, R.A. & GONTEN, W.D.v. (1972) : Productivity of vertically fractured wells prior to stabilized flow. J. Petrol. Technol., **24** : 807 - 811. SPE Paper, **3631** (1971).

MOSCHOVIDIS, Z.A. (1987) : Interpretation of pressure decline for minifrac treatments initiated at the interface of two formations. SPE Paper, **16188**.

MOSLOW, T.F. & TILLMAN, R.W. (1986) : Sedimentary facies and reservoir characteristics of Frontier Formation sandstones, Southwestern Wyoming. In : SPENCER, C.W. & MAST, R.F. (Editors), Geology of tight gas reservoirs. Amer. Assoc. Petrol. Geol., Stud. Geol., **24** : 271 - 295.

MOUSLI, N.A., RAGHAVAN, R., CINCO-LEY, H. & SAMANIEGO, F.V. (1982) : The influence of vertical fractures intercepting active and observation wells on interference tests. Soc. Petrol. Engin. J., **22** : 933 - 944. SPE Paper, **9346** (1980).

MOWER, L.N. & SOVERALL, D.G. (1980) : Applied technology meets Trinidad sand control challenge. SPE Paper, **9422**.

MUCH, M. (1987) : Proppant evaluation test improved. Oil Gas J., 06.04.**1987** : 33 - 36.
MUCH, M.G. & PENNY, G.S. (1987) : Long-term performance of proppants under simulated reservoir conditions. SPE/DOE Paper, **16415**.
MUECKE, T.W. (1974) : Factors influencing the deterioration of plastic sand consolidation treatments. J. Petrol. Technol., **26** : 157 - 166. SPE Paper, **4354** (1973).
MUECKE, T.W. (1978) : Formation fines and factors controlling their movement in porous media. SPE Paper, **7007**. J. Petrol. Technol., **31** (1979) : 144 - 150.
MUKHERJEE, H. & ECONOMIDES, M.J. (1988) : A parametric comparison of horizontal and vertical well performance. SPE Paper, **18303**.
MULLINS, L.D., BALDWIN, W.F. & BERRY, P.M. (1974) : Surface flowline sand detection. SPE Paper, **5152**.
MUNGAN, N. (1965) : Permeability reduction through changes in pH and salinity. SPE Paper, **1283**.
MURE, E. (1987) : Frigg. In : SPENCER, A.M. et al. (Editors), Geology of the Norwegian oil and gas fields, 203 - 214; Graham & Trotman, London/England.
MURER, H. (1981) : Methoden und Erfahrungen der Sandbekämpfung in den Ölfeldern Matzen - Pirawarth - Hochleiten. Erdoel-Erdgas, **97** : 357 - 365.
MURER, H. & STRAUCH, E. (1980) : Sidetrack saves lost oil despite unique problems. World Oil, **1980**/9.
MURPHEY, J.R. (1980) : Gelled water epoxy sand consolidation system. U.S. Patent, **4,199,484** (20.04.1980), **4,216,829** (12.08.1980).
MURPHEY, J.R., BILA, V.J. & TOTTY, K. (1974) : Sand consolidation systems placed with water. SPE Paper, **5031**.
MURPHEY, J.R., ROLL, D.L. & WONG, L. (1985) : Resin-coated sand slurries for repair of damaged liners. SPE Paper, **13649**.
MURPHY, D.L. & CARNEY, M.L. (1977) : Massive frac - a second look. Proc. MHF Symp., 57 - 75; Oklahoma Univ., 02.1977.
MURPHY, H.D. & BROWN, D.W. (1976) : Investigation of down-hole hydraulic-fracture geometry and connection conditions with flow testing. SPE Paper, **6093**.
MURPHY, H.D. & FEHLER, M.C. (1986) : Hydraulic fracturing of jointed formations. SPE Paper, **14088**.
MURPHY, H.D., LAWTON, R.G., TESTER, J.W., POTTER, R.M., BROWN, D.W. & AAMODT, R.L. (1977) : Preliminary assessment of a geothermal energy reservoir formed by hydraulic fracturing. Soc. Petrol. Engin. J., **17** : 317 - 326. SPE Paper, **6093**.
MURPHY, H.D. & PINE, R.J. (1984) : Dendritic hydraulic fracturing in geothermal energy reservoirs. ASME Petrol. Divis. Conference, New Orleans/Louisiana, 02.1984.
MURPHY, H.D. & TESTER, J.W. (1979) : Heat production from a hot dry rock geothermal reservoir formed by hydraulic fracturing - a comparison of field and theoretical results. SPE Paper, **8265**.
MURPHY, P.J. (1988) : Performance of horizontal wells in the Helder field. SPE Paper, **18340**.
MURPHY, W.B. & JUCH, A.H. (1960) : Pin-point sand-fracturing - a method of simultaneous injection into selected sands. J. Petrol. Technol., **12** : 21 - 24.
MUSKAT, M. (1949) : Physical principles of oil production. McGraw Hill, New York.
MYERS, G.D. (1988) : Practical pulsed neutron spectroscopy logging with a high resolution gamma ray detector. Trans. 29th Soc. Prof. Well Log Analysts Annual Logging Symp., San Antonio/Texas, 05. - 08.06.1988.
MYUNG, J.L. (1976) : Fracture investigation of the Devonian Shale using geophysical well logging techniques. SPE Paper, **6368**.
MYUNG, J.L. & CALDWELL, J. (1971) : Fracture investigation by wireline logging methods. Proc. 10th Annual Ontario Petrol. Inst. Inc., Paper **9**, Toronto/Ontario, 24. - 27.10.1971.
NADERI, N.C. (1984) : Discussion of experience in gravel packing long, perforated intervals in deviated holes. J. Petrol. Technol., **36** : p. 487. SPE Paper, **12886**.
NAJURIETA, H.L. (1979) : Interference and pulse testing in uniformly fractured reservoirs. SPE Paper, **8283**.

NAJURIETA, H.L., ROBLES, O.O. & EDWARDS, D.P. (1986) : Interference well testing to predict early water break-through in naturally fractured reservoirs. SPE Paper, **15636**.
NAKORNTHAP, K. & EVANS, R.D. (1986) : Numerical simulation and reservoir characterization of multiphase fluid flow in naturally fractured formations. SPE Paper, **15130**.
NALL, A., CAMPBELL, P. & BONEY, C. (1983) : Fracture treatments in the Abo Formation, Pecos Slope field. Proc. 30th Annual Southwestern Petroleum Short Course, Lubbock/Texas, 27. - 28.04.1983, pp. 50 - 56.
NARAGHI, M. (1988) : OPEC`s current $ 18/barrel price policy : its rationale, member country economic consequences, and future implications. SPE Paper, **17455**.
NARENDRAN, V.M. & CLEARY, M.P. (1983) : Analysis of growth and interaction of multiple hydraulic fractures. SPE Paper, **12272**.
NAYBERG, T.M. & LINAFELTER, R.L. (1984) : Controlling cement circulation loss to both high-permeability and fractured formations. SPE Paper, **12905**.
NAYLOR, J.D. & FERRIES, N.R. (1988) : Managing gas supplies in the deregulated markets of the 1980`s. SPE Paper, **17747**.
NEAL, E.A., PARMLEY, E.L. & COLPAYS, P.J. (1977) : Oxide ceramic proppants for treatment of deep well fractures. SPE Paper, **6816**.
NEAL, M.R. (1983) : Gravel pack evaluation. J. Petrol. Technol., **35** : 1611 - 1616. SPE Paper, **11232** (1982).
NEAL, M.R. & CARROLL, J.F. (1985) : A quantitative approach to gravel pack evaluation. J. Petrol. Technol., **37** : 1035 - 1040. SPE Paper, **12477** (1984).
NEAL, R.L. & REED, M.E. (1981) : The impact of taxes on exploration economics in the USA. SPE Paper, **9559**.
NEARING, T.R. & STARTZMAN, R.A. (1988) : Effects of stimulation/completion practices on Eastern Devonian shale well productivity. SPE Paper, **18553**.
NEASHAM, J.W. (1977 a) : The morphology of dispersed clay in sandstone reservoirs and its effect on sandstone shaliness, pore space and fluid flow properties. SPE Paper, **6858**.
NEASHAM, J.W. (1977 b) : Applications of scanning electron microscopy to the characterization of hydrocarbon-bearing rocks. Scanning Electron Microscopy, **1977**/3 : 101 - 108.
NECTOUX, A. (1987) : Equilibrium gas oil drainage : velocity, gravitational and compositional effects - North Sea field applications. 4th Europ. Symp. Enhanced Oil Recovery, Modern Reservoir Management Session III, Hamburg/Germany FRG.
NEHMER, W.L. (1988) : Viscoelastic gravel-pack carrier fluid. SPE Paper, **17168**.
NEILL, B.E. et al. (1977) : The use of photon logs to evaluate gravel packing. SPE Paper, **6532**.
NEILL, G.H., DOBBS, J.B., PRUITT, G.T. & CRAWFORD, H.R. (1964) : Field and laboratory results of carbon dioxide and nitrogen in well stimulation. J. Petrol. Technol., **16** : 244 - 248.
NELSON, A.C. & SMITH, D.M. (1969) : A new development in gravel packing. SPE Paper, **2745**.
NELSON, P.H., MAGNUSSON, K.A. & RACHIELE, R. (1983) : Applications of borehole geophysics at an experimental waste storage site. Geophys. Prospect., **30** : p. 910 ff.
NELSON, R.A. (1981) : Significance of fracture sets associated with stylolite zones. Bull. Amer. Assoc. Petrol. Geol., **65** : 2417 - 2425.
NELSON, R.A. (1985) : Geologic analysis of naturally fractured reservoirs. Contrib. Petrol. Geol. Engin., **1** : 320 pp.; Gulf Publ., Houston/Texas.
NELSON, R.A. (1987) : Fractured reservoirs : turning knowledge into practice. J. Petrol. Technol., **39** : 407 - 456. SPE Paper, **16470**.
NELSON, R.A. (1988) : Do workers and managers view fractured reservoirs differently? Amer. Assoc. Petrol. Geol., Annual Convention, Houston/Texas, 20. - 23.03.1988.
NELSON, R.A., LENOX, L.C. & WARD, B.J. (1987) : Oriented core : its use, error, and uncertainty. Bull. Amer. Assoc. Petrol. Geol., **71** : 357 - 368.
NEMAT-NASSER, S. & OHTSUBO, H. (1978) : Fluid flow and heat transfer through hydraulically induced fractures in hot, dry rock masses. J. Press. Vessel

Techn., **100** : 277 - 284.
NERETNIEKS, I. & RASMUSON, A. (1984) : An approach to modeling radionuclide migration in a medium with strongly varying velocity and block sizes along the flow path. Water Resour. Res., **20** : 1823 - 1836.
NEWBERRY, B.M., NELSON, R.F. & AHMED, U. (1985) : Prediction of vertical hydraulic fracture migration using compressional and shear wave slowness. SPE/DOE Paper, **13895**.
NEWBERRY, M.E. & BARKER, K.M. (1985) : Formation damage prevention through the control of paraffin and asphaltene deposition. SPE Paper, **13796**.
NEWENDORP, P.D. (1975 a) : Petroleum exploration risk analysis methods. Decis. Anal. Petrol. Explor.; **7**, Petrol. Publ., Tulsa/Oklahoma.
NEWENDORP, P.D. (1975 b) : Risk analysis using simulation techniques. Decis. Anal. Petrol. Explor.; **8**, Petrol. Publ., Tulsa/Oklahoma.
NEWENDORP, P.D. (1983) : A strategy for implementing risk analysis. SPE Paper, **11299**. J. Petrol. Technol., **36** (1984) : 1791 - 1796.
NEWMAN, G.H. (1981) : The effect of water chemistry on the laboratory compression and permeability characteristics of some North Sea chalks. SPE Paper, **10203**. J. Petrol. Technol., **35** (1983) : 976 - 980.
NGHIEM, L.X. (1983) : Modeling infinite-conductivity vertical fractures by using source and sink terms. Soc. Petrol. Engin. J., **23** : 633 - 644. SPE Paper, **10507** (1982).
NGHIEM, L.X., FORSYTH, P.A. & BEHIE, A. (1984) : A fully implicit hydraulic fracture model. J. Petrol. Technol., **36** : 1191 - 1198. SPE Paper, **10506** (1982).
NGUYEN, H.X. & LARSON, D.B. (1983) : Fracture height containment by creating an artificial barrier with a new additive. SPE Paper, **12061**.
NGUYEN, T.V. (1986) : Experimental study of non-Darcy flow through perforations. SPE Paper, **15473**.
NGUYEN, T.V. & STEVENS, C.E. (1988) : The use of inert gas radioactive tracers for steam injection profiling. SPE Paper, **17419**.
NIELSEN, R.M. & KOHLHAAS, C.A. (1979) : Acoustic and biaxial measurement of rock properties for interpretation of logs for design of well-completion operations. SPE Paper, **8238**.
NIEMEYER, B.L. & REINART, M.R. (1986) : Hydraulic fracturing of a moderate permeability reservoir, Kuparuk River unit. SPE Paper, **15507**.
NIERODE, D.E. (1983) : Comparison of hydraulic fracture design methods to observed field results. SPE Paper, **12059**. J. Petrol. Technol., **37** (1985) : 1831 - 1839.
NIERODE, D.E. (1987) : Hydraulically fractured gas well productivity modeling. SPE Paper, **16897**.
NIERODE, D.E. & KRUK, K.F. (1973) : An evaluation of acid fluid loss additives, retarded acids, and acidized fracture conductivity. SPE Paper, **4549**.
NINI, C.J. & OWEN, G.W. (1983) : Successful high angle gravel packing techniques. SPE Paper, **12105**.
NOBLETT, B.R. & FERTL, W.H. (1987) : Circumferential acoustic log finds vertical fractures. World Oil, **206** (1987/11).
NOBLETT, B.R., FERTL, W.H. & GUY, J.O. (1987) : Recent advances in fracture evaluation. SPE Paper, **16226**.
NOLTE, K.G. (1979) : Determination of fracture parameters from fracturing pressure decline. SPE Paper, **8341**.
NOLTE, K.G. (1982) : Fracture design considerations based on pressure analysis. SPE Paper, **10911**. SPE Product. Engin., **3** (1988) : 22 - 30.
NOLTE, K.G. (1984 a) : Determination of proppant and fluid schedules from fracturing pressure decline. SPE Paper, **13278**. SPE Product. Engin., **1** (1986) : 255 - 265.
NOLTE, K.G. (1984 b) : A general analysis of fracturing pressure decline. SPE Paper, **12941**.
NOLTE, K.G. (1986) : A general analysis of fracturing pressure decline with application to three models. SPE Form. Eval., **1** : 571 - 583. SPE Paper, **12941** (1984).
NOLTE, K.G. (1988 a) : Principles for fracture design based on pressure analysis. SPE Product. Engin., **3** : 22 - 30. SPE Paper, **10911** (1982).

NOLTE, K.G. (1988 b) : Application of fracture design based on pressure analysis. SPE Product. Engin., **3** : 31 - 42.
NOLTE, K.G. (1988 c) : Fluid flow considerations in hydraulic fracturing. SPE Paper, **18537**.
NOLTE, K.G. & SMITH, M.B. (1981) : Interpretation of fracturing pressures. J. Petrol. Technol., **33** : 1767 - 1775. SPE Paper, **8297** (1979).
NOMAN, R. & ARCHER, J.S. (1987) : The effect of pore structure on non-Darcy gas flow in some low-permeability reservoir rocks. SPE/DOE Paper, **16400**.
NOMAN, R., SHRIMANKER, N. & ARCHER, J.S. (1985) : Estimation of the coefficient of inertial resistance in high-rate gas wells. SPE Paper, **14207**.
NORBURY, I. (1987) : Hod. In : SPENCER, A.M. et al. (Editors), Geology of the Norwegian oil and gas fields, 107 - 116; Graham & Trotman, London/England.
NORDGREN, R.P. (1972) : Propagation of a vertical hydraulic fracture. SPE Paper, **3009**. Soc. Petrol. Engin. J., **12** : 306 - 314.
NORDGREN, R.P. (1977) : Strength of well completions. Proc. 18th U.S. Symp. Rock Mechan., 22. - 24.06.1977.
NORENG, Ø. (1987) : Norway and OPEC. Rivalry or joint action? The oil strategy for a less and less marginal producer. In : Norway, OPEC and IEA - global dialogue and price stabilizing measures in the oil market? Intern. Polit., **1987**/5-6, Norwegian Institute of International Affairs.
NORMAN, L.R. & CONWAY, M.W. (1983) : High temperature gelled acid, laboratory development and field results. SPE Paper, **11510**.
NORMAN, L.R., CONWAY, M.W. & WILSON, J.M. (1981) : Temperature-stable acid-gelling polymers : laboratory evaluation and field results. SPE Paper, **10260**. J. Petrol. Technol., **36** (1984) : 2011 - 2018.
NORMAN, M.E., CIPOLLA, C.L. & WEBB, M.L. (1983) : The application of manufactured proppants in moderately permeable oil reservoirs. SPE Paper, **12357**.
NORMAN, M.E. & FAST, C.R. (1985) : Proppant monograph. General Abrasive Division, Dresser Industries, Houston/Texas.
NORTH, F.K. (1985) : Petroleum geology. 607 pp.; Allen & Unwin, Winchester/Massachusetts. ISBN 0-04-533004-1.
NORTHCUTT, J.C., ROBERTSON, C.J. & HANNAH, R.R. (1987) : State-of-the-art fracture stimulation of the Upper Morrow Formation in the Anadarko Basin. SPE Paper, **16913**, **18260** (1988).
NORTHROP, D.A. (1988) : Insights into natural gas production from low permeability reservoirs. SPE Paper, **17706**.
NORTHROP, D.A. & FROHNE, K.H. (1988) : Insights and contributions from the multiwell experiment : a field laboratory in tight sandstone reservoirs. SPE Paper, **18286**.
NORTHROP, D.A., SATTLER, A.R. & WESTHUSING, J.K. (1983) : Multi-well experiment : a field laboratory for tight gas sands. SPE/DOE Paper, **11646**.
NORTON, M.G. & HOFFMAN, S.J. (1982) : The use of foam in stimulating fractured California reservoirs. SPE Paper, **10769**.
NORTON-ALCOA PROPPANTS (1985) : Norton proppant products. Brochure; Norton-Alcoa Proppants, Dallas/Texas.
NORVIK, H. (1988) : The European gas market. Offshore Northern Seas Conference, Stavanger/Norway, 23. - 26.08.1988.
NOVOTNY, E.J. (1976) : Prediction of stimulation from acid fracturing treatments using finite fracture conductivity. SPE Paper, **6123**. J. Petrol. Technol., **29** (1977). SPE Reprint Series, Well Completions, **5 a**/II (1978) : 126 - 134.
NOVOTNY, E.J. (1977) : Proppant transport. SPE Paper, **6813**. SPE Reprint Series, Well Completions, **5 a**/II (1978) : 208 - 219.
NOVOTNY, R.J. & MATSON, R.P. (1975) : Laboratory observations of gravel placement techniques. SPE Paper, **5659**.
NUMBERE, D.T., JOSEPH, J. & ZOUKAGHE, M. (1985) : Transient pressure behavior in a low-permeability reservoir due to an irregular vertical fracture. SPE/DOE Paper, **13868**.
NURMI, R.D. (1985) : Eolian sandstone reservoirs : bedding facies and production geology modelling. SPE Paper, **14172**, **13048** (1984).
NURMI, R.D., CHARARA, M. & WATFA, M. (1987) : New techniques for the evaluation of Middle East fractured reservoirs. SPE Paper, **16807**.

NURMI, R., WATERHOUSE, M. & AKBAR, M. (1988) : Analysis and modeling of fractures in Middle East reservoirs. Amer. Assoc. Petrol. Geol., Annual Convention, Houston/Texas, 20. - 23.03.1988.

NYDEGGER, G.L., RICE, D.D. & BROWN, C.A. (1979) : Development of shallow gas reserves in low-permeability reservoirs of late Cretaceous age, Bowdoin Dome area, North-Central Montana. SPE Paper, **7945**. J. Petrol. Technol., **32** (1980) : 2111 - 2120.

NYSTAD, A.N. (1981) : Economic analysis of the North Sea oil and gas region. SPE Paper, **9566**. J. Petrol. Technol., **33** : 2515 - 2527.

ODD, R.A. (1982) : Economics of oil and gas production in North America : a benchmark for Europe. EUR Paper, **302**.

ODEH, A.S. (1976) : Pseudo-steady state flow capacity of wells with limited entry, and with an altered zone around the wellbore. SPE Paper, **6132**.

ODEH, A.S. & COHEN, M.F. (1988) : A proposed technique for simulation of viscous fingering in one-dimensional miscible flow. SPE Paper, **17116, 17087** (1987).

OEHME, W. (1986) : Energiewirtschaft im Wandel. Erdöl-Erdgas-Kohle, **102** : 430 - 432.

OEN, P.M., ENGELL-JENSEN, M. & BARENDREGT, A.A. (1988) : Skjold field, Danish North Sea : early evaluations of oil recovery through water imbibition in a fractured reservoir. SPE Reser. Engin., **3** : 17 - 22. SPE Paper, **15569** (1986).

OFFICE OF TECHNOLOGY ASSESSMENT (1985) : U.S. natural gas availability - gas supply through the year 2000. US. Congress Office of Technology Assessment Rept., **OTA-E-245** : 252 pp.

OFFSHORE ENGINEER (1984) : A big orange out to stimulate the North Sea oil juices. Offshore Engin., **1984**/8 : p. 47.

OFFSHORE ENGINEER (1985 a) : Oil and gas activity Northwest European continental shelf. Offshore Engin., **1985**/9 : 219 - 248.

OFFSHORE ENGINEER (1985 b) : Phillips examines Ekofisk waterflood extension. Offshore Engin., **1985**/9 : 92 - 94.

OFFSHORE ENGINEER (1985 c) : Oil and gas activity Northwest European continental shelf. Offshore Engin., **1985**/9 : 219 - 248.

OFFSHORE ENGINEER (1986 a) : V-block pipeline will cast wider net. Industry tightens belt as oil prices fall. Offshore Engin., **1986**/3 : 17 - 18.

OFFSHORE ENGINEER (1986 b) : Patient German probing brings scant offshore reward. Offshore Engin., **1986**/8 : 108 - 109.

OFFSHORE ENGINEER (1986 c) : Towards 2000 - detailing the developments. Offshore Engin., **1986**/1 : 38 - 52.

OFFSHORE ENGINEER (1986 d) : Britain offshore. Offshore Engin., **1986** : 26 - 28.

OFFSHORE ENGINEER (1986 e) : Morecambe`s winter warmer moves on a stage. Looking forward to forty years of peak shaving. Offshore Engin., **1986**/10 : 78 - 85.

OFFSHORE ENGINEER (1986 f) : Oil price best for every one at $ 15 - 30. Offshore Engin., **1986**/10 : p. 17.

OFFSHORE ENGINEER (1986 g) : Rospo Mare - Elf keeps faith with horizontal wells. Offshore Engin., **1986**/9 : 57 - 58.

OFFSHORE ENGINEER (1988) : North Sea drillers to break new ground. Offshore Engin., **1988**/4 : p. 14.

OFFSHORE OIL INTERNATIONAL (1988) : A 1988-89 recession would reduce the demand for oil say U.S. experts. Offshore Oil International, **1988**/4 : p. 9.

OGBE, D.O. & BRIGHAM, W.E. (1984 a) : A model for interference testing with wellbore storage and skin effects at both wells. SPE Paper, **13253**.

OGBE, D.O. & BRIGHAM, W.E. (1984 b) : Pulse testing with wellbore storage and skin effects. SPE Paper, **12780**. SPE Form. Eval., **2** (1987) : 29 - 42.

OHNO, K., NANBA, T. & HORNE, .N. (1985) : Analysis of an interwell tracer test in a depleted heavy oil reservoir. SPE Paper, **13672**.

OIL DAILY (1988 a) : Preserving market control seen Saudi`s major goal. Saudis rejected accord to keep market control. Oil Daily, 04.05.**1988** (OTC Edition) : 1 - 2.

OIL DAILY (1988 b) : Major U.K. North Sea projects gain approval. Oil Daily, 04.05.**1988** (OTC Edition) : 1, 9.

OIL DAILY (1988 c) : Consumer group claims steep gas reserves drop. Oil Daily, 04.05.**1988** (OTC Edition) : p. 10
OIL DAILY (1988 d) : Saudi cite effort to disrupt output pact. OPEC meetings in Vienna, major`s earnings captures spotlight. Oil Daily, 02.05.**1988** (OTC Edition) : 1 - 2.
OIL DAILY (1988 e) : Gasoline prices continue to rise. U.S. prices up despite OPEC indecision. Oil Daily, 02.05.**1988** (OTC Edition) : 1 - 2.
OIL DAILY (1988 f) : Opec maneuverings take the spotlight as oil prices continue drop. Oil Daily, 03.10.**1988** : p. 6.
OILFIELD PUBLICATIONS (1986) : The North Sea oil and gas activity and concession map 1986. Oilfield Publ., Ledbury/England.
OIL GAS JOURNAL (1965) : New "beads" help acidizing, fracturing. Oil Gas J., 20.08.**1965** : p. 52 ff.
OIL GAS JOURNAL (1984) : Annual drilling report. Oil Gas J., 01.10.**1984** : 103 - 161.
OIL GAS JOURNAL (1986) : Industry survey illustrates gain in U.S. well costs. Oil Gas J., 22./29.12.**1986** : p. 22.
OIL GAS JOURNAL (1987 a) : U.S.S.R. tries nuclear stimulation in East Siberia. Oil Gas J., 17.08.**1987** : p. 20.
OIL GAS JOURNAL (1987 b) : U.S.S.R. fails to produce big oil in Baltic Sea. Oil Gas J., 03.08.**1987** : p. 61.
OIL GAS JOURNAL (1987 c) : OPEC talks to center on third quarter output. Oil Gas J., 22.06.**1987** : p. 21.
OIL GAS JOURNAL (1987 d) : Soviets claim oil boost with nuclear explosions. Oil Gas J., 22.06.**1987** : p. 22.
OIL GAS JOURNAL (1987 e) : New oil market wild card : recession. Oil Gas J., 01.06.**1987** : 58 - 59.
OIL GAS JOURNAL (1987 f) : Fear of another 1986 may inspire restraint needed in oil market. Oil Gas J., 11.05.**1987** : p. 15.
OIL GAS JOURNAL (1987 g) : Oil price uncertainty catapults financial managers into top slots. Oil Gas J., 30.03.**1987** : 25 - 27.
OIL GAS JOURNAL (1987 h) : Oil price slide in 1986 trims Soviet trade value. Oil Gas J., 09.02.**1987** : p. 50.
OIL GAS JOURNAL (1987 i) : Controls eased on U.S. goods exports to Soviets. Oil Gas J., 26.01.**1987** : 29 - 30.
OIL GAS JOURNAL (1987 k) : New tool measures fracture direction, distance. Oil Gas J., 19.01.**1987** : 22 - 23.
OIL GAS JOURNAL (1987 l) : U.S.S.R. imports more oil and gas equipment. Oil Gas J., 31.08.**1987** : 15 - 16.
OIL GAS JOURNAL (1987 m) : Reserves hikes show OPEC still holds control of oil market. Oil Gas J., 28.12.**1987** : p. 13.
OIL GAS JOURNAL (1987 n) : OPEC`s hollow Vienna accord dashes hopes for oil price gains. Oil Gas J., 21.12.**1987** : p. 13.
OIL GAS JOURNAL (1987 o) : Western Siberia will be a key to Soviet plans to boost hydrocarbon production. Oil Gas J., 07.12.**1987** : 33 - 40.
OIL GAS JOURNAL (1987 p) : Relaxed restrictions improve prospects for U.S. sales, joint ventures in U.S.S.R. Oil Gas J., 07.12.**1987** : 42 - 46.
OIL GAS JOURNAL (1988 a) : Soviets` oil and gas production soars. Oil Gas J., 30.05.**1988** : 26 - 27.
OIL GAS JOURNAL (1988 b) : Amoco claims records in N. Sea high angle well. Oil Gas J., 11.07.**1988** : p. 21.
OIL GAS JOURNAL (1988 c) : New crude oil price drop need not to be disaster for U.S. drilling. Oil Gas J., 25.08.**1988** : p. 21.
OIL GAS JOURNAL (1988 d) : OGJ Newsletter. Oil Gas J., 03.10.**1988** : 4 - 6.
OIL GAS JOURNAL (1988 e) : Mixed oil market signals spur divergent analyses. Oil Gas J., 03.10.**1988** : 31 - 32.
OIL GAS JOURNAL (1988 f) : U.S.S.R. strives to hike gas exports. Oil Gas J., 20.06.**1988** : 24 - 25.
OIL GAS JOURNAL (1988 g) : Non-OPEC alliance attempt falls short. Oil Gas J., 09.05.**1988** : p. 20.
OILMAN (1985 a) : A record year for drilling. Oilman, **1985**/2 : 32 - 35.
OILMAN (1985 b) : Stimulation in demand. Oilman, **1985**/9 : p. 28.

OILMAN (1987 a) : Shadows sour Southern Basin gas. Oilman, **1987**/5 : 20 - 22.
OILMAN (1987 b) : Complex trade routes to the USSR. Oilman, **1987**/6 : p. 33.
OILMAN (1987 c) : E. Europe : built on the Soviet model. Oilman, **1987**/6 : 31 - 32.
OILMAN (1987 d) : Amoco`s Valhall sinks into chalk alongside neighbouring Ekofisk. Oilman, **1987**/1 : p. 10.
OILMAN (1987 e) : Sandstone`s hidden treasures. Leading consultants are carrying out evaluation of Upper Carboniferous prospects. Oilman, **1987**/6 : p. 50.
OLAREWAJU, J.S. & LEE, W.J. (1988) : A study of pressure behavior of layered and dual-porosity reservoirs in the presence of skin, wellbore storage, and phase segregation. SPE Paper, **17302**.
OLESEN, J.R., MAHDAVI, M., STEINMAN, D.K. & YVER, J.P. (1987) : Dual-burst thermal decay time logging overview and examples. SPE Paper, **15716**.
OLIVE, C. (1988) : Fracture orientation determination in the Sandhills (McKnight) field, Crane County, Texas. Amer. Assoc. Petrol. Geol., Annual Convention, Houston/Texas, 20. - 23.03.1988.
OLIVER, D.S. (1984) : Pressure transients caused by fracturing. SPE Paper, **12735**.
OLIVER, D.W. & FERTL, W.H. (1986) : Recent advances in pulsed neutron logging. CIM Paper, **86-37-19**.
OLIVER, D.W., WYATT, D.F. & SMITH, H.D. (1986) : Reservoir monitoring with the thermal multigate decay log. SPE Paper, **14137**.
OLSEN, T.N. & DEBONIS, P. (1988) : Real time optimization of fracturing fluid viscosity to meet changing fracture conditions. SPE Paper, **17540**.
ONUR, M., REYNOLDS, A.C. & RAGHAVAN, R. (1986) : Interference testing in commingled multilayered reservoirs. SPE Paper, **15132**.
ONUR, M., SERRA, K.V. & REYNOLDS, A.C. (1988) : Analysis of pressure buildup data obtained at a well located in a multi-well system. SPE Paper, **18123**.
ONYIA, E.C. (1987) : Geology drilling log - a computer database system for drilling simulation. SPE Drill. Engin., **2** : 27 - 36.
ONYIA, E.C. (1988) : Relationships between formation strength, drilling strength, and electric log properties. SPE Paper, **18166**.
OPDAL, S.T. & KOSSACK, C.A. (1988) : Recovery of condensate from a heterogeneous reservoir by the injection of a slug of methane followed by nitrogen. SPE Paper, **18265**.
ORASIANU, T.O. (1986) : Romania : analysis of recent offshore operations. Oil Gas J., **84**/7 : 94 - 95.
ORASIANU, T.O. (1987) : Soviet oil industry is set new targets. Oilman, **1987**/6 : 28 - 30.
ORTIZ, J. & McLANE, J.E. (1984) : Low-pH methanol : an alternative for stimulation in water-sensitive tight, dirty sandstones. SPE Paper, **12502**. SPE Product. Engin., **1** (1986) : 195 - 202.
OSBORNE, M.W., McLEOD, H.O. & SCHROEDER, H.D. (1981) : The analysis and control of hydraulic fracturing problems. SPE/DOE Paper, **9868**.
O`SHEA, P. & MURPHY, W.O. (1982) : GRI program for tight gas sands research. SPE/DOE Paper, **10808**.
O`SHEA, P., ROSENBERG, J.I. & BRASHEAR, J.P. (1982) : An analytical review of the National Petroleum Council Study of tight gas reservoirs. SPE Paper, **11256**.
OSMAN, M.E. (1987) : Worldwide oil and gas supply and demand. SPE Paper, **16292**.
OSTENSEN, R.W. (1983) : Microcrack permeability in tight gas sandstone. Soc. Petrol. Engin. J., **23** : 919 - 927. SPE Paper, **10924** (1982).
O`SULLIVAN, M.J., BODVARSSON, G.S., PRUESS, K. & BLAKELEY. M.R. (1985) : Fluid and heat flow in gas-rich geothermal reservoirs. Soc. Petrol. Engin. J., **25** : 215 - 226. SPE Paper, **12102** (1983).
OTERO, C., SAEZ, A.E. & RUSINEK, I. (1987) : Effective permeabilities for heterogeneous reservoirs. 4th Europ. Symp. Enhanced Oil Recovery, Poster, Hamburg/Germany FRG.
OTUOMAGIE, R. & MENZIE, D.E. (1976) : Analysis of pressure buildup in an infinite two-layered oil reservoir without crossflow. SPE Paper, **6130**.
OTTO, S.B. (1986) : Low crude prices could trigger serious gas shortages. World Oil, **1986**/9 : 25 - 29.

OUSTERHOUT, R.S. & HALL, C.D. (1961) : Reduction of friction loss in fracturing operations. J. Petrol. Technol., **13** : 217 - 222.
OVERBEY, W.K., YOST, A.B. II & WILKINS, D.A. (1988) : Inducing multiple hydraulic fractures from a horizontal wellbore. SPE Paper, **18249**.
OVERBEY, W.K., YOST, L.E. & YOST, A.B. II (1988) : Analysis of natural fractures observed by borehole video camera in a horizontal well. SPE Paper, **17760**.
OWEN, L.B. (1982) : Fracture orientation analysis by the solid earth tidal strain method. SPE Paper, **11070**.
OWEN, L.B., TORONTO, T.W. & PETERSON, R.E. (1988) : Reliability of anelastic strain recovery estimates for stress orientation in the Travis Peak Formation, Harrison County, Texas. SPE Paper, **18165**.
OWENS, W.W. & ARCHER, D.L. (1971) : The effect of rock wettability on oil-water relative permeability relationships. J. Petrol. Technol., **23** : 873 - 878. SPE Paper, **3034** (1970).
OWUSU, L.A. & ERSHAGHI, I. (1986) : Estimation of reserves and recovery performance in highly complex naturally fractured reservoirs. SPE Paper, **15935**.
OYENEYIN, M.B. (1985) : Formation sand size prediction from well-log correlations. SPE Paper, **16245**.
OYENEYIN, M.B. (1987 a) : Computer programs help pick best gravelpack design. Oil Gas J., 02.03.**1987**.
OYENEYIN, M.B. (1987 b) : Formation sand size prediction from well-log correlations. SPE Paper, **16245**.
OYENEYIN, M.B. (1988) : Numerical analysis of the effects of gravel packing on gas well productivity. SPE Paper, **17170**.
OYENEYIN, M.B. & PEDEN, J.M. (1985) : Factors to consider in the effective gravel packing of deviated wells. SPE Paper, **13916**.
OZKAN, E., OHAERI, C.U. & RAGHAVAN, R. (1987) : Unsteady flow to a gas well produced at a constant pressure in a fractured reservoir. SPE Form. Eval., **2** : 186 - 200. SPE Paper, **9902** (1981).
OZKAN, E. & RAGHAVAN, R. (1988) : Performance of horizontal wells subject to bottom water drive. SPE Paper, **18545**.
OZKAN, E., RAGHAVAN, R. & JOSHI, S.D. (1987) : Horizontal well pressure analysis. SPE Paper, **16378**.
OZKAYA, I. (1984) : Computer simulation of hydraulic fracturing in shales - influences on primary migration. J. Petrol. Technol., **36** : 826 - 828. SPE Paper, **11455** (1983).
OZKAYA, I. (1986) : Analysis of natural hydraulic fracturing of shales during sedimentation. SPE Product. Engin., **1** : 191 - 194. SPE Paper, **13343** (1984).
OZKAYA, I. (1987) : Geological and petrophysical factors that control excess field pressures of isolated sand lenses within shales. SPE Paper, **15746**.
PABLEY, A.S., EWING, B.C. & CALLAWAY, R.E. (1982) : Performance of crosslinked hydrochloric acid in the Rocky Mountains region. SPE Paper, **10877**.
PABLEY, A.S. & HOLCOMB, D.L. (1980) : A new stimulation technique : high strength cross linked acid. SPE Paper, **9241**.
PABLEY, A.S. & HOLCOMB, D.L. (1981) : Crosslinked acid gels offer advantages. Oil Gas J., 28.09.**1981** : 286 - 291.
PABLEY, A.S. & HOLCOMB, D.L. (1982) : Crosslinked high-strength hydrochloric acid gel system - Rocky Mountain case histories. J. Petrol. Technol., **34** : 2080 - 2086. SPE Paper, **10622**.
PACCALONI, G. (1979) : Matrix stimulation planning. Oil Gas J., 09.11.**1979** : 155 - 160; 26.11.**1979** : 61 - 65.
PAI, V.J. (1982) : Predicting, preventing and remedying hydraulic fracturing screen outs. Texas Techn. Univ. Short Course, 21. - 22.04.1982.
PAI, V.J. & GARBIS, S.J. (1983 a) : MHF treatment design. Oil Gas J., 25.07.**1983** : 135 - 140; 08.08.**1983** : 126 - 128; 22.08.**1983** : 144 - 146; 05.09.**1983** : 140 - 142.
PAI, V.J. & GARBIS, S.J. (1983 b) : Review of the completion practices in the Morrow Formation in Eddy, Chaves and Lea Counties of Southeast New Mexico. Proc. 30th Southwestern Petroleum Short Course, Texas Techn. Univ., Lubbock/Texas, 27. - 28.04.1983, 153 - 159. SPE Paper, **1335** (1982).
PAI, V.J., GARBIS, S.J. & HALL, H. (1983) : MHF - from concept to execution.

Proc. 30th Annual Southwestern Petroleum Short Course, Texas Techn. Univ., Lubbock/Texas, 27. - 28.04.1983, 100 - 121.

PAILLET, F.L. (1980) : Acoustic propagation in the vicinity of fractures which intersect a fluid-filled borehole. Trans. 21st Soc. Prof. Well Log Analysts Annual Logging Symp., Paper **DD**, Lafayette/Louisiana, 08. - 11.07.1980.

PAILLET, F.L. (1981) : A comparison of fracture characterization techniques applied to near-vertical fractures in a limestone reservoir. Trans. 22nd Soc. Prof. Well Log Analysts Annual Logging Symp., Paper **XX**, Mexico City/Mexico, 23. - 26.06.1981.

PAILLET, F.L. (1983 a) : Acoustic characterization of fracture permeability at Chalk River, Ontario, Canada. Canad. Geotechn. J., **20** : 468 - 476.

PAILLET, F.L. (1983 b) : Acoustic character of hydraulic fractures in granite. Proc. 24th U.S. Rock Mechan. Symp., 327 - 334, College Station/Texas.

PAILLET, F.L. (1985 a) : Applications of borehole-acoustic methods in rock mechanics. Proc. 26th U.S. Rock Mechan. Symp., **1** : 207 - 220; Balkema, Boston.

PAILLET, F.L. (1985 b) : Problems in fractured reservoir evaluation and possible routes to their solutions. Log Analyst, **26**/6 : 26 - 41.

PAILLET, F.L. (1987) : Experimental verification of acoustic waveform and VSP tube wave measurements of fracture permeability. Trans. 28th Soc. Prof. Well Log Analysts Annual Logging Symp., London/England, 29.06. - 02.07.1987.

PAILLET, F.L., CHENG, C.H., HESS, A.E. & HARDIN, E.L. (1986) : Comparison of fracture permeability estimates based on tube-wave generation in vertical seismic profiles, acoustic waveform-log attenuation, and pumping-test analysis. National Water Well Assoc. Conference and Exposition, Denver/Colorado, 398 - 416.

PAILLET, F.L. & HESS, A.E. (1986) : Geophysical well log analysis of fractured crystalline rocks at East Bull Lake, Ontario, Canada. U.S. Geol. Surv., Water Resour. Invest. Rept., **86-4052** : 37 pp.

PAILLET, F.L., HESS, A.E., CHENG, C.H. & HARDIN, E.L. (1987) : Characterization of fracture permeability with high-resolution vertical flow measurements during borehole pumping. Ground Water, **25** : 28 - 40.

PAILLET, F., HSIEH, P. & CHENG, C.H. (1987) : Experimental verification of acoustic waveform and vertical seismic profile measurements of fracture permeability. Trans. 28th Soc. Prof. Well Log Analysts Annual Logging Symp., Paper **OO**, London/England, 29.06. - 02.07.1987.

PAILLET, F.L., KEYS, W.S. & HESS, A.E. (1985) : Effects of lithology on acoustic televiewer-log quality and fracture interpretation. Trans. 26th Soc. Prof. Well Log Analysts Annual Logging Symp., Paper **JJJ**, Dallas/Texas, 17. - 20.06.1985.

PAILLET, F.L. & KIM, K. (1985) : Distribution and acoustic properties of borehole wall breakouts in deeply buried basalt. J. Geophys. Res., **1985**.

PAILLET, F.L. & WHITE, J.E. (1982) : Acoustic modes of propagation in the borehole and their relationship to rock properties. Geophysics, **47** : 1215 - 1228.

PALLATT, N., WILSON, J. & McHARDY, B. (1984) : The relationship between permeability and the morphology of diagenetic illite in reservoir rocks. J. Petrol. Technol., **36** : 2225 - 2227. SPE Paper, **12788** (1983).

PALMER, I.D. & CARROLL, H.B. (1981) : A stochastic model of fracture propagation : initial results. SPE/DOE Paper, **9880**.

PALMER, I.D. & CARROLL, H.B. (1982) : Three-dimensional hydraulic fracture propagation in the presence of stress variations. SPE/DOE Paper, **10849**. Soc. Petrol. Engin. J., **23** (1983) : 870 - 878.

PALMER, I.D. & CARROLL, H.B. (1983) : Numerical solution for height of elongated hydraulic fractures with leakoff. SPE/DOE Paper, **11627**.

PALMER, I.D. & CRAIG, H.R. (1984) : Modeling of asymmetric vertical growth in elongated hydraulic fractures and application to first MWX stimulation. SPE/DOE/GRI Paper, **12879**.

PALMER, I.D. & LUISKUTTY, C.T. (1985 a) : A model of the hydraulic fracturing process for elongated vertical fractures and comparison of results with other models. A comparison of highly elongated fracture models of variable heights. SPE/DOE Paper, **13864**.

PALMER, I.D. & LUISKUTTY, C.T. (1985 b) : Comparison of hydraulic fracture models for highly elongated fractures of variable height. J. Energy Res. Technol., **1985**.
PALMER, I.D. & VEATCH, R.W. (1987) : Abnormally high fracturing pressures in step-rate tests. SPE Paper, **16902**.
PALMER, I.D. & VEATCH, R.W. (1988) : Uplifts and tilts at earth`s surface induced by pressure transients associated with five hydraulic fracture treatments. SPE Paper, **18538**.
PALUMBO, S., GIACCA, D., ZUCCA, G., FERRARI, M., GRAZIOSI, M. & PIROVANO, P. (1989) : The development of potassium cellulosic polymers and their contribution to the inhibition of hydratable clays. SPE Paper, **18477**.
PAPADOPOULOS, J.M., NARENDRAN, V.M. & CLEARY, M.P. (1983) : Laboratory simulations of hydraulic fracturing. SPE/DOE Paper, **11618**.
PARKER, C.D. (1980) : MHF problems. SPE Paper, **9066**.
PARKER, C.D. (1981) : Logistic and operational considerations for massive hydraulic fracturing. J. Petrol. Technol., **33** : 1189 - 1194. SPE Paper, **9066**.
PARKER, M.A., ADAMS, R.W. & LIANKUI, L. (1986) : Stimulation studies of Chinese formations. SPE Paper, **14091**.
PARKER, M.A. & McDANIEL, B.W. (1987) : Fracturing treatment design improved by conductivity measurements under in-situ conditions. SPE Paper, **16901**.
PARMESWEAR, R. & MAEREFAT, N.L. (1986) : A comparison of methods for the representation of three-phase relative permeability data. SPE Paper, **15061**.
PARRISH, R.L., STEVENS, A.L. & TURNER, T.F. (1981) : A true in-situ fracturing experiment - final results. J. Petrol. Technol., **33** : 1297 - 1304.
PARROT, D.I. & LONG, M.G. (1979) : A case history of massive hydraulic refracturing in the tight Muddy "J" Formation. SPE Paper, **7936**.
PARSONS, R.S. & FINCHER, R.W. (1987) : Short-radius lateral drilling : a completion alternative. SPE Paper, **15943**.
PASINI, J. P. III & OVERBEY, W.K. (1969) : Natural and induced fracture systems and their application to petroleum production. SPE Paper, **2565**.
PASTERNACK, E.S. & GOODWILL, W.P. (1983) : Applications of digital borehole televiewer logging. Trans. 24th Soc. Prof. Well Log Analysts Annual Logging Symp., Paper **X**, Calgary/Alberta, 27. - 30.06.1983.
PATTILLO, P.D. & SMITH, M.B. (1982) : The effect of formation-flow on the integrity of perforated casing. SPE Paper, **11123**.
PAULS, R.W., VENDITTO, J., CHISHOLM, P., HOLTMYER, M. & GREGORCZYK, W. (1985) : Successful stimulation of the Olmos Formation using oil-base fluids and high-proppant concentrations. SPE Paper, **13817**.
PAULUS, C.A. (1987) : Leveraged buyouts and the petroleum industry. SPE Paper, **16301**.
PEACEMAN, D.W. (1976) : Convection in fractured reservoirs - numerical calculation of convection in a vertical fissure, including the effect of matrix-fissure transfer. SPE Paper, **5523** (1975).
PEACEMAN, D.W. (1983) : Interpretation of well-block pressures in numerical reservoir simulation with nonsquare grid blocks and anisotropic permeability. Soc. Petrol. Engin. J., **23** : 531 - 543. SPE Paper, **10528** (1982).
PEAKE, W.T. (1984) : Hydraulic fracture and steam stimulation of a diatomite reservoir. SPE Paper, **12737**.
PEARCE, K.W. (1983) : Application of high strength proppants in relatively shallow and hard formations. Proc. 30th Southwestern Petroleum Short Course, Texas Techn. Univ., Lubbock/Texas, 27. - 28.04.1983, pp. 153 - 159.
PEARCE, R.M. (1979) : Evaluation of fracture treatments using tracer and temperature surveys. SPE Paper, **7910**.
PEARSON, C.F. (1981) : The relationship between microseismicity and high pore pressures during hydraulic fracturing stimulation experiments in low permeability granitic rocks. J. Geophys. Res., **86** (B 9) : 7855 - 7864.
PEARSON, C.M. (1987) : Development and application of an operator`s stimulation monitoring system. SPE Paper, **16903**.
PEARSON, C.M., ABOU-SAYED, I.S. & NGUYEN, N.B. (1985) : Parameters affecting fracture design and their measurement during treatments. SPE Paper, **13798**.
PEARSON, C.M., LYNCH, K.W., SCHMIDT, J.H. & McCASLIN, N.F. (1988) : Improvement of massive hydraulic fracture treatments in the Cotton Valley Formation of

East Texas. SPE Paper, **17714**.
PEDEN, J.M. & LUO, Y. (1987) : Settling velocity of variously shaped particles in drilling and fracturing fluids. SPE Drill. Engin., **2** : 337 - 343. SPE Paper, **16243** (1986).
PEDEN, J.M., OYENEYIN, M.B. & RUSSELL, J. (1982) : Prediction model for optimum gravel packing efficiency in highly deviated wells. EUR Paper, **312**. SPE Paper, **12806** (1984).
PEDEN, J.M., OYENEYIN, M.B., RUSSELL, J., LIU, Z. & GUITARD, F.A. (1982 a) : Laboratory studies on the effectiveness of gravel packing perforations and screen-casing annulus in highly deviated wells. EUR Paper, **311**. SPE Paper, **12805** (1984).
PEDEN, J.M., RUSSELL, J. & OYENEYIN, M.B. (1984) : A numerical approach to the design of a gravel pack for effective sand control in deviated wells. SPE Paper, **13084**.
PEDEN, J.M., RUSSELL, J. & OYENEYIN, M.B. (1985) : Design of an effective gravel pack for sand control : a numerical approach. SPE Paper, **13647**.
PEDEN, J.M. & YASSIN, A.A.M. (1986 a) : The effect of perforation conditions upon maximum sand-free production rate in friable conditions. 6th Offshore Southeast Asia Conference and Exhibition, 368 - 378, Singapore, 26. - 31.01. 1986.
PEDEN, J.M. & YASSIN, A.A.M. (1986 b) : The determination of optimum completion and production conditions for sand free oil production. SPE Paper, **15406**.
PEETERS, M. & HARTLEY, R. (1984) : Induced fracture height detection from wireline logs. Trans. 25th Soc. Prof. Well Log Analysts Annual Logging Symp., Paper **GGG**, New Orleans/Louisiana, 10. - 13.06.1984.
PEKOT, L.J. & GERSIB, G.A. (1987) : Ekofisk. In : SPENCER, A.M. et al. (Editors), Geology of the Norwegian oil and gas fields, 73 - 88; Graham & Trotman, London/England.
PELLEY, W. (1982) : Banking, energy, and the Federal role. SPE Paper, **11268**.
PELLEY, W.E. & CONSTABLE, R.W. (1982) : Evaluation of reservoir risks and economic assumptions in bank lending. SPE Paper, **11100**.
PEMPER, R.R., FLECKER, M.J., McWHIRTER, V.C. & OLIVER, D.W. (1988) : Hydraulic fracture evaluation with multiple radioactive tracers. Geophysics, **53**.
PENBERTHY, W.L. (1985) : Gravel placement through perforations and perforation cleaning for gravel packing. SPE Paper, **14161**. J. Petrol. Technol., **40** (1988) : 229 - 236.
PENBERTHY, W.L. & COPE, B.J. (1979) : Design and productivity of gravel-packed completions. SPE Paper, **8428**. J. Petrol. Technol., **32** (1980) : 1670 - 1686.
PENBERTHY, W.L., SHAUGHNESSY, C.M., GRUESBECK, C. & SALATHIEL, W.M. (1978) : Sand consolidation pre-flush dynamics. J. Petrol. Technol., **30** : 845 - 850. SPE Paper, **6804** (1977).
PENCE, S.A. (1975) : Evaluating formation damage in low permeability sandstone. SPE Paper, **5638**.
PENDLETON, L.E. & RAMESH, B. (1985) : An innovative method of drilling horizontal boreholes. Heavy oil and oil sands, Techn. Symp., Calgary/Alberta, 20.02.1985.
PENG, C.P., YANOSIK, J.L. & STEPHENSON, R.E. (1987) : A generalized compositional model for naturally fractured reservoirs. SPE Paper, **16340**.
PENNEBAKER, E.S. & WOODY, R.T. (1977) : The temperature-sound log and borehole density scans for problem wells. SPE Paper, **6782**.
PENNY, G.S. (1982) : Non-damaging fluid-loss additives for use in hydraulic fracturing of gas wells. SPE Paper, **10659**.
PENNY, G.S. (1987) : An evaluation of the effects of environmental conditions and fracturing fluids upon the long-term conductivity of proppants. SPE Paper, **16900**.
PENNY, G.S., CONWAY, M.W. & LEE, W. (1985) : Control and modelling of fluid leakoff during hydraulic fracturing. J. Petrol. Technol., **37** : 1071 - 1080. SPE Paper, **12486** (1984).
PENNY, G.S. & FORD, W.G. (1985) : Influence of downhole conditions on the leakoff properties of fracturing fluids. SPE Paper, **14016**. SPE Product. Engin., **3** (1988) : 43 - 51.
PENNY, G.S., SOLIMAN, M.Y., CONWAY, M.W. & BRISCOE, J.E. (1983) : Enhanced load

water-recovery technique improves stimulation results. SPE Paper, **12149**.
PENNY, G.S., THOMAS, B.D. & SOLIMAN, M.Y. (1983) : Enhanced load water recovery technique improves stimulation results in the Appalachian area. SPE Paper, **12314**.
PEREZ, G. & KELKAR, B.G. (1988) : A new method to predict two-phase pressure drop across perforations. SPE Paper, **18248**.
PERKINS, C.A. (1988) : Oil price instability coulds tubular outlook. World Oil, **207** (1988/10) : 82 - 84.
PERKINS, T.K. & GONZALEZ, J.A. (1984) : Changes in earth stresses around a wellbore caused by radially symmetrical pressure and temperature gradients. Soc. Petrol. Engin. J., **24** : 129 - 140. SPE Paper, **10080** (1981).
PERKINS, T.K. & GONZALEZ, J.A. (1985) : The effect of thermoelastic stresses on injection well fracturing. Soc. Petrol. Engin. J., **25** : 78 - 88. SPE Paper, **11332** (1982).
PERKINS, T.K. & KERN, L.R. (1961) : Widths of hydraulic fractures. J. Petrol. Technol., **13** : 937 - 949. SPE Reprint Series, Well Completion, **5** (1970) : 265 - 277.
PERKINS, T.K. & KRECH, W.W. (1968) : The energy balance concept of hydraulic fracturing. Soc. Petrol. Engin. J., **8** : 1 - 12.
PERKINS, T.K. & WEINGARTEN, J.S. (1988) : Stability and failure of spherical cavities in unconsolidated sand and weakly consolidated rock. SPE Paper, **18244**.
PERLMAN, W. (1986) : Coal seam fracing method. U.S. Patent, **4,566,539** (28.01. 1986). Canad. Patent, **1,222,942** (16.06.1987).
PERRIER, R. & QUIBLIER, J. (1978) : Orientation des fractures observés sur carottes à l`aide de la pentagemétrie. Rev. Inst. Franç. Pétrol., **28** : 963 - 974.
PERRY, G. & SMITH, G. (1980) : Natural completions. Overcoming the damage caused by drilling and perforating. Drilling-DCW, 01.04.1980.
PERTHUIS, H. (1989) : Weak or slowly generating HF solutions : which choice for sandstone acidizing? SPE Paper, **18469**.
PESLJAK, J.A. (1985) : Calculation of circular cracks in hydraulic fracturing. In : Dynamics of multi-phase media, **34** : 188 - 193 (in Russian); Novosibirsk.
PETAK, K.R., SOLIMAN, M.Y. & ANDERSON, M.F. (1986) : Type curves for analyzing naturally fractured reservoirs. SPE Paper, **15638**.
PETERS, F.W. & STOUT, C.M. (1977) : Clay stabilization during fracturing treatments with hydrolyzable zirconium salts. J. Petrol. Technol., **29** : 187 - 194. SPE Paper, **5687** (1975).
PETERSON, R.E. & KOHOUT, J. (1983) : An approximation of continuity of lenticular Mesaverde sandstone lenses utilizing close-well correlations, Piceance Basin, Northwestern Colorado. SPE/DOE Paper, **11610**.
PETHRICK, W.D., SENNHAUSER, E.S. & HARDING, T.G. (1986) : Numerical modelling optimization of cyclic steam stimulation in Cold Lake oil sands. CIM Paper, **86-37-21**.
PETRIE, T.A. (1983) : Recent petroleum mergers and acquisitions : an oil analyst`s perspective. SPE Paper, **11663**.
PETROLEUM ENGINEER (1987) : Ekofisk lift is second heaviest on record. Petrol. Engin. Intern., **1987**/8 : 10 - 14.
PETROLEUM ENGINEER (1988) : Conoco would double horizontal drilling record. Petrol. Engin. Intern., **1988**/5 : p. 10. Oil Gas J., 21.03.**1988** : p. 102.
PETROLEUM REVIEW (1987) : The pressures on the $ and OPEC oil. Petrol. Rev., **1987**/12 : 10 - 16.
PETROLEUM REVIEW (1988 a) : Sheikh Yamani suggests a dialogue on price. Petrol. Rev., **1988**/4 : p. 42.
PETROLEUM REVIEW (1988 b) : North Sea activity on rising trend. Petrol. Rev., **1988**/2 : p. 44.
PETROLEUM REVIEW (1988 c) : Regulate to avert a 4th price disaster. Petrol. Rev., **1988**/2 : 13 - 14.
PETROLEUM REVIEW (1988 d) : Conoco takes an optimistic view on the future of Southern Basin gas. Petrol. Rev., **1988**/10 : 4 - 6.
PETROLEUM REVIEW (1988 e) : First lateral well in North Sea completed. Petrol.

Rev., **1988**/10 : p. 8.
PETRYK, R.P. & GARUK, B.W. (1979) : Fracture acidizing with foamed acid : success in the D-1 carbonate. CIM Paper, **79-30-37**.
PETZET, G.A. (1982) : High sand fracs hike output in Oklahoma wells. Oil Gas J., 11.10.**1982** : p. 90.
PETZET, G.A. (1987) : Canadians combine steam drive, horizontal well. Oil Gas J., 01.06.**1987** : p. 22.
PETZET, G.A. (1988) : Operators make wide use of horizontal drilling technology. Oil Gas J., 11.04.**1988** : 15 - 17.
PHILIPP, W. & REINICKE, K.M. (1982) : Zur Entstehung und Erschließung der Gasprovinz Osthannover. Erdoel-Erdgas, **98** : 85 - 90.
PHILLIPS, A.M. & ANDERSON, R.W. (1985) : Use of proppant selection models to optimize fracturing treatment designs in low-permeability reservoirs. SPE/DOE Paper, **13855**.
PHILLIPS, A.M., COUCHMAN, D.D. & WILKE, J.G. (1987) : Successful field application of high-temperature rheology of CO2 foam fracturing fluids. SPE/DOE Paper, **16416**.
PHILLIPS, A.M. & WILSON, W.J. (1984 a) : Improved drainage of sand pack enhances fracturing fluid recovery and increases production. SPE Paper, **12924**.
PHILLIPS, A.M. & WILSON, W.J. (1984 b) : New stimulation technique improves frac fluid recovery and increases production. SPE Paper, **13374**.
PHILLIPS, A.M. & WILSON, W.J. (1985) : Optimization of fracture fluid recovery and production through the use of a new stimulation technique. SPE Paper, **13815**.
PICKENS, T.B. (1983) : Oil and gas industry : challenges for the 1980`s. SPE Paper, **11666**.
PICKETT, G.R. & REYNOLDS, E.B. (1969) : Evaluation of fractured reservoirs. Soc. Petrol. Engin. J., **9** : 28 - 38. SPE Paper, **1866** (1967).
PIERCE, A.E., VELA, S. & KOONCE, K.T. (1974) : Determination of the compass orientation and length of hydraulic fractures by pulse testing. SPE Paper, **5132**.
PIERCE, D., CROCKER, C., KLEMENTICH, E. & McFATHRIDGE, D. (1983) : Hydraulic fracturing the deep Smackover Formation of the Johns field, Rankin county, MS. SPE Paper, **11933**.
PILEHVARI, A. & CLARK, P.E. (1985) : Rheology of hydraulic fracturing fluids : wall slip during viscosity measurement. J. Petrol. Technol., **37** : 1840 - 1846. SPE Paper, **12029** (1983).
PINE, R.J. (1983) : Pressure transient analysis for large scale hydraulic injections in the Carnmenellis granite, England. Proc. 9th Workshop Geothermal Reserv. Engin., Stanford Univ., Stanford/California, 12.1983, pp. 1 - 9.
PINE, R.J. (1987) : The design and observed effects of medium viscosity, large volume fluid injection for HDR development in jointed rock. Workshop on forced fluid flow through strong fractured rock masses, Garchy (Nievre)/France, 12. - 15.04.1987, pp. 1 - 7.
PINE, R.J., BARIA, R., PEARSON, R.A., KWAKWA, K. & McCARTNEY, R. (1986) : A technical summry of phase 2 B of the Camborne School of Mines HDR project, 1983 - 1986. EEC/US Workshop on hot dry rock geothermal energy, Brussels/Belgium, 05.1986.
PINE, R.J. & BATCHELOR, A.S. (1982) : In-situ stresses and jointing in the Carnmenellis granite and the implications for hydraulic behaviour. Camborne School Min. J., **1982**.
PINE, R.J. & BATCHELOR, A.S. (1984) : Downward growth of hydraulic stimulation by shearing in jointed rock. Intern. J. Rock Mechan., Miner. Sci., Geomechan. Abstr., **21**/5 : 249 - 263.
PINE, R.J. & LEDINGHAM, P. (1983) : In-situ hydraulic parameters for the Carnmenellis granite hot dry rock geothermal energy research reservoir. SPE Paper, **12020** (1983). J. Petrol. Technol., **36** (1984) : 1982 - 1990.
PIRSON, S.J. (1983) : Geologic well log analysis. 3rd edition, 476 pp.; Gulf Publ., Houston/Texas.
PIRSON, S.J. (1987) : How to map fracture development from well logs. World Oil, **1967**/3 : 106 - 114.
PITMAN, J.K., ANDERS, D.E., FOUCH, T.D. & NICHOLS, D.J. (1986) : Hydrocarbon

potential of nonmarine Upper Cretaceous and Lower Tertiary rocks, Eastern Uinta Basin, Utah. In : SPENCER, C.W. & MAST, R.F. (Editors), Geology of tight gas reservoirs. Amer. Assoc. Petrol. Geol., Stud. Geol., **24** : 235 - 252.

PITTMAN, E.D. (1979) : Porosity, diagenesis and productive capability of sandstone reservoirs. In : SCHOLLE, P.A. & SCHLUGER, P.R. (Editors), Aspects of diagenesis. Soc. Econ. Paleont. Miner., Spec. Publ., **26** : 159 - 173.

PITTMAN, E.D. & THOMAS, J.B. (1979) : Some applications of scanning electron microscopy to the study of reservoir rock. J. Petrol. Technol., **31** : 1375 - 1380. SPE Paper, **7550** (1978).

PITTS, J.P. (1988) : Largest frac proves successful. Mobil taps Wilcox Sand. Newpaper article.

PLAHN, S.V., STARTZMAN, R.A. & WATTENBARGER, R.A. (1987) : A method for predicting horizontal well performance in solution-gas-drive reservoirs. SPE Paper, **16201**.

PLEIN, E. (1975) : Der deutsche Nordsee-Schelf aus geologischer Sicht. Erdoel-Erdgas, **91** : p. 233.

PLEIN, E. (1978) : Rotliegend-Ablagerungen im Norddeutschen Becken. Z. dt. geol. Ges. **129** : 71 - 97.

PLUMB, R.A. (1988) : Fracture characterization from well logs. Amer. Assoc. Petrol. Geol., Annual Convention, Houston/Texas, 20. - 23.03.1988.

PLUMB, R.A., BRIE, A. & HSU, K. (1985) : Fracture detection and evaluation using new wireline methods. Proc. 26th U.S. Symp. Rock Mechan., Rapid City/South Dakota, p. 227 ff.

PLUMB, R.A. & HICKMAN, S.H. (1985 a) : Stress induced borehole elongation : a comparison between the four-arm dipmeter and the borehole televiewer in the Auburn geothermal well. J. Geophys. Res., **90** : 5513 - 5521.

PLUMB, R.A. & HICKMAN, S.H. (1985 b) : The correspondence between stress induced borehole elongation determined by the four-arm dipmeter and the Auburn geothermal well. J. Geophys. Res., **90**.

PLUMB, R.A. & HORNBY, B. (1988) : New fracture characterization methods : example integrating wellbore images and borehole Stoneley wave analysis. SPE Paper, **18374**.

PLUMB, R.A. & LUTHI, S.M. (1986) : Application of borehole images to geological modelling of an eolian reservoir. SPE Paper, **15487**.

POE, B.D. & JENNINGS, J.W. (1988) : A review of the methods of estimating stabilized absolute open-flow potential of gas wells. SPE Paper, **17308**.

POINTE, P.R.la, BELFIELD, W.C. & HELWIG, J.A. (1984) : Analysis of fracturing and fluid-flow characteristics of the Monterey Formation, Santa Barbara Channel, CA. SPE Paper, **12734**.

POLIKAR, M., ALI, S.M.F. & PUTTAGUNTA, V.R. (1988) : High-temperature relative permeabilities for Athabasca oil sands. SPE Paper, **17424**.

POLLARD, D.D. (1976) : On the form and stability of open hydraulic fractures in the earth´s crust. Geophys. Res. Letters, **3** : 513 - 516.

POLLARD, D.D. & HOLZHAUSEN, G.R. (1979) : On the mechanical interaction between a fluid-filled fracture and the earth´s surface. Tectonophys., **53** : 27 - 57.

POLLASTRO, R.M. & SCHOLLE, P.A. (1986) : Exploration and development of hydrocarbons from low-permeability chalks - an example from the Upper Cretaceous Niobrara Formation, Rocky Mountain region. In : SPENCER, C.W. & MAST, R.F. (Editors), Geology of tight gas reservoirs. Amer. Assoc. Petrol. Geol., Stud. Geol., **24** : 129 - 142.

POOLLEN, H.K.v. (1957 a) : Productivity vs. permeability damage in hydraulically produced fractures. API Drill. Product. Pract., **1957** : 103 - 110.

POOLLEN, H.K.v. (1957 b) : Do fracture fluids damage productivity? Oil Gas J., 05.**1957**.

POOLLEN, H.K.v. & MALONE, W.T. (1959) : A stimulation - sand control method. SPE Paper, **1321-G**.

POOLLEN, H.K.v., TINSLEY, J.M. & SAUNDERS, C.D. (1958) : Hydraulic fracturing - fracture flow capacity vs. well productivity. Trans. AIME, **213** : 91 - 95. SPE Paper, **890-G** (1957).

POPE, C.D., WILES, T.J. & PIERCE, B.R. (1987) : Curable resin-coated sand con-

trols proppant flowback. SPE Paper, **16209**.
POPESCU, B. & HODGSHON, S. (1987) : World-survey of petroleum activity in 1986. Petrol. Econ., **54**/8 : 302 - 304.
POSTON, S.W. & CHEN, H.Y. (1987) : The simultaneous determination of formation compressibility and gas-in-place in abnormally pressured reservoirs. SPE Paper, **16227**.
POSTON, S.W., TURNER, S.E., TOBOLA, D.P. & BARGER, B.R. (1986) : Development of a two-stage sand consolidation technique. SPE Paper, **14814**.
POTTER, G.F. (1984) : Formation damage of brine corrosion inhibitors. SPE Paper, **12495**.
POTTER, J.M. & DIBBLE, W.E. (1983) : Formation damage due to colloid plugging. SPE Paper, **11801**.
POTTER, J.M. & DIBBLE, W.E. (1985) : Chemical aspects of iron colloid plugging in quartz sands and implications for formation damage. J. Petrol. Technol., **37** : 1682 - 1688. SPE Paper, **11801** (1983).
POTTER, N. (1987) : Shadows sour Southern Basin gas. Oilman, **1987**/5 : 20 - 33.
POULSEN, D.K. (1988) : A comprehensive theoretical treatment of fracturing fluid loss. SPE Paper, **18262**.
POULSEN, D.K. & LEE, W.S. (1984) : Fracture design with time- and temperature-dependent fluid properties. SPE Paper, **12483**.
POULSEN, D.K. & SOLIMAN, M.Y. (1987) : A procedure for optimal hydraulic fracturing treatment design. SPE Paper, **15940**.
POUPON, A., HOYLE, W.R. & SCHMIDT, A.W. (1970) : Log analysis in formations with complex lithologies. SPE Paper, **2925**.
POWELL, J.P. & JOHNSTON, K.H. (1960) : Waterflood fracturing pays off. IPAA Monthly, **1960**/10 : p. 22 ff.
POWER, D.V., SCHUSTER, C.L., HAY, R. & TWOMBLEY, J. (1976) : Detection of hydraulic fracture orientation and dimensions in cased wells. J. Petrol. Technol., **28** : 1116 - 1124. SPE Paper, **5626** (1975).
POWERS, L.W. & STEVENSON, W.M. (1987) : Perceptions : the key to oil prices. SPE Paper, **16838**.
POWERS, M.L., DODSON, C.J., GHASSEMI, F. & MOORE, J.S. (1985) : Commercial application of steamflooding in an oilfield comprising multiple thin-sand reservoirs. J. Petrol. Technol., **37** : 1707 - 1715. SPE Paper, **13035** (1984).
PRACHNER, W. (1976) : Evaluation of completion damage to high permeability non-consolidated formations. SPE Paper, **5711**.
PRADO, L.R., KRUYSDIJK, C.v., NIKO, H., SANCEVIC, Z.A. & RODRIGUEZ, R. (1988) : Falloff testing a waterflood-induced fractured well in Western Venezuela. SPE Paper, **18142**.
PRADO, L.R. & PRAT, G.d. (1987) : An analytical solution for unsteady liquid flow in a reservoir with a uniformly fractured zone around the well. SPE/DOE Paper, **16395**.
PRASAD, R.K. (1985) : Fracturing improves productivity of a damaged well. SPE Paper, **14710**.
PRASAD, R.K. (1986) : Series of well tests help determine remedial operations. SPE Paper, **14709**.
PRASAD, R.K. (1987) : Abnormally pressured gas reservoirs : case studies. SPE Paper, **16861**.
PRAST, W.G. (1980) : The role of American service companies in the Mexican petroleum industry. Proc. 109th Annual Meeting of the Council of Economics, Las Vegas/Nevada, 24. - 29.02.1980, pp. 17 - 20.
PRAT, G.da, MANUCCI, J., PRADO, L. & MILLAN, E. (1984) : Use of pressure transient testing to evaluate fractured reservoirs in Western Venezuela. SPE Paper, **13054**.
PRATER, R. (1968) : Hydraulic fracturing. U.S. Patent, **3,372,752** (12.03.1968).
PRATS, M. (1961) : Effect of vertical fractures on reservoir behaviour - incompressible fluid case. Soc. Petrol. Engin. J., **1** : 105- 118.
PRATS, M. (1981) : Effect of burial history on the subsurface horizontal stresses of formations having different material properties. Soc. Petrol. Engin. J., **21** : 658 - 662. SPE Paper, **9017**.
PRATS, M. (1983) : Interpretation of pulse tests in reservoirs with crossflow between contiguous layers. SPE Paper, **11963**. SPE Form. Eval., **1** (1986) : 511

- 520.
PRATS, M., HAZEBROEK, P. & STRICKLER, W.F. (1962) : Effect of vertical fractures on reservoir behaviour - compressible fluid case. Soc. Petrol. Engin. J., **2** : 87 - 94.
PRATS, M. & LEVINE, J.S. (1963) : Effect of vertical fractures on reservoir behaviour results on oil and gas flow. SPE Paper, **593**.
PREVEDEL, B. (1985 a) : Berkhöpen 2001 - Horizontalbohrprojekt. Erdoel-Erdgas, **101** : 377 - 383.
PREVEDEL, B. (1985 b) : How one operator drilled horizontally through a salt dome. World Oil, **1985**/12 : 69 - 76; 80.
PREVEDEL, B. (1987 a) : New techniques in horizontal and drainhole drilling optimization : Part 1 - Lehrte 41, lateral drilling project. SPE Paper, **15694**.
PREVEDEL, B. (1987 b) : New techniques in horizontal and drainhole drilling optimization : Part 2 - Georgsdorf 503, horizontal drilling project. SPE Paper, **17080**.
PRICE, R.S. (1986) : Government policy and international gas trade. SPE Paper, **14630**.
PRIEST, M.A. (1988) : Injection of radioisotopes at wellhead improves fracturing operations. SPE Paper, **17464**.
PRIISHOLM, S., NIELSEN, B.L. & HASLUND, O. (1987) : Fines migration, blocking and clay swelling of potential geothermal sandstone reservoirs, Denmark. SPE Form. Eval., **2** : 168 - 178. SPE Paper, **15199** (1985).
PRIJAMBODO, R., RAGHAVAN, R. & REYNOLDS, A.C. (1985) : Well test analysis for wells producing layered reservoirs with crossflow. Soc. Petrol. Engin. J., **25** : 380 - 396. SPE Paper, **10262** (1981).
PRO-TECHNICS (1987) : See as you´ve never seen before. Brochure; Pro-Technics International, Houston/Texas.
PROYER, G. (1980) : Thermodynamische Gesichtspunkte bei der Planung von Dampfinjektionsbohrungen. Erdoel-Erdgas, **96** : 444 - 456.
PROYER, G., CHAZITEODOROU, G., MUSS, H. & ROSSKAMP, M. (1983) : Results of a steam-drive pilot project in the Ruehlertwist field, Federal Republic of Germany. SPE Paper, **11956**. J. Petrol. Technol., **37** (1985) : 284 - 294.
PRUDHOMME, R.K., ELLIS, S., CONSTIEN, V. & KNOLL, S. (1988) : Reproducible rheological measurements on crosslinked fracturing fluids. SPE Paper, **18210**.
PRUESS, K. (1983) : Heat transfer in fractured geothermal reservoirs with boiling. Water Resources Res., **19** : 201 - 208.
PRUESS, K. (1985) : A quantitative model of vapor-dominated geothermal reservoirs as heat pipes in fractured porous rock. Trans. Geotherm. Resour. Counc., **9** (II) : 353 - 361.
PRUESS, K. & BODVARSSON, G.S. (1984) : Thermal effects of reinjection in geothermal reservoirs with major vertical fractures. J. Petrol. Technol., **36** : 1567 - 1578. SPE Paper, **12099** (1983).
PRUESS, K. & NARASIMHAN, T.N. (1982) : On fluid reserves and the production of super-heated steam from fractured, vapor-dominated geothermal reservoirs. J. Geophys. Res., **87** (B 11) : 9329 - 9339.
PRUESS, K. & NARASIMHAN, T.N. (1985) : A practical method for modeling fluid and heat flow in fractured porous media. Soc. Petrol. Engin. J., **25** : 14 - 26. SPE Paper, **10509** (1982).
PRUITT, G.T., SIMMONS, C.M., NEILL, G.H. & CRAWFORD, H.R. (1965) : A method to minimize cost of pumping fluids containing friction reducing additives. J. Petrol. Technol., **17** : 641 - 646. SPE Paper, **997** (1964).
PUGH, G.E. (1982) : Drilling of three horizontal hole pattern, Ft. McMurray, Alberta. 33rd Annual Technical Meeting of the Petroleum Society of CIM, Calgary/Alberta, 06. - 09.06.1982. 2nd Intern. Conf. on the future of heavy crude and tar sands, **894-902**.
PUGH, T.D., McDANIEL, B.W. & SEGLEM, R.L. (1977) : A new fracturing technique for Dean Sand. SPE Paper, **6378**.
PURCELL, W.R. (1949) : Capillary pressures - their measurement using mercury and their calculation of permeability therefrom. Trans. AIME, **186** : 39 - 48.
PURSELL, D.A., HOLDITCH, S.A. & BLAKELEY, D. (1988) : Laboratory investigation of inertial flow in high strength fracture proppants. SPE Paper, **18319**.
PYE, D.S. & SMITH, W.A. (1973) : Fluid loss additive seriously reduces fracture

proppant conductivity and formation permeability. SPE Paper, **4680** (1973), **5114** (1975).
QADEER, S., DEHGHANI, K., OGBE, D.O. & OSTERMANN, R.D. (1988) : Correcting oil/ water relative permeability data for capillary end effect in displacement experiments. SPE Paper, **17423**.
QI, Y.F. & ZHANG, X. (1986) : Parameter estimation in a naturally fractured reservoir by method of cybernetics. SPE Paper, **14864**.
QUINLAN, M. (1988) : North Sea oilfield survey - new concern over oil prices. Petrol. Econ., **55**/4 : 111 - 116.
QUIREIN, J.A., VIGNE, J.la & CHAPMAN, S. (1987) : Enhancement to the pulsed-neutron gamma spectroscopy interpretation process. Trans. 28th Soc. Prof. Well Log Analysts Annual Logging Symp., London/England, 29.06. - 02.07.1987.
QUITZAU, R. & BASSIOUNI, Z. (1981) : The possible impact of the geopressure energy resource on conventional oil and gas explorations. SPE Paper, **10281**.
QUONG, R., OWEN, L.B. & LOCKE, F.E. (1982) : Technical potential of methods for methane extraction from geopressured-geothermal fluids at high temperature and pressure. J. Petrol. Technol., **34** : 504 - 510. SPE Paper, **9469** (1980).
RAGHAVAN, R., CADY, G.V. & RAMEY, H.J. (1972) : Well-test analysis for vertically fractured wells. J. Petrol. Technol., **24** : 1014 - 1020. SPE Paper, **3013** (1970).
RAGHAVAN, R. & OHAERI, C.U. (1981) : Unsteady flow to a well produced at constant pressure in a fractured reservoir. SPE Paper, **9902**. SPE Form. Eval., **2** (1987) : 186 - 200.
RAGHAVAN, R., PRIJAMBODO, R. & REYNOLDS, A.C. (1985) : Well test analysis for wells producing layered reservoirs with crossflow. Soc. Petrol. Engin. J., **25** : 380 - 396.
RAGHAVAN, R., TOPALOGLU, H.N., COBB, W.M. & RAMEY, H.J. (1974) : Well-test analysis for wells producing from two commingled zones of unequal thickness. J. Petrol. Technol., **26** : 1035 - 1043. SPE Paper, **4559** (1973).
RAGHAVAN, R., URAIET, A. & THOMAS, G.W. (1978) : Vertical fracture height : effect on transient flow behavior. Soc. Petrol. Engin. J., **18** : 265 - 277. SPE Paper, **6016** (1976).
RAHIM, Z. & LEE, W.J. (1986) : An iterative technique to determine reservoir and fracture properties using buildup data for hydraulically fractured wells. SPE Paper, **15929**.
RAIBLE, C.J. & GALL, B.L. (1985) : Laboratory formation damage studies of Western tight gas sands. SPE/DOE Paper, **13903**.
RAKOP, K.C. (1985) : Laboratory analysis to determine the embedment behaviour of Zirprop 20/40 proppant and intermediate sintered 20/40 proppant. TerraTek Research Repts., **TR-85-20.08.84** and **TR-85-14.05.85**.
RAMBOW, F.H.K. (1984) : The borehole televiewer : some field examples. Trans. 25th Soc. Prof. Well Log Analysts Annual Logging Symp., Paper **C**, New Orleans/Louisiana, 10. - 13.06.1984.
RAMBOW, F.H.K. (1985) : Unique logging tool "sees" into wellbore. World Oil, **1985**/7 : 79 - 88.
RAMEY, H.J. (1975 a) : Non-Darcy flow and wellbore storage effects in pressure buildup and drawdown of gas wells. SPE Reprint Series, **9**.
RAMEY, H.J. (1975 b) : Interference analysis for anisotropic formations - a case history. SPE Paper, **5319**.
RAMEY, H.J. & GRINGARTEN, A.C. (1975) : Effect of high volume vertical fractures on geothermal steam well behavior. 2nd United Nations Symp. Use Developm. Geotherm. Energy, San Francisco 05.1975.
RAMEY, H.J., KRUGER, P. & RAGHAVAN, R. (1973) : Explosive stimulation of hydrothermal reservoirs. In : KRUGER, P. & OTTE, C. (Editors), Geothermal Energy, 231 - 249; Stanford Univ. Press, Stanford/California.
RAMSAY, H.J. (1983) : Project financing for the small independent company. SPE Paper, **11289**.
RANDALL, R.R., GRAY, T., CRAIK, G. & HOPKINSON, E.C. (1985) : A new digital multiscale pulsed neutron logging system. SPE Paper, **14461**. SPE Form. Eval., **2** (1987) : 395 - 400.
RANDOL, W.L. (1986) : Opec presents an oil price war. Starring Saudi Arabia as Rambo. First Boston Research, Spec. Rept., 05.02.**1986** : 10 pp.

RANDOLPH, P.L. (1974) : Massive stimulation effort may double U.S. gas supply. World Oil, **1974**/11 : 81 - 84, 178 - 179.
RANDOLPH, P.L. (1977) : Natural gas from geopressured aquifers? SPE Paper, **6826**.
RANDOLPH, P.L. (1983) : Porosity and permeability of Mesaverde Sandstone core from the U.S. DOE multiwell experiment, Garfield county, Colorado. SPE/DOE Paper, **11765**.
RANDOLPH, P.L., SOEDER, D.J. & CHOWDIAH, P. (1984) : Porosity and permeability of tight sands. SPE/DOE/GRI Paper, **12836**.
RANDOLPH, P.L., SOEDER, D.J. & CHOWDIAH, P. (1985) : Effects of water and stress upon permeabilities to gas of paludal and coastal sands. Institute of Gas Technology, Topical Rept., DOE/MC, **20342-1838**, 02.1985.
RANKIN, M. (1982) : Risk stopping rules, and the economic viability of exploratory drilling programs. SPE Paper, **11098**.
RANNEY, L. (1941) : Drilling wells horizontally. Oil Weekly, 20.01.**1941** : 12 - 14.
RAO, H.V.S.G., ADVANI, S.H., CHANG, P., LEE, S.C. & DEAN, S.L. (1979) : In situ stress determination based on fracture responses associated with coring operations. Proc. 20th U.S. Symp. Rock Mechan., 683 - 690; Austin/Texas.
RASHID, O.M. (1984) : Packerless single-selective gravel-packed completions designed to improve recovery at Cognac Field. SPE Paper, **13129**. J. Petrol. Technol., **37** (1985) : 1819 - 1830.
RASMUS, J.C. (1981) : Fluid evaluation of fractured hydrocarbon intervals in the Twin Creek Formation of the Yellow Creek field. SPE/DOE Paper, **9857**.
RAVENNE, C., GALLI, A., MATHIEU, Y., MONTADERT, L. & RUDXIBWICZ, J.L. (1987) : Heterogeneities and geometry of sedimentary bodies in a fluvio-deltaic reservoir. SPE Paper, **16752**.
RAWLINGS, H.E. (1958) : Use of hydraulic fracturing equipment for formation sand control. J. Petrol. Technol., **10**/5 : 29 - 32.
RAY, E.O. (1976) : Devonian shale development in eastern Kentucky. In : Natural gas from unconventional geologic sources, pp. 100 - 111, National Academy of Science, Washington/D.C.
RAYMOND, L.R. & BINDER, G.G. (1967) : Productivity of wells in vertically fractured, damaged formations. J. Petrol. Technol., **19** : 120 - 130. SPE Paper, **1454** (1966).
REDDY, G.S., PROWSE, P.R. & REDFORD, D.A. (1980) : Physical modelling and numerical simulation of in situ recovery of bitumen from oil sands by steam injection. 31st Annual Technical Meeting of the Petroleum Society of CIM, Calgary/Alberta, 25. - 28.05.1980.
REED, M.G. (1972) : Stabilization of formation clays with hydroxy-aluminum solutions. J. Petrol. Technol., **24** : 860 - 864. SPE Paper, **3694** (1971).
REED, M.G. (1977) : Formation permeability damage by mica alteration and carbonate dissolution. J. Petrol. Technol., **29** : 1056 - 1060. SPE Paper, **6009** (1976).
REED, M.G. (1980) : Gravel pack and formation sandstone dissolution during steam injection. J. Petrol. Technol., **32** : 941 - 949. SPE Paper, **8424** (1979).
REED, M.G. & COPPEL, C.P. (1972) : Sand stabilization with hydroxy aluminum solutions. SPE Paper, **4186**.
REEVES, S.R., WALLACE, J.A. & BEAVERS, W.M. (1987) : The influence of reservoir properties and geologic setting on coalbed fracturing and production : a preliminary evaluation of the TEAM project. SPE/DOE Paper, **16423**.
REGALBUTO, J.A. & RIGGS, R.S. (1985) : High differential pressure, radial flow characteristics of gun perforations. SPE Paper, **14319**.
REGALBUTO, J.A. & RIGGS, R.S. (1986) : Underbalanced perforation characteristics as affected by differential pressure. OTC Paper, **5245**. SPE Paper, **15816**. SPE Product. Engin., **3** (1988) : 83 - 88.
REHM, W. (1987) : Horizontal drilling applied in slim holes. Petrol. Engin. Intern., **1987**/2 : 24 - 28.
REID, A.G. (1981) : Oil exploration possibilities in China. Proc. 110th Annual Meeting of the Council of Economics, Chicago/Illinois, 22. - 26.02.1981, pp. 135 - 140.

REIDENBACH, V.G., HARRIS, P.C., LEE, Y.N. & LORD, D.L. (1986) : Rheological study of foam fracturing fluids using nitrogen and carbon dioxide. SPE Product. Engin., **1** : 31 - 43. SPE Paper, **12026** (1983).

REILEY, R.H., STAGG, T.O. & VIDRINE, J.G. (1988) : Horizontal well case histories in Alaska. SPE Paper, **17570**.

REINICKE, K.M., BRINKMANN, F.W., SCHWARZ, H. & HUENI, G. (1985) : Interpretation of buildup data obtained from MHF wells in Northern Germany. J. Petrol. Technol., **37** : 2173 - 2183. SPE/DOE Paper, **11605** (1983).

REINICKE, K.M., TRENEL, K., HERBST, E., KEMPGEN, M. & LINZ, E. (1988 a) : Anpassung der Erdölfeldaktivitäten an niedrige Erlöse. Erdöl-Erdgas-Kohle, **104** : 263 - 270.

REINICKE, K.M., TRENEL, K., HERBST, E., KEMPGEN, M. & LINZ, E. (1988 b) : Optimizing oilfield operation. SPE Paper, **17552**.

REINICKE, K. & WIETHOFF, B. (1987) : BEB-operated thermal projects : comparison of performance to forecast. 4th Europ. Symp. Enhanced Oil Recovery, Thermal Recovery Session III, Hamburg/Germany FRG.

REISS, L.H. (1981) : Reservoir engineering aspects of fractured formations. 112 pp.; Gulf Publ., Houston/Texas.

REISS, L.H. (1985) : Horizontal wells production, after five years. SPE Paper, **14338**. J. Petrol. Technol., **39** (1987) : 1411 - 1416.

REISS, L.H., JOURDAN, A. & GIGER, F. (1984) : Horizontal wells for the first time offshore : deep and prolific. SPE Paper, **13024**.

REISS, L.H., JOURDAN, A.P.L., GIGER, F.M. & ARMESSEN, P.A. (1984) : Offshore and onshore European horizontal wells. OTC Paper, **4791**. SPE Paper, **13328**.

REISTLE, C.E. (1945) : U.S. Patent, **2,368,424** (30.01.1945).

REN, N.K. & HUDSON, P.J. (1985) : Predicting the in-situ state of stress using differential wave velocity analysis. Proc. 26th U.S. Rock Mechan. Symp., Rapid City/South Dakota.

REN, N.K. & ROEGIERS, J.C. (1983) : Differential strain curve analysis: a new method for determining the pre-existing in-situ stress state from rock core measurements. Proc. 5th Intern. Congr. Rock Mechan., **2** : 117 - 127; Balkema, Netherlands.

REN, S.Q. & XIONG, H.J. (1986) : Effective of temperature field and common ion effect on effective acid penetration in fracture systems. SPE Paper, **14852**.

RENSBURG, W.C.J.v. & MALIK, K.A. (1987) : Tracking changes in oil exploration and development technology and their economic implications. SPE Paper, **17021**.

RENSVOLD, R.F. (1982) : Applications of sand control technology in thermal recovery systems. CIM Paper, **82-33-69**.

RENSVOLD, R.F. (1983) : Sand consolidation resins - their stability in hot brine. Soc. Petrol. Engin. J., **23** : 238 - 248. SPE Paper, **10653** (1982).

RENSVOLD, R.F. & DECKER, L.R. (1978) : Full scale gravel packing model studies. Paper **EUR 39**, European Offshore Petroleum Conference and Exhibition, London/England, 24. - 27.10.1978.

RENTSCH, H.C. & MES, M.J. (1988) : Measurement of Ekofisk subsidence. OTC Paper, **5619**.

RESTINE, J.L. (1983) : Effect of preheating on Kern River steam drive. J. Petrol. Technol., **35** : 523 - 529. SPE Paper, **10750**.

REVIERE, R.H. & WU, C.H. (1986) : An economic evaluation of waterflood infill drilling in nine Texas waterflood units. SPE Paper, **15037**.

REYNOLDS, A.C., CHANG, W.L., YEH, N. & RAGHAVAN, R. (1985) : Wellbore pressure response in naturally fractured reservoirs. J. Petrol. Technol., **37** : 908 - 920. SPE Paper, **12012** (1983).

REYNOLDS, J.J., SCOTT, J.B., POPHAM, J.L. & COFFER, H.F. (1961) : Hydraulic fracture - field test to determine areal extent and orientation. J Petrol. Technol., **13** : 371 - 376.

REYNOLDS, S.M. (1987) : Photogeology : unique system for detecting more and better drillable prospects. Oil Gas J., 30.03.**1987** : 120 - 124.

RIAL, R.M. (1984) : 3D thermal stimulation using a horizontal wellbore for steamflooding. SPE Paper, **13076**.

RICE, D.D. & SHURR, G.W. (1978) : Potential for major natural gas resources in shallow, low-permeability reservoirs of the Northern Great Plains - an as-

sessment of their natural resources. 24th Annual Conf. Montana Geol. Soc., Guidebook, Williston Basin Symp., 265 - 281. Amer Assoc. Petrol. Geol. Bull., **64** (1980) : 969 - 987.
RICHARDSON, A.S. (1985) : Slant drilling of Valhalla waterflood project. CIM Paper, **85-36-60**.
RICHARDSON, E.A. & HAMBY, T.W. (1970) : Consolidation of silty sands with an epoxy resin overflush process. J. Petrol. Technol., **22** : 1103 - 1108. SPE Paper, **2189** (1968). SPE Reprint Series, Well Completions, **5** : 196 - 201.
RICHARDSON, J.G., SANGREE, J.B. & SNEIDER, R.M. (1986 a) : Coastal barrier reservoirs. SPE Paper, **15777**. J. Petrol. Technol., **40** (1988) : 1121 - 1122.
RICHARDSON, J.G., SANGREE, J.B. & SNEIDER, R.M. (1986 b) : Sand-rich deltas. SPE Paper, **15799**.
RICHARDSON, J.G., SANGREE, J.B. & SNEIDER, R.M. (1986 c) : Meandering stream reservoirs. SPE Paper, **15781**. J. Petrol. Technol., **39** (1987) : 1501 - 1502.
RICHARDSON, J.G., SANGREE, J.B. & SNEIDER, R.M. (1986 d) : Braided streams. SPE Paper, **15782**. J. Petrol. Technol., **39** (1987) : 1499 - 1500.
RICHARDSON, J.G., SANGREE, J.B. & SNEIDER, R.M. (1986 e) : Permeability distributions in reservoirs. SPE Paper, **15785**. J. Petrol. Technol., **39** (1987) : 1197 - 1199.
RICHARDSON, J.B., SANGREE, J.B. & SNEIDER, R.M. (1988) : Aeolian dunes. J. Petrol. Technol., **40** : 11 - 12. SPE Paper, **15783** (1986).
RIECKMANN, M. (1970) : Untersuchungen von Turbulenzerscheinungen beim Fließen von Gasen durch Speichergesteine unter Berücksichtigung der Gleitströmung. Erdoel-Erdgas, **86** : 36 - 51.
RIECKMANN, M. (1971) : Untersuchungen über die effektiven Permeabilitäten der Sandsteine des Rotliegenden und des Karbons. Erdoel-Erdgas, **87** : 2 - 8.
RIECKMANN, M. & PUSCH, G. (1984) : Neuere Entwicklungen auf dem Gebiet der Lagerstättentechnik. Erdoel-Erdgas, **100** : 132 - 138, 214 - 222.
RIEPE, L., PAPE, H. & SCHOPPER, J.R. (1985) : Detection of fractures and microfissures in granites by logs. Trans. 26th Soc. Prof. Well Log Analysts Annual Logging Symp., Paper **P**; Dallas/Texas, 17. - 20.06.1985.
RIESE, A.C., SCHECHTER, R.S. & RIESE, W.C. (1988) : Origins and mechanisms of formation damage. Amer. Assoc. Petrol. Geol., Annual Convention, Houston/Texas, 20. - 23.03.1988.
RIETMAN, N.D. & KIDD, R.D. (1977) : Improvement of ultimate recovery of natural gas by realistic gas prices. SPE Paper, **6636**.
RIGASSI, D.A. (1985) : Eastern Europe. The USSR has production trouble. World Oil, **1985**/8 : 90 - 100.
RIGASSI, D.A. (1986) : Eastern Europe. USSR aims to improve production. World Oil, **1986**/8 : 71 - 74.
RIGASSI, D.A. (1987) : Eastern Europe. Gorbachev spurs oil recovery. World Oil, **1987**/8 : 62 - 64.
RIGHTMIRE, C.T. & CHOATE, R. (1986) : Coal-bed methane and tight gas sands interrelationships. In : SPENCER, C.W. & MAST, R.F. (Editors), Geology of tight gas reservoirs. Amer. Assoc. Petrol. Geol., Stud. Geol., **24** : 87 - 110.
RIGHTMIRE, C.T., EDDY, G.E. & KIRR, J.N. (1984) : Coalbed methane resources of the United States. AAPG Stud. Geol., **17** : 378 pp.
RIKE, J.L. (1970) : Review of sand consolidaton experience in South Louisiana. SPE Reprint Series, Well Completions, **5** : 176 - 181. J. Petrol. Technol., **18** (1966) : 545 - 550. SPE Paper, **1238** (1965).
RIKE, J.L. (1975) : Shortcomings of present sand control methods suggest new fail-safe concept. Oil Gas J., **1975**/2 : 97 - 102.
RILEY, E.A. (1959) : Hydraulic fracturing in waterflood operations in Kermit, Cherrykirk and Pecos Valley fields. J. Petrol. Technol., **11** : 15 - 22.
RISNES, R., BRATLI, R.K. & HORSRUD, P. (1982 a) : Sand stresses around a wellbore. Soc. Petrol. Engin. J., **22** : 883 - 898. SPE Paper, **9650** (1981).
RISNES, R., BRATLI, R.K. & HORSRUD, P. (1982 b) : Sand arching - a case study. EUR Paper, **310**. SPE Paper, **12948** (1984).
RIXE, F.H., FAST, C.R. & HOWARD, G.C. (1963) : Selection of propping agents for hydraulic fracturing. API Drill. Product. Pract., **1963** : 138 - 143.
ROBERTS, C.N. (1981) : Fracture optimization in a tight gas play : Muddy "J"

Formation, Wattenberg field, Colorado. SPE/DOE Paper, **9851**.
ROBERTS, G. & CRAMPIN, S. (1986) : Shear-wave polarizations in a hot dry rock geothermal reservoir : anisotropic effects of fractures. Intern. J. Rock Mechan., Min. Sci., Geomechan. Abstr., **23** : 291 - 302.
ROBERTS, J.D., MATHIESON, A.S. & HAMPSON, J.M. (1987) : Statfjord. In : SPENCER, A.M. et al. (Editors), Geology of the Norwegian oil and gas fields, 319 - 340; Graham & Trotman, London/England.
ROBERTS, L.D. & GUIN, J.A. (1974) : A new method for predicting acid penetration distance. SPE Paper, **5155**. SPE Reprint Series, Well Completions, **5 a**/ II (1978) : 116 - 125.
ROBERTS, M.L. (1984) : Crude oil prices - the direction? J. Petrol. Technol., **36** : 811 - 816. SPE Paper, **11314** (1983).
ROBERTS, M.L. (1986) : OPEC and Adam Smith. SPE Paper, **14637**.
ROBERTS, M.L. (1987) : OPEC and Adam Smith : Part II. SPE Paper, **16288**.
ROBINSON, A.E. (1981) : Facies types and reservoir quality of the Rotliegendes sandstone, North Sea. SPE Paper, **10303**.
ROBINSON, B.A. & TESTER, J.W. (1984) : Dispersed fluid flow in fractured reservoir : an analysis of tracer-determined residence time distributions. J. Geophys. Res., (B 12) **89** : 10374 - 10384.
ROBINSON, B.A., TESTER, J.W. & BROWN, L. (1984) : Reservoir sizing using inert and chemically reactive tracers. SPE Paper, **13147**. SPE Form. Eval., **3** (1988) : 227 - 234.
ROBINSON, B.M., HOLDITCH, S.A. & LEE, W.J. (1983) : A case study of the Wilcox (Lobo) Trend in Webb and Zapata Counties, Texas. SPE/DOE Paper, **11600**. J. Petrol. Technol., **38** (1986) : 1355 - 1364.
ROBINSON, B.M., HOLDITCH, S.A. & WHITEHEAD, W.S. (1986) : Minimizing damage to a propped fracture by controlled flowback procedures. SPE Paper, **15250**. J. Petrol. Technol., **40** (1988) : 753 - 759.
ROBINSON, J.C. (1985) : Practical onsite inspection and testing improves fracture treatment quality. SPE Paper, **14435**.
RODRIGUEZ, F., HORNE, R.N. & CINCO-LEY, H. (1984 a) : Partially penetrating fractures : pressure transient analysis of an infinite conductivity fracture. SPE Paper, **12743**.
RODRIGUEZ, F., HORNE, R.N. & CINCO-LEY, H. (1984 b) : Partially-penetrating vertical fractures : pressure transient analysis of a finite-conductivity fracture. SPE Paper, **13057**.
ROEGIERS, J.C. (1974) : The development and evaluation of a field method for in situ stress determination using hydraulic fracturing. Ph. D. Thesis, Univ. of Minnesota, Minneapolis/Minnesota.
ROEGIERS, J.C. (1987) : Elements of rock mechanics. In : ECONOMIDES, M.J. & NOLTE, K.G. (Editors), Reservoir stimulation. Schlumberger Educational Services, 2-1 - 2-21; Dowell Schlumberger Document, **TSL-2612**; Schlumberger Document, **SMP-7018**; Houston/Texas.
ROEGIERS, J.C., BEN NACEUR, K. & TOUBOUL, E. (1986) : Numerical investigation of the effects of fluid rheological properties on three-dimensional fracture propagation. CIM Paper, **86-37-26**.
ROEGIERS, J.C. & DETOURNAY, E. (1988) : Considerations on failure initiation in inclined boreholes. 29th U.S. Rock Mechan. Symp., Univ. of Minnesota, Minneapolis/Minnesota, 06.1988.
ROEGIERS, J.C. & ISHIJIMA, Y. (1983) : A coupled fracturing model and its application to hydraulic fracturing. SPE Paper, **12311**.
ROEGIERS, J.C., McLENNAN, J.D. & MURPHY, D.L. (1982) : Influence of preexisting discontinuities on the hydraulic fracturing propagation process. Japan-United States First Joint Seminar on Hydraulic Fracturing and Geothermal Energy, Tokyo/Japan, 08.11.1982.
ROEGIERS, J.C. & WILES, T.D. (1981) : Hydraulic fracturing of rock masses - influence of preexisting discontinuities. Univ. of Toronto, Dept. Civil Engin. Publ., **81-04**. ISBN 0-7727-2.
ROEHL, E.A. & HACKETT, J.L. (1982) : A laboratory technique for screening shale swelling inhibitors. SPE Paper, **11117**.
ROEMERSHAUSER, A.E. & HAWKINS, M.F. (1955) : The effect of slant hole, drainhole, and lateral hole drilling on well productivity. J. Petrol. Technol.,

7 : 11 - 14.
ROEST, J.A. (1986) : Pulse testing reveals poor lateral and vertical continuity in a reservoir consisting of distributary-channel sands. SPE Paper, **15613**.
ROGERS, E.B. (1971) : Sand control in oil and gas wells. Oil Gas J., **1971**/11.
ROGERS, E.B. (1976) : Successful well stimulation program has revitalized a California oil field. SPE Paper, **5699**.
ROGERS, R.E., VEATCH, R.W. & NOLTE, K.G. (1984) : Pipe viscosimeter study of fracturing fluid rheology. Soc. Petrol. Engin. J., **24** : 575 - 581. SPE Paper, **10258** (1981).
ROHRET, M.T. & JONES, T.C. (1978) : Stimulation of the Niobrara Formation using foamed methanol-water. SPE Paper, **7174**.
ROLL, D.L., HIMES, R., EWERT, D.P. & DOERKSEN, J. (1986) : Effects of pumping equipment on sand-laden slurries. SPE Paper, **15071**. SPE Product. Engin., **2** (1987) : 291 - 304.
ROODHART, L.P. (1983) : Fracturing fluids : fluid loss measurements under dynamic conditions. SPE Paper, **11900**. Soc. Petrol. Engin. J., **25** (1985) : 629 - 636.
ROODHART, L.P. (1985) : Proppant settling in non-Newtonian fracturing fluids. SPE/DOE Paper, **13905**.
ROODHART, L.P. & DAVIES, D.R. (1987) : Polymer emulsion : the revival of a fracturing fluid. SPE/DOE Paper, **16413**.
ROODHART, L., KUIPER, T.O. & DAVIES, D.R. (1986) : Proppant rock impairment during hydraulic fracturing. SPE Paper, **15629**.
ROSATO, N.D., BENNETT, C.O., REYNOLDS, A.C. & RAGHAVAN, R. (1981) : Analysis of short-time buildup data for finite conductivity fractures. SPE/DOE Paper, **9890**. J. Petrol. Technol., **34** (1982) : 2413 - 2422.
ROSBACO, J.A. (1983) : Investment evaluations in inflationary and unstable scenarios. SPE Paper, **11302**.
ROSE, R.E. (1988) : New techniques lead to better completions in Red Fork and other low-permeability formations. SPE Paper, **17305**.
ROSE, R.E. & DOHERTY, M.G. (1982) : Data collection instrumentation and interpretation for geopressured aquifer well tests. SPE/DOE Paper, **10835**.
ROSE, W. (1982) : Permeability anisotropy in low permeability formations. SPE/DOE Paper, **10810**.
ROSE, W. (1983) : A note on the role played by sediment bedding in causing permeability anisotropy. J. Petrol. Technol., **35** : 330 - 332. SPE Paper, **10916**.
ROSENBERG, J.I., O`SHEA, P., MERCER, J., MORRA, F. & BRASHEAR, J.P. (1983) : A sensitivity analysis of the NPC study of tight gas. SPE Paper, **11645**.
ROSENE, R.B. & SHUMAKER, E.F. (1971) : Viscous fluids provide improved results from hydraulic fracturing treatments. SPE Paper, **3347**.
ROSEPILER, M.H. (1979) : Determination of principal stresses and the confinement of hydraulic fractures in Cotton Valley. SPE Paper, **8405**.
ROSEPILER, M.J. (1981) : Calculation and significance of water saturations in low porosity shaly gas sands. Trans. 22nd Soc. Prof. Well Log Analysts Annual Logging Symp., Paper **000**.
ROSS, S.E. (1984) : Analysis of pressure transient data obtained in hydraulically fractured wells : a field example. SPE Paper, **12736**.
ROSS, W.M., PIERSON, N.O. & COULTER, A.W. (1963) : Matrix acidizing corrects formation damage in sandstones. Petrol. Engin. Intern., **35** : p. 12 ff.
ROSSEN, R.H. & SHEN, E.I. (1987) : Simulation of gas/oil drainage and water/oil imbibition in naturally fractured reservoirs. SPE Paper, **16982**.
ROSSKAMP, M. (1982) : Untertagekomplettierung von Produktions- und Injektionsbohrungen in den Dampfflutgebieten Emlichheim und Rühlertwist. Erdoel-Erdgas, **98** : 317 - 323.
ROSSUM, B.v. (1975) : Aspects of the geology and appraisal/development of the Groningen gas field. Erdoel-Erdgas, **91** : 254 - 256.
ROWLEY, J.C., PETTIT, R.A., HENDRON, R.H., SINCLAIR, A.R. & NICHOLSON, R.W. (1983) : Fracturing operations in a dry geothermal reservoir. SPE Paper, **12100**.
ROWSE, J.G. (1987) : World energy markets : coping with instability. Proc. 9th Intern. Conf. of the Intern. Assoc. Energy Econom., 06. - 08.07.1987, Calgary/Alberta.

ROYCE, T.N., BECK, L.M. & RICKARDS, A.R. (1984) : Rheological characteristics of adjustable crosslinked fracturing fluids. SPE Paper, **13178**.

RUDDY, I., ANDERSON, M.A., PATTILLO, P.D., BISHLAWI, M. & FOGED, N. (1988) : Rock compressibility, compaction, and subsidence in a high-porosity chalk reservoir : a case study of Valhall field. SPE Paper, **18278**.

RUEDRICH, R.A., PERKINS, T.K. & O`BRIEN, D.E. (1974) : Precise joint length determination using a multiple casing locator tool. SPE Paper, **5087**.

RUHOVETS, N. & FERTL, W.H. (1982) : Volumes, types, and distribution of clay minerals in reservoir rocks based on well logs. SPE/DOE Paper, **10796**.

RUMMEL, F. (1985) : Manipulation der Frac-Ausbreitung in geklüftetem Fels. Statusbericht zum Forschungsvorhaben, **03E-6075-A** : 49 - 51. Programm Energieforschung und Energietechnologien. Statusreport 1985 Geotechnik und Lagerstätten (KFA Jülich).

RUMMEL, F., HEUSER, U., MÜLLER, W. & OTTO, F. (1986) : Basisdaten und Modelluntersuchungen zur Rißausbreitung bei Stimulationsversuchen in geringpermeablen Sedimenten der norddeutschen Erdgasträgerformationen. DGMK-Forschungsbericht, **167-4** : 137 pp.

RUMMEL, F., HEUSER, U., MÜLLER, W. & OTTO, F. (1987) : Hydraulic fracturing - Basisdaten und Modelluntersuchungen zur Rißausbreitung. Erdöl-Erdgas-Kohle, **103** : 29 - 30.

RUMMO, G.J. (1982) : Crosslinking delay achieved with gel additive. Oil Gas J., 13.09.**1982** : 84 - 89.

RUNGE, H.C. (1986) : Die derzeitige Ölpreisentwicklung und ihre Auswirkung auf die Exploration und Produktion von Erdöl und Erdgas. Erdöl-Erdgas-Kohle, **102** : 323 - 327. DVGI Spring Meeting, Celle/Germany FRG, 30.05.1986.

RUOTSALA, J.E. & WILLIAMS, R.T. (1982) : Reflection seismology as an exploration tool for fractured zones in gas-bearing shale - Cottageville, West Virginia : a case study. SPE/DOE Paper, **10792**.

RUSSELL, D.G. & TRUITT, N.E. (1964) : Transient pressure behavior in vertically fractured reservoirs. J. Petrol. Technol., **16** : 1159 - 1170.

RUSSELL, F.G. & COADY, A.B. (1982) : Gas-permeation process economically recovers CO2 from heavily concentrated streams. Oil Gas J., 28.06.**1982** : 126 - 134.

RUZYLA, K. (1984) : Characterization of pore space by quantitative image analysis. SPE Paper, **13133**. SPE Form. Eval., **1** (1986) : 389 - 398.

RYAN, W.M. (1976) : Remote sensing fracture study - West Virginia and Southeastern Kentucky. Proc. 7th Appalachian Petrol. Geol. Symp., Morgantown/West Virginia, 01. - 04.03.1976, pp. 94 ff.

SACUTA, A., NGUYEN, D.M. & KISSEL, G.A. (1986) : Stability of nickel-coated sands as gravel pack material for thermal wells. SPE Paper, **15058**.

SAFER, A.E. (1977) : World oil - challenges and opportunities. SPE Paper, **6337**.

SAGE, B.H. & LACEY, W.N. (1942) : Effectiveness of gravel screens. Trans. AIME, **146** : 89 - 106.

SAGEEV, A. (1986 a) : Multiple interference testing in the Ohaaki geothermal field. SPE Paper, **15122**. SPE Form. Eval., **3** (1988) : 429 - 437.

SAGEEV, A. (1986 b) : Interference testing detecting an impermeable or compressible region. SPE Paper, **15585**. SPE Form. Eval., **3** (1988) : 420 - 428.

SAHUQUET, B.C. & FERRIER, J. (1980) : Steam drive pilot in a fractured carbonate reservoir : Lacq superieur field. SPE Paper, **9453**. J. Petrol. Technol., **24** (1982) : 873 - 880.

SAHUQUET, B. & SITBON, A. (1987) : Recent results of steamdrive development in Lacq Superieur Reservoir. 4th Europ. Symp. Enhanced Oil Recovery, Thermal Recovery Session III, Hamburg/Germany FRG.

SALAMA, S.Y. (1988) : Natural gas supply and federal regulations : causes and effects. SPE Paper, **17745**.

SALAMY, S.P., SARADJI, B.S., OKOYE, C.O., MERCER, J.C. & YOST, A.B. II (1987) : Recovery efficiency aspects of horizontal well drilling in Devonian shale. SPE/DOE Paper, **16411**.

SALDANA, M.A., BRIGHAM, W.E. & RAMEY, H.J. (1986) : Wellbore inertial and frictional effects on drillstem tests from naturally-fractured reservoirs. SPE Paper, **15478**.

SALZ, L.B. (1977) : Relationship between fracture propagation pressure and pore

pressure. SPE Paper, **6870**.
SAMANIEGO, F.V. & CINCO-LEY, H. (1983) : Pressure transient analysis for naturally fractured gas reservoirs : field case. SPE Paper, **12010**.
SAMHABER, R. (1977) : Sandbekämpfung in Erdölförderbohrungen - Erfahrungen in der Bundesrepublik Deutschland und Österreich. 26. DGMK Main Meeting, Berlin/Germany FRG, 05. - 06.10.1977.
SAMPATH, K. & KEIGHIN, C.W. (1982) : Factors affecting gas slippage in tight sandstones of Cretaceous age in the Uinta Basin. J. Petrol. Technol., **34** : 2715 - 2720. SPE Paper, **9872** (1981).
SAMPSON, L.E. & MARCOU, J.A. (1988) : Interference test in a fractured carbonate : a Lisburne case history. SPE Paper, **18138**.
SAMUELS, G. (1980) : An evaluation of the geopressured energy resource of Louisiana and Texas. SPE/DOE Paper, **8948**.
SANDAL, H. & RAMEY, H.J. (1978) : Interference testing with wellbore storage and skin effect at the produced well. SPE Paper, **7454**.
SANDY, J., WIGGINS, M. & VENDITTO, J. (1986) : Crosslinked DGG gels carry high proppant loads in low-temperature applications. Oil Gas J., 01.09.**1986** : 52 - 54.
SANGREE, J.B. (1969) : What you should know to analyze core fractures. World Oil, **168**/4 : 69 - 72.
SANTROL (1975) : Super sand, a new high-strength, low-cost proppant. Brochure; Santrol, Houston/Texas.
SANTROL (1985) : Long term stability of resin coated proppants. Brochure, 01.1985; Santrol, Houston/Texas.
SANTROL (1986 a) : Dual coating makes Santrol`s Super HS proppant twice as effective. Brochure, **STL 85562/1M/2-86**; Santrol, Houston/Texas.
SANTROL (1986 b) : Better sand control with resin coated gravel. Brochure, 09.1986; Santrol, Houston/Texas.
SARABIAN, A.R. (1980) : An analysis of pressure transient behavior in an infinite two-layered heterogeneous system with a vertical fracture. SPE Paper, **9045**.
SARATHI, P.S., ROARK, D. & STRYCKER, A.R. (1988) : Light oil steamflooding : a laboratory study. SPE Paper, **17447**.
SARDA, J.P. (1981) : Use of high-strength ceramic beads for propping deep hydraulic fractures. J. Petrol. Technol., **33** : 55 - 56. SPE Paper, **7572** (1978).
SARDA, J.P. (1984) : Liaison des puits par fracturation hydraulique en vue de la gazéification souterraine du charbon : le problème de la direction de fracture. Bull. Soc. Géol. France, **26**.
SARDA, J.P., PERREAU, P.J. & DEFLANDRE, J.P. (1988) : Acoustic emission interpretation for estimating hydraulic fracture extent. SPE Paper, **17723**, **18192**.
SARDA, J.P., WITTRISCH, C., PERREAU, P. & DEFLANDRE, J.P. (1987) : Downhole frac monitoring system developed. Oil Gas J., 06.04.**1987** : 40 - 46.
SARKAR, A.K. & SHARMA, M.M. (1988) : An experimental investigation of fines migration in two-phase flow. SPE Paper, **17437**.
SARMA, H.K. (1986) : Viscous fingering : one of the main factors behind poor flood efficiencies in petroleum reservoirs. Powder Technol., **48** : 39 - 49.
SATMAN, A. (1988) : Pressure transient analysis of a composite naturally fractured reservoir. SPE Paper, **18587**.
SATTLER, A.R., GALL, B.L., RAIBLE, C.J. & GILL, P.J. (1988) : Frac fluid systems for naturally fractured tight gas sandstones : a general case study from multiwell experiment stimulations. SPE Paper, **17717**.
SATTLER, A.R., HECKES, A.A. & CLARK, J.A. (1984) : Pressure core measurements in tight sandstone lenses during the multiwell experiment. SPE/DOE/GRI Paper, **12853**.
SATTLER, A.R., HUDSON, P.J., RAIBLE, C.J., GALL, B.L. & MALONEY, D.R. (1986) : Laboratory studies for the design and analysis of hydraulic fracture stimulations in lenticular, tight gas reservoirs. SPE Paper, **15245**.
SATTLER, A.R., RAIBLE, C.J. & GALL, B.L. (1985) : Integration of laboratory and field data for insight on the multiwell experiment paludal stimulation. SPE/DOE Paper, **13891**.

SATTLER, A.R., RAIBLE, C.J., GALL, B.L. & GILL, P.J. (1988) : Frac fluid systems for naturally fractured tight gas sandstones : a general case study from multiwell experiment stimulations. SPE Paper, **17717**.
SAUCIER, R.J. (1974) : Considerations in gravel pack design. J. Petrol. Technol., **26** : 205 - 212. SPE Paper, **4030** (1972). SPE Reprint Series, Well Completions, **5 a**/II (1978) : 50 - 57.
SAUCIER, R.J. (1988) : A field development program in hydraulic fracture stimulation. SPE Paper, **18257**.
SAUCIER, R.J. & LANDS, J.F. (1977) : Simulated in-situ perforation tests. SPE Paper, **6758**.
SAUNDERS, L.W. & McKENZIE, H.L. (1979) : Performance review of phenolic-resin gravel packing. SPE Paper, **8425**. J. Petrol. Technol., **33** (1981) : 221 - 228.
SAUNDERS, R.J. & SHIRAZI, M. (1986) : The role of the World Bank in gas development in developing countries. SPE Paper, **14632**.
SAWDON, W.A. (1937) : Gravel packing oil wells. Petrol. Engin., **1937**.
SAWOLO, N., KRUEGER, R.F. & MALY, G.P. (1983) : Case history of Yakin field : its development and sand control. J. Petrol. Technol., **35** : 23 - 30. SPE Paper, **10452** (1982).
SAWYER, W.K., LOCKE, C.D. & OVERBEY, W.K. (1974) : Simulation of finite-capacity vertical fracture in a gas reservoir. SPE Paper, **4593**.
SCANLON, T. (1988) : A new forum for Middle East analysis. Petrol. Rev., **1988**/2 : 30 - 31.
SCHAEFER, G.T. & POWERS, L.W. (1984) : Impact of rapidly changing economic environment on petroleum exploration and acquisition. SPE Paper, **12021**.
SCHAIBLE, D.F., AKPAN, B. & AYOUB, J.A. (1986) : Identification, evaluation, and treatment of formation damage, offshore Louisiana. SPE Paper, **14820**.
SCHECHTER, R.S. (1982) : A description of chemical precipitation mechanisms and their role in formation damage during stimulation by hydrochloric acid. SPE Paper, **10625**. J. Petrol. Technol., **34** : 2097 - 2112.
SCHENEWERK, P.A., KNAPP, R.M. & FERTL, W.H (1982) : The accuracy of pulsed neutron capture logs for residual oil saturation determinations. SPE Paper, **11148**.
SCHENK, L. & PUIG, F. (1983) : Aspects of compaction/subsidence in the Bolivar Coast heavy oil fields, highlighted by performance data of the M-6 project area. Proc. 1982 Forum on subsidence due to fluid withdrawals, 109 - 120; Checotah/Oklahoma.
SCHERUBEL, G.A. & CROWE, C.W. (1978) : Foamed acid, a new concept in fracture acidizing. SPE Paper, **7568**.
SCHETTLER, P.D., PARMLEY, C.R. & LEE, W.J. (1987) : Gas storage and transport in the Devonian shales. SPE Paper, **17070**.
SCHEUERMAN, R.F. (1984) : A new look at gravel pack carrier fluid properties. SPE Paper, **12476**. SPE Product. Engin., **1** (1986) : 9 - 16.
SCHEUERMAN, R.F. (1988) : A buffer-regulated HF acid for sandstone acidizing to 550 oF. SPE Product. Engin., 3 : 15 - 21. SPE Paper, **13563** (1985).
SCHINDEWOLF, E. (1985) : Erdgas in Wärmemarkt - Chancen und Risiken bis zum Jahr 2000. Talk BGW/DVGW-Arbeitstagung, Landesgruppe Niedersachsen/Bremen, Hameln/Germany FRG, 26.05.1985.
SCHLESINGER, B. (1982) : Outlook for U.S. gas energy, demand, and competitiveness. SPE Paper, **14632**.
SCHLOTTMAN, B.W., MILLER, W.K. II & LUEDERS, R.K. (1981) : Massive hydraulic fracture design for the East Texas Cotton Valley Sands. SPE Paper, **10133**.
SCHLUMBERGER (1972) : Log interpretation. Volume I - Principles. 113 pp.; New York.
SCHLUMBERGER (1974) : Log interpretation. Volume II - Applications. 116 pp.; New York.
SCHLUMBERGER (1982 a) : Dipmeter interpretation. Volume I - Fundamentals. 61 pp.; New York.
SCHLUMBERGER (1982 b) : Well evaluation developments Continental Europe. 296 pp.
SCHLUMBERGER (1983) : Stratigraphic high resolution dipmeter tool. Schlumberger Marketing Document, **M-086300** : 23 pp.; Houston/Texas.
SCHLUMBERGER (1985) : Array-sonic service. Schlumberger Brochure, **SMP-9050**;

Schlumberger Marketing Document, **M-089050**; Houston/Texas.
SCHLUMBERGER (1986 a) : Schlumberger`s FracHite Log. A major advance in hydraulic fracturing design technology. Brochure; Houston/Texas.
SCHLUMBERGER (1986 b) : Formation MicroScanner service. Schlumberger Brochure, **SMP-9100**; Schlumberger Marketing Document, **M-090033** and **M-089100** : 37 pp.; Houston/Texas.
SCHLUMBERGER (1987 a) : Fracture detection with logs. Techn. Rev., **37** : 22 - 34.
SCHLUMBERGER (1987 b) : Array sonic service. Schlumberger Brochure, **SMP-5060**; Houston/Texas.
SCHLUMBERGER (1987 c) : The emergence of geochemical well logging. Techn. Rev., **35**/2 : 27 - 35.
SCHLUMBERGER (1988 a) : Avoid a void. Schlumberger`s one-trip gravel pack evaluation and repair service. Schlumberger Brochure, **SMP 5048**; Houston/ Texas.
SCHMIDT, F. (1987) : Die Zukunft der Energieversorgung aus der Sicht der Mineralölindustrie. Erdöl-Erdgas-Kohle, **103** : 255 - 257.
SCHMIDT, R.A., BOADE, R.R. & BASS, R.C. (1979) : A new perspective on well shooting - the behavior of contained explosions and deflagrations. SPE Paper, **8346**. J. Petrol. Technol., **33** (1981) : 1305 - 1311.
SCHMIDT, R.A., WARPINSKI, N.R. & COOPER, P.W. (1980) : In-situ evaluation of several tailored-pulse well-shooting concepts. SPE Paper, **8934**.
SCHMIDT, S.H., CAUDLE, B.H. & MILLER, M.A. (1986) : Gas well decline analysis incorporating the real gas behaviour and non-Darcy flow. SPE Paper, **15521**.
SCHMIDT, V. & McDONALD, D.A. (1979 a) : Texture and recognition of secondary porosity in sandstones. In : SCHOLLE, P.A. & SCHLUGER, P.R. (Editors), Aspects of diagenesis. Soc. Econ. Paleont. Miner., Spec. Publ., **26** : 209 - 225; Tulsa/Oklahoma.
SCHMIDT, V. & McDONALD, D.A. (1979 b) : The role of secondary porosity in the course of sandstone diagenesis. In : SCHOLLE, P.A. & SCHLUGER, P.R. (Editors), Aspects of diagenesis. Soc. Econ. Paleont. Miner., Spec. Publ., **26** : 175 - 207; Tulsa/Oklahoma.
SCHÖNEICH, H. (1981) : Die Entwicklung der Erdöl- und Erdgasexploration in der Nordsee. Erdoel-Erdgas, **97** : 278 - 288.
SCHÖNEICH, H. (1983) : Erdöl und Erdgas in der Nordsee - Entwicklung von Exploration und Produktion 1982/83. Erdoel-Erdgas, **99** : 359 - 369.
SCHOLL, E.S. (1986) : Future manpower requirements in a dramatically changing petroleum industry. SPE Paper, **15444**.
SCHOLLE, P.A. (1977) : Current oil and gas production from North American Upper Cretaceous chalks. U.S. Geol. Surv. Circ., **767** : 51 pp.
SCHOLS, R.S. (1983) : Special topics on hydraulic fracturing. Seminar given at BEB in Hannover/Germany FRG, Unpubl. Rept., 10.1983.
SCHOLS, R.S. & VISSER, W. (1974) : Proppant bank build-up in a vertical fracture without fluid-loss. SPE Paper, **4834**.
SCHOONBEEK, J.B. (1976) : Land subsidence as a result of natural gas extraction in the province of Groningen. SPE Paper, **5751**.
SCHRANK, J.A. & HUNT, E. (1981) : Improved reservoir evaluation with the SEM. Scanning Electron Microscopy, **1981** : 573 - 578.
SCHRAUFNAGEL, R.A. (1987) : Multiple coal seam project. Quart. Rev. of Methane from Coal Seams Technol., **4** : 33 - 42.
SCHRAUFNAGEL, R.A. & LAMBERT, S.W. (1988) : Multiple coal seam project. Quart. Rev. of Methane from Coal Seams Technol., **5**/3-4 : 33 - 44; **6**/1 : 29 - 37.
SCHRIDER, L.A. (1985) : An assessment for development drilling in the Clinton Sandstone, a low permeable oil and gas reservoir in Northeastern Ohio. SPE Paper, **13939**.
SCHRIDER, L.A., OVERBEY, W.K., KOMAR, C.A. & PASINI, J. III (1977) : Natural gas from Eastern U.S. shales. SPE Paper, **6841**.
SCHRIDER, L.A. & WISE, R.L. (1980) : Potential new sources of natural gas. J. Petrol. Technol., **32** : 703 - 716. SPE Paper, **7628** (1978).
SCHRIETER, F.E. & SHAW, M.S. (1978) : Use of fine salt as a fluid loss material in acid fracturing stimulation treatments. SPE Paper, **7570**.
SCHROEDER, D.E. (1987) : Gravel pack studies in a full scale, high pressure

wellbore model. SPE Paper, **16890**.
SCHRÖDER, L. (1980) : Die Erdöl- und Erdgasexploration in der Bundesrepublik Deutschland 1979. Erdoel-Erdgas, **96** : 253 - 264.
SCHRÖDER, L. (1985) : Erdöl- und Erdgasexploration in der Bundesrepublik Deutschland 1984. Erdoel-Erdgas, **101** : 219 - 230.
SCHRÖDER, L. (1986) : Erdöl- und Erdgasexploration in der Bundesrepublik Deutschland 1985. Erdöl-Erdgas-Kohle, **102** : 438 - 448.
SCHRÖDER, L. (1987) : Erdöl- und Erdgasexploration in der Bundesrepublik Deutschland 1986. Erdöl-Erdgas-Kohle, **103** : 314 - 324.
SCHRÖDER, L. (1988) : Erdöl- und Erdgasexploration in der Bundesrepublik Deutschland 1987. Erdöl-Erdgas-Kohle, **104** : 299 - 308.
SCHRÖDER, L. & SCHÖNEICH, H. (1986) : International map of natural gas fields in Europe 1 : 2 500 000, Explanatory Notes. 2nd. edition, 175 pp.; Hannover/ Germany FRG.
SCHROEDER, R.H. & TUCKER, M.J. (1974) : Evaluation of completion practices for improved sand control success and longevity. SPE Paper, **5027**.
SCHUBERT, E. (1987) : Chancen und Risiken der derzeitigen Erdöl- und Erdgas-exploration. Erdöl-Erdgas-Kohle, **103** : 302 - 304.
SCHUCK, L.Z. & KEECH, T.W. (1975) : Monitoring acoustic emission from propagating fractures in petroleum reservoir rocks. Proc. 1st Conf. Acoustic Emission and Microseismic Activity in Geologic Structure and Materials, 309 - 338, Pennsylvan. State Univ., 06.1975.
SCHULTE, W.M. (1984) : Production from a hydraulically fractured well with well inflow limited to part of the fracture height. SPE/DOE/GRI Paper, **12882**. SPE Product. Engin., **1** (1986) : 333 - 343.
SCHULTE, W.M. & VRIES, A.S.d. (1985) : In-situ combustion in naturally fractured heavy oil reservoirs. Soc. Petrol. Engin. J., **25** : 67 - 77. SPE/DOE Paper, **10723** (1982).
SCHULTZ, W.E., SMITH, H.D., VERBOUT, J.L., BRIDGES, J.R. & GARCIA, G. (1983) : Experimental basis for a new borehole corrected pulsed neutron capture logging system (TMD). Trans. 24th Soc. Prof. Well Log Analysts Annual Logging Symp., Paper **CC**, Calgary/Alberta, 27. - 30.06.1983.
SCHUMACHER, J.P. & DICKERMAN, R.P. (1986) : Subzero hydraulic fracturing : a field case study, Lisburne Carbonate reservoir, Prudhoe Bay Unit, North Slope, Alaska. SPE Paper, **15068**.
SCHURTZ, G.C., BREINER, W.G. & COMEAUX, B.G. (1975) : New thru-tubing gravel pack technique. SPE Paper, **5660**.
SCHUSTER, C.L. (1978) : Detection within the wellbore of seismic signals created by hydraulic fracturing. SPE Paper, **7448**.
SCHWARTZ, D.H. (1969) : Successful sand control design for high rate oil and water wells. J. Petrol. Technol., **21** : 1193 - 1198. SPE Paper, **2330** (1968). SPE Reprint Series, Well Completion, **5** (1970) : 182 - 187.
SCHWARZ, H. & SCHÖBER, K. (1982) : Die Erschließung tiefer, geringpermeabler Gaslagerstätten durch großvolumige Frac-Behandlungen. gwf-gas/erdgas, **123** : 271 - 278.
SCOTT, J.N. (1986) : The politics of energy policy : the corporate perspective. SPE Paper, **14648**.
SCOTT, J.O. (1963) : The effect of vertical fractures on transient pressure behavior of wells. J. Petrol. Technol., **15** : 1365 - 1369.
SCOTT, J.O. (1979) : Application of a new method for determining flow characteristics of fractured wells in tight sands. SPE Paper, **7931**.
SCOTT, T.A. (1986) : Perforation breakdown improves Permian Basin stimulation results. SPE Paper, **15015**.
SCOTT, W.W. (1928) : Methods of completing wells in Gulf Coast fields. API Fall Meeting, Galveston/Texas, 05. - 06.09.1928.
SCOTTISH PETROLEUM ANNUAL (1987) : Scottish petroleum annual. 140 pp.; Aberdeen Petroleum Publ., Aberdeen/Scotland. ISBN 0-9508220-5-1.
SCRIMGEOUR, J.A. & SHEPHERD, C.E. (1988) : Maintening Balmoral field productivity while controlling sand production with gravel packs. SPE Paper, **18382**.
SEANARD, K.C. (1986) : Underbalanced perforating in a closed system. SPE Paper, **14828**.
SEARS, G.F. & PHILLIPS, N.V. (1987) : Fractured reservoir evaluation using Mon-

te Carlo techniques. J. Petrol. Technol., **39** : 71 - 76. SPE Paper, **13777** (1985).

SECCOMBE, J.C. & ANDERSON, G.E. (1982) : Selection of a fracture proppant in a tight gas field, bauxite vs. sand, Wamsutter area, Wyoming. SPE/DOE Paper, **10827**.

SECCOMBE, J.C. & SAKASHITA, B.J. (1985) : Preliminary economic assessment of the commercial potential for deep coalbed methane production from the Red Mountain Unit, Piceance Basin, Colorado. Gas Research Institute Rept., **GRI**-85/0318; Chicago/Illinois.

SEETHARAM, R.V. & DEANS, H.A. (1986) : Simulator models for single-well tracer test parameter estimation and model selection in heterogeneous formations. SPE Paper, **15435**.

SEIDEL, W.R. & STAHL, E.J. (1972) : Gas well stimulation with a viscous water-base fracturing fluid. J. Petrol. Technol., **24** : 1385 - 1390. SPE Paper, **3502** (1971).

SEKERA, P. (1986) : Oil prices 1921 to 2121 (past, present and future). CIM Paper, **86-37-59**.

SENECAL, J.E. & MULHERN, M.E. (1985) : Monterey Formation analysis : state-of-the-art interpretation of lithology and identification of fractures and fluid type in fine-grained, mineralogically variable rocks. Trans. 26th Soc. Prof. Well Log Analysts Annual Logging Symp., Paper **KKK**, Dallas/Texas, 17. - 20.06.1985.

SENER, I. (1986) : Relative diffusion flux : a practical concept for characterizing fracture to matrix gas transfer : application to a fractured reservoir model. SPE Paper, **15522**.

SENGUPTA, S.K., HAYATDAVOUDI, A., TIAB, J.O., KAIRA, S.K., LEBLANC, J.L. & SCHLUNTZ, E.K. (1982) : Effect of flow rate and rheology on shear strength of migrating formation fines due to flow of pseudoplastic fluids. SPE Paper, **10669**.

SEPR (1983) : Zirprop 126. The unique light high strength proppant for hydraulic fracturing. 47 pp.; Paris/France.

SERRA, K.V., REYNOLDS, A.C. & RAGHAVAN, R. (1983) : New pressure transient analysis methods for naturally fractured reservoirs. J. Petrol. Technol., **35** : 1902 - 1914, 2271 - 2283. SPE Paper, **10780** (1982).

SERRA, O. (1985) : Fundamentals of well-log interpretation. **1** : The acquisition of logging data. Developm. Petrol. Sci., **15 A** : 424 pp.; Elsevier, Amsterdam/Netherlands.

SERRA, O. (1986) : Advanced interpretation of wireline logs. 295 pp.; Schlumberger.

SERRA, O. (1987) : Fundamentals of well-log interpretation. **2** : The interpretation of logging data. Developm. Petrol. Sci., **15 B** : 684 pp.; Elsevier, Amsterdam/Netherlands.

SERRA, O., BALDWIN, J. & QUIREIN, J. (1980) : Theory, interpretation, and practical applications of natural gamma ray spectroscopy. Trans. 21st Soc. Prof. Well Log Analysts Annual Logging Symp., Paper **Q**, Lafayette/Louisiana, 08. - 11.07.1980.

SETSER, G.G. (1981) : Fracture detection by circumferential propagation of acoustic energy. SPE Paper, **10204**.

SETTARI, A. (1979) : Simulation of the hydraulic fracturing processes. SPE Paper, **7693**.

SETTARI, A. (1980) : Simulation of hydraulic fracturing processes. Soc. Petrol. Engin. J., **20** : 487 - 500. SPE Paper, **7693** (1979).

SETTARI, A. (1982) : Development and testing of a pseudo-three-dimensional model of hydraulic fracture geometry (P3DH). SPE Paper, **10505**. SPE Product. Engin., **1** (1986) : 449 - 466.

SETTARI, A. (1983 a) : Quantitative analysis of factors controlling vertical fracture growth (containment). SPE Paper, **11629**.

SETTARI, A. (1983 b) : A new general model of fluid loss in hydraulic fracturing. SPE Paper, **11625**. Soc. Petrol. Engin. J., **25** (1985) : 491 - 501.

SETTARI, A. (1985) : Quantitative analysis of factors influencing vertical and lateral fracture growth. SPE/DOE Paper, **13862**. SPE Product. Engin., **3** (1988) : 310 - 322.

SETTARI, A. (1988 a) : General model of fluid flow (leak-off) from fractures induced in injection operations. SPE Paper, **18197**.
SETTARI, A. (1988 b) : Modelling of fracture and deformation processes in oil sands. 4th UNITAR/UNDP Conference on heavy crude and tar sands, Edmonton/Alberta.
SETTARI, A. & CLEARY, M.P. (1982) : Three-dimensional simulation of hydraulic fracturing. SPE Paper, **10504**. J. Petrol. Technol., **36** (1984) : 1177 - 1190.
SETTARI, A. & CLEARY, M.P. (1986) : Development and testing of a pseudo-three dimensional model of hydraulic fracture geometry (P3DH). SPE Product. Engin., **1** : 449 - 466. SPE Paper, **10505** (1982).
SETTARI, A., KRY, R.P. & YEE, C.T. (1988) : Coupling of fluid flow and soil behaviour to model injection into unconsolidated oil sands. CIM Paper, **88-39-72**.
SETTARI, A. & PRICE, H.S. (1980) : Simulation of hydraulic fracturing in low-permeability reservoirs. SPE Paper, **8939**. Soc. Petrol. Engin. J., **24** (1984) : 141 - 152.
SETTARI, A., PUCHYR, P.J. & BACHMAN, R.C. (1987) : Partially decoupled modeling of hydraulic fracturing processes. SPE Paper, **16031**.
SETTARI, A. & RAISBECK, J.M. (1981) : Analysis and numerical modeling of hydraulic fracturing during cyclic steam stimulation in oil sands. J. Petrol. Technol., **33** : 2201 - 2212. SPE Paper, **9078**.
SEVOUGIAN, S.D., SCHECHTER, R.S. & SEPEHRNOORI, K. (1987) : Optimization of vertical acid fractures in steady-state flow. SPE Paper, **16252**.
SHAFFER, R.J., THORPE, R.K., INGRAFFEA, A.R. & HEUZE, F.E. (1984) : Numerical and physical studies of fluid-driven fracture propagation in jointed rock. SPE/DOE/GRI Paper, **12881**.
SHAH, S.N. (1982) : Proppant settling correlations for non-Newtonian fluids under static and dynamic conditions. Soc. Petrol. Engin. J., **22** : 164 - 170, 556. SPE Paper, **9330** (1980).
SHAH, S.N. (1984) : Correlations predict friction pressures of fracturing gels. Oil Gas J., 16.01.**1984** : 92 - 98.
SHAH, S.N. (1986) : Proppant-settling correlations for non-Newtonian fluids. SPE Product. Engin., **1** : 446 - 448. SPE Paper, **13835** (1984).
SHAH, S.N. & LEE, Y.N. (1986) : Friction pressures of proppant-laden hydraulic fracturing fluids. SPE Product. Engin., **1** : 437 - 445. SPE Paper, **13836** (1984).
SHAH, S.N., LEE, Y.N. & JENSEN, D.G. (1985) : Frac treatment quality improvement with field rheology unit. Oil Gas J., 04.02.**1985** : 47 - 51.
SHAH, S.N., SMITH, D.R. & DONALDSON, D.E. (1983) : Ultra high sand concentration in hydraulic fracturing - a case history. SPE Paper, **11578**.
SHAH, S.H. & WATTERS, L.T. (1986) : Time and shear effects on rheological properties of crosslinked fluids - an evaluation method. SPE Product. Engin., **1** : 55 - 61. SPE Paper, **12923** (1984).
SHANKS, R.T., KWON, B.S., VRIES, M.R.d. & WICHMAN, P.A. (1976) : A review of fracture detection with well logs. Soc. Petrol. Engin. AIME Publ., **6159** : 1 - 17.
SHARABI, H. (1988) : The next Arab decade. Alternative futures. 347 pp.; Westview Press, Boulder/Colorado; Mansell Publ., London/England. ISBN 0-8133-7440-5.
SHARMA, B., HONARPOUR, M., SZPAKIEWICZ, M. & SCHATZINGER, R. (1987) : Critical heterogeneities in a barrier bar deposit and their influences on waterflood and chemical EOR operations : Part I. SPE Paper, **16749**.
SHARMA, G.D., PATIL, S.L., KAMATH, V.A. & GODBOLE, S.P. (1988) : The potential of natural gas in the Alaskan Arctic. SPE Paper, **17456**.
SHARMA, M.M. & YORTSOS, Y.C. (1986) : Permeability impairment due to fines migration in sandstones. SPE Paper, **14819**.
SHARMA, M.M., YORTSOS, Y.C. & HANDY, L.L. (1985) : Release and deposition of clays in sandstones. SPE Paper, **13672**.
SHAUGHNESSY, C.M. & KUNZE, K.R. (1981) : Understanding sandstone acidizing leads to improved field practices. J. Petrol. Technol., **33** : 1196 - 1202. SPE Paper, **9388** (1980).

SHAUGHNESSY, C.M. & SALATHIEL, W.M. (1977) : Method of consolidating subterranean formations using diluted resins. U.S. Patent, **3,976,140**.

SHAUGHNESSY, C.M., SALATHIEL, W.M. & PENBERTHY, W.L. (1977) : A new, low-viscosity epoxy sand consolidation process. SPE Paper, **6803**. J. Petrol. Technol., **30** (1978) : 1805 - 1812. SPE Reprint Series, Well Completions, **5 a**/II (1978) : 95 - 102.

SHAW, J.S., GATENS, J.M. III, LANCASTER, D.E., LEE, W.J., AVARY, K.L. & TERRY, D.P. (1986) : Reservoir and stimulation analysis of Devonian Shale gas field. SPE Paper, **15938**.

SHAW, M.N. & SUTHERLAND, W.G. (1988) : Innovative approaches to drilling high angle wells in Bass Strait. IADC/SPE Paper, **17233**.

SHEA, C. & BUCHER, D.H. (1982) : Foam fracturing of Upper Devonian Benson Formation in Central West Virginia. SPE/DOE Paper, **10826**.

SHELL (1979) : WISPER - Acid fracturing through widely spaced perforations. Shell Intern. Research Maatsch., European Patent, **0,005,874**.

SHELLEY, R.F. & McGOWEN, J.M. (1986) : Pump-in test correlation predicts proppant placement. SPE Paper, **15151**.

SHERRARD, D.W., BRICE, B.W. & MACDONALD, D.G. (1986) : Application of horizontal wells at Prudhoe Bay. SPE Paper, **15376**. J. Petrol. Technol., **39** (1987) : 1417 - 1425.

SHI, G.C. & CHA, B.C.K. (1974) : A fracture criterion for three-dimensional crack problems. Engin. Fract. Mechan., **6** : 699 - 723.

SHIELDS, C. (1974) : The dipmeter used to recognize and correlate depositional environment. Trans. 3rd Europ. Symp., Soc. Prof. Well Log Analysts, Paper **H**, London/England.

SHLYAPOBERSKY, J. (1985) : Energy analysis of hydraulic fracturing. Proc. 26th U.S. Rock Mechan. Symp., Rapid City/South Dakota, 26. - 28.06.1985, pp. 539 - 546. In : ASHWORTH, E. (Editor), Research & engineering applications in rock masses; Balkema, Rotterdam/Boston.

SHLYAPOBERSKY, J. (1987) : Field measurements of fracture toughness of rock by hydraulic fracturing in light of damage mechanics. 24th Annual Meeting of the Society of Engineering Science, Salt Lake City/Utah, 09.1987.

SHLYAPOBERSKY, J. (1988) : Hydraulic fracture determination of minimum in-situ stress from extended pressure falloff and from fracture reopening and flowback tests. 2nd Intern. Workshop on Hydraulic Fracturing Stress Measurements, Minneapolis/Minnesota, 06.1988.

SHLYAPOBERSKY, J., WALHAUG, W.W., SHEFFIELD, R.E. & HUCKABEE, P.T. (1988) : Field determination of fracturing parameters for overpressure calibrated design of hydraulic fracturing. SPE Paper, **18195**.

SHLYAPOBERSKY, J., WONG, G.K. & WALHAUG, W.W. (1988) : Overpressure calibrated design of hydraulic fracture stimulations. SPE Paper, **18194**.

SHORT, J.A. (1978) : Massive frac treatments tapping tight gas sands in Uinta Basin. Oil Gas J., 01.**1978** : 71 - 77.

SHREDER, L.A. (1984) : Case history of acid stimulation in the Fryburg Formation in Western North Dakota. SPE Paper, **12921**. SPE Product. Engin., **1** (1986) : 405 - 410.

SHRYOCK, S.G. (1980) : Gravel packing studies in a full scale, deviated model wellbore. SPE Paper, **9421**. J. Petrol. Technol., **35** (1983) : 603 - 609.

SHRYOCK, S.G., AHMAD, S., MELOY, J.M. & KENT, J.W. (1988) : Gravel packing for the Duri steamflood. OSEA Paper, **88164**.

SHRYOCK, S.G., DUNLAP, R.G. & MILLHONE, R.S. (1979) : Preliminary results from full-scale gravel-packing studies. J. Petrol. Technol., **31** : 669 - 675. SPE Paper, **7123** (1978).

SHRYOCK, S.G. & MILLHONE, R.S. (1979) : Gravel packing studies in a full-scale, vertical model wellbore - Progress Report 2. J. Petrol. Technol., **31** : 1137 - 1143. SPE Paper, **8429**.

SHUCK, L.Z. (1977) : Method for selectively orienting induced fractures in subterranean earth formations. U.S. Patent, **4,005,750** (01.02.1977).

SHURTZ, G.C. & BREINER, W.G. (1975) : New thru-tubing gravel pack techniques. SPE Paper, **5660**.

SIBBIT, A.M. & FAIVRE, O. (1985) : The dual laterolog response in fractured rocks. Trans. 26th Soc. Prof. Well Log Analysts Annual Logging Symp., Paper

T, Dallas/Texas, 17. - 20.06.1985.
SIEGFRIED, R. & SIMMONS, G. (1978) : Characterization of oriented cracks with differential strain analysis. J. Geophys. Res., **83** : 1269 - 1278.
SIEVERT, J.A., WAHL, H.A., CLARK, P.E. & HARKIN, M.W. (1981) : Prop transport in a large vertical symposium. SPE/DOE Paper, **9865**.
SILVA, R.J., NIKO, H., BERGH, J.N.v.d. & SANCEVIC, Z.A. (1988) : Numerical modeling of gas and water injection in a geologically complex volative oil reservoir in the Lake Maracaibo area, Venezuela. SPE Paper, **18322**.
SIMMONS, A.L. (1983) : STRESSFRAC - a new dimension in well treatment. SPE Paper, **11827**.
SIMMONS, E.L. & RAU, W.E. (1988) : Predicting deepwater fracture pressures : a proposal. SPE Paper, **18025**.
SIMMONS, G., SIEGFRIED, R.W. & FEVES, M. (1974) : Differential strain analysis : a new method for examining cracks in rocks. J. Geophys. Res., **79** : 4383 - 4385.
SIMOES, W.C. (1985) : Mergers in the petroleum industry. CIM Paper, **85-36-20**.
SIMON, D.E. & COON, R.M. (1976) : Evaluation of fluid pH effects on low permeability sandstones. SPE Paper, **6010**.
SIMON, D.E., COULTER, G.R., KING, G.E. & HOLMAN, G.B. (1982) : North Sea chalk completions - a laboratory study. J. Petrol. Technol., **34** : 2531 - 2536. SPE Paper, **10395** (1981).
SIMON, D.E., KAUL, F.W. & CULBERTSON, J.N. (1977) : Anadarko Basin Morrow-Springer sandstone stimulation study. SPE Paper, **6757**.
SIMON, D.E. & UNDERWOOD, P. (1977) : Morrow stimulation in Southeast New Mexico. SPE Paper, **6375**.
SIMONSON, E.R., ABOU-SAYED, A.S. & CLIFTON, R.J. (1978) : Containment of massive hydraulic fractures. SPE Paper, **6089**. Soc. Petrol. Engin. J., **18** : 27 - 32.
SINAL, M.L. & LANCASTER, G. (1987) : Liquid CO2 fracturing : advantages and limitations. J. Canad. Petrol. Technol., **1987**/9-10 : 16 - 30.
SINCLAIR, A.R. (1971) : The effects of heat transfer in deep well fracturing. J. Petrol. Technol., **23** : 1484 - 1492. SPE Paper, **3011** (1970). SPE Reprint Series, Well Completions, **5 a**/II (1978) : 237 - 245.
SINCLAIR, A.R. (1980) : Geothermal fracture stimulation technology. **1** : Fracturing proppants and their properties (**1980**/7); **2** : High temperature proppant testing, 53 pp. (**1980**/7). **4** : Proppant analysis at geothermal conditions (**1981**/1). Geothermal Reservoir Well Stimulation Program, U.S. Dept. Energy, Contract, **DE-AC-32-79AL 10563**.
SINCLAIR, A.R. & GRAHAM, J.W. (1977) : Resin coated gravel is used to help control sand. Oil Gas J., 18.04.**1977** : 58 - 60.
SINCLAIR, A.R. & GRAHAM, J.W. (1978 a) : A new proppant for hydraulic fracturing. Paper **78-PET-34**, ASME Technol. Conference, Houston/Texas, 05. - 08.11.1978.
SINCLAIR, A.R. & GRAHAM, J.W. (1978 b) : An effective method of sand control. SPE Paper, **7004**.
SINCLAIR, A.R., GRAHAM, J.W. & SINCLAIR, C.P. (1983) : Improved well stimulation with resin-coated proppants. SPE Paper, **11579**.
SINCLAIR, A.R., PITTARD, F.J. & HANOLD, R.J. (1980) : Geothermal well stimulation. Geothermal Resources Council Trans., **4** : 423 - 426.
SINCLAIR, A.R., TERRY, W.M. & KIEL, O.M. (1974) : Polymer emulsion fracturing. J. Petrol. Technol., **26** : 731 - 738. SPE Paper, **4675** (1973).
SINGH, P.K. & AGARWAL, R.G. (1988) : Two-step rate test : a new procedure for determining formation parting pressure. SPE Paper, **18141**.
SINGH, P.K., AGARWAL, R.G. & KRASE, L.D. (1988) : Systematic design and analysis of step rate tests to determine formation parting pressure. SPE Paper, **16798**.
SINGHAI, A.K. & CARD, C.C. (1988) : Monitoring of steam stimulation in the McMurray formation, Athabasca deposit, Alberta. J. Petrol. Technol., **40** : 483 - 490. SPE Paper, **14215** (1985).
SINHA, B.K. (1976) : Fluid leakoff under dynamic and static conditions utilizing the same equipment. SPE Paper, **6126**.
SINHA, B.K. & ELBEL, J.L. (1983) : Stimulation treatment design and evaluation

of old producing wells. SPE Paper, **11552**.
SINHA, B.K. & FURLONG, K.D. (1979) : Gas well deliverability prediction - for hydraulically fractured (vertical) wells in tight reservoirs. SPE Paper, **7948**.
SINHA, B.K. & SCHLITTLER, J.E. (1977) : Analysis of pressure fall-off data from wells penetrating highly dipping formations and their stimulation treatment design. SPE Paper, **6886**.
SINOR, J.E. (1986) : Long-term cycles and the prediction of petroleum demand. J. Petrol. Technol., **38** : 669 - 672.
SINSON, C.M., OGBE, D.O. & DEHGHANI, K. (1988) : Optimization of well stimulation strategies in oil and gas fields. SPE Paper, **17792**.
SKAGGS, B.C. (1984) : Transport efficiency of high-density gravel packing slurries. SPE Paper, **12480**.
SKIDMORE, T.D. & WILLINGHAM, J.R. (1976) : Acid fracture stimulation of the Ekofisk. SPE Paper, **6125**.
SLATON, M. (1988) : Puddle pack solves the hole problem. Southwest Oil World, **36**/9 : 14 - 16.
SLATTER, T.D., RUCKER, J.R. & CRISP, E.L. (1986) : Natural gas stimulation in tight clay-bearing sandstone using foamed CO2 as hydraulic fracturing media. SPE Paper, **15238**.
SLOAT, B.F. (1988) : Field test results with alkaline-potassium solutions to permanently stabilize clays and lower residual oil saturations. SPE Paper, **17537**.
SLOAT, B.F. & LARSEN, D. (1984) : How to stabilize clays and improve injectivity. SPE Paper, **12927**.
SLUSSER, M.L. & RIECKMANN, M. (1976) : Fracturing low permeable gas reservoirs. Erdoel-Erdgas, **92** : 70 - 76.
SMALL, G.P. (1985) : Steam-injection profile control using limited-entry perforations. SPE Paper, **13607**. SPE Product. Engin., **1** (1986) : 388 - 394.
SMITH, A.F., SMITH, T.E. & MONSHAUSEN, T. (1988) : Ekofisk subsidence : conceptual and design considerations along the road to jacking. OTC Paper, **5652**.
SMITH, C.F., CROWE, C.W. & NOLAN, T.J. (1968) : Secondary deposition of iron compounds following acid treatments. SPE Paper, **2358**. J. Petrol. Technol., **21** (1969) : 1121 - 1129.
SMITH, C.F., DOLLARHIDE, F.E. & BYTH, N.J. (1978) : Acid corrosion inhibitors - are we getting what we need? J. Petrol. Technol., **30** : 737 - 747. SPE Paper, **5644** (1975).
SMITH, C.F. & HENDRICKSON, A.R. (1965) : Hydrofluoric acid stimulation of sandstone reservoirs. J. Petrol. Technol., **17** : 215 - 222. SPE Reprint Series, Well Completion, **5** (1970) : 211 - 218.
SMITH, C.F. & MURPHY, D.L. (1974) : Field evaluation of gas well stimulation methods - Sutton County, Texas. SPE Paper, **4800**.
SMITH, C.F., PAVLICH, J.P. & SLOVINSKY, R.L. (1964) : Potassium, calcium treatments inhibit clay swelling. Oil Gas J., 30.11.**1964** : 80 - 83.
SMITH, C.F., ROSS, W.M. & HENDRICKSON, A.R. (1965) : Hydrofluoric acid stimulation - developments for field application. SPE Paper, **1284**.
SMITH, C.W., BASS, R.C. & TYLER, L.D. (1978) : Puff n`tuff, a residual stress-gas fracturing experiment. Proc. 19th U.S. Rock Mechan. Symp., p. 7.
SMITH, D.J. (1988) : Project management of subsidence and Ekofisk jacking project. OTC Paper, **5655**.
SMITH, F.M. (1968) : Method of permeably consolidating incompetent sands with a heat-curable resin. U.S. Patent, **3,373,812** (19.03.1968).
SMITH, G.E. (1986) : Fluid flow and sand production in heavy oil reservoirs under solution gas drive. SPE Paper, **15094**. SPE Product. Engin., **3** (1988) : 169 - 180.
SMITH, H.D., ARNOLD, D.M. & PEELMAN, H.E. (1983) : Applications of a new borehole corrected pulsed neutron capture logging system (TMD). Trans. 24th Soc. Prof. Well Log Analysts Annual Logging Symp., Paper **DD**, Calgary/Alberta, 27. - 30.06.1983.
SMITH, H.D., ROBBINS, C.A., ARNOLD, D.M., GADEKEN, L.L. & DEATON, J.G. (1983) : A multi-function compensated spectral natural gamma ray logging system. SPE Paper, **12050**.

SMITH, J.E. (1965) : Hydraulic fracture treatment design. SPE Paper, **1286**.
SMITH, J.E. (1973) : Effect of incomplete fracture fill up at the well-bore on productivity ratio. SPE Paper, **4677**.
SMITH, K.T. & SCHMID, J.C. (1984) : Application of an alternate stimulation method for the Ferguson Sandstone in the Powder River Basin. SPE Paper, **12902**.
SMITH, M.A., DAWSON, J. & SCOGGINS, D. (1983) : High temperature, crosslinked high strength acid system. Proc. 30th Annual Southwestern Petroleum Short Course, Texas Techn. Univ., Lubbock/Texas, 27. - 28.04.1983, pp. 163 - 167.
SMITH, M.B. (1979) : Effect of fracture azimuth on production with application to the Wattenberg gas field. SPE Paper, **8298**.
SMITH, M.B. (1981) : Stimulation design for short precise hydraulic fractures - mHF. SPE Paper, **10313**. Soc. Petrol. Engin. J., **25** (1985) : 371 - 379.
SMITH, M.B. (1987) : The role of pressure analysis. SPE Continuing Education, Hydraulic Fracturing Seminar, London/England, 01.12.1987.
SMITH, M.B., HOLMAN, G.B., FAST, C.R. & COVLIN, R.J. (1978) : The azimuth of deep penetrating fractures in the Wattenberg field. J. Petrol. Technol., **30** : 185 - 193. SPE Paper, **6092**.
SMITH, M.B., LOGAN, J.M. & WOOD, M.D. (1980) : Fracture azimuth - a shallow experiment. J. Energy Resourc., **102** : 99 - 105. Trans. ASME, **102** : 99 - 105.
SMITH, M.B., ROSENBERG, R.J. & BOWEN, J.F. (1982) : Fracture width - design vs. measurement. SPE Paper, **10965**.
SMITH, M.B., MILLER, W.K. II & HAGA, J. (1987) : Tip screenout fracturing : a technique for soft, unstable formations. SPE Product. Engin., **2** : 95 - 103. SPE Paper, **13273** (1984).
SMITH, M.B., REN, N.K., SORRELS, G.G. & TEUFEL, L.W. (1985) : A comprehensive fracture diagnostics experiment : Part II - Comparison of seven fracture azimuth measurements. SPE/DOE Paper, **13894**. SPE Product. Engin., **1** (1986) : 423 - 431.
SMITH, M.J., CLARK, G.J., HENDRICKSON, A.R. & EGBOGAH, E.O. (1986) : Acidization of dirty sandstones with buffered HF acid systems. SPE Paper, **14826**.
SMITH, P.J. & BROWN, C.E. (1984) : The interpretation of tracer test response from heterogeneous reservoirs. SPE Paper, **13262**.
SMITH, R.C., DOBKINS, T.A., SMITH, M.B. & PATTILLO, P.D. (1978) : Interpretation guidelines for after frac temperature logs. SPE Paper, **7559**.
SMITH, R.C. & STEFFENSON, R.J. (1973) : Improved interpretation guidelines for temperature profiles in water injection wells. SPE Paper, **4649**.
SMITH, R.C. & STEFFENSON, R.J. (1975) : Interpretation of temperature profiles in water-injection wells. J. Petrol. Technol., **27** : 777 - 784. SPE Paper, **7220**.
SMITH, R.H. & POLK, A.C. (1947) : Use of plastics in consolidating loose sands in wells. Trans. AIME, **170** : 243 - 249.
SMITH, R.K., SAWYER, W.K. & ESPOSITO, P.R. (1981) : The effect of anisotropy and vertical fracture penetration by production decline curves for low permeability gas wells. SPE Paper, **10369**.
SMITH, T.K. (1969) : Sand consolidation through production tubing. Oil Gas J., 21.04.**1969**.
SMITS, R.M.M. & WAAL, J.A.d. (1988) : A comparison between the pressure-lag model and the rate-type model for the prediction of reservoir compaction and surface subsidence. SPE Form. Eval., **3** : 357 - 363. SPE Paper, **16389** (1986).
SMITS, R.M.M., WAAL, J.D.d. & KOOTEN, J.F.C. (1988) : Prediction of abrupt reservoir compaction and surface subsidence caused by pore collapse in carbonates. SPE Form. Eval., **3** : 340 - 346. SPE Paper, **15642** (1986).
SMOLARCHUK, P. & DILL, W. (1986) : Iron control in fracturing and acidizing operations. CIM Paper, **86-37-28**.
SNEDDON, I.N. & LOWENGRUB, M. (1969) : Crack problems in the classical theory of elasticity. Wiley, New York.
SNEIDER, R.M., KING, H.R., HAWKES, H.E. & DAVIS, T.K. (1981) : Methods for detection and characterization of reservoir rock, Deep Basin Gas area, Western Canada. SPE Paper, **10072**.
SNOW, S.E. & HOUGH, E.V. (1988) : Field and laboratory experience in stimulating Ekofisk area North Sea chalk reservoirs. SPE Paper, **18225**.
SOBBI, F.A. & STEWART, G. (1988) : Well test interpretation for naturally frac-

tured reservoirs. SPE Paper, **18173**.
SOBOCINSKI, D.P., LESKOVA, H. & GREEBE, F. (1984) : Simulating a steamflood at the Georgsdorf field, Federal Republic of Germany. J. Petrol. Technol., **36** : 1952 - 1964. SPE Paper, **11951** (1983).
SOBOTKA, S.P. (1986) : Financing mergers and acquisitions. SPE Paper, **14642**.
SOEDER, D.J. (1988) : Porosity and permeability of Eastern Devonian gas shale. SPE Form. Eval., **3** : 116 - 124. SPE Paper, **15213** (1986).
SOEDER, D.J. & CHOWDIAH, P. (1988) : Comparison of pore geometry in high- and low-permeability sandstones, Travis Peak Formation, East Texas. SPE Paper, **17729**.
SOEDER, D.J. & DOHERTY, M.G. (1983) : The effects of laboratory drying techniques on the permeability of tight sandstone core. SPE/DOE Paper, **11622**. SPE Form. Eval., **1** (1986) : 16 - 22.
SOEDER, D.J. & RANDOLPH, P.L. (1987) : Porosity, permeability, and pore structure of the tight Mesaverde Sandstone, Piceance Basin, Colorado. SPE Form. Eval., **2** : 129 - 136. SPE Paper, **13134** (1984).
SOKOLL, R.E. & ROGERS, A.K. (1982) : Short interval gravel packing technique. SPE Paper, **10668**.
SOLBERG, P., LOCKNER, D. & BYERLEE, J.D. (1980) : Hydraulic fracturing in granite under geothermal conditions. Intern. J. Rock Mechan., Min. Sci., Geomechan. Abstr., **17** : 25 - 33.
SOLIMAN, M.Y. (1983) : Modifications to production increase calculation for a hydraulically fractured well. J. Petrol. Technol., **35** : 170 - 172.
SOLIMAN, M.Y. (1986 a) : Technique for considering fluid compressibility and temperature changes in mini-frac analysis. SPE Paper, **15370**.
SOLIMAN, M.Y. (1986 b) : Fracture conductivity distribution studied. Oil Gas J., 10.02.**1986** : 89 - 93.
SOLIMAN, M.Y. & HUNT, J.L. (1985) : Effect of fracturing fluid and its cleanup on well performance. SPE Paper, **14514**.
SOLIMAN, M.Y., HUNT, J.L. & EL RABAA, A.W. (1988) : On fracturing horizontal wells. SPE Paper, **18542**.
SOLIMAN, M.Y., VENDITTO, J.J. & SLUSHER, G.L. (1984) : Evaluating fractured well performance using type curves. SPE Paper, **12598**.
SOLLEE, S.S. (1985) : Gravel pack logging experiments. SPE Paper, **14163**.
SOLHEIM, O. (1987) : The oil market, OPEC and Norwegian foreign policy from the point of view of an oil company. In : Norway, OPEC and IEA - global dialogue and price stabilizing measures in the oil market? Intern. Polit., **1987**/5-6, Norwegian Institute of International Affairs.
SOLUM, J.R. (1984) : A new technique in sand control using liner vibration with gravel packing. SPE Paper, **12479**.
SOLUM, J.R. (1986) : Evaluation of gravel blending and pumping techniques for accurate control and protection of the gravel during the gravel packing operation. SPE Paper, **14125**.
SOLUM, J.R. & RAMEZANI, R. (1987) : Design and application of downhole pump protectors for sand control with submersible pumping. SPE Artificial Lift Workshop, Houston/Texas, 22. - 24.04.1987.
SOLUM, J.R., RAMEZANI, R. & MIMS, M.G. (1988) : New technology for gravel packing through perforation tubes. SPE Paper, **17495**.
SOMERTON, W.H., CHEN, S.P., SCHUH, M.J. & YUEN, J.P. (1984) : Screening tests to evaluate formation damage in EOR operations. SPE Paper, **12500**.
SOMMER, J. (1987) : CO2-Frac Fehndorf 2T - first MHF treatment with CO2-foam in Europe. Erdöl-Erdgas-Kohle, **103** : 22 - 26 (in German).
SOMMER, J. (1988) : Carbon dioxide frac Fehndorf 2 T - first MHF treatment with CO2 foam in Europe. Oil Gas, **14**/1 : 16 - 19.
SONI, Y. & HARMON, R.A. (1986) : Simulation of the Saner Ranch fracture-assisted steamflood pilot. J. Canad. Petrol. Technol., **1986**/1-2 : 57 - 70. J. Petrol. Technol., **39** (1987) : 684 - 696. SPE Paper, **13036** (1984). CIM Paper, **85-36-04** (1985).
SONIER, F. (1987) : A new simulator for naturally fractured reservoirs. SPE Paper, **16006**.
SOOKPRASONG, A. (1987) : Determination of fracture conductivity and length. SPE Continuing Education, Hydraulic Fracturing Seminar, London/England,

01.12.1987.
SOOKPRASONG, P.A. & ELBEL, J.L. (1985) : Hydraulic fracture evaluation in a geopressured reservoir. SPE Paper, **13816**.
SORRELLS, G.G., GRANT, L., MAHRER, K.D., MAUK, F.J., MULCAHY, C.C., STUMP, B.W. & SWANSON, J.G. (1986) : Hydraulic fracture diagnostics research. Gas Research Institute, Annual Rept., **GRI-86/0283** : 105 pp.
SORRELS, G.G. & MULCAHY, C.C. (1986) : Advances in the microseismic method of hydraulic fracture azimuth estimation. SPE Paper, **15216**. In Focus - Tight Gas Sands, **3** : 14 - 24.
SPARLIN, D.D. (1967) : A new sand control technique for old sand problems. API Paper, **906-12-H**.
SPARLIN, D.D. (1969) : Fight sand with sand - a realistic approach to gravel packing. SPE Paper, **2649**.
SPARLIN, D.D. (1971) : Pressure-packing technique controls unconsolidated sand. Oil Gas J., 18.10.**1971**.
SPARLIN, D.D. (1974) : Sand and gravel - a study of their permeabilities. SPE Paper, **4772**.
SPARLIN, D.D. (1987) : Gravel packing innovations : what`s new? Drill. Magaz., 01.11.**1987** : 17 - 20.
SPARLIN, D.D. & BOND, W.C. (1969) : Method of placing gravel rocks. U.S. Patent, **3,443,637** (23.05.1969).
SPARLIN, D.D. & COPELAND, C.T. (1972) : Pressure packing with concentrated gravel slurry. SPE Paper, **4033**. SPE Reprint Series, Well Completions, **5 a**/II (1978) : 58 - 68.
SPARLIN, D.D. & HAGEN, R. (1983) : The new intermediate strength. Drilling, **1983**/6.
SPARLIN, D.D. & HAGEN, R.W. (1985) : Gravel packs require planning. Oil Gas J., 05.08.**1985** : 87 - 90.
SPARLIN, D.D. & HAGEN, R.W. (1986) : Soluble fluid-loss additives can reduce well productivities and prevent complete gravel placement. SPE Paper, **14817**. SPE Product. Engin., **3** (1988) : 63 - 68.
SPATSCHEK, H.R. (1987) : Die internationale Ölindustrie - Probleme und Aufgaben in Gegenwart und Zukunft. Leobener Bergmannstag 1987, Leoben/Austria, 22.-26.09.1987. Erdöl-Erdgas-Kohle, **103** : 343 - 344.
SPE OF AIME (1975) : General index to publications of the Society of Petroleum Engineers of AIME 1967 - 1974. Henry L. Doherty Series. 255 pp.; Dallas/Texas.
SPE OF AIME (1982) : General index to publications of the Society of Petroleum Engineers of AIME 1975 - 80. Henry L. Doherty Series. 147 pp.; Dallas/Texas.
SPE (1986) : General index to publications of the Society of Petroleum Engineers 1981 - 85. Henry L. Doherty Series. 233 pp.; Richardson/Texas.
SPENCER, A.M. (1981) : Astrofrac - a safe cost effective explosive fracturing method for tight formations. Proc. Intern. Gas Research Conf., Los Angeles, 28.09. - 01.10.1981.
SPENCER, A.M. et al. (1987) : Geology of the Norwegian oil and gas fields. 493 pp.; Graham & Trotman, London/England. ISBN 0-86010-908-9.
SPENCER, C.W. (1983) : Geologic aspects of tight gas reservoirs in the Rocky Mountain region. SPE/DOE Paper, **11647**. J. Petrol. Technol., **37** (1985) : 1308 - 1314.
SPENCER, C.W. (1984) : Overpressured tight gas reservoirs in the Pinedale anticline area, Sublette County, Wyoming. In : LAW, B.E. (Editor), Geological characteristics of low-permeability Upper Cretaceous and lower Tertiary rocks in the Pinedale anticline area, Sublette County, Wyoming. U.S. Geol. Surv. Open-File Rept., **84-753** : 51 - 59.
SPENCER, C.W. & LAW, B.E. (1981) : Overpressured, low-permeability gas reservoirs in Green River, Washakie, and Great Divide basins, southwestern Wyoming. Bull. Amer. Assoc. Petrol. Geol., **65** : p. 569.
SPENCER, C.W. & MAST, R.F. (1986) : Geology of tight gas reservoirs. Amer. Assoc. Petrol. Geol., Stud. Geol., **24** : 299 pp.; Tulsa/Oklahoma.
SPREUX, A., GEORGES, C. & LESSI, J. (1988) : Most problems in horizontal completions are resolved. Oil Gas J., 13.06.**1988** : 48 - 52.
SPREUX, A., LOUIS, A. & ROCCA, M. (1987) : Logging horizontal wells - field

practice for various techniques. SPE Paper, **16565**. J. Petrol. Technol., **40** (1988) : 1352 - 1354.
SPRINGER, F.P. & MÜLLER, R. (1985) : Lösungsmöglichkeiten zur Beseitigung von Überschußwasser bei Dampfflutprojekten. Erdoel-Erdgas, **101** : 56 - 62.
SPURLOCK, J.W. & DEMSKI, D.B. (1972) : A new approach to the sand control problem - a multi-layer, wire-wrapped sand screen. SPE Paper, **4014**.
STACEY, R.E. & THAIN, I.A. (1983) : 25 years of operation at Wairakei geothermal power station. Proc. 5th New Zealand Geothermal Workshop, 79 - 90; Auckland/New Zealand.
STADALMAN, J.R., NOVOTNY, R.J. & HOUCHIN, L.R. (1985) : Understanding wellbore pressures during changing well conditions improves sand control longevity. SPE Paper, **14160**.
STAHMER, A.M. (1979) : Viel Erdöl - zu wenig Exploration. Erdoel-Erdgas, **95** : 255 - 259.
STALDER, P.I. (1973) : Influence of crystallographic habit and aggregate structure of authigenic clay minerals on sandstone permeability. Geol. Mijnb., **52** : 217 - 220.
STALKUP, F.I. & EBANKS, W.J. (1986) : Permeability variation in a sandstone barrier island-tidal channel-tidal delta complex, Ferron Sandstone (Lower Cretaceous), Central Utah. SPE Paper, **15532**.
STANDARD OIL PROPPANTS (1986 a) : Long term proppant stability - recent laboratory testing. Unpubl. Rept., 05.1986; Standard Oil Proppants, Dallas/Texas.
STANDARD OIL PROPPANTS (1986 b) : A prop for all reasons. Brochure; Standard Oil Proppants, Dallas/Texas.
STANG, H.R. & SONI, Y. (1987) : Saner Ranch pilot test of fracture-assisted steamflood technology. J. Petrol. Technol., **39** : 684 - 696. SPE Paper, **13036** (1984).
STÄUBLE, A.J. & MILIUS, G. (1970) : Geology of Groningen gas field, Netherlands. In : HALBOUTY, M.T. (Editor), Geology of giant petroleum fields. Amer. Assoc. Petrol. Geol. Mem., **14** : 359 - 369; Tulsa/Oklahoma.
STAUFER, T.R. (1988) : Down stream integration by OPEC states : the compelling competitive edge. SPE Paper, **18513**.
STEANSON, R.E., ELBEL, J.L. & WENDORFF, C.L. (1979) : How to select a propping agent for hydraulic fracturing treatments. Petrol. Engin., **1979** : 82 - 92.
STEENBOCK, A. (1985) : Neue Drainhole-Bohrtechnik. Erdoel-Erdgas, **101** : 80 - 84.
STEENBOCK, A. & MÜLLER-LINK, D. (1987) : Alternative drainhole drilling system; theoretical background and practical experience. Offshore Northern Seas Conference, Aberdeen/Scotland, 08. - 11.09.1987.
STEFANSSON, V. (1981) : The Krafla geothermal field, Northeast Iceland. In : RYBACK, L. & MUFFLER, L.J.P. (Editors), Geothermal systems, principles and case histories. 273 - 294; Wiley, New York.
STEIN, N. (1975) : Mechanical properties of friable sands from conventional log data. SPE Paper, **5500**. J. Petrol. Technol., **28** (1976) : 757 - 763.
STEIN, N. (1976) : Use of conventional well log data to select gravel pack materials for sand control. SPE Paper, **6175**.
STEIN, N. (1983) : Designing gravel packs for changing well conditions. World Oil, 01.02.**1983** : 41 - 47.
STEIN, N. & HILCHIE, D.W. (1972) : Estimating the maximum production rates possible from friable sandstones without using sand control. J. Petrol. Technol., **24** : 1157 - 1160. SPE Paper, **3439** (1971).
STEIN, N., ODEH, A.S. & JONES, L.G. (1974) : Estimating maximum sand-free production rates from friable sands for different well completion geometries. J. Petrol. Technol., **26** : 1156 - 1158. SPE Paper, **4534** (1973).
STEINMAN, D.K., ADOLPH, R.A., MAHDAVI, M. & PREEG, W.E. (1988) : Dual-burst thermal decay time logging principles. SPE Form. Eval., **3** : 377 - 385. SPE Paper, **15437** (1986).
STEINMANN, K. (1985) : Die westeuropäische Erdgasversorgung und das Pipelinesystem. Erdoel-Erdgas, **101** : 182 - 187.
STENA (1986) : Well servicing stimulation and workover. Brochure; Stena, Gothenburg/Sweden.
STEPHEN-PYE, C. & SMITH, W.A. (1973) : Fluid loss additive seriously reduces

fracture proppant conductivity and formation permeability. SPE Paper, **4680**.
STEPHENS, R.W. & MARTINS, P.J. (1985) : Hydraulic fracturing - twenty years of North Sea experience. Offshore Europe Conference, Aberdeen/Scotland.
STERN, J.P. (1986) : Soviet oil and gas exports to the West : commercial transaction or security threat? Gower Publ., Hampshire/England.
STEWART, C.D. & WILLIAMSON, D.R. (1988) : Horizontal drilling aspects of the Helder field redevelopment. OTC Paper, **5792**. SPE Paper, **17886**.
STEWART, D. (1986) : Secured oil and gas lending in bankruptcies. SPE Paper, **14644**.
STEWART, G., GUPTA, A. & WESTAWAY, P.J. (1984) : Interpretation of interference tests in a reservoir with sealing and partially communicating faults. SPE Paper, **12967**.
STEWART, G., WITTMAN, M.J. & GOLF-RACHT, T.v. (1981) : The application of the repeat formation tester to the analysis of naturally fractured reservoirs. SPE Paper, **10181**.
STILES, R. (1985) : Gravel packing in deviated holes. Gulf Coast Oil World, **4**/ 11 (1985/9) : 28 - 29.
STILES, R.F., COLOMB, G.T. & FARLEY, D.L. (1986) : Development of a gravity-assisted gravel pack system. SPE Paper, **15409**.
STIM-LAB (1986) : The investigation of effects of environmental conditions and fracturing fluids upon fracture conductivity and proppants. Phase I and II (incl. continuation). Unpubl. Repts., 06.1986 and 10.1986; Stim-Lab.
STOLLER, H.M. (1985) : A perspective on tailored pulse loading : a new approach to oil and gas well stimulation. SPE/DOE Paper, **13860**.
STORMONT, D.H. (1935) : Increasing drainage of oil into the well by drainhole drilling. Oil Gas J., 17.08.**1935** : 105 - 108.
STORY, J.C. (1986) : International crude oil pricing decision-making : the quest for market stability. SPE Paper, **14626**.
STOUT, C.M. & PETERS, F.W. (1976) : Preventing formation damage during fracturing of the Clinton Formation. SPE Paper, **6128**.
STRACKE, M.L. (1987) : Success in offshore gravel pack methods. SPE Paper, **16212**.
STRAMP, R.L. (1980) : The use of horizontal drainholes in the Empire Abo Unit. SPE Paper, **9221**.
STRANG, D.L. & NORTON, J.L. (1983) : Foamed sand provides improved stimulation results from the Devonian shale. SPE Paper, **12312**.
STRELTSOVA, T.D. (1983) : Well pressure behavior of a naturally fractured reservoir. Soc. Petrol. Engin. J., **23** : 769 - 780. SPE Paper, **10782** (1982).
STRELTSOVA, T.D. & McKINLEY, R.M. (1984) : Effect of flow time duration on buildup pattern for reservoirs with heterogeneous properties. Soc. Petrol. Engin. J., **24** : 294 - 306. SPE Paper, **11140** (1982).
STRICKLAND, F.G. (1985) : Reasons for production decline in the diatomite, Belridge oil field : a rock mechanics view. J. Petrol. Technol., **37** : 521 - 526. SPE Paper, **10773** (1982).
STRICKLAND, F.G. & REN, N.K. (1980) : Use of differential strain curve analysis in predicting in-situ stress state for deep wells. Proc. 21st U.S. Rock Mechan. Symp., Rolla/Missouri.
STRICKLAND, F.G. & REN, N.K. (1981) : Predicting the in-situ stress for deep wells using the differential strain curve analysis. SPE/DOE Paper, **8954**.
STRICKLAND, W.T., RICHARDSON, E.A., HAMBY, T.W. & TORREST, R.S. (1975) : A metal-plating process for sand consolidation. Soc. Petrol. Engin. J., **15** : 209 - 216. SPE Paper, **4958** (1974).
STRIEGLER, J.H. (1982) : ARCO finishes fourth horizontal drainhole. Oil Gas J., **1982** : 55 - 61.
STRIGHT, D.H. & GORDON, J.I. (1983) : Decline curve analysis in fractured low permeability gas wells in the Piceance Basin. SPE/DOE Paper, **11640**.
STROBL, W. & PLACKETT, W.G. (1984) : Cavity and gravel-pack logging in Groningen gas wells. SPE Paper, **12998**.
STROHM, P.J., MANTOOTH, M.A. & PRIESTER, C.L.d. (1970) : Controlled injection of sand consolidation plastic. SPE Reprint Series, Well Completions, **5** : 188 - 195. SPE Paper, **1485** (1966). J. Petrol. Technol., **19** (1967) : 487 - 494.
STRUBHAR, M.K., FITCH, J.L. & GLENN, E.E. (1974) : Multiple vertical fractures

from an inclined wellbore - a field experiment. SPE Paper, **5115**. J. Petrol. Technol., **27** (1975) : 641 - 647. SPE Reprint Series, Well Completions, **5 a**/II (1978) : 260 - 266.
STRUBHAR, M.K., FITCH, J.L. & MEDLIN, W.L. (1980) : Demonstration of massive hydraulic fracturing, Piceance Basin, Colorado. SPE Paper, **9336**.
STRUBHAR, M.K., FITCH, J.L. & UHRI, D.C. (1977) : Simplified performance prediction method for vertically fractured gas wells. SPE Paper, **6021**.
STRUBHAR, M.K. & GLENN, E.E. (1974) : Method of creating a plurality of features from a deviated well. U.S. Patent, **3,835,928**.
STRUBHAR, M.K. & JENNINGS, A.R. (1987) : Trends in massive hydraulic fracturing - a recent example. DGMK Spring Meeting, Celle/Germany FRG, 22.05.1987.
STRUBHAR, M.K., MEDLIN, W.L., NABI, S.M. & ANDREANI, F.S. (1984) : Fracturing results in diatomaceous earth formations, South Belridge field, California. J. Petrol. Technol., **36** : 495 - 502. SPE Paper, **10966** (1982).
SUAU, J. & GARTNER, J. (1979) : Fracture detection from the logs. Trans. 6th European Symp., Soc. Prof. Well Log Analysts, London/England, 03.1979. Log Analyst, **1980**/3-4 : 3 - 13.
SULAK, R.M. & DANIELSEN, J. (1988) : Reservoir aspects of Ekofisk subsidence. OTC Paper, **5618**. SPE Paper, **17852**.
SULLIVAN, J. (1982) : A method of evaluating economic risk for an offshore development project. EUR Paper, **316**.
SULLIVAN, R.B., LEE, W.J. & HOLDITCH, S.A. (1987) : Pressure transient response in multilayer gas reservoirs containing hydraulic fractures. SPE/DOE Paper, **16399**.
SULLIVAN, S.A., POSTON, S.W. & PIPER, L.D. (1988) : Using short-term pressure buildup tests for reserves estimation in tight gas reservoirs. SPE Paper, **17707**.
SUMAN, G.O. (1972) : Perforations - a prime source of well performance problems. J. Petrol. Technol., **24** : 399 - 411. SPE Paper, **3445** (1971).
SUMAN, G.O. (1975 a) : Sand control handbook. World Oil Reprint Series, **12** : 57 pp.
SUMAN, G.O. (1975 b) : Unconsolidated sand stabilization through wellbore stress state control. SPE Paper, **5717**.
SUMAN, G.O., ELLIS, R.C. & SNYDER, R.E. (1983) : Sand control handbook. 2nd edition, 89 pp.; Gulf Publ., Houston/Texas. ISBN 0-87201-793-1.
SUMAN, G.O. & SNYDER, R.E. (1982) : Sand control design criteria and gravel packing concepts. SPE Paper, **10031**.
SUN, R.J. (1969) : Theoretical size of hydraulically induced horizontal fractures and corresponding surface uplift in an idealized medium. J. Geophys. Res., **74** : 5995 - 6011.
SUN, R.J. & MONGAN, C.E. (1974) : Hydraulic fracturing in shale at West Valley, New York - a study of bedding-plane fractures induced in shale for waste disposal. U.S. Geol. Surv., Open-File Rept., **74-365**.
SUNG, W. & ERTEKIN, T. (1987) : Performance comparison of vertical and horizontal hydraulic fractures and horizontal boreholes in low permeability reservoirs: a numerical study. SPE/DOE Paper, **16407**.
SUNG, W., ERTEKIN, T. & SCHWERER, F.C. (1986) : The development, testing and application of a comprehensive coal seam degasification model. SPE Paper, **15247**.
SUTTON, G.D. (1976) : Uses and results of a liquid friction reducer in acidizing treatments. 23rd Annual Southwestern Petroleum Short Course, Texas Techn. Univ., Lubbock/Texas, 22. - 23.04.1976.
SUTTON, G.D. & LASATER, R.M. (1972) : Aspects of acid additive selection in sandstone acidizing. SPE Paper, **4114**.
SWAAN, A.d. (1982) : A three-phase model for fractured reservoirs presenting fluid segregation. SPE Paper, **10510**.
SWANSON, G.S., GOTTSCHLING, J.C. & MEEKEN, R. (1980) : An analysis of treating pressure and well production phenomena resulting from fracturing treatments in California`s Belridge and Lost Hills fields. SPE Paper, **9263**.
SWANSON, G.S. & MEEKEN, R.B. (1981) : An analysis of fracturing pressures in the South Belridge and Lost Hills fields. SPE Paper, **9935**.
SWANSON, P.L. & SPETZLER, H. (1984) : Ultrasonic probing of the fracture pro-

cess zone in rock using surface waves. Proc. 25th U.S. Rock Mechan. Symp., 67 - 76.

SWANSON, R.K., BERNARD, W.J. & OSOBA, J.S. (1986) : A summary of the geothermal and methane production potential of U.S. Gulf Coast geopressured zones from test well data. J. Petrol. Technol., **38** : 1365 - 1370. SPE Paper, **13090** (1984).

SWANSON, V.F. (1967) : The development of formula for direct determination of free settling velocity of any size particle. Trans. ASME, **1967** : 160 - 166.

SWARTZ, G. & AHMED, U. (1982) : Optimization of hydraulic fracturing techniques for Eastern Devonian shales. SPE/DOE Paper, **10848**.

SWEENEY, J.L. & WEYANT, J.P. (1980) : The energy modeling forum`s U.S. oil and gas supply study. Proc. 109th Annual Meeting of the Council of Economics, Las Vegas/Nevada, 24. - 29.02.1980, pp. 78 - 84.

SWIFT, R.P. & KUSUBOV, A.S. (1980) : Technique for studying multiple fractures produced at intermediate loading rates. Proc. 21st U.S. Rock Mechan. Symp., 682 - 690.

SWIFT, R.P. & KUSUBOV, A.S. (1981) : Multiple fracturing of boreholes using tailored-pulse loading. SPE/DOE Paper, **9892**. Soc. Petrol. Engin. J., **22** (1982) : 923 - 932.

SWIFT, S.C. & BROWN, L. (1976) : Interference testing for reservoir definition - the state of the art. SPE Paper, **5809**.

SYDANSK, R.D. (1984) : Stabilizing clays with potassium hydroxide. J. Petrol. Technol., **36** : 1366 - 1374. SPE Paper, **11721** (1983).

SYLTE, J.E., HALLENBECK, L.D. & THOMAS, L.K. (1988) : Ekofisk formation pilot waterflood. SPE Paper, **18276**.

SYLVESTER, N.D. & CHEN, H.L. (1988) : An improved cyclic steam stimulation model for pressure-depleted reservoirs. SPE Paper, **17420**.

TAHA, R., HILL, A.D. & SEPEHRNOORI, K. (1986) : Simulation of sandstone-matrix acidizing in heterogeneous reservoirs. J. Petrol. Technol., **38** : 753 - 767. SPE Paper, **13218** (1984).

TAHA, R.M., HILL, A.D. & SEPEHRNOORI, K. (1987) : Sandstone acidizing treatment using a generalized model. SPE Paper, **16885**.

TALEBI, S. & CORNET, F.H. (1985) : Seismo-acoustic activity generated by fluid injections in a granitic rock mass. 4th Conf. Acoustic Emission and Microseismic Activity, Pennsylvania State Univ., University Park/Pennsylvania, 22. - 24.10.1985.

TAM INTERNATIONAL (1986) : TAM-J heavy-duty inflatable packers for well testing, workovers, and production. Tam International Brochure, **TOP 5M 4/87**.

TAN, H.C., McGOWEN, J.M., LEE, W.S. & SOLIMAN, M.Y. (1988) : Field application of minifrac analysis to improve fracturing treatment design. SPE Paper, **17463**.

TANG, Z., LIU, H. & JIANG, C. (1986) : Technical measures for improving oilfield development by waterflooding in a large multi-layered heterogeneous sandstone reservoir - realization of high and stable output in Daqing oilfield. SPE Paper, **14860**.

TANNICH, J.D. (1975) : Liquid removal from hydraulically fractured gas wells. J. Petrol. Technol., **27** : 1309 - 1317. SPE Paper, **5113** (1974).

TANNICH, J.D. & NIERODE, D.E. (1986) : The effect of vertical fractures on gas well productivity. SPE Paper, **15902**.

TANSOE, A., KING, G.E. & HOLMAN, G.B. (1986) : Valhall - production from high porosity chalk. Offshore Northern Seas, Reservoir Engineering Conference, Stavanger/Norway, Paper **R 3** : 15 pp.

TARIKI, A. (1985) : Global oil markets under stress. SPE Paper, **14193**.

TARIQ, S.M. (1987) : Evaluation of flow characteristics of perforations including nonlinear effects with the finite-element method. SPE Product. Engin., **2** : 104 - 112.

TARIQ, S.M., ICHARA, M.J. & AYESTARAN, L. (1985) : Performance of perforated completions in the presence of common heterogeneities : anisotropy, laminations, or natural fractures. SPE Paper, **14320**.

TAUSCH, G.H. & CORLEY, C.B. (1958) : Sand exclusion in oil and gas wells. API Drill. Product. Pract., **1958** : 66 - 82.

TAYLOR, T.J. (1983) : Interpretation and application of borehole televiewer surveys. Trans. 24th Soc. Prof. Well Log Analysts Annual Logging Symp., Paper **QQ**, Calgary/Alberta, 27. - 30.06.1983.

TECLAW, C.E. (1988) : Natural gas : a strategic fuel for the 1990`s. Symposium on Energy, Finance and Taxation Policies, Washington/D.C., 19. - 20.09.1988.

TEK, M.R. (1979) : The challenge of worldwide natural gas matching supply limited markets to market limited supplies. SPE Paper, **8290**.

TEMPLETON, C.C., RICHARDSON, E.A., KARNES, G.T. & LYBARGER, J.H. (1975) : Self-generating mud acid. J. Petrol. Technol., **27** : 1199 - 1203. SPE Paper, **5153** (1974).

TEPLITZ, A.J. & HASSEBROEK, W.E. (1946) : An investigation of oil well cementing. API Drill. Product. Pract., **1946** : p. 76 ff.

TERWILLIGER, P.L., SMITH, F.M. & GOODWIN, R.J. (1964) : Warm-air coking - a new completion method for unconsolidated sands. J. Petrol. Technol., **16** : 367 - 371.

TESLJUK, E.W. (1957) : Die Anwendung radioaktiver "körniger" Isotope bei der hydraulischen Sprengung von Schichten auf den Feldern der Krasnodarneft. Z. angew. Geol., **3** : 79 - 81.

TESTER, J.N., BIVENS, R.L. & POTTER, R.M. (1982) : Interwell tracer analysis of a hydraulically fractured granitic geothermal reservoir. Soc. Petrol. Engin. J., **22** : 537 - 554. SPE Paper, **8270** (1979).

TESTER, J.W., MURPHY, H.D., GRIGSBY, G.O., POTTER, R.M. & ROBINSON, B.A. (1986) : Fractured geothermal reservoir growth induced by heat extraction. SPE Paper, **15124**.

TEUFEL, L.W. (1979) : An experimental study of hydraulic fracture propagation in layered rocks. Ph.D. Thesis, Texas Univ., College Station/Texas, 08.1979.

TEUFEL, L.W. (1981) : Strain relaxation method for predicting hydraulic fracture azimuth from oriented core. SPE/DOE Paper, **9836**.

TEUFEL, L.W. (1982) : Prediction of hydraulic fracture azimuth from anelastic strain recovery measurements of oriented core. In : GOODMAN, R.E. & HUGHS, F.F. (Editors), Issues in rock mechanics; Berkeley/California. Proc. 23rd U.S. Rock Mechan. Symp., 238 - 246; New York.

TEUFEL, L.W. (1983) : Determination of in-situ stresses from anelastic strain recovery measurements of oriented core : applications to hydraulic fracturing treatment design. SPE Paper, **11649**.

TEUFEL, L.W. (1985 a) : Determination of in-situ stress and prediction of hydraulic fracture azimuth at the Mounds Experiment by the anelastic strain recovery method. SPE/DOE Paper, **13896**.

TEUFEL, L.W. (1985 b) : Insights into the relationship between wellbore breakouts, natural fractures and in situ stress. Proc. 16th U.S. Rock Mechan. Symp., Rapid City/South Dakota, 26. - 28.06.1985, pp. 1199 - 1206.

TEUFEL, L.W. (1985 c) : In situ stress measurements in inclined holes in the North Sea : application to water flooding and enhanced oil recovery. SPE Paper, **13986**.

TEUFEL, LW. (1988) : Microcrack genesis during anelastic strain recovery of core : implications to petrophysical properties. SPE Paper, **18168**.

TEUFEL, L.W. & BLANTON, T.L. (1985) : In situ stress determination from wellbore elongation measurements. SPE Paper, **13877**.

TEUFEL, L.W. & CLARK, J.A. (1981) : Hydraulic fracture propagation in layered rock : experimental studies of fracture containment. SPE/DOE Paper, **9878**. Soc. Petrol. Engin. J., **24** (1984) : 19 - 32.

TEUFEL, L.W., HART, C.M., SATTLER, A.R. & CLARK, J.A. (1984) : Determination of hydraulic fracture azimuth by geophysical, geological and oriented-core methods at the Multiwell Experiment site, Rifle, Co. SPE Paper, **13226**.

TEUFEL, L.W. & WARPINSKI, N.R. (1983) : In-situ stress variations and hydraulic fracture propagation in layered rock - observations from a mineback experiment. Proc. 5th Intern. Congr. Rock Mechan., Melbourne/Australia.

TEUFEL, L.W. & WARPINSKI, N.R. (1984) : Determination of in situ stress from anelastic strain recovery measurements of oriented core : comparison to hydraulic fracture stress measurements in the Rollins Sandstone, Piceance Basin, Colorado. Proc. 25th U.S. Rock Mechan. Symp., Evanston/Illinois, 25. - 27.06.1984, pp. 176 - 185.

TEUFEL, L.W. & WARPINSKI, N.R. (1988) : Influence of natural fractures on hy-

draulic fracture propagation. SPE Paper, **18134**.
TEWKSBURY, J.B. (1977) : Long range forecasting of foreign oil price - technique and results. SPE Paper, **6340**.
THACKER, W.C. & LAVELLE, J.W. (1978) : Stability of settling of suspended sediments. Phys. Fluids, **21** : 291 - 292.
THIERCELIN, M.J., BEN NACEUR, K. & LEMANCZYK, Z.R. (1985) : Simulation of three-dimensional propagation of a vertical hydraulic fracture. SPE/DOE Paper, **13861**.
THIERCELIN, M.J., JEFFREY, R.G. & BEN NACEUR, K. (1987) : The influence of fracture toughness on the geometry of hydraulic fractures. SPE Paper, **16431**.
THIERCELIN, M.J. & LEMANCZYK, Z.R. (1983) : The effect of stress gradient on the height of vertical hydraulic fractures. SPE/DOE Paper, **11626**. SPE Product. Engin., **1** (1986) : 245 - 254.
THOMAS, D.C., HSING, H. & MENZIE, D.E. (1984) : Evaluation of core damage caused by oil-based drilling and coring fluids. SPE Paper, **13097**.
THOMAS, G.E., SONNENBERG, F.P., LANGENBAHN, W.E. & MURPHY, D.L. (1988) : Better well siting through paleotopography. World Oil, **207** (1988/9) : 57 - 62.
THOMAS, I.P. & JOHNSON, C.E. (1976) : Perforating resin consolidated zones for capacity improvement. SPE Paper, **5697**.
THOMAS, J.B. (1978) : Diagenetic sequences in low-permeability argillaceous sandstones. J. Geol. Soc., **135** : 93 - 100.
THOMAS, L.K., DIXON, T.N., EVANS, C.E. & VIENOT, M.E. (1987) : Ekofisk waterflood pilot. J. Petrol. Technol., **39** : 221 - 232. SPE Paper, **13120** (1984).
THOMAS, L.K., DIXON, T.N. & PIERSON, R.G. (1983) : Fractured reservoir simulation. Soc. Petrol. Engin. J., **23** : 42 - 54. SPE Paper, **9305** (1980).
THOMAS, R.D. & WARD, D.C. (1972) : Effect of overburden pressure and water saturation on gas permeability of tight sandstone cores. J. Petrol. Technol., **24** : 120 - 124. SPE Paper, **3634** (1971).
THOMAS, R.L. & CROWE, C.W. (1978 a) : New chemical treatment provides stimulation and clay control in sandstone formations. SPE Paper, **7012**.
THOMAS, R.L. & CROWE, C.W. (1978 b) : Matrix treatment employs new acid system for stimulation and control of fines migration in sandstone formations. SPE Paper, **7566**. J. Petrol. Technol., **33** (1981) : 1491 - 1499.
THOMAS, R.L. & CROWE, C.W. (1978 c) : Single-stage chemical treatment provides stimulation and clay control in sandstone formations. SPE Paper, **7124**.
THOMAS, R.L., CROWE, C.W. & SIMPSON, B.E. (1976) : Effect of chemical treatment upon formation clays as revealed by improved SEM technique. SPE Paper, **6007**.
THOMAS, R.L. & ELBEL, J.L. (1979) : The use of viscosity stabilizers in high temperature fracturing. SPE Paper, **8344**.
THOMAS, R.L. & ROOT, R.L. (1979) : The use of viscosity stabilizers in high temperature fracturing. SPE Paper, **8344**.
THOMAS, S.A., YOST, M.E. & CATHEY, S.R. (1987) : Silica removal from steamflood-produced water : South Texas Tar Sands pilot. SPE Product. Engin., **2** : 131 - 136. SPE Paper, **13021** (1984).
THOMAS, W.A. (1986) : North Sea field developments : historic costs and future trends. J. Petrol. Technol., **38** : 1211 - 1220. SPE Paper, **12984** (1984).
THOMPSON, J.C. (1977) : The theory, methods of application and results of the use of 100-mesh sand in stimulating Permian Basin formations. SPE Paper, **6374**.
THOMPSON, J.K. (1981) : Use of constant pressure, finite capacity type curves for performance prediction of fractured wells in low-permeability reservoirs. SPE/DOE Paper, **9839**.
THOMPSON, R.G. & SINGLETON, F.D. (1988) : A world oil crisis is near at hand. World Oil, **207** (1988/4) : 75 - 82.
THOMPSON, R.S. (1988) : The impact of the new tax law on internal cash flow generation. SPE Paper, **16309**.
THON, A. (1985) : Das Aufsuchen von Erdöl- und Erdgaslagerstätten im Paläozoikum auf dem Gebiet der CSSR. Erdoel-Erdgas, **101** : 275 - 278.
THORNE, B.J. & MORRIS, H.E. (1987) : Advances in borehole seismic fracture diagnostics. SPE/DOE Paper, **16405**.
THORNE, B.J. & MORRIS, H.E. (1988) : An assessment of borehole seismic fracture diagnostics. SPE Paper, **18193**.

THORNOCK, R.A. & COVLIN, R.J. (1977) : Stimulation and development of the Upper Cretaceous sands - Spindle field, Colorado. SPE Paper, **6705**.

THORNTON, S.D. & RADKE, C.J. (1985) : Dissolution and condensation kinetics of silica in alkaline solution. SPE Paper, **13601**.

TIAB, D. & PUTHIGAI, S.K. (1988) : Pressure-derivative type curves for vertically fractured wells. SPE Form. Eval., **3** : 156 - 158. SPE Paper, **11028** (1982).

TILLMAN, J.E. (1983) : Development of fracture-enhanced reservoirs in Upper Devonian sands and shales. SPE Paper, **12301**.

TIMKO, D.J. & FERTL, W.H. (1971) : Relationship between hydrocarbon accumulations and geopressure and its economic significance. J. Petrol. Technol., **23** : 923 - 931. SPE Paper, **2990** (1970).

TIMUR, A., HEMPKINS, W.B. & WEINBRANDT, R.M. (1971) : Scanning electron microscope study of pore systems in rocks. J. Geophys. Res., **76** : 4932 - 4947.

TINDELL, W.A., NEAL, J.K. & HUNTER, J.C. (1981) : Evolution of fracturing the Cotton Valley Sands in Oak Fill field. J. Petrol. Technol., **33** : 799 - 807. SPE Paper, **9067** (1980).

TINER, R.L., STAHL, E.J. & MALONE, W.T. (1974) : Developments in fluids to reduce potential damage from fracture treatments. SPE Paper, **4790**.

TINSLEY, J.M., LASATER, R.M. & KNOX, J.A. (1967) : Design techniques for chemical fracture-squeeze treatments. J. Petrol. Technol., **19** : 1493 - 1499. SPE Paper, **1771**.

TINSLEY, J.M. & WILLIAMS, J.R. (1975) : A new method for providing increased fracture conductivity and improving stimulation results. J. Petrol. Technol., **27** : 1319 - 1325. SPE Paper, **4676** (1973).

TINSLEY, J.M., WILLIAMS, J.R., TINER, R.L. & MALONE, W.T. (1969) : Vertical fracture height - its effect on steady-state production increase. SPE Paper, **1900** (1967). J. Petrol. Technol., **21** : 633 - 638. SPE Reprint Series, Well Completion, **5** (1970) : 278 - 283.

TINSON, J.K. et al. (1982) : A method for selecting potential infill locations in the East Texas Cotton Valley tight gas play. SPE Paper, **11022**.

TIPPIE, D.B. (1973) : The effect of producing rate on the formation of arches in unconsolidated sands. Thesis, **T-1575**; Colorado School of Mines, Golden/ Colorado.

TIPPIE, D.B. & KOHLHAAS, C.A. (1973) : Effect of flow rate on stability of unconsolidated production sands. SPE Paper, **4533**.

TIPPIE, D.B. & KOHLHAAS, C.A. (1974) : Variations of skin damage with flow rate associated with sand flow or stability in unconsolidated-sand reservoirs. SPE Paper, **4886**.

TIRANT, P.le & DUPUIS, M. (1967) : Dimensions of the fractures obtained by hydraulic fracturing of oil-bearing formations. Rev. Inst. Franç. Pétrol., **1967** : 44 - 98.

TIRANT, P.le, GAY, L. et al. (1972) : Manuel de fracturation hydraulique. Edition Technip, Paris/France.

TIXIER, M.P., LOVELESS, G.W. & ANDERSON, R.A. (1973) : Estimation of formation strength from the mechanical-properties log. SPE Paper, **4532**. J. Petrol. Technol., **27** (1975) : 283 - 293. SPE Reprint Series, Well Completions, **5 a**/ II (1978) : 103 - 113.

TODD, A.C., SOMERVILLE, J.E. & SCOTT, G. (1984) : The application of depth of formation damage measurements in predicting water injectivity decline. SPE Paper, **12498**.

TODD, B.J. & BRADLEY, D.J. (1986) : Perforation geometry and skin effects on well productivity. SPE Paper, **15029**.

TOMA, P. & LIVESEY, D.B. (1986) : A new sand control filter for thermal recovery wells. SPE Paper, **15057**. SPE Product. Engin., **3** (1988) : 249 - 257.

TOMLINSON, H.L., BYRD, A.C. & BERG, C.F.v. (1988) : Fracturing process control and automation. SPE Paper, **17315**.

TORREST, R.S. (1975) : Electroless nickel sand consolidation : laboratory studies of cementation and calculated effect on well productivity. Soc. Petrol. Engin. J., **15** : 203 - 208. SPE Paper, **4964**.

TORREST, R.S. (1982 a) : Deposit buildup during gravel packing with viscous polymer solutions and water. SPE Paper, **10772**. J. Petrol. Technol., **35** (1983)

: 325 - 328.
TORREST, R.S. (1982 b) : The influence of placement conditions on the response of gravel packs to production. SPE Paper, **11011**.
TORREST, R.S. (1982 c) : The flow of viscous polymer solutions for gravel packing through porous media. SPE Paper, **11010**.
TORREST, R.S. (1982 d) : Injectivity testing for gravel pack quality - why not? SPE Paper, **11344**.
TORREST, R.S. & MARCANO, F. (1986) : Gel surface filtration from polymer solutions during gravel packing. SPE Paper, **15546**.
TORREST, R.S. & SAVAGE, R.W. (1975) : Particle collection from vertical suspension flows through small side ports - a correlation. Canad. J. Chem. Engin., **53** : p. 699.
TORREY, P.D. (1940) : Progress in squeeze cementing application and technique. Oil Weekly, 29.07.**1940**.
TORSAETER, O. (1984) : An experimental study of water imbibition in chalk from the Ekofisk field. SPE/DOE Paper, **12688**.
TORTIKE, W.S. & ALI, S.M.F. (1989) : Fracture-based cyclic steam stimulation of tar sands. SPE Paper, **18430**.
TORY, E.M. & PICKARD, D.K. (1977) : A three-parameter Markov model for sedimentation. Canad. J. Chem. Engin., **55** : 655 - 665.
TOUBOUL, E., BEN NACEUR, K. & THIERCELIN, M. (1986) : Variational methods in the simulation of three-dimensional fracture propagation. Proc. 27th U.S. Rock Mechan. Symp., 659 - 668, Tuscaloosa/Alabama.
TRACY, P.B. (1988) : Lateral drilling technology tested on in-situ coal gasification project. IADC/SPE Paper, **17237**.
TRAHAN, G.J. & SPIES, R.J. (1981) : A technique for making high rate completions in partially pressure depleted, unconsolidated reservoirs. SPE Paper, **10174**.
TRAVIS, B.J. (1981) : The KRAK model of fluid driven fracture propagation in permeable rock. SPE/DOE Paper, **9881**.
TREADWAY, B.R., BRANDT, H. & PARKER, P.H. (1966) : Studies of a new process to consolidate oil sands with plastics. J. Petrol. Technol., **18** : 1537 - 1543. SPE Paper, **1612**.
TREVITS, M.A., HANSON, M.E. & WARD, V.L. (1982) : Methane drainage : identification and evaluation of the parameters controlling induced fracture geometry. SPE/DOE Paper, **10820**.
TRIELOFF, H.J. (1983) : Kohlendioxid zur Ausbeutesteigerung in Erdöllagerstätten. Erdöl und Kohle - Erdgas, **36** : p. 495. DGMK-Forschungsbericht, **202-1** : 240 pp.
TSANG, Y.W. & WITHERSPOON, P.A. (1983) : The dependence of fracture mechanical and fluid flow properties on fracture roughness and sample size. J. Geophys. Res., **88** : 2359 - 2366.
TSAY, F.S. & FANG, J.H. (1986) : A core-log study of the Hartselle Sandstone in North Alabama with emphasis in log interpretation problems in tight gas sands. SPE Paper, **15211**.
TUCKER, R.L. (1979) : Practical pressure analysis in evaluation of proppant selection for the low permeability, highly geopressured reservoirs of the McAllen Ranch (Vicksburg) field. SPE Paper, **7925**.
TUNN, W. (1971) : Stimulationen bei tiefen Gasbohrungen. Erdoel-Erdgas, **87** : 202 - 208.
TWEEDIE, J.A. & KENNY, G.M. (1982) : North Sea development risk : a historical analysis. EUR Paper, **317**.
TYLER, L.D. & VOLLENDORF, W.C. (1975) : Physical observations and mapping of cracks resulting from hydraulic fracturing in-situ stress measurements. SPE Paper, **5542**.
TYLER, N. (1987) : Depositional setting and production characteristics of low permeability sandstones, Upper Cretaceous Olmos Formation, South Texas. SPE/DOE Paper, **16424**.
TYLER, N. & FINLEY, R.J. (1988) : Reservoir architecture : a critical element in extended conventional recovery of mobile oil in heterogeneous reservoirs. SPE Paper, **18135**.
TYLER, T.N., METZGER, R.R. & TWYFORD, L.R. (1984) : Analysis and treatment of

formation damage at Prudhoe Bay, AK. SPE Paper, **12471**. J. Petrol. Technol., **37** (1985) : 1010 - 1018.
TYSSEE, D.A. & VETTER, O.J. (1981) : Chemical characterization problems of water-soluble polymers. Soc. Petrol. Engin. J., **21** : 721 - 730.
UBANI, E.A. & EVANS, R.D. (1982) : Non-Darcy compressible flow of real gases in propped fractures. SPE Paper, **11101**.
UBANI, E.A. & RAY, R.M. (1988) : On the performance of noncontinuous tight gas sands. SPE Paper, **17728**.
UDELL, K.S. & LOFY, J.D. (1985) : Permeability reduction of unconsolidated media due to stress-induced silica dissolution. SPE Paper, **13655**.
UDGAARD, N.M. (1987) : Norway between the IEA and OPEC : oil prices and foreign policy. In : Norway, OPEC and IEA - global dialogue and price stabilizing measures in the oil market? Intern. Polit., **1987**/5-6; Norwegian Institute of International Affairs.
UHRI, D.C. (1987) : Stimulation of earth formations surrounding a deviated wellbore by sequential hydraulic fracturing. U.S. Patent, **4,687,061** (18.08. 1987).
UKOOA (1984) : Potential oil and gas production from the UK offshore in the year 2000. UK Offshore Operators Association, London/England, 09.1984.
ULDRICH, D.O. & ERSHAGHI, I. (1979) : A method for estimating the interporosity flow parameter in naturally fractured reservoirs. Soc. Petrol. Engin. J., **19** : 324 - 332. SPE Paper, **7142** (1978).
UNDERDOWN, D.R. & DAS, K. (1982) : New proppant for deep hydraulic fracturing. SPE Paper, **10889**. J. Petrol. Technol., **37** (1985) : 98 - 104.
UNDERDOWN, D.R. & DAS, K. (1985) : Stability of gravel-packing materials for thermal wells. J. Petrol. Technol., **37** : 2006 - 2012. SPE Paper, **11793** (1983).
UNDERDOWN, D.R., DAS, K. & NGUYEN, H. (1984) : Gravel packing highly deviated wells with a crosslinked polymer system. SPE Paper, **12481**. J. Petrol. Technol., **37** (1985) : 2197 - 2202.
UNDERDOWN, D.R., DAY, J.C. & SPARLIN, D.D. (1980) : A plastic pre-coated gravel for controlling formation sand. SPE Paper, **8801**.
URAIET, A., RAGHAVAN, R. & THOMAS, G.W. (1976) : Vertical fracture height : effect on transient flow behavior. SPE Paper, **6016**.
URAIET, A., RAGHAVAN, R. & THOMAS, G.W. (1977) : Determination of the orientation of a vertical fracture by interference tests. J. Petrol. Technol., **29** : 73 - 80. SPE Paper, **5845** (1976).
USATCHOV, P.M. (1986) : Hydraulic fracturing of reservoirs. 165 pp. (in Russian); Nedra, Moscow/USSR.
VAIROGS, J., HEARN, C.L., DAREING, D.W. & RHOADES, V.W. (1971) : Effect of rock stress on gas production from low-permeability reservoirs. J. Petrol. Technol., **23** : 1161 - 1167. SPE Paper, **3001** (1970).
VALAIS, M. (1987) : The outlook for European gas equilibrium, supply and demand to 2000. IFA Enlarged Meeting on Raw Materials, Sevilla/Spain, 21. - 22.10. 1987.
VALENTIN, E.P., VALAIS, M.R. & CHAMPLON, D.J. (1988) : Gas production outlook and technological progress : European problematics and answers. SPE Paper, **17718**.
VANDAMME, L.M., JEFFREY, R.G. & CURRAN, J.H. (1986 a) : Pressure distribution in three-dimensional hydraulic fractures. SPE Paper, **15265**. SPE Product. Engin., **3** (1988) : 181 - 186.
VANDAMME, L.M., JEFFREY, R.G. & CURRAN, J.H. (1986 b) : Effects of three-dimensionalization on a hydraulic fracture pressure profile. Proc. 27th U.S. Rock Mechan. Symp., 580 - 590; Tuscaloosa/Alabama.
VANORSDALE, C.R. (1987) : Evaluation of Devonian shale gas reservoirs. SPE Product. Engin., **2** : 209 - 216. SPE Paper, **14446** (1985).
VARGAS, L.d. & OHLSON, M. (1986) : Viscosity breakers for a sand pack completion fluid. Guidrapak II. SPE Paper, **15195**.
VARZI, M. (1986) : The outlook for oil prices in 1987. Talk Gastech 86 Conference, 12th Intern. LNG/LPG Conference, Hamburg/Germany FRG, 25. - 28.11. 1986.
VASILEV, M.L., BANDALL, R.R., OLIVER, D.W. & FERTL, W.H. (1987) : Applications

of a new pulsed neutron capture log to Middle East reservoirs. SPE Paper, **15717**.

VAZIRI, H. (1986) : Mechanics of fluid and sand production from oil sands reservoirs. CIM Paper, **86-37-75**.

VEATCH, R.W. (1977) : Computer studies of a method for determining in-situ fracture flow capacity in low-permeability gas formations. SPE Paper, **6839**.

VEATCH, R.W. (1981) : A brief survey of the technology challenge to improve recovery from tight gas reservoirs. Proc. ASME 16th Intersoc. Energy Convers. Conference, Atlanta/Georgia.

VEATCH, R.W. (1983) : Overview of current hydraulic fracturing design and treatment technology - Part 1, Part 2. J. Petrol. Technol., **35** : 677 - 687, 853 - 864. SPE Paper, **10039** (1982), **11922**.

VEATCH, R.W. (1986) : Economics of fracturing : some methods, examples, and case studies. SPE Paper, **15509**.

VEATCH, R.W. & BAKER, O. (1983) : How technology and price affect U.S. tight gas potential. Petrol. Engin. Intern., **1983**/1 : 84 - 96.

VEATCH, R.W. & CROWELL, R.F. (1982) : Joint research/operations programs accelerate massive hydraulic fracturing technology. J. Petrol. Technol., **34** : 2763 - 2775. SPE Paper, **9337** (1980).

VEATCH, R.W. & MOSCHOVIDIS, Z.A. (1986) : An overview of recent advances in hydraulic fracturing technology. SPE Paper, **14085**.

VEEN, F.R.v. (1975) : Geology of the Leman gas-field. In : WOODLAND, A.W. (Editor), Petroleum and the continental shelf of North-West Europe, **1** : 223 - 231; Applied Science Publ., London/England.

VELEY, C.D. (1969) : How hydrolyzable metal ions react with clays to control formation water sensitivity. J. Petrol. Technol., **21** : 1111 - 1118. SPE Paper, **2188** (1968).

VELEY, C.D. & CASEY, G. (1973) : In-place clay stabilization cuts formation sand entry. World Oil, **1973**/6 : 52 - 54.

VELEY, C.D. & JOLY, G. (1987) : Reduction of water influx in fractured reservoirs. SPE Paper, **15774**.

VENDITTO, J.J., CHISHOLM, P., WIGGINS, M. & CONWAY, M. (1986) : Refracturing completed wells can be cost effective. Petrol. Engin. Intern., **1986**/8 : 26 - 34.

VENTURA, J.L. (1985) : Slator Ranch fracture optimization study. J. Petrol. Technol., **37** : 1251 - 1262. SPE Paper, **12816** (1983).

VERBEEK, C.M.J. (1982) : Analysis of production tests of hydraulically fractured wells in a tight solution gas-drive reservoir. SPE Paper, **11084**.

VERITY, R.V. & MORRIS, C.W. (1981) : Fracture stimulation of Union`s Baca well 23. Geotherm. Resourc. Counc. Trans., **5**.

VETTER, O.J. & CRICHLOW, H.B. (1979) : Radioactive tracer adsorption chromatography in geothermal reservoirs. SPE Paper, **7960**.

VETTER, O.J., LANKFORD, S., NILSSEN, T. & SHELTON, M. (1988) : Well stimulations and scale inhibitors. SPE Paper, **17284**.

VICANEK, J. (1985) : Erdölförderung durch horizontale Bohrungen. Erdoel-Erdgas, **101** : 85 - 90.

VIELVOYE, R. (1986) : Offshore Northern Europe. Oil Gas J., 30.06.**1986** : 47 - 73.

VIELVOYE, R. (1987 a) : OPEC restraint on production strengthens crude oil prices. Oil Gas J., 06.07.**1987** : p. 25.

VIELVOYE, R. (1987 b) : OPEC actions boost crude prices toward $ 18/bbl target. Oil Gas J., 12.01.**1987** : 15 - 17.

VIELVOYE, R. (1987 c) : Recovery in North Sea drilling yields string of oil and gas strikes. Oil Gas J., 05.10.**1987** : 21 - 23.

VIELVOYE, R. (1987 d) : Outlook for field development growing brighter in North Sea. Oil Gas J., 07.12.**1987** : 19 - 22.

VIELVOYE, R. (1988 a) : Operators remain busy with gas development work in U.K. North Sea. Oil Gas J., 15.08.**1988** : 33 - 39.

VIELVOYE, R. (1988 b) : Oil price drop possible unless OPEC cuts flow. Oil Gas J., 04.07.**1988** : 12 - 14.

VIELVOYE, R. (1988 c) : Advances open marginal offshore sectors. Oil Gas J., 02.05.**1988** : 33 - 35.

VISSER, W. & DOMSELAAR, H.R.v. (1974) : Effect of interference of fractured wells on obtainable productivity improvements. KSEPL Research Summary, **1974**/5; Rijswijk/Netherlands.
VITTORATOS, E. (1986) : Interpretation of temperature profiles from the steam-stimulated Cold Lake reservoir. SPE Paper, **15050**.
VLIS, A.C.v.d., DUNS, H. & FERNANDEZ-LUQUE, R. (1979) : Increasing well productivity in tight chalk reservoirs. Proc. 10th World Petrol. Congr., **3** : 71 - 78, Panel Discussion 7, Paper 3; Bucharest/Romania.
VLIS, A.C.v.d., HAAFKENS, R., SCHIPPER, B.A. & VISSER, W. (1975) : Criteria for proppant placement and fracture conductivity. SPE Paper, **5637**.
VOEGELE, M.D., ABOU-SAYED, A.S. & JONES, A.H. (1983) : Optimization of stimulation design through the use of in-situ stress determination. J. Petrol. Technol., **35** : 1071 - 1081. SPE Paper, **10308** (1981).
VOEGELE, M.D. & JONES, A.H. (1980) : A wireline hydraulic fracturing tool for determination of in-situ stress contrasts. SPE/DOE Paper, **8937**.
VOGE, E.A. & CRAFTON, J.W. (1988) : Carbonate acidizing : a waterflood approach for modeling wormhole growth. SPE Paper, **17496, 17518**.
VOGEL, C.B. & HEROLZ, R.A. (1981) : The CAD - a circumferential acoustic device for well logging. J. Petrol. Technol., **33** : 1985 - 1987. SPE Paper, **6819** (1977).
VOIGHT, B. (1968) : Determination of the virgin state of stress in the vicinity of a borehole from measurements of a partial anelastic strain tensor in drill cores. Felsmechan. Ingen. Geol., **6** : 201 - 215.
VOLK, L.J., GALL, B.L., RAIBLE, C.J. & CARROLL, H.B. (1983) : A method for evaluation of formation damage due to fracturing fluids. SPE/DOE Paper, **11638**.
VOLK, L.J., RAIBLE, C.J., CARROLL, H.B. & SPEARS, J.S. (1981) : Embedment of high strength proppant into low-permeability reservoir rock. SPE/DOE Paper, **9867**.
VONDE, T.R. (1980) : Specialized pumping techniques applied to a very low-gravity sand laden crude : Cat Canyon field, California. SPE Paper, **8900**. J. Petrol. Technol., **34** (1982) : p. 1951 ff.
VONGVUTHIPORNCHAI, S. & RAGHAVAN, R. (1987) : Pressure falloff behavior in vertically fractured wells : non-Newtonian power-law fluids. SPE Form. Eval., **2** : 573 - 589. SPE Paper, **13058** (1984).
VOSPER, K.W & STEVENS, B.G. (1982) : Development options for small gas fields. EUR Paper, **344**.
VRIELINK, H.J. & HIPPMAN, A. (1988) : Optimization of slant well drilling in the Lindbergh field. SPE Paper, **17529**.
VRIEZEN, P.B., SPIJKER, A. & VLIS, A.C.v.d. (1975) : Erosion of perforation tunnels in gas wells. SPE Paper, **5661**.
WAAL, J.A.d. & SMITS, R.M.M. (1988) : Prediction of reservoir compaction and surface subsidence : field application of a new model. SPE Form. Eval., **3** : 347 - 356. SPE Paper, **14214** (1985).
WADE, J.E. (1967) : Techniques for treating and completing water injection wells in California. J. Petrol. Technol., **19** : 49 - 53. SPE Paper, **1543** (1966).
WAGES, P.E. (1982) : Interpretation of post-fracture temperature surveys. SPE Paper, **11189**.
WAGNER, O.R. (1977) : The use of tracers in diagnosing interwell reservoir heterogeneities - field results. J. Petrol. Technol., **29** : 1410 - 1416. SPE Paper, **6046** (1976).
WAGNER, O.R., BAKER, L.E. & SCOTT, G.R. (1974) : The design and implementation of multiple tracer program for multifluid, multiwell injection projects. SPE Paper, **5125**.
WAHL, H.A. (1956) : An electrolytic model study of the effects of horizontal fractures on well productivity. M. Sc. Thesis, Univ. of Oklahoma, Norman/ Oklahoma.
WAHL, H.A. (1965) : Horizontal fracture design based on propped fracture area. J. Petrol. Technol., **17** : 723 - 731. SPE Paper, **976** (1964).
WAHL, H.A. & CAMPBELL, J.M. (1963) : Sand movement in horizontal fractures. J. Petrol. Technol., **15** : 1239 - 1246.
WAHL, J.S., TITTMAN, J., ALGER, R.P. & JOHNSTONE, C.W. (1964) : The dual spa-

cing formation density log. SPE Paper, **989**.
WAHLMEIER, M.A. & ANDREWS, P.W. (1988) : Mechanics of gravel placement and packing : a design and evaluation approach. SPE Product. Engin., **3** : 69 - 82. SPE Paper, **16181** (1986).
WALBE, K. (1986) : Utilization of the borehole television camera in conjunction with regular open and cased-hole logging in the Devonian shale of the Appalachian Basin. SPE Paper, **15610**.
WALDGREN, D. & CORRIGAN, A.F. (1985) : An engineering and geological review of the problems encountered in simulating naturally fractured reservoirs. SPE Paper, **13717**.
WALDORF, D.M. (1965) : Effect of steam on permeabilities of water-sensitive formations. SPE Paper, **1119**.
WALKER, M.L., FREDRICKSON, S.E. & NORMAN, L.R. (1984) : Heated acid for improved stimulation results. SPE Paper, **13371**.
WALKER, T.A. (1962) : Fracture zones vary acoustic signal amplitudes. World Oil, **1962**/6 : 135 - 142.
WALKUP, G.W. & HORNE, R.N. (1985) : Characterization of tracer retention processes and their effect on tracer transport in fractured geothermal reservoirs. SPE Paper, **13610**.
WALLING, R. & STABLER, F. (1978) : New applications of proven sand technology in high-angle holes. SPE Paper, **7000**.
WALLS, J.D. (1981) : Tight gas sands : permeability, pore structure and clay. SPE/DOE Paper, **9871**. J. Petrol. Technol., **34** (1982) : 2708 - 2714.
WALLS, J.D. (1982) : Measurement of fluid salinity effects on tight gas sands with a computer-controlled permeameter. SPE Paper, **11092**.
WALLS, J.D. & AMAEFULE, J.O. (1985) : Capillary pressure and permeability relationships in tight gas sands. SPE/DOE Paper, **13879**.
WALLS, J.D., NUR, A.M. & BOURBIE, T. (1982) : Effects of pressure and water saturation on gas permeability in tight sands. J. Petrol. Technol., **34** : 930 - 936. SPE Paper, **9378** (1980).
WALSER, D.W. (1985) : Field study of a new high-temperature fracturing fluid in South Texas. SPE Paper, **13813**. SPE Product. Engin., **3** (1988) : 187 - 191.
WALSH, J.B. (1981) : Effect of pore pressure and confining stress on fracture permeability. Intern. J. Rock Mechan., Min. Sci., Geomechan. Abstr., **18** : 429 - 435.
WALSH, M.D. & HOLDER, D.J. (1988) : Inflatable packers : production application. SPE Paper, **17443**.
WALSH, M.P., LAKE, L.W. & SCHECHTER, R.S. (1982) : A description of chemical precipitation mechanisms and their role in formation damage during stimulation by hydrofluoric acid. SPE Paper, **10625**. J. Petrol. Technol., **34** (1982) : 2097 - 2112.
WANG, J.J. & CLIFTON, R.J. (1989) : Influence of elastic modulus and fracture toughness on hydraulic fracturing simulation. SPE Paper, **18454**.
WANNER, J.J., CHESNUT, D.A. & COX, D.O. (1979) : Gas production from tight Mesa Verde sands in Wyoming - a field case history. SPE Paper, **7934**.
WARD, J.S. & MORROW, N.R. (1985) : Capillary pressures and gas relative permeabilities of low-permeability sandstone. SPE/DOE Paper, **13882**. SPE Form. Eval., **2** (1987) : 345 - 356.
WARD, V.L. (1984) : Nitrogen and carbon dioxide in the oil field : stimulation and completion applications. SPE Paper, **12594**. SPE Product. Engin., **1** (1986) : 275 - 278.
WARDLAW, N.C. & CASSAN, J.P. (1978) : Estimation of recovery efficiency by visual observation of pore systems in reservoir rocks. Bull. Canad. Petrol. Geol., **26** : 572 - 585.
WARDLAW, N.C. & TAYLOR, R.P. (1976) : Mercury capillary pressure curves and the interpretation of pore structure and capillary behavior in reservoir rocks. Bull. Canad. Petrol. Geol., **24** : 225 - 262.
WAREMBOURG, P. (1980) : Professional design imperative in fracturing. Western Oil Reporter, **1980**.
WAREMBOURG, P. (1988) : New analysis technology for determining fracture parameters from calibration treatments : case histories, Rocky Mountain area. SPE Paper, **17502**.

WAREMBOURG, P.A., KLINGENSMITH, E.A., HODGES, J.E. & ERDLE, J.E. (1985) : Fracture stimulation design and evaluation. SPE Paper, **14379**.
WARNOCK, W.E., HARRIS, P.C. & KING, D.S. (1985) : Successful field applications of CO2-foam fracturing fluids in the Arkansas-Louisiana-Texas region. J. Petrol. Technol., **37** : 80 - 88. SPE Paper, **11932** (1983).
WARPINSKI, N.R. (1983 a) : Measurement of width and pressure in a propagating hydraulic fracture. SPE/DOE Paper, **11648**. Soc. Petrol. Engin. J., **25** (1985) : 46 - 54.
WARPINSKI, N.R. (1983 b) : Investigation on the accuracy and reliability of in-situ stress measurements using hydraulic fracturing in perforated cased holes. Proc. 24th U.S. Rock Mechan. Symp., 773 - 786, College Station/Texas.
WARPINSKI, N.R. (1986) : Elastic and viscoelastic calculations of stress in sedimentary basins. SPE Paper, **15243**.
WARPINSKI, N.R. (1988 a) : Dual leakoff behavior in hydraulic fracturing of tight, lenticular gas sands. SPE Paper, **18259**.
WARPINSKI, N.R. (1988 b) : Determining the minimum in situ stresses from hydraulic fracturing through perforations. Proc. 2nd Intern. Workshop on Hydraulic Fracture Stress Measurements, Minneapolis/Minnesota, 06.1988.
WARPINSKI, N.R. & BRANAGAN, P.T. (1988) : Altered-stress fracturing. SPE Paper, **17533**.
WARPINSKI, N.R., BRANAGAN, P.T., SATTLER, A.R., CIPOLLA, C.L., LORENZ, J.C. & THORNE, B.J. (1988) : A case study of a stimulation experiment in a fluvial, tight sandstone gas reservoir. SPE Paper, **18258**.
WARPINSKI, N.R., BRANAGAN, P.T., SATTLER, A.R., LORENZ, J.C., NORTHROP, D.A., MANN, R.L. & FROHNE, K.H. (1985) : Fracturing and testing case study of paludal, tight, lenticular gas sands. SPE/DOE Paper, **13876**. SPE Form. Eval., **2** (1987) : 535 - 545.
WARPINSKI, N.R., BRANAGAN, P. & WILMER, R. (1985) : In-situ stress measurements at U.S. DOE`s Multiwell Experiment Site, Mesaverde Group, Rifle, Colorado. J. Petrol. Technol., **37** : 527 - 536. SPE Paper, **12142** (1983).
WARPINSKI, N.R., CLARK, J.A., SCHMIDT, R.A. & HUDDLE, C.W. (1982) : Laboratory investigations on the effect of in-situ stresses on hydraulic fracture containment. Soc. Petrol. Engin. J., **22** : 333 - 340. SPE/DOE Paper, **9834** (1981).
WARPINSKI, N.R., SCHMIDT, R.A., COOPER, P.W., WALLING, H.C. & NORTHROP, D.A. (1979) : High-energy gas frac : multiple fracturing in a wellbore. Proc. 20th U.S. Rock Mechan. Symp., pp. 143 - 152.
WARPINSKI, N.R., SCHMIDT, R.A. & NORTHROP, D.A. (1980) : In-situ stresses : the predominant influence on fracture containment. SPE/DOE Paper, **8932**. J. Petrol. Technol., **34** (1982) : 653 - 664.
WARPINSKI, N.R. & TEUFEL, L.W. (1984) : Influence of geological discontinuities on hydraulic fracture propagation. SPE Paper, **13224**. J. Petrol. Technol., **39** (1987) : 209 - 220, p. 999.
WARPINSKI, N.R. & TEUFEL, L.W. (1986) : Viscoelastic constitutive model for determining in situ stress magnitudes from anelastic strain recovery of core. SPE Paper, **15368**.
WARPINSKI, N.R. & TEUFEL, L.W. (1987) : In-situ stresses in low-permeability, nonmarine rocks. SPE/DOE Paper, **16402**.
WARREN, J.E. & ROOT, P.J. (1963) : The behaviour of naturally fractured reservoirs. Soc. Petrol. Engin. J., **3** : 245 - 255.
WATERHOUSE, M., CHARARA, M. & NURMI, R. (1987) : Focus on fractures. Middle East Well Eval. Rev., **3** : 14 - 27; Schlumberger, Houston/Texas.
WATERS, A.B. (1980) : Stimulation of hydrocarbon reservoirs - an introduction to hydraulic fracturing. Industrial Minerals, **1980**/10 : 57 - 65. Industrial Minerals Conference, Session "Minerals and chemicals in drilling muds - the 80s and beyond", Houston/Texas.
WATERS, A.B. (1981) : Hydraulic fracturing - what is it? J. Petrol. Technol., **33** : p. 1416 ff.
WATKINS, D.R., KALFAYAN, L.J. & BLASER, S.M. (1987) : Cyclic steam stimulation in a tight clay-rich reservoir. SPE Paper, **16336**.
WATKINS, D.R., KALFAYAN, L.J., WATANABE, D.J. & HOLM, J.A. (1985) : Preventing gravel pack and formation dissolution during steam injection. SPE Paper,

13660. SPE Product. Engin., **1** (1986) : 471 - 477.
WATKINS, D.R. & WATANABE, D.J. (1984) : Method of inhibiting silica dissolution during injection of steam into a reservoir. U.S. Patent, **4,475,595**.
WATKINS, E.K., WENDORFF, C.L. & AINLEY, B.R. (1983) : A new crosslinked foamed fracturing fluid. SPE Paper, **12027**.
WATKINS, H. (1982) : Higher sand concentrations in East Texas. SPE Paper, **10913**.
WATSON, A.T., KERIG, P.D., RICHMOND, P.C. & TAO, T.M. (1986) : An improved model for estimating relative permeability from displacement experiments. SPE Paper, **15064**.
WATSON, R.W. & ERTEKIN, T. (1988) : The effect of the steep temperature gradient on relative permeability measurements. SPE Paper, **17505**.
WATSON, S.C., BENSON, G.R. & FILLO, G.J. (1985) : Liquid propellant stimulation : two case studies in shallow oil and gas fields. SPE Paper, **14516**.
WATSON, S.C., BENSON, G.R. & FILLO, G.J. (1986) : Liquid propellant stimulation : case studies in shallow Appalachian Basin wells. SPE Product. Engin., **1** : 203 - 212. SPE Paper, **13376** (1984).
WATTENBERGER, R.A. & RAMEY, H.J. (1969) : Well test interpretation of vertically fractured gas wells. J. Petrol. Technol., **21** : 625 - 632. SPE Paper, **2155** (1968).
WATTS, N.L. (1983) : Microfractures in chalks of Albuskjell field, Norwegian sector, North Sea : possible origin and distribution. Bull. Amer. Assoc. Petrol. Geol., **67** : 201 - 234.
WAY, S.C. (1987) : Hydrologic characterization of coal seams for optimal dewatering and methane drainage. Quart. Rev. of Methane from Coal Seams Technol., **4** : 42 - 43.
WAY, S.C. (1988) : Hydrologic characterization of coal seams for optimal dewatering and methane drainage. Quart. Rev. of Methane from Coal Seams Technol., **5**/3-4 : 44 - 45.
WEAVER, J.D. & MORGAN, J.A. (1988) : Furan resin process replaces workovers in gas storage reservoirs. SPE Paper, **17742**.
WEBER, K.J. (1982) : Influence of common sedimentary structures on fluid flow in reservoir models. J. Petrol. Technol., **34** : 665 - 672. SPE Paper, **9247** (1980).
WEBER, R., PUSCH, G. & MÜLLER, T. (1987) : Entölungsprozess in geschichteten Trägern. Einfluß von Fließbarrieren. Erdöl und Kohle - Erdgas - Petrochemie, **40** : 467 - 474.
WEBSTER, K. (1977) : Current completion practices in tight reservoirs. SPE Paper, **6379**.
WEBSTER, K.R., GOINS, W.C. & BERRY, S.C. (1970) : A continuous multistage fracturing technique. SPE Reprint Series, Well Completions, **5** : 284 - 290. J. Petrol. Technol., **17** (1965) : 619 - 625. SPE Paper, **977** (1964).
WEEKS, S.G. (1974) : Formation damage or limited penetration? Test well shooting may give a clue. J. Petrol. Technol., **26** : 979 - 985. SPE Paper, **4794**.
WEEKS, S.G. (1979) : Revised completion procedures for S.W. Texas low-strength, low-permeability, geopressured deep Wilcox Sands. SPE Paper, **7913**.
WEFA (1988) : "Boom or bust" : the economy and the oil market. Wharton Econometric Forecasting Associates, Rept., Winter 1988.
WEI, K.K., MORROW, N.R. & BROWER, K.R. (1986) : Effect of fluid, confining pressure, and temperature on absolute permeabilities of low-permeability sandstones. SPE Form. Eval., **1** : 413 - 423. SPE Paper, **13093** (1984).
WEIGHTMAN, G. & ROBERTS, T. (1987) : Suitcase-size data recording system improves fracturing operations. SPE Paper, **17042**.
WEIMER, R.J. & SONNENBERG, S.A. (1982) : Wattenberg field, paleostructure - stratigraphic trap, Denver Basin, Colorado. Oil Gas J., 22.03.**1982** : 204 - 210.
WEIMER, R.J. & SONNENBERG, S.A. (1983) : Codell Sandstone, a new exploration play, Denver Basin. Oil Gas J., 30.05.**1983** : 119 - 125.
WEIMER, R.J., SONNENBERG, S.A. & YOUNG, G.B.C. (1986) : Wattenberg field, Denver Basin, Colorado. In : SPENCER, C.W. & MAST, R.F. (Editors), Geology of tight gas reservoirs. Amer. Assoc. Petrol. Geol., Stud. Geol., **24** : 143 - 164.

WEINBRANDT, R.M., RAMEY, H.J. & CASSE, F.J. (1975) : The effect of temperature on relative and absolute permeability of sandstones. Soc. Petrol. Engin. J., **15** : 376 - 384. SPE Paper, **4142** (1972).
WEIRICH, J.B., ZALESKI, T.E. & MULCAHY, P.M. (1987) : Perforating the horizontal well : designs and techniques prove successful. SPE Paper, **16929**.
WEIRICH, J.B., ZALESKI, T.E. & TYLER, S.L. (1988) : One-trip multiple-zone gravel packing technique for low pressure, shallow wells. SPE Paper, **17590**.
WEISSENBURGER, K.W., MORITA, N., MARTIN, A.J. & WHITFILL, D.L. (1987) : The engineering approach to sand production prediction. SPE Paper, **16892**.
WELEX (1986) : Full wave sonic. Brochure; Welex, Houston/Texas.
WELEX (1988) : Compensated spectral natural gamma logging. Welex Brochure, **EL-1021R**; Houston/Texas.
WENANDE, B. (1987) : A review of the first horizontal well - Denmark. Offshore Drilling Conference, London/England, 02. - 03.12.1987.
WENDEL, D.J., KUNKEL, L.A. & SWANSON, G.S. (1988) : Waterflood potential of diatomite : new laboratory methods. SPE Paper, **17439**.
WENDORFF, C.L. (1978) : Frac sand control - a must for good frac treatments. ASME Meeting, Petrol. Divis., Houston 11.1978.
WENDORFF, C.L. (1982) : Fracture porosimeter - a new tool for determining fracture conductivity and downhole stress. SPE Paper, **10963**.
WENDORFF, C.L. & AINLEY, B.R. (1981) : Massive hydraulic fracturing of high temperature wells with stable frac foams. SPE Paper, **10257**.
WENDORFF, C.L., BURKE, R.F. & GARVIS, S.J. (1979) : The effect sand handling equipment has on prop sand quality. SPE Paper, **8402**.
WENDORFF, C.L. & EARL, R.B. (1983) : Foam fracturing laboratory. SPE Paper, **12026**.
WERES, O., JUN, C.H. & TSAO, L. (1986) : New three-phase equilibrium model (oil/gas/brine) used to interpret production of liquid hydrocarbons from a geopressured gas well. SPE Paper, **15084**.
WEST, J. (1987) : Independent operators begin to see first signs of a gradual recovery. Oil Gas J., 12.10.**1987** : 33 - 39.
WEST, T.S. (1951) : Gravel pack completions for exclusion of water and gas. Trans. AIME, **192** : 183 - 190.
WESTAWAY, P., HERTZOG, R. & PLASEK, R.E. (1980) : The gamma spectrometer tool inelastic and capture gamma-ray spectroscopy for reservoir analysis. SPE Paper, **9461**. Soc. Petrol. Engin. J., **23** (1983) : p. 553 ff.
WESTCOTT, P.A. (1985) : Present and future application of multiple radial fracturing techniques to tight reservoirs. SPE Paper, **13859**.
WESTERHAUSEN, H. (1980) : Aussichten der weltweiten Exploration auf Erdöl und Erdgas. Erdoel-Erdgas, **96** : 44 - 50.
WESTERMARK, R.V. & SCHOLES, P. (1987) : Fracture identification in open hole using data from cement evaluation tools. SPE Paper, **16387**.
WESTERN ATLAS (1988 a) : Sand production analysis. Rept.; Western Atlas, Houston/Texas.
WESTERN ATLAS (1988 b) : PRISM Precision radioactive isotope spectral measurement. Brochure; Western Atlas, Houston/Texas.
WESTERN OIL WORLD (1988) : Horizontal drilling holds promise. Western Oil World, **45**/9 : 26 - 27.
WESTERN PETROLEUM (1984) : Proppants. 2nd. edition; Western Petroleum, Fort Worth/Texas.
WHEELER, J.A. (1969) : Analytical calculations of heat transfer from fractures. SPE Paper, **2494**.
WHEELER, V.J., PARSONS, T.V., CONCHIE, S.J. & DURHAM, B. (1985) : Application of radioactive tracers in oil reservoir water flood studies. SPE Paper, **13985**.
WHITE, B., WALTER, T. & DIEBOLD, J. (1965) : A proven gas-well completion technique for higher deliverability. J. Petrol. Technol., **17** : 647 - 656. SPE Paper, **1002** (1964).
WHITE, J. & DANIEL, G. (1980) : Design of MHF treatments in the Cotton Valley. SPE Paper, **9065**.
WHITE, J.L. & DANIEL, E.F. (1981) : Key factors in MHF design. J. Petrol. Technol., **33** : 1501 - 1512. SPE Paper, **9065** (1980).

WHITEHEAD, W.S., HUNT, E.R., FINLEY, R.J. & HOLDITCH, S.A. (1986) : In-situ stresses: a comparison between log-derived values and actual field-measured values in the Travis Peak Formation of East Texas. SPE Paper, **15209**.

WHITEHEAD, W.S., HUNT, E.R. & HOLDITCH, S.A. (1987) : The effects of lithology and reservoir pressure on the in-situ stresses in the Waskom (Travis Peak) field. SPE/DOE Paper, **16403**.

WHITSITT, N.F. & DYSART, G.R. (1970) : The effect of temperature on stimulation design. J. Petrol. Technol., **22** : 493 - 502. SPE Paper, **2497** (1969).

WHITTING, J.J. (1979) : Natural gas supply demand strategy in the U.K. SPE Paper, **8293**.

WIBORG, R. & JEWHURST, J. (1986) : Ekofisk subsidence - detailed and solutions assessed. Oil Gas J., 02.**1986** : 47 - 51.

WICHMANN, P.A., McWHIRTER, V.C. & HOPKINSON, E.C. (1975) : Field results of the natural gamma ray spectralog. Trans. 16th Soc. Prof. Well Log Analysts Annual Logging Symp., Paper **O**, New Orleans/Louisiana, 04. - 07.06.1975.

WIEDERHORN, S.M. & BOLZ, L.H. (1970) : Stress corrosion static fatigue of glass. J. Amer. Ceram. Soc., **53** : p. 543 ff.

WIERCZEYKO, E. (1985) : Seismische und geoelektrische Frac-Ortung von der Erdoberfläche aus. Statusbericht zum Forschungsvorhaben **03E-2079-A** : 223 - 232. Programm Energieforschung und Energietechnologien, Statusreport 1985 Geotechnik und Lagerstätten (KFA Jülich).

WIJESINGHE, A.M. (1985) : Green`s functions for solving unsteady flow problems in naturally fractured reservoirs with arbitrary fracture connectivity : Part I - Theory. SPE Paper, **13626**.

WIJHESINGHE, A.M. & CULHAM, W.E. (1984) : Single-well pressure testing solutions for naturally fractured reservoirs with arbitrary fracture connectivity. SPE Paper, **13055**.

WIJESINGHE, A.M. & KECECIOGLU, I. (1985) : Analysis of interference pressure tests in naturally fractured reservoirs with macroscopic fracture and pore system permeabilities and unsteady interporosity flow. SPE Paper, **14522**.

WIJESINGHE, A.M. & KECECIOGLU, I. (1986 a) : Green`s functions for solving unsteady flow problems in naturally fractured reservoirs with arbitrary fracture connectivity : Part II - Applications. SPE Paper, **15113**.

WIJESINGHE, A.M. & KECECIOGLU, I. (1986 b) : Vertical interference pressure testing across a low-permeability zone with unsteady crossflow. SPE Paper, **15583**.

WILEY, C.B. (1981) : Success of a high-friction diverting gel in acid stimulation of a carbonate reservoir - Cornell Unit, San Andres field. J. Petrol. Technol., **33** : 2196 - 2200.

WILEY, R. (1980) : Borehole televiewer - revisited. Trans. 21st Soc. Prof. Well Log Analysts Annual Logging Symp., Paper **HH**, Lafayette/Louisiana, 08. - 11.07.1980.

WILEY, R. & COCANOWER, R.D. (1975) : A quantitative technique for determining injectivity profiles using radioactive tracers. SPE Paper, **5513**.

WILKIRSON, J.P., SMITH, J.H., STAGG, T.O. & WALTERS, D.A. (1986) : Horizontal drilling techniques at Prudhoe Bay, Alaska. SPE Paper, **15372**.

WILLIAMS, B. (1988 a) : OPEC ventures downstream : industry threat or stability aid? Oil Gas J., 16.05.**1988** : 15 - 17.

WILLIAMS, B. (1988 b) : Oil price rise revives profitability of OGJ independent group in `87. Oil Gas J., 25.04.**1988** : 15 - 17.

WILLIAMS, B.B. (1970) : Fluid loss from hydraulically induced fractures. J. Petrol. Technol., **22** : 882 - 888. SPE Paper, **2769** (1969). SPE Reprint Series, Well Completions, **5** : 302 - 308.

WILLIAMS, B.B. (1975) : Hydrofluoric acid reaction with sandstone formations. J. Engin. Industr., **1975**/2 : 252 - 258.

WILLIAMS, B.B., ELLIOTT, L.S. & WEAVER, R.H. (1972) : Productivity of inside casing gravel-pack completions. J. Petrol. Technol., **24** : 419 - 425. SPE Paper, **3590** (1971).

WILLIAMS, B.B., GIDLEY, J.L. & SCHECHTER, R.S. (1979) : Acidizing fundamentals. Soc. Petrol. Engin. Monograph Series, **6** : 124 pp. ISBN 0-89520-205-0.

WILLIAMS, B.B. & NIERODE, D.E. (1971) : Design of acid fracturing treatments. SPE Paper, **3720**. J Petrol. Technol., **24** (1972) : 849 - 859. SPE Reprint

Series, **5 a**/II (1978) : 135 - 145.
WILLIAMS, B.B., NIETO, G., GRAHAM, H.L. & LEIBACH, R.E. (1973) : A staged fracturing treatment for multisand intervals. J. Petrol. Technol., **25** : 897 - 904. SPE Paper, **4093** (1972).
WILLIAMS, B.B. & WHITELEY, M.E. (1971) : Hydrofluoric acid reaction with a porous sandstone. SPE Paper, **3112**. Soc. Petrol. Engin. J., **11** : 306 - 314. SPE Reprint Series, Well Completions, **5 a**/II (1978) : 154 - 162.
WILLIAMS, D.A., WOODROOF, R.A. & BOX, P.C. (1982) : High-temperature frac job successful at geothermal well. Oil Gas J., **1982**/3 : 141 - 145.
WILLIAMS, D.B., SHERRARD, D.W. & LIN, C.Y. (1987) : The impacts on waterflood management of inducing fractures in injection wells in the Prudhoe Bay oil field. SPE Paper, **16358**.
WILLIAMS, J.L. (1977) : The influence of induced vertical fractures on overall reservoir performance. SPE Paper, **6381**.
WILLIAMS, L.H. & UNDERDOWN, D.R. (1980) : New polymer offers effective permanent clay stabilization treatment. SPE Paper, **8797**. J. Petrol. Technol., **33** (1981) : 1211 - 1217.
WILLIAMS, R.E., ROWLEY, J.C., NEUDECKER, J.W. & BRITTENHAM, T.L. (1979) : Equipment for drilling and fracturing hot granite wells. SPE Paper, **8268**.
WILLIAMS, R.L. & McCARTHY, J. (1987) : Using multiple radioactive tracers to optimize stimulation designs. SPE Paper, **16383**.
WILLIAMSON, C.D. & ALLENSON, S.J. (1989) : A new non-damaging particulate fluid loss additive. SPE Paper, **18474**.
WILSEY, L.E. & BEARDEN, W.G. (1954) : Reservoir fracturing - a method of oil recovery from extremely low permeability formations. Trans. AIME, **201** : 169 - 175.
WILSON, B., NELSON, C. & STADLEMAN, J. (1983) : Resin-coated sand slurry pack gas deliverabilities : field and laboratory results. SPE Paper, **11984**.
WILSON, J.H., STALLCUP, B.S. & GLAZE, O.H. (1978) : Gravel packing in the High Island field : a study of sand control in high rate wells. SPE Paper, **7593**.
WILSON, L. (1956) : Some aspects of sand-oil fracturing in long heterogeneous sand sections. J. Petrol. Technol., **8** : 12 - 16.
WILSON, M.A. & SABINS, F.L. (1987) : A laboratory investigation of cementing horizontal wells. SPE Paper, **16928**.
WILSON, M.D. (1981) : Origins of clays controlling permeability in tight gas sands. SPE/DOE Paper, **9843**. J. Petrol. Technol., **34** (1982) : 2871 - 2876.
WILSON, M.D. & PITTMAN, E.D. (1977) : Authigenic clay in sandstones : recognition and influence on reservoir properties and paleoenvironmental analysis. J. Sedim. Petrol., **47** : 3 - 31.
WILSON, M.J. & KISER, S.C. (1987) : Synergistic approach for siting and design for injection of hazardous liquid wastes : case study in Western San Joaquin Valley, Kern County, California. SPE Paper, **16327**.
WILSON, R.C. & WILLIS, D.N. (1986) : Successful high angle drilling in the Statfjord field. SPE Paper, **15465**.
WILSON, S.C. (1978) : Foamed acid stimulation of the San Andres : an area study. Petrol. Engin. Intern., **1978**/5 : 90 - 98.
WINKLE, W.v. & MIGNOTTE, H.X. (1963) : Implosive shock treatment of injection wells. SPE Paper, **737**.
WINNOCK, E. & PONTALIER, Y. (1970) : Lacq gas field, France. In : HALBOUTY, M.T. (Editor), Geology of giant petroleum fields. Amer. Assoc. Petrol. Geol. Mem., **14** : 370 - 387; Tulsa/Oklahoma.
WINTERBURN, R. (1947) : Control of unconsolidated sands in Wilmington oil field. API Drill. Product. Pract., **1947** : 63 - 79.
WINTERS, W.J., LUCIA, F.de & JOHNSON, R. (1988) : Drilling a large medium radius horizontal hole in hard rock. SPE Paper, **17528**.
WITHJACK, E.M. (1985) : Analysis of naturally fractured reservoirs with bottom-water drive : Nido A and B fields, offshore Northwest Palawan, Philippines. J. Petrol. Technol., **37** : 1481 - 1490. SPE Paper, **12019** (1983).
WITT, W.d. (1986) : Devonian gas-bearing shales in the Appalachian Basin. In : SPENCER, C.W. & MAST, R.F. (Editors), Geology of tight gas reservoirs. Amer. Assoc. Petrol. Geol., Stud. Geol., **24** : 1 - 8; Tulsa/Oklahoma.
WITTERHOLT, E.J. & TIXIER, M.P. (1972) : Temperature logging in injection

wells. SPE Paper, **4022**.
WITTRISCH, C. (1987) : Direction of hydraulic fracture - Simfrac. Trans. 28th Soc. Prof. Well Log Analysts Annual Logging Symp., London/England, 29.06. - 02.07.1987.
WITTRISCH, C. & SARDA, J.P. (1987) : Simfrac - method for mapping hydraulic fractures. Trans. 28th Soc. Prof. Well Log Analysts Annual Logging Symp., Paper **RR**, London/England, 29.06. - 02.07.1987.
WOJTANOWICZ, A.K., KRILOV, Z. & LANGLINAIS, J.P. (1987) : Study on the effect of pore blocking mechanisms on formation damage. SPE Paper, **16233**.
WOLFF, R.G., BREDEHOEFT, J.D., KEYS, W.S. & SHUTER, E. (1974) : Tectonic stress determinations, Northern Piceance Creek Basin, Colorado. In : Energy resources of the Piceance Creek Basin, Colorado, pp. 193 - 197; Rocky Mountain Assoc. Geol., Denver/Colorado.
WONG, D.W., HARRINGTON, A.B. & CINCO-LEY, H. (1984) : Application of the pressure derivative function in the pressure transient testing of fractured wells. SPE Paper, **1305**. SPE Form. Eval., **1** (1986) : 470 - 480.
WONG, J.F. (1987) : China technology, equipment market grows as oil, gas sectors keep expanding. Oil Gas J., 07.12.**1987** : 46 - 49.
WONG, S.K. (1986) : Analysis and implications of in-situ stress changes during steam stimulation of Cold Lake oil sands. SPE Paper, **15173**. SPE Reser. Engin., **3** (1988) : 55 - 81.
WONG, S.K. & CLEARY, M.P. (1985) : Numerical simulation of unsteady fluid flow and propagation of a circular hydraulic fracture. Intern. J. Numer. Anal. Meth. Geomechan., **9** : 1 - 14.
WONG, S.W. (1970) : Effect of liquid saturation on turbulence factors for gas-liquid systems. J. Canad. Petrol. Technol., **1970** : p. 274 ff.
WOOD, D.B. & JUNKIN, G. (1970) : Stresses and displacements around hydraulically fractured wells. SPE Paper, **3030**.
WOOD, D.C. (1979) : The effect of sand size on arch stability in unconsolidated sands. Thesis, **T-2165**; Colorado School of Mines, Golden/Colorado.
WOOD, M.D. (1981) : Method of determining change in the subsurface structure due to application of fluid pressure to the earth. U.S. Patent, **4,271,696** (09.06.1981).
WOOD, M.D., PARKIN, C.W., YOTAM, R., HANSON, M.E., SMITH, M.B., ABBOTT, B.L., COX, D. & O`SHEA, P. (1983) : Fracture proppant mapping using surface superconducting magnetometers. SPE/DOE Paper, **11612**.
WOOD, M.D., POLLARD, D.D. & RALEIGH, C.B. (1976) : Determination of in-situ geometry of hydraulically generated fractures using tiltmeters. SPE Paper, **6091**.
WOODROOF, R.A. & ANDERSON, R.W. (1977) : Synthetic polymer friction reducers can cause formation damage. SPE Paper, **6812**.
WOODROOF, R.A., BAKER, J.R. & JENKINS, R.A. (1975) : Corrosion inhibition of hydrochloric-hydrofluoric acid/mutual solvent mixtures at elevated temperatures. SPE Paper, **5645**.
WOODS, T.J. (1987) : The effects of current low drilling levels on Lower-48 oil and gas production. SPE Paper, **16844**.
WOODS, T.J. & HOLTBERG, P.D. (1986) : Hydrocarbon activity in an era of low oil prices. SPE Paper, **15355**.
WOOLEY, G.R. & PRACHNER, W. (1984) : Reservoir compaction loads on casings and gravel-pack screens. SPE Paper, **13088**.
WORLD OIL (1975 a) : East Texas gas field gets first massive frac. World Oil, **1975**/5 : 98 - 100.
WORLD OIL (1975 b) : Completion method reduces effect of formation damage. World Oil, **1975**/11.
WORLD OIL (1985) : Western Europe. North Sea keeps up hearty pace. World Oil, **201** (15.08.**1985**) : 73 - 89.
WORLD OIL (1988) : Some discipline, please. World Oil, **207** (1988/8) : 31 - 34.
WOZNIAK, K.C. (1988) : Pulsed-neutron gas saturation analysis : applications and limitations in thermal recovery projects, Kern River field, CA. SPE Paper, **18148**.
WRIGHTSMAN, G.G. (1949) : Method for consolidation of sands. U.S. Patent, **2,476,015** (12.07.1949).

WU, P.T. (1984) : Comparison of digital and analog techniques for determining sonic velocity from borehole acoustic waveforms. SPE Paper, **13286**.
WU, Y.S. & PRUESS, K. (1986) : A multiple-porosity method for simulation of naturally fractured petroleum reservoirs. SPE Paper, **15129**. SPE Reser. Engin., **3** (1988) : 327 - 336.
WYATT, D.F. & SMITH, H.D. (1987) : Reservoir monitoring with the thermal multigate decay log. SPE Form. Eval., **2** : 401 - 411. SPE Paper, **14137** (1986).
WYMAN, R.E. (1985) : Gas recovery from tight sands. SPE Paper, **13940**.
WYMAN, R.E., HOLDITCH, S.A. & RANDOLPH, P.L. (1979) : Analyses of an Elmworth hydraulic fracture - Alberta, Canada. SPE Paper, **7935**. J. Petrol. Technol., **32** (1980) : 1621 - 1630.
WYMAN, S. (1978) : The small independent producer`s role in the rapidly changing oil and gas industry. SPE Paper, **7626**.
XUE, P.H. (1986) : A point bar facies reservoir model - semi-communicated sandbody. SPE Paper, **14837**.
YAGER, D. (1985) : Esso drills world`s longest horizontal well. The Roughneck, **1985**/1.
YASSIN, A.A.M. & PEDEN, J.M. (1987) : Sand stability predictions on a microcomputer. SPE Paper, **16504**.
YASSIN, A.A.M. & PEDEN, J.M. (1988) : The effect of formation damage upon maximum sandfree production rate in friable sandstone formations. SPE Paper, **17144**.
YEW, C.H. & CHIOU, Y.J. (1983) : The effects of in-situ stresses and layer properties on the containment of a hydraulic fracture. SPE Paper, **12332**.
YEW, C.H. & LI, Y. (1987) : Fracturing of a deviated well. SPE Paper, **16930**.
YEW, C.H. & LODDE, P. (1983) : Propagation of a hydraulically induced fracture in layered medium. SPE Paper, **11870**.
YILDIZ, T. & LANGLINAIS, J.P. (1988) : Calculation of pressure losses across gravel packs. SPE Paper, **17167**.
YOST, A.B. II, FROHNE, K.H., KOMAR, C.A. & AMERI, S. (1982) : Techniques to determine natural and induced fracture relationships in Devonian shale. J. Petrol. Technol., **34** : 1371 - 1377. SPE Paper, **9271** (1980).
YOST, A.B. II, OVERBEY, W.K., SALAMY, S.P., OKOYE, C.O. & SARADJI, B.S. (1987) : Devonian shale horizontal well : rationale for wellsite selection and well design. SPE/DOE Paper, **16410**.
YOST, A.B. II, OVERBEY, W.K., WILKINS, D.A. & LOCKE, C.D. (1988) : Hydraulic fracturing of a horizontal well in a naturally fractured reservoir : case study for multiple fracture design. SPE Paper, **17759**.
YOUNG, B.M. (1967) : Consolidation technique in clayey sand boosts flow rate. Oil Gas J., 19.09.**1967**.
YOUNG, B.M. (1971) : Method of improving strength and stability of sand consolidations made with resin systems. U.S. Patent, **3,625,287** (07.12.1971).
YOUNG, B.M., COOK, J.G. & DONALDSON, A.L. (1969) : Here`s a new sand-pack method. Oil Gas J., 10.03.**1969**.
YOUNG, B.M., McLAUGHLIN, H.C. & BORCHARDT, J.K. (1979) : Clay stabilization agents - their effectiveness in high temperature steam. SPE Paper, **7895**. J. Petrol. Technol., **32** (1980) : 2121 - 2131.
YOUNG, C., BARKER, D.B. & CLARK, H.C. (1984) : Field tests of the stem-induced explosive fracturing technique. SPE/DOE/GRI Paper, **12840**. SPE Product. Engin., **1** (1986) : 266 - 274.
YOUNG, W.S. & ZALESKI, T.E. (1985) : Procedural considerations associated with tubing-conveyed underbalanced perforating. SPE Paper, **13646**.
YOUSEF, O.M. & EVANS, R.D. (1986) : Type-curve techniques for hydraulically fractured lenticular formations. SPE Paper, **15159**.
YU, M. (1988): World crude oil prices - prices as paid in national currencies. Petrol. Econ., **55**/4 : 135 - 137.
YURKEWICZ, G.A. (1986) : The economic impact of take-or-pay settlements. SPE Paper, **14643**.
YUSTER, S.T. & CALHOUN, J.C. (1945) : Pressure parting of formations in water flood operations. Oil Weekly, 12.03.**1945**, 19.03.**1945**.
ZAHNER, R.L. & CRAFTON, J.W. (1985) : A review of the long-term performance of hydraulic fracture treatments in the Muddy "J" Sand, Wattenberg field. A re-

view of the optimization of hydraulic fracture design for the Muddy "J" Sand, Wattenberg field. SPE/DOE Paper, **13873**.

ZAITOUN, A., KOHLER, N. & GUERRINI, Y. (1989) : Improved polyacrylamide treatments for water control in producing wells. SPE Paper, **18501**.

ZALESKI, T.E. & DONOVAN, J.F. (1986) : Gravel packing operations from a floating vessel. SPE Paper, **14816**.

ZALESKI, T.E. & SPATZ, E. (1988) : Horizontal completions challenge for industry. Oil Gas J., 02.05.**1988** : 58 - 70.

ZALESKI, T.E. & WEIRICH, J.B. (1988) : A new approach to underbalanced perforating design. SPE Paper, **18246**.

ZEMANEK, J., ANGONA, F.A., WILLIAMS, D.M. & CALDWELL, R.L. (1984) : Continuous acoustic shear wave logging. Trans. 25th Soc. Prof. Well Log Analysts Annual Logging Symp., Paper **U**, Dallas/Texas, 17. - 20.06.1985.

ZEMANEK, J., CALDWELL, R.L., GLENN, E.E., HOLCOMB, S.V., NORTON, L.J. & STRAUS, A.J.D. (1969) : The borehole televiewer - a new logging concept for fracture location and other types of borehole inspection. J. Petrol. Technol., **21** : 762 - 774. Geophysics, **35** (1970) : 254 - 269. SPE Paper, **2402** (1968).

ZHELTOV, Y.P. & KHRISTIANOVICH, S.A. (1955) : Hydraulic fracture of an oil-bearing bed. Izvest. Akad. Nauk SSSR, OTN No. 5, **1955** : 3 - 41 (in Russian).

ZHOU, W., QI, W. & XIAO, T. (1988) : Development in the limited-entry fracturing technique. SPE Paper, **17834**.

ZHU, Y.D. (1986) : Well testing analysis in a double porosity reservoir. SPE Paper, **14867**.

ZHUANG, H. (1982) : Interference testing and pulse testing in the Kenli carbonate oil pool - a case history. SPE Paper, **10581**. J. Petrol. Technol., **36** (1984) : 1009 - 1017.

ZHUVAGIN, I.G. & AKCHASYANOV, Y.A. (1960) : New methods of controlling results of hydraulic fracturing by means of radioactive tracers. Neft. Khoz., **38**/6 : p. 7 ff.

ZIARA, B.A., MANCILLAS, G. & MATSON, R.P. (1977) : Ekofisk stimulation program. Petrol. Engin. Intern., **1977**/6 : 39 - 50.

ZIEGLER, P.A. (1982) : Geological atlas of Western and Central Europe. 130 pp., Shell Internationale Petroleum Maatschappij B.V., The Hague/Netherlands (distributed by Elsevier, Amsterdam/Netherlands).

ZIEGLER, V.M. (1988) : Injection well testing in a light oil steamflood, Bueno Vista Hills field, California. SPE Paper, **18140**.

ZIGRANGE, D.J. & SYLVESTER, N.D. (1981) : An explicit equation for particle settling velocities in solid-liquid systems. Al. Chem. Engin. J., **27** : 1043 - 1044.

ZIGRYE, J.L., OSBORNE, M.W. & WESTBROOK, G.H. (1986) : Field analysis and flow loop tests diagnose problems with delayed action crosslinked fracturing fluid systems. SPE Paper, **15633**.

ZIGRYE, J.L., WHITFILL, D.L. & SIEVERT, J.A. (1985) : Fluid-loss control differences of crosslinked and linear fracturing fluids. J. Petrol. Technol., **37** : 315 - 320. SPE Paper, **12153** (1983).

ZOBACK, M.D. & HAIMSON, B.C. (1982) : Status of the hydraulic fracturing method for in-situ stress measurements. Proc. 23rd U.S. Rock Mechan. Symp., 143 - 157.

ZOBACK, M.D., MOOS, D., MASTIN, L. & ANDERSON, R.N. (1985) : Wellbore breakouts and in situ stress. J. Geophys. Res., **90** : 5523 - 5530.

ZOBACK, M.D. & POLLARD, D. (1978) : Hydraulic fracture propagation and interpretation of pressure-time records for in-situ stress determinations. Proc. 19th U.S. Rock Mechan. Symp., Stateline/Nevada, pp. 14 - 22.

ZOBACK, M.D., ZOBACK, M.L. & SEEBURGER, D. (1981) : State of stress in the conterminous U.S. and its effect on fractures. SPE Paper, **10335**.

ZUBER, M.D., KUUSKRAA, V.A. & SAWYER, W.K. (1988) : Optimizing well spacing and hydraulic fracture design for economical recovery of coalbed methane. SPE Paper, **17726**.

ZUBER, M.D., LEE, W.J. & GATENS, J.M. III (1987) : Effect of stimulation on the performance of Devonian shale gas wells. SPE Product. Engin., **2** : 250 - 256. SPE Paper, **14508** (1985).

ZUBER, M.D., REEVES, S.R., JONES, A.H. & SCHRAUFNAGEL, R.A. (1988) : Under-

standing variability in coalbed methane well performance : a case study of three wells in the Oak Grove field, Alabama. SPE Paper, **18251**.
ZWICKY, R.W. (1984) : Discussion of price forecasting and project evaluation in the 1980`s. J. Petrol. Technol., **36** : p. 1499. SPE Paper, **13347**.
ZWOLLE, S. & DAVIES, D.R. (1983) : Gravel packing sand quality - a quantitative study. J. Petrol. Technol., **35** : 1042 - 1050. SPE Paper, **10660** (1982).

Addendum

AHMED, U., MARKELEY, M. & PLUMB, D. (1989) : Enhanced in-situ stress profiling using minifrac, core, and sonic logging data. SPE Paper, **19004**.
BARREE, R.D. (1989) : A new look at tip screenout behavior using a 3-D fracture simulation with coupled proppant transport. SPE Paper, **18955**.
BEN NACEUR, K. (1989) : Design and evaluation of acid fracturing treatments. SPE Paper, **18978**. Analysis of the mechanisms affecting pressure decline in hydraulic fracturing. SPE Paper, **18999**.
CORNELL, F.L. (1989) : Engineering improvements for Red Fork fracturing. SPE Paper, **18963**.
HAAS, M., McFALL, K. & COATES, J.M. (1989) : Technology advances unlock the potential of the tight Frontier Formation in the Moxa Arch. SPE Paper, **18962**.
HARMS, W.M. & CURTICE, R.J. (1989) : Viscosity measurement throughout frac job gives increased gel control during continuous mixing. SPE Paper, **18968**.
HEUZE, F.E., SHAFFER, R.J., BLAIR, S.C. & THORPE, R.K. (1989) : Understanding the effect of natural fractures on the hydrofracture stimulation of gas reservoirs. SPE Paper, **18956**.
HUNT, W.C. III & SHU, W.R. (1989) : Controlled pulse fracturing for well stimulation. SPE Paper, **18972**.
JEFFREY, R. (1989) : The combined effect of fluid lag and fracture toughness on hydraulic fracture propagation. SPE Paper, **18957**.
KEMP, F. (1989) : Study of Nordgren`s equation of hydraulic fracturing. SPE Paper, **18959**.
LEE, T.S., ADVANI, S.H., AVASTHI, J.M. & LEE, J.K. (1989) : Characteristic time concept associated with hydraulic fracture configuration evolution and optimization. SPE Paper, **19000**.
MOORE, P.J.R. (1989) : Barque and Clipper - well test analysis in low permeability fractured gas reservoirs. SPE Paper, **18966**.
MIGLICCO, T.P., MOSCHOVIDIS, Z.A. & VEATCH, R.W. (1989) : Closure stress profiling and 3D fracture simulations of the staged field experiment well no. 2, Upper Travis Peak Formation, North Appleby field. SPE Paper, **18967**.
NILSON, G.J., KELSO, B. & HAAS, M. (1989) : A comparative study of future production of possible conventional and unconventional gas, Piceance Basin, Colorado. SPE Paper, **18961**.
NOLTE, K.G. & ECONOMIDES, M.J. (1989) : Fracture length determination and implications for treatment design. SPE Paper, **18979**.
OKOYE, C.U. & OSHINUGA, A. (1989) : Pressure transient behavior in multilayer vertically fractured gas reservoirs with finite conductivity. SPE Paper, **18960**.
SETTARI, A. & WONG, S. (1989) : Modeling of fracture propagation in layered reservoirs. SPE Paper, **18958**.
SHAH, S.N. & LORD, D.L. (1989) : Hydraulic fracture slurry transport in horizontal pipes. SPE Paper, **18994**.
STREETER, R. & HUNT, E. (1989) : Prefracture analysis of Red Mountain tight gas sands : an integrated analysis of open-hole and cased-hole logs. SPE Paper, **19003**.
VEEKEN, C.A.M. & DAVIES, D.R. (1989) : Limited communication between hydraulic fracture and (deviated) wellbore. SPE Paper, **18962**.
WORLOW, D.W. & HOLDITCH, S.A. (1989) : Rheologic measurements of a crosslinked fracture fluid under shear conditions expected during a fracture treatment. SPE Paper, **18970**.
WINSLOW, D., ZIMMERMAN, C., HINKIE, R. & LOCKMAN, R. (1989) : Selection of tools for stimulation in horizontal cased hole. SPE Paper, **18995**.
ZAMMERILLI, A.M. (1989) : A simulation study of horizontal, high-angle and vertical wells in Eastern Devonian Shale. SPE Paper, **18998**.

Index